D0989840

Atlas of the Flora of the Great Plains 

ATLAS *of the* Flora of the Great Plains

THE GREAT PLAINS FLORA ASSOCIATION

R. L. McGregor, Coordinator
University of Kansas

T. M. Barkley, Editor
Kansas State University

William T. Barker
North Dakota State University

Ralph E. Brooks
The University of Kansas

Steven P. Churchill
University of Nebraska—Lincoln

Robert B. Kaul
University of Nebraska—Lincoln

Ole A. Kolstad
Kearney State College

David M. Sutherland
University of Nebraska at Omaha

Theodore Van Bruggen
University of South Dakota

Ronald R. Weedon
Chadron State College

James S. Wilson
Emporia Kansas State College

THE IOWA STATE UNIVERSITY PRESS / AMES 1977

Contribution No. 1283-B, Division of Biology
Kansas Agricultural Experiment Station
Kansas State University

Contribution No. 658, Department of Botany
North Dakota Agricultural Experiment Station
North Dakota State University

Composed and printed by
The Iowa State University Press
First edition, 1977

Library of Congress Cataloging in Publication Data

Great Plains Flora Association.
Atlas of the Flora of the Great Plains.

Precursor volume to a Flora of the Great Plains.
"Contribution no. 1283-B, Division of Biology, Kansas Agricultural Experiment Station, Kansas State University."
"Contribution no. 658, Department of Botany, North Dakota Agricultural Experiment Station, North Dakota State University."
Bibliography: p.
Includes index.
1. Botany—Great Plains—Maps. I. Barker, William T. II. Barkley, Theodore Mitchell, 1934-
III. Great Plains Flora Association. Flora of the Great Plains. IV. Title.
G1421.J9G7 1976 912'.1'581978 76-54301
ISBN 0-8138-0135-4

Contents

Introduction

THIS ATLAS provides distributional information for the vascular plants growing in the Great Plains of North America. The area covered extends from the Canadian border south through the Texas panhandle, and from the Rocky Mountain uplift east to the beginnings of continuous woodland. The entire states of Kansas, Nebraska, South Dakota, and North Dakota are included; plus the eastern portions of Montana, Wyoming, and Colorado; the northeast sixth of New Mexico; the Texas panhandle; the northwest half of Oklahoma; and the western border areas of Iowa, Missouri, and Minnesota. The Great Plains are geographically and floristically coherent, and they have become an area of intensive and extensive farming and ranching.

The Great Plains have received inadequate botanical attention, for the only treatise to deal with the flora of the region is the *Flora of the Prairies and Plains of Central North America* by P. A. Rydberg (1932). His work is still useful, but it is rapidly becoming relevant merely as a historical document. It was based upon a rather limited number of specimens, mostly in the herbarium of the New York Botanical Garden, and it used a very narrow systematic concept that is now obsolete: Rydberg recognized large numbers of minor morphological variants as distinct species; subsequent research has proved that notion untenable.

Renewed floristic studies in the Great Plains began some 20 years ago with R. L. McGregor at the University of Kansas, who with his students and colleagues has collected more than 135,000 specimens from the region. Dr. McGregor and eight other taxonomists from the Midwest founded the Great Plains Flora Association (GPFA) in the spring of 1973 for the purpose of creating this atlas and ultimately a detailed *Flora of the Great Plains.* Two additional botanists have since been added to the GPFA membership on the basis of their interest and contributions to the project.

This atlas is a prime-source document of biological data. We trust that it will be useful to persons interested in the biology of the Great Plains, as well as to those with special interests in environmental matters. The urgency for publication of the atlas at this time is two-fold. First, the agricultural success of the Great Plains depends in some measure on the sensible manipulation of the native flora or its derivatives, which makes accurate knowledge of the flora and its distribution essential.

Second, ecological concerns demand a readily useable and reliable catalog of plant distributions to aid in writing environmental impact statements, monitoring the effects of pollution and designing land-use formulae. Certainly, the flora of a region is a major component of the broad but pre-eminent complex of interactions between human activities and the natural environment.

This atlas is a precursor of a *Flora of the Great Plains* that will include keys, descriptions, illustrations, ecological information, and the properly aligned nomenclature for each of the approximately 3,000 taxa occurring in the area.

Sources and Presentation of the Data

The distribution maps were derived by plotting the collection locales for the pertinent specimens in the regional herbaria; thus, each dot on each map is based upon specimen evidence. Absence of a dot does not preclude the possibility that the plant occurs there. Data from the literature have been incorporated for the eastern edges of the region. Since the counties of the Great Plains region are relatively small (30-40 miles square), usually no more than one dot is spotted on the map for a county, even when there are numerous collections. However, for the very large counties (for example, Cherry County, Nebraska) several collections may be dotted on the map, provided the dots do not overlap to the point of confusion.

The following herbaria have been surveyed by the Great Plains Flora Association members for the preparation of this atlas:

The University of Kansas, Lawrence
Emporia Kansas State College, Emporia
Kansas State University, Manhattan
University of Nebraska—Lincoln
University of Nebraska at Omaha
Kearney State College, Kearney, Nebraska
Chadron State College, Chadron, Nebraska
University of South Dakota, Vermillion
North Dakota State University, Fargo

In addition, the herbaria of the University of Oklahoma, Norman, and Texas Tech University, Lubbock, were surveyed by Steve Stephens of the University of Kansas. Pertinent distribution maps from the herbarium of the University of Minnesota, St. Paul, were supplied by Dr. Gerald Ownbey, curator.

Data for Iowa were largely derived from the *Flora of Southwestern Iowa* by Marcus J. Fay and the *Flora of Northwestern Iowa* by Jack Carter; both of which are floristic dissertations done at the University of Iowa, Iowa City. In addition, the Iowa State University herbarium, Ames, has been surveyed.

Data for Missouri are derived primarily from the *Flora of Missouri* by Julian Steyermark (1963).

The line on the distribution maps signifies the boundary of the Great Plains as covered by this atlas. Specimens from outside the region but belonging to species within the Great Plains Flora are dotted when convenient to do so. However, the dots occurring outside the range should not be construed as an accurate depiction of the species distribution.

A distribution map is provided for each entity that naturally occurs within the Great Plains, with some exceptions. For sake of space-economy, entities that have highly restricted distributions are cited in the List of Infrequent Great Plains Taxa, but are not accorded a map. Among this category are the purely occasional weeds that are rarely collected in the region, and entities that may be common elsewhere but barely enter our range and are seldom thought of as members of the Great Plains flora (e.g., *Sassafras*). Inasmuch as the eastern boundary for the region is imperfectly defined, there are numerous plants in this category, viz., plants of the eastern woodlands with small localized populations in the Great Plains flora, but which are not plants of the grasslands.

The maps are sequentially numbered, and each map is identified with the proper Latin name and by family. Family names and numbers are provided at the first appearance of a family; succeeding pages of maps referable to a particular family have the family name given only in parentheses at the top of the page. Genera are alphabetical within each family and species are alphabetical within each genus. Families are listed and circumscribed according to the scheme of Cronquist (1968), and the families of flowering plants are sequentially numbered as an aid to reference. A family sequence list precedes the map section of this atlas.

Several families not represented in the map section are present in the List of Infrequent Great Plains Taxa. These families are numbered in the family sequence list; thus there are occasional gaps in the family numbers in the map section.

The List of Infrequent Great Plains Taxa also follows the Cronquist scheme. States are cited according to Postal Service abbreviations; states and counties within each state are listed alphabetically.

Colloquial names are provided in the atlas where appropriate. They were derived from numerous regional source books and from our own experience. We urge the use of generic names as common names for species not supplied with colloquial names.

In spite of the intensive field program of the past two decades, our knowledge of the flora of the Great Plains remains incomplete. Natural vegetation is a dynamic thing, with new species constantly being introduced, and other species passing from the flora. We anticipate that the observer of the Great Plains flora will find species not included in the atlas, and well-known plants will be found in new localities. We welcome specimens to document the occurrence of new species and range extensions. Specimens may be referred to any of the GPFA members listed on the title page, and particularly to the University of Kansas Herbarium, Lawrence, KS 66044, in care of Dr. R. L. McGregor.

Nomenclature and Taxonomic Concepts

While the purpose of this atlas is to document plant distributions, it is *de facto* a checklist for the vascular plants of the region. This may imply that species concepts and nomenclatural ambiguities have been resolved, but this is not the case. Such matters must await the writing of the *Flora of the Great Plains.* The taxonomic concept is generally "conservative" and akin to that used in the *Manual of Vascular Plants* by Gleason and Cronquist (1963), but with considerable leniency in admitting infraspecific taxa.

Acknowledgments

It is a pleasure to acknowledge the labors and commitment of H. A. "Steve" Stephens, naturalist, author, and plant collector formerly associated with the University of Kansas Herbarium. He has worked in every county in the region covered by this atlas, and in the past 10 years he has made nearly 90,000 collections. To him we owe a particular debt of gratitude. We also acknowledge JoAnn Luehring, who has served as the principal editorial assistant in this project and is the secretary of the GPFA, and Karen West, who patiently plotted the map distributions. Finally, we members of the GPFA wish to thank the administrations of our home institutions for supporting this project. Our special acknowledgments go to the following agencies: The State Biological Survey of Kansas, the Kansas and North Dakota Agricultural Experiment Stations, the University of Nebraska Research Council, the Nebraska State Museum, the Chadron State College Research Council, and the Kearney State College Research Services Council.

T. M. Barkley
Editor, GPFA

August, 1976

References

Carter, Jack. 1960. "Flora of Northwestern Iowa." Dissertation, University of Iowa, Iowa City.

Cronquist, Arthur. 1968. *The Evolution and Classification of Flowering Plants.* Houghton Mifflin, Boston.

Fay, Marcus J. 1953. "Flora of Southwestern Iowa." Dissertation, University of Iowa, Iowa City.

Gleason, Henry A. and Arthur Cronquist. 1963. *Manual of Vascular Plants of Northwestern United States and Adjacent Canada.* Van Nostrand and Co., Princeton.

Rydberg, Per Axel. 1932. *Flora of the Prairies and Plains of Central North America.* The New York Botanical Garden, New York.

Steyermark, Julian A. 1963. *Flora of Missouri.* The Iowa State University Press, Ames.

List of Families in Sequence 

[Colloquial names are in parentheses. Families marked with an asterisk occur only in the list of infrequent taxa]

VASCULAR CRYPTOGAMS
*Lycopodiaceae (Clubmoss)
Selaginellaceae (Selaginella)
Isoetaceae (Quillwort)
Equisetaceae (Horsetail)
Ophioglossaceae (Adder's-tongue)
*Osmundaceae (Osmunda)
Polypodiaceae (Fern)
Marsileaceae (Marsilea)
Salviniaceae (Salvinia)

Division PINOPHYTA (Conifers)
Pinaceae (Pine)
Cupressaceae (Cypress)
Ephedraceae (Ephedra)

Division MAGNOLIOPHYTA
Class Magnoliopsida (Dicots)

1. Annonaceae (Custard-apple)
*2. Lauraceae (Laurel)
*3. Saururaceae (Lizard's-tail)
4. Aristolochiaceae (Birthwort)
5. Nymphaeaceae (Water Lily)
5a. Nelumbonaceae
6. Ceratophyllaceae (Hornwort)
7. Ranunculaceae (Buttercup)
8. Berberidaceae (Barberry)
9. Menispermaceae (Moonseed)
10. Papaveraceae (Poppy)
11. Fumariaceae (Dutchman's Breeches)
12. Platanaceae (Sycamore)
13. Ulmaceae (Elm)
14. Moraceae (Mulberry)
15. Cannabaceae (Hemp)
16. Urticaceae (Nettle)
17. Juglandaceae (Walnut)
18. Fagaceae (Beech)
19. Betulaceae (Birch)
20. Phytolaccaceae (Pokeweed)
21. Nyctaginaceae (Four-O'Clock)
22. Cactaceae (Cactus)
23. Aizoaceae (Carpetweed)
24. Caryophyllaceae (Pink)
25. Portulaceae (Purslane)
26. Chenopodiaceae (Goosefoot)
27. Amaranthaceae (Pigweed)
28. Polygonaceae (Buckwheat)
*29. Plumbaginaceae (Leadwort)
30. Elatinaceae (Waterwort)
31. Hypericaceae (St. John's-wort)
32. Tiliaceae (Linden)
33. Malvaceae (Mallow)
*34. Droseraceae (Sundew)
35. Violaceae (Violet)
36. Passifloraceae (Passion-flower)
37. Cistaceae (Rockrose)
38. Tamaricaceae (Tamarisk)
39. Loasaceae (Sand Lily)
40. Cucurbitaceae (Gourd)
41. Salicaceae (Willow)
42. Capparaceae (Caper)
43. Brassicaceae (Mustard)
*44. Resedaceae (Reseda)
45. Ericaceae (Heath)
46. Pyrolaceae (Wintergreen)
47. Monotropaceae (Indian Pipe)
48. Sapotaceae (Sapodilla)
49. Ebenaceae (Ebony)
50. Primulaceae (Primrose)
51. Crassulaceae (Stonecrop)
52. Saxifragaceae (Saxifrage)

53. Rosaceae (Rose)
54a. Fabaceae-Mimosoideae (Bean)
54b. Fabaceae-Caesalpinoideae (Bean)
54c. Fabaceae-Papilionoideae (Bean)
55. Haloragaceae (Water Milfoil)
56. Hippuridaceae (Mare's-tail)
57. Lythraceae (Loosestrife)
*58. Thymelaeaceae (Leatherwood)
59. Onagraceae (Evening Primrose)
*60. Melastomataceae (Melastome)
61. Elaeagnaceae (Russian Olive)
62. Cornaceae (Dogwood)
*62a. Garryaceae (Garrya)
63. Santalaceae (Sandalwood)
64. Loranthaceae (Mistletoe)
65. Rafflesiaceae
66. Celastraceae (Staff-tree)
67. Aquifoliaceae (Holly)
68. Euphorbiaceae (Spurge)
69. Rhamnaceae (Buckthorn)
70. Vitaceae (Grape)
71. Staphyleaceae (Bladdernut)
72. Sapindaceae (Soapberry)
73. Hippocastanaceae (Buckeye)
74. Aceraceae (Maple)
75. Anacardiaceae (Cashew)
76. Simaroubaceae (Quassia)
77. Rutaceae (Rue)
78. Zygophyllaceae (Caltrop)
79. Oxalidaceae (Wood Sorrel)
80. Geraniaceae (Geranium)
81. Balsaminaceae (Touch-me-not)
82. Linaceae (Flax)
83. Polygalaceae (Milkwort)
84. Araliaceae (Ginseng)
85. Apiaceae (Parsley)
86. Gentianaceae (Gentian)
87. Apocynaceae (Dogbane)
88. Asclepiadaceae (Milkweed)
89. Solanaceae (Nightshade)
90. Convolvulaceae (Convolvulus)
91. Cuscutaceae (Dodder)
*92. Menyanthaceae (Buckbean)
93. Polemoniaceae (Polemonium)
94. Hydrophyllaceae (Waterleaf)
95. Boraginaceae (Borage)
96. Callitrichaceae (Water Starwort)
97. Verbenaceae (Vervain)
98. Phrymaceae (Lopseed)
99. Lamiaceae (Mint)
100. Plantaginaceae (Plantain)
101. Oleaceae (Olive)
102. Scrophulariaceae (Figwort)
103. Orobanchaceae (Broomrape)
104. Bignoniaceae (Bignonia)
105. Acanthaceae (Acanthus)
106. Pedaliaceae
107. Lentibulariaceae (Bladderwort)
108. Campanulaceae (Bellflower)
109. Rubiaceae (Madder)
110. Caprifoliaceae (Honeysuckle)
*111. Adoxaceae (Adoxa)
112. Valerianaceae (Valerian)
113. Dipsacaceae (Teasel)
114. Asteraceae (Sunflower)

Division MAGNOLIOPHYTA
Class Liliopsida (Monocots)

*115. Butomaceae (Butomus)
116. Alismataceae (Water Plantain)
117. Hydrocharitaceae (Frog's-bit)
*118. Scheuchzeriaceae
119. Juncaginaceae (Arrowgrass)
120. Najadaceae (Naiad)
121. Potamogetonaceae (Pondweed)
122. Ruppiaceae (Ditchgrass)
123. Zannichelliaceae (Horned Pondweed)
124. Commelinaceae (Spiderwort)
125. Juncaceae (Rush)
126. Cyperaceae (Sedge)
127. Poaceae (Grass)
128. Sparganiaceae (Bur-reed)
129. Typhaceae (Cattail)
130. Araceae (Arum)
131. Lemnaceae (Duckweed)
132. Pontederiaceae (Pickerelweed)
133. Liliaceae (Lily)
134. Iridaceae (Iris)
135. Dioscoreaceae (Yam)
136. Orchidaceae (Orchid)

Distribution Map Showing County Names 

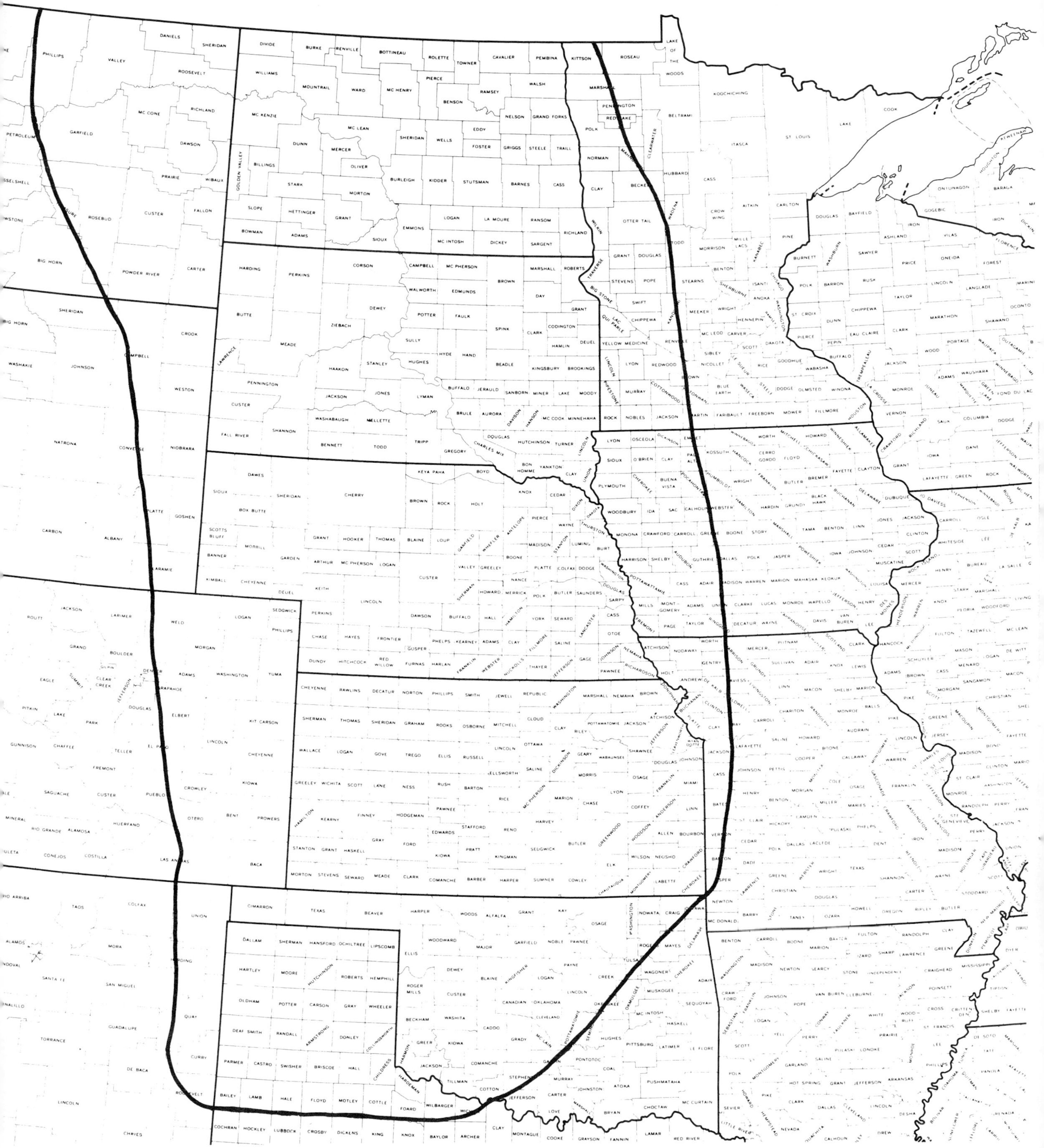

SECTION ONE

Distribution Maps

Selaginellaceae (Selaginella Family)

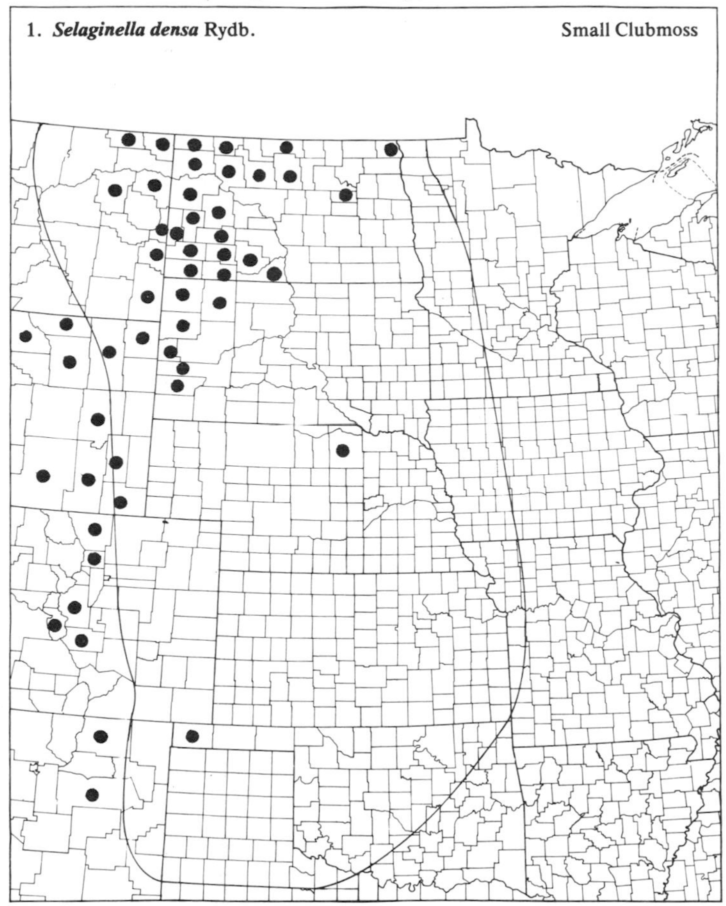

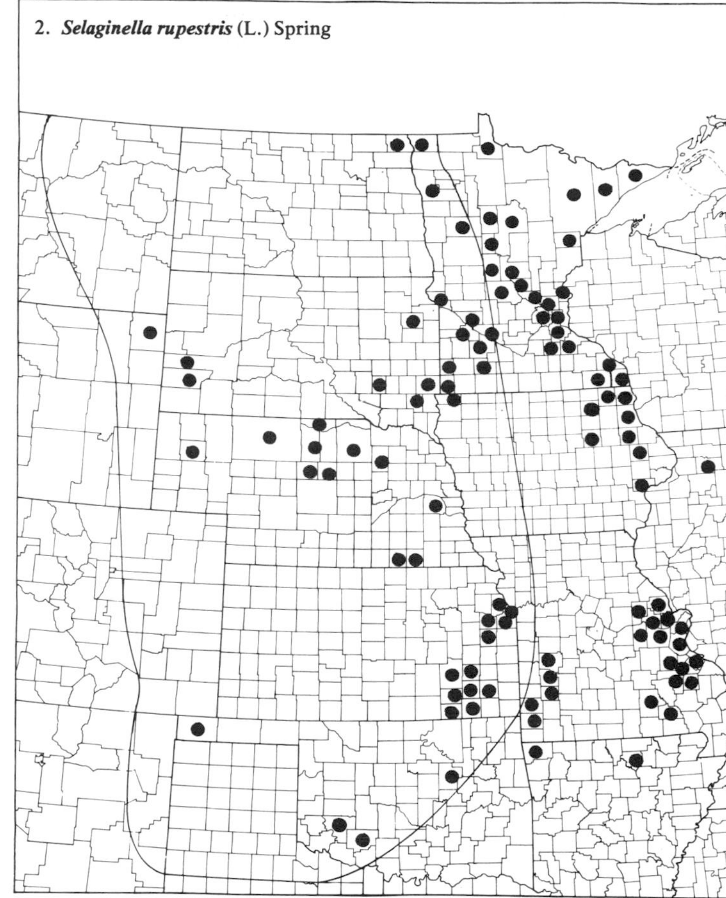

Isoetaceae (Quillwort Family)

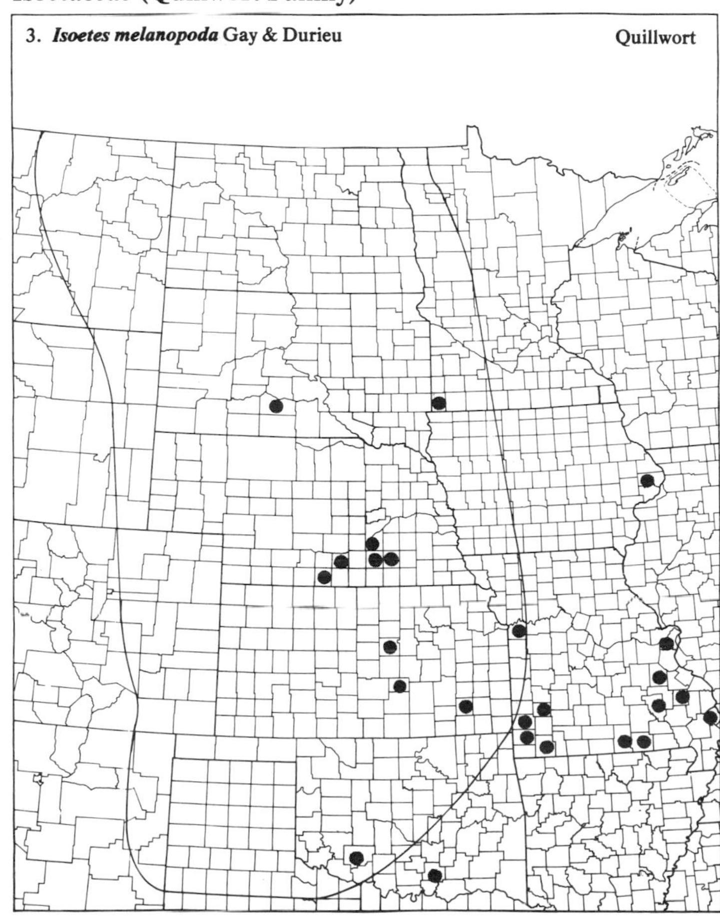

Equisetaceae (Horsetail Family)

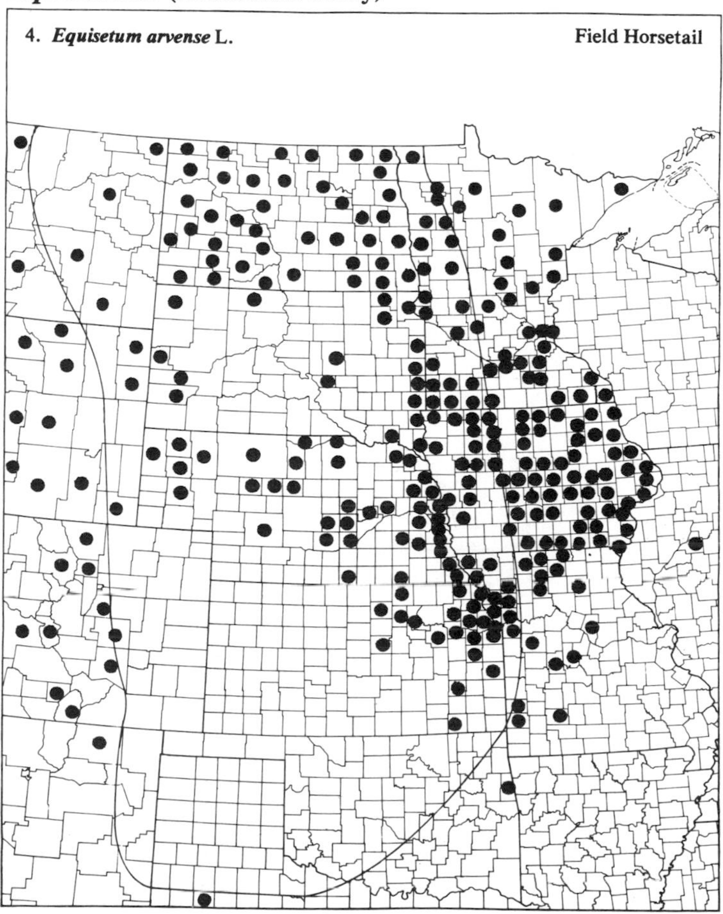

(Equisetaceae)

5. ***Equisetum X. ferrissii*** Clute

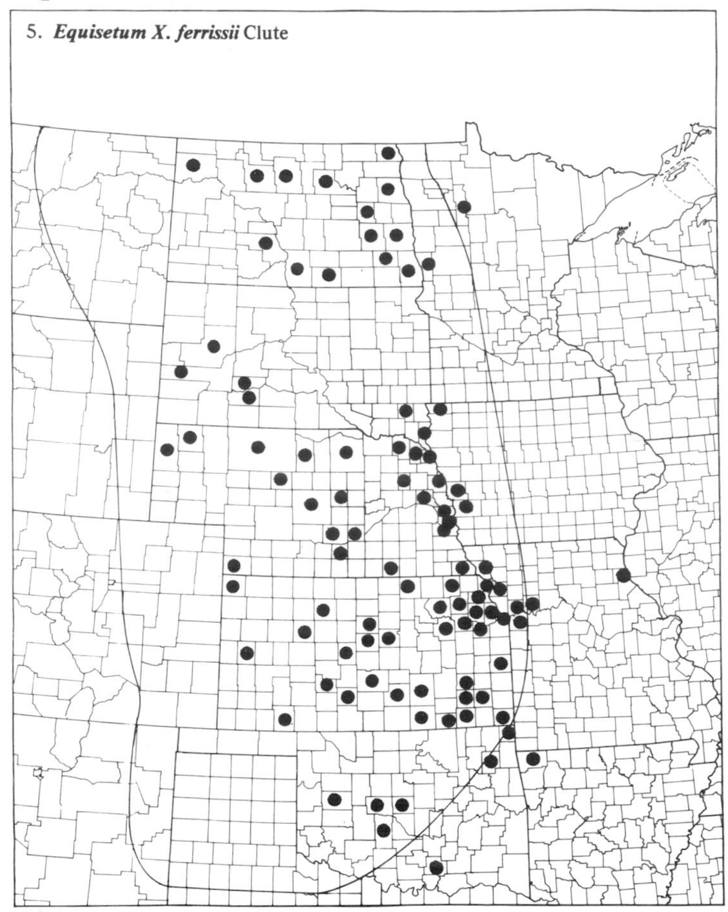

6. ***Equisetum fluviatile*** L. Water Horsetail

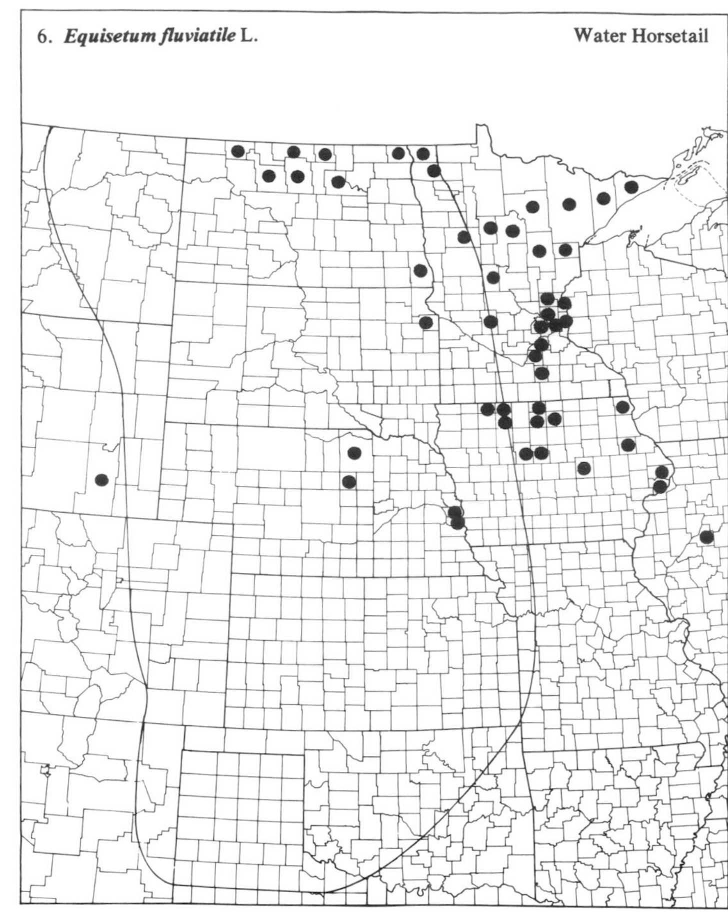

7. ***Equisetum hyemale*** L. Scouring Rush

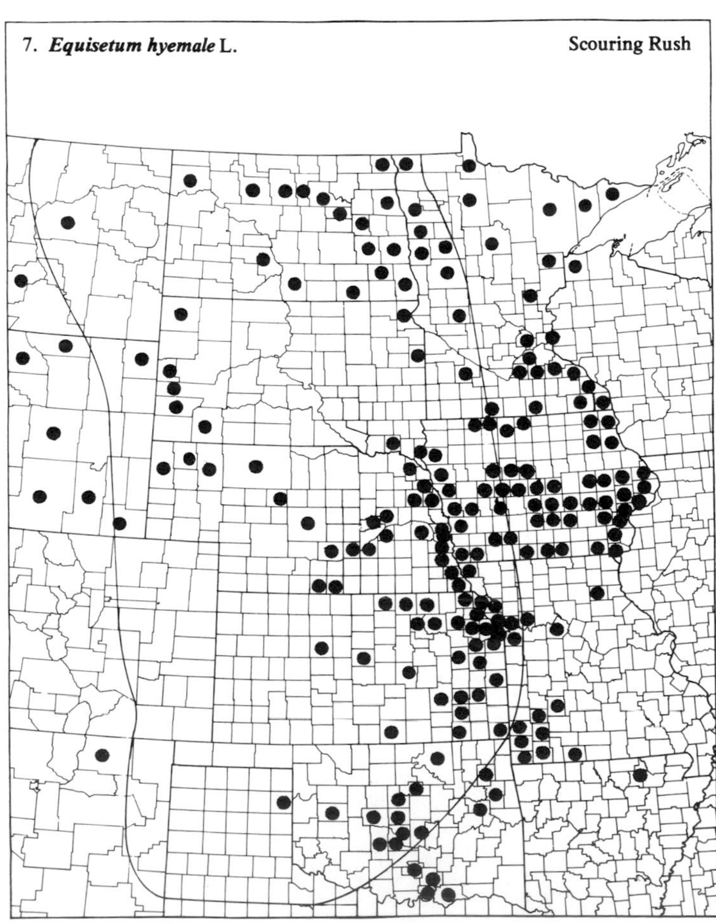

8. ***Equisetum laevigatum*** A. Br. Smooth Horsetail

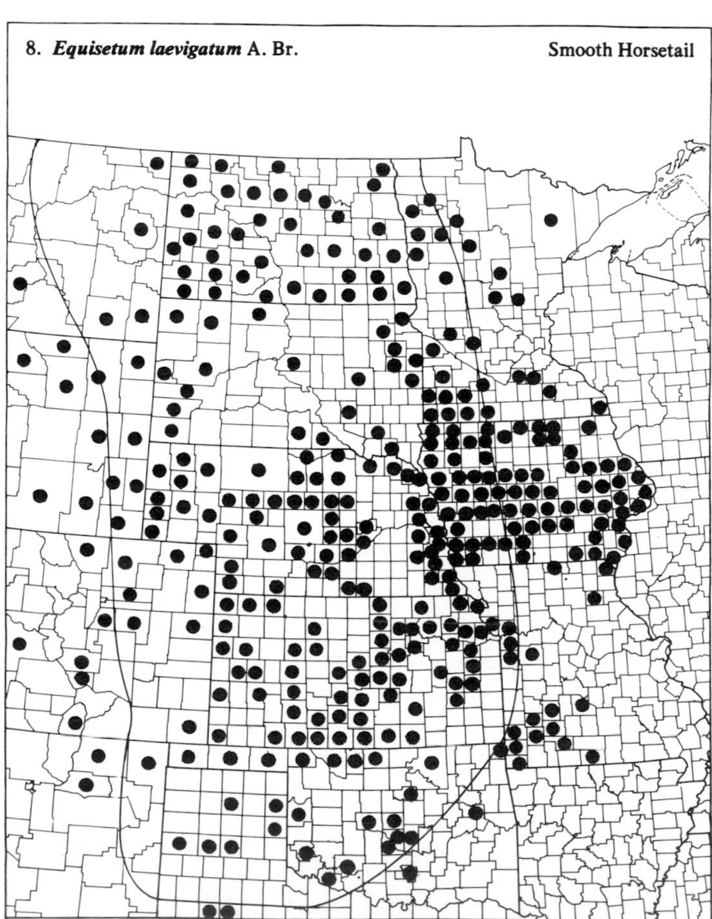

(Equisetaceae)

9. ***Equisetum pratense*** Ehrh. Meadow Horsetail

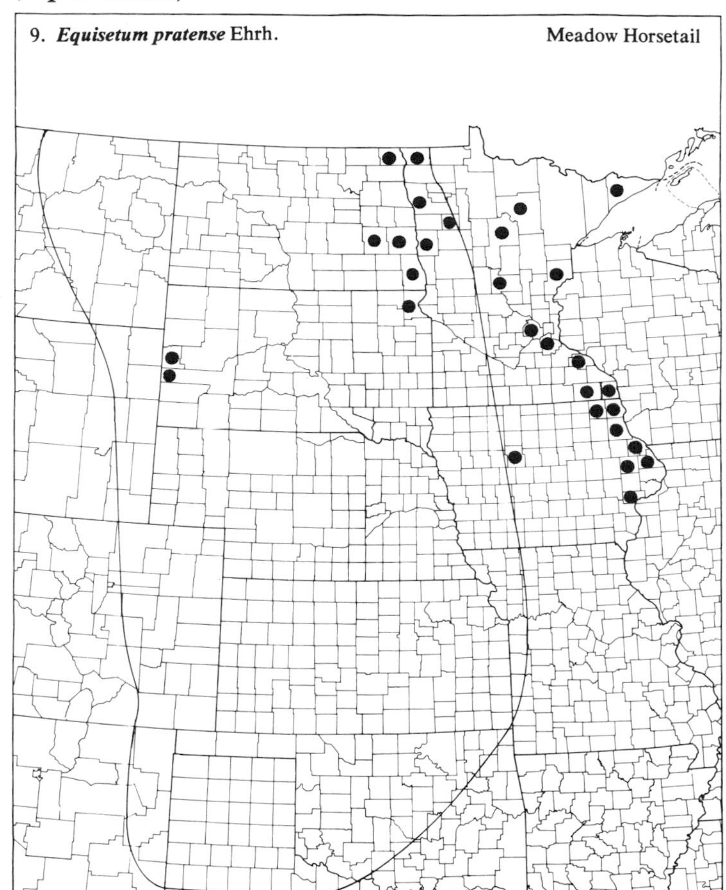

Ophioglossaceae (Adder's-tongue Family)

10. ***Botrychium dissectum*** Spreng. var. ***obliquum*** (Muhl.) Clute

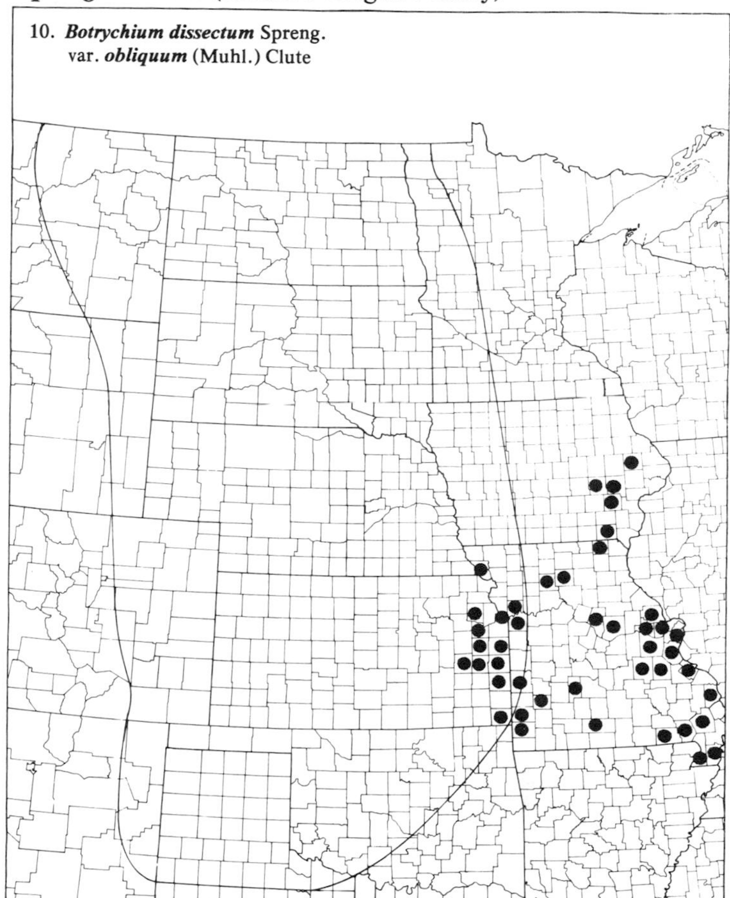

11. ***Botrychium virginianum*** (L.) Sw. Rattlesnake Fern

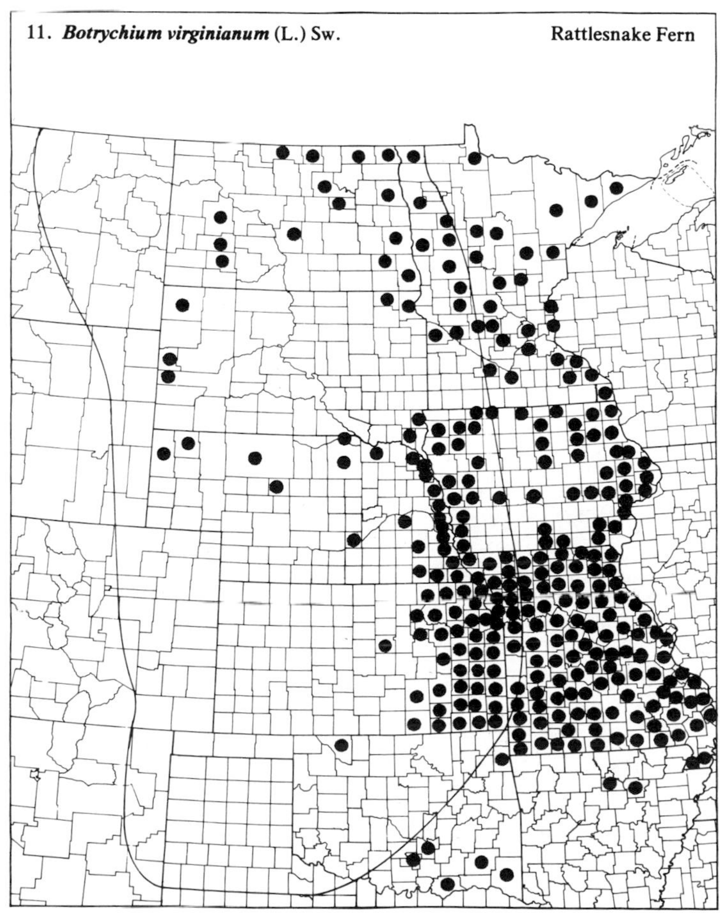

12. ***Ophioglossum engelmannii*** Prantl Adder's-tongue

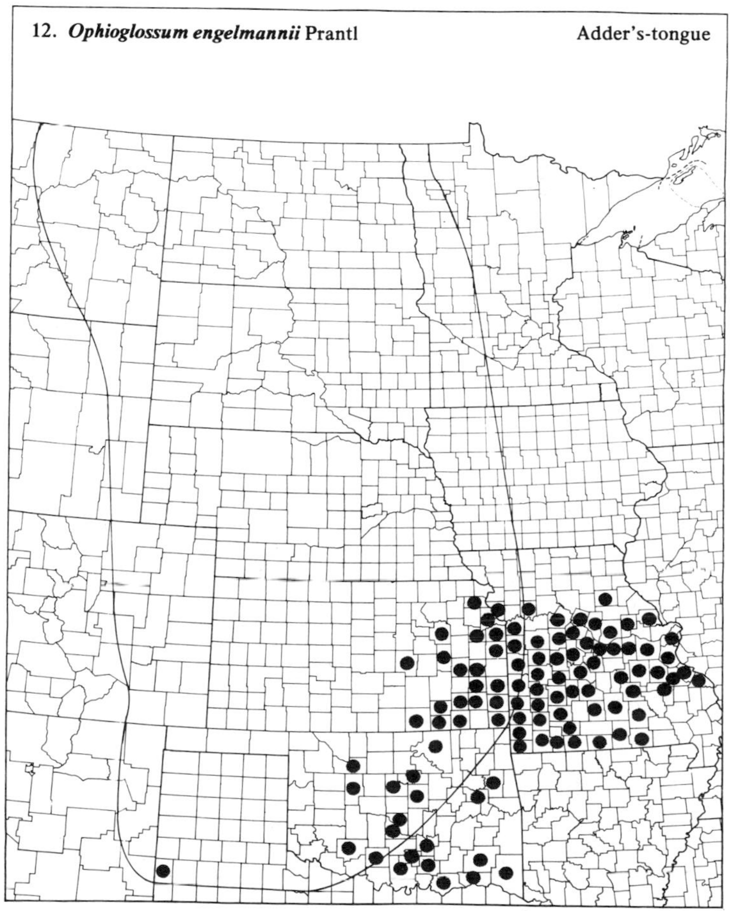

Polypodiaceae (Fern Family)

13. ***Adiantum capillus-veneris*** L. — Venus'-hair Fern

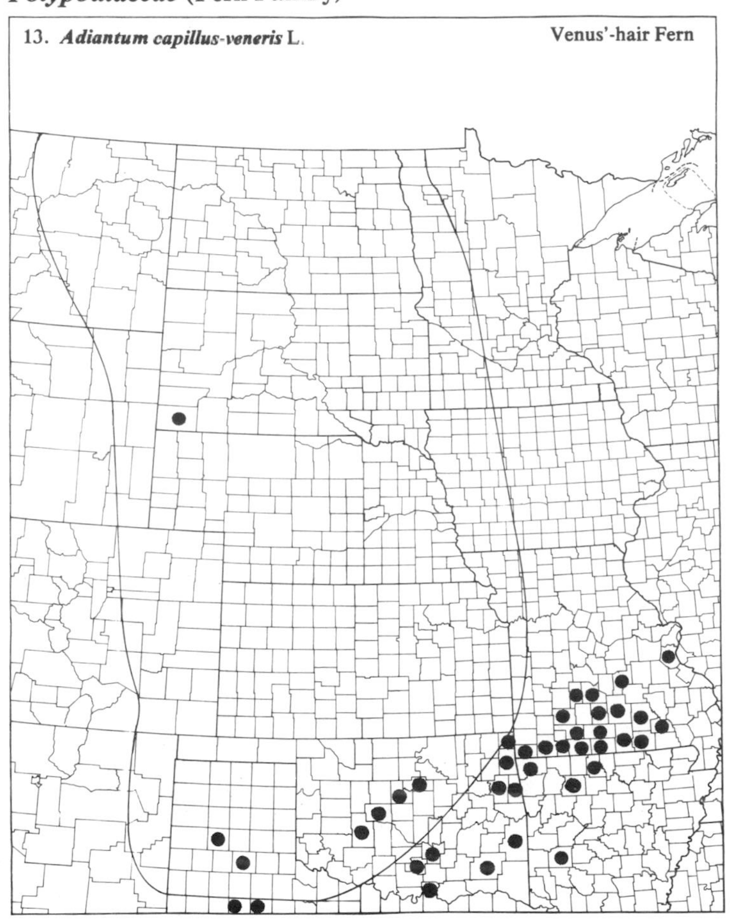

14. ***Adiantum pedatum*** L. — Maidenhair Fern

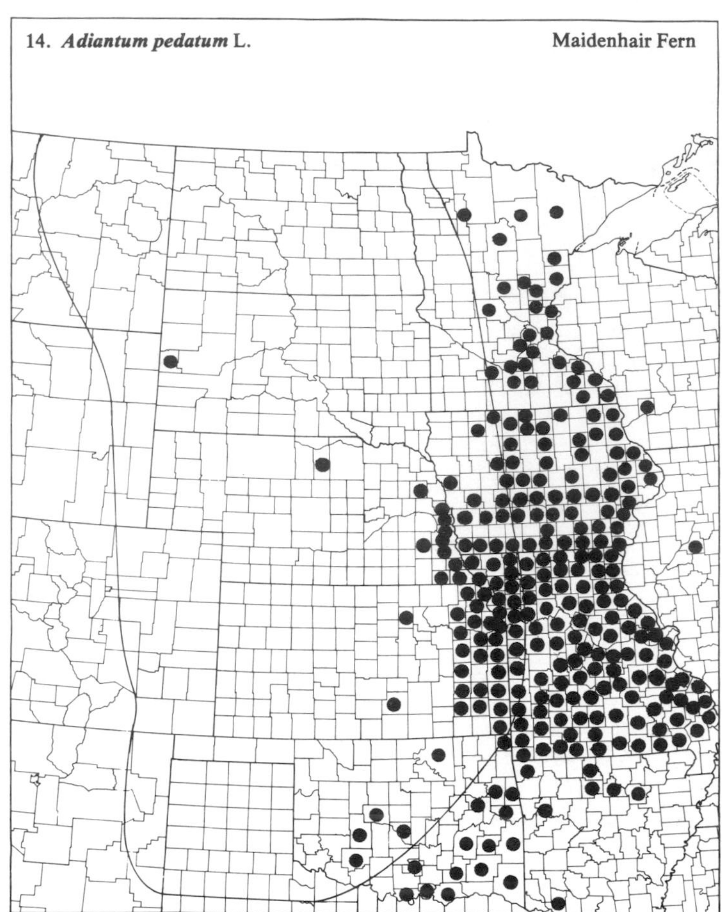

15. ***Asplenium platyneuron*** (L.) Oakes ex D.C. Eat. — Ebony Spleenwort

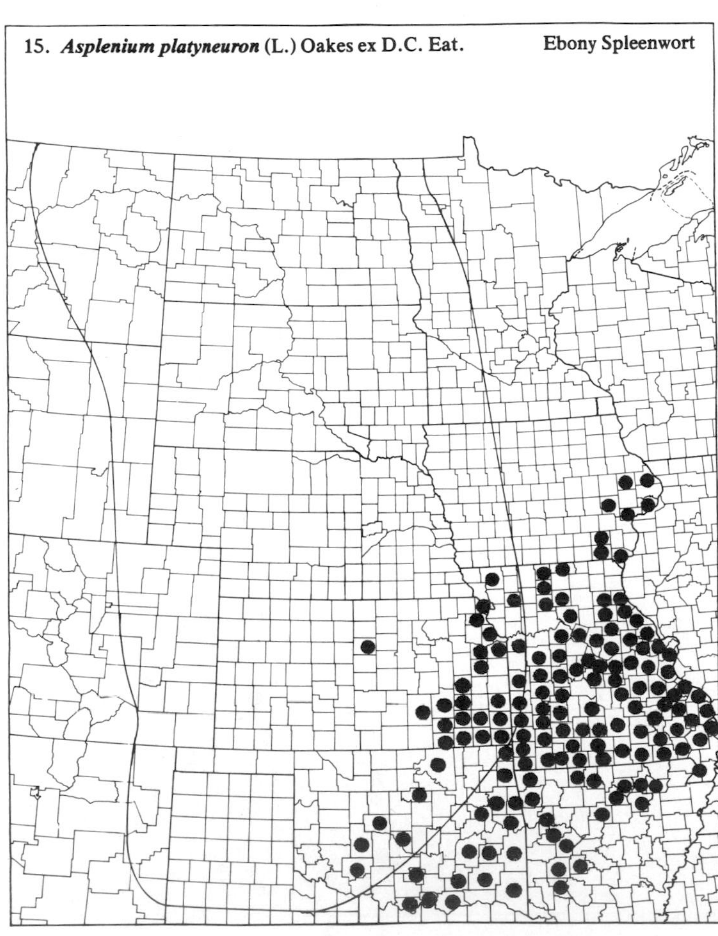

16. ***Asplenium resiliens*** Kunze — Black-stem Spleenwort

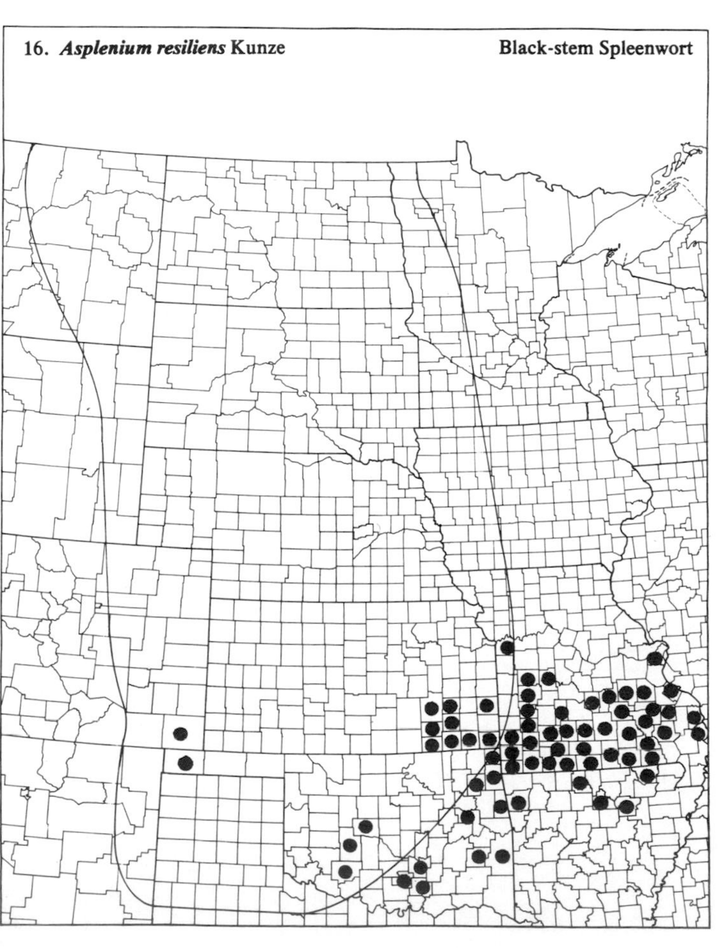

(Polypodiaceae)

17. ***Asplenium trichomanes*** L. — Maidenhair Spleenwort

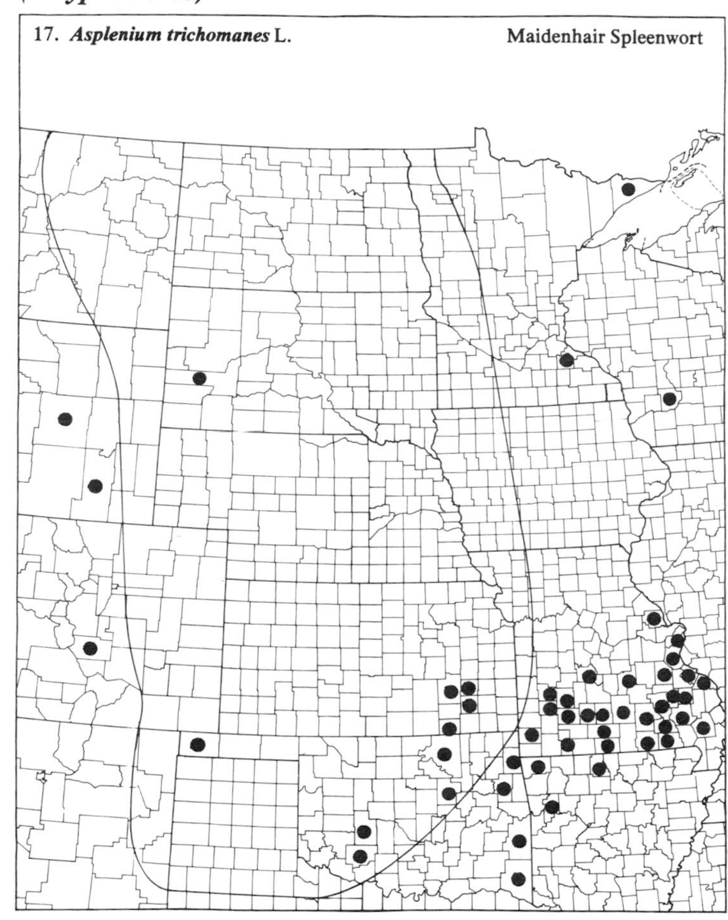

18. ***Athyrium filix-femina*** (L.) Roth — Lady Fern

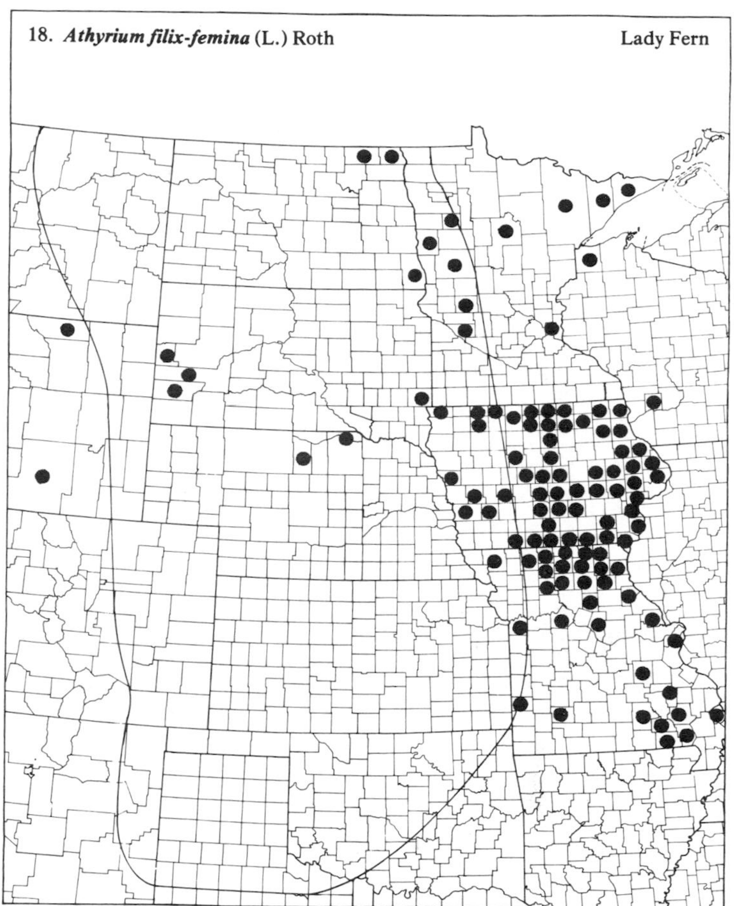

19. ***Camptosorus rhizophyllus*** (L.) Link — Walking Fern

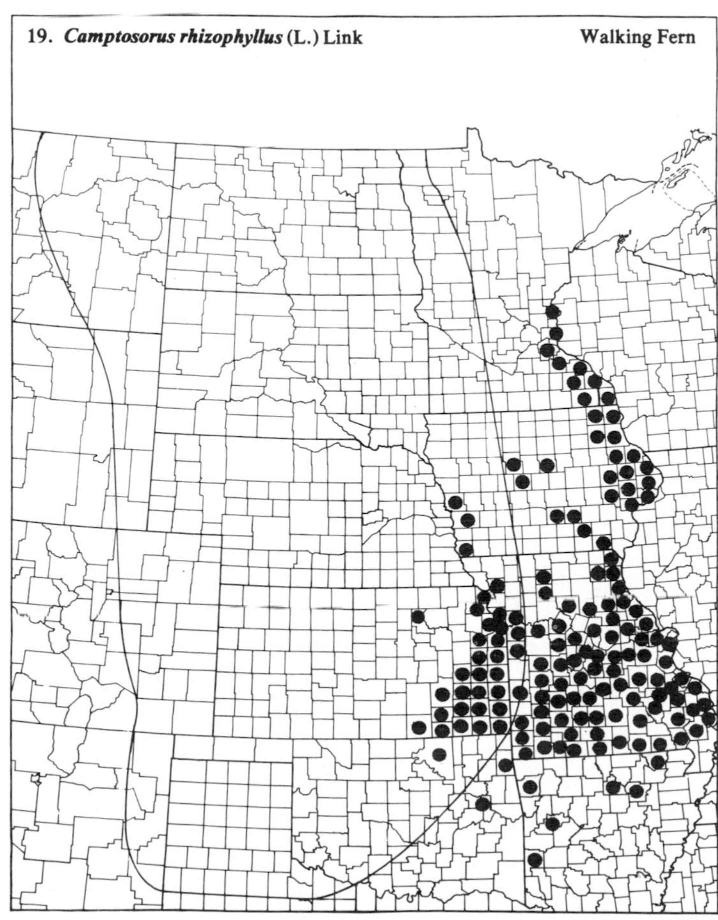

20. ***Cheilanthes eatonii*** Baker in Hook. & Baker

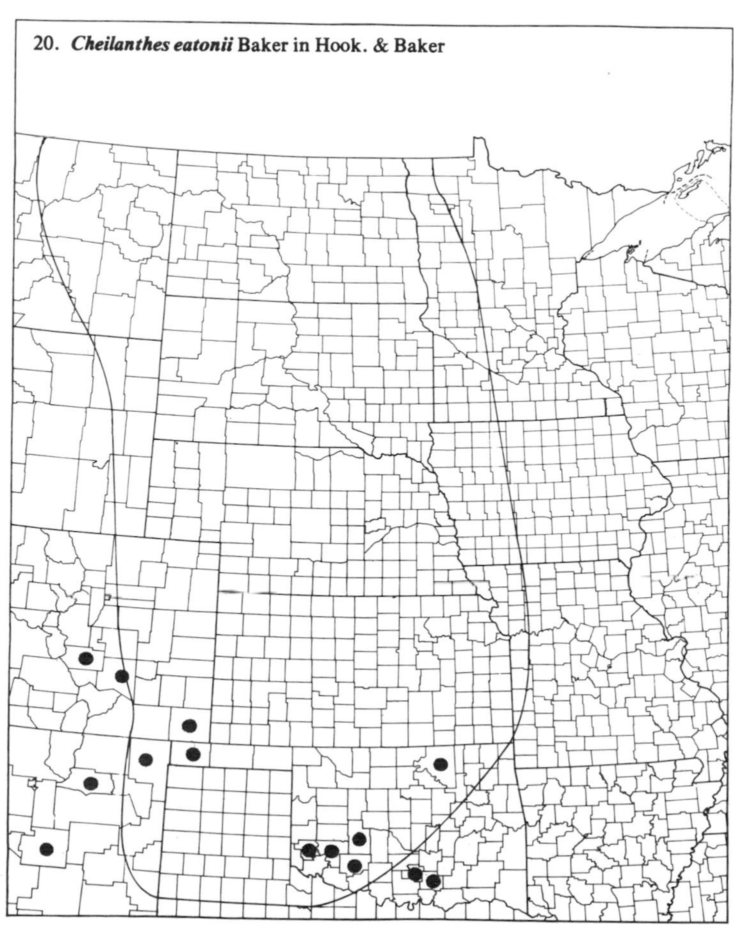

(Polypodiaceae)

21. ***Cheilanthes feei*** Moore — Lip Fern

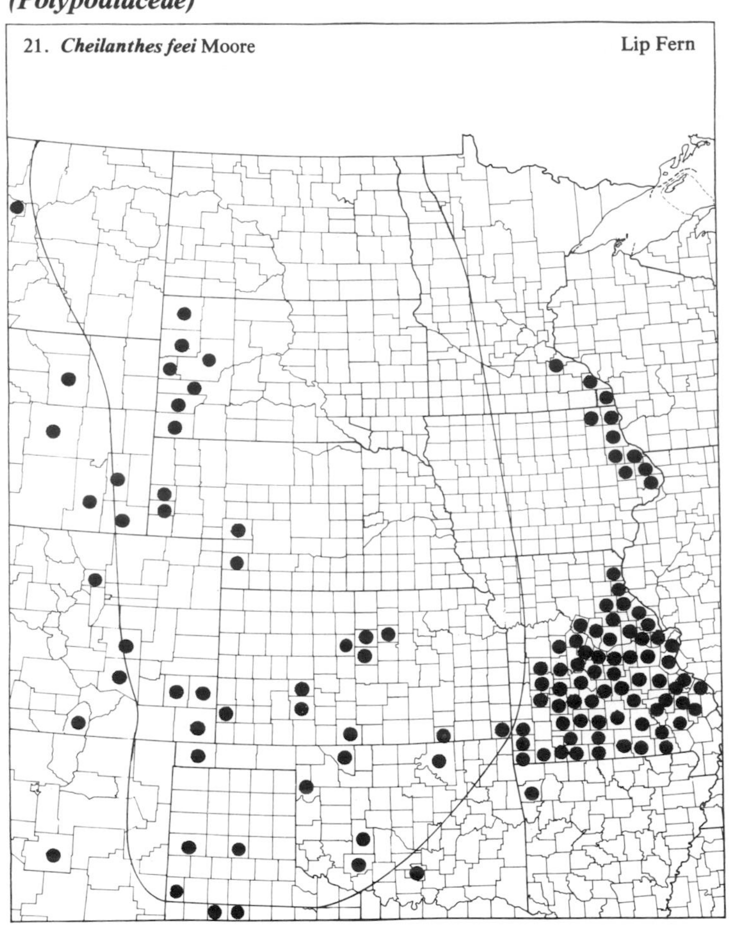

22. ***Cheilanthes lanosa*** D.C. Eat. in Torr. — Woolly Lip Fern

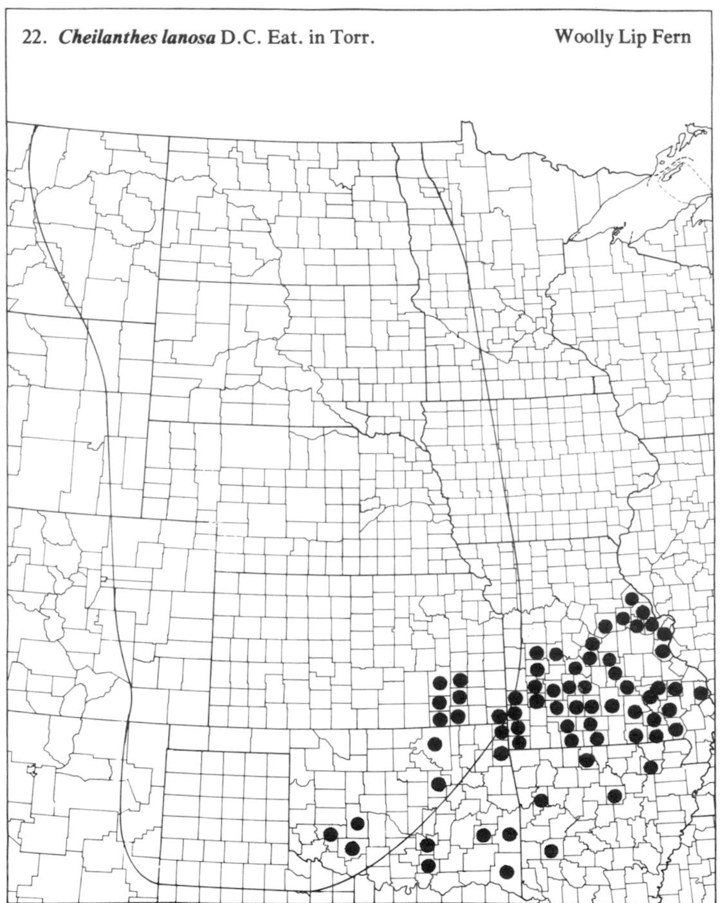

23. ***Cystopteris fragilis*** (L.) Bernh. — Common Bladder Fern

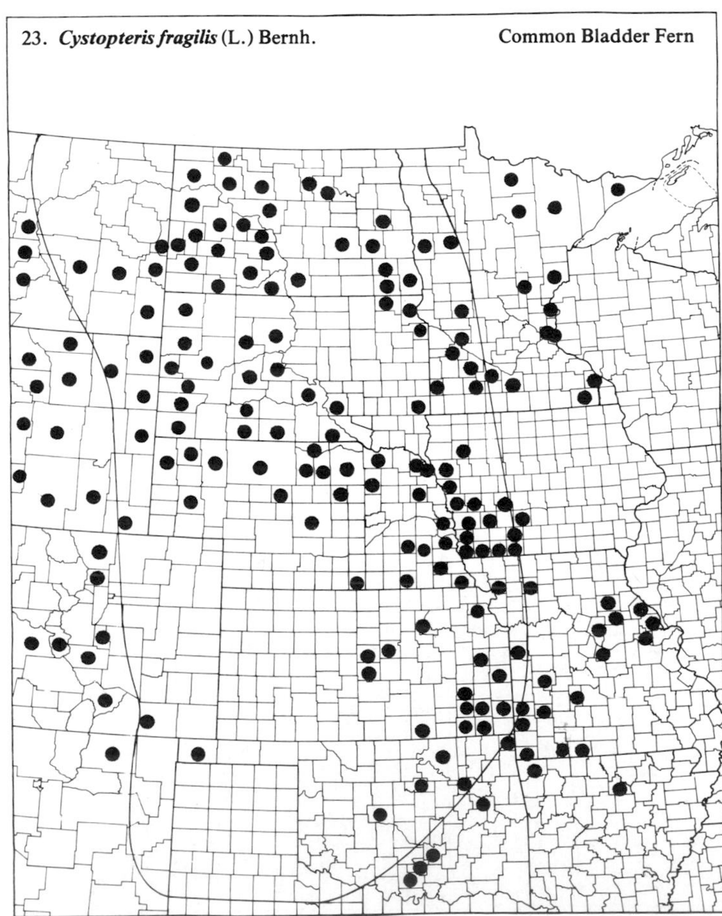

24. ***Cystopteris protrusa*** (Weath.) Blasd.

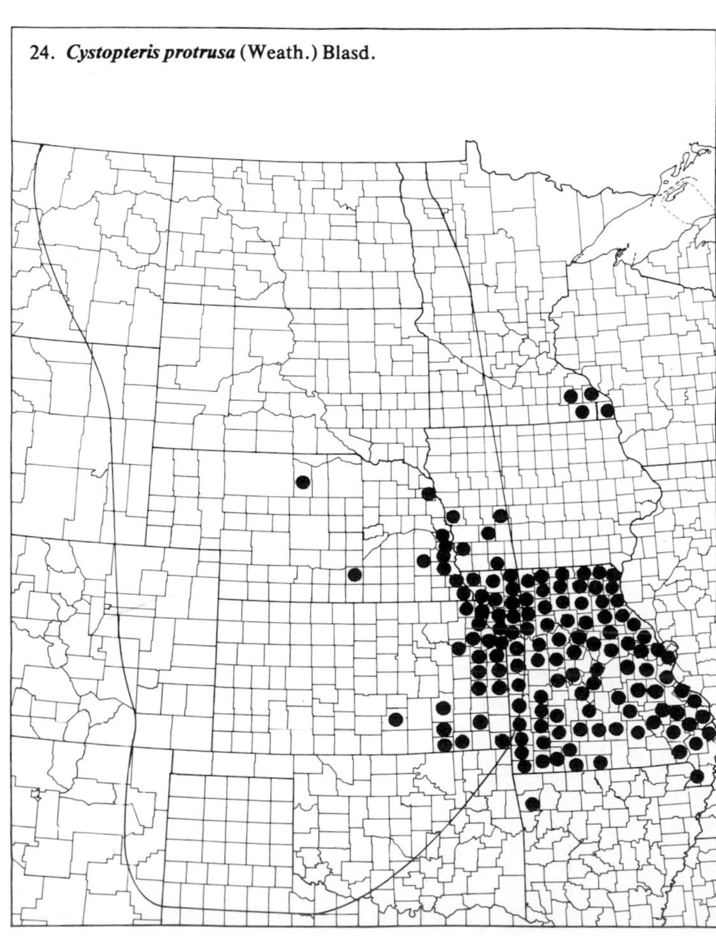

(Polypodiaceae)

25. ***Cystopteris tennesseensis*** Shaver

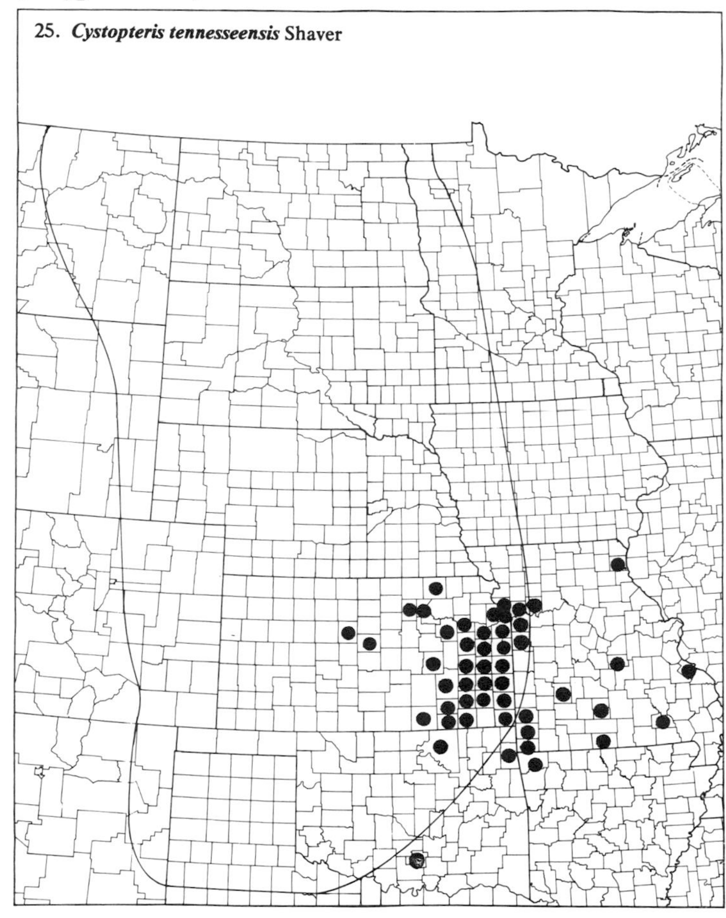

26. ***Dryopteris carthusiana*** (Vill.) H.P. Fuchs — Wood Fern

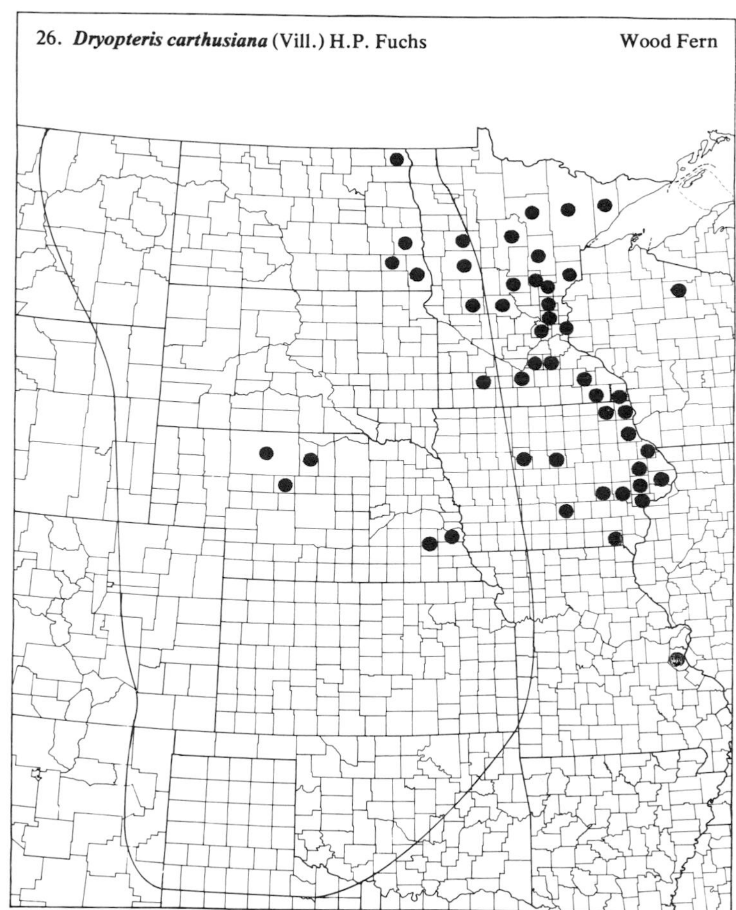

27. ***Dryopteris cristata*** (L.) Gray — Crested Wood Fern

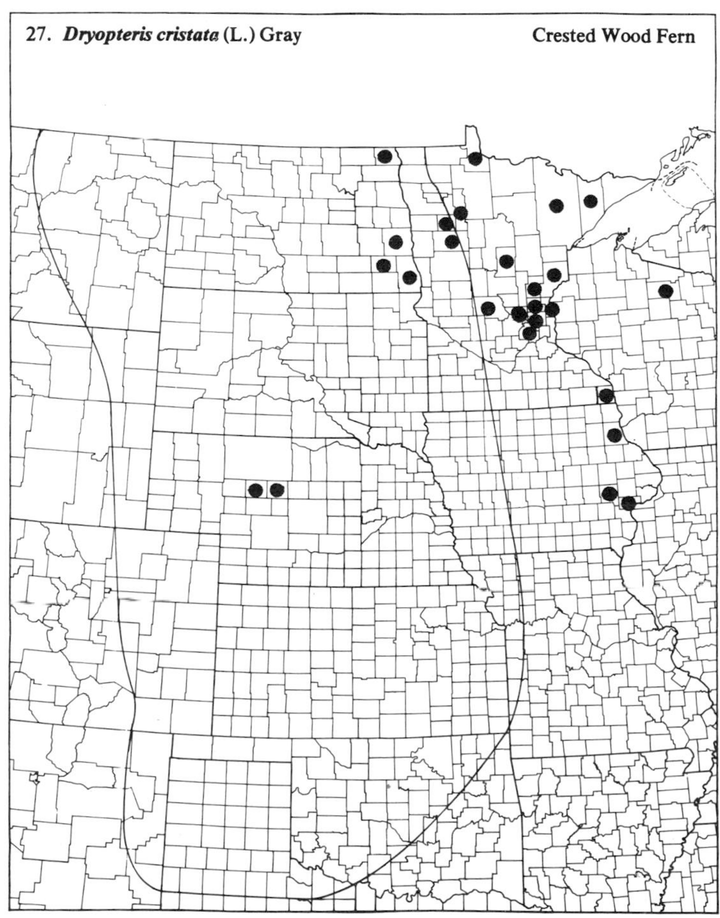

28. ***Dryopteris filix-mas*** (L.) Schott — Shield Fern

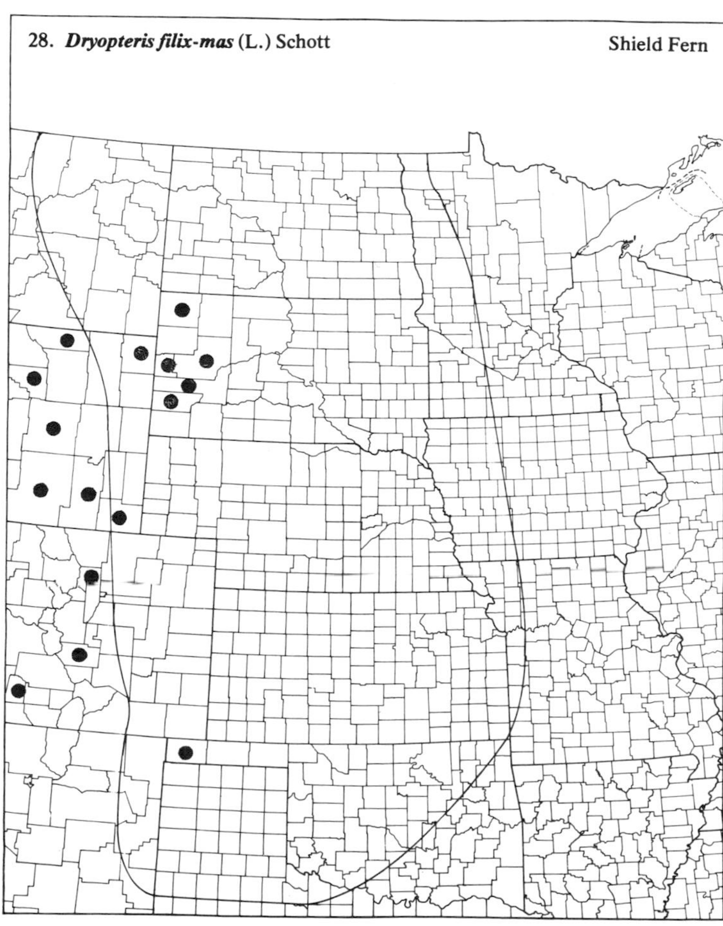

(Polypodiaceae)

29. ***Dryopteris marginalis*** (L.) Gray — Evergreen Wood Fern

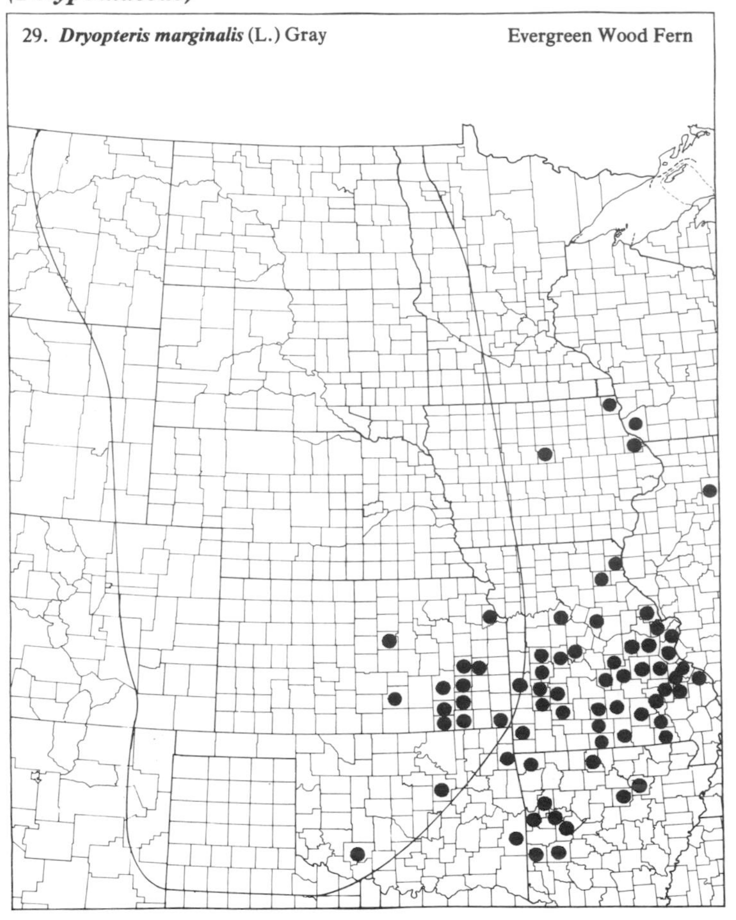

30. ***Matteuccia struthiopteris*** (L.) Todaro — Ostrich Fern

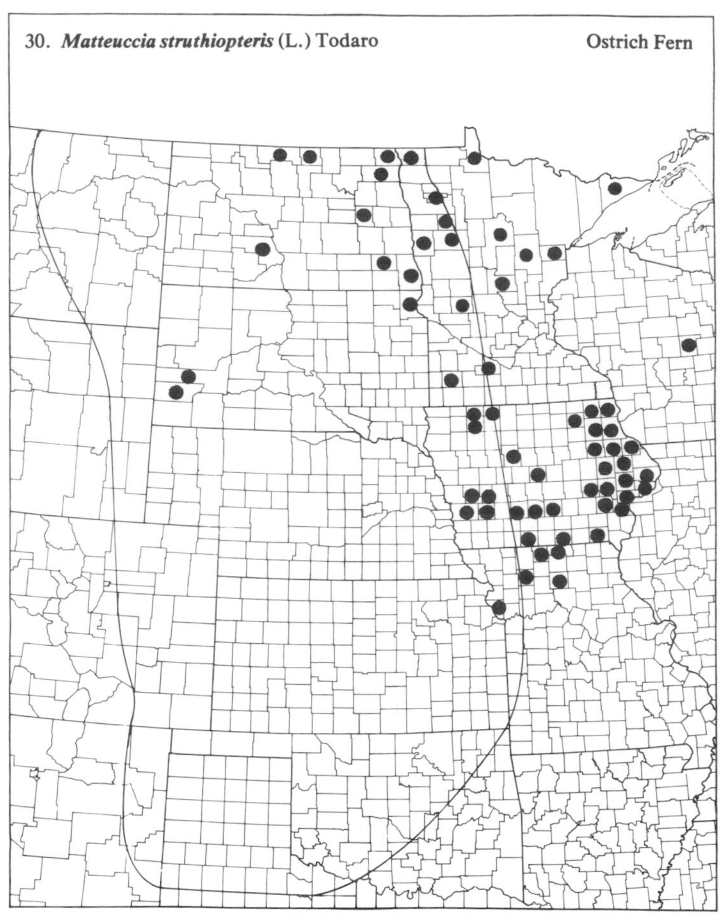

31. ***Notholaena dealbata*** (Pursh) Kunze — Cloak Fern

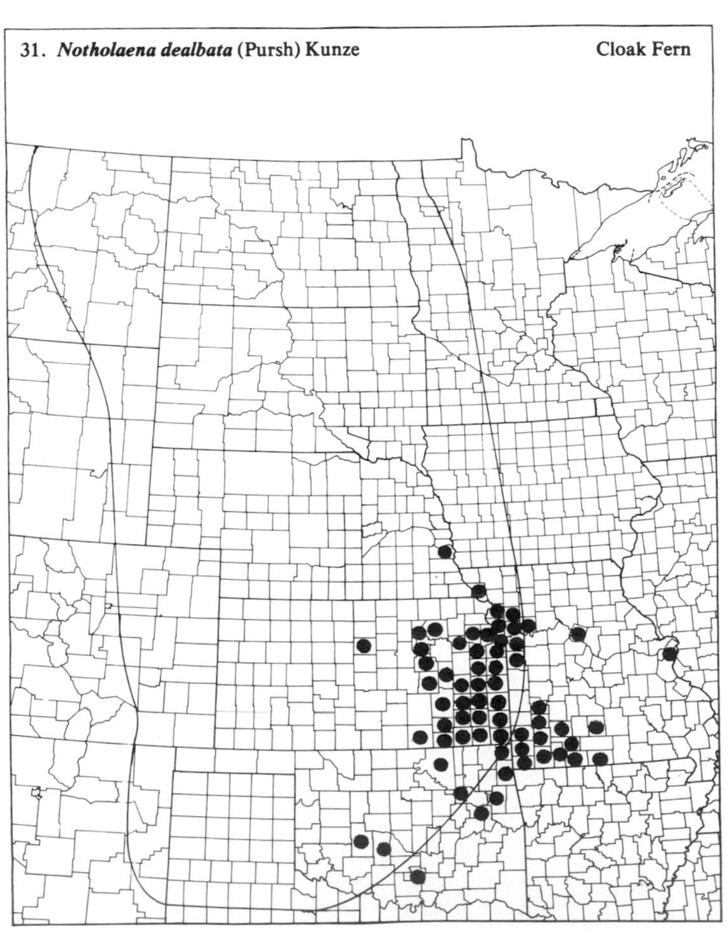

32. ***Onoclea sensibilis*** L. — Sensitive Fern

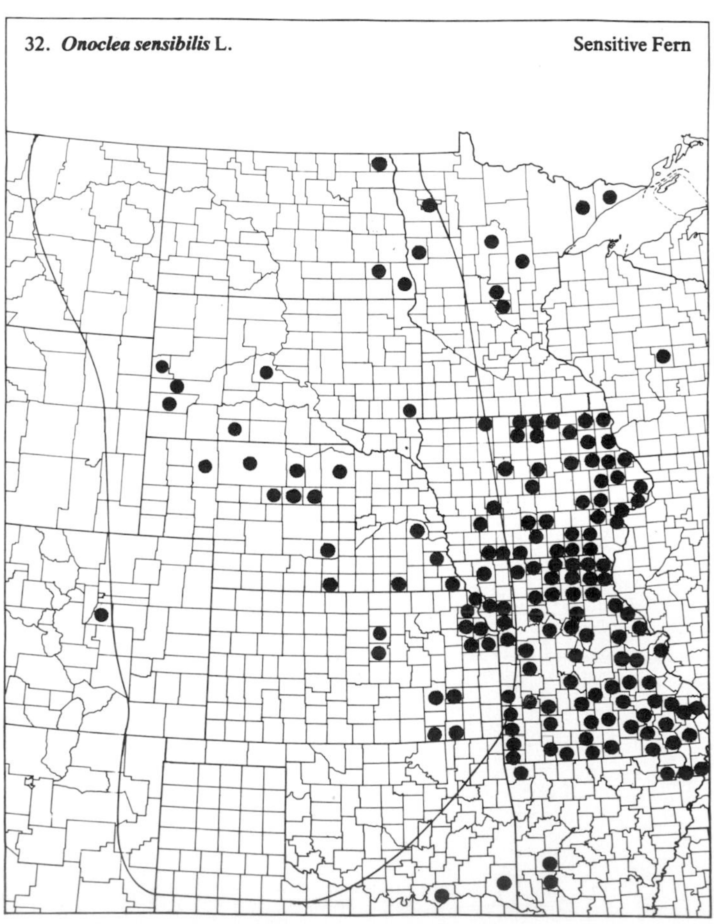

(Polypodiaceae)

33. ***Pellaea atropurpurea*** (L.) Link — Purple Cliff-brake

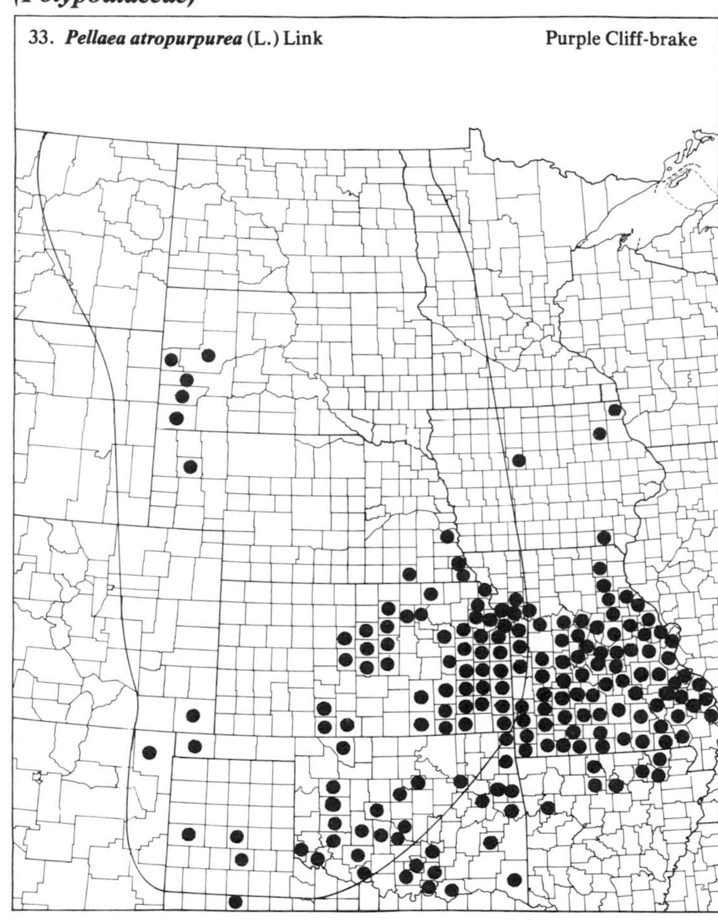

34. ***Pellaea glabella*** Mett. ex Kuhn var. ***glabella*** — Smooth Cliff-brake

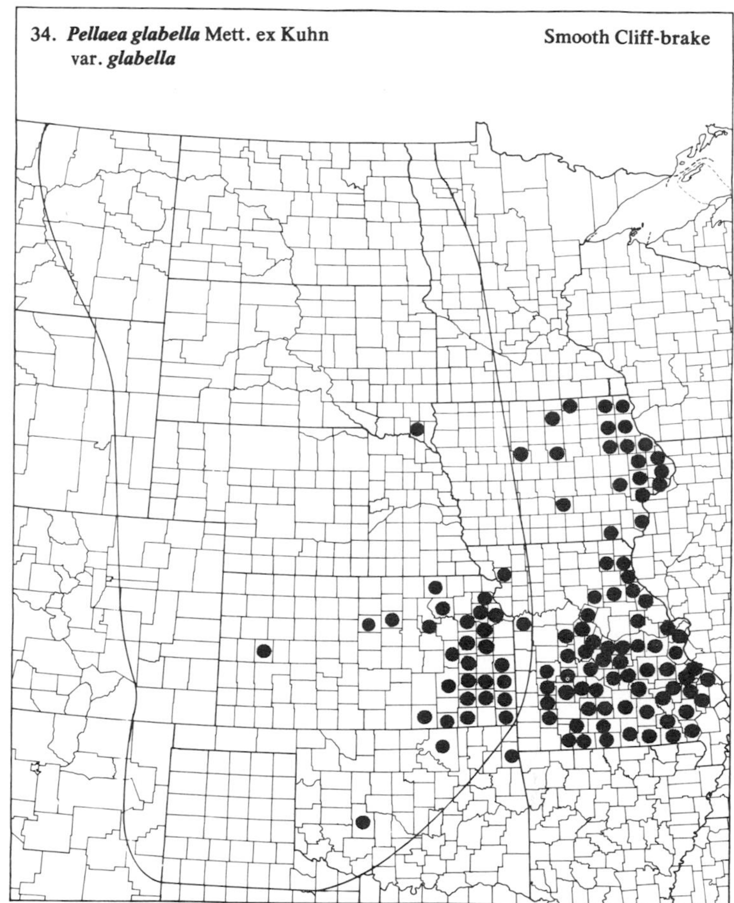

35. ***Pellaea glabella*** Mett. ex Kuhn var. ***occidentalis*** (E.Nels.) Butt. — Smooth Cliff-brake

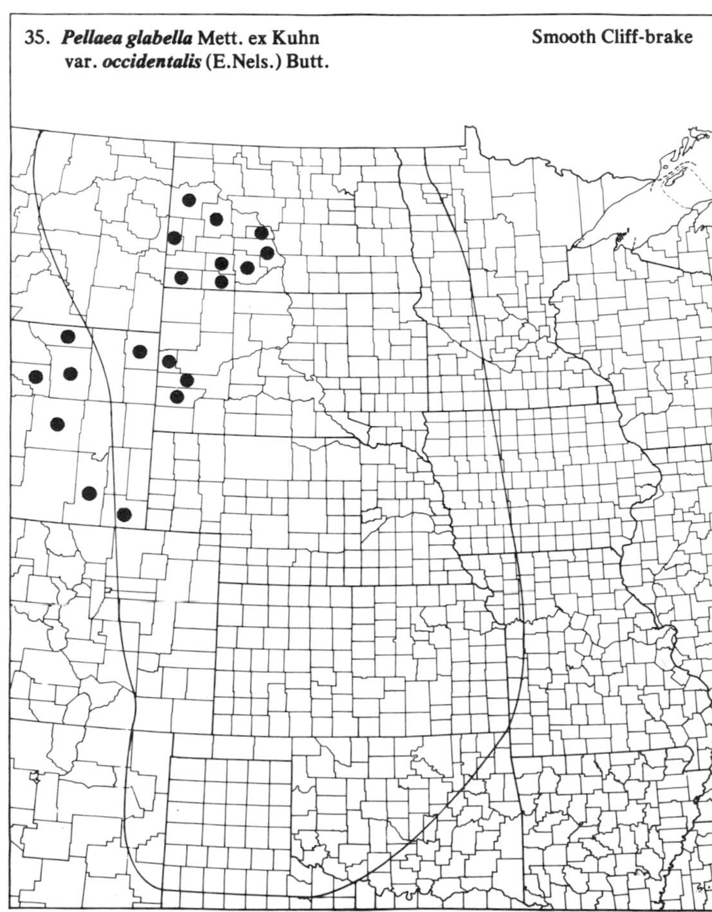

36. ***Polystichum acrostichoides*** (Michx.) Schott — Christmas Fern

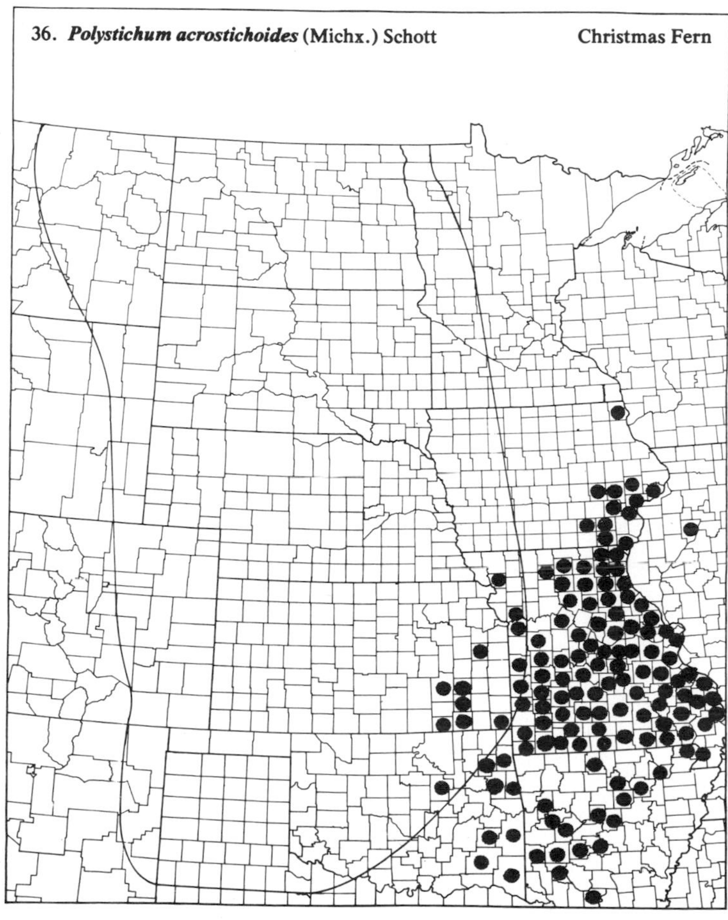

(Polypodiaceae)

37. ***Pteridium aquilinum*** (L.) Kuhn var. ***latiusculum*** (Desv.) Underw. ex Heller — Common Bracken

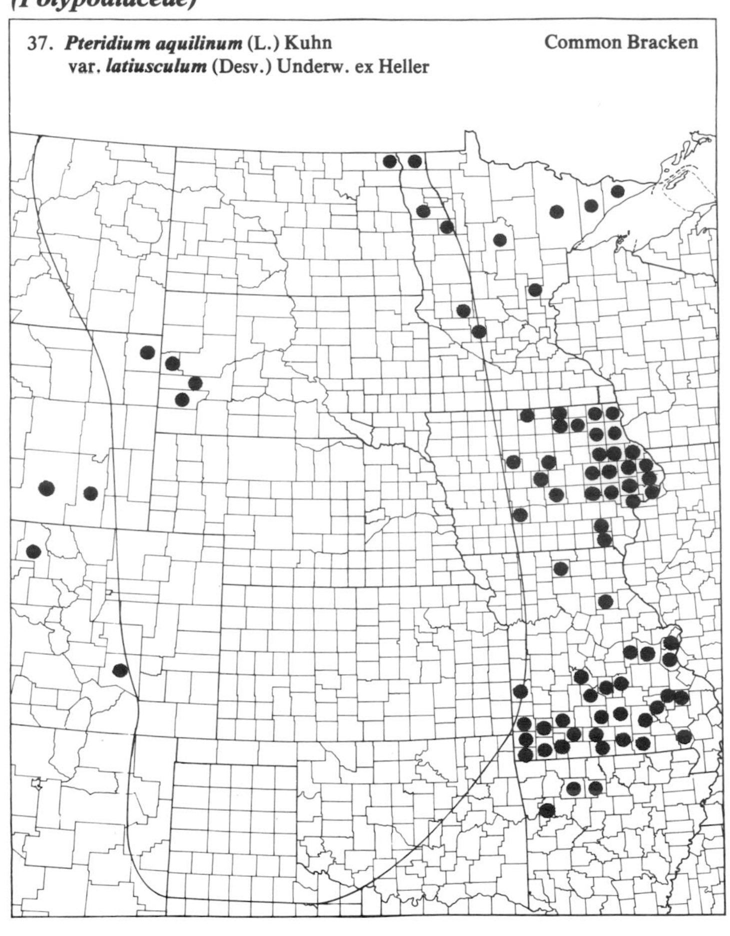

38. ***Thelypteris palustris*** Schott — Marsh Fern

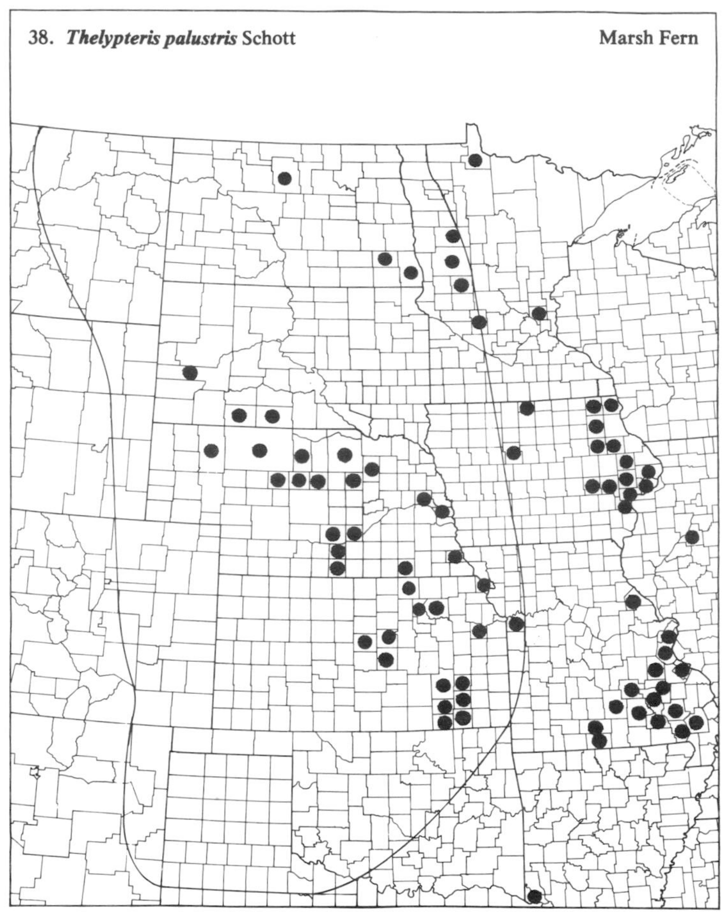

39. ***Woodsia obtusa*** (Spreng.) Torr. — Blunt-lobed Woodsia

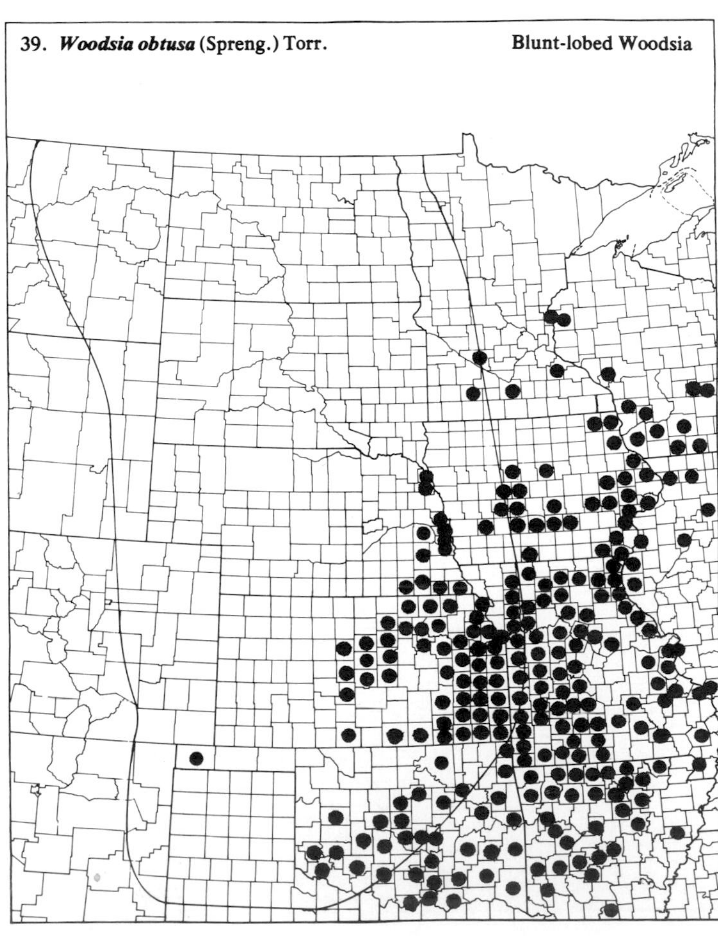

40. ***Woodsia oregana*** D.C. Eat. — Oregon Woodsia

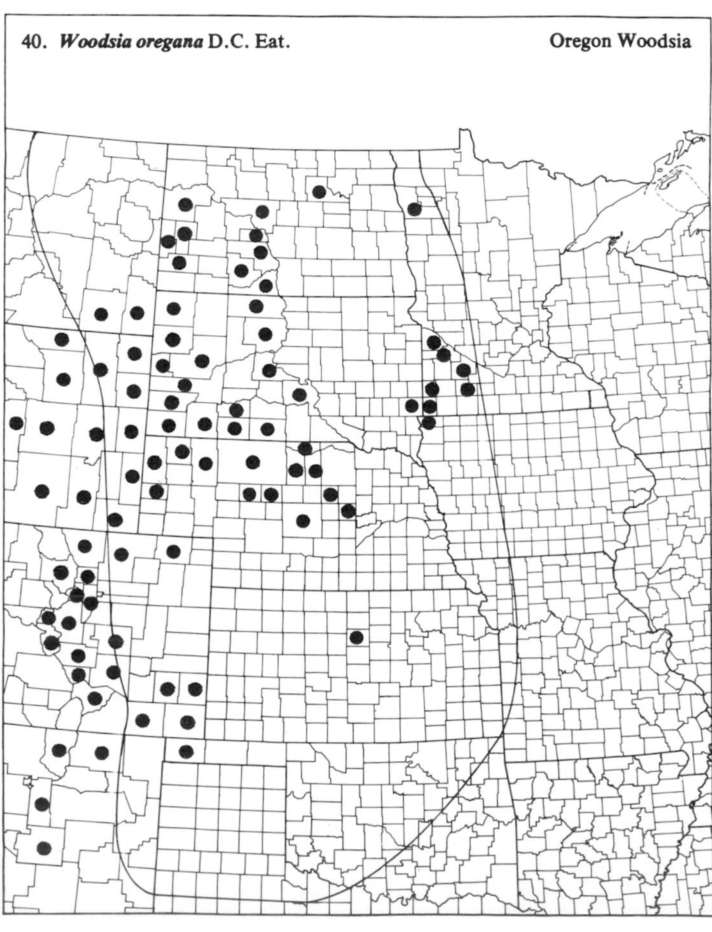

Marsileaceae (Marsilea Family)

41. ***Marsilea vestita*** Grev. & Hook. — Pepperwort

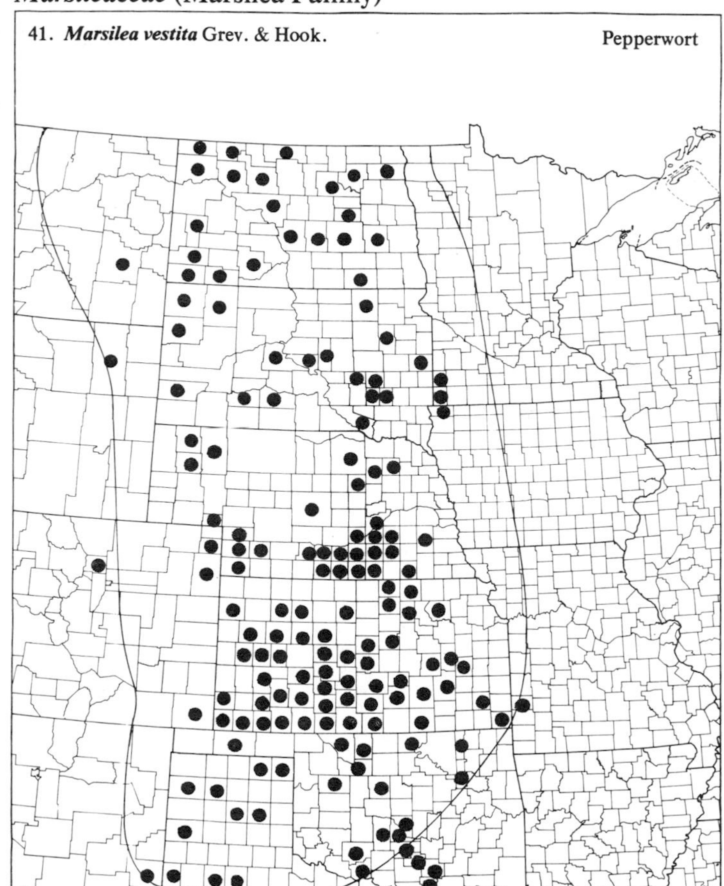

Salviniaceae (Salvinia Family)

42. ***Azolla mexicana*** Presl.

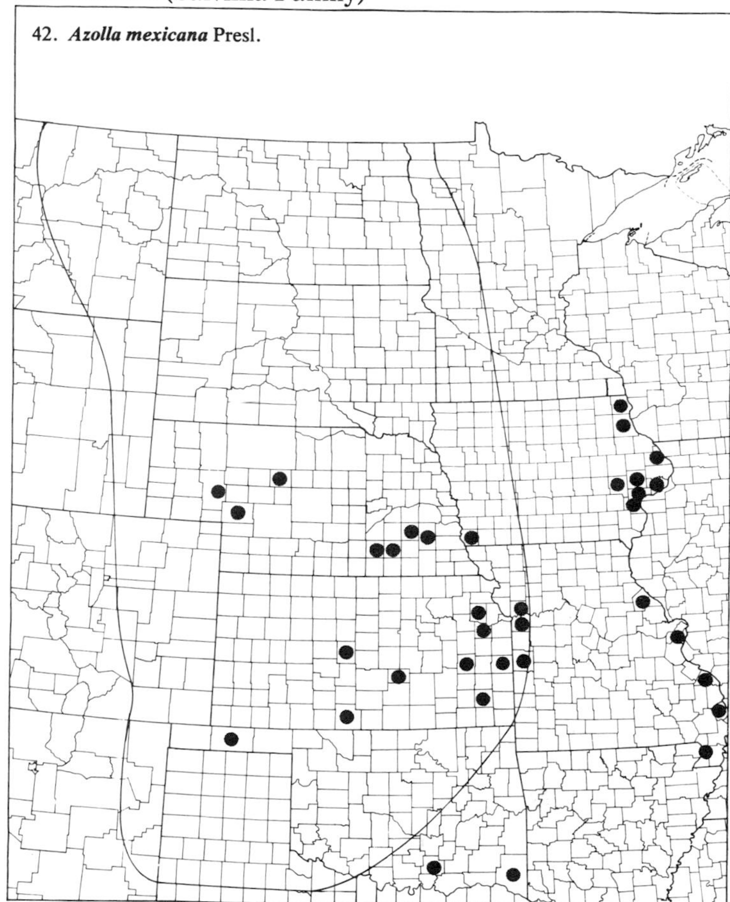

Pinaceae (Pine Family)

43. ***Picea glauca*** (Moench) Voss — White Spruce

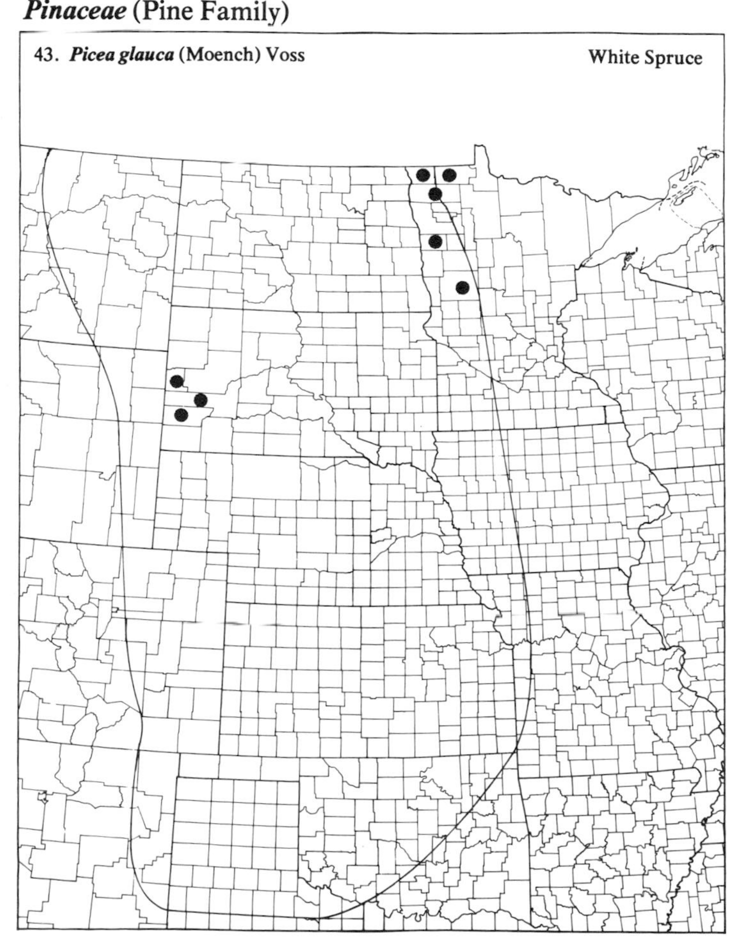

44. ***Pinus edulis*** Engelm. — Pinyon Pine

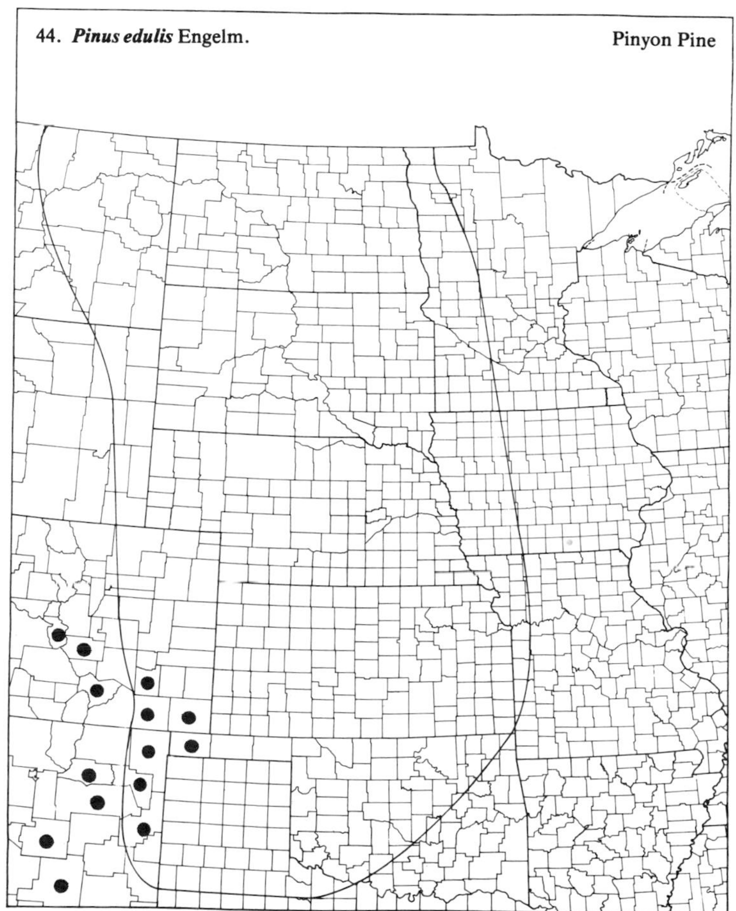

(Pinaceae)

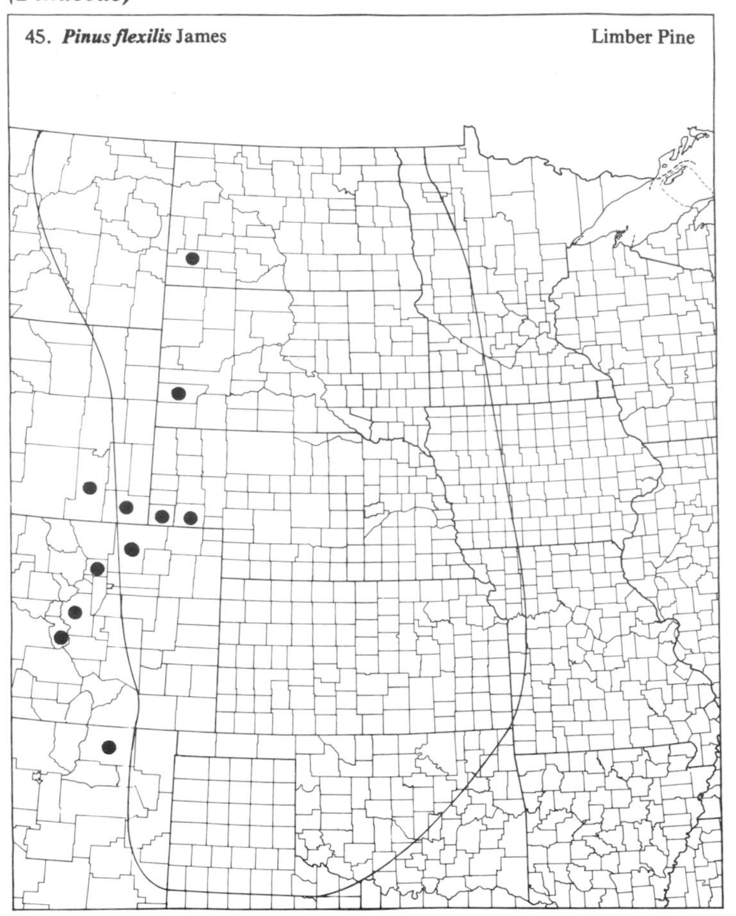

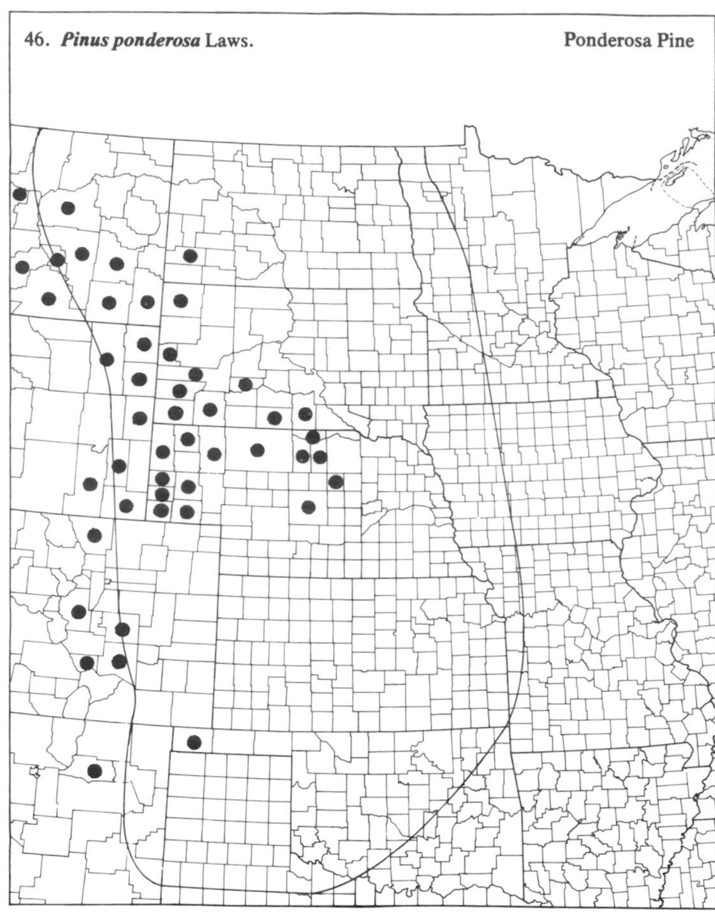

Cupressaceae (Cypress Family)

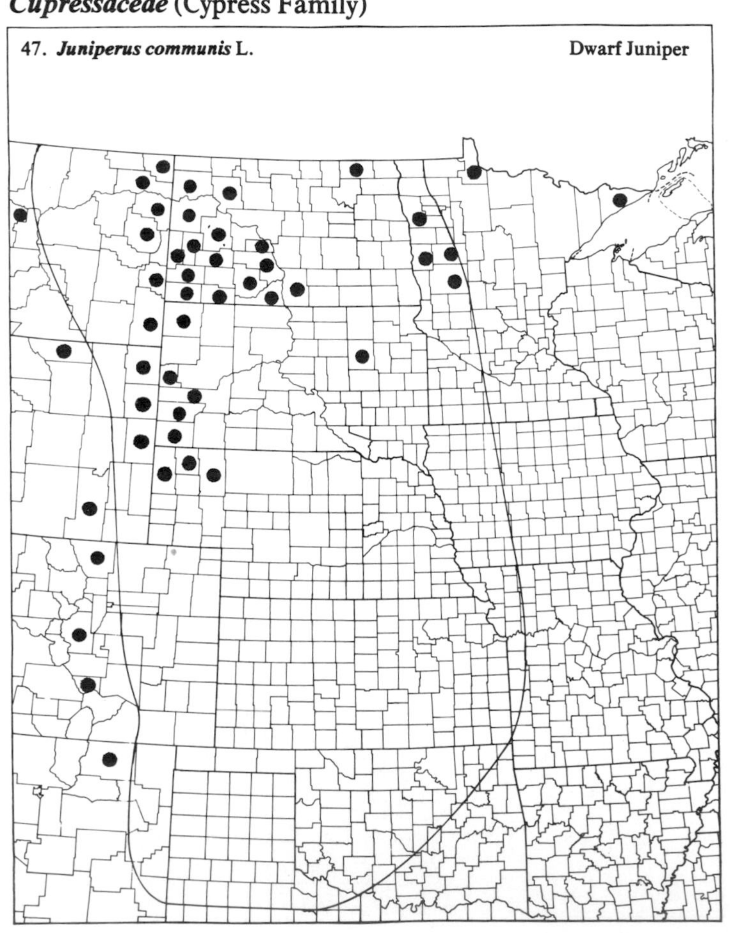

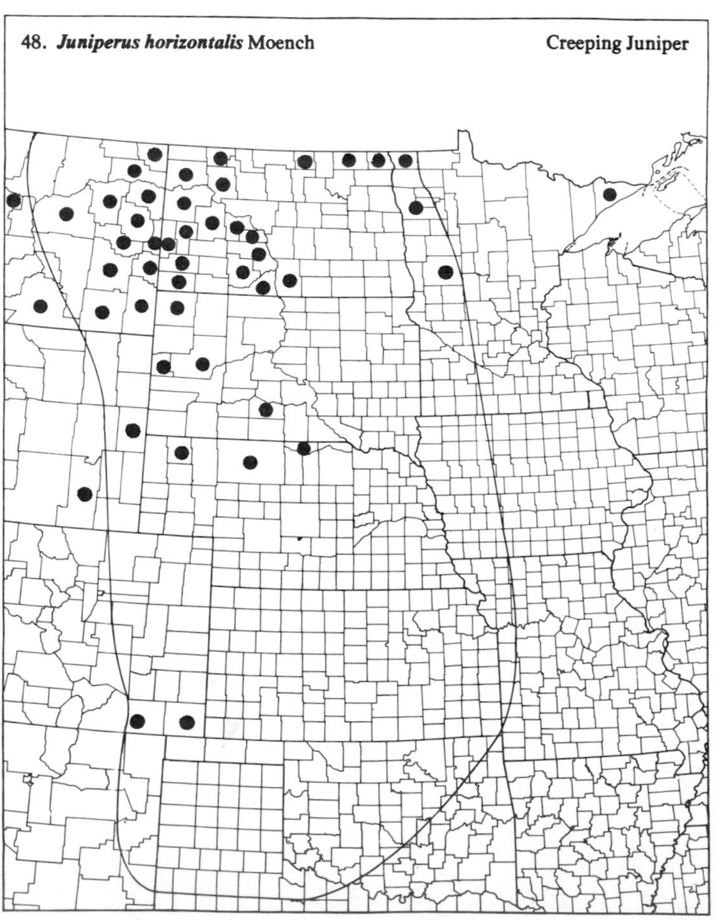

(Cupressaceae)

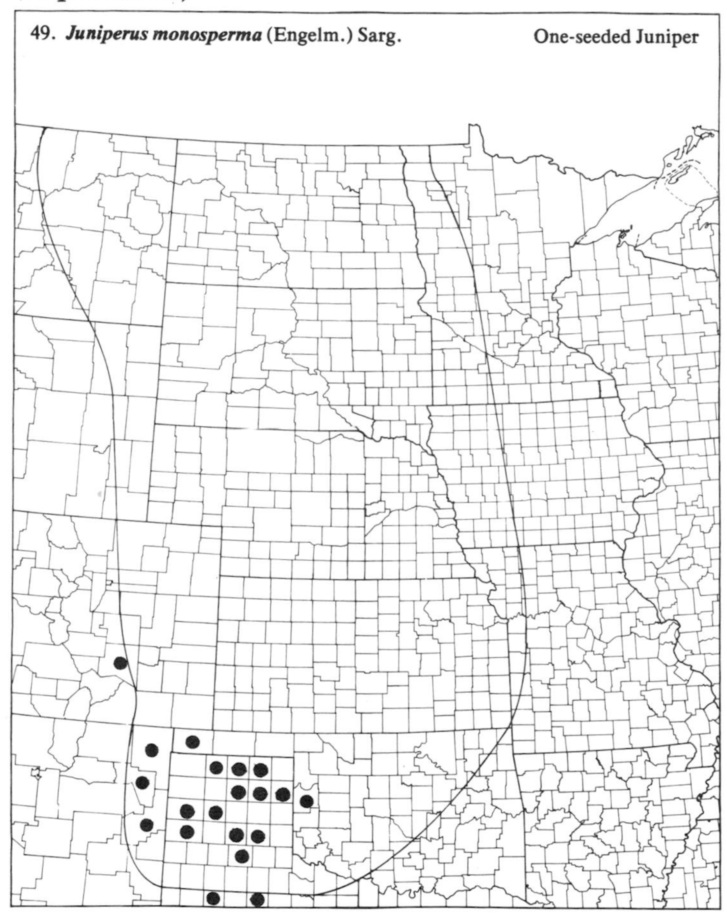

49. ***Juniperus monosperma*** (Engelm.) Sarg. One-seeded Juniper

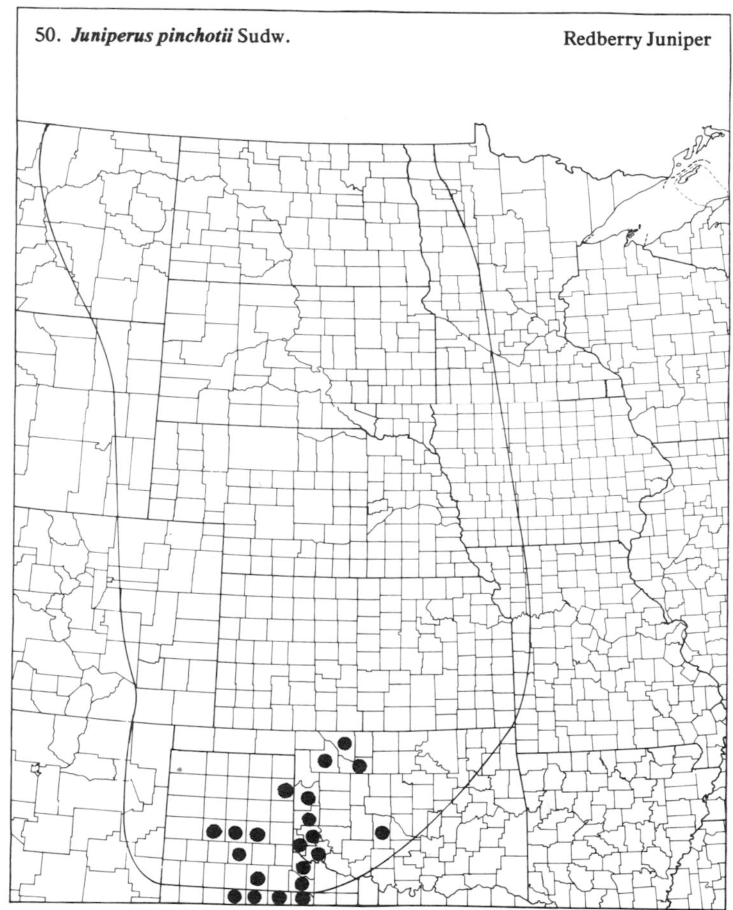

50. ***Juniperus pinchotii*** Sudw. Redberry Juniper

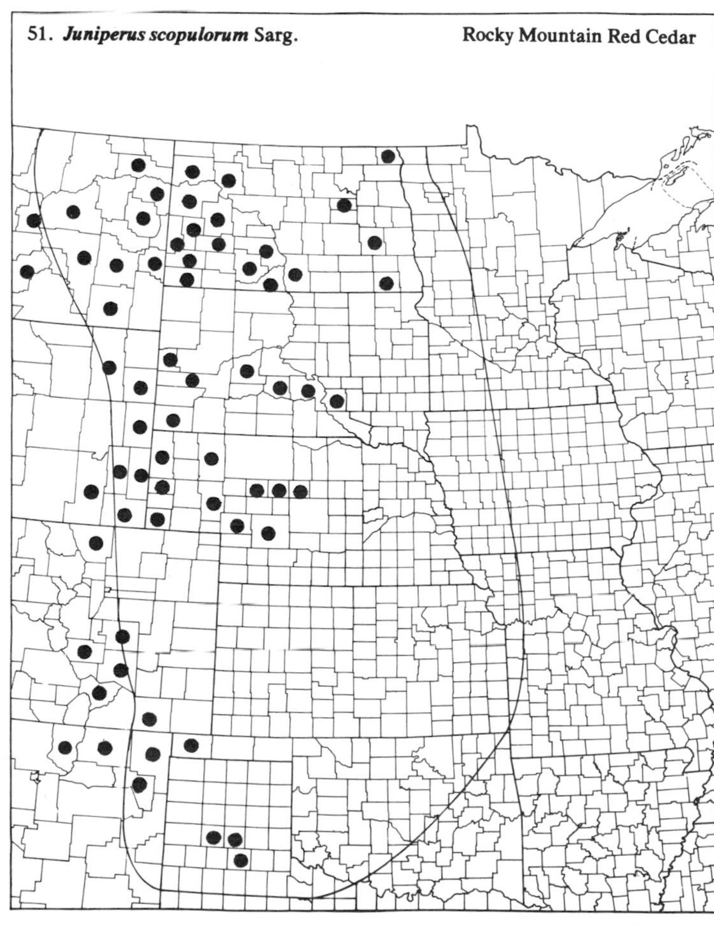

51. ***Juniperus scopulorum*** Sarg. Rocky Mountain Red Cedar

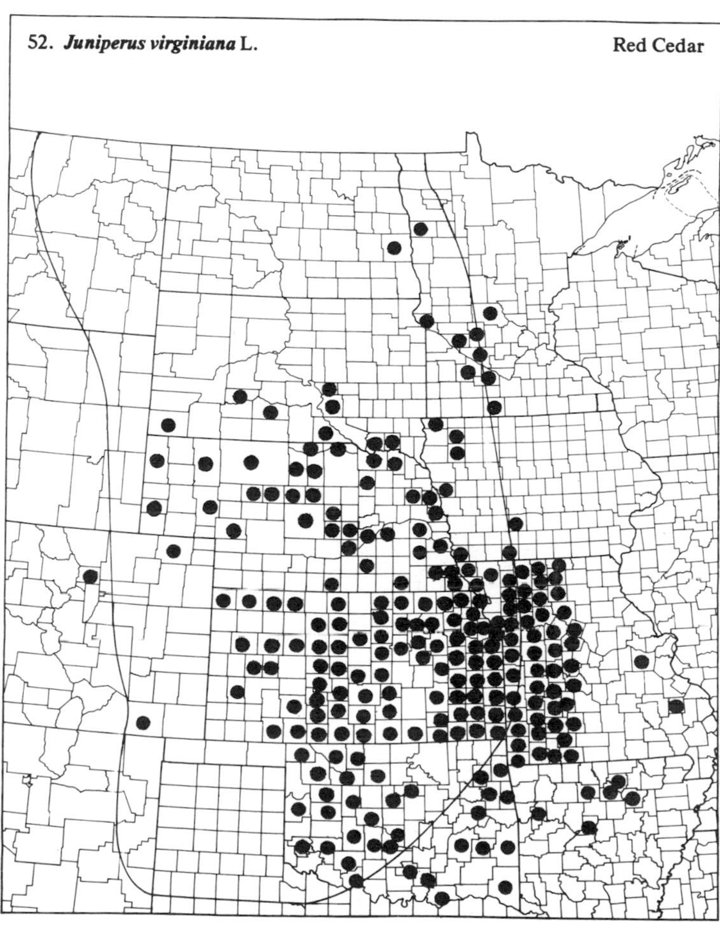

52. ***Juniperus virginiana*** L. Red Cedar

Ephedraceae (Ephedra Family)

53. ***Ephedra antisyphilitica*** C.A. Mey. — Ephedra

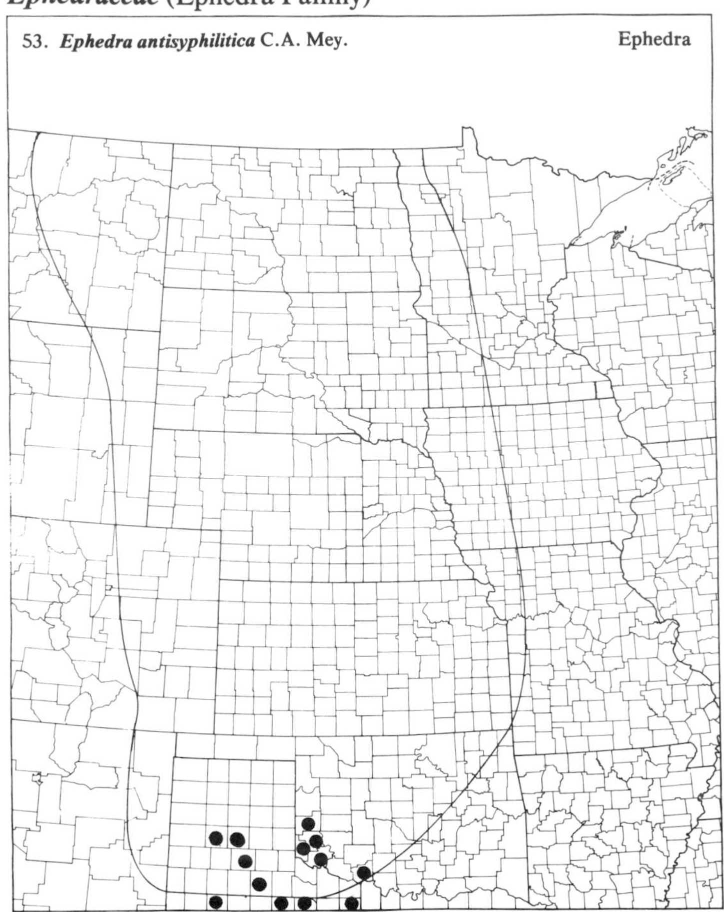

54. ***Ephedra torreyana*** Wats. — Torrey Ephedra

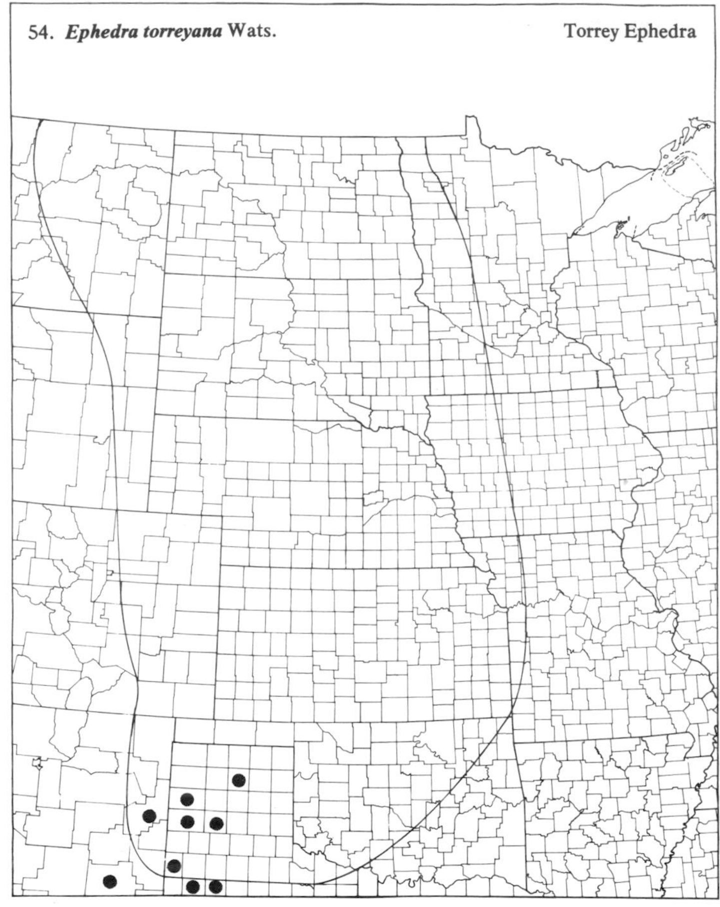

1. *Annonaceae* (Custard-apple Family)

55. ***Asimina triloba*** (L.) Dun. — Pawpaw

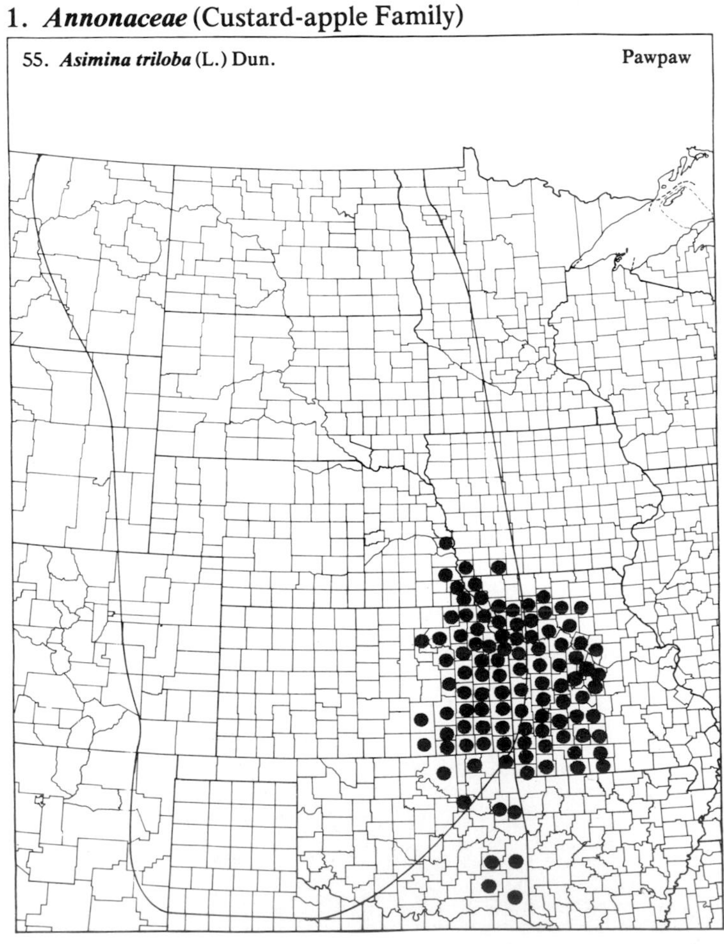

4. *Aristolochiaceae* (Birthwort Family)

56. ***Asarum canadense*** L. — Wild Ginger

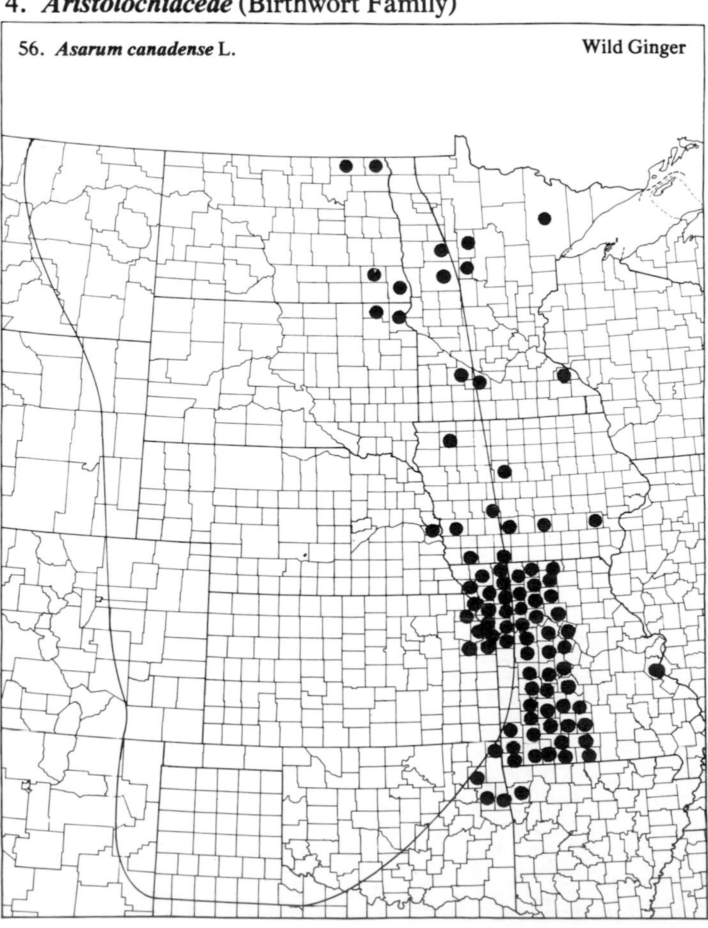

(Aristolochiaceae)

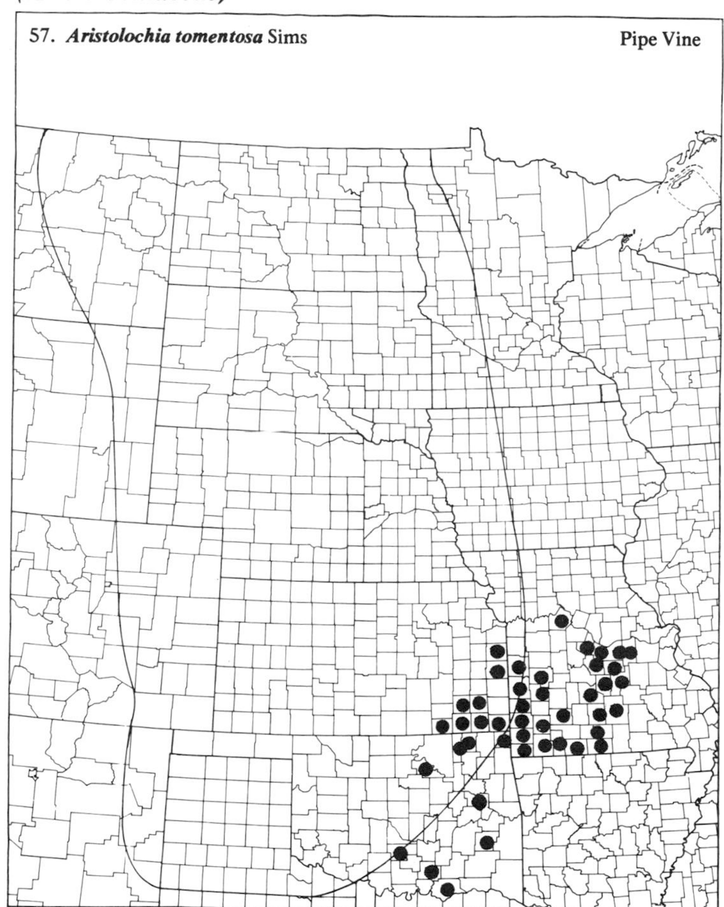

5. *Nymphaeaceae* (Water Lily Family)

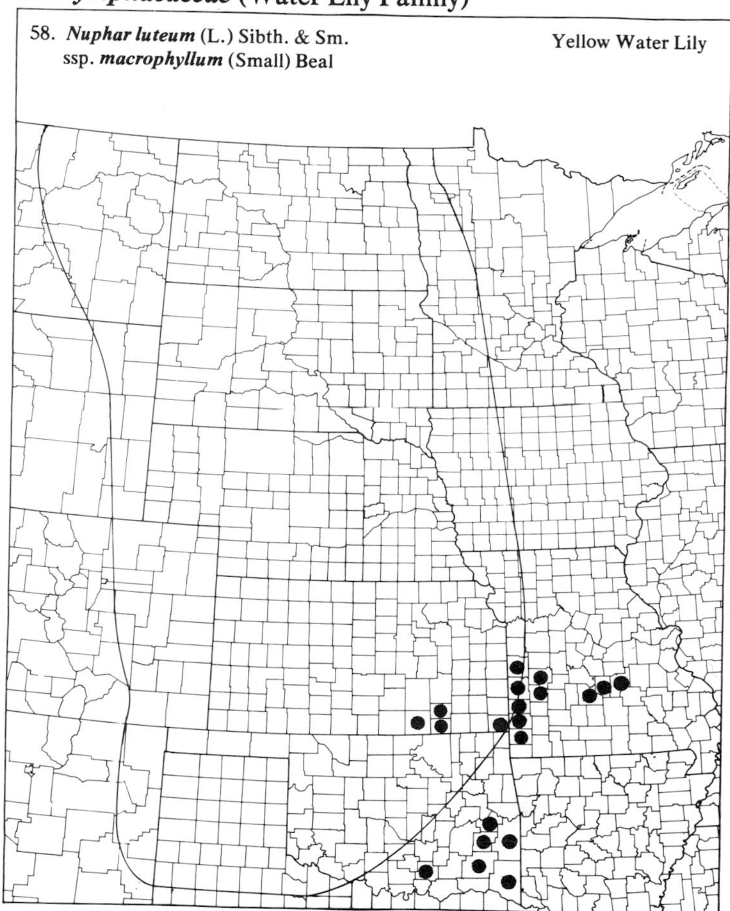

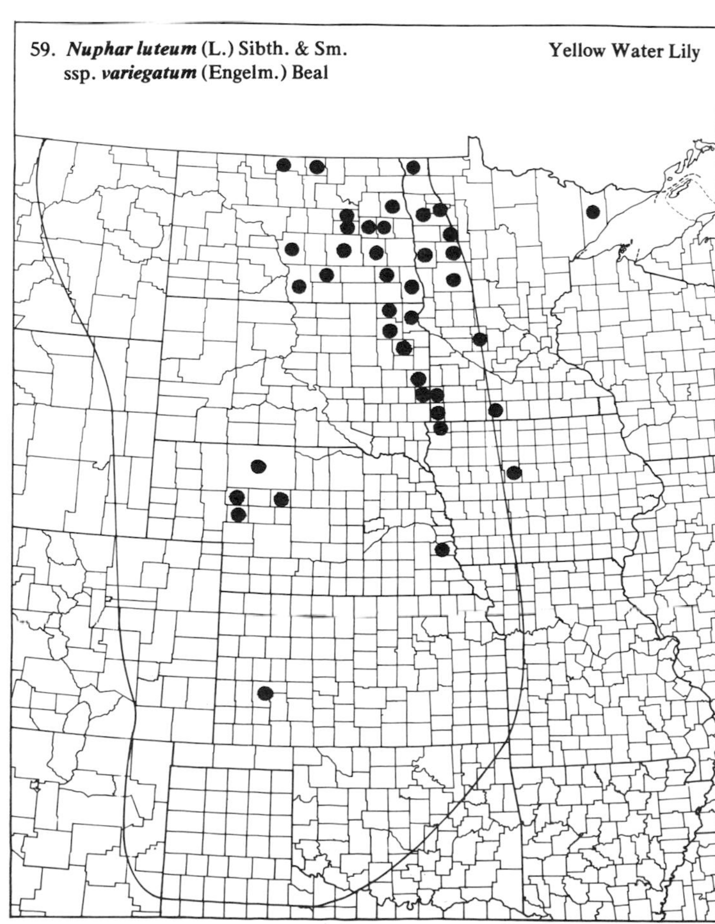

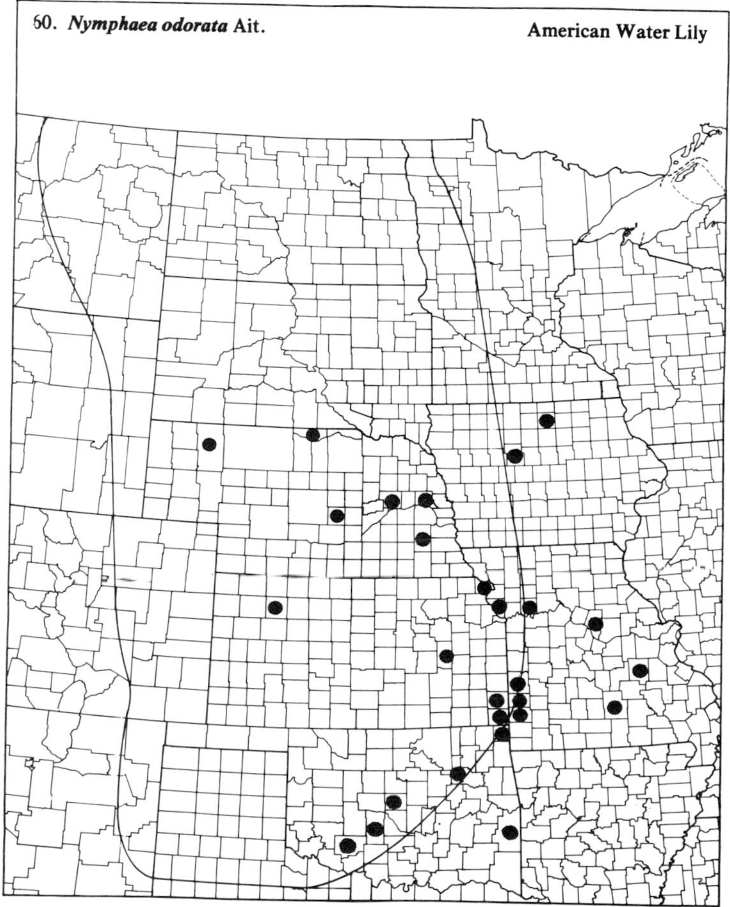

(Nymphaeaceae)

61. ***Nymphaea tuberosa*** Paine — White Water Lily

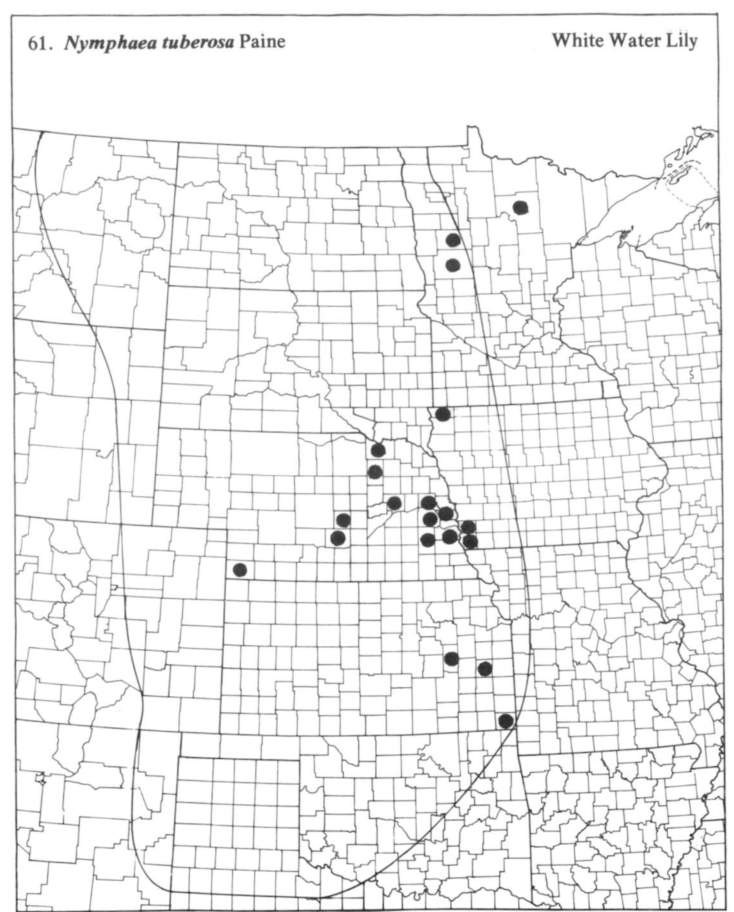

5a. *Nelumbonaceae*

62. ***Nelumbo lutea*** (Willd.) Pers. — American Lotus

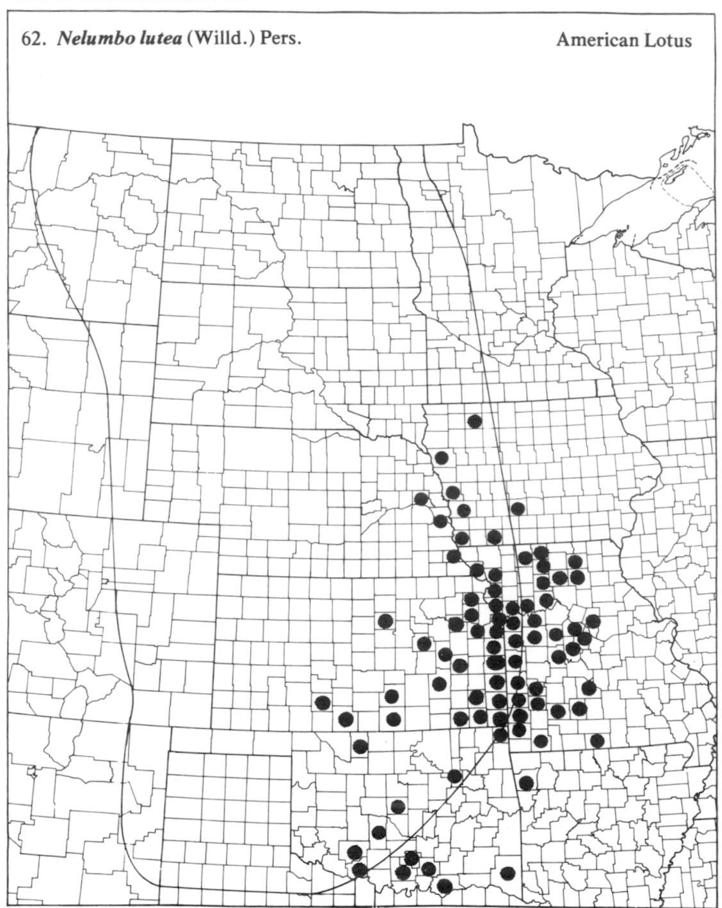

6. *Ceratophyllaceae* (Hornwort Family)

63. ***Ceratophyllum demersum*** L. — Hornwort

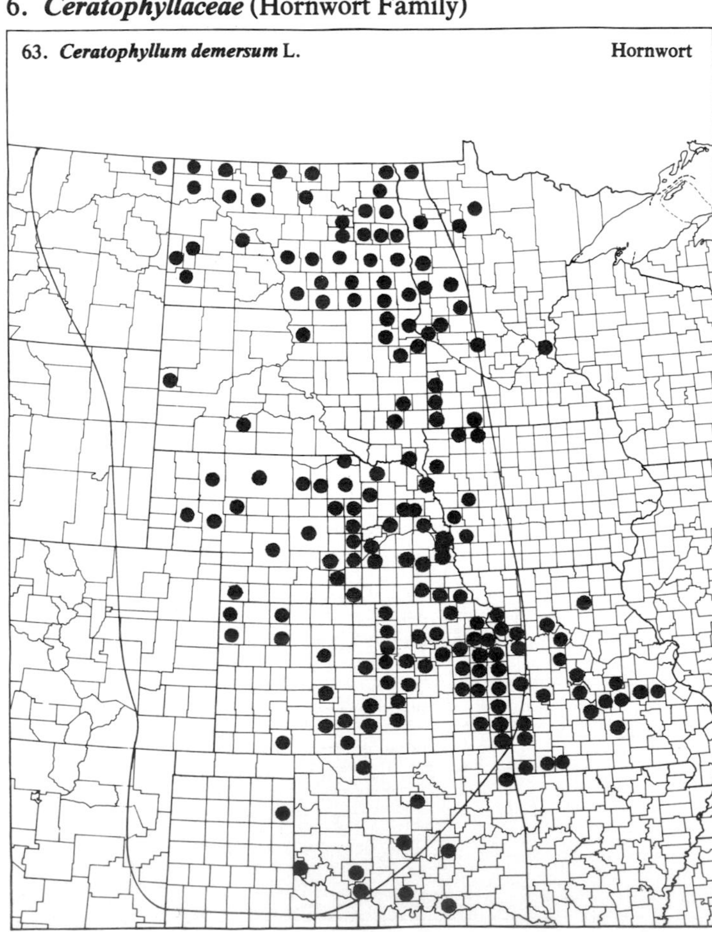

7. *Ranunculaceae* (Buttercup Family)

64. ***Aconitum columbianum*** Nutt. — Monkshood

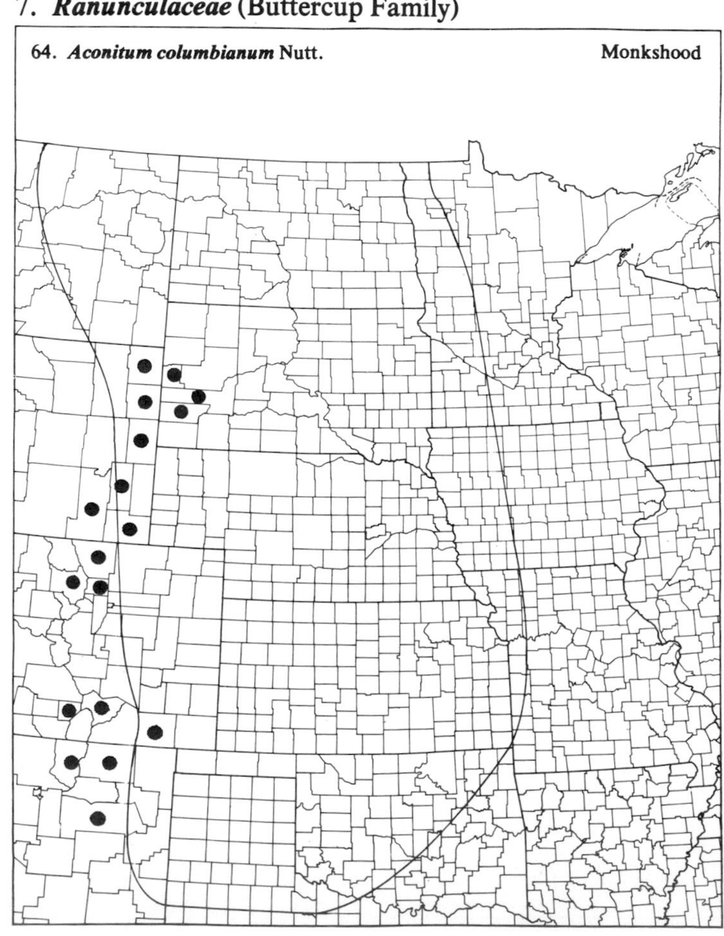

(Ranunculaceae)

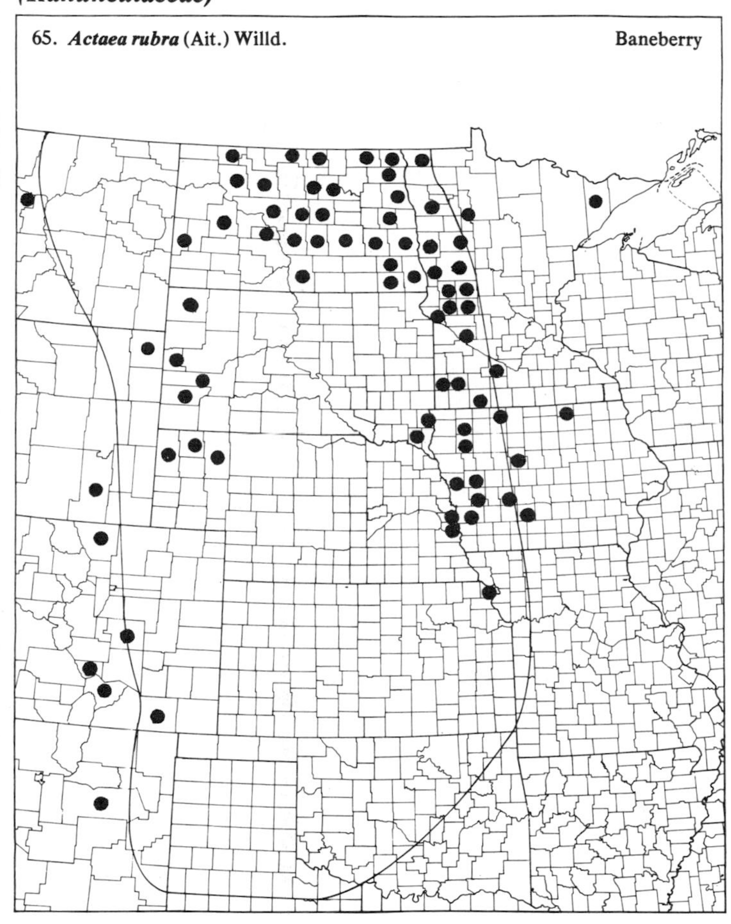

65. ***Actaea rubra*** (Ait.) Willd. Baneberry

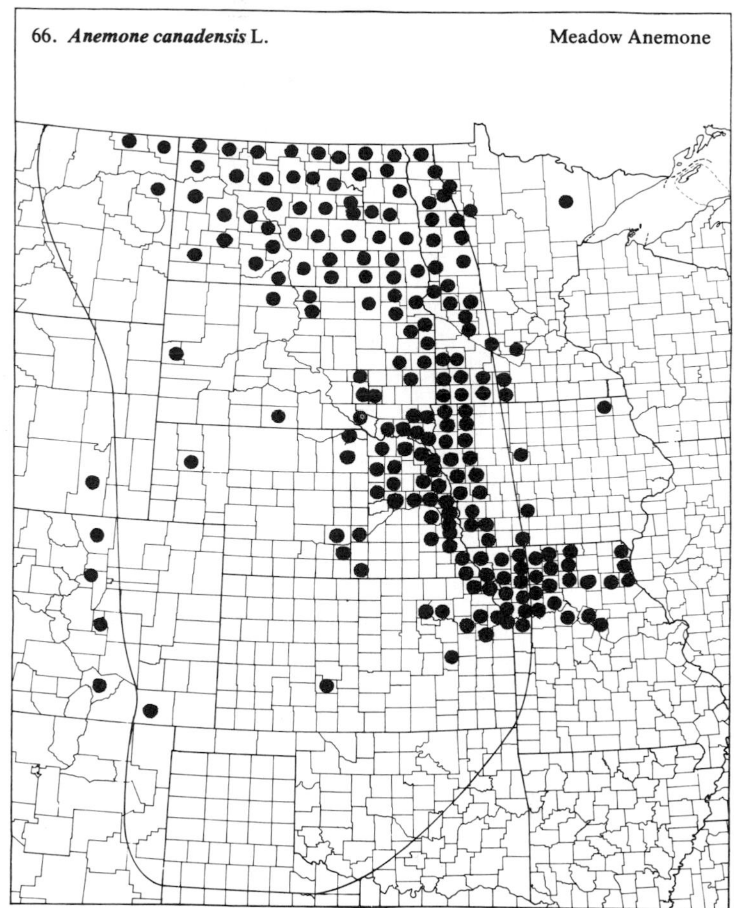

66. ***Anemone canadensis*** L. Meadow Anemone

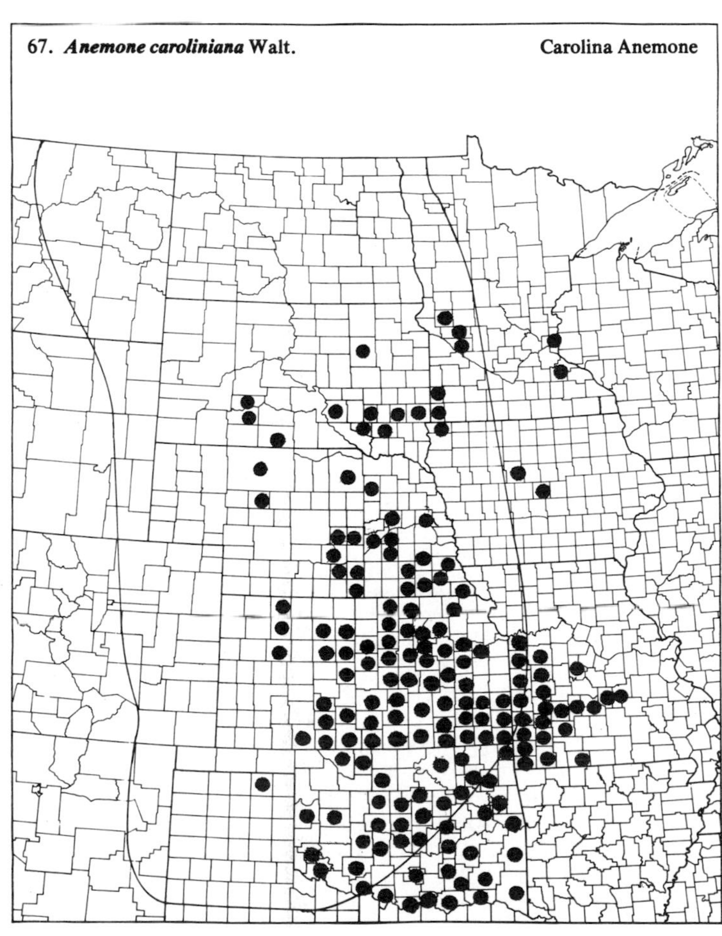

67. ***Anemone caroliniana*** Walt. Carolina Anemone

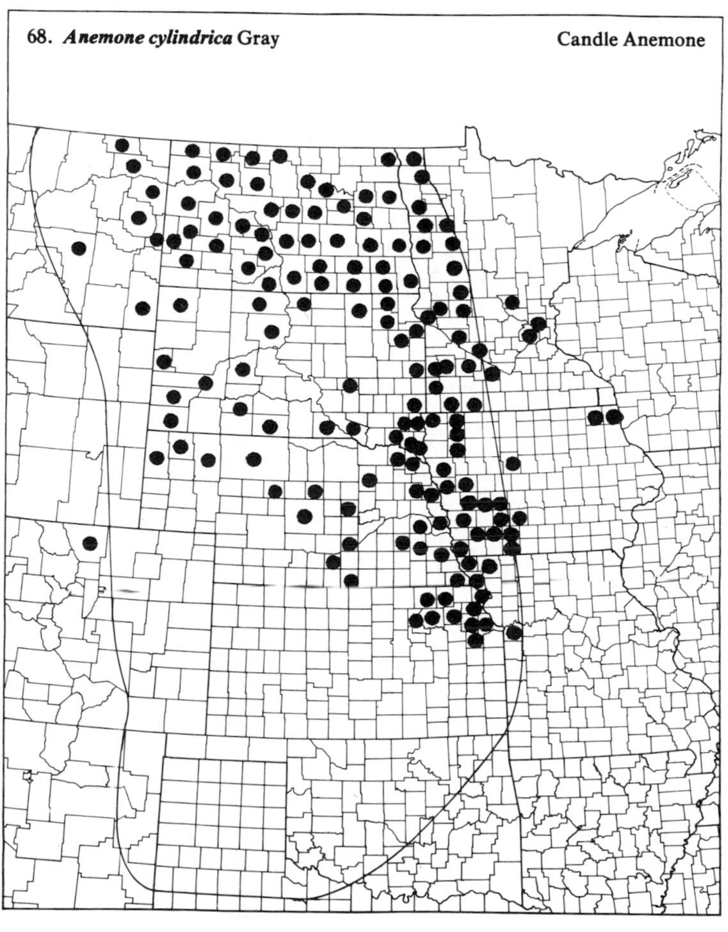

68. ***Anemone cylindrica*** Gray Candle Anemone

(Ranunculaceae)

69. ***Anemone decapetala*** Ard. Tenpetal Anemone

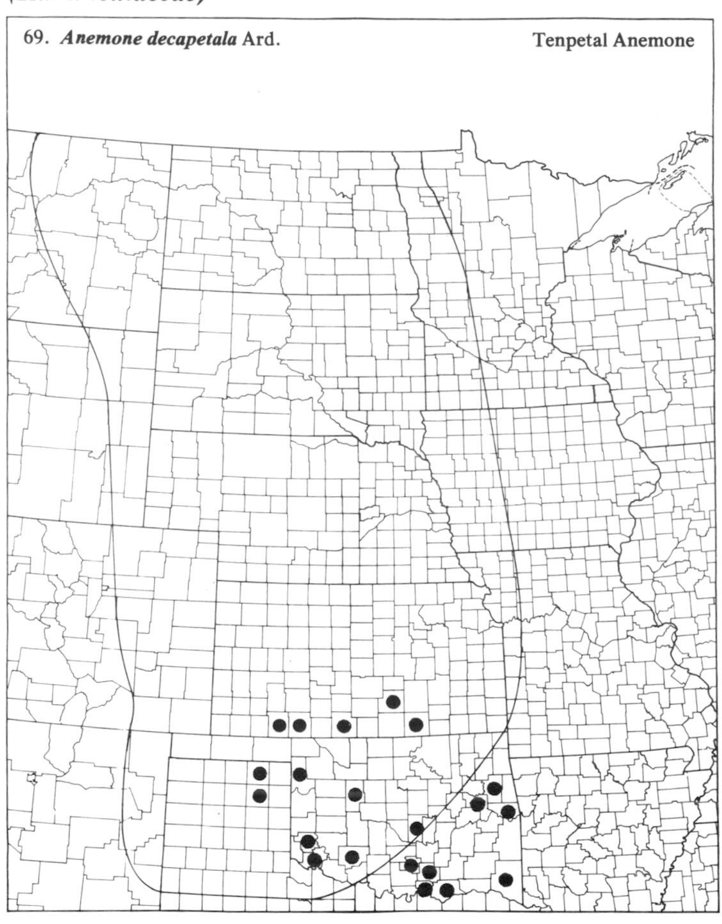

70. ***Anemone multifida*** Poir.

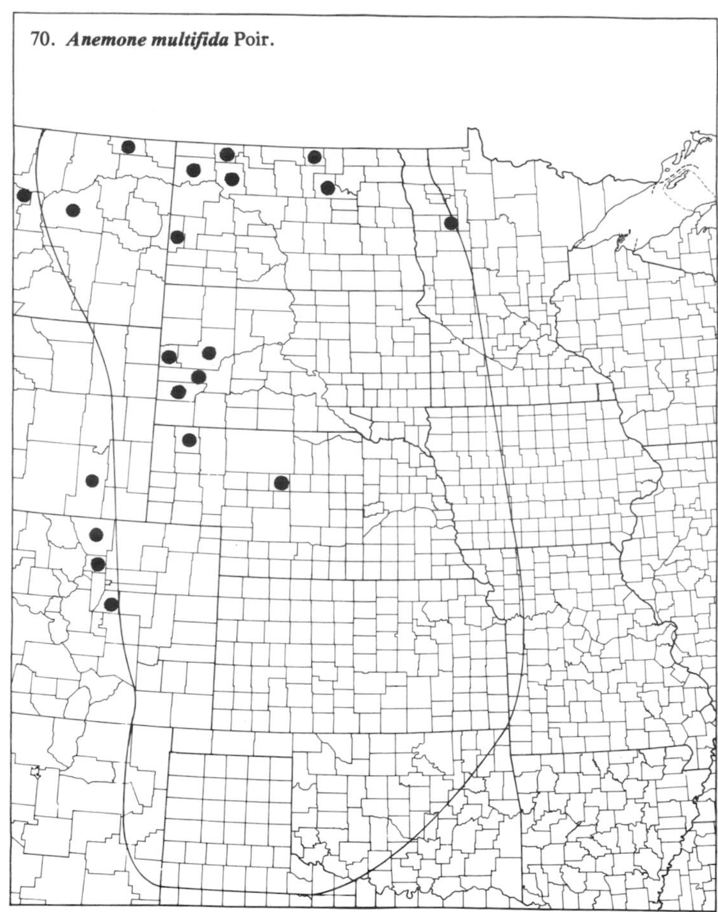

71. ***Anemone patens*** L. Pasque-flower

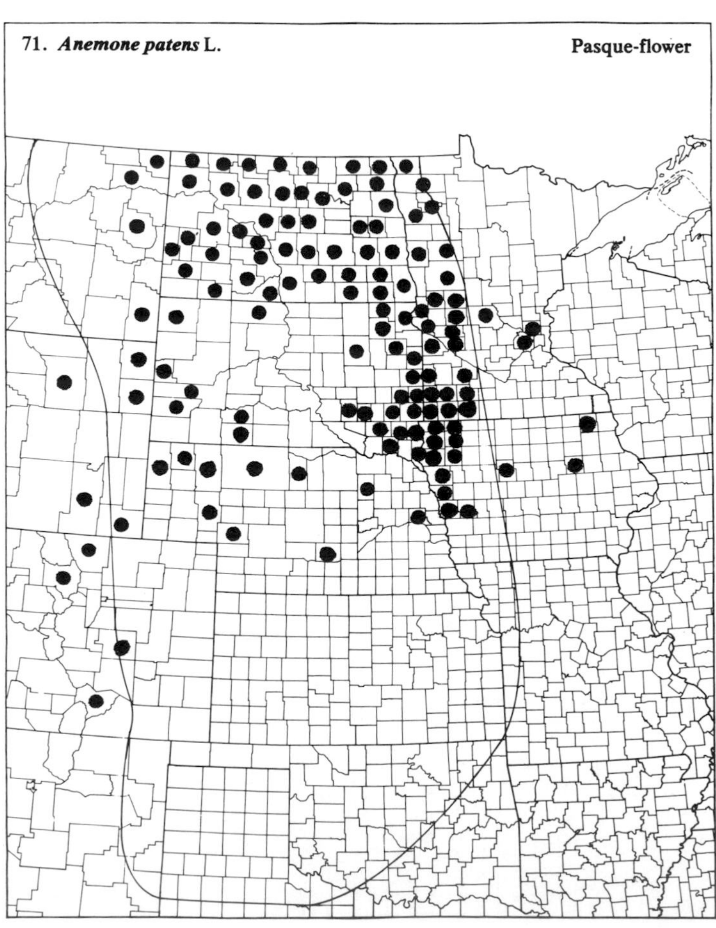

72. ***Anemone quinquefolia*** L. Wood Anemone

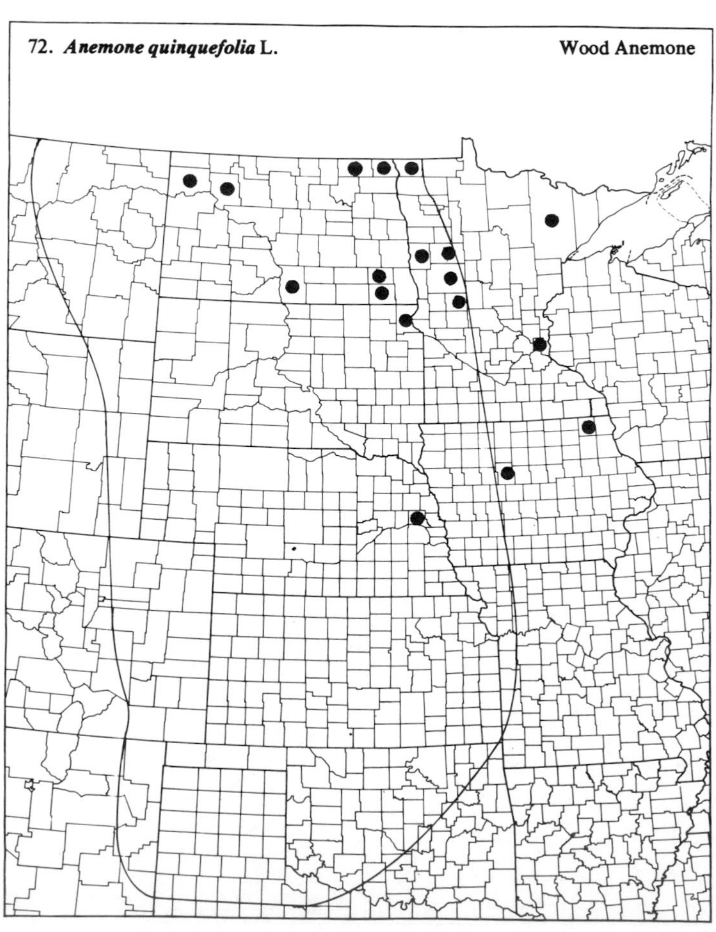

(Ranunculaceae)

73. ***Anemone virginiana*** L. Tall Anemone

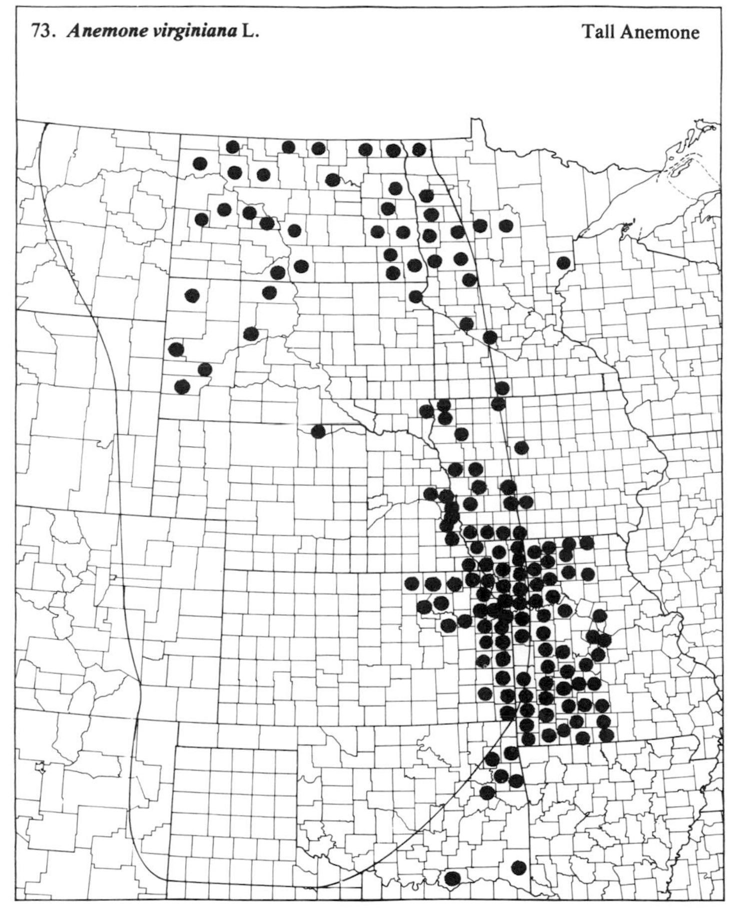

74. ***Anemonella thalictroides*** (L.) Spach Rue Anemone

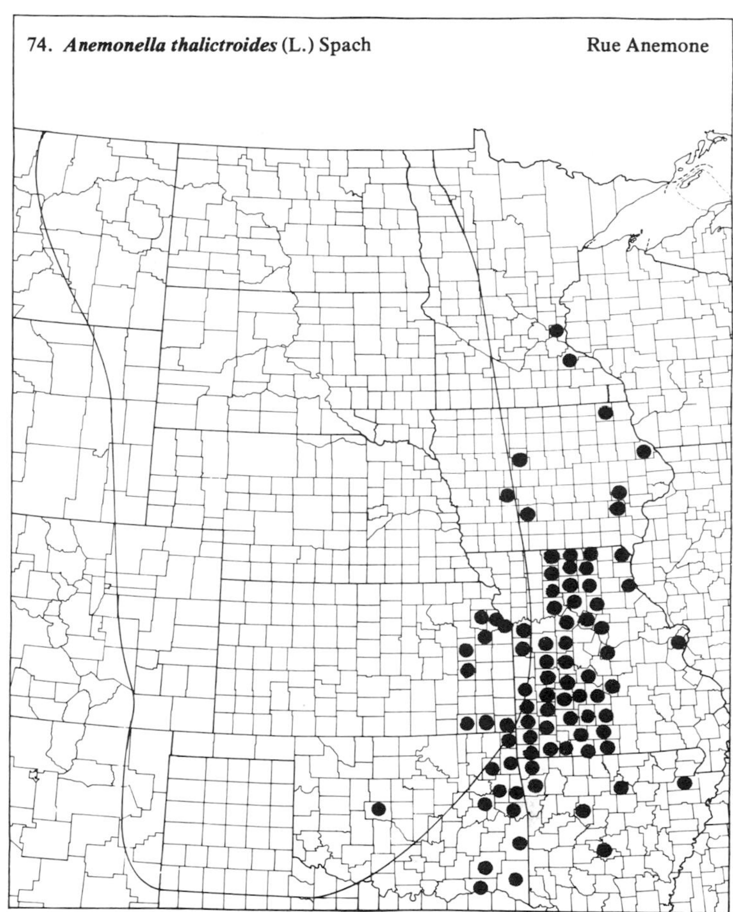

75. ***Aquilegia canadensis*** L. Wild Columbine

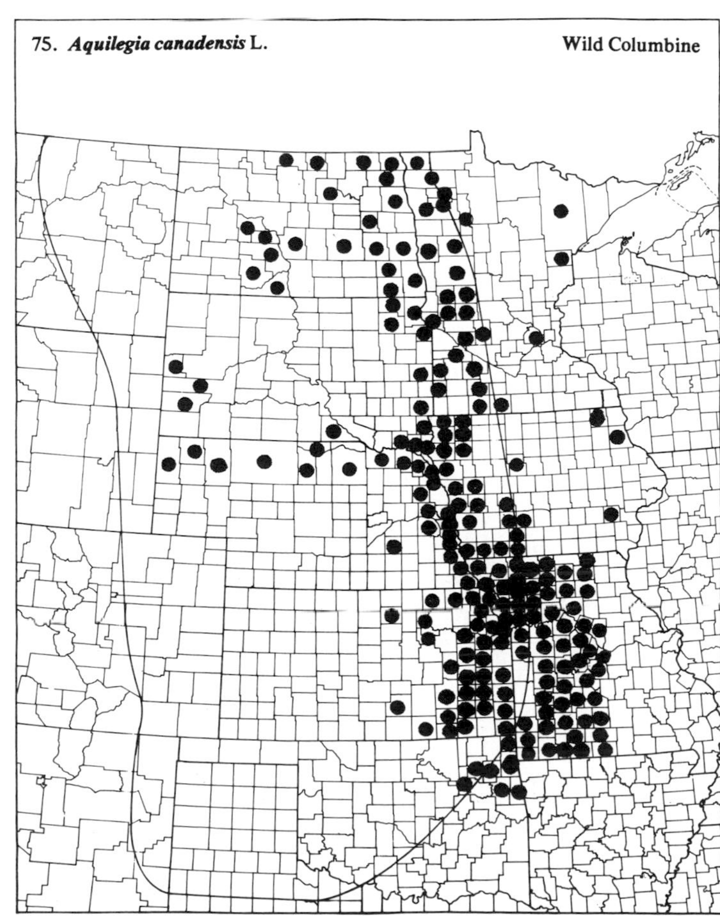

76. ***Caltha palustris*** L. Marsh Marigold

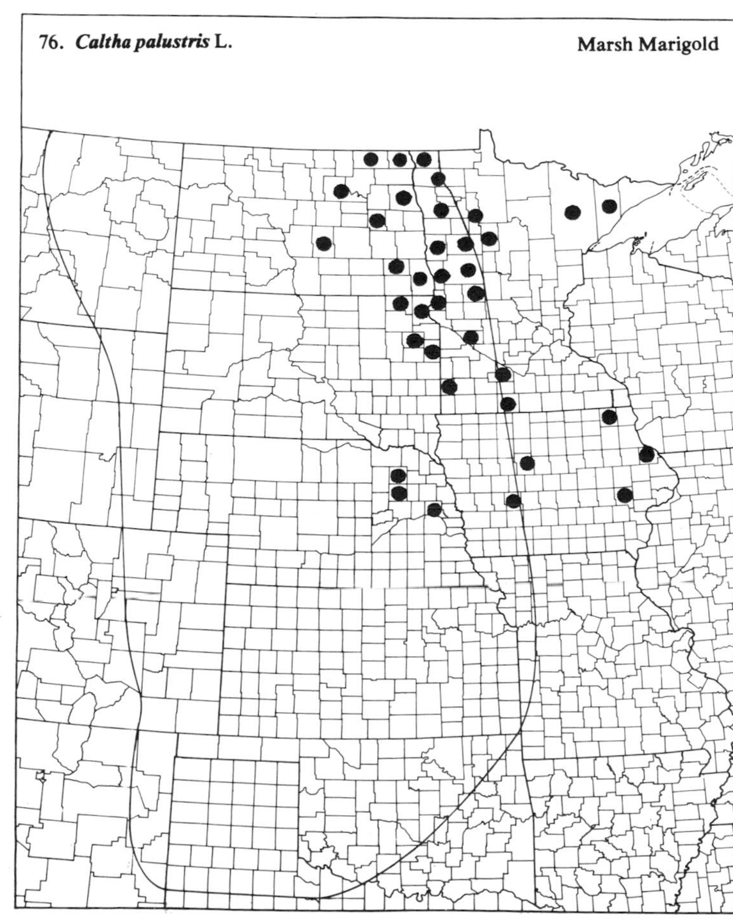

(Ranunculaceae)

77. ***Clematis fremontii*** Wats. — Fremont's Clematis

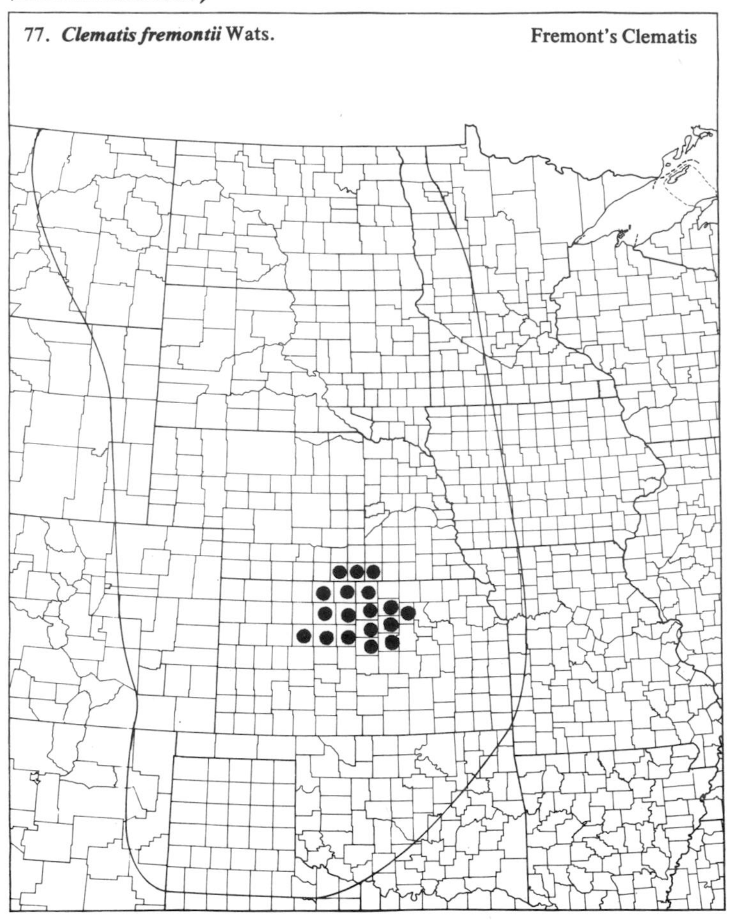

78. ***Clematis ligusticifolia*** Nutt. — Western Clematis

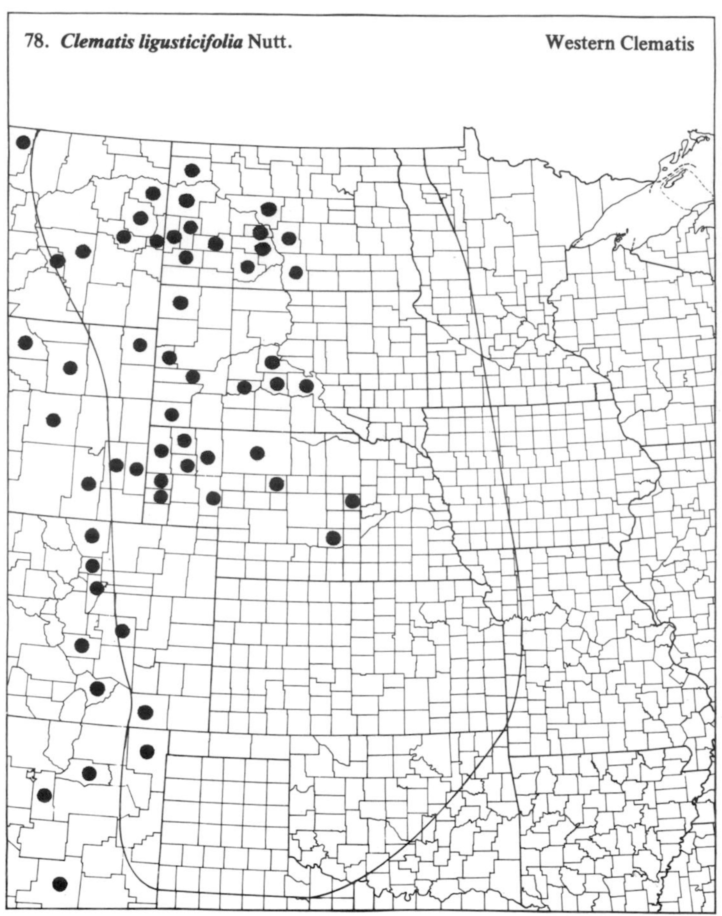

79. ***Clematis pitcheri*** T. & G. — Pitcher's Clematis

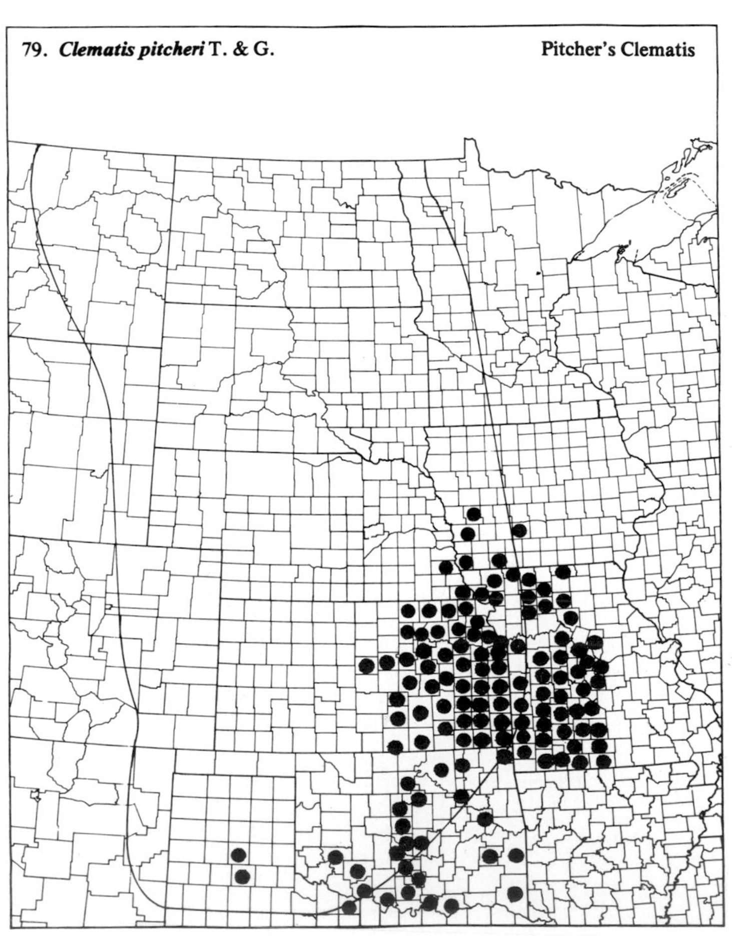

80. ***Clematis virginiana*** L. — Virgin's Bower

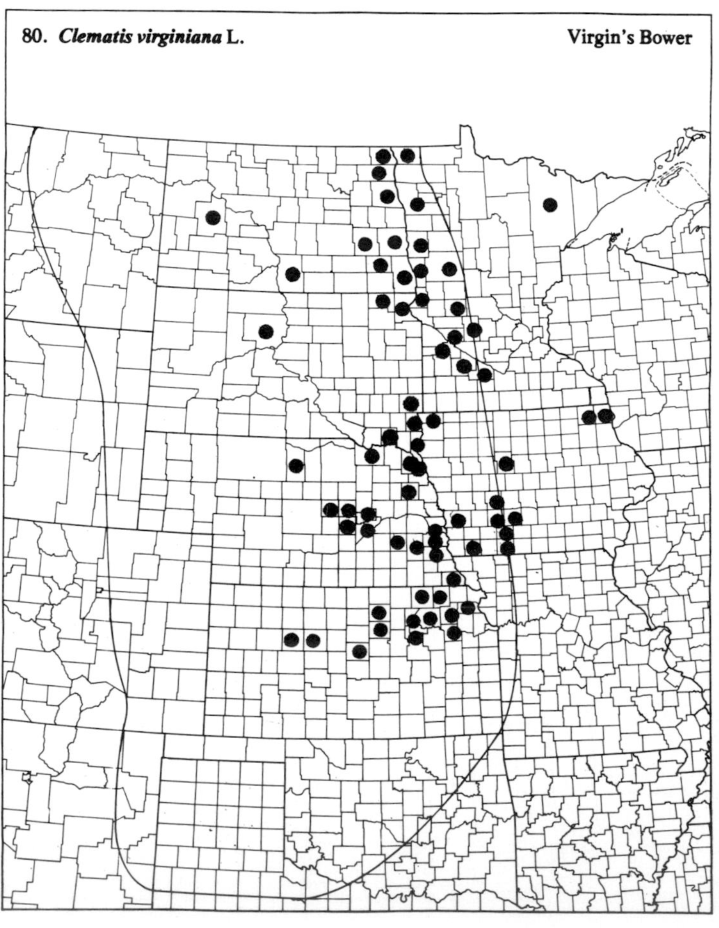

(Ranunculaceae)

81. ***Delphinium ajacis*** L. Rocket Larkspur

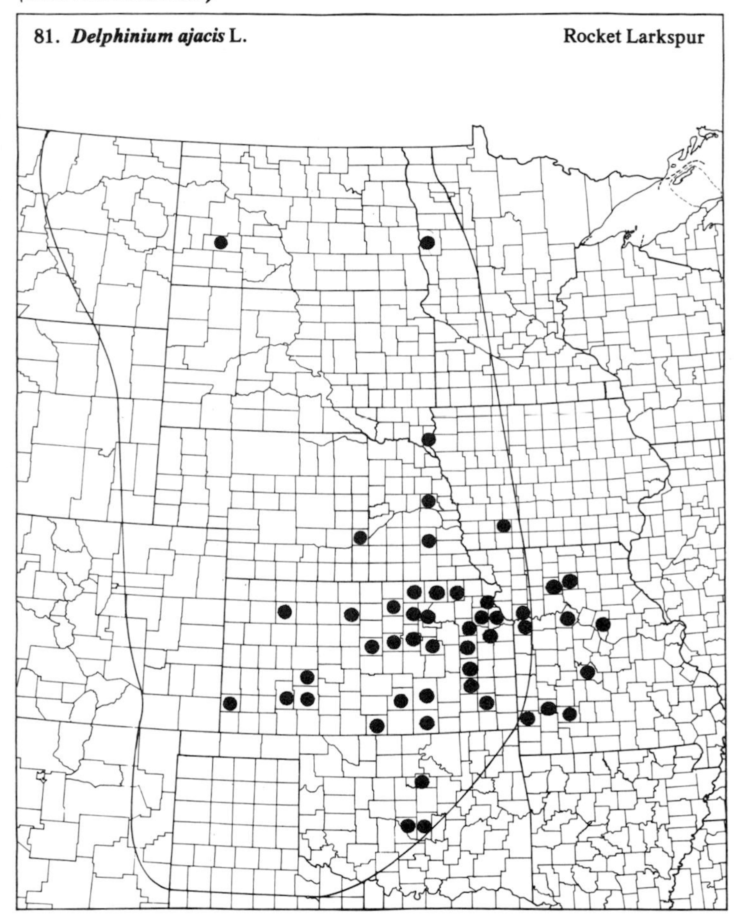

82. ***Delphinium bicolor*** Nutt. Little Larkspur

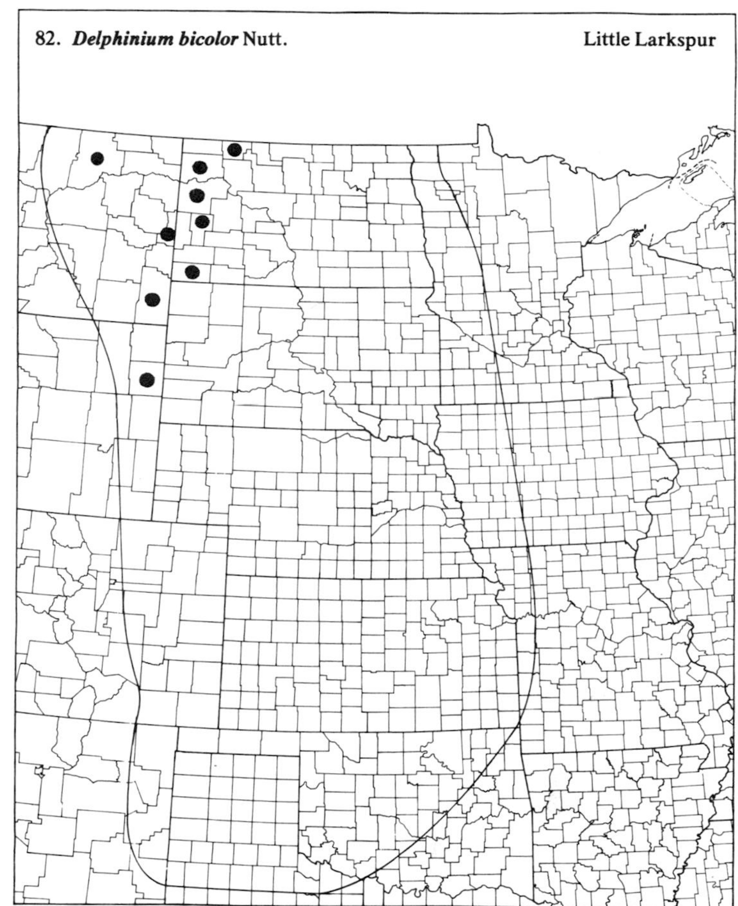

83. ***Delphinium nuttallianum*** Pritz. ex Walp. Blue Larkspur

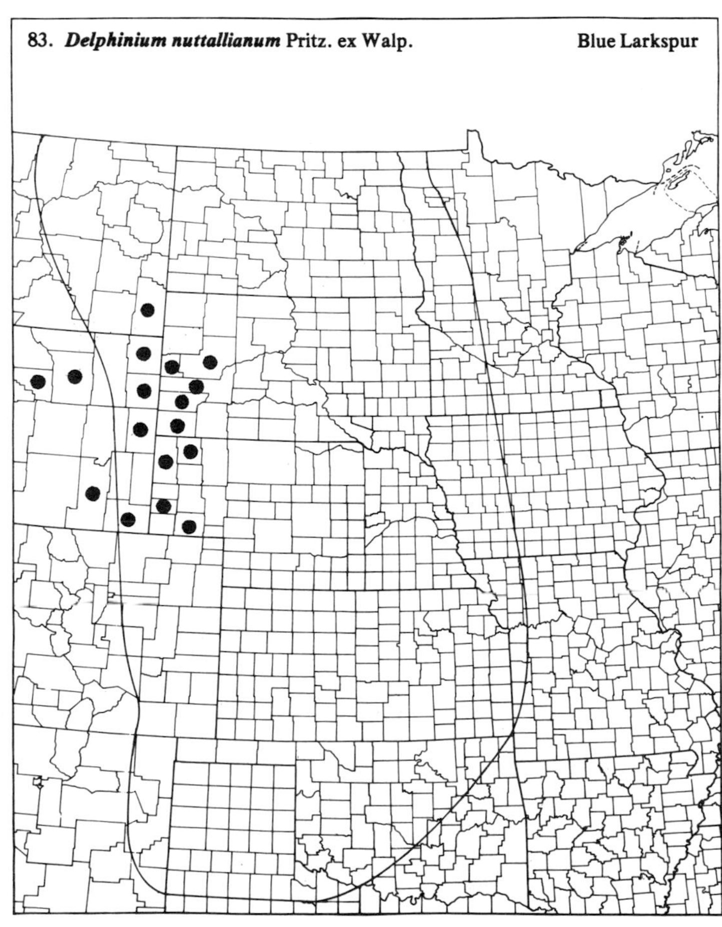

84. ***Delphinium tricorne*** Michx. Dwarf Larkspur

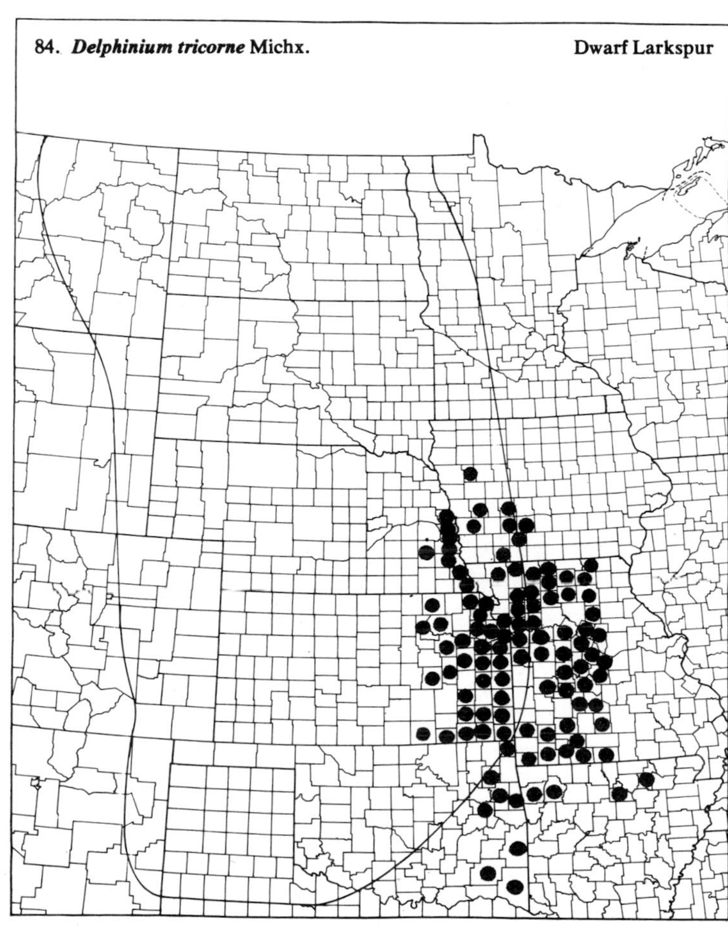

(Ranunculaceae)

85. ***Delphinium virescens*** Nutt. Prairie Larkspur

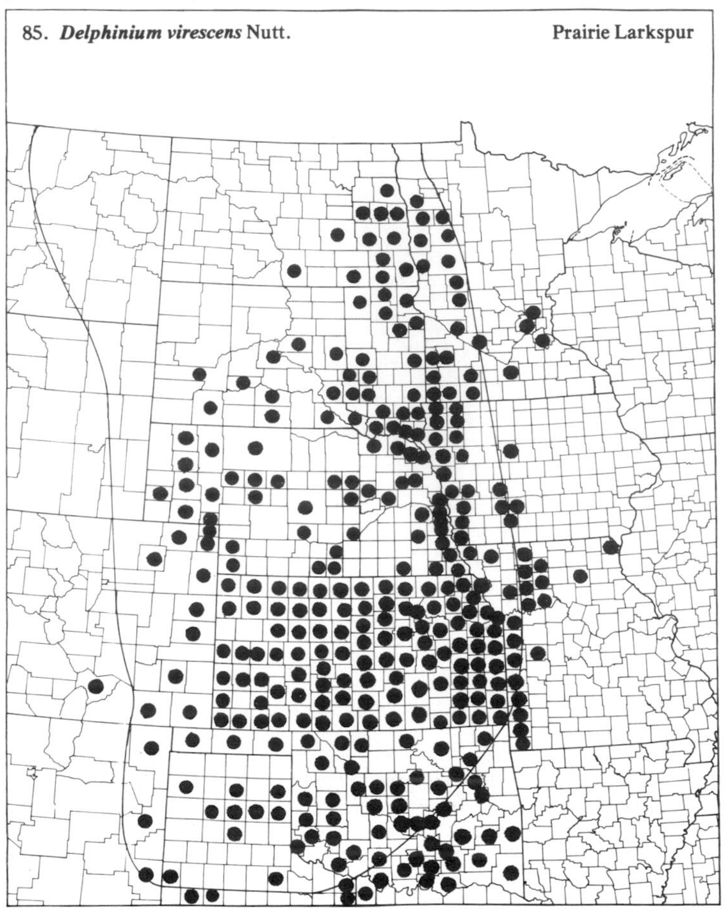

86. ***Isopyrum biternatum*** (Raf.) T. & G. False Rue Anemone

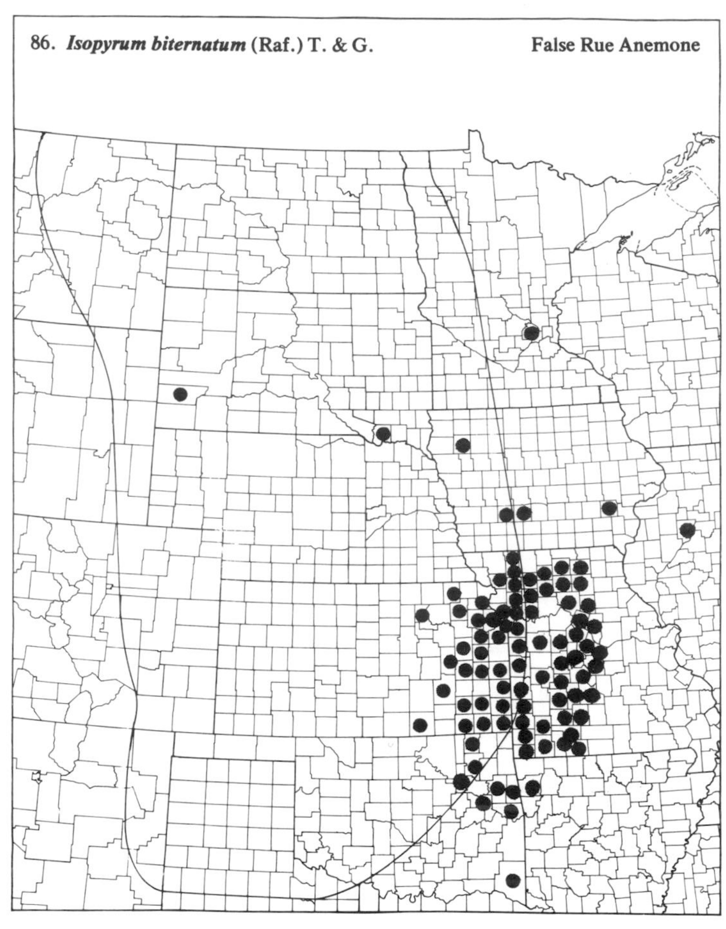

87. ***Myosurus minimus*** L. Mousetail

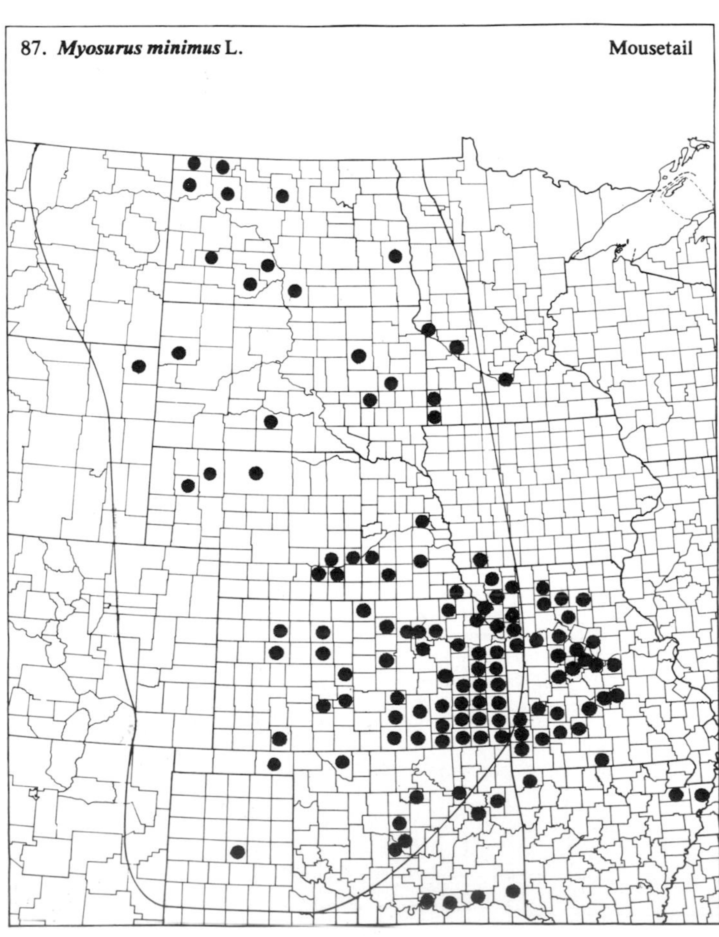

88. ***Ranunculus abortivus*** L. Early Wood Buttercup

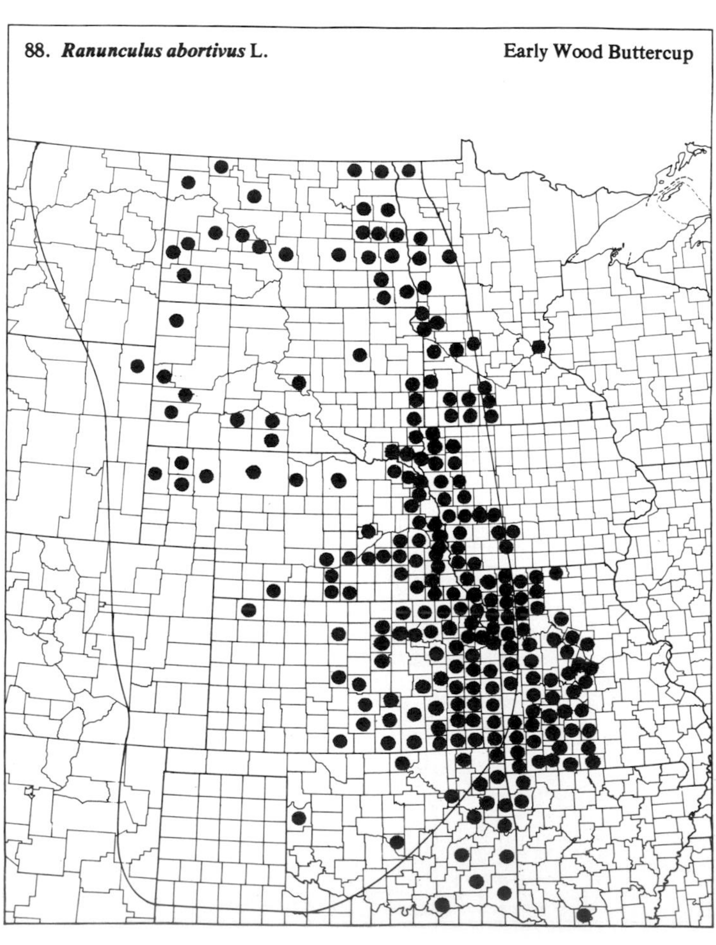

(Ranunculaceae)

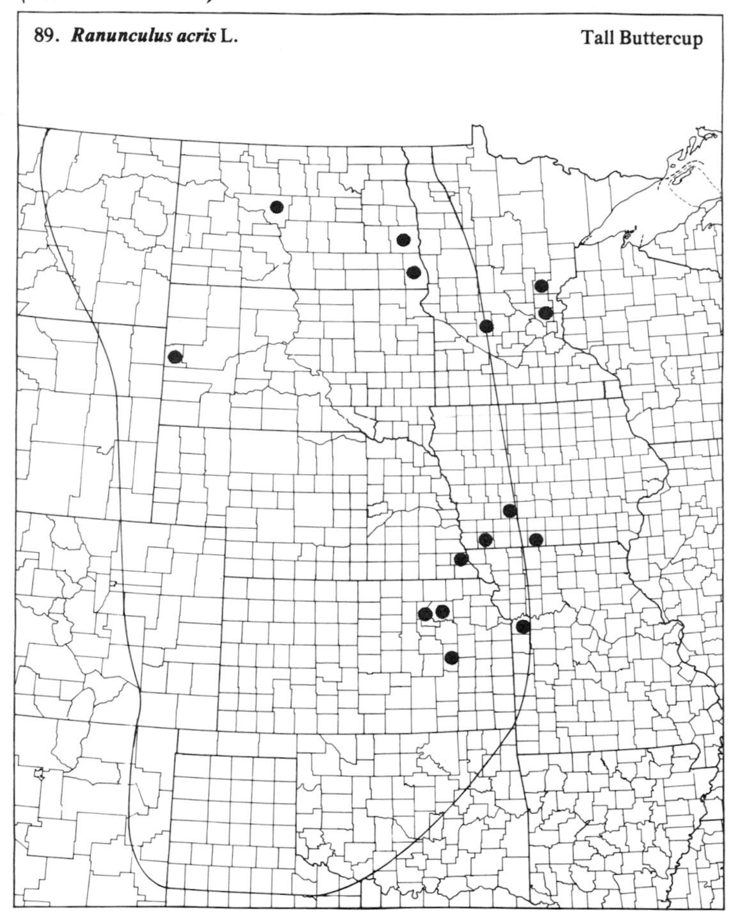

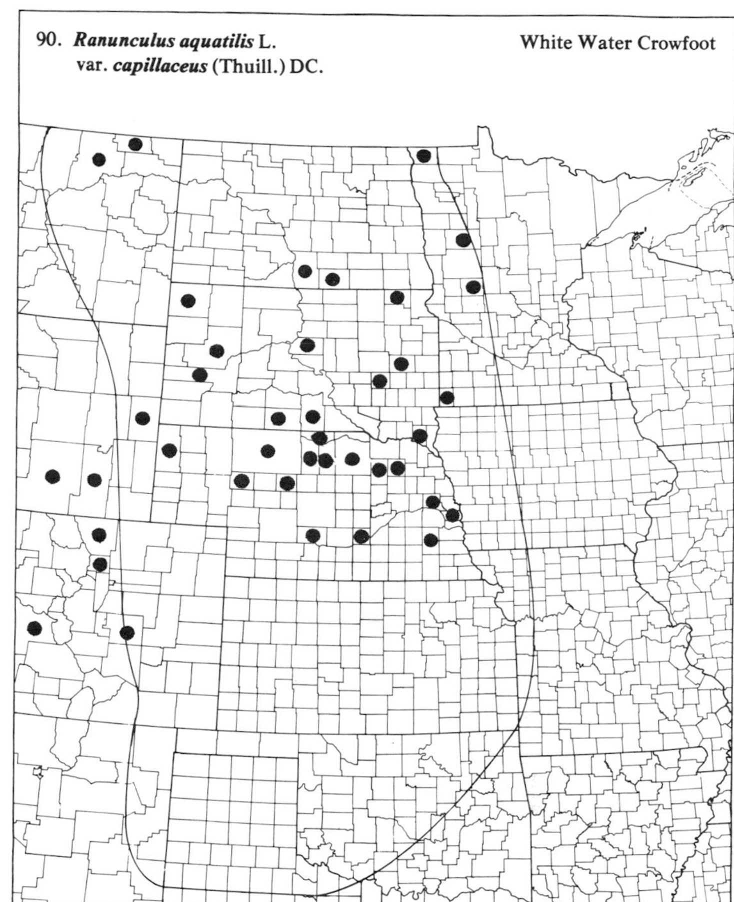

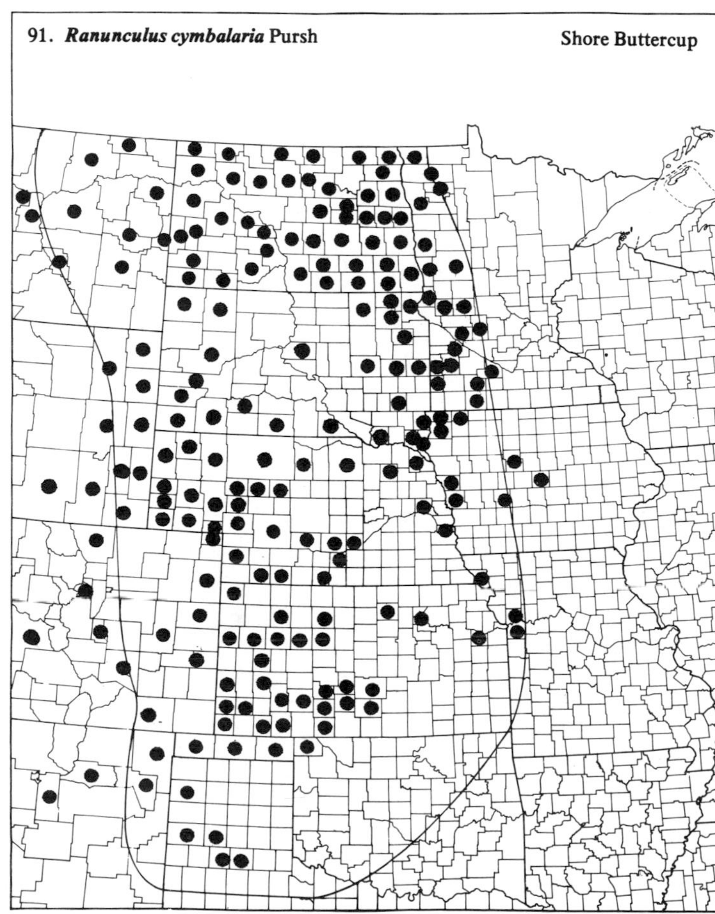

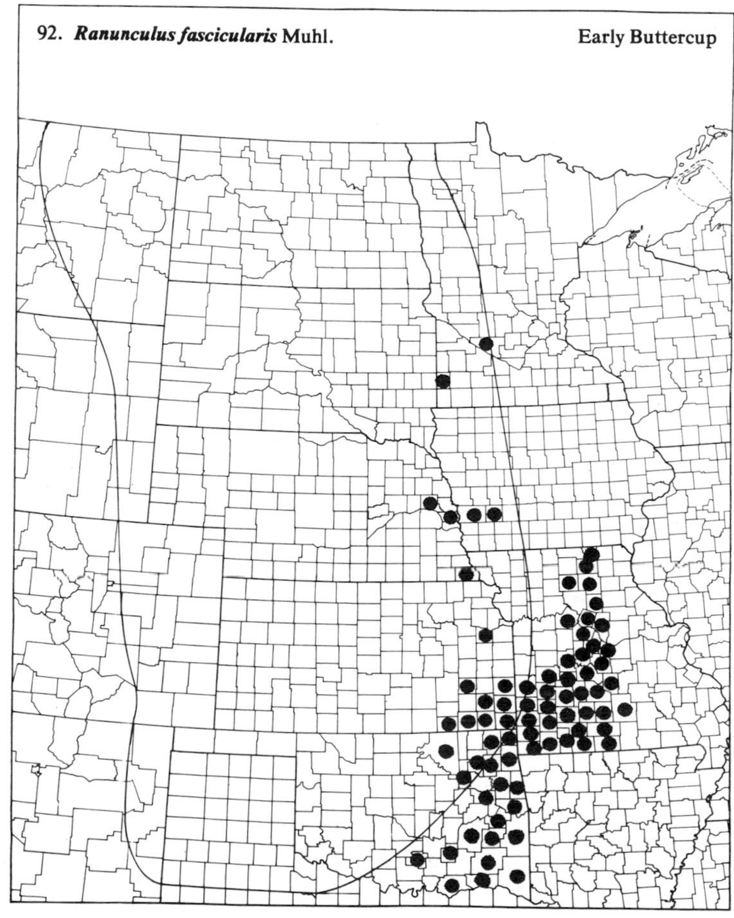

(Ranunculaceae)

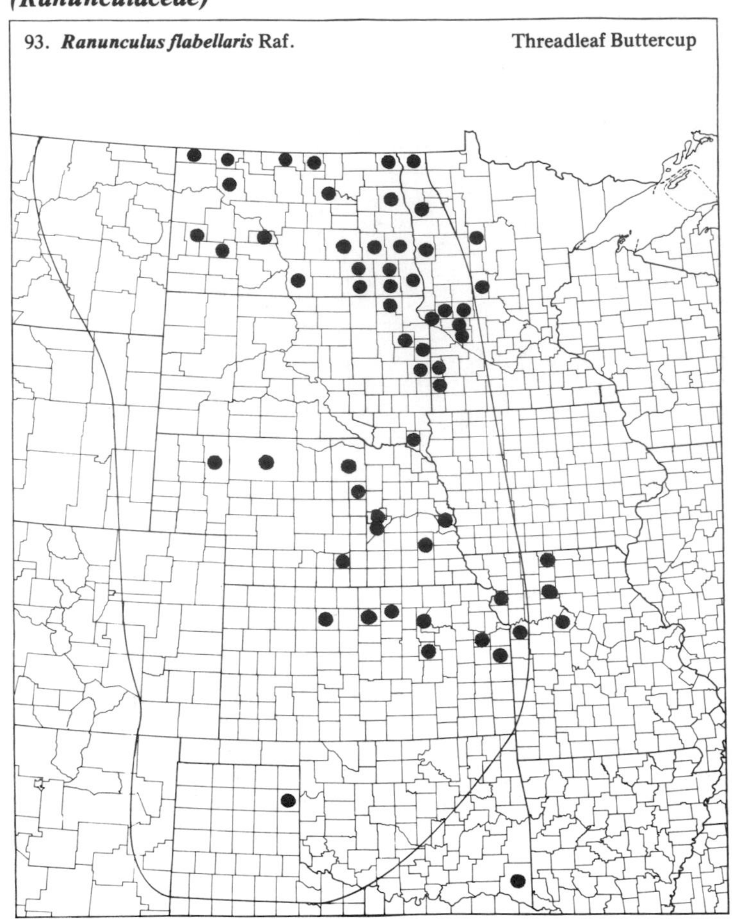

93. ***Ranunculus flabellaris*** Raf. Threadleaf Buttercup

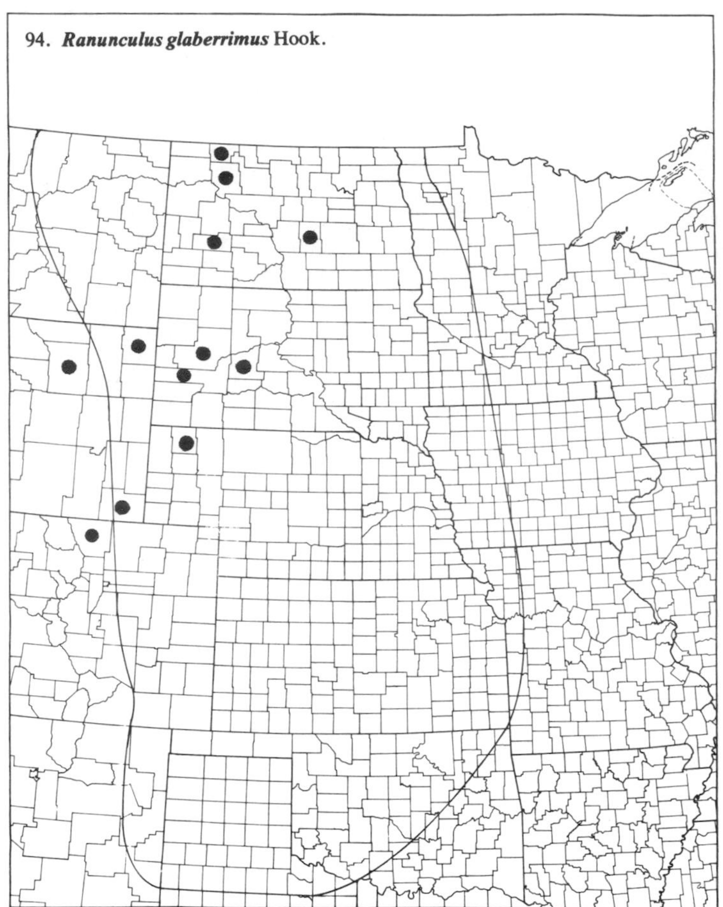

94. ***Ranunculus glaberrimus*** Hook.

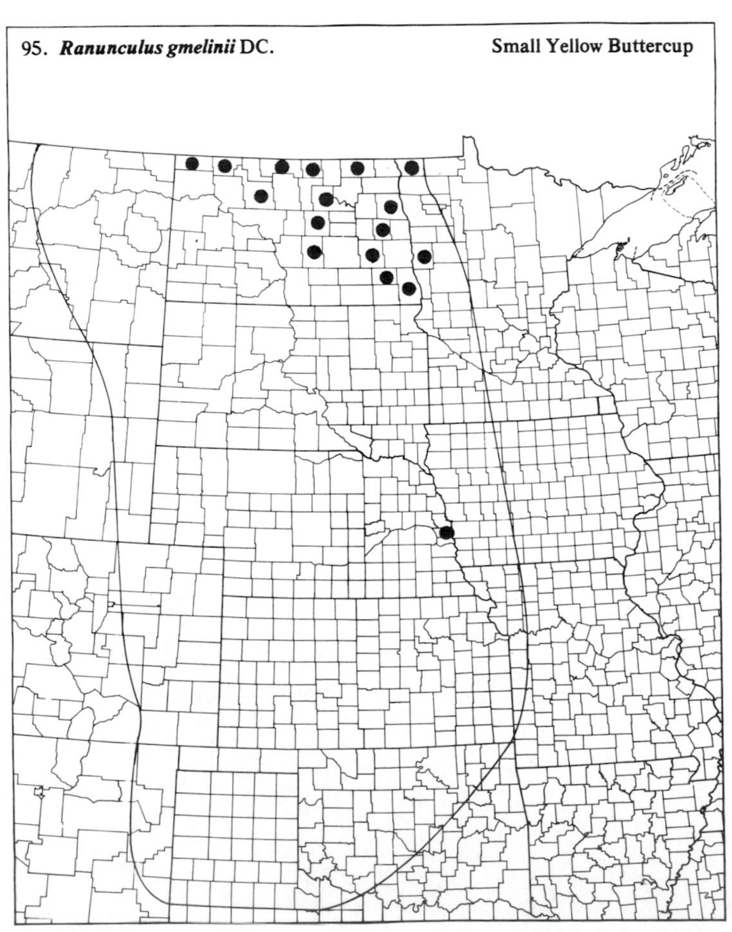

95. ***Ranunculus gmelinii*** DC. Small Yellow Buttercup

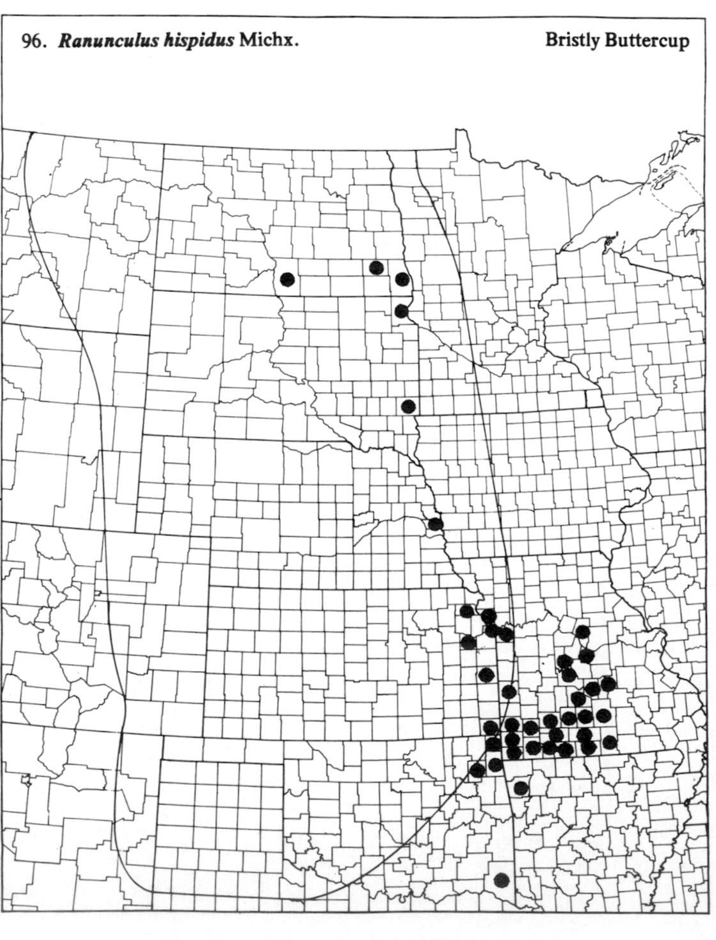

96. ***Ranunculus hispidus*** Michx. Bristly Buttercup

(Ranunculaceae)

97. ***Ranunculus longirostris*** Godr. White Water Crowfoot

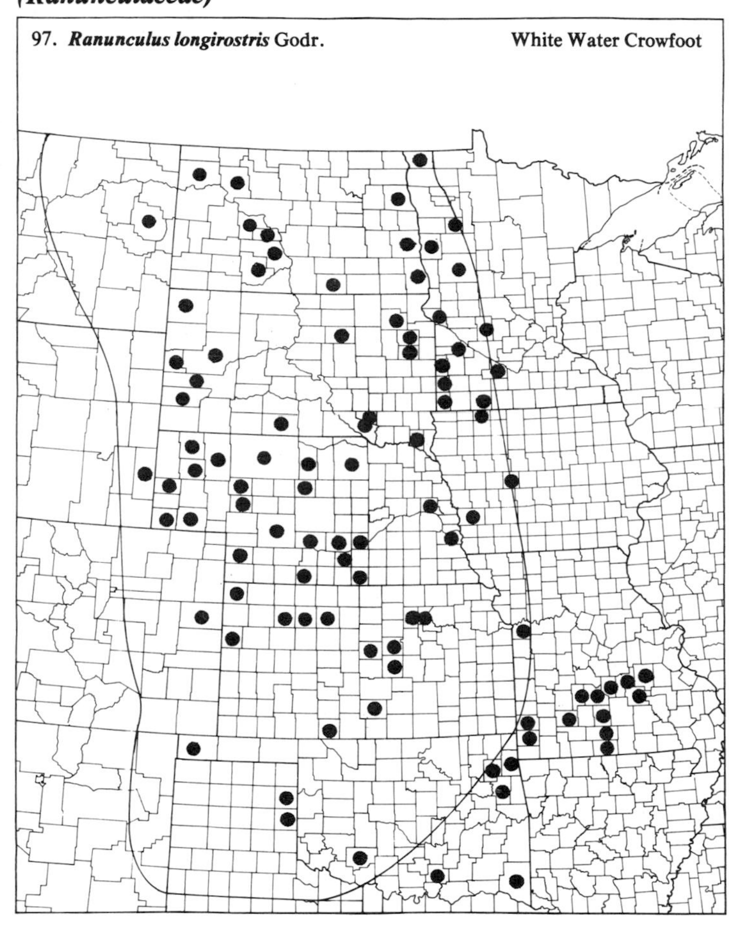

98. ***Ranunculus macounii*** Britt. Macoun's Buttercup

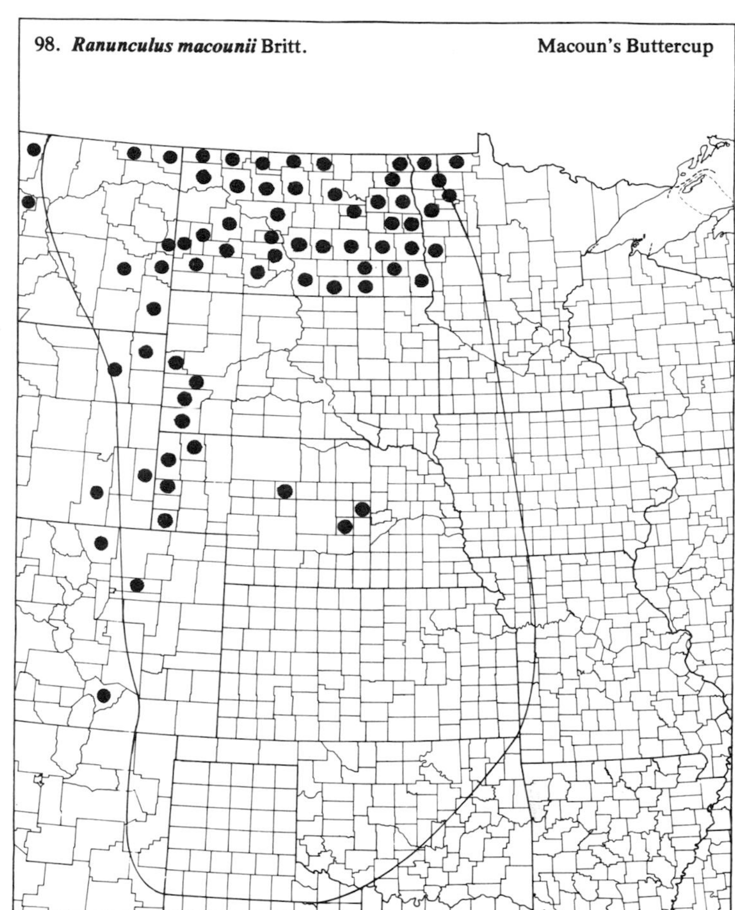

99. ***Ranunculus micranthus*** Nutt.

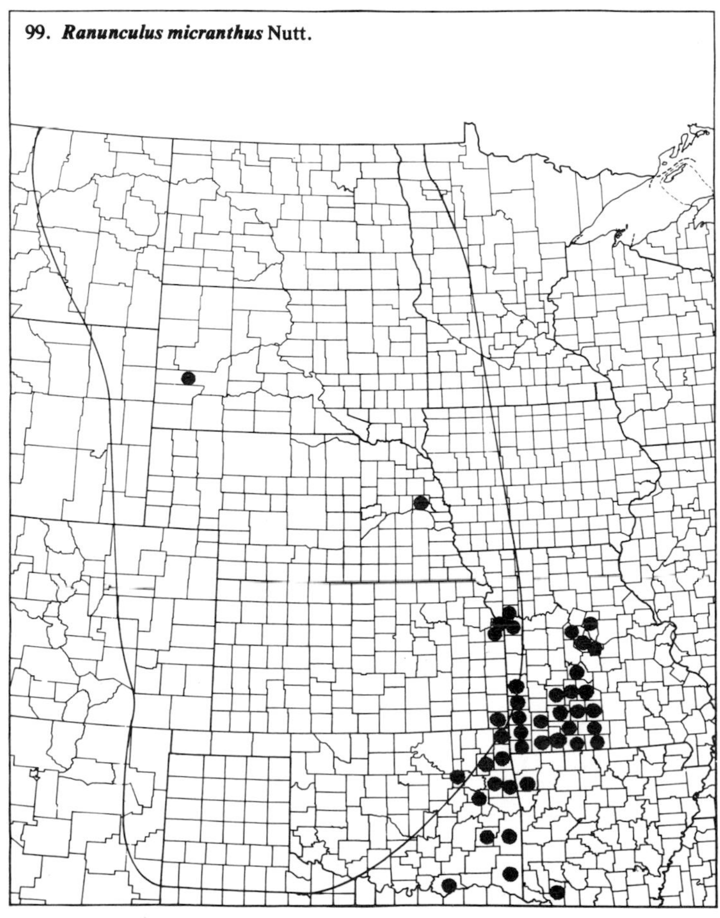

100. ***Ranunculus pensylvanicus*** L. Bristly Crowfoot

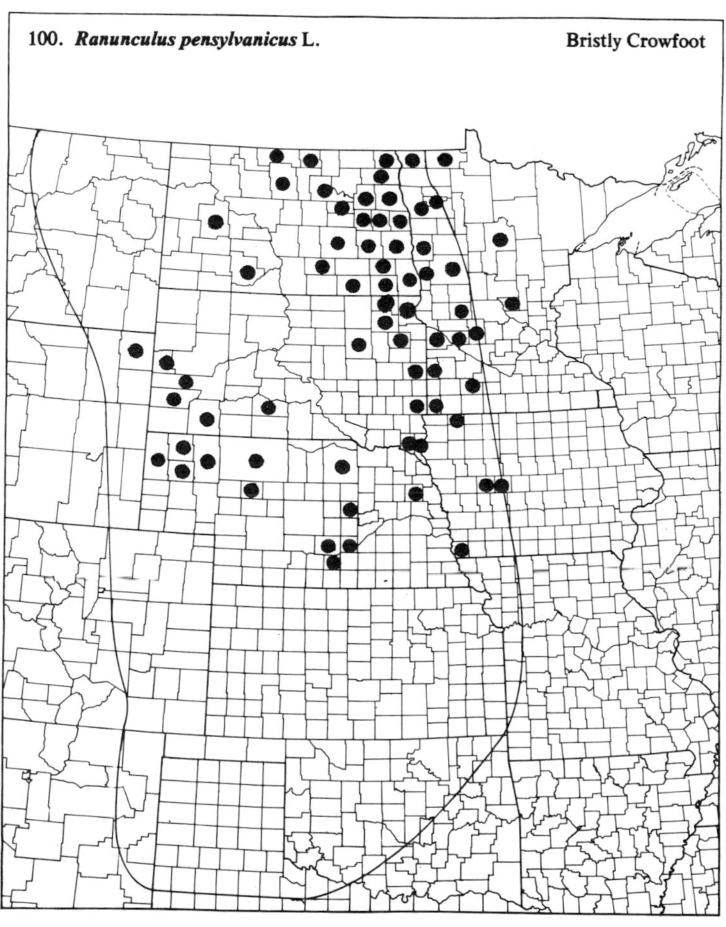

(Ranunculaceae)

101. **Ranunculus recurvatus** Poir. Hooked Buttercup

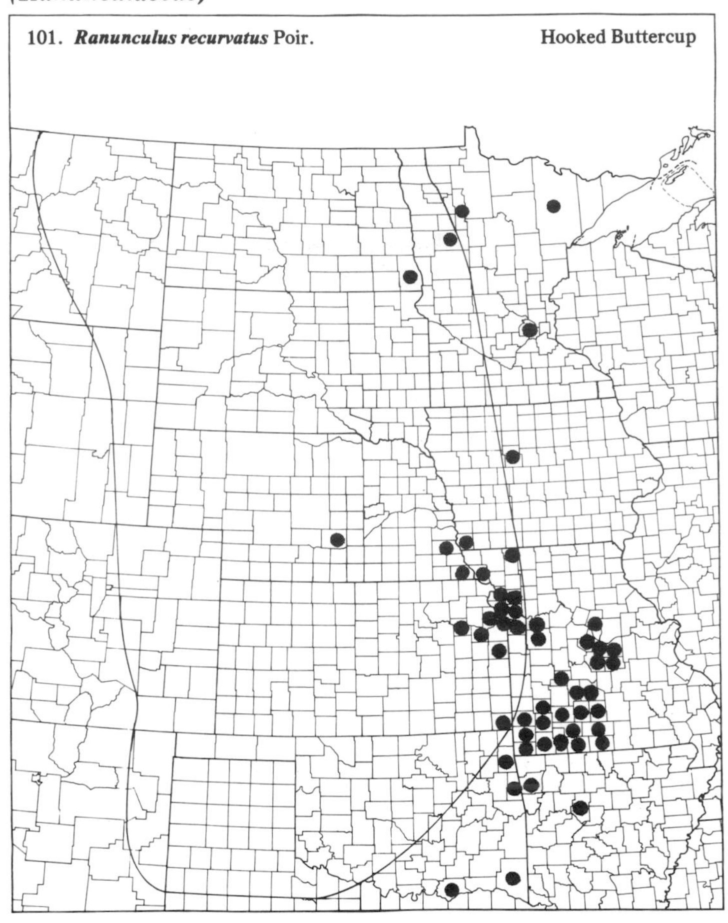

102. **Ranunculus rhomboideus** Goldie Prairie Buttercup

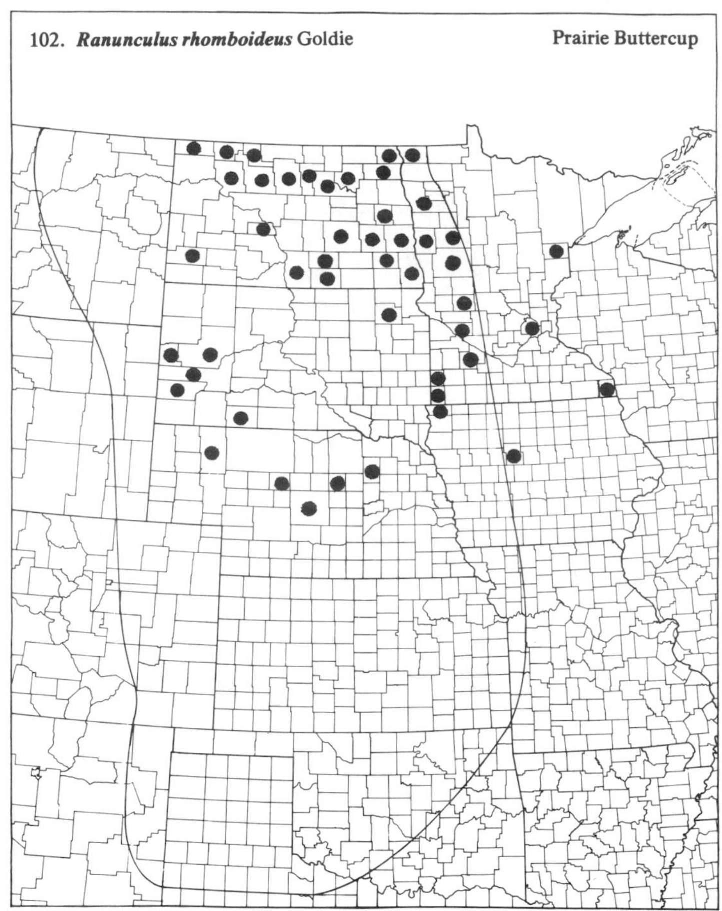

103. **Ranunculus sceleratus** L. Cursed Crowfoot

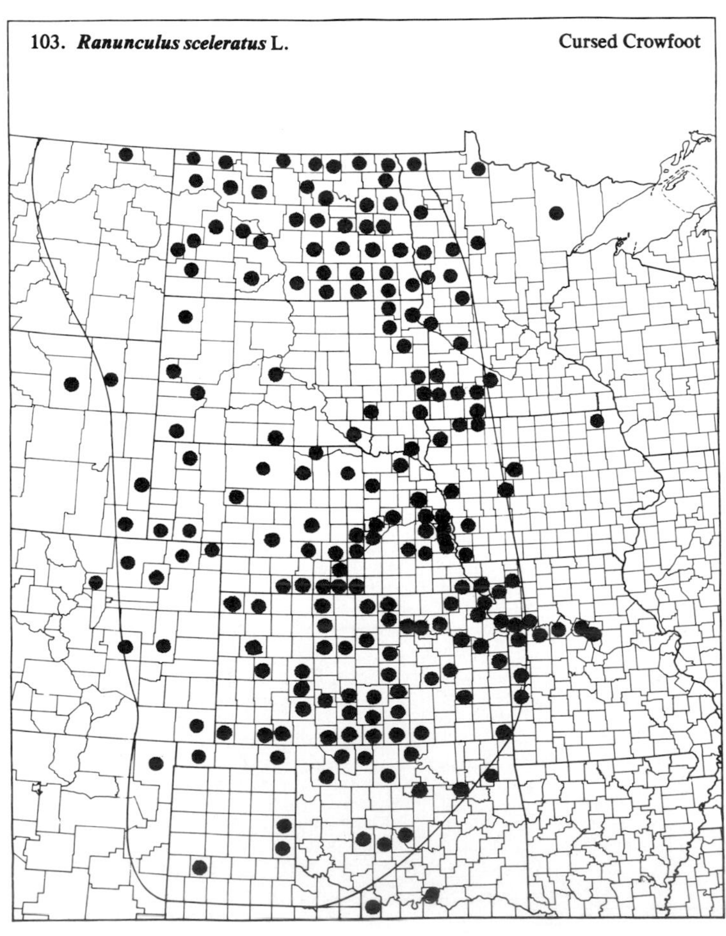

104. **Ranunculus septentrionalis** Poir. Marsh Buttercup

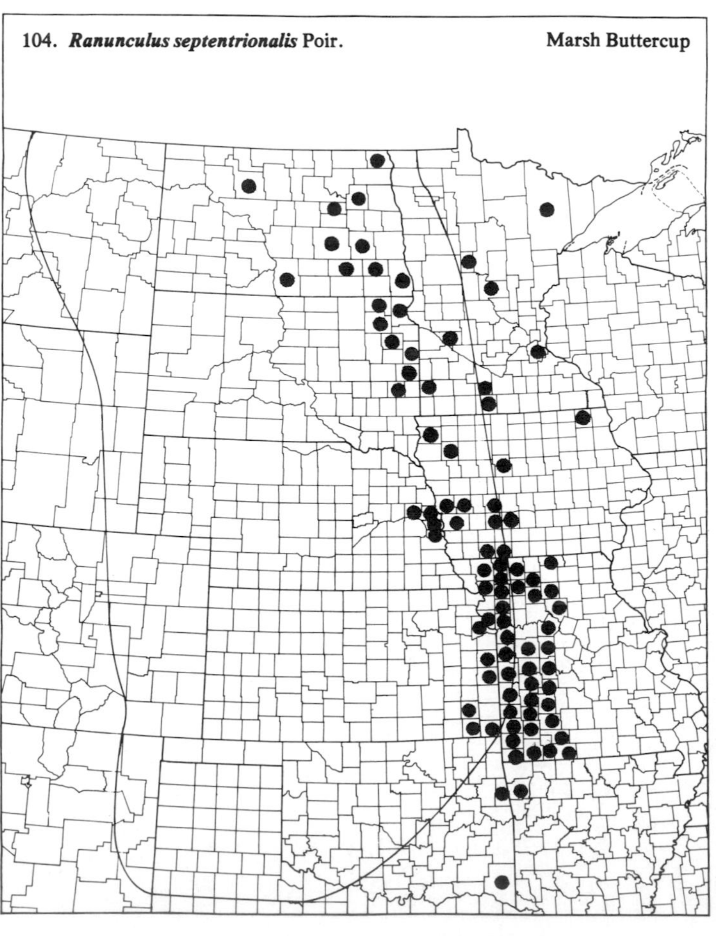

(Ranunculaceae)

105. ***Ranunculus subrigidus*** Drew — White Water Crowfoot

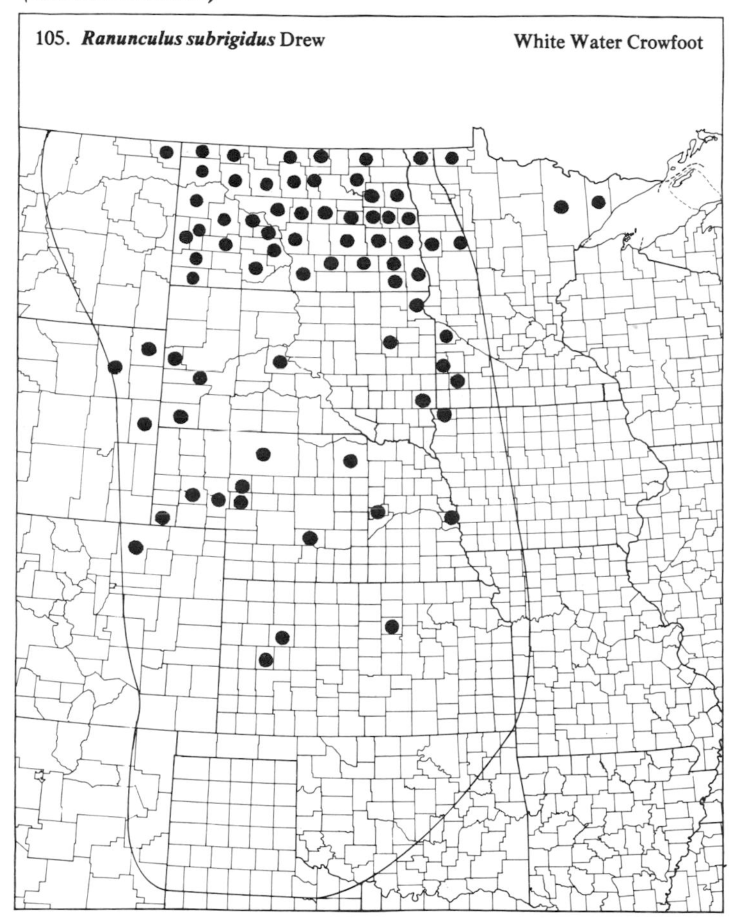

106. ***Thalictrum dasycarpum*** Fisch. & Lall. — Purple Meadowrue

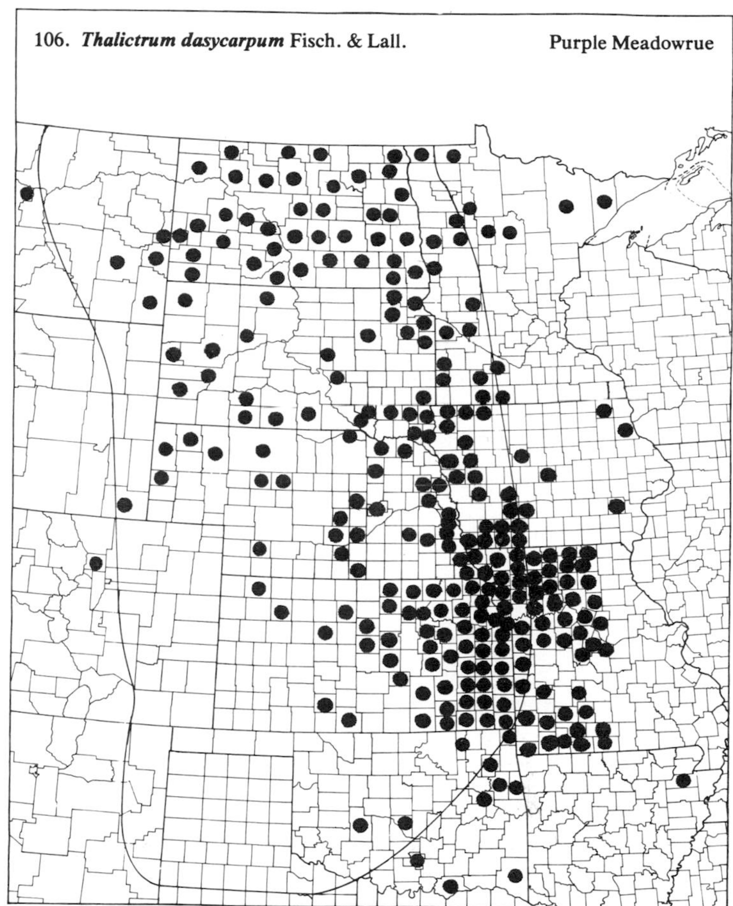

107. ***Thalictrum dioicum*** L. — Quicksilver-weed

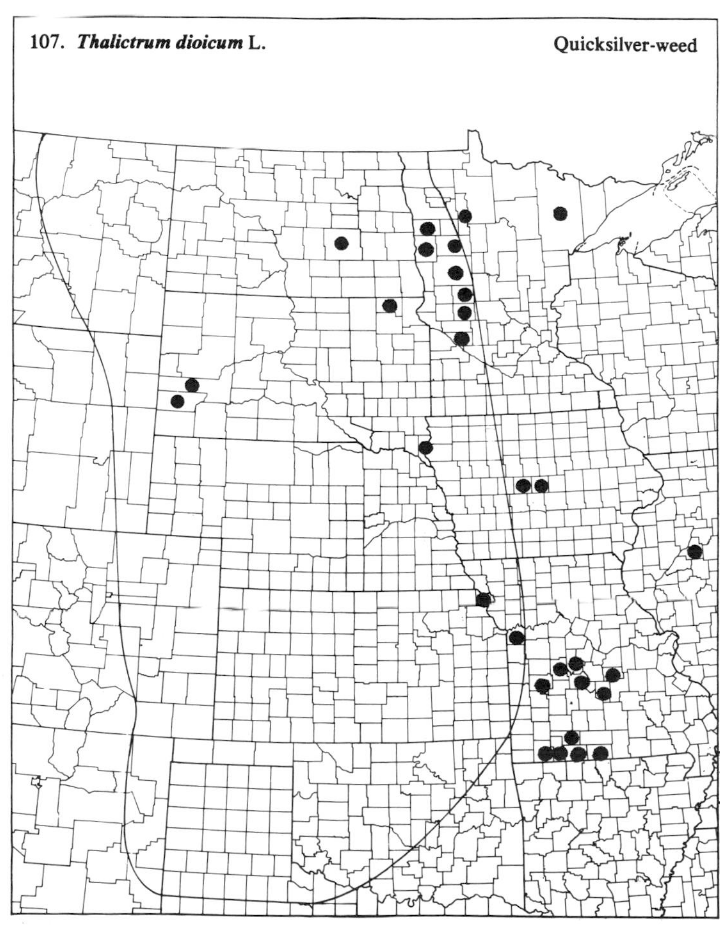

108. ***Thalictrum venulosum*** Trel. — Early Meadowrue

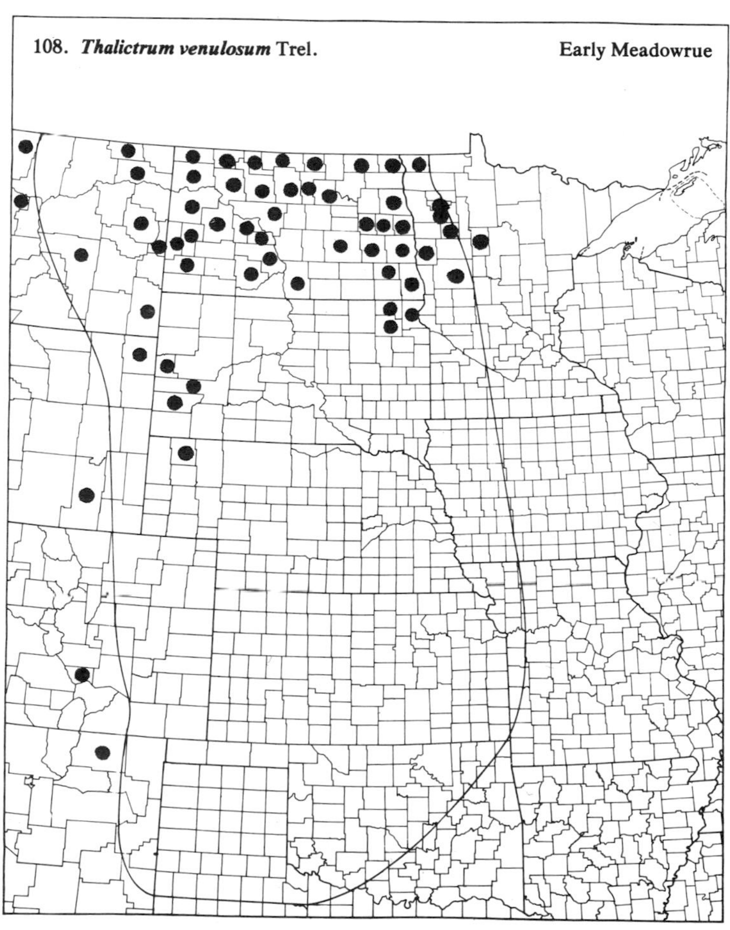

8. *Berberidaceae* (Barberry Family)

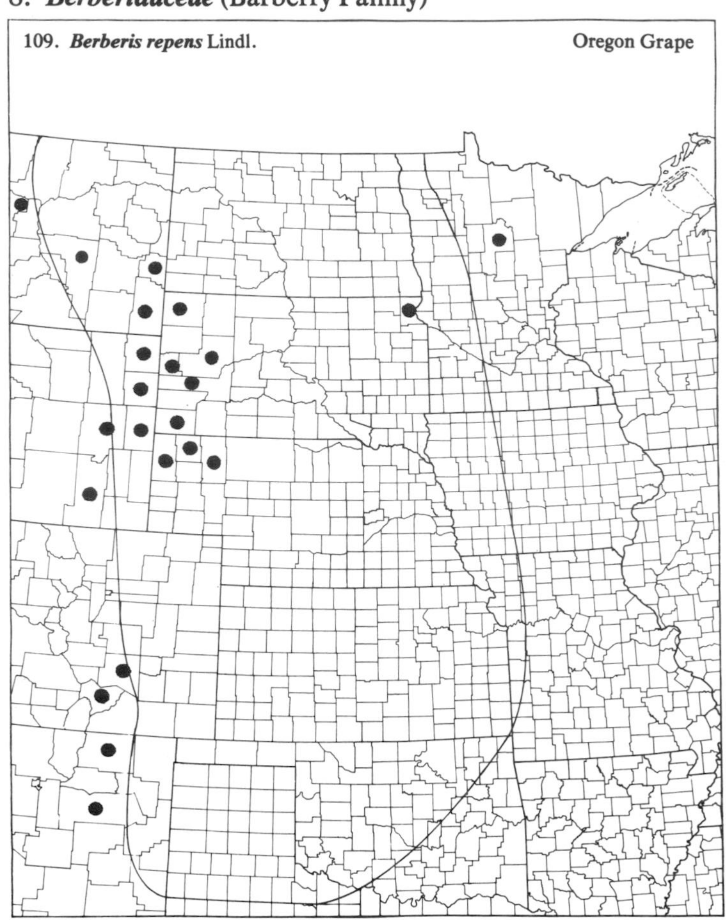

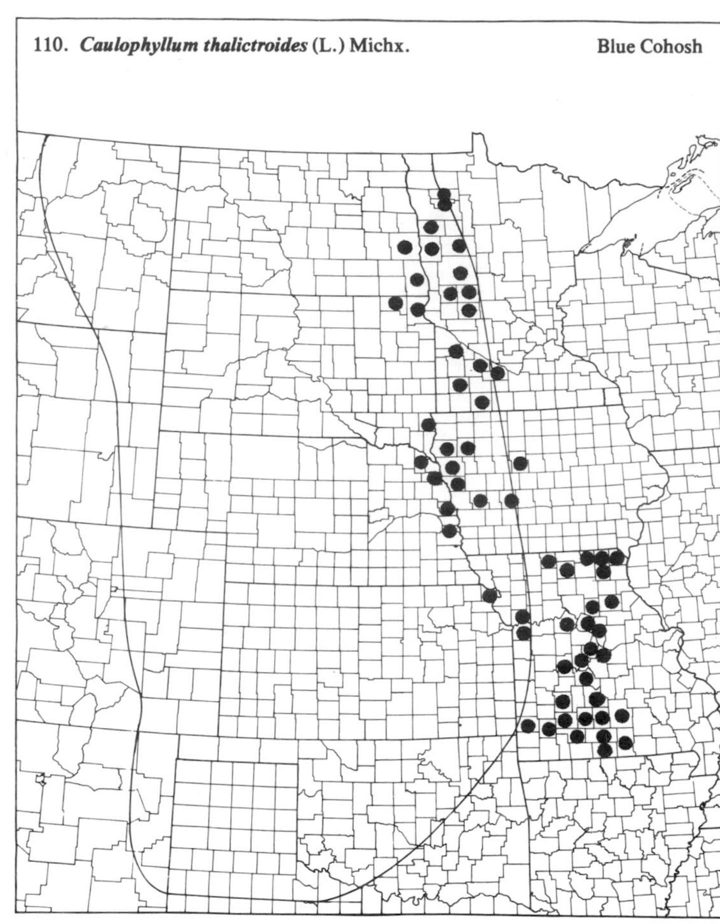

9. *Menispermaceae* (Moonseed Family)

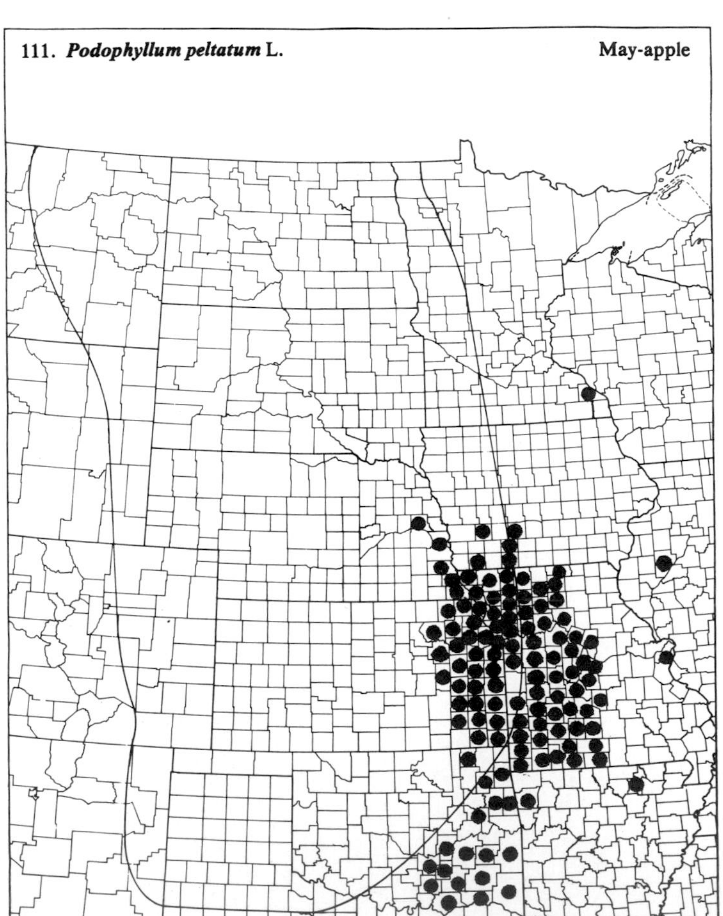

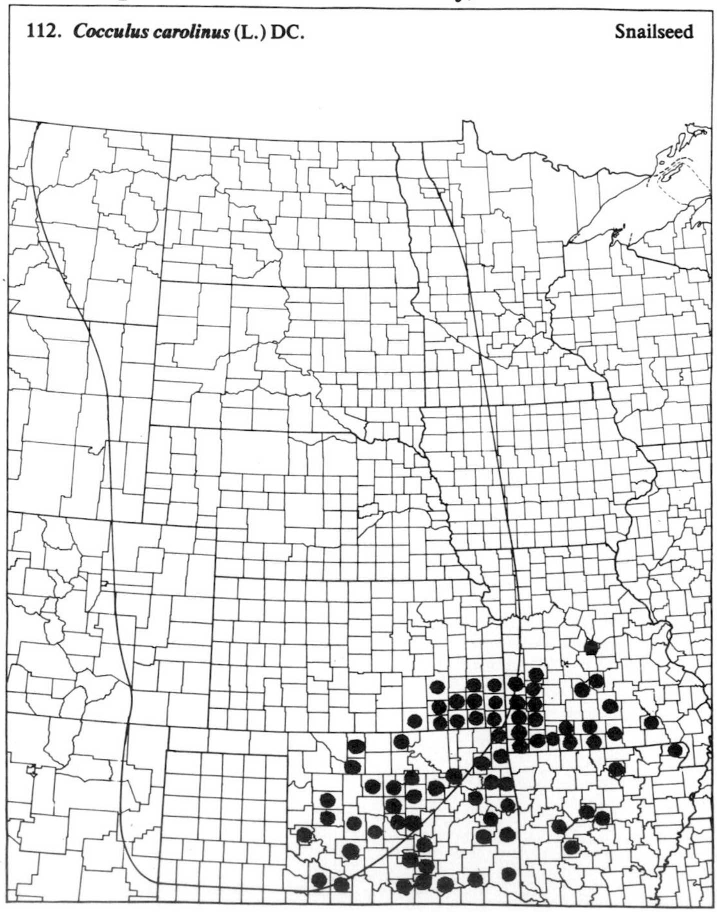

(Menispermaceae)

113. ***Menispermum canadense*** L. Moonseed

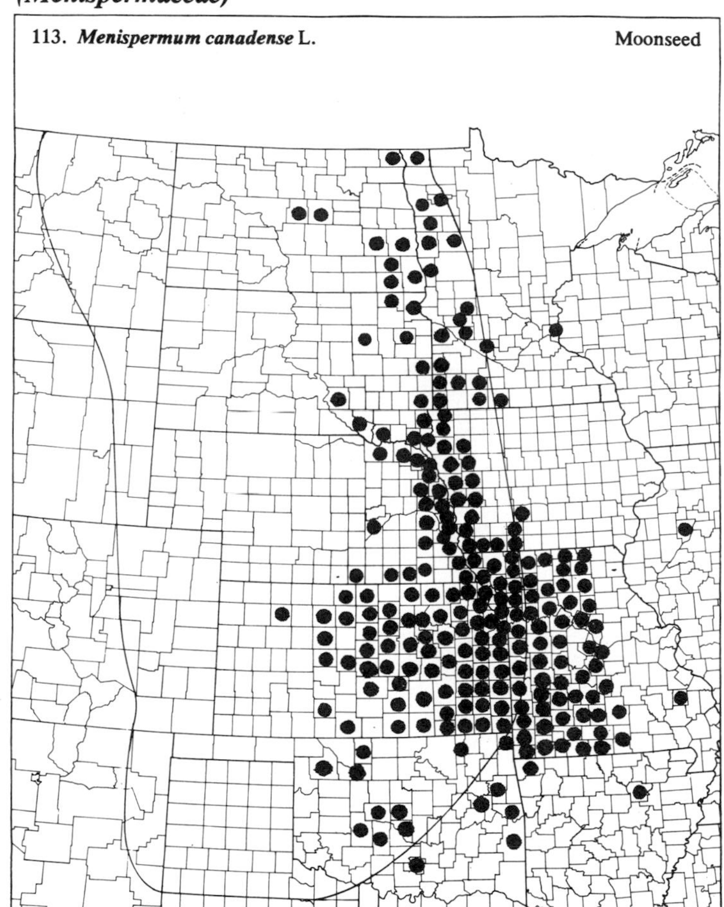

10. *Papaveraceae* (Poppy Family)

114. ***Argemone polyanthemos*** (Fedde) G.Ownbey Prickly Poppy

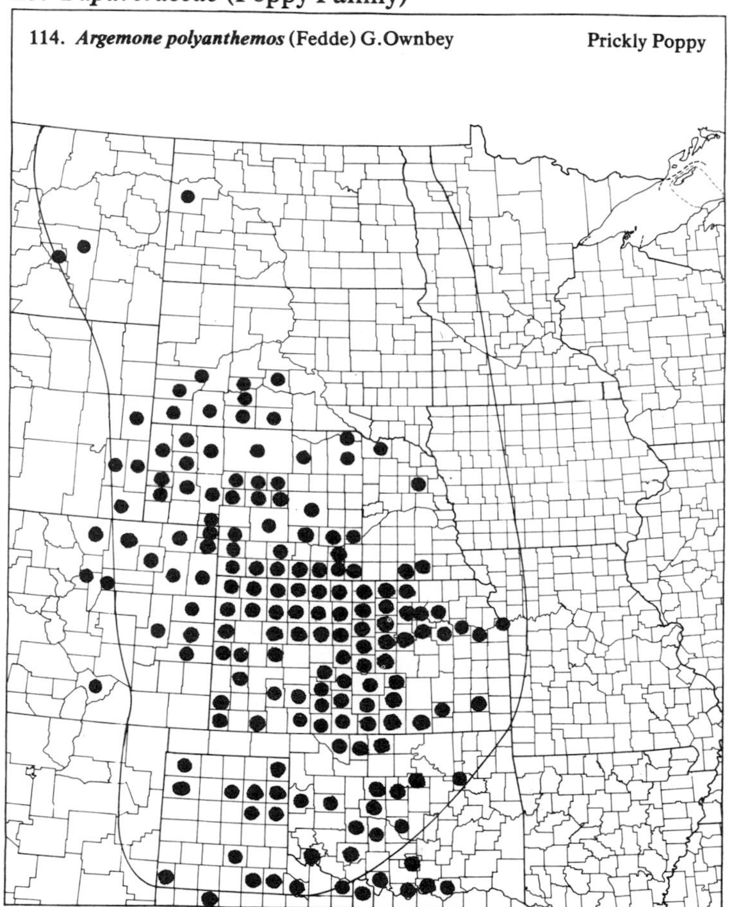

115. ***Argemone squarrosa*** Greene Hedgehog Prickly Poppy

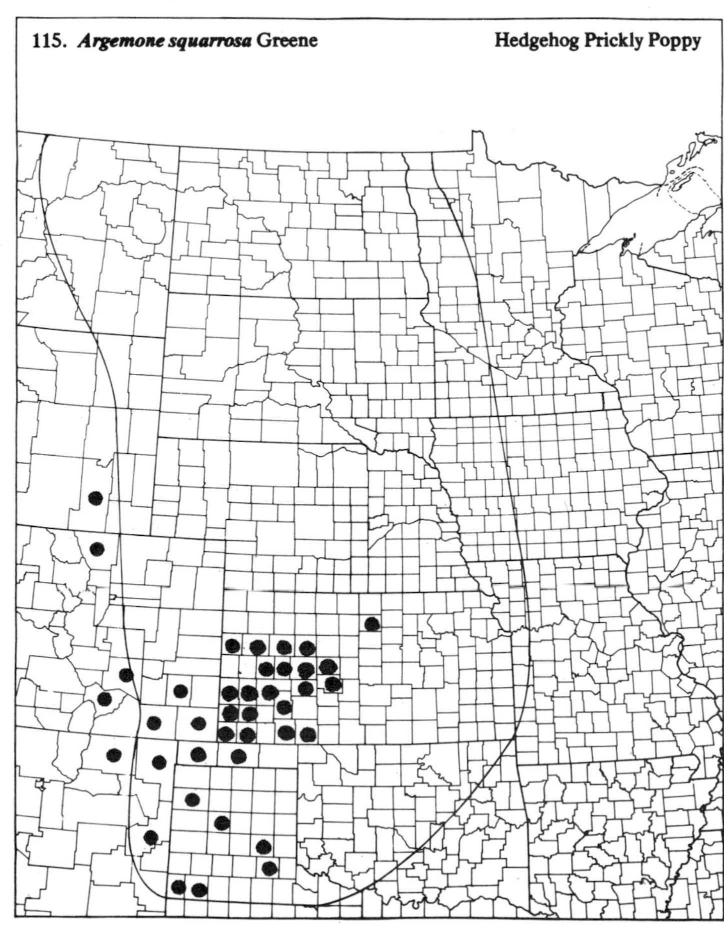

116. ***Sanguinaria canadensis*** L. Bloodroot

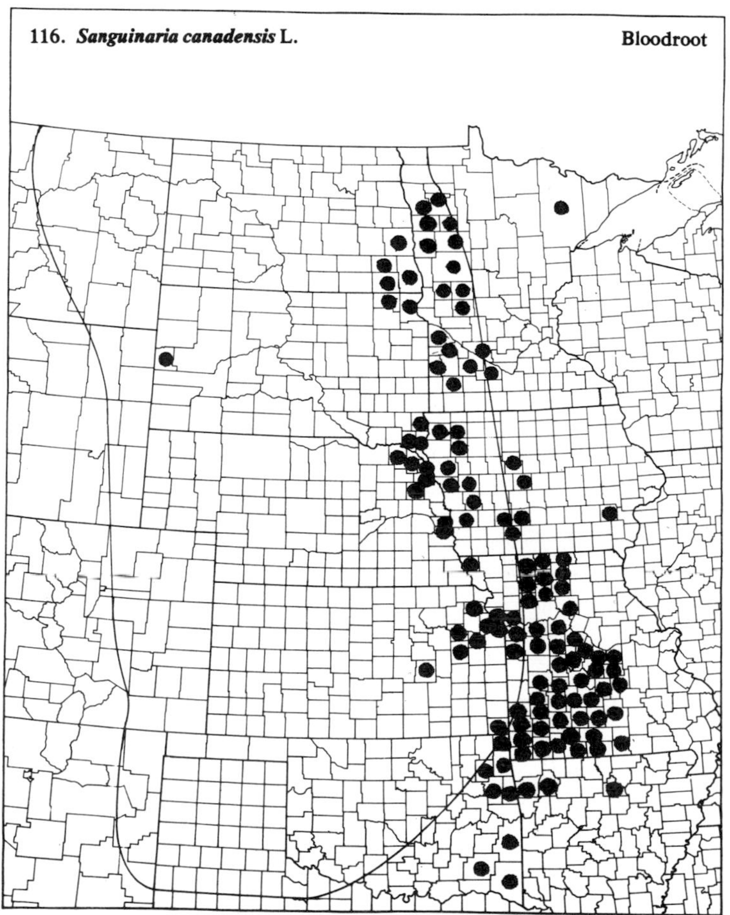

11. *Fumariaceae* (Dutchman's Breeches Family)

117. ***Corydalis aurea*** Willd. var. ***aurea*** — Golden Corydalis

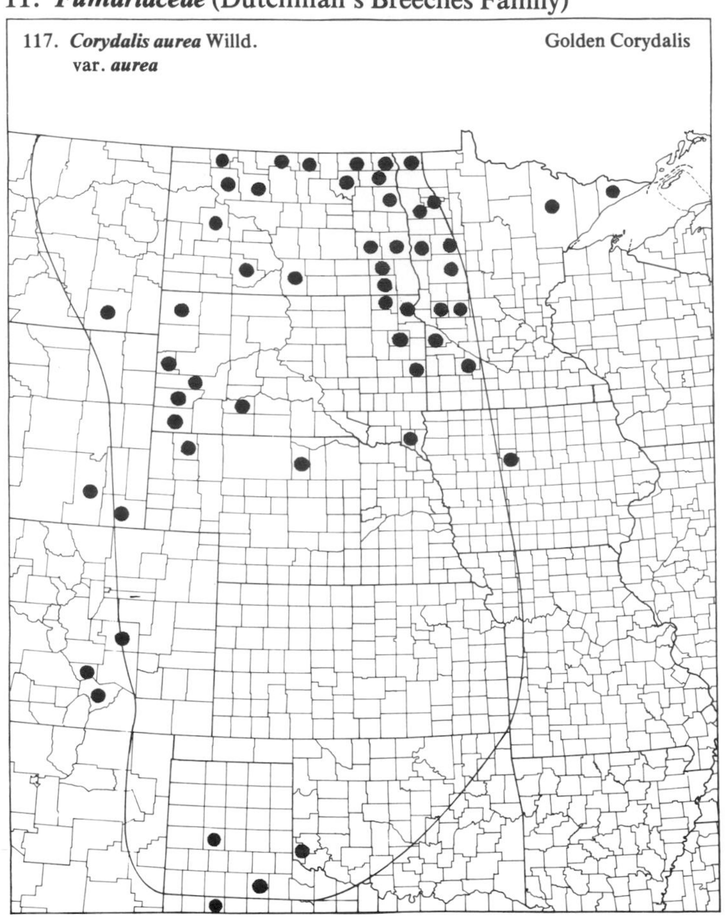

118. ***Corydalis aurea*** Willd. var. ***occidentalis*** Engelm. — Golden Corydalis

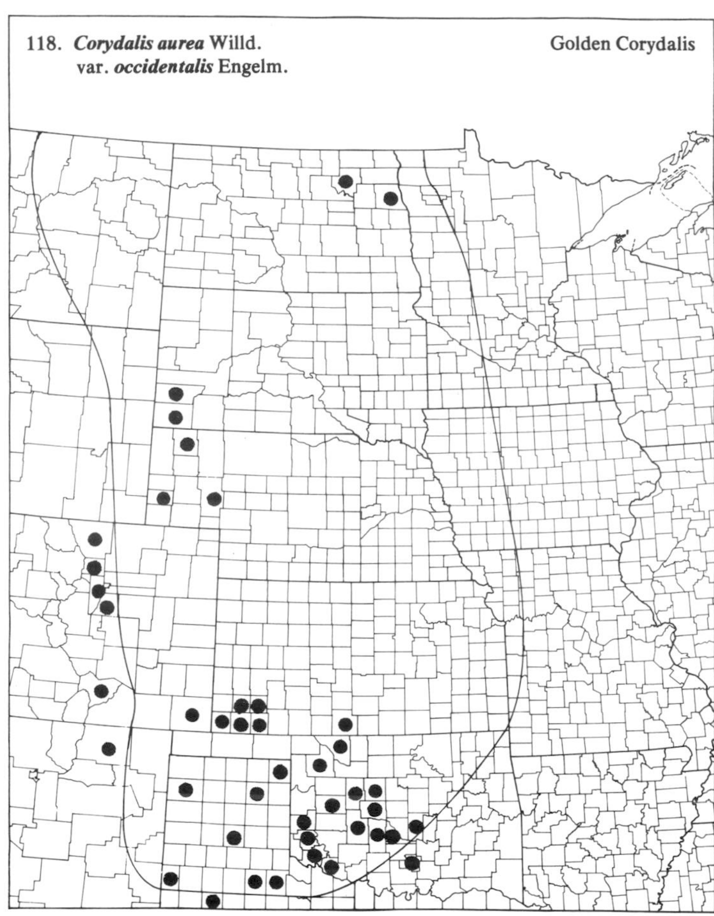

119. ***Corydalis crystallina*** Engelm. — Mealy Corydalis

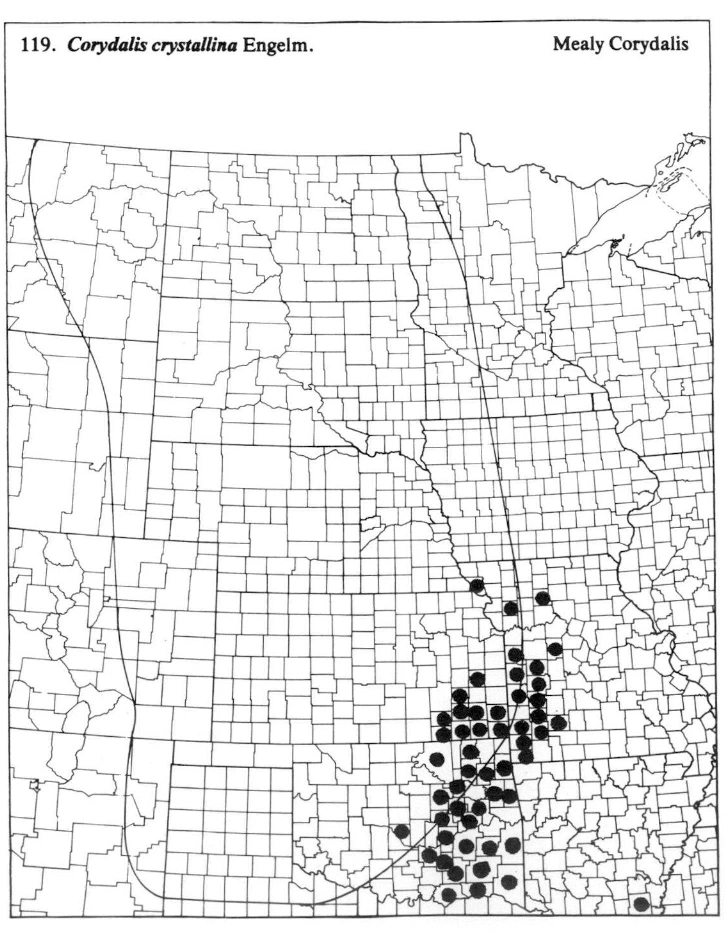

120. ***Corydalis curvisiliqua*** Engelm. ssp. ***grandibracteata*** (Fedde) G. Ownbey

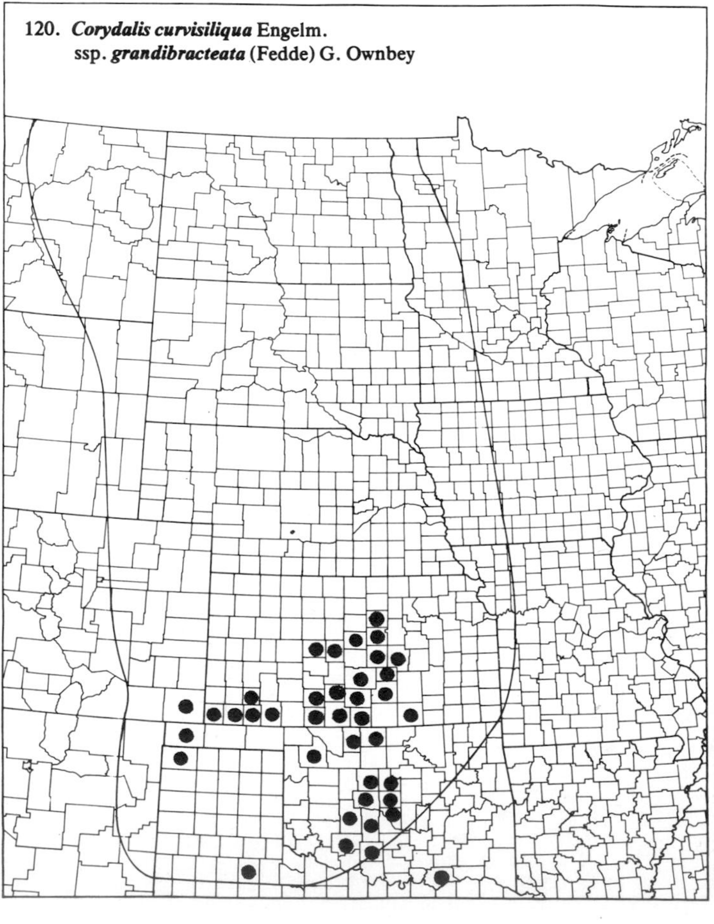

(Fumariaceae)

121. ***Corydalis flavula*** (Raf.) DC. — Yellow Harlequin

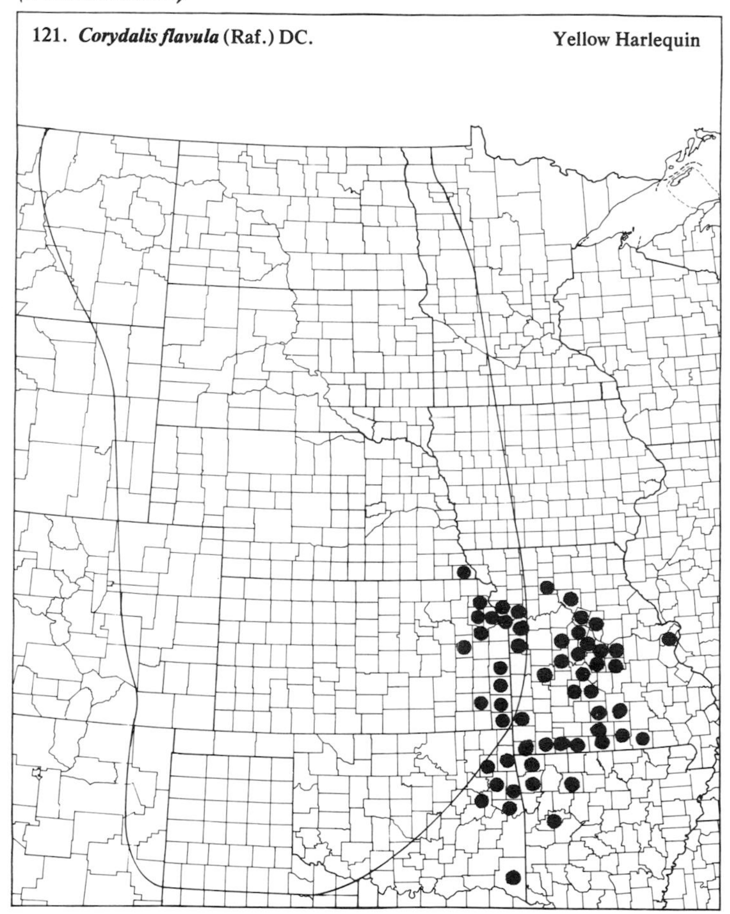

122. ***Corydalis micrantha*** (Engelm.) Gray — Slender Fumewort

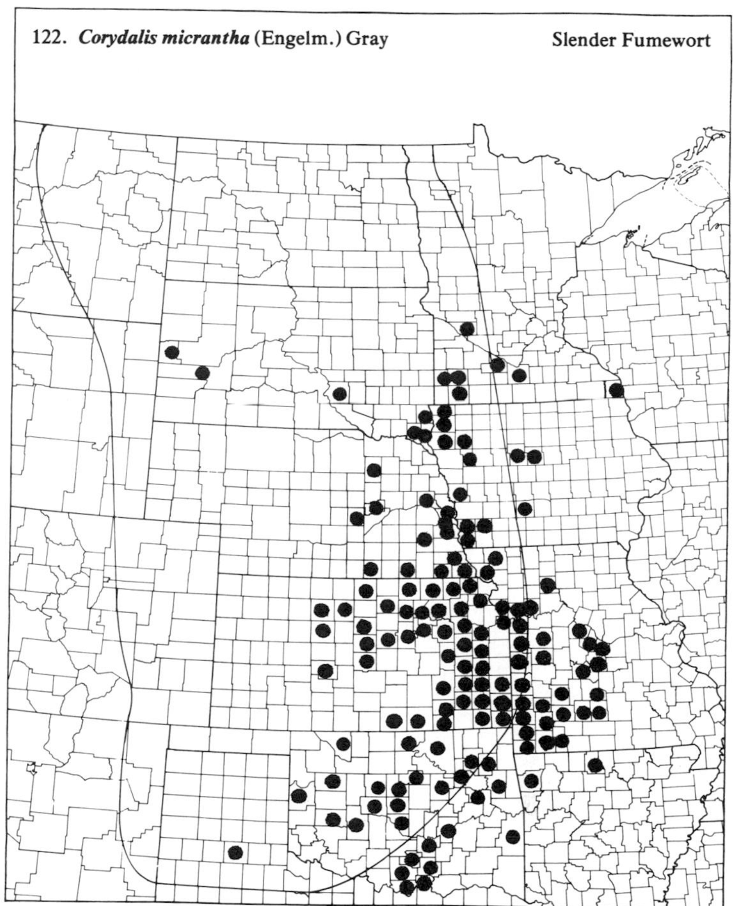

123. ***Dicentra cucullaria*** (L.) Bernh. — Dutchman's Breeches

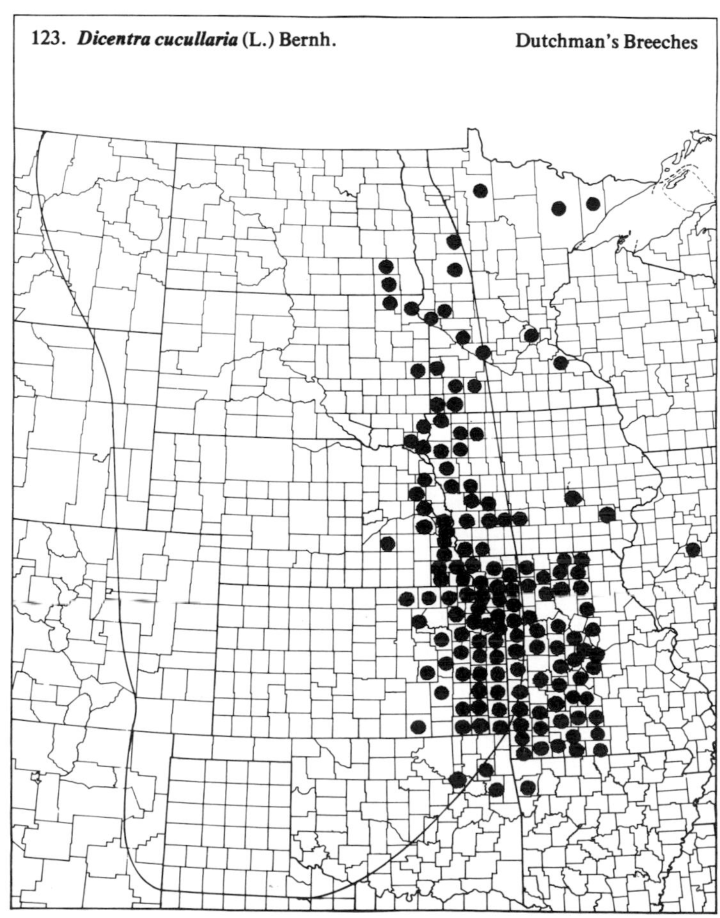

124. ***Fumaria officinalis*** L. — Common Fumitory

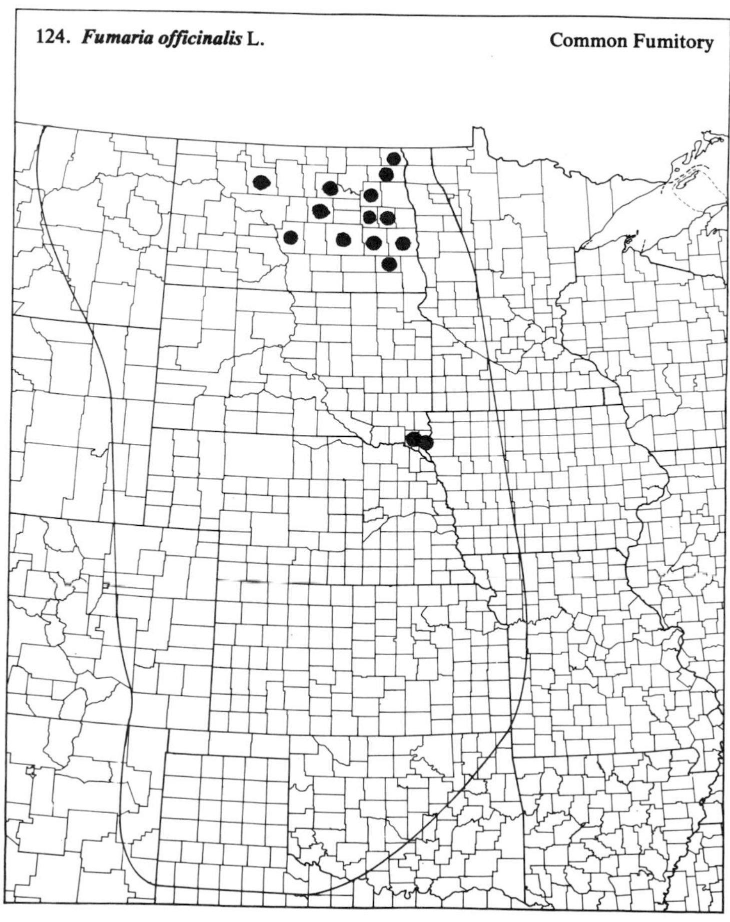

12. *Platanaceae* (Sycamore Family)

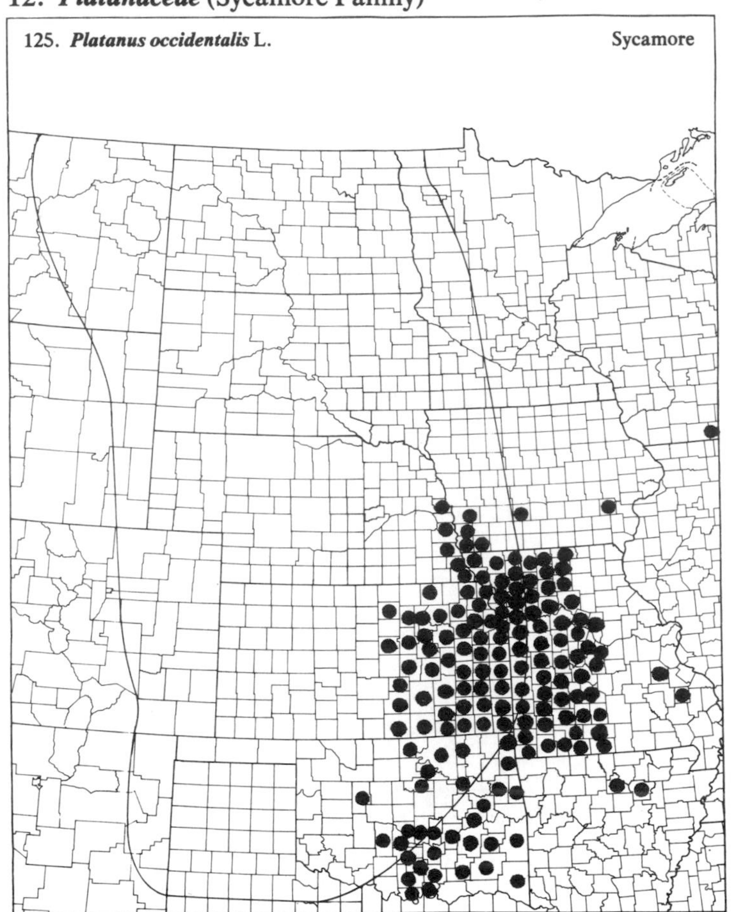

13. *Ulmaceae* (Elm Family)

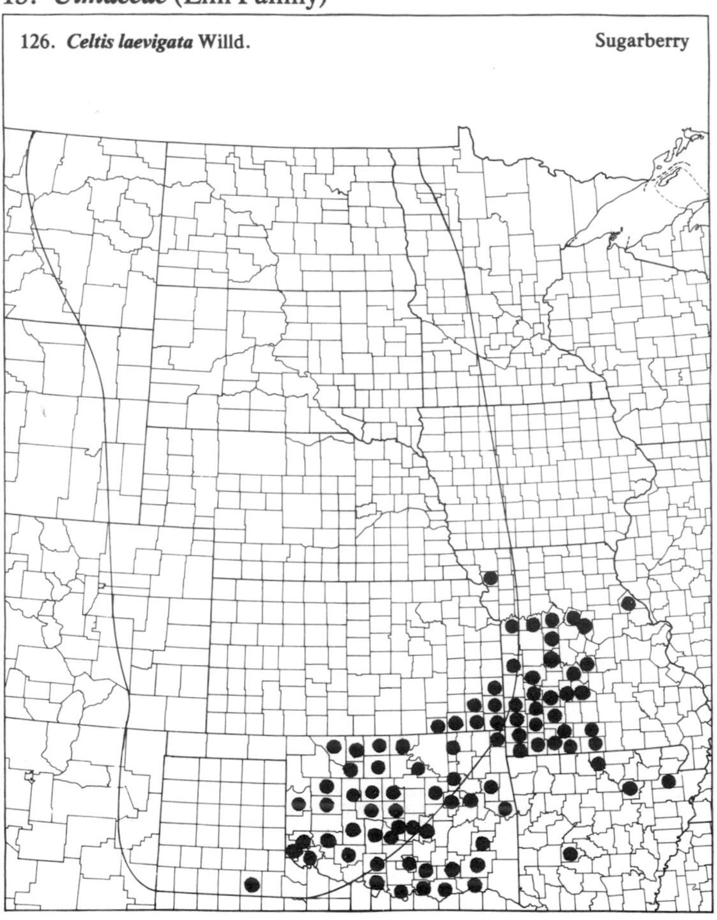

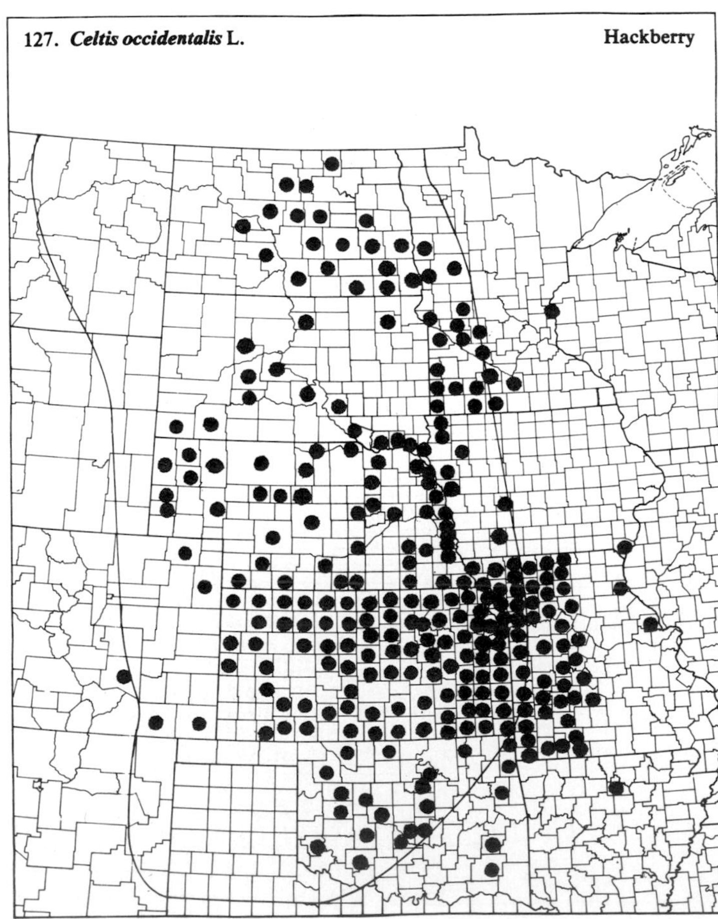

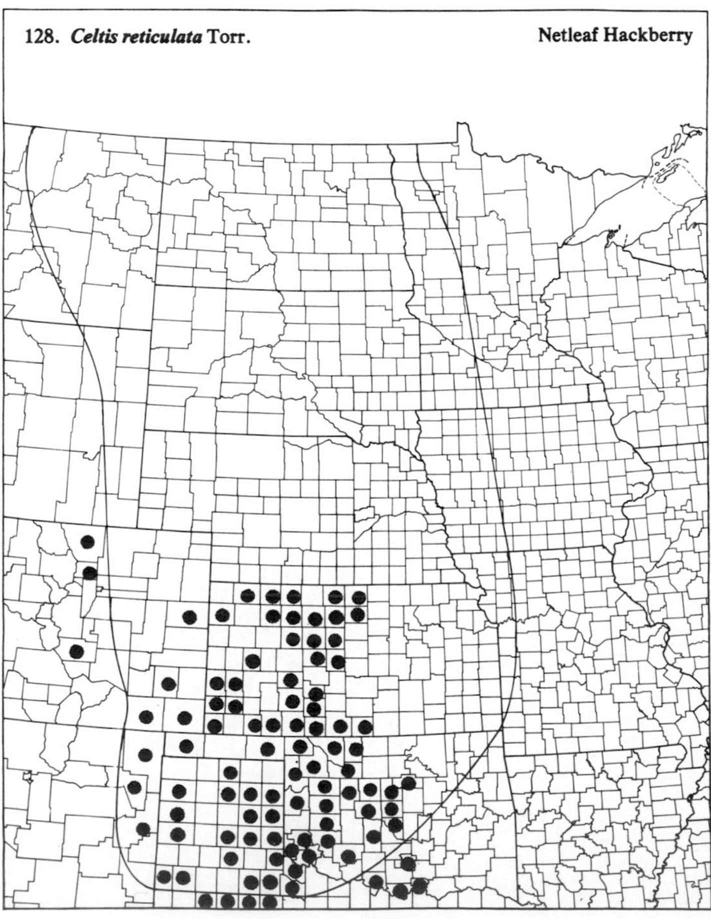

(Ulmaceae)

129. ***Celtis tenuifolia*** Nutt. — Dwarf Hackberry

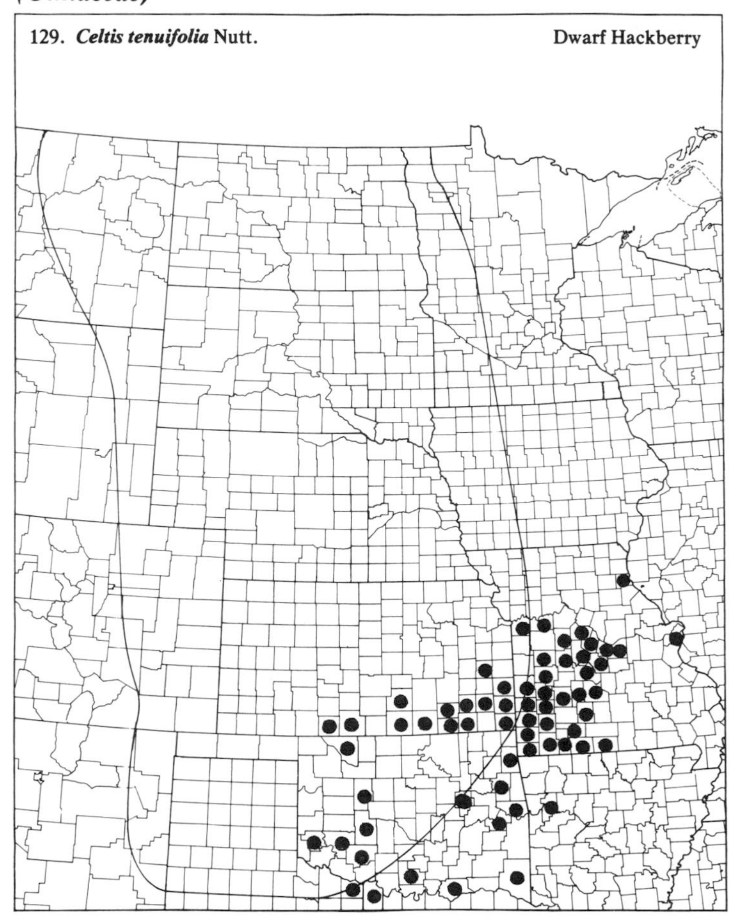

130. ***Ulmus americana*** L. — American Elm

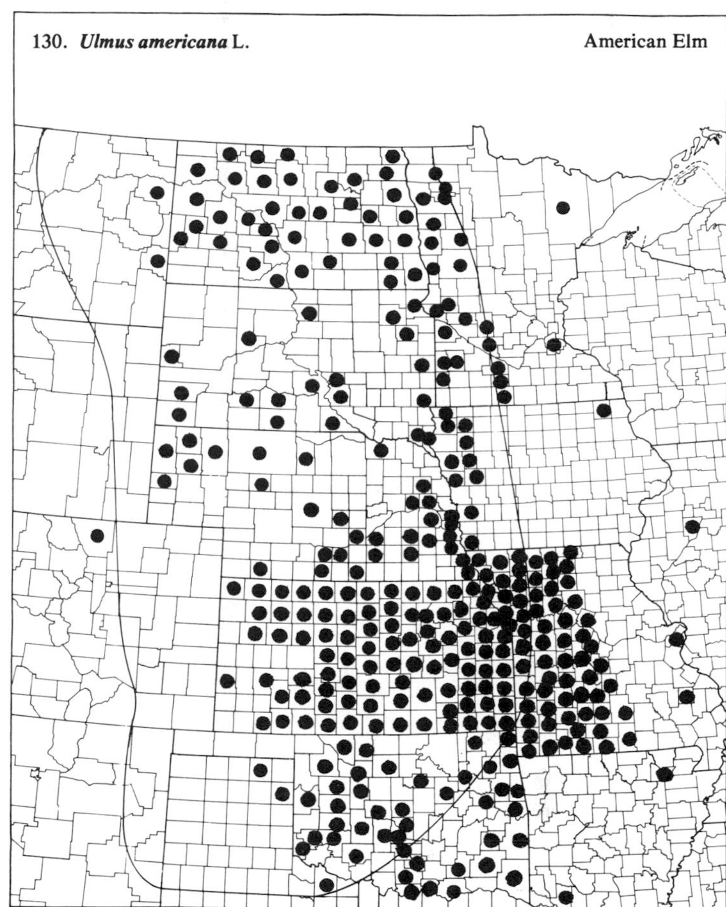

131. ***Ulmus pumila*** L. — Siberian Elm

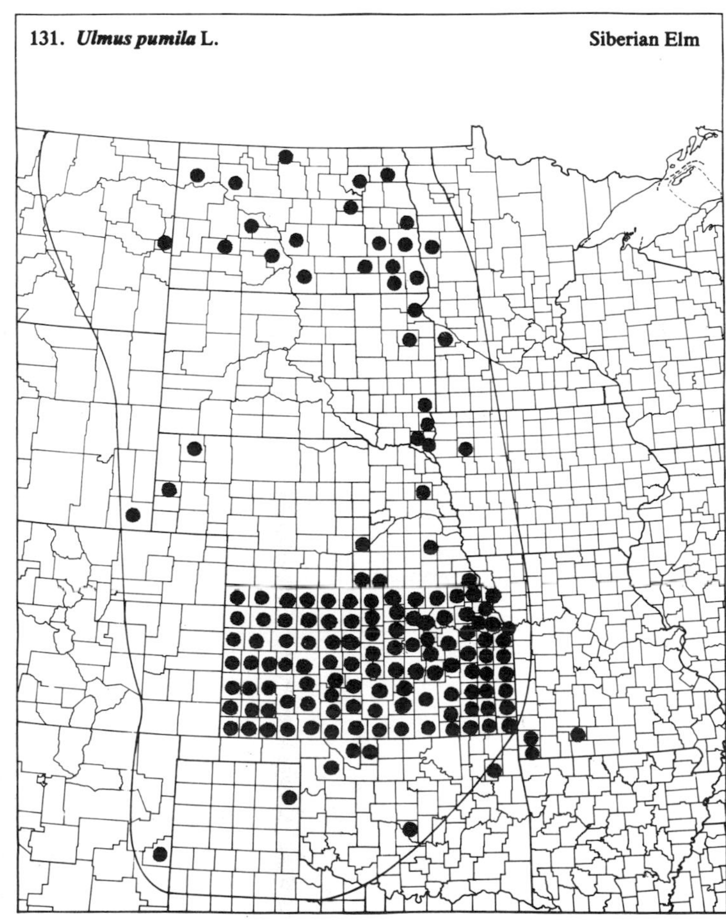

132. ***Ulmus rubra*** Muhl. — Slippery Elm

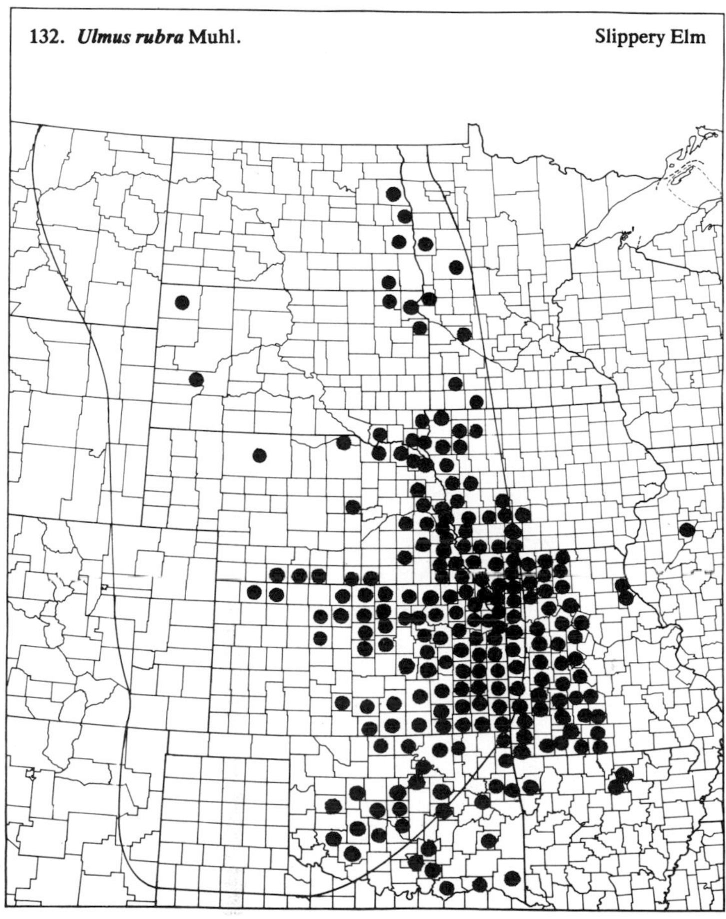

(Ulmaceae)

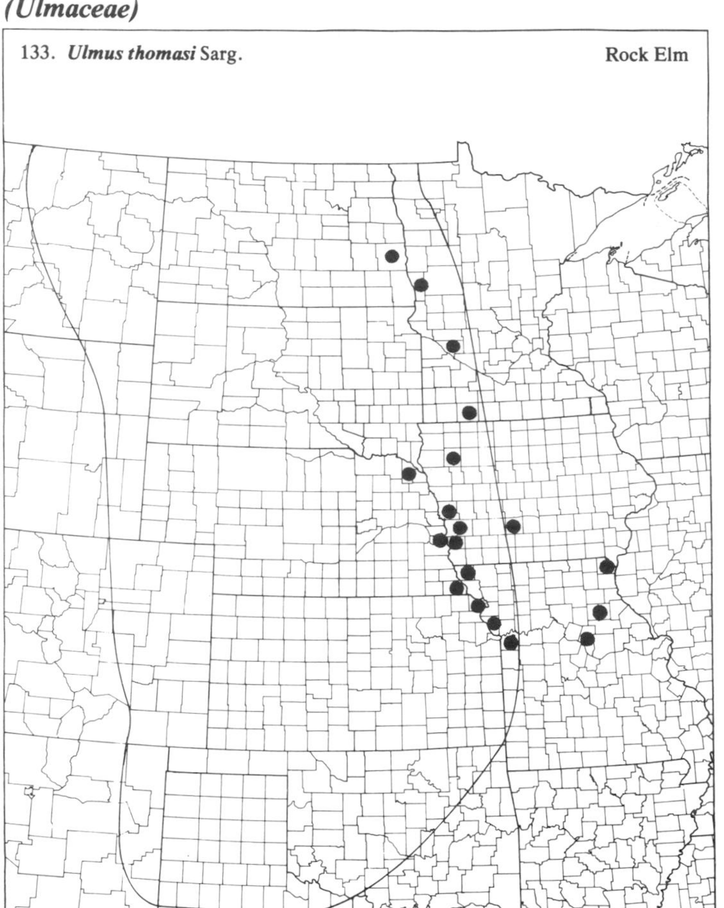

133. ***Ulmus thomasi*** Sarg. Rock Elm

14. *Moraceae* (Mulberry Family)

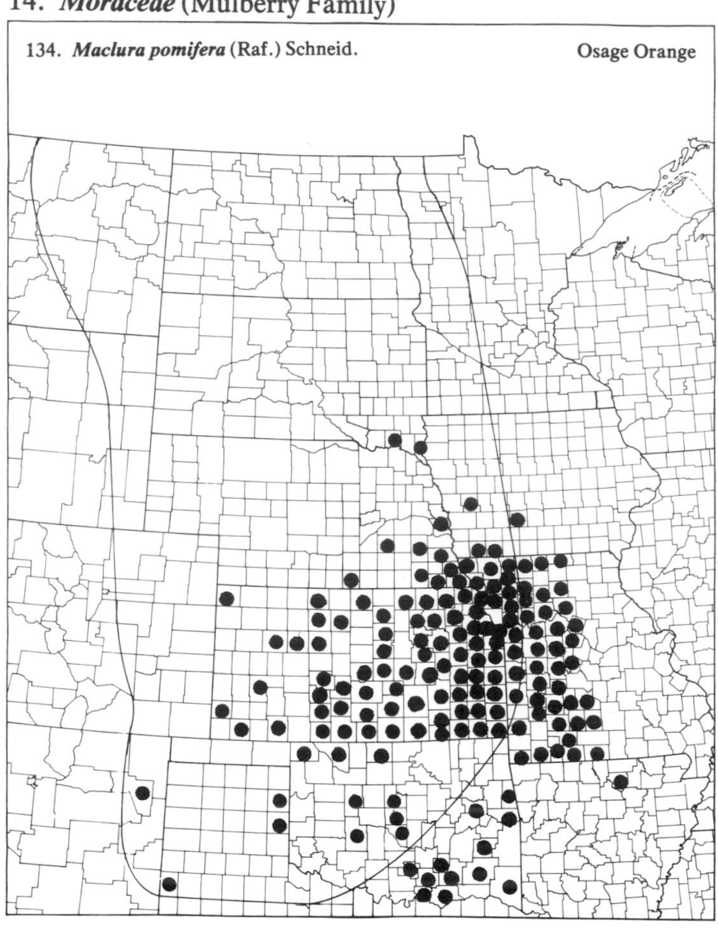

134. ***Maclura pomifera*** (Raf.) Schneid. Osage Orange

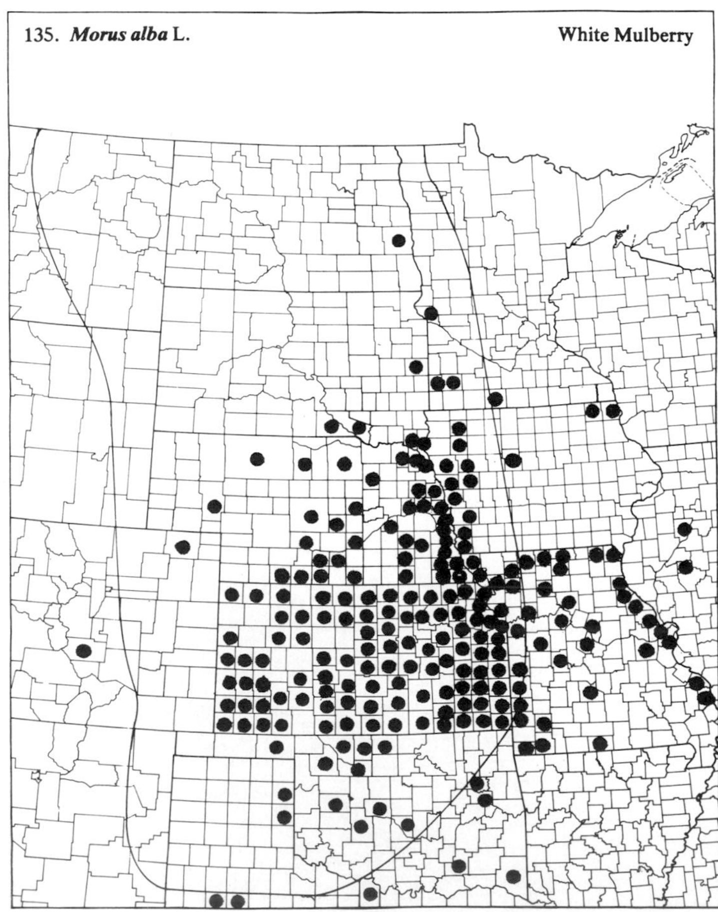

135. ***Morus alba*** L. White Mulberry

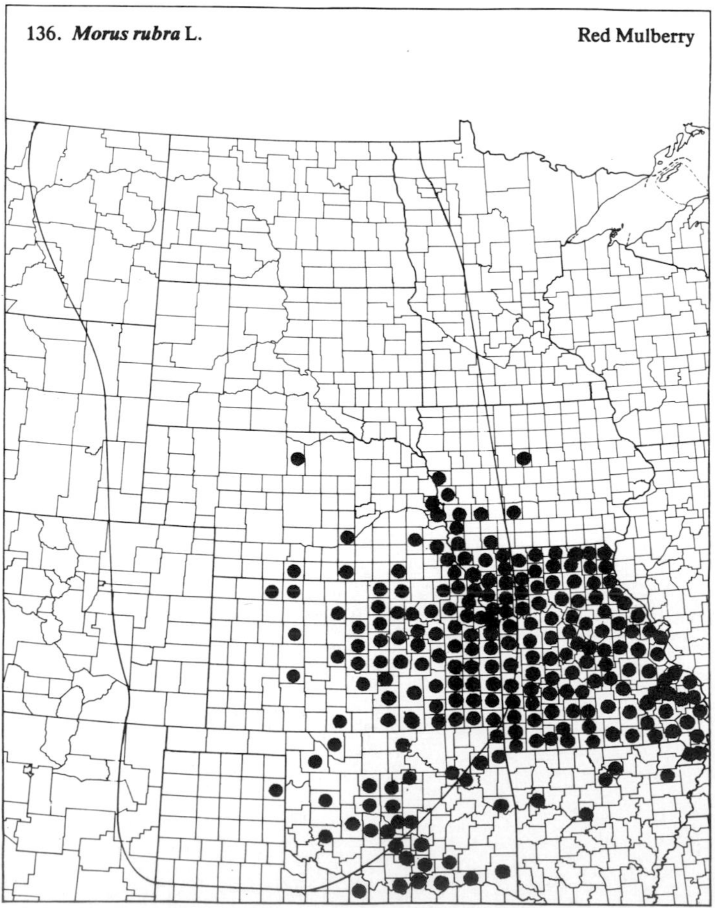

136. ***Morus rubra*** L. Red Mulberry

15. *Cannabaceae* (Hemp Family)

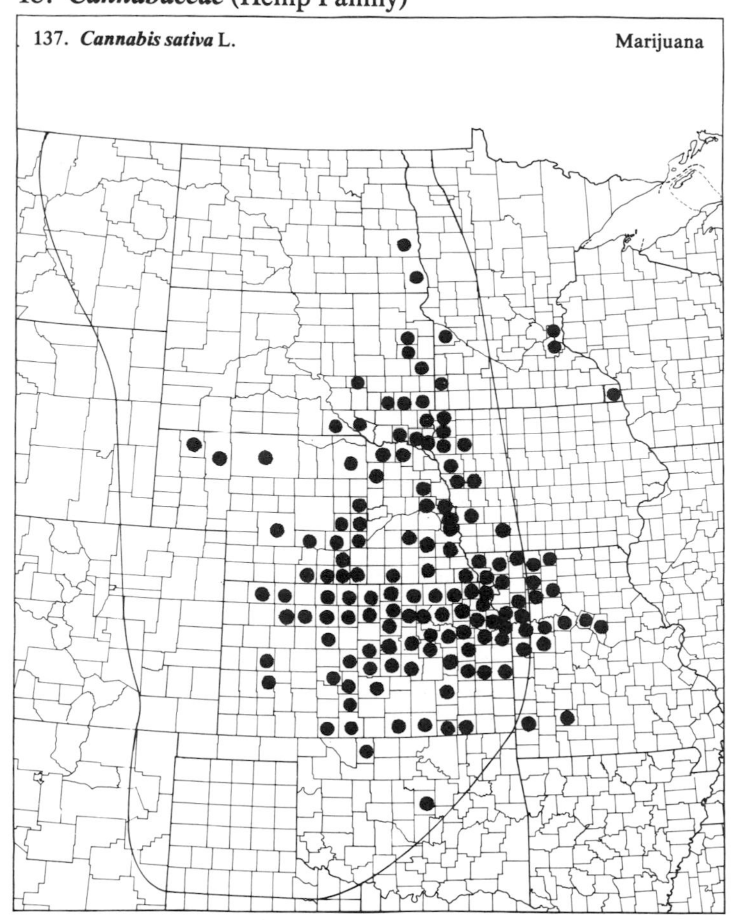

137. ***Cannabis sativa*** L. — Marijuana

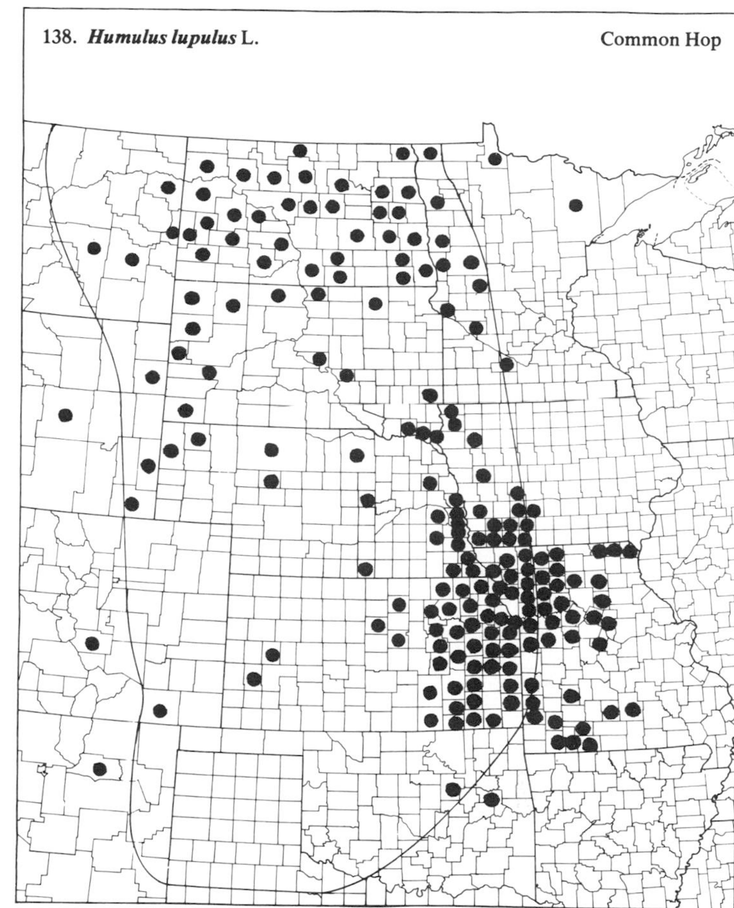

138. ***Humulus lupulus*** L. — Common Hop

16. *Urticaceae* (Nettle Family)

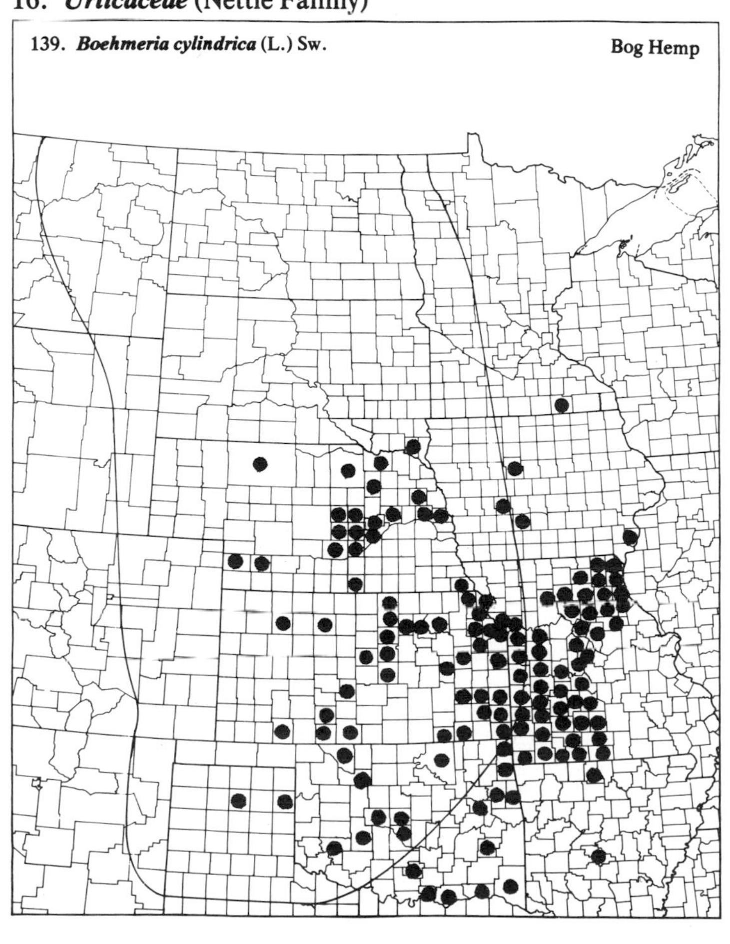

139. ***Boehmeria cylindrica*** (L.) Sw. — Bog Hemp

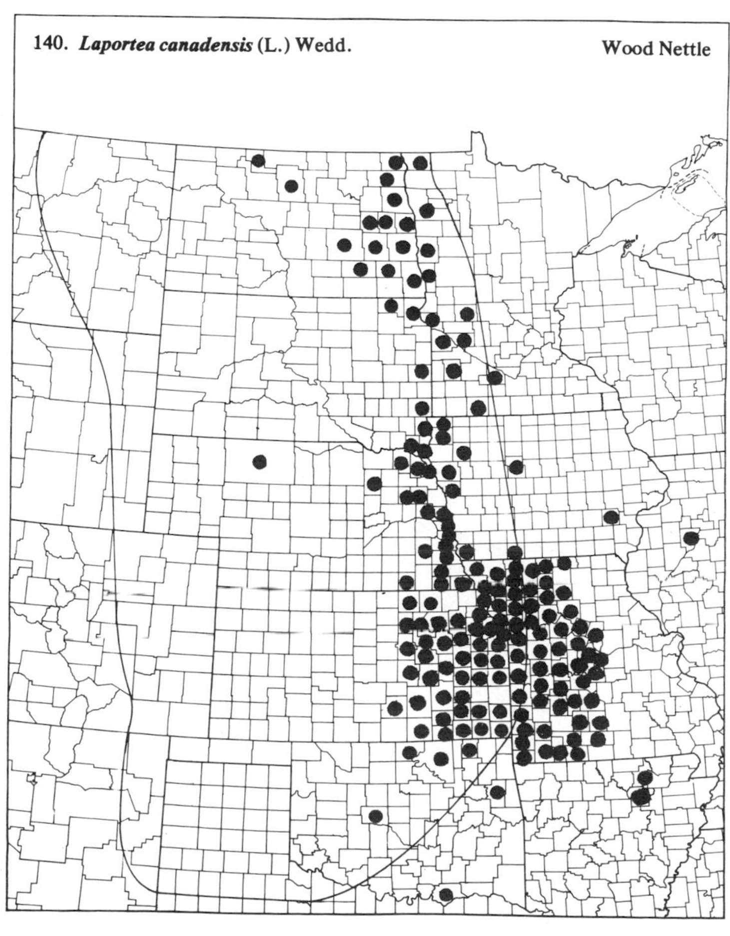

140. ***Laportea canadensis*** (L.) Wedd. — Wood Nettle

(Urticaceae)

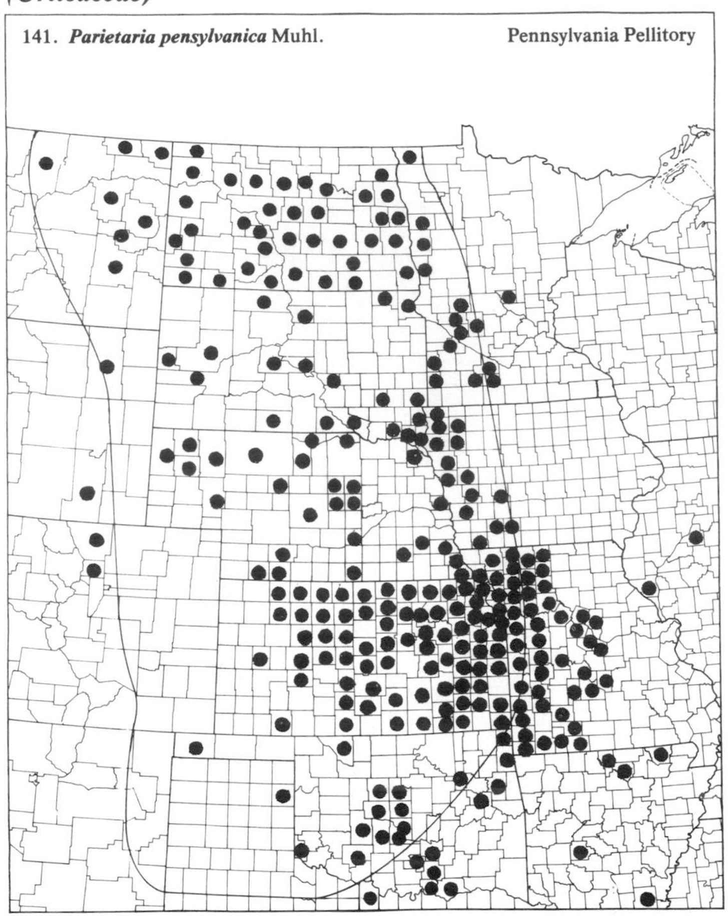

141. ***Parietaria pensylvanica*** Muhl. Pennsylvania Pellitory

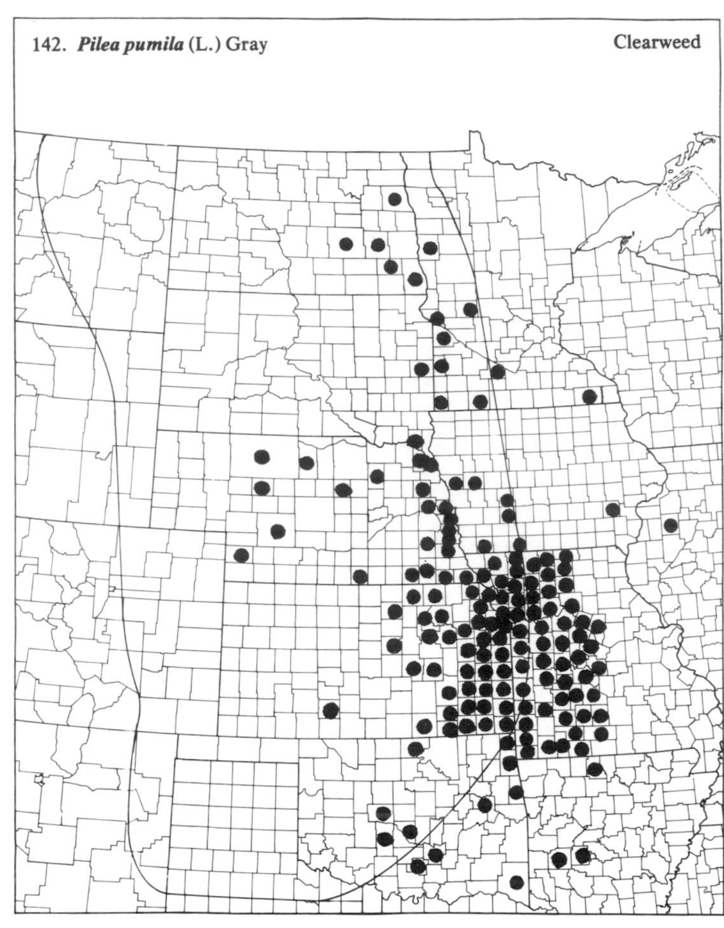

142. ***Pilea pumila*** (L.) Gray Clearweed

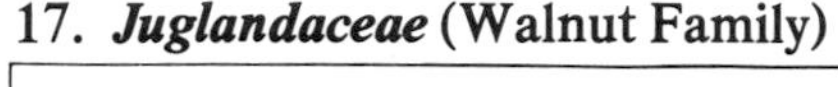

17. *Juglandaceae* (Walnut Family)

143. ***Urtica dioica*** L. ssp. ***gracilis*** (Ait.) Seland. Stinging Nettle

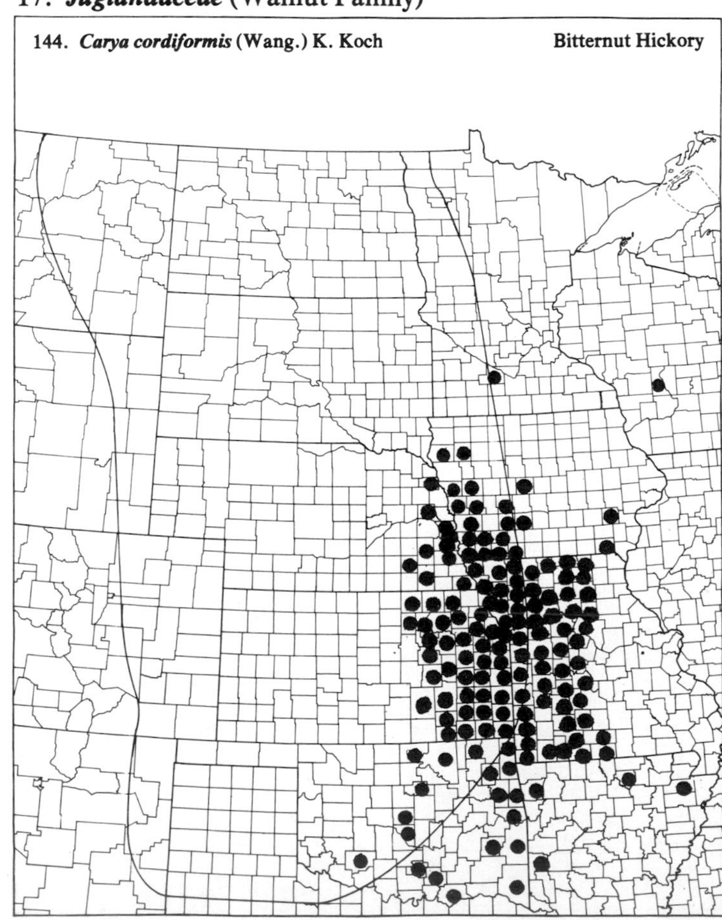

144. ***Carya cordiformis*** (Wang.) K. Koch Bitternut Hickory

(Juglandaceae)

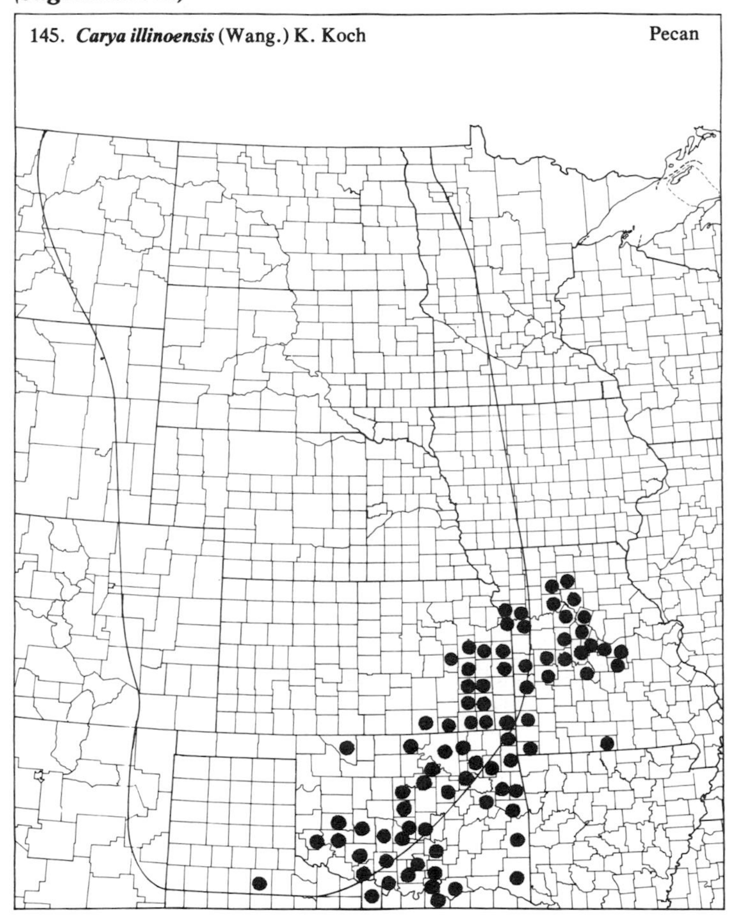

145. ***Carya illinoensis*** (Wang.) K. Koch — Pecan

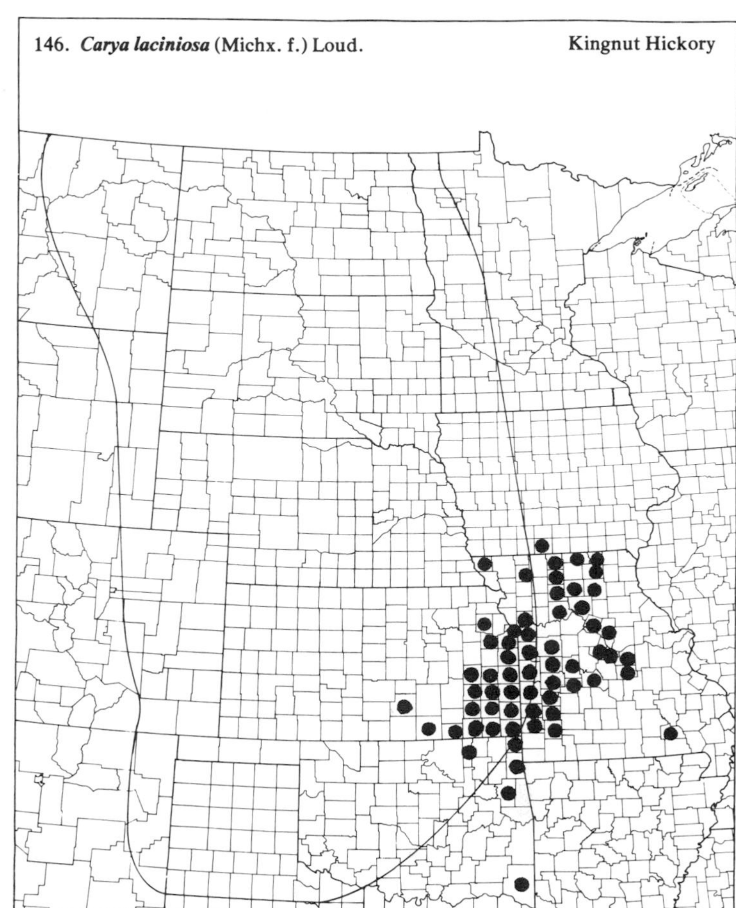

146. ***Carya laciniosa*** (Michx. f.) Loud. — Kingnut Hickory

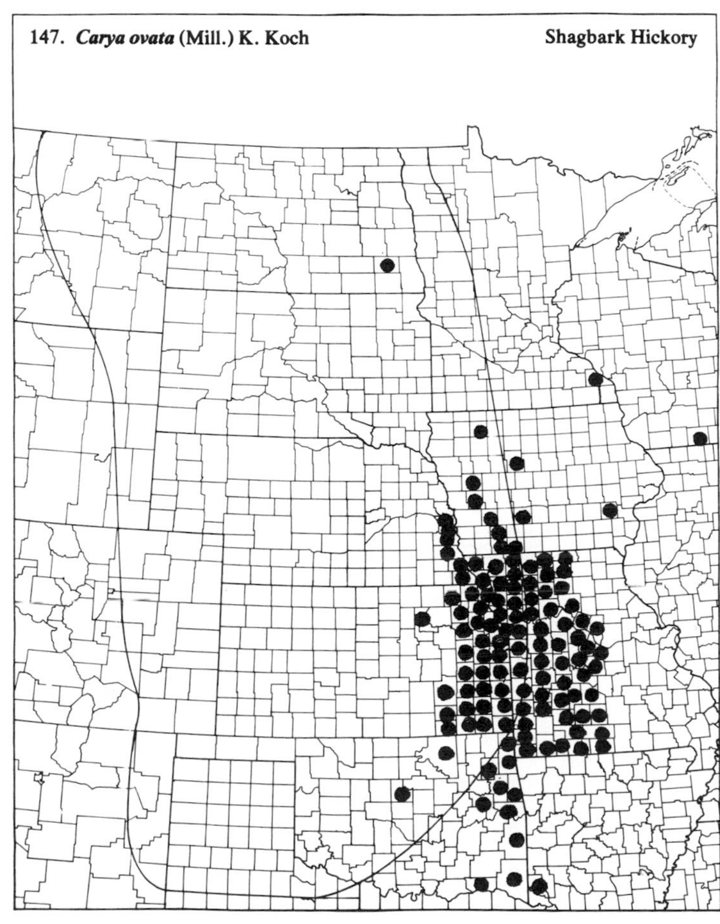

147. ***Carya ovata*** (Mill.) K. Koch — Shagbark Hickory

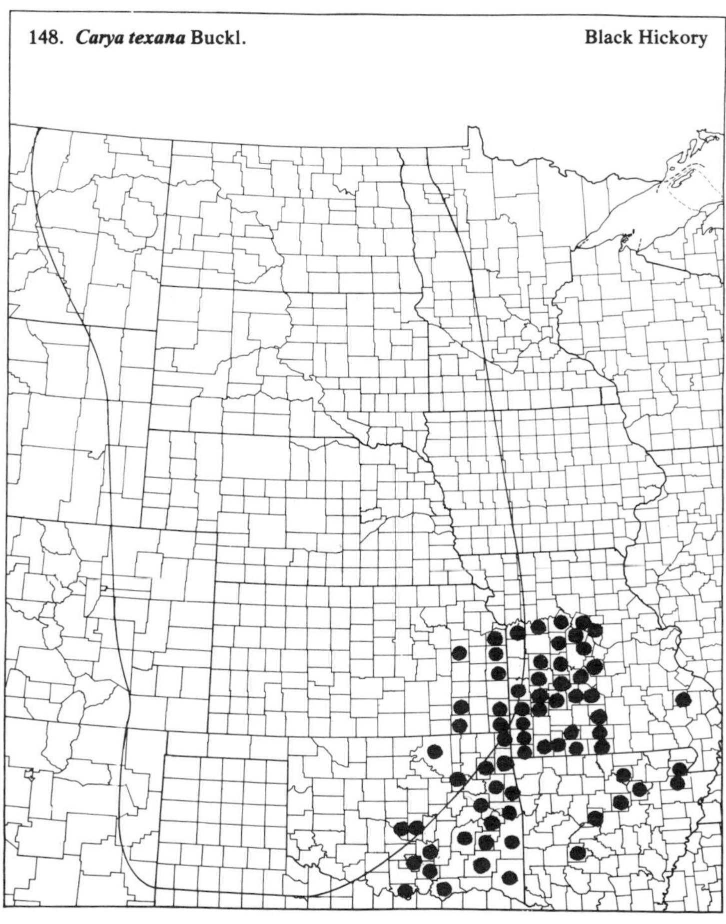

148. ***Carya texana*** Buckl. — Black Hickory

(Juglandaceae)

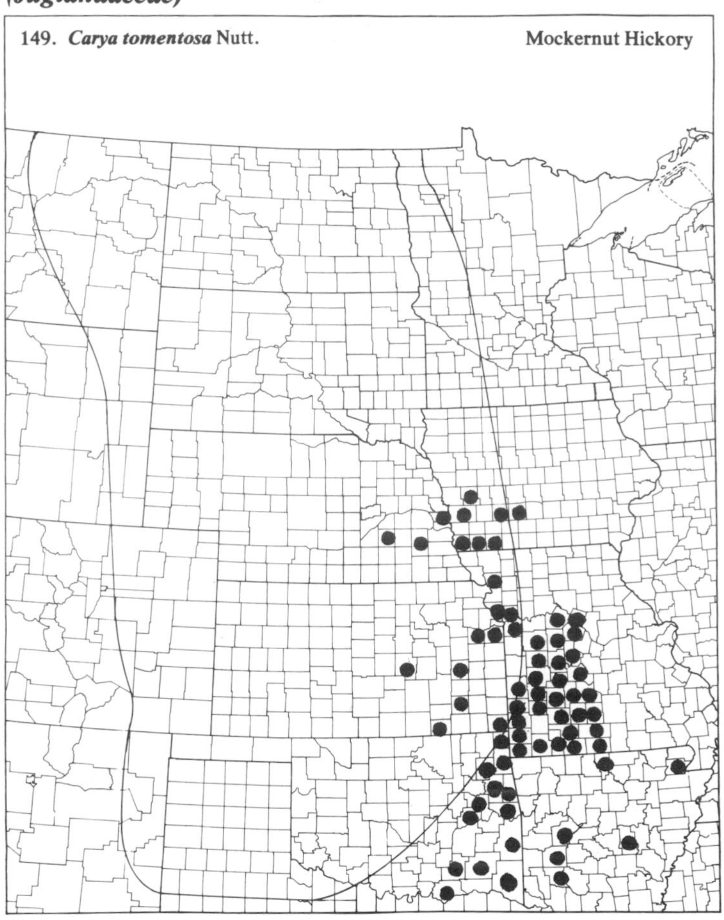

149. ***Carya tomentosa*** Nutt. Mockernut Hickory

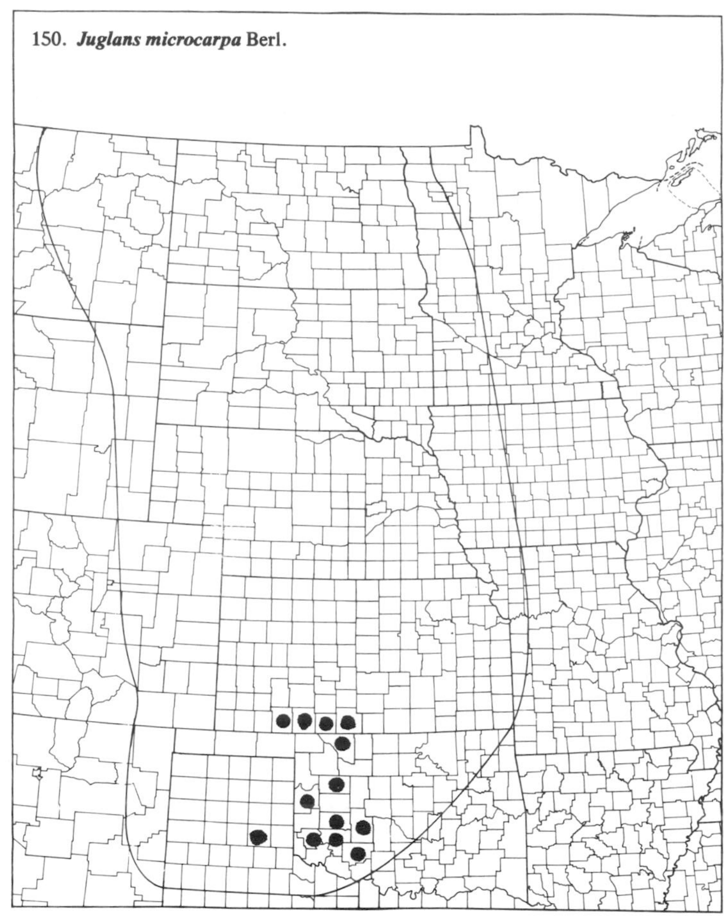

150. ***Juglans microcarpa*** Berl.

18. *Fagaceae* (Beech Family)

151. ***Juglans nigra*** L. Black Walnut

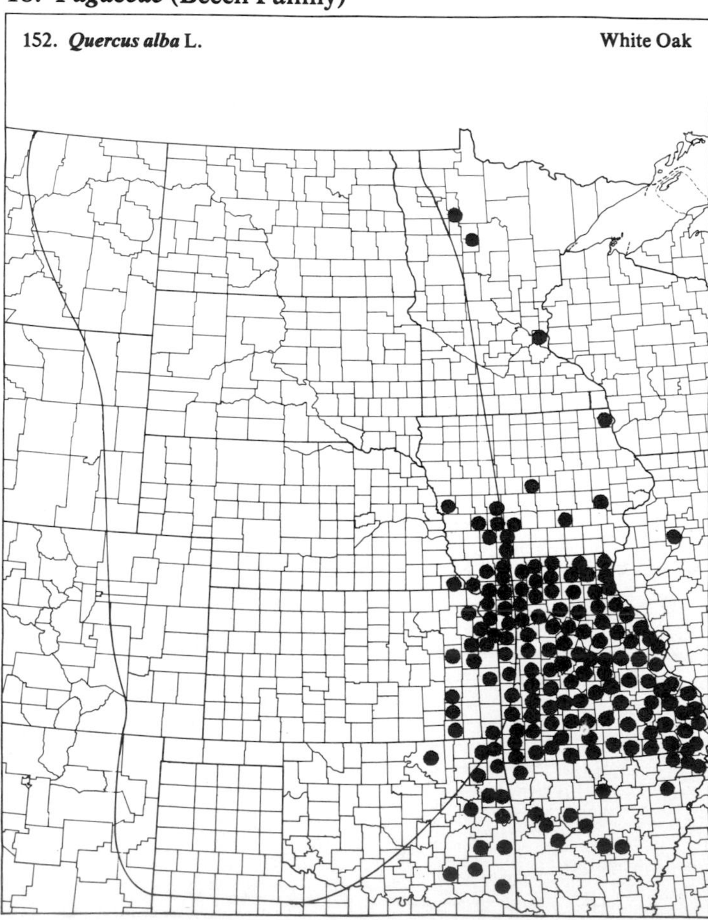

152. ***Quercus alba*** L. White Oak

(Fagaceae)

153. ***Quercus borealis*** Michx. f. var. ***maxima*** (Marsh.) Ashe — Red Oak

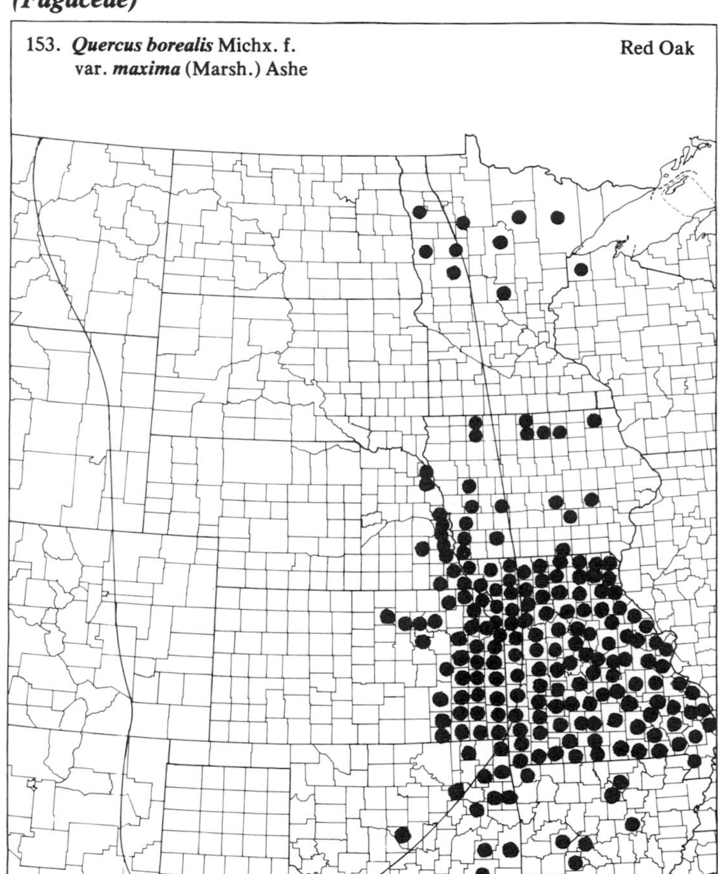

154. ***Quercus havardii*** Rydb.

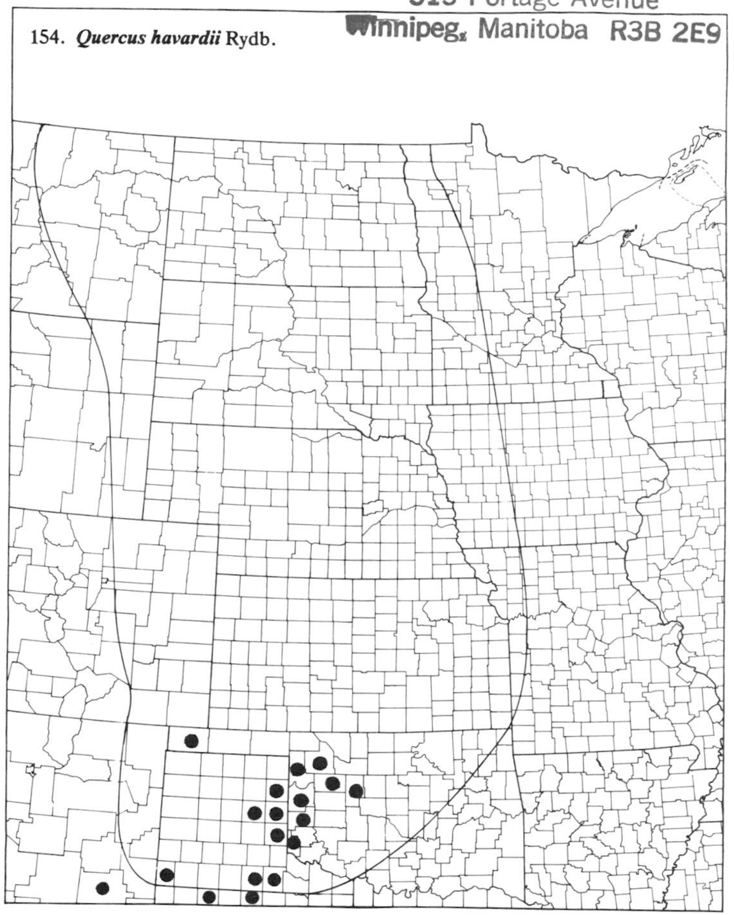

155. ***Quercus imbricaria*** Michx. — Shingle Oak

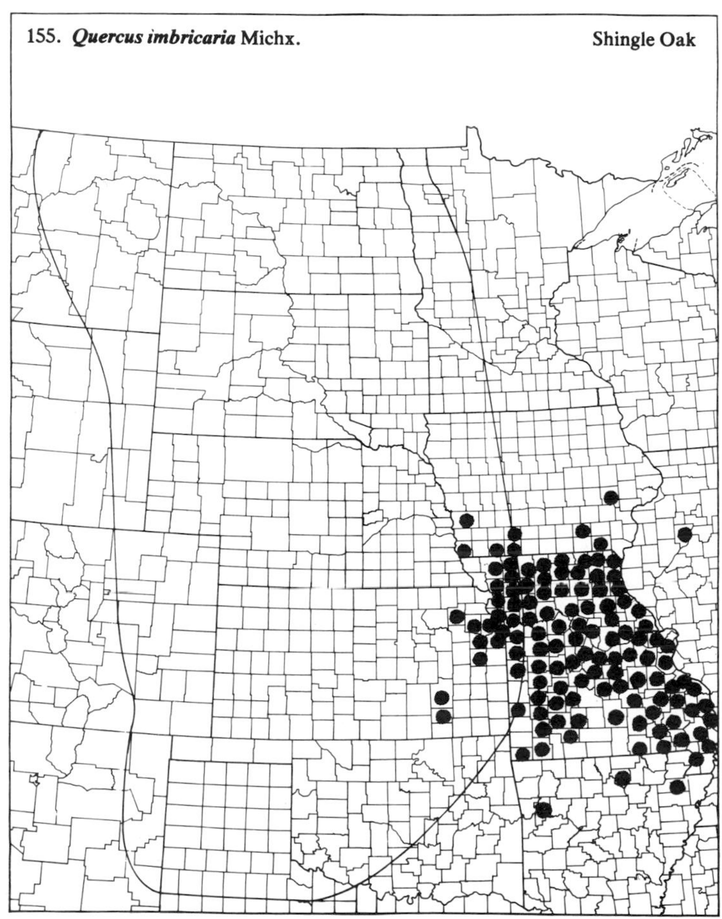

156. ***Quercus macrocarpa*** Michx. — Bur Oak

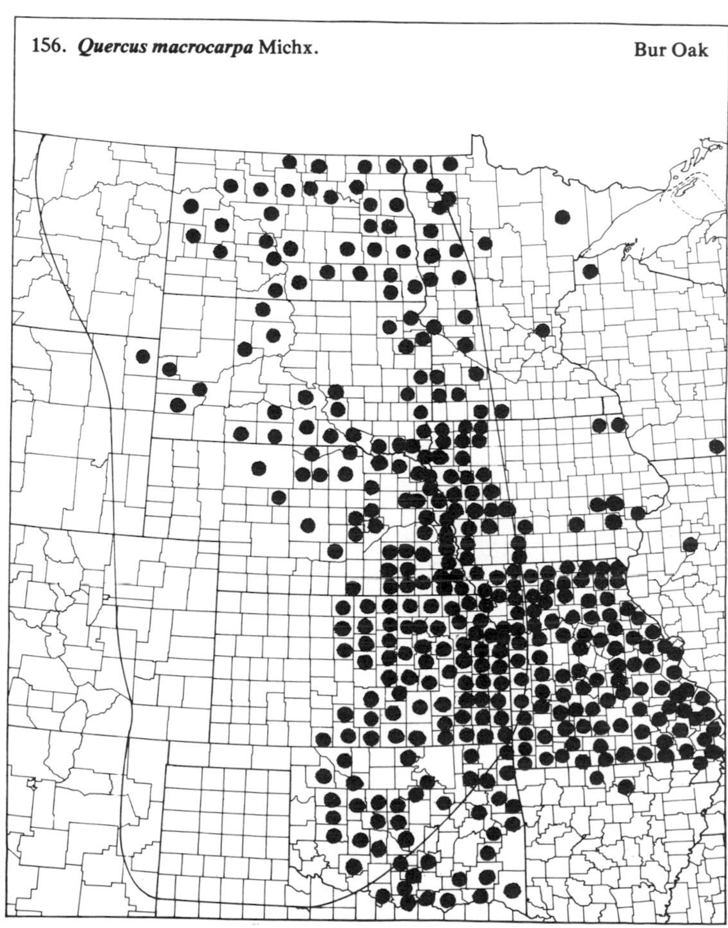

(Fagaceae)

157. ***Quercus marilandica*** Muenchh. Black Jack Oak

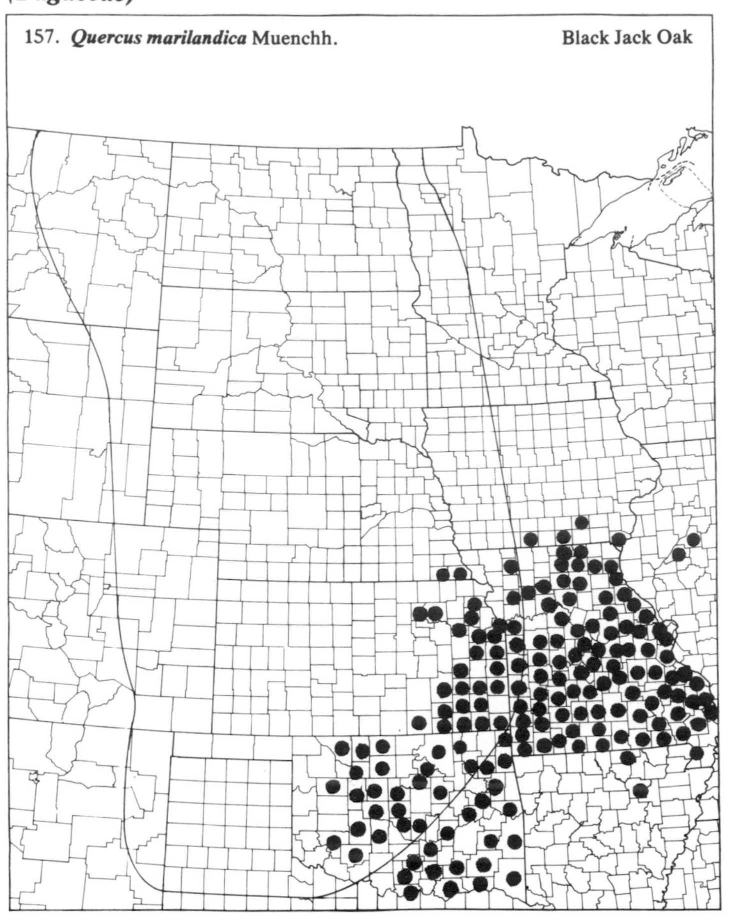

158. ***Quercus mohriana*** Buckl.

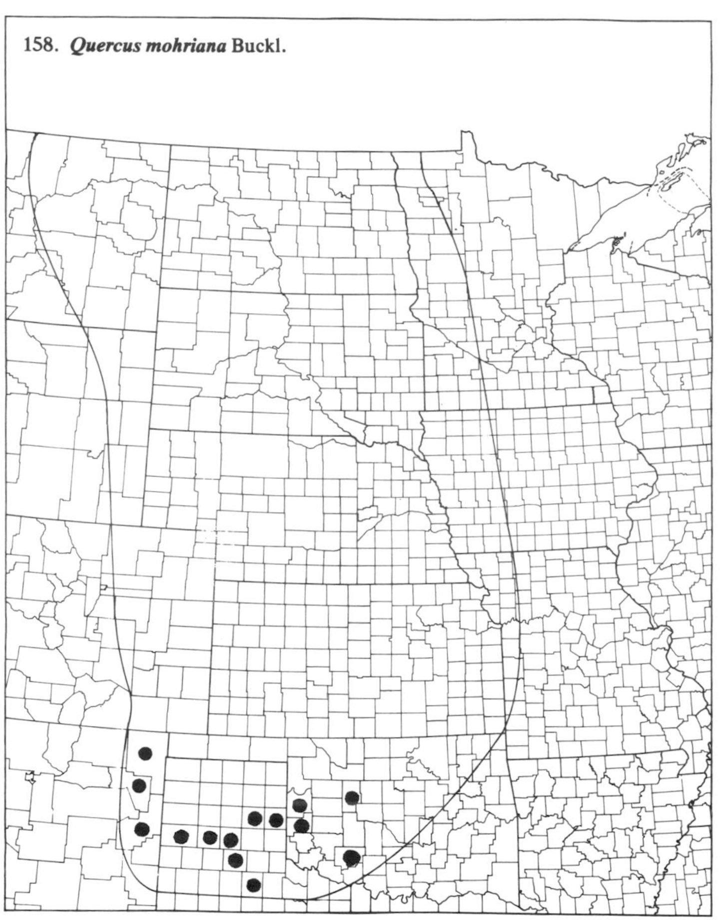

159. ***Quercus muhlenbergii*** Engelm. Chinquapin Oak

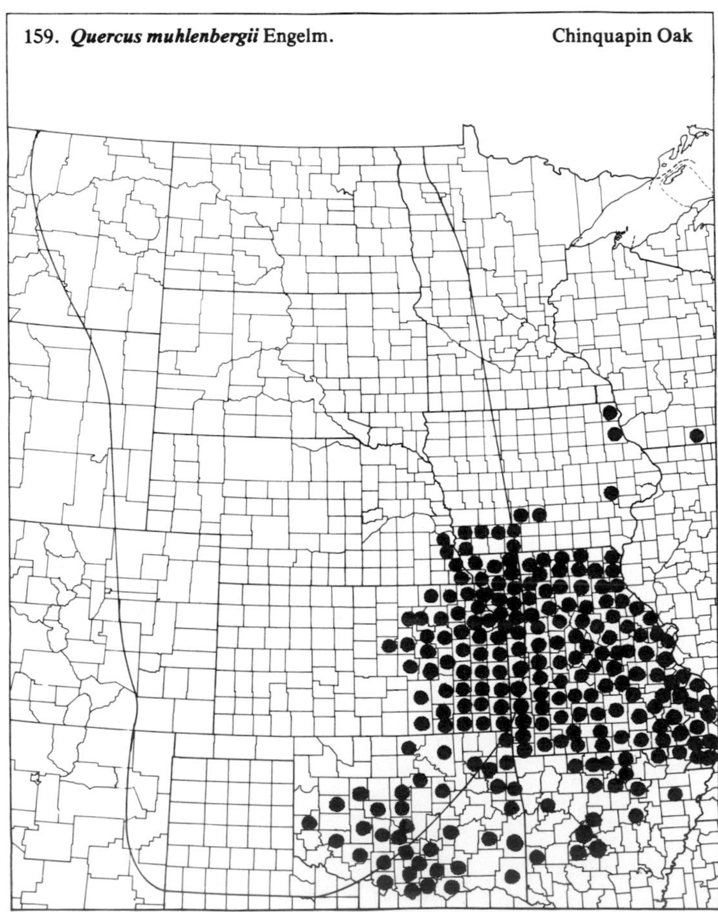

160. ***Quercus palustris*** Muenchh. Pin Oak

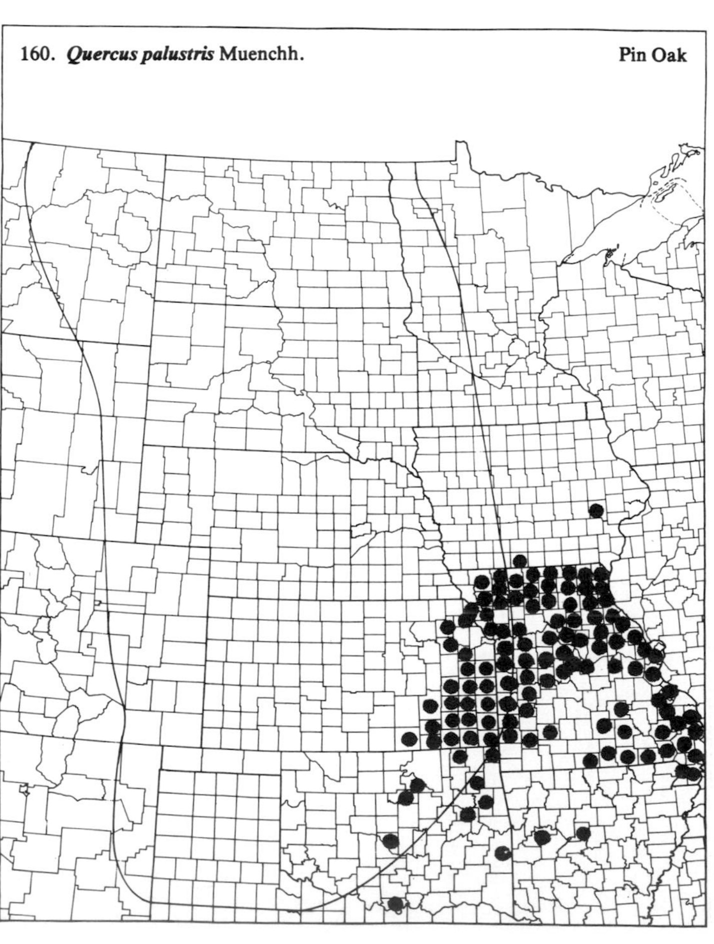

(Fagaceae)

161. ***Quercus prinoides*** Willd. Dwarf Chinquapin Oak

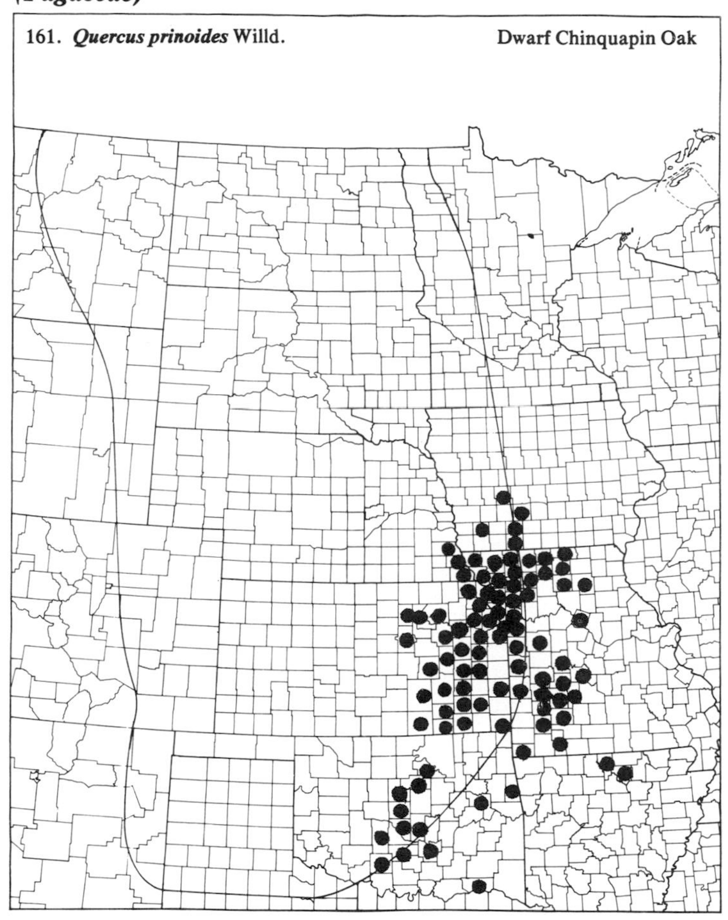

162. ***Quercus shumardii*** Buckl. Shumard's Red Oak

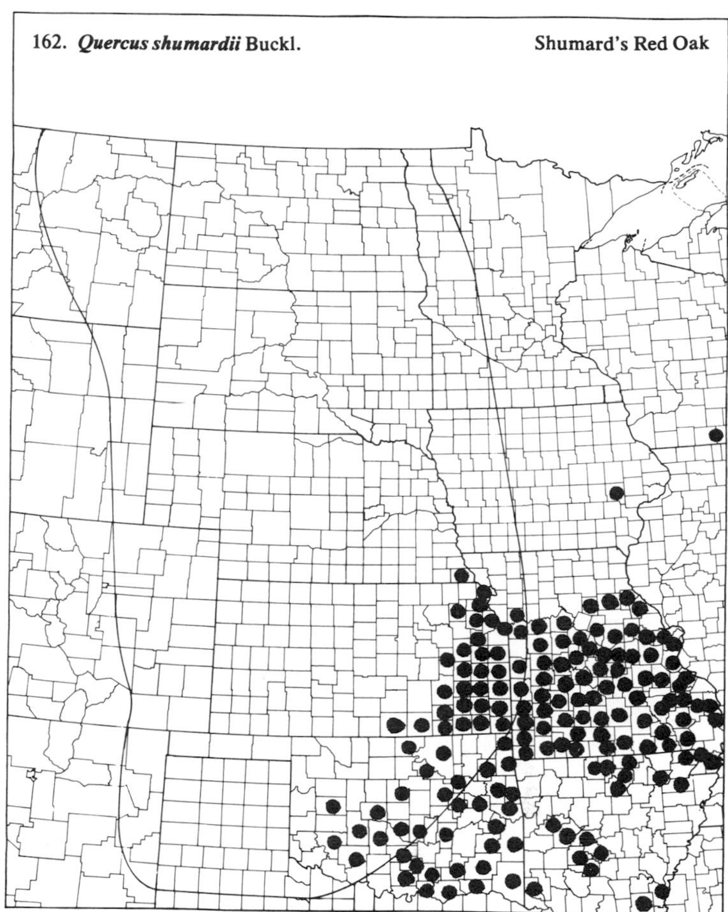

163. ***Quercus stellata*** Wang. Post Oak

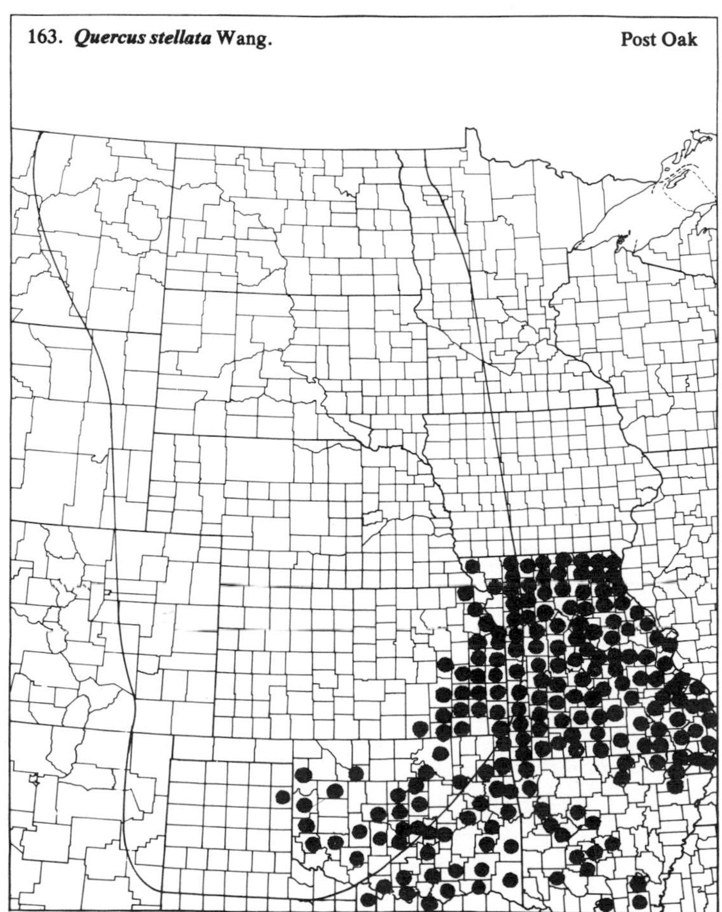

164. ***Quercus velutina*** Lam. Black Oak

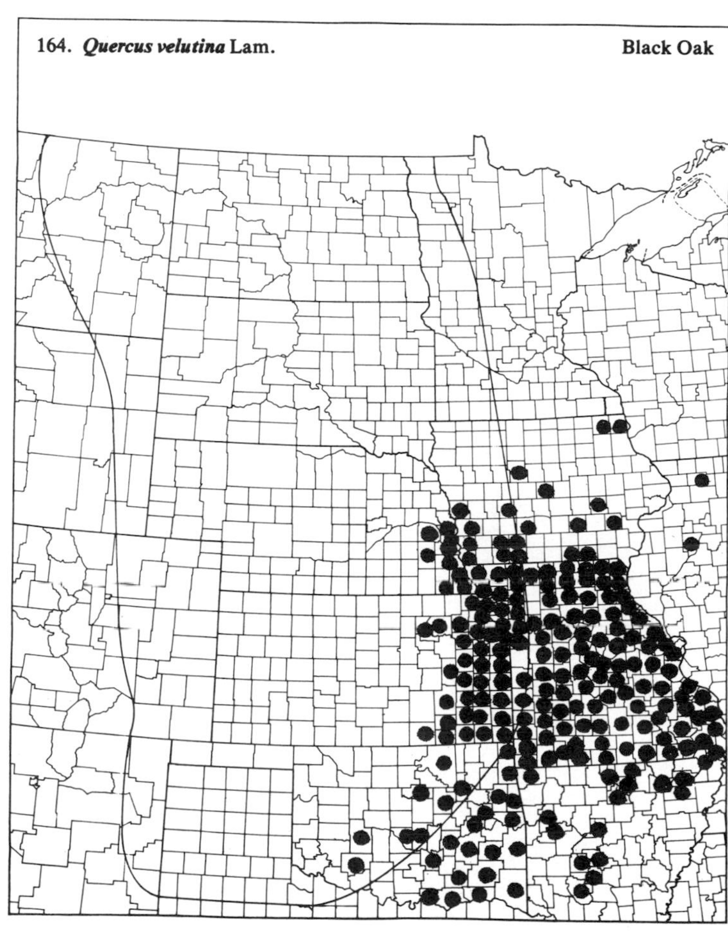

19. *Betulaceae* (Birch Family)

165. ***Alnus incana*** (L.) Moench ssp. ***rugosa*** (DuRoi) Clausen — Speckled Alder

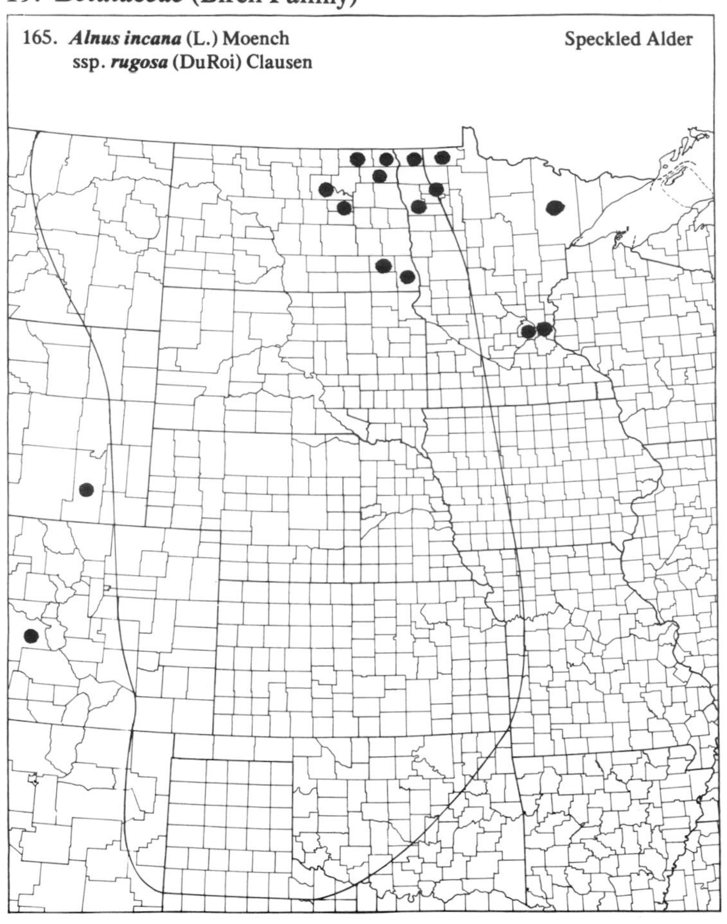

166. ***Betula glandulosa*** Michx. var. ***glandulifera*** Regel — Dwarf Birch

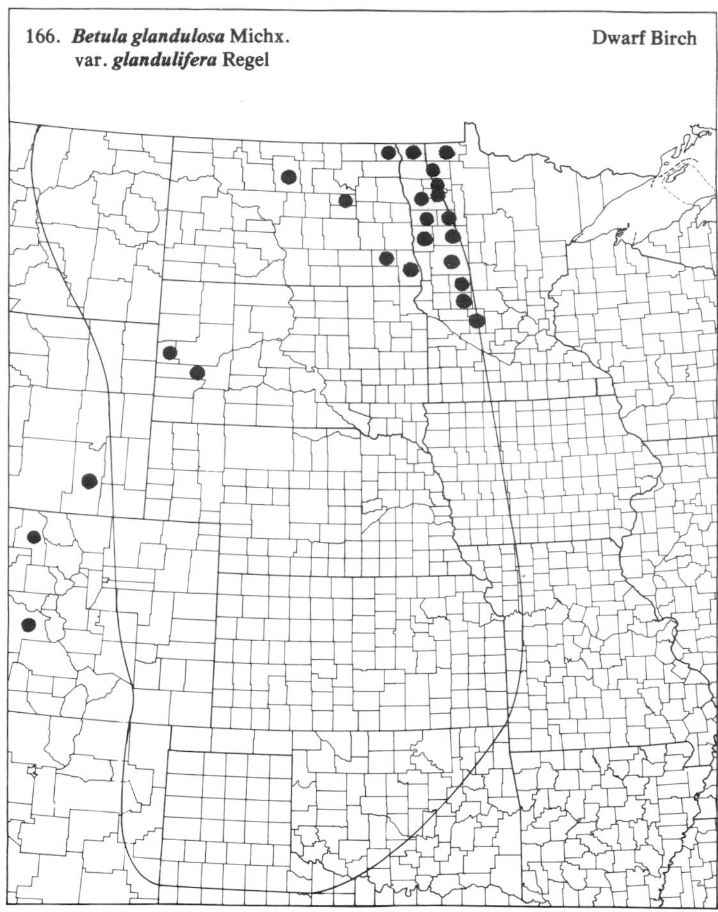

167. ***Betula nigra*** L. — River Birch

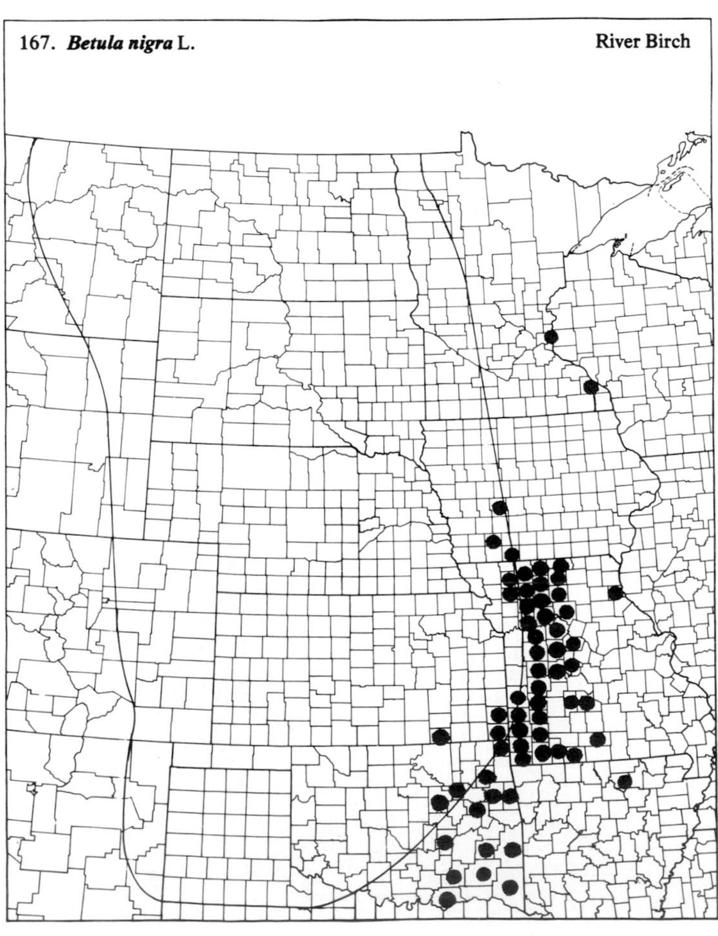

168. ***Betula occidentalis*** Hook. — Mountain Birch

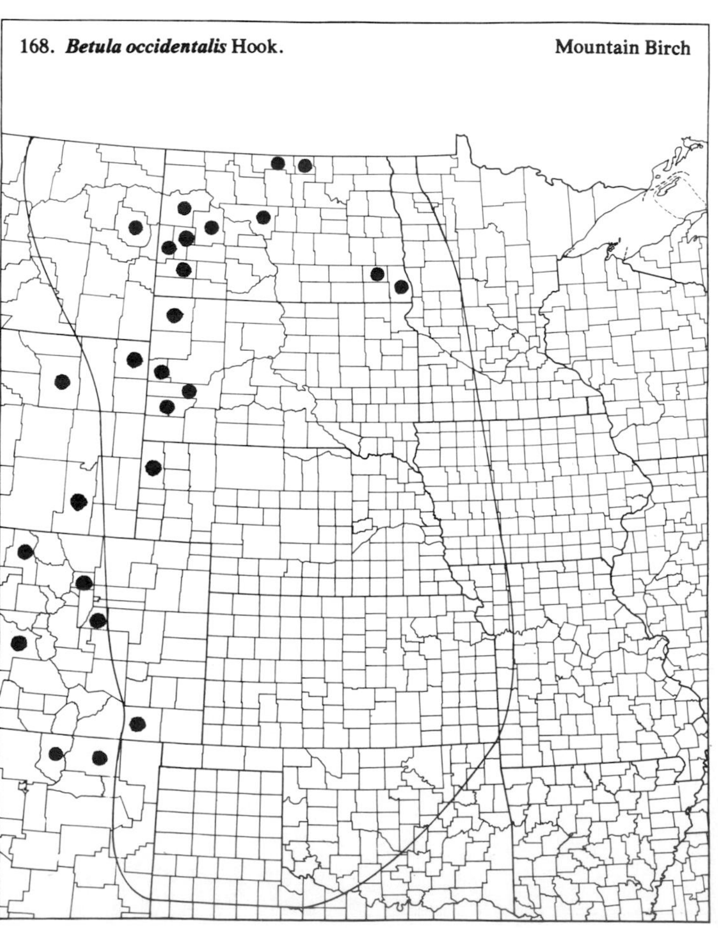

(Betulaceae)

169. ***Betula papyrifera*** Marsh. — Paper Birch

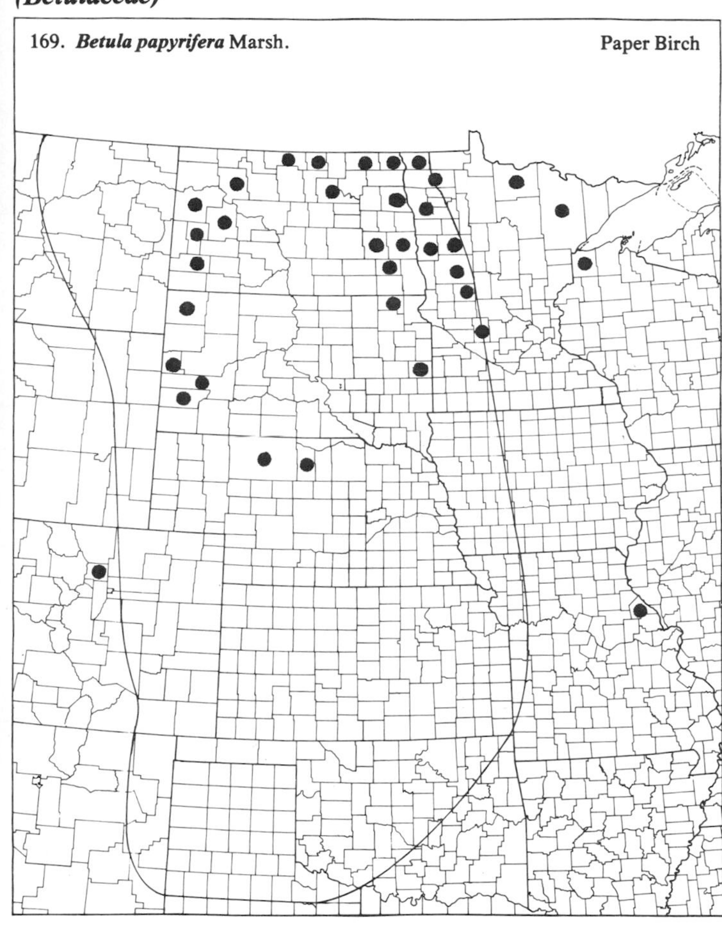

170. ***Corylus americana*** Walt. — American Hazelnut

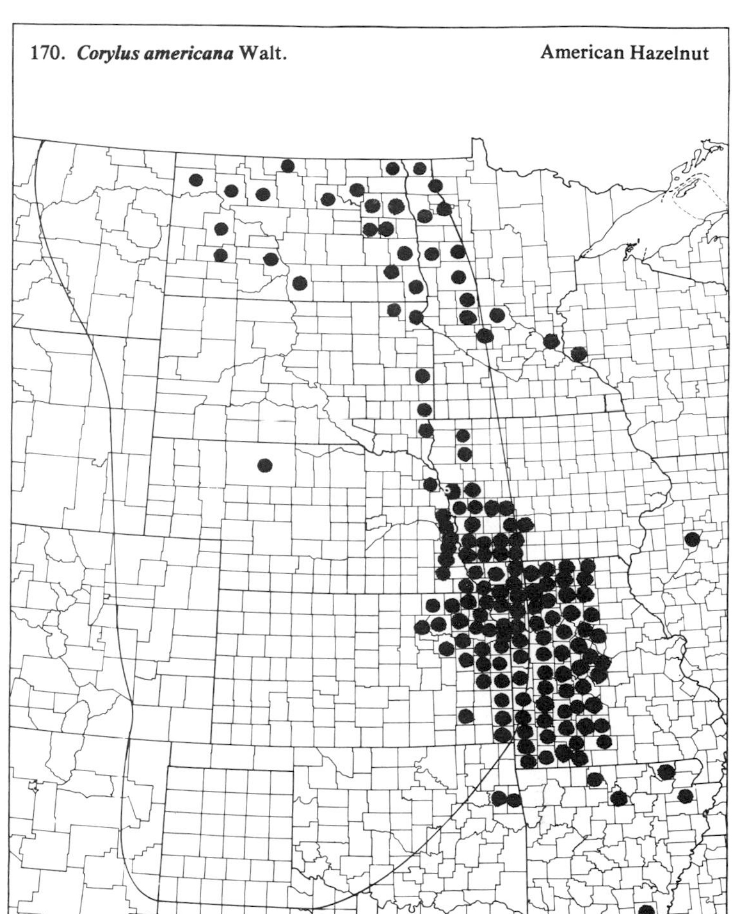

171. ***Corylus cornuta*** Marsh. — Beaked Hazelnut

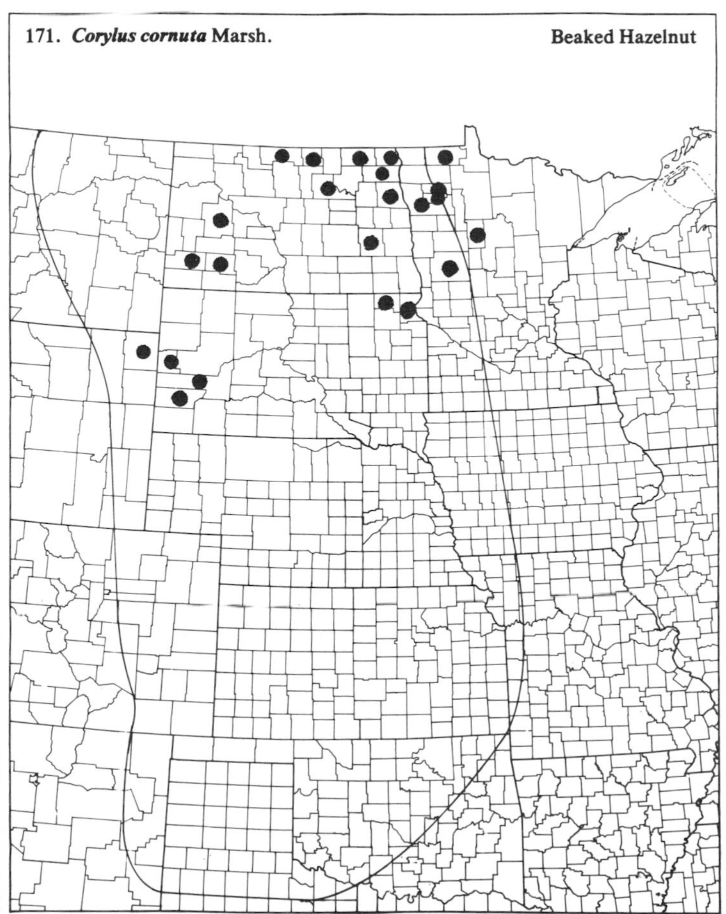

172. ***Ostrya virginiana*** (Mill.) K. Koch — Hop-hornbeam

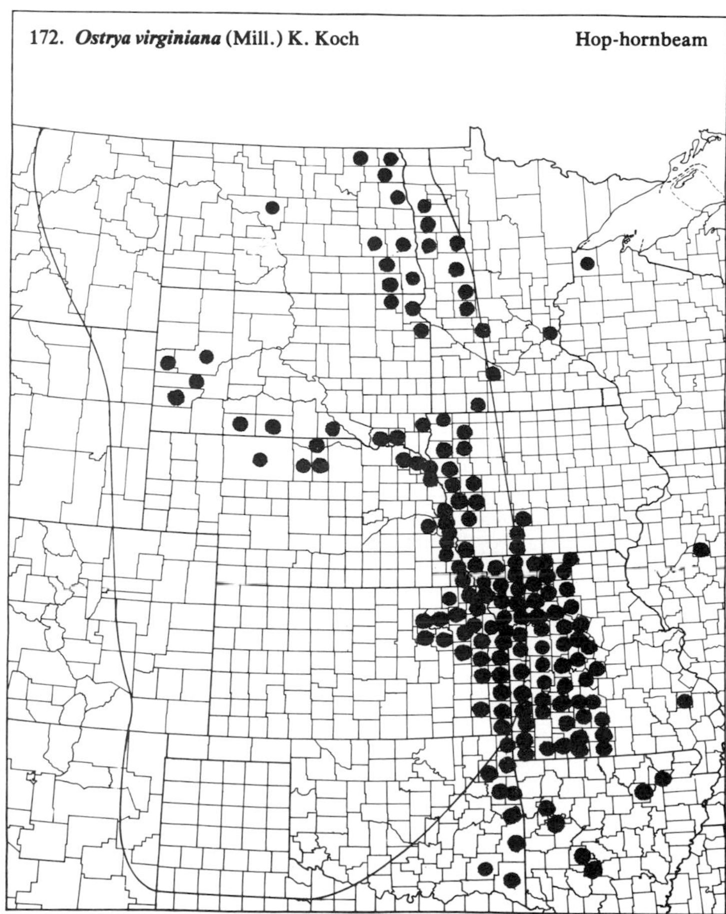

20. *Phytolaccaceae* (Pokeweed Family)

173. ***Phytolacca americana*** L. Pokeweed

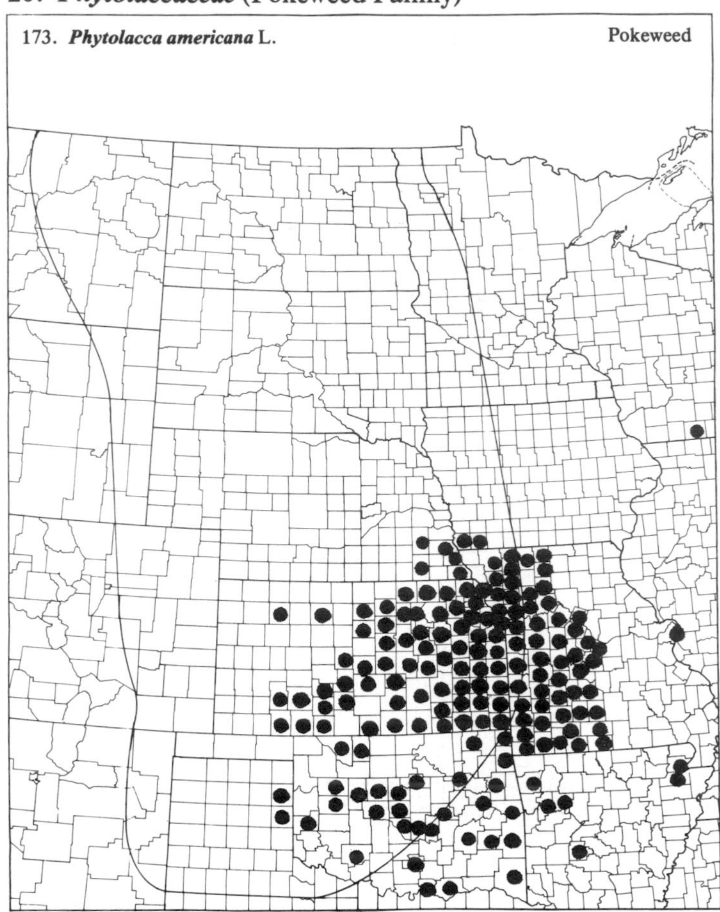

21. *Nyctaginaceae* (Four-O'Clock Family)

174. ***Abronia fragrans*** Nutt. Sweet Sand Verbena

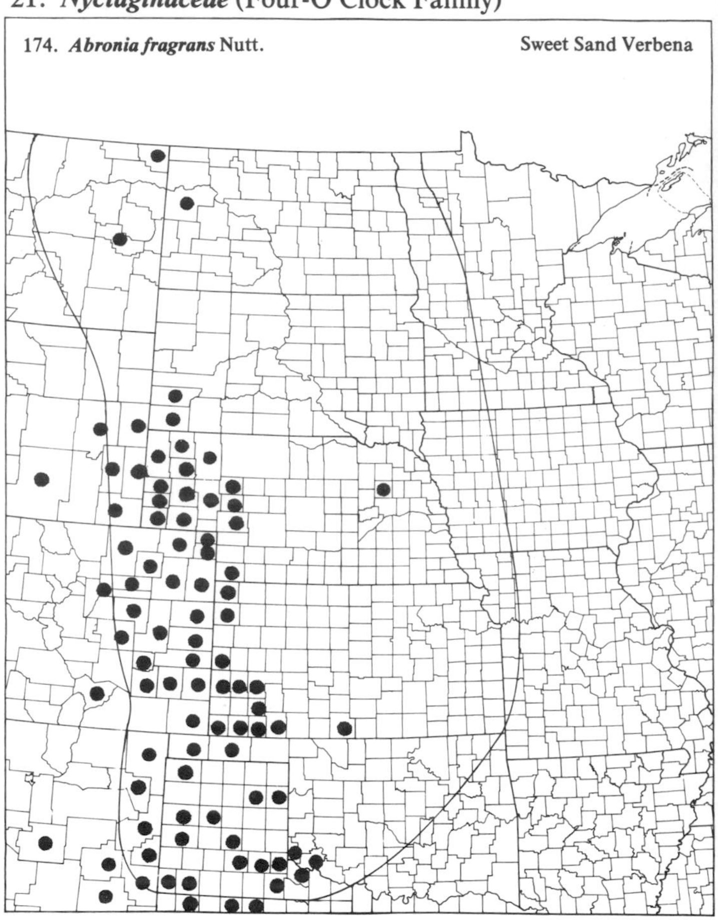

175. ***Mirabilis albida*** (Walt.) Heimerl. White Four-O'Clock

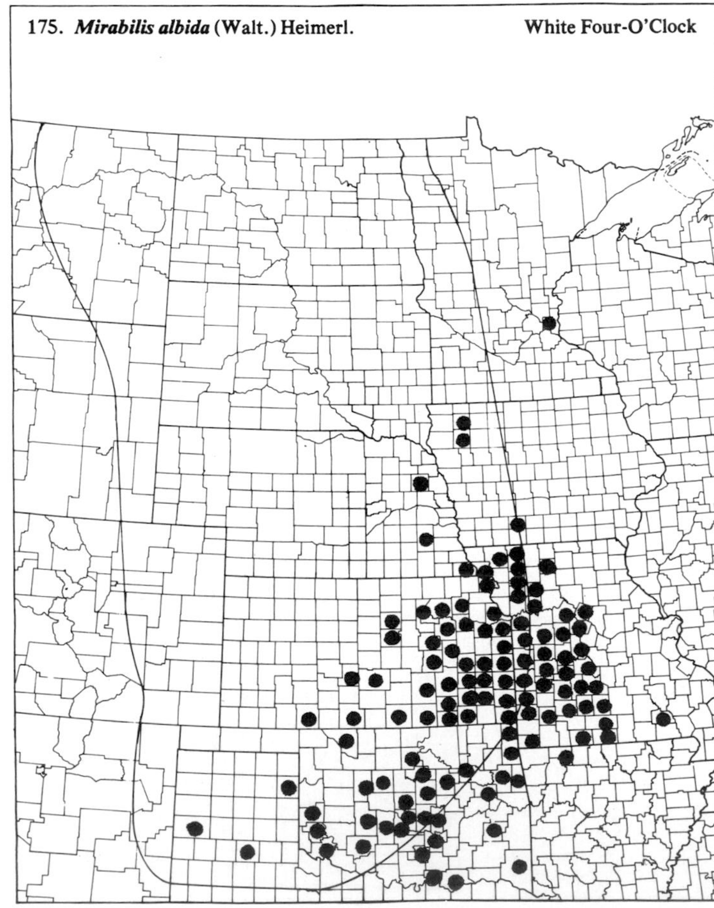

176. ***Mirabilis carletonii*** (Standl.) Standl. Carleton Four-O'Clock

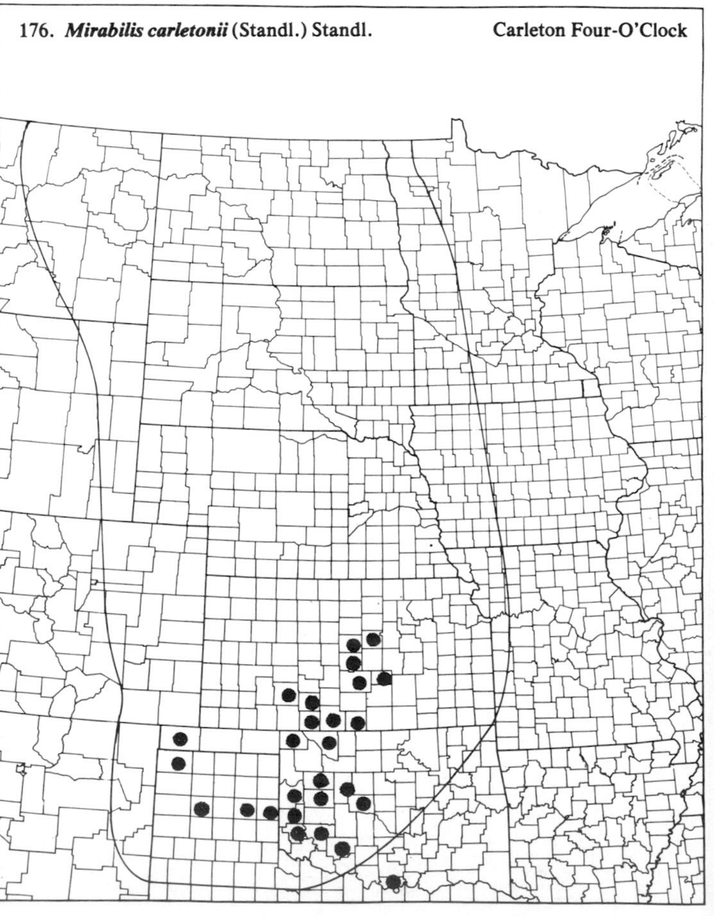

(Nyctaginaceae)

177. ***Mirabilis exaltata*** (Standl.) Standl.

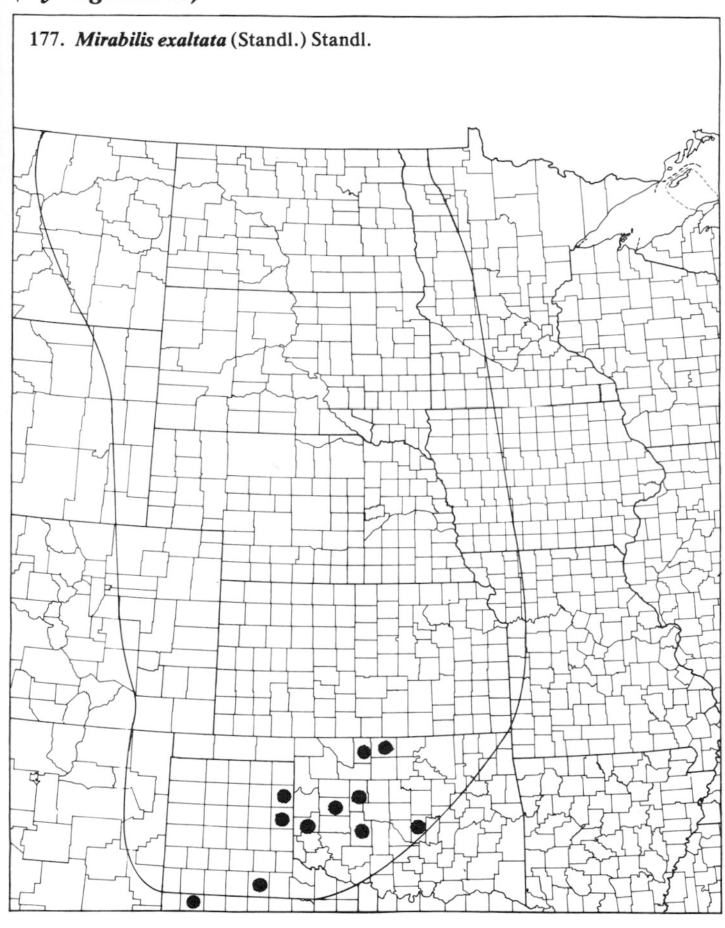

178. ***Mirabilis gausapoides*** (Standl.) Standl.

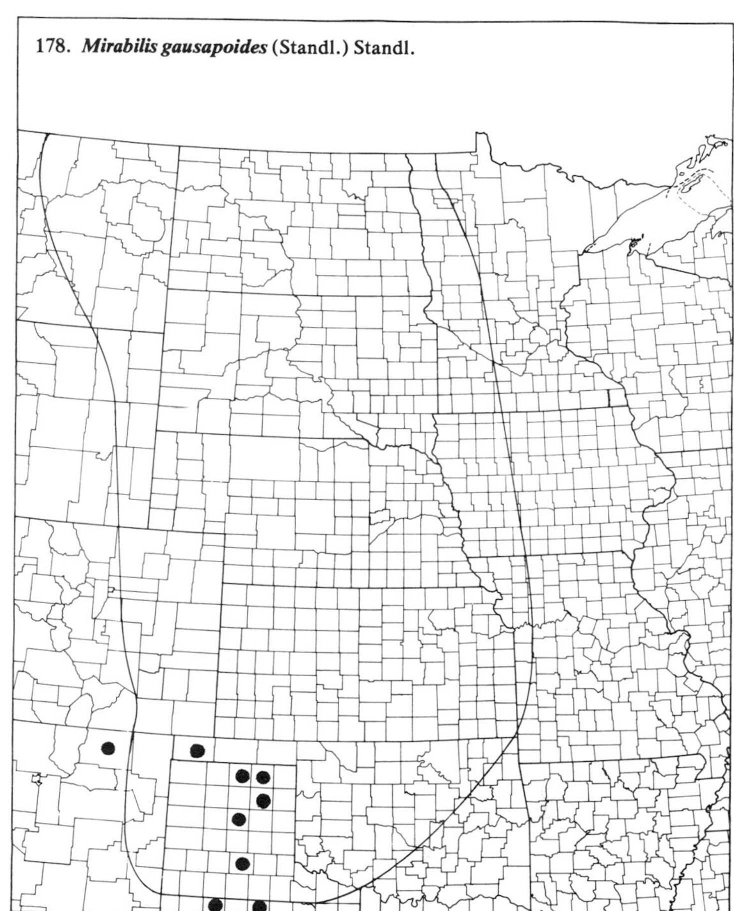

179. ***Mirabilis glabra*** (Wats.) Standl.

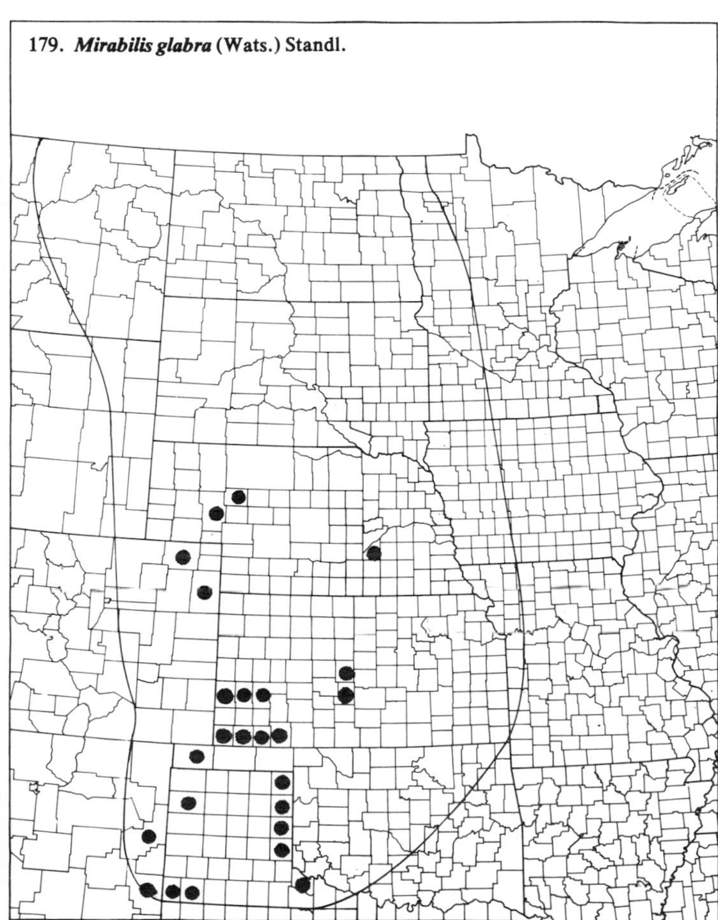

180. ***Mirabilis hirsuta*** (Pursh) MacM. Hairy Four-O'Clock

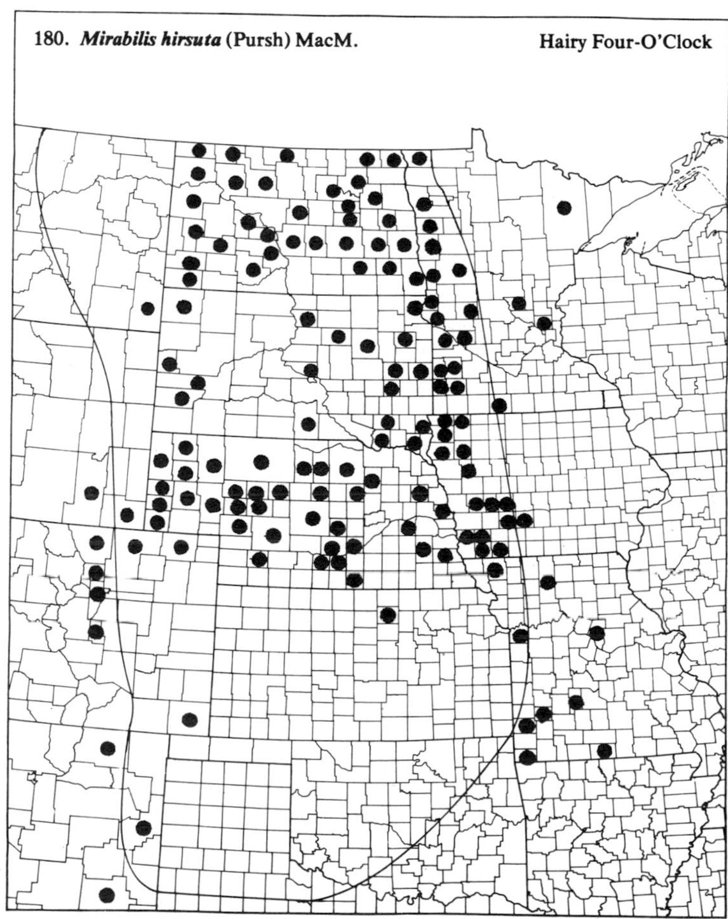

(Nyctaginaceae)

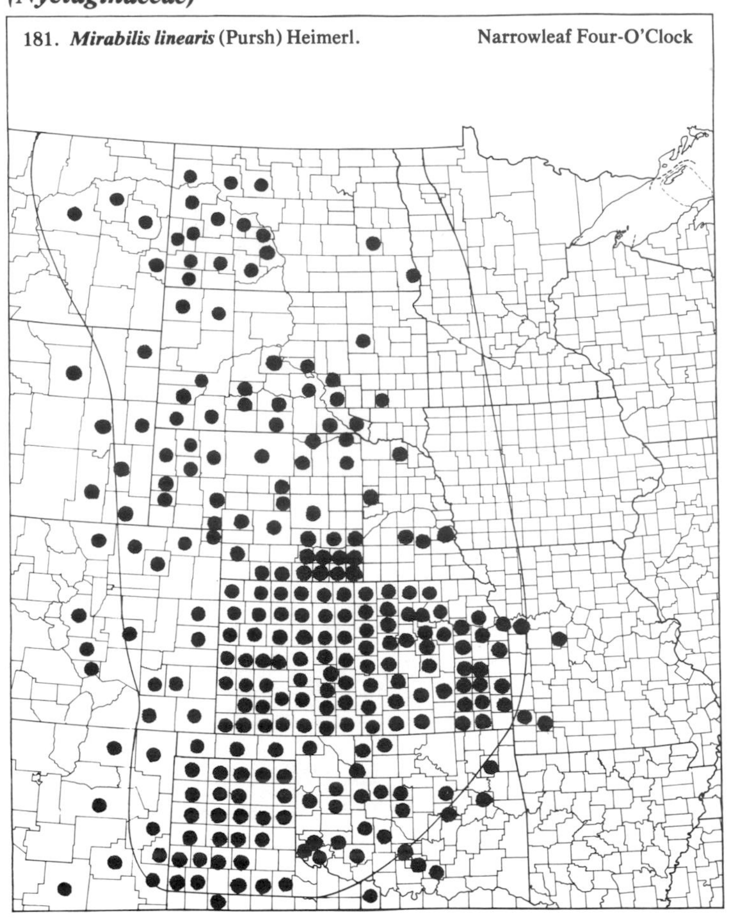

181. ***Mirabilis linearis*** (Pursh) Heimerl. Narrowleaf Four-O'Clock

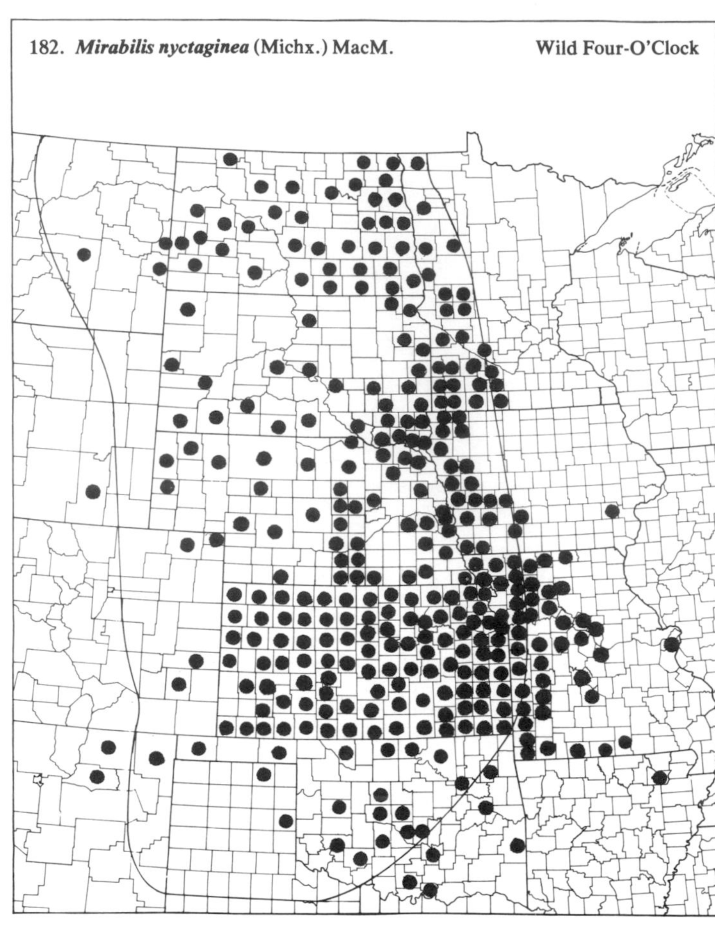

182. ***Mirabilis nyctaginea*** (Michx.) MacM. Wild Four-O'Clock

183. ***Tripterocalyx micranthus*** (Torr.) Hook. Sand Puffs

22. *Cactaceae* (Cactus Family)

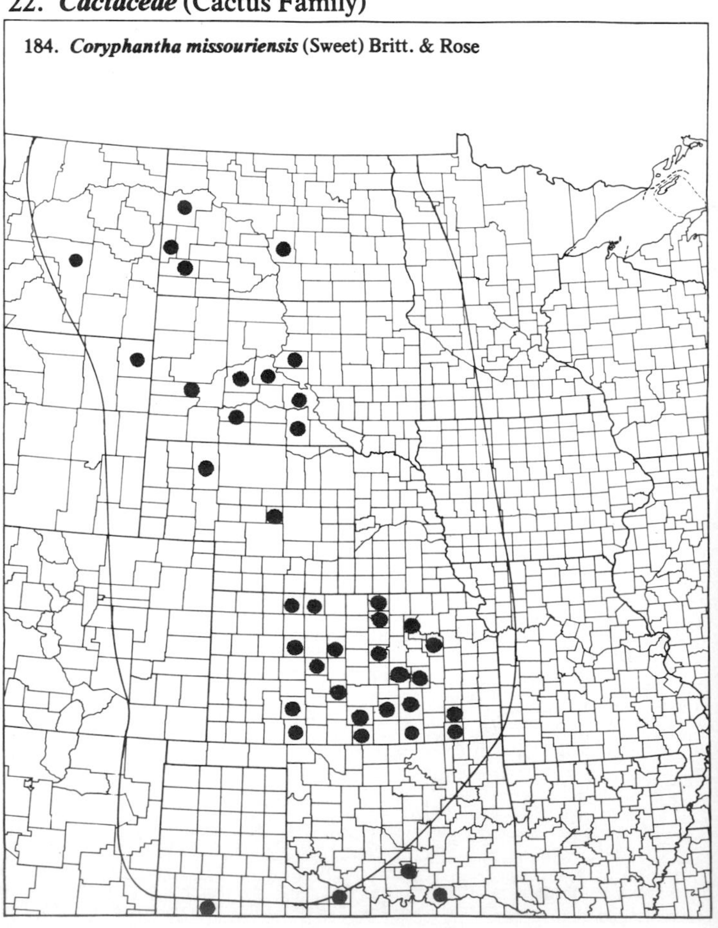

184. ***Coryphantha missouriensis*** (Sweet) Britt. & Rose

(Cactaceae)

185. *Coryphantha vivipara* (Nutt.) Britt. & Rose — Pincushion Cactus

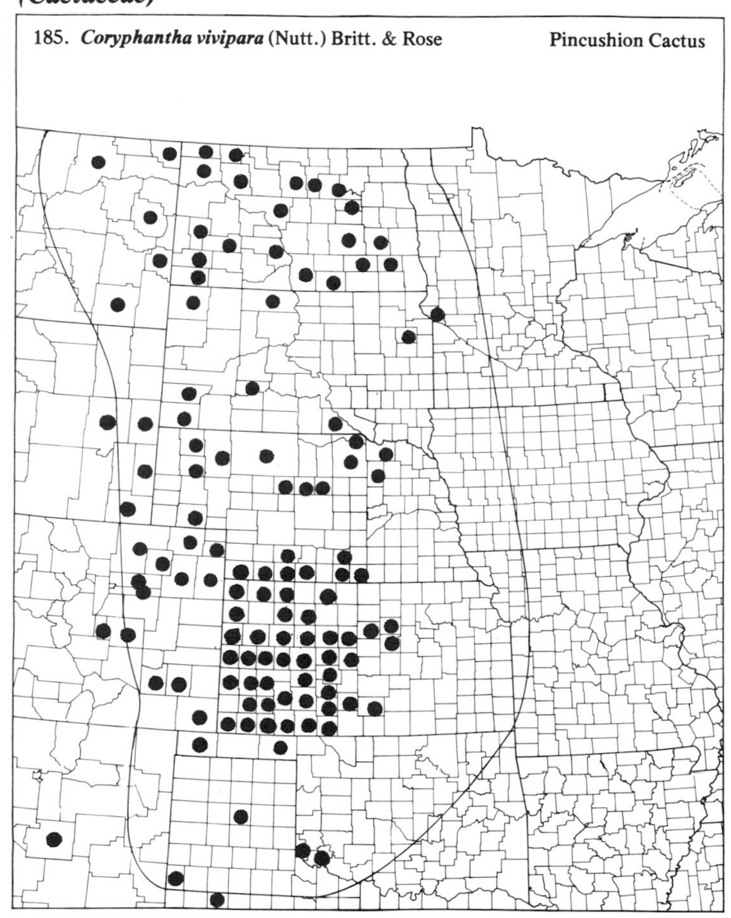

186. *Echinocereus reichenbachii* (Terscheck) Haage var. *albispinus* (Lahman) L. Benson — Lace Echinocereus

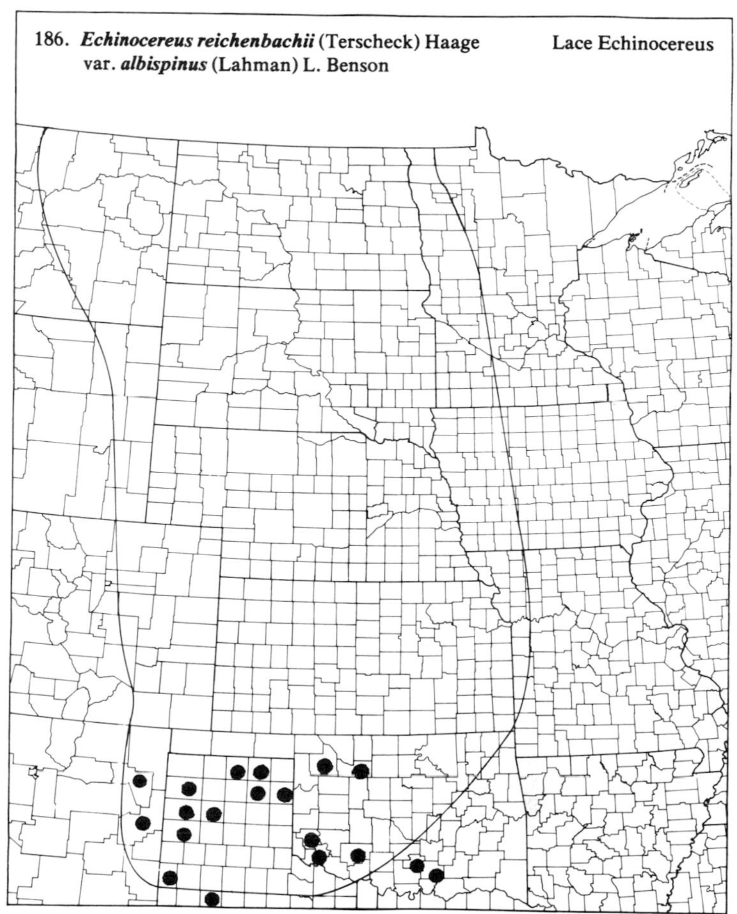

187. *Echinocereus viridiflorus* Engelm. — Hedgehog Cactus

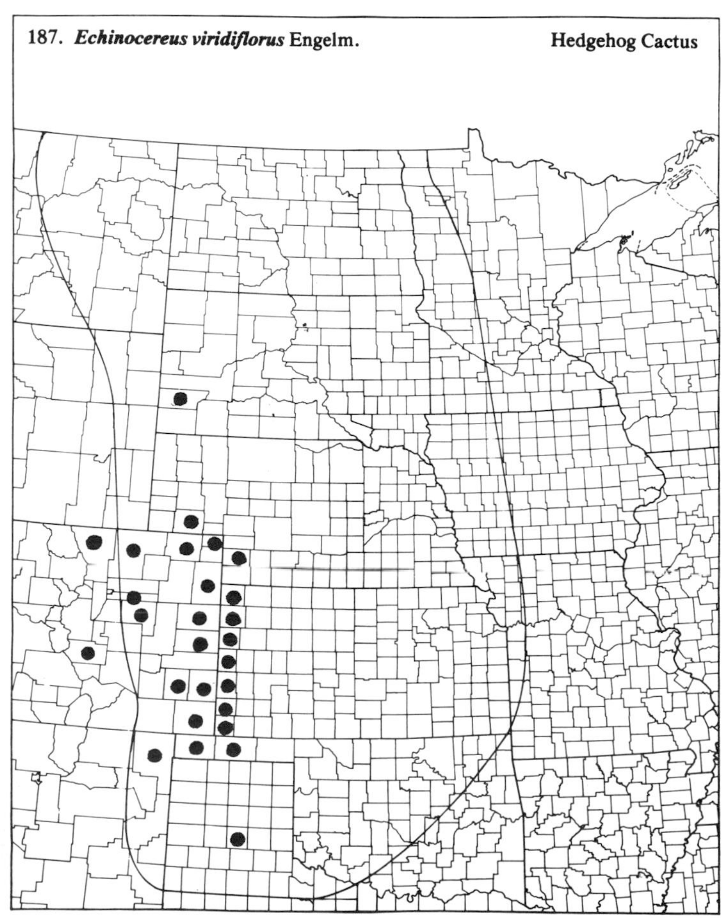

188. *Opuntia davisii* Engelm. & Bigel

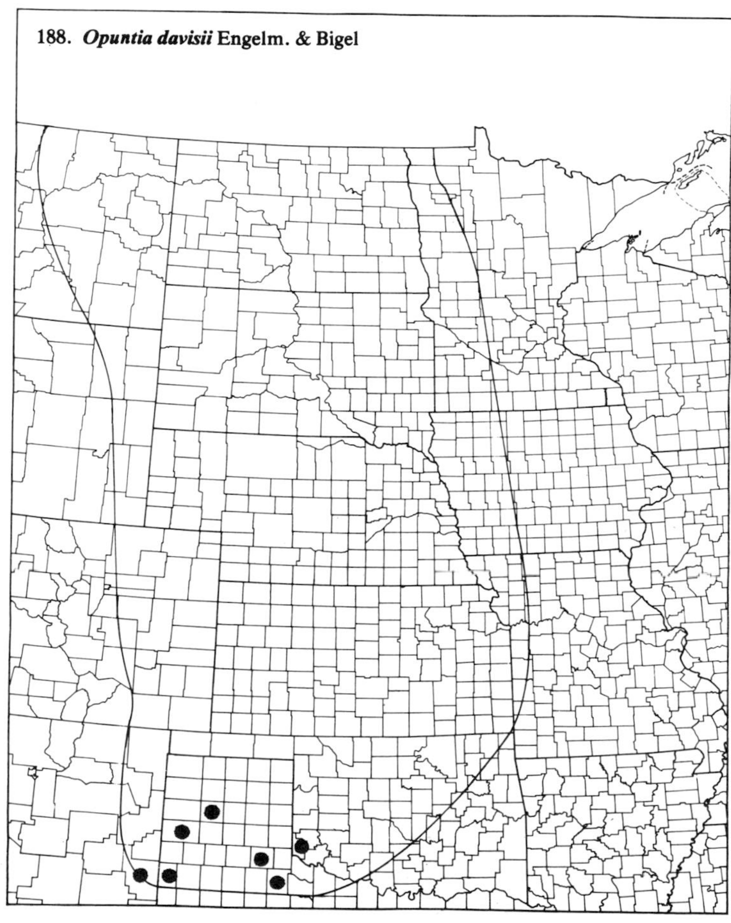

(Cactaceae)

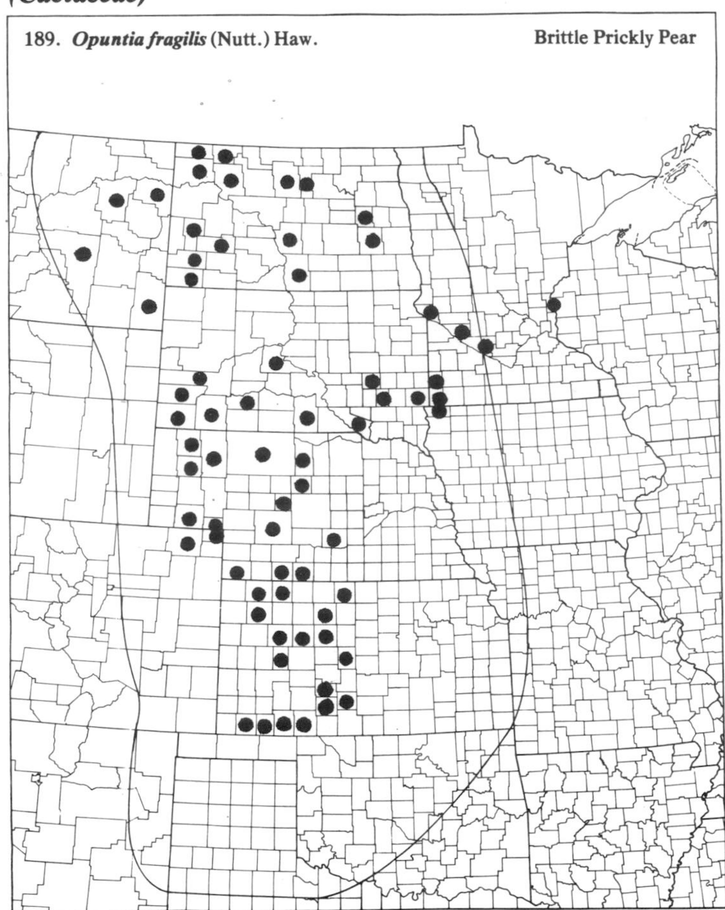
189. *Opuntia fragilis* (Nutt.) Haw.
Brittle Prickly Pear

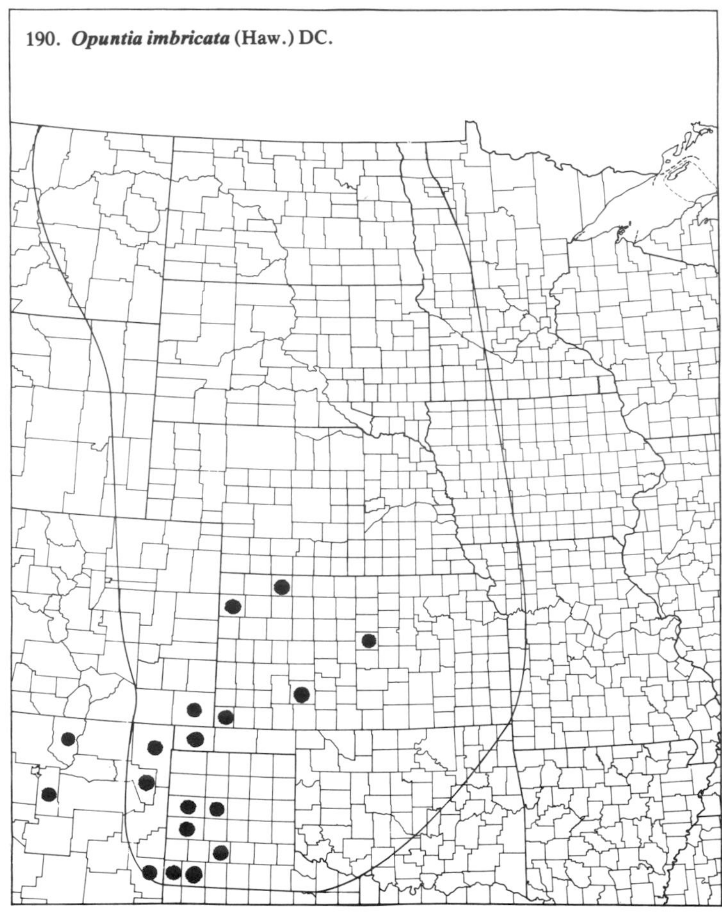
190. *Opuntia imbricata* (Haw.) DC.

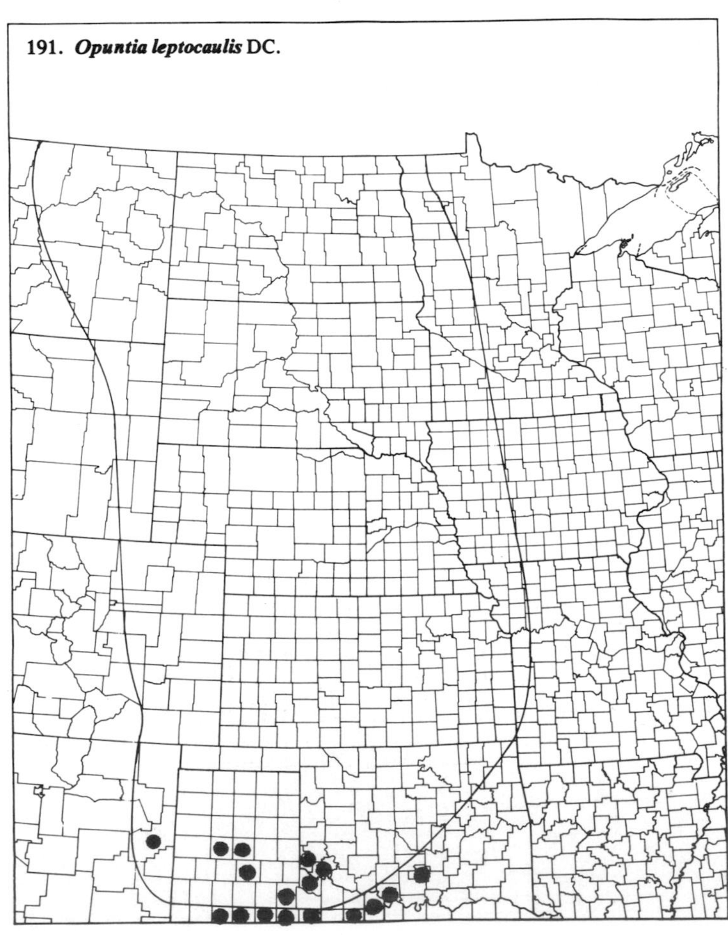
191. *Opuntia leptocaulis* DC.

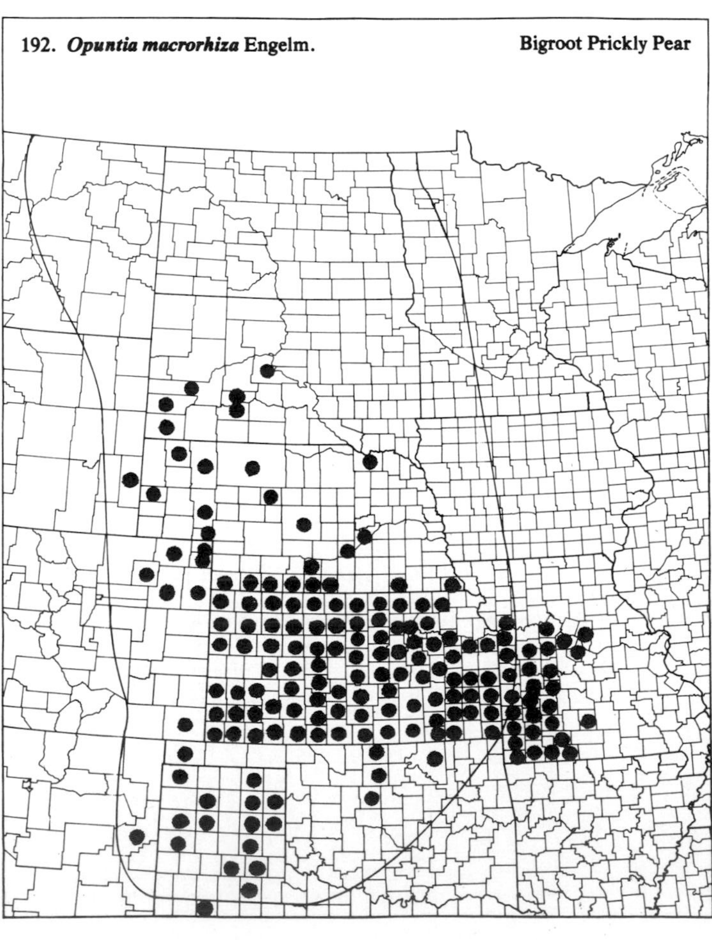
192. *Opuntia macrorhiza* Engelm.
Bigroot Prickly Pear

(Cactaceae)

193. ***Opuntia phaeacantha*** Engelm.

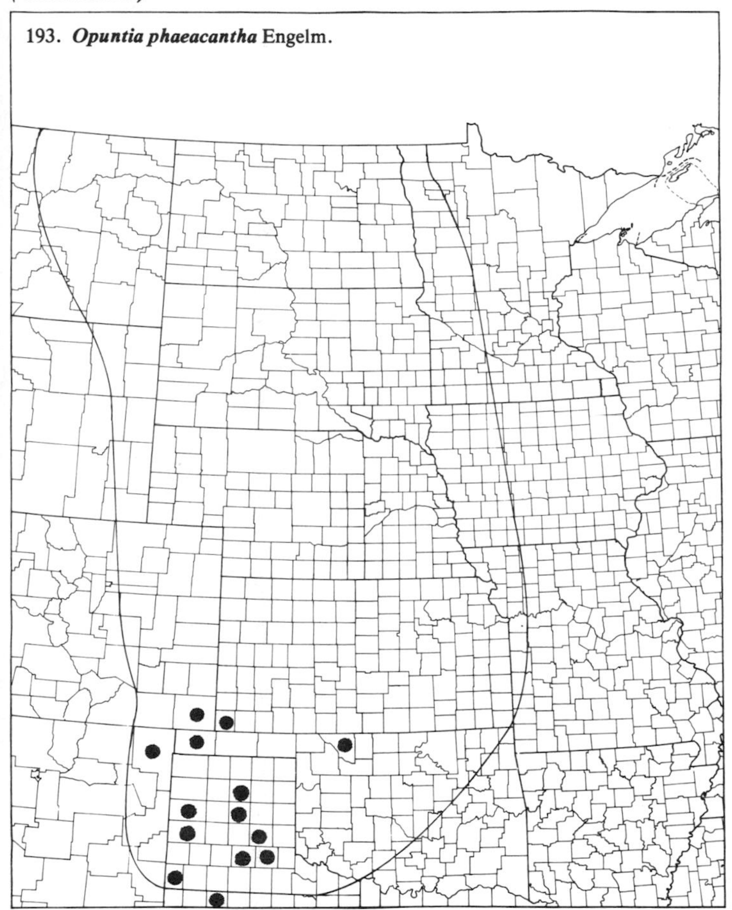

194. ***Opuntia polyacantha*** Haw. Plains Prickly Pear

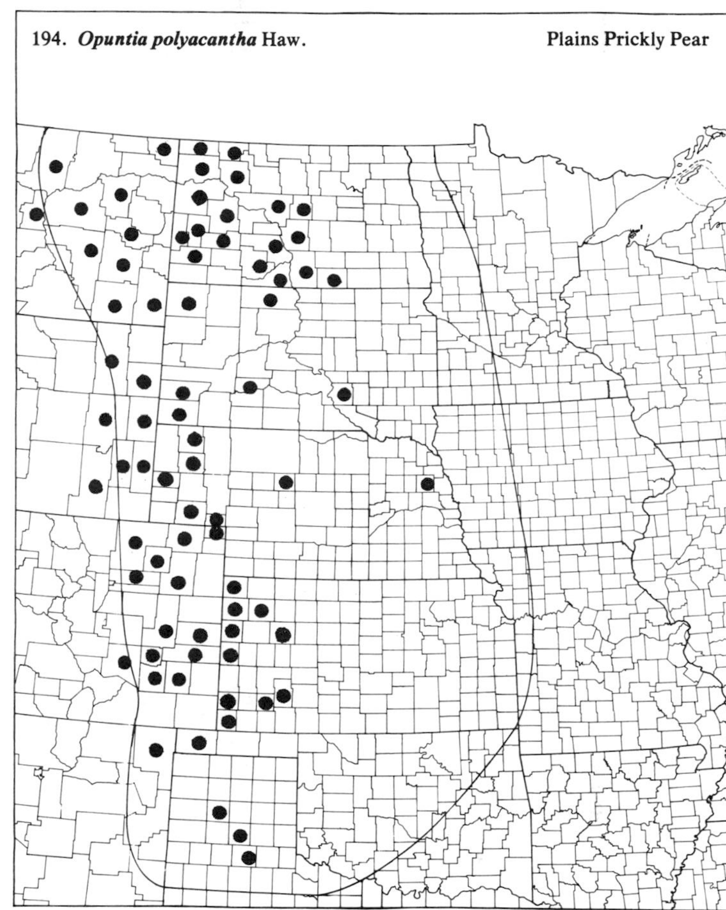

23. *Aizoaceae* (Carpetweed Family)

195. ***Mollugo verticillata*** L. Carpetweed

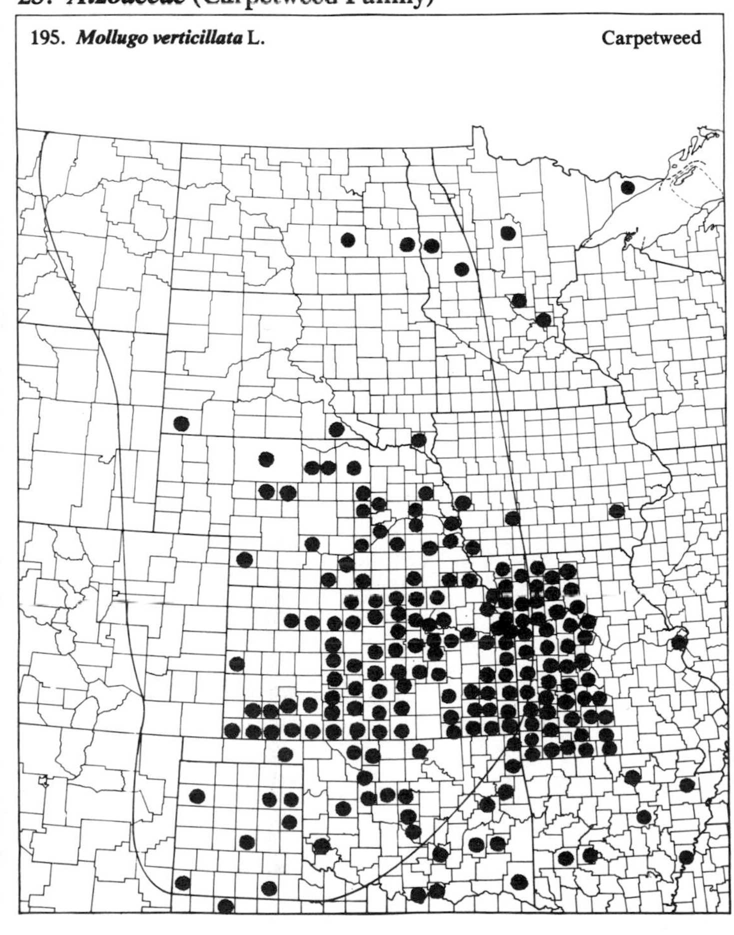

196. ***Sesuvium verrucosum*** Raf.

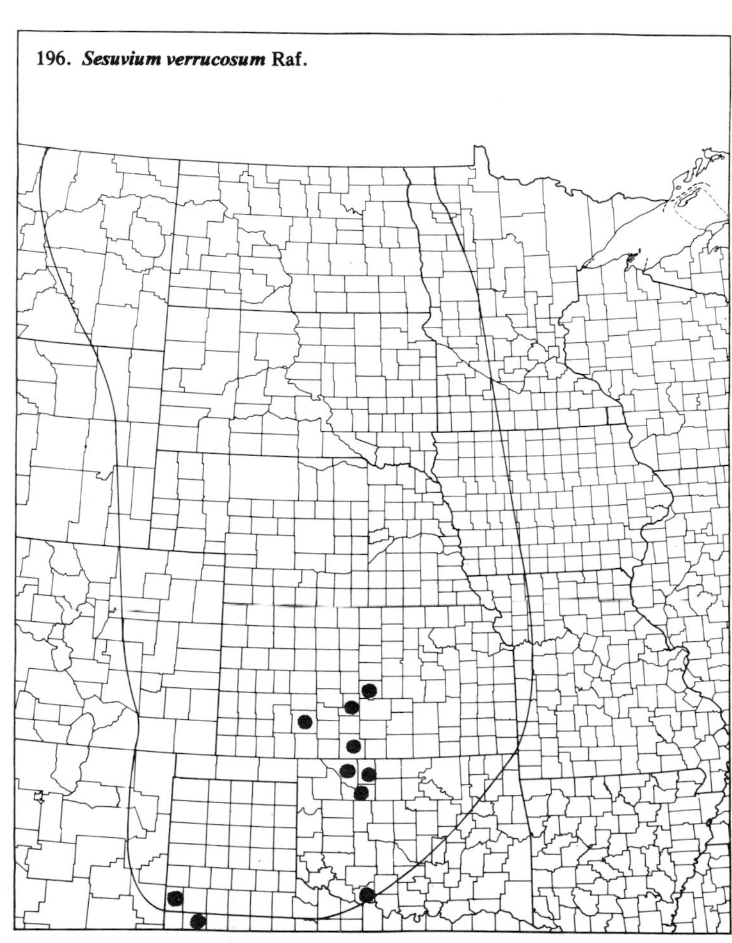

24. *Caryophyllaceae* (Pink Family)

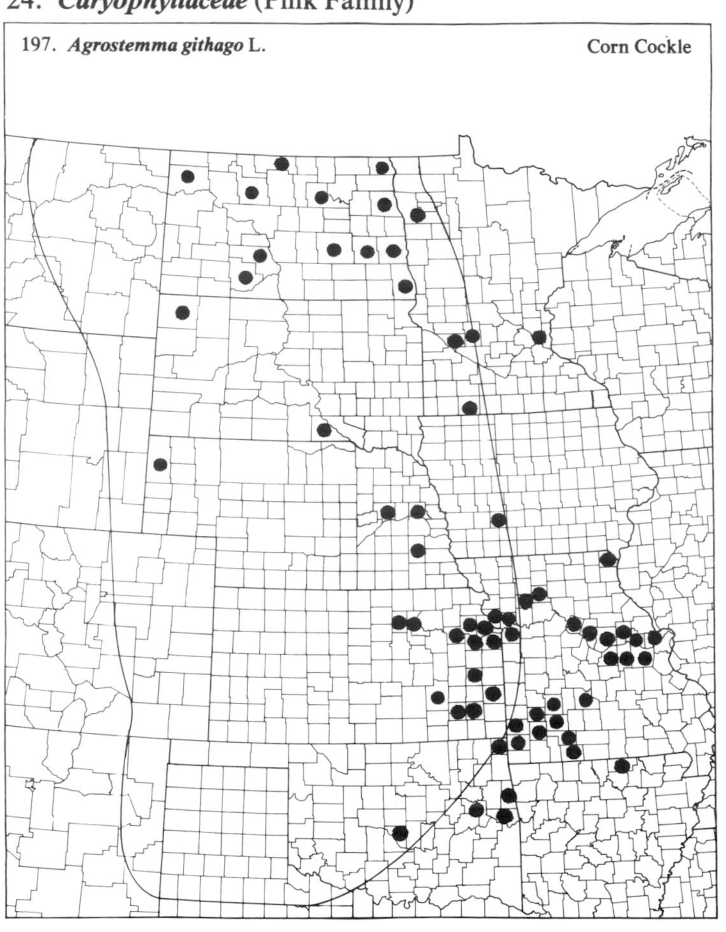
197. ***Agrostemma githago*** L. Corn Cockle

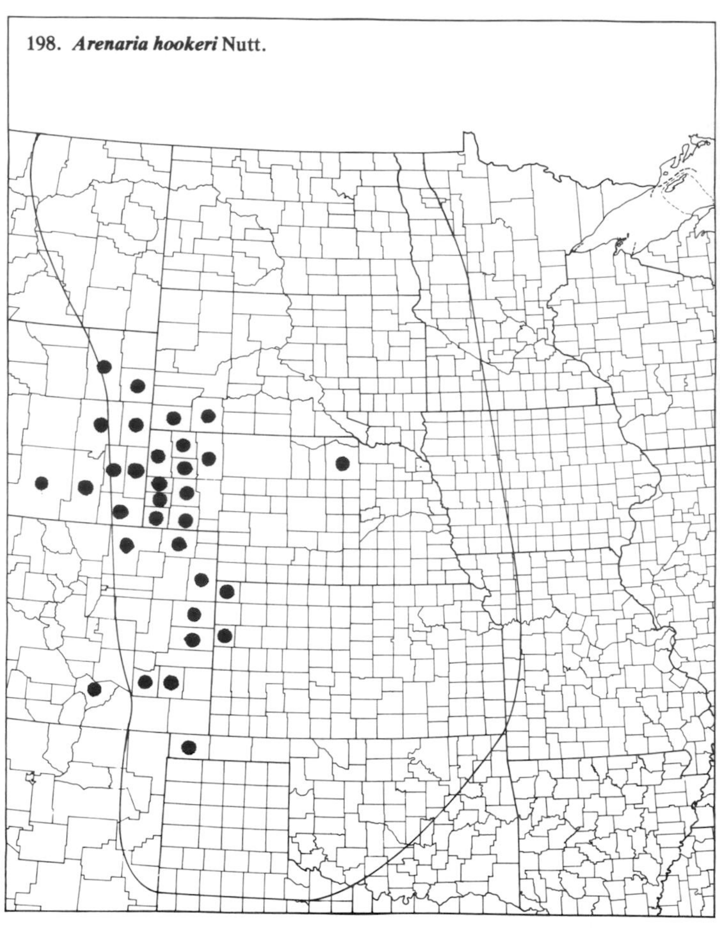
198. ***Arenaria hookeri*** Nutt.

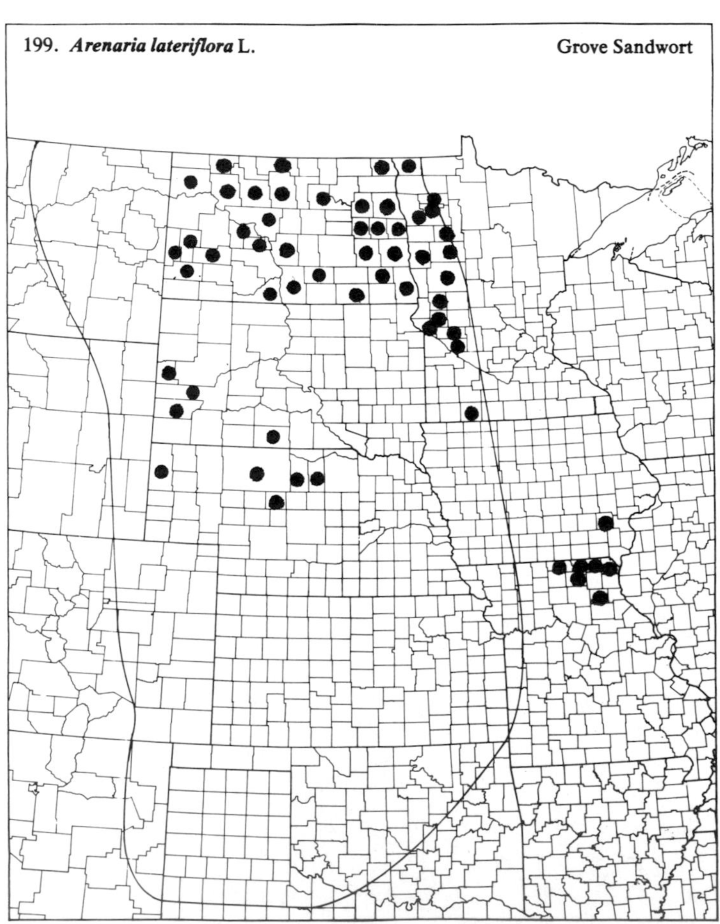
199. ***Arenaria lateriflora*** L. Grove Sandwort

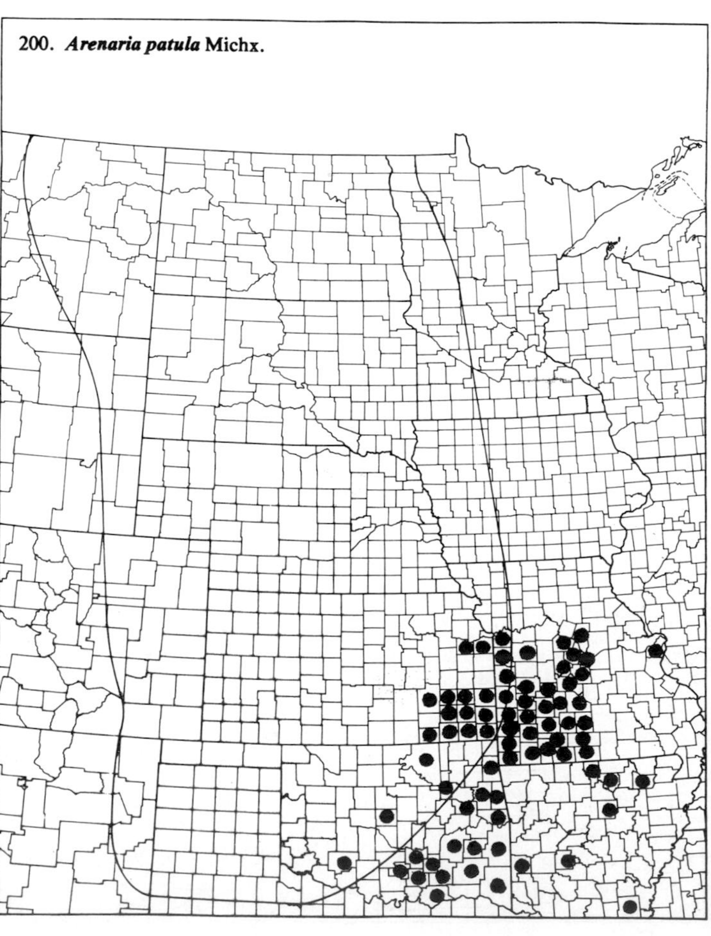
200. ***Arenaria patula*** Michx.

(Caryophyllaceae)

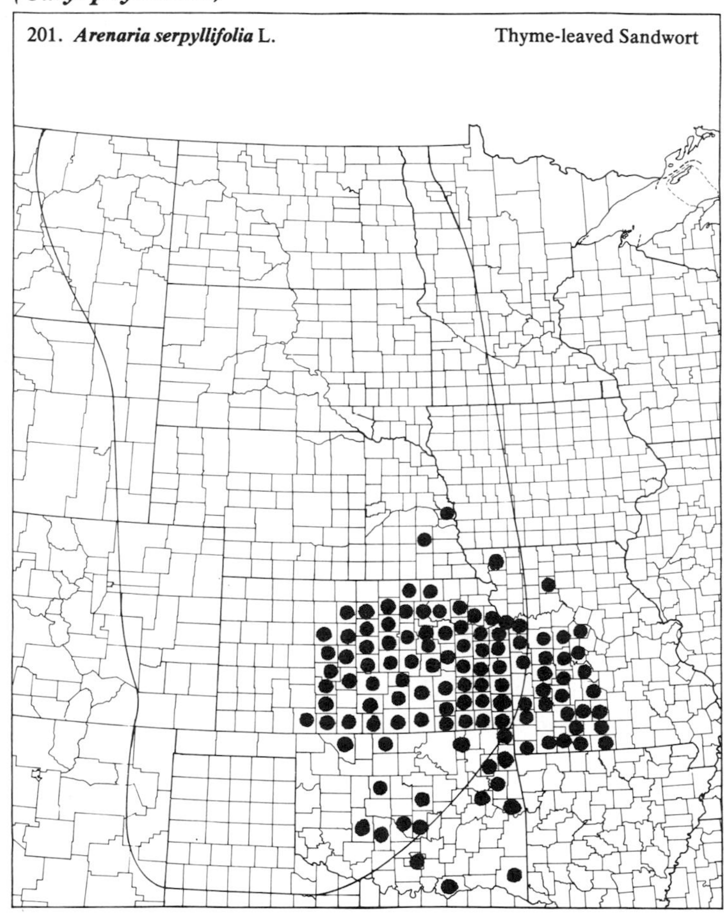

201. ***Arenaria serpyllifolia*** L. Thyme-leaved Sandwort

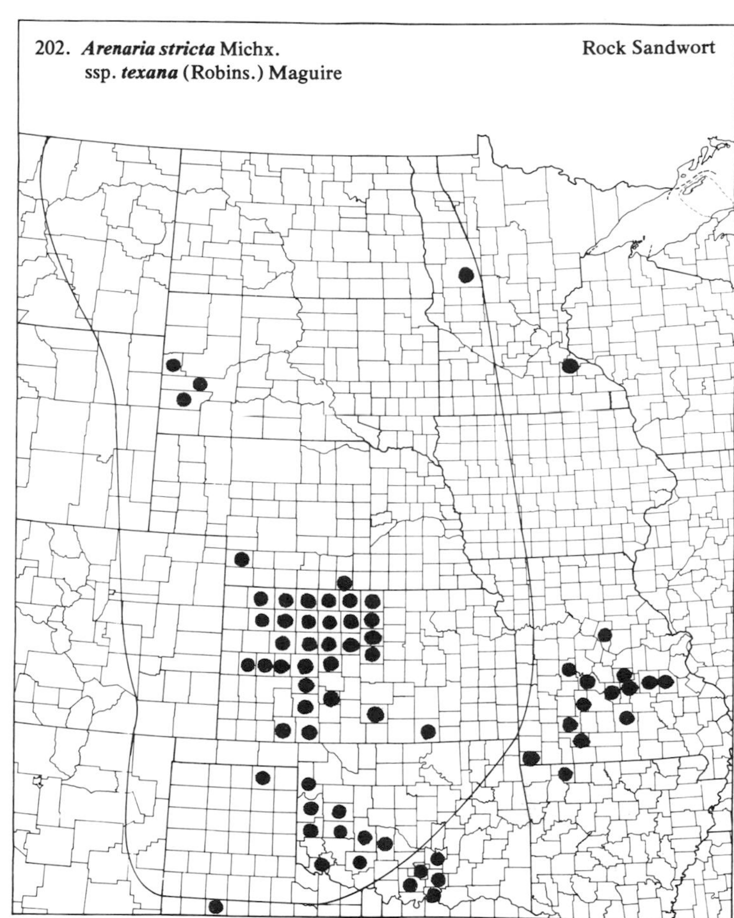

202. ***Arenaria stricta*** Michx. ssp. ***texana*** (Robins.) Maguire Rock Sandwort

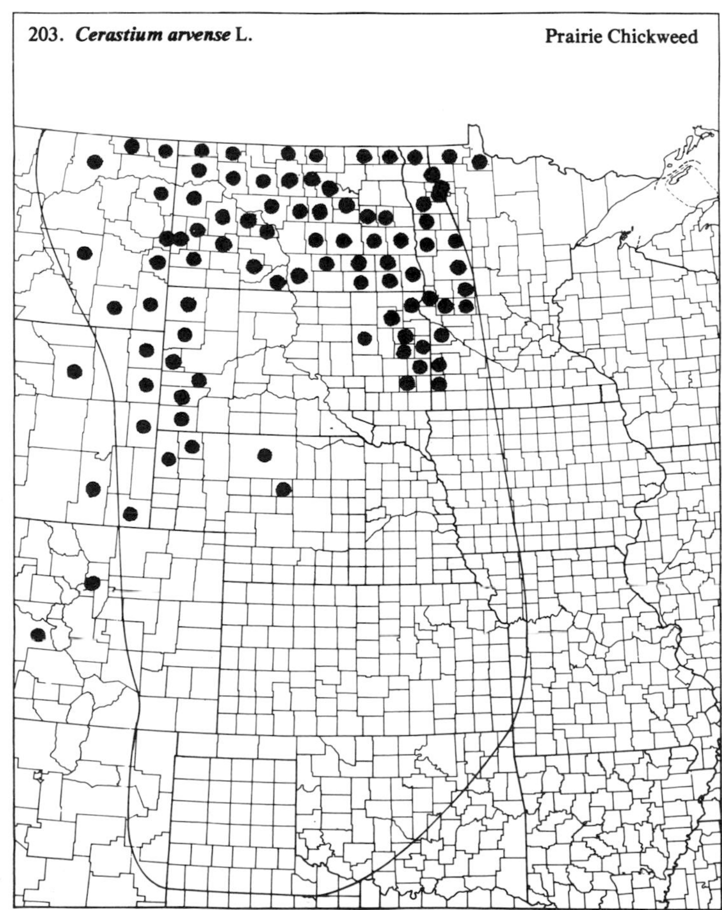

203. ***Cerastium arvense*** L. Prairie Chickweed

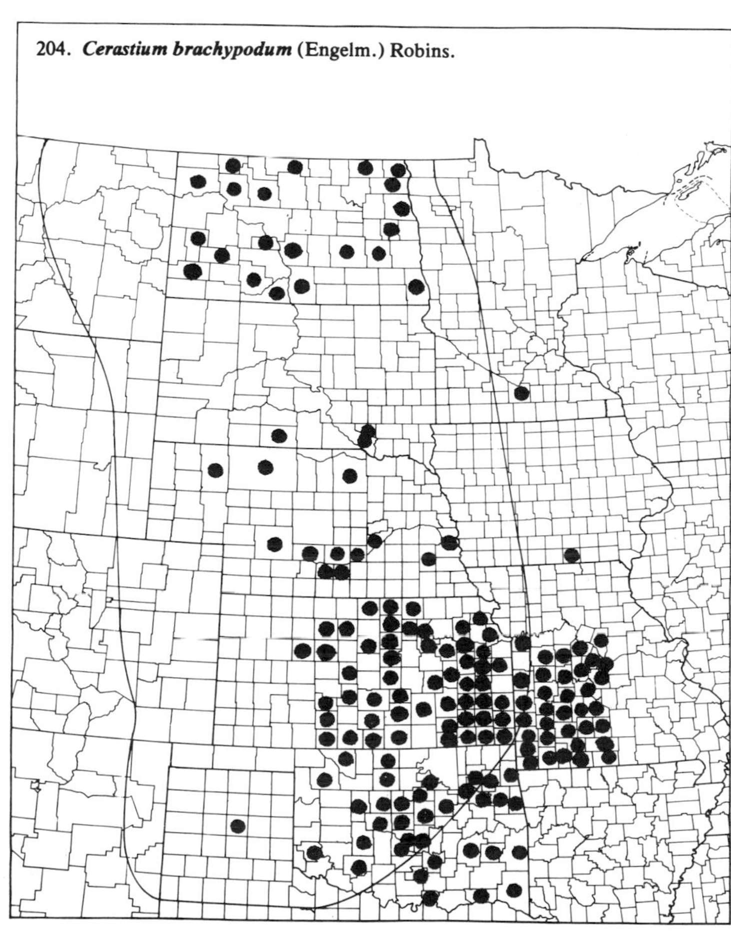

204. ***Cerastium brachypodum*** (Engelm.) Robins.

(Caryophyllaceae)

205. ***Cerastium glomeratum*** Thuill.

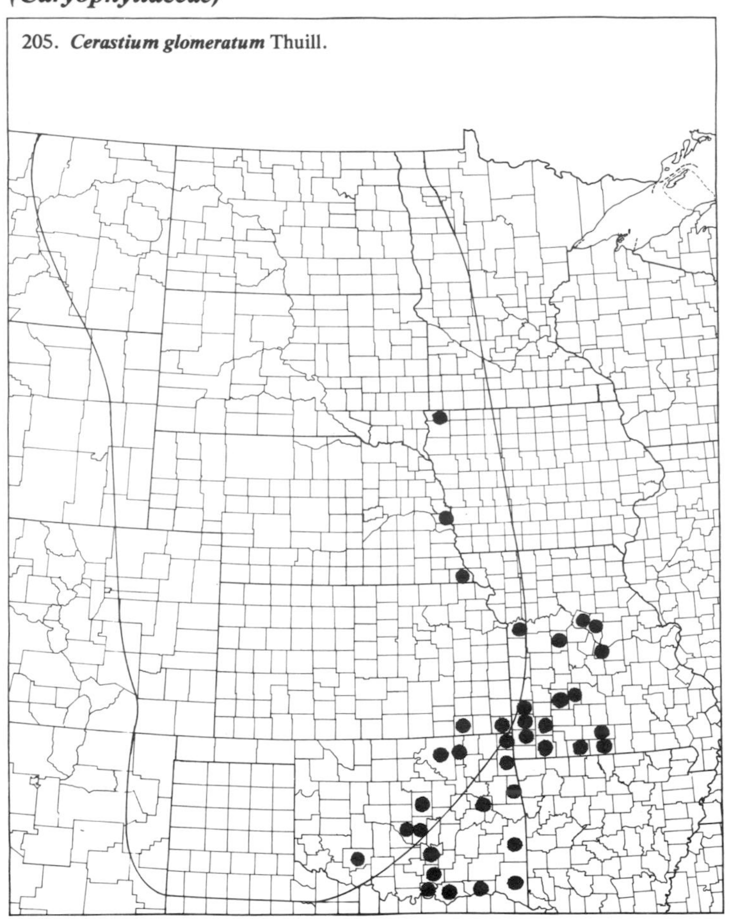

206. ***Cerastium nutans*** Raf. Powderhorn Cerastium

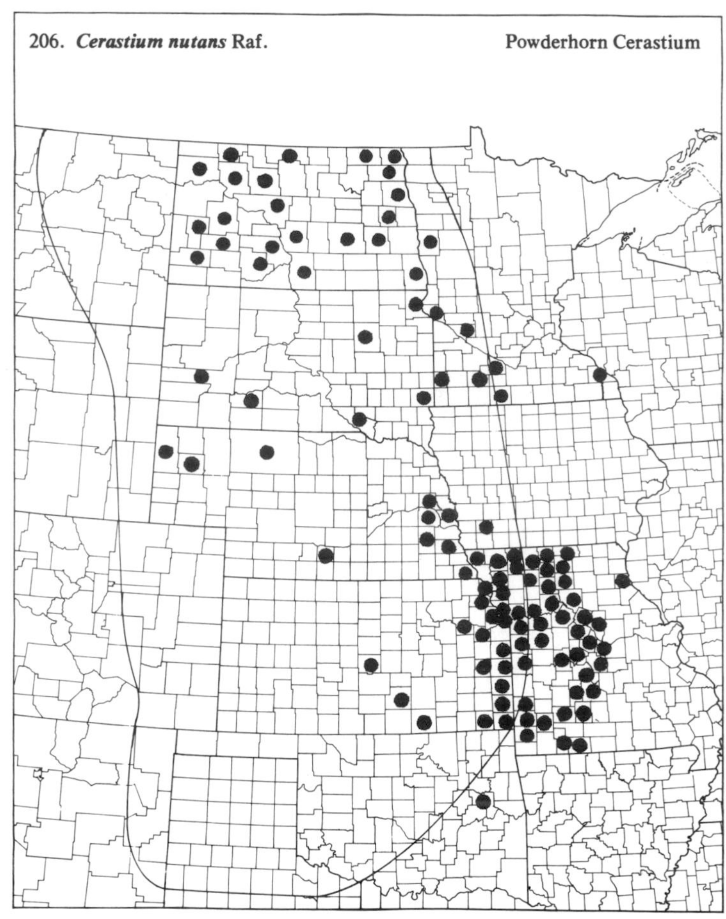

207. ***Cerastium vulgatum*** L. Mouse-ear Chickweed

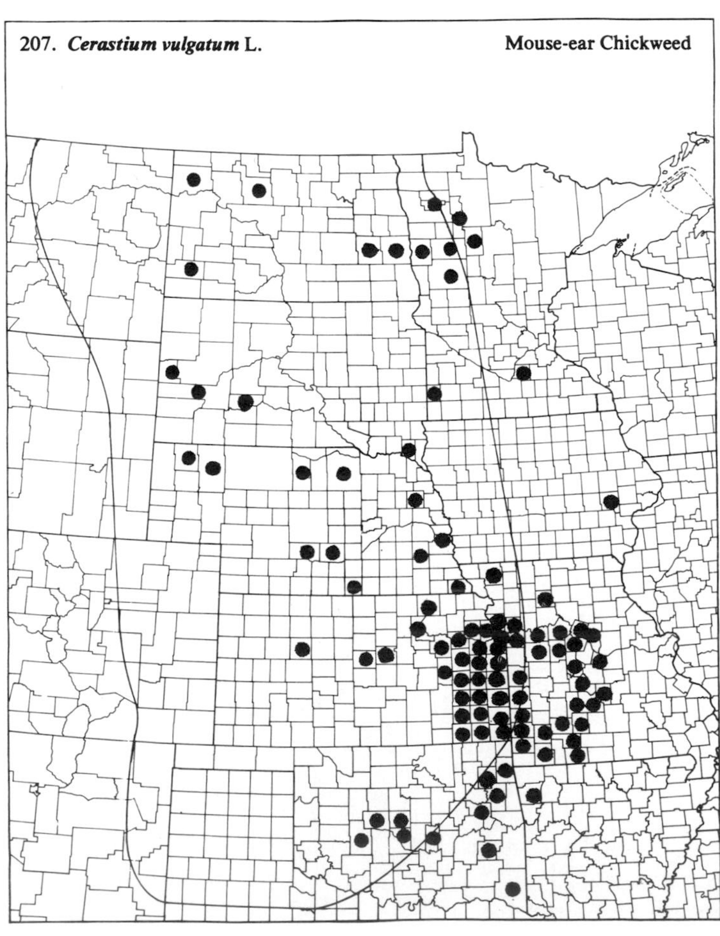

208. ***Dianthus armeria*** L. Deptford Pink

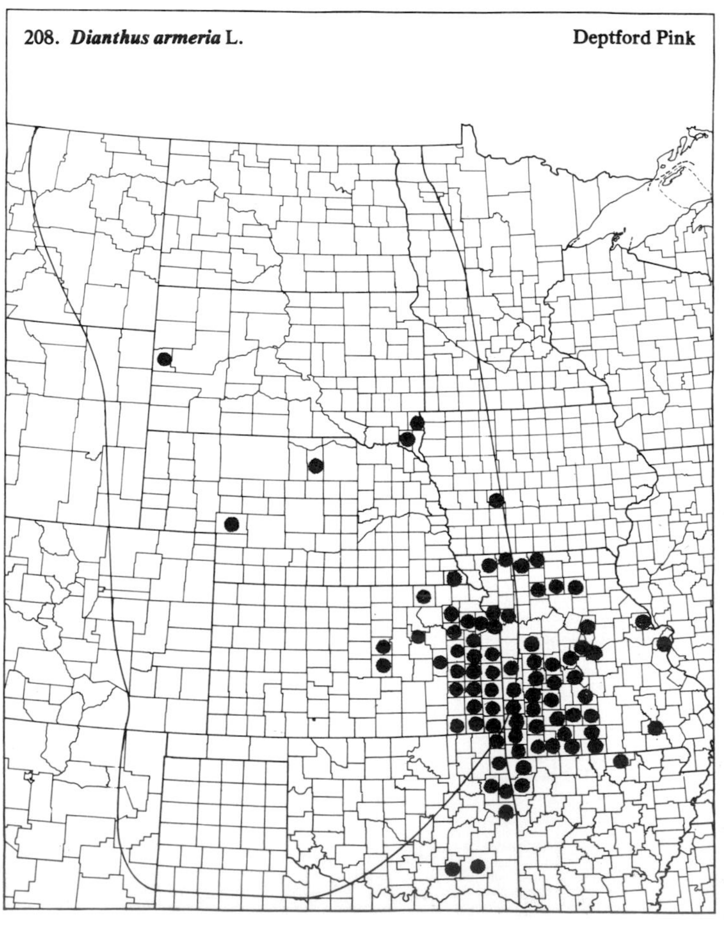

(Caryophyllaceae)

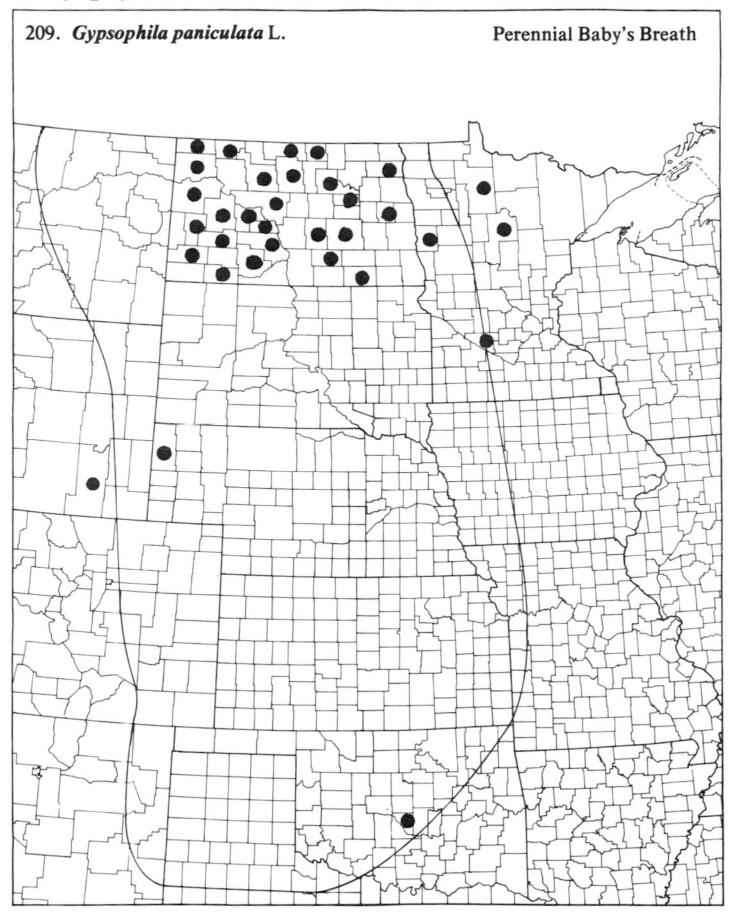

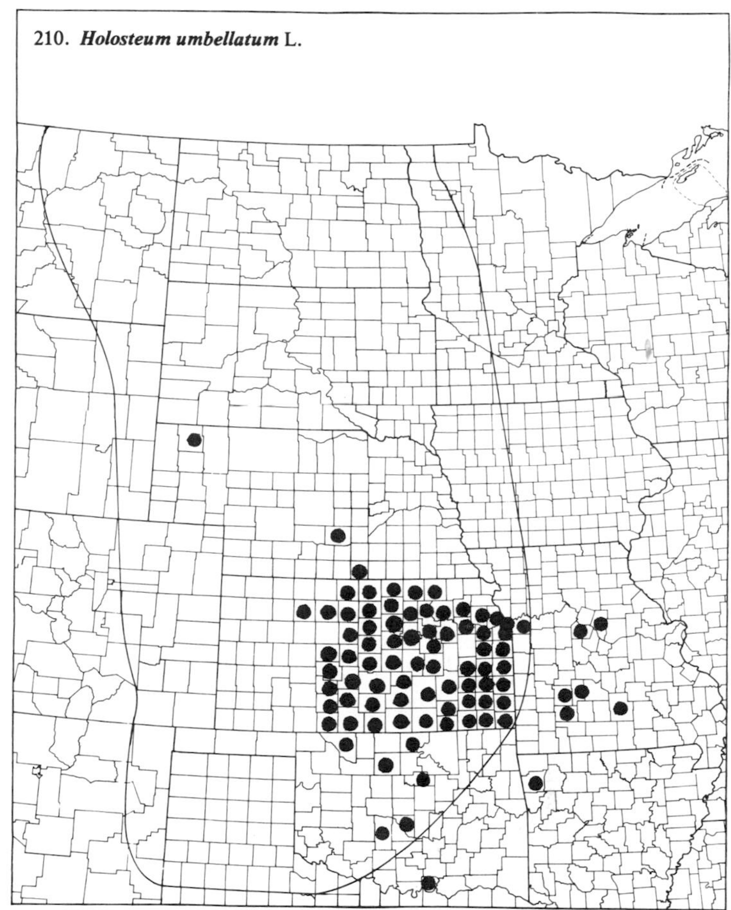

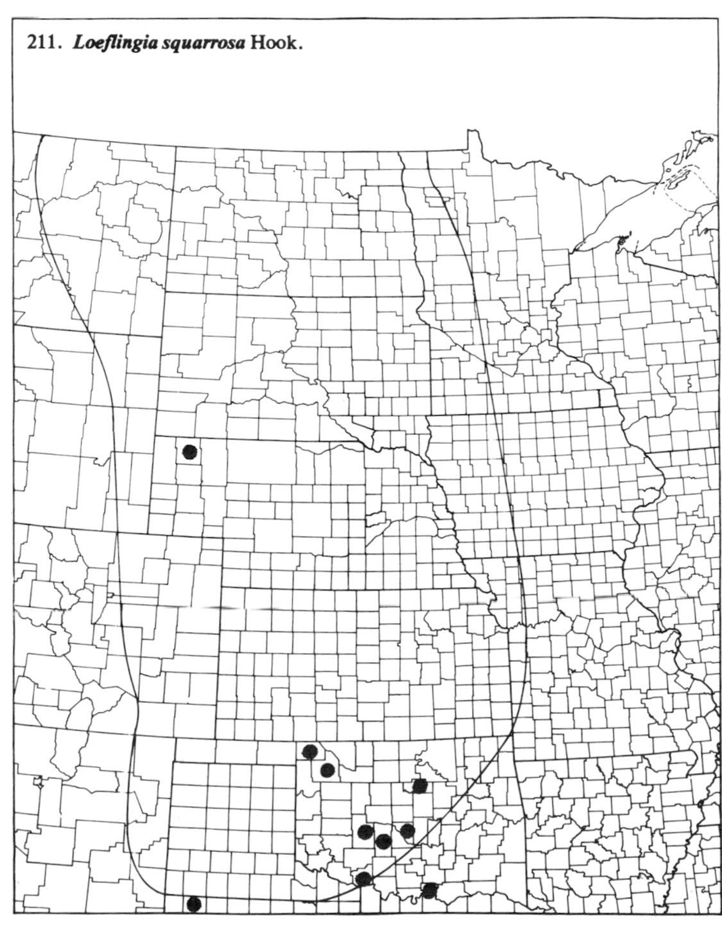

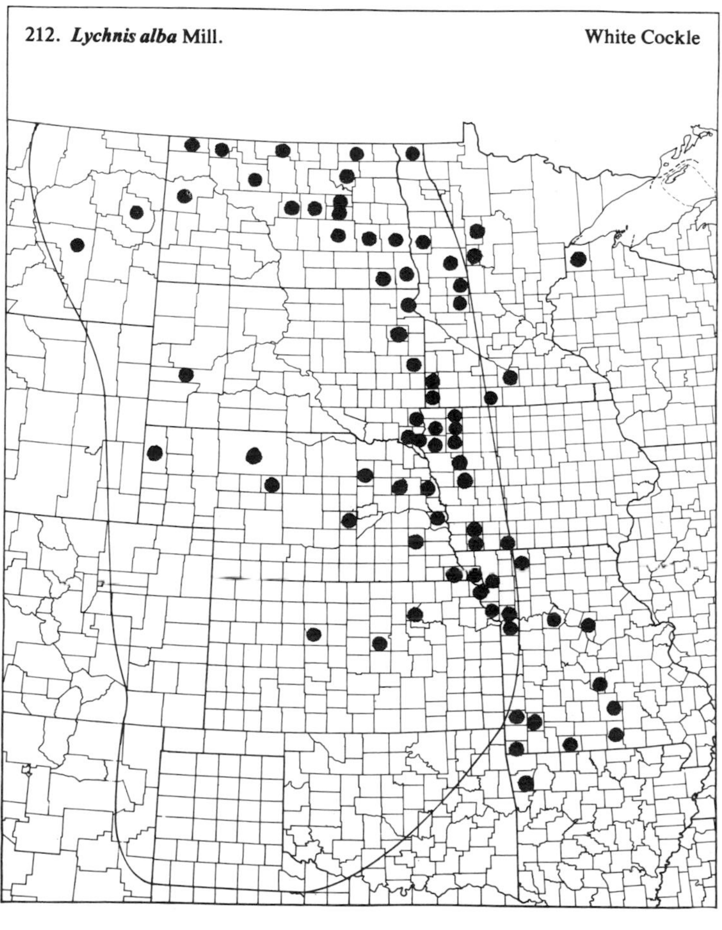

(Caryophyllaceae)

213. ***Lychnis drummondii*** (Hook.) Wats.

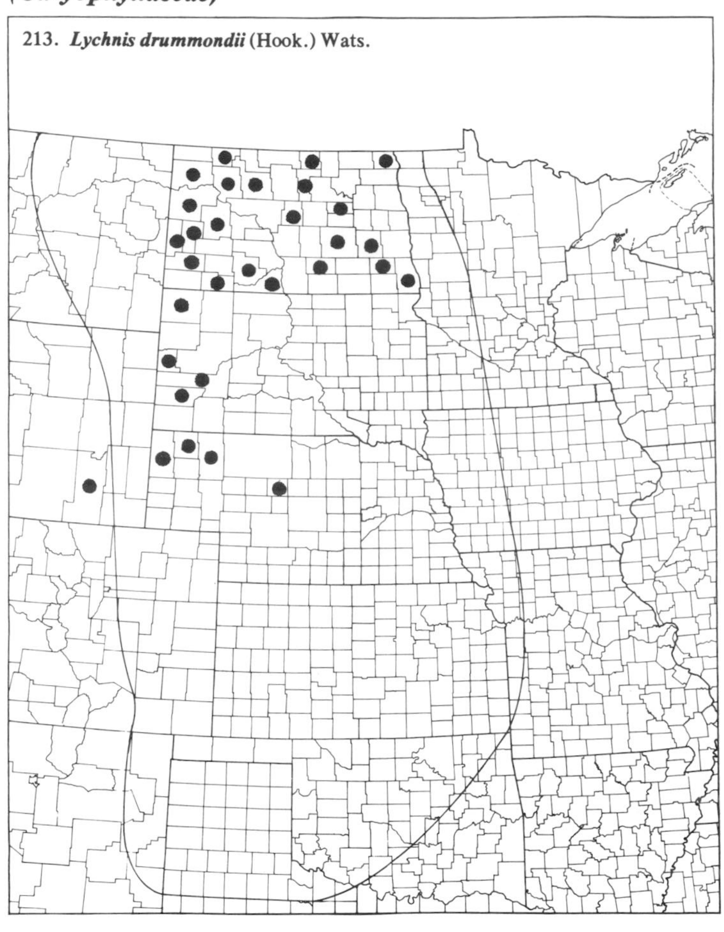

214. ***Paronychia canadensis*** (L.) Wood — Forked Chickweed

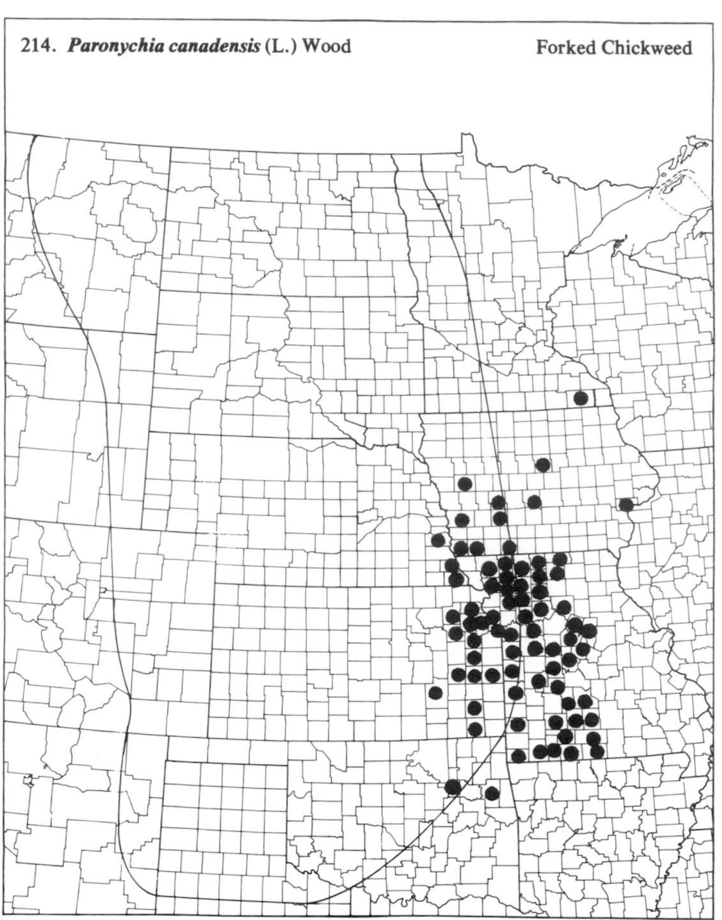

215. ***Paronychia depressa*** Nutt.

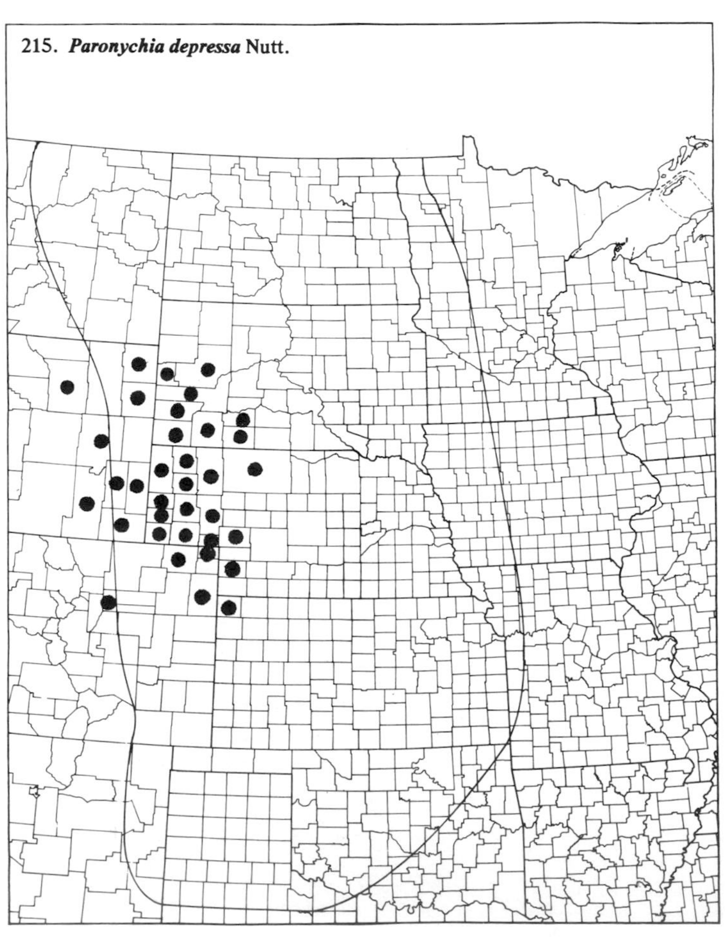

216. ***Paronychia fastigiata*** (Raf.) Fern. — Forked Chickweed

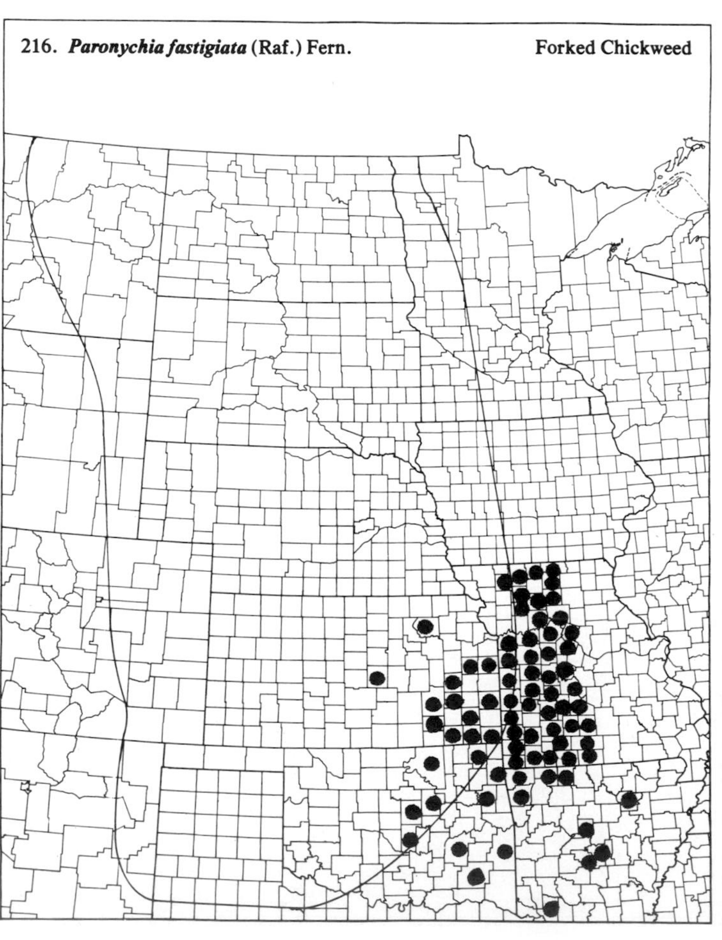

(Caryophyllaceae)

217. ***Paronychia jamesii*** T. & G. — James Nailwort

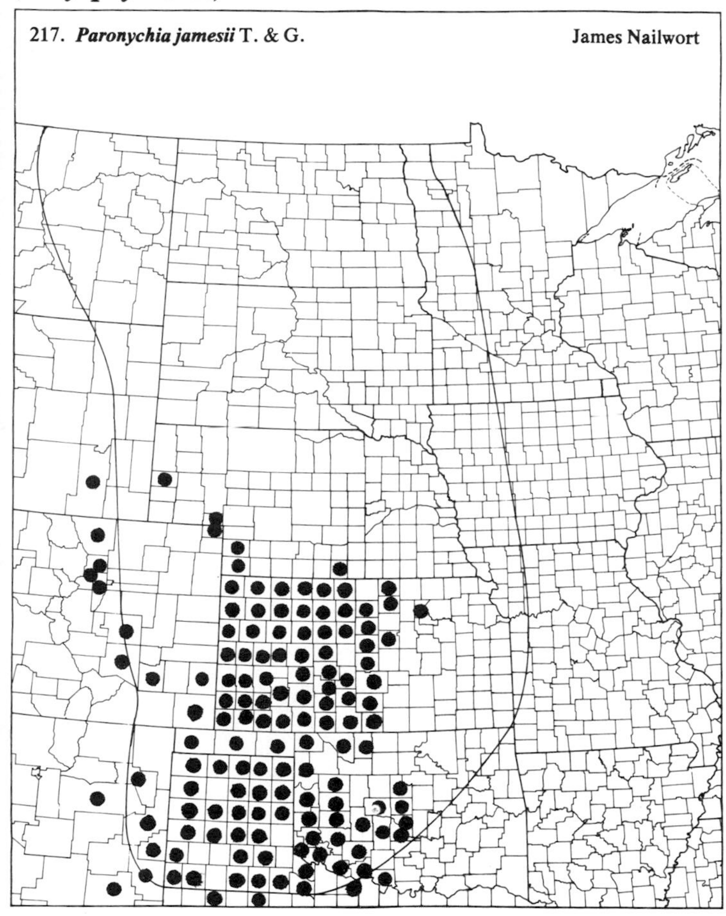

218. ***Paronychia sessiliflora*** Nutt. — Whitlowwort

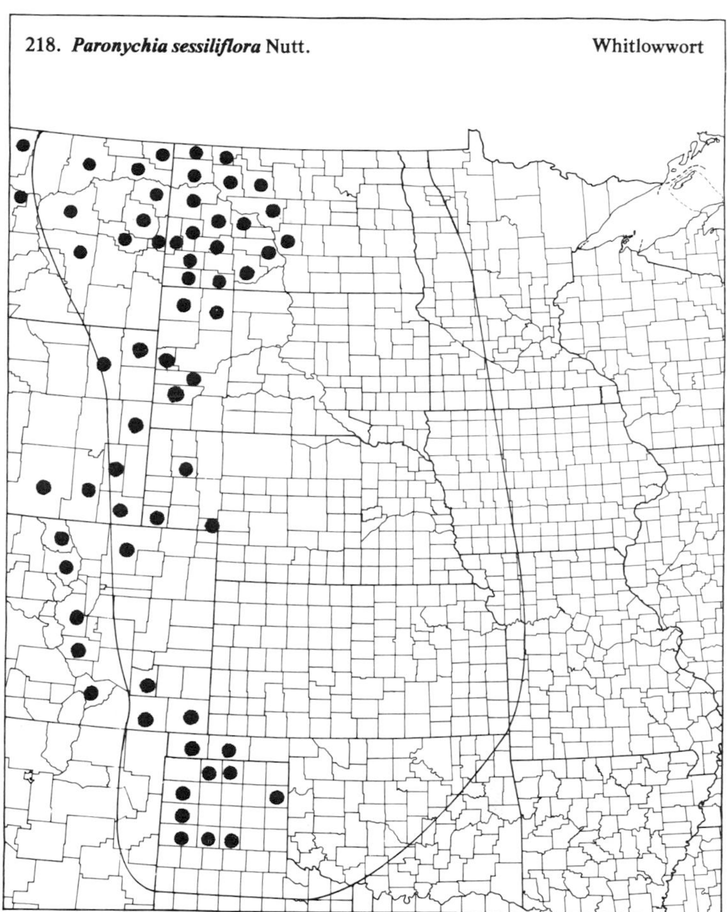

219. ***Sagina decumbens*** (Ell.) T. & G. — Trailing Pearlwort

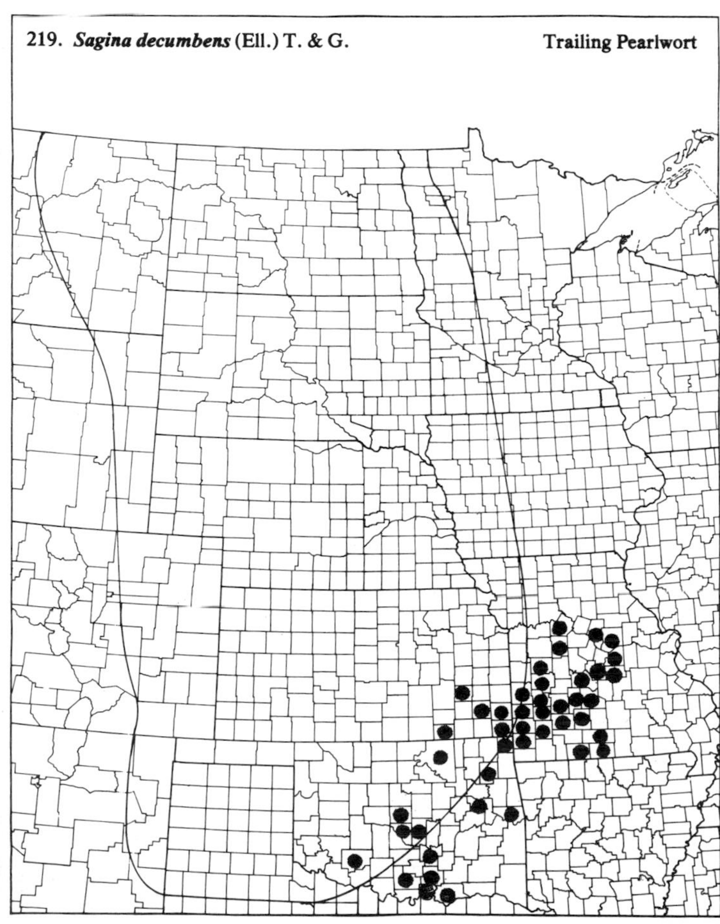

220. ***Saponaria officinalis*** L. — Bouncing Bet

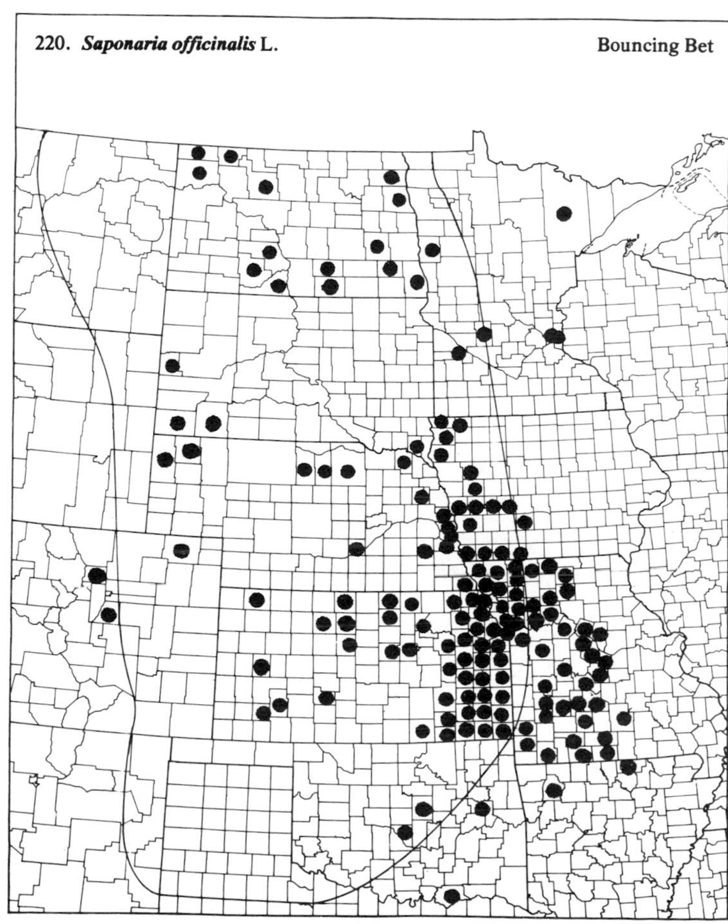

(Caryophyllaceae)

221. ***Silene antirrhina*** L. Sleepy Catchfly

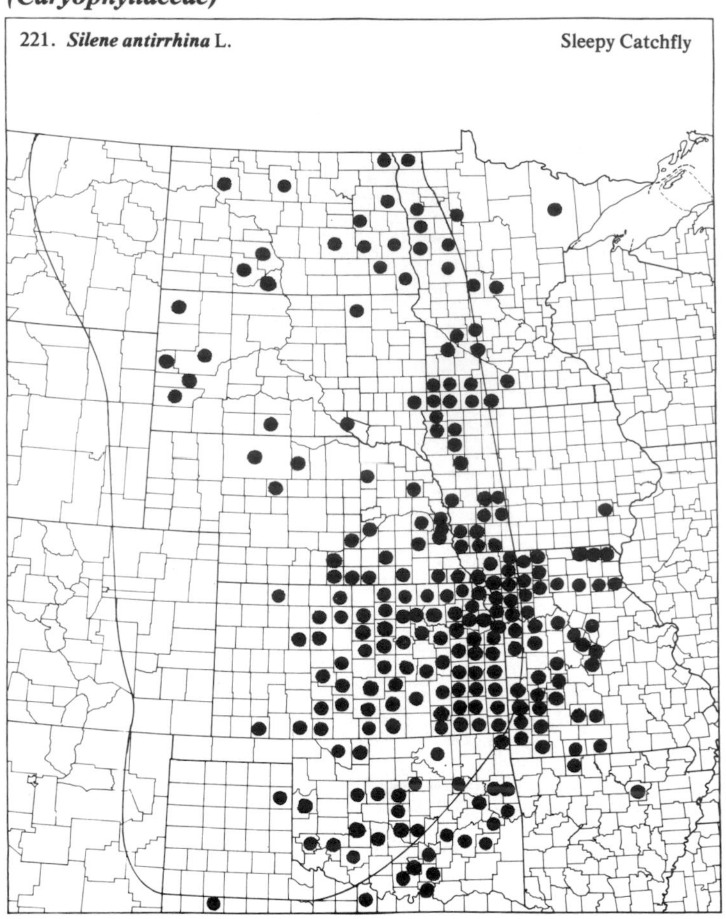

222. ***Silene cserei*** Baumg. Smooth Catchfly

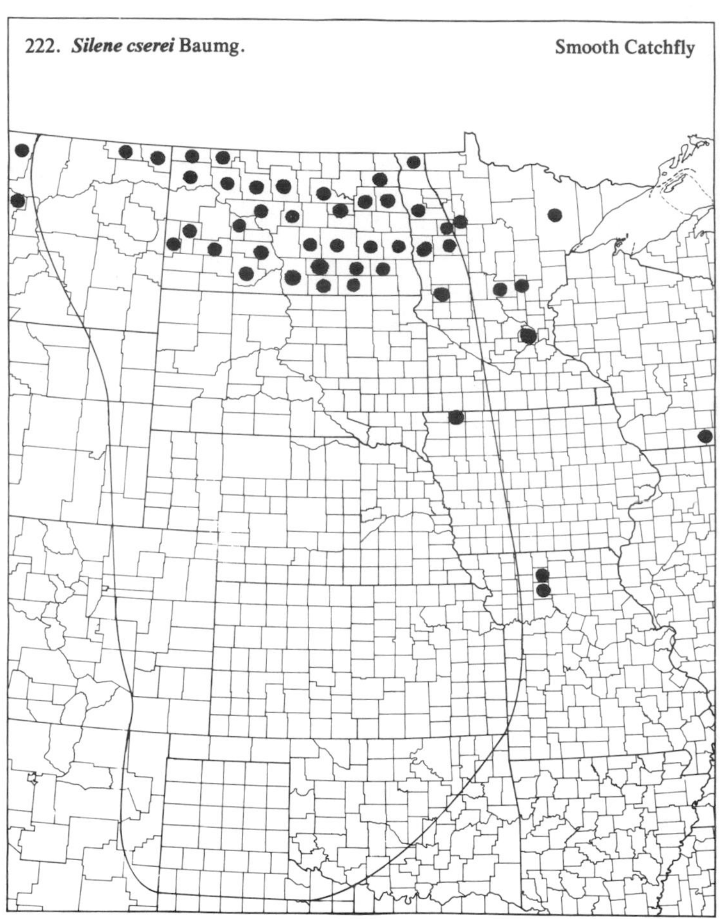

223. ***Silene cucubalus*** Wibel Bladder Campion

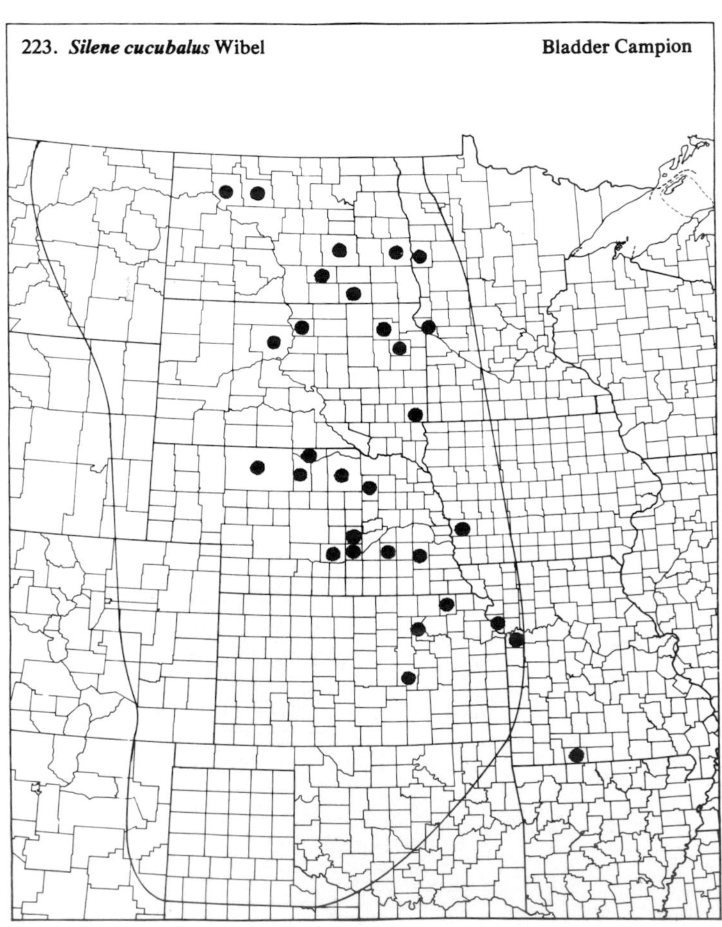

224. ***Silene dichotoma*** Ehrh. Forked Catchfly

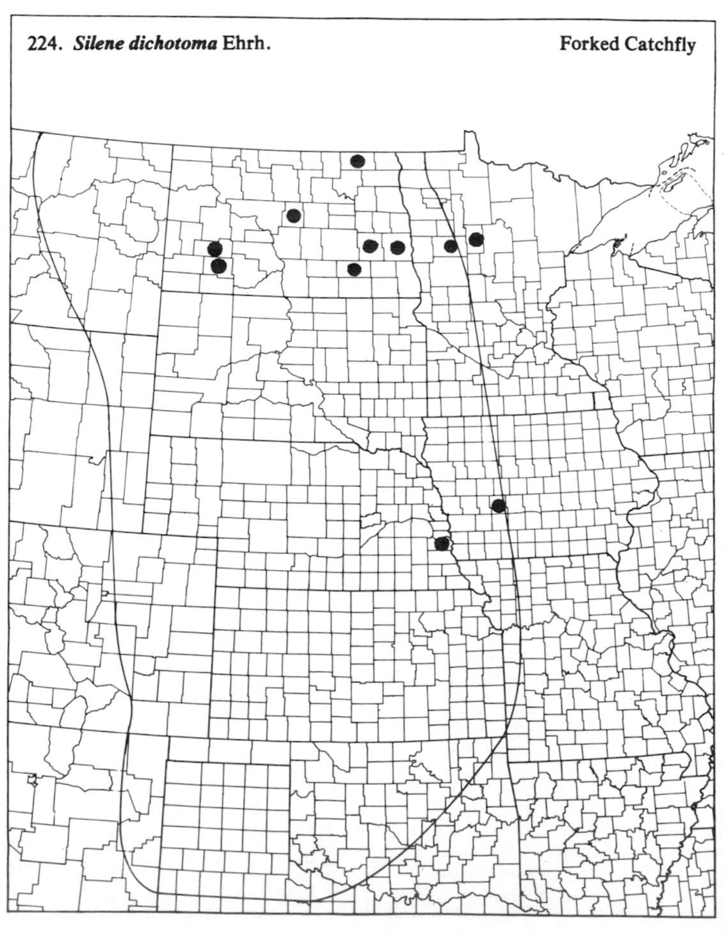

(Caryophyllaceae)

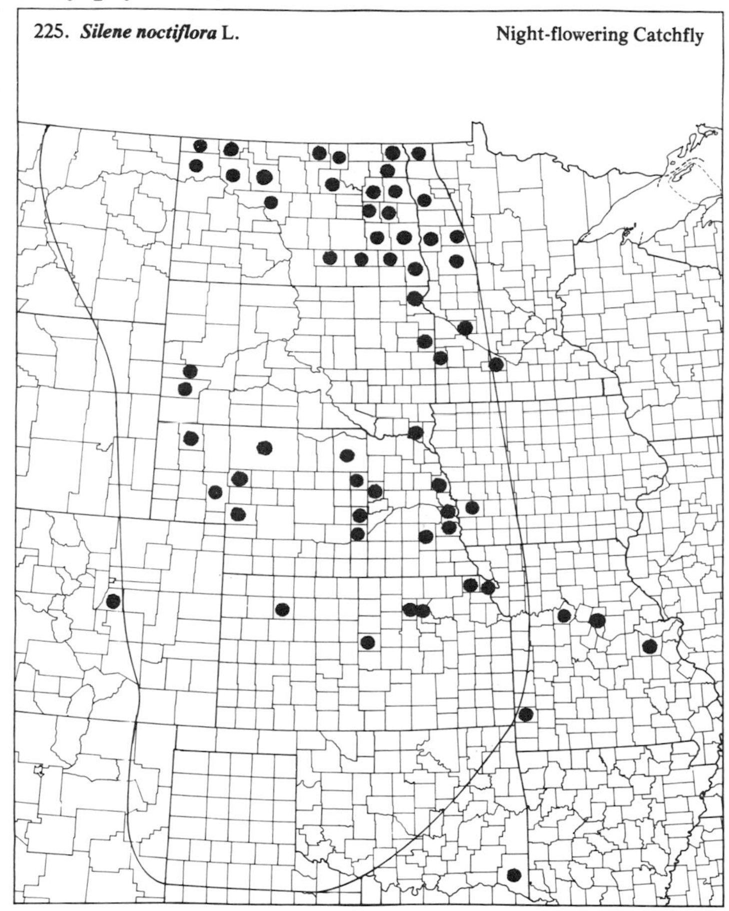
225. ***Silene noctiflora*** L. Night-flowering Catchfly

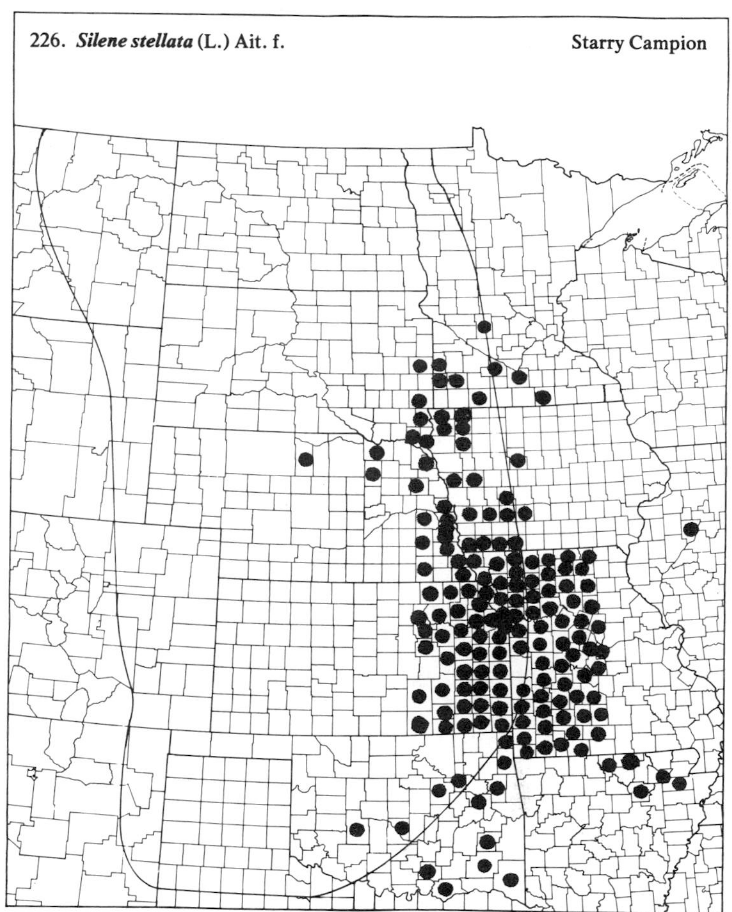
226. ***Silene stellata*** (L.) Ait. f. Starry Campion

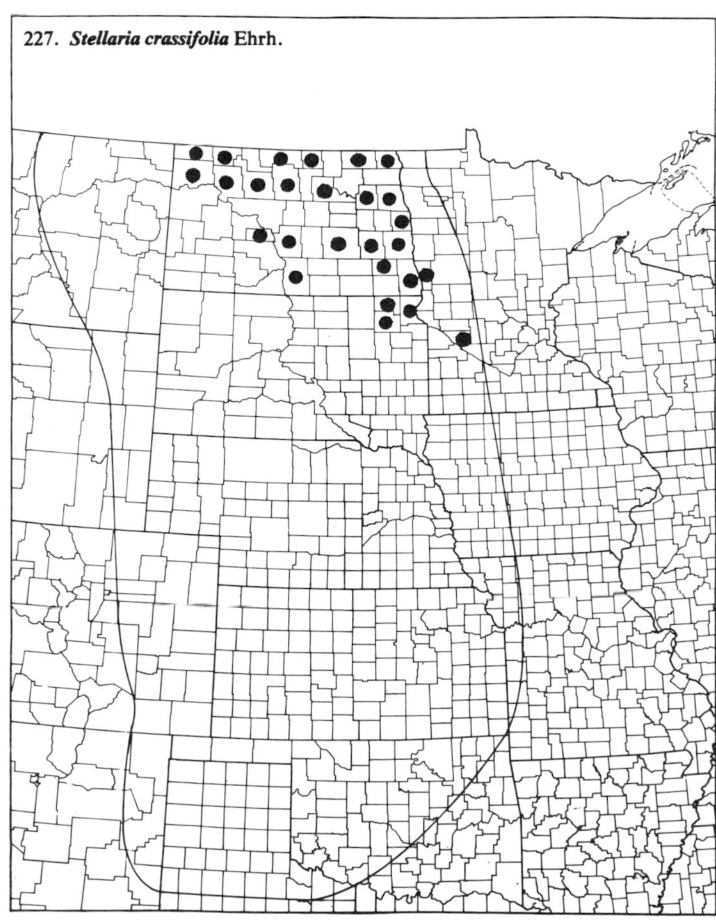
227. ***Stellaria crassifolia*** Ehrh.

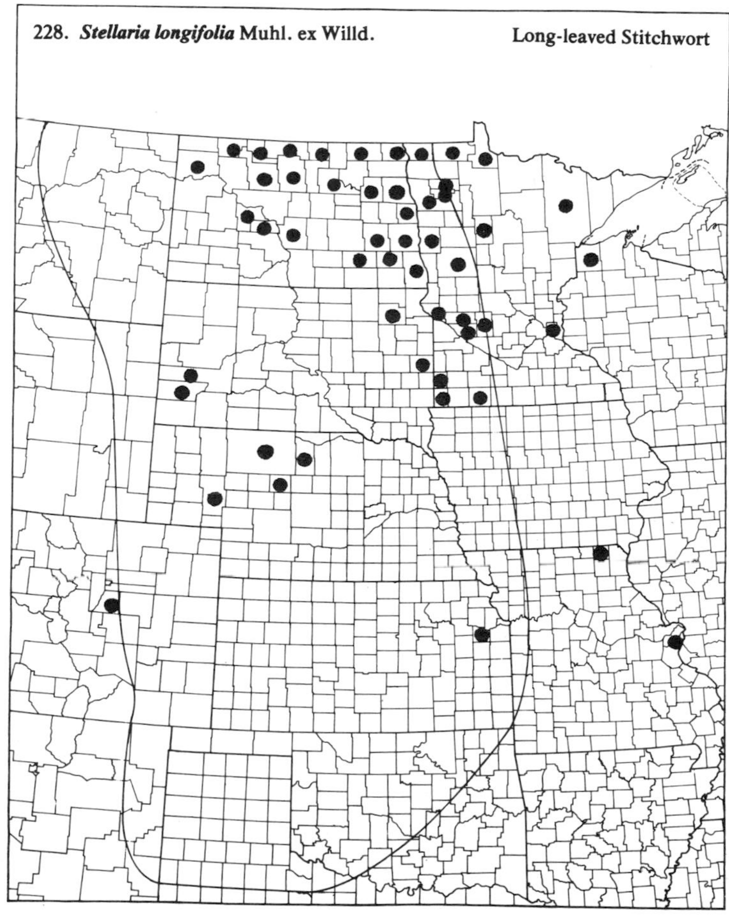
228. ***Stellaria longifolia*** Muhl. ex Willd. Long-leaved Stitchwort

(Caryophyllaceae)

229. ***Stellaria longipes*** Goldie

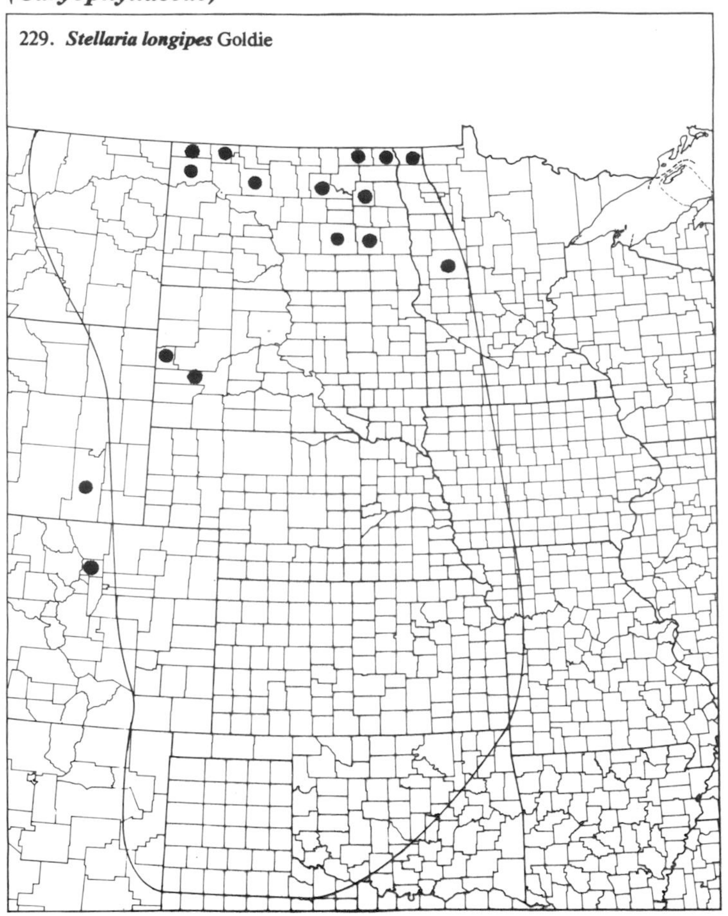

230. ***Stellaria media*** (L.) Cyr. Common Chickweed

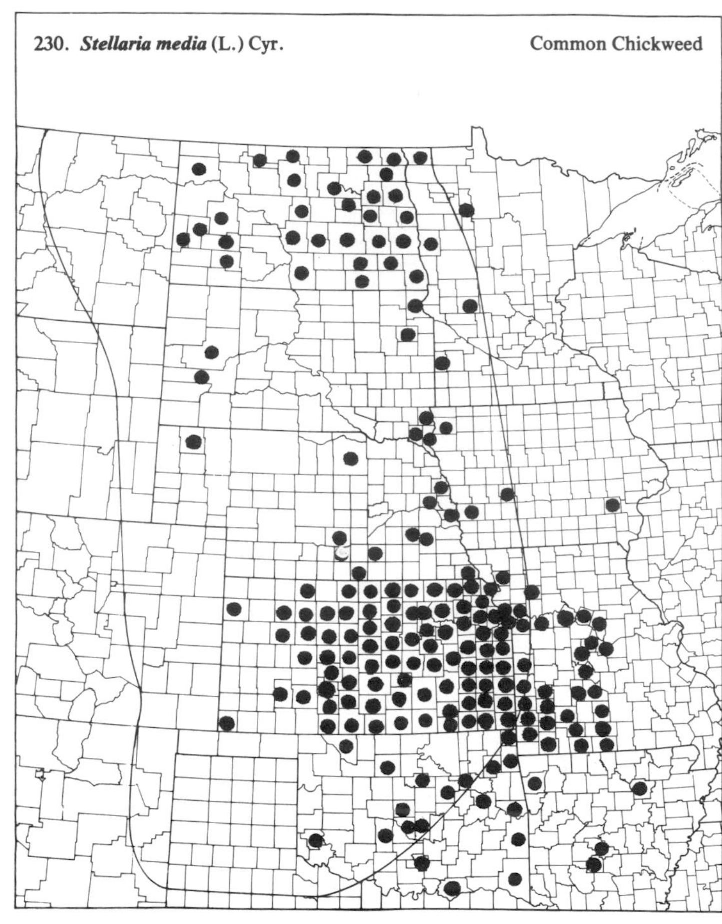

231. ***Vaccaria segetalis*** (Neck.) Gke. Cowherb

25. *Portulacaceae* (Purslane Family)

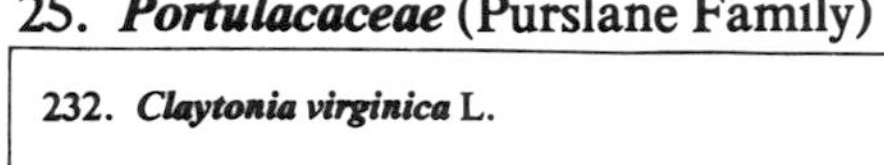

232. ***Claytonia virginica*** L. Virginia Springbeauty

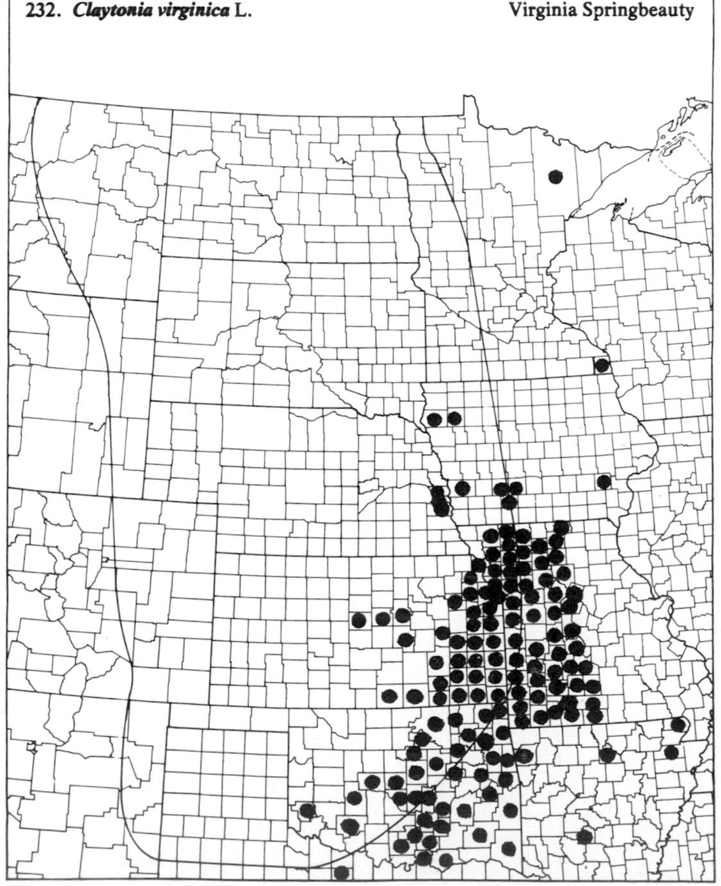

(Portulacaceae)

233. ***Portulaca mundula*** I.M. Johnst.

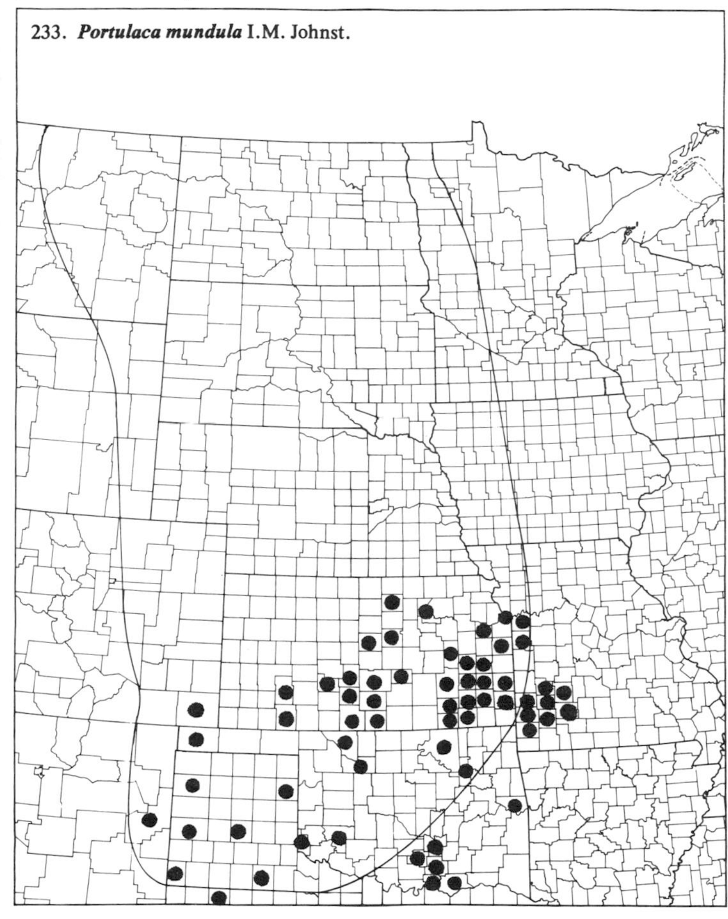

234. ***Portulaca oleracea*** L. Common Purslane

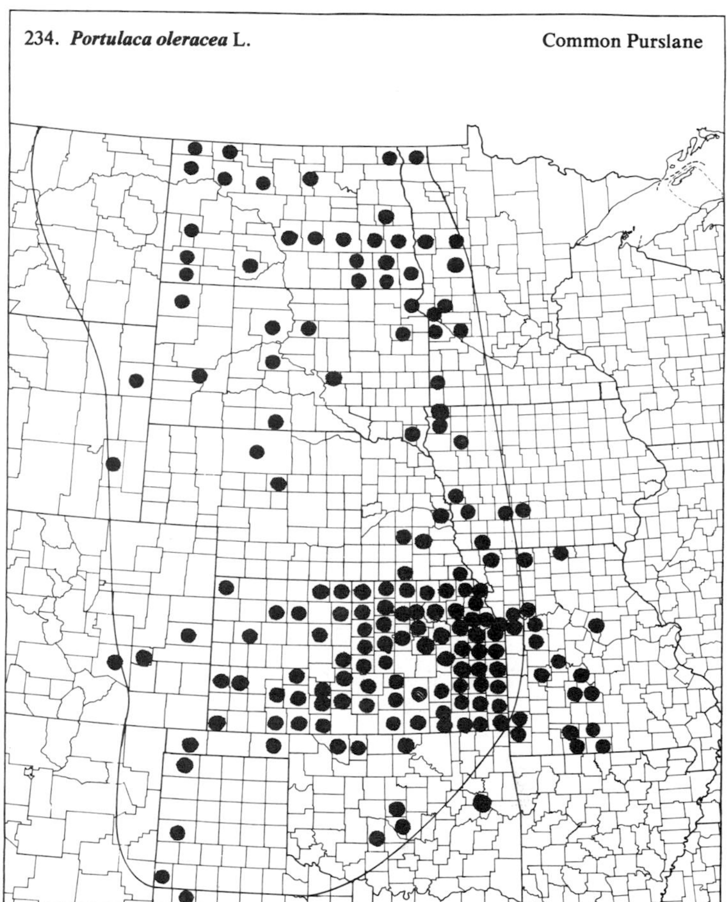

235. ***Portulaca parvula*** Gray Slenderleaf Purslane

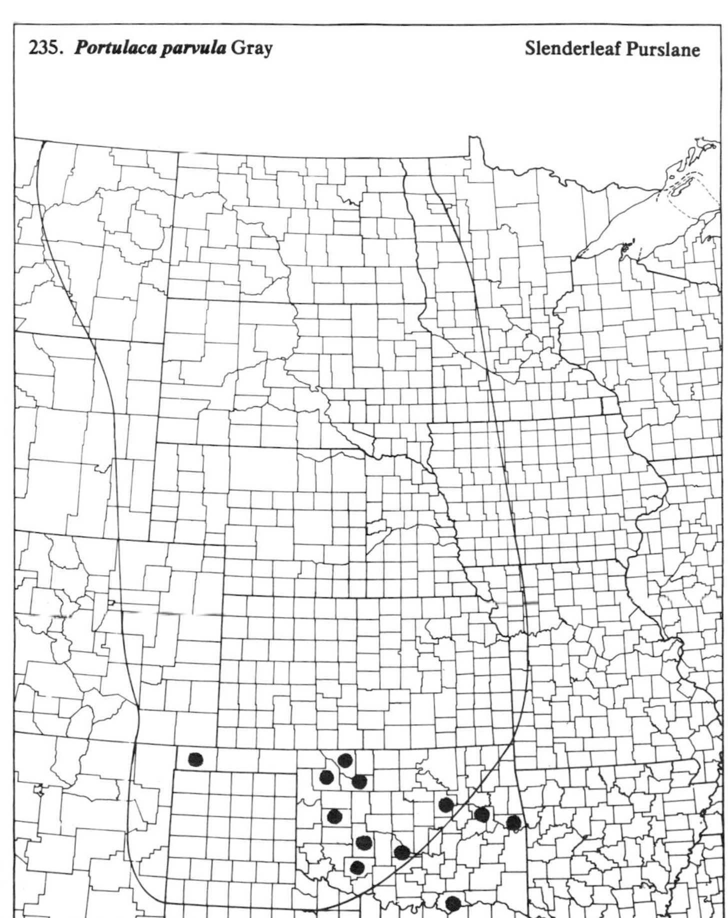

236. ***Portulaca retusa*** Engelm.

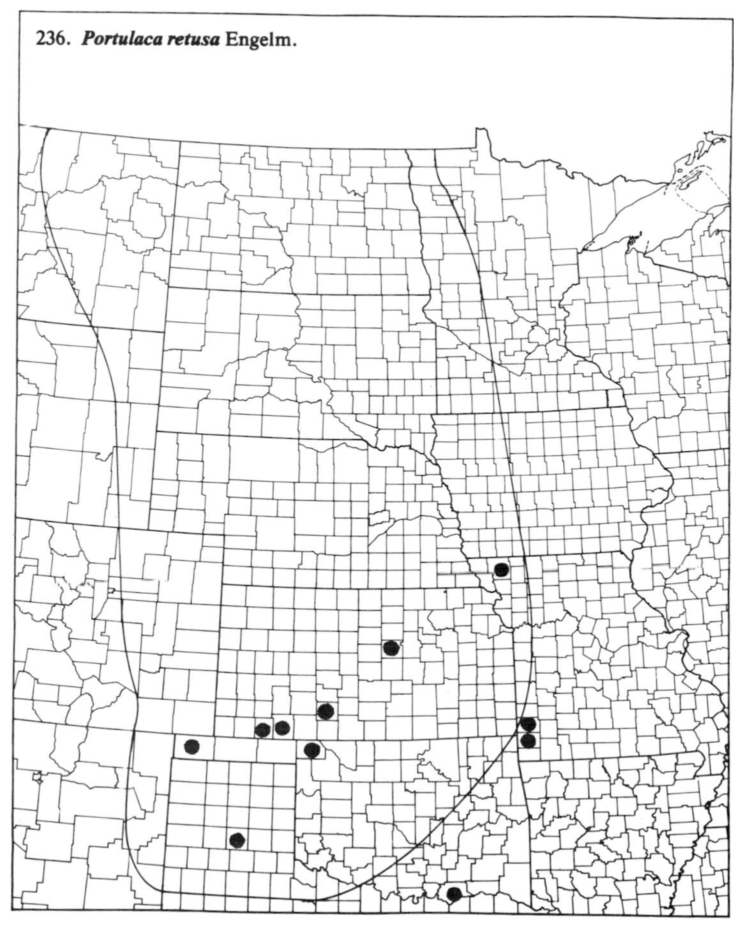

(Portulacaceae)

237. ***Talinum calycinum*** Engelm. — Rockpink Fameflower

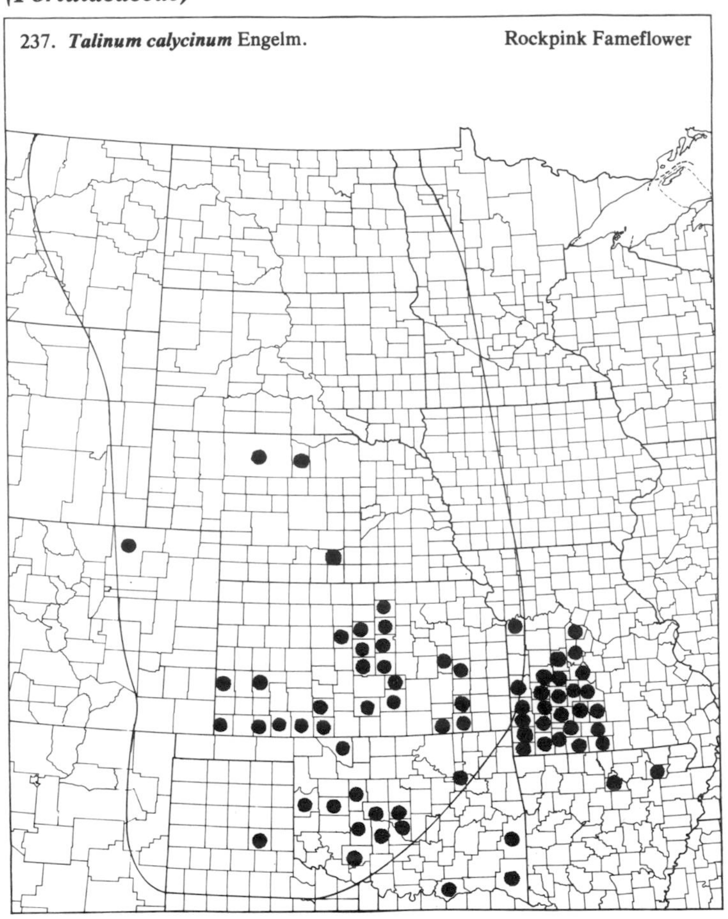

238. ***Talinum parviflorum*** Nutt. — Prairie Fameflower

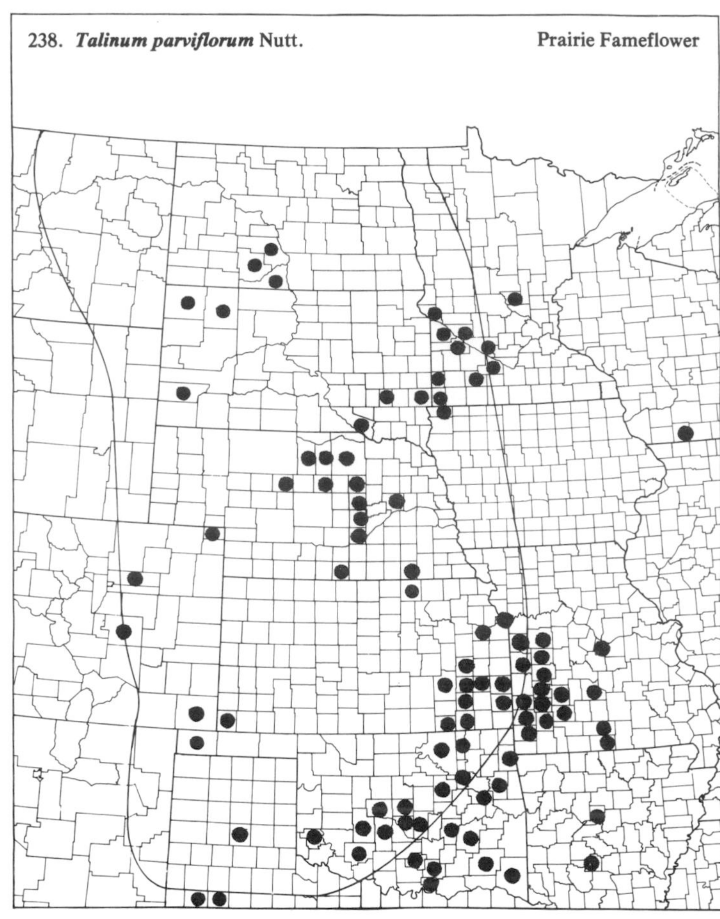

26. *Chenopodiaceae* (Goosefoot Family)

239. ***Atriplex argentea*** Nutt. ssp. ***argentea*** — Silverscale Saltbush

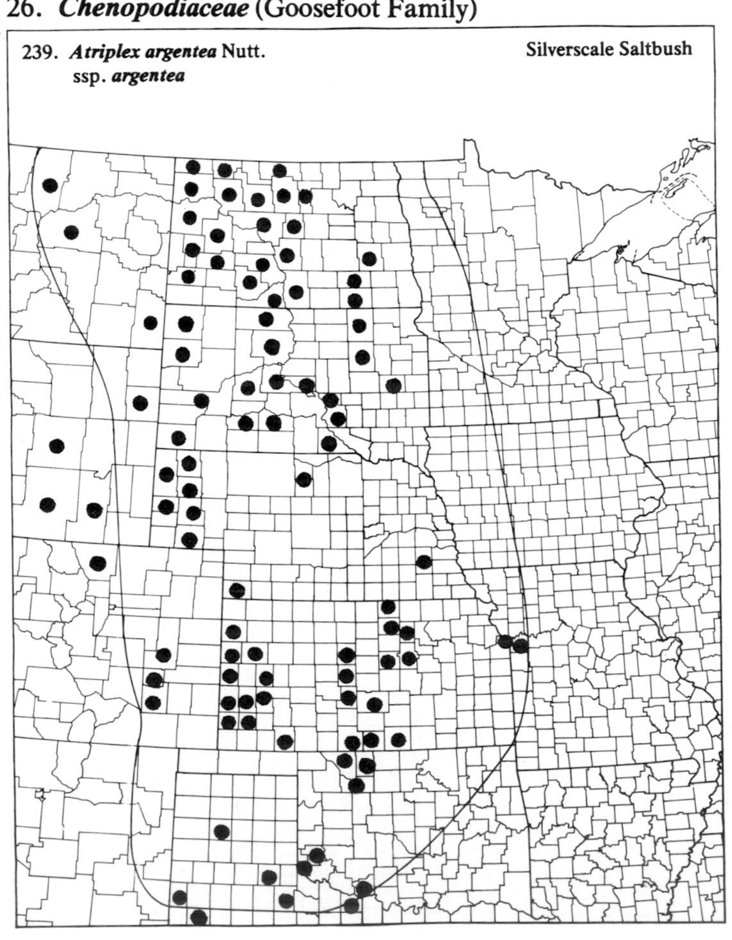

240. ***Atriplex canescens*** (Pursh) Nutt. — Shadscale

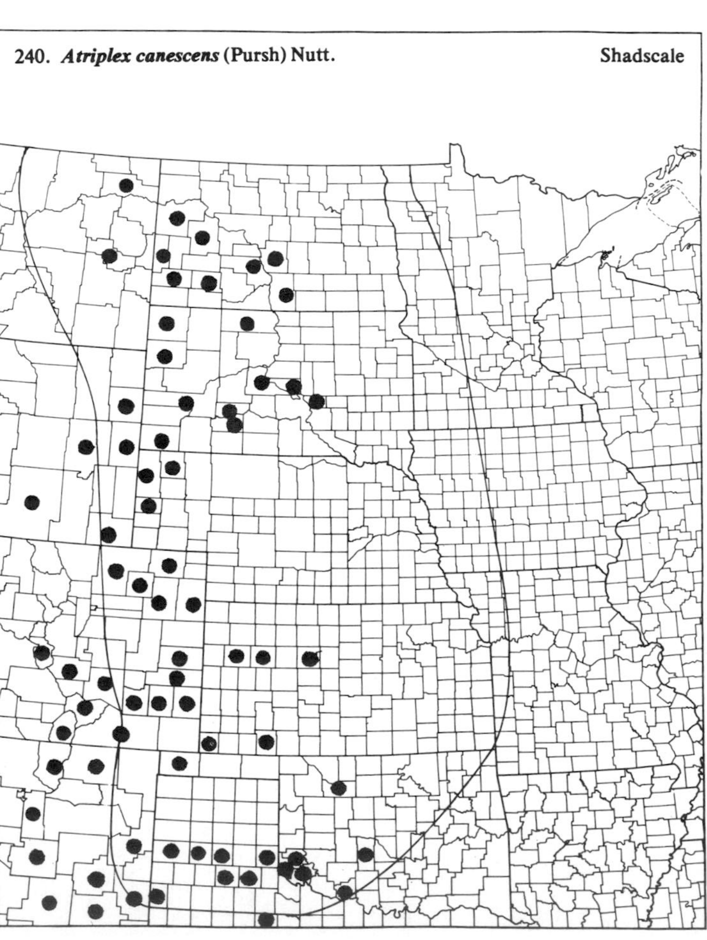

(Chenopodiaceae)

241. ***Atriplex confertifolia*** (Torr. & Frem.) Wats. Spiny Saltbush

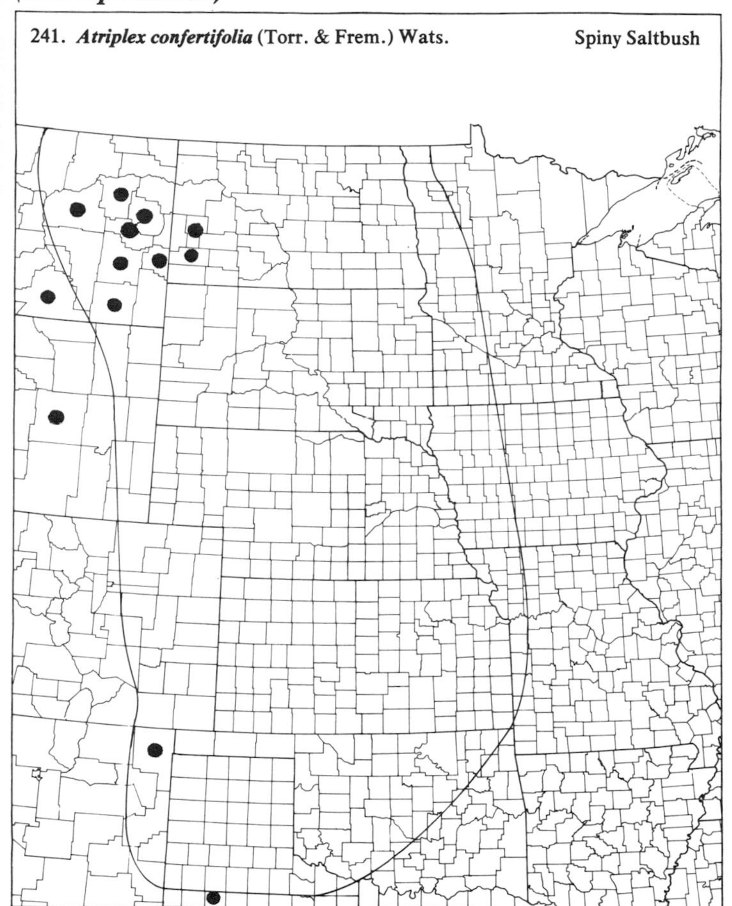

242. ***Atriplex dioica*** (Nutt.) Macbr. Rillscale

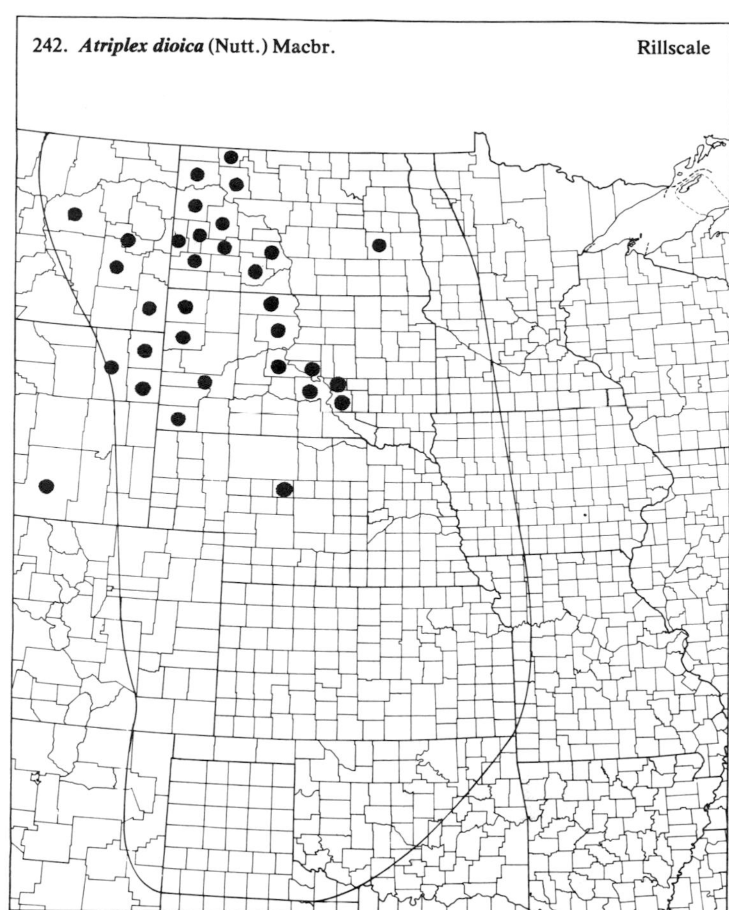

243. ***Atriplex heterosperma*** Bunge

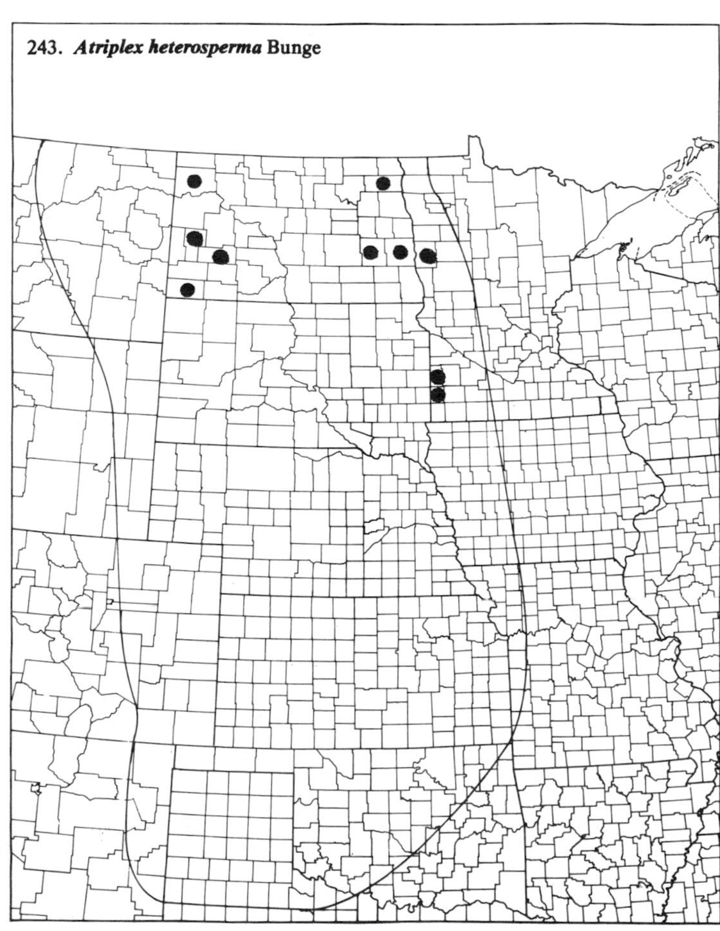

244. ***Atriplex hortensis*** L. Garden Orach

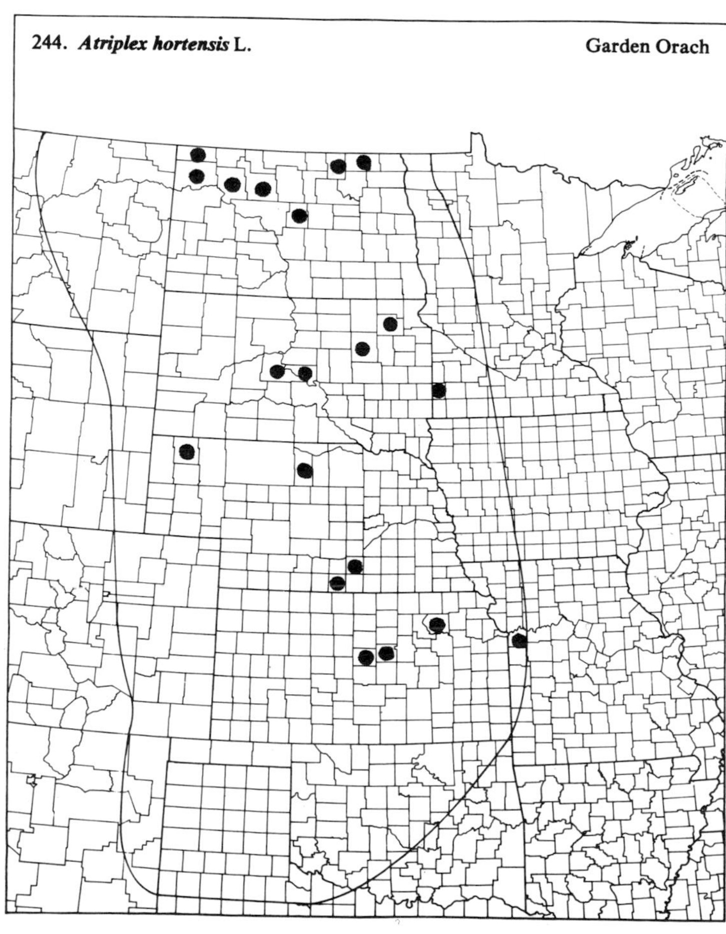

(Chenopodiaceae)

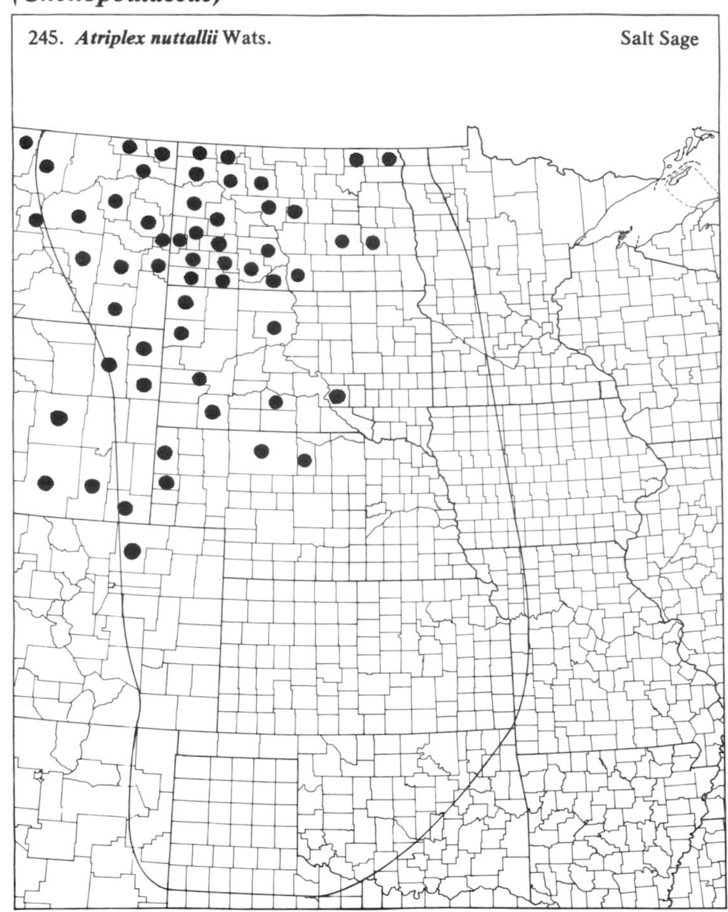
245. ***Atriplex nuttallii*** Wats. Salt Sage

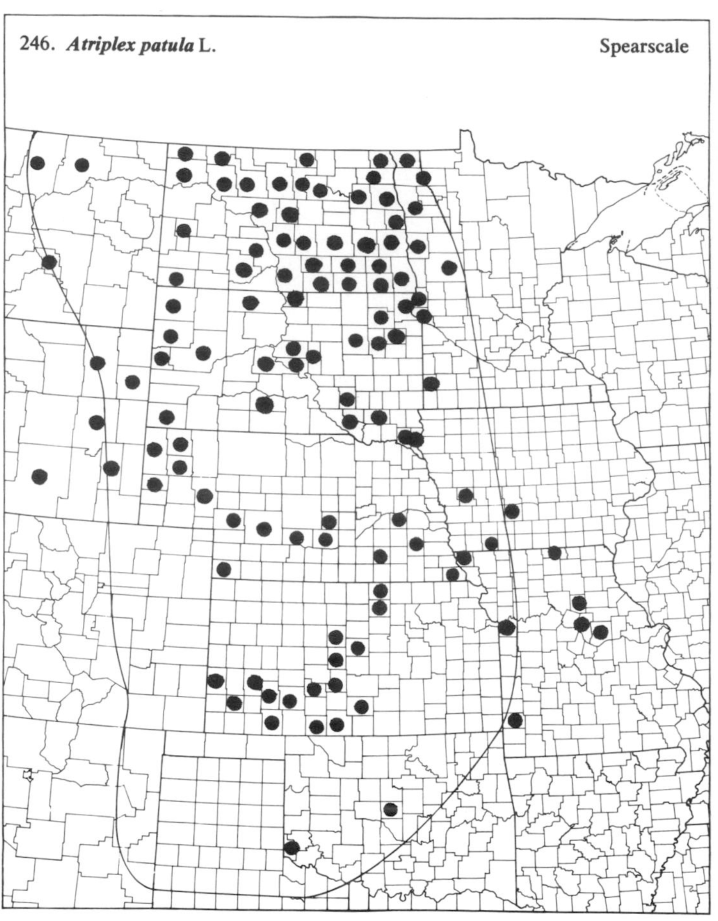
246. ***Atriplex patula*** L. Spearscale

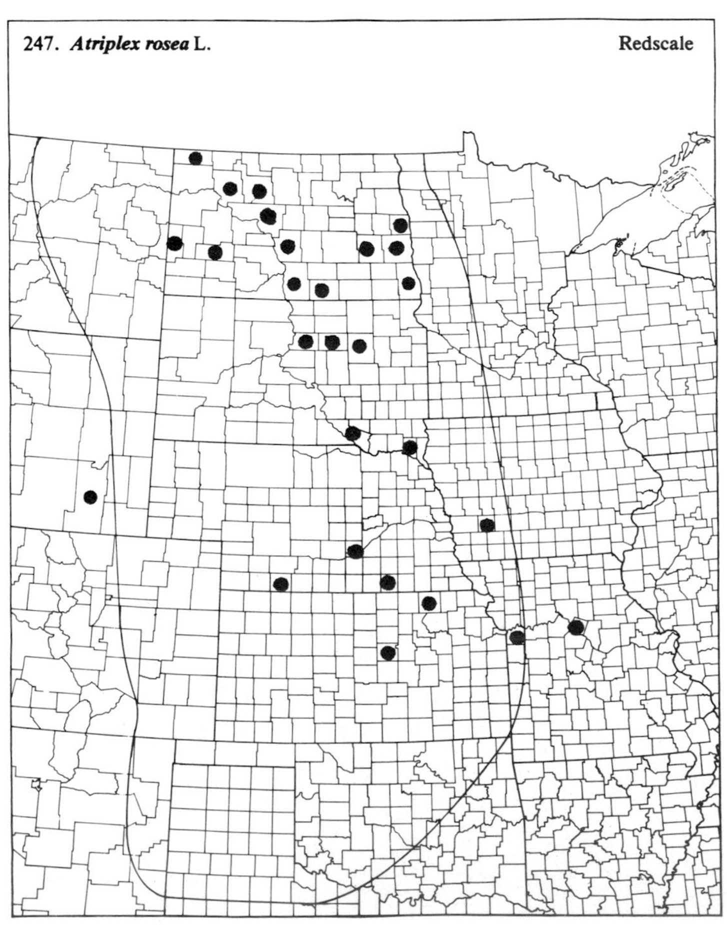
247. ***Atriplex rosea*** L. Redscale

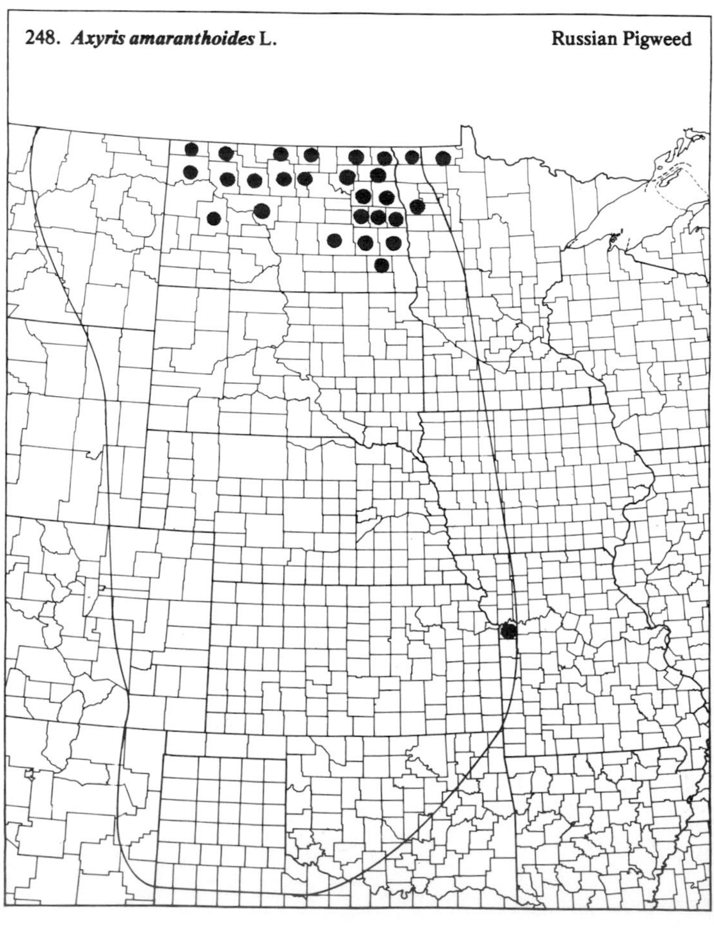
248. ***Axyris amaranthoides*** L. Russian Pigweed

(Chenopodiaceae)

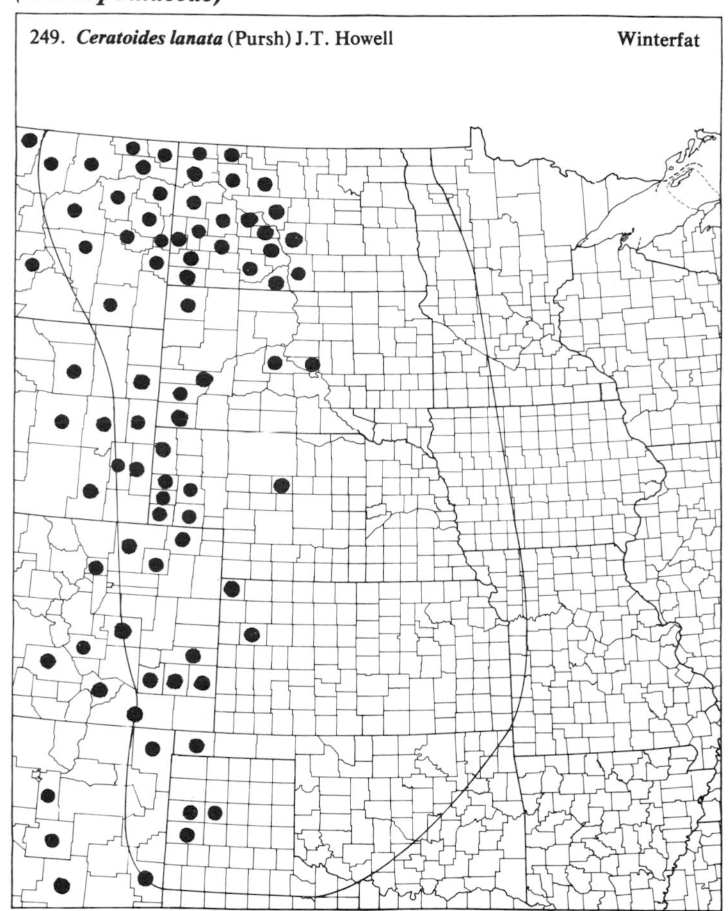

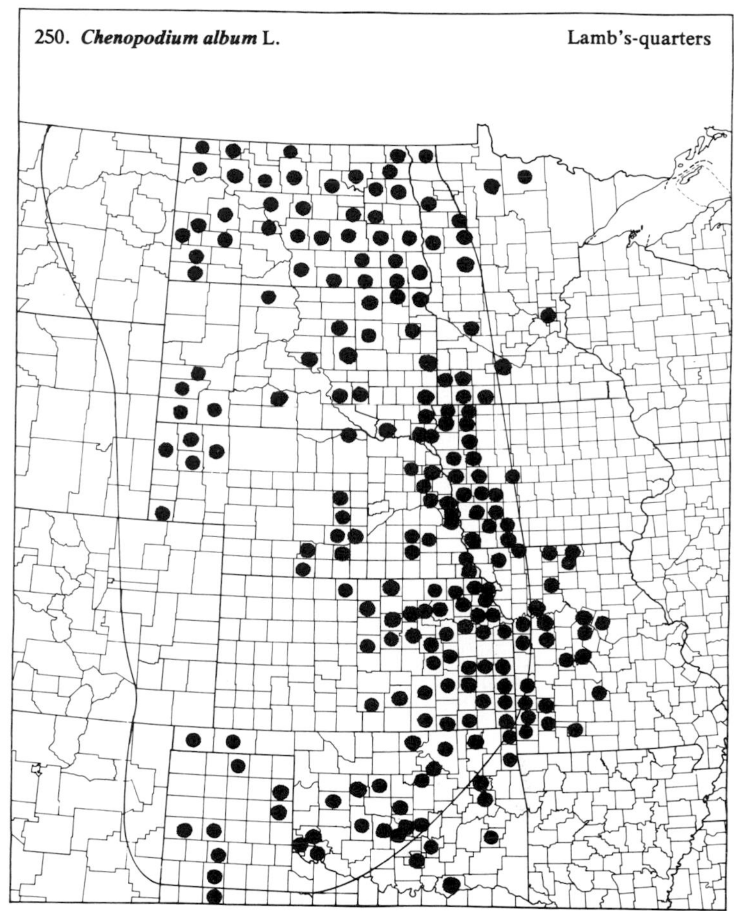

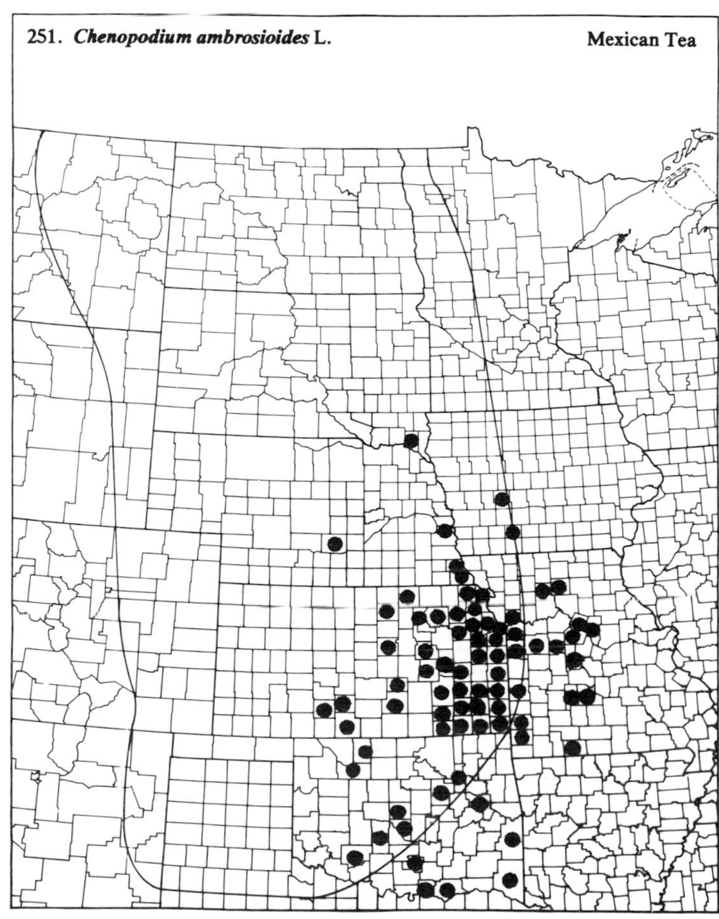

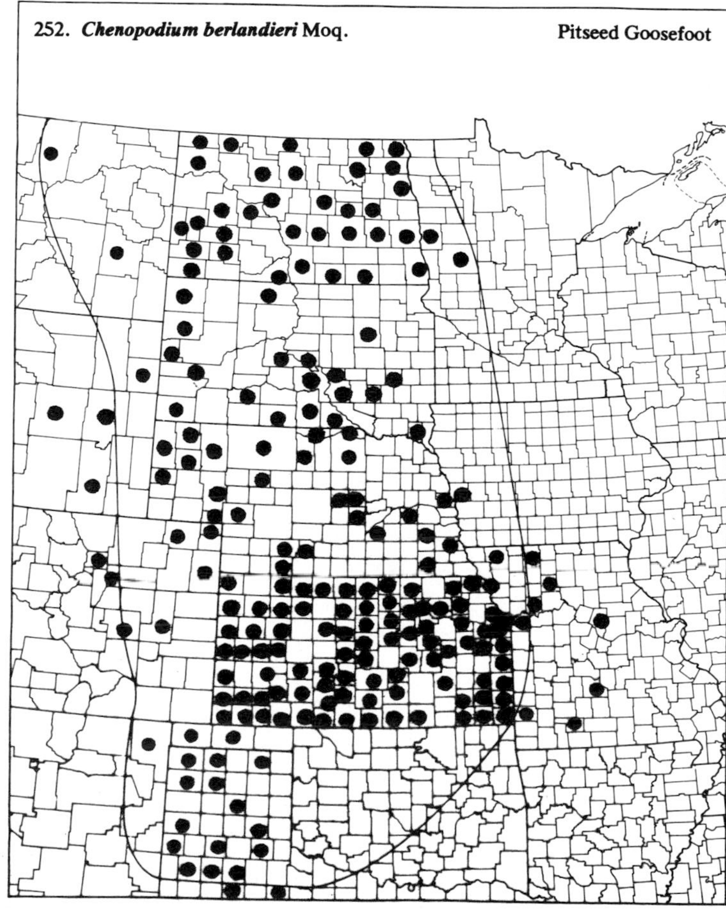

(Chenopodiaceae)

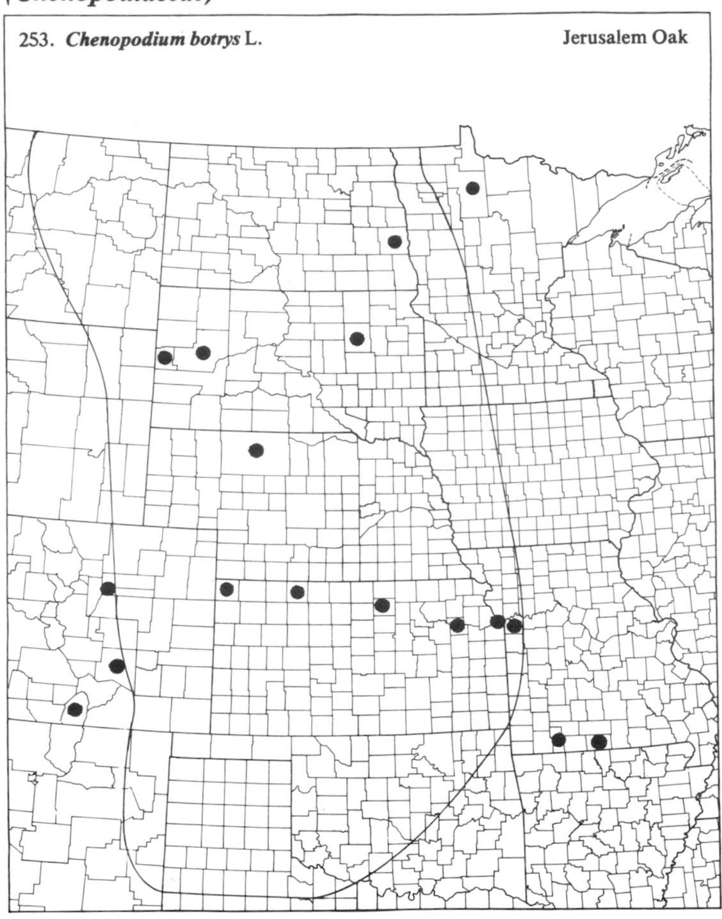

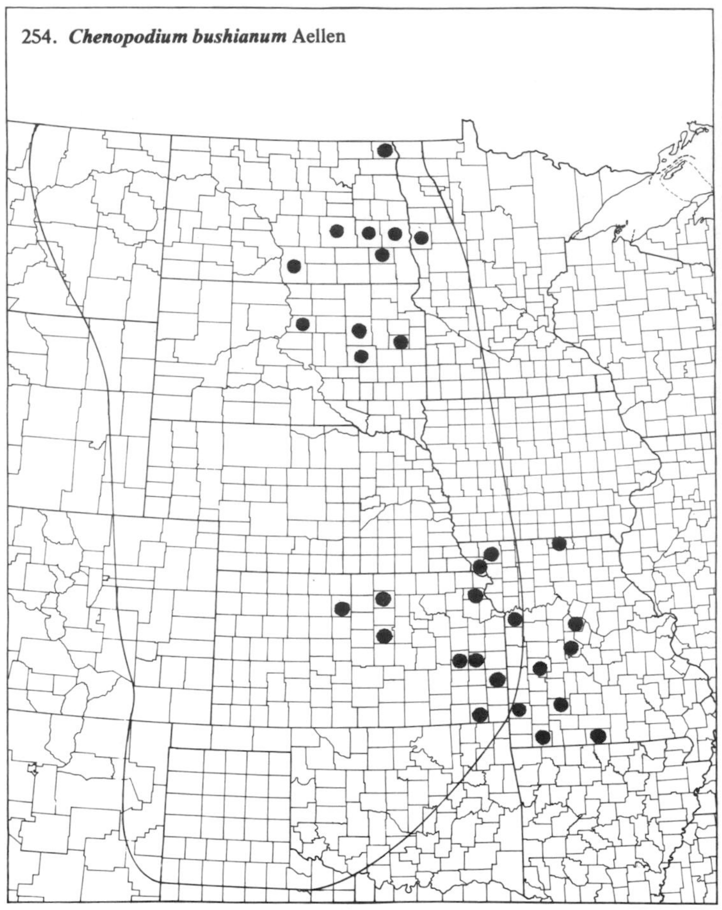

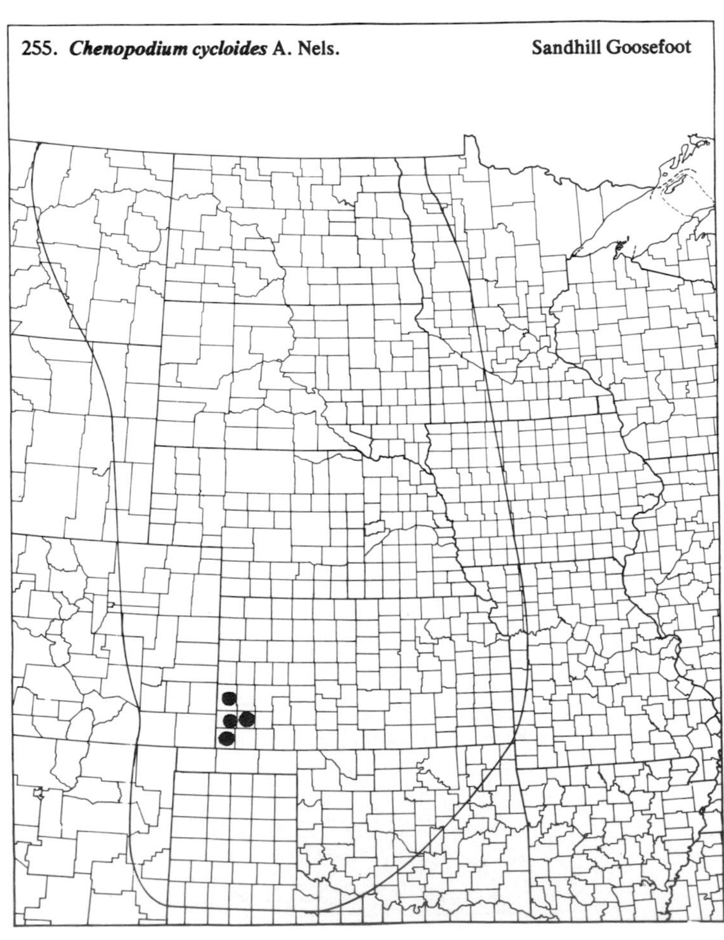

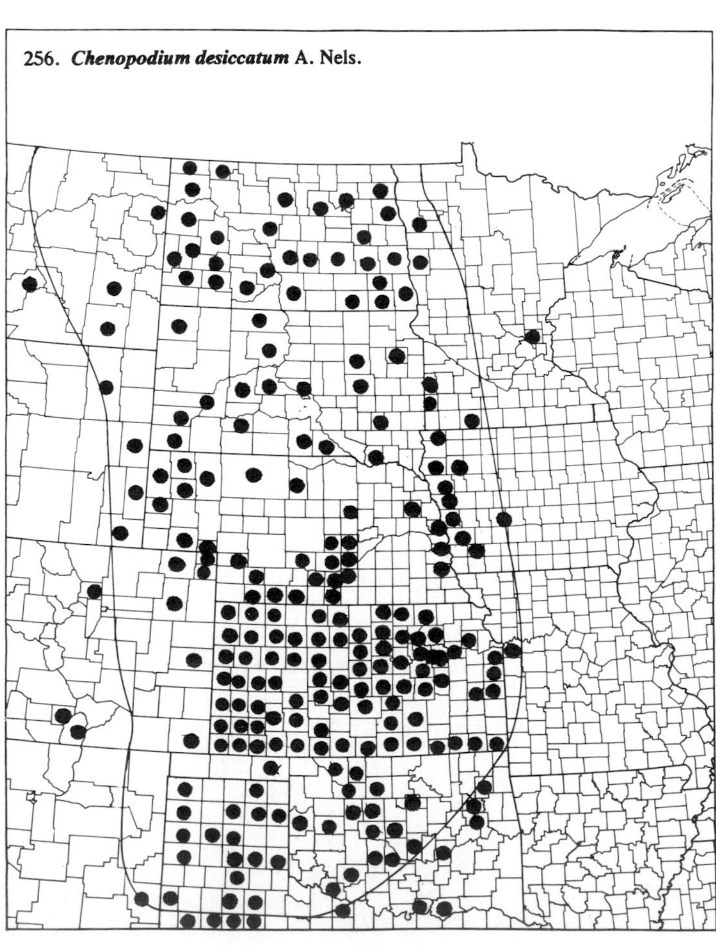

(Chenopodiaceae)

257. ***Chenopodium fremontii*** Wats. Fremont Goosefoot

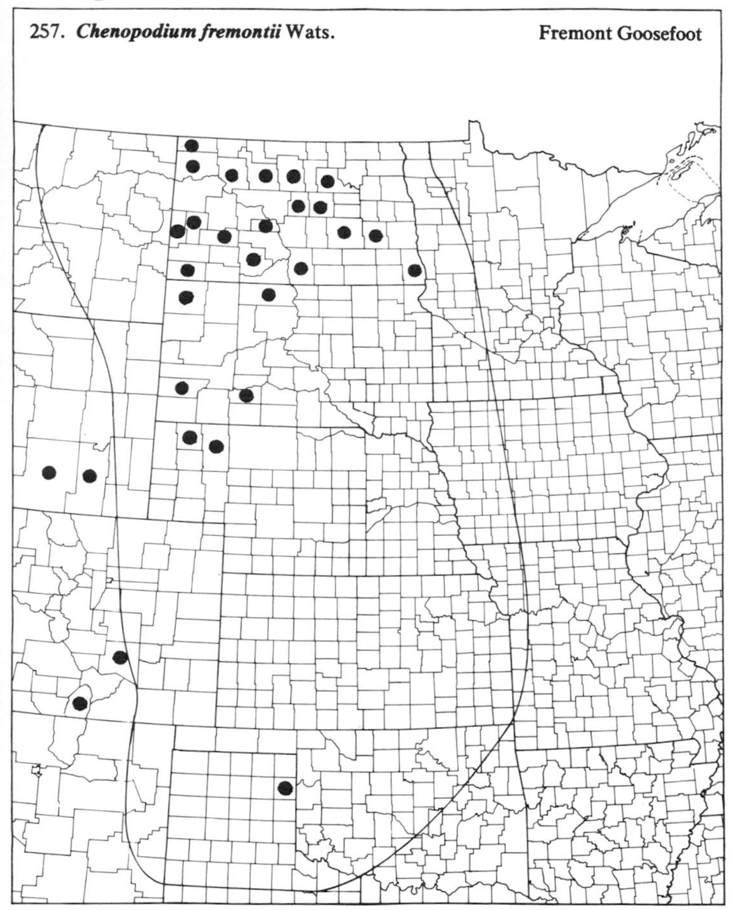

258. ***Chenopodium glaucum*** L. Oak-leaved Goosefoot

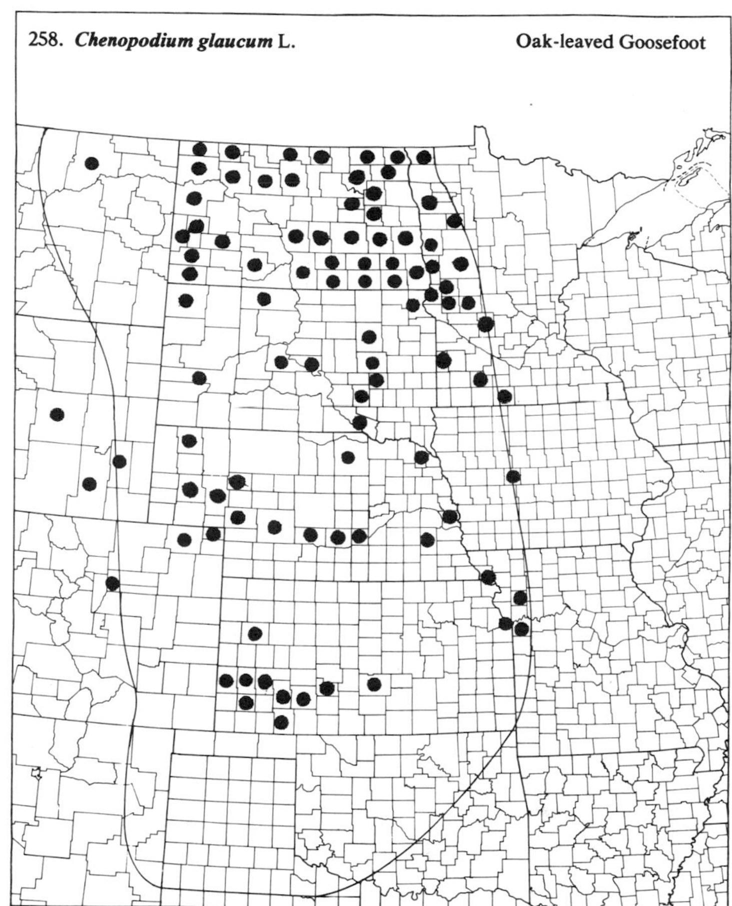

259. ***Chenopodium hybridum*** L. Maple-leaved Goosefoot

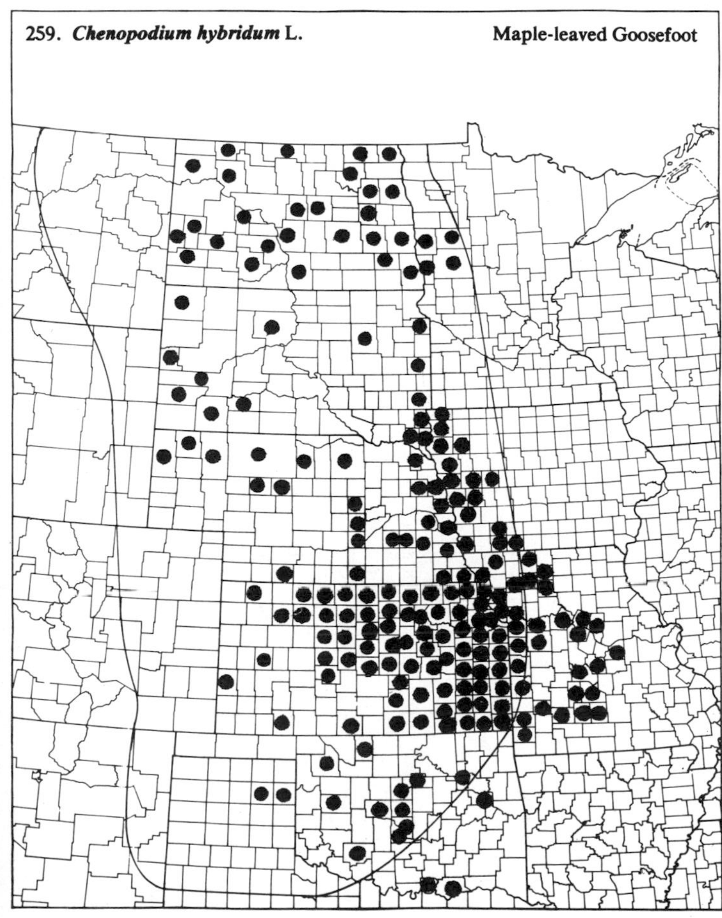

260. ***Chenopodium incanum*** (Wats.) Heller

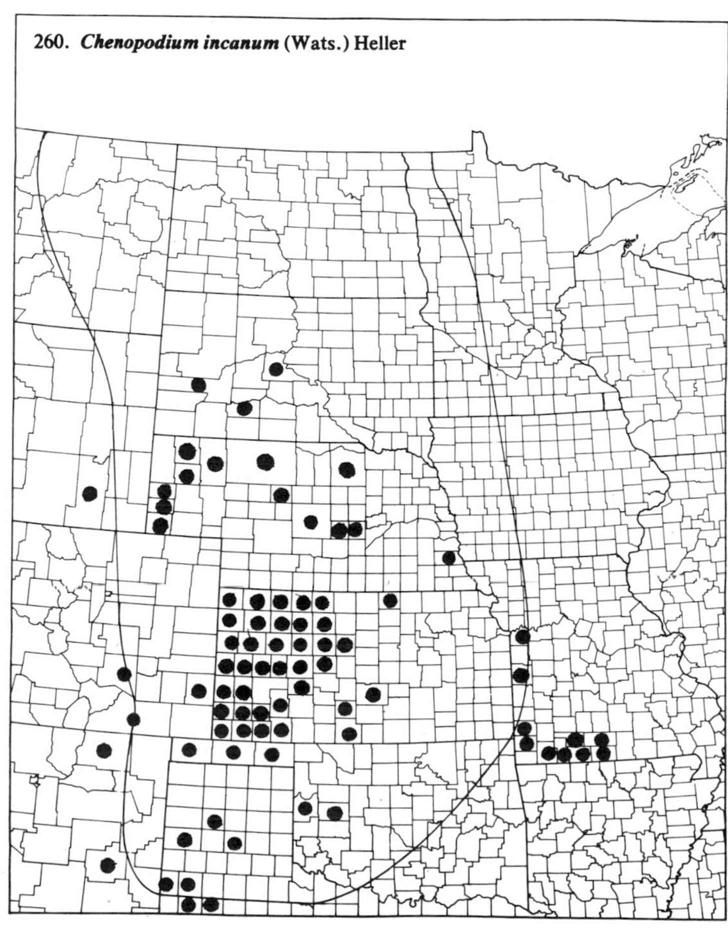

(Chenopodiaceae)

261. ***Chenopodium missouriense*** Aellen

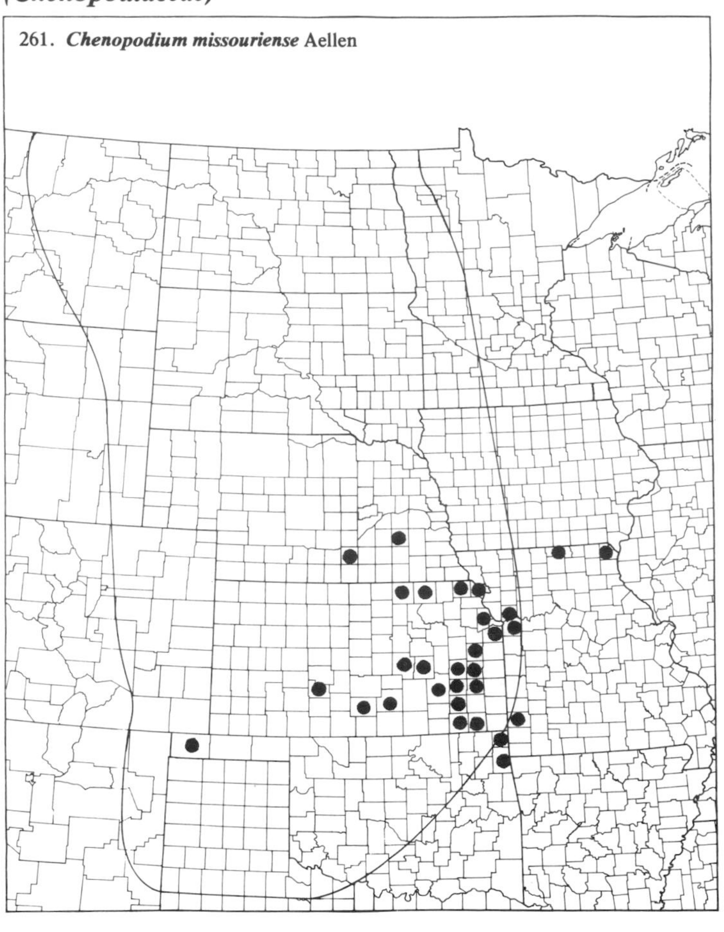

262. ***Chenopodium pallescens*** Standl.

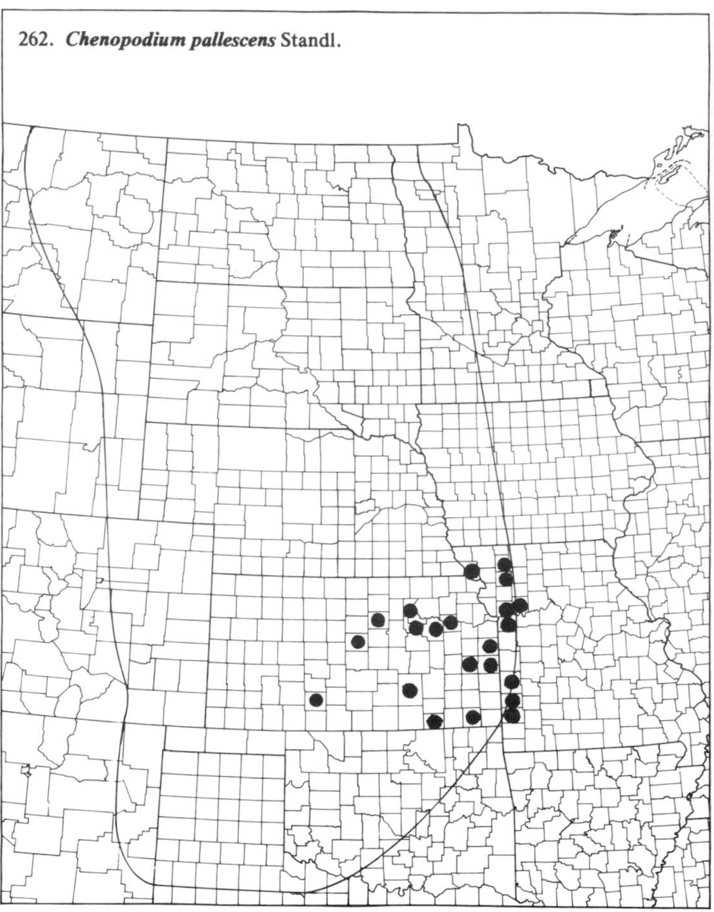

263. ***Chenopodium rubrum*** L. Alkali Blite

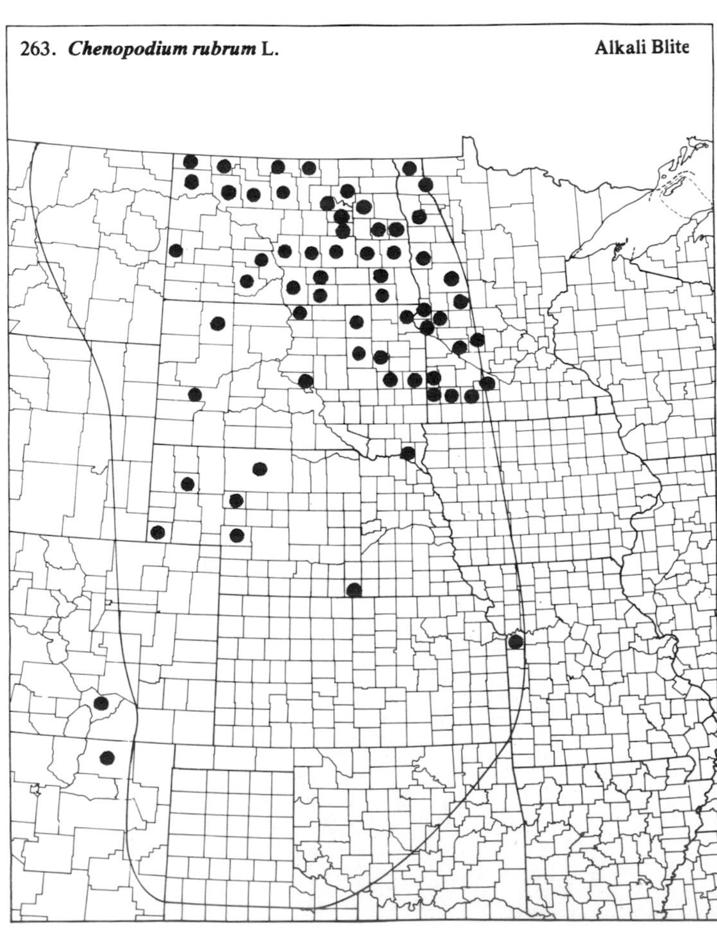

264. ***Chenopodium standleyanum*** Aellen

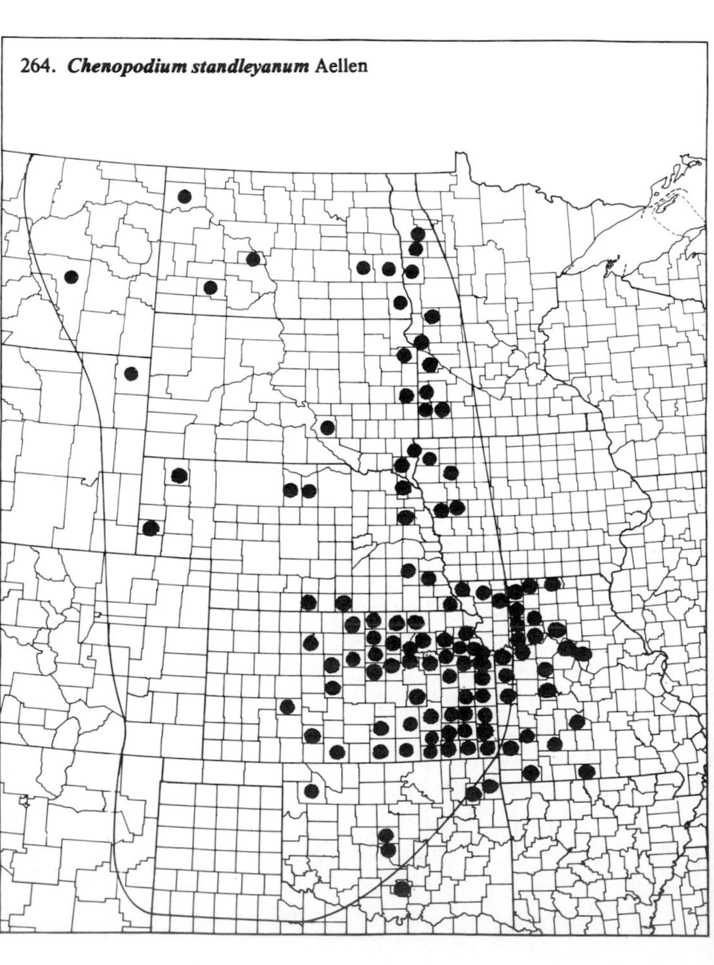

(Chenopodiaceae)

265. ***Chenopodium strictum*** Roth

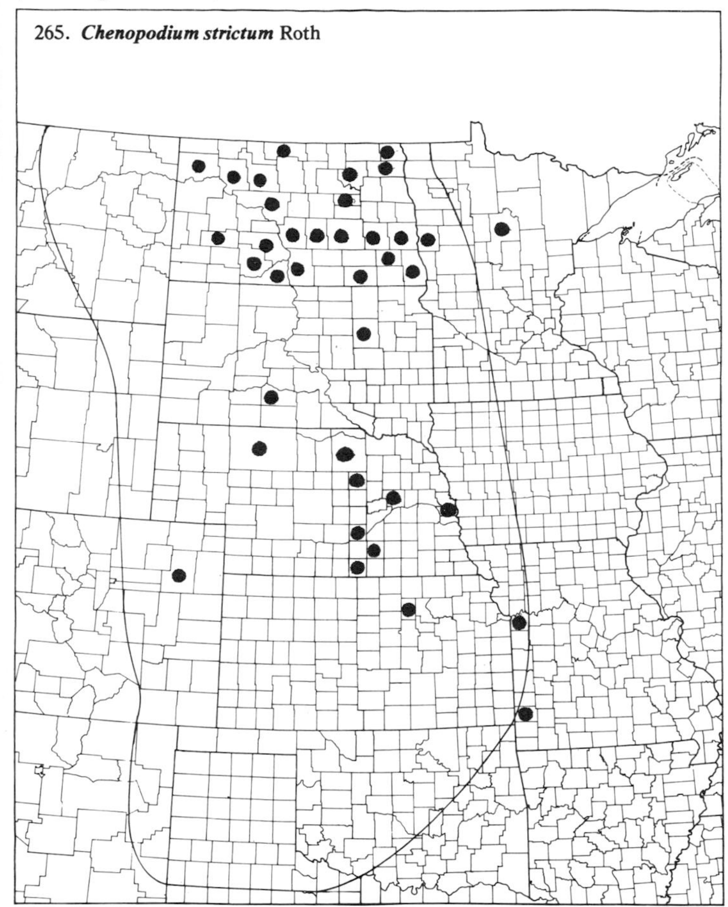

266. ***Chenopodium watsonii*** A. Nels.

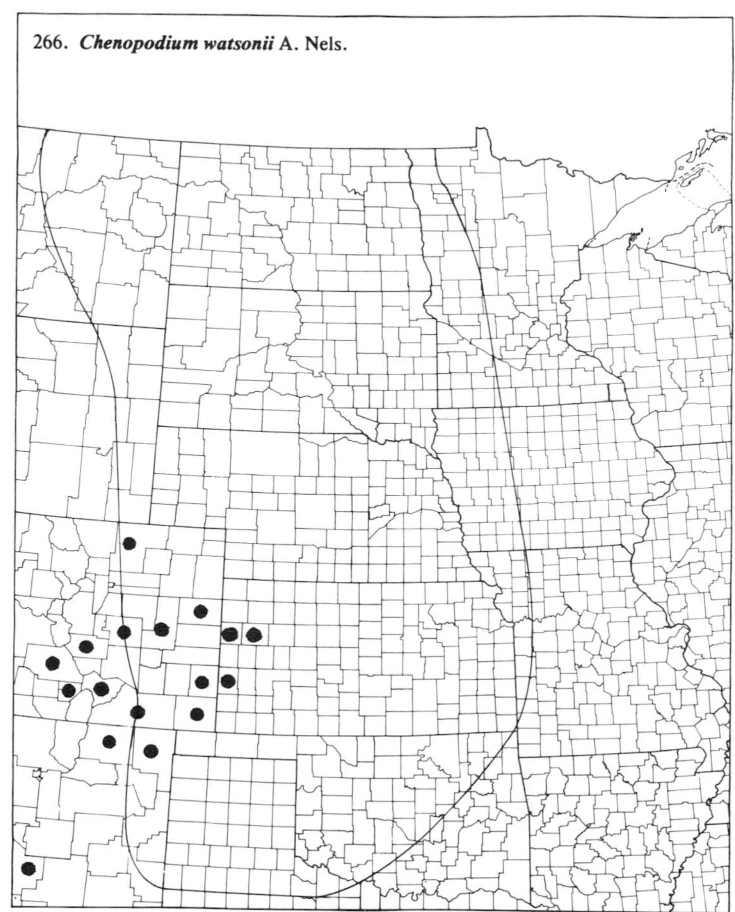

267. ***Corispermum hyssopifolium*** L. Bugseed

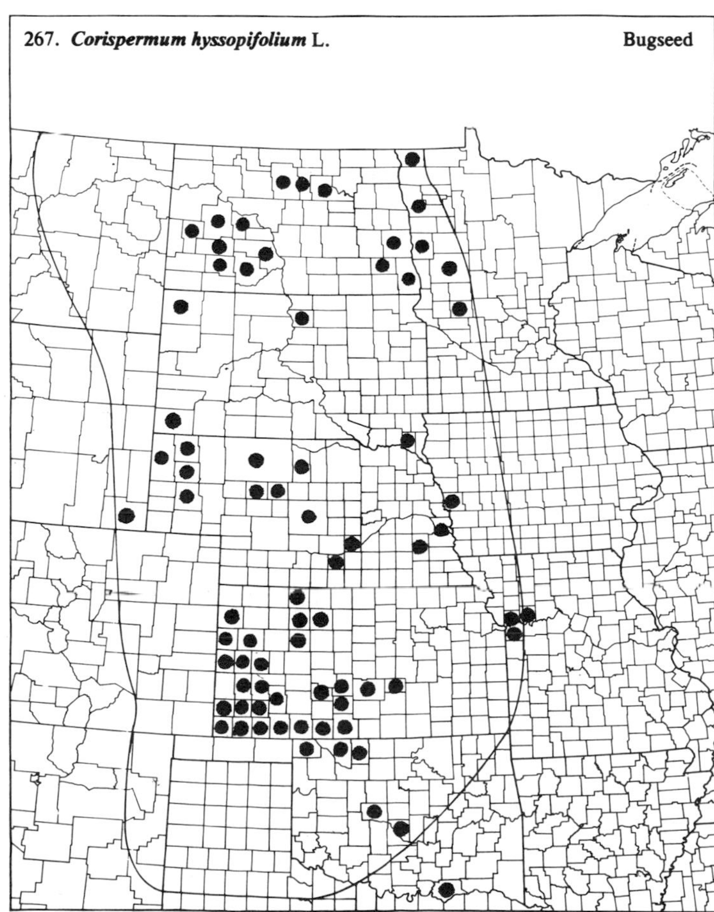

268. ***Corispermum nitidum*** Kit. Bugseed

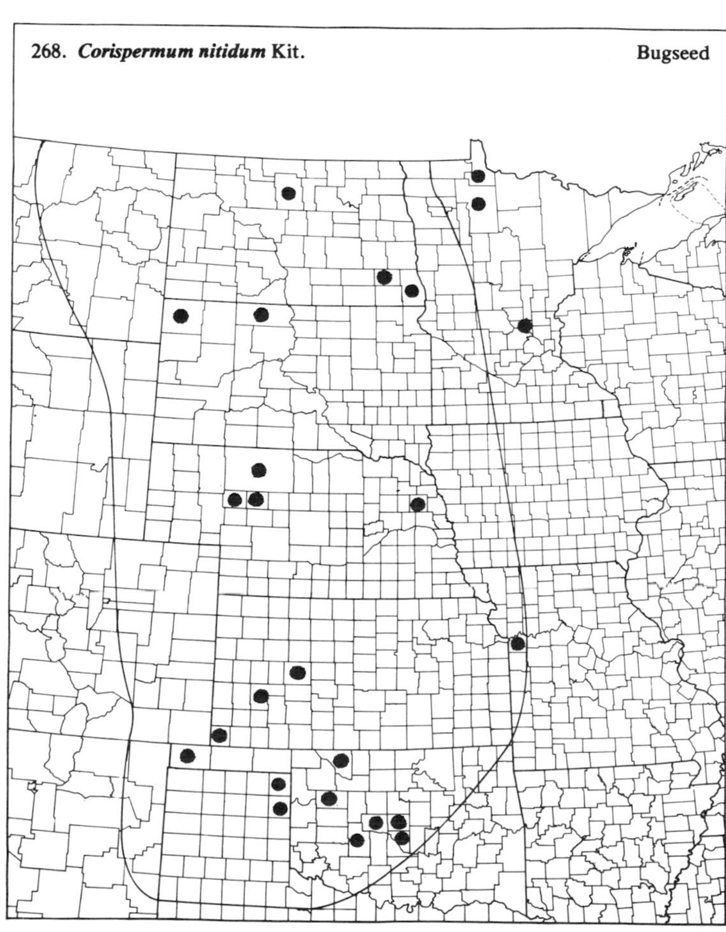

(Chenopodiaceae)

269. **Cycloloma atriplicifolium** (Spreng.) Coult. Winged Pigweed

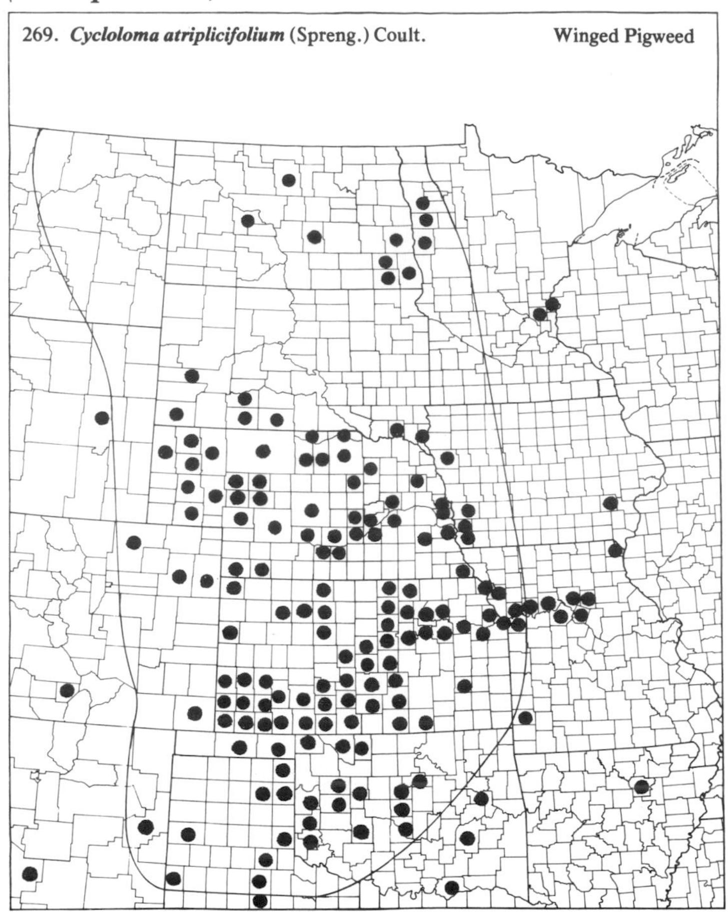

270. **Kochia scoparia** (L.) Schrad. Kochia

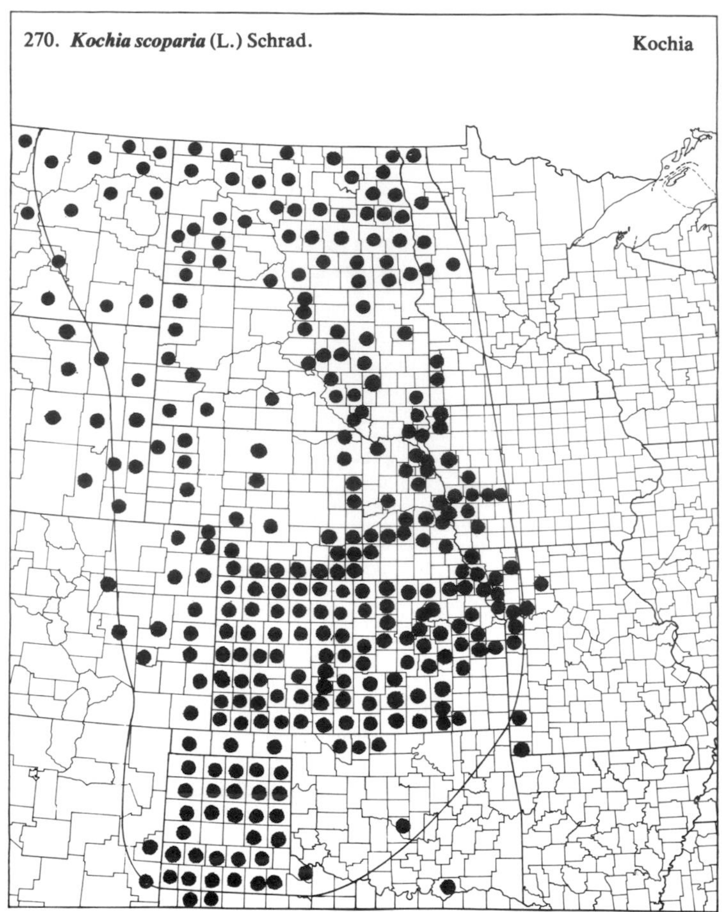

271. **Monolepis nuttalliana** (Schult.) Greene Povertyweed

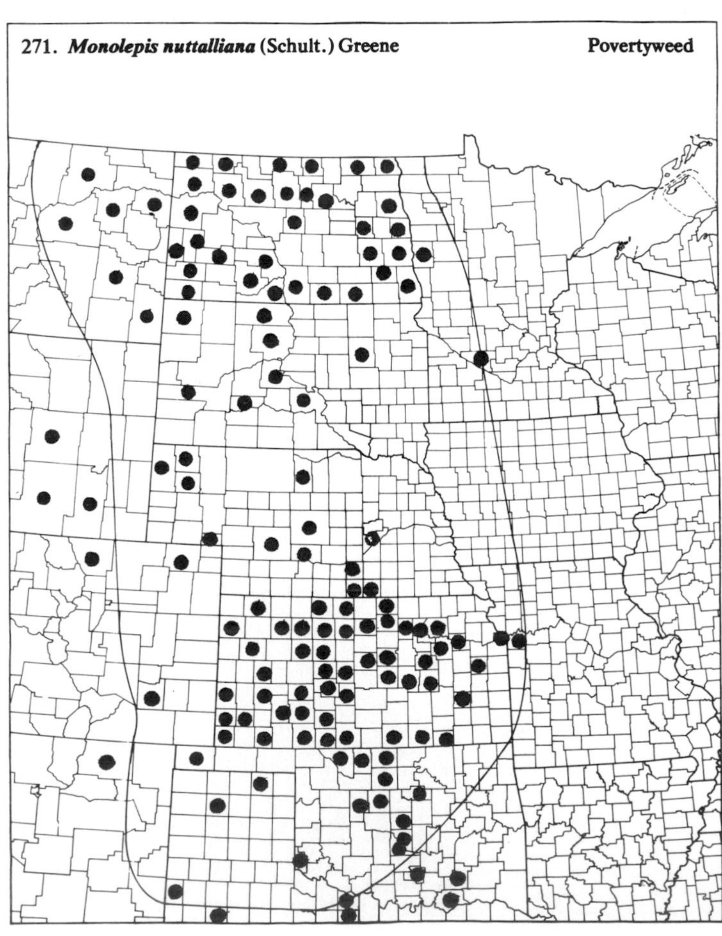

272. **Salicornia rubra** A. Nels. Saltwort

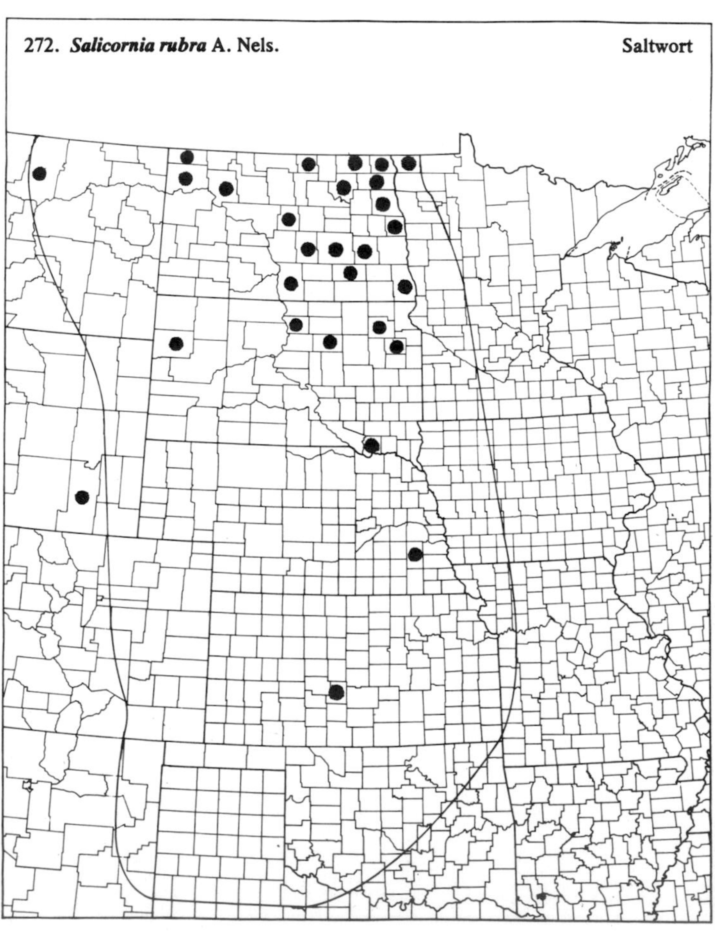

(Chenopodiaceae)

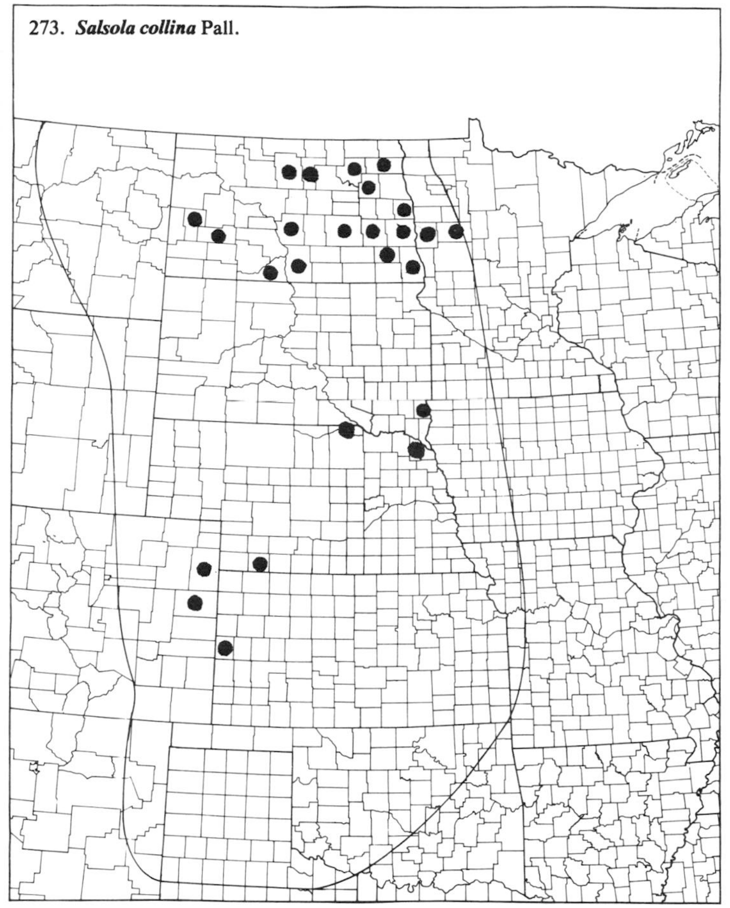

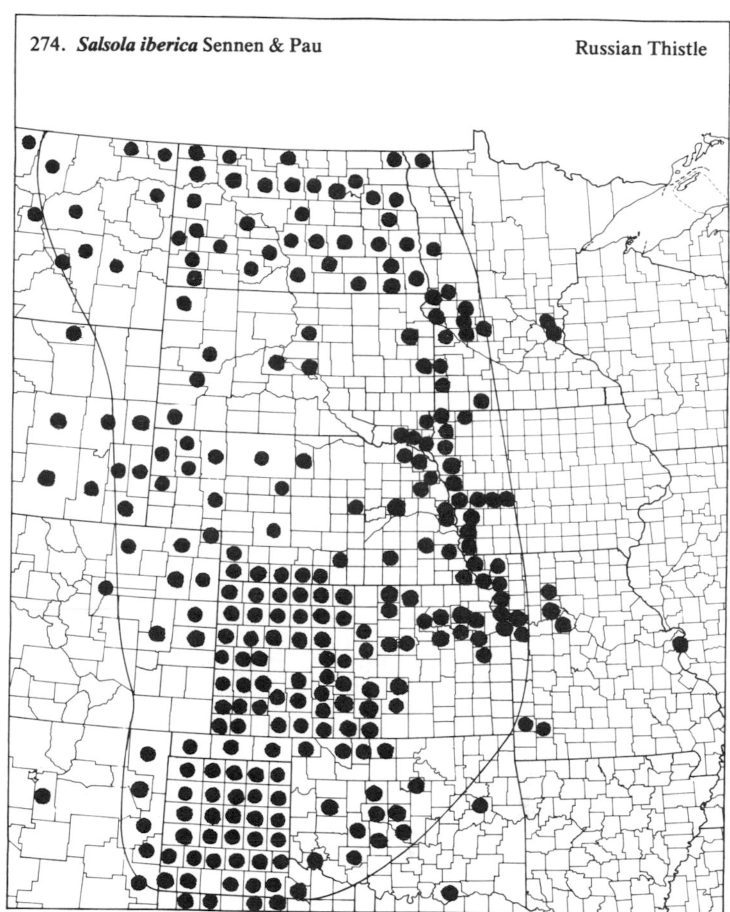

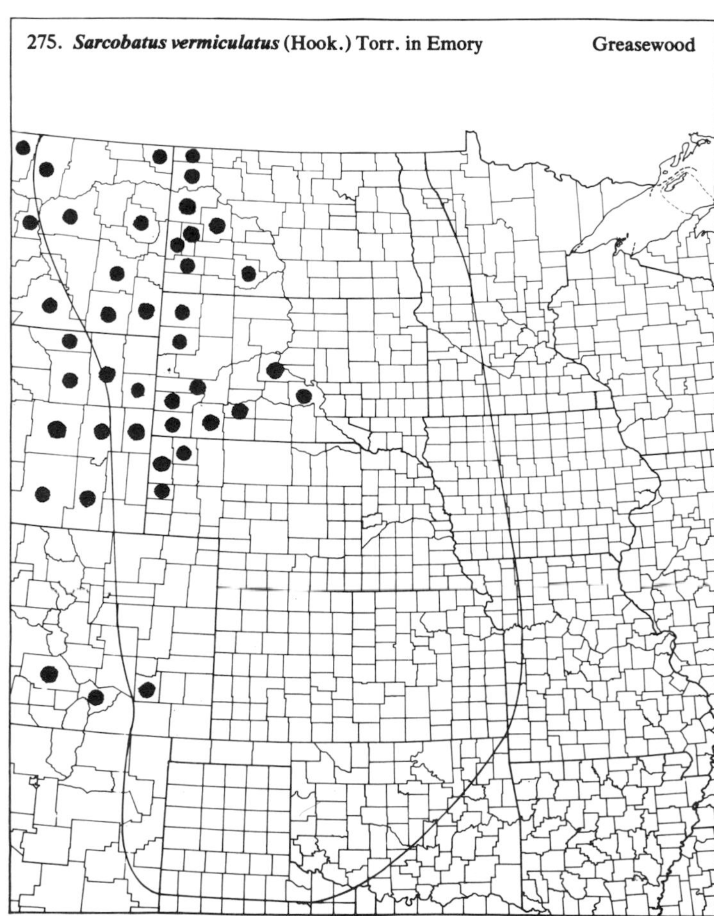

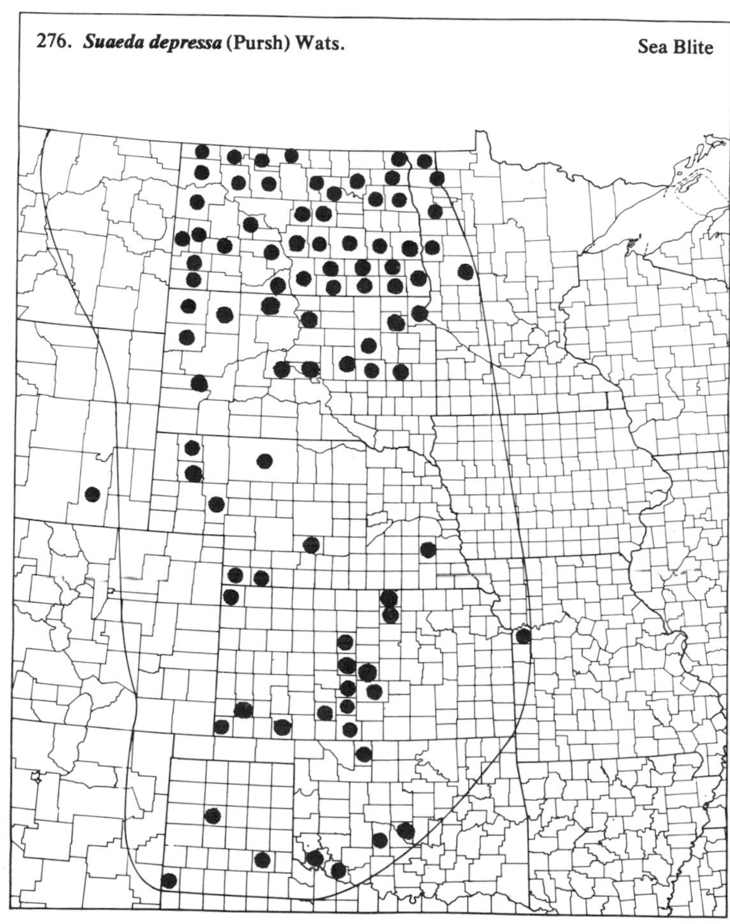

(Chenopodiaceae)

277. ***Suaeda intermedia*** Wats.

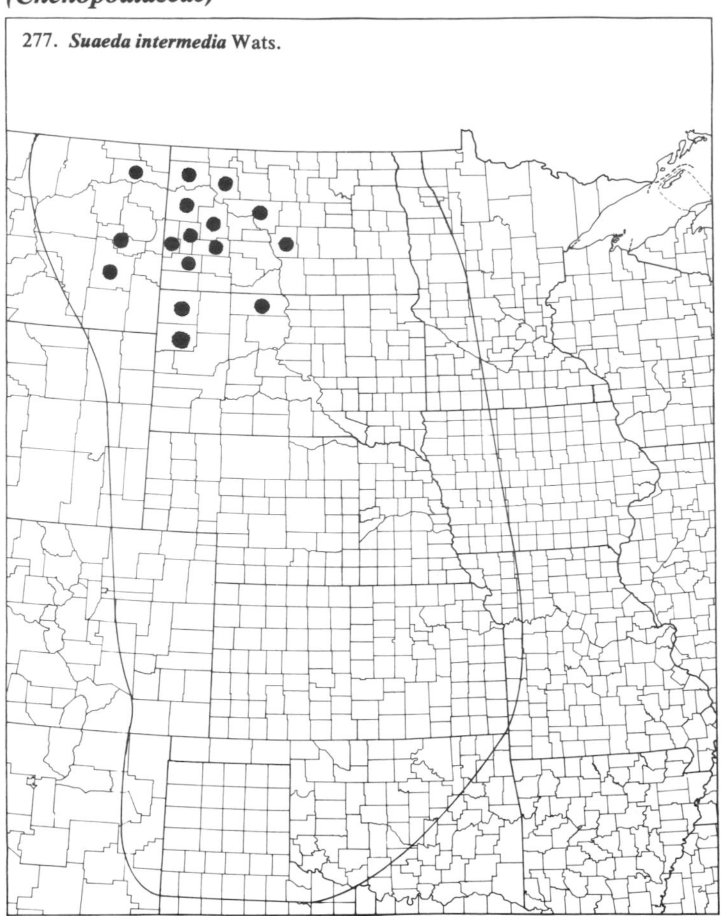

278. ***Suckleya suckleyana*** (Torr.) Rydb.

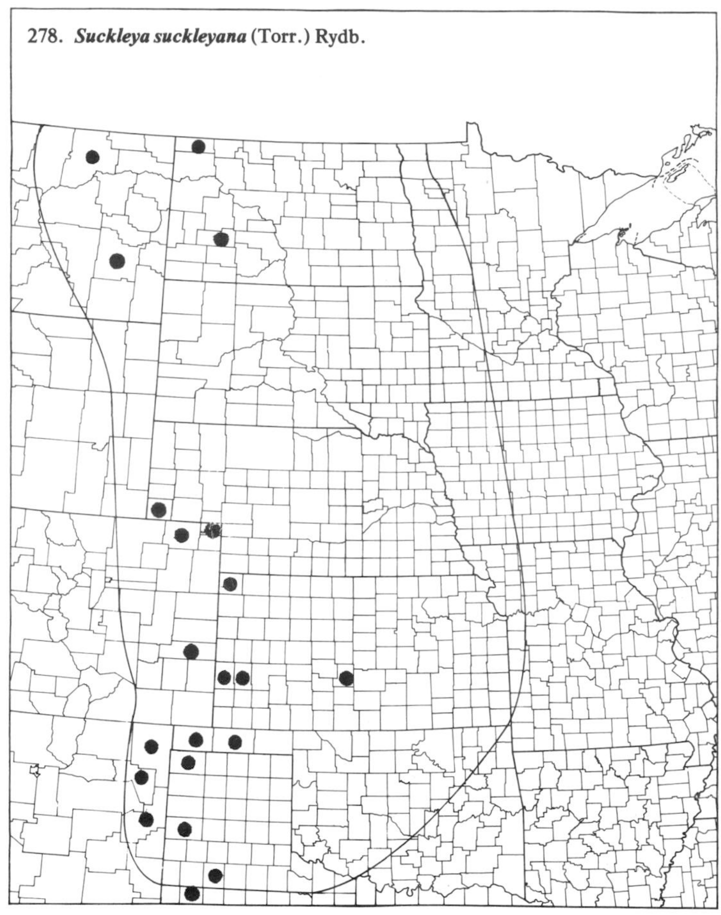

27. *Amaranthaceae* (Pigweed Family)

279. ***Amaranthus albus*** L. Tumbleweed

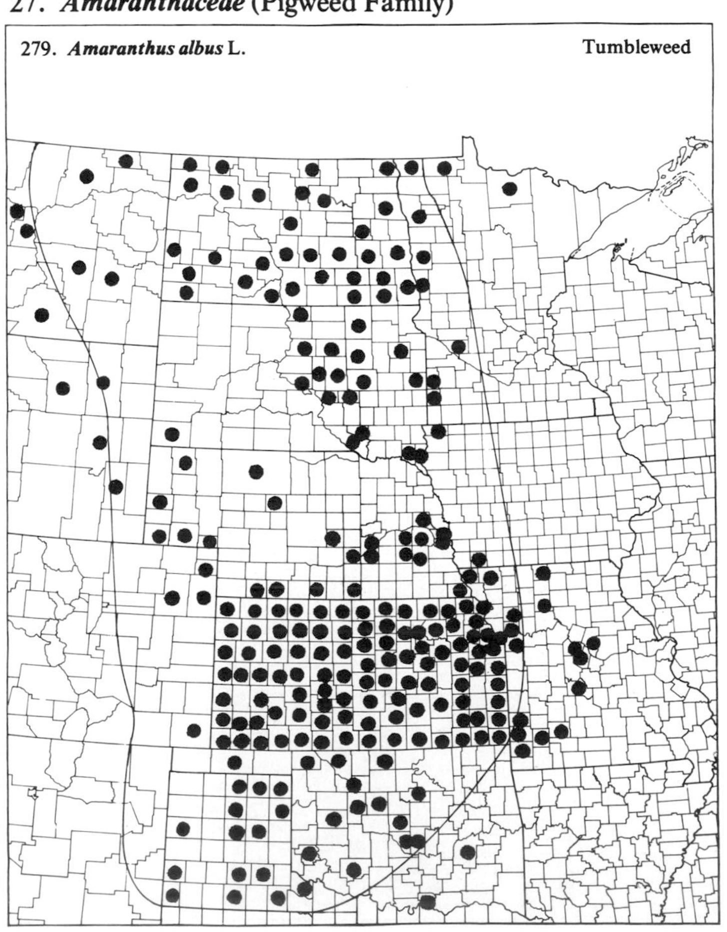

280. ***Amaranthus arenicola*** I.M. Johnst.

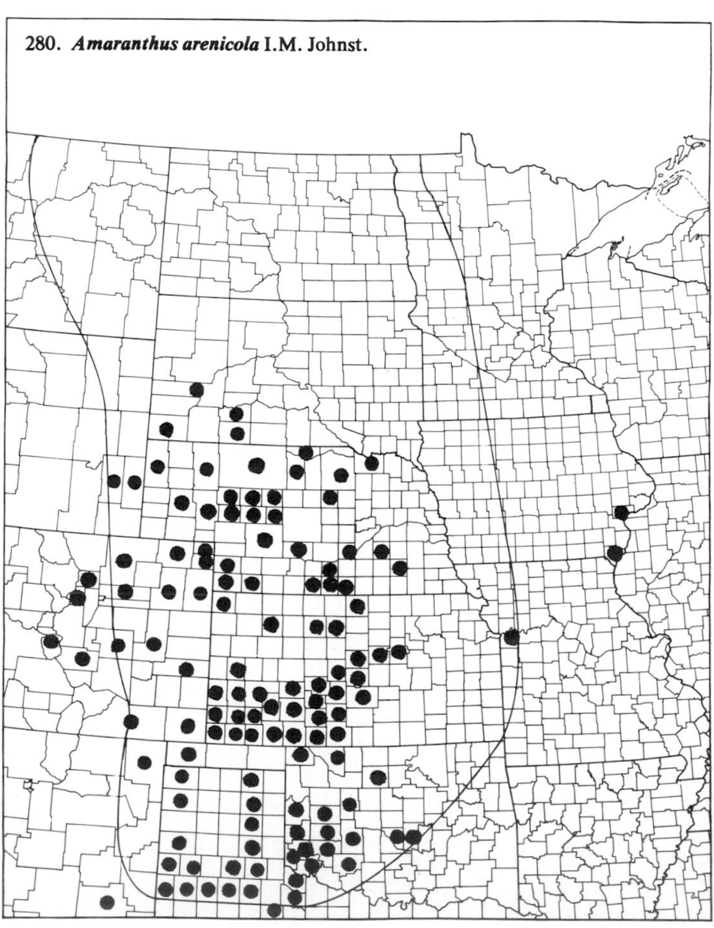

(Amaranthaceae)

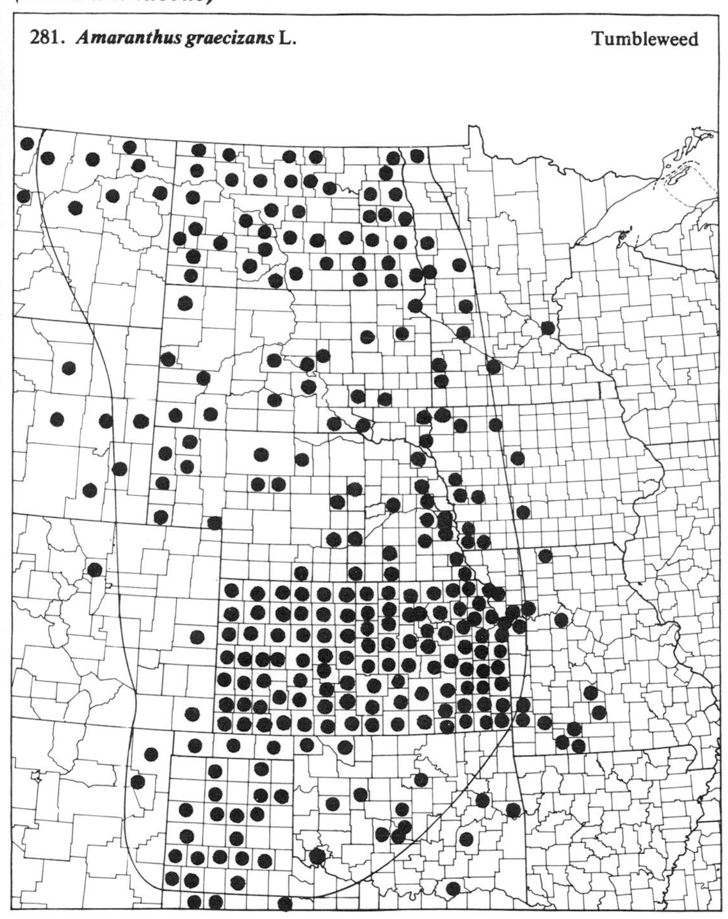

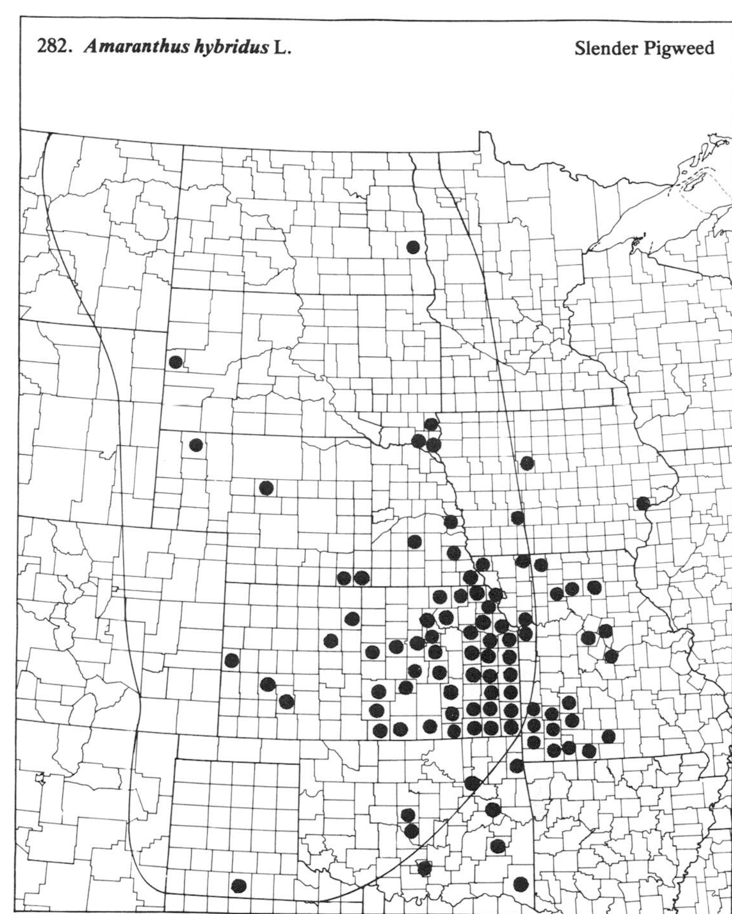

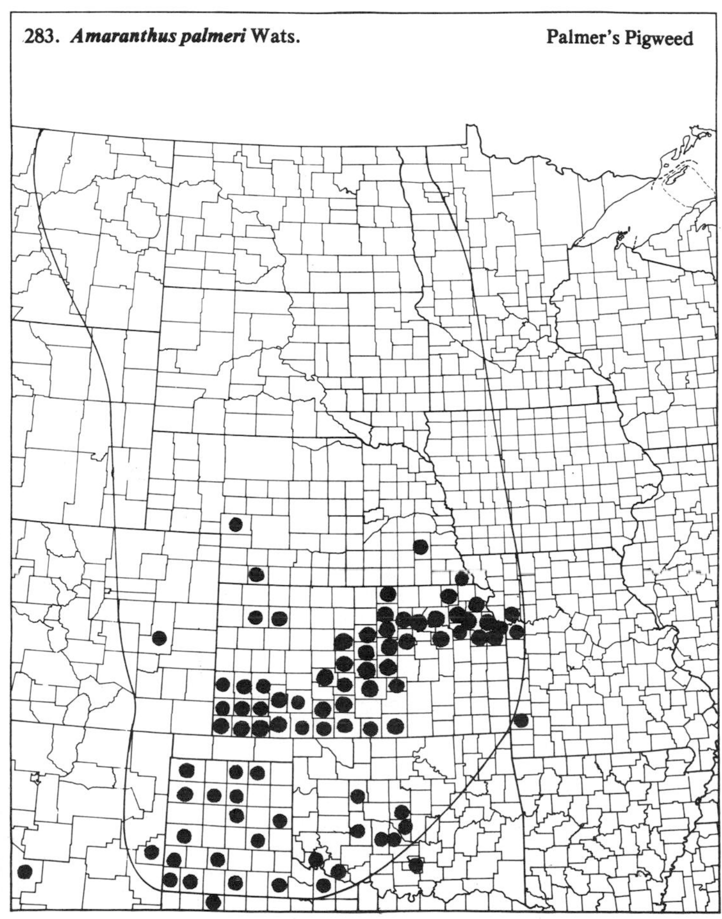

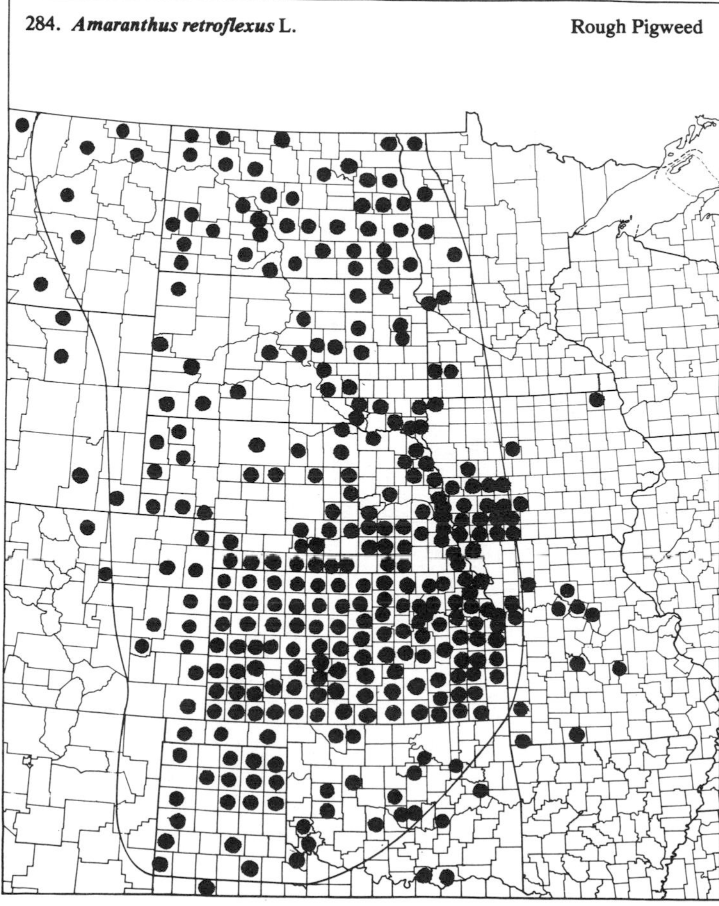

(Amaranthaceae)

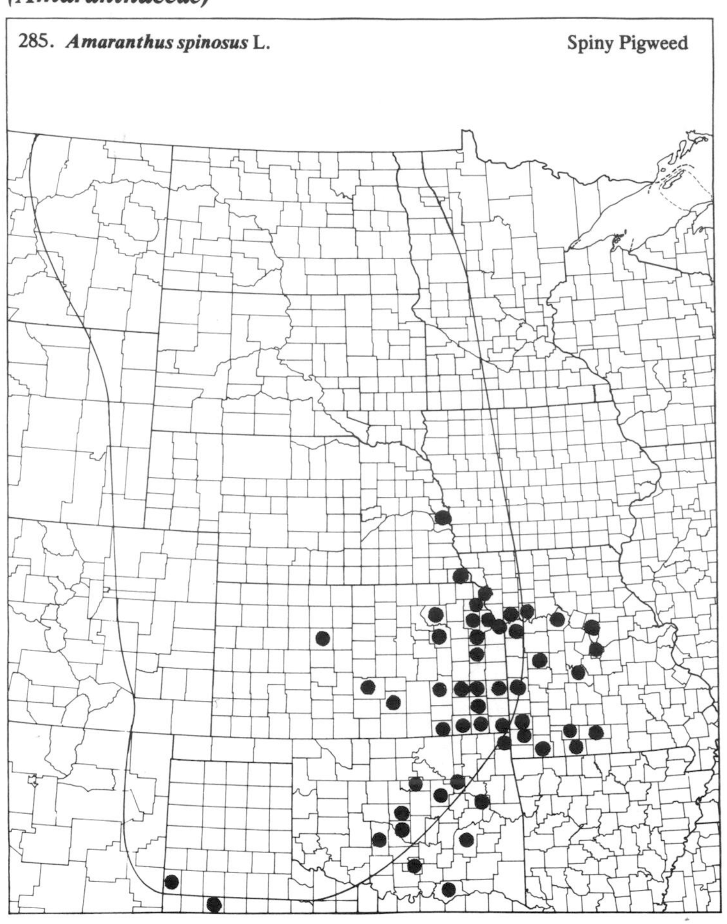

285. ***Amaranthus spinosus*** L. Spiny Pigweed

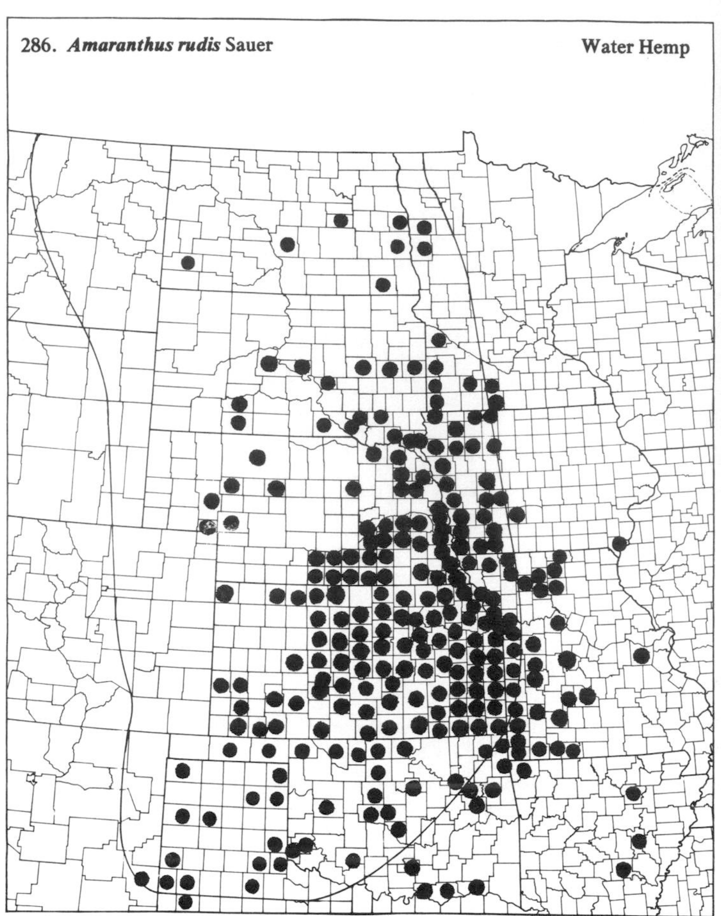

286. ***Amaranthus rudis*** Sauer Water Hemp

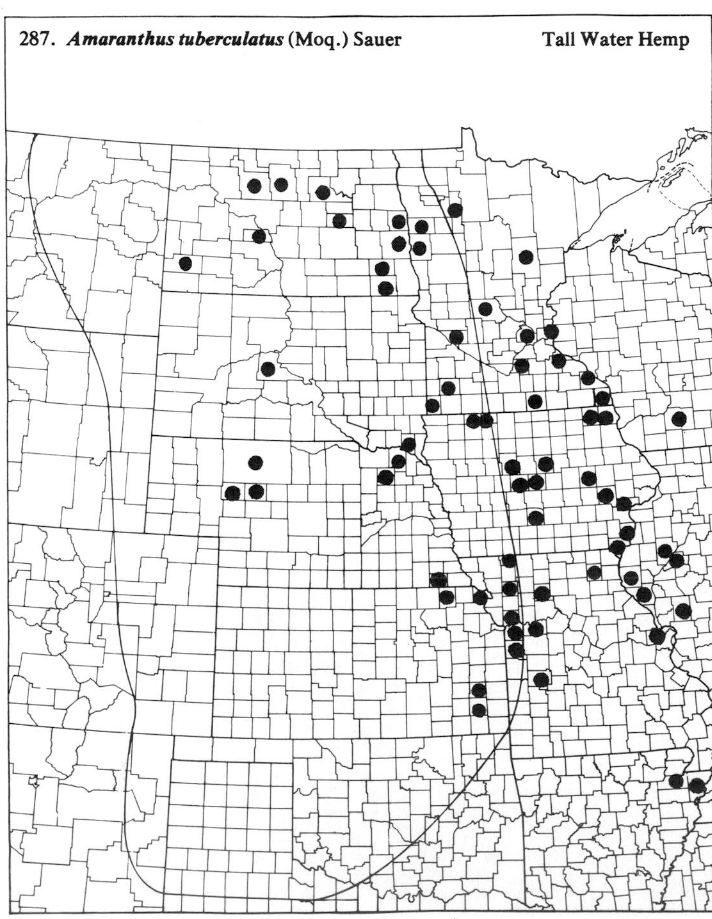

287. ***Amaranthus tuberculatus*** (Moq.) Sauer Tall Water Hemp

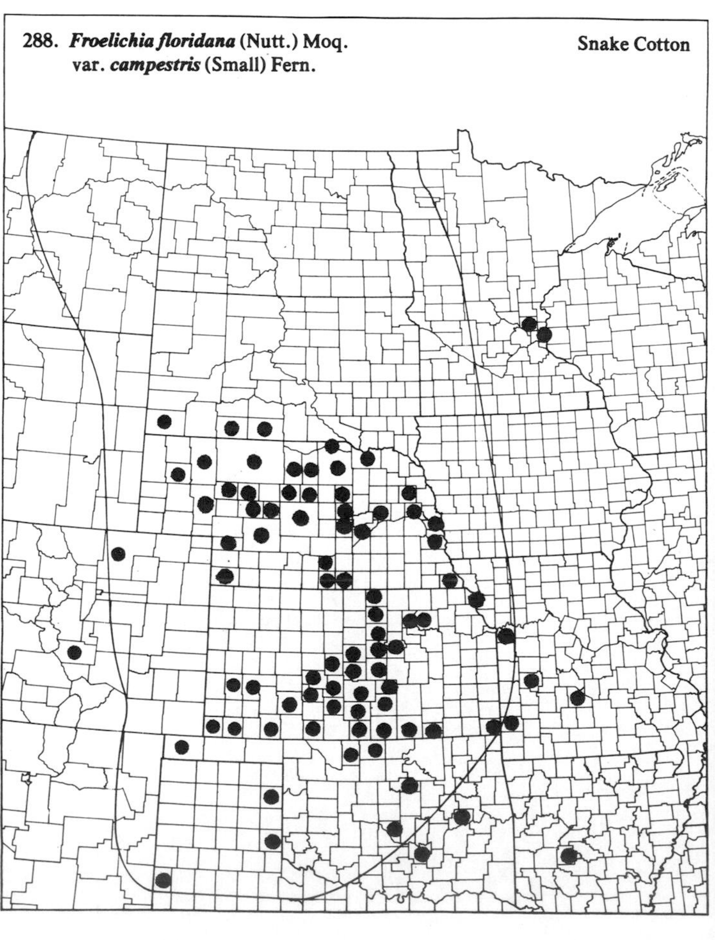

288. ***Froelichia floridana*** (Nutt.) Moq. var. ***campestris*** (Small) Fern. Snake Cotton

(Amaranthaceae)

289. ***Froelichia gracilis*** (Hook.) Moq. Cottonweed

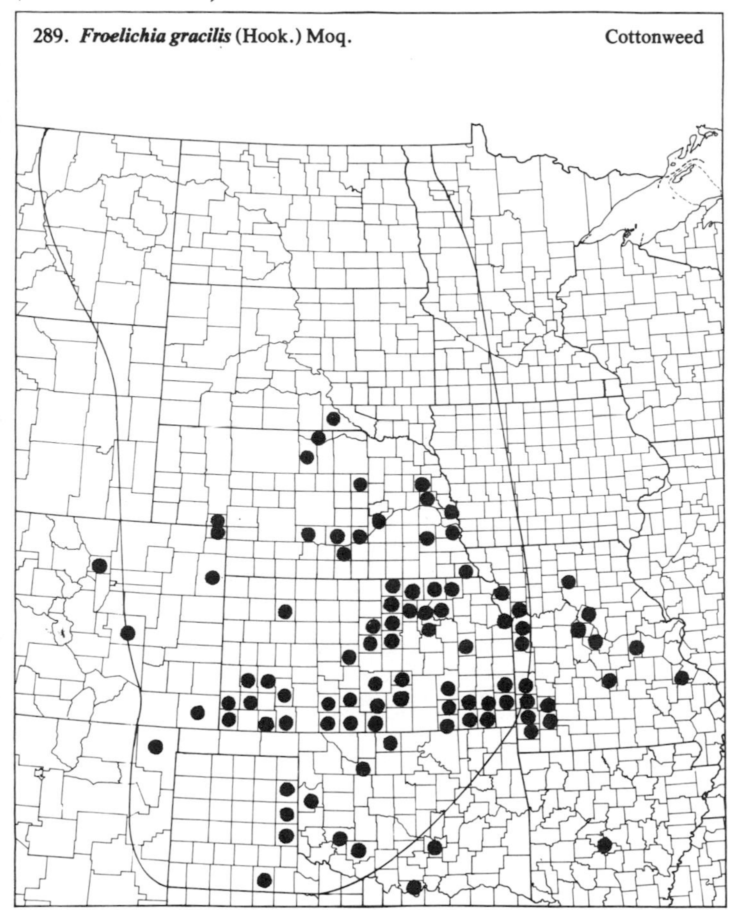

290. ***Iresine rhizomatosa*** Standl.

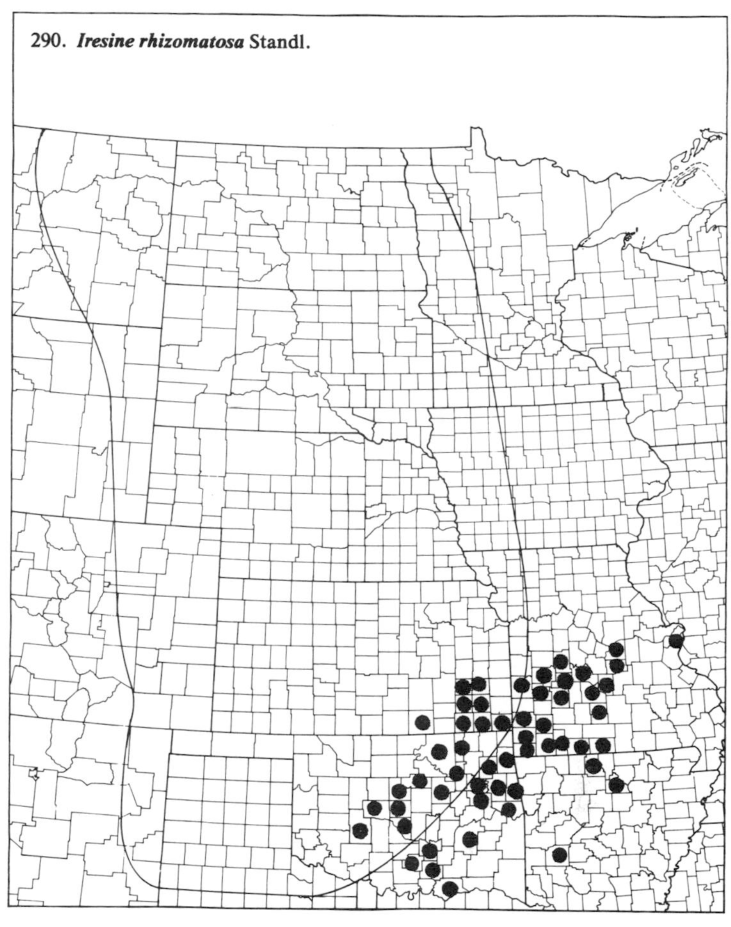

28. *Polygonaceae* (Buckwheat Family)

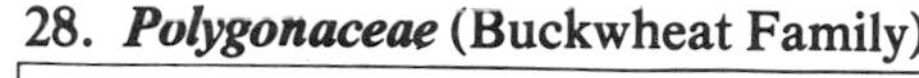

291. ***Tidestromia lanuginosa*** (Nutt.) Standl. Woolly Tidestromia

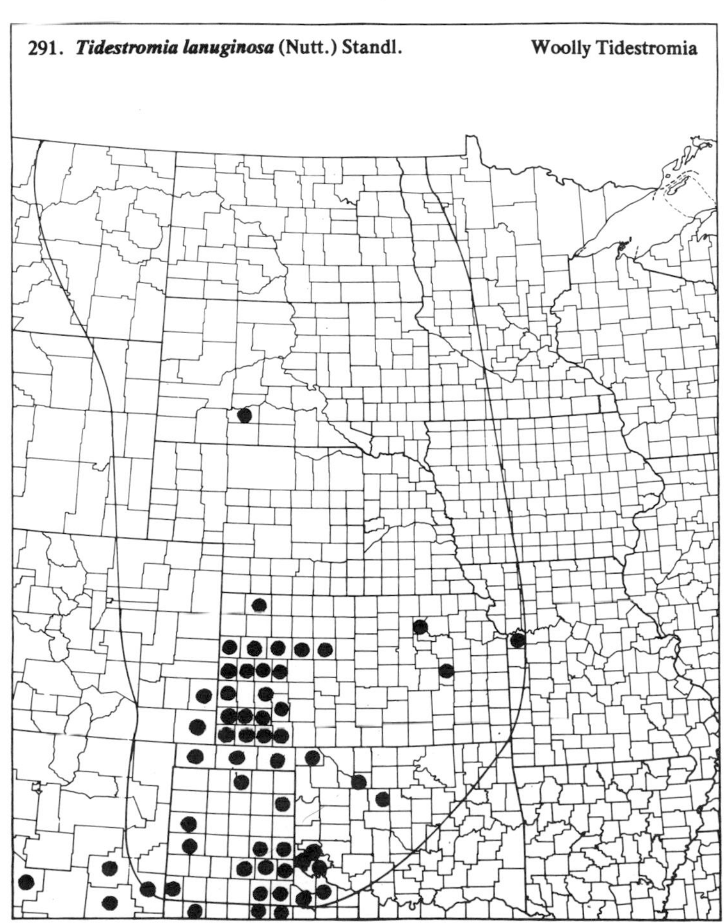

292. ***Eriogonum alatum*** Torr. var. ***alatum***

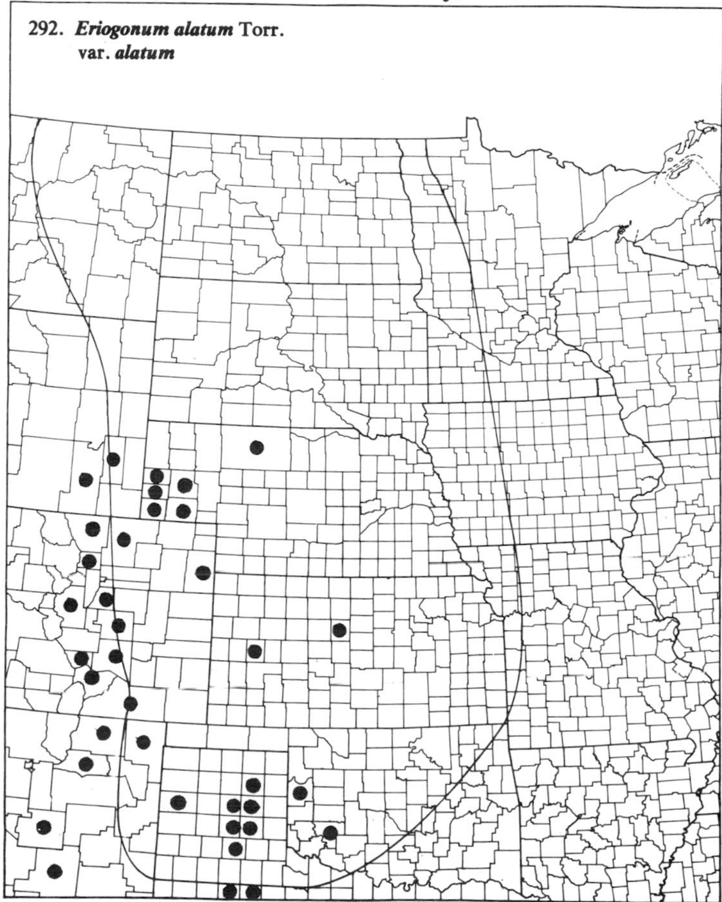

(Polygonaceae)

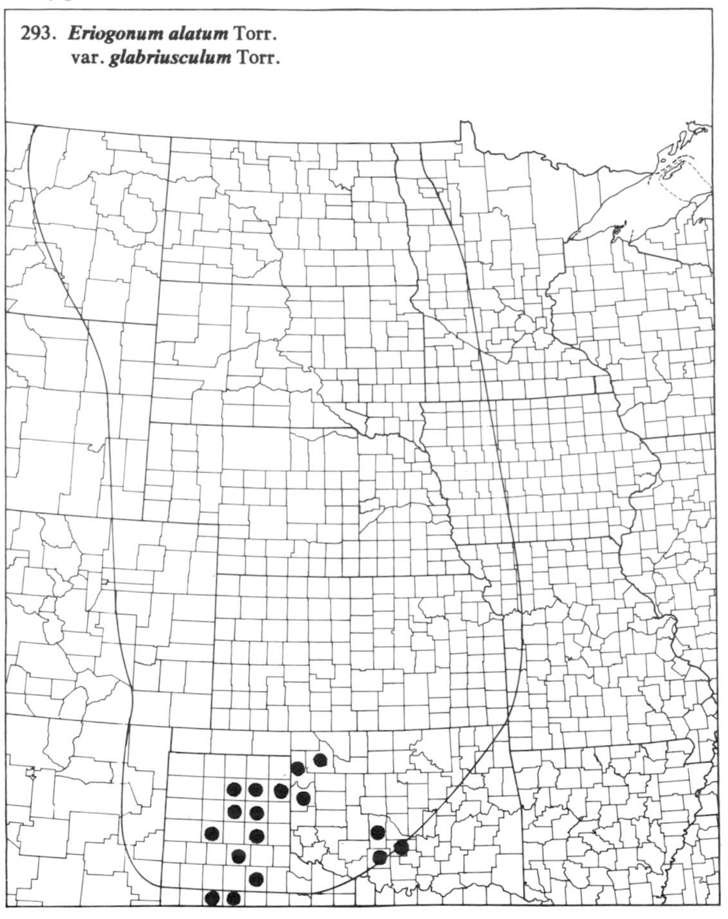

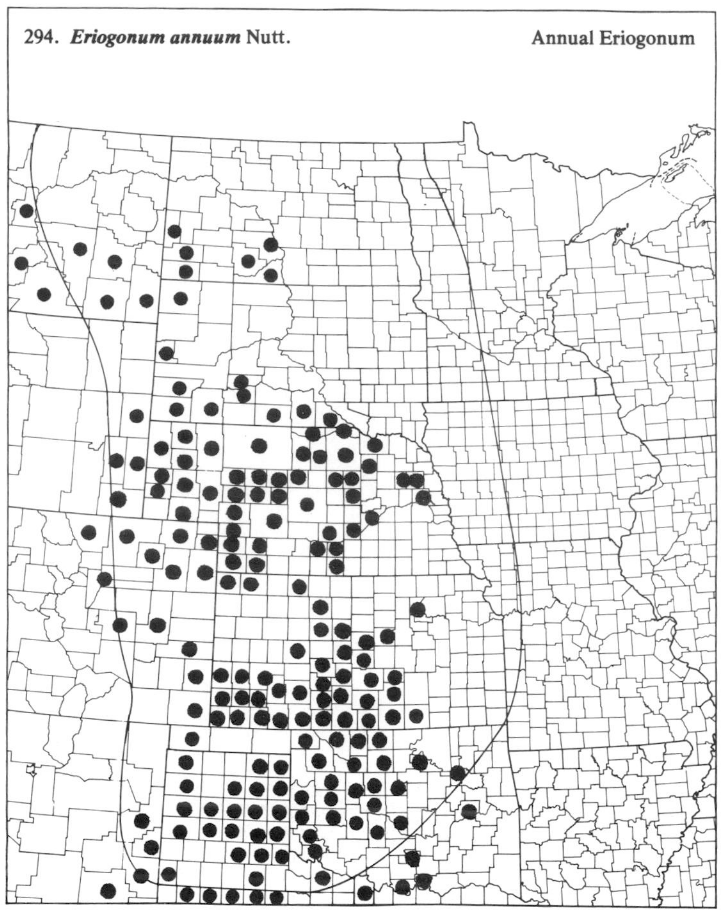

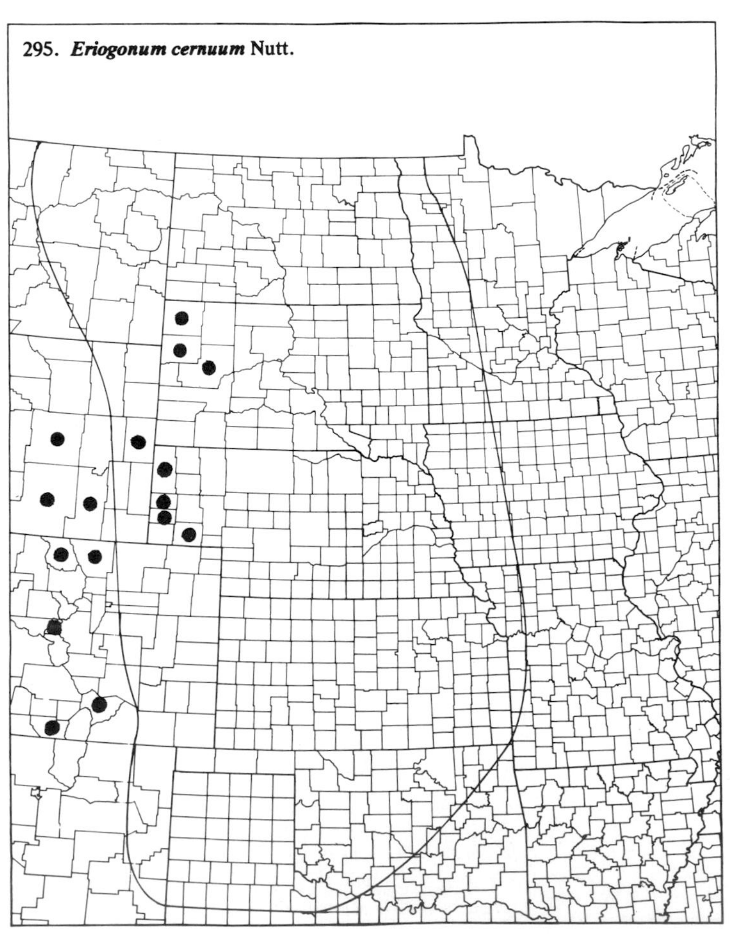

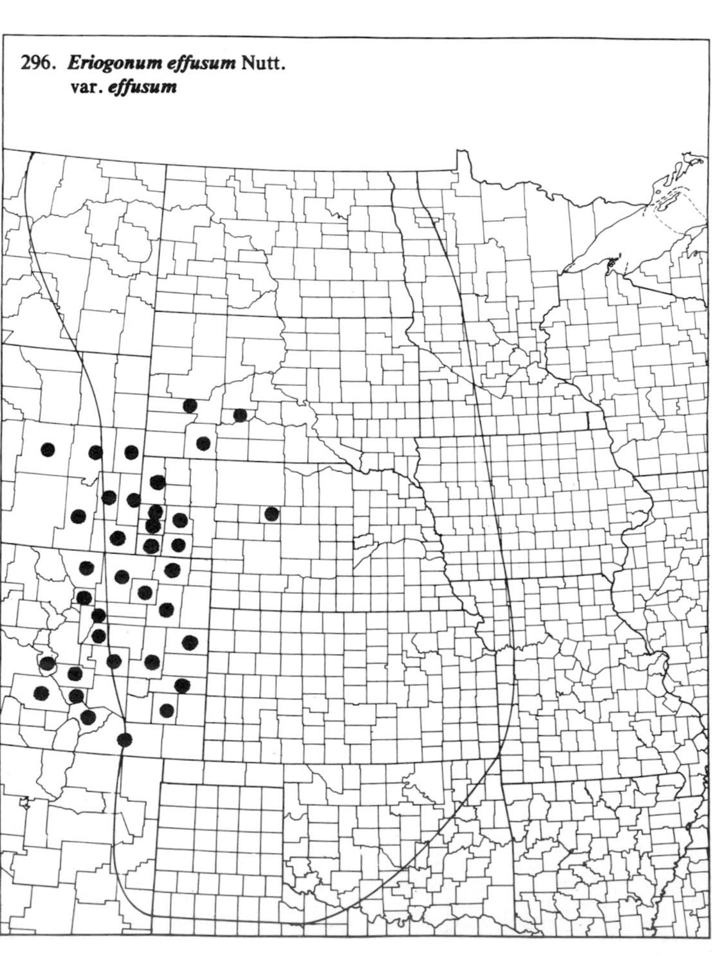

(Polygonaceae)

297. ***Eriogonum effusum*** Nutt. var. ***rosmarinoides*** Benth. in DC.

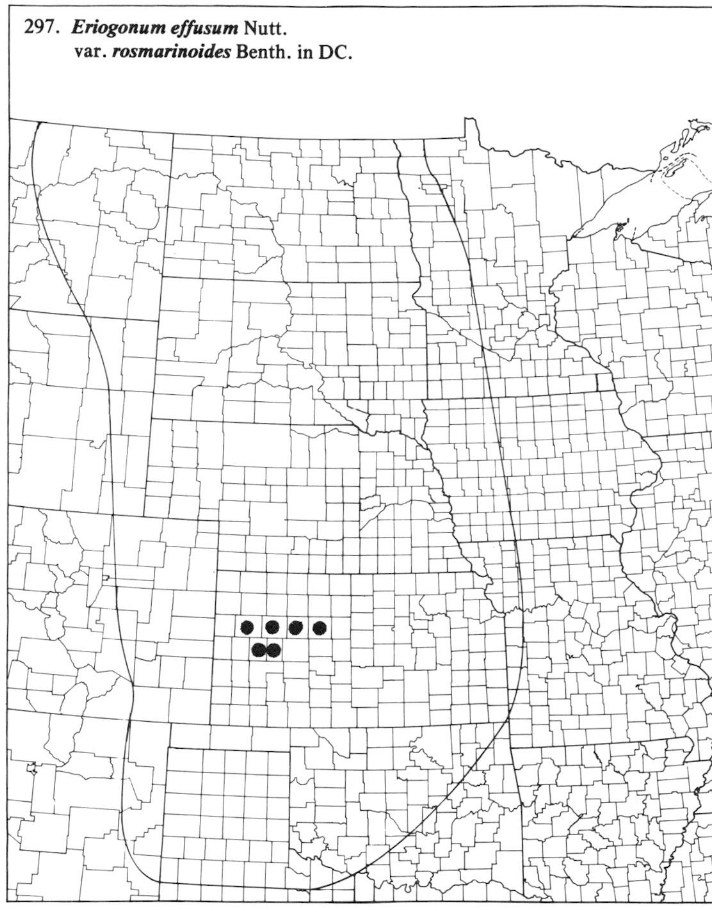

298. ***Eriogonum flavum*** Nutt.

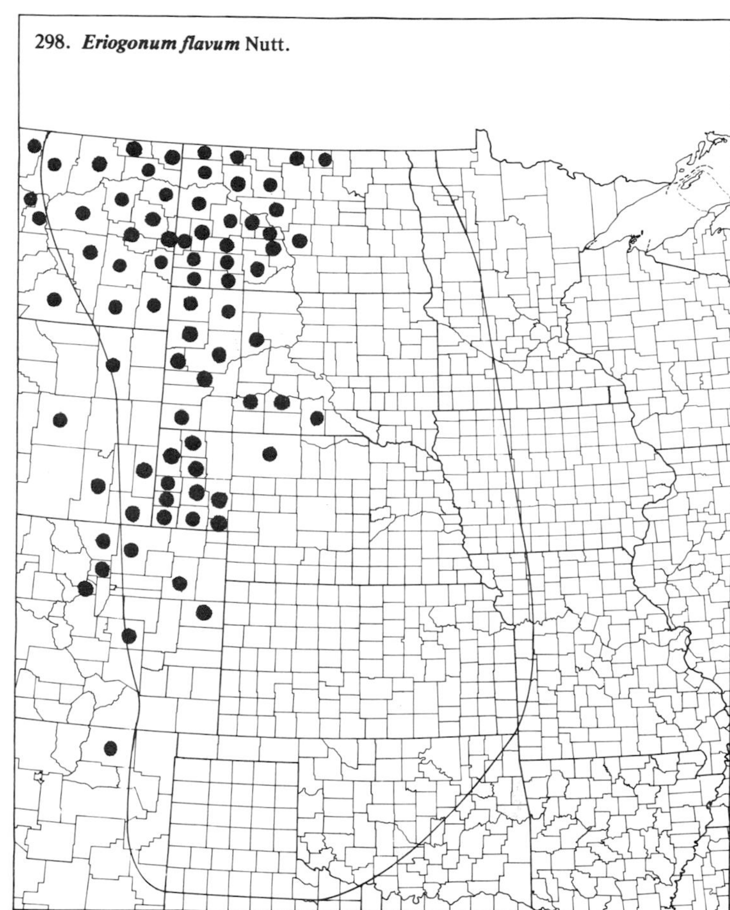

299. ***Eriogonum jamesii*** Benth.

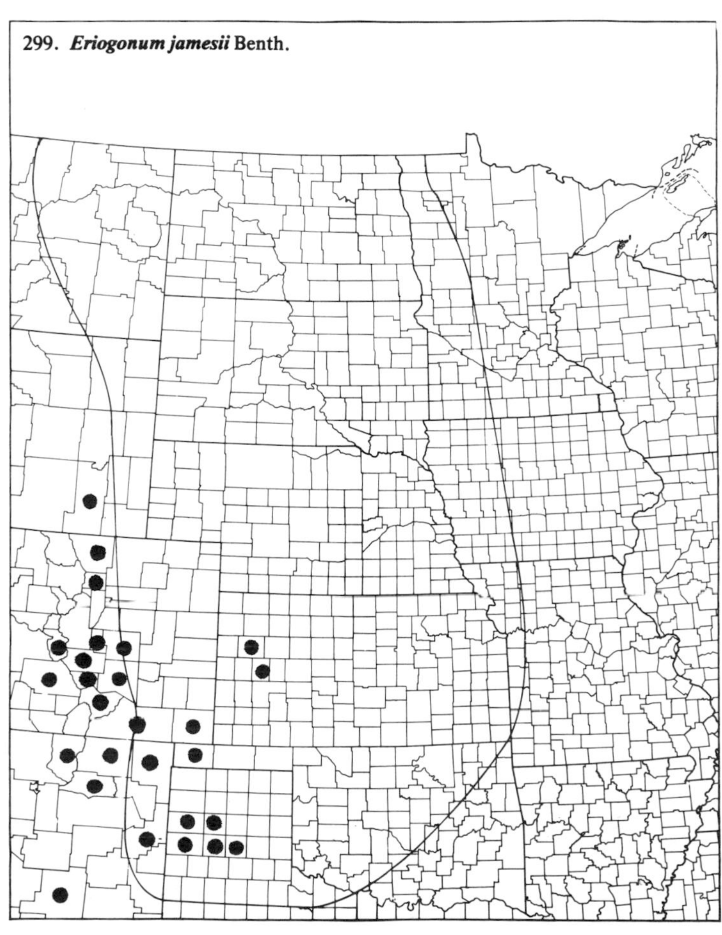

300. ***Eriogonum lachnogynum*** Torr.

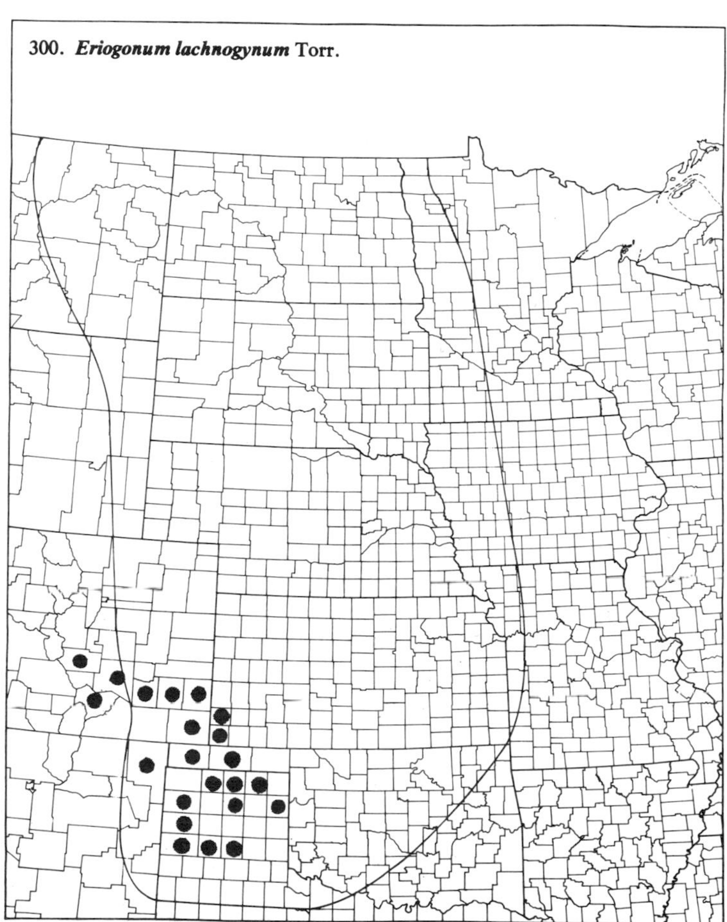

(Polygonaceae)

301. ***Eriogonum longifolium*** Nutt. var. ***lindheimeri*** Gand. — Longleaf Eriogonum

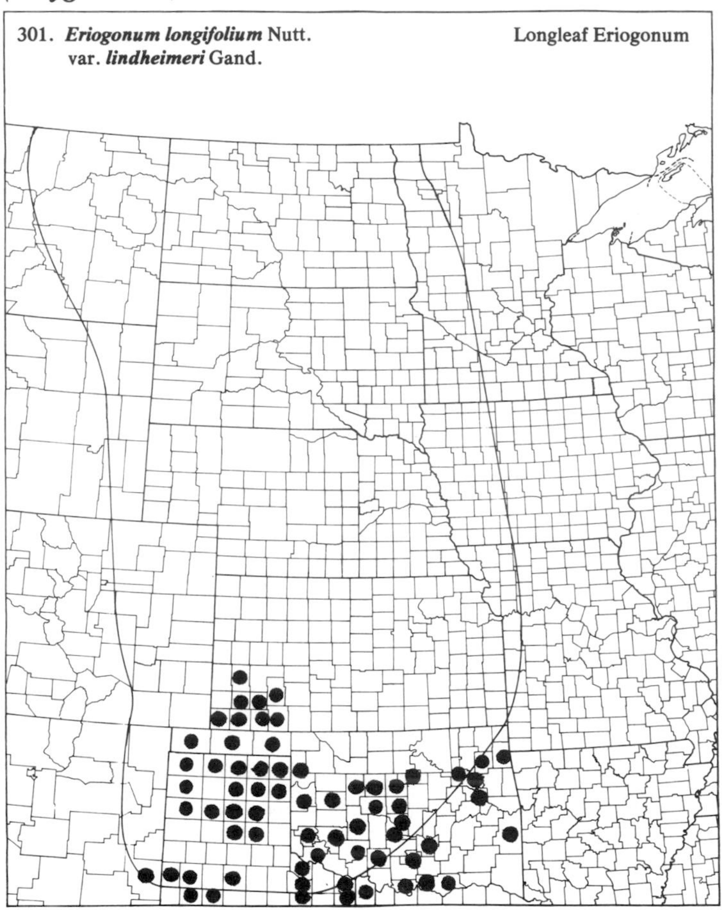

302. ***Eriogonum pauciflorum*** Pursh var. ***pauciflorum***

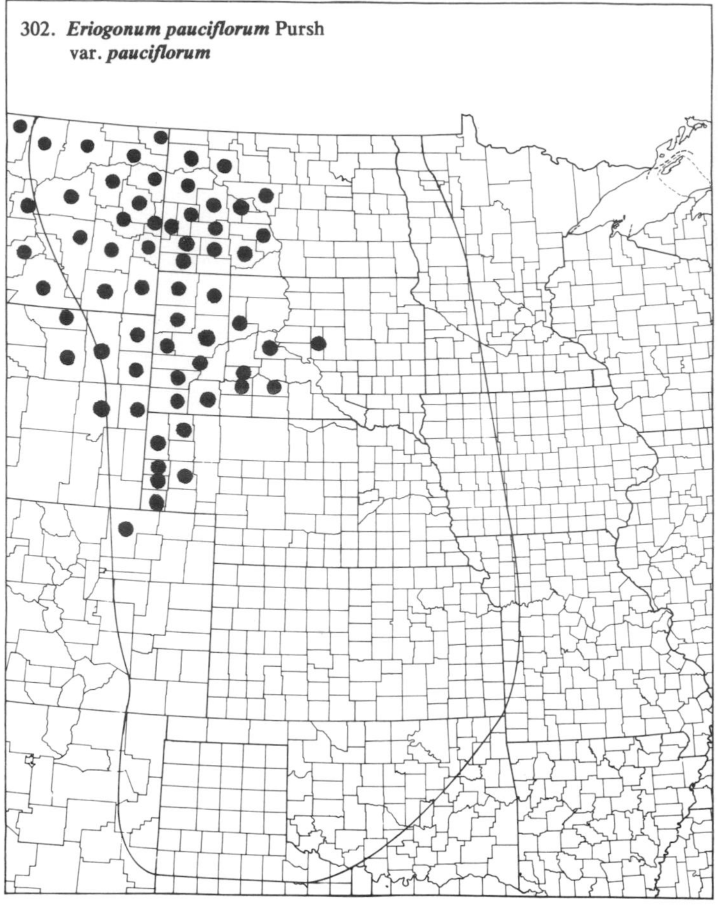

303. ***Eriogonum pauciflorum*** Pursh var. ***gnaphaloides*** (Benth.) Reveal

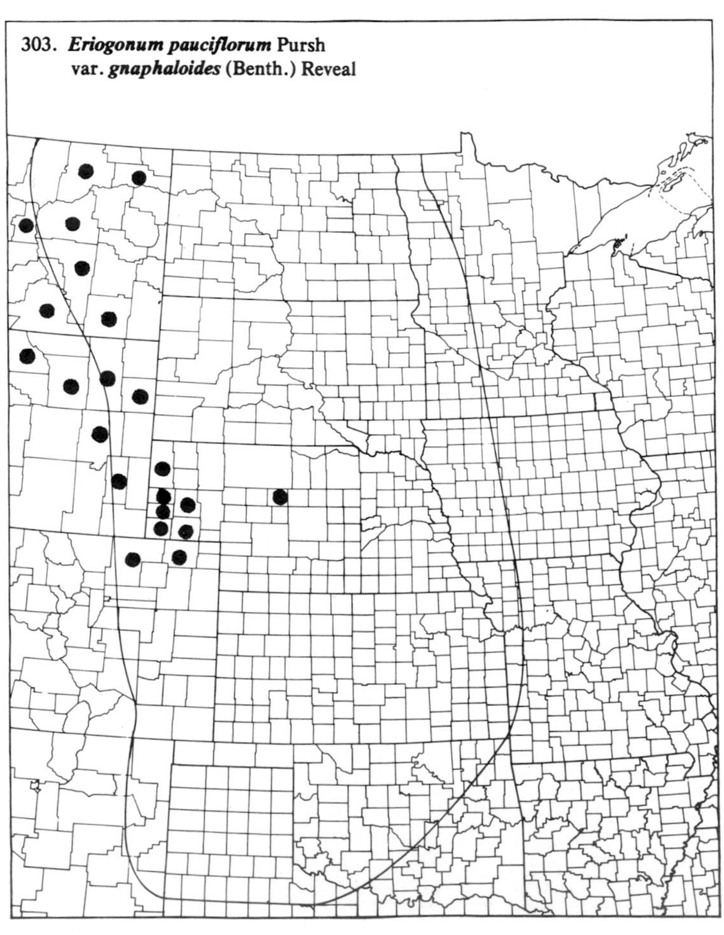

304. ***Eriogonum visheri*** A. Nels.

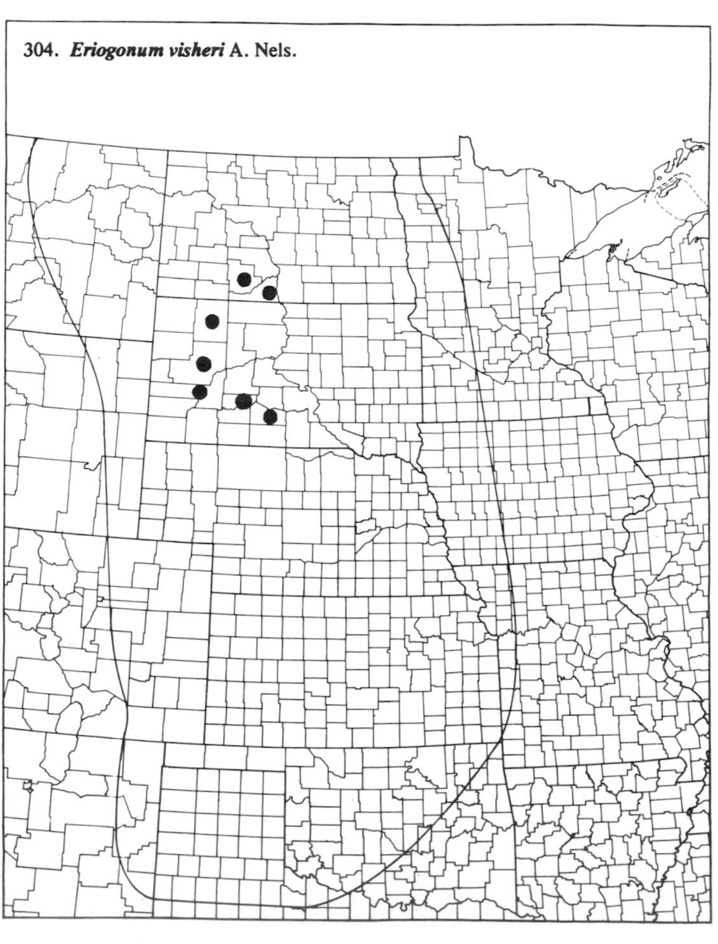

(Polygonaceae)

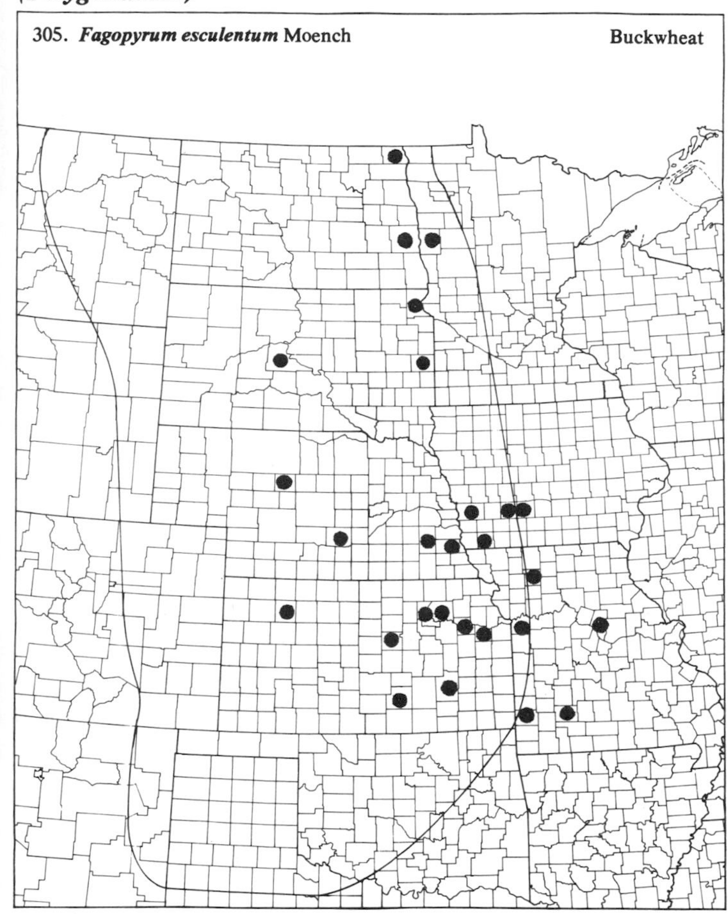

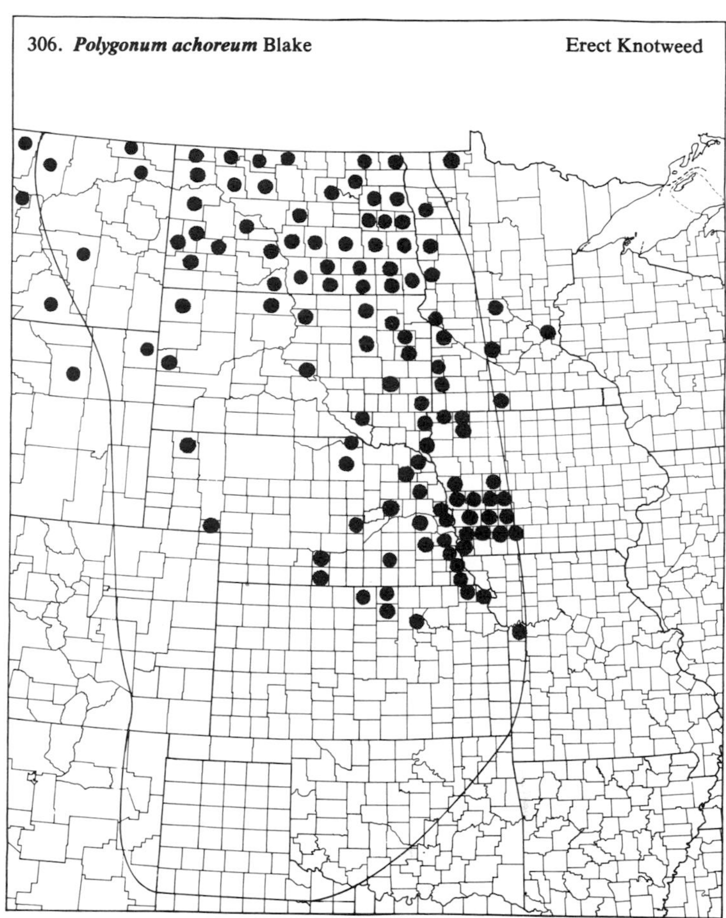

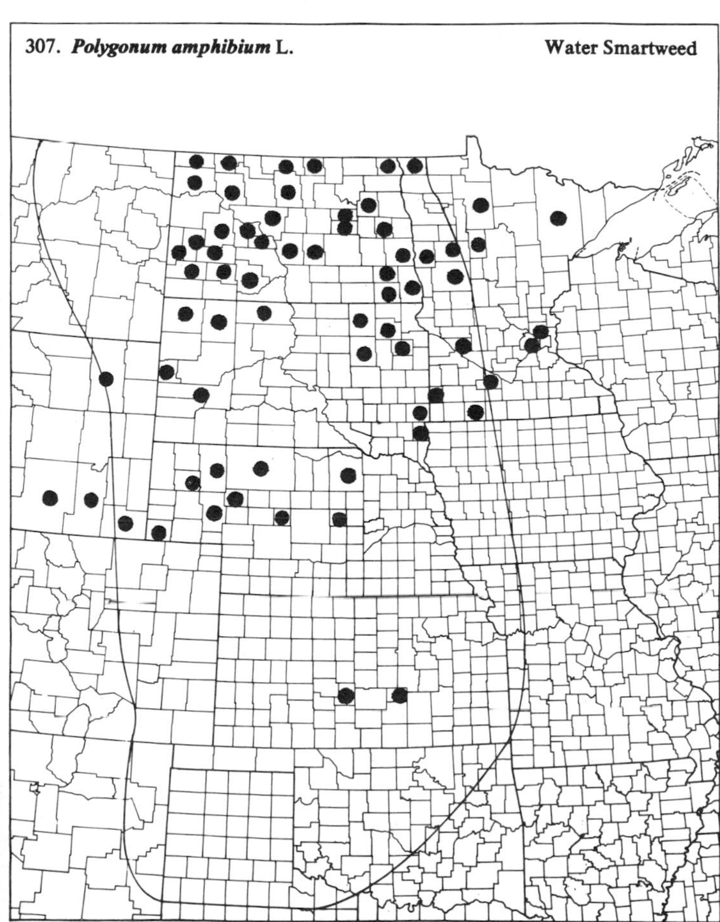

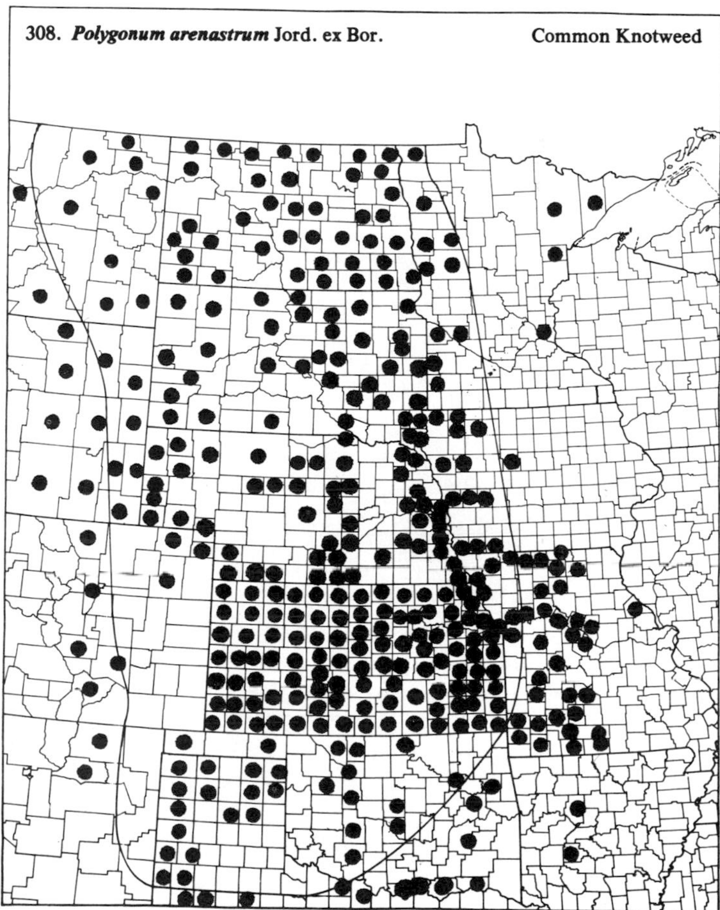

(Polygonaceae)

309. **Polygonum bicorne** Raf.

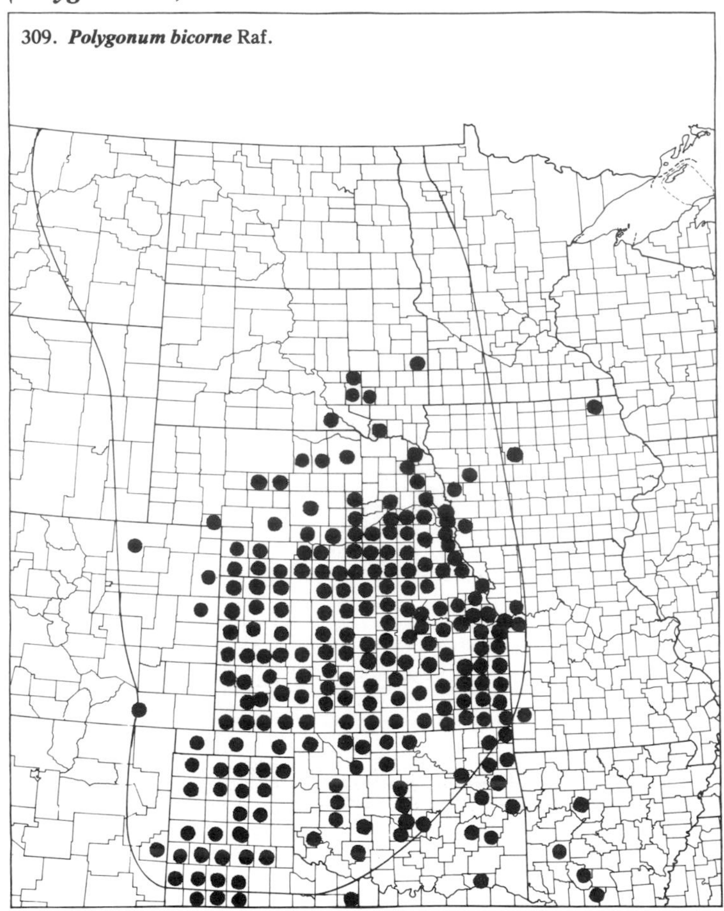

310. **Polygonum coccineum** Muhl. Swamp Smartweed

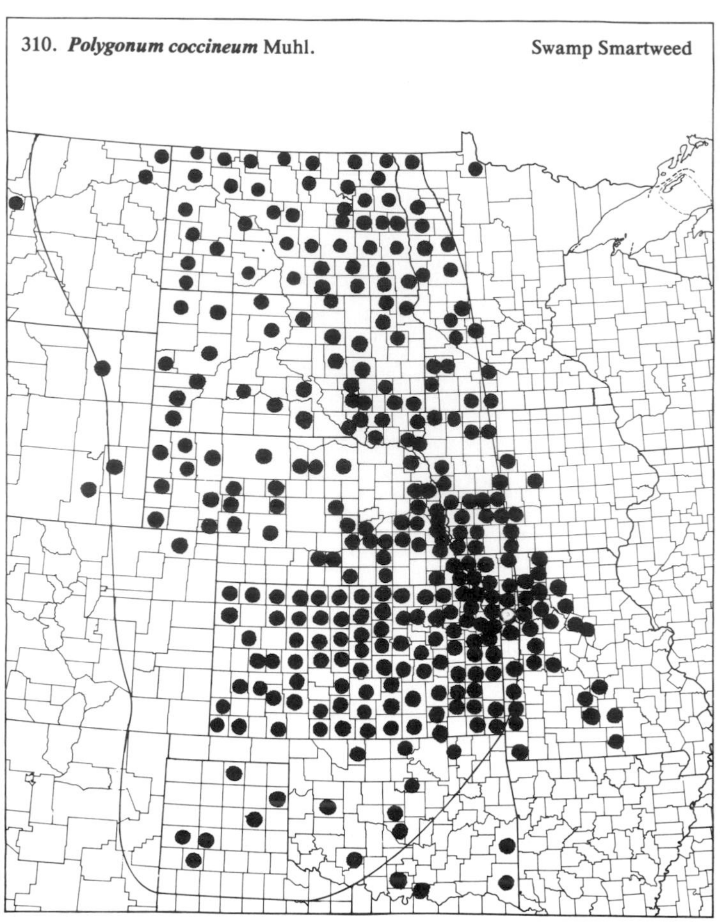

311. **Polygonum convolvulus** L. Wild Buckwheat

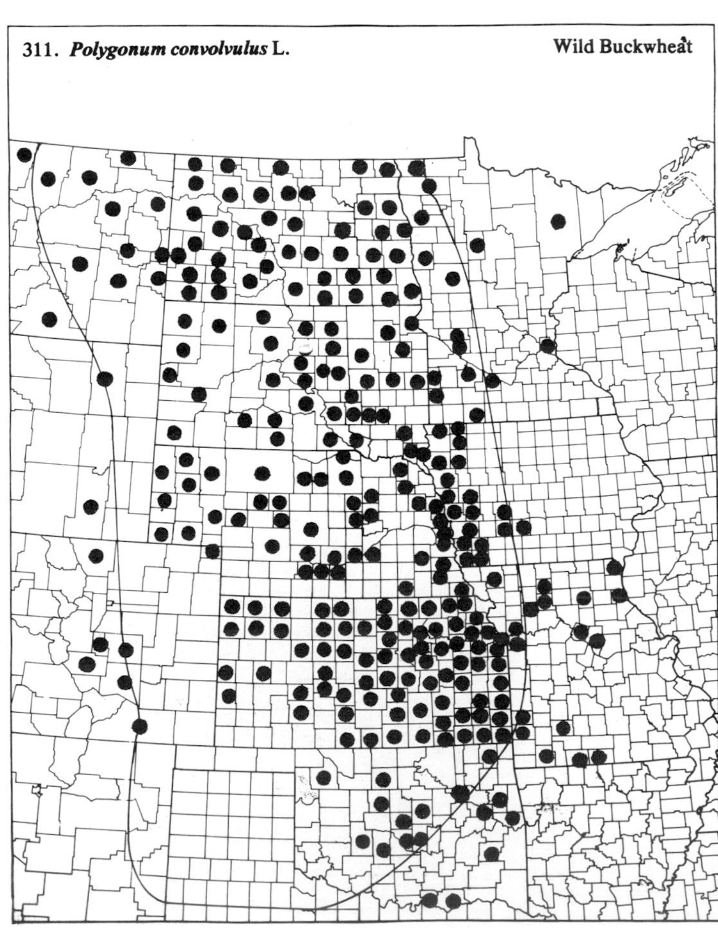

312. **Polygonum douglasii** Greene

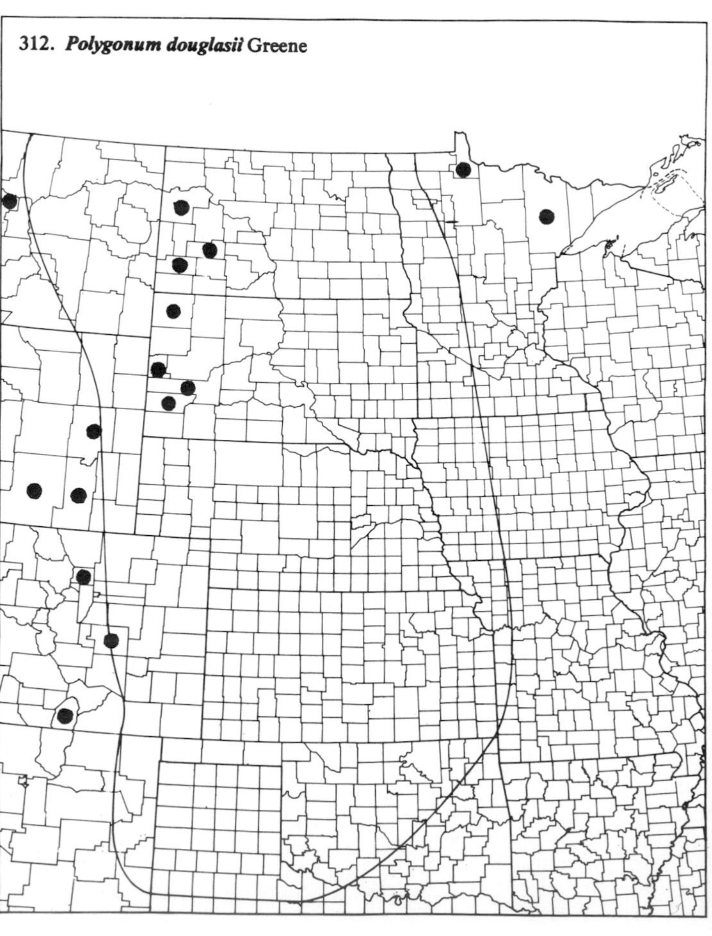

(Polygonaceae)

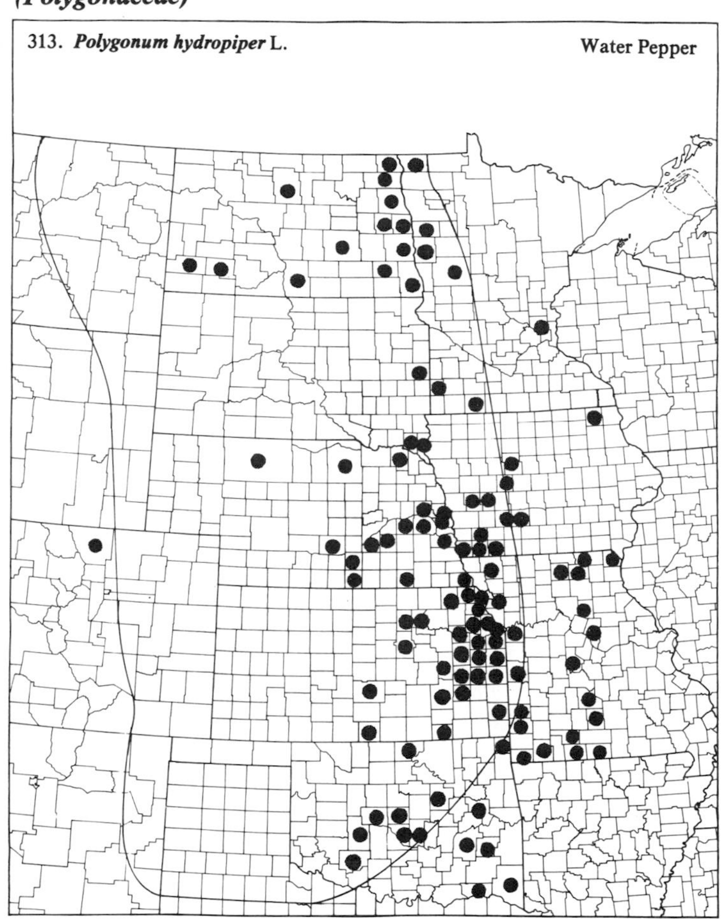

313. ***Polygonum hydropiper*** L. — Water Pepper

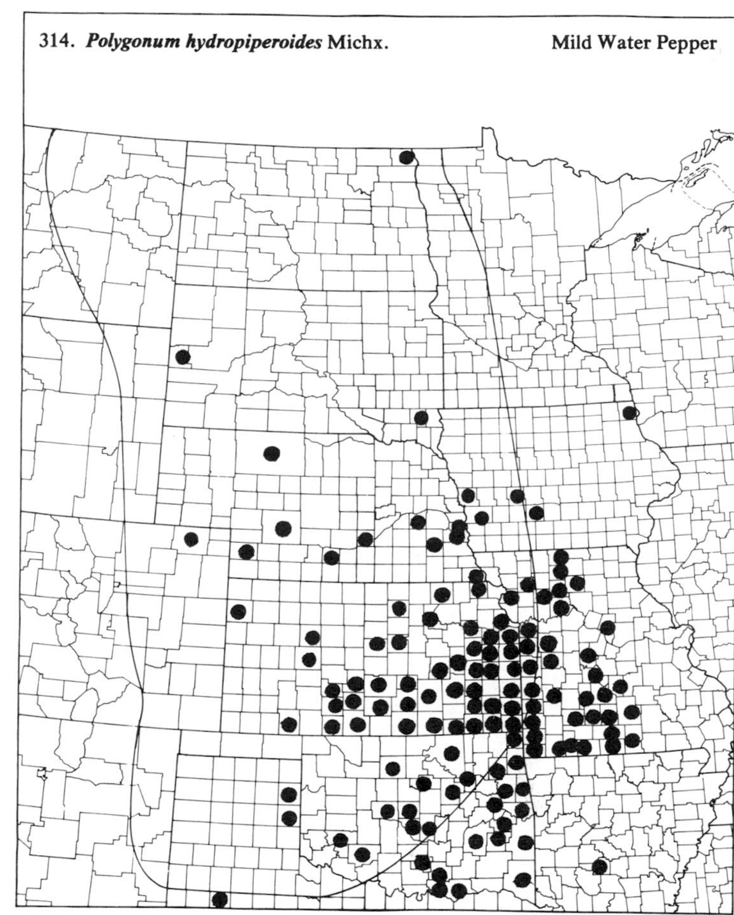

314. ***Polygonum hydropiperoides*** Michx. — Mild Water Pepper

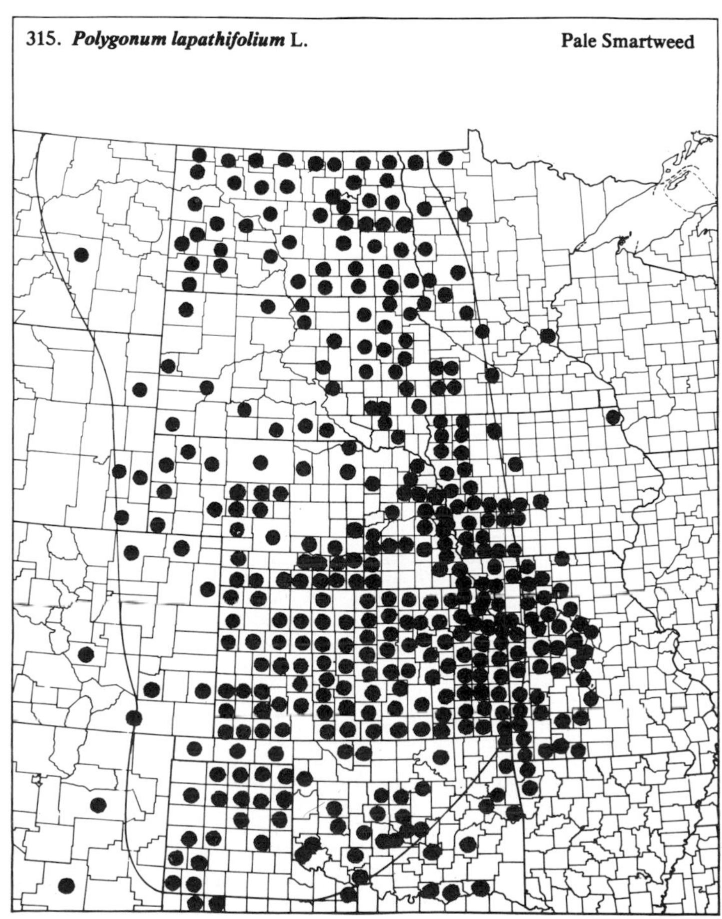

315. ***Polygonum lapathifolium*** L. — Pale Smartweed

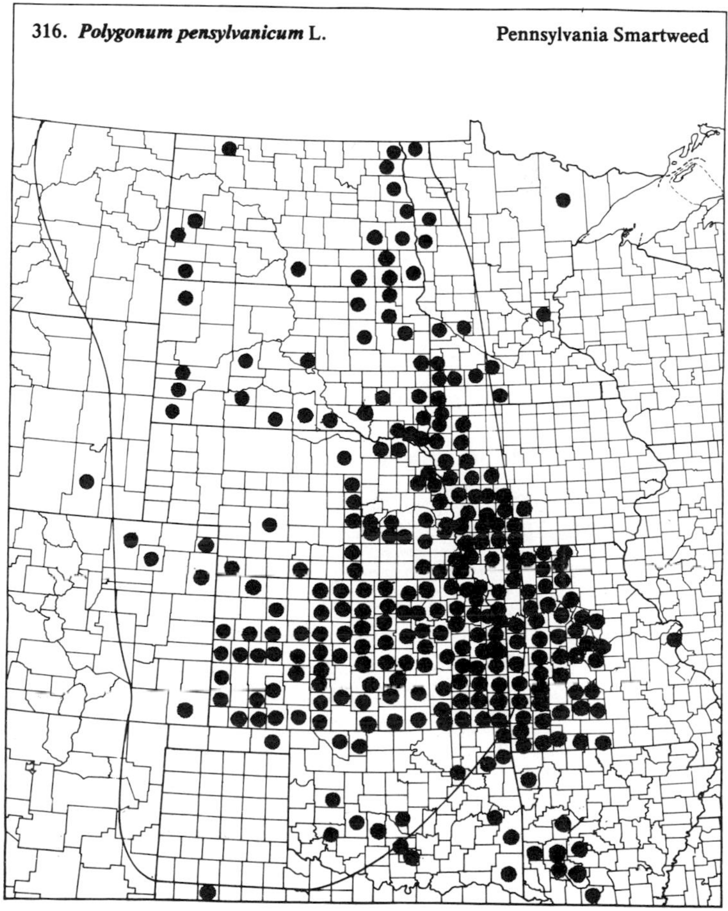

316. ***Polygonum pensylvanicum*** L. — Pennsylvania Smartweed

(Polygonaceae)

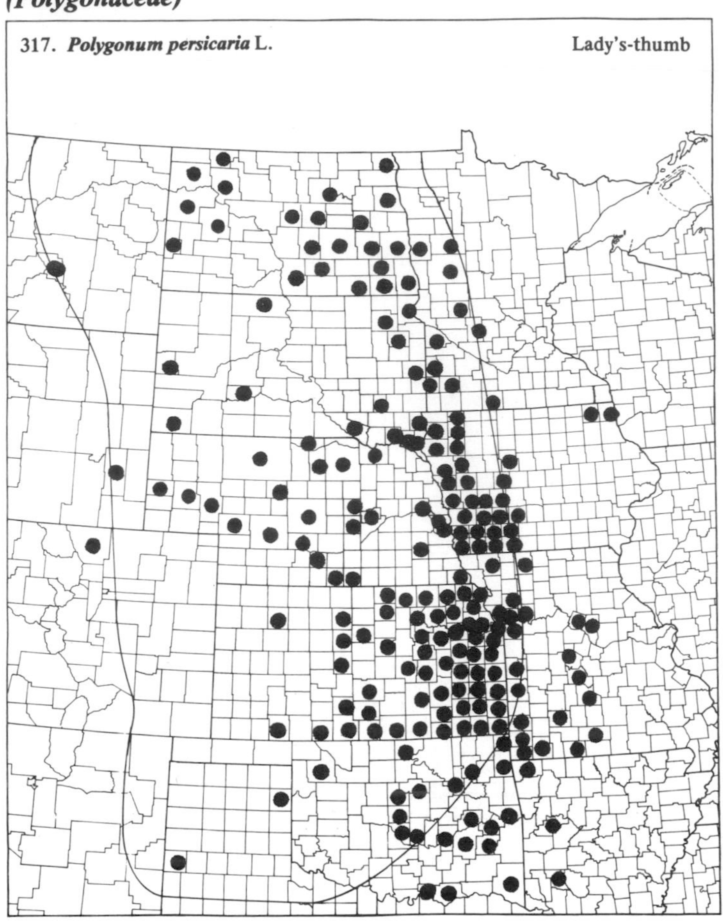

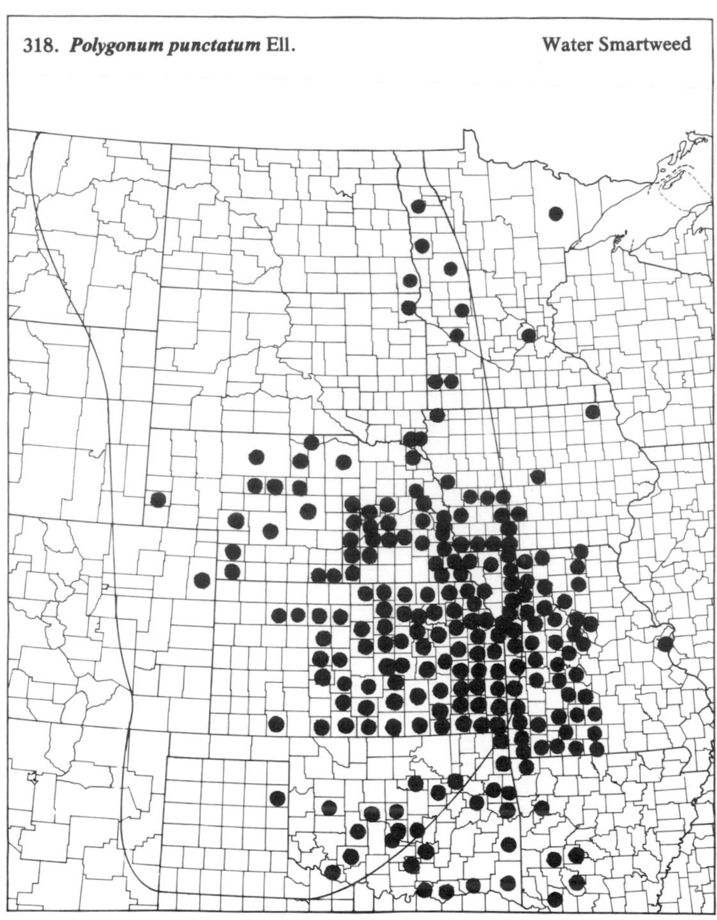

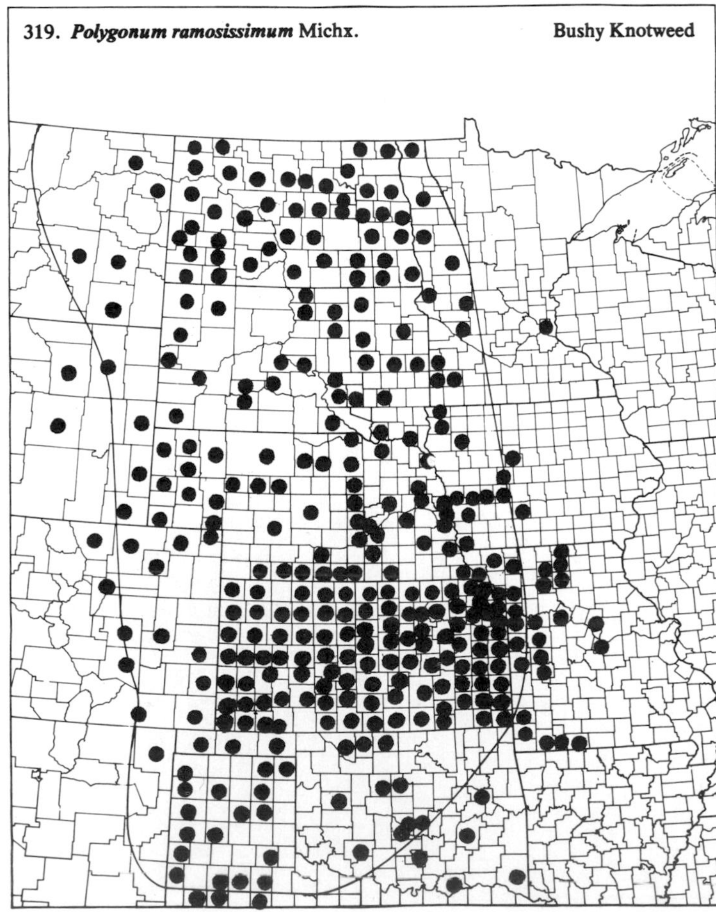

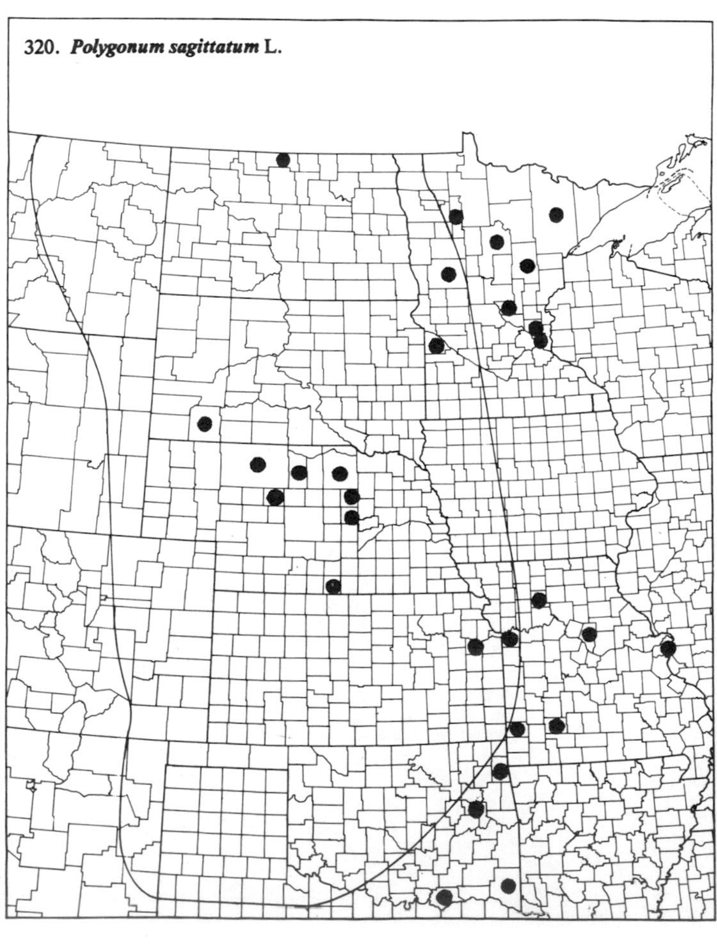

(Polygonaceae)

321. ***Polygonum scandens*** L. Climbing False Buckwheat

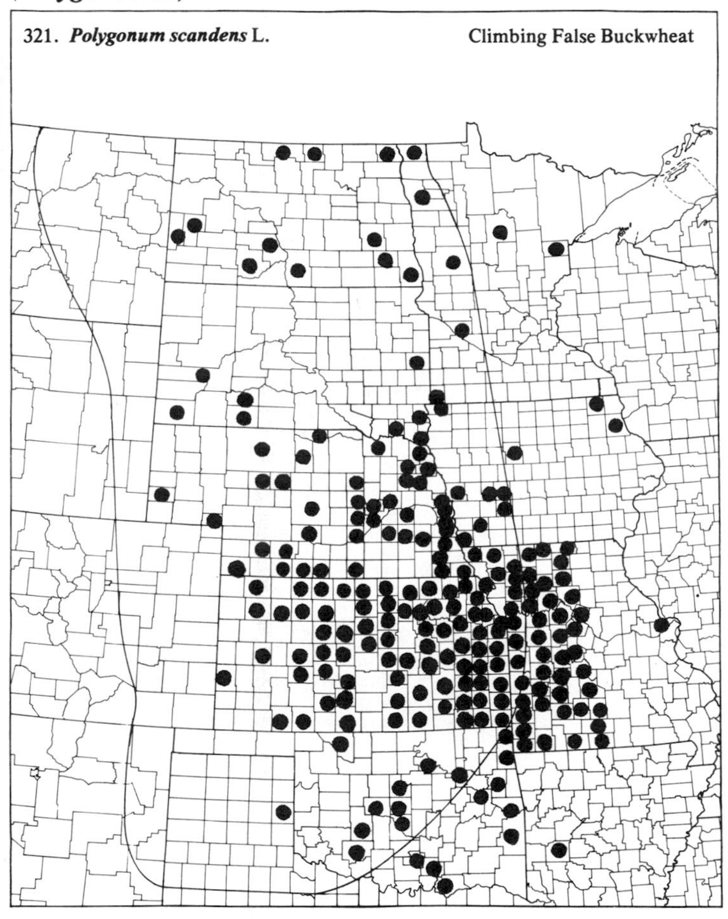

322. ***Polygonum tenue*** Michx. Slender Knotweed

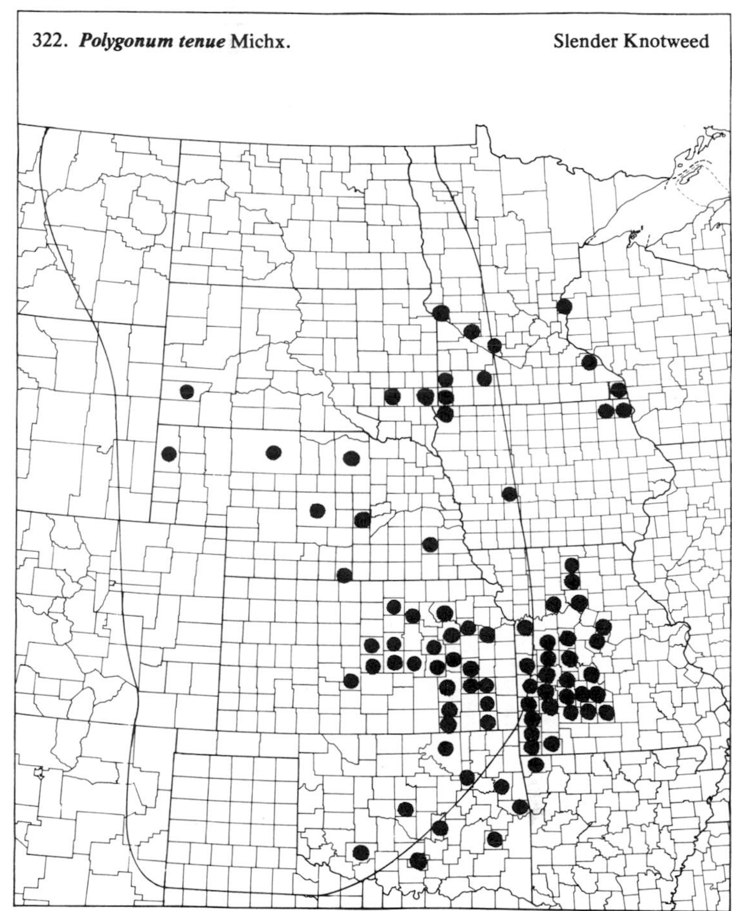

323. ***Polygonum virginianum*** L.

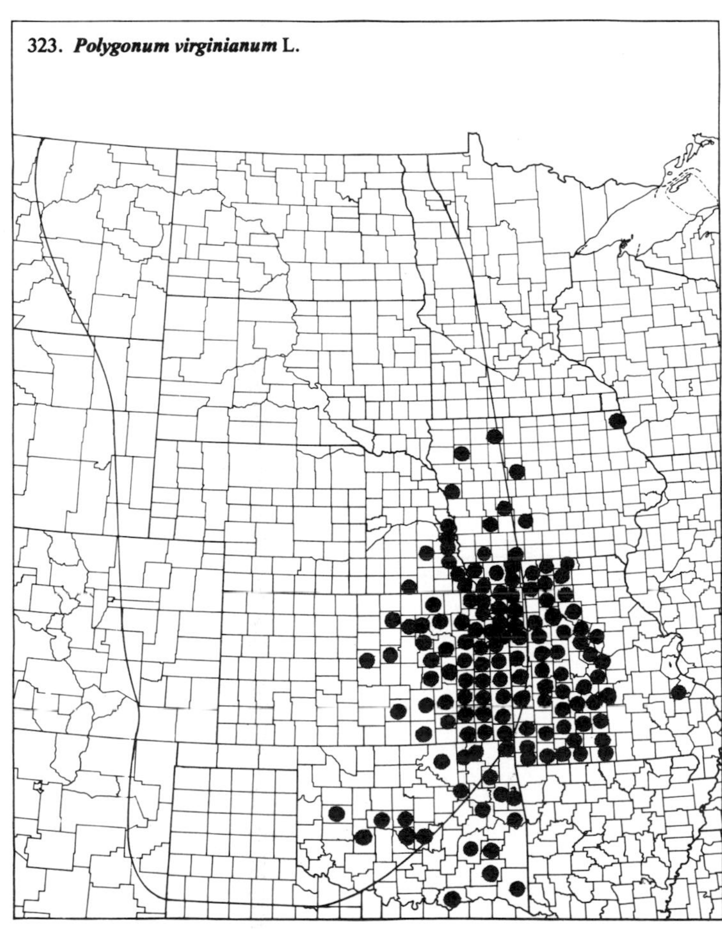

324. ***Rumex acetosella*** L. Sheep Sorrel

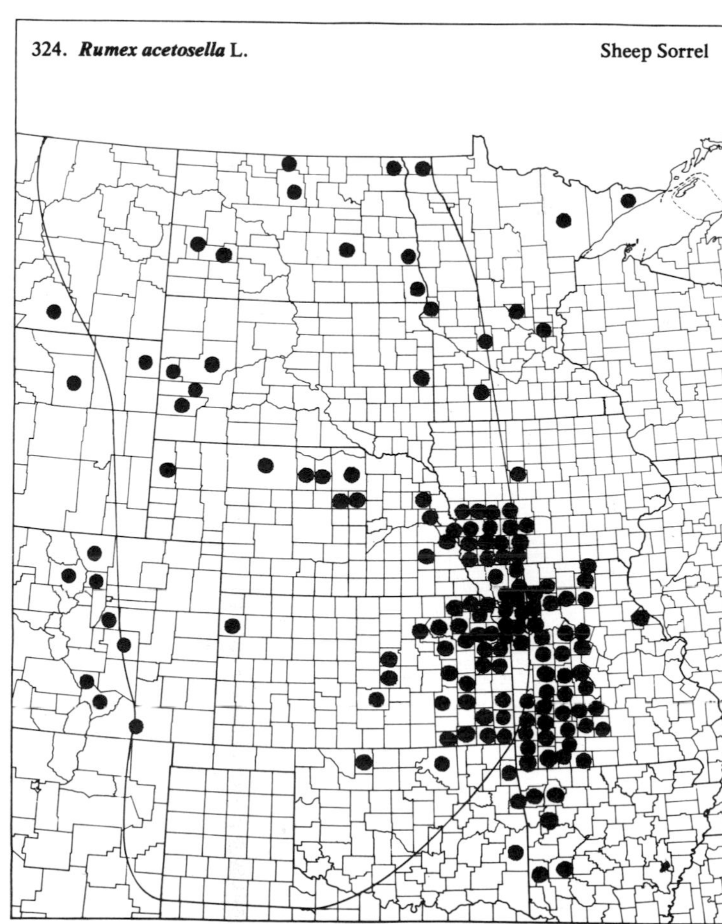

(Polygonaceae)

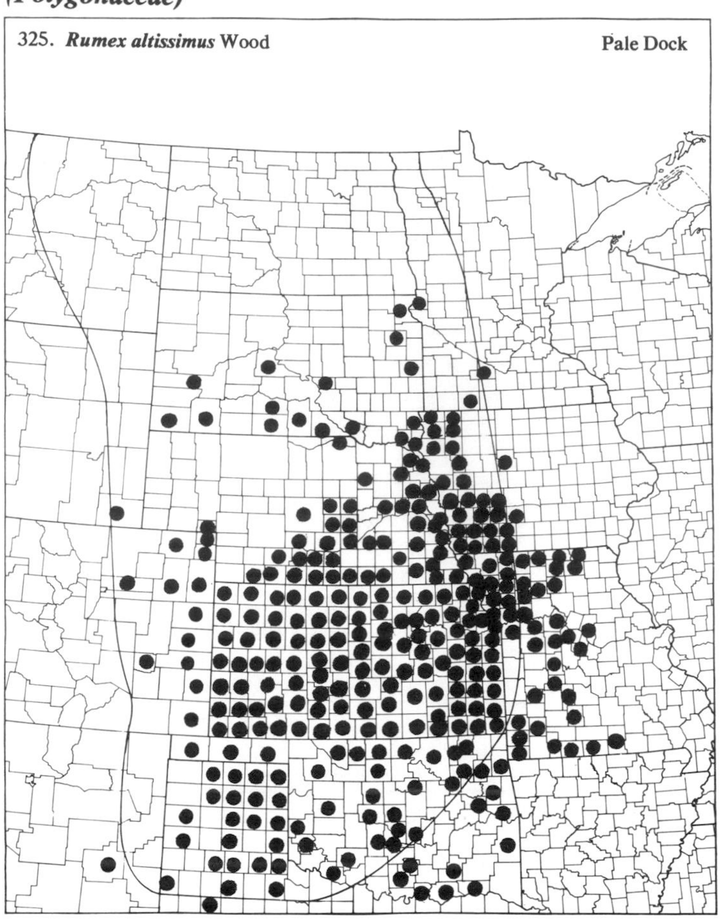

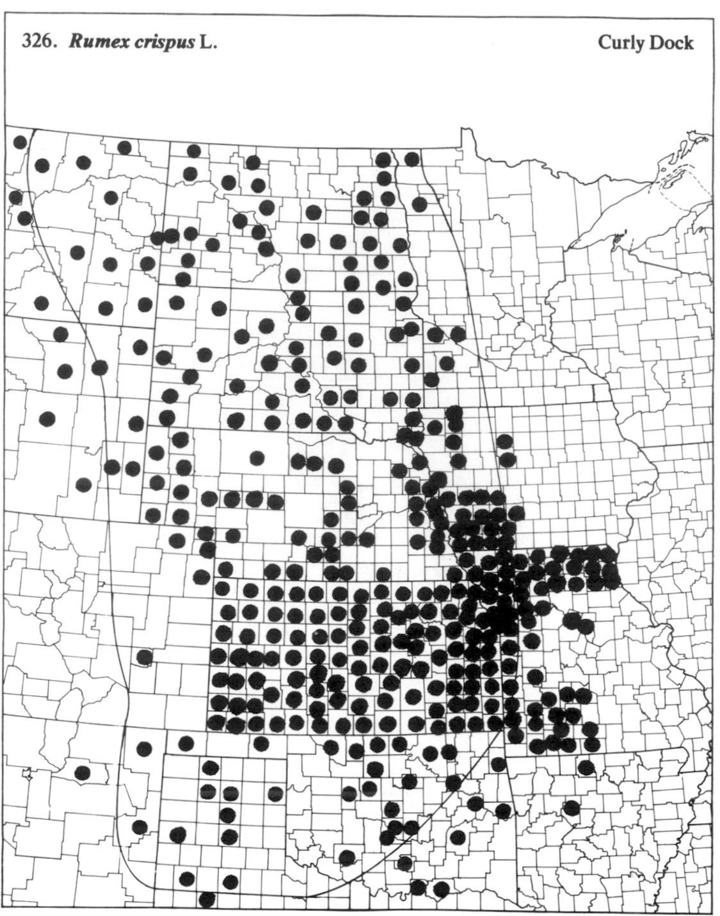

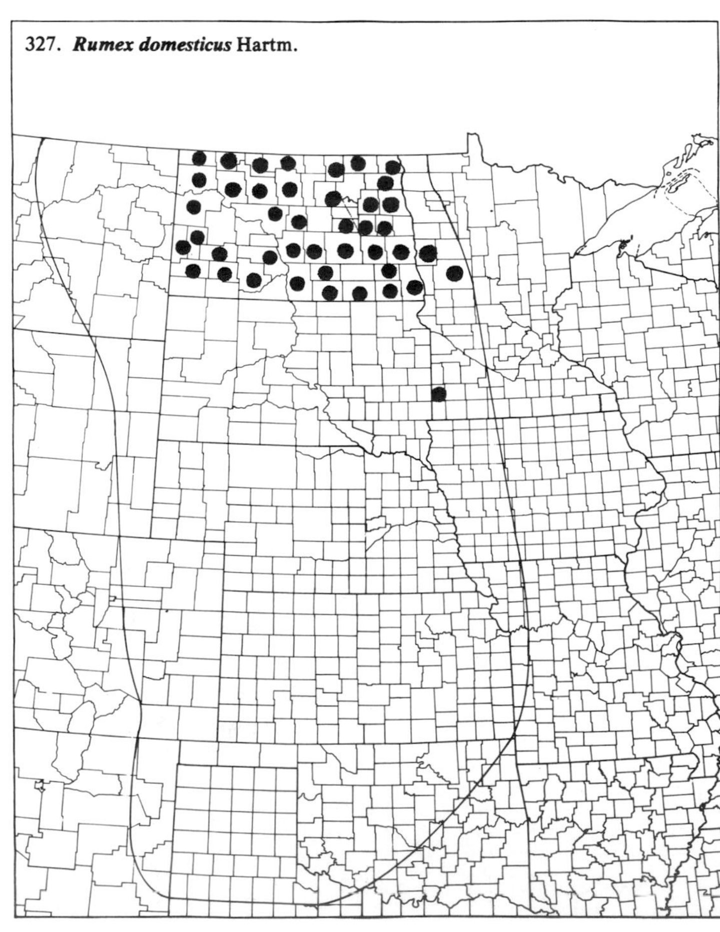

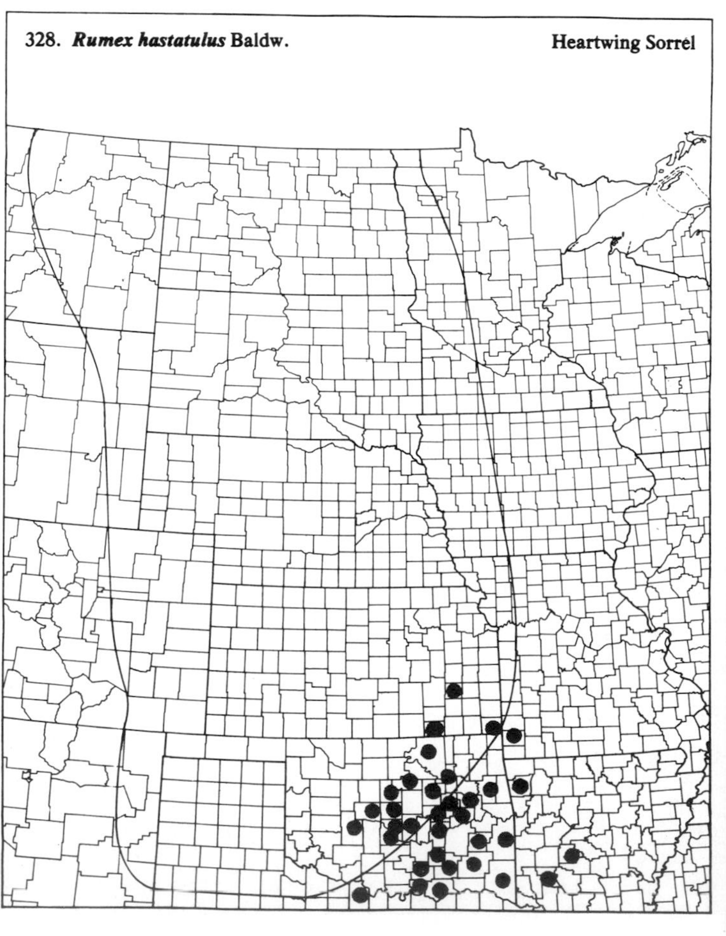

(Polygonaceae)

329. ***Rumex hymenosepalus*** Torr.

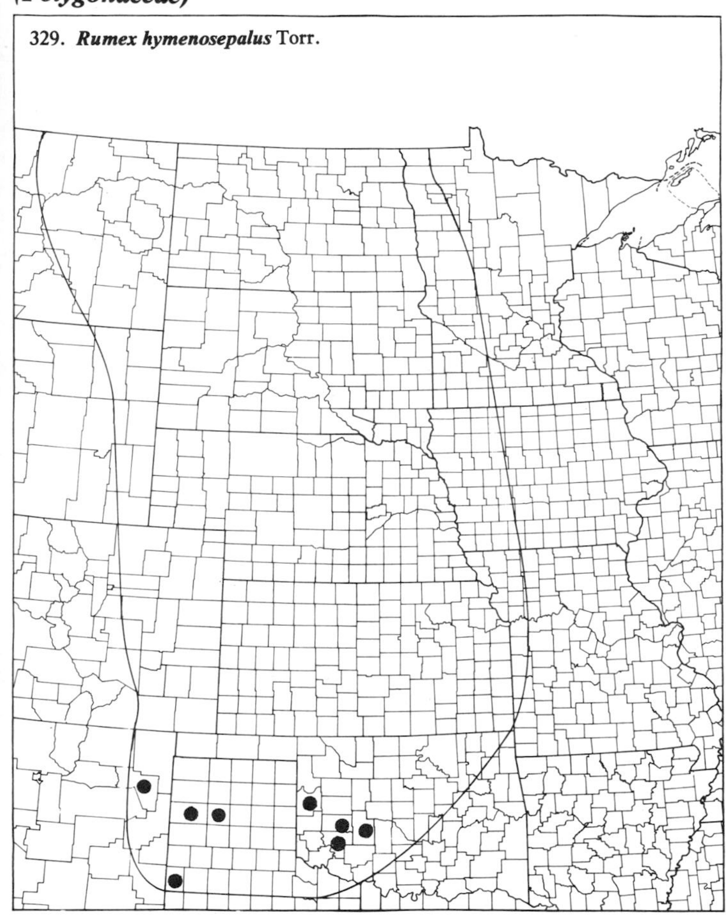

330. ***Rumex maritimus*** L. var. ***fueginus*** (Phil.) Dusen — Golden Dock

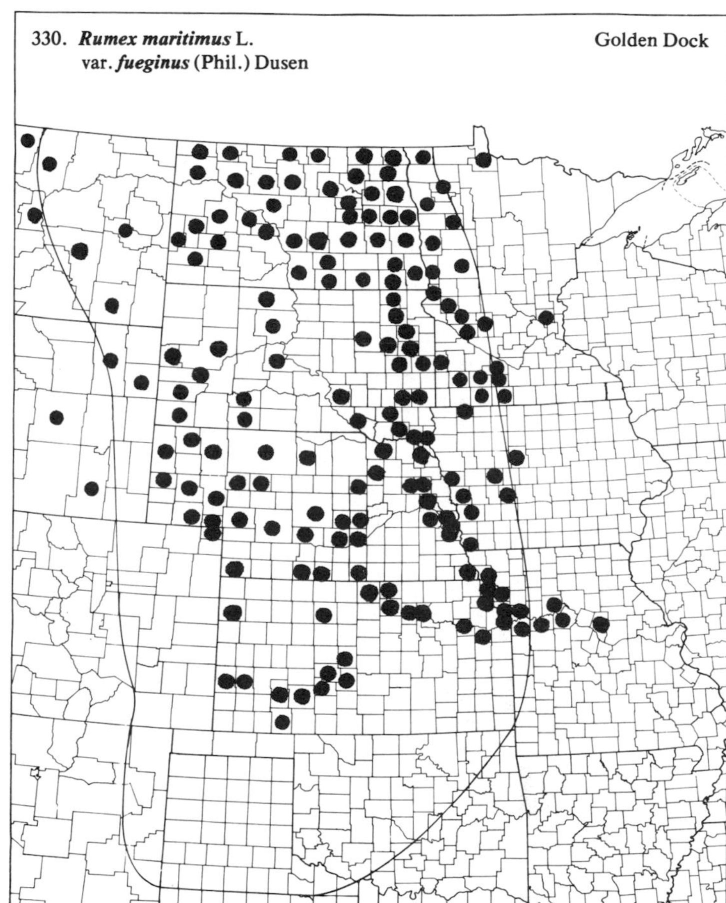

331. ***Rumex mexicanus*** Meisn. — Willow-leaved Dock

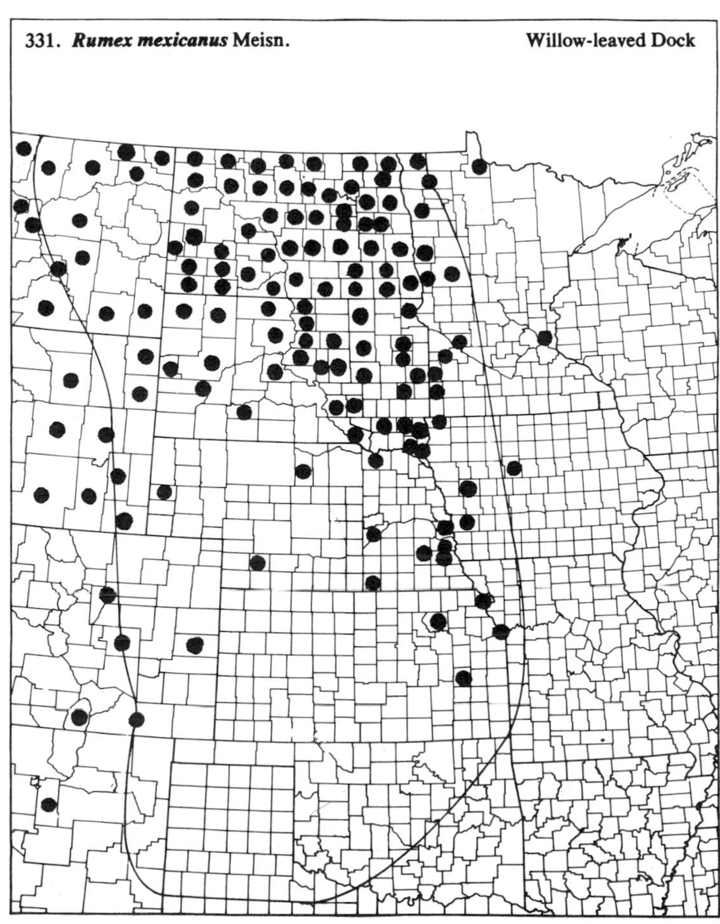

332. ***Rumex obtusifolius*** L. — Bitter Dock

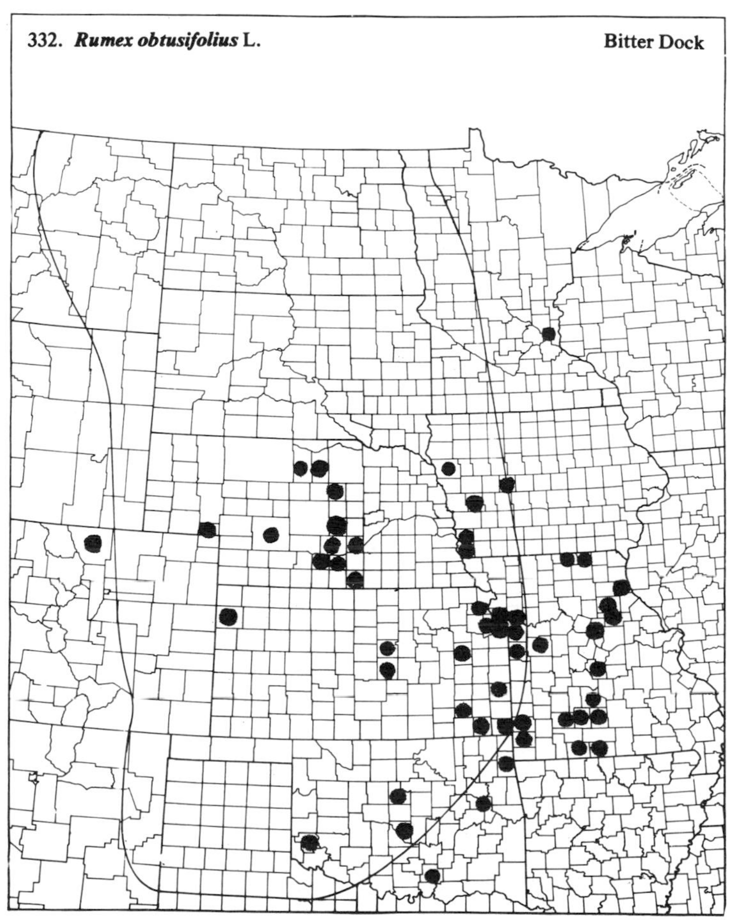

(Polygonaceae)

333. **Rumex occidentalis** Wats. — Western Dock

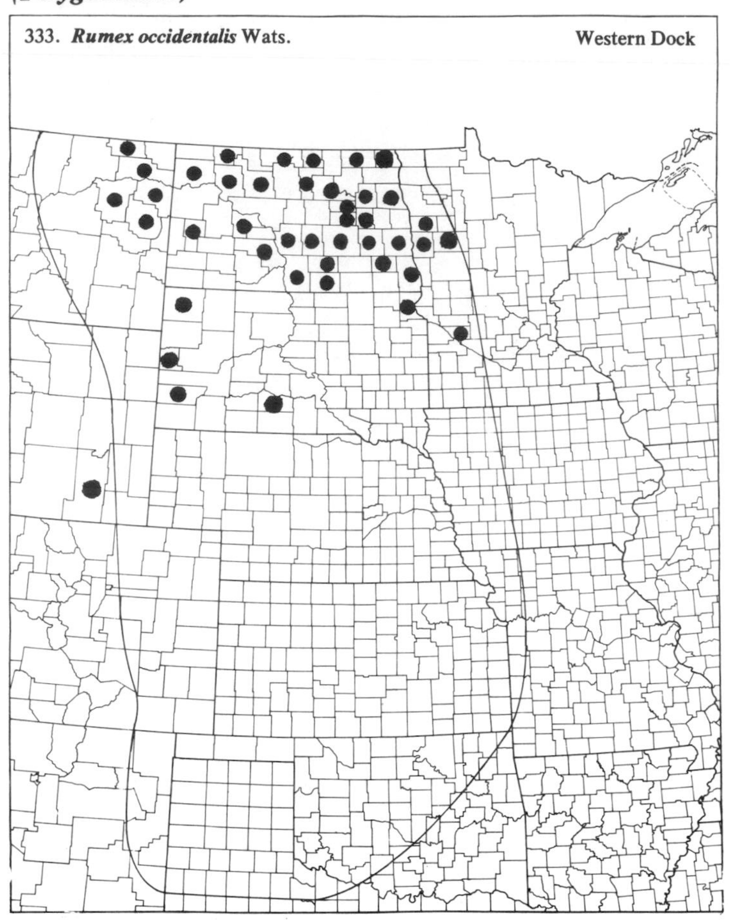

334. **Rumex orbiculatus** Gray — Great Water Dock

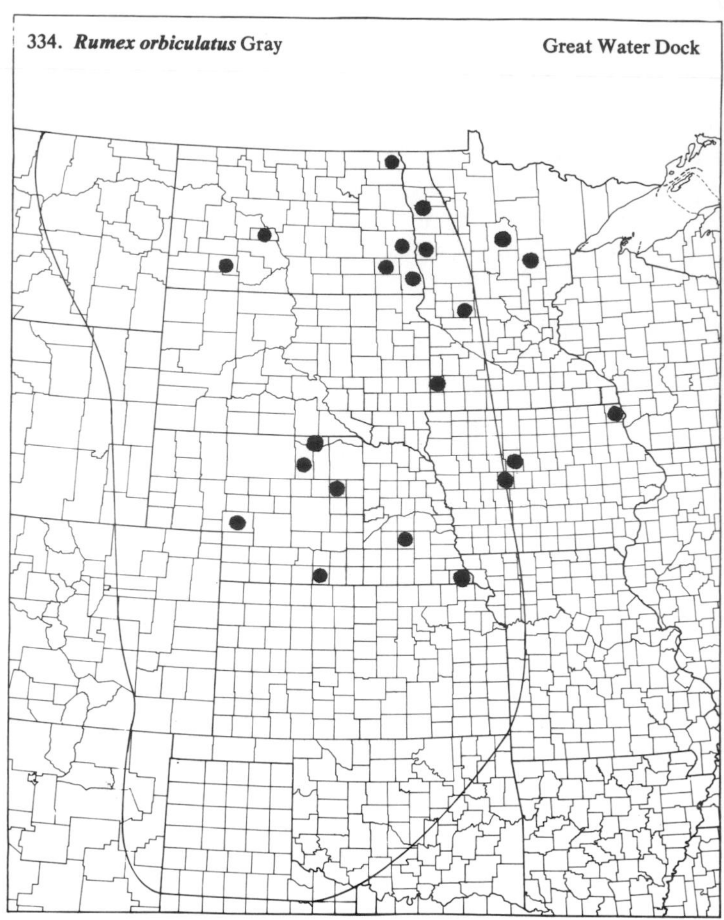

335. **Rumex patientia** L. — Patience Dock

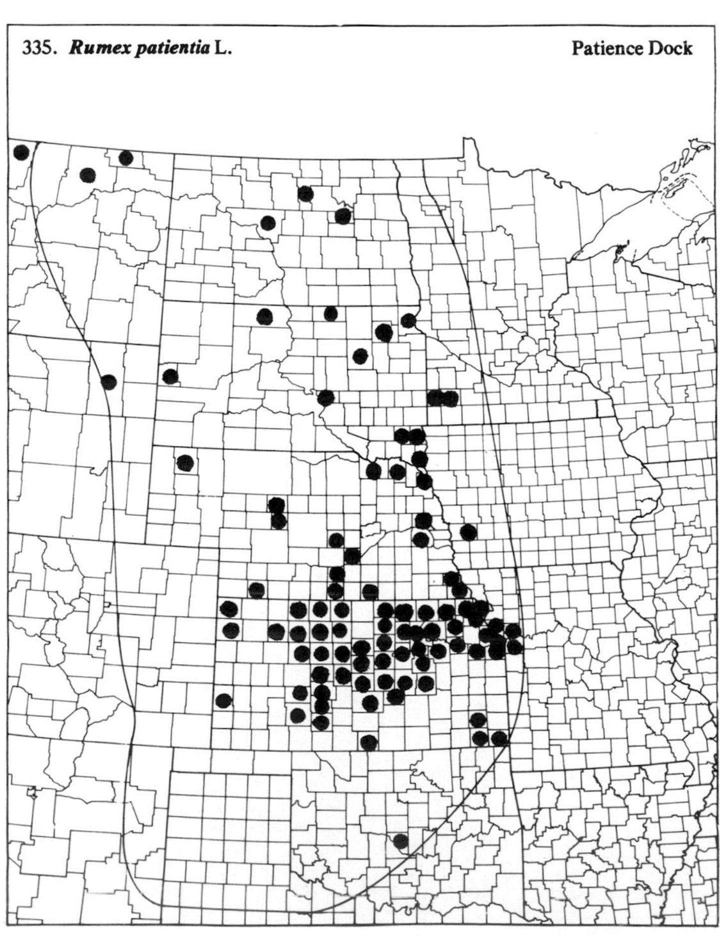

336. **Rumex stenophyllus** Ledeb.

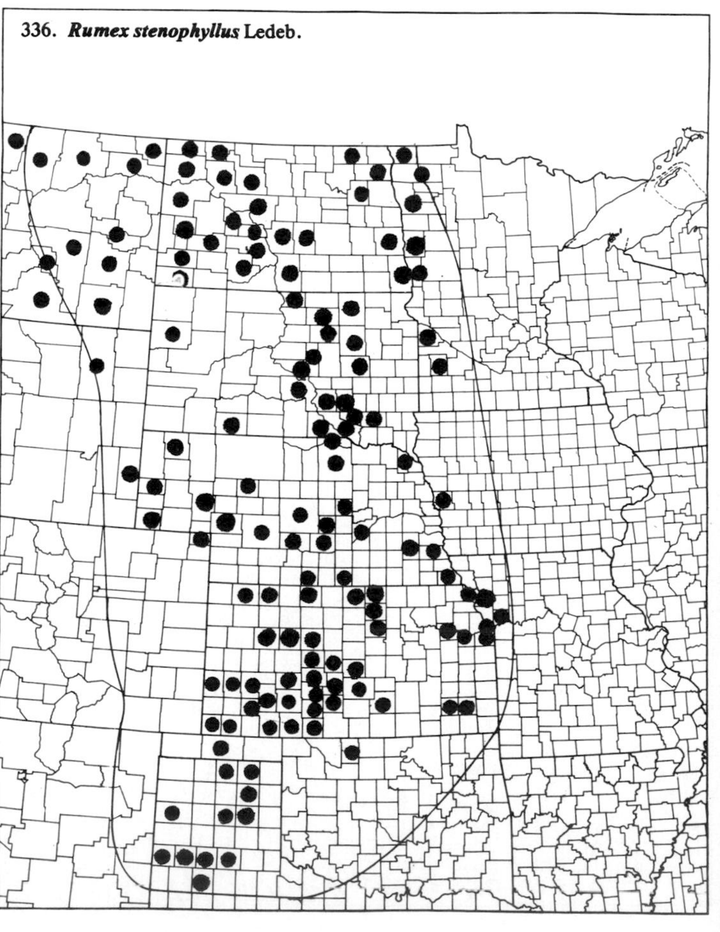

(Polygonaceae)

337. ***Rumex venosus*** Pursh — Veined Dock

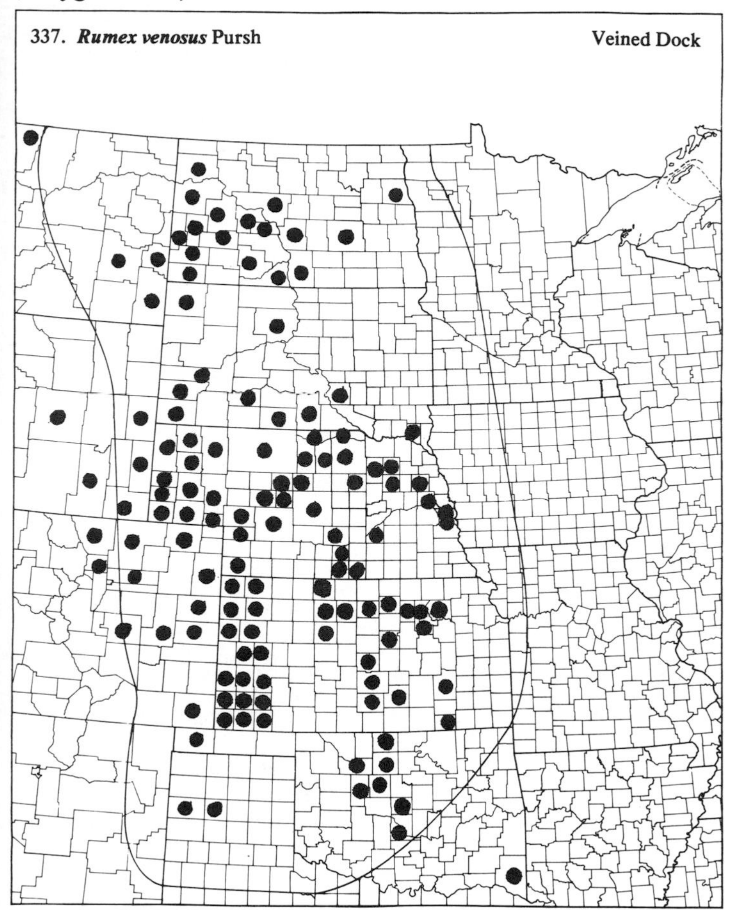

338. ***Rumex verticillatus*** L. — Water Dock

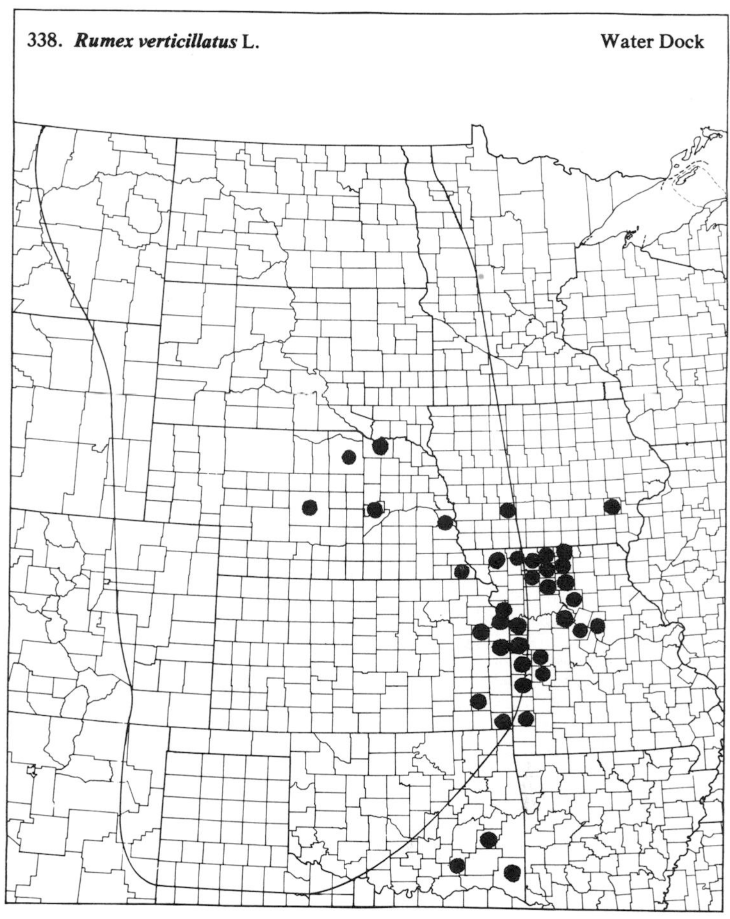

30. *Elatinaceae* (Waterwort Family)

339. ***Bergia texana*** (Hook.) Seubert

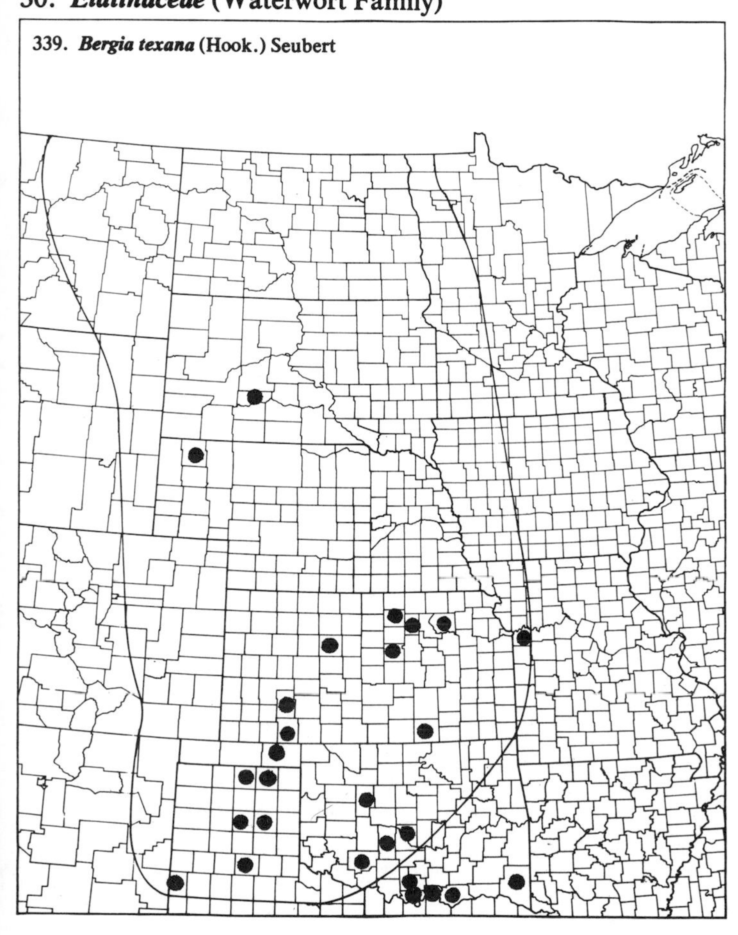

340. ***Elatine triandra*** Schkuhr — Waterwort

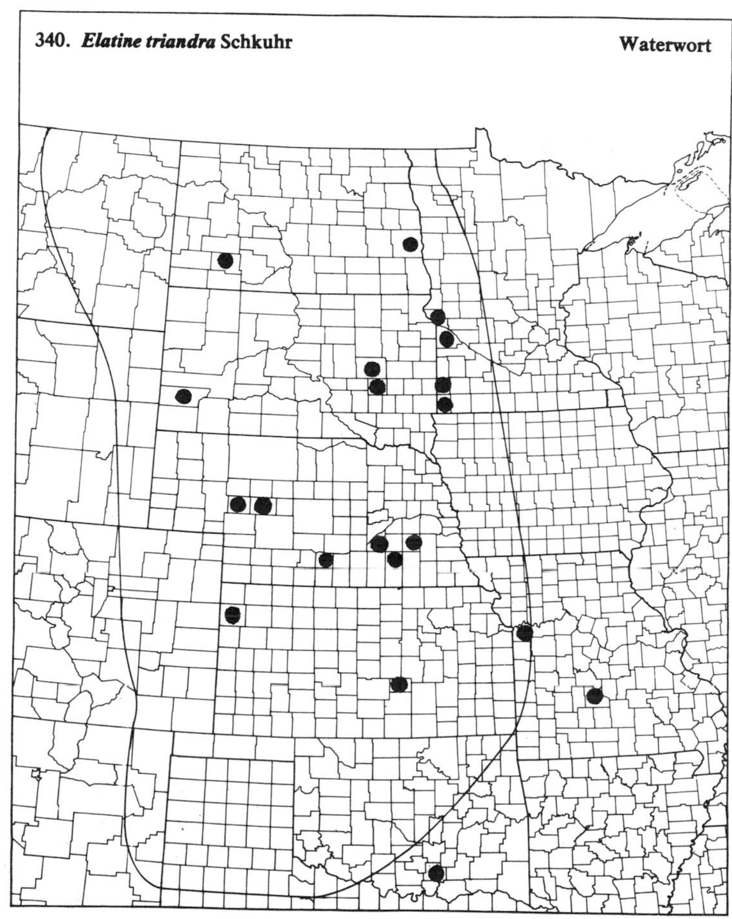

31. *Hypericaceae* (St. John's-wort Family)

341. *Ascyrum hypericoides* L. var. *multicaule* (Michx.) Fern. — St. Andrew's Cross

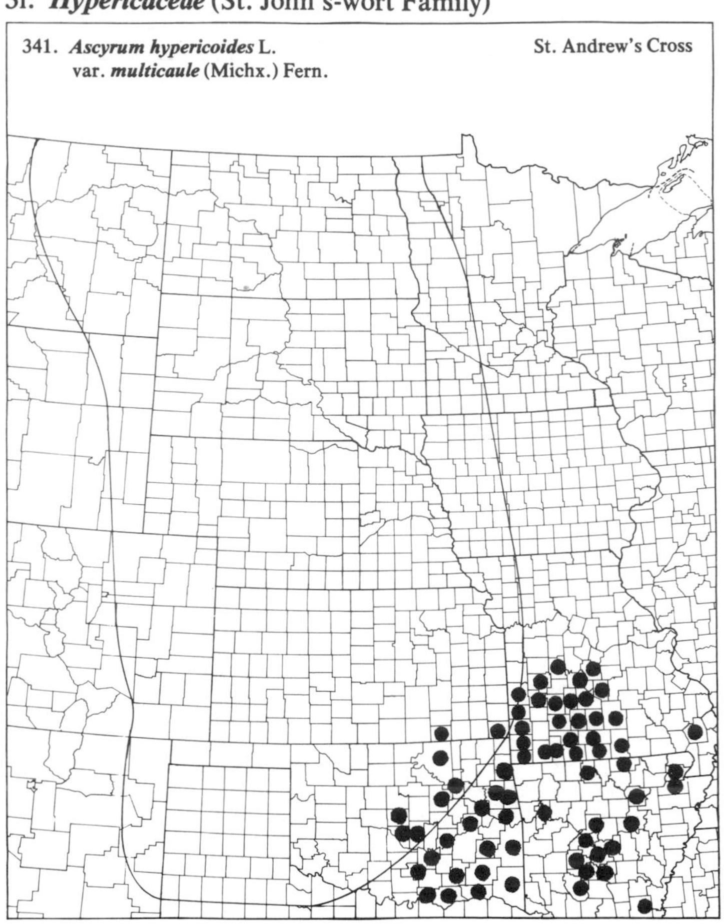

342. *Hypericum drummondii* (Grev. & Hook.) T. & G. — Nits-and-lice

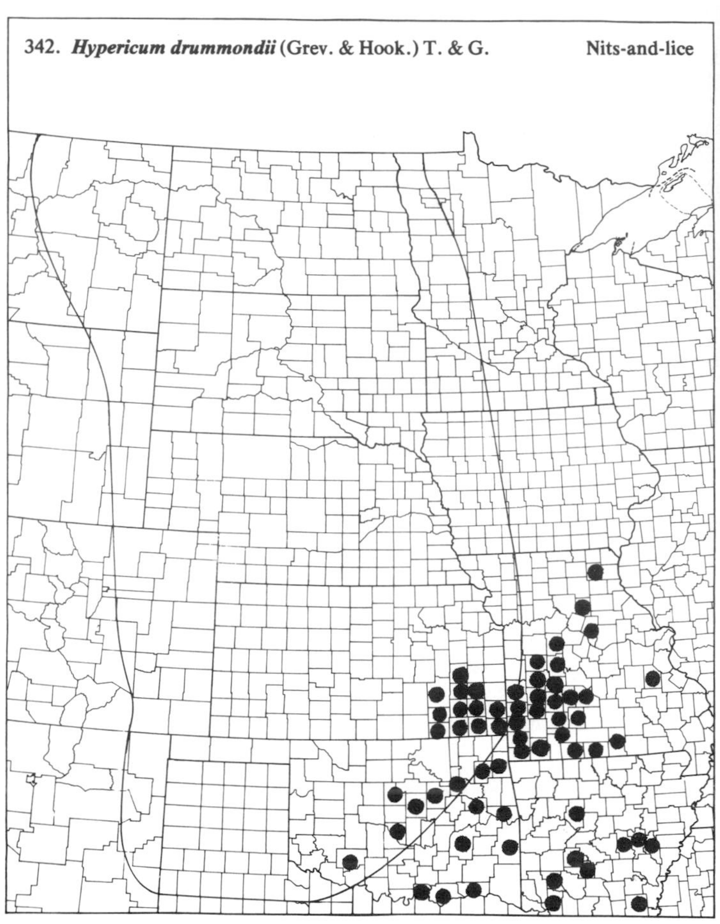

343. *Hypericum majus* (Gray) Britt. — Greater St. John's-wort

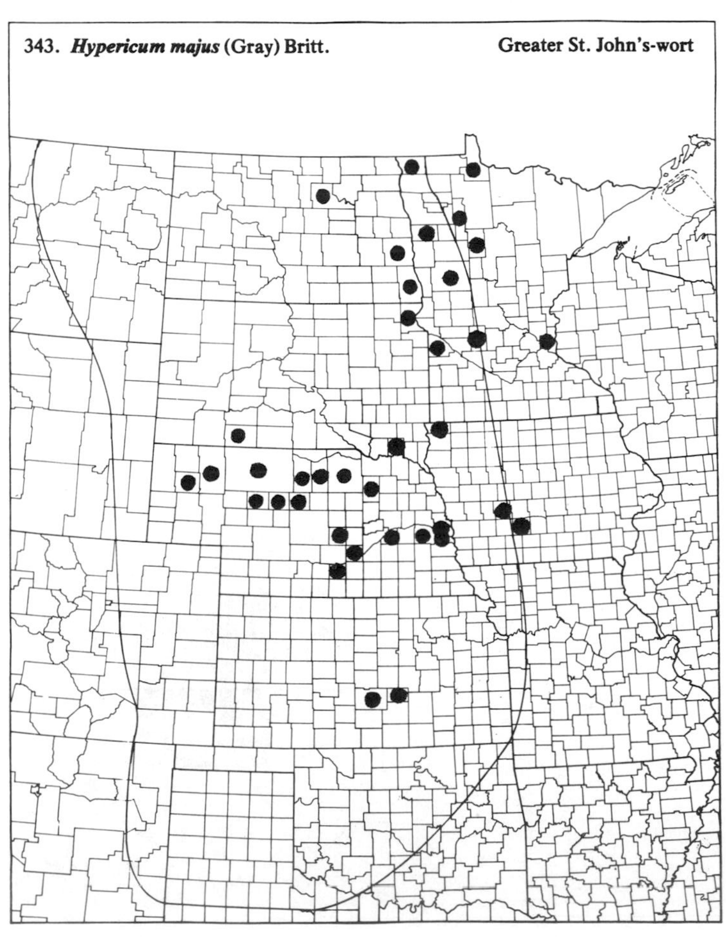

344. *Hypericum mutilum* L. — Dwarf St. John's-wort

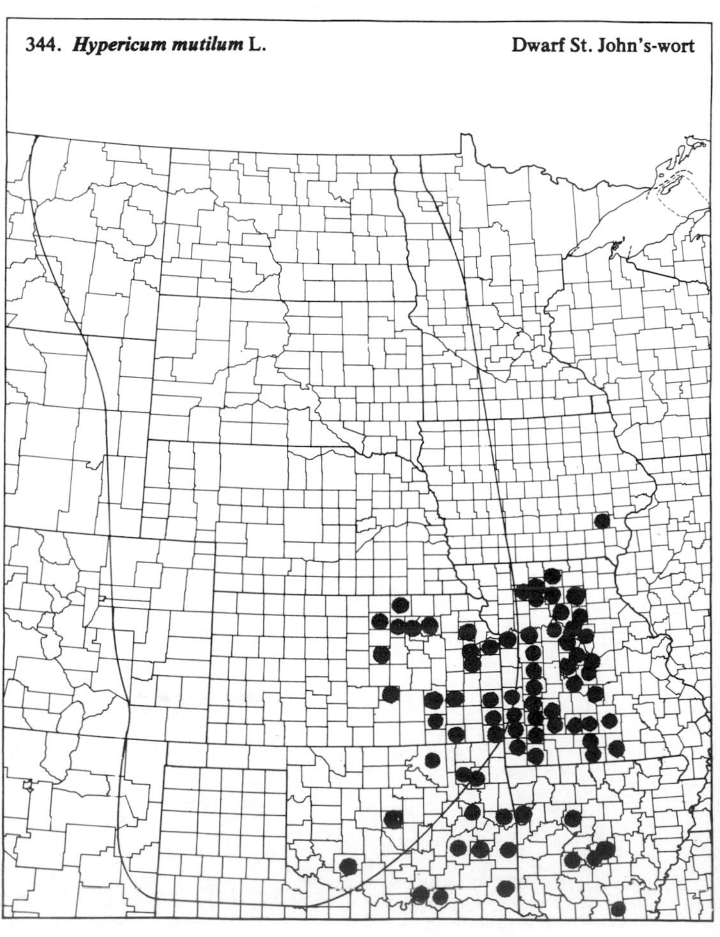

(Hypericaceae)

345. ***Hypericum perforatum*** L. Common St. John's-wort

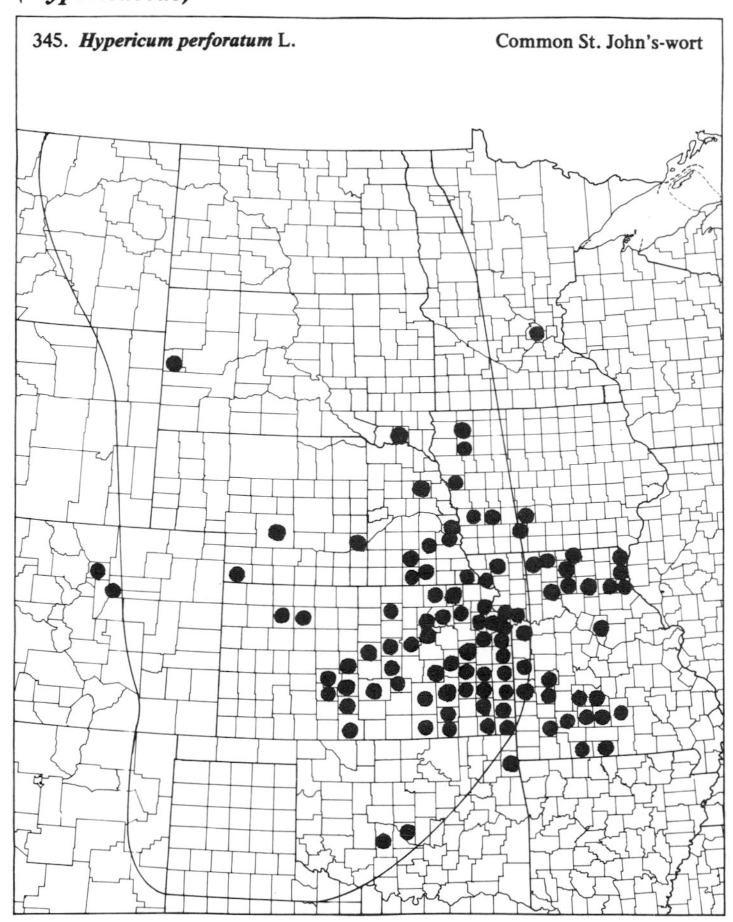

346. ***Hypericum punctatum*** Lam. Spotted St. John's-wort

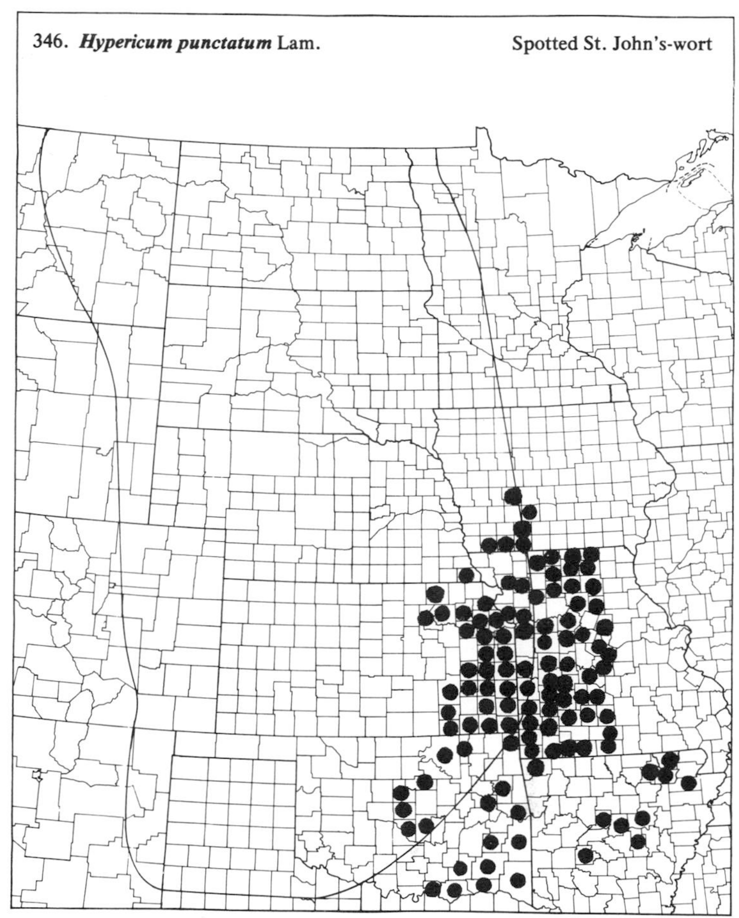

347. ***Hypericum sphaerocarpum*** Michx. Roundfruit St. John's-wort

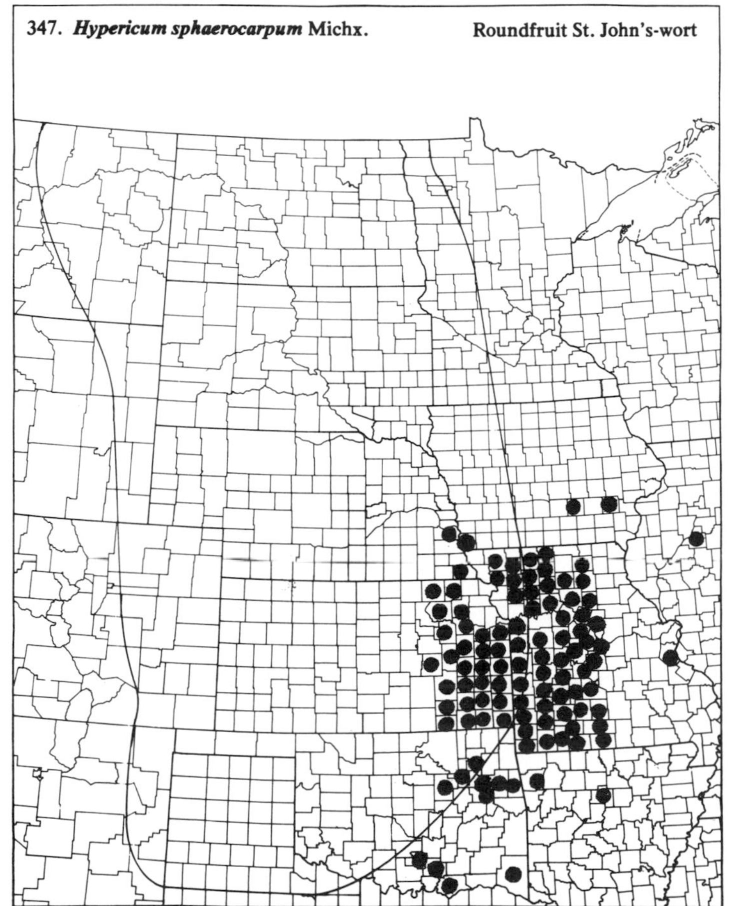

32. *Tiliaceae* (Linden Family)

348. ***Tilia americana*** L. American Linden

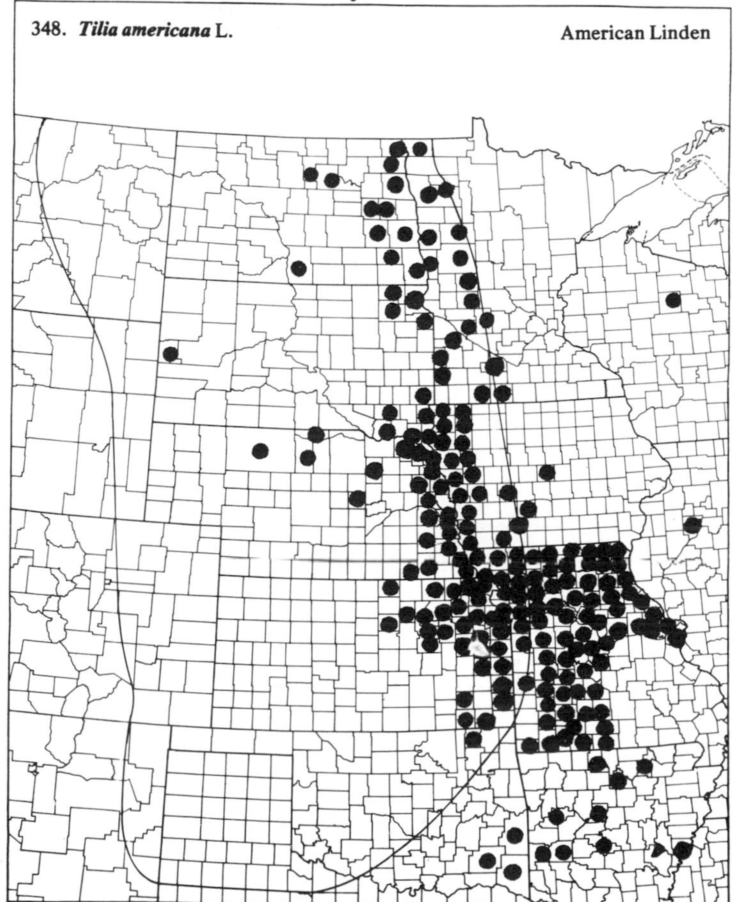

33. *Malvaceae* (Mallow Family)

349. ***Abutilon theophrasti*** Medic. Velvet-leaf

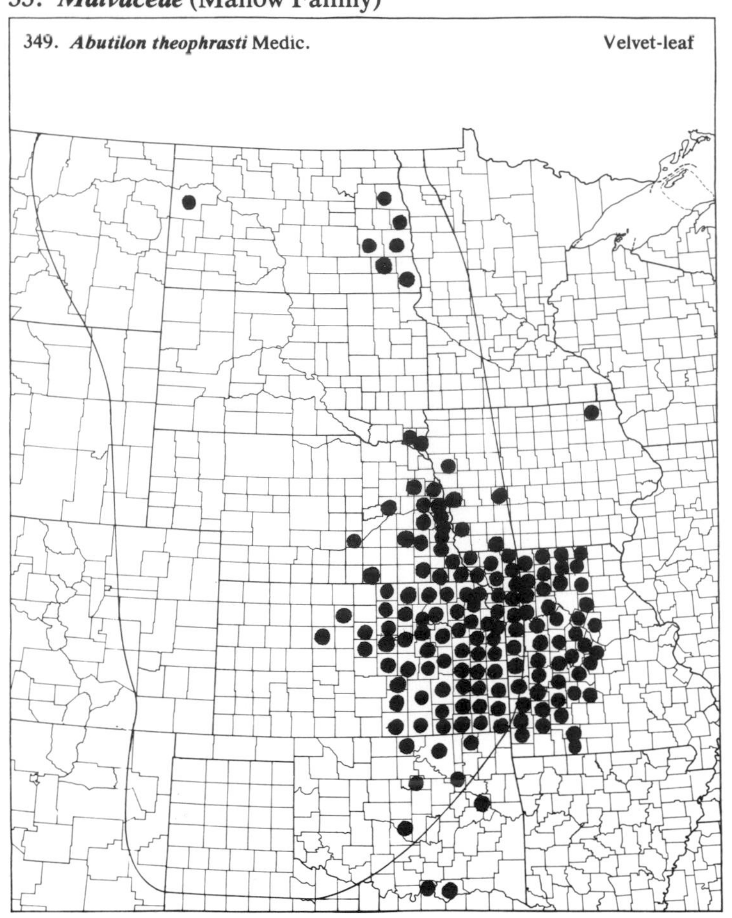

350. ***Althaea rosea*** (L.) Cav. Hollyhock

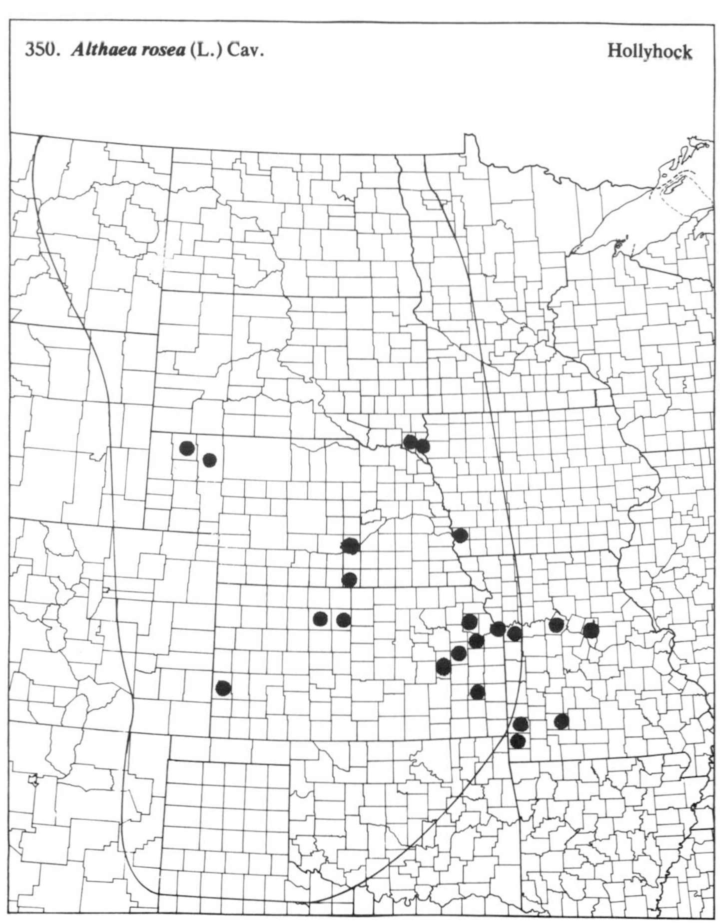

351. ***Callirhoe alcaeoides*** (Michx.) Gray Pink Poppy Mallow

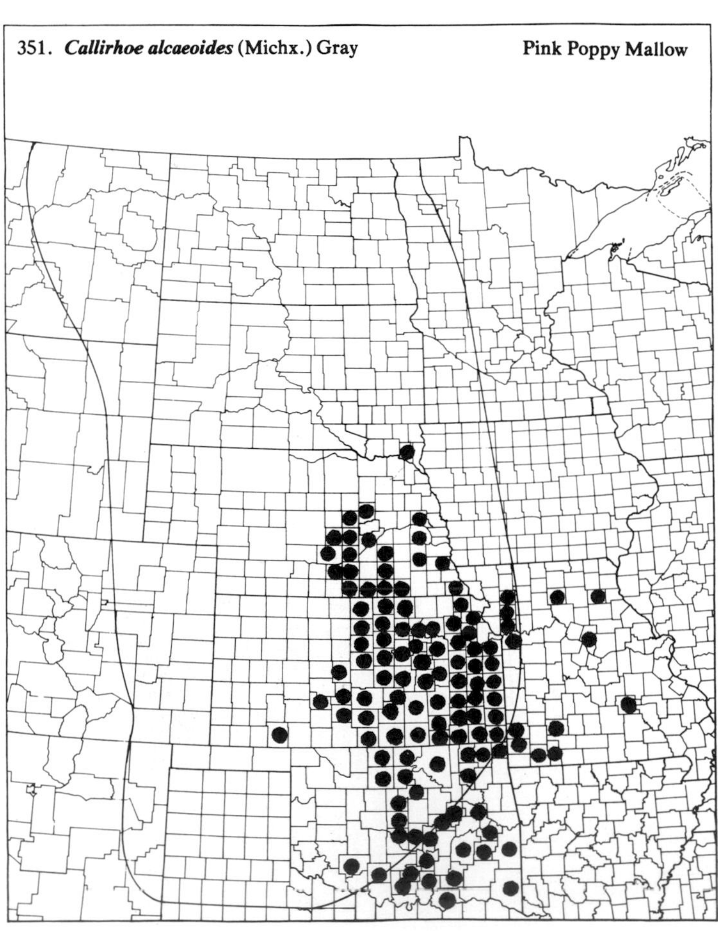

352. ***Callirhoe involucrata*** (T. & G.) Gray Purple Poppy Mallow

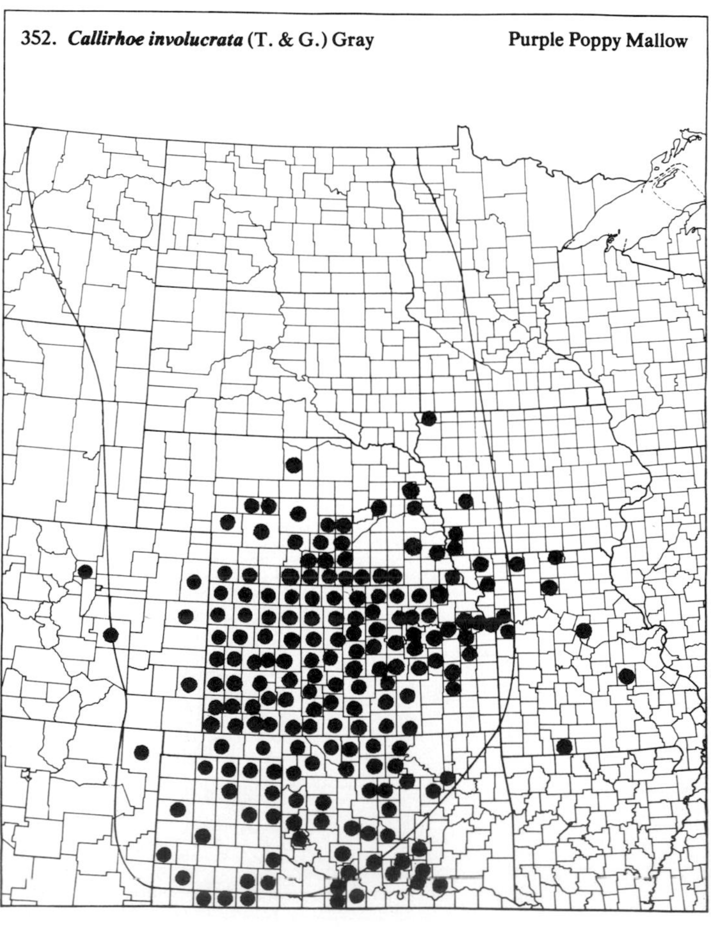

(Malvaceae)

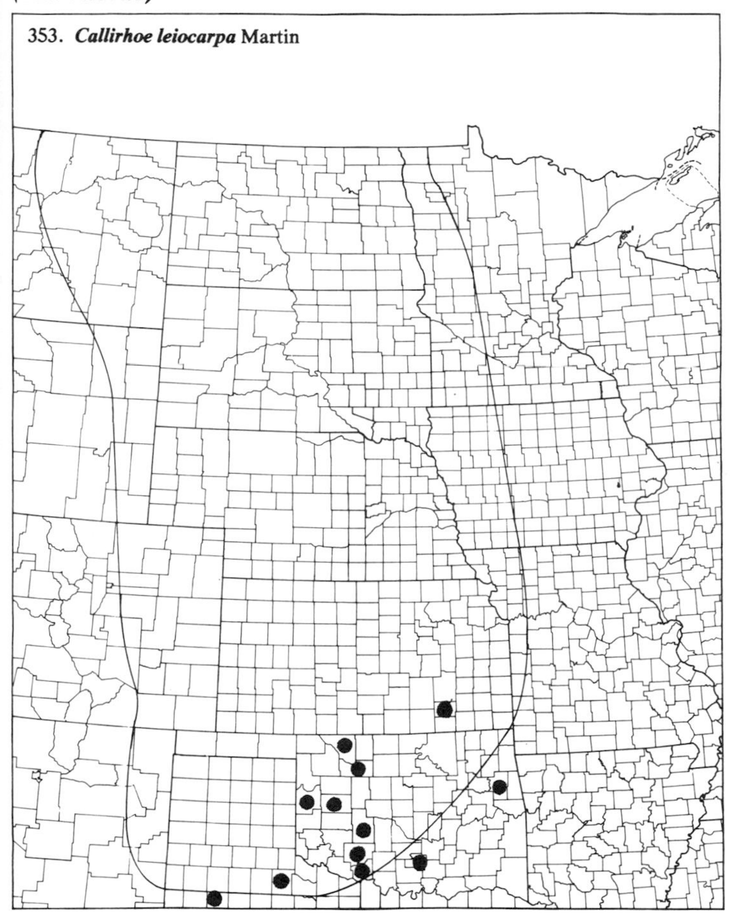
353. **Callirhoe leiocarpa** Martin

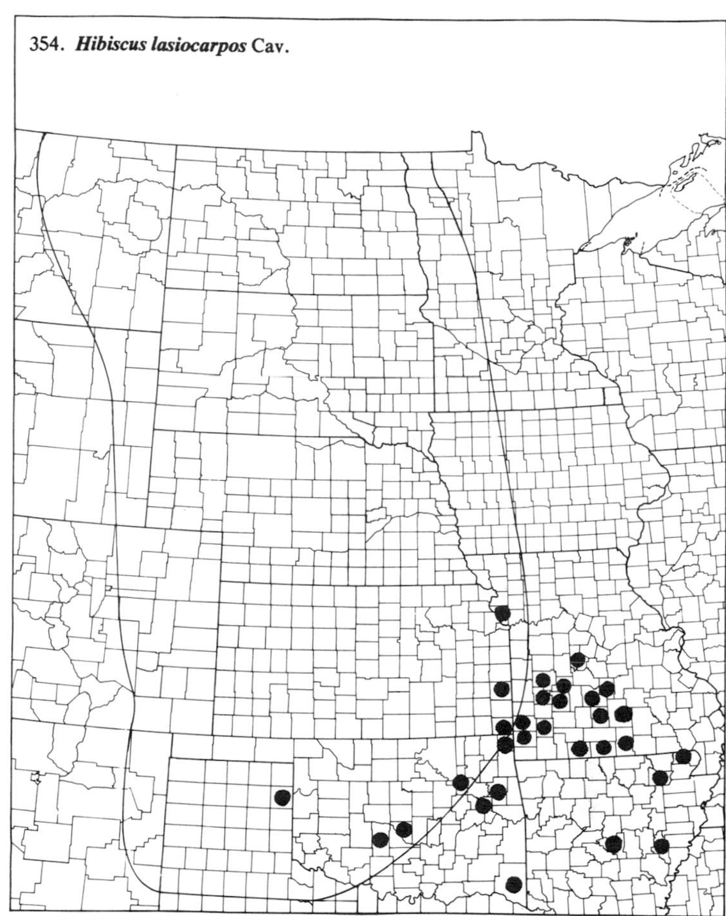
354. **Hibiscus lasiocarpos** Cav.

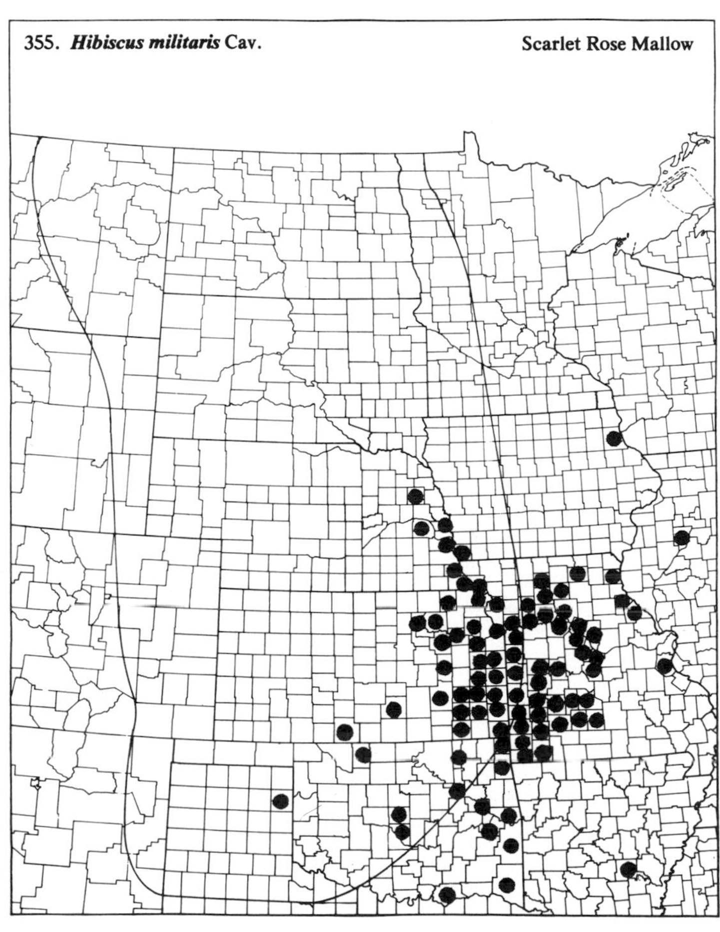
355. **Hibiscus militaris** Cav.
Scarlet Rose Mallow

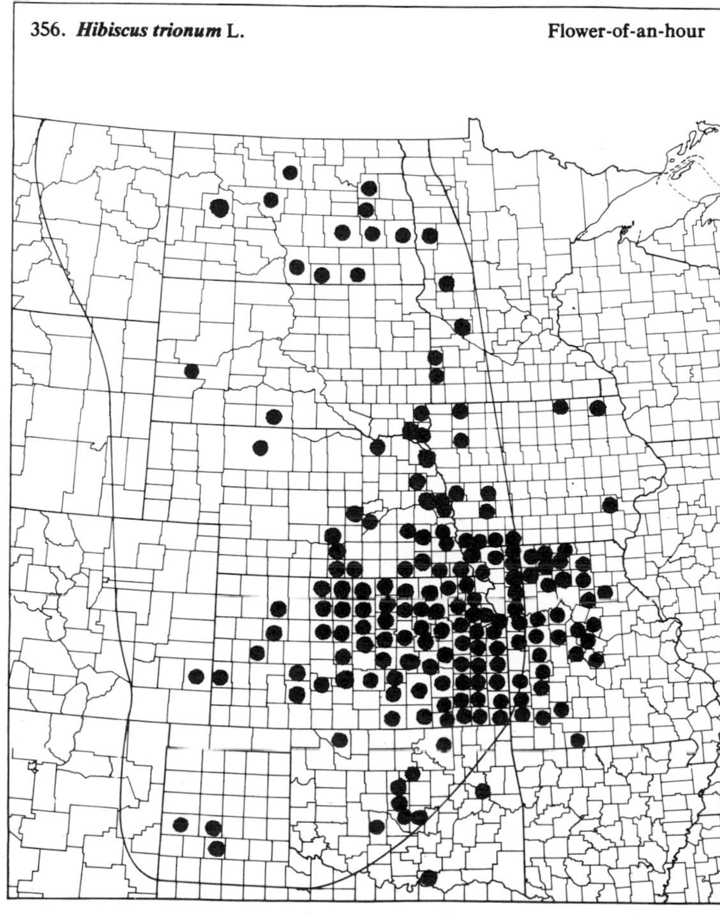
356. **Hibiscus trionum** L.
Flower-of-an-hour

(Malvaceae)

357. ***Malva neglecta*** Wallr. Common Mallow

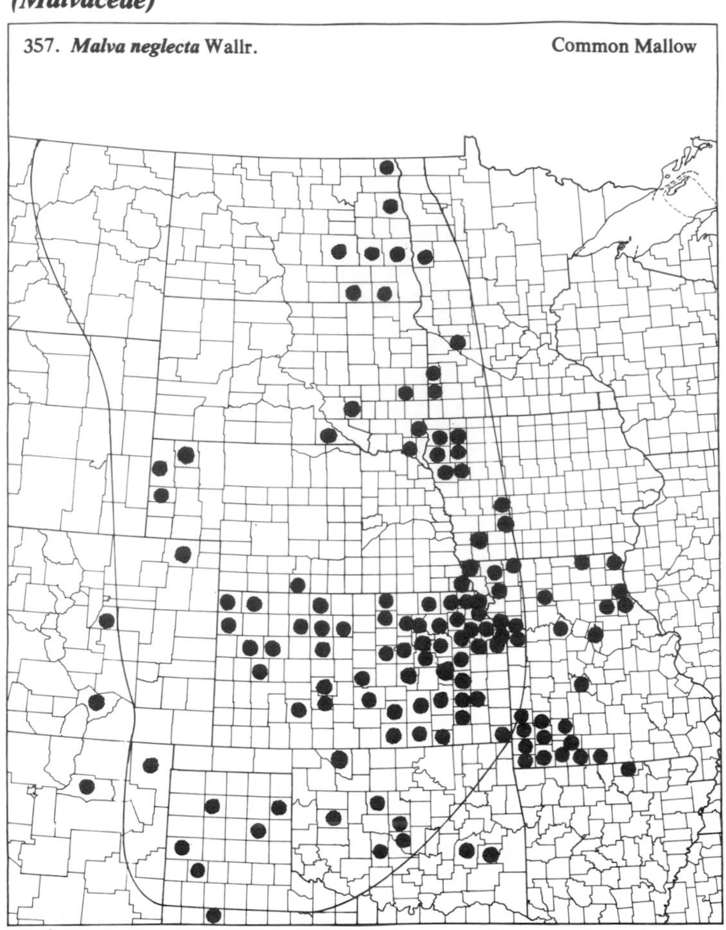

358. ***Malva parviflora*** L. Small-fruited Mallow

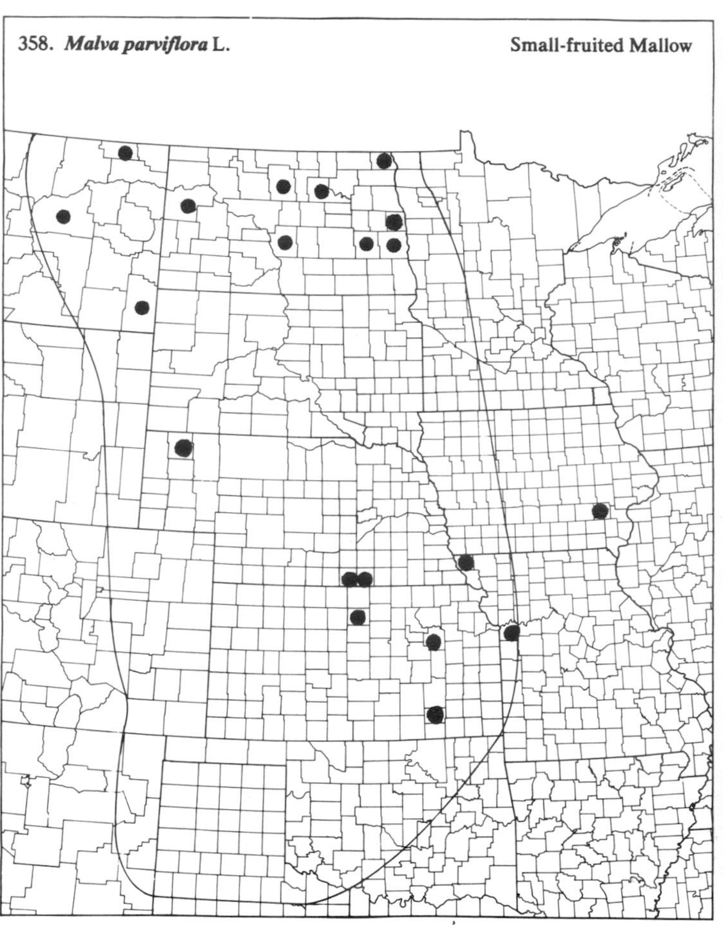

359. ***Malva rotundifolia*** L. Common Mallow

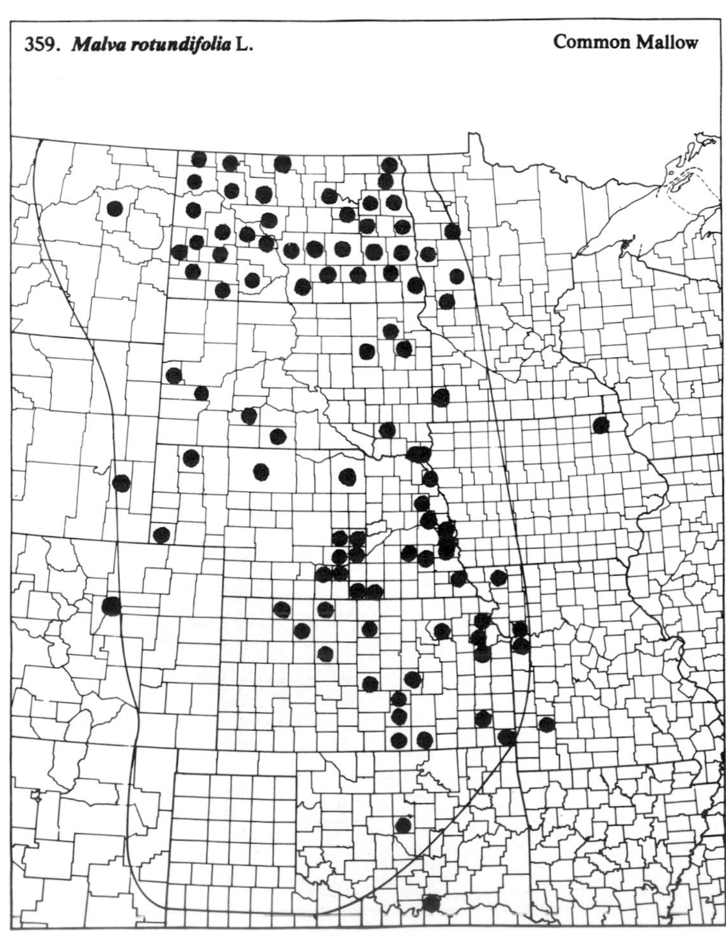

360. ***Malva sylvestris*** L. High Mallow

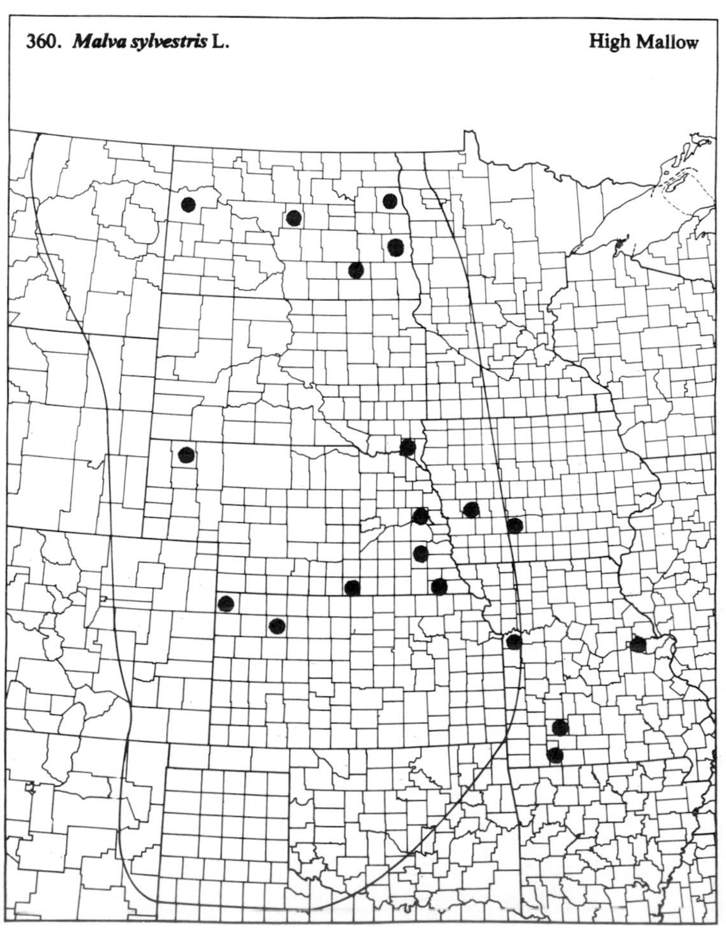

(Malvaceae)

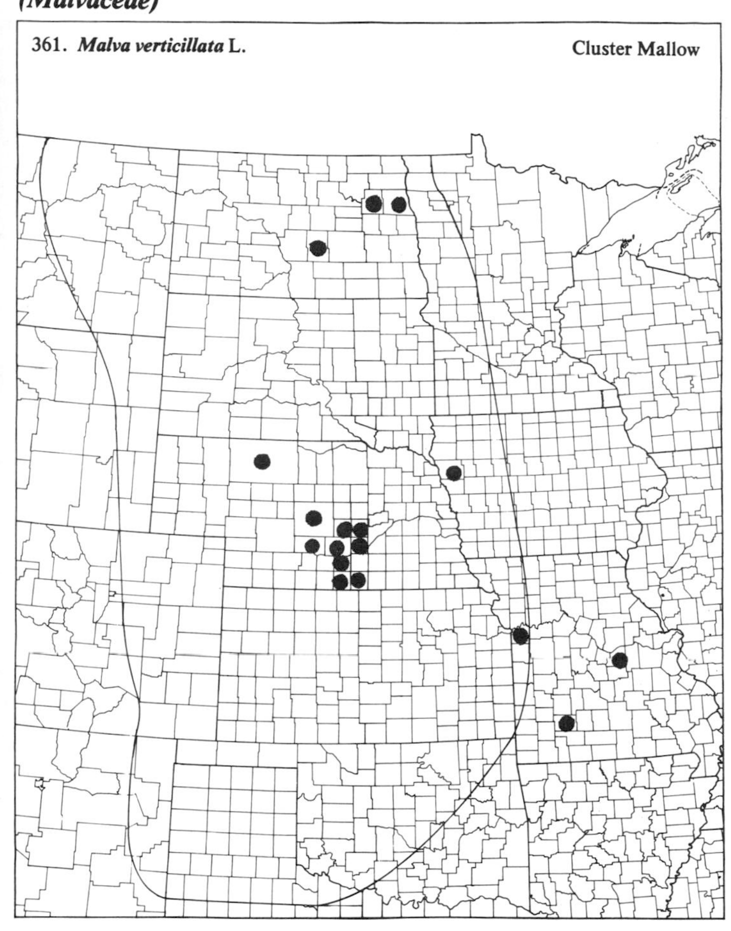

361. ***Malva verticillata*** L. Cluster Mallow

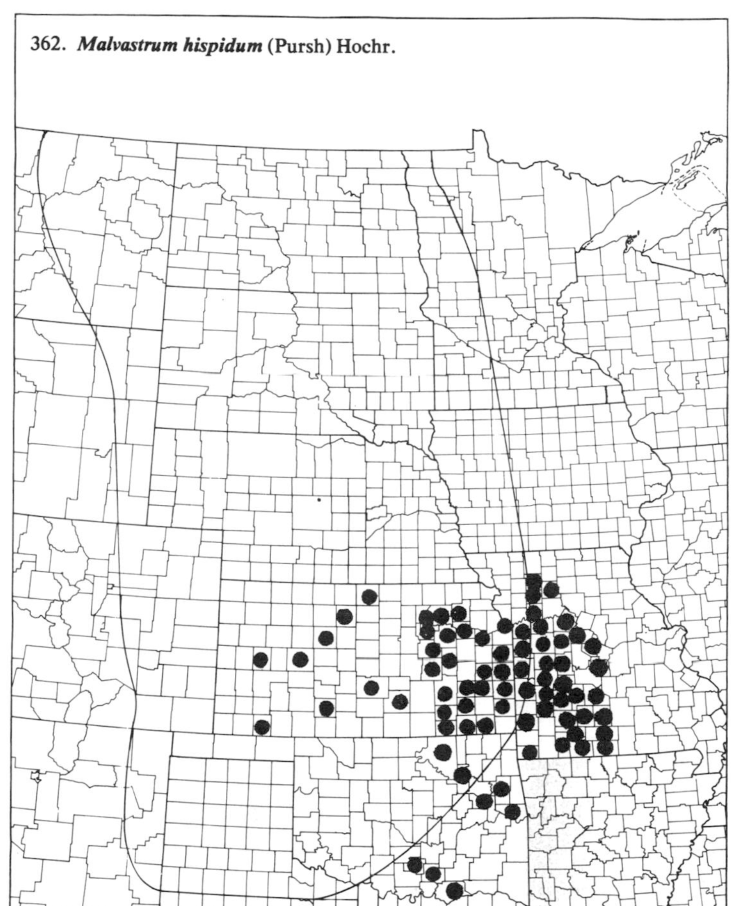

362. ***Malvastrum hispidum*** (Pursh) Hochr.

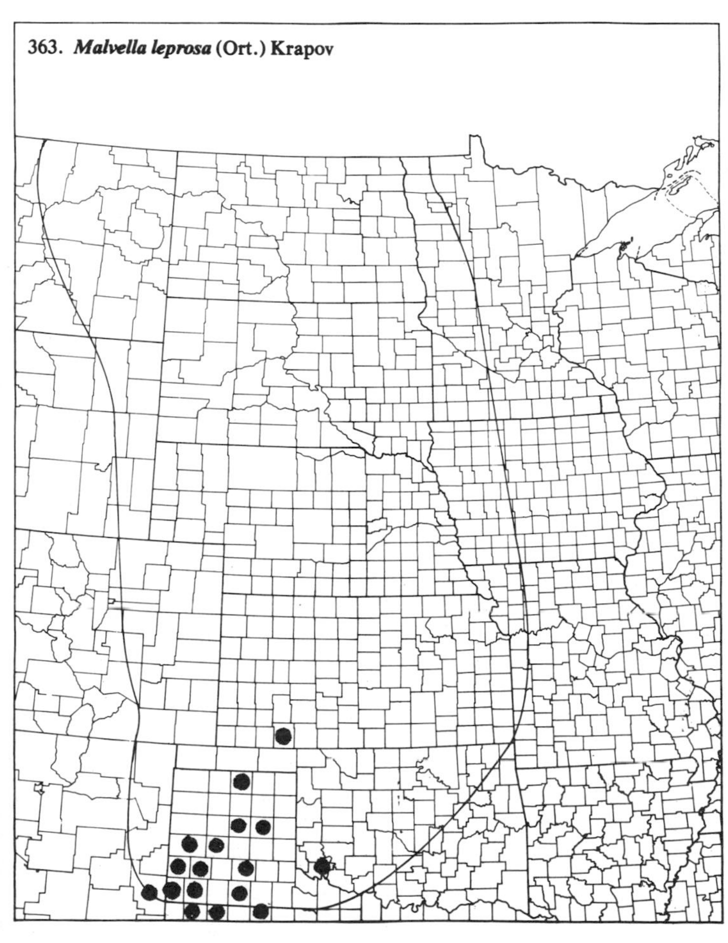

363. ***Malvella leprosa*** (Ort.) Krapov

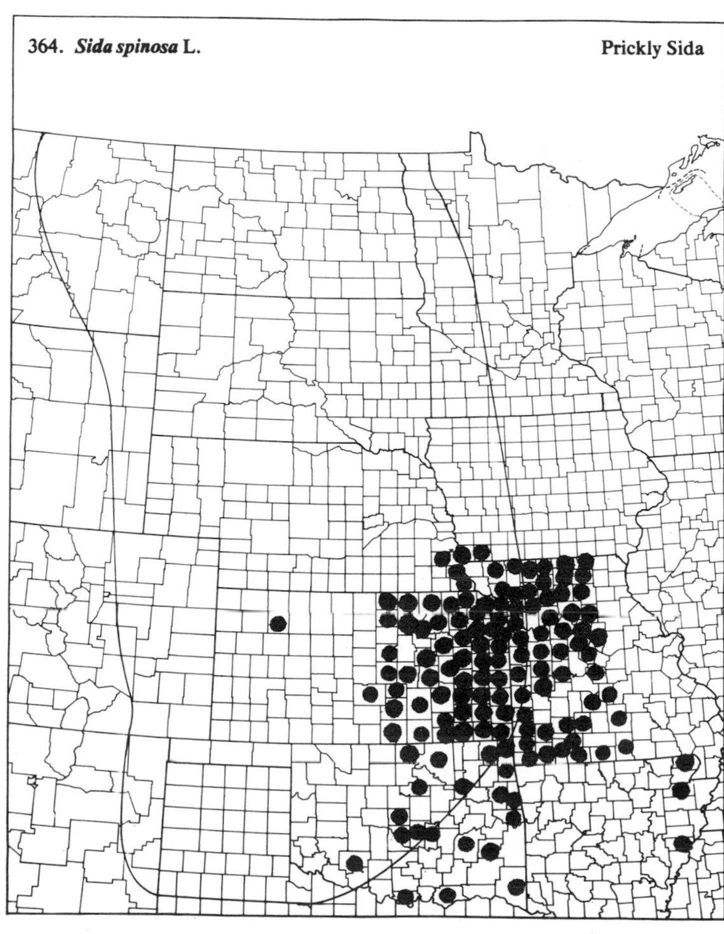

364. ***Sida spinosa*** L. Prickly Sida

(Malvaceae)

365. ***Sphaeralcea angustifolia*** (Cav.) D. Don var. ***cuspidata*** Gray — Narrowleaf Globe Mallow

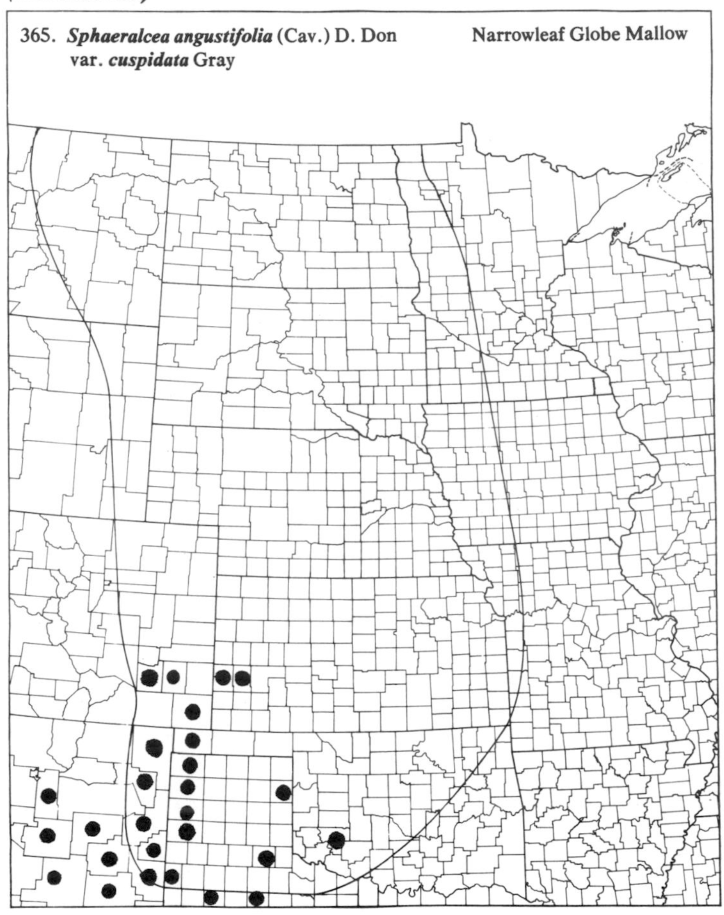

366. ***Sphaeralcea coccinea*** (Pursh) Rydb. — Red False Mallow

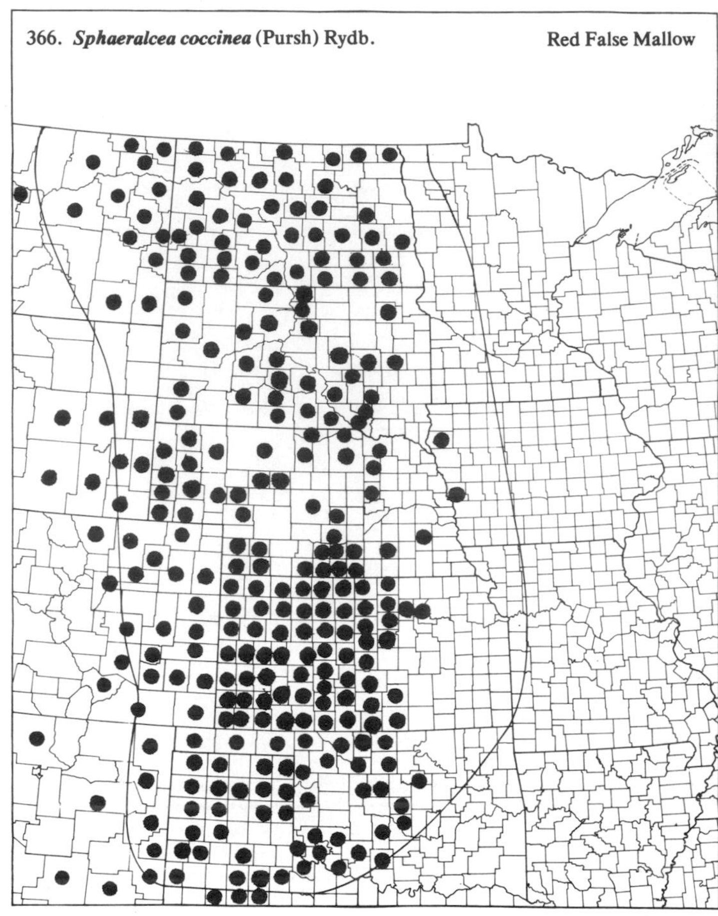

35. *Violaceae* (Violet Family)

367. ***Hybanthus concolor*** (T.F. Forst.) Spreng. — Green Violet

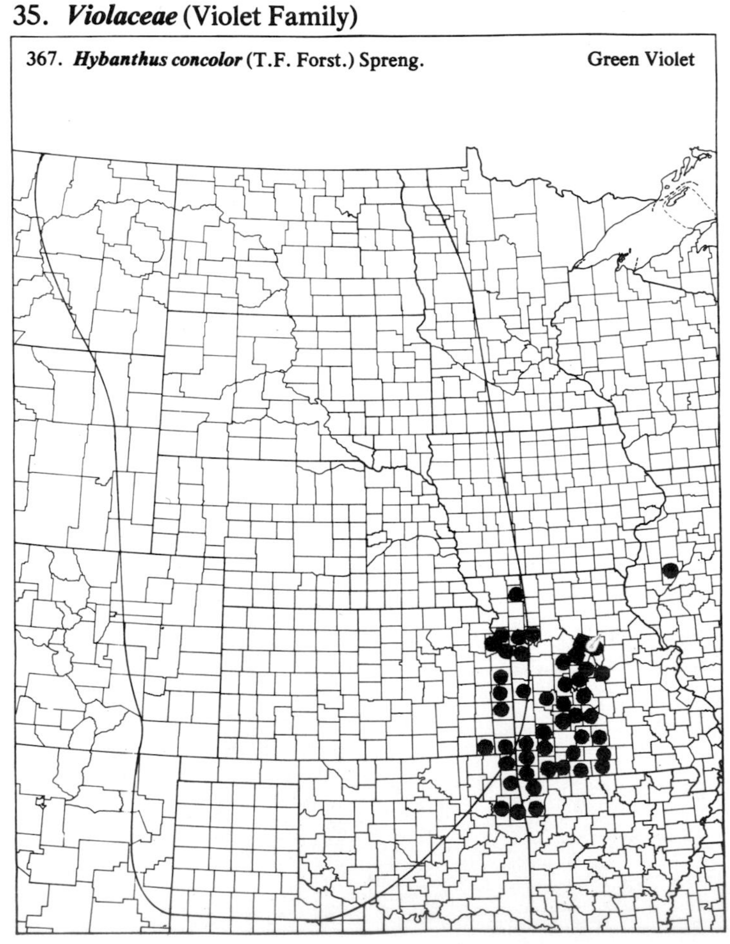

368. ***Hybanthus verticillatus*** (Ort.) Baill. — North American Calceolaria

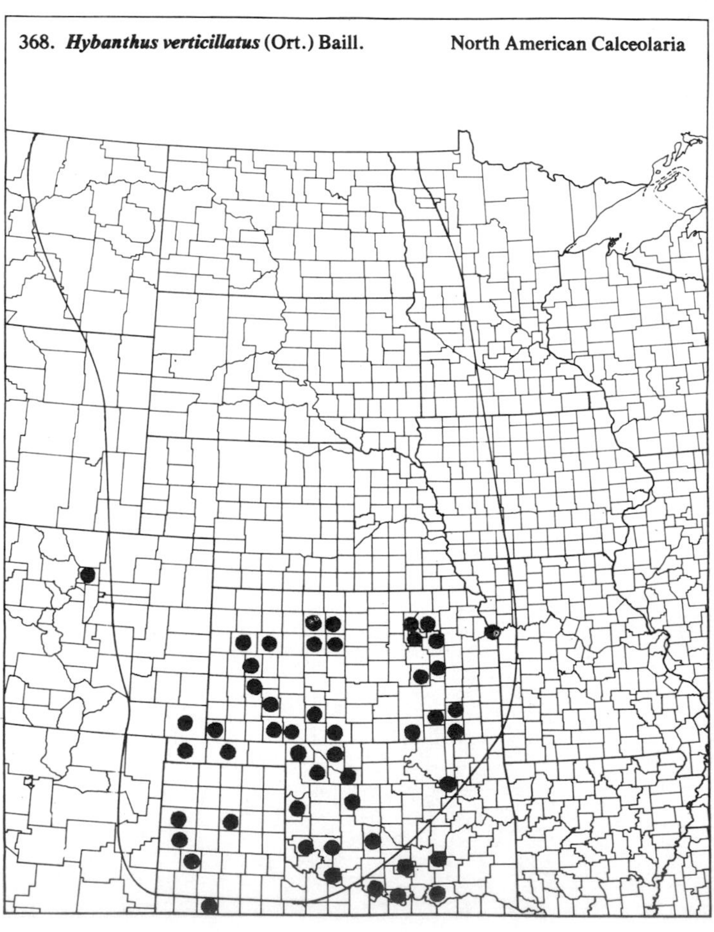

(Violaceae)

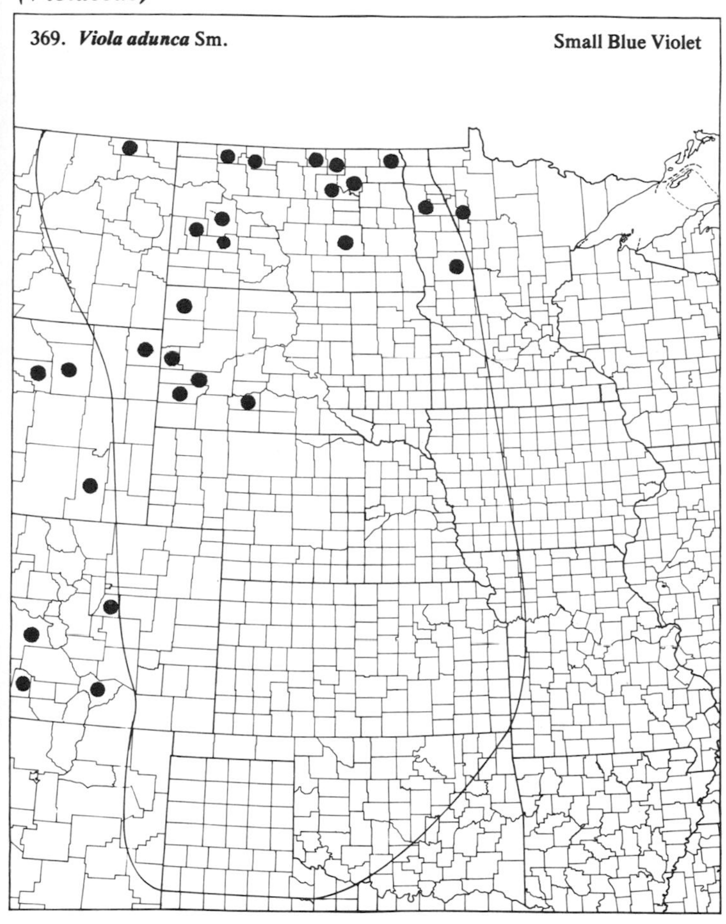
369. *Viola adunca* Sm.
Small Blue Violet

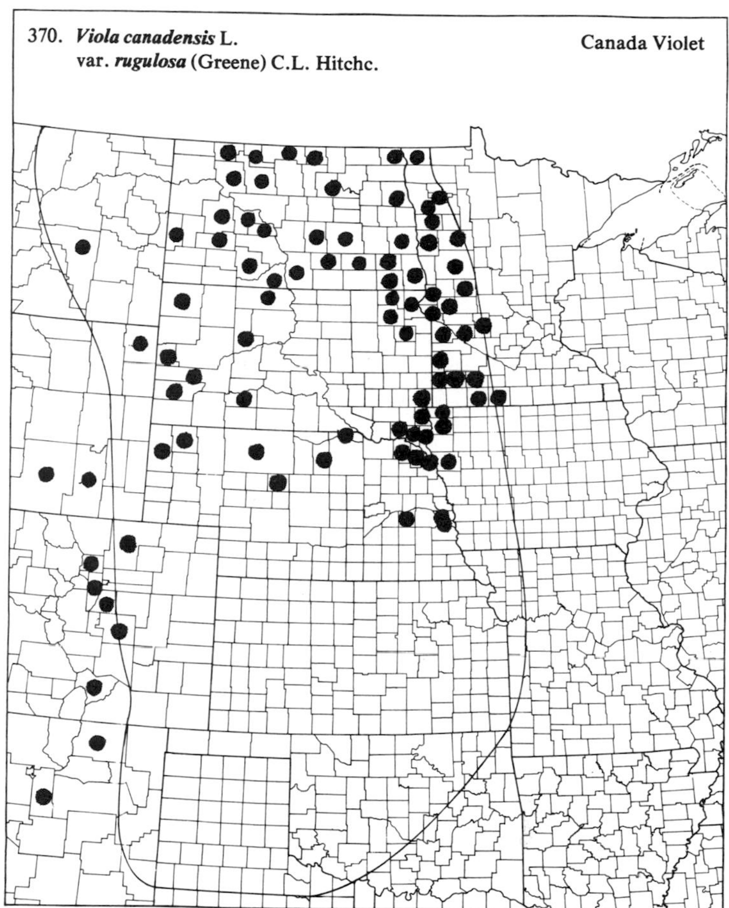
370. *Viola canadensis* L.
var. *rugulosa* (Greene) C.L. Hitchc.
Canada Violet

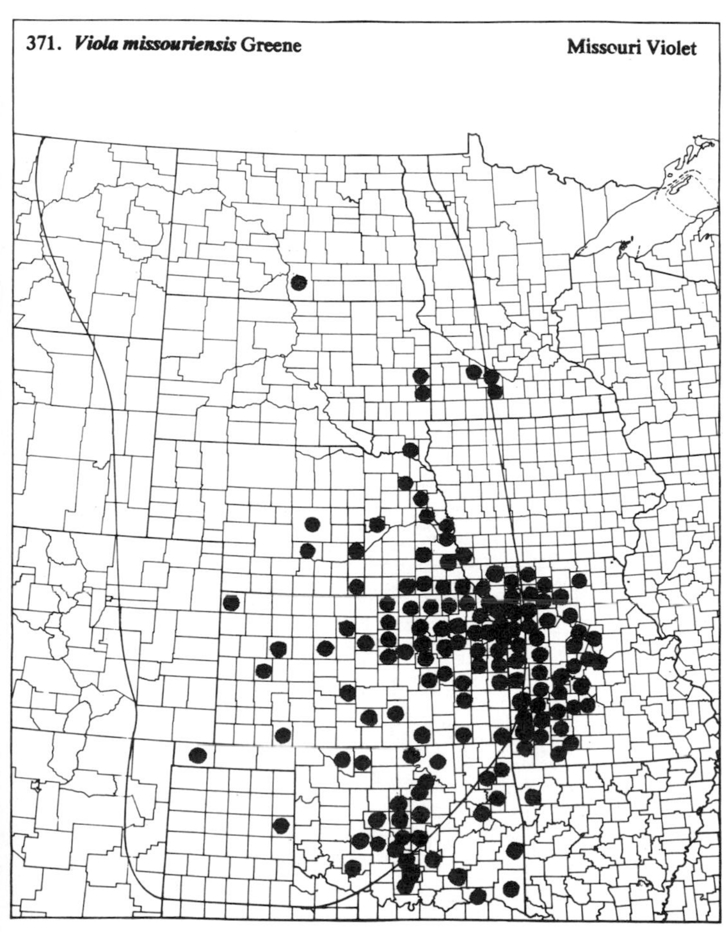
371. *Viola missouriensis* Greene
Missouri Violet

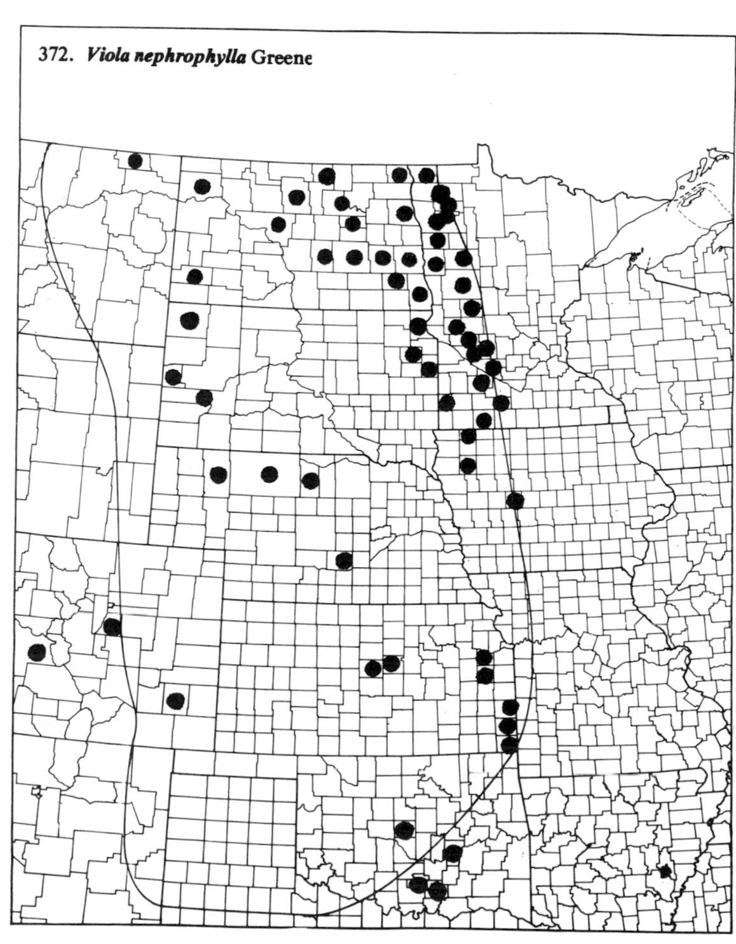
372. *Viola nephrophylla* Greene

(Violaceae)

373. ***Viola nuttallii*** Pursh — Yellow Prairie Violet

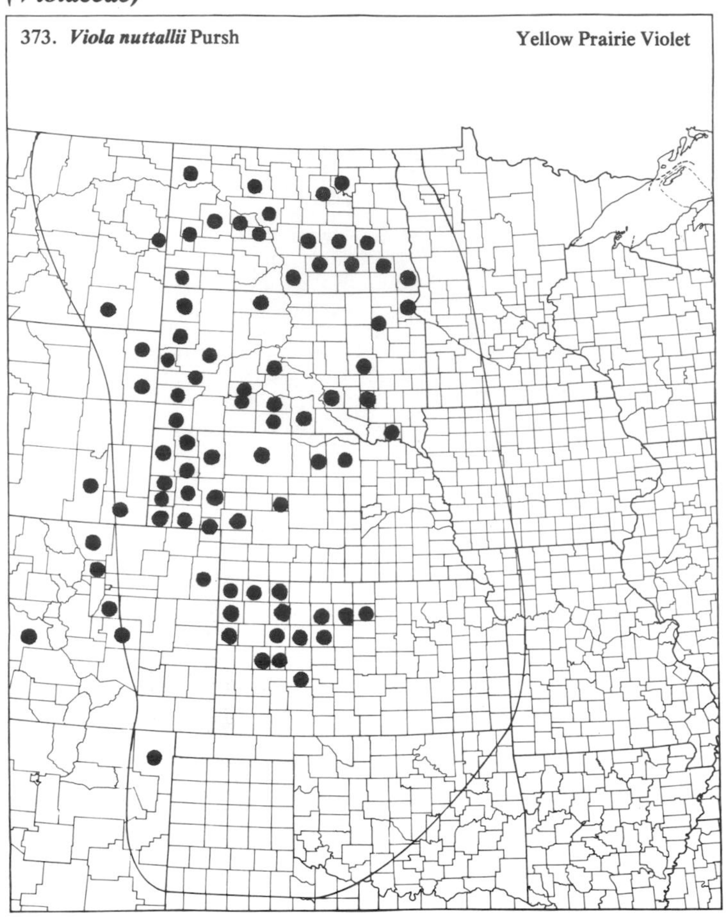

374. ***Viola pedata*** L. — Pansy Violet

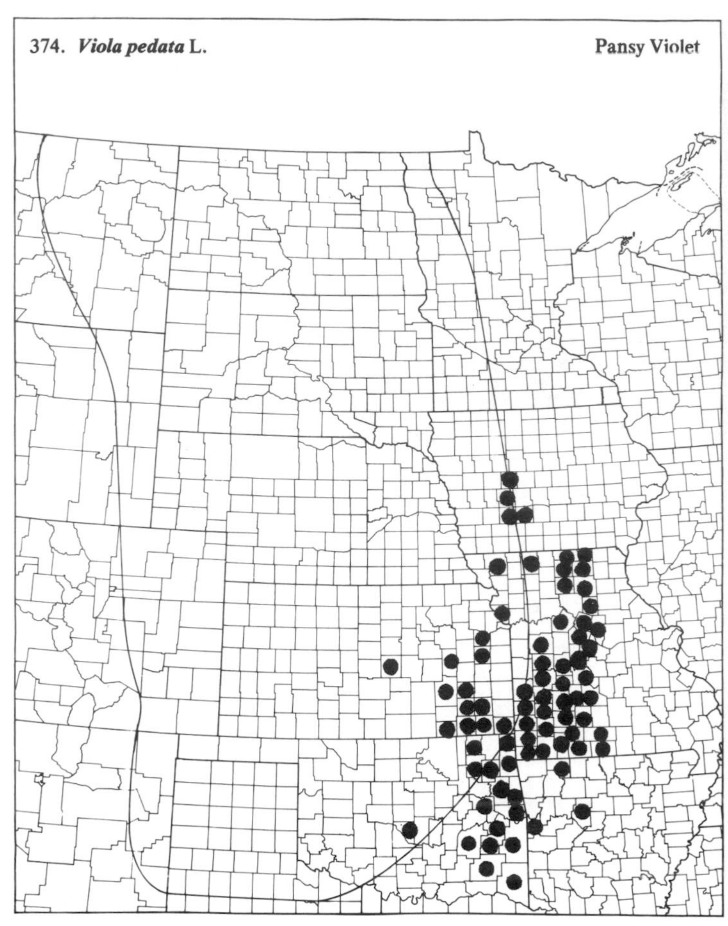

375. ***Viola pedatifida*** G. Don — Prairie Violet

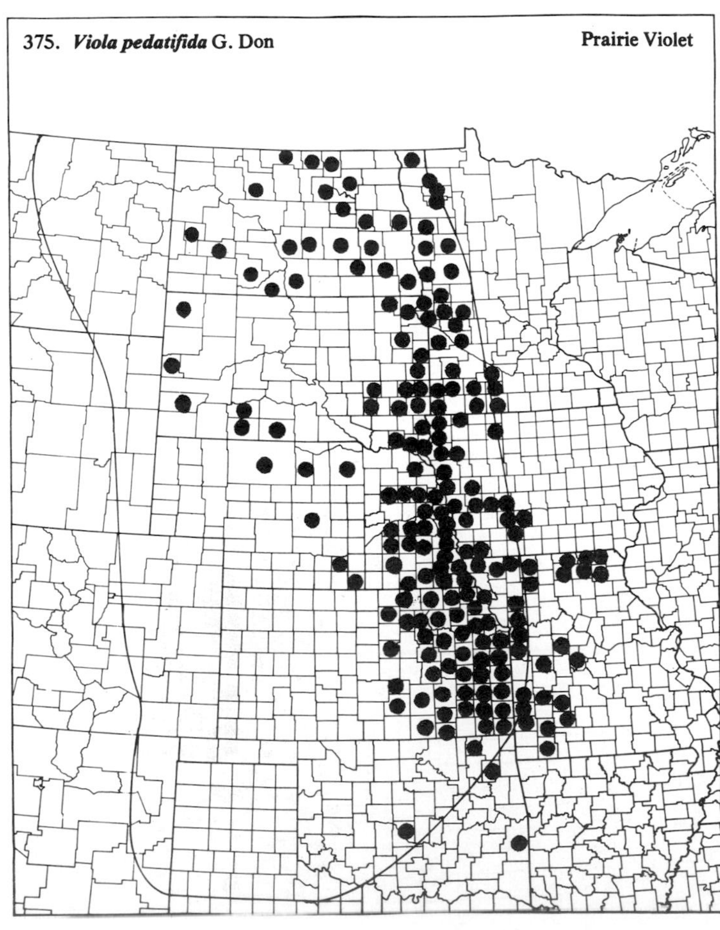

376. ***Viola pratincola*** Greene — Meadow Violet

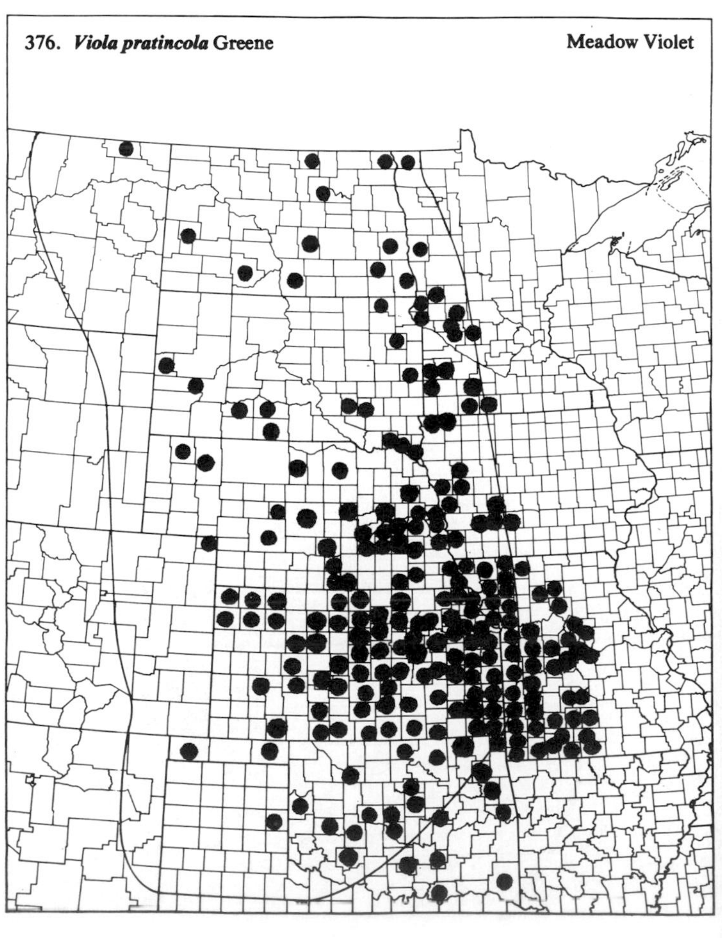

(Violaceae)

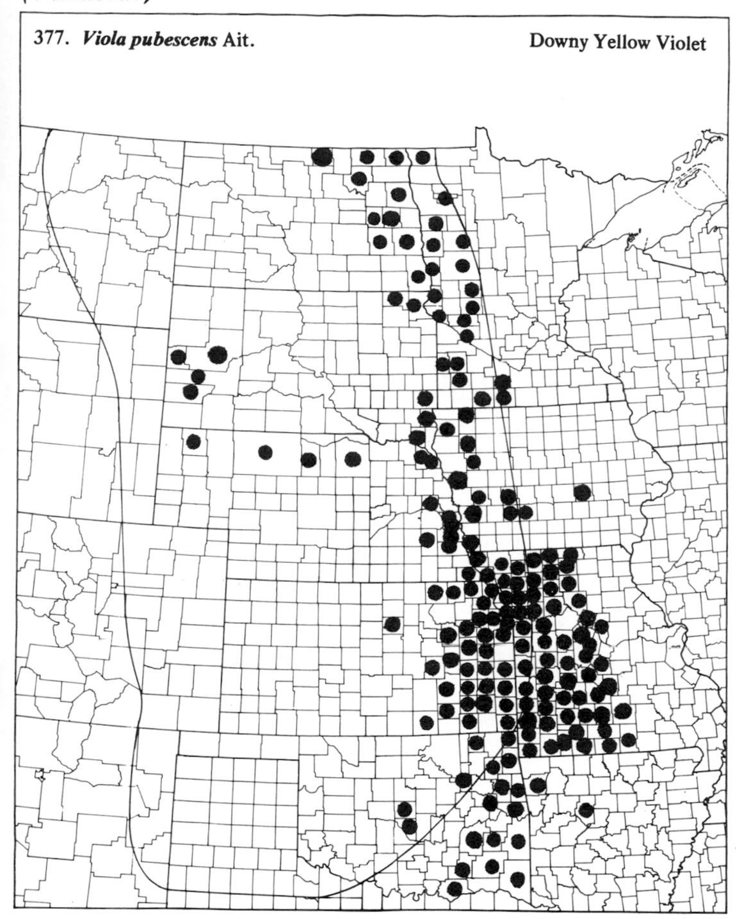
377. Viola pubescens Ait.
Downy Yellow Violet

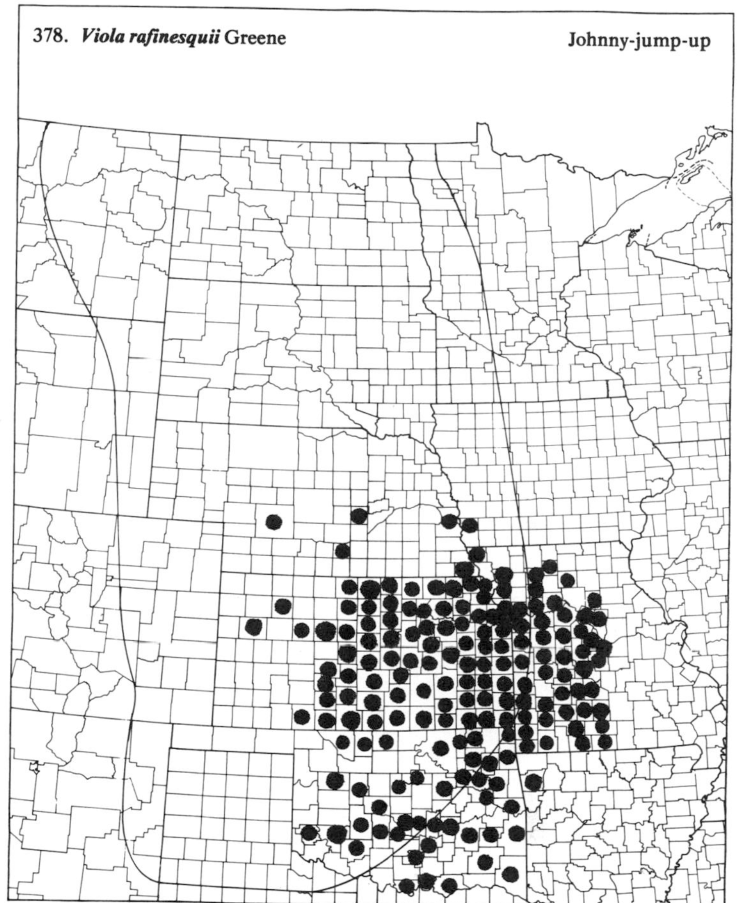
378. Viola rafinesquii Greene
Johnny-jump-up

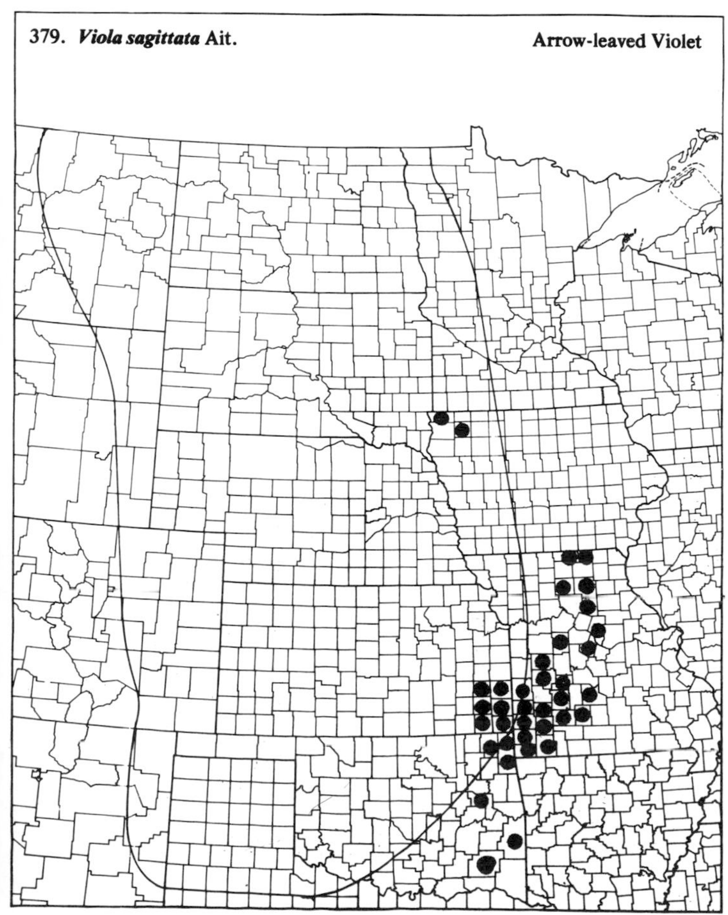
379. Viola sagittata Ait.
Arrow-leaved Violet

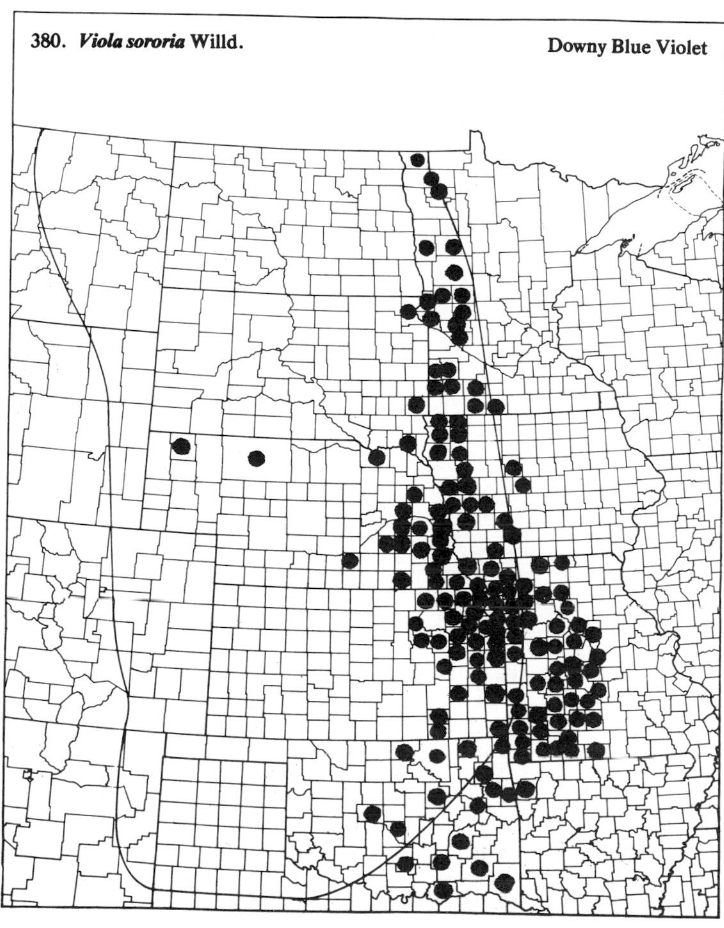
380. Viola sororia Willd.
Downy Blue Violet

(Violaceae)

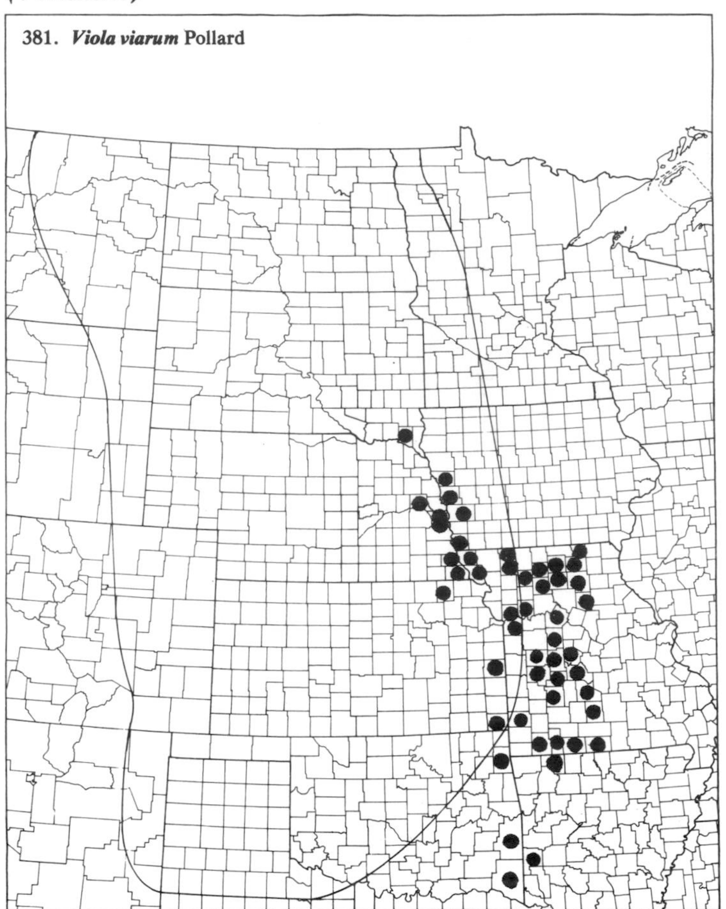

381. ***Viola viarum*** Pollard

36. *Passifloraceae* (Passion-flower Family)

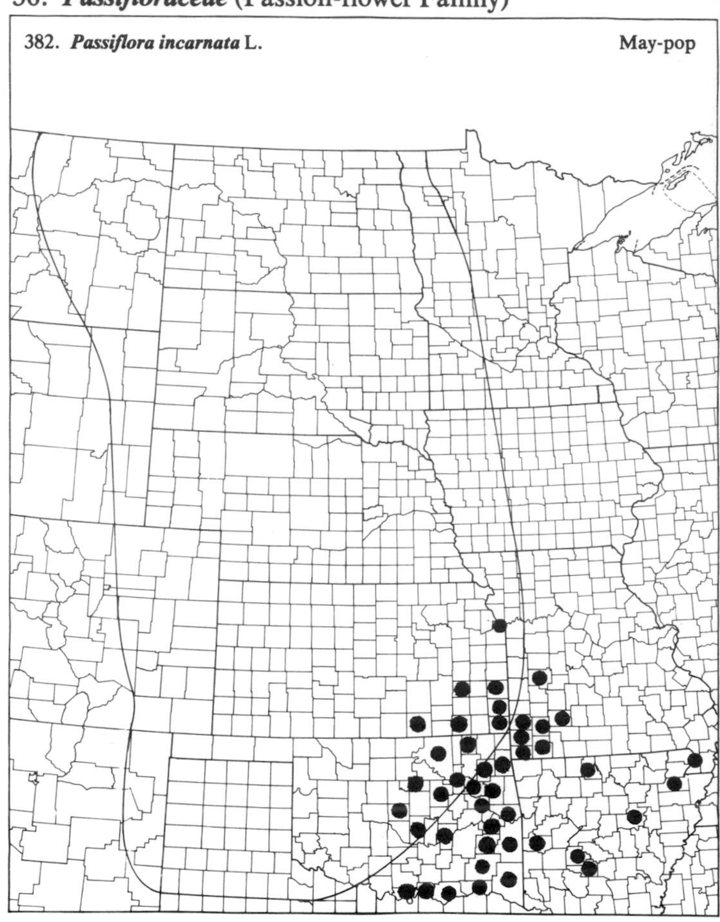

382. ***Passiflora incarnata*** L. May-pop

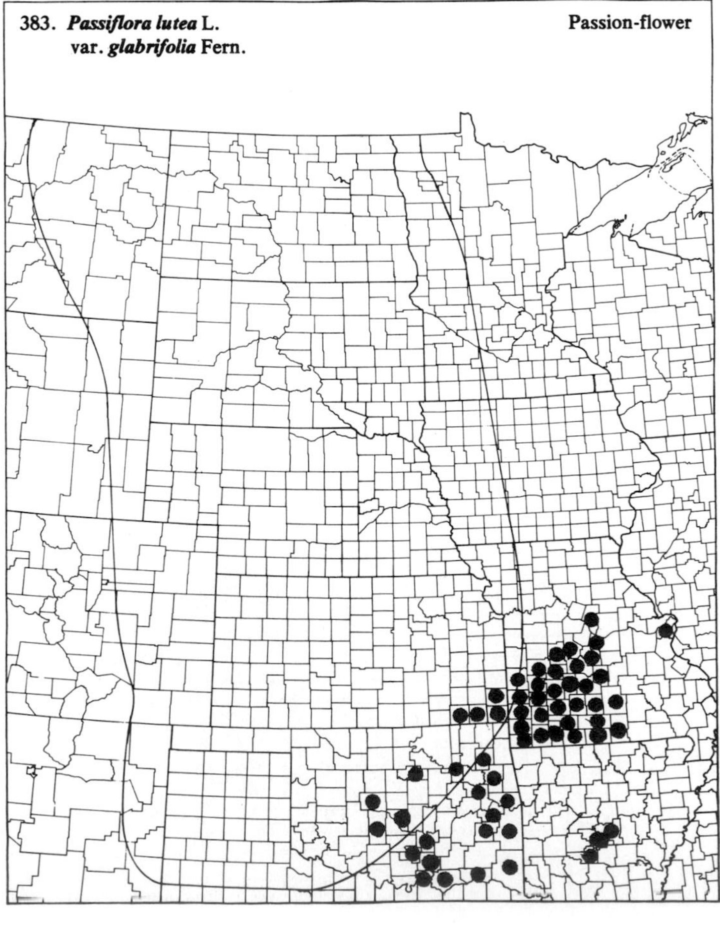

383. ***Passiflora lutea*** L. var. ***glabrifolia*** Fern. Passion-flower

37. *Cistaceae* (Rockrose Family)

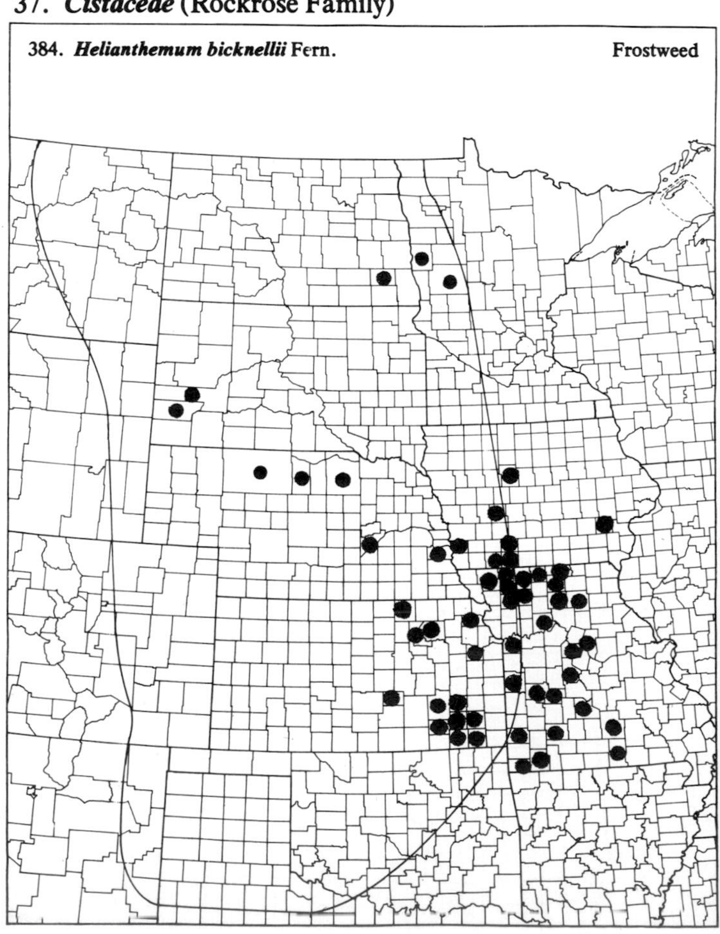

384. ***Helianthemum bicknellii*** Fern. Frostweed

(Cistaceae)

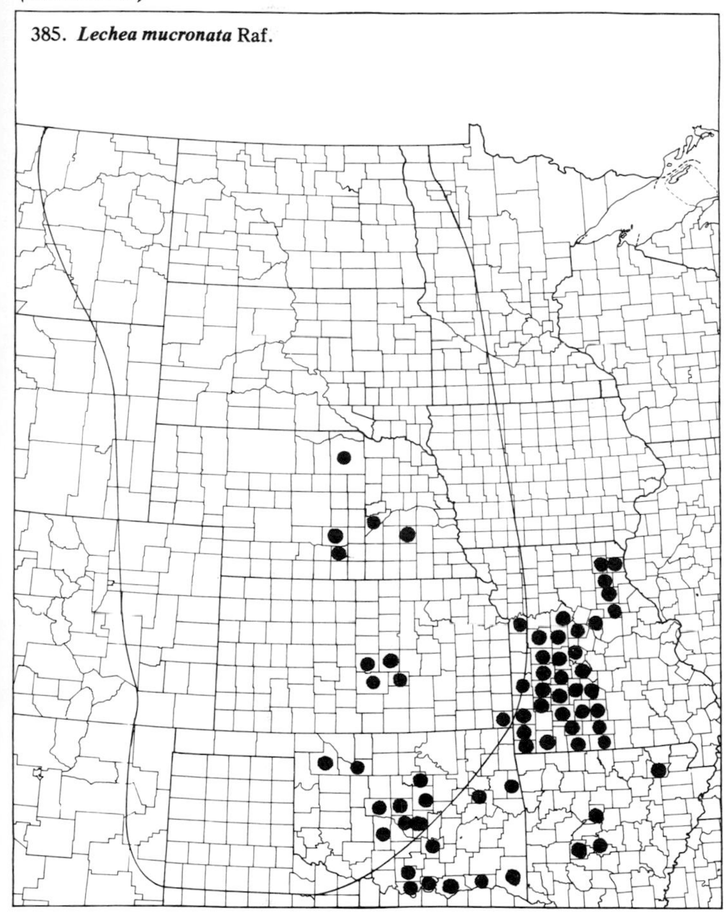

385. ***Lechea mucronata*** Raf.

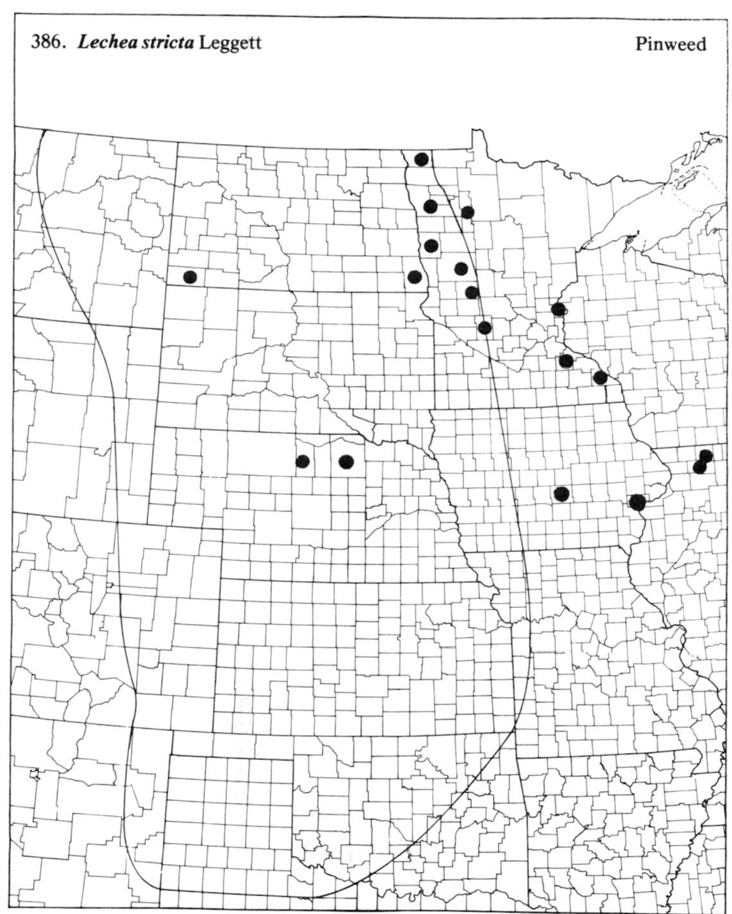

386. ***Lechea stricta*** Leggett — Pinweed

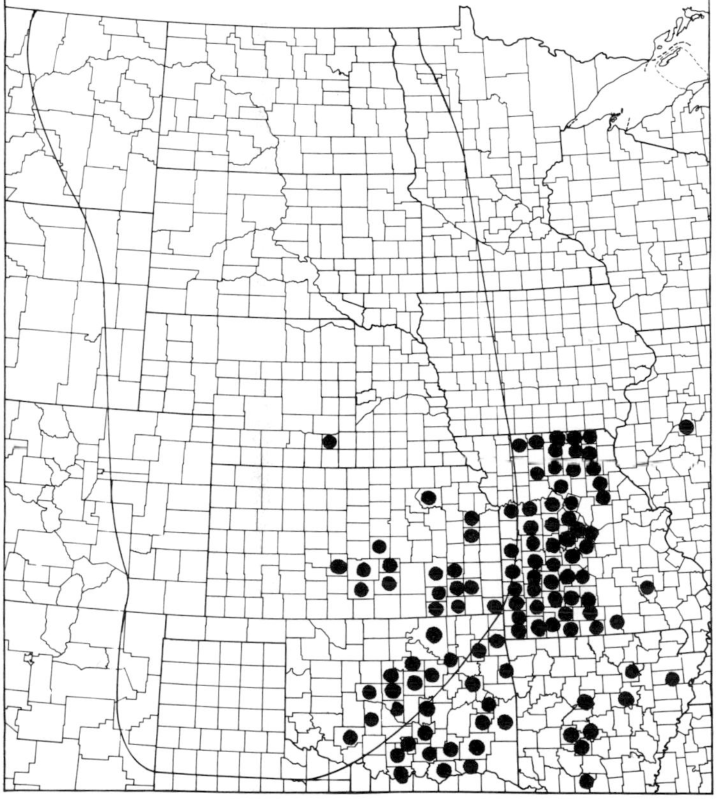

38. *Tamaricaceae* (Tamarisk Family)

387. ***Lechea tenuifolia*** Michx. — Pinweed

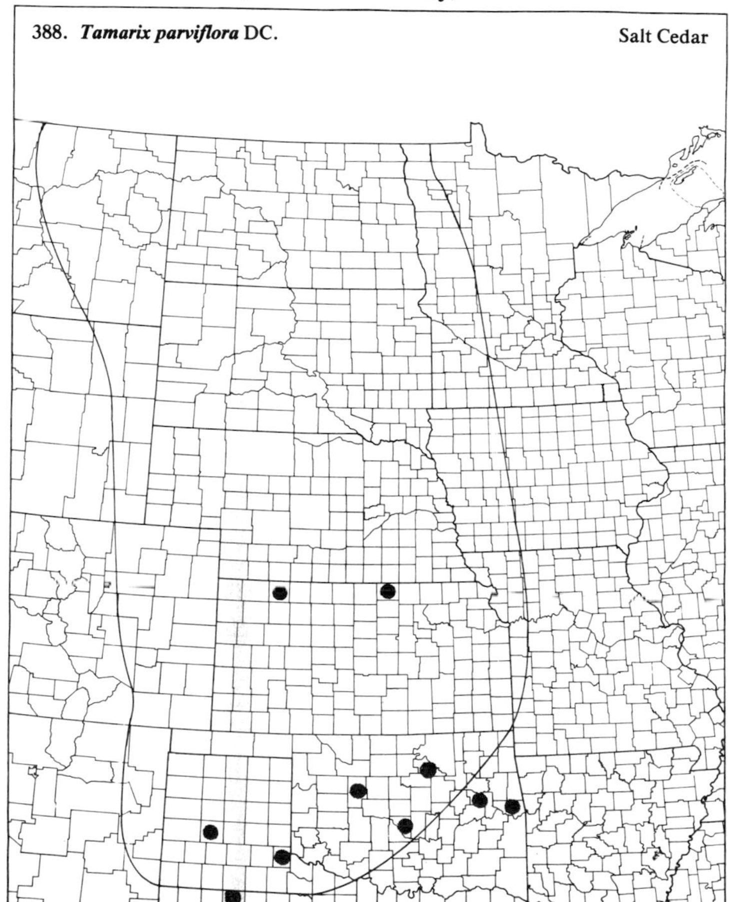

388. ***Tamarix parviflora*** DC. — Salt Cedar

(Tamaricaceae)

389. ***Tamarix ramosissima*** Ledeb. Salt Cedar

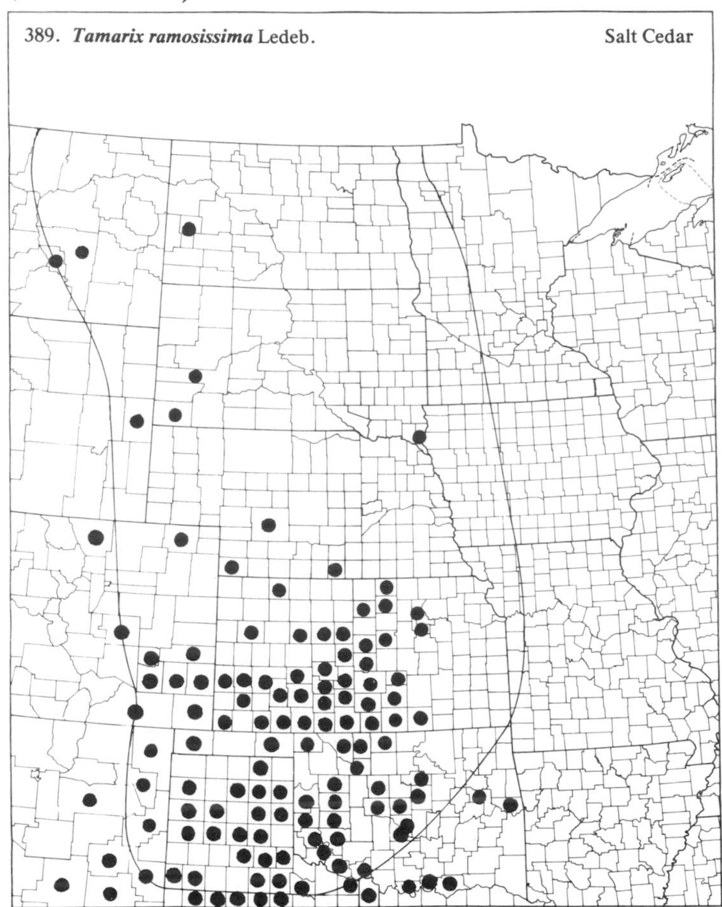

39. *Loasaceae* (Sand Lily Family)

390. ***Mentzelia decapetala*** (Pursh) Urban & Gilg Sand Lily

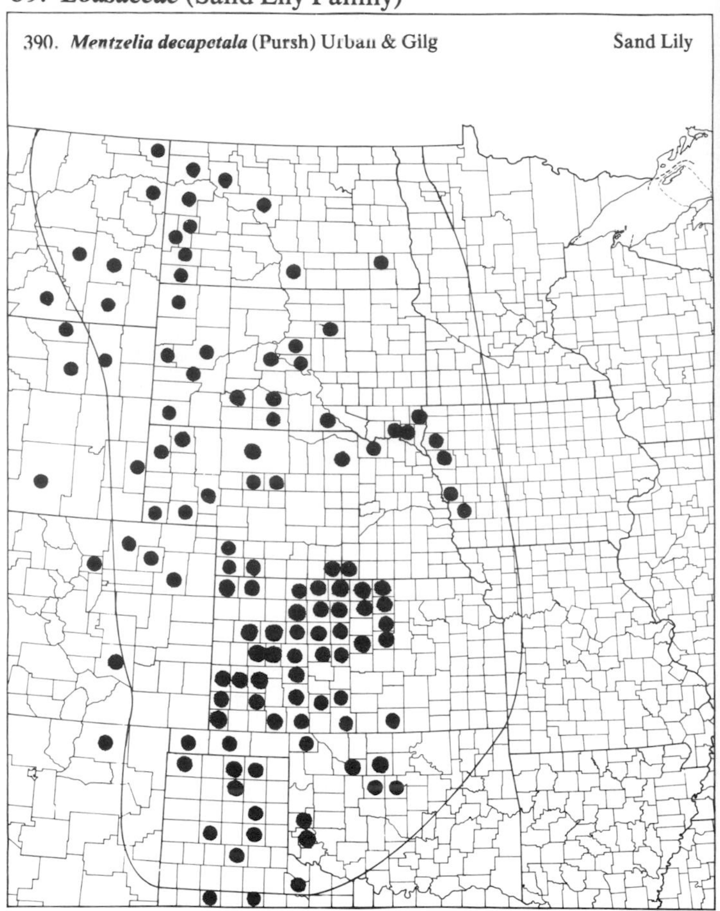

391. ***Mentzelia multiflora*** (Nutt.) Gray

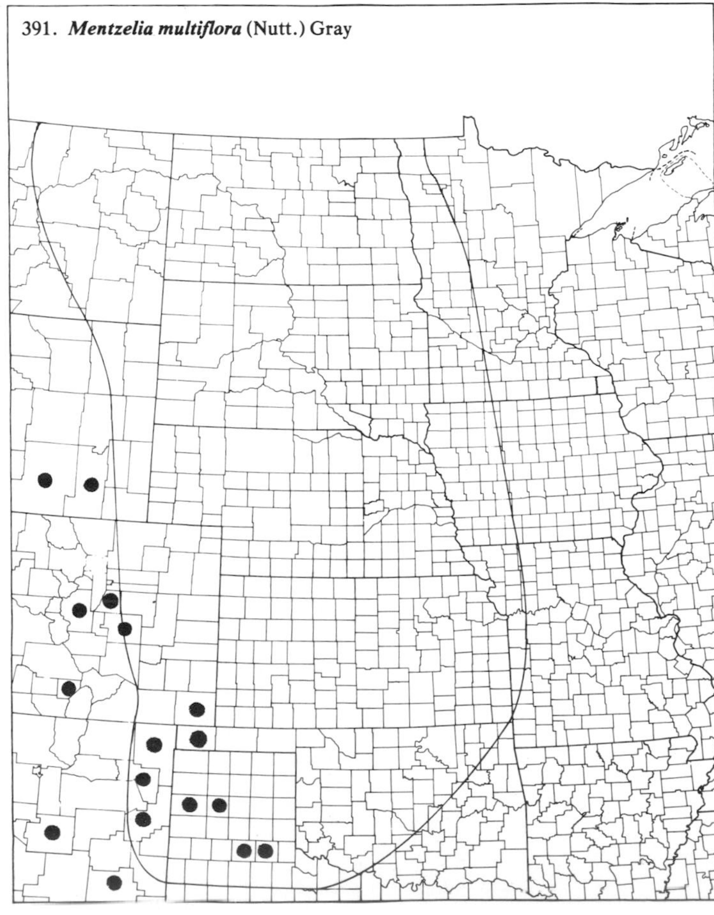

392. ***Mentzelia nuda*** (Pursh) T. & G. Bractless Mentzelia

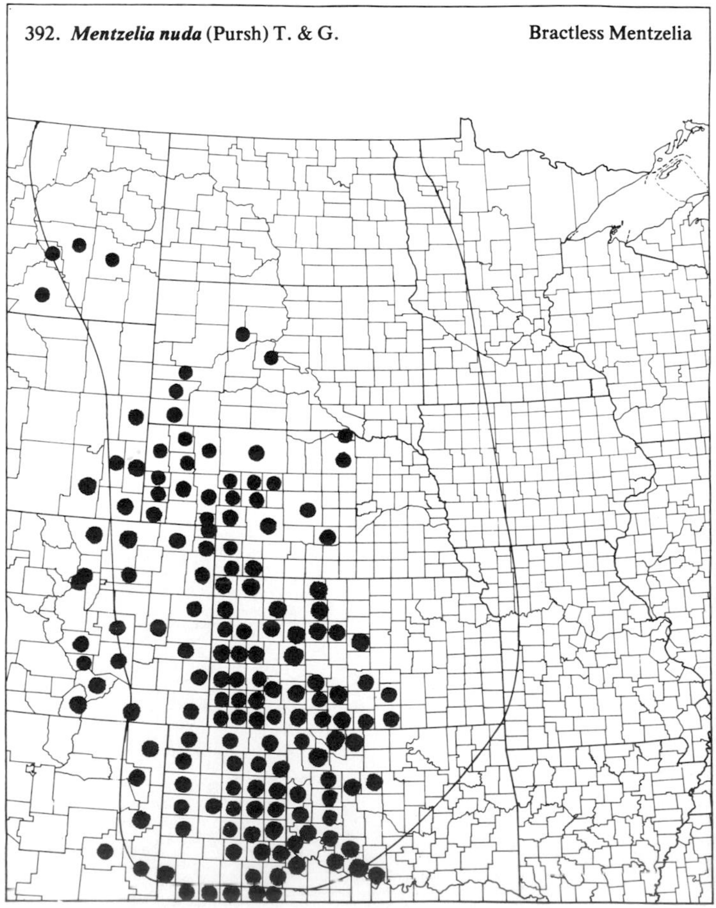

(Loasaceae)

393. ***Mentzelia oligosperma*** Nutt. Stickleaf Mentzelia

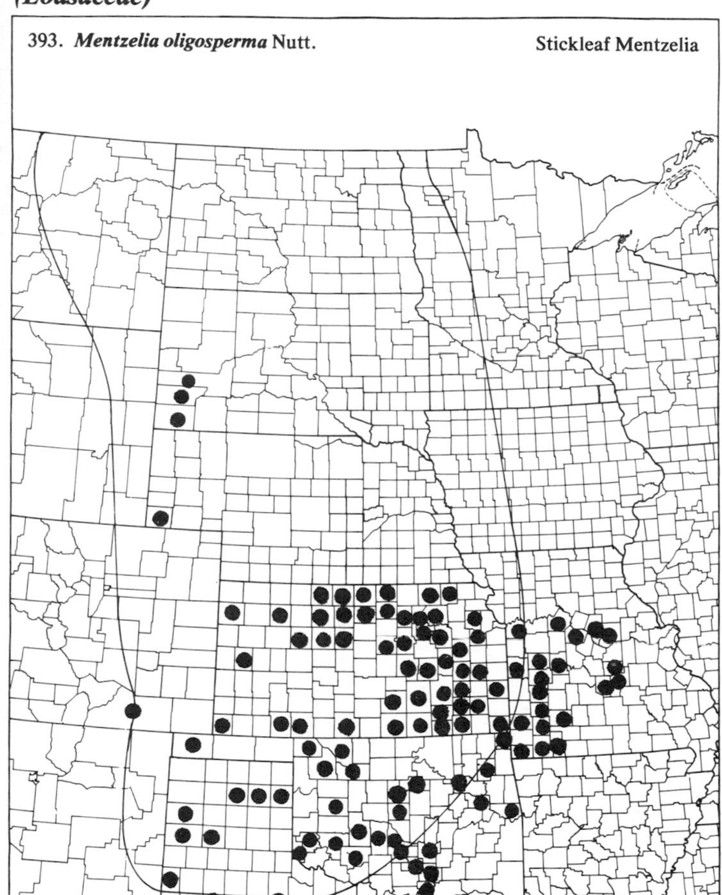

40. *Cucurbitaceae* (Gourd Family)

394. ***Cucurbita foetidissima*** H.B.K. Buffalo Gourd

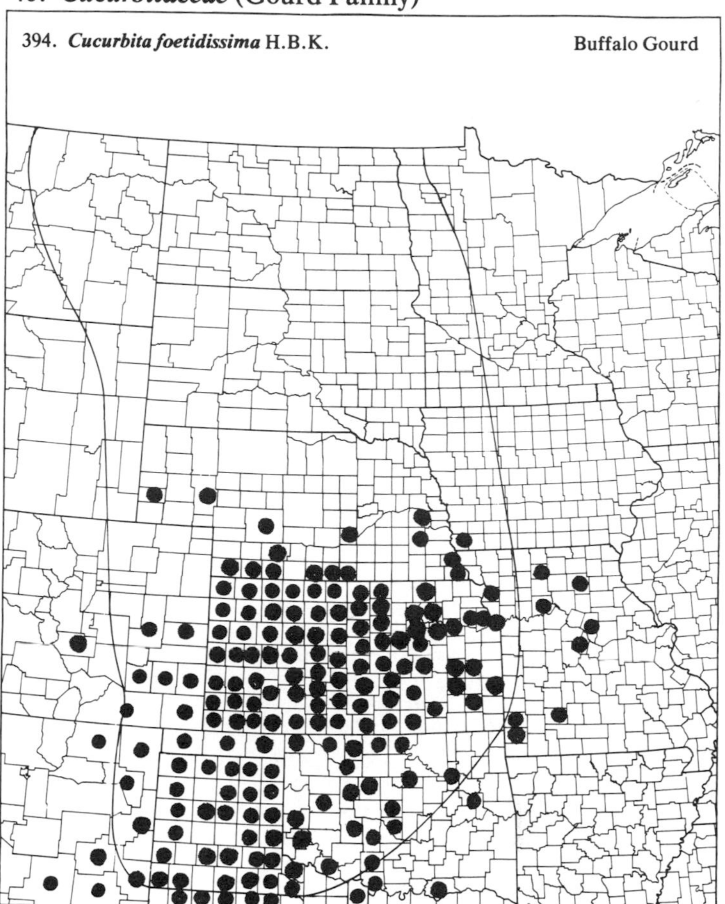

395. ***Cyclanthera dissecta*** (T. & G.) Arn. Cyclanthera

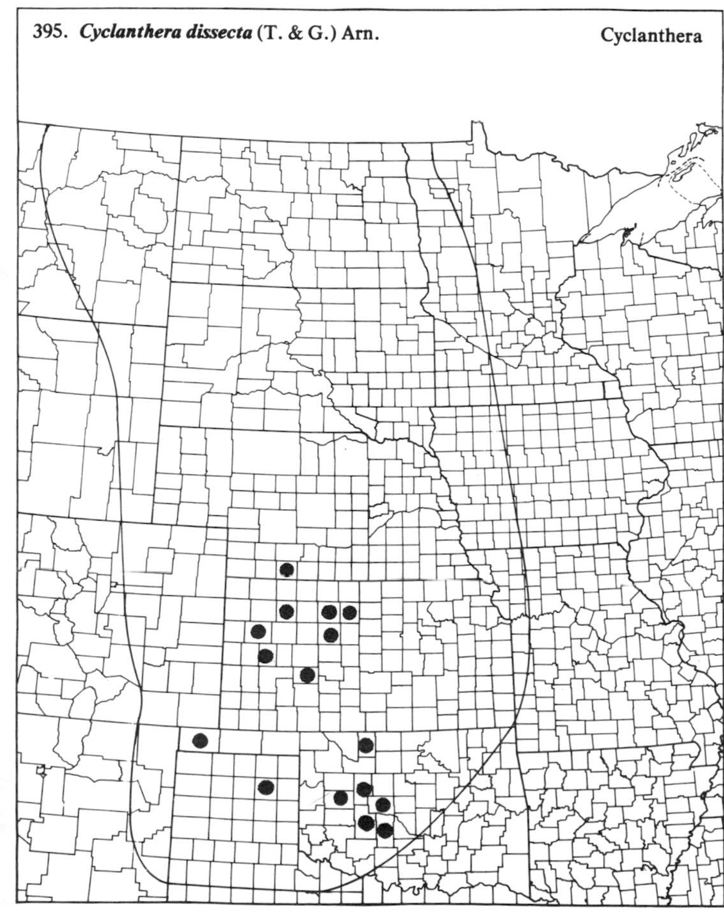

396. ***Echinocystis lobata*** (Michx.) T. & G. Wild Cucumber

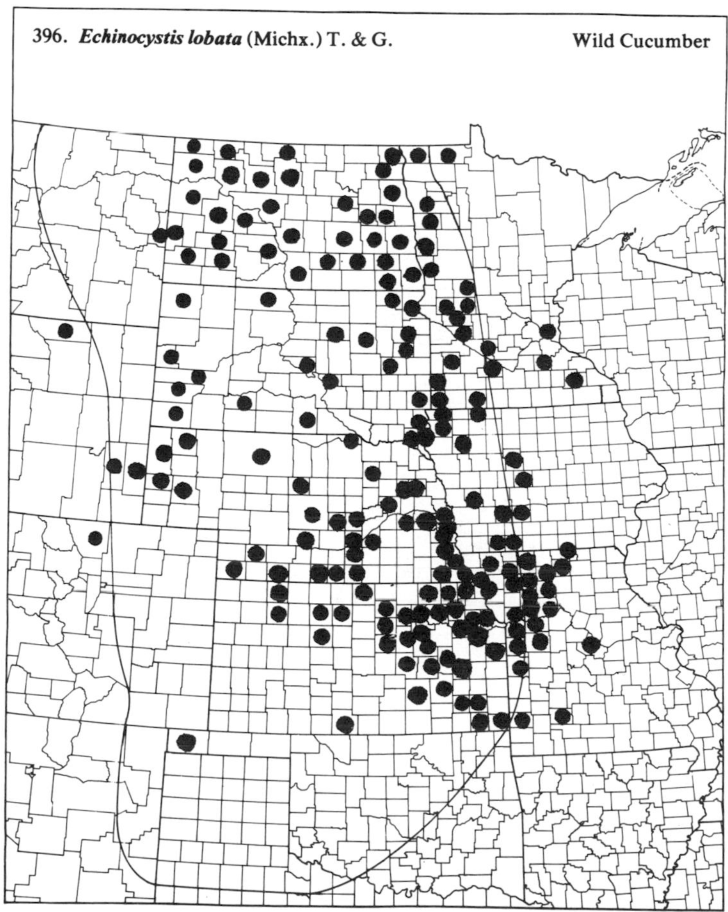

(Cucurbitaceae)

397. ***Melothria pendula*** L. Creeping Cucumber

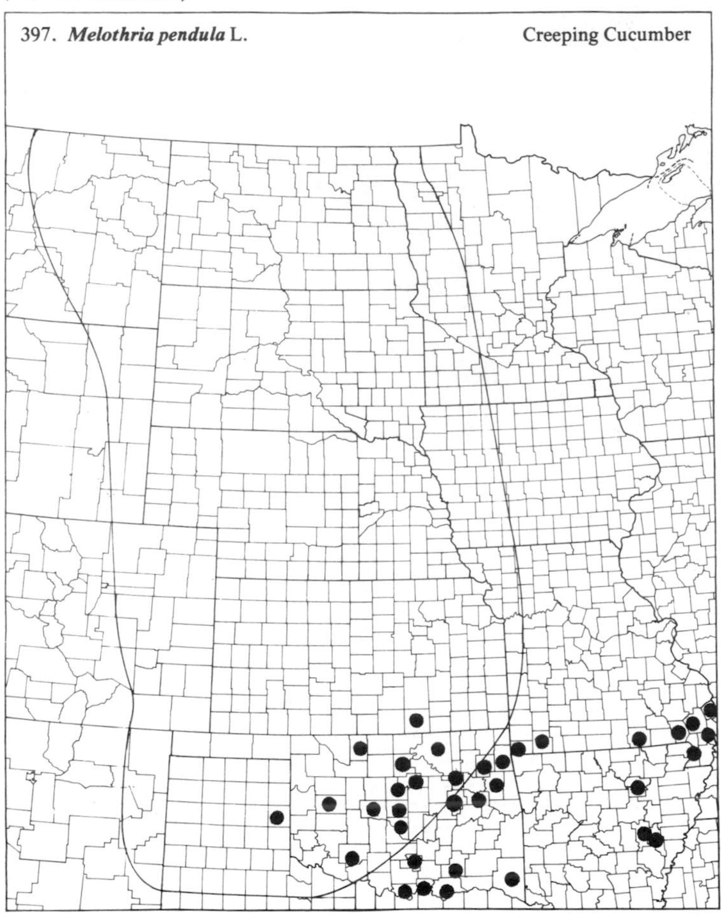

398. ***Sicyos angulatus*** L. Bur Cucumber

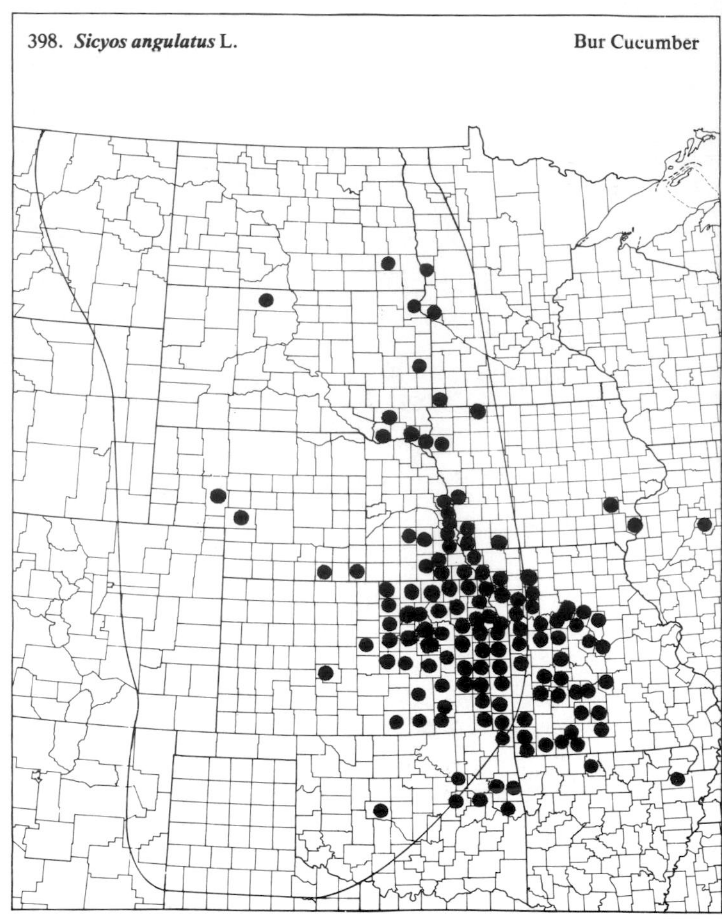

41. *Salicaceae* (Willow Family)

399. ***Populus acuminata*** Rydb. Smooth-barked Cottonwood

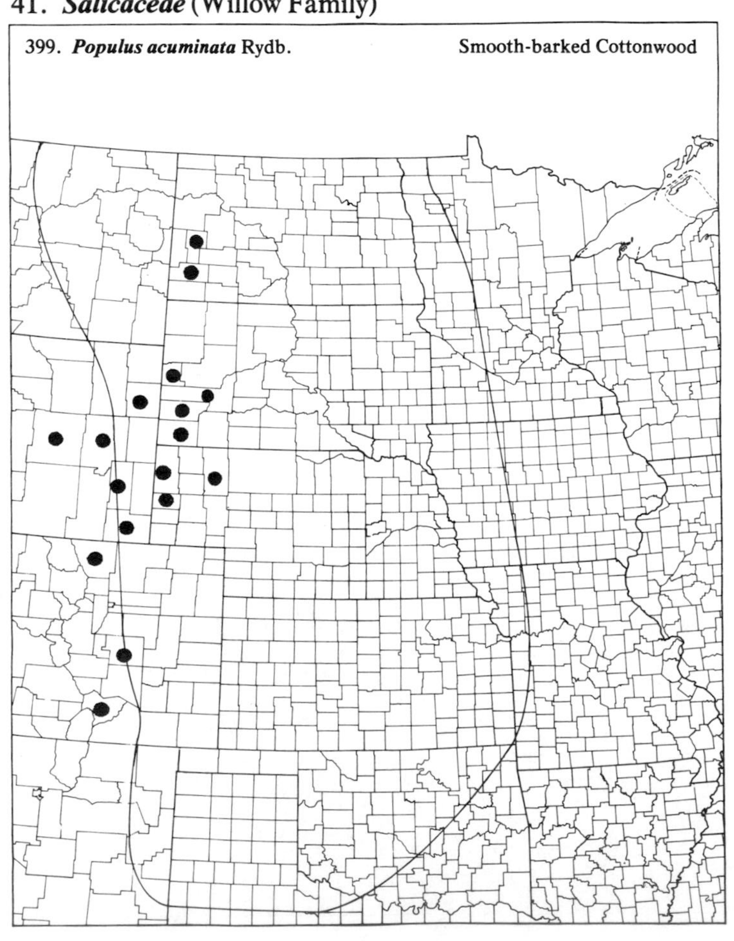

400. ***Populus alba*** L. Silver Poplar

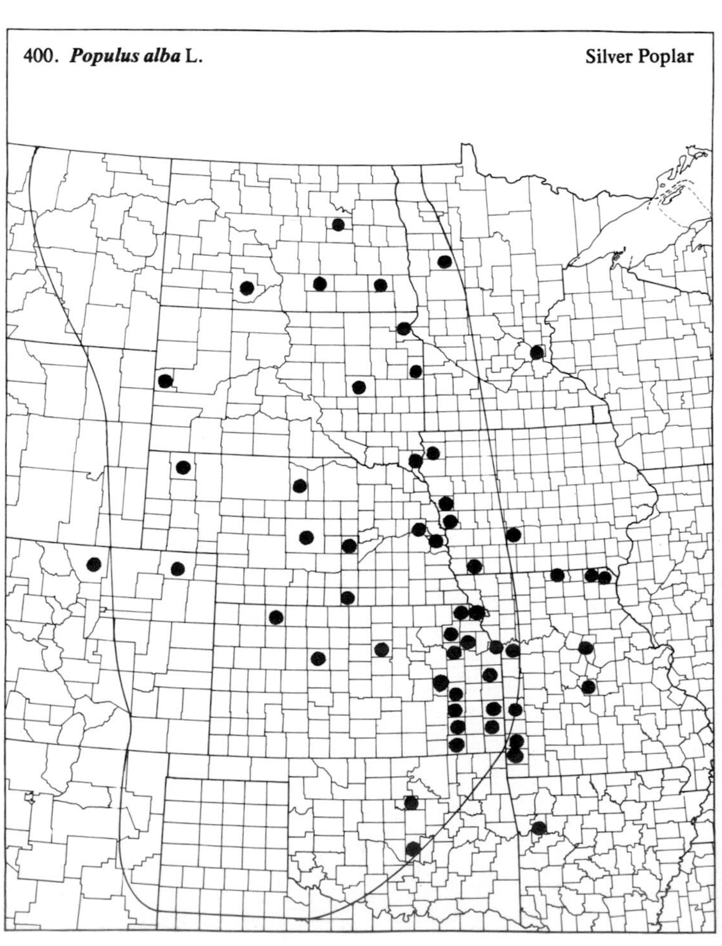

(Salicaceae)

401. ***Populus balsamifera*** L. Balsam Poplar

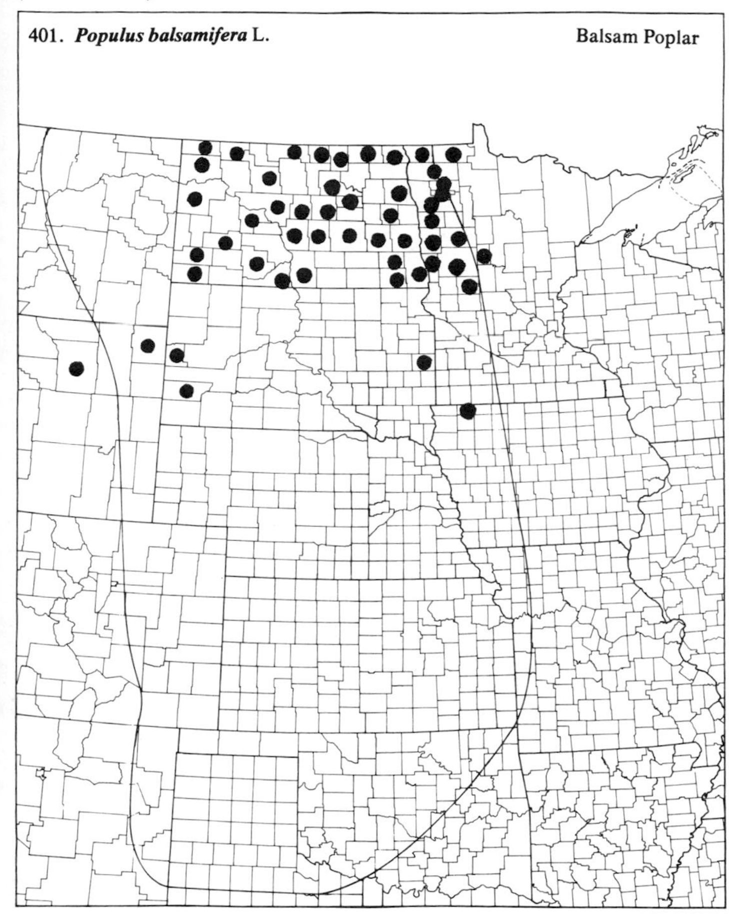

402. ***Populus deltoides*** Marsh. var. ***deltoides*** Cottonwood

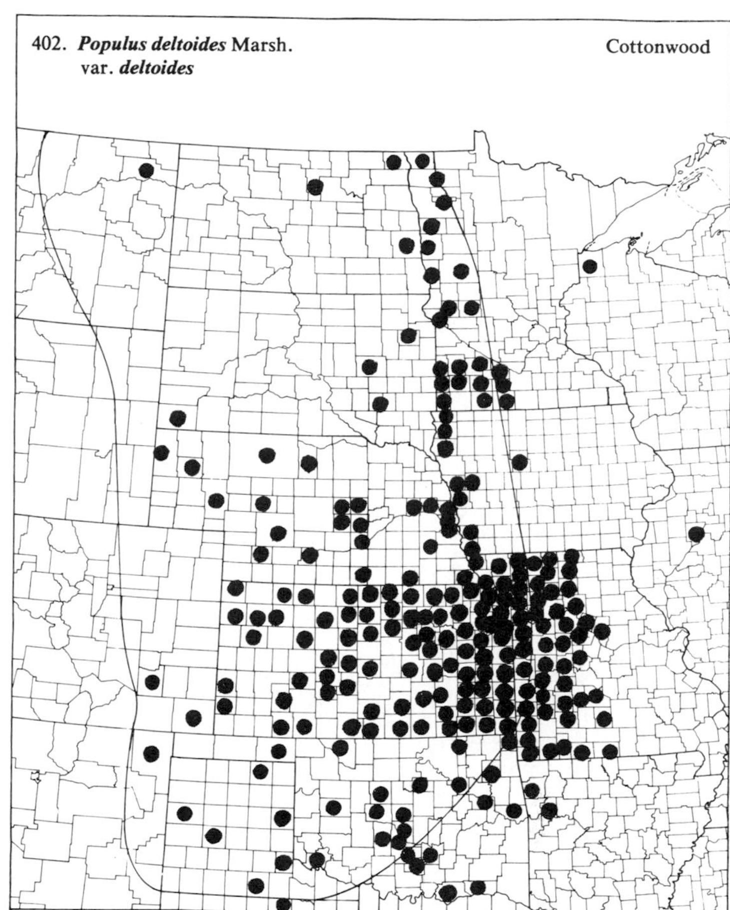

403. ***Populus deltoides*** Marsh. var. ***occidentalis*** Rydb. Cottonwood

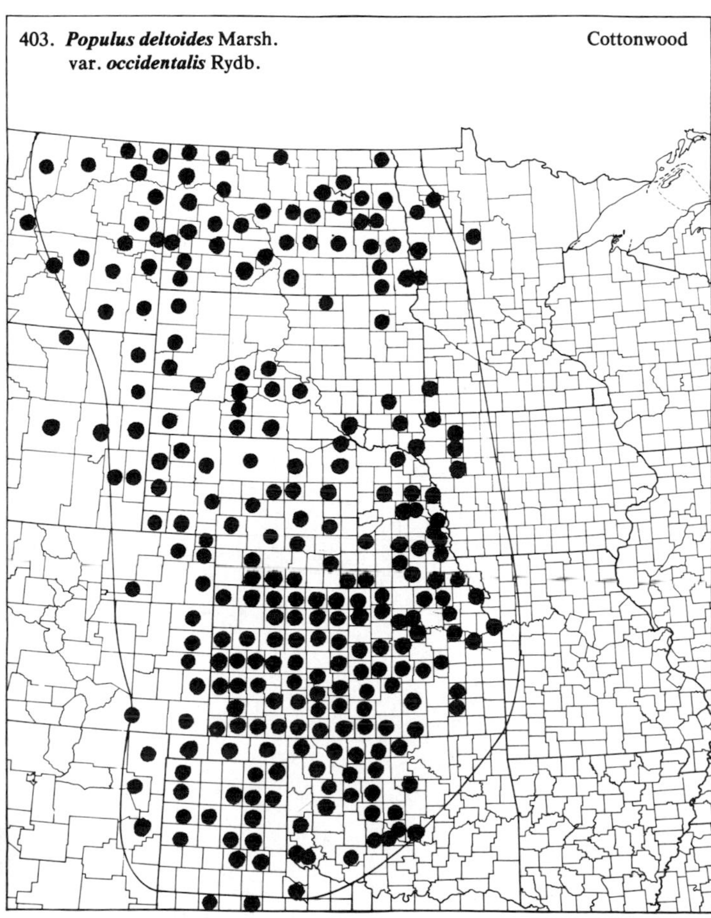

404. ***Populus tremuloides*** Michx. Quaking Aspen

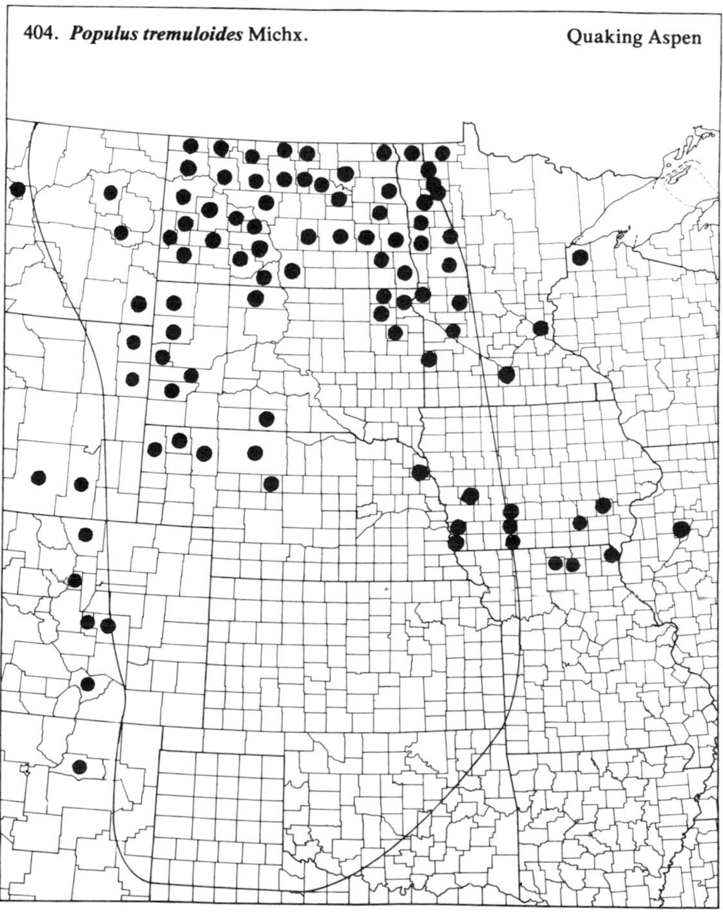

(Salicaceae)

405. ***Salix alba*** L. White Willow

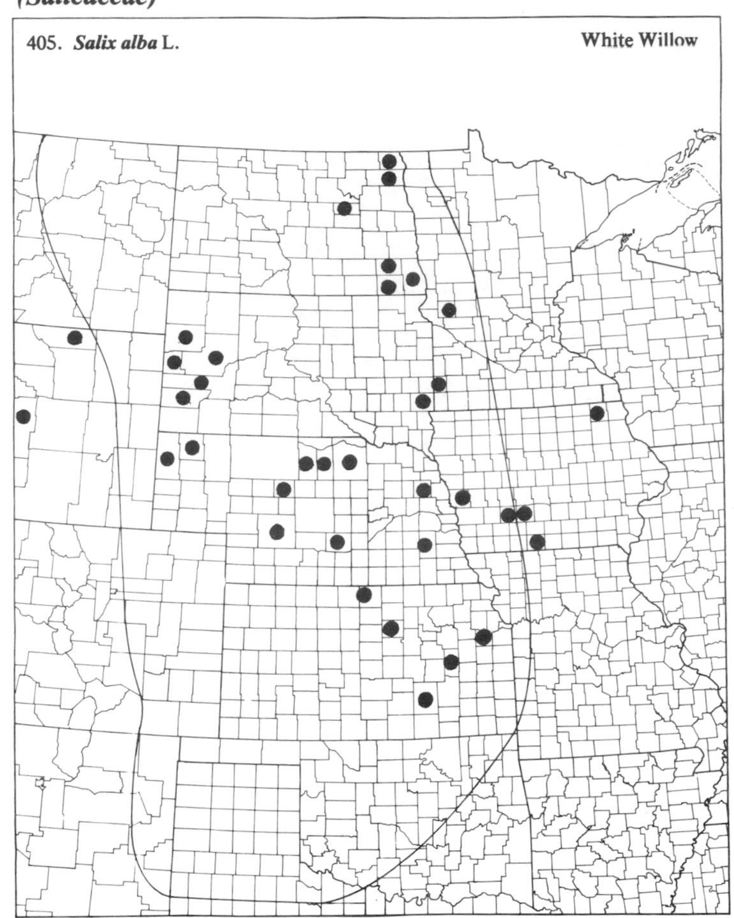

406. ***Salix amygdaloides*** Anderss. Peach-leaved Willow

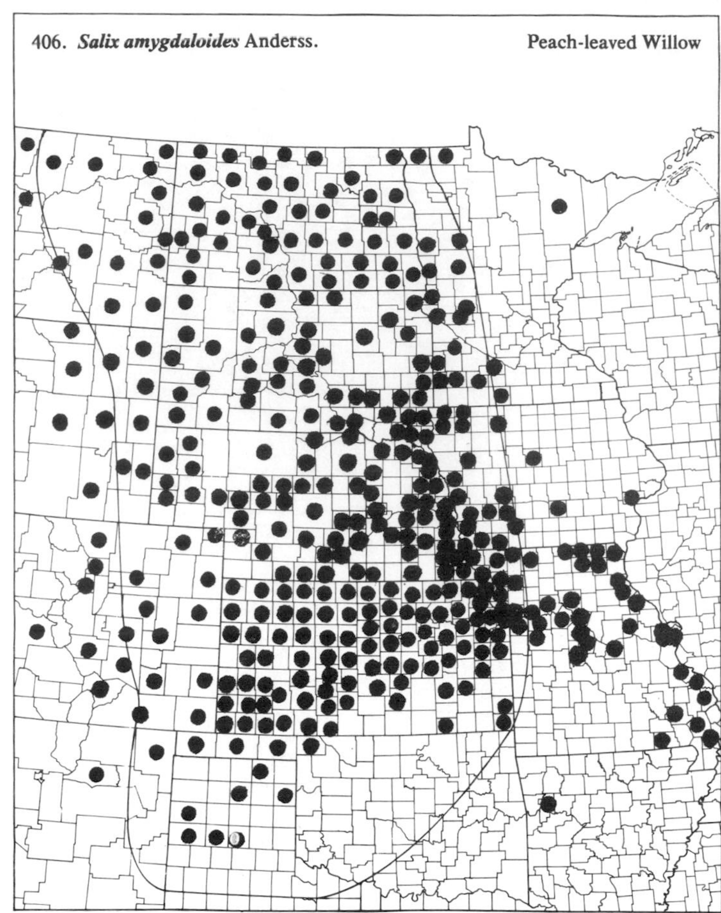

407. ***Salix bebbiana*** Sarg. Long-beaked Willow

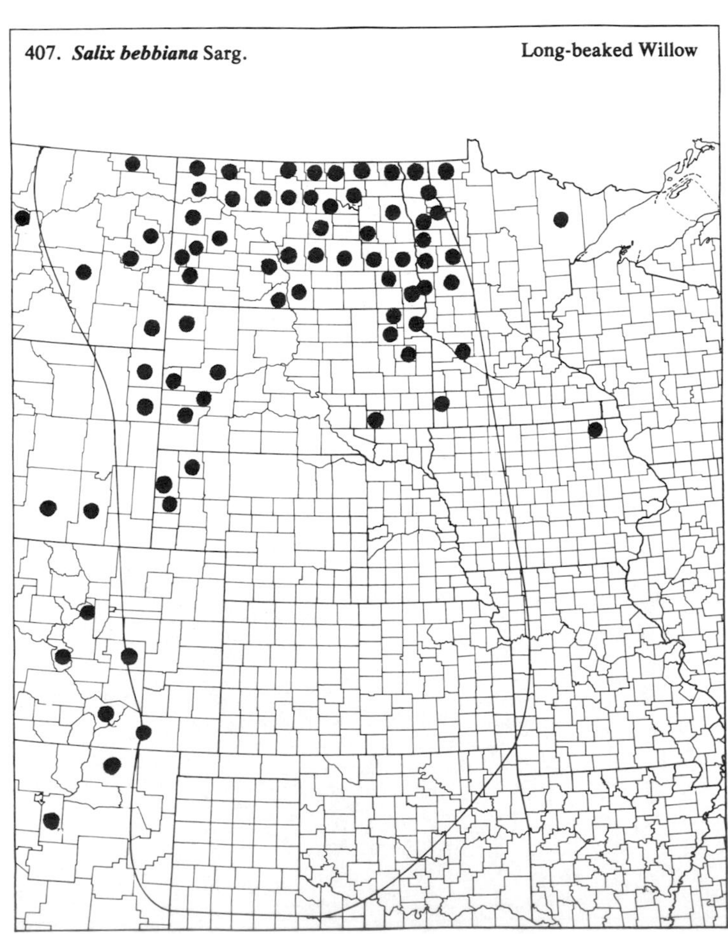

408. ***Salix candida*** Flugge Hoary Willow

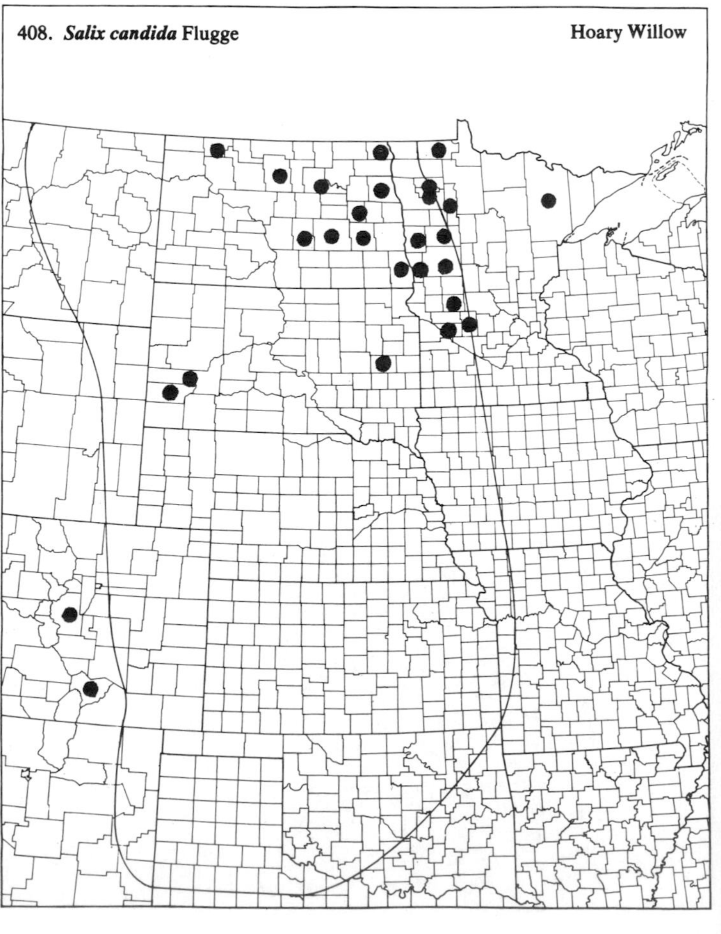

(Salicaceae)

409. ***Salix caroliniana*** Michx. Carolina Willow

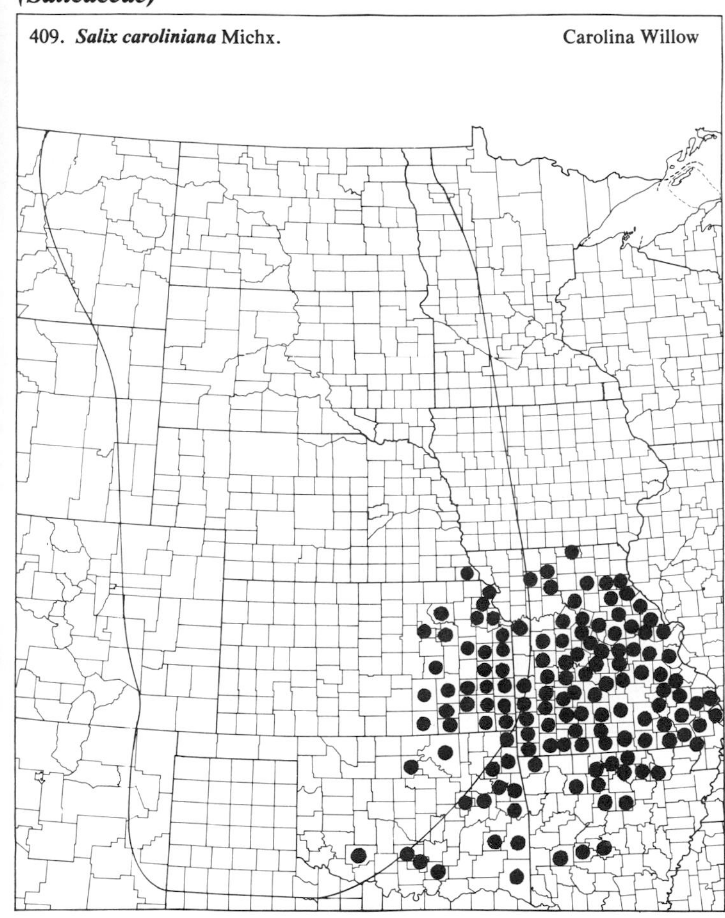

410. ***Salix discolor*** Muhl. Large Pussy Willow

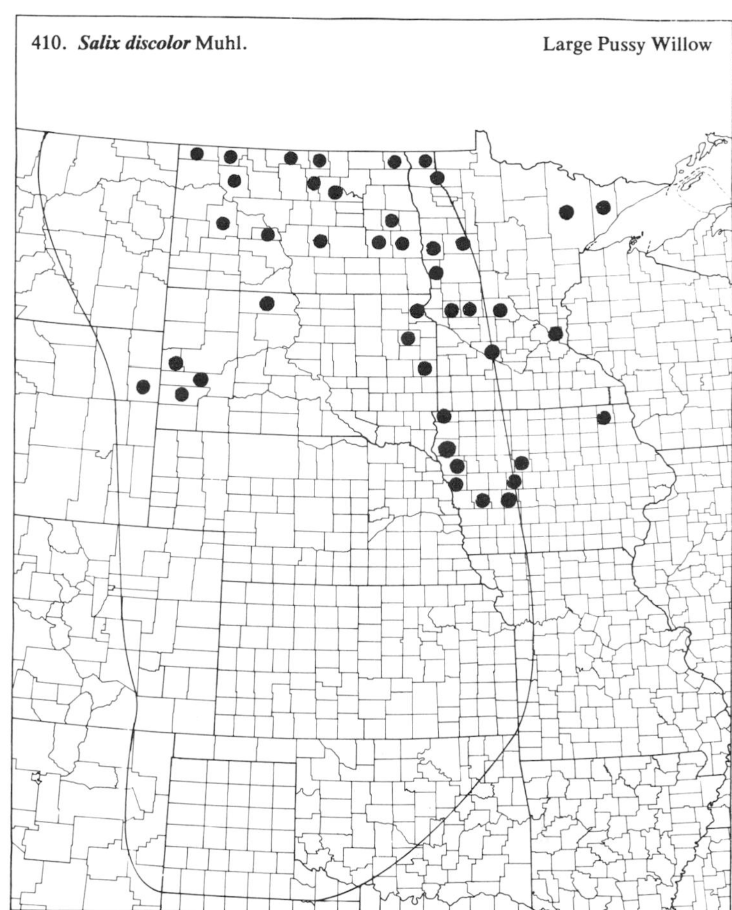

411. ***Salix exigua*** Nutt.
ssp. ***exigua*** Coyote Willow

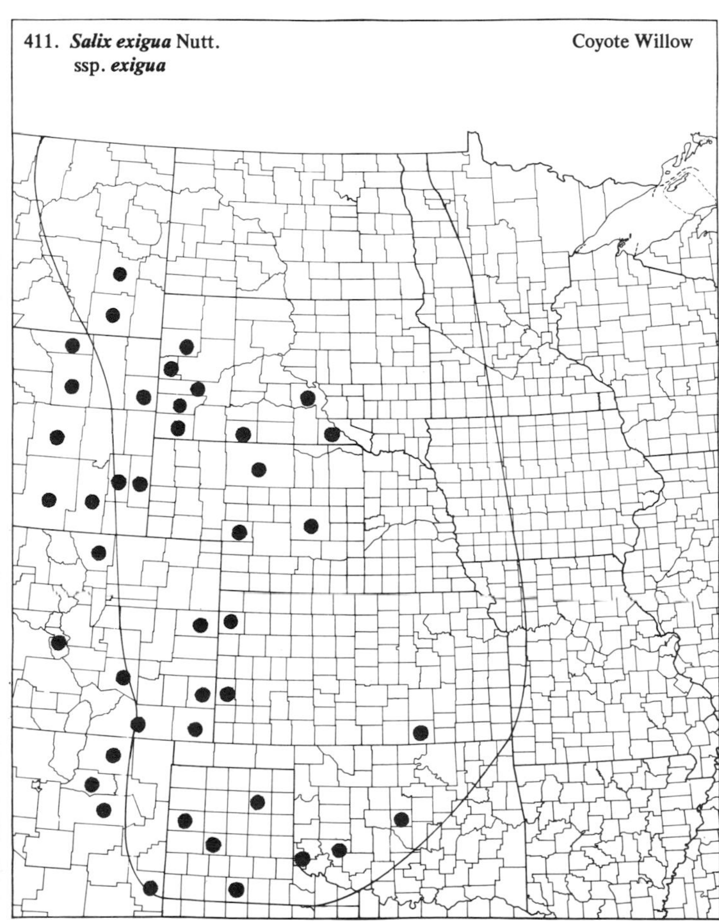

412. ***Salix exigua*** Nutt.
ssp. ***interior*** (Rowlee) Cronq. Coyote Willow

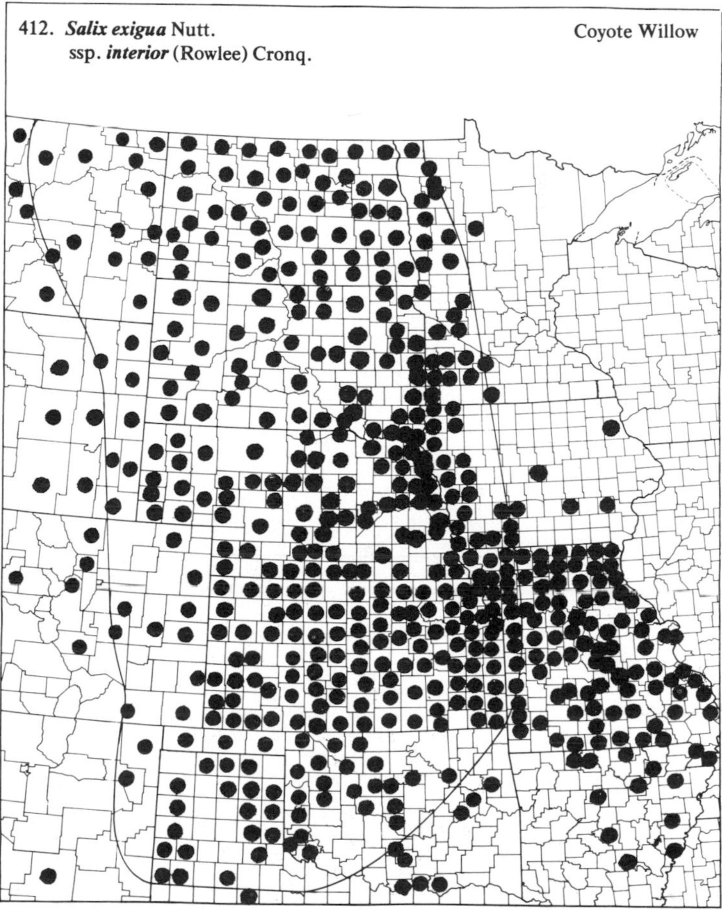

(Salicaceae)

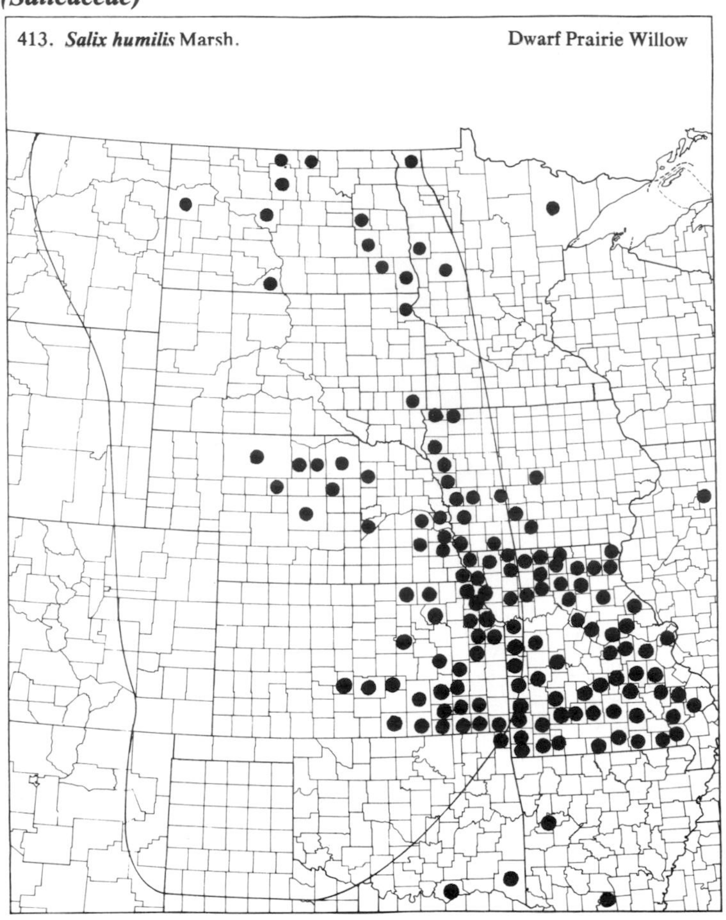
413. *Salix humilis* Marsh.
Dwarf Prairie Willow

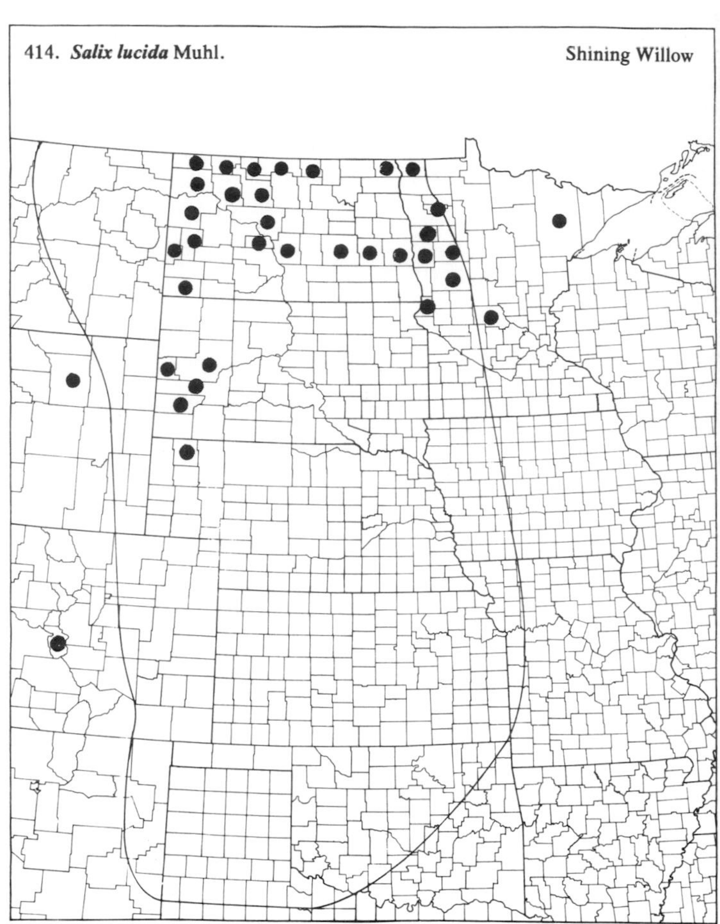
414. *Salix lucida* Muhl.
Shining Willow

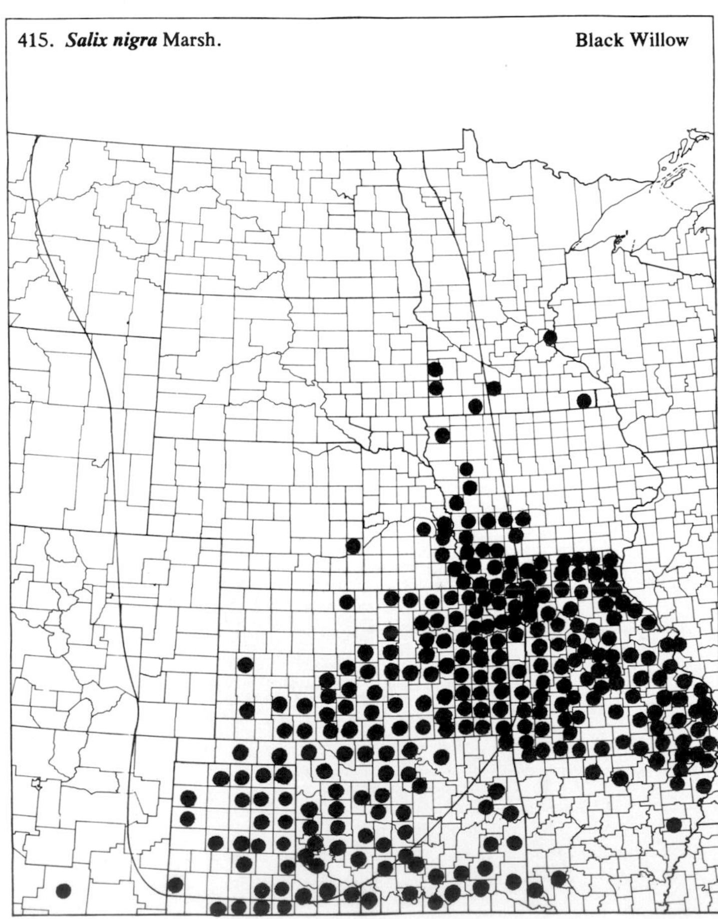
415. *Salix nigra* Marsh.
Black Willow

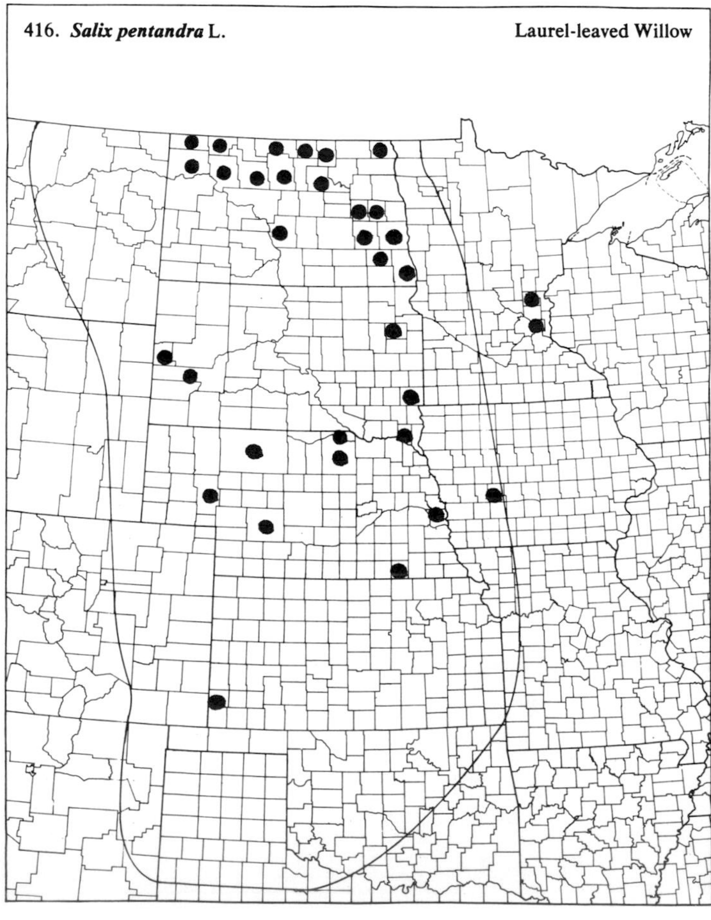
416. *Salix pentandra* L.
Laurel-leaved Willow

(Salicaceae)

417. *Salix petiolaris* Sm. Meadow Willow

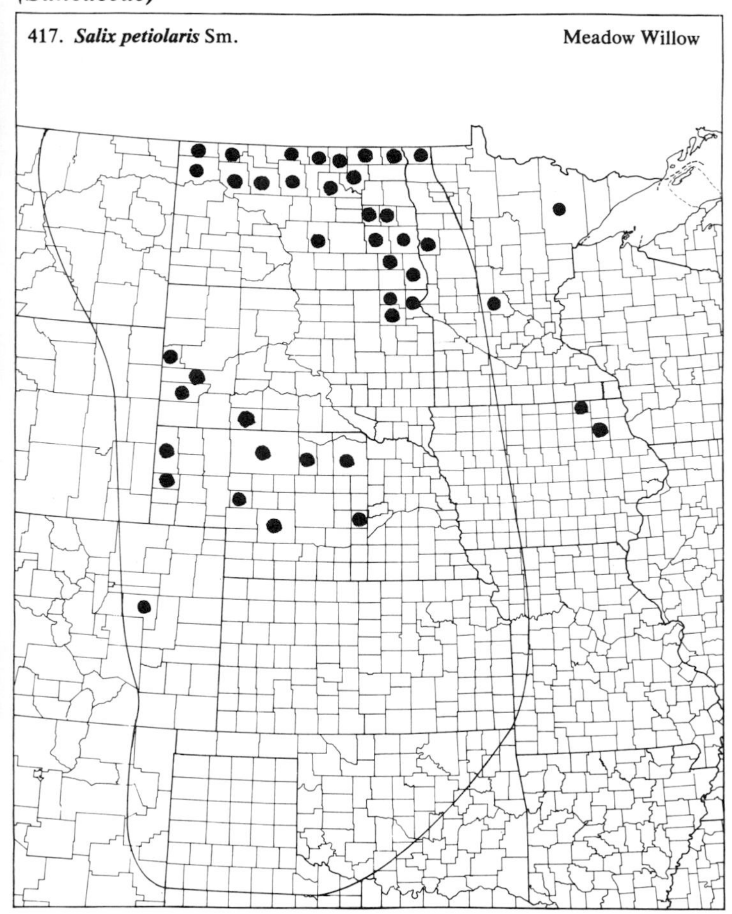

418. *Salix rigida* Muhl. var. *rigida* Diamond Willow

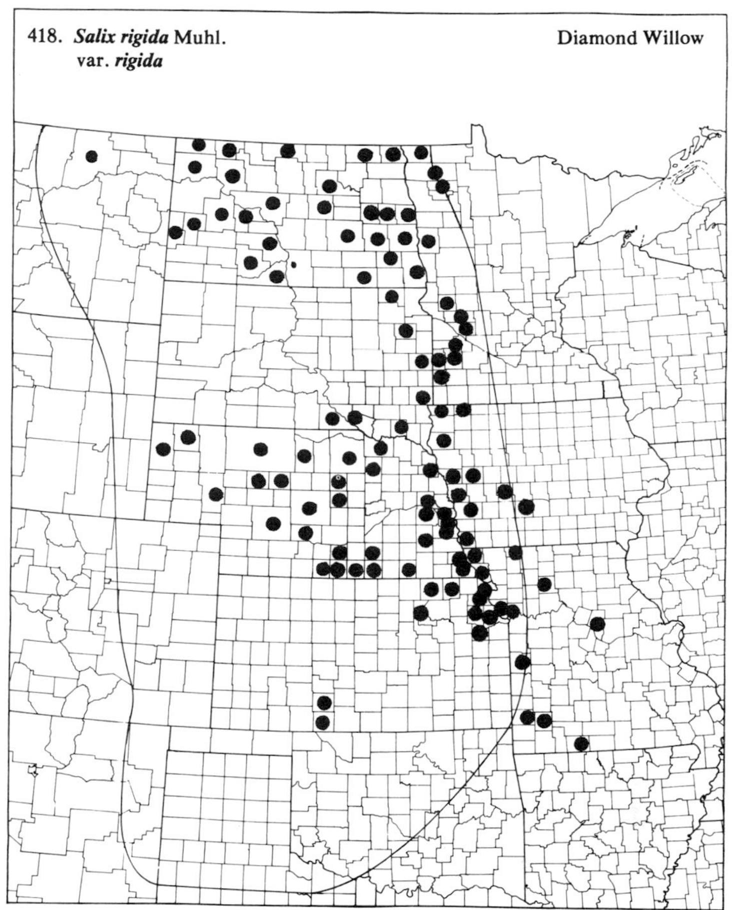

419. *Salix rigida* Muhl. var. *watsonii* (Bebb) Cronq. Diamond Willow

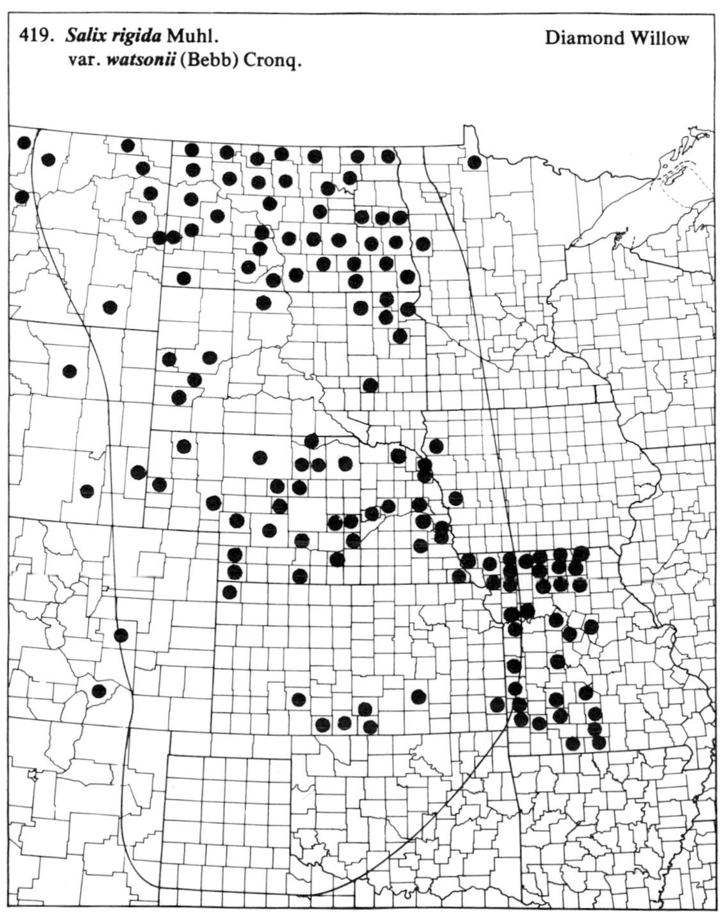

420. *Salix serissima* (Bailey) Fern. Autumn Willow

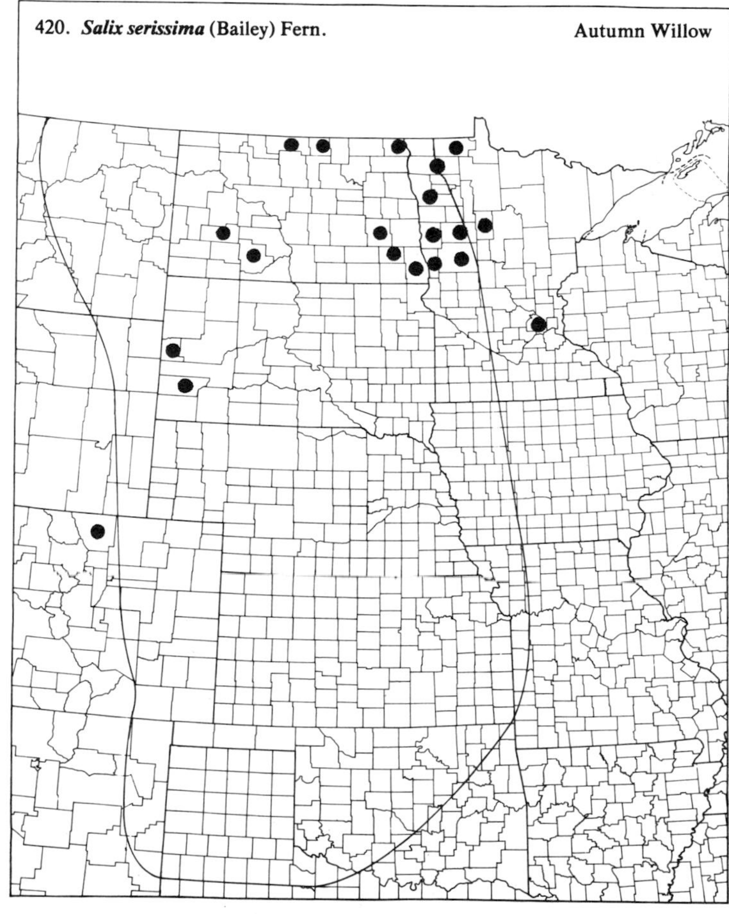

42. *Capparaceae* (Caper Family)

421. *Cleome serrulata* Pursh — Rocky Mountain Beeplant

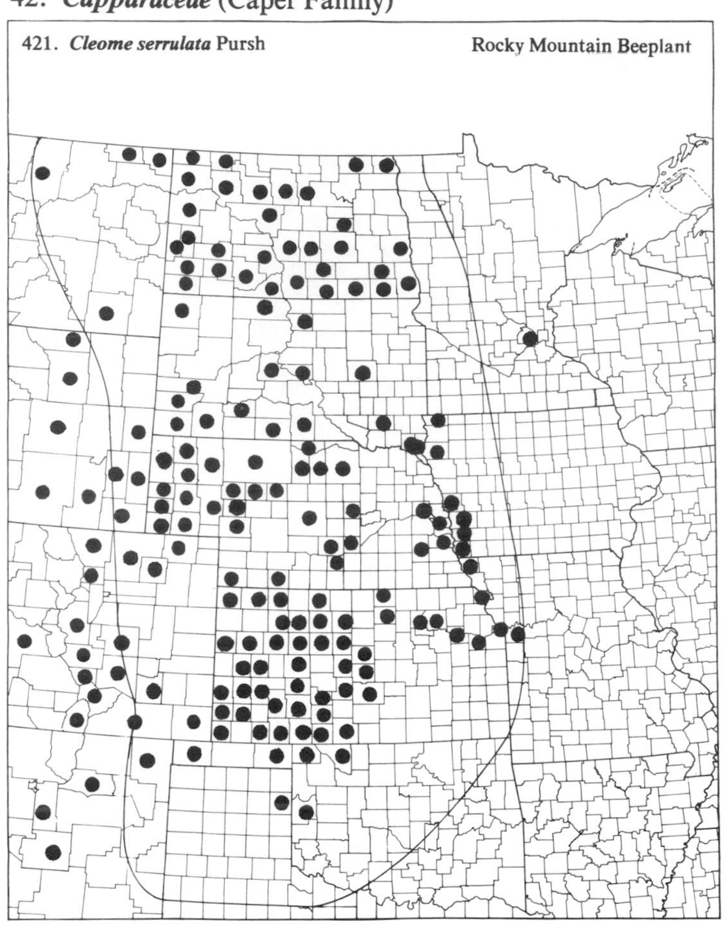

422. *Cleomella angustifolia* Torr. — Cleomella

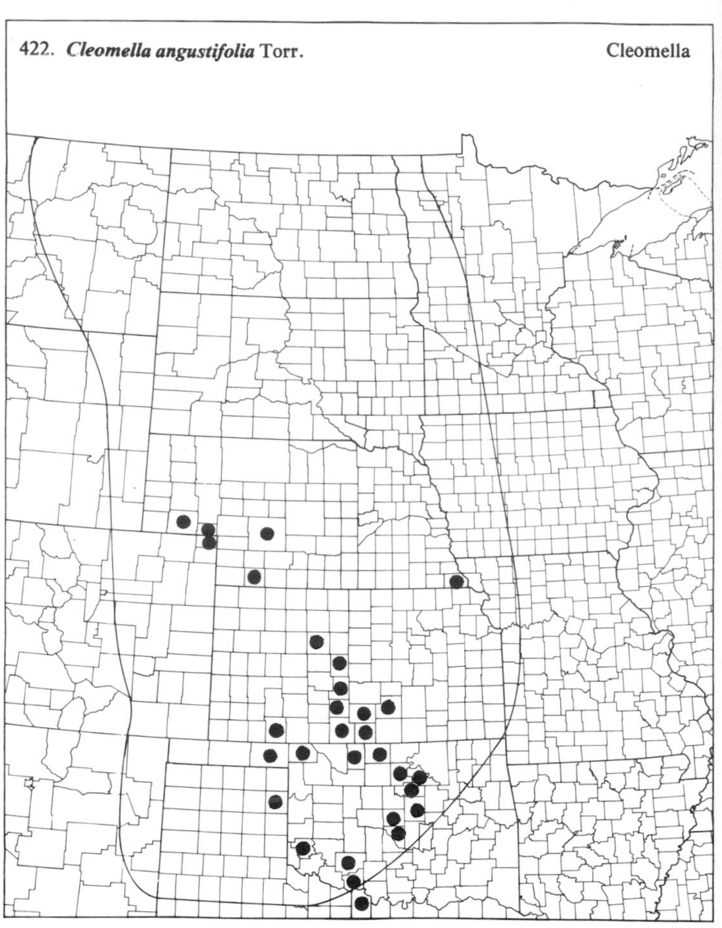

423. *Cristatella jamesii* T. & G. — Cristatella

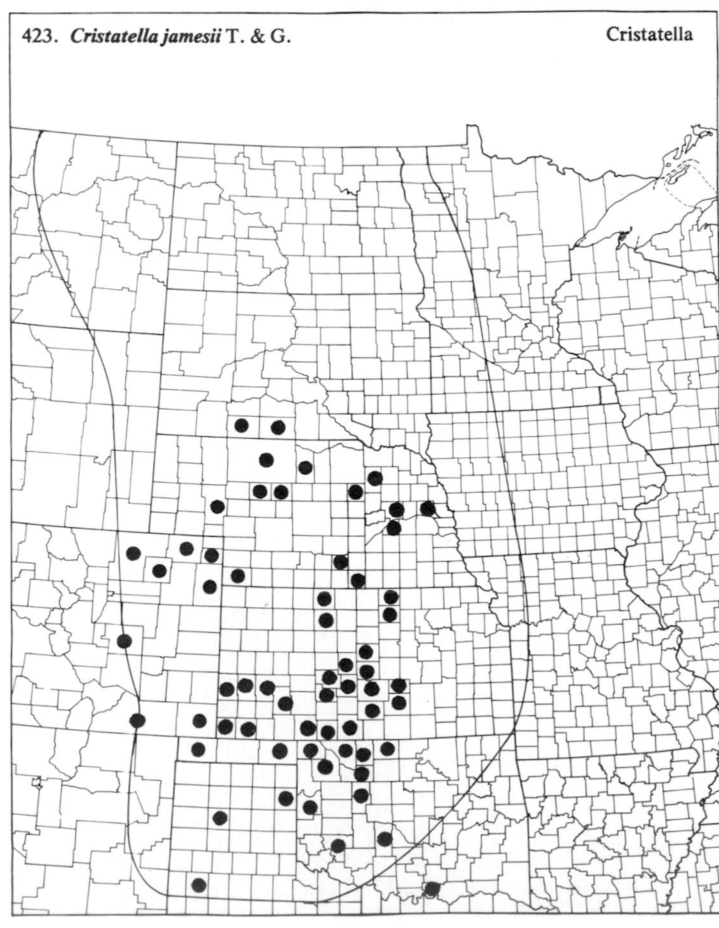

424. *Polanisia dodecandra* (L.) DC. ssp. *trachysperma* (T. & G.) Iltis — Clammyweed

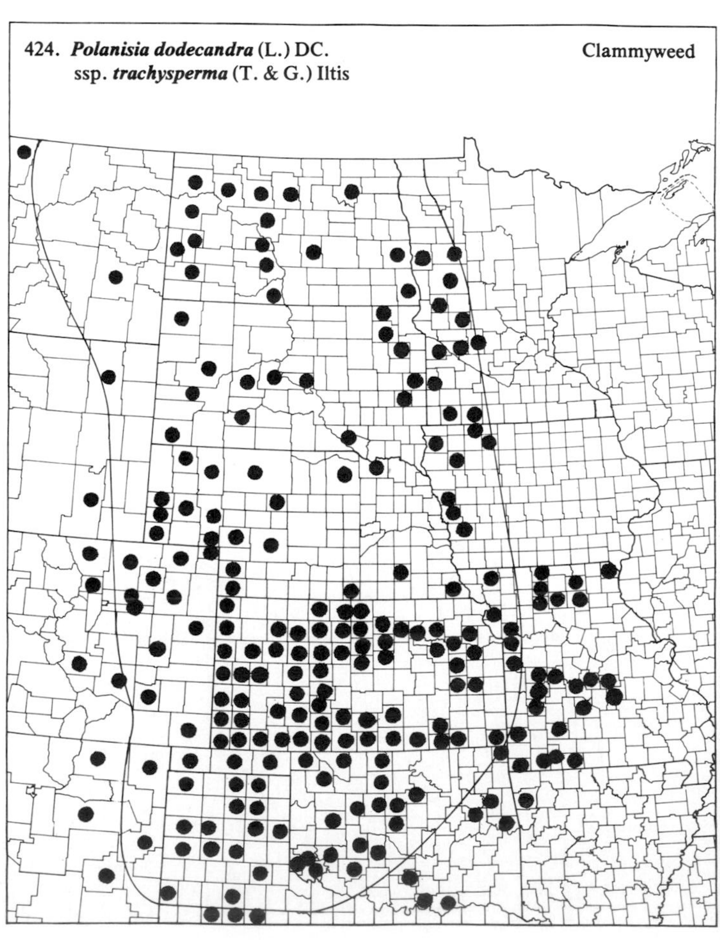

43. *Brassicaceae* (Mustard Family)

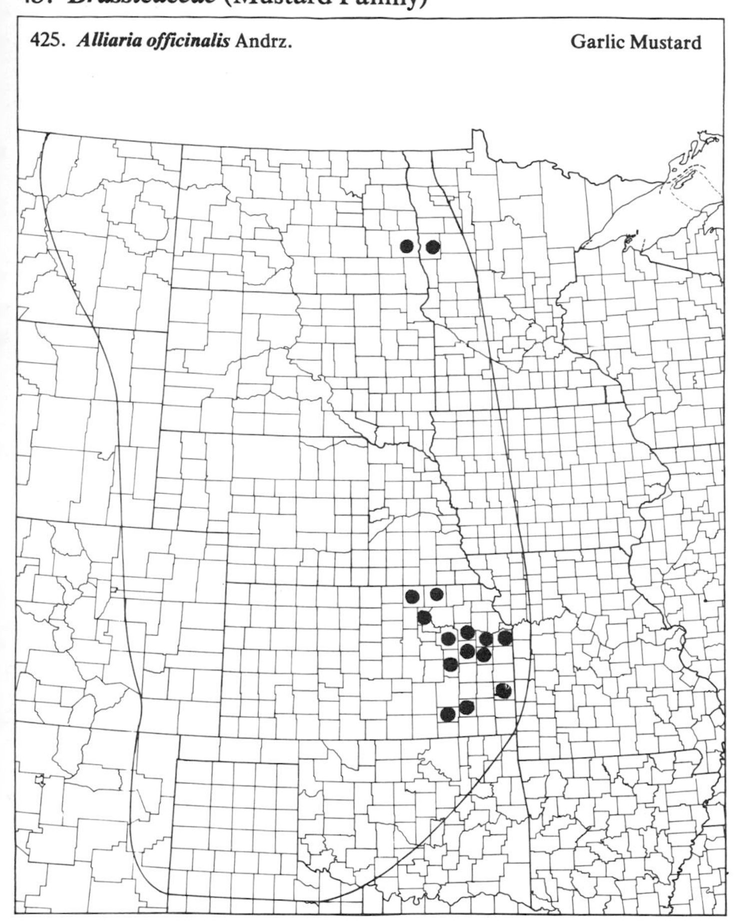

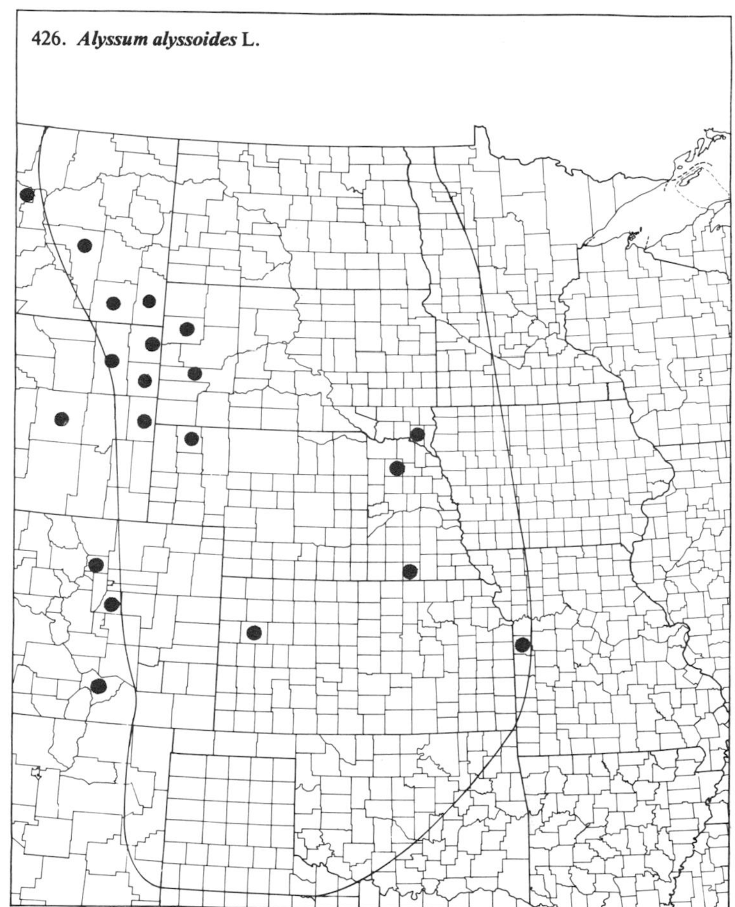

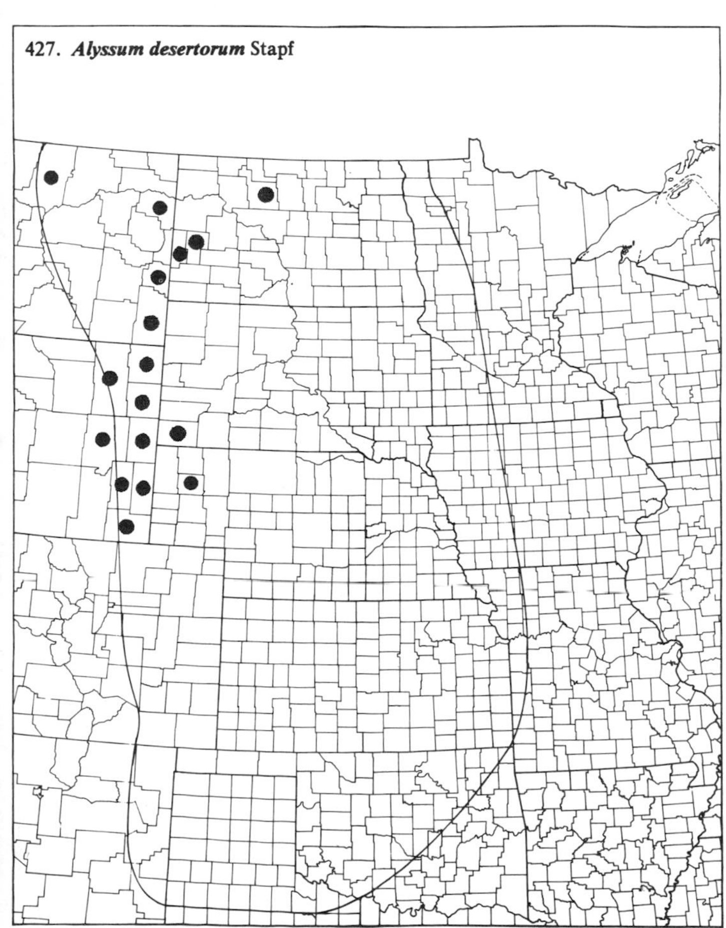

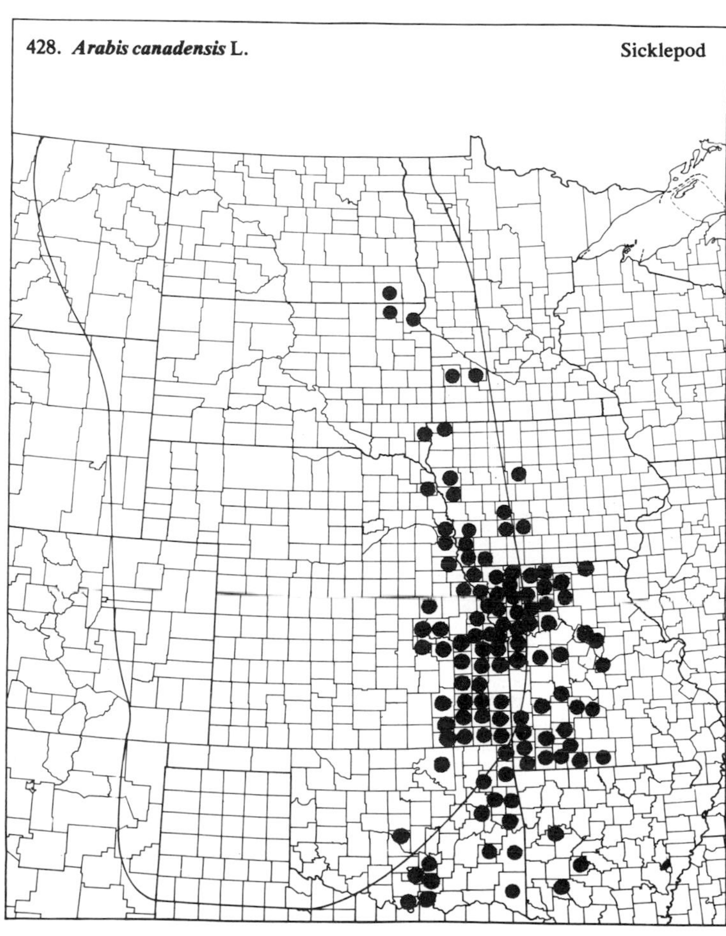

(Brassicaceae)

429. ***Arabis divaricata*** A. Nels.

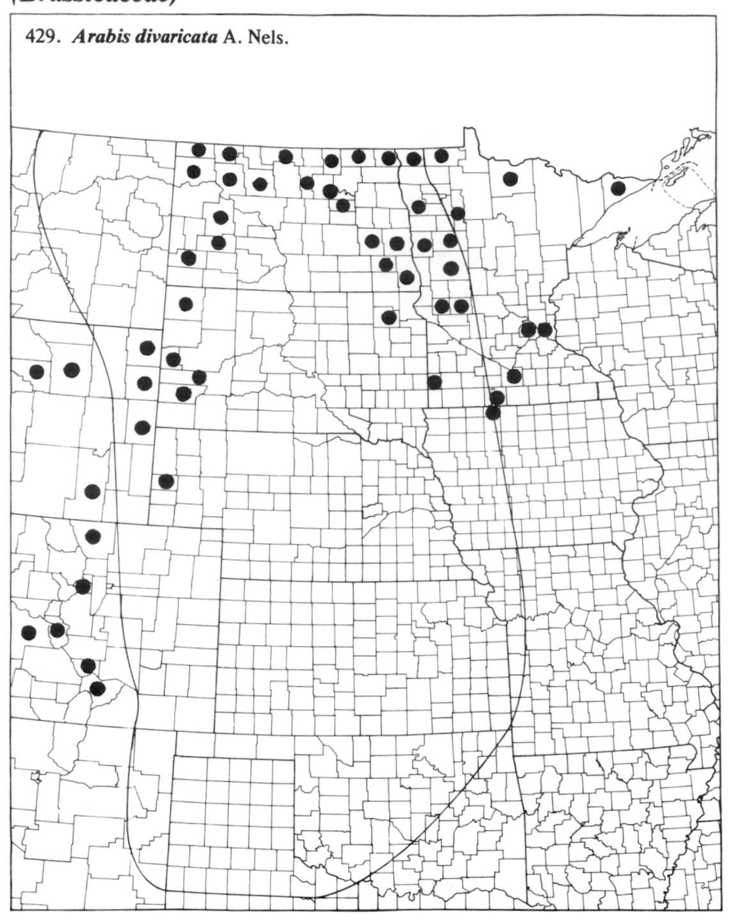

430. ***Arabis drummondii*** Gray

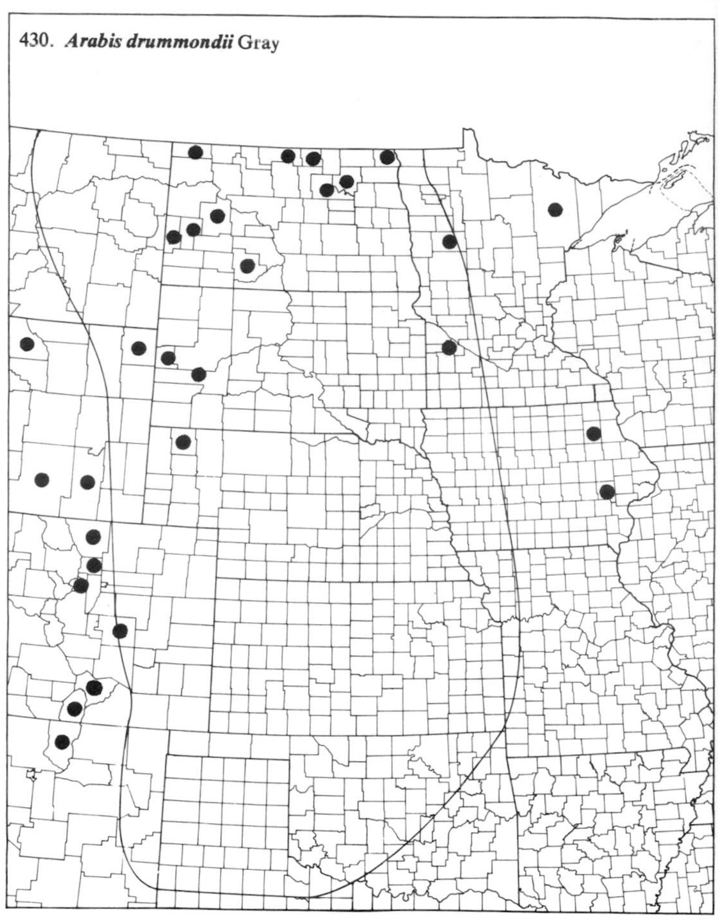

431. ***Arabis glabra*** (L.) Bernh. Tower Mustard

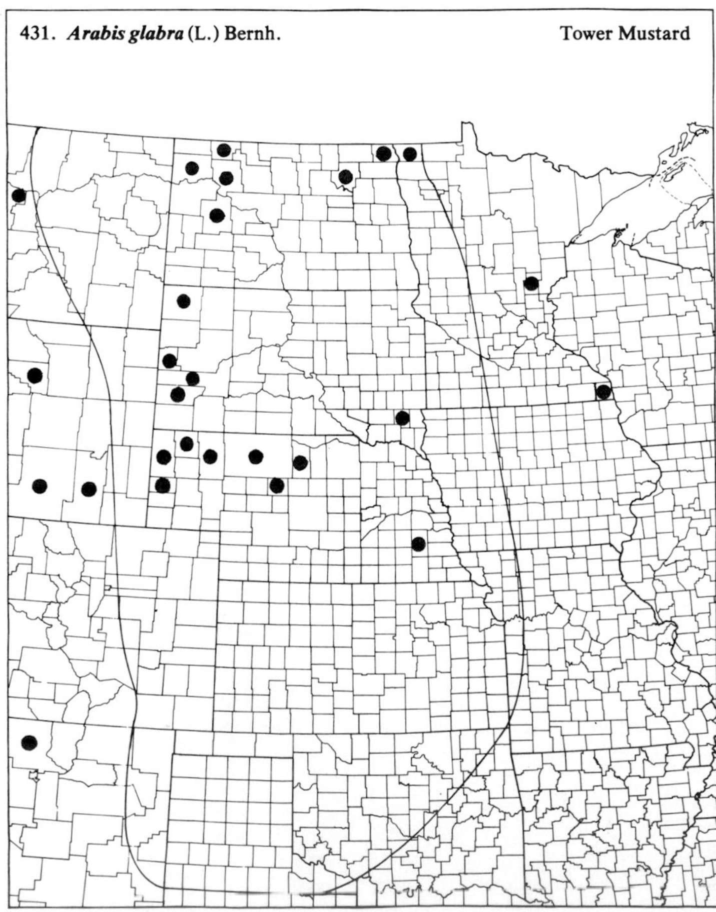

432. ***Arabis hirsuta*** (L.) Scop. var. ***pycnocarpa*** (M. Hopk.) Roll. Rock Cress

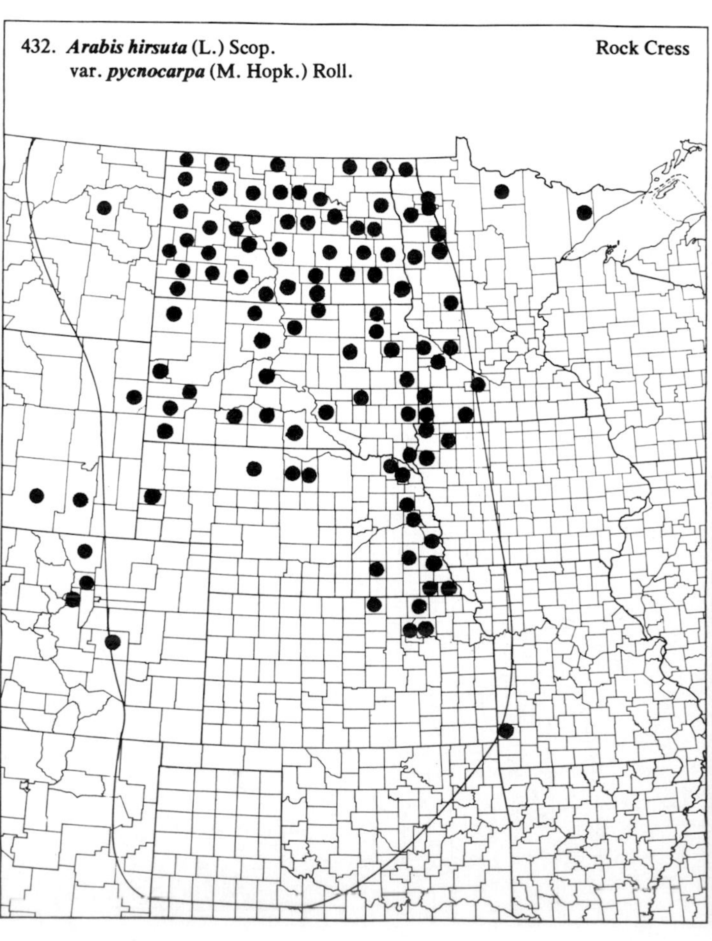

(Brassicaceae)

433. ***Arabis holboellii*** Hornem. var. ***collinsii*** (Fern.) Roll. — Rock Cress

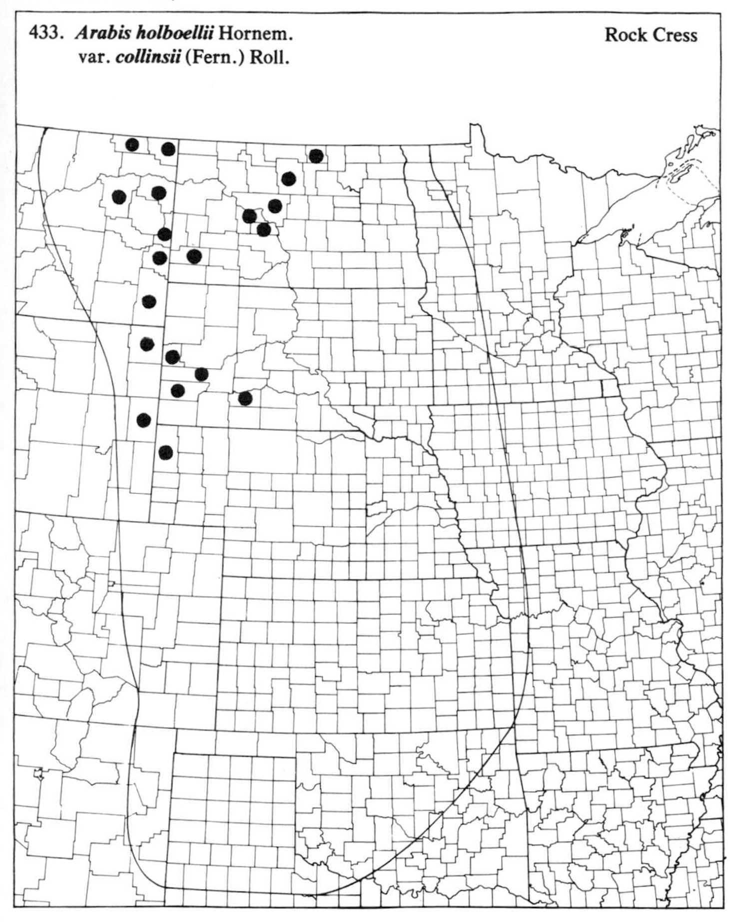

434. ***Arabis holboellii*** Hornem. var. ***pinetorum*** (Tidest.) Roll. — Rock Cress

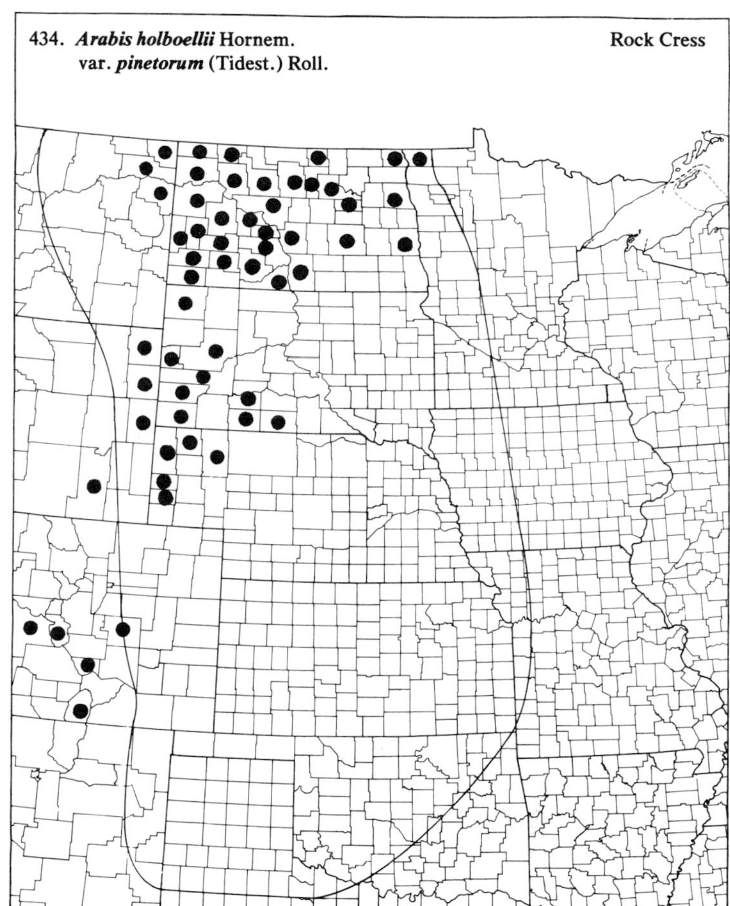

435. ***Arabis shortii*** (Fern.) Gl. — Rock Cress

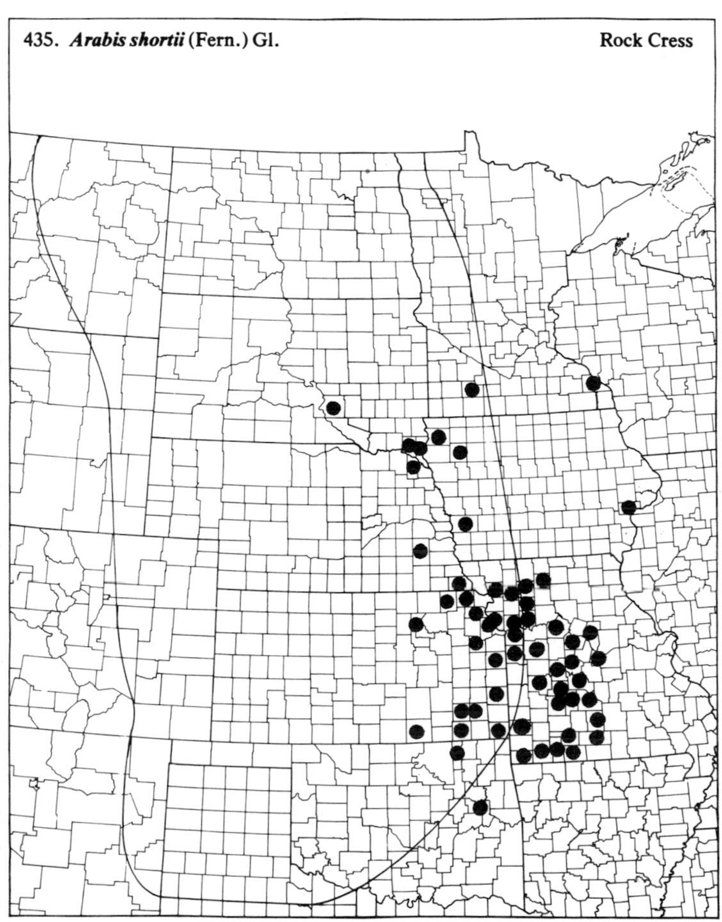

436. ***Arabis virginica*** (L.) Poir. — Rock Cress

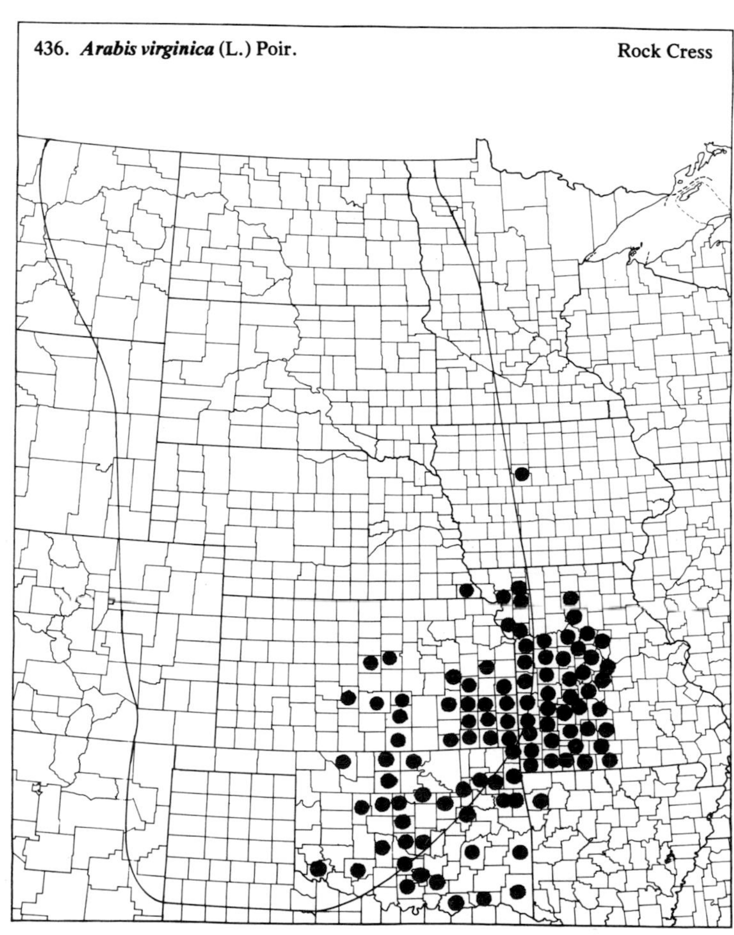

(Brassicaceae)

437. ***Armoracia rusticana*** Gaertn., Mey., & Schreb. Horseradish

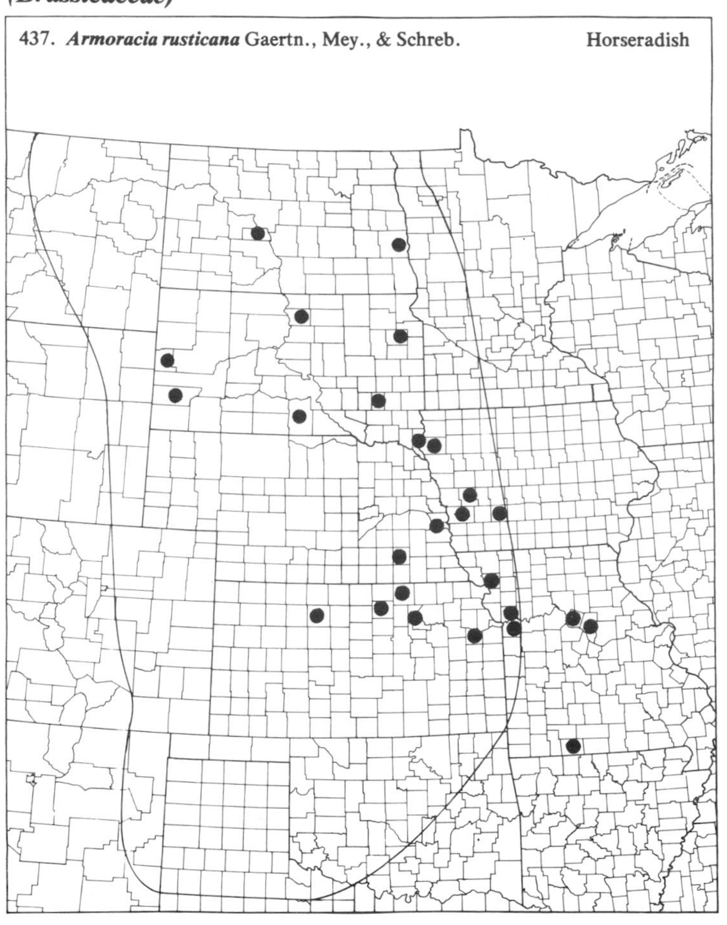

438. ***Barbarea vulgaris*** R. Br. Winter Cress

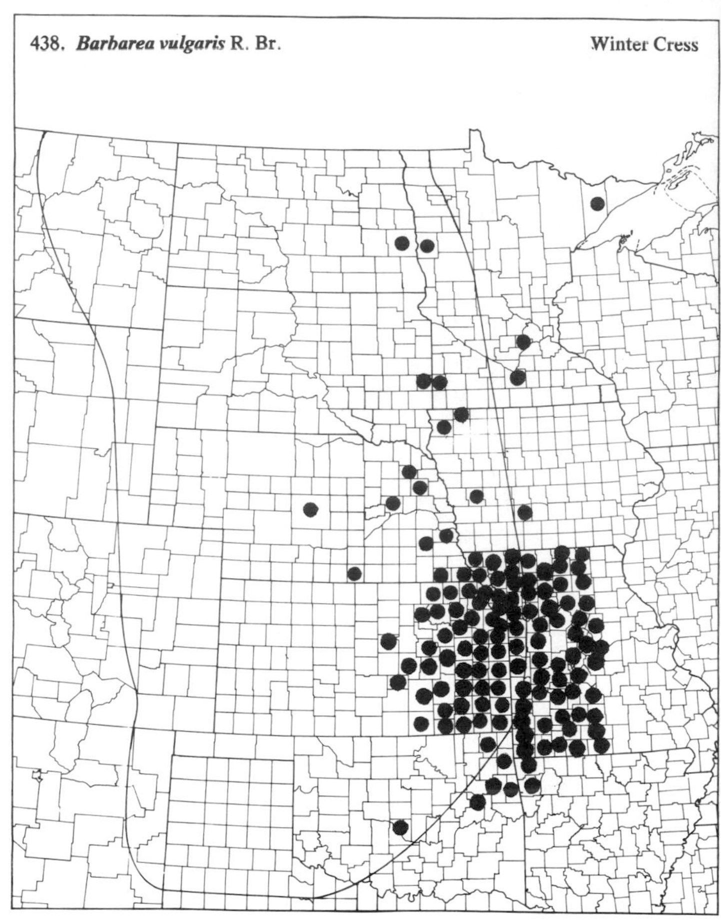

439. ***Berteroa incana*** (L.) DC. Hoary False Alyssum

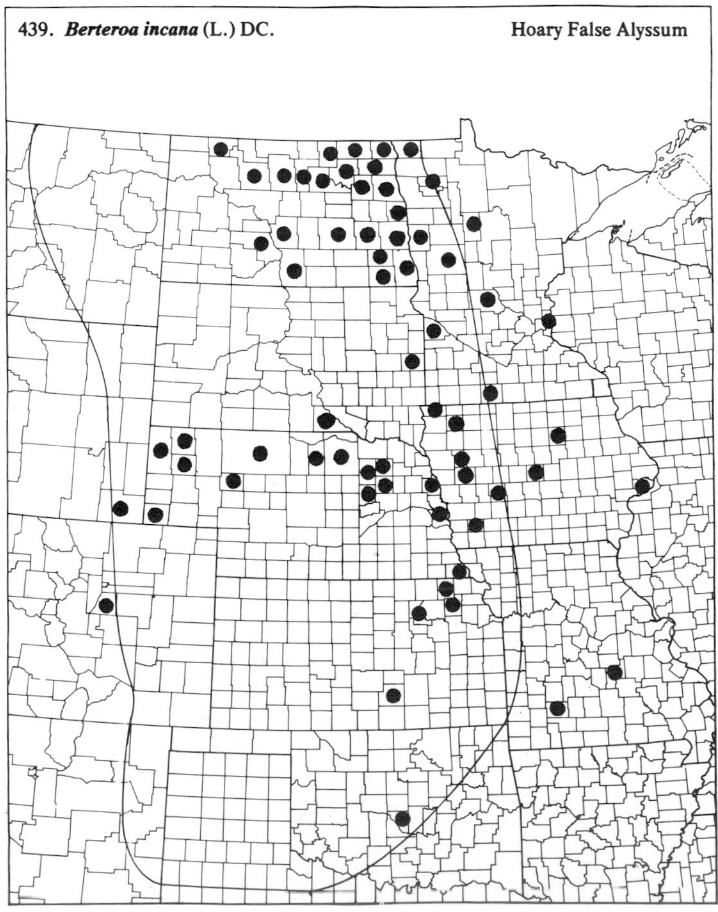

440. ***Brassica campestris*** L. Wild Turnip

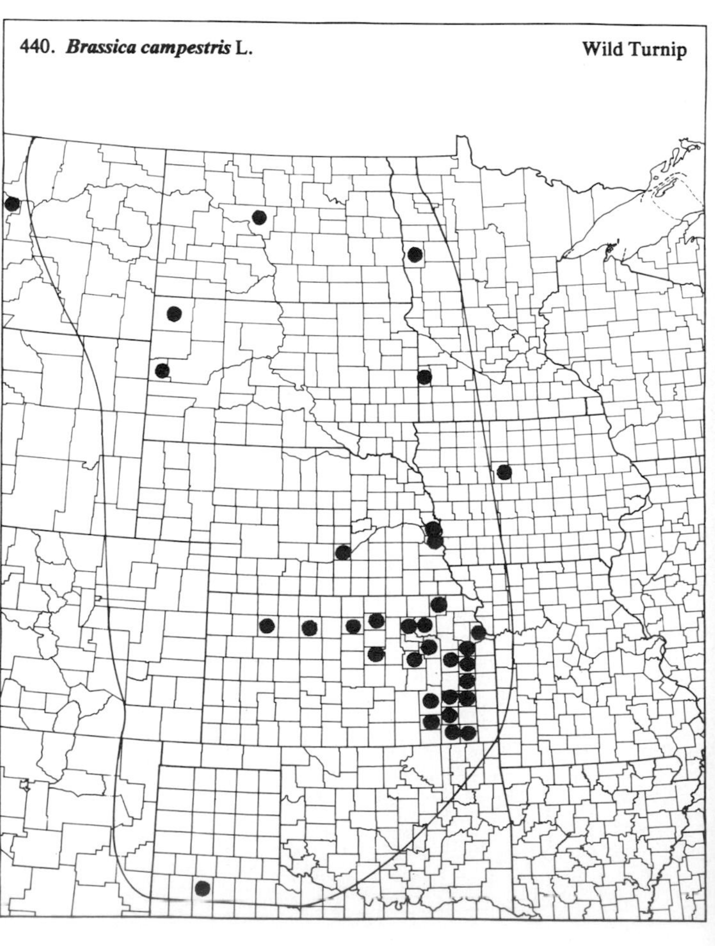

(Brassicaceae)

441. **Brassica hirta** Moench — White Mustard

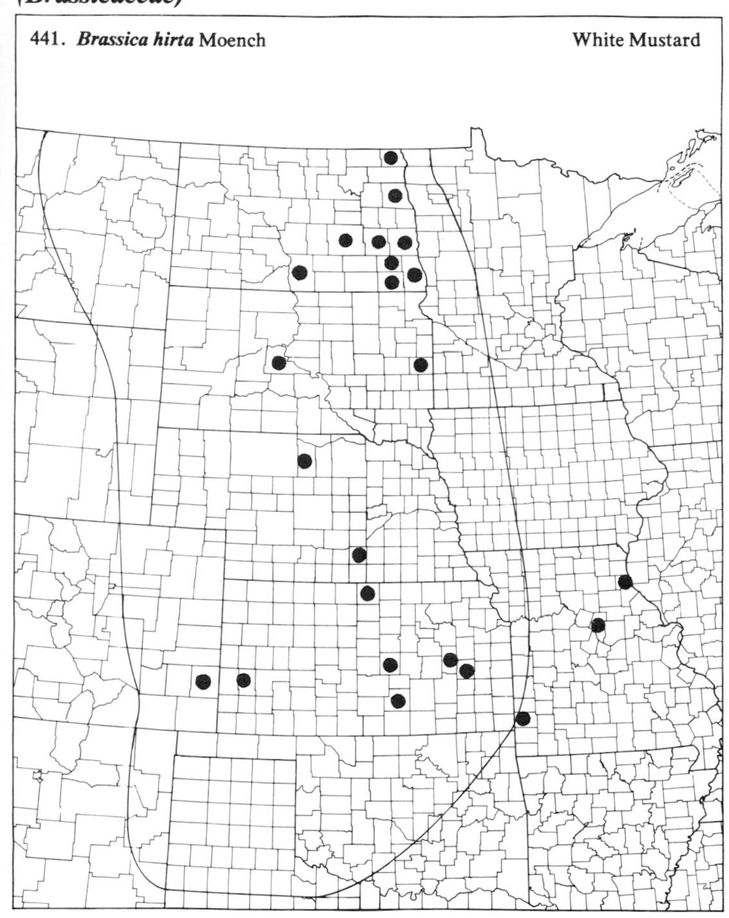

442. **Brassica juncea** (L.) Coss. — Indian Mustard

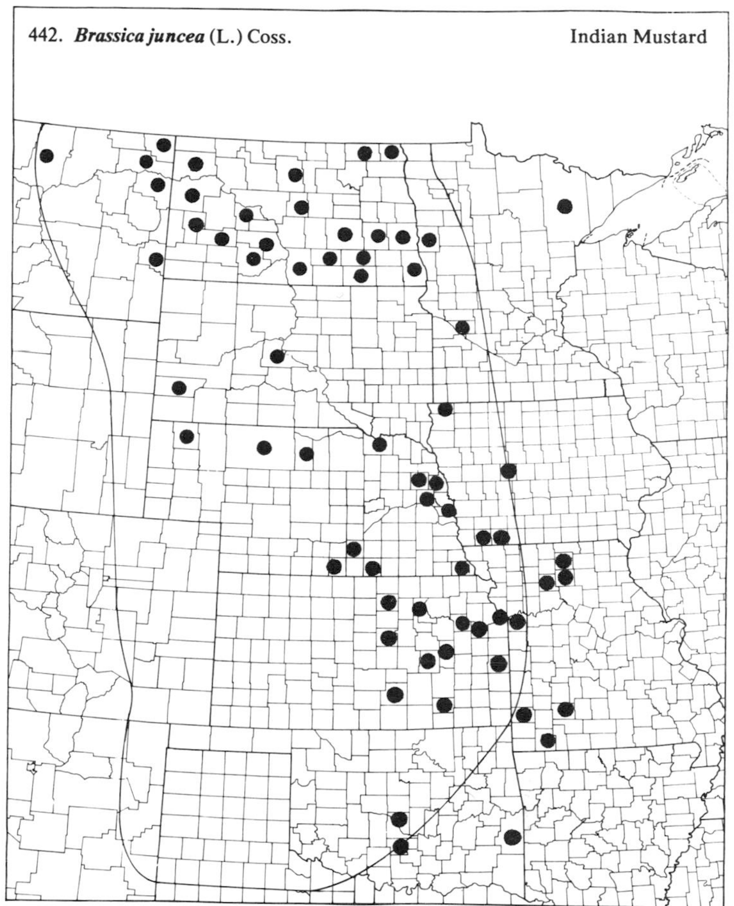

443. **Brassica kaber** (DC.) Wheeler — Charlock

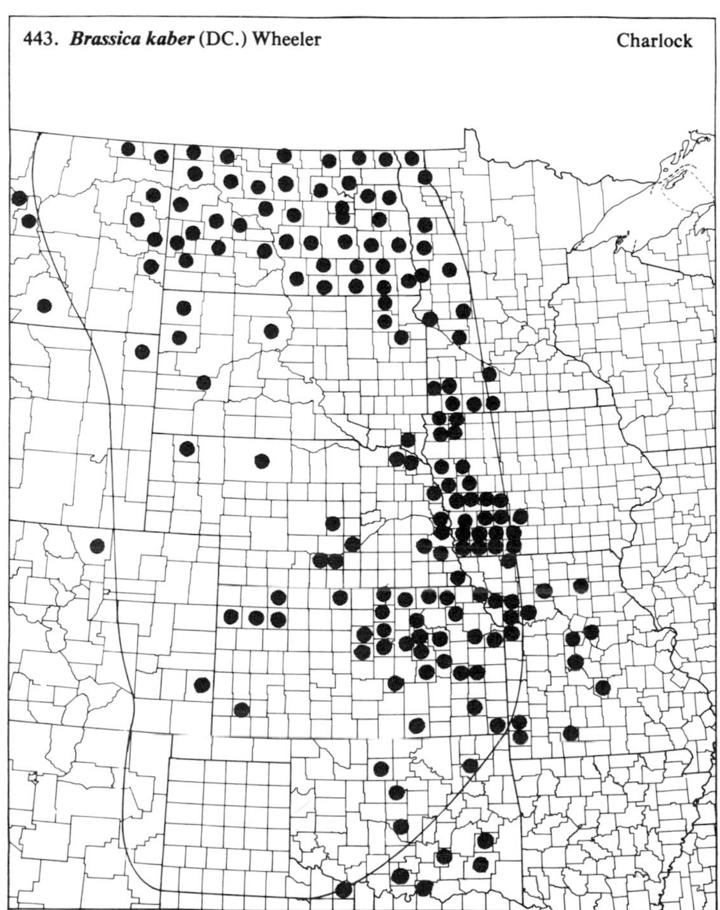

444. **Brassica nigra** (L.) Koch — Black Mustard

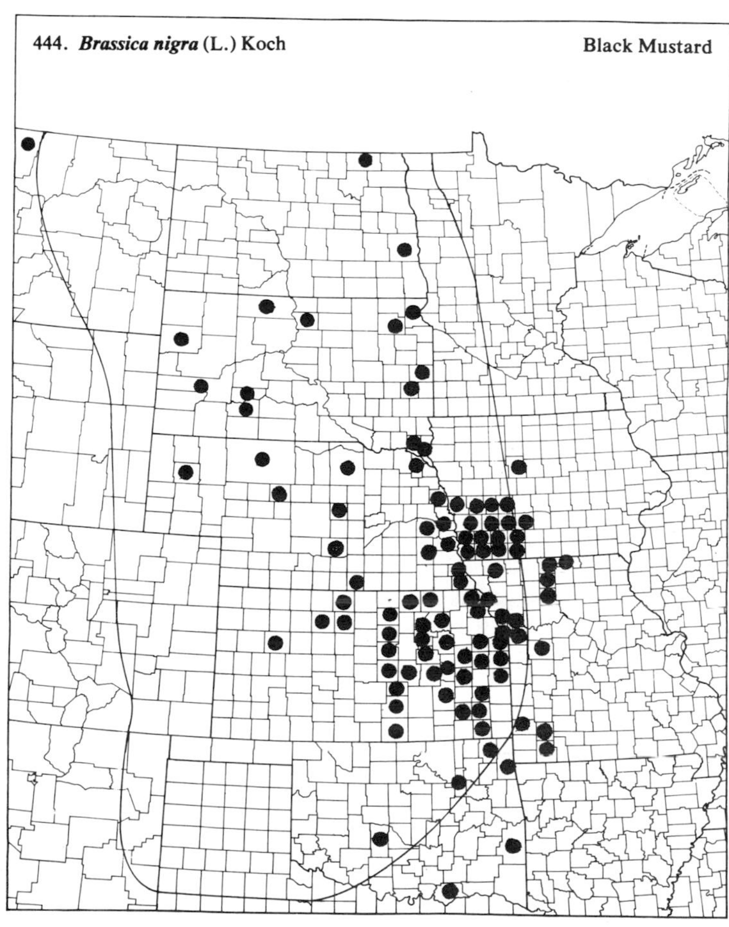

(Brassicaceae)

445. ***Camelina microcarpa*** Andrz. Small-seeded False Flax

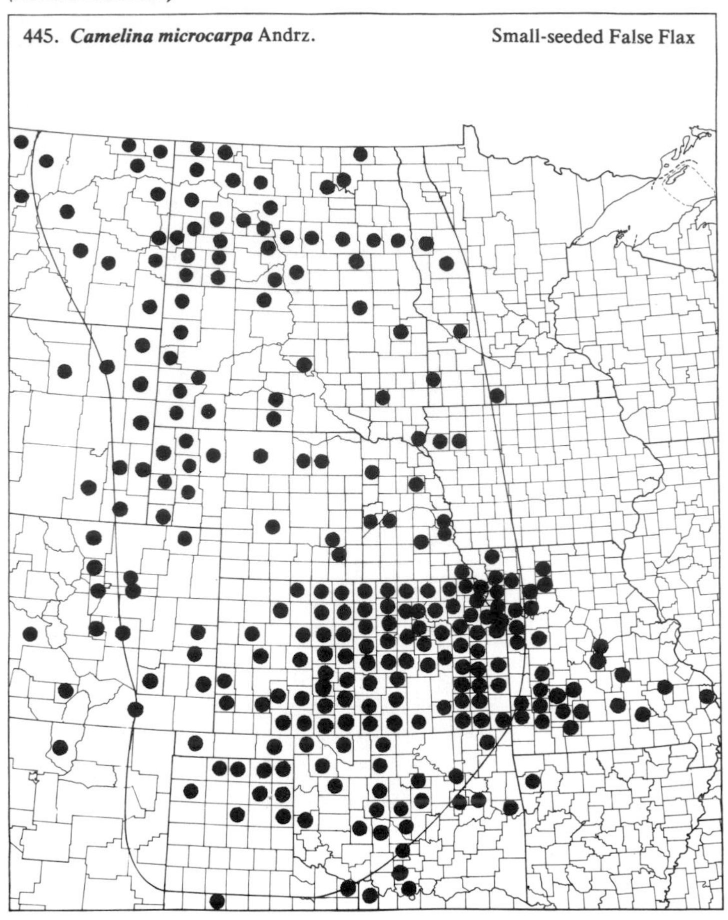

446. ***Camelina sativa*** (L.) Crantz Gold-of-pleasure

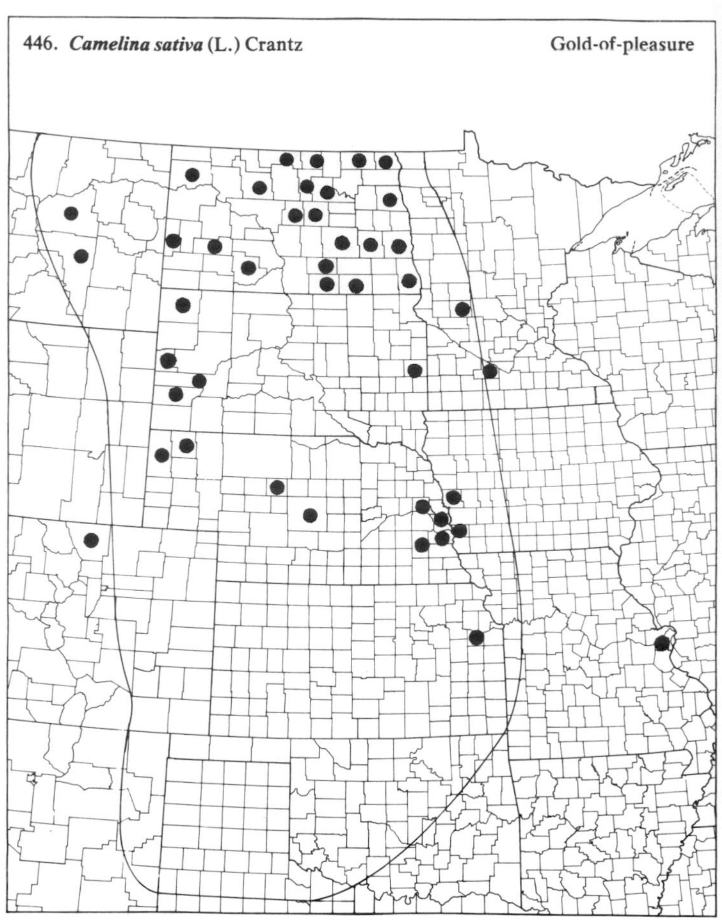

447. ***Capsella bursa-pastoris*** (L.) Medic. Shepherd's Purse

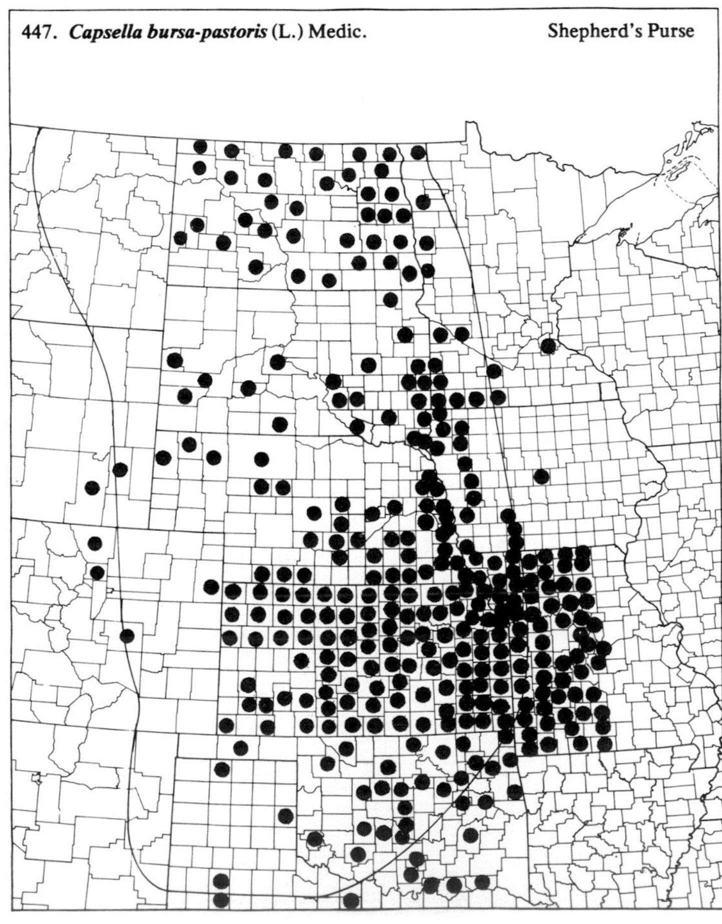

448. ***Cardamine bulbosa*** (Schreb.) B.S.P. Spring Cress

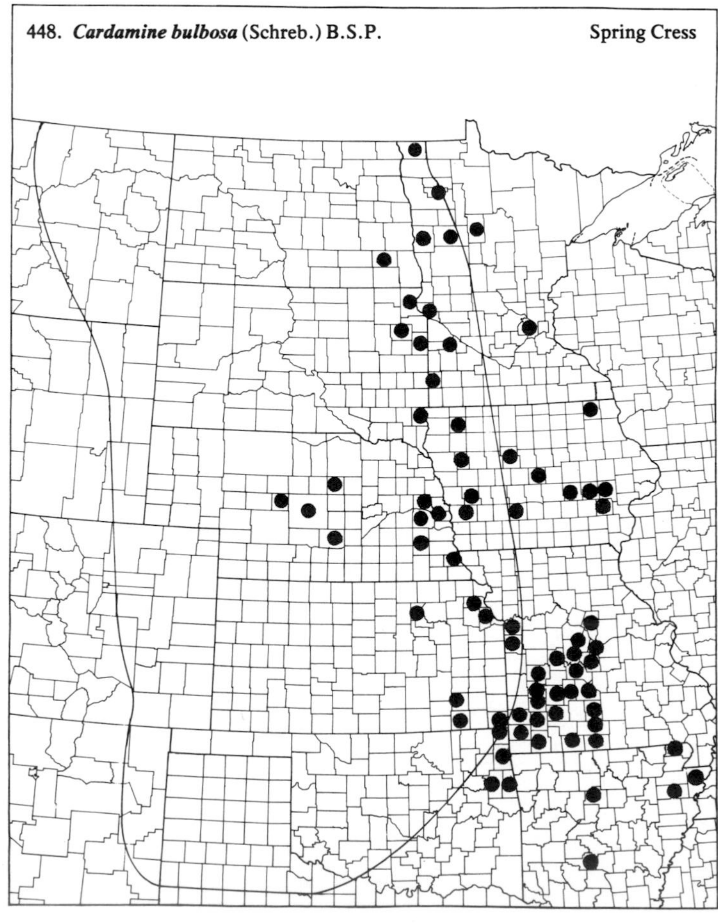

(Brassicaceae)

449. ***Cardamine parviflora*** L. var. ***arenicola*** (Britt.) Schulz — Smallflower Bitter Cress

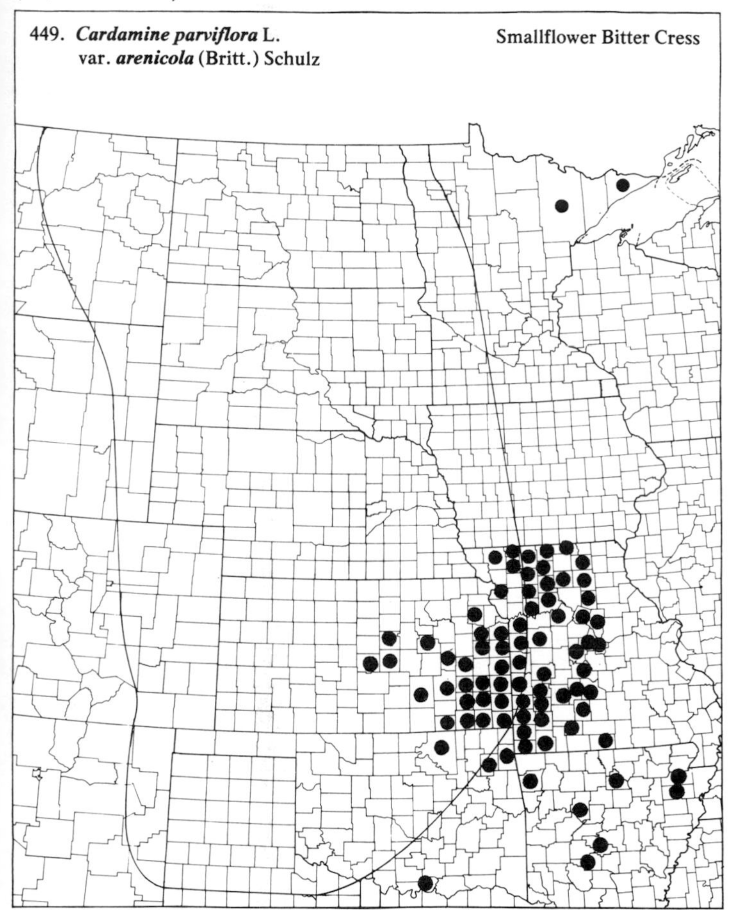

450. ***Cardamine pensylvanica*** Muhl. — Bitter Cress

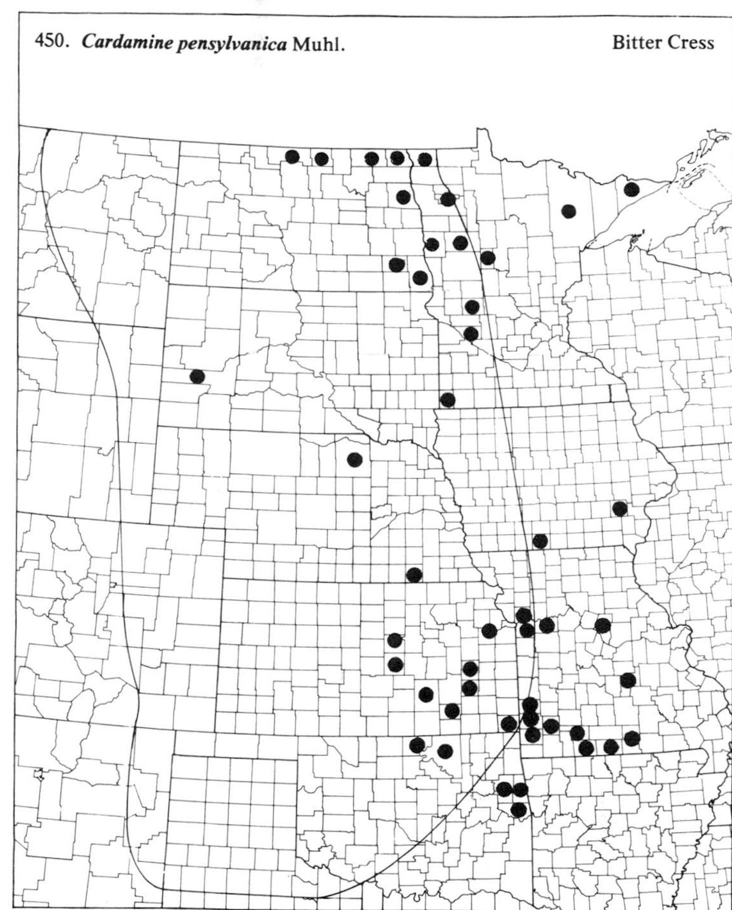

451. ***Cardaria draba*** (L.) Desv. — Hoary Cress

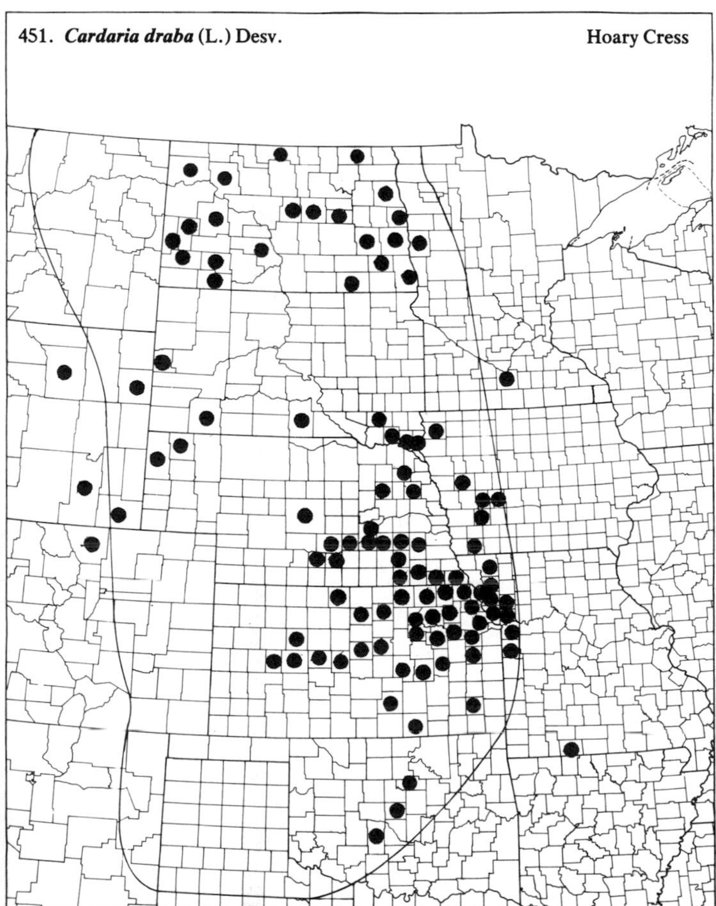

452. ***Chorispora tenella*** (Pall.) DC. — Blue Mustard

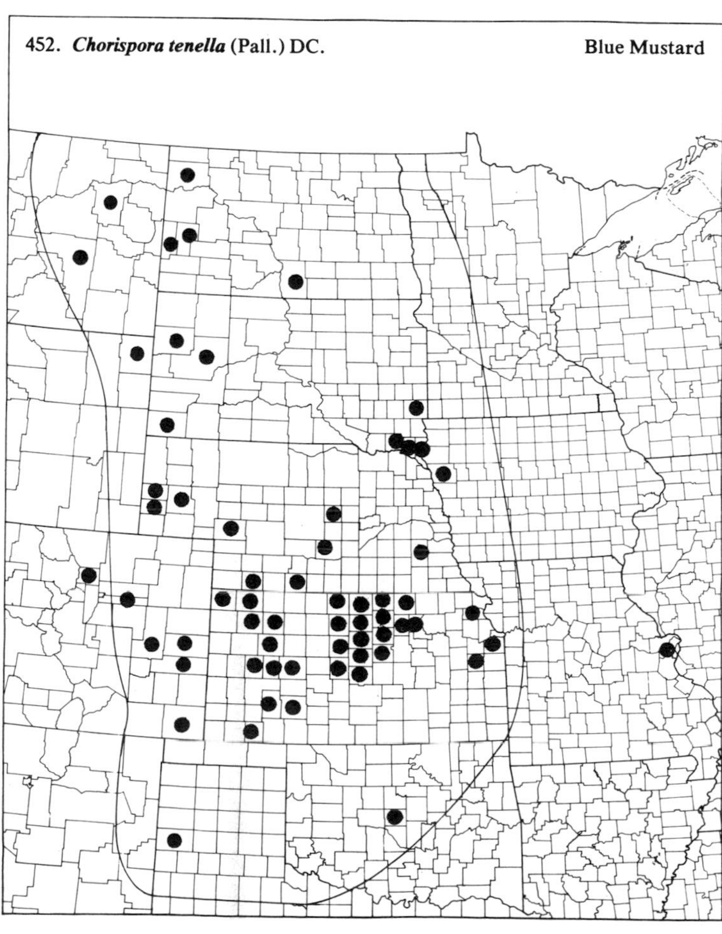

(Brassicaceae)

453. ***Conringia orientalis*** (L.) Dum. — Hare's-ear Mustard

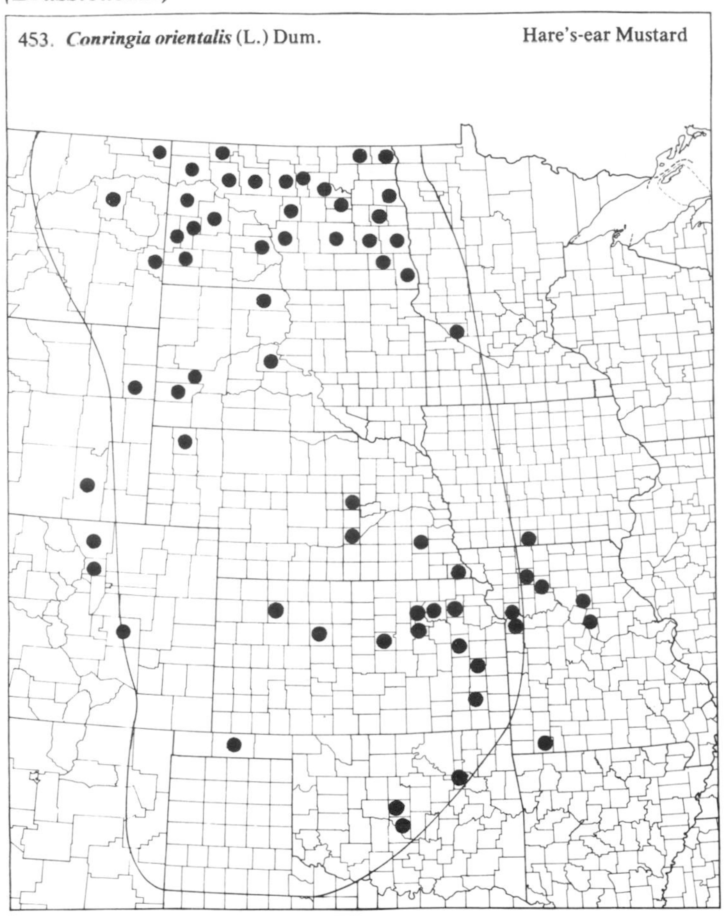

454. ***Dentaria laciniata*** Muhl. — Toothwort

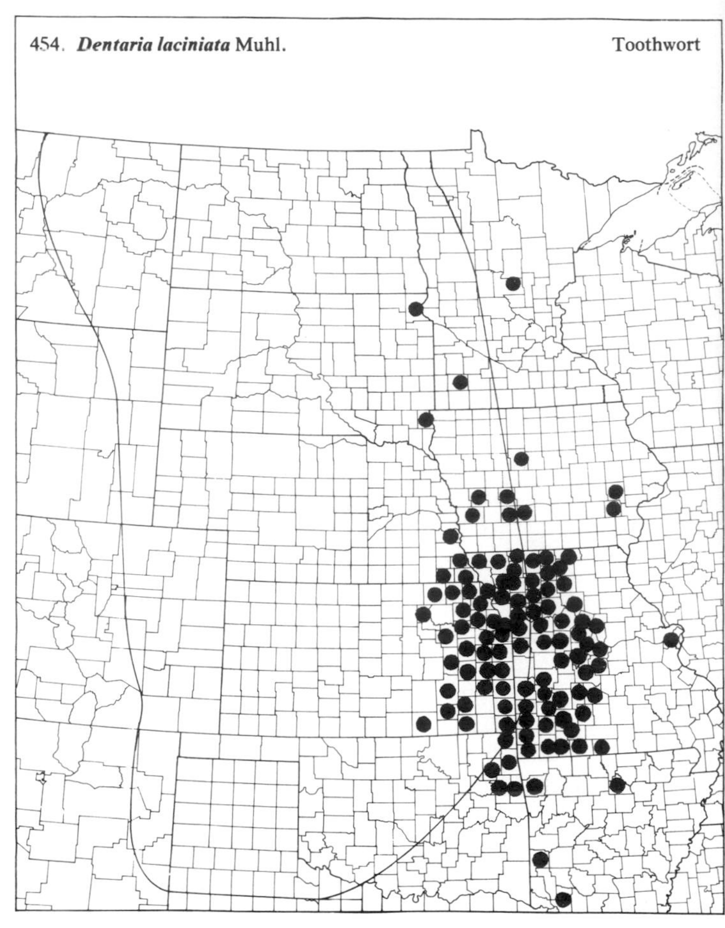

455. ***Descurainia pinnata*** (Walt.) Britt. var. ***brachycarpa*** (Richards.) Fern. — Tansy Mustard

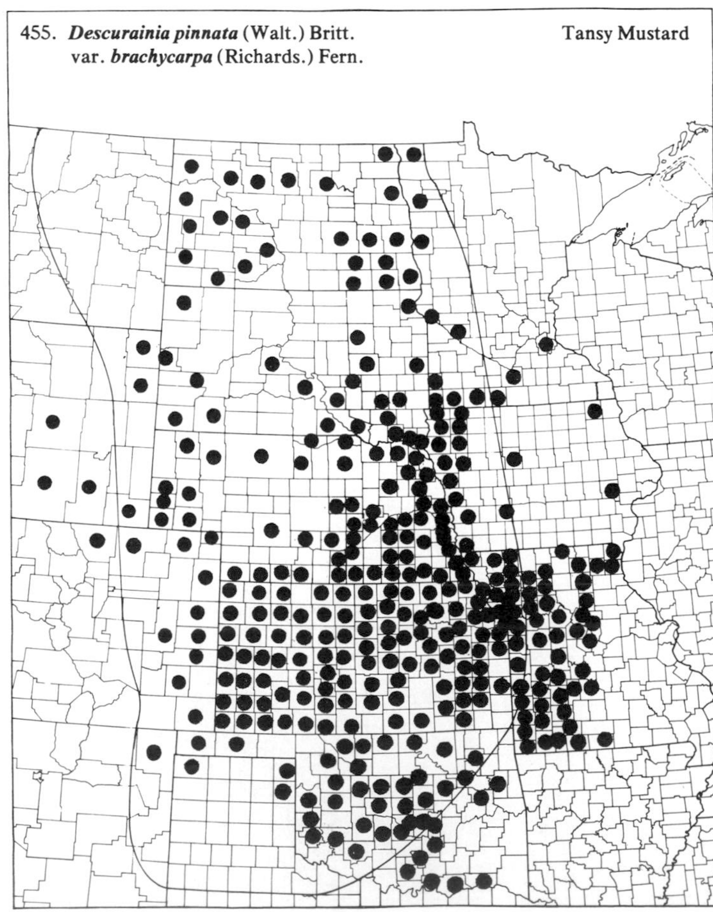

456. ***Descurainia pinnata*** (Walt.) Britt. var. ***osmiarum*** (Cockll.) Shinners — Tansy Mustard

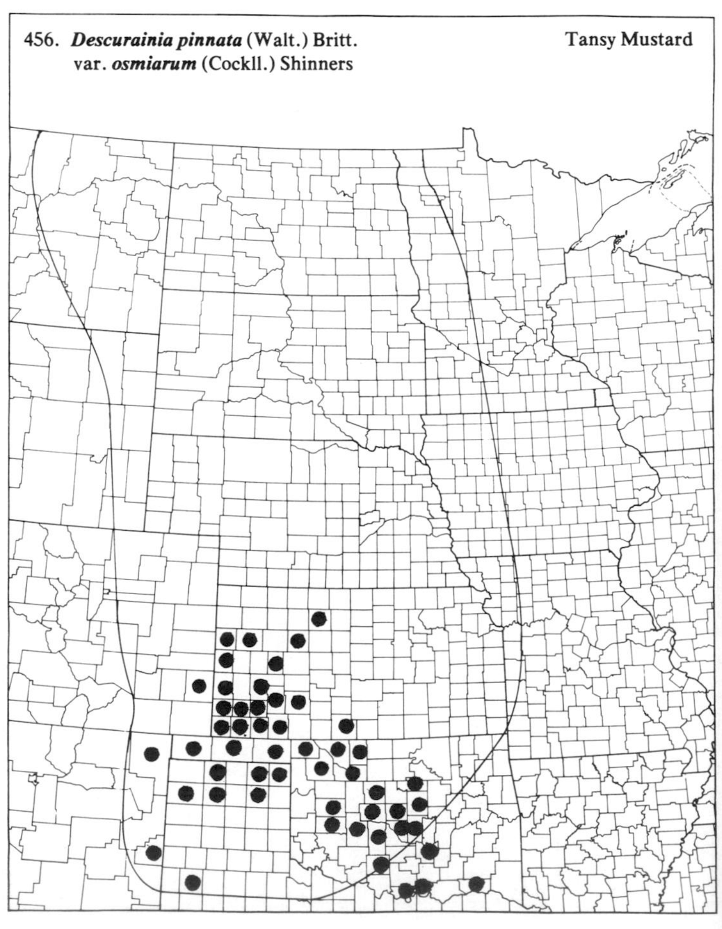

(Brassicaceae)

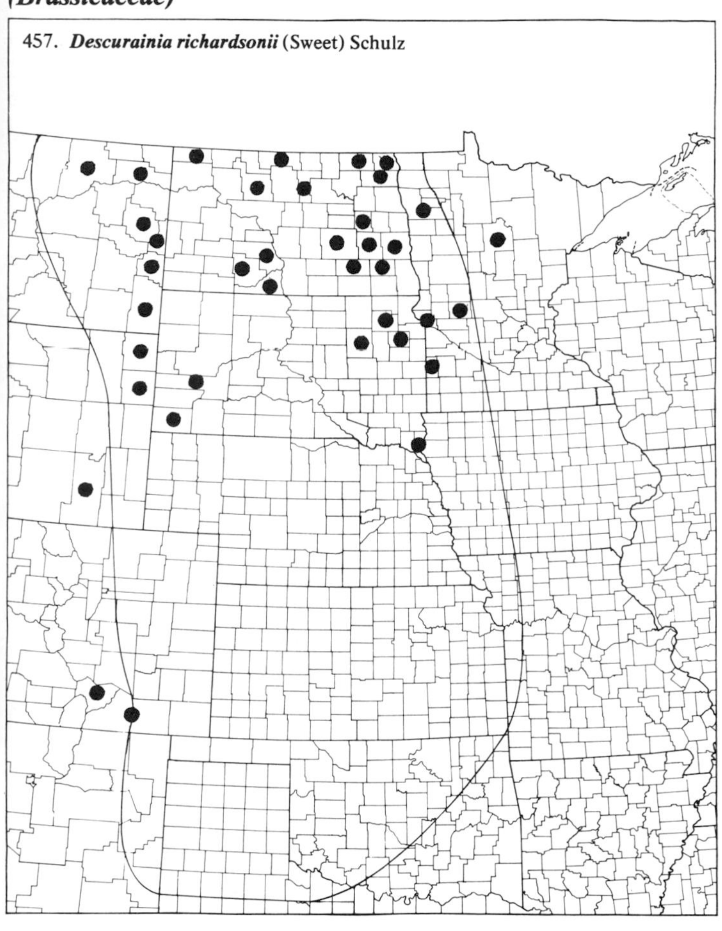

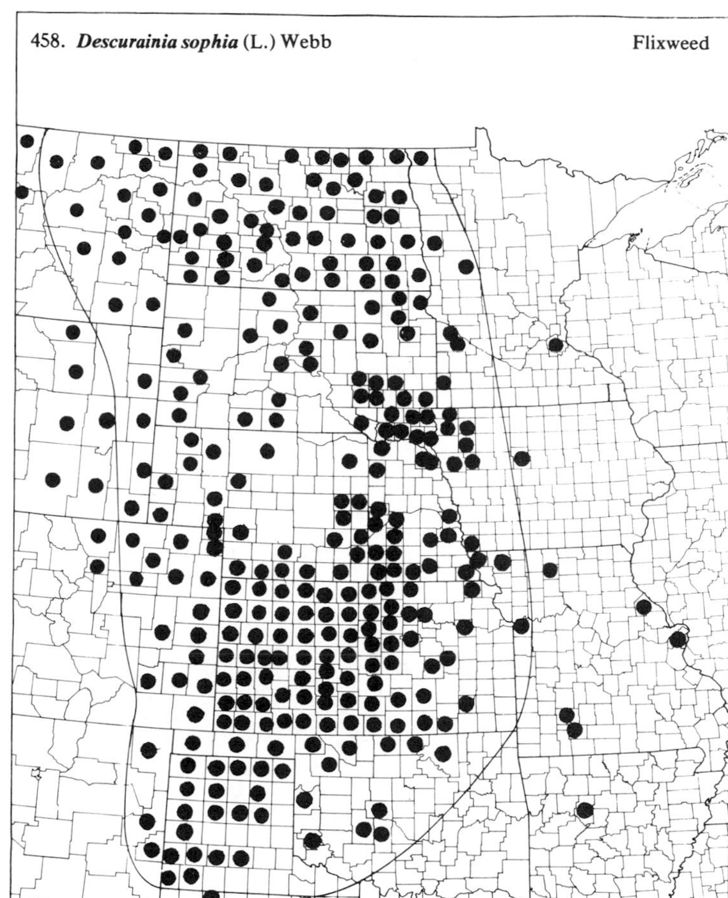

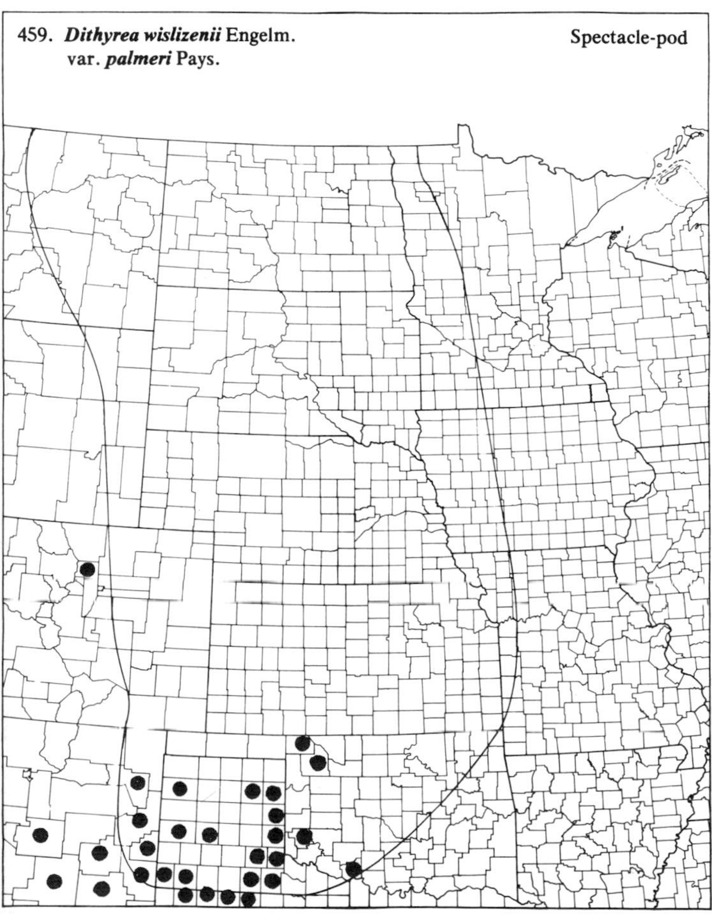

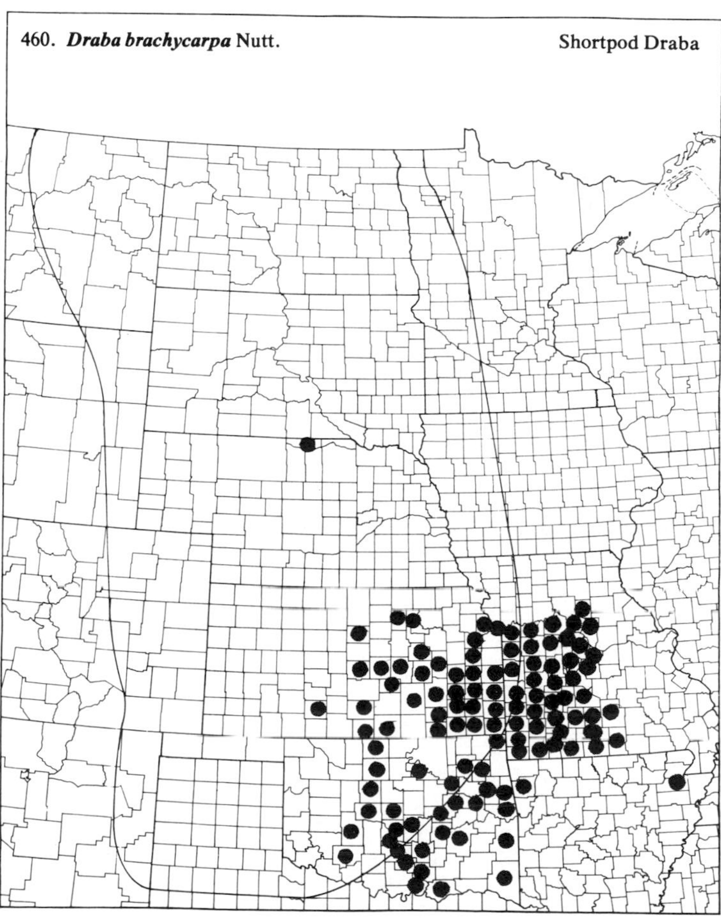

(Brassicaceae)

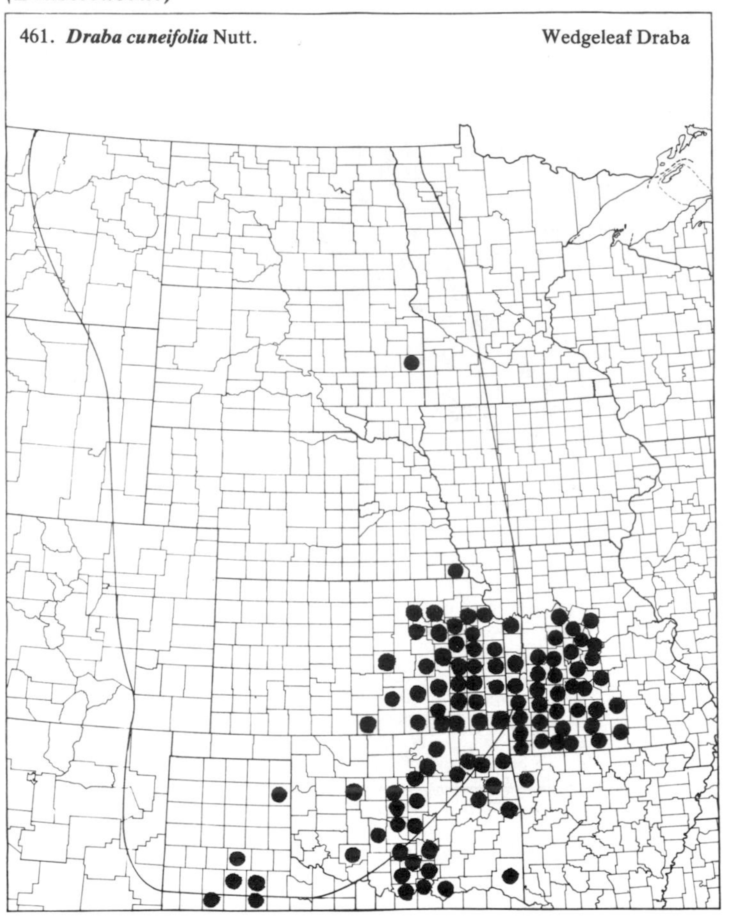

461. ***Draba cuneifolia*** Nutt. Wedgeleaf Draba

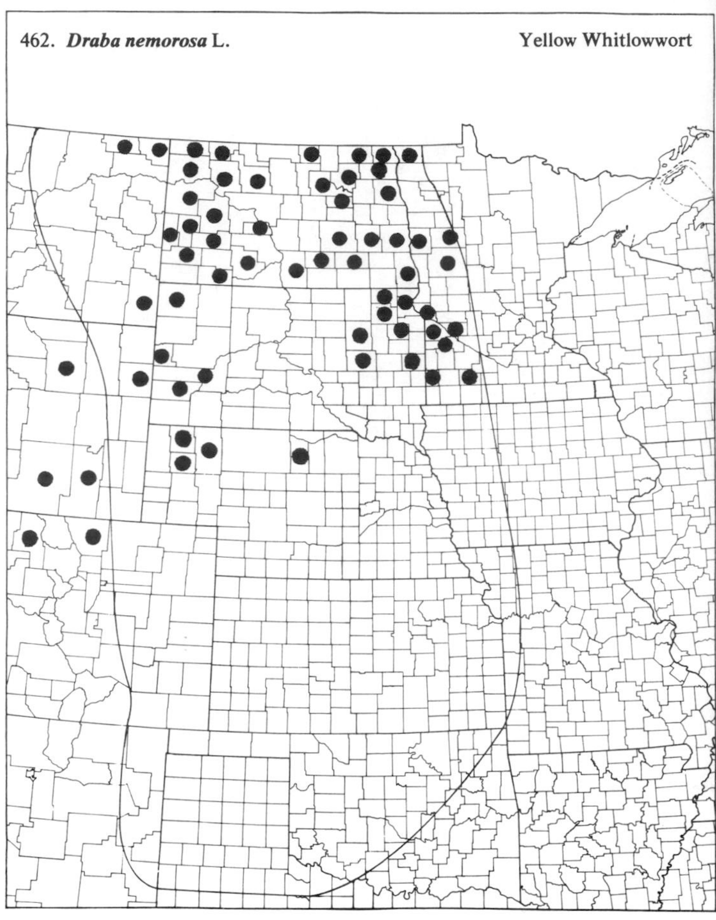

462. ***Draba nemorosa*** L. Yellow Whitlowwort

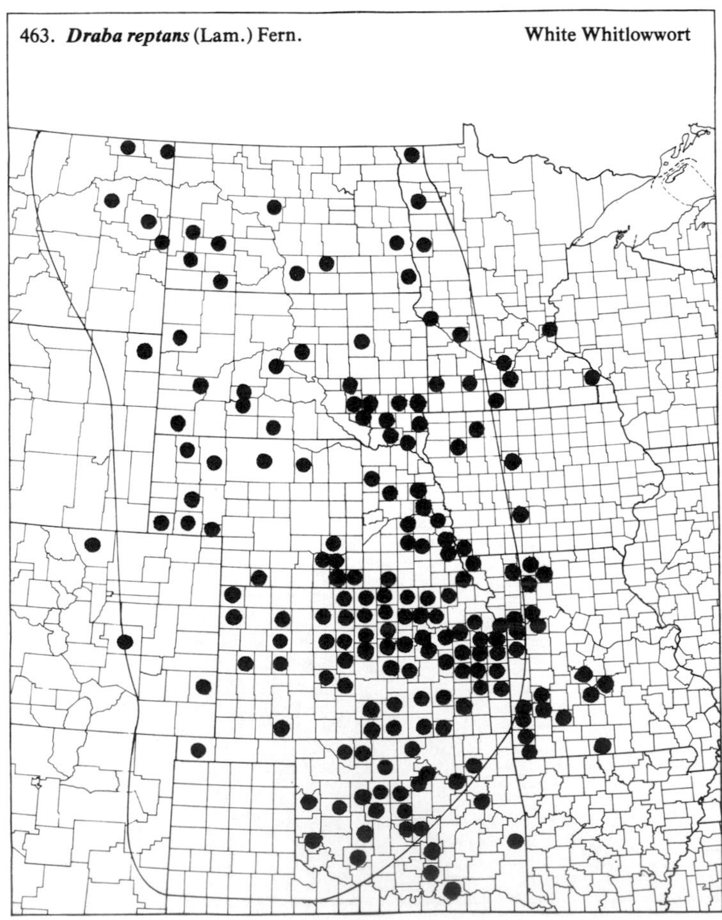

463. ***Draba reptans*** (Lam.) Fern. White Whitlowwort

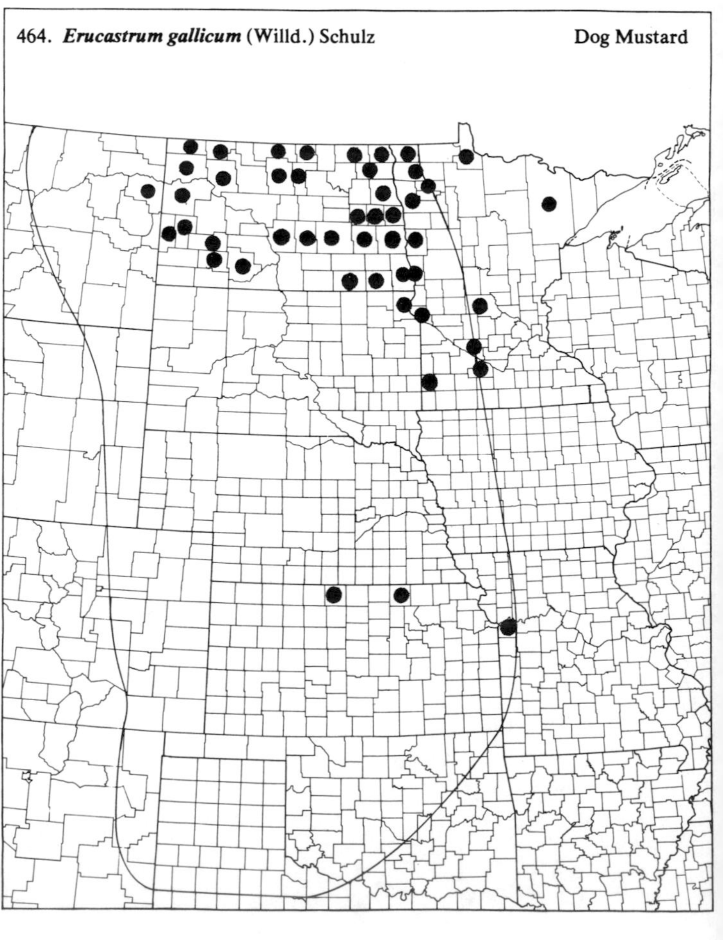

464. ***Erucastrum gallicum*** (Willd.) Schulz Dog Mustard

(Brassicaceae)

465. ***Erysimum asperum*** (Nutt.) DC. Western Wallflower

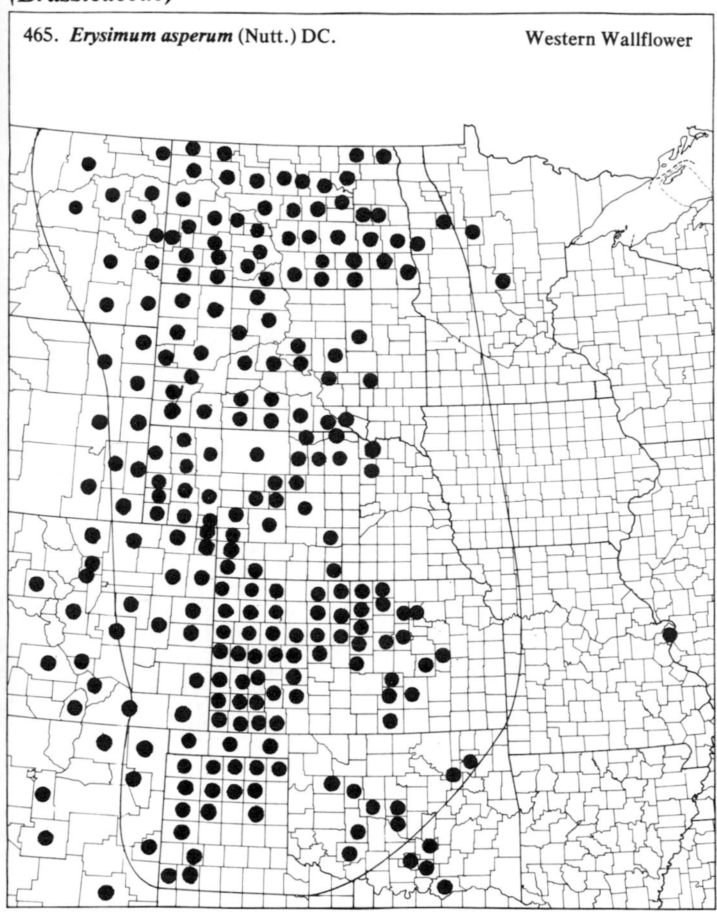

466. ***Erysimum cheiranthoides*** L. Wormseed Wallflower

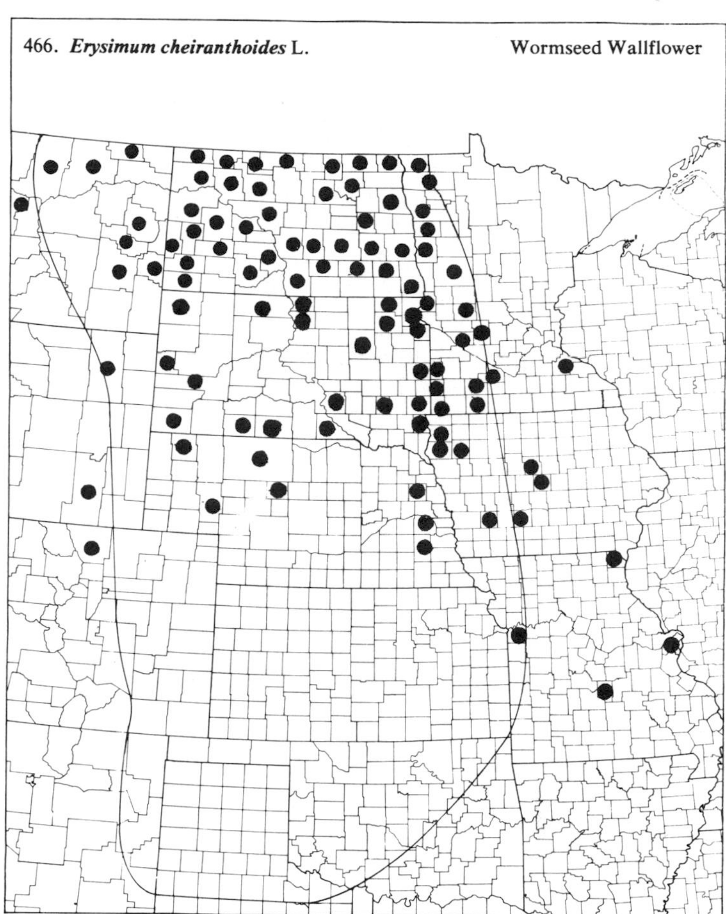

467. ***Erysimum inconspicuum*** (Wats.) MacM. Smallflower Wallflower

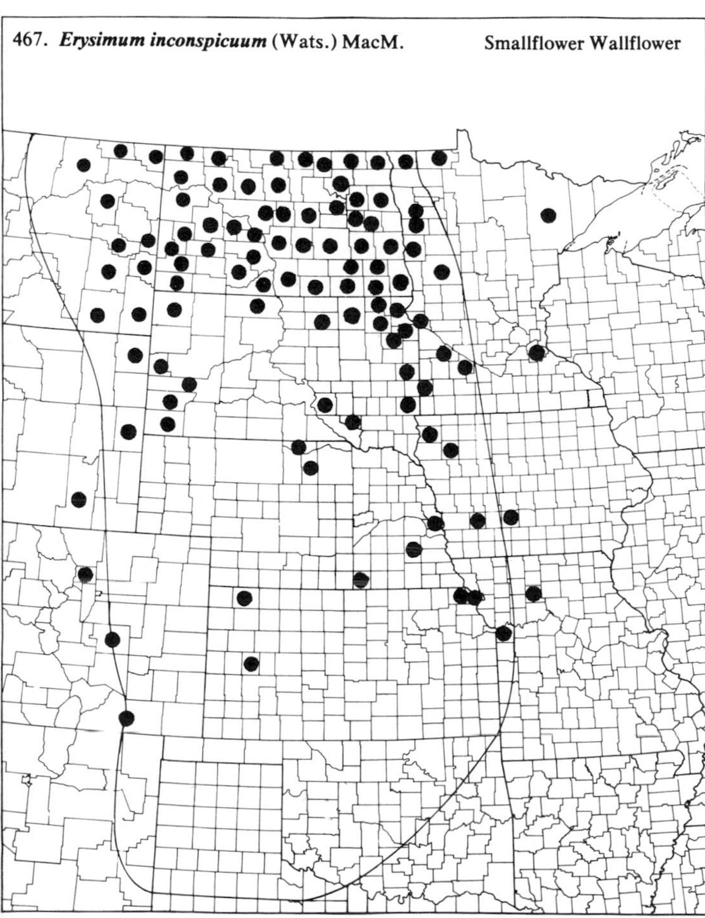

468. ***Erysimum repandum*** L. Bushy Wallflower

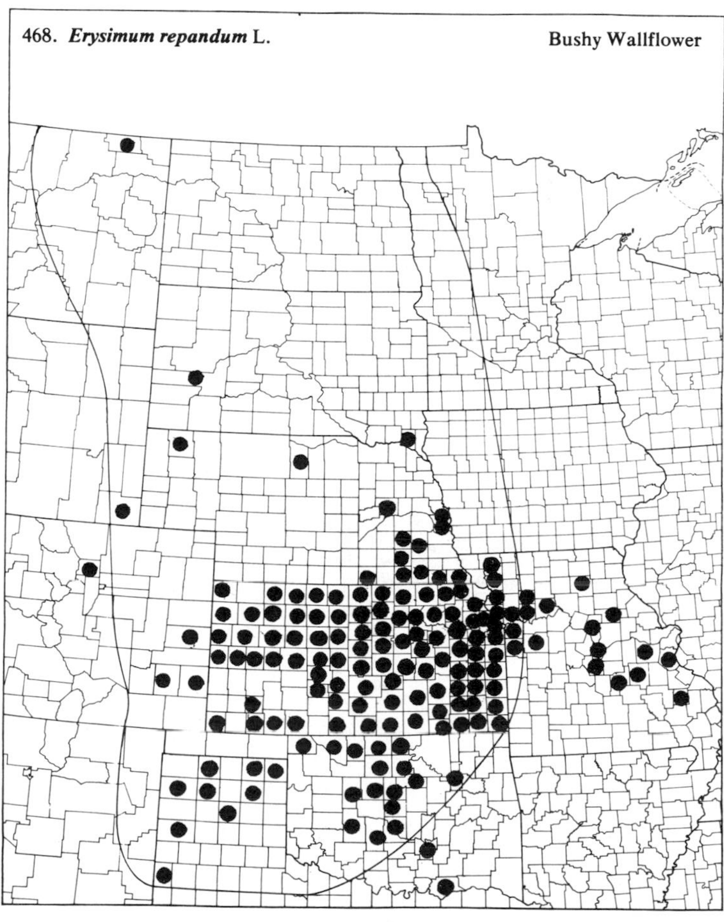

(Brassicaceae)

469. ***Hesperis matronalis*** L. — Dame's Rocket

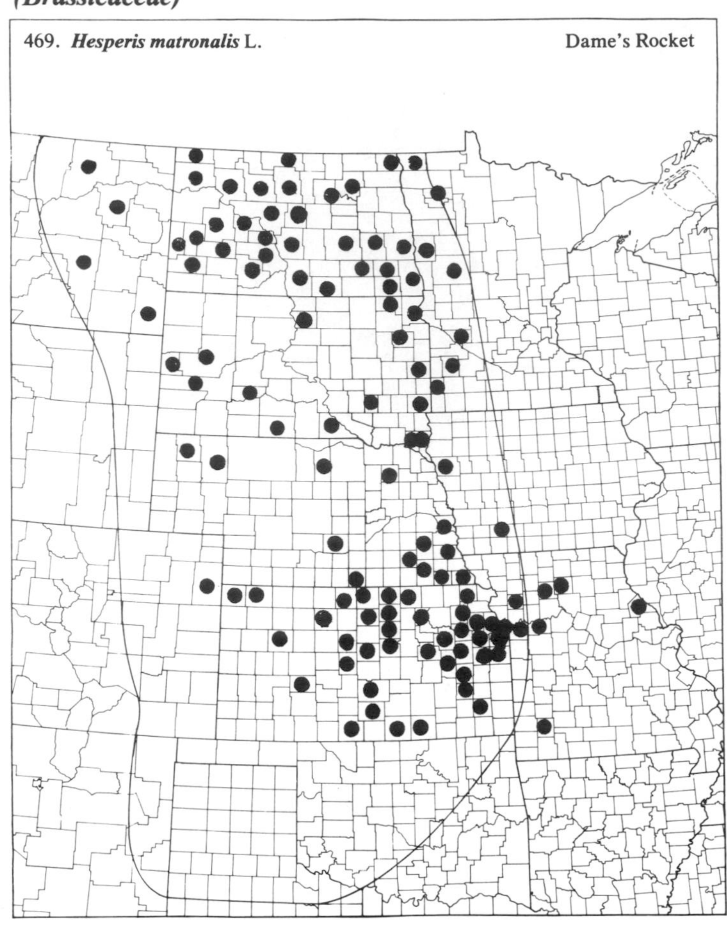

470. ***Iodanthus pinnatifidus*** (Michx.) Steud. — Purple Rocket

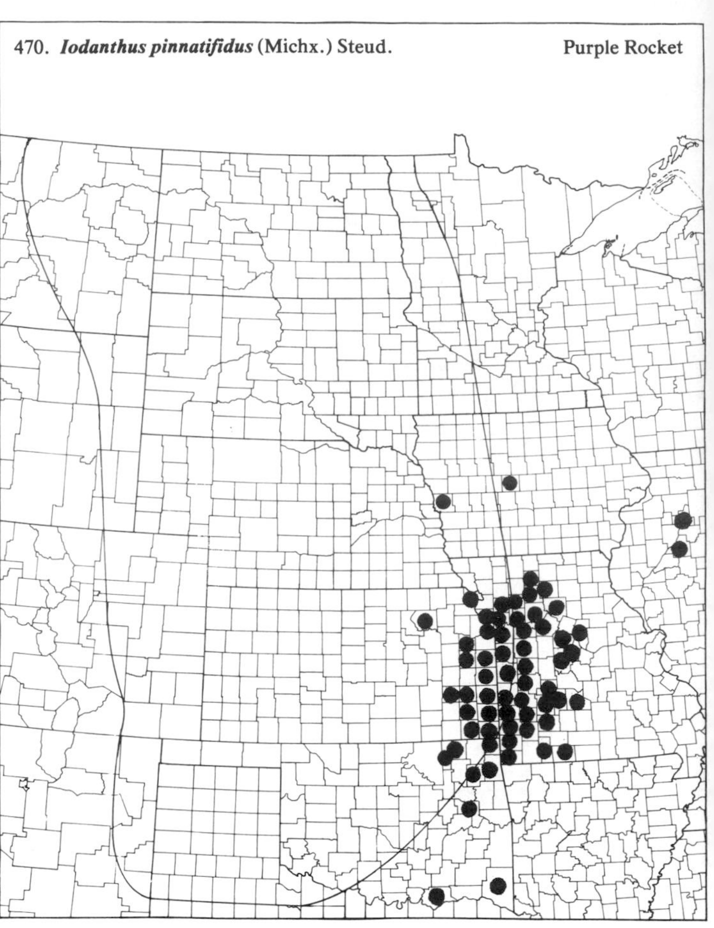

471. ***Lepidium campestre*** (L.) R. Br. — Field Peppergrass

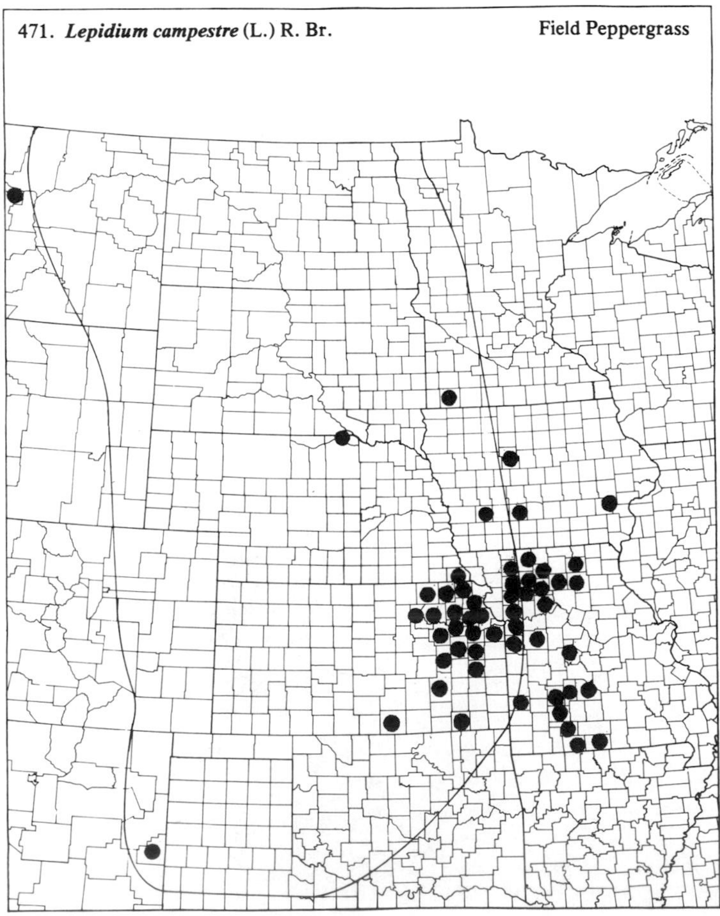

472. ***Lepidium densiflorum*** Schrad. — Peppergrass

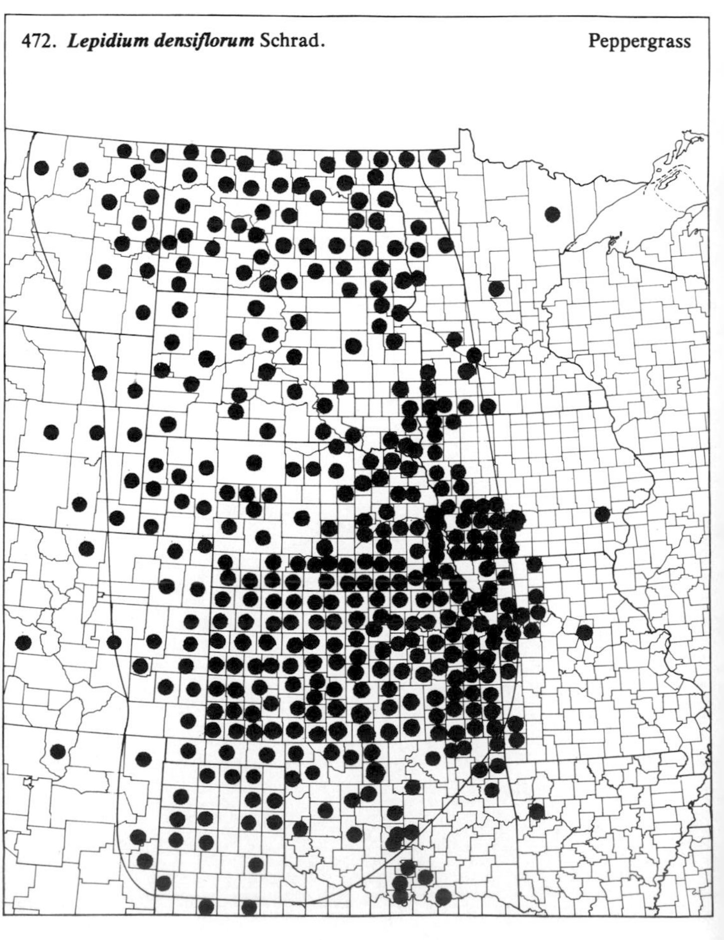

(Brassicaceae)

473. **Lepidium oblongum** Small

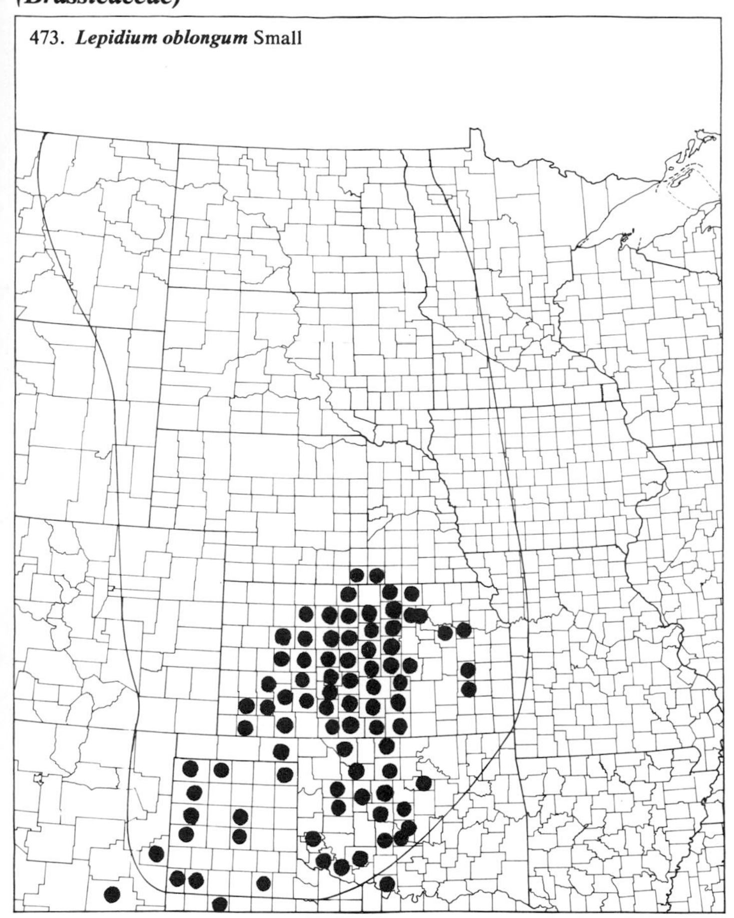

474. **Lepidium perfoliatum** L. Clasping Peppergrass

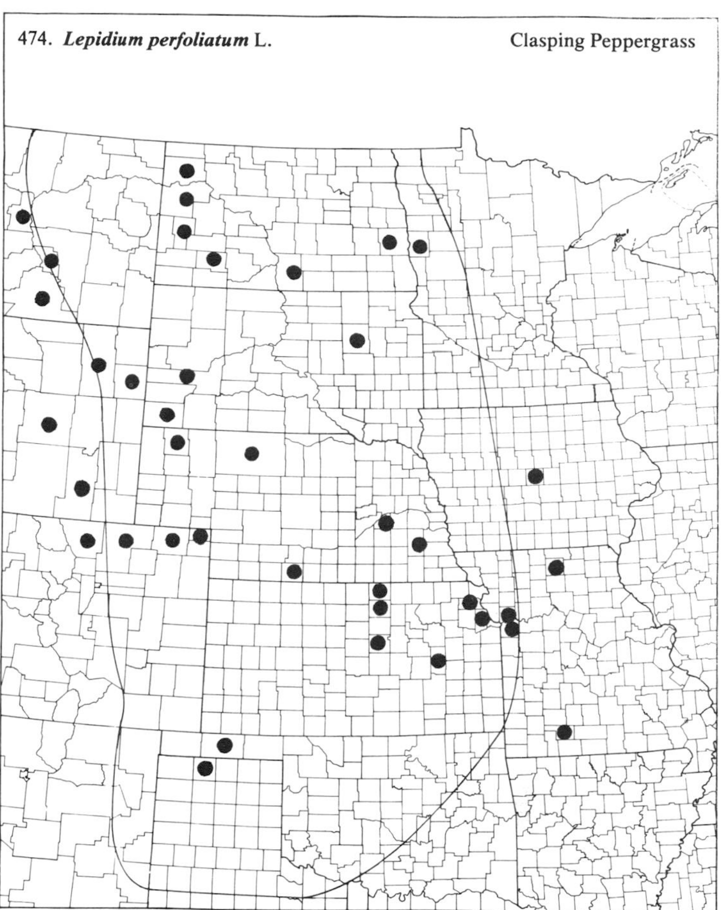

475. **Lepidium ramosissimum** A. Nels. Bushy Peppergrass

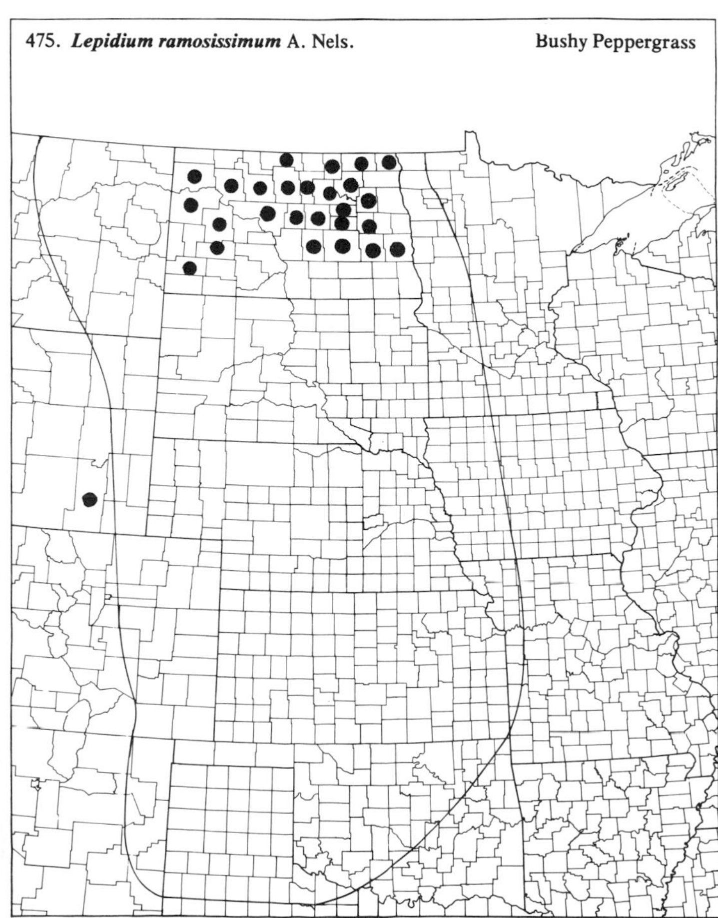

476. **Lepidium virginicum** L. Virginia Peppergrass

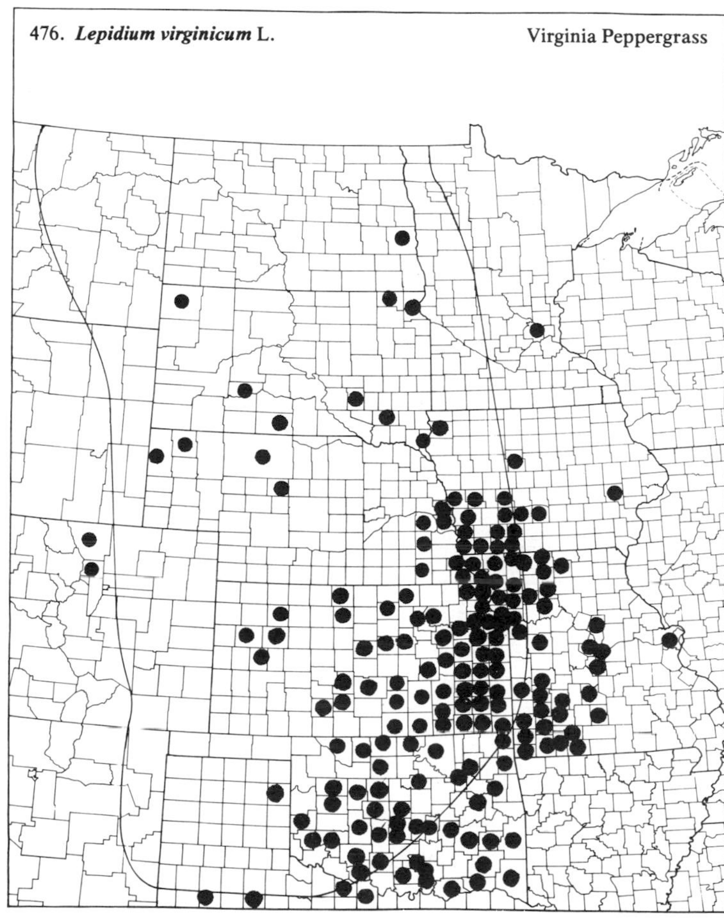

(Brassicaceae)

477. ***Lesquerella alpina*** (Nutt.) Wats.

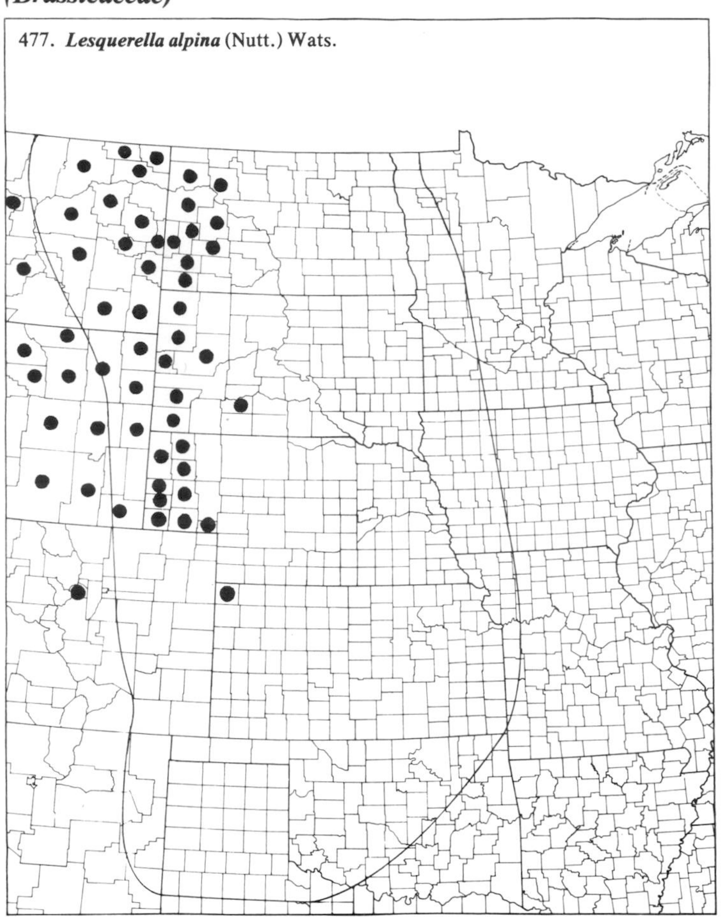

478. ***Lesquerella arenosa*** (Richards.) Rydb. var. ***arenosa***

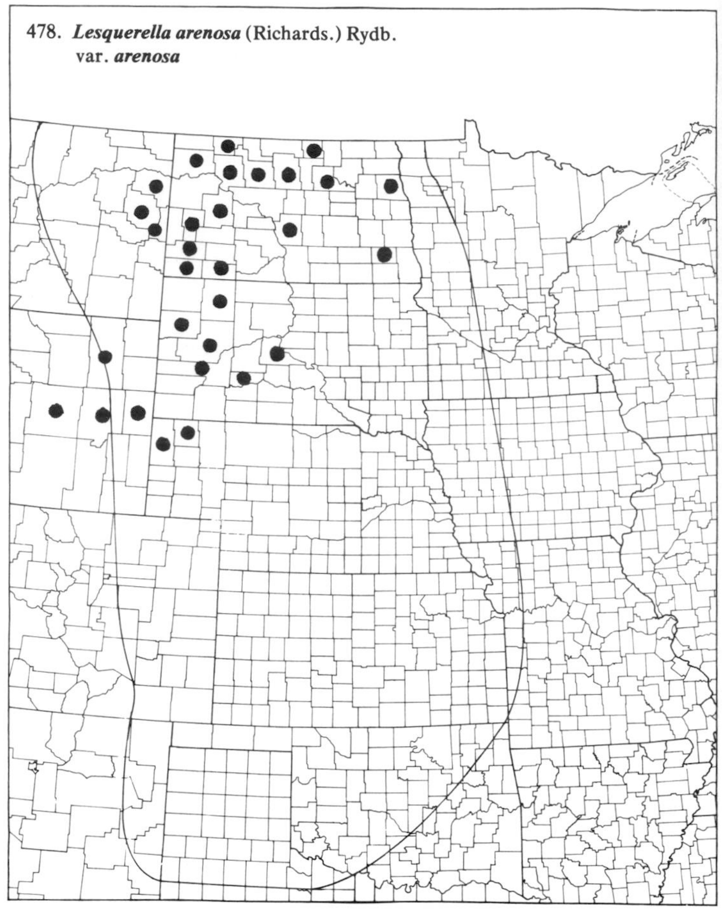

479. ***Lesquerella arenosa*** (Richards.) Rydb. var. ***argillosa*** Roll. & Shaw

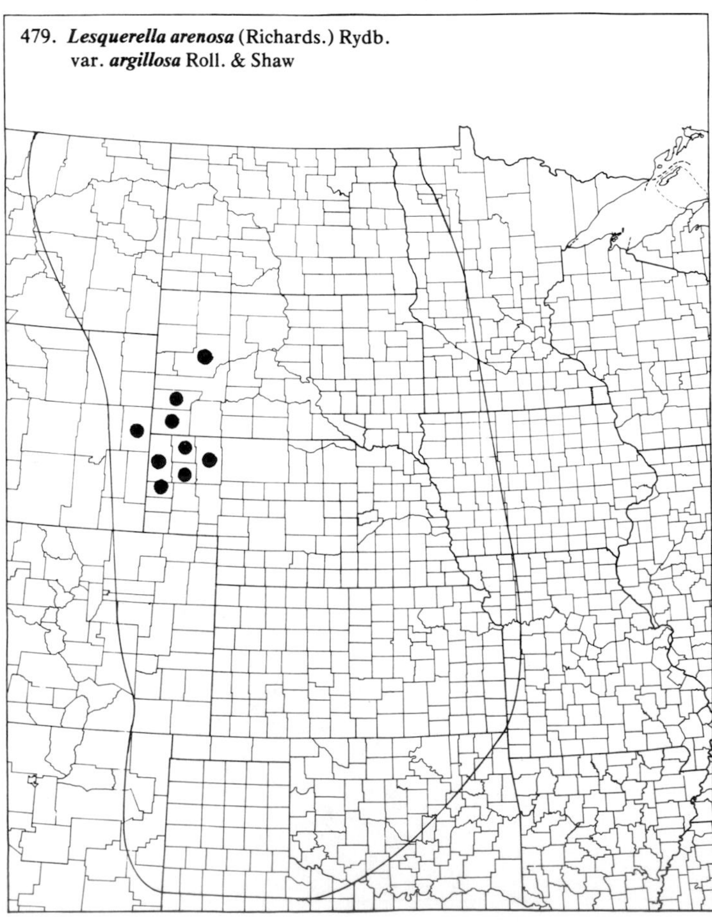

480. ***Lesquerella auriculata*** (Engelm. & Gray) Wats.

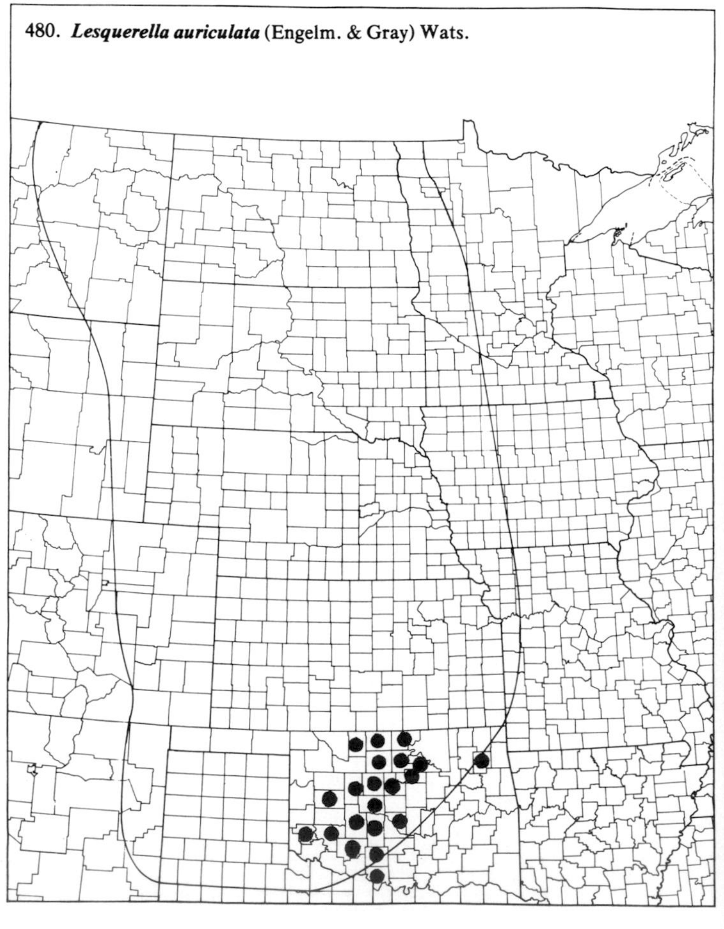

(Brassicaceae)

481. ***Lesquerella fendleri*** (Gray) Wats.

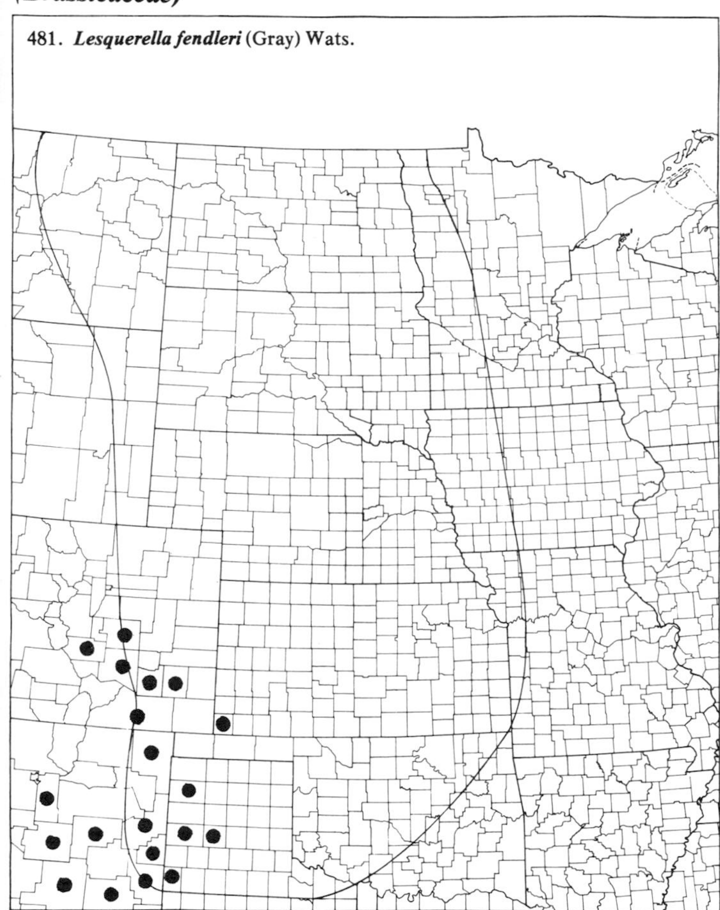

482. ***Lesquerella gordonii*** (Gray) Wats.

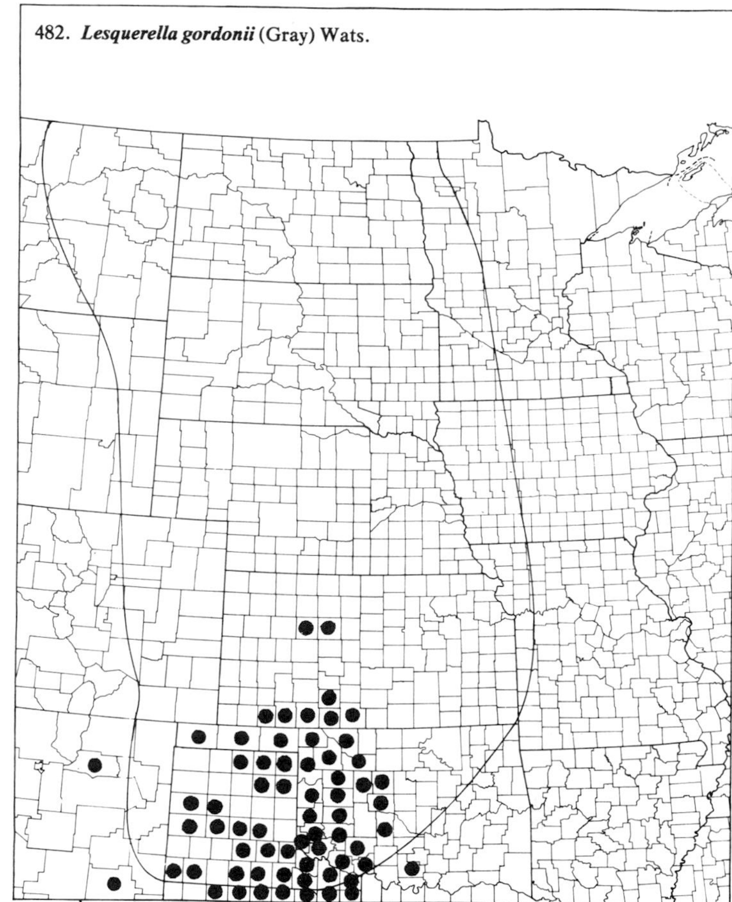

483. ***Lesquerella gracilis*** (Hook.) Wats. ssp. ***nuttallii*** (T. & G.) Roll. & Shaw — Spreading Bladderpod

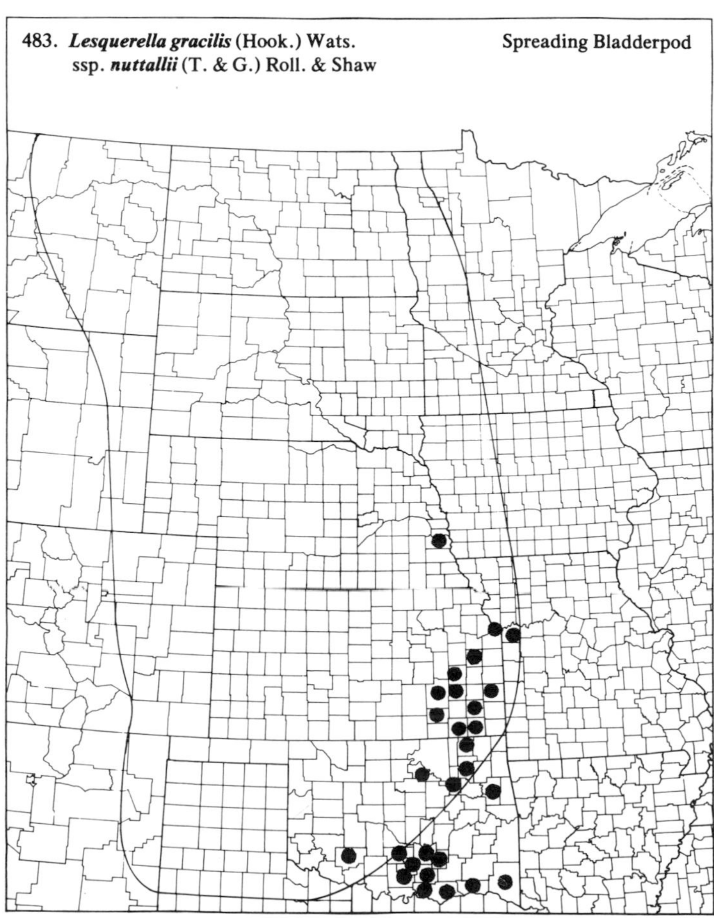

484. ***Lesquerella ludoviciana*** (Nutt.) Wats. — Bladderpod

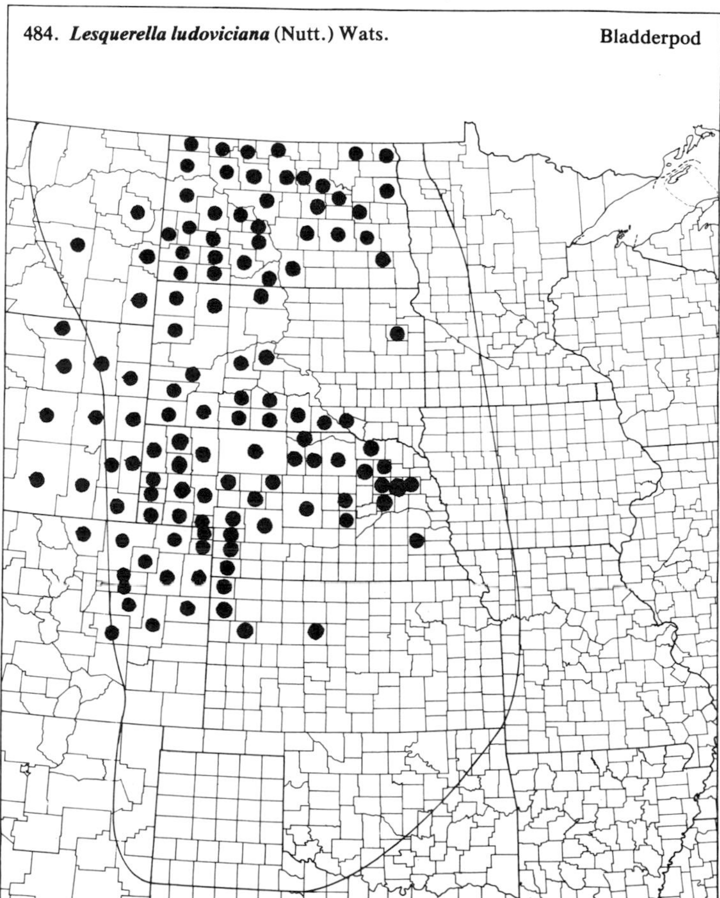

(Brassicaceae)

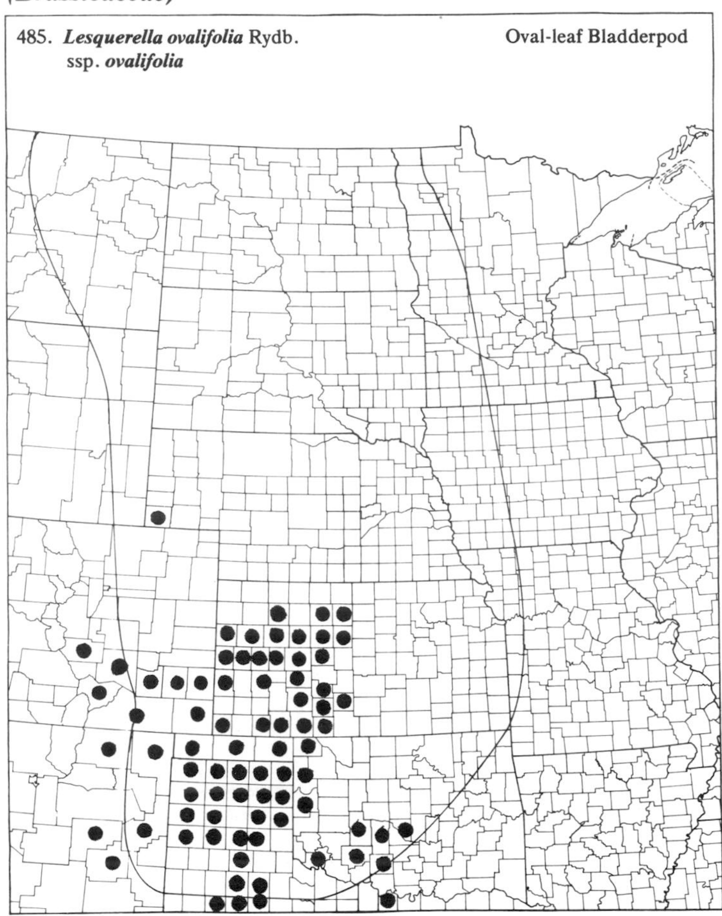

485. ***Lesquerella ovalifolia*** Rydb. ssp. ***ovalifolia*** — Oval-leaf Bladderpod

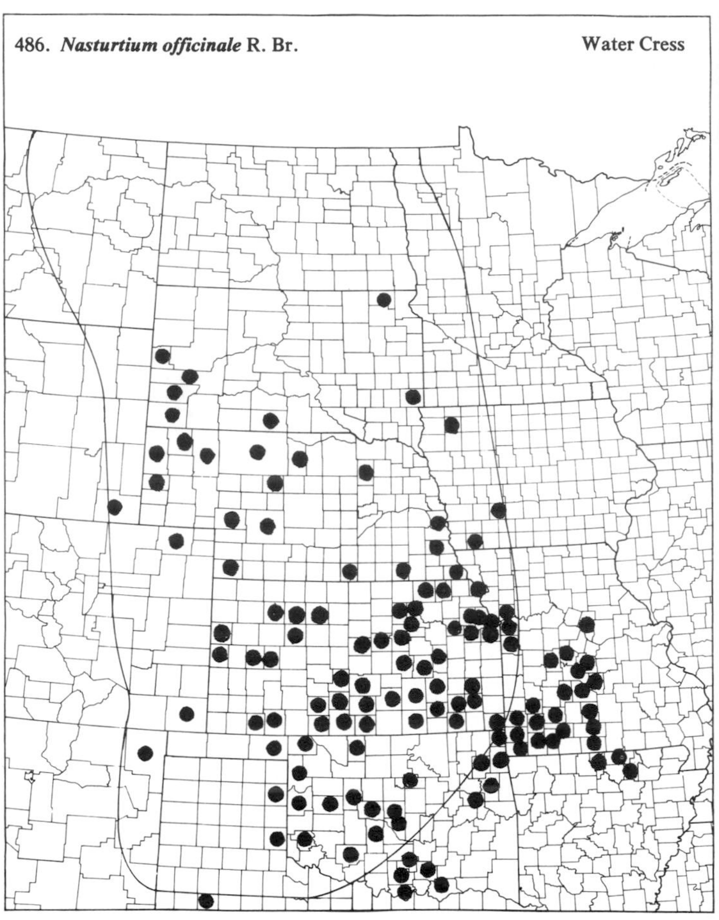

486. ***Nasturtium officinale*** R. Br. — Water Cress

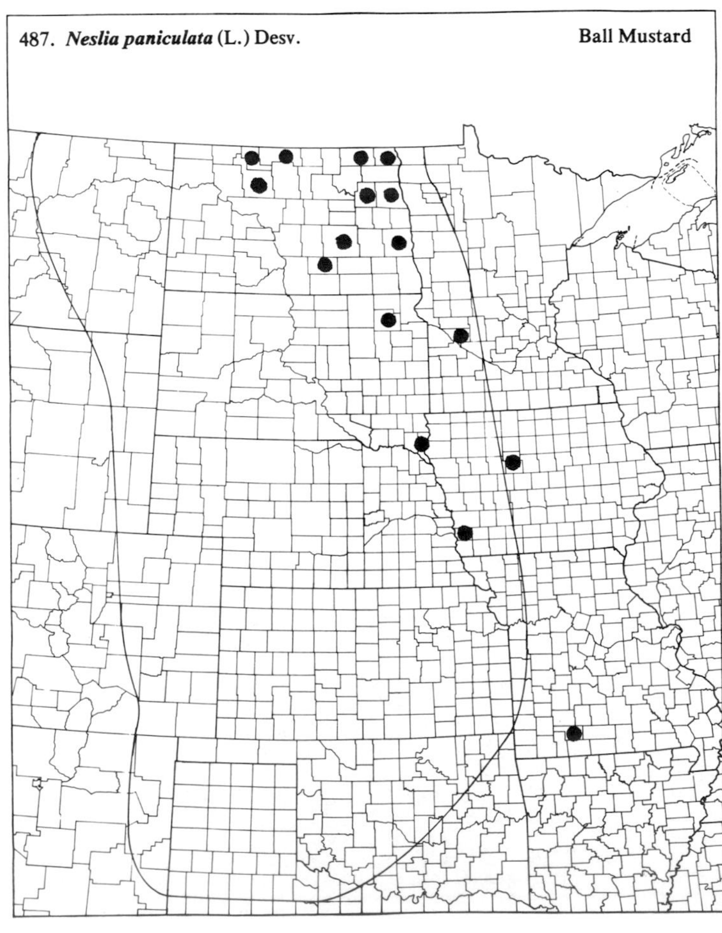

487. ***Neslia paniculata*** (L.) Desv. — Ball Mustard

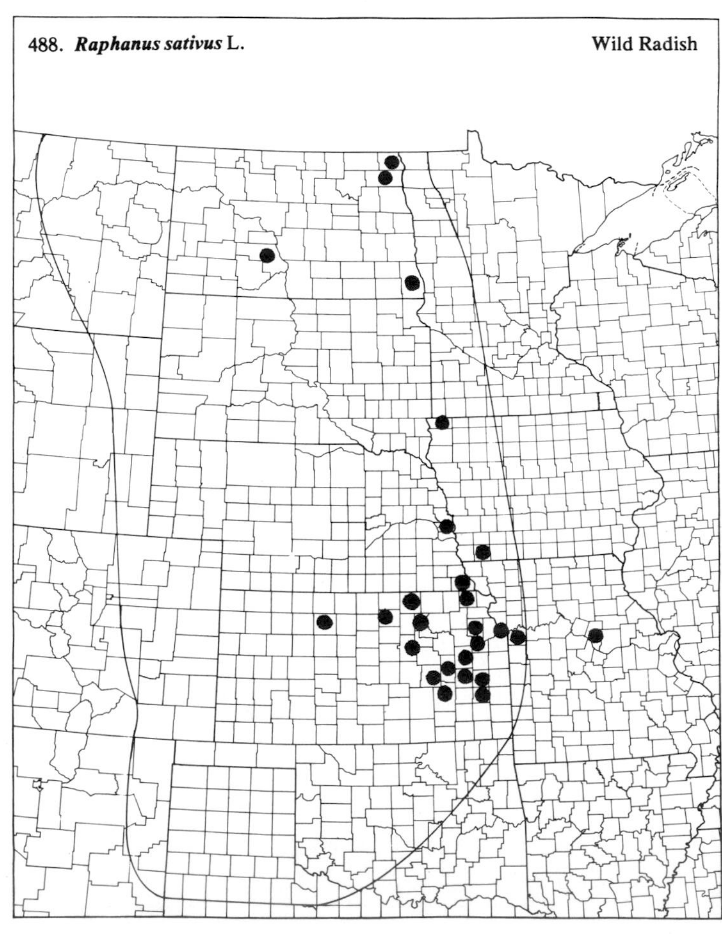

488. ***Raphanus sativus*** L. — Wild Radish

(Brassicaceae)

489. ***Rorippa palustris*** (L.) Bess.
ssp. ***glabra*** (Schulz) Stuckey
var. ***fernaldiana*** (Butt. & Abbe) Stuckey

Bog Yellow Cress

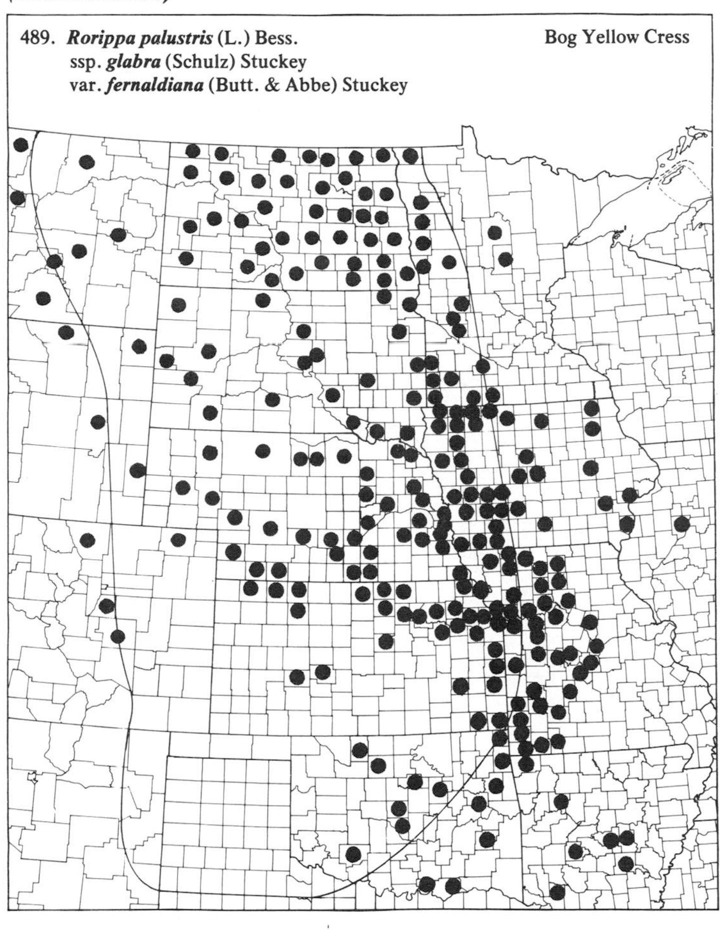

490. ***Rorippa palustris*** (L.) Bess.
ssp. ***hispida*** (Desv.) Jonsell
var. ***hispida***

Bog Yellow Cress

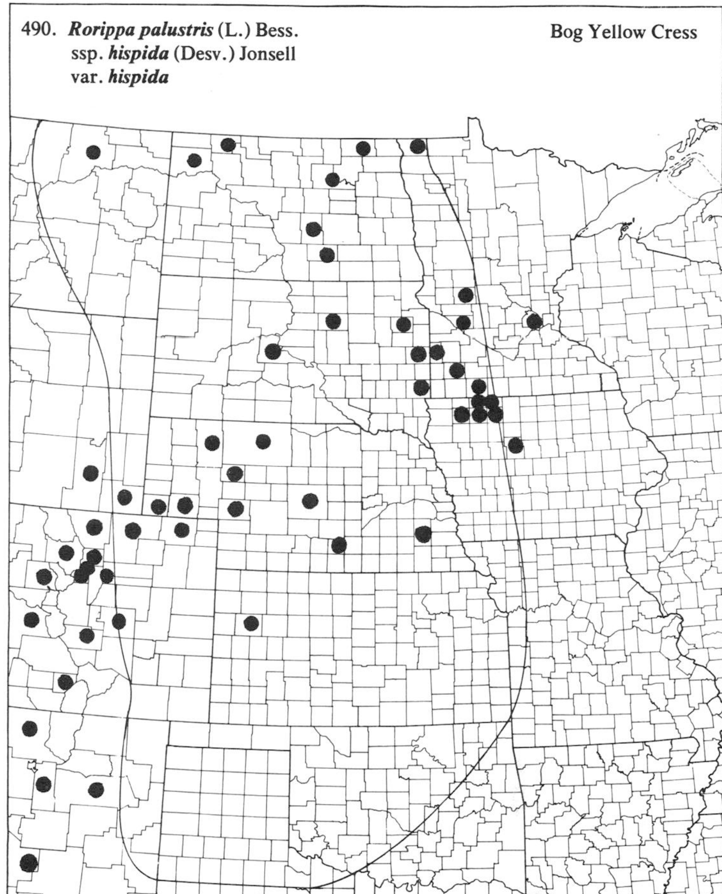

491. ***Rorippa sessiliflora*** (Nutt.) Hitchc.

Yellow Cress

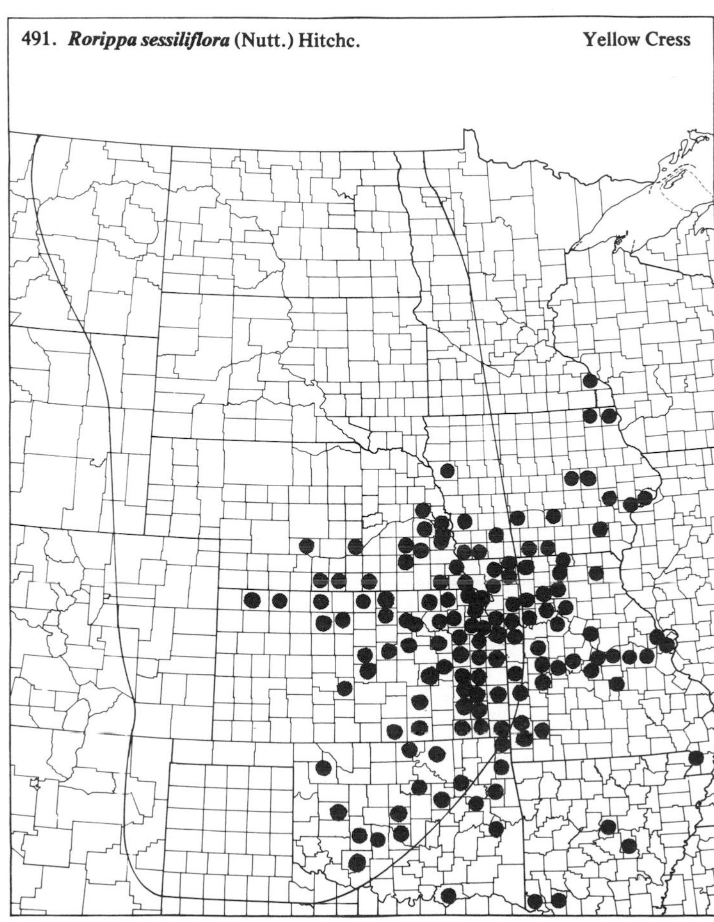

492. ***Rorippa sinuata*** (Nutt.) Hitchc.

Spreading Yellow Cress

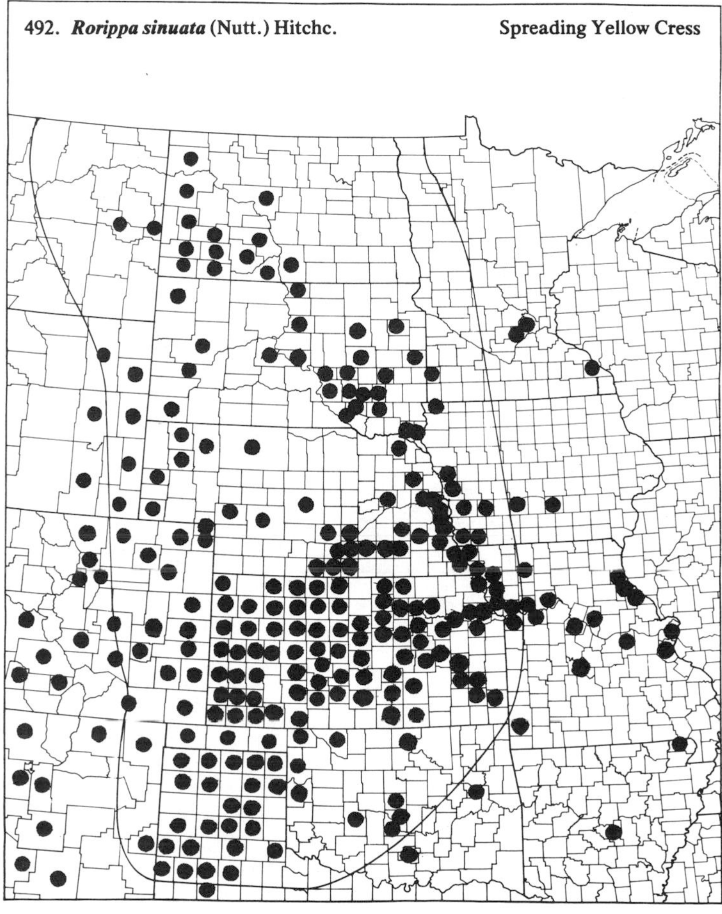

(Brassicaceae)

493. ***Rorippa truncata*** (Jeps.) Stuckey

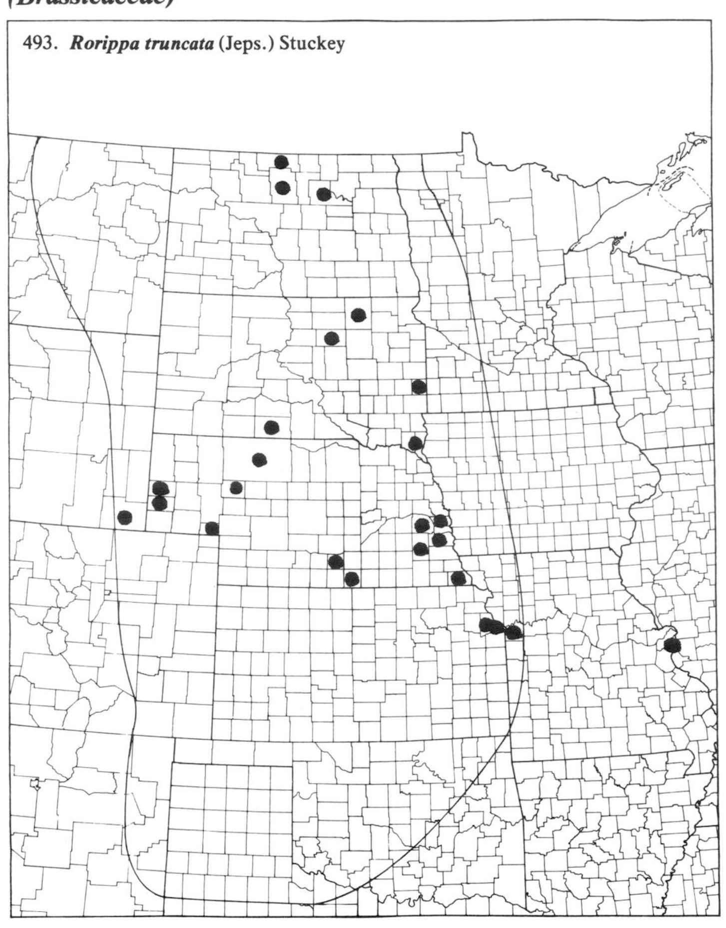

494. ***Selenia aurea*** Nutt. Golden Selenia

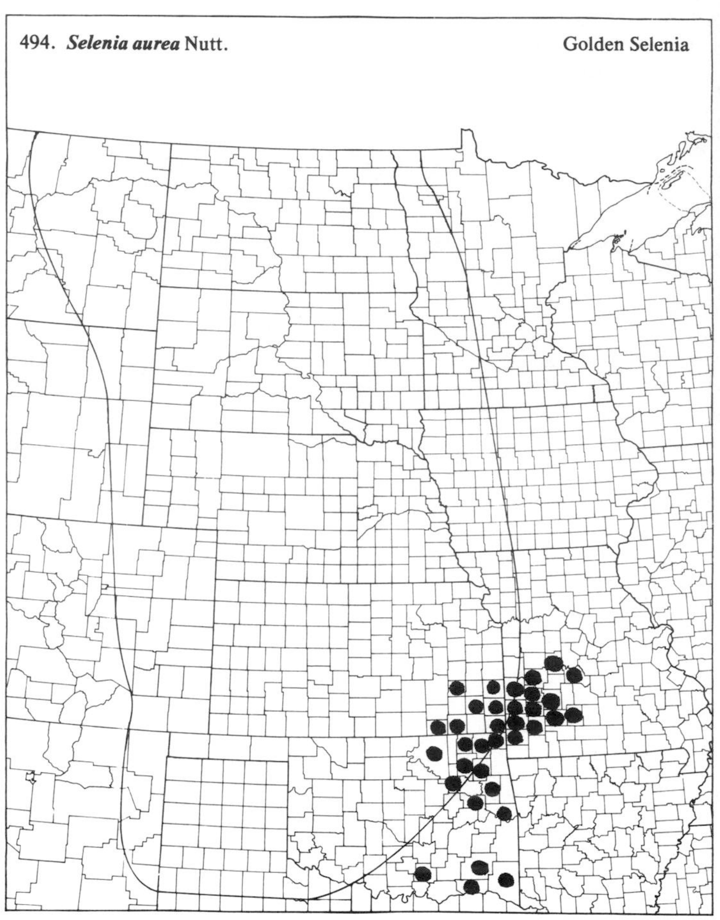

495. ***Sisymbrium altissimum*** L. Tumbling Mustard

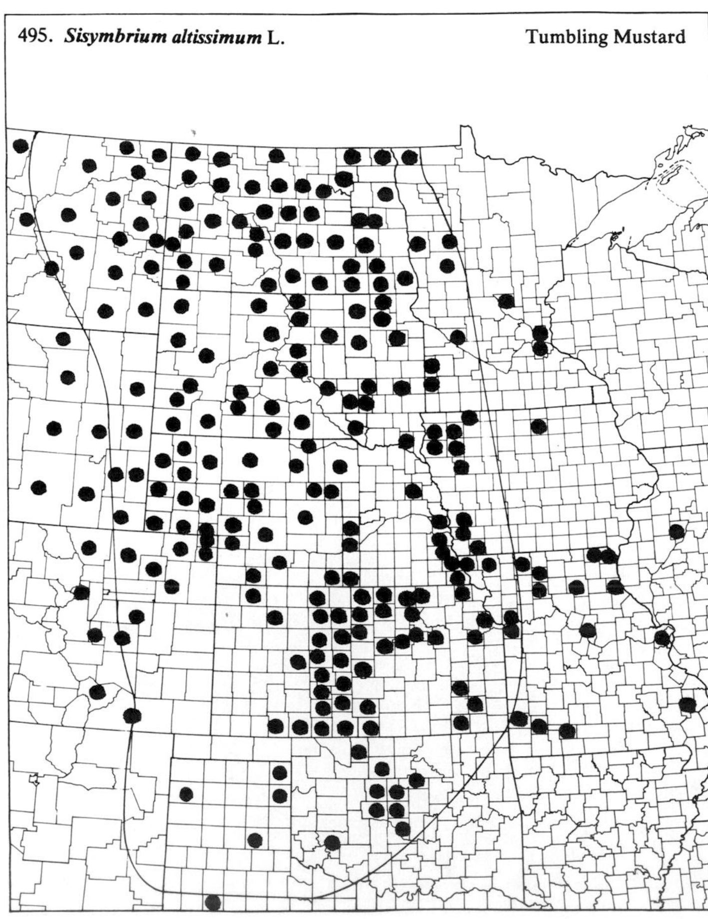

496. ***Sisymbrium loeselii*** L. Tall Hedge Mustard

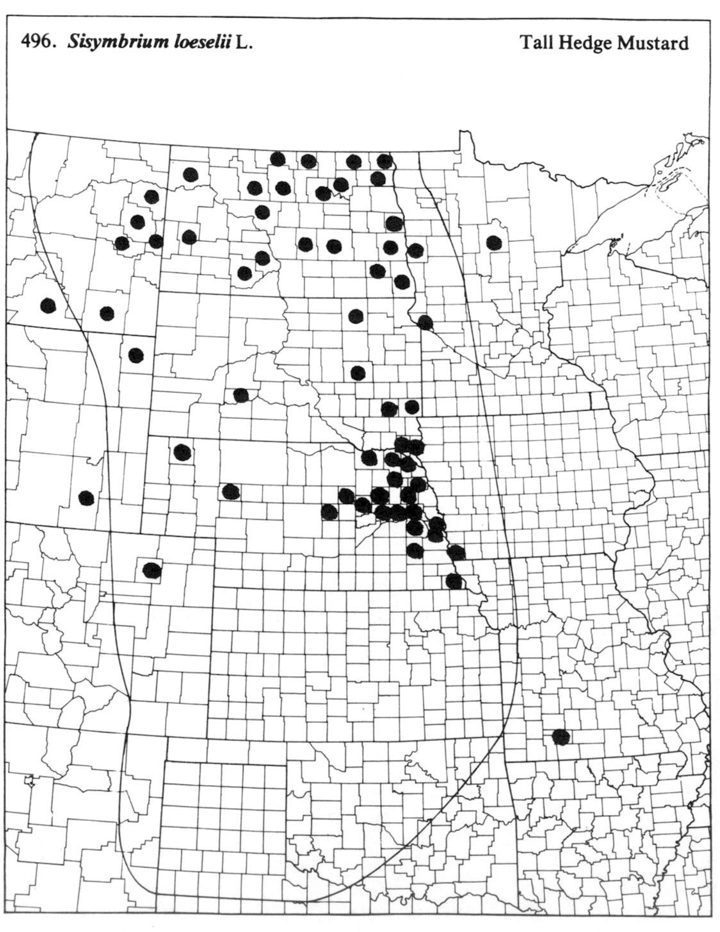

(Brassicaceae)

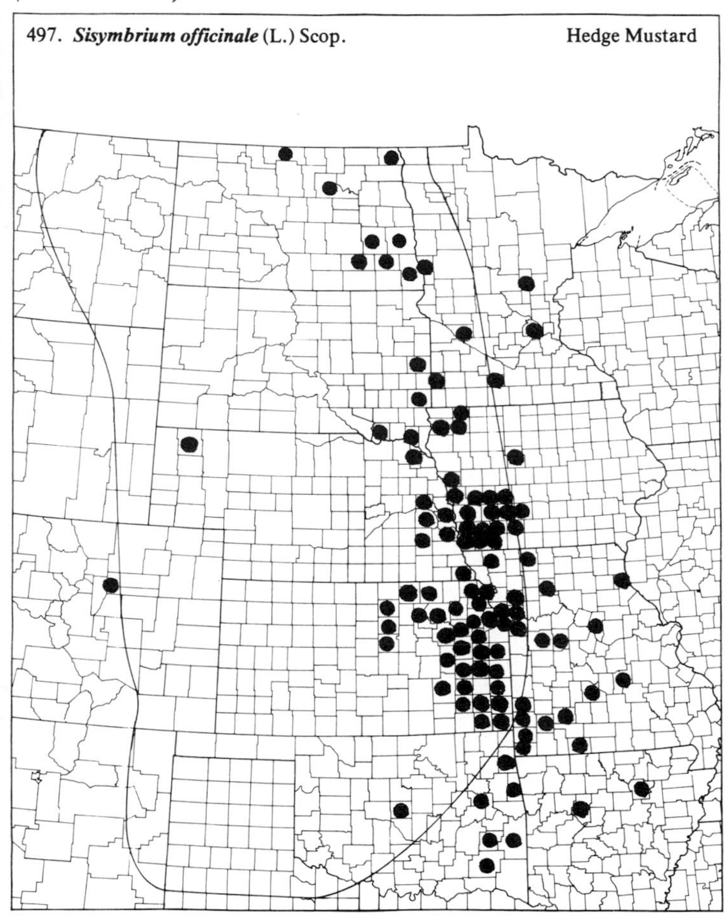

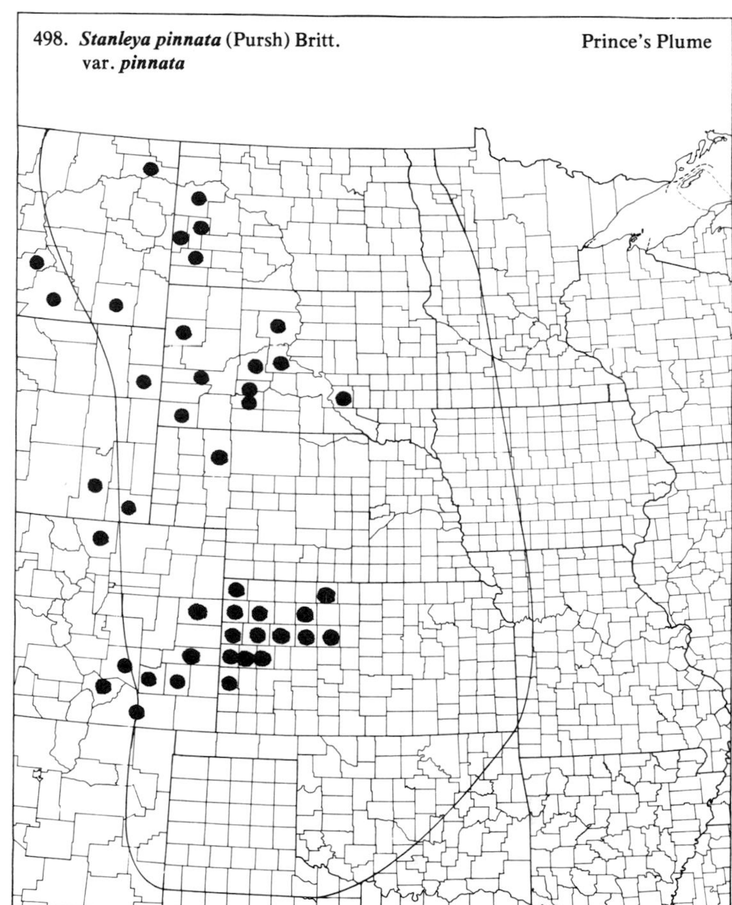

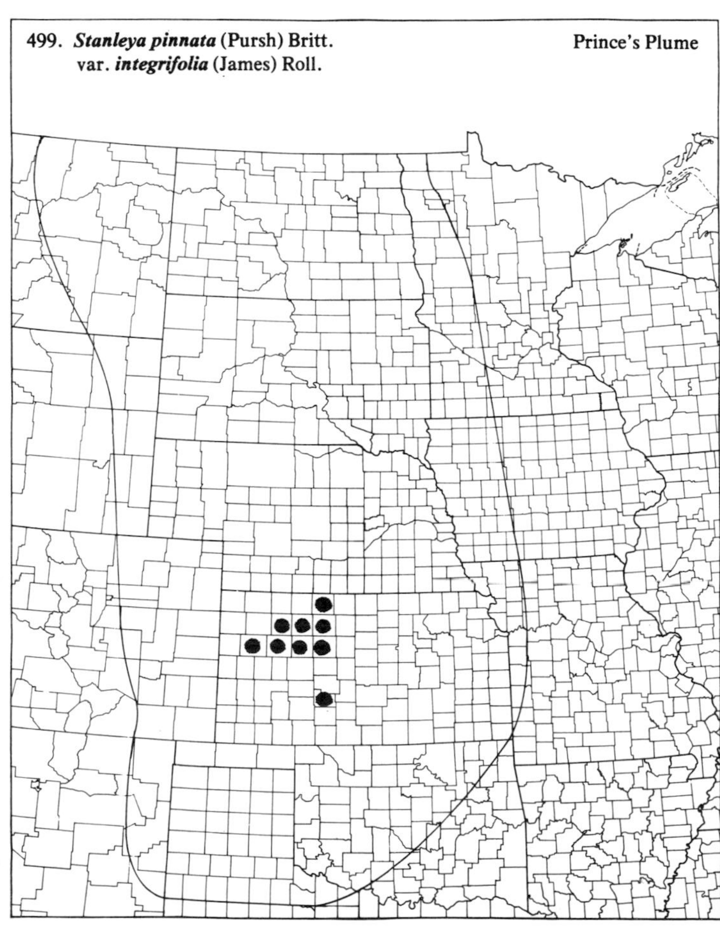

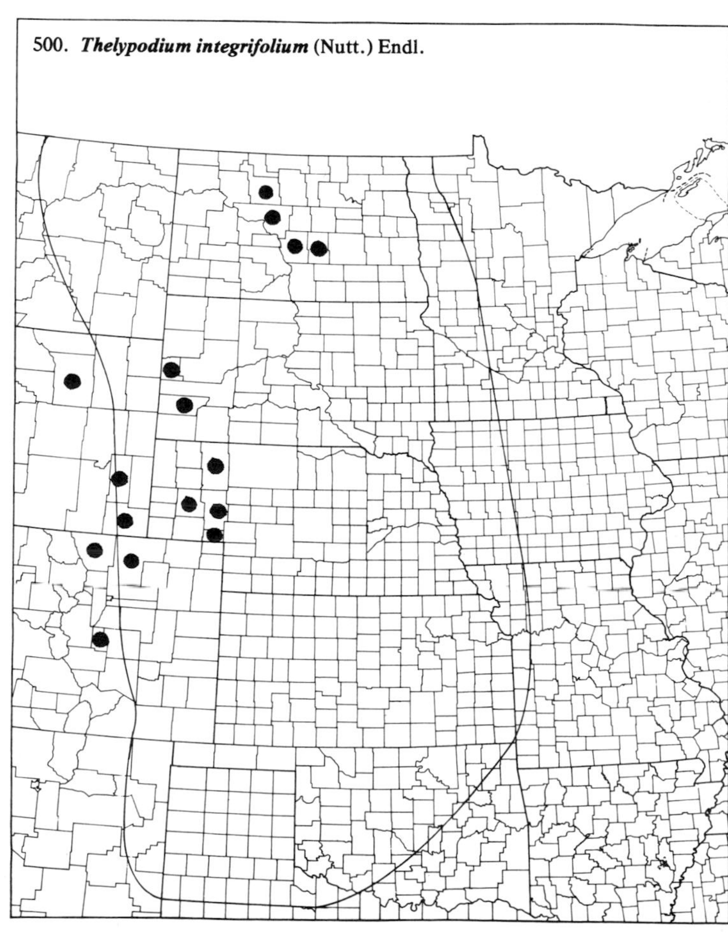

(Brassicaceae)

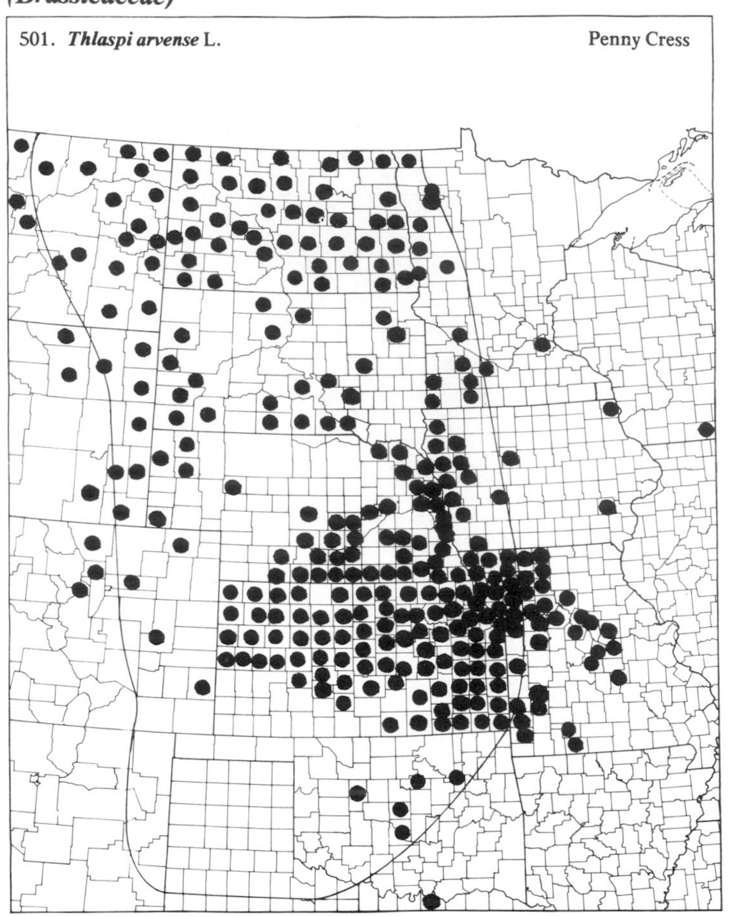

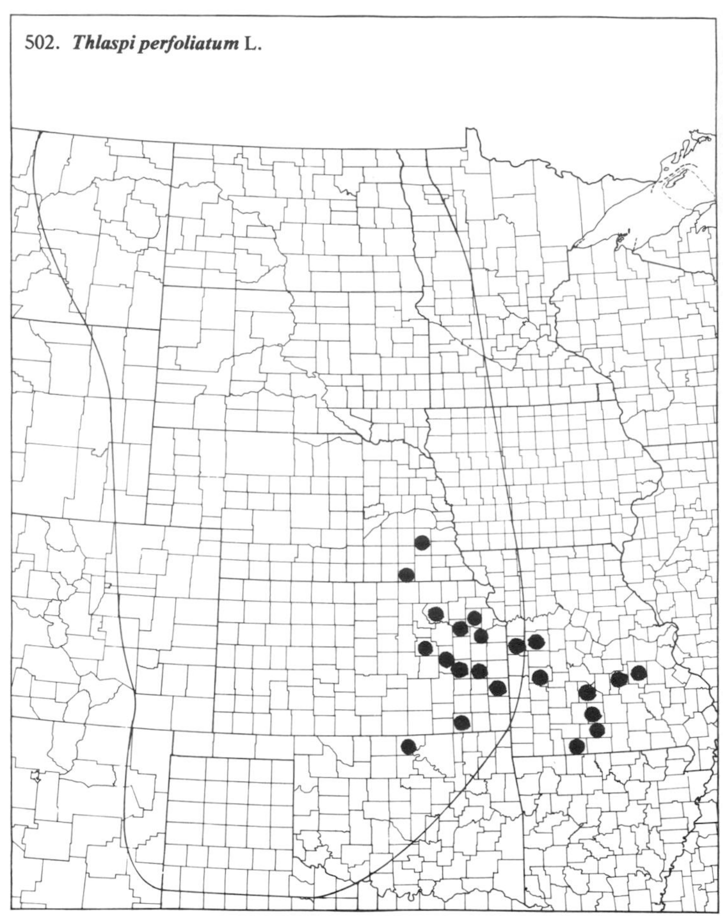

45. *Ericaceae* (Heath Family)

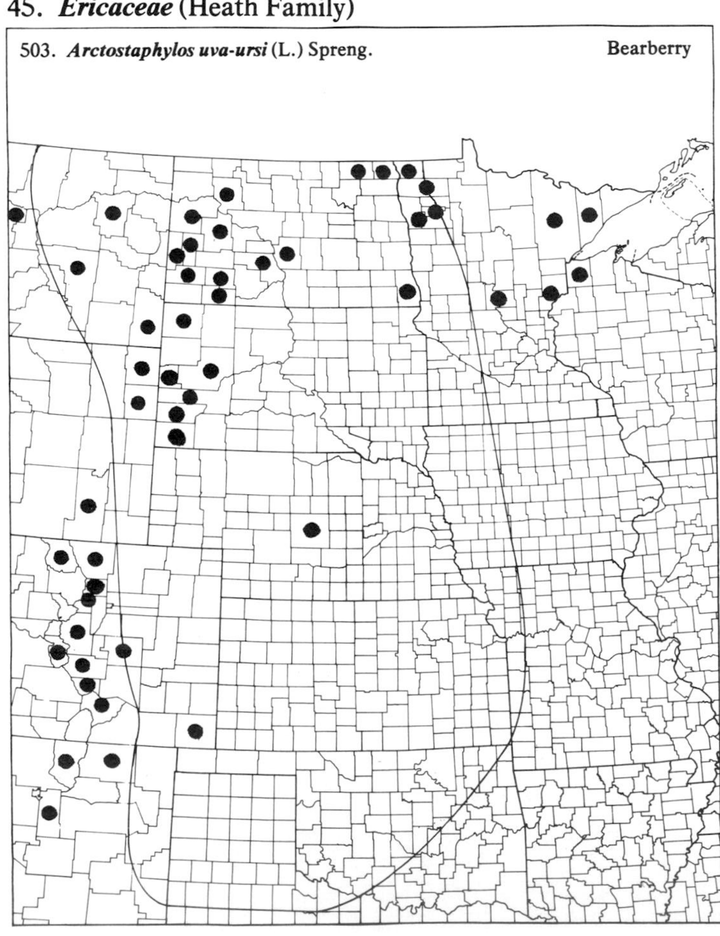

46. *Pyrolaceae* (Wintergreen Family)

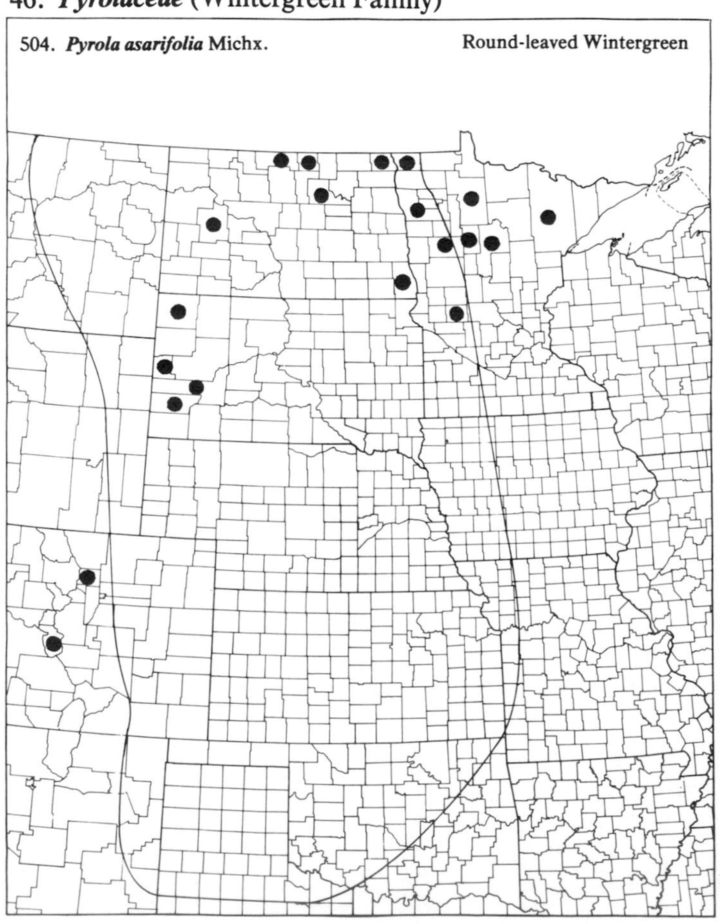

(Pyrolaceae)

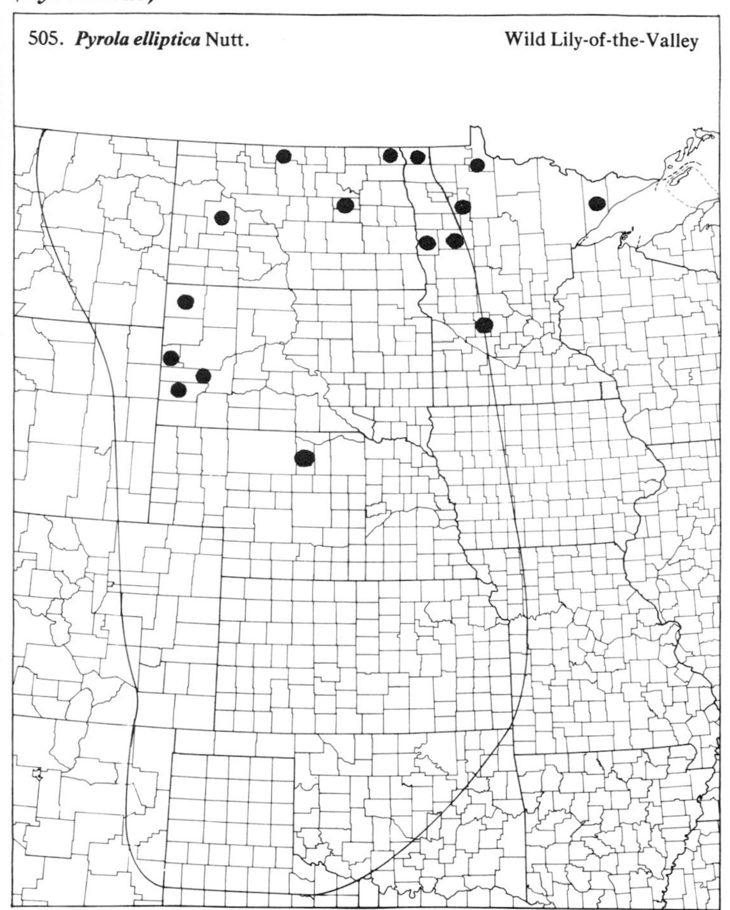

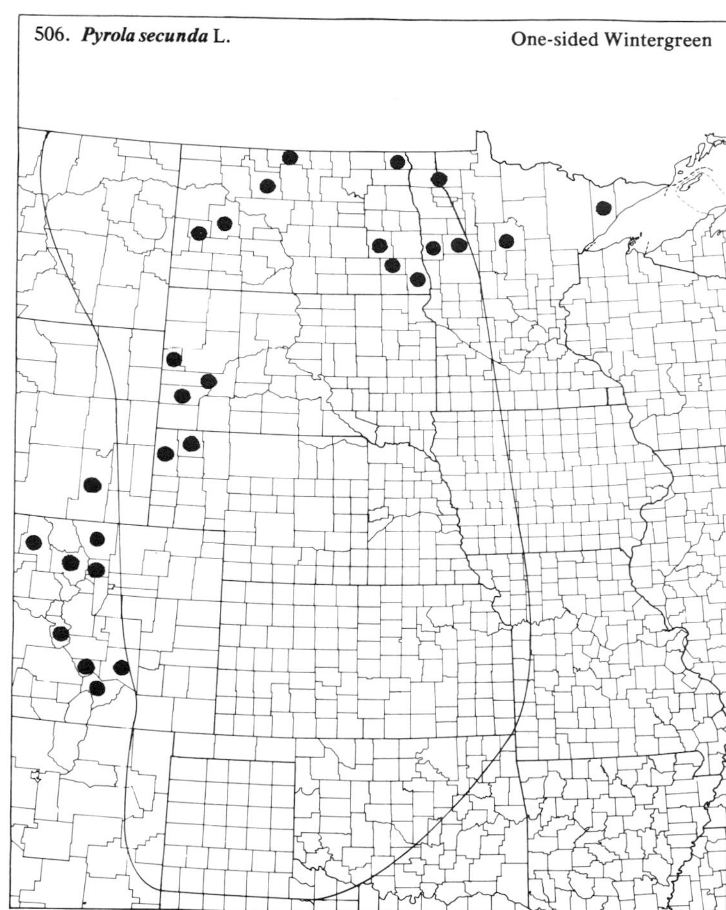

47. *Monotropaceae* (Indian Pipe Family)

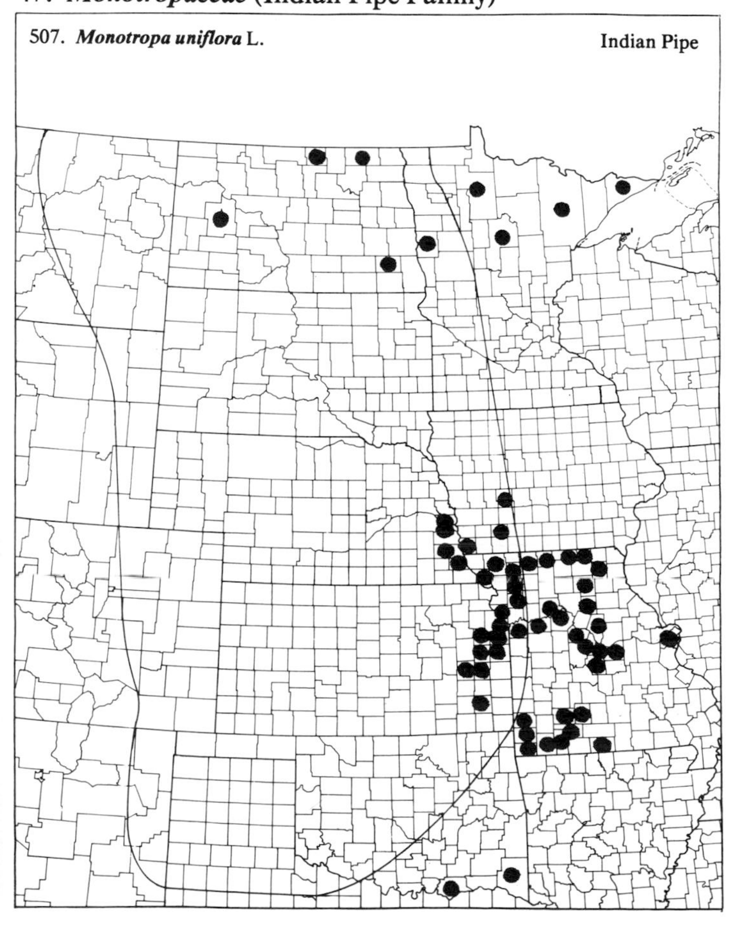

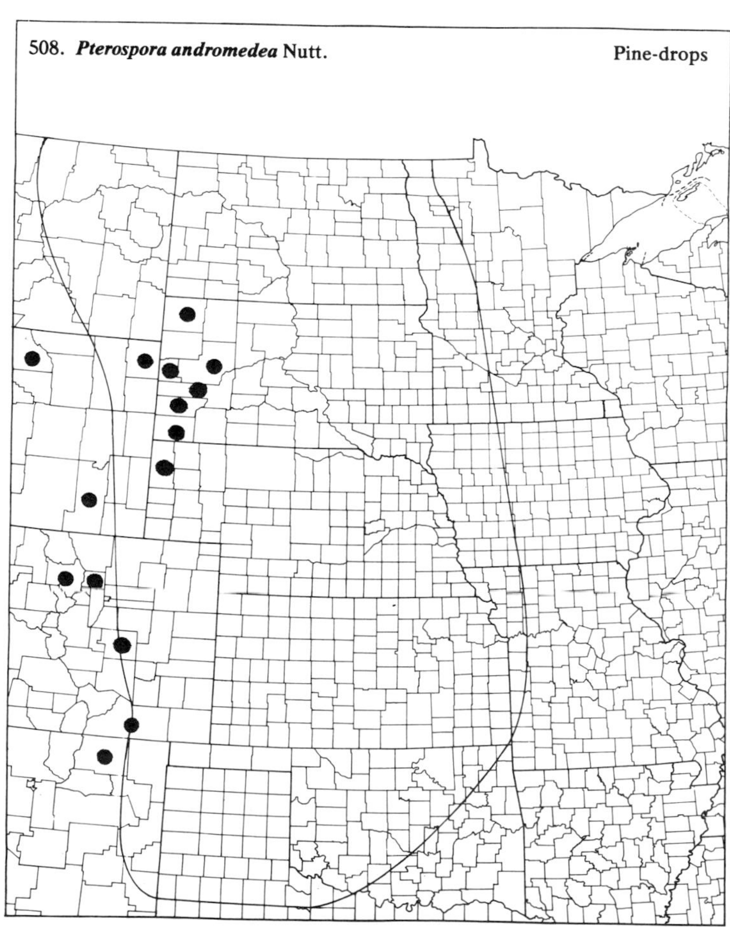

48. *Sapotaceae* (Sapodilla Family)

509. ***Bumelia lanuginosa*** (Michx.) Pers. var. ***oblongifolia*** (Nutt.) Clark — Woolly Buckthorn

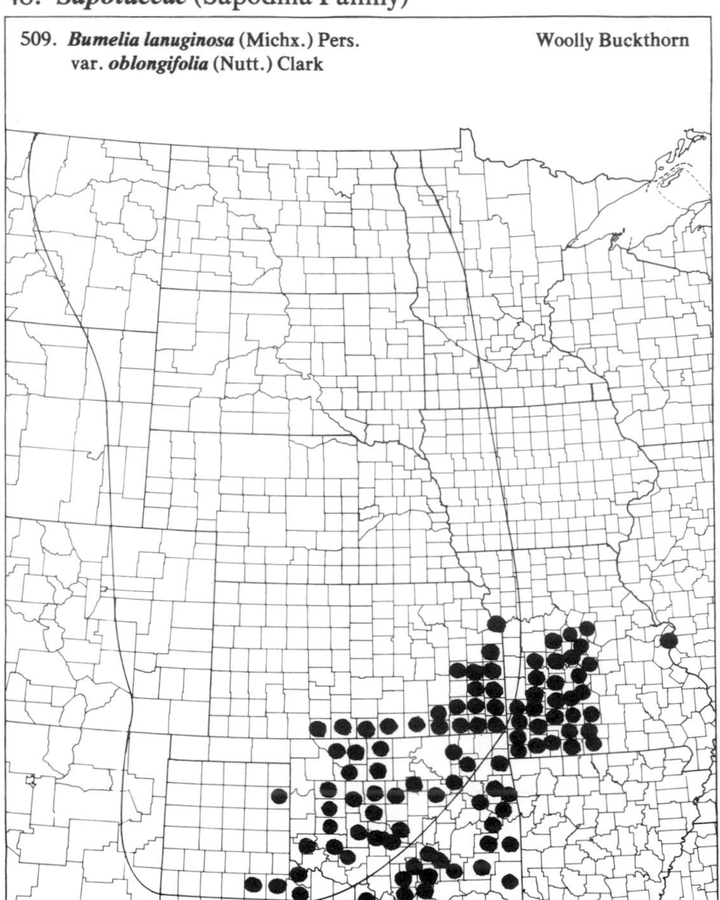

49. *Ebenaceae* (Ebony Family)

510. ***Diospyros virginiana*** L. — Persimmon

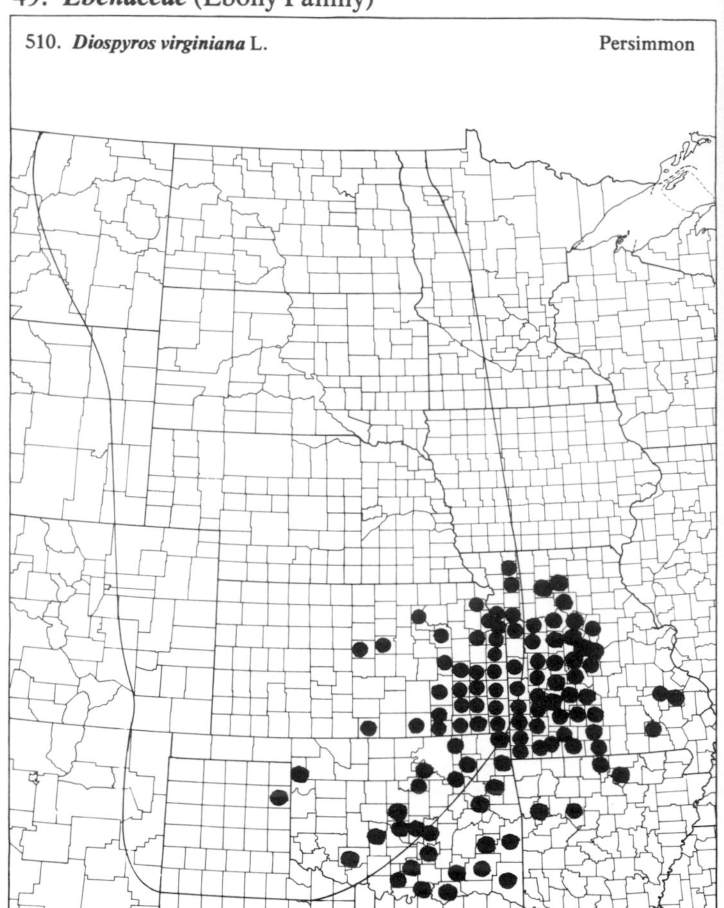

50. *Primulaceae* (Primrose Family)

511. ***Anagallis arvensis*** L. — Scarlet Pimpernel

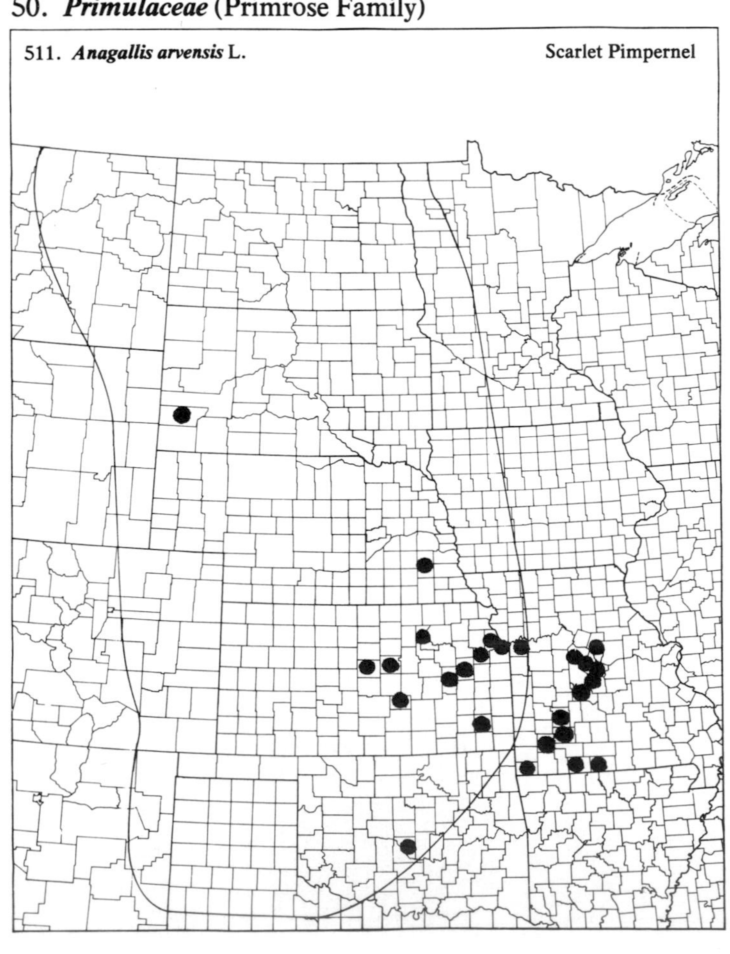

512. ***Androsace occidentalis*** Pursh — Western Rock Jasmine

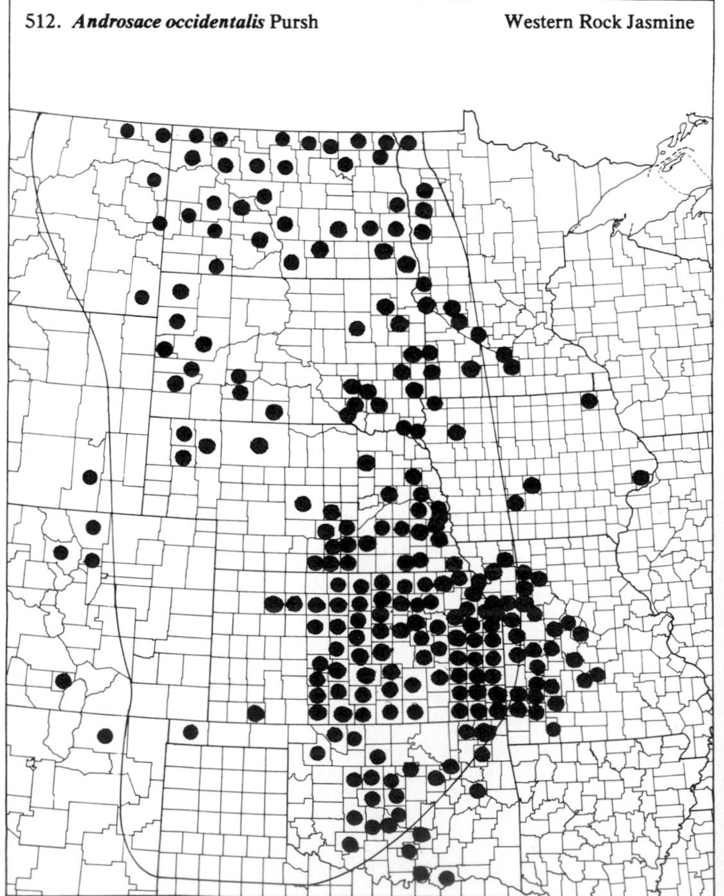

(Primulaceae)

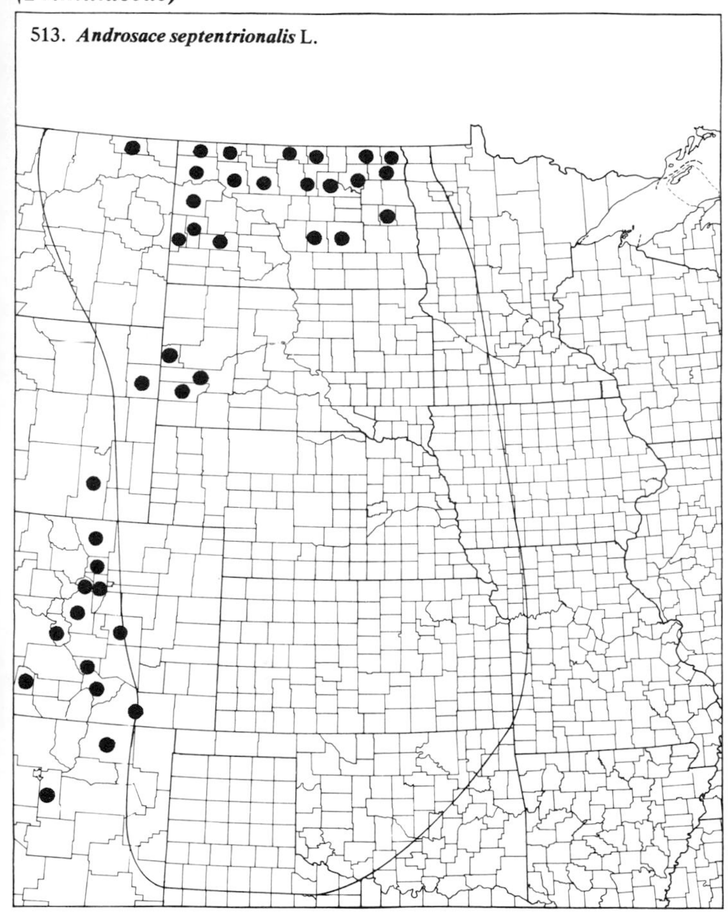

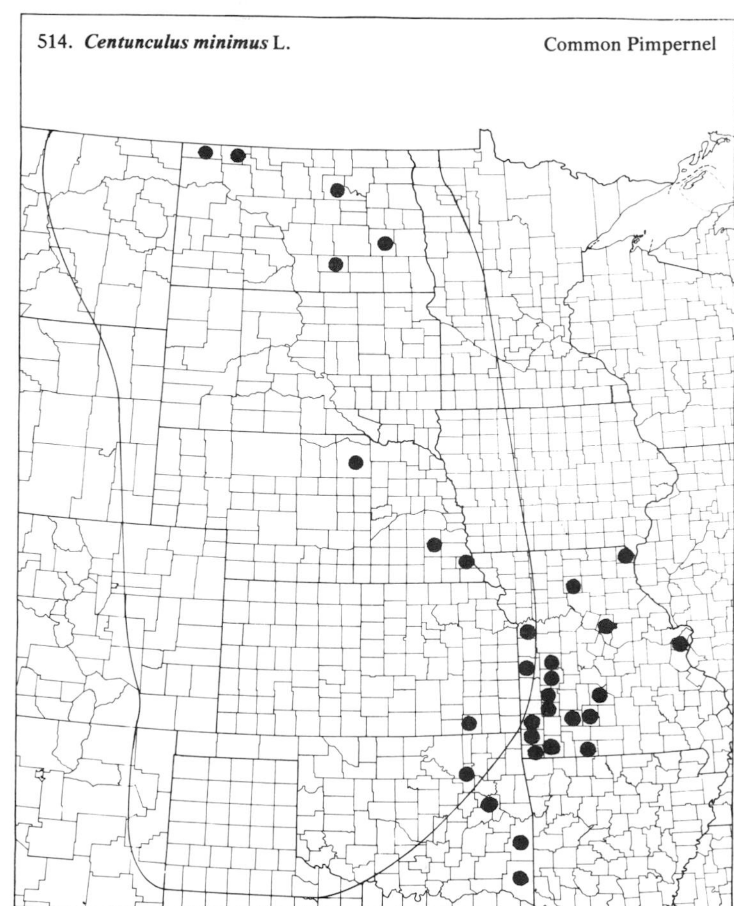

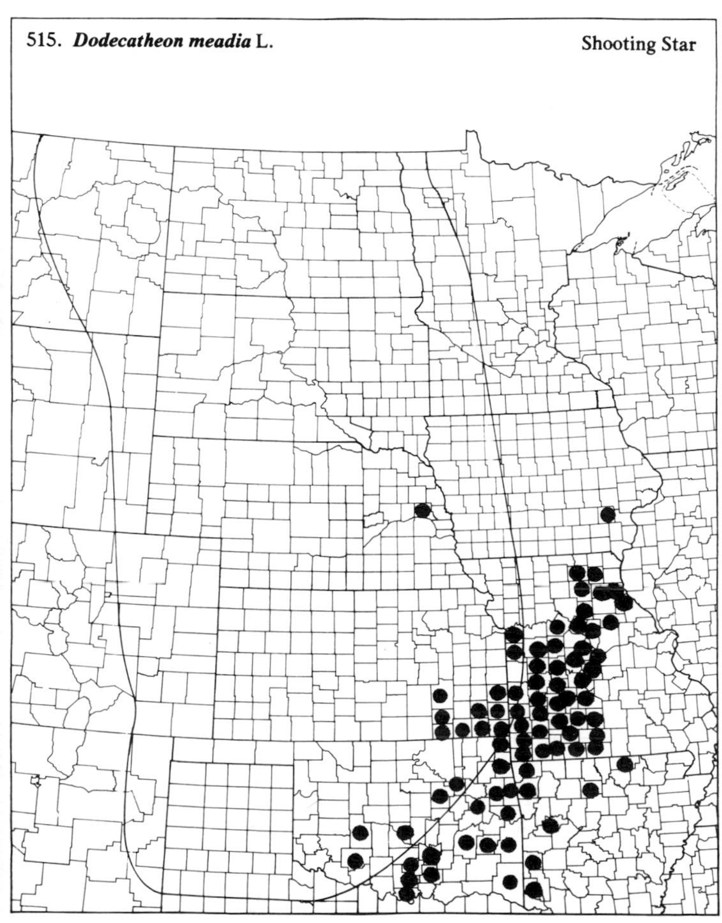

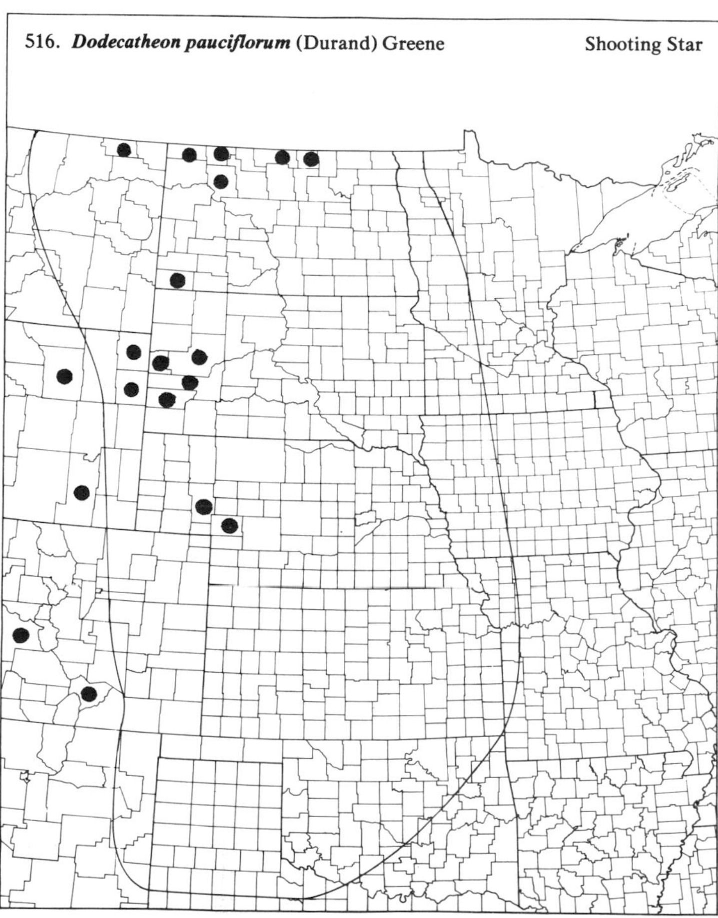

(Primulaceae)

517. ***Glaux maritima*** L. Sea Milkwort

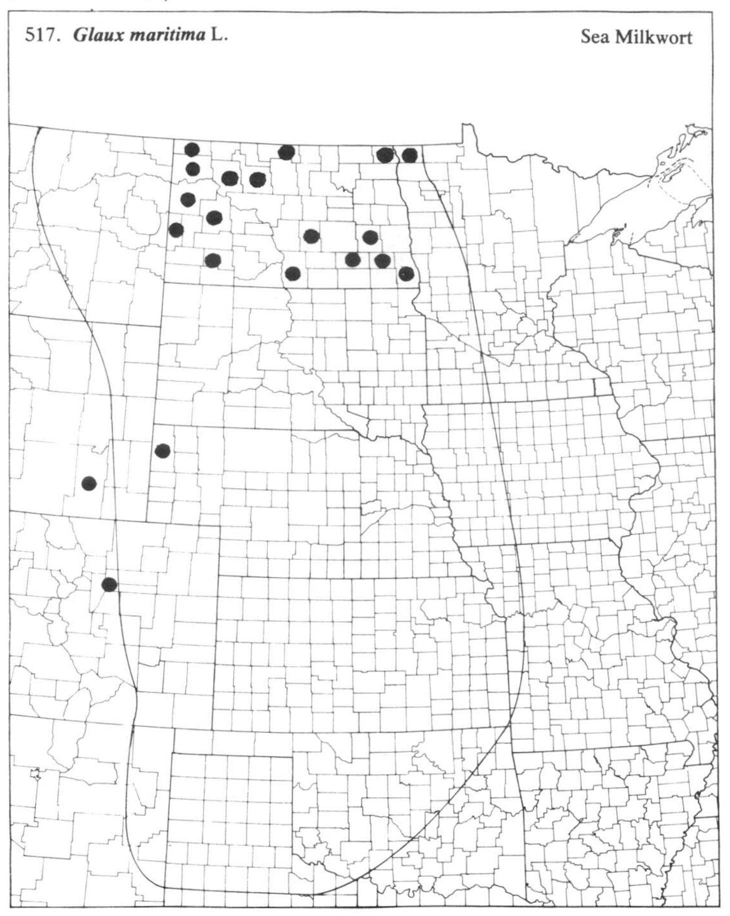

518. ***Lysimachia ciliata*** L. Fringed Loosestrife

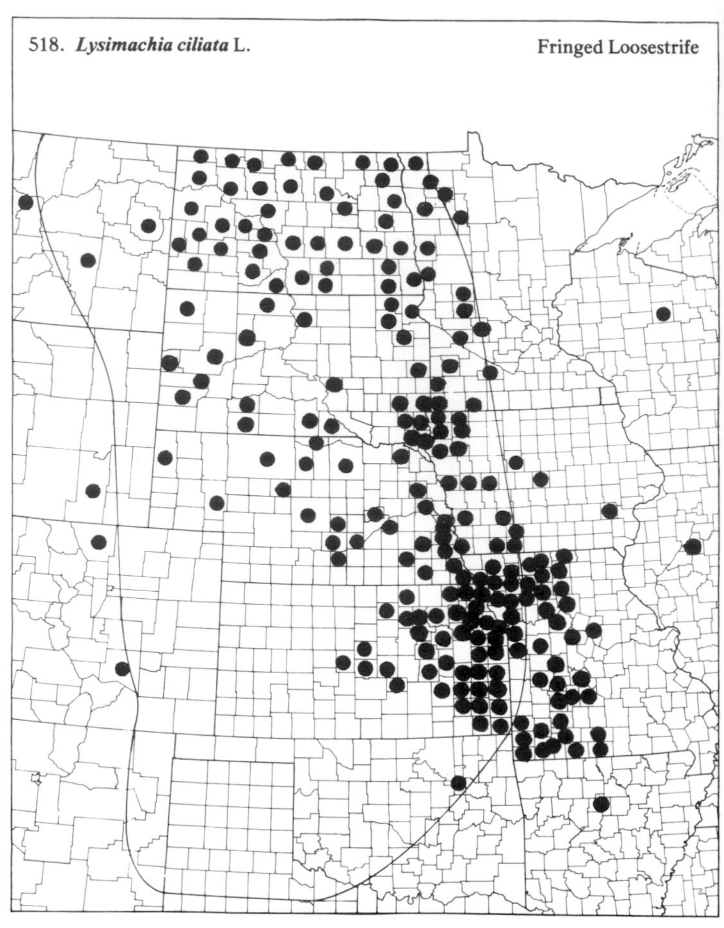

519. ***Lysimachia hybrida*** Michx. Loosestrife

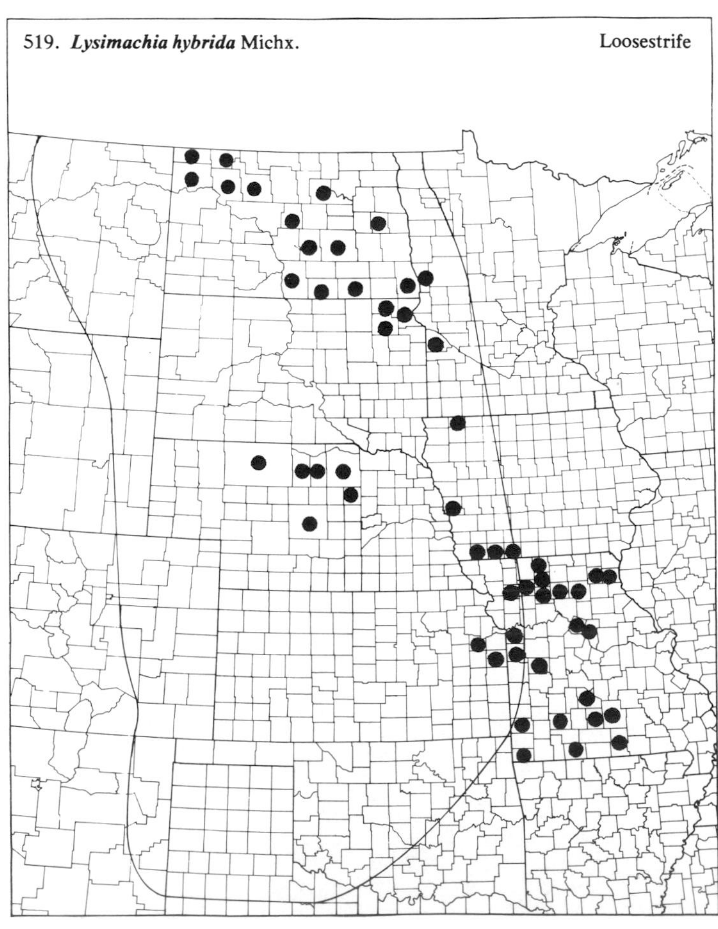

520. ***Lysimachia lanceolata*** Walt. Loosestrife

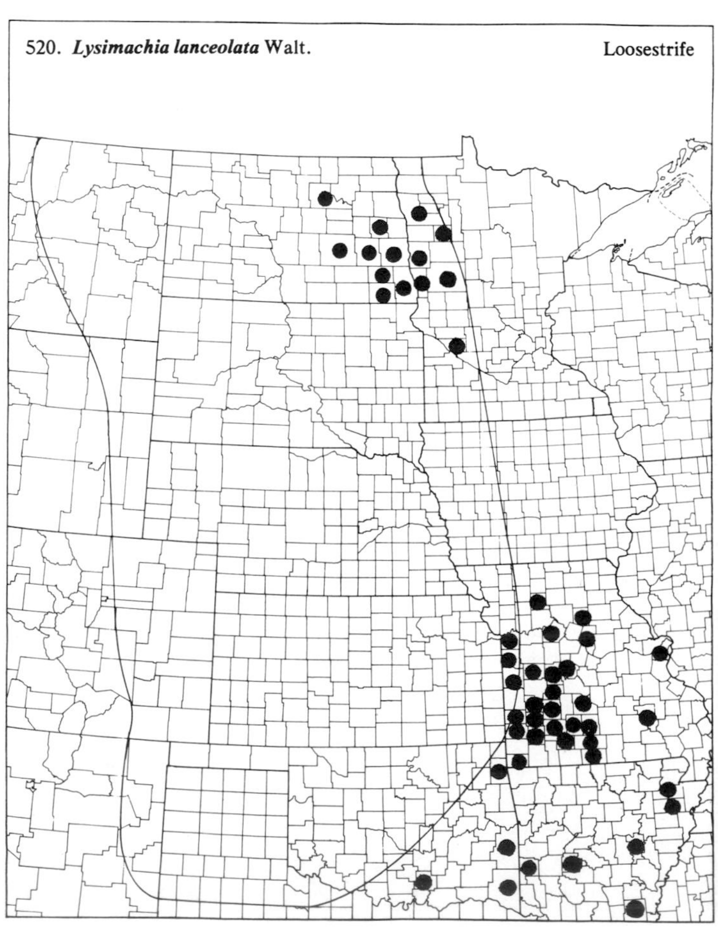

(Primulaceae)

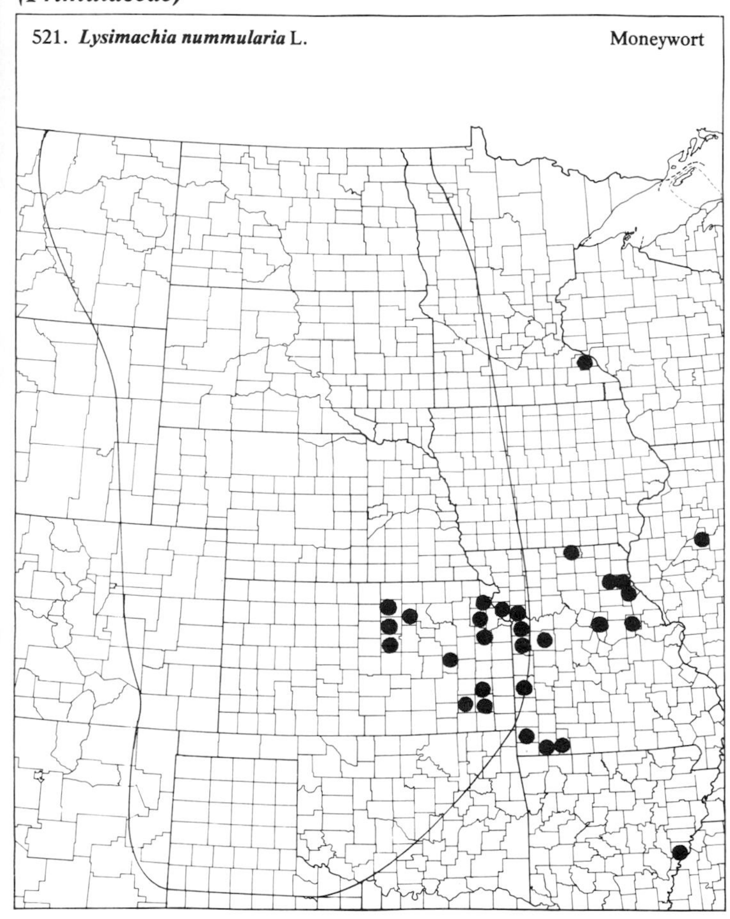

521. ***Lysimachia nummularia*** L. Moneywort

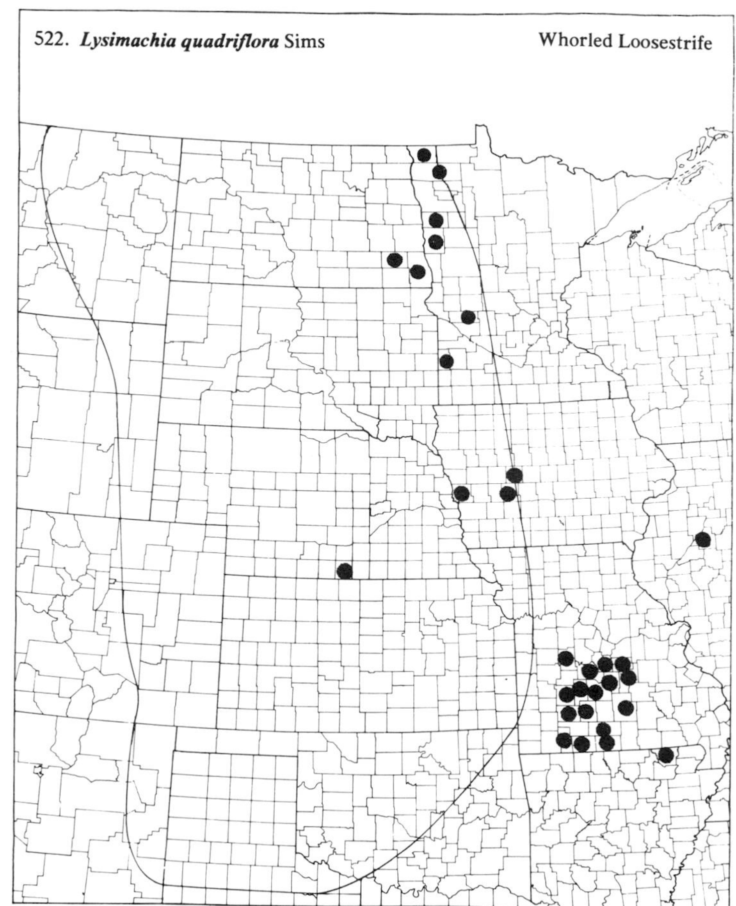

522. ***Lysimachia quadriflora*** Sims Whorled Loosestrife

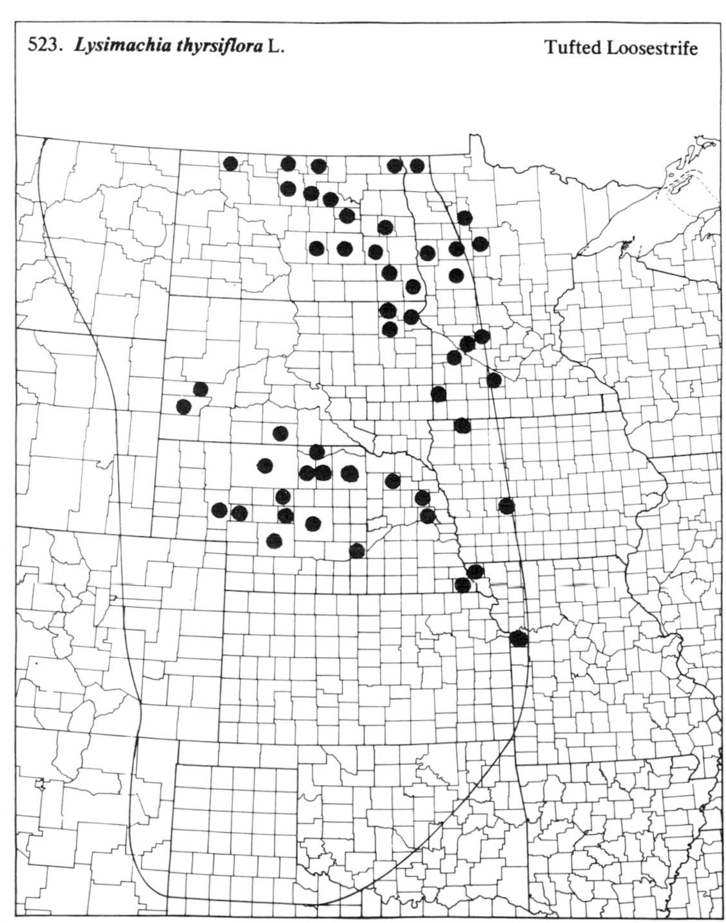

523. ***Lysimachia thyrsiflora*** L. Tufted Loosestrife

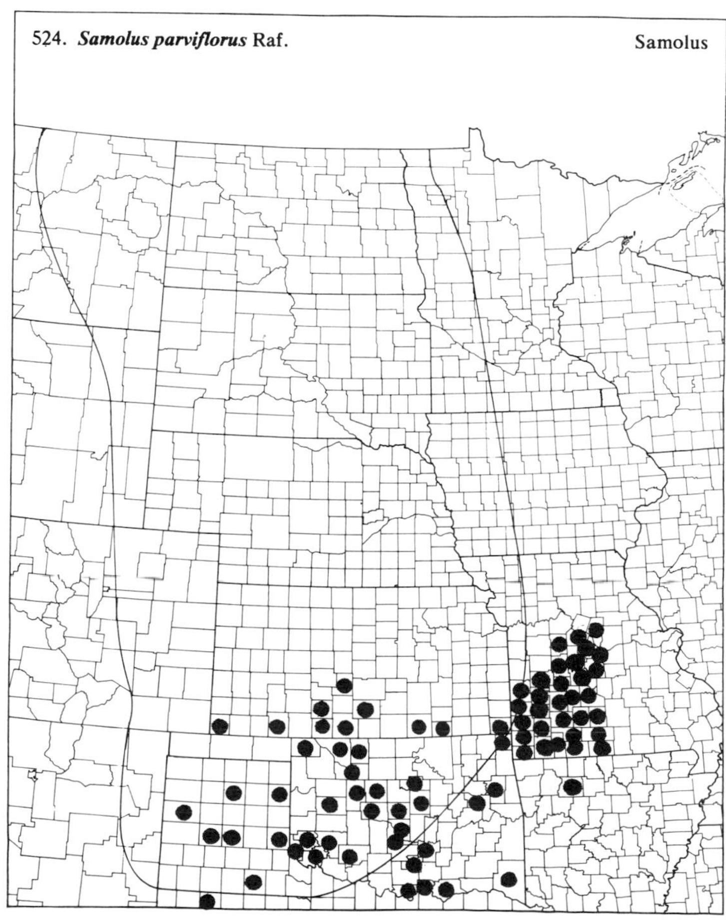

524. ***Samolus parviflorus*** Raf. Samolus

51. *Crassulaceae* (Stonecrop Family)

525. ***Penthorum sedoides*** L. Ditch Stonecrop

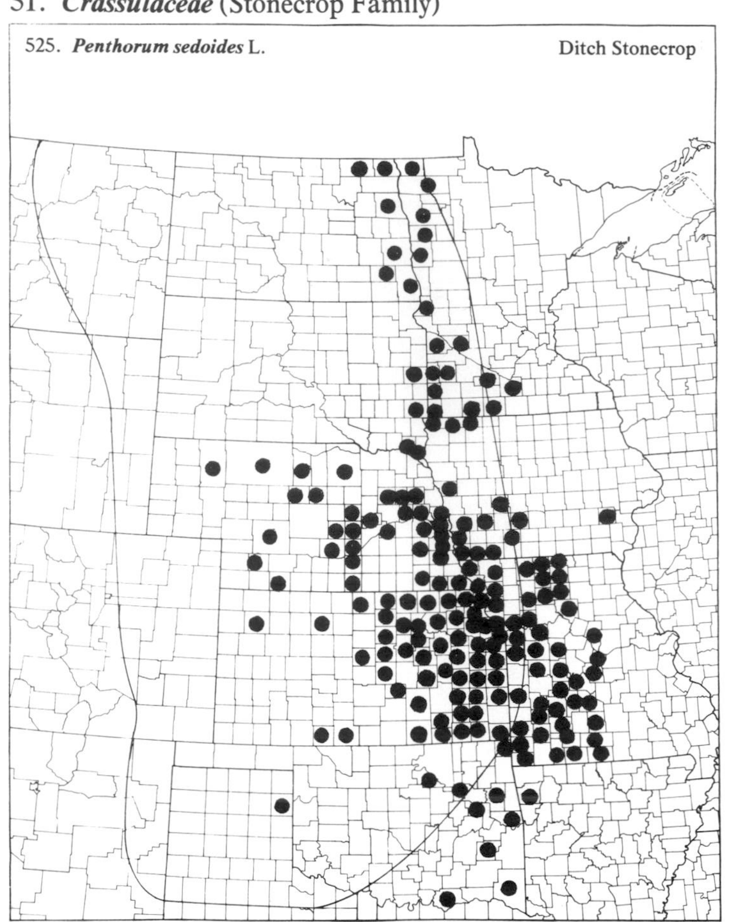

526. ***Sedum lanceolatum*** Torr. Stonecrop

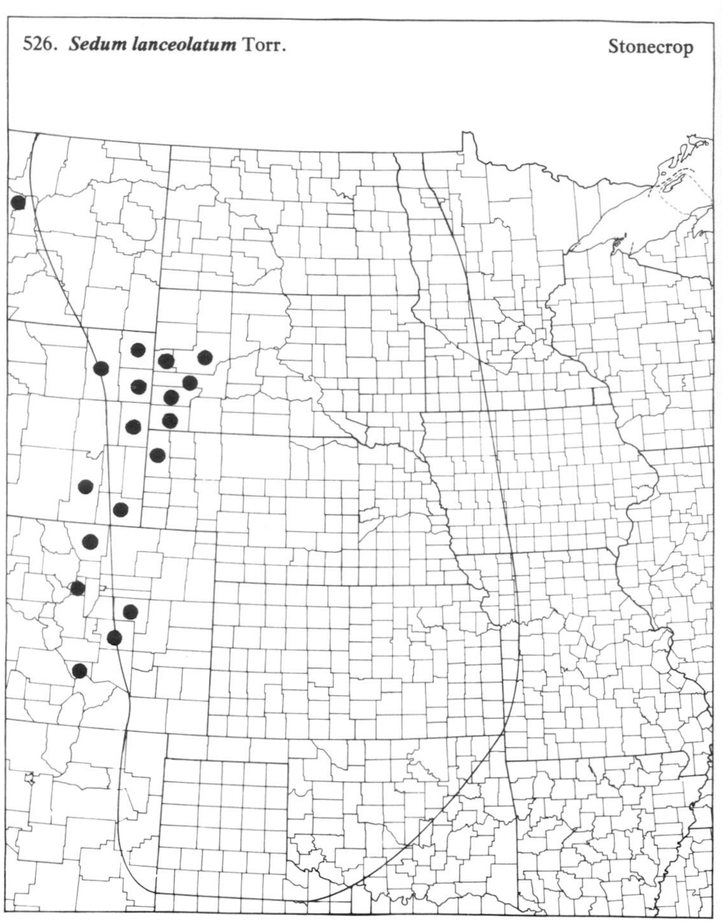

527. ***Sedum nuttallianum*** Raf. Stonecrop

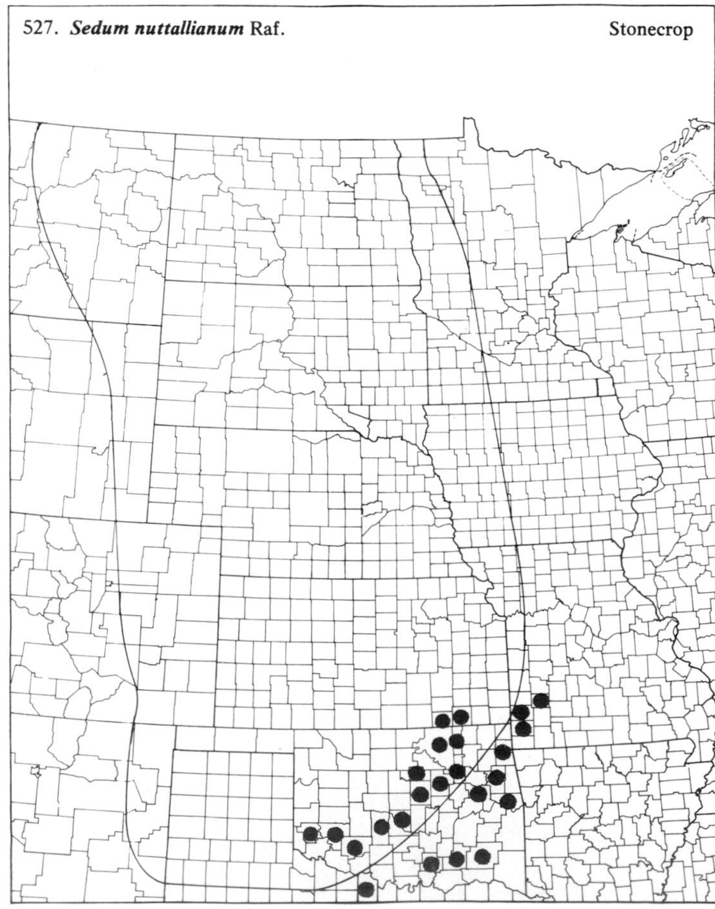

528. ***Sedum pulchellum*** Michx. Stonecrop

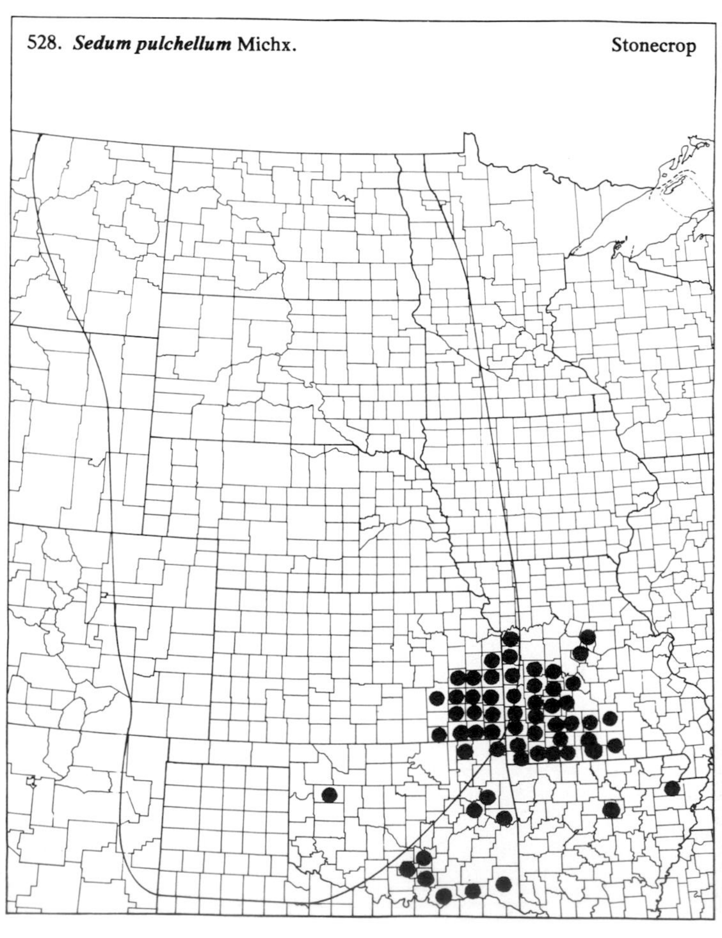

52. *Saxifragaceae* (Saxifrage Family)

529. ***Heuchera hirsuticaulis*** (Wheeler) Rydb. Alumroot

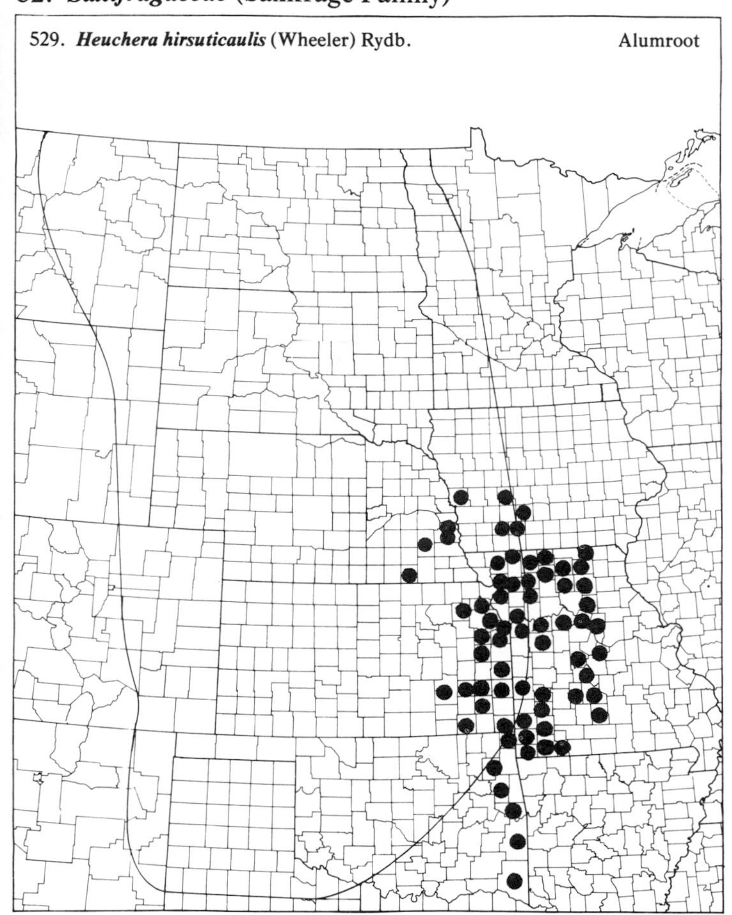

530. ***Heuchera richardsonii*** R. Br. Alumroot

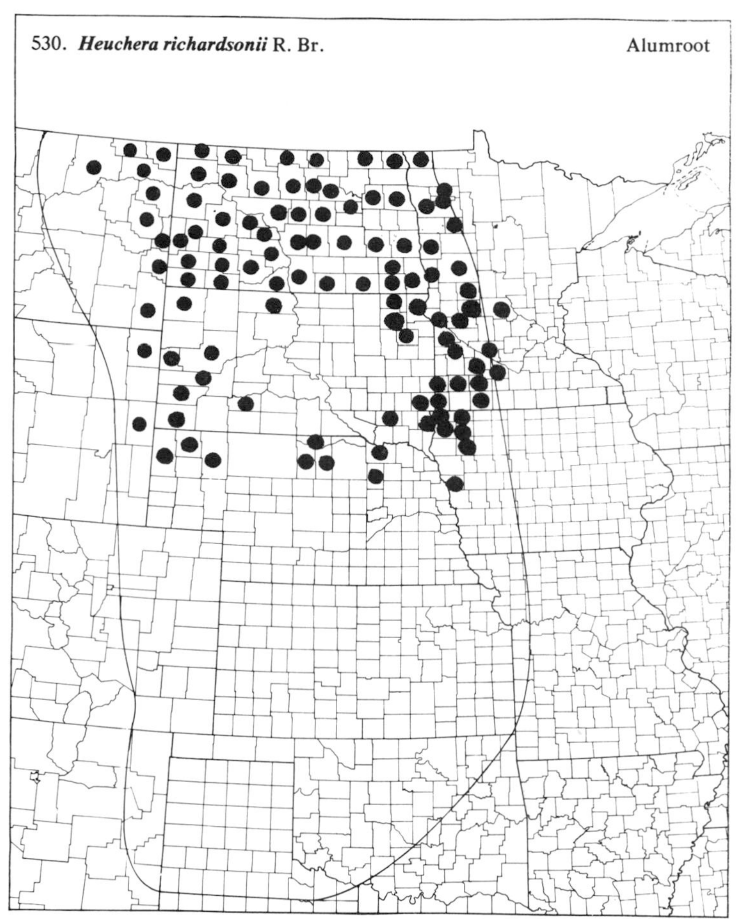

531. ***Mitella nuda*** L. Bishop's Cap

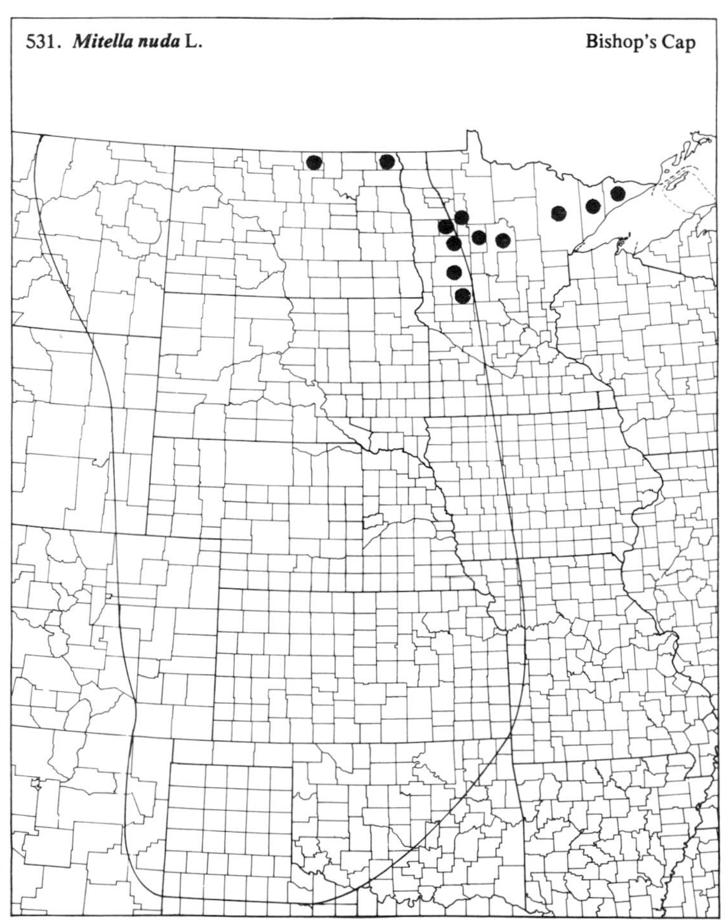

532. ***Parnassia glauca*** Raf. Grass-of-Parnassus

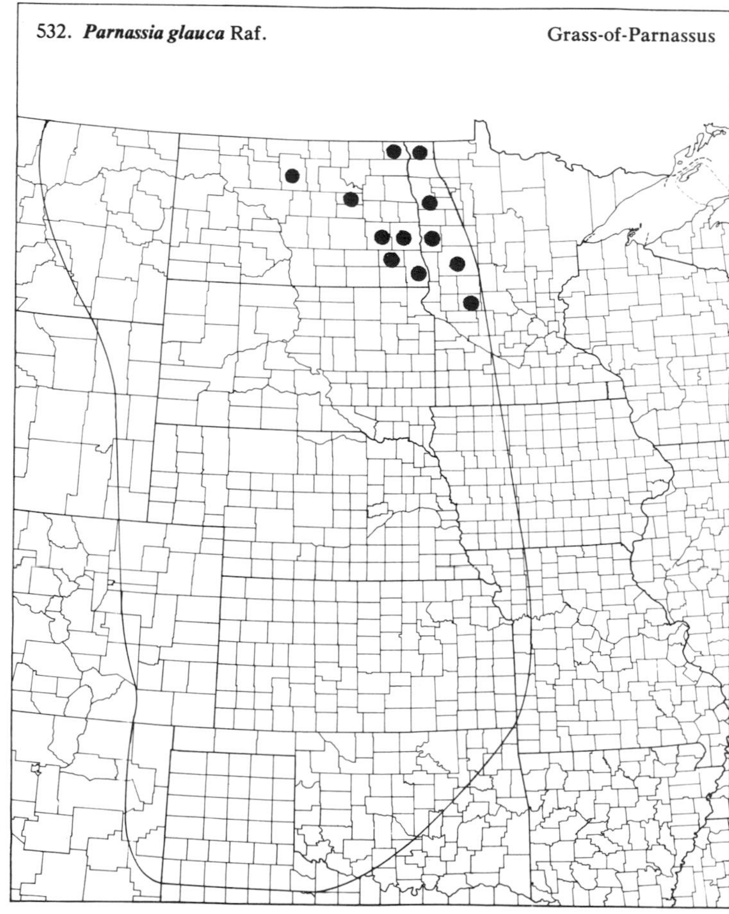

(Saxifragaceae)

533. ***Parnassia palustris*** L. Northern Grass-of-Parnassus

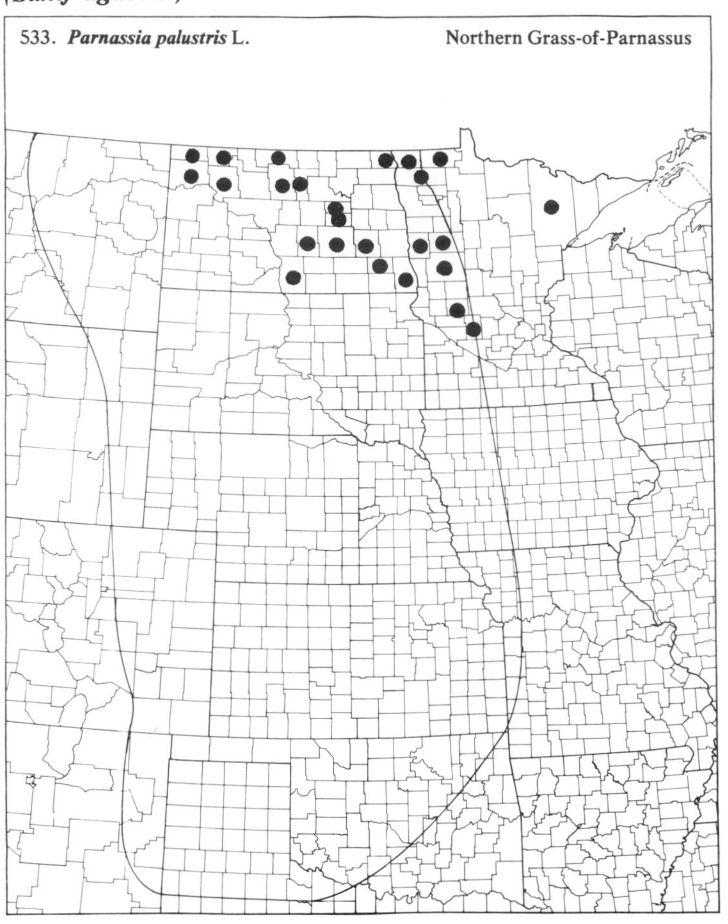

534. ***Ribes americanum*** Mill. Wild Black Currant

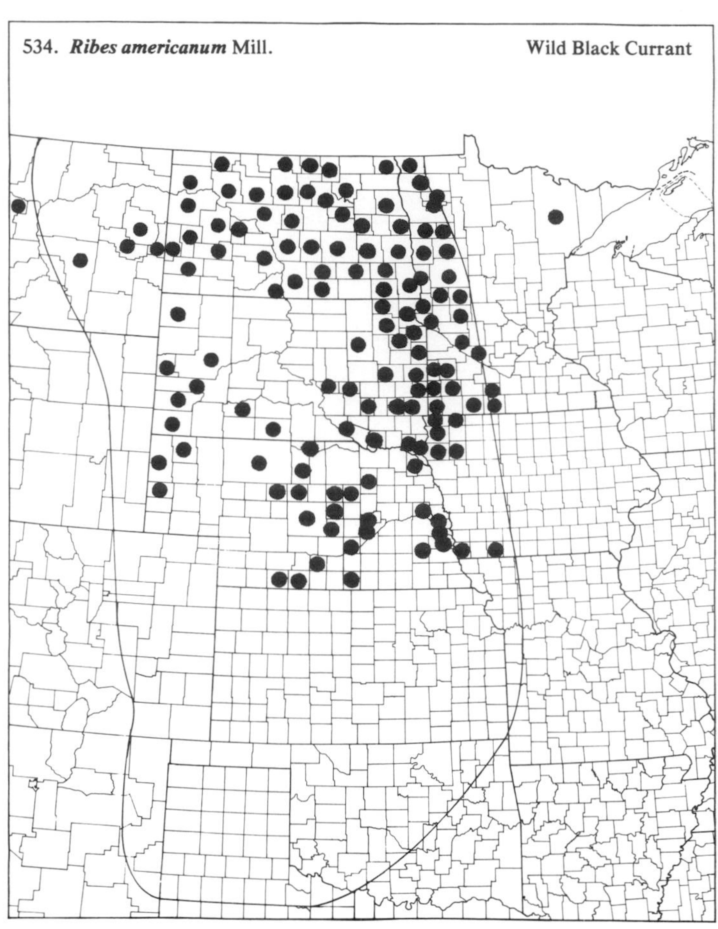

535. ***Ribes cereum*** Dougl. var. ***inebrians*** (Lindl.) C.L. Hitchc. Western Red Currant

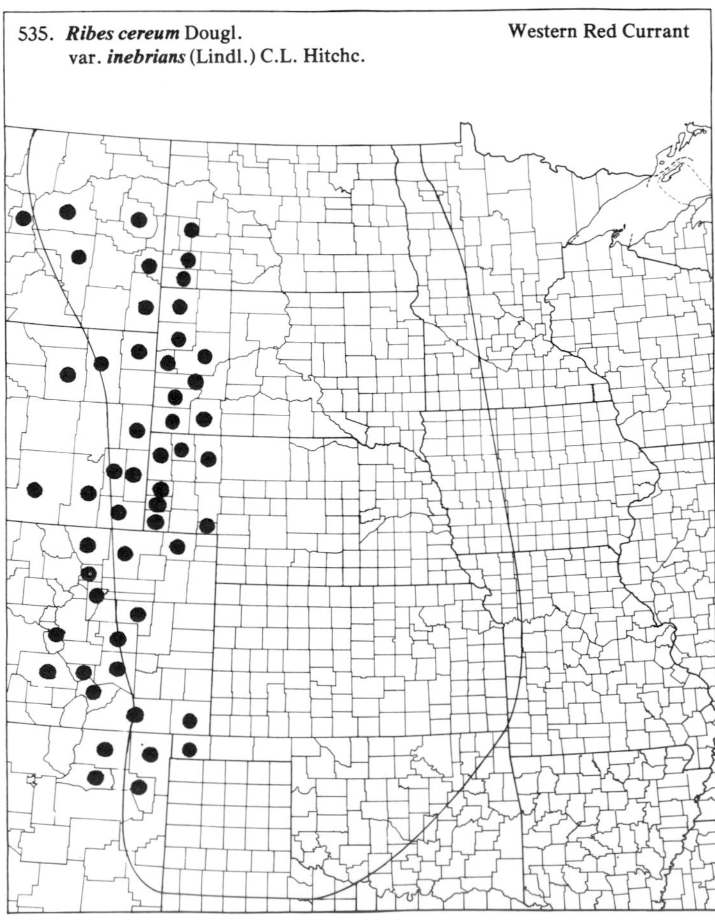

536. ***Ribes cynosbati*** L. Dogberry

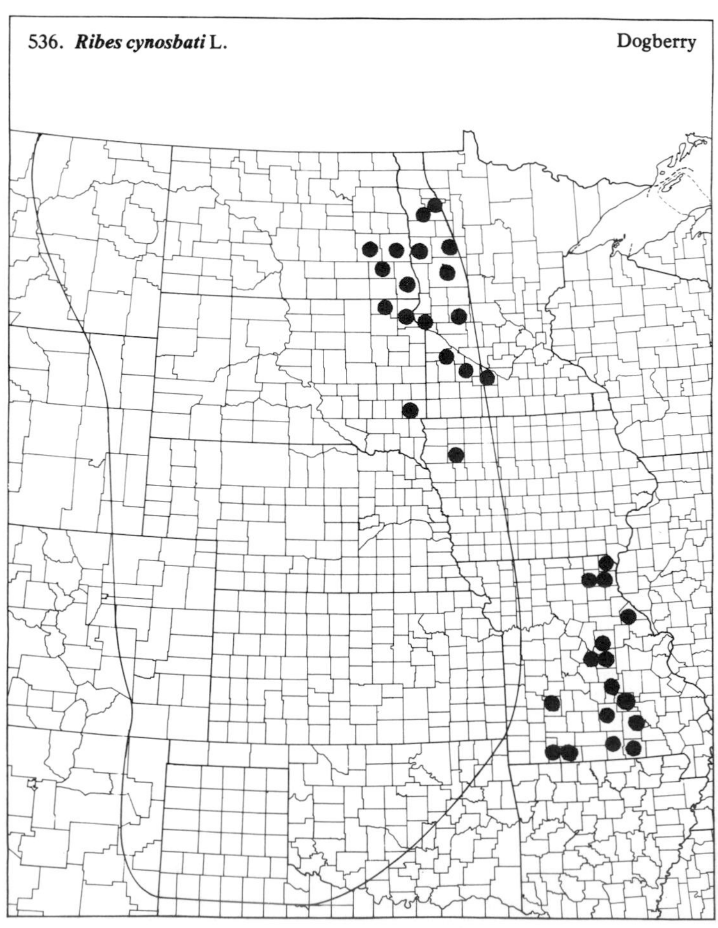

(Saxifragaceae)

537. ***Ribes hirtellum*** Michx.

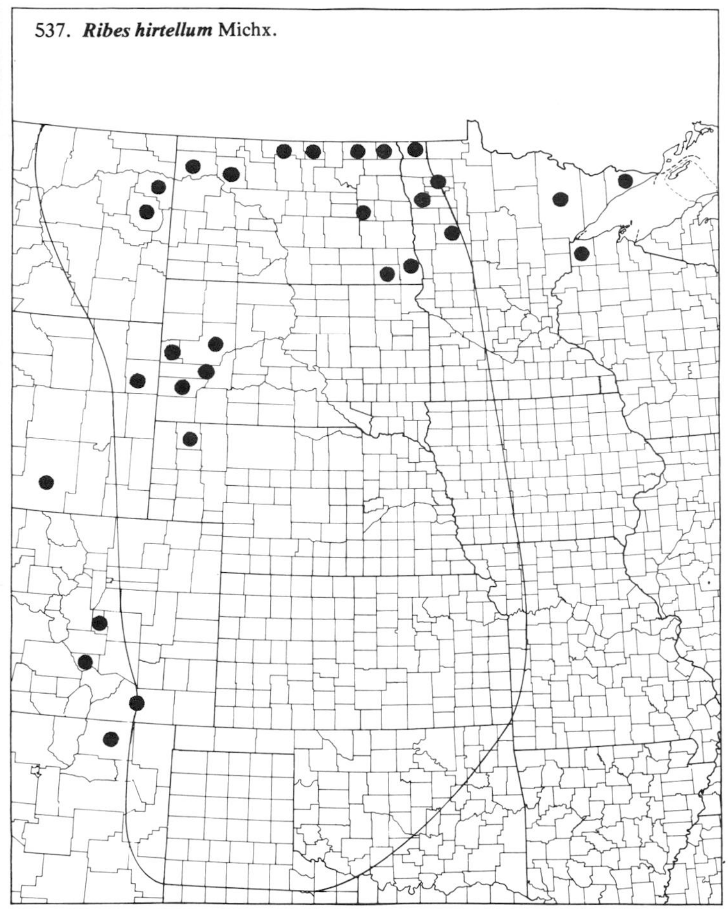

538. ***Ribes missouriense*** Nutt. Missouri Gooseberry

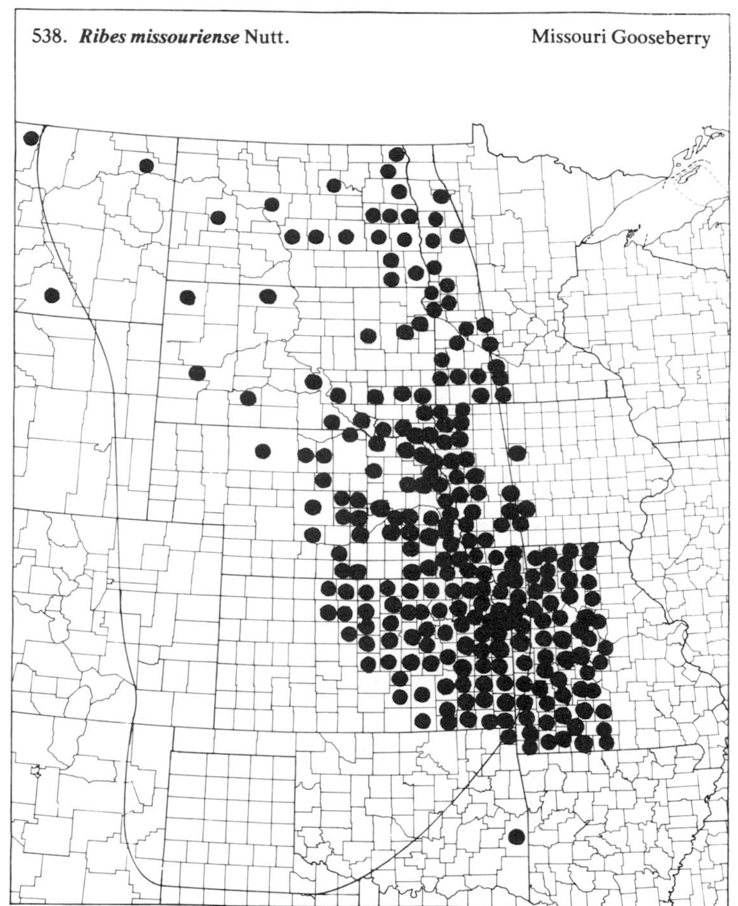

539. ***Ribes odoratum*** Wendl. f. Buffalo Currant

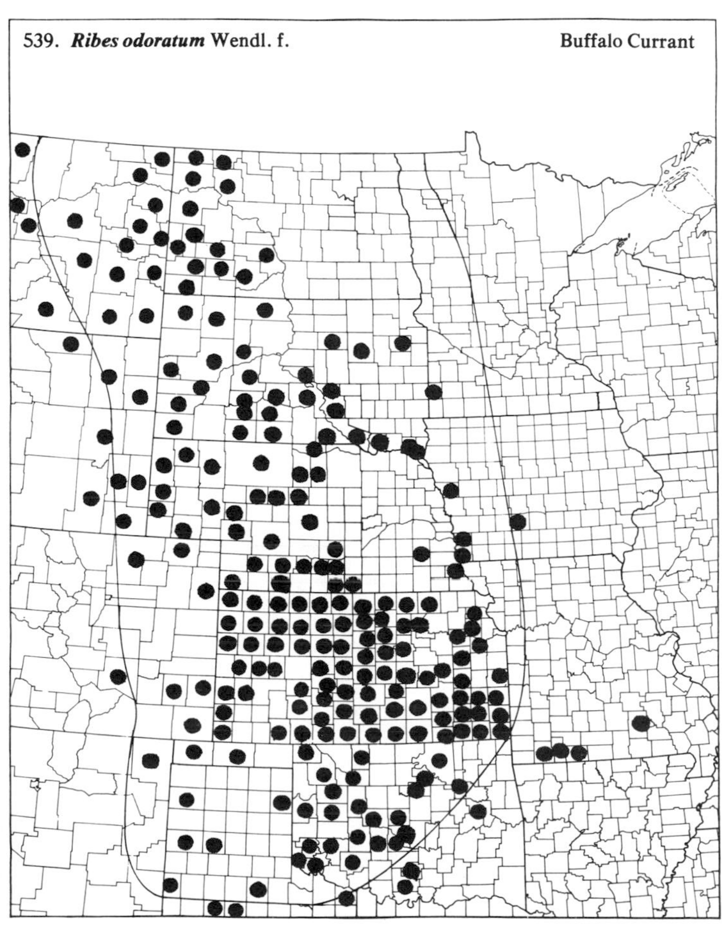

540. ***Ribes oxyacanthoides*** L.

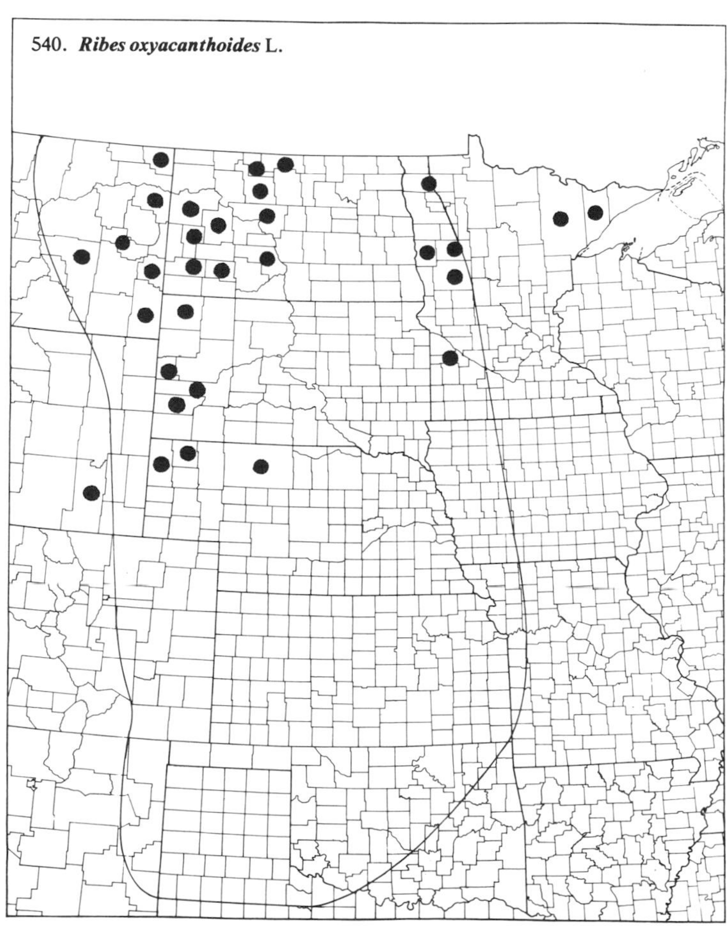

(Saxifragaceae)

541. ***Ribes setosum*** Lindl. Bristly Gooseberry

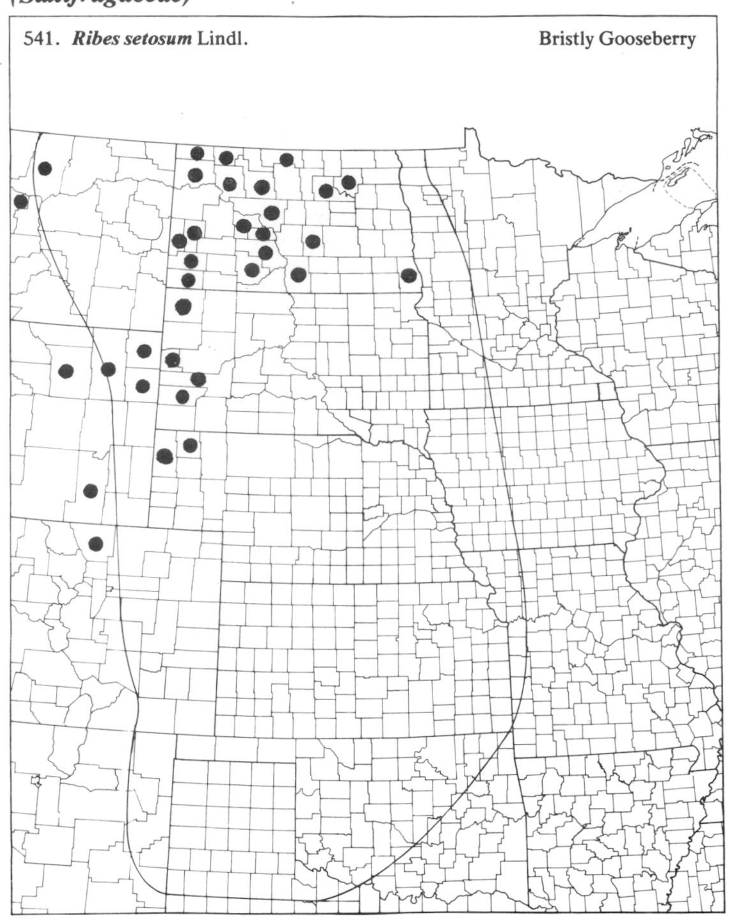

542. ***Ribes triste*** Pall. Swamp Currant

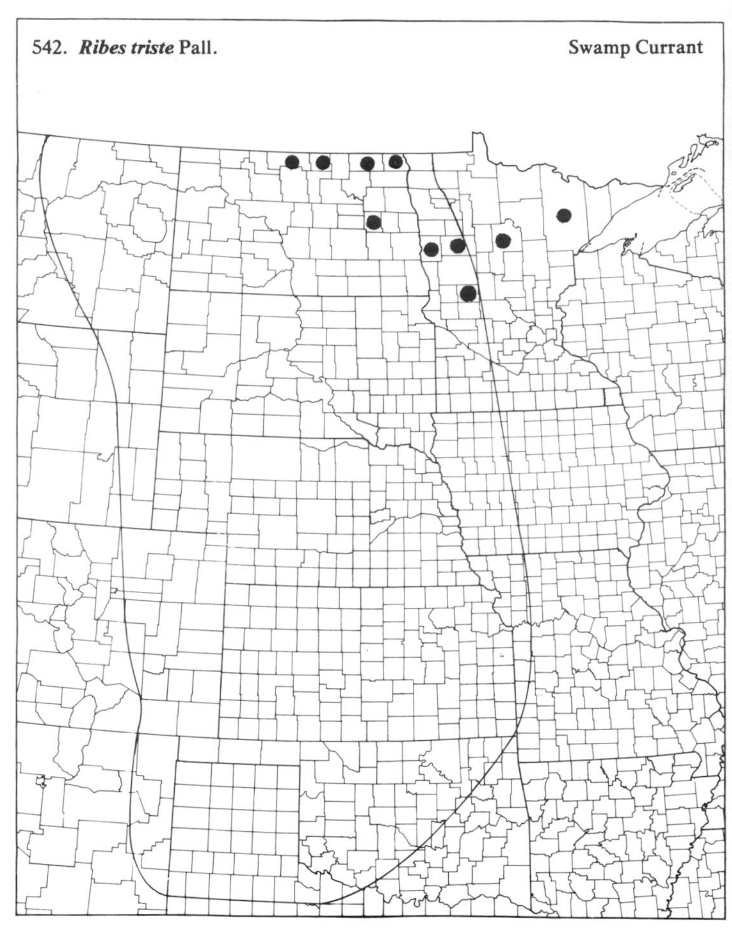

53. *Rosaceae* (Rose Family)

543. ***Agrimonia gryposepala*** Wallr. Agrimony

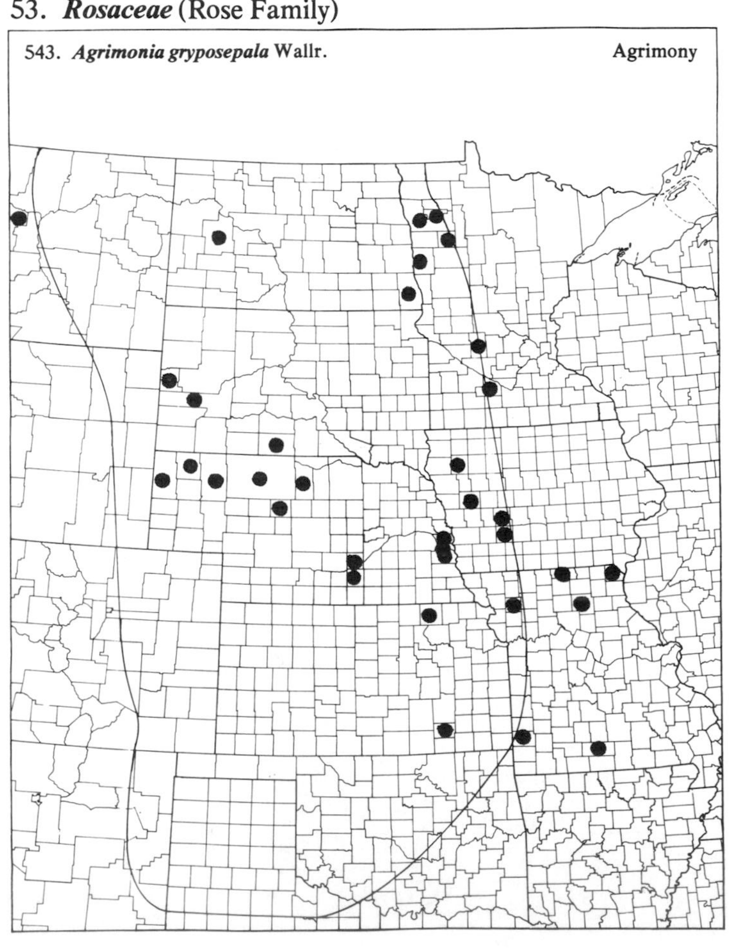

544. ***Agrimonia parviflora*** Ait. Many-flowered Agrimony

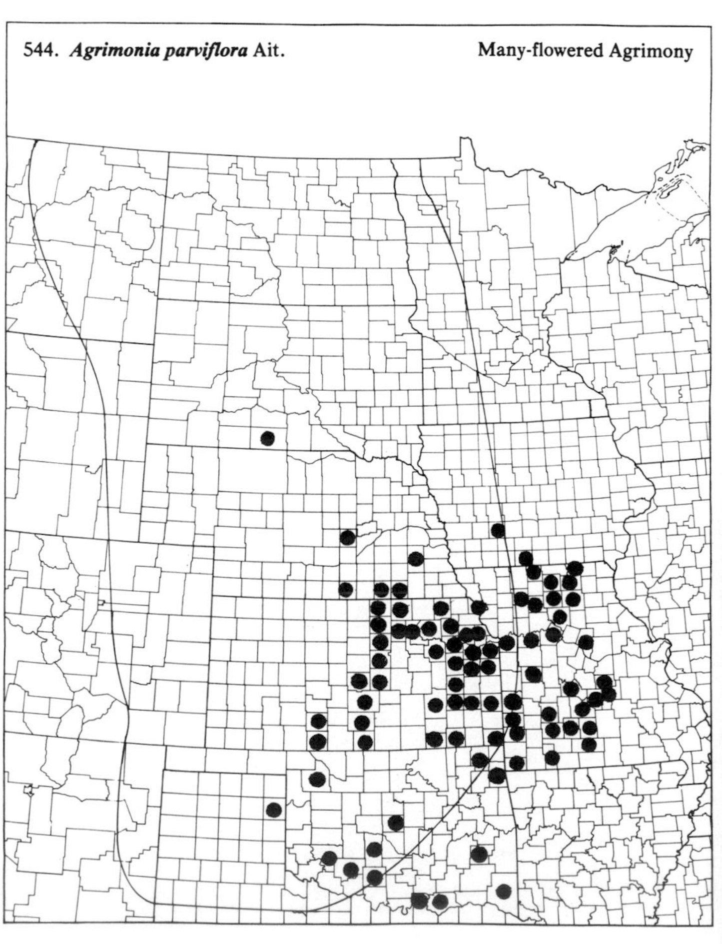

(Rosaceae)

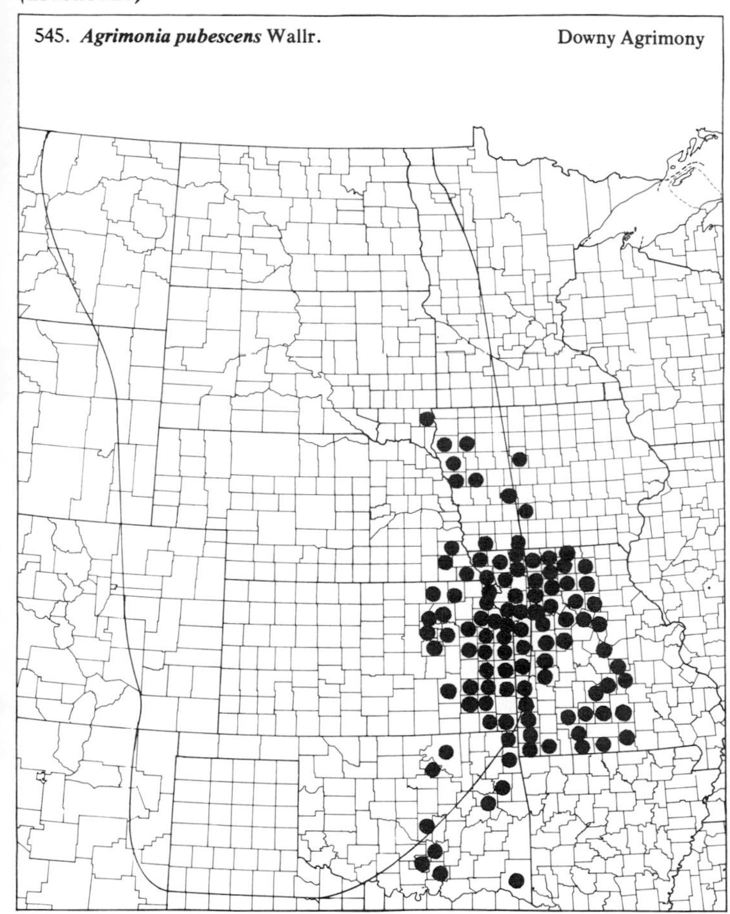
545. *Agrimonia pubescens* Wallr.
Downy Agrimony

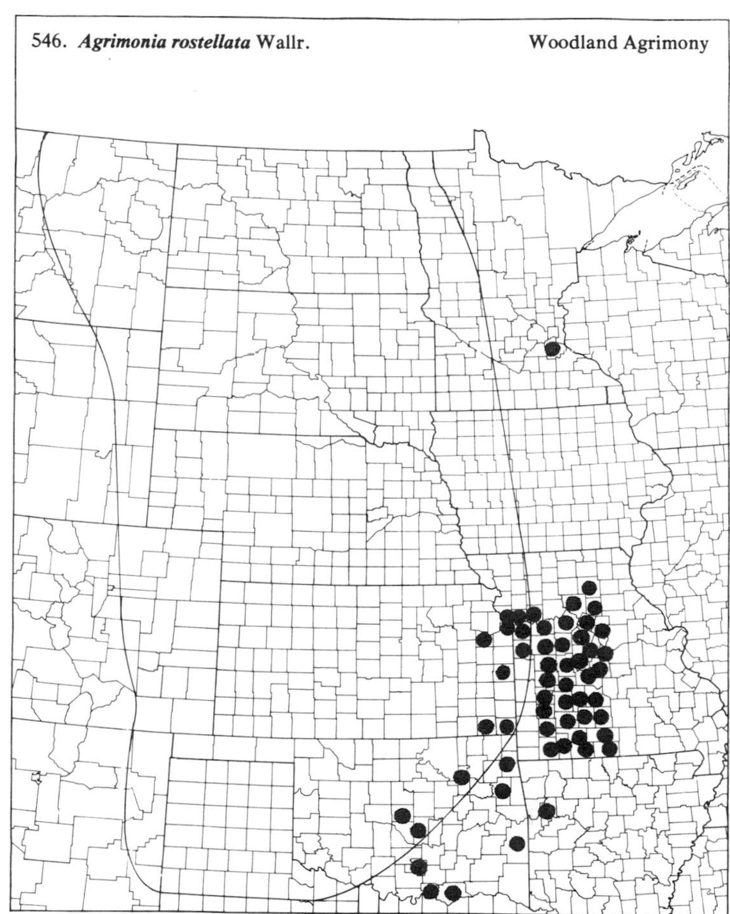
546. *Agrimonia rostellata* Wallr.
Woodland Agrimony

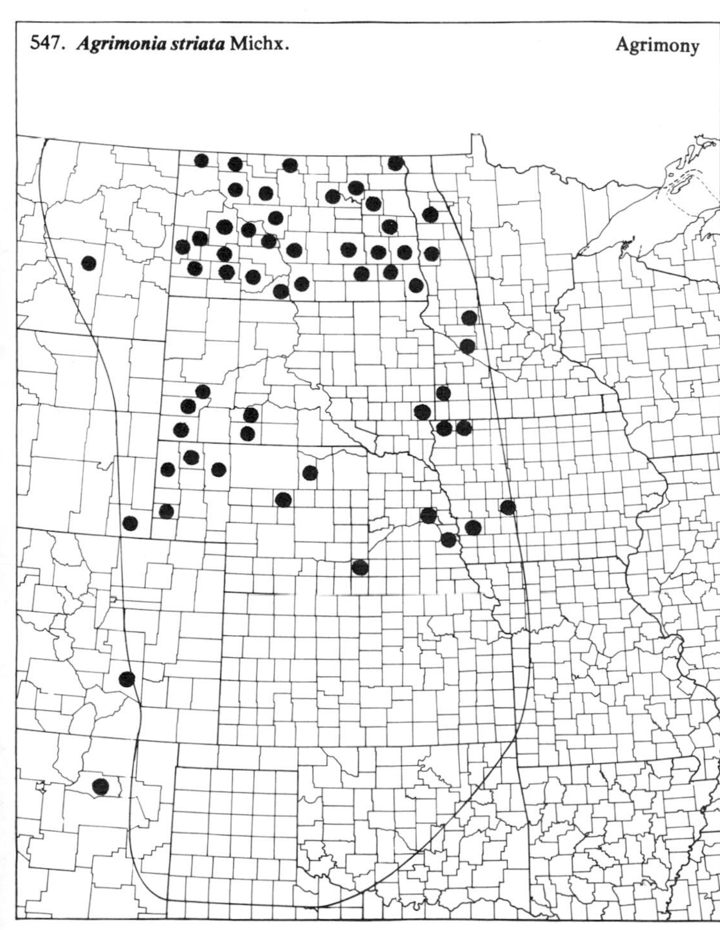
547. *Agrimonia striata* Michx.
Agrimony

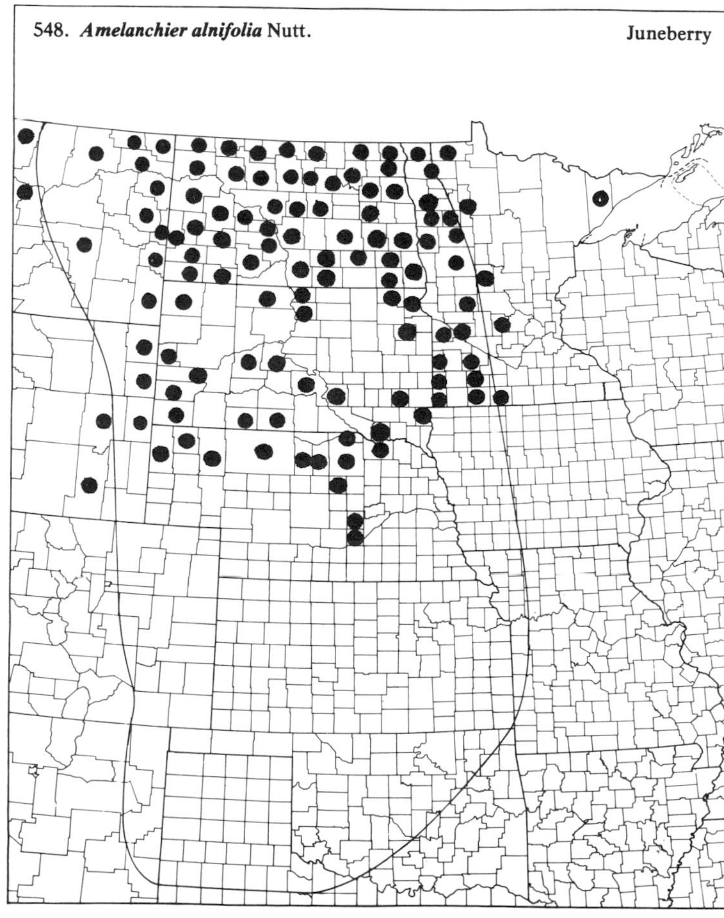
548. *Amelanchier alnifolia* Nutt.
Juneberry

(Rosaceae)

549. ***Amelanchier arborea*** (Michx. f.) Fern. Juneberry

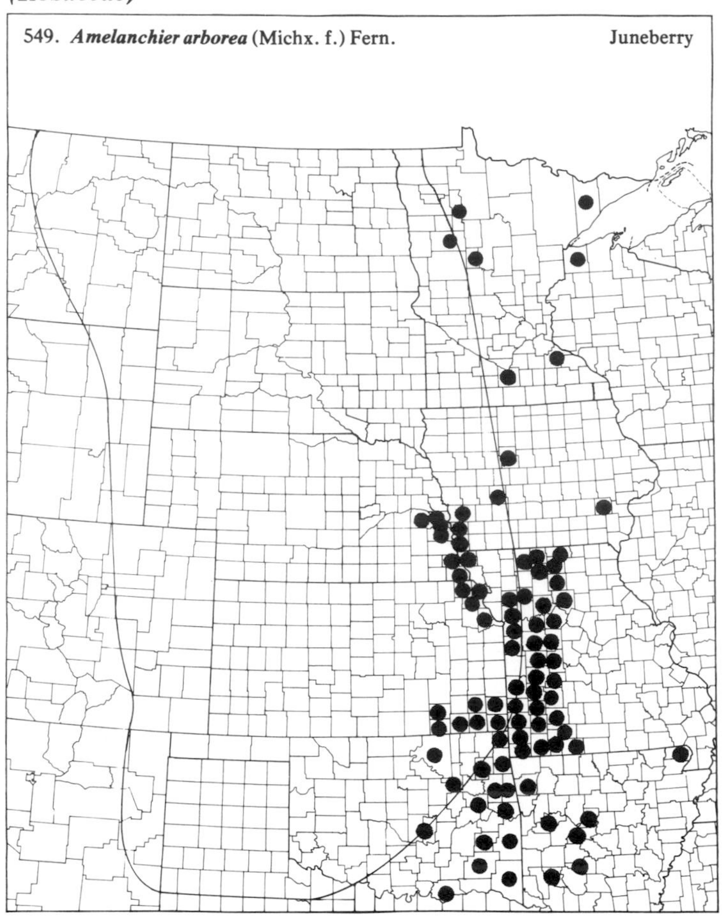

550. ***Amelanchier sanguinea*** (Pursh) DC.

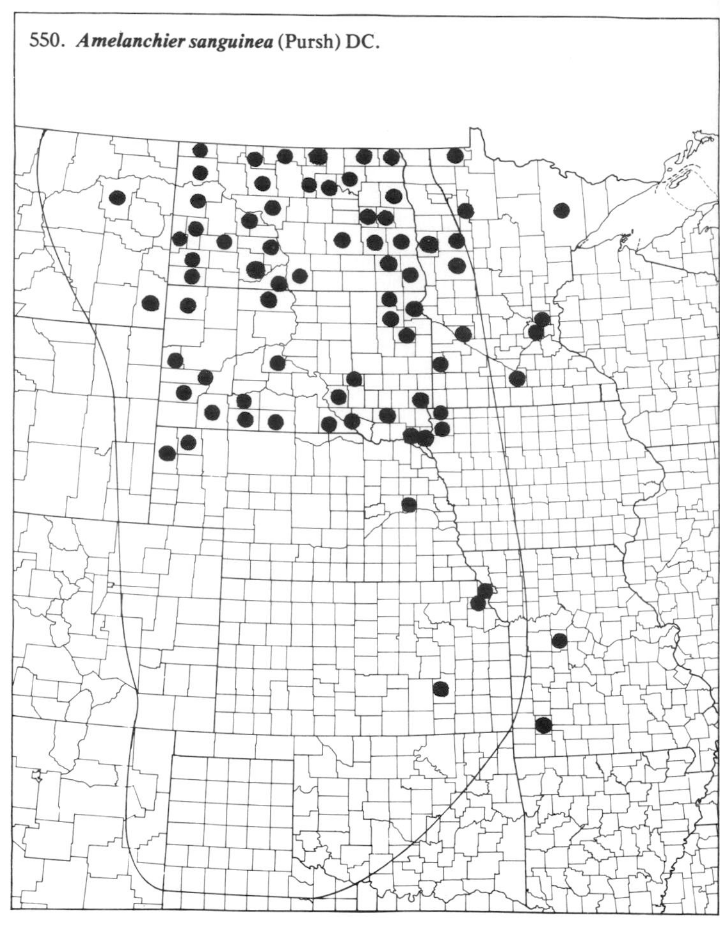

551. ***Cercocarpus montanus*** Raf. Mountain Mahogany

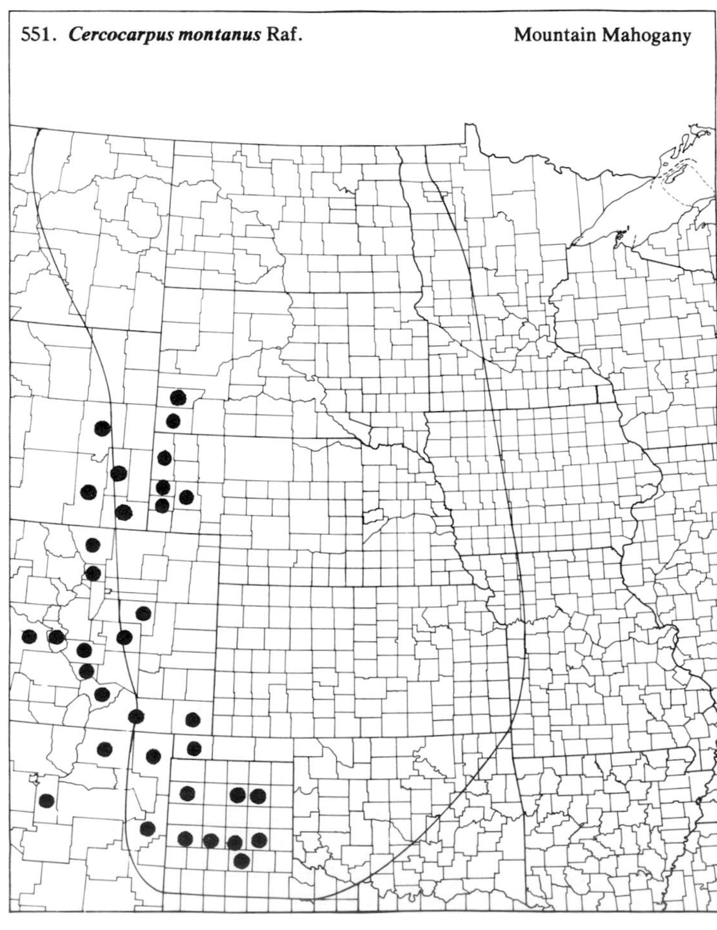

552. ***Chamaerhodos erecta*** (L.) Bunge var. ***parviflora*** (Nutt.) C.L. Hitchc. Little Rose

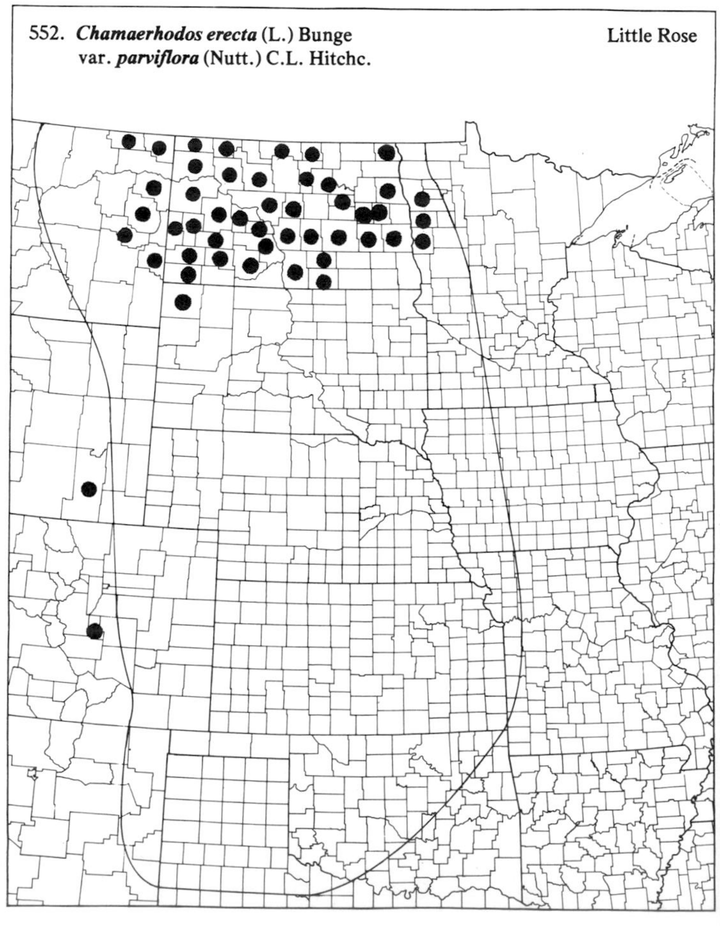

(Rosaceae)

553. ***Crataegus calpodendron*** (Ehrh.) Medic. Hawthorn

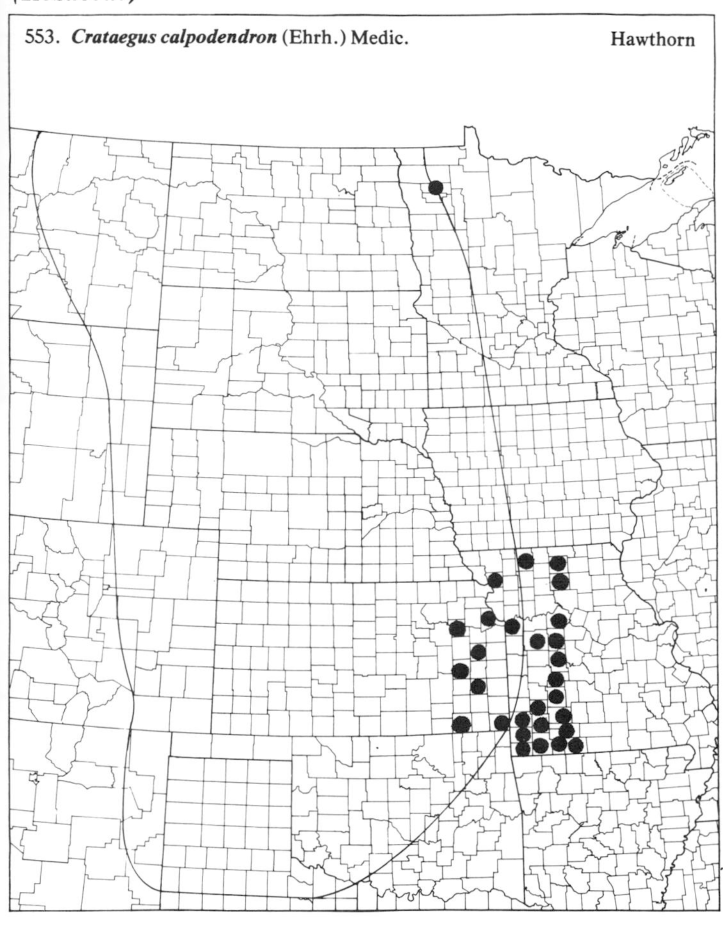

554. ***Crataegus chrysocarpa*** Ashe

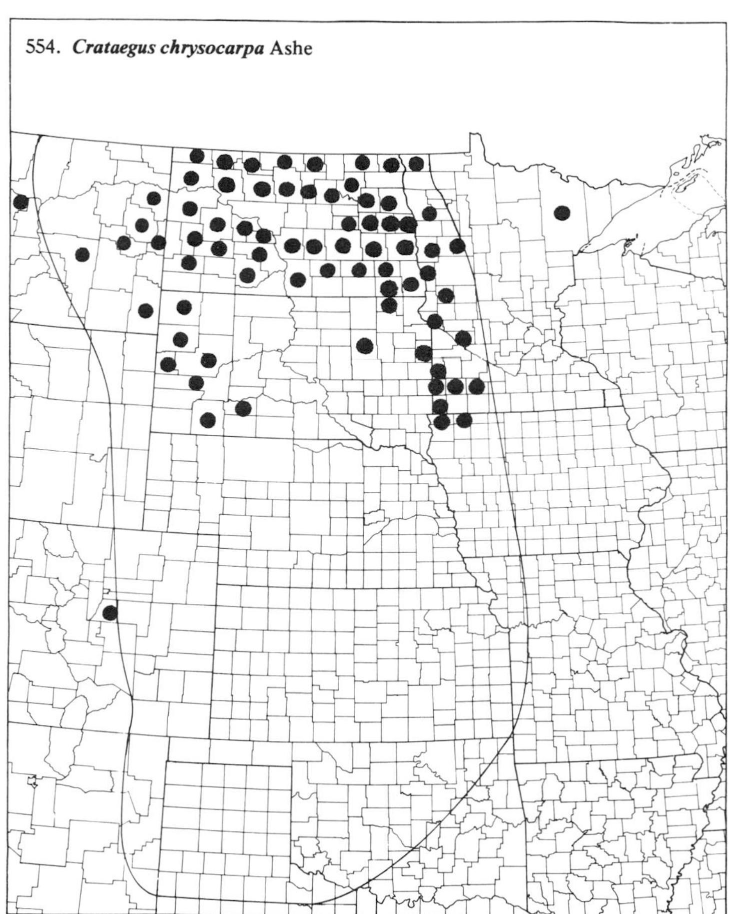

555. ***Crataegus crus-galli*** L. Cockspur Hawthorn

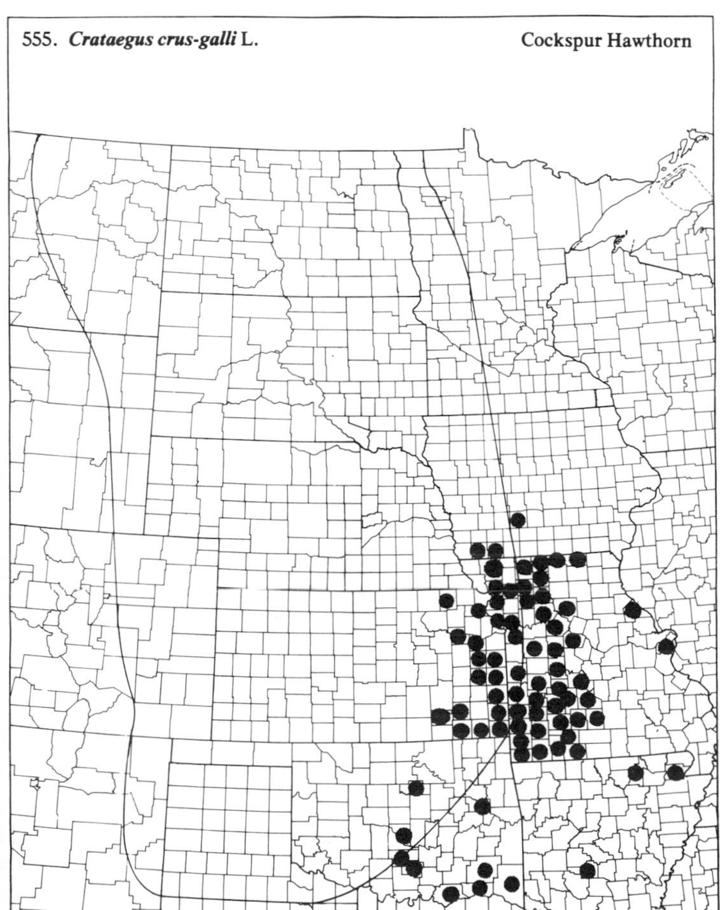

556. ***Crataegus mollis*** (T. & G.) Scheele Downy Hawthorn

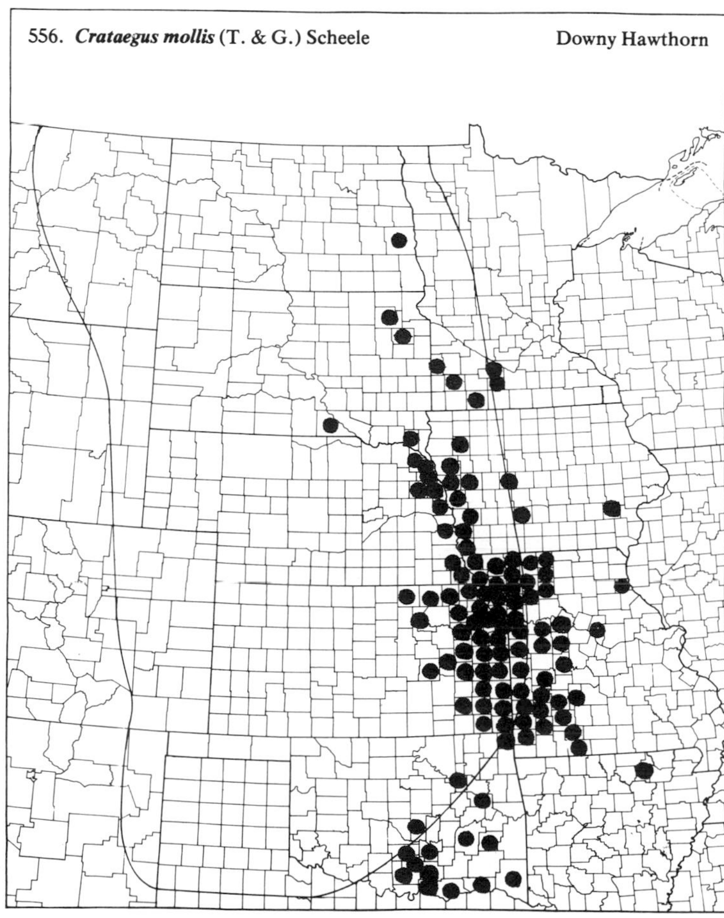

(Rosaceae)

557. ***Crataegus succulenta*** Link var. ***succulenta*** — Hawthorn

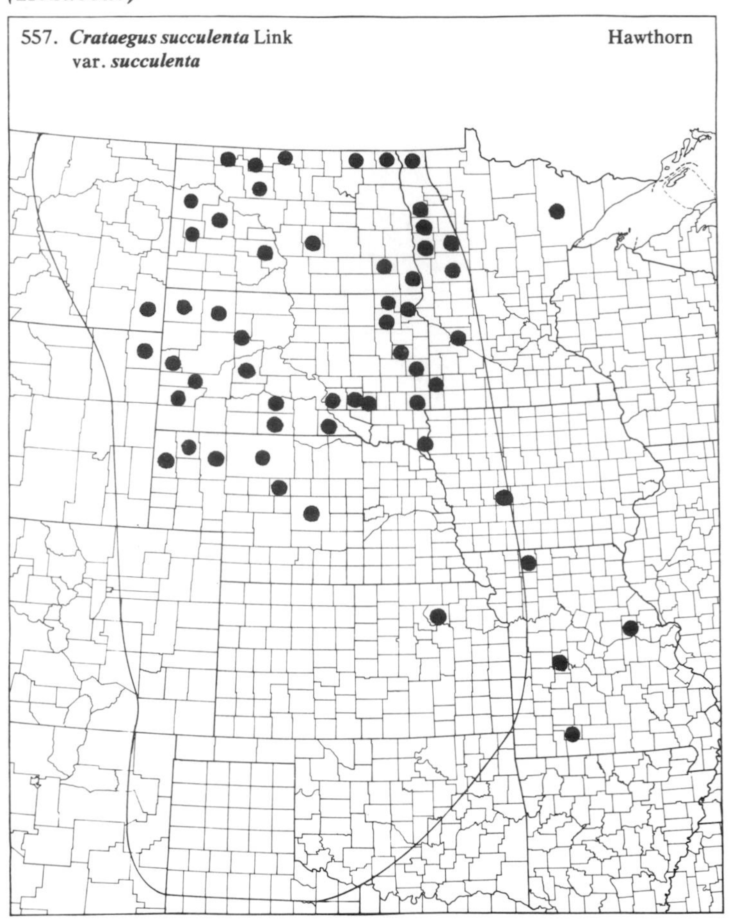

558. ***Crataegus succulenta*** Link var. ***pertomentosa*** (Ashe) E.J. Palm. — Hawthorn

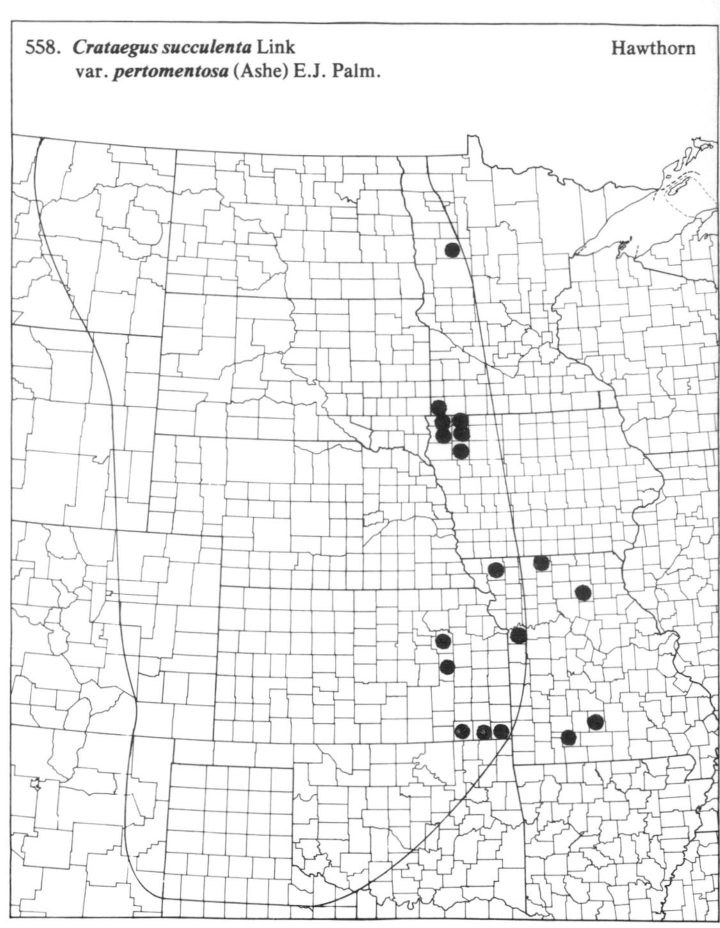

559. ***Crataegus viridis*** L. — Hawthorn

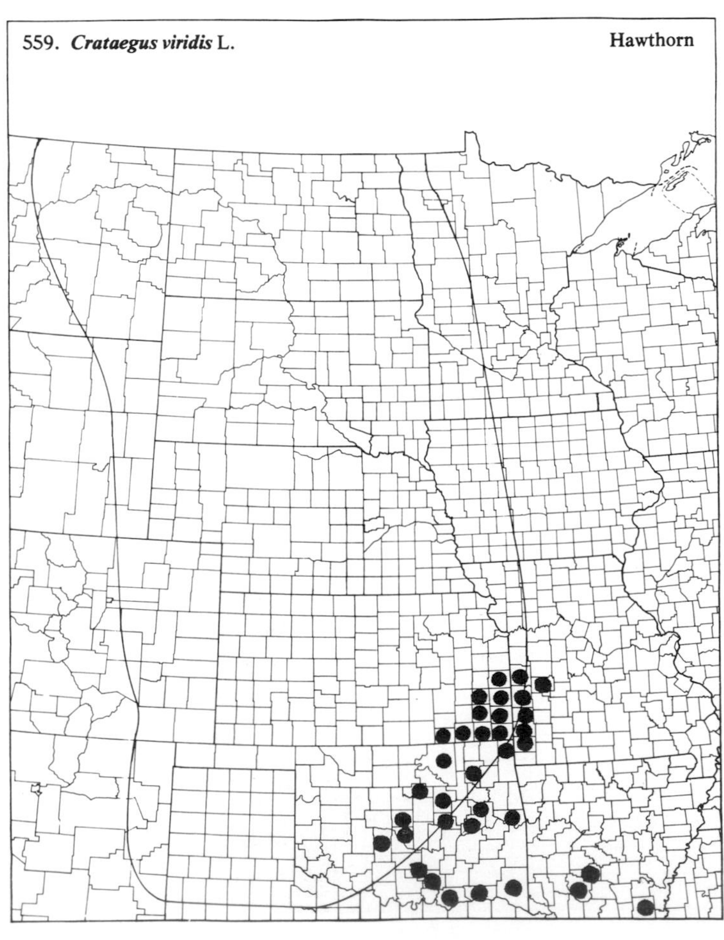

560. ***Fragaria vesca*** L. var. ***americana*** Porter — Wood Strawberry

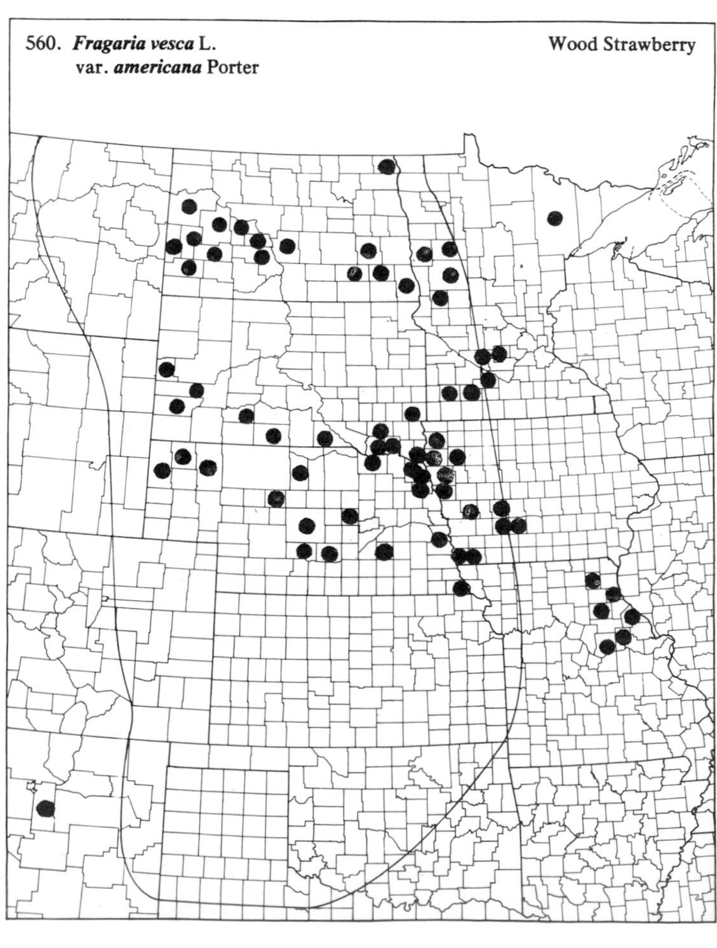

(Rosaceae)

561. ***Fragaria virginiana*** Duchn. var. ***glauca*** Wats. — Wild Strawberry

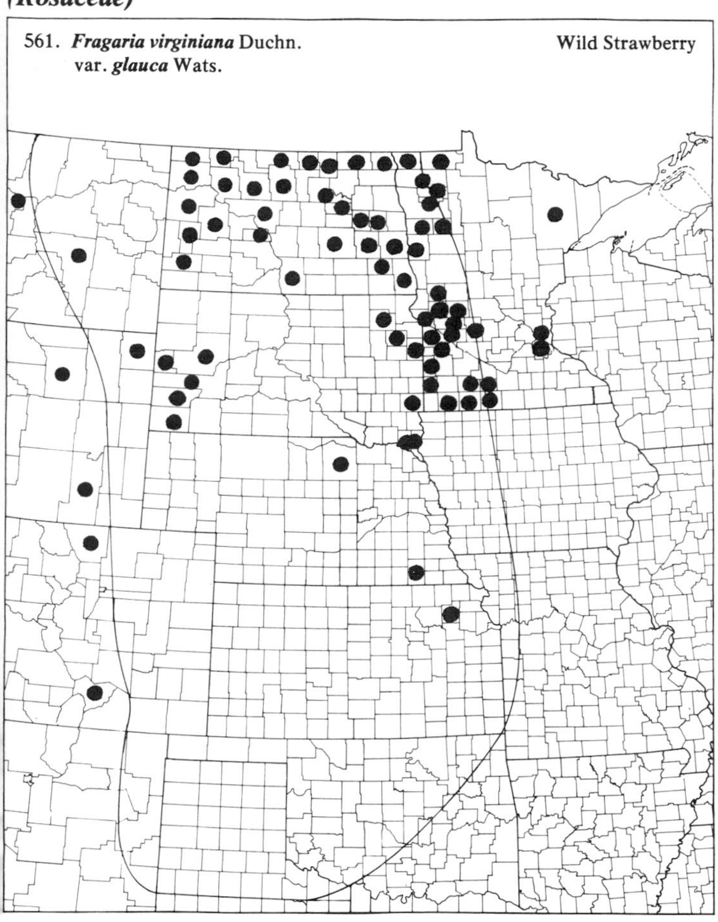

562. ***Fragaria virginiana*** Duchn. var. ***illinoensis*** (Prince) Gray — Wild Strawberry

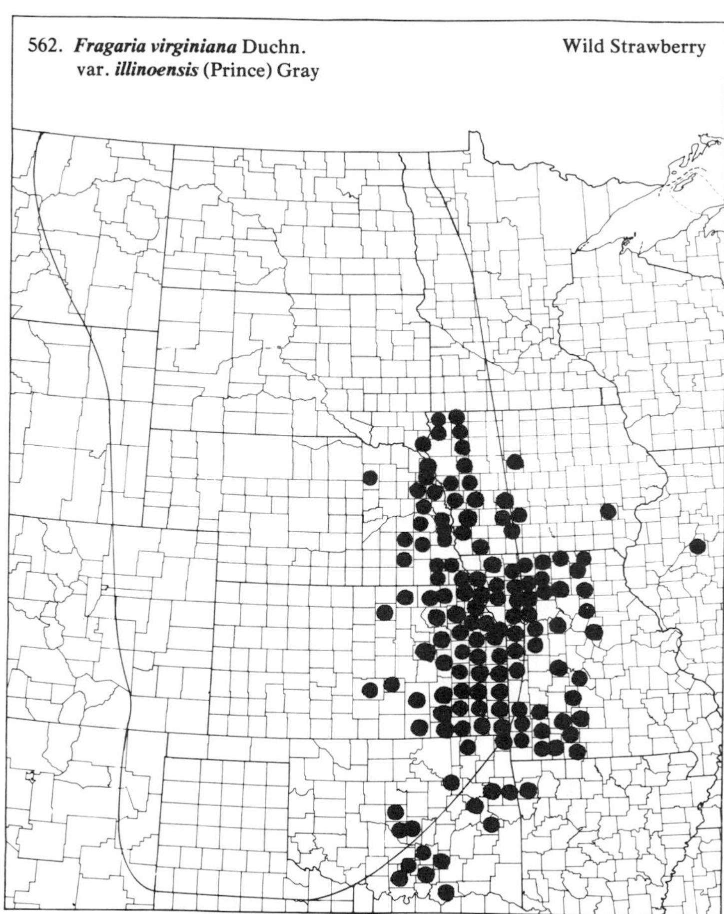

563. ***Geum aleppicum*** Jacq. var. ***strictum*** (Ait.) Fern. — Yellow Avens

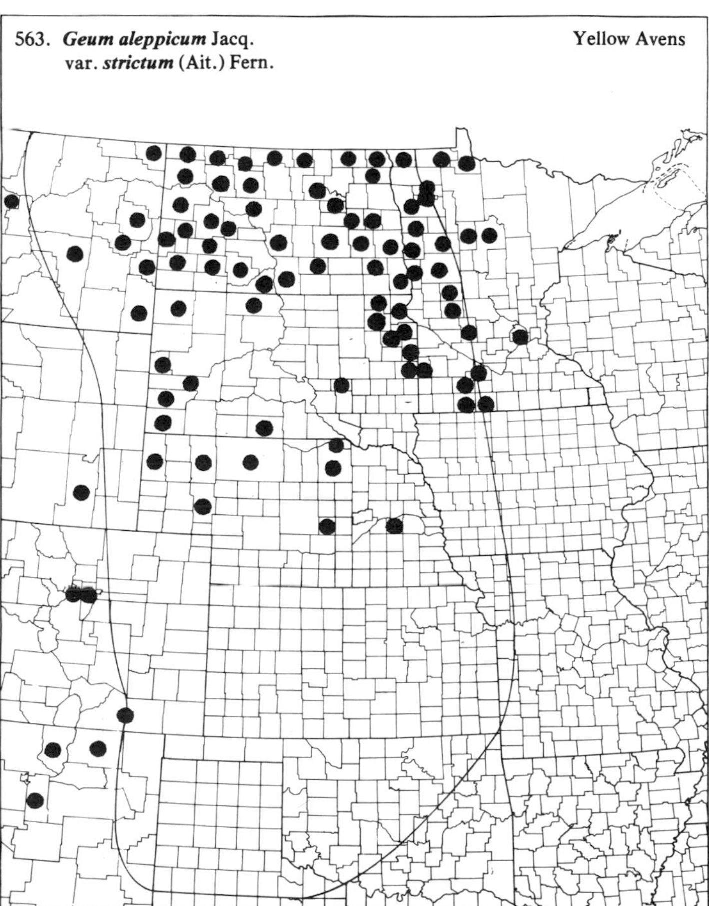

564. ***Geum canadense*** Jacq. — White Avens

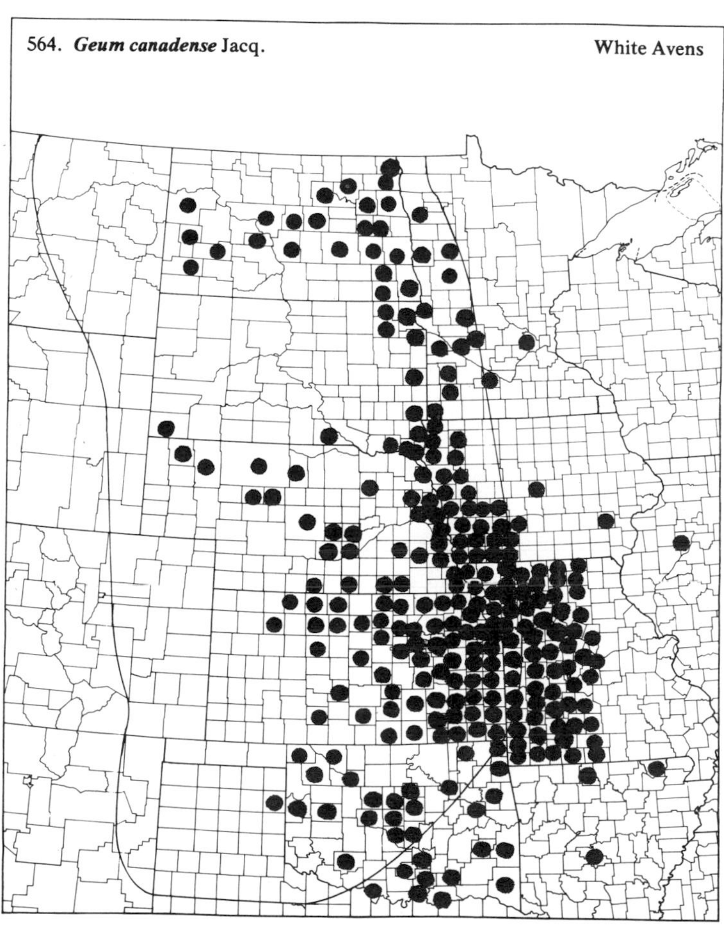

(Rosaceae)

565. **Geum macrophyllum** Willd. — Large-leaved Avens

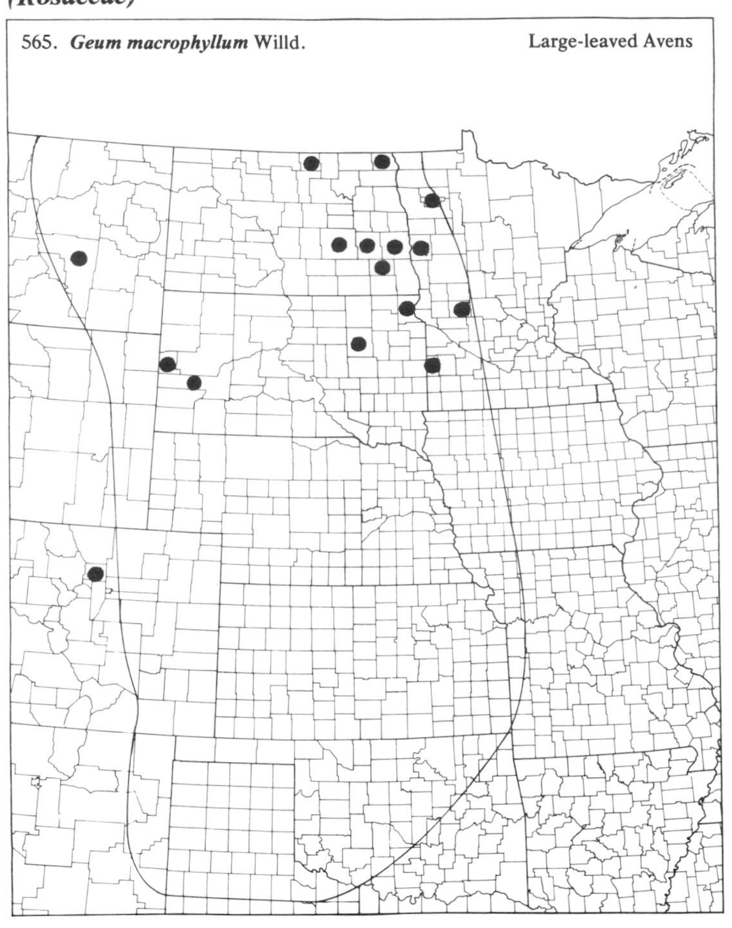

566. **Geum triflorum** Pursh — Purple Avens

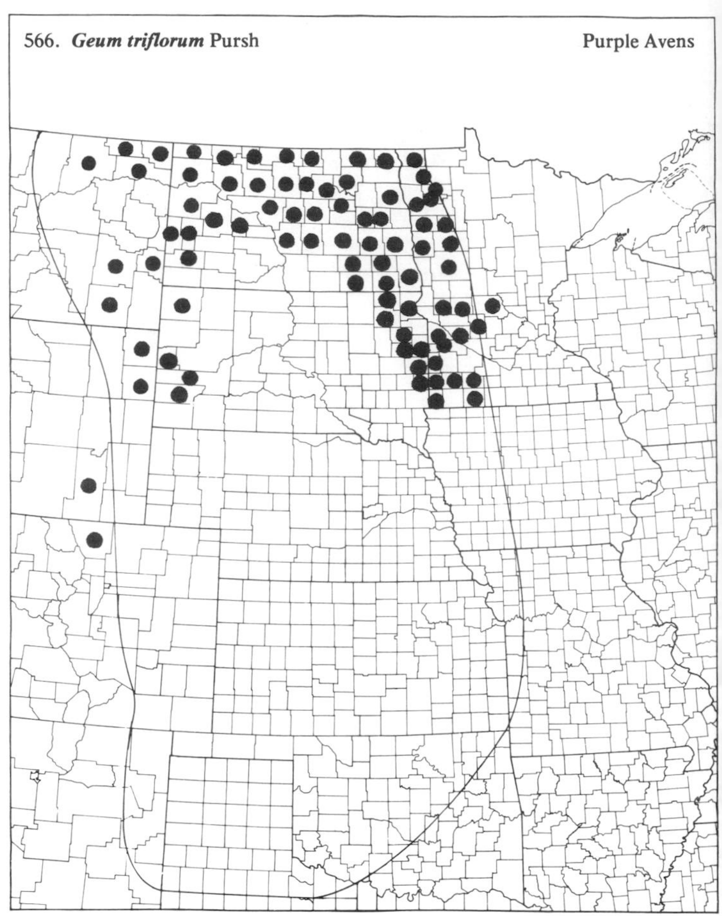

567. **Geum vernum** (Raf.) T. & G. — Heartleaf Avens

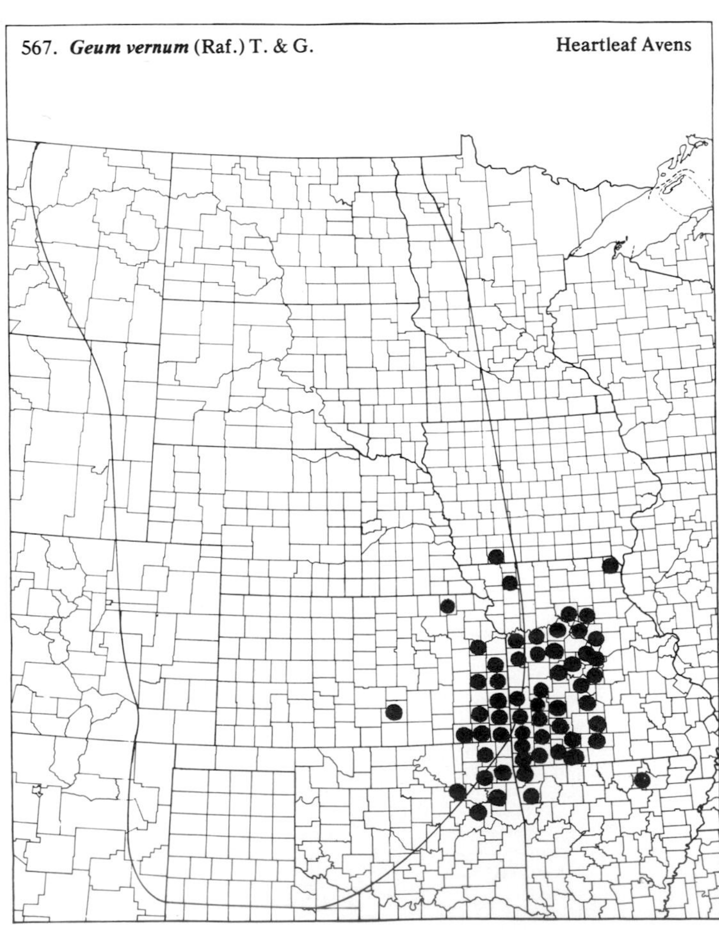

568. **Physocarpus opulifolius** (L.) Maxim. — Ninebark

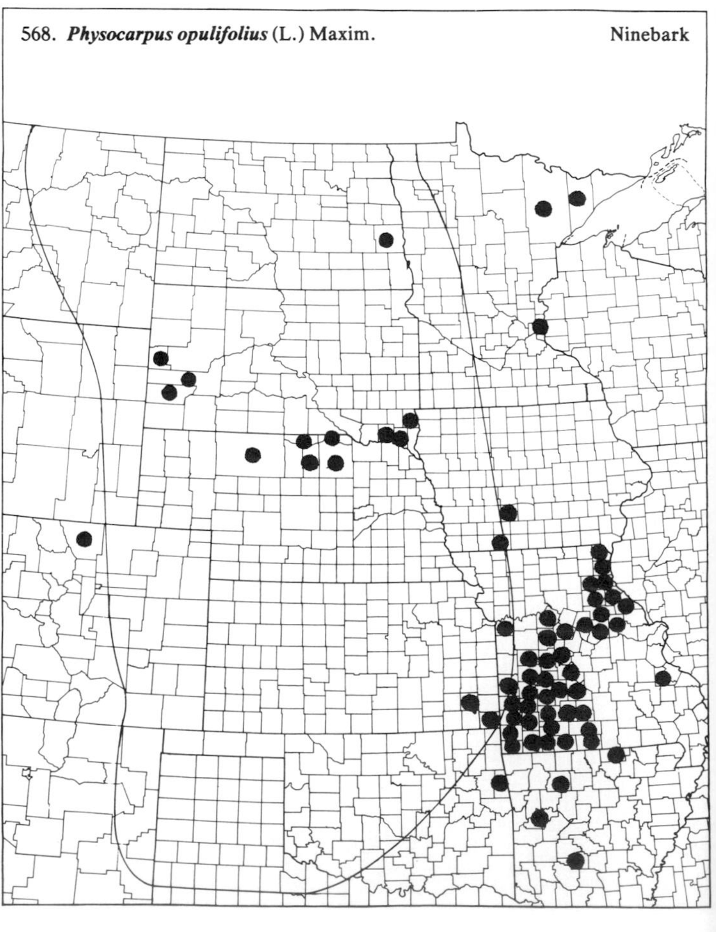

(Rosaceae)

569. ***Potentilla anserina*** L. Silverweed

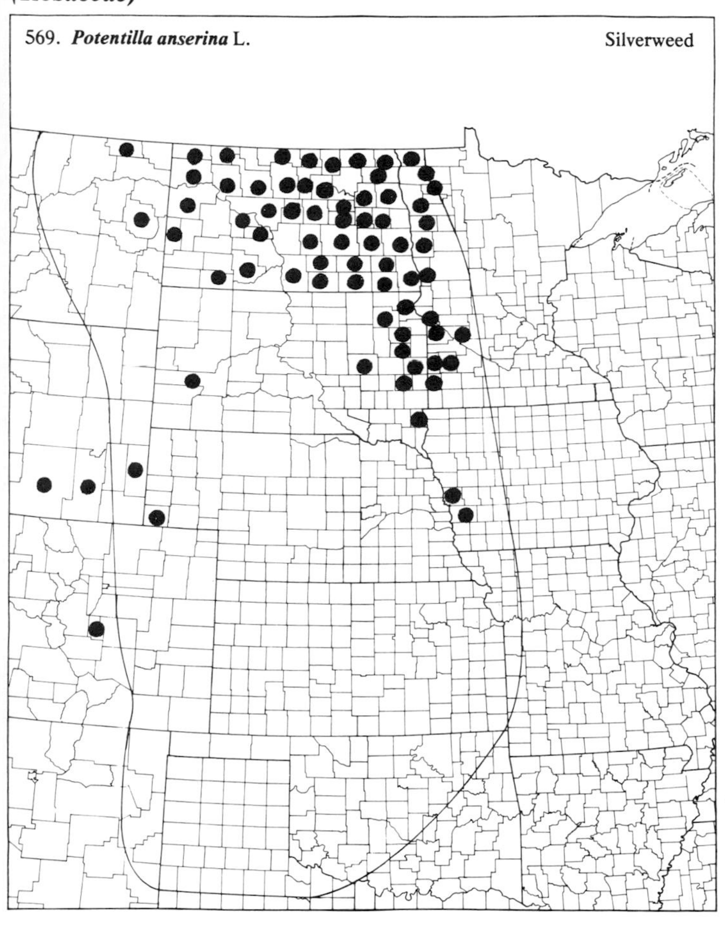

570. ***Potentilla argentea*** L. Silvery Cinquefoil

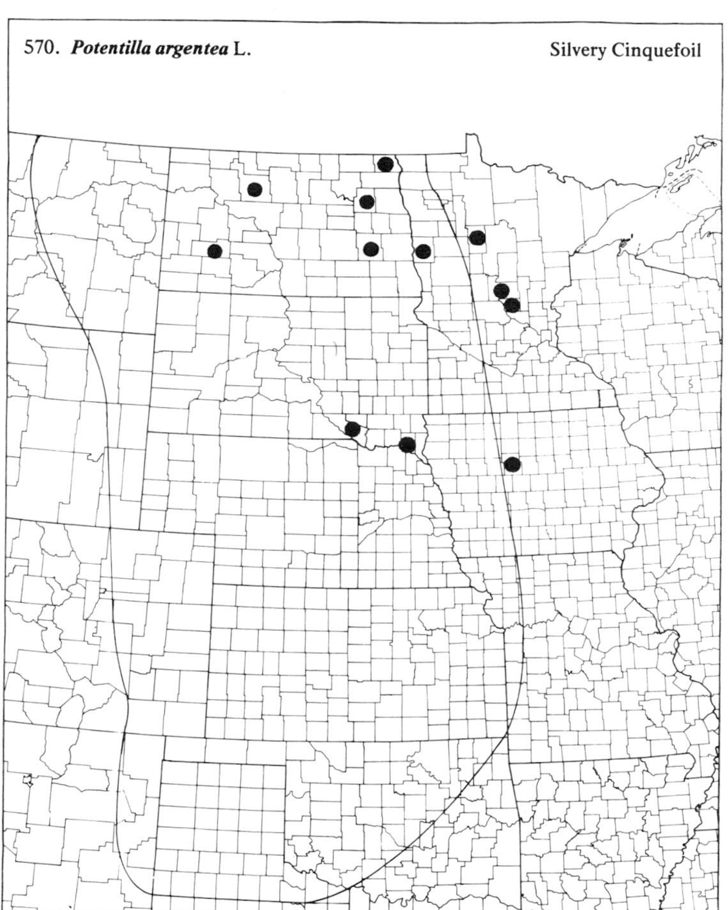

571. ***Potentilla arguta*** Pursh Tall Cinquefoil

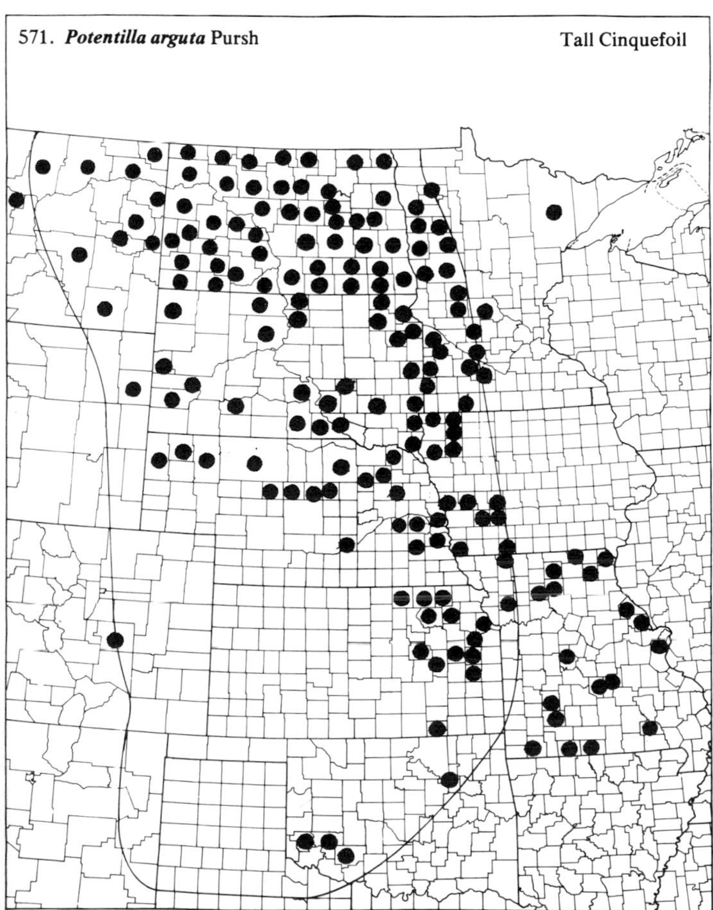

572. ***Potentilla concinna*** Richards.

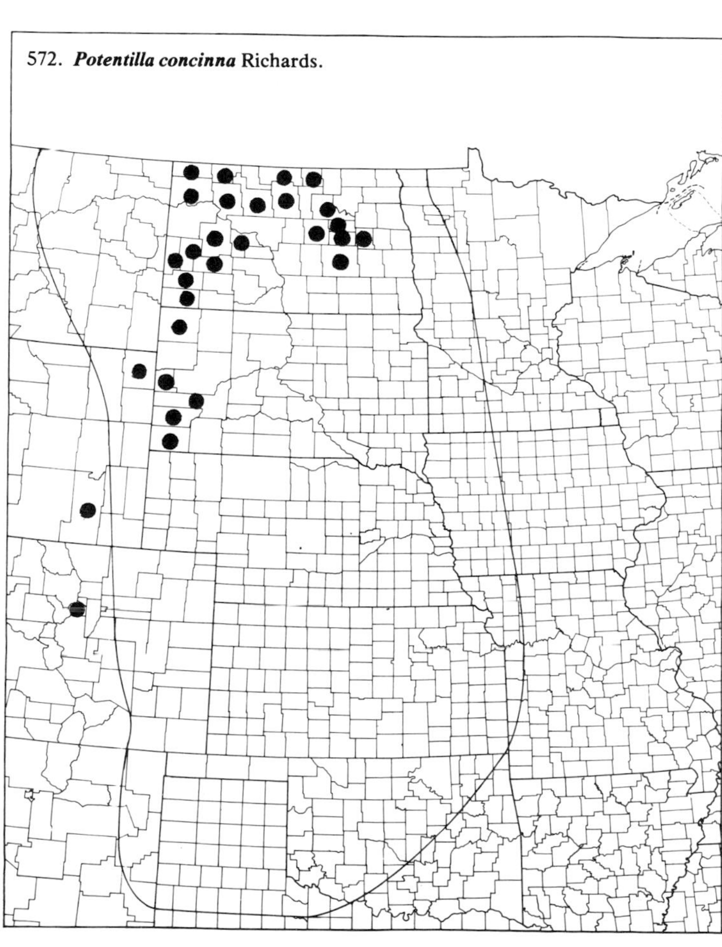

(Rosaceae)

573. ***Potentilla fruticosa*** L. Shrubby Cinquefoil

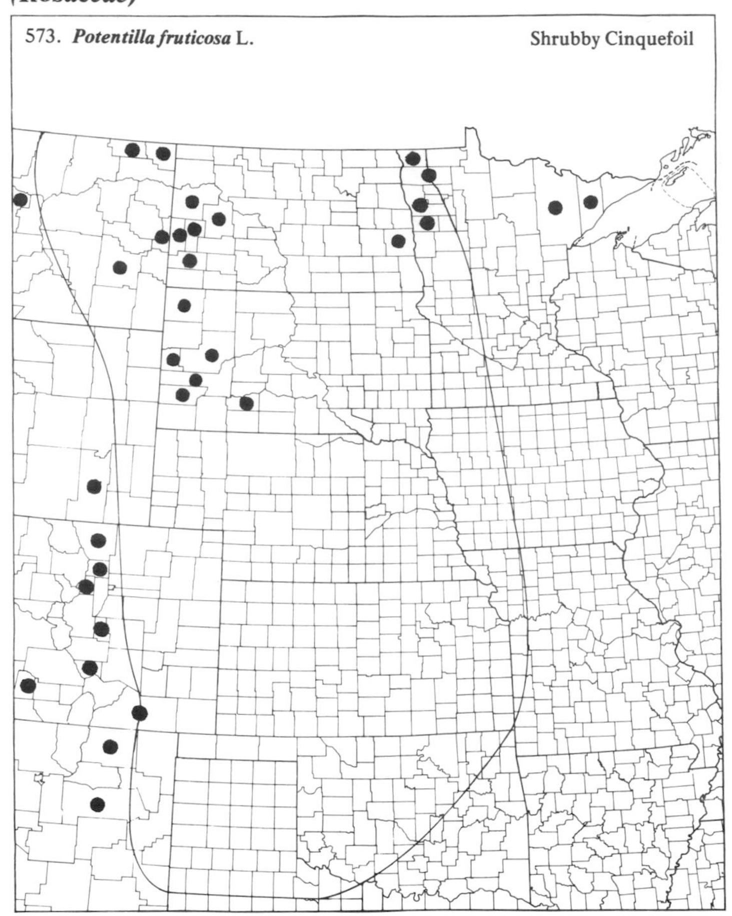

574. ***Potentilla gracilis*** Dougl.

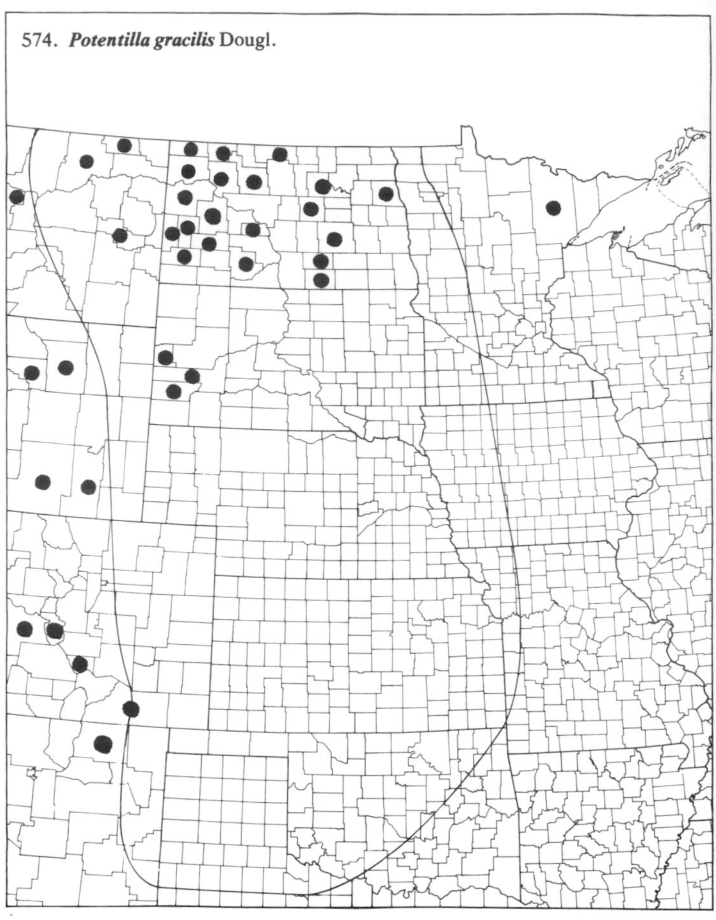

575. ***Potentilla hippiana*** Lehm.

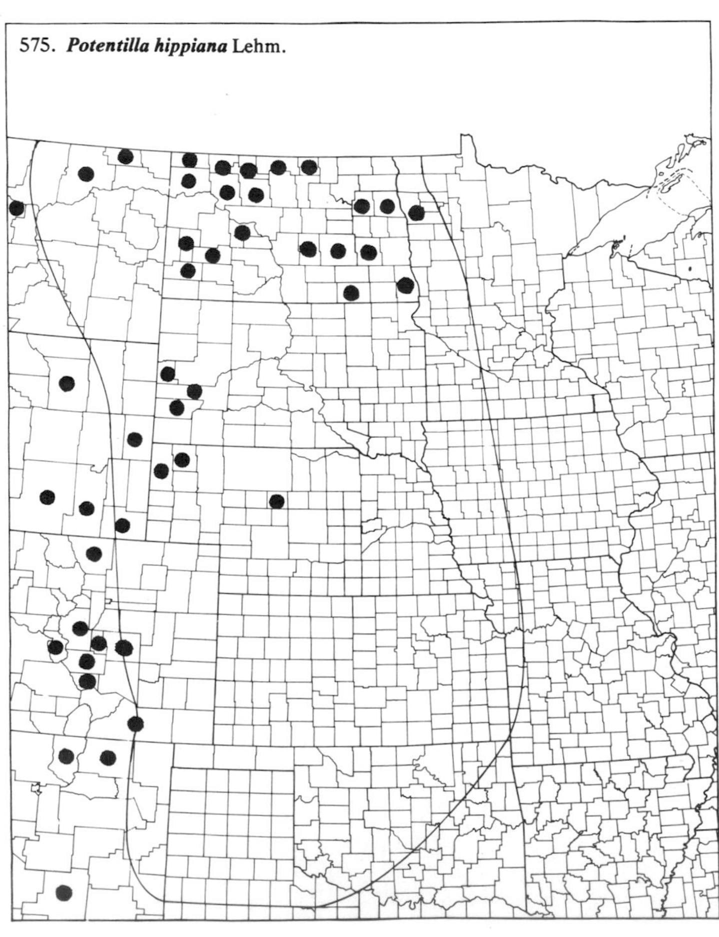

576. ***Potentilla norvegica*** L. Strawberryweed

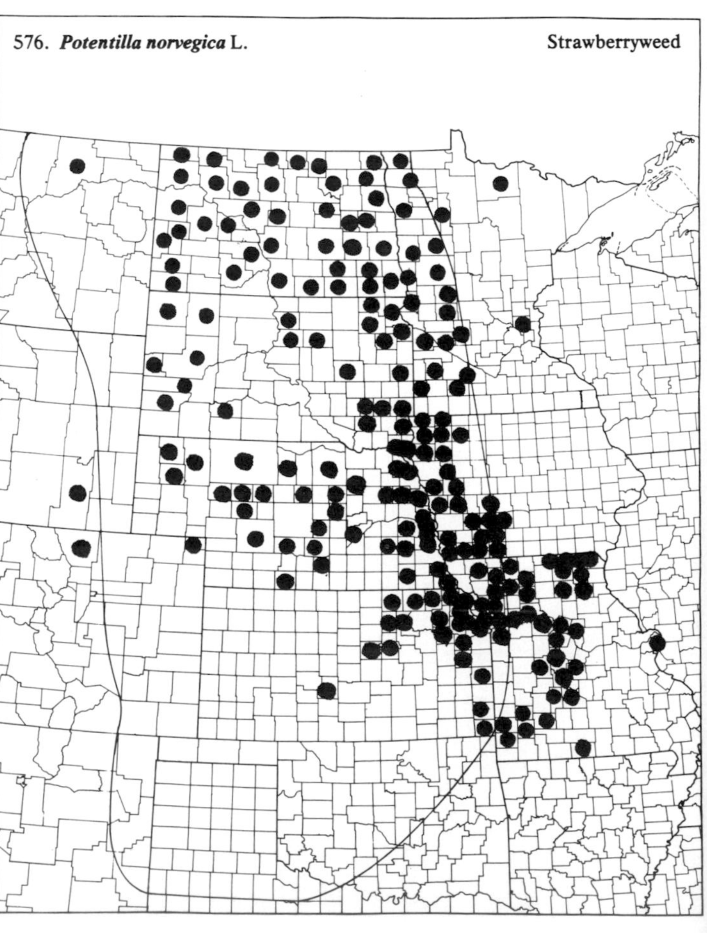

(Rosaceae)

577. **Potentilla paradoxa** Nutt. Bushy Cinquefoil

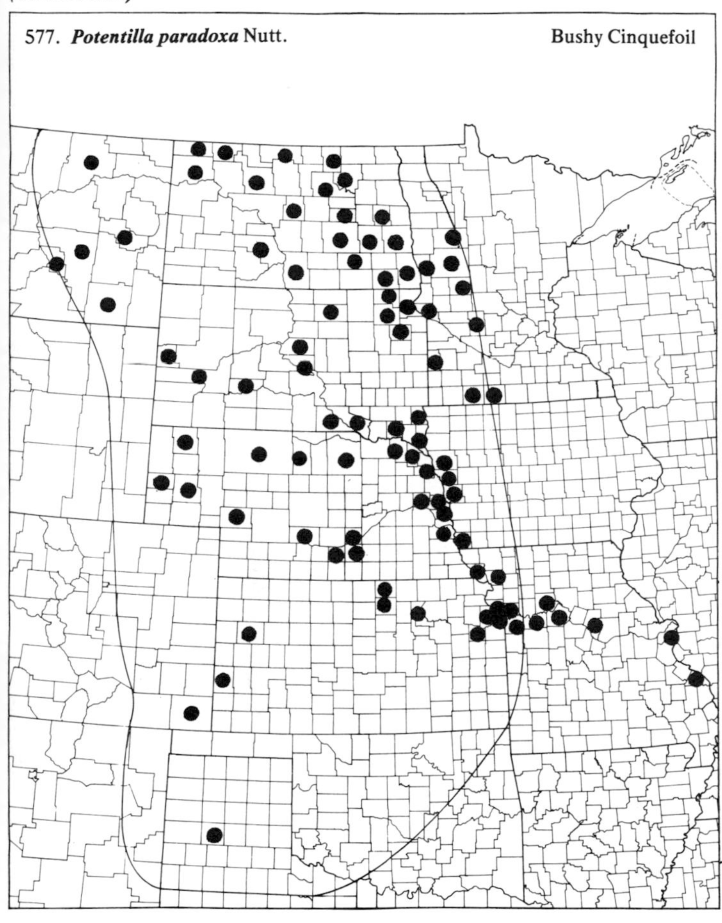

578. **Potentilla pensylvanica** L.

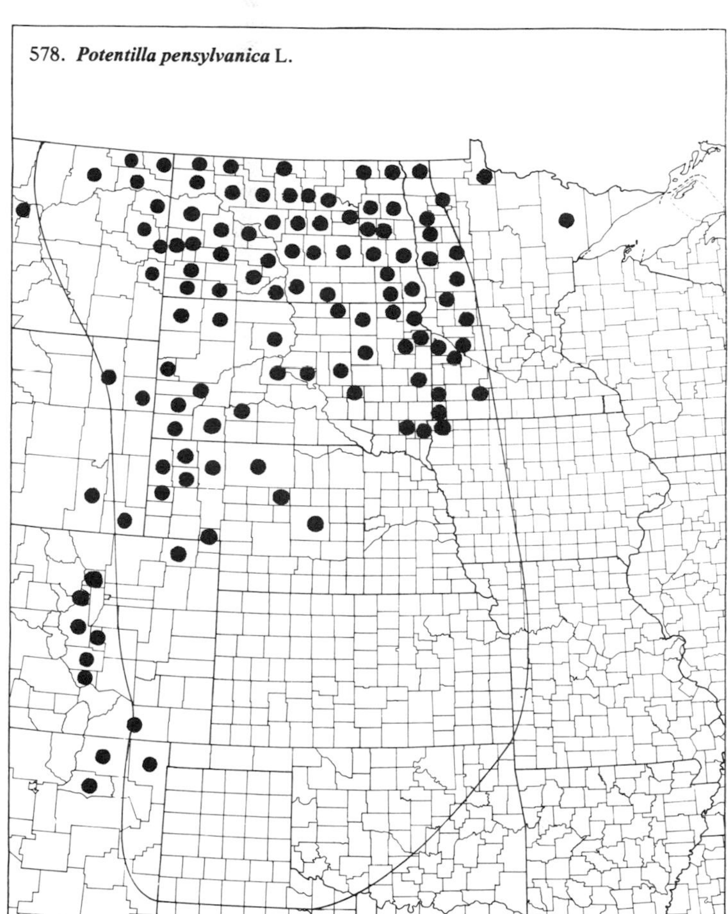

579. **Potentilla recta** L. Sulphur Cinquefoil

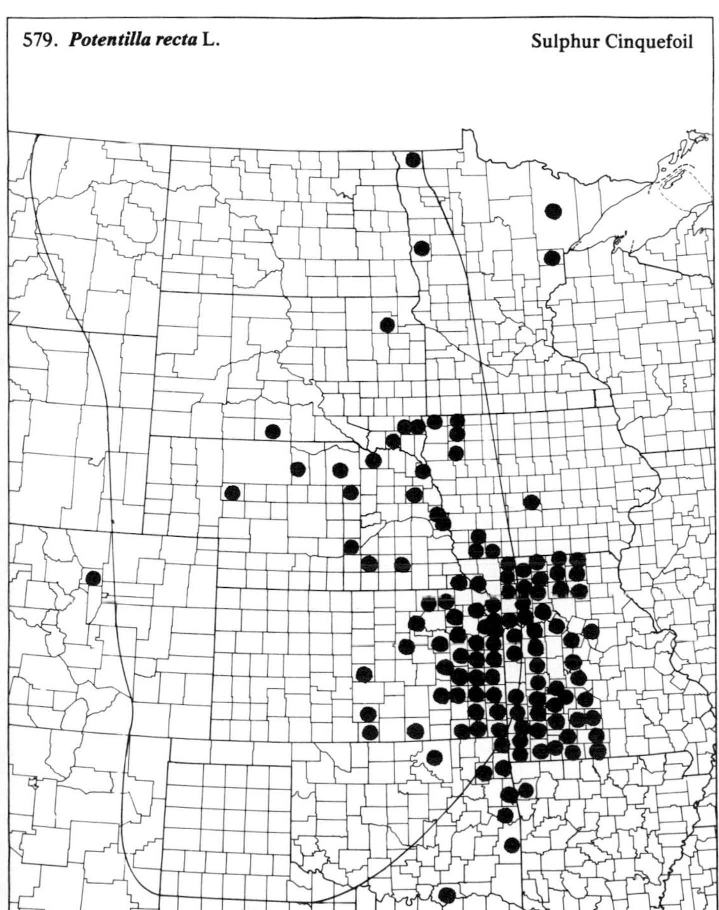

580. **Potentilla rivalis** Nutt. Brook Cinquefoil

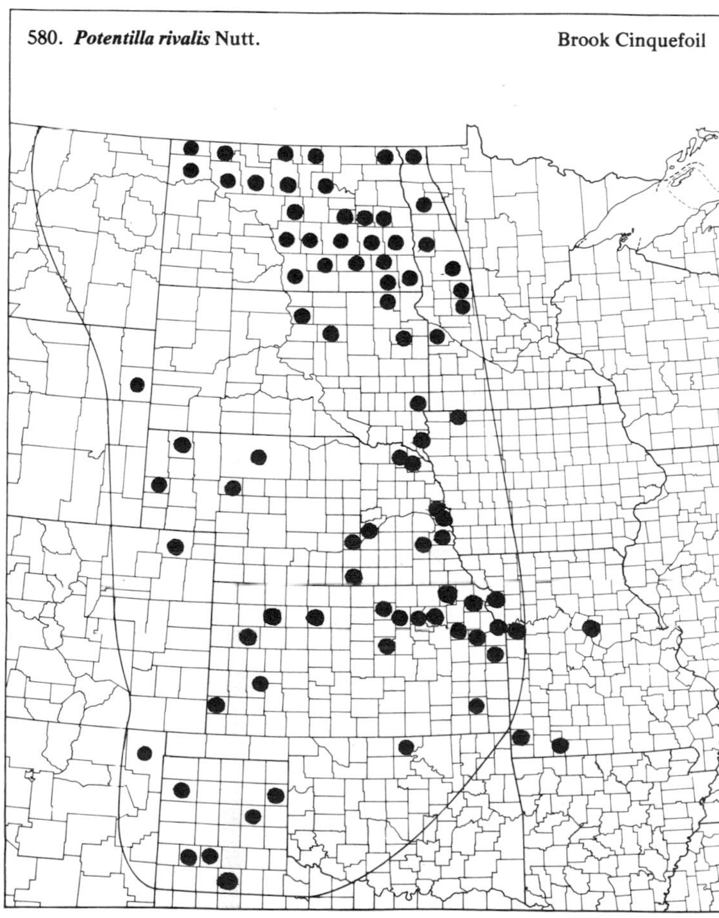

(Rosaceae)

581. ***Potentilla simplex*** Michx. Old-field Cinquefoil

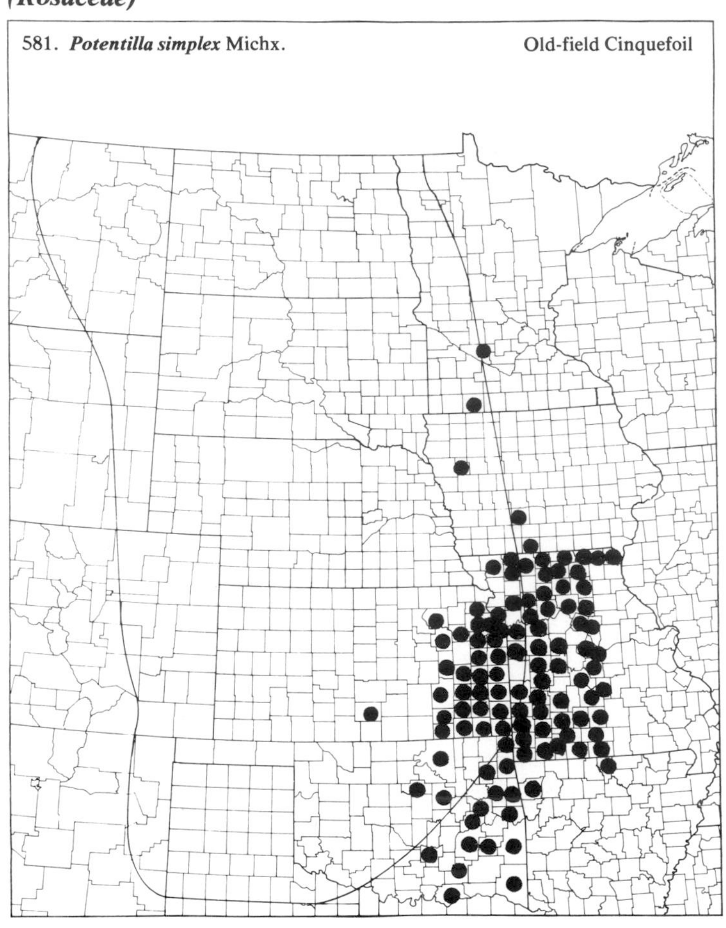

582. ***Prunus americana*** Marsh. Wild Plum

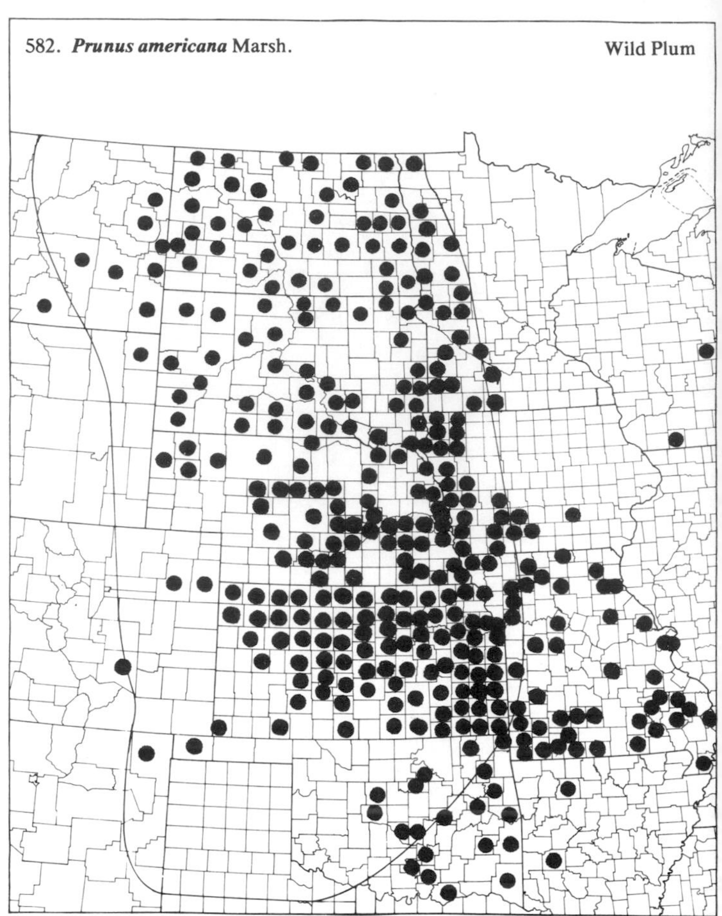

583. ***Prunus angustifolia*** Marsh. Chickasaw Plum

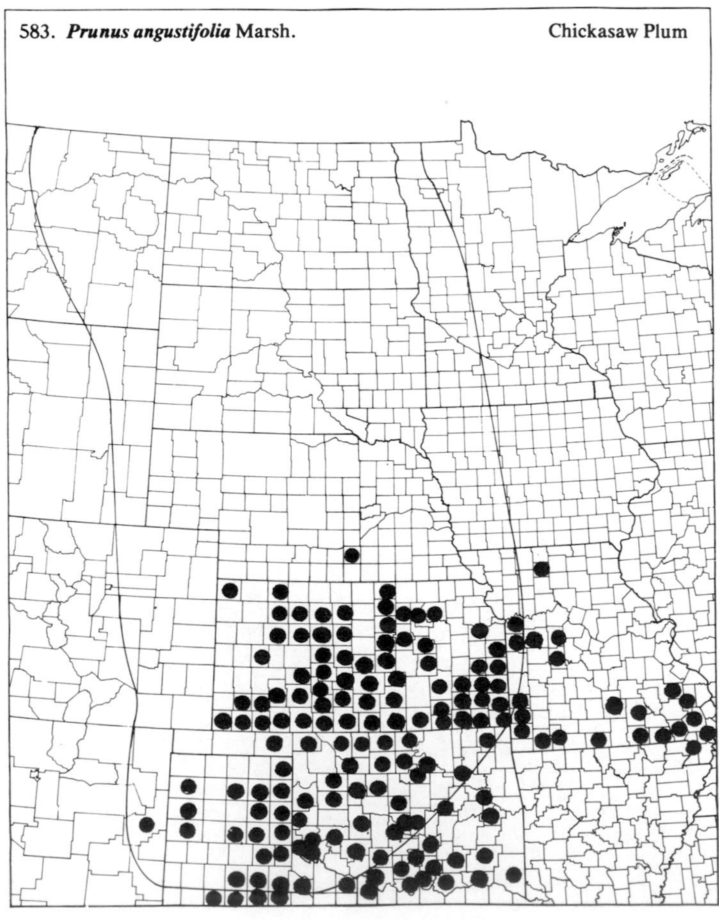

584. ***Prunus besseyi*** Bailey Sand Cherry

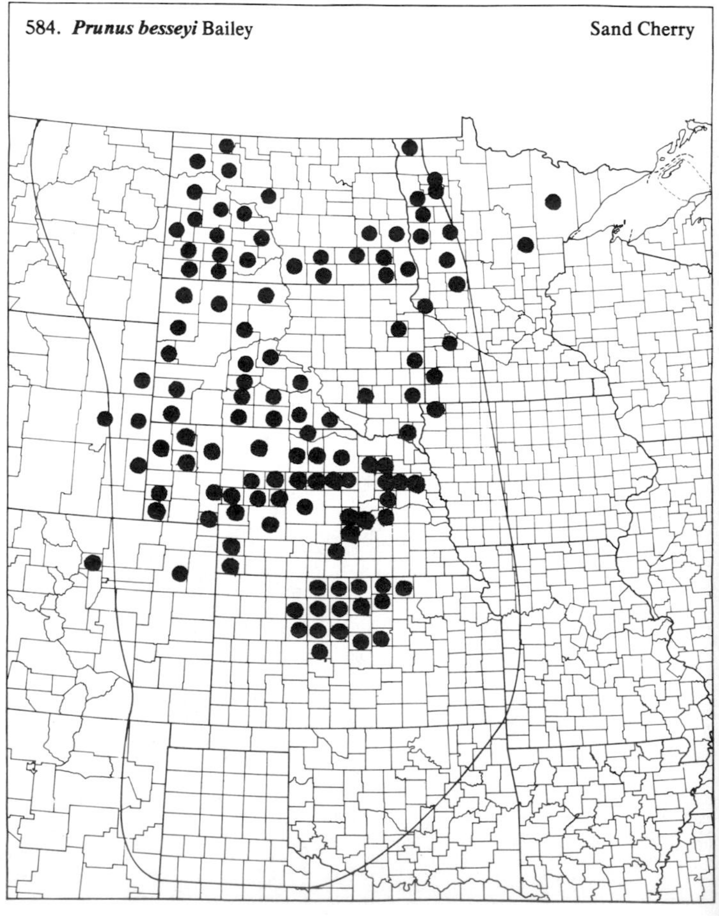

(Rosaceae)

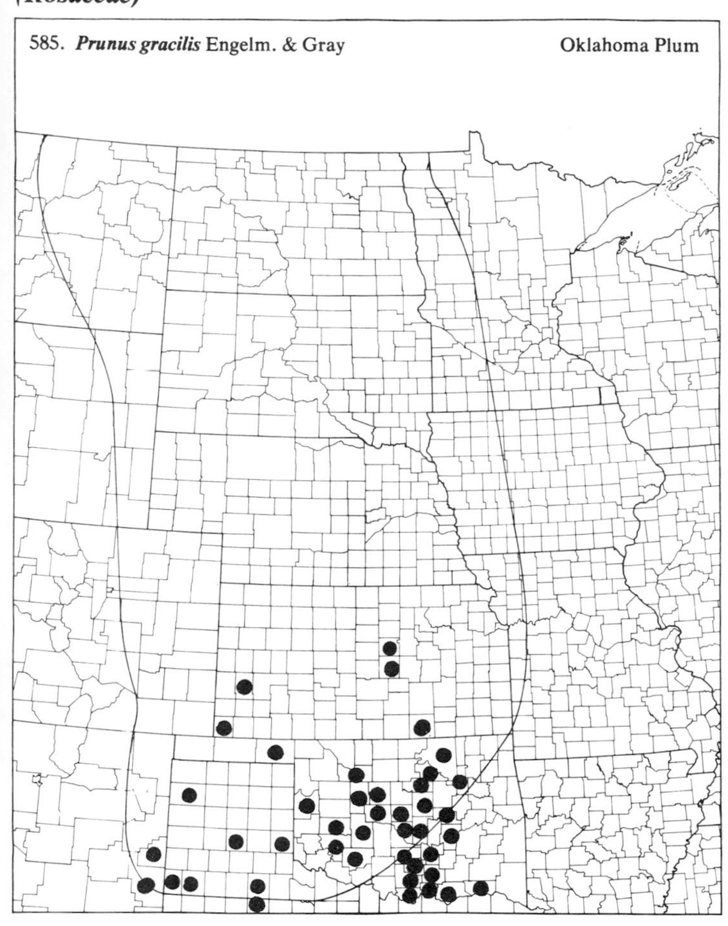
585. Prunus gracilis Engelm. & Gray
Oklahoma Plum

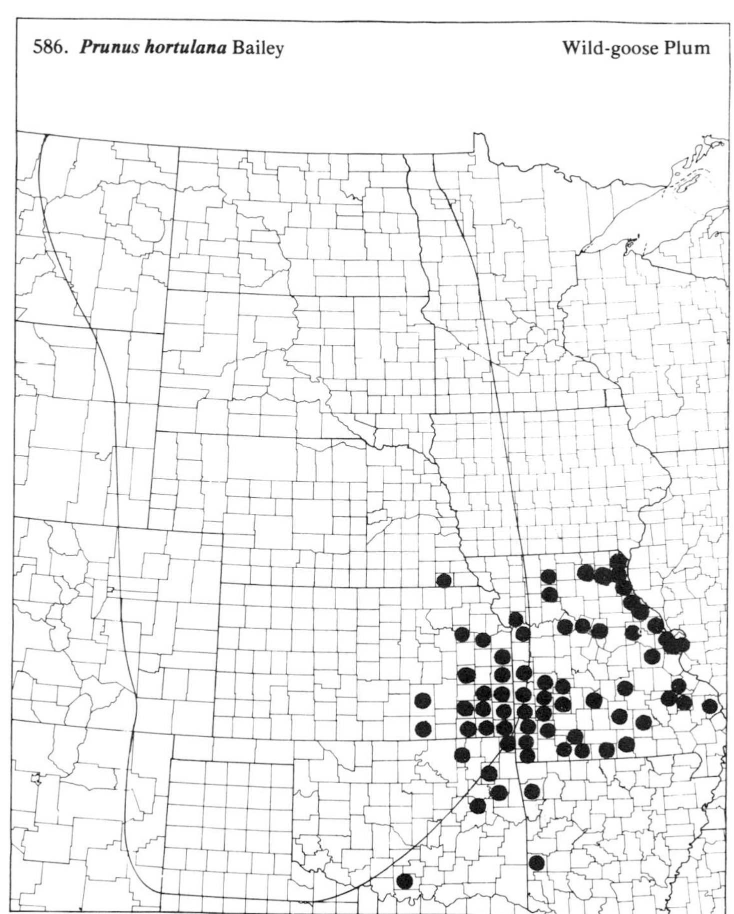
586. Prunus hortulana Bailey
Wild-goose Plum

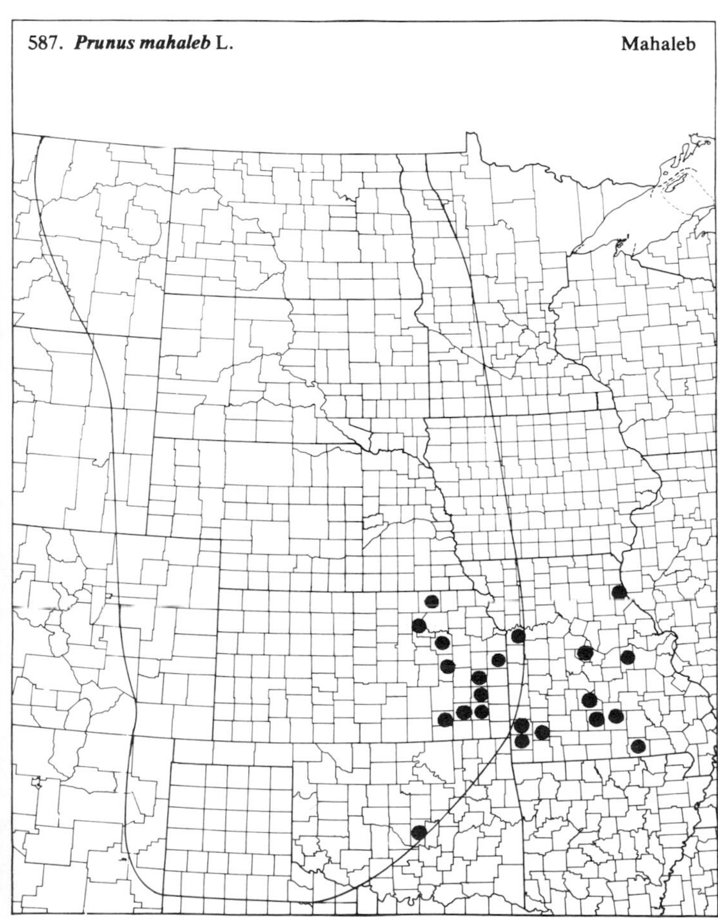
587. Prunus mahaleb L.
Mahaleb

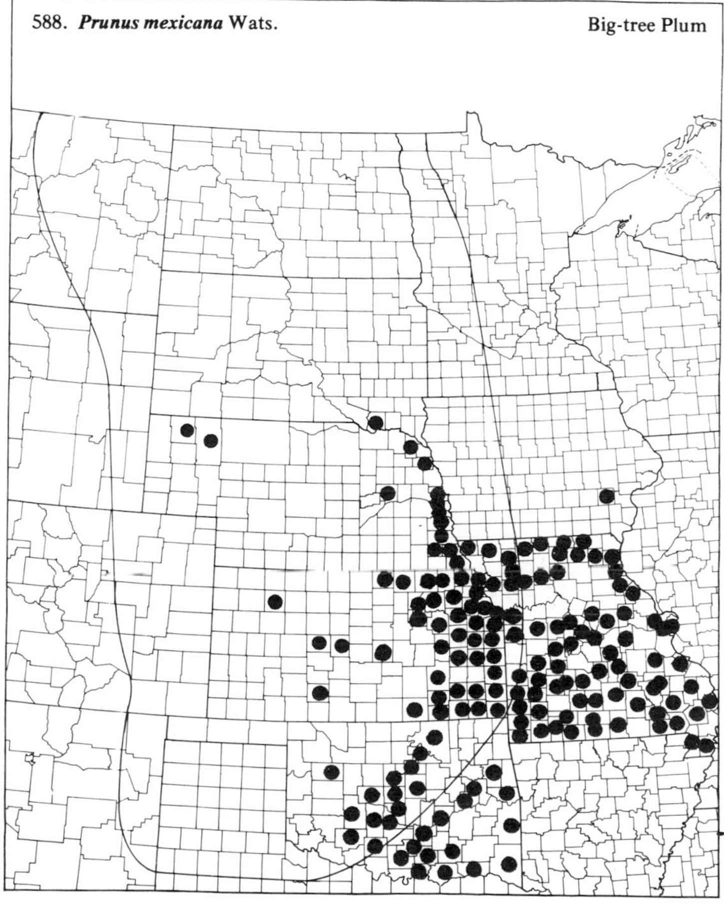
588. Prunus mexicana Wats.
Big-tree Plum

(Rosaceae)

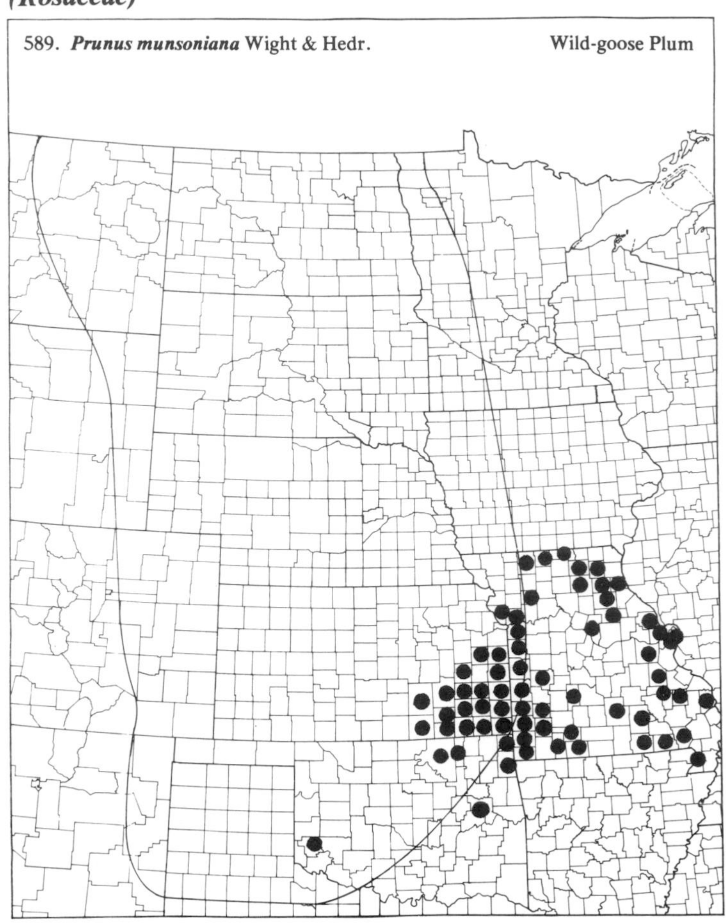

589. ***Prunus munsoniana*** Wight & Hedr. Wild-goose Plum

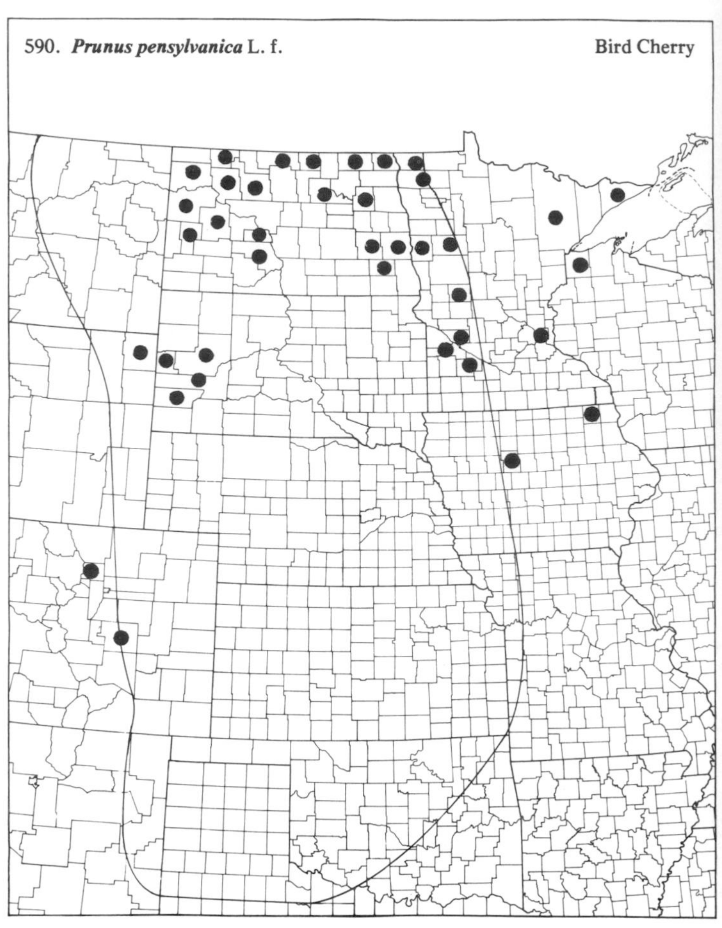

590. ***Prunus pensylvanica*** L. f. Bird Cherry

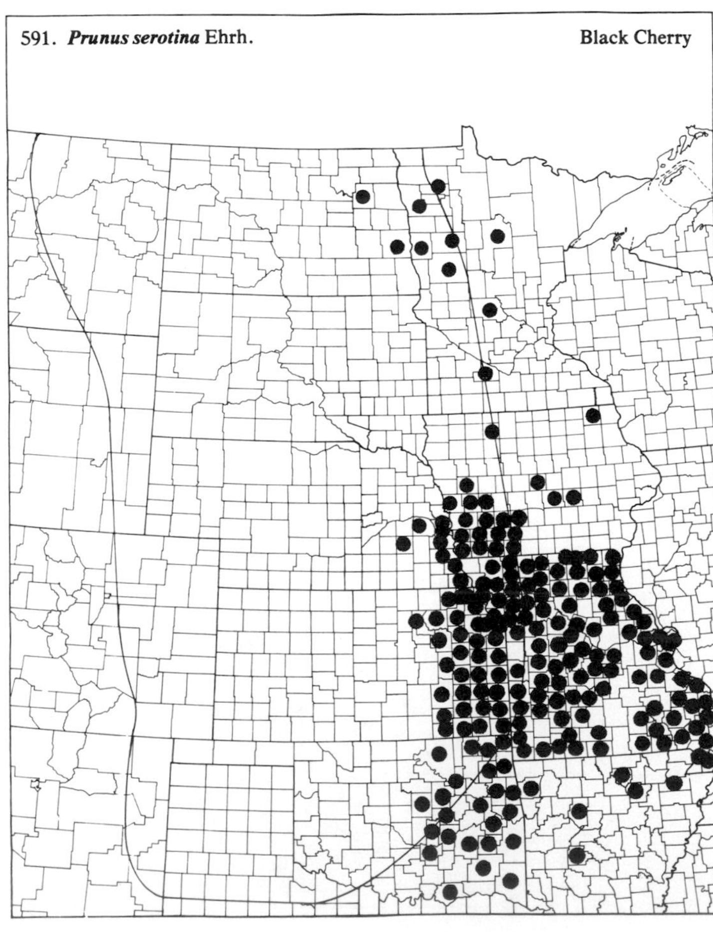

591. ***Prunus serotina*** Ehrh. Black Cherry

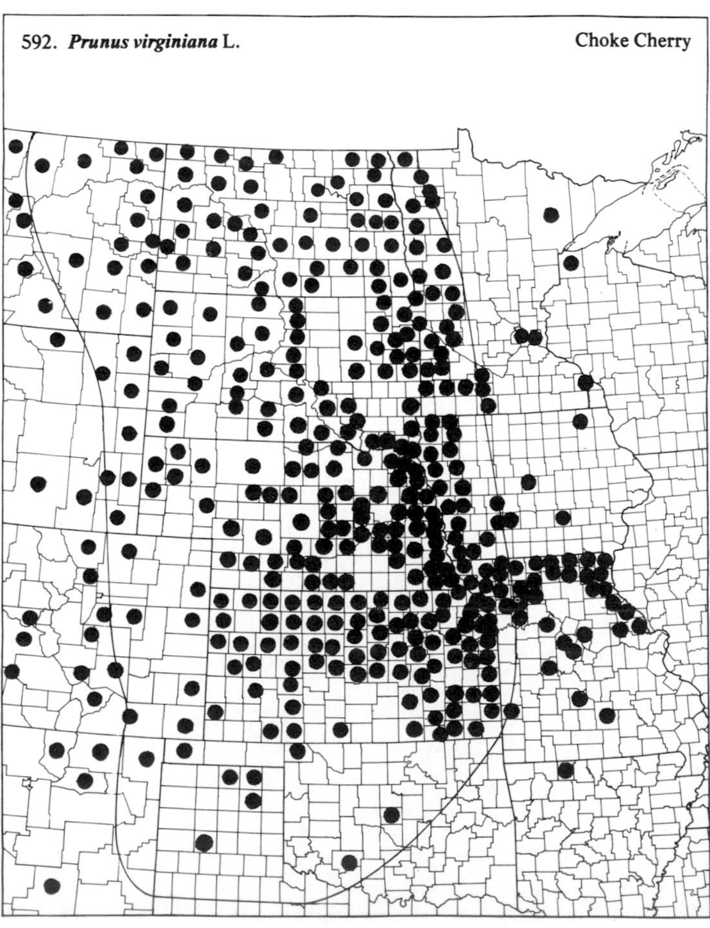

592. ***Prunus virginiana*** L. Choke Cherry

(Rosaceae)

593. ***Pyrus ioensis*** (Wood) Bailey — Wild Crabapple

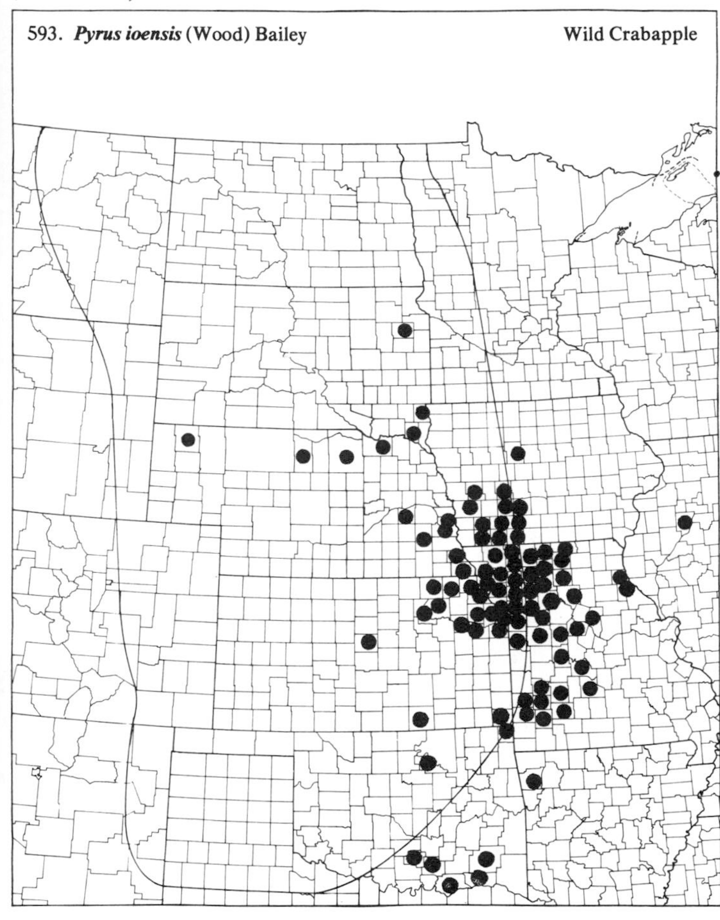

594. ***Rosa acicularis*** Lindl. ssp. ***sayi*** (Schwein.) W.H. Lewis — Prickly Wild Rose

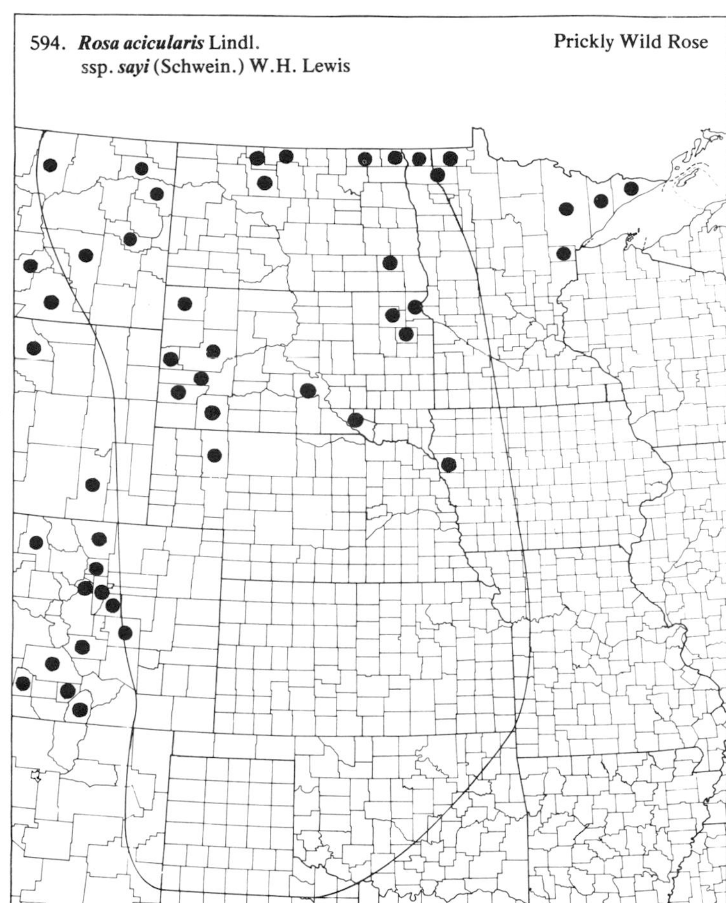

595. ***Rosa arkansana*** Porter — Prairie Wild Rose

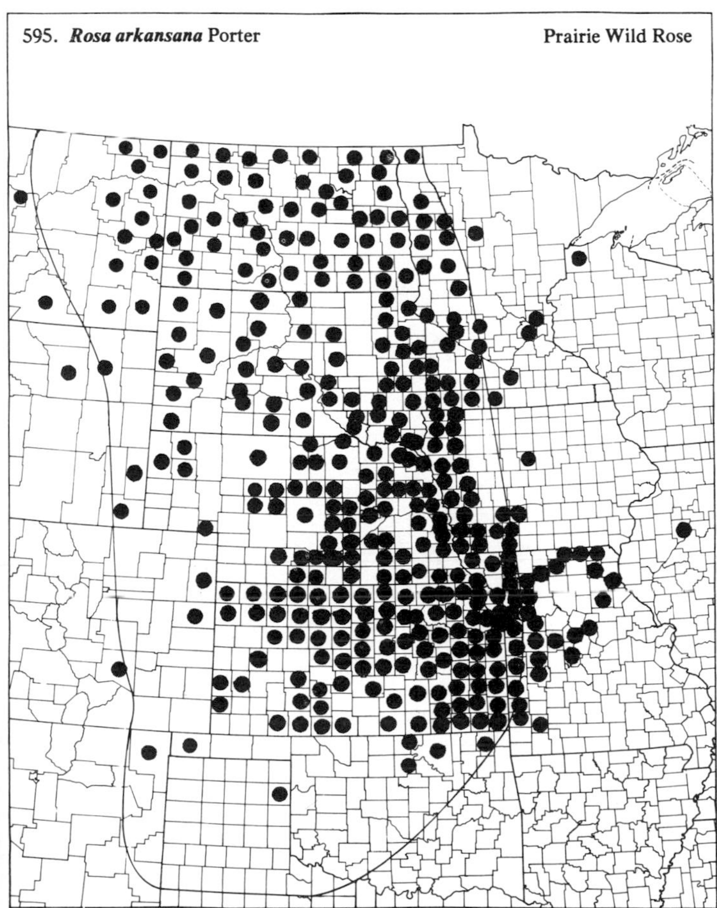

596. ***Rosa blanda*** Ait. — Smooth Wild Rose

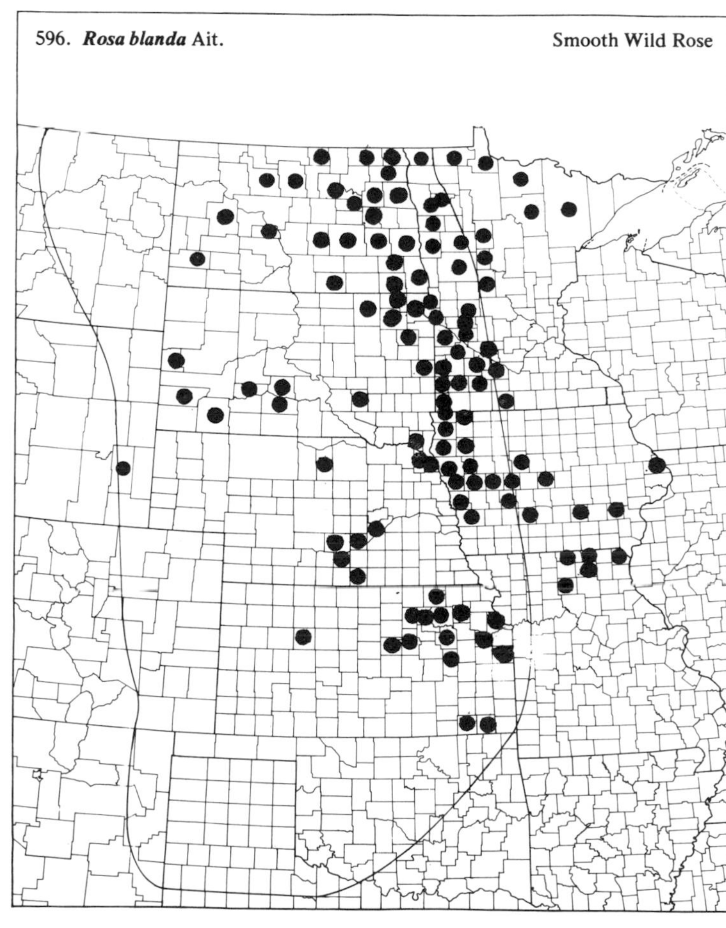

(Rosaceae)

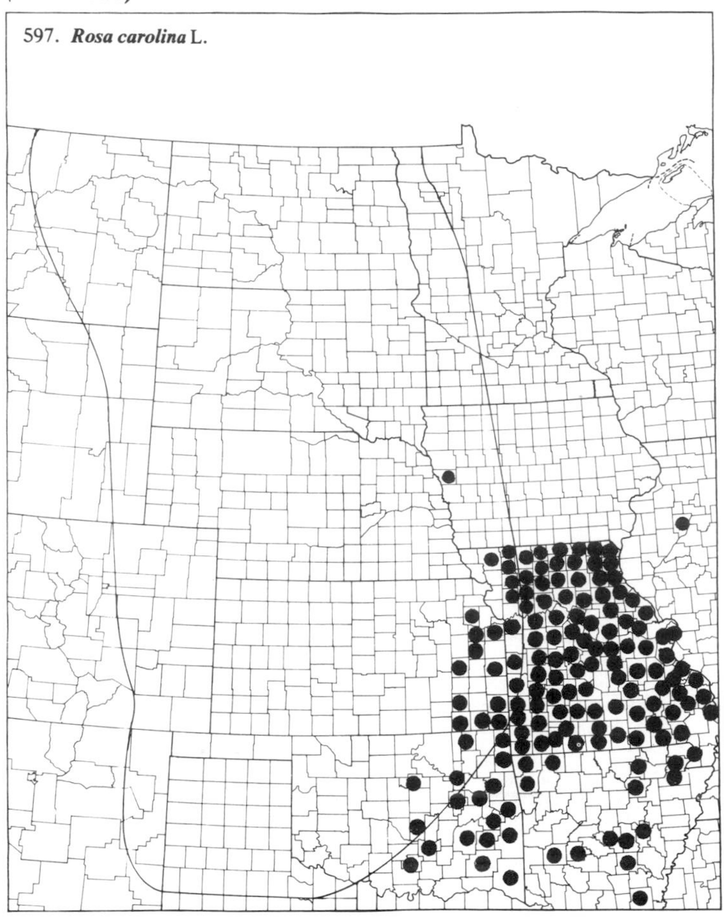
597. **Rosa carolina** L.

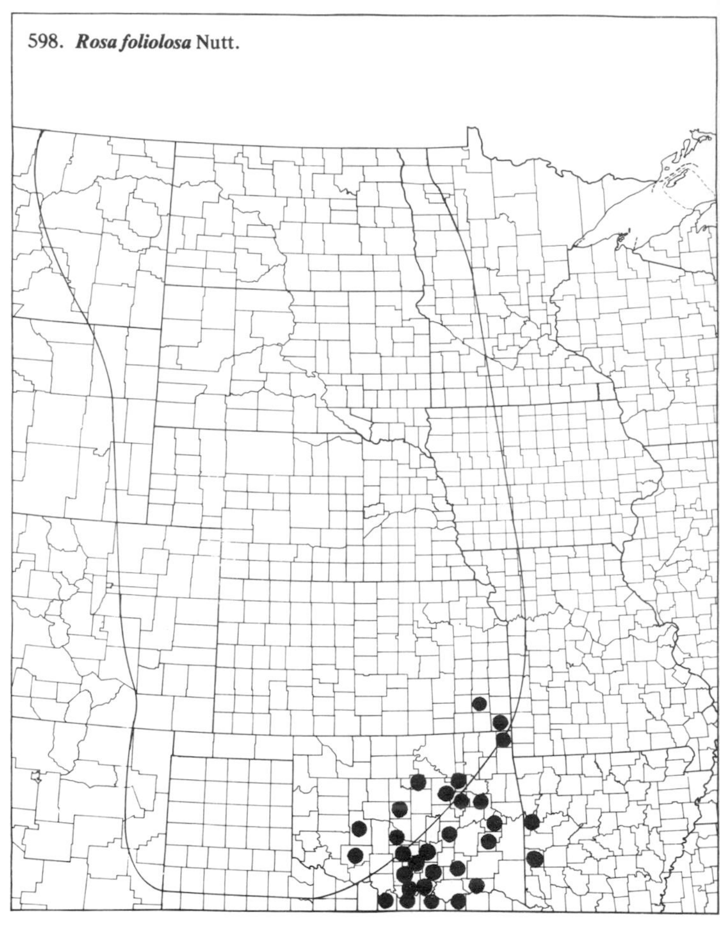
598. **Rosa foliolosa** Nutt.

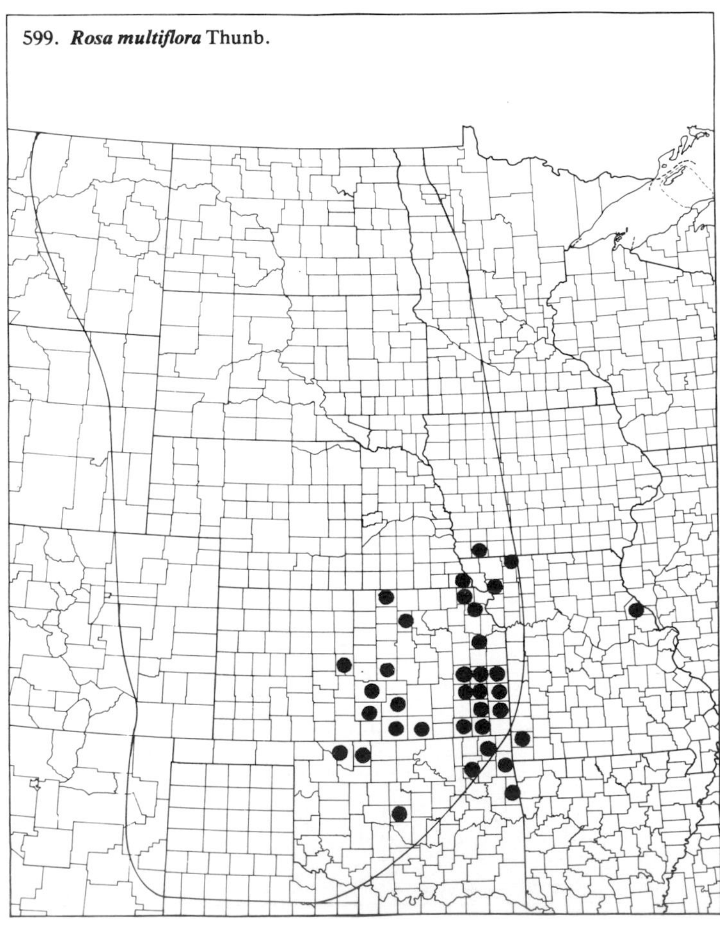
599. **Rosa multiflora** Thunb.

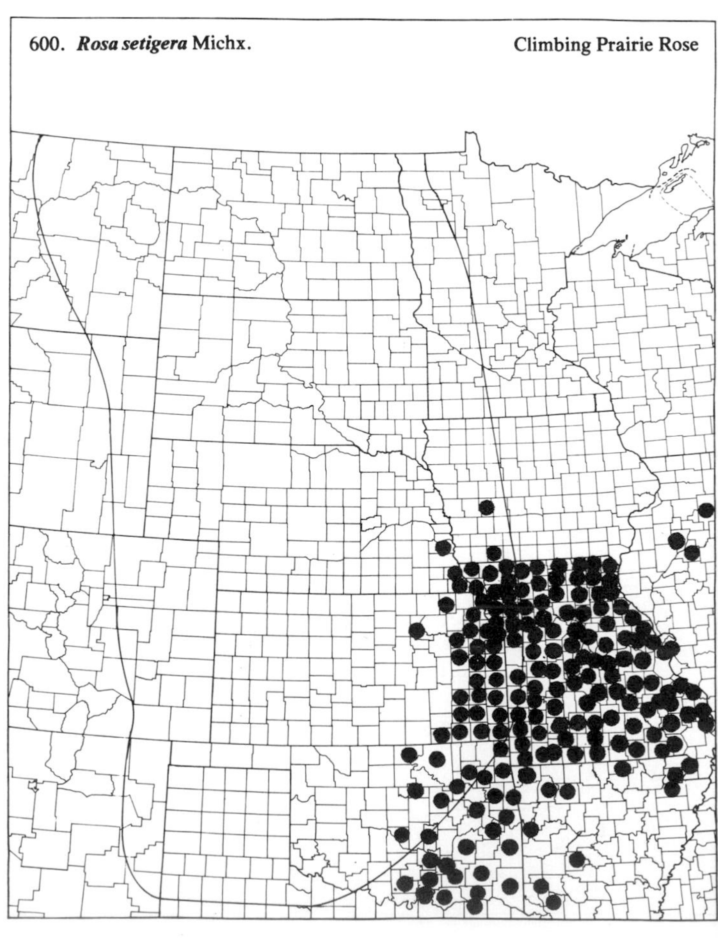
600. **Rosa setigera** Michx.
Climbing Prairie Rose

(Rosaceae)

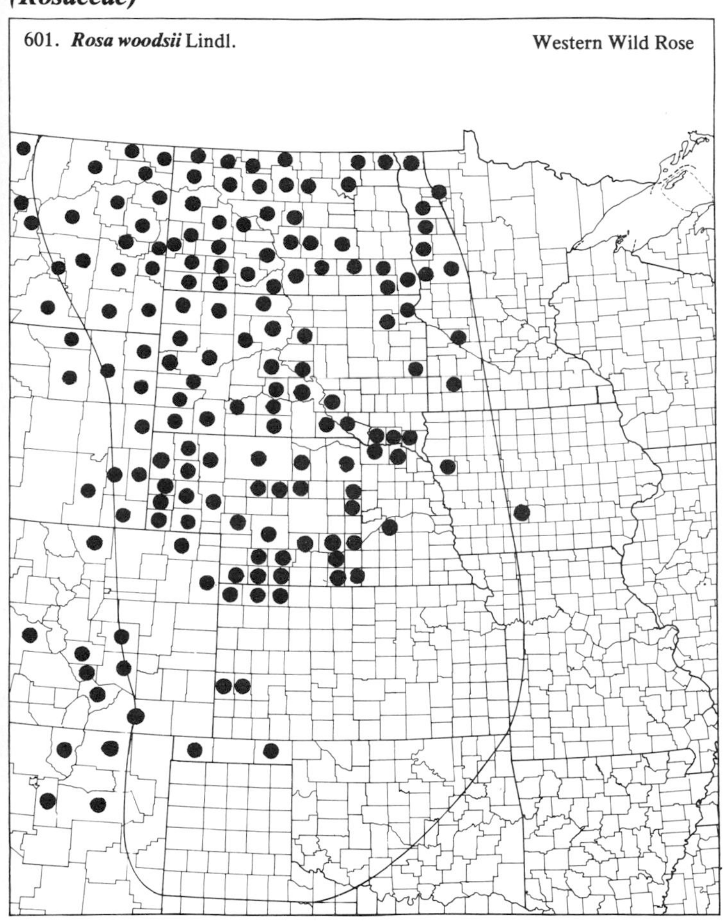
601. Rosa woodsii Lindl.
Western Wild Rose

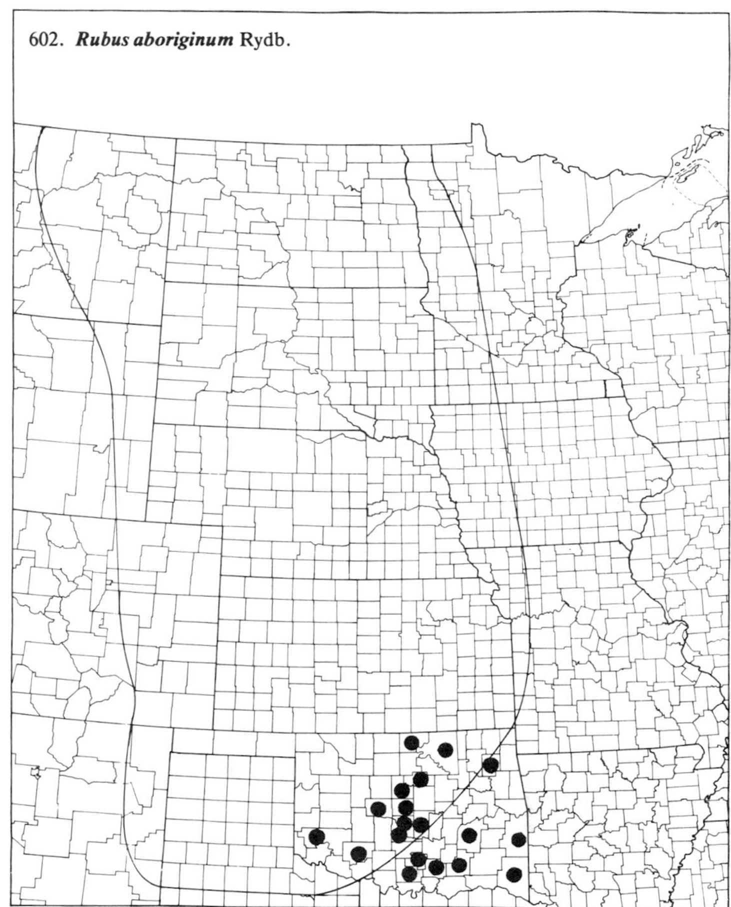
602. Rubus aboriginum Rydb.

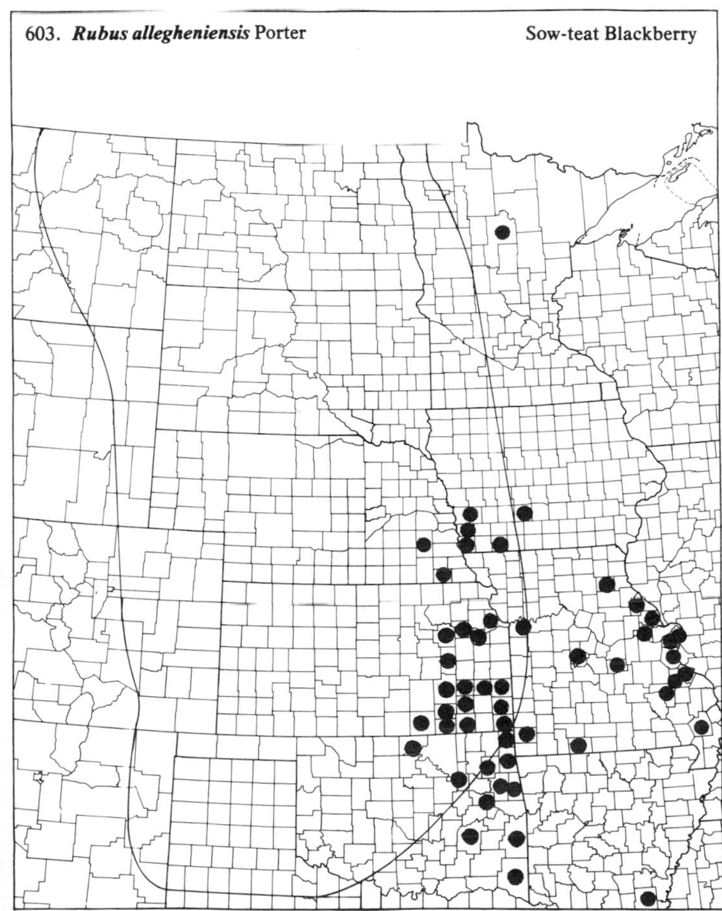
603. Rubus allegheniensis Porter
Sow-teat Blackberry

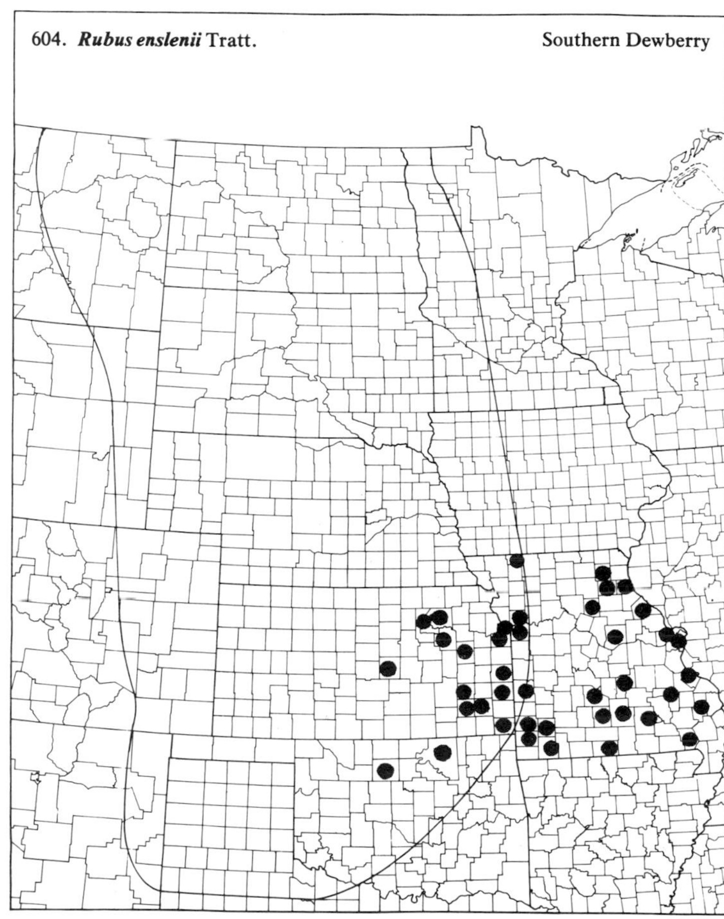
604. Rubus enslenii Tratt.
Southern Dewberry

(Rosaceae)

605. ***Rubus flagellaris*** Willd. Dewberry

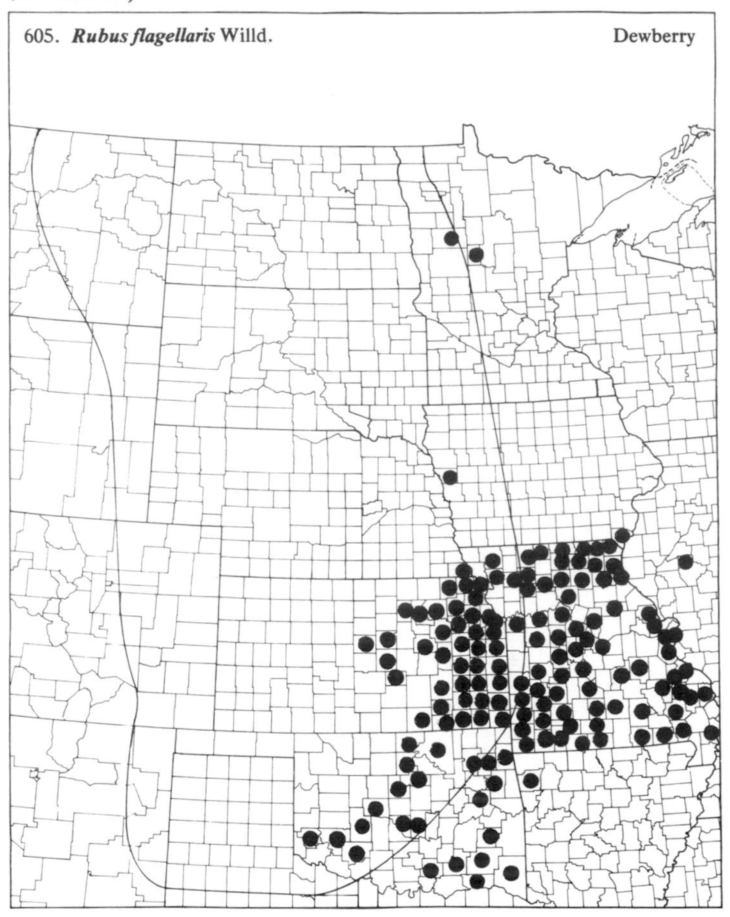

606. ***Rubus idaeus*** L. ssp. ***sachalinensis*** (Levl.) Focke Red Raspberry

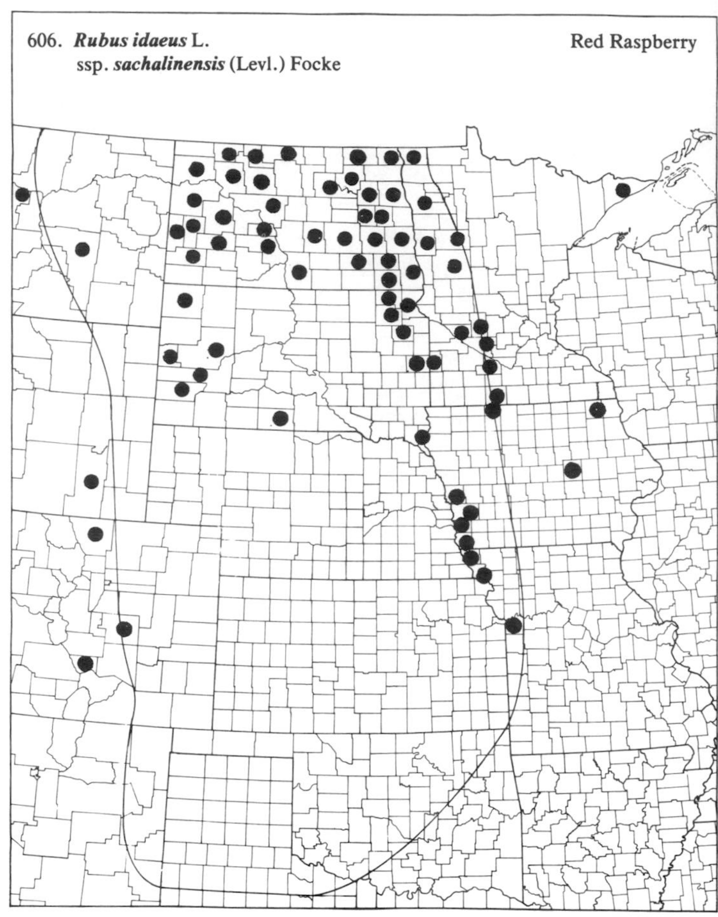

607. ***Rubus occidentalis*** L. Black Raspberry

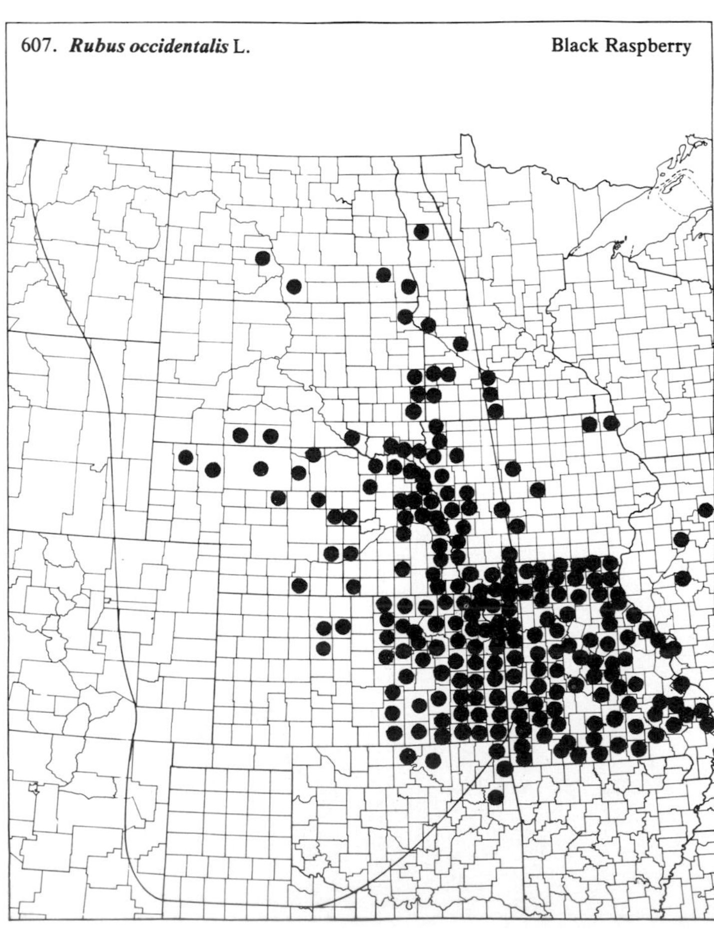

608. ***Rubus ostryifolius*** Rydb. Highbush Blackberry

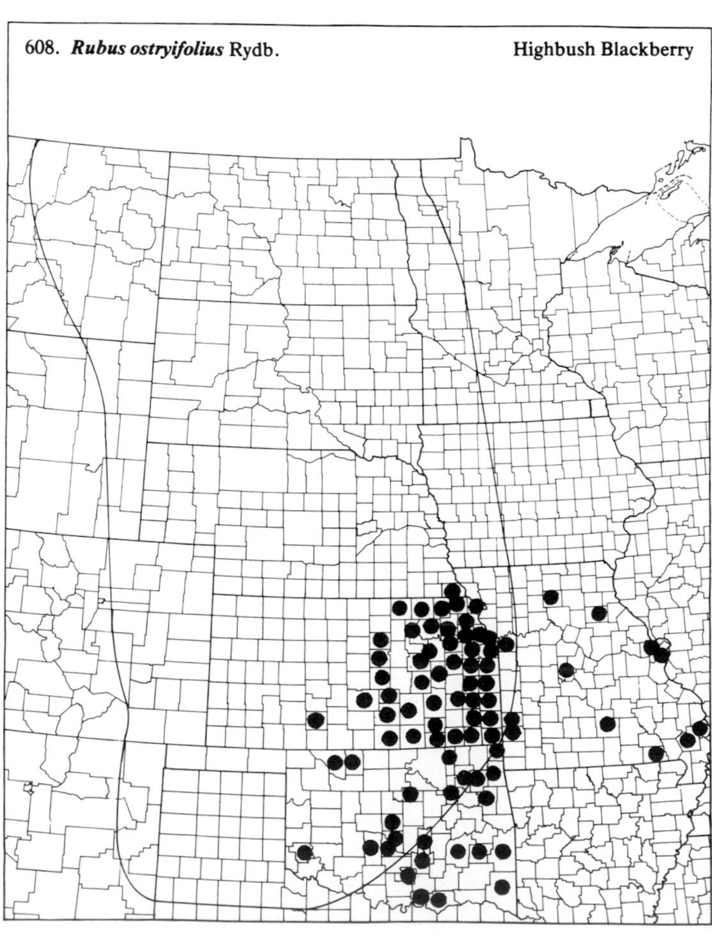

(Rosaceae)

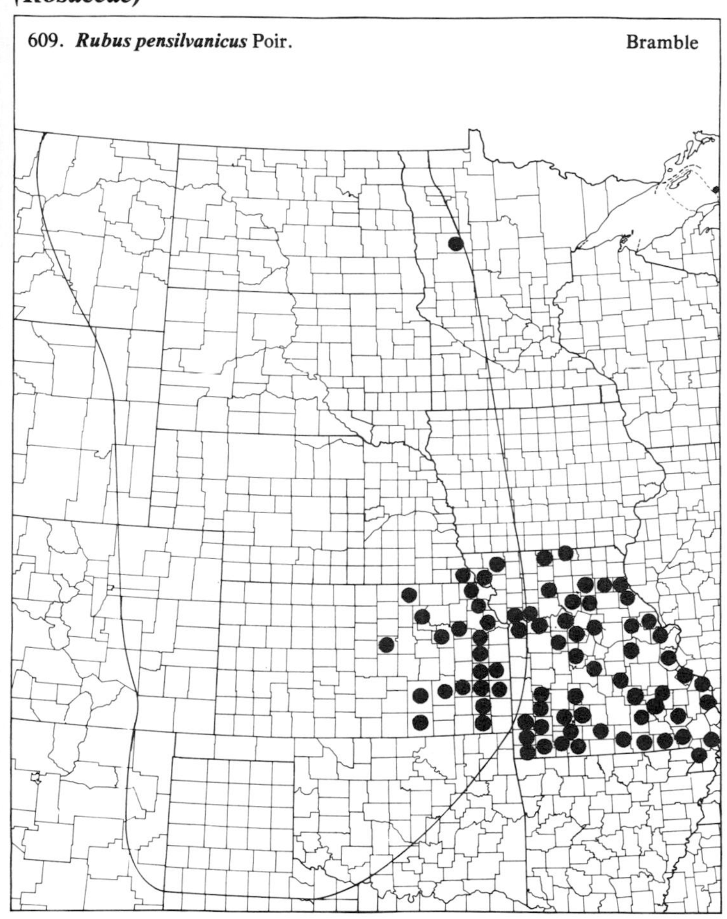

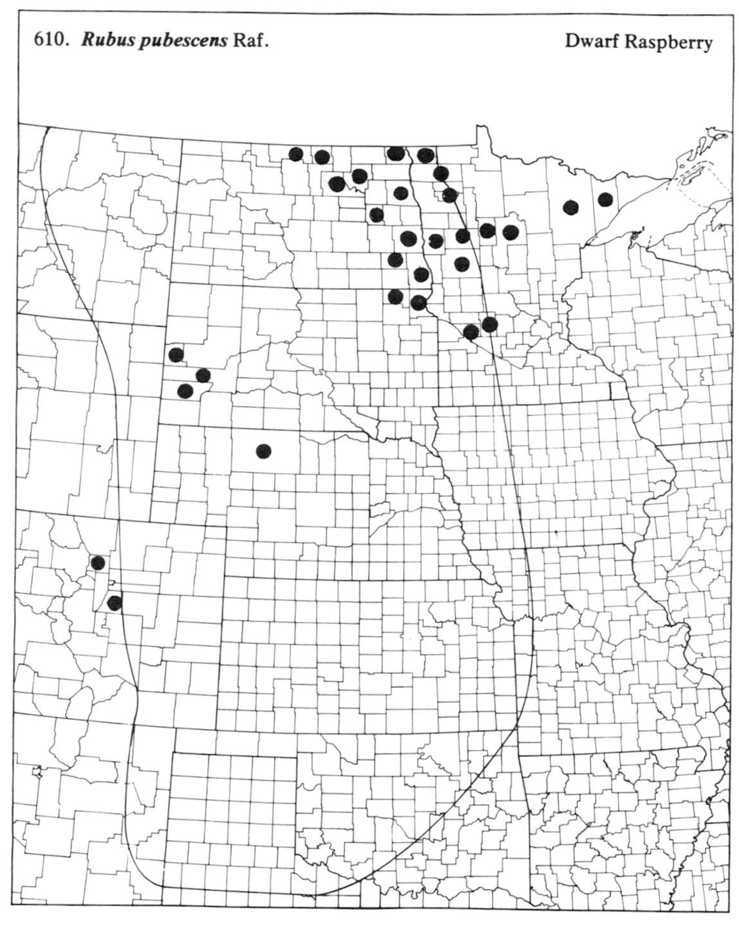

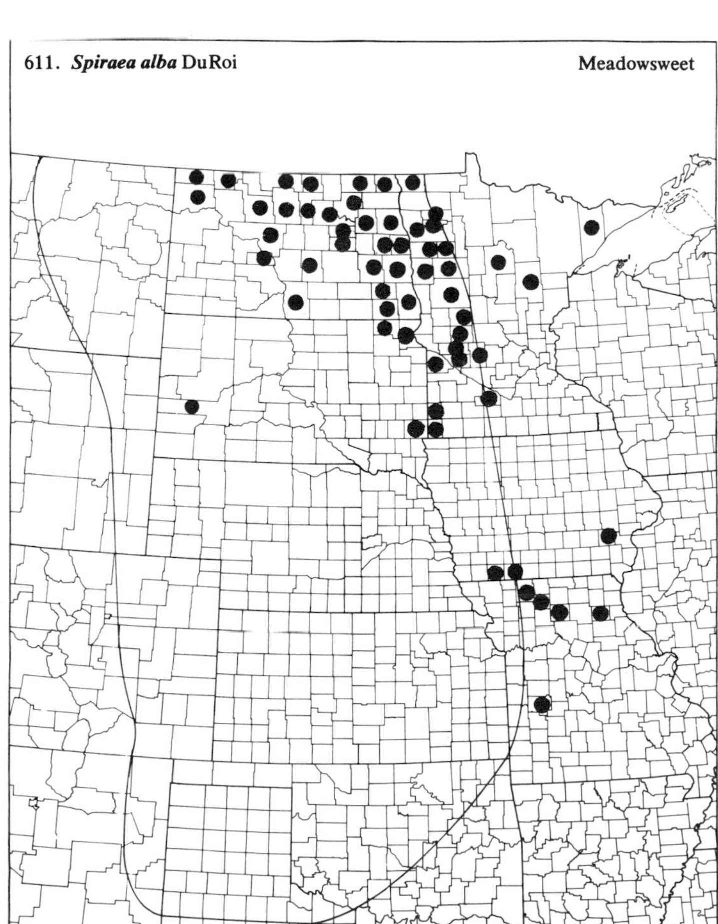

54a. *Fabaceae* (Bean Family), Subfamily *Mimosoideae*

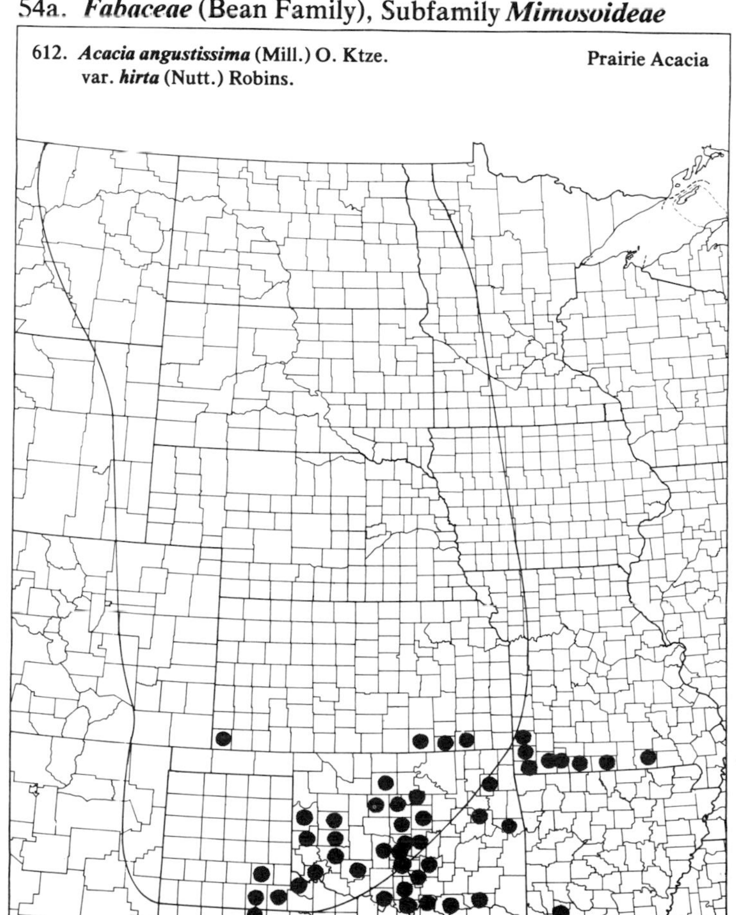

(Fabaceae-Mimosoideae)

613. **Desmanthus cooleyi** (Eat.) Trel. Bundleflower

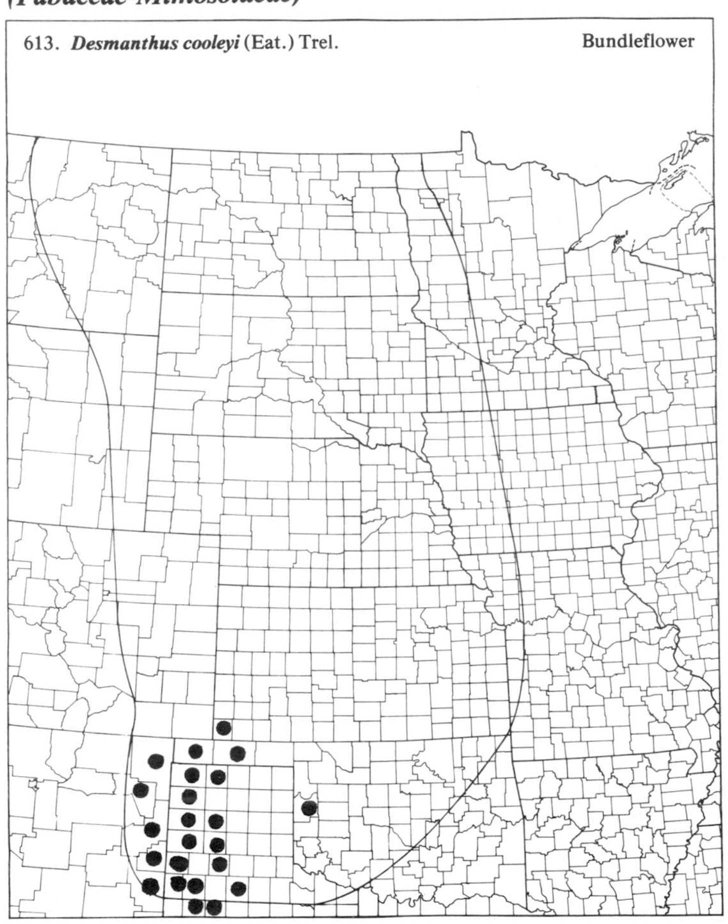

614. **Desmanthus illinoensis** (Michx.) MacM. Bundleflower

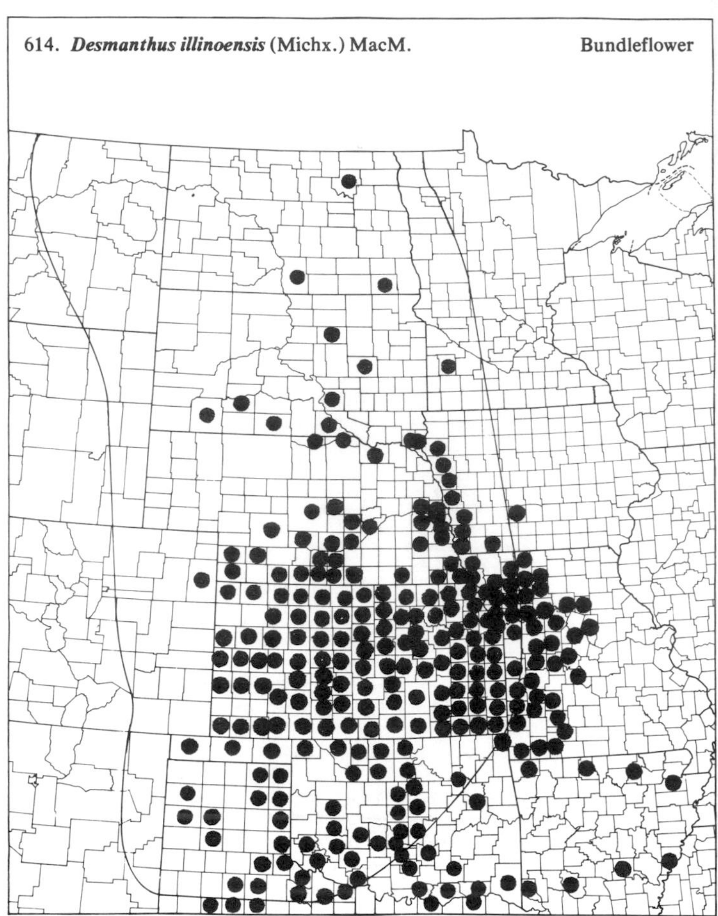

615. **Desmanthus leptolobus** T. & G. Bundleflower

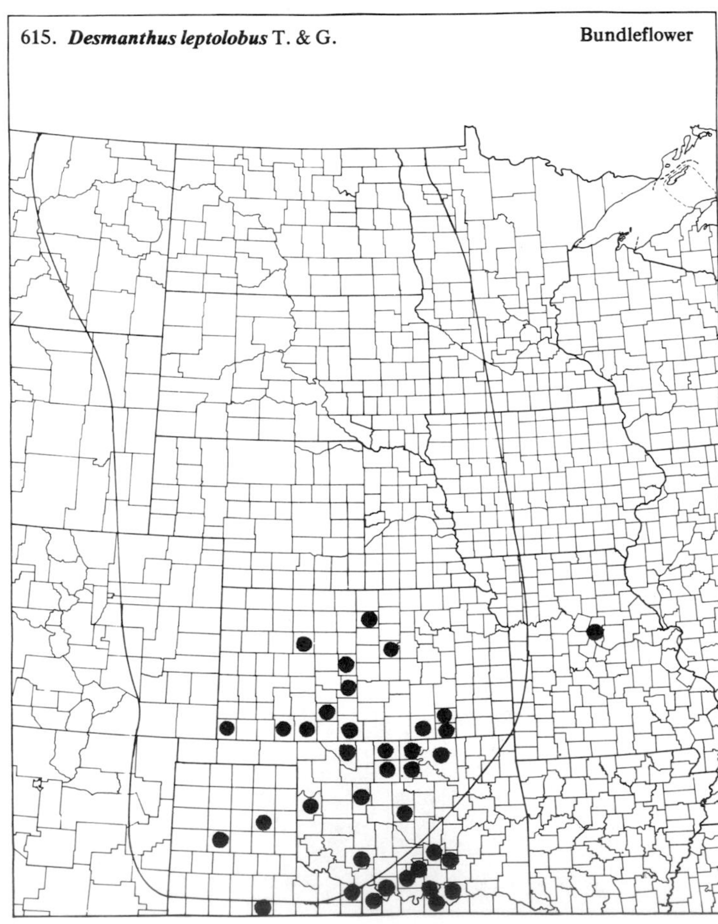

616. **Mimosa borealis** Gray Mimosa

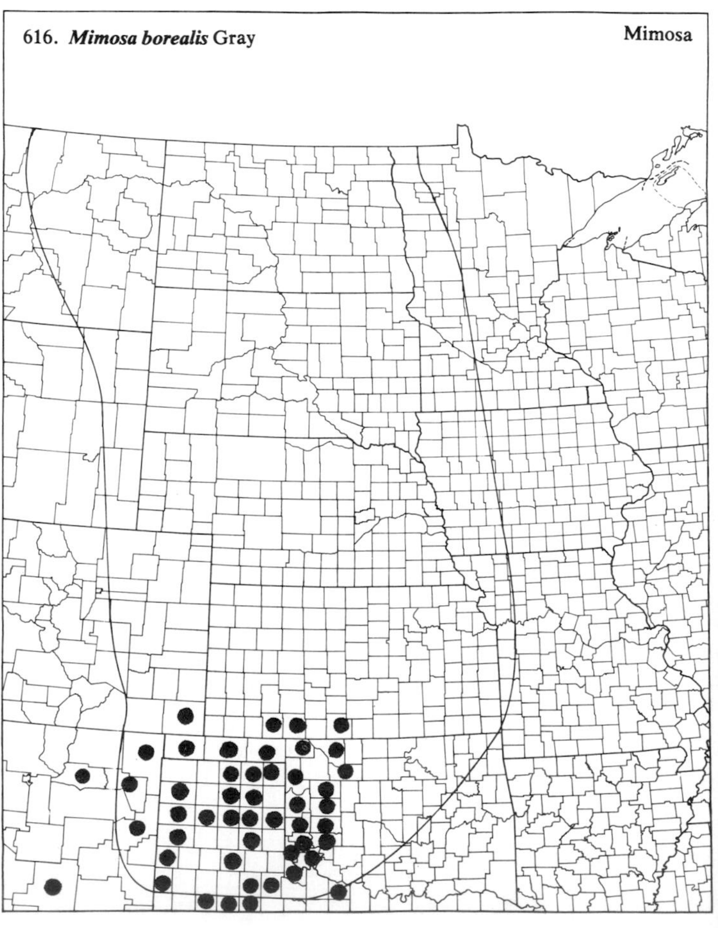

(Fabaceae-Mimosoideae)

617. ***Prosopis glandulosa*** Torr. Mesquite

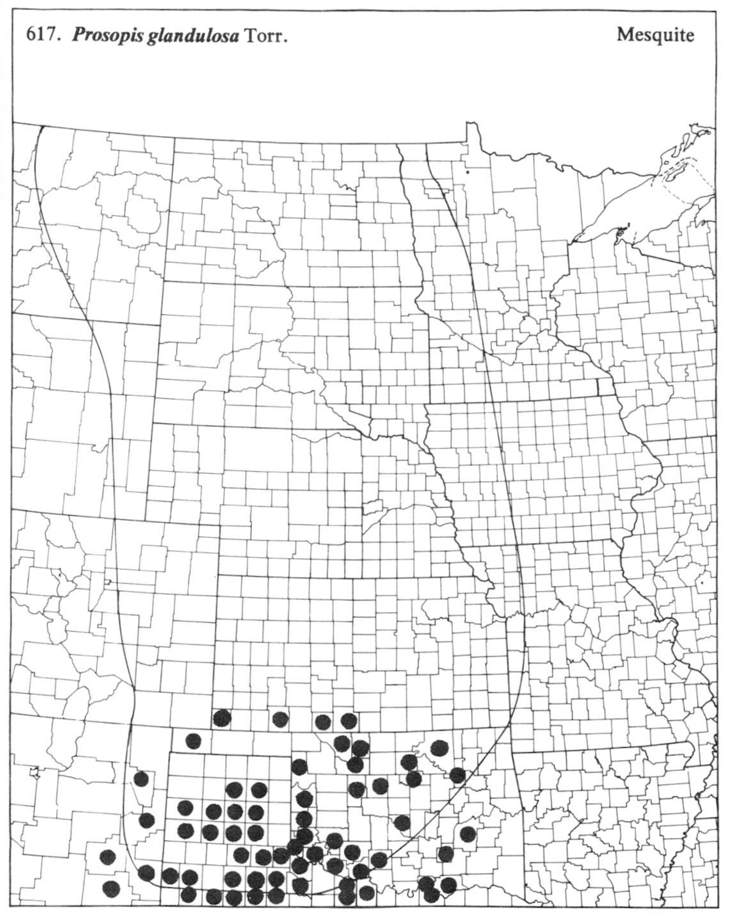

618. ***Schrankia nuttallii*** (DC.) Standl. Sensitive Briar

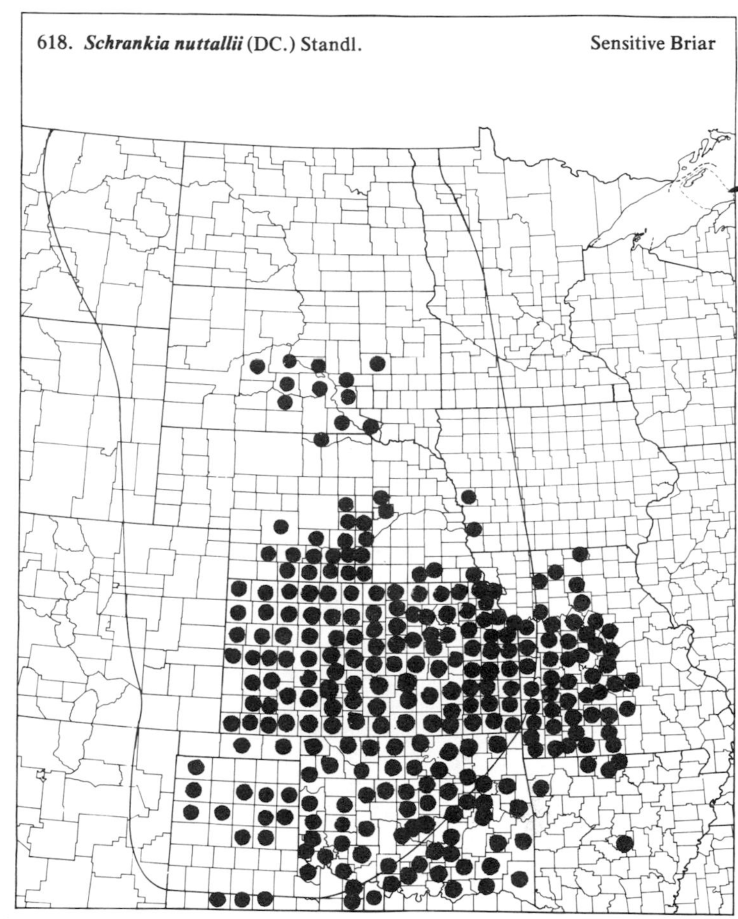

619. ***Schrankia occidentalis*** (W. & S.) Standl. Sensitive Briar

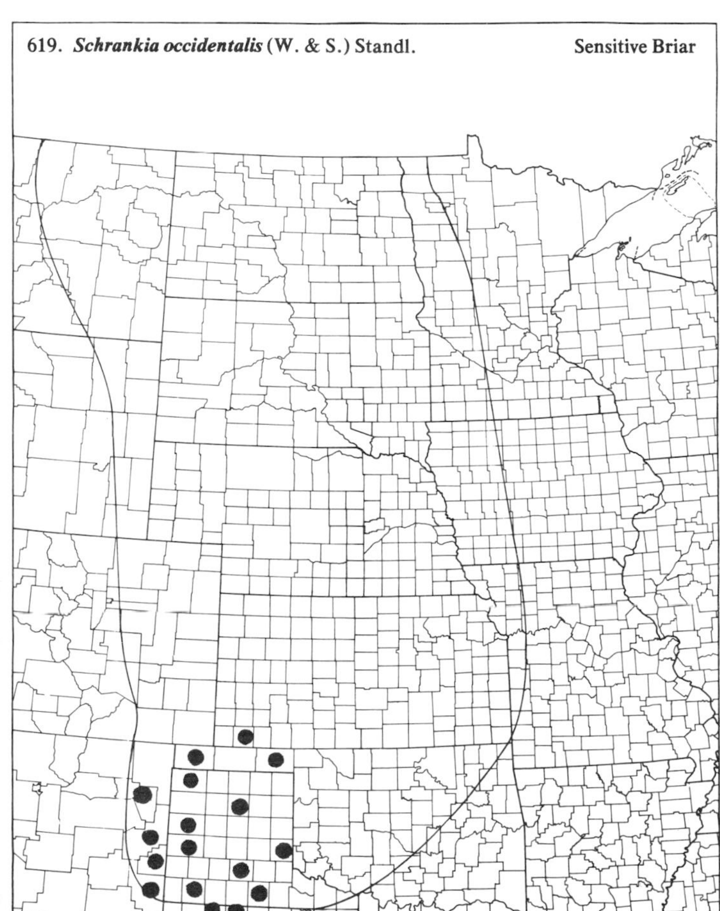

54b. *Fabaceae* (Bean Family), Subfamily *Caesalpinoideae*

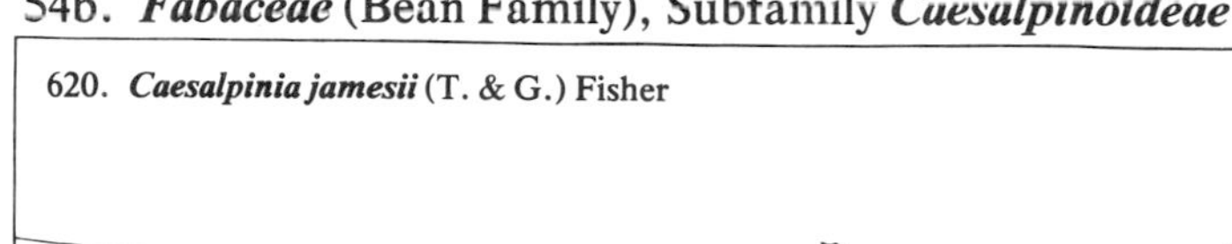

620. ***Caesalpinia jamesii*** (T. & G.) Fisher

(Fabaceae-Caesalpinoideae)

621. ***Cassia fasciculata*** Michx. — Partridge Pea

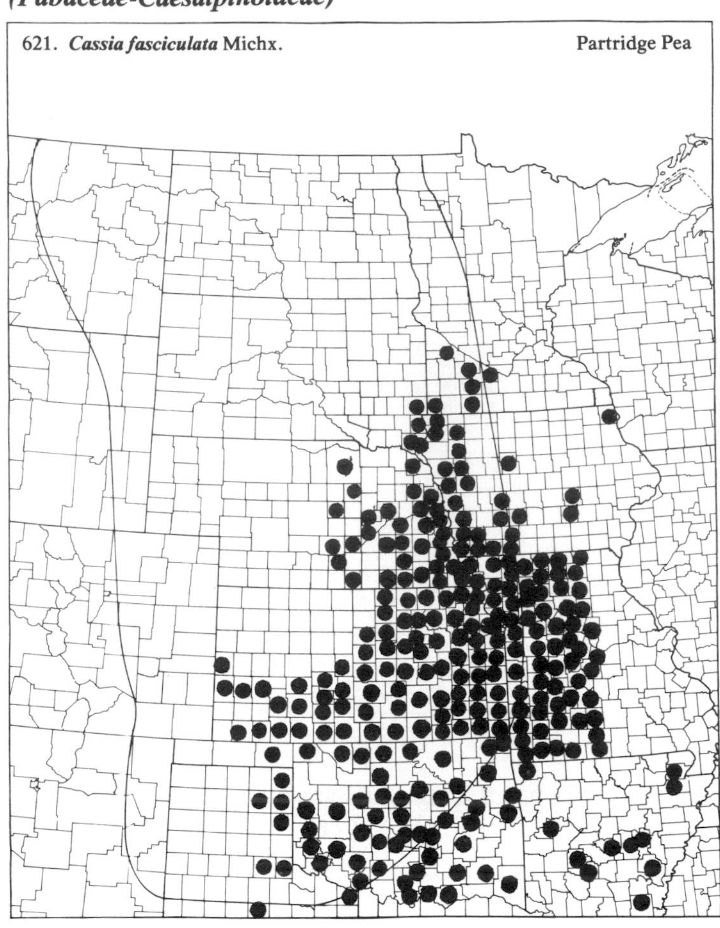

622. ***Cassia marilandica*** L. — Wild Senna

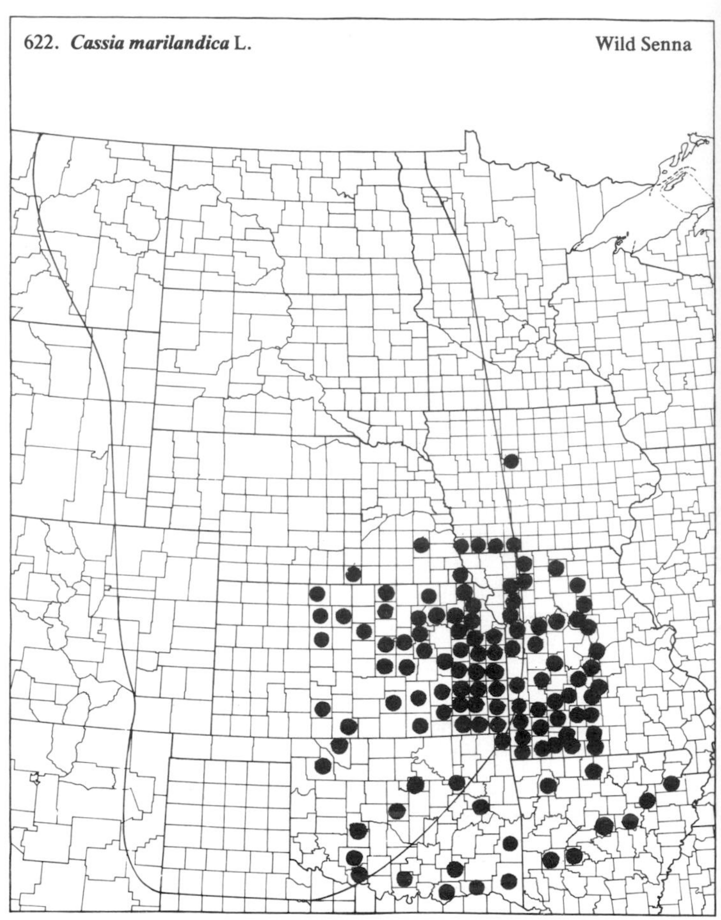

623. ***Cassia nictitans*** L. — Sensitive Pea

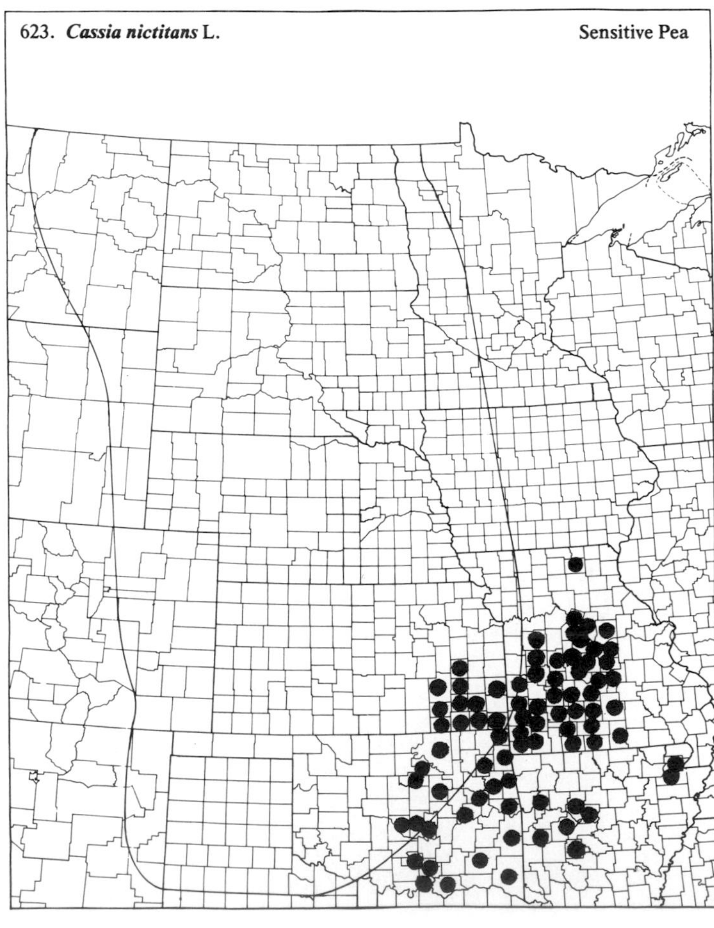

624. ***Cercis canadensis*** L. — Redbud

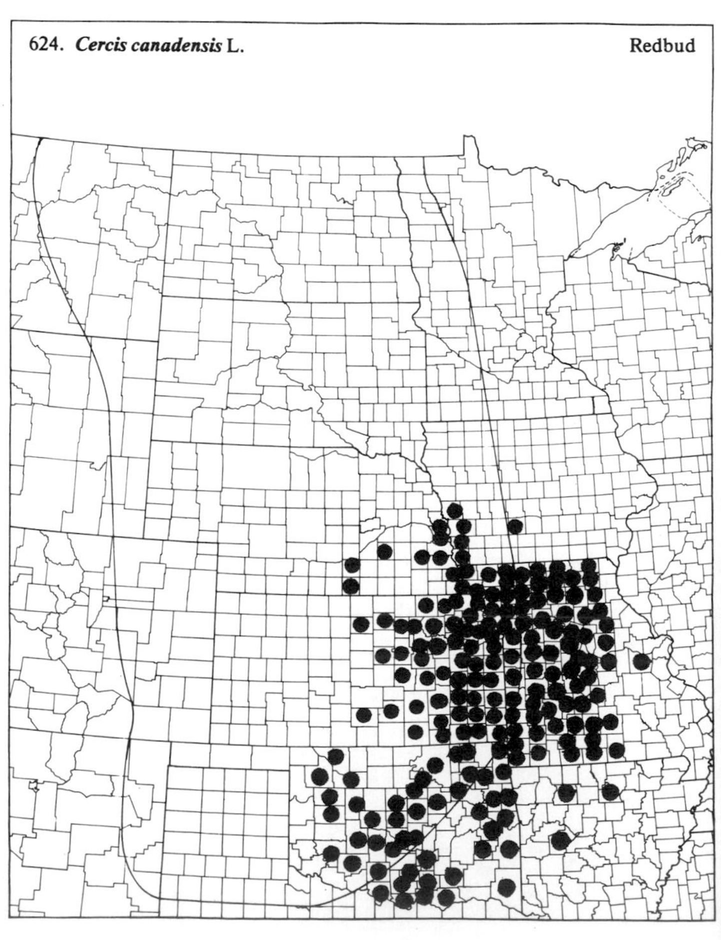

(Fabaceae-Caesalpinoideae)

625. **Gleditsia triacanthos** L. Honey Locust

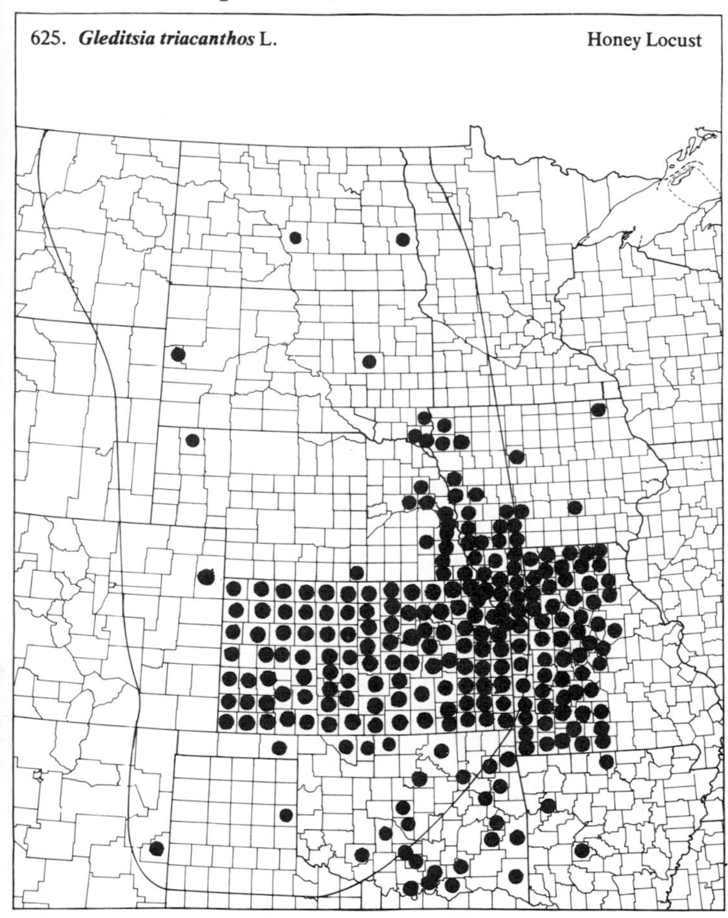

626. **Gymnocladus dioica** (L.) K. Koch Kentucky Coffee Tree

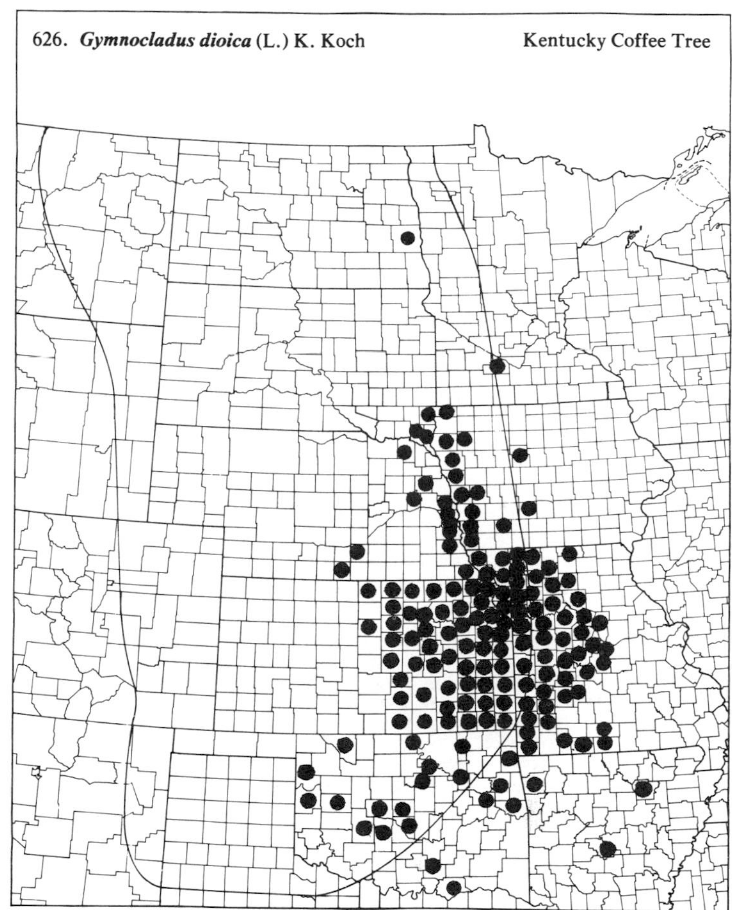

627. **Hoffmanseggia glauca** (Ort.) Eifert Pig Nut

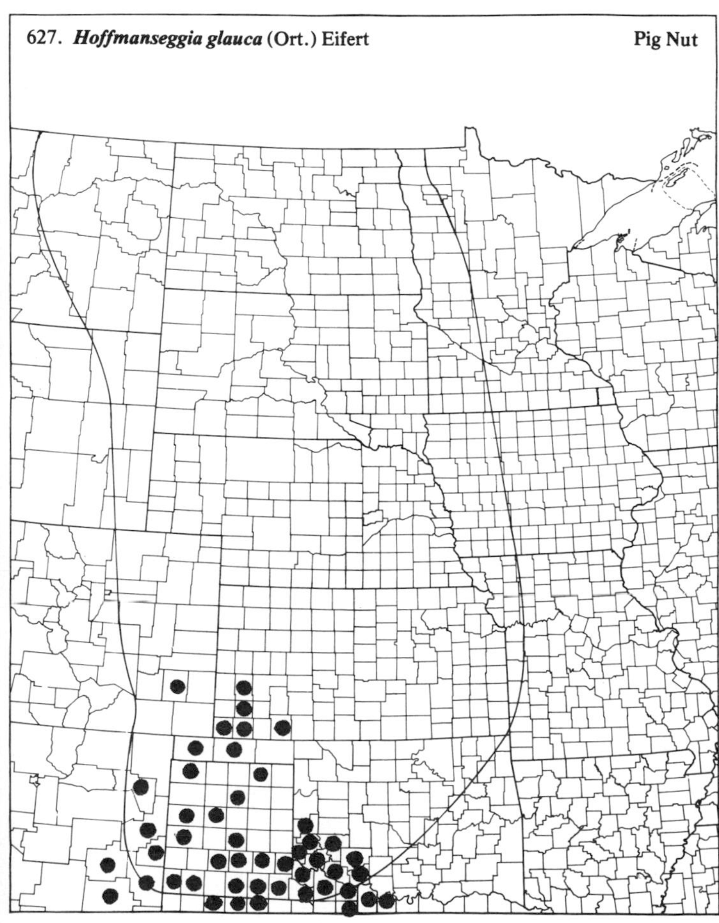

628. **Krameria lanceolata** Torr. Ratany

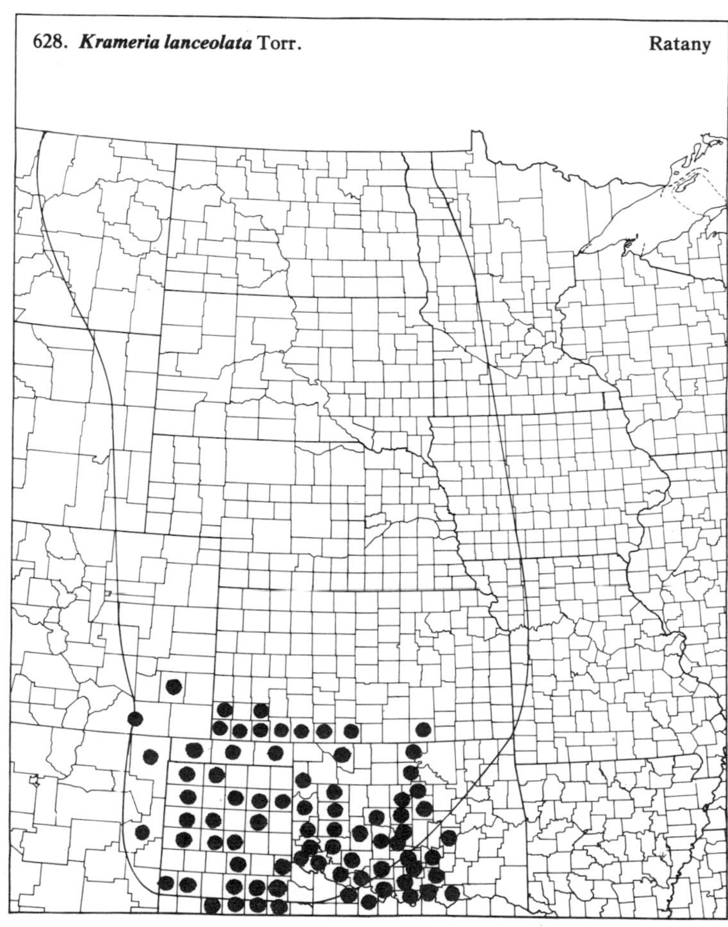

54c. ***Fabaceae*** (Bean Family), Subfamily ***Papilionoideae***

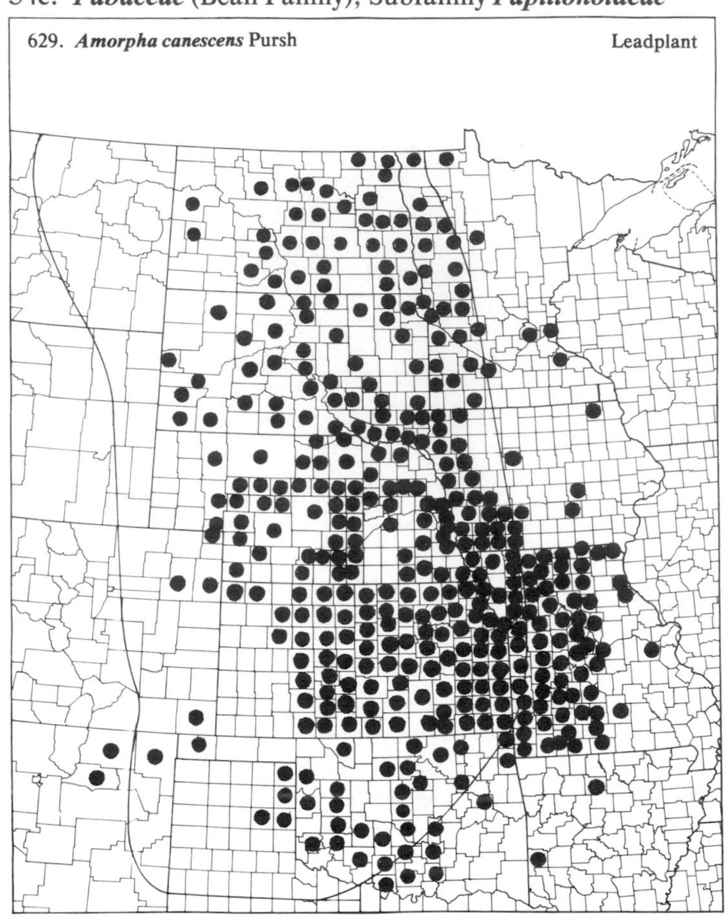
629. ***Amorpha canescens*** Pursh — Leadplant

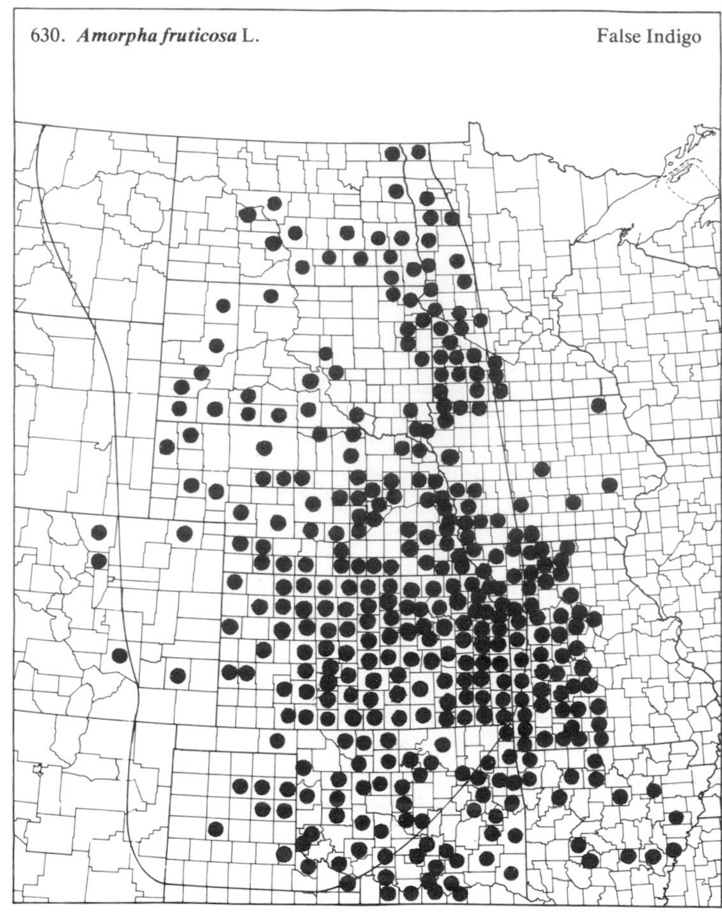
630. ***Amorpha fruticosa*** L. — False Indigo

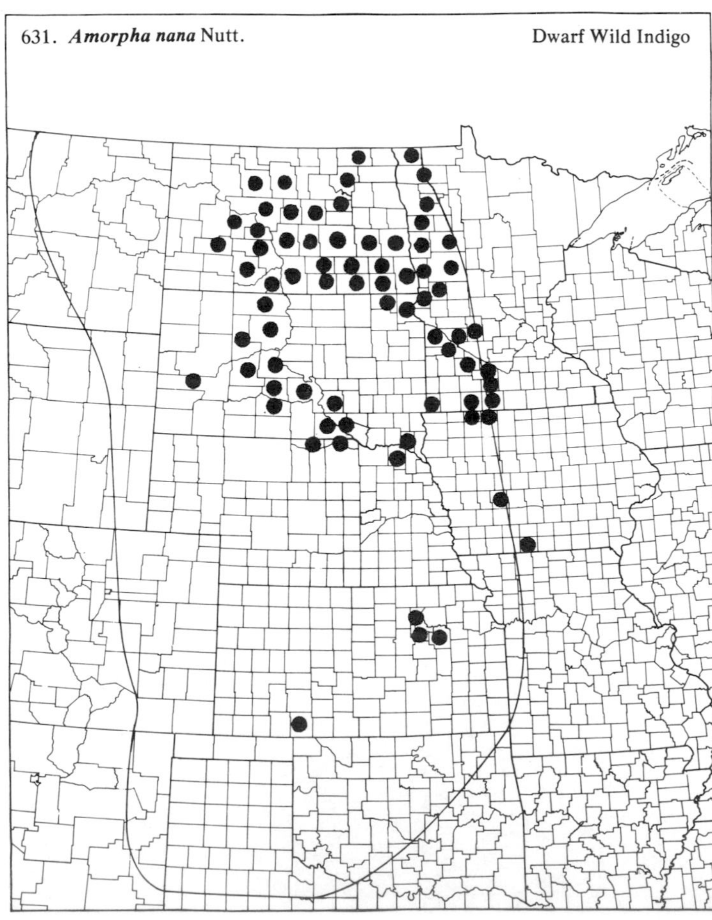
631. ***Amorpha nana*** Nutt. — Dwarf Wild Indigo

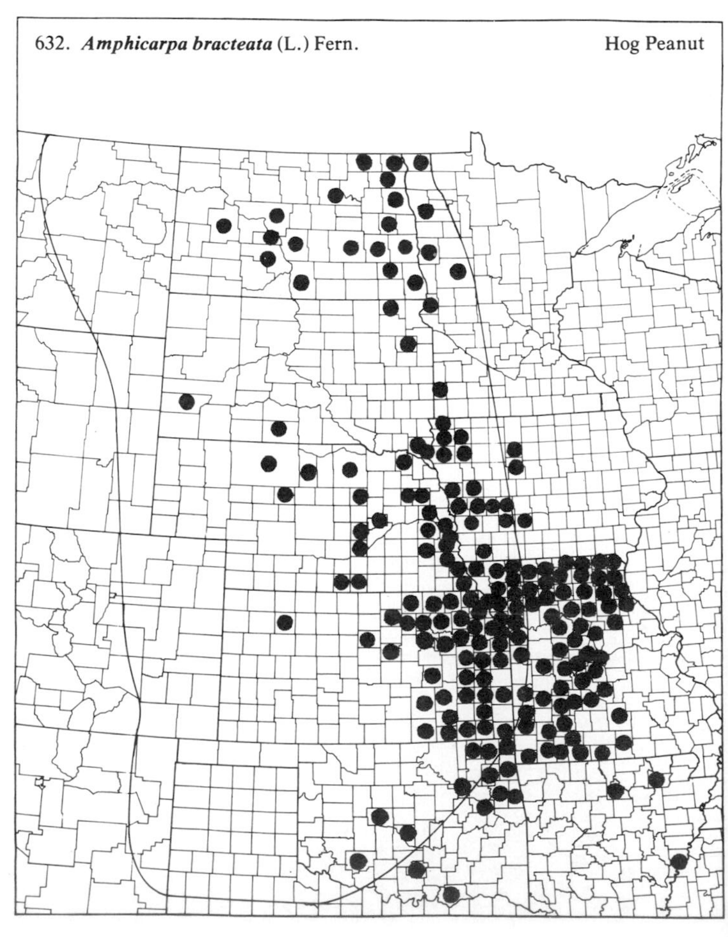
632. ***Amphicarpa bracteata*** (L.) Fern. — Hog Peanut

(Fabaceae-Papilionoideae)

633. *Apios americana* Medic. Ground Nut

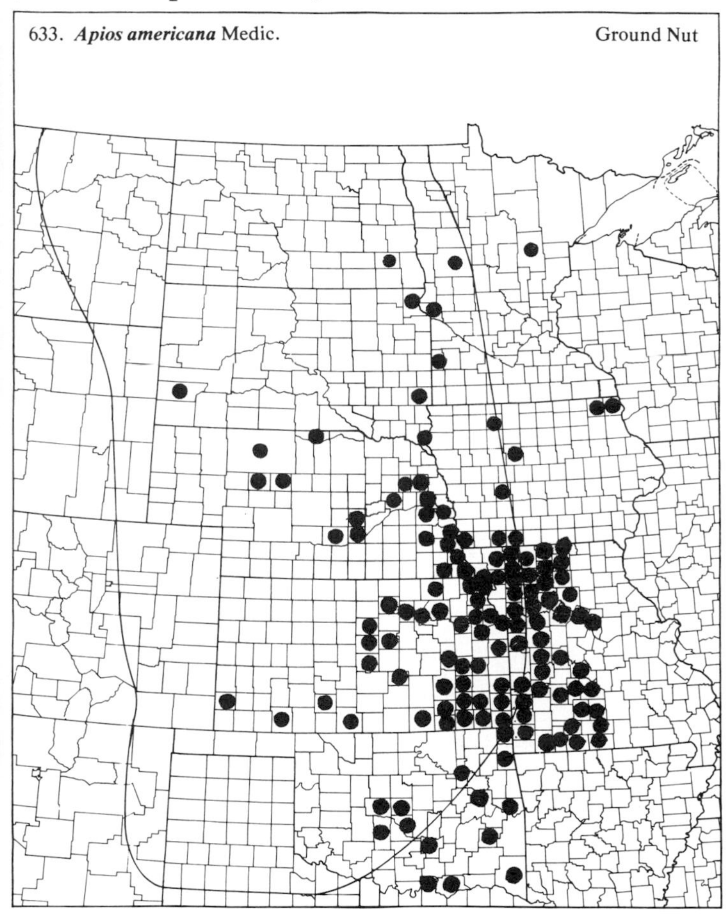

634. *Astragalus aboriginum* Richards.

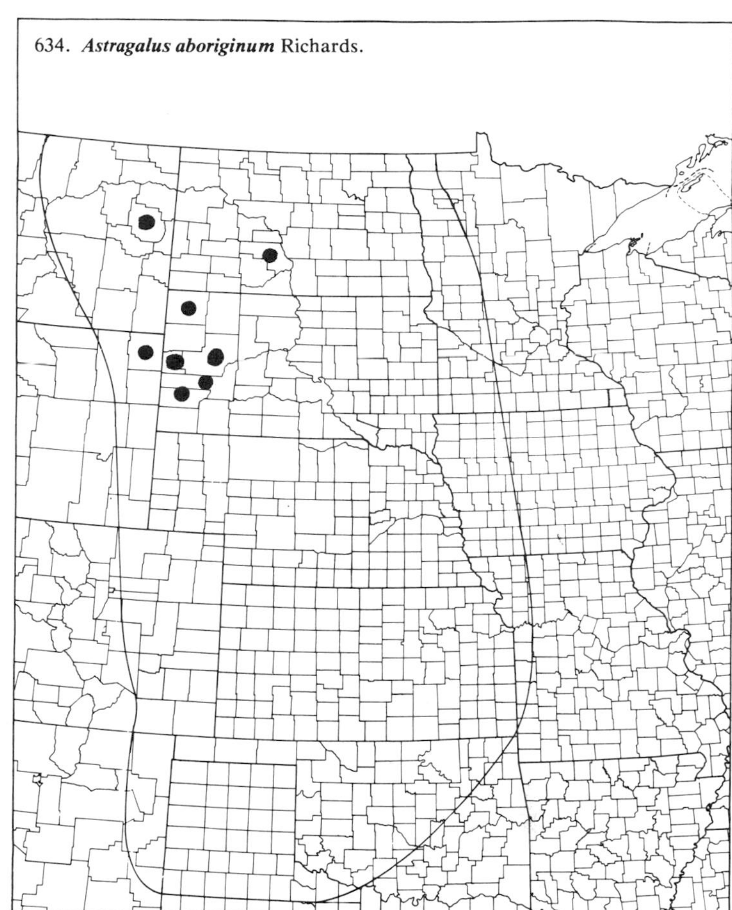

635. *Astragalus adsurgens* Pall. var. *robustior* Hook.

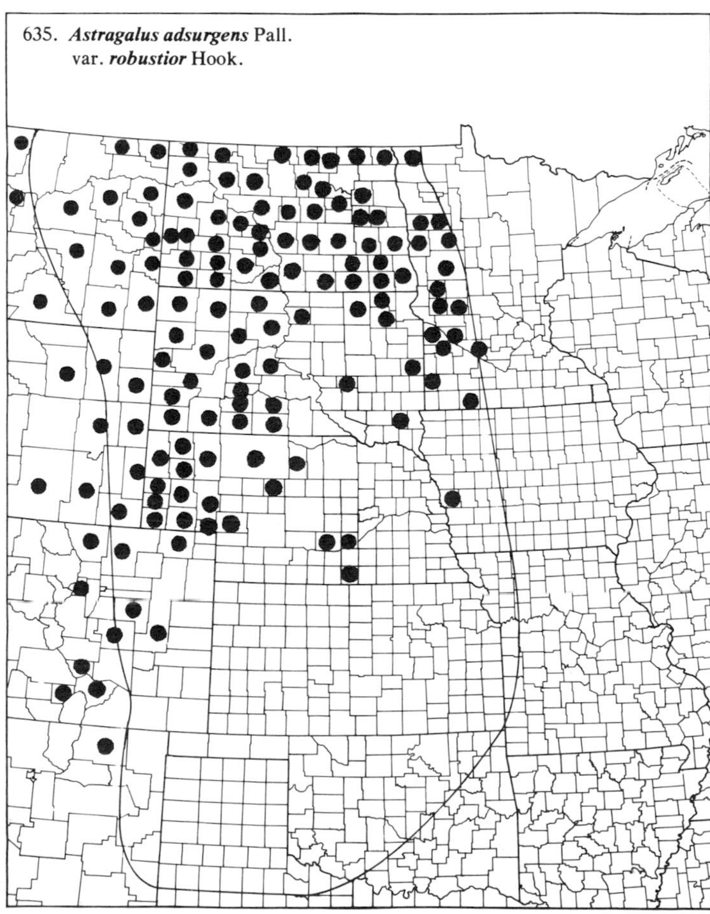

636. *Astragalus agrestis* Dougl. ex D. Don

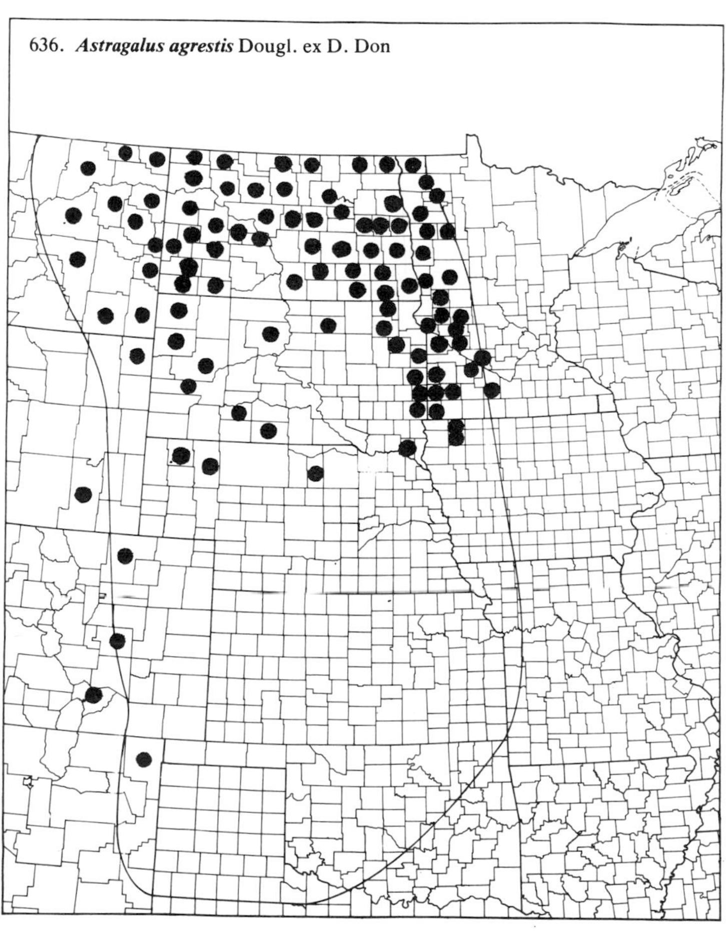

(Fabaceae-Papilionoideae)

637. ***Astragalus bisulcatus*** (Hook.) Gray — Two-grooved Milkvetch

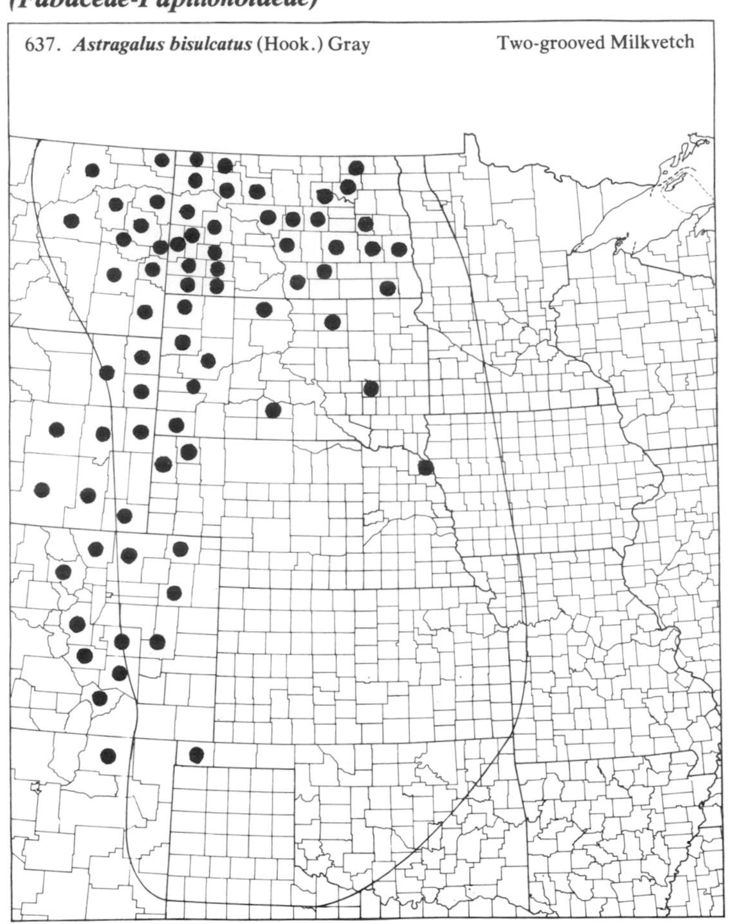

638. ***Astragalus canadensis*** L. — Canada Milkvetch

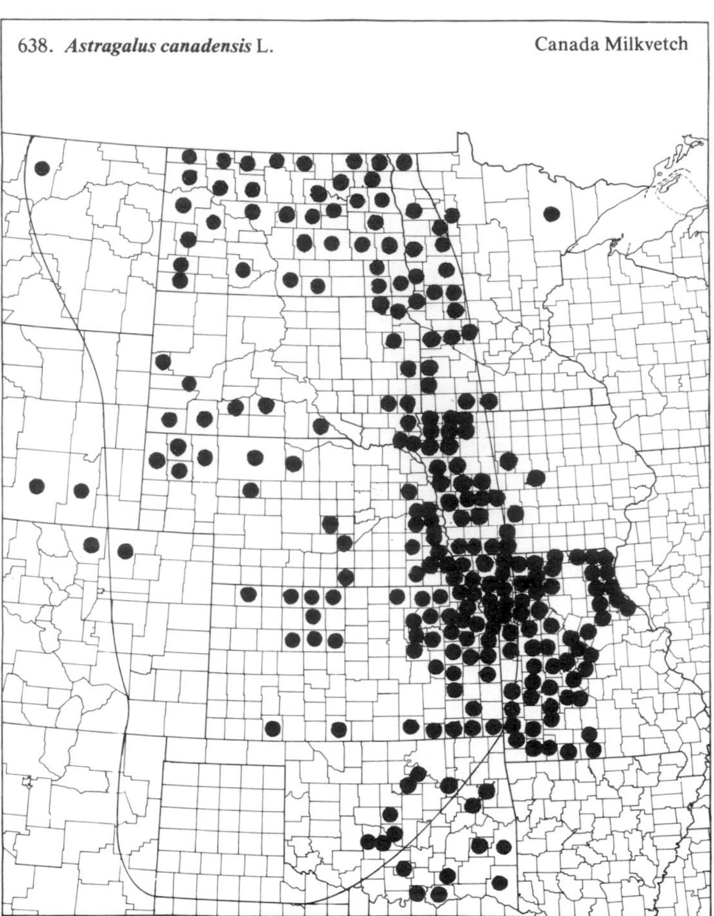

639. ***Astragalus ceramicus*** Sheld. var. ***filifolius*** (Gray) Herm.

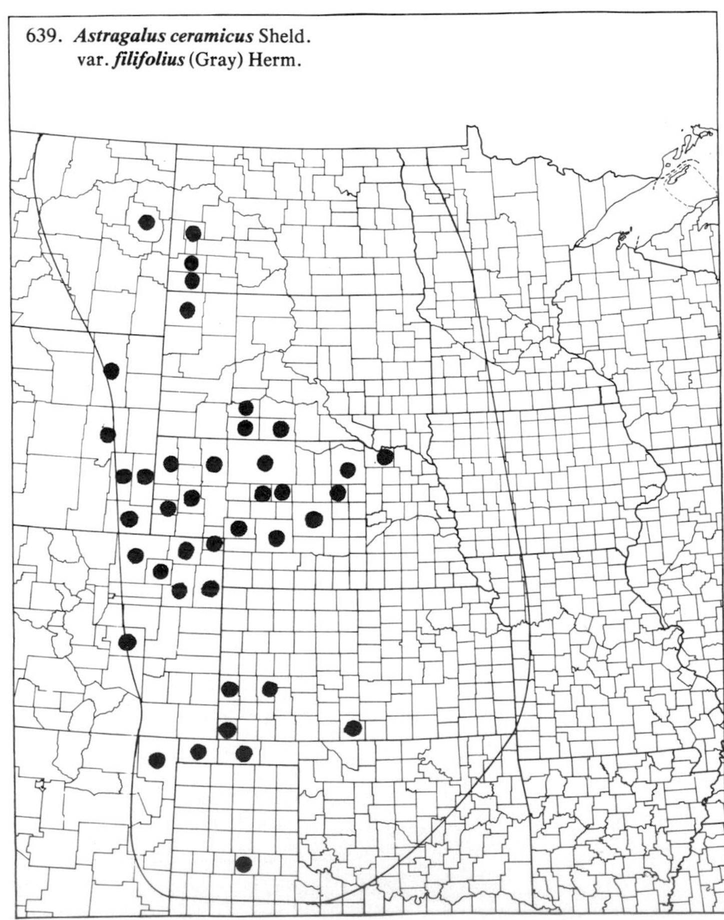

640. ***Astragalus crassicarpus*** Nutt. var. ***crassicarpus***

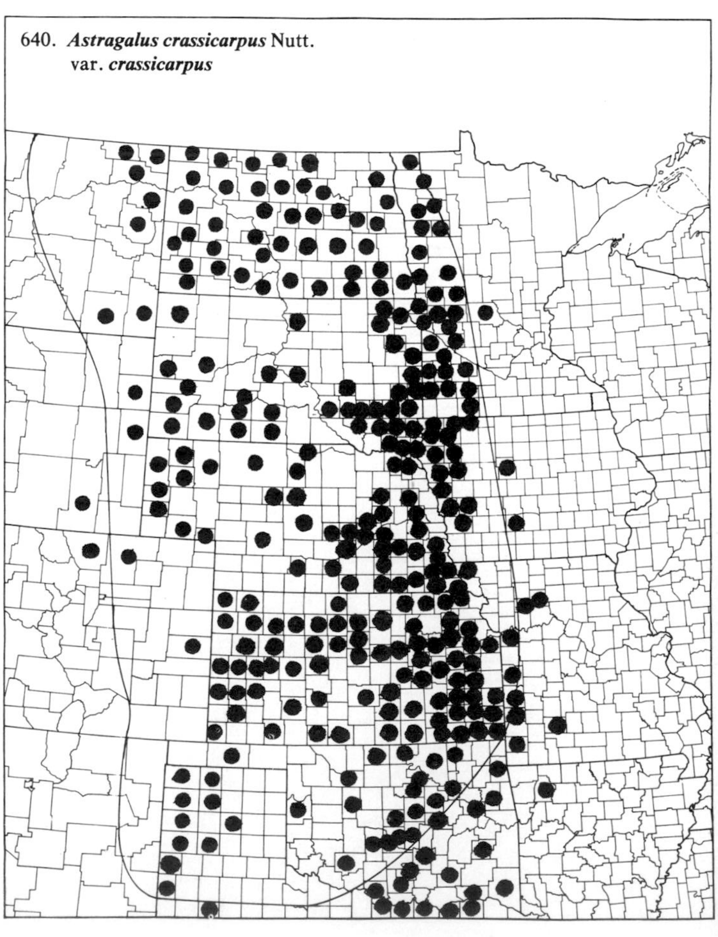

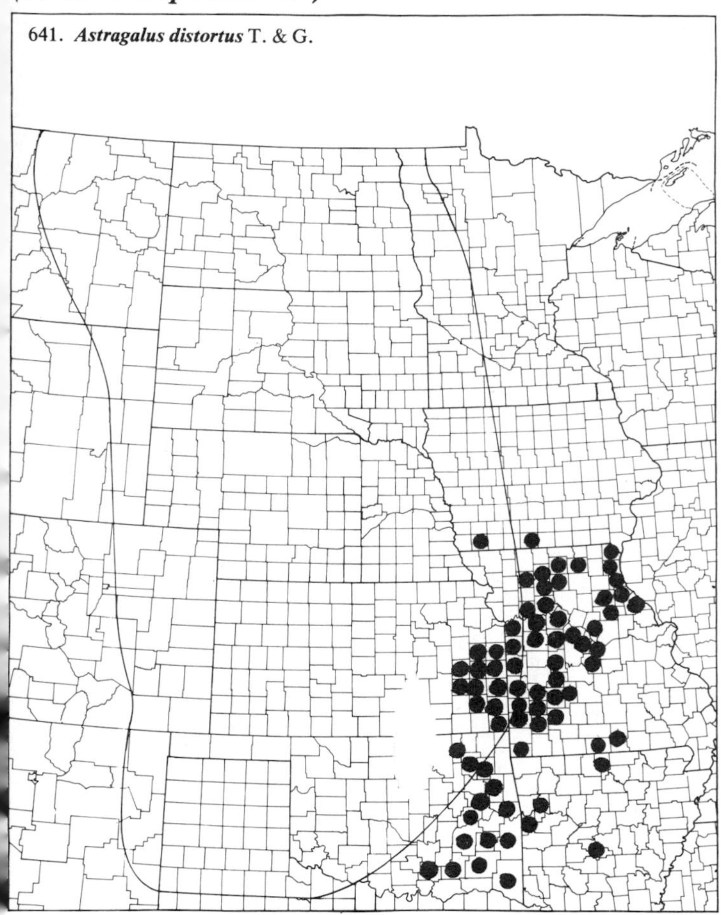
641. ***Astragalus distortus*** T. & G.

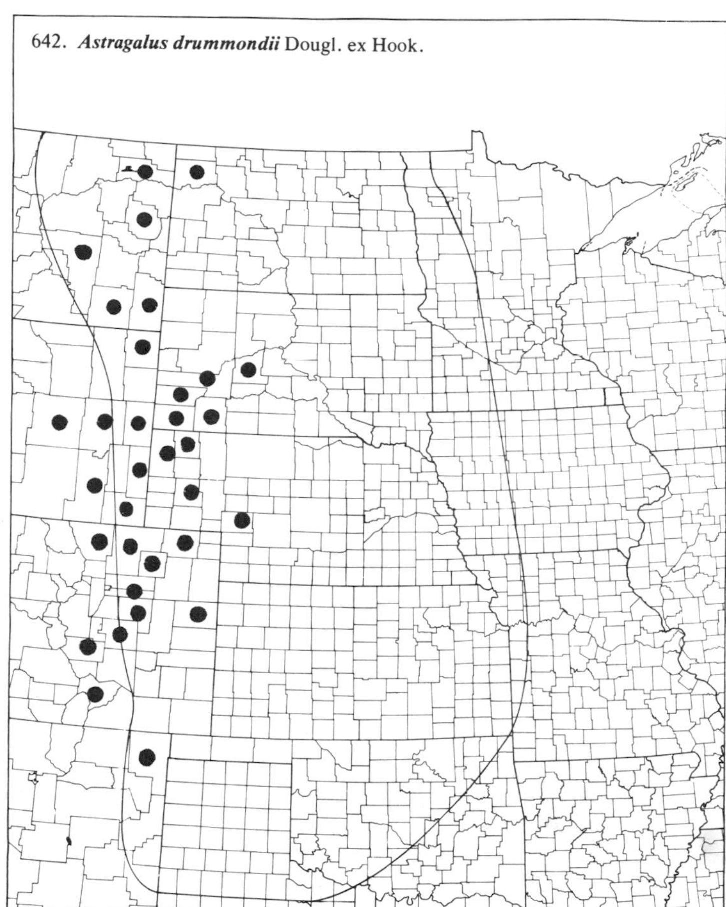
642. ***Astragalus drummondii*** Dougl. ex Hook.

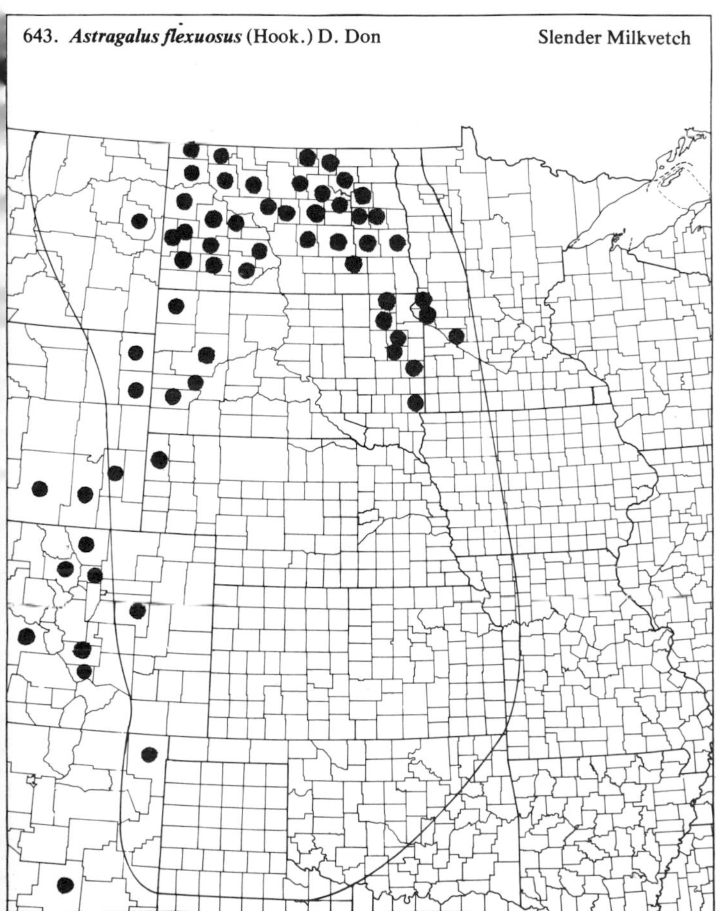
643. ***Astragalus flexuosus*** (Hook.) D. Don   Slender Milkvetch

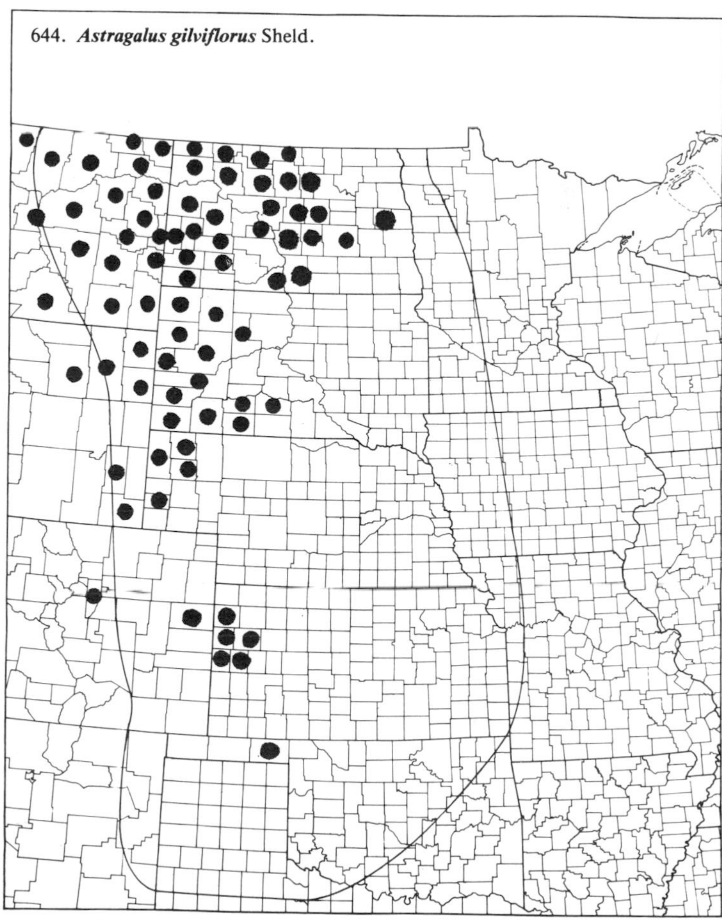
644. ***Astragalus gilviflorus*** Sheld.

(Fabaceae-Papilionoideae)

645. *Astragalus gracilis* Nutt. Slender Milkvetch

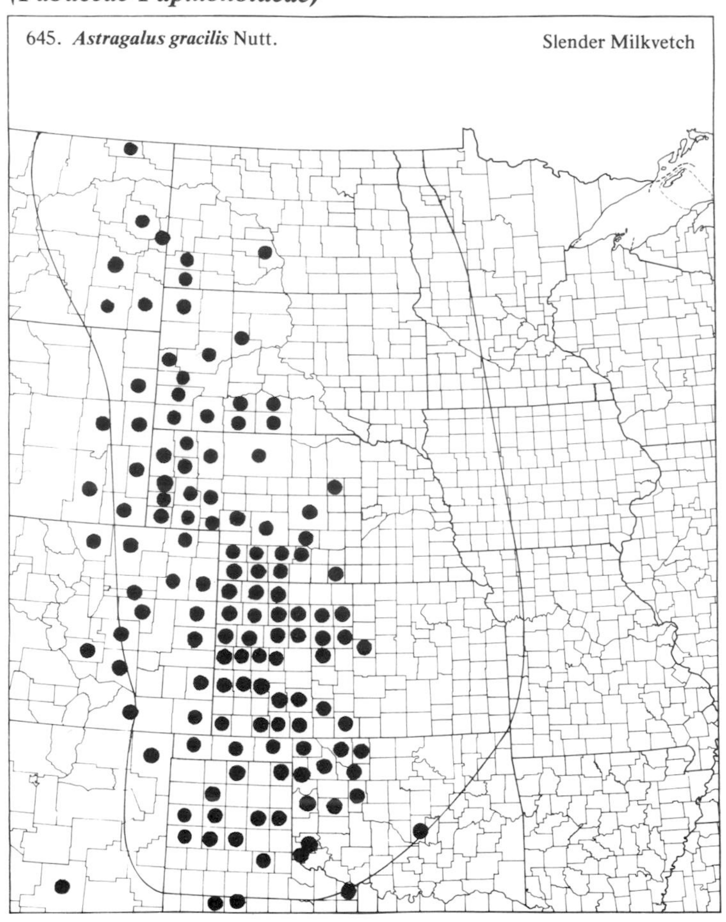

646. *Astragalus hyalinus* M.E. Jones

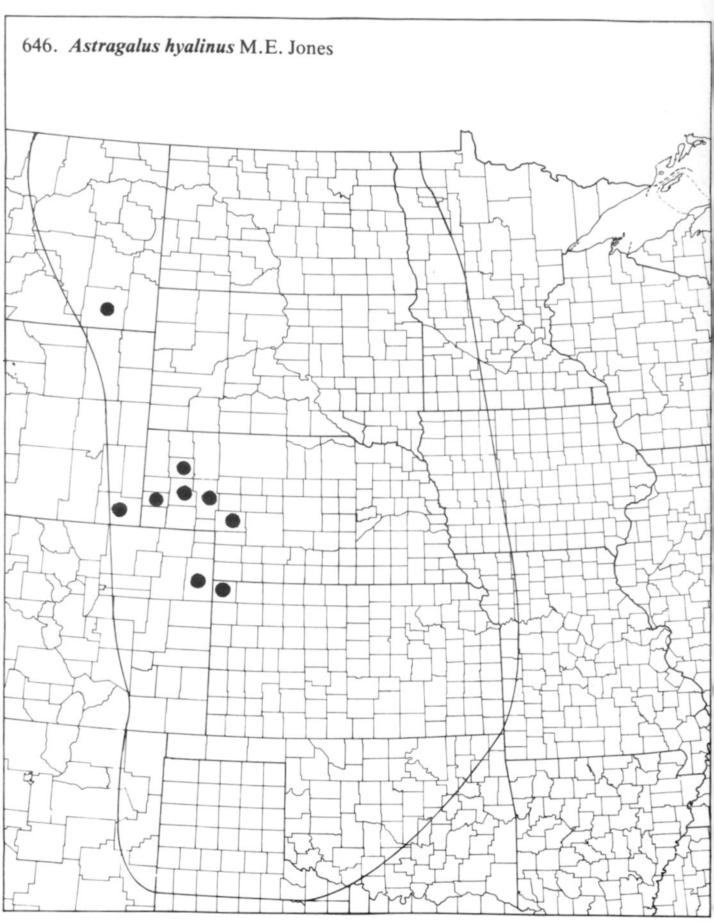

647. *Astragalus kentrophyta* Gray

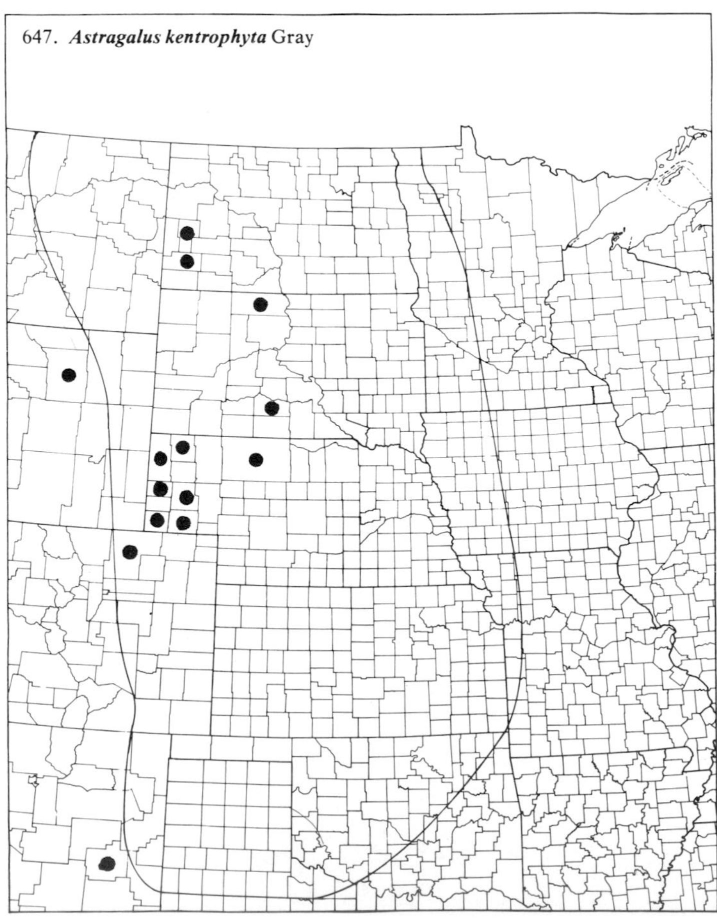

648. *Astragalus lindheimeri* Engelm. & Gray

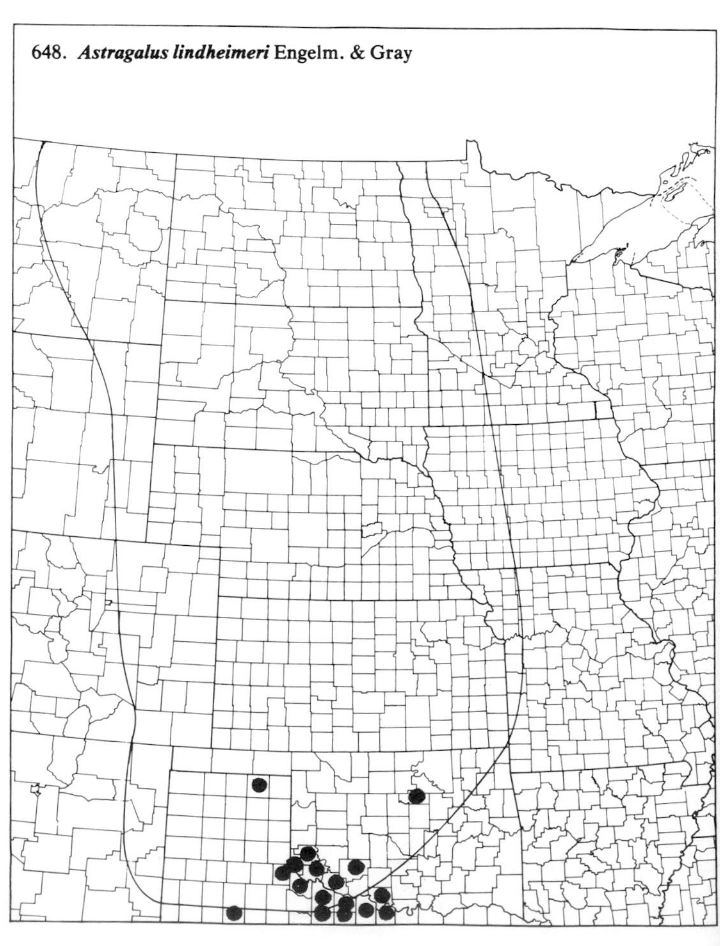

(Fabaceae-Papilionoideae)

649. ***Astragalus lotiflorus*** Hook. Lotus Milkvetch

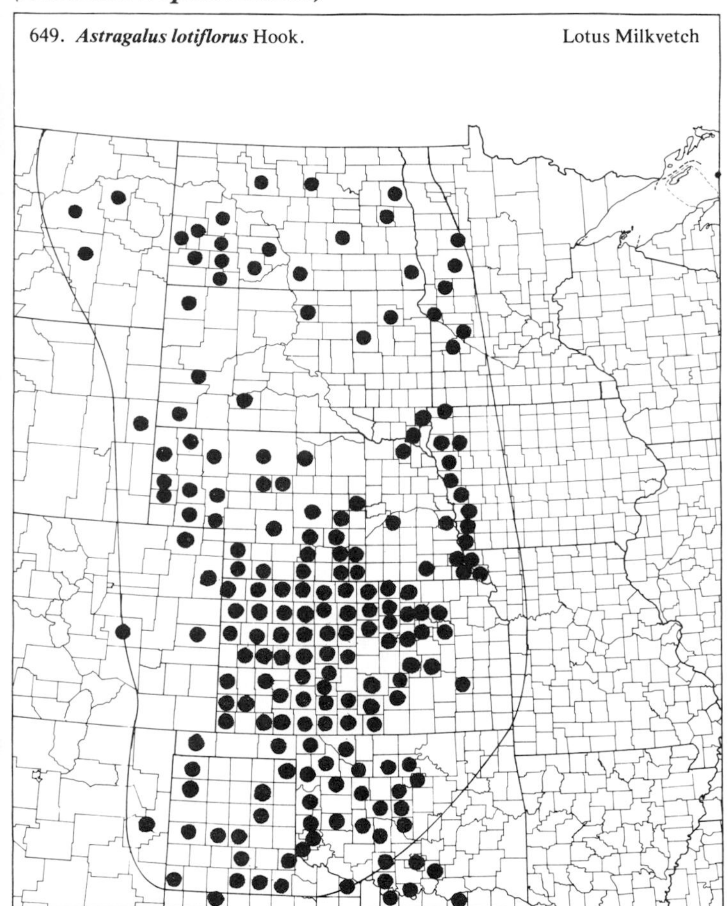

650. ***Astragalus missouriensis*** Nutt. Missouri Milkvetch

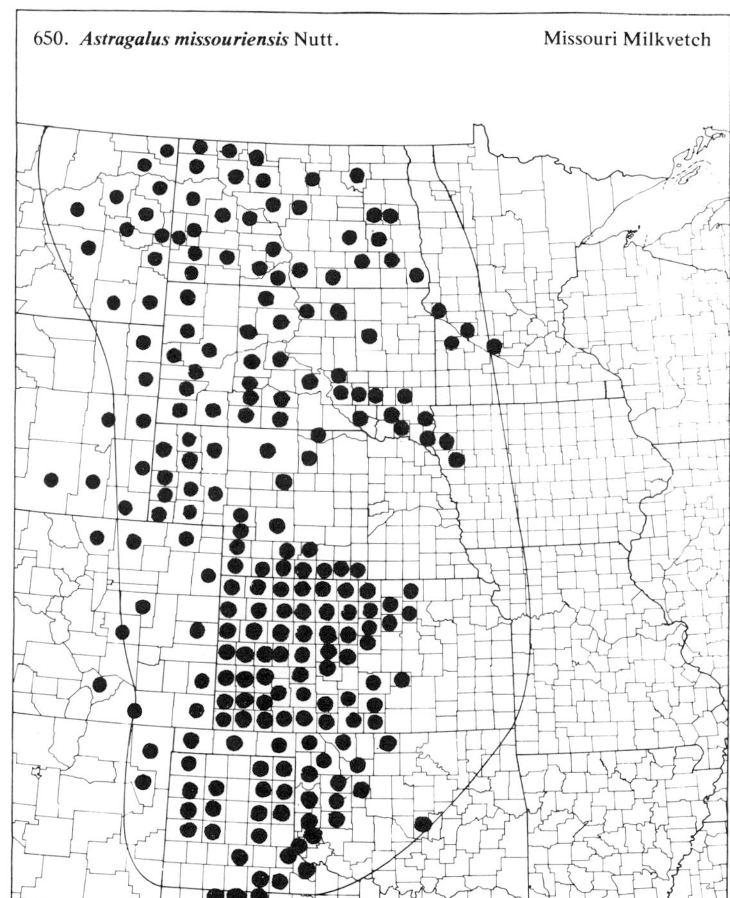

651. ***Astragalus mollissimus*** Torr. Woolly Loco

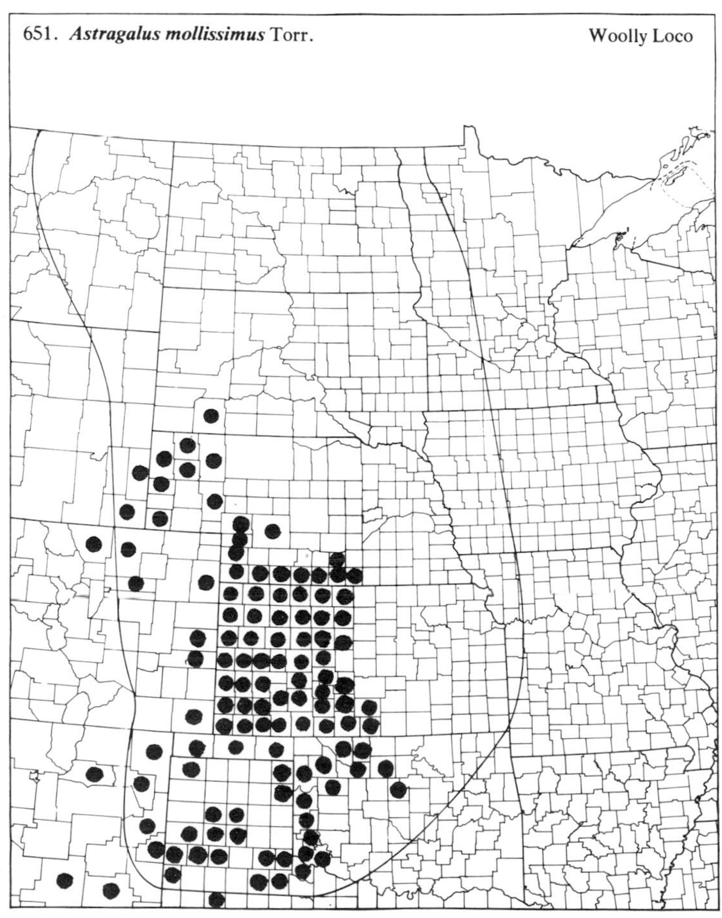

652. ***Astragalus nuttallianus*** DC.

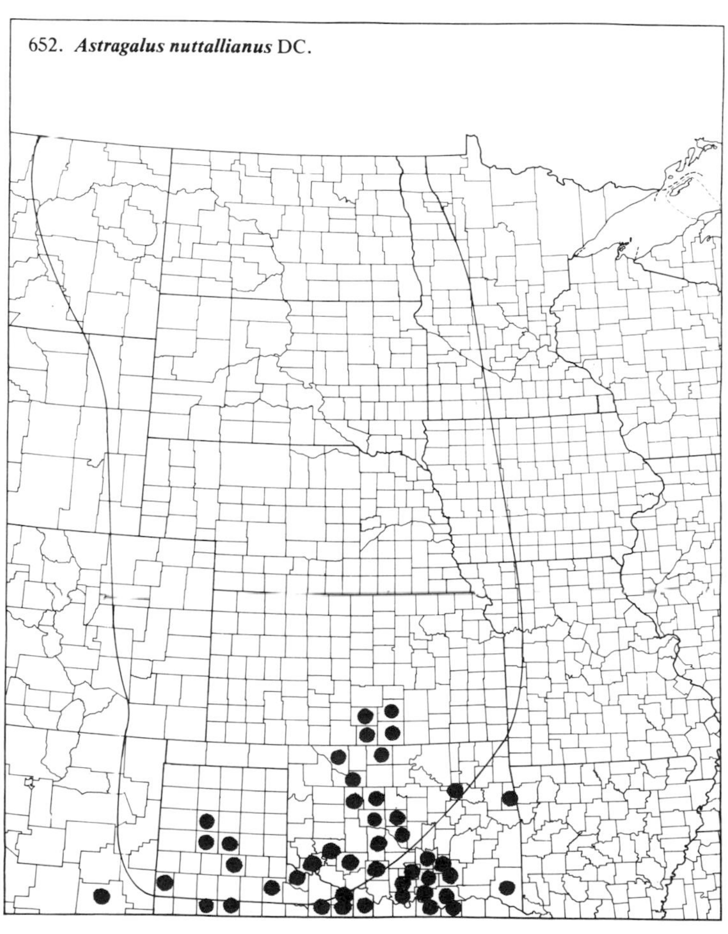

(Fabaceae-Papilionoideae)

653. ***Astragalus pectinatus*** Dougl. ex G. Don — Narrow-leaved Poisonvetch

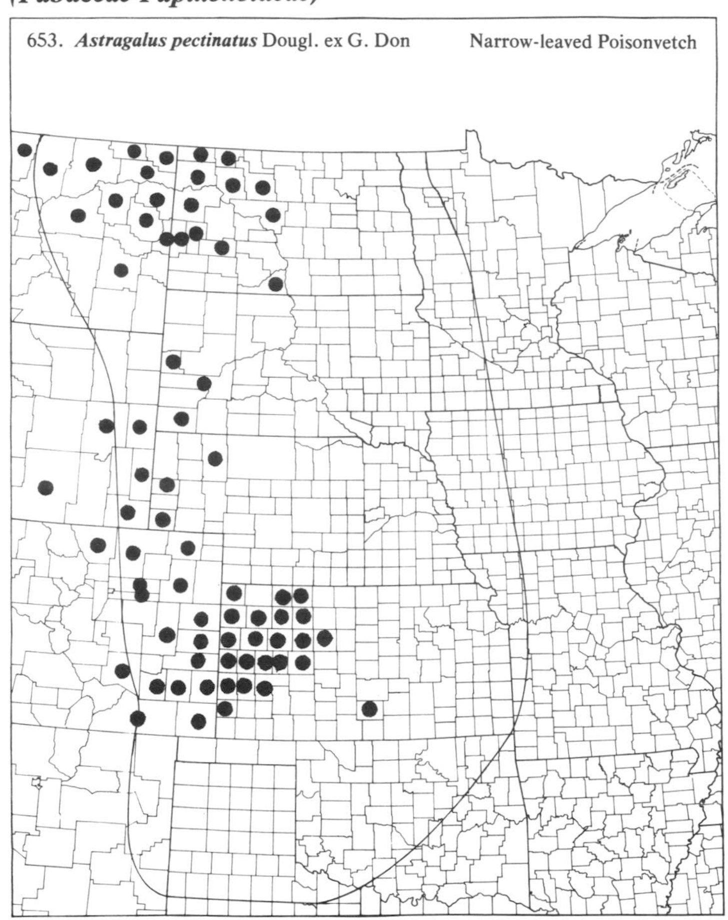

654. ***Astragalus plattensis*** Nutt. ex T. & G. — Ground Plum

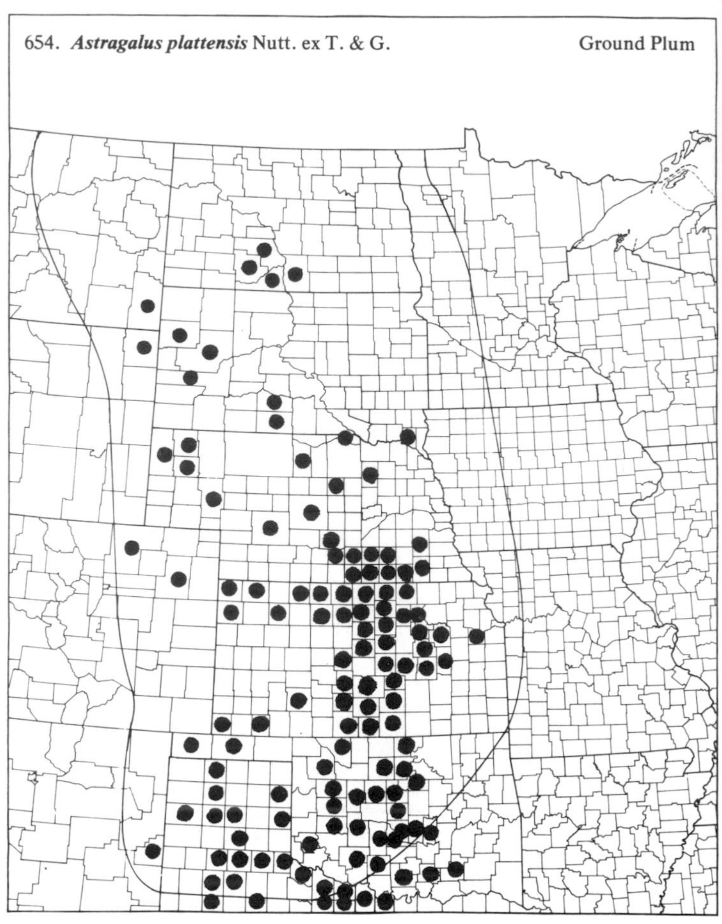

655. ***Astragalus purshii*** Dougl. ex Hook.

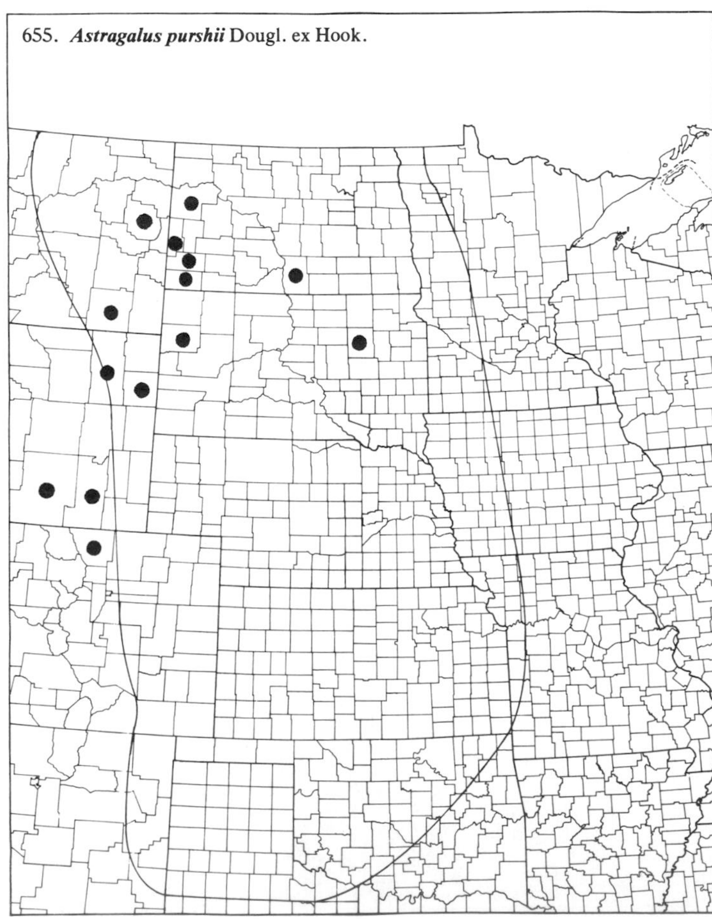

656. ***Astragalus racemosus*** Pursh — Creamy Poisonvetch

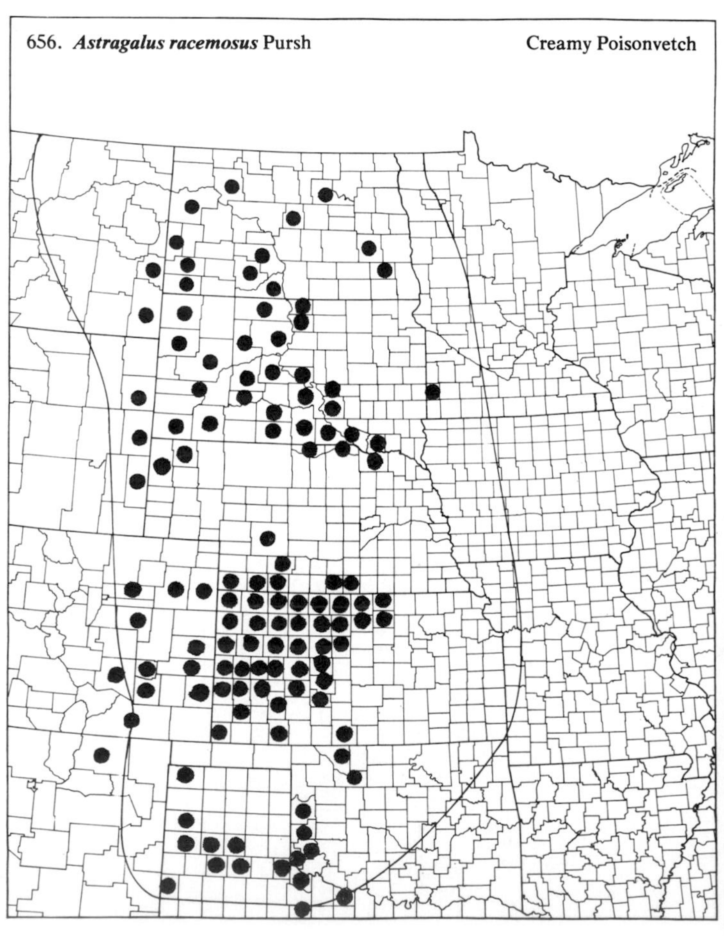

(Fabaceae-Papilionoideae)

657. **Astragalus sericoleucus** Gray

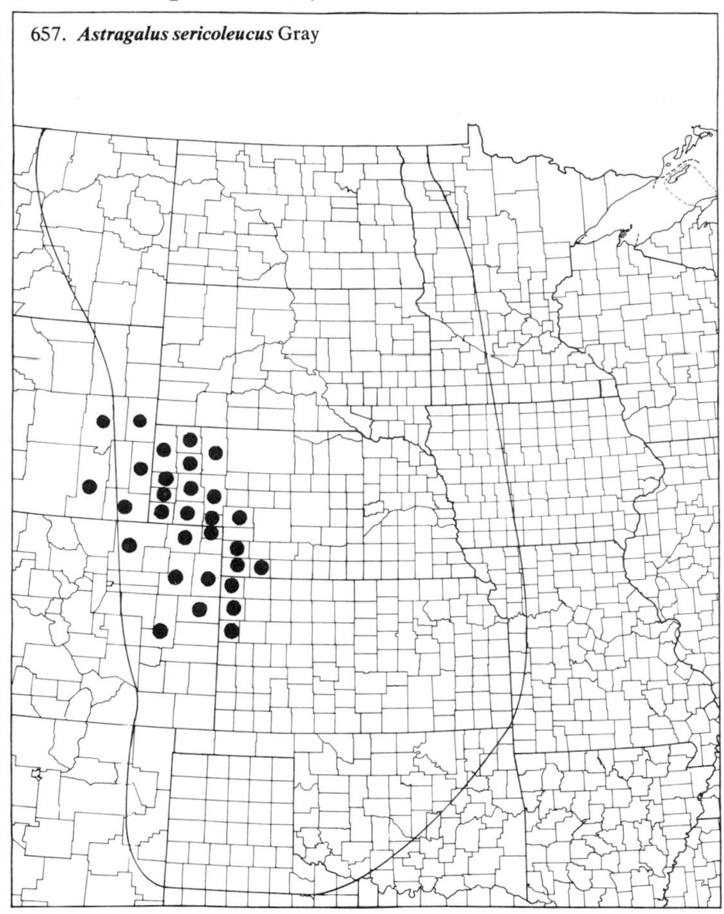

658. **Astragalus spatulatus** Sheld. Tufted Milkvetch

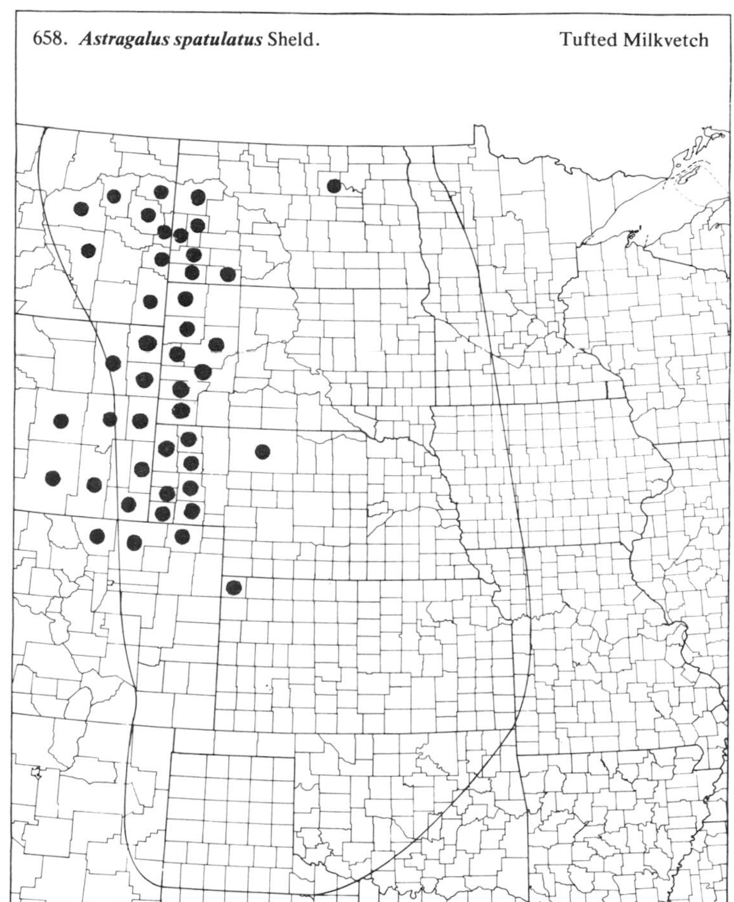

659. **Astragalus tenellus** Pursh Loose-flowered Milkvetch

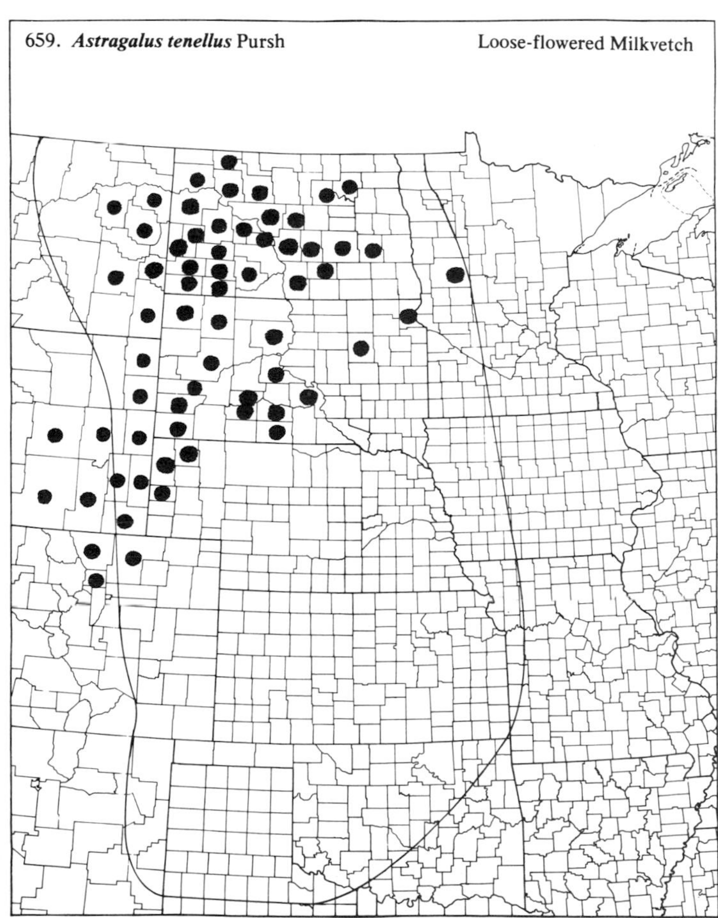

660. **Astragalus vexilliflexus** Sheld.

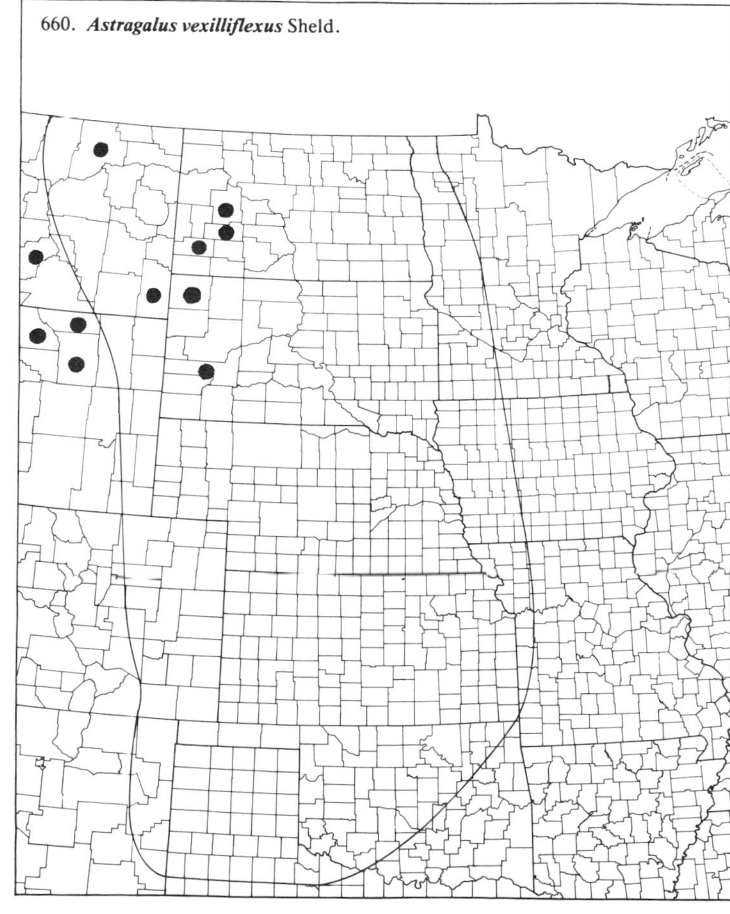

(Fabaceae-Papilionoideae)

661. ***Baptisia australis*** (L.) R. Br. var. ***minor*** (Lehm.) Fern. — Blue False Indigo

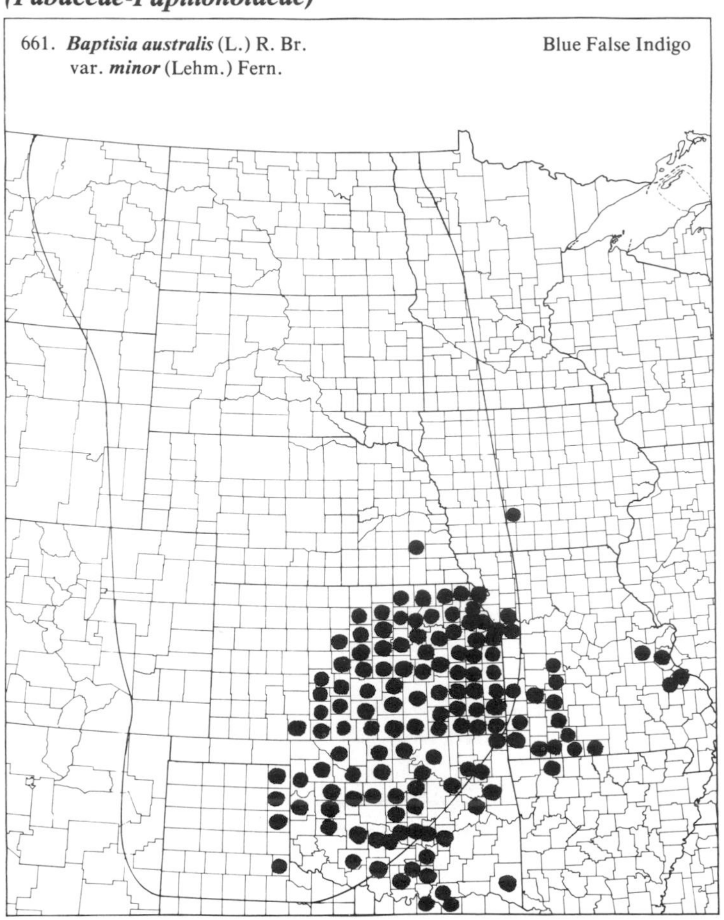

662. ***Baptisia leucantha*** T. & G. — Wild Indigo

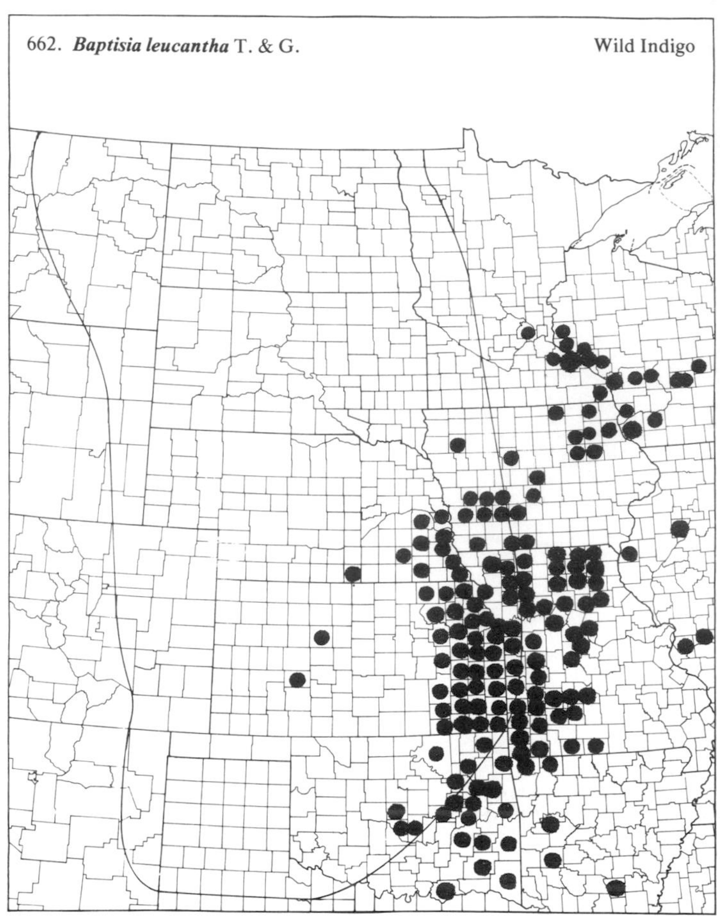

663. ***Baptisia leucophaea*** Nutt. — Plains Wild Indigo

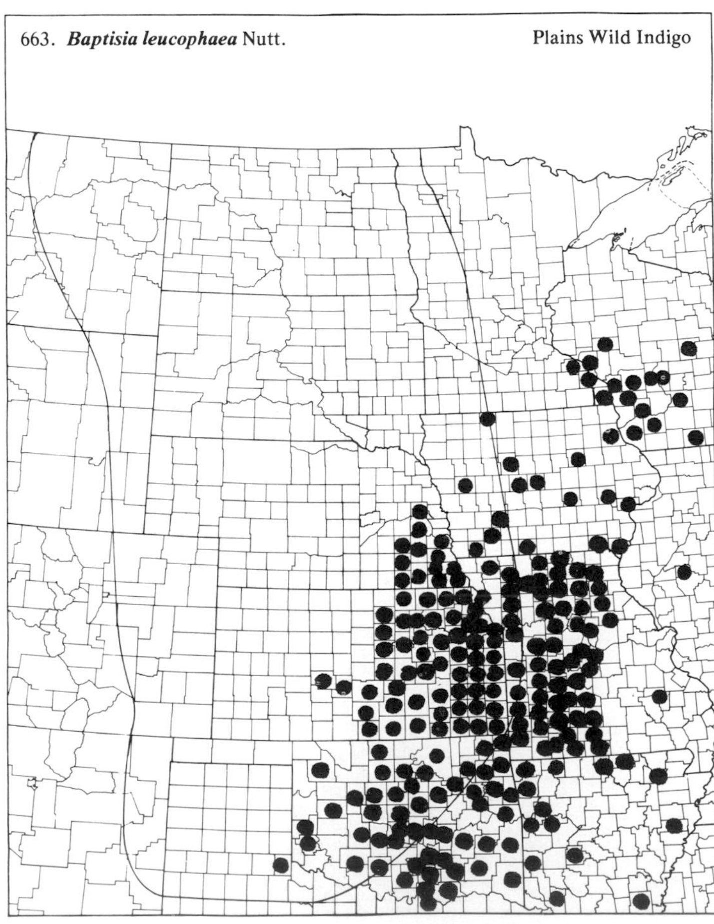

664. ***Caragana arborescens*** Lam. — Pea Tree

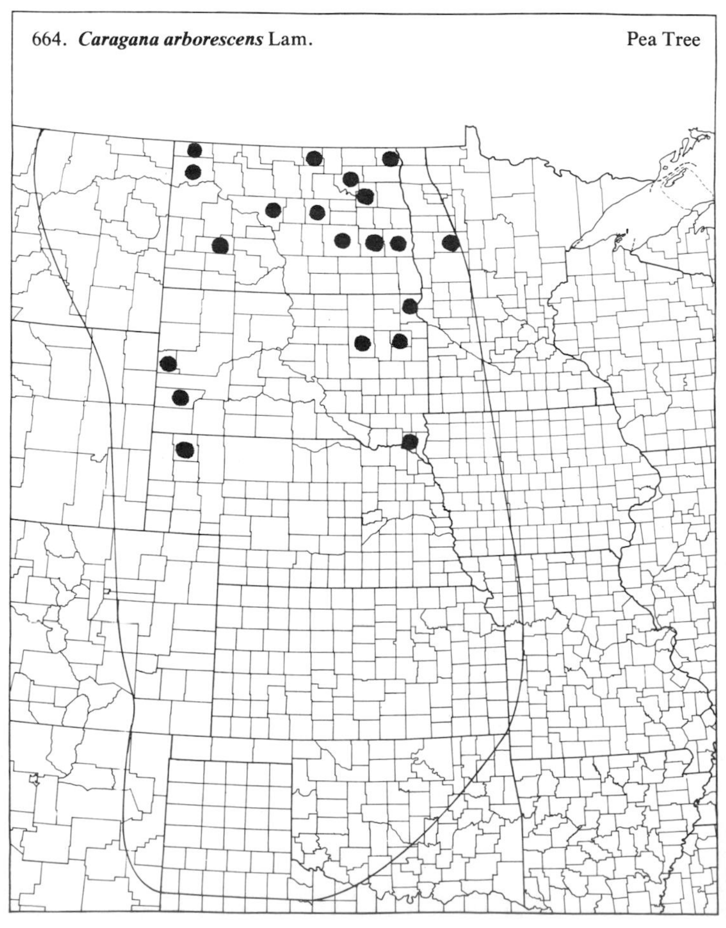

(Fabaceae-Papilionoideae)

665. ***Coronilla varia*** L. Crown Vetch

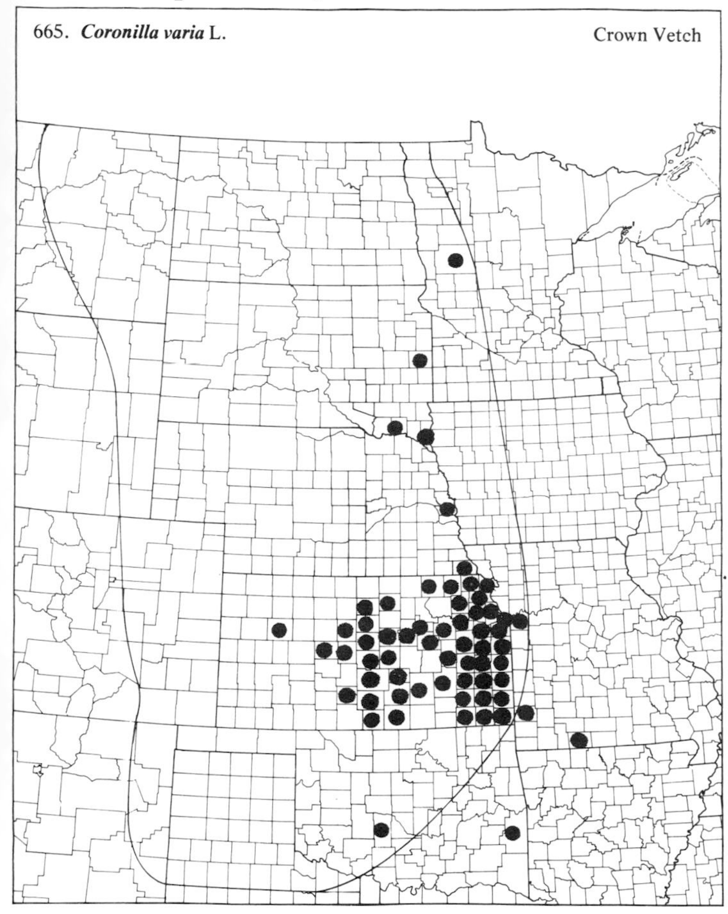

666. ***Crotalaria sagittalis*** L. Rattlebox

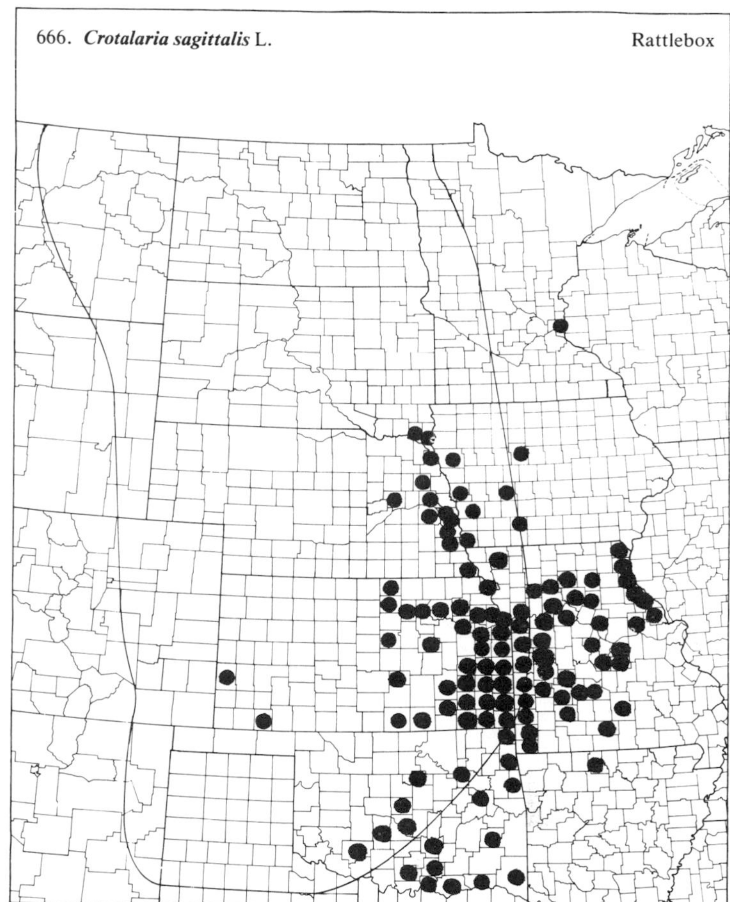

667. ***Dalea aurea*** Nutt. Silktop Dalea

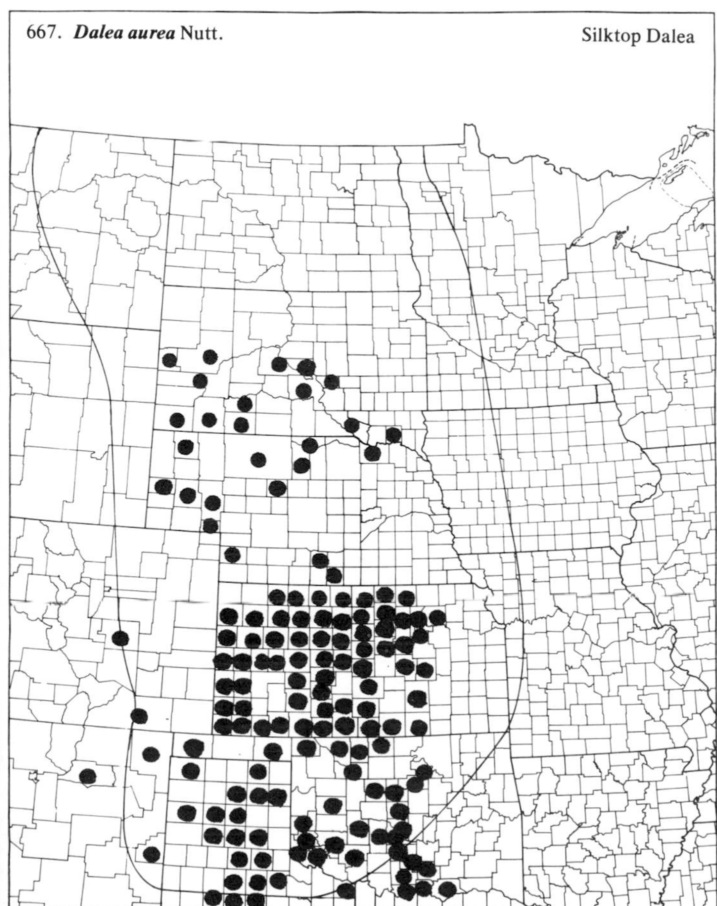

668. ***Dalea enneandra*** Nutt. Nineanther Dalea

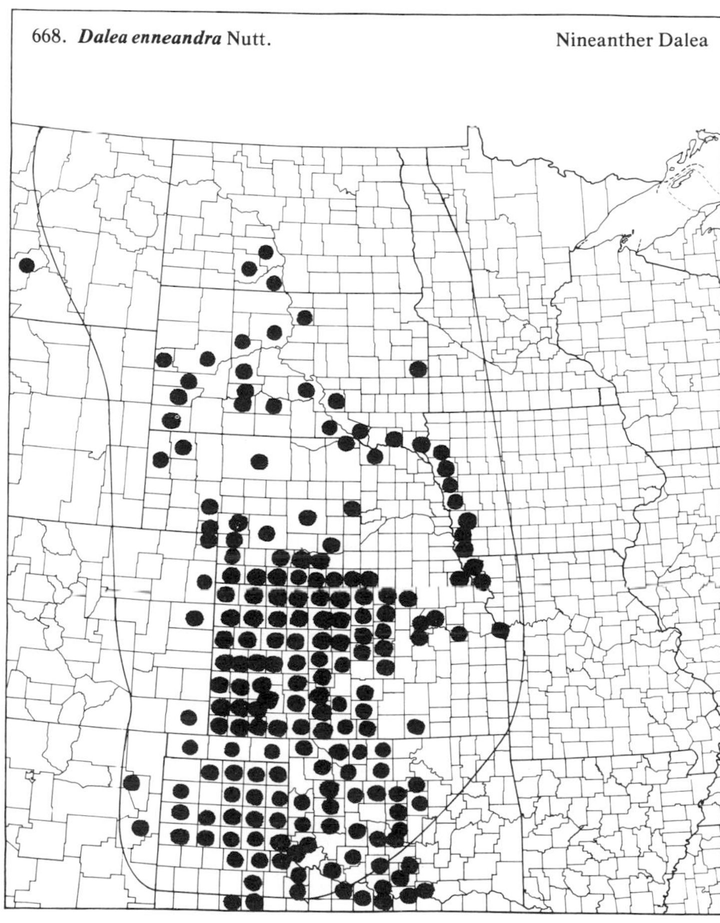

(Fabaceae-Papilionoideae)

669. **Dalea formosa** Torr.

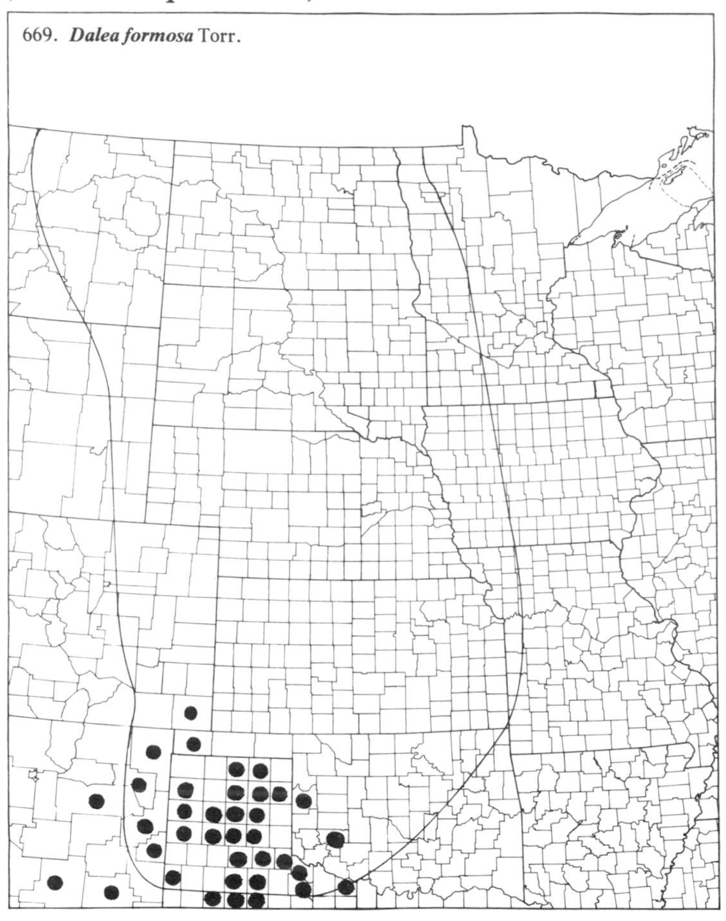

670. **Dalea jamesii** (Torr.) T. & G. James' Dalea

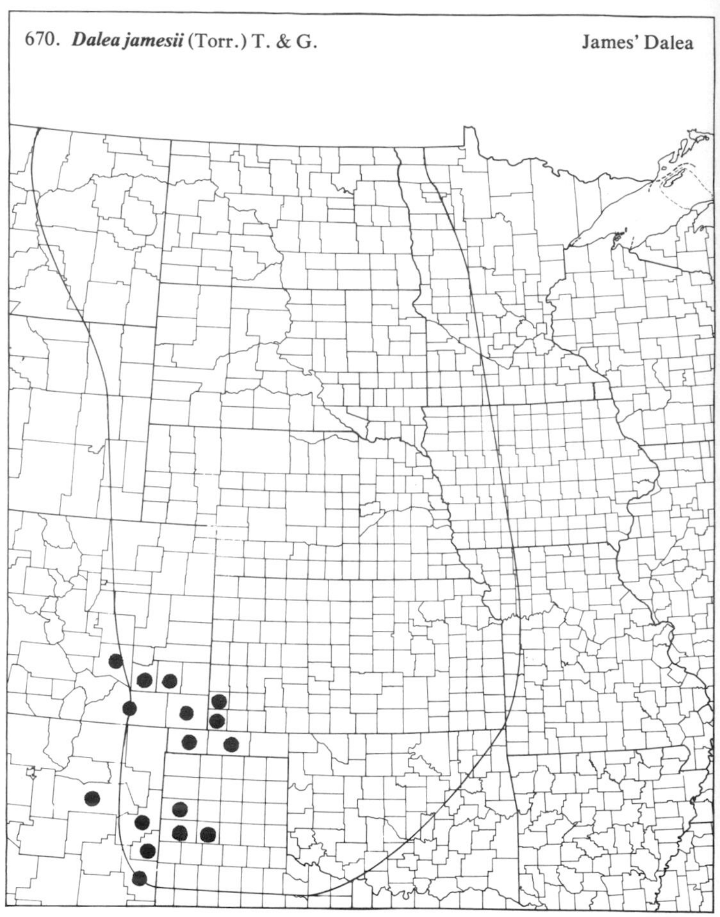

671. **Dalea lanata** Spreng. Woolly Dalea

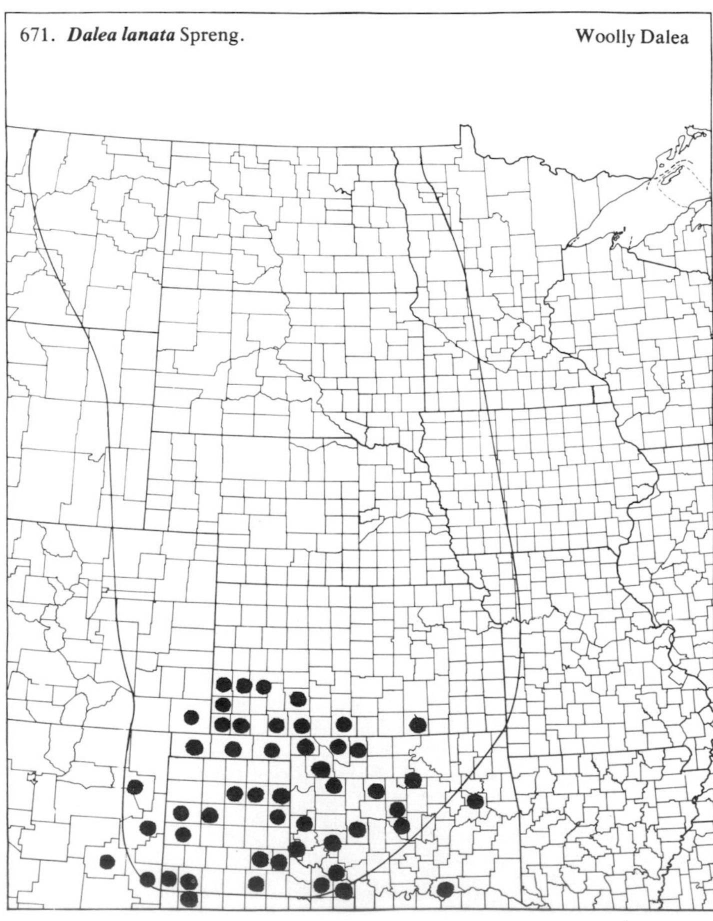

672. **Dalea leporina** (Ait.) Bullock

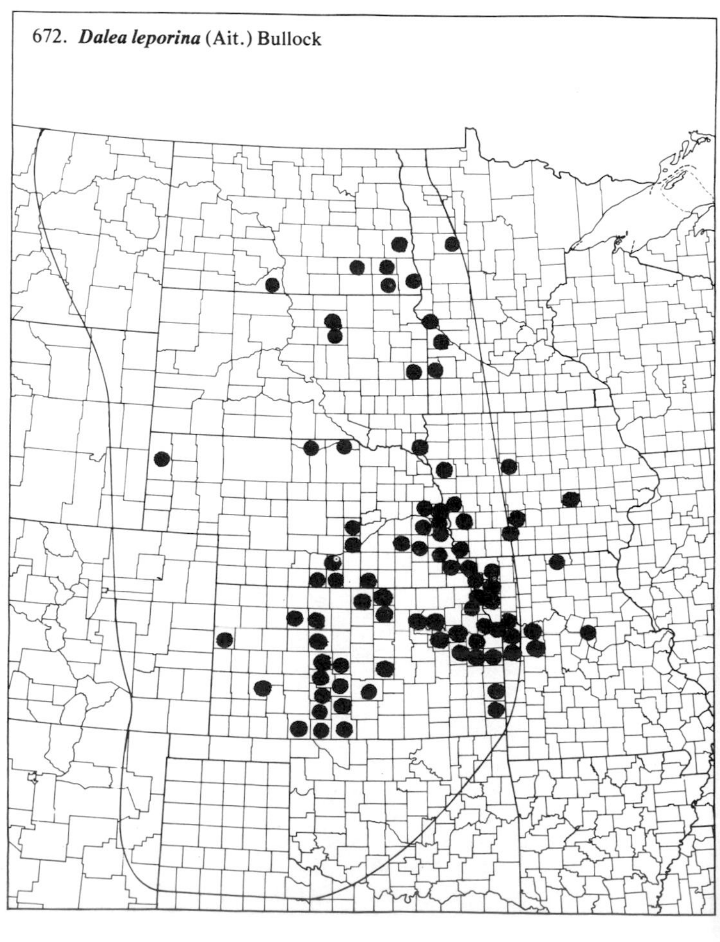

(Fabaceae-Papilionoideae)

673. ***Dalea nana*** Torr. Dwarf Dalea

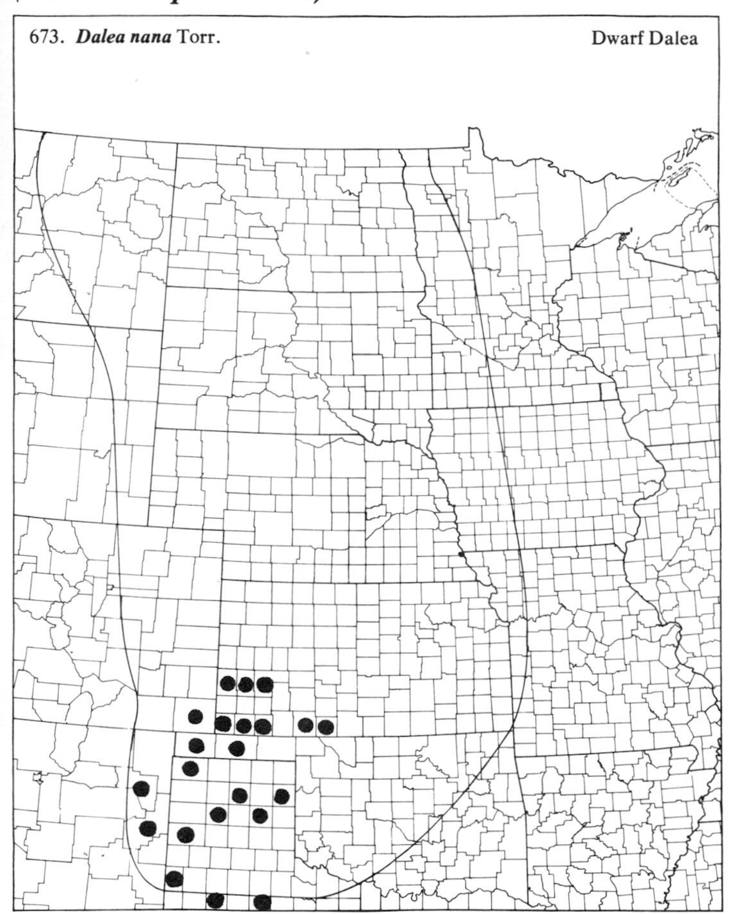

674. ***Desmodium canadense*** (L.) DC. Canada Tickclover

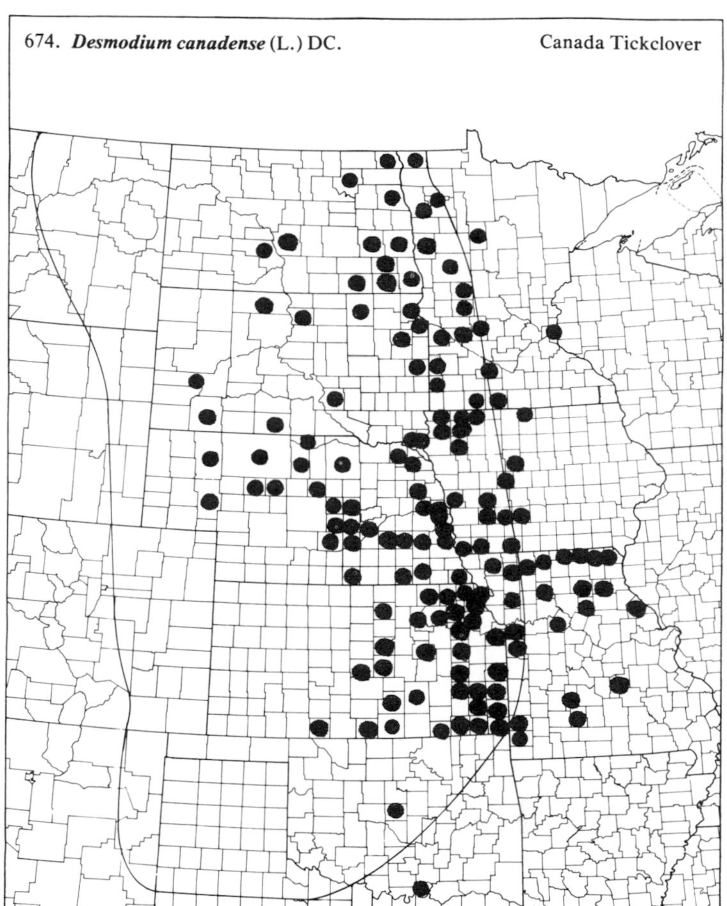

675. ***Desmodium canescens*** (L.) DC. Hoary Tickclover

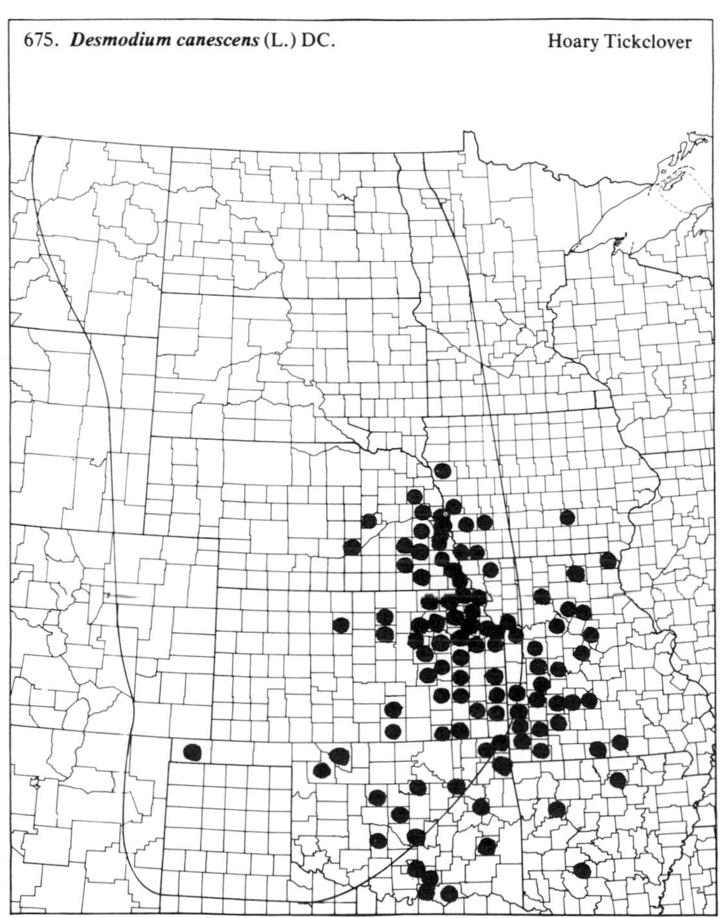

676. ***Desmodium ciliare*** (Willd.) DC.

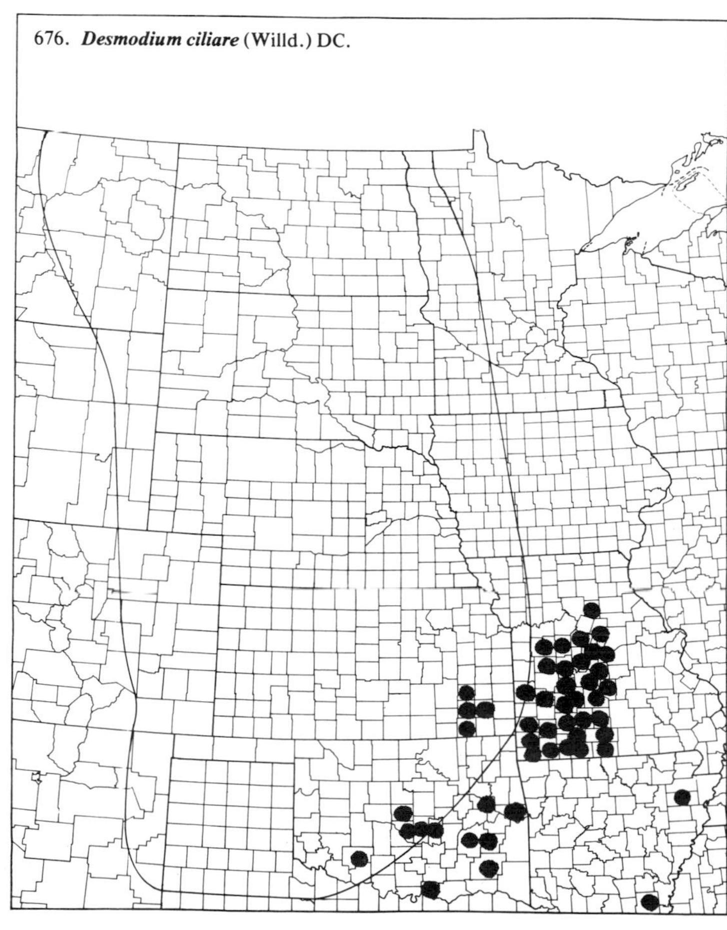

(Fabaceae-Papilionoideae)

677. **Desmodium cuspidatum** (Muhl.) Loud. — Longleaf Tickclover

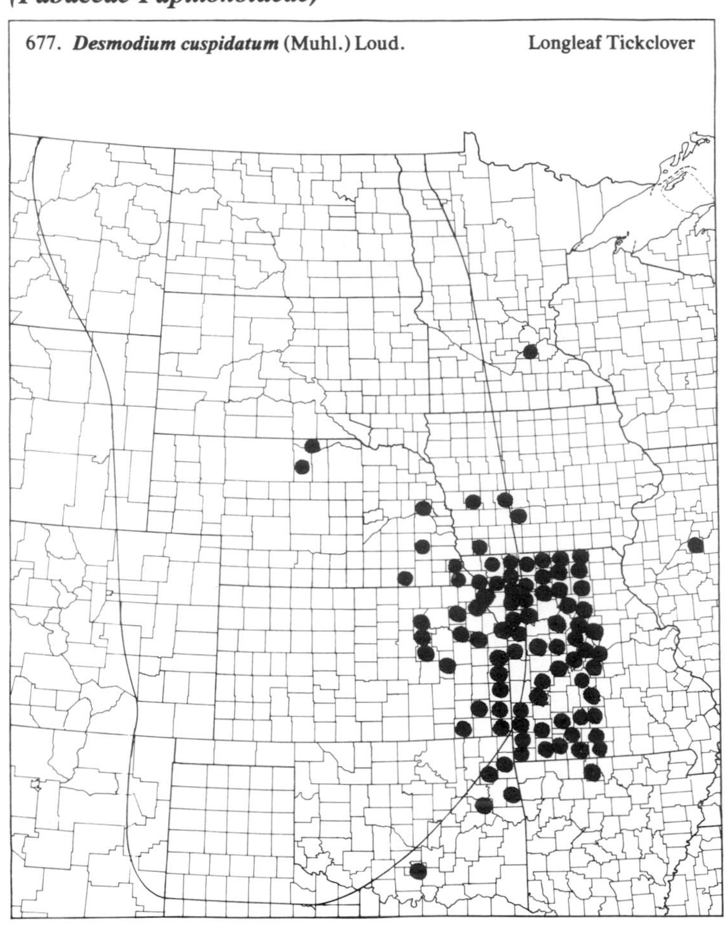

678. **Desmodium glutinosum** (Muhl.) Wood — Large-flowered Tickclover

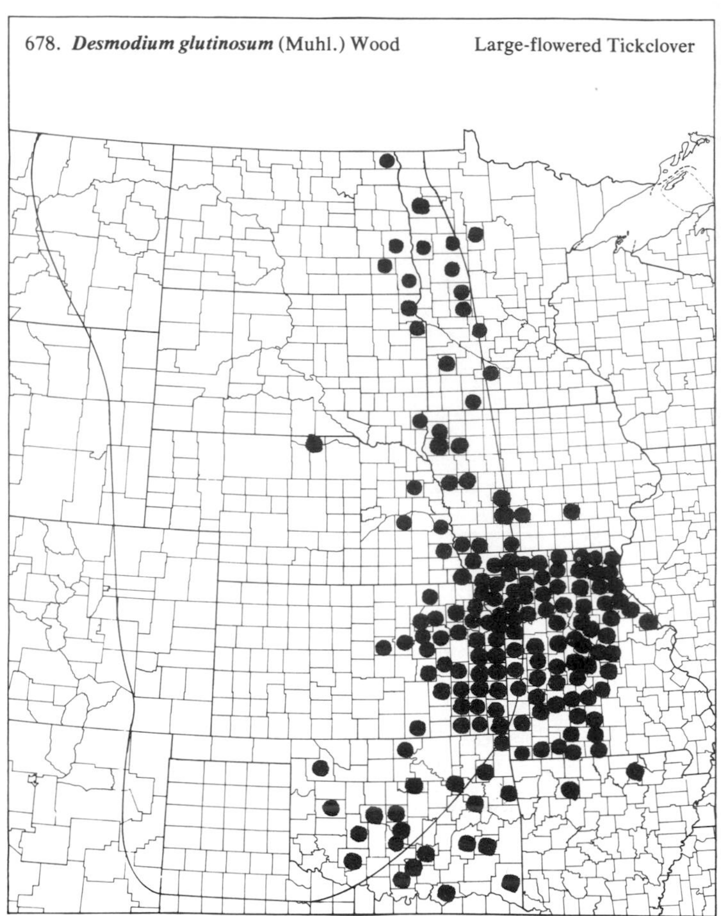

679. **Desmodium illinoense** Gray — Illinois Tickclover

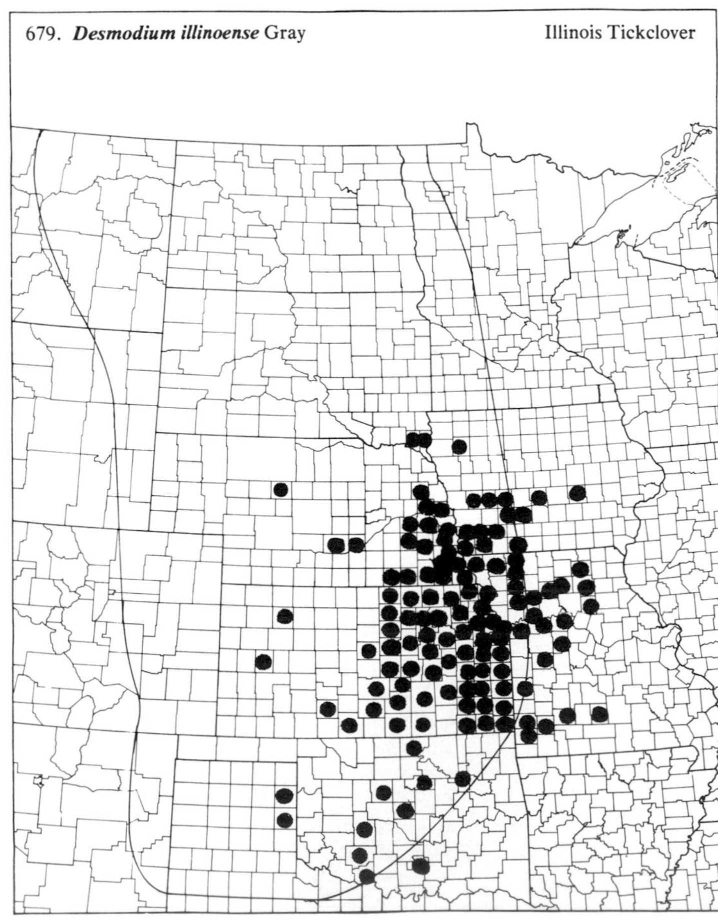

680. **Desmodium paniculatum** (L.) DC. — Panicled Tickclover

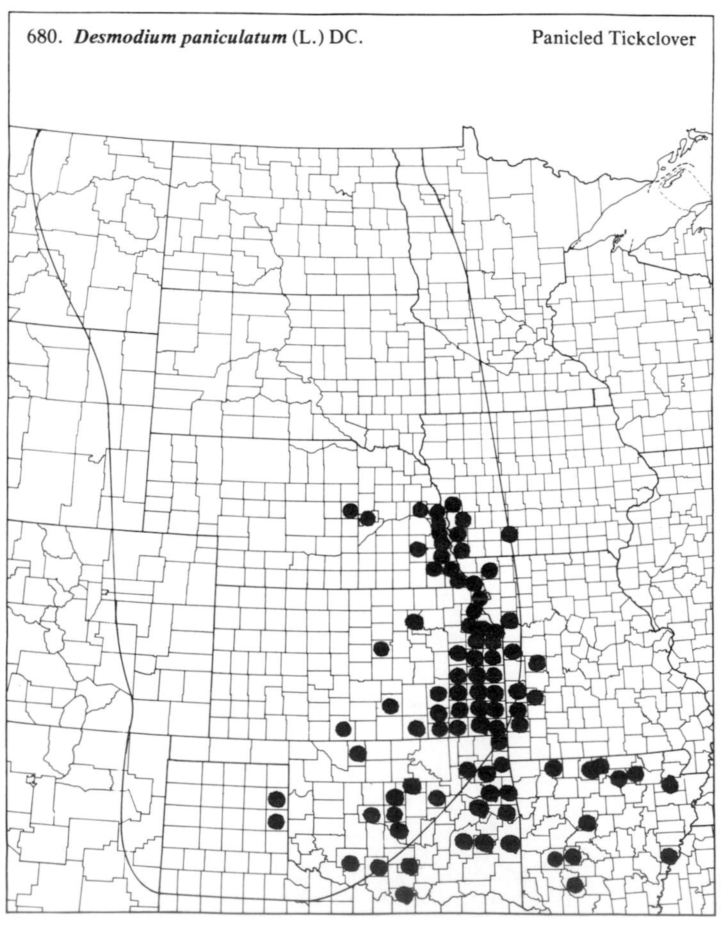

(Fabaceae-Papilionoideae)

681. ***Desmodium sessilifolium*** (Torr.) T. & G. Sessile Tickclover

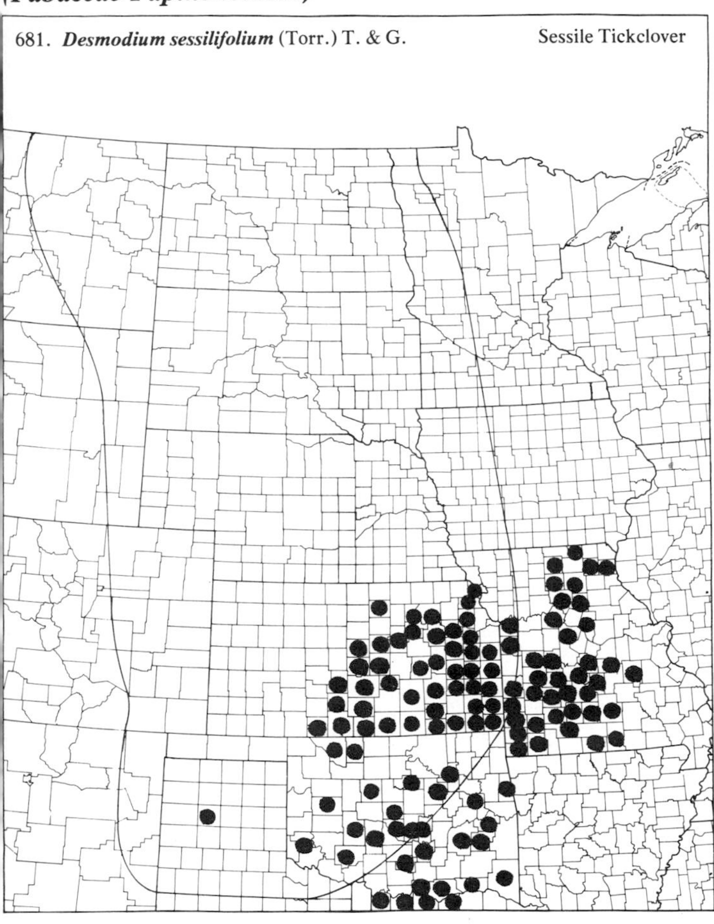

682. ***Glycyrrhiza lepidota*** Pursh Wild Licorice

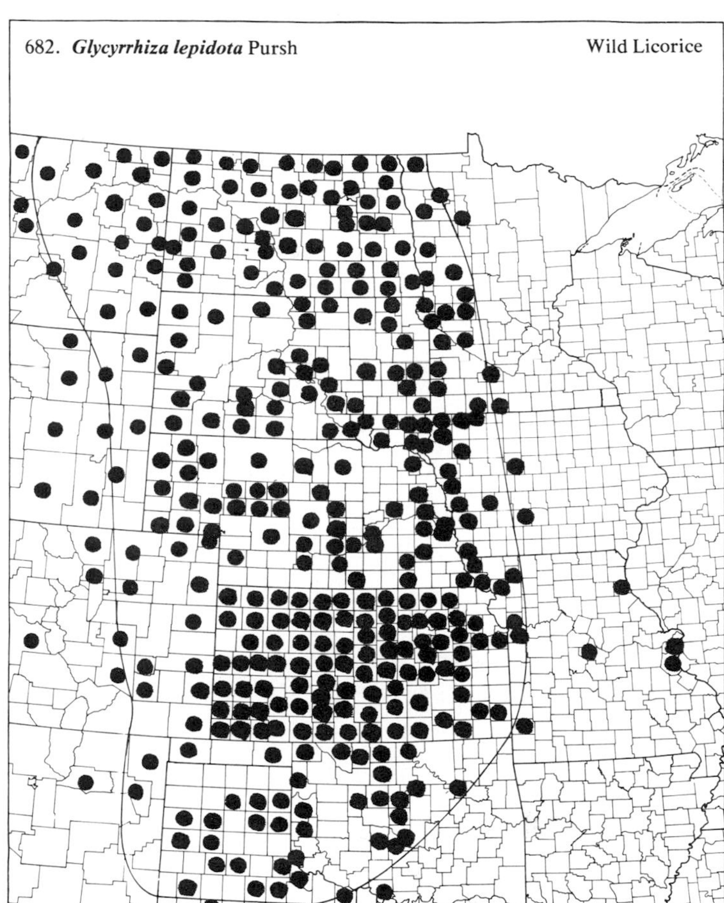

683. ***Hedysarum boreale*** Nutt. Sweet Vetch

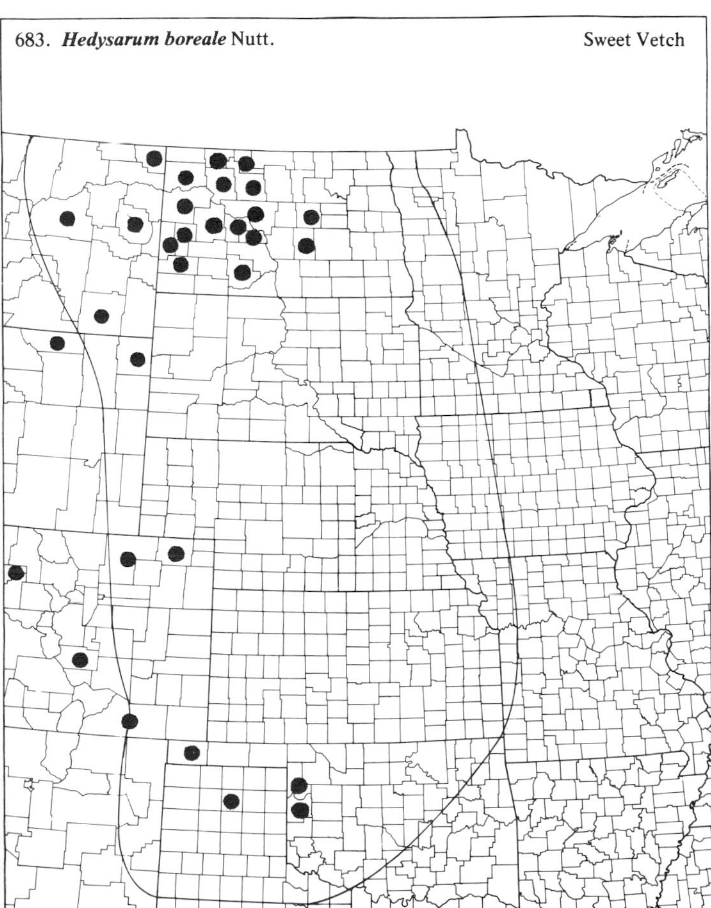

684. ***Indigofera miniata*** Ort. var. ***leptosepala*** (Nutt.) B.L. Turner

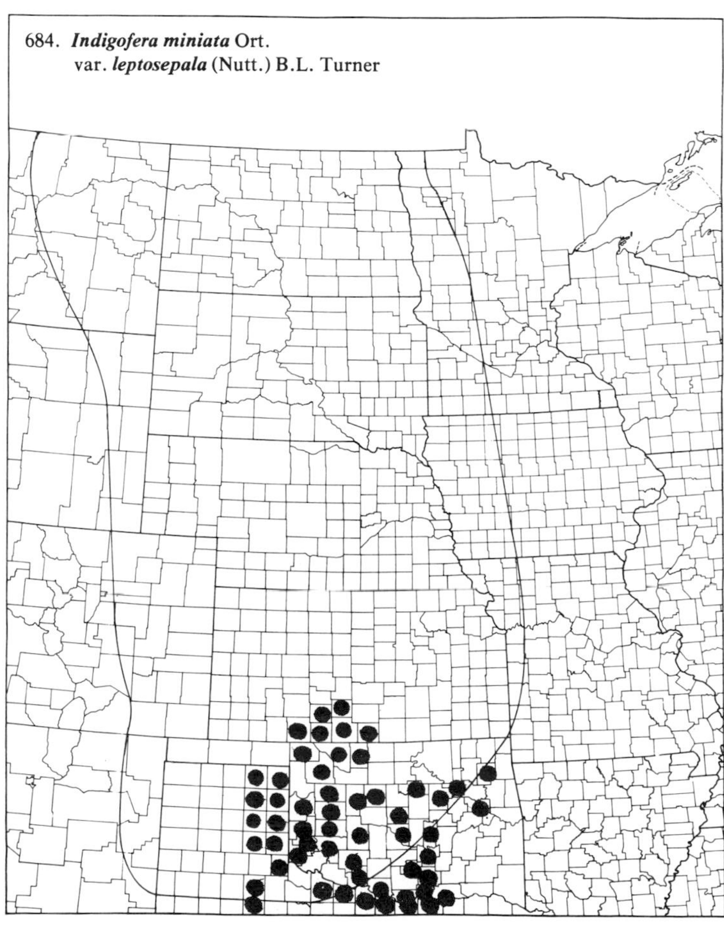

(Fabaceae-Papilionoideae)

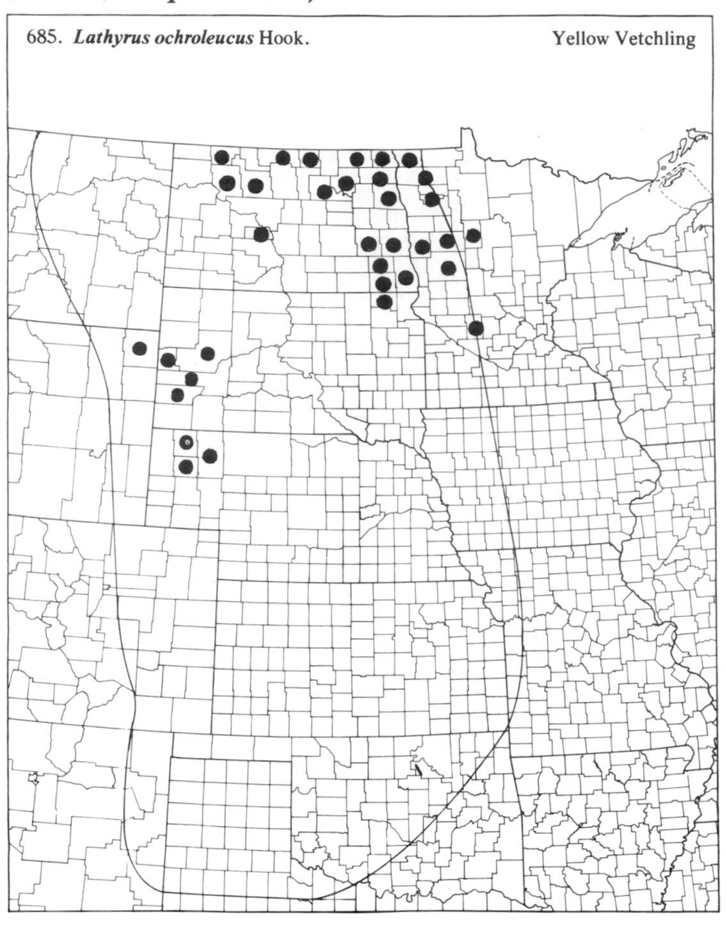

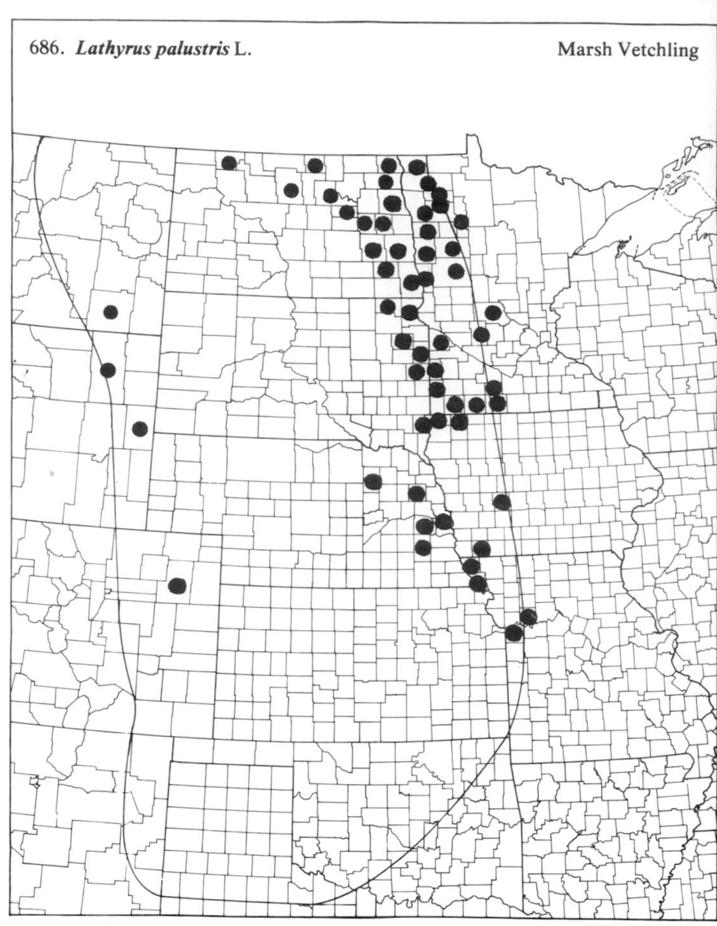

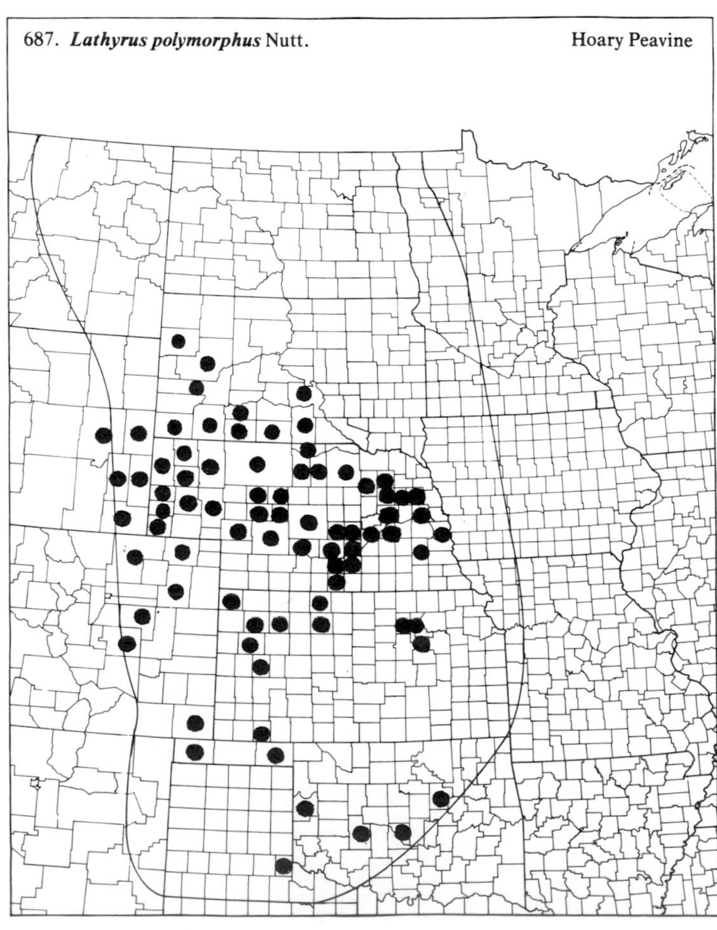

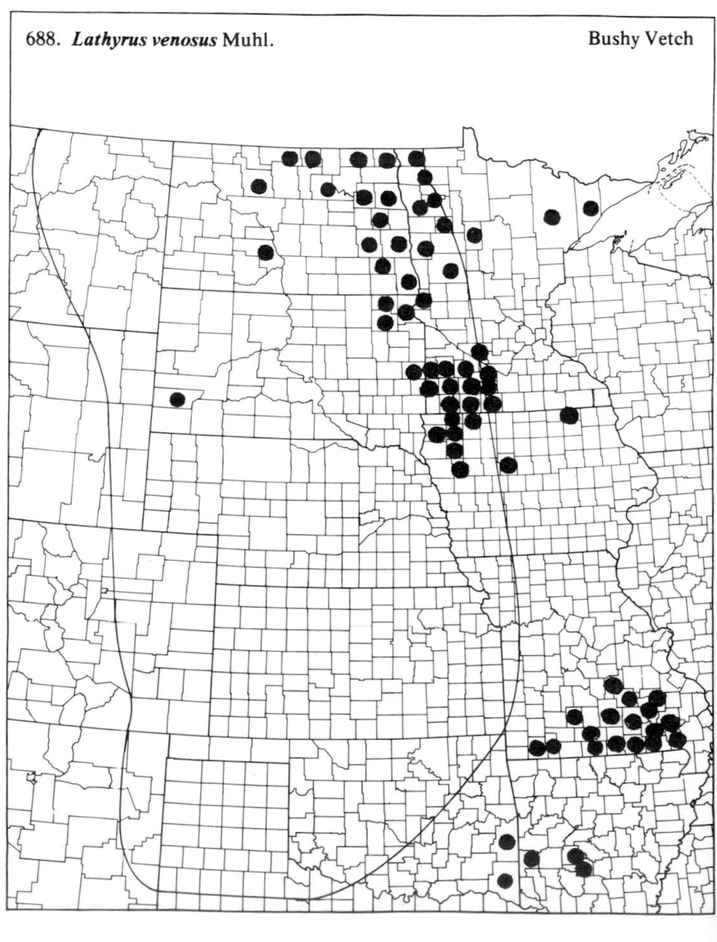

(Fabaceae-Papilionoideae)

689. ***Lespedeza capitata*** Michx. Bush Clover

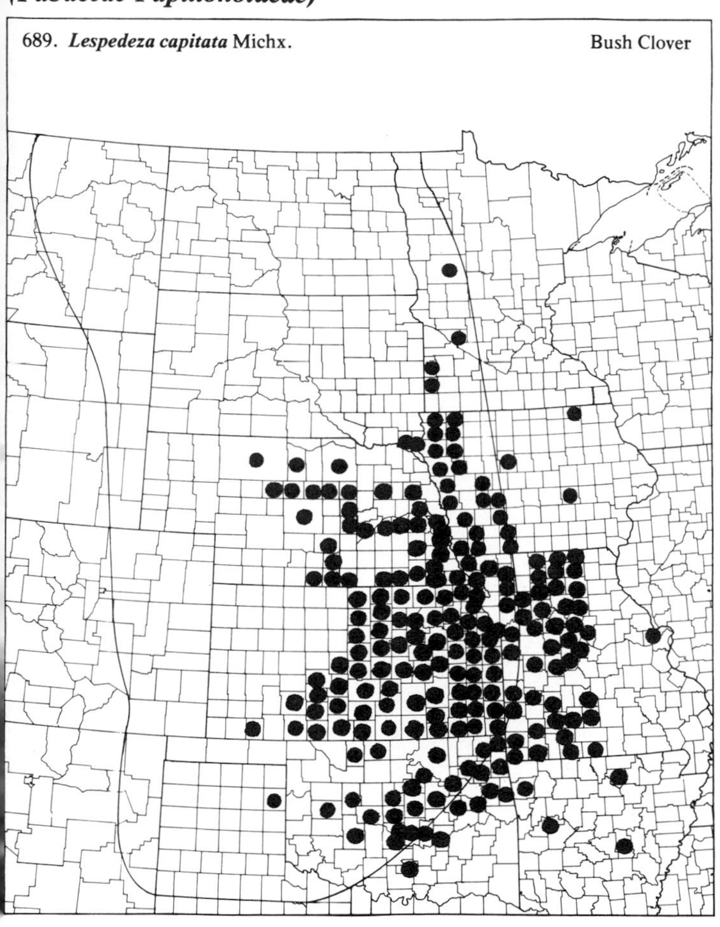

690. ***Lespedeza cuneata*** (Dumont) G. Don Sericea Lespedeza

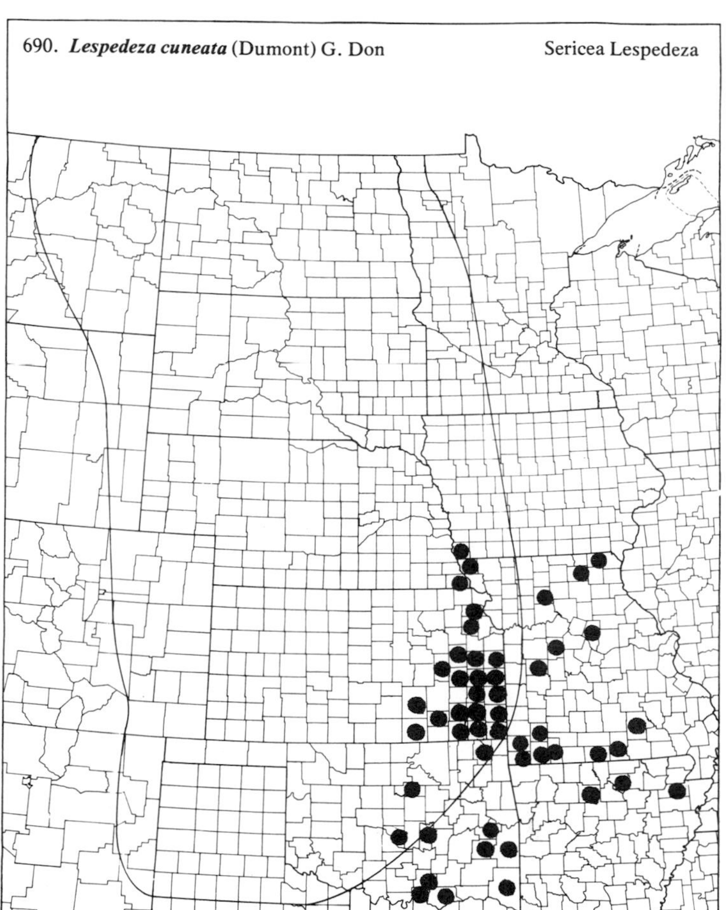

691. ***Lespedeza procumbens*** Michx. Trailing Lespedeza

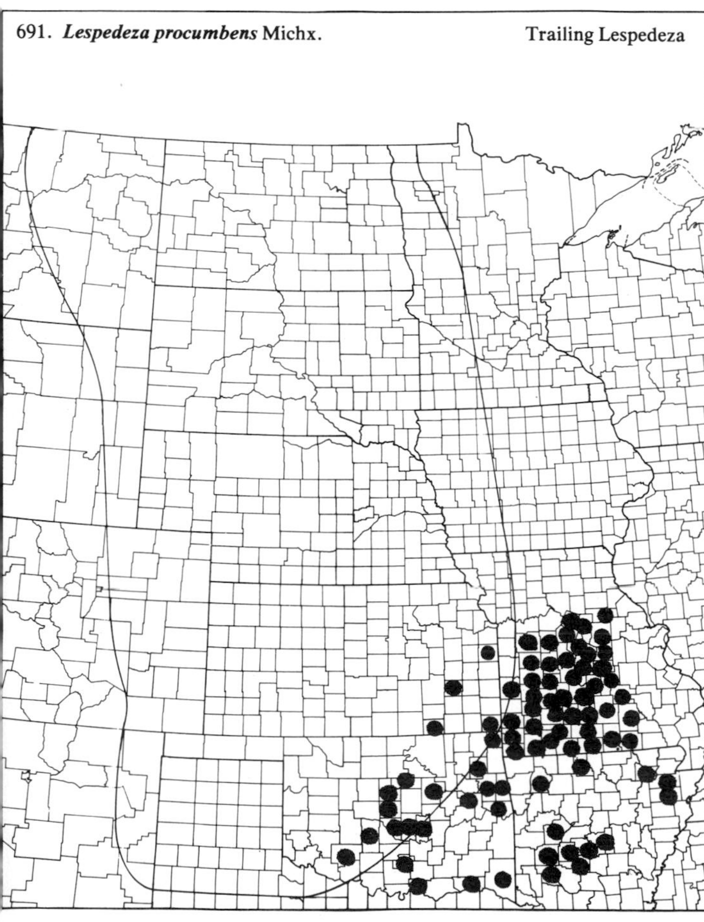

692. ***Lespedeza repens*** (L.) Bart. Creeping Lespedeza

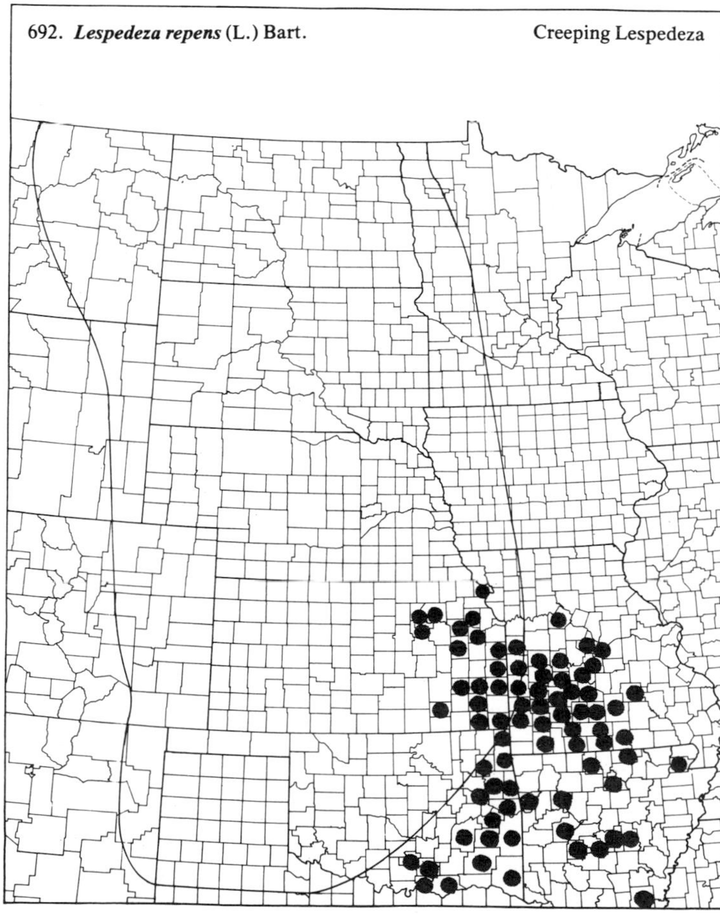

(Fabaceae-Papilionoideae)

693. *Lespedeza stipulacea* Maxim. Korean Clover

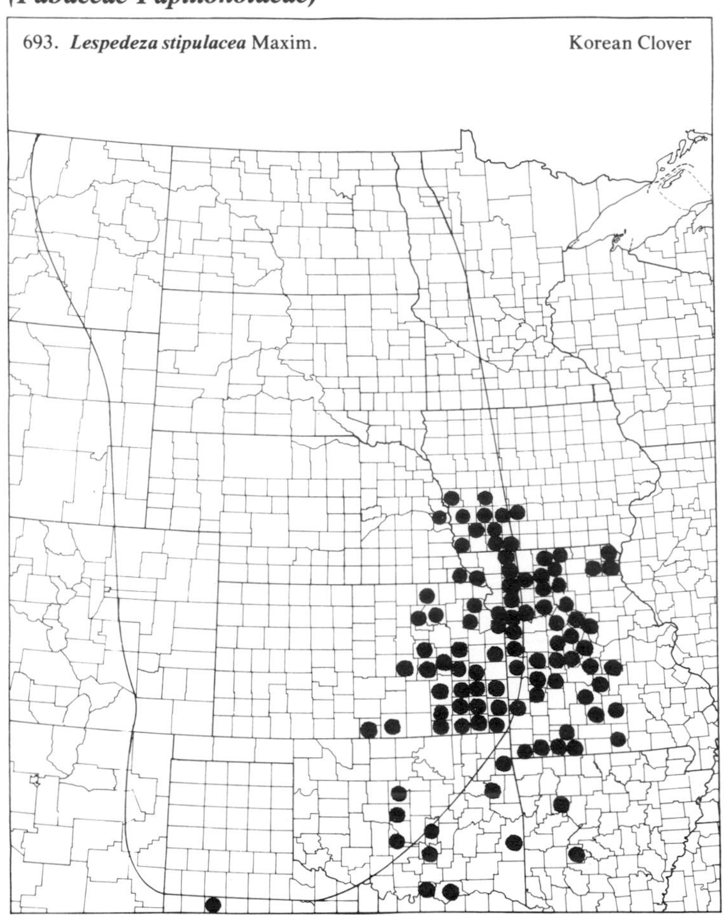

694. *Lespedeza striata* (Thunb.) H. & A. Japanese Clover

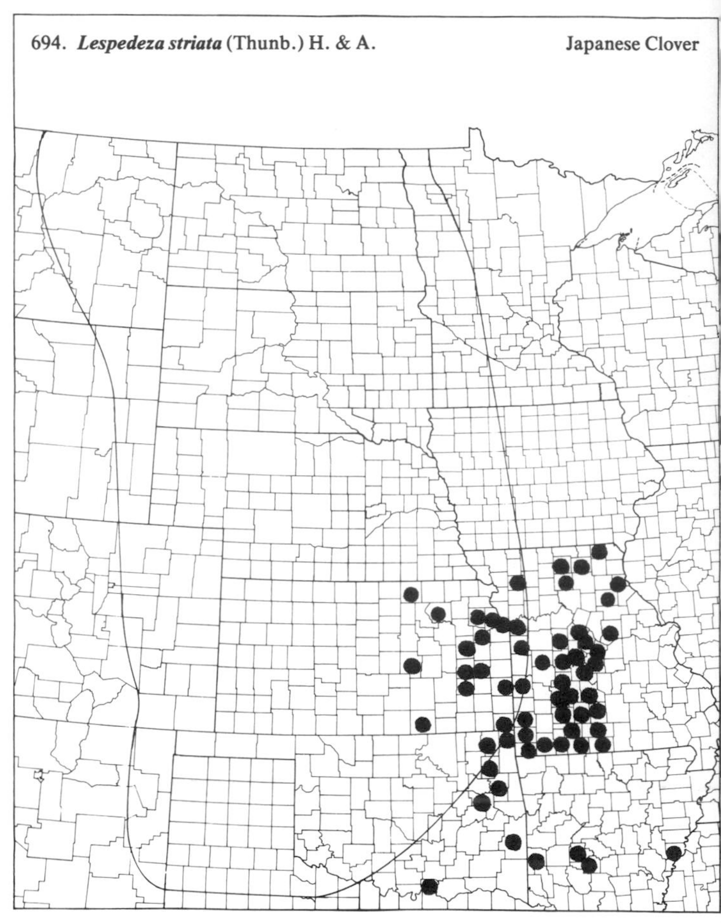

695. *Lespedeza stuevei* Nutt. Stuve's Lespedeza

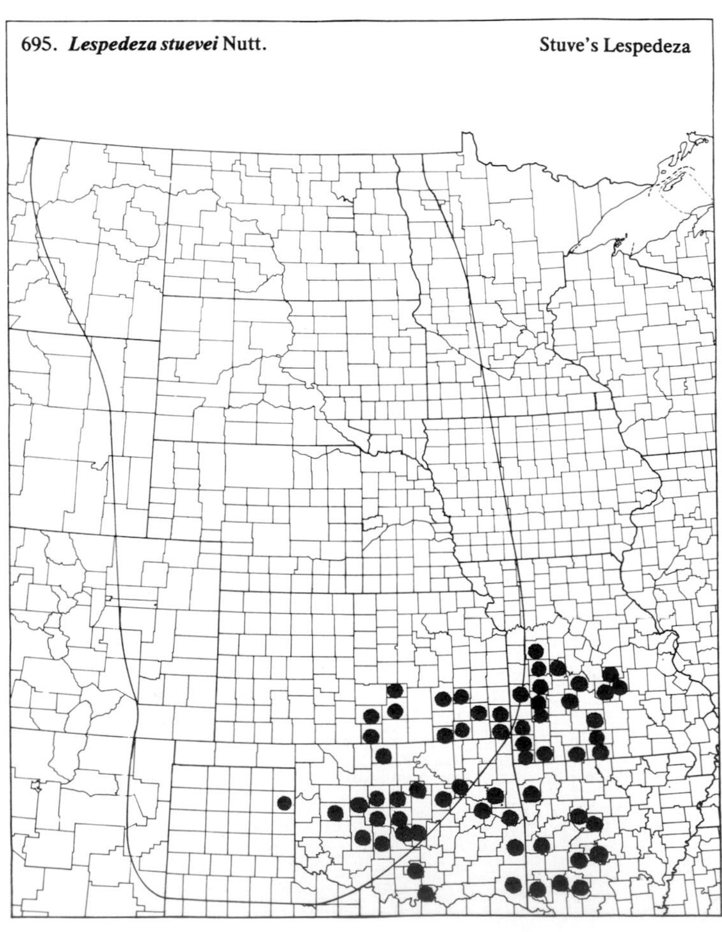

696. *Lespedeza violacea* (L.) Pers. Violet Lespedeza

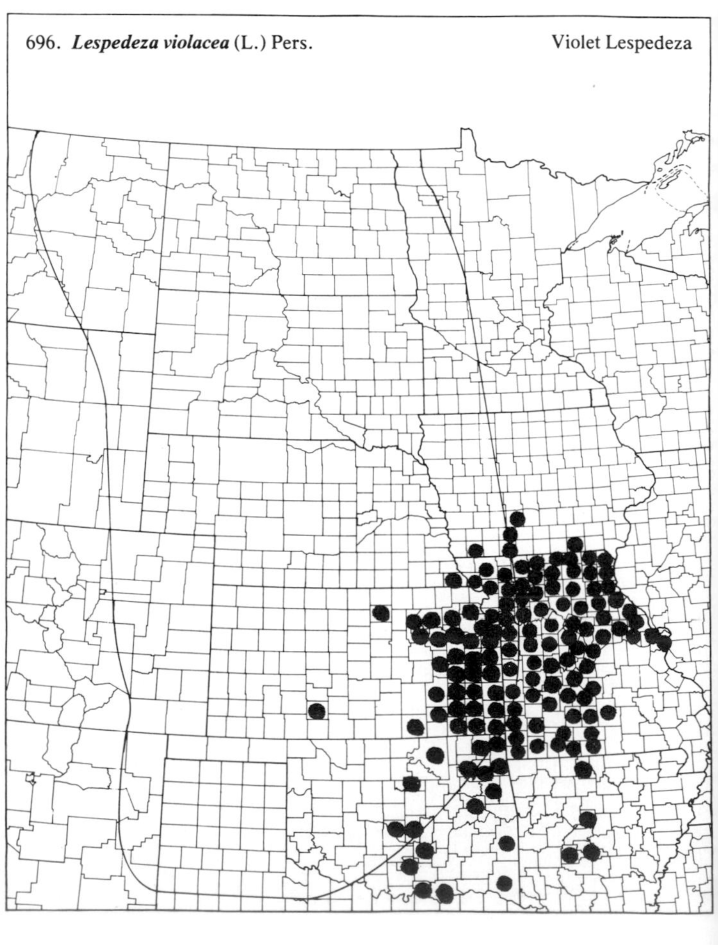

(Fabaceae-Papilionoideae)

697. **Lespedeza virginica** (L.) Britt. — Slender Lespedeza

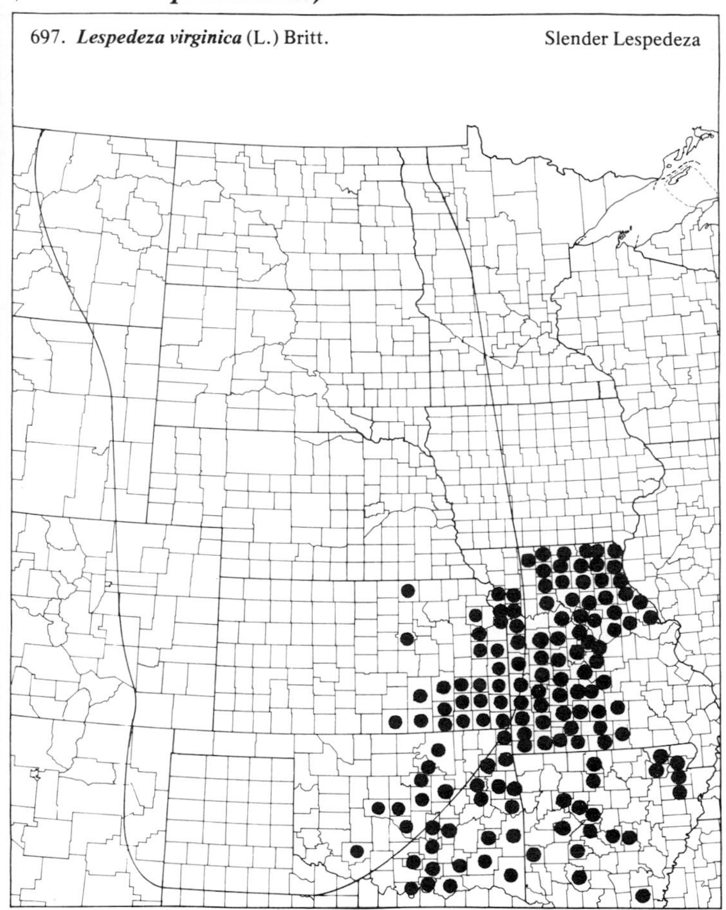

698. **Lotus corniculatus** L. — Bird's-foot Trefoil

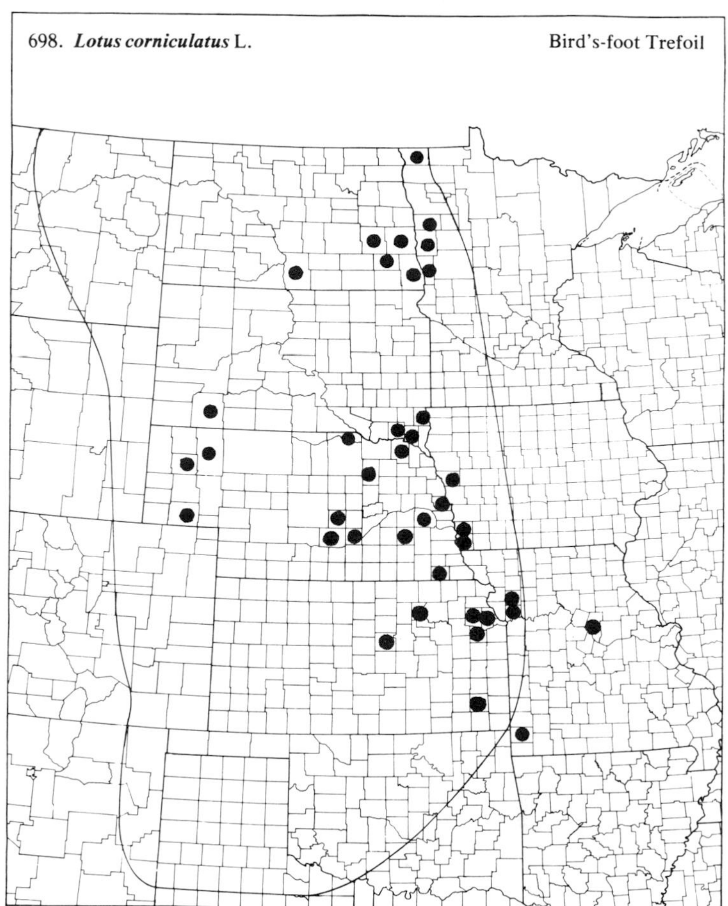

699. **Lotus purshianus** Clem. & Clem.

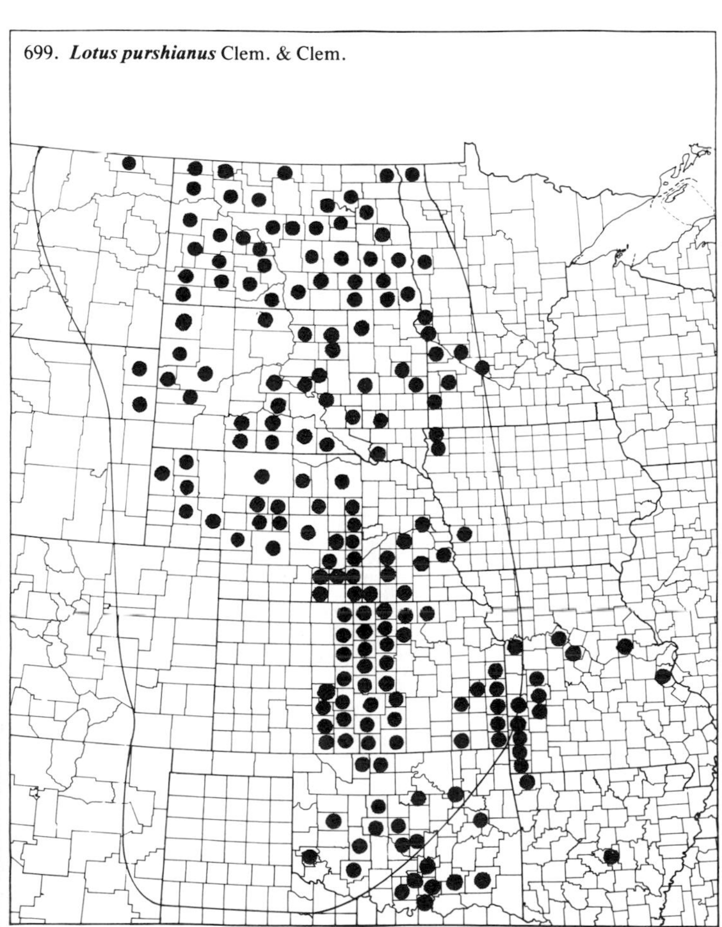

700. **Lupinus argenteus** Pursh var. **argenteus** — Silvery Lupine

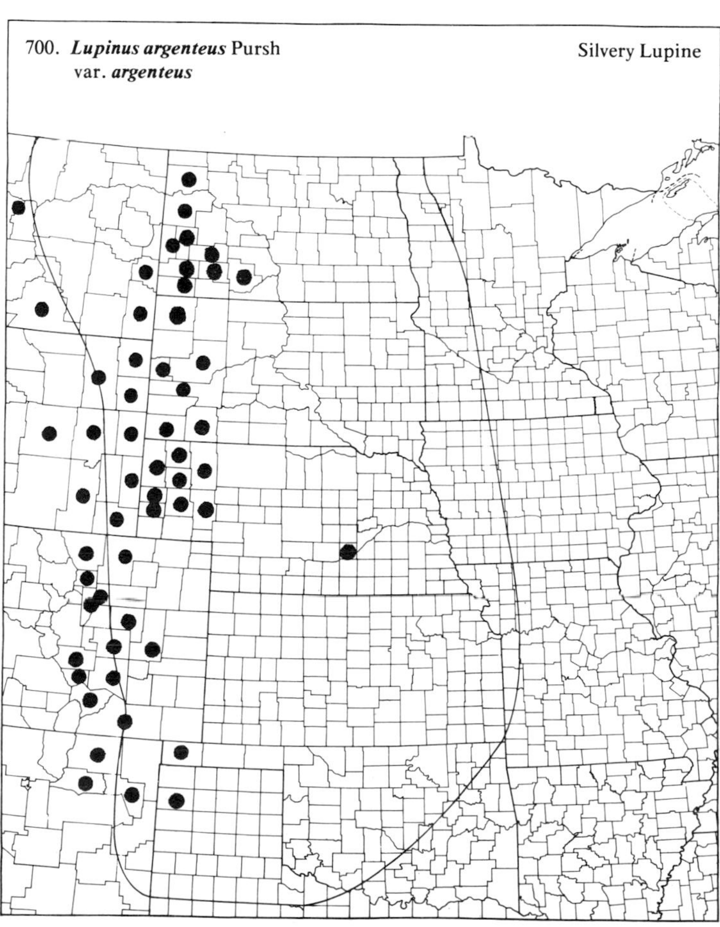

(Fabaceae-Papilionoideae)

701. ***Lupinus argenteus*** Pursh var. ***parviflorus*** (Nutt.) C.L. Hitchc. — Silvery Lupine

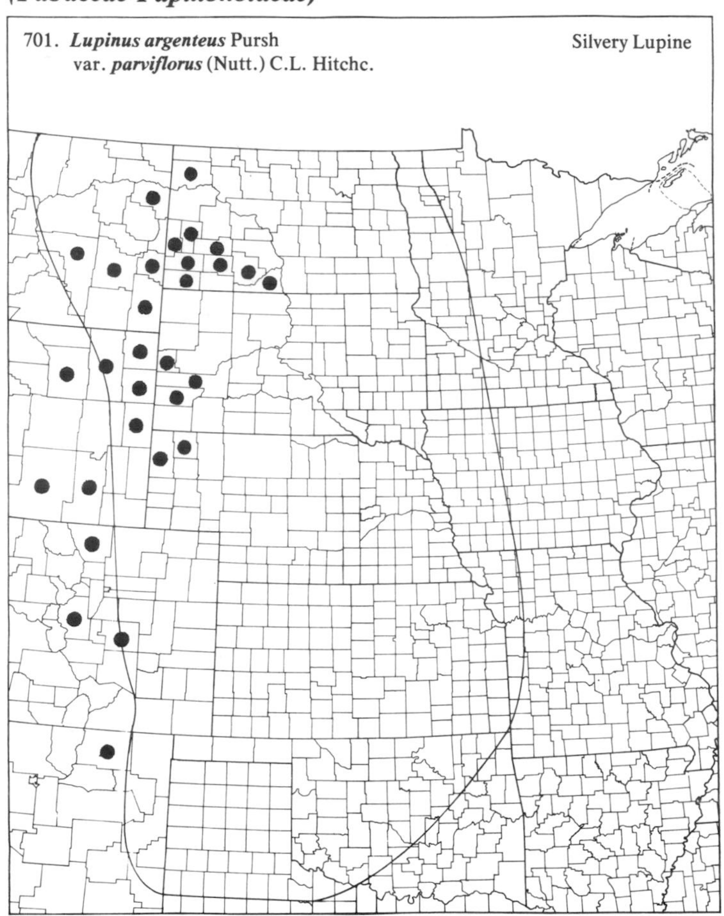

702. ***Lupinus pusillus*** Pursh — Small Lupine

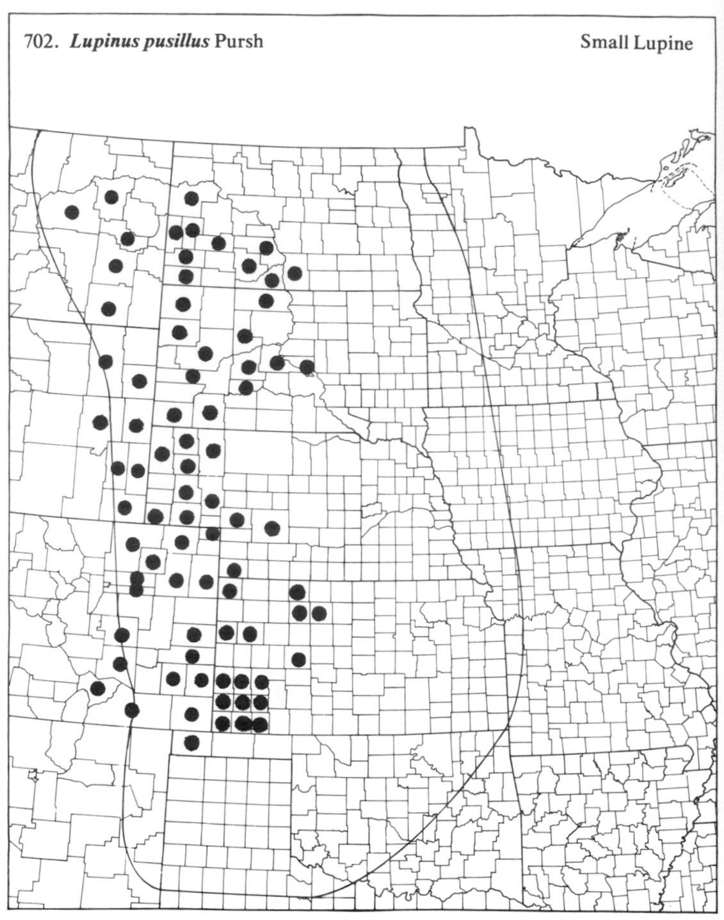

703. ***Medicago falcata*** L. — Yellow Alfalfa

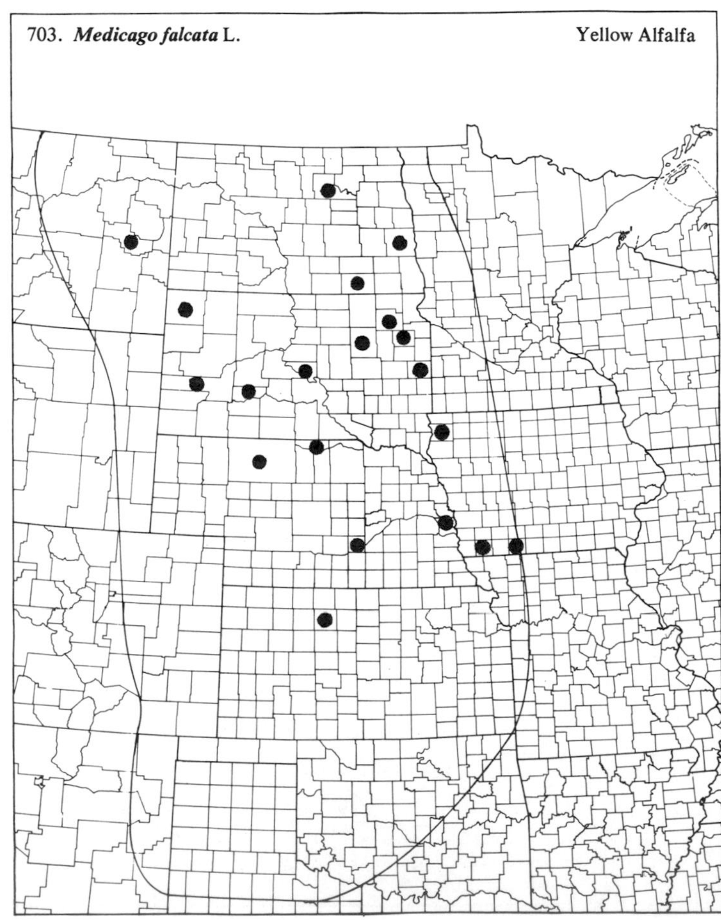

704. ***Medicago lupulina*** L. — Black Medick

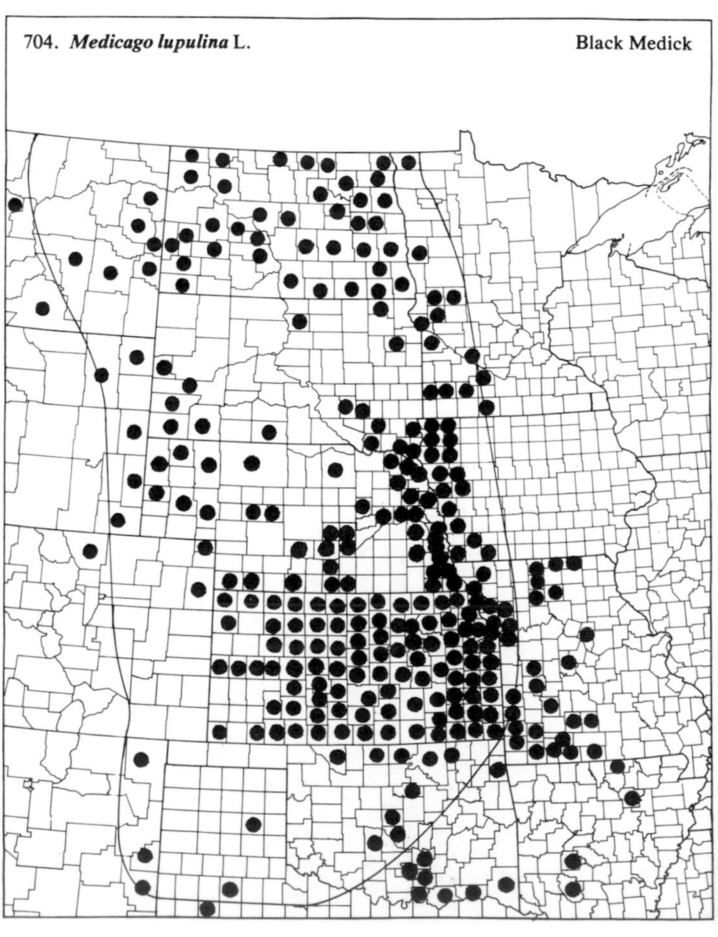

(Fabaceae-Papilionoideae)

705. ***Medicago sativa*** L. Alfalfa

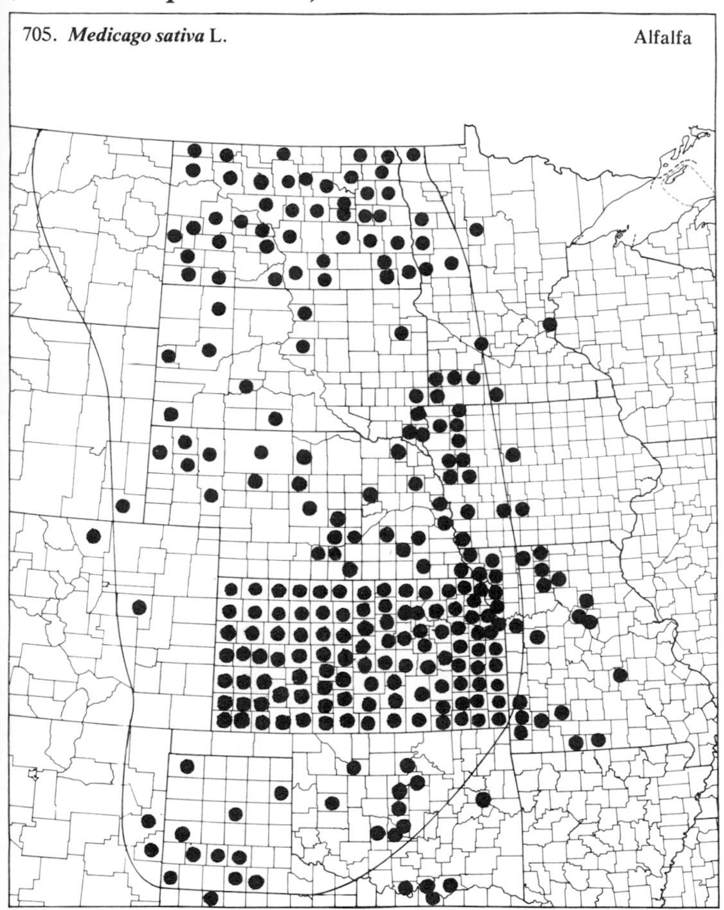

706. ***Melilotus albus*** Desr. White Sweet Clover

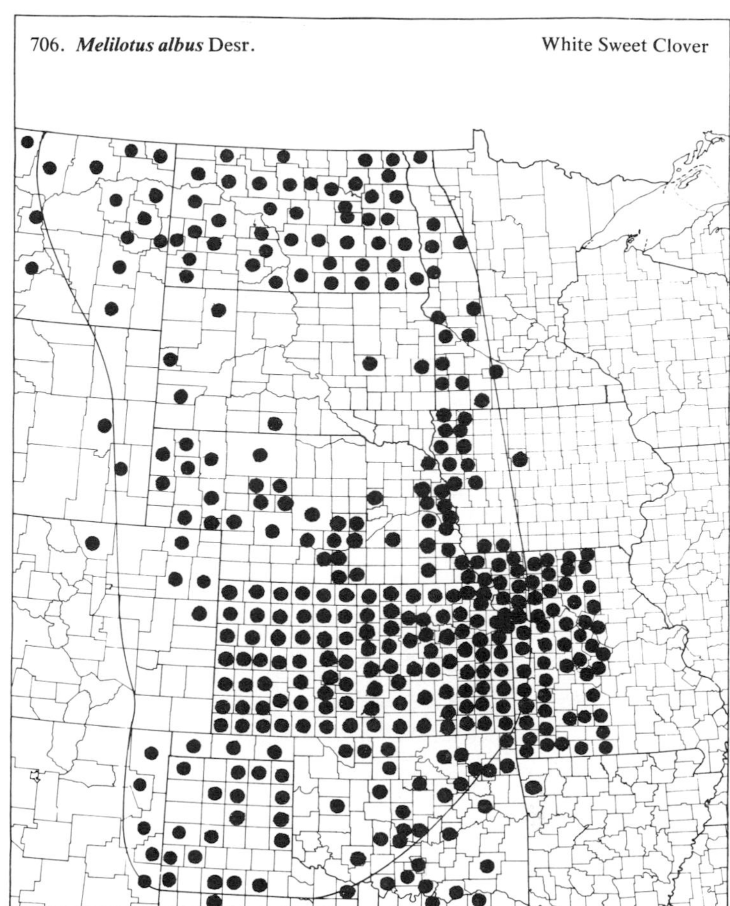

707. ***Melilotus officinalis*** (L.) Lam. Yellow Sweet Clover

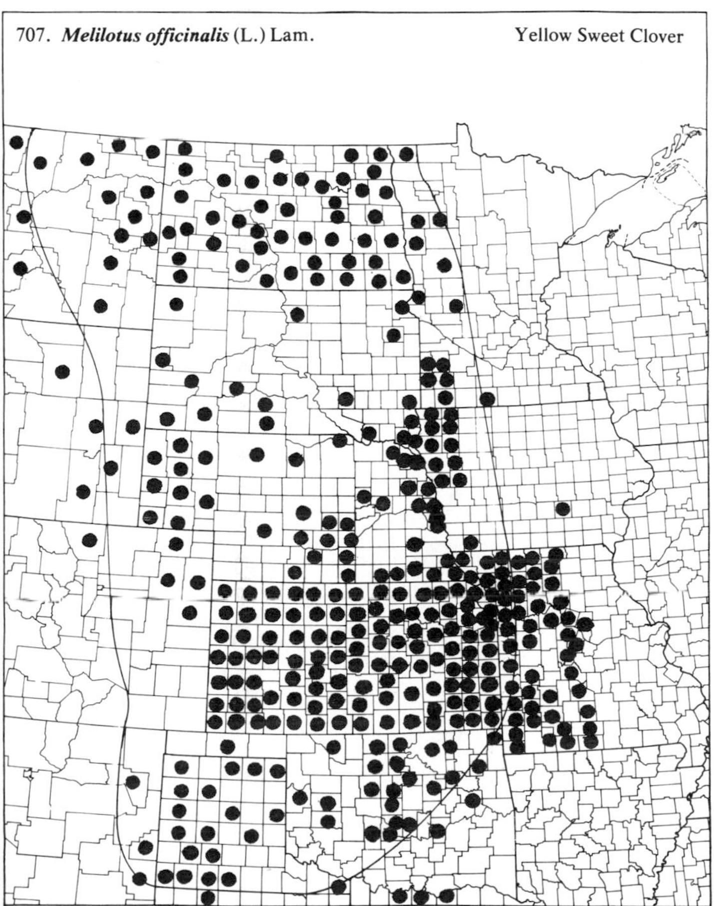

708. ***Oxytropis campestris*** (L.) DC. var. ***gracilis*** (A. Nels.) Barneby

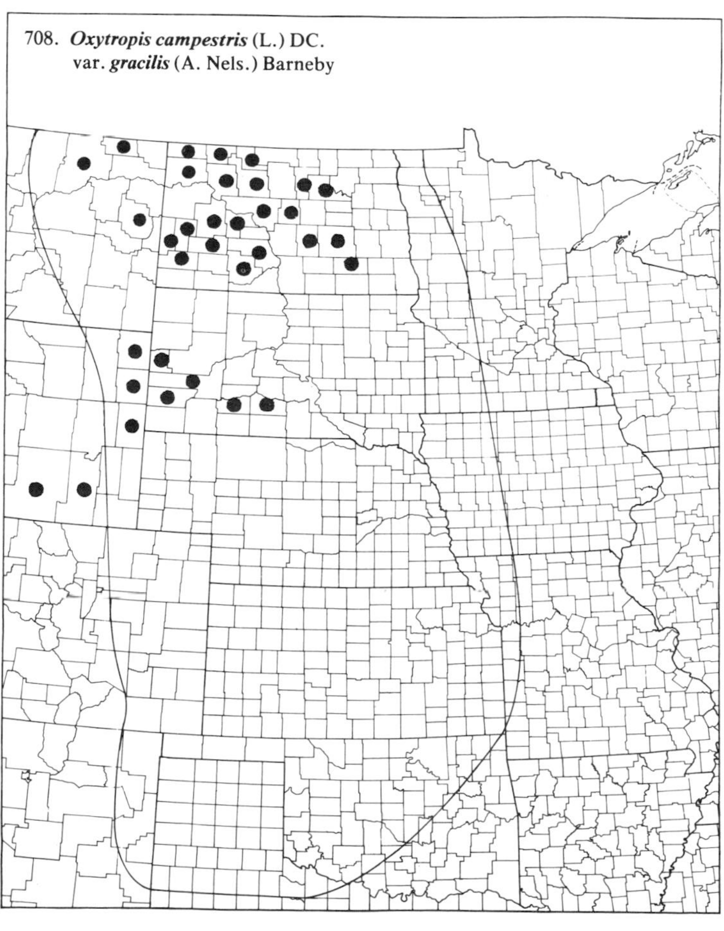

(Fabaceae-Papilionoideae)

709. *Oxytropis lambertii* Pursh — Purple Locoweed

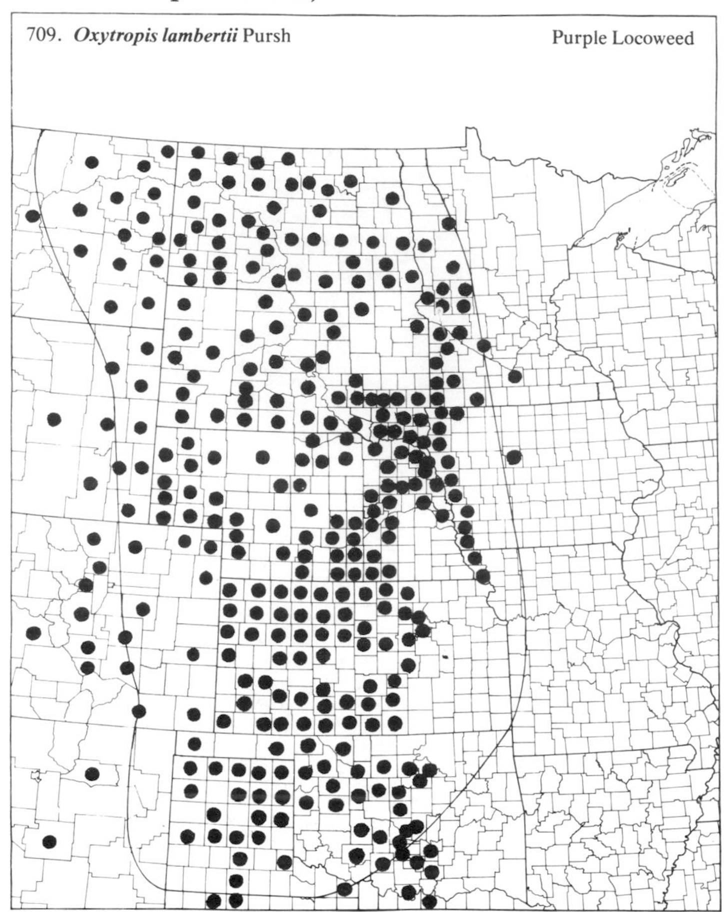

710. *Oxytropis multiceps* Nutt.

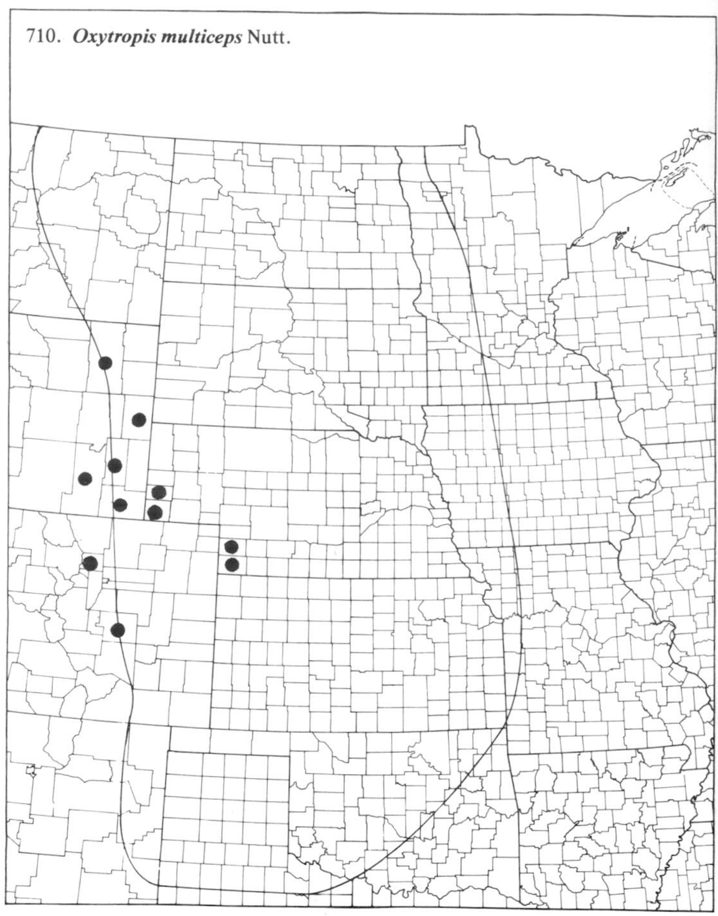

711. *Oxytropis sericea* Nutt. — White Locoweed

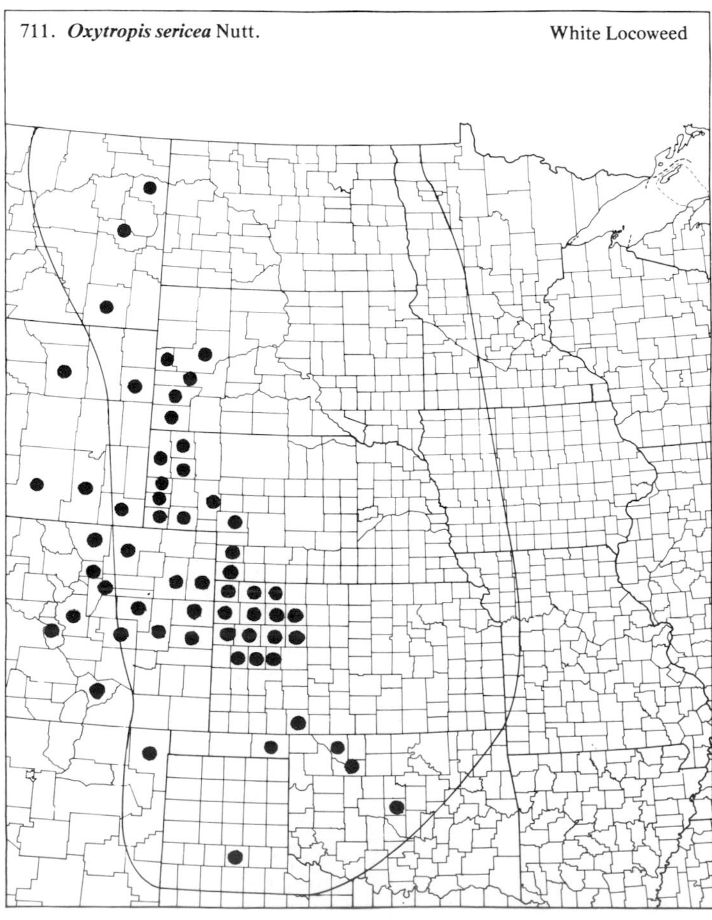

712. *Oxytropis splendens* Dougl. — Showy Locoweed

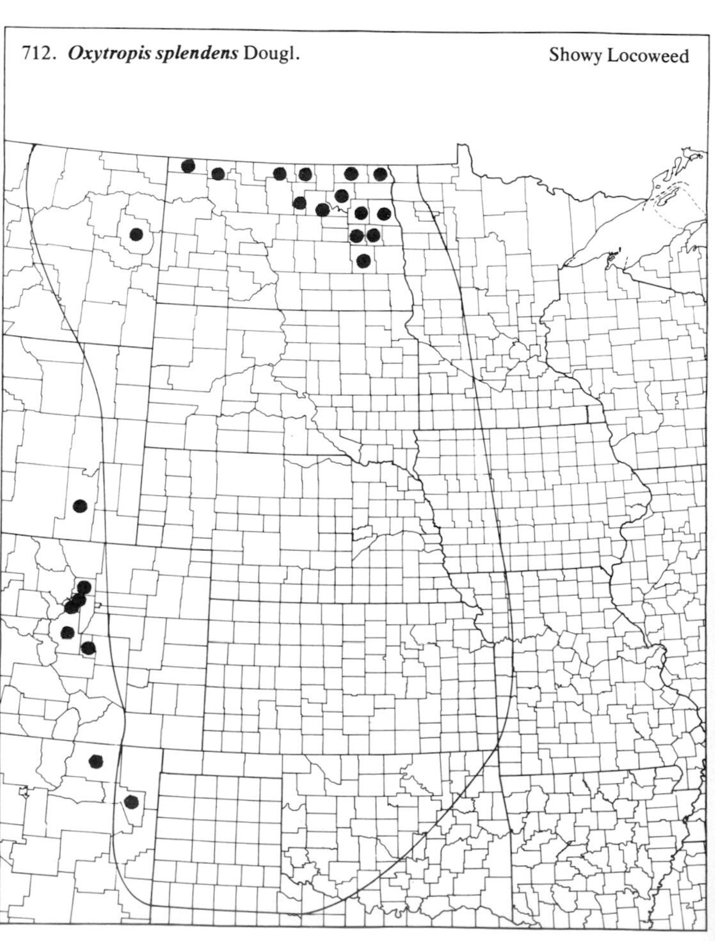

(Fabaceae-Papilionoideae)

713. **Petalostemon arenicola** Wemple

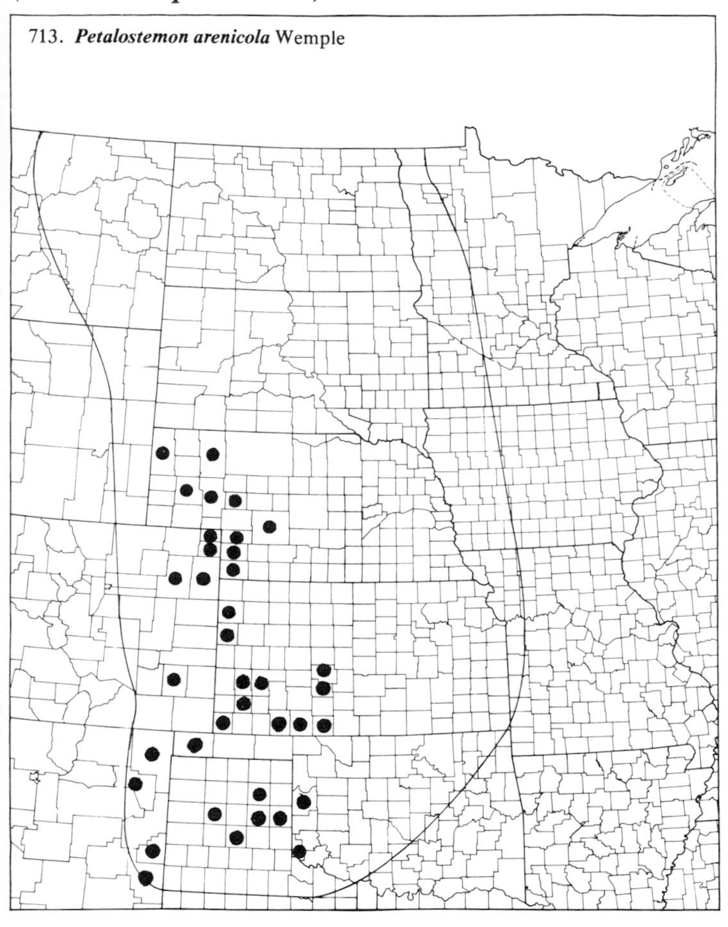

714. **Petalostemon candidum** (Willd.) Michx. White Prairie Clover

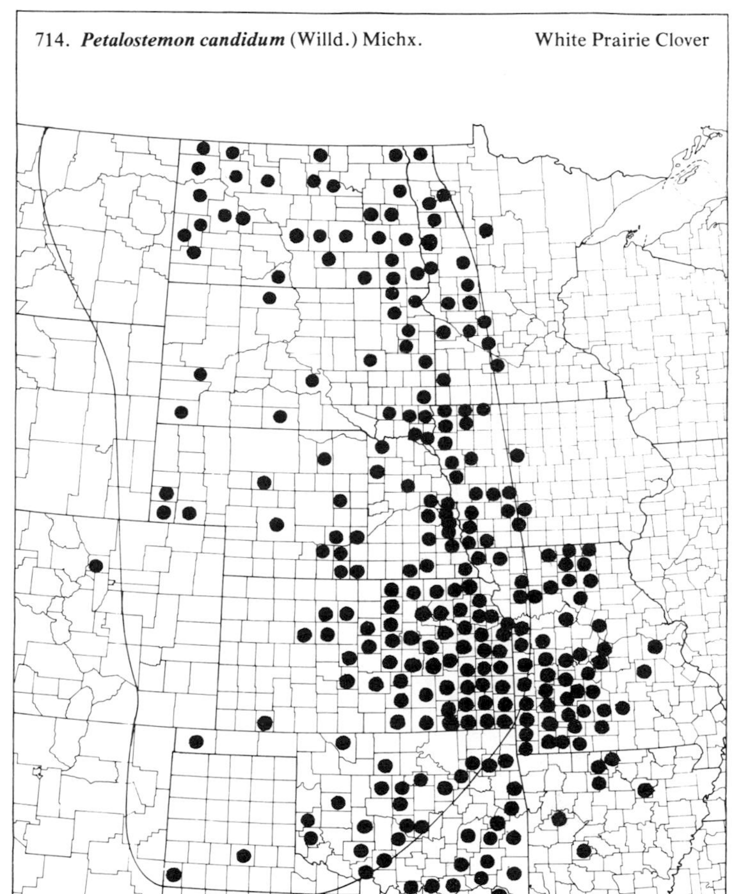

715. **Petalostemon compactum** (Spreng.) Swezey Compact Prairie Clover

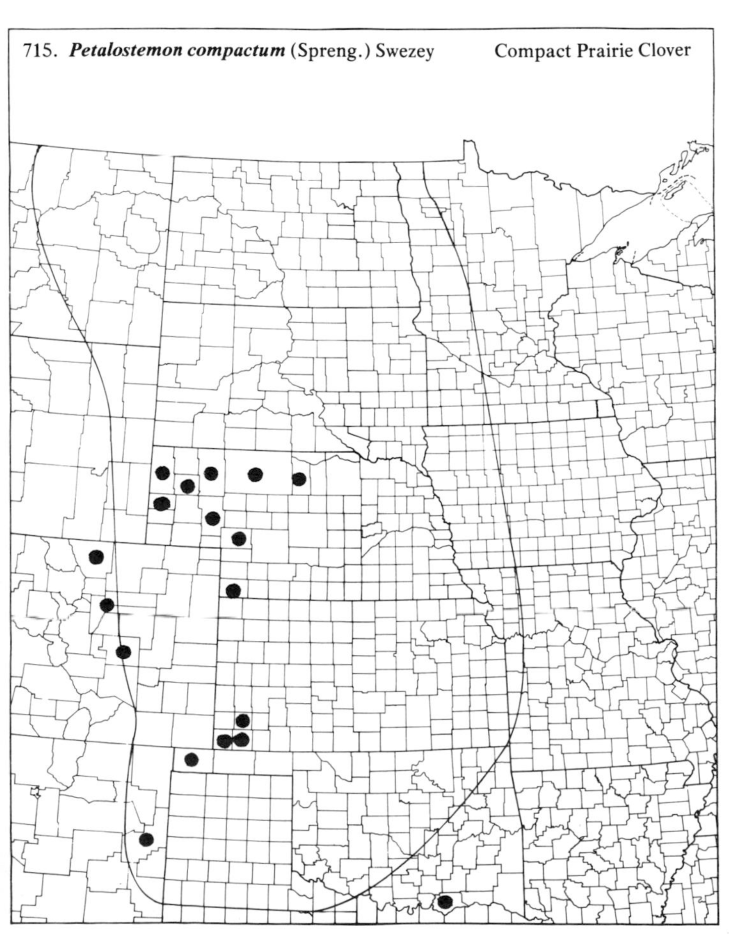

716. **Petalostemon multiflorum** Nutt. Round-headed Prairie Clover

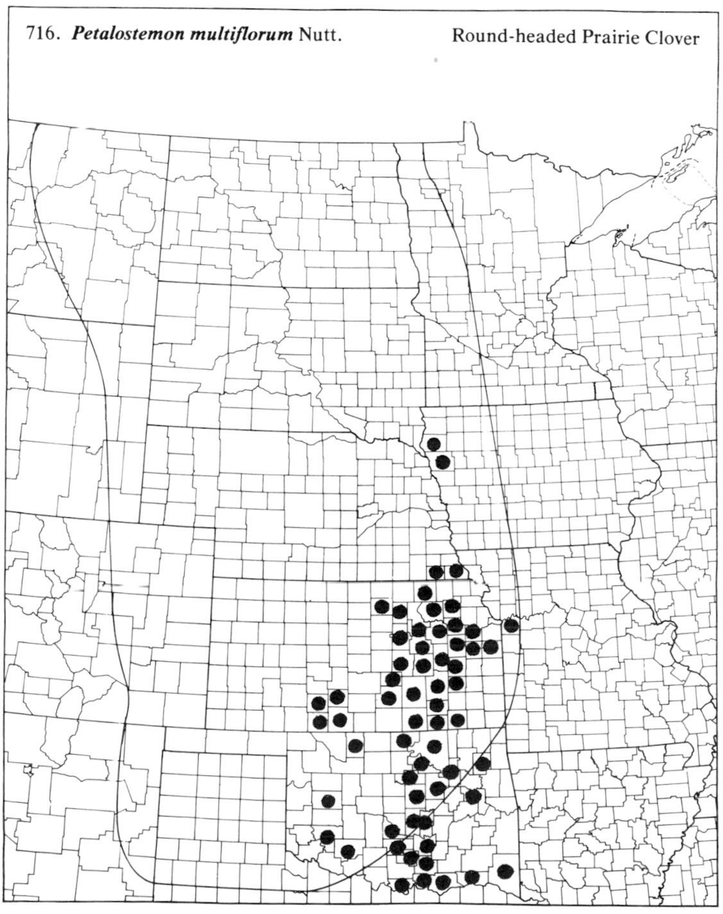

(Fabaceae-Papilionoideae)

717. ***Petalostemon occidentale*** (Gray) Fern. Western Prairie Clover

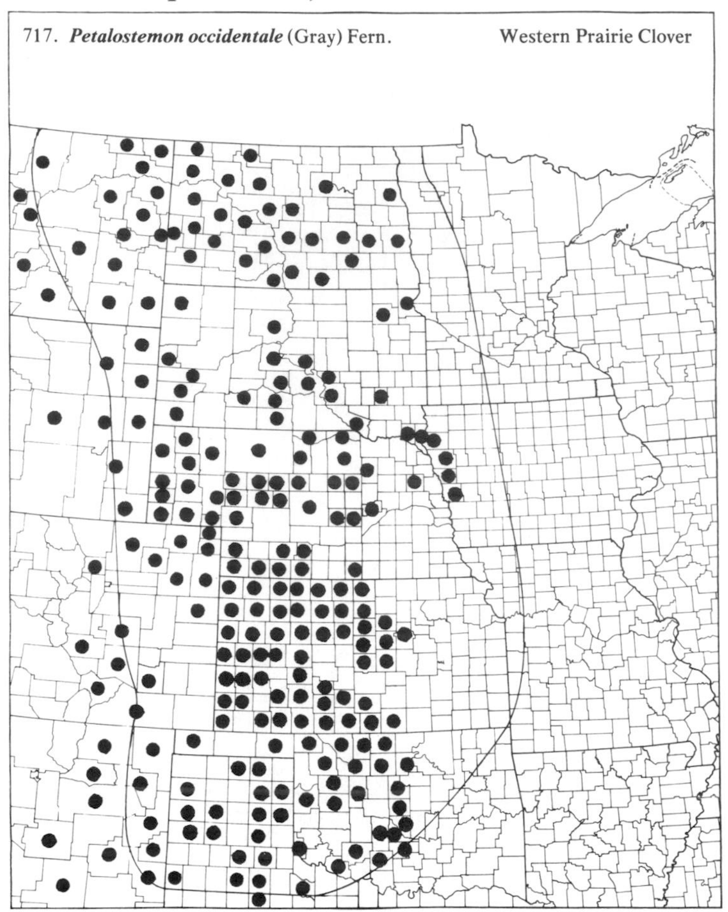

718. ***Petalostemon purpureum*** (Vent.) Rydb. Purple Prairie Clover

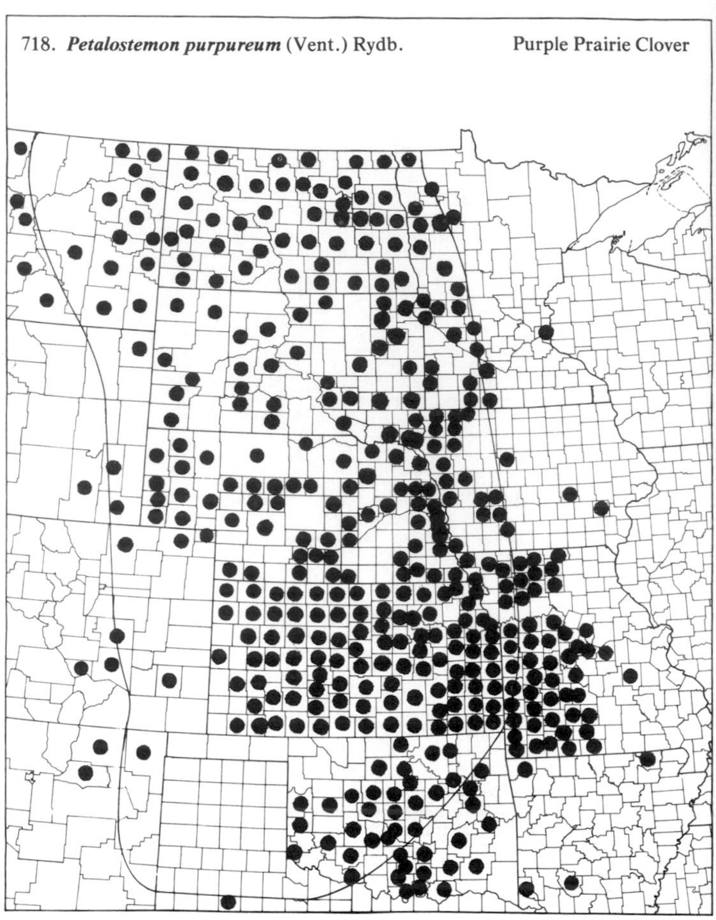

719. ***Petalostemon tenuifolium*** Gray Slimleaf Prairie Clover

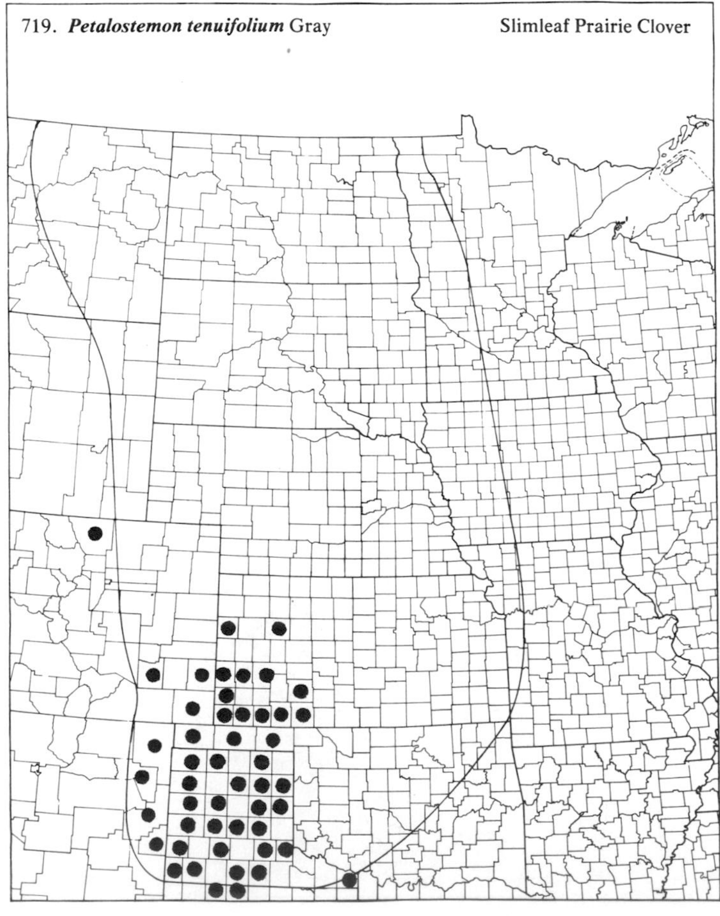

720. ***Petalostemon villosum*** Nutt. Silky Prairie Clover

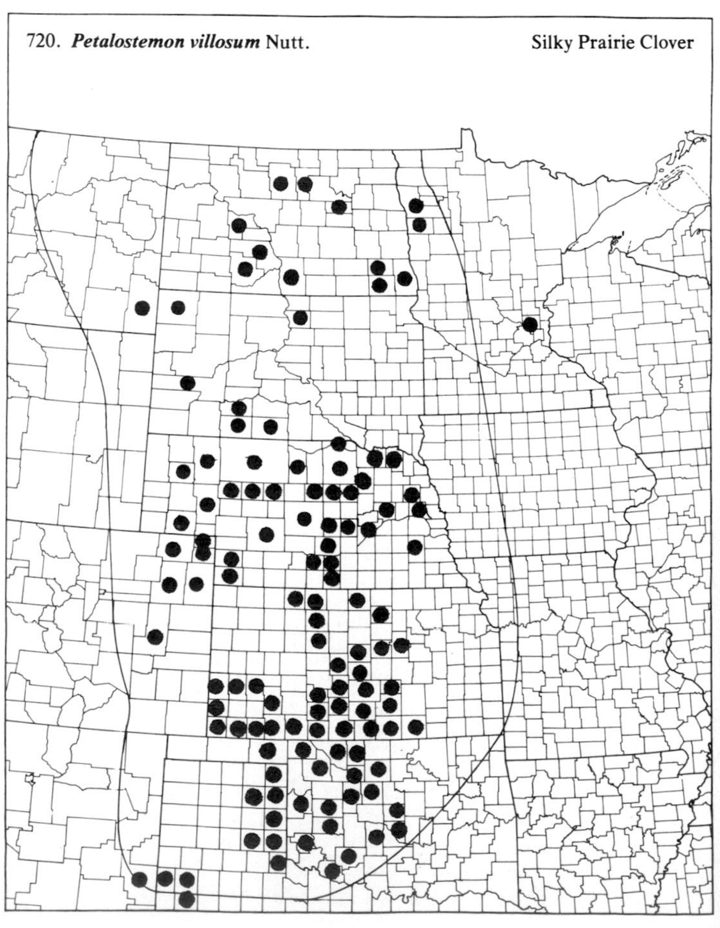

(Fabaceae-Papilionoideae)

721. ***Psoralea argophylla*** Pursh — Silver-leaf Scurf Pea

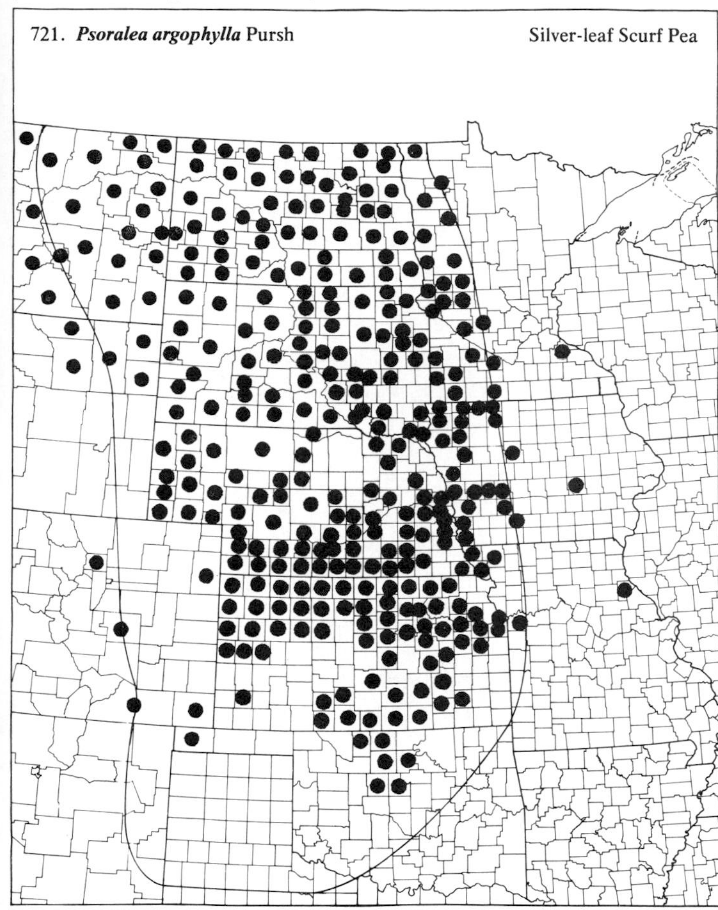

722. ***Psoralea cuspidata*** Pursh

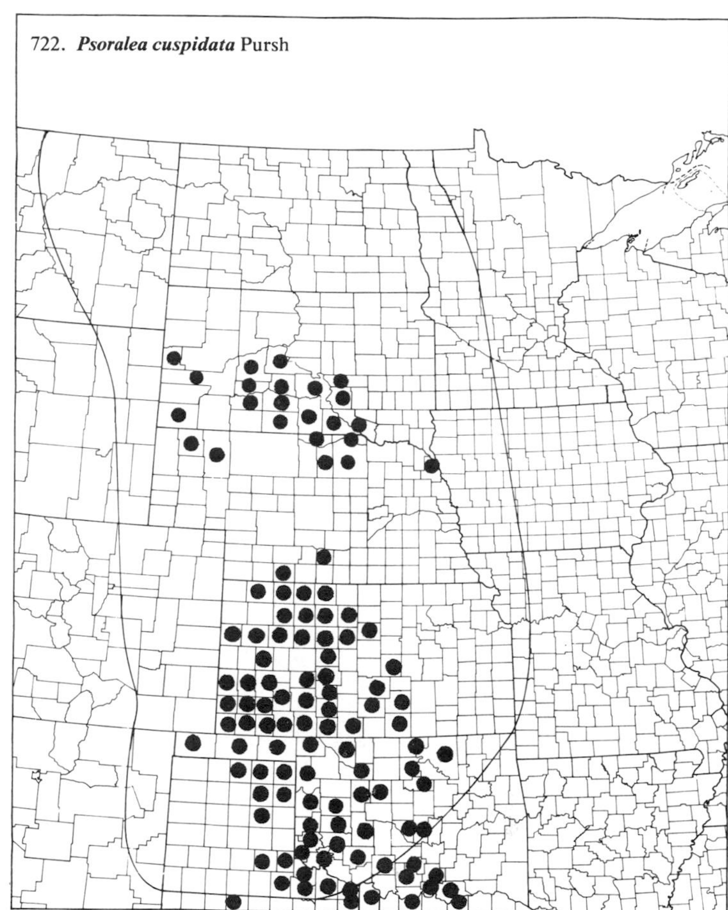

723. ***Psoralea digitata*** Nutt. — Palm-leaved Scurf Pea

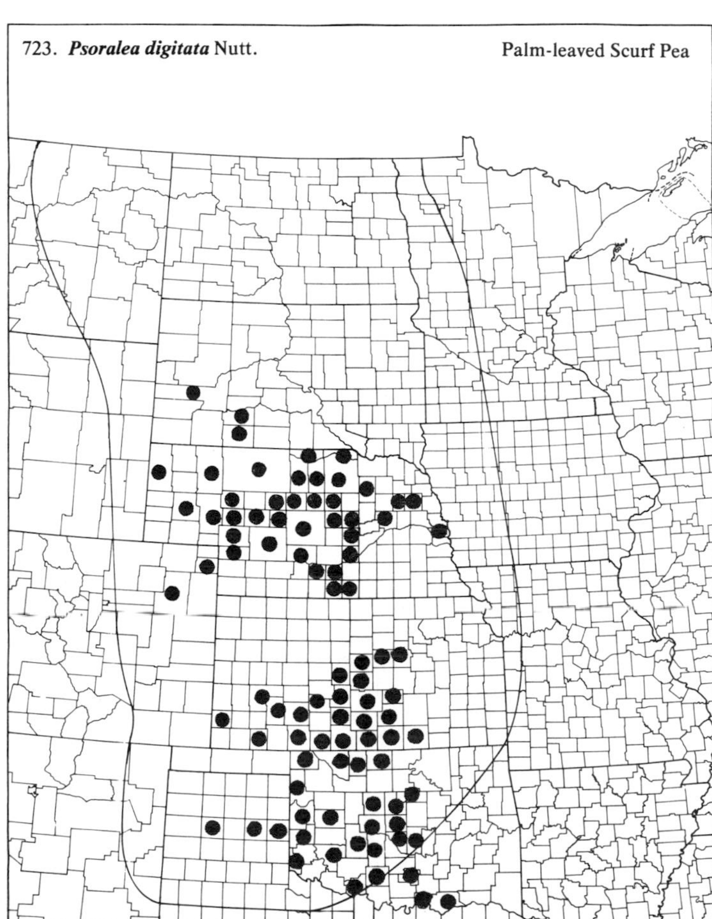

724. ***Psoralea esculenta*** Pursh — Breadroot Scurf Pea

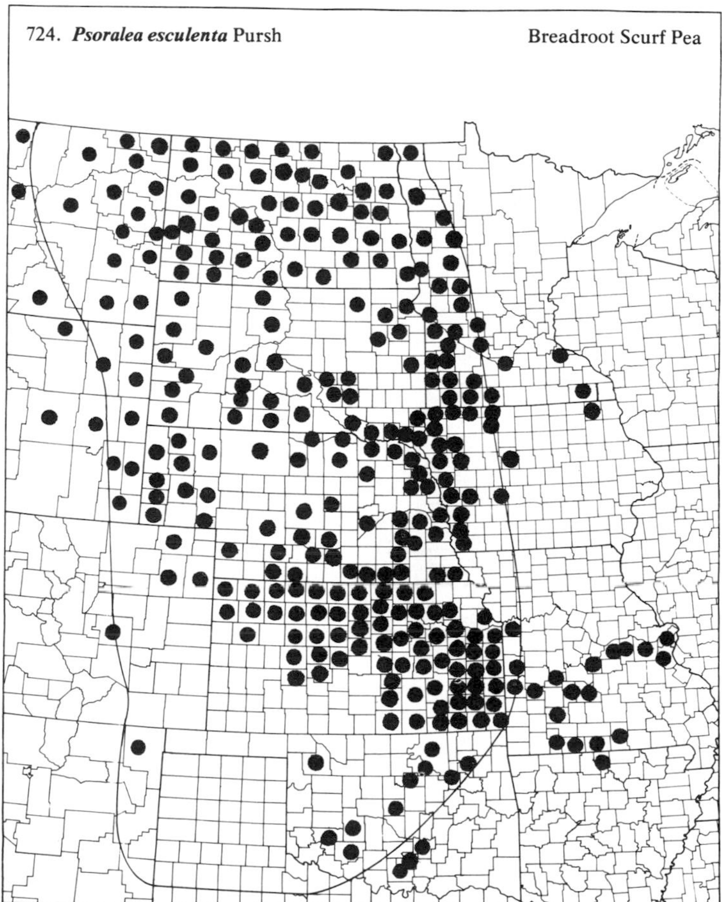

(Fabaceae-Papilionoideae)

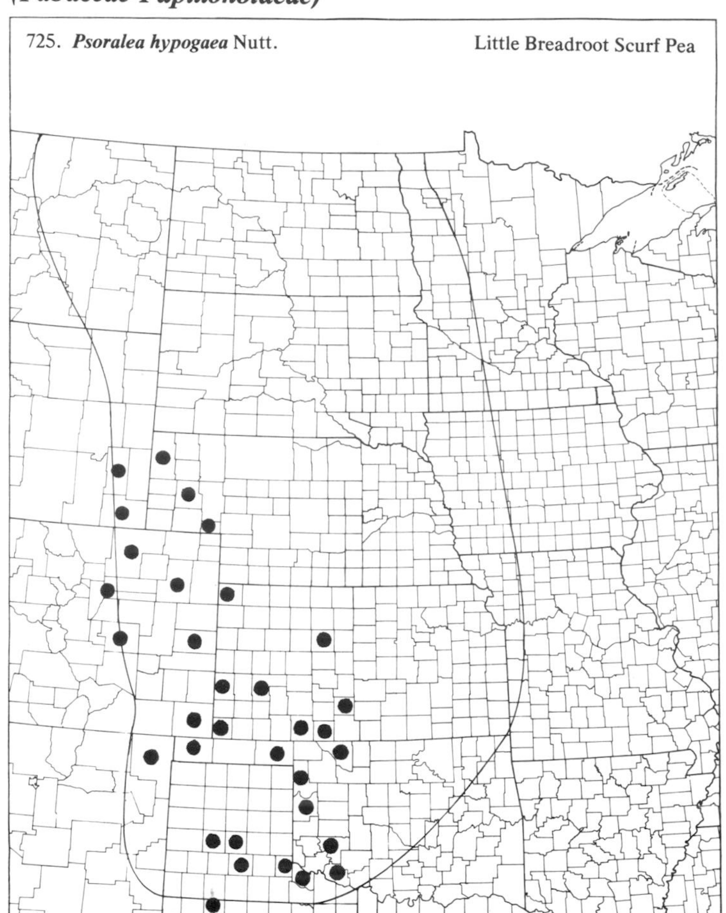

725. ***Psoralea hypogaea*** Nutt. Little Breadroot Scurf Pea

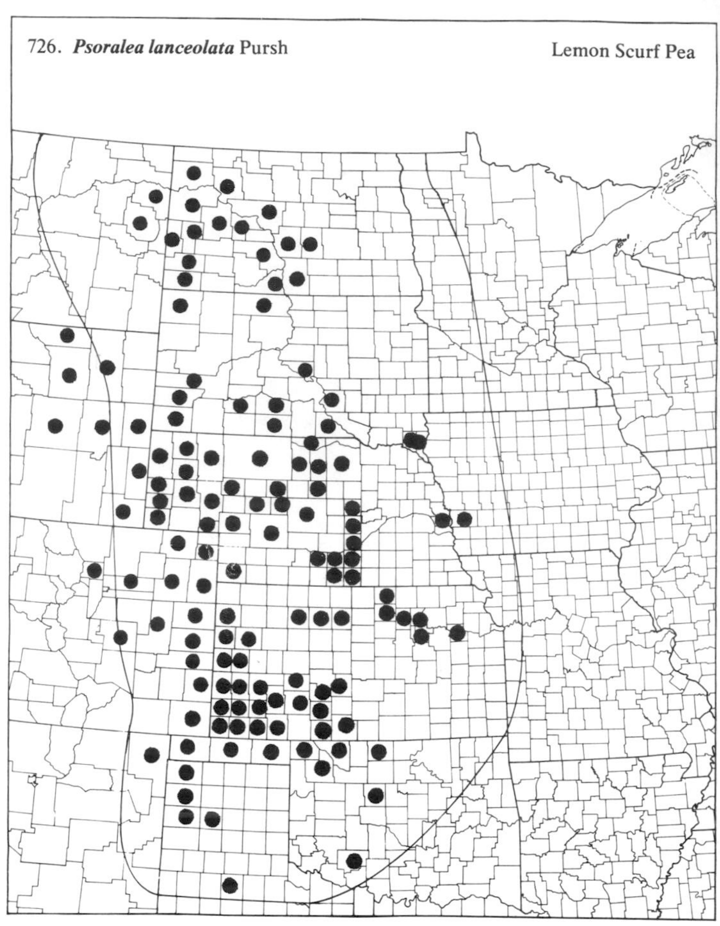

726. ***Psoralea lanceolata*** Pursh Lemon Scurf Pea

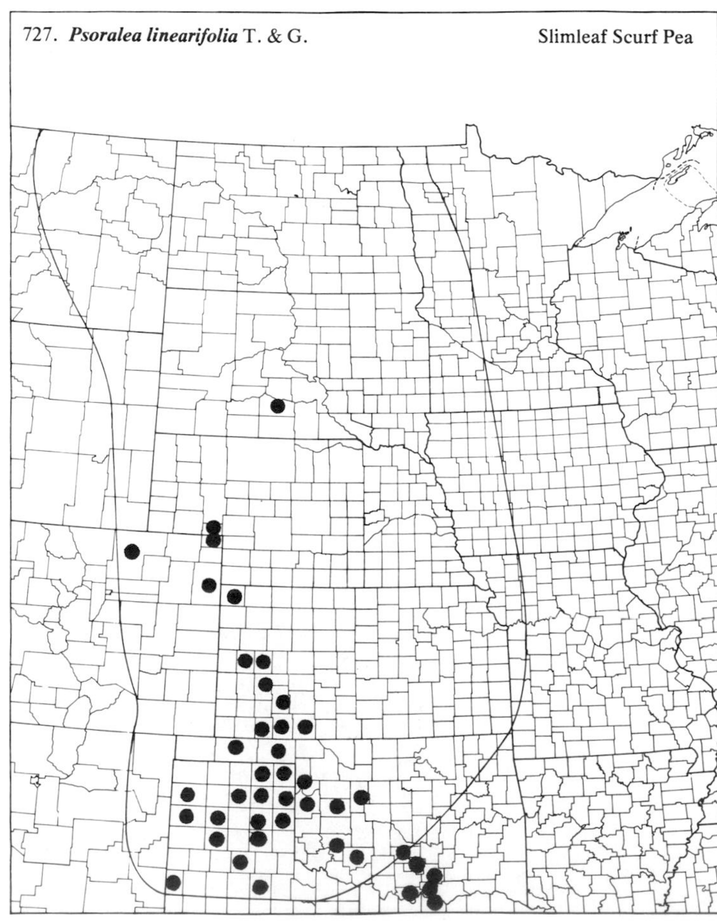

727. ***Psoralea linearifolia*** T. & G. Slimleaf Scurf Pea

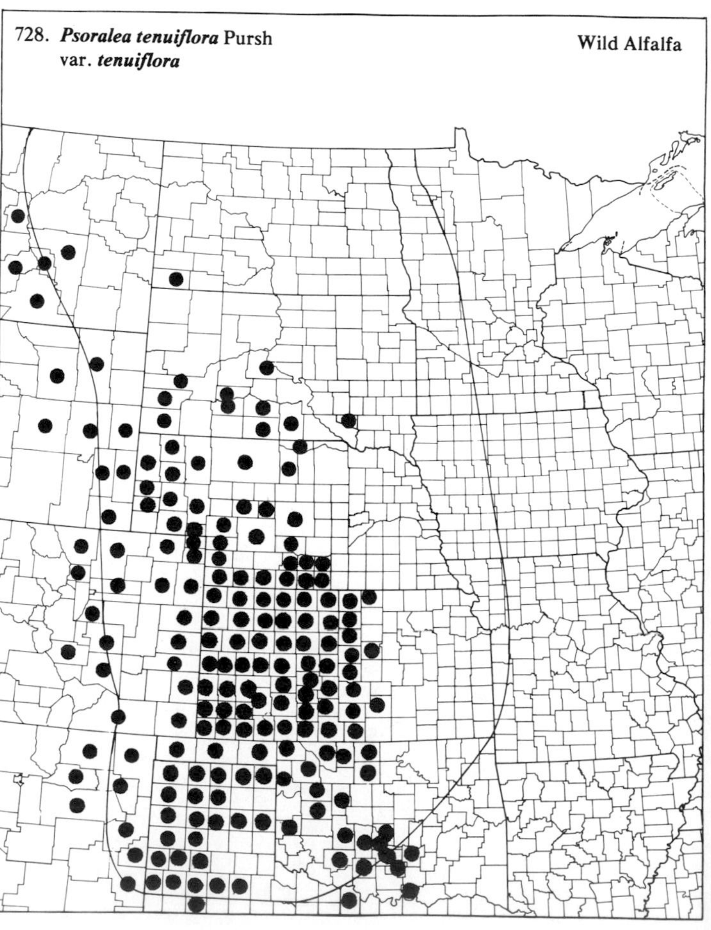

728. ***Psoralea tenuiflora*** Pursh var. ***tenuiflora*** **Wild Alfalfa**

(Fabaceae-Papilionoideae)

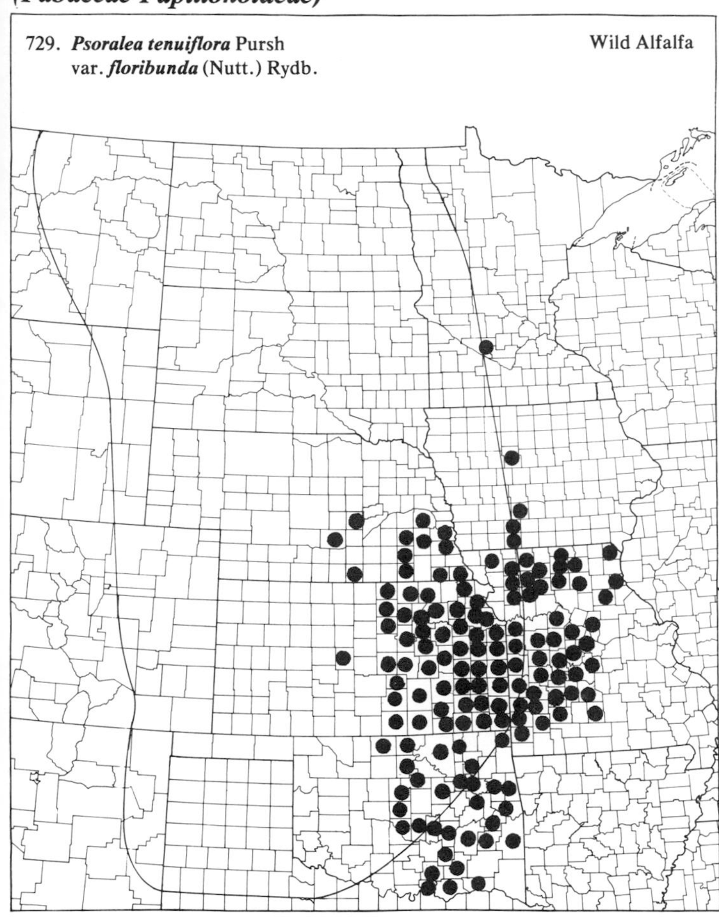

729. ***Psoralea tenuiflora*** Pursh var. ***floribunda*** (Nutt.) Rydb. — Wild Alfalfa

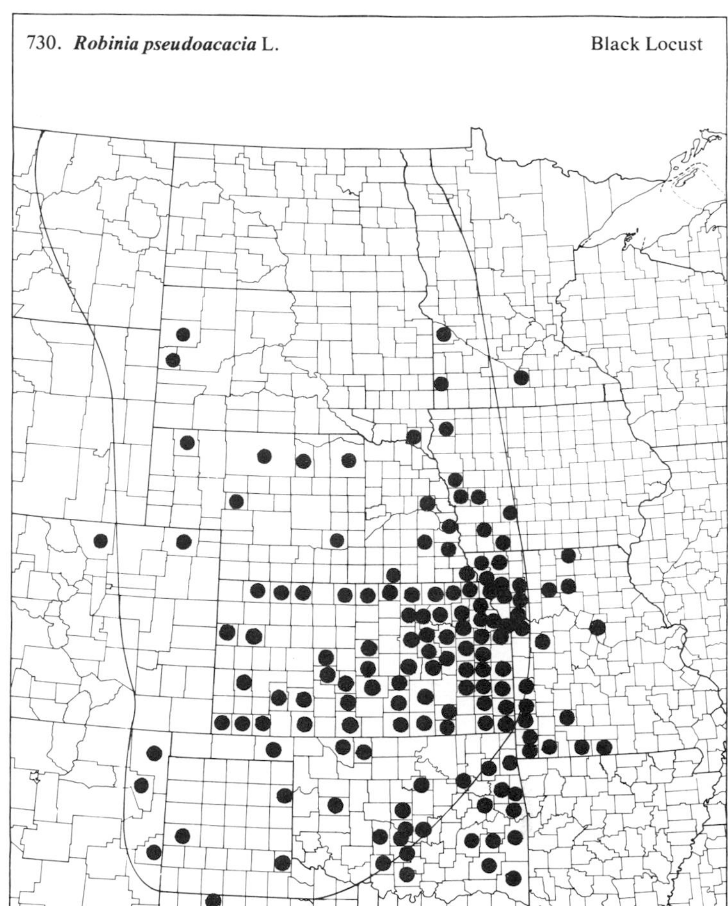

730. ***Robinia pseudoacacia*** L. — Black Locust

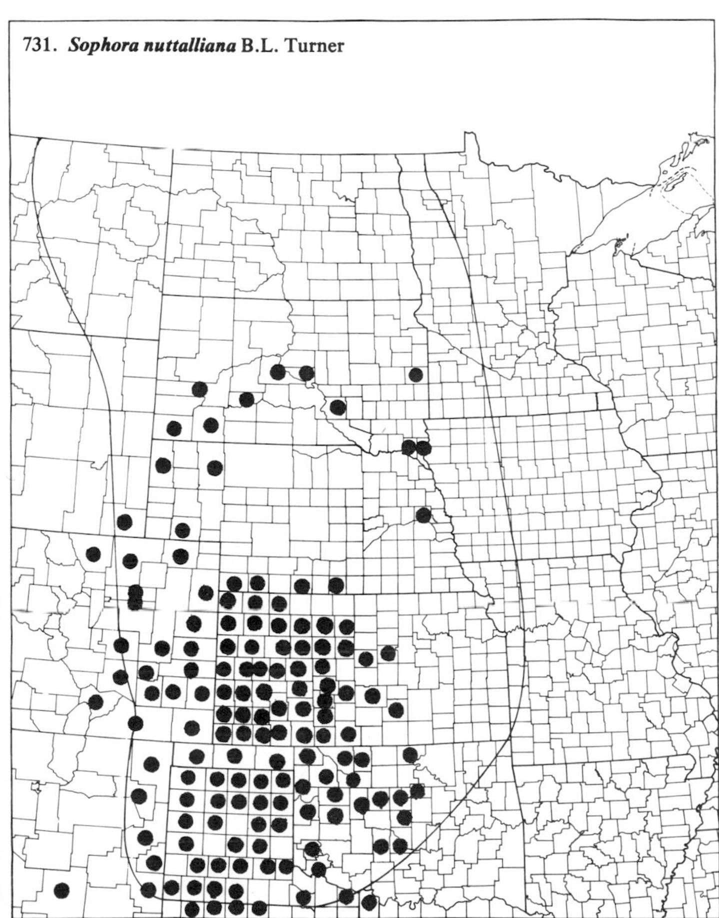

731. ***Sophora nuttalliana*** B.L. Turner

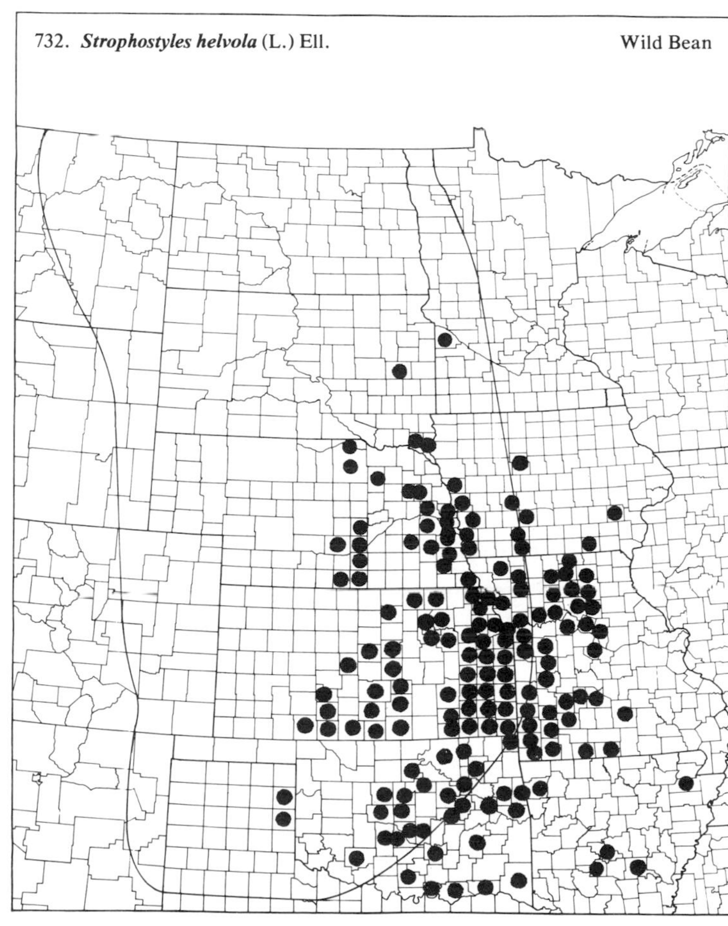

732. ***Strophostyles helvola*** (L.) Ell. — Wild Bean

(Fabaceae-Papilionoideae)

733. ***Strophostyles leiosperma*** (T. & G.) Piper — Smoothseed Wild Bean

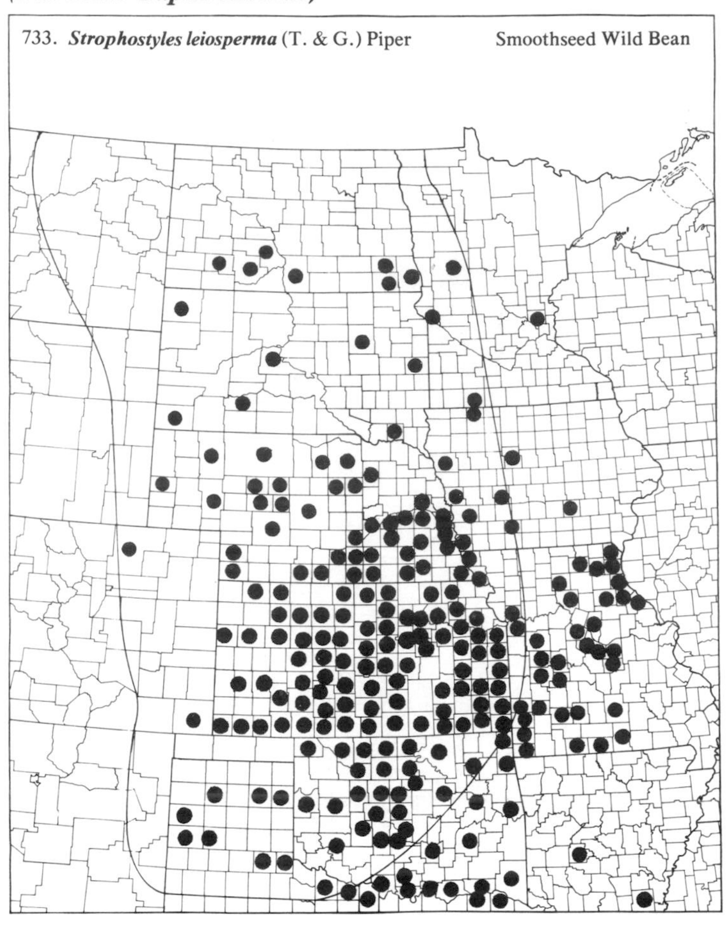

734. ***Stylosanthes biflora*** (L.) B.S.P. — Pencil-flower

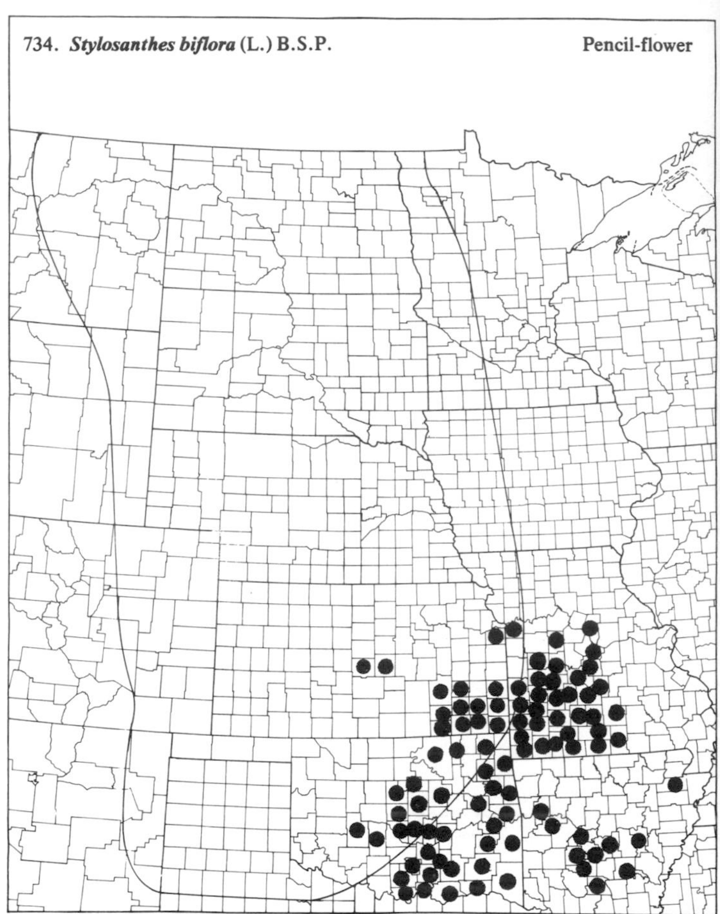

735. ***Tephrosia virginiana*** (L.) Pers. — Tephrosia

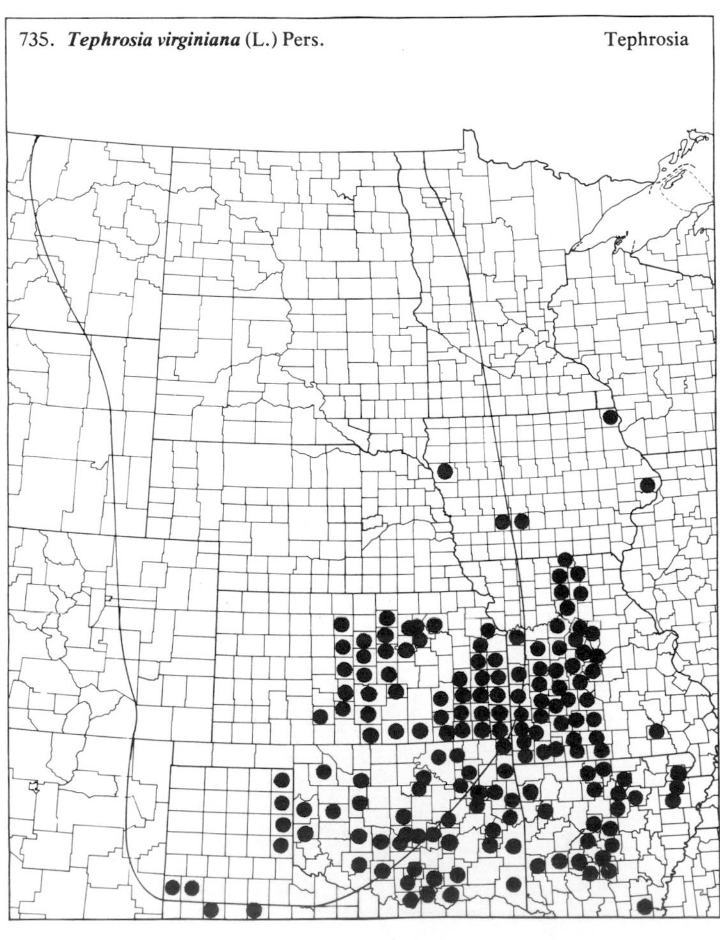

736. ***Thermopsis rhombifolia*** Nutt. — Golden Pea

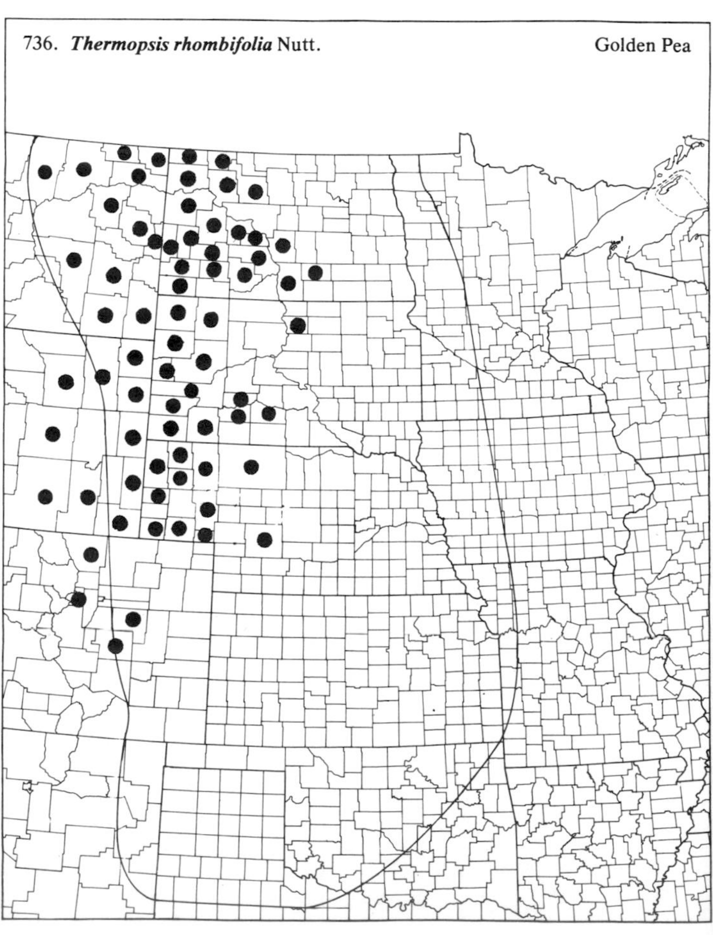

(Fabaceae-Papilionoideae)

737. ***Trifolium campestre*** Schreb. Plains Clover

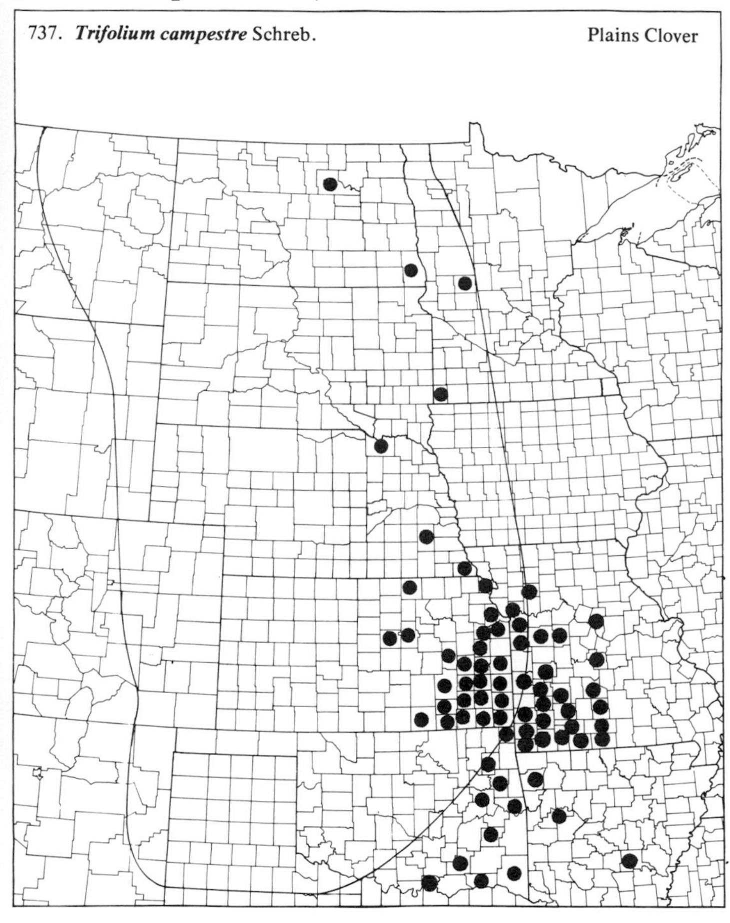

738. ***Trifolium dubium*** Sibth. Small Hop-clover

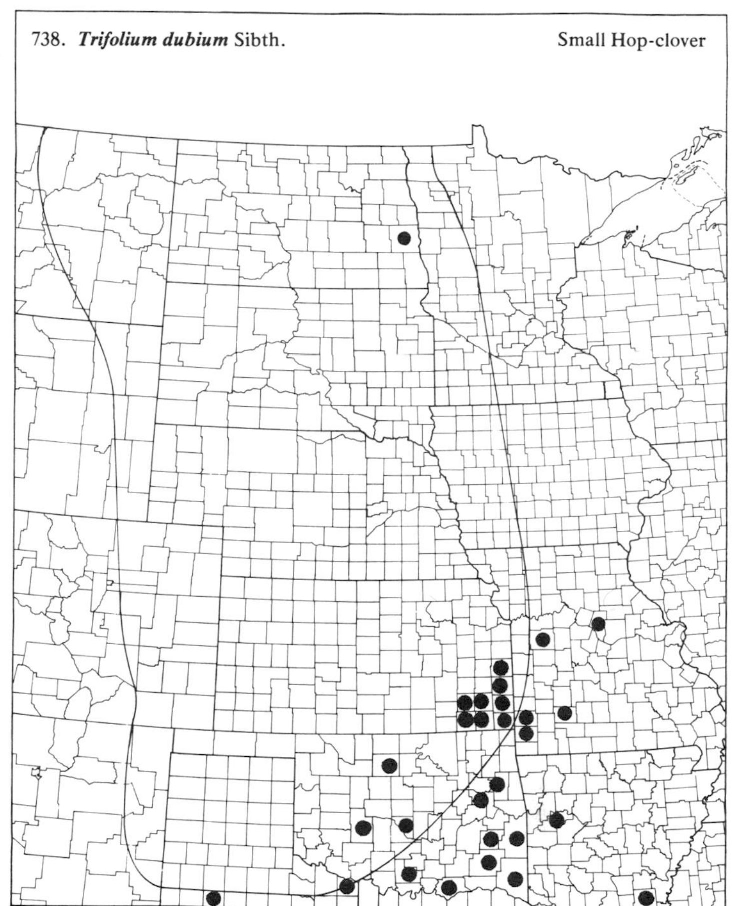

739. ***Trifolium hybridum*** L. Alsike Clover

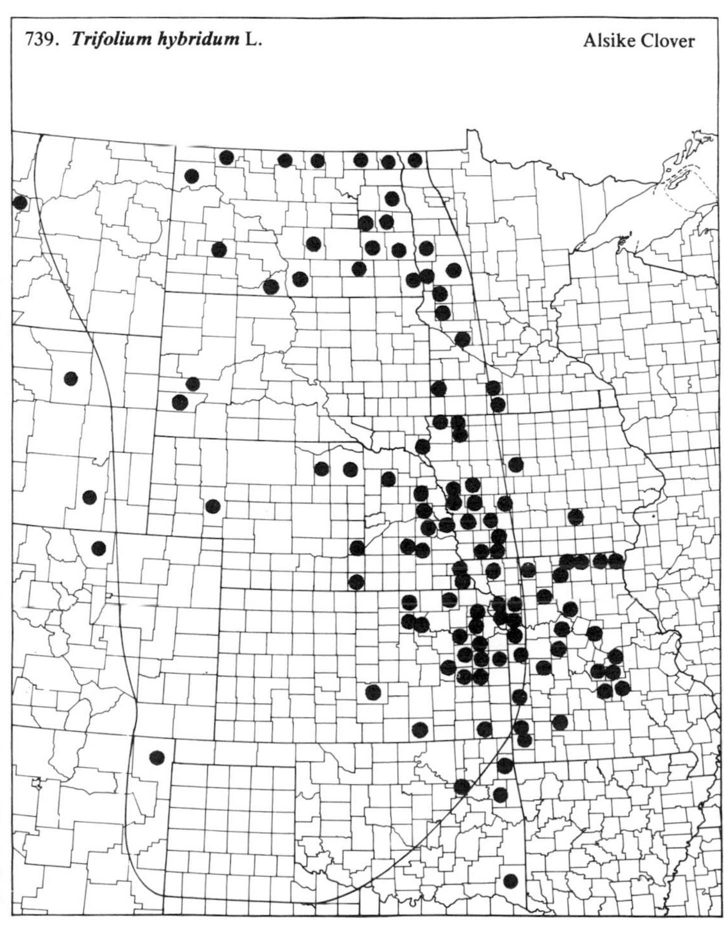

740. ***Trifolium pratense*** L. Red Clover

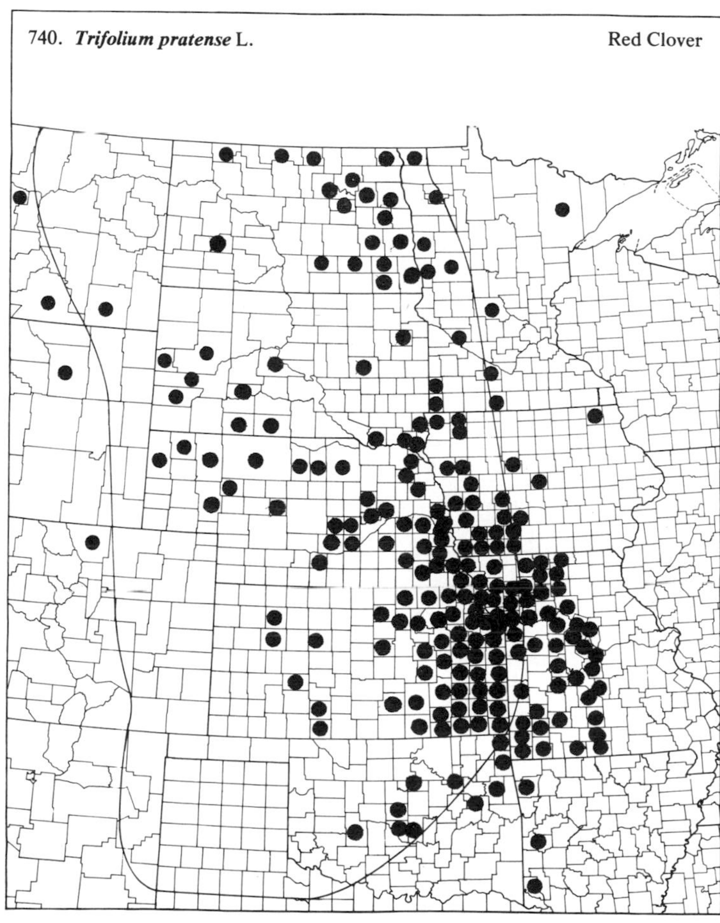

(Fabaceae-Papilionoideae)

741. ***Trifolium reflexum*** L. Buffalo Clover

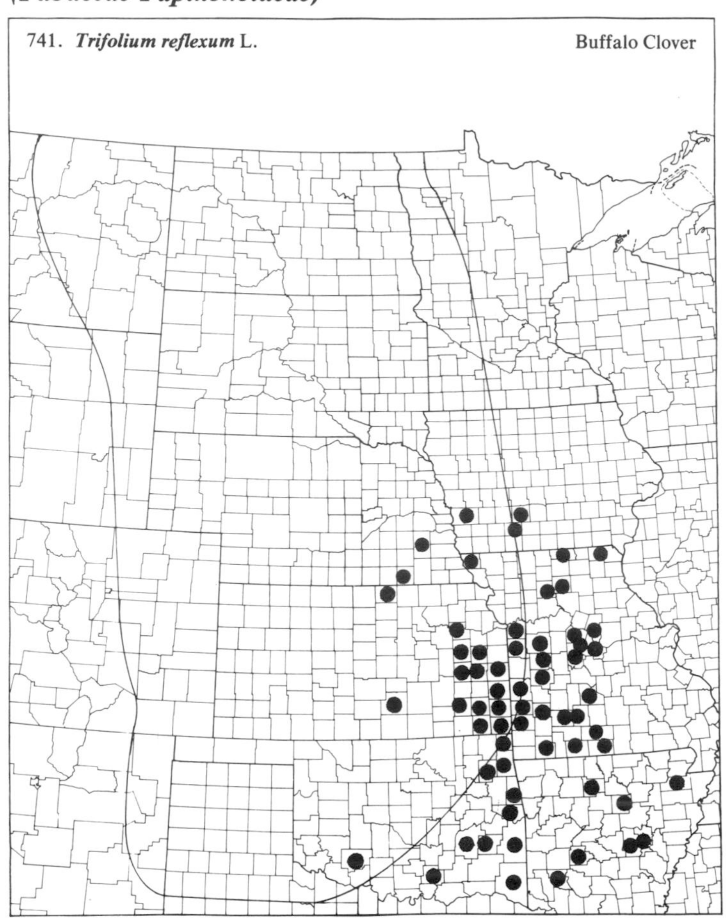

742. ***Trifolium repens*** L. White Clover

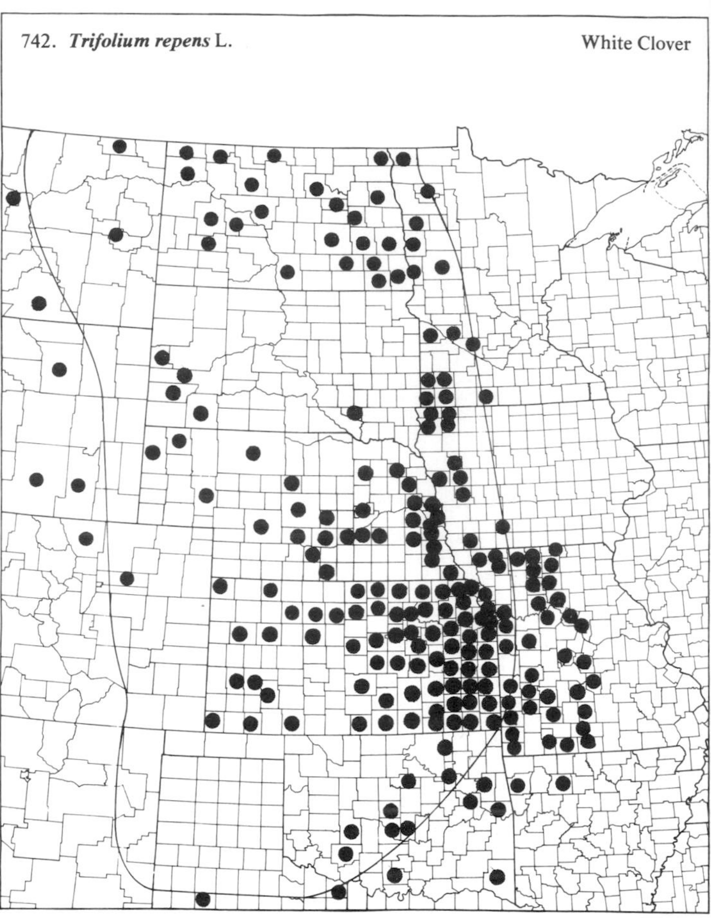

743. ***Vicia americana*** Muhl. var. ***americana*** American Vetch

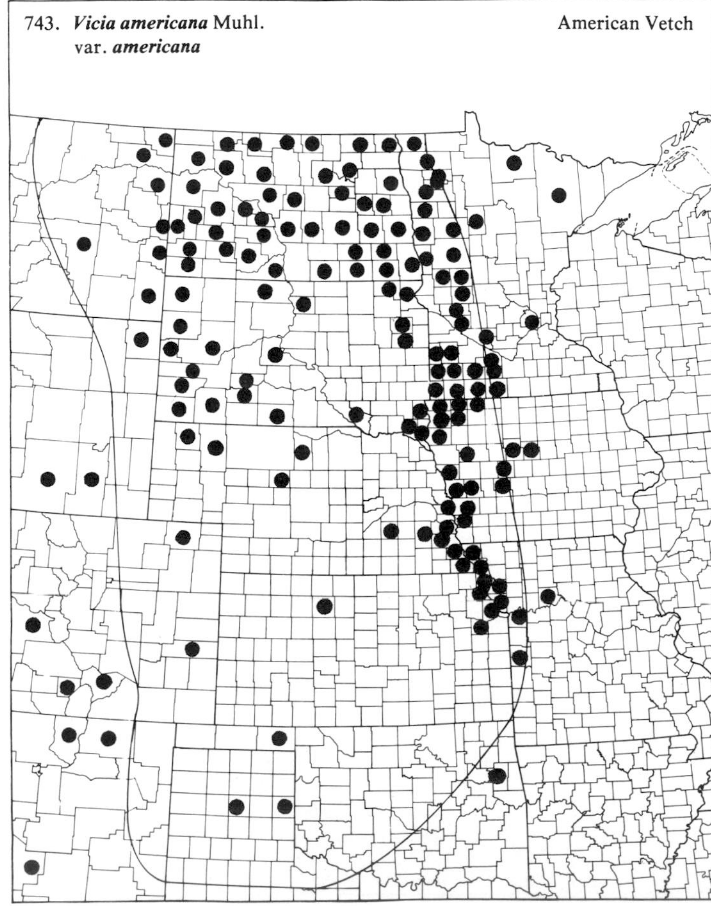

744. ***Vicia americana*** Muhl. var. ***minor*** Hook. American Vetch

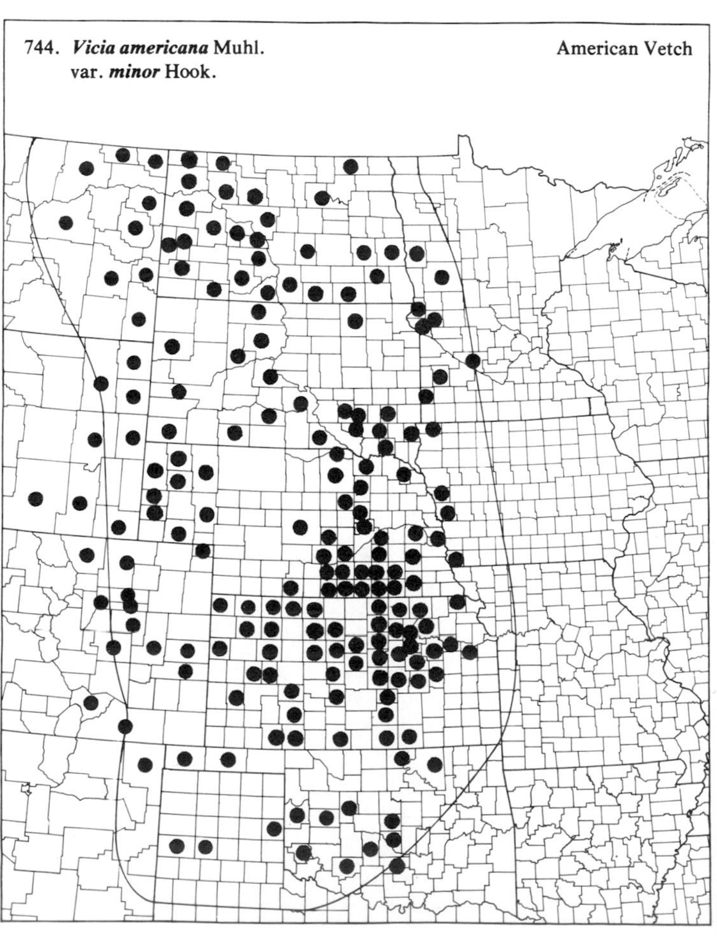

(Fabaceae-Papilionoideae)

745. ***Vicia dasycarpa*** Ten. Woollypod Vetch

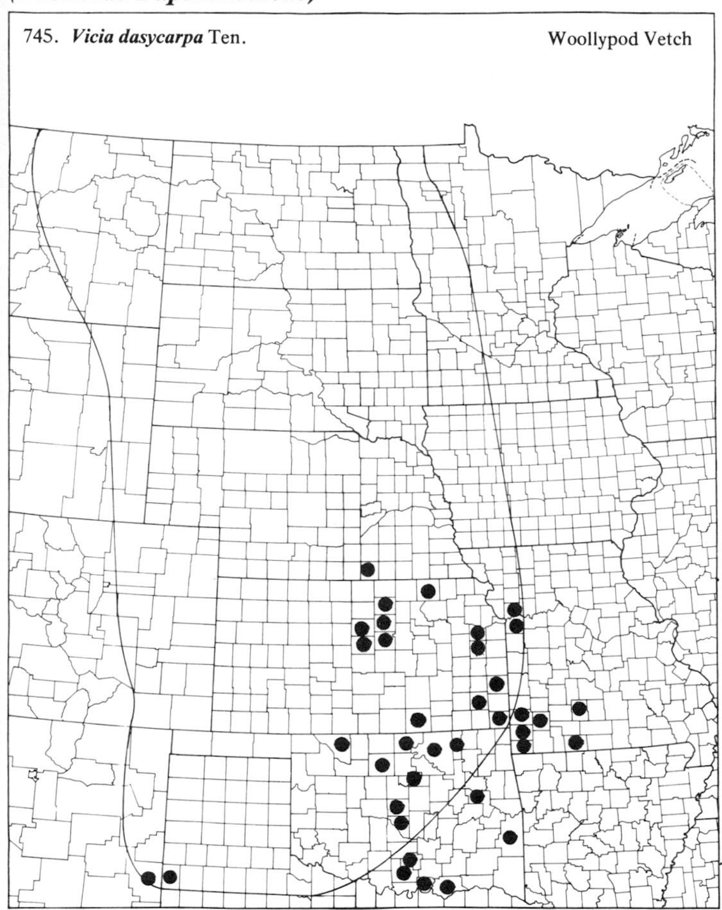

746. ***Vicia ludoviciana*** Nutt. Louisiana Vetch

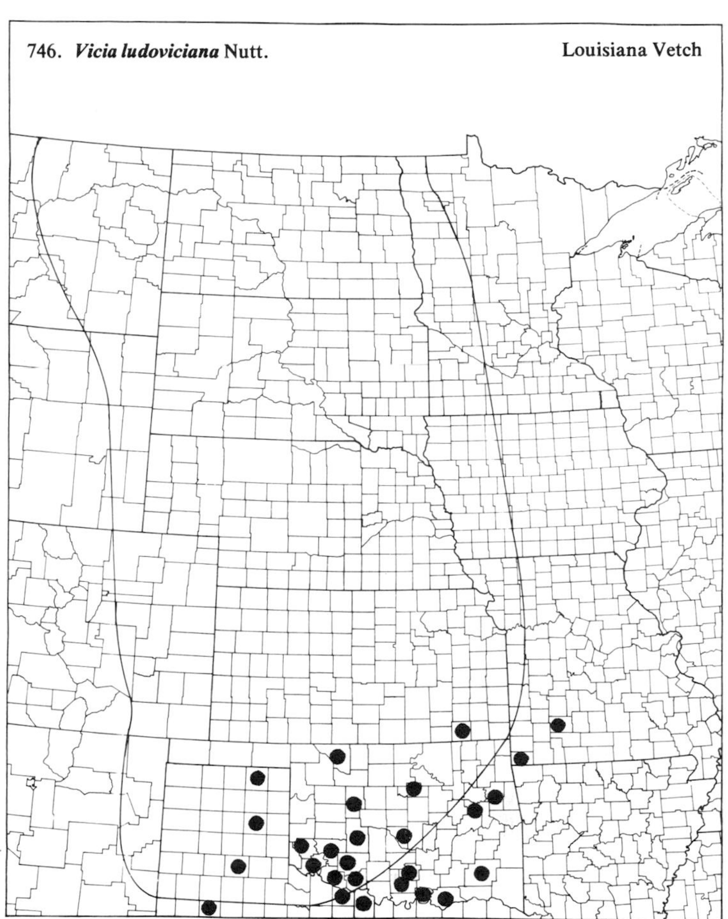

55. *Haloragaceae* (Water Milfoil Family)

747. ***Vicia villosa*** Roth Hairy Vetch

748. ***Myriophyllum heterophyllum*** Michx. Water Milfoil

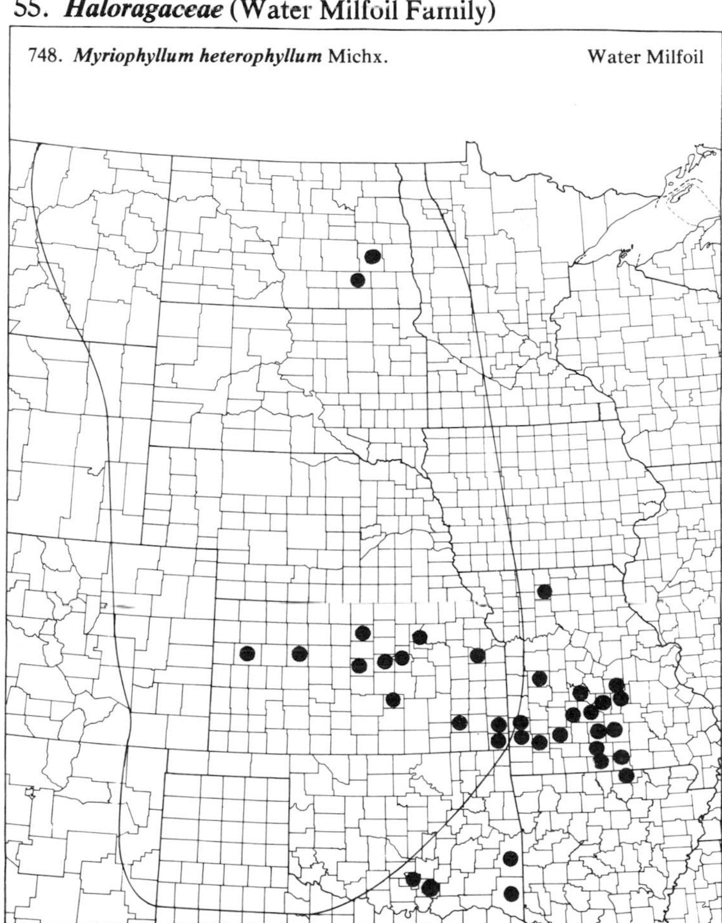

(Haloragaceae)

749. ***Myriophyllum pinnatum*** (Walt.) B.S.P. Water Milfoil

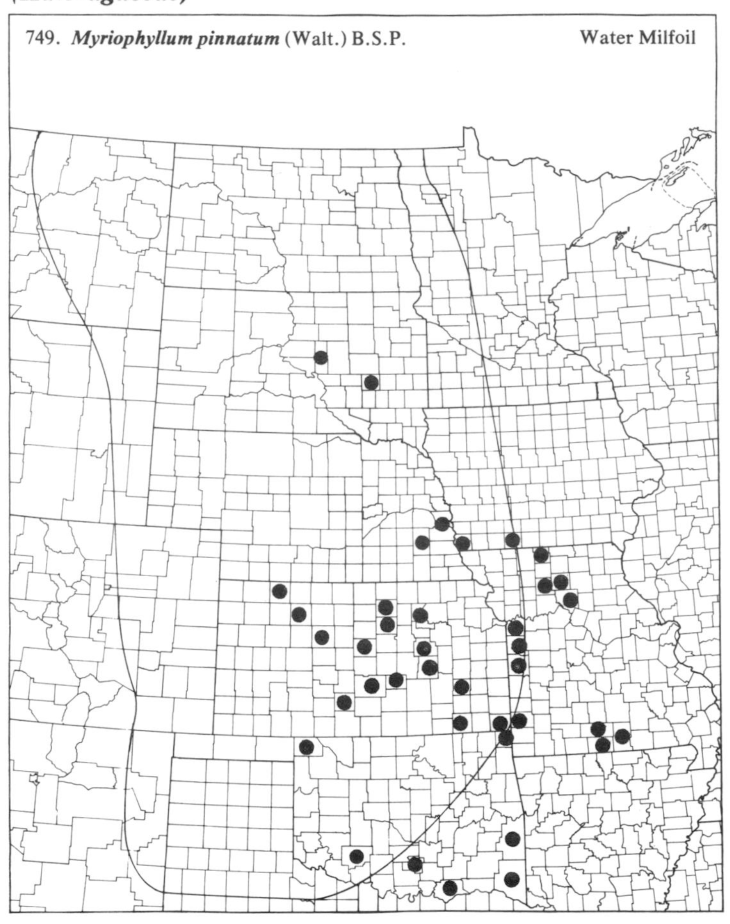

750. ***Myriophyllum spicatum*** L. var. ***exalbescens*** (Fern.) Jeps. Water Milfoil

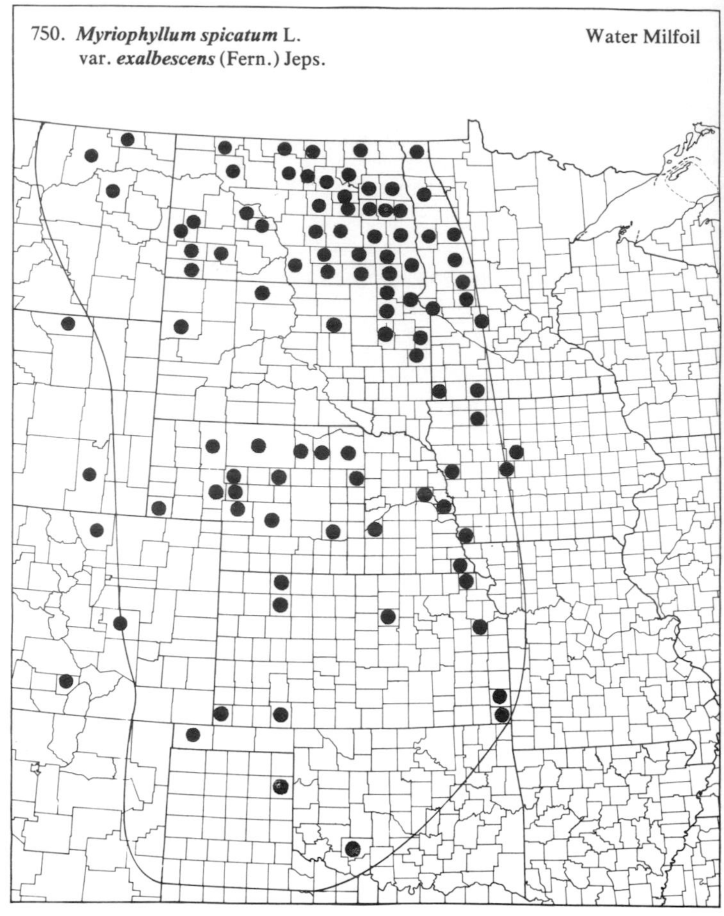

751. ***Myriophyllum verticillatum*** L. Water Milfoil

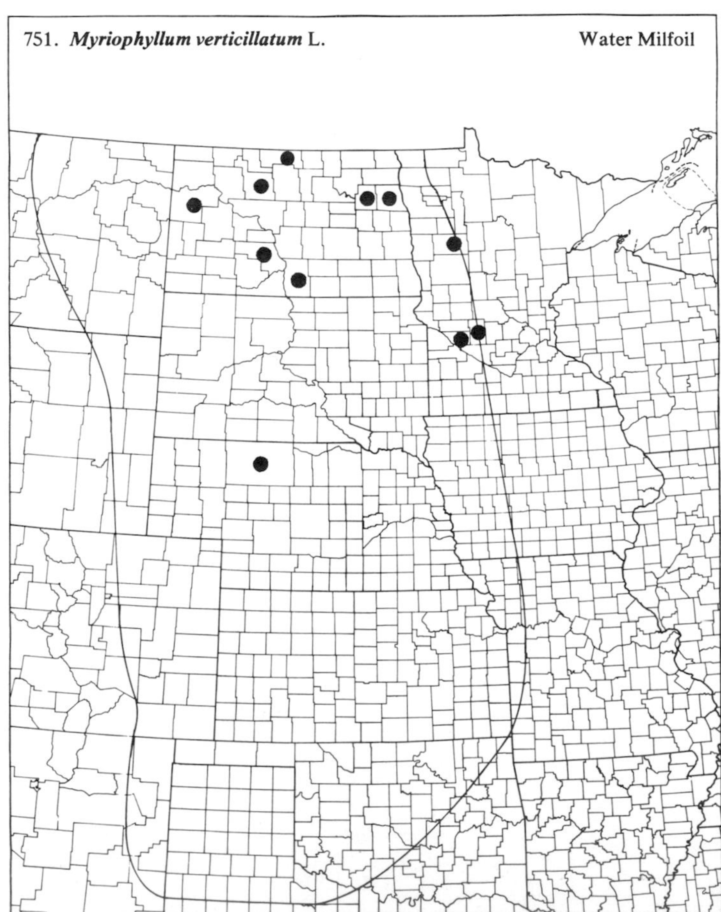

56. *Hippuridaceae* (Mare's-tail Family)

752. ***Hippuris vulgaris*** L. Mare's-tail

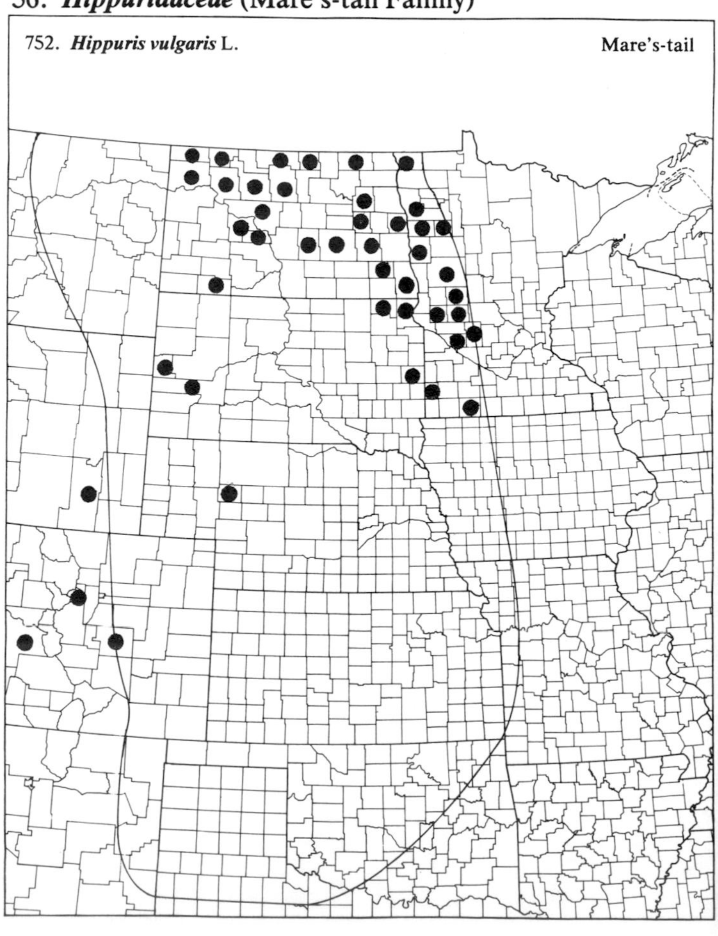

57. *Lythraceae* (Loosestrife Family)

753. ***Ammannia auriculata*** Willd. — Earleaf Ammannia

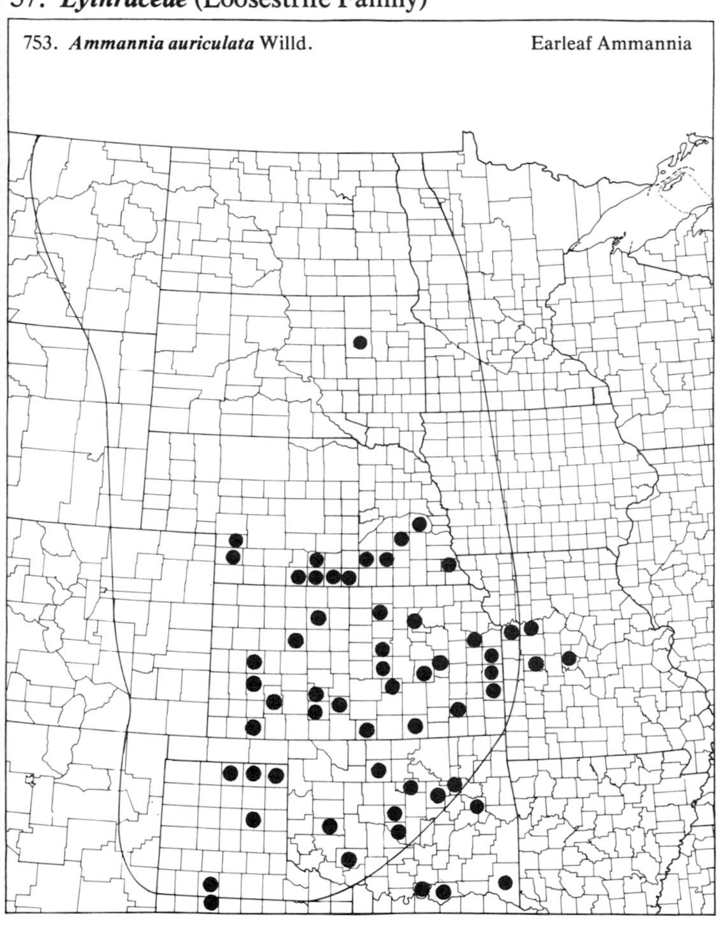

754. ***Ammannia coccinea*** Rottb. — Tooth-cup

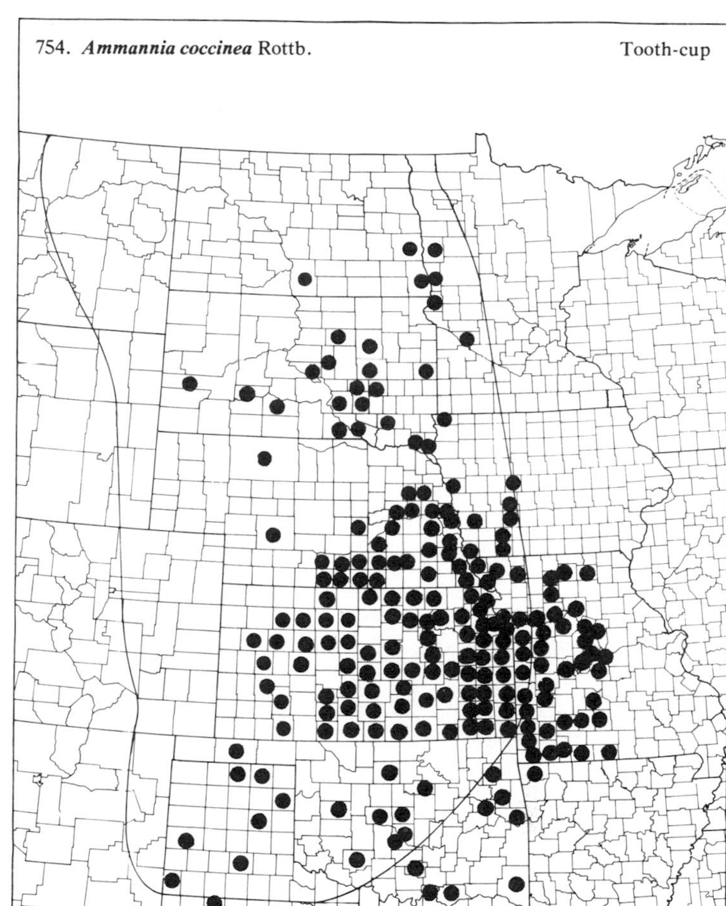

755. ***Cuphea viscosissima*** Jacq. — Blue Waxweed

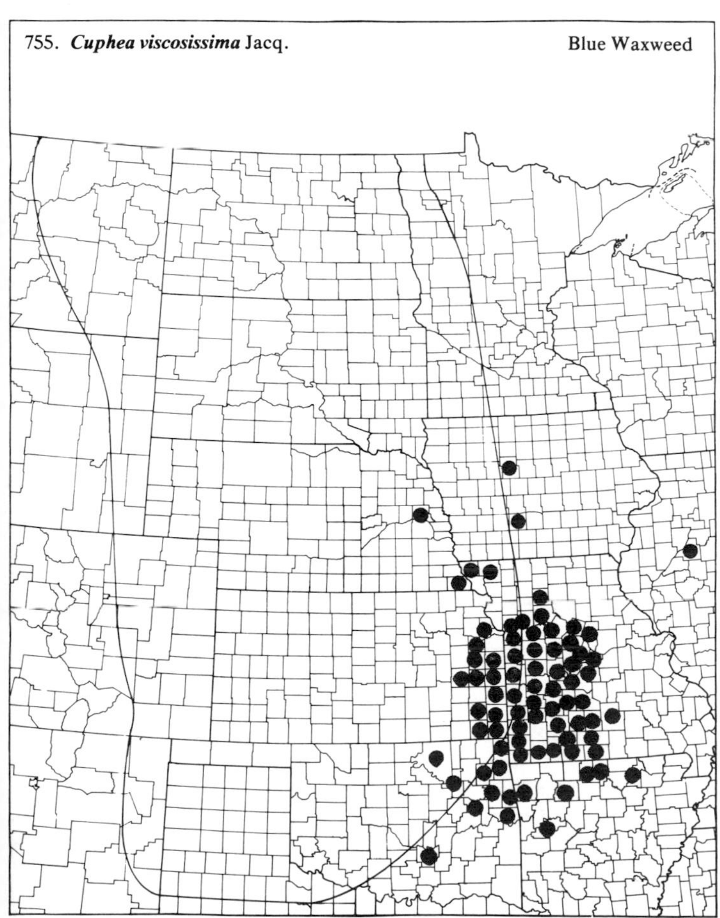

756. ***Lythrum californicum*** T. & G. — Loosestrife

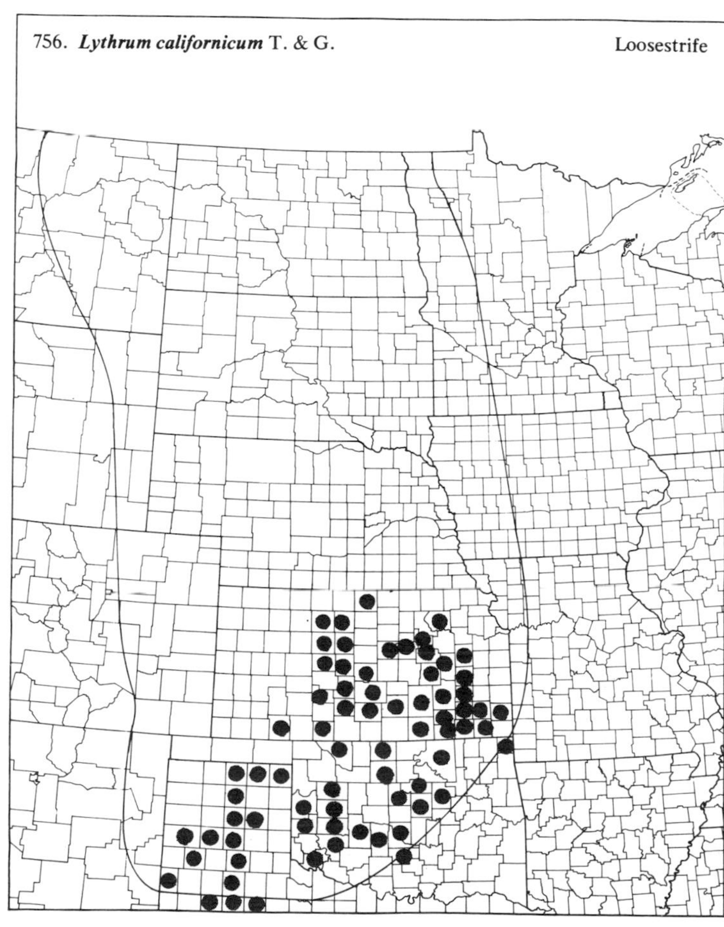

(Lythraceae)

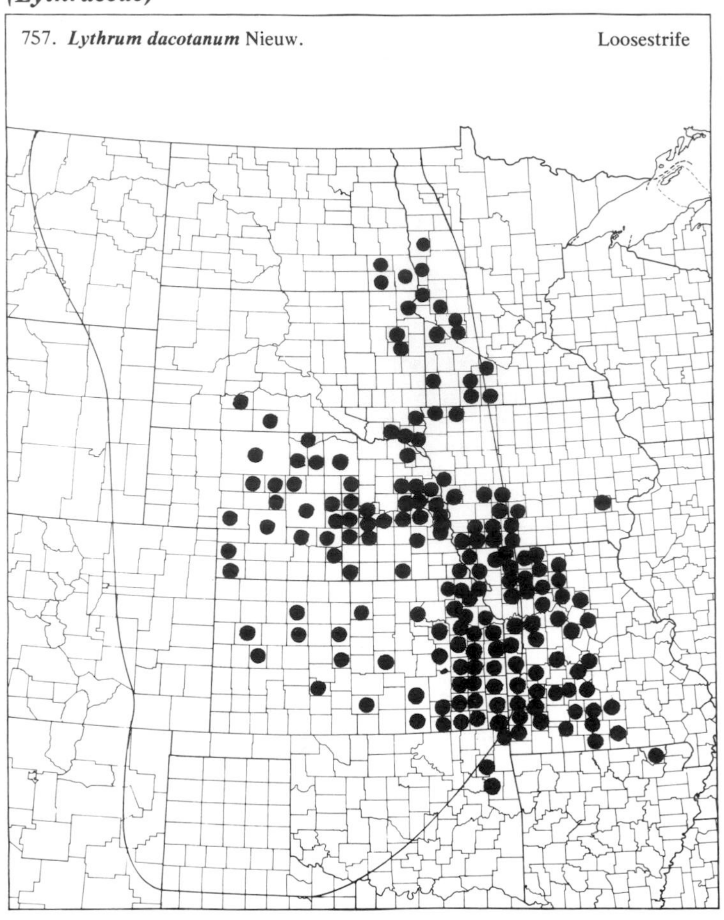

757. ***Lythrum dacotanum*** Nieuw. Loosestrife

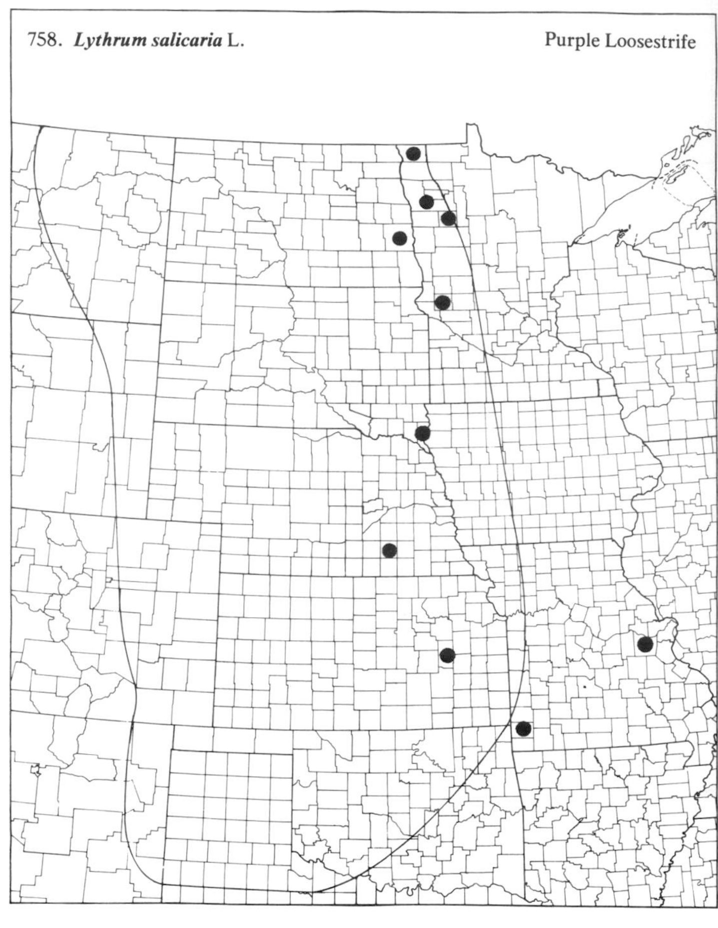

758. ***Lythrum salicaria*** L. Purple Loosestrife

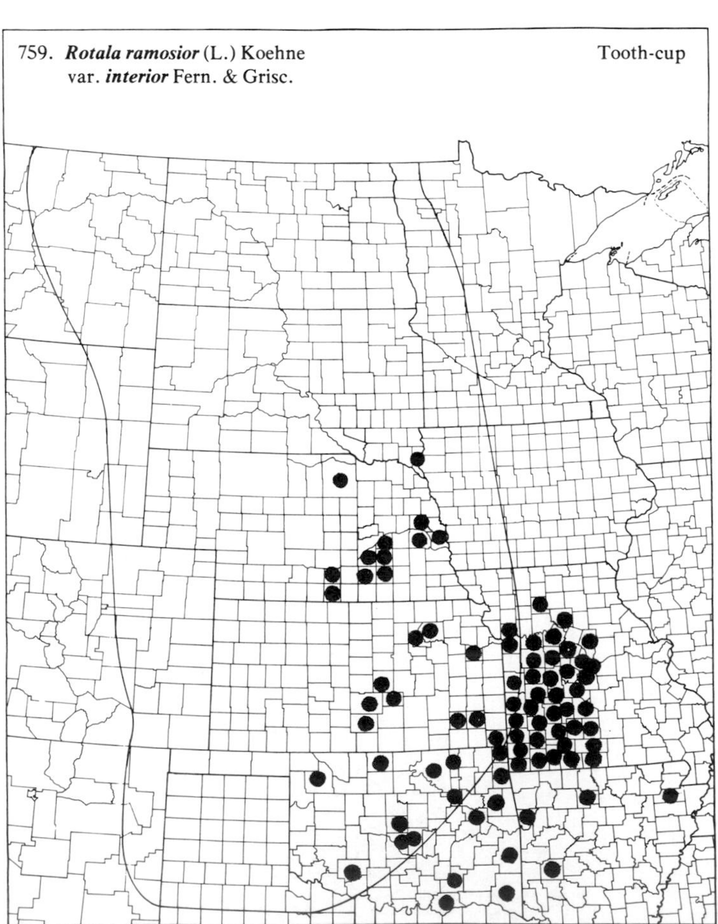

759. ***Rotala ramosior*** (L.) Koehne var. ***interior*** Fern. & Grisc. Tooth-cup

59. *Onagraceae* (Evening Primrose Family)

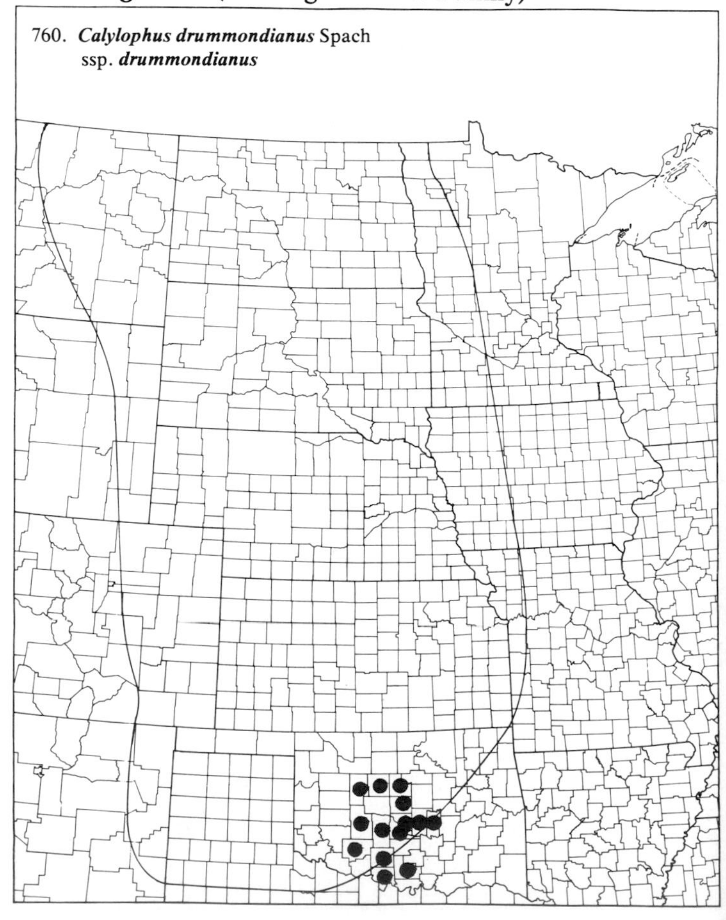

760. ***Calylophus drummondianus*** Spach ssp. ***drummondianus***

(Onagraceae)

761. ***Calylophus drummondianus*** Spach
ssp. ***berlandieri*** (Spach) Towner & Raven

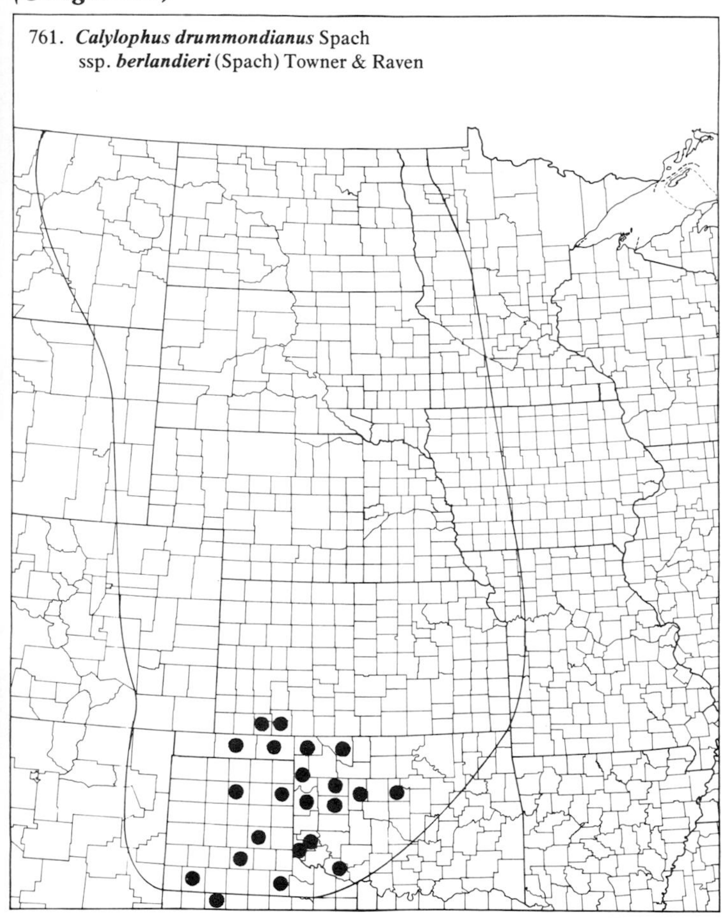

762. ***Calylophus hartwegii*** (Benth.) Raven
ssp. ***fendleri*** (Gray) Towner & Raven

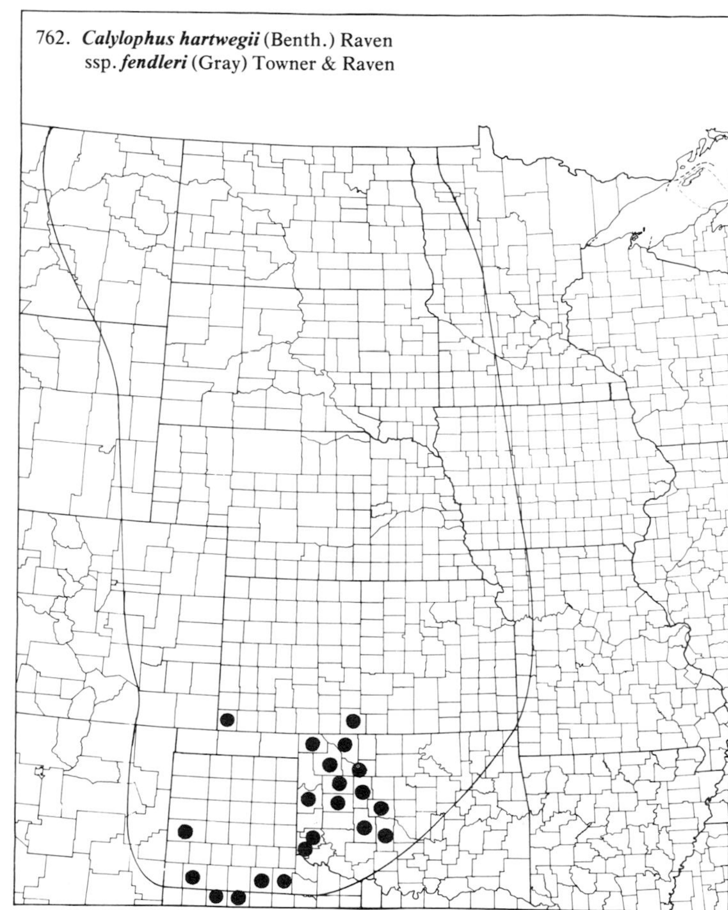

763. ***Calylophus hartwegii*** (Benth.) Raven
ssp. ***lavandulifolius*** (T. & G.) Towner & Raven

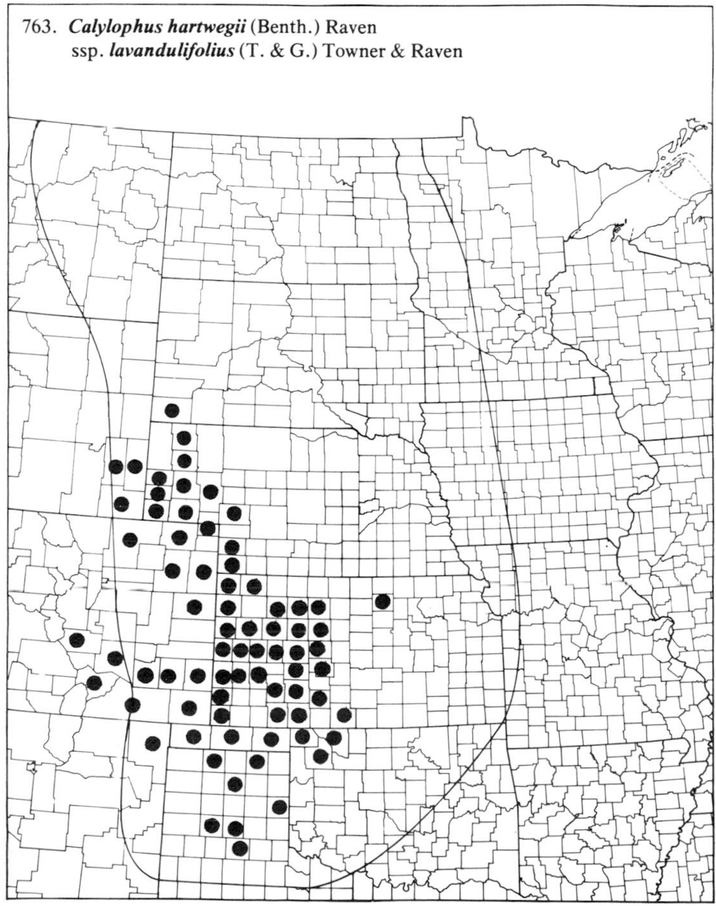

764. ***Calylophus hartwegii*** (Benth.) Raven
ssp. ***pubescens*** (Gray) Towner & Raven

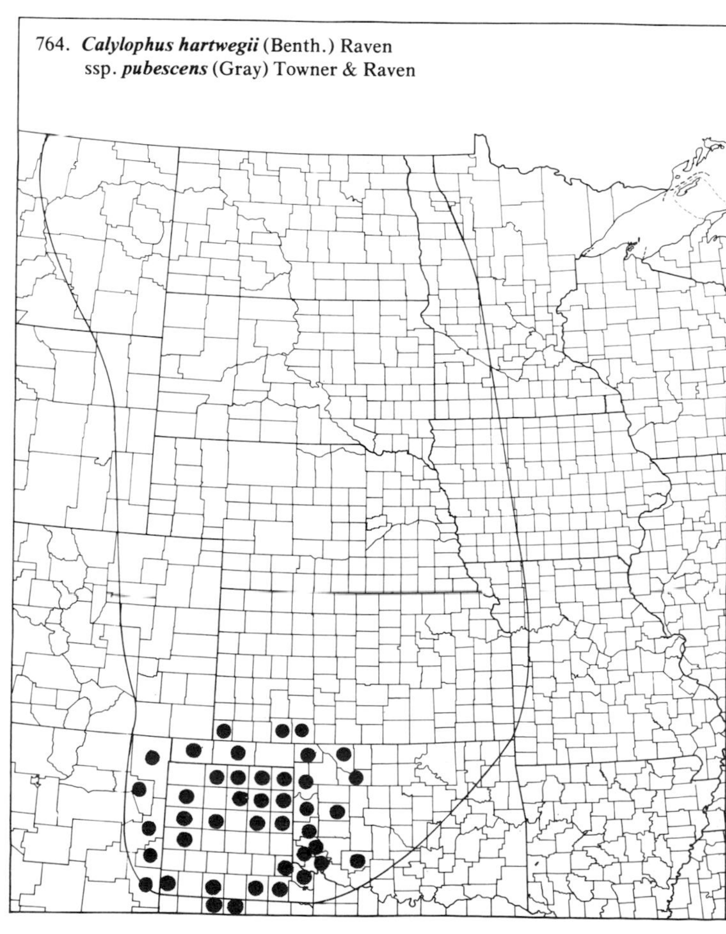

(Onagraceae)

765. ***Calylophus serrulatus*** (Nutt.) Raven — Yellow Evening Primrose

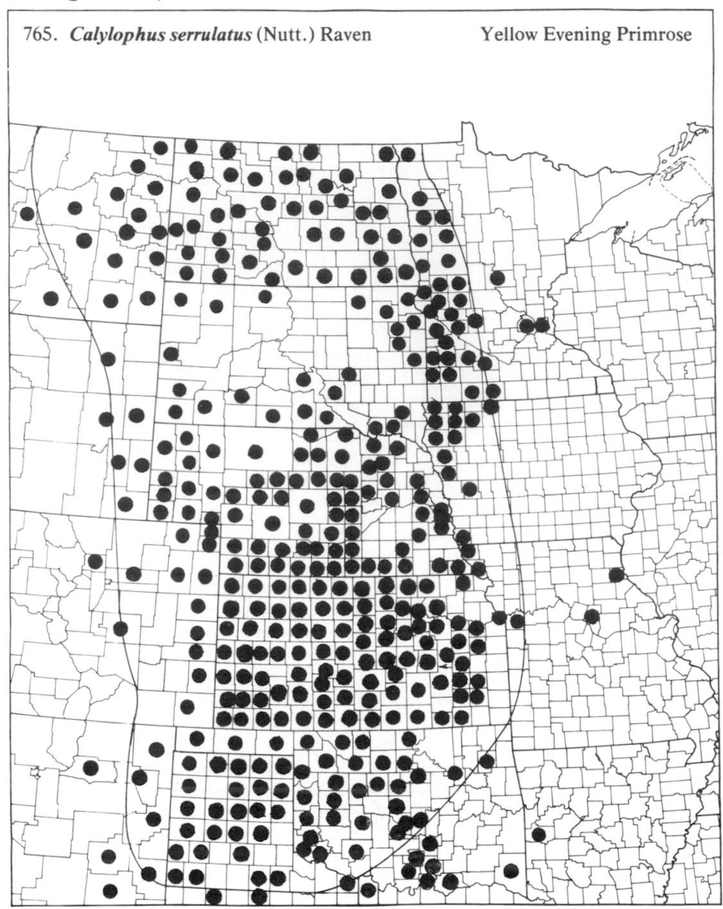

766. ***Circaea alpina*** L.

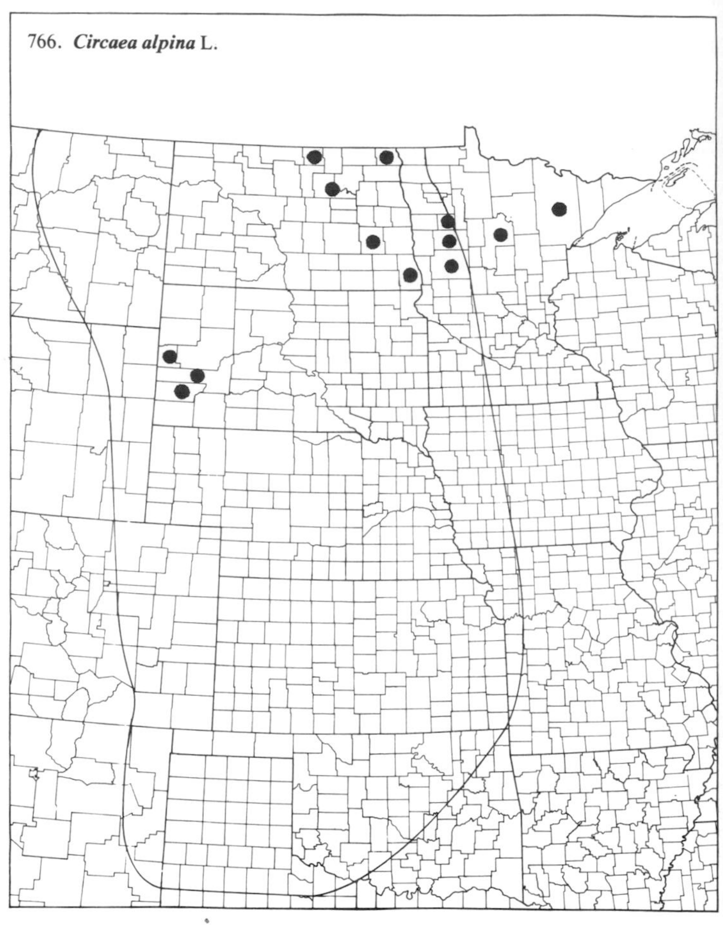

767. ***Circaea lutetiana*** L. ssp. ***canadensis*** (L.) Asch. & Magnus — Enchanter's Nightshade

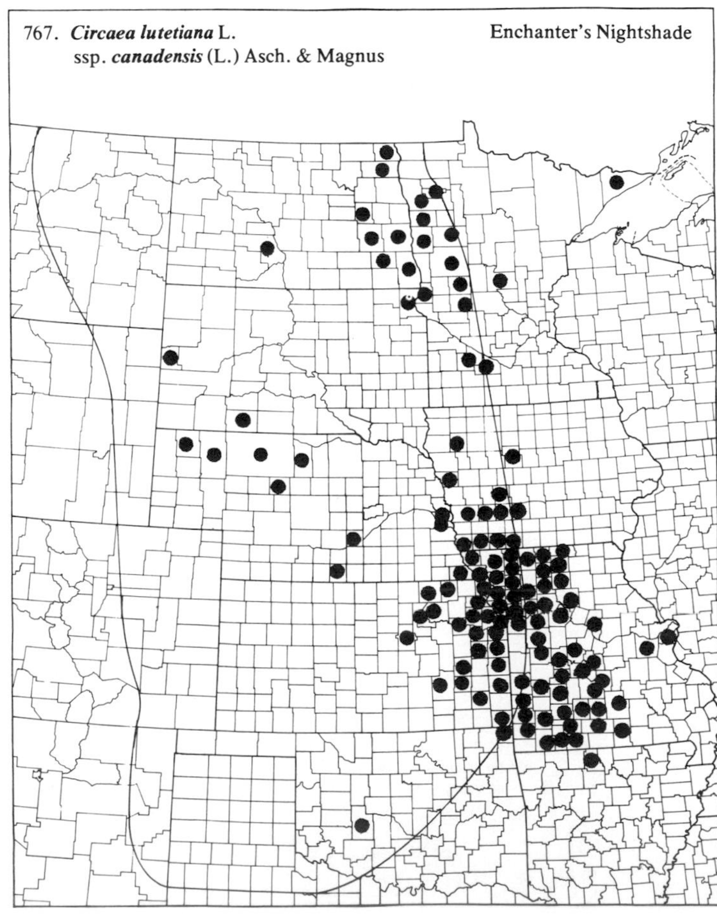

768. ***Epilobium adenocaulon*** Hausskn. — Willow Herb

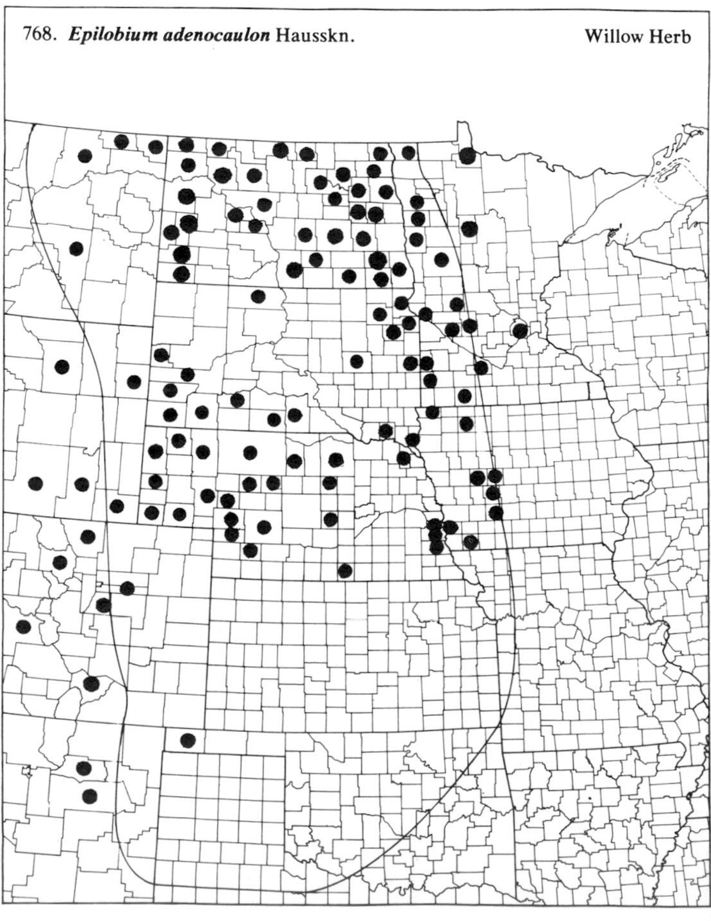

(Onagraceae)

769. ***Epilobium angustifolium*** L. Fireweed

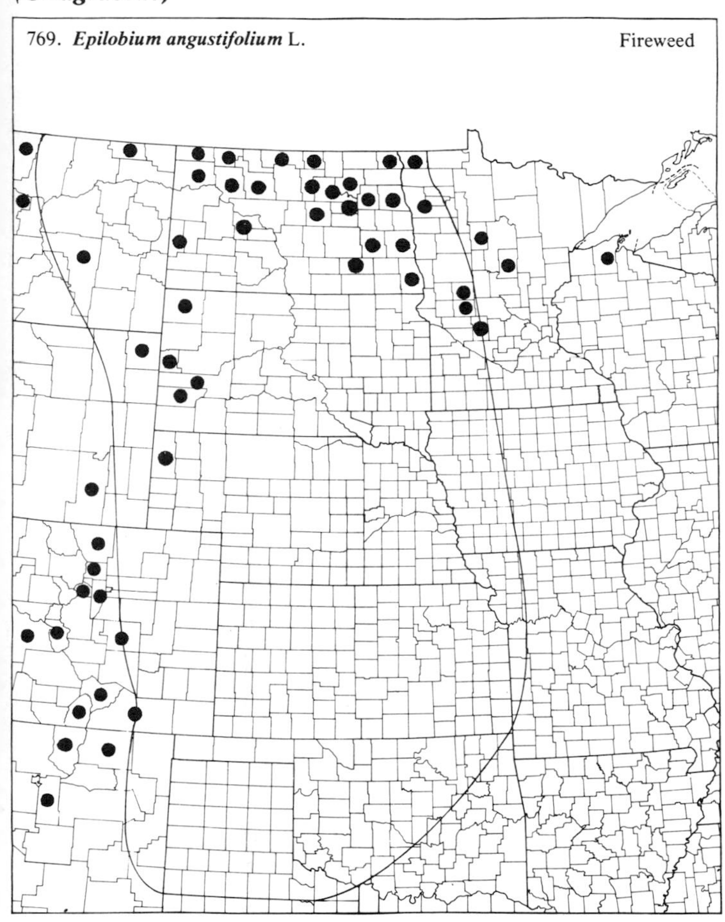

770. ***Epilobium coloratum*** Biehler Purple-leaved Willow Herb

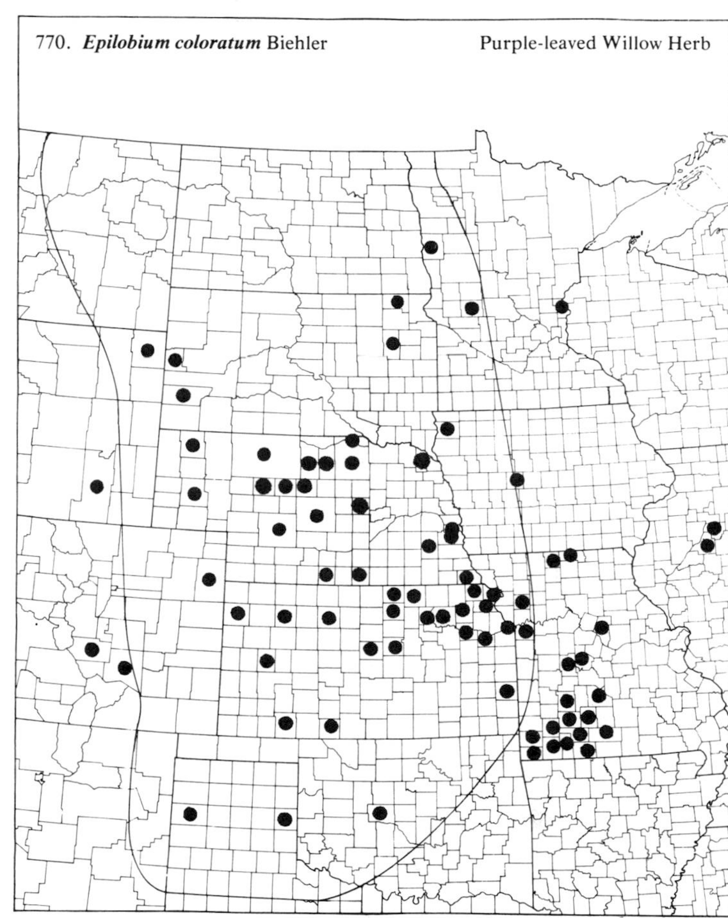

771. ***Epilobium leptophyllum*** Raf. Narrow-leaved Willow Herb

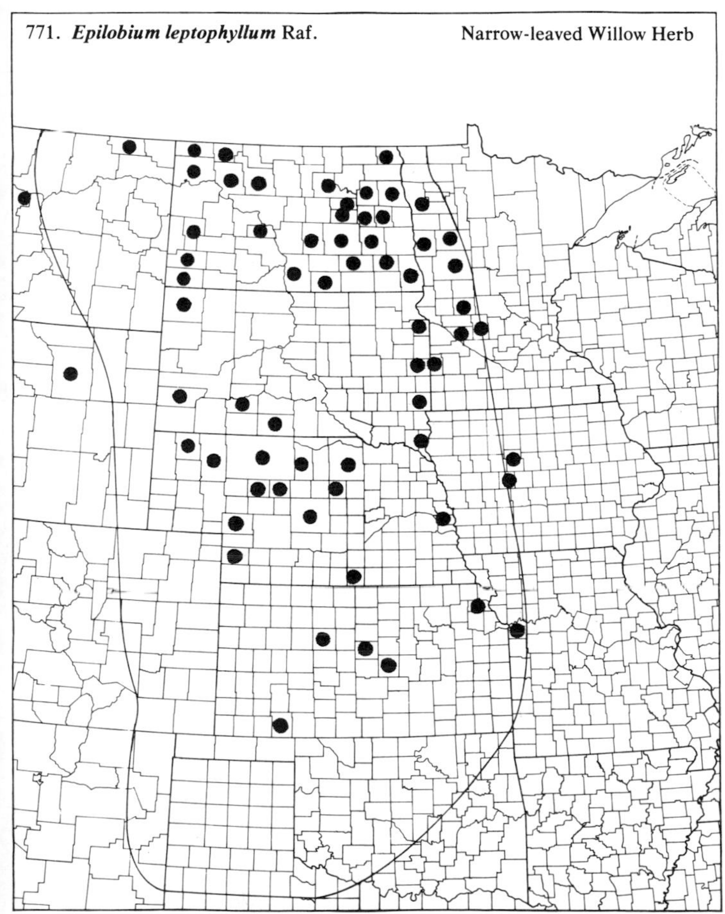

772. ***Epilobium paniculatum*** Nutt. Willow Herb

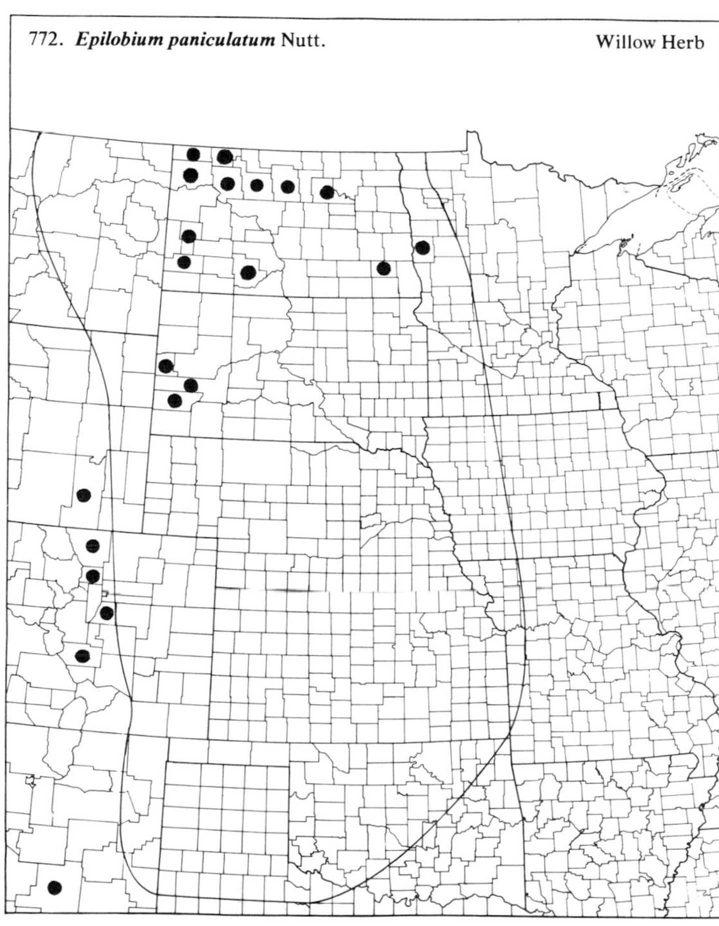

(Onagraceae)

773. ***Gaura coccinea*** Pursh — Scarlet Gaura

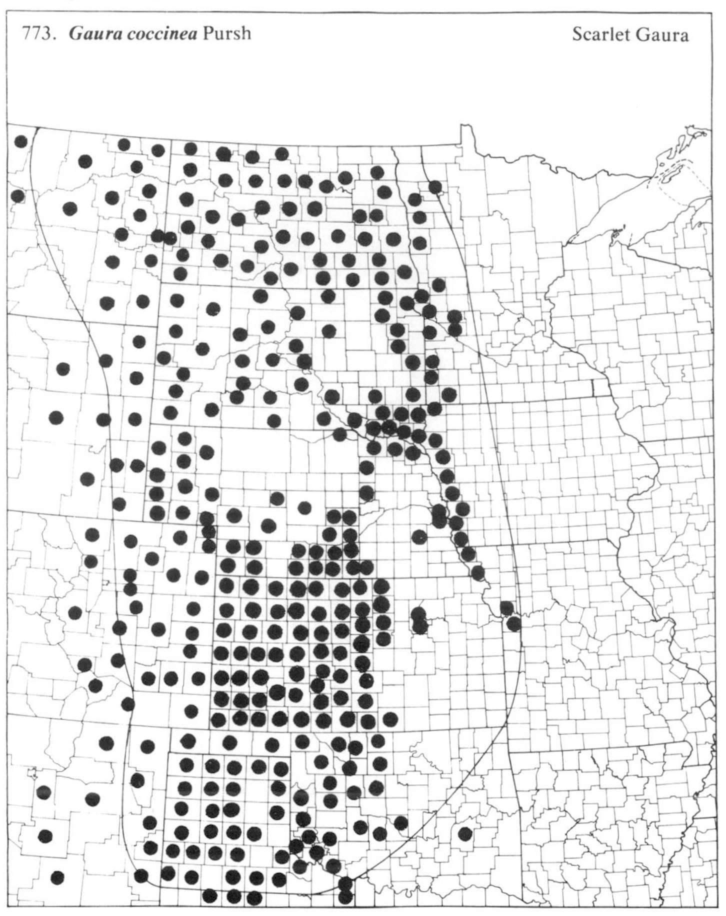

774. ***Gaura longiflora*** Spach

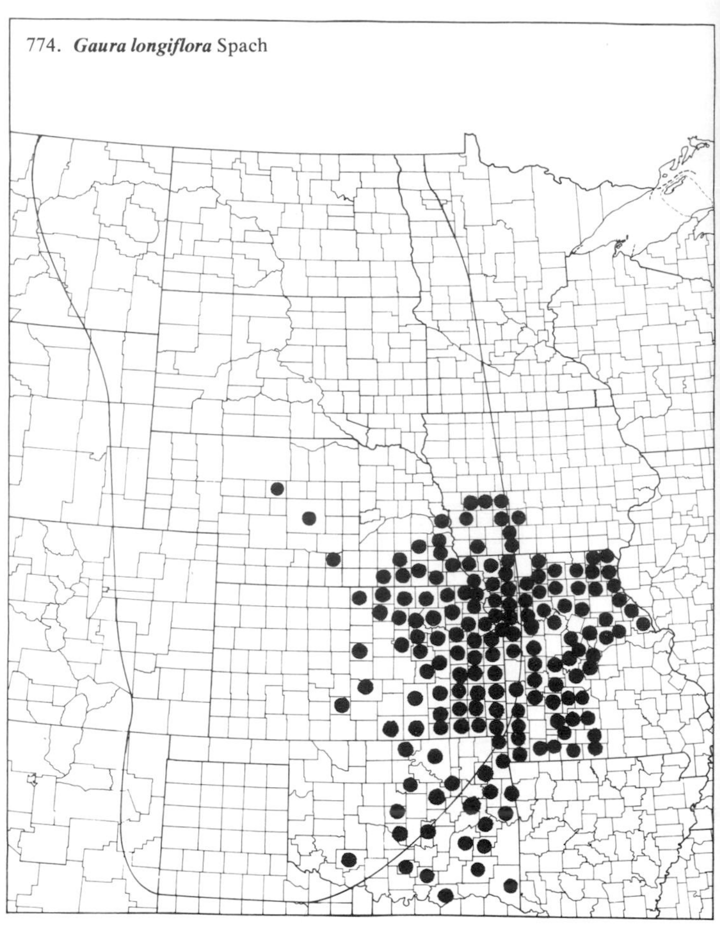

775. ***Gaura parviflora*** Dougl. — Velvety Gaura

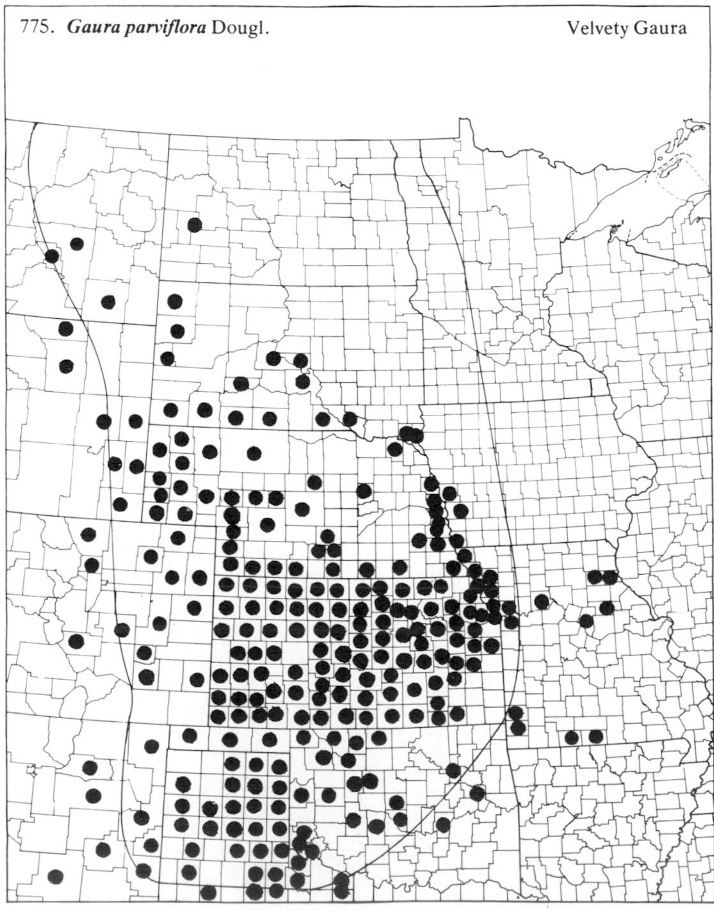

776. ***Gaura sinuata*** Nutt.

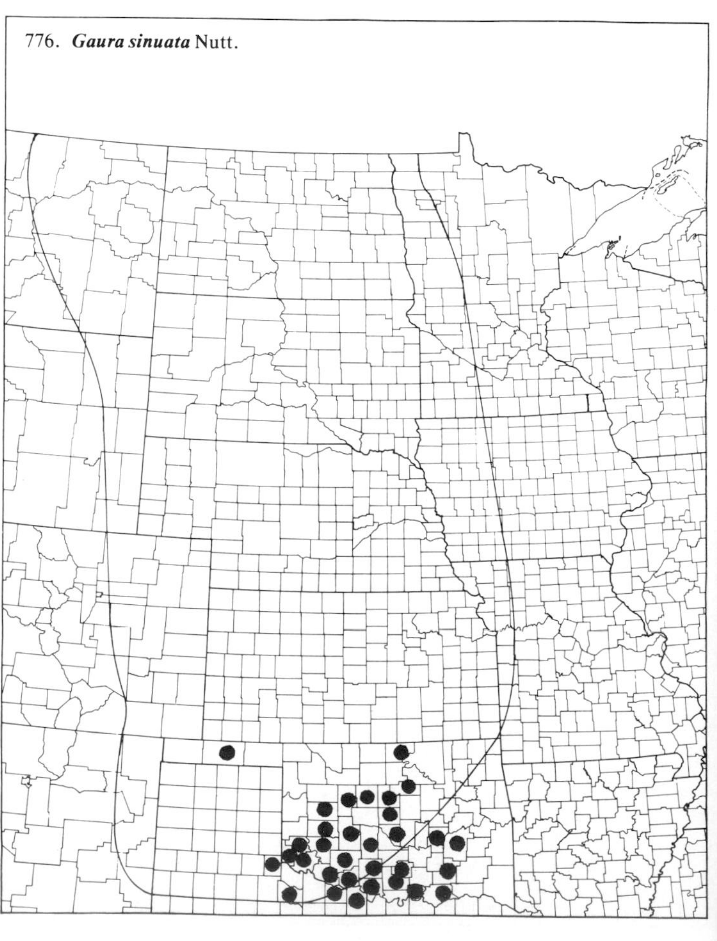

(Onagraceae)

777. ***Gaura suffulta*** Engelm. ex Gray

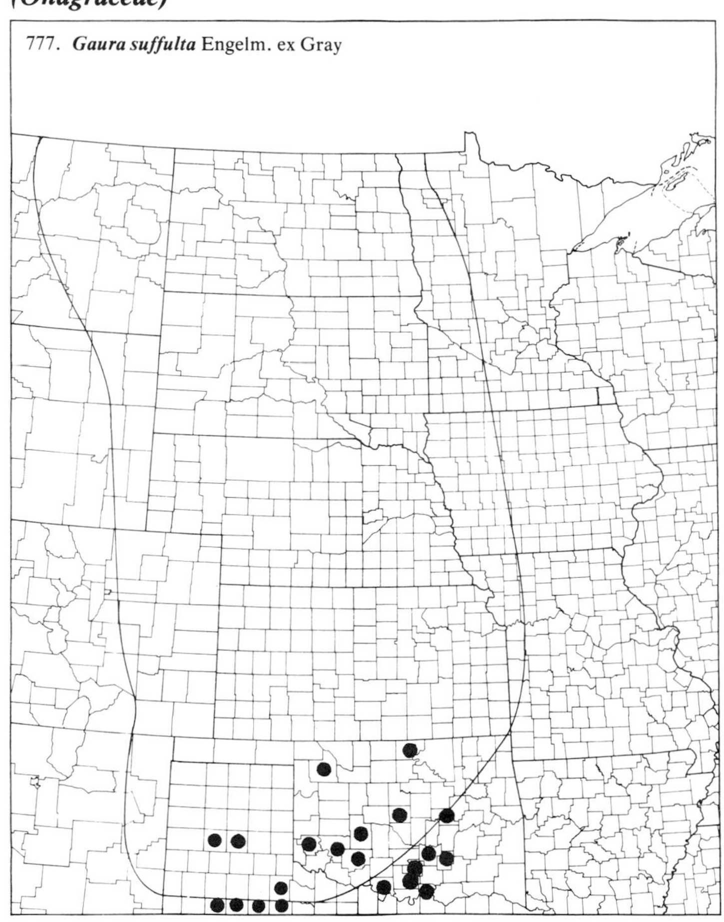

778. ***Gaura triangulata*** Buckl.

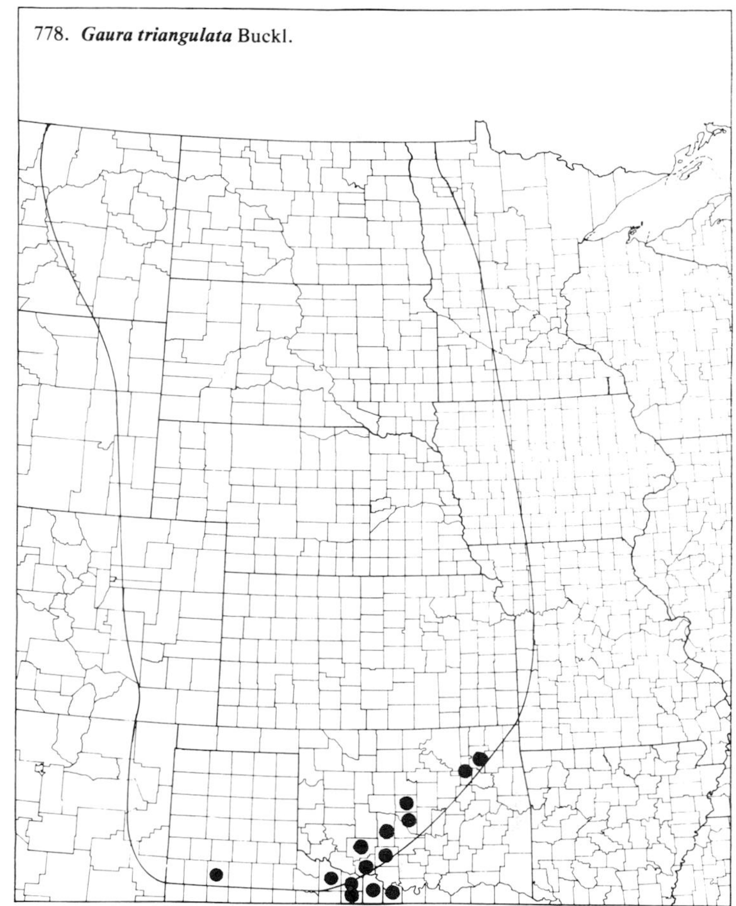

779. ***Gaura villosa*** Torr. — Hairy Gaura

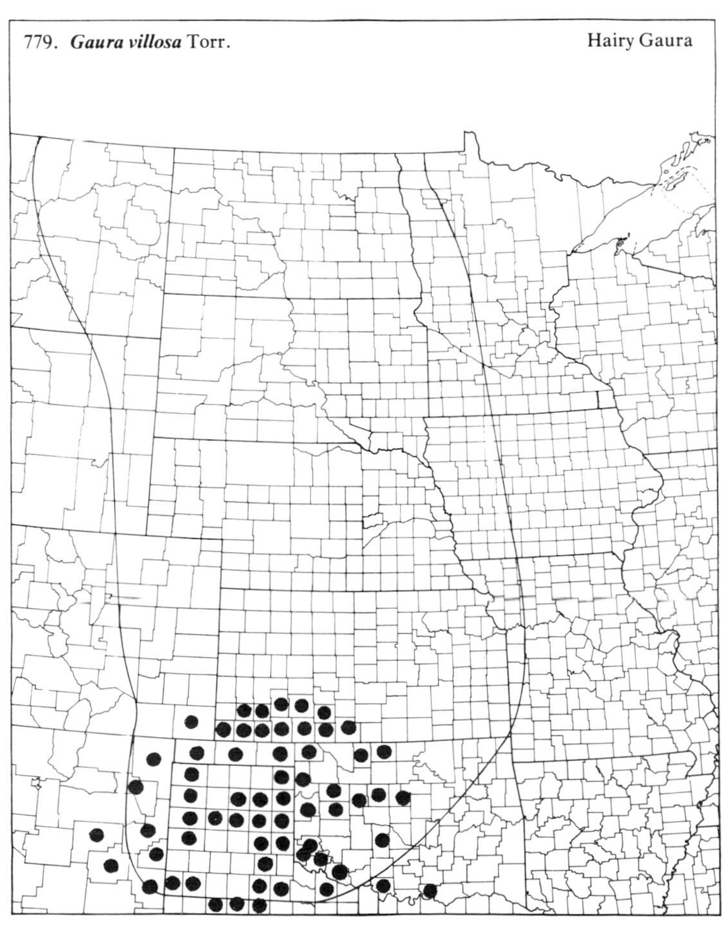

780. ***Ludwigia alternifolia*** L. var. ***pubescens*** Palm. & Steyerm. — Bushy Seedbox

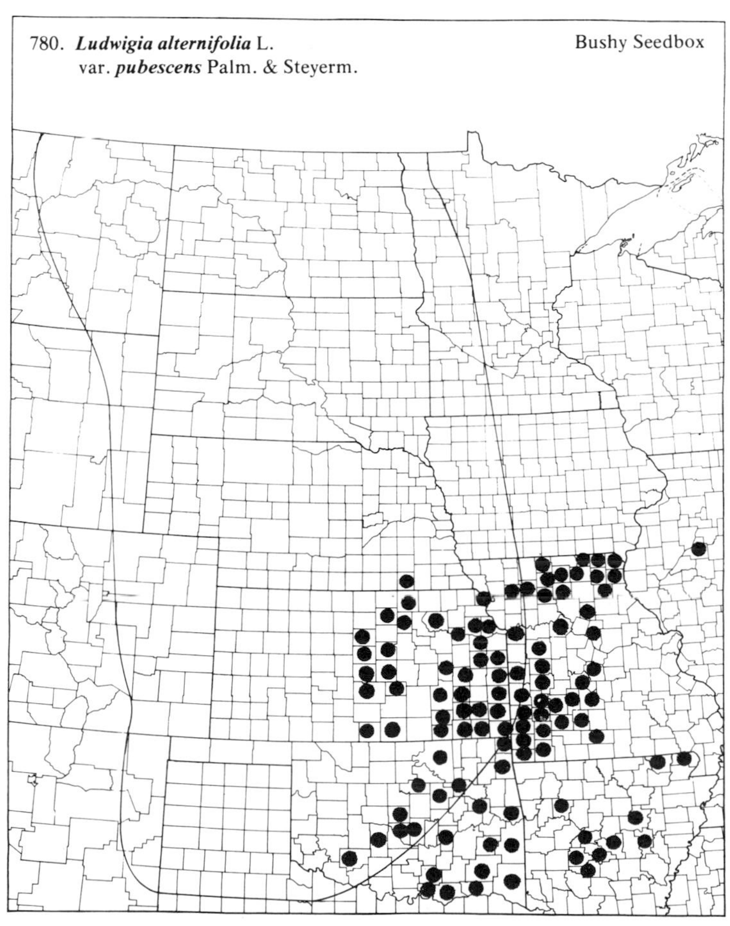

(Onagraceae)

781. ***Ludwigia palustris*** (L.) Ell. Water Purslane

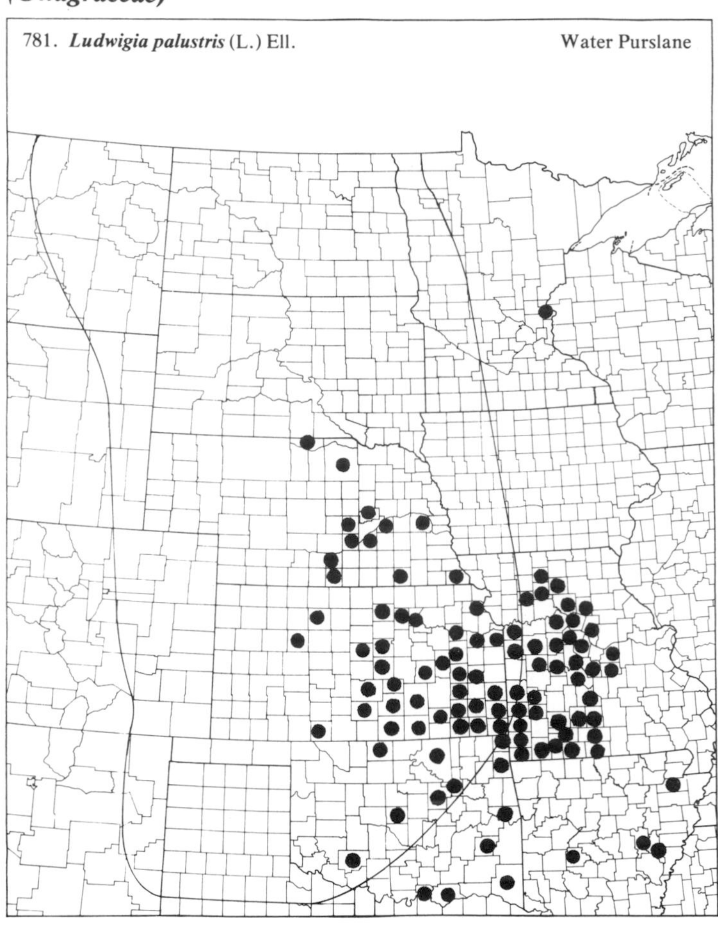

782. ***Ludwigia peploides*** (H.B.K.) Raven ssp. ***glabrescens*** (O. Ktze.) Raven

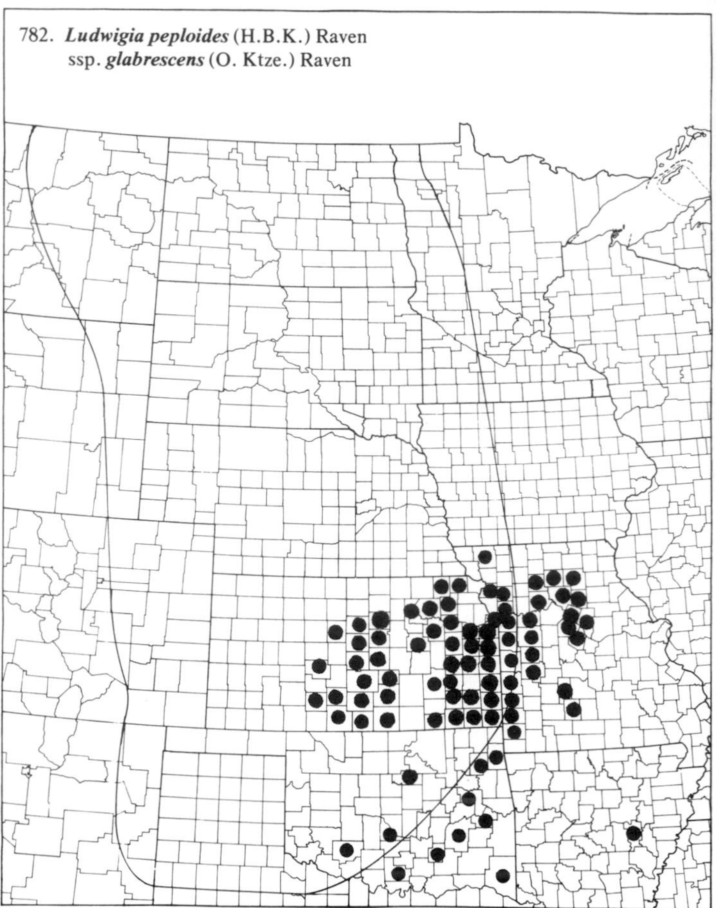

783. ***Ludwigia polycarpa*** Short & Peter Many-seeded Seedbox

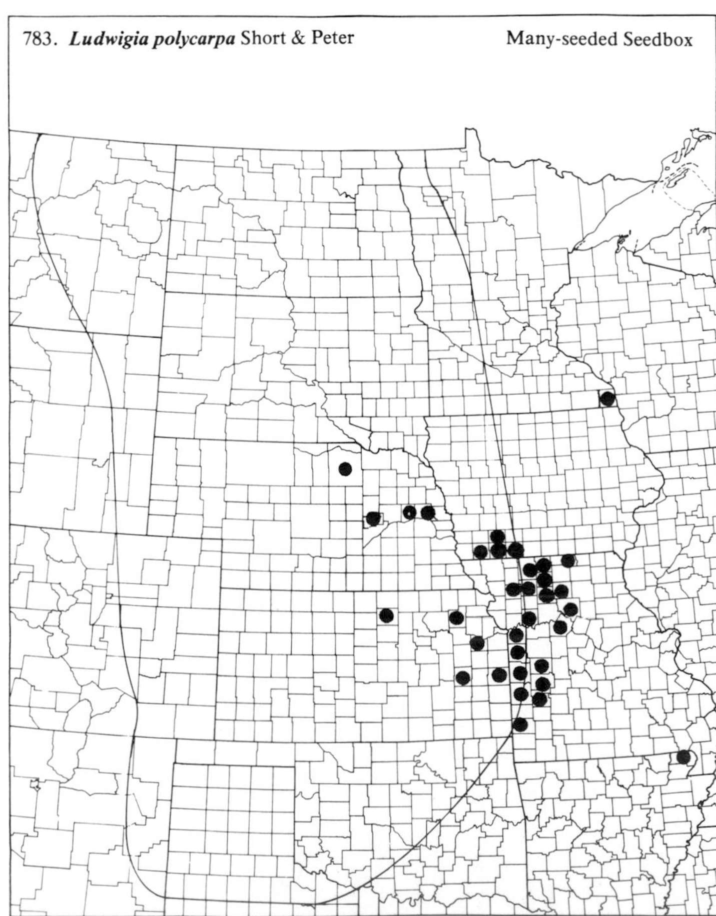

784. ***Oenothera albicaulis*** Pursh Pale Evening Primrose

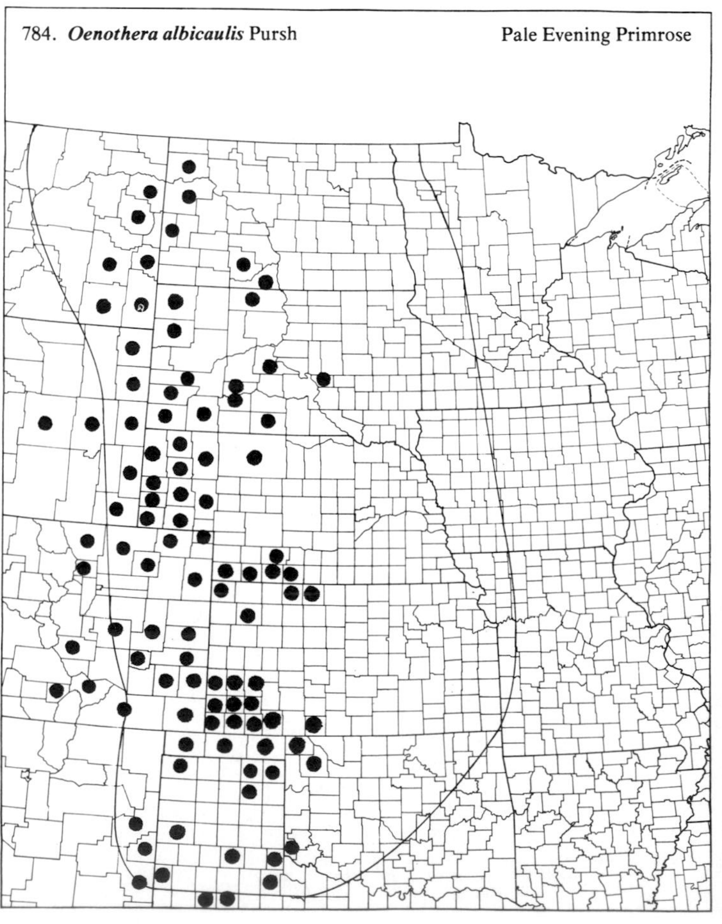

(Onagraceae)

785. ***Oenothera biennis*** L. ssp. ***centralis*** Munz — Common Evening Primrose

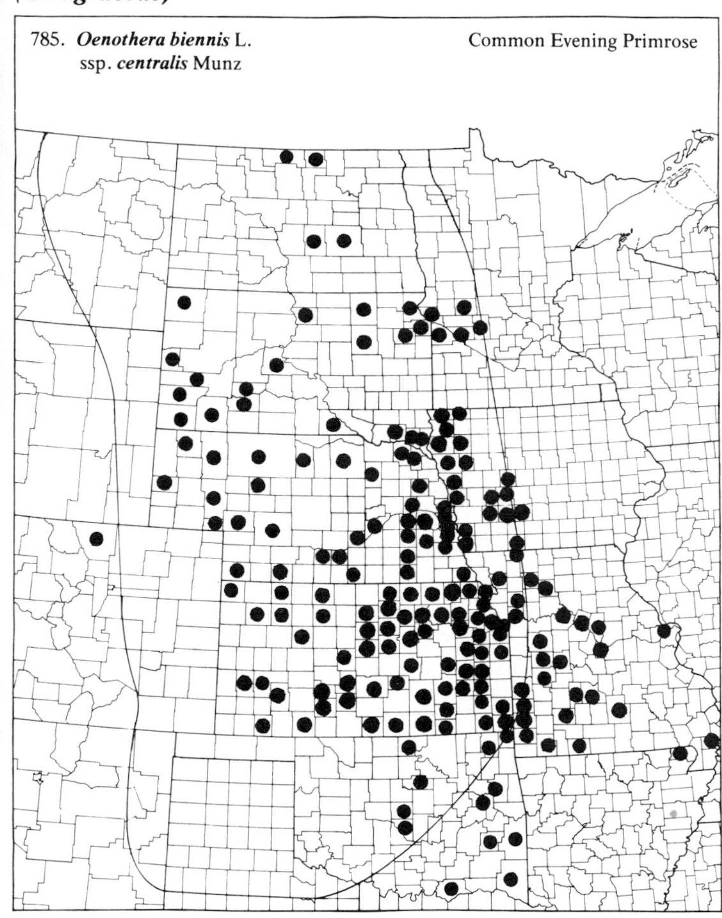

786. ***Oenothera caespitosa*** Nutt. ssp. ***caespitosa*** — Gumbo Lily

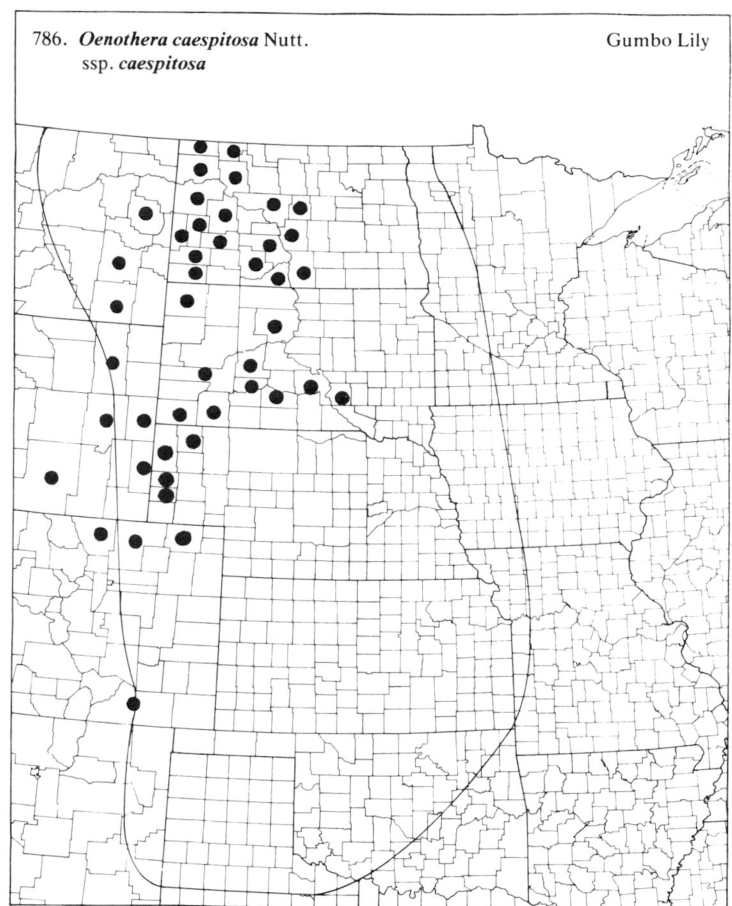

787. ***Oenothera caespitosa*** Nutt. ssp. ***montana*** (Nutt.) Munz — Gumbo Lily

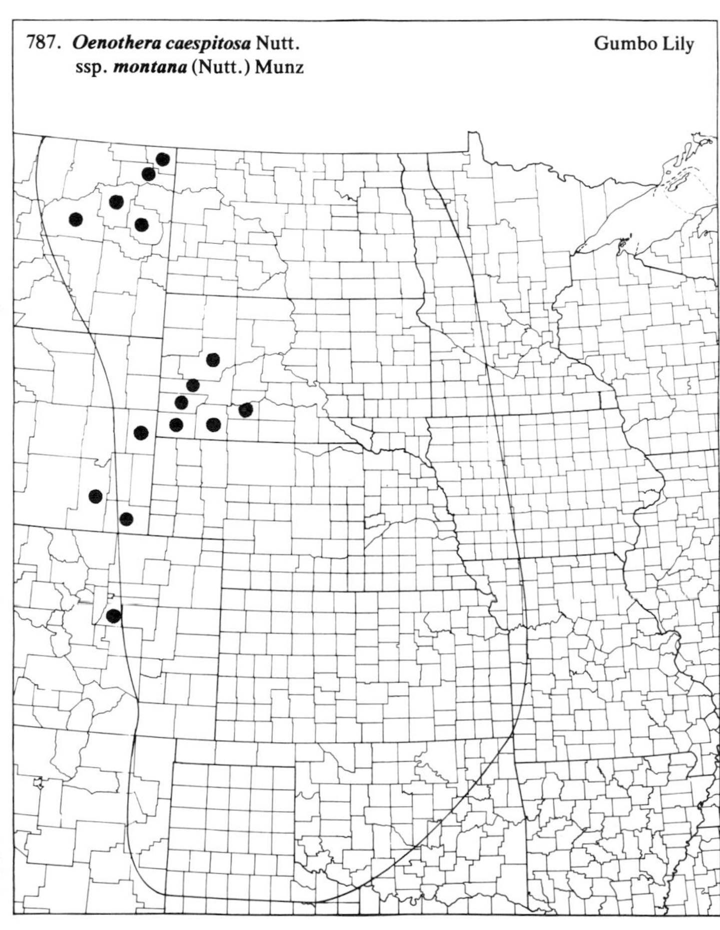

788. ***Oenothera canescens*** Torr. & Frem. — Beakpod Evening Primrose

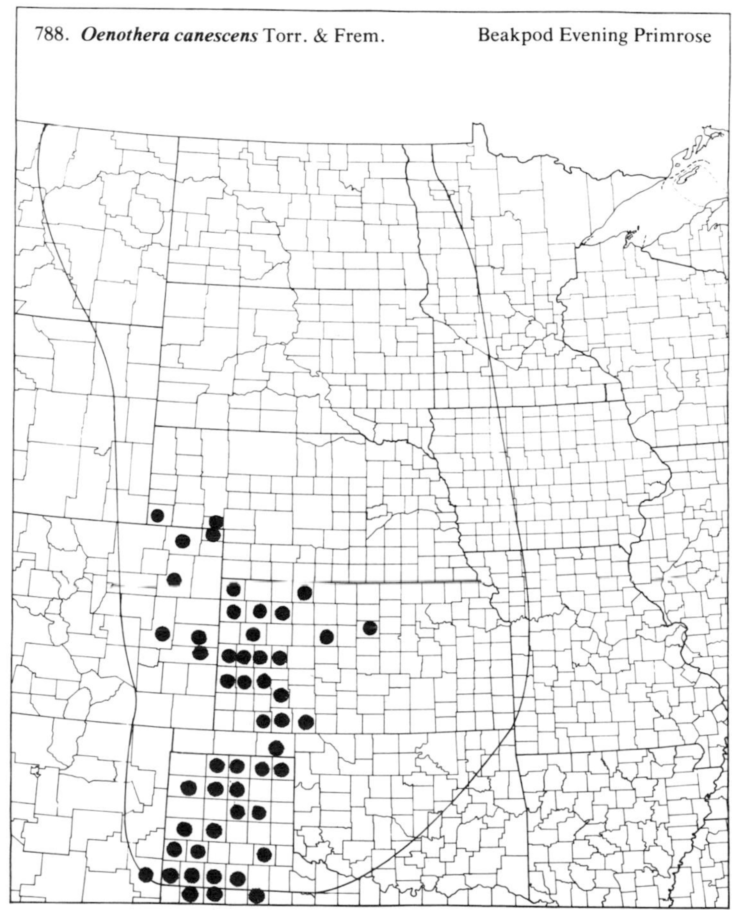

(Onagraceae)

789. ***Oenothera coronopifolia*** T. & G. Combleaf Evening Primrose

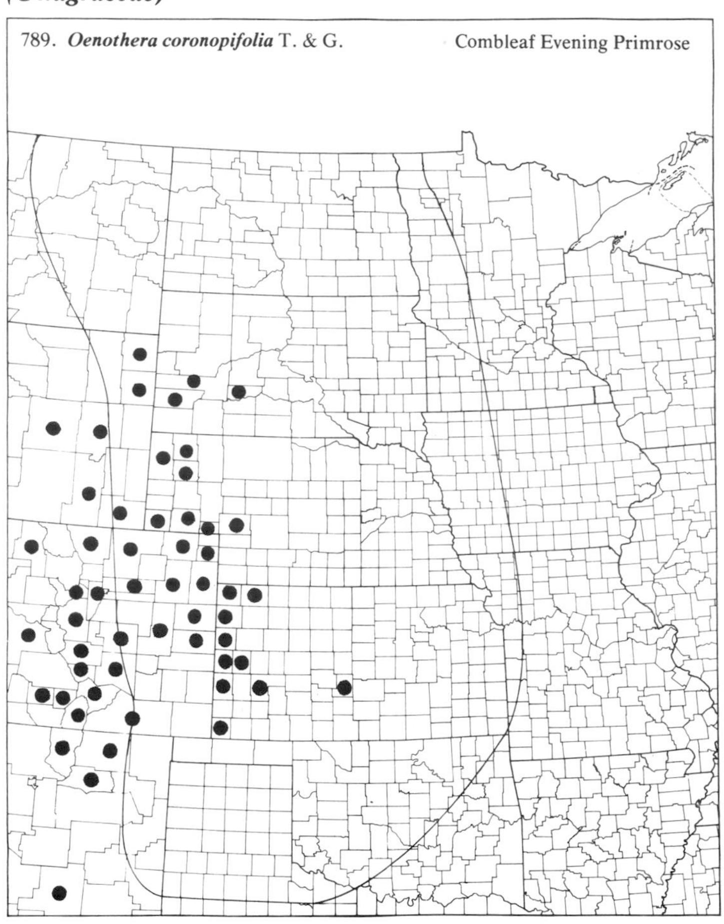

790. ***Oenothera engelmannii*** (Small) Munz

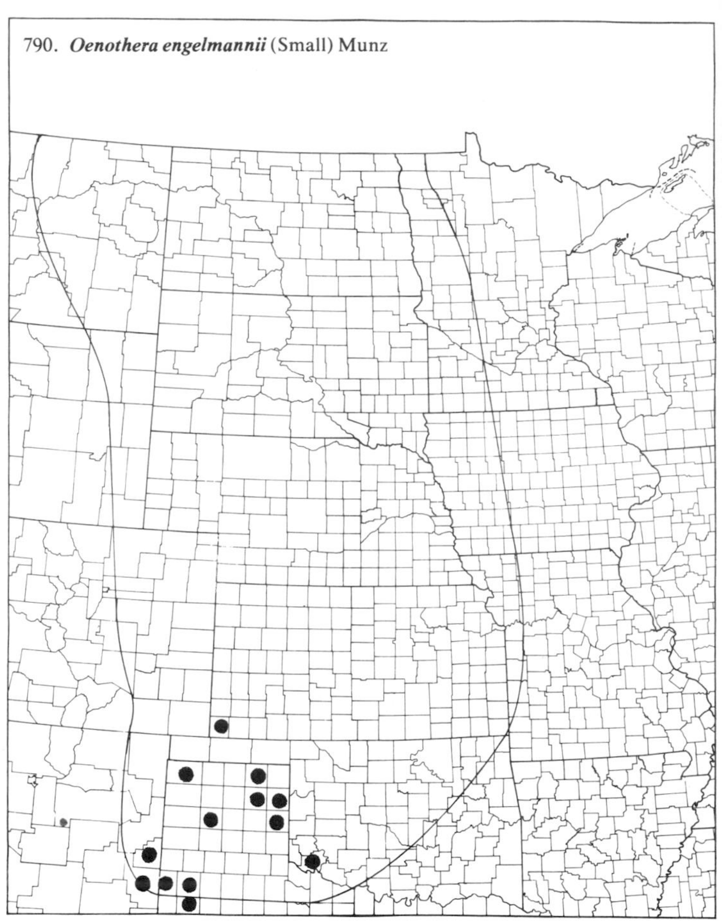

791. ***Oenothera flava*** (A. Nels.) Garrett

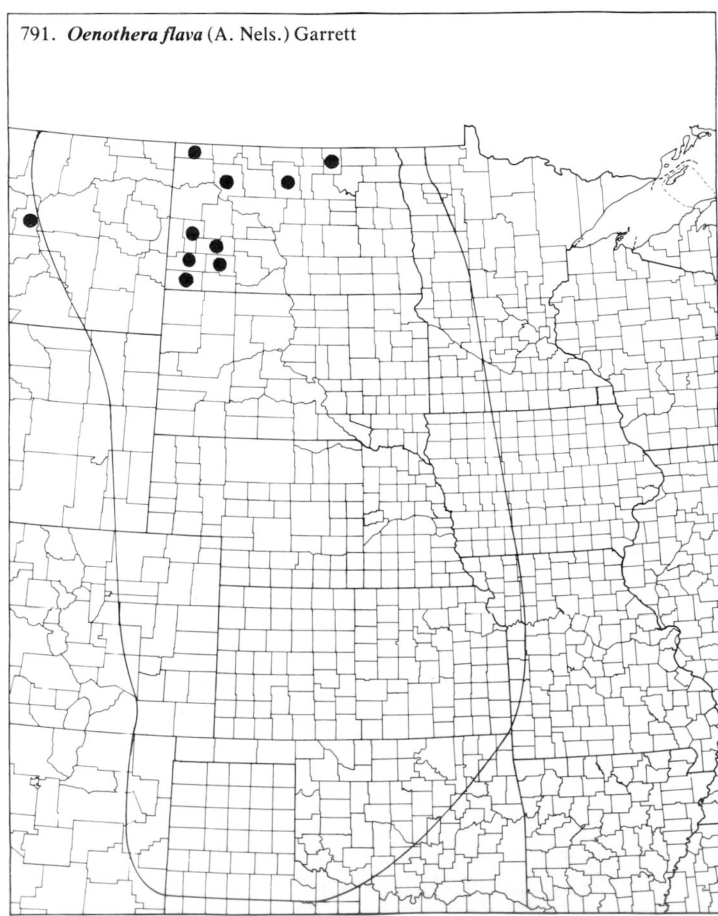

792. ***Oenothera fremontii*** Wats. Fremont's Evening Primrose

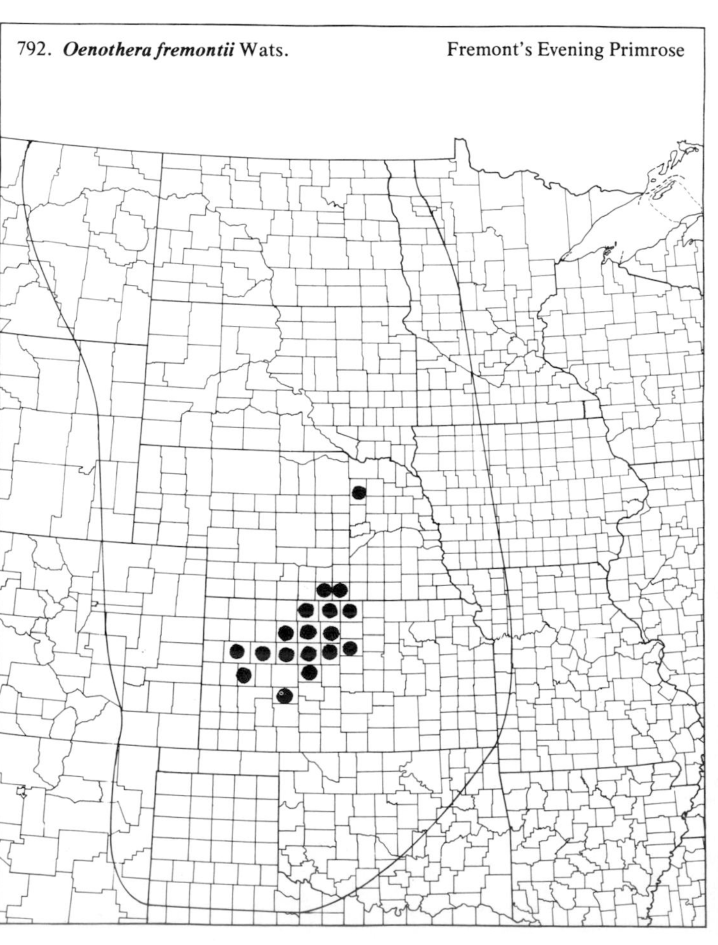

(Onagraceae)

793. ***Oenothera grandis*** (Britt.) Smyth

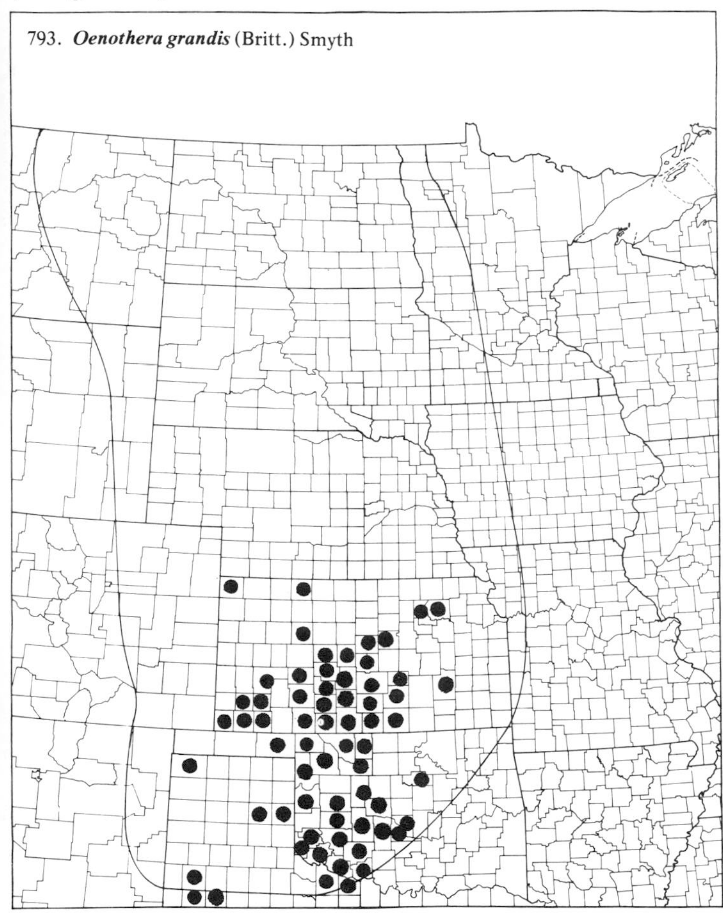

794. ***Oenothera hookeri*** T. & G. ssp. ***hirsutissima*** (Gray) Munz — Hooker's Evening Primrose

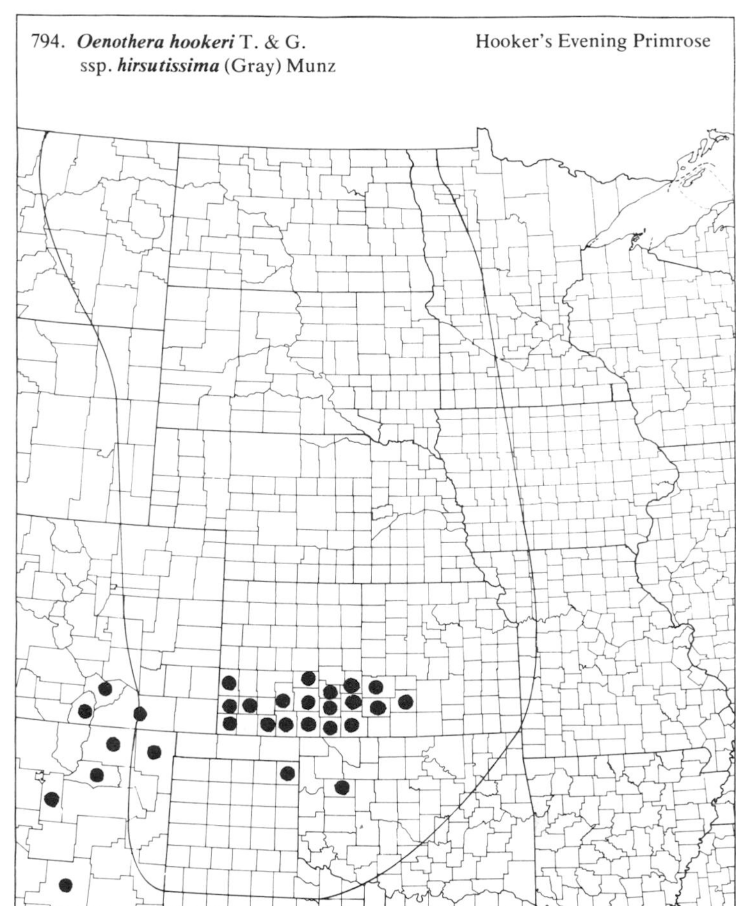

795. ***Oenothera jamesii*** T. & G.

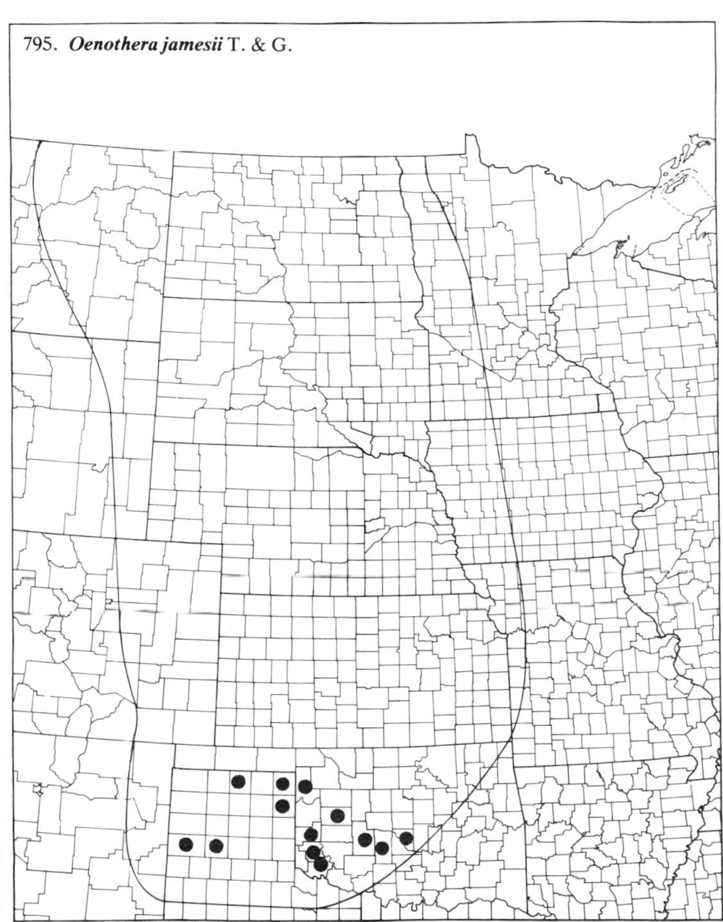

796. ***Oenothera laciniata*** Hill — Cutleaf Evening Primrose

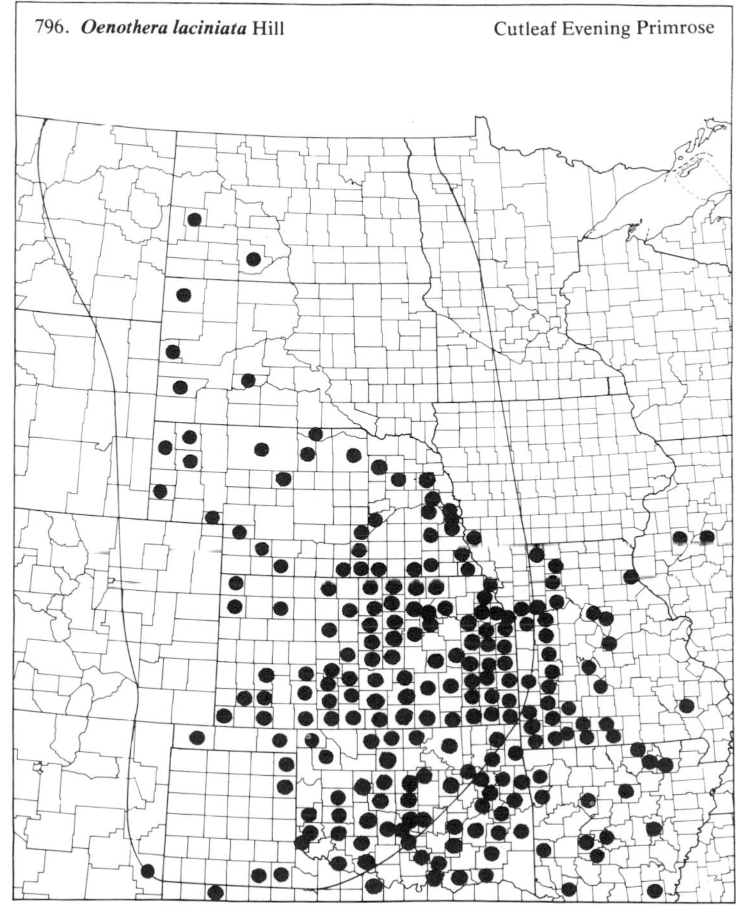

(Onagraceae)

797. ***Oenothera linifolia*** Nutt. Narrow-leaved Evening Primrose

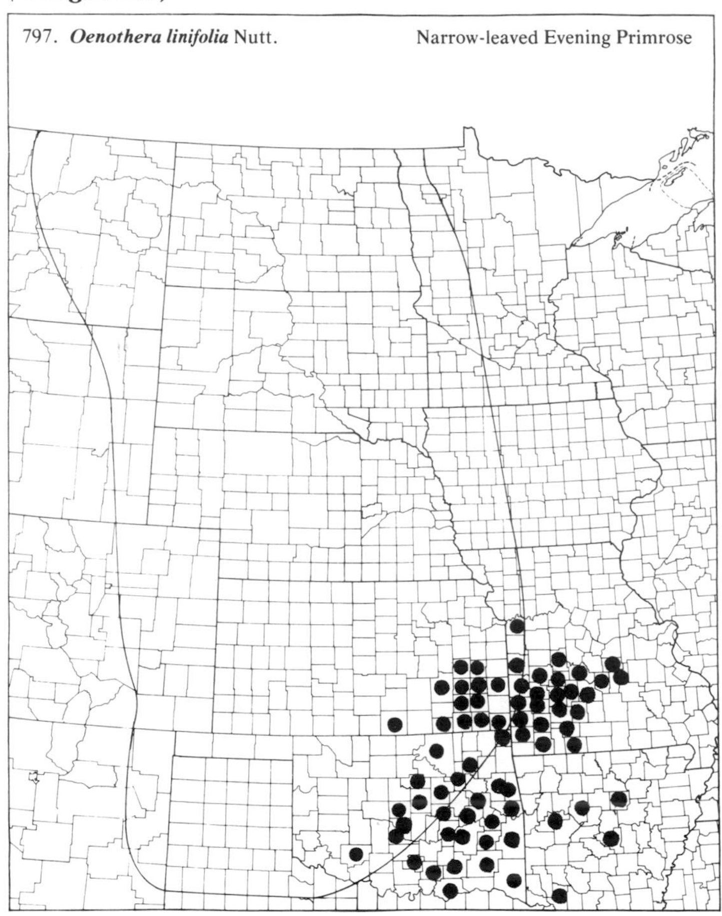

798. ***Oenothera macrocarpa*** Nutt.
var. ***macrocarpa***

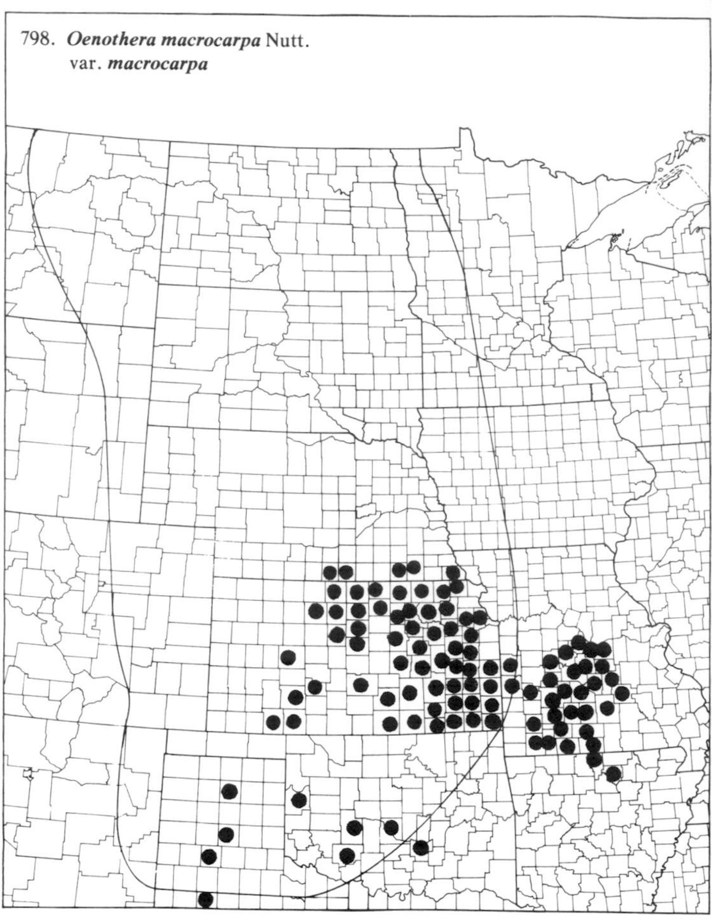

799. ***Oenothera macrocarpa*** Nutt.
var. ***incana*** (Gray) Reveal

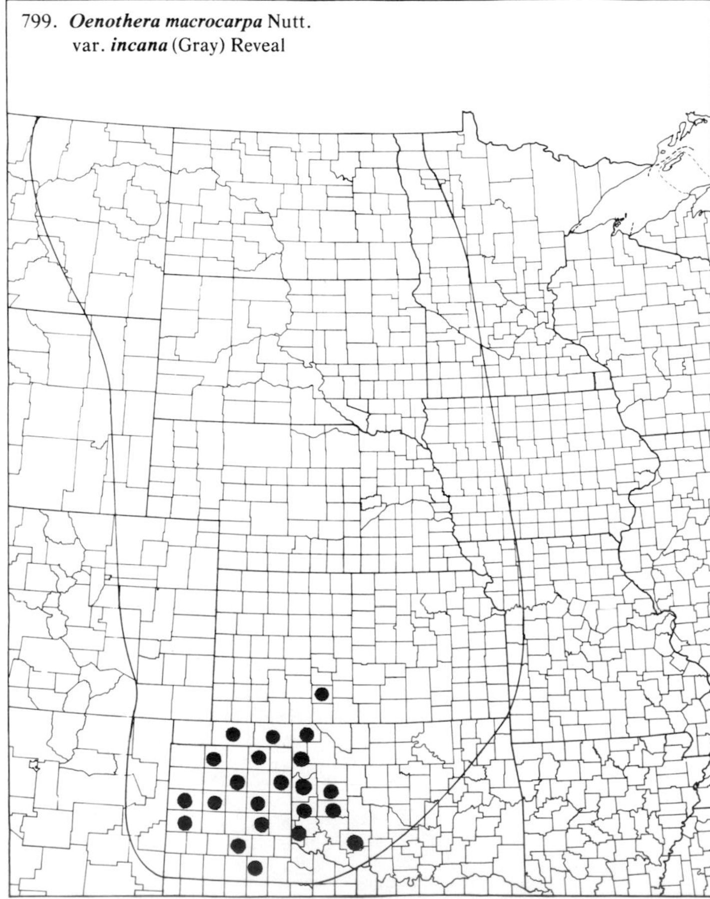

800. ***Oenothera macrocarpa*** Nutt.
var. ***oklahomensis*** (Nort.) Reveal

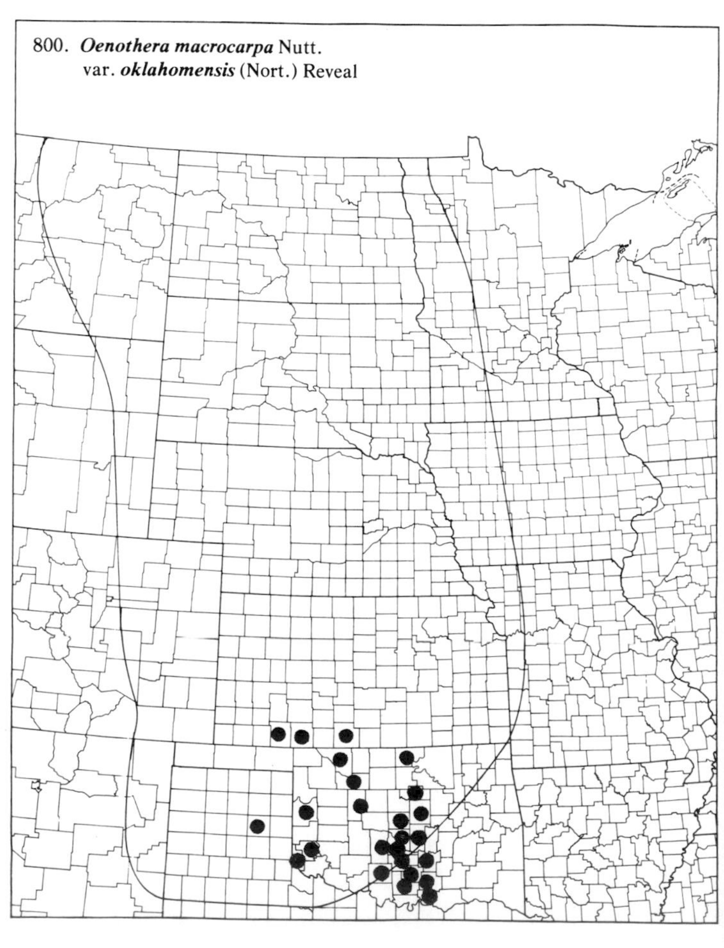

(Onagraceae)

801. ***Oenothera nuttallii*** Sweet — White-stemmed Evening Primrose

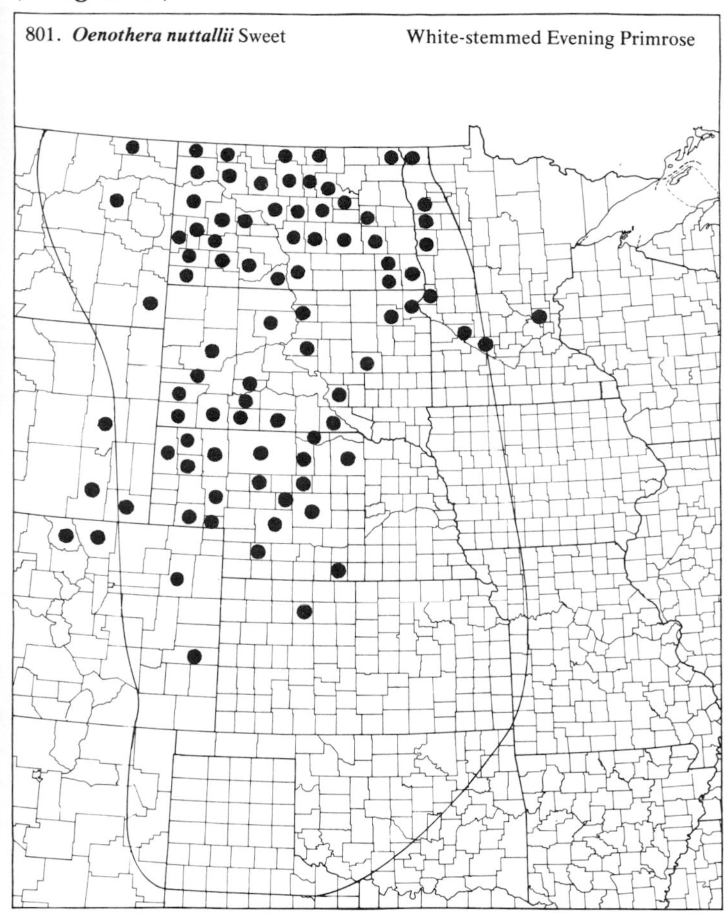

802. ***Oenothera pallida*** Lindl. ssp. ***latifolia*** (Rydb.) Munz

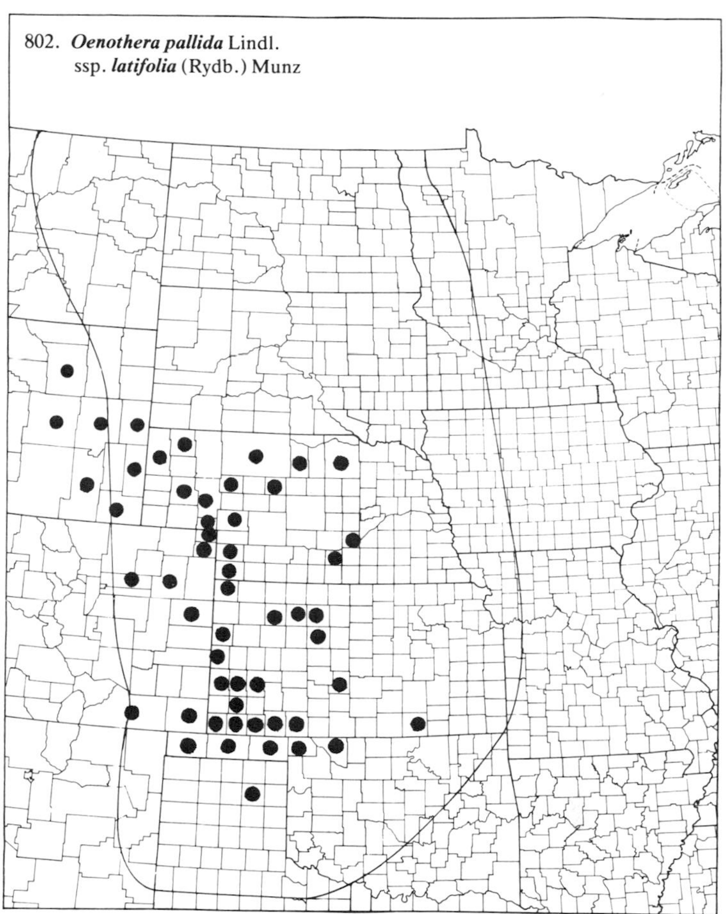

803. ***Oenothera rhombipetala*** Nutt. — Fourpoint Evening Primrose

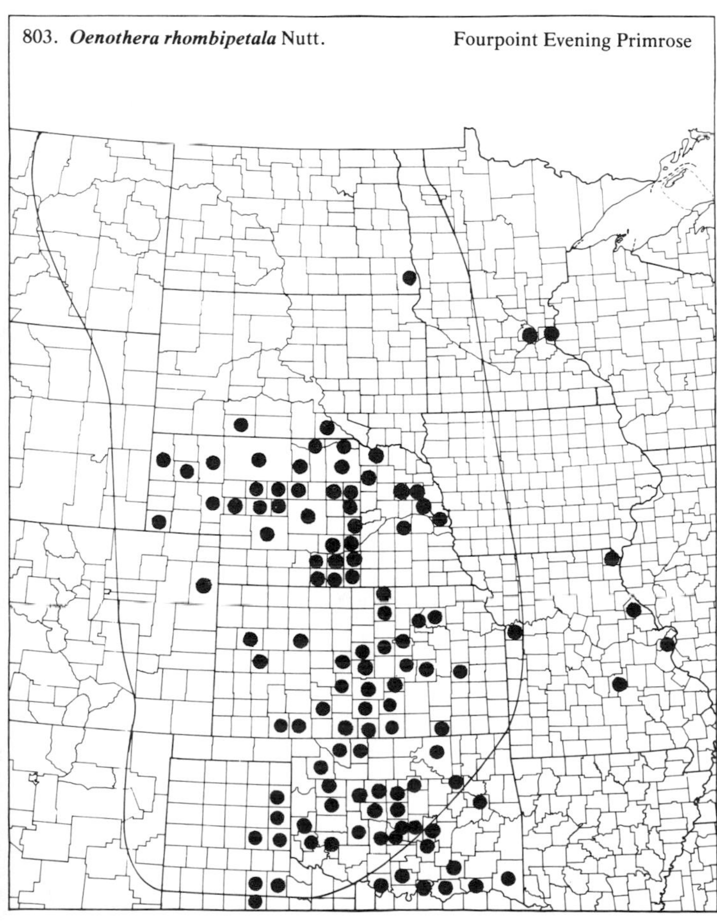

804. ***Oenothera speciosa*** Nutt. — White Evening Primrose

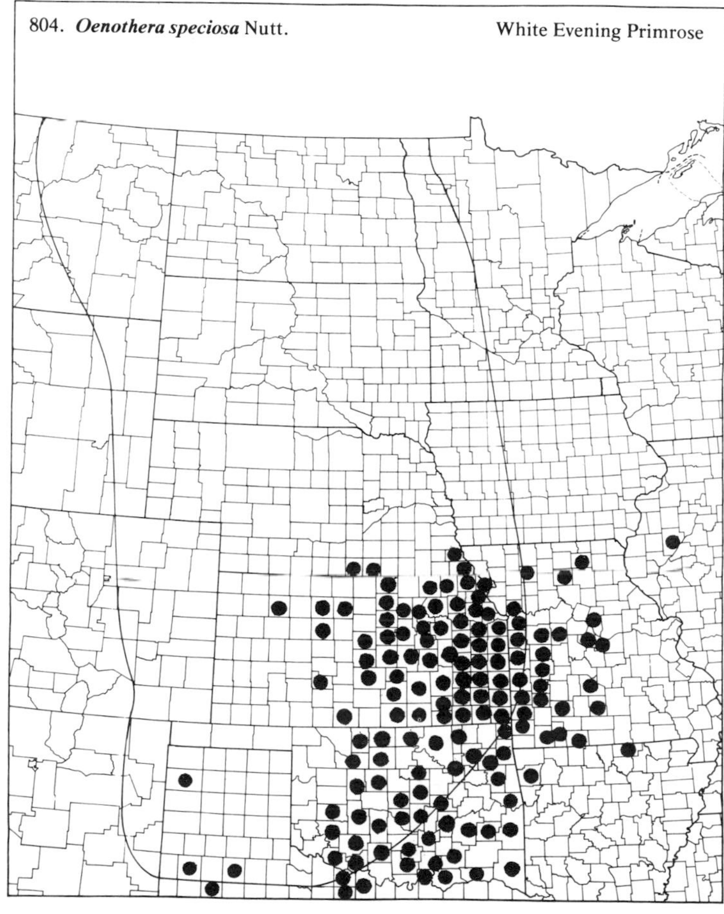

(Onagraceae)

805. ***Oenothera strigosa*** (Rydb.) Mack. & Bush ssp. ***strigosa*** — Common Evening Primrose

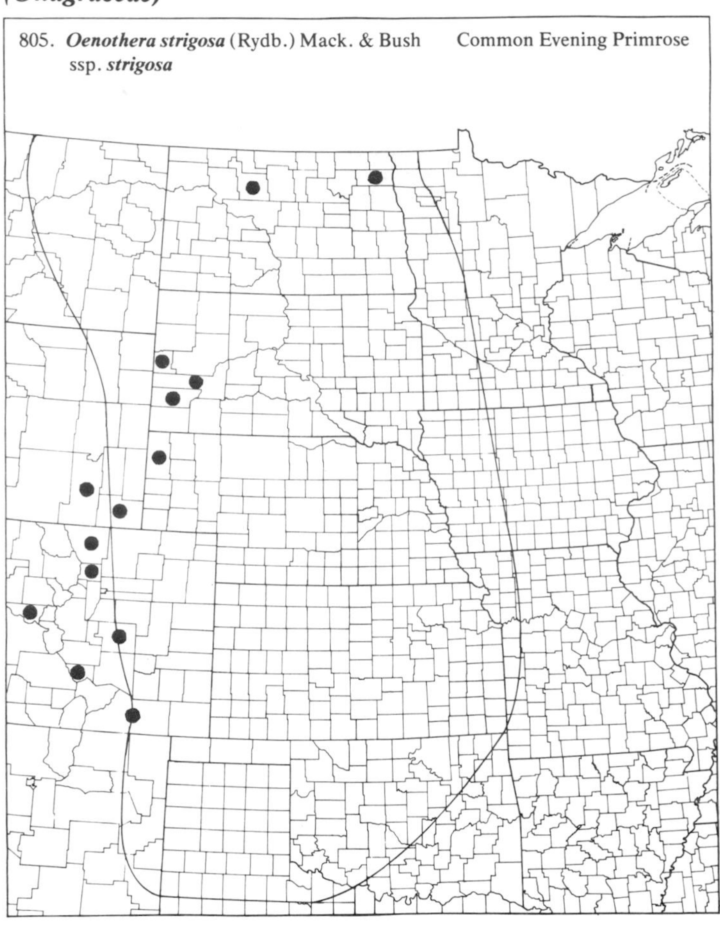

806. ***Oenothera strigosa*** (Rydb.) Mack. & Bush ssp. ***canovirens*** (Steel) Munz — Common Evening Primrose

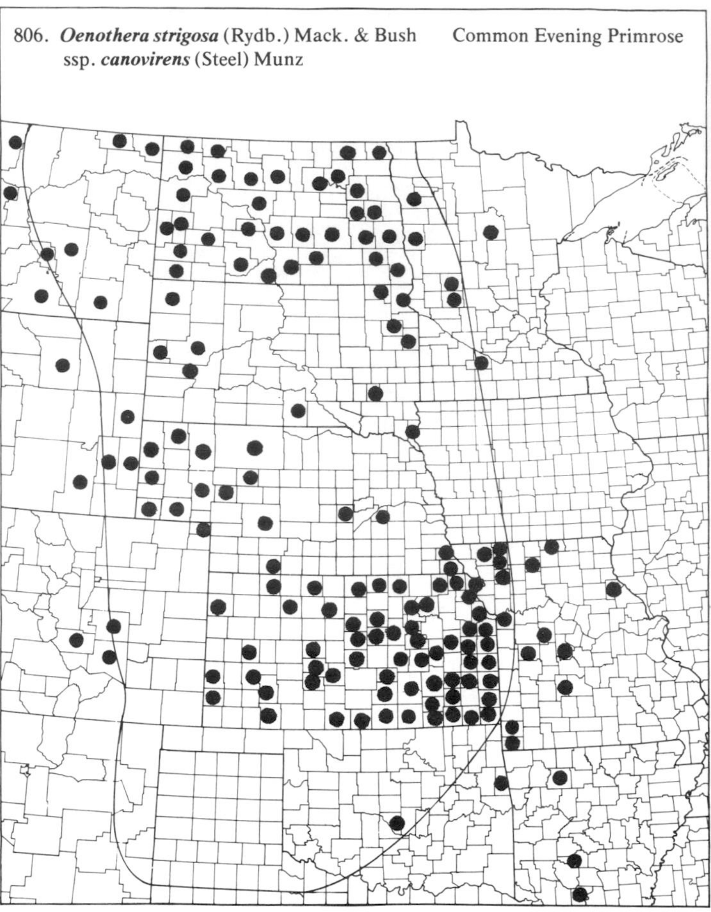

807. ***Oenothera triloba*** Nutt. — Stemless Evening Primrose

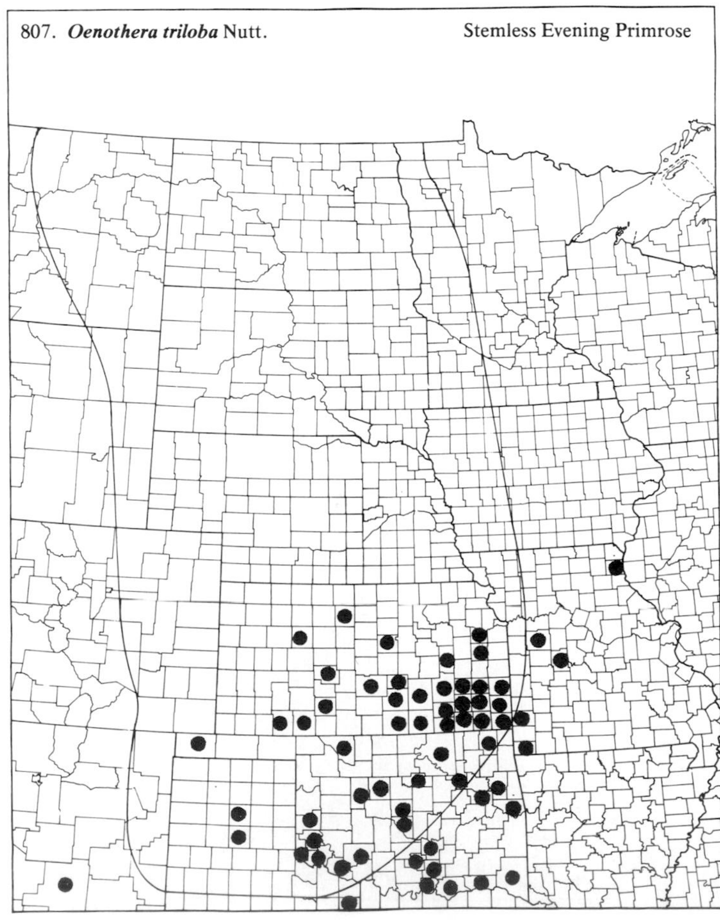

808. ***Stenosiphon linifolius*** (Nutt.) Heynh. — Stenosiphon

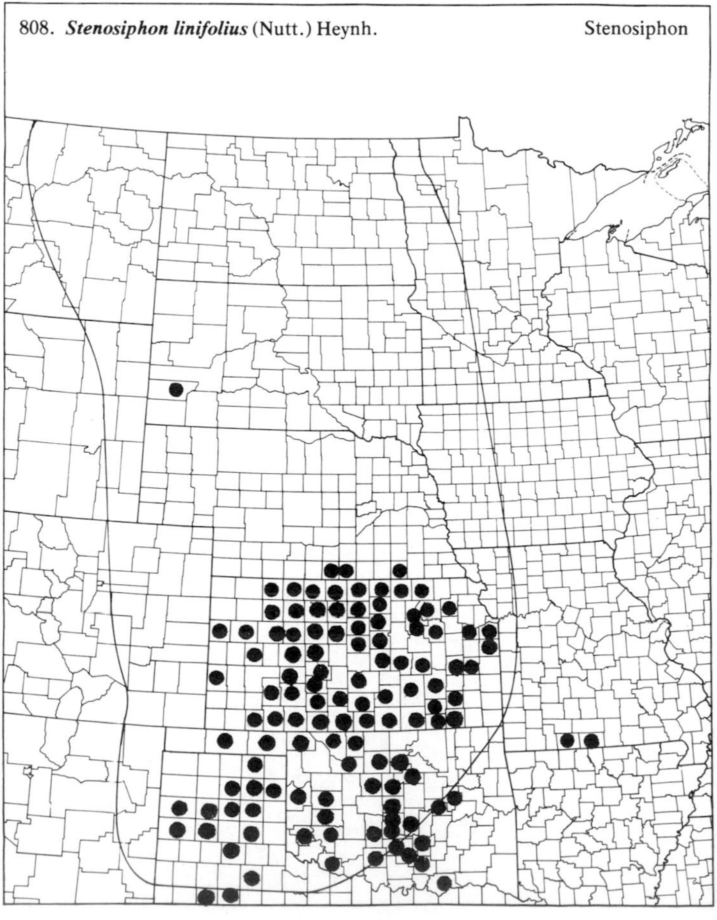

61. *Elaeagnaceae* (Russian Olive Family)

809. *Elaeagnus angustifolia* L. Russian Olive

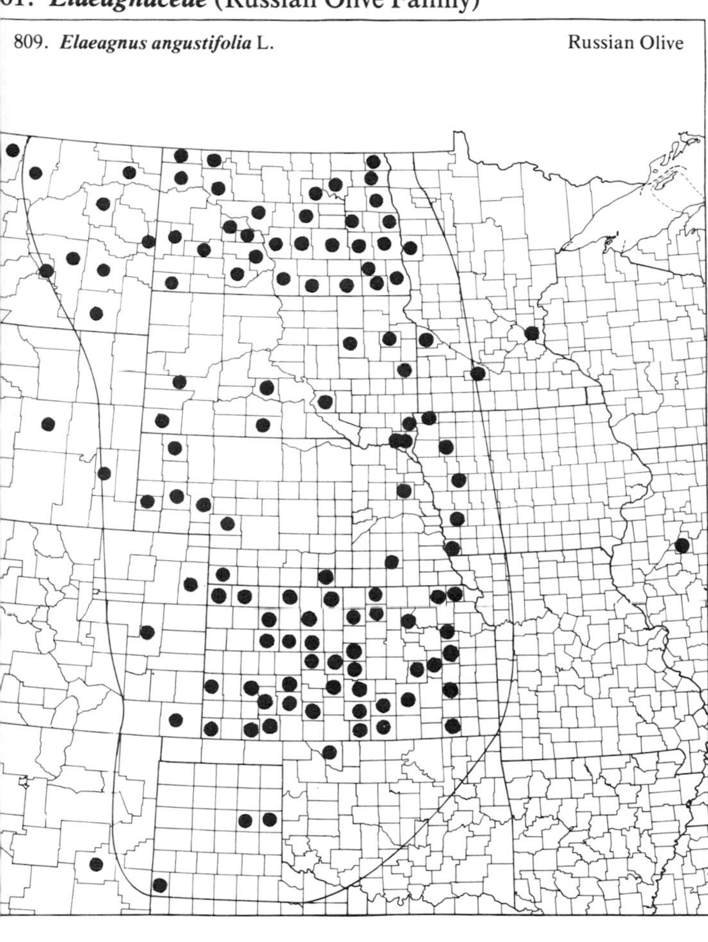

810. *Elaeagnus commutata* Bernh. Silverberry

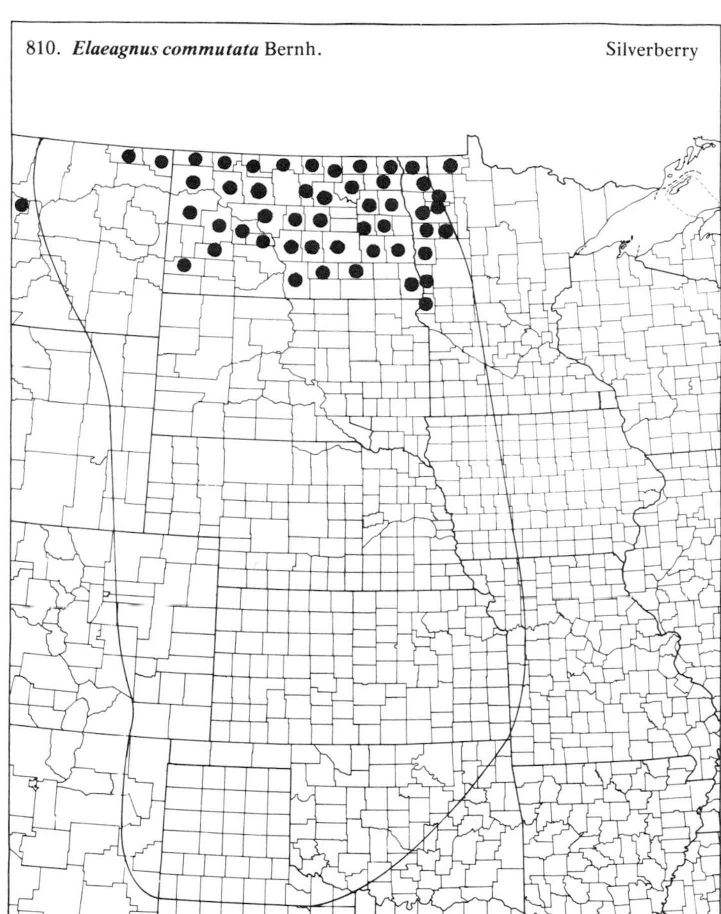

811. *Shepherdia argentea* (Pursh) Nutt. Buffaloberry

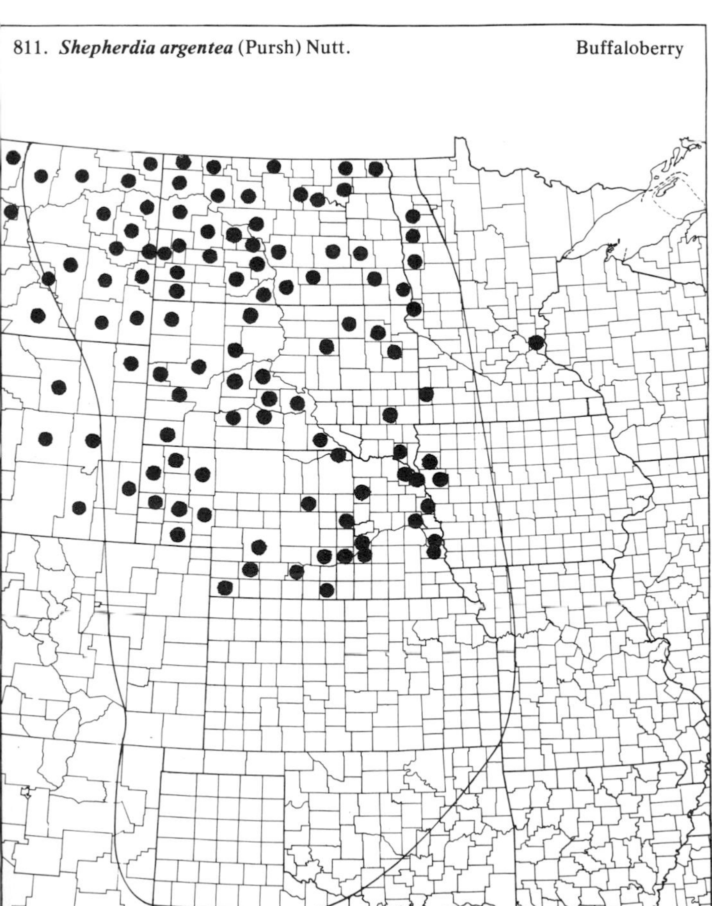

812. *Shepherdia canadensis* (L.) Nutt. Rabbitberry

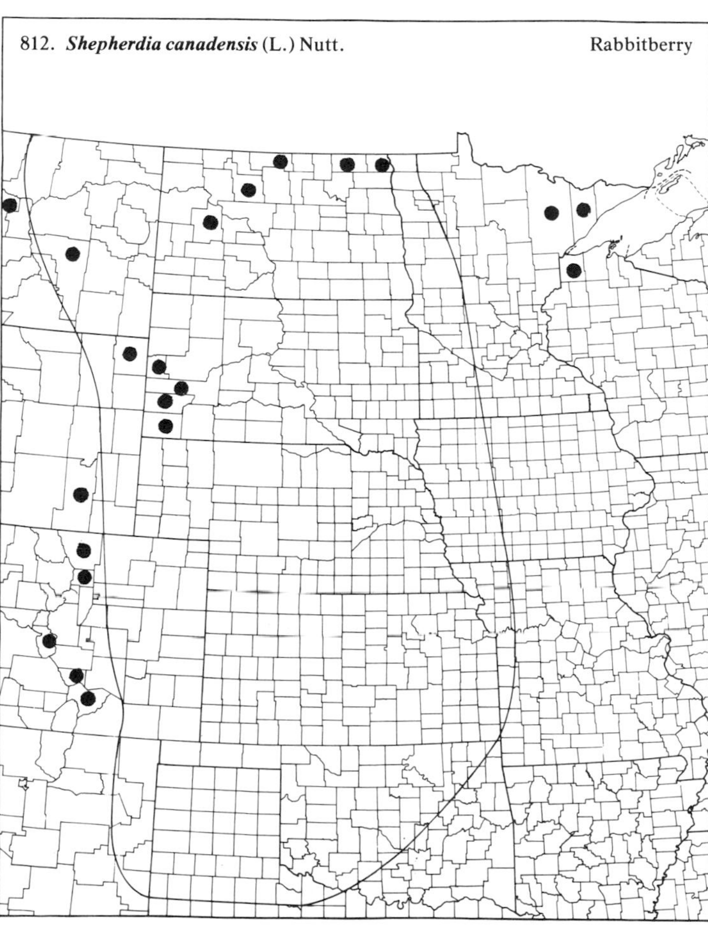

62. *Cornaceae* (Dogwood Family)

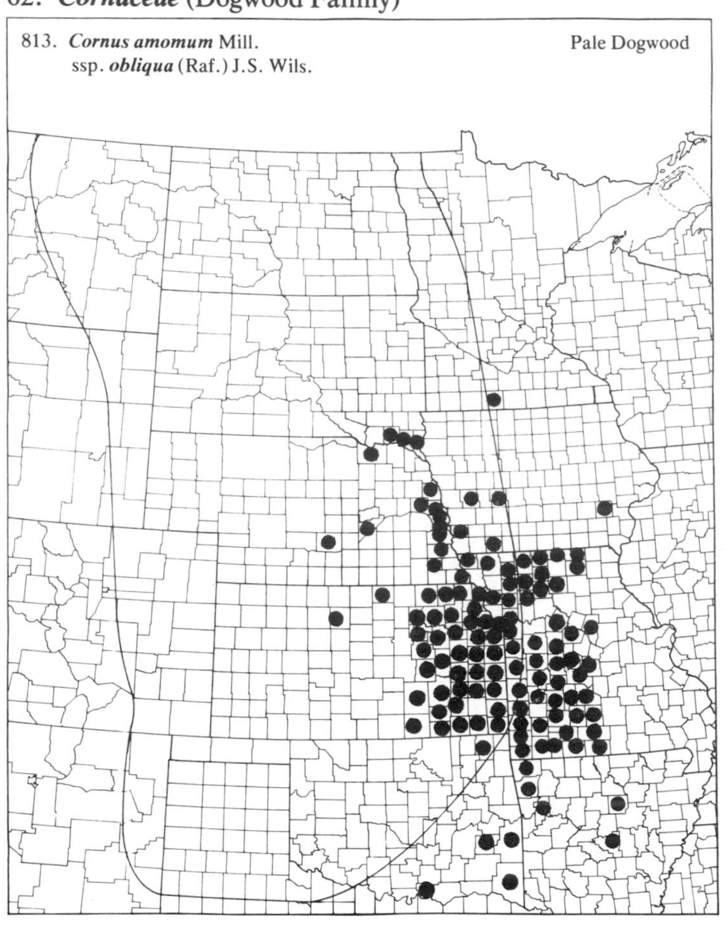

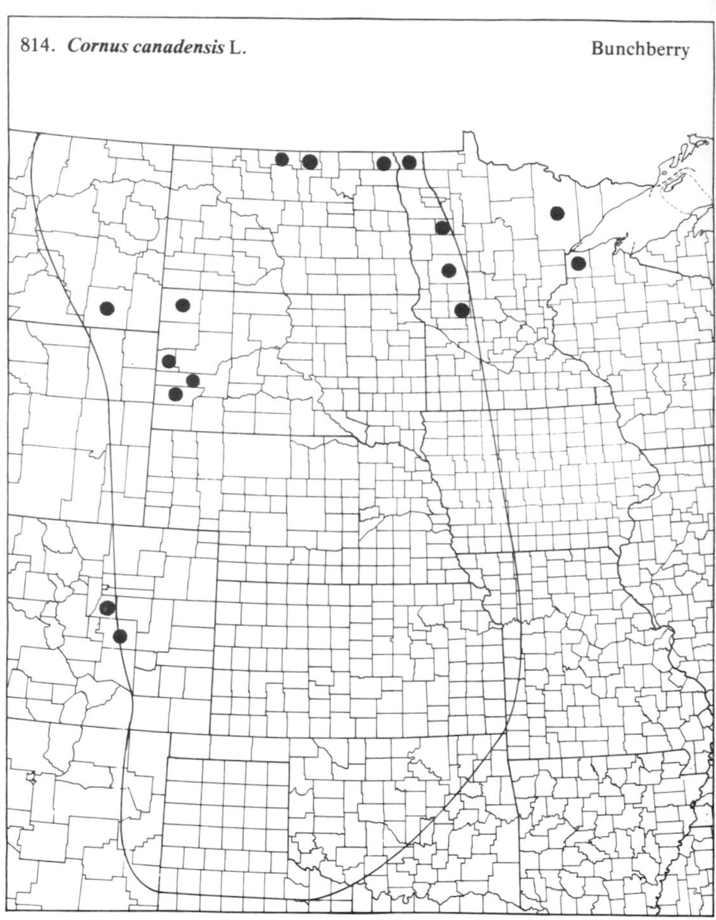

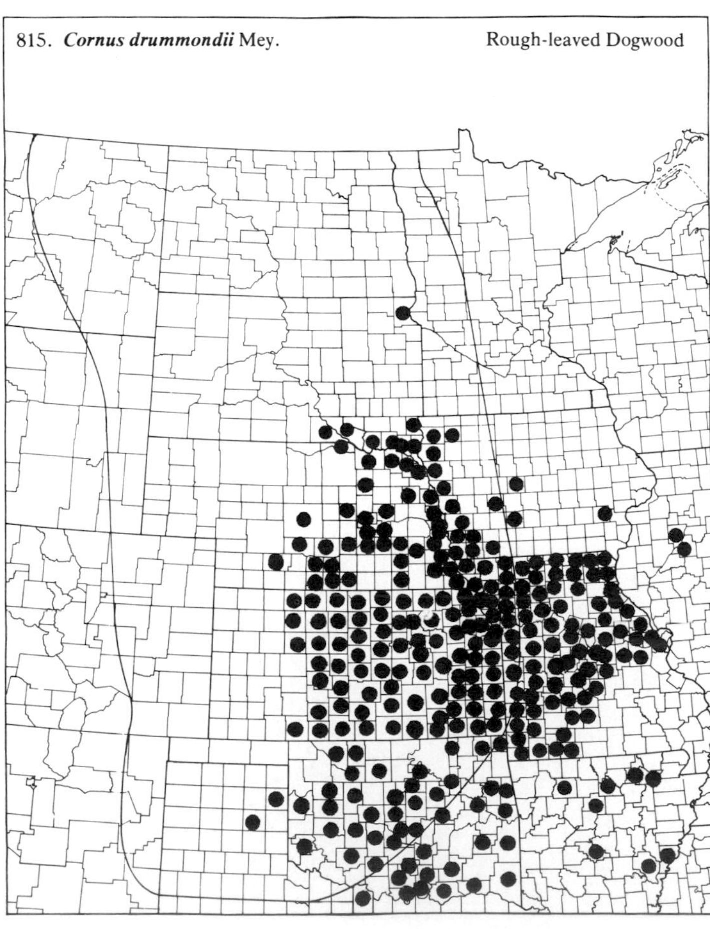

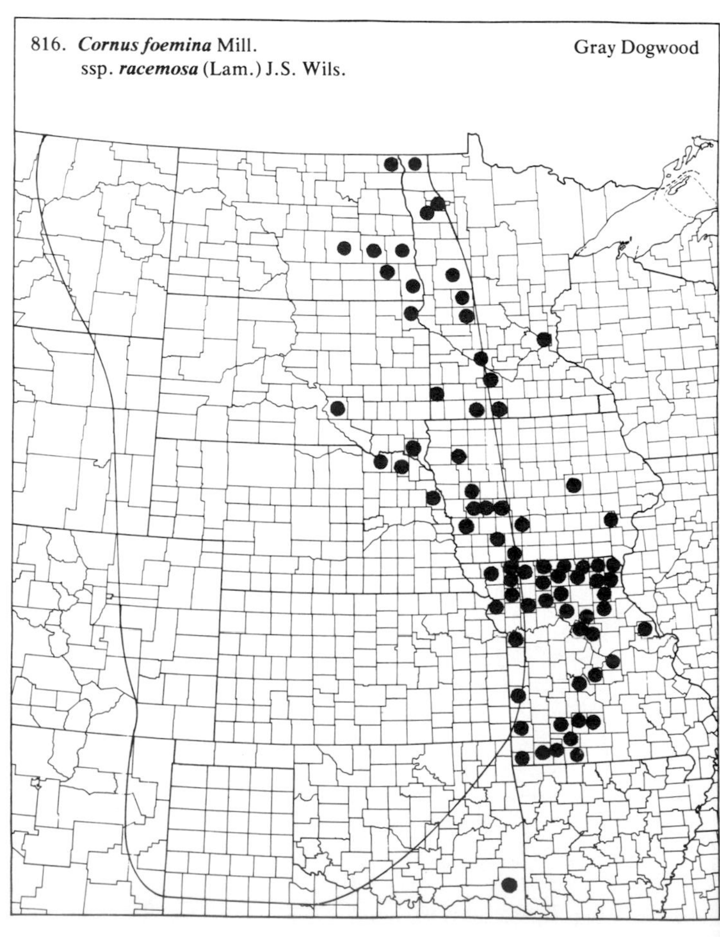

(Cornaceae)

817. ***Cornus stolonifera*** Michx. Red Osier

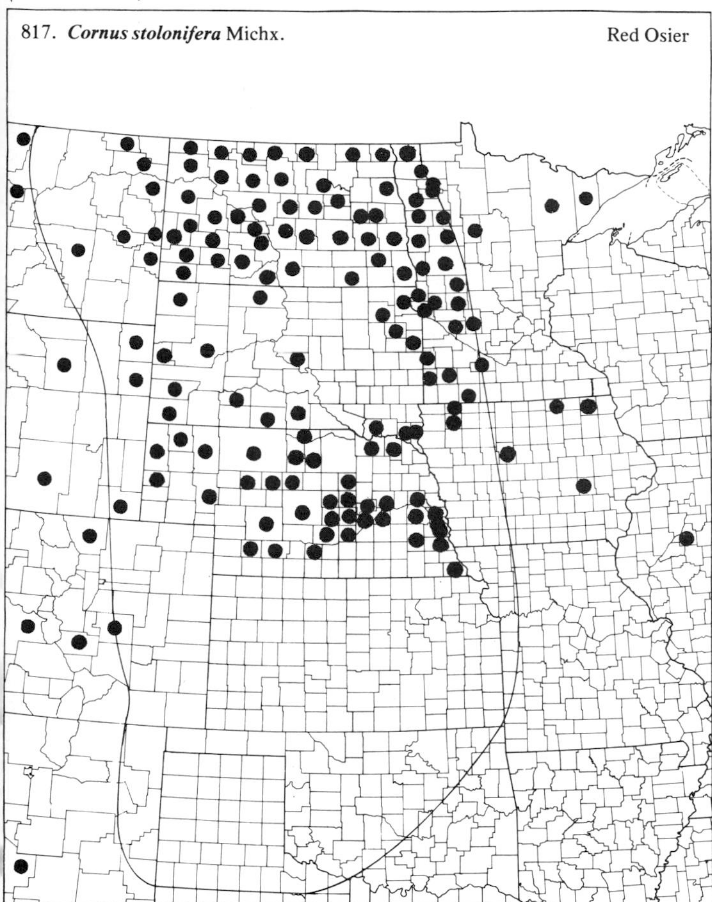

63. *Santalaceae* (Sandalwood Family)

818. ***Comandra umbellata*** (L.) Nutt. ssp. ***umbellata*** Bastard Toadflax

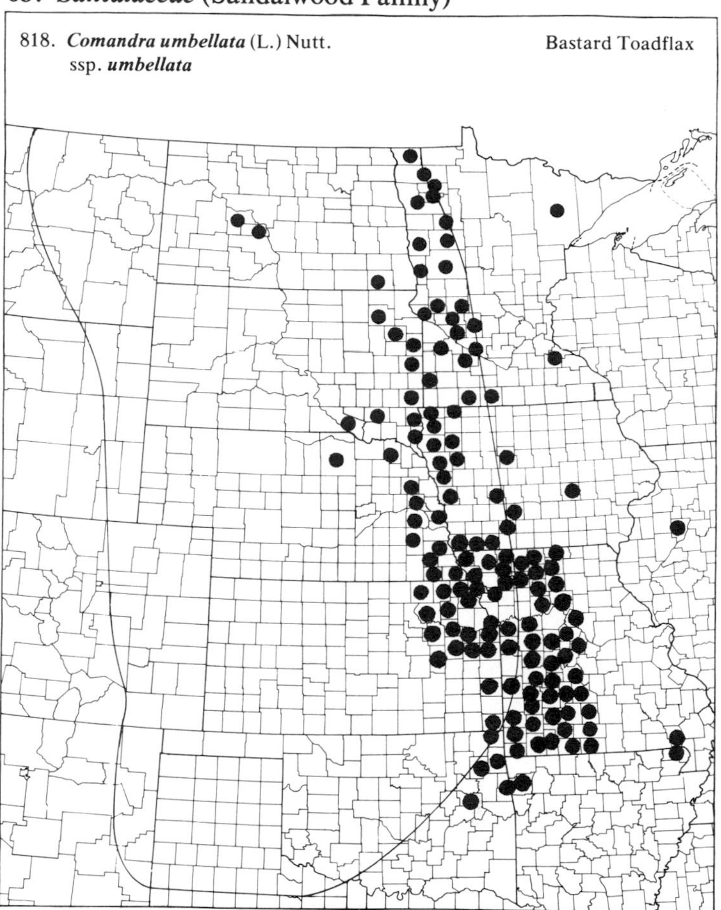

819. ***Comandra umbellata*** (L.) Nutt. ssp. ***pallida*** (A. DC.) Piehl. Bastard Toadflax

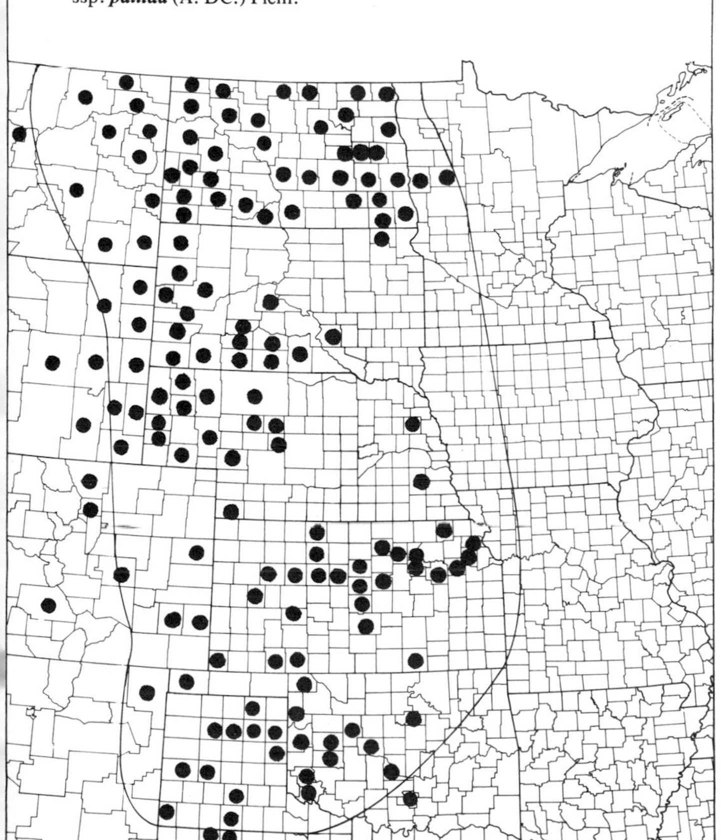

64. *Loranthaceae* (Mistletoe Family)

820. ***Phoradendron tomentosum*** (DC.) Gray

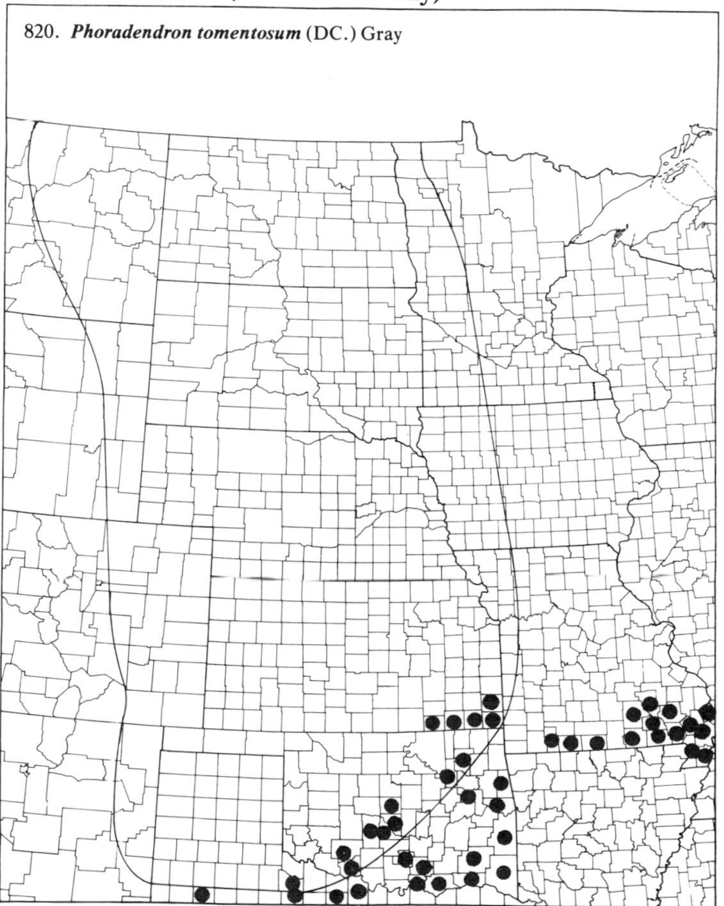

65. *Rafflesiaceae*

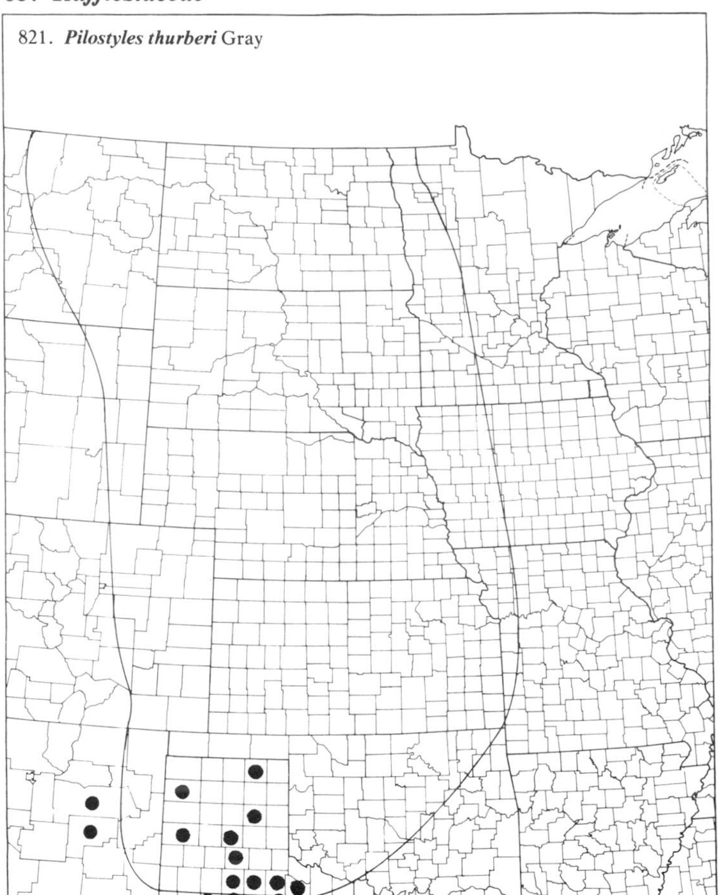
821. ***Pilostyles thurberi*** Gray

66. *Celastraceae* (Staff-tree Family)

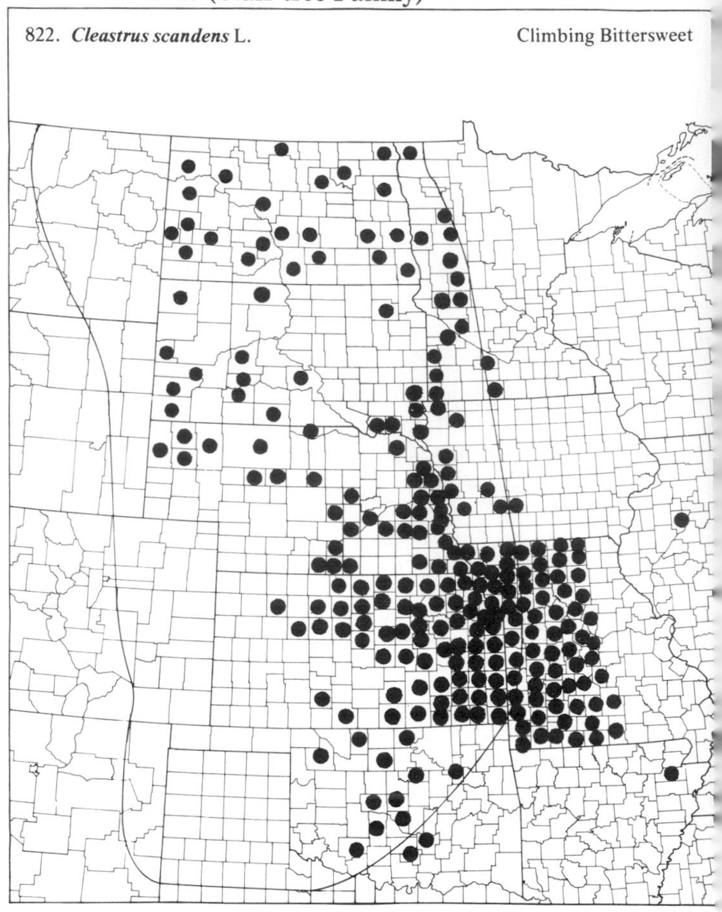
822. ***Cleastrus scandens*** L. Climbing Bittersweet

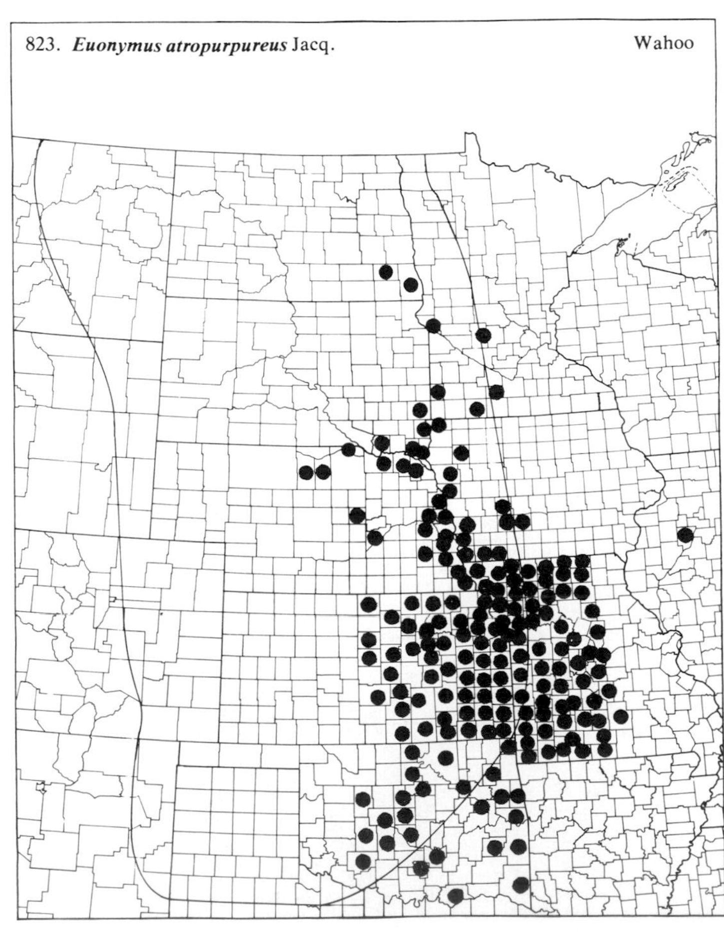
823. ***Euonymus atropurpureus*** Jacq. Wahoo

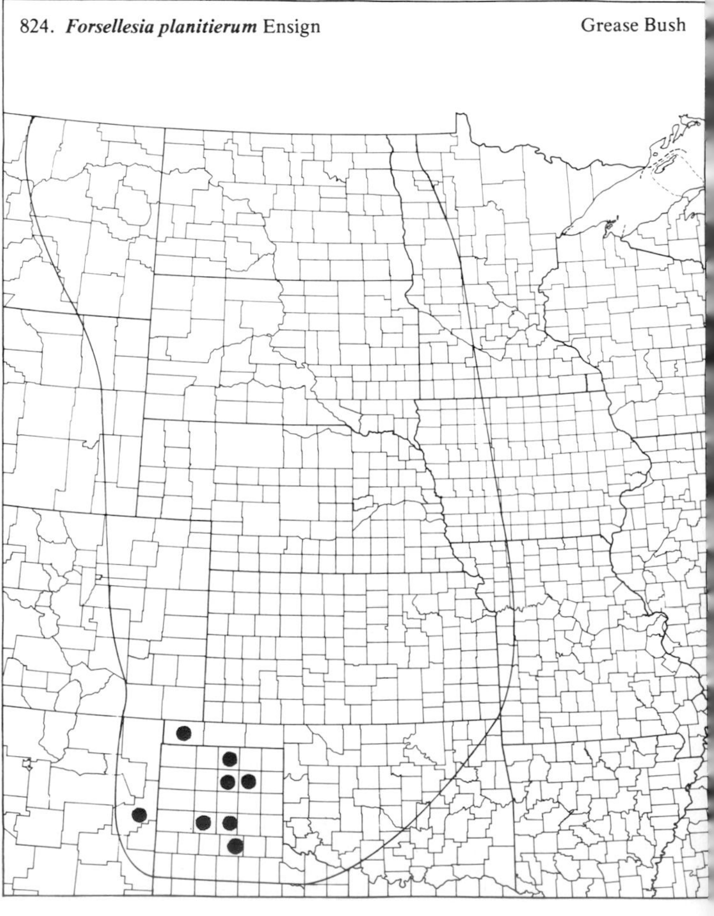
824. ***Forsellesia planitierum*** Ensign Grease Bush

67. *Aquifoliaceae* (Holly Family)

825. ***Ilex decidua*** Walt. Deciduous Holly

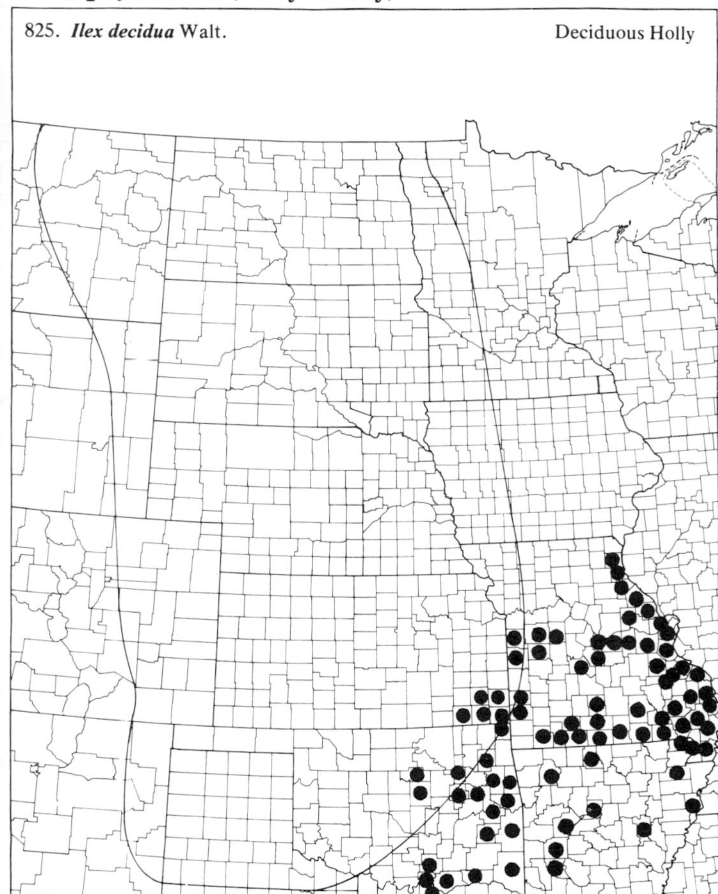

68. *Euphorbiaceae* (Spurge Family)

826. ***Acalypha monococca*** (Engelm.) Mill.

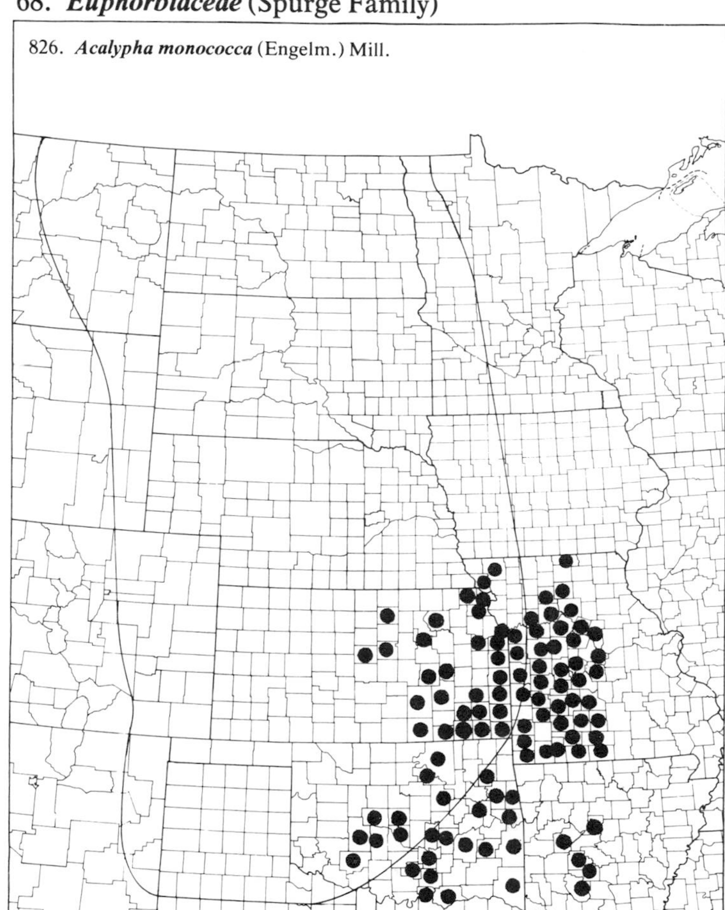

827. ***Acalypha ostryaefolia*** Ridd. Three-seeded Mercury

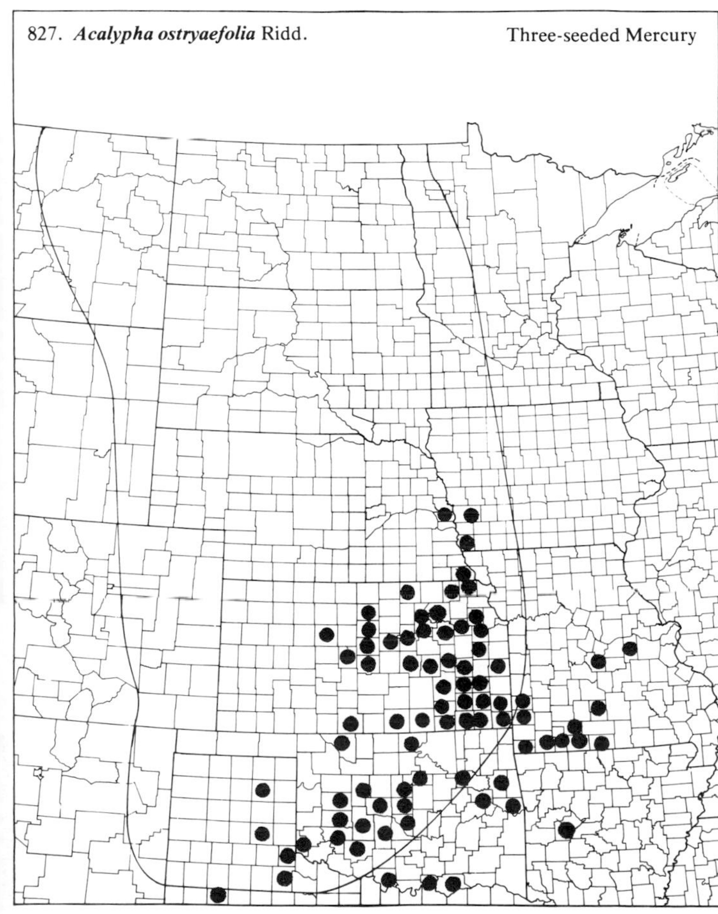

828. ***Acalypha rhomboidea*** Raf. Rhombic Copperleaf

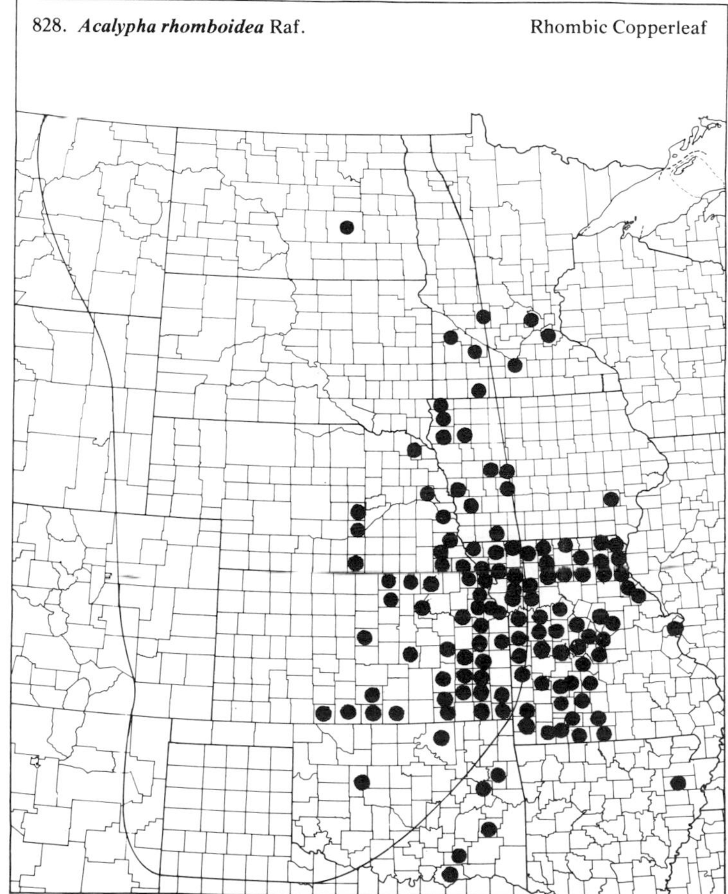

(Euphorbiaceae)

829. ***Acalypha virginica*** L. — Three-seeded Mercury

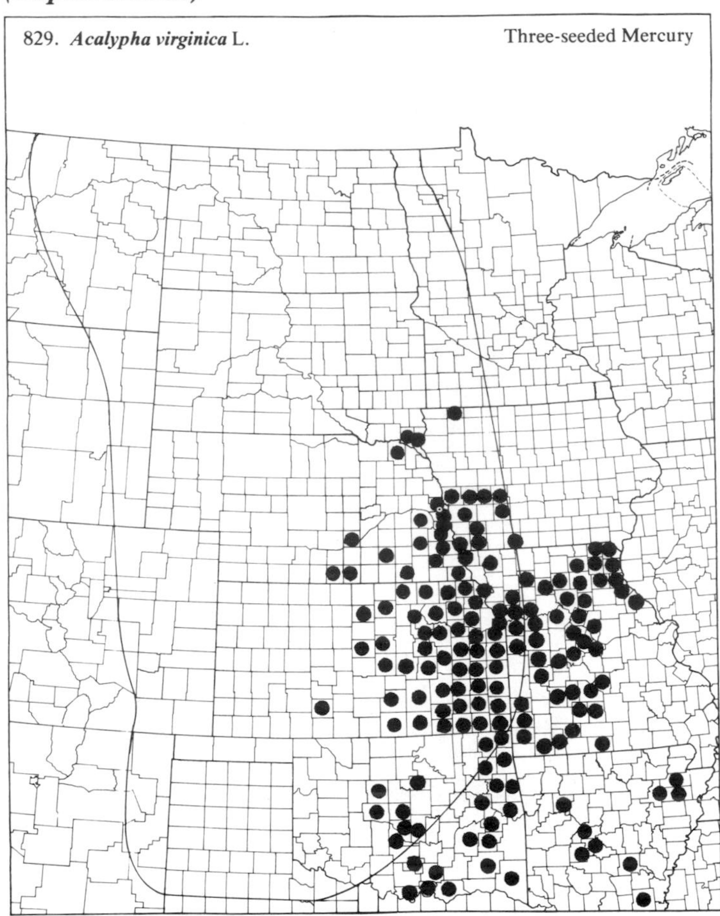

830. ***Argythamnia humilis*** (Engelm. & Gray) Muell. Arg. var. ***humilis*** — Wild Mercury

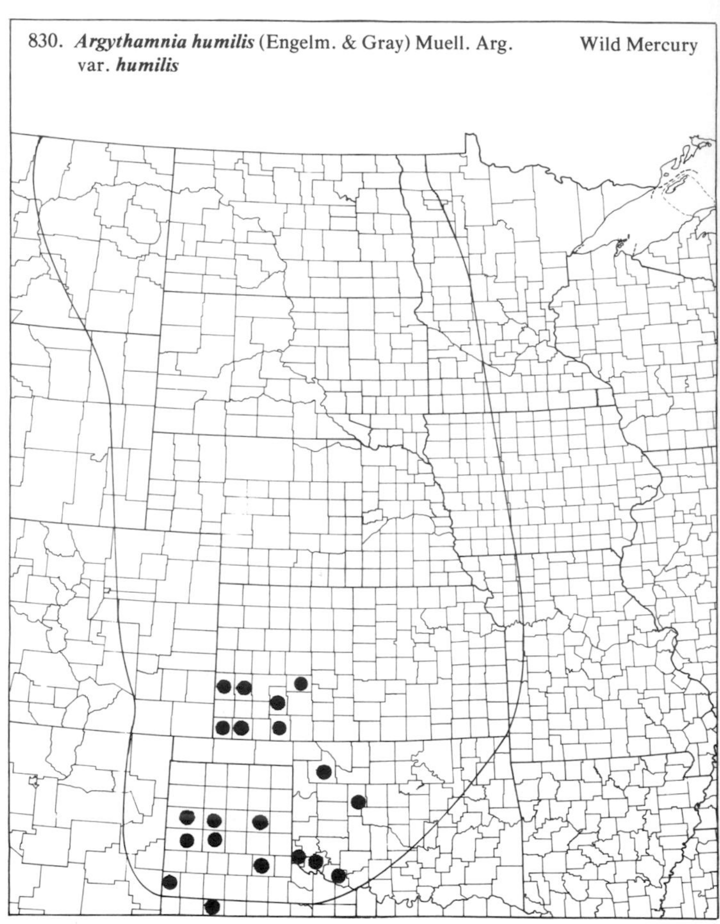

831. ***Argythamnia mercurialina*** (Nutt.) Muell. Arg. — Wild Mercury

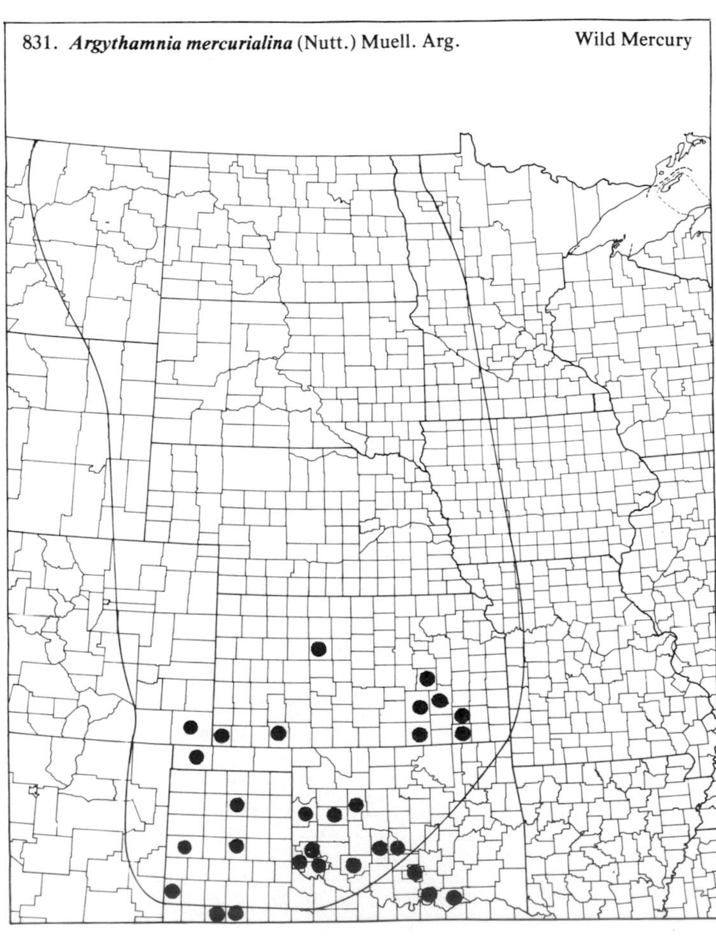

832. ***Cnidoscolus texanus*** (Muell. Arg.) Small — Bull Nettle

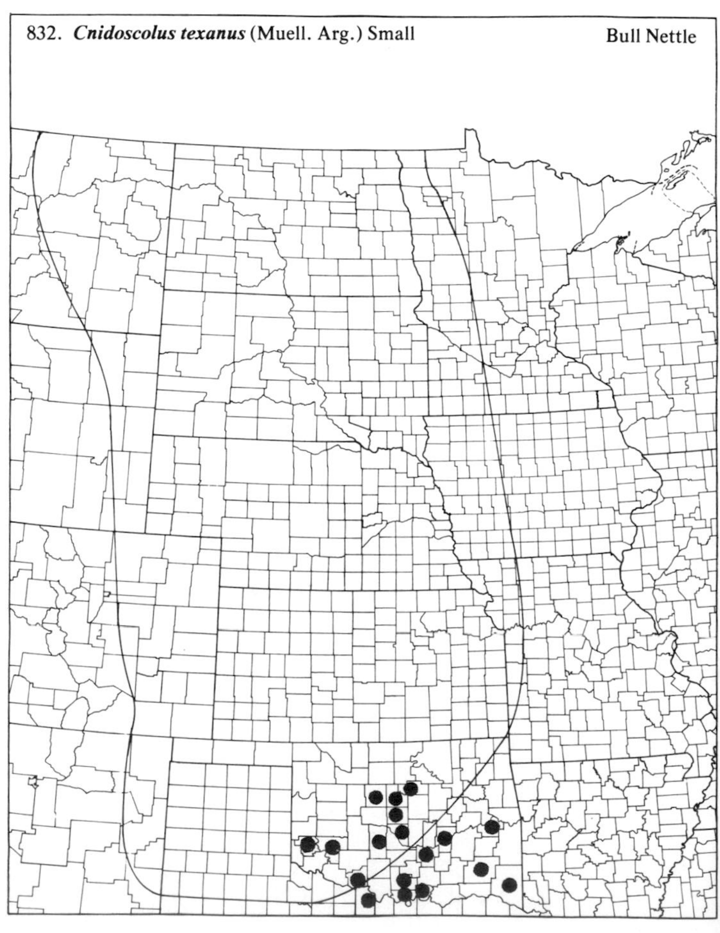

(Euphorbiaceae)

833. *Croton capitatus* Michx. var. *capitatus* — Woolly Croton

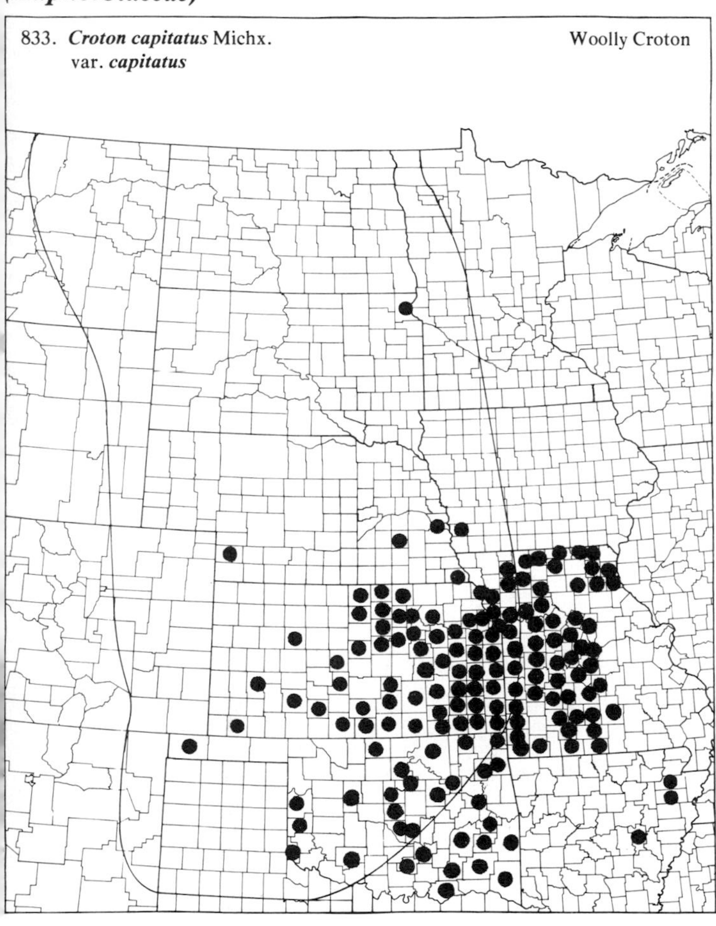

834. *Croton glandulosus* L. var. *septentrionalis* Muell. Arg. — Tropic Croton

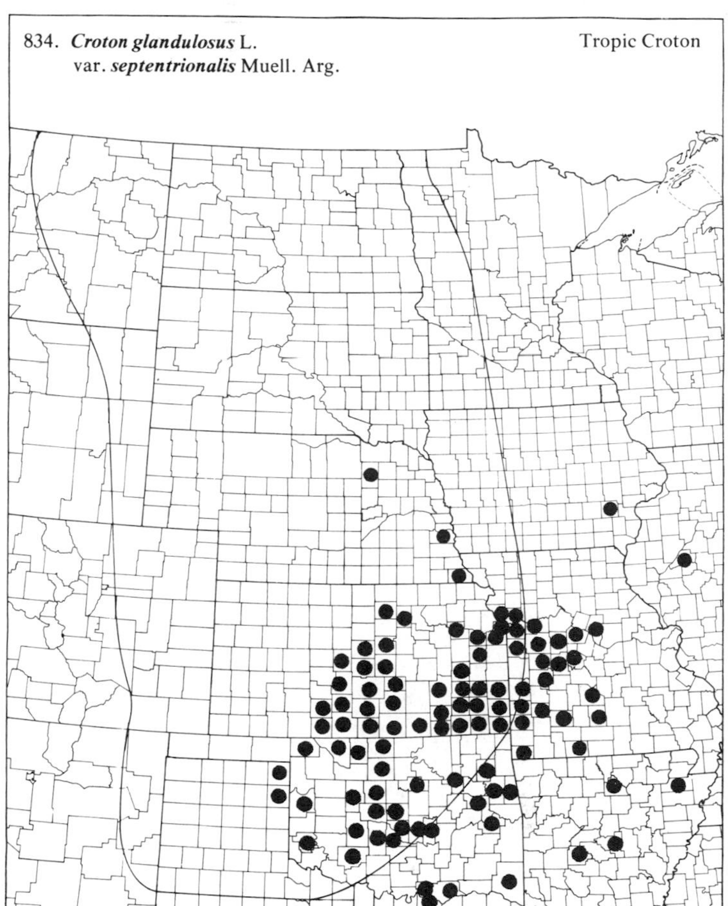

835. *Croton lindheimerianus* Scheele — Three-seeded Croton

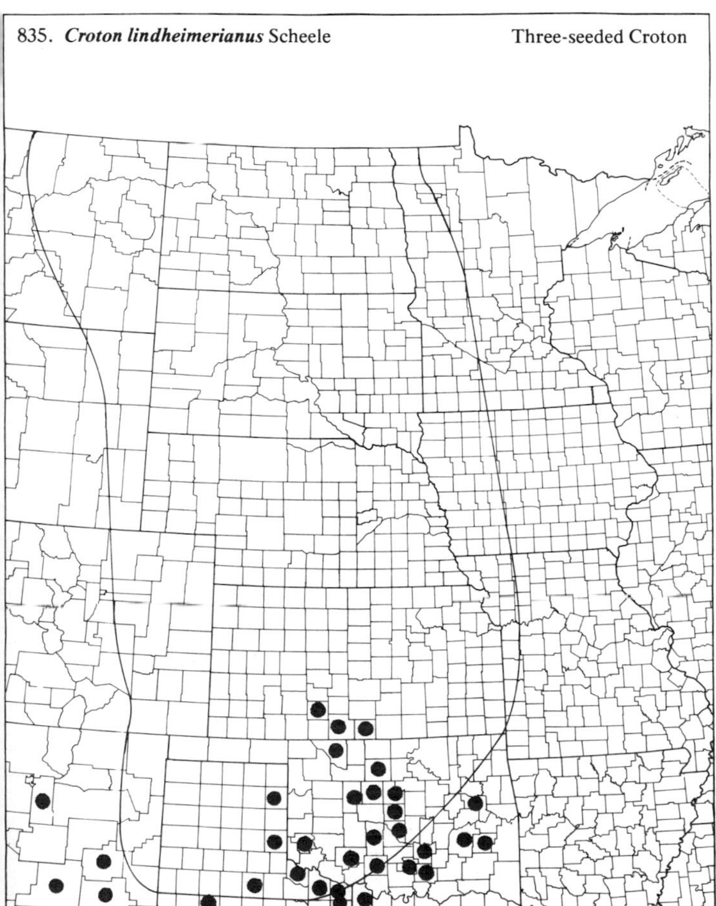

836. *Croton monanthogynus* Michx. — One-seeded Croton

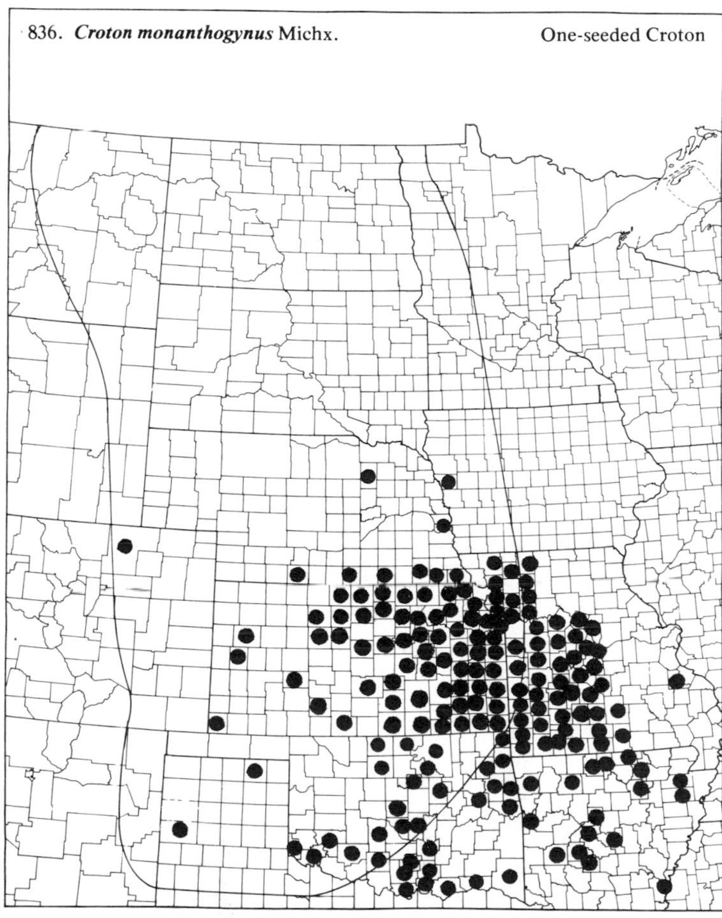

(Euphorbiaceae)

837. ***Croton texenis*** (Klotzsch) Muell. Arg. Texas Croton

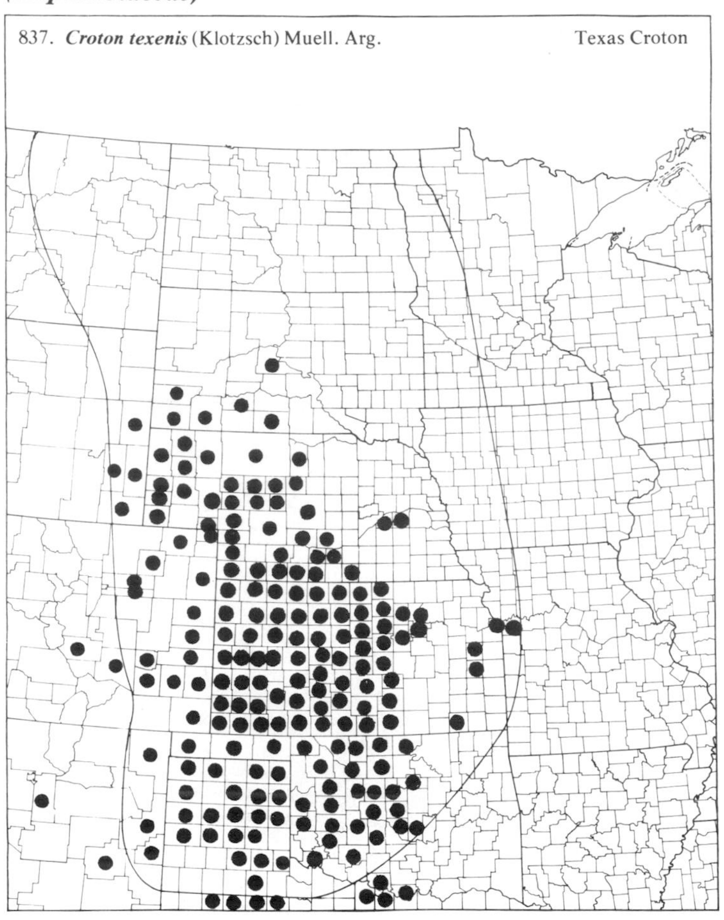

838. ***Crotonopsis elliptica*** Willd. Crotonopsis

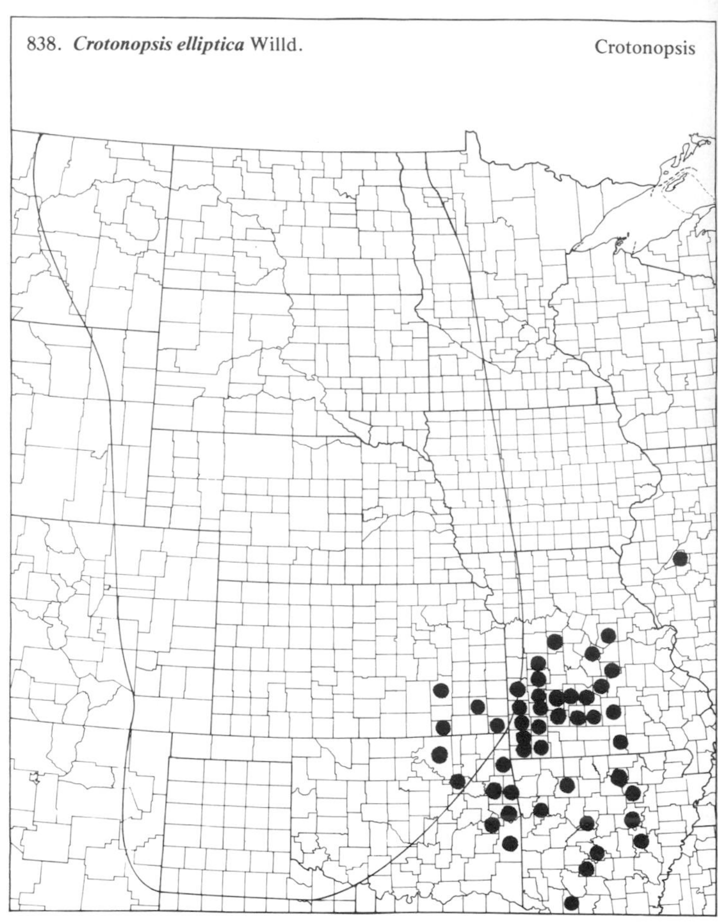

839. ***Euphorbia albomarginata*** T. & G.

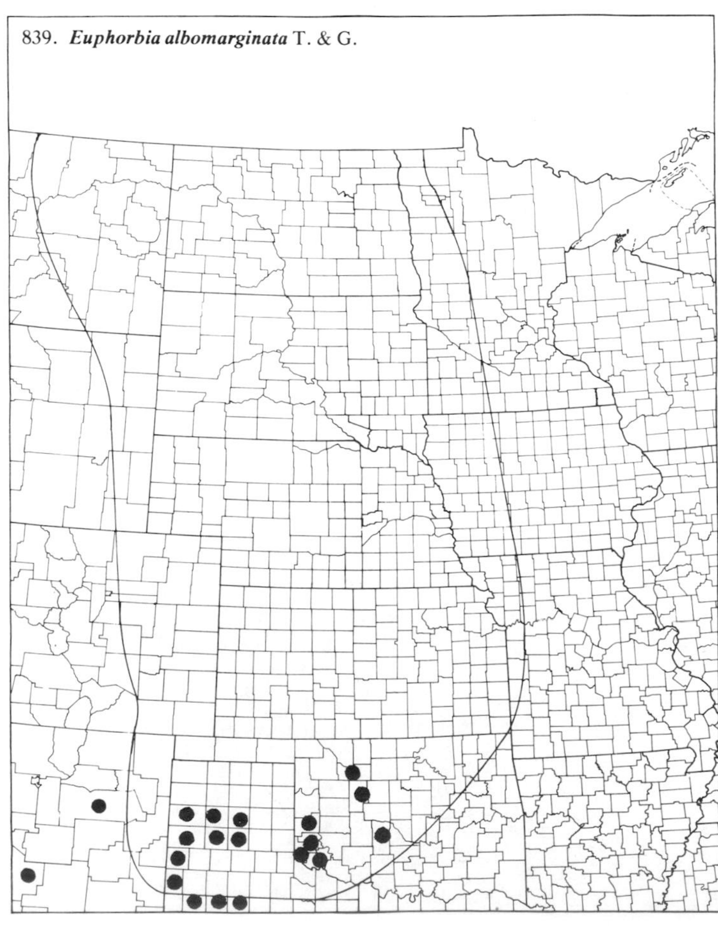

840. ***Euphorbia corollata*** L. Flowering Spurge

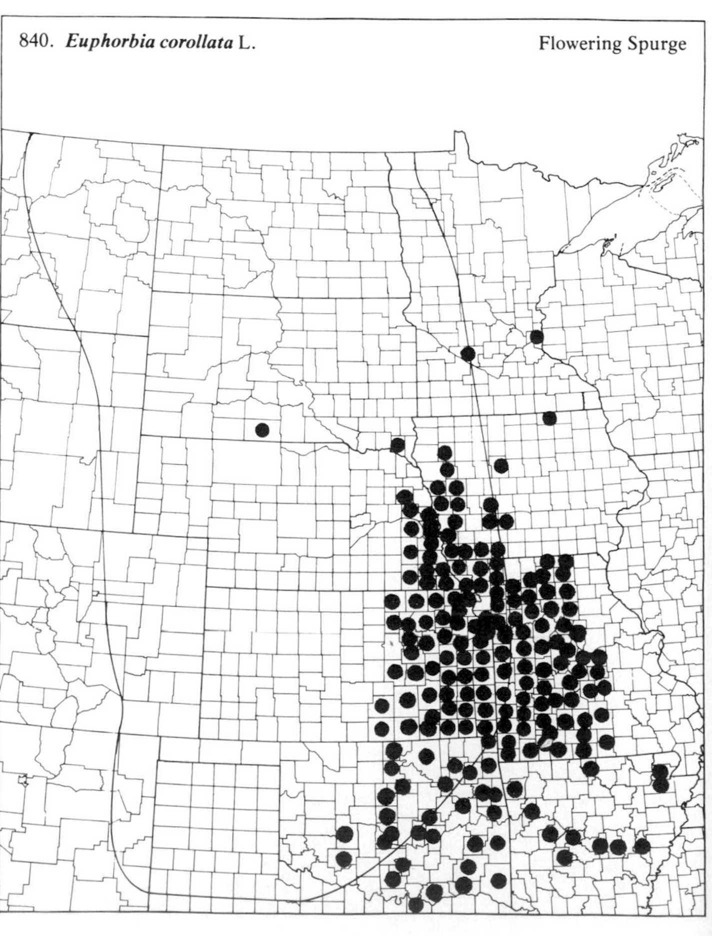

(Euphorbiaceae)

841. ***Euphorbia cyathophora*** Murr. Fire-on-the-mountain

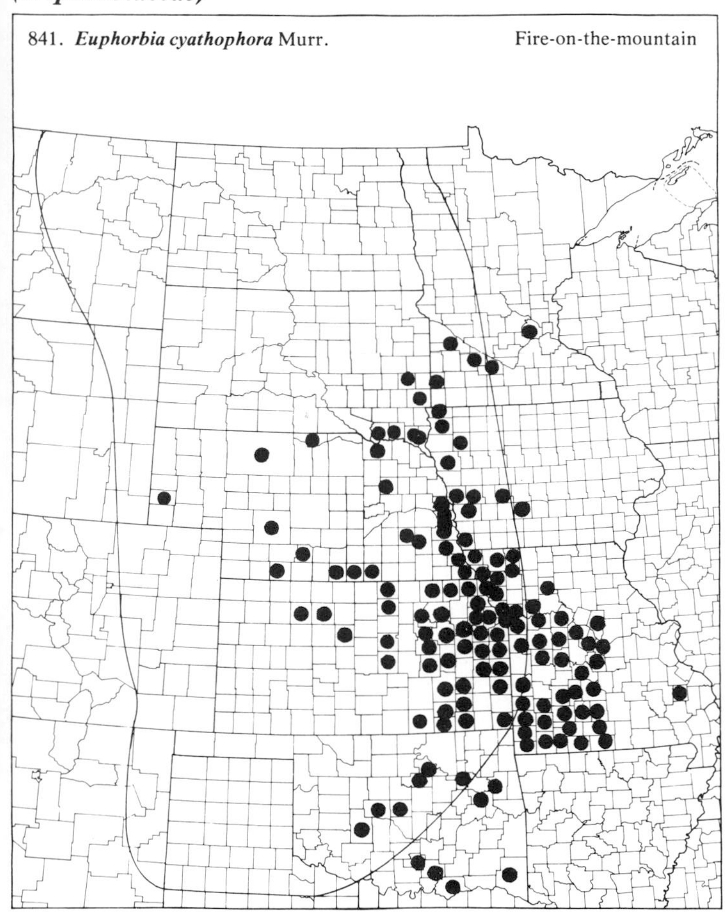

842. ***Euphorbia cyparissias*** L. Cypress Spurge

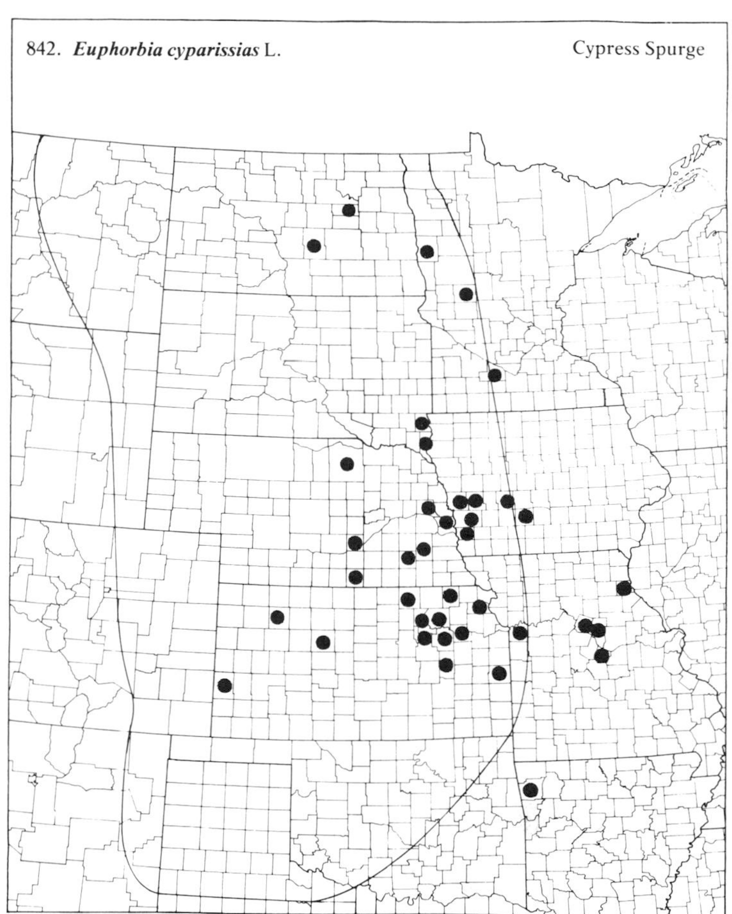

843. ***Euphorbia dentata*** Michx. Toothed Spurge

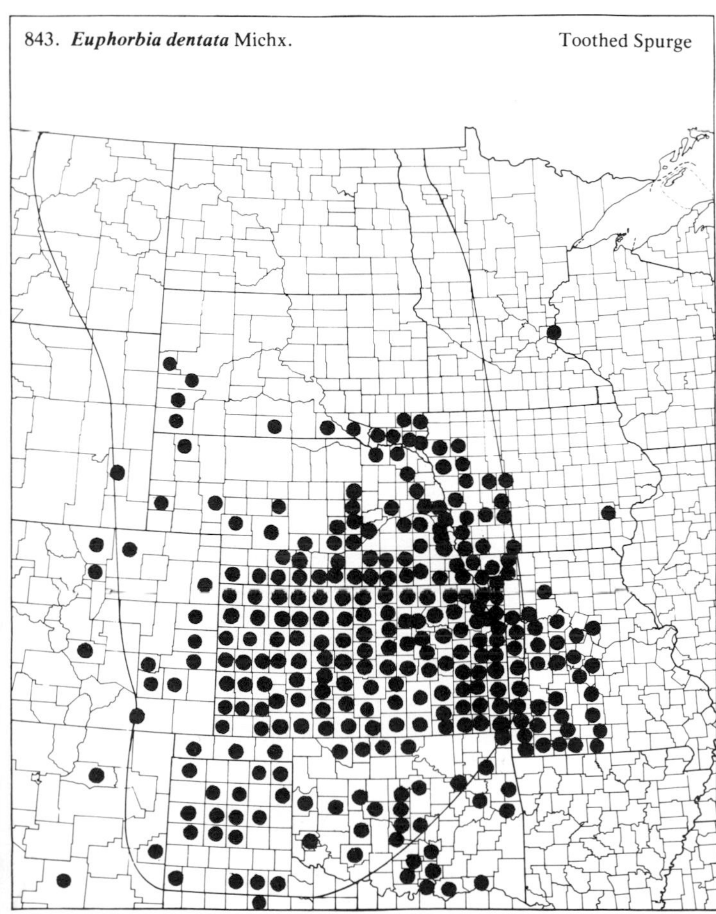

844. ***Euphorbia fendleri*** T. & G. Fendler's Euphorbia

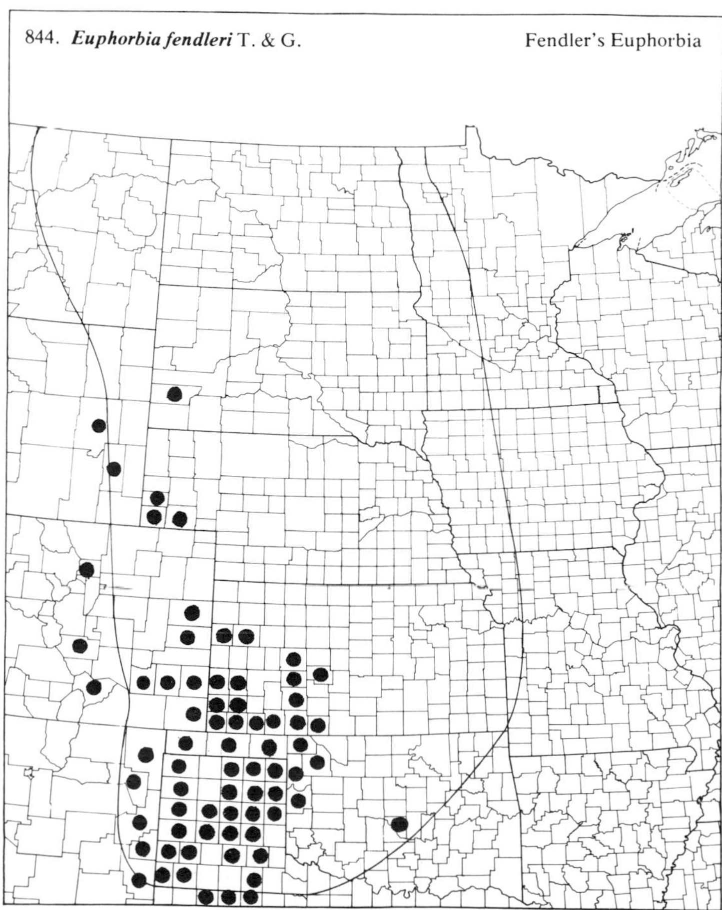

(Euphorbiaceae)

845. **Euphorbia geyeri** Engelm. — Geyer's Spurge

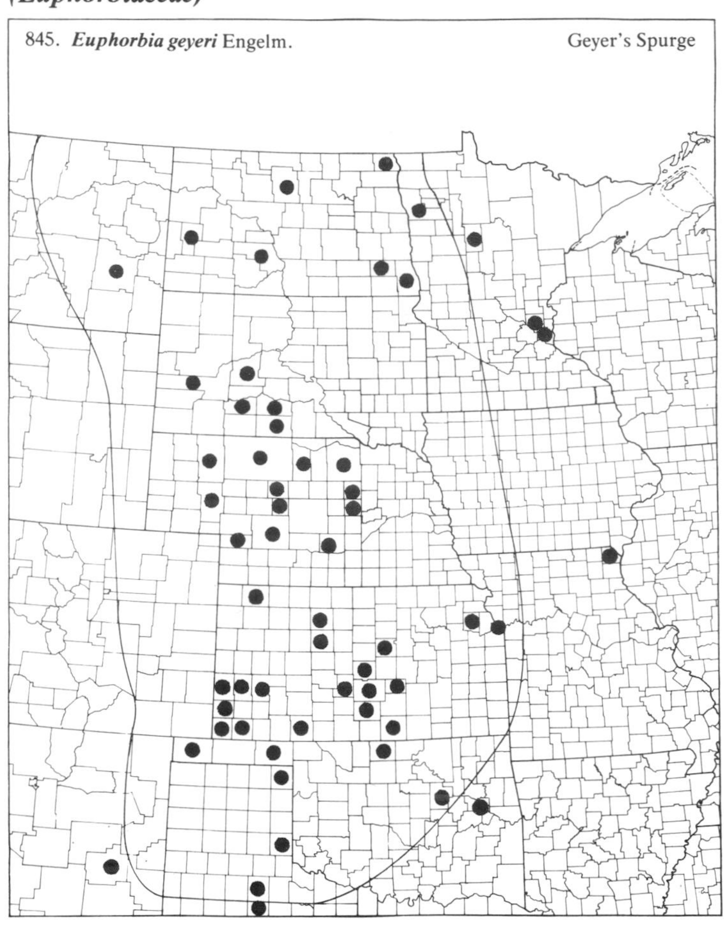

846. **Euphorbia glyptosperma** Engelm. — Ridge-seeded Spurge

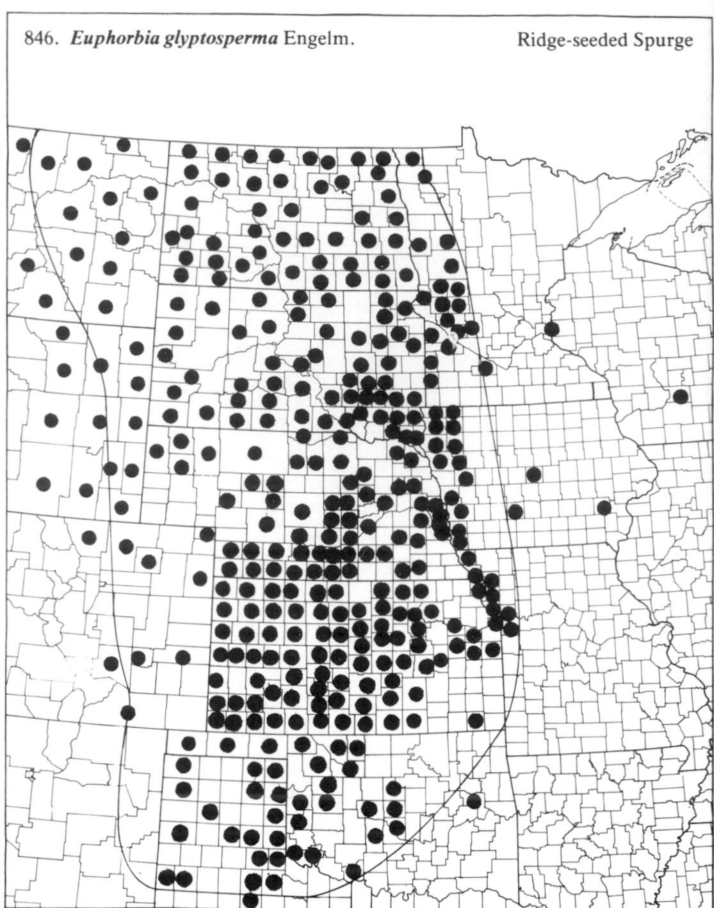

847. **Euphorbia hexagona** Nutt. — Six-angled Spurge

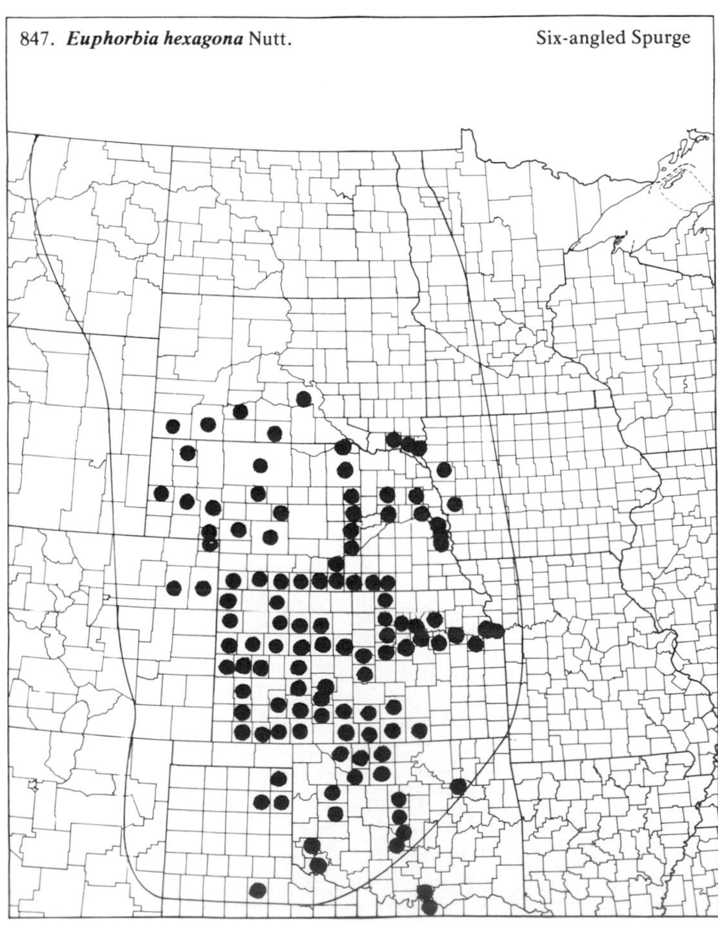

848. **Euphorbia humistrata** Engelm. — Spreading Euphorbia

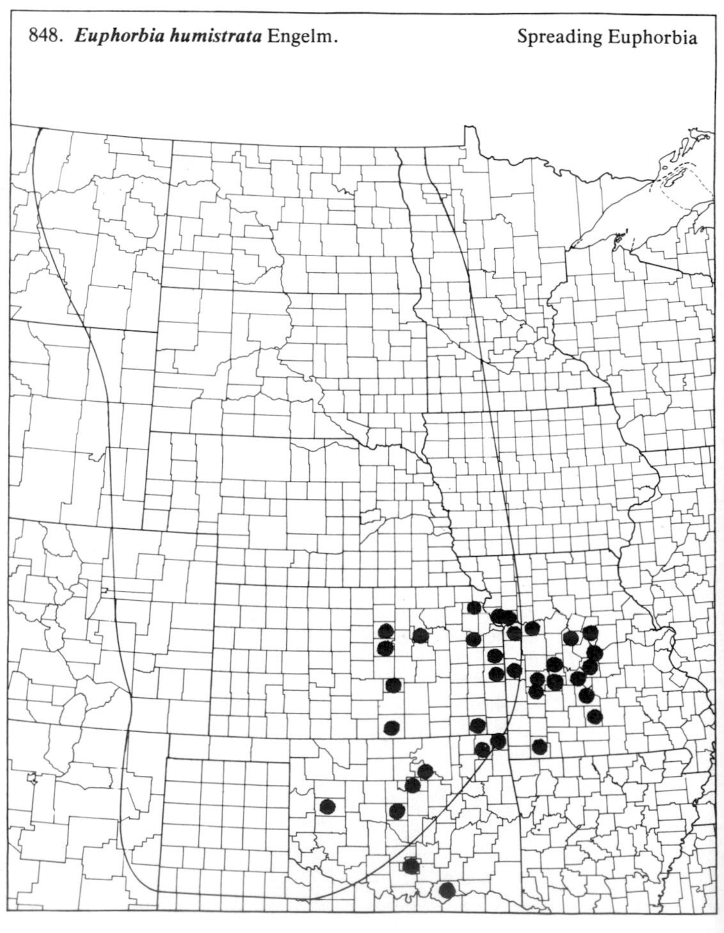

(Euphorbiaceae)

849. ***Euphorbia lata*** Engelm. Hoary Euphorbia

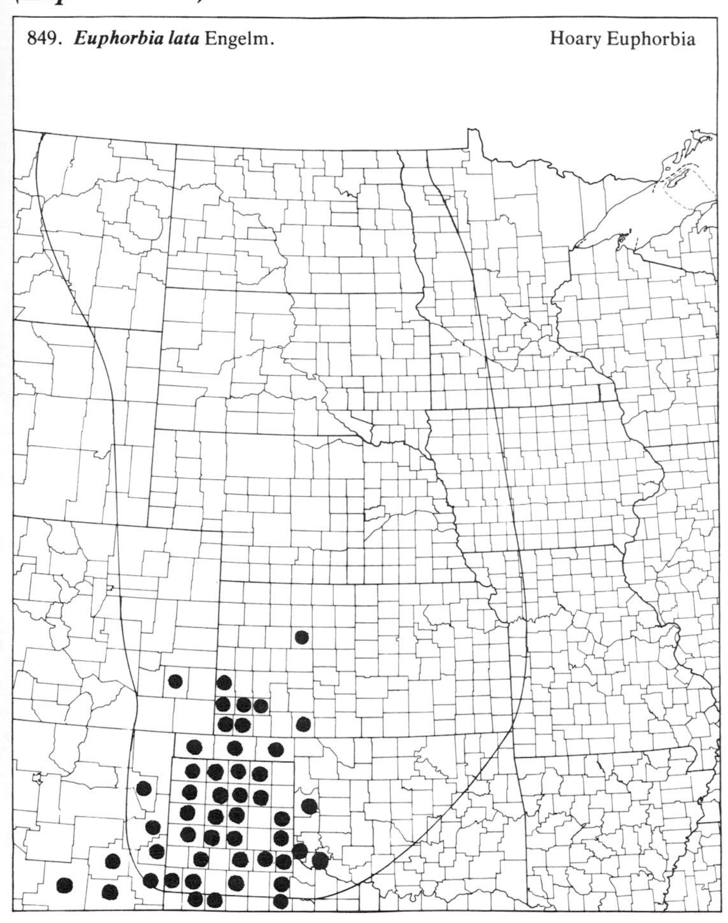

850. ***Euphorbia maculata*** L. Spotted Spurge

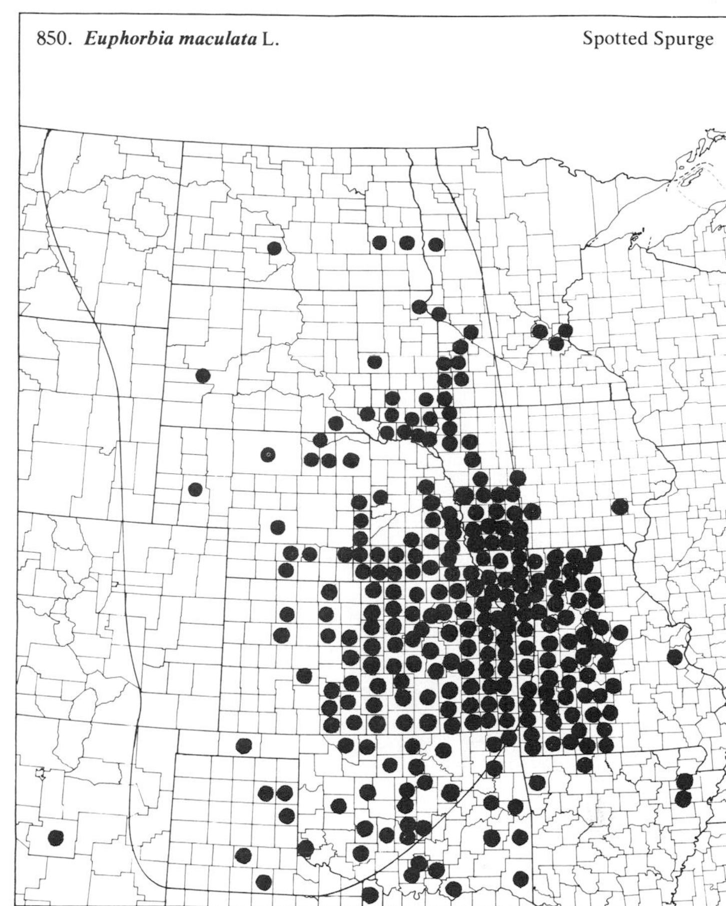

851. ***Euphorbia marginata*** Pursh Snow-on-the-mountain

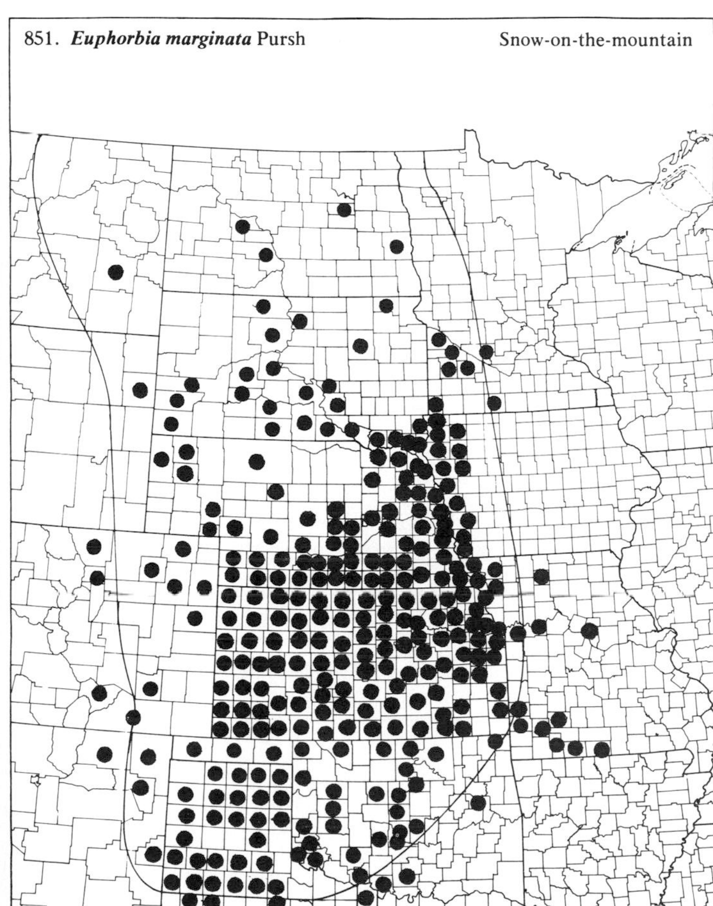

852. ***Euphorbia missurica*** Raf. Missouri Spurge

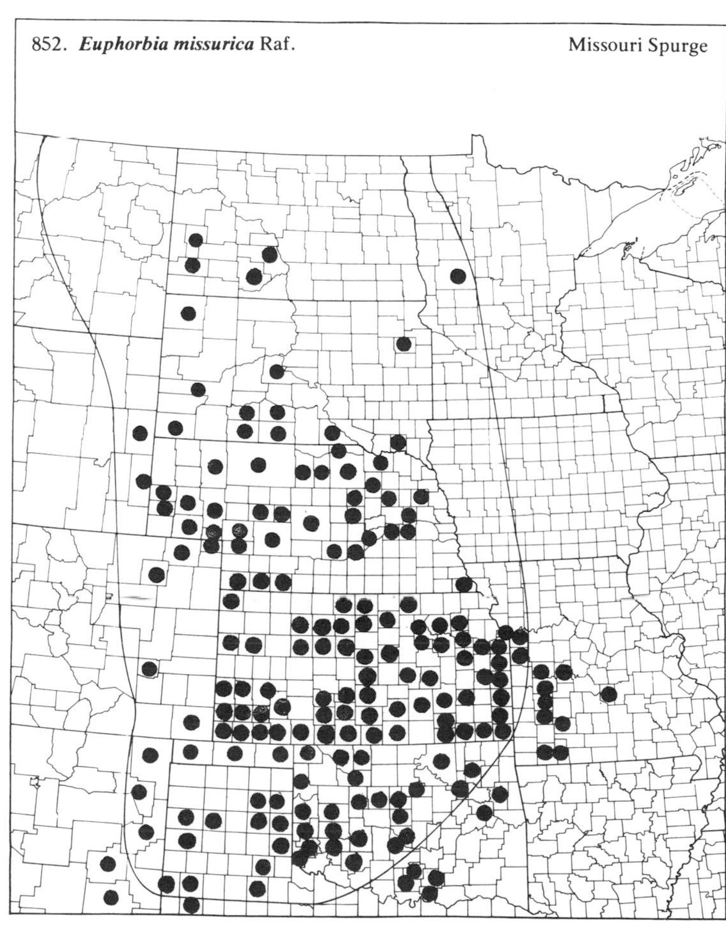

(Euphorbiaceae)

853. **Euphorbia nutans** Lag.

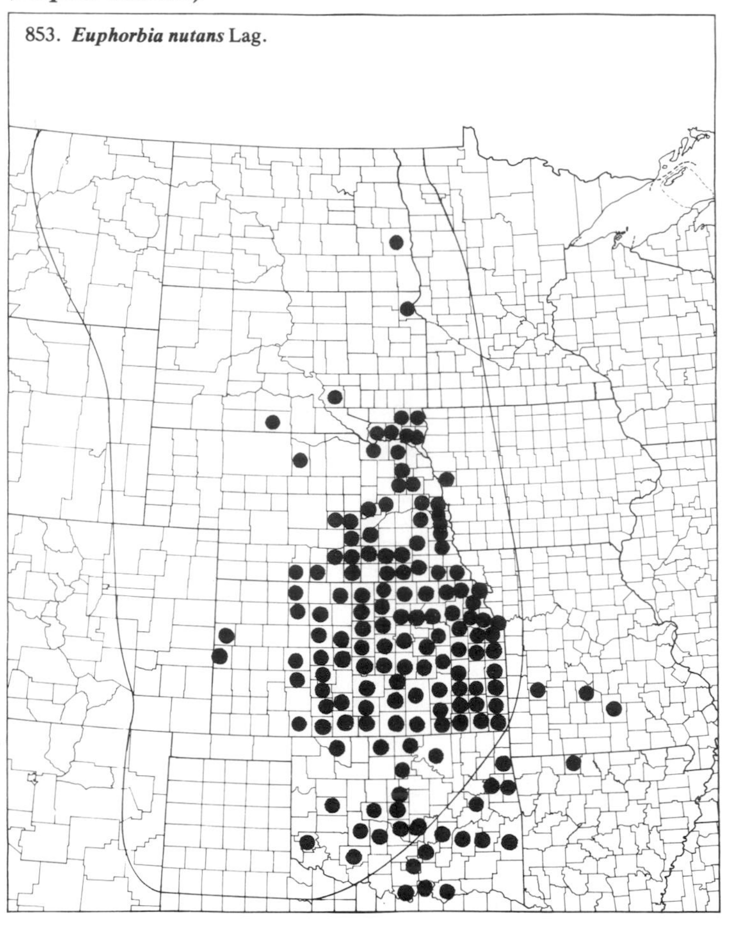

854. **Euphorbia podperae** Croizat — Leafy Spurge

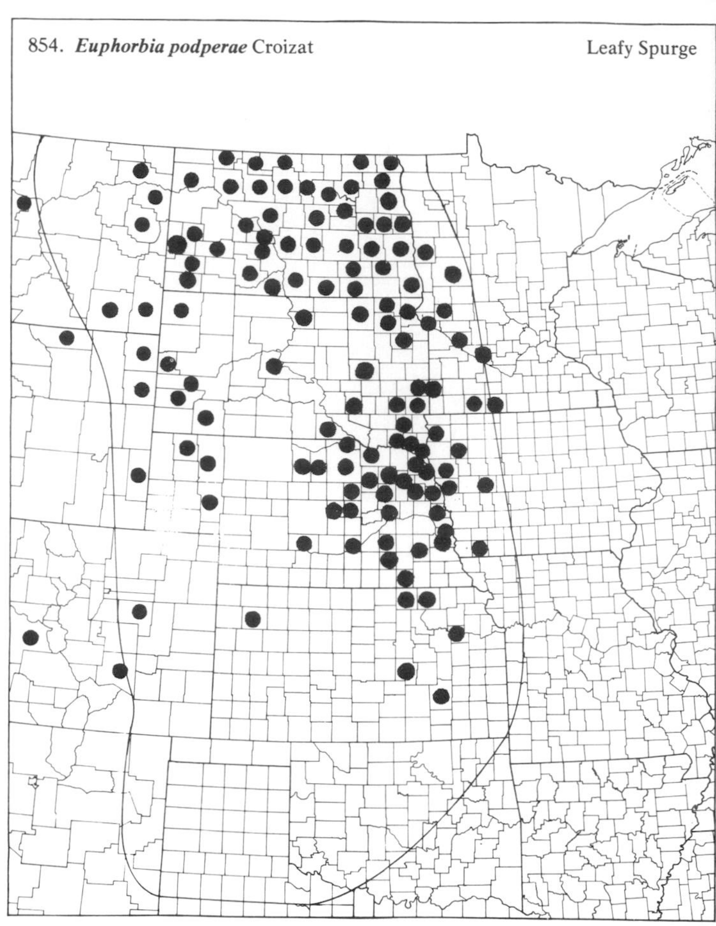

855. **Euphorbia prostrata** Ait.

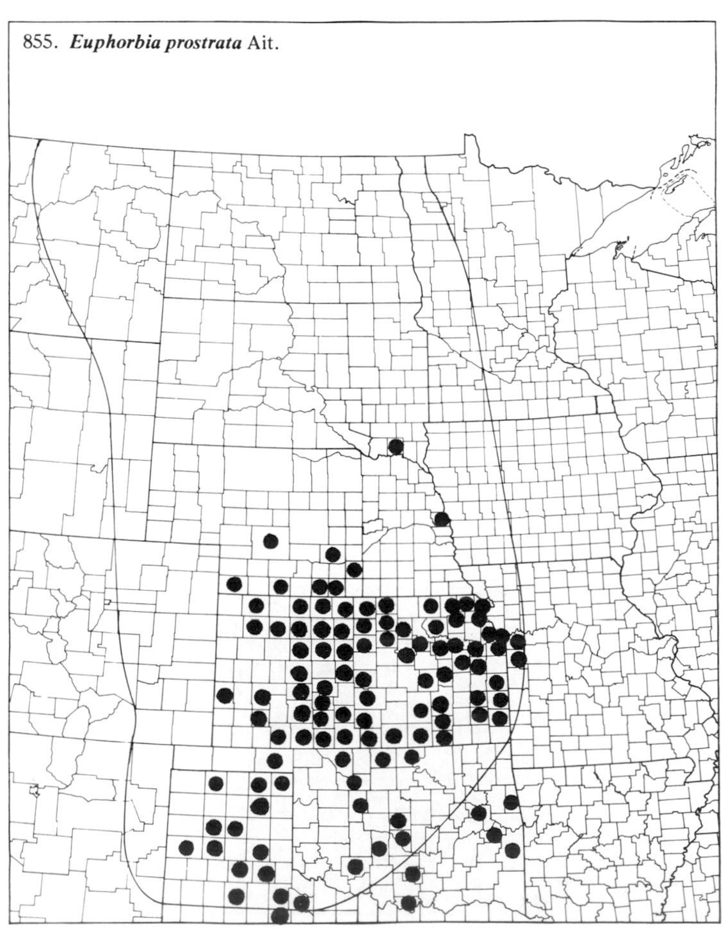

856. **Euphorbia robusta** (Engelm.) Small

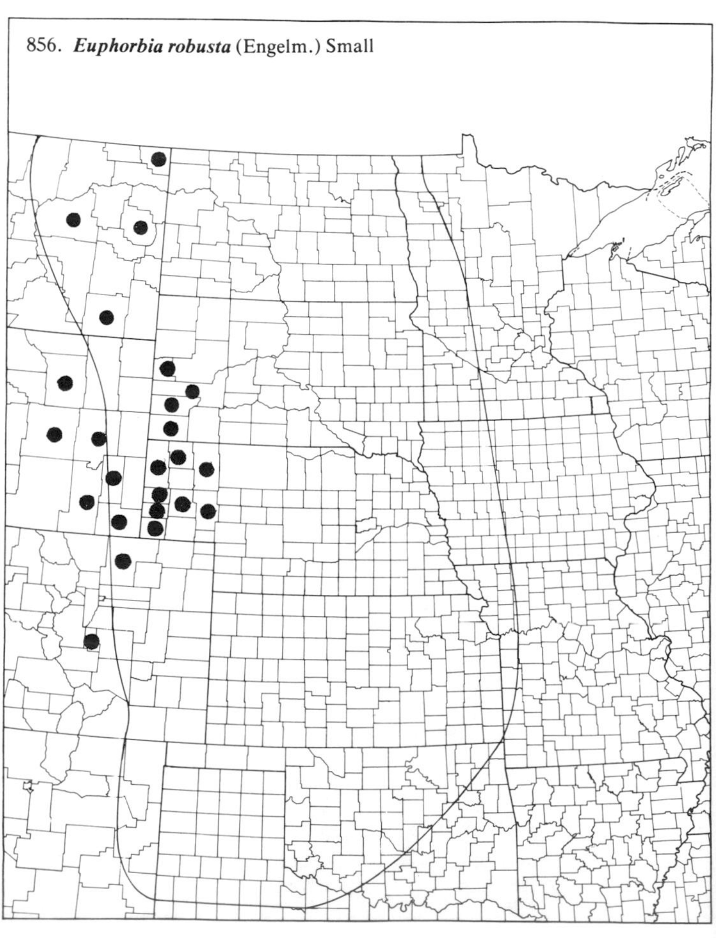

(Euphorbiaceae)

857. ***Euphorbia serpens*** H.B.K. Round-leaved Spurge

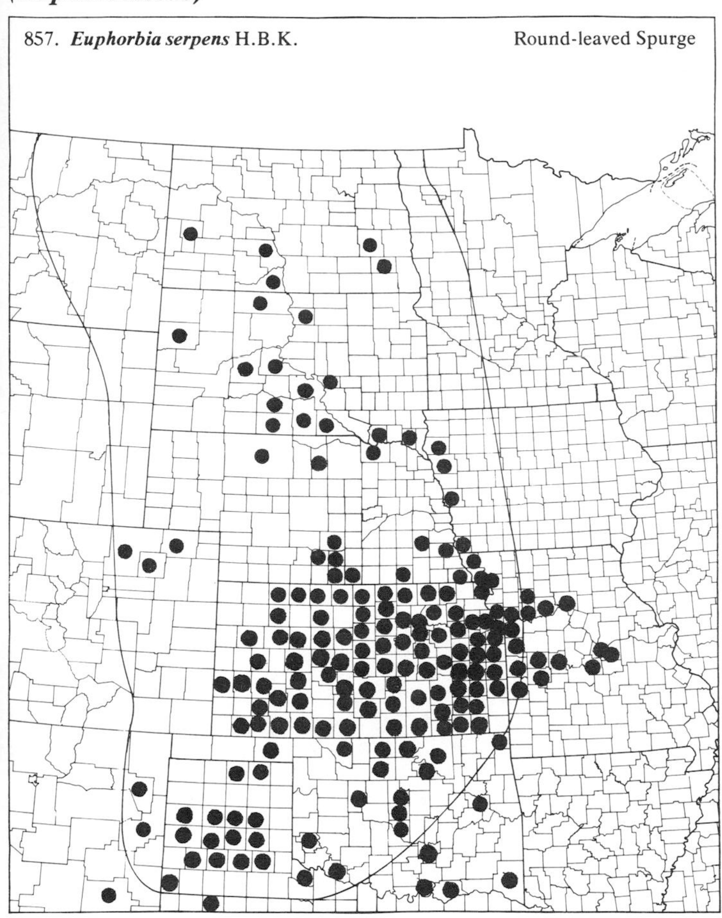

858. ***Euphorbia serpyllifolia*** Pers. Thyme-leaved Spurge

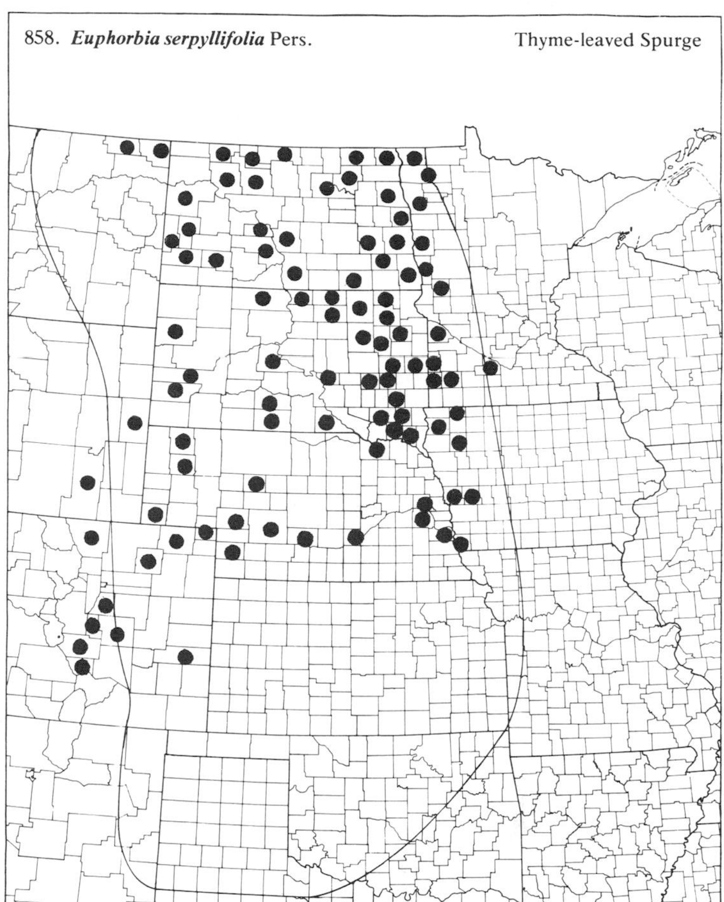

859. ***Euphorbia spathulata*** Lam.

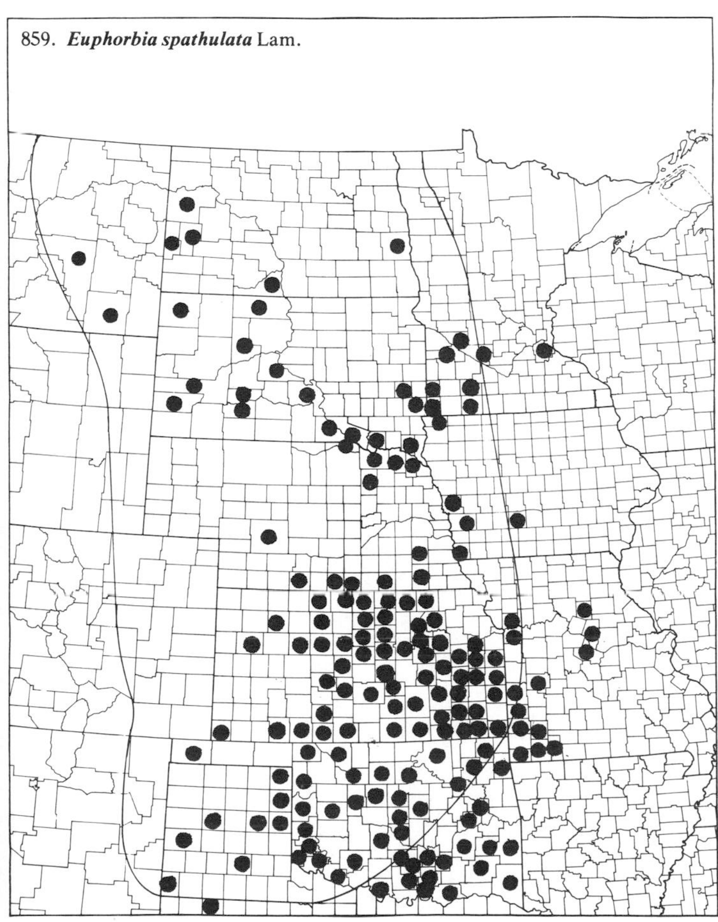

860. ***Euphorbia stictospora*** Engelm. Mat Spurge

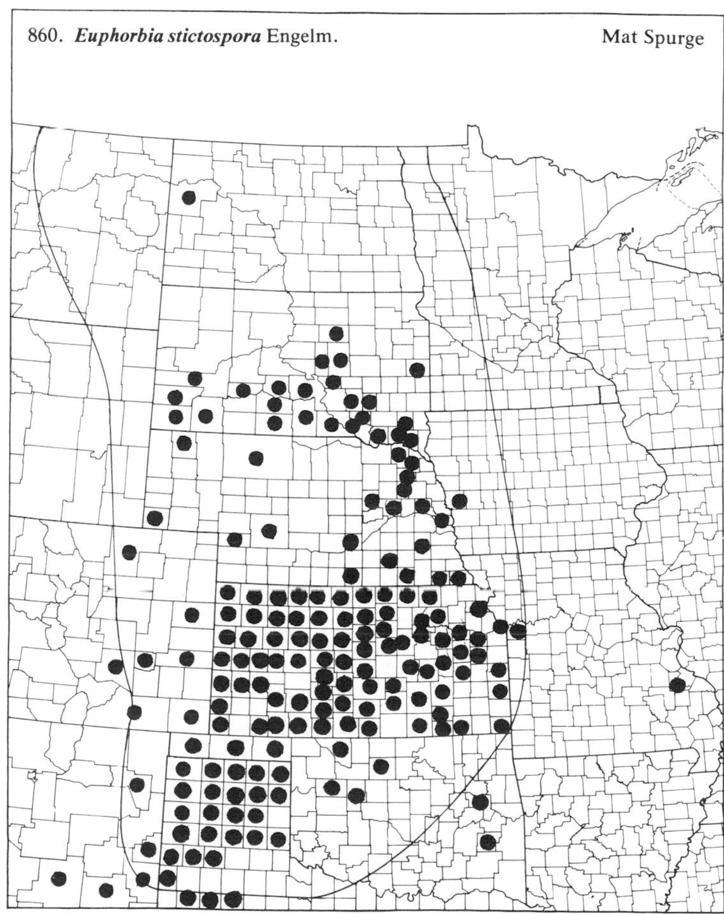

(Euphorbiaceae)

861. ***Reverchonia arenaria*** Gray

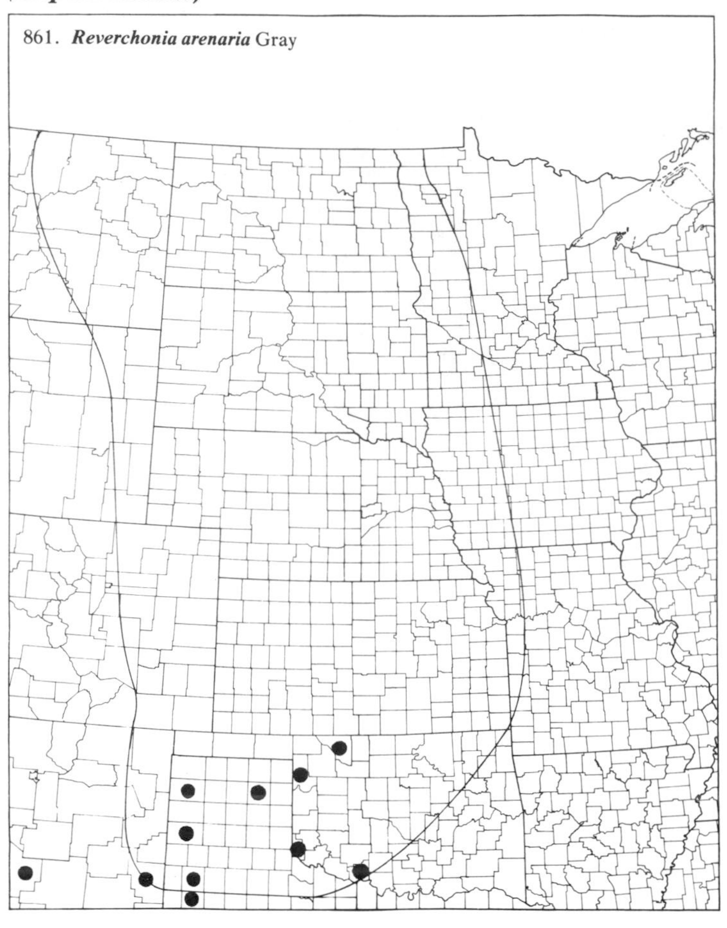

862. ***Stillingia sylvatica*** L. Queen's-delight

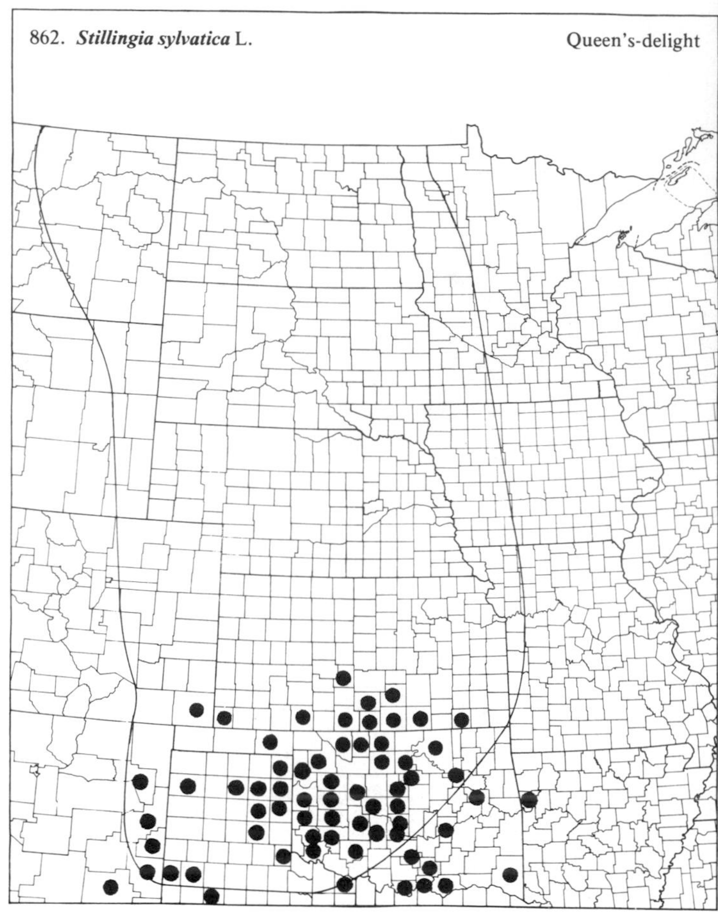

863. ***Tragia betonicifolia*** Nutt.

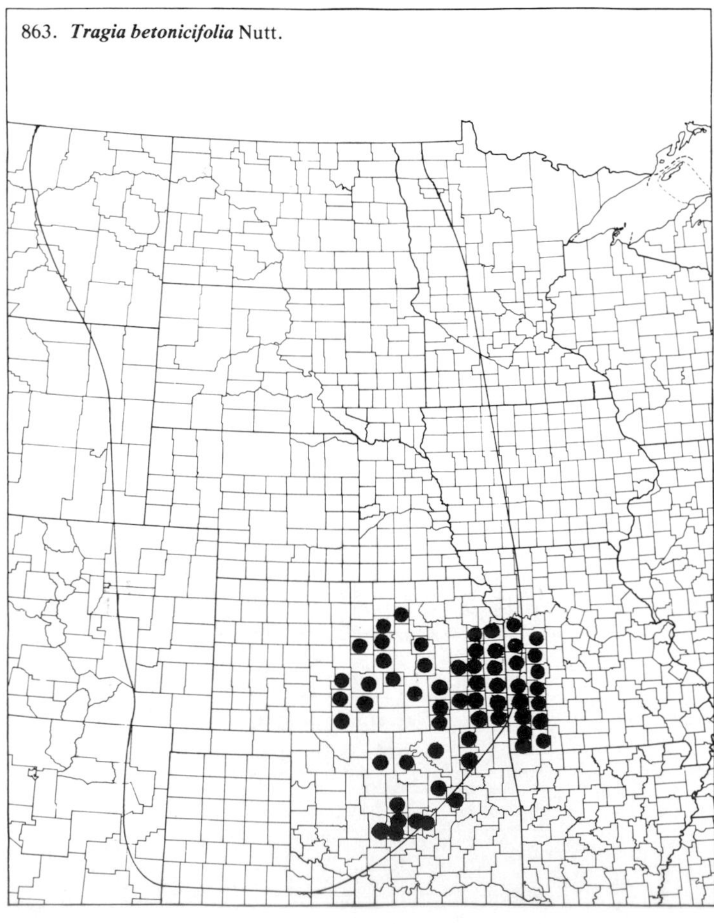

864. ***Tragia ramosa*** Torr.

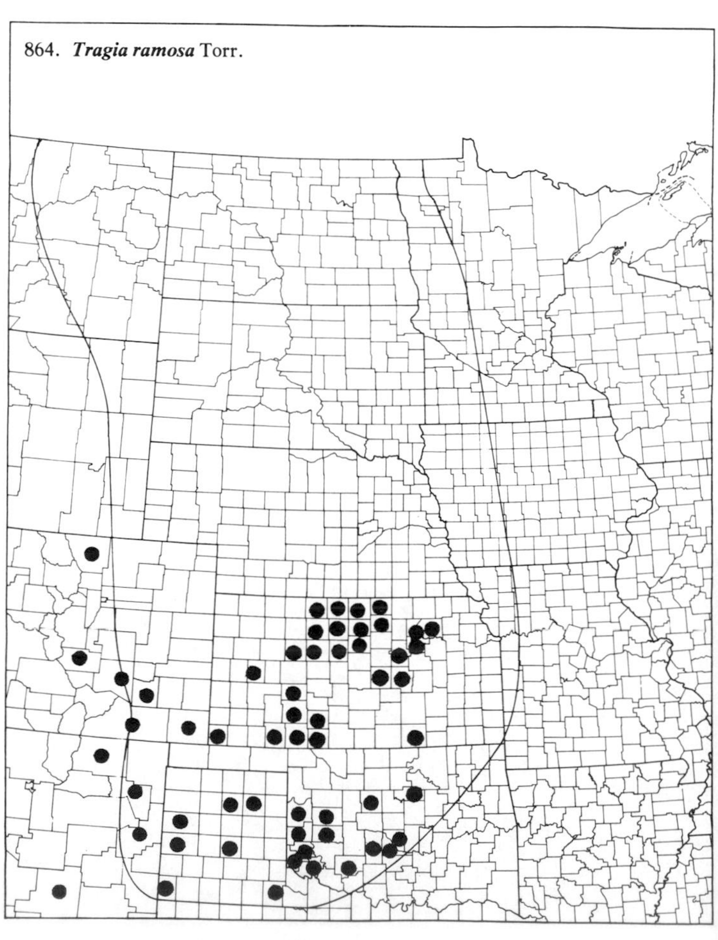

69. *Rhamnaceae* (Buckthorn Family)

865. *Ceanothus americanus* L. var. *pitcheri* T. & G. — New Jersey Tea

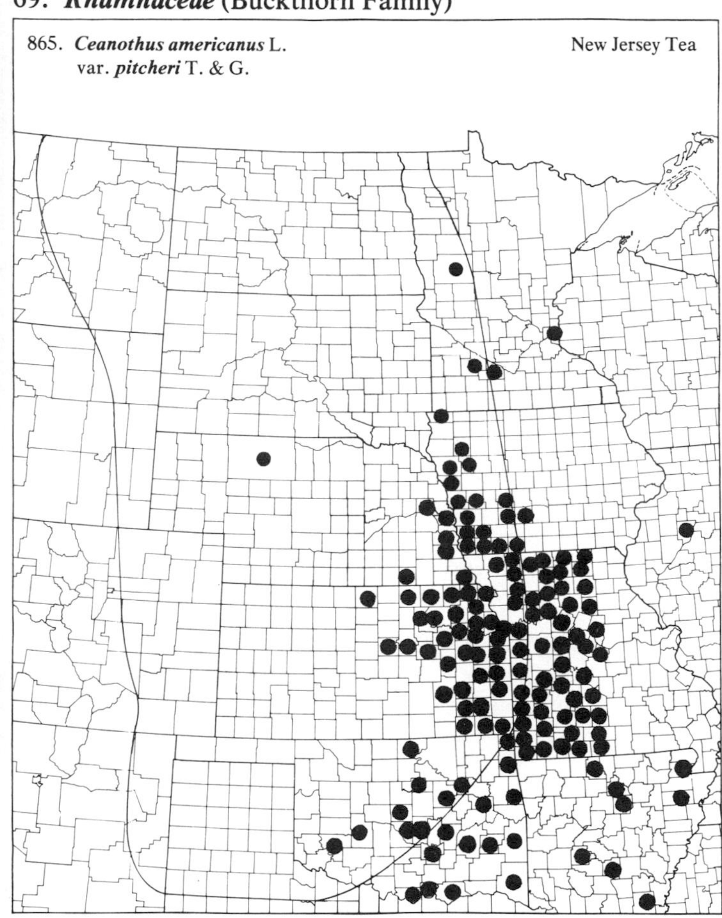

866. *Ceanothus herbaceus* Raf. var. *pubescens* (T. & G.) Shinners — New Jersey Tea

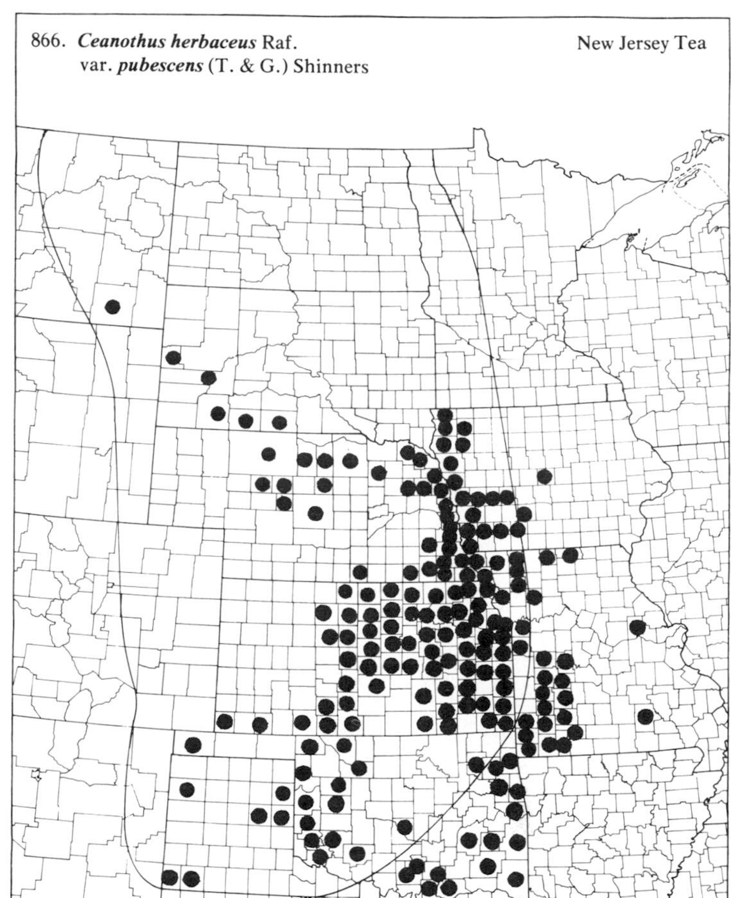

867. *Rhamnus alnifolia* L'Her. — Alder Buckthorn

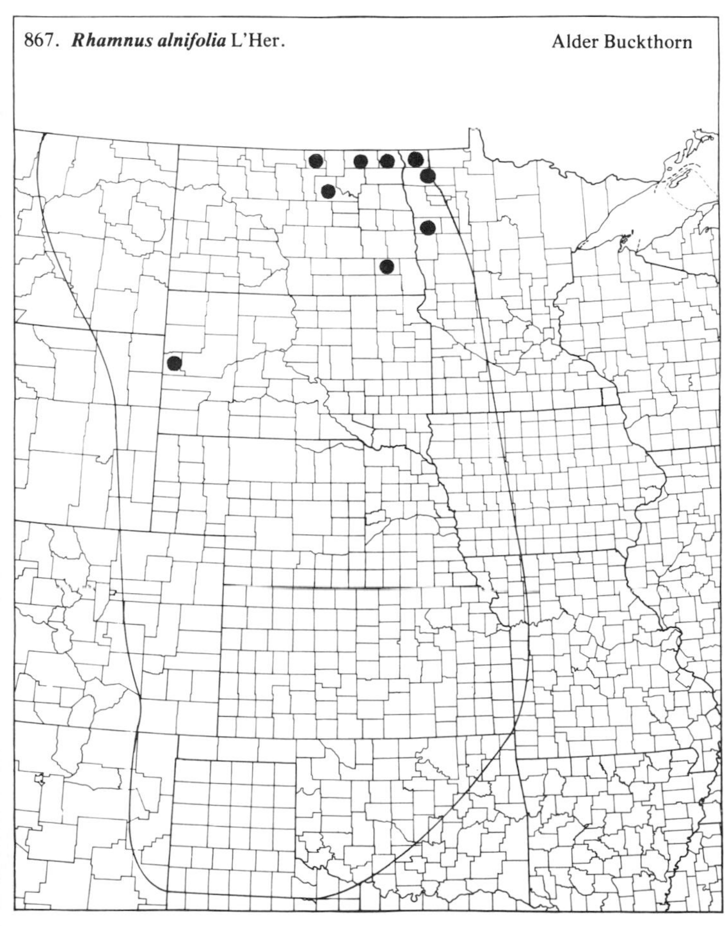

868. *Rhamnus catharticus* L. — Common Buckthorn

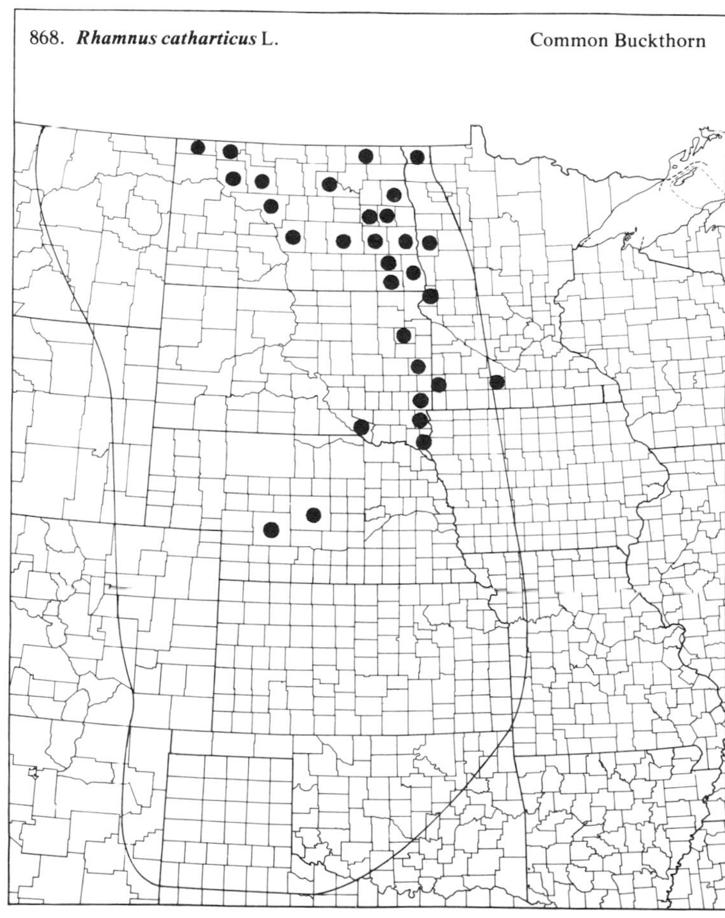

(Rhamnaceae)

869. ***Rhamnus lanceolata*** Pursh var. ***glabratus*** Gl. — Lance-leaved Buckthorn

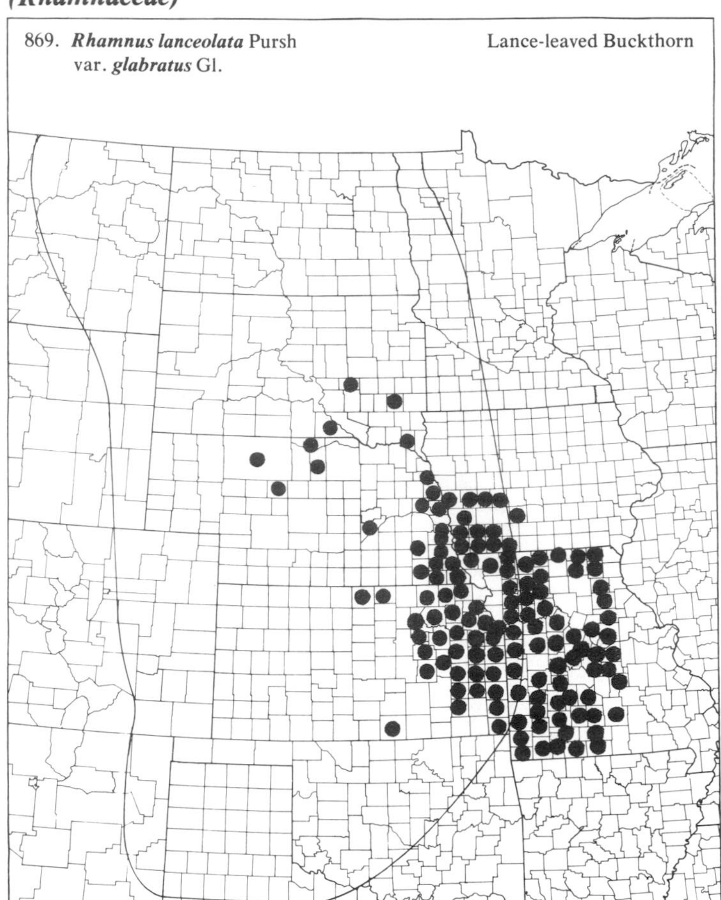

70. *Vitaceae* (Grape Family)

870. ***Ampelopsis cordata*** Michx. — Raccoon Grape

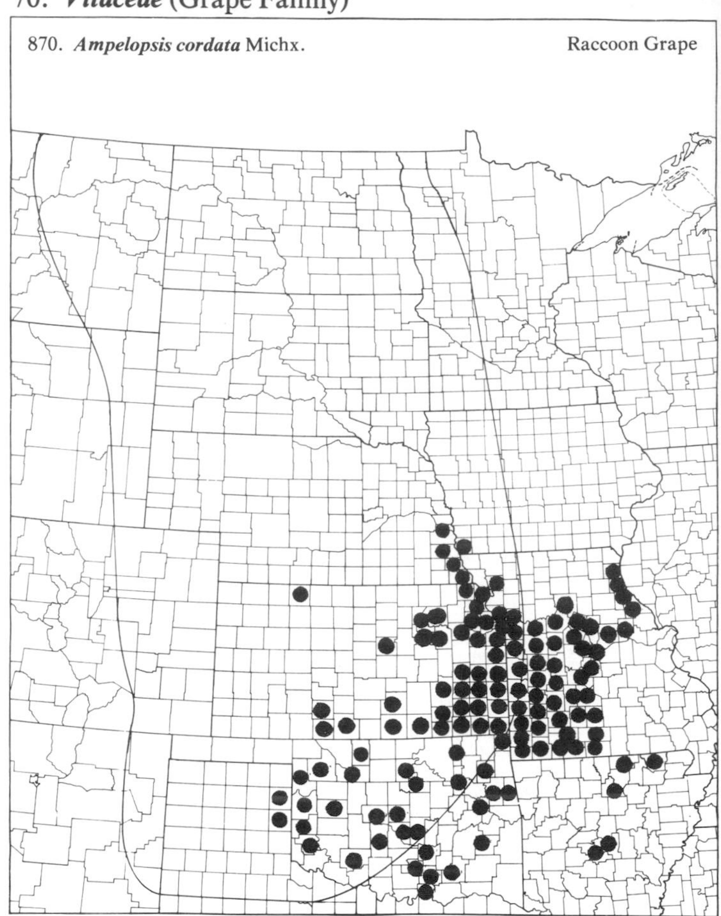

871. ***Cissus incisa*** (Nutt.) Des Moul. — Possum Grape

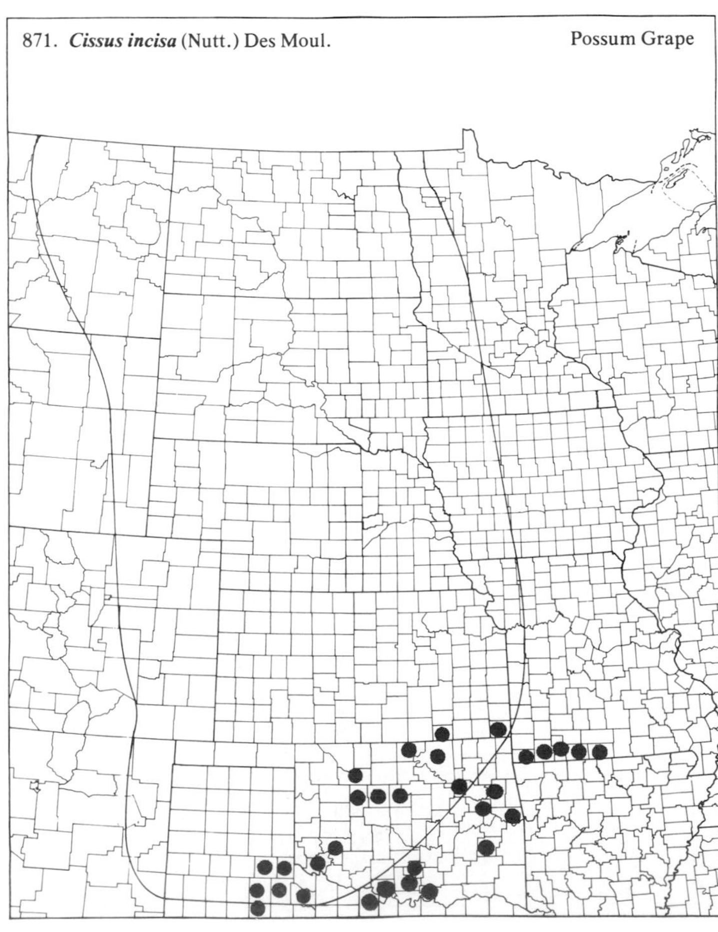

872. ***Parthenocissus quinquefolia*** (L.) Planch. — Virginia Creeper

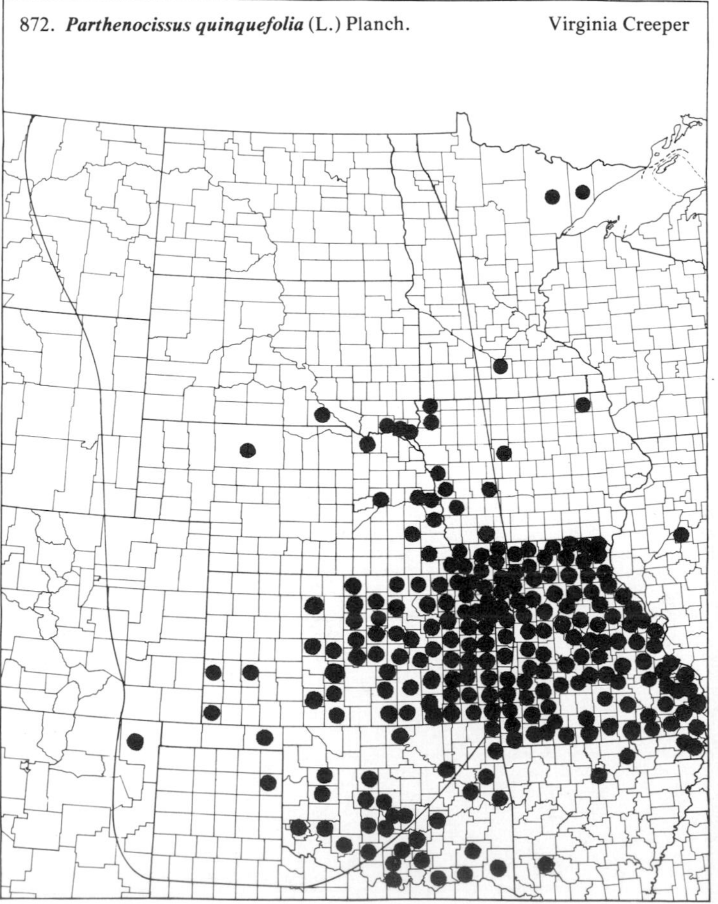

(Vitaceae)

873. ***Parthenocissus vitacea*** (Knerr) Hitchc. Woodbine

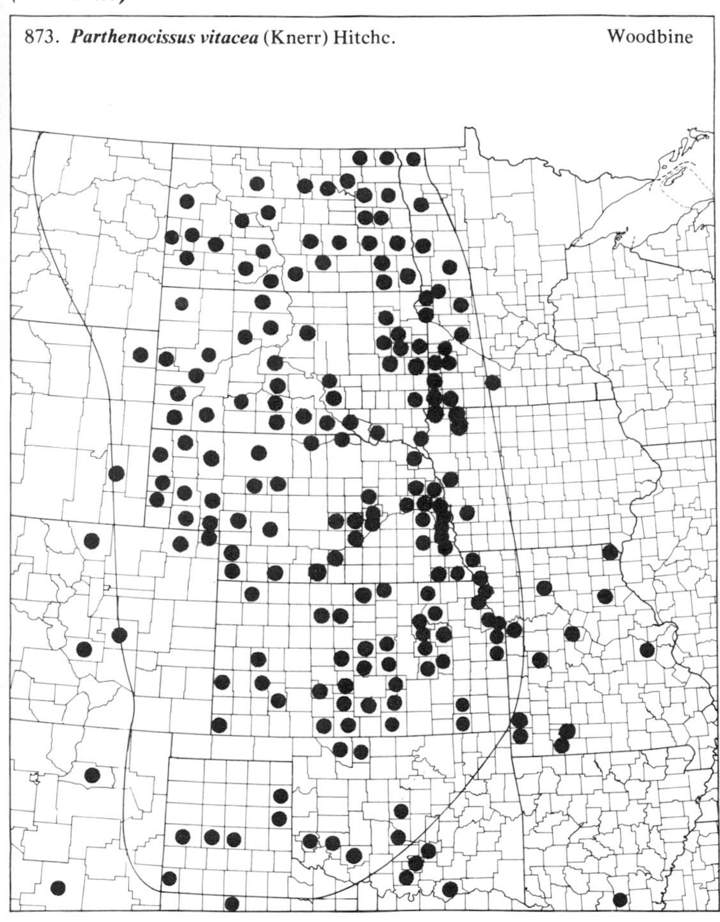

874. ***Vitis acerifolia*** Raf. Bush Grape

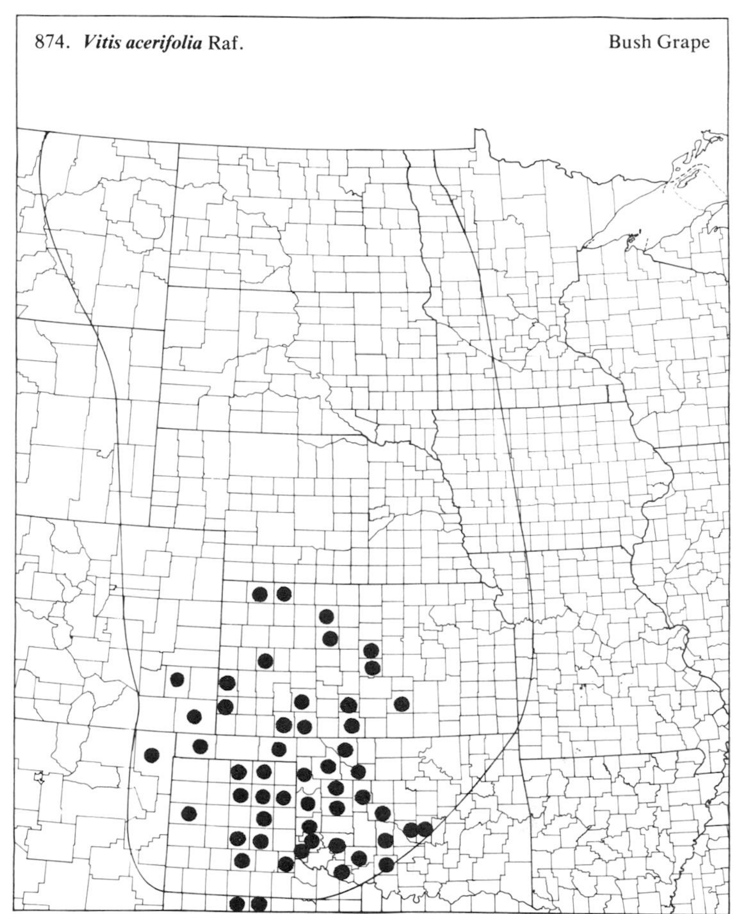

875. ***Vitis aestivalis*** Michx. Pigeon Grape

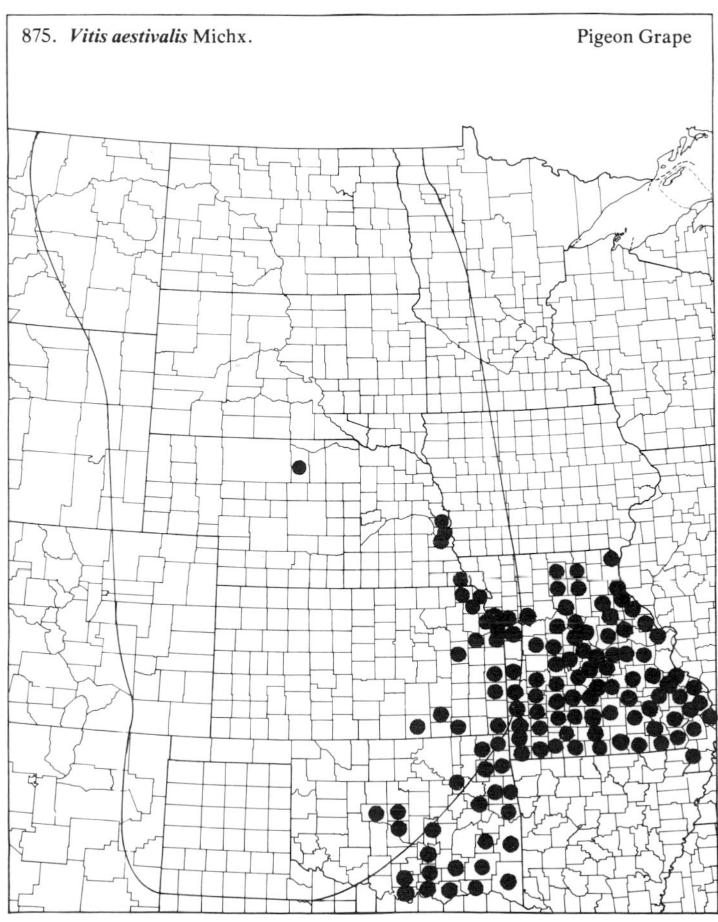

876. ***Vitis cinerea*** Engelm. Graybark Grape

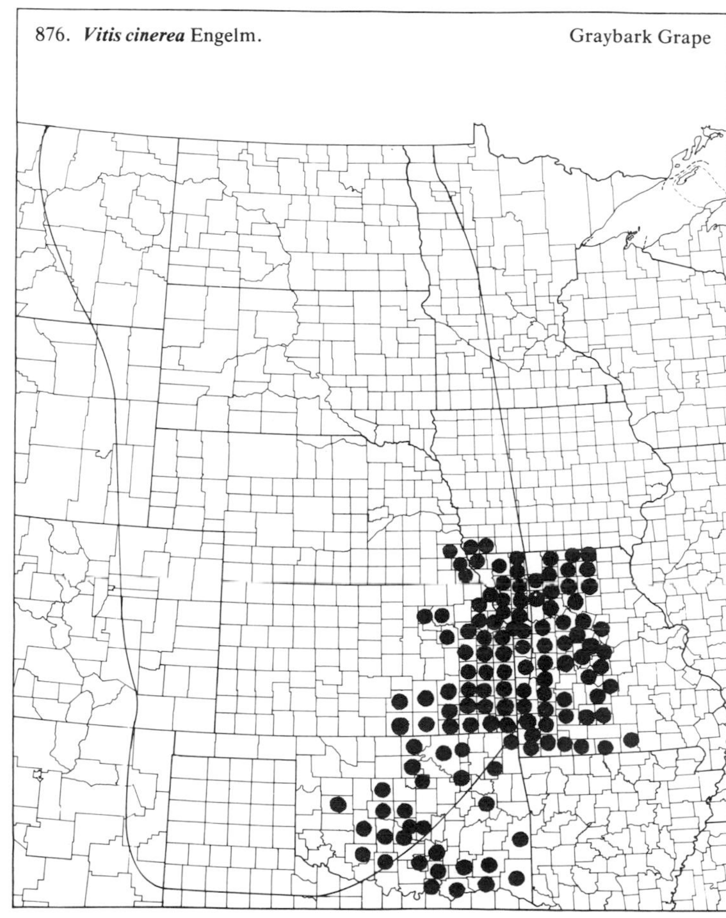

(Vitaceae)

877. ***Vitis riparia*** Michx. River-bank Grape

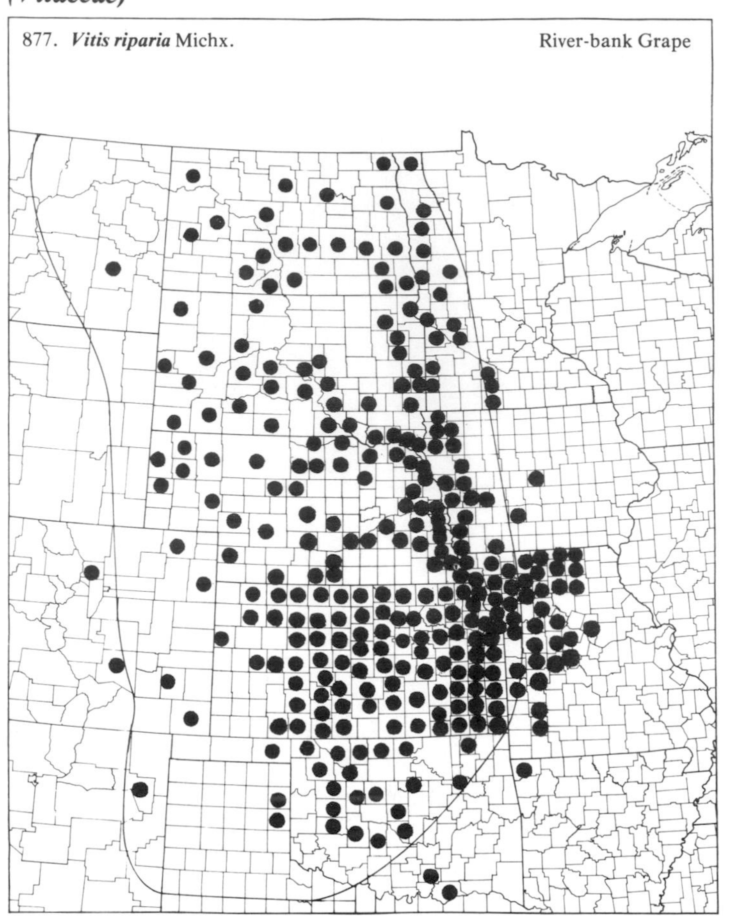

878. ***Vitis vulpina*** L. Winter Grape

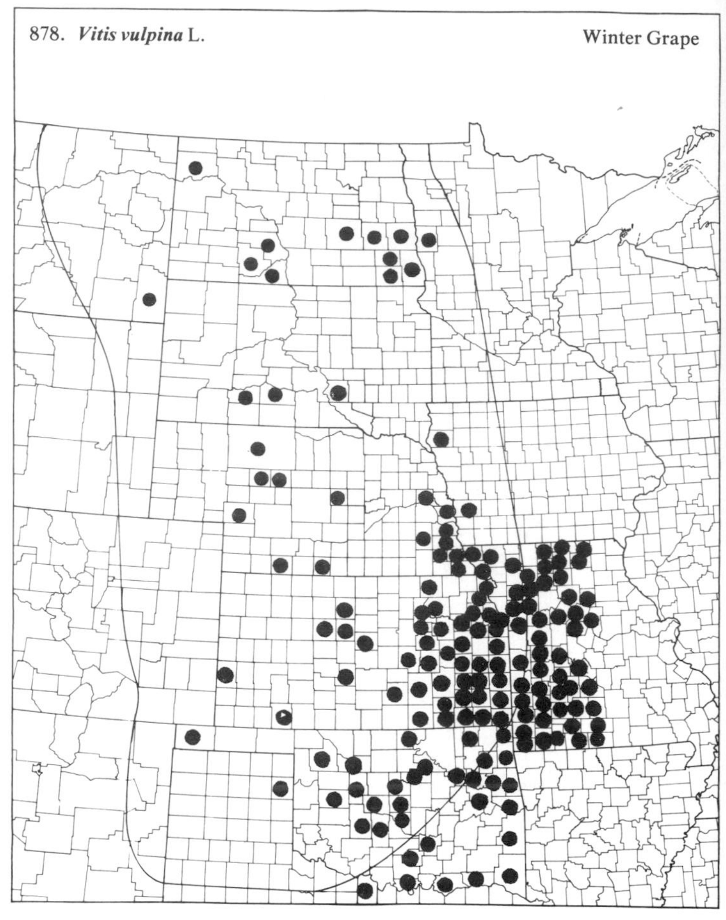

71. *Staphyleaceae* (Bladdernut Family)

879. ***Staphylea trifolia*** L. Bladdernut

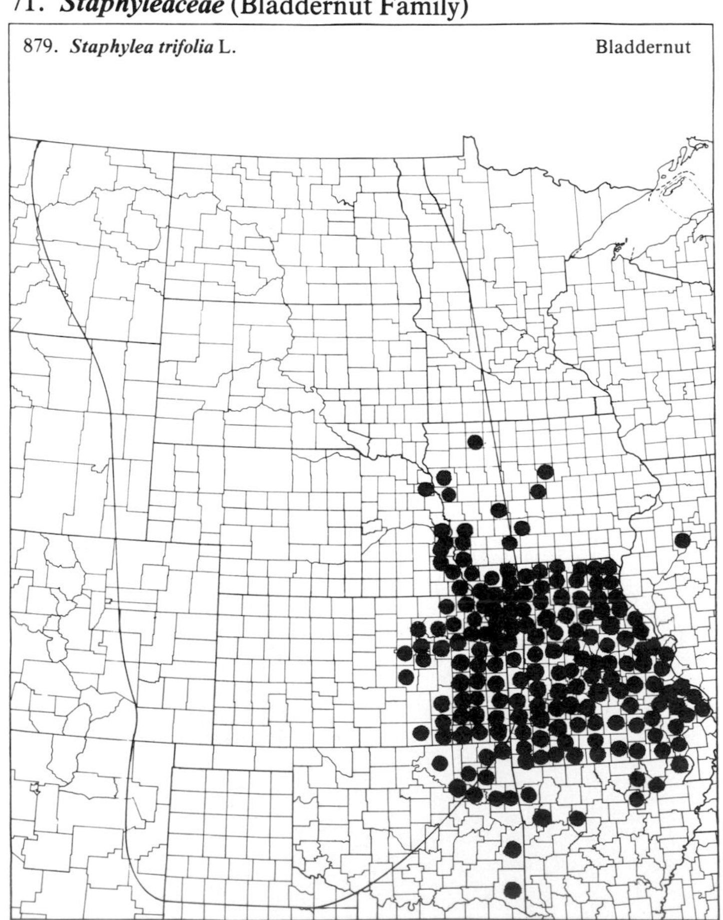

72. *Sapindaceae* (Soapberry Family)

880. ***Cardiospermum halicacabum*** L.

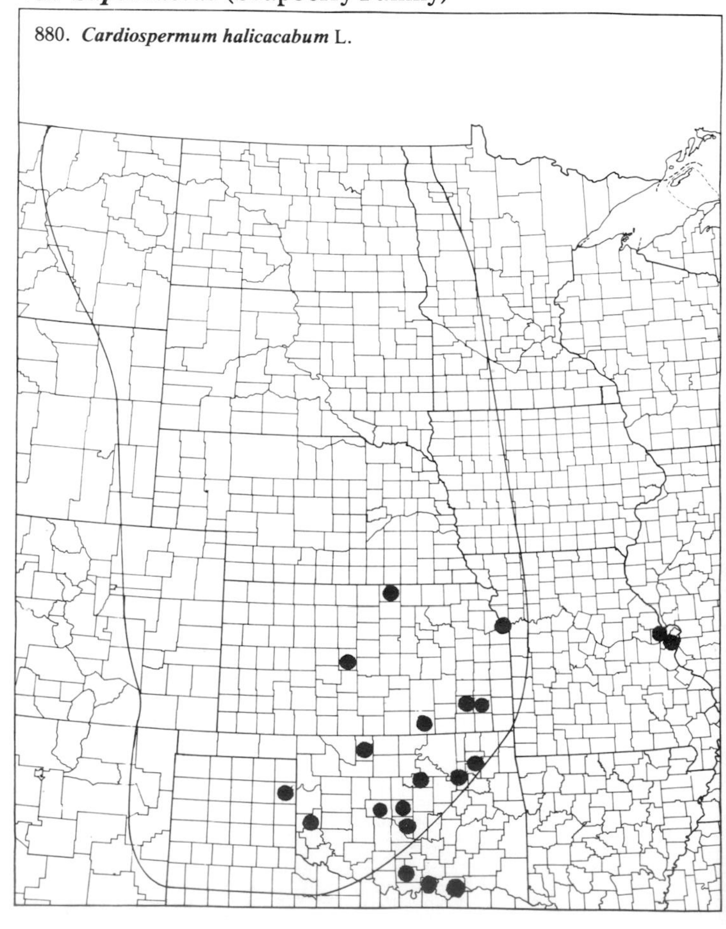

(Sapindaceae)

881. ***Sapindus drummondii*** H. & A. — Soapberry

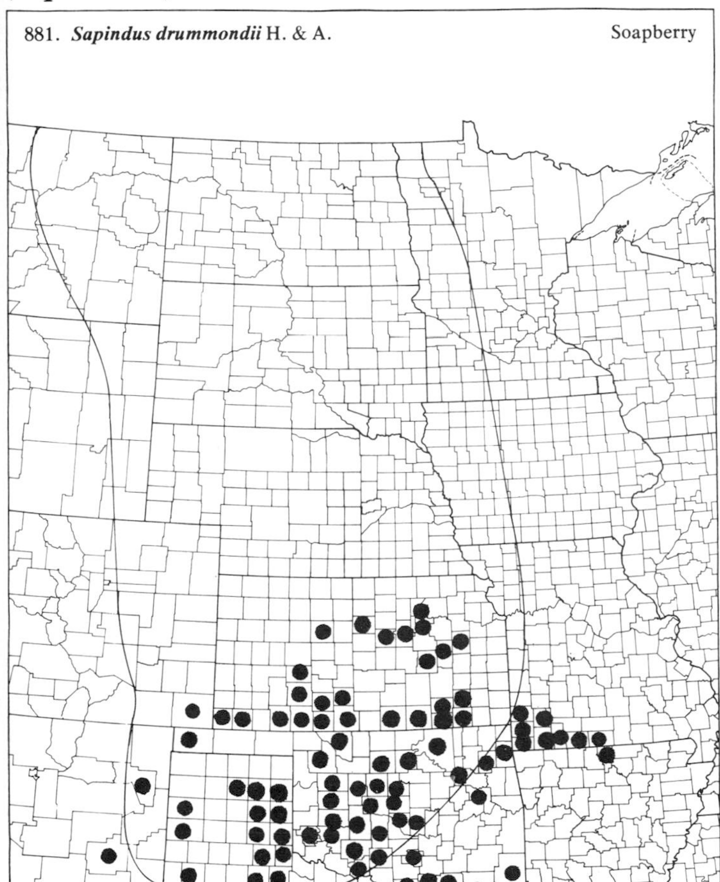

73. *Hippocastanaceae* (Buckeye Family)

882. ***Aesculus glabra*** Willd. var. ***arguta*** (Buckl.) Robins. — Western Buckeye

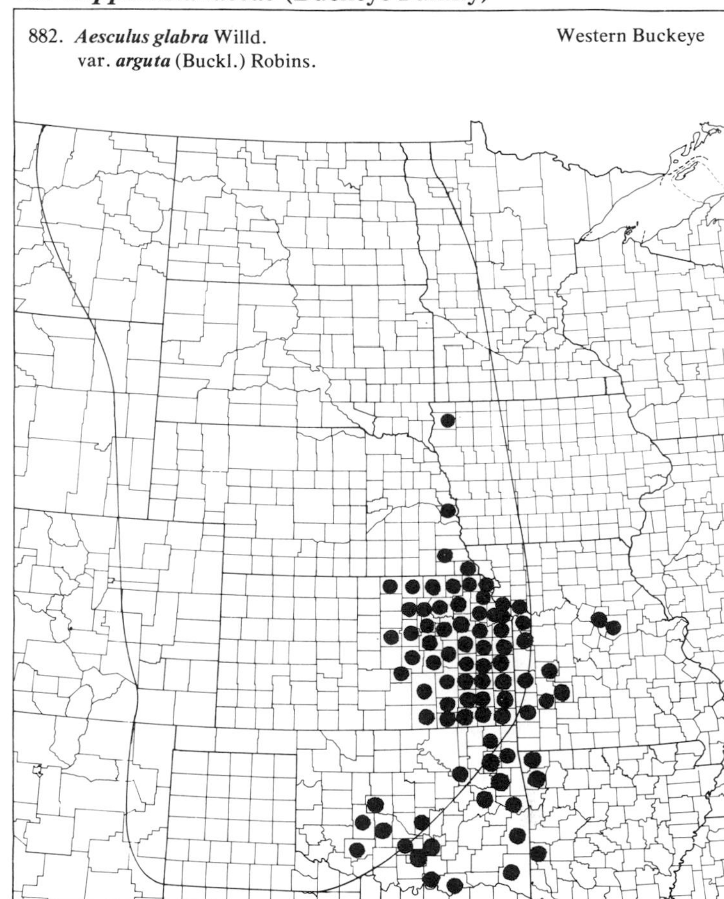

74. *Aceraceae* (Maple Family)

883. ***Acer negundo*** L. — Box Elder

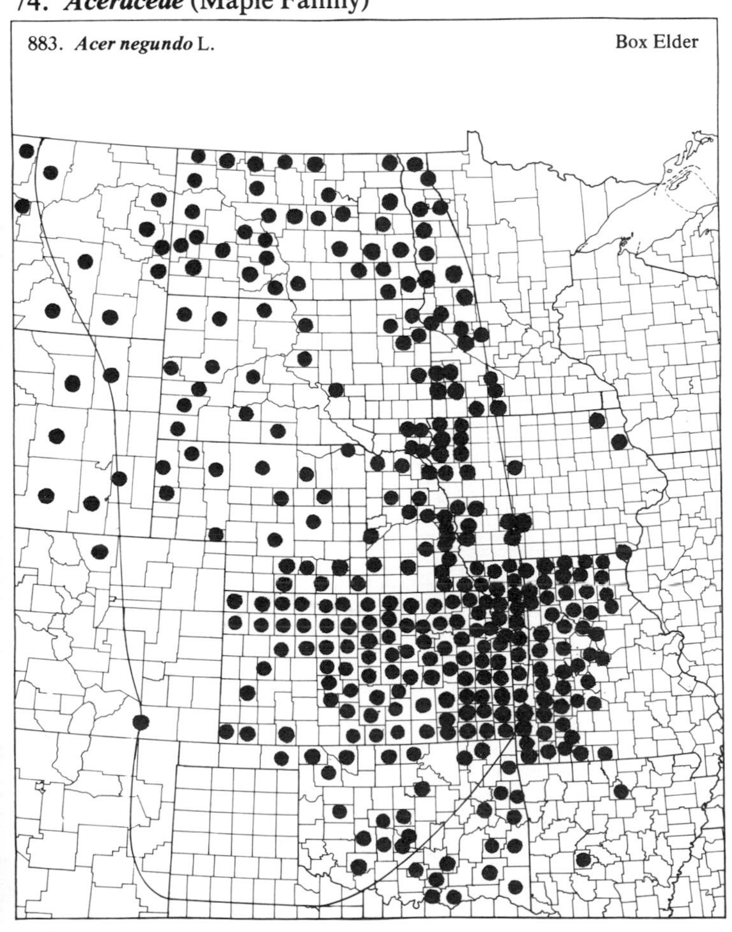

884. ***Acer saccharinum*** L. — Silver Maple

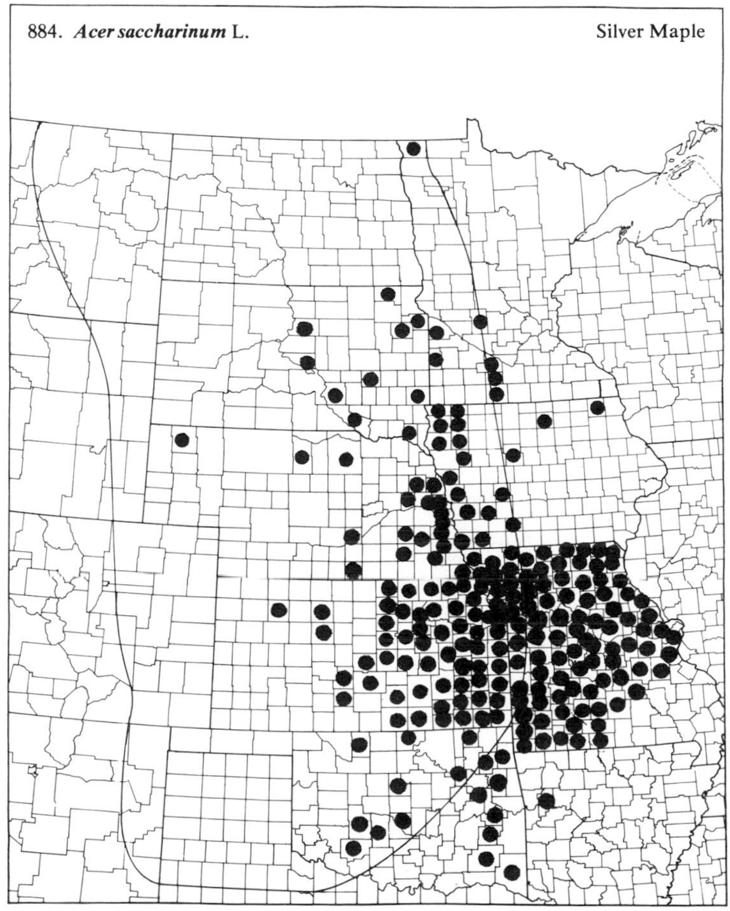

75. *Anacardiaceae* (Cashew Family)

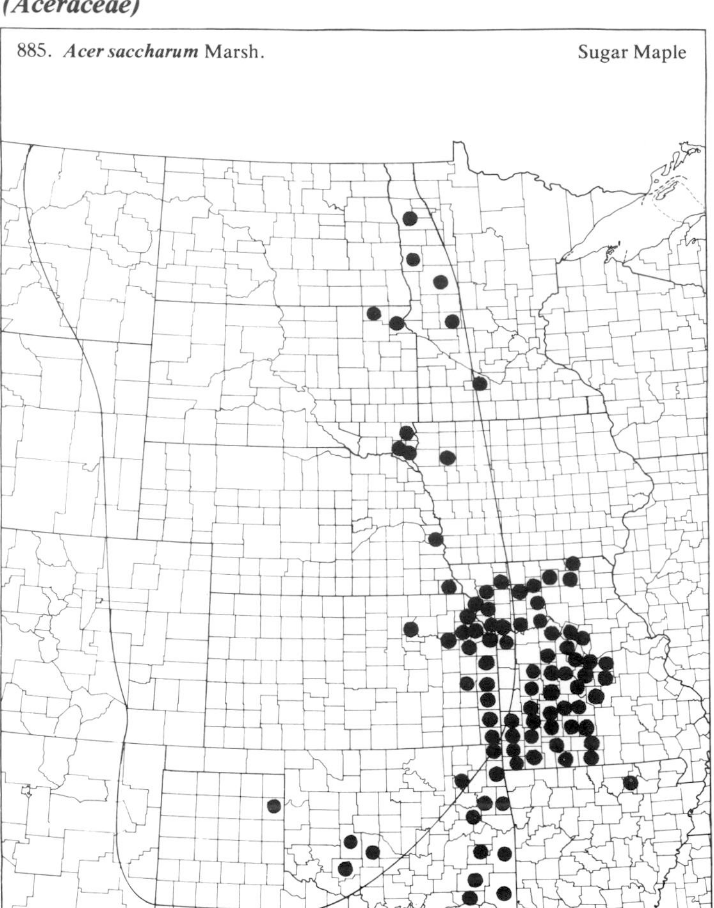

885. *Acer saccharum* Marsh. Sugar Maple

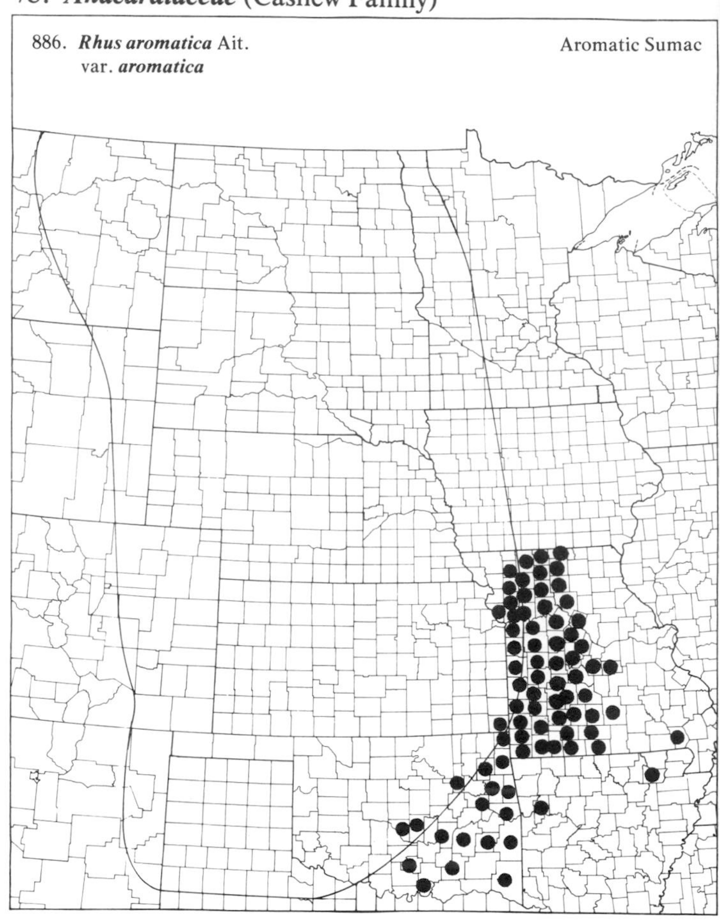

886. ***Rhus aromatica*** Ait. var. ***aromatica*** Aromatic Sumac

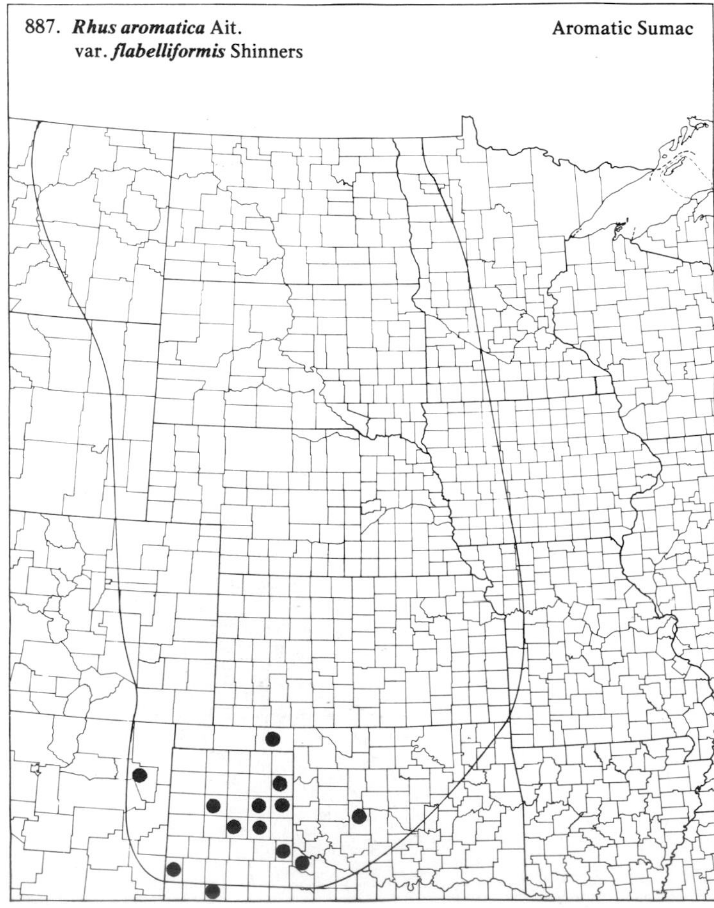

887. ***Rhus aromatica*** Ait. var. ***flabelliformis*** Shinners Aromatic Sumac

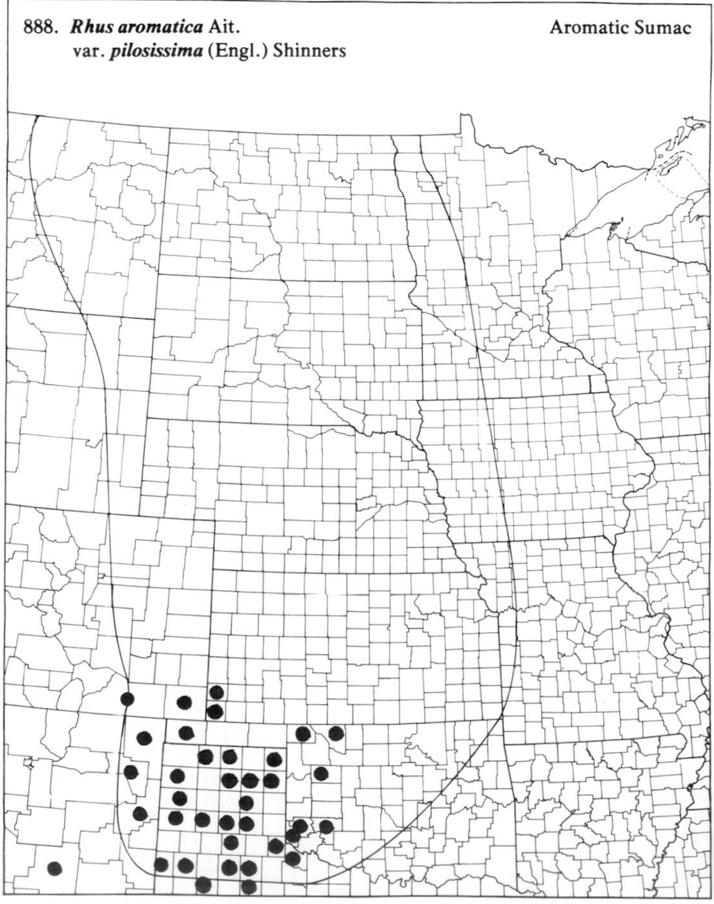

888. ***Rhus aromatica*** Ait. var. ***pilosissima*** (Engl.) Shinners Aromatic Sumac

(Anacardiaceae)

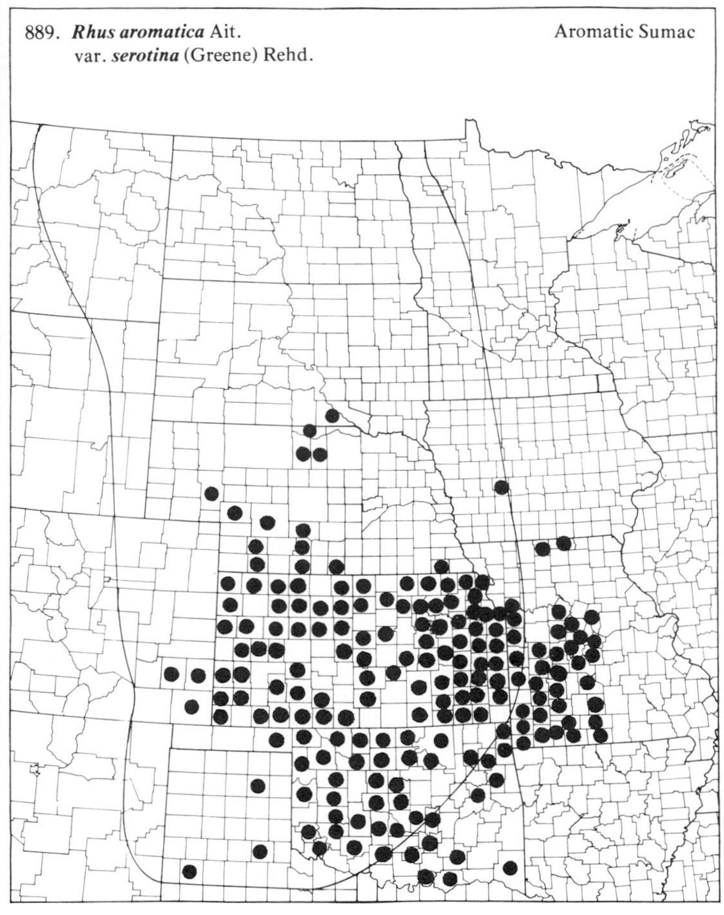

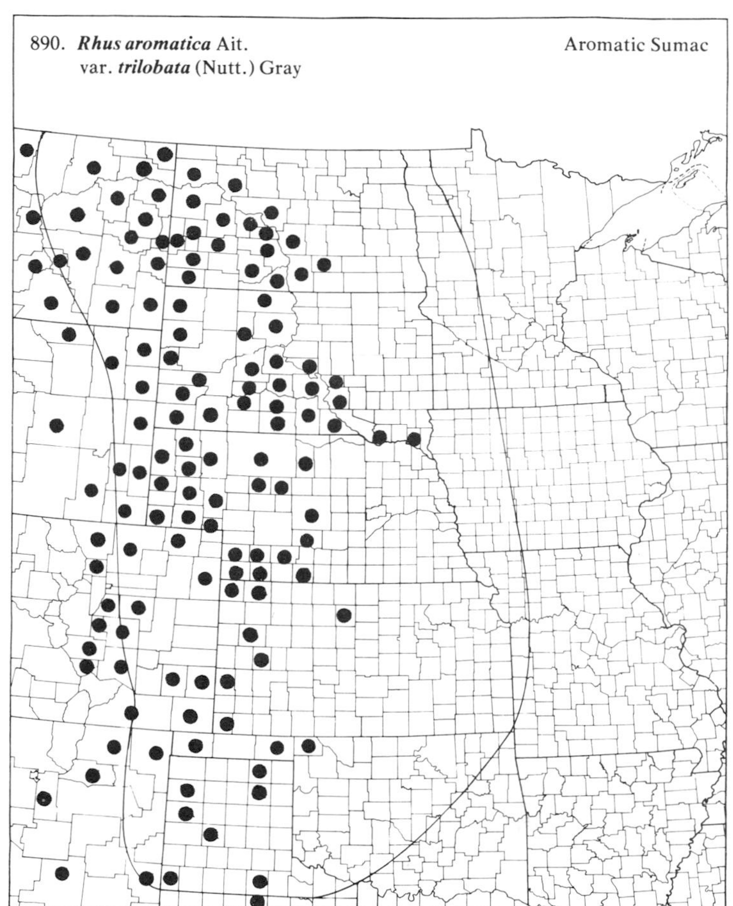

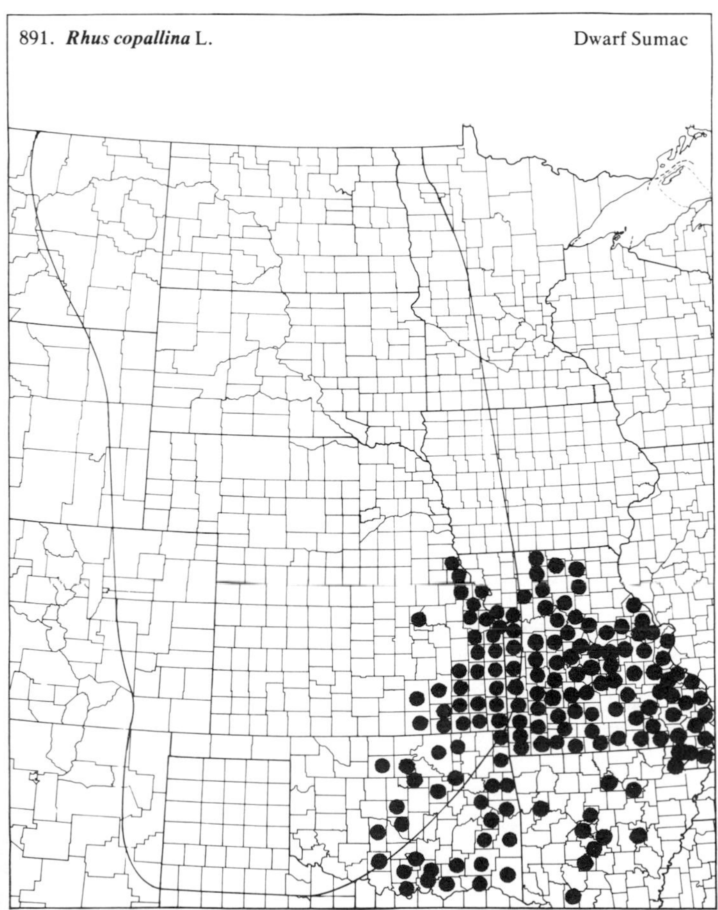

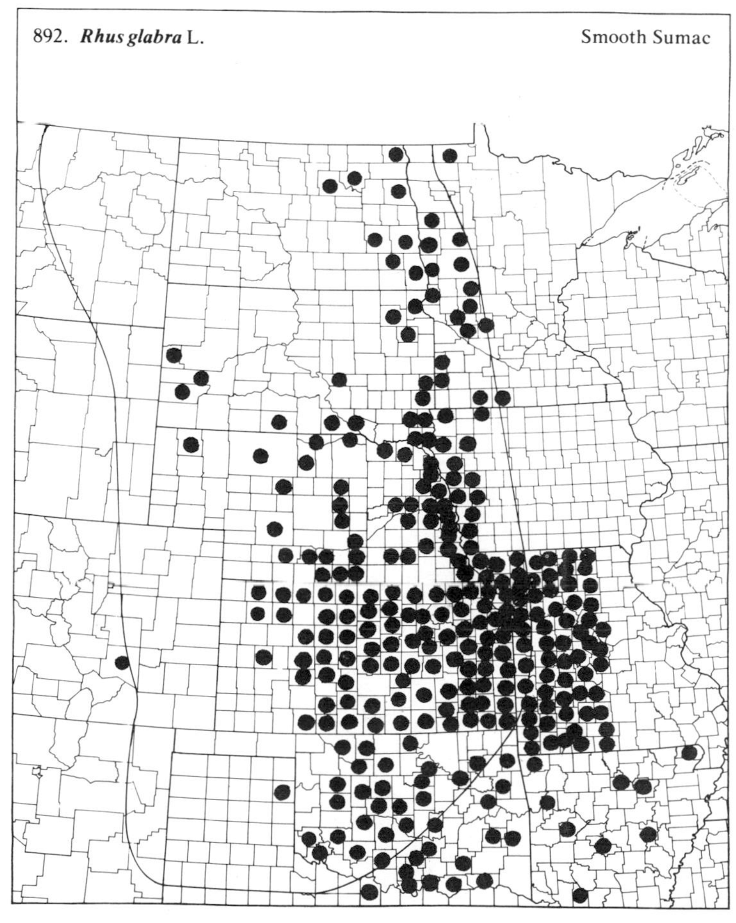

(Anacardiaceae)

893. ***Rhus microphylla*** Engelm.

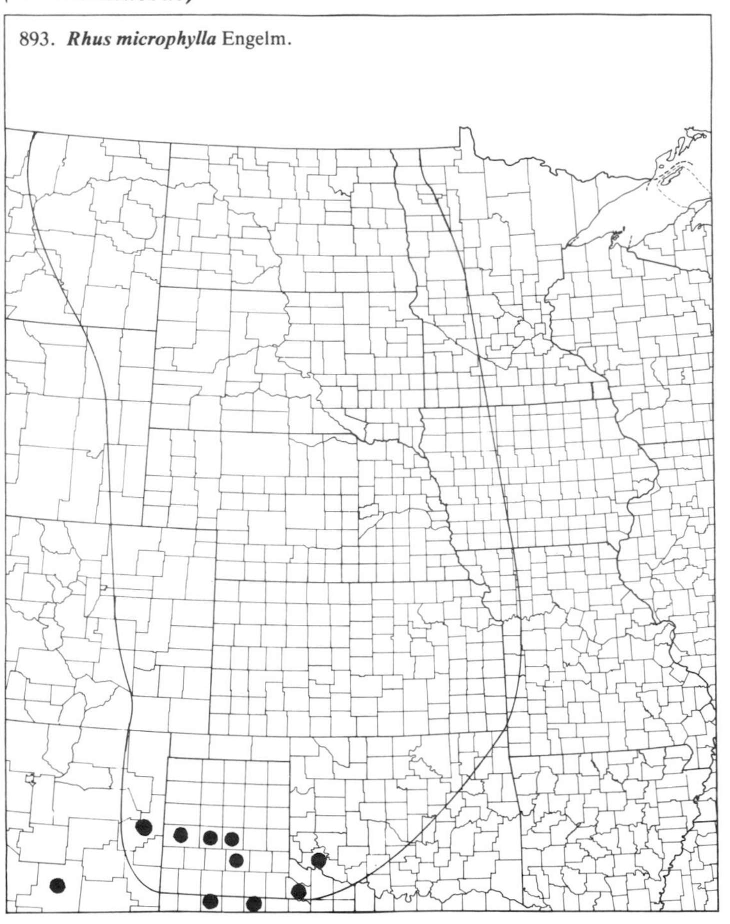

894. ***Toxicodendron radicans*** (L.) O. Ktze. ssp. ***negundo*** (Greene) Gillis. Poison Ivy

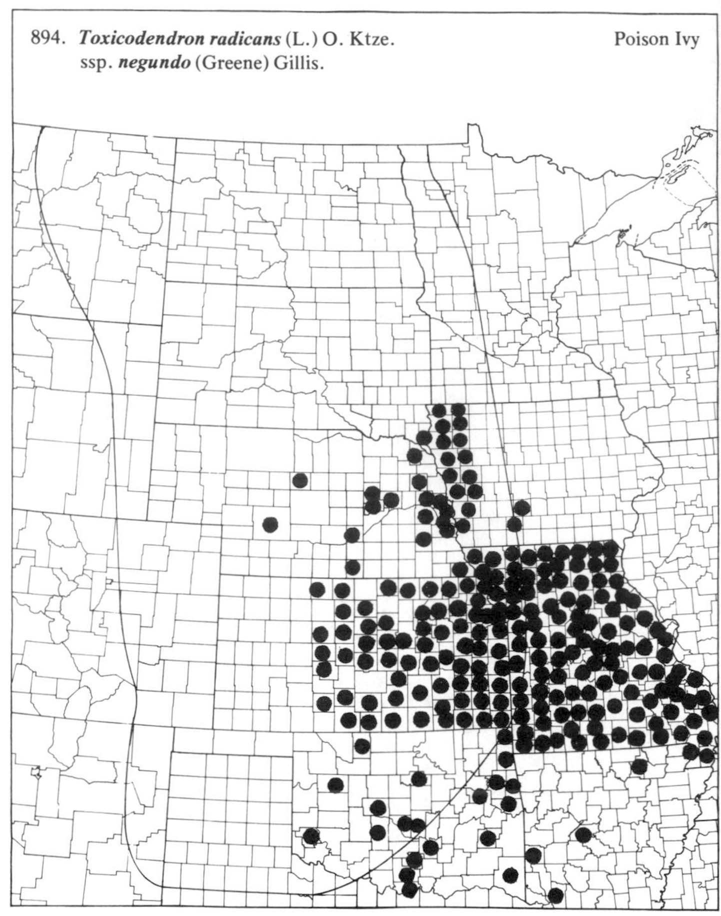

76. *Simaroubaceae* (Quassia Family)

895. ***Toxicodendron rydbergii*** (Small ex Rydb.) Greene Poison Ivy

896. ***Ailanthus altissima*** (Mill.) Swingle Tree of Heaven

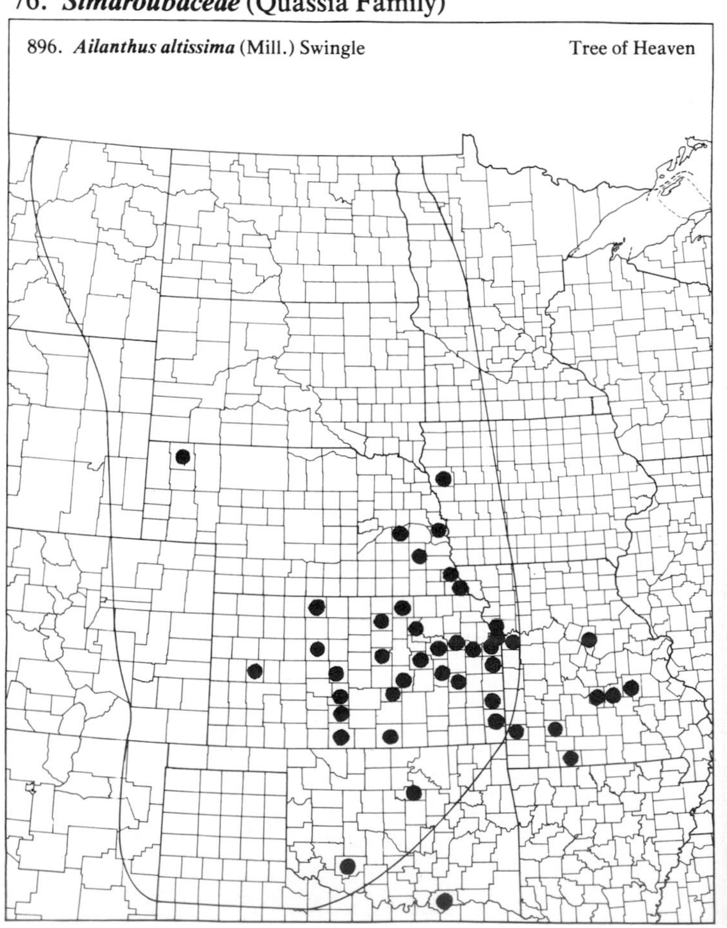

77. ***Rutaceae*** (Rue Family)

897. ***Ptelea trifoliata*** L. Hop Tree

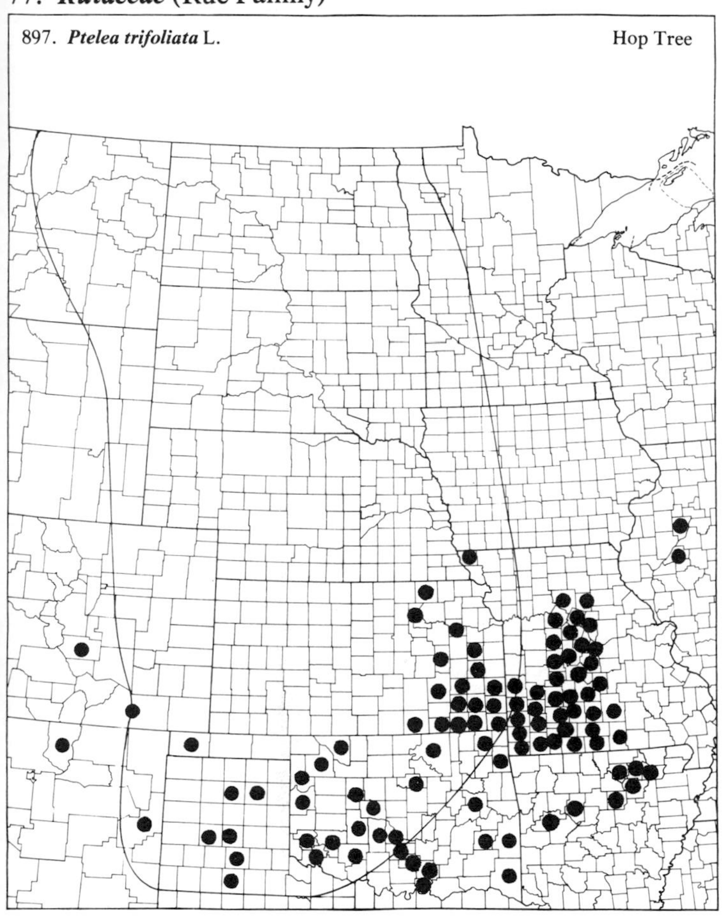

898. ***Zanthoxylum americanum*** Mill. Prickly Ash

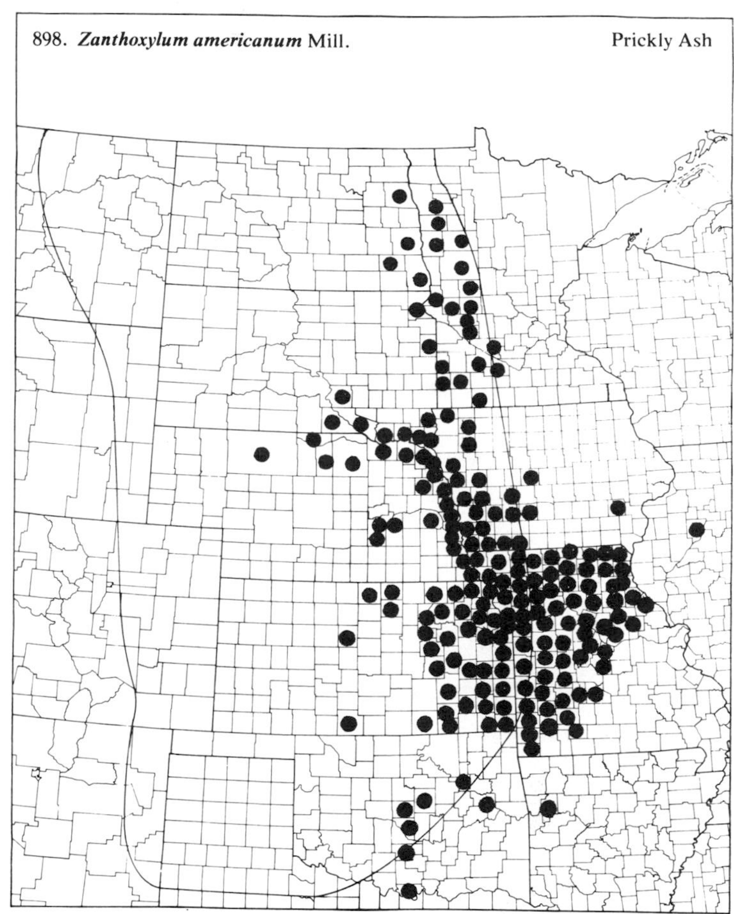

78. ***Zygophyllaceae*** (Caltrop Family)

899. ***Kallstroemia parviflora*** Nort.

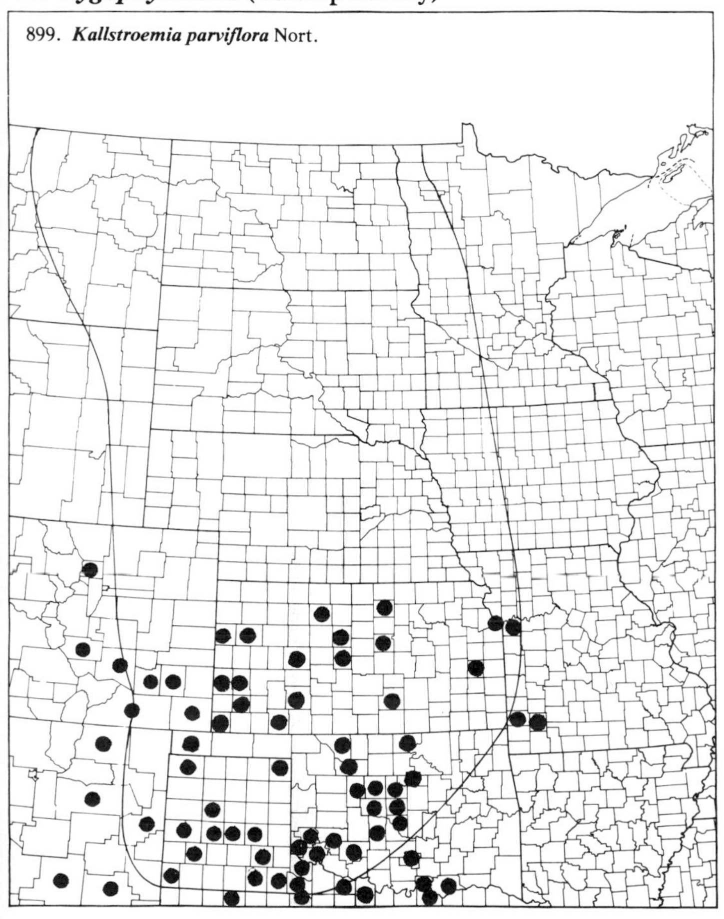

900. ***Tribulus terrestris*** L. Puncture Vine

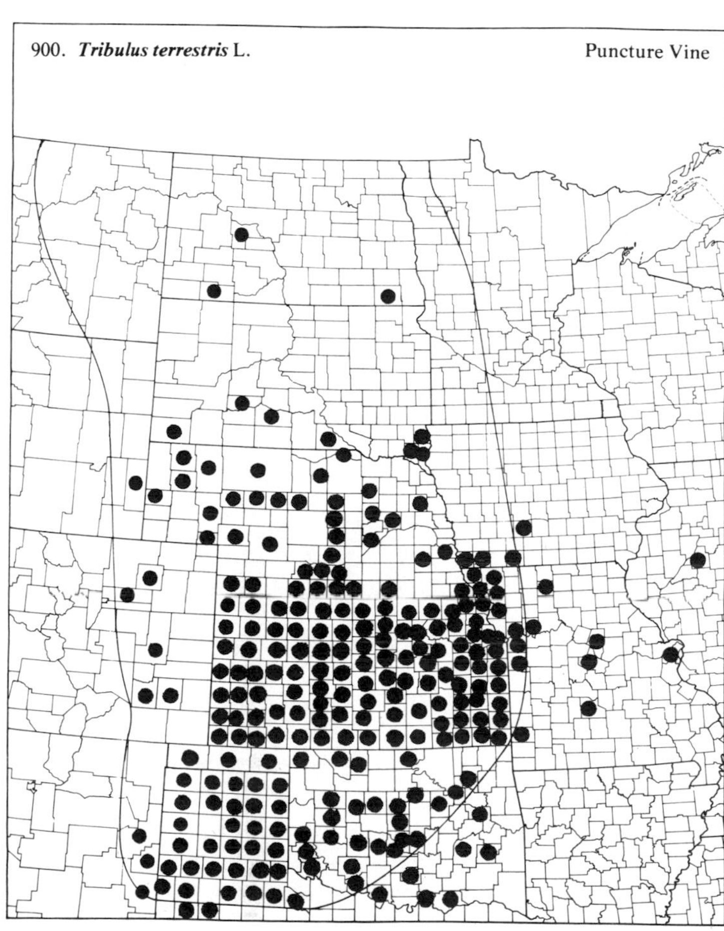

79. *Oxalidaceae* (Wood Sorrel Family)

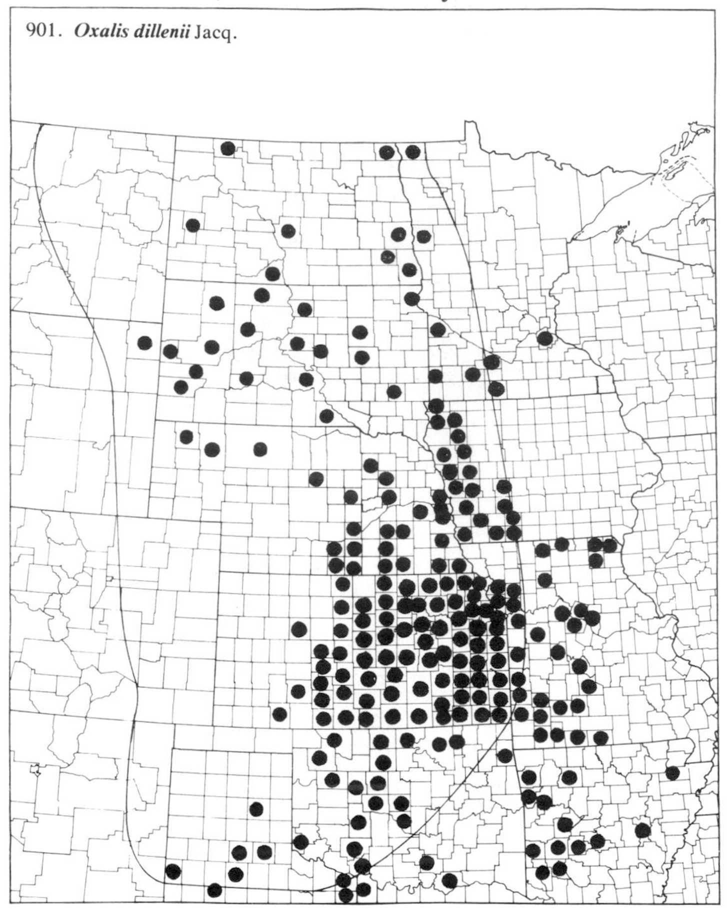
901. *Oxalis dillenii* Jacq.

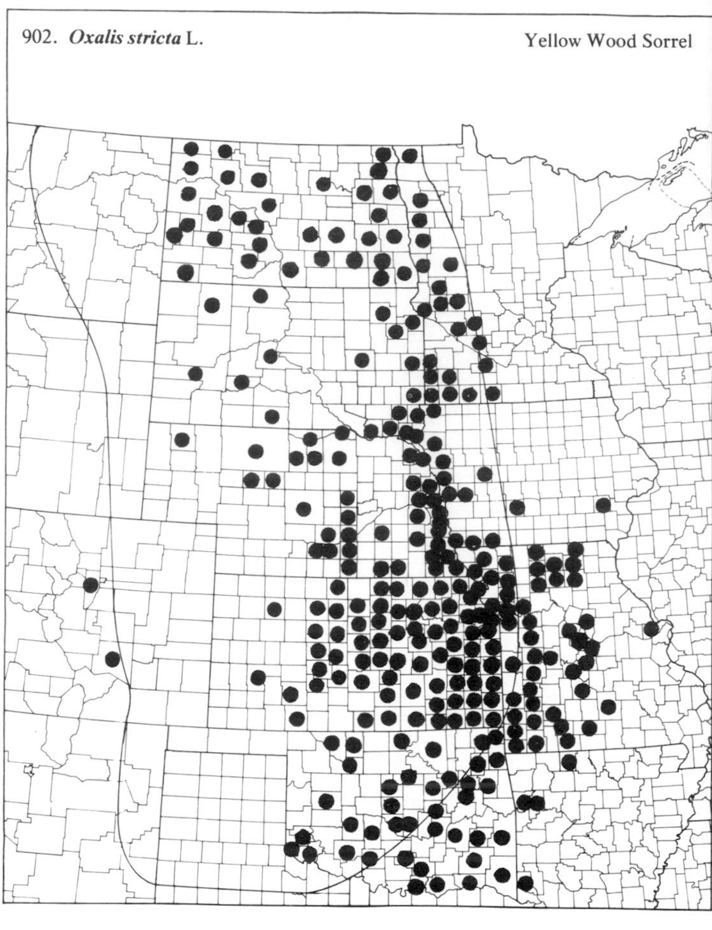
902. *Oxalis stricta* L. Yellow Wood Sorrel

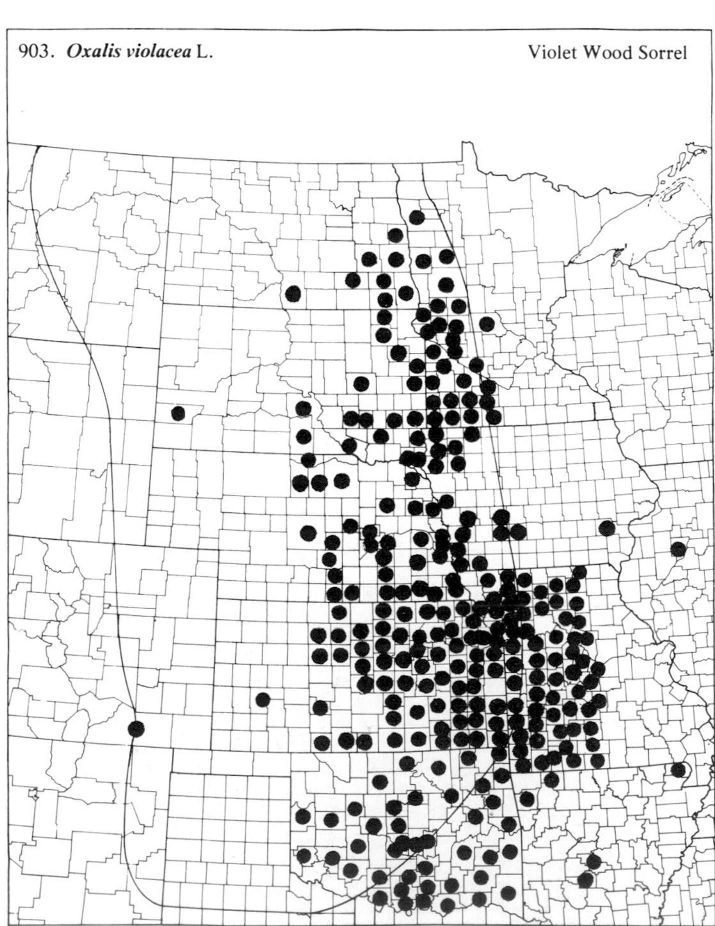
903. *Oxalis violacea* L. Violet Wood Sorrel

80. *Geraniaceae* (Geranium Family)

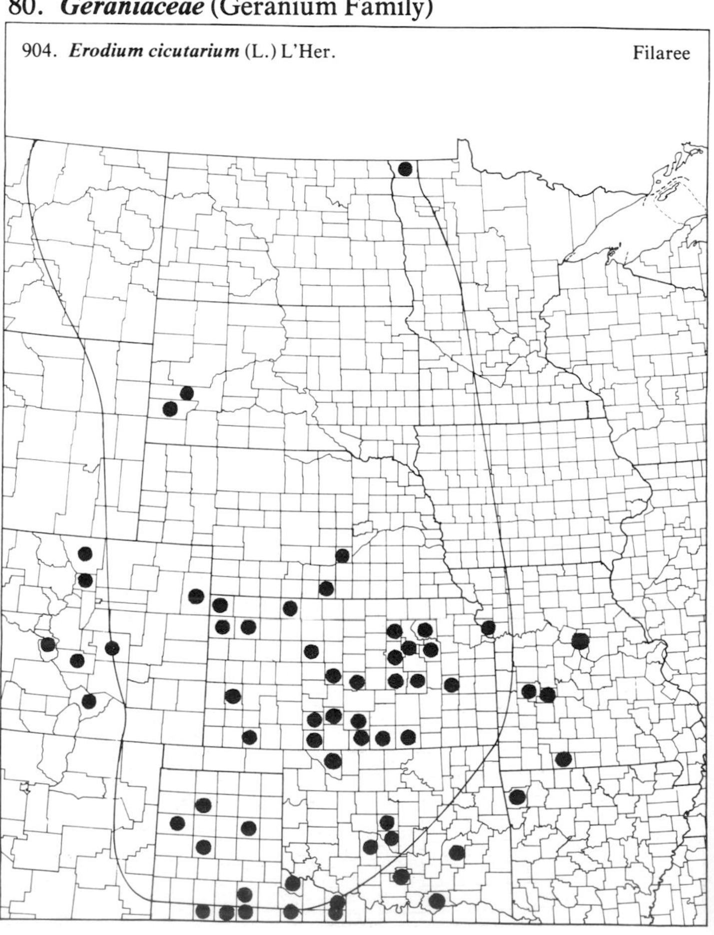
904. *Erodium cicutarium* (L.) L'Her. Filaree

(Geraniaceae)

905. ***Geranium bicknellii*** Britt. Bicknell's Cranesbill

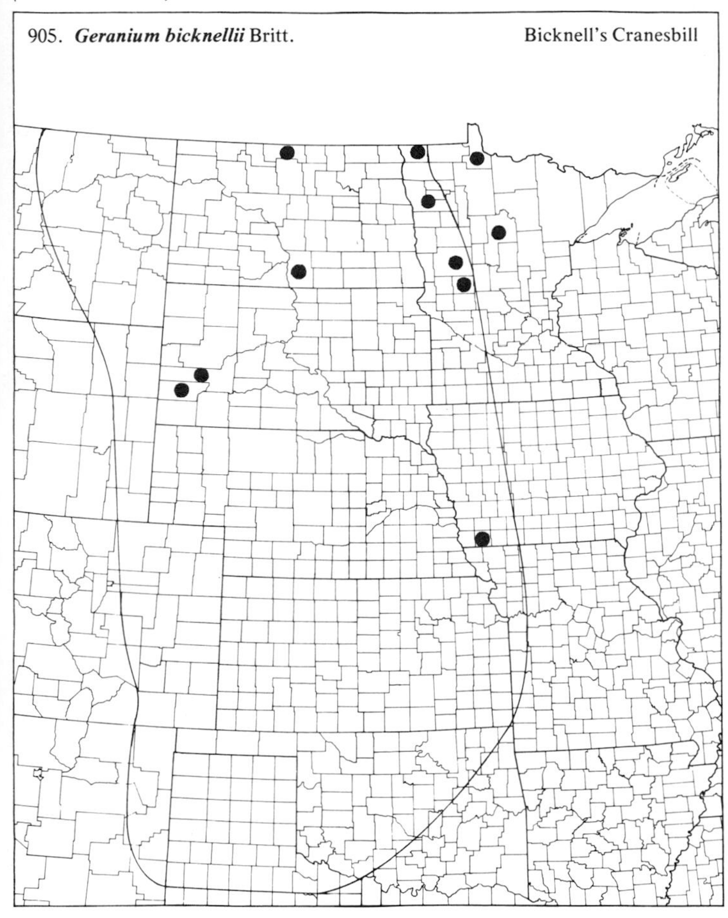

906. ***Geranium carolinianum*** L. Carolina Cranesbill

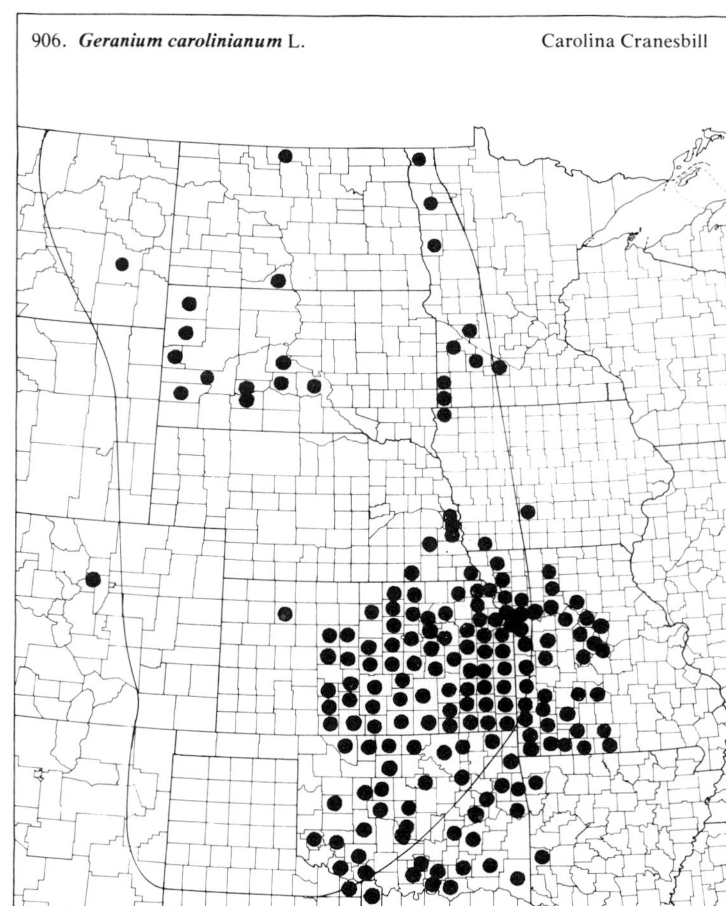

907. ***Geranium maculatum*** L. Wild Cranesbill

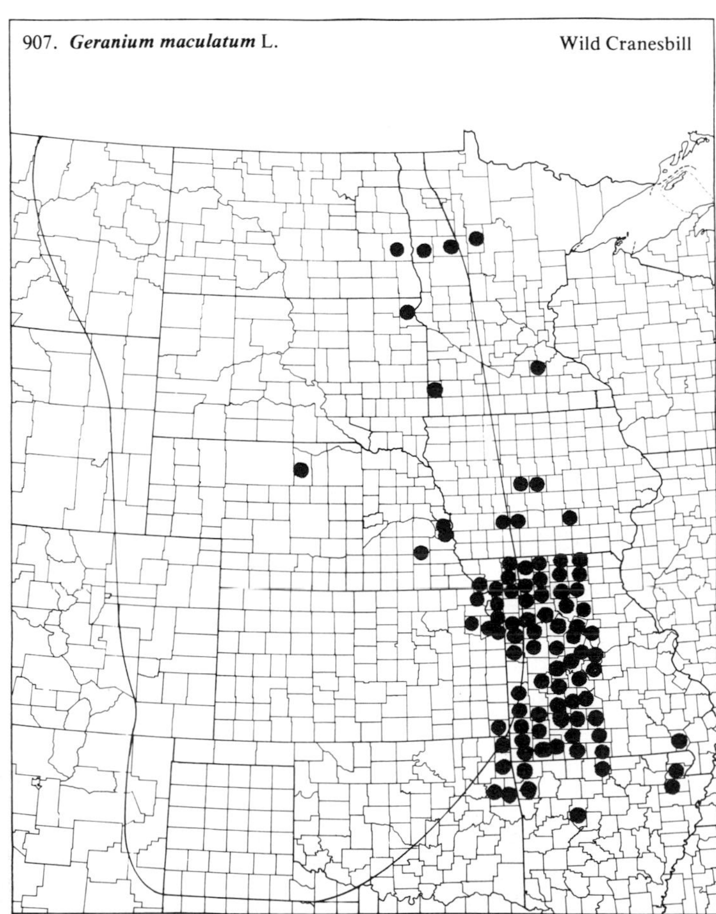

908. ***Geranium pusillum*** L. Small Cranesbill

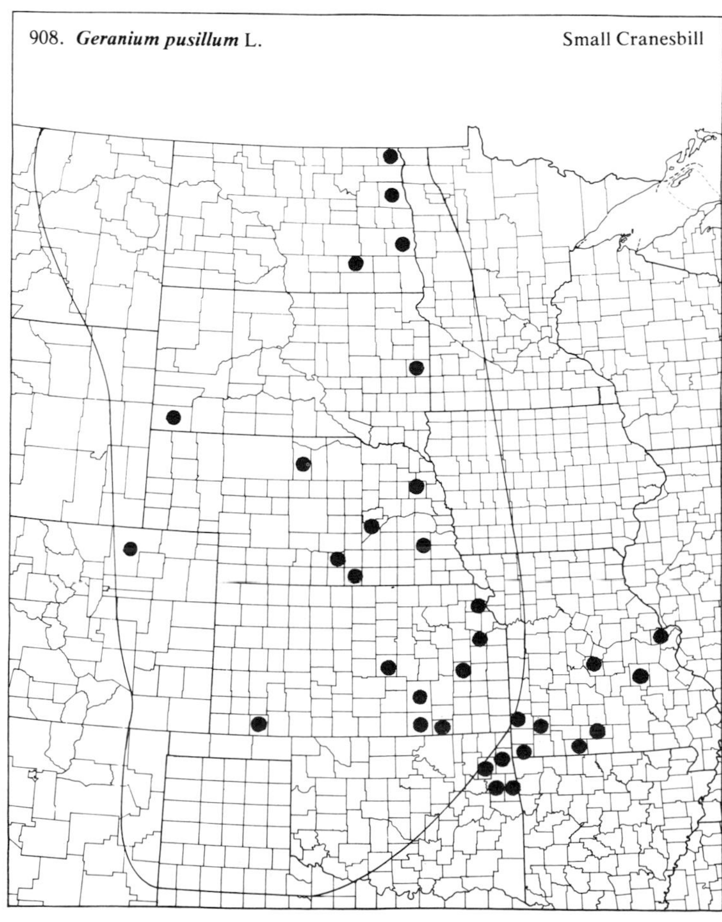

81. *Balsaminaceae* (Touch-me-not Family)

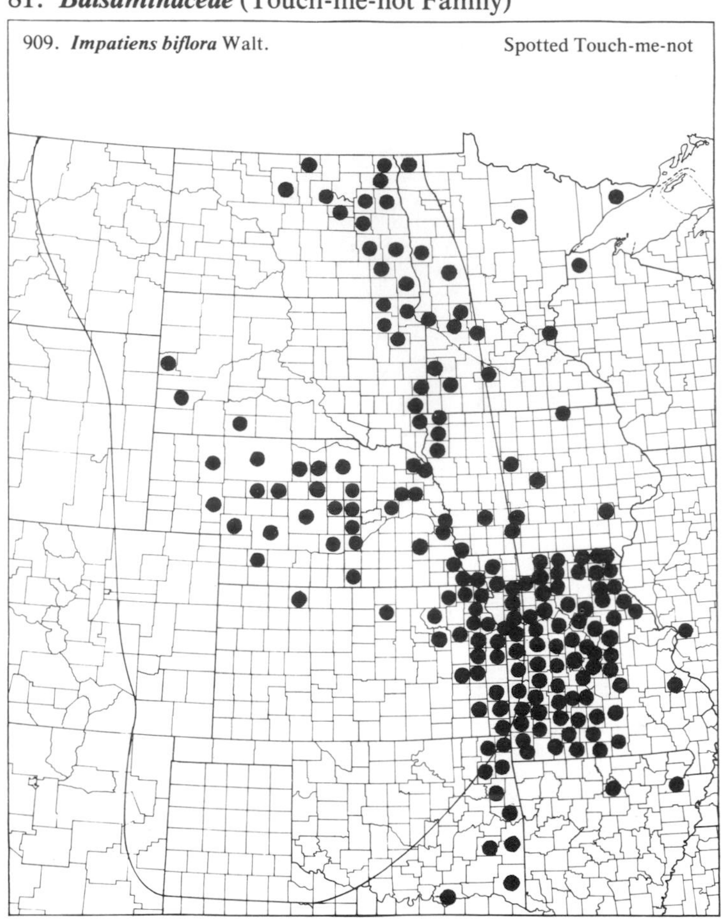

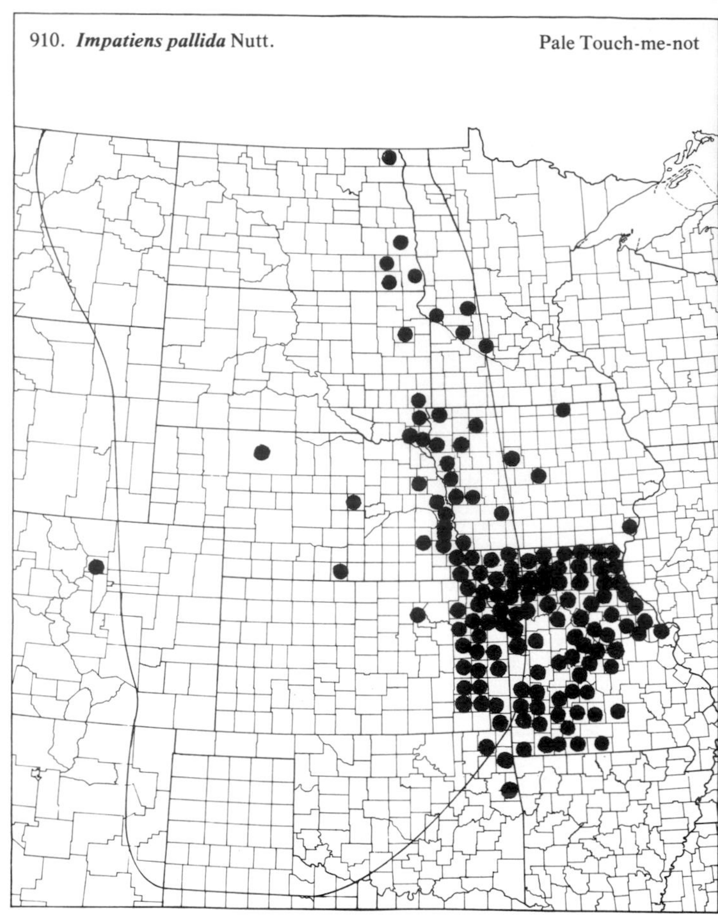

82. *Linaceae* (Flax Family)

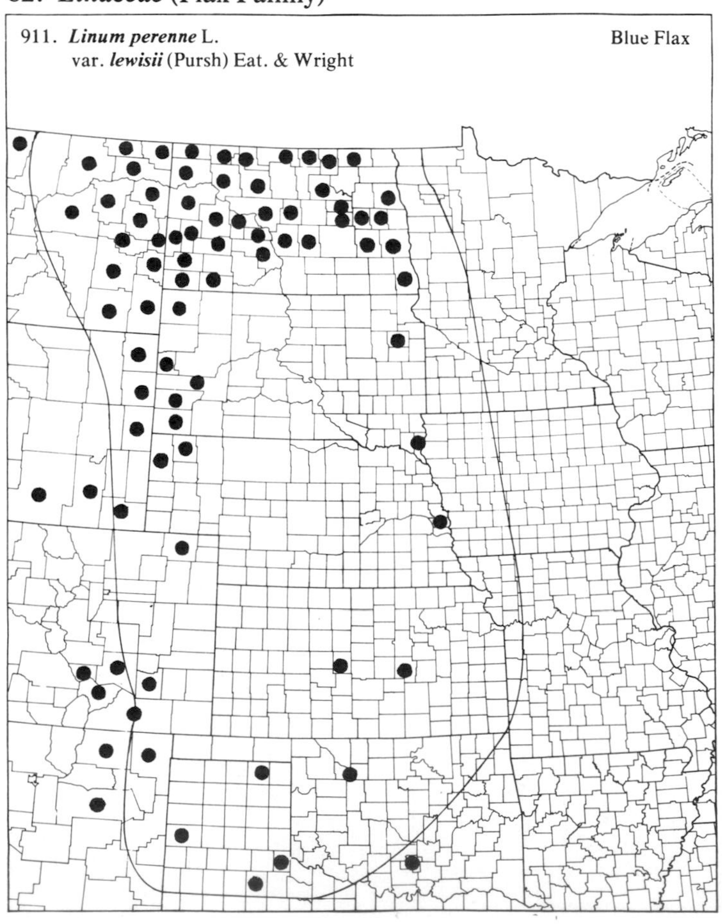

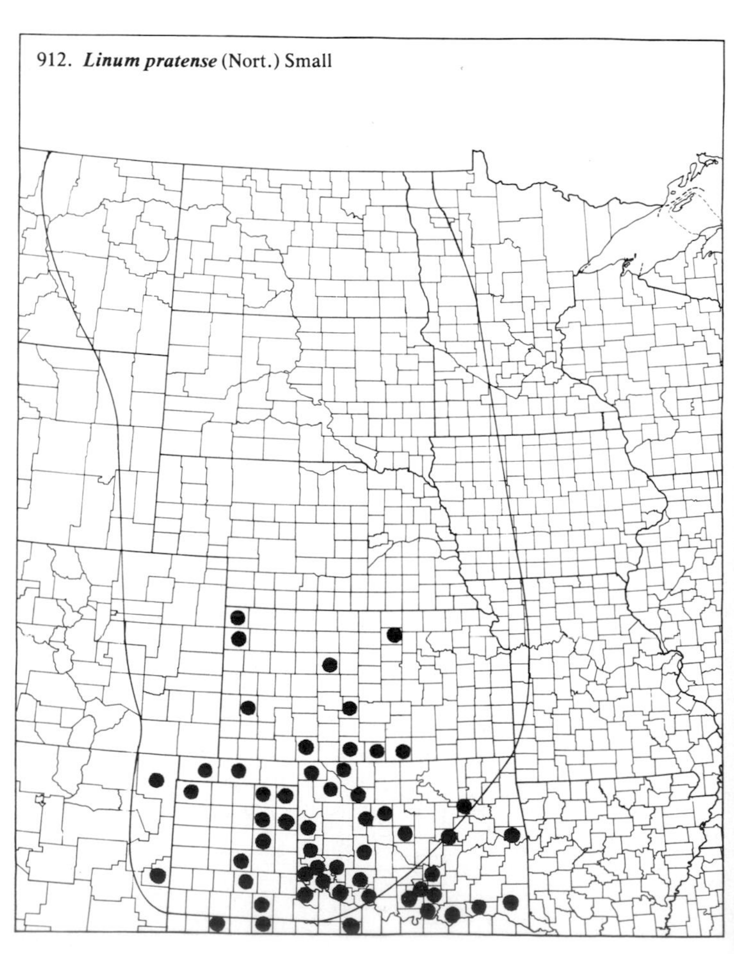

(Linaceae)

913. ***Linum rigidum*** Pursh var. ***rigidum*** — Stiffstem Flax

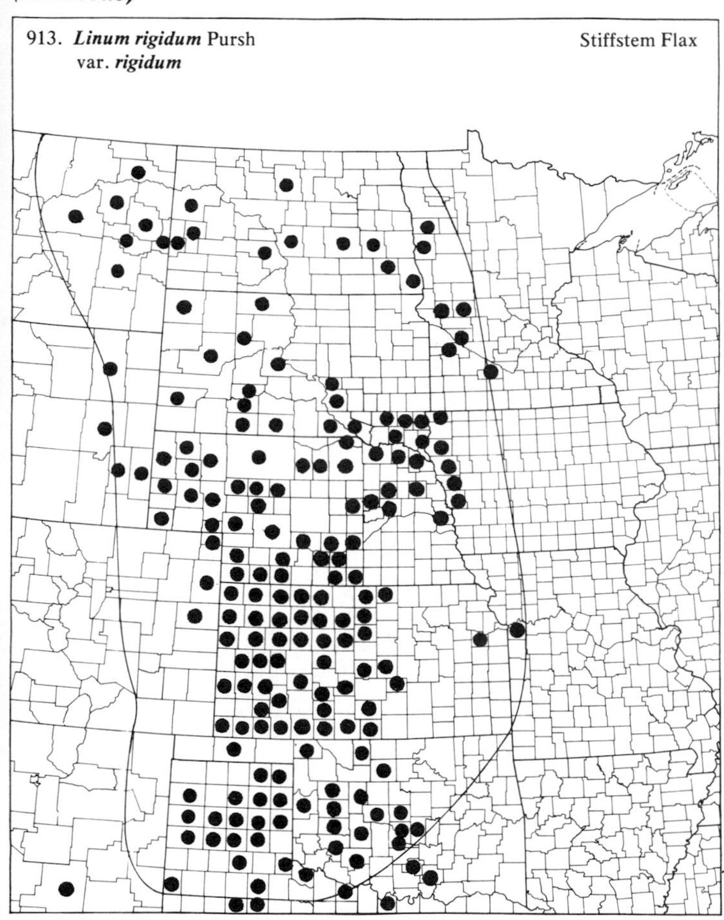

914. ***Linum rigidum*** Pursh var. ***berlandieri*** (Hook.) T. & G. — Stiffstem Flax

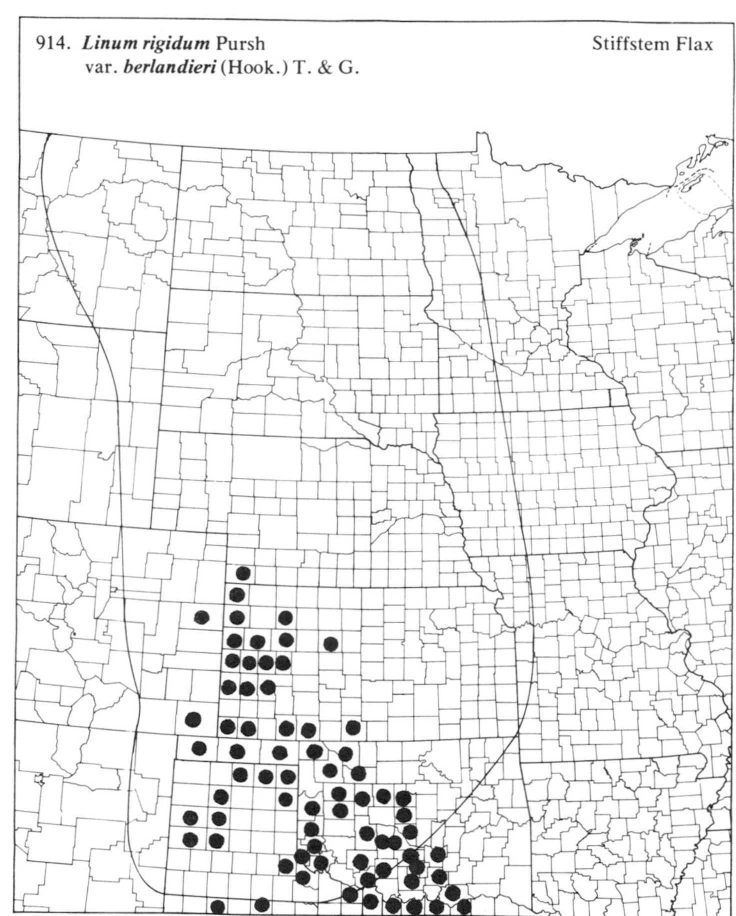

915. ***Linum rigidum*** Pursh var. ***compactum*** (A. Nels.) Rogers — Stiffstem Flax

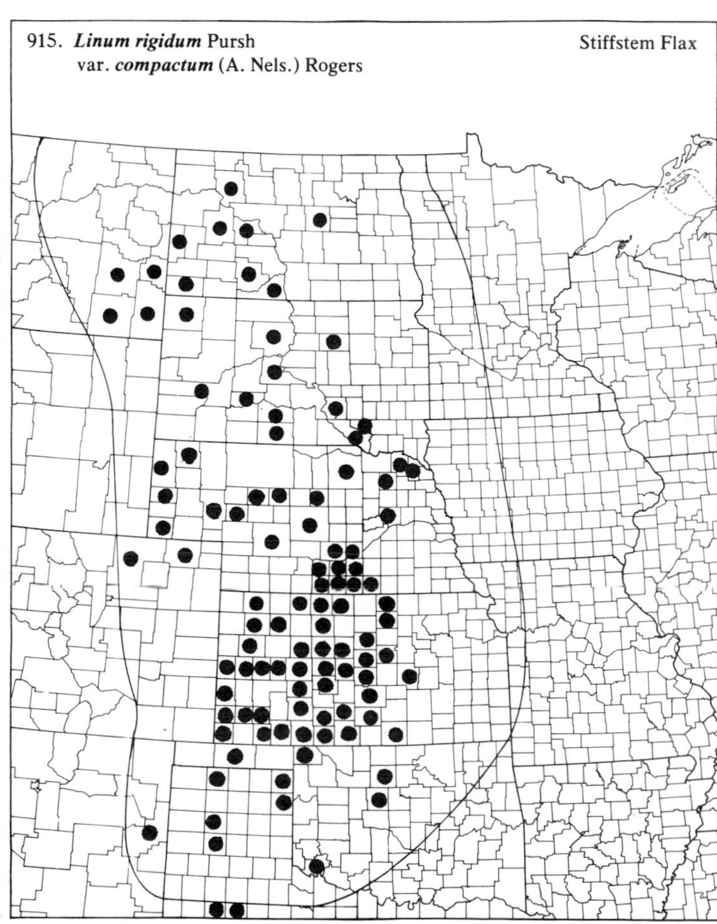

916. ***Linum sulcatum*** Ridd. — Grooved Flax

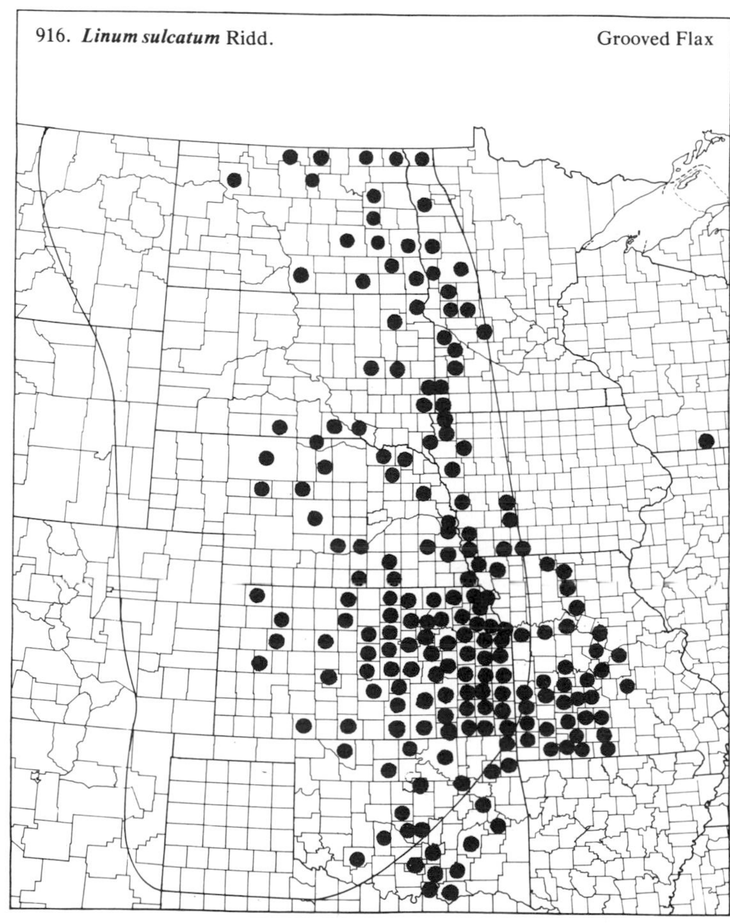

(Linaceae)

917. ***Linum usitatissimum*** L. Common Flax

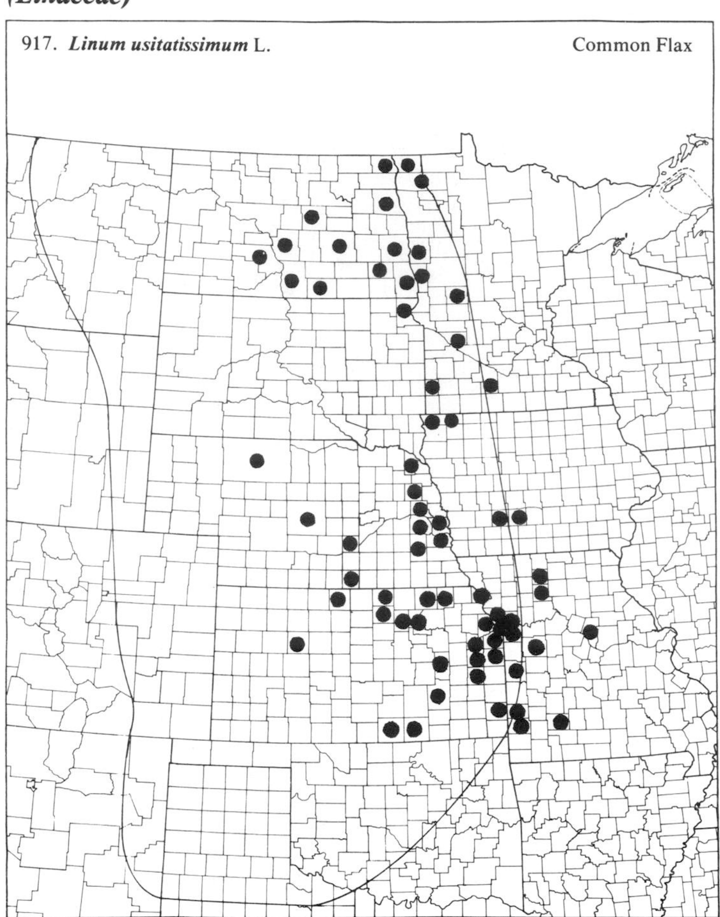

83. *Polygalaceae* (Milkwort Family)

918. ***Polygala alba*** Nutt. White Milkwort

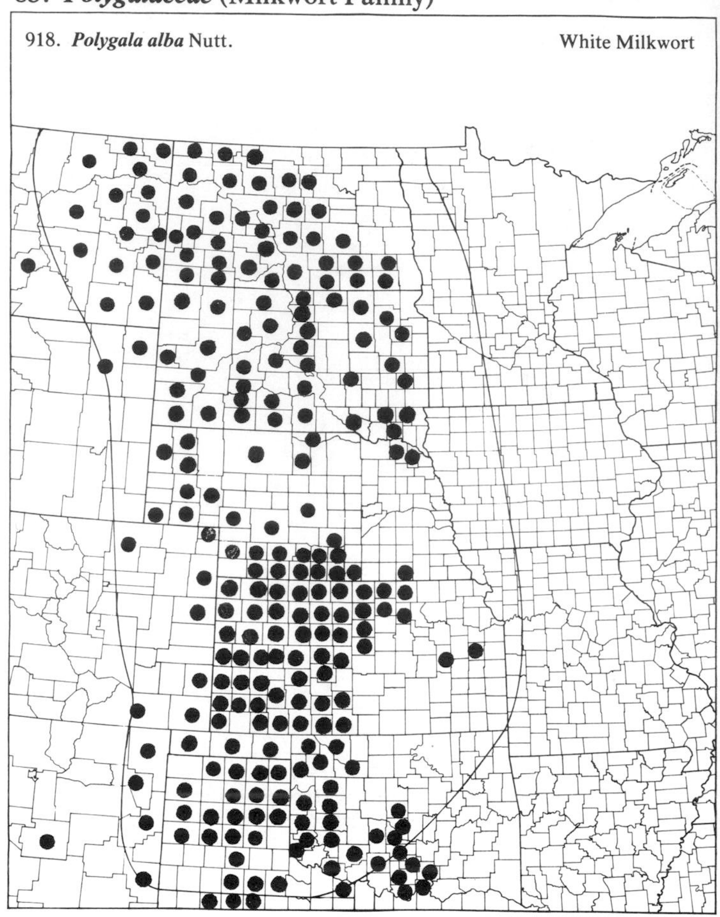

919. ***Polygala incarnata*** L.

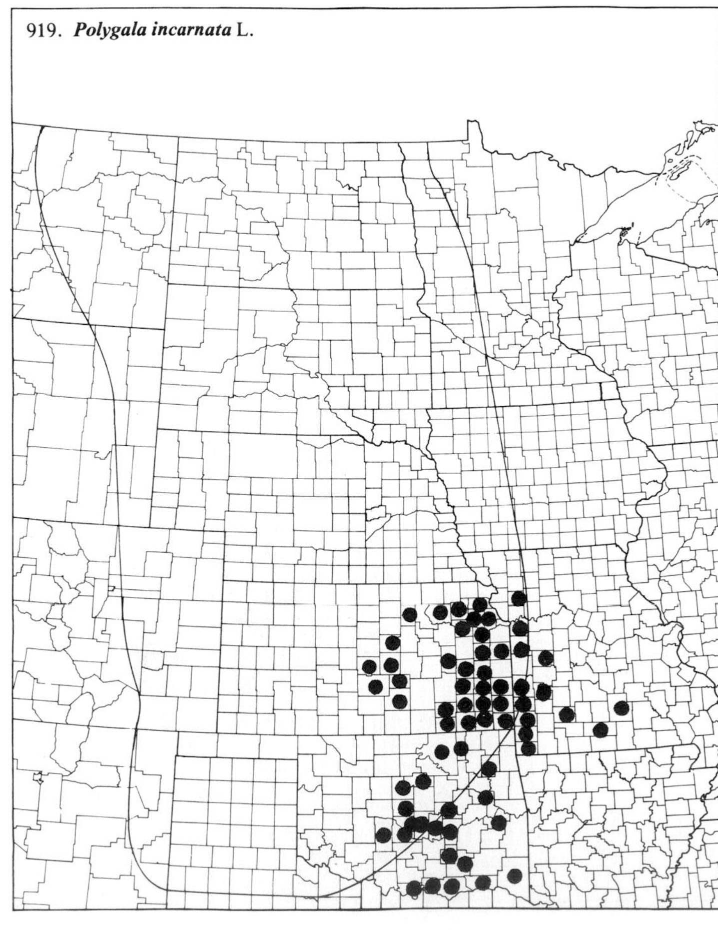

920. ***Polygala sanguinea*** L. Blood Polygala

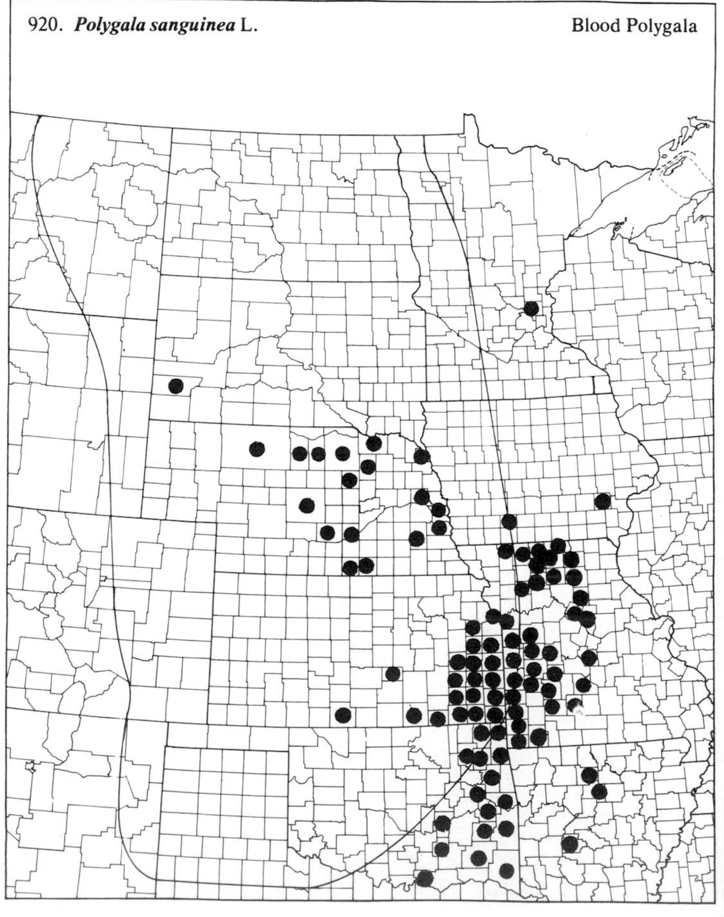

(Polygalaceae)

921. **Polygala senega** L. Seneca Snakeroot

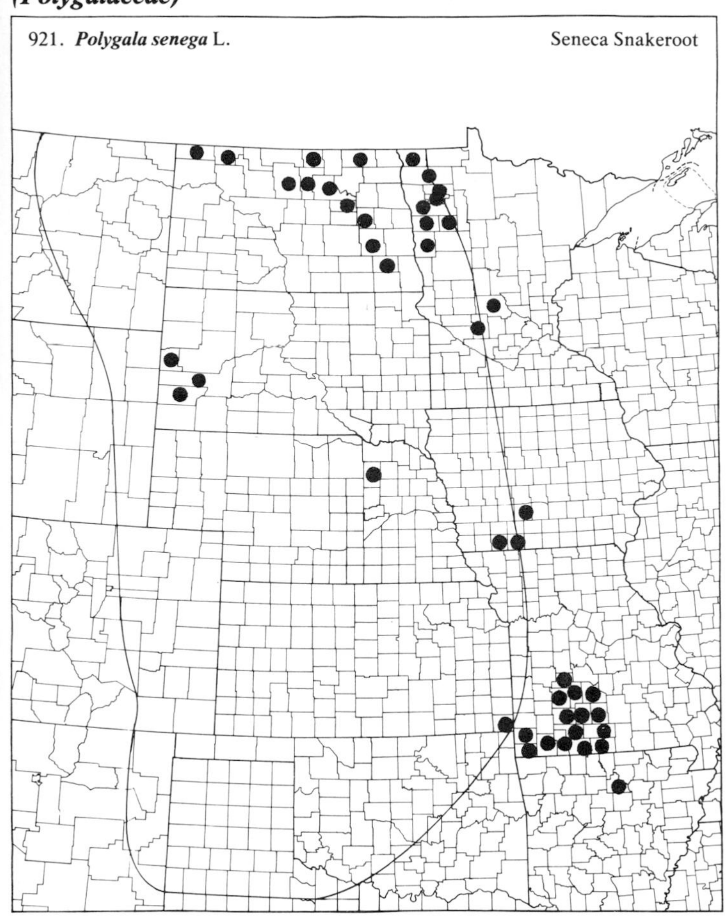

922. **Polygala verticillata** L. Whorled Milkwort

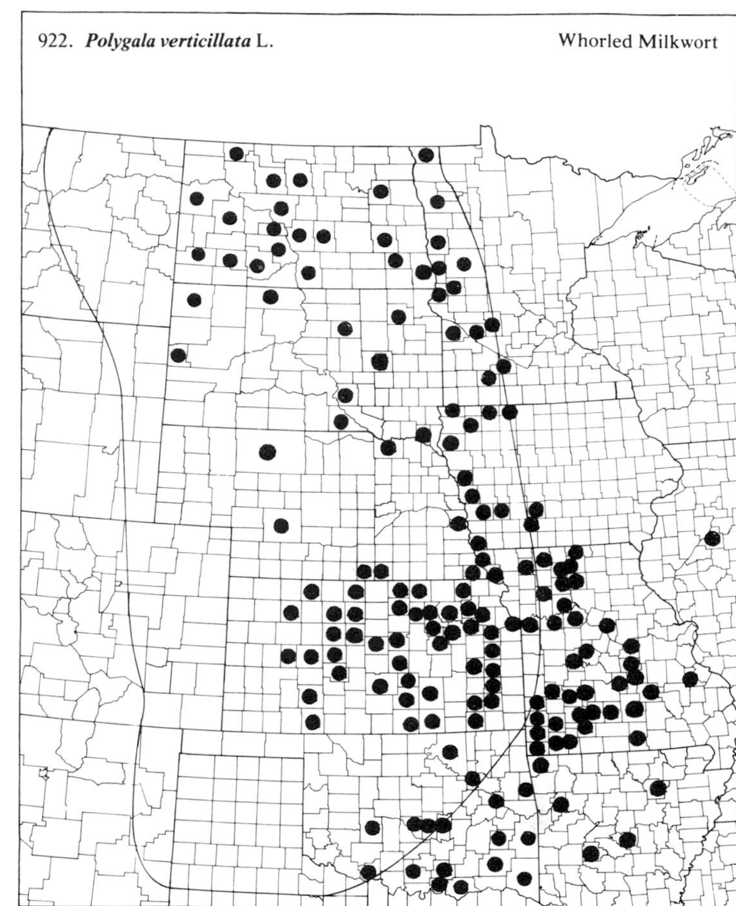

84. *Araliaceae* (Ginseng Family)

923. **Aralia nudicaulis** L. Wild Sarsaparilla

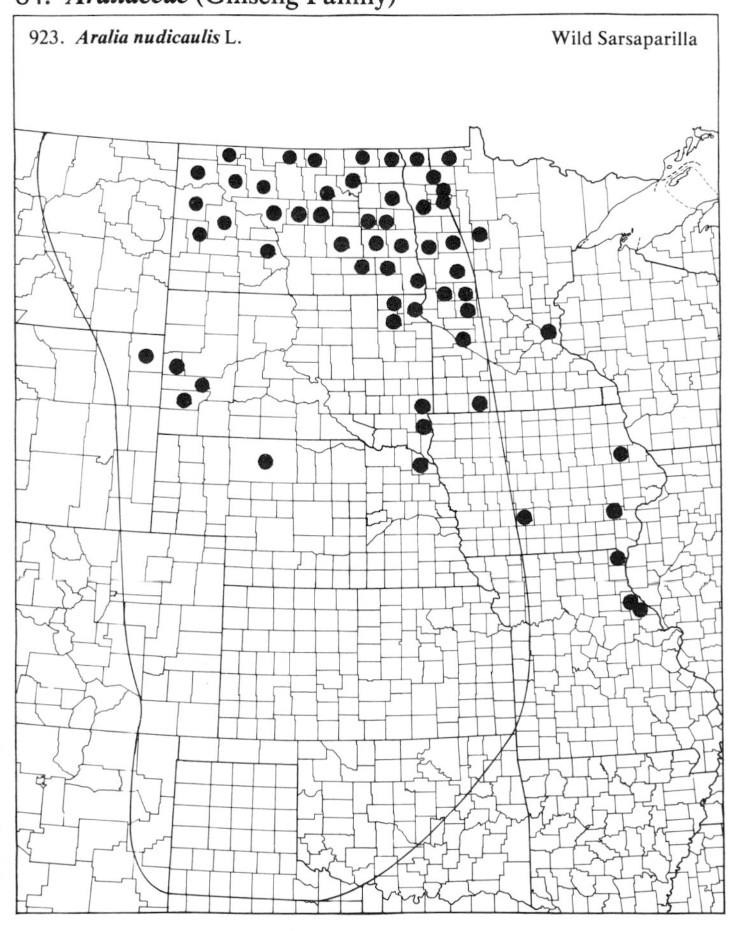

924. **Aralia racemosa** L. Spikenard

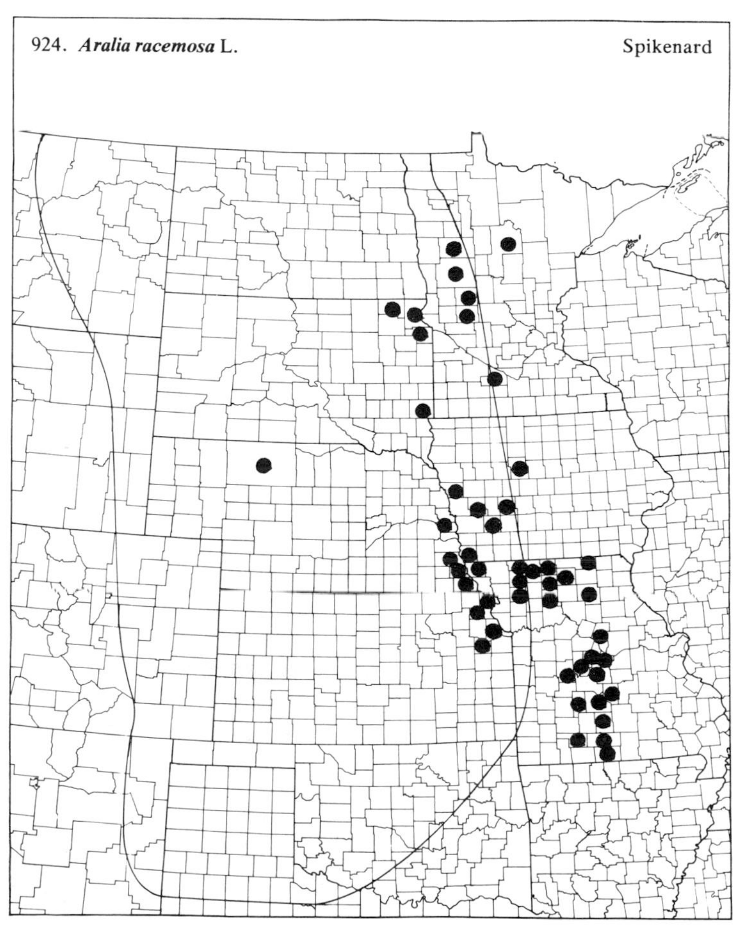

(Araliaceae)

925. ***Panax quinquefolium*** L. Ginseng

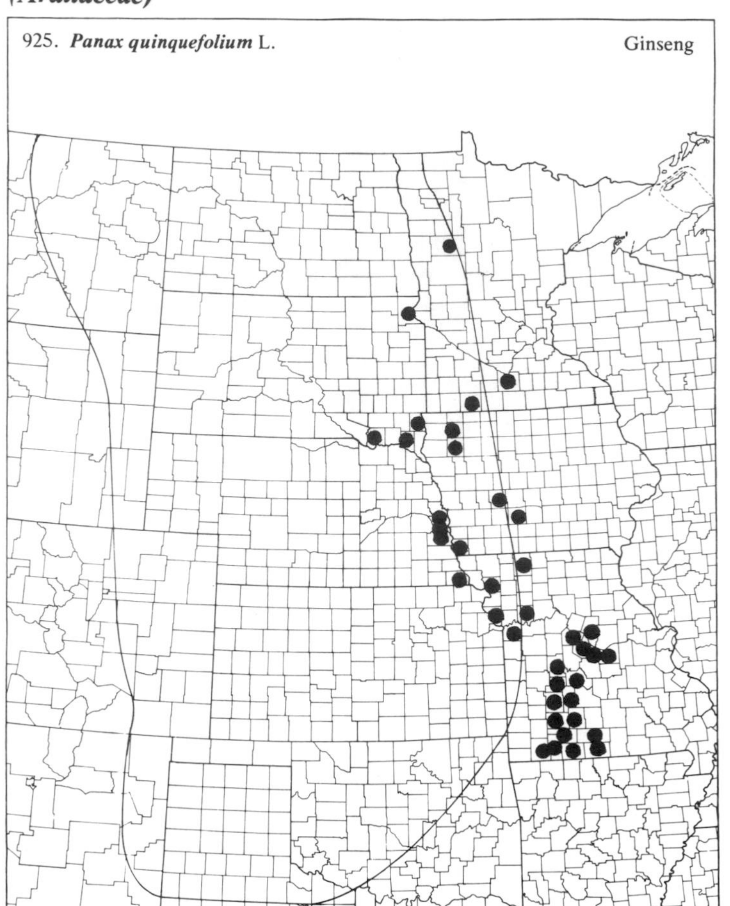

85. *Apiaceae* (Parsley Family)

926. ***Ammoselinum popei*** T. & G.

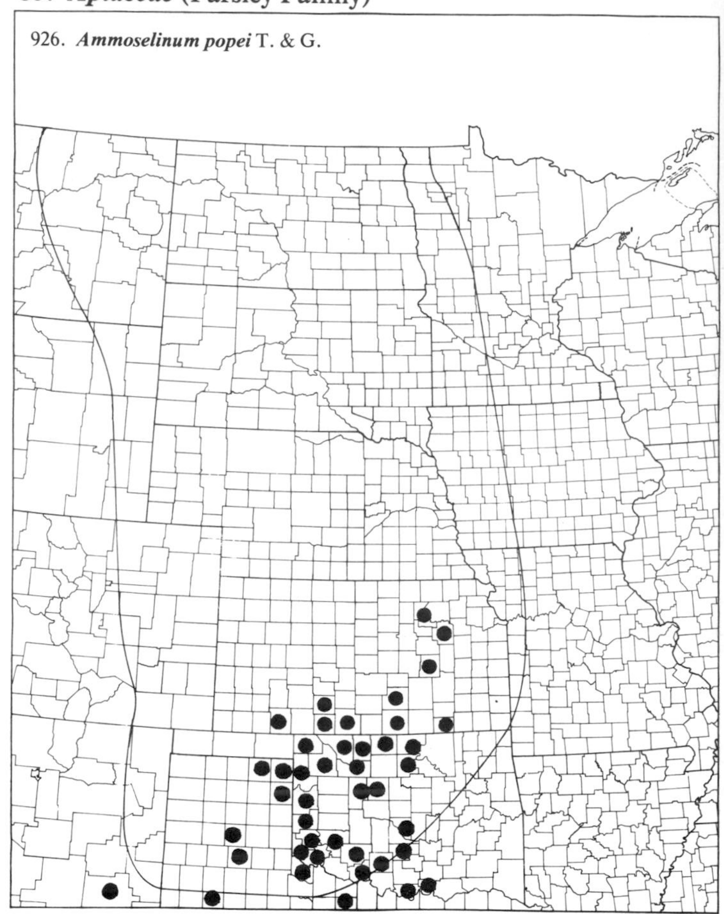

927. ***Anethum graveolens*** L. Dill

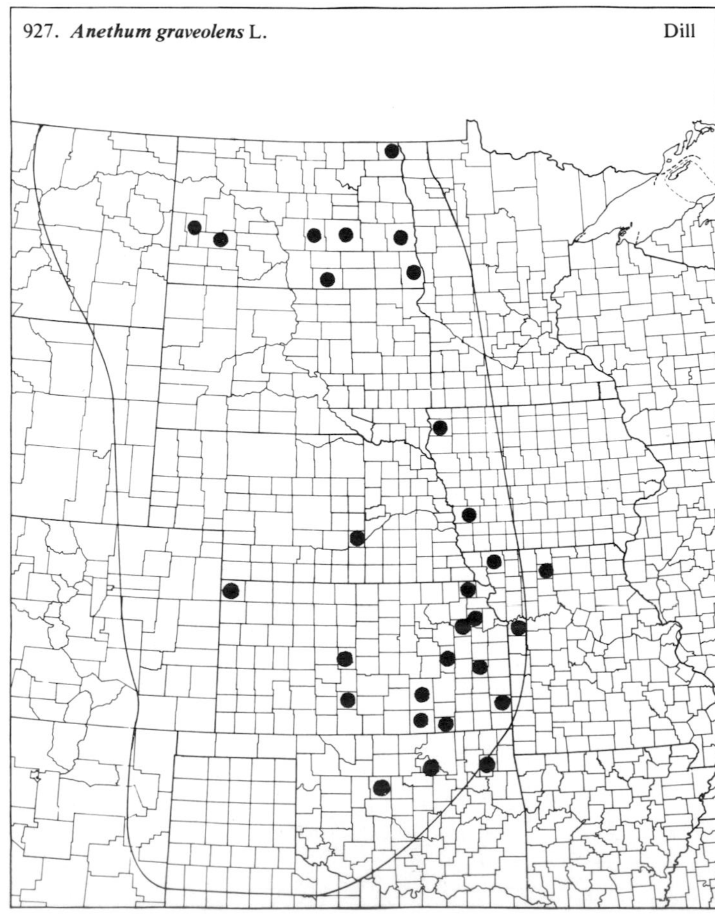

928. ***Berula erecta*** (Huds.) Cov. var. ***incisum*** (Torr.) Cronq. Water Parsnip

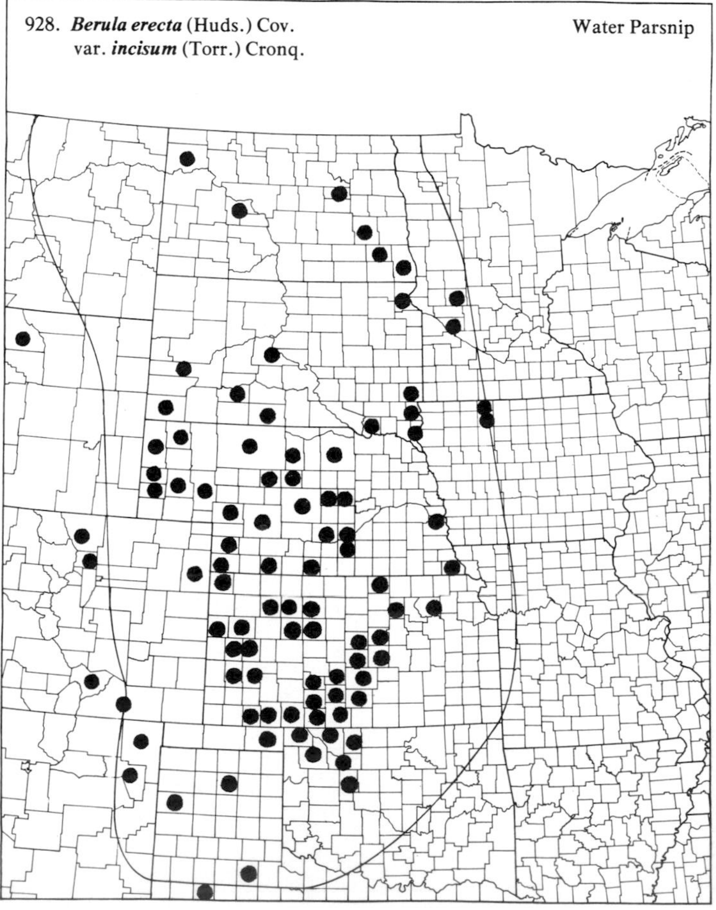

(Apiaceae)

937. ***Cymopterus montanus*** Nutt. — Mountain Corkwing

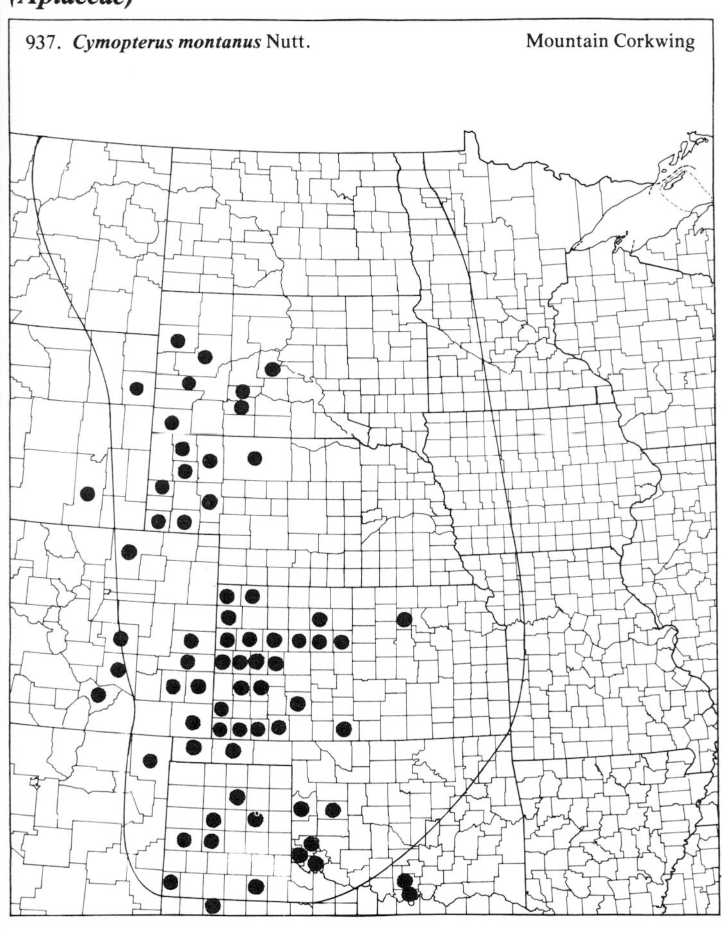

938. ***Daucus carota*** L. — Wild Carrot

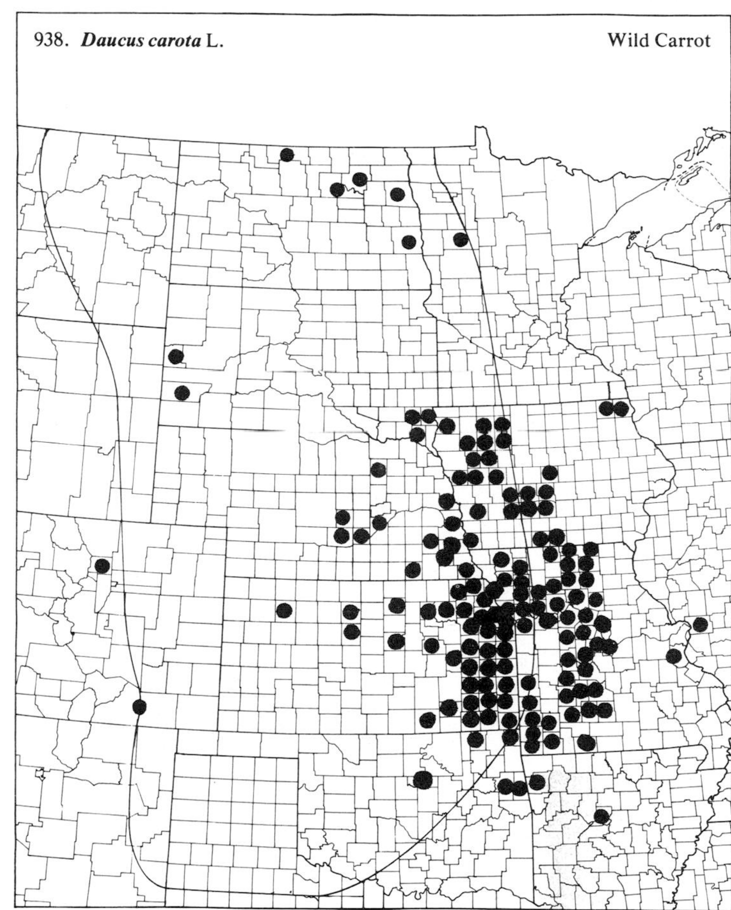

939. ***Daucus pusillus*** Michx. — Southwestern Carrot

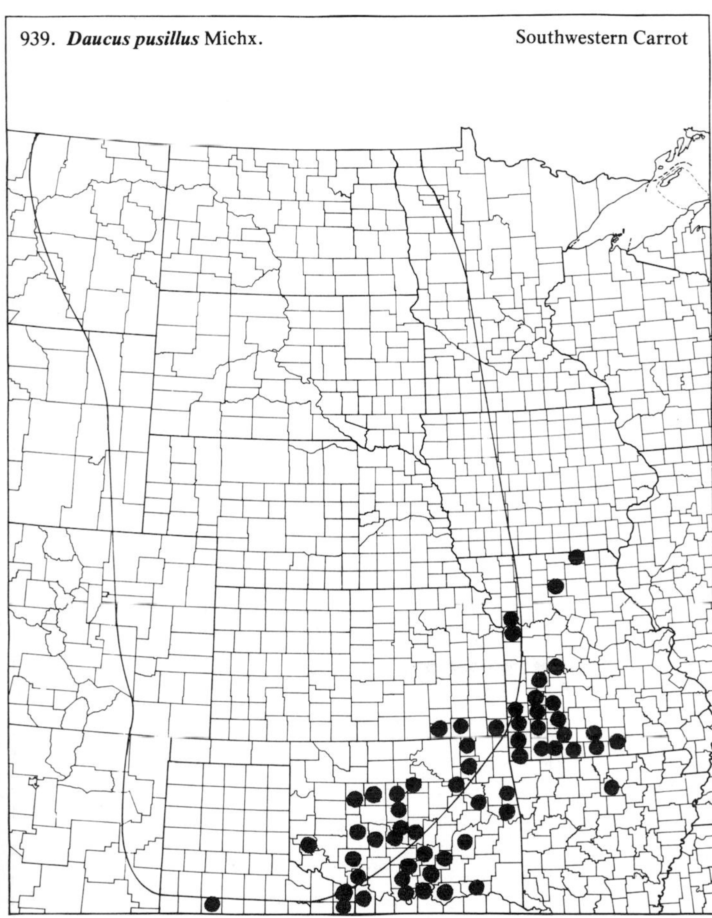

940. ***Eryngium leavenworthii*** T. & G. — Leavenworth Eryngo

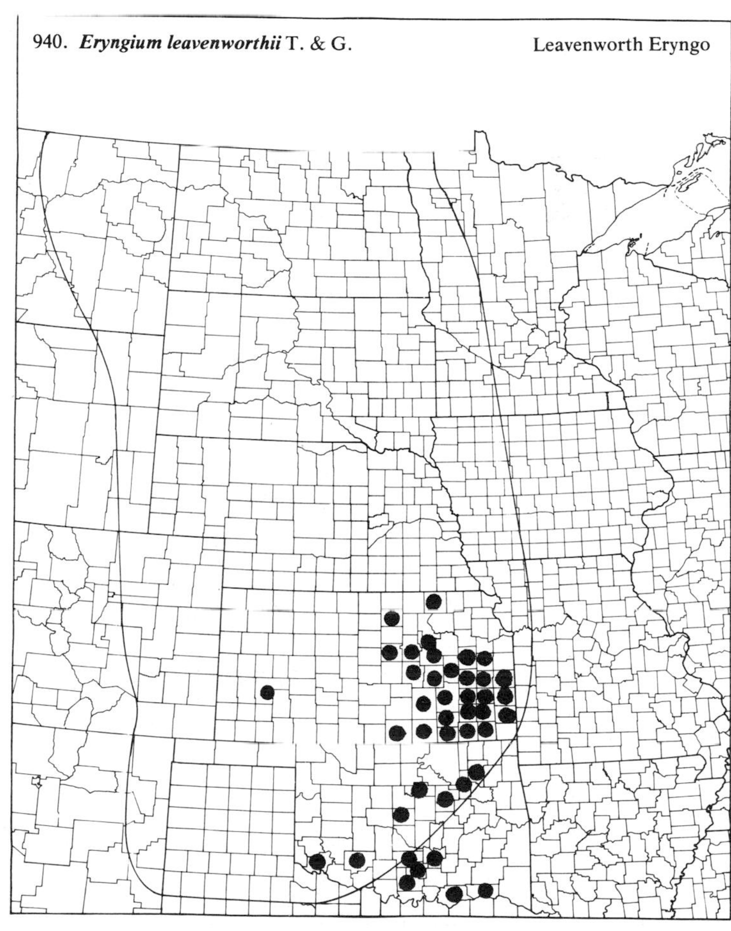

(Apiaceae)

941. ***Eryngium yuccifolium*** Michx. Button Snakeroot

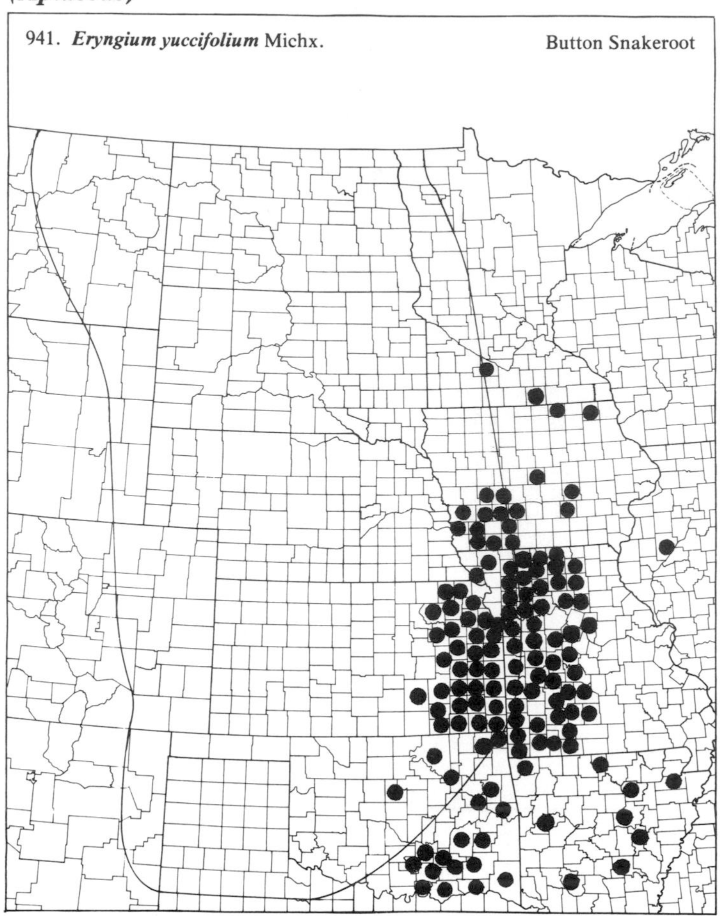

942. ***Eurytaenia texana*** T. & G.

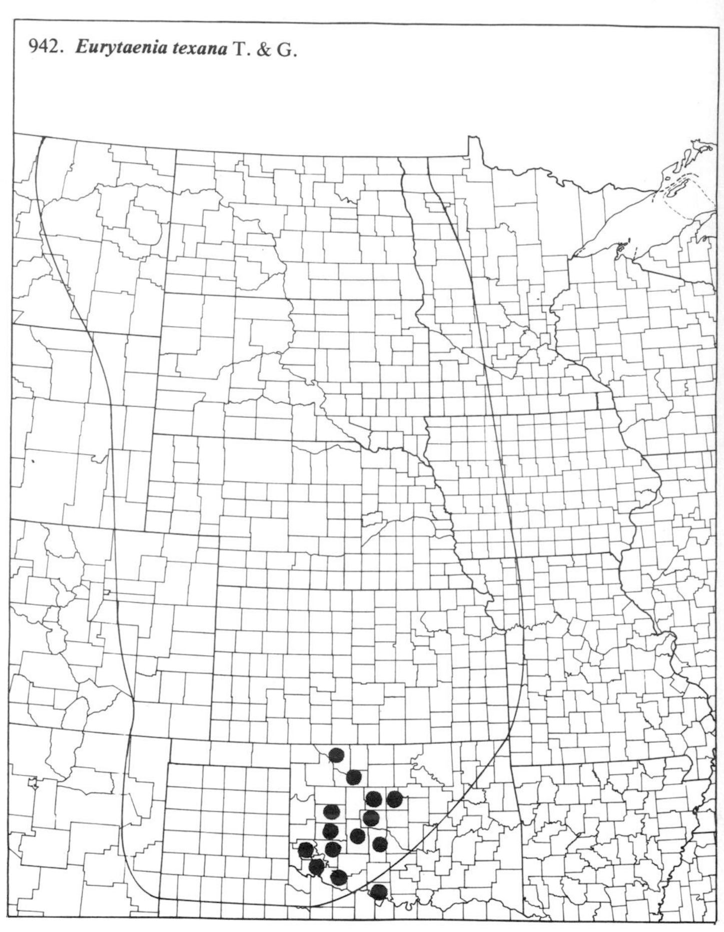

943. ***Heracleum sphondylium*** L. ssp. ***montanum*** (Schleicher) Briq. Eltrot

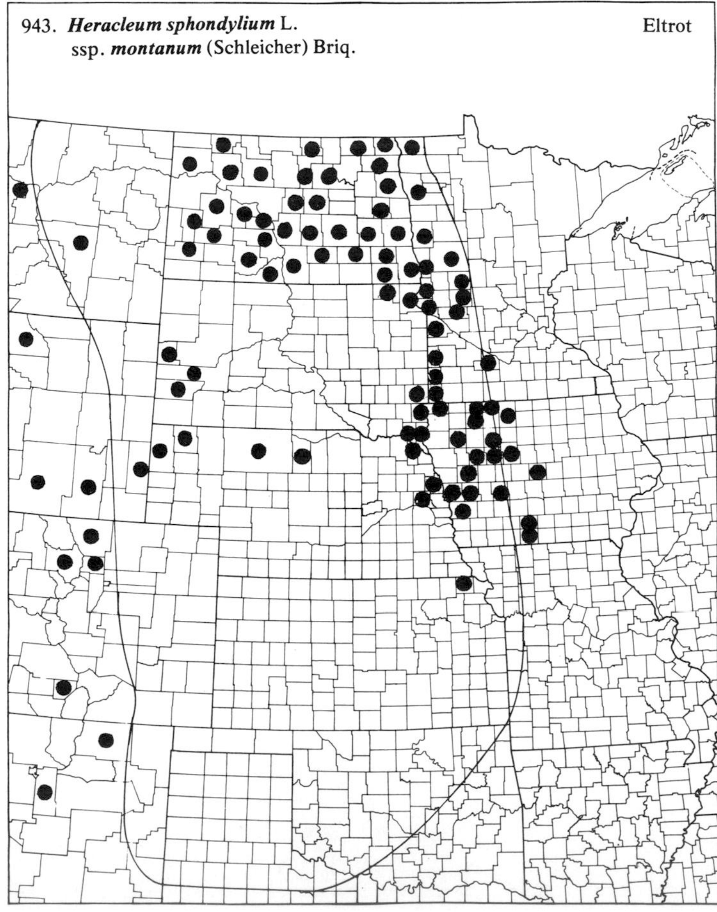

944. ***Limnosciadium pinnatum*** (DC.) Math. & Const.

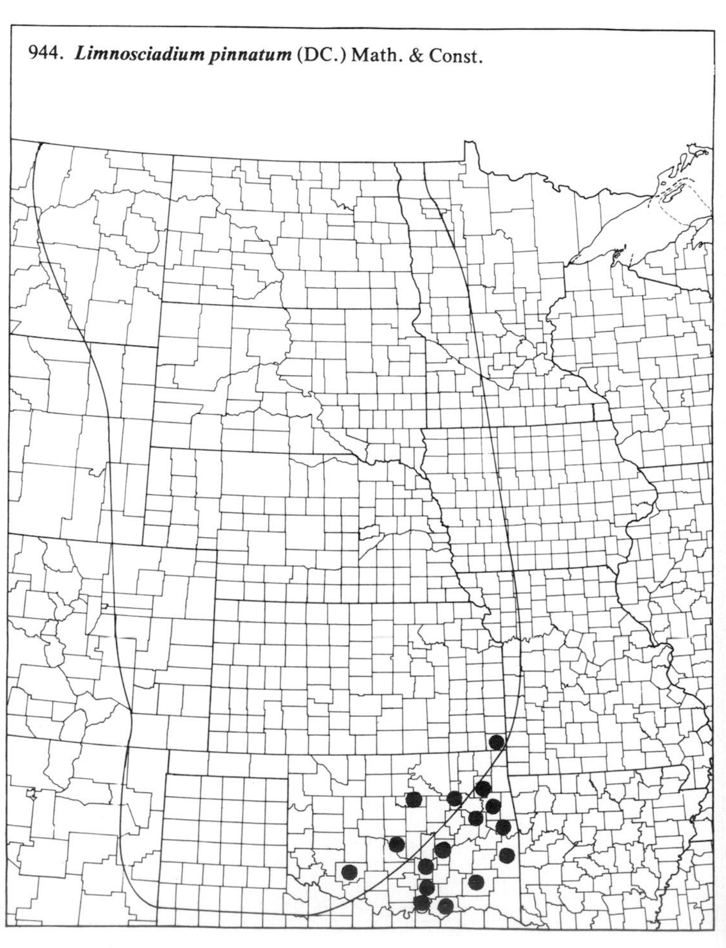

(Apiaceae)

945. ***Lomatium foeniculaceum*** (Nutt.) Coult. & Rose var. ***foeniculaceum*** Wild Parsley

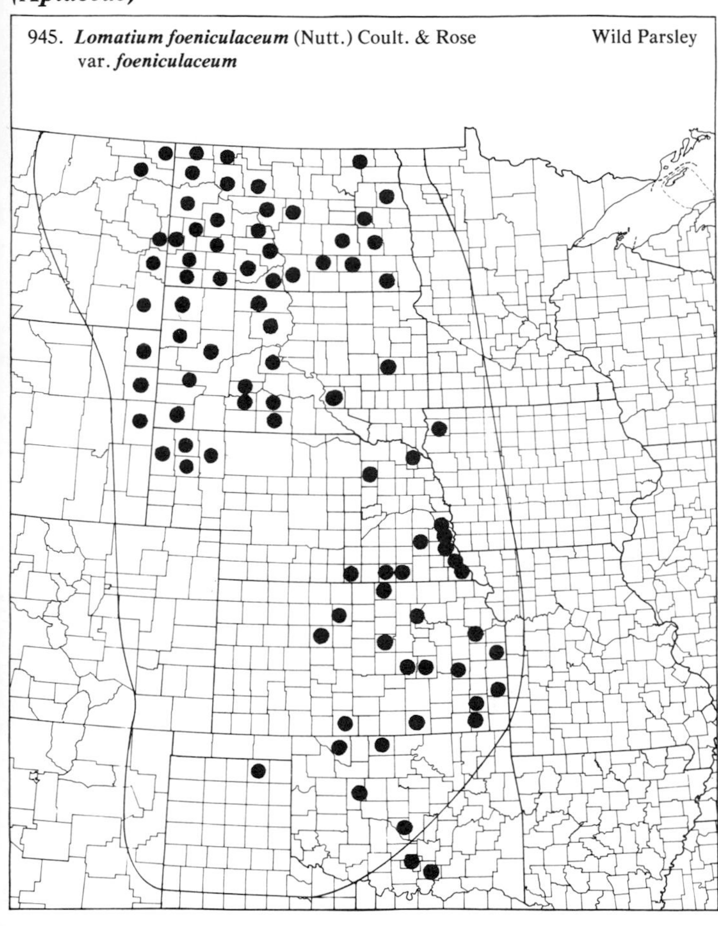

946. ***Lomatium foeniculaceum*** (Nutt.) Coult. & Rose var. ***daucifolium*** (T. & G.) Cronq. Wild Parsley

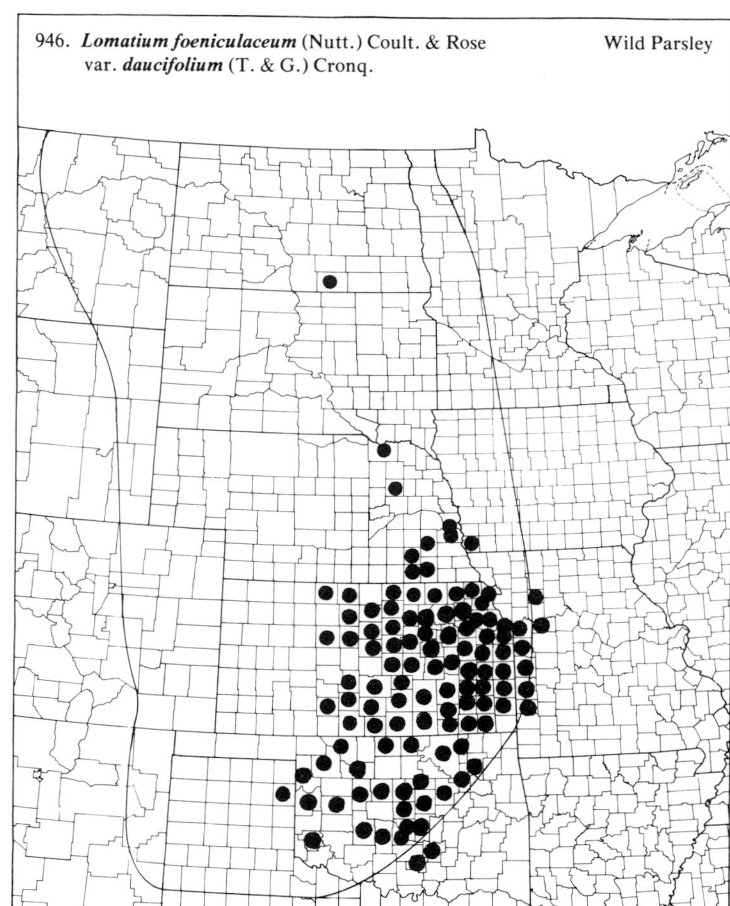

947. ***Lomatium macrocarpa*** (Nutt.) Coult. & Rose

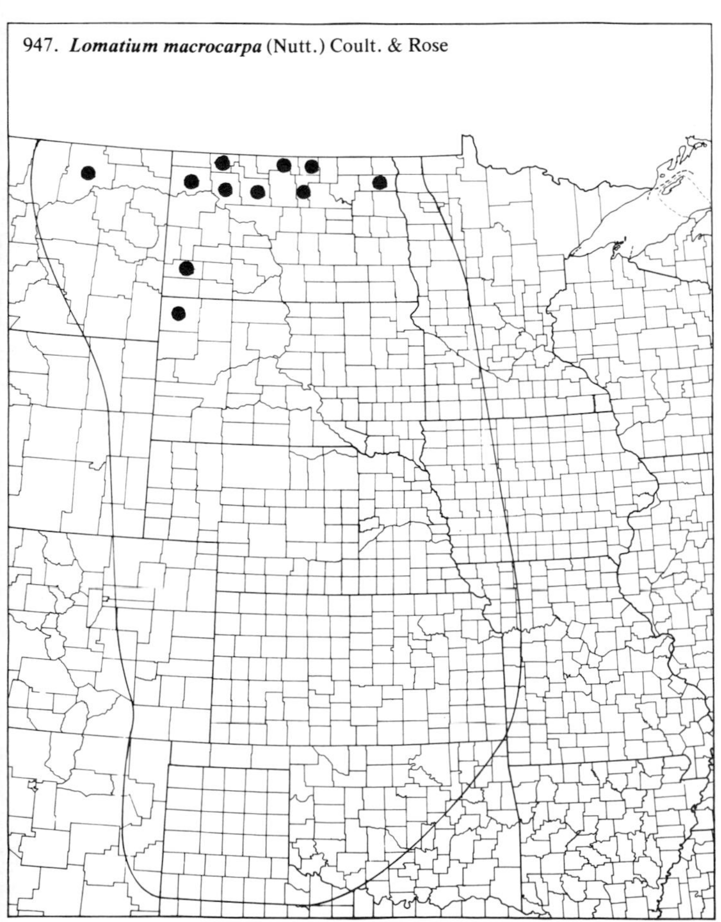

948. ***Lomatium orientale*** Coult. & Rose Wild Parsley

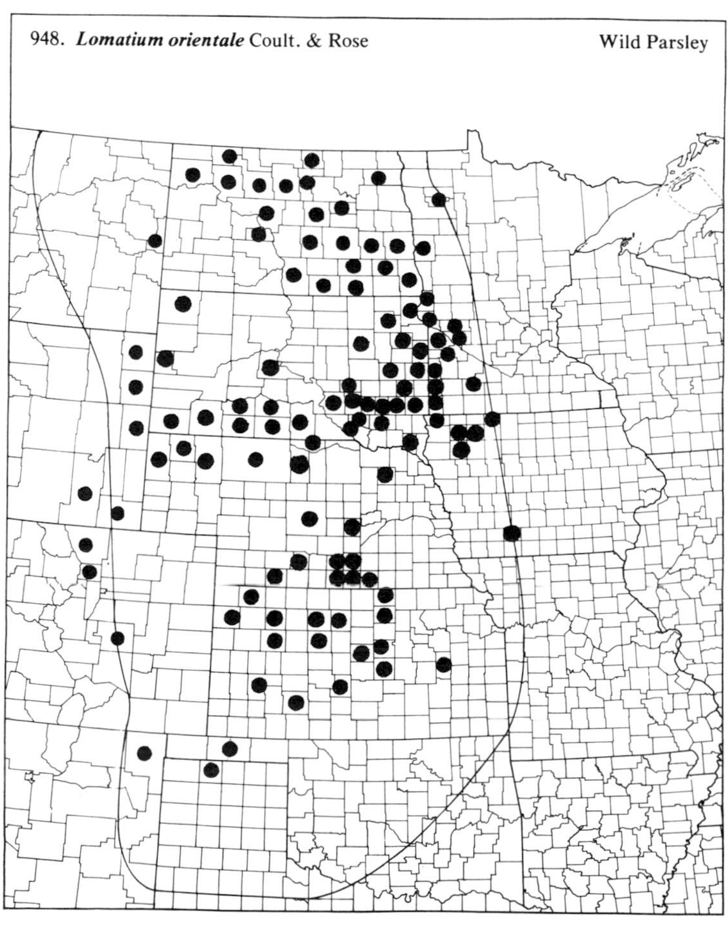

(Apiaceae)

949. ***Musineon divaricatum*** (Pursh) Nutt. ex T. & G.

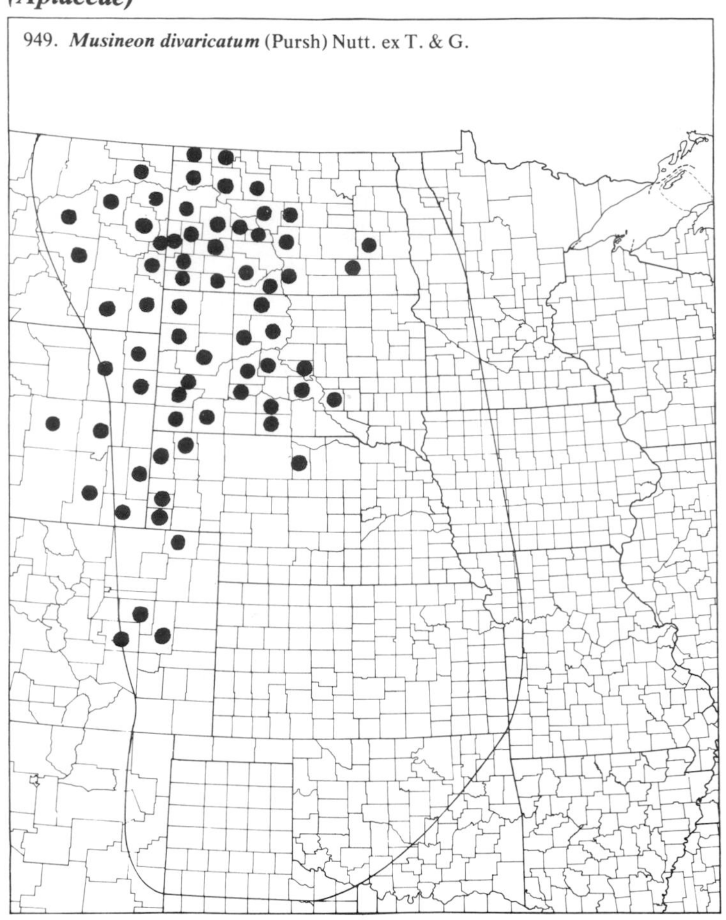

950. ***Musineon tenuifolium*** Nutt.

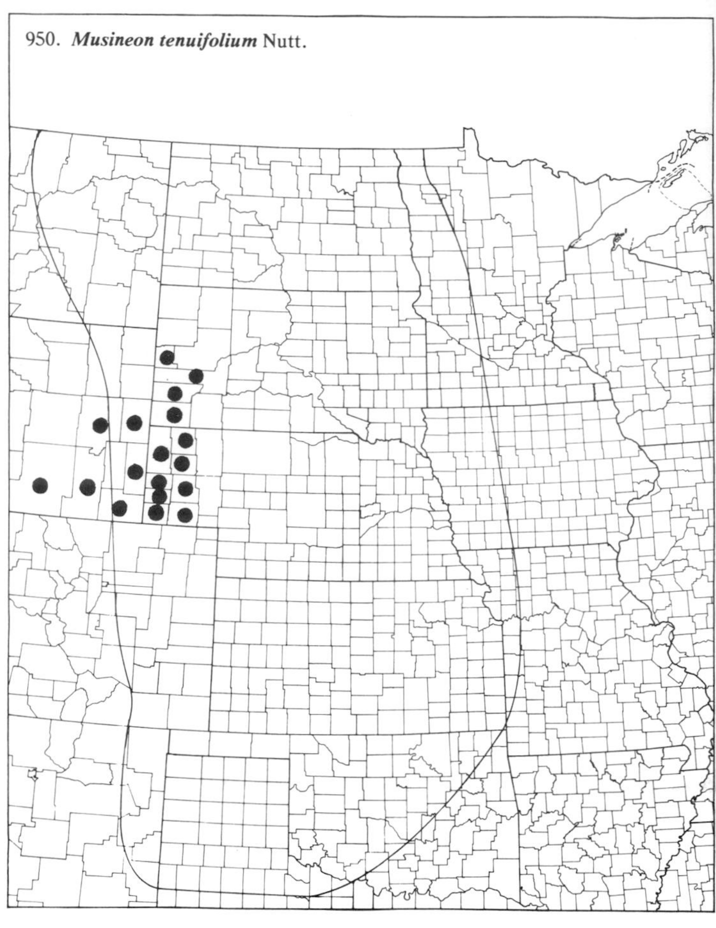

951. ***Osmorhiza claytonii*** (Michx.) Clarke — Sweet Jarvil

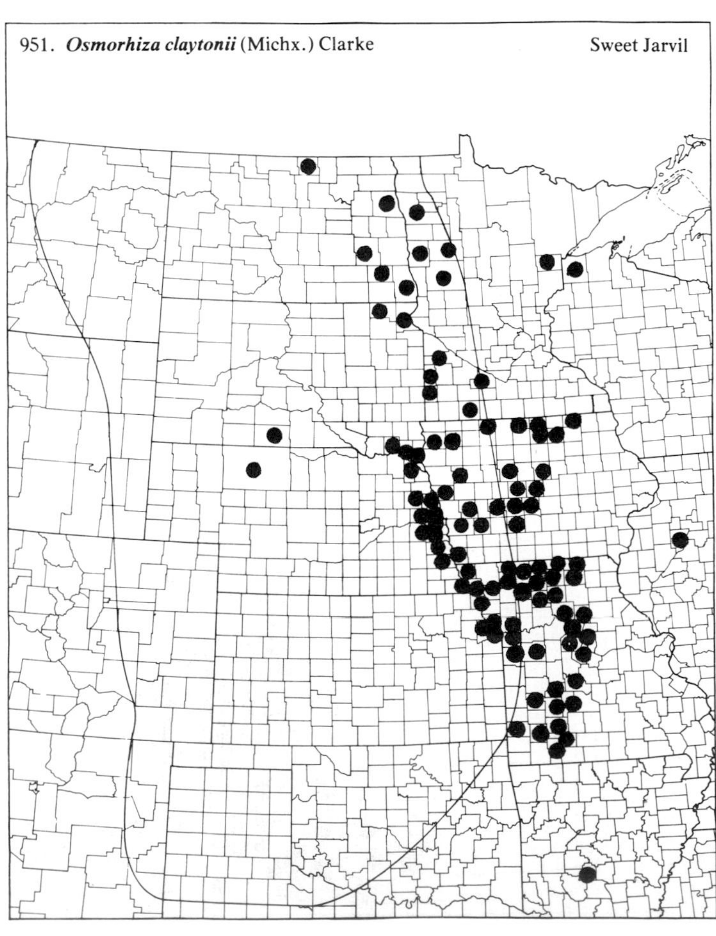

952. ***Osmorhiza longistylis*** (Torr.) DC. var. ***longistylis*** — Aniseroot

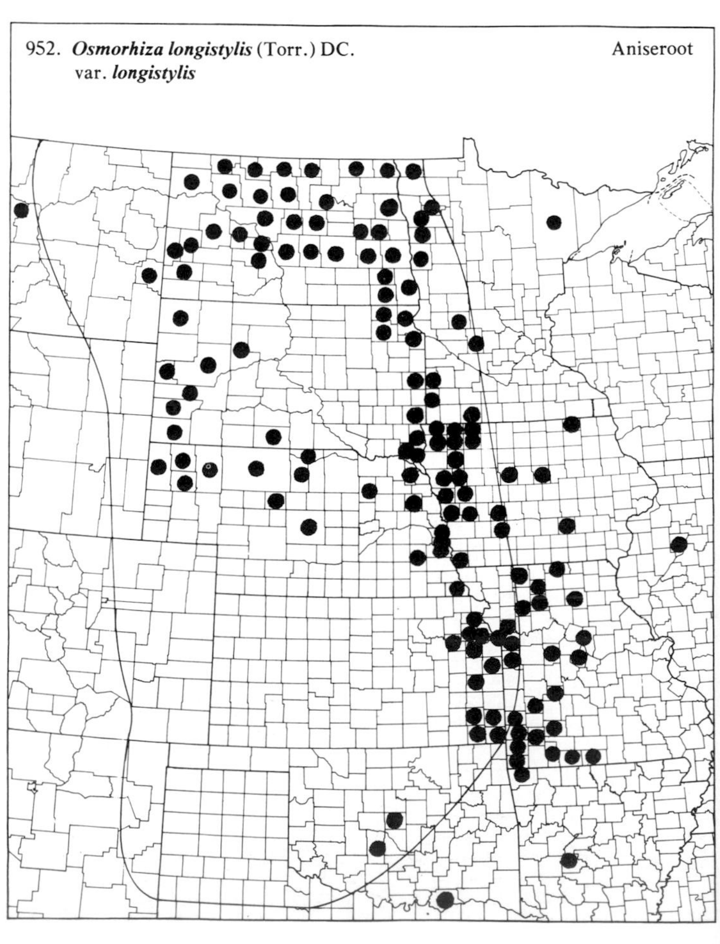

(Apiaceae)

953. ***Osmorhiza longistylis*** (Torr.) DC. var. ***villicaulis*** Fern. — Aniseroot

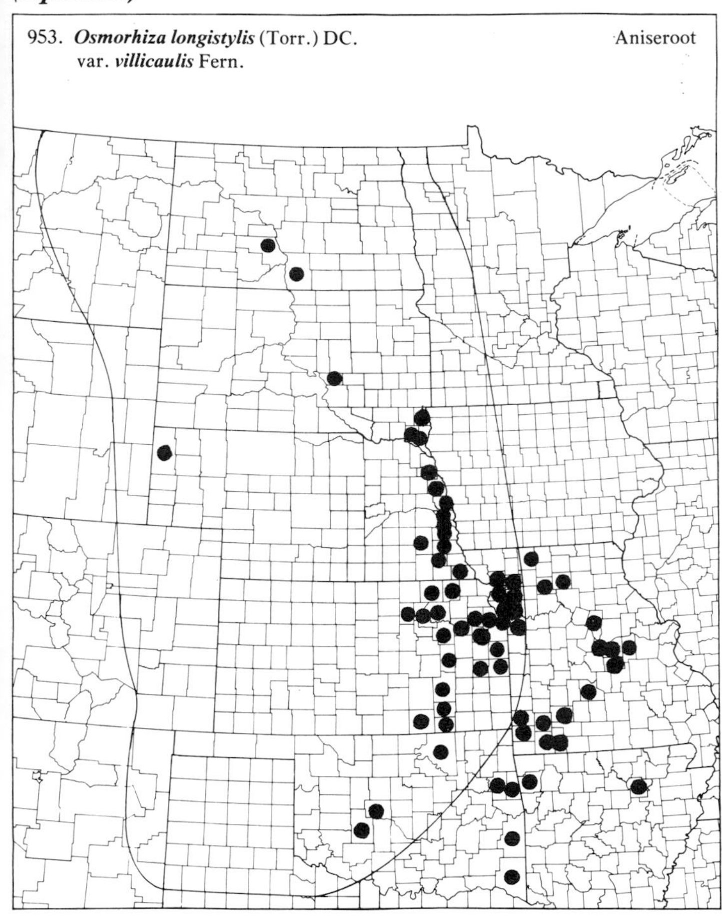

954. ***Pastinaca sativa*** L. — Wild Parsnip

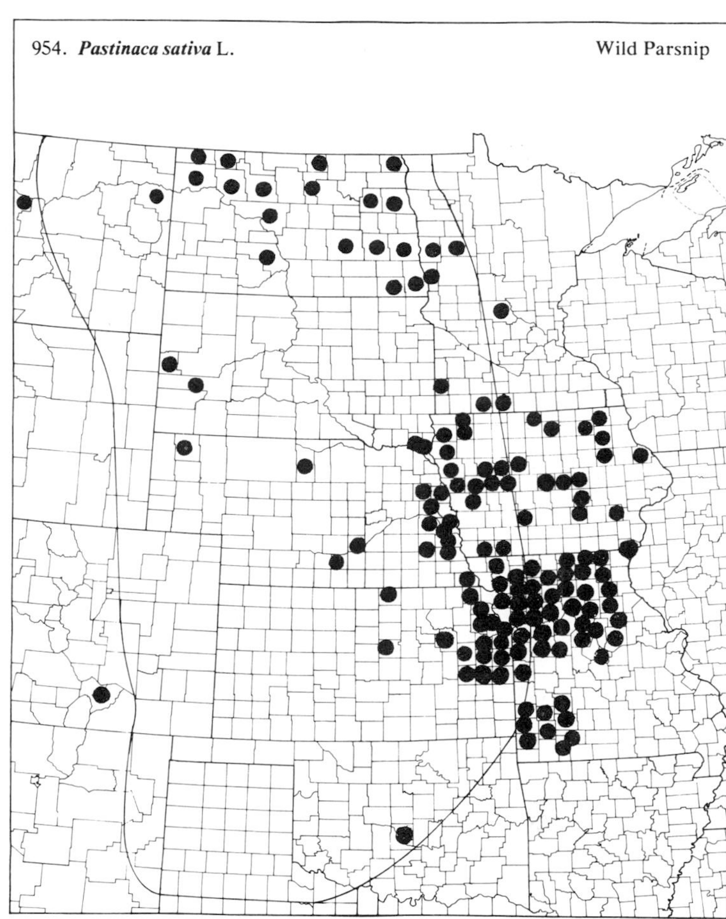

955. ***Perideridia americana*** (Nutt.) Reichb. — Eulophus

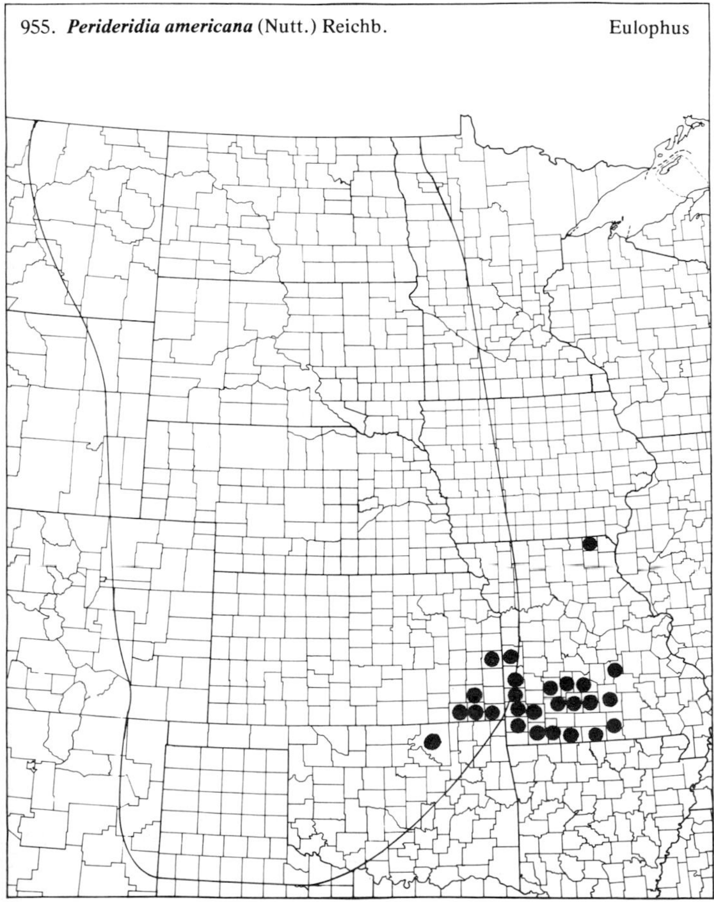

956. ***Polytaenia nuttallii*** DC. — Prairie Parsley

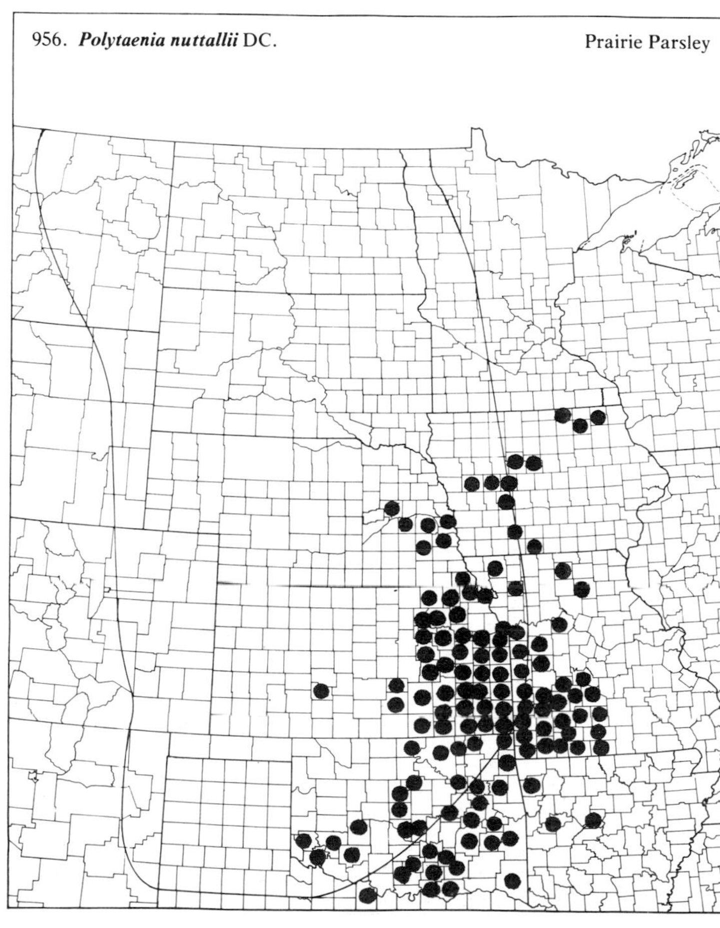

(Apiaceae)

957. ***Ptilimnium nuttallii*** (DC.) Britt. Mock Bishopweed

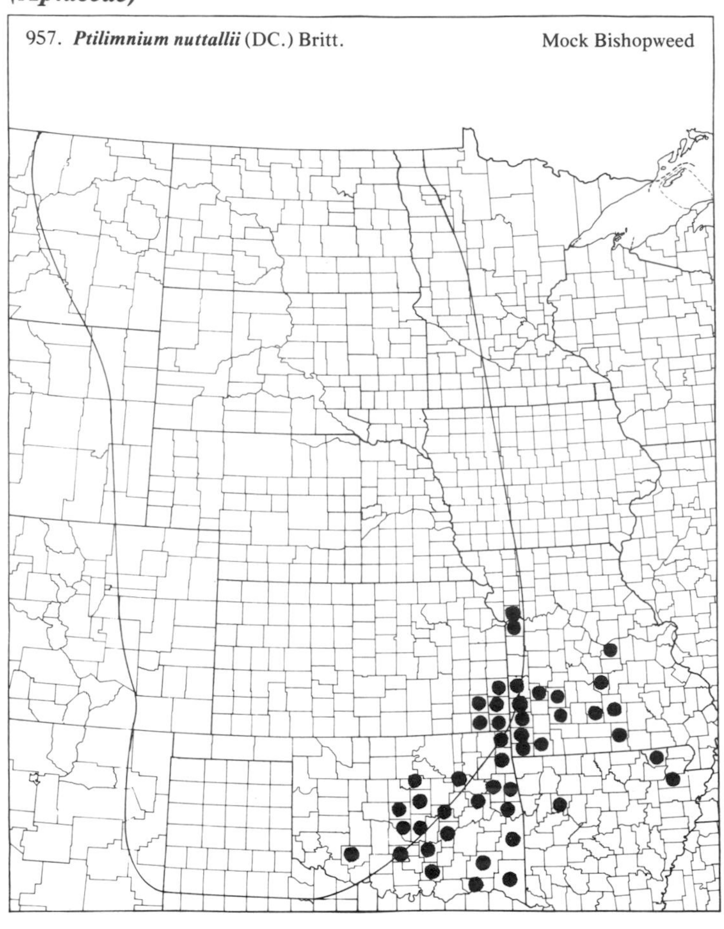

958. ***Sanicula canadensis*** L. Canada Sanicle

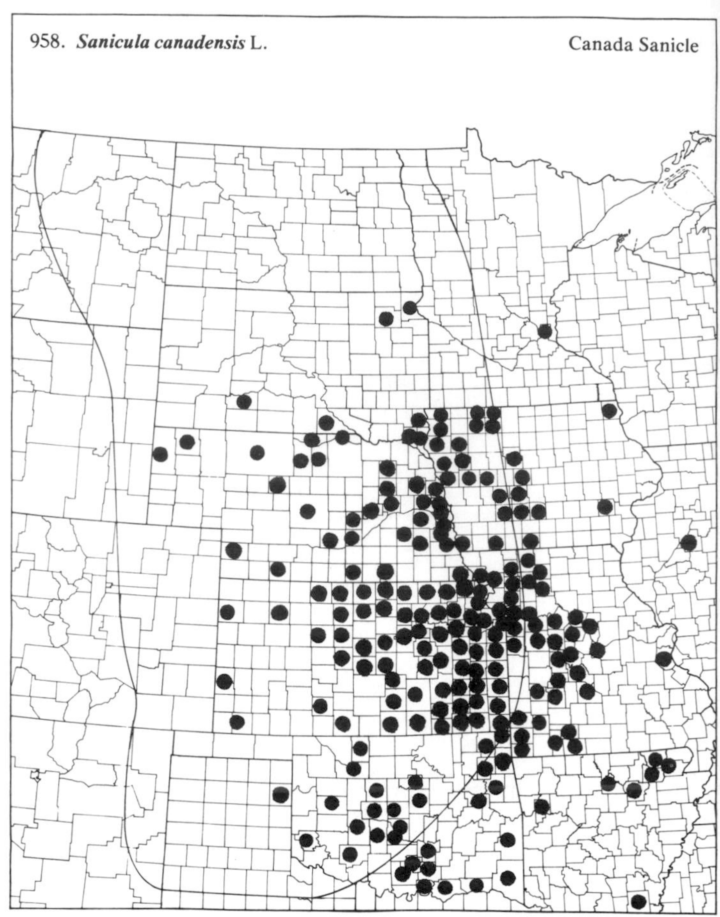

959. ***Sanicula gregaria*** Bickn. Cluster Sanicle

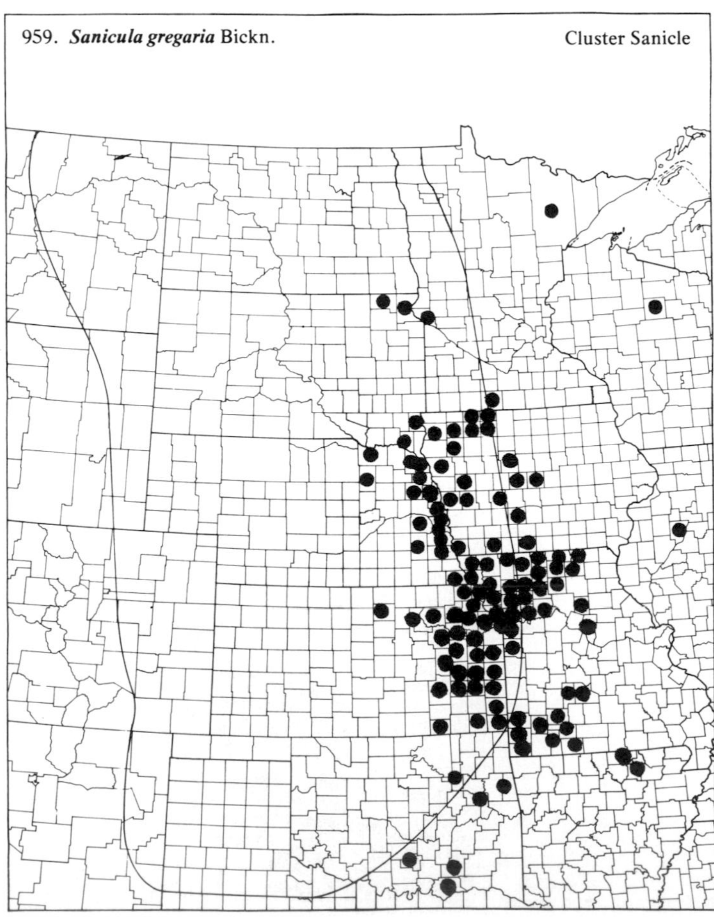

960. ***Sanicula marilandica*** L. Black Snakeroot

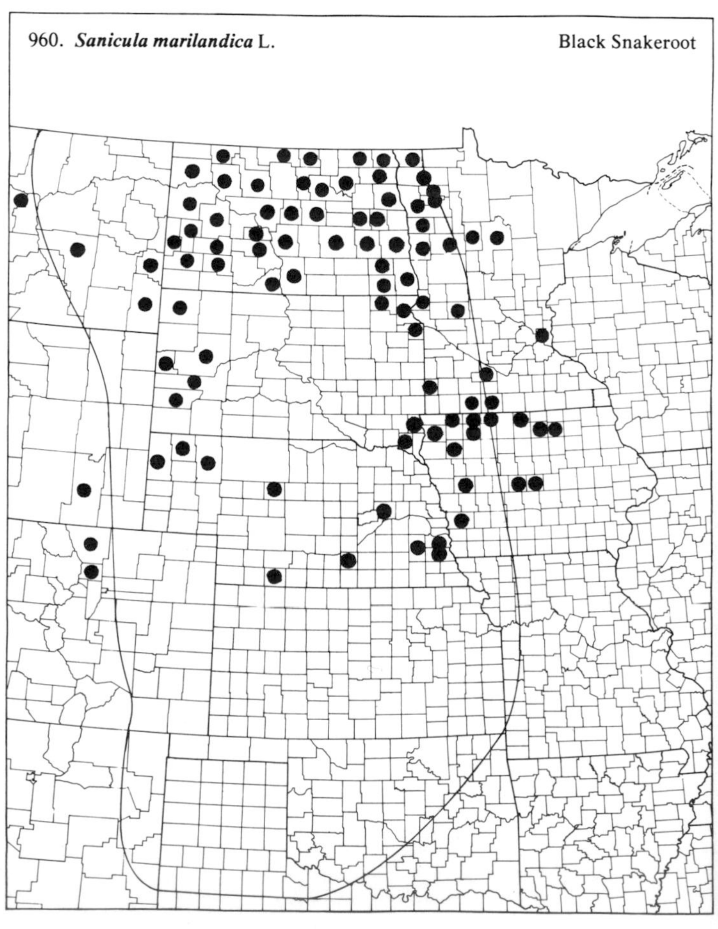

(Apiaceae)

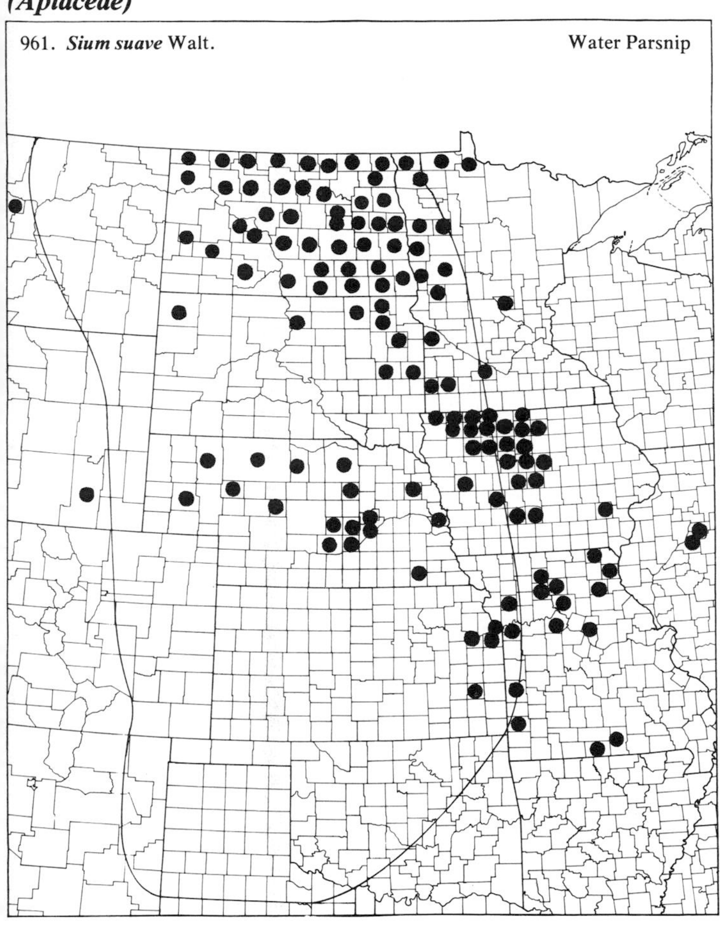

961. ***Sium suave*** Walt. Water Parsnip

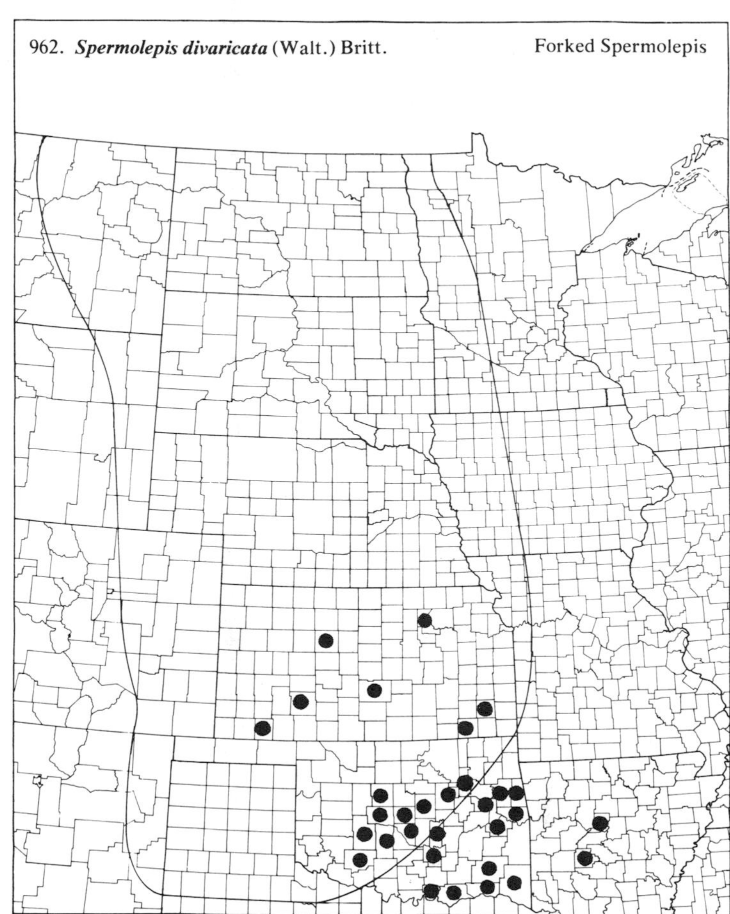

962. ***Spermolepis divaricata*** (Walt.) Britt. Forked Spermolepis

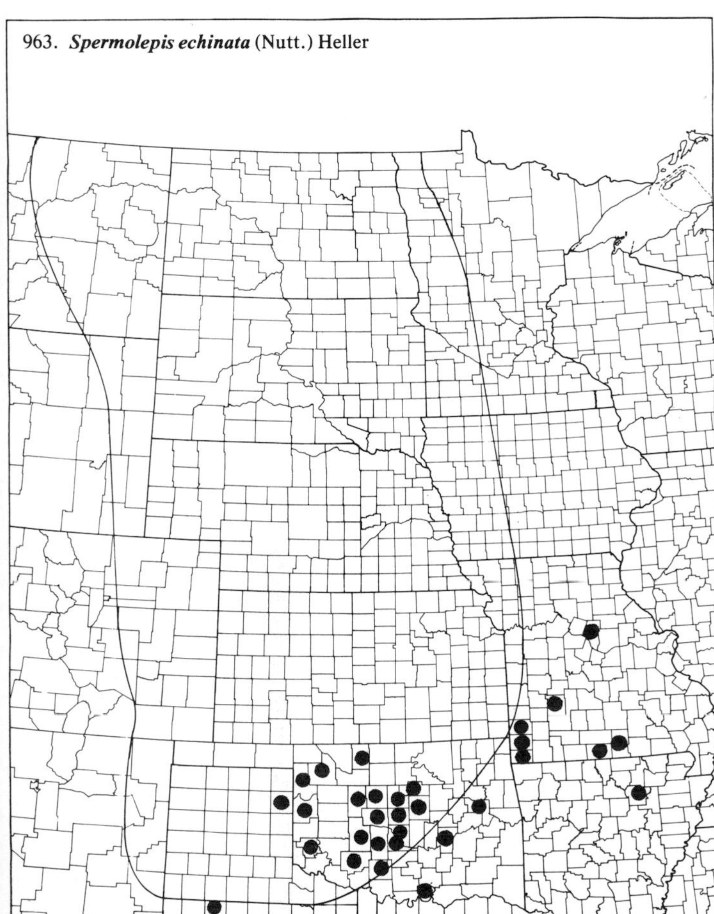

963. ***Spermolepis echinata*** (Nutt.) Heller

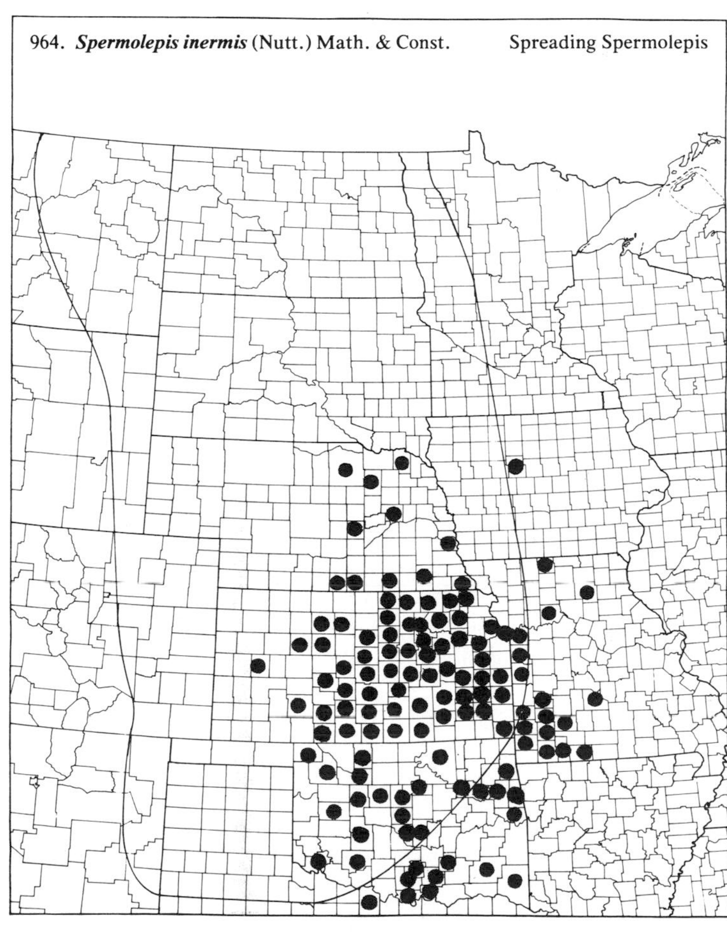

964. ***Spermolepis inermis*** (Nutt.) Math. & Const. Spreading Spermolepis

(Apiaceae)

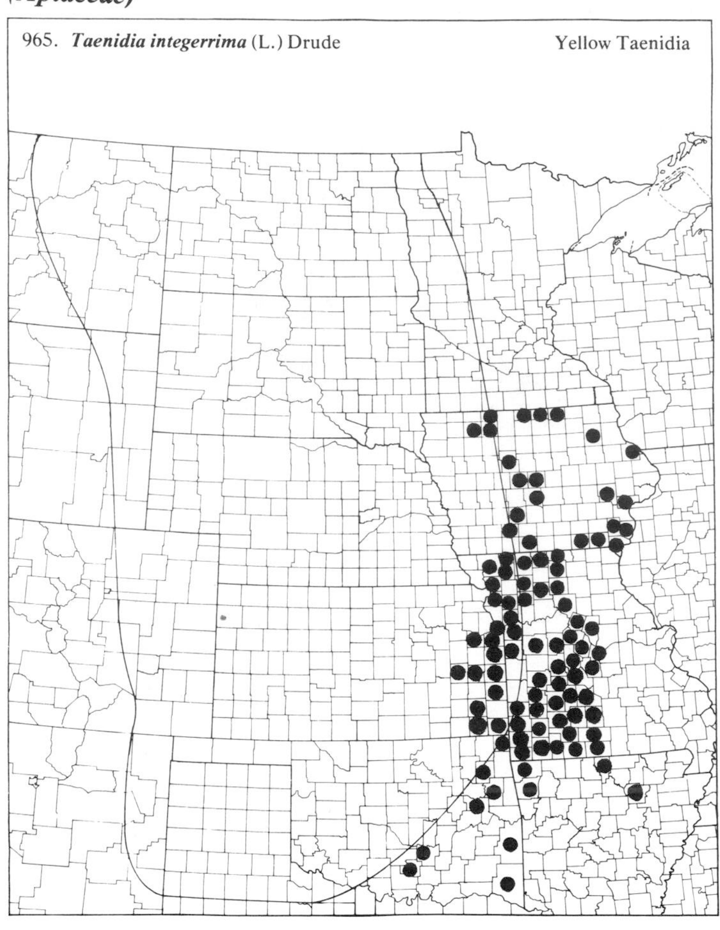
965. **Taenidia integerrima** (L.) Drude Yellow Taenidia

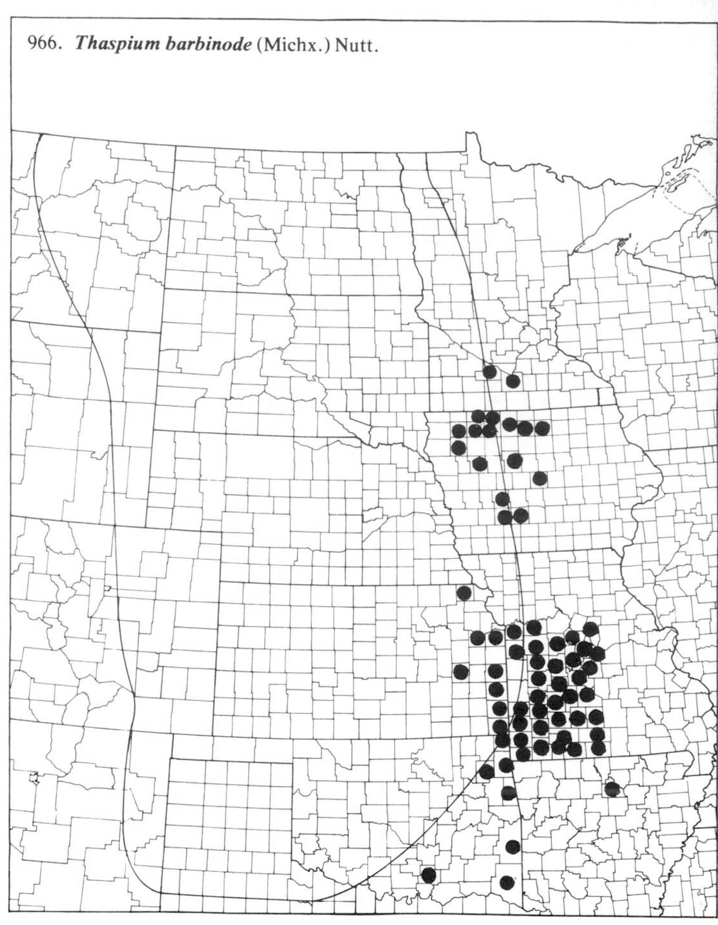
966. **Thaspium barbinode** (Michx.) Nutt.

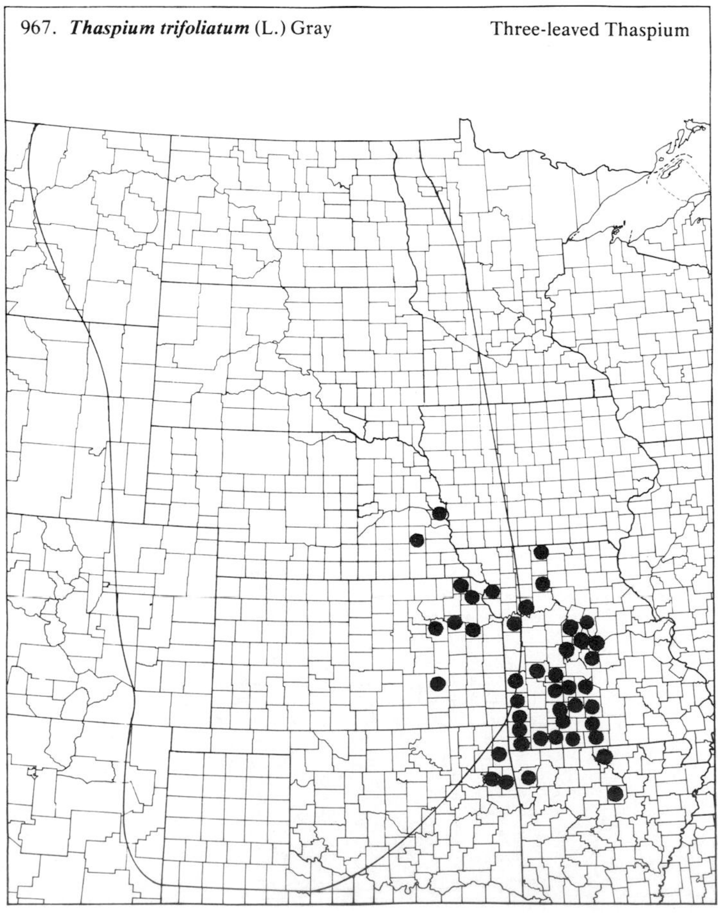
967. **Thaspium trifoliatum** (L.) Gray Three-leaved Thaspium

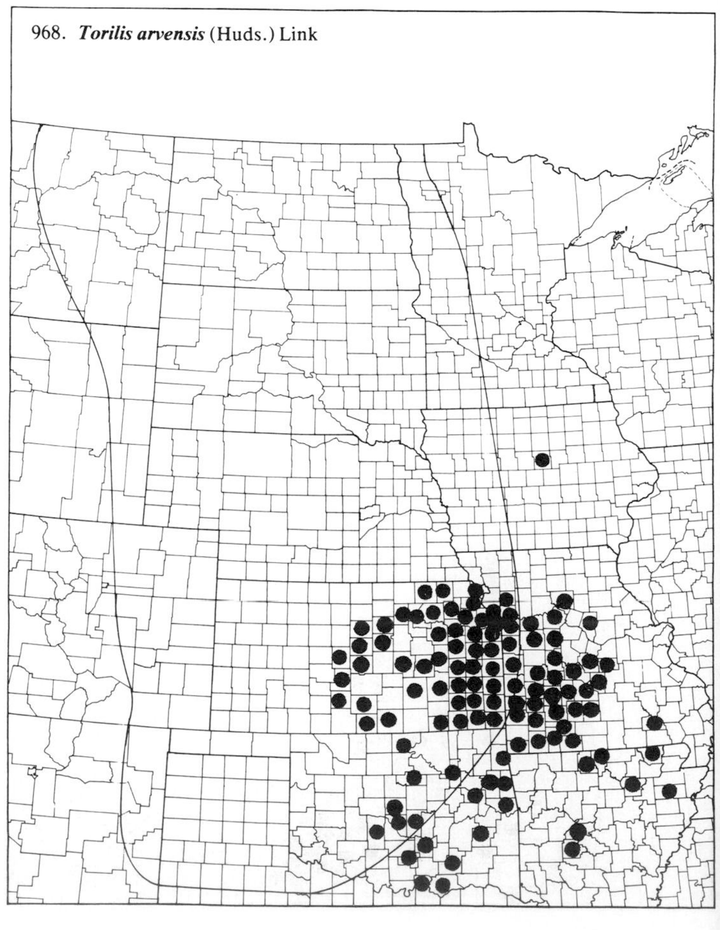
968. **Torilis arvensis** (Huds.) Link

(Apiaceae)

969. ***Zizia aptera*** (Gray) Fern. Meadow Parsnip

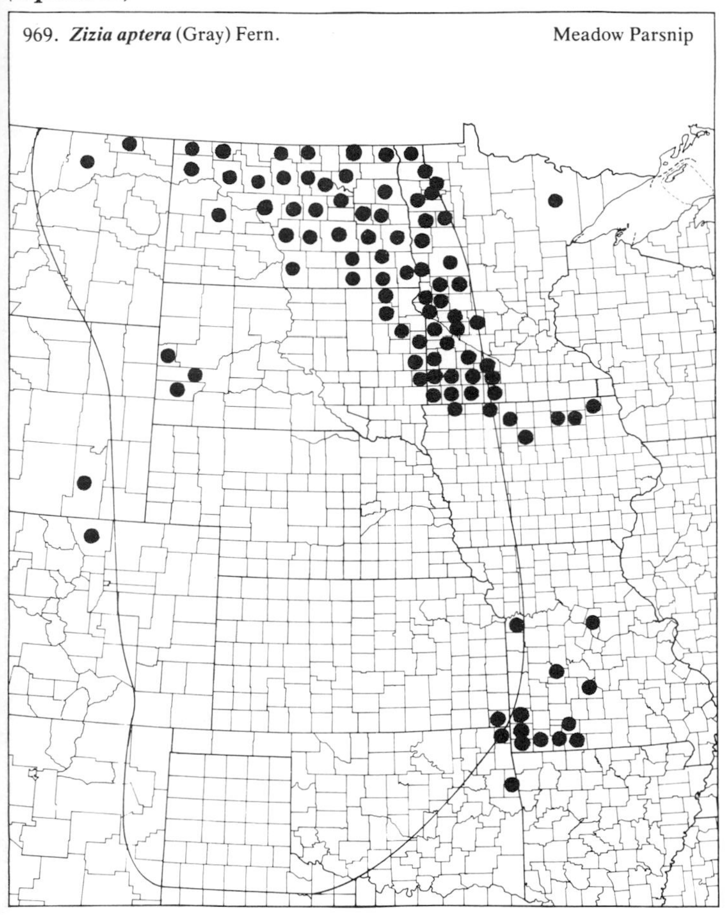

970. ***Zizia aurea*** (L.) Koch Golden Alexanders

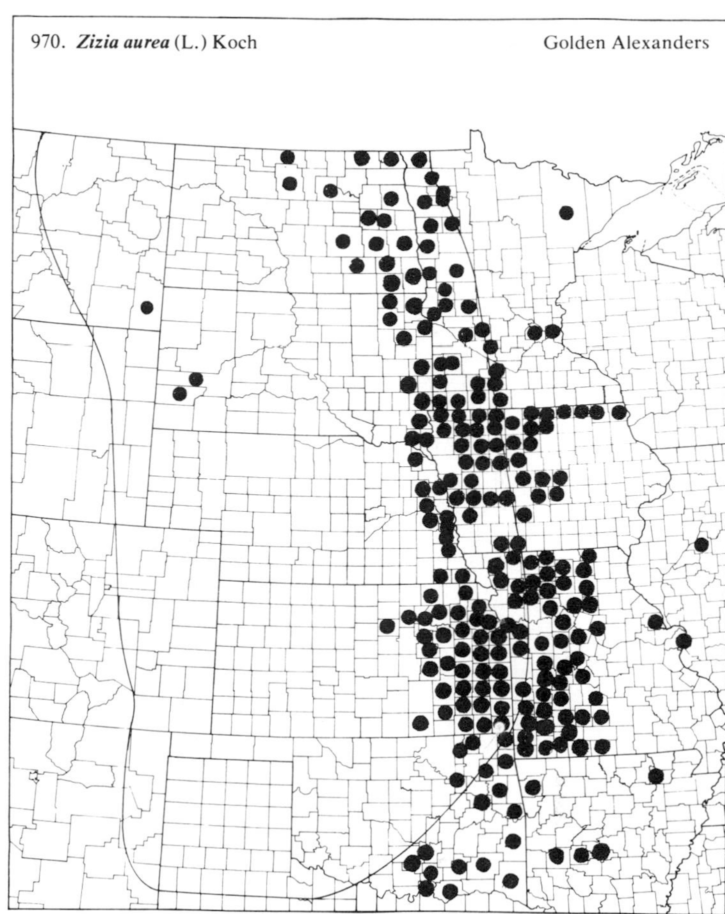

86. *Gentianaceae* (Gentian Family)

971. ***Eustoma grandiflorum*** (Raf.) Shinners Prairie Gentian

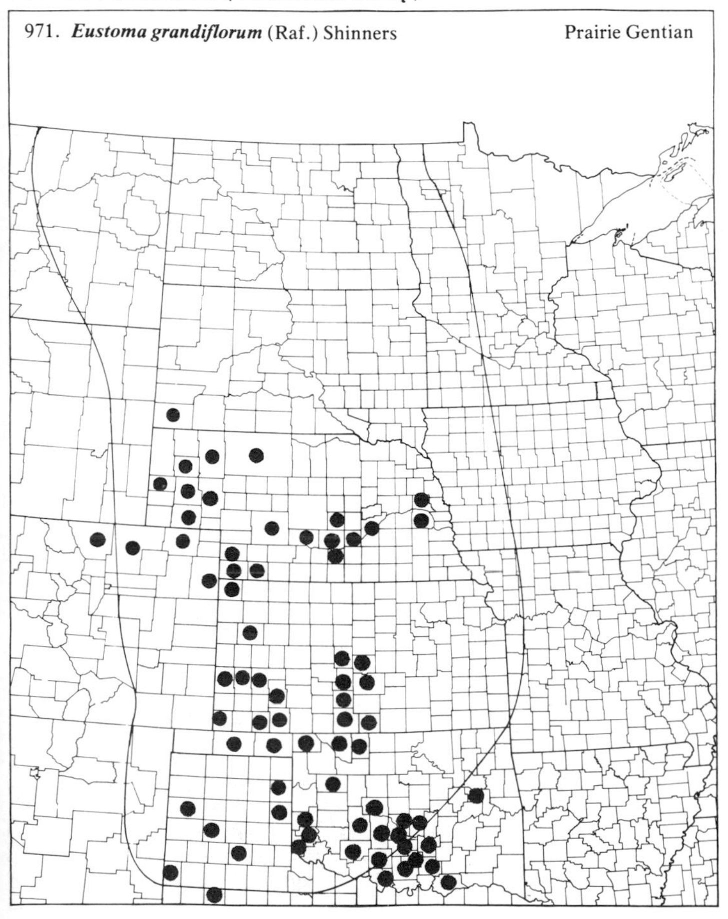

972. ***Gentiana affinis*** Griseb. Northern Gentian

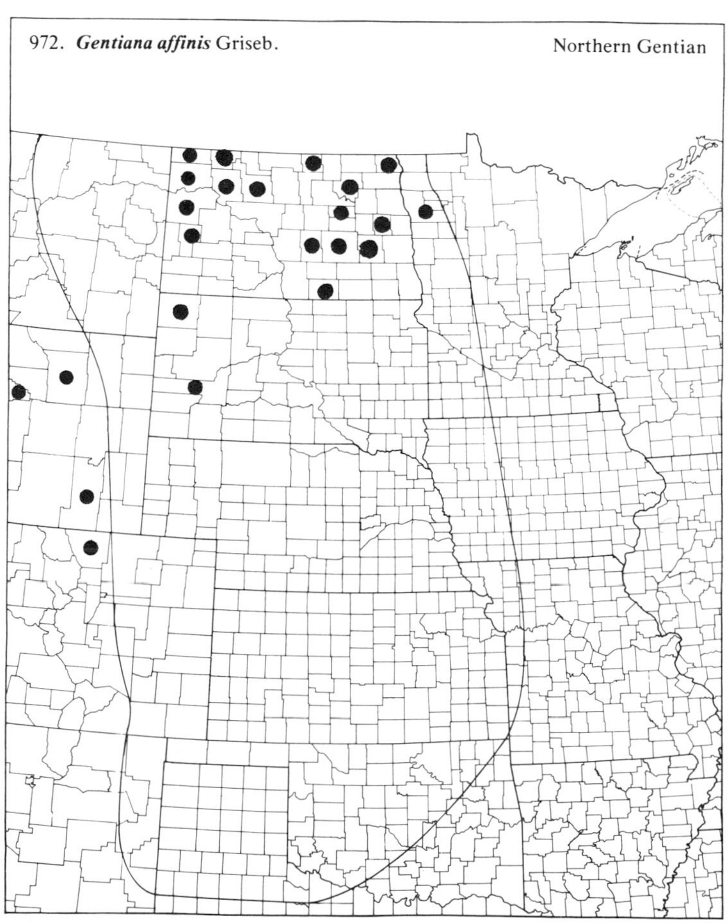

(Gentianaceae)

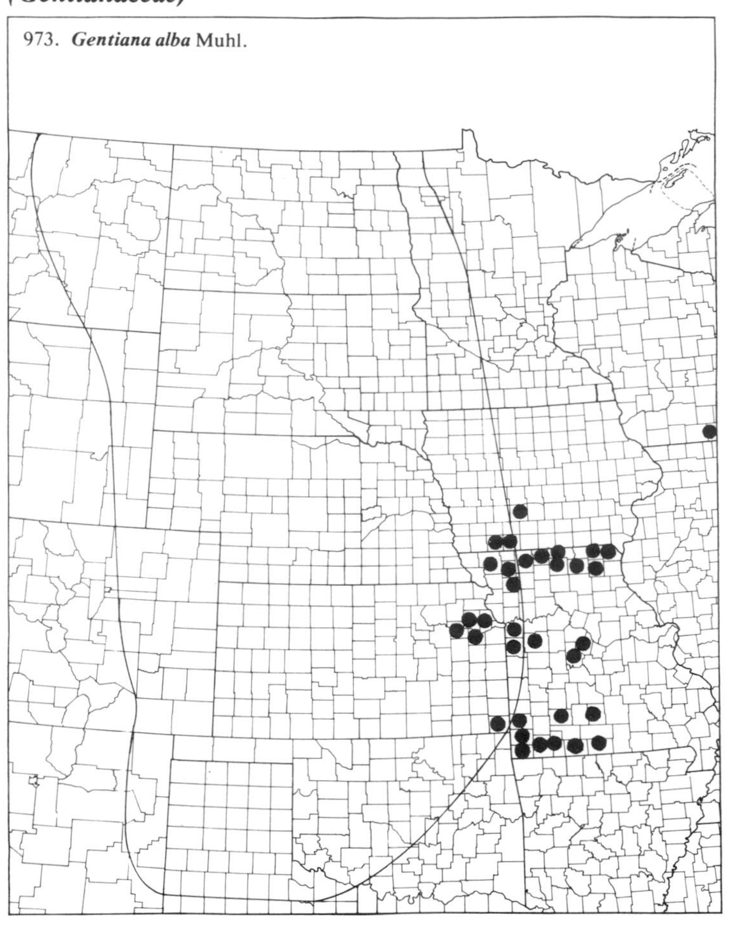
973. ***Gentiana alba*** Muhl.

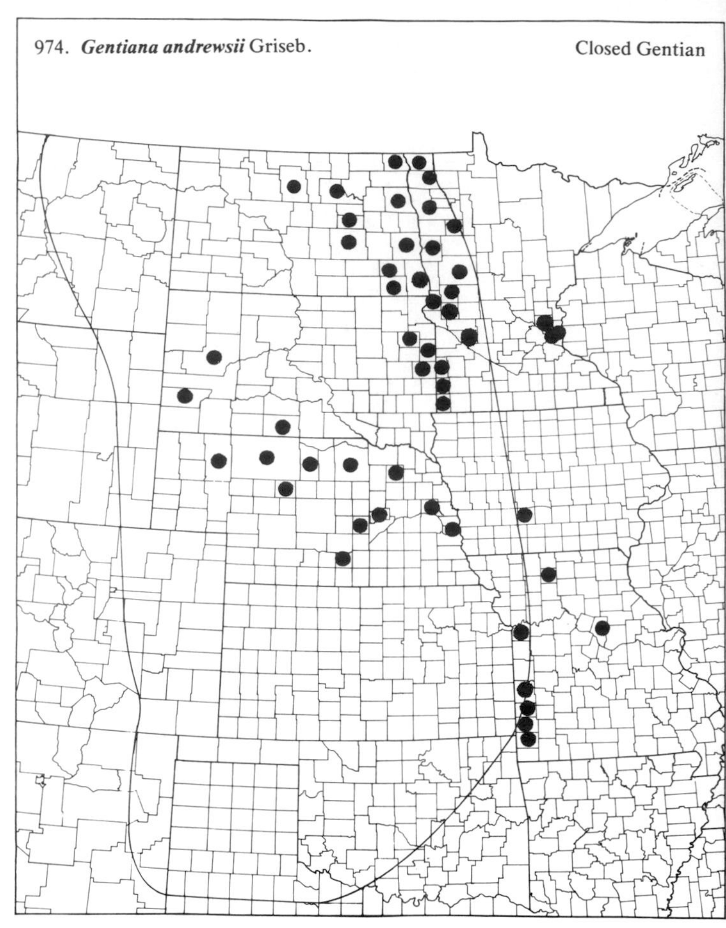
974. ***Gentiana andrewsii*** Griseb. Closed Gentian

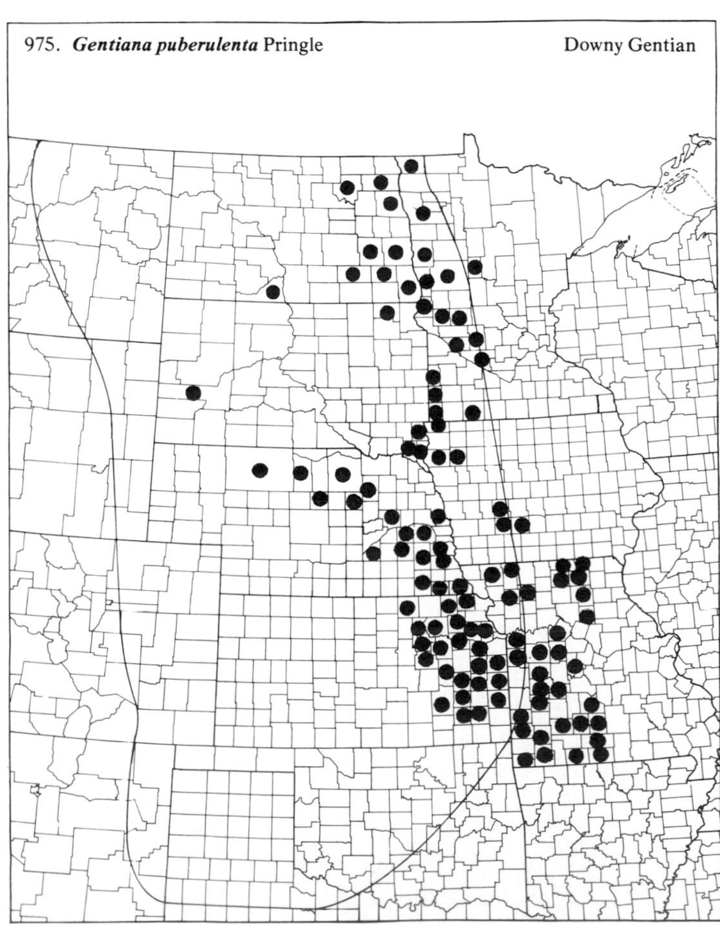
975. ***Gentiana puberulenta*** Pringle Downy Gentian

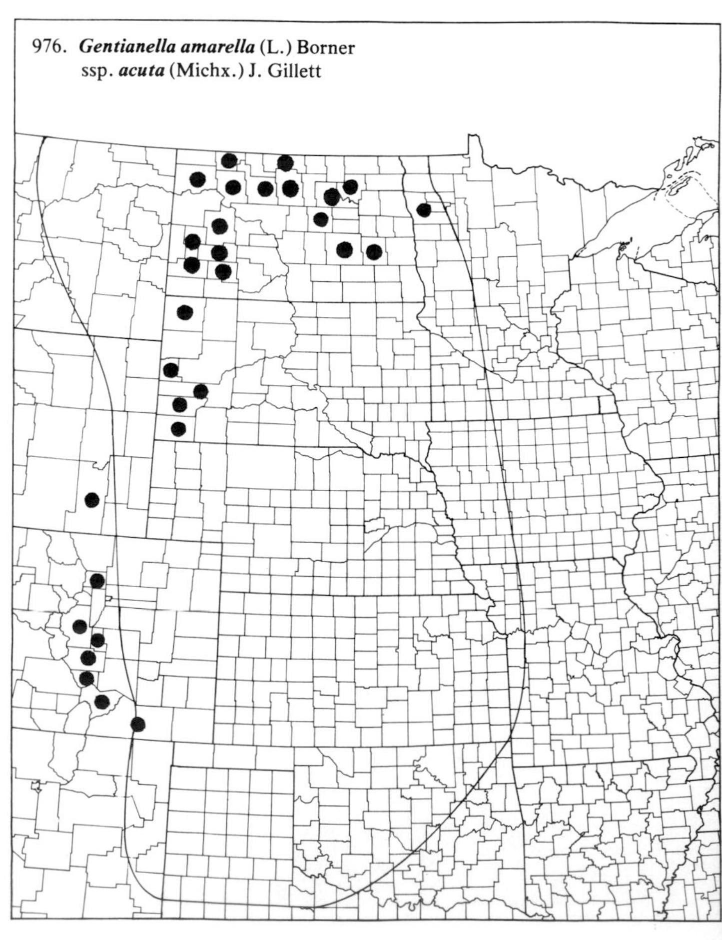
976. ***Gentianella amarella*** (L.) Borner ssp. ***acuta*** (Michx.) J. Gillett

977. ***Gentianopsis crinita*** (Froel.) Ma

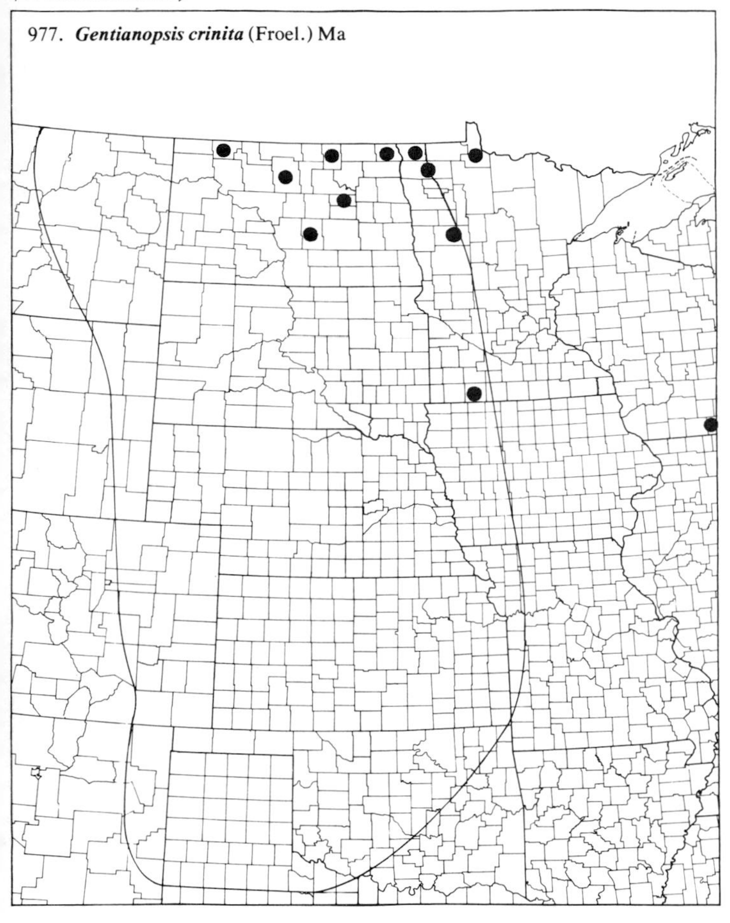

978. ***Halenia deflexa*** (Sm.) Griseb. Spurred Gentian

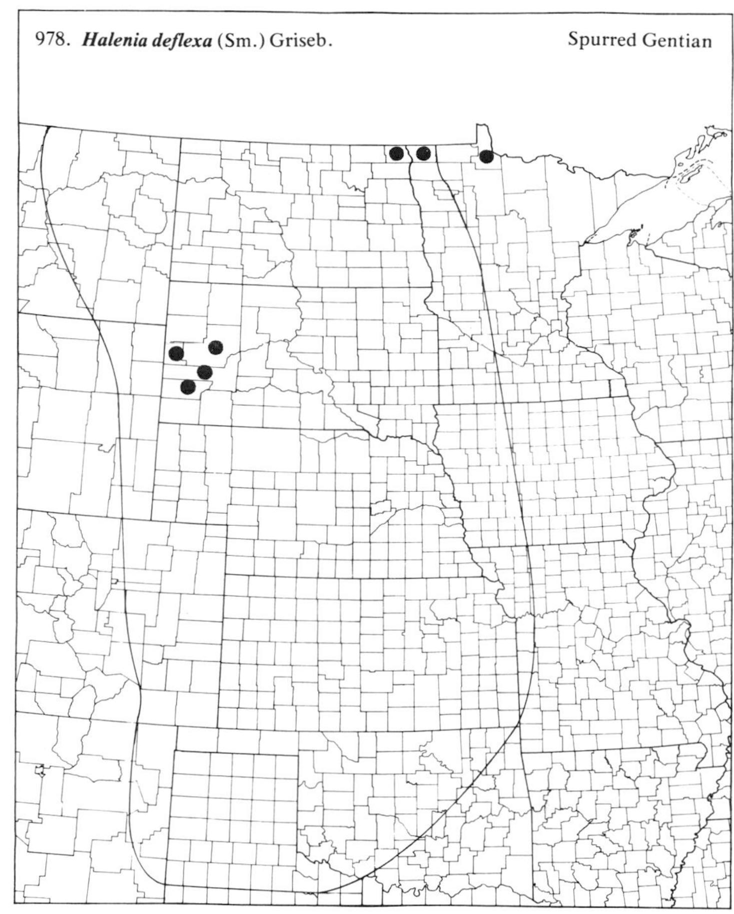

979. ***Sabatia campestris*** Nutt. Prairie Rose Gentian

87. *Apocynaceae* (Dogbane Family)

980. *Apocynum androsaemifolium* L. Spreading Dogbane

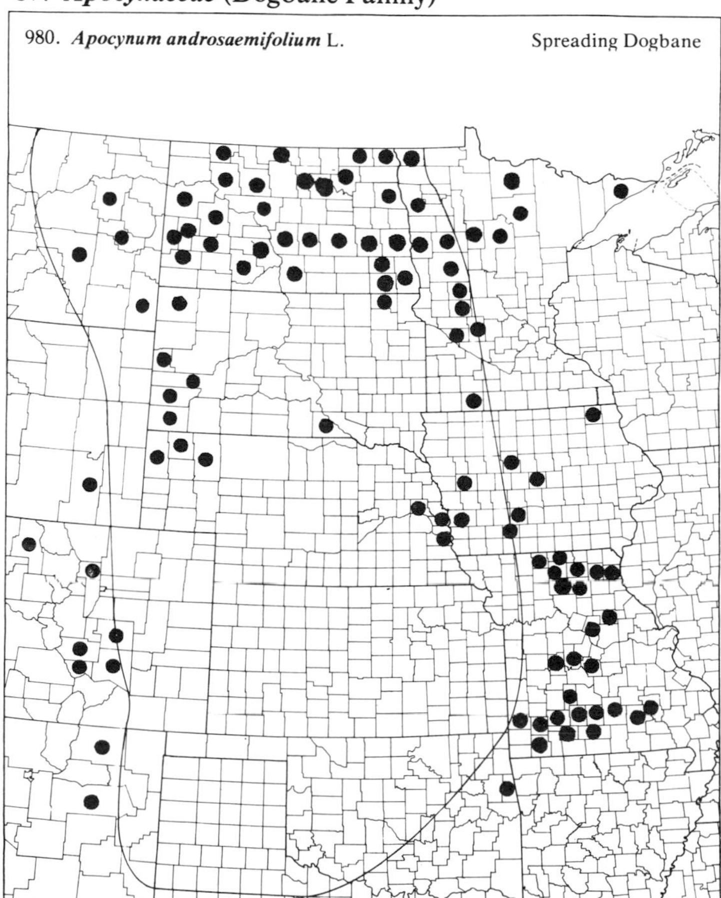

(Apocynaceae)

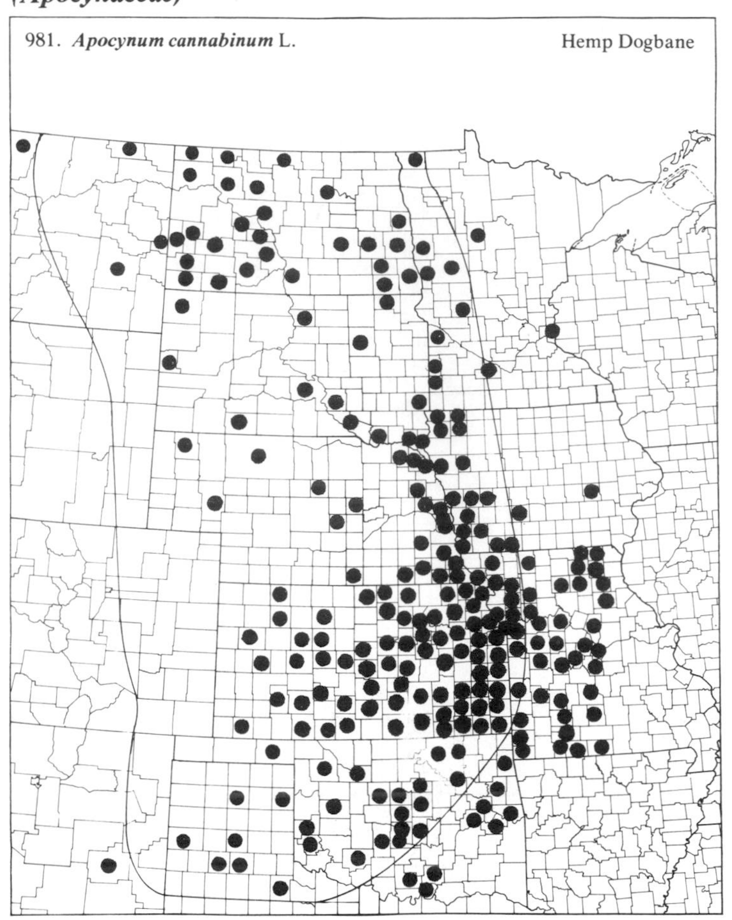

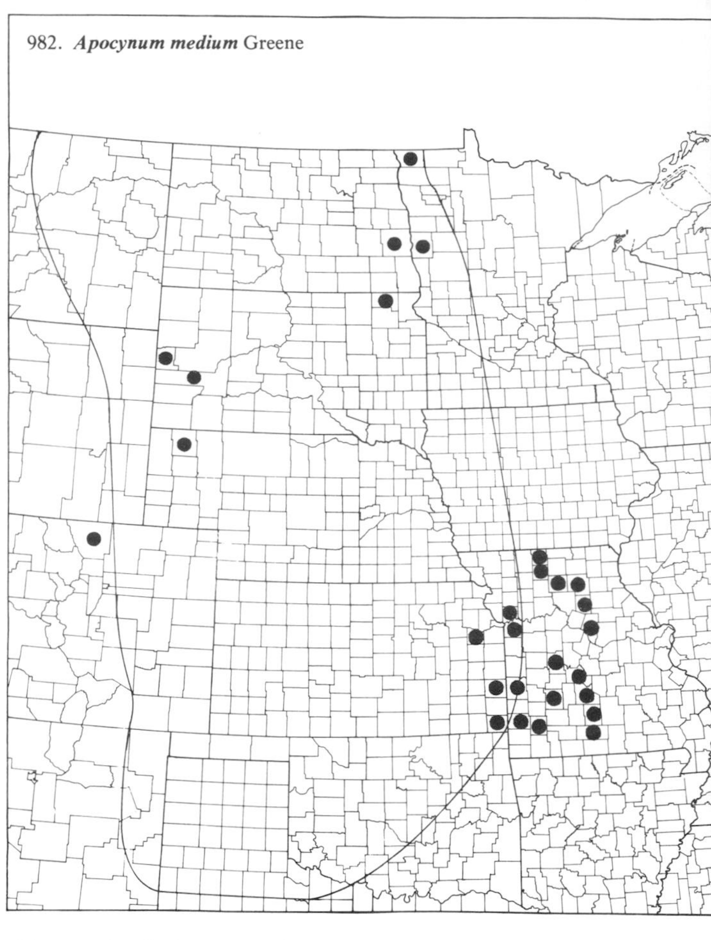

88. *Asclepiadaceae* (Milkweed Family)

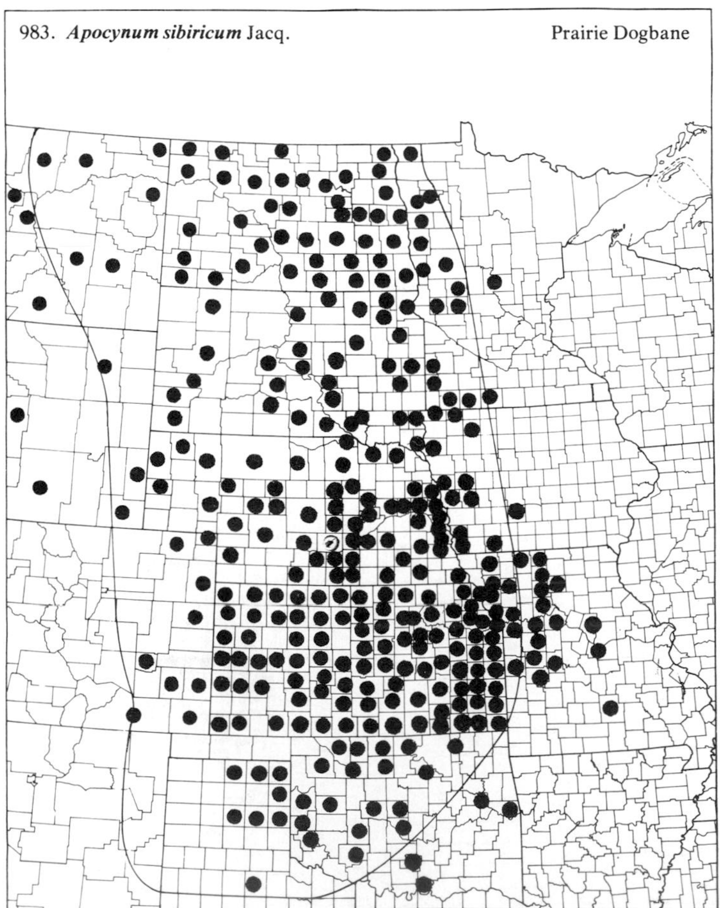

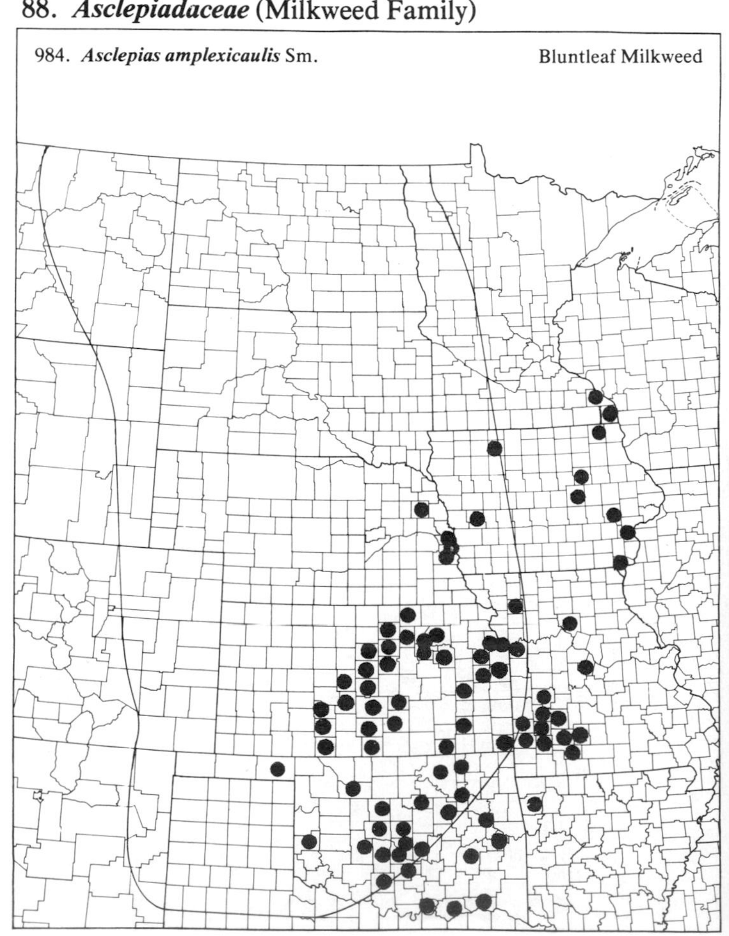

(Asclepiadaceae)

985. ***Asclepias arenaria*** Torr. Sand Milkweed

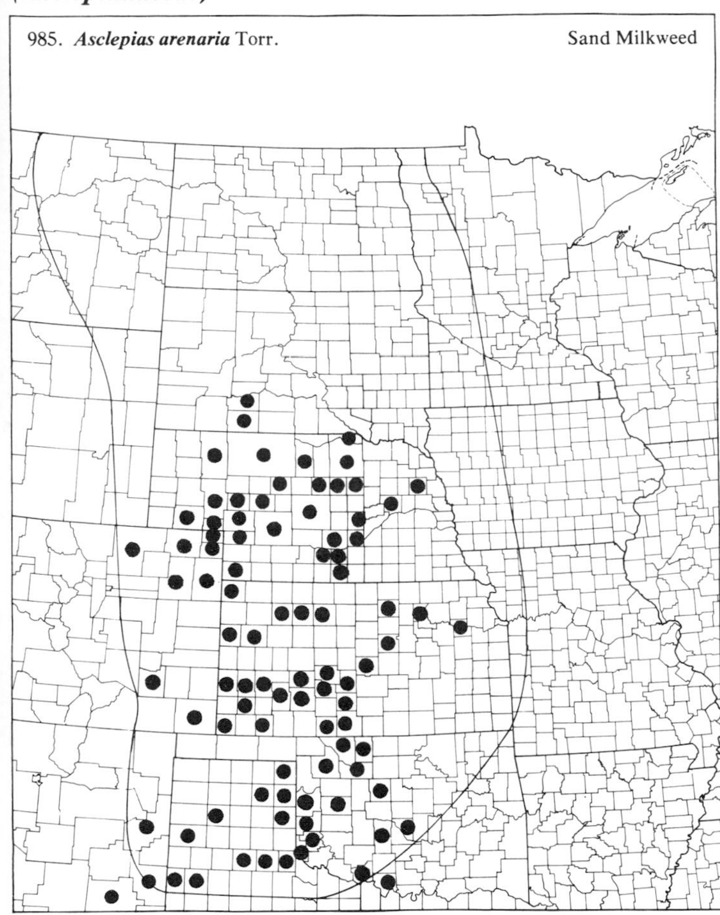

986. ***Asclepias asperula*** (Dcne.) Woods. var. ***decumbens*** (Nutt.) Shinners

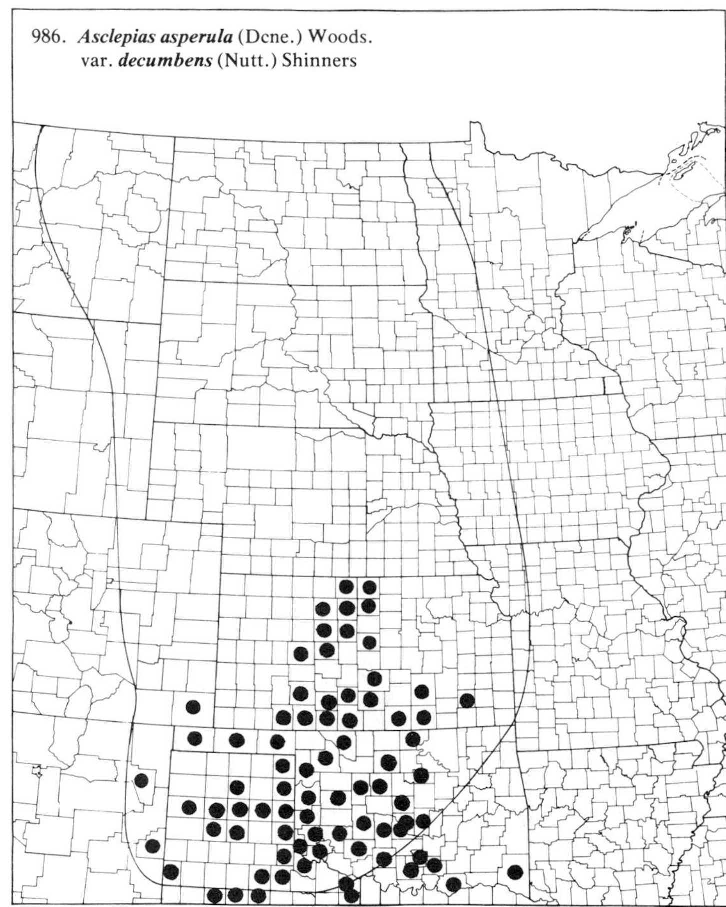

987. ***Asclepias engelmanniana*** Woods.

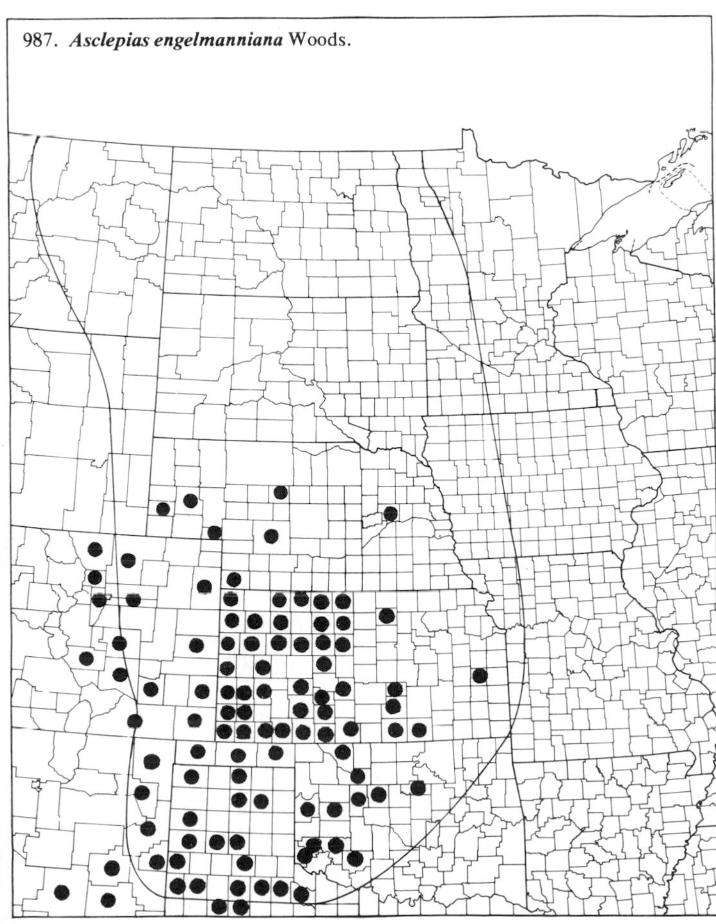

988. ***Asclepias hirtella*** (Penn.) Woods. Prairie Milkweed

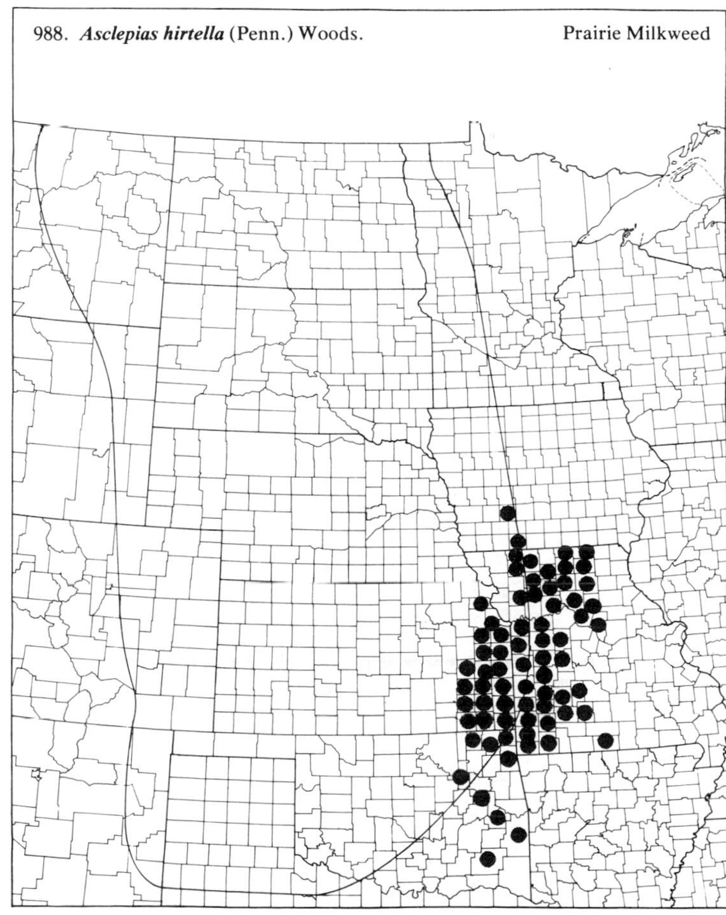

(Asclepiadaceae)

989. ***Asclepias incarnata*** L. Swamp Milkweed

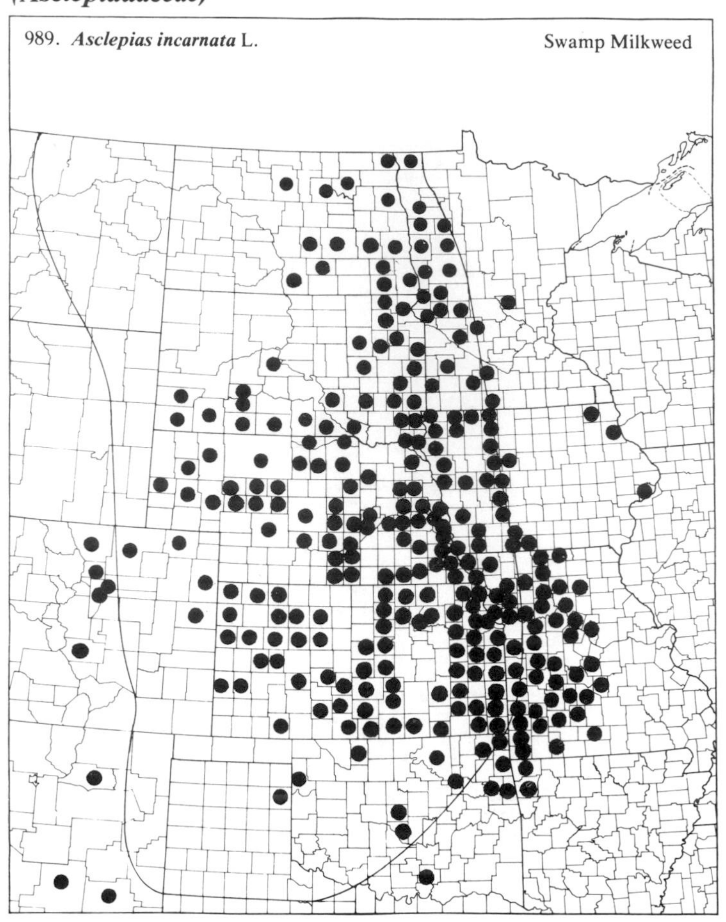

990. ***Asclepias lanuginosa*** Nutt. Woolly Milkweed

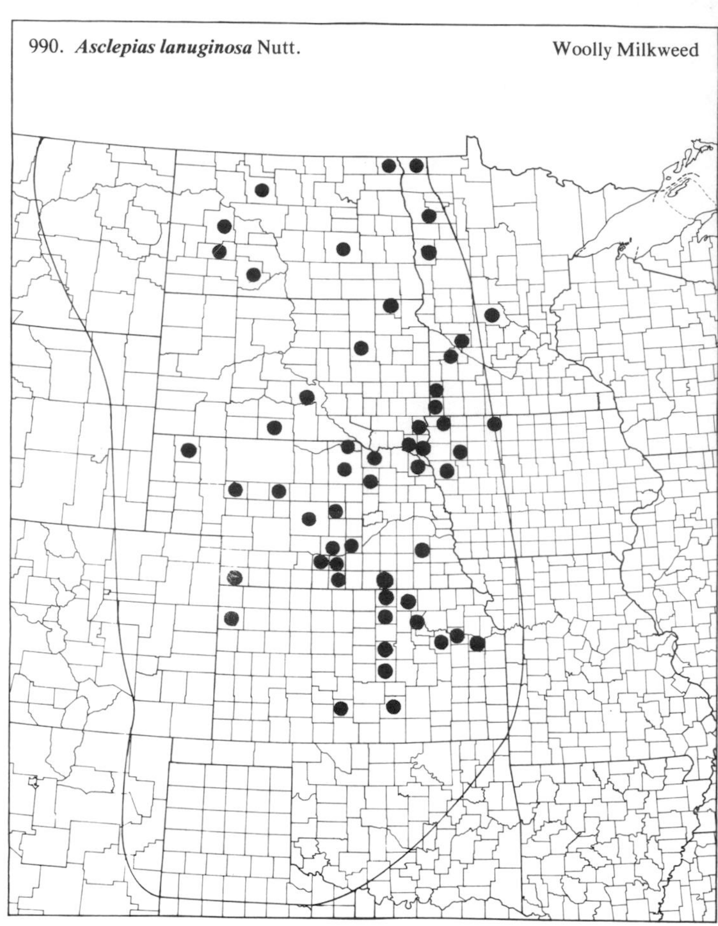

991. ***Asclepias latifolia*** (Torr.) Raf. Broadleaf Milkweed

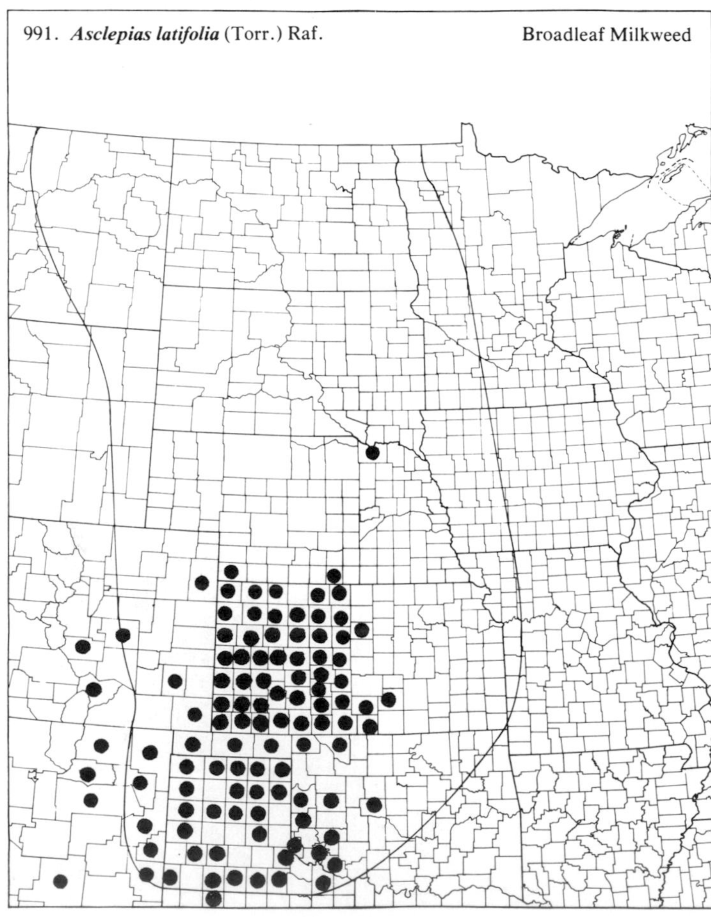

992. ***Asclepias meadii*** Torr.

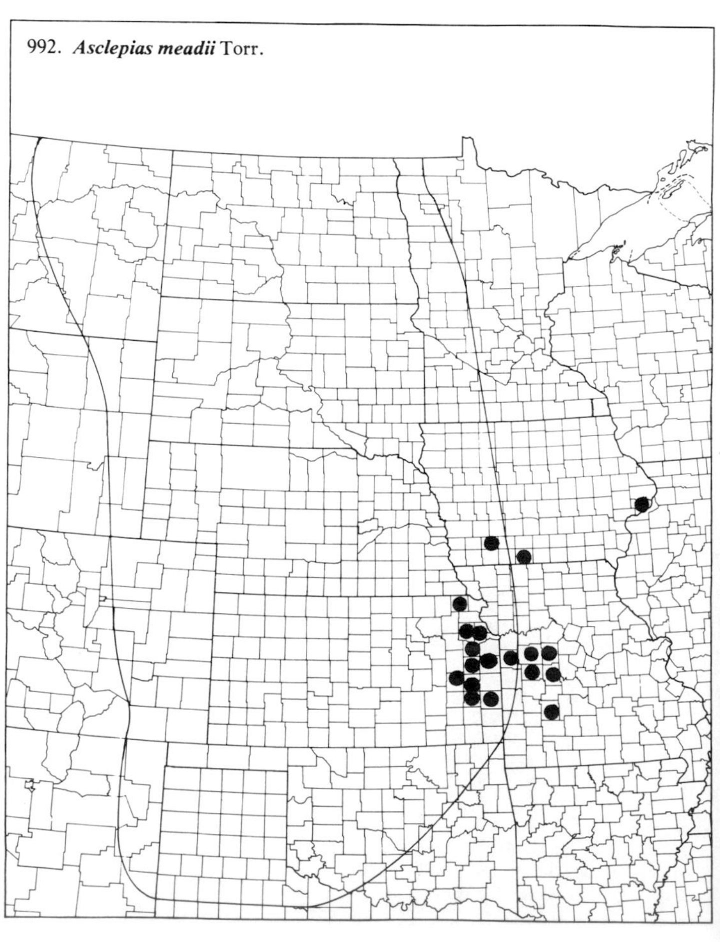

(Asclepiadaceae)

993. ***Asclepias ovalifolia*** Dcne.

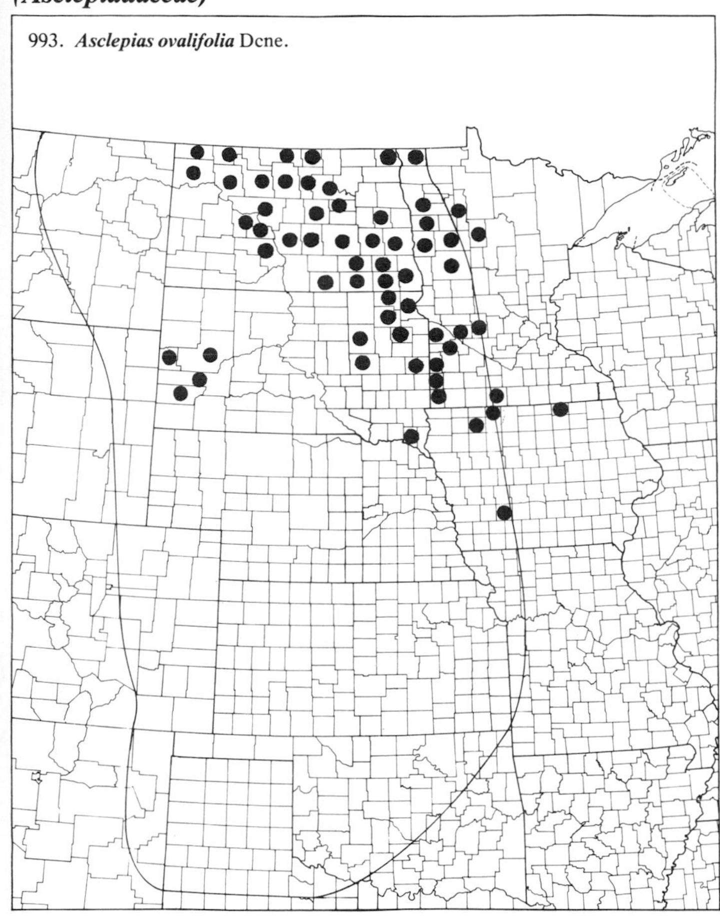

994. ***Asclepias pumila*** (Gray) Vail — Dwarf Milkweed

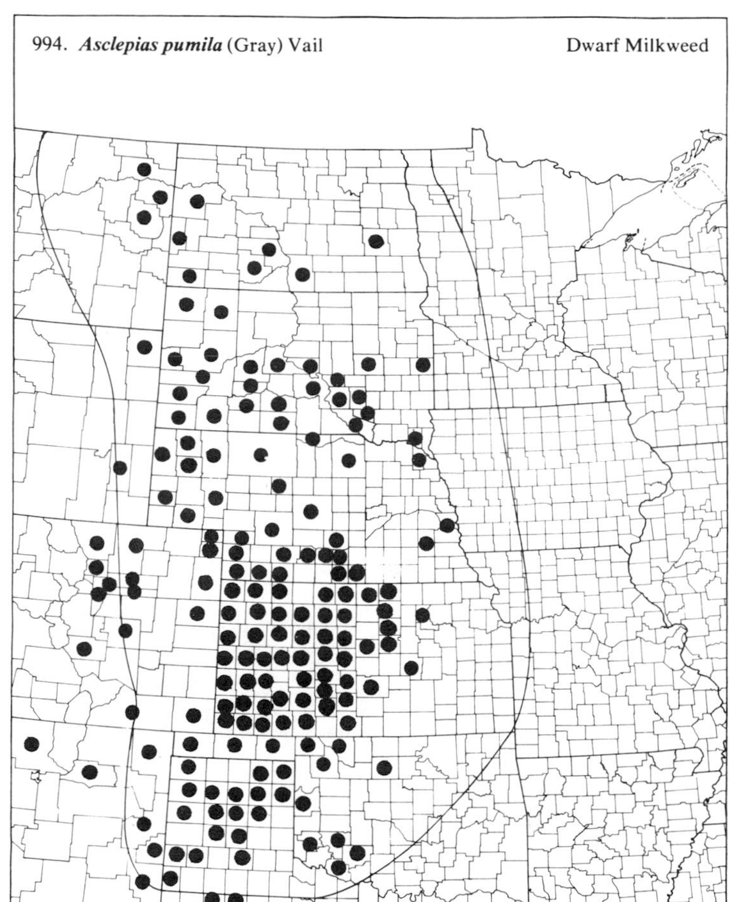

995. ***Asclepias purpurascens*** L. — Purple Milkweed

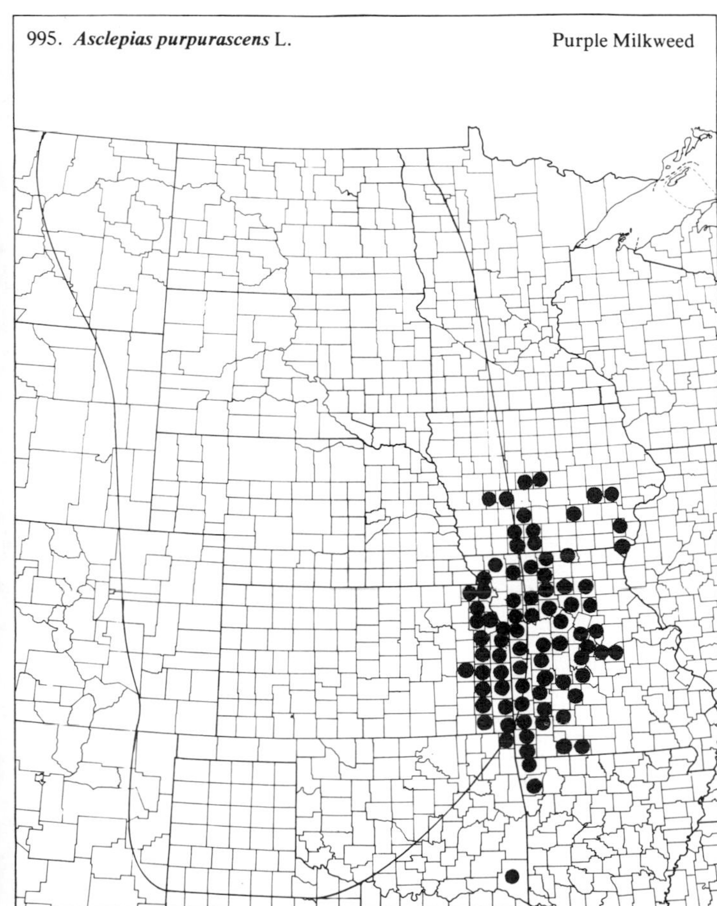

996. ***Asclepias speciosa*** Torr. — Showy Milkweed

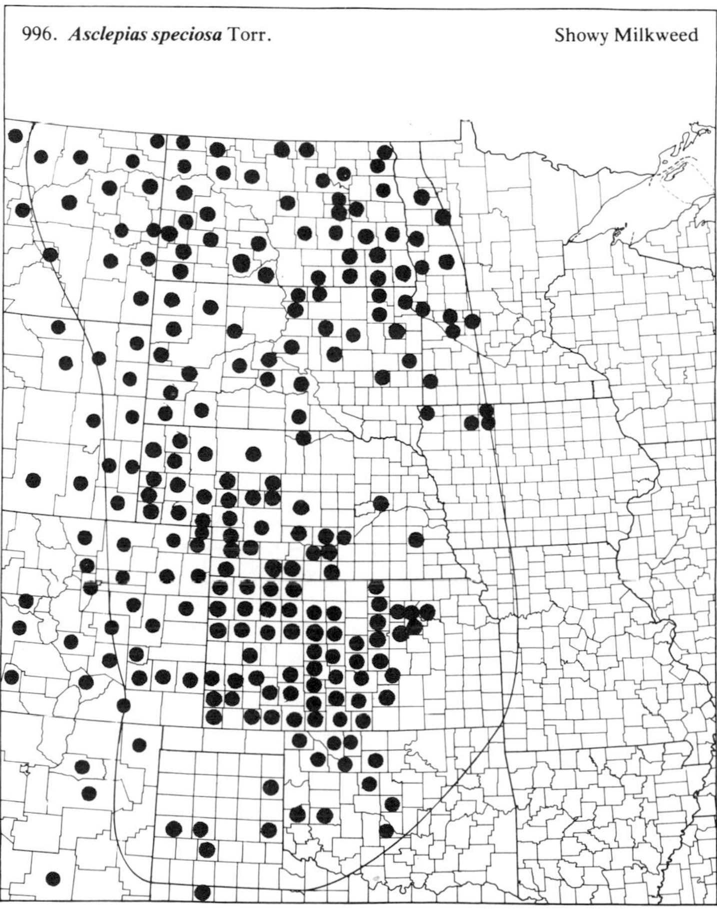

(Asclepiadaceae)

997. ***Asclepias stenophylla*** Gray — Narrow-leaved Milkweed

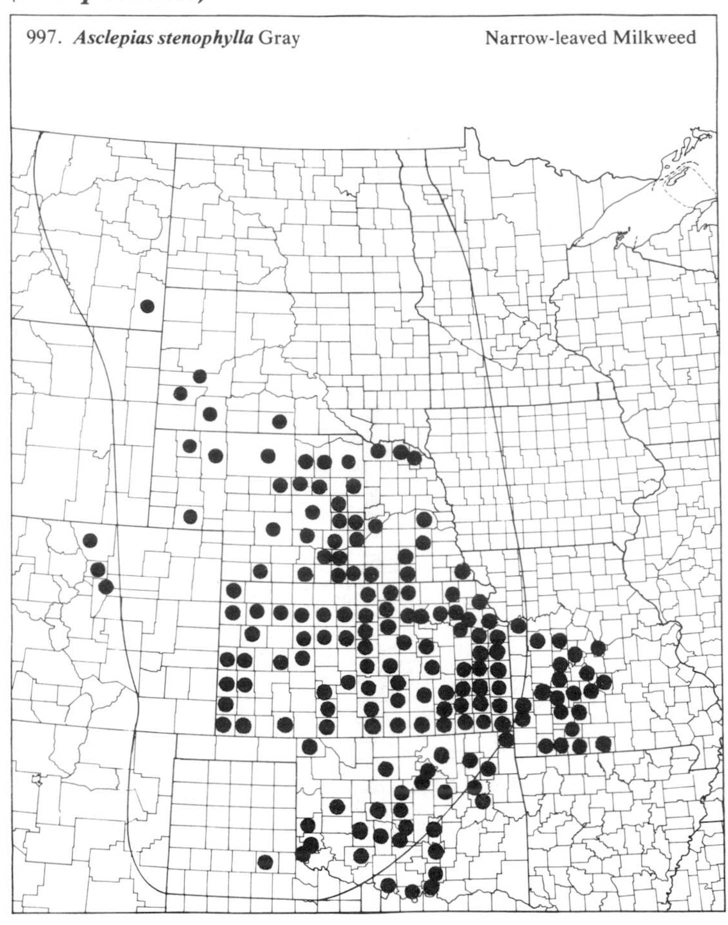

998. ***Asclepias subverticillata*** (Gray) Vail — Poison Milkweed

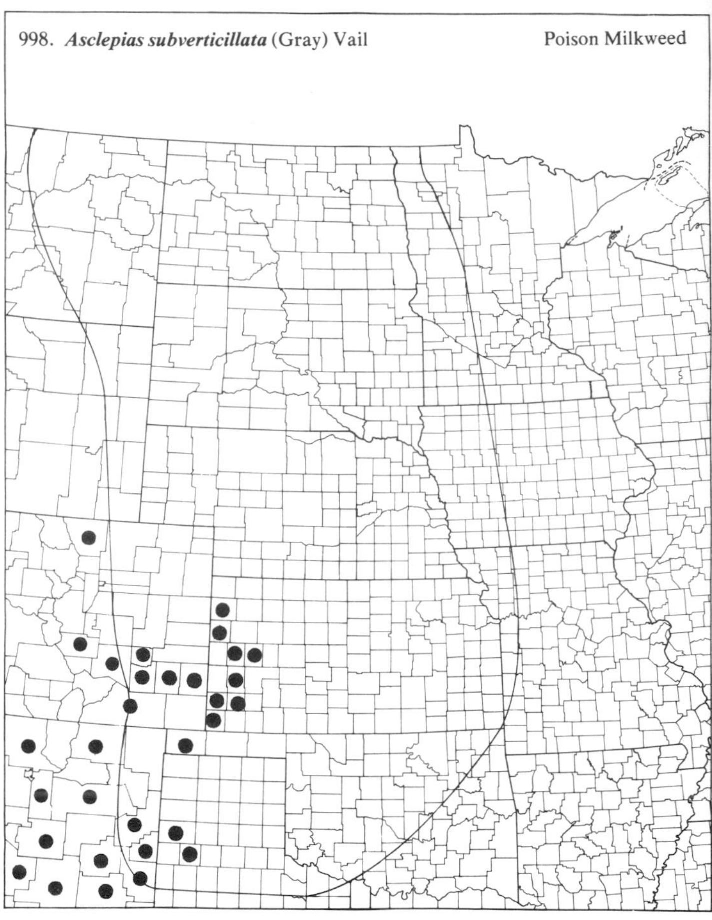

999. ***Asclepias sullivantii*** Engelm. — Smooth Milkweed

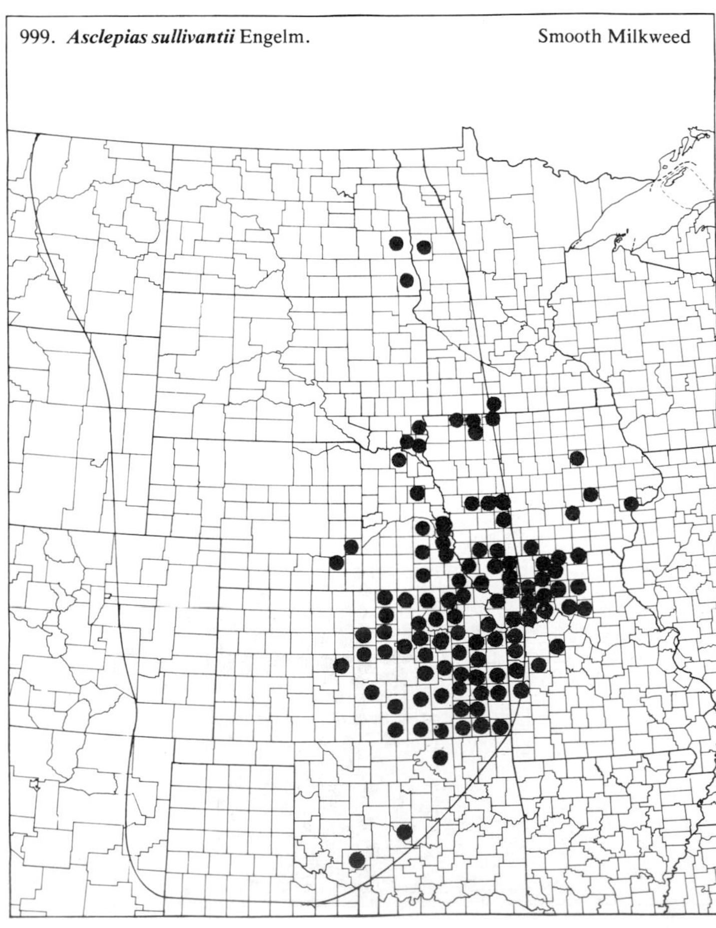

1000. ***Asclepias syriaca*** L. — Common Milkweed

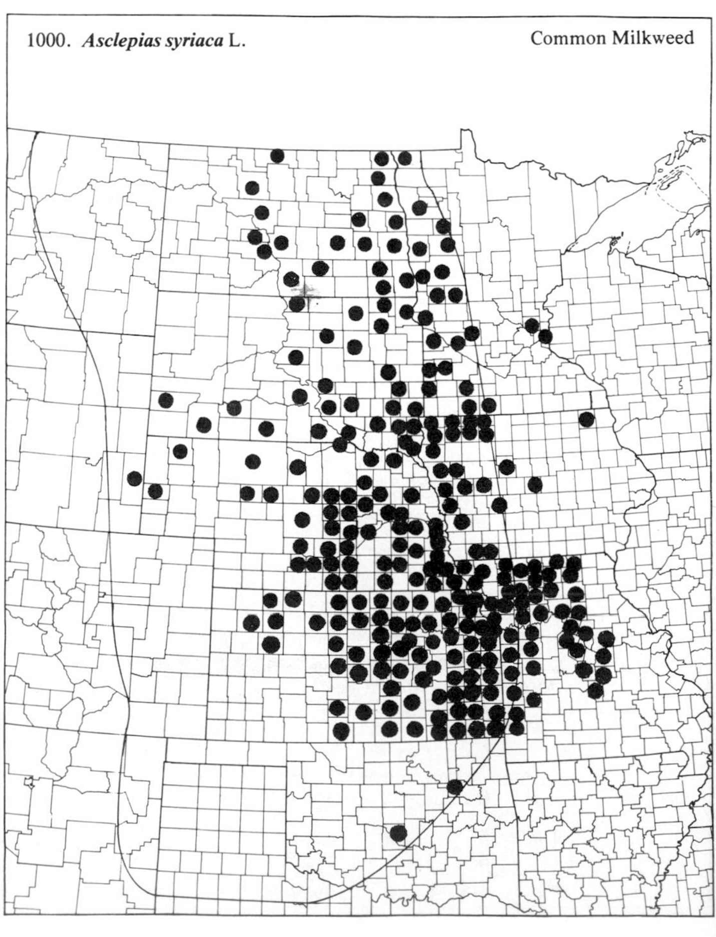

(Asclepiadaceae)

1001. *Asclepias tuberosa* L. ssp. *interior* Woods. — Butterfly Milkweed

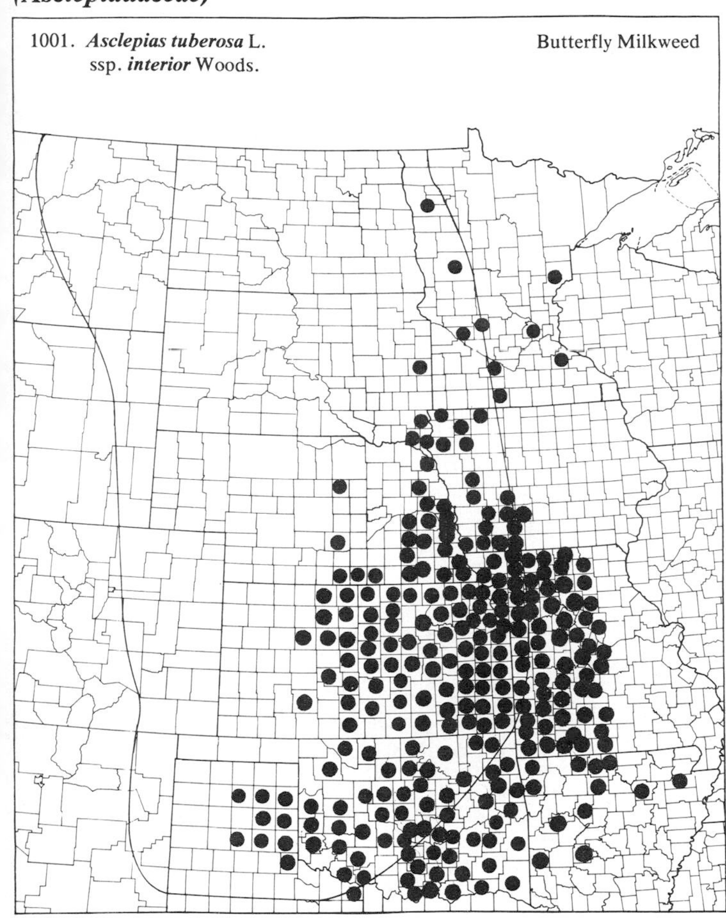

1002. *Asclepias verticillata* L. — Whorled Milkweed

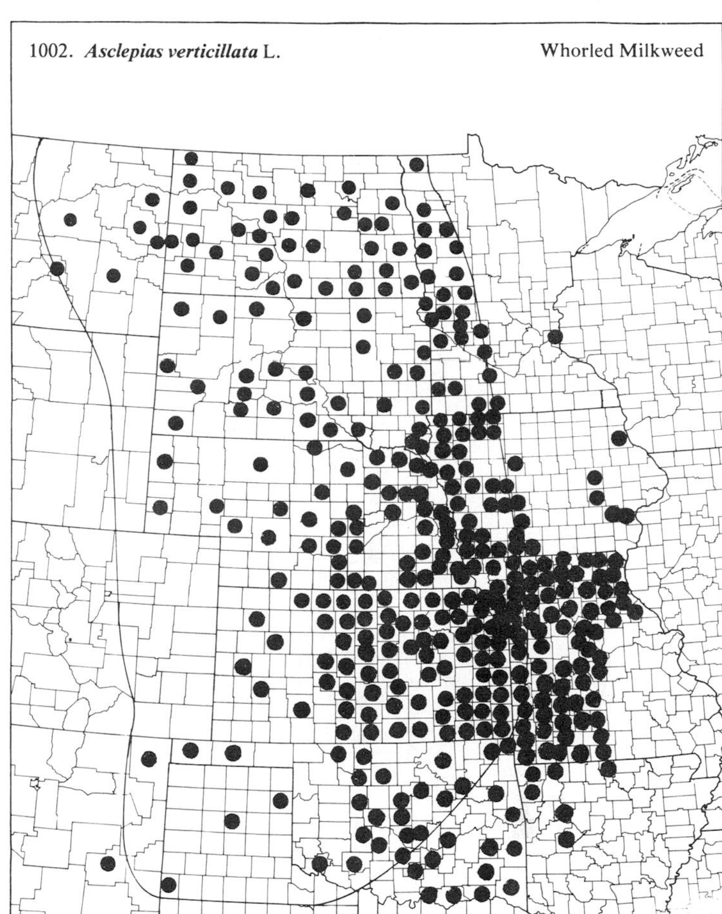

1003. *Asclepias veridiflora* Raf. — Green Milkweed

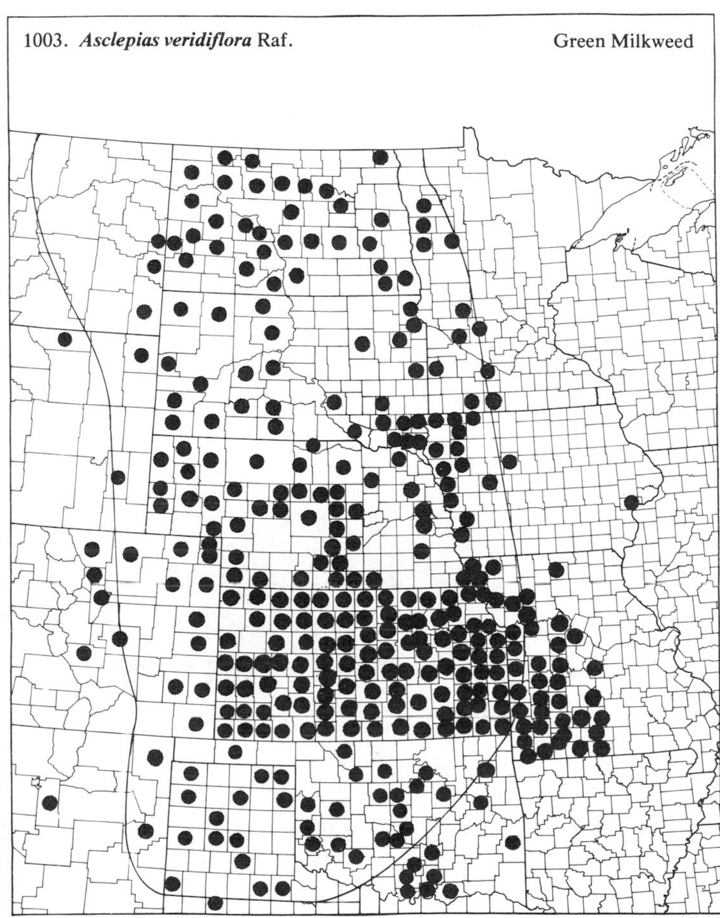

1004. *Asclepias viridis* Walt.

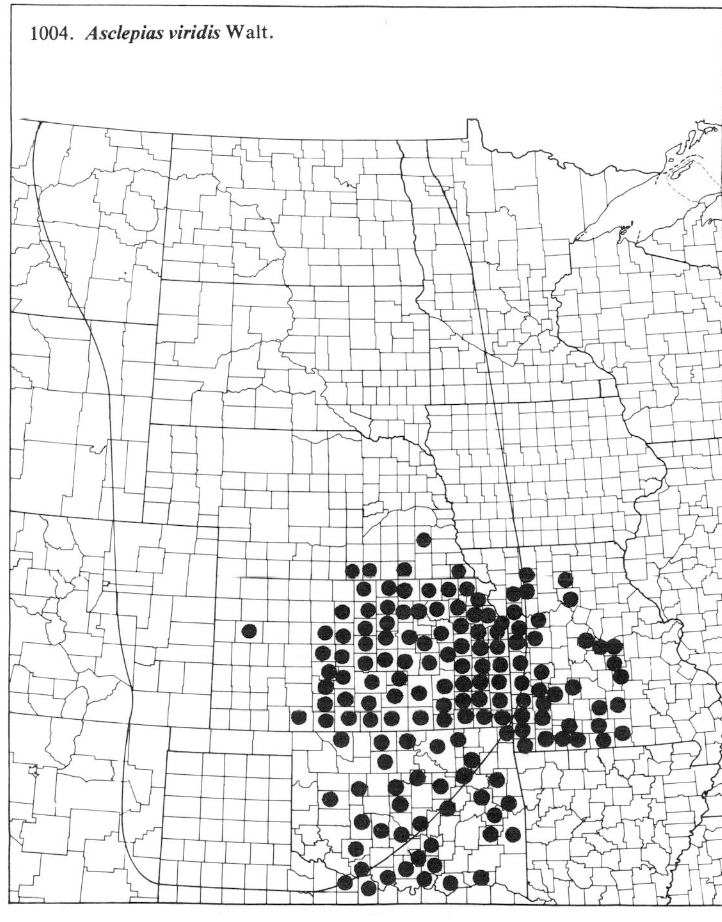

(Asclepiadaceae)

1005. ***Cynanchum laeve*** (Michx.) Pers. Sand Vine

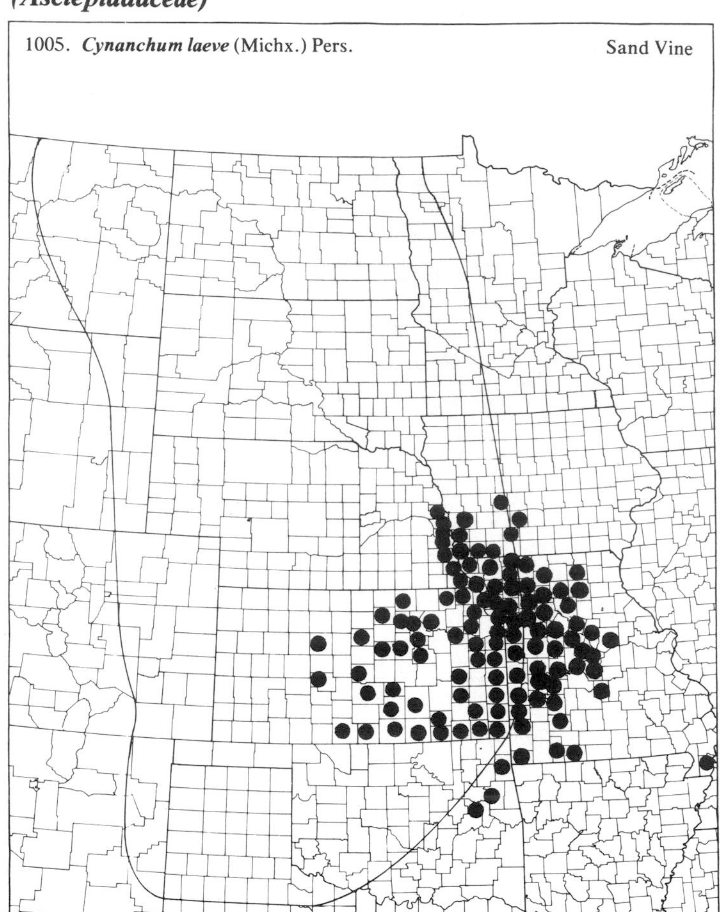

89. *Solanaceae* (Nightshade Family)

1006. ***Chamaesaracha conioides*** (Moric. ex Dun.) Britt. Chamaesaracha

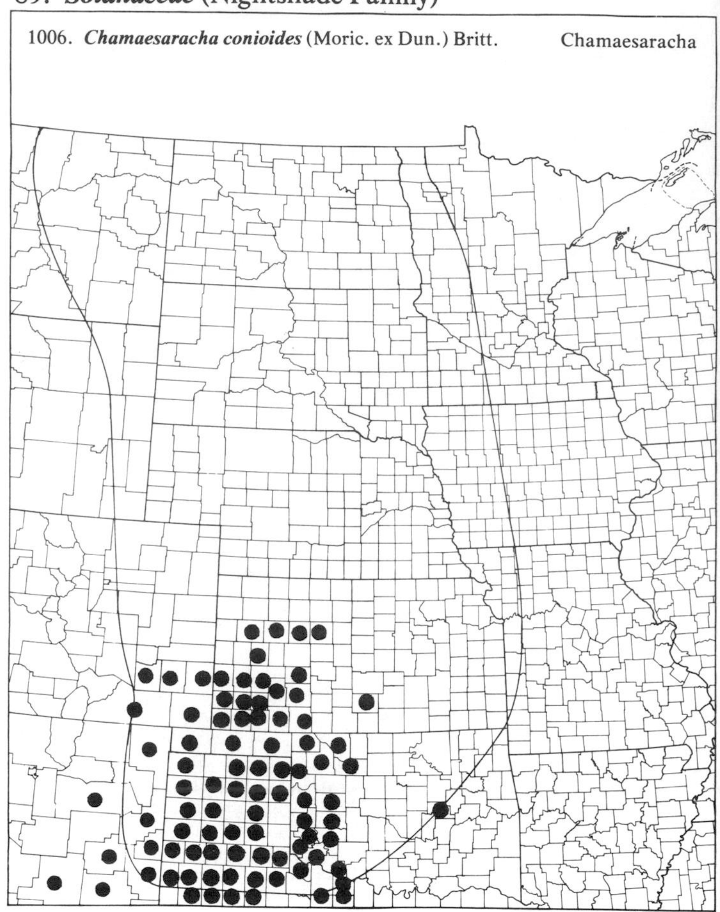

1007. ***Chamaesaracha coronopus*** (Dun.) Gray Green False Nightshade

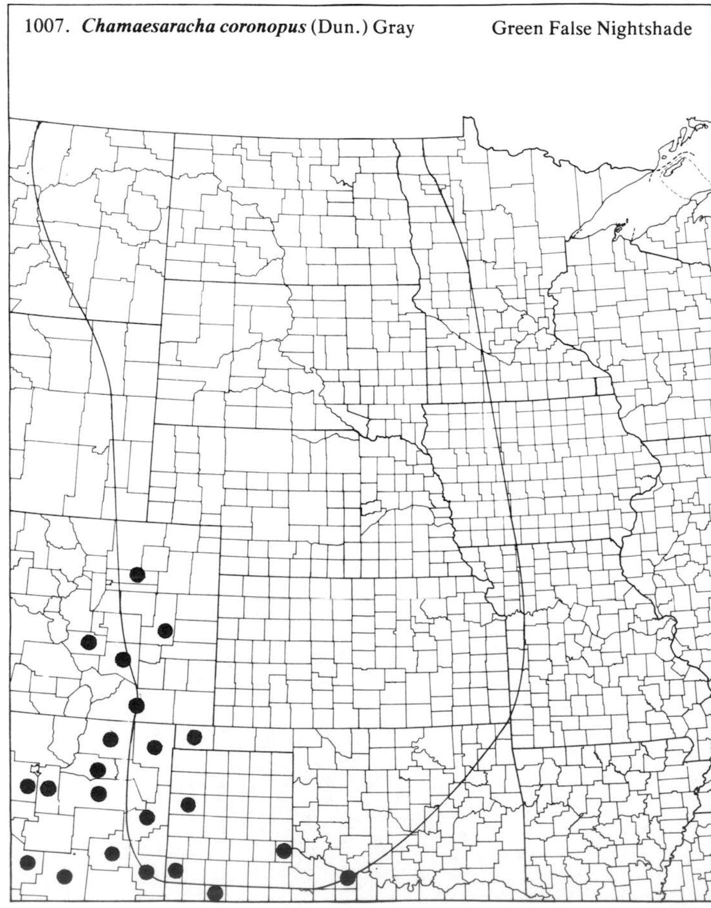

1008. ***Datura stramonium*** L. Jimsonweed

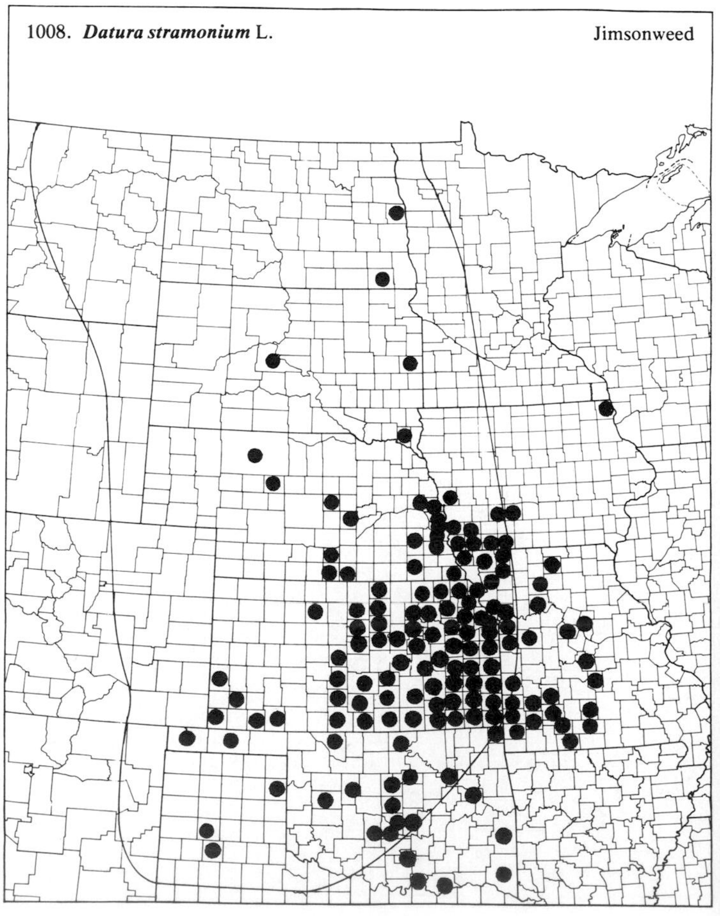

(Solanaceae)

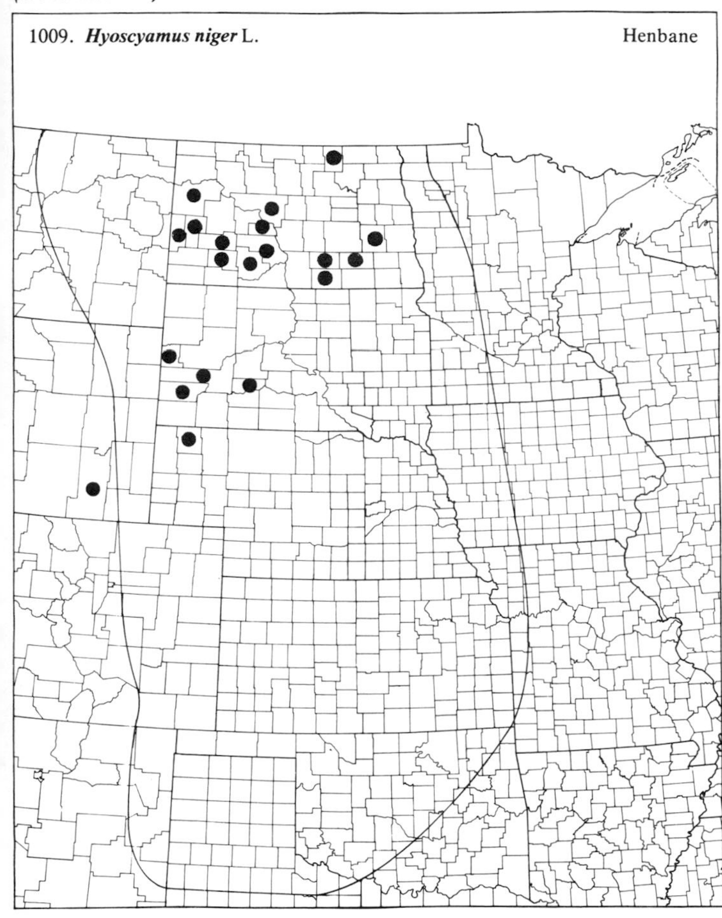
1009. Hyoscyamus niger L.
Henbane

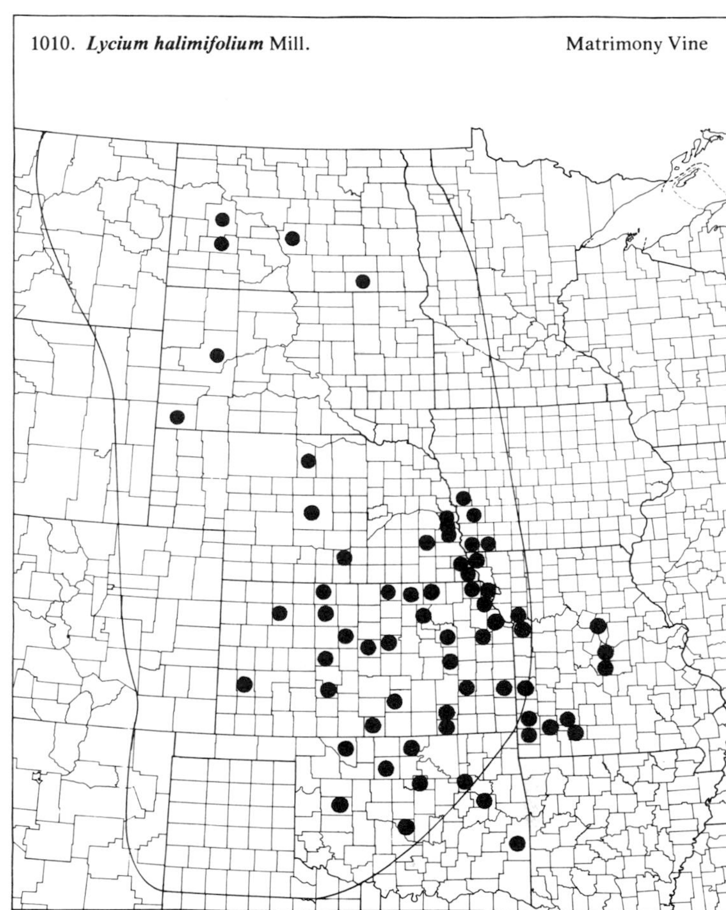
1010. Lycium halimifolium Mill.
Matrimony Vine

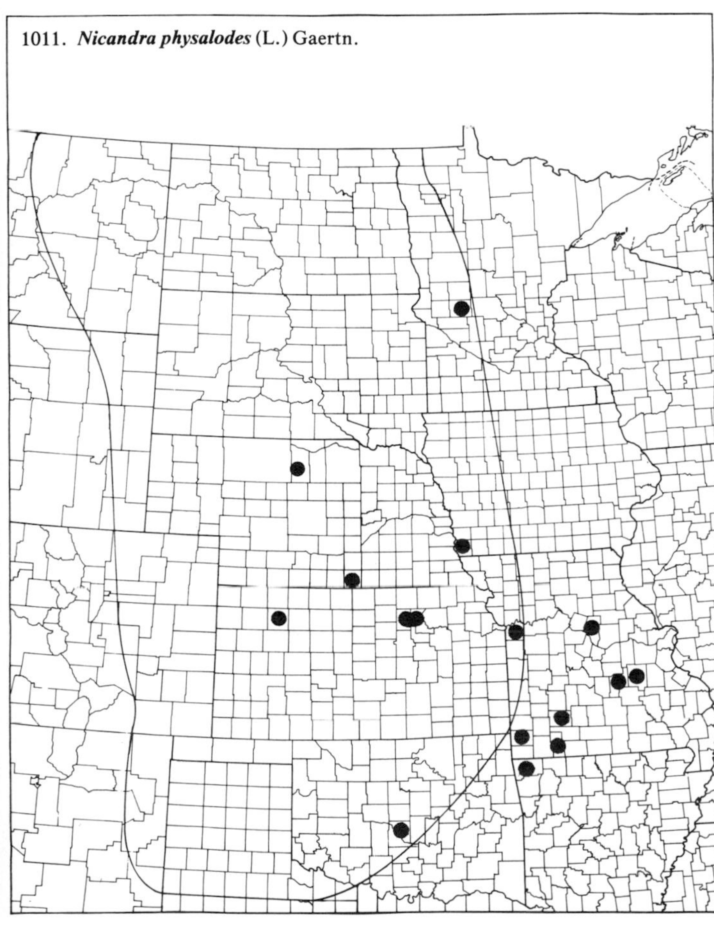
1011. Nicandra physalodes (L.) Gaertn.

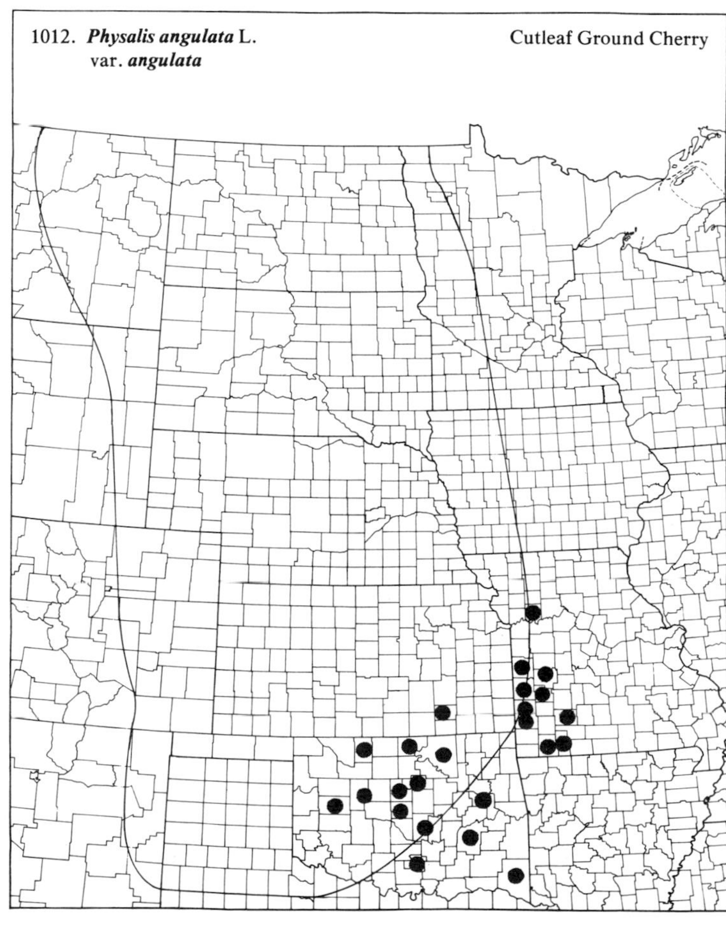
1012. Physalis angulata L.
var. angulata
Cutleaf Ground Cherry

(Solanaceae)

1013. ***Physalis angulata*** L. var. ***pendula*** (Rydb.) Waterfall — Cutleaf Ground Cherry

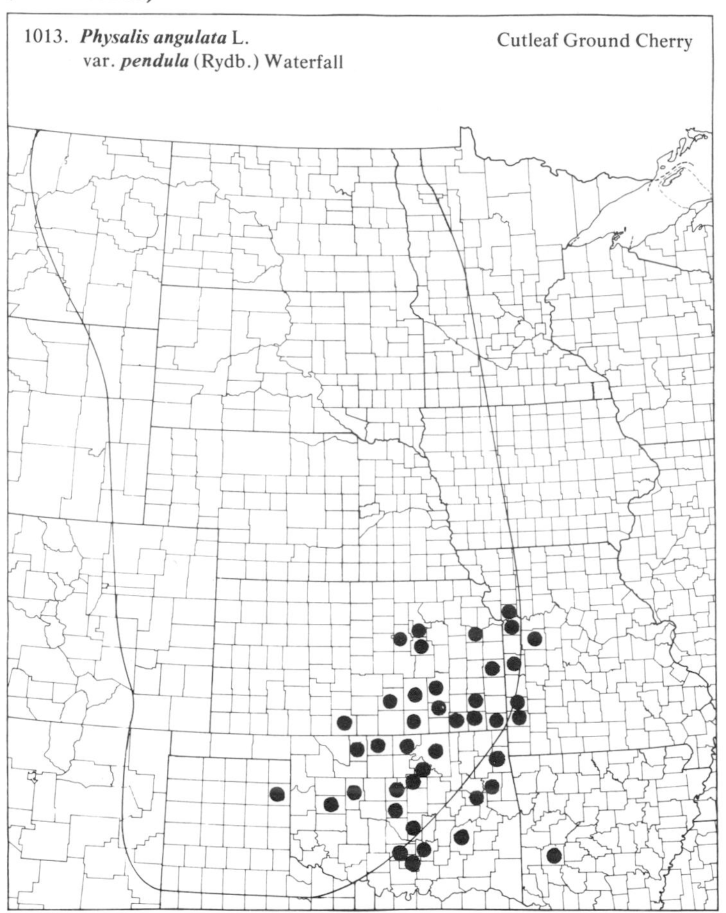

1014. ***Physalis hederaefolia*** Gray var. ***comata*** (Rydb.) Waterfall

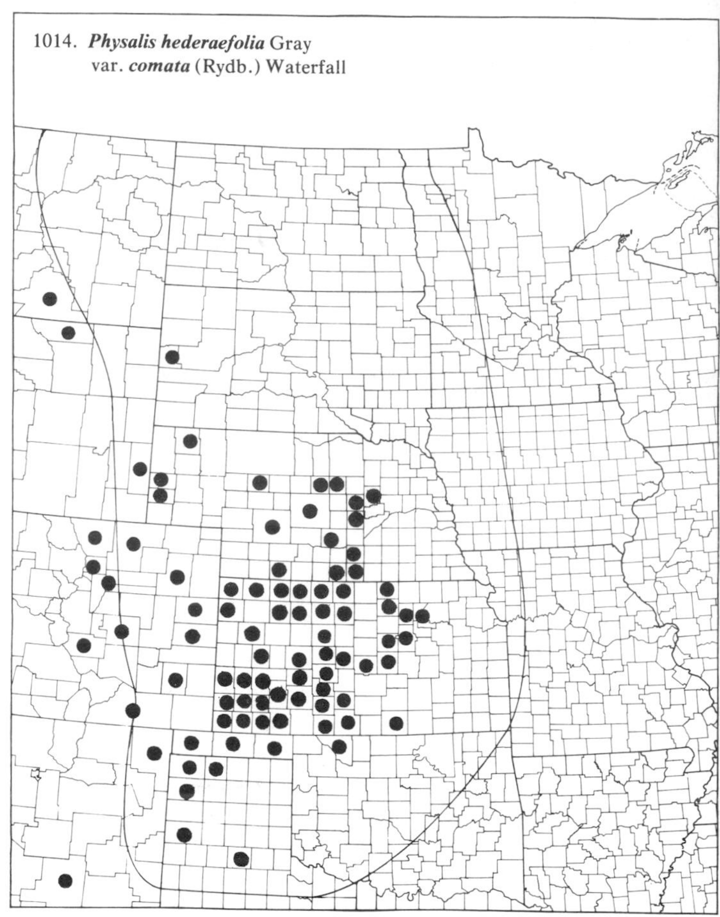

1015. ***Physalis heterophylla*** Nees — Clammy Ground Cherry

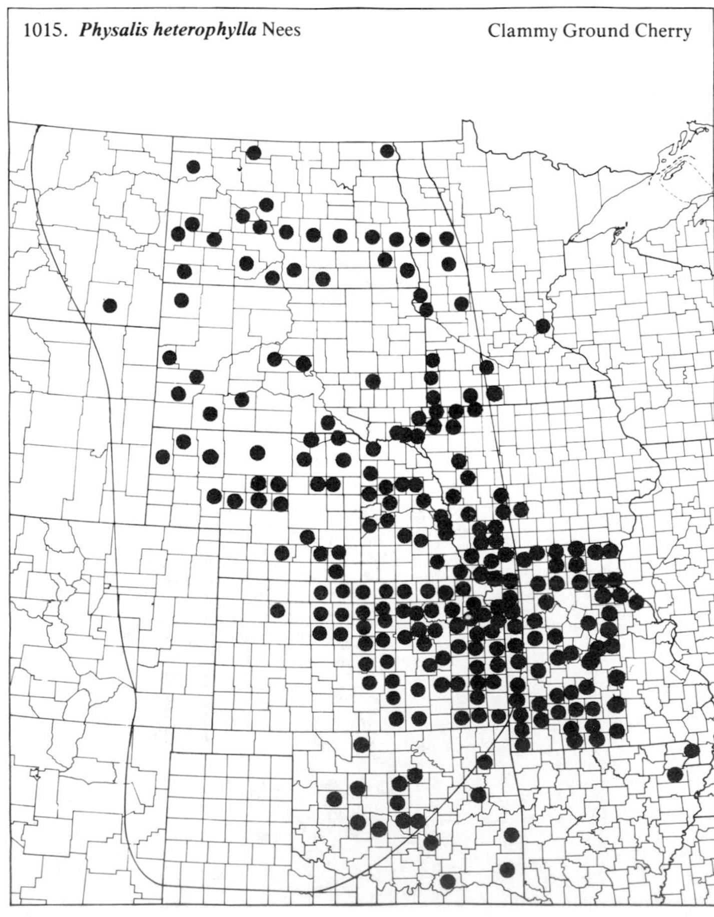

1016. ***Physalis lobata*** Torr. — Chinese Lantern

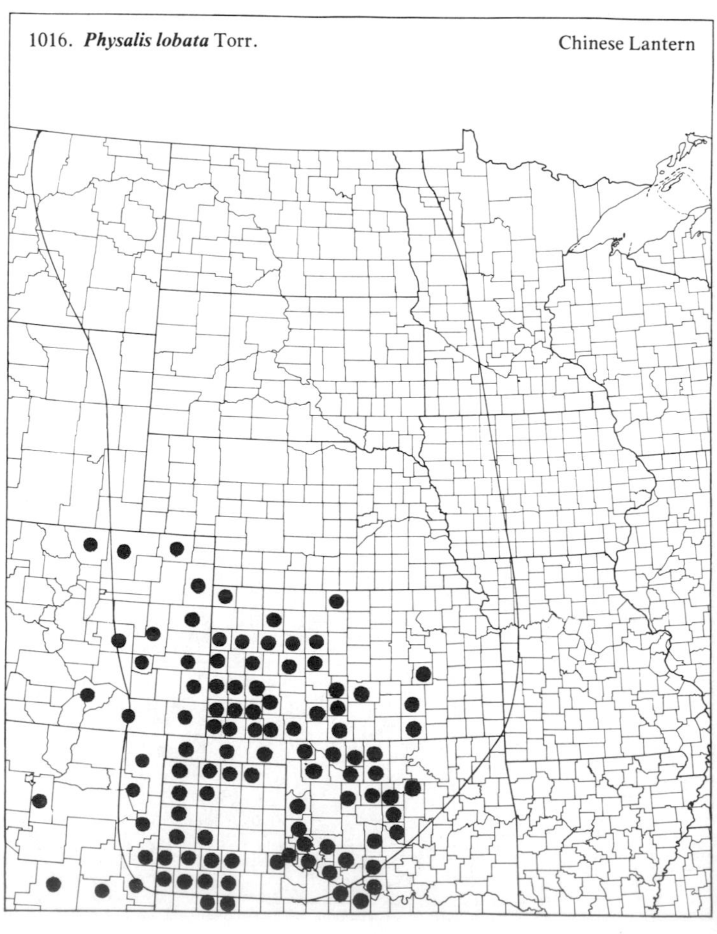

(Solanaceae)

1017. ***Physalis pubescens*** L. var. ***integrifolia*** (Dun.) Waterfall — Downy Ground Cherry

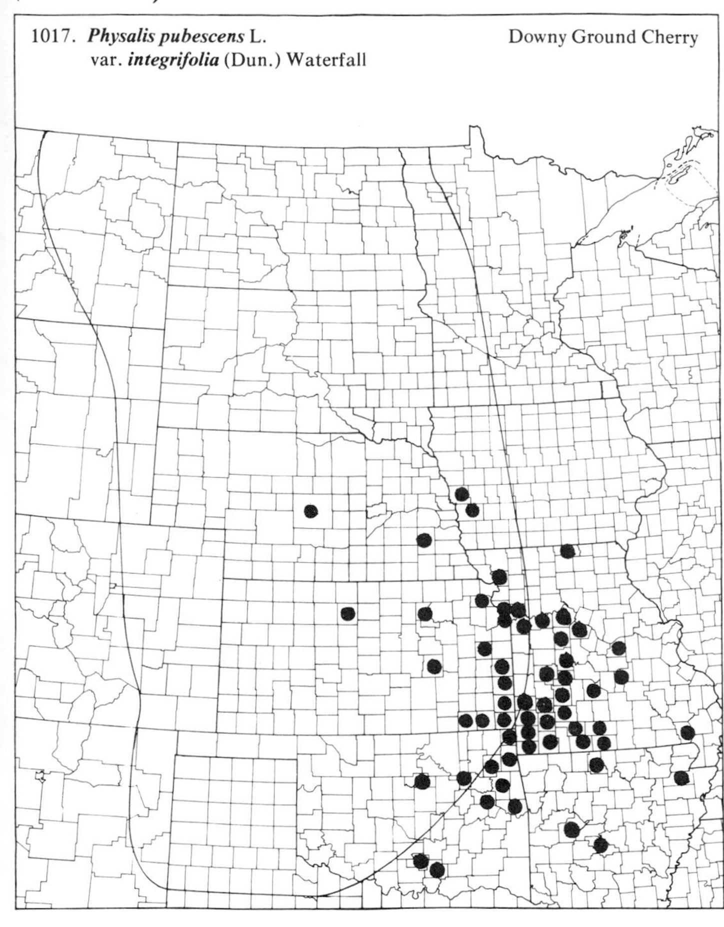

1018. ***Physalis pubescens*** L. var. ***missouriensis*** (Mack. & Bush) Waterfall — Downy Ground Cherry

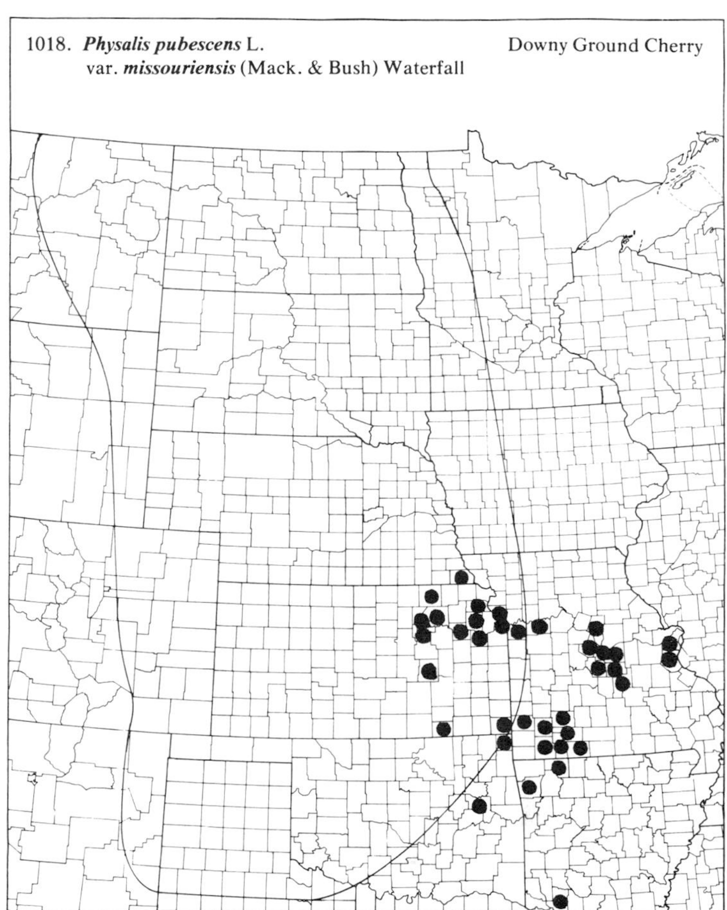

1019. ***Physalis pumila*** Nutt. — Prairie Ground Cherry

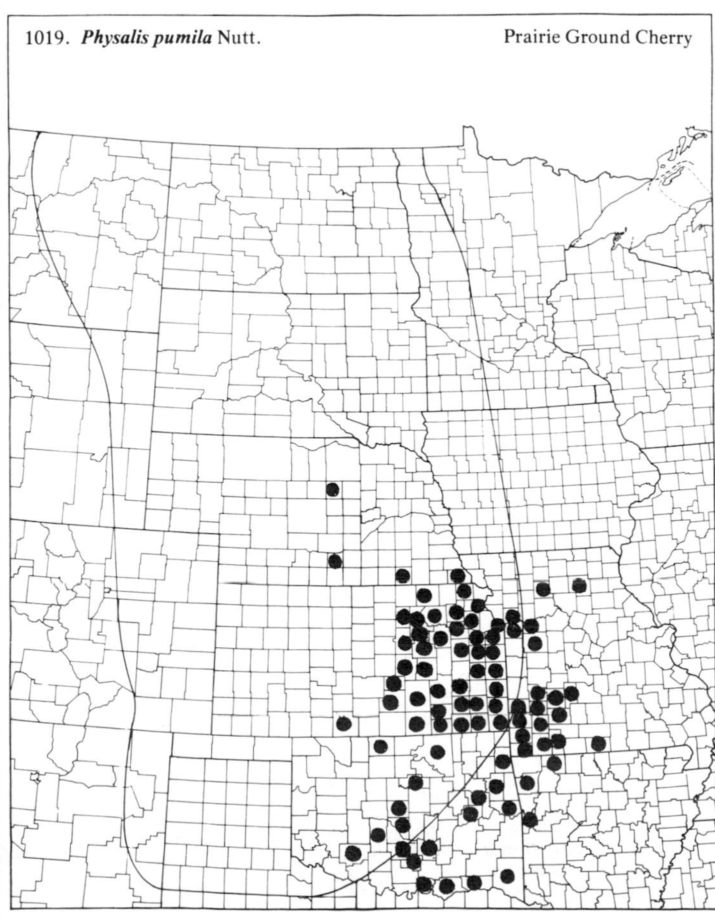

1020. ***Physalis virginiana*** Mill. var. ***virginiana*** — Ground Cherry

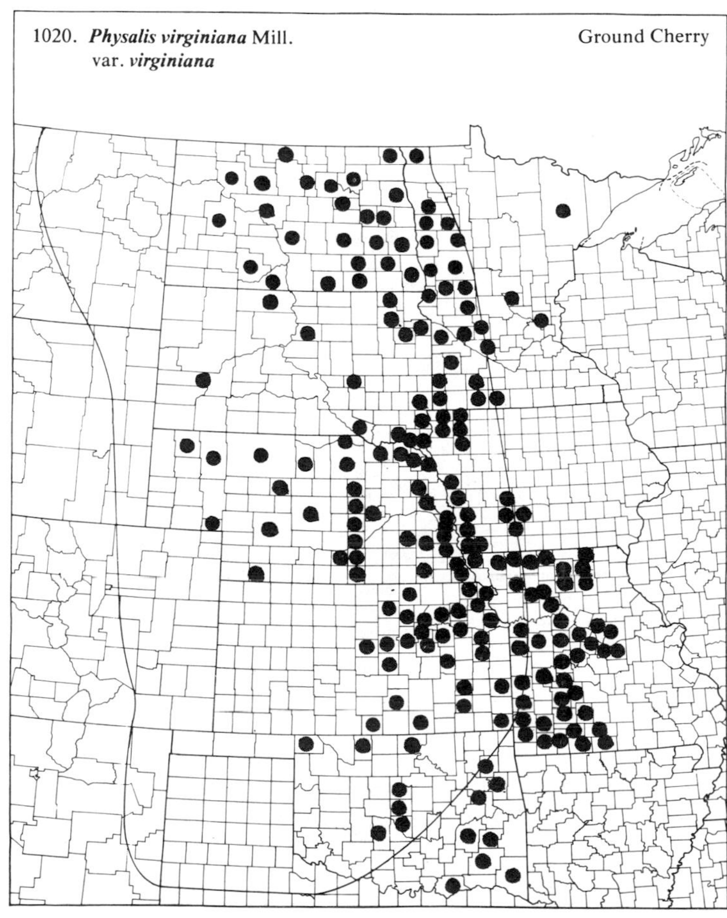

(Solanaceae)

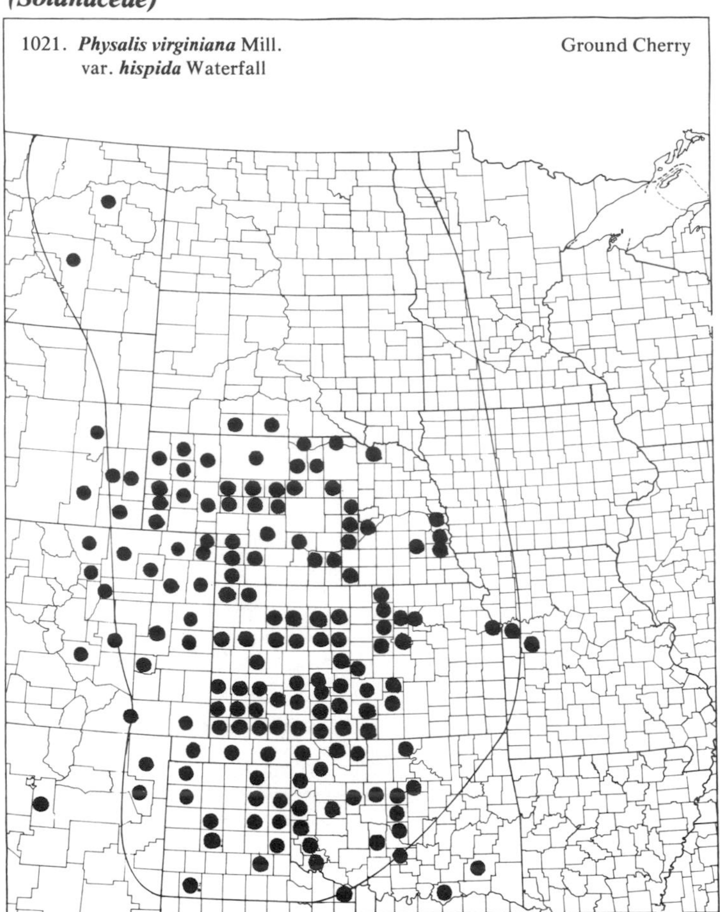

1021. ***Physalis virginiana*** Mill. var. ***hispida*** Waterfall — Ground Cherry

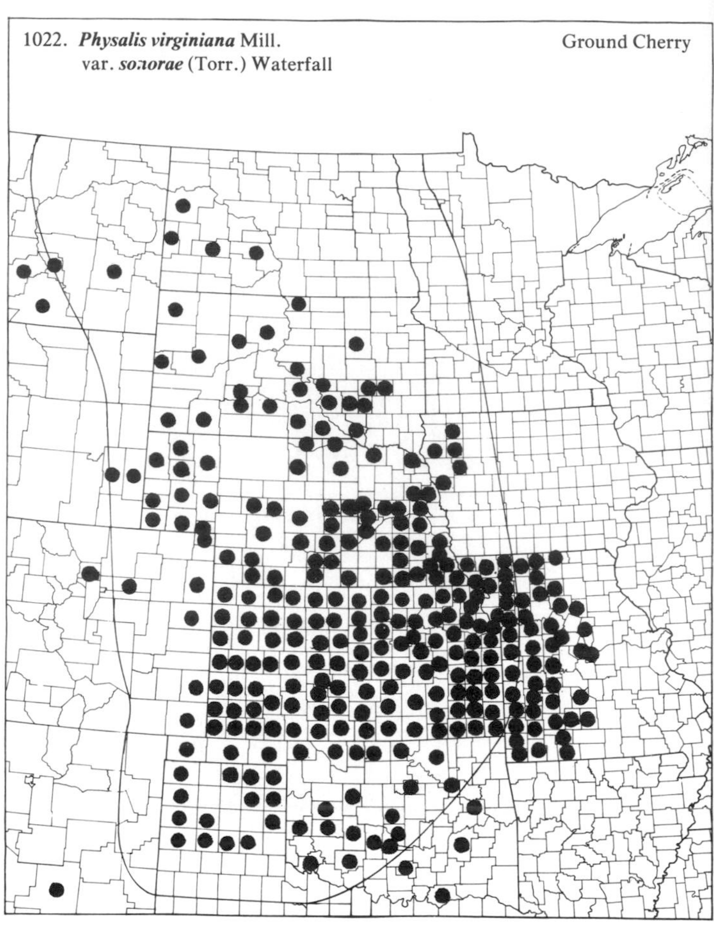

1022. ***Physalis virginiana*** Mill. var. ***sonorae*** (Torr.) Waterfall — Ground Cherry

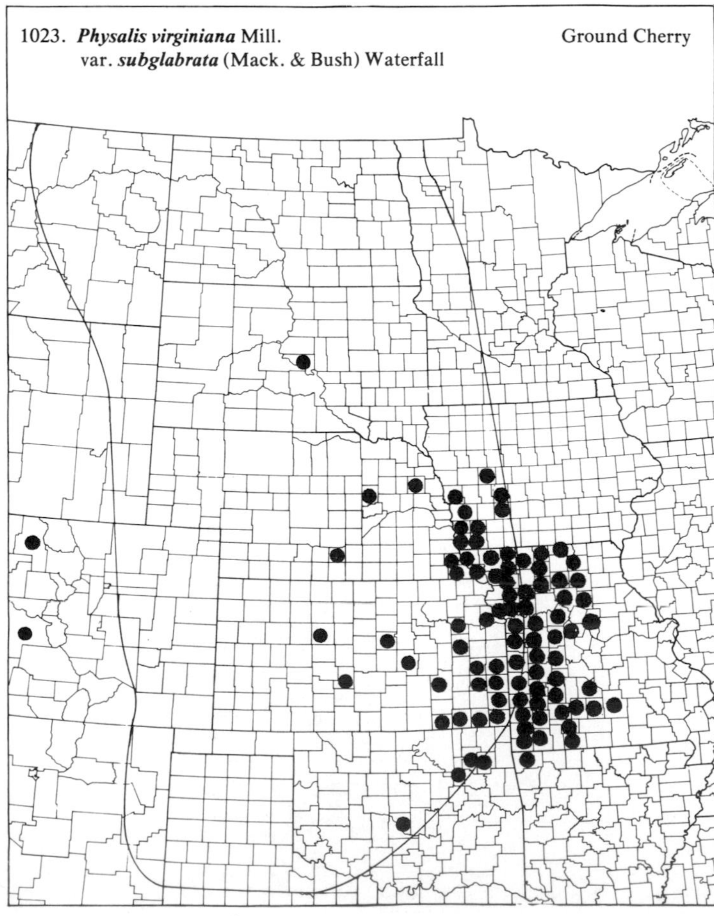

1023. ***Physalis virginiana*** Mill. var. ***subglabrata*** (Mack. & Bush) Waterfall — Ground Cherry

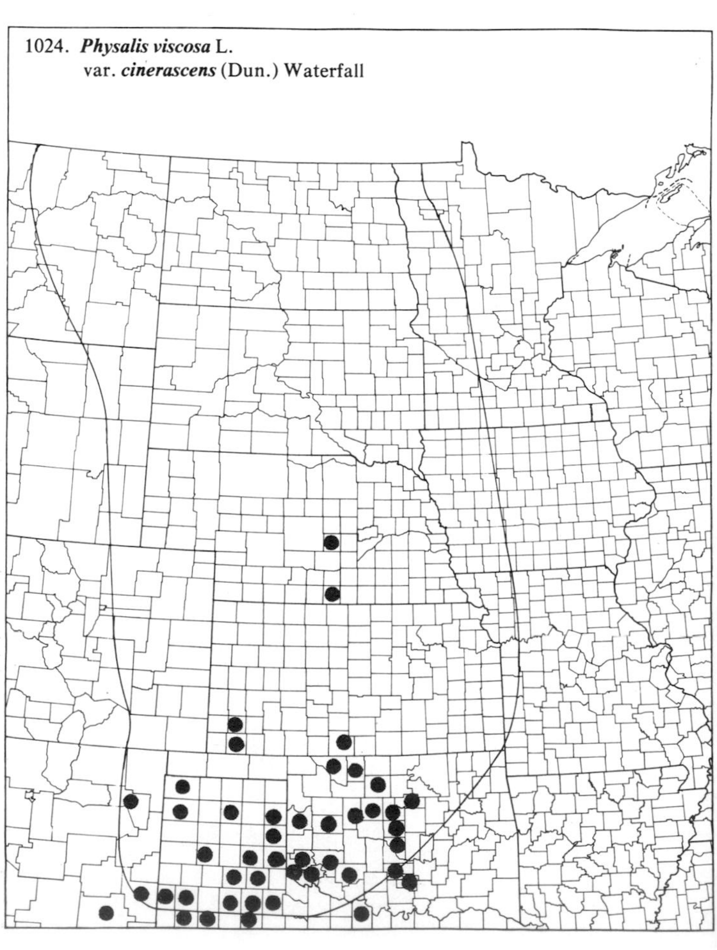

1024. ***Physalis viscosa*** L. var. ***cinerascens*** (Dun.) Waterfall

(Solanaceae)

1025. **Solanum americanum** Mill. Black Nightshade

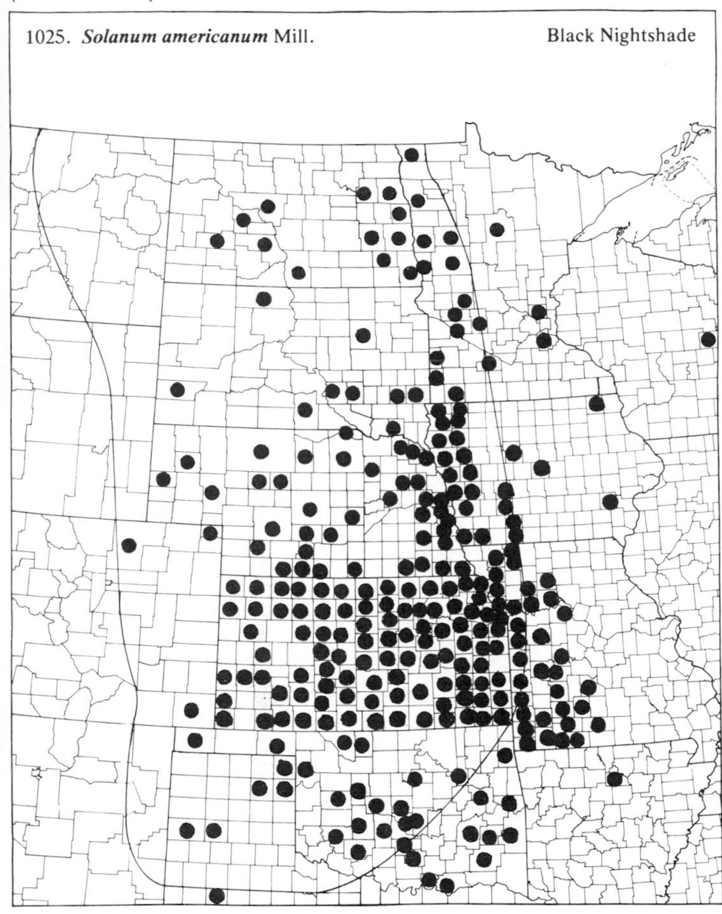

1026. **Solanum carolinense** L. Horse Nettle

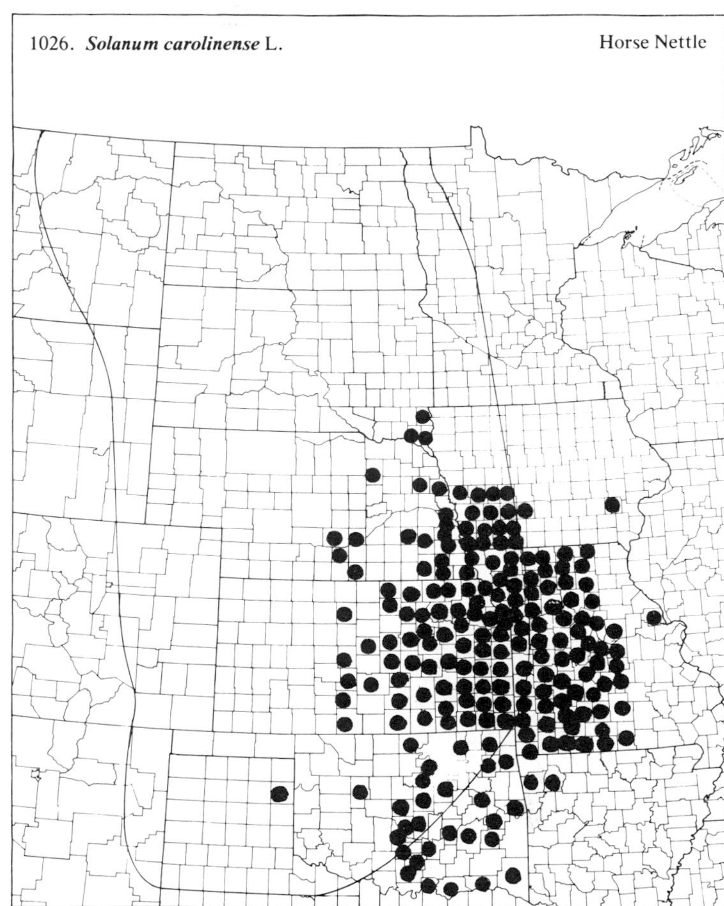

1027. **Solanum dimidiatum** Raf.

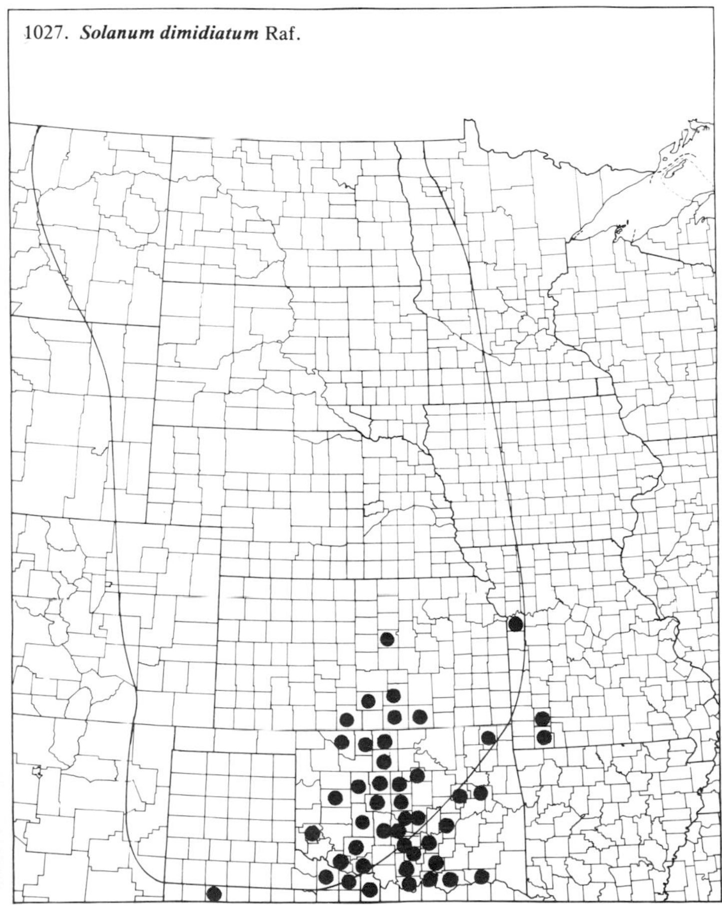

1028. **Solanum dulcamara** L. Bittersweet

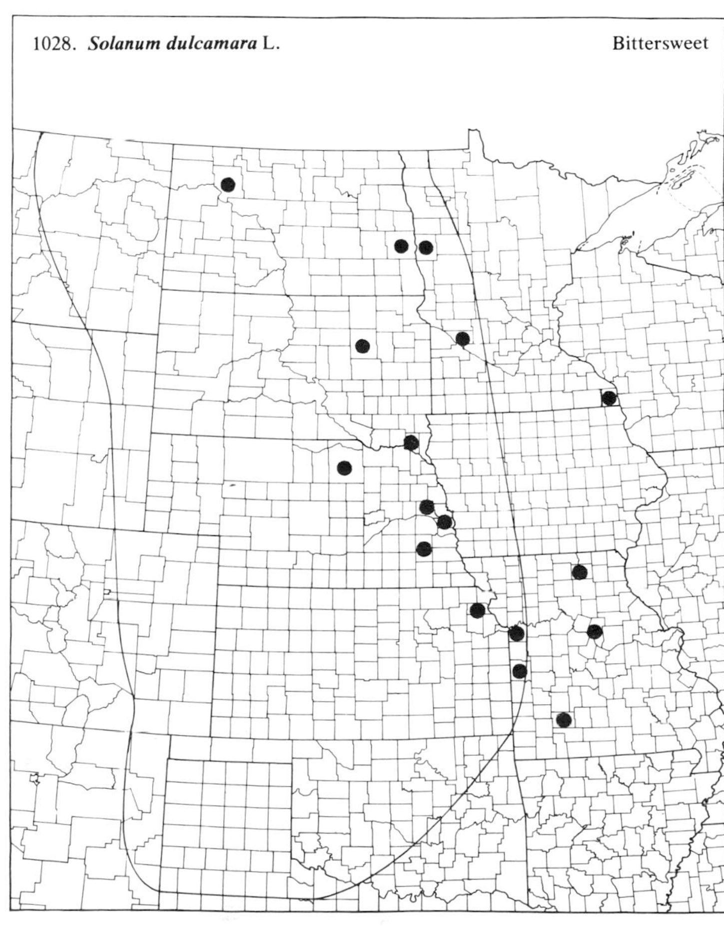

(Solanaceae)

1029. ***Solanum elaeagnifolium*** Cav. Silverleaf Nightshade

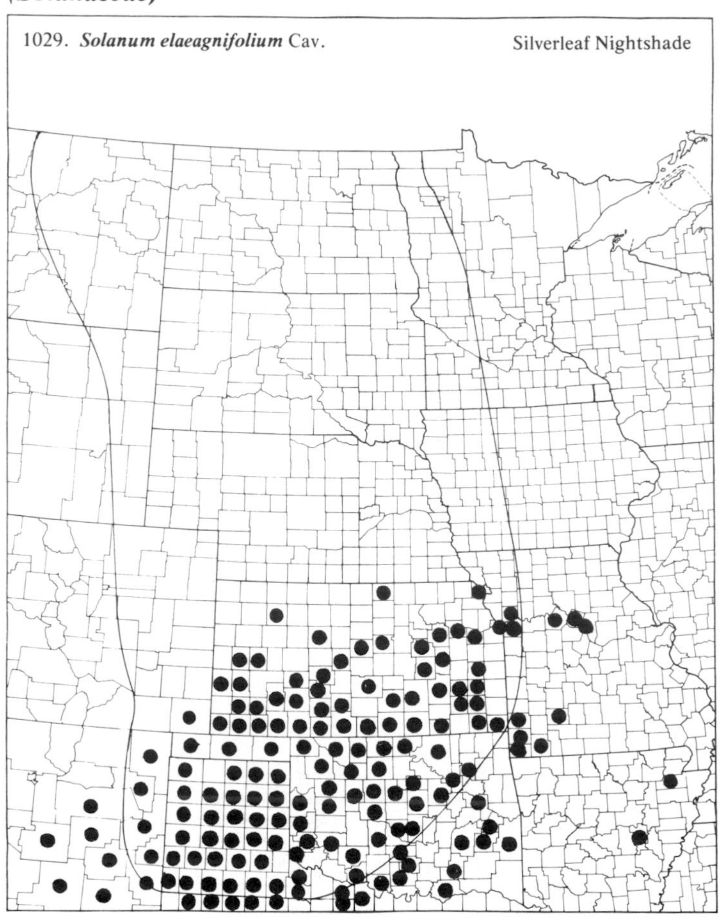

1030. ***Solanum rostratum*** Dun. Buffalo Bur

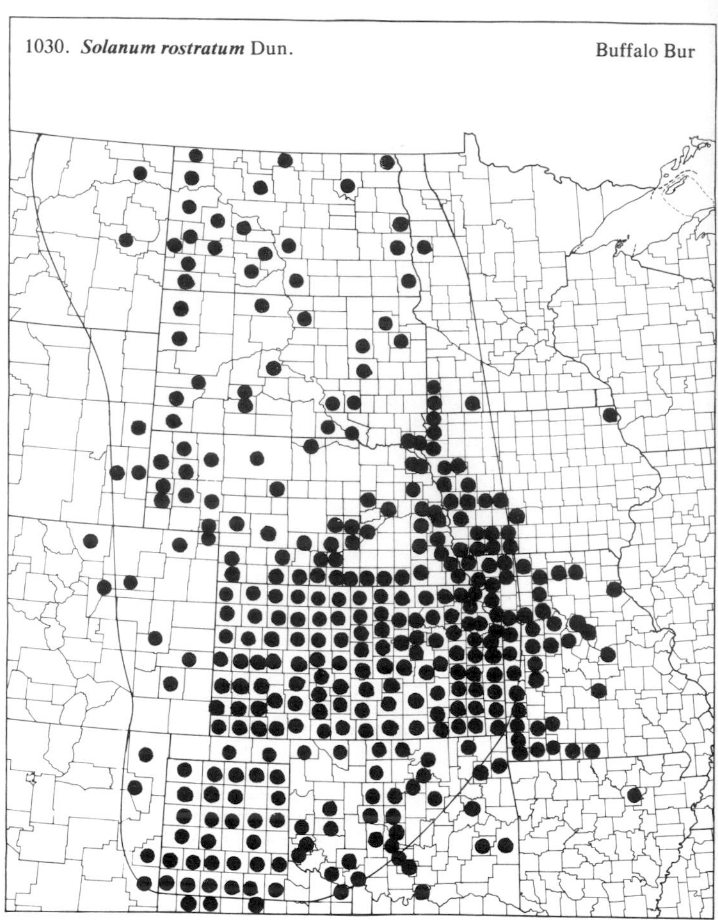

1031. ***Solanum triflorum*** Nutt. Cut-leaved Nightshade

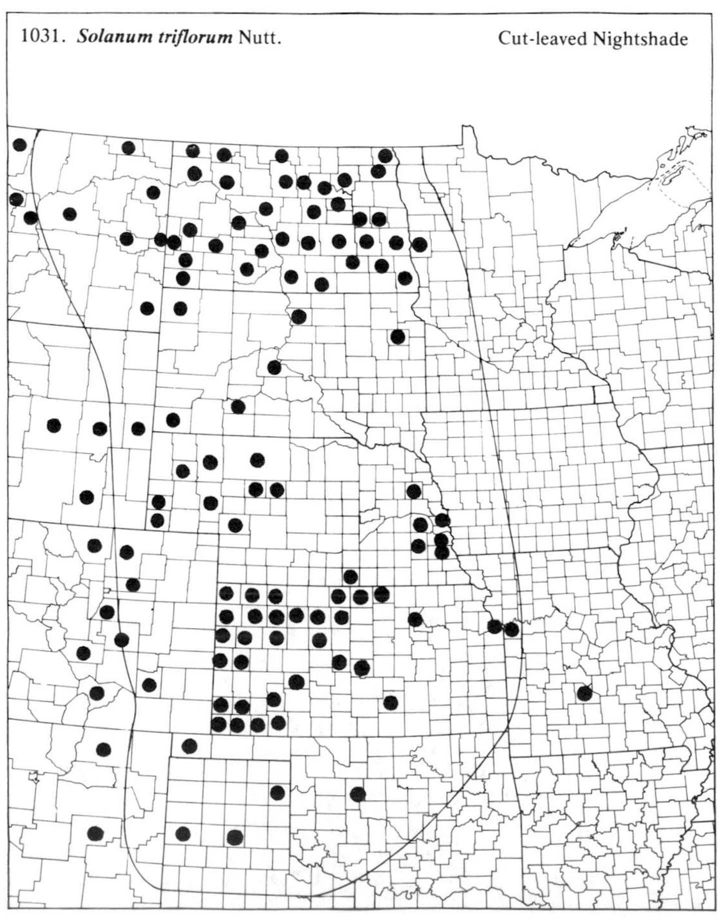

1032. ***Solanum villosum*** Mill. Hairy Nightshade

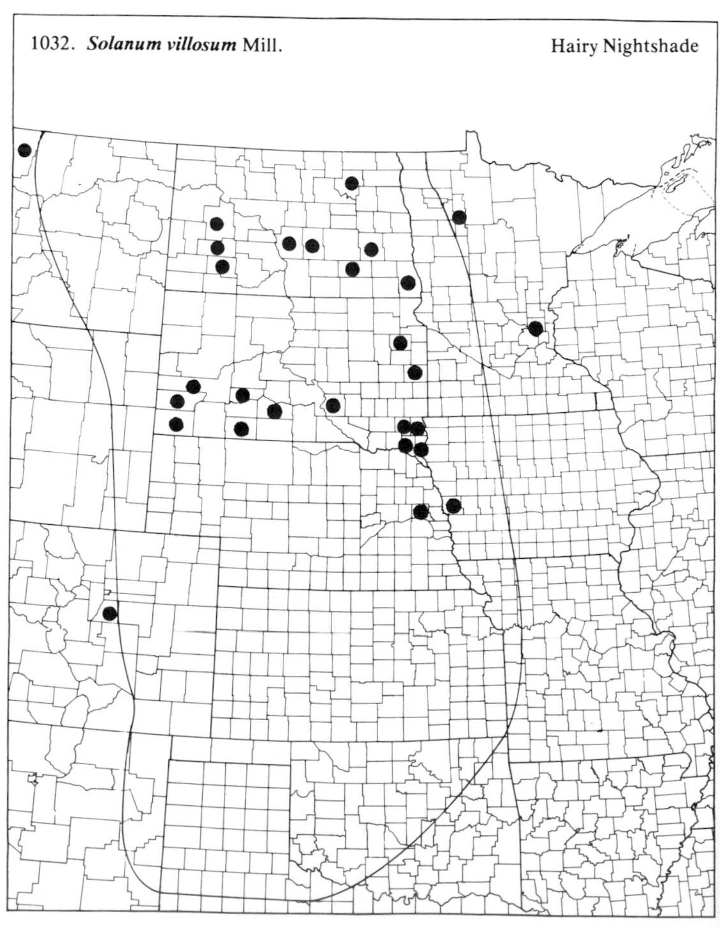

90. *Convolvulaceae* (Convolvulus Family)

1033. ***Convolvulus arvensis*** L. — Field Bindweed

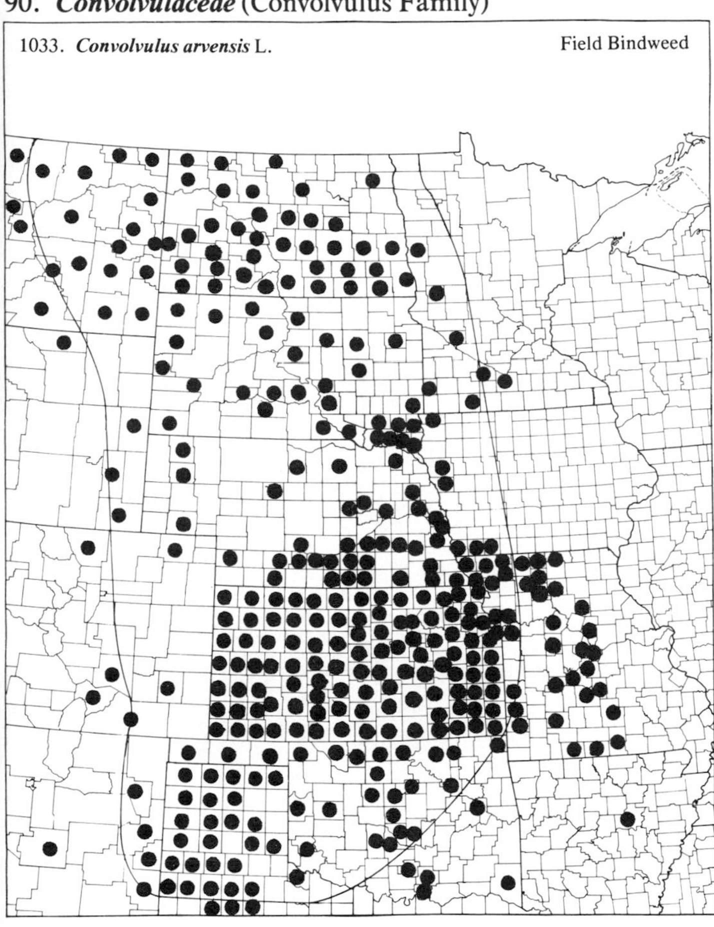

1034. ***Convolvulus equitans*** Benth.

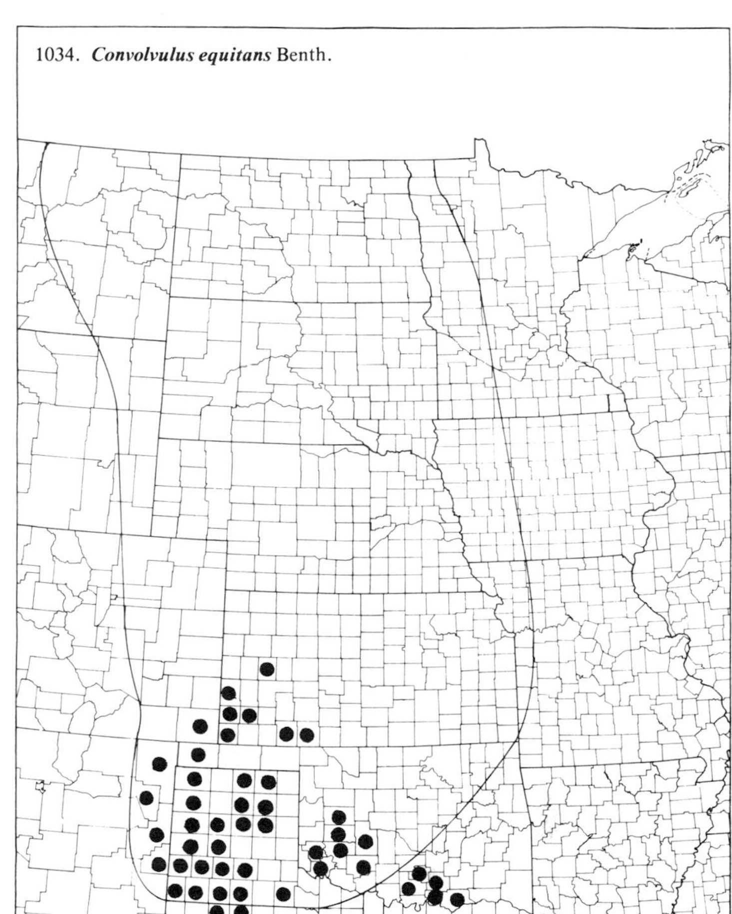

1035. ***Convolvulus sepium*** L. — Hedge Bindweed

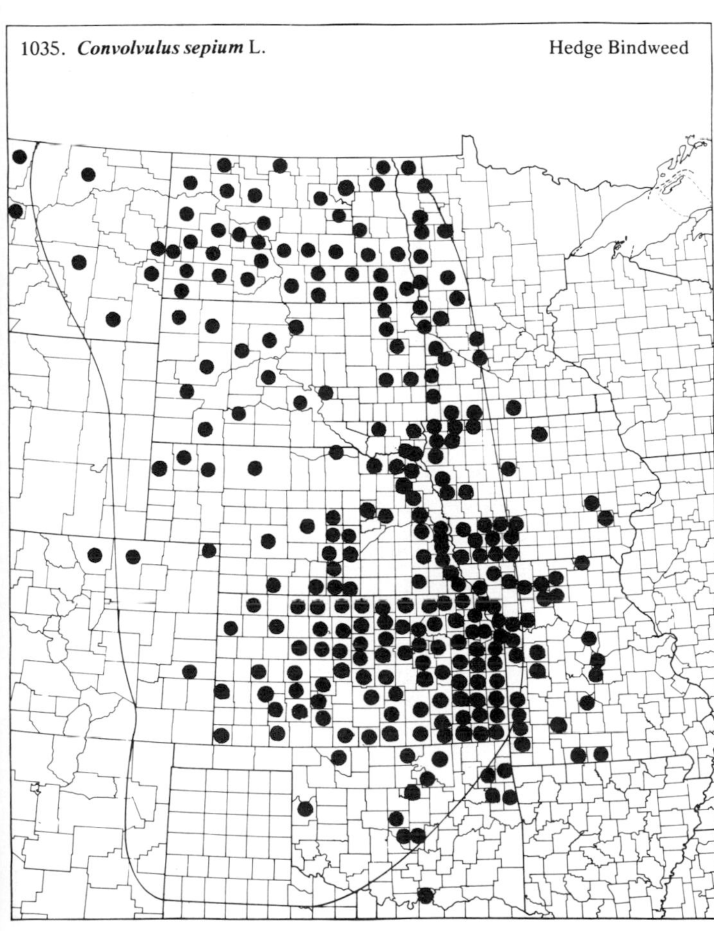

1036. ***Evolvulus nuttallianus*** R. & S. — Nuttall Evolvulus

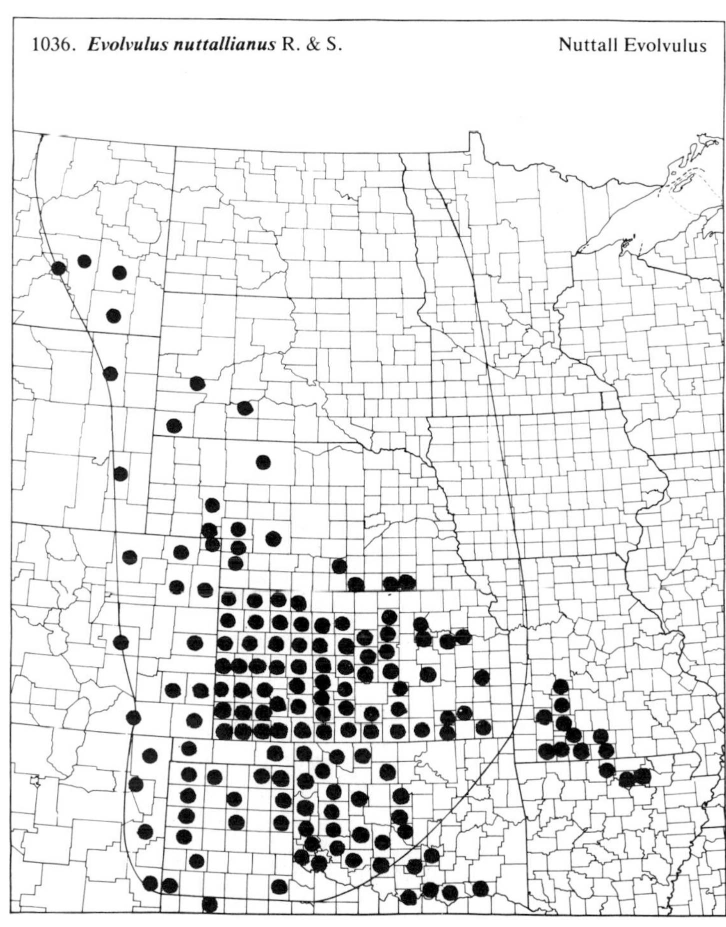

(Convolvulaceae)

1037. ***Ipomoea coccinea*** L. Red Morning-glory

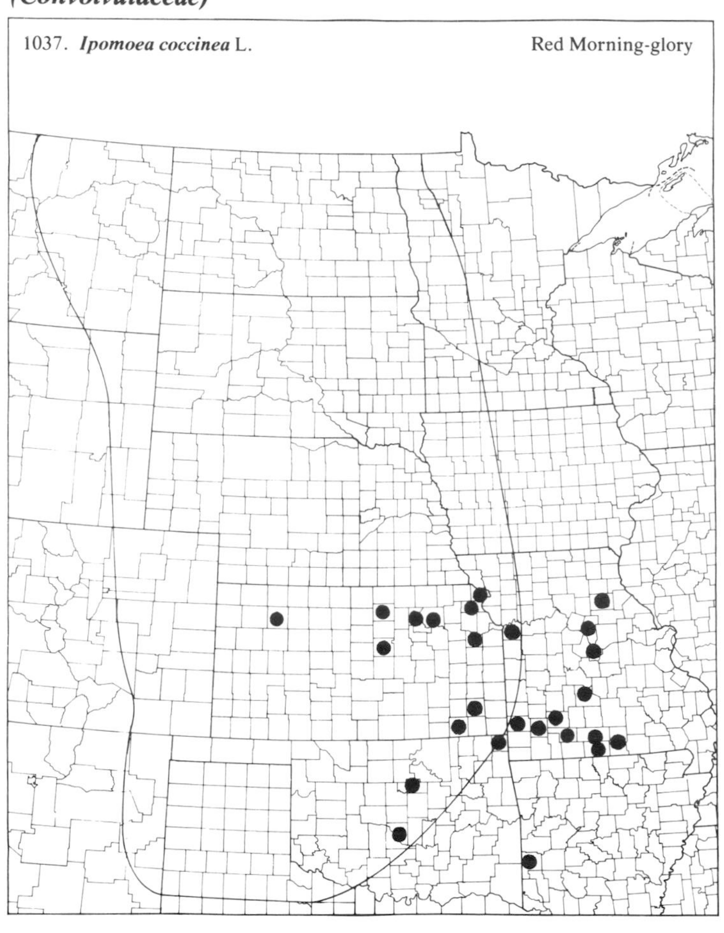

1038. ***Ipomoea hederacea*** (L.) Jacq. Ivyleaf Morning-glory

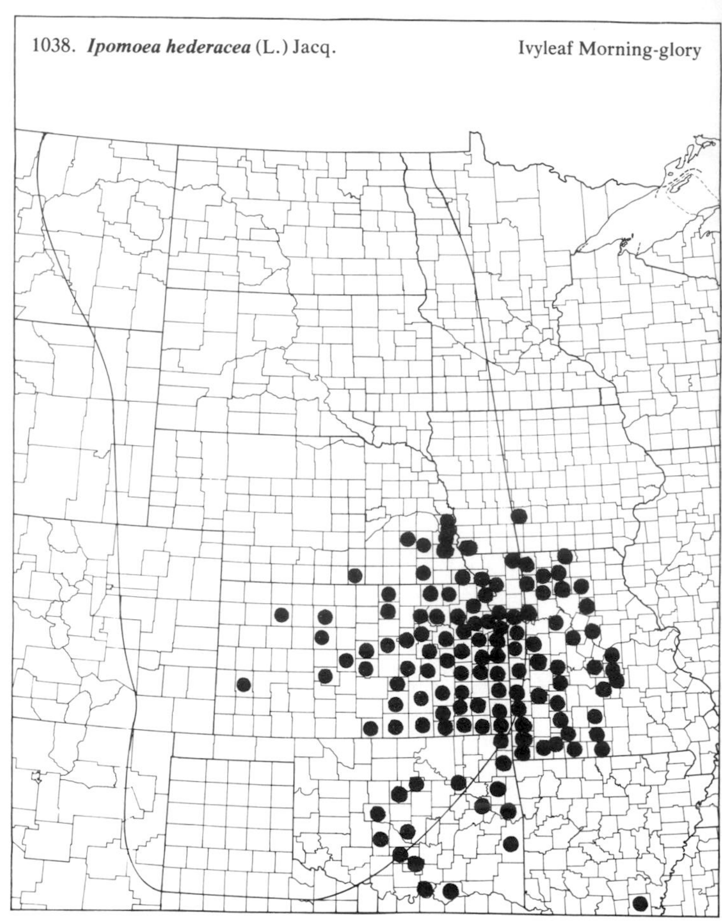

1039. ***Ipomoea lacunosa*** L. White Morning-glory

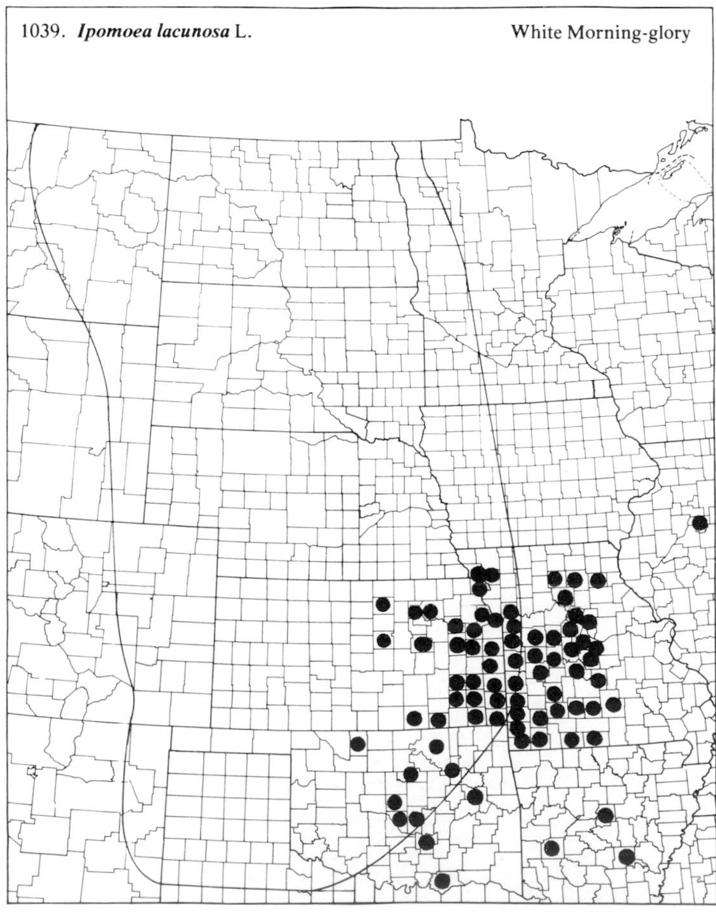

1040. ***Ipomoea leptophylla*** Torr. Bush Morning-glory

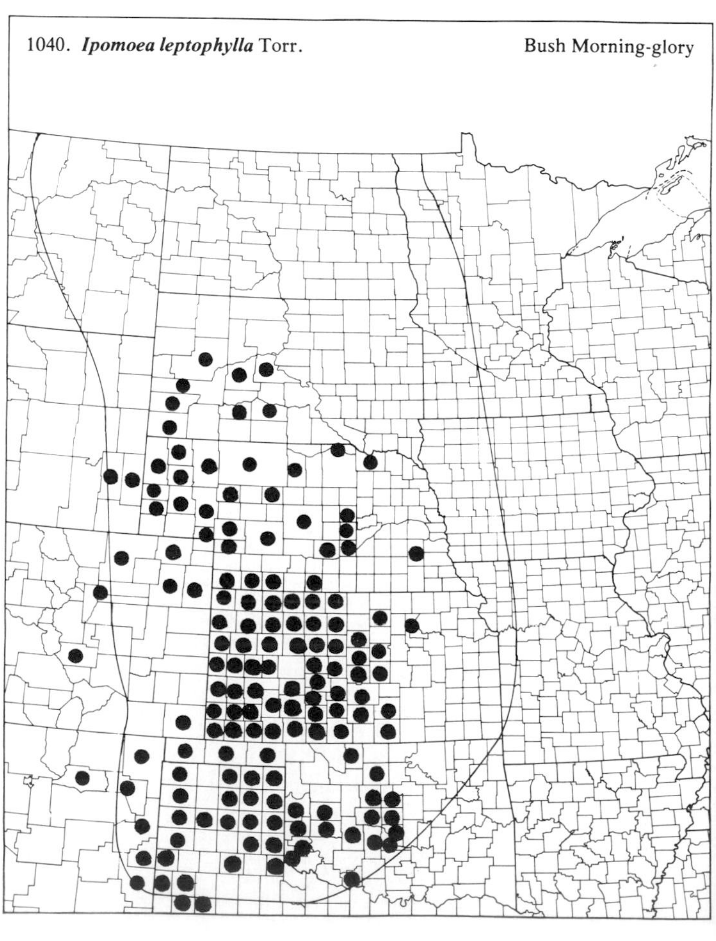

(Convolvulaceae)

1041. ***Ipomoea pandurata*** (L.) Mey. — Bigroot Morning-glory

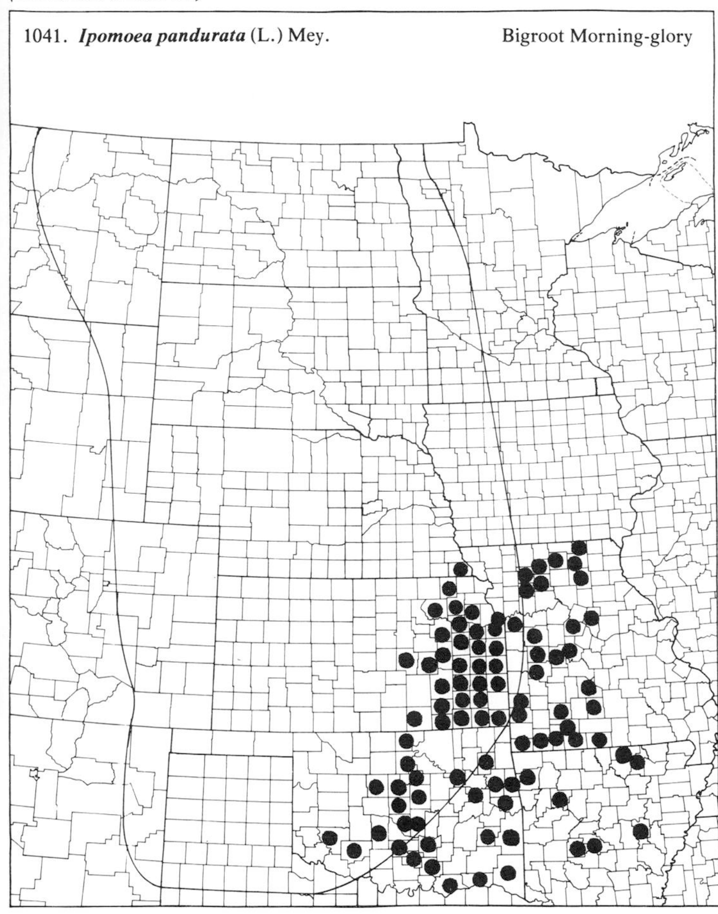

1042. ***Ipomoea purpurea*** (L.) Roth — Common Morning-glory

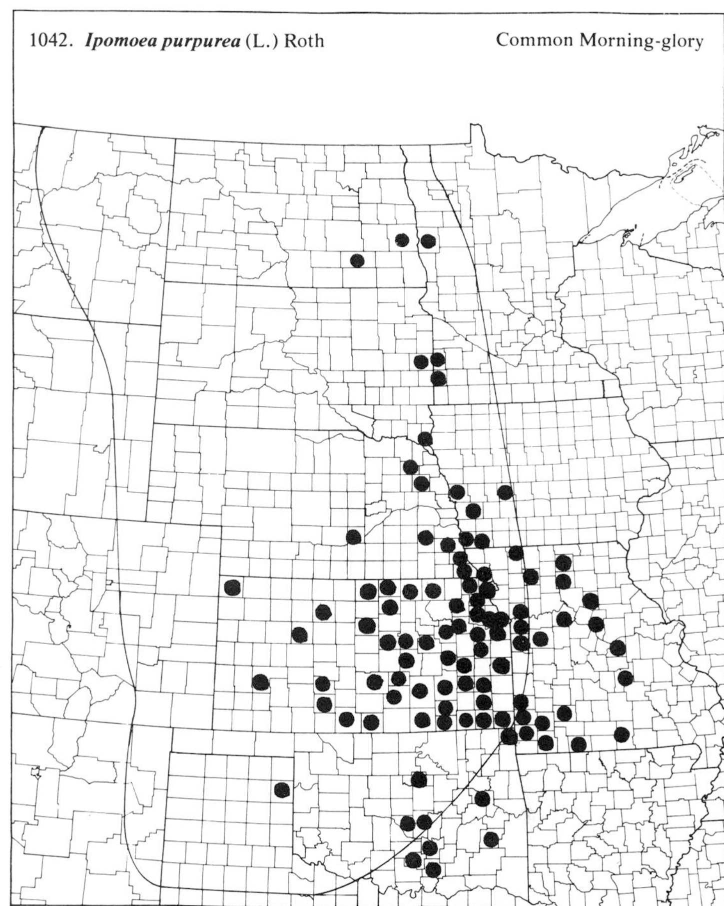

1043. ***Ipomoea shumardiana*** (Torr.) Shinners

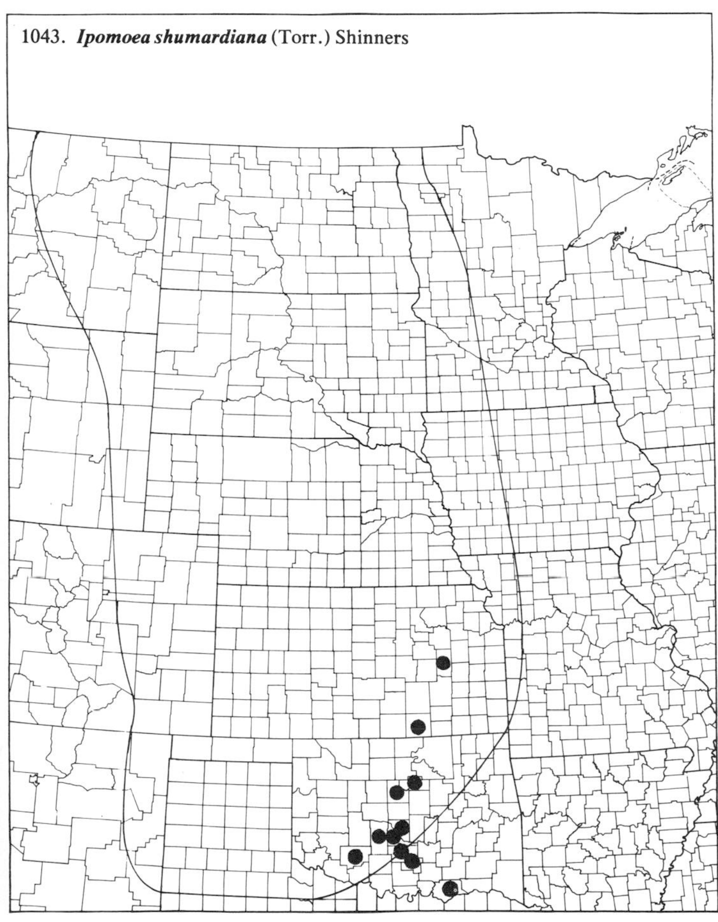

1044. ***Stylisma pickeringii*** (Torr.) Gray var. ***pattersonii*** (Fern. & Schub.) Myint

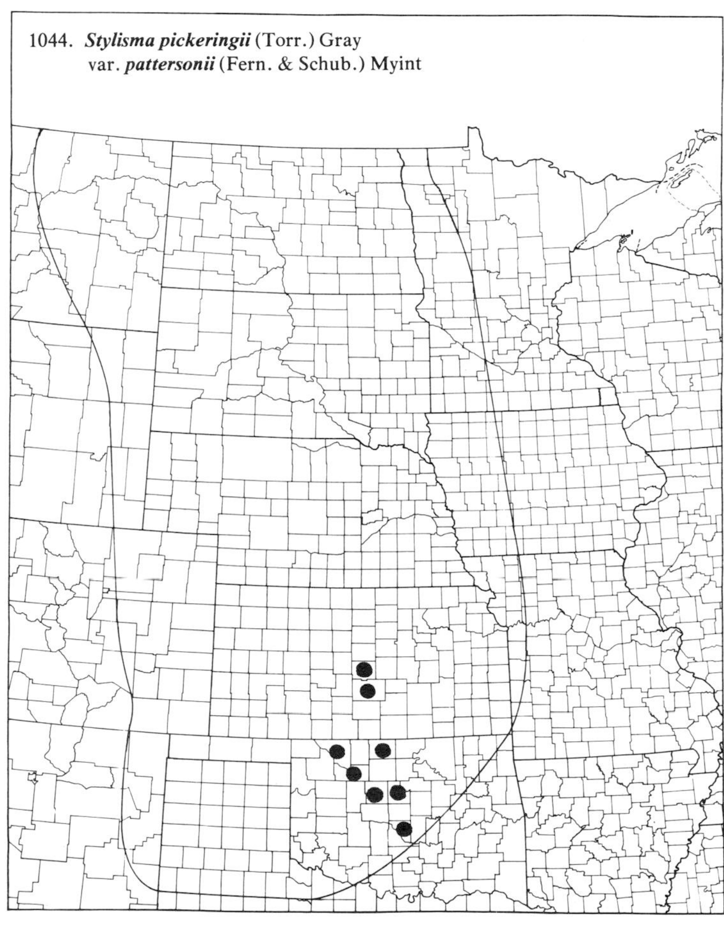

91. *Cuscutaceae* (Dodder Family)

1045. ***Cuscuta cephalanthi*** Engelm. Buttonbush Dodder

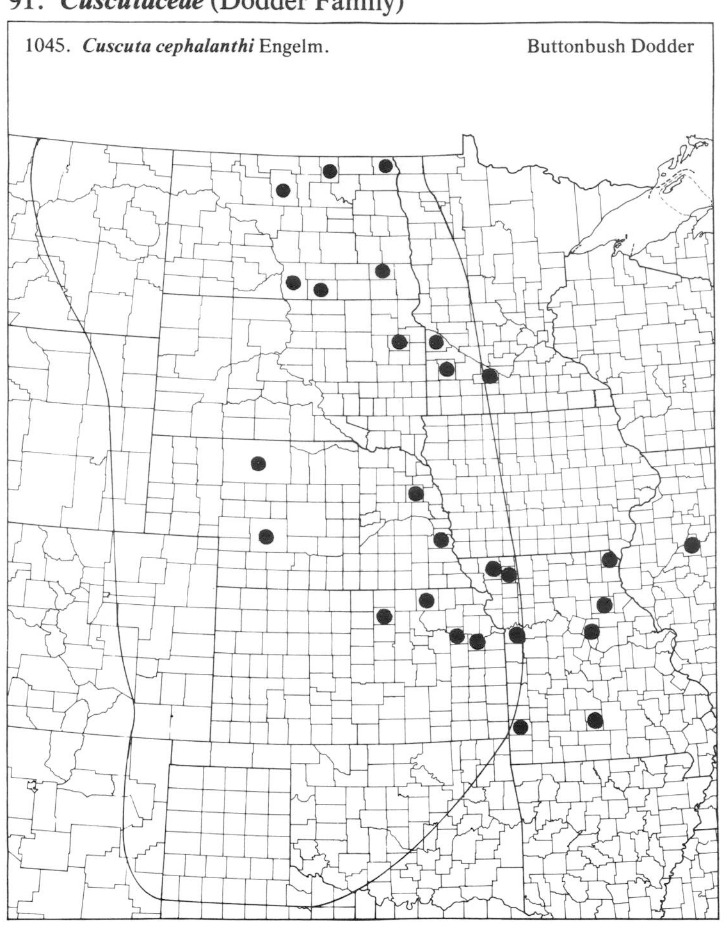

1046. ***Cuscuta compacta*** Juss.

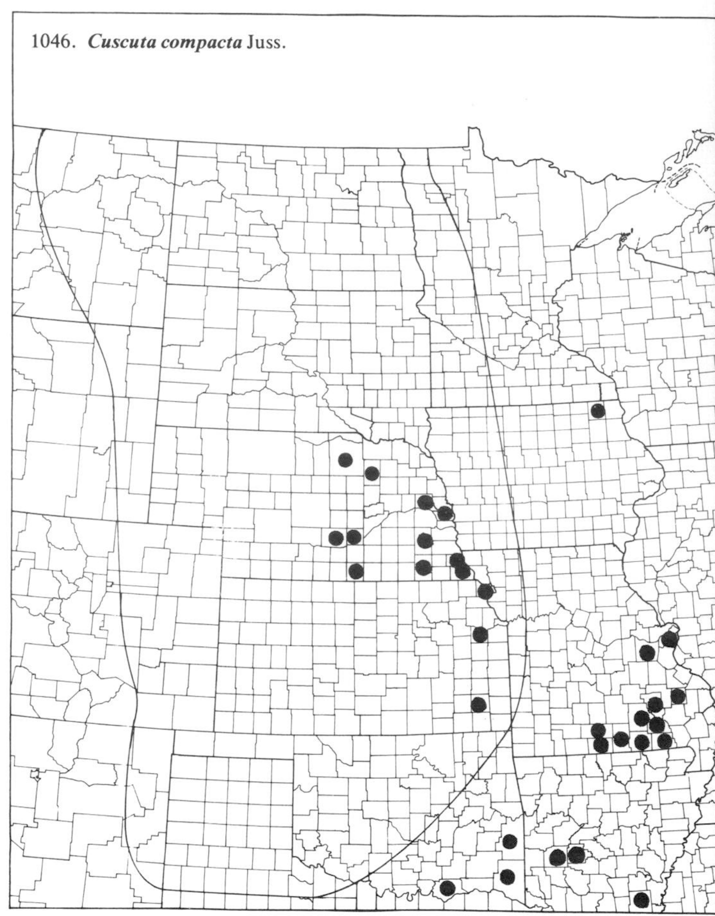

1047. ***Cuscuta coryli*** Engelm. Hazel Dodder

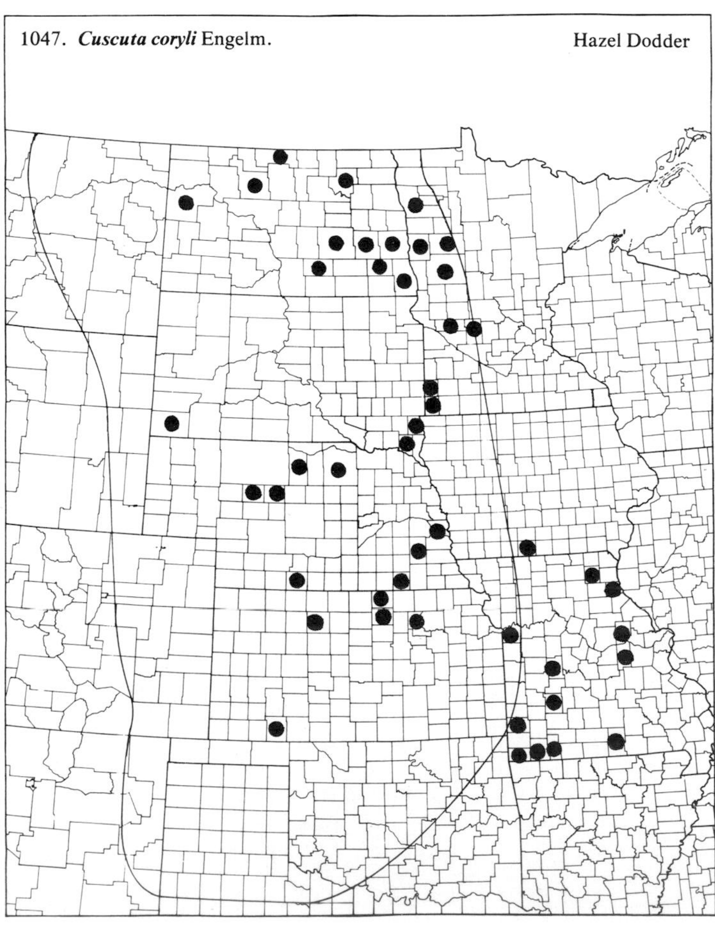

1048. ***Cuscuta cuspidata*** Engelm. Cusp Dodder

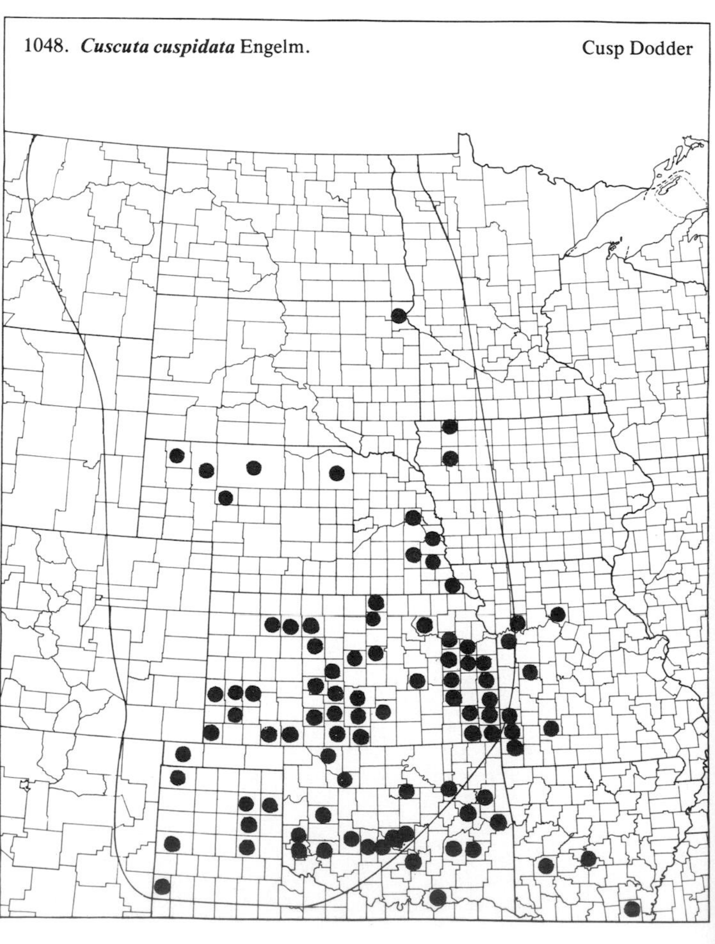

(Cuscutaceae)

1049. ***Cuscuta glomerata*** Choisy — Cluster Dodder

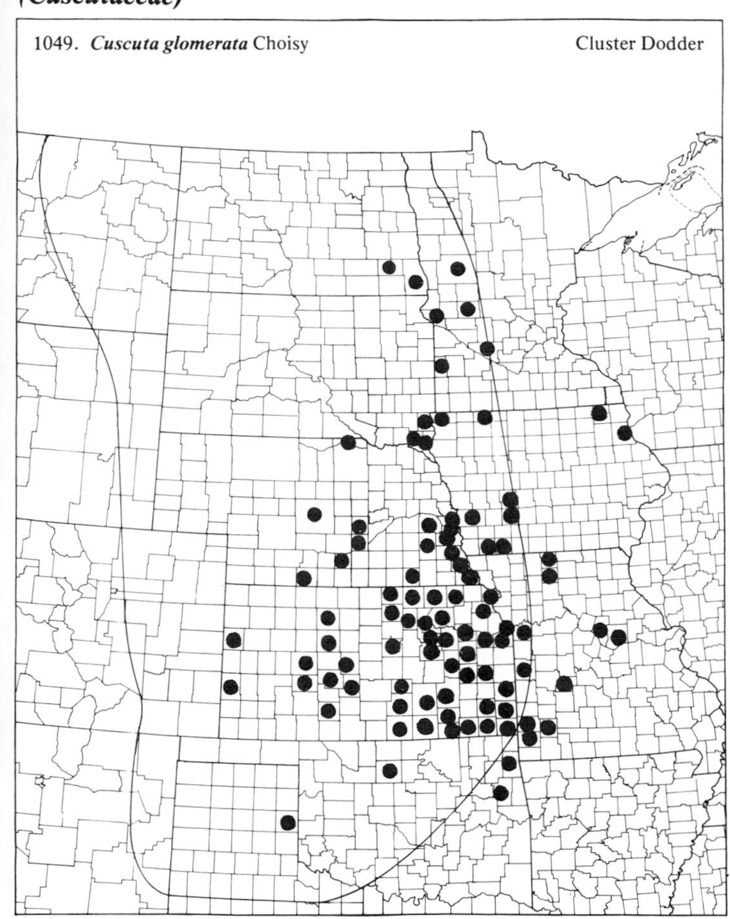

1050. ***Cuscuta gronovii*** Willd. — Gronovius' Dodder

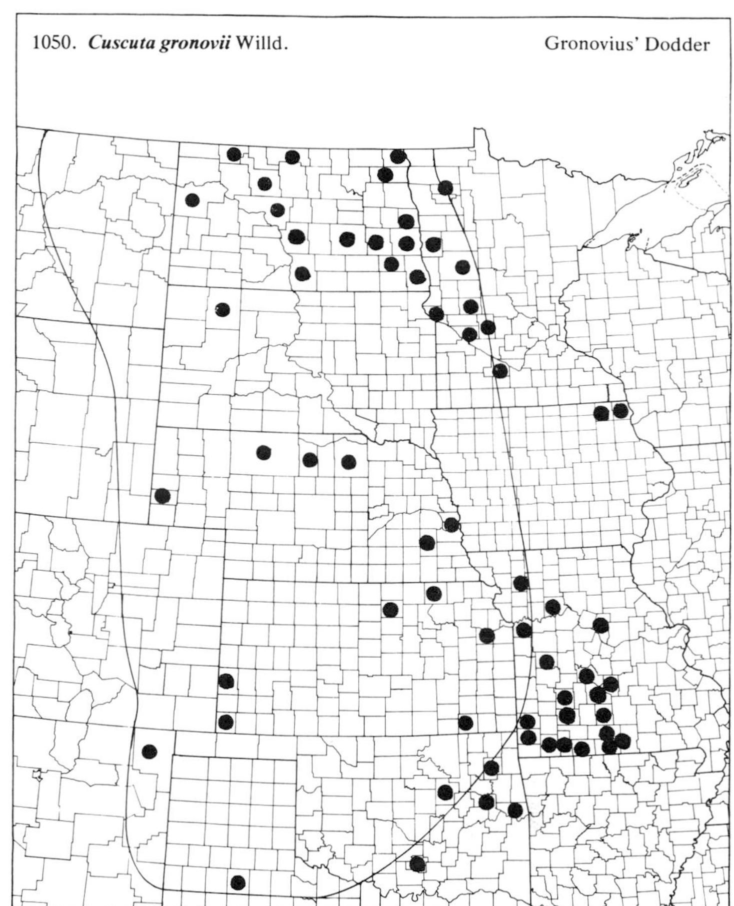

1051. ***Cuscuta indecora*** Choisy — Large Alfalfa Dodder

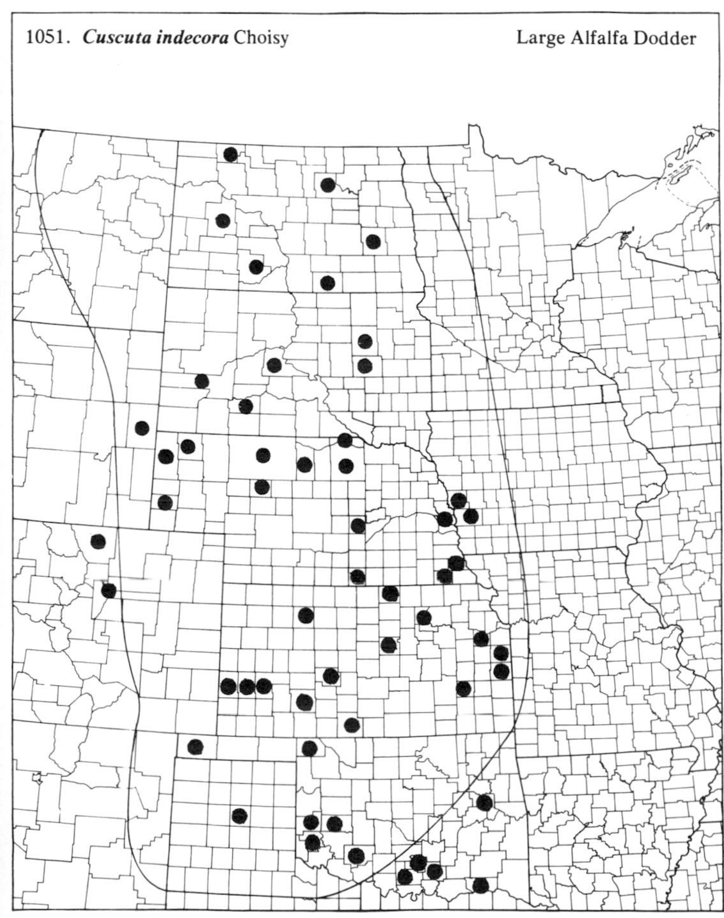

1052. ***Cuscuta pentagona*** Engelm. — Field Dodder

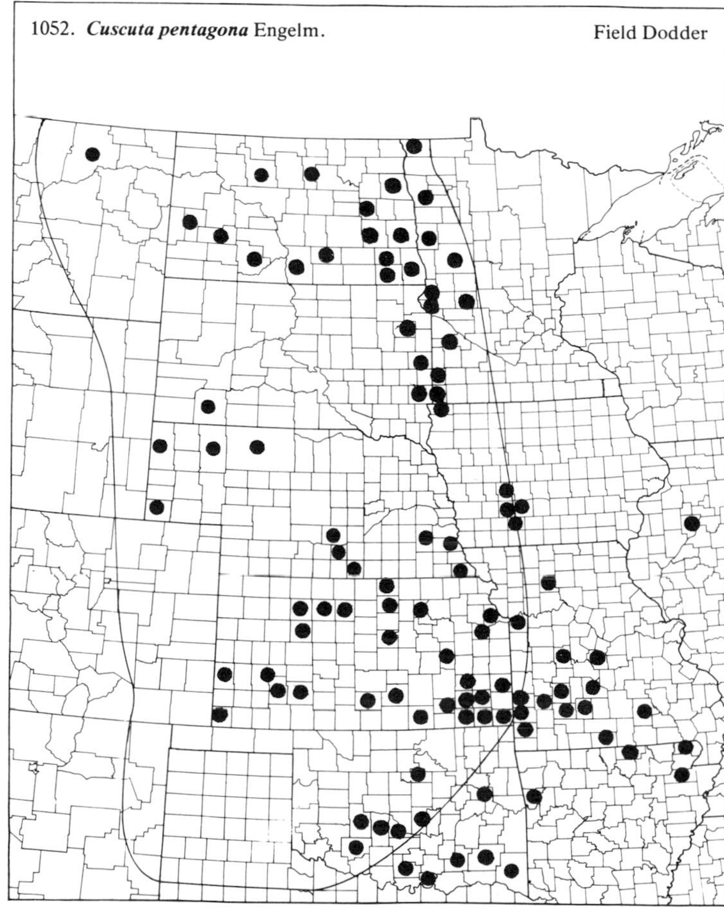

(Cuscutaceae)

1053. *Cuscuta polygonorum* Engelm. Smartweed Dodder

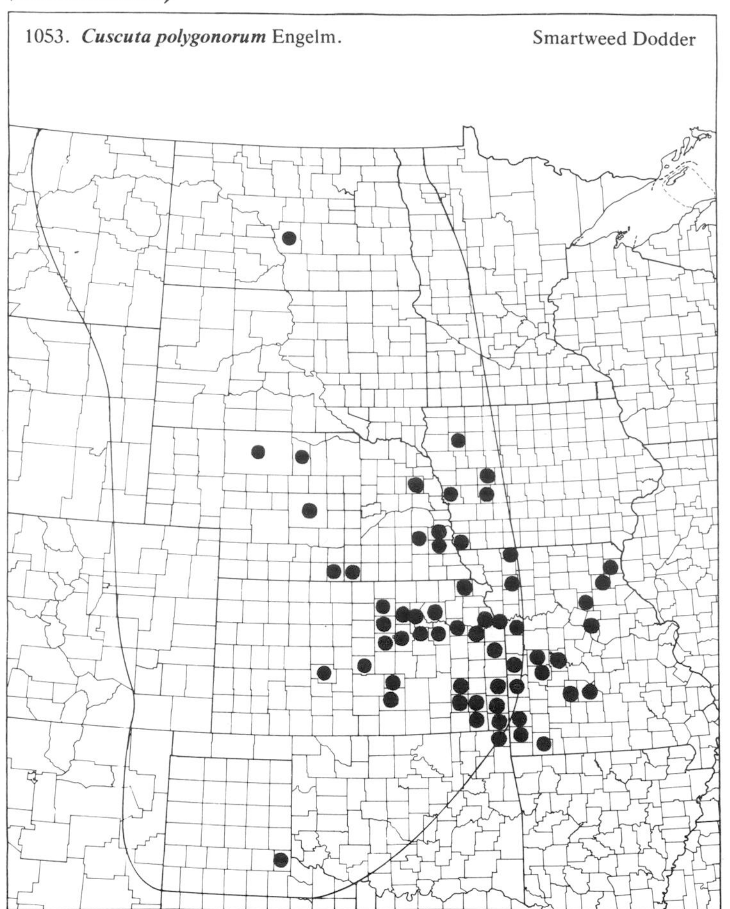

1054. *Cuscuta umbrosa* Hook.

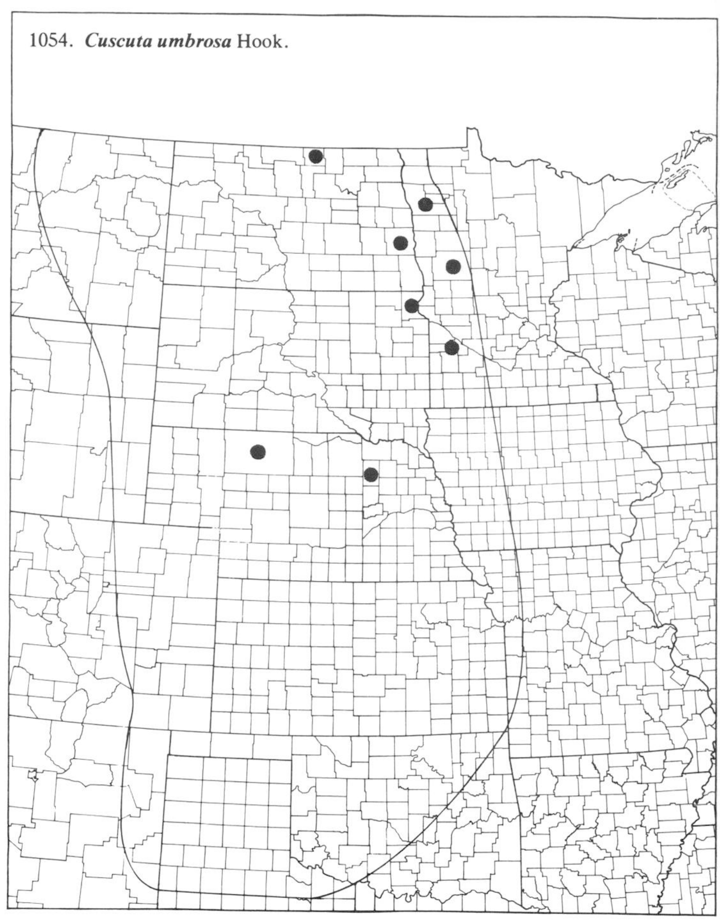

93. *Polemoniaceae* (Polemonium Family)

1055. *Collomia linearis* Nutt. Collomia

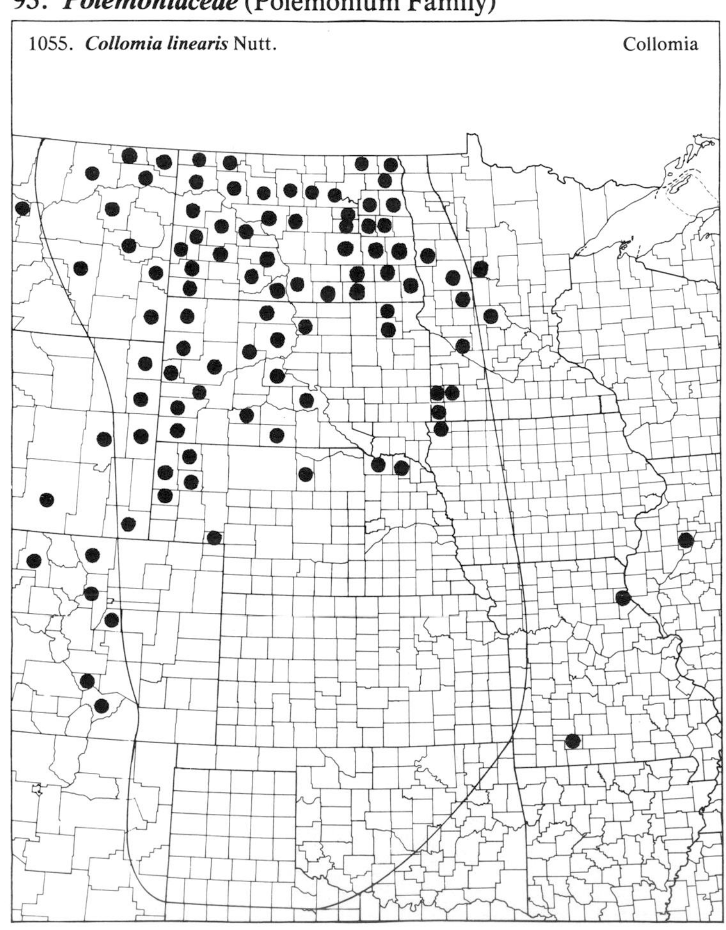

1056. *Gilia rigidula* Benth. ssp. *acerosa* (Gray) Wherry

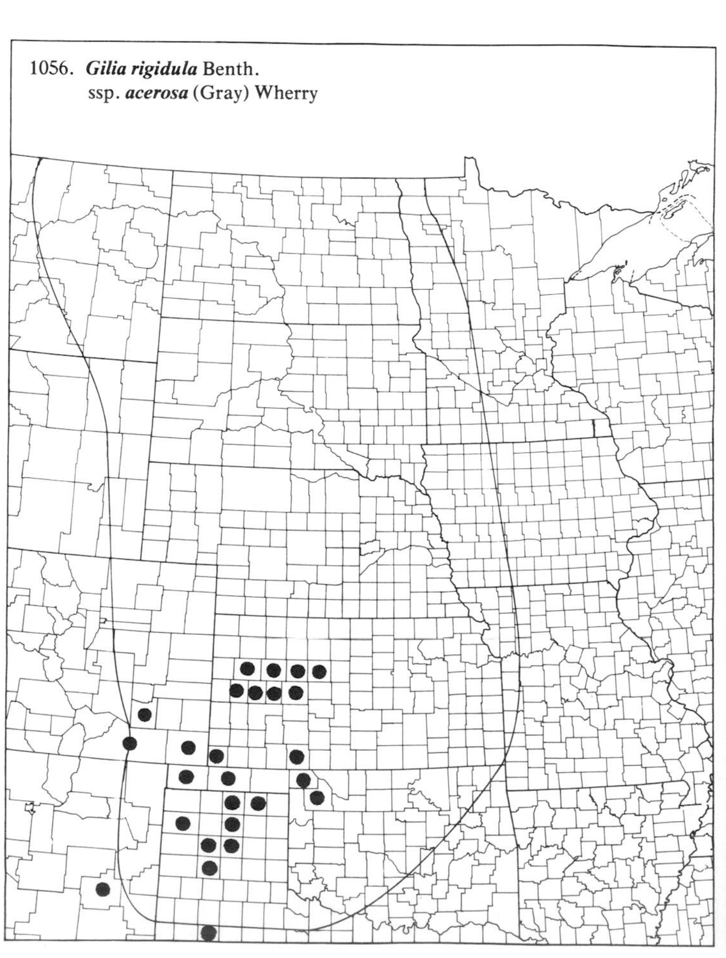

(Polemoniaceae)

1057. ***Gilia spicata*** Nutt. — Spike Gilia

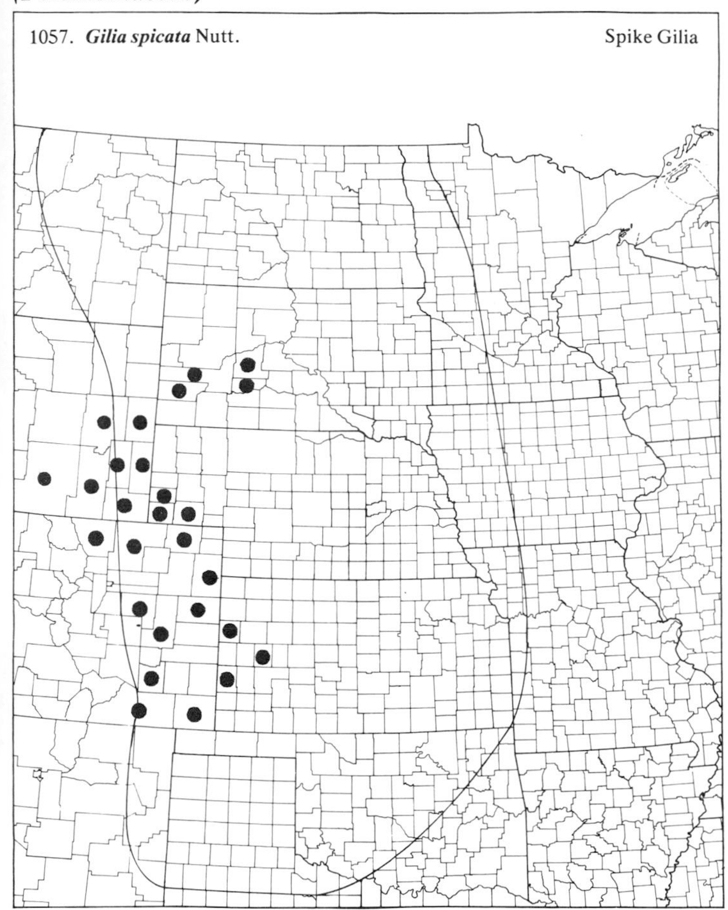

1058. ***Ipomopsis congesta*** (Hook.) V. Grant — Ball-head

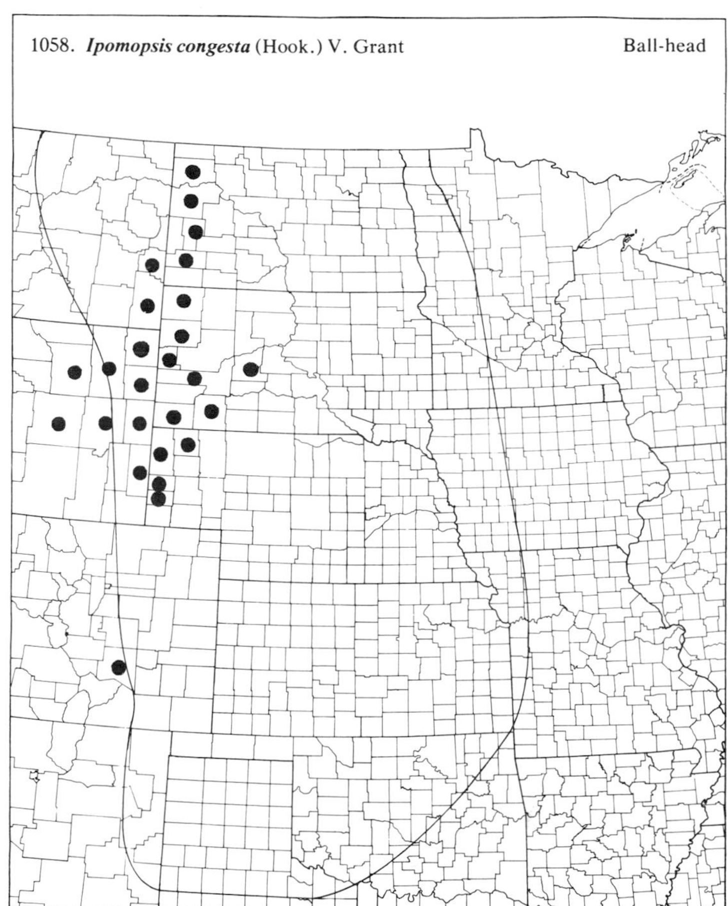

1059. ***Ipomopsis laxiflora*** (Coult.) V. Grant

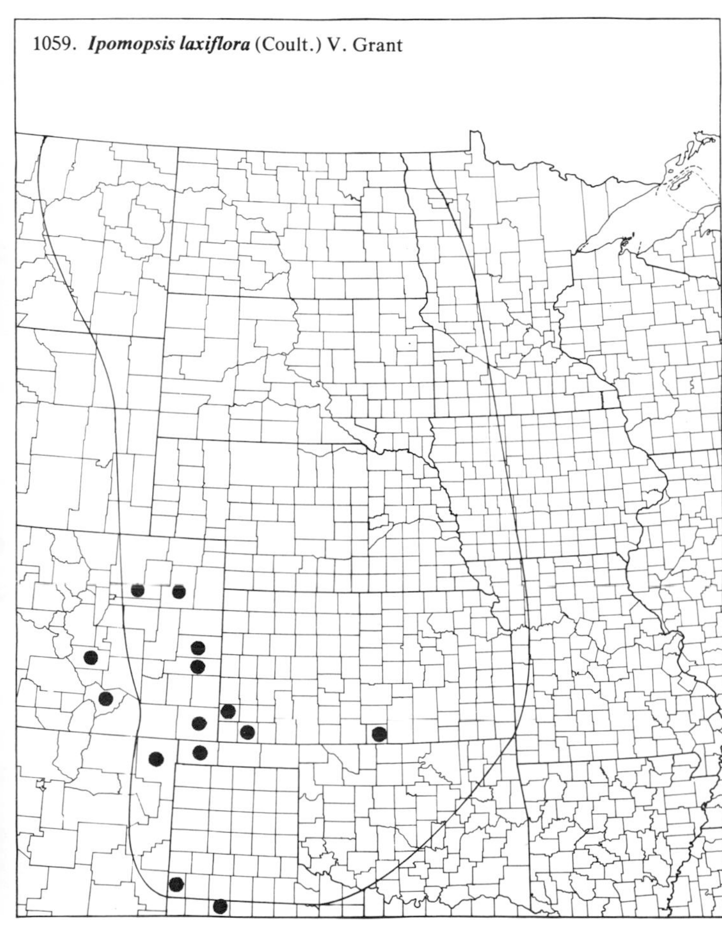

1060. ***Ipomopsis longiflora*** (Torr.) V. Grant

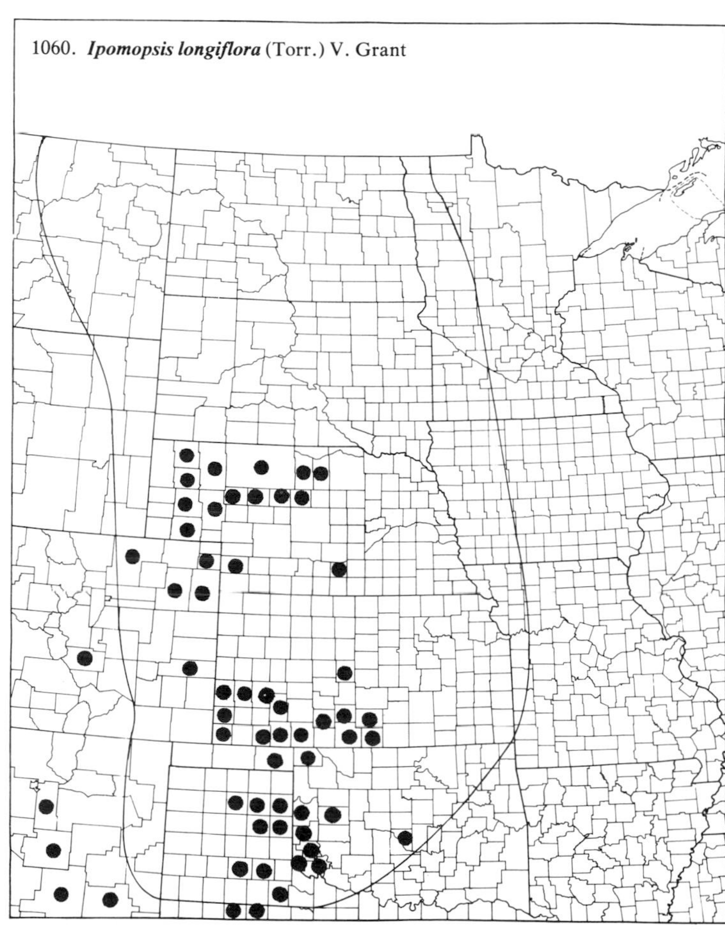

(Polemoniaceae)

1061. ***Microsteris gracilis*** (Hook.) Greene var. ***humilior*** (Hook.) Cronq.

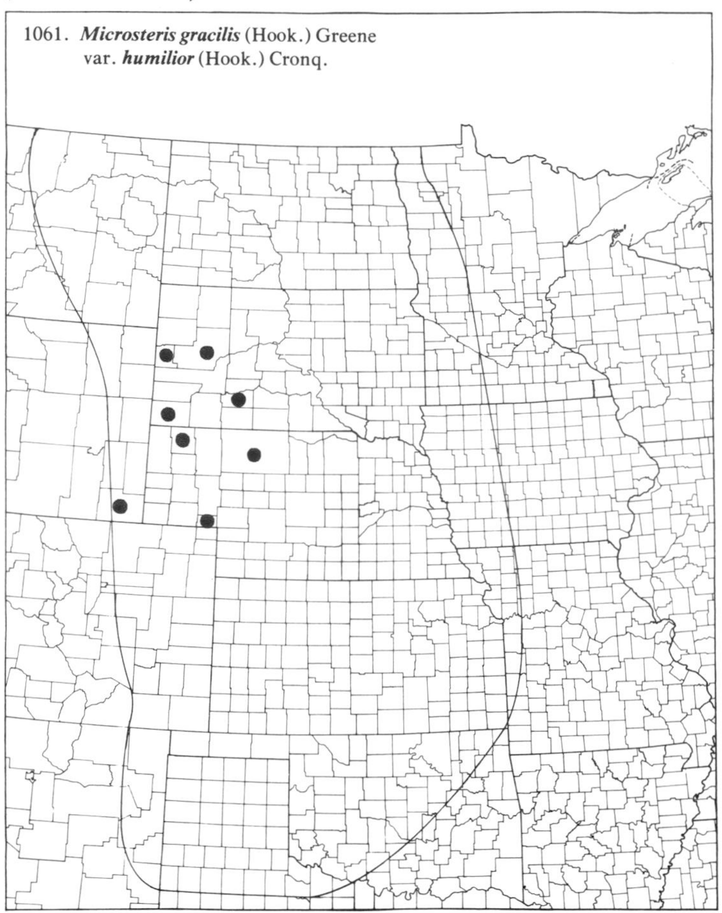

1062. ***Navarretia intertexta*** (Benth.) Hook. var. ***propinqua*** (Suksd.) Brand

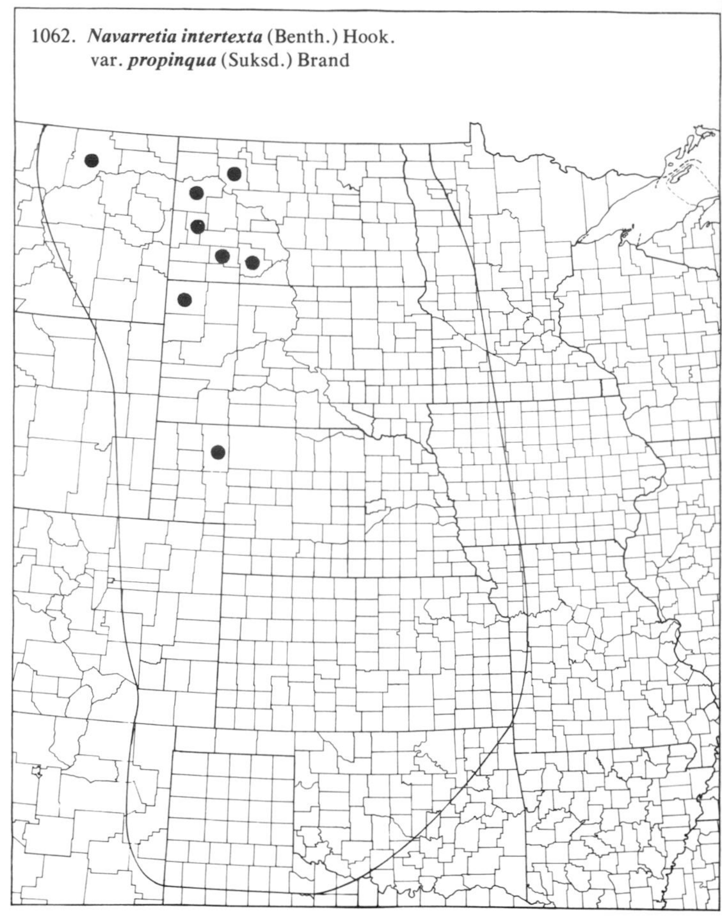

1063. ***Phlox alyssifolia*** Greene

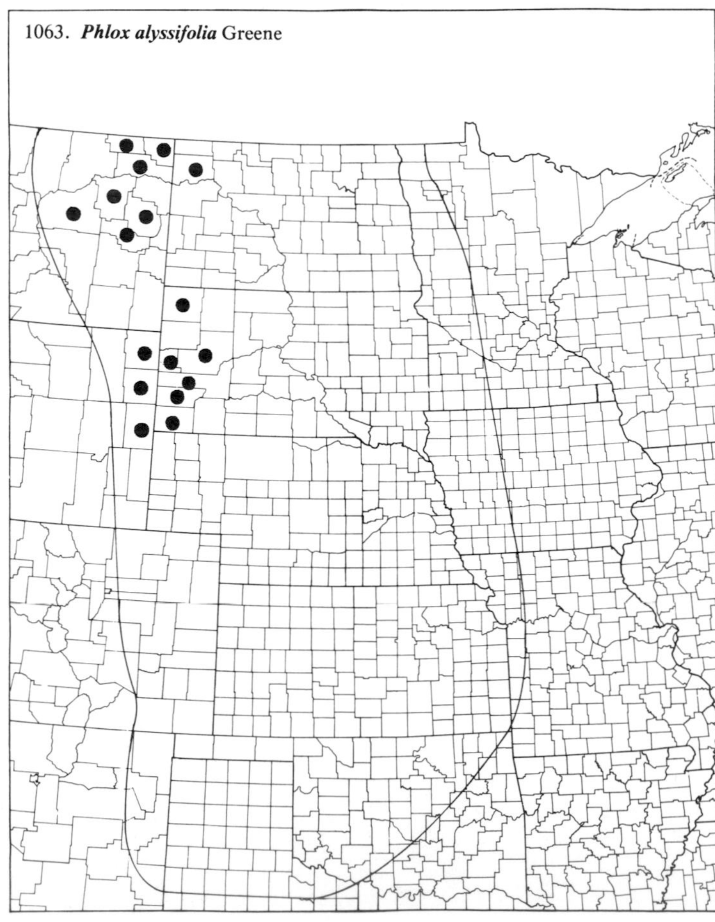

1064. ***Phlox andicola*** Nutt. Moss Phlox

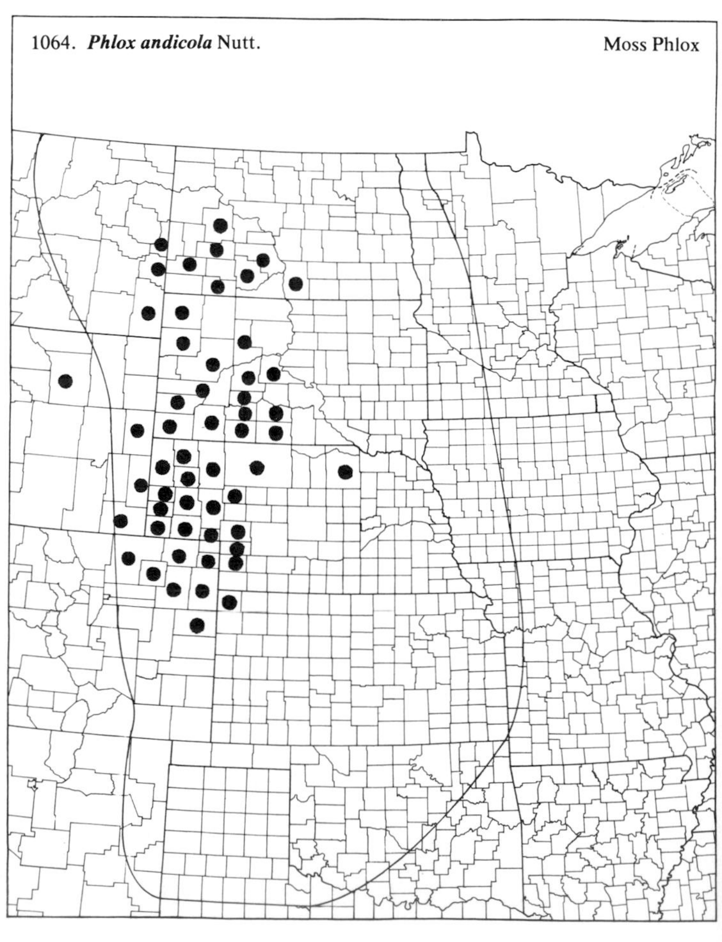

(Polemoniaceae)

1065. ***Phlox divaricata*** L. ssp. ***laphamii*** (Wood) Wherry — Blue Phlox

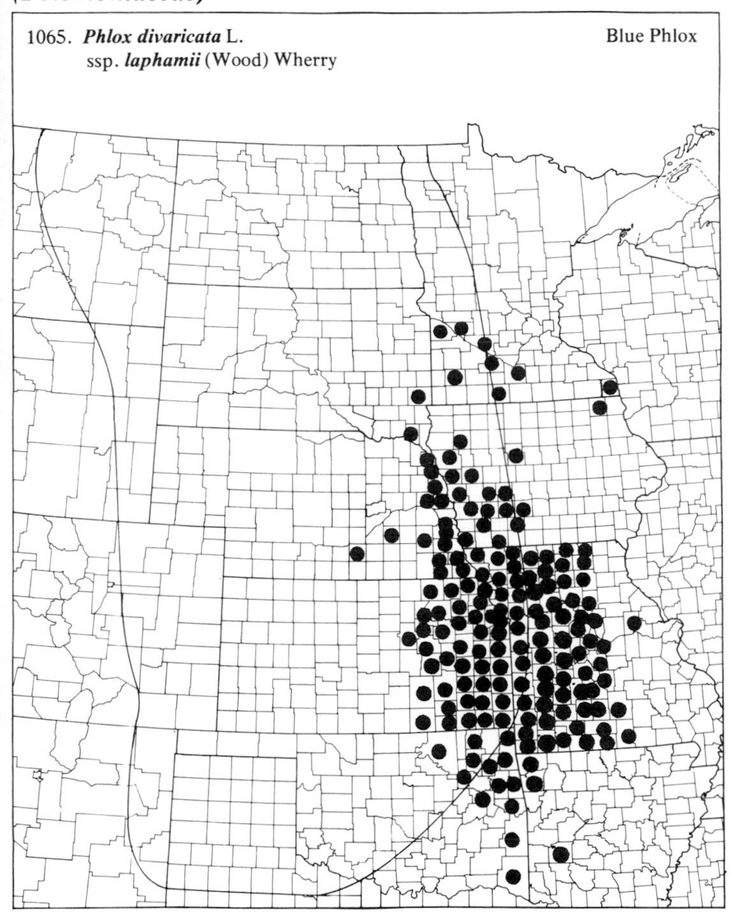

1066. ***Phlox hoodii*** Rich. — Hood's Phlox

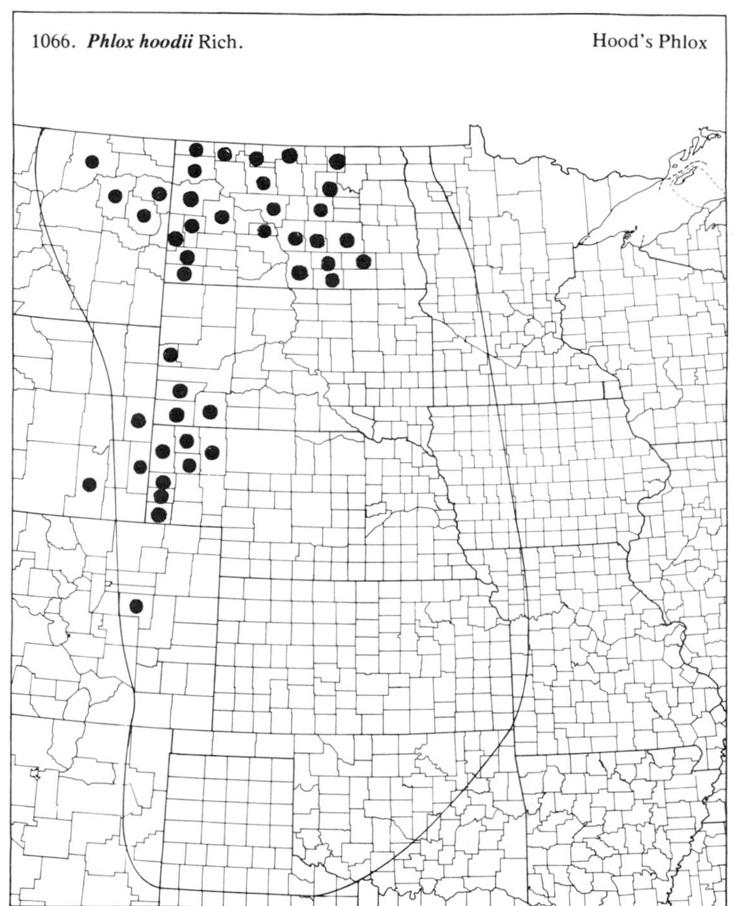

1067. ***Phlox oklahomensis*** Wherry

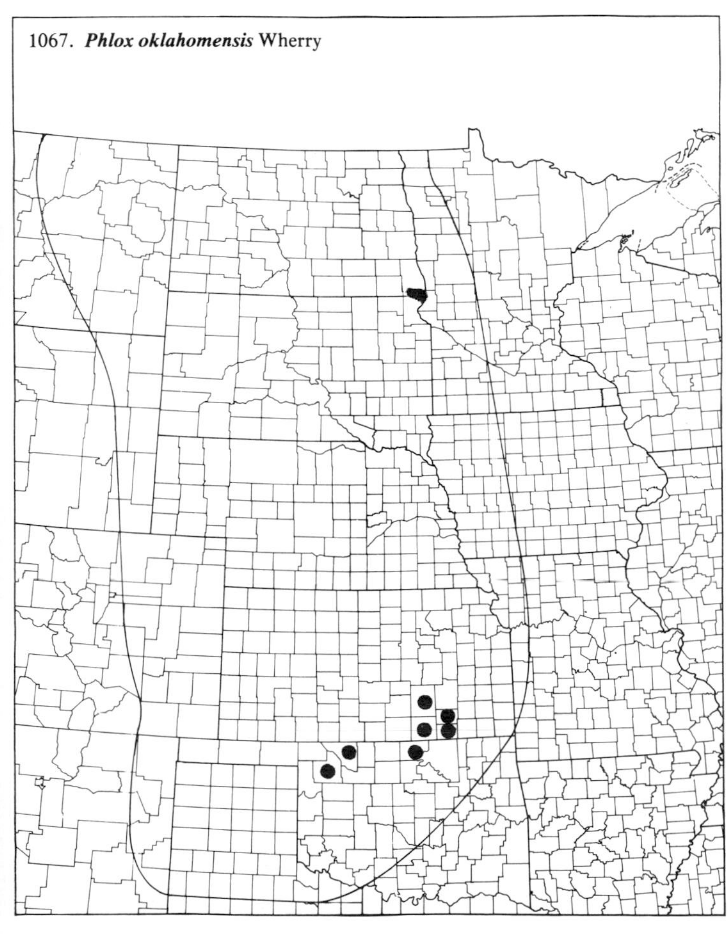

1068. ***Phlox paniculata*** L. — Fall Phlox

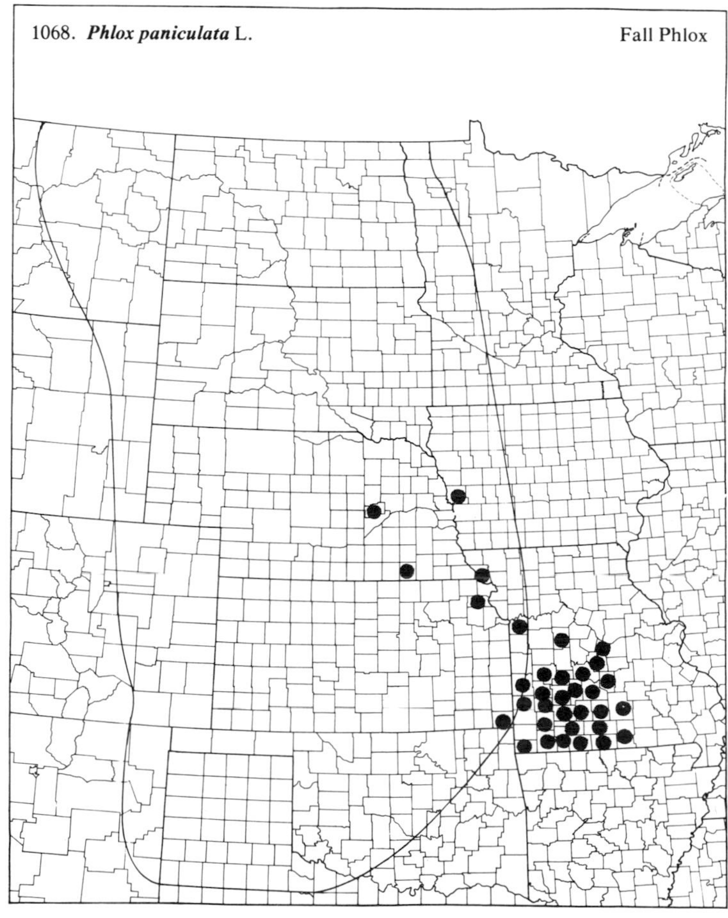

(Polemoniaceae)

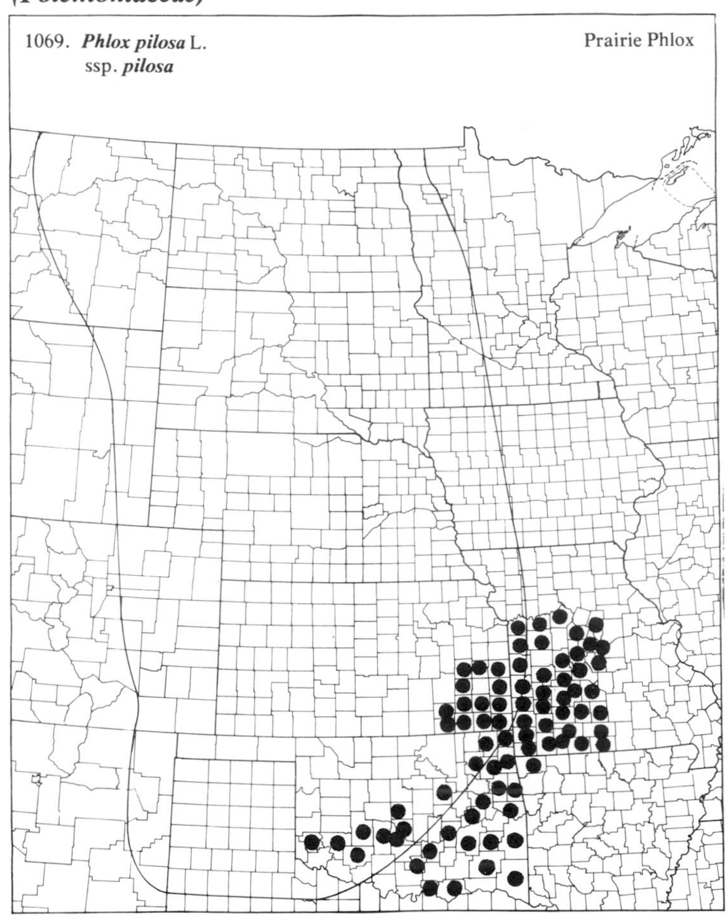

1069. ***Phlox pilosa*** L.
ssp. ***pilosa*** — Prairie Phlox

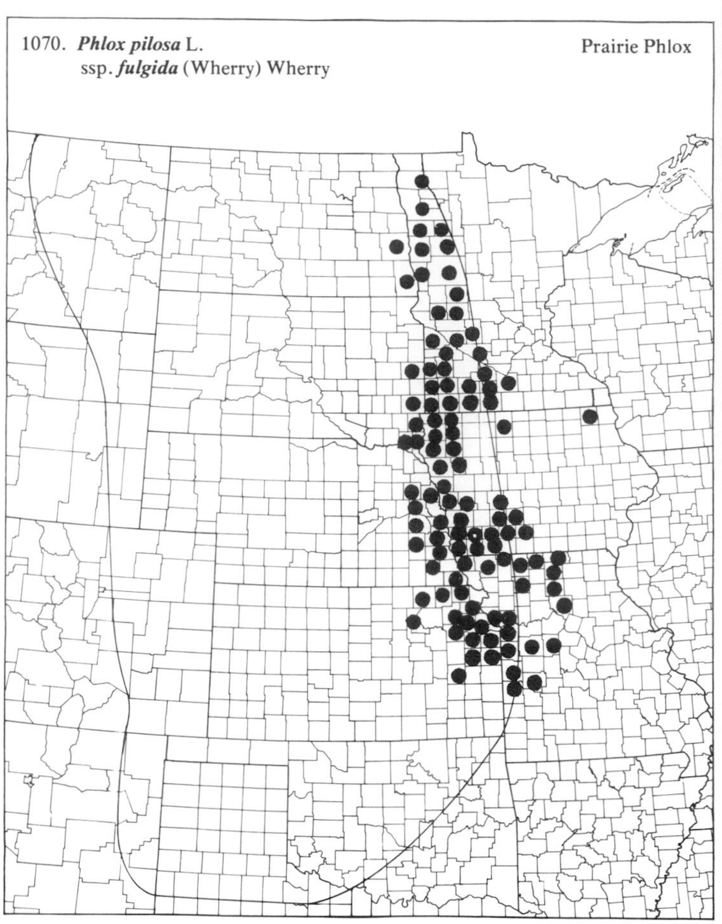

1070. ***Phlox pilosa*** L.
ssp. ***fulgida*** (Wherry) Wherry — Prairie Phlox

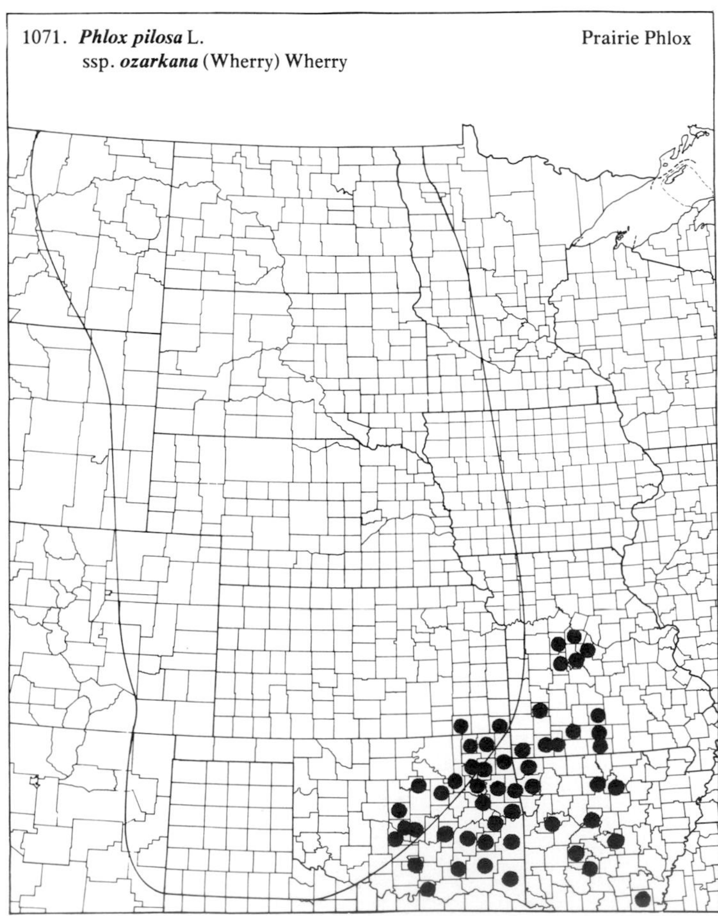

1071. ***Phlox pilosa*** L.
ssp. ***ozarkana*** (Wherry) Wherry — Prairie Phlox

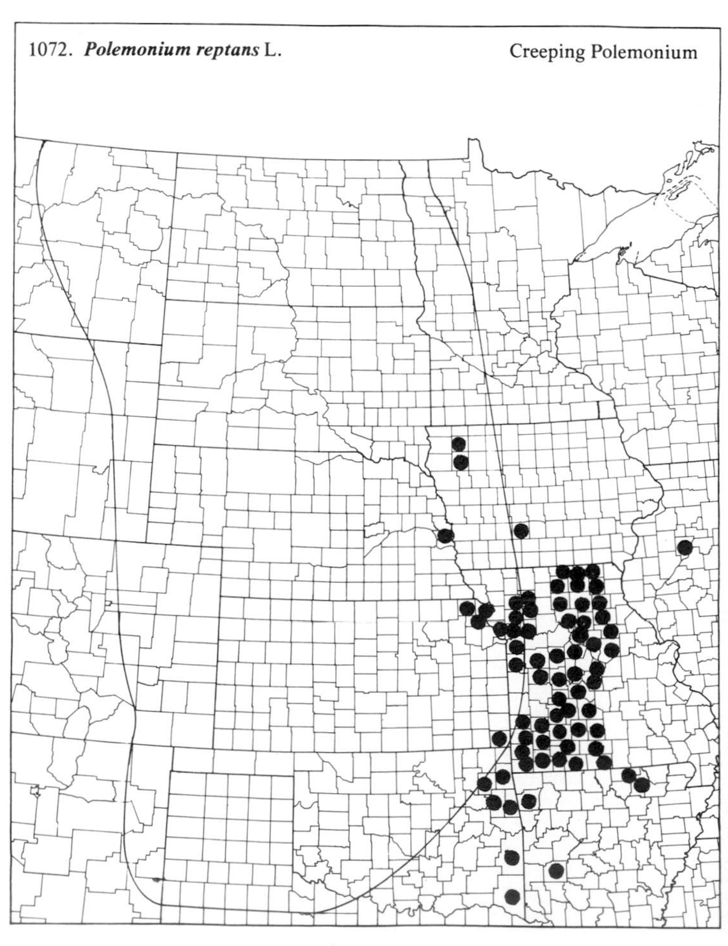

1072. ***Polemonium reptans*** L. — Creeping Polemonium

94. *Hydrophyllaceae* (Waterleaf Family)

1073. *Ellisia nyctelea* L. Waterpod

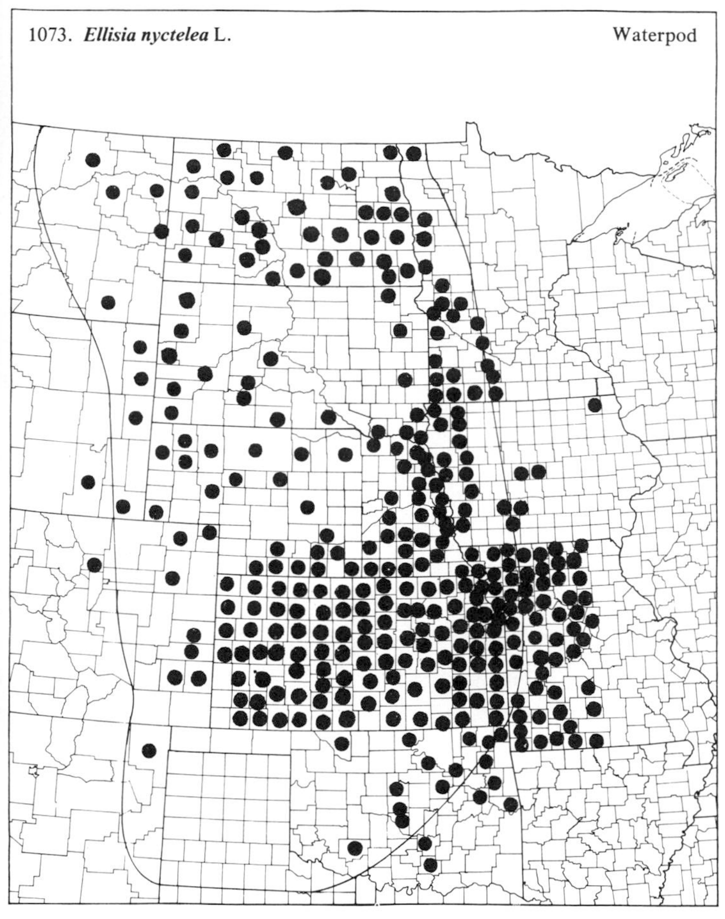

1074. *Hydrophyllum appendiculatum* Michx. Notchbract Waterleaf

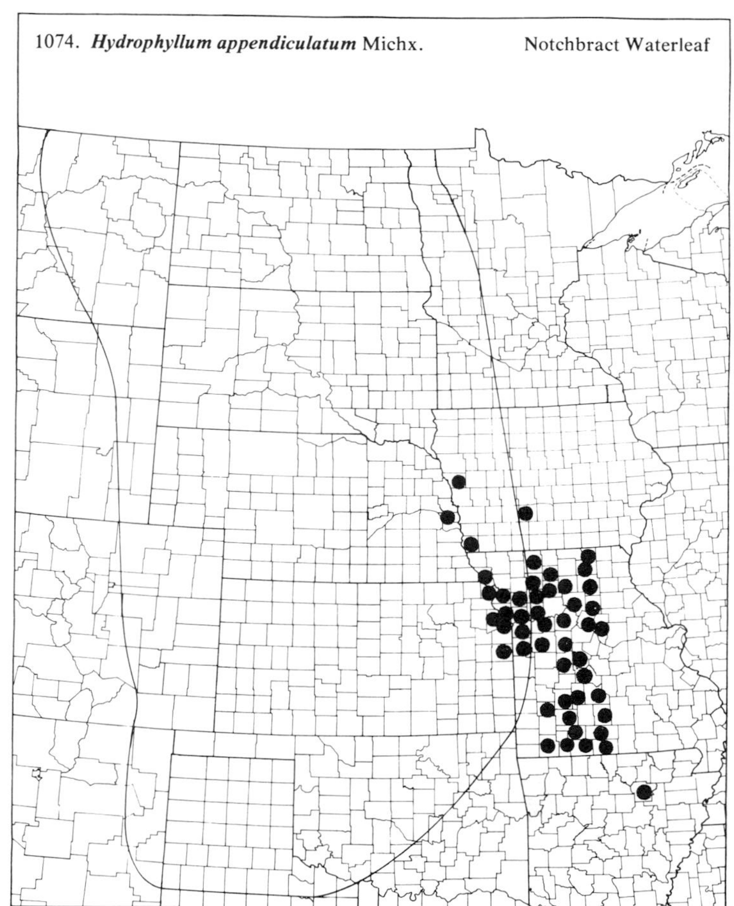

1075. *Hydrophyllum virginianum* L. Waterleaf

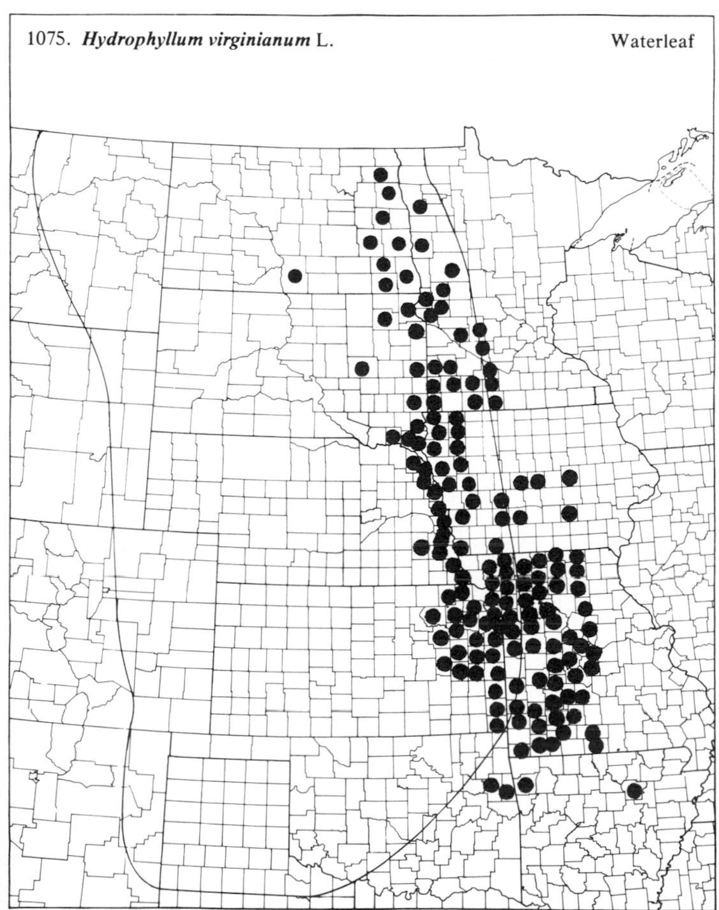

1076. *Nama hispidum* Gray

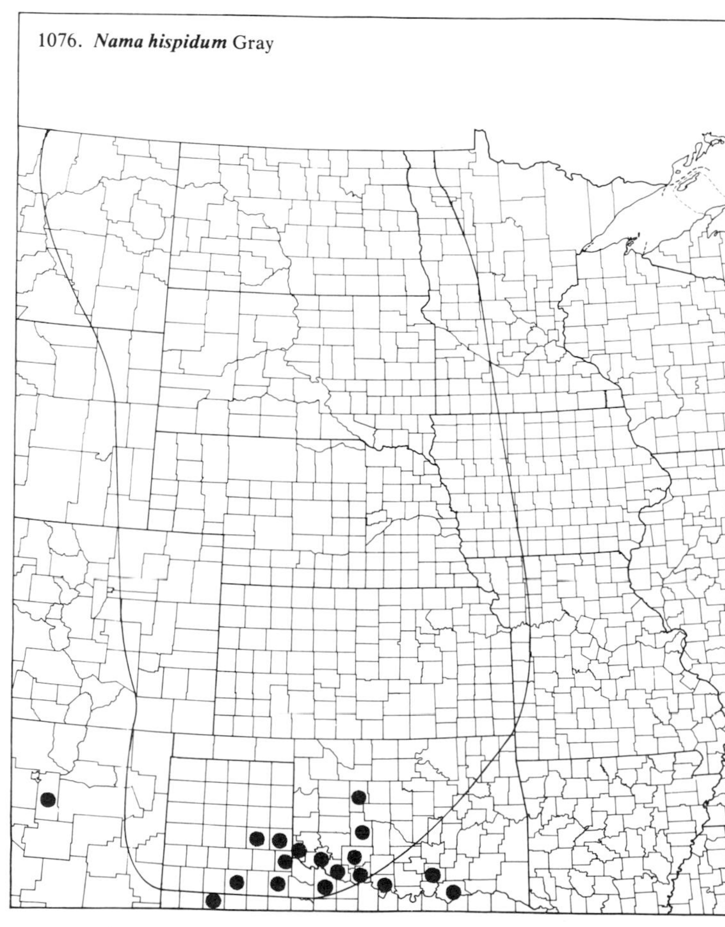

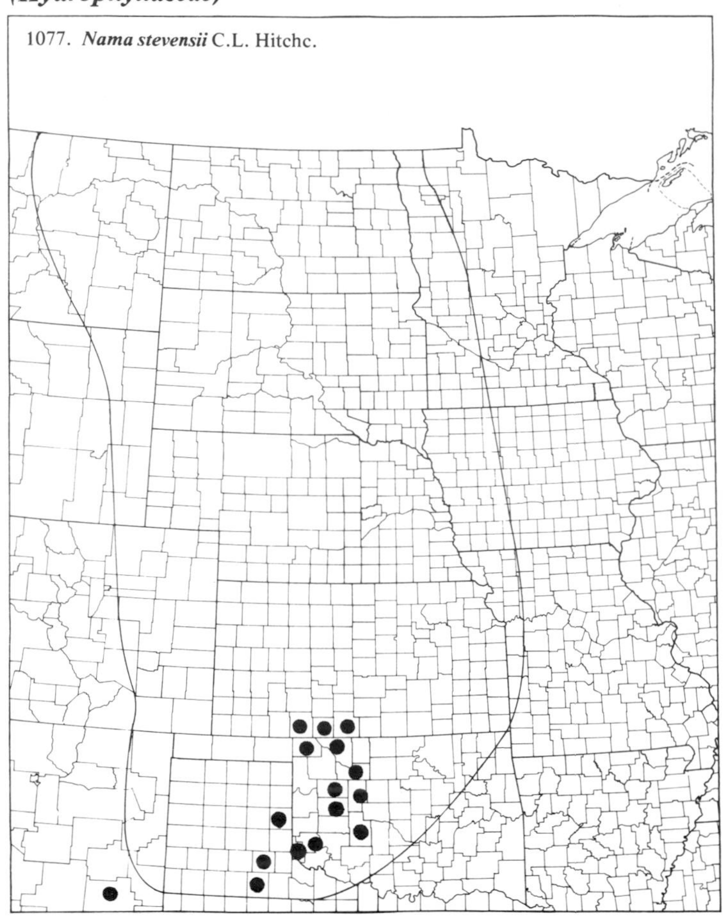

1077. ***Nama stevensii*** C.L. Hitchc.

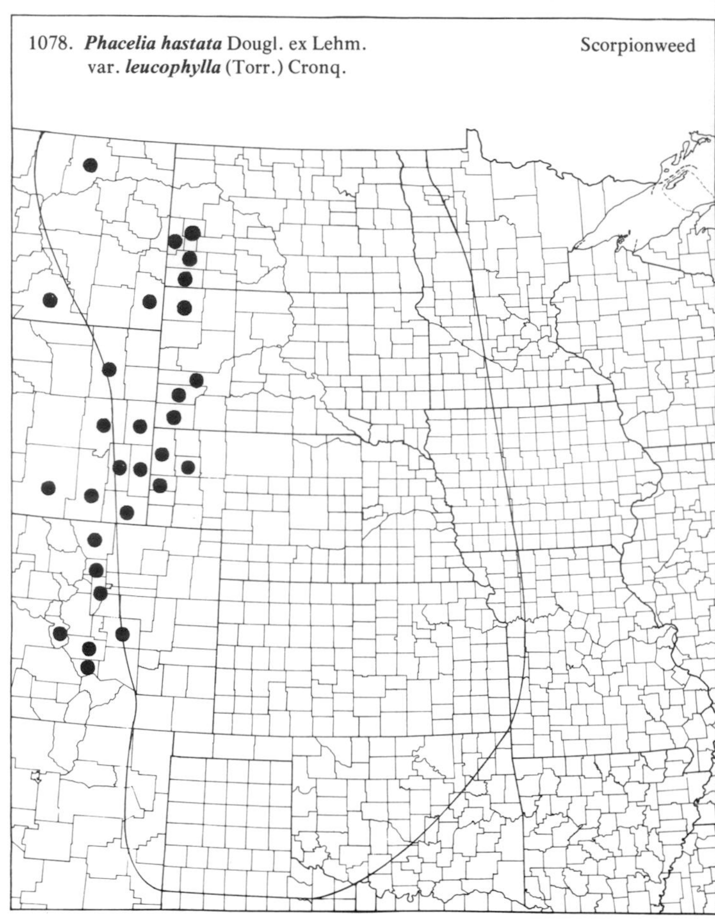

1078. ***Phacelia hastata*** Dougl. ex Lehm. var. ***leucophylla*** (Torr.) Cronq. Scorpionweed

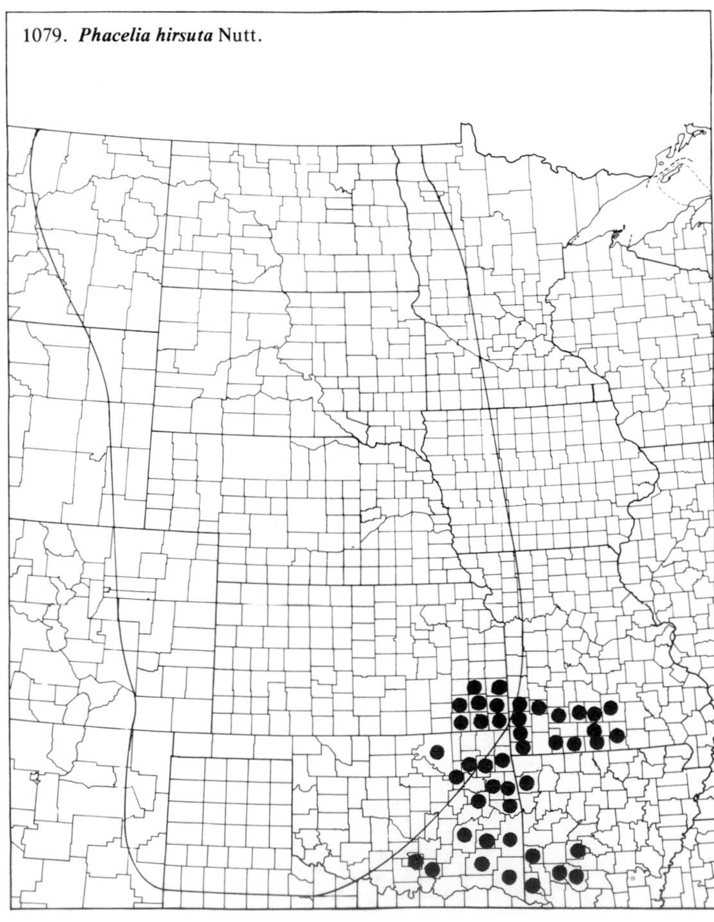

1079. ***Phacelia hirsuta*** Nutt.

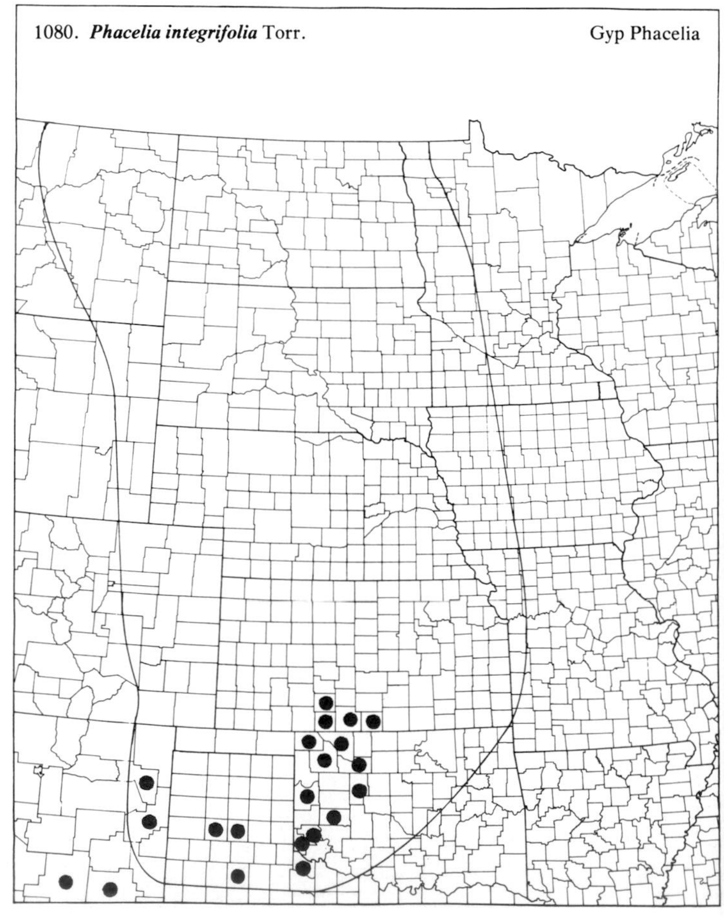

1080. ***Phacelia integrifolia*** Torr. Gyp Phacelia

95. *Boraginaceae* (Borage Family)

1081. *Cryptantha cana* (A. Nels.) Pays.

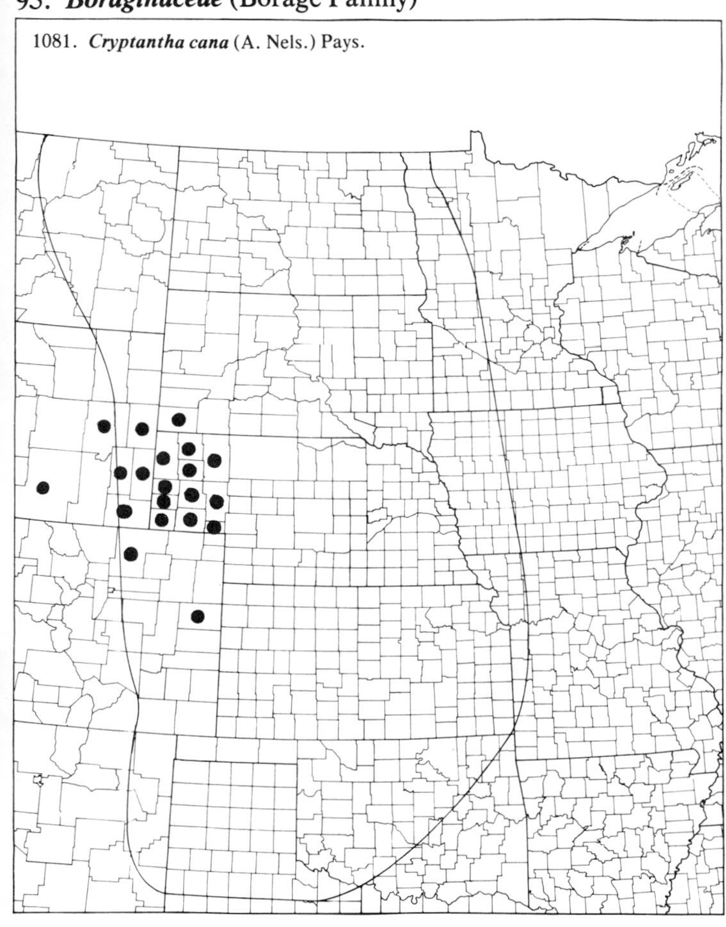

1082. *Cryptantha celosioides* (Eastw.) Pays.

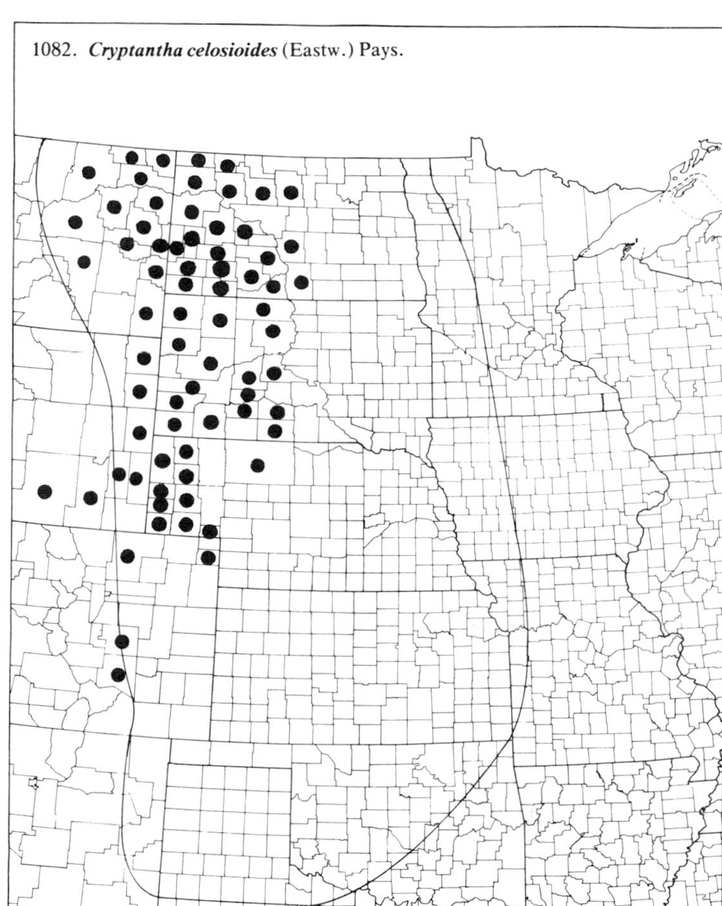

1083. *Cryptantha crassisepala* (T. & G.) Greene Thicksepal Cryptantha

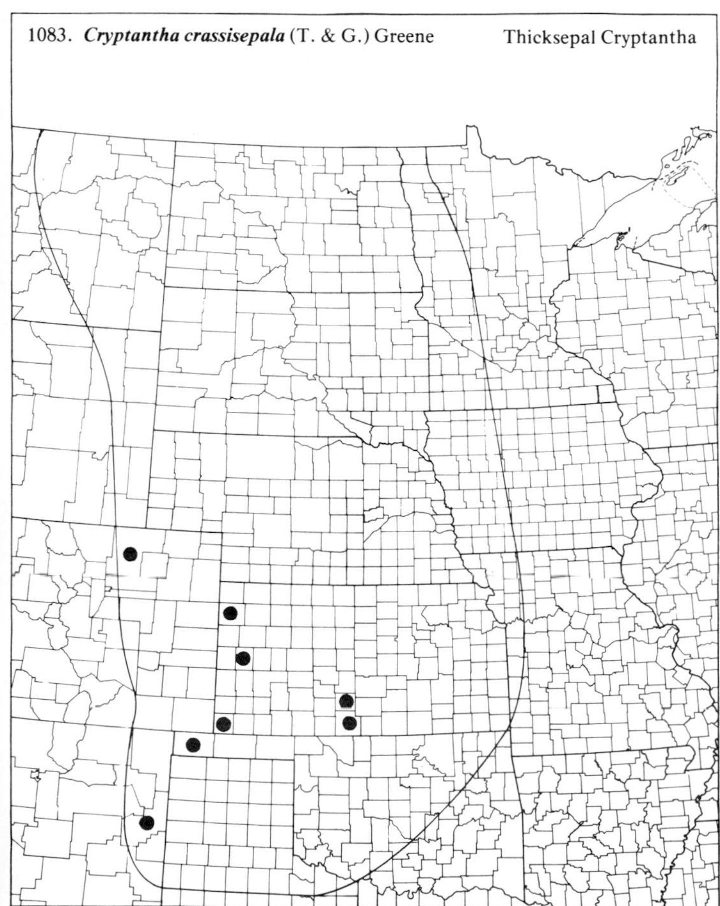

1084. *Cryptantha fendleri* (Gray) Greene

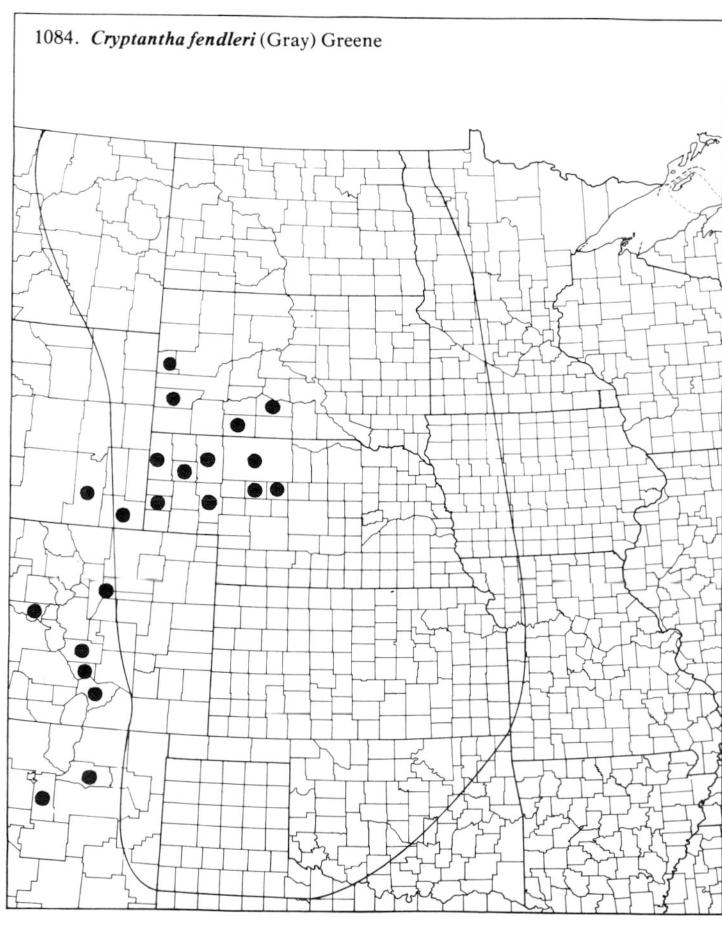

(Boraginaceae)

1085. ***Cryptantha jamesii*** (Torr.) Pays. James' Cryptantha

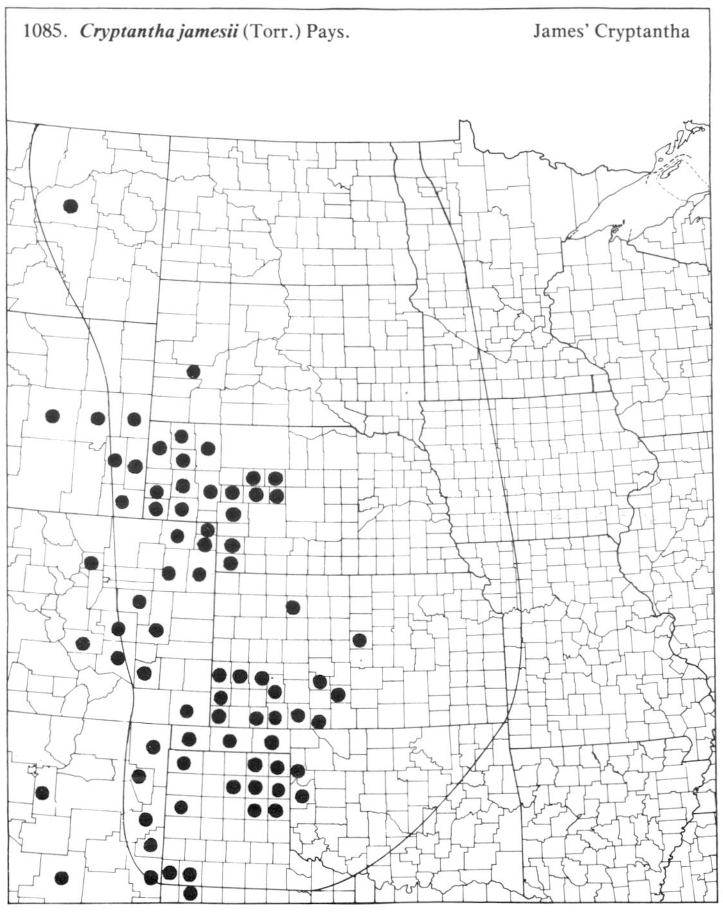

1086. ***Cryptantha minima*** Rydb.

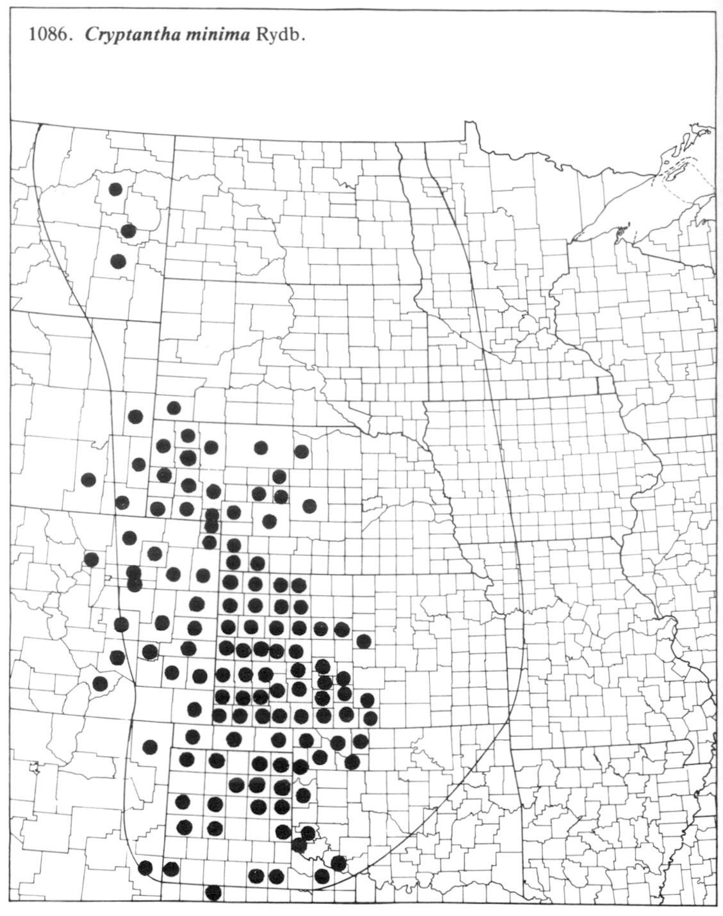

1087. ***Cryptantha thyrsiflora*** (Greene) Pays.

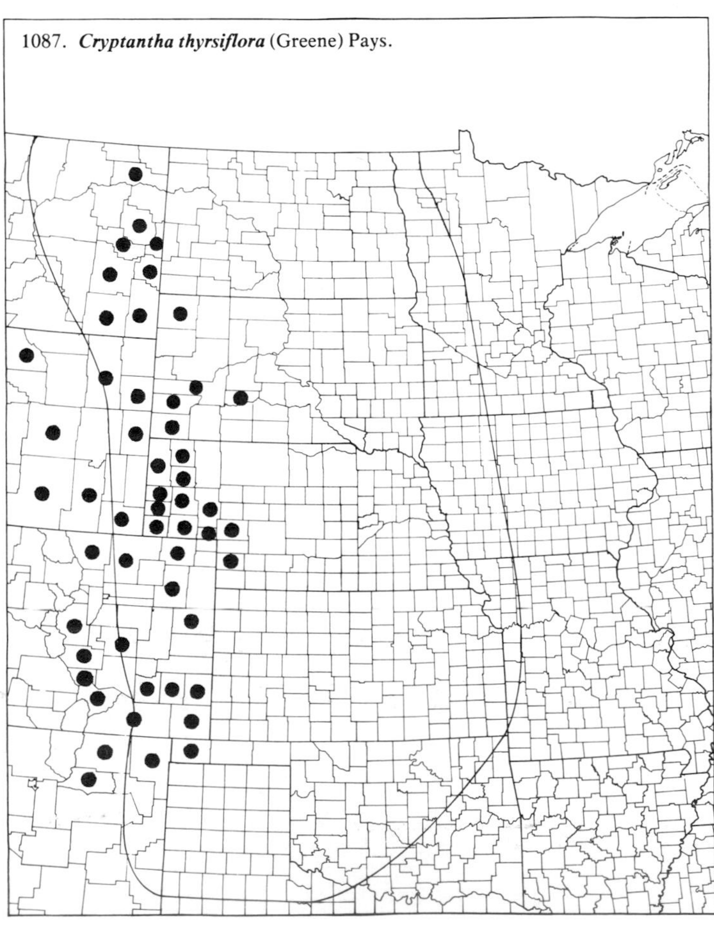

1088. ***Cynoglossum officinale*** L. Hound's Tongue

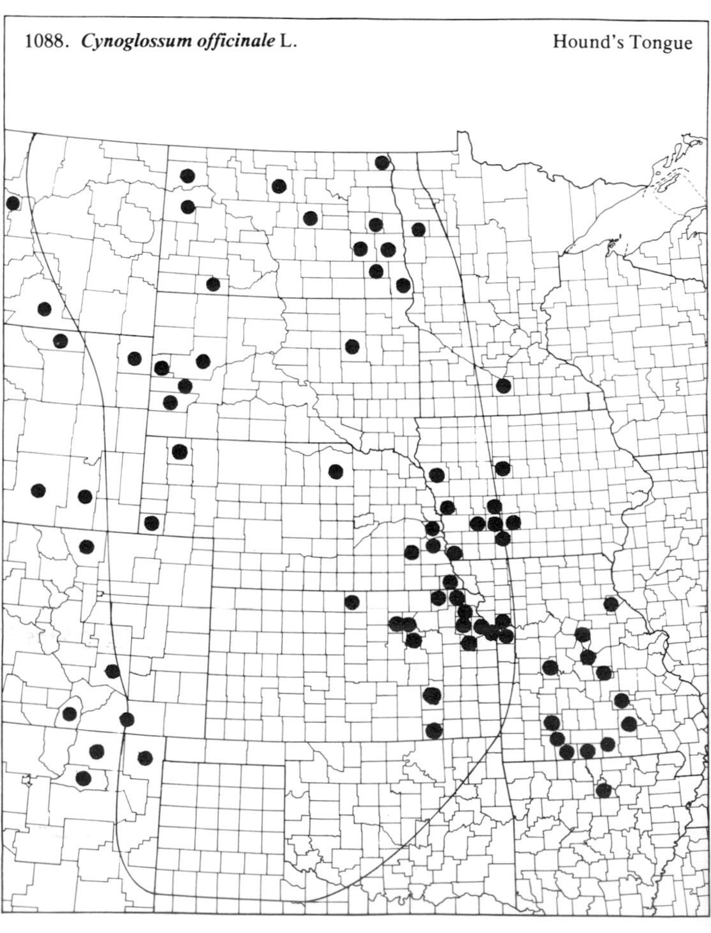

(Boraginaceae)

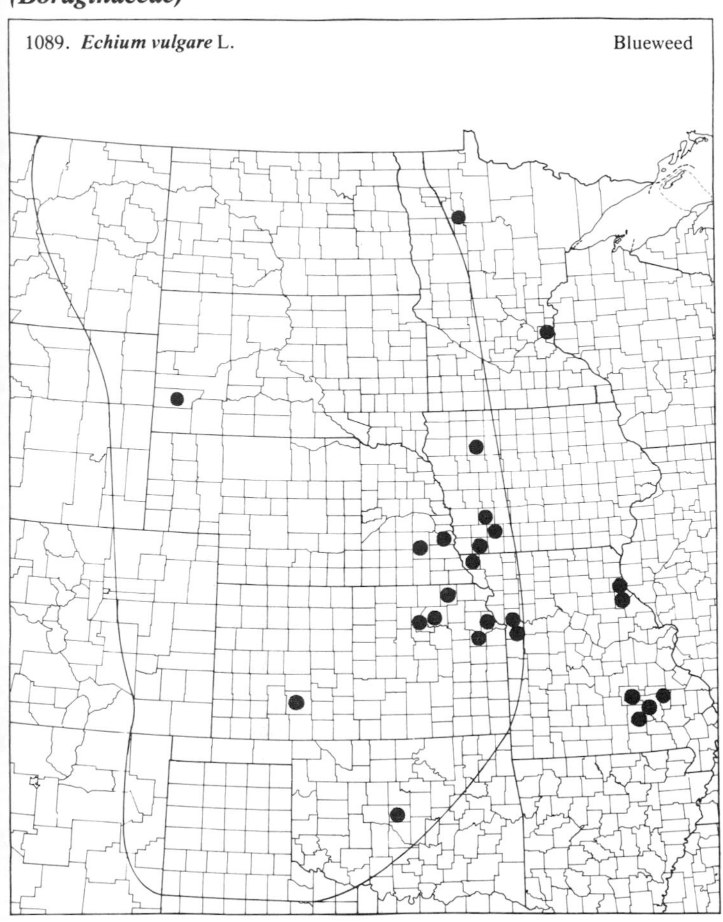

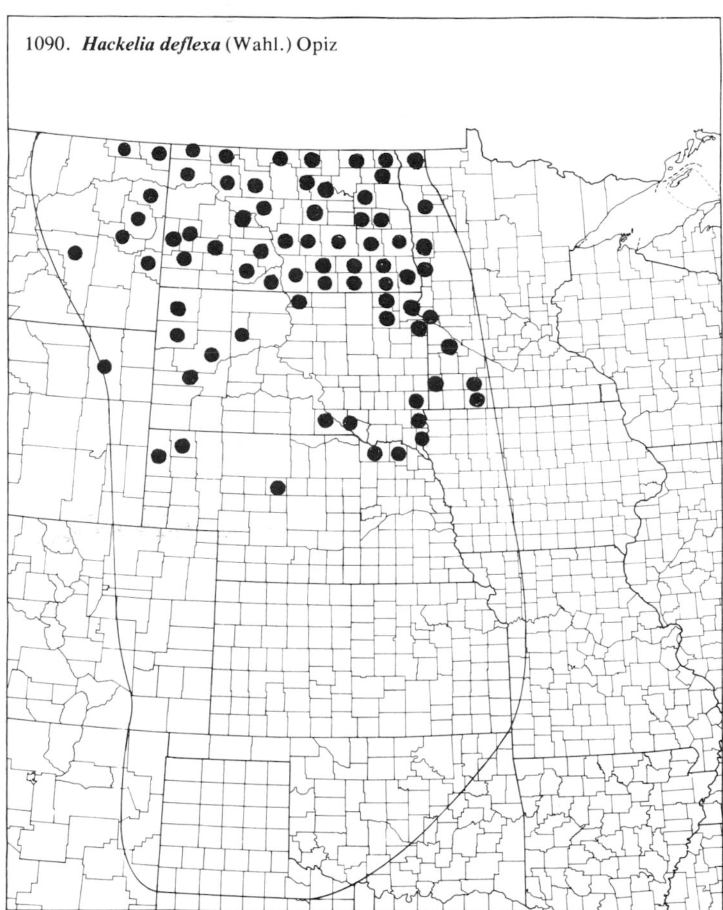

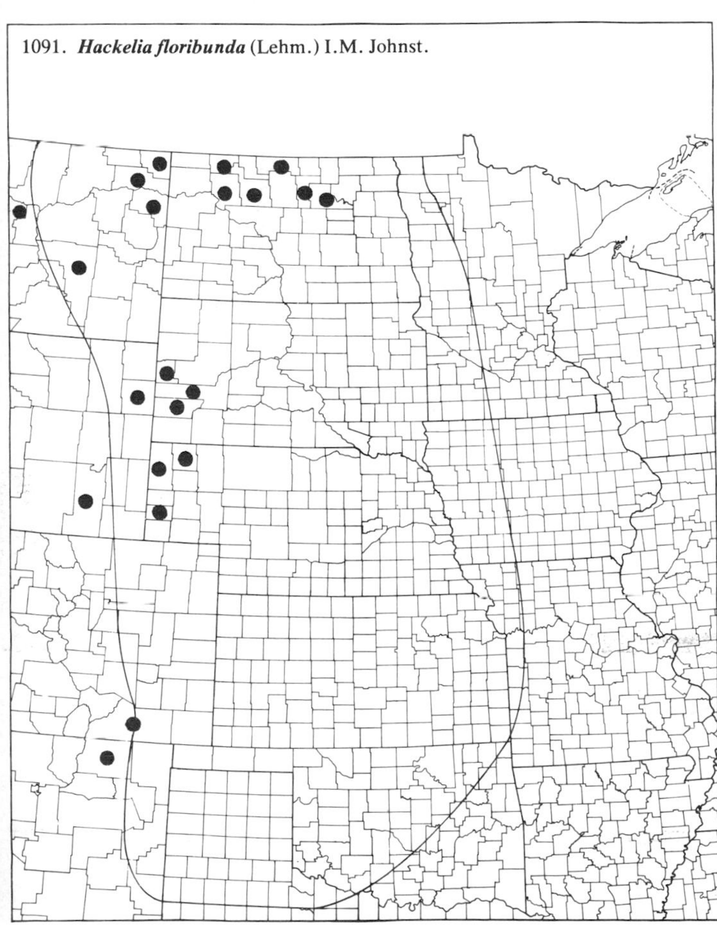

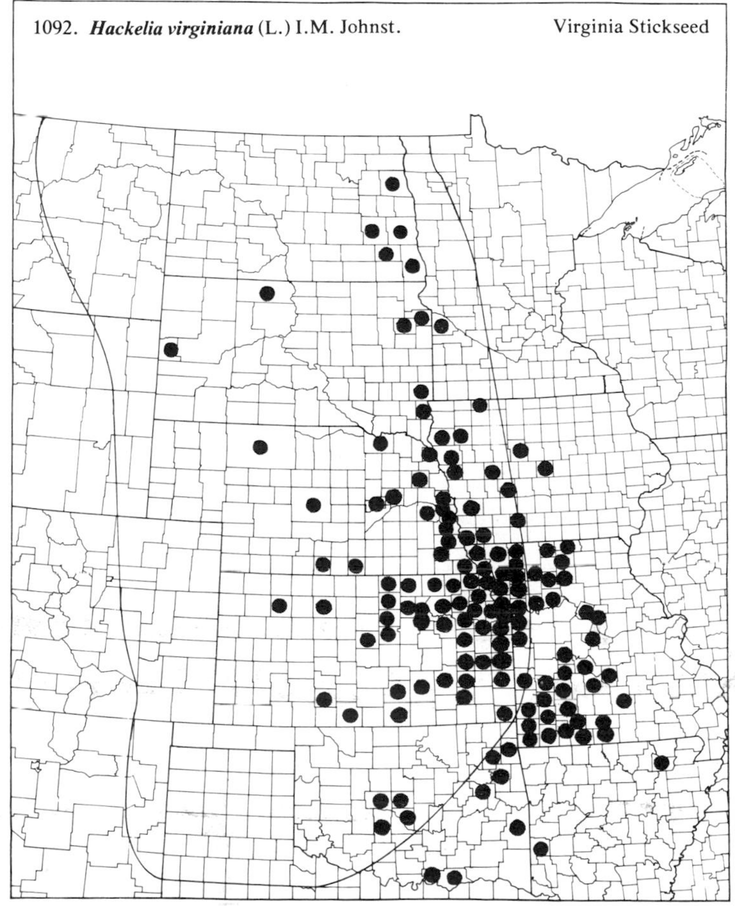

(Boraginaceae)

1093. ***Heliotropium convolvulaceum*** (Nutt.) Gray — Wild Heliotrope

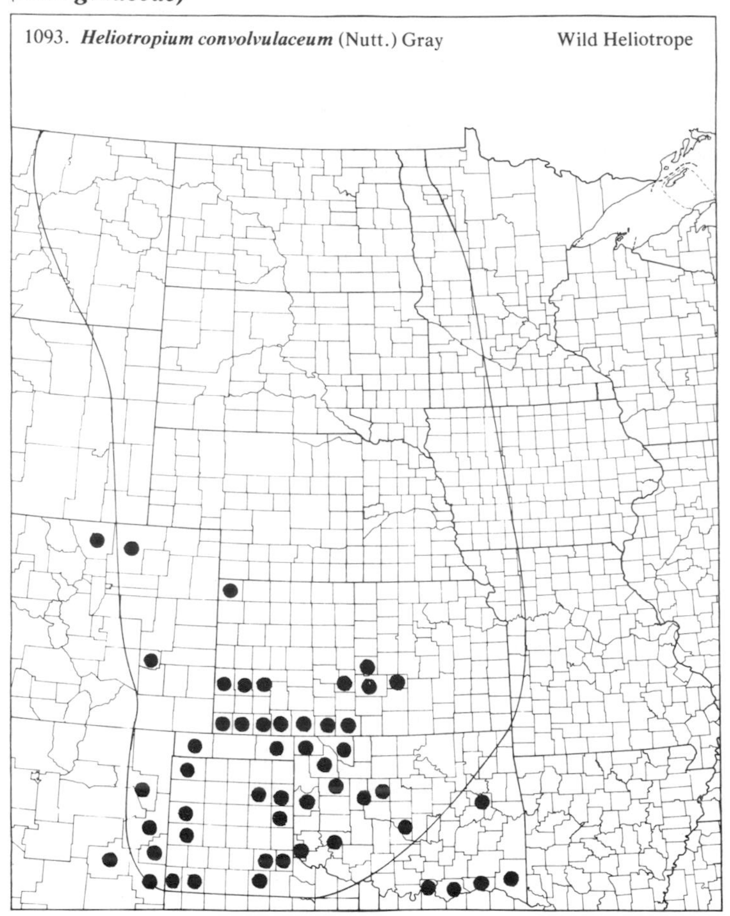

1094. ***Heliotropium curassavicum*** L. — Seaside Heliotrope

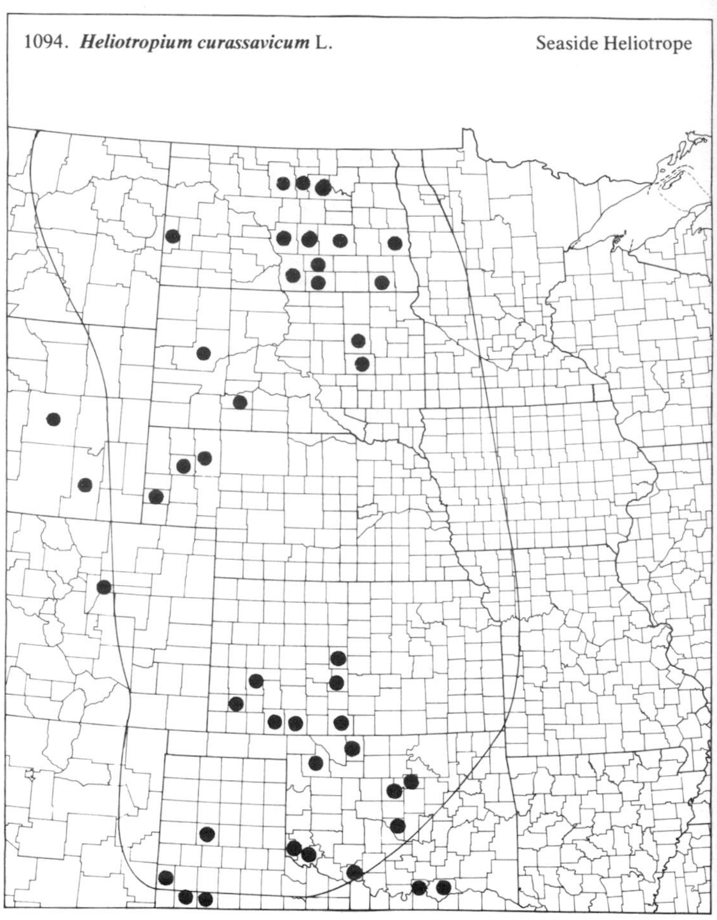

1095. ***Heliotropium tenellum*** (Nutt.) Torr. — Pasture Heliotrope

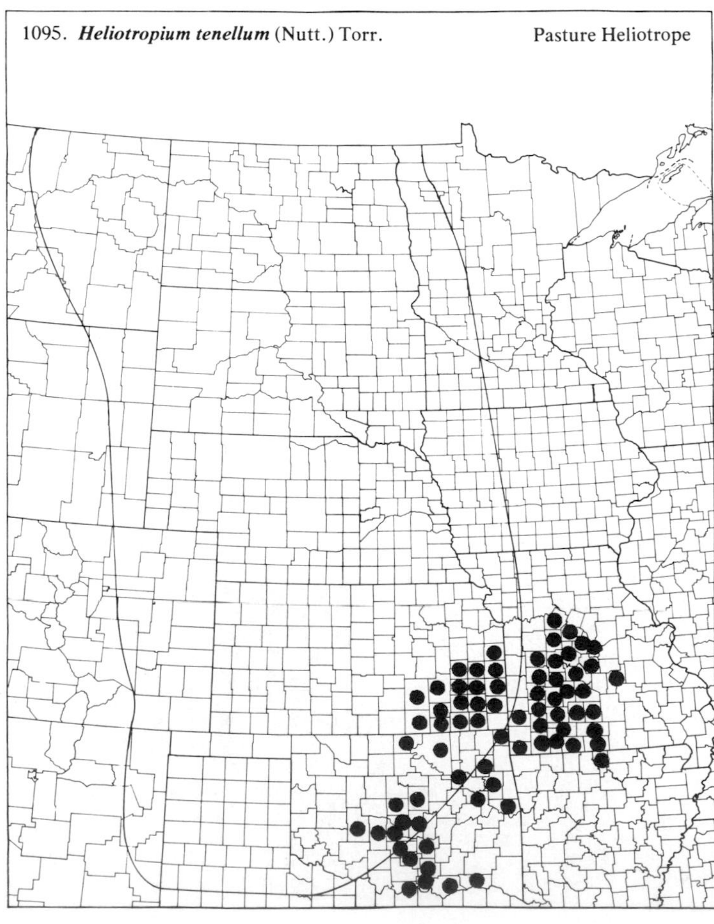

1096. ***Lappula cenchrusoides*** A. Nels.

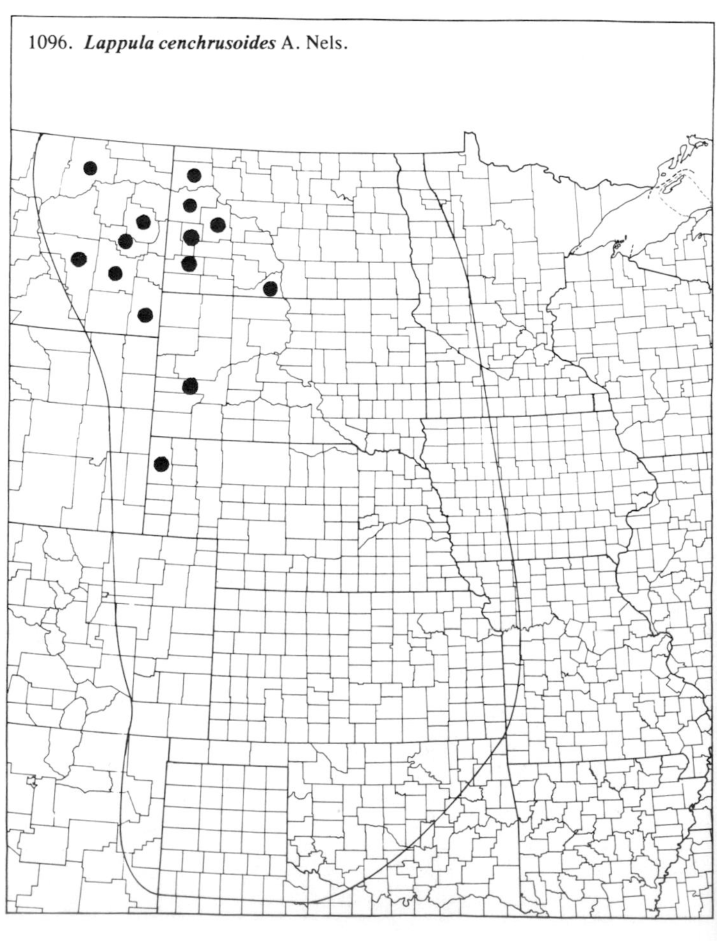

(Boraginaceae)

1097. ***Lappula echinata*** Gilib. — Blue Stickseed

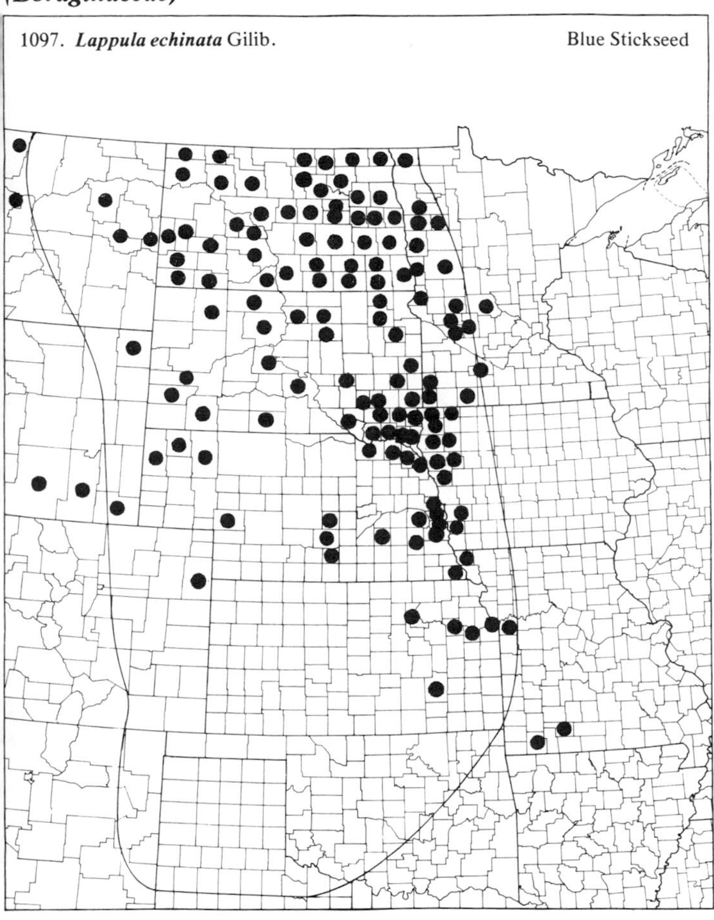

1098. ***Lappula redowskii*** (Hornem.) Greene — Low Stickseed

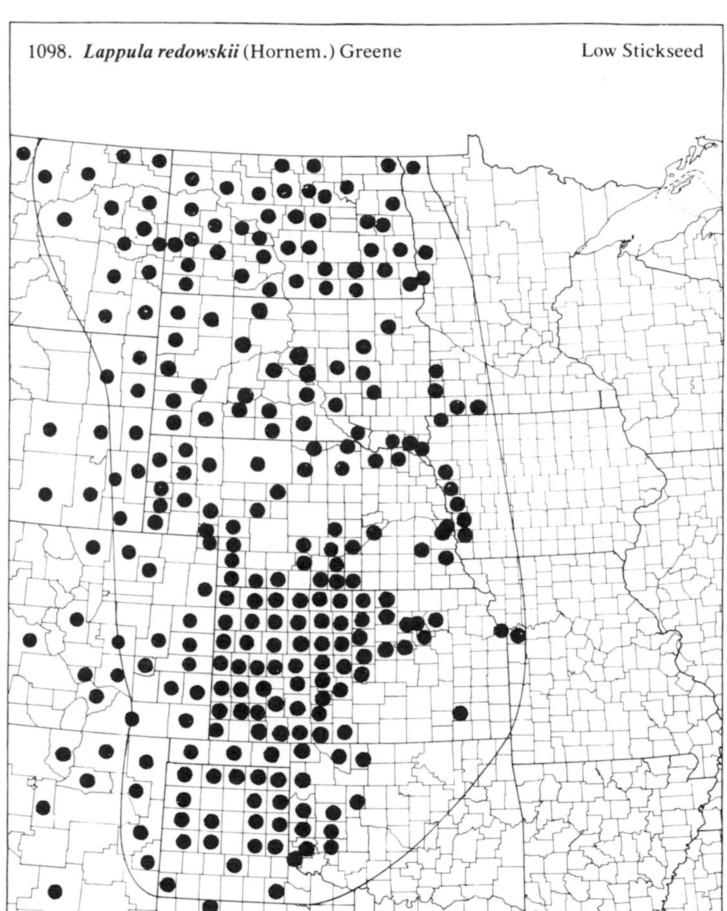

1099. ***Lappula texana*** (Scheele) Britt. — Cupseed Stickseed

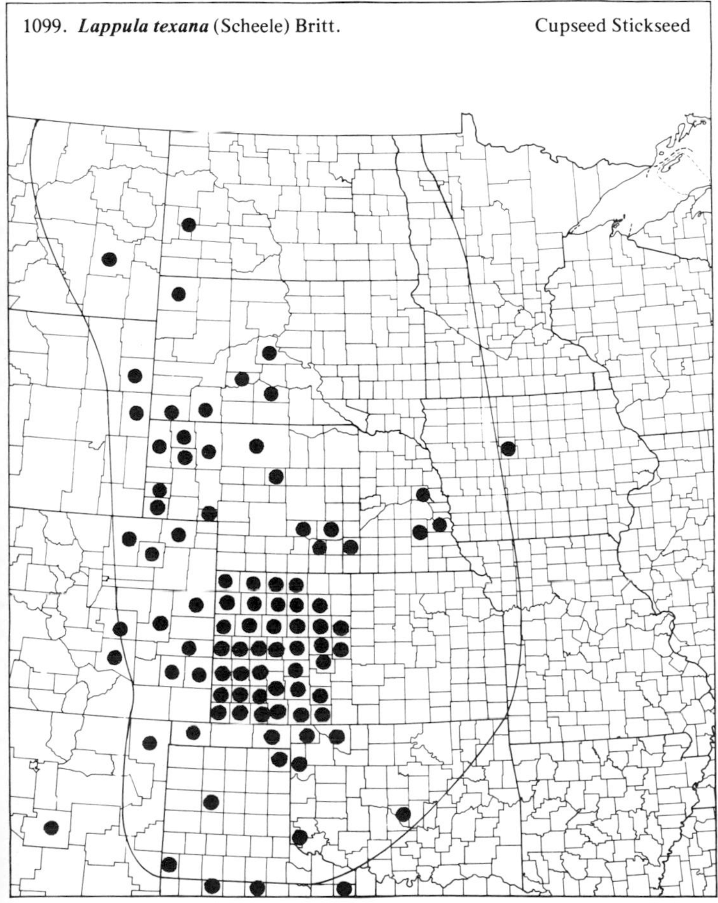

1100. ***Lithospermum arvense*** L. — Corn Gromwell

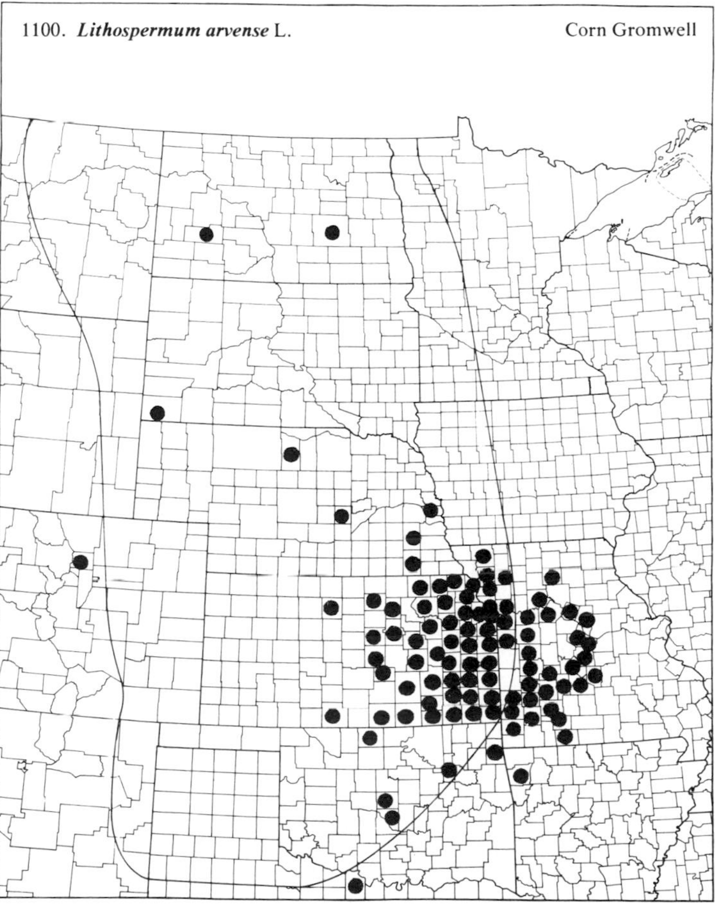

(Boraginaceae)

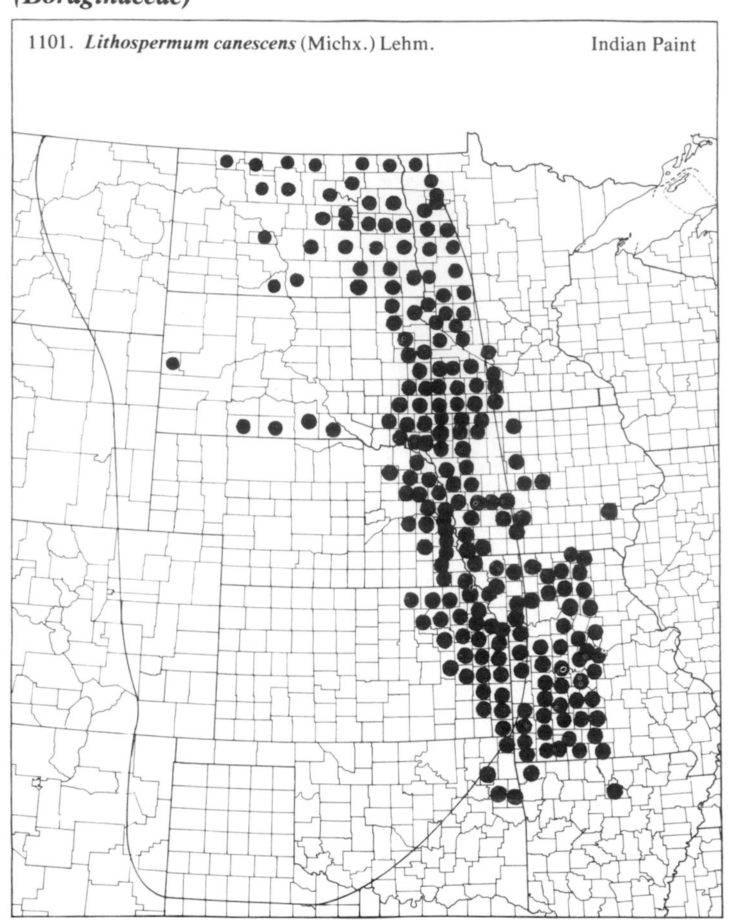

1101. ***Lithospermum canescens*** (Michx.) Lehm. Indian Paint

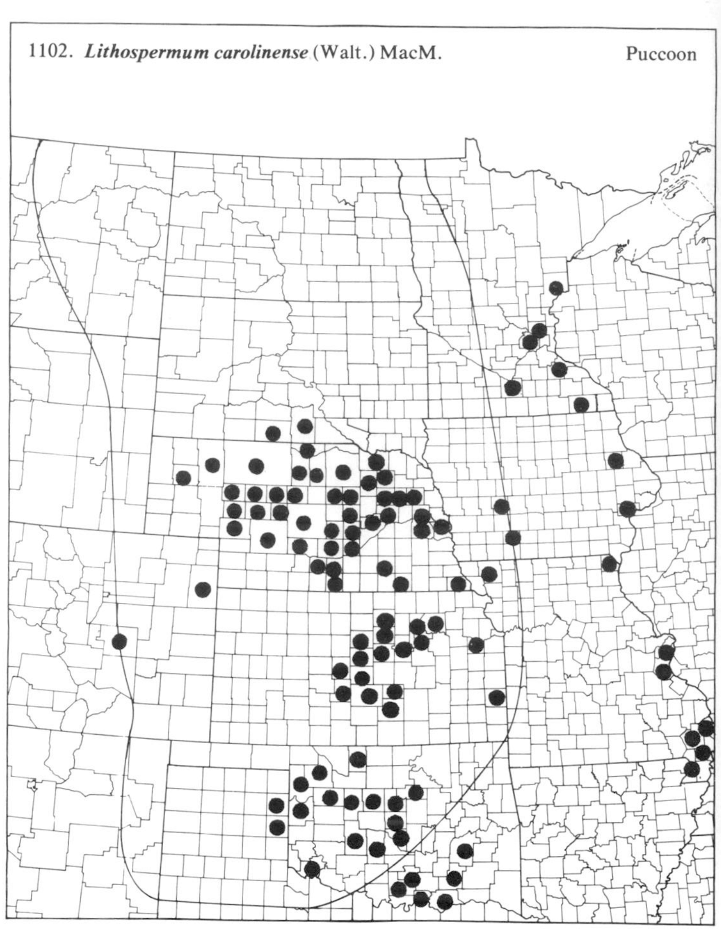

1102. ***Lithospermum carolinense*** (Walt.) MacM. Puccoon

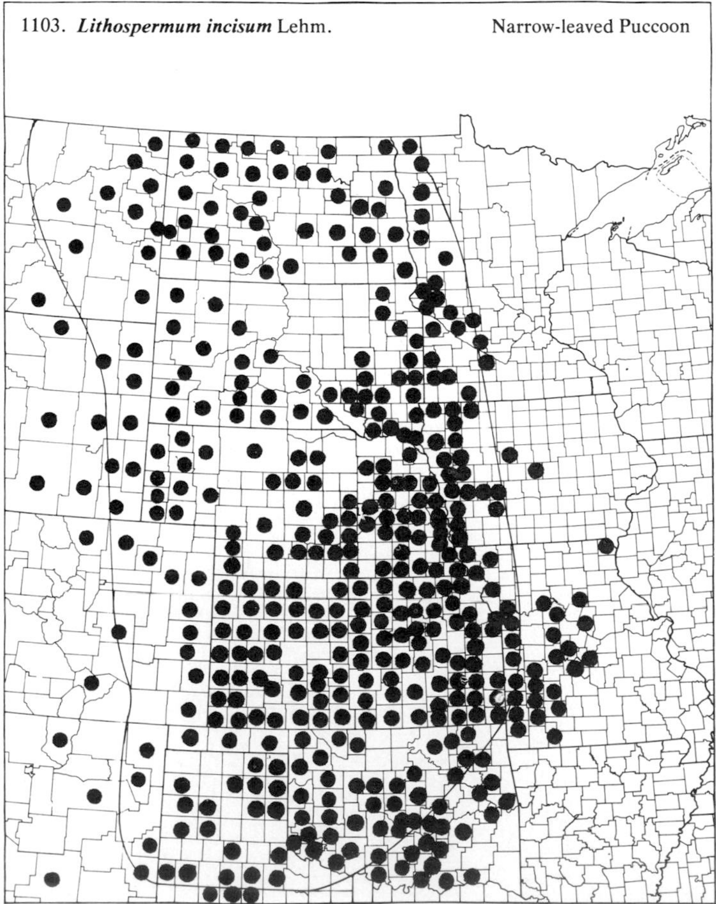

1103. ***Lithospermum incisum*** Lehm. Narrow-leaved Puccoon

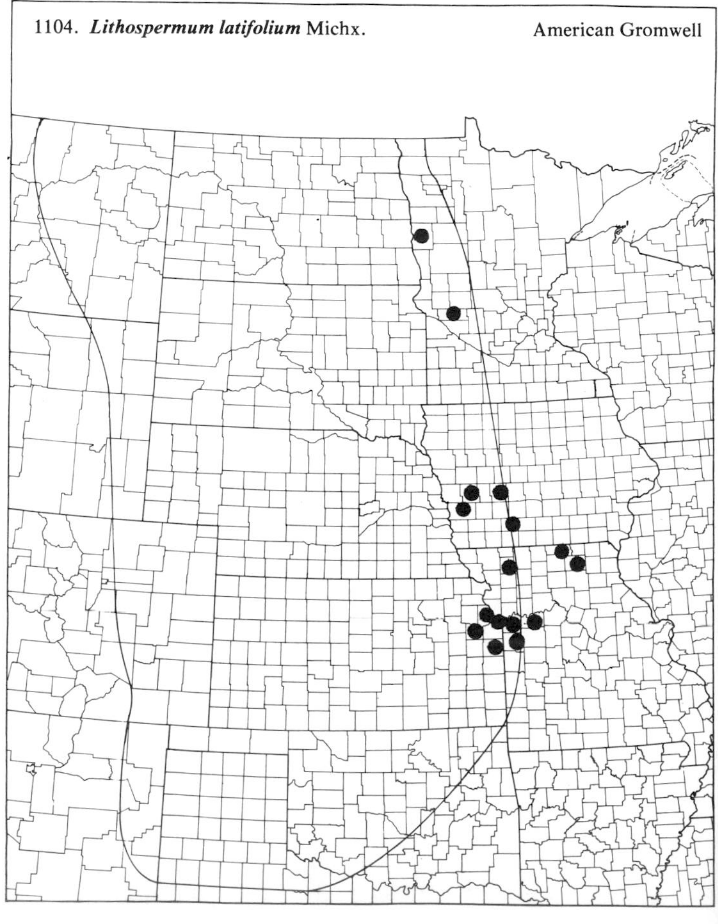

1104. ***Lithospermum latifolium*** Michx. American Gromwell

(Boraginaceae)

1105. **Mertensia lanceolata** (Pursh) A. DC. Wild Forget-me-not

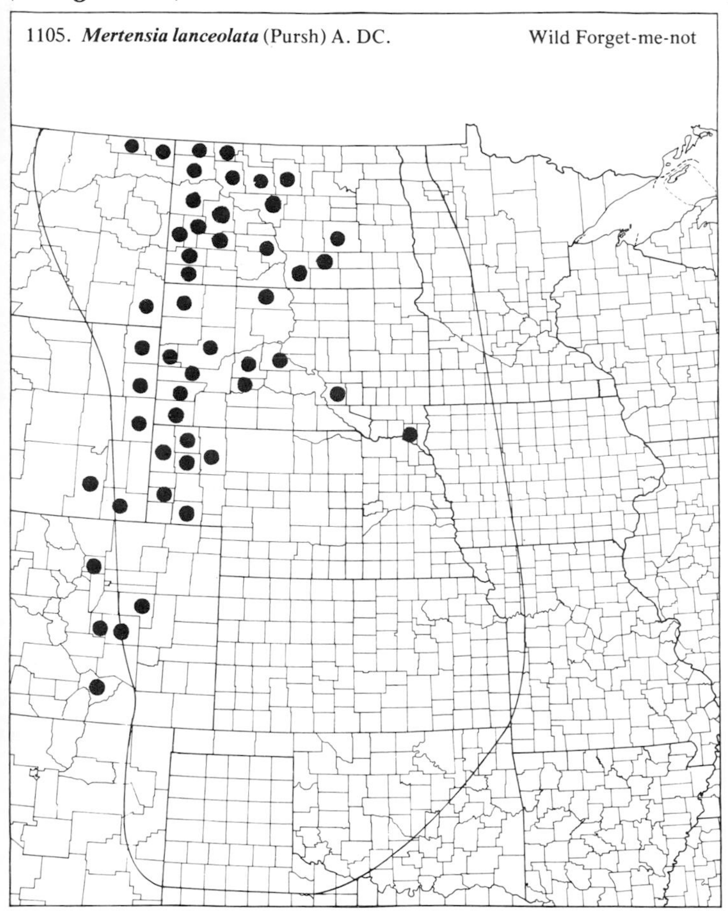

1106. **Mertensia oblongifolia** (Nutt.) G. Don

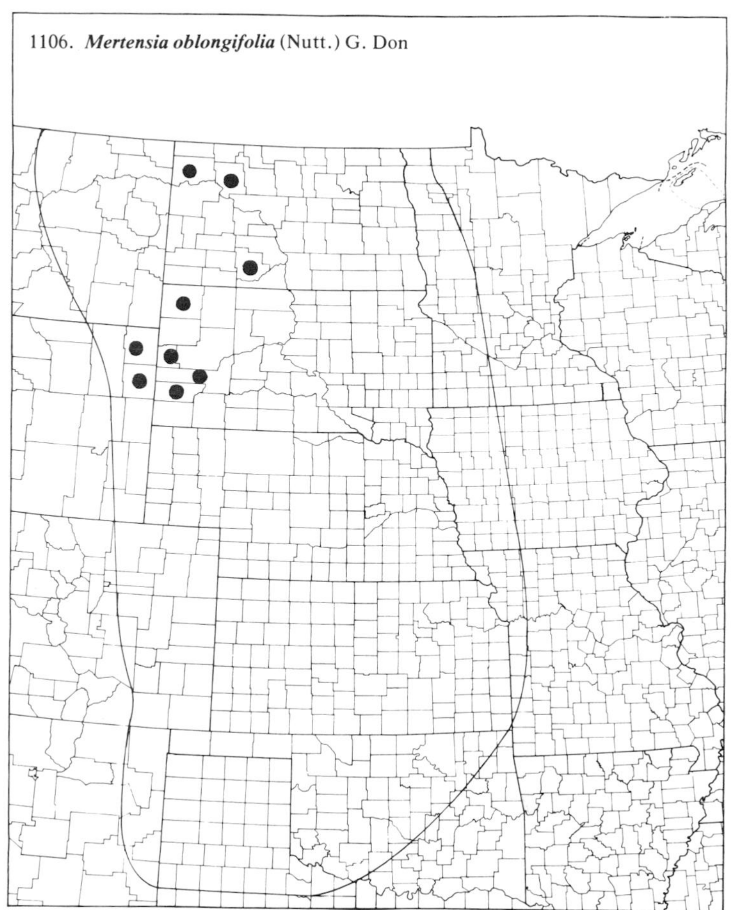

1107. **Myosotis verna** Nutt. Virginia Forget-me-not

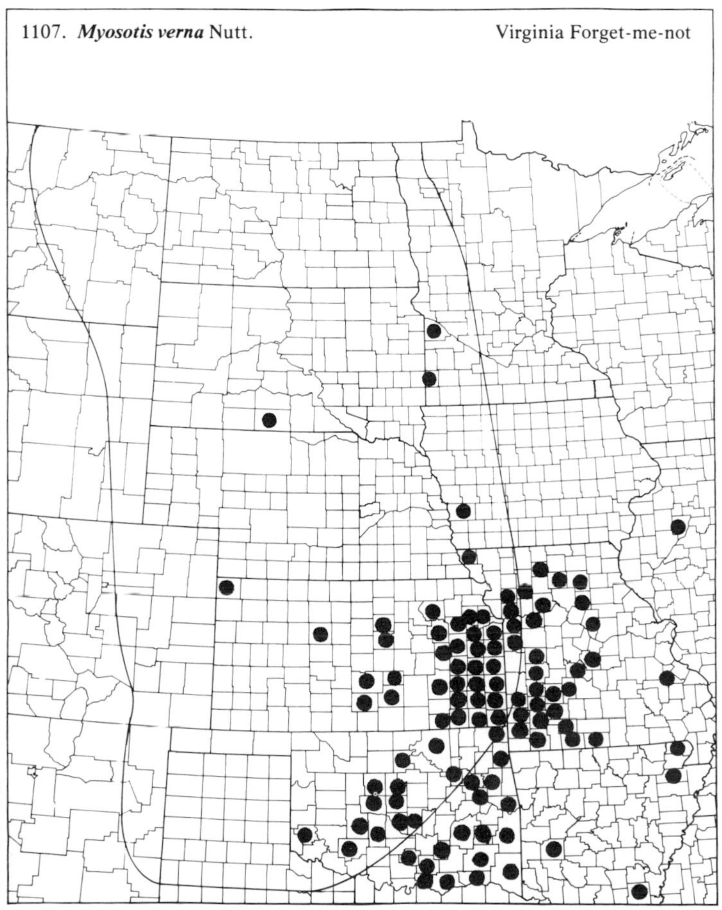

1108. **Onosmodium molle** Michx. var. **hispidissimum** (Mack.) Cronq. False Gromwell

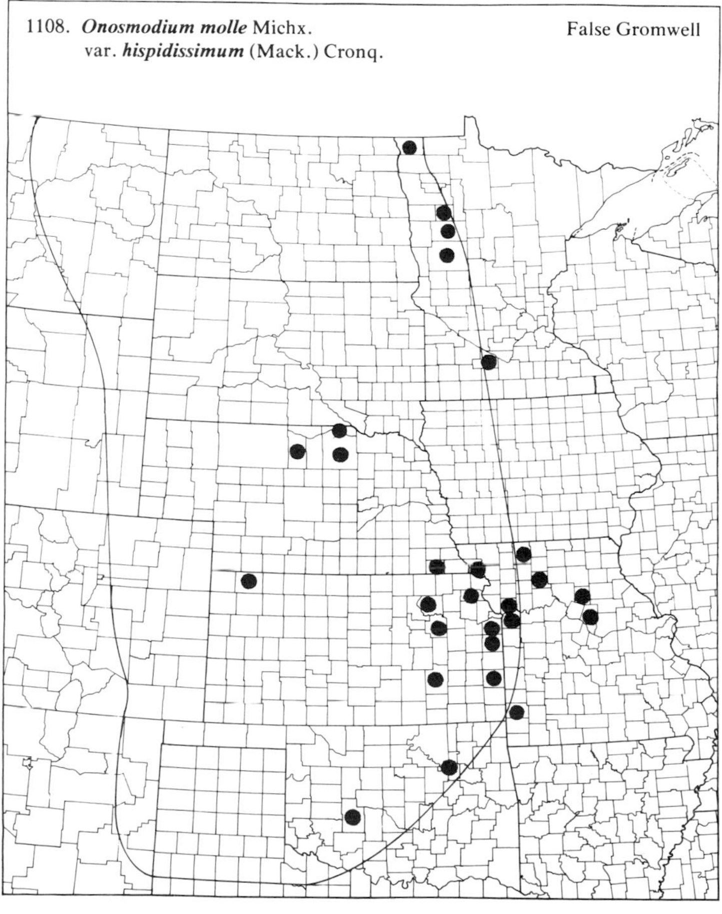

(Boraginaceae)

1109. *Onosmodium molle* Michx. var. *occidentale* (Mack.) I.M. Johnst. — False Gromwell

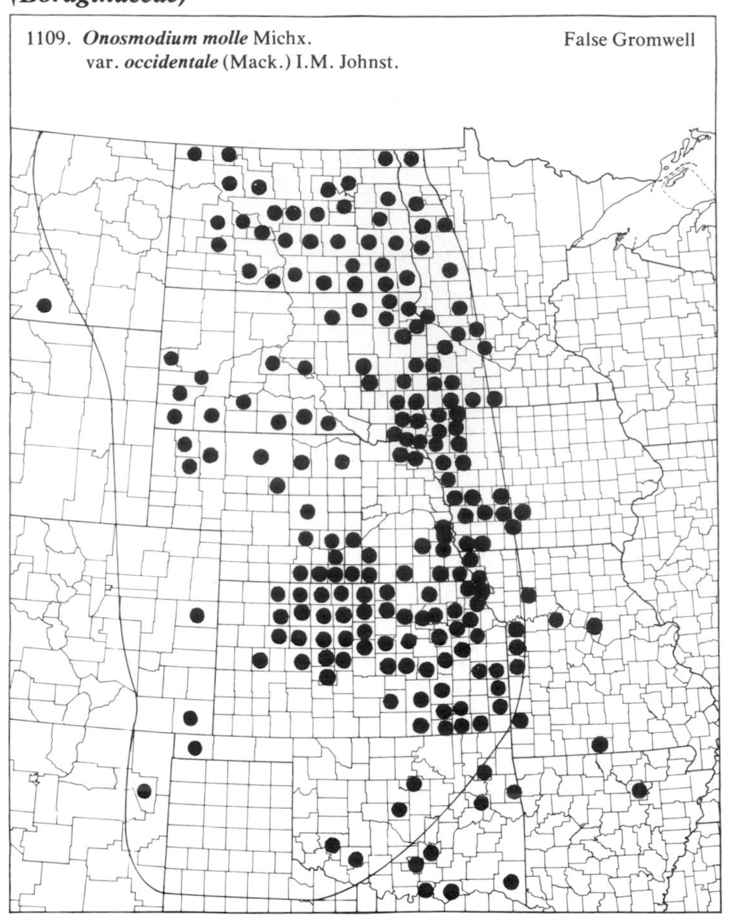

1110. *Plagiobothrys scouleri* (H. & A.) I.M. Johnst. — Popcorn-flower

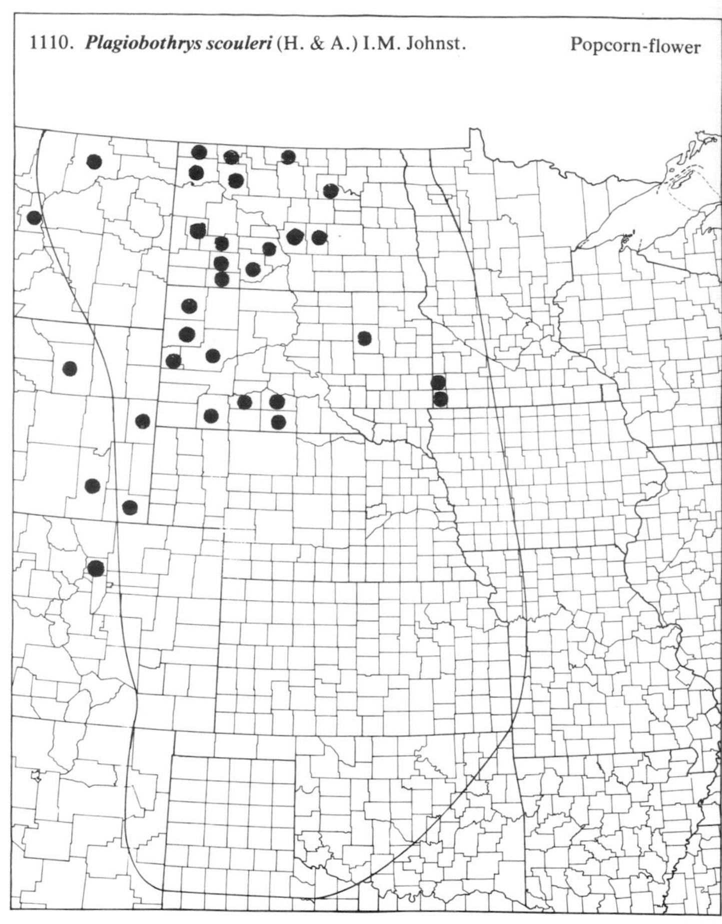

96. *Callitrichaceae* (Water Starwort Family)

1111. *Callitriche hermaphroditica* L. — Water Starwort

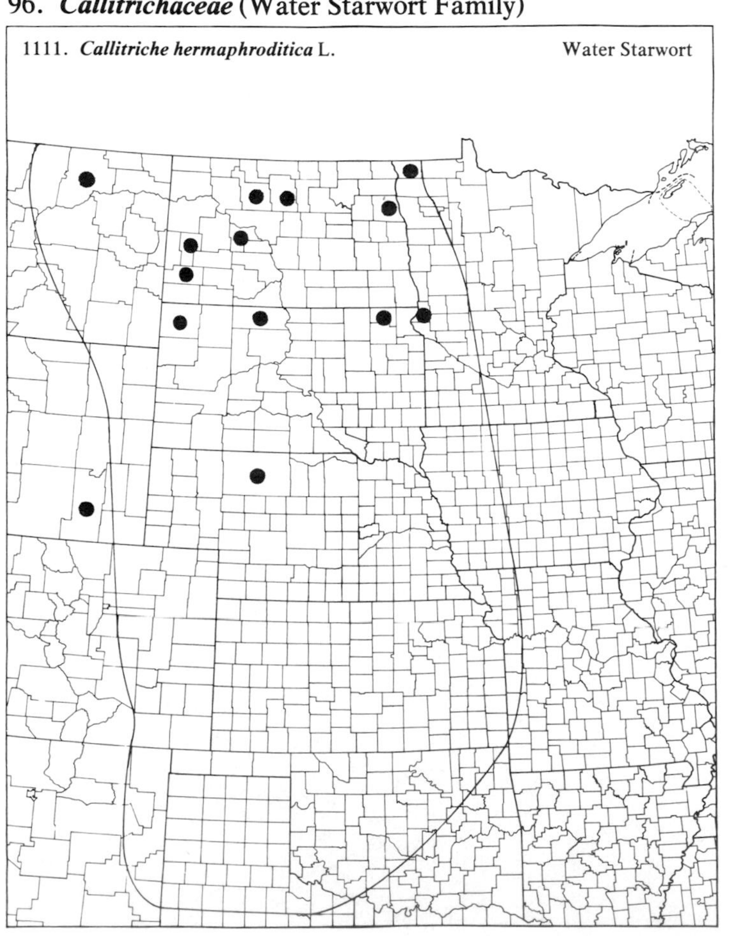

1112. *Callitriche heterophylla* Pursh — Water Starwort

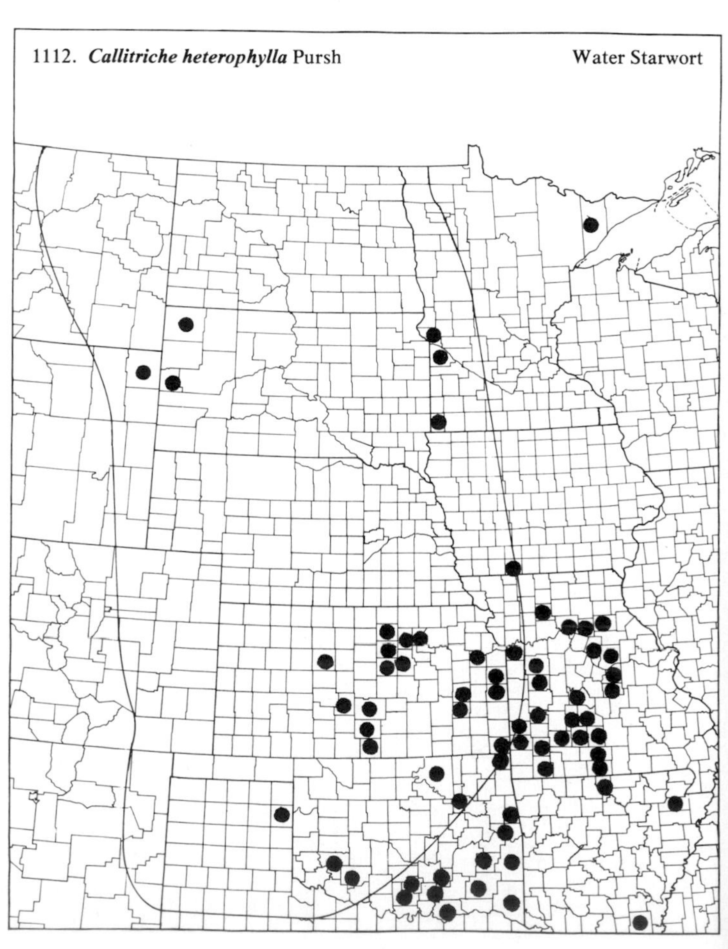

(Callitrichaceae)

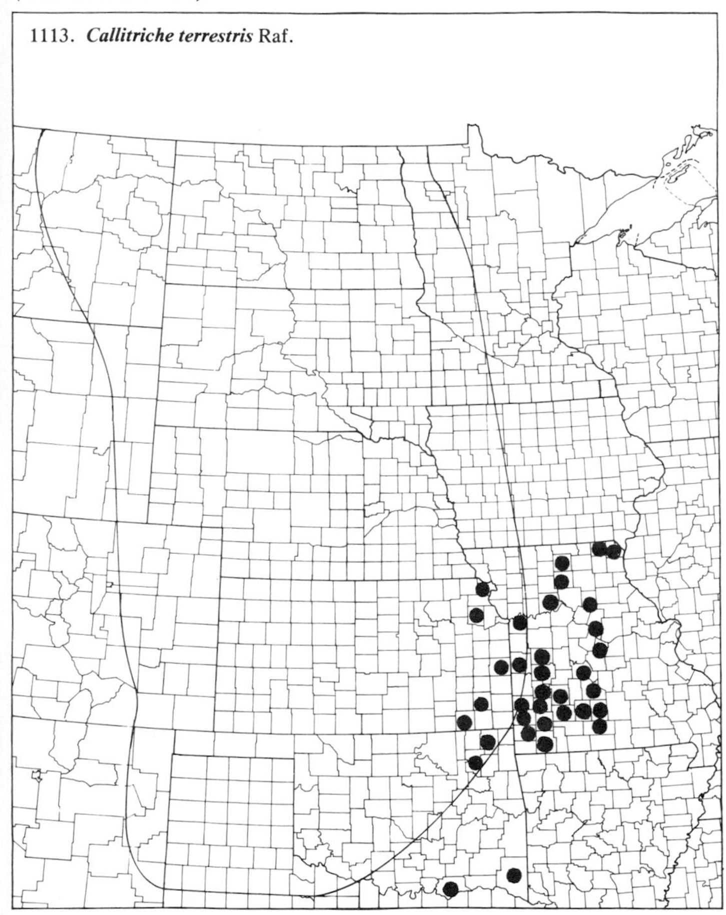
1113. *Callitriche terrestris* Raf.

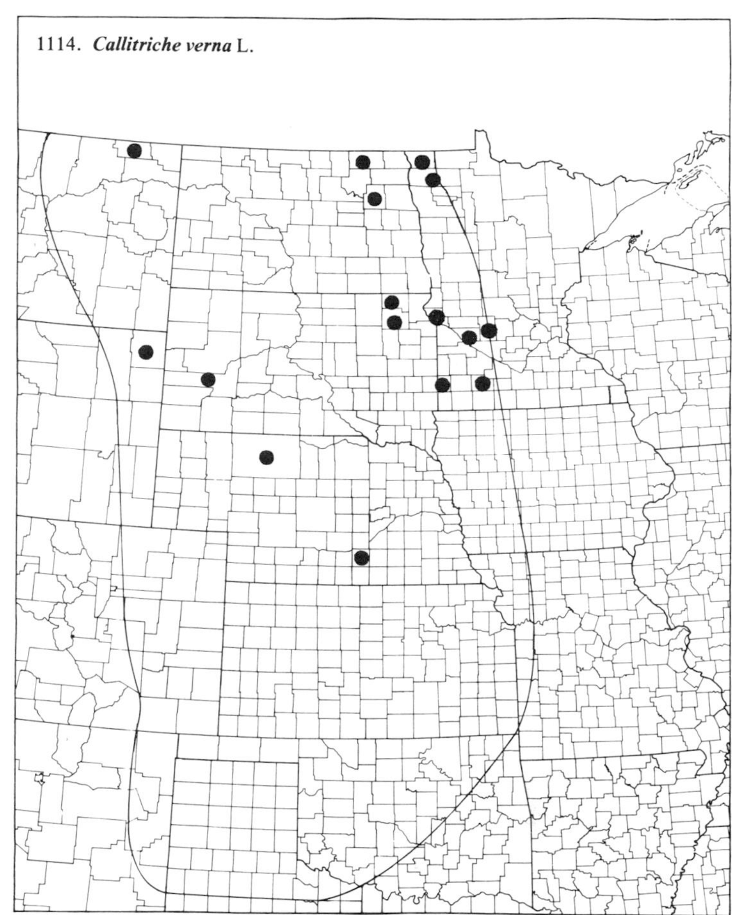
1114. *Callitriche verna* L.

97. *Verbenaceae* (Vervain Family)

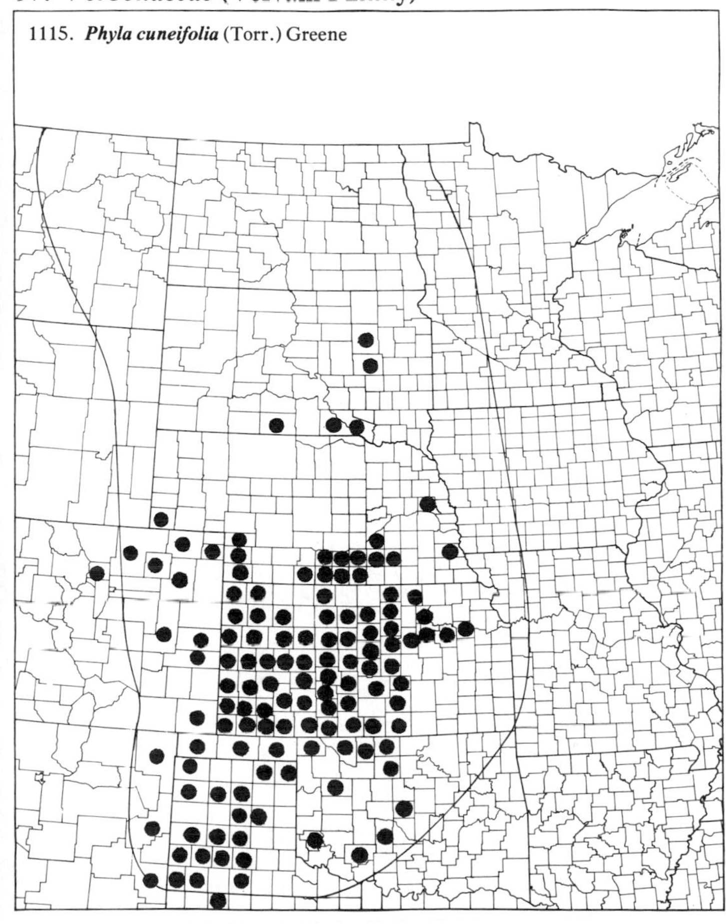
1115. *Phyla cuneifolia* (Torr.) Greene

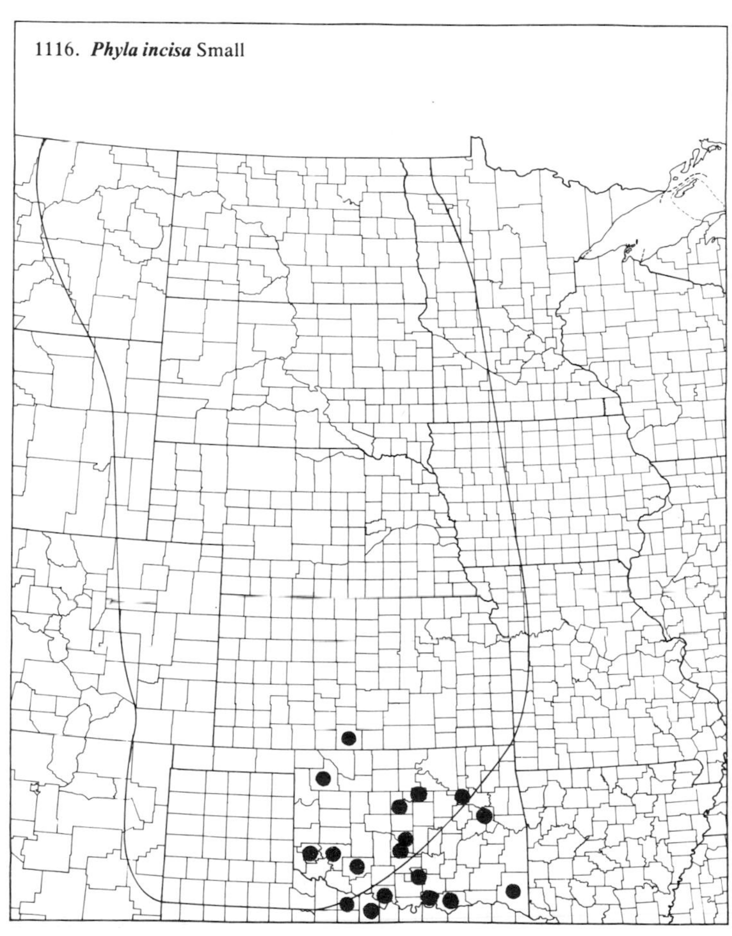
1116. *Phyla incisa* Small

(Verbenaceae)

1117. ***Phyla lanceolata*** (Michx.) Greene — Fog Fruit

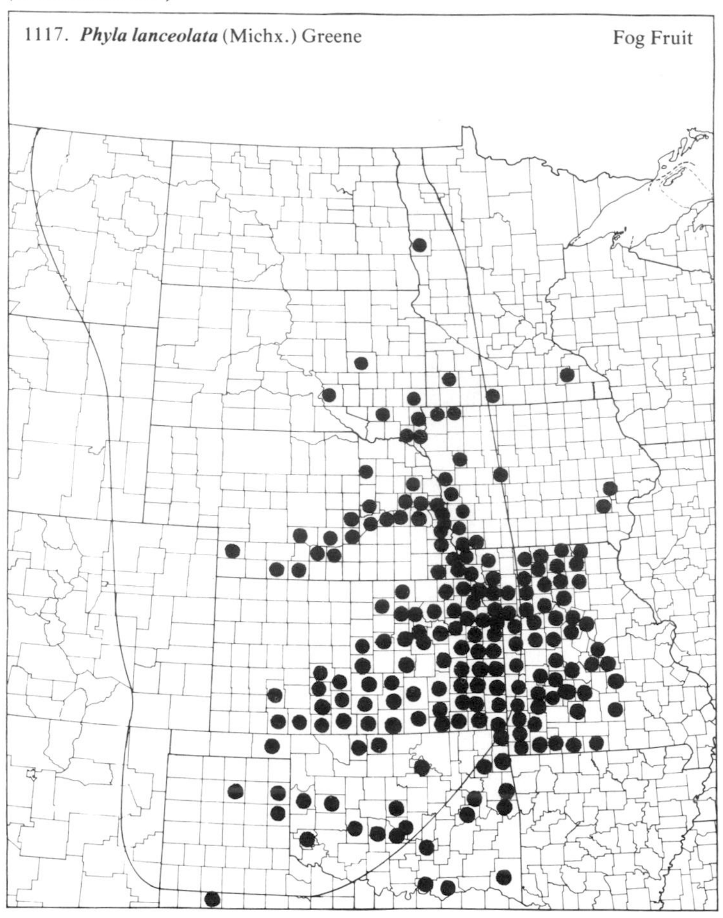

1118. ***Verbena ambrosifolia*** Rydb.

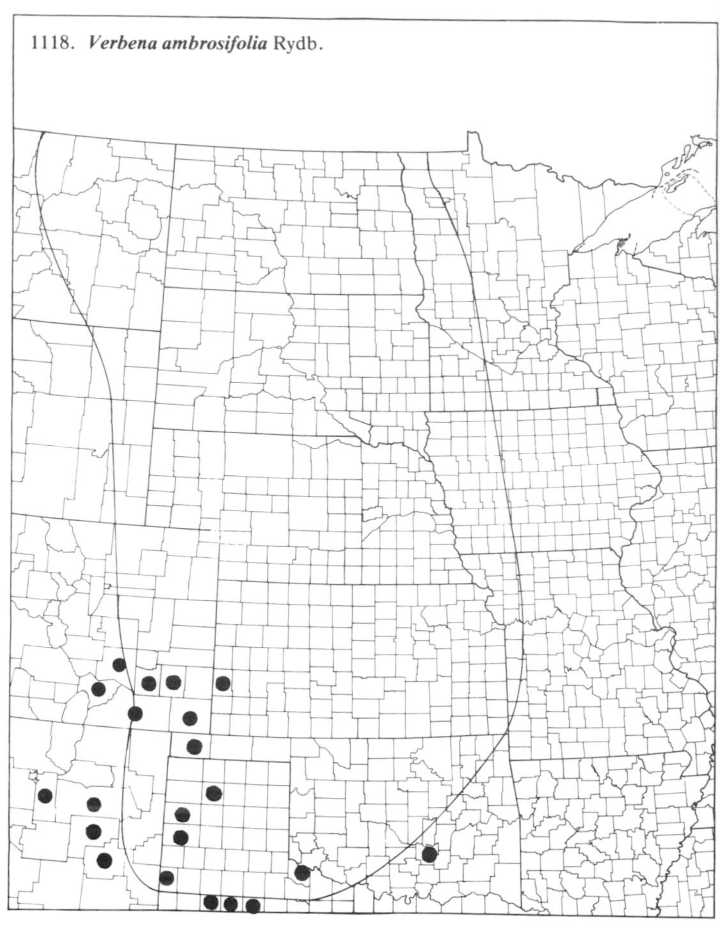

1119. ***Verbena bipinnatifida*** Nutt. — Vervain

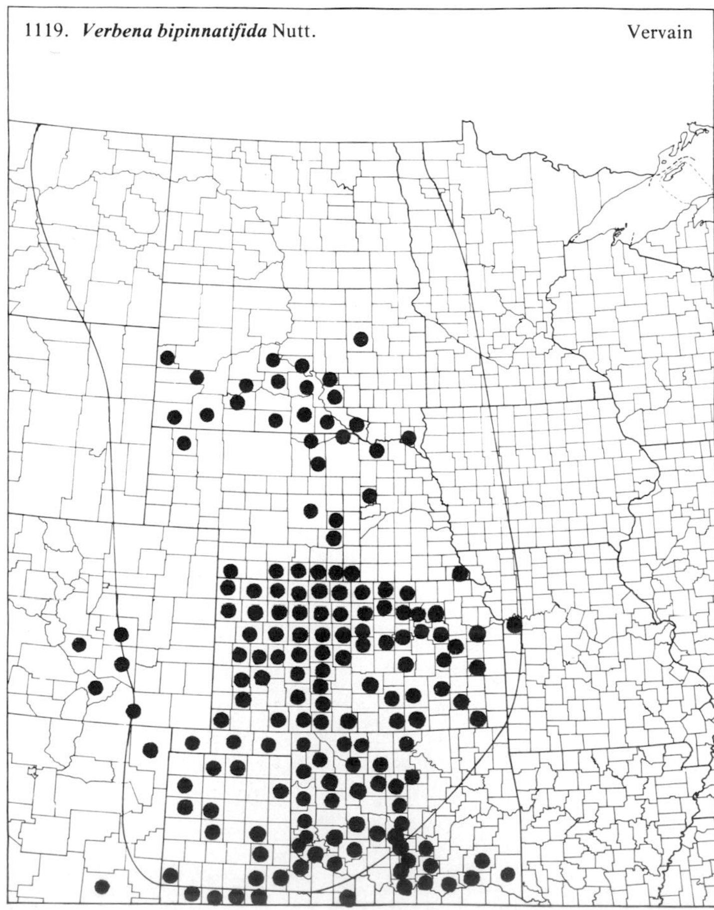

1120. ***Verbena bracteata*** Lag. & Rodr. — Bracted Vervain

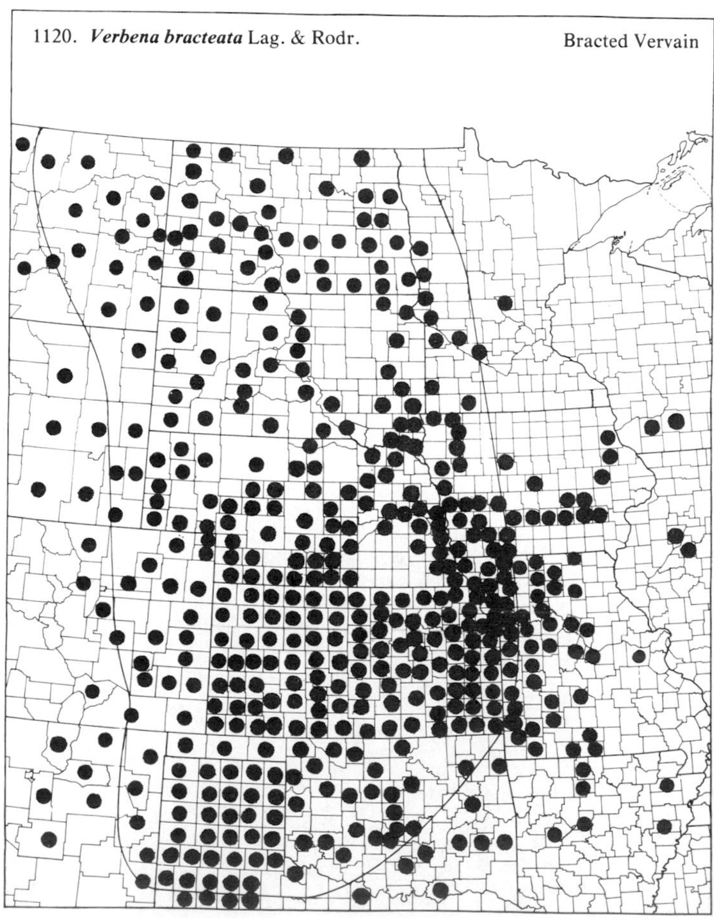

(Verbenaceae)

1121. ***Verbena canadensis*** (L.) Britt. Rose Vervain

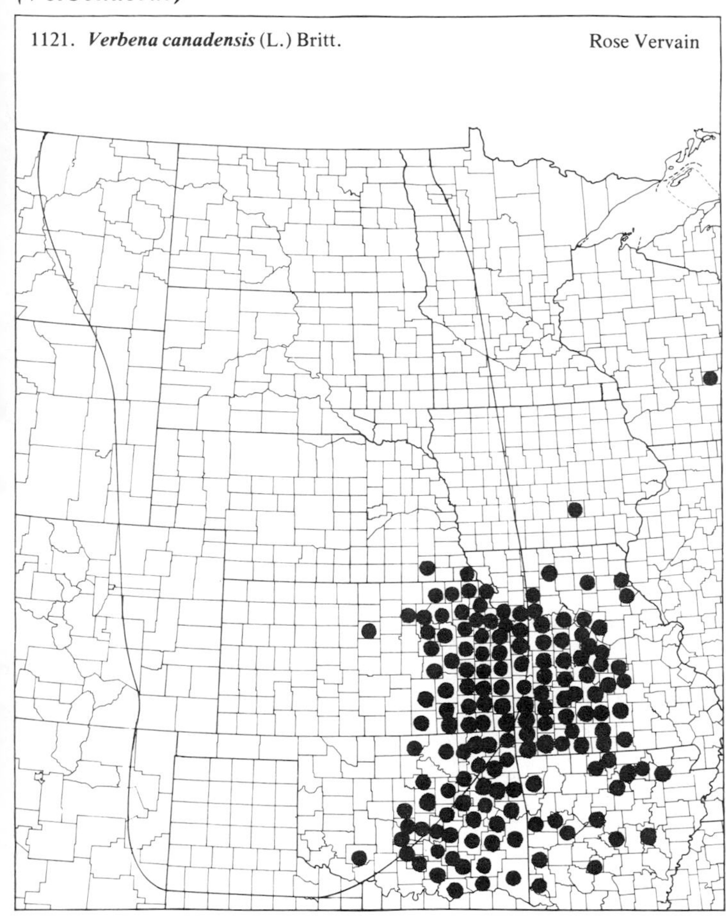

1122. ***Verbena hastata*** L. Blue Vervain

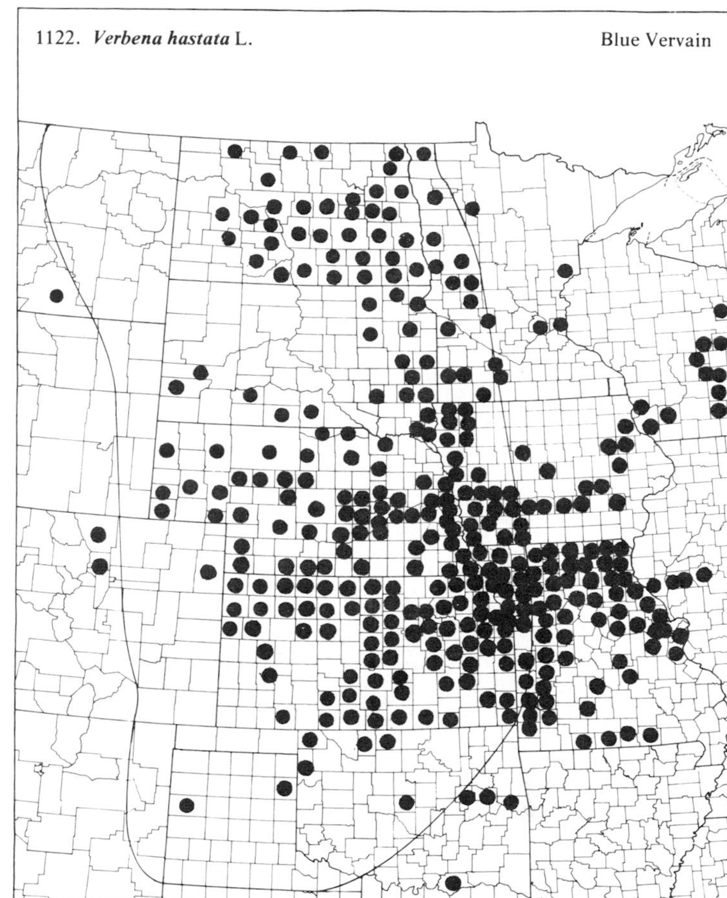

1123. ***Verbena pumila*** Rydb.

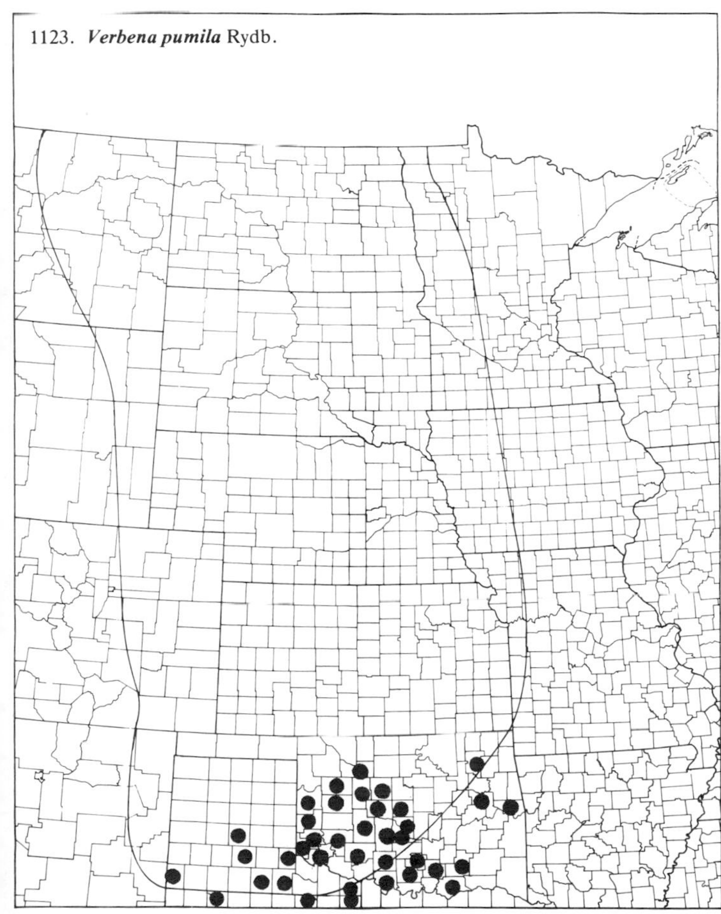

1124. ***Verbena simplex*** Lehm. Narrow-leaved Verbena

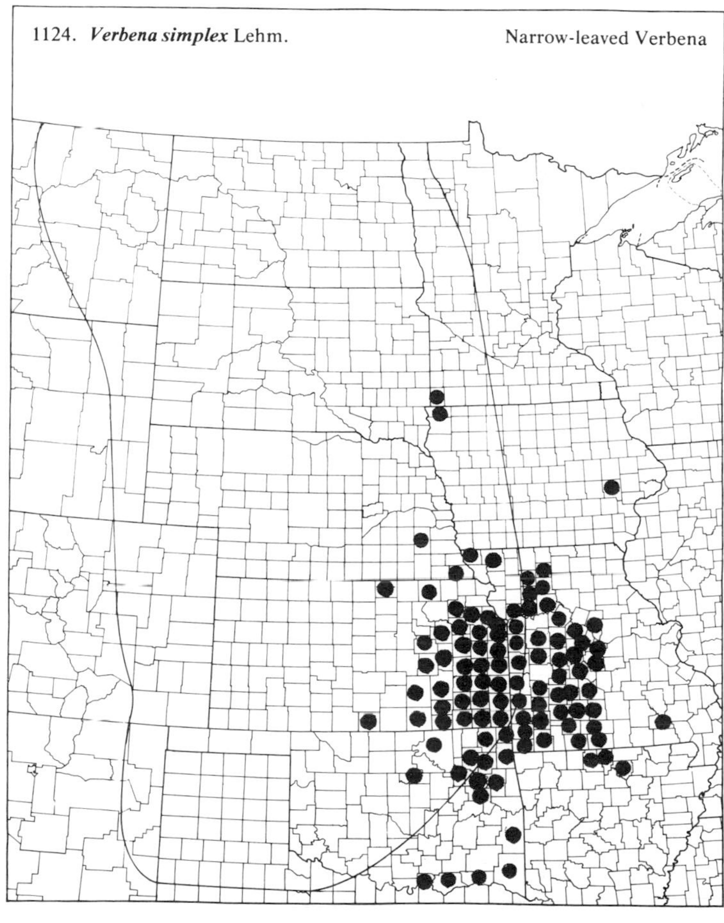

(Verbenaceae)

1125. ***Verbena stricta*** Vent. Hoary Vervain

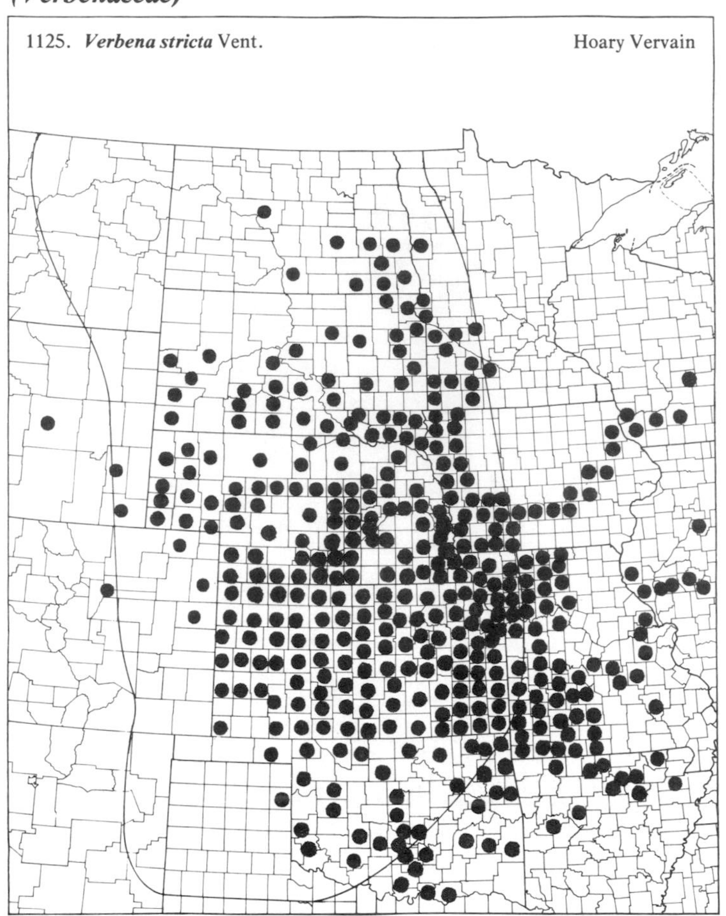

1126. ***Verbena urticifolia*** L. Nettle-leaved Vervain

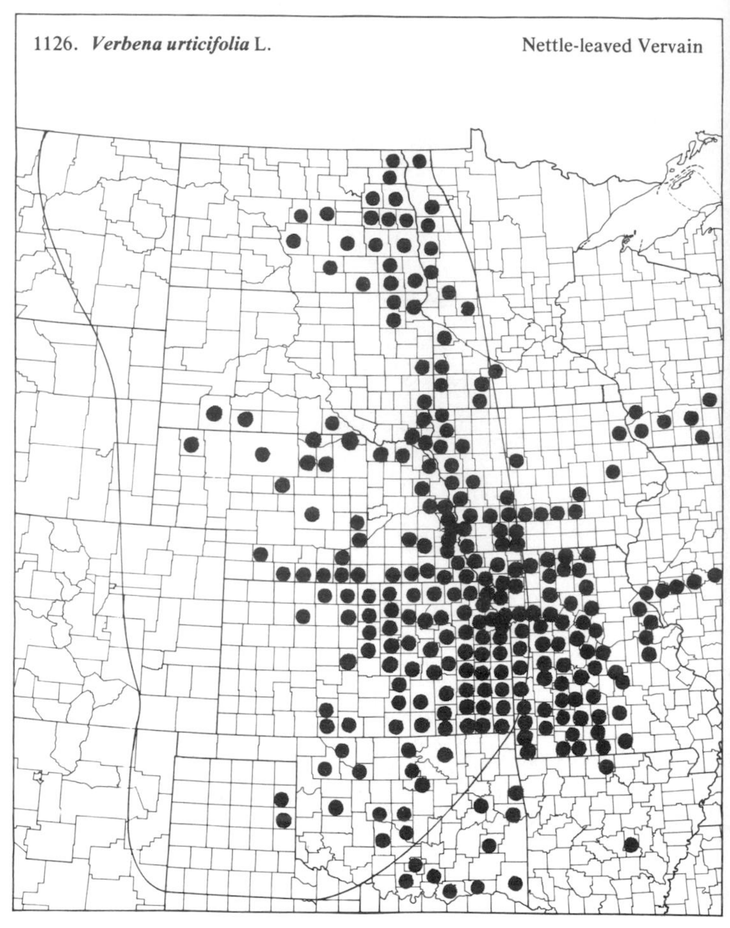

98. ***Phrymaceae*** (Lopseed Family)

1127. ***Phryma leptostachya*** L. Lopseed

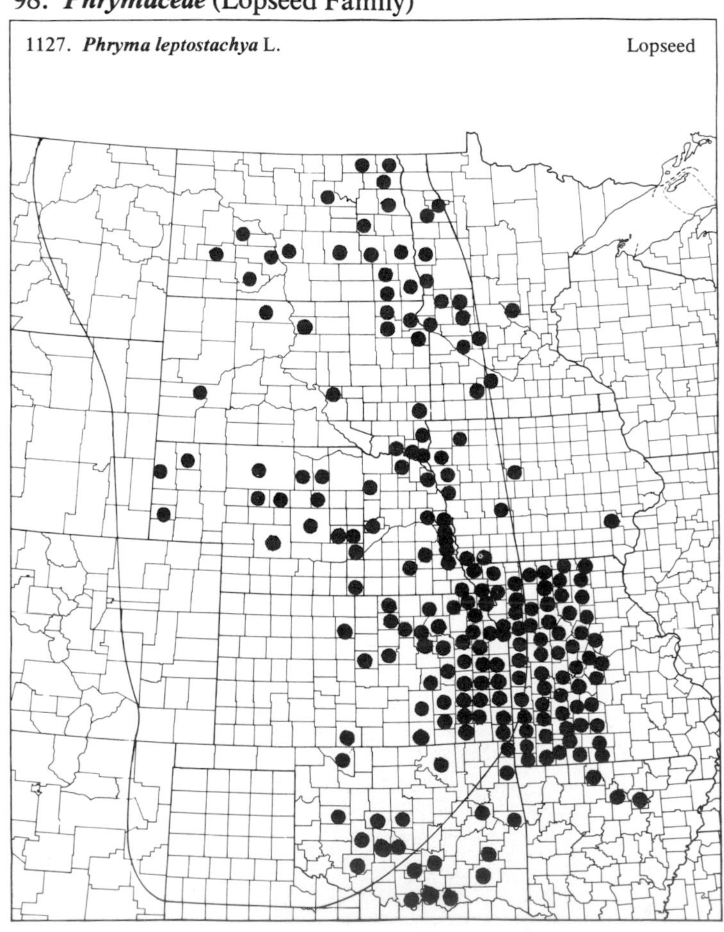

99. ***Lamiaceae*** (Mint Family)

1128. ***Agastache foeniculum*** (Pursh) O. Ktze. Lavender Hyssop

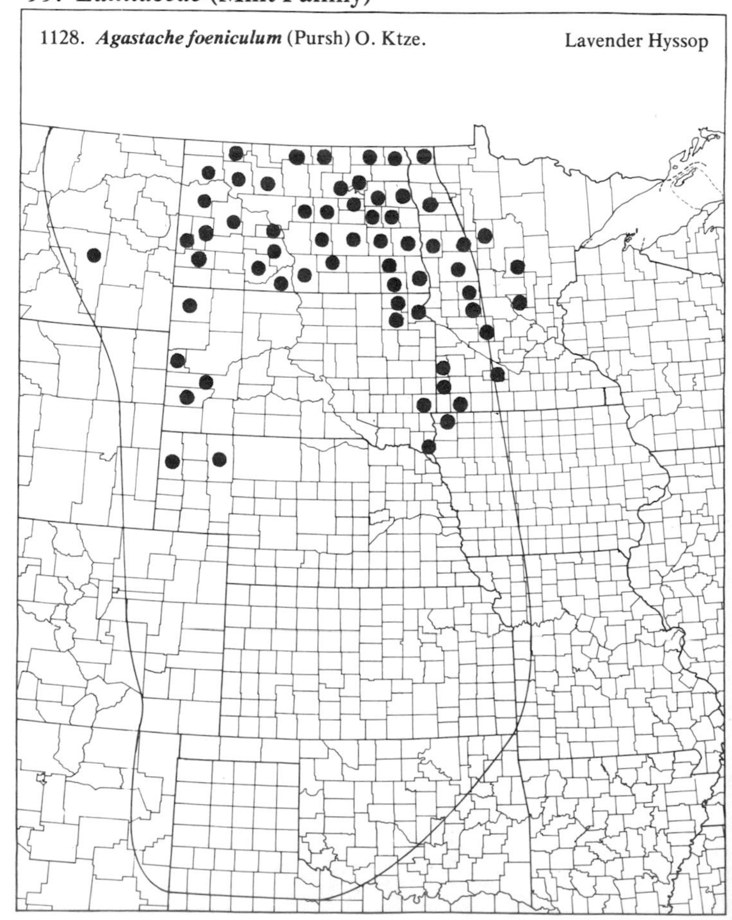

(Lamiaceae)

1129. ***Agastache nepetoides*** (L.) O. Ktze. Catnip Giant Hyssop

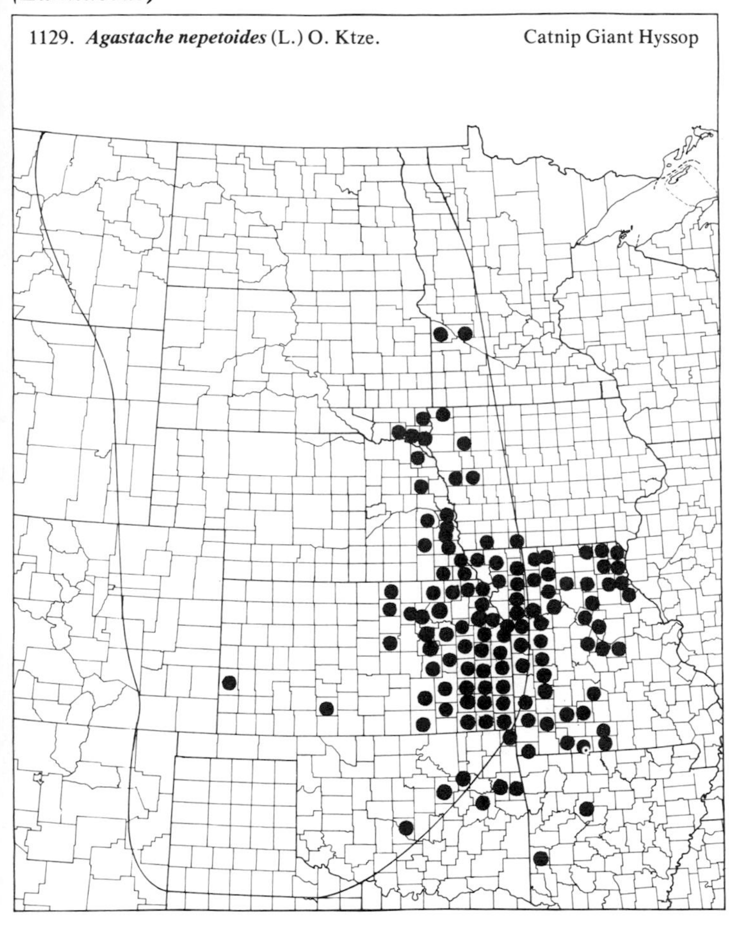

1130. ***Agastache scrophulariaefolia*** (Willd.) O. Ktze. Purple Giant Hyssop

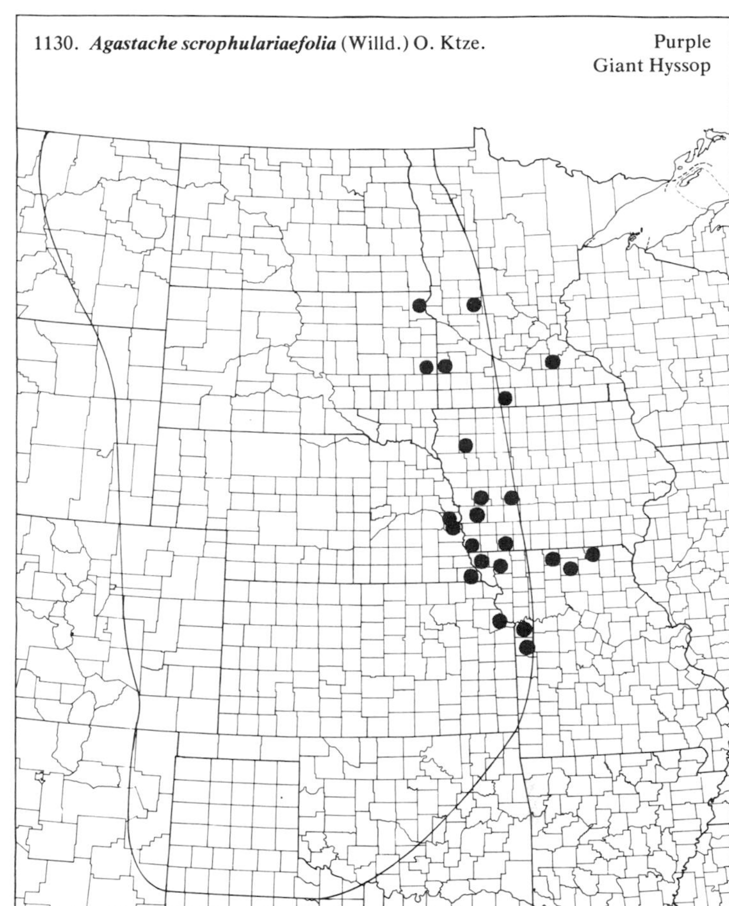

1131. ***Blephilia hirsuta*** (Pursh) Benth. Wood Mint

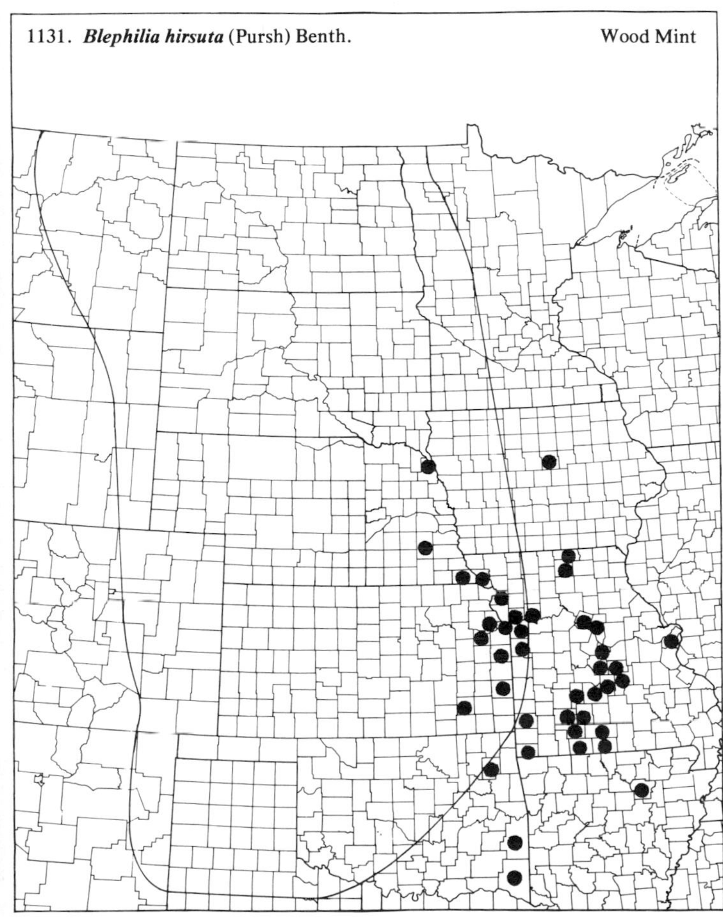

1132. ***Dracocephalum parviflorum*** Nutt. Dragonhead

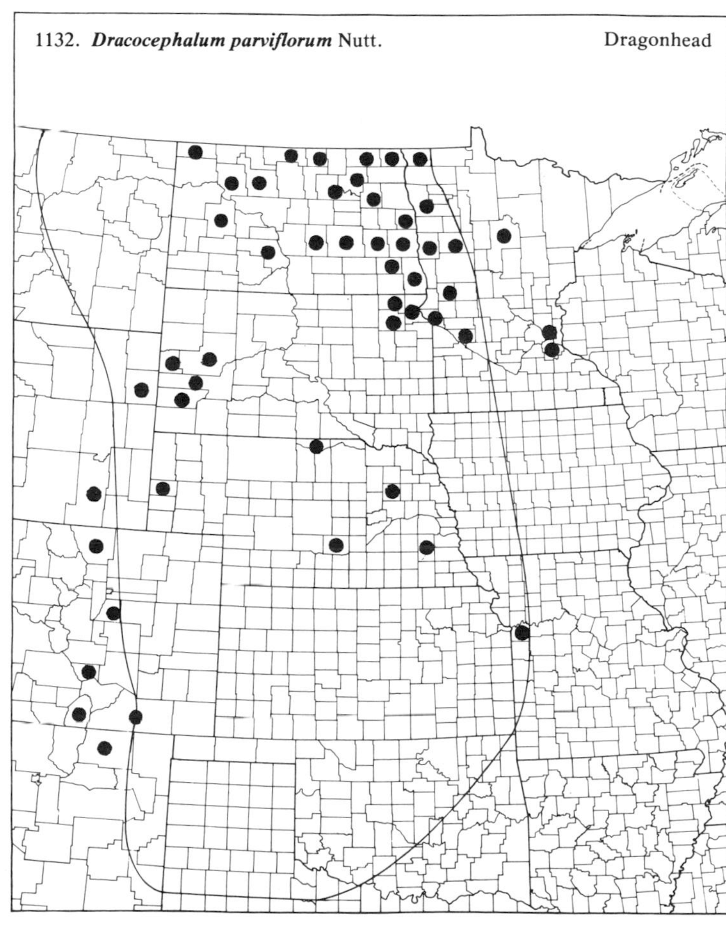

(Lamiaceae)

1133. *Galeopsis tetrahit* L. Hemp Nettle

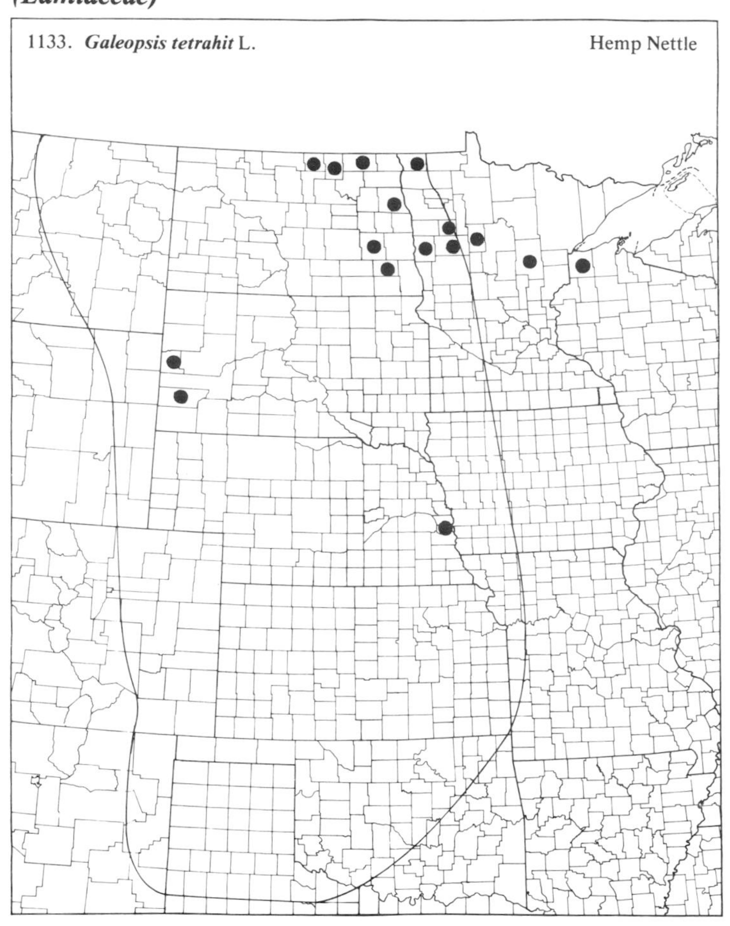

1134. *Glecoma hederacea* L. Ground Ivy

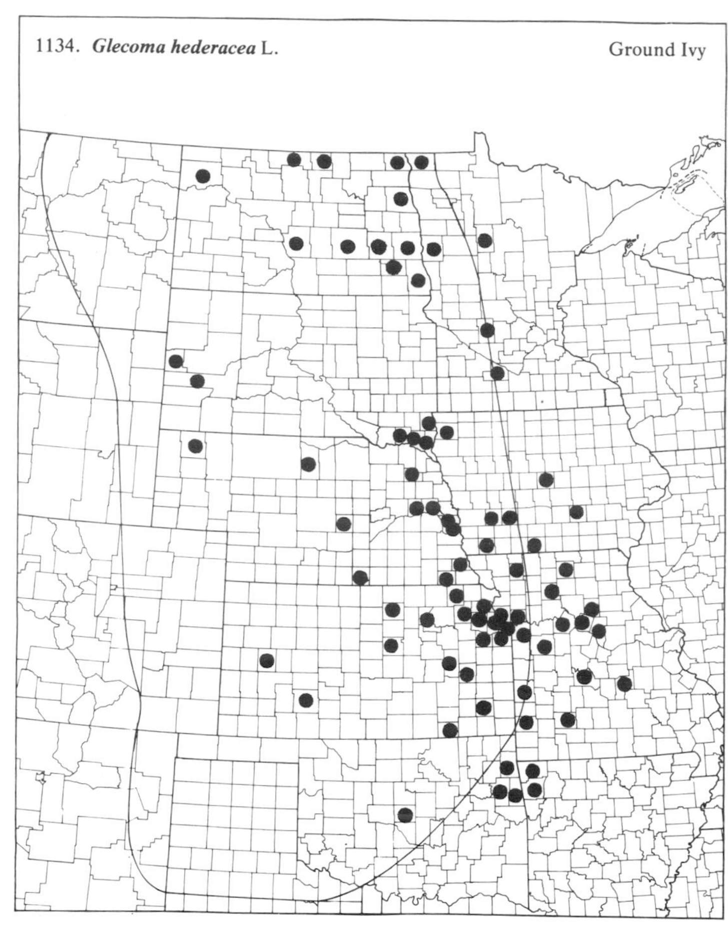

1135. *Hedeoma drummondii* Benth. Drummond False Pennyroyal

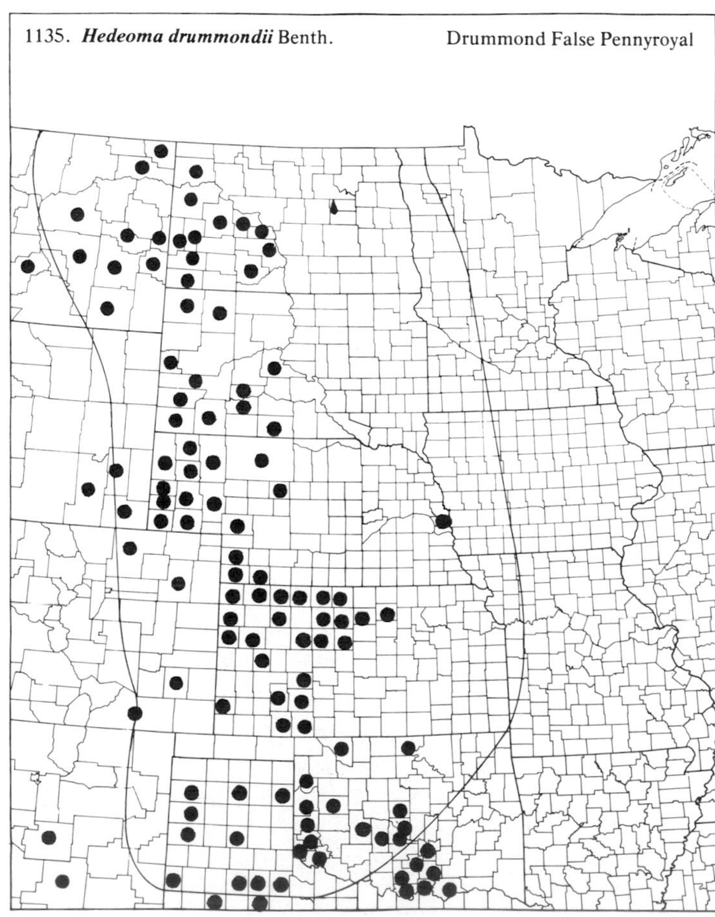

1136. *Hedeoma hispida* Pursh Rough Pennyroyal

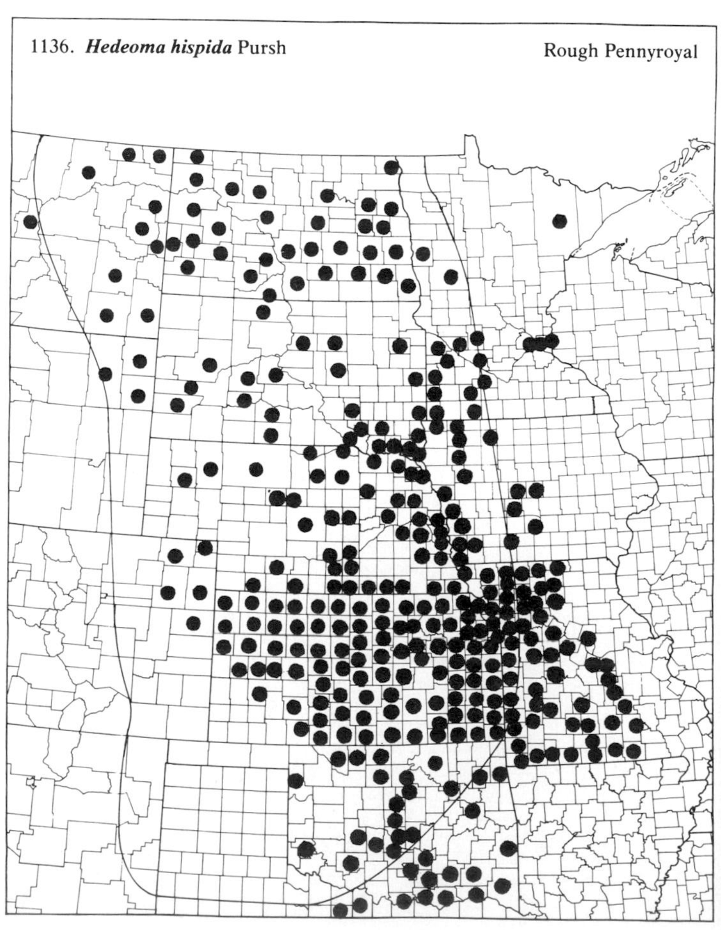

(Lamiaceae)

1137. **Hedeoma pulegioides** (L.) Pers. Pennyroyal

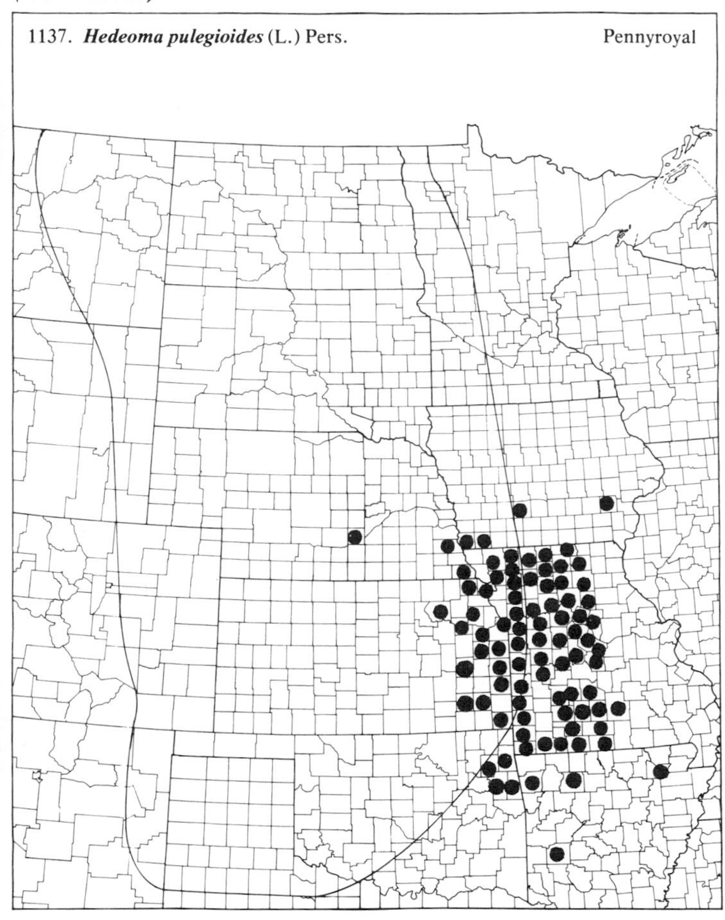

1138. **Isanthus brachiatus** (L.) B.S.P. False Pennyroyal

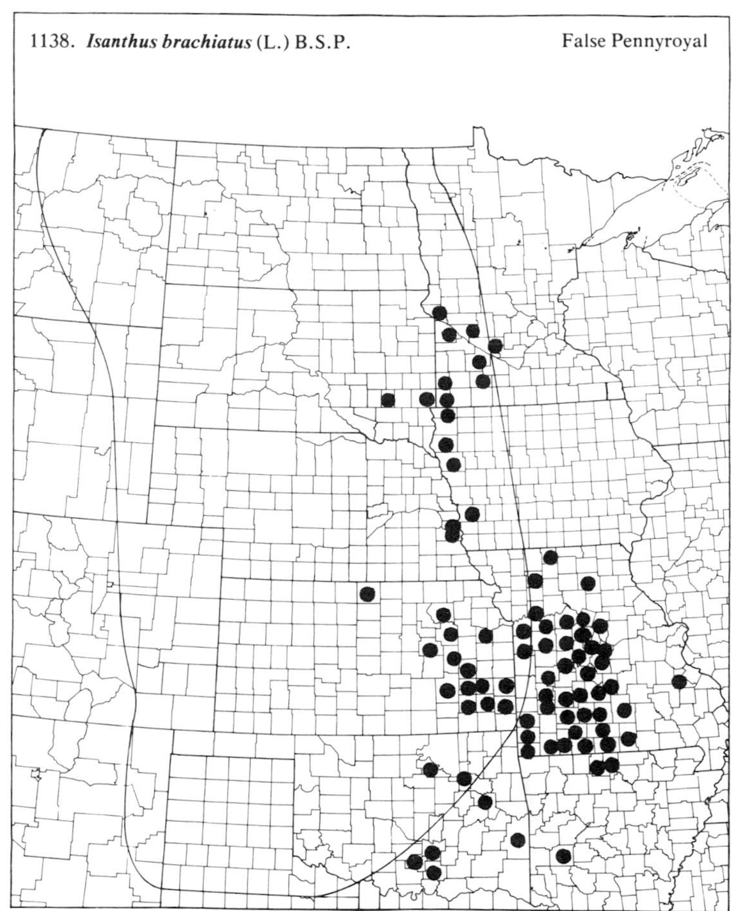

1139. **Lamium amplexicaule** L. Henbit

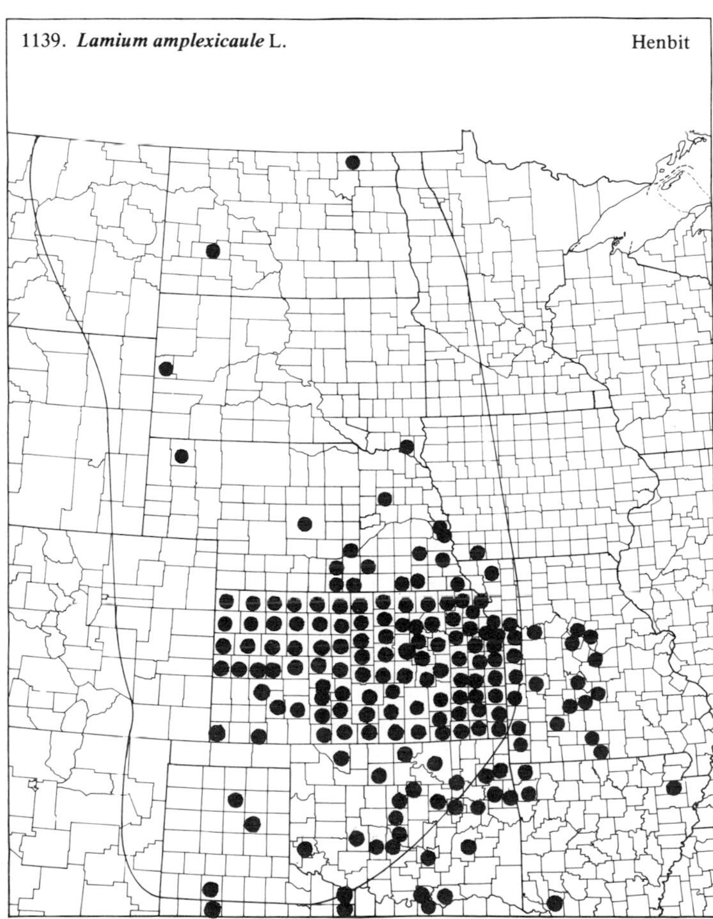

1140. **Lamium purpureum** L. Purple Dead Nettle

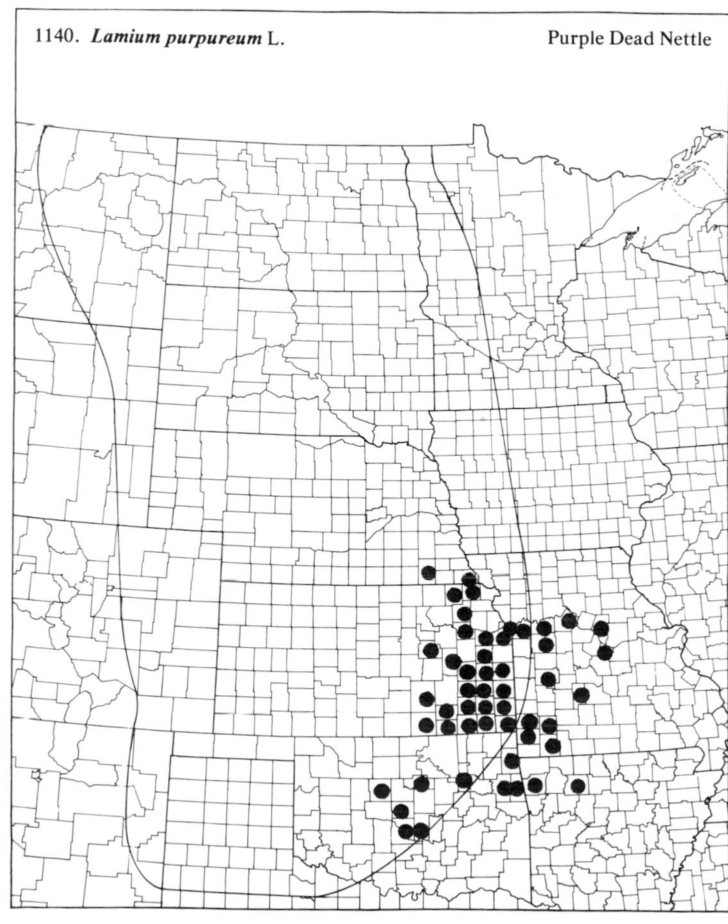

(Lamiaceae)

1141. **Leonurus cardiaca** L. Common Motherwort

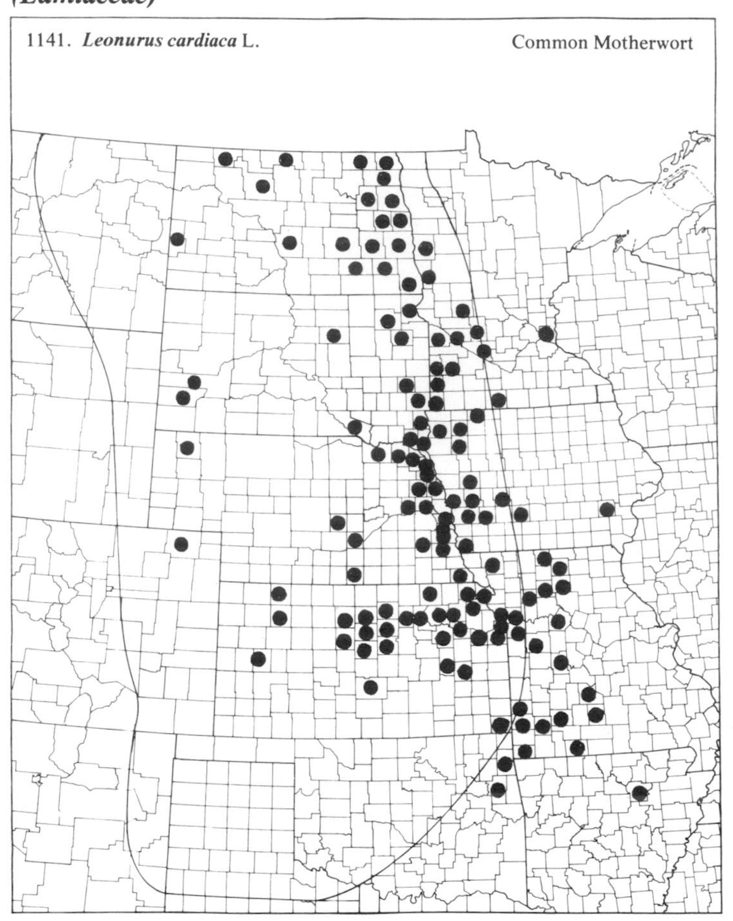

1142. **Lycopus americanus** Muhl. American Bugleweed

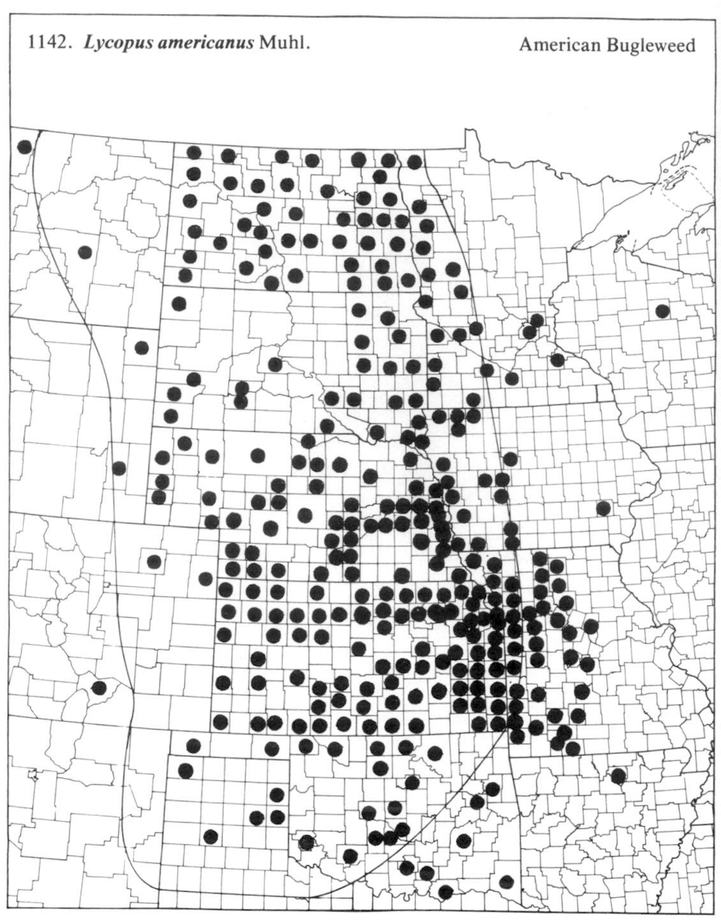

1143. **Lycopus asper** Greene Rough Bugleweed

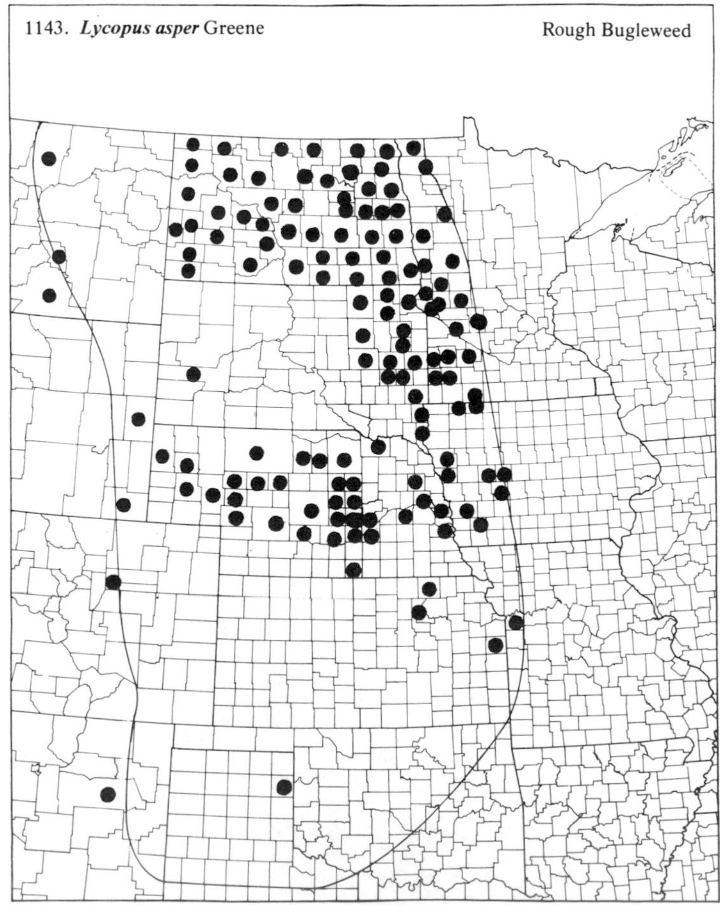

1144. **Lycopus uniflorus** Michx. Bugleweed

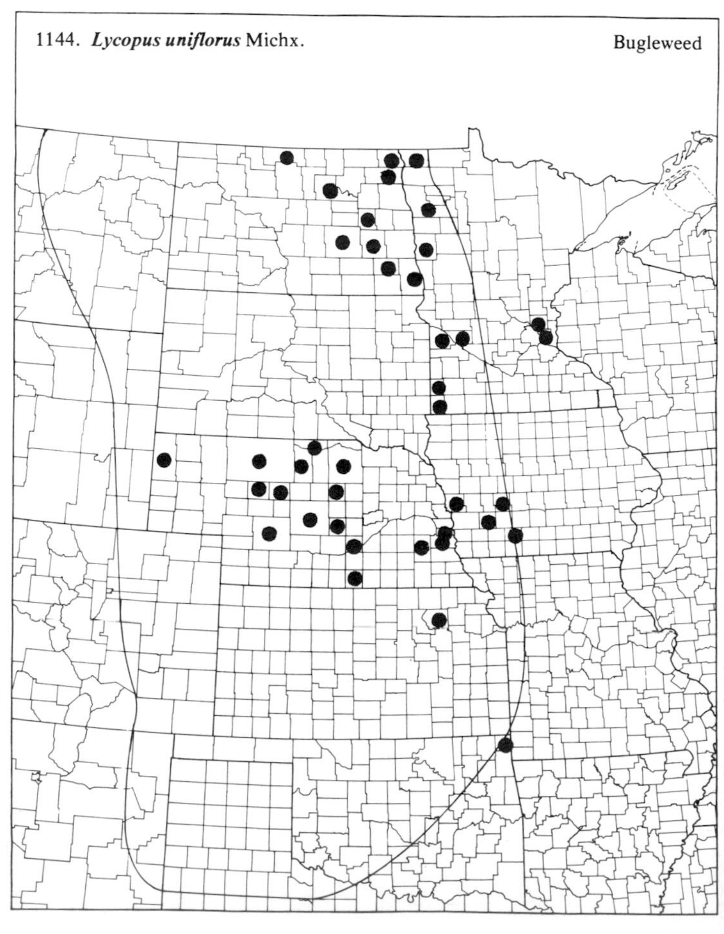

(Lamiaceae)

1145. ***Lycopus virginicus*** L. Virginia Bugleweed

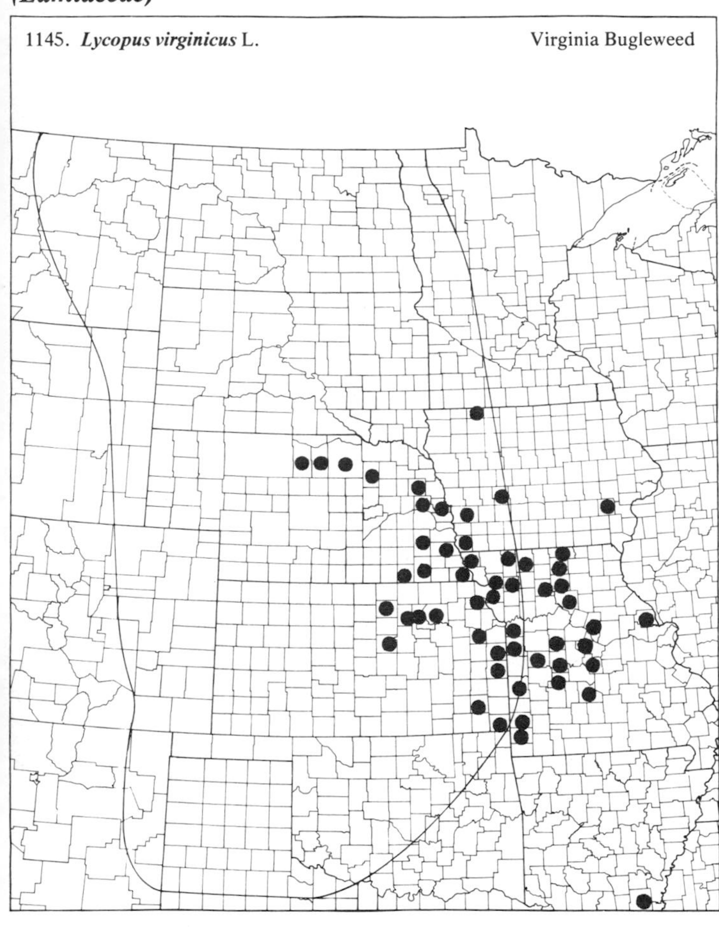

1146. ***Marrubium vulgare*** L. Common Horehound

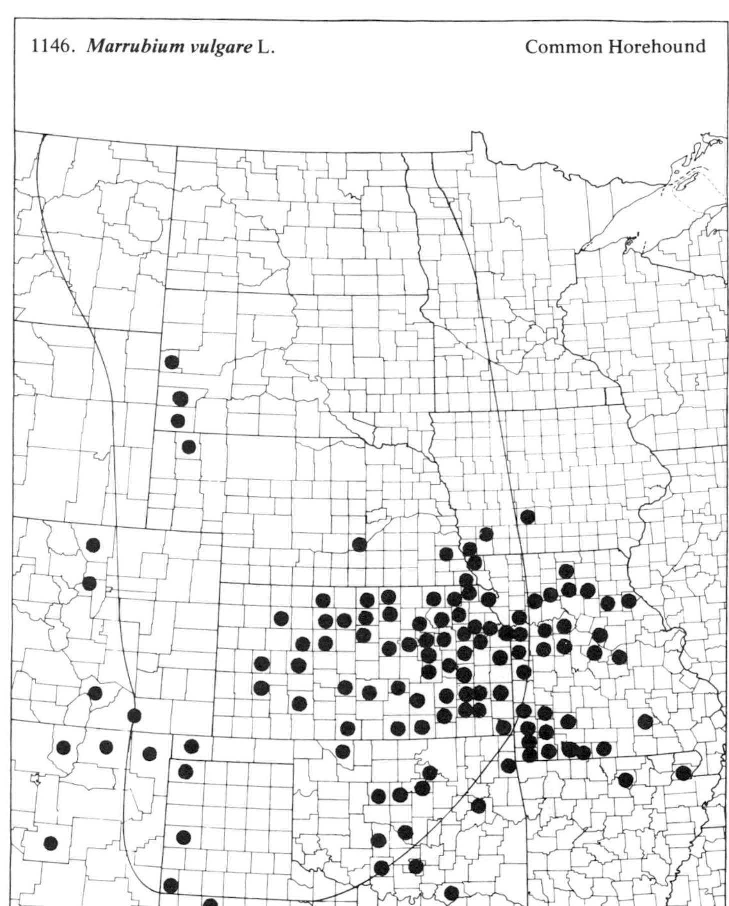

1147. ***Mentha arvensis*** L. Field Mint

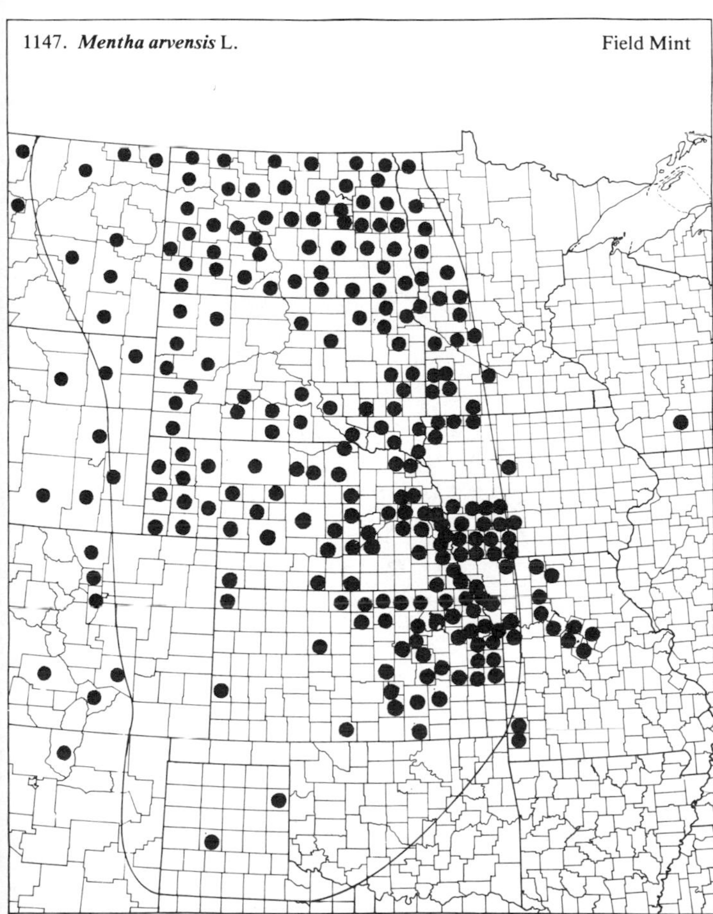

1148. ***Monarda citriodora*** Cerv. Lemon Beebalm

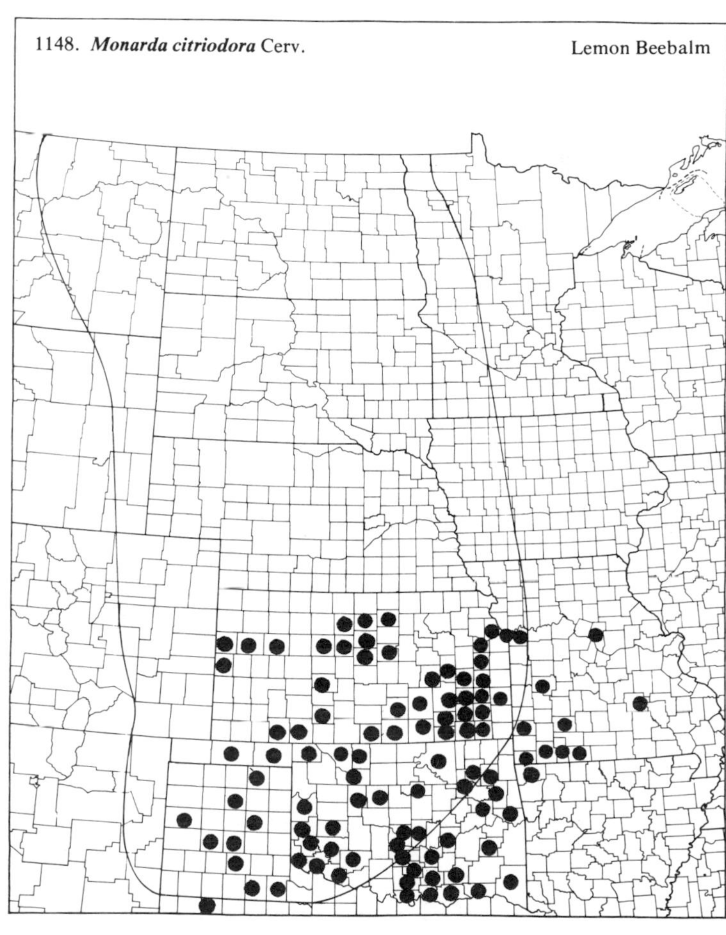

(Lamiaceae)

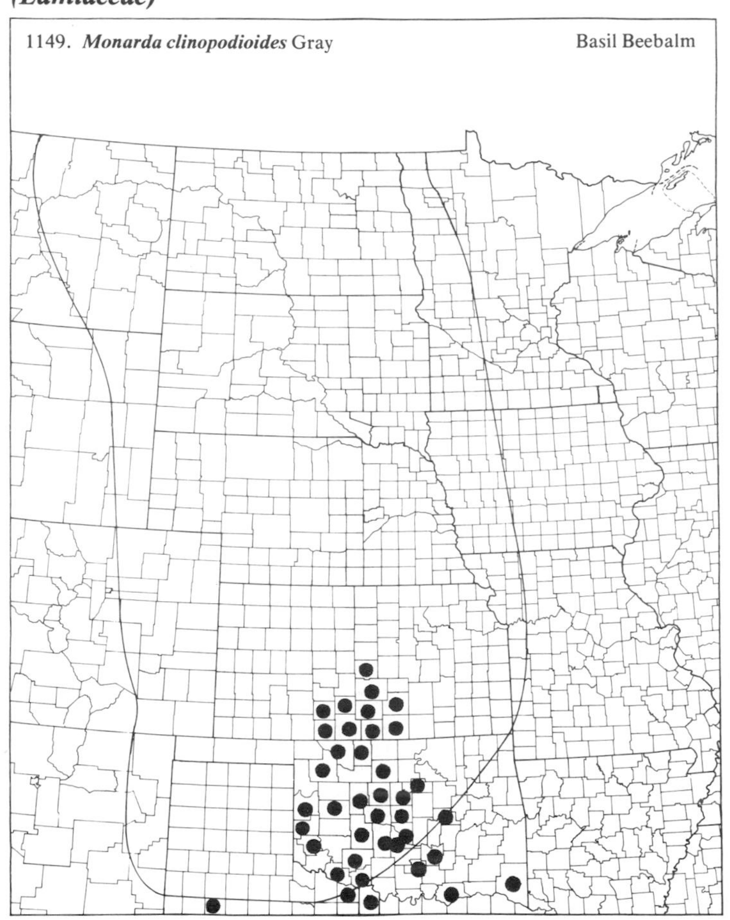
1149. Monarda clinopodioides Gray
Basil Beebalm

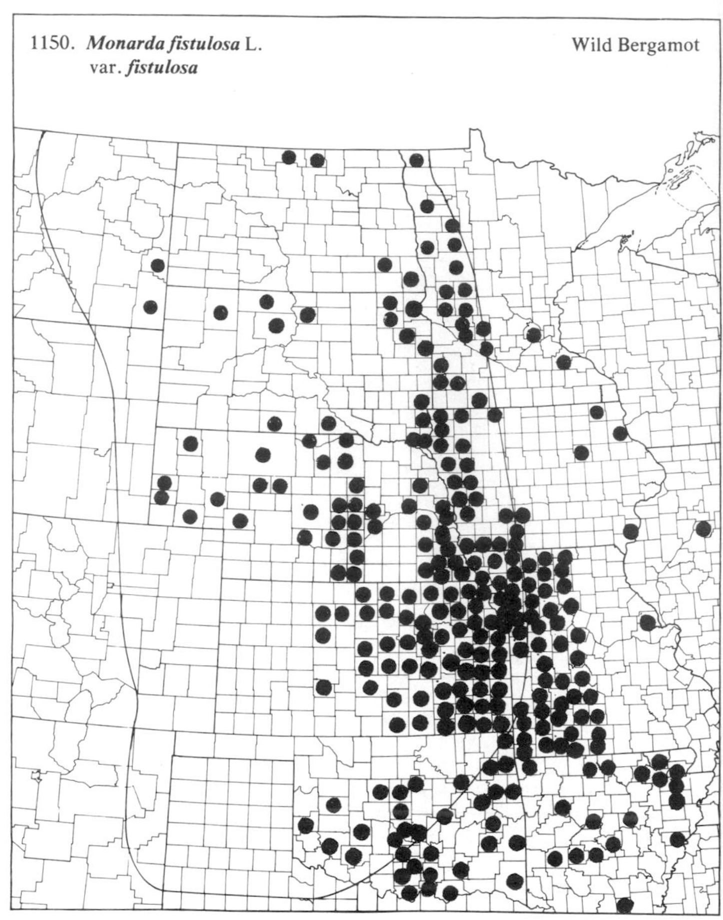
1150. Monarda fistulosa L.
var. fistulosa
Wild Bergamot

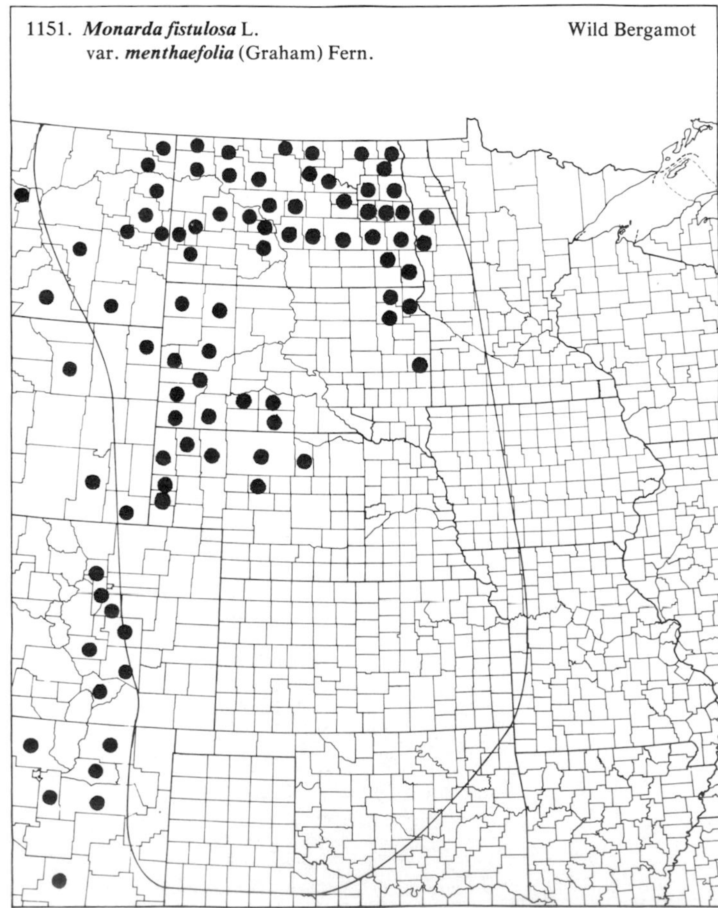
1151. Monarda fistulosa L.
var. menthaefolia (Graham) Fern.
Wild Bergamot

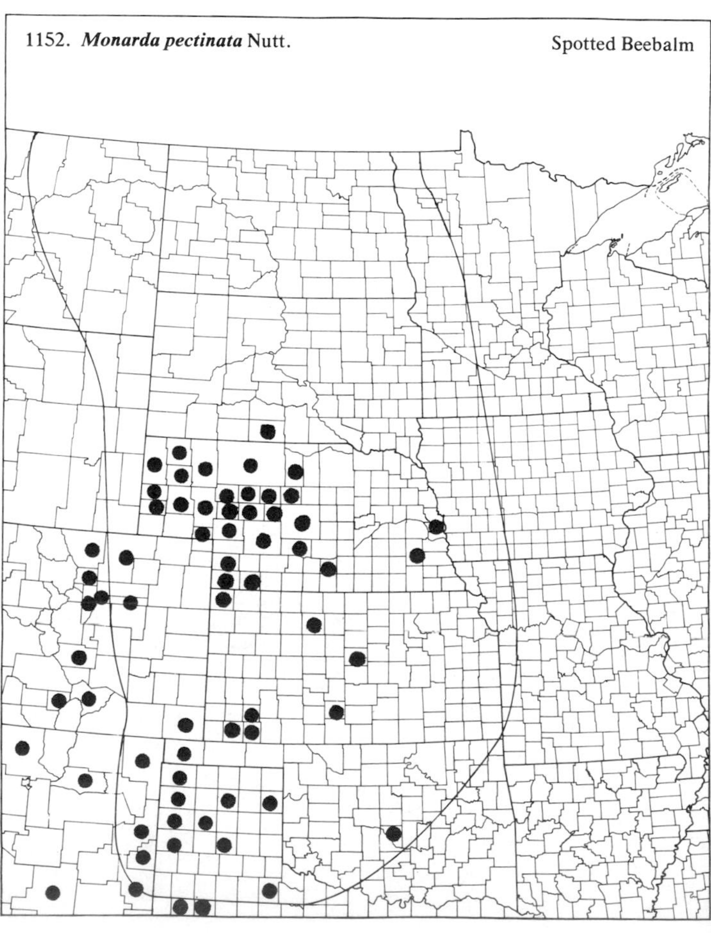
1152. Monarda pectinata Nutt.
Spotted Beebalm

(Lamiaceae)

1153. ***Monarda punctata*** L. ssp. ***occidentalis*** Epl. — Horse Mint

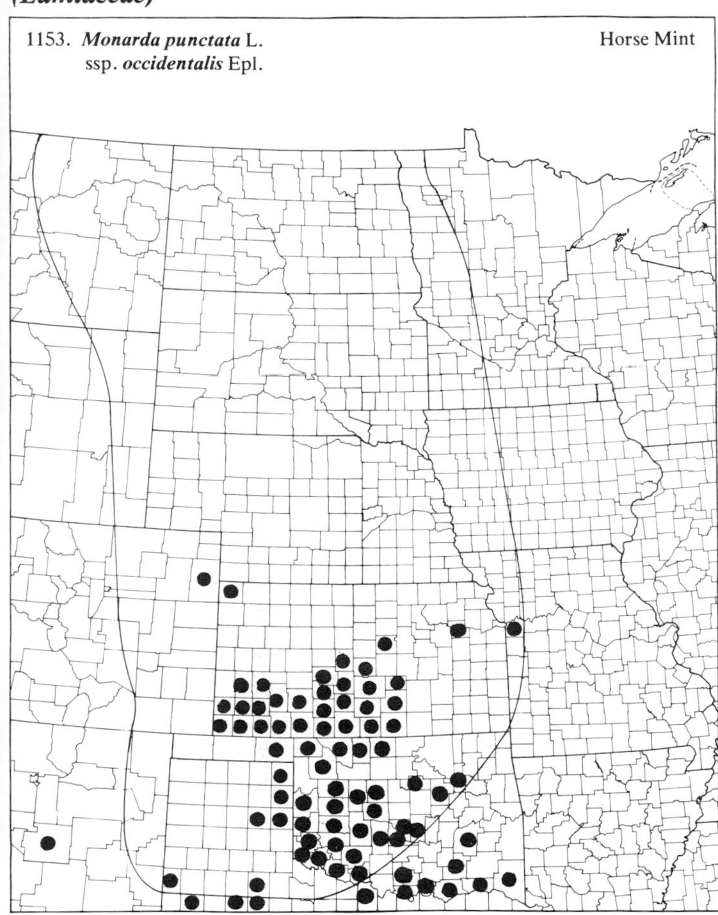

1154. ***Nepeta cataria*** L. — Catnip

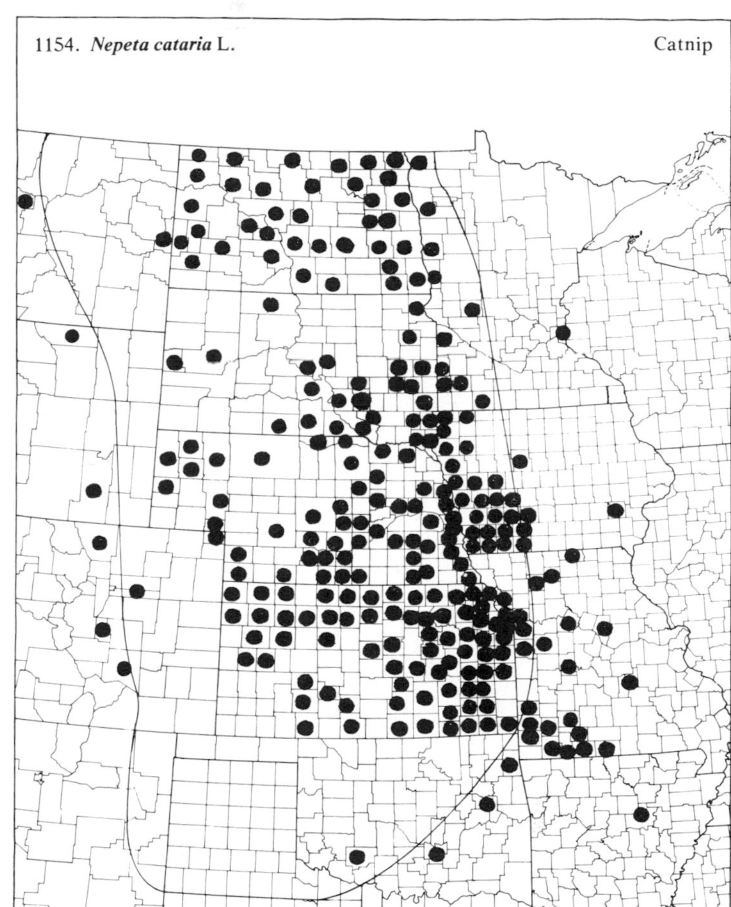

1155. ***Perilla frutescens*** (L.) Britt. — Common Perilla

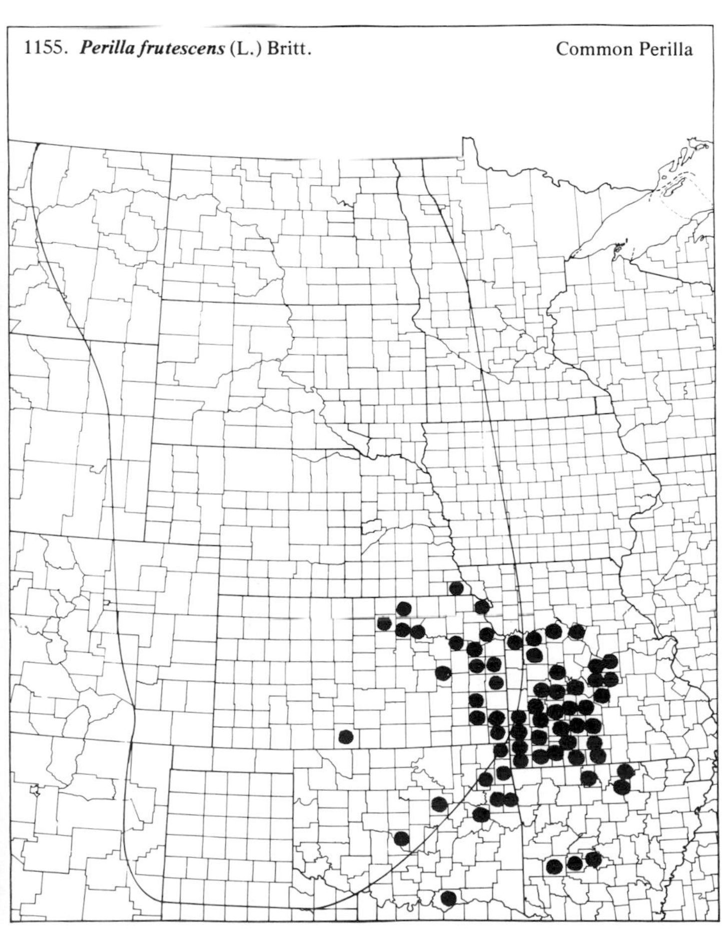

1156. ***Physostegia intermedia*** (Nutt.) Gray — Intermediate Lionsheart

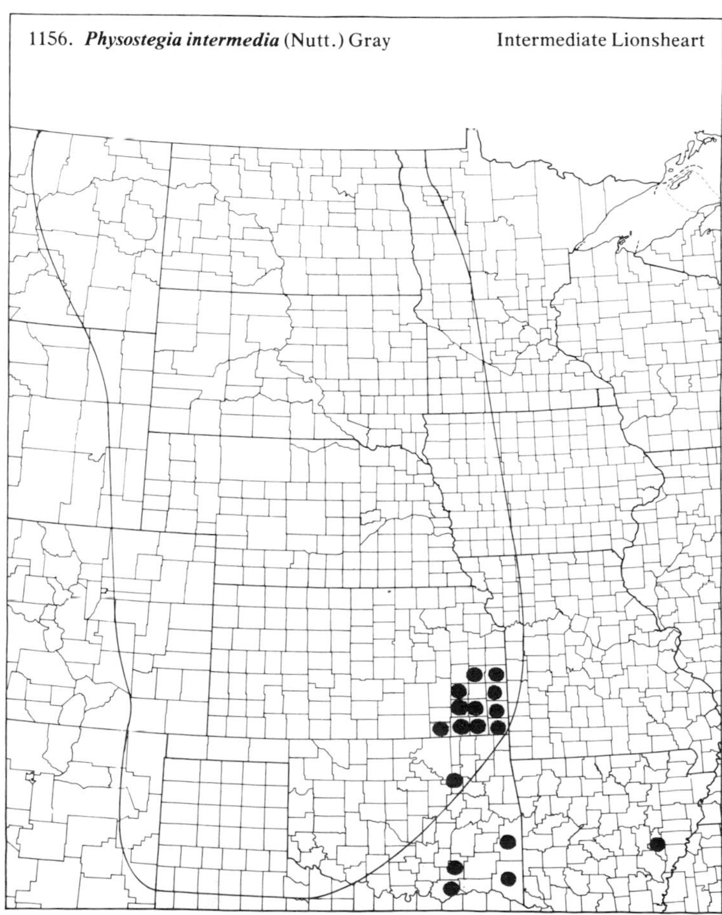

(Lamiaceae)

1157. ***Physostegia parviflora*** Nutt. Obedient Plant

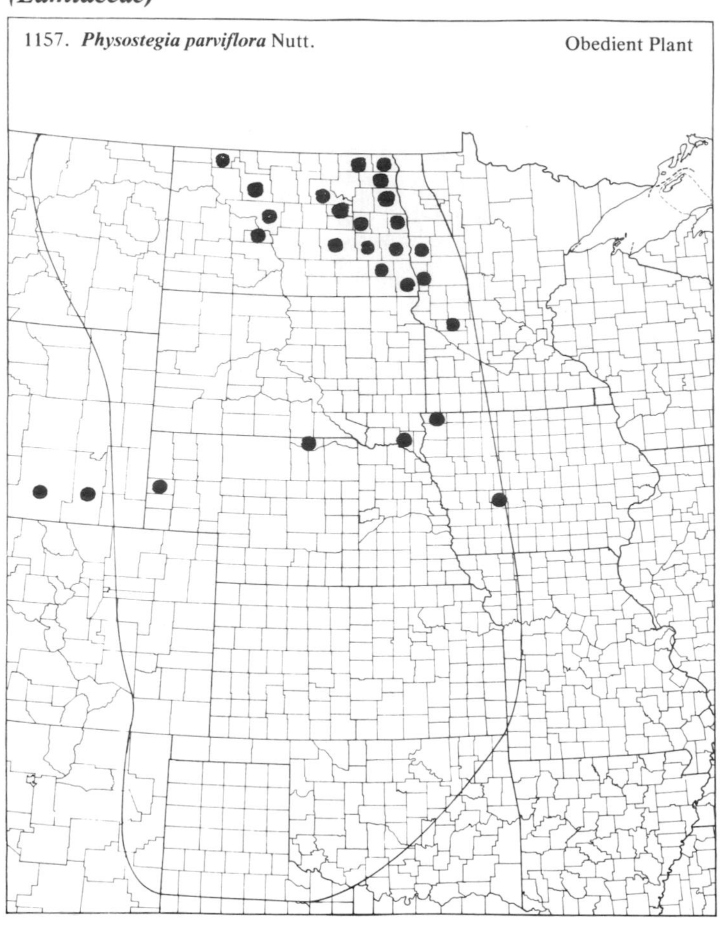

1158. ***Physostegia virginiana*** (L.) Benth. False Dragonhead

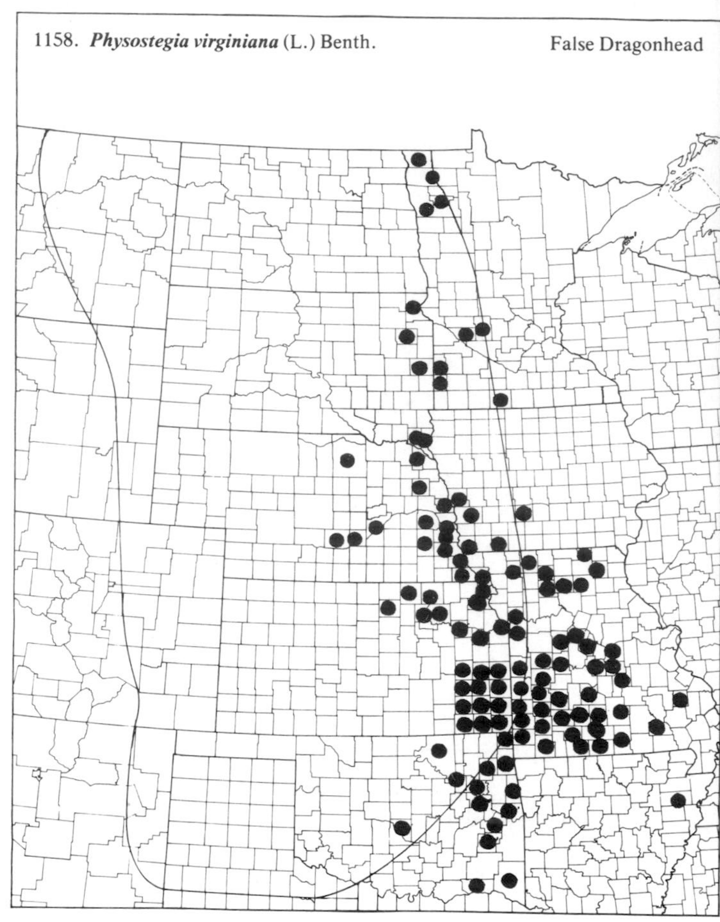

1159. ***Prunella vulgaris*** L. Selfheal

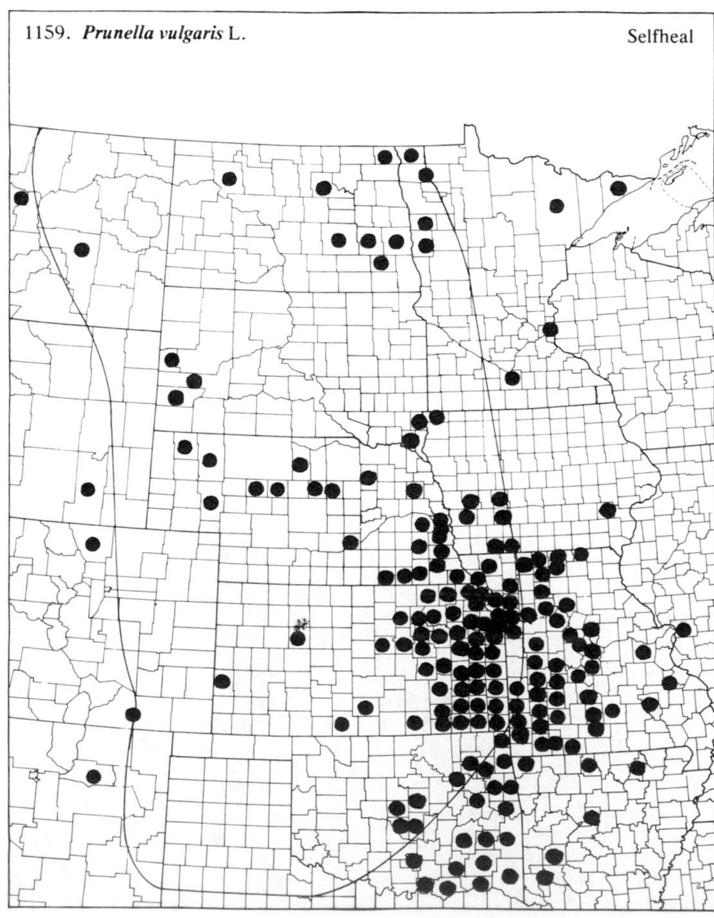

1160. ***Pycnanthemum pilosum*** Nutt. Woods Mountain Mint

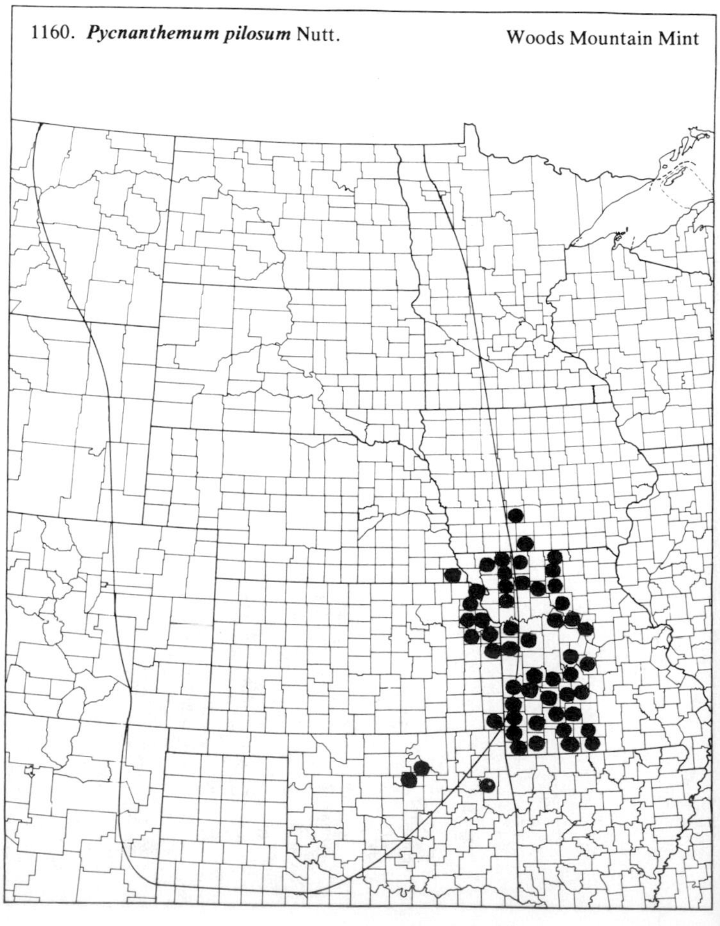

(Lamiaceae)

1161. ***Pycnanthemum tenuifolium*** Schrad.

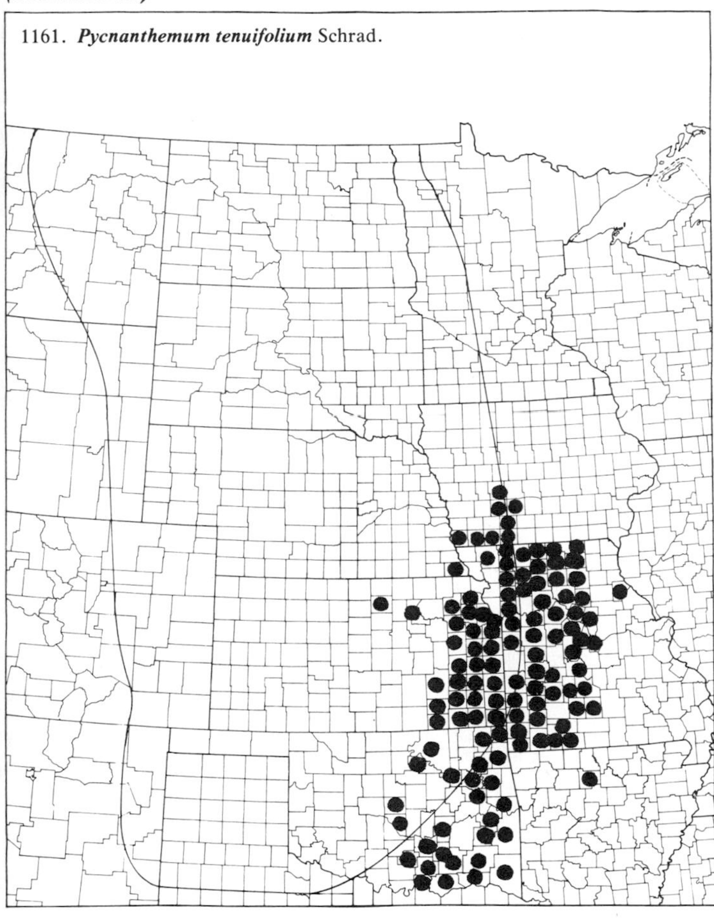

1162. ***Pycnanthemum virginianum*** (L.) Durand & Jacks. Mountain Mint

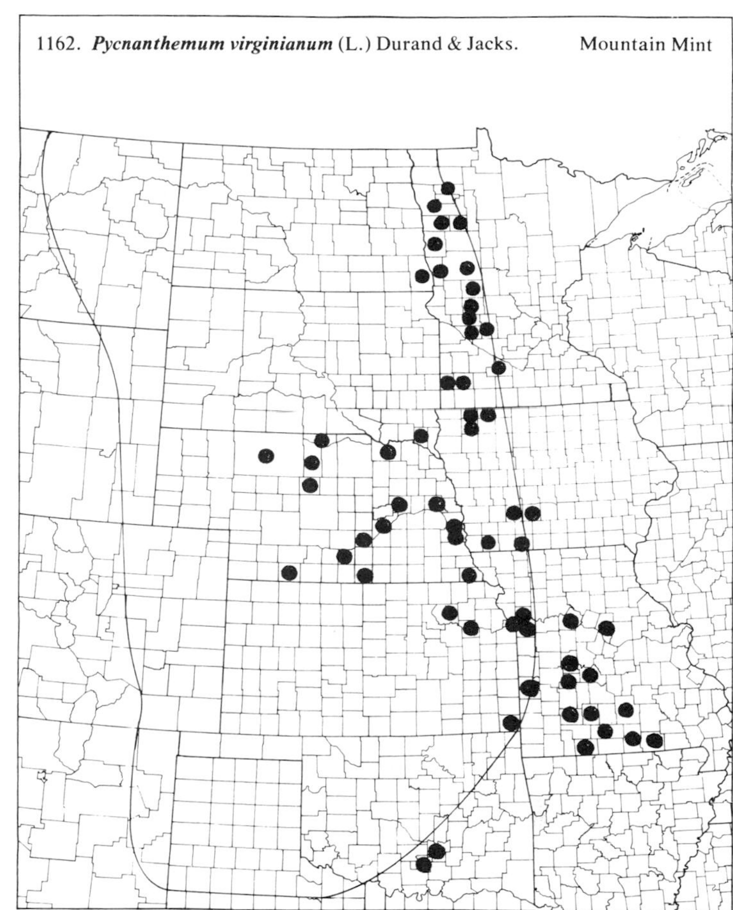

1163. ***Salvia pitcheri*** Torr. Pitcher's Sage

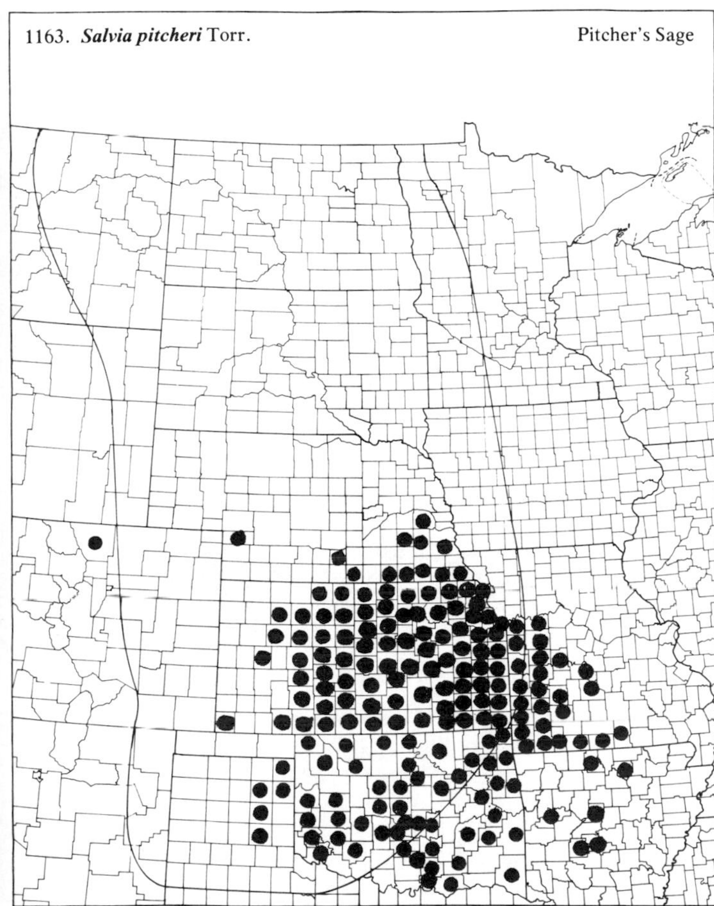

1164. ***Salvia reflexa*** Hornem. Lance-leaved Sage

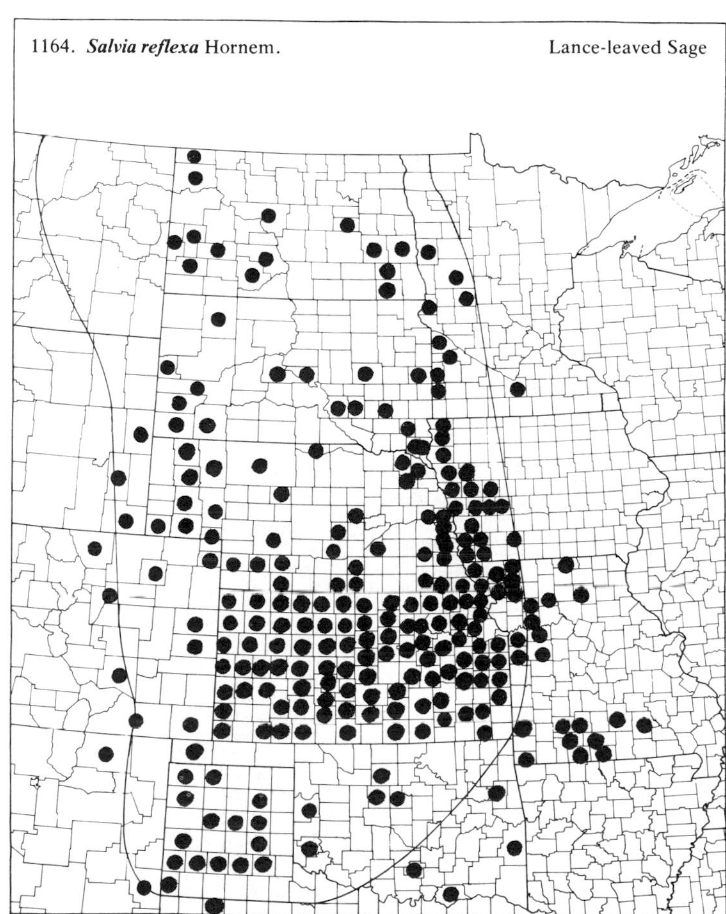

(Lamiaceae)

1165. ***Scutellaria brittonii*** Porter

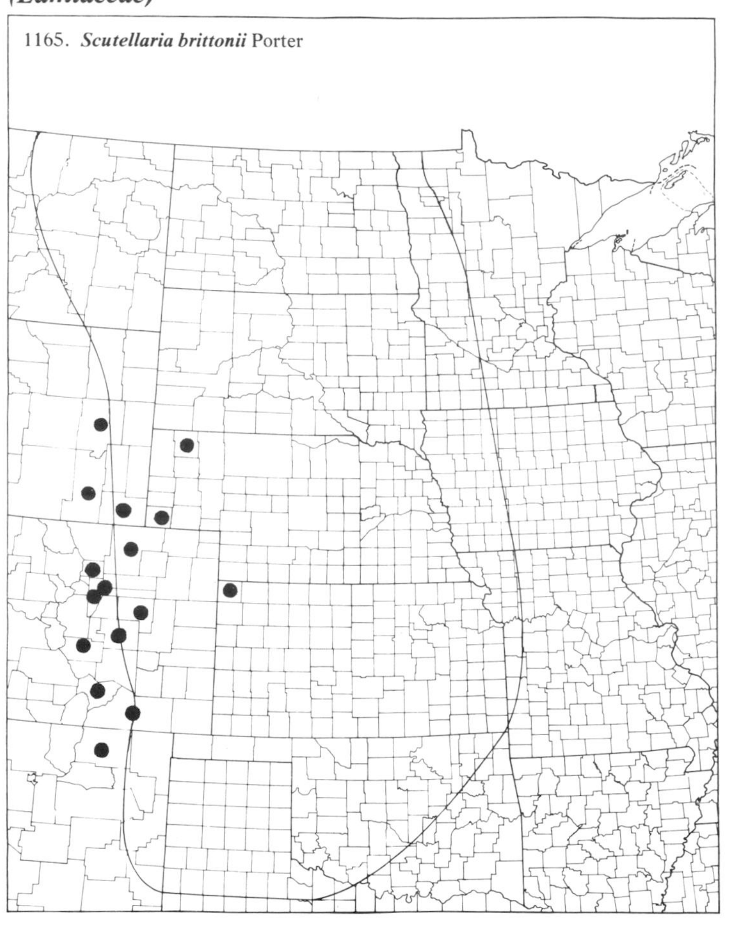

1166. ***Scutellaria drummondii*** Benth.

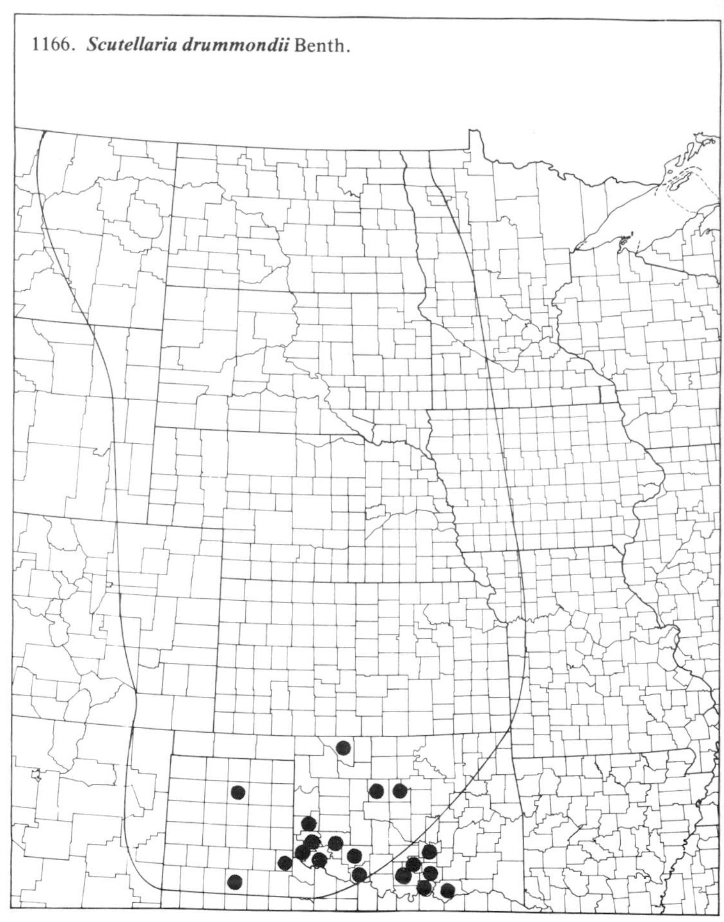

1167. ***Scutellaria galericulata*** L. — Marsh Skullcap

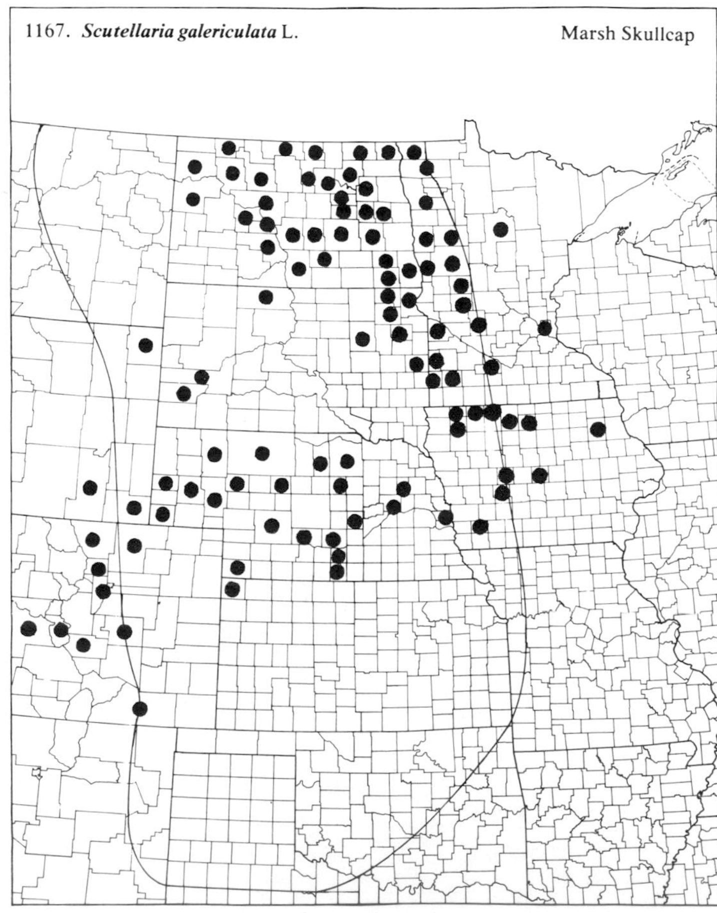

1168. ***Scutellaria lateriflora*** L. — Blue Skullcap

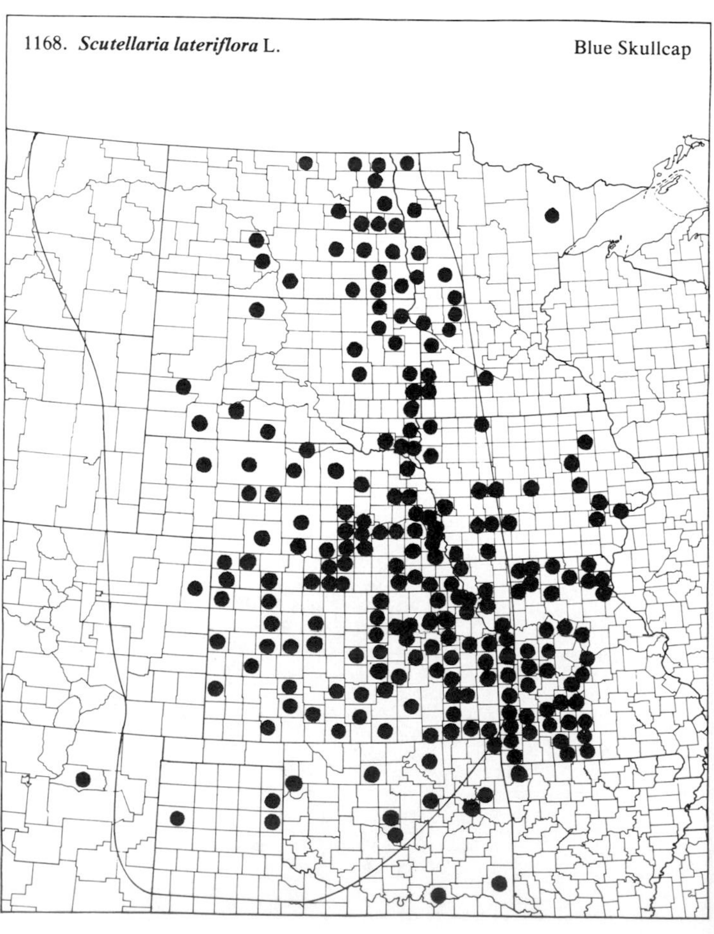

(Lamiaceae)

1169. ***Scutellaria ovata*** Hill — Eggleaf Skullcap

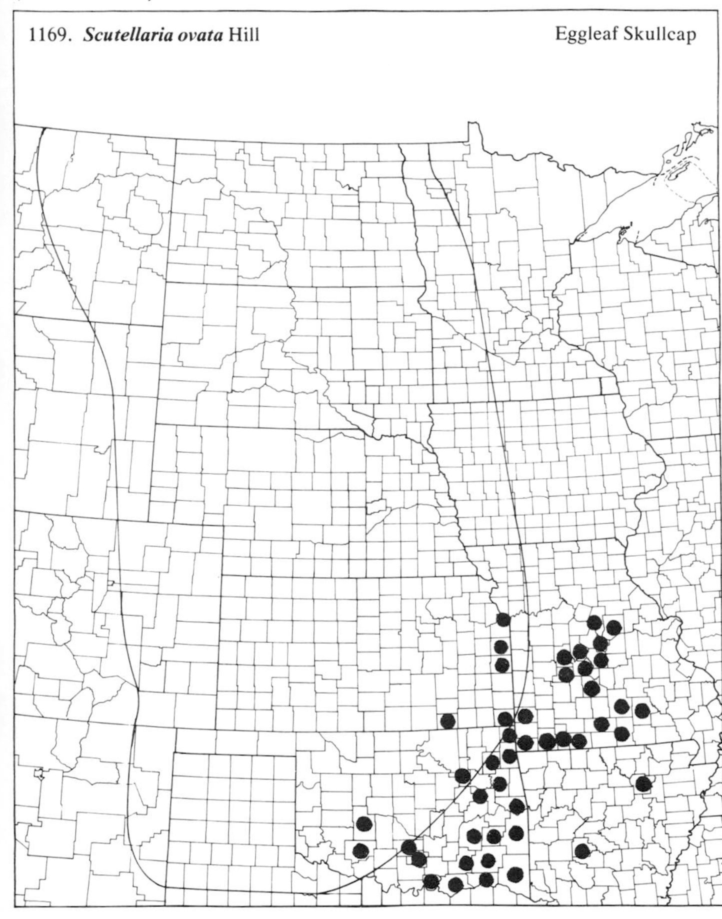

1170. ***Scutellaria parvula*** Michx. var. ***parvula*** — Small Skullcap

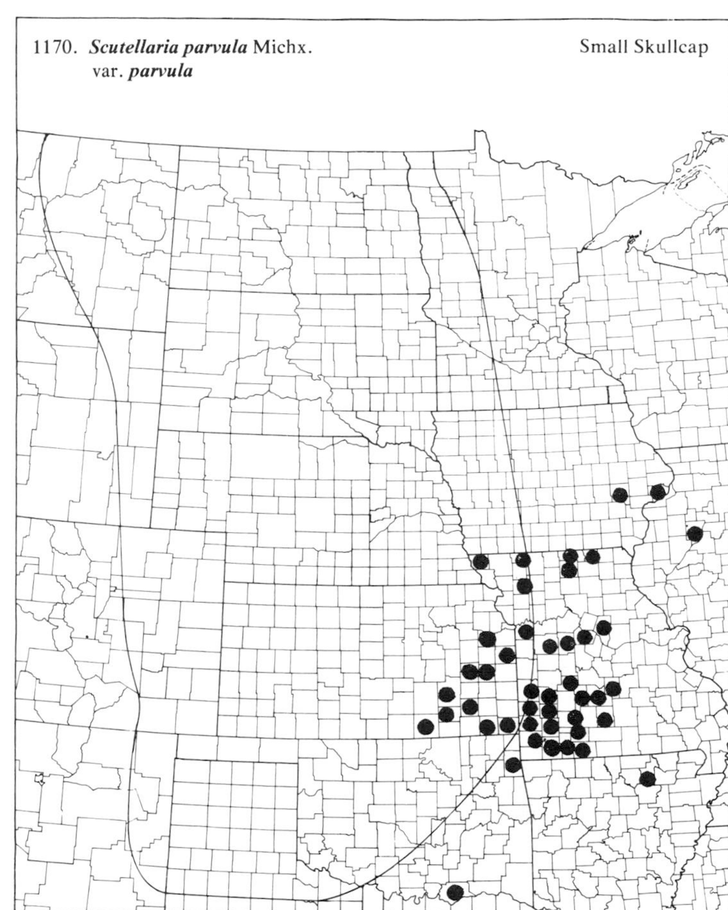

1171. ***Scutellaria parvula*** Michx. var. ***australis*** Fassett — Southern Small Skullcap

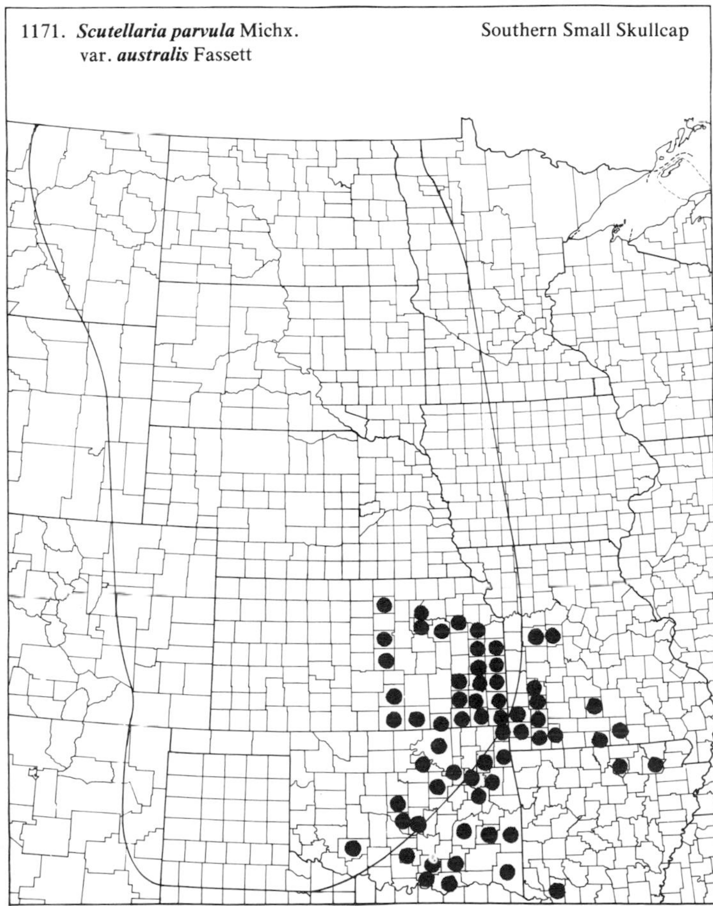

1172. ***Scutellaria parvula*** Michx. var. ***leonardi*** (Epl.) Fern. — Leonard Small Skullcap

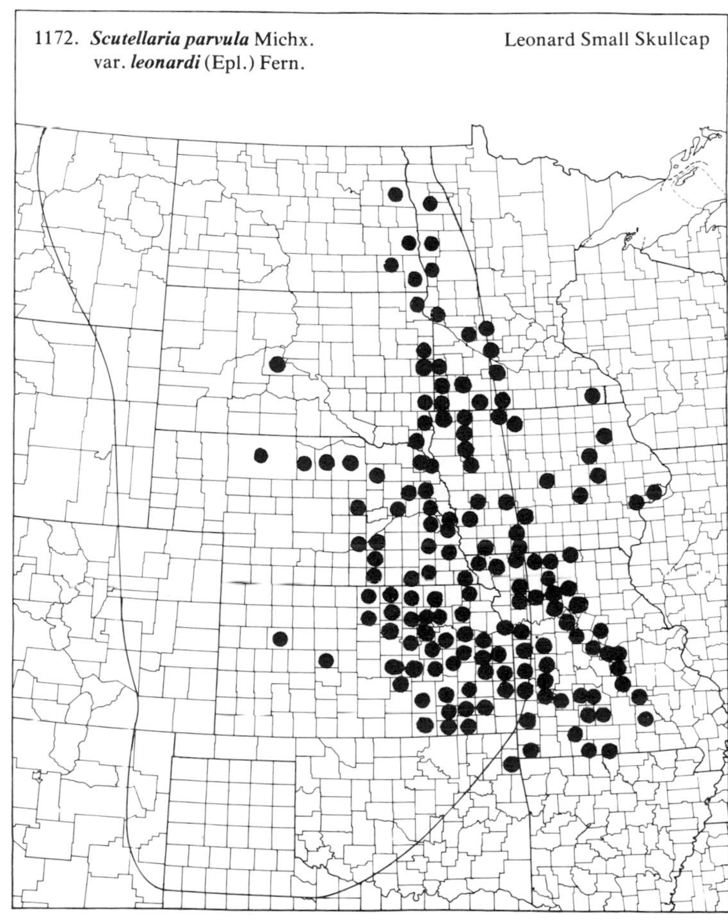

(Lamiaceae)

1173. ***Scutellaria resinosa*** Torr. Resinous Skullcap

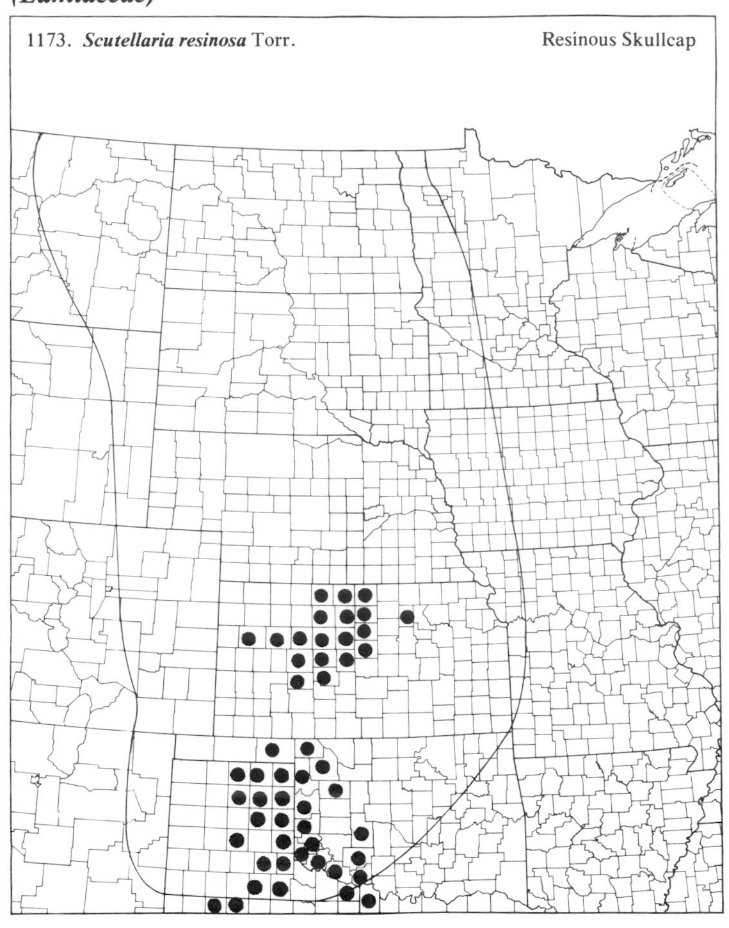

1174. ***Scutellaria wrightii*** Gray

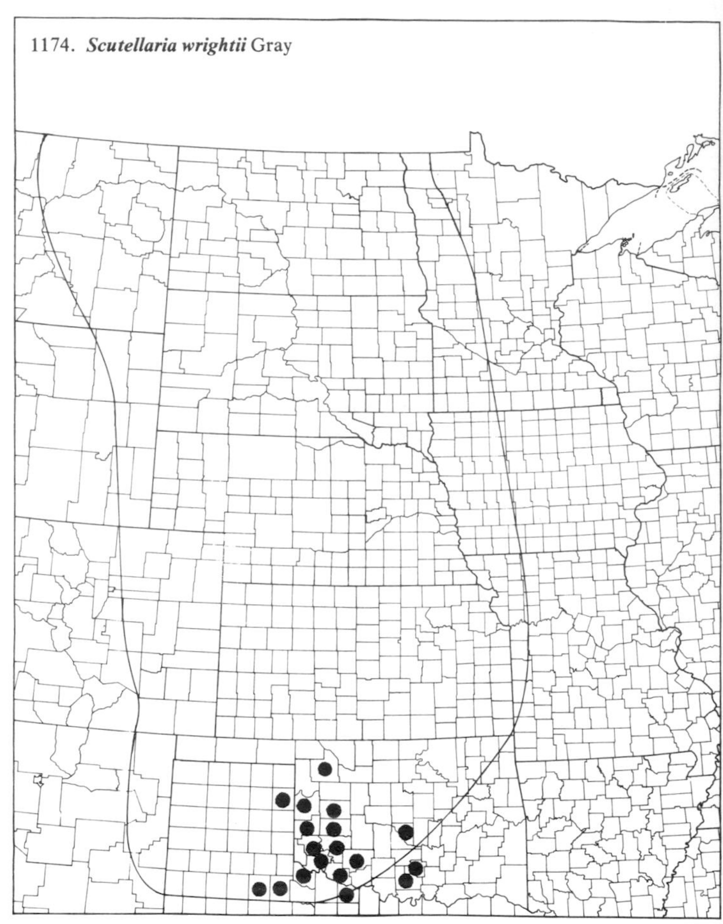

1175. ***Stachys palustris*** L. var. ***pilosa*** (Nutt.) Fern. Hedge Nettle

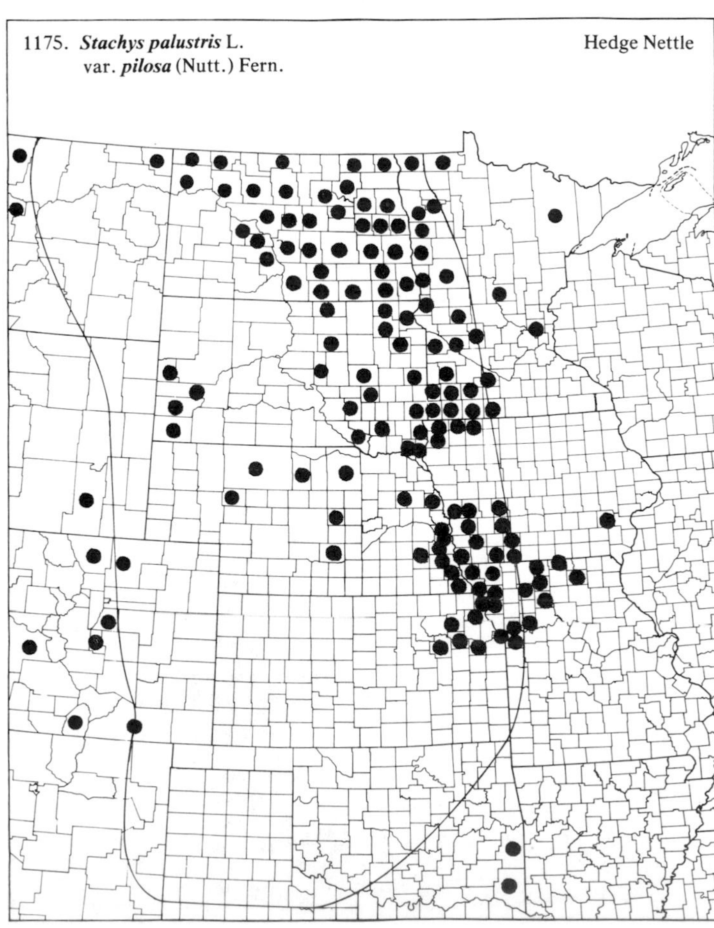

1176. ***Stachys tenuifolia*** Willd. Slenderleaf Betony

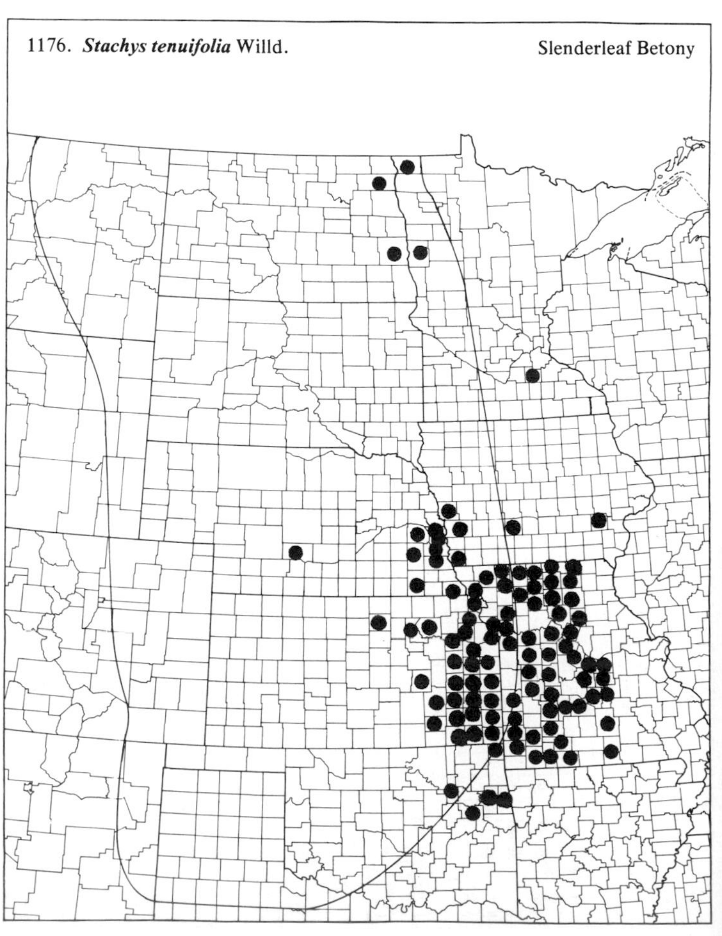

(Lamiaceae)

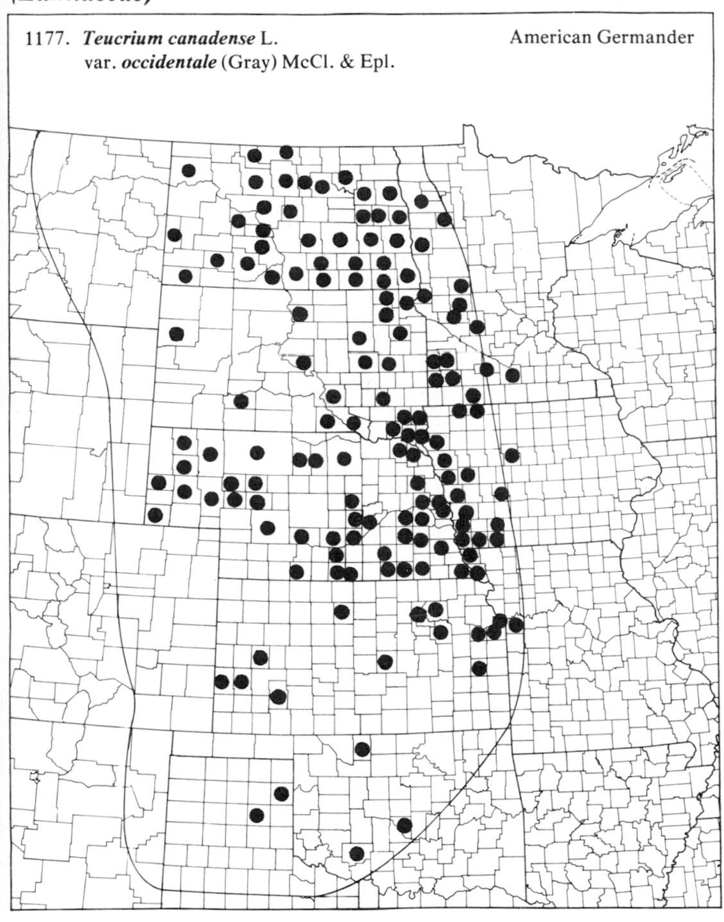

1177. ***Teucrium canadense*** L. var. ***occidentale*** (Gray) McCl. & Epl. — American Germander

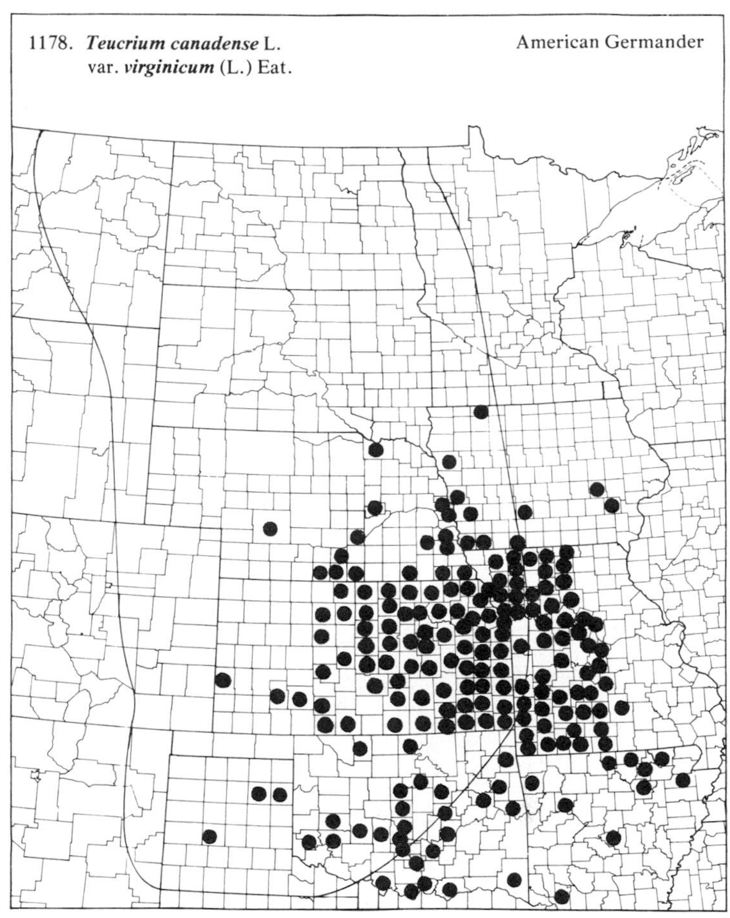

1178. ***Teucrium canadense*** L. var. ***virginicum*** (L.) Eat. — American Germander

1179. ***Teucrium laciniatum*** Torr. — Cutleaf Germander

100. *Plantaginaceae* (Plantain Family)

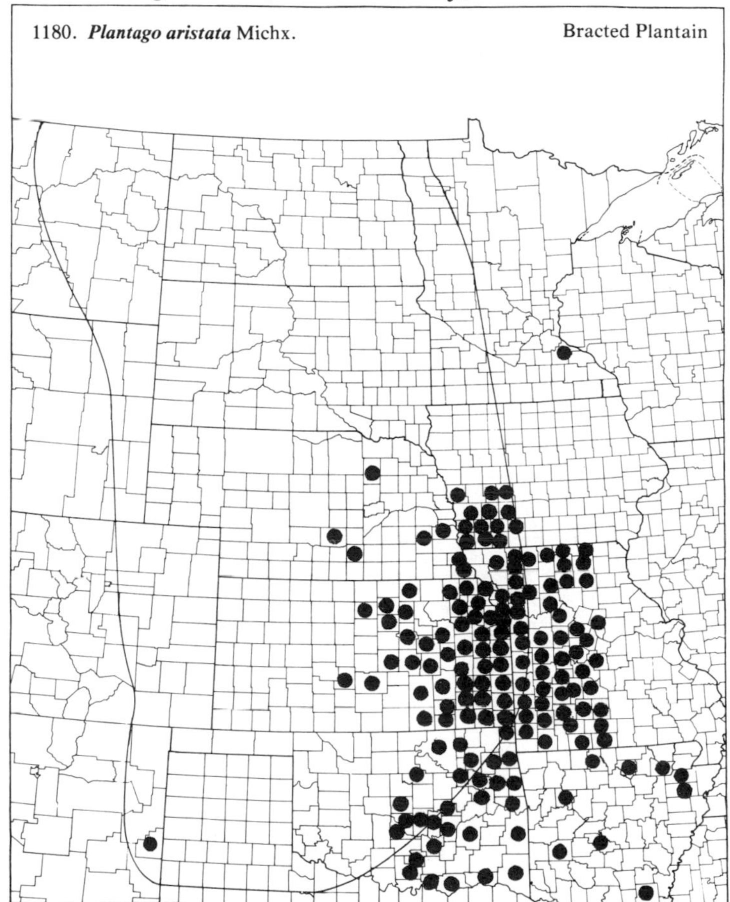

1180. ***Plantago aristata*** Michx. — Bracted Plantain

(Plantaginaceae)

1181. ***Plantago elongata*** Pursh

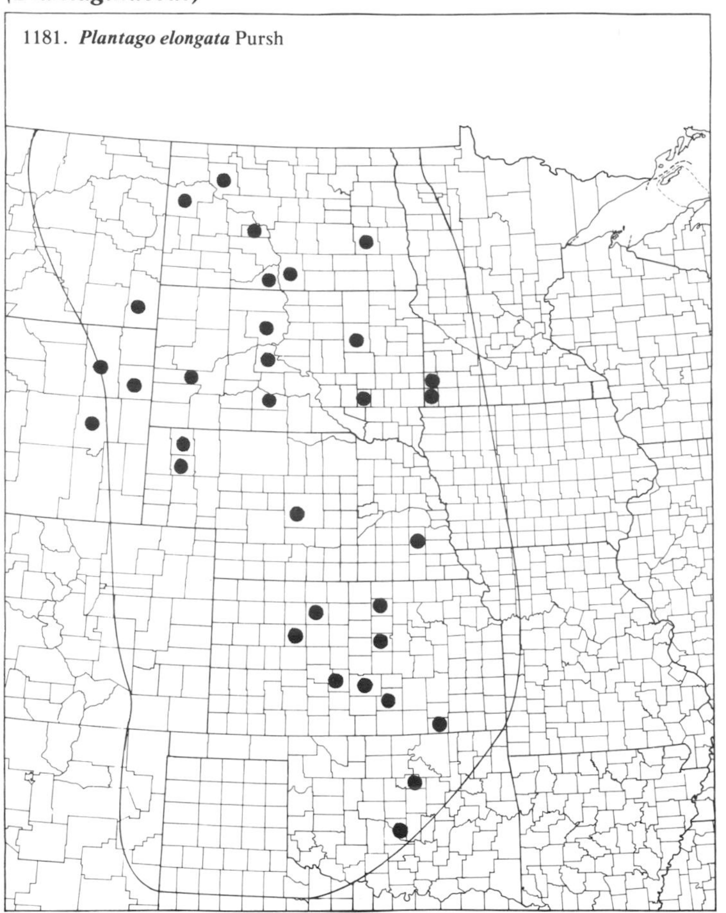

1182. ***Plantago eriopoda*** Torr. — Alkali Plantain

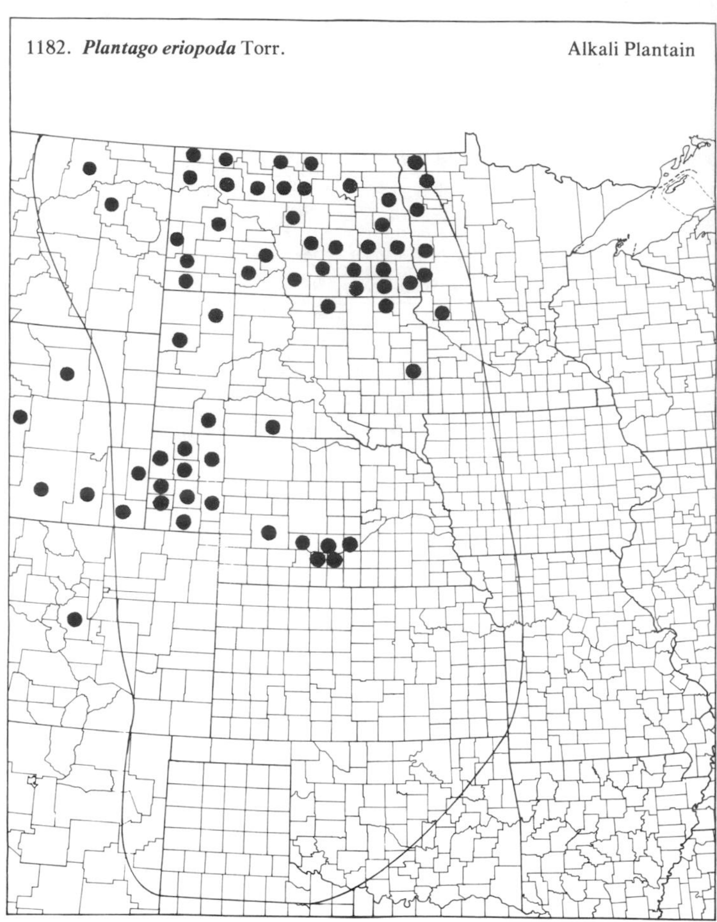

1183. ***Plantago lanceolata*** L. — English Plantain

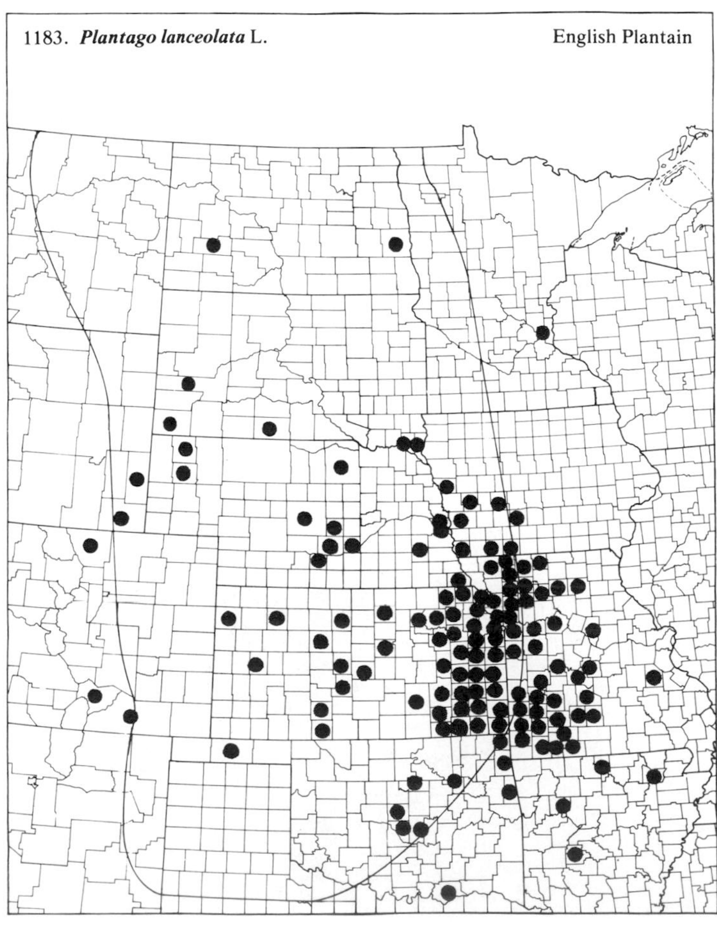

1184. ***Plantago major*** L. — Common Plantain

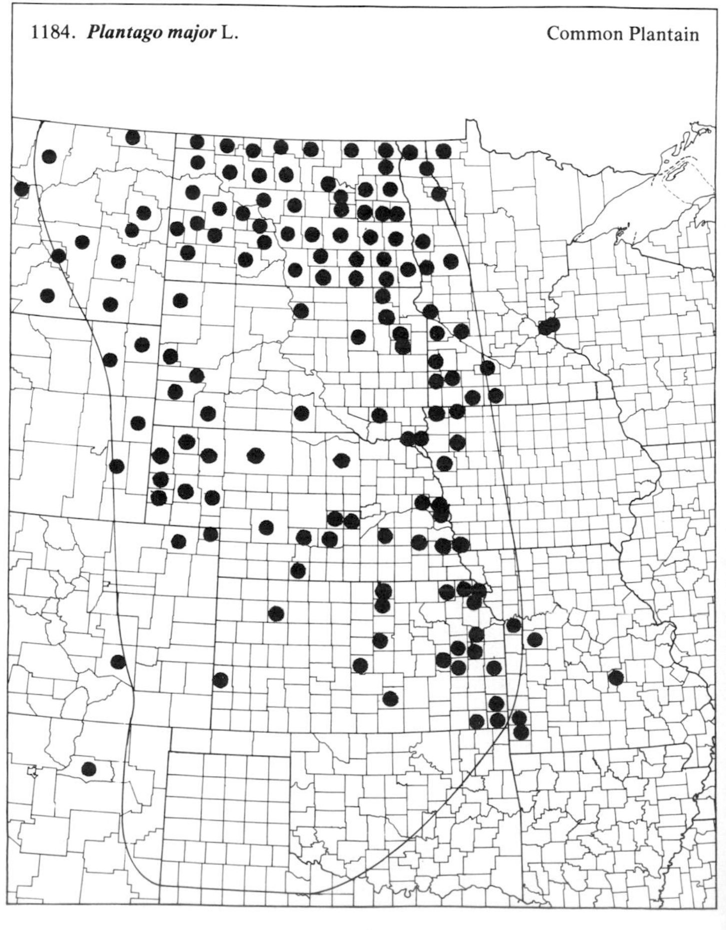

(Plantaginaceae)

1185. **Plantago patagonica** Jacq. var. **patagonica** — Buckhorn

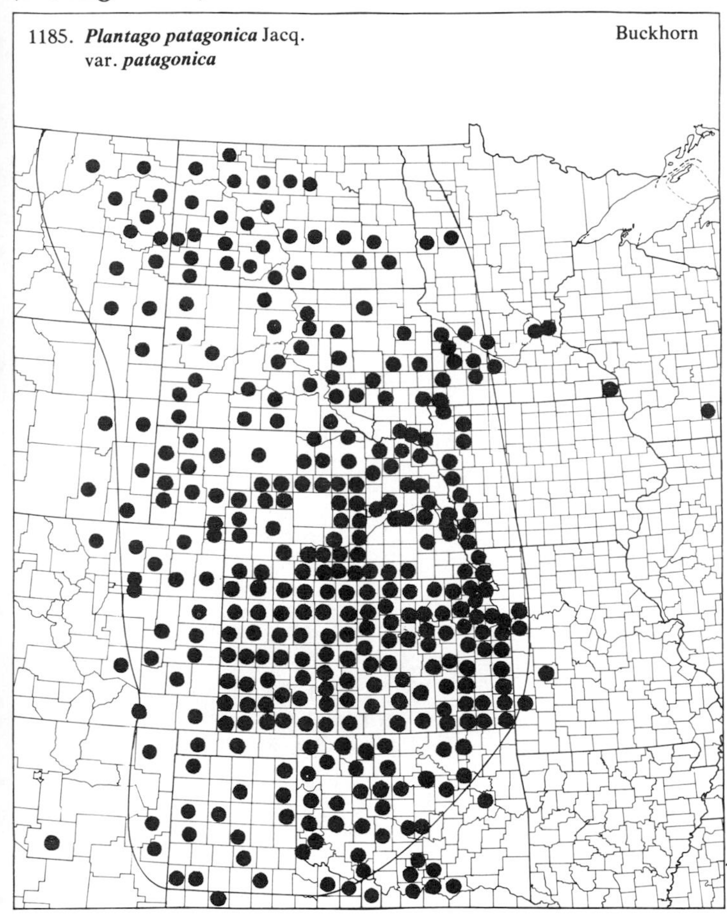

1186. **Plantago patagonica** Jacq. var. **spinulosa** (Dcne.) Gray — Buckhorn

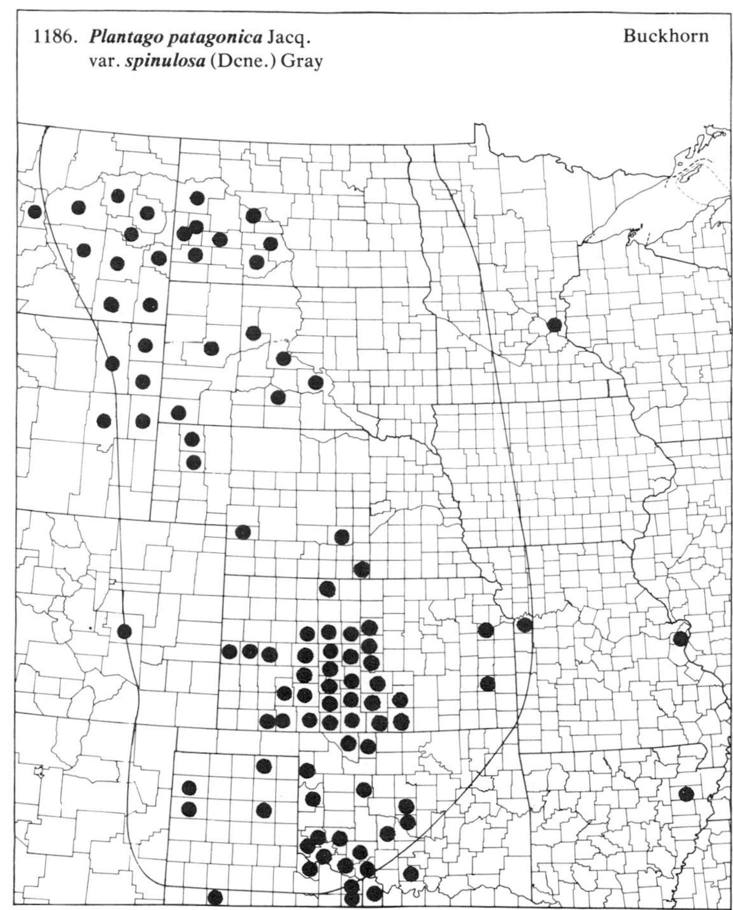

1187. **Plantago pusilla** Nutt. — Slender Plantain

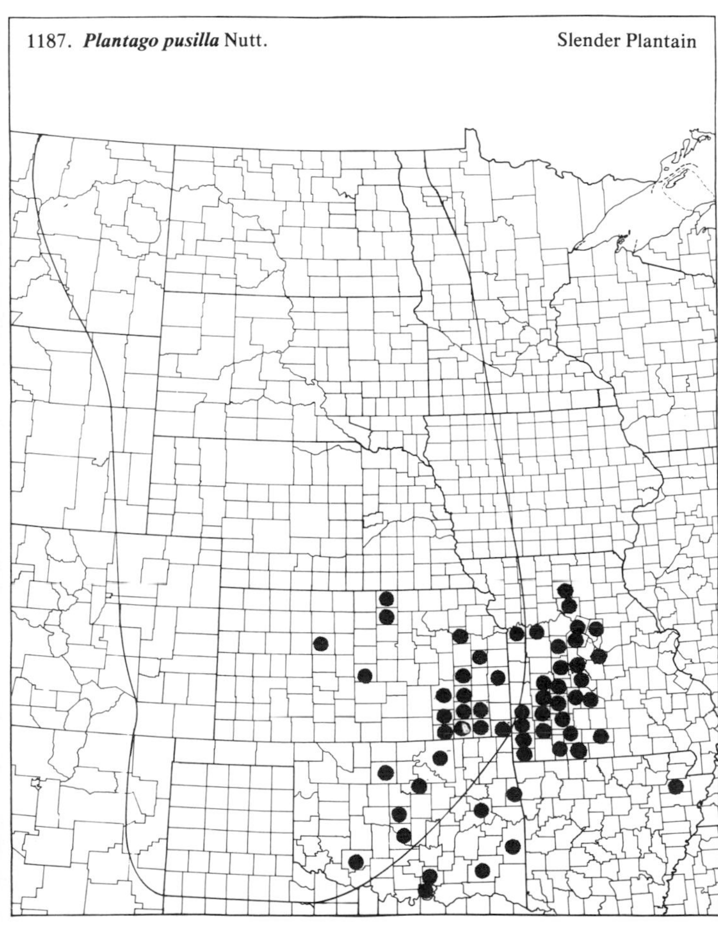

1188. **Plantago rhodosperma** Dcne. — Red-seeded Plantain

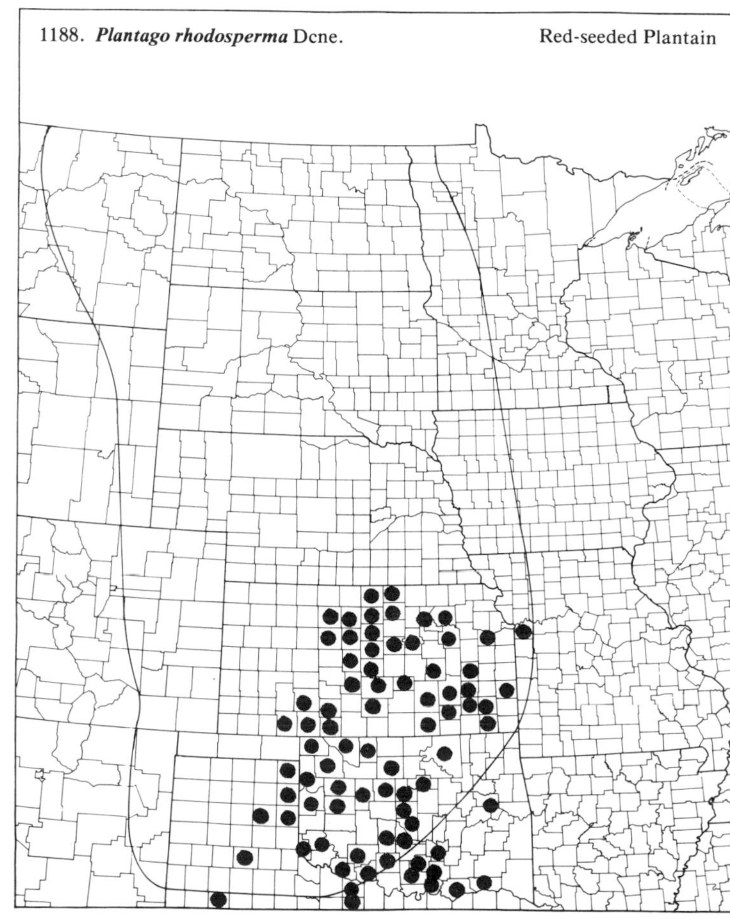

(Plantaginaceae)

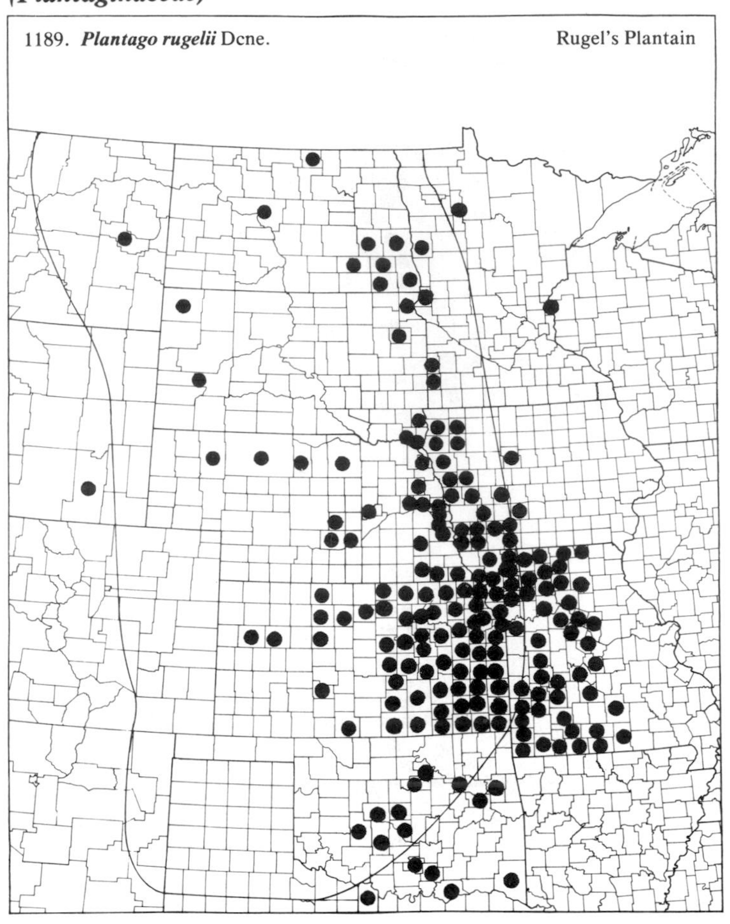

1189. ***Plantago rugelii*** Dcne. Rugel's Plantain

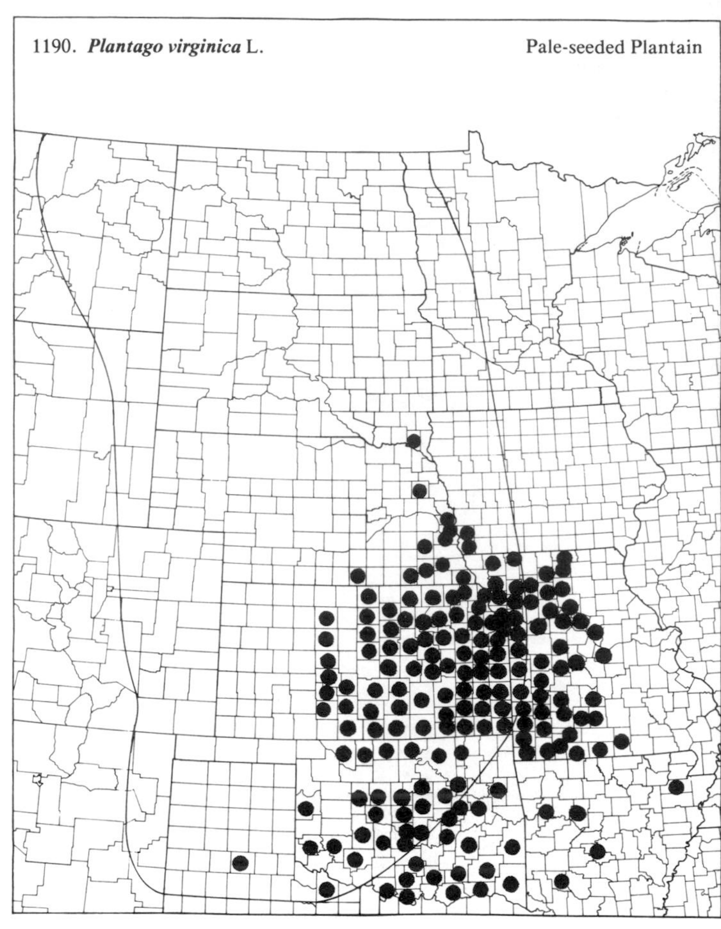

1190. ***Plantago virginica*** L. Pale-seeded Plantain

101. *Oleaceae* (Olive Family)

1191. ***Plantago wrightiana*** Dcne.

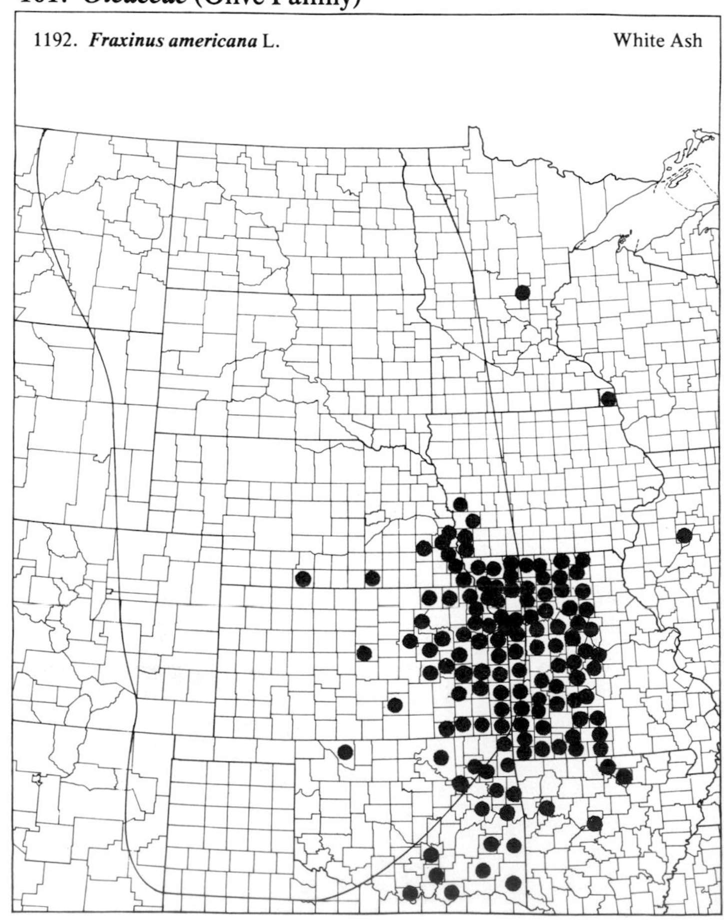

1192. ***Fraxinus americana*** L. White Ash

(Oleaceae)

1193. ***Fraxinus pennsylvanica*** Marsh. var. ***pennsylvanica*** Red Ash

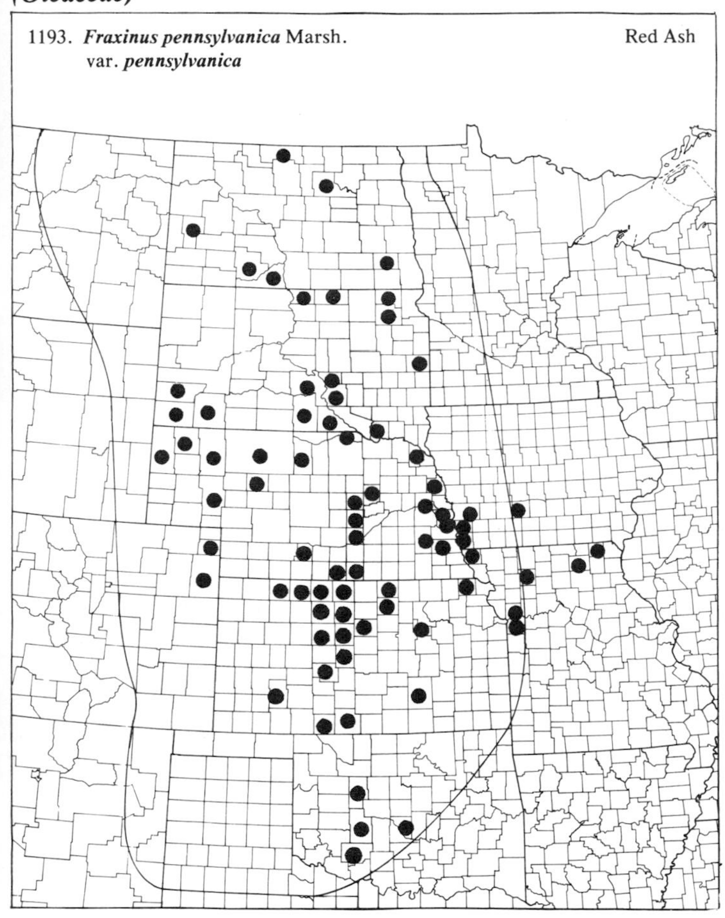

1194. ***Fraxinus pennsylvanica*** Marsh. var. ***subintegerrima*** (Vahl) Fern. Green Ash

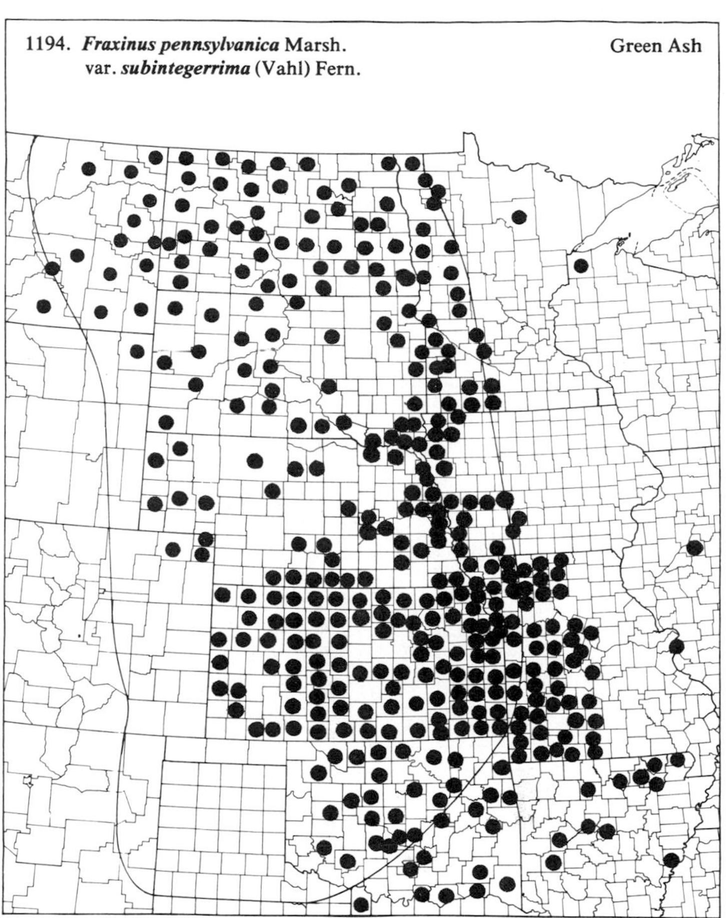

1195. ***Fraxinus quadrangulata*** Michx. Blue Ash

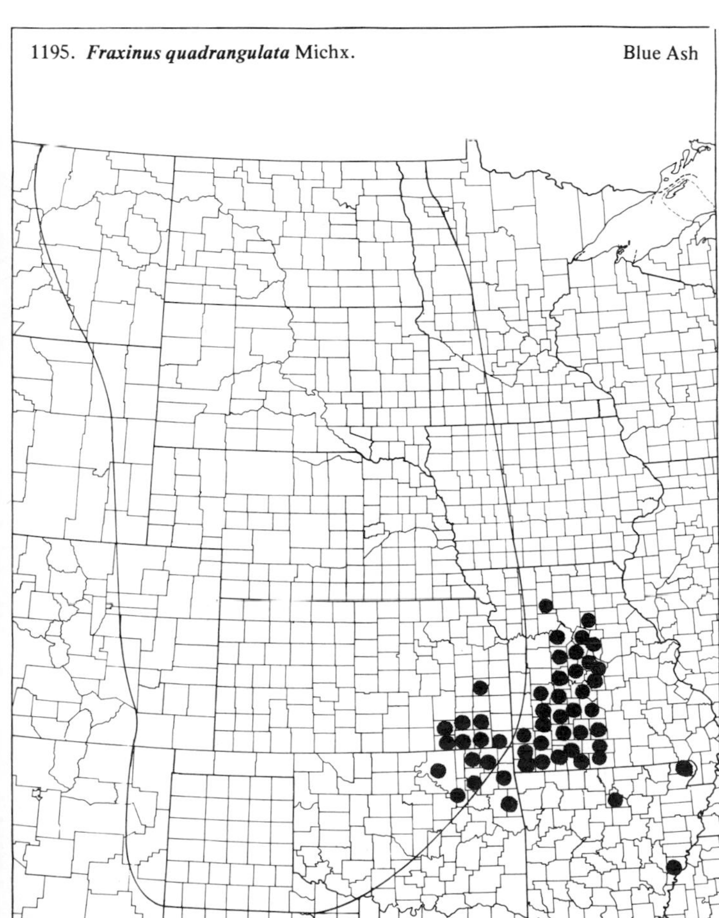

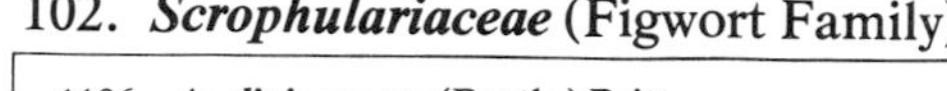

102. *Scrophulariaceae* (Figwort Family)

1196. ***Agalinis aspera*** (Benth.) Britt.

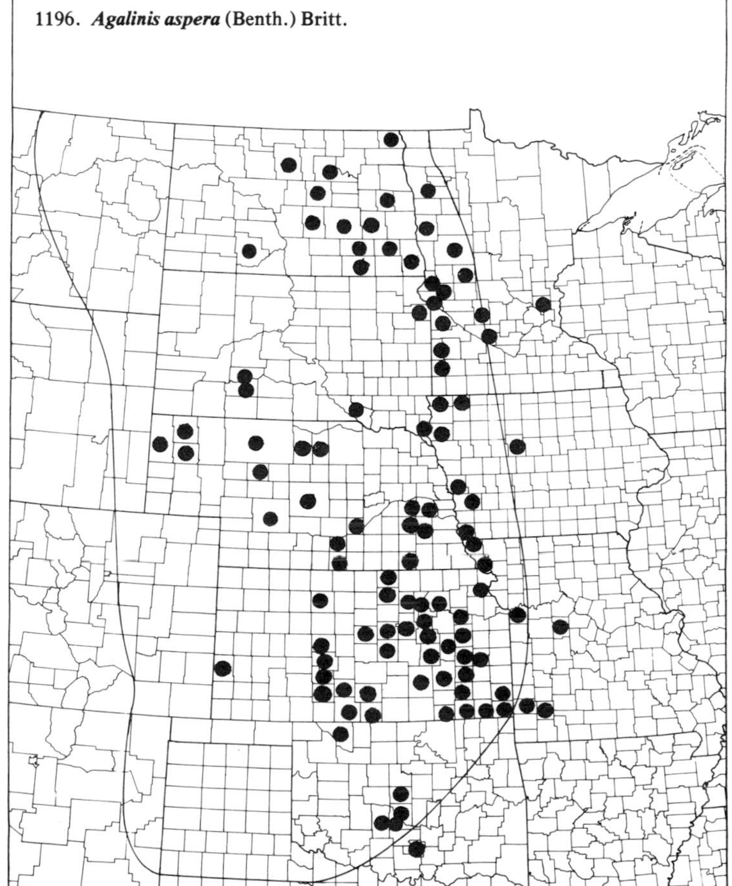

(Scrophulariaceae)

1197. ***Agalinis fasciculata*** (Ell.) Raf.

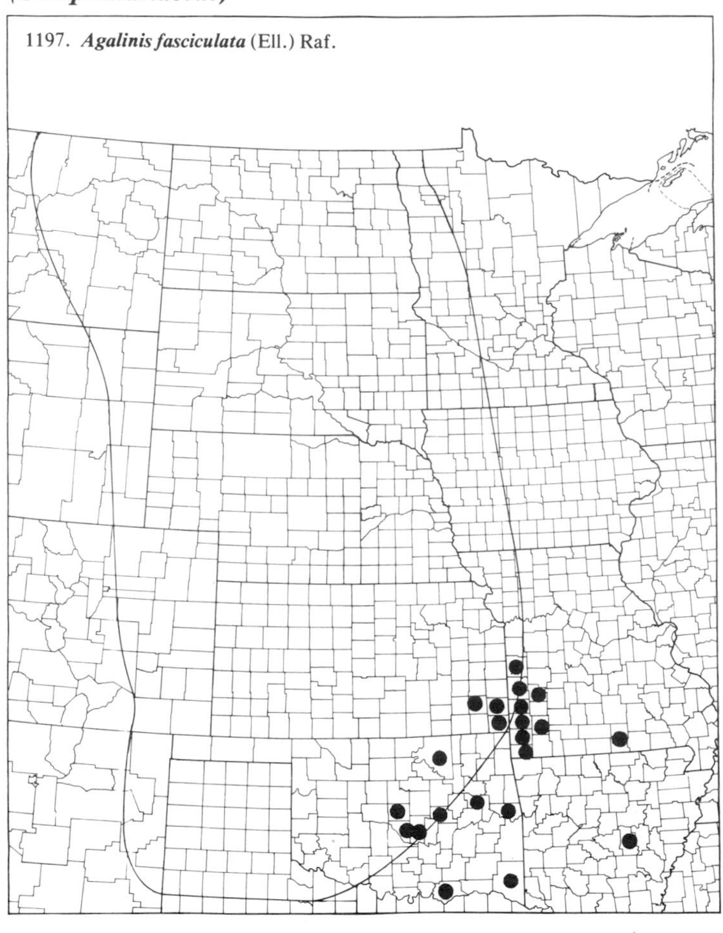

1198. ***Agalinis gattingerii*** (Small) Small

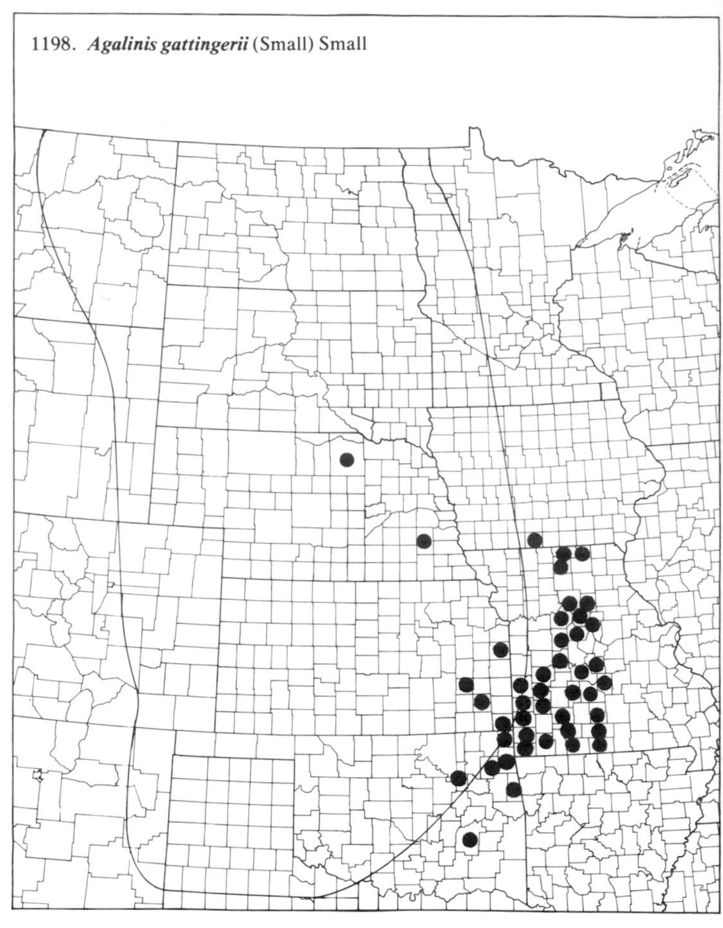

1199. ***Agalinis heterophylla*** (Nutt.) Small

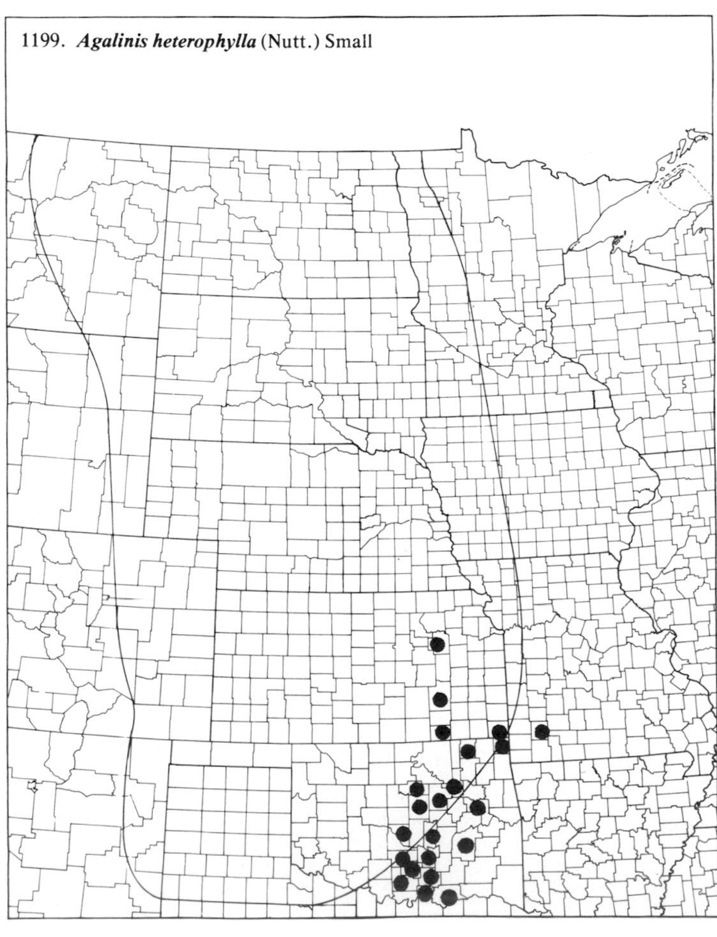

1200. ***Agalinis purpurea*** (L.) Penn.

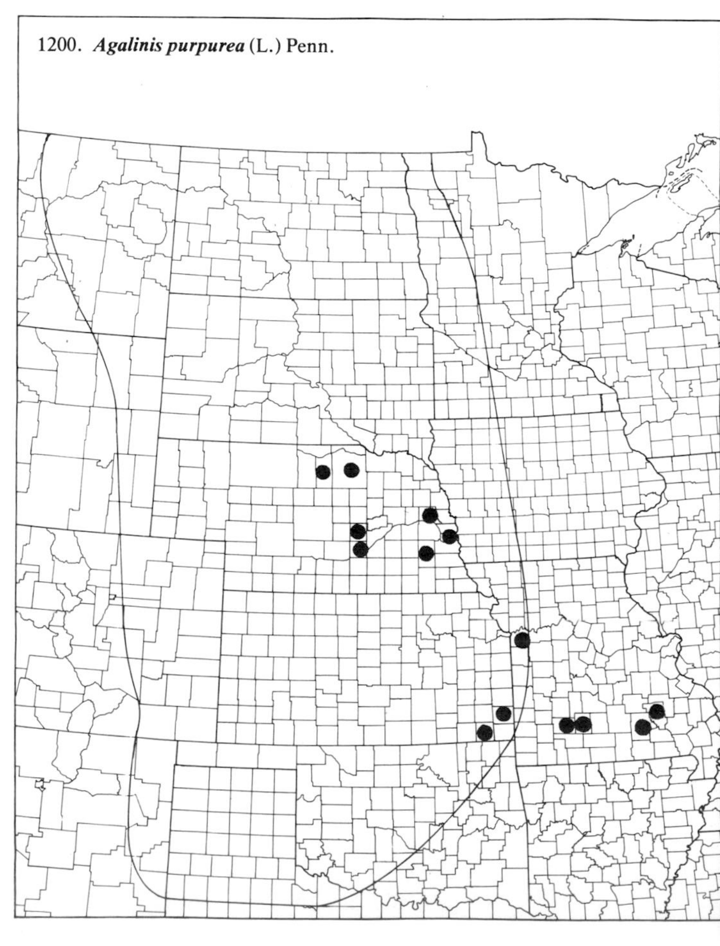

(Scrophulariaceae)

1201. ***Agalinis skinneriana*** (Wood) Britt.

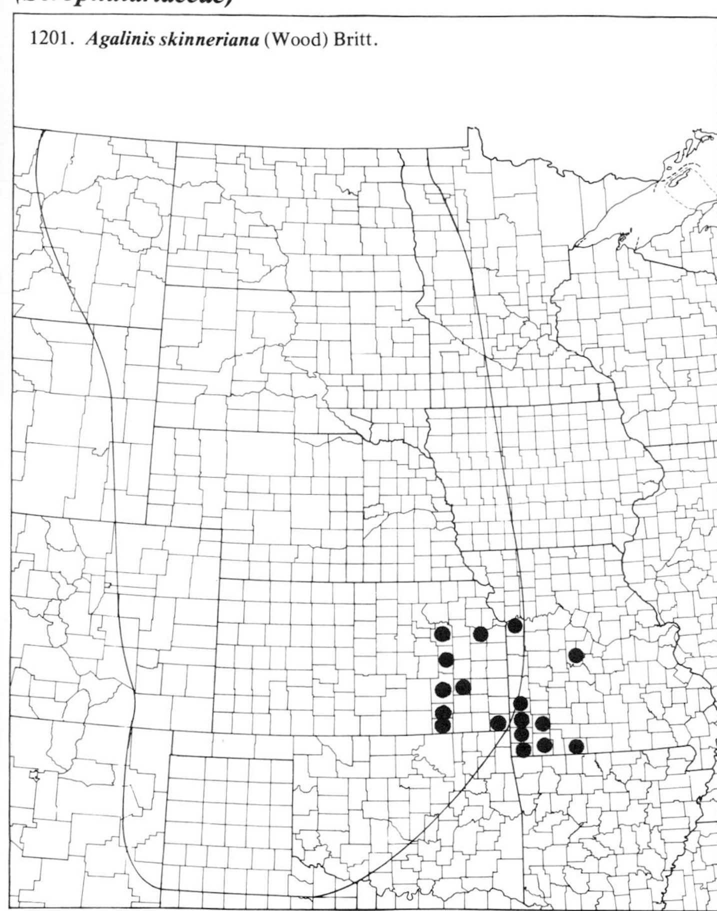

1202. ***Agalinis tenuifolia*** (Vahl) Raf.

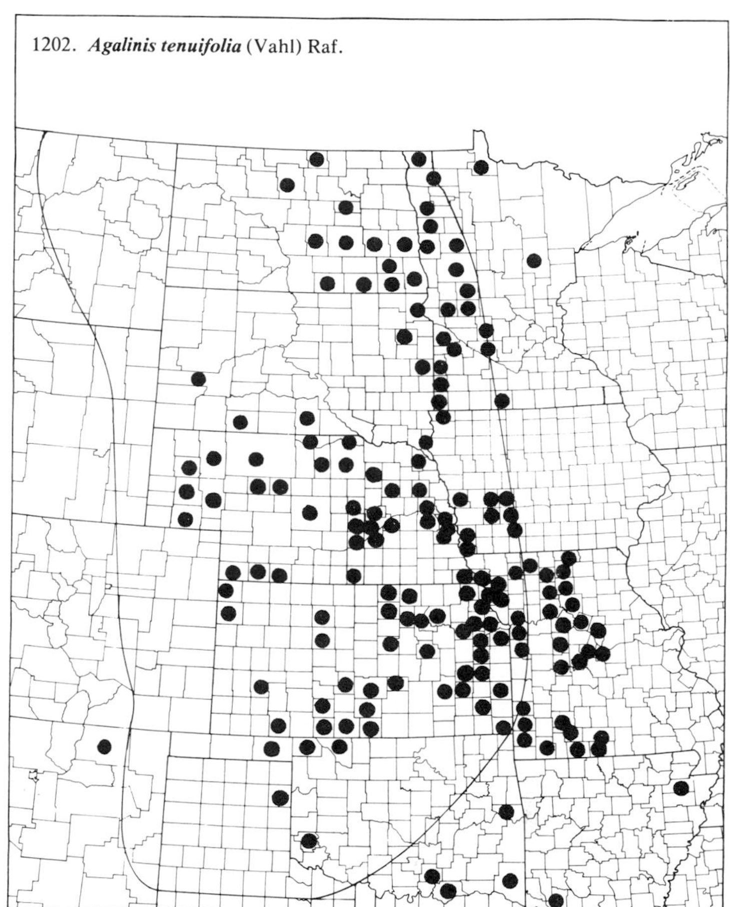

1203. ***Aureolaria grandiflora*** (Benth.) Penn. var. ***cinerea*** Penn. Bigflower Gerardia

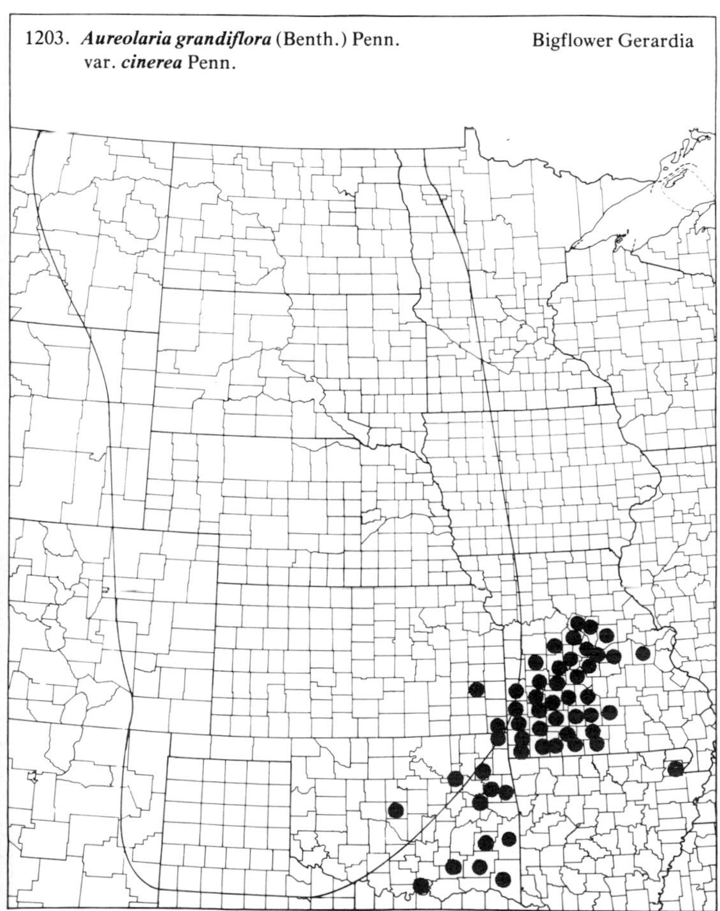

1204. ***Bacopa rotundifolia*** (Michx.) Wettst. Water Hyssop

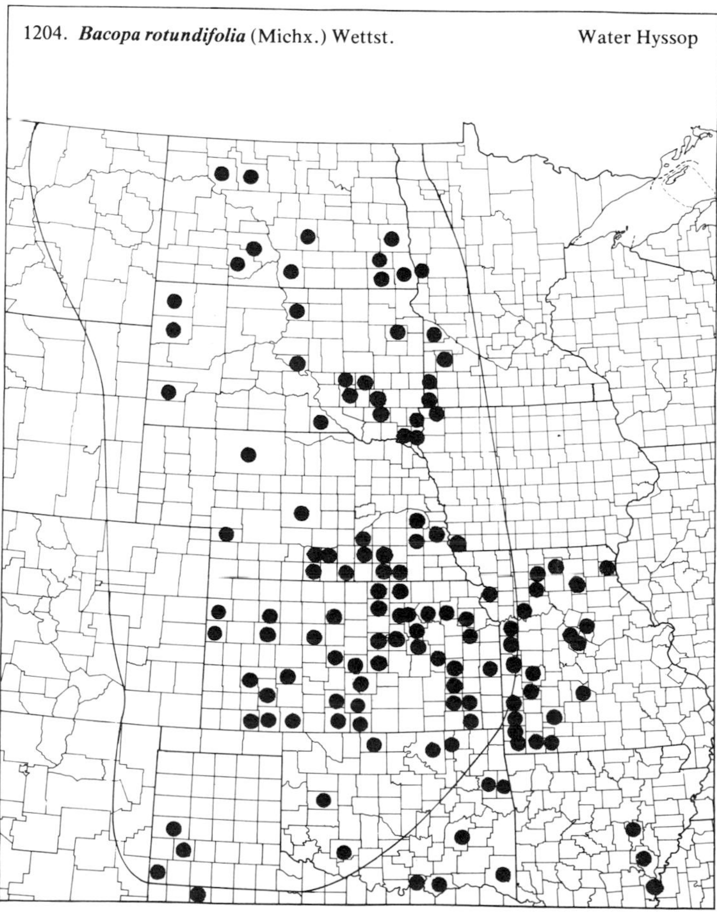

(Scrophulariaceae)

1205. ***Besseya wyomingensis*** (A. Nels.) Rydb. Kitten-tails

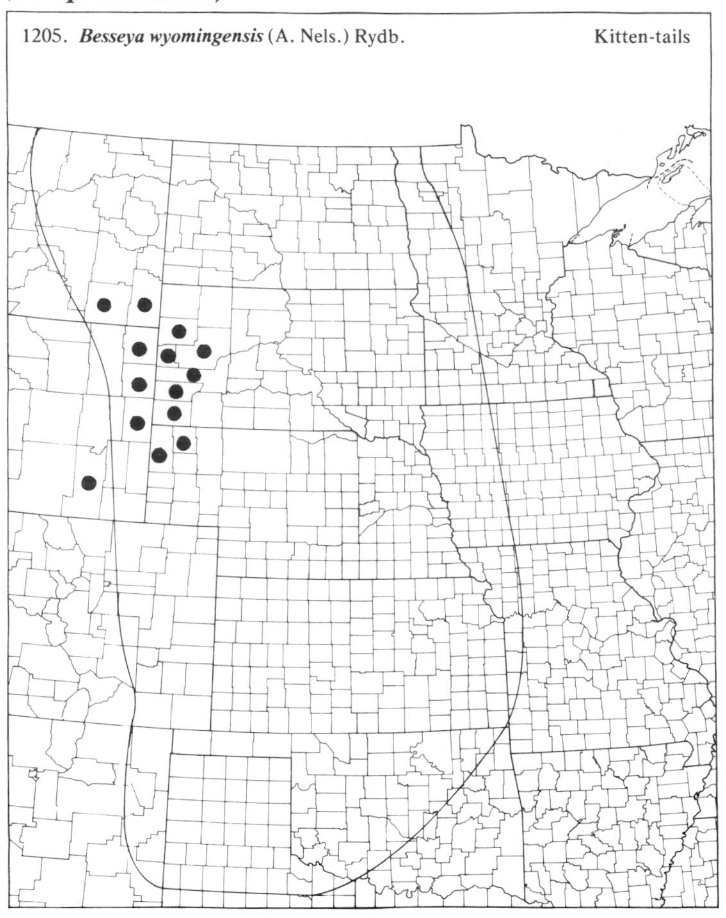

1206. ***Buchnera americana*** L. American Bluehearts

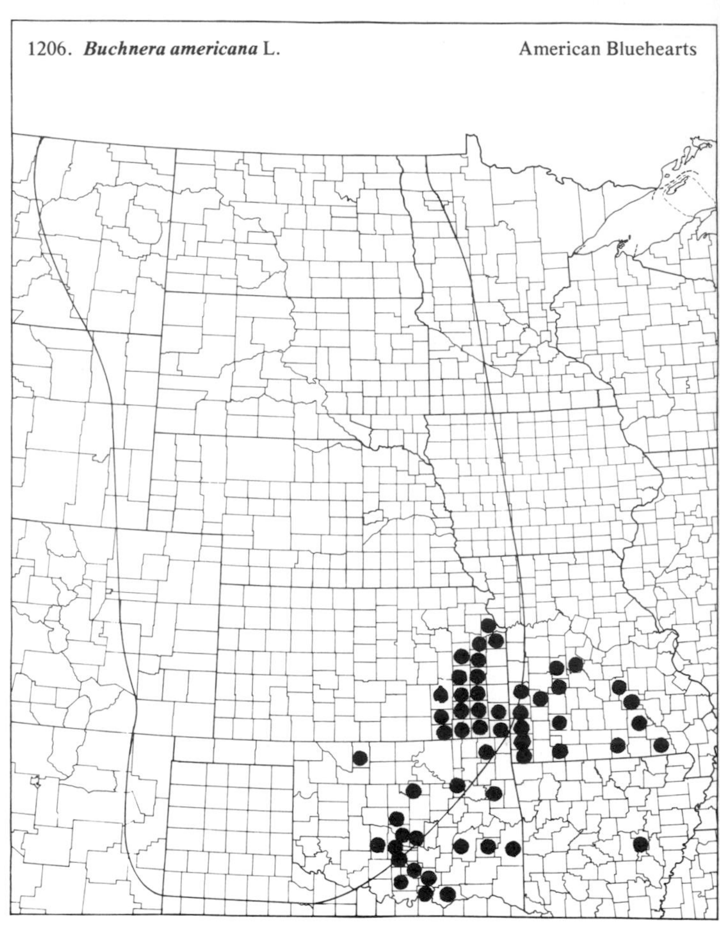

1207. ***Castilleja coccinea*** (L.) Spreng. Indian Paintbrush

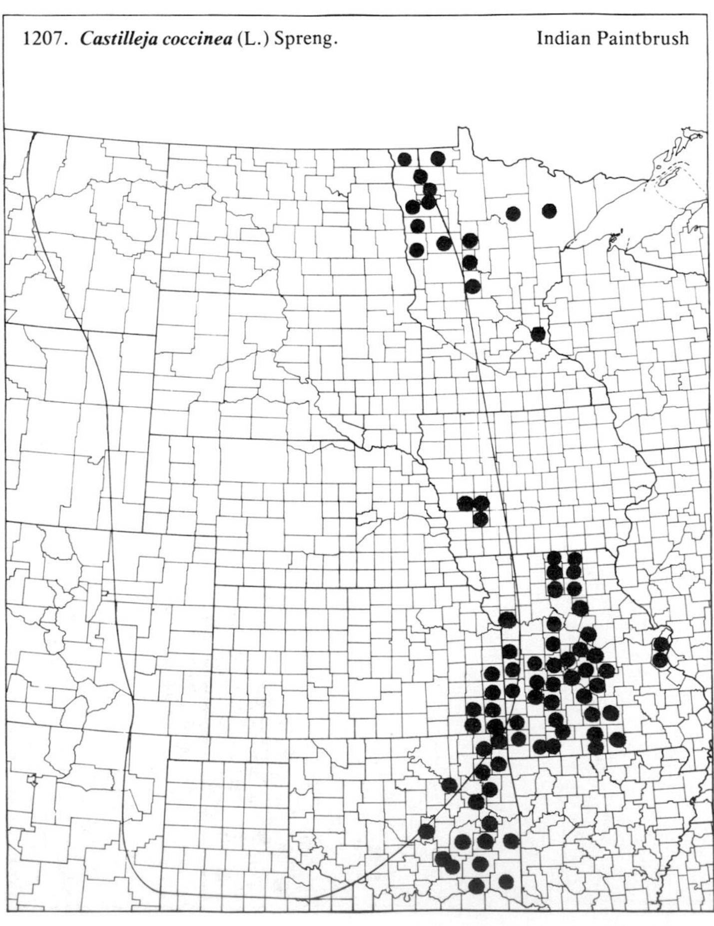

1208. ***Castilleja purpurea*** (Nutt.) G. Don var. ***citrina*** (Penn.) Shinners Citron Paintbrush

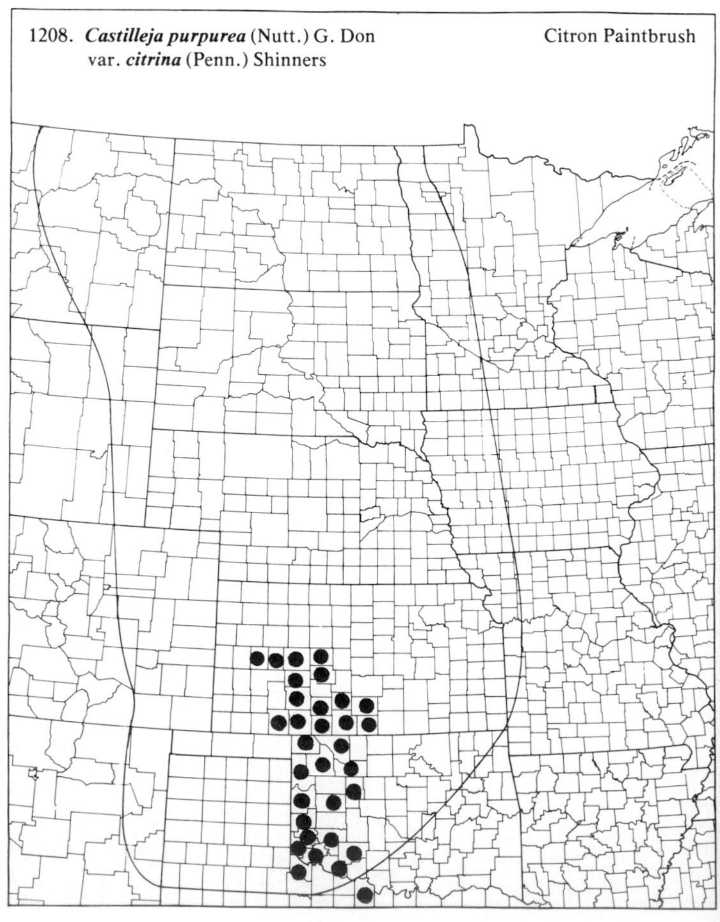

(Scrophulariaceae)

1209. ***Castilleja sessiliflora*** Pursh — Downy Paintbrush

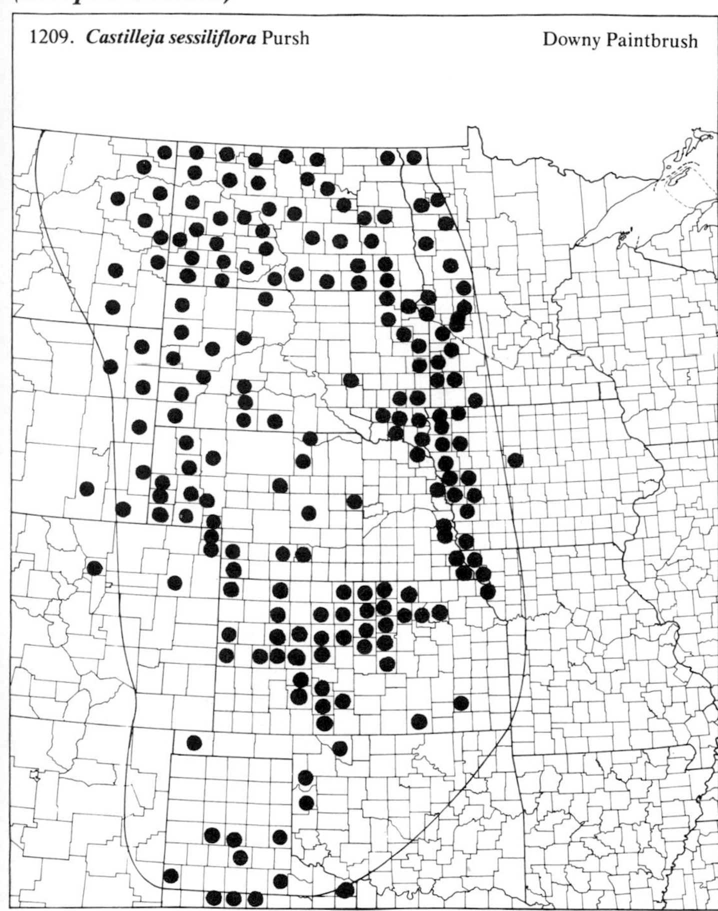

1210. ***Chaenorrhinum minus*** (L.) Lange

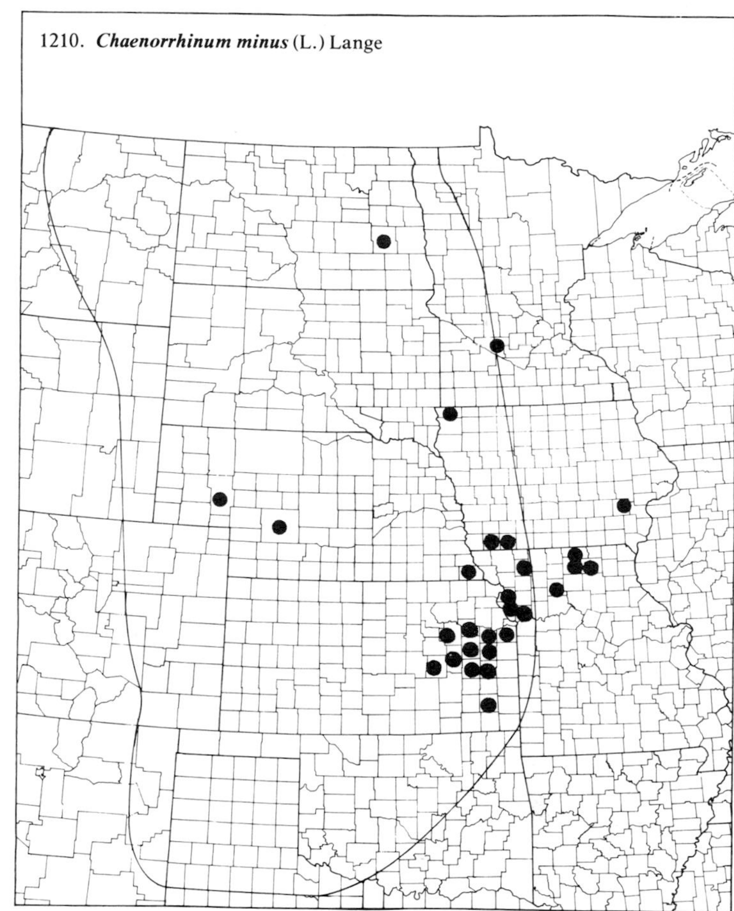

1211. ***Collinsia parviflora*** Lindl. — Blue Lips

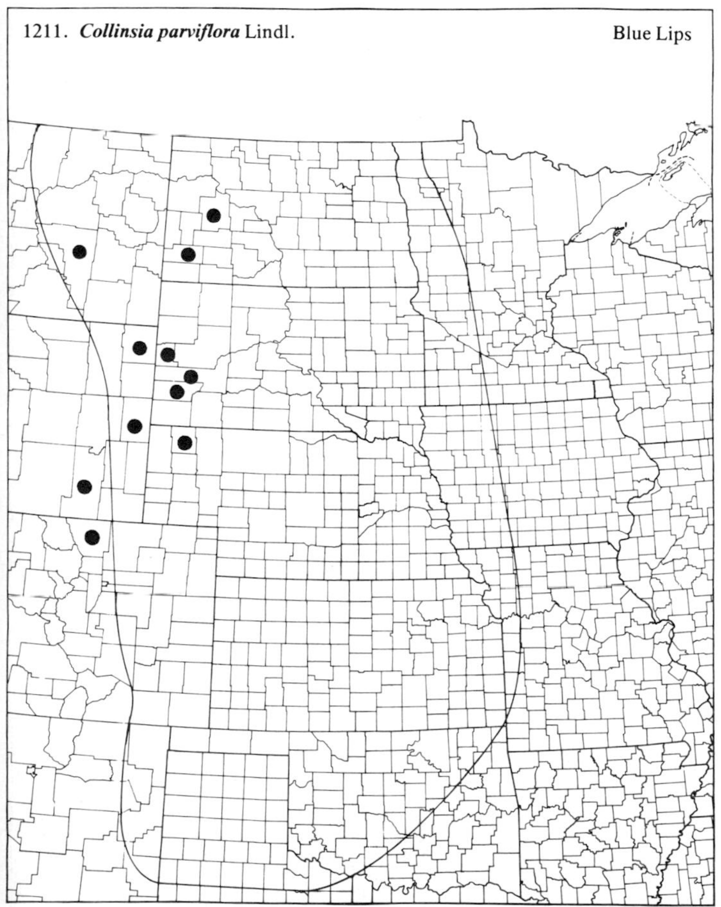

1212. ***Collinsia violacea*** Nutt. — Collinsia

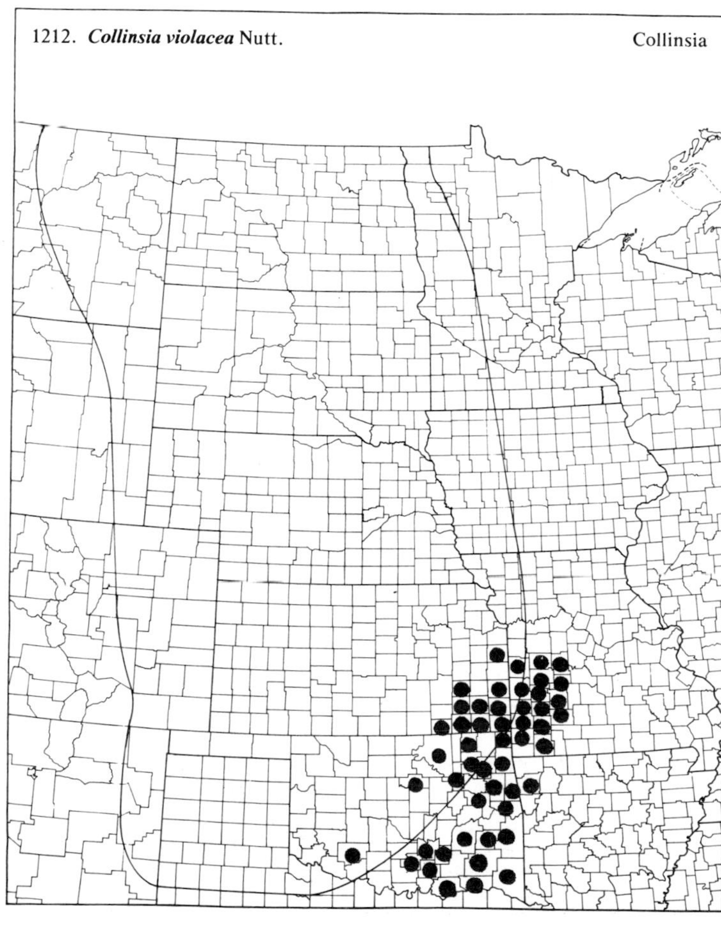

(Scrophulariaceae)

1213. ***Gratiola neglecta*** Torr. Hedge Hyssop

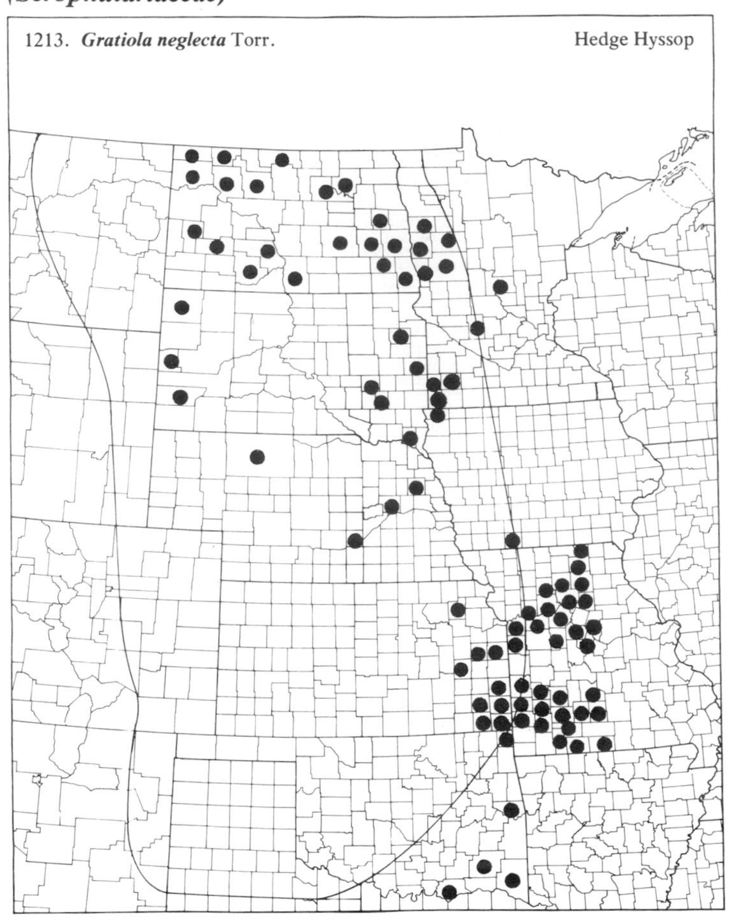

1214. ***Gratiola virginiana*** L. Virginia Hedge Hyssop

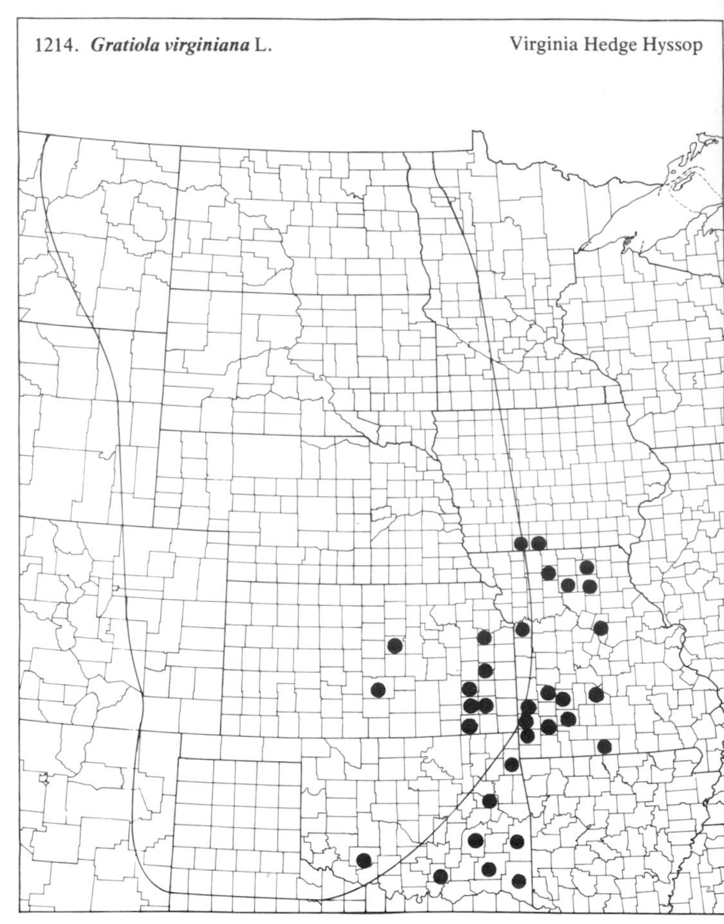

1215. ***Leucospora multifida*** (Michx.) Nutt. Leucospora

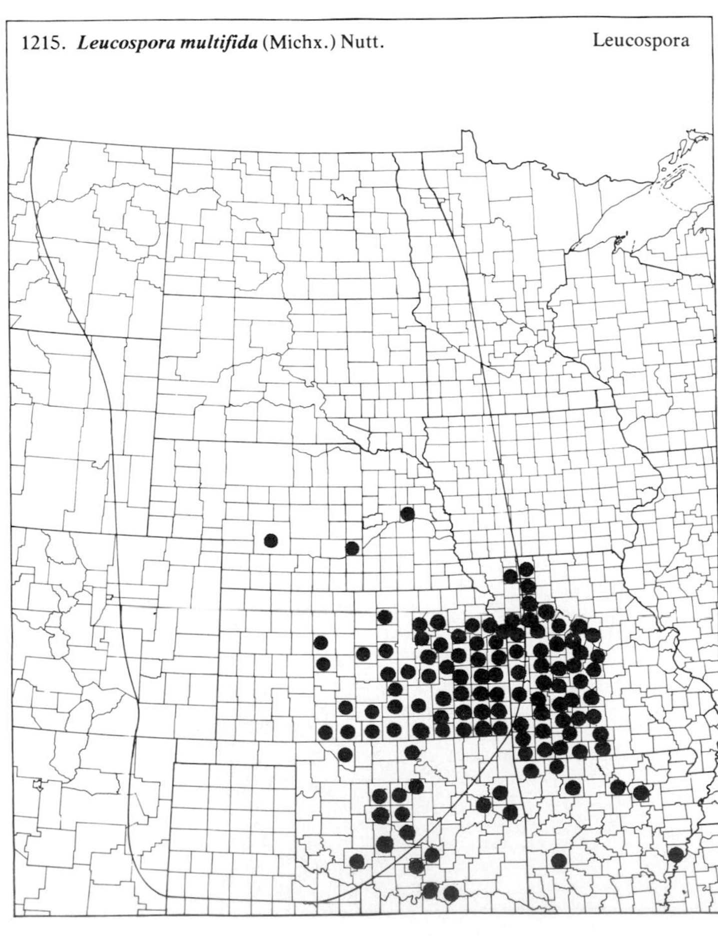

1216. ***Limosella aquatica*** L. Mudwort

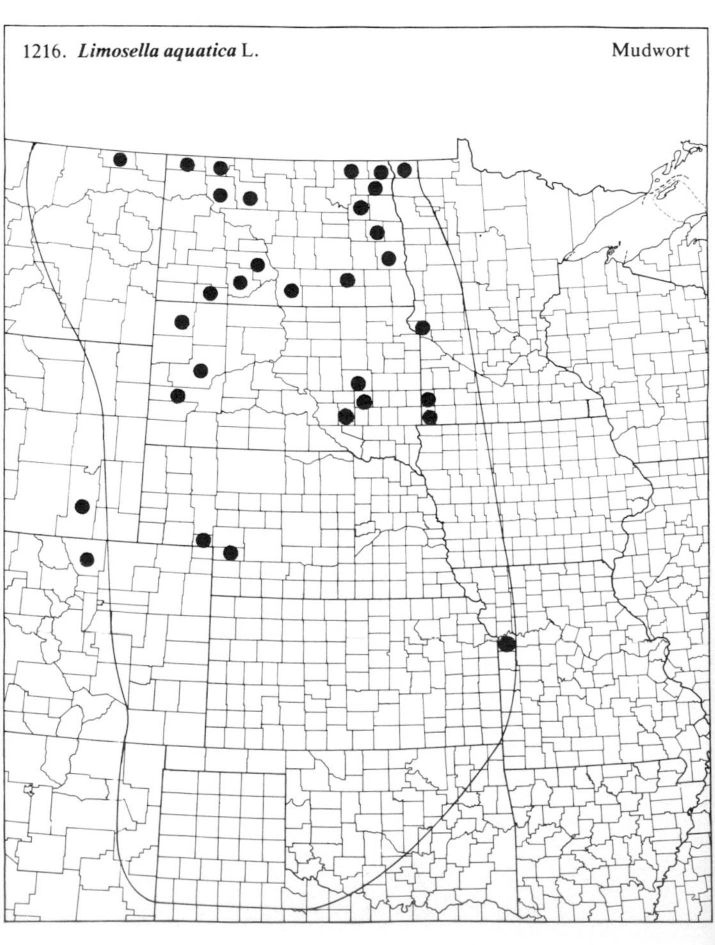

(Scrophulariaceae)

1217. ***Linaria canadensis*** (L.) Dumont var. ***texana*** (Scheele) Penn. — Old-field Toadflax

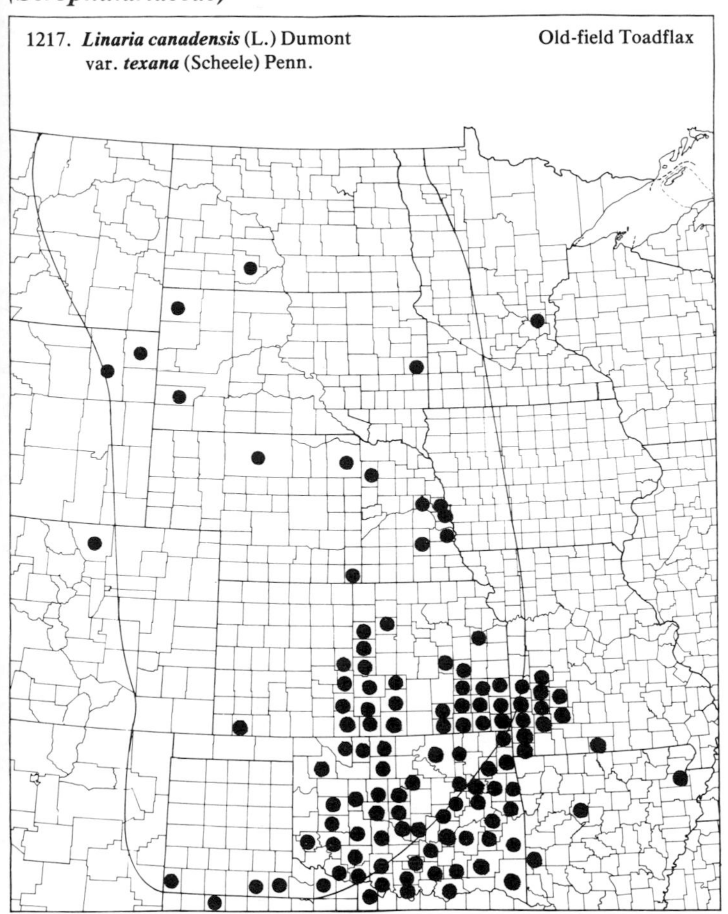

1218. ***Linaria dalmatica*** (L.) Mill. — Toadflax

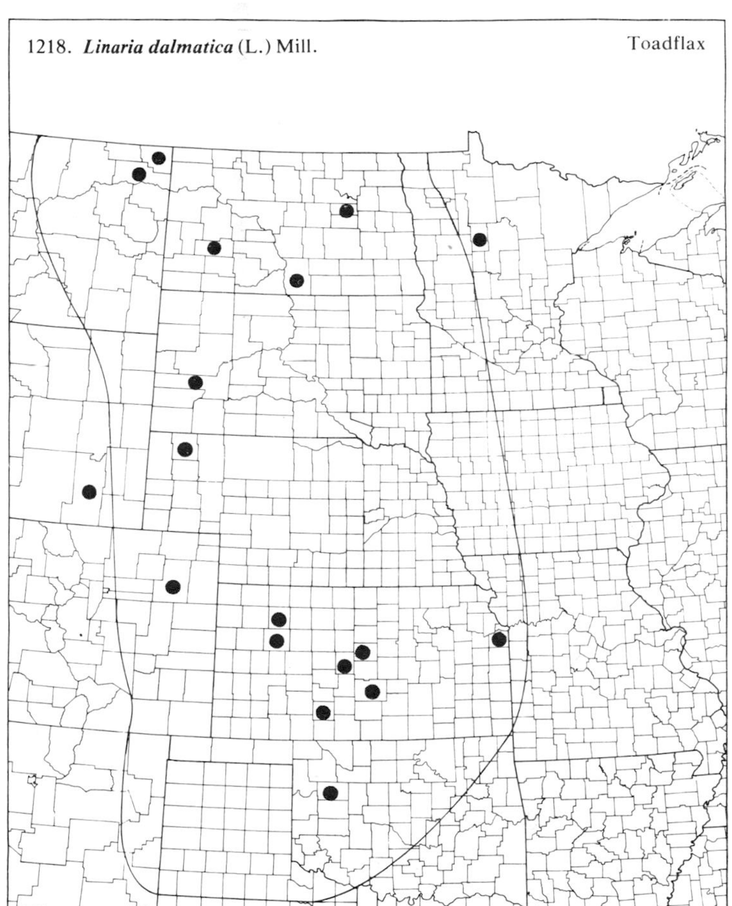

1219. ***Linaria vulgaris*** Hill — Butter-and-eggs

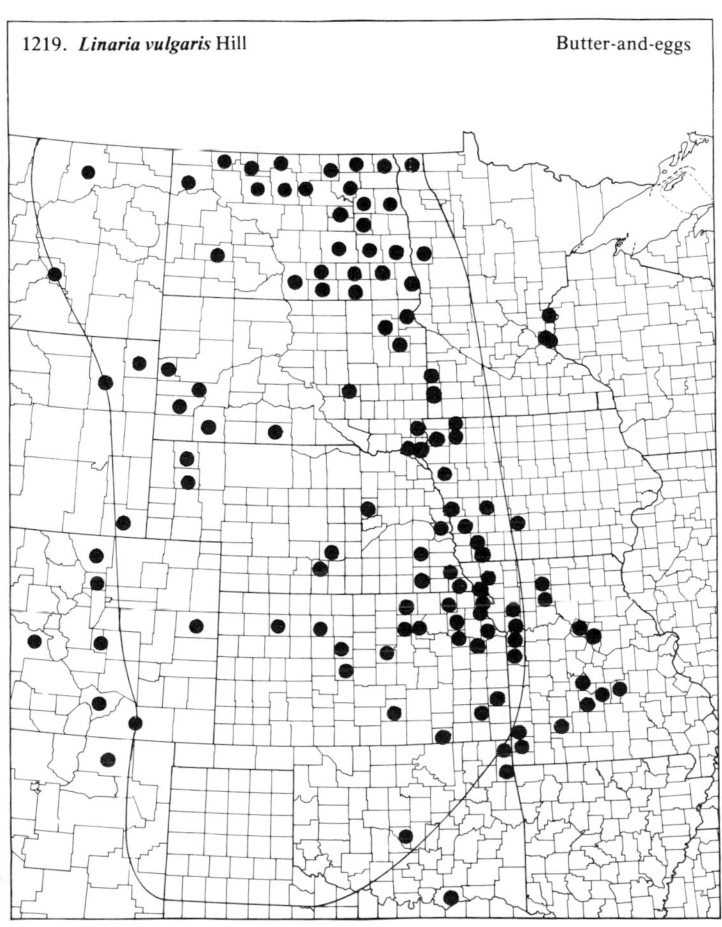

1220. ***Lindernia anagallidea*** (Michx.) Penn. — False Pimpernel

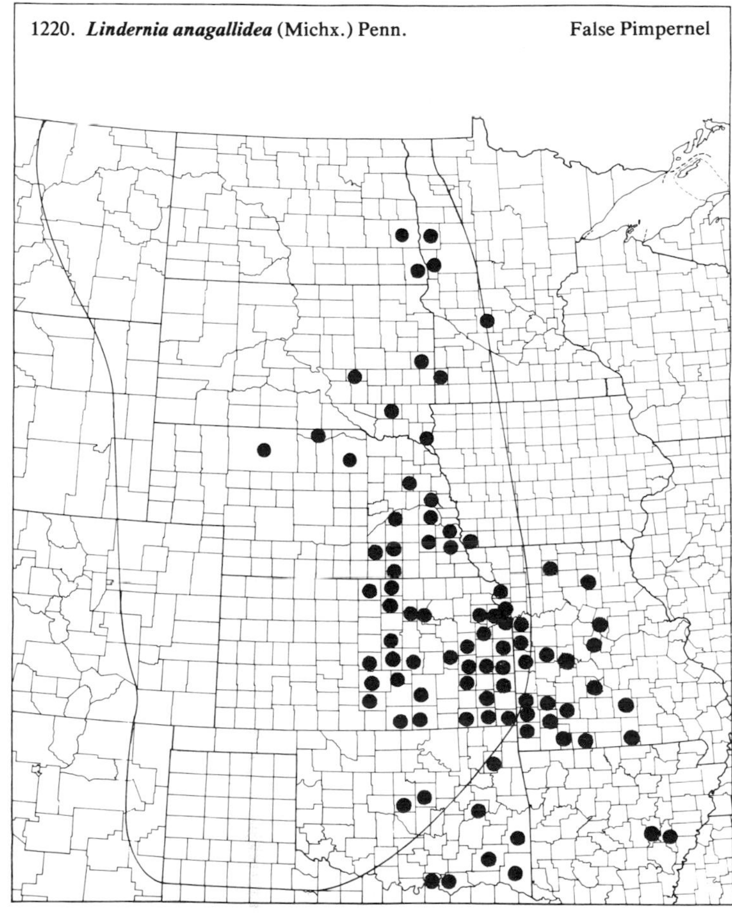

(Scrophulariaceae)

1221. ***Lindernia dubia*** (L.) Penn. False Pimpernel

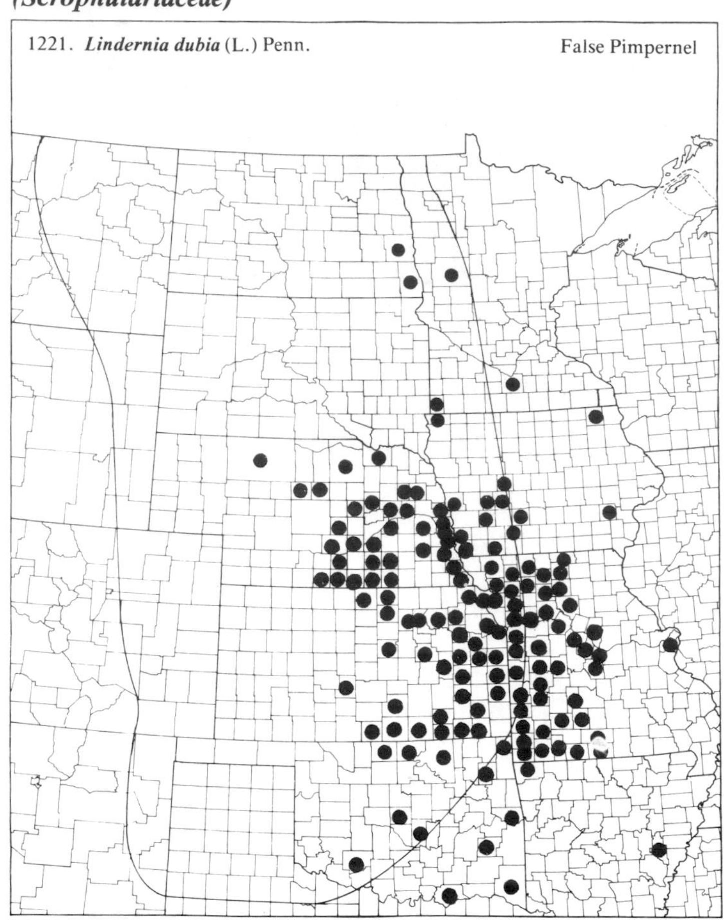

1222. ***Mimulus alatus*** Ait. Sharpwing Monkey-flower

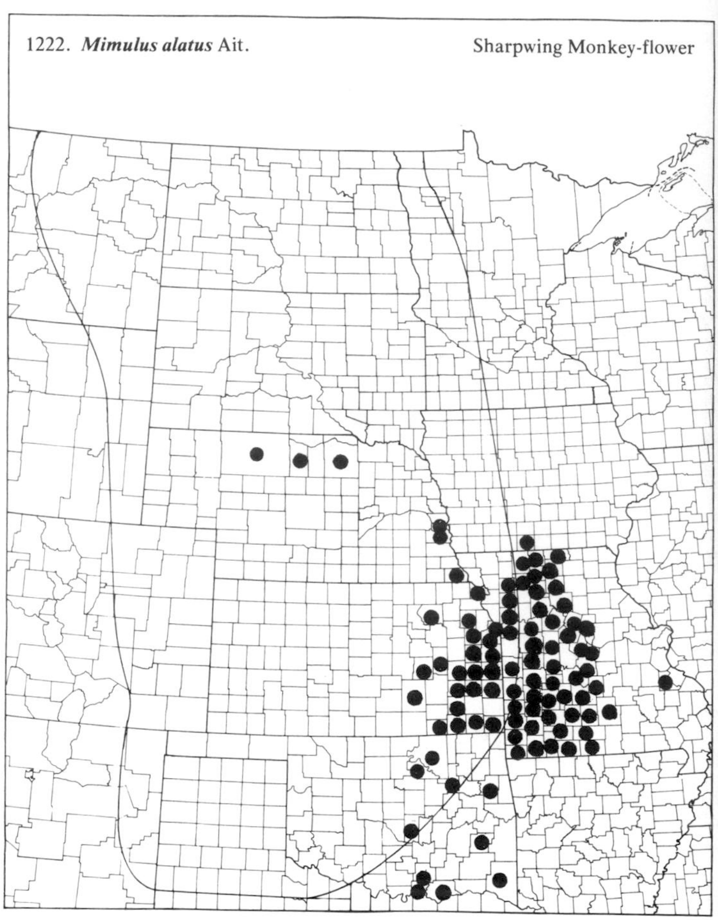

1223. ***Mimulus glabratus*** H.B.K. var. ***fremontii*** (Benth.) Grant Roundleaf Monkey-flower

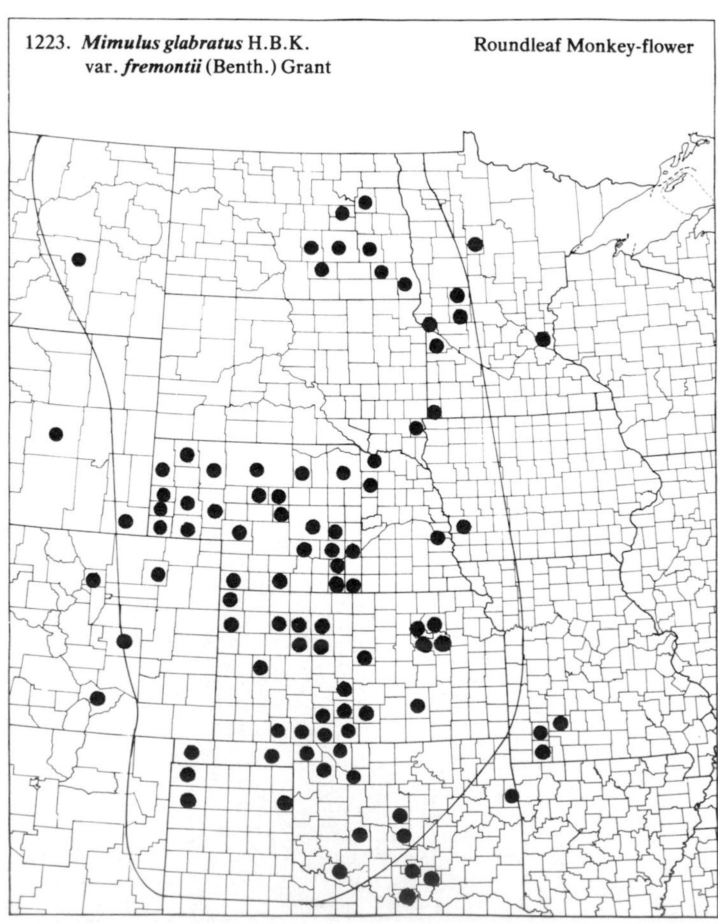

1224. ***Mimulus guttatus*** DC. Monkey-flower

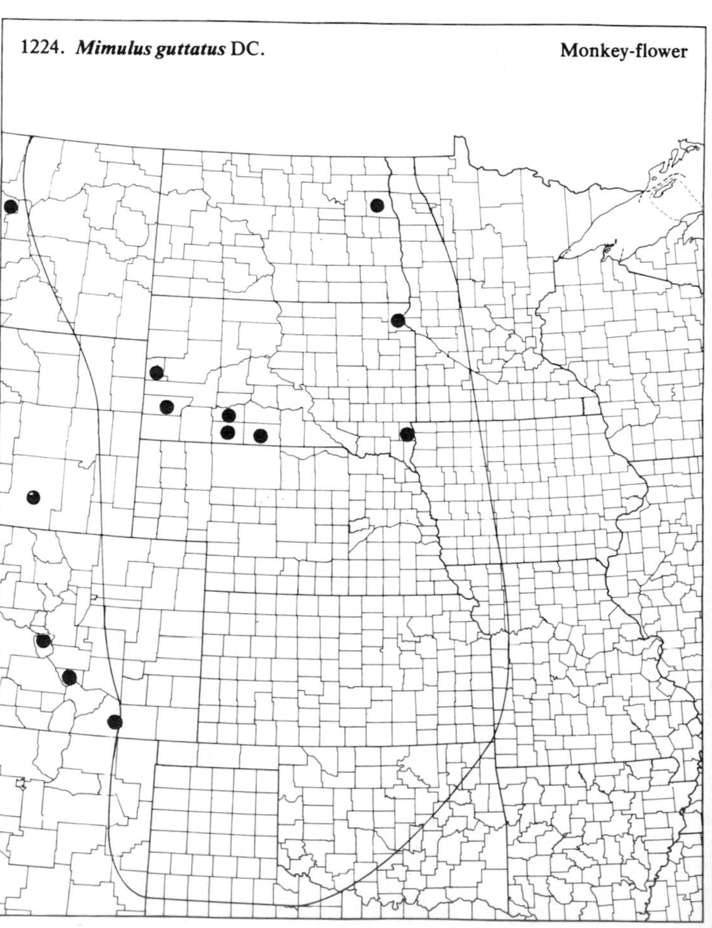

(Scrophulariaceae)

1225. ***Mimulus ringens*** L. Alleghany Monkey-flower

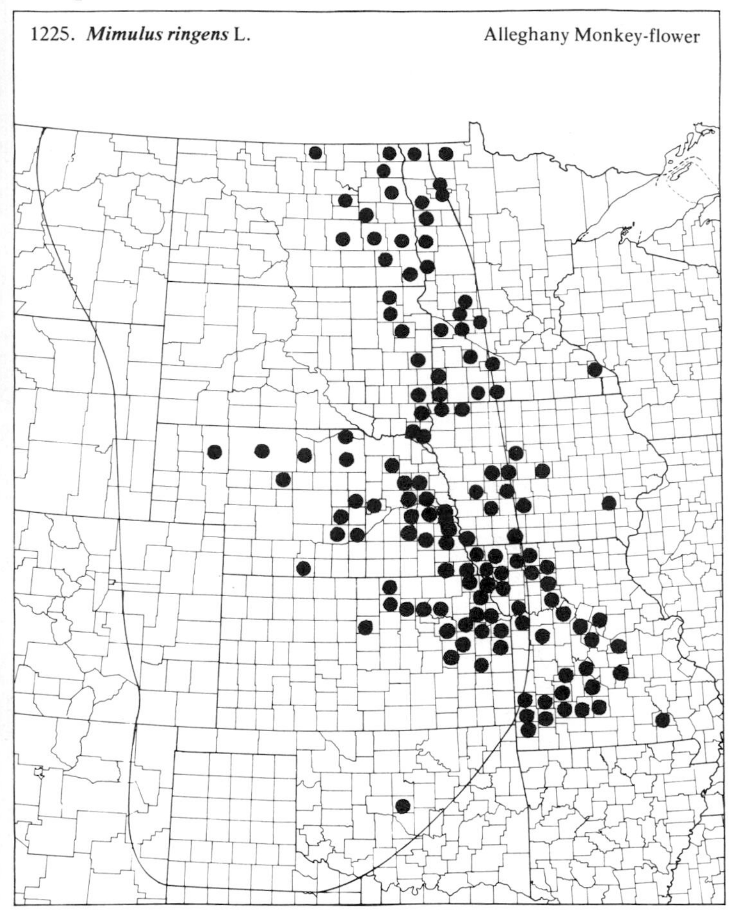

1226. ***Orthocarpus luteus*** Nutt. Owl Clover

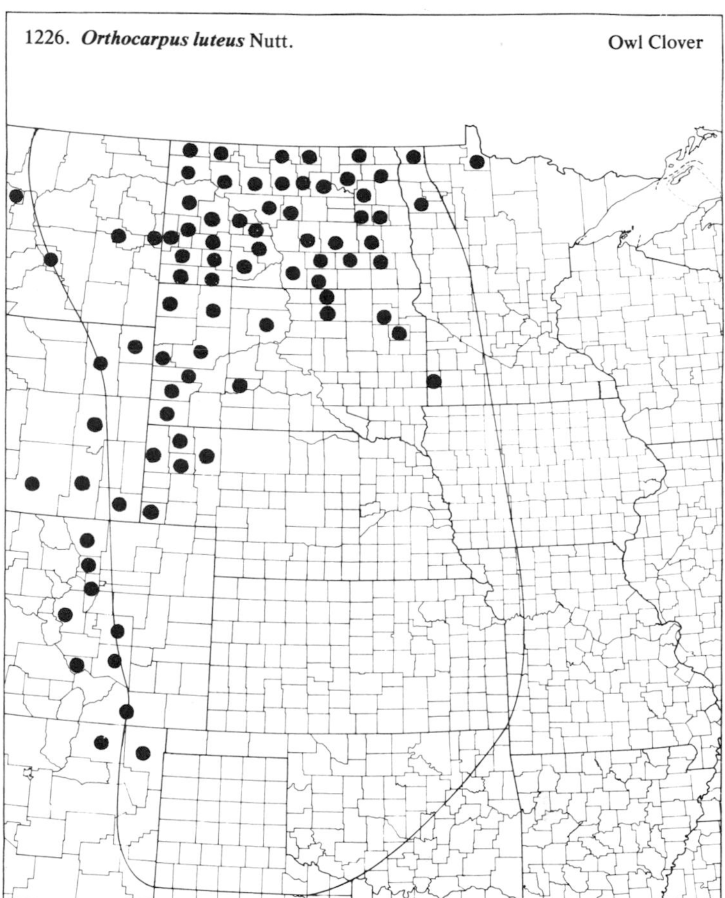

1227. ***Pedicularis canadensis*** L. Lousewort

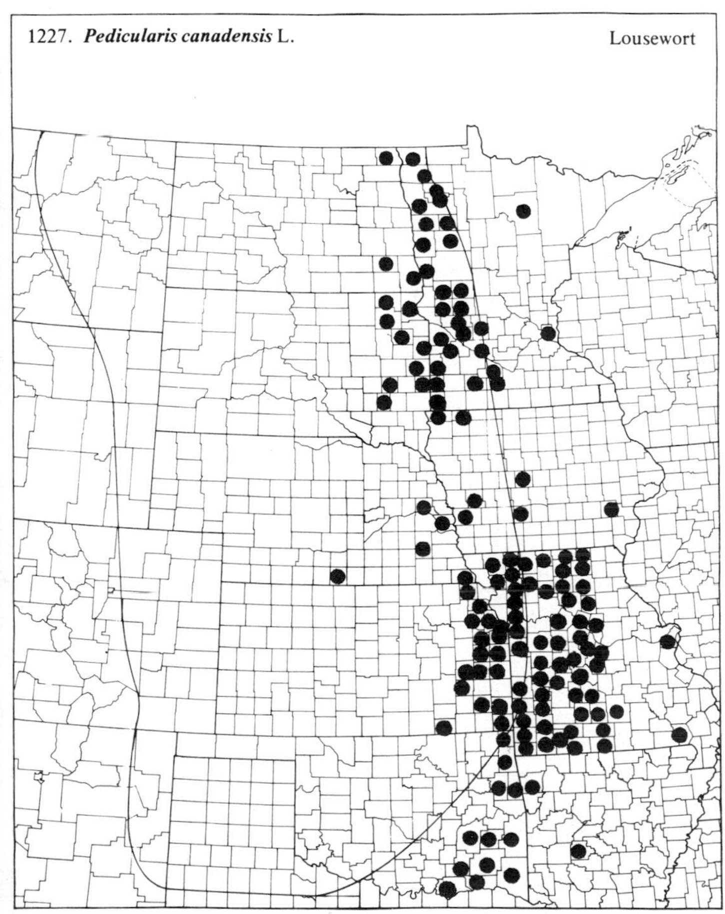

1228. ***Pedicularis lanceolata*** Michx.

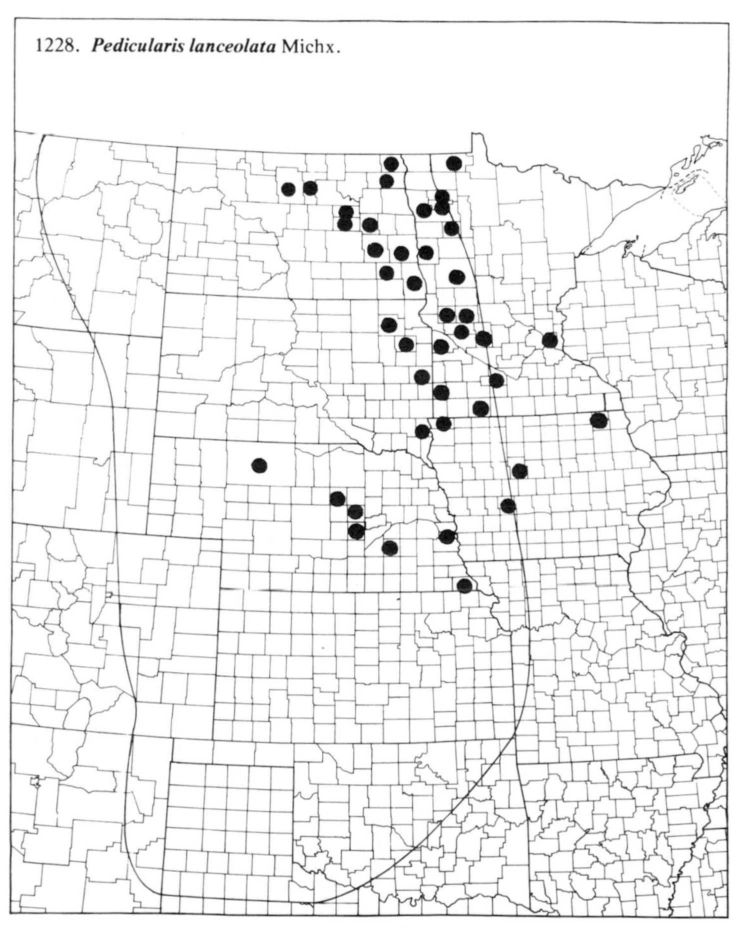

(Scrophulariaceae)

1229. ***Penstemon albidus*** Nutt. White Beardtongue

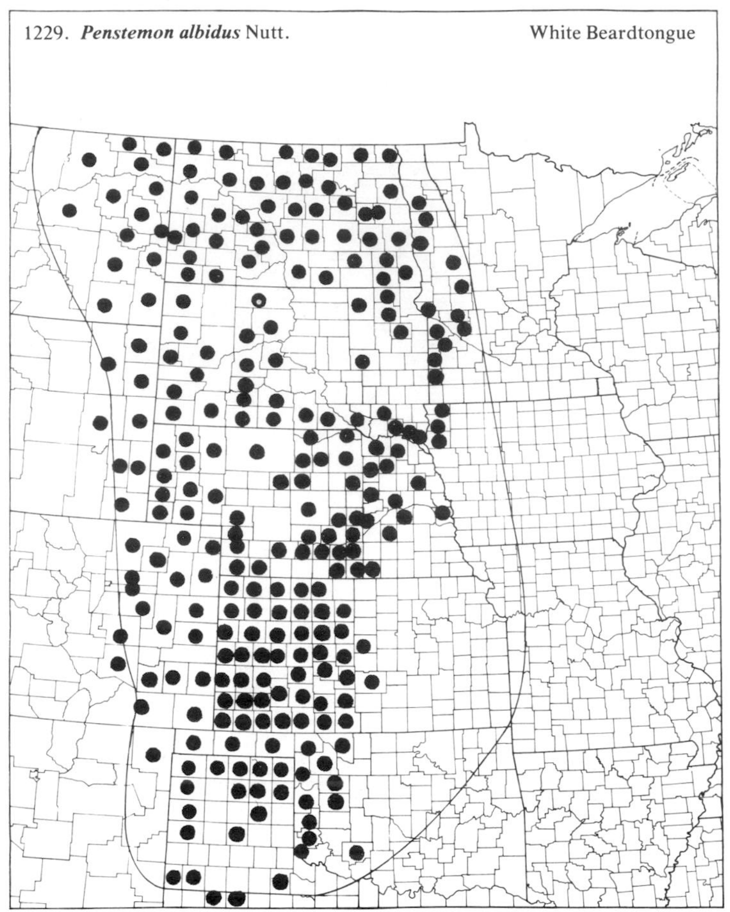

1230. ***Penstemon ambiguus*** Torr. Gilia Penstemon

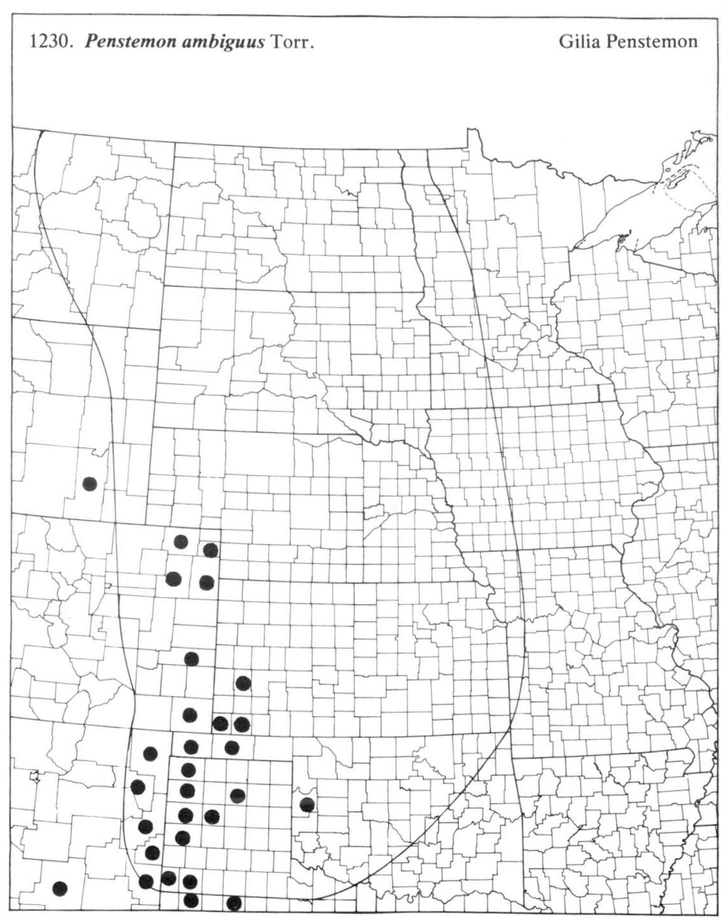

1231. ***Penstemon angustifolius*** Nutt. ssp. ***angustifolius*** Narrow Beardtongue

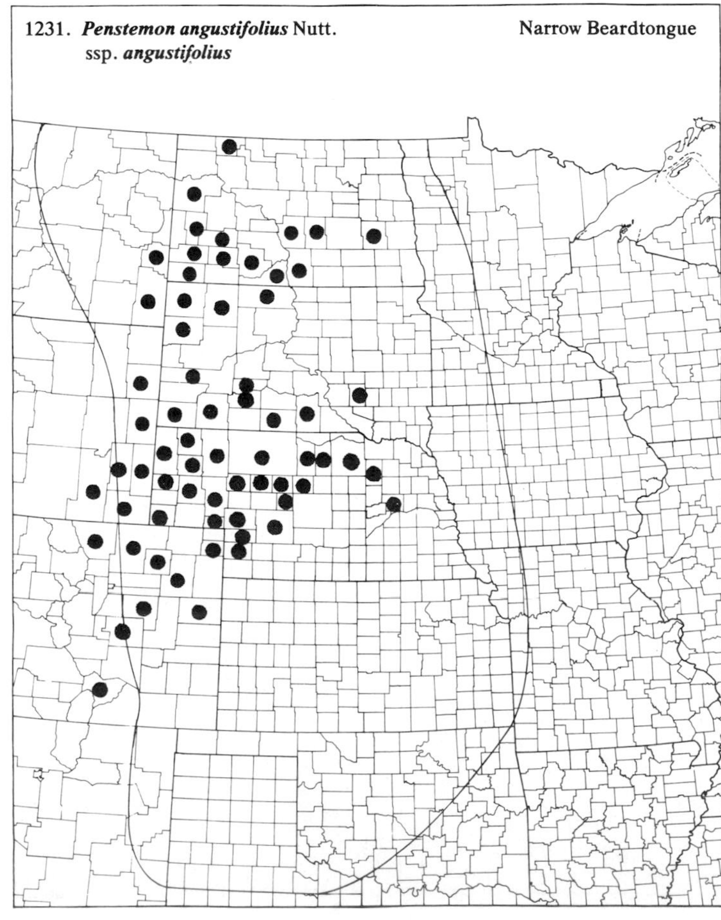

1232. ***Penstemon angustifolius*** Nutt. ssp. ***caudatus*** (Heller) Keck Narrow Beardtongue

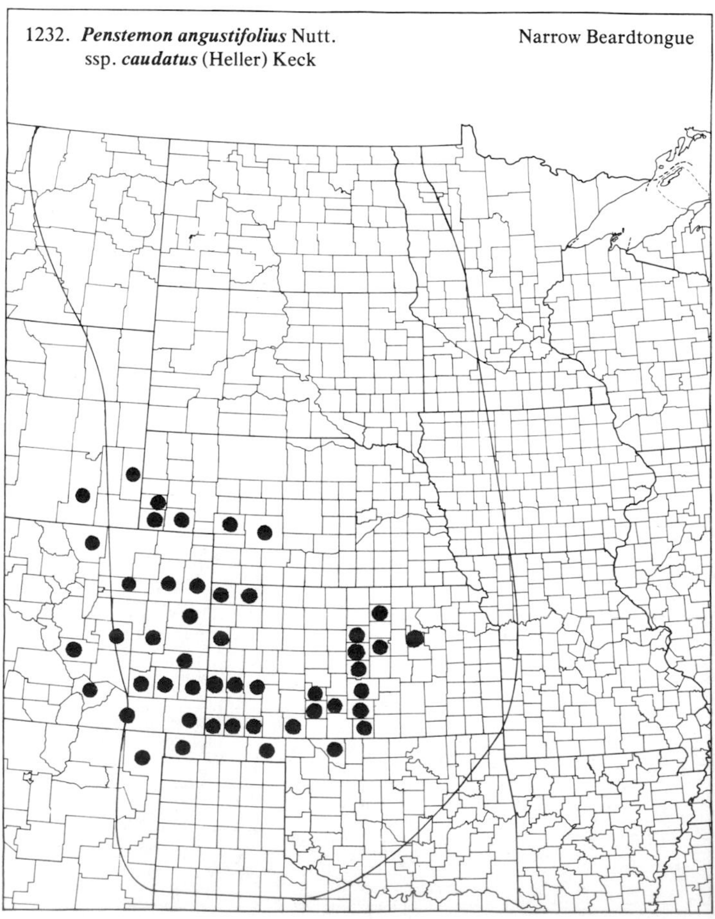

(Scrophulariaceae)

1233. ***Penstemon buckleyi*** Penn. Buckley's Penstemon

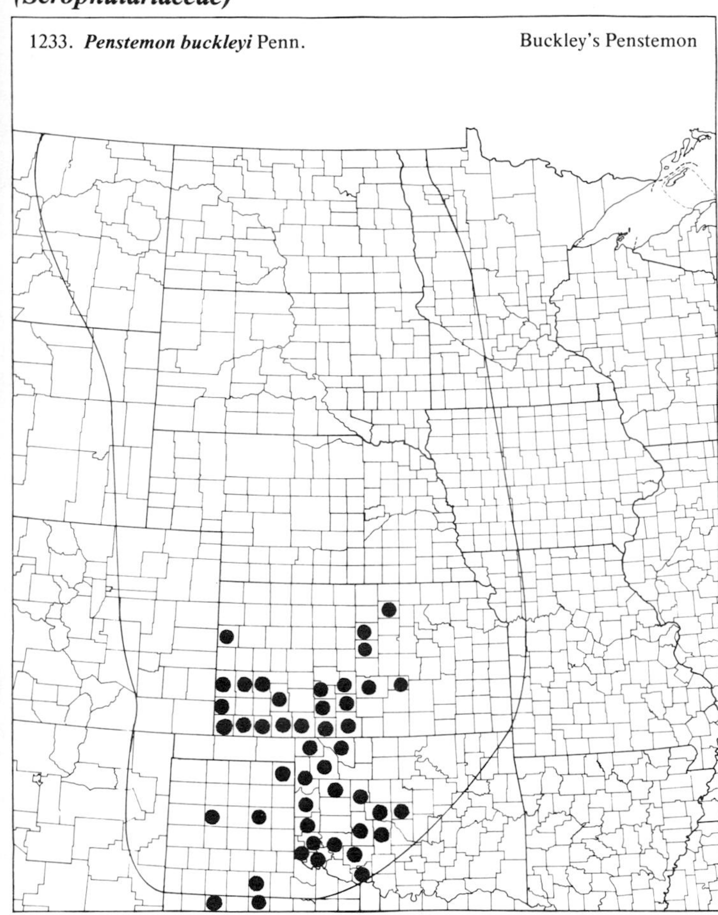

1234. ***Penstemon cobaea*** Nutt. Cobaea Penstemon

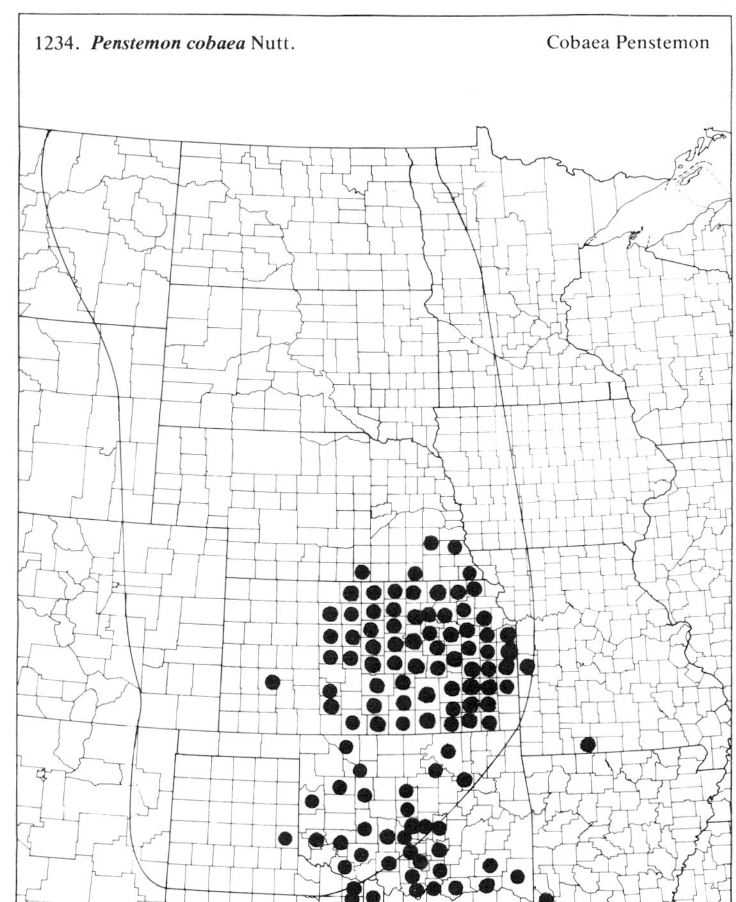

1235. ***Penstemon digitalis*** Nutt. Smooth Penstemon

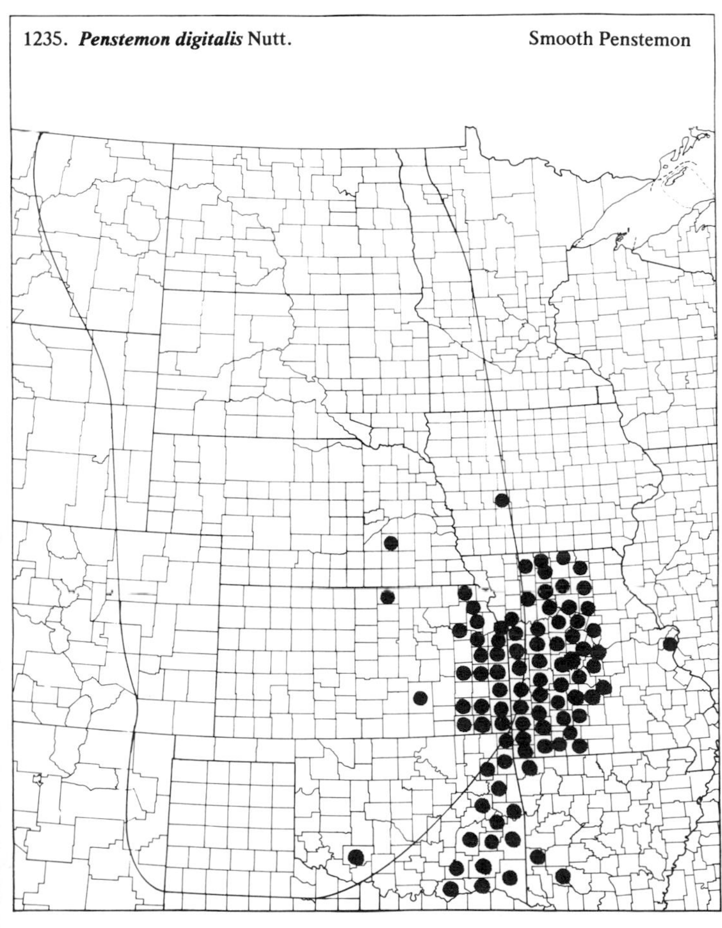

1236. ***Penstemon eriantherus*** Pursh Crested Beardtongue

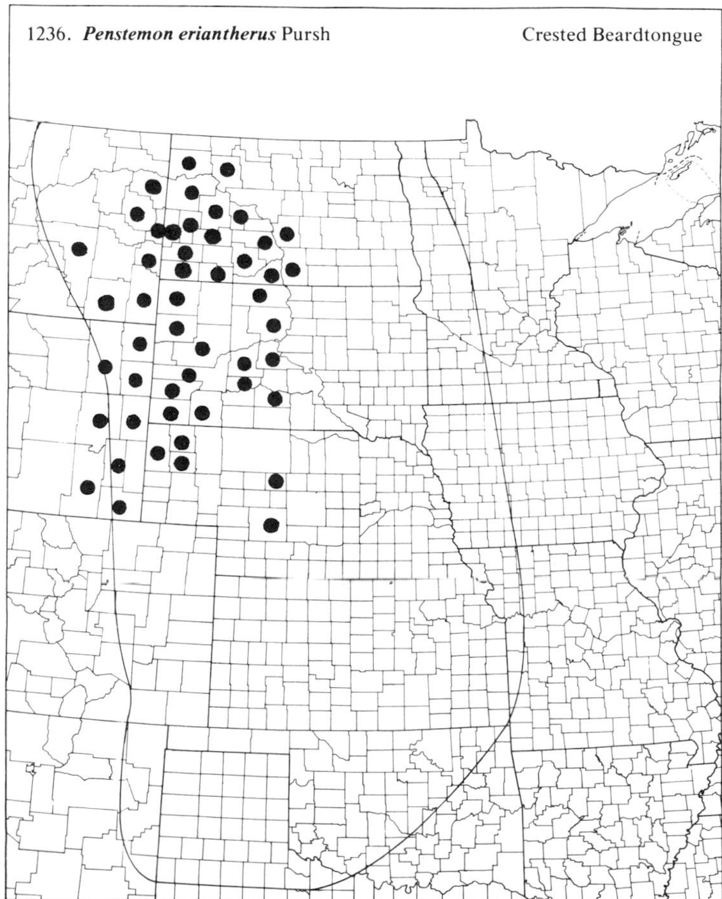

(Scrophulariaceae)

1237. ***Penstemon fendleri*** T. & G.

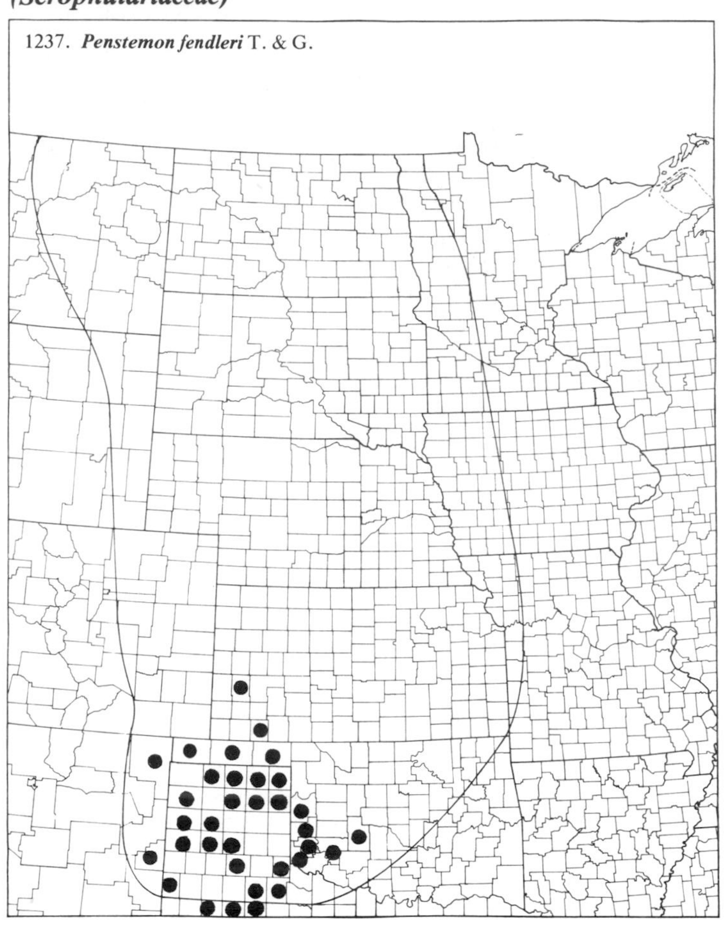

1238. ***Penstemon glaber*** Pursh

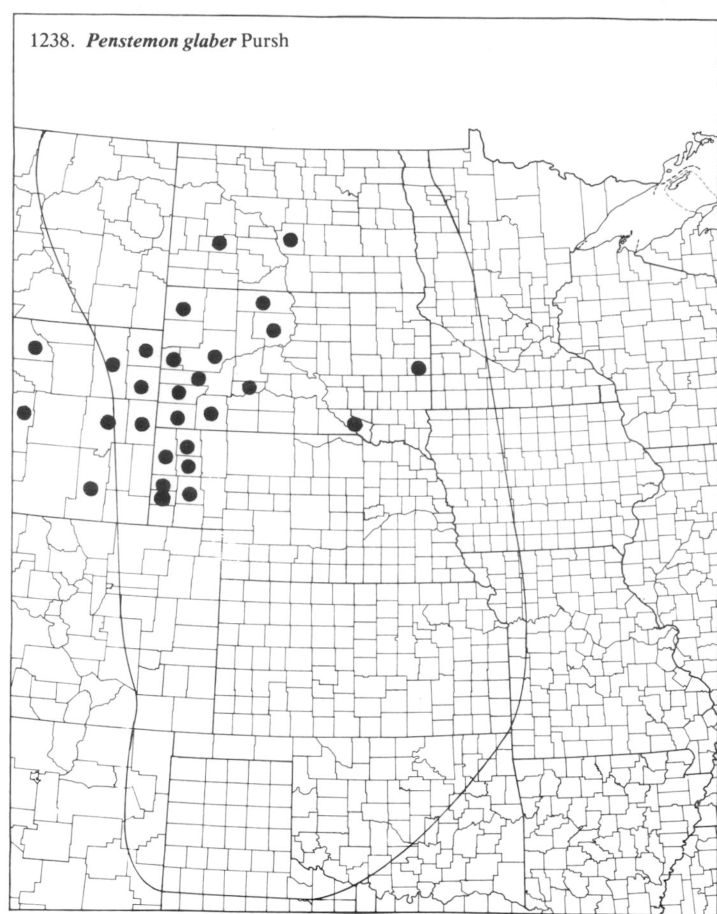

1239. ***Penstemon gracilis*** Nutt. Slender Beardtongue

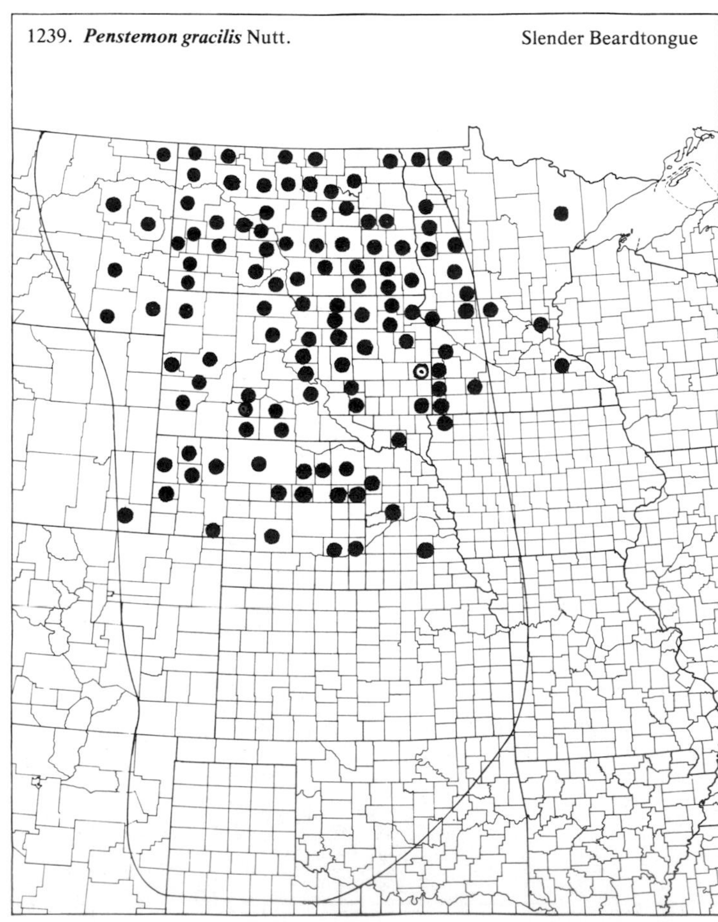

1240. ***Penstemon grandiflorus*** Nutt. Large Beardtongue

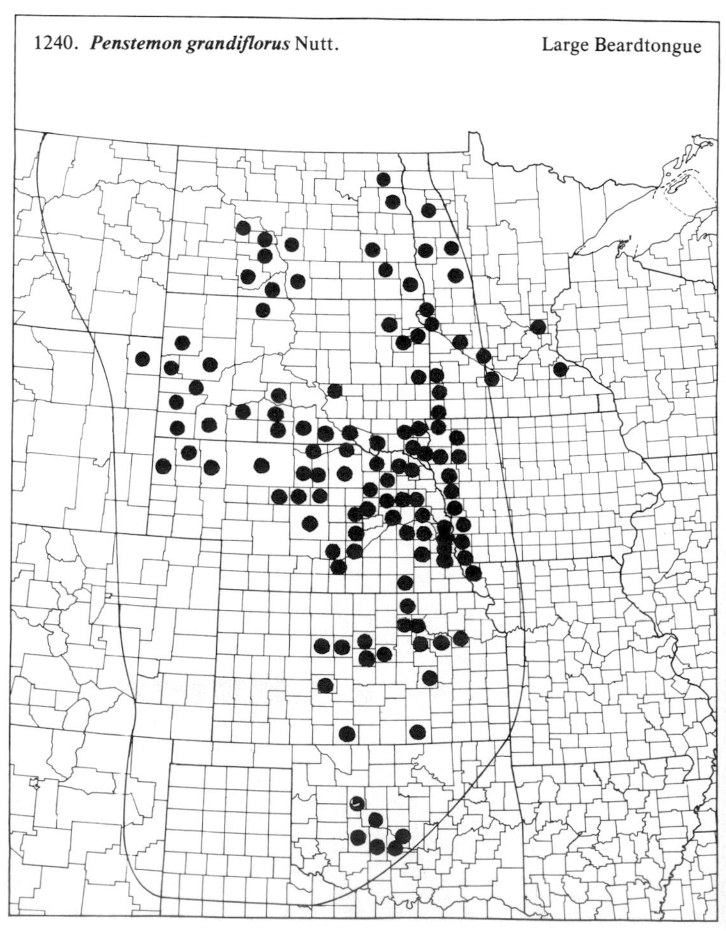

(Scrophulariaceae)

1241. **Penstemon laxiflorus** Penn.

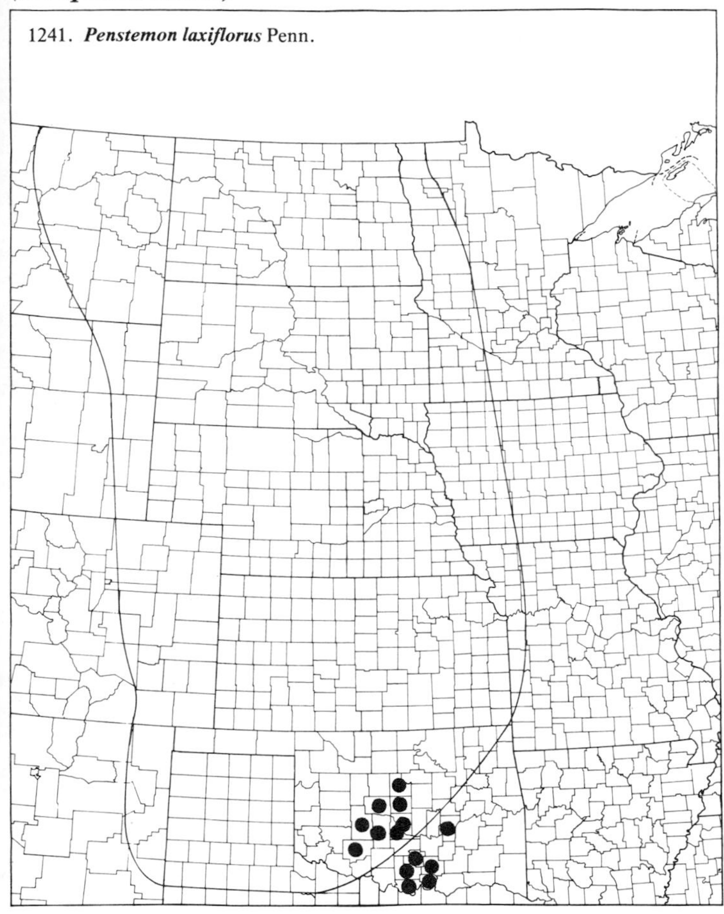

1242. **Penstemon nitidus** Dougl.

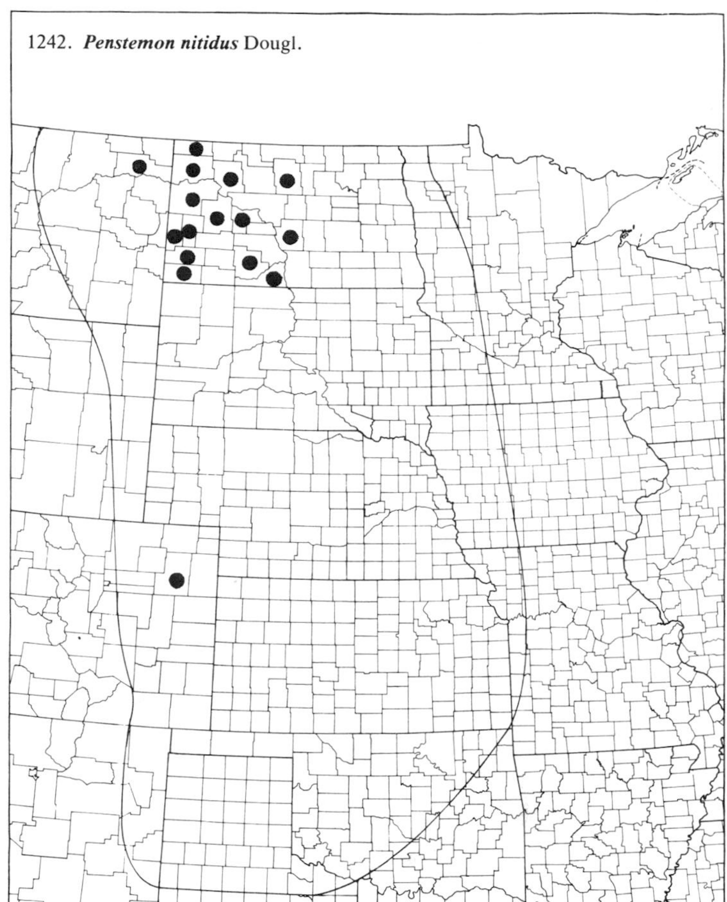

1243. **Penstemon tubaeflorus** Nutt. Tube Penstemon

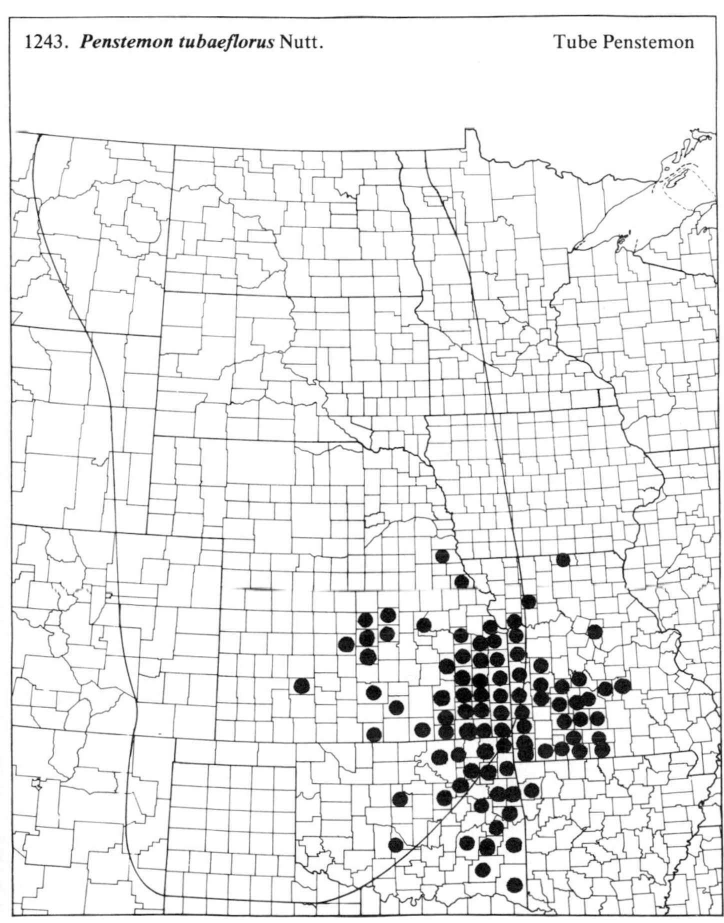

1244. **Scrophularia lanceolata** Pursh Figwort

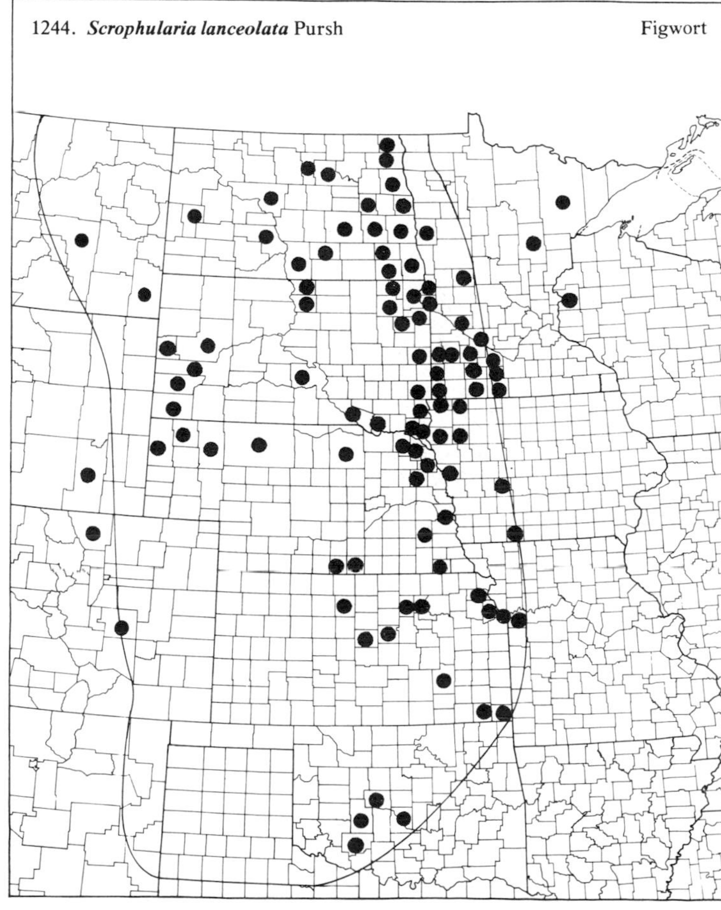

(Scrophulariaceae)

1245. ***Scrophularia marilandica*** L. Maryland Figwort

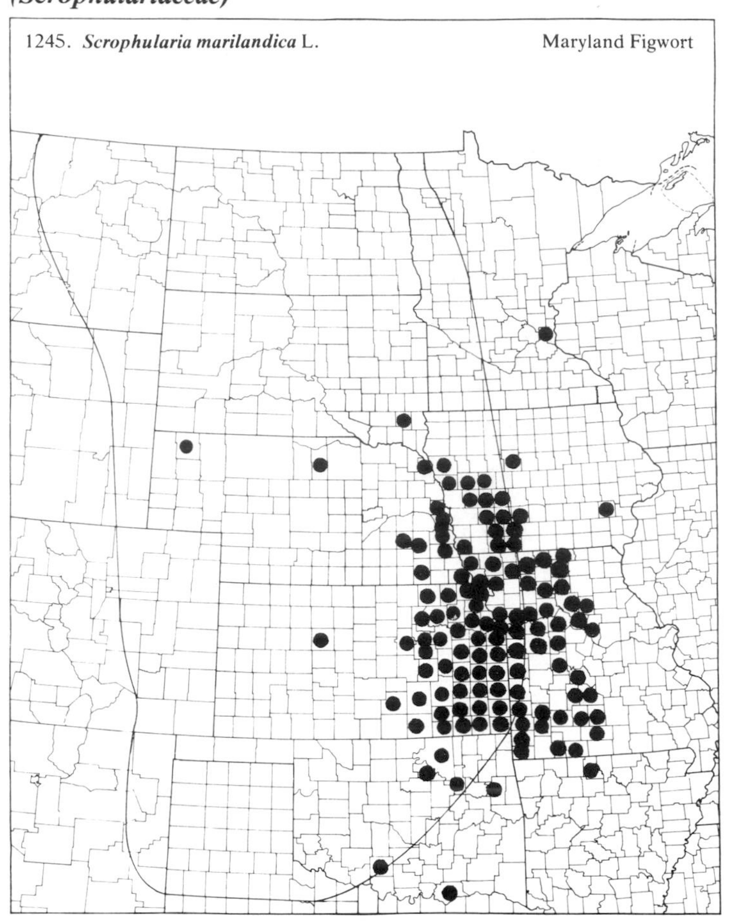

1246. ***Seymeria macrophylla*** Nutt. Mullein Foxglove

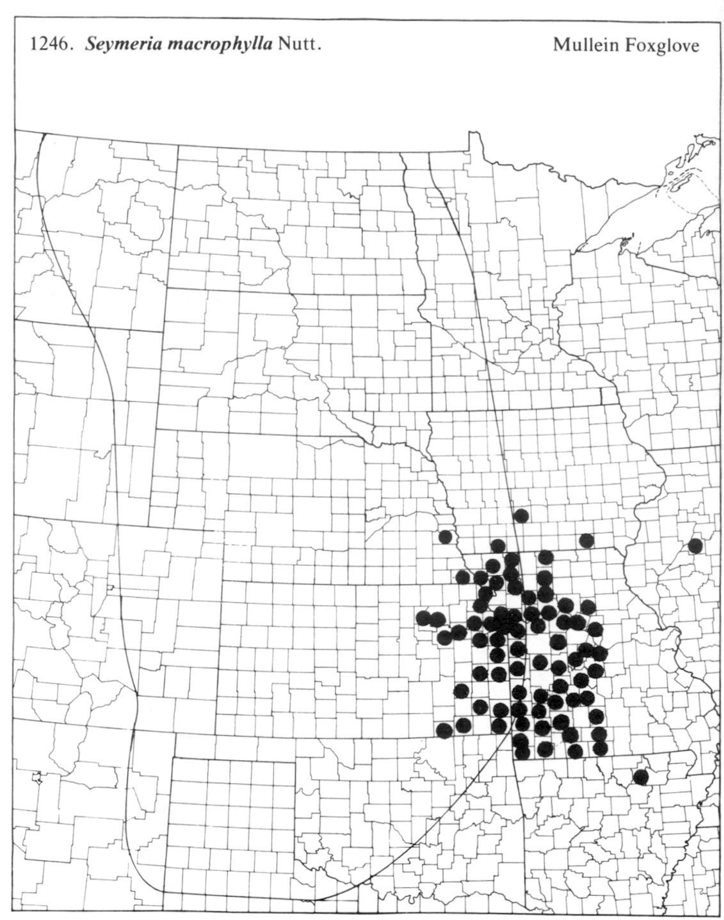

1247. ***Tomanthera auriculata*** (Michx.) Raf. Earleaf Gerardia

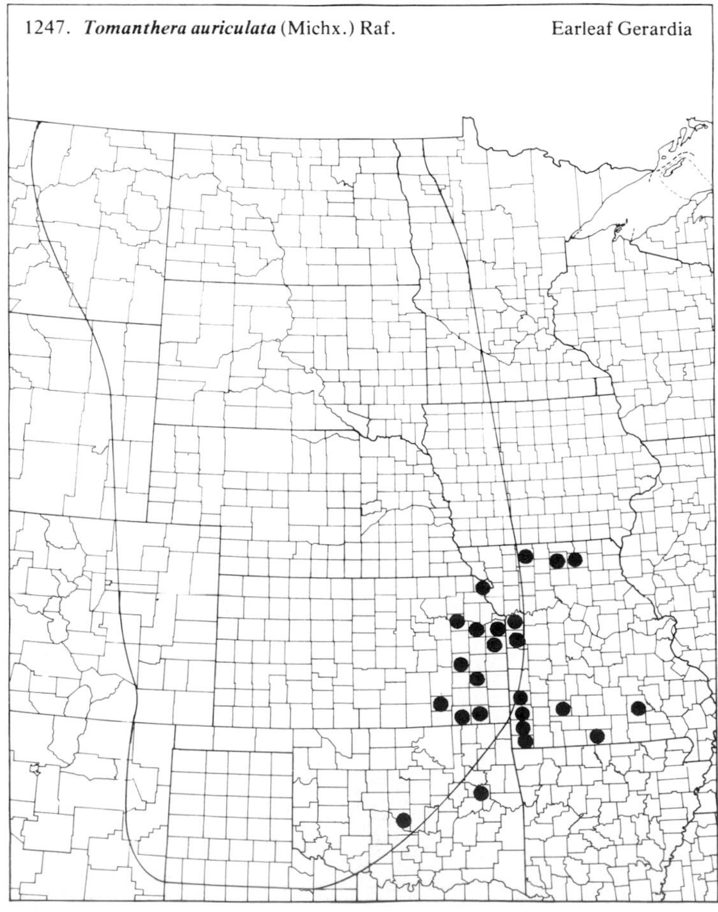

1248. ***Tomanthera densiflora*** Benth. Fineleaf Gerardia

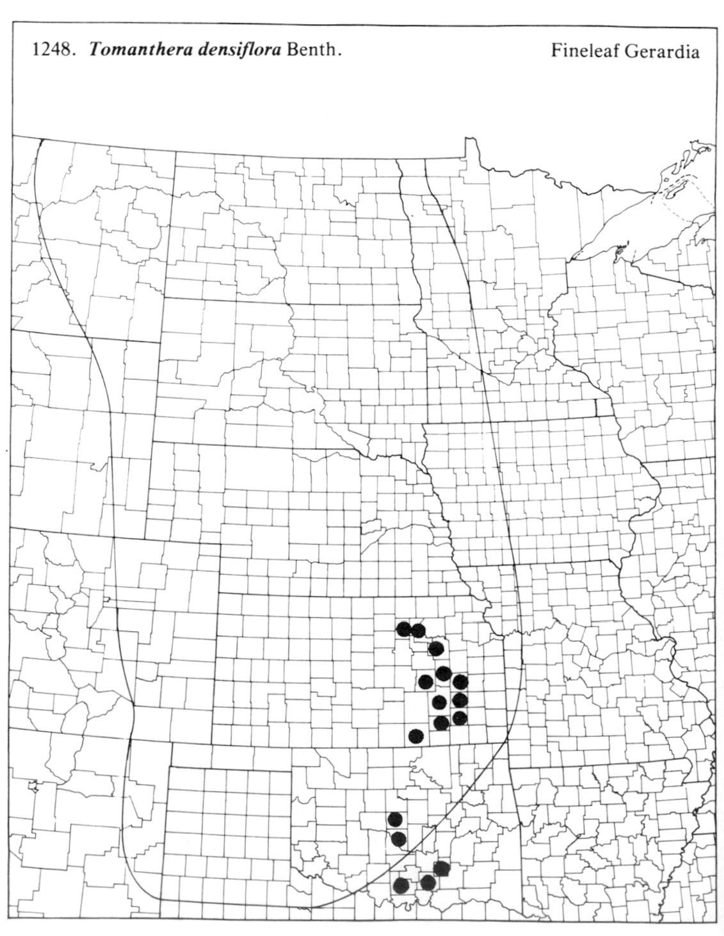

(Scrophulariaceae)

1249. ***Verbascum blattaria*** L. Moth Mullein

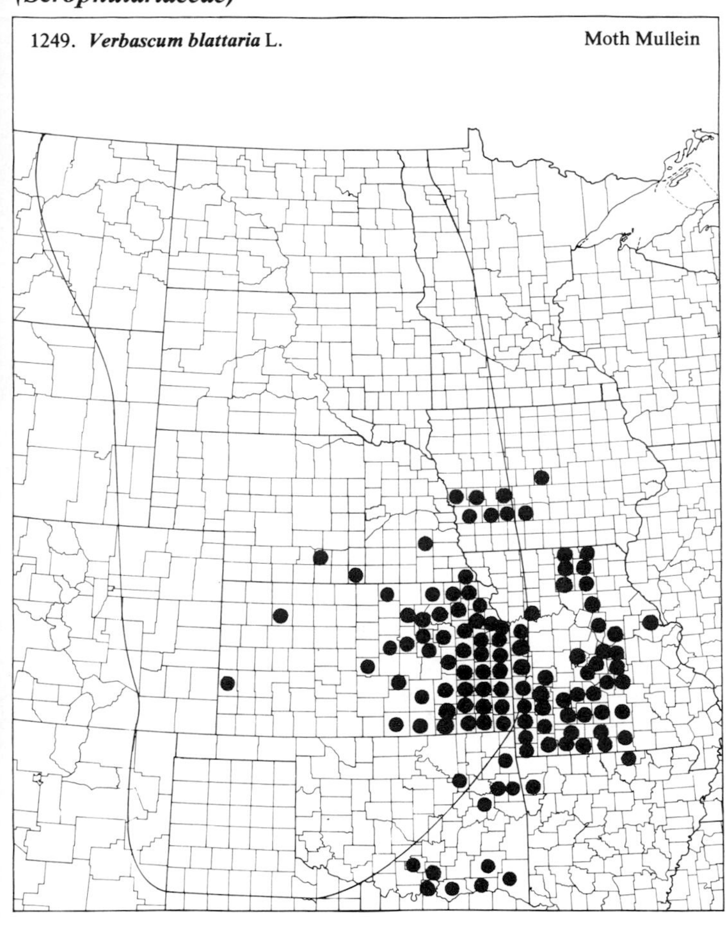

1250. ***Verbascum thapsus*** L. Common Mullein

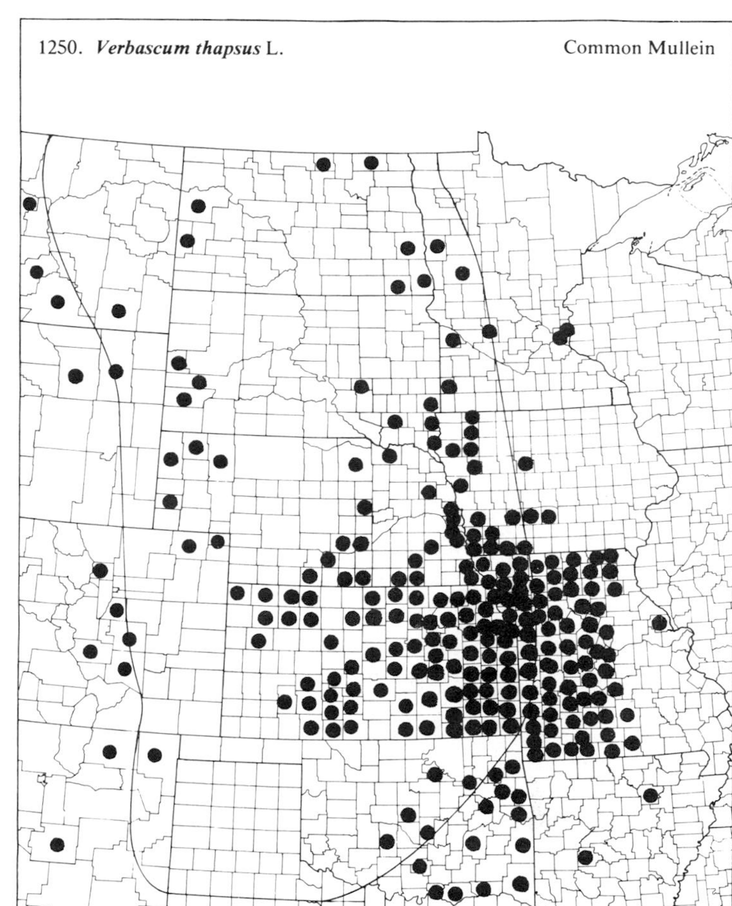

1251. ***Veronica agrestis*** L. Field Speedwell

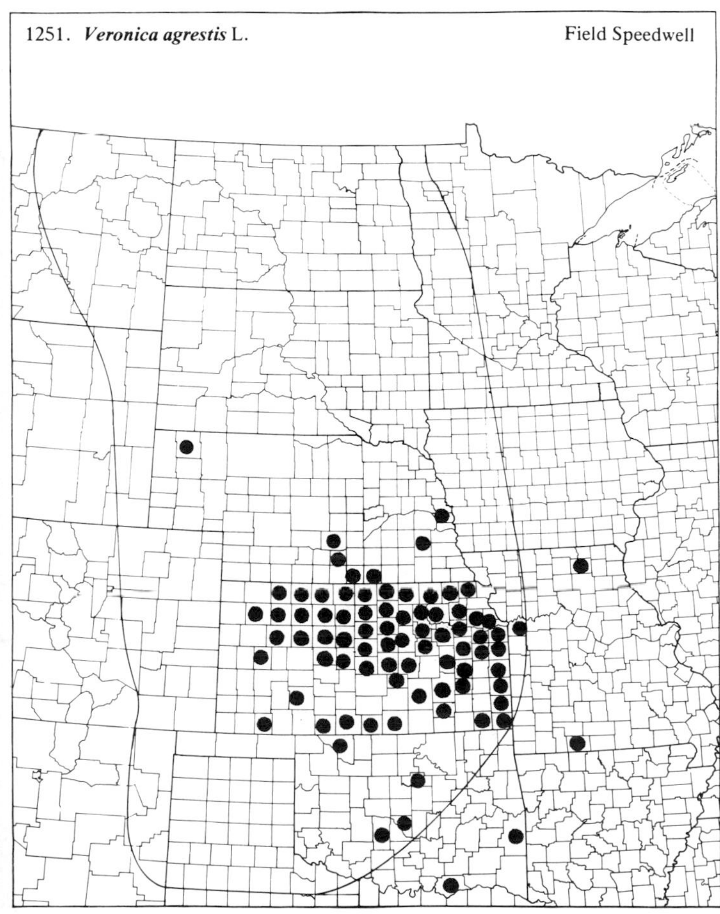

1252. ***Veronica americana*** (Raf.) Schwein. Brooklime

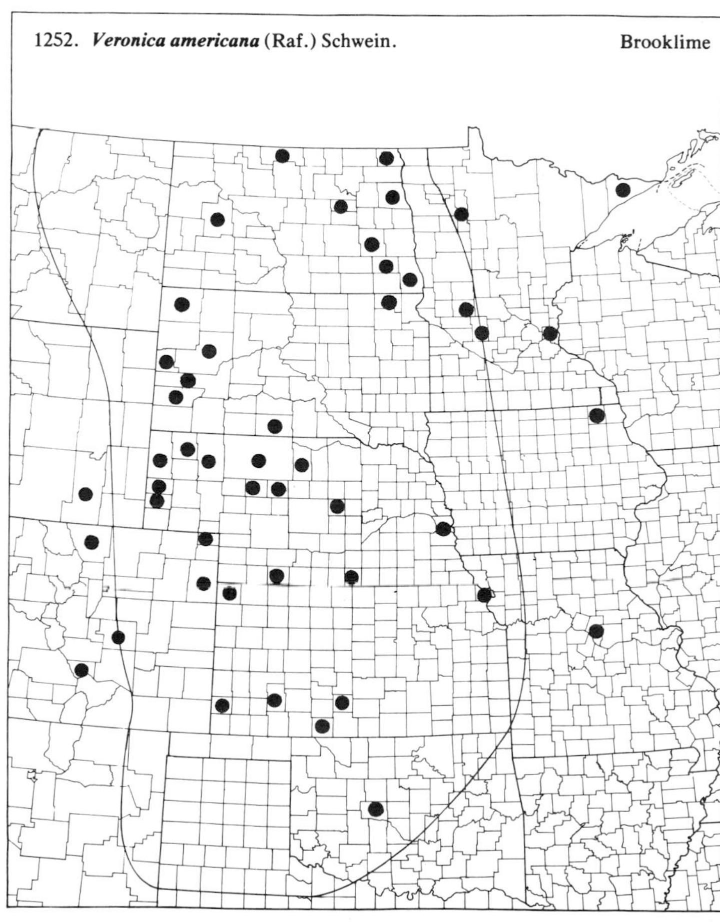

(Scrophulariaceae)

1253. ***Veronica anagallis-aquatica*** L. Water Speedwell

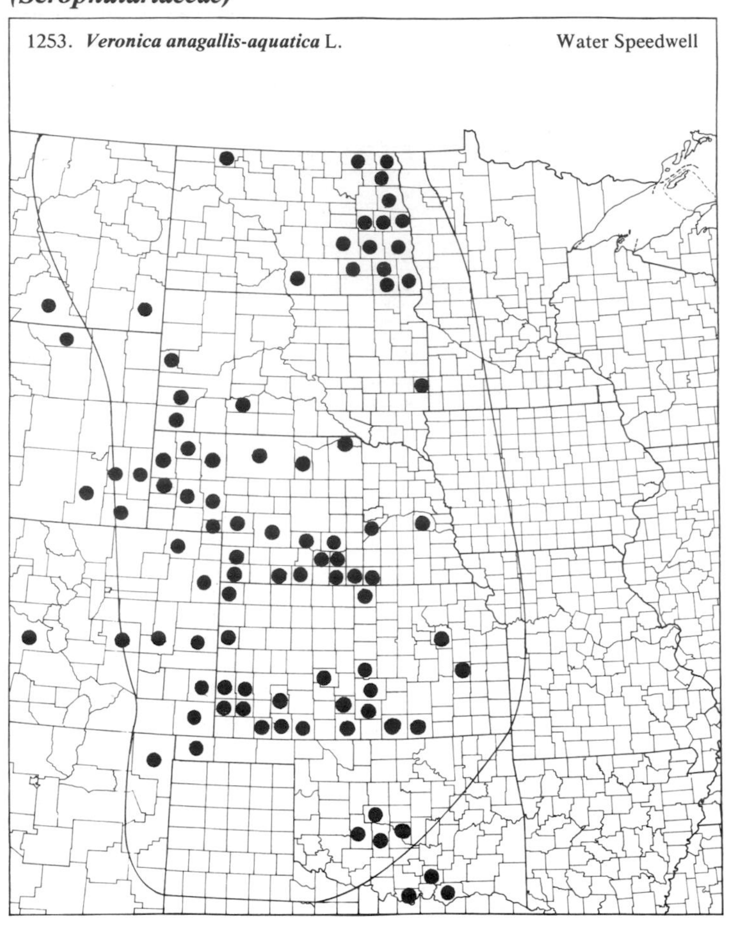

1254. ***Veronica arvensis*** L. Corn Speedwell

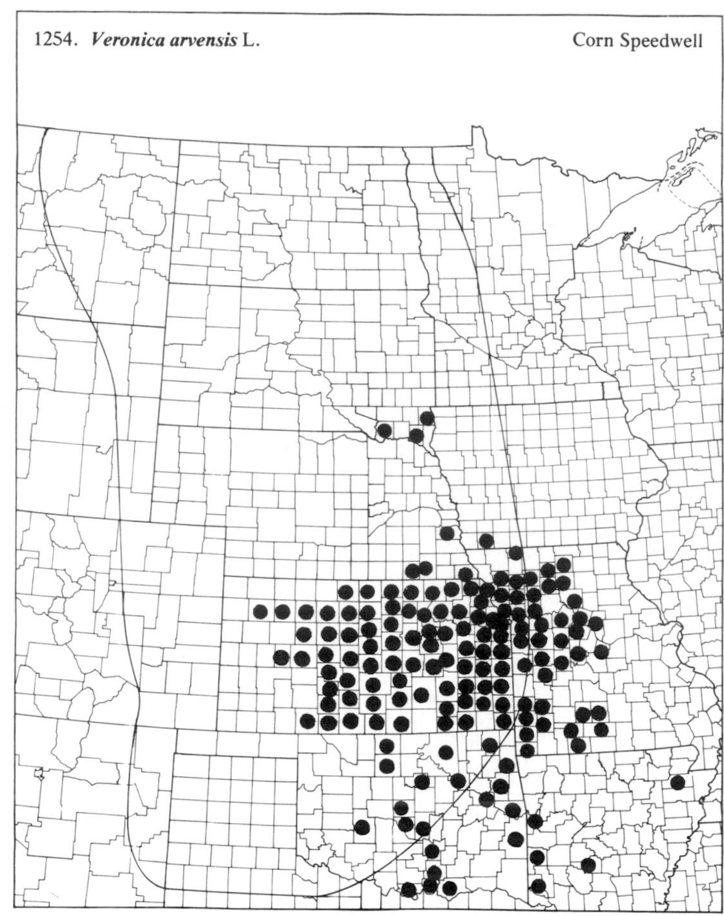

1255. ***Veronica catenata*** Penn.

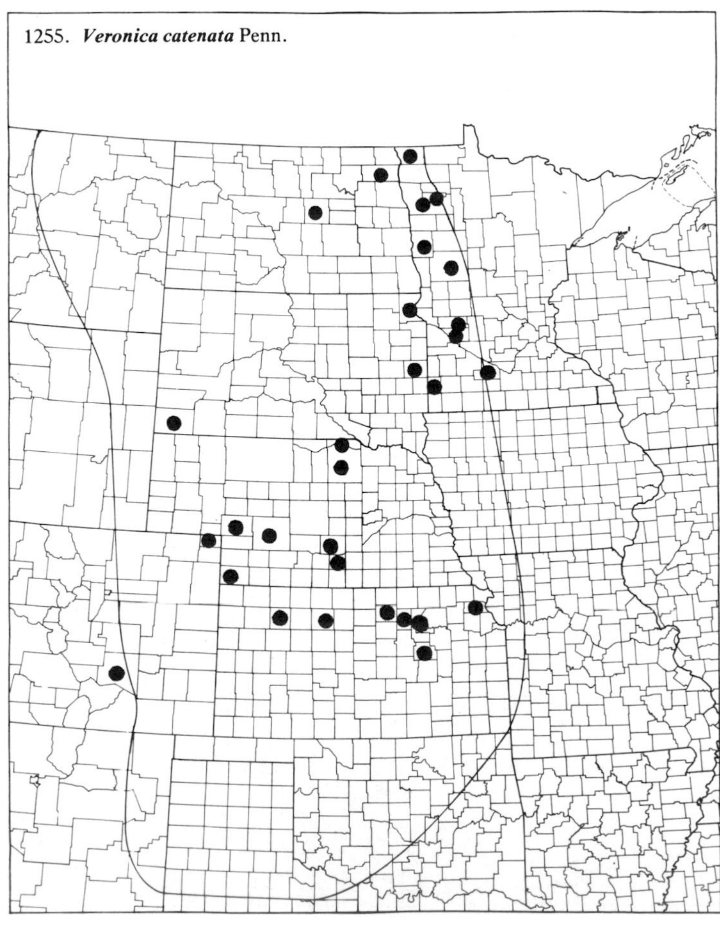

1256. ***Veronica peregrina*** L. var. ***peregrina*** Purslane Speedwell

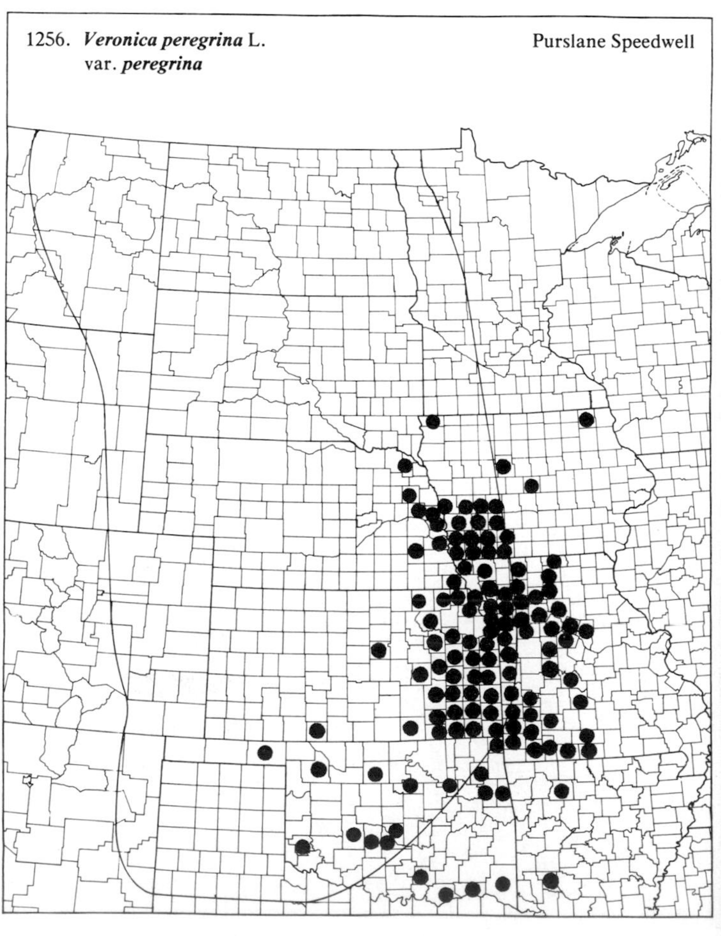

(Scrophulariaceae)

1257. ***Veronica peregrina*** L. var. ***xalapensis*** (H.B.K.) St. John & Warren — Purslane Speedwell

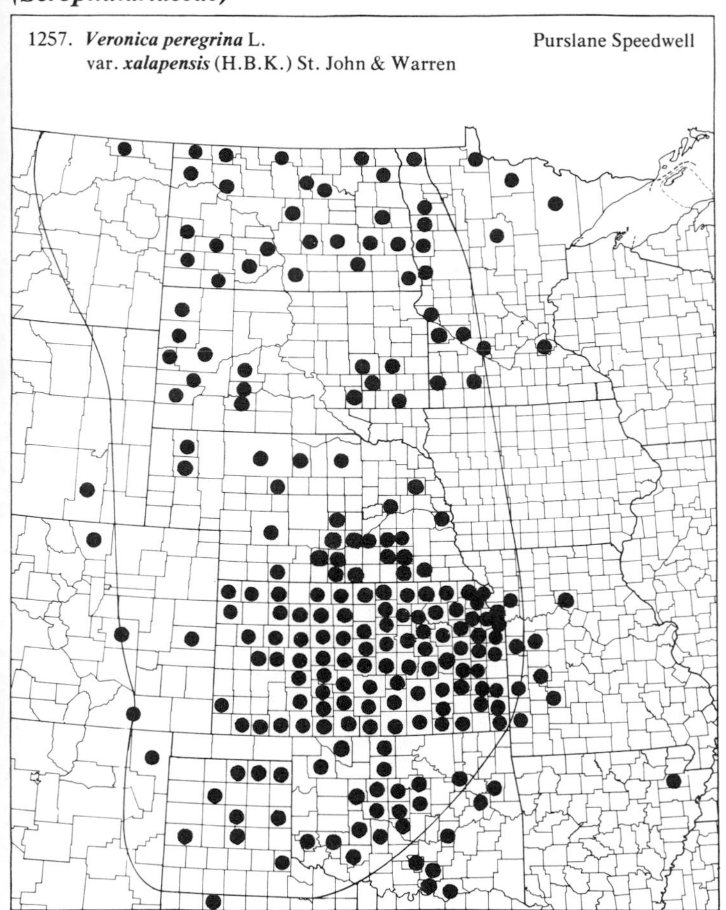

1258. ***Veronica scutellata*** L. — Marsh Speedwell

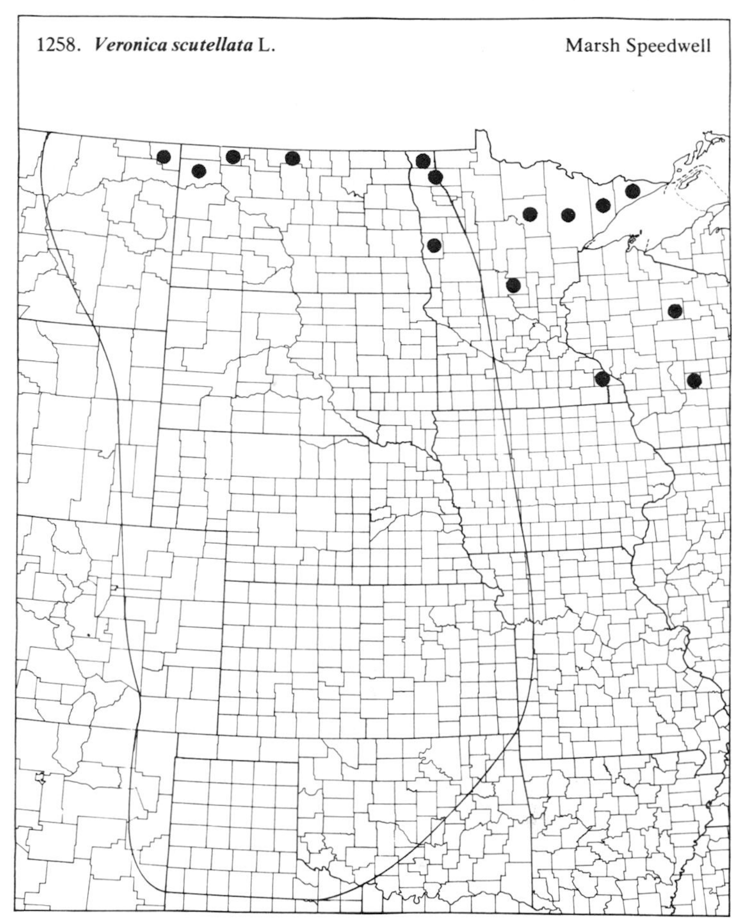

1259. ***Veronicastrum virginicum*** (L.) Farw. — Culver's-root

103. *Orobanchaceae* (Broomrape Family)

1260. ***Orobanche fasciculata*** Nutt.

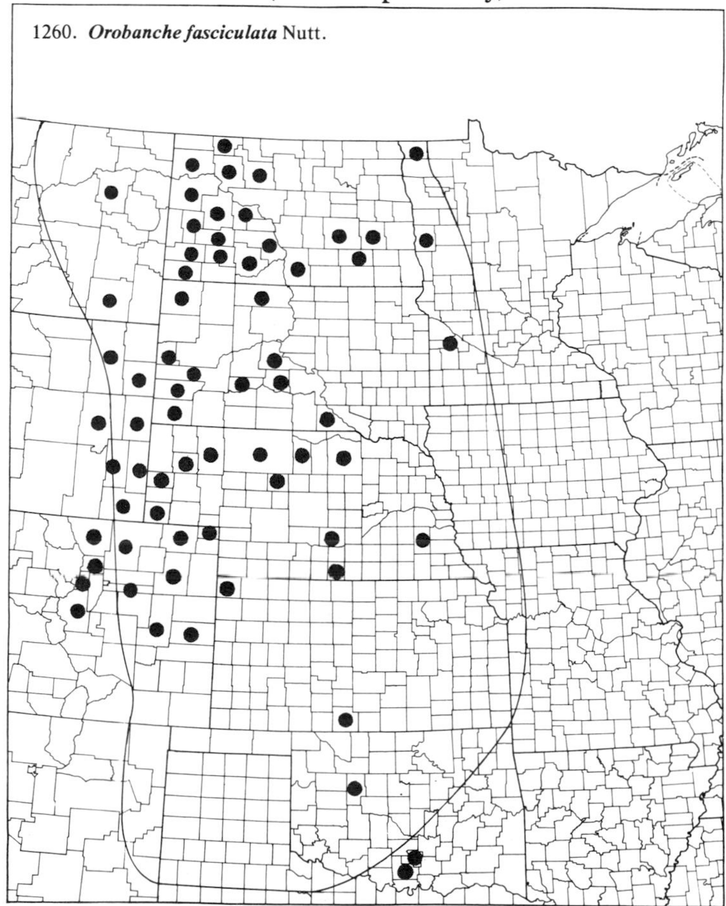

(Orobanchaceae)

1261. ***Orobanche ludoviciana*** Nutt. Broomrape

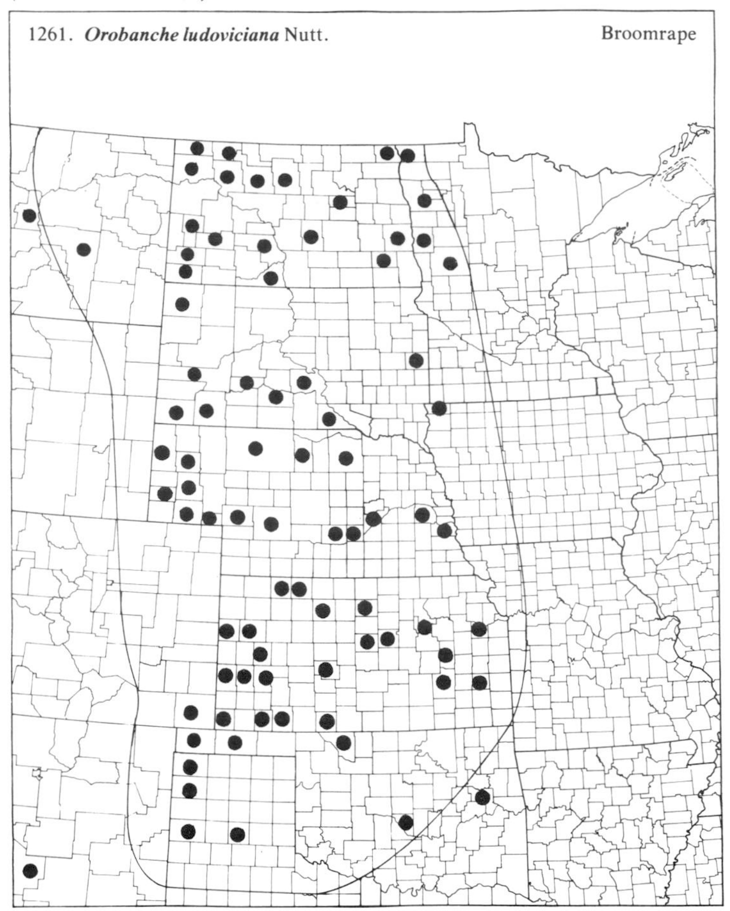

1262. ***Orobanche multiflora*** Nutt.

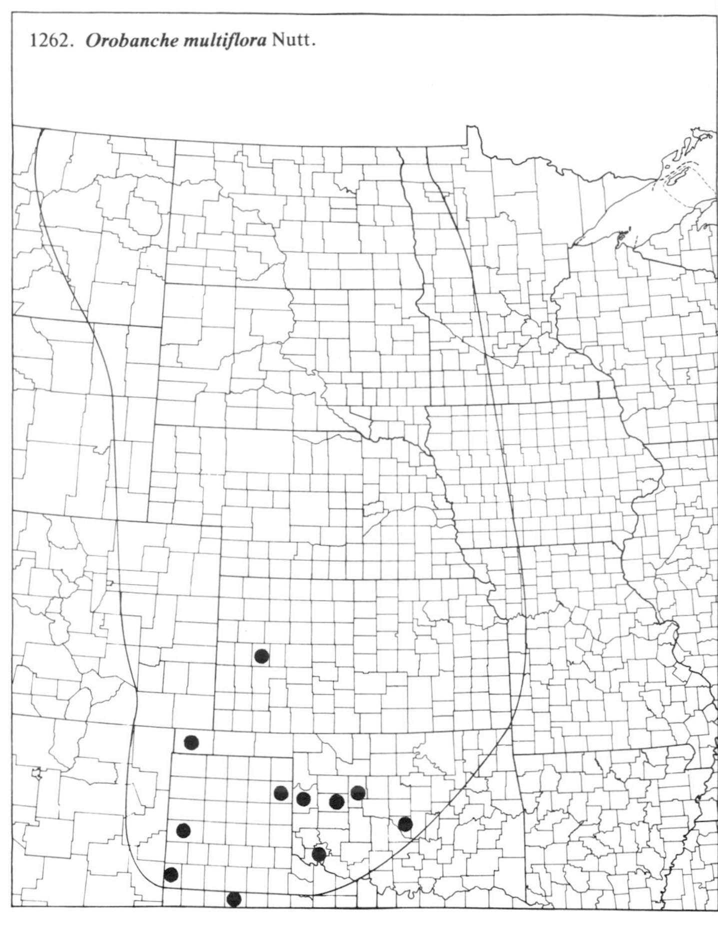

1263. ***Orobanche uniflora*** L. One-flowered Cancerroot

104. ***Bignoniaceae*** (Bignonia Family)

1264. ***Campsis radicans*** (L.) Seem. Trumpet Vine

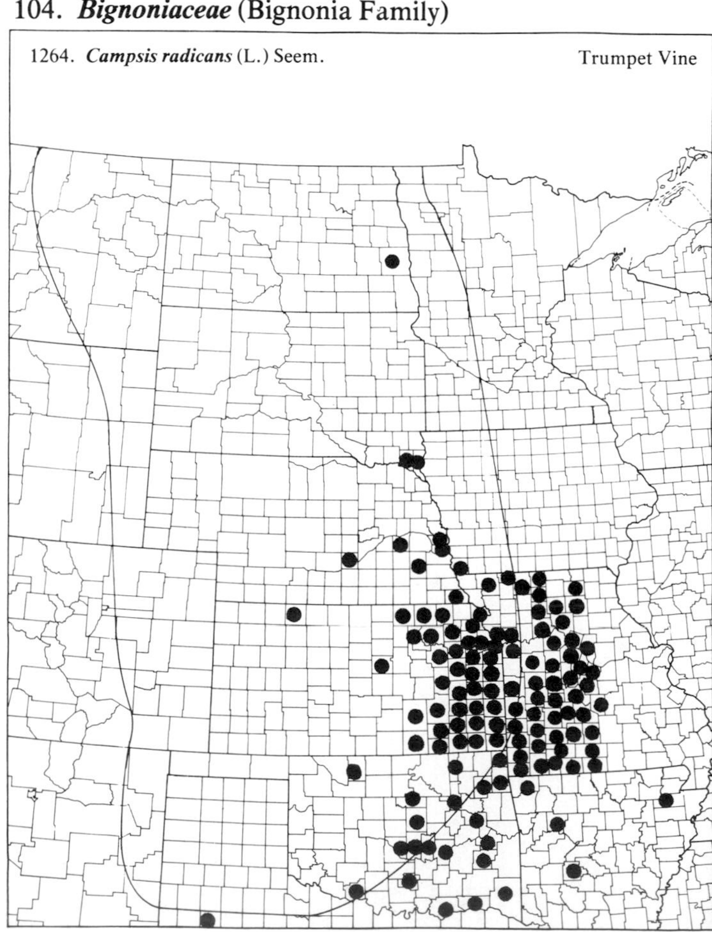

(Bignoniaceae)

1265. ***Catalpa speciosa*** Warder — Catalpa

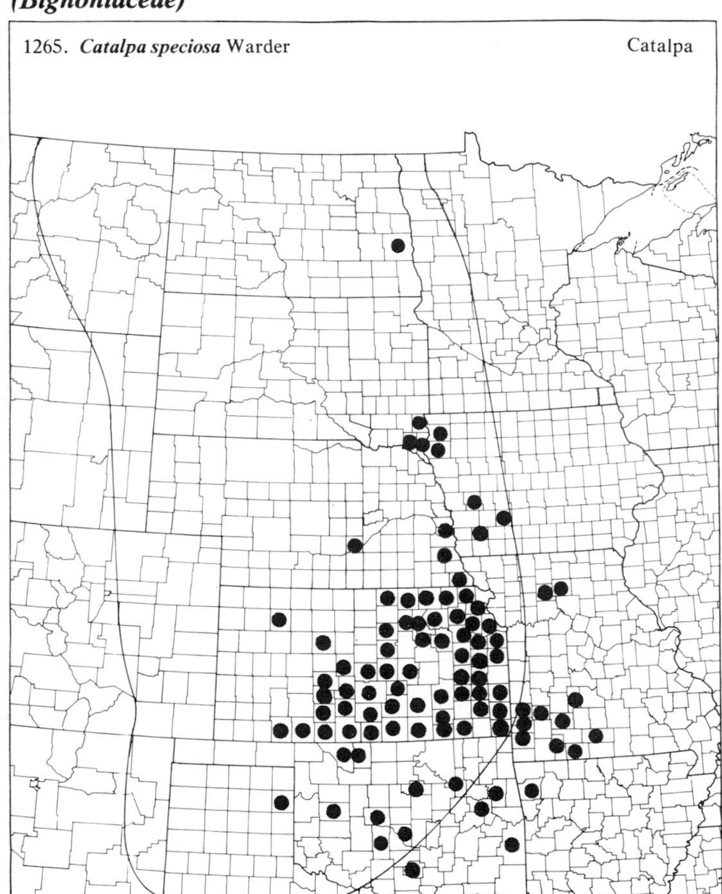

105. *Acanthaceae* (Acanthus Family)

1266. ***Dicliptera brachiata*** (Pursh) Spreng. — Dicliptera

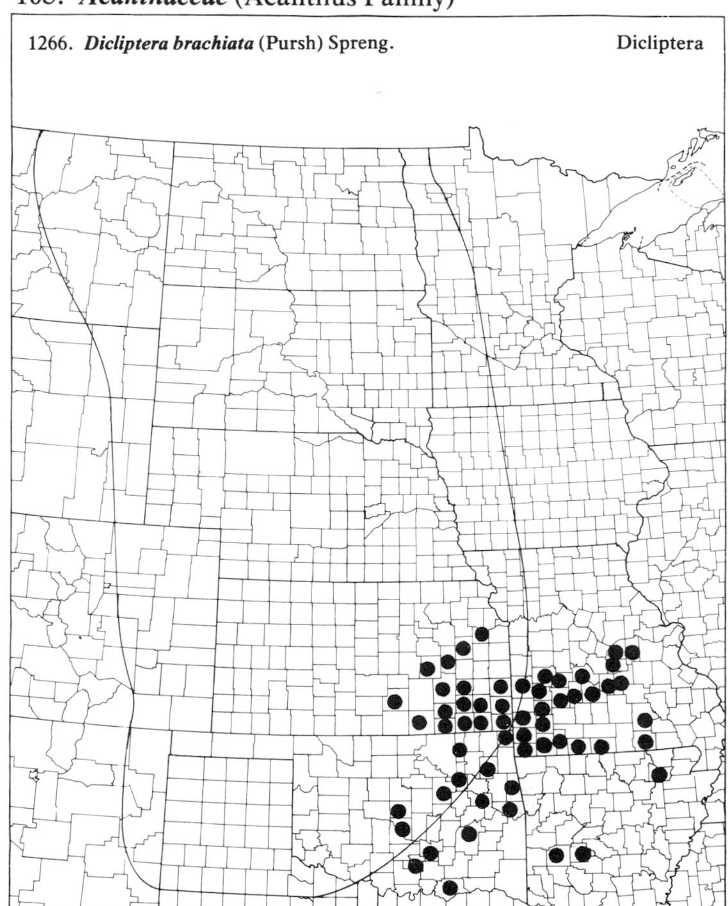

1267. ***Justicia americana*** (L.) Vahl — American Dianthera

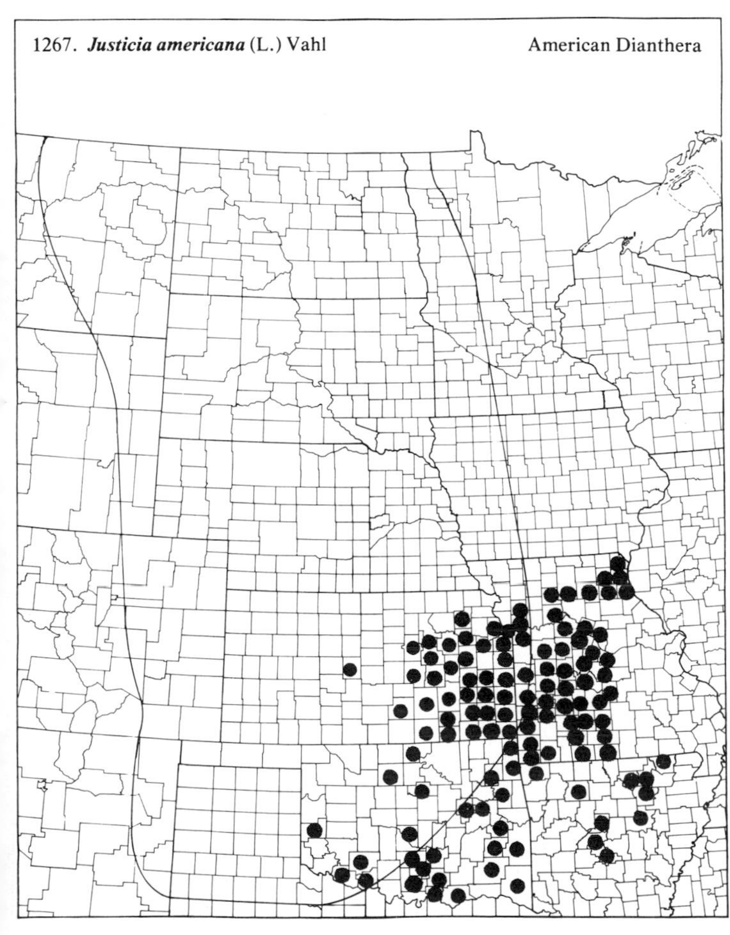

1268. ***Ruellia humilis*** Nutt. — Fringeleaf Ruellia

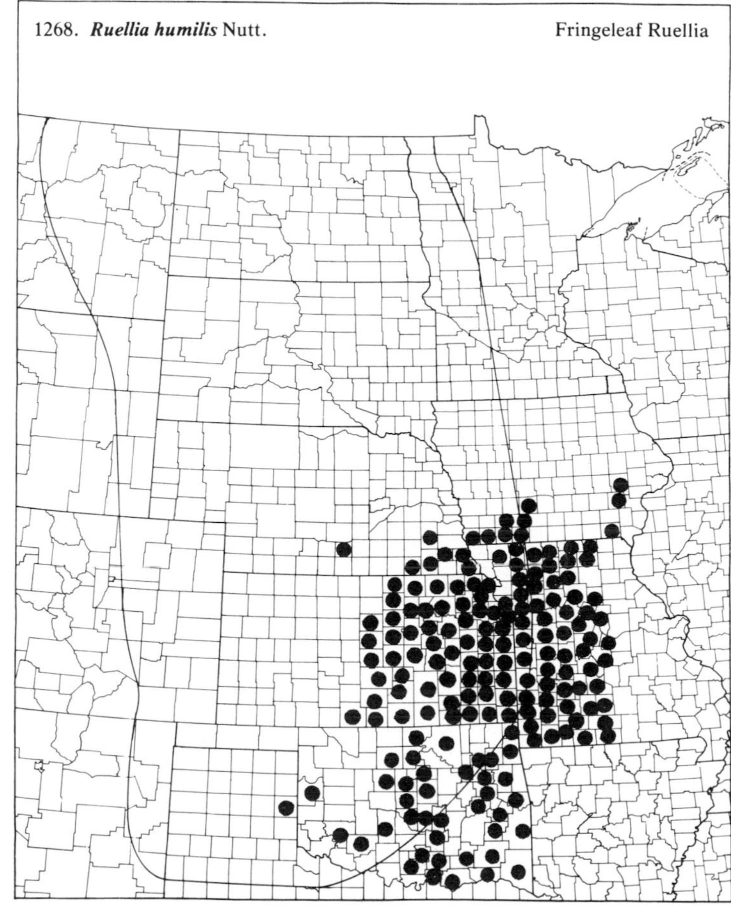

(Acanthaceae)

1269. ***Ruellia strepens*** L. Limestone Ruellia

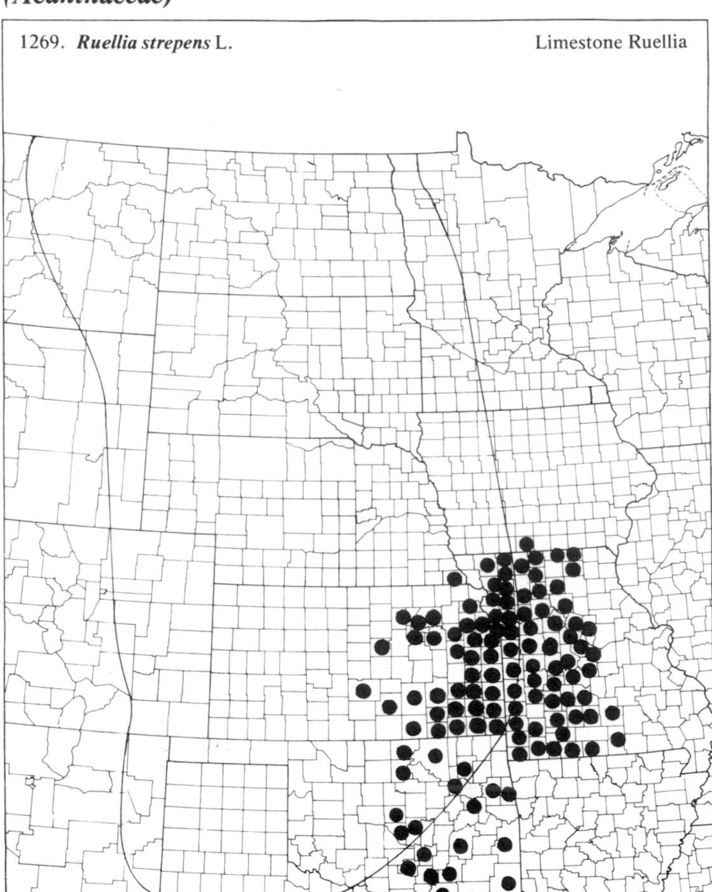

106. *Pedaliaceae*

1270. ***Proboscidea louisianica*** (Mill.) Thell. Devil's Claw

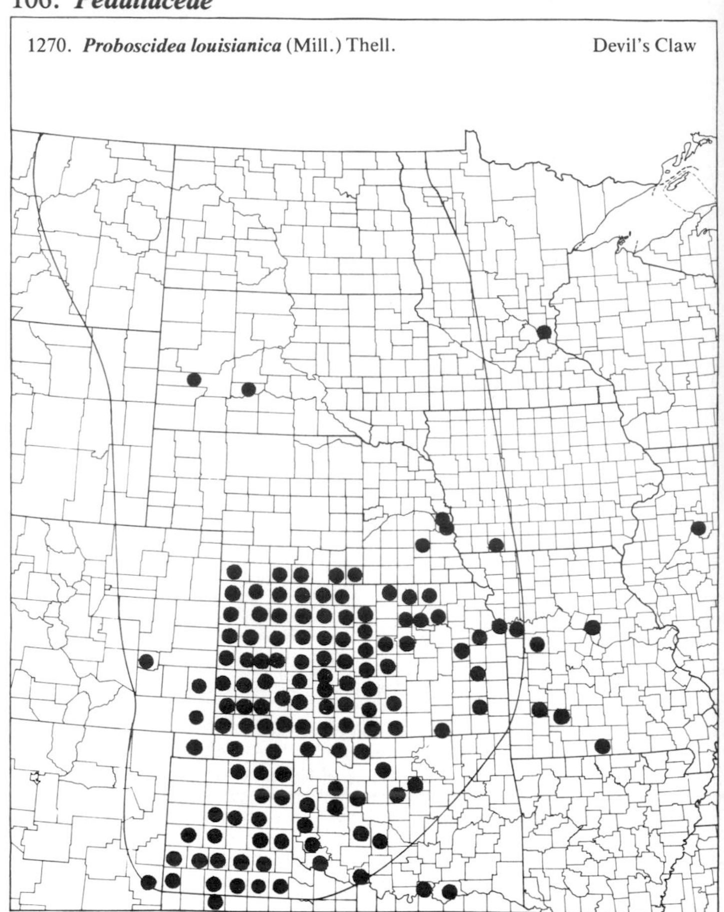

107. *Lentibulariaceae* (Bladderwort Family)

1271. ***Utricularia gibba*** L. Bladderwort

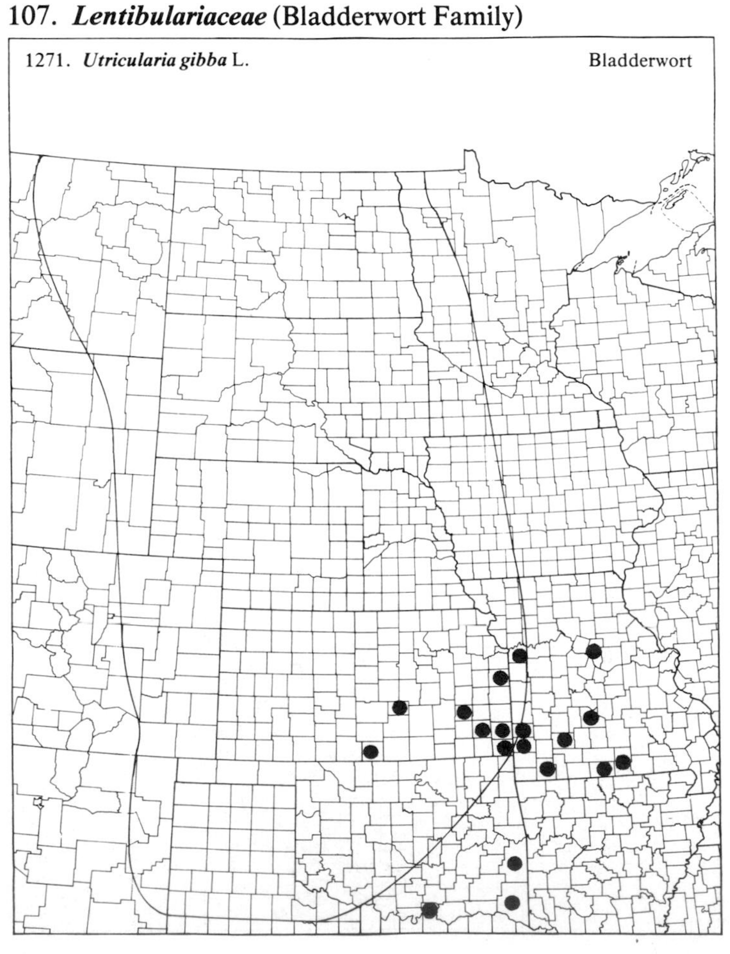

1272. ***Utricularia minor*** L. Bladderwort

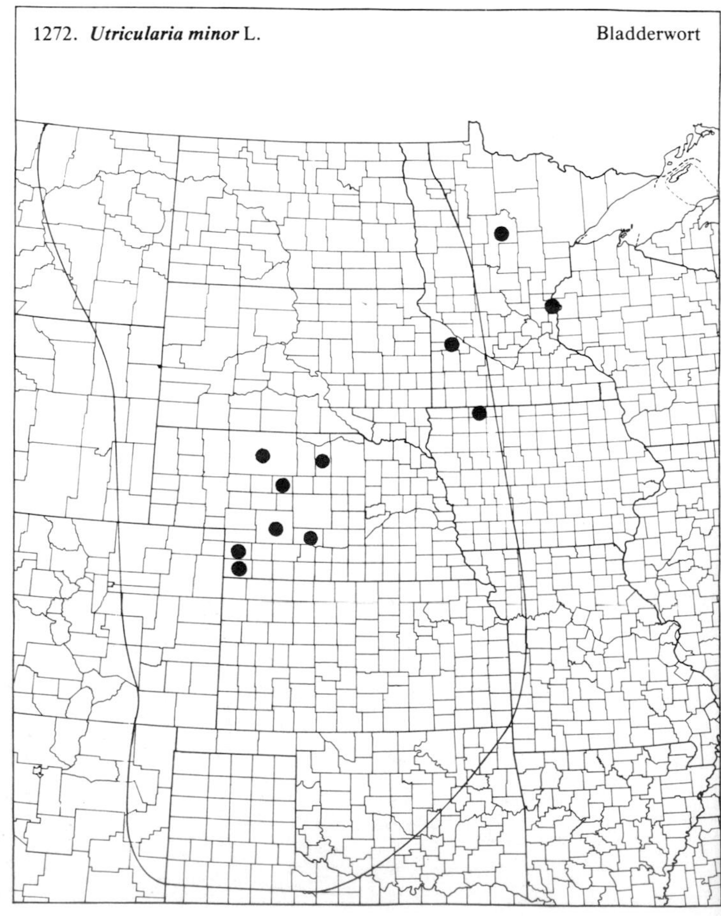

(Lentibulariaceae)

108. *Campanulaceae* (Bellflower Family)

1273. ***Utricularia vulgaris*** L. Bladderwort

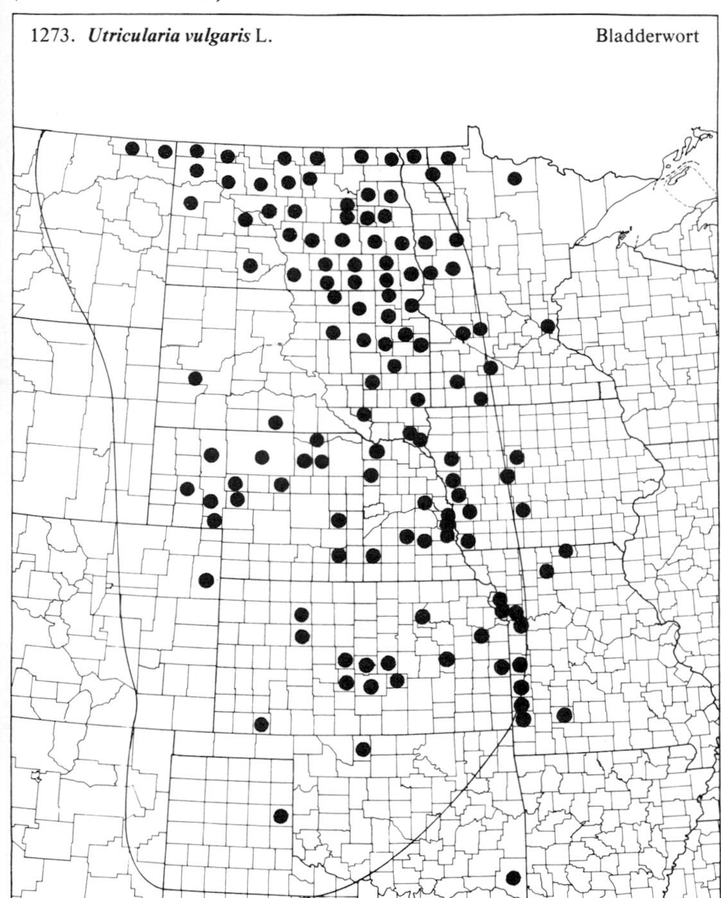

1274. ***Campanula americana*** L. Tall Bellflower

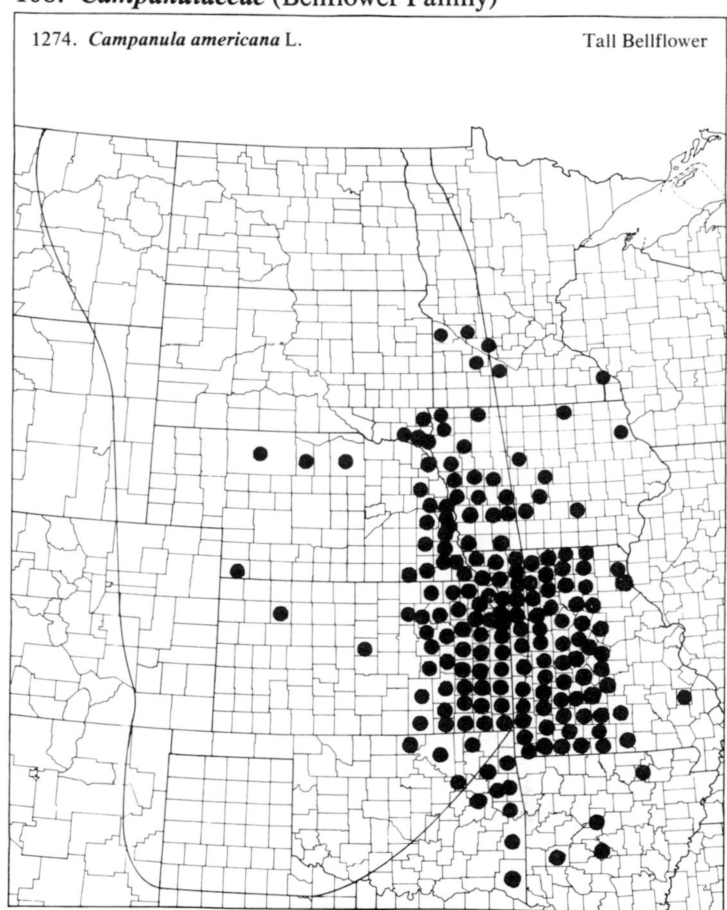

1275. ***Campanula aparinoides*** Pursh Marsh Bellflower

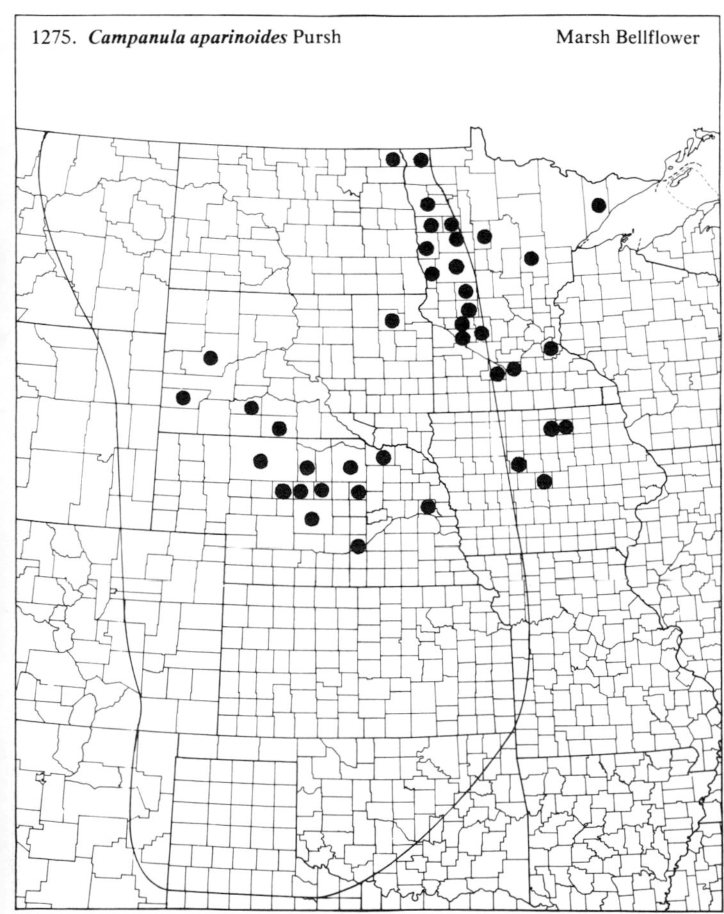

1276. ***Campanula rapunculoides*** L. Creeping Bellflower

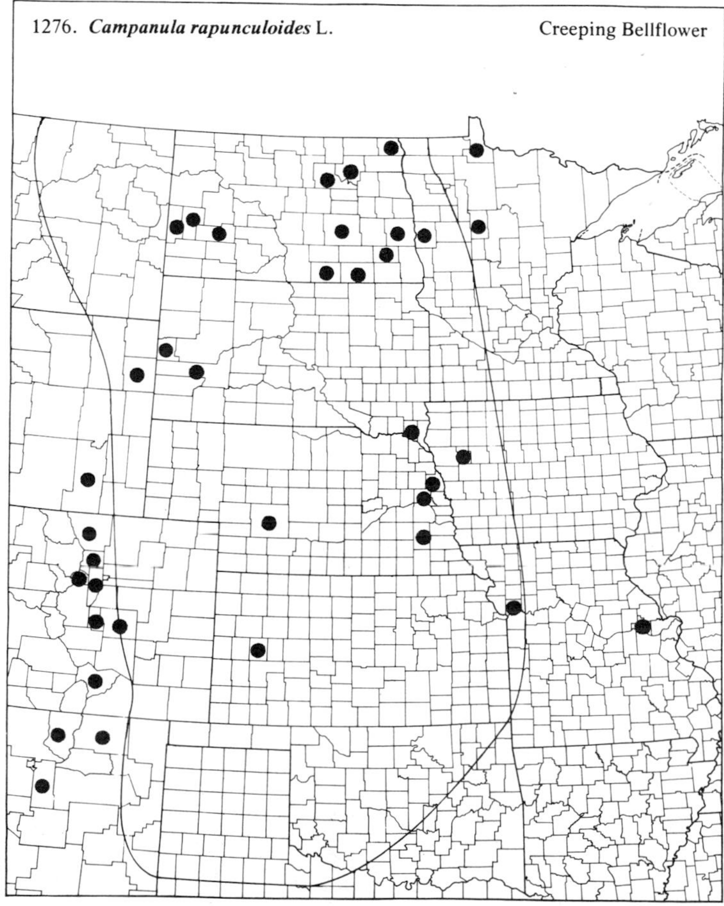

(Campanulaceae)

1277. ***Campanula rotundifolia*** L. Harebell

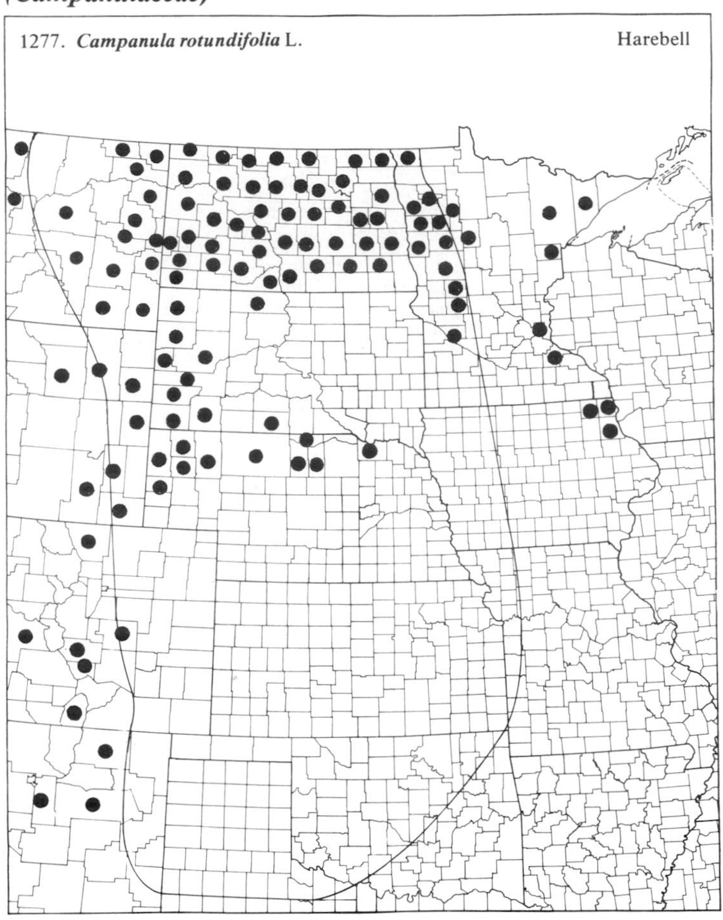

1278. ***Lobelia cardinalis*** L. Cardinal-flower

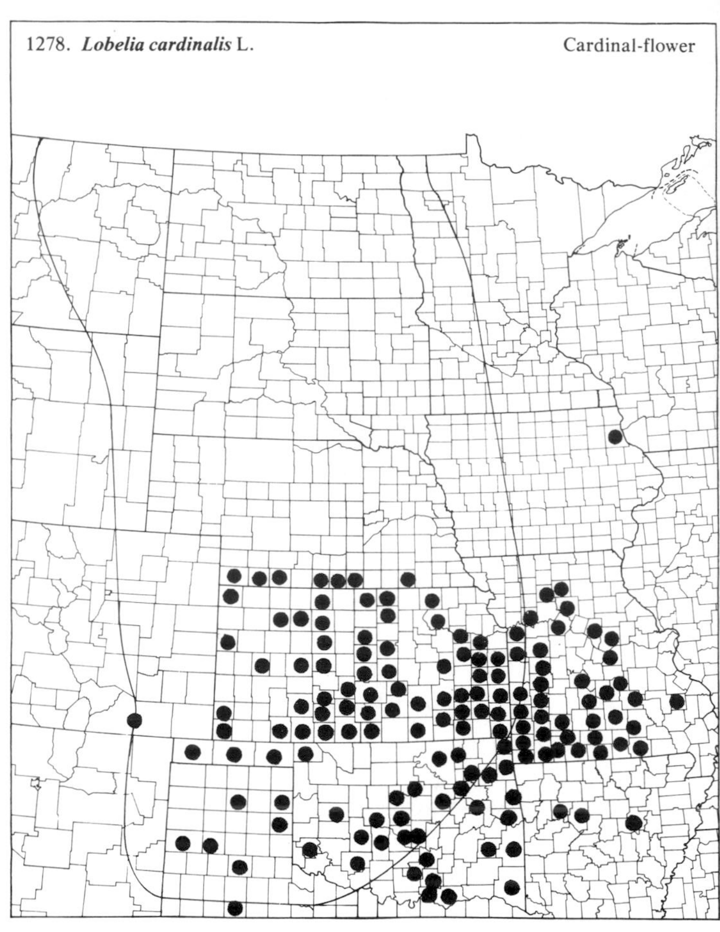

1279. ***Lobelia inflata*** L. Indian Tobacco

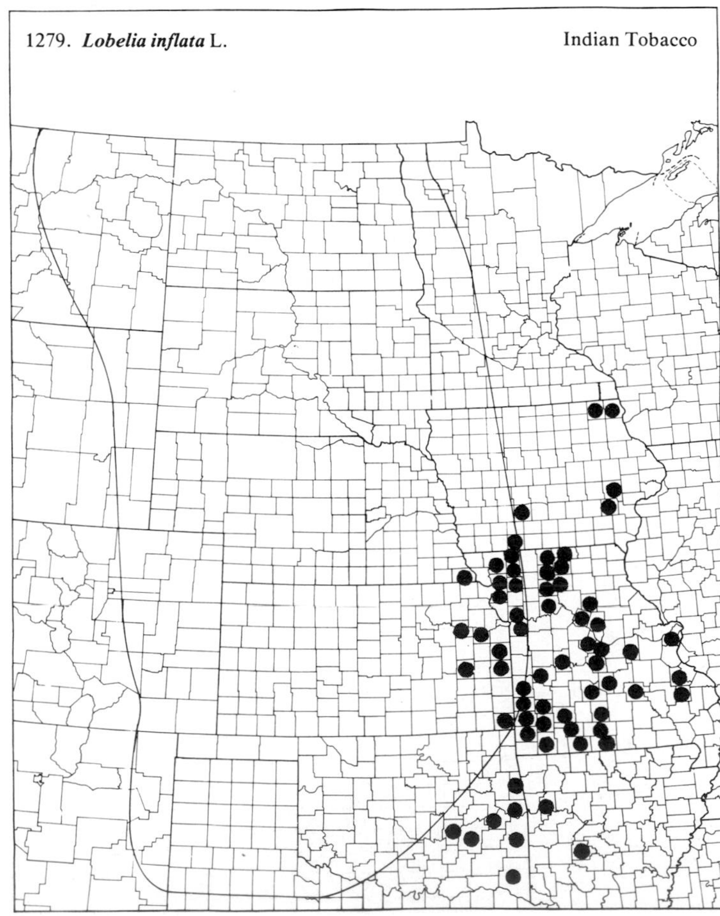

1280. ***Lobelia kalmii*** L. Kalm's Lobelia

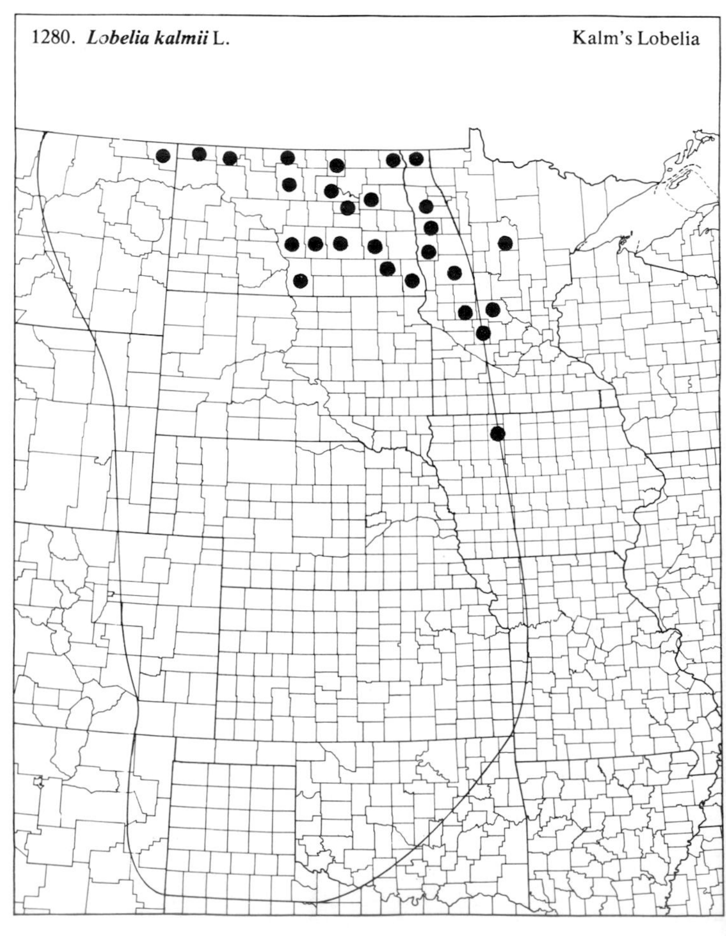

(Campanulaceae)

1281. ***Lobelia siphilitica*** L. Blue Cardinal-flower

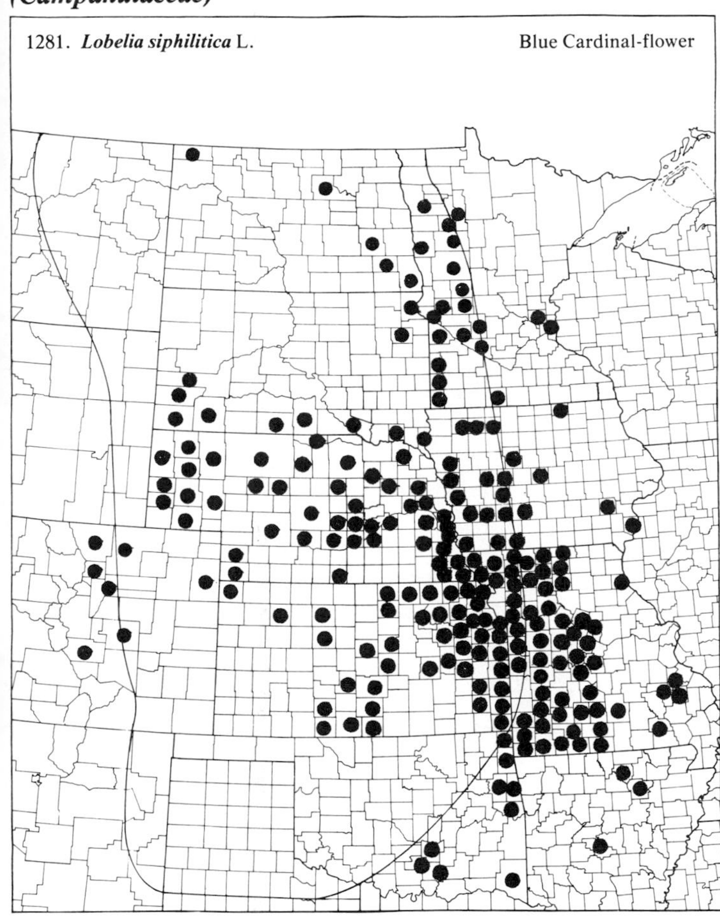

1282. ***Lobelia spicata*** Lam. Pale-spike Lobelia

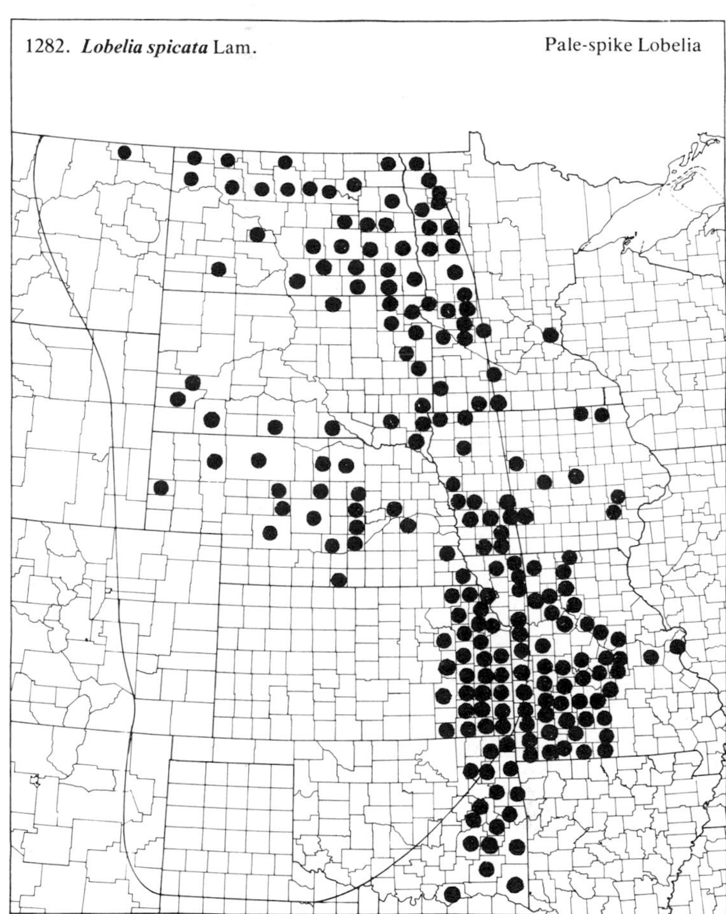

1283. ***Triodanis biflora*** (R. & P.) Greene

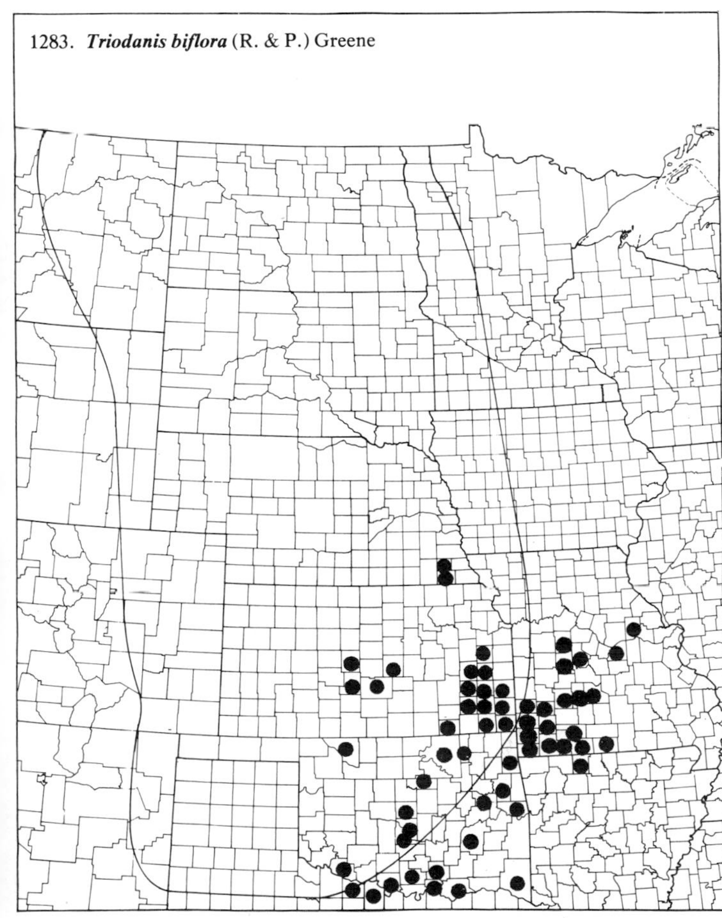

1284. ***Triodanis holzingeri*** McVaugh

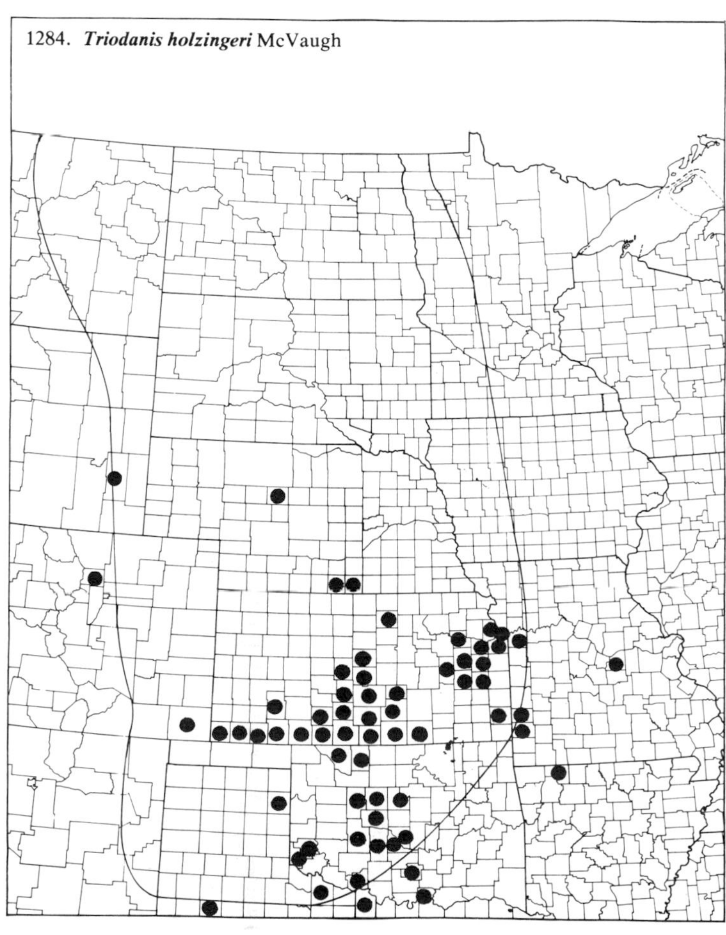

(Campanulaceae)

1285. ***Triodanis leptocarpa*** (Nutt.) Nieuw.

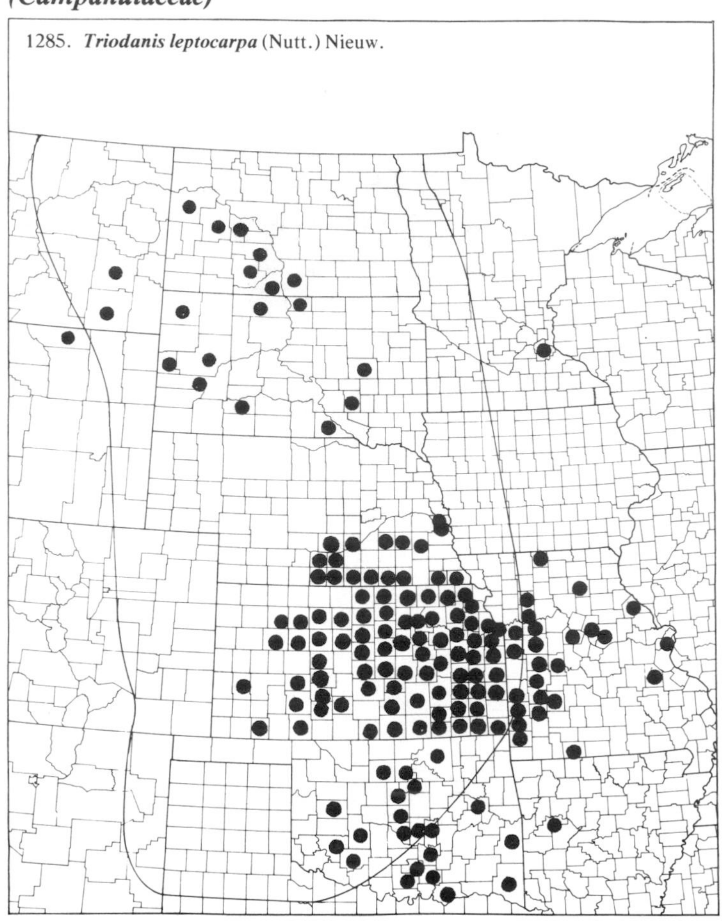

1286. ***Triodanis perfoliata*** (L.) Nieuw.

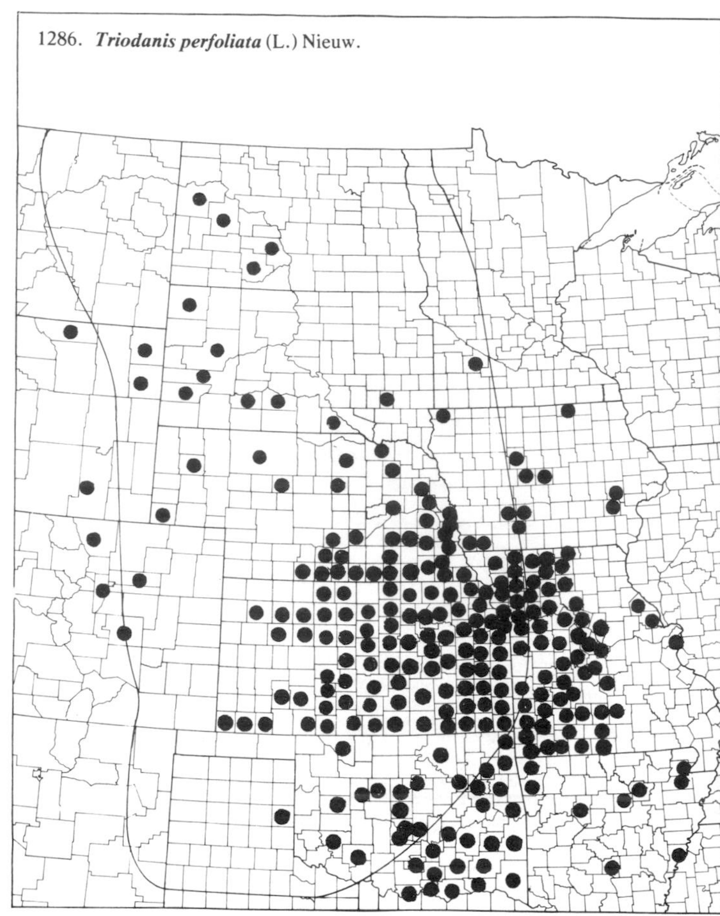

109. ***Rubiaceae*** (Madder Family)

1287. ***Cephalanthus occidentalis*** L.

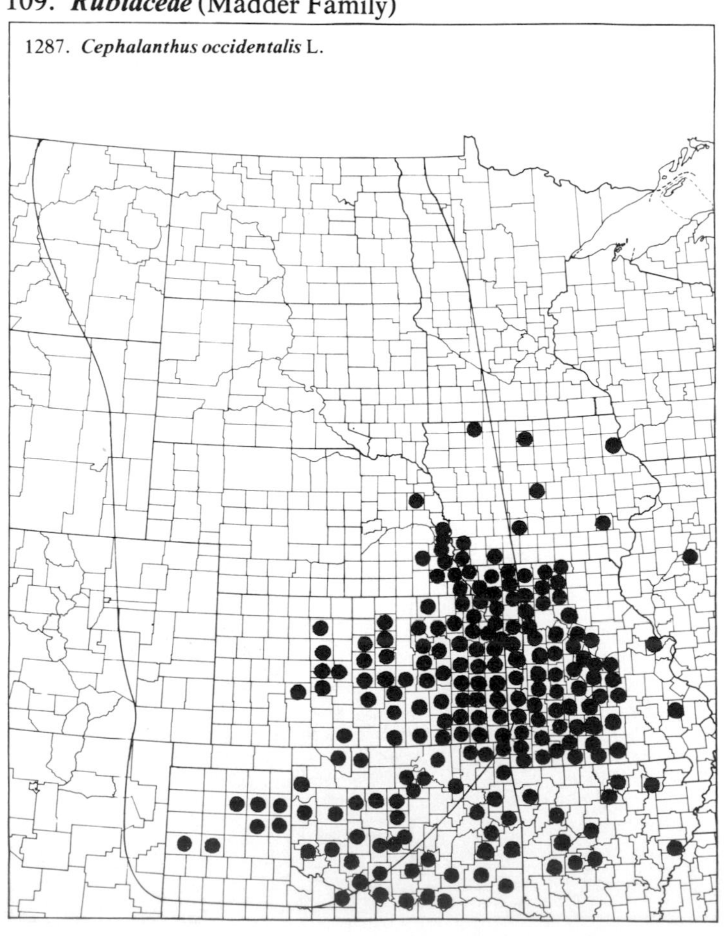

1288. ***Diodia teres*** Walt. Buttonweed

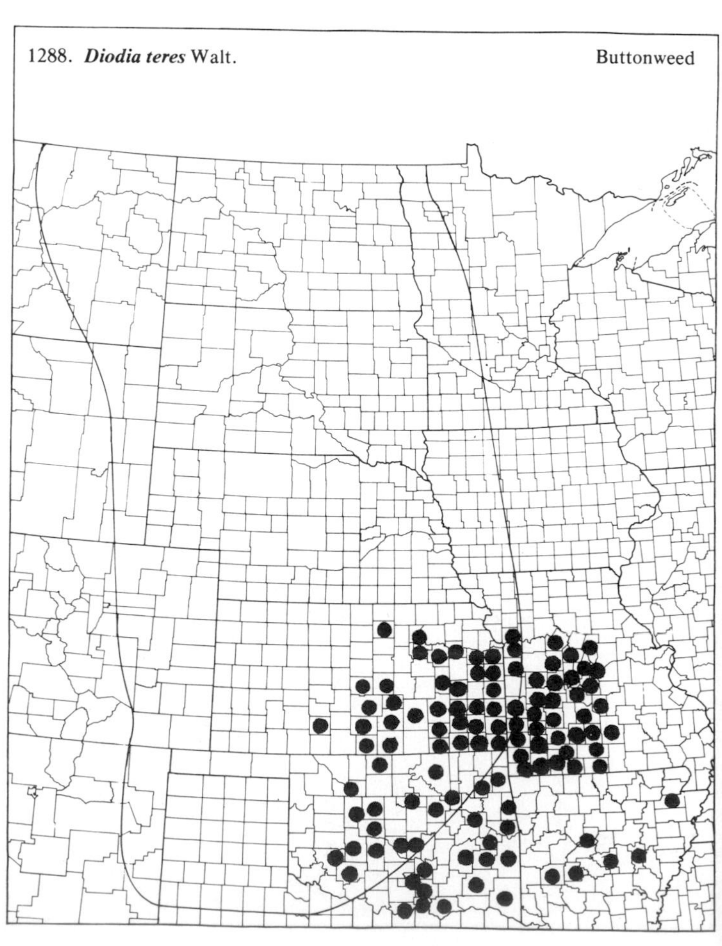

(Rubiaceae)

1289. **Galium aparine** L. Catchweed Bedstraw

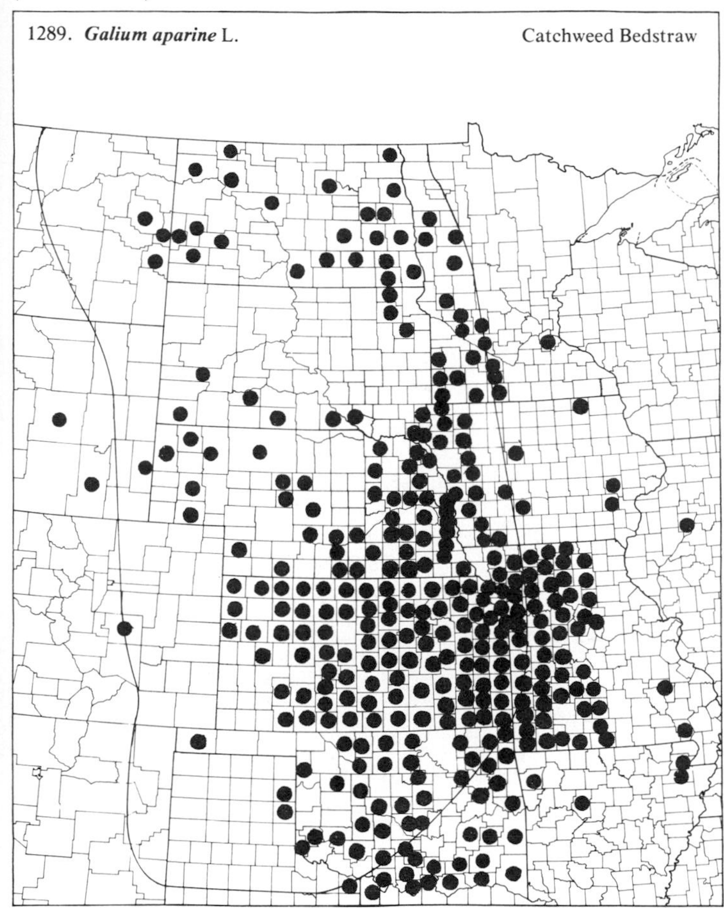

1290. **Galium boreale** L. Northern Bedstraw

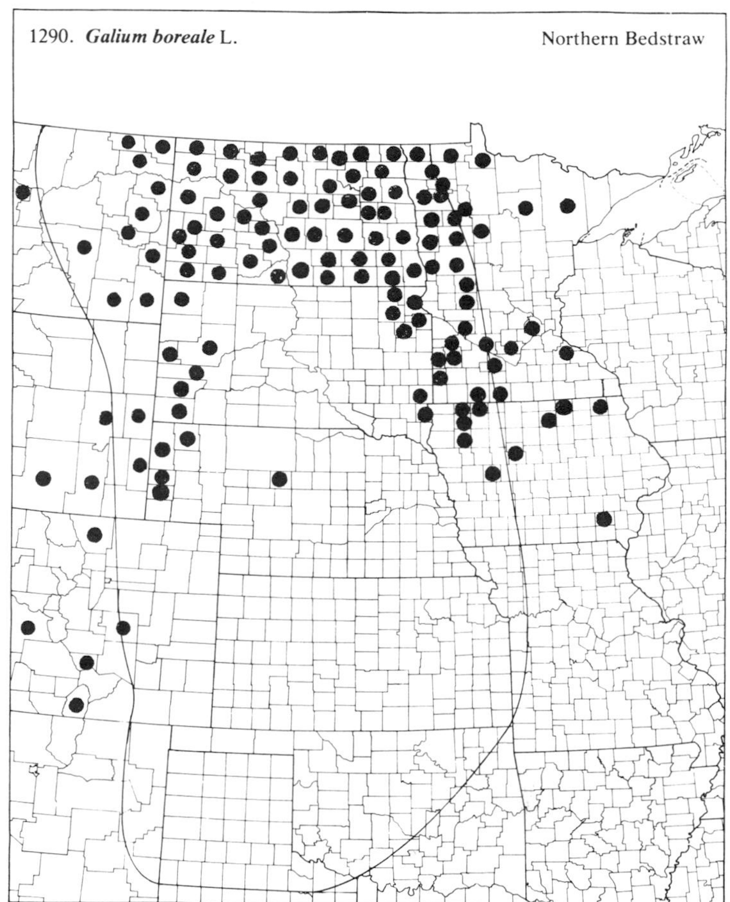

1291. **Galium circaezans** Michx. Woods Bedstraw

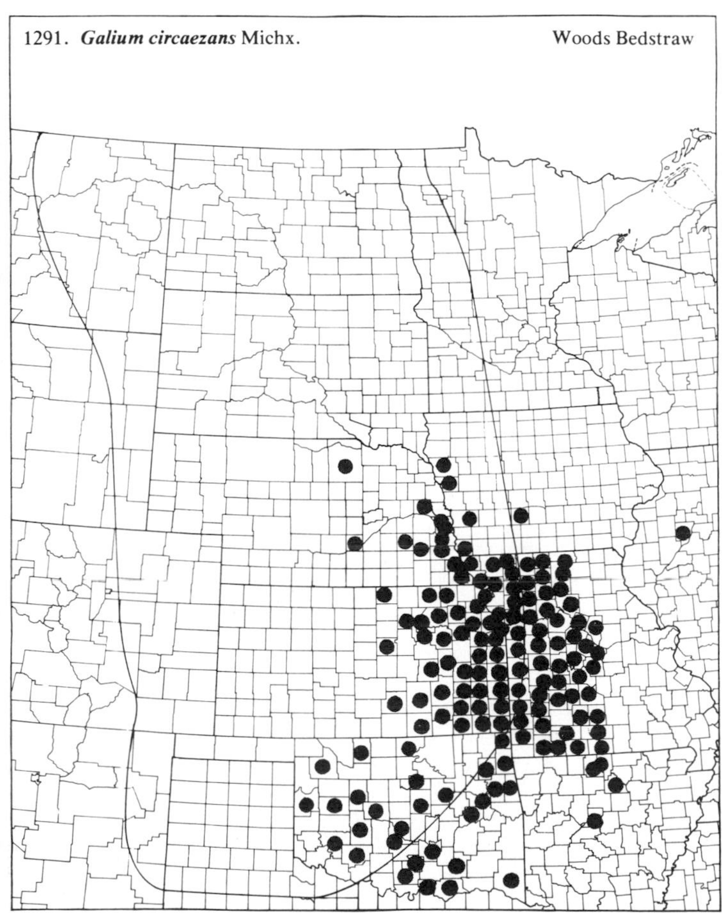

1292. **Galium concinnum** T. & G. Shining Bedstraw

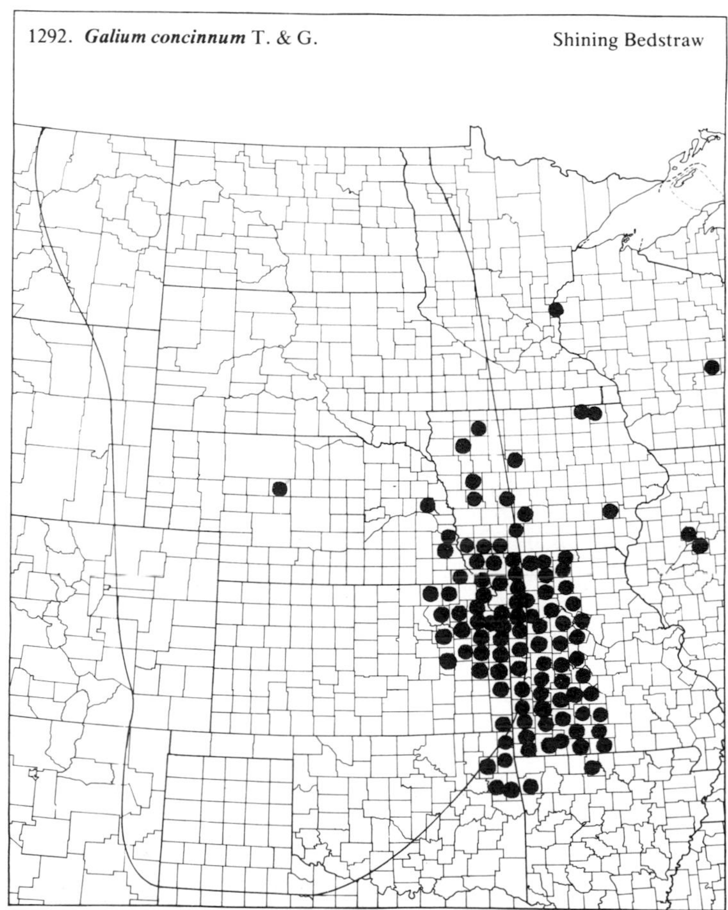

(Rubiaceae)

1293. ***Galium obtusum*** Bigel. Bluntleaf Bedstraw

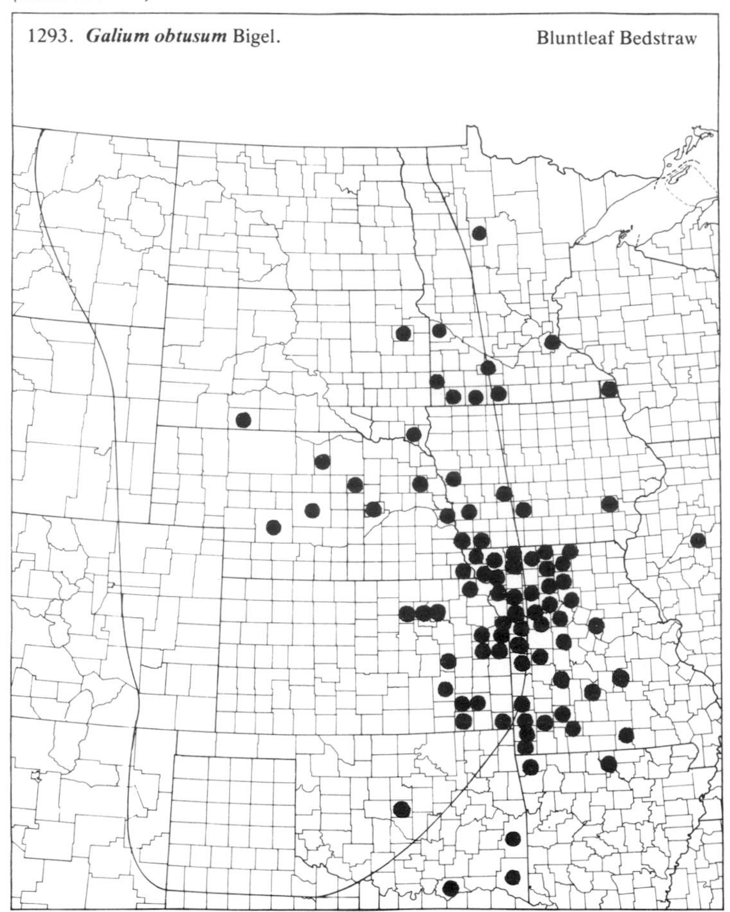

1294. ***Galium pilosum*** Ait. Hairy Bedstraw

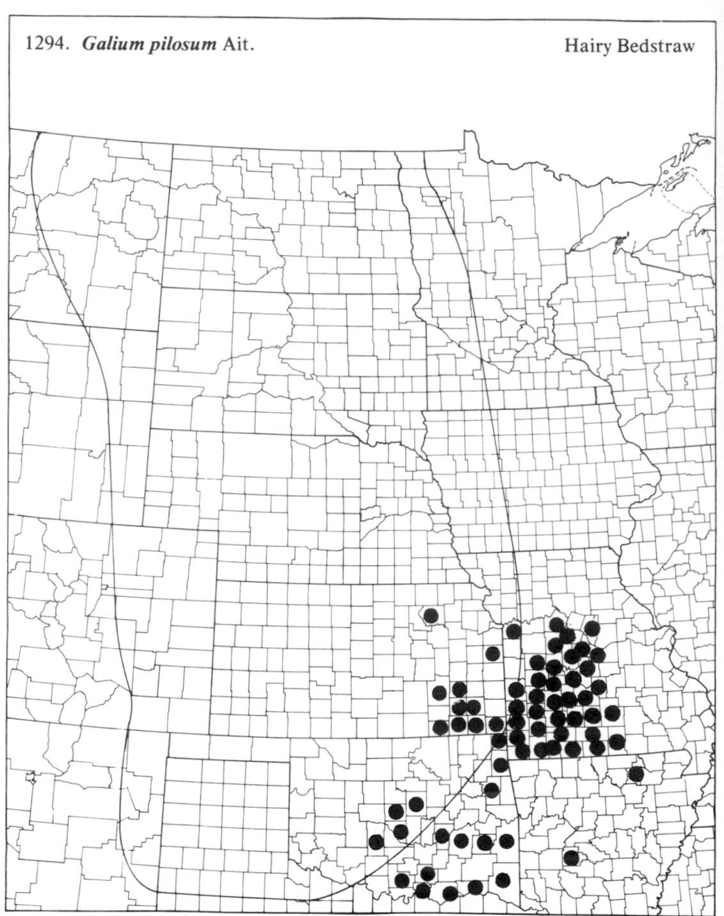

1295. ***Galium trifidum*** L. Small Bedstraw

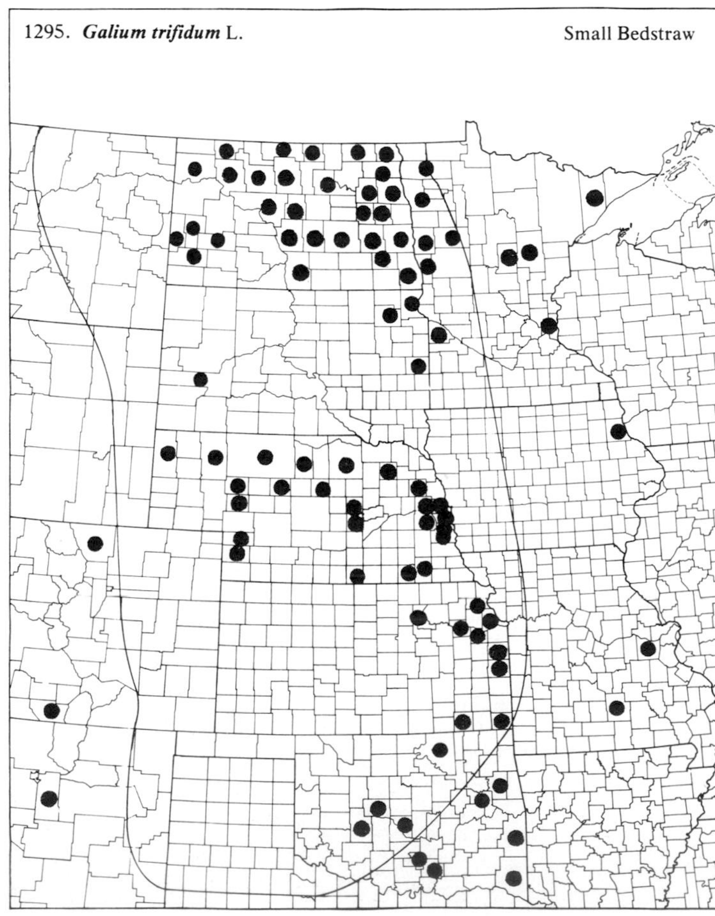

1296. ***Galium triflorum*** Michx. Sweet-scented Bedstraw

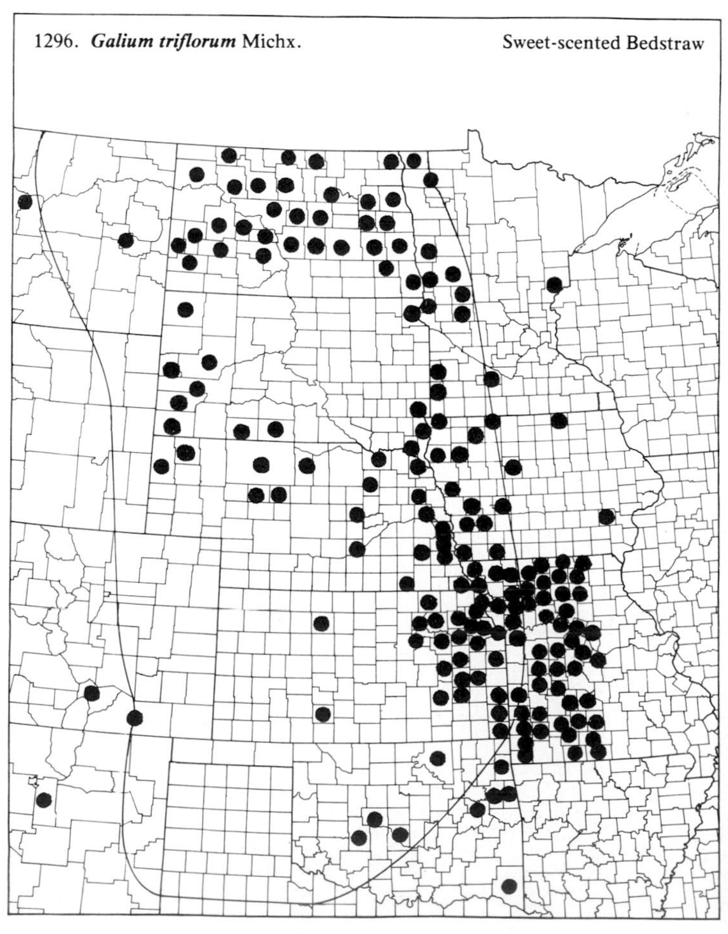

(Rubiaceae)

1297. ***Galium verum*** L. Yellow Bedstraw

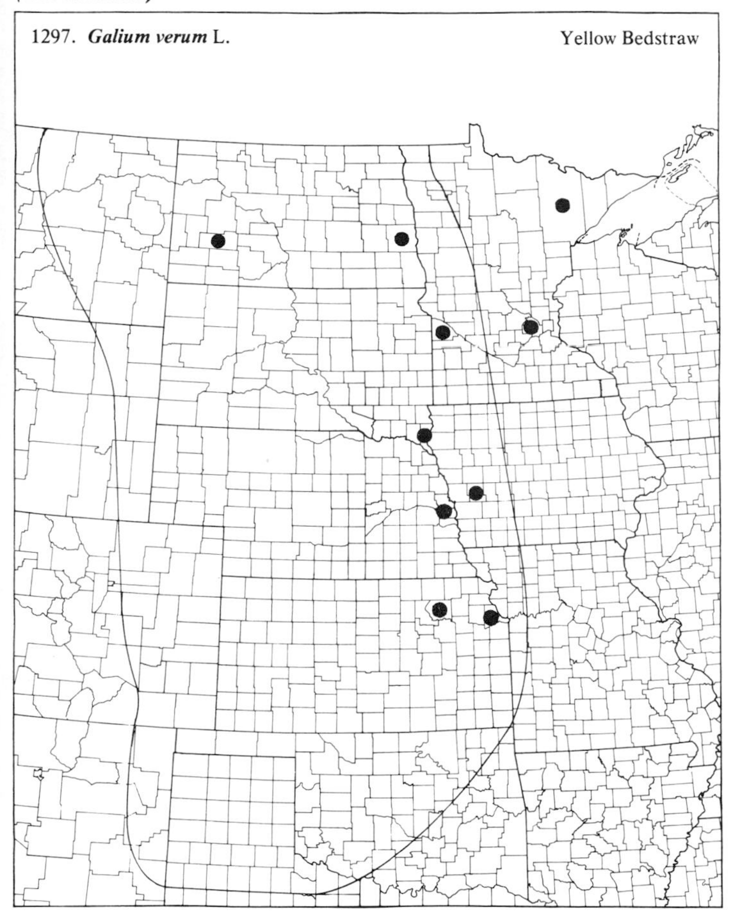

1298. ***Galium virgatum*** Nutt.

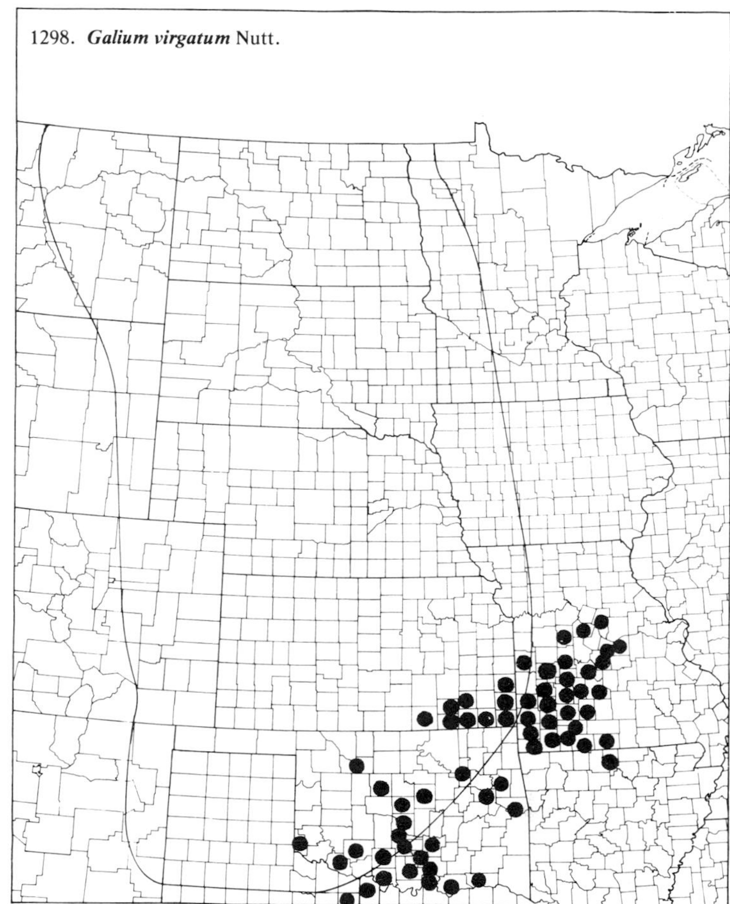

1299. ***Hedyotis crassifolia*** Raf. Small Bluets

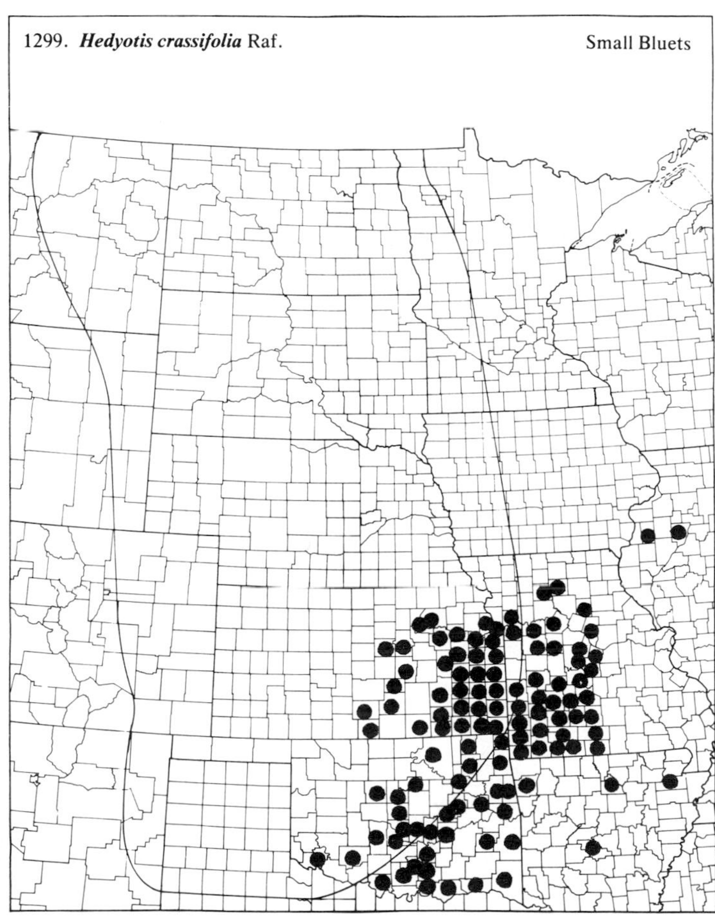

1300. ***Hedyotis longifolia*** (Gaertn.) Hook. Bluets

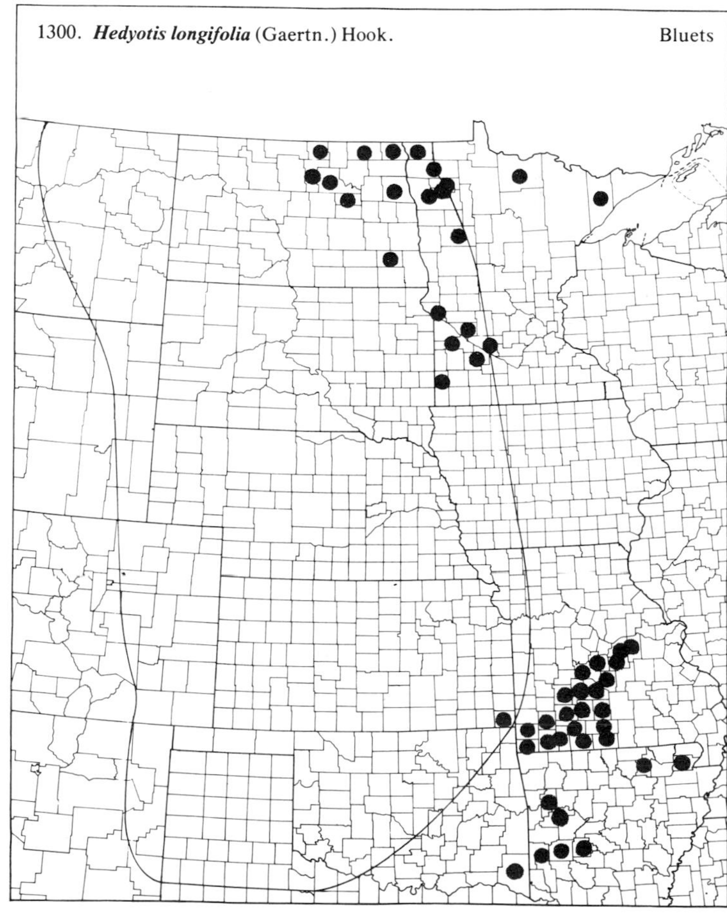

(Rubiaceae)

1301. ***Hedyotis nigricans*** (Lam.) Fosb. Narrowleaf Bluets

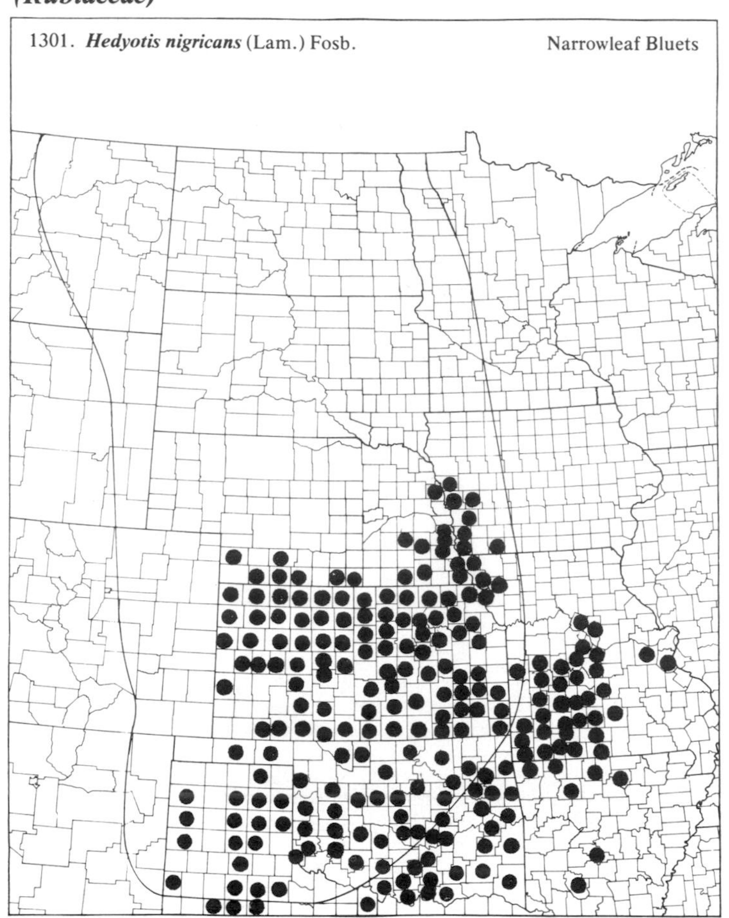

1302. ***Spermacoce glabra*** Michx. Buttonweed

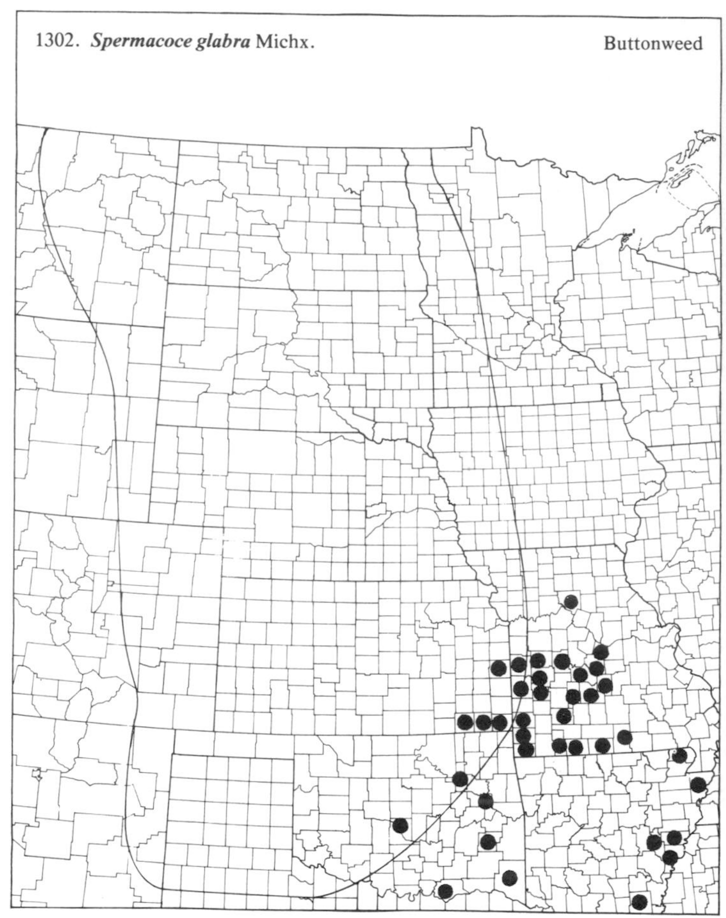

110. *Caprifoliaceae* (Honeysuckle Family)

1303. ***Linnaea borealis*** L. Twinflower

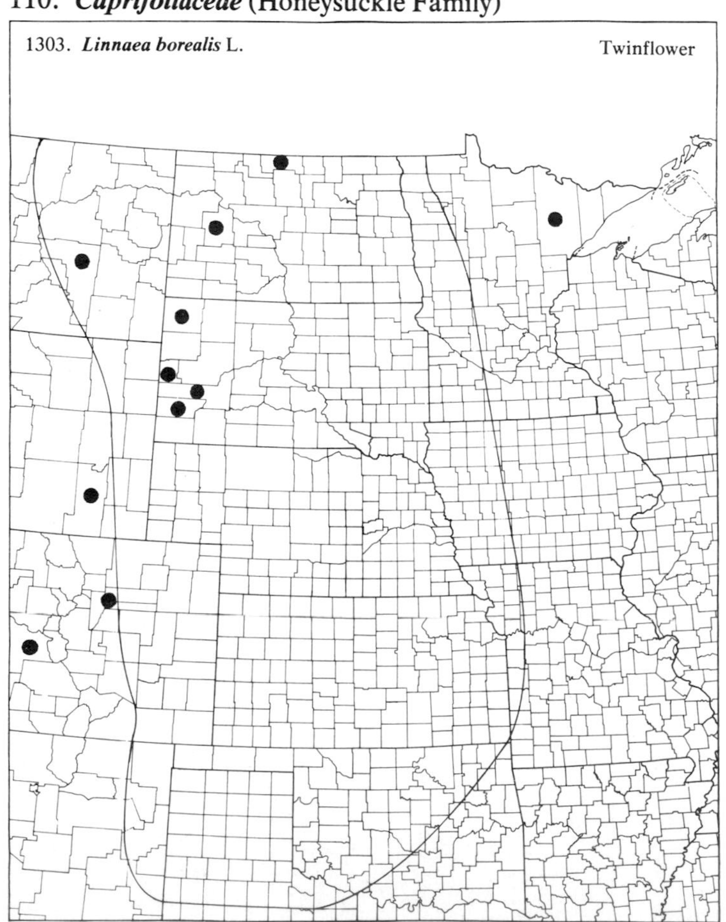

1304. ***Lonicera albiflora*** T. & G.

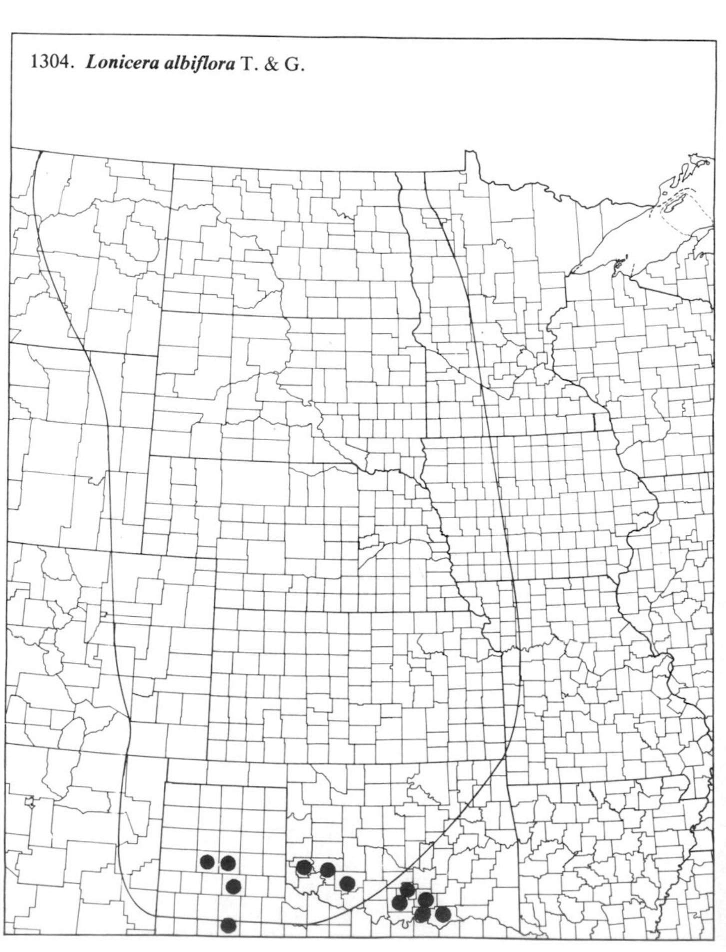

(Caprifoliaceae)

1305. **Lonicera dioica** L. var. **glaucescens** (Rydb.) Butters — Wild Honeysuckle

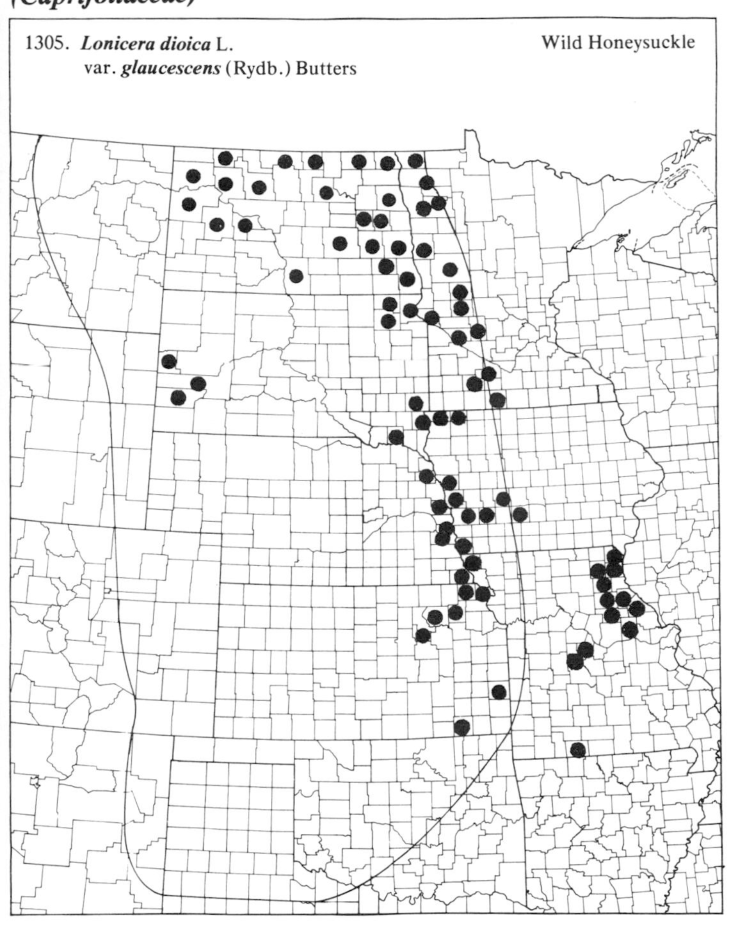

1306. **Lonicera japonica** Thunb. — Japanese Honeysuckle

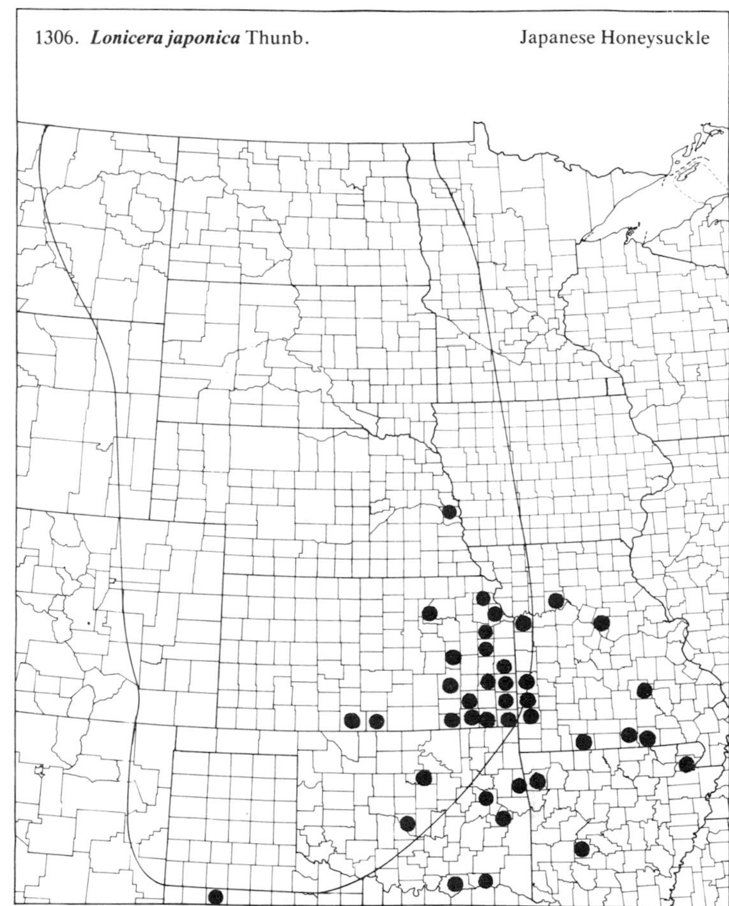

1307. **Lonicera prolifera** (Kirchn.) Rehd. — Grape Honeysuckle

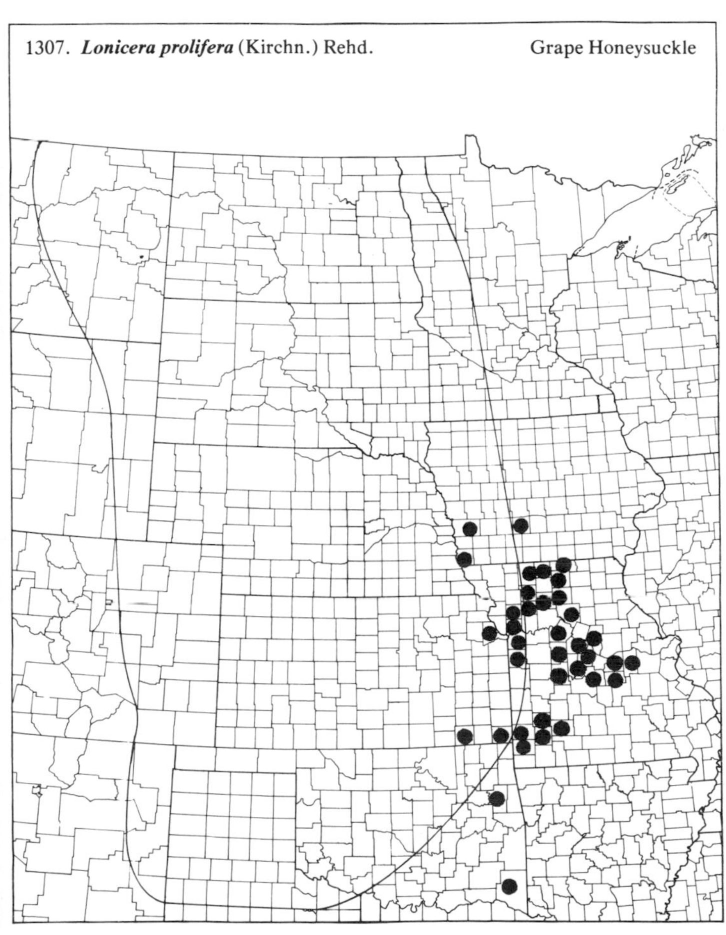

1308. **Lonicera tatarica** L. — Tatarian Honeysuckle

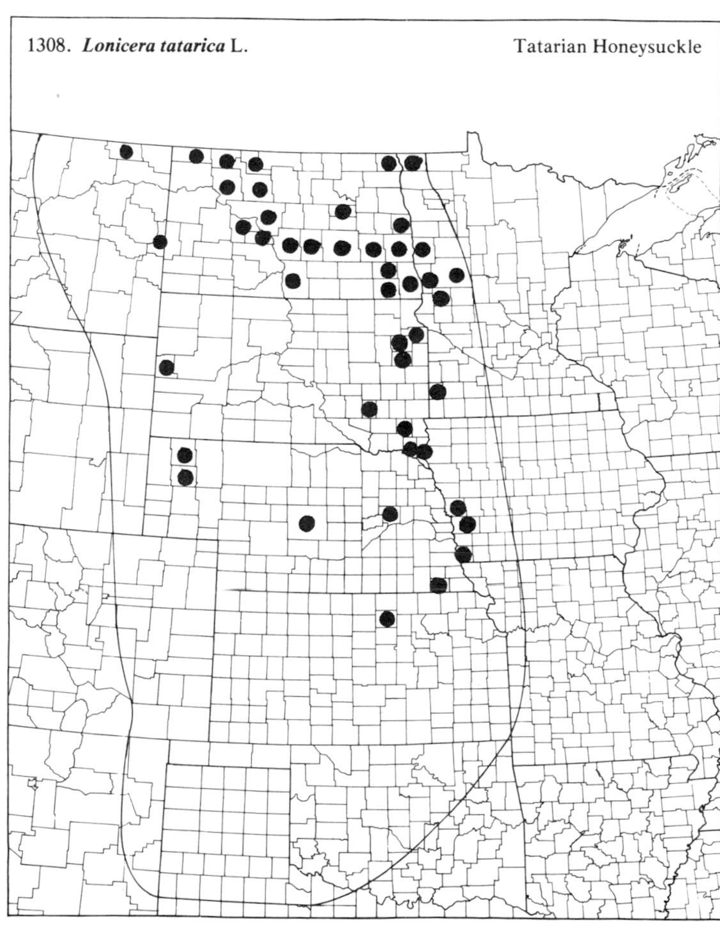

(Caprifoliaceae)

1309. ***Sambucus canadensis*** L. Elderberry

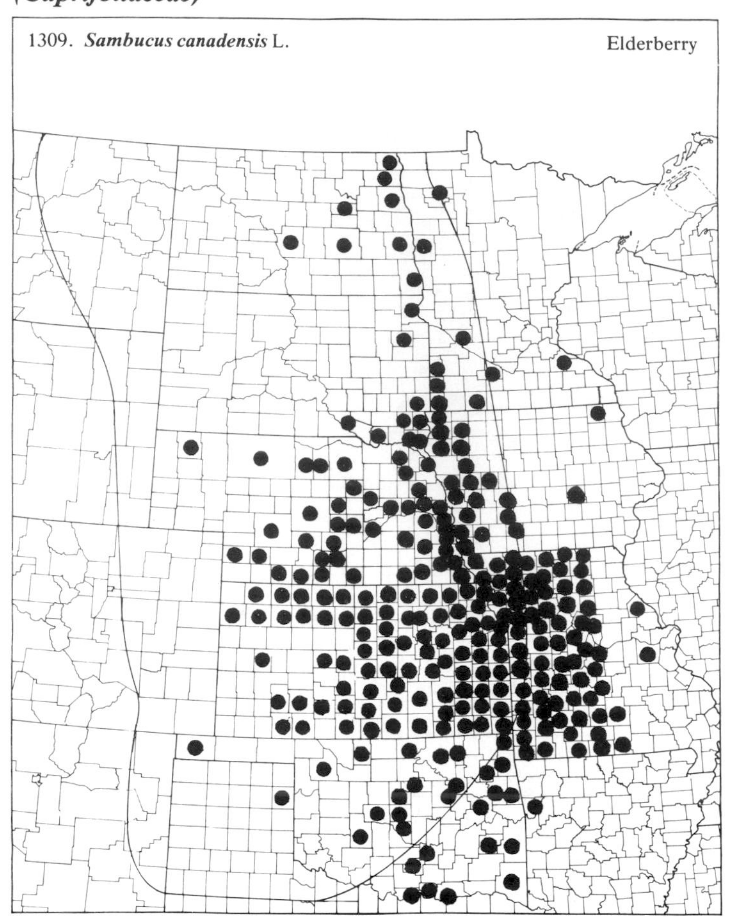

1310. ***Sambucus racemosa*** L. ssp. ***pubens*** (Michx.) House Stinking Elderberry

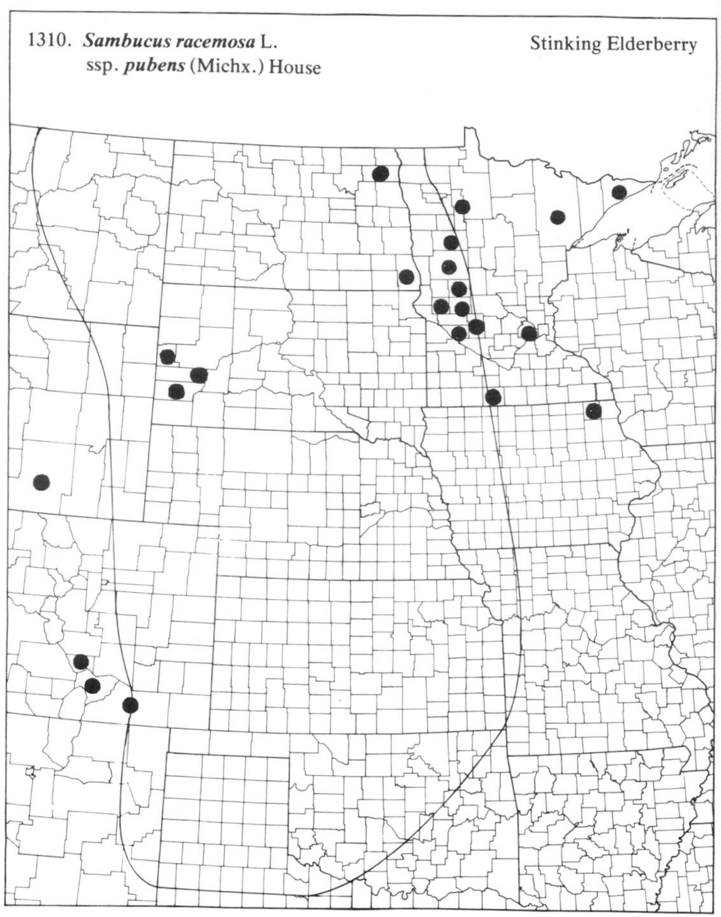

1311. ***Symphoricarpos albus*** (L.) Blake Snowberry

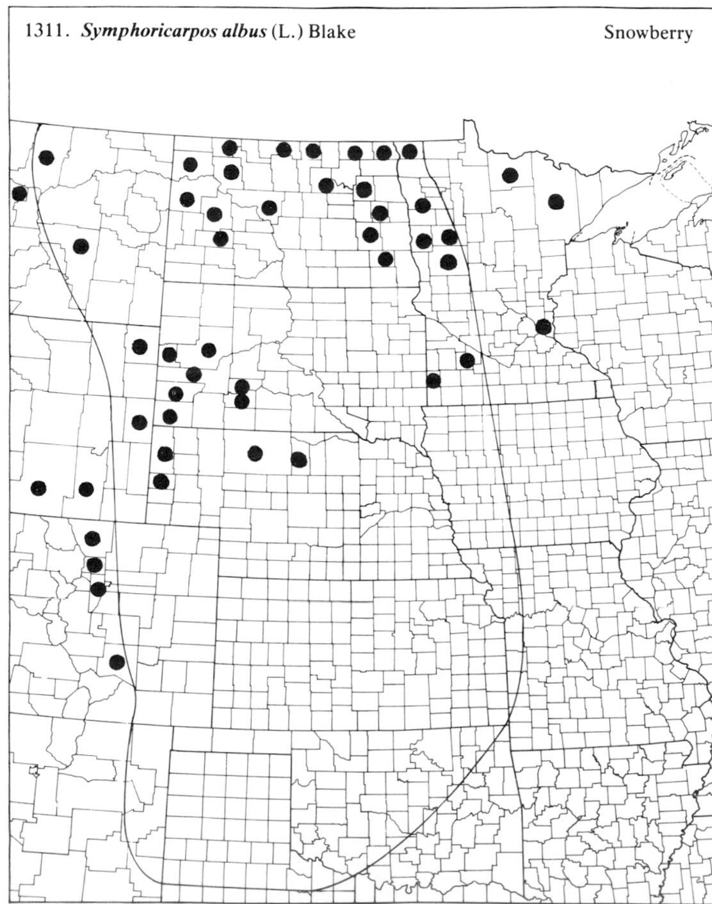

1312. ***Symphoricarpos occidentalis*** Hook. Western Snowberry

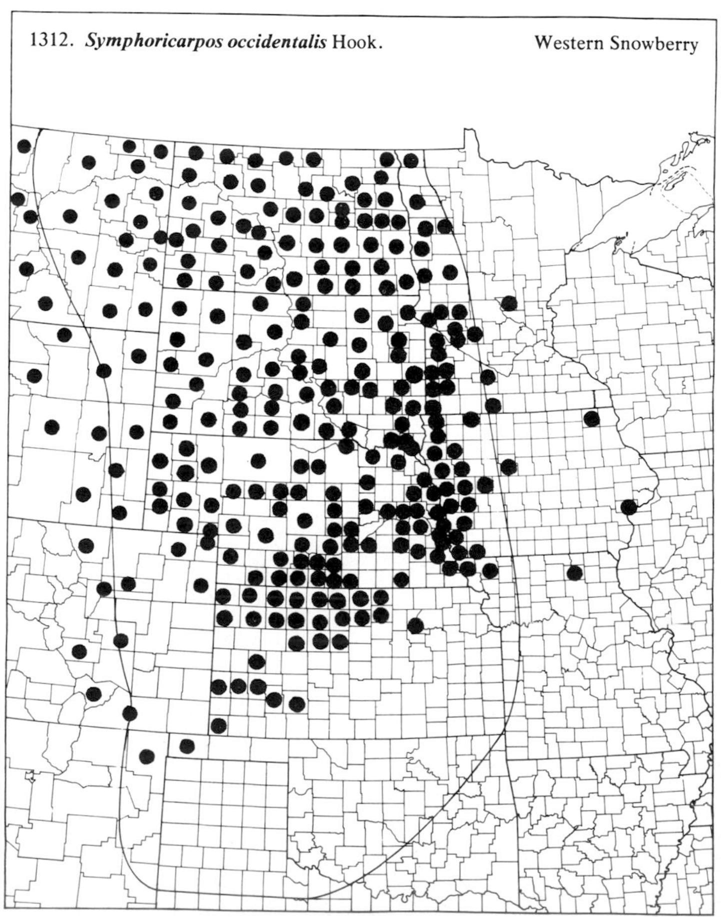

(Caprifoliaceae)

1313. *Symphoricarpos orbiculatus* Moench — Buckbrush

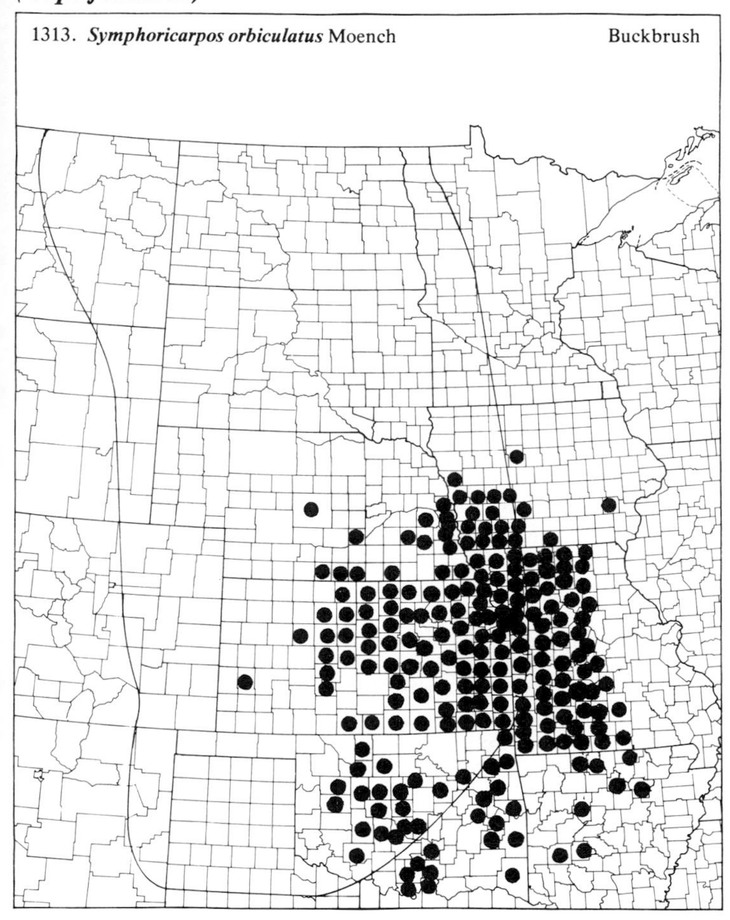

1314. *Triosteum aurantiacum* Bickn. var. *illinoense* (Wieg.) Palm. & Steyerm. — Orange Horse Gentian

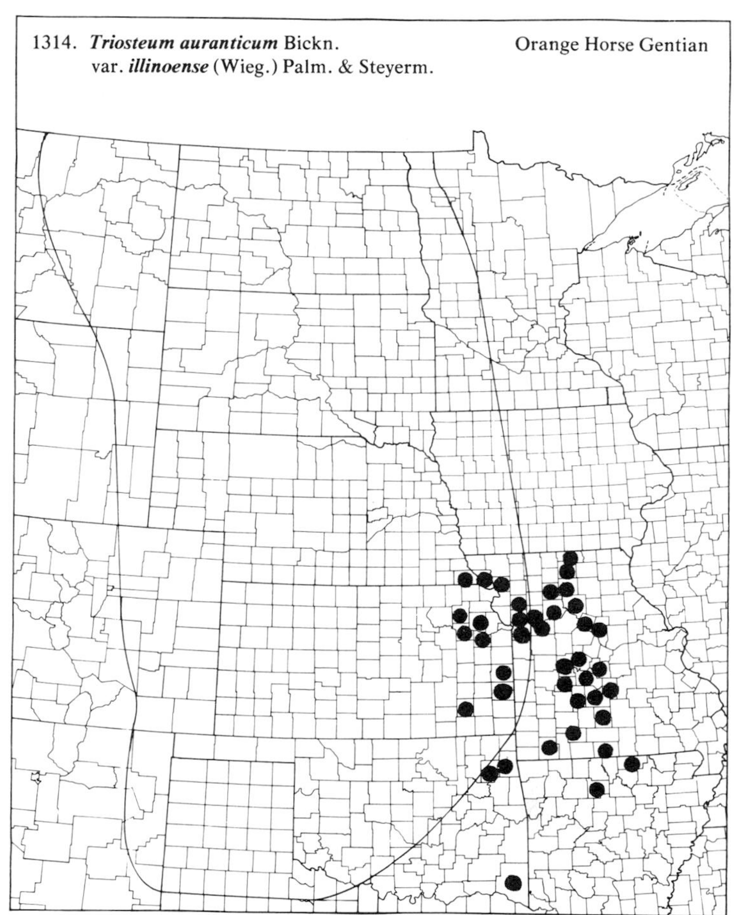

1315. *Triosteum perfoliatum* L. — Horse Gentian

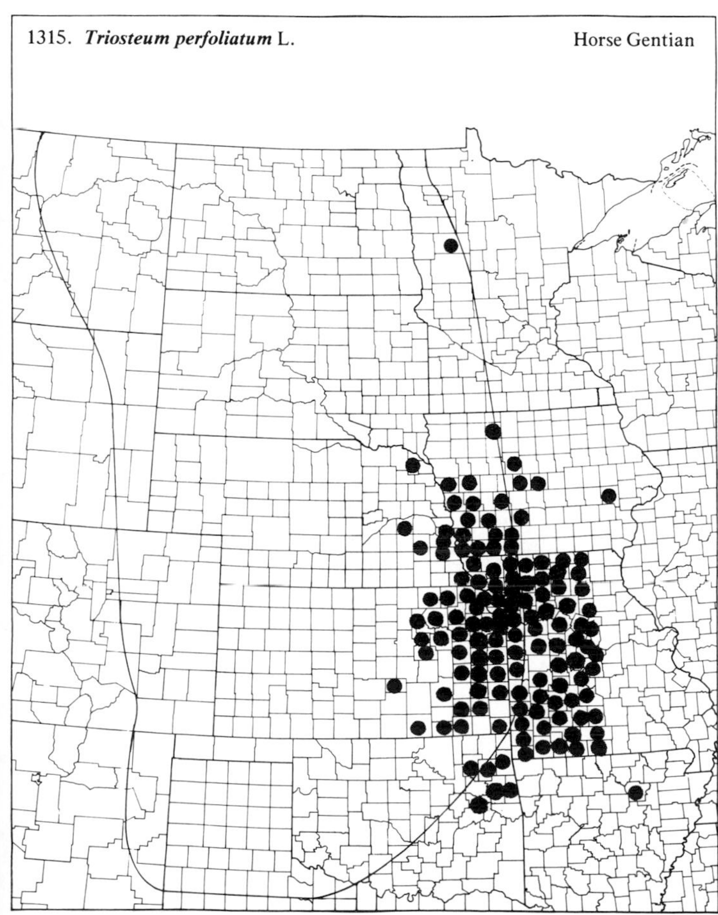

1316. *Viburnum lentago* L. — Nannyberry

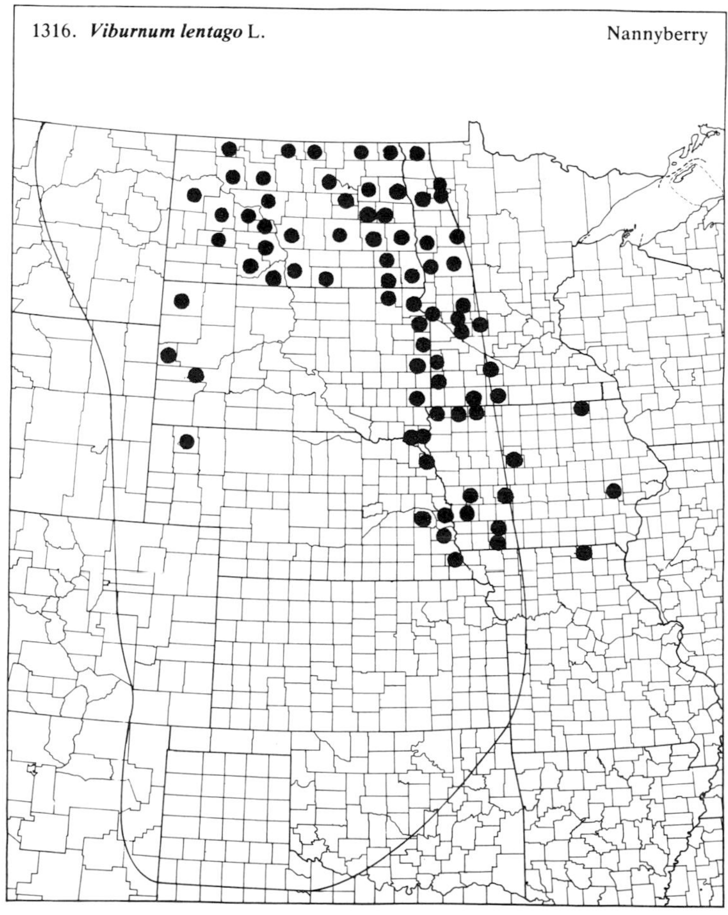

(Caprifoliaceae)

1317. ***Viburnum opulus*** L. var. ***americanum*** Ait. — Highbush Cranberry

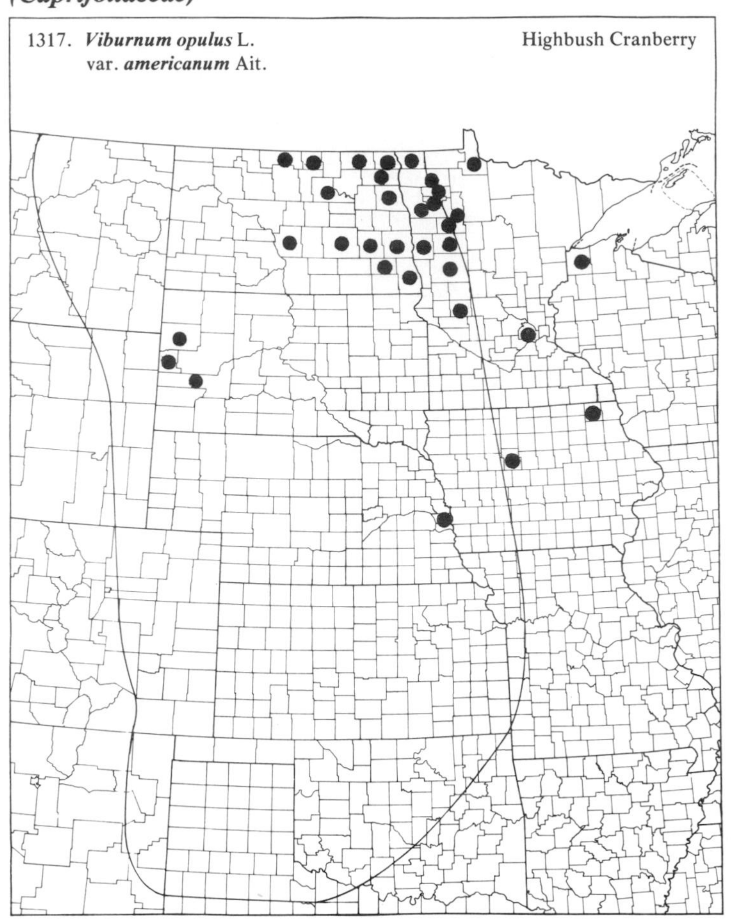

1318. ***Viburnum prunifolium*** L. — Black Haw

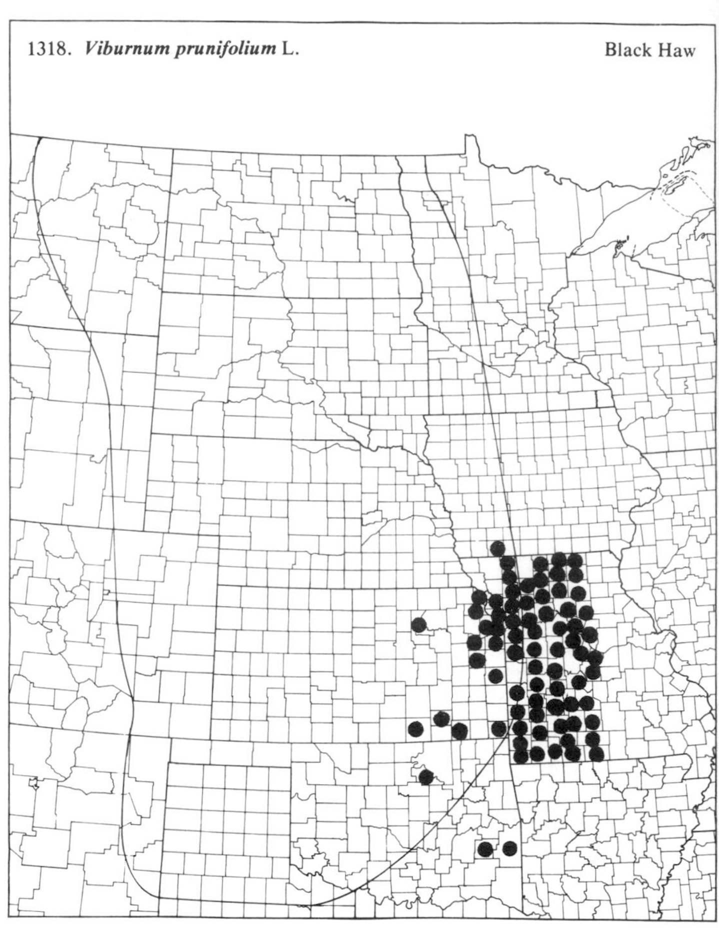

1319. ***Viburnum rafinesquianum*** Schult. var. ***affine*** (Bush) House — Downy Arrowwood

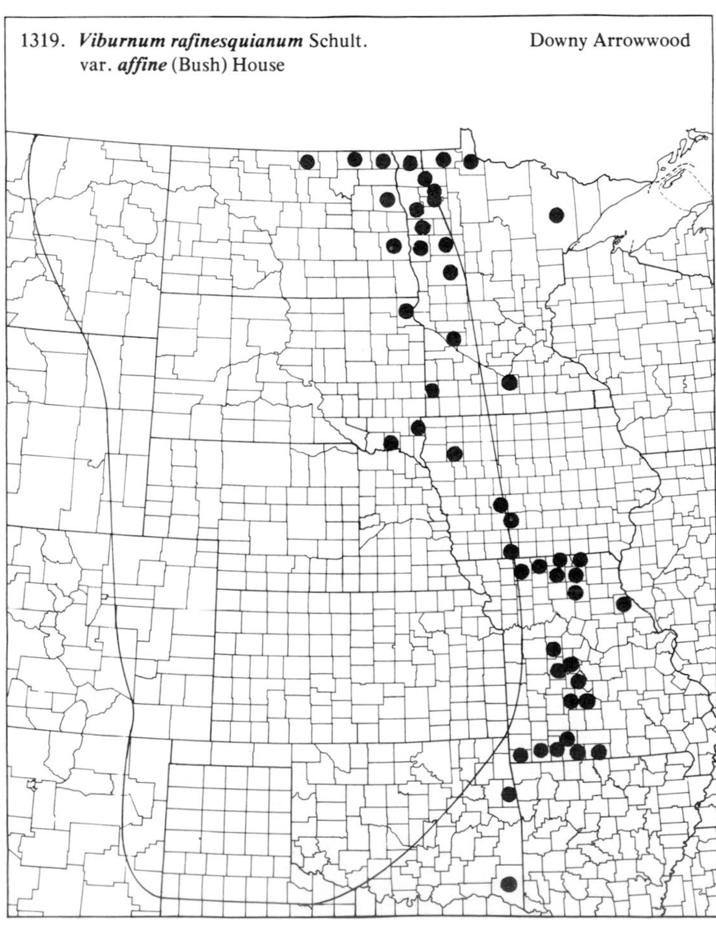

1320. ***Viburnum rufidulum*** Raf. — Southern Black Haw

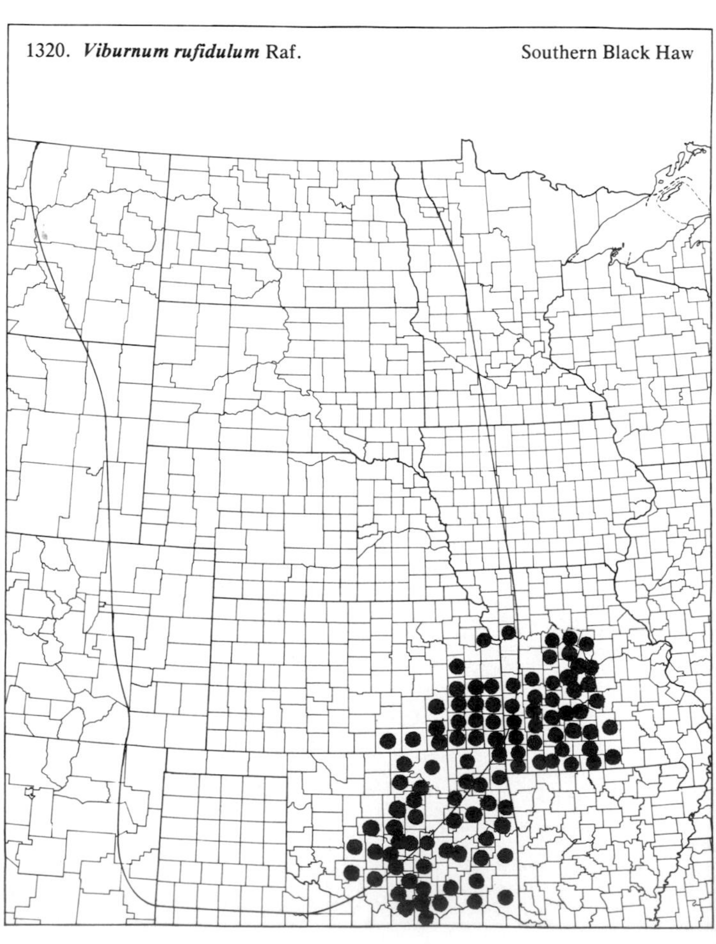

112. *Valerianaceae* (Valerian Family)

1321. ***Valerianella radiata*** (L.) Dufr. — Corn-salad

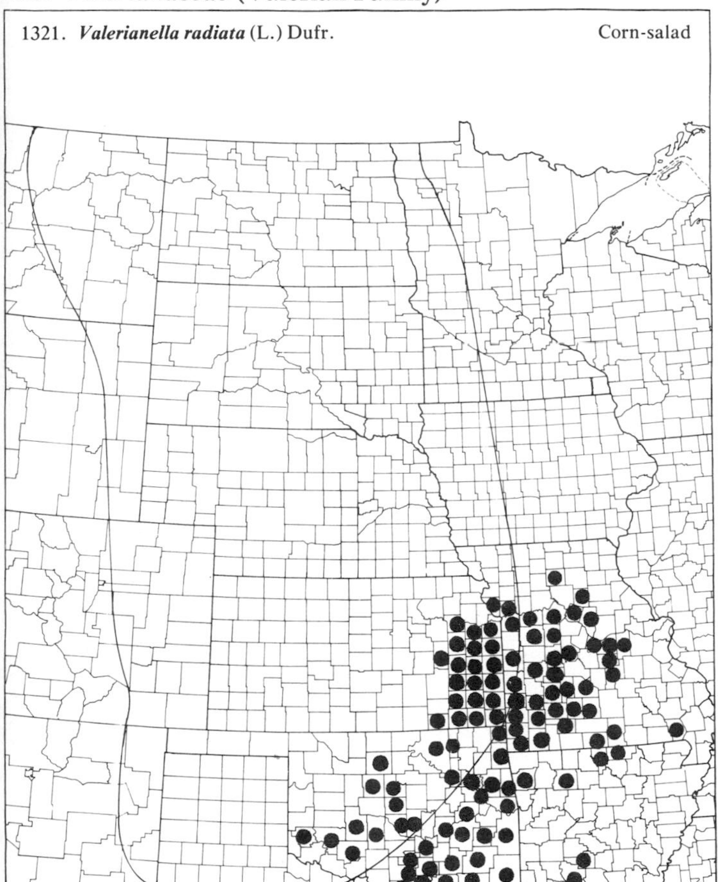

113. *Dipsacaceae* (Teasel Family)

1322. ***Dipsacus sylvestris*** Huds. — Teasel

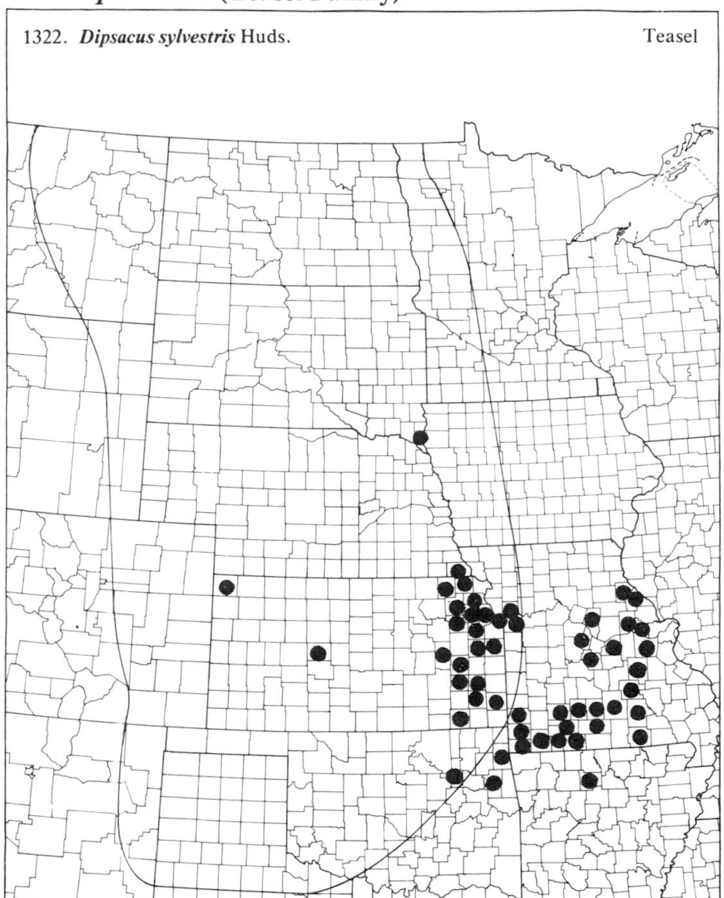

114. *Asteraceae* (Sunflower Family)

1323. ***Achillea millefolium*** L. ssp. ***lanulosa*** (Nutt.) Piper — Yarrow

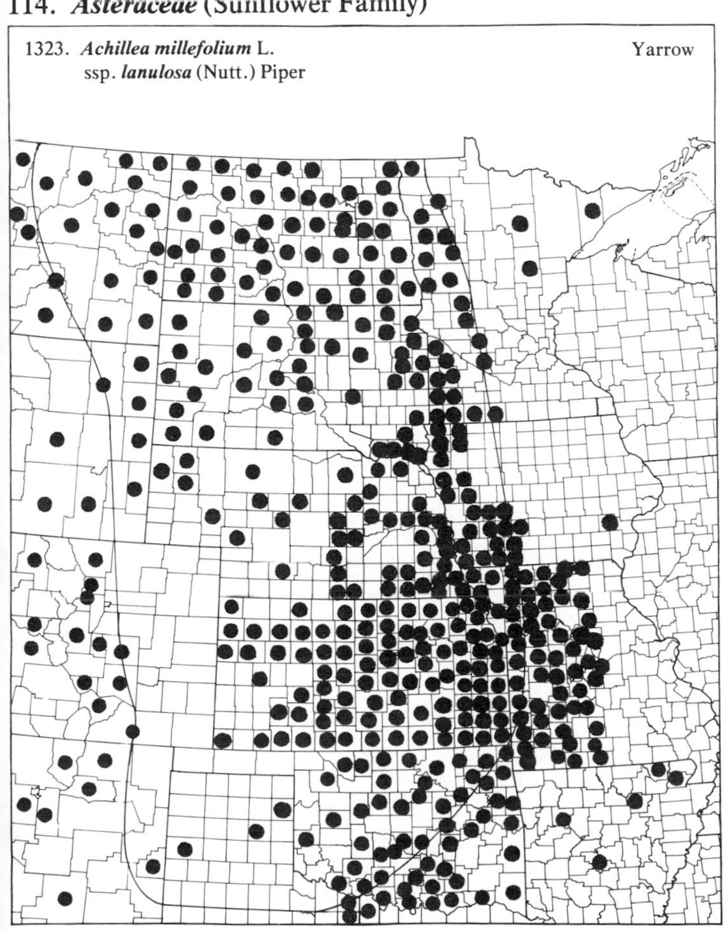

1324. ***Agoseris glauca*** (Pursh) Dietr. — False Dandelion

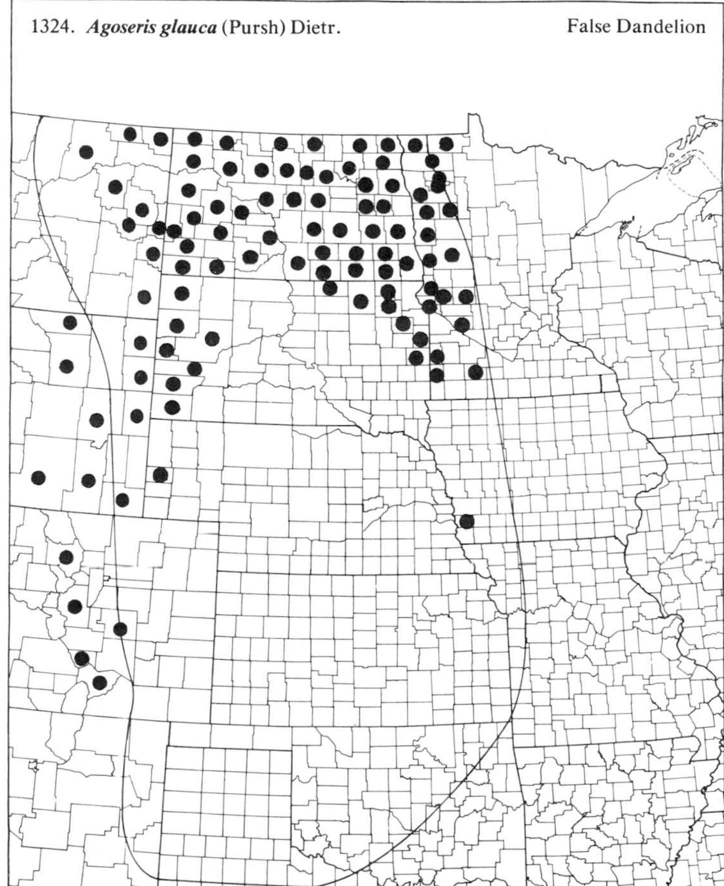

(Asteraceae)

1325. ***Ambrosia acanthicarpa*** Hook. Annual Bursage

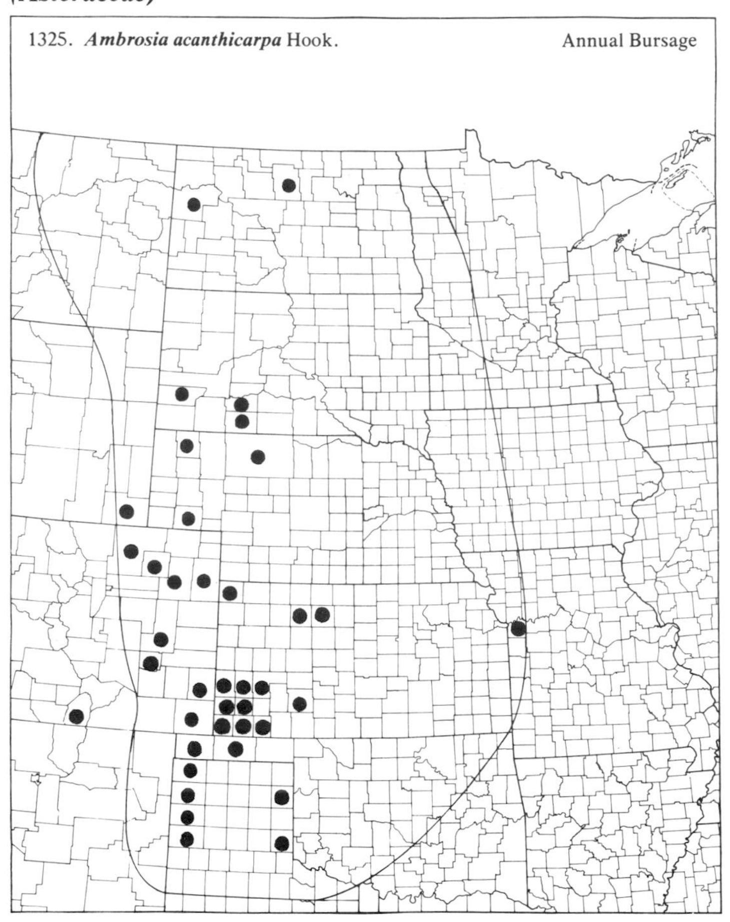

1326. ***Ambrosia artemisiifolia*** L. Common Ragweed

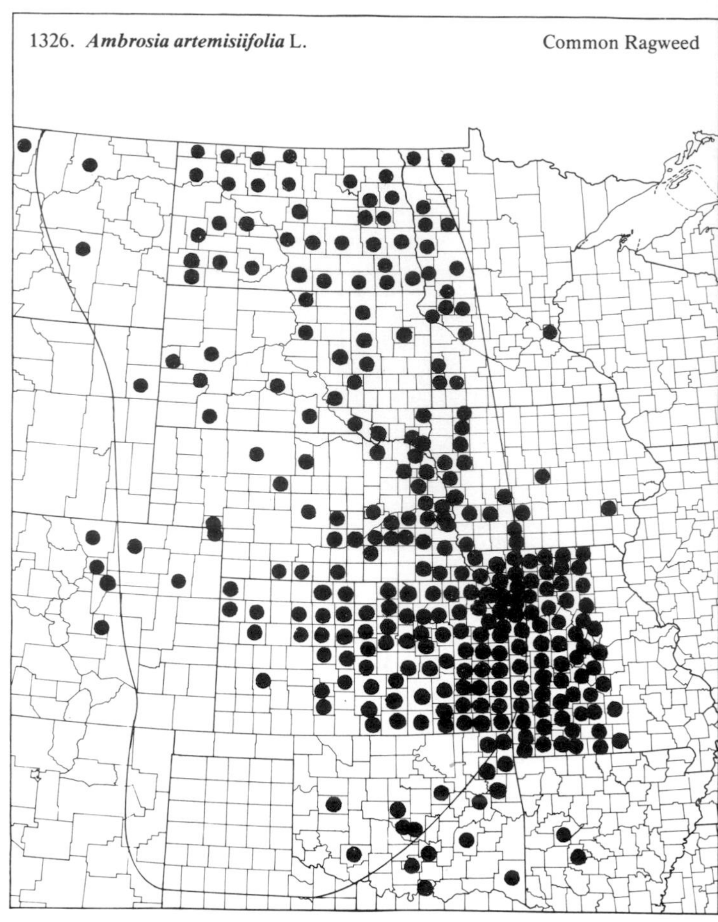

1327. ***Ambrosia bidentata*** Michx. Ragweed

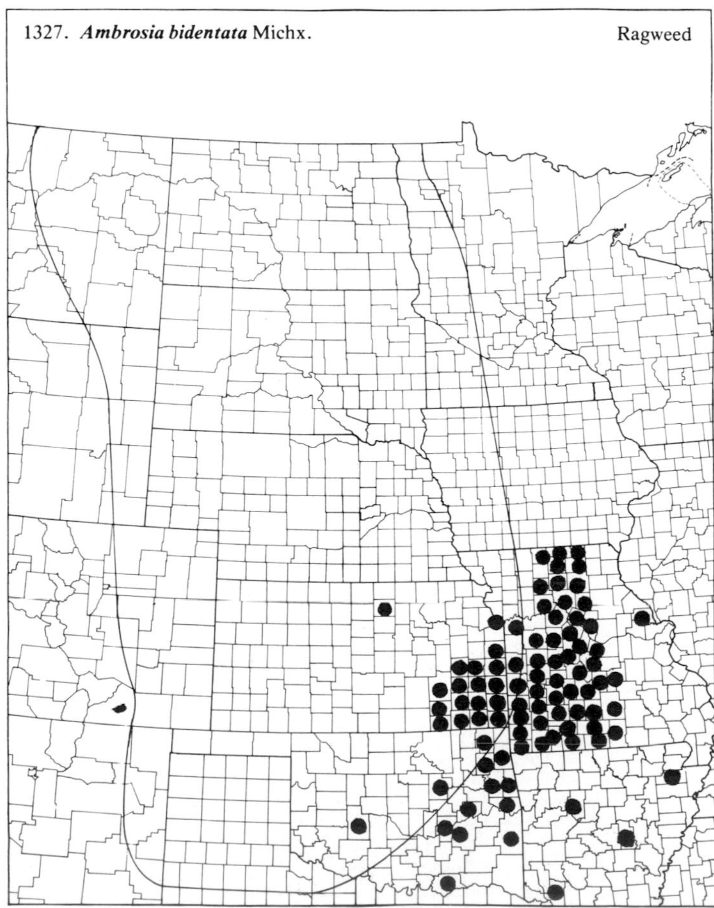

1328. ***Ambrosia confertiflora*** DC.

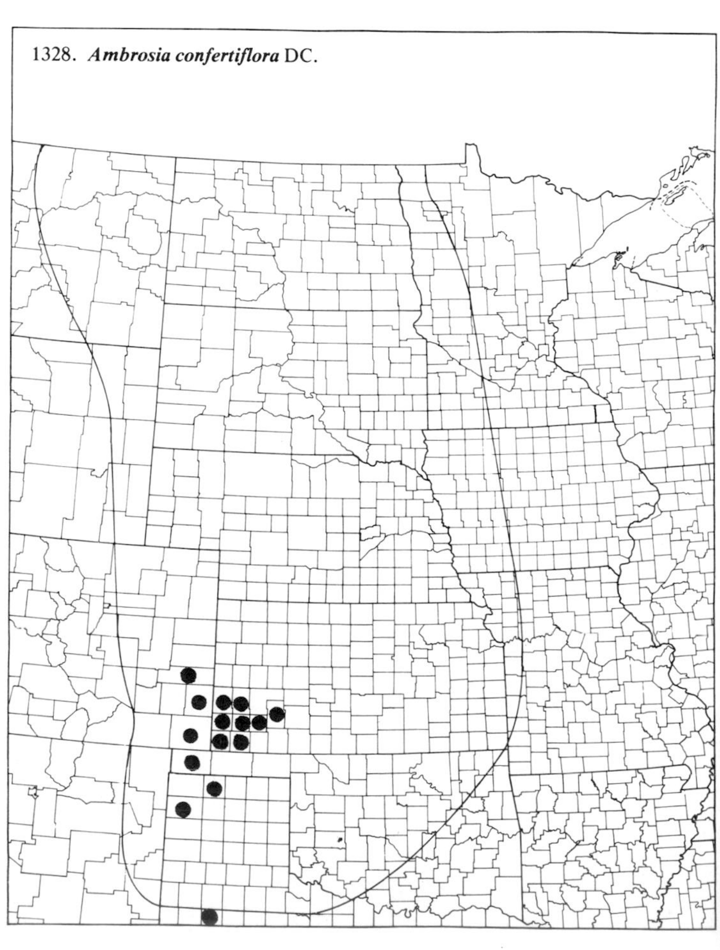

(Asteraceae)

1329. ***Ambrosia grayi*** (A. Nels.) Shinners — Bur Ragweed

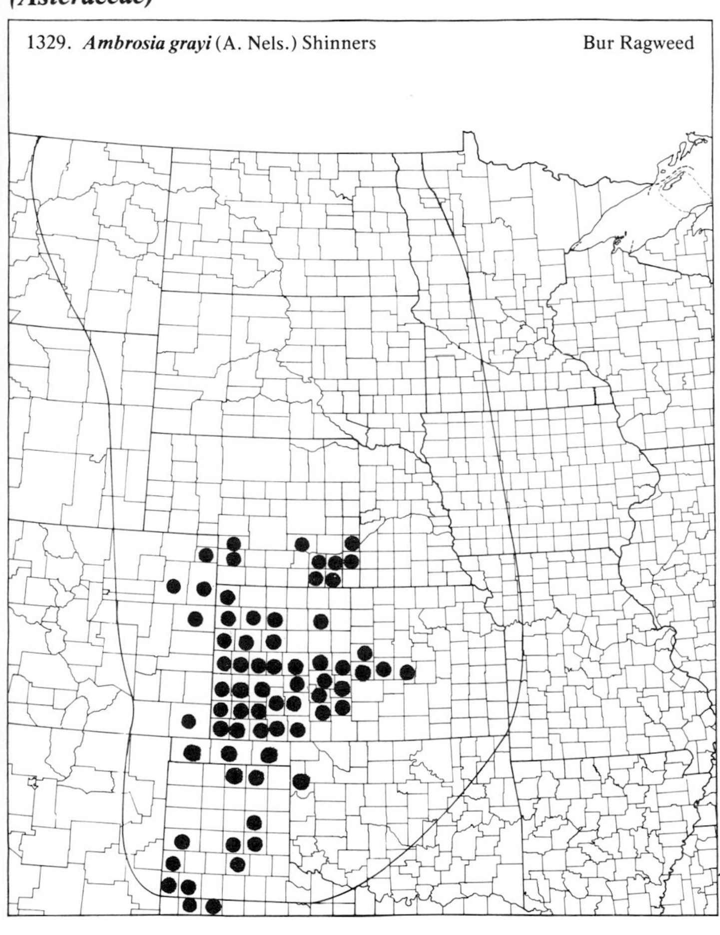

1330. ***Ambrosia psilostachya*** DC. — Western Ragweed

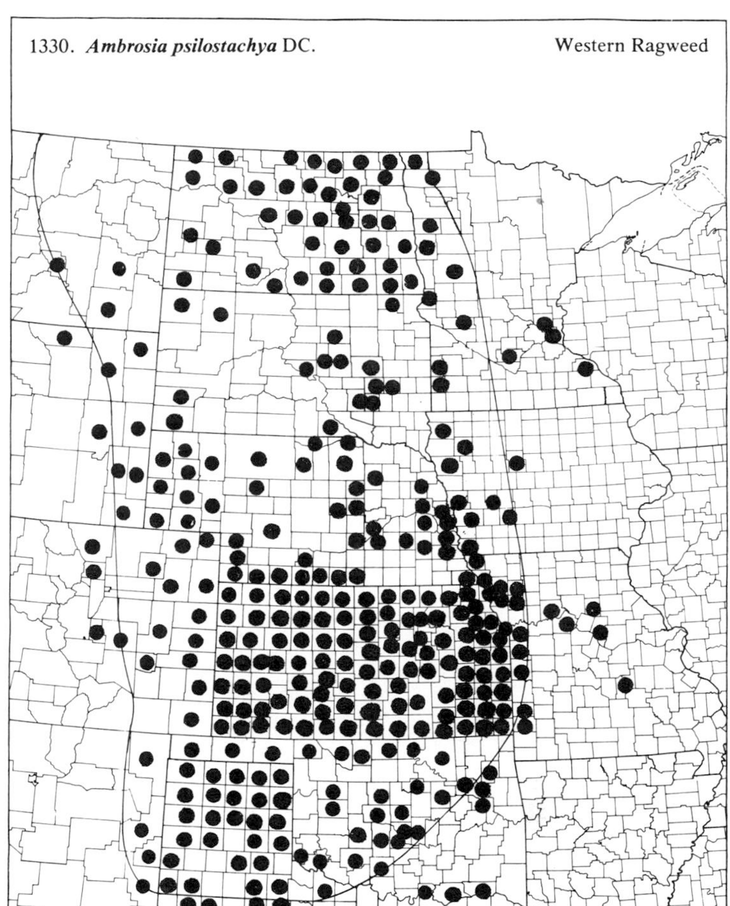

1331. ***Ambrosia tomentosa*** Nutt. — Perennial Bursage

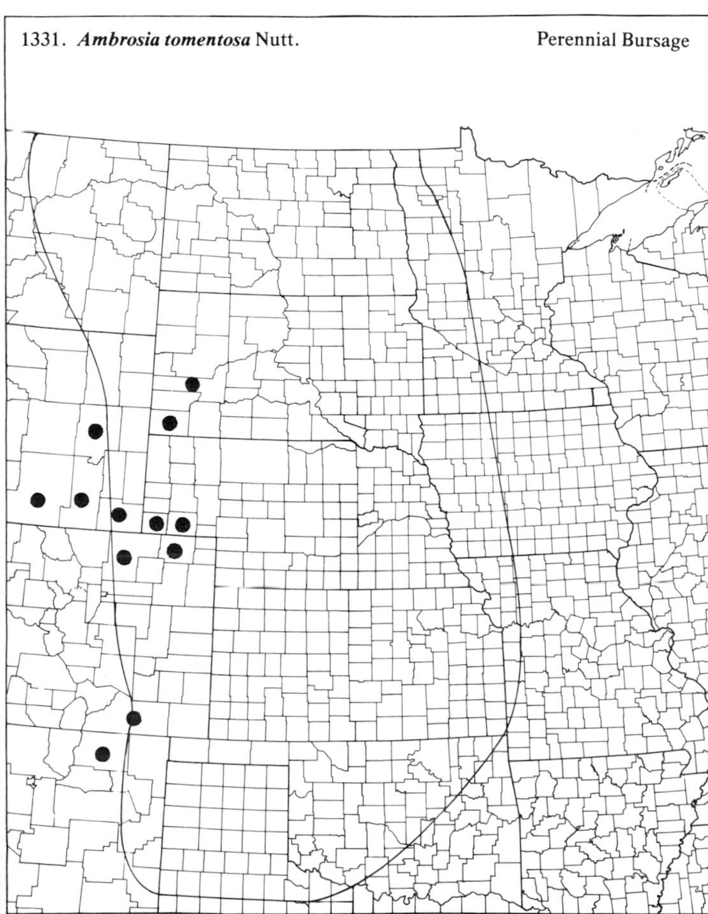

1332. ***Ambrosia trifida*** L. — Giant Ragweed

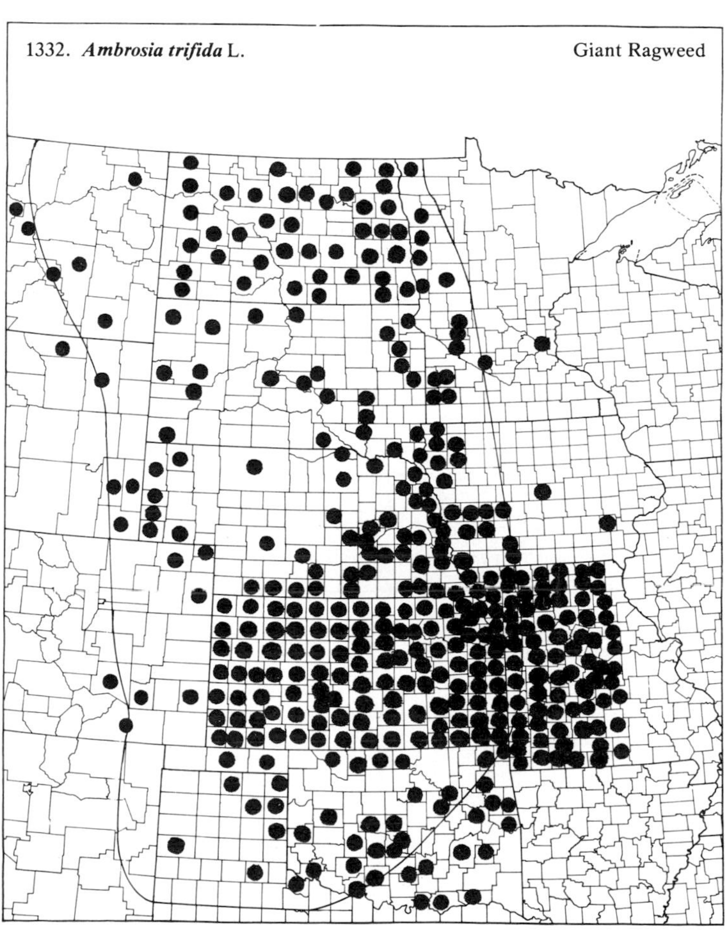

(Asteraceae)

1333. ***Anaphalis margaritacea*** (L.) Benth. & Hook. Pearly Everlasting

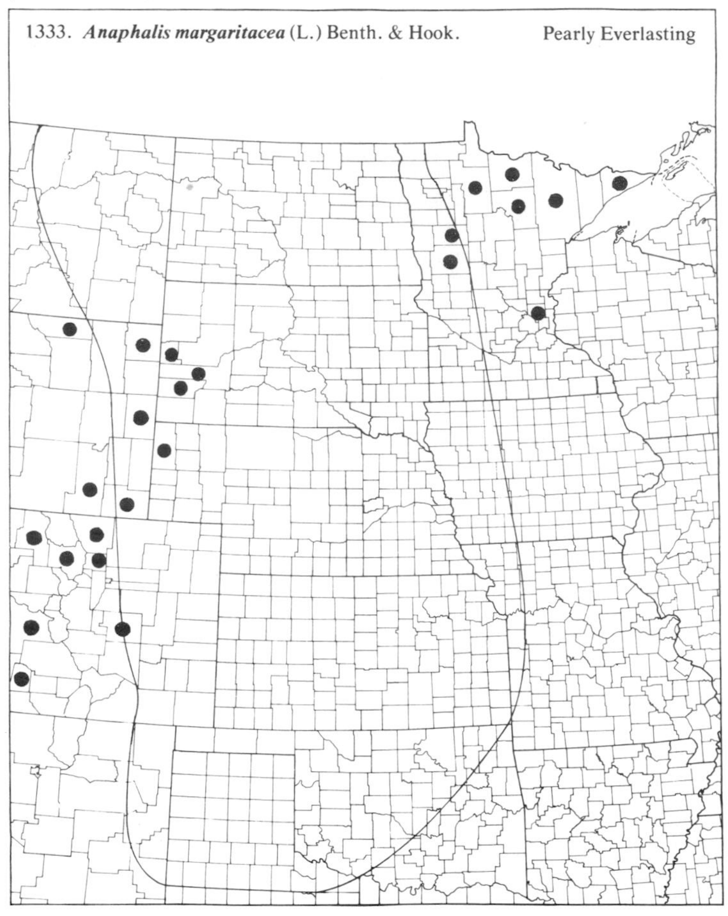

1334. ***Antennaria neglecta*** Greene Field Pussytoes

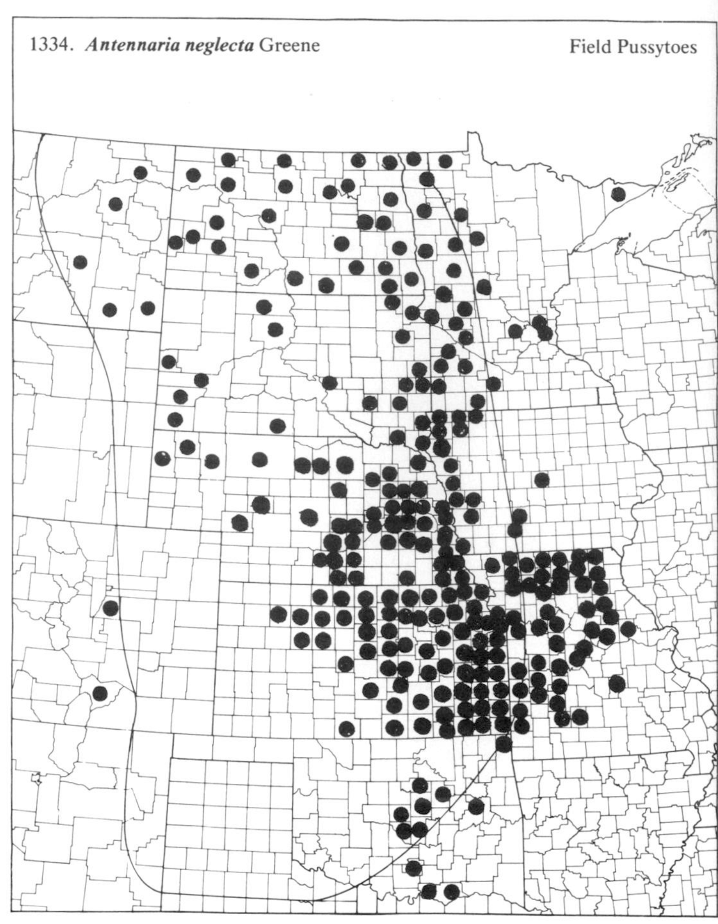

1335. ***Antennaria parvifolia*** Nutt. Pussytoes

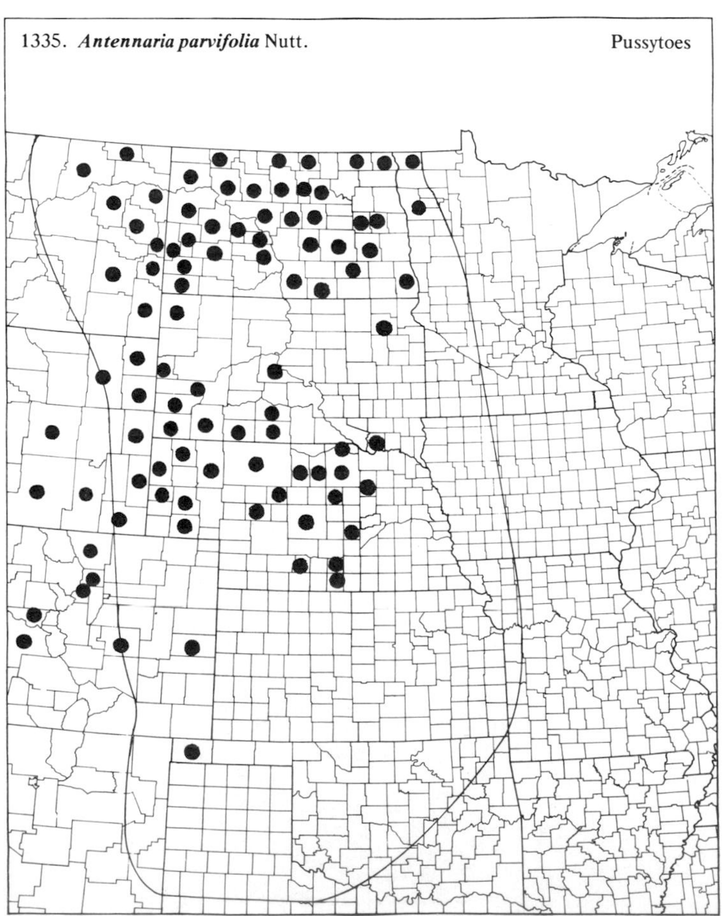

1336. ***Antennaria plantaginifolia*** (L.) Richards. Plainleaf Pussytoes

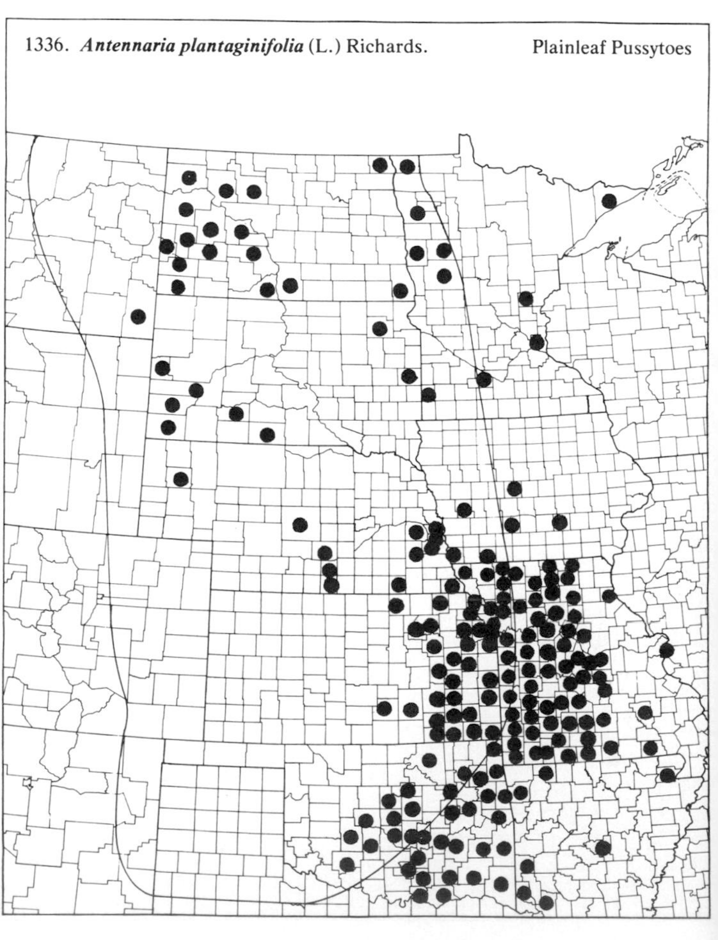

(Asteraceae)

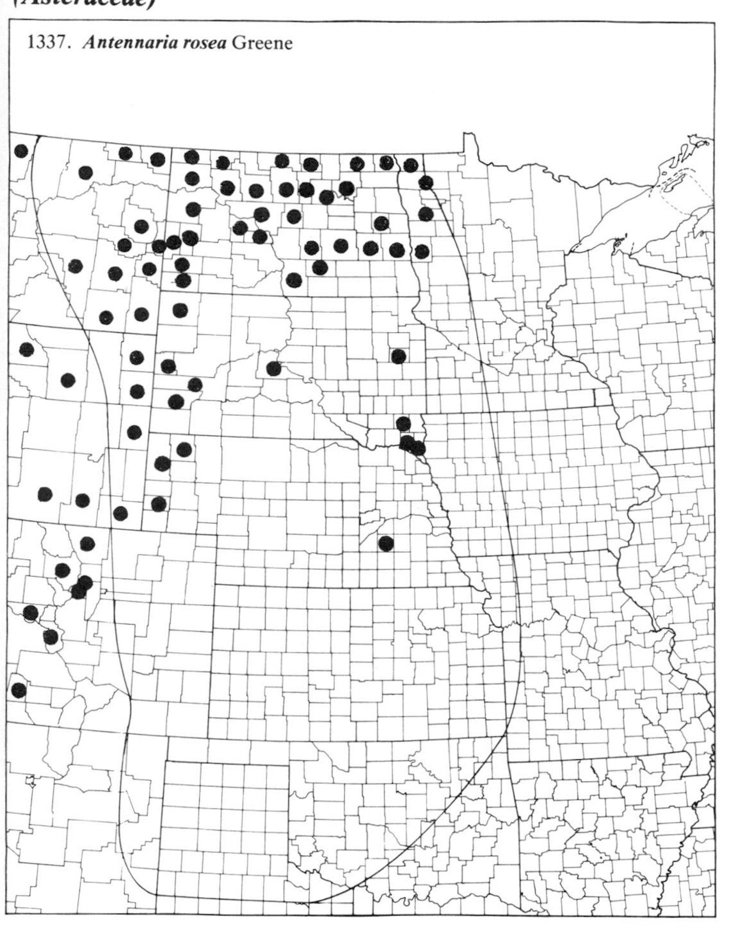

1337. ***Antennaria rosea*** Greene

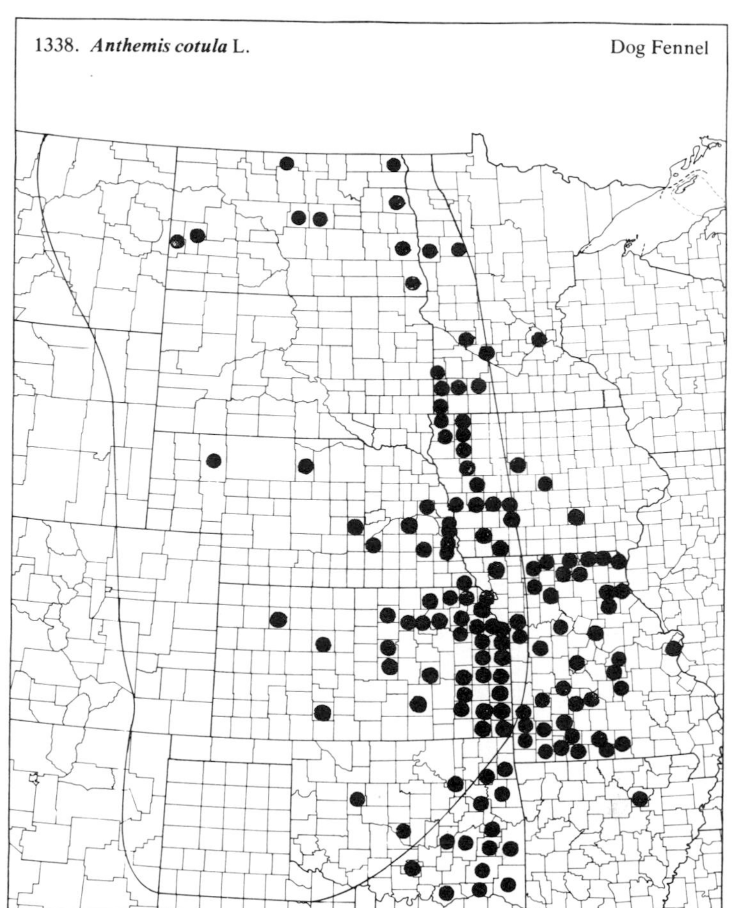

1338. ***Anthemis cotula*** L. Dog Fennel

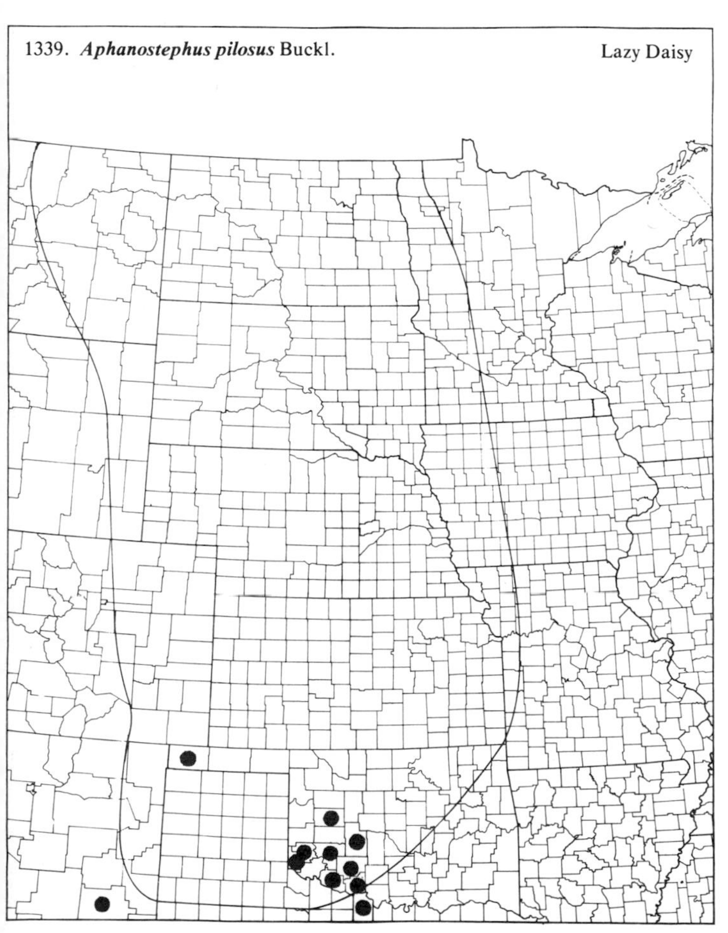

1339. ***Aphanostephus pilosus*** Buckl. Lazy Daisy

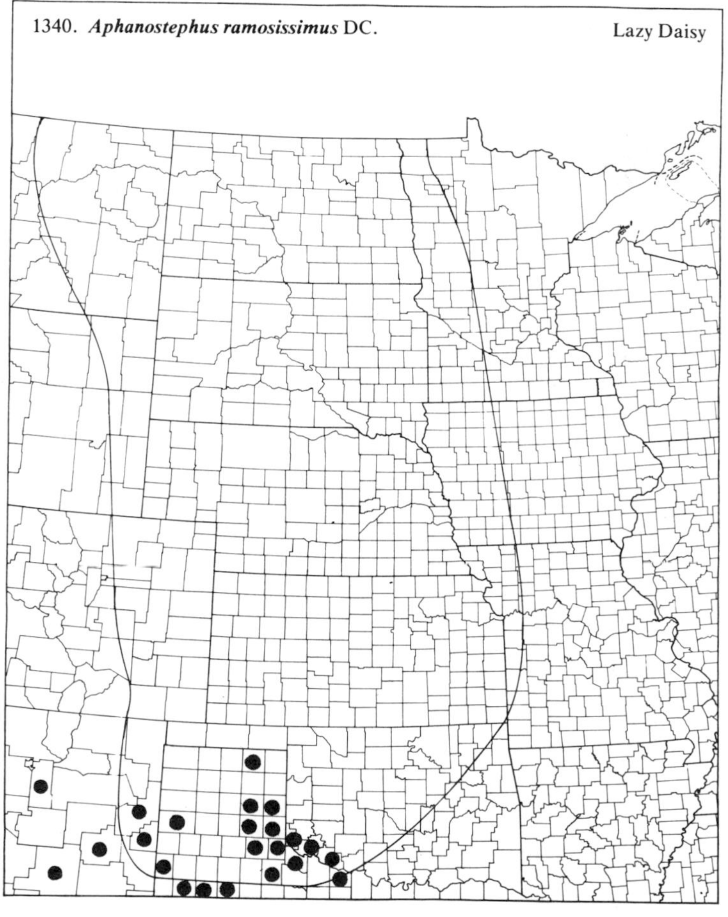

1340. ***Aphanostephus ramosissimus*** DC. Lazy Daisy

(Asteraceae)

1341. ***Aphanostephus skirrobasis*** (DC.) Trel. Lazy Daisy

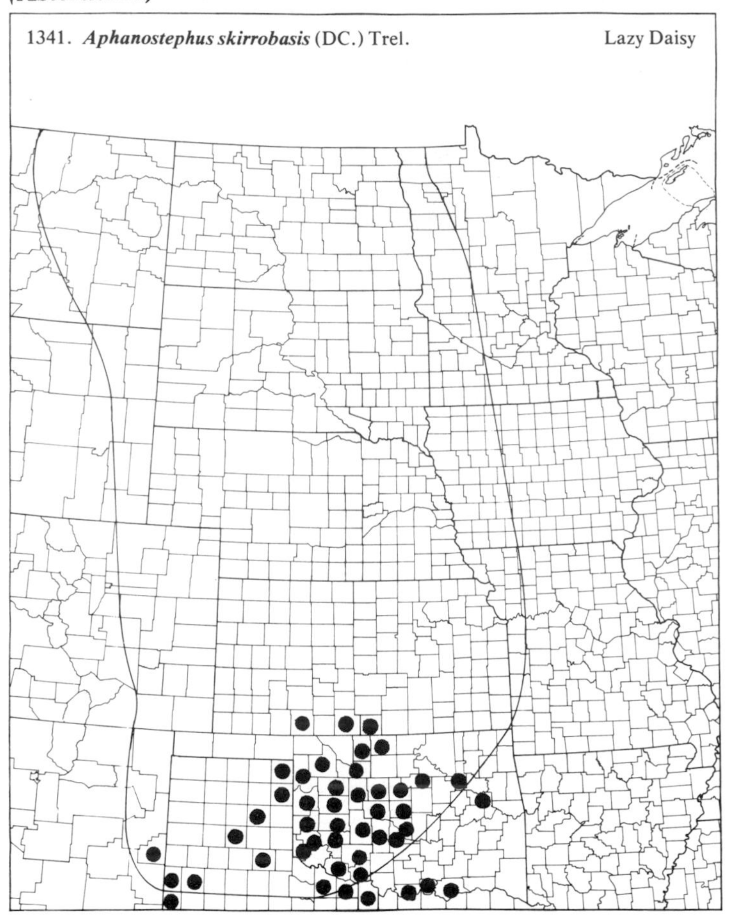

1342. ***Arctium minus*** Schkuhr Common Burdock

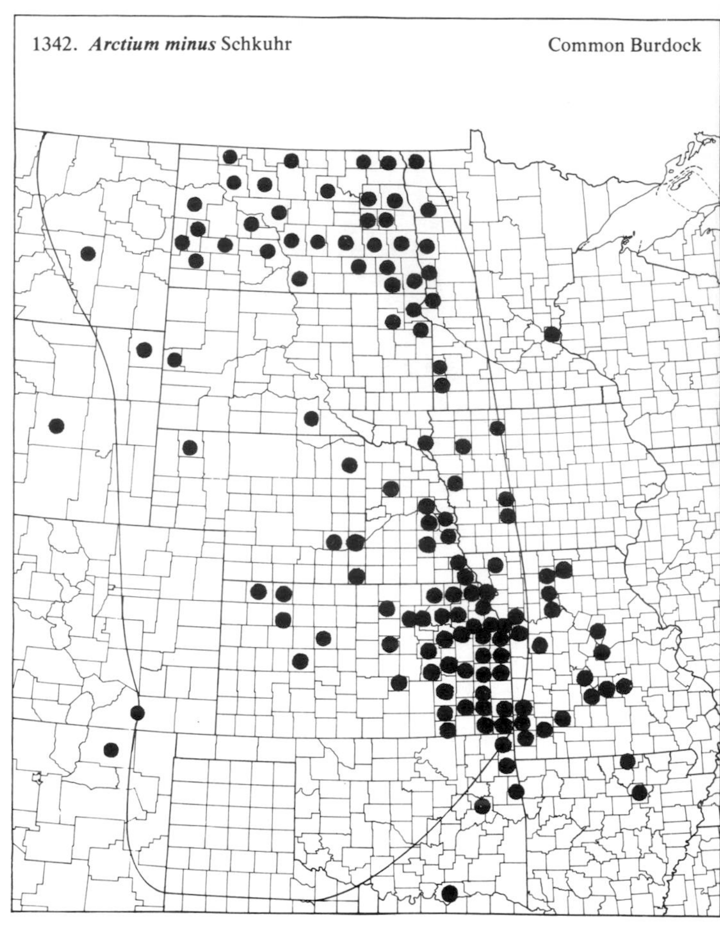

1343. ***Arnica fulgens*** Pursh Arnica

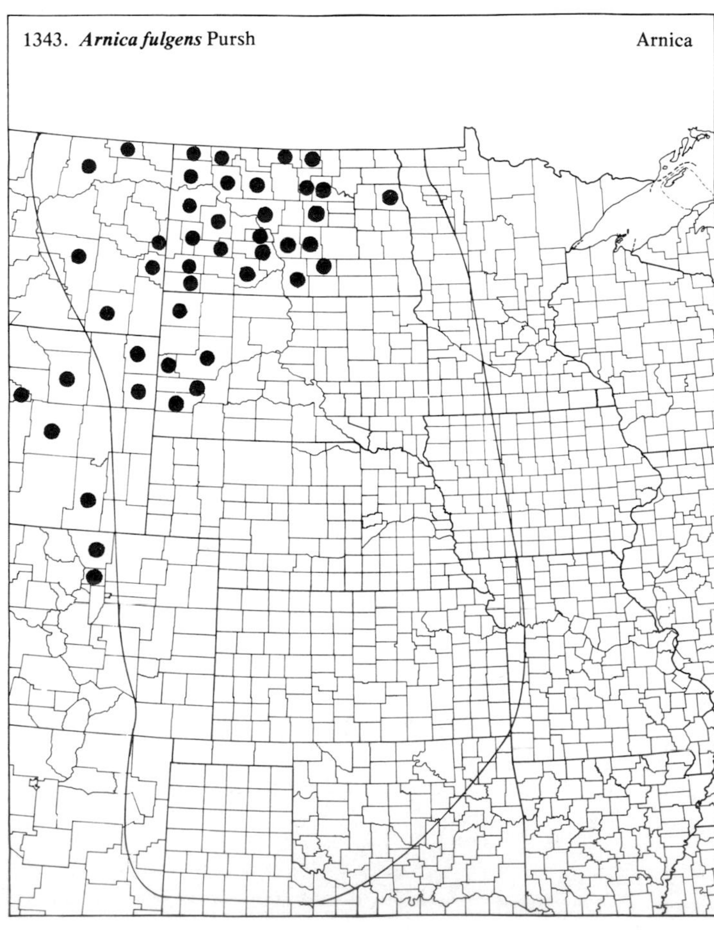

1344. ***Artemisia absinthium*** L. Wormwood

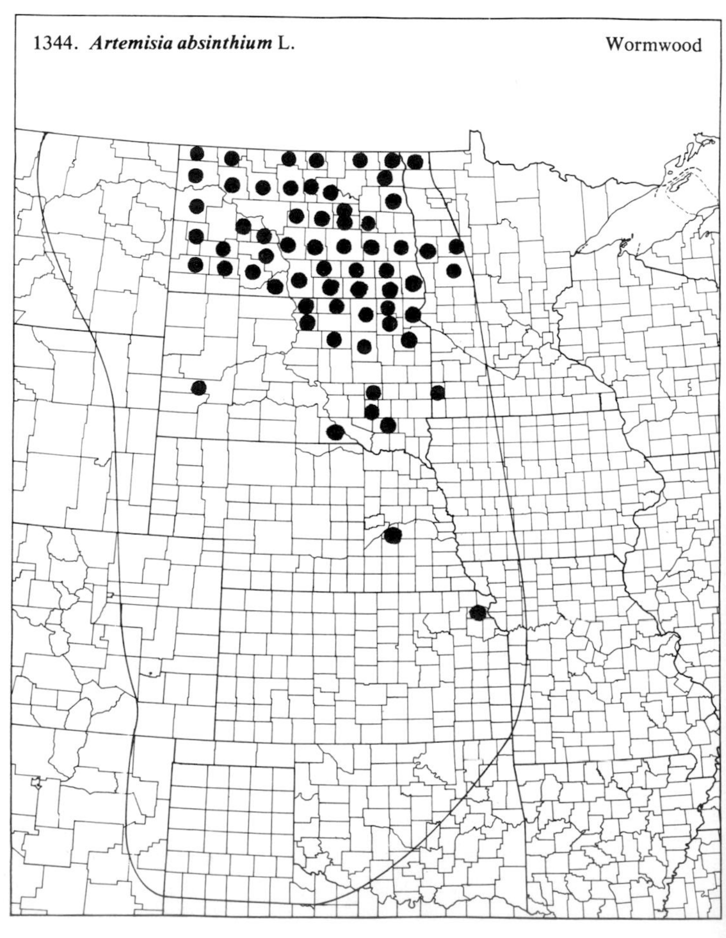

(Asteraceae)

1345. ***Artemisia biennis*** Willd. Biennial Wormwood

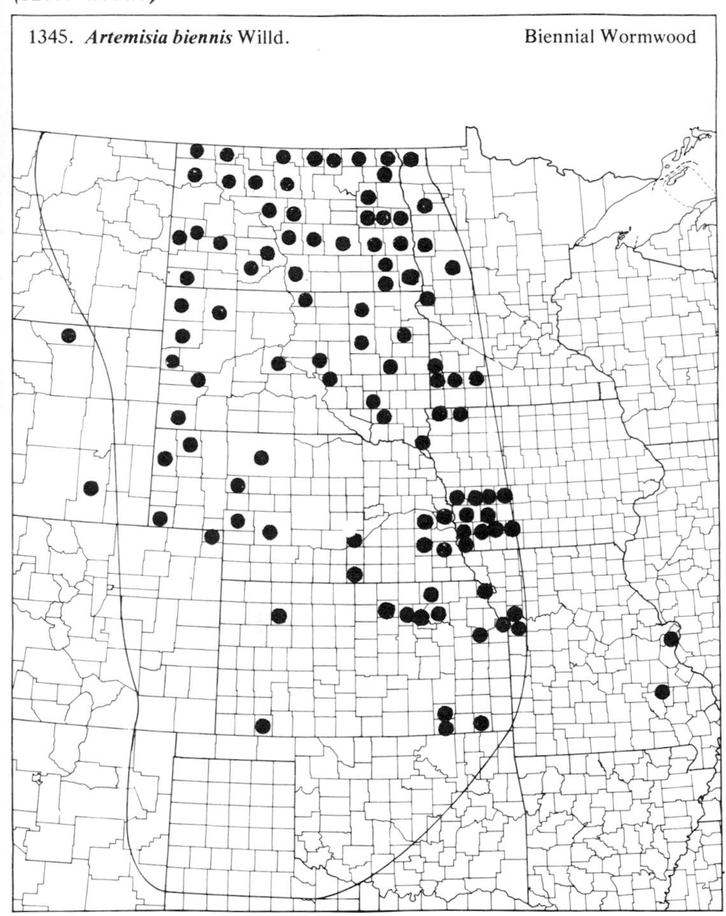

1346. ***Artemisia campestris*** L. ssp. ***caudata*** (Michx.) Hall & Clem. Western Sagebrush

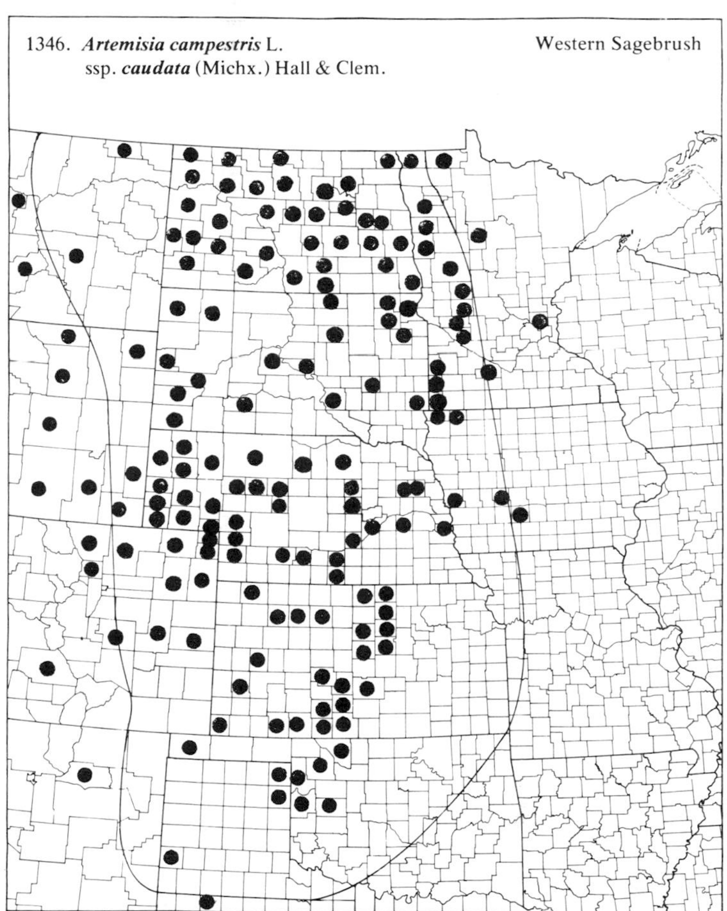

1347. ***Artemisia cana*** Pursh Dwarf Sagebrush

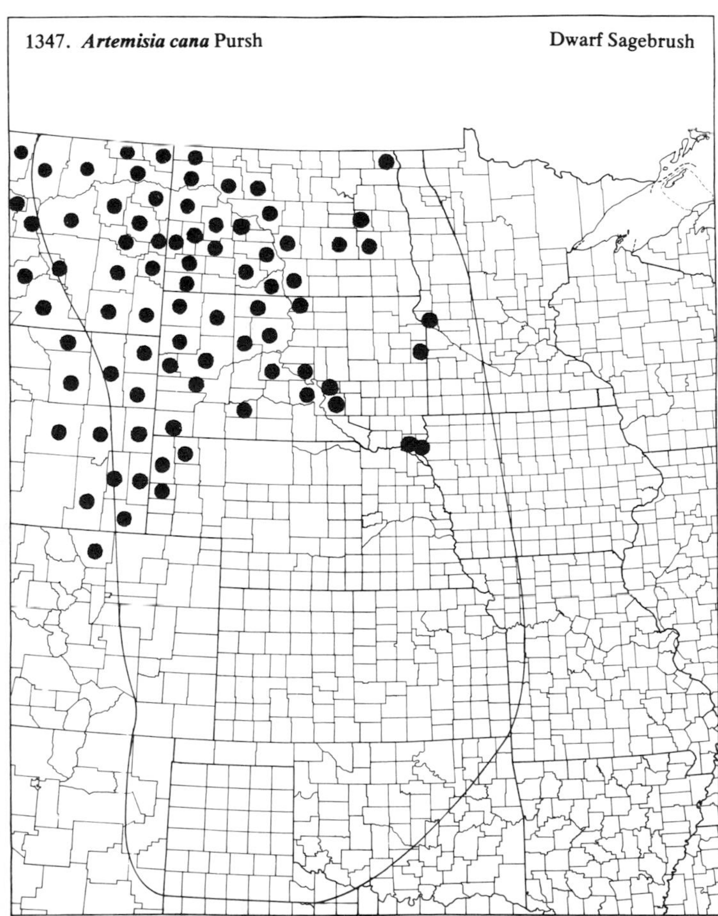

1348. ***Artemisia carruthii*** Wood

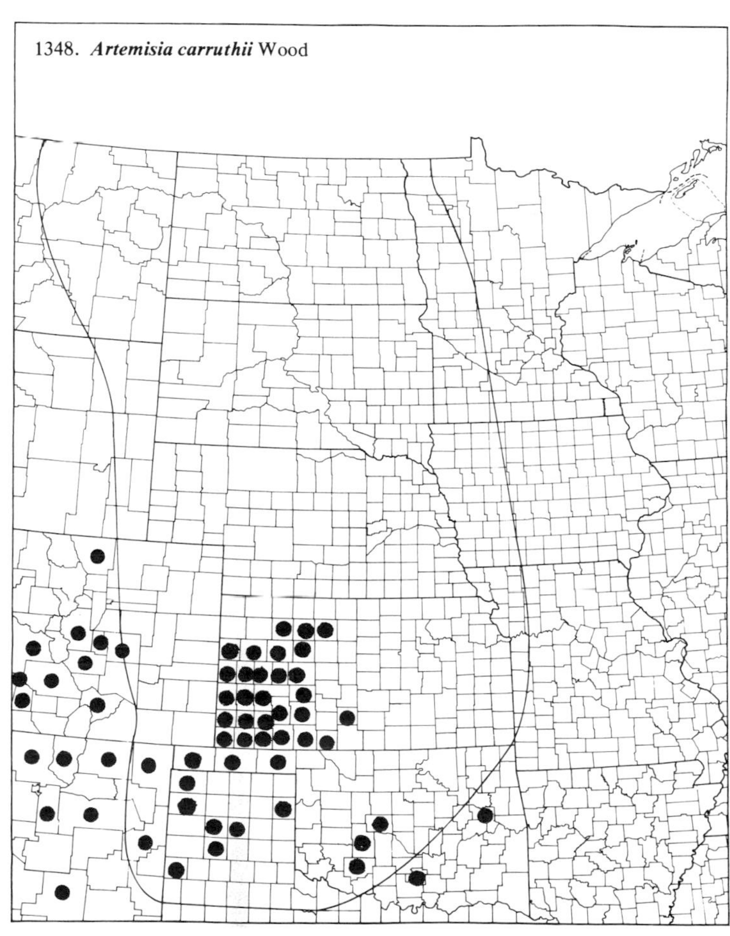

(Asteraceae)

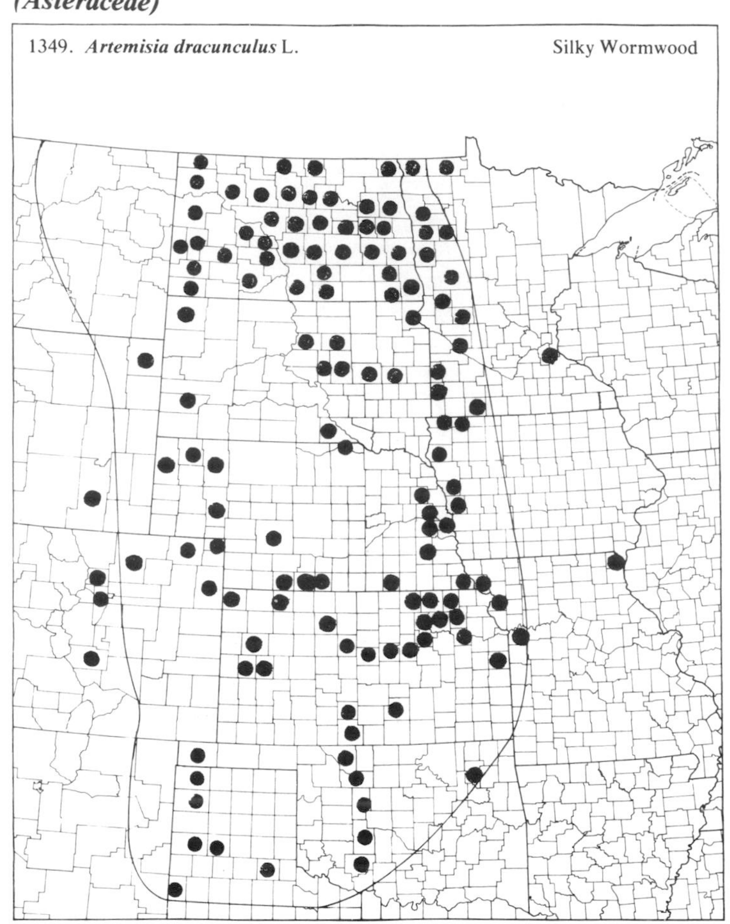
1349. *Artemisia dracunculus* L. Silky Wormwood

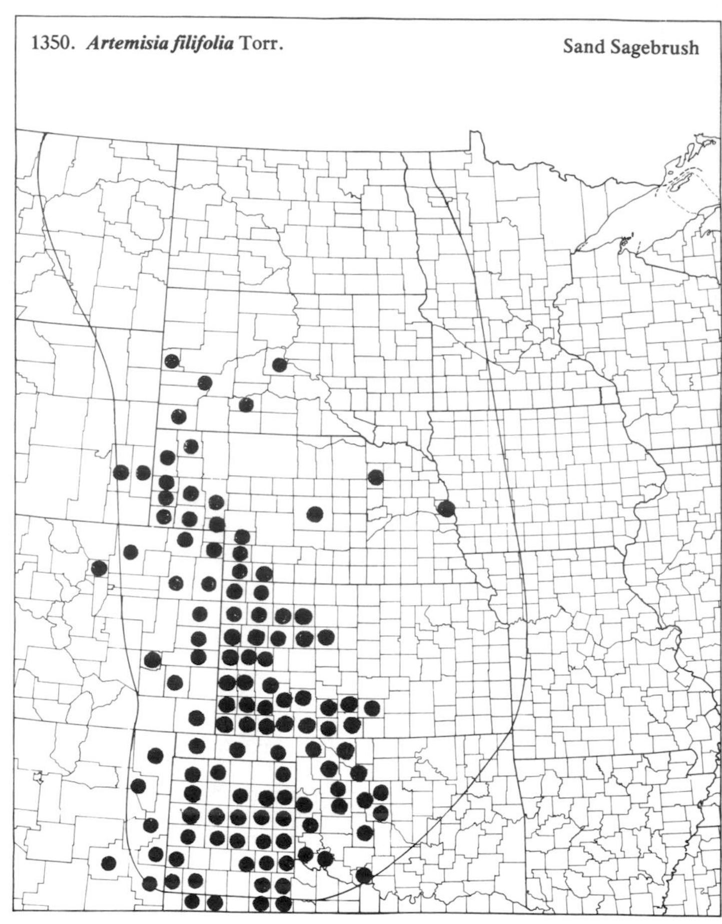
1350. *Artemisia filifolia* Torr. Sand Sagebrush

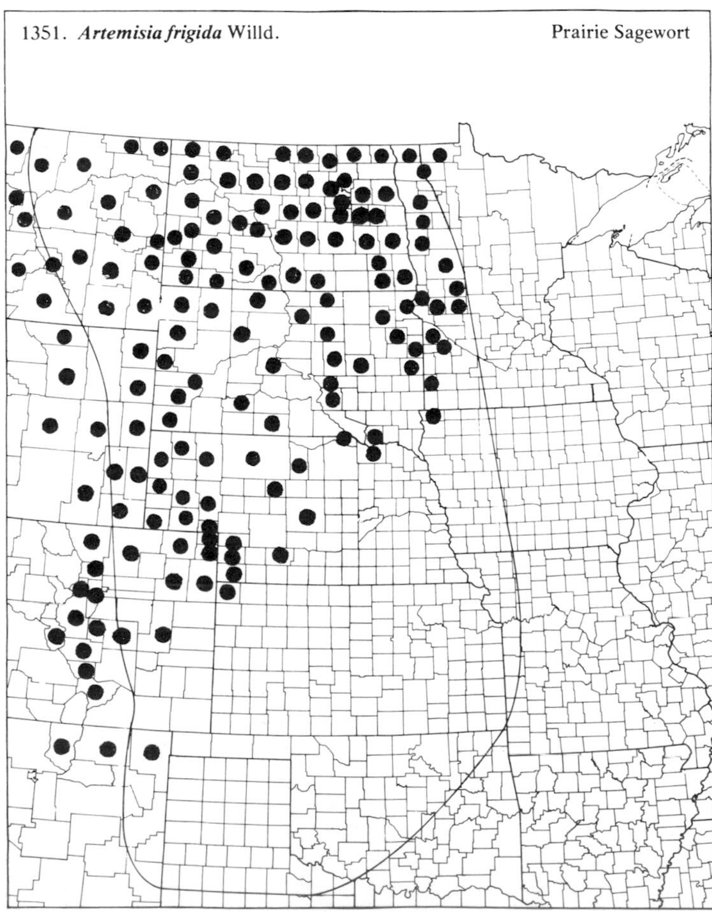
1351. *Artemisia frigida* Willd. Prairie Sagewort

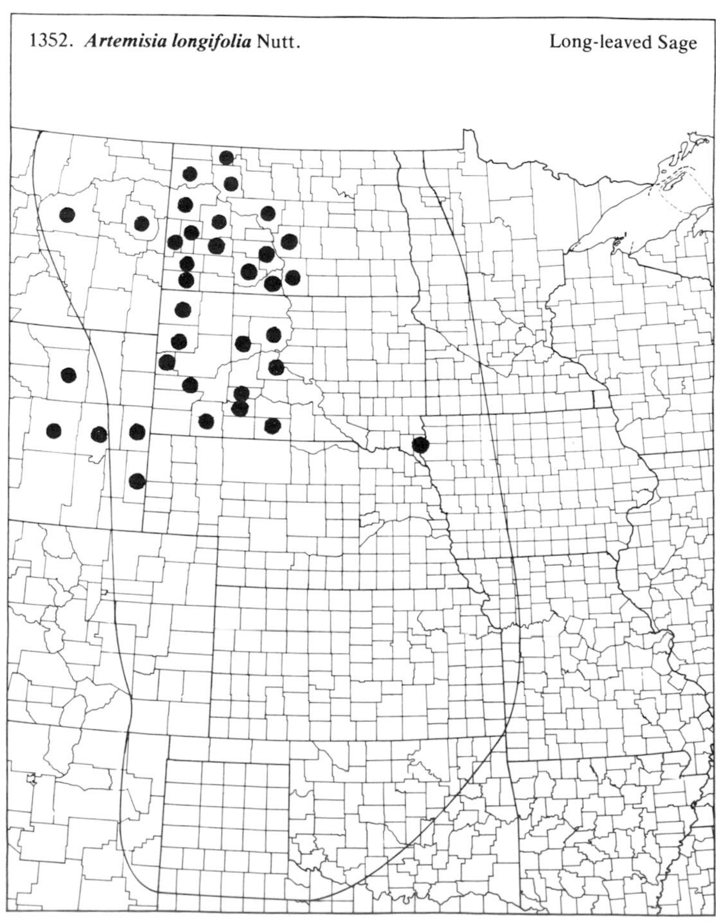
1352. *Artemisia longifolia* Nutt. Long-leaved Sage

(Asteraceae)

1353. *Artemisia ludoviciana* Nutt. var. *ludoviciana* — White Sage

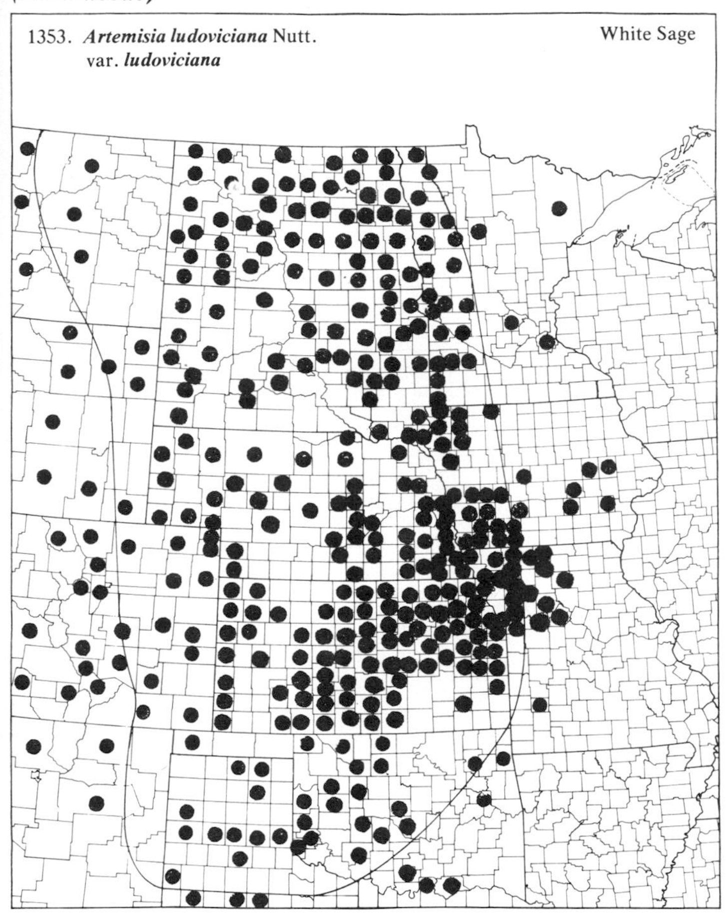

1354. *Artemisia ludoviciana* Nutt. var. *mexicana* (Willd.) Fern. — White Sage

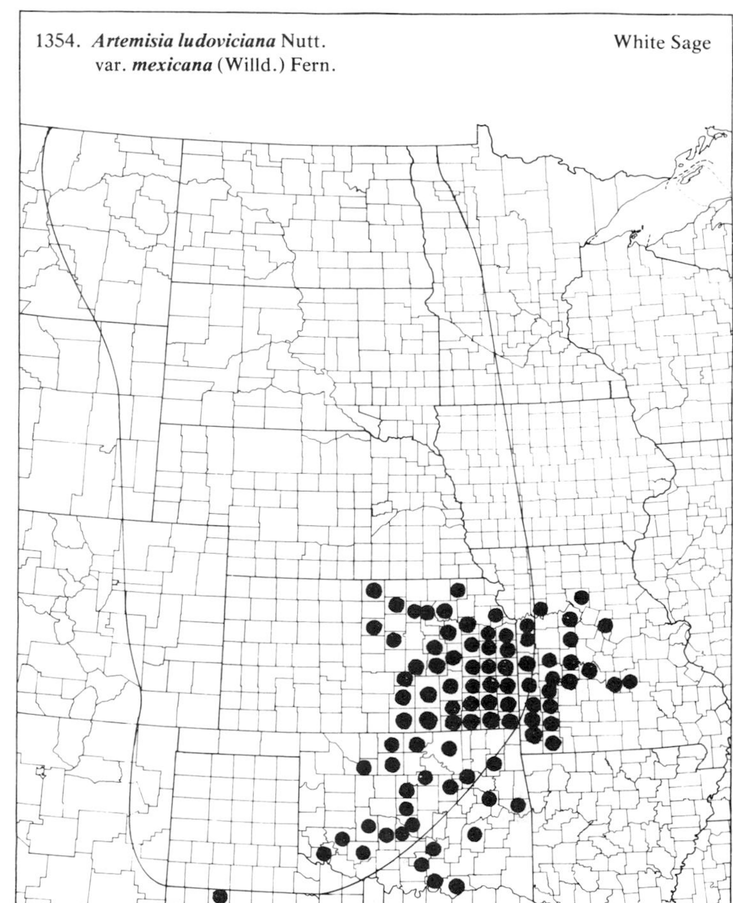

1355. *Artemisia tridentata* Nutt. — Big Sagebrush

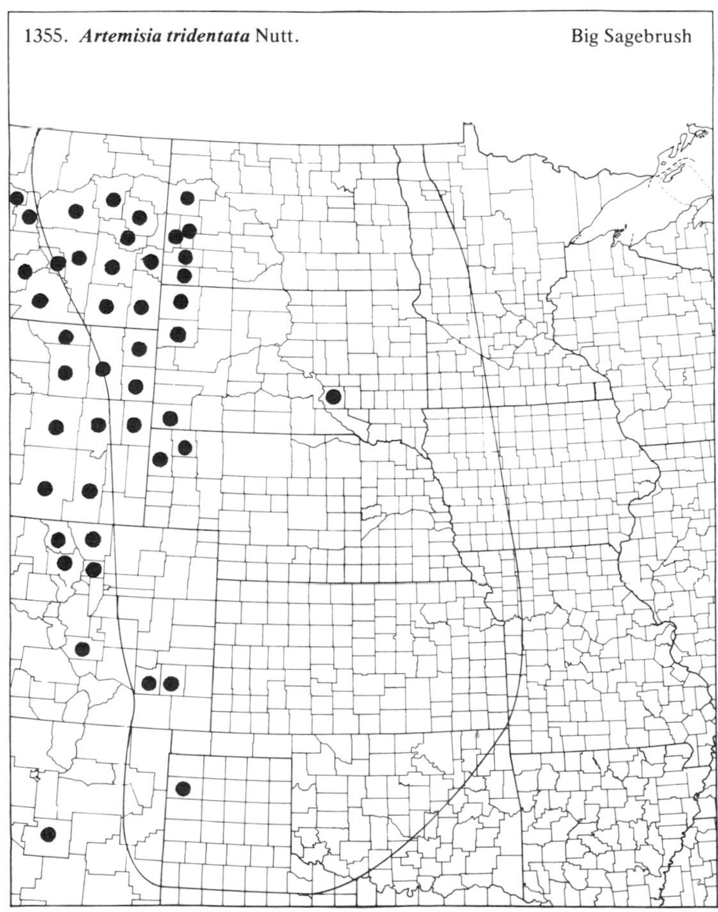

1356. *Aster azureus* Lindl. — Azure Aster

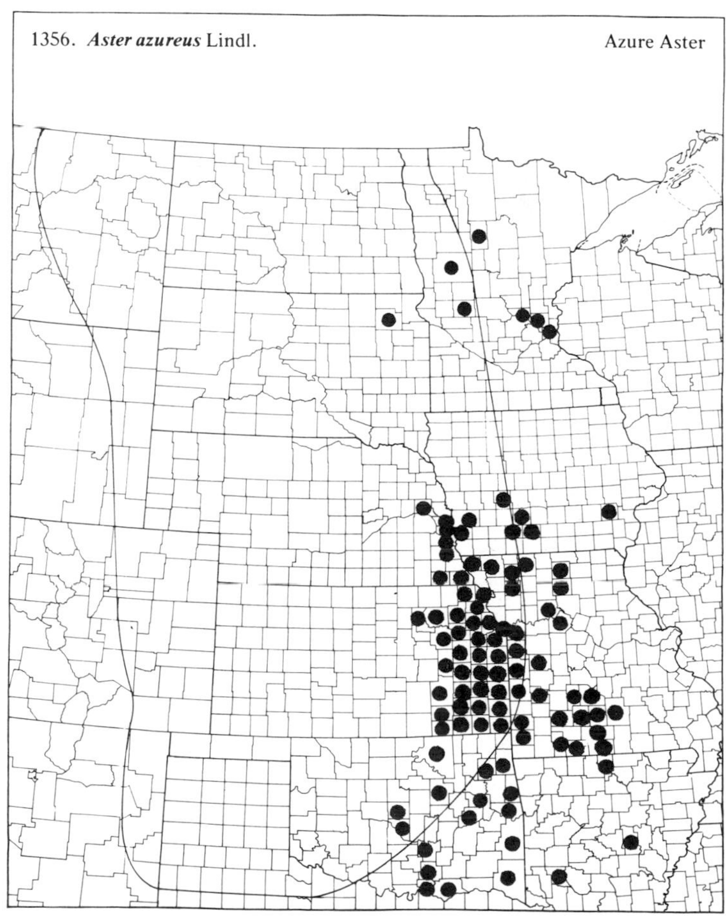

(Asteraceae)

1357. ***Aster brachyactis*** Blake — Rayless Aster

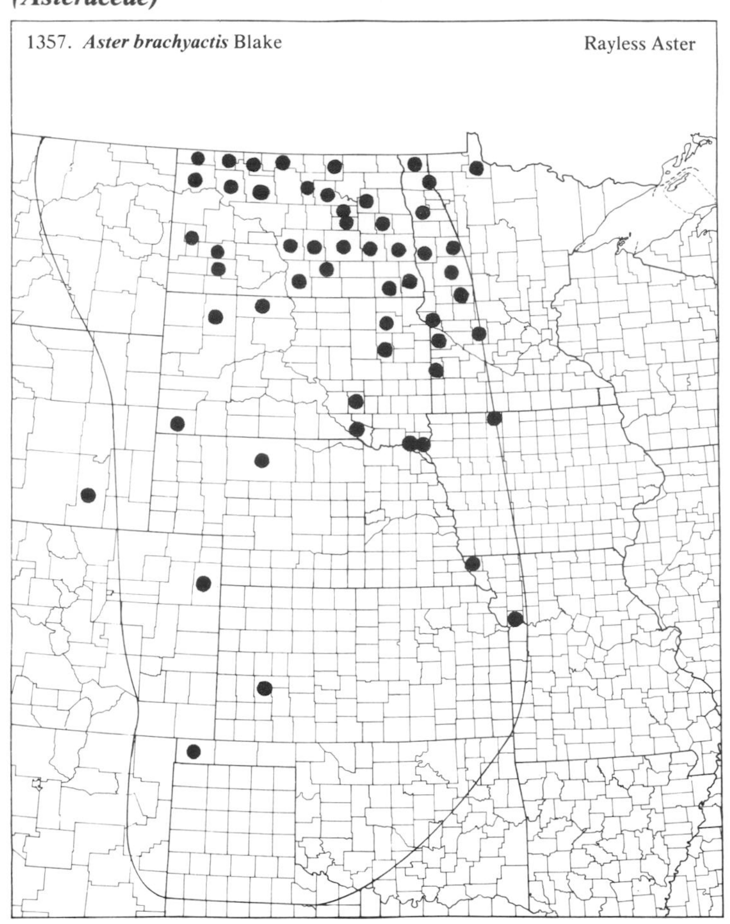

1358. ***Aster ciliolatus*** Lindl.

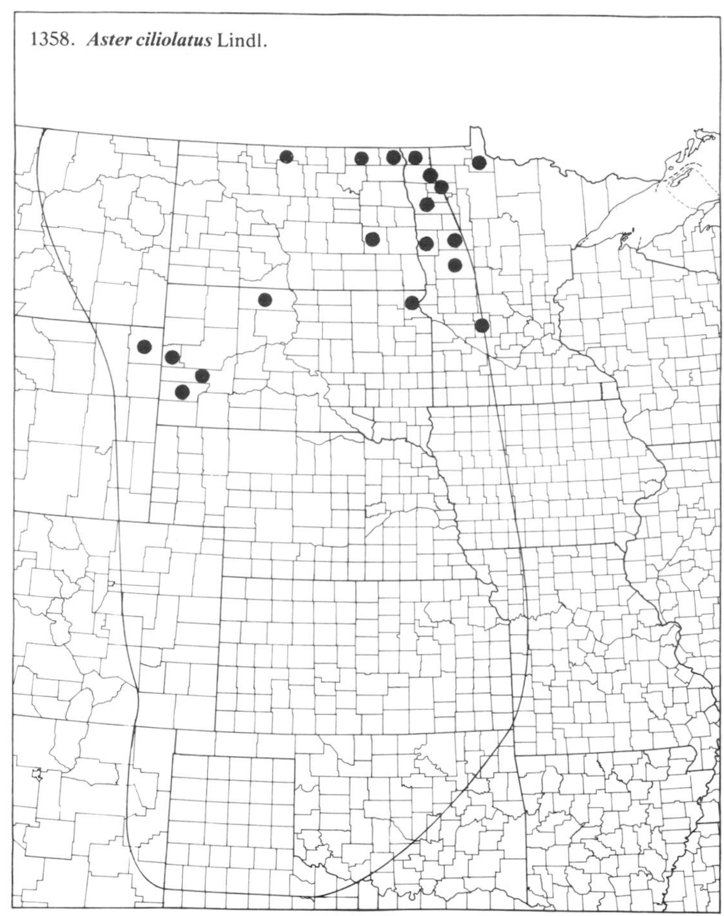

1359. ***Aster drummondii*** Lindl. — Drummond Aster

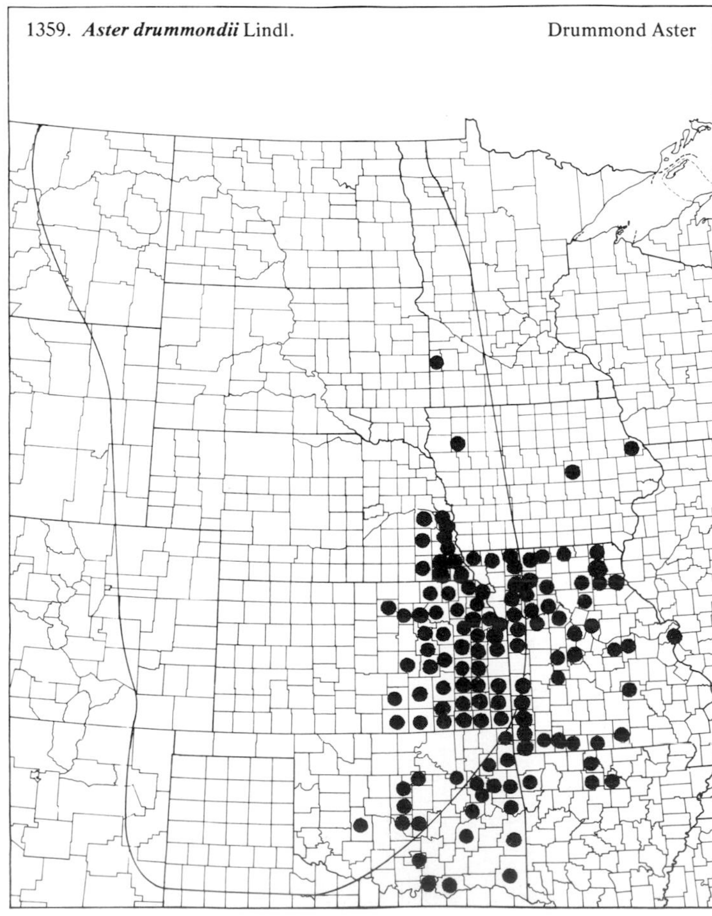

1360. ***Aster ericoides*** L. — White Aster

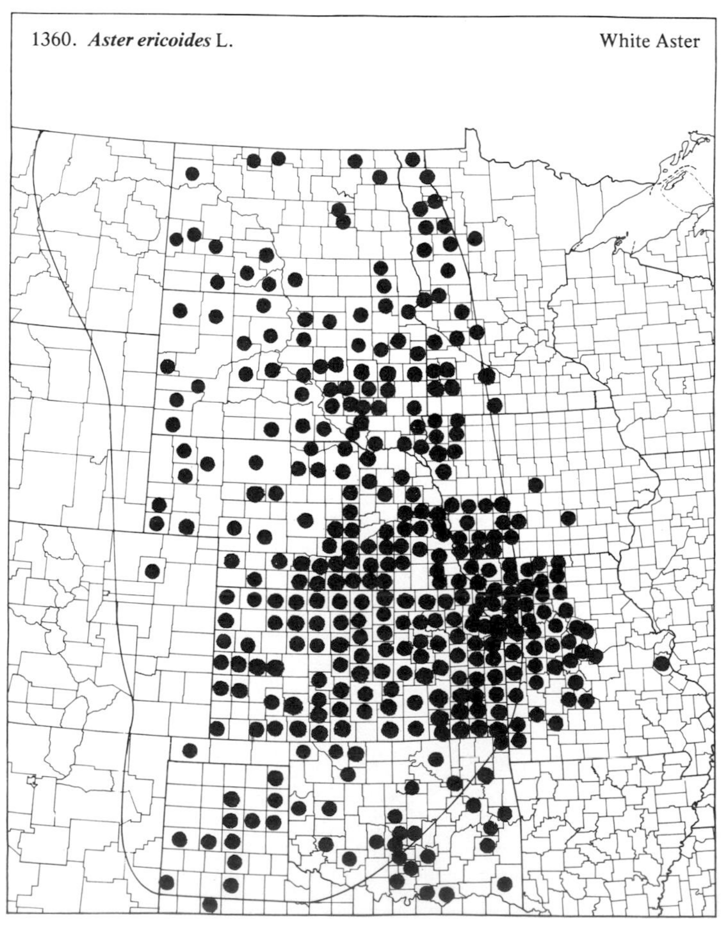

(Asteraceae)

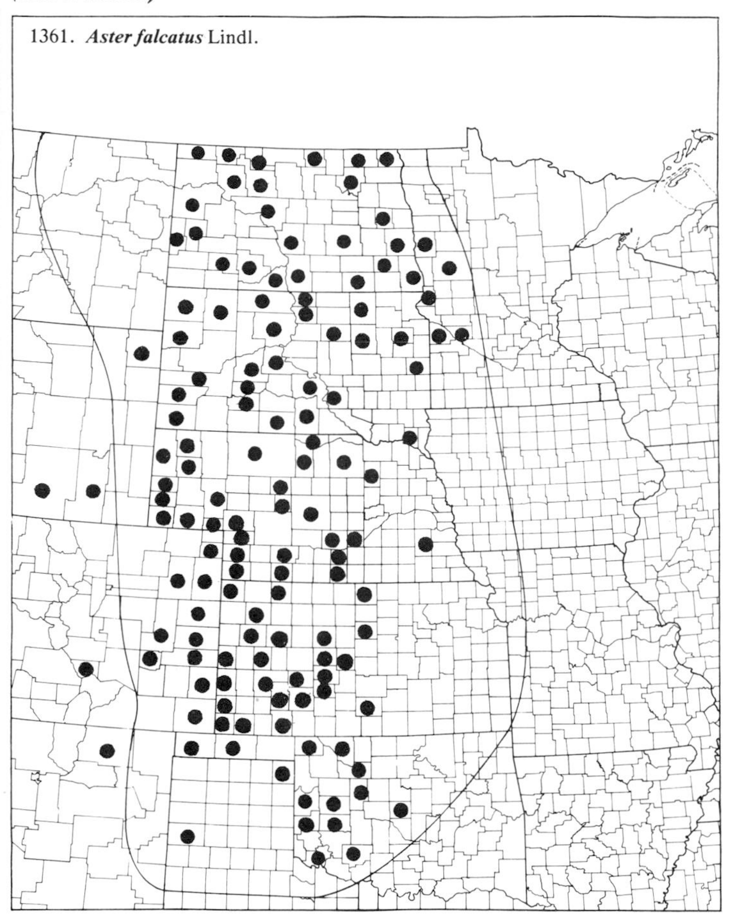
1361. Aster falcatus Lindl.

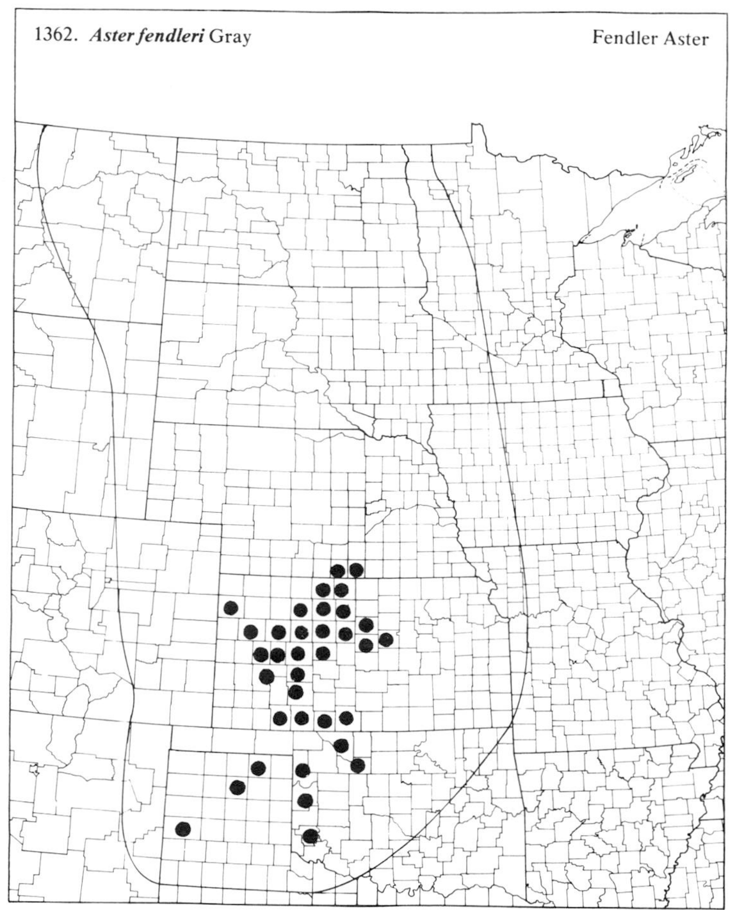
1362. Aster fendleri Gray
Fendler Aster

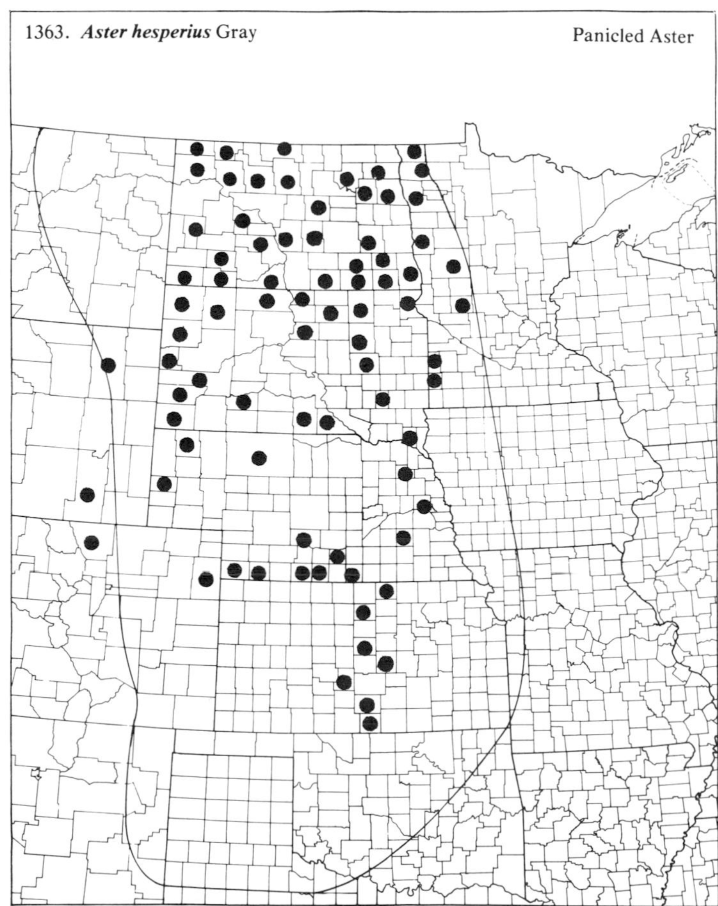
1363. Aster hesperius Gray
Panicled Aster

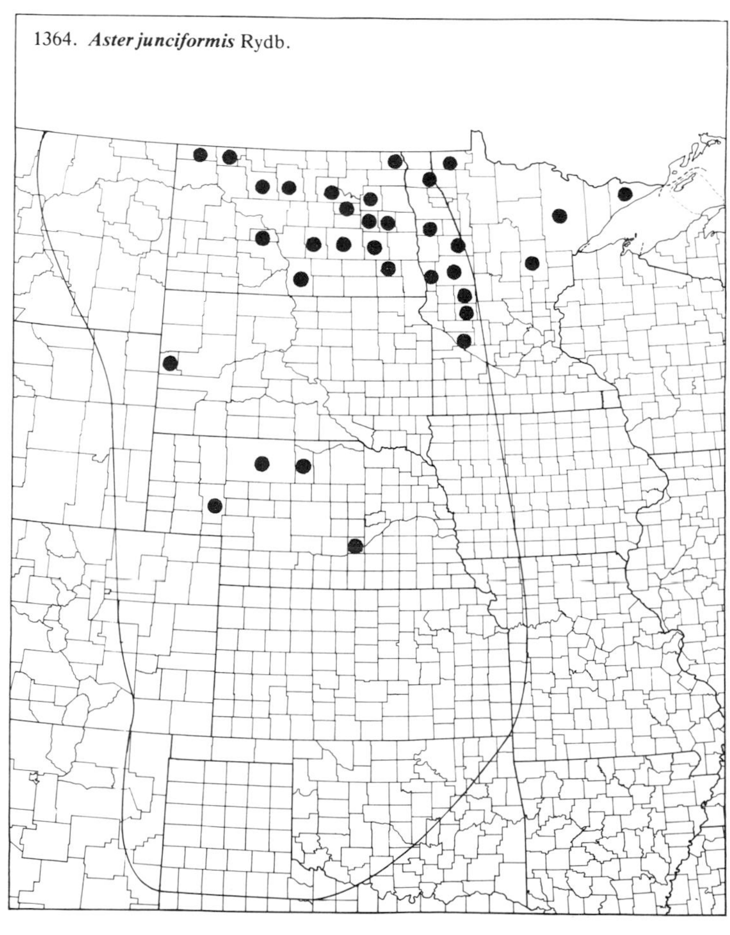
1364. Aster junciformis Rydb.

(Asteraceae)

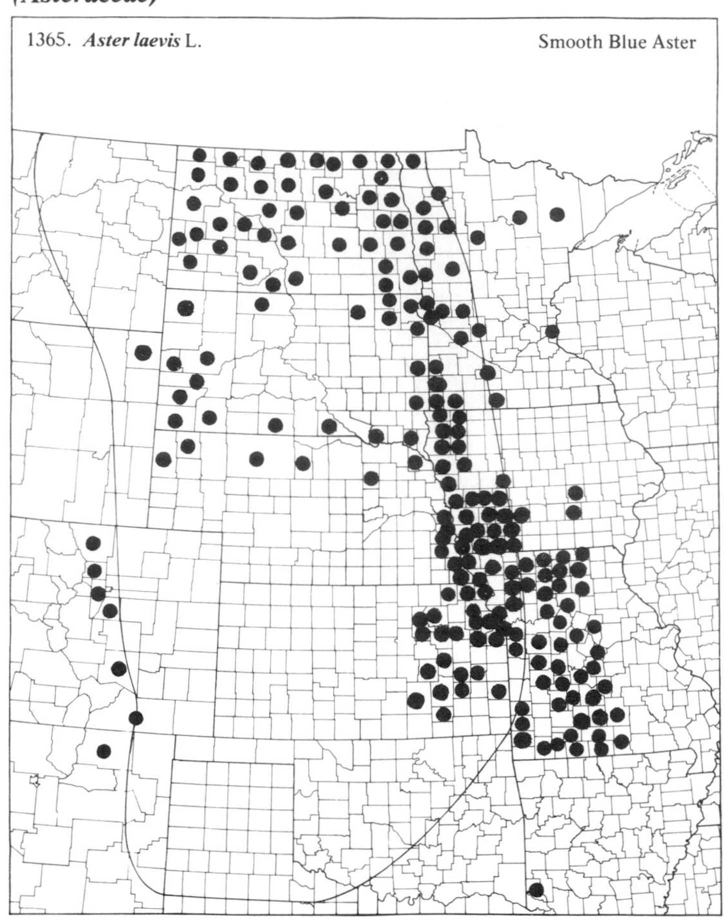
1365. **Aster laevis** L.
Smooth Blue Aster

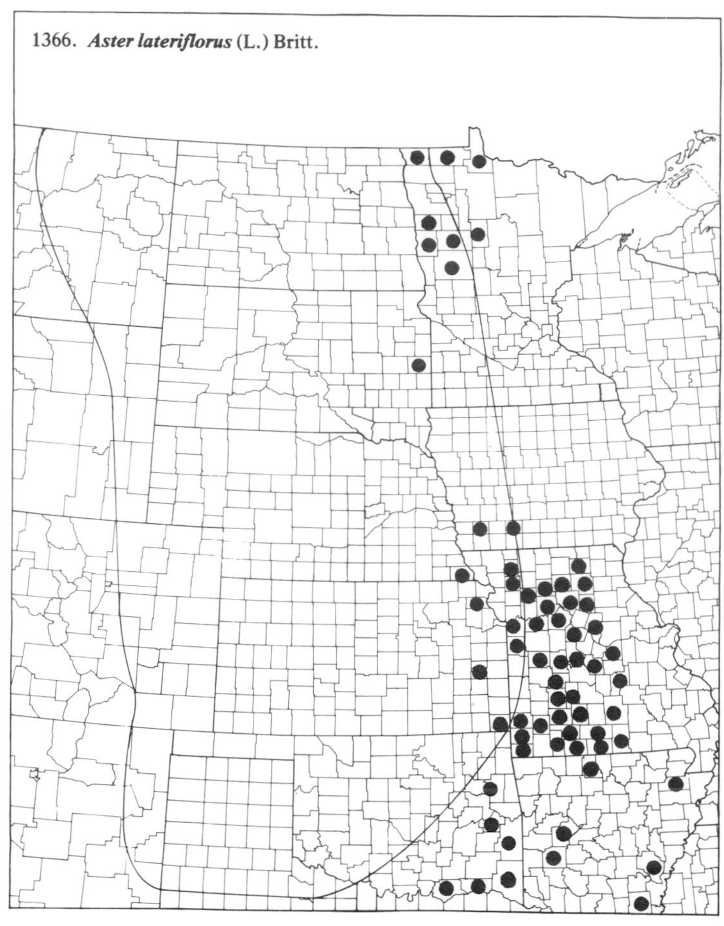
1366. **Aster lateriflorus** (L.) Britt.

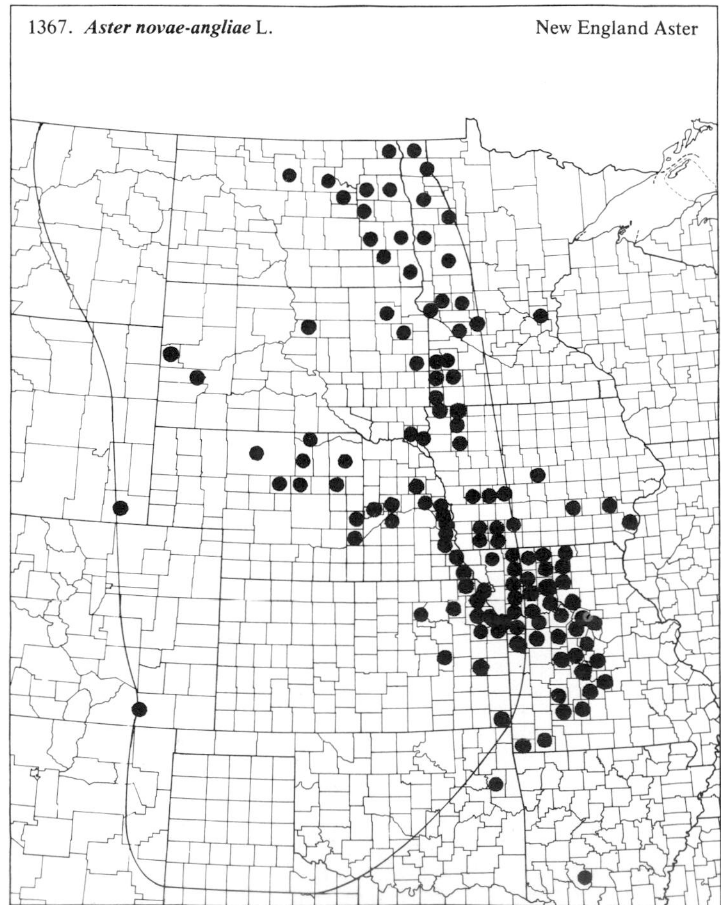
1367. **Aster novae-angliae** L.
New England Aster

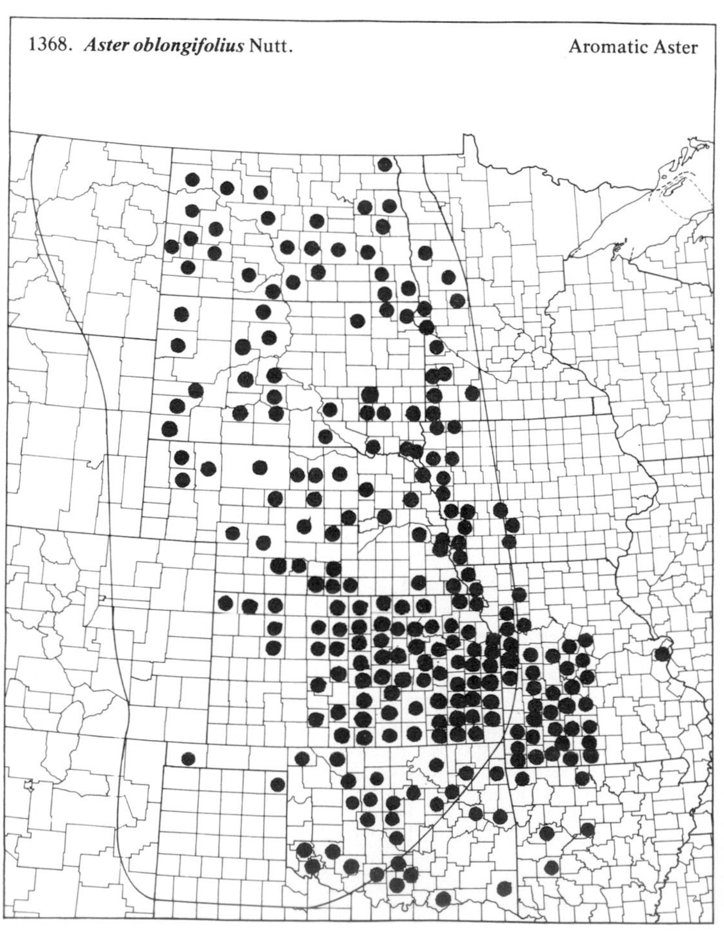
1368. **Aster oblongifolius** Nutt.
Aromatic Aster

(Asteraceae)

1369. ***Aster ontarionis*** Wieg.

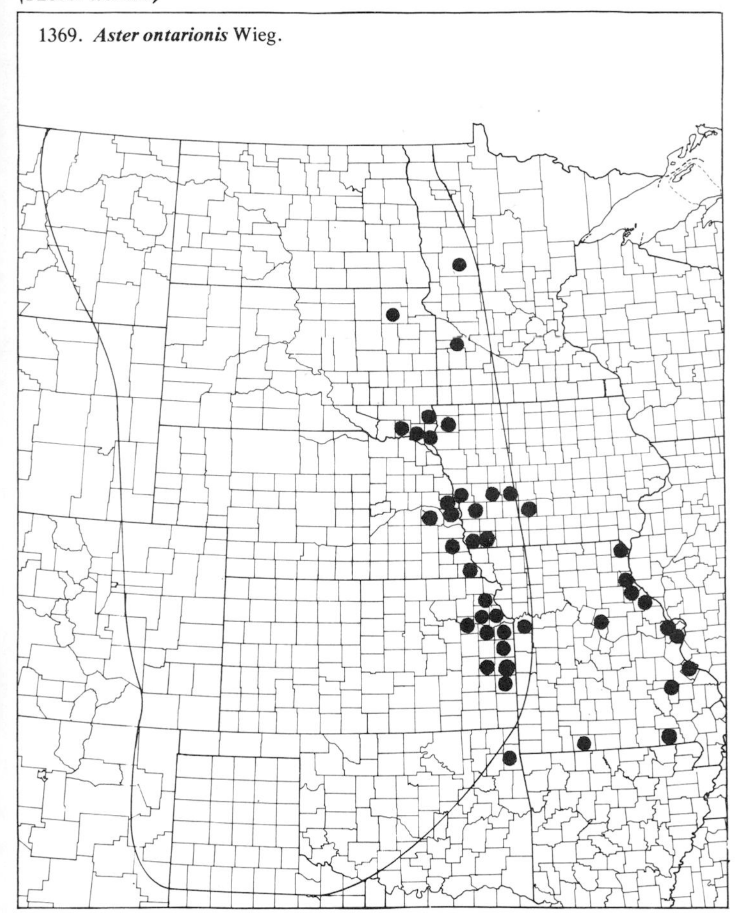

1370. ***Aster pansus*** (Blake) Cronq.

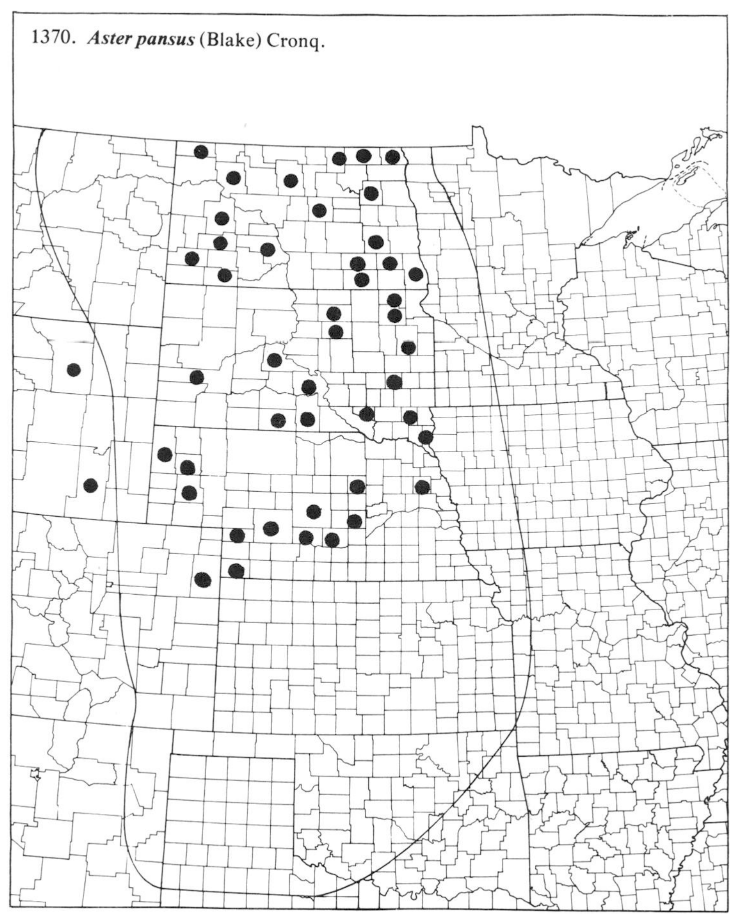

1371. ***Aster parviceps*** (Burgess) Mack. & Bush

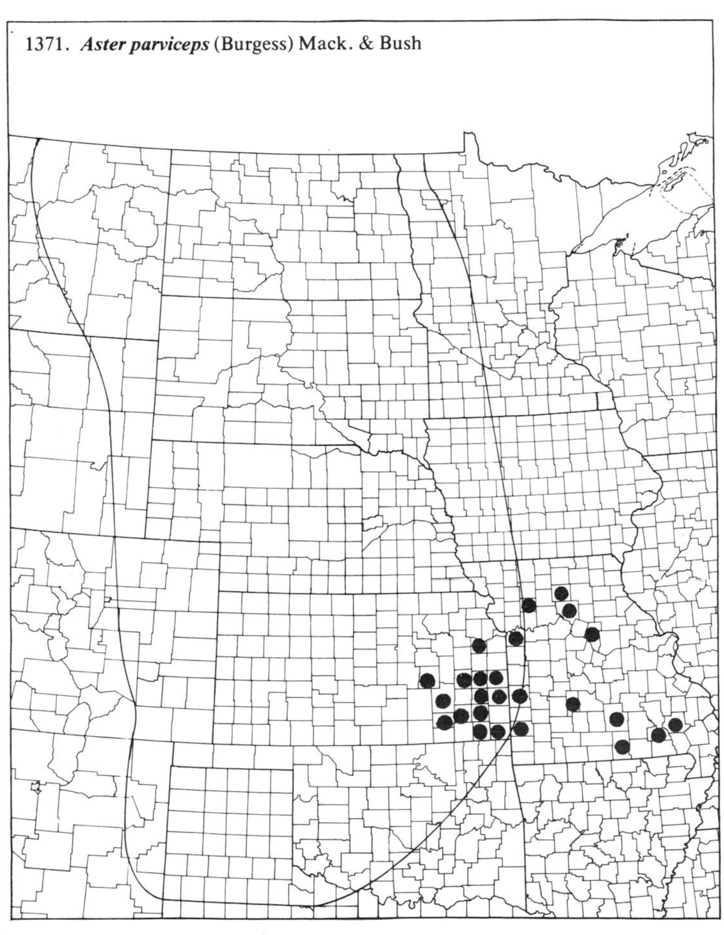

1372. ***Aster patens*** Ait.
var. ***patentissimus*** (Lindl.) T. & G.

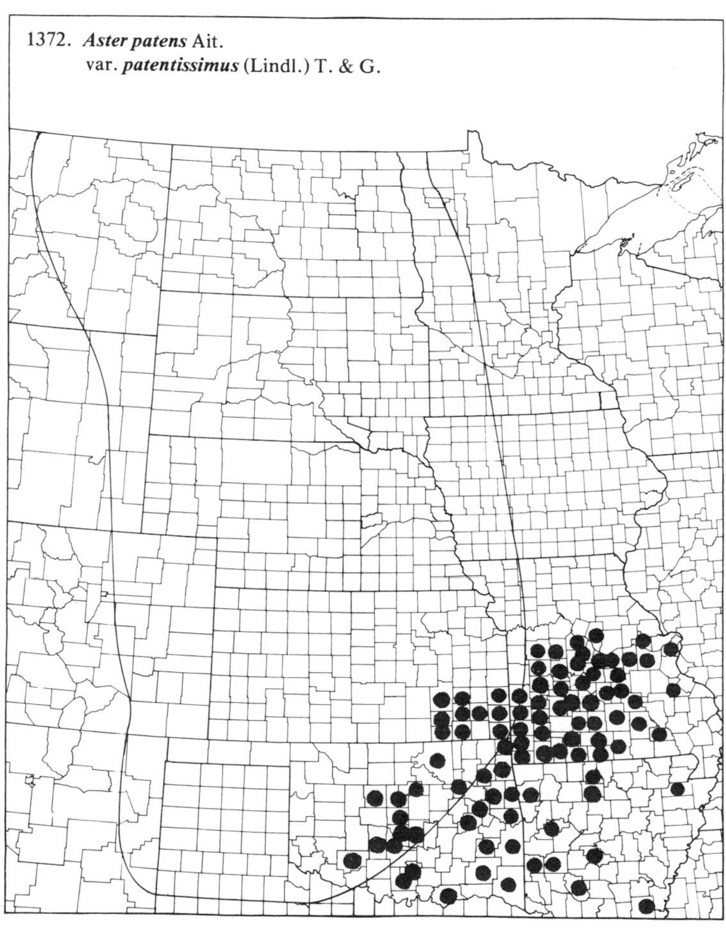

(Asteraceae)

1373. ***Aster pauciflorus*** Nutt. Few-flowered Aster

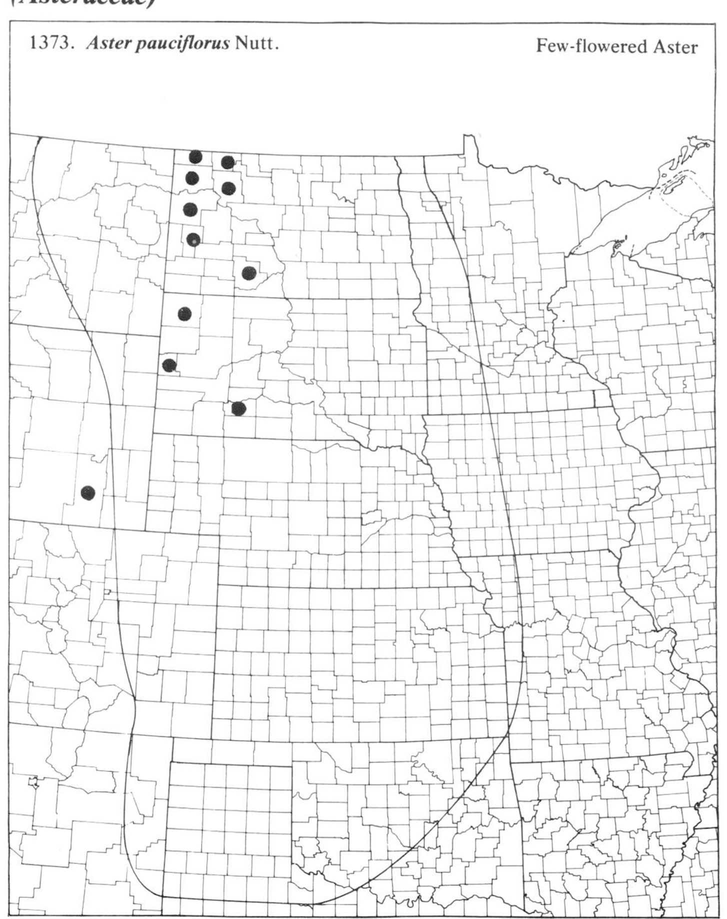

1374. ***Aster pilosus*** Willd.

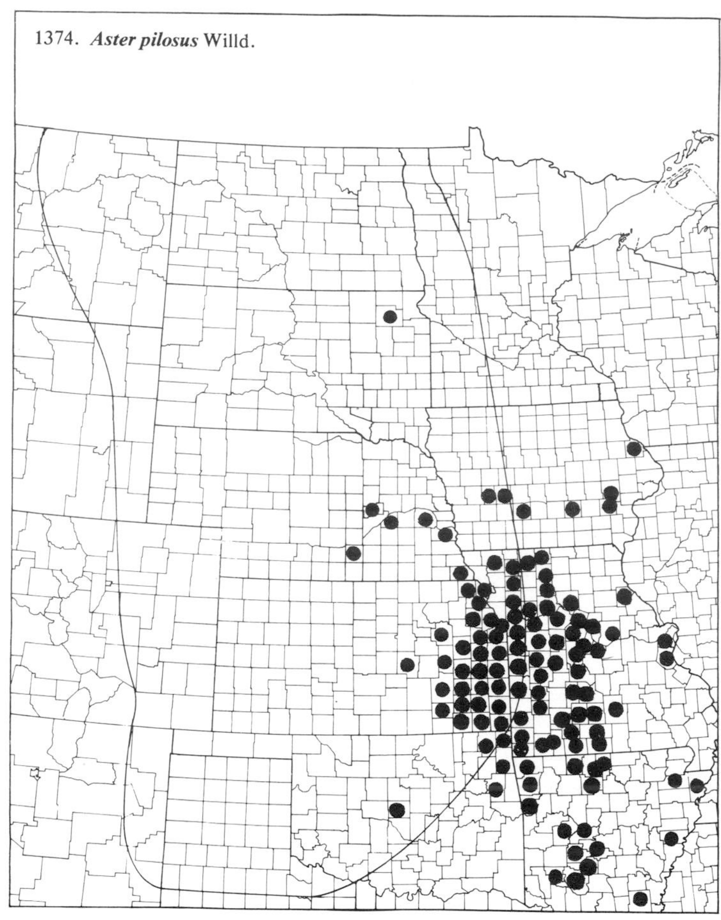

1375. ***Aster praealtus*** Poir. var. ***praealtus*** Willowleaf Aster

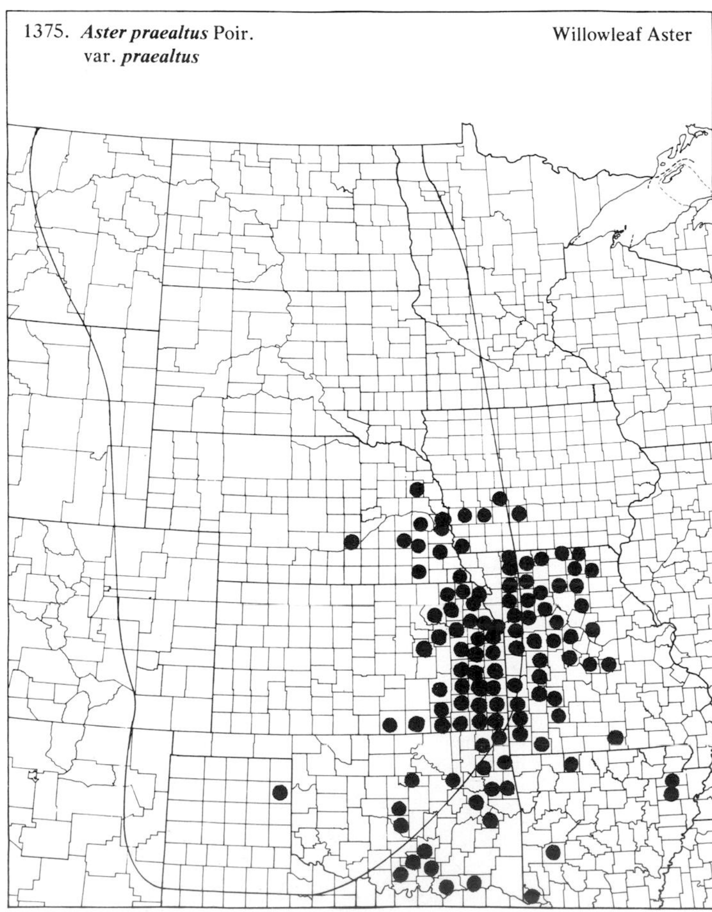

1376. ***Aster praealtus*** Poir. var. ***nebraskensis*** (Britt.) Wieg. Willowleaf Aster

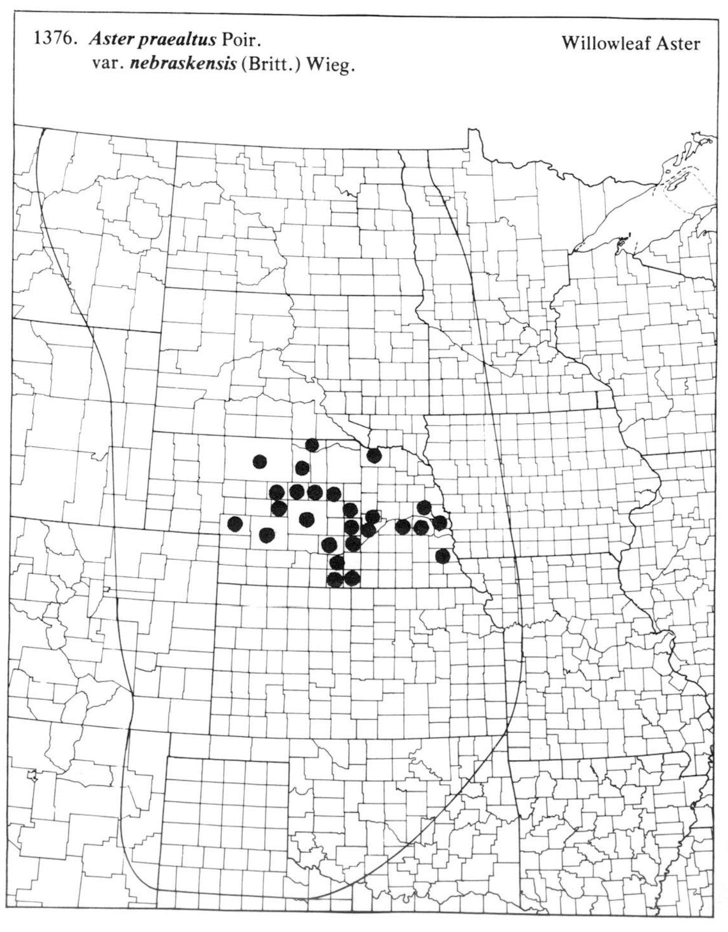

(Asteraceae)

1377. ***Aster ptarmicoides*** (Nees) T. & G. Sneezewort Aster

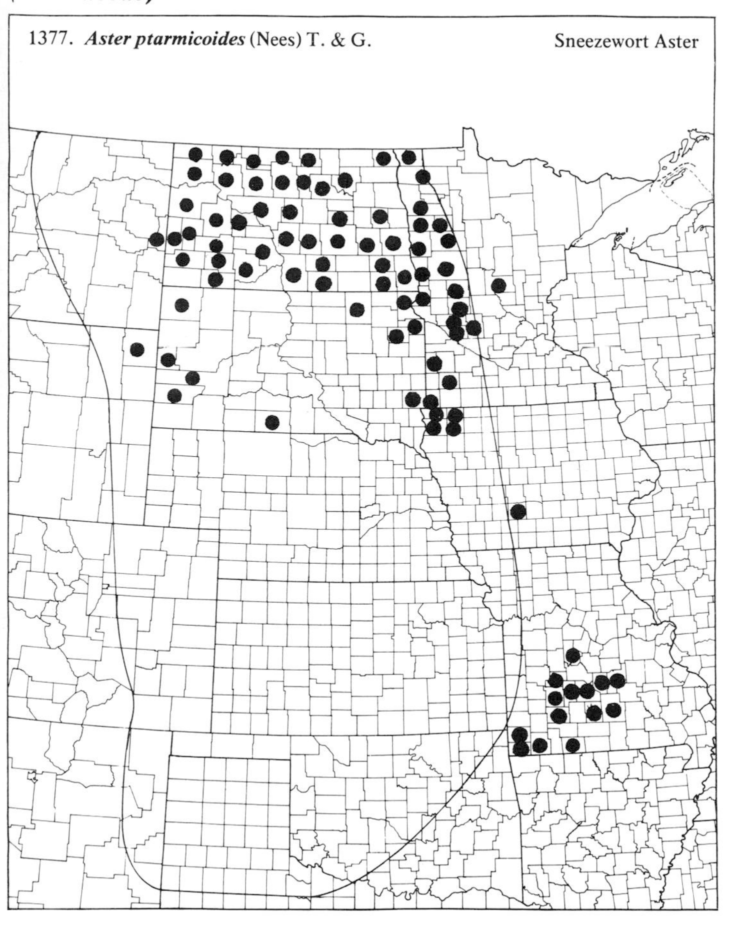

1378. ***Aster pubentior*** Cronq.

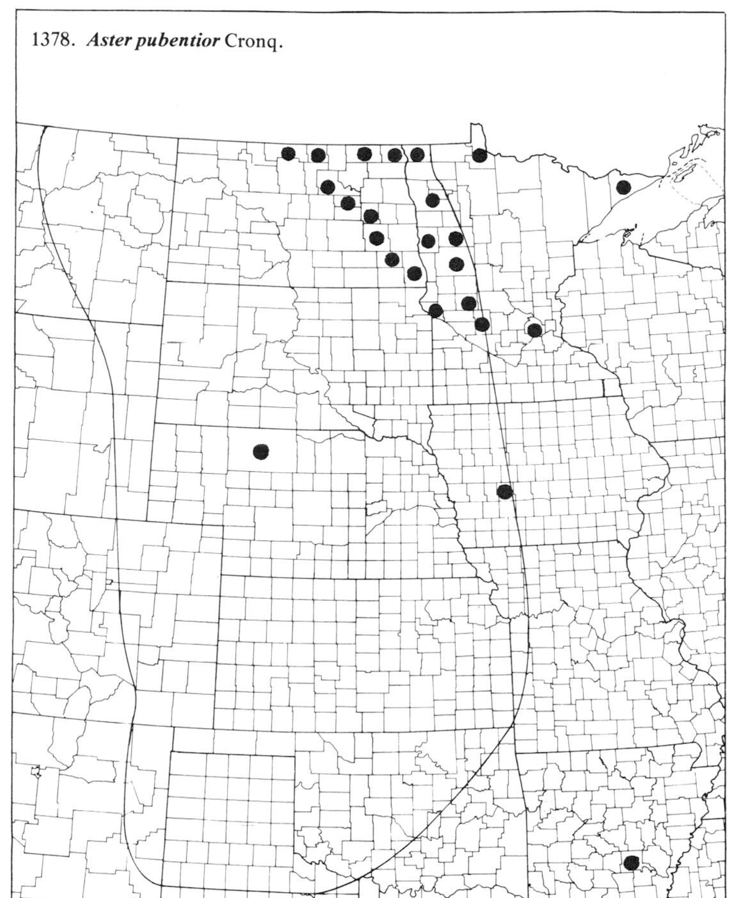

1379. ***Aster puniceus*** L. Swamp Aster

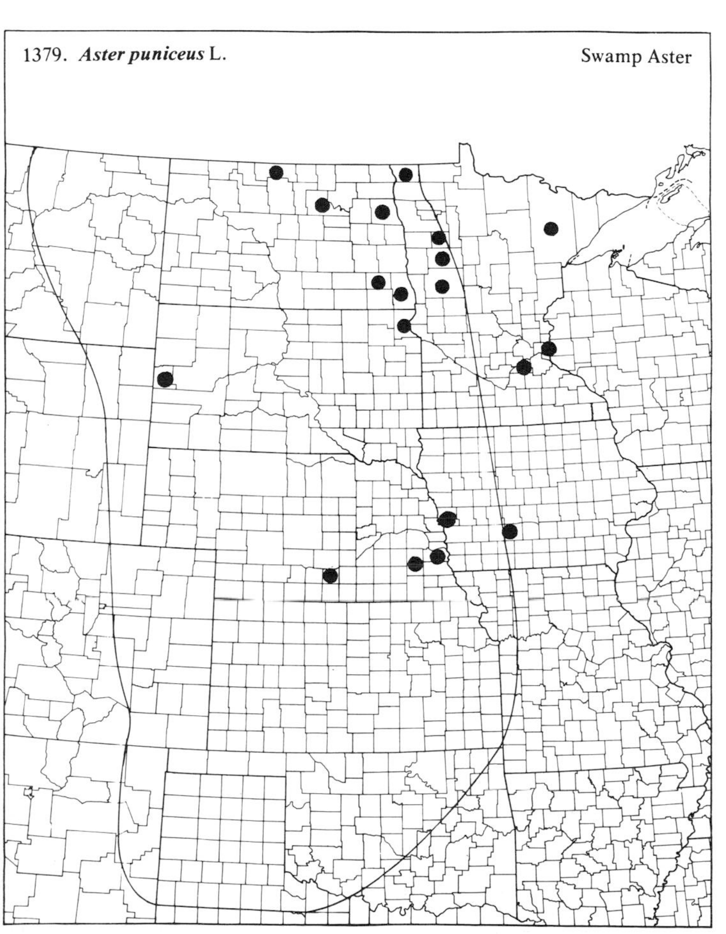

1380. ***Aster sagittifolius*** Willd. Arrow-leaved Aster

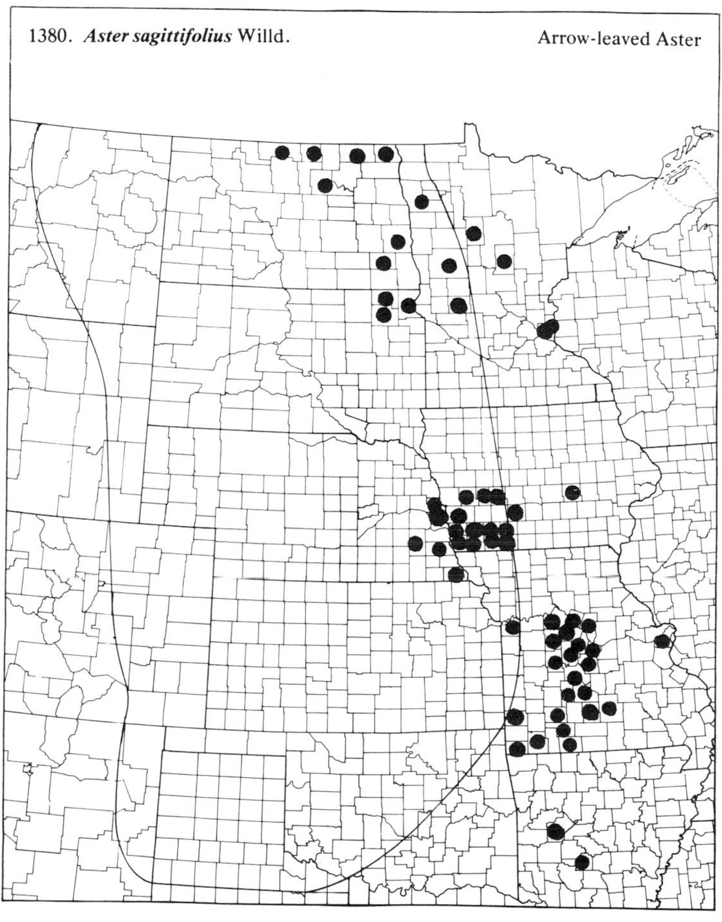

(Asteraceae)

1381. ***Aster sericeus*** Vent. Silky Aster

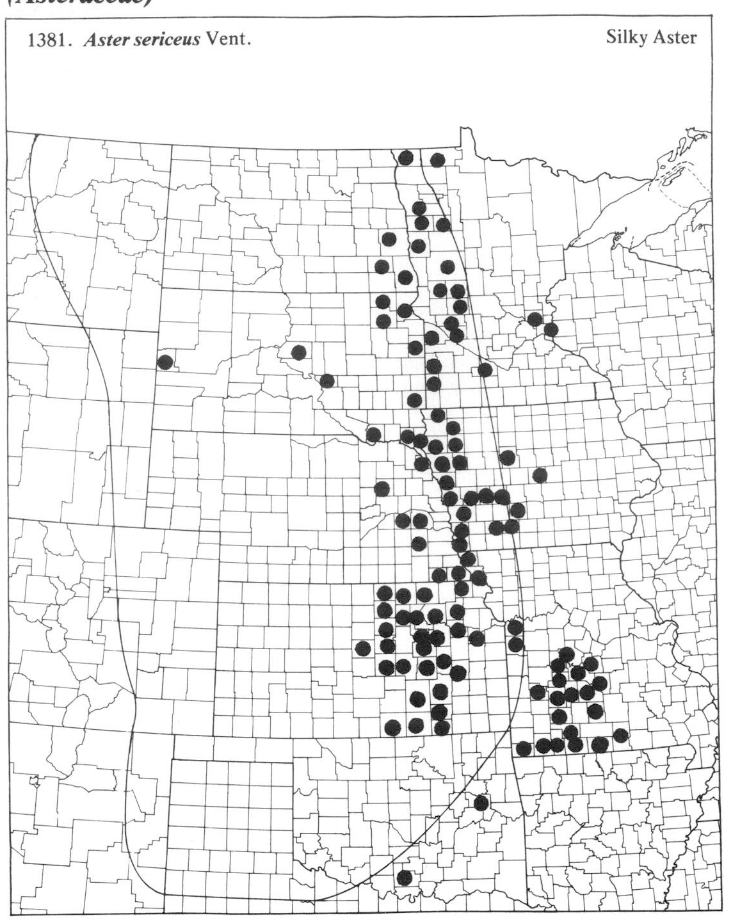

1382. ***Aster simplex*** Willd. var. ***simplex*** Panicled Aster

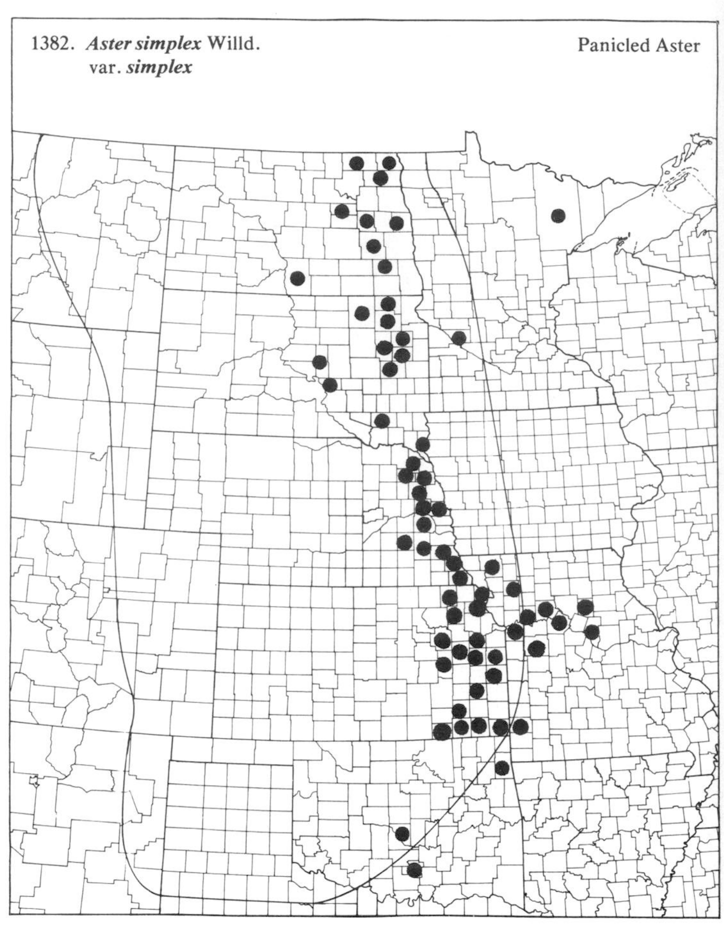

1383. ***Aster simplex*** Willd. var. ***interior*** (Wieg.) Cronq. Panicled Aster

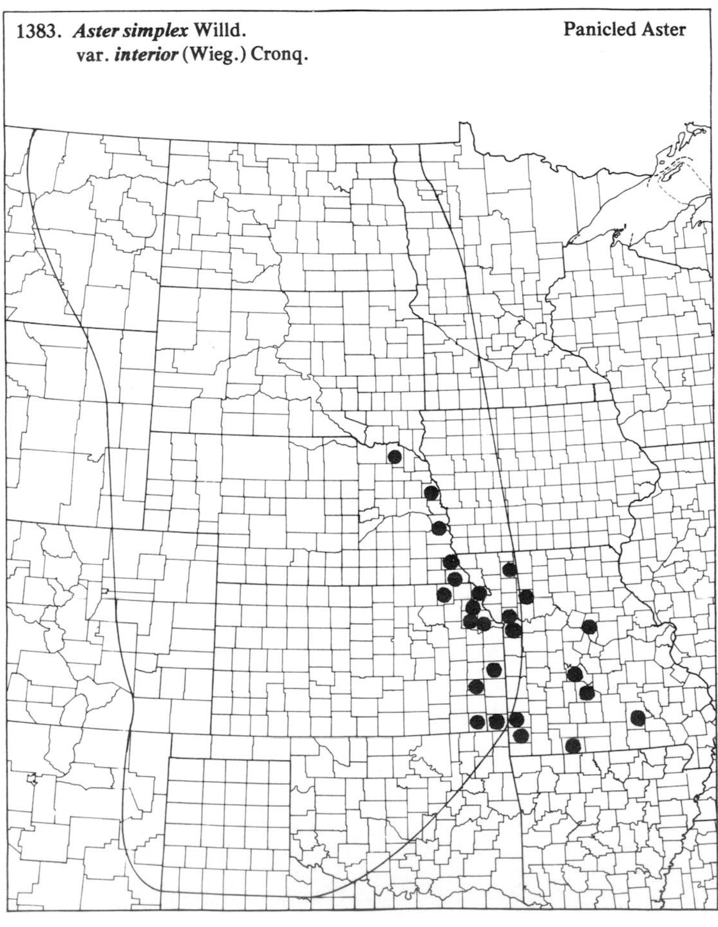

1384. ***Aster simplex*** Willd. var. ***ramosissimus*** (T. & G.) Cronq. Panicled Aster

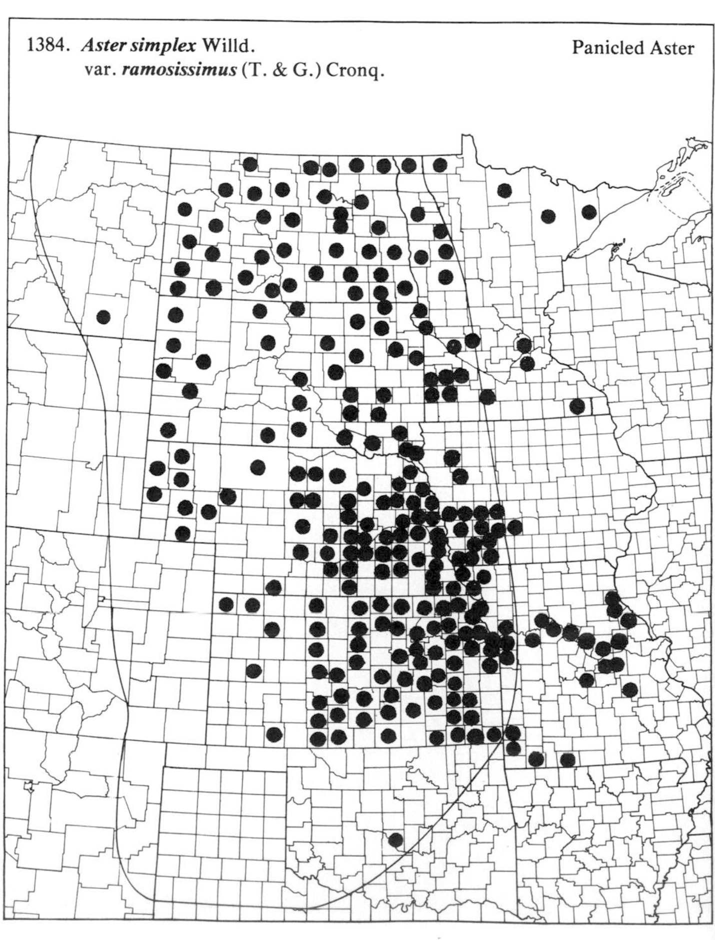

(Asteraceae)

1385. ***Aster subulatus*** Michx. var. ***ligulatus*** Shinners

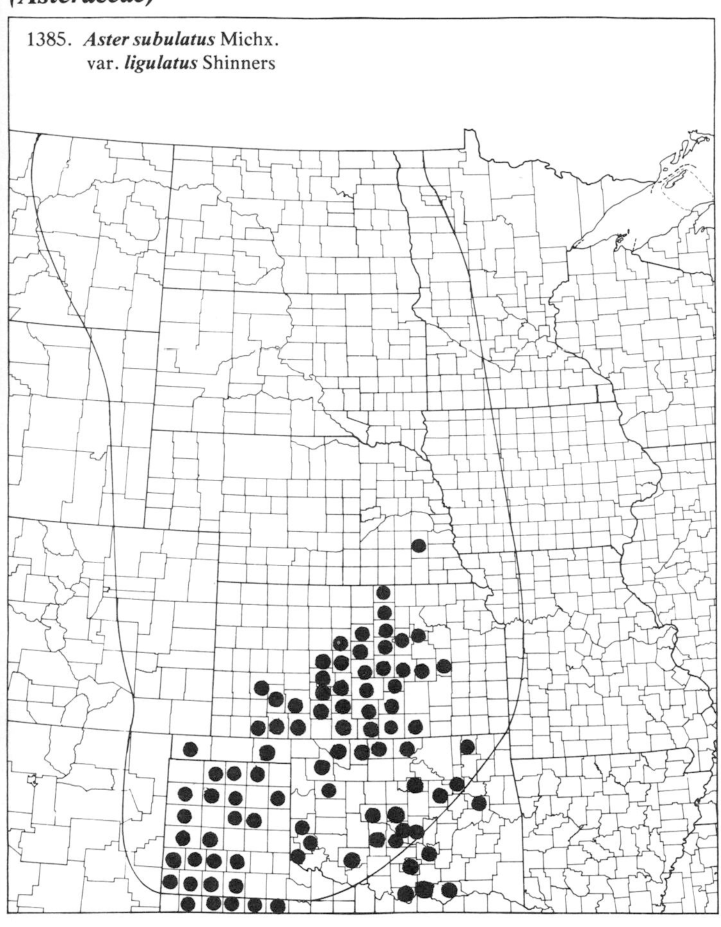

1386. ***Astranthium integrifolium*** (Michx.) Nutt. ssp. ***ciliatum*** (Raf.) DeJong — Western Daisy

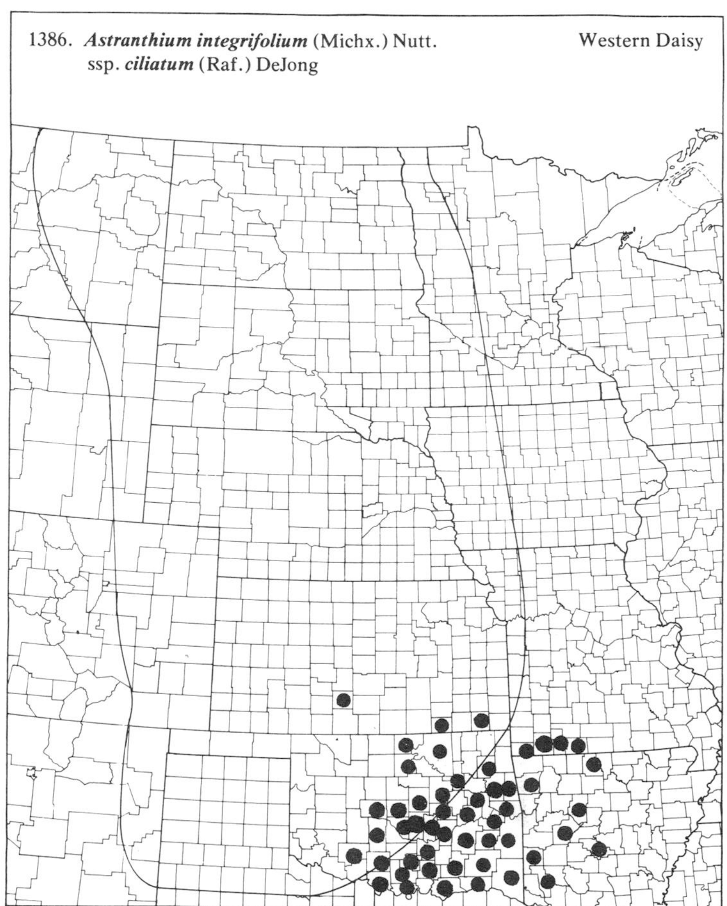

1387. ***Baccharis salicina*** T. & G. — Willow Baccharis

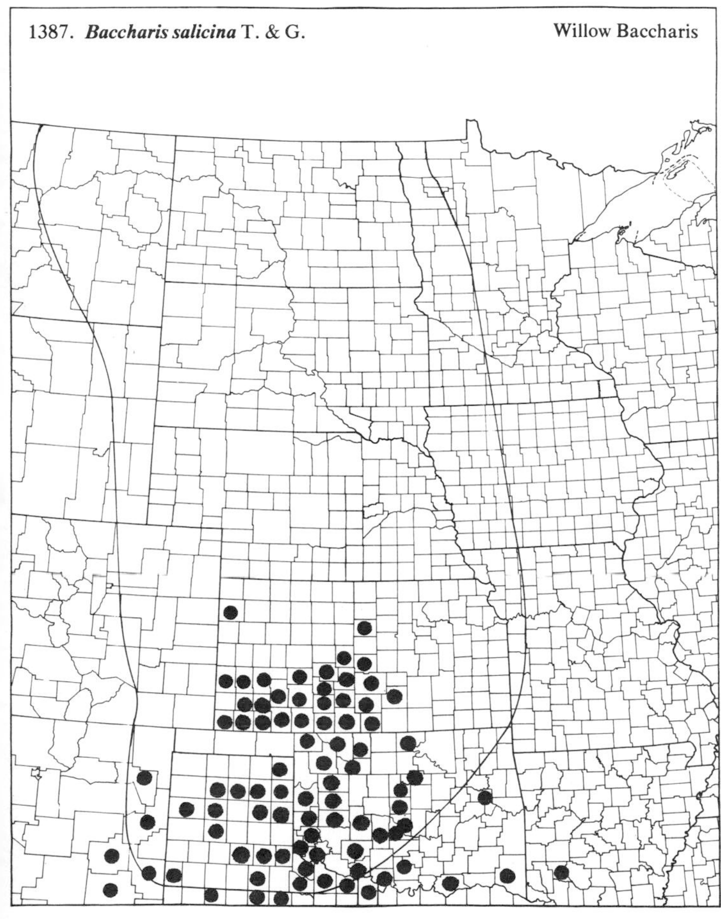

1388. ***Baccharis wrightii*** Gray

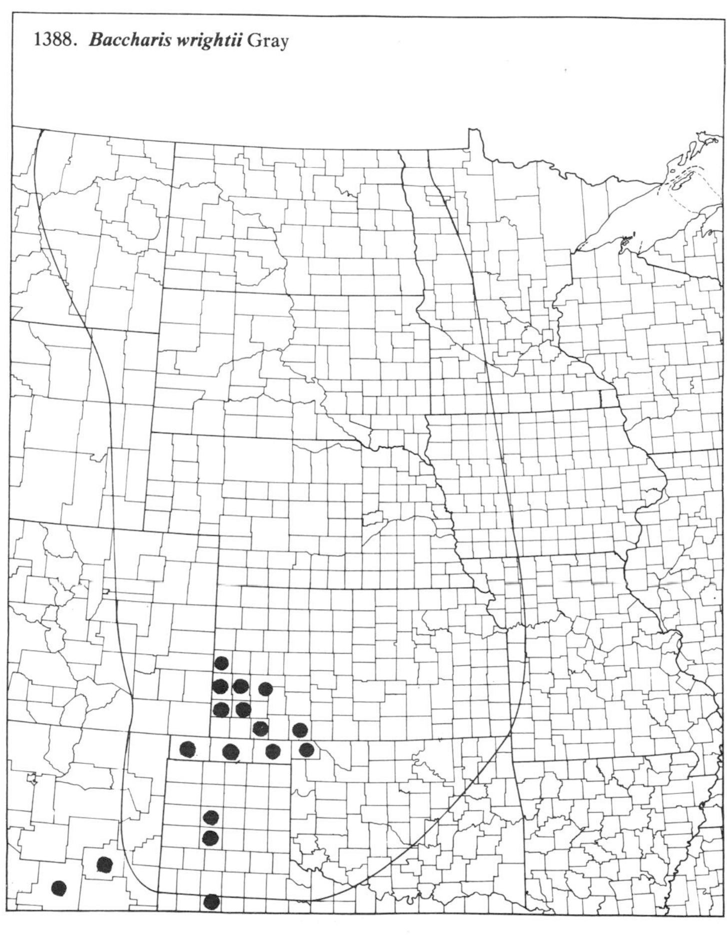

(Asteraceae)

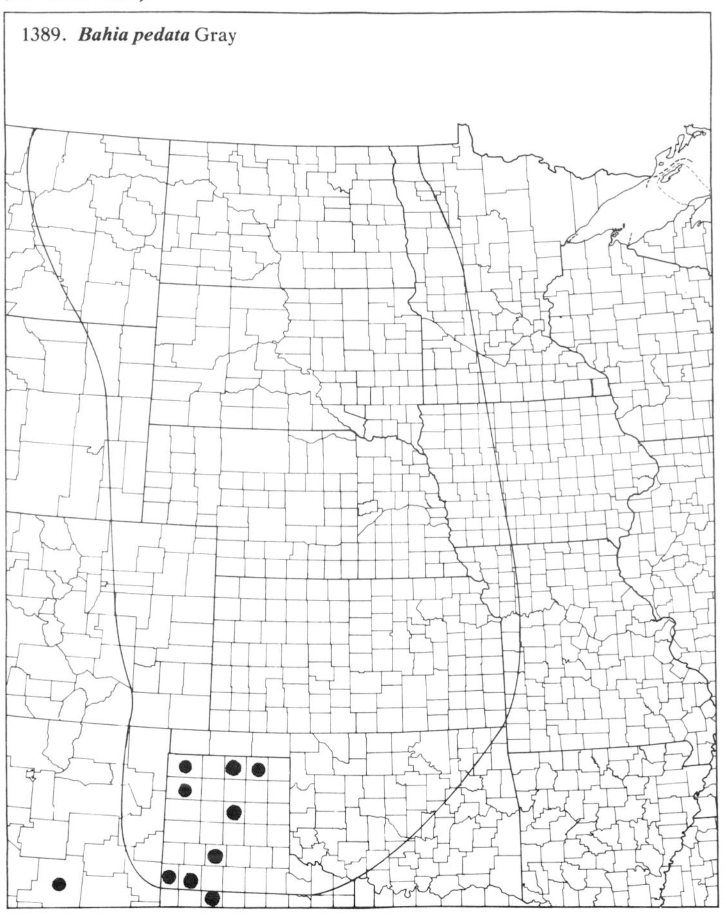
1389. ***Bahia pedata*** Gray

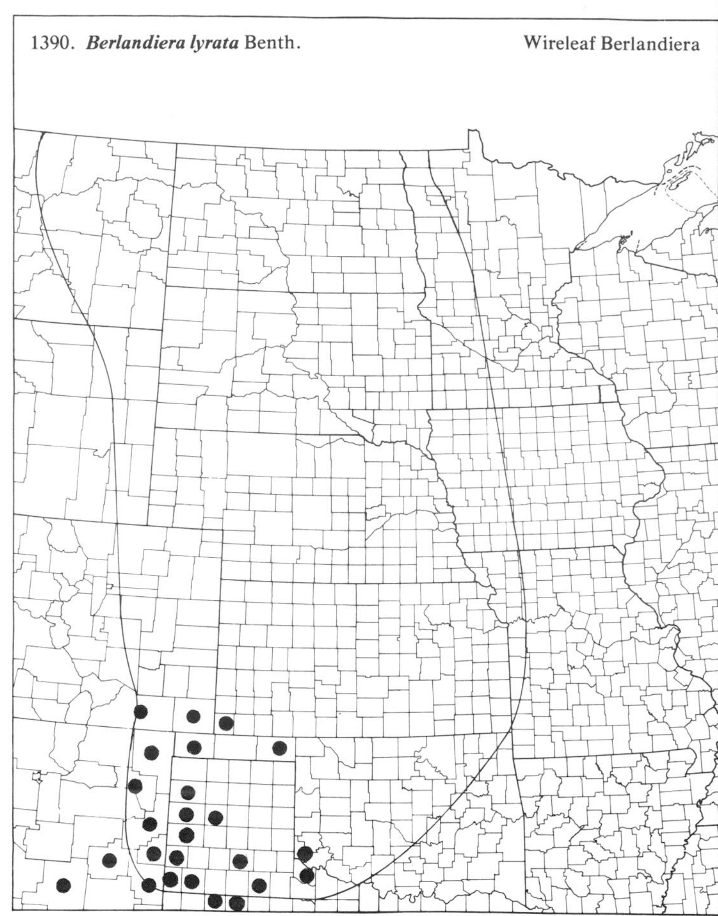
1390. ***Berlandiera lyrata*** Benth. Wireleaf Berlandiera

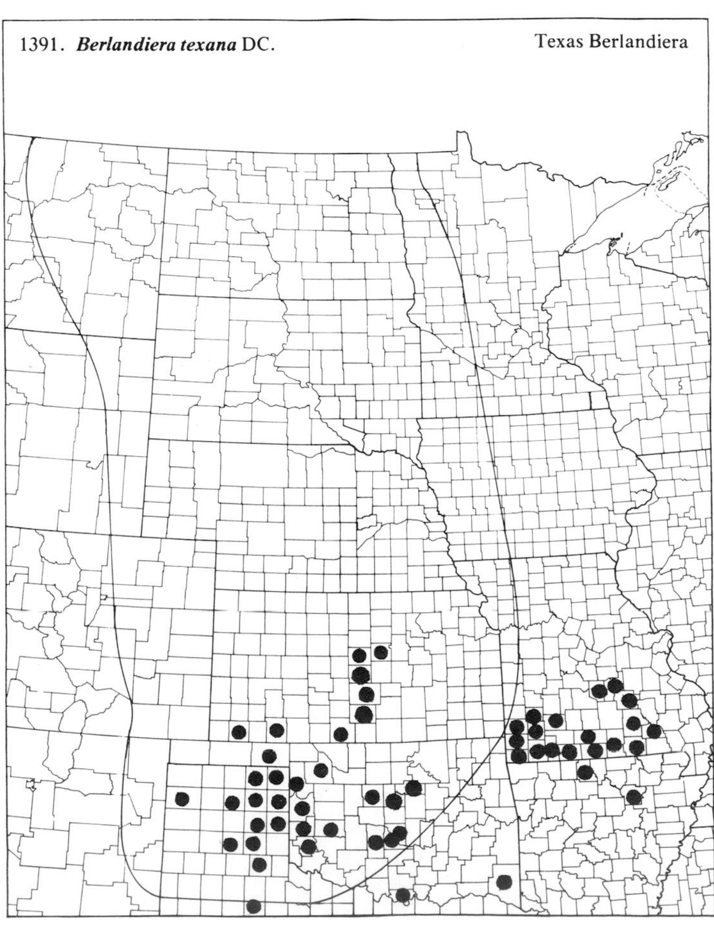
1391. ***Berlandiera texana*** DC. Texas Berlandiera

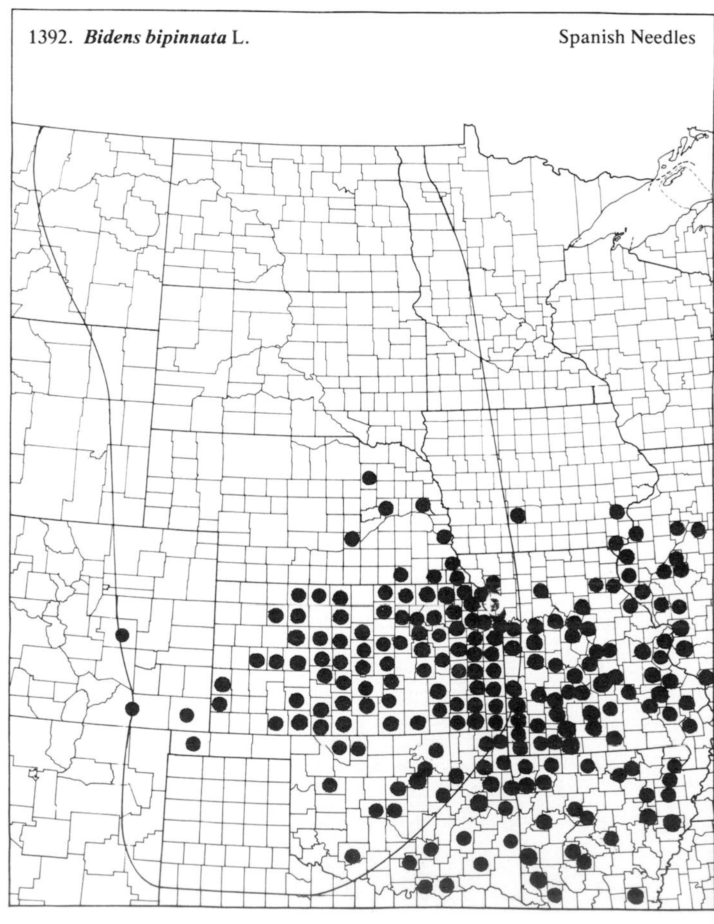
1392. ***Bidens bipinnata*** L. Spanish Needles

(Asteraceae)

1393. **Bidens cernua** L. Nodding Beggarticks

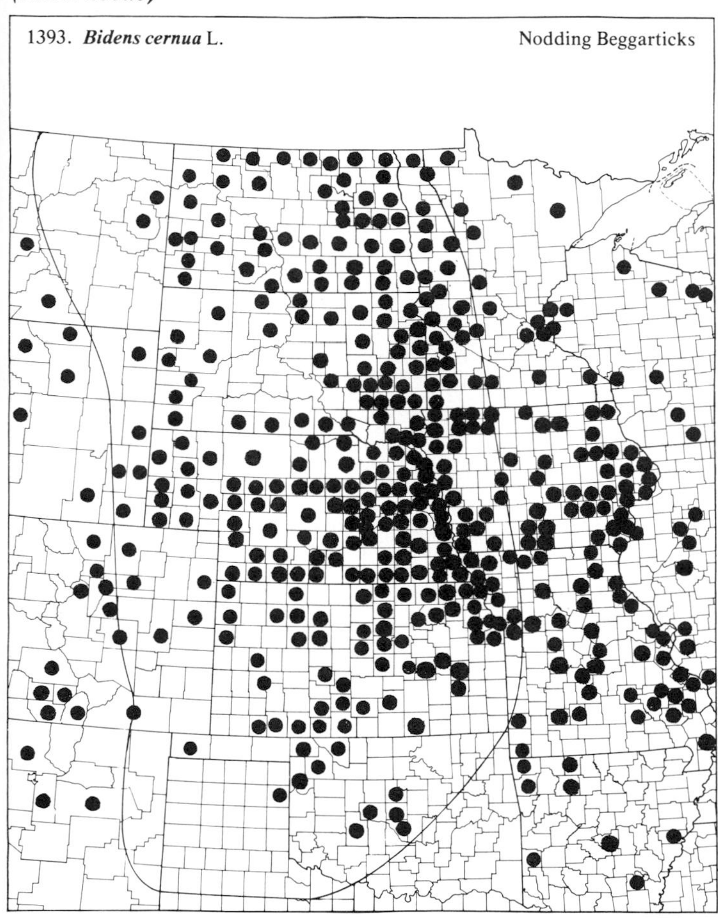

1394. **Bidens comosa** (Gray) Wieg. Beggarticks

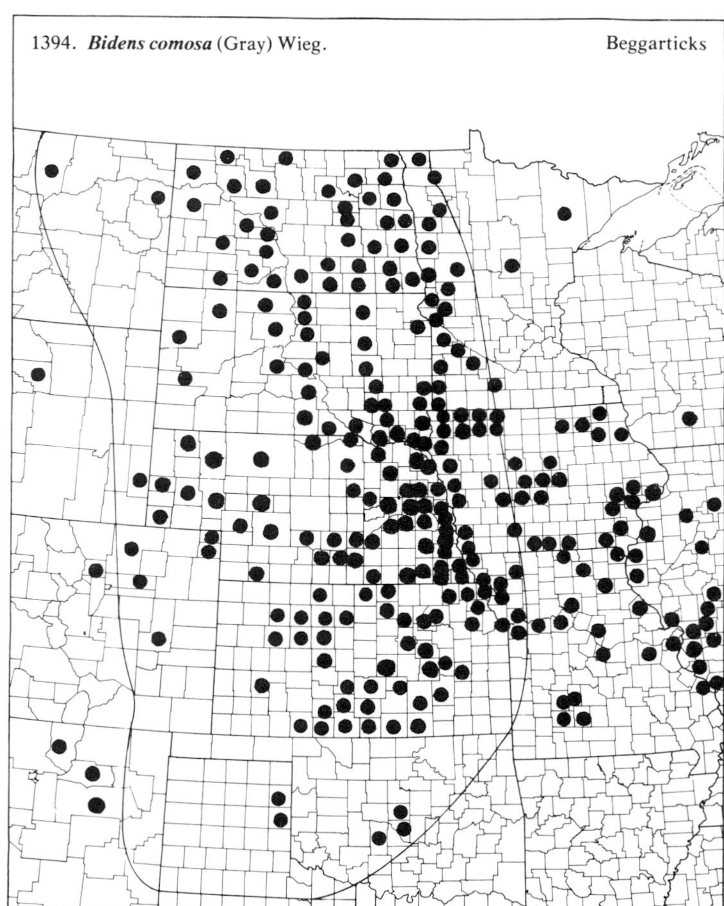

1395. **Bidens connata** Muhl. Sticktight

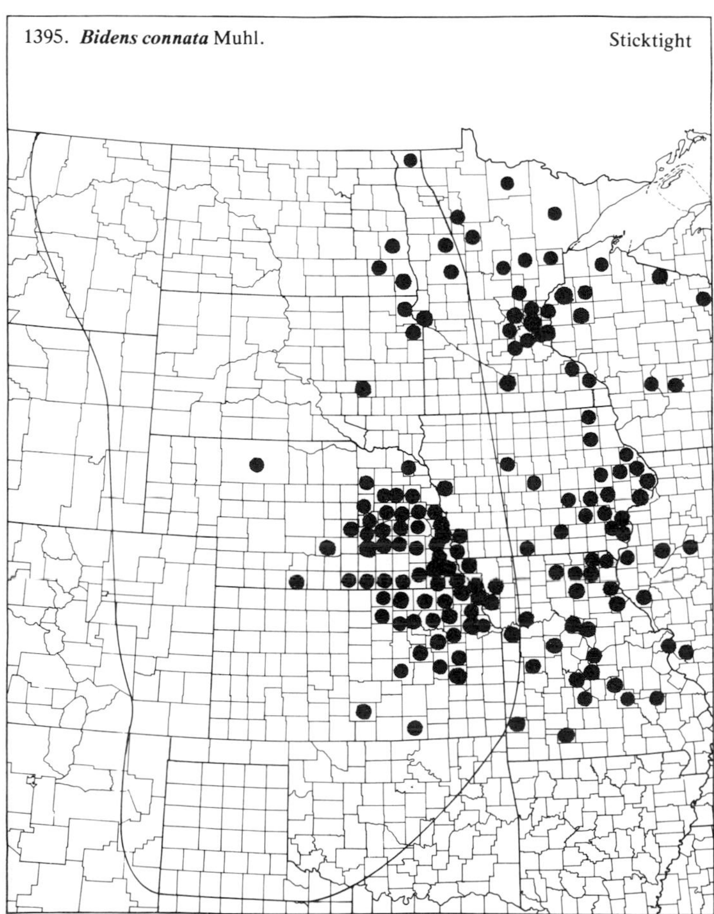

1396. **Bidens coronata** (L.) Britt. Tickseed Sunflower

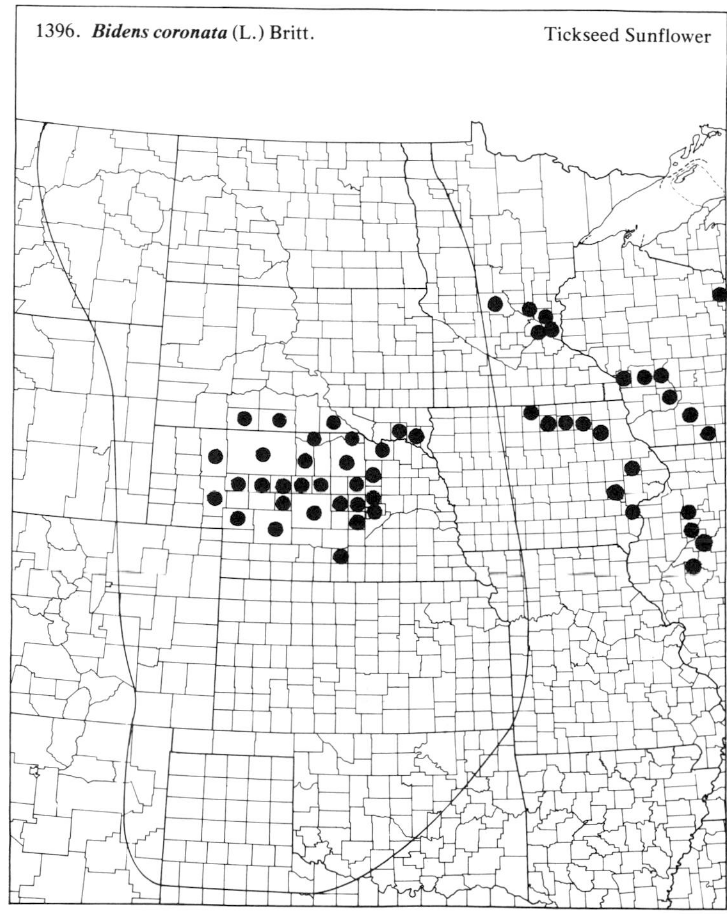

(Asteraceae)

1397. ***Bidens frondosa*** L. Beggarticks

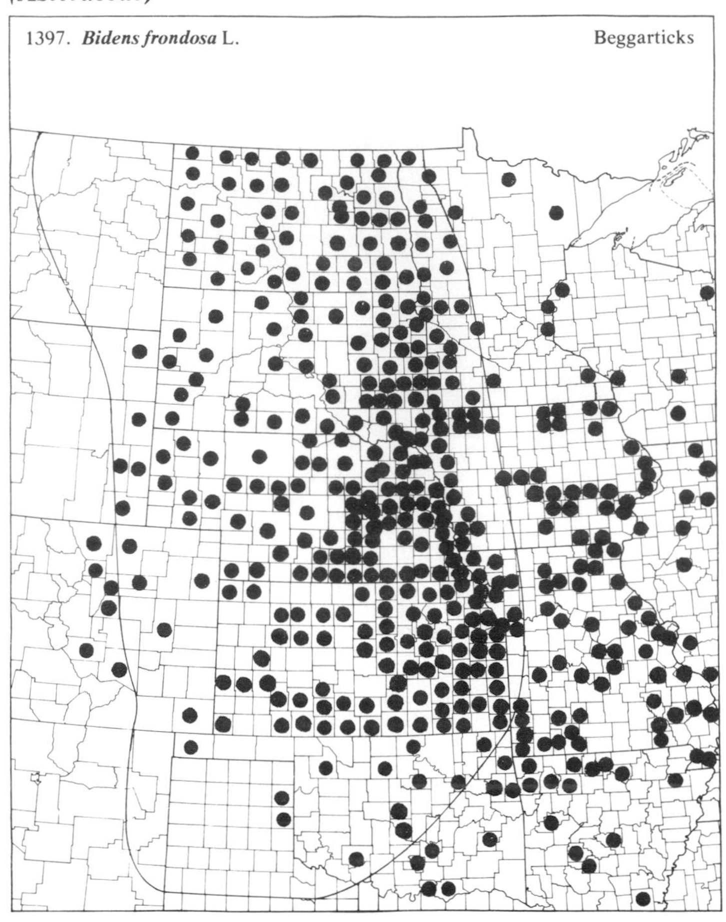

1398. ***Bidens polylepis*** Blake Coreopsis Beggarticks

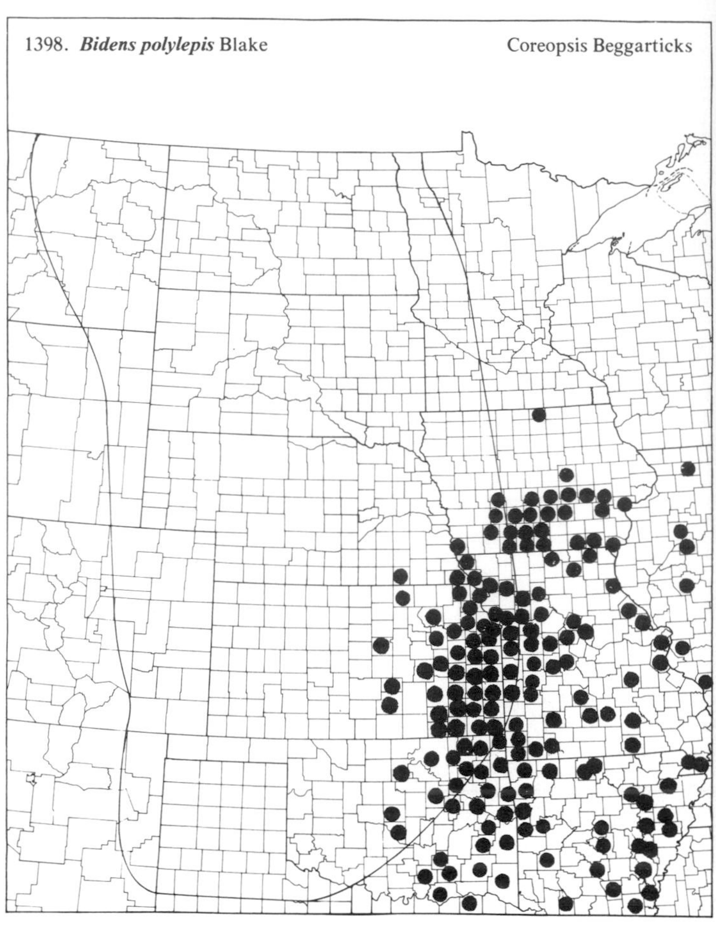

1399. ***Bidens vulgata*** Greene Beggarticks

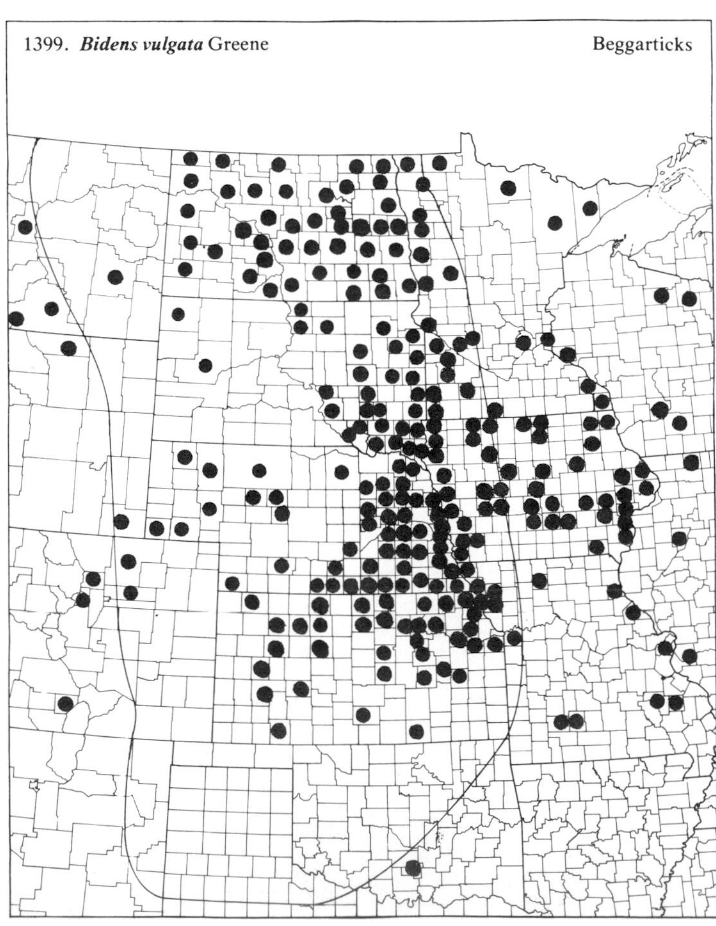

1400. ***Boltonia asteroides*** (L.) L'Her. var. ***latisquama*** (Gray) Cronq. Violet Boltonia

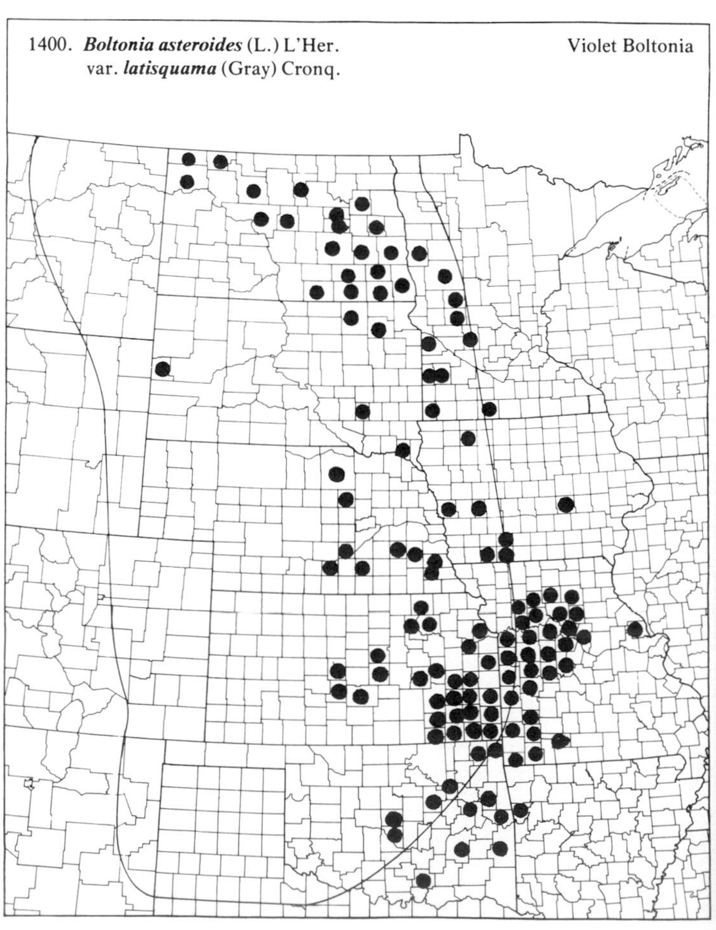

(Asteraceae)

1401. ***Boltonia asteroides*** (L.) L'Her. var. ***recognita*** (Fern. & Grisc.) Cronq. — White Boltonia

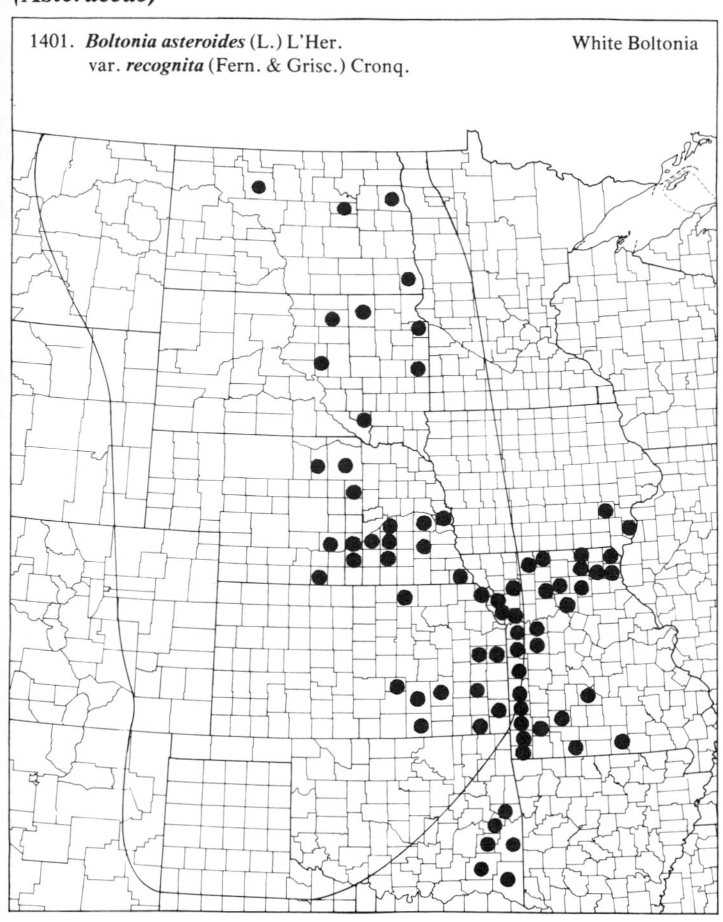

1402. ***Brickellia grandiflora*** (Hook.) Nutt. — Brickellia

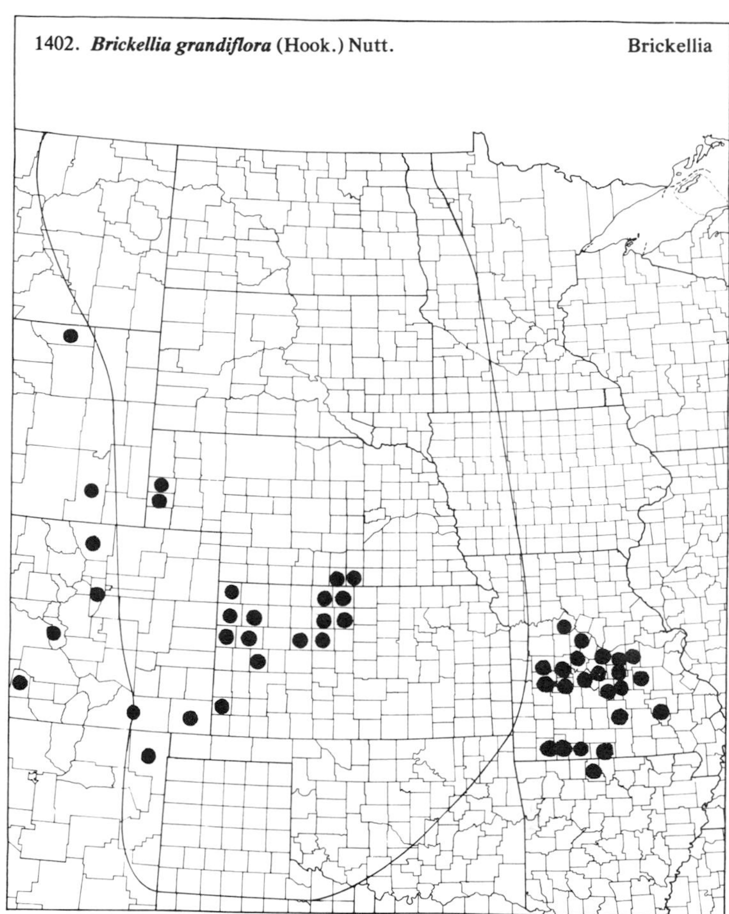

1403. ***Cacalia atriplicifolia*** L. — Pale Indian Plantain

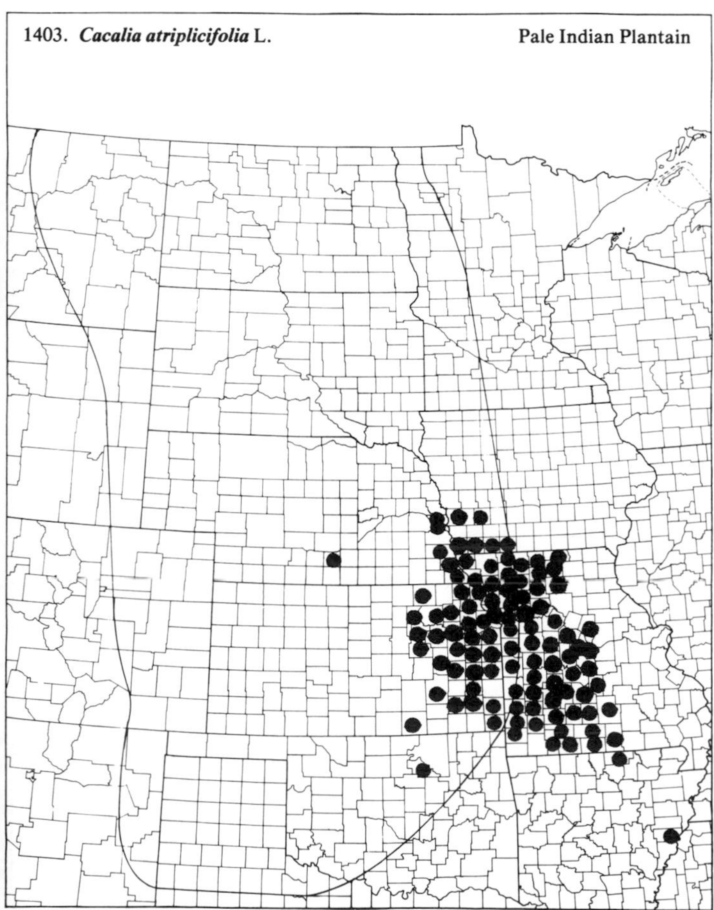

1404. ***Cacalia tuberosa*** Nutt. — Indian Plantain

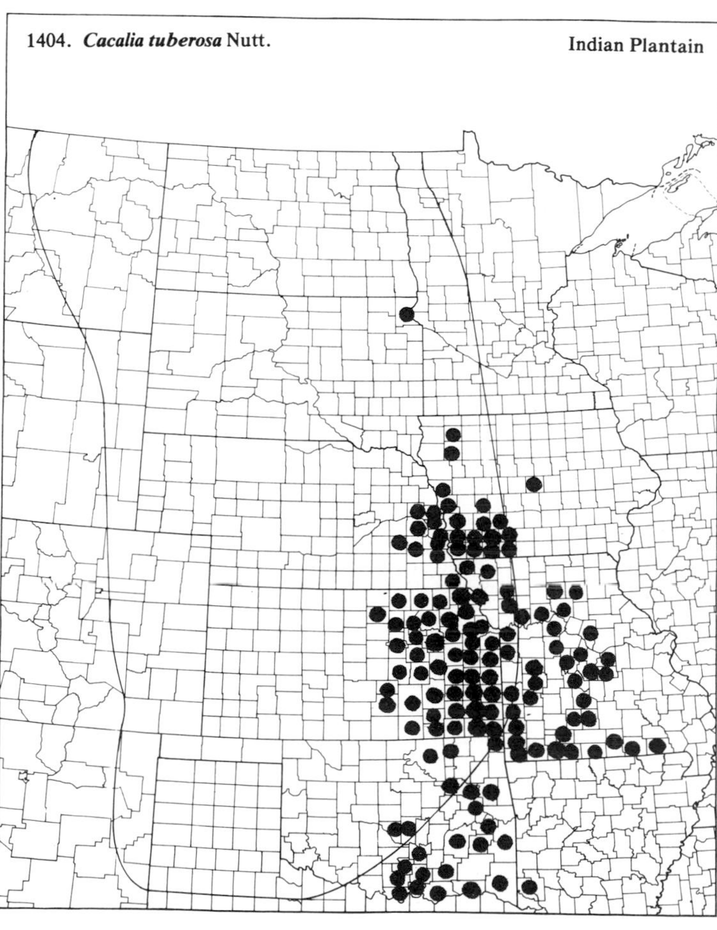

(Asteraceae)

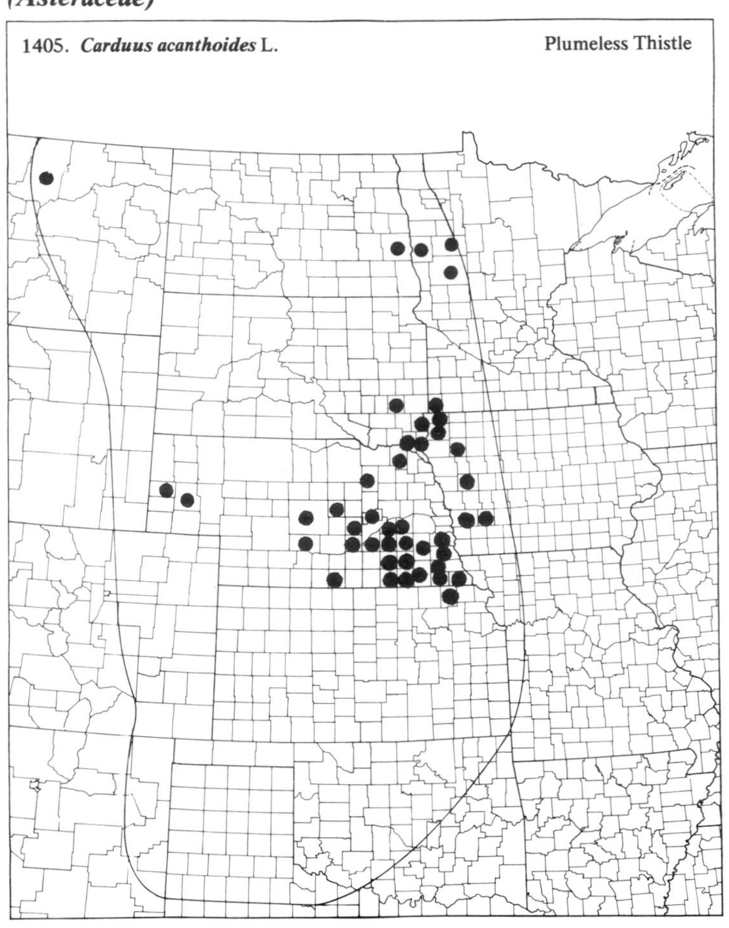
1405. ***Carduus acanthoides*** L. Plumeless Thistle

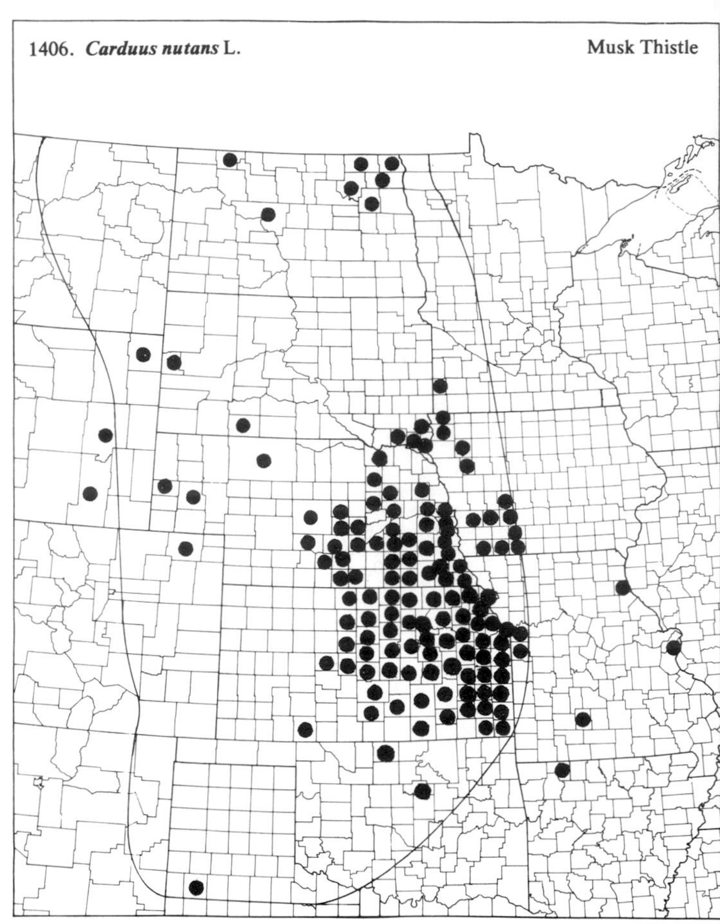
1406. ***Carduus nutans*** L. Musk Thistle

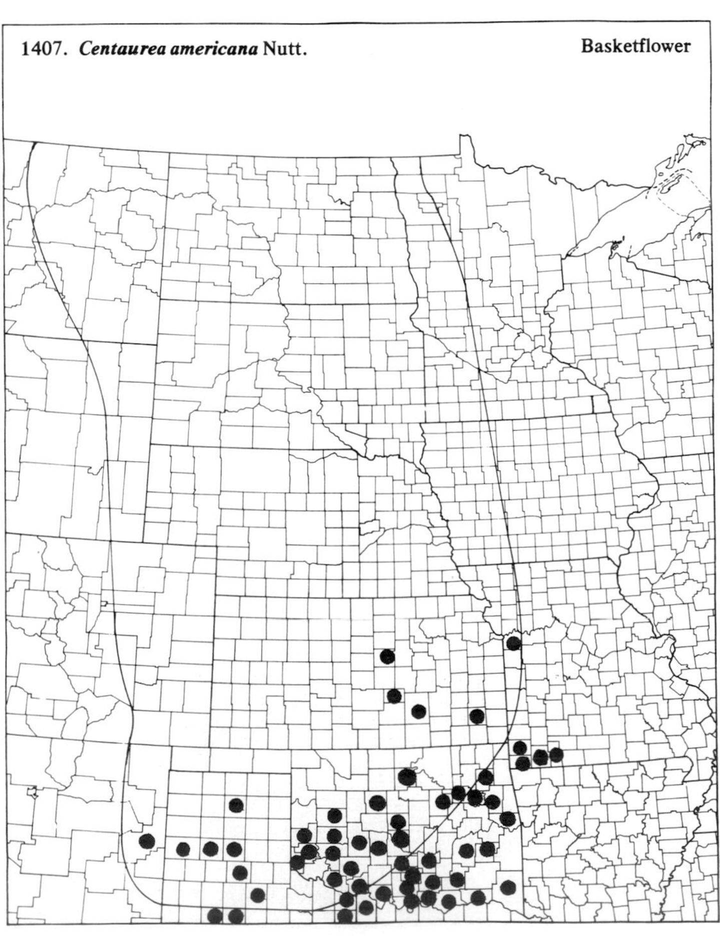
1407. ***Centaurea americana*** Nutt. Basketflower

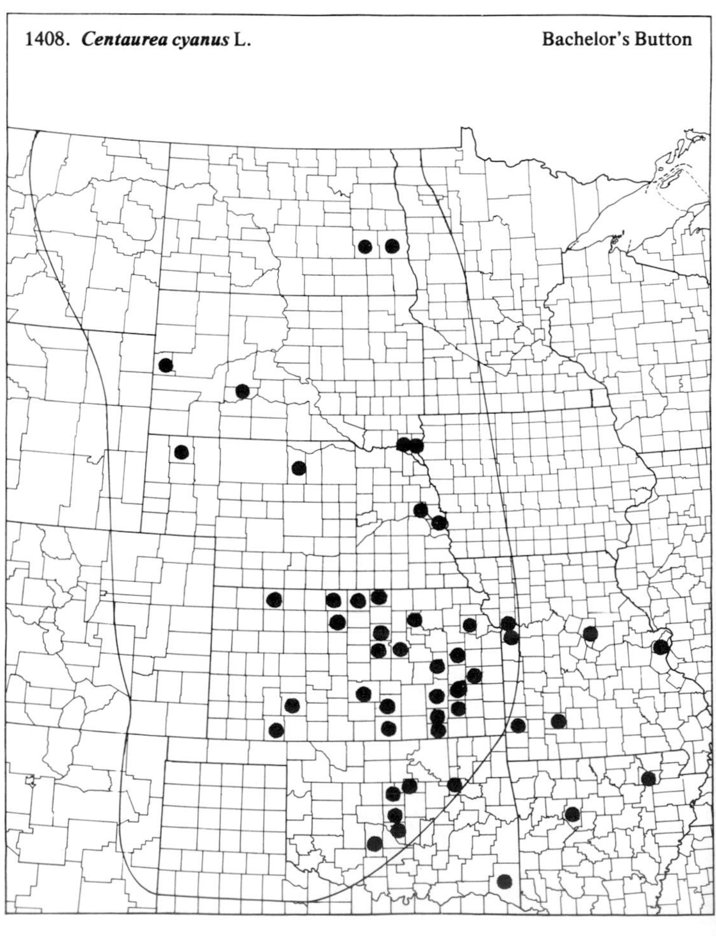
1408. ***Centaurea cyanus*** L. Bachelor's Button

(Asteraceae)

1409. ***Centaurea maculosa*** Lam. Spotted Knapweed

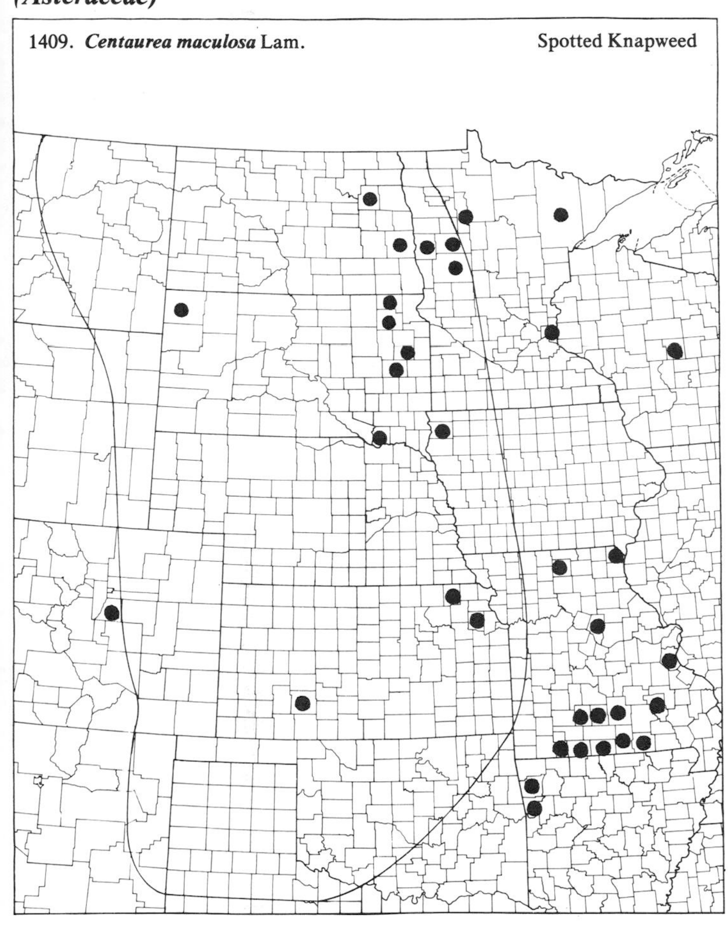

1410. ***Centaurea repens*** L. Russian Knapweed

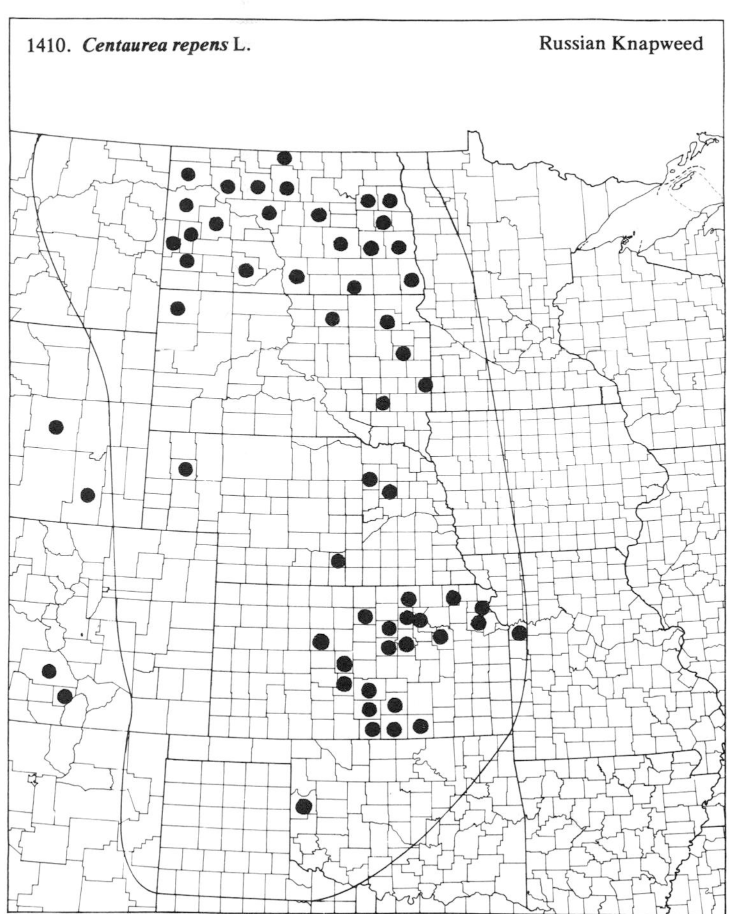

1411. ***Centaurea solstitialis*** L. Barnaby's Thistle

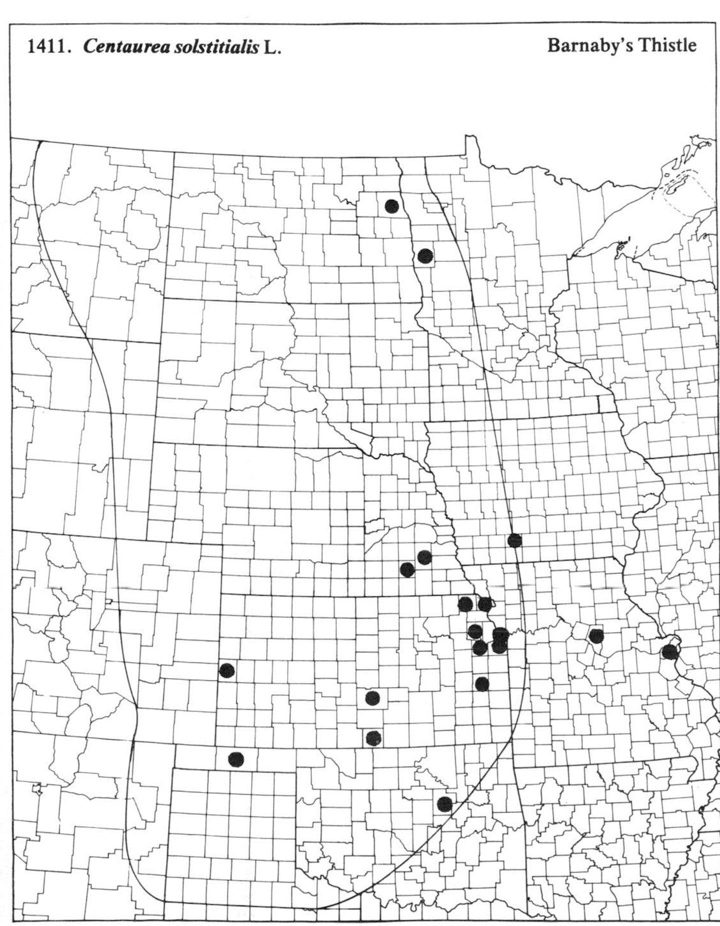

1412. ***Chaetopappa asteroides*** (Nutt.) DC.

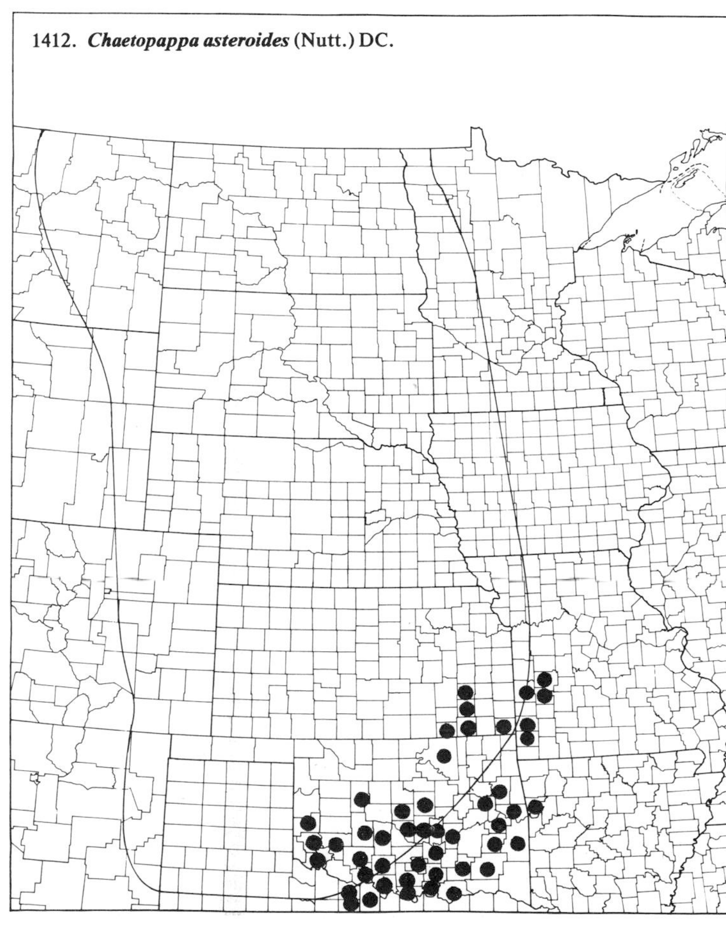

(Asteraceae)

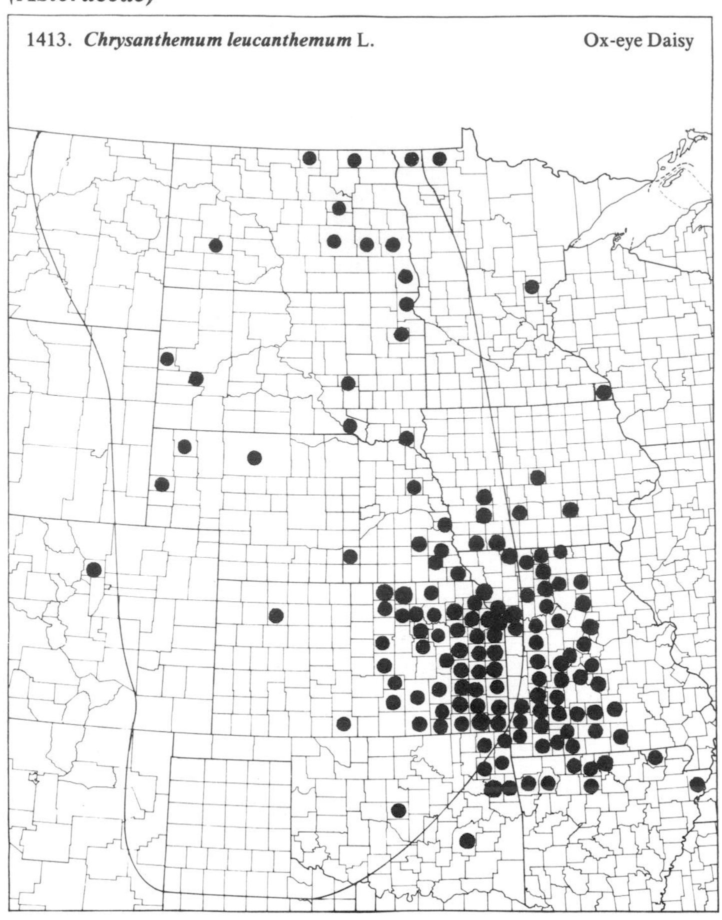
1413. ***Chrysanthemum leucanthemum*** L. — Ox-eye Daisy

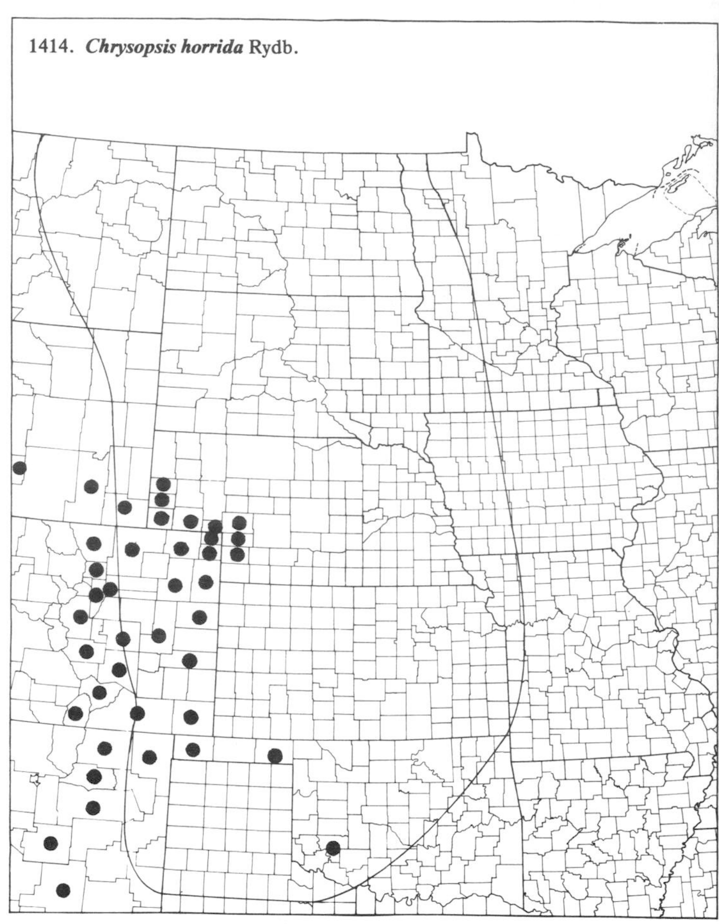
1414. ***Chrysopsis horrida*** Rydb.

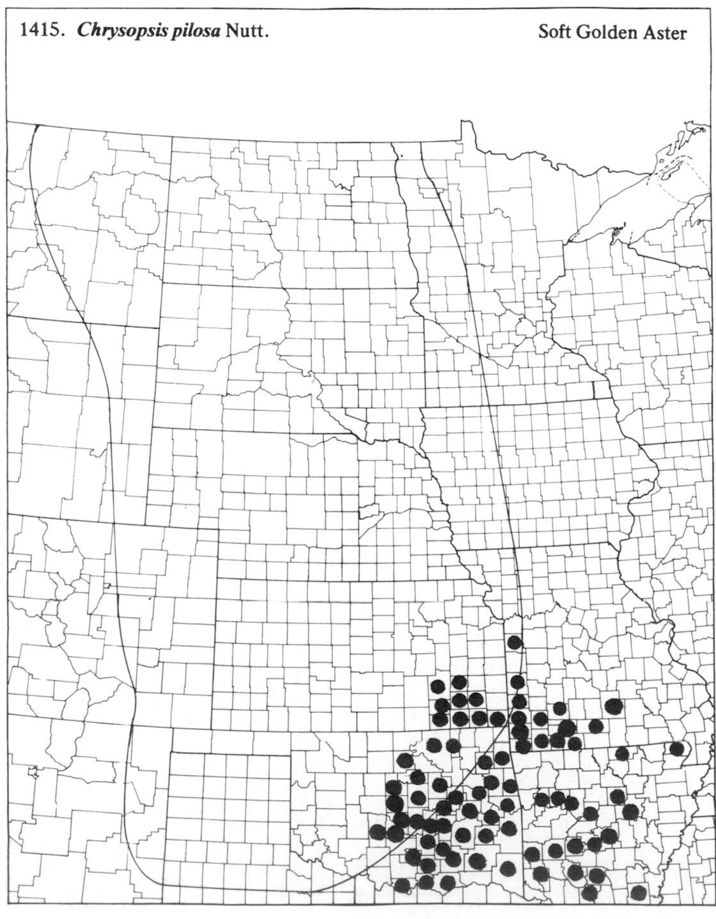
1415. ***Chrysopsis pilosa*** Nutt. — Soft Golden Aster

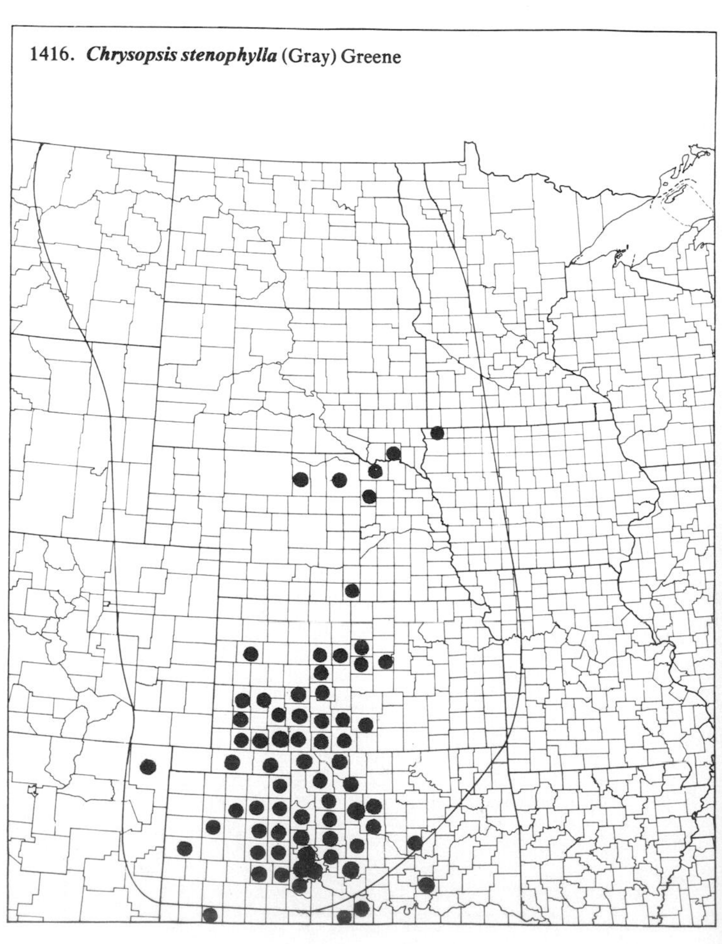
1416. ***Chrysopsis stenophylla*** (Gray) Greene

(Asteraceae)

1417. ***Chrysopsis villosa*** (Pursh) Nutt.
var. ***villosa***
Golden Aster

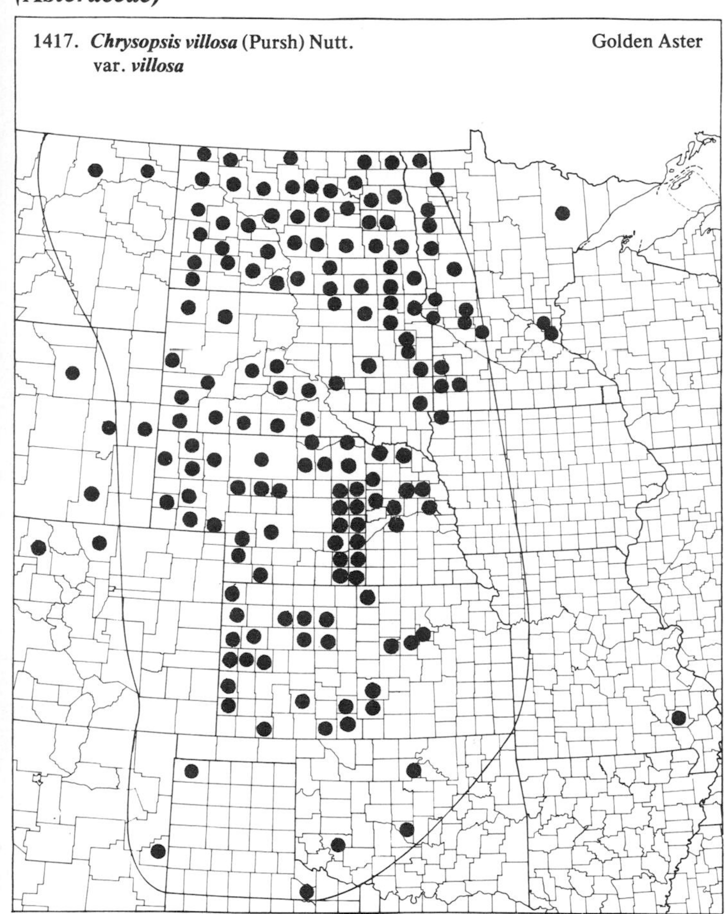

1418. ***Chrysopsis villosa*** (Pursh) Nutt.
var. ***angustifolia*** (Rydb.) Cronq.
Golden Aster

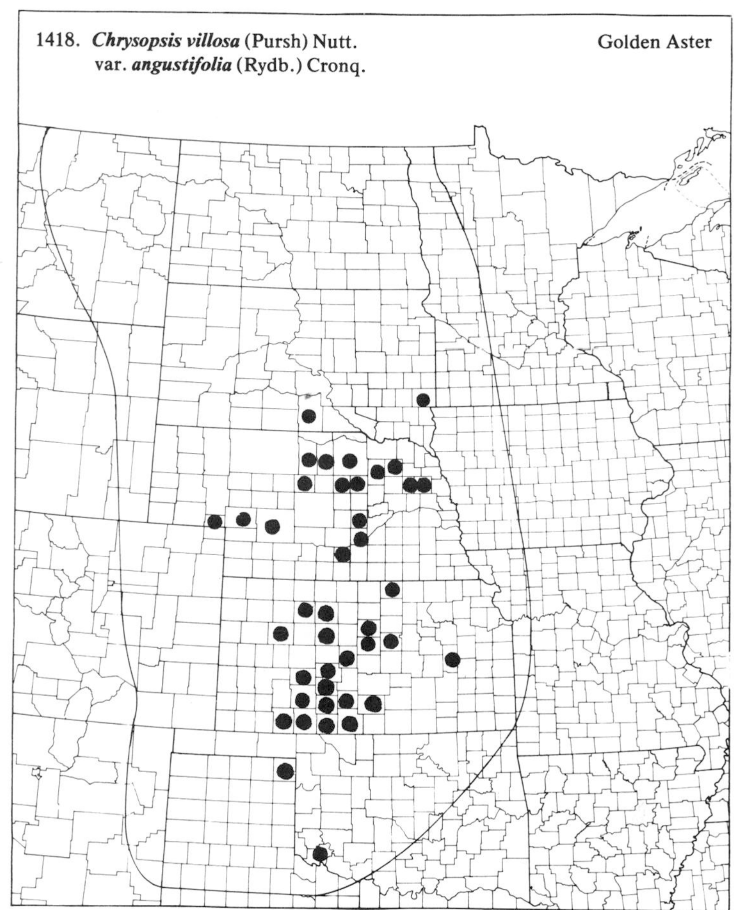

1419. ***Chrysopsis villosa*** (Pursh) Nutt.
var. ***canescens*** (DC.) Gray
Golden Aster

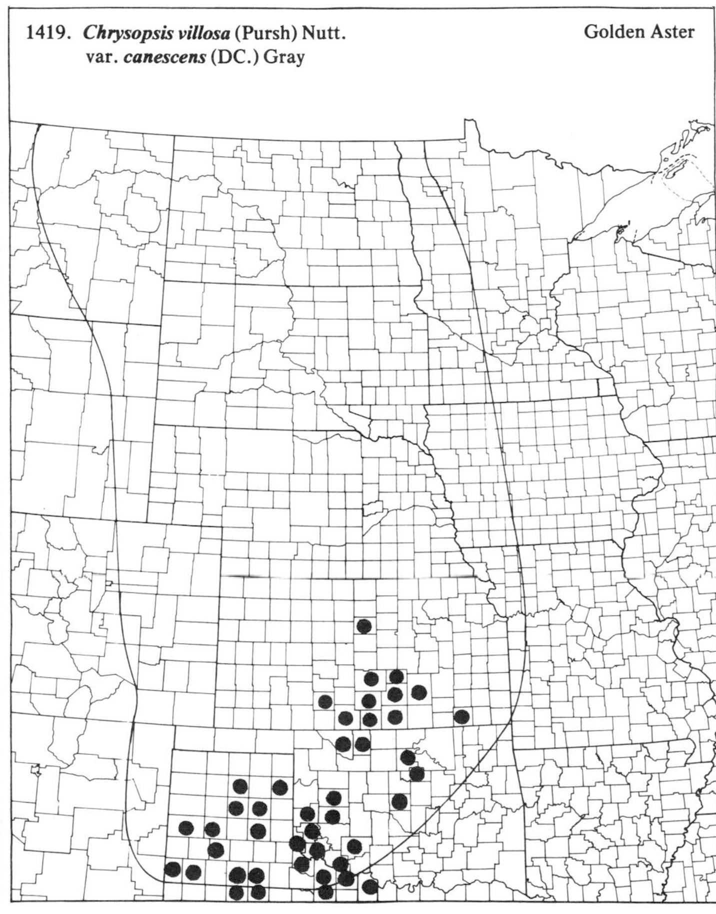

1420. ***Chrysopsis villosa*** (Pursh) Nutt.
var. ***foliosa*** (Nutt.) DC. Eat.
Golden Aster

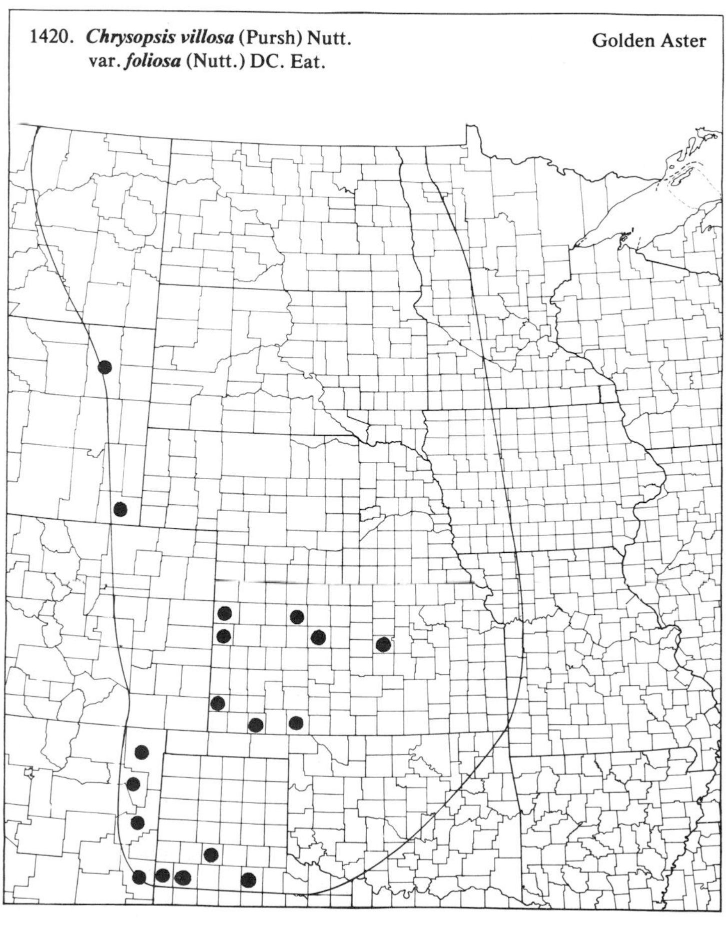

(Asteraceae)

1421. ***Chrysopsis villosa*** (Pursh) Nutt. var. ***hispida*** (Hook.) Gray — Golden Aster

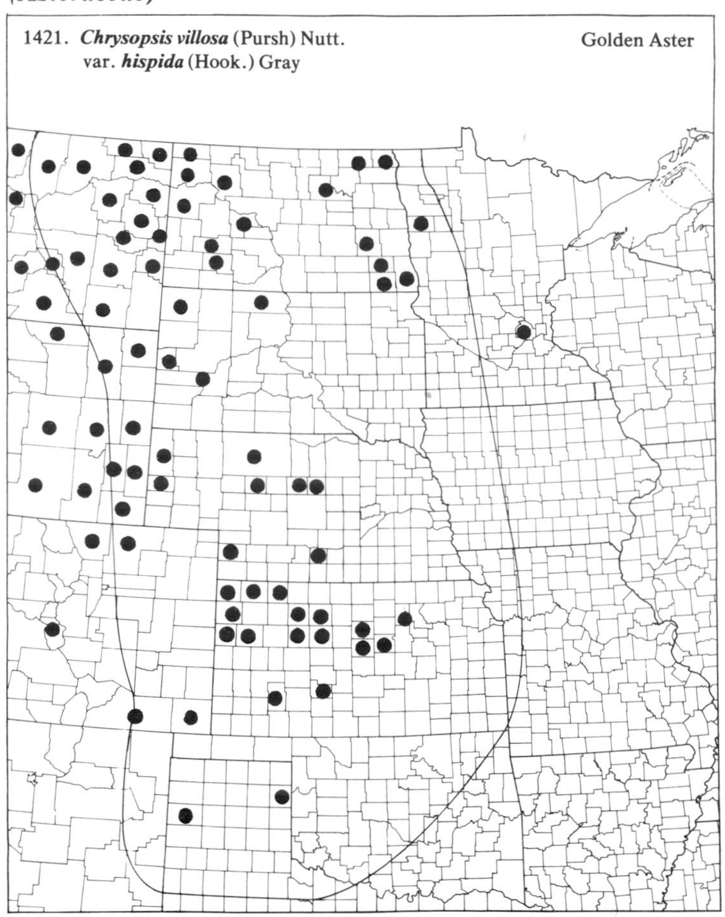

1422. ***Chrysothamnus nauseosus*** (Pall.) Britt. ssp. ***nauseosus*** — Rabbit Brush

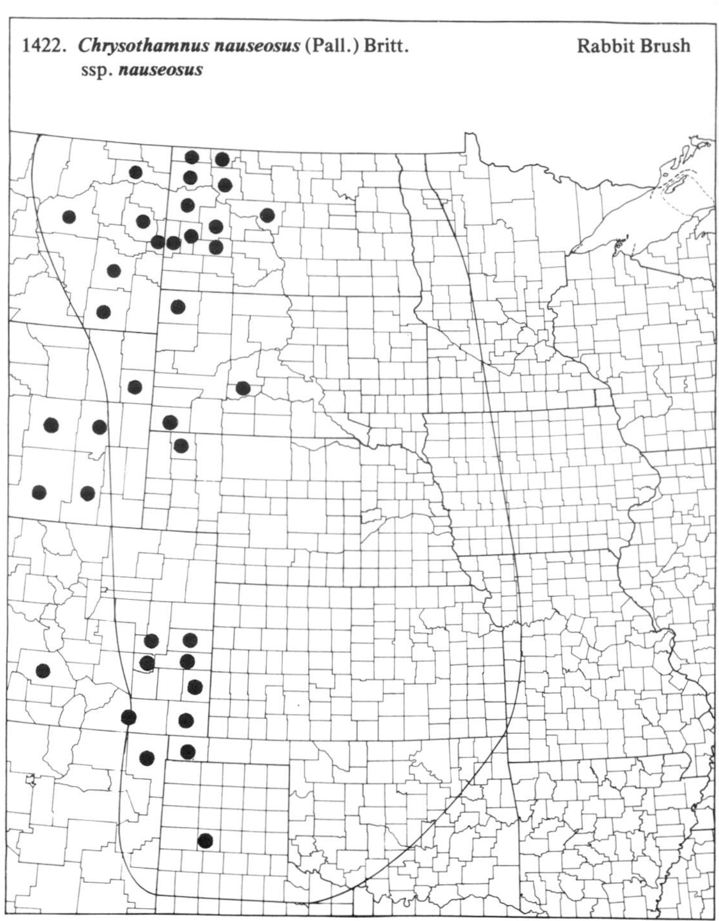

1423. ***Chrysothamnus nauseosus*** (Pall.) Britt. ssp. ***graveolens*** (Nutt.) Piper — Rabbit Brush

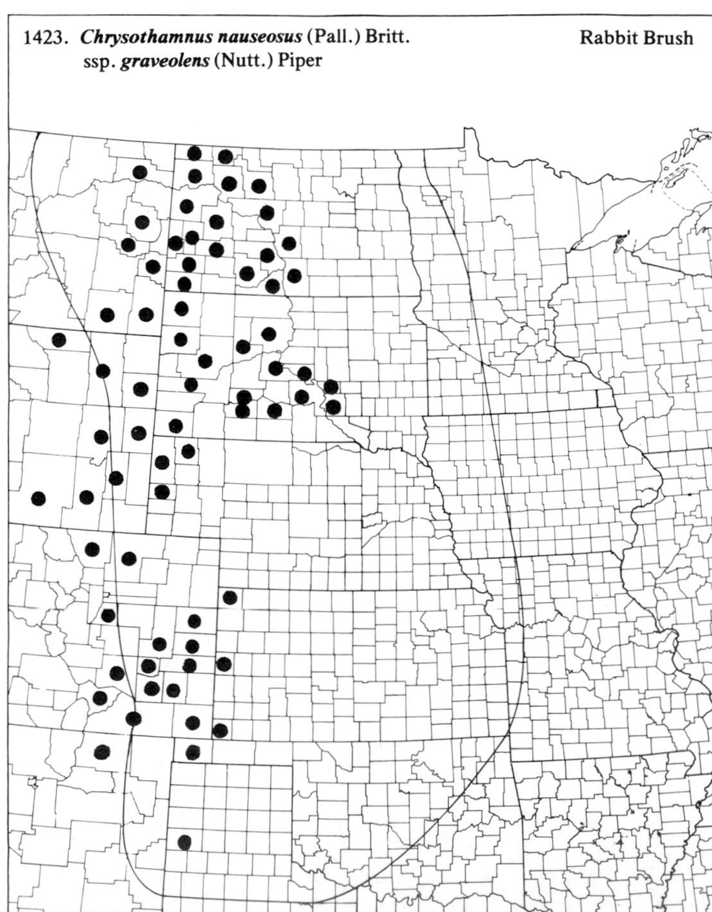

1424. ***Chrysothamnus pulchellus*** (Gray) Greene

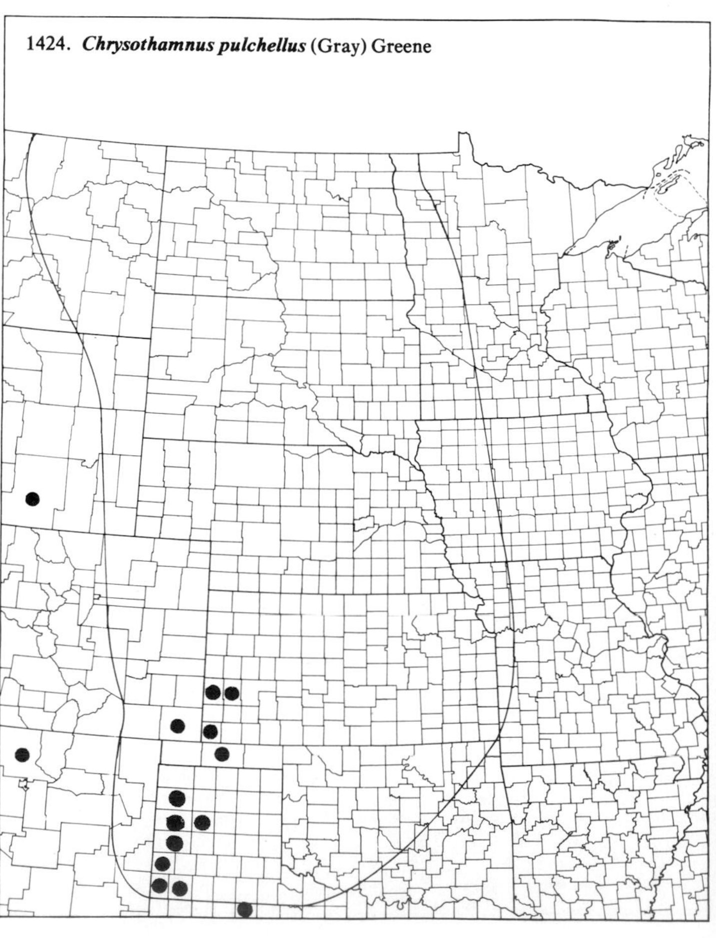

(Asteraceae)

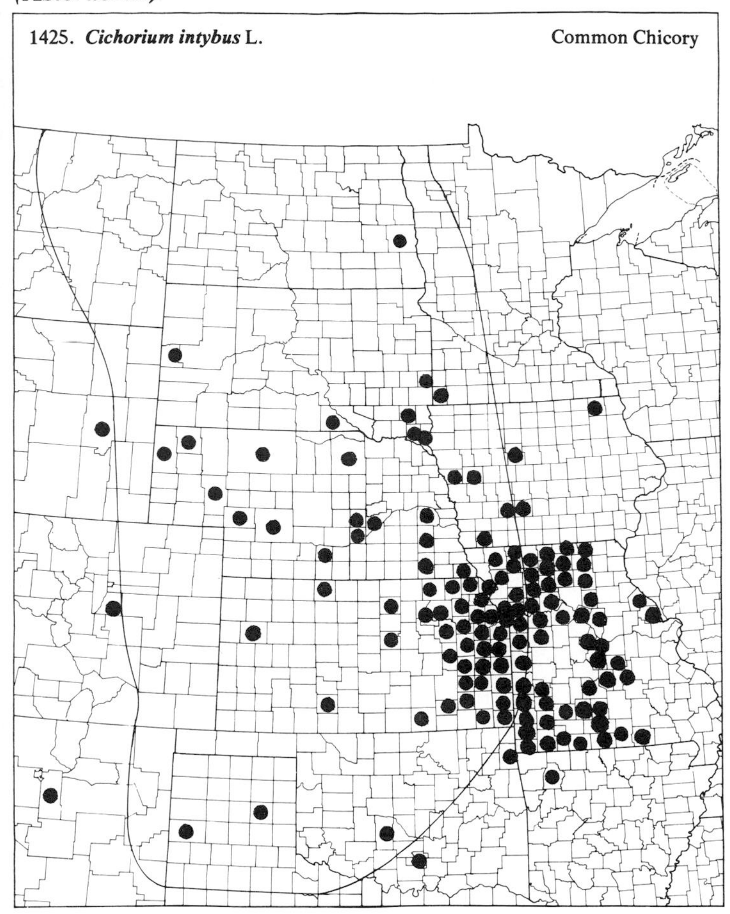
1425. ***Cichorium intybus*** L. Common Chicory

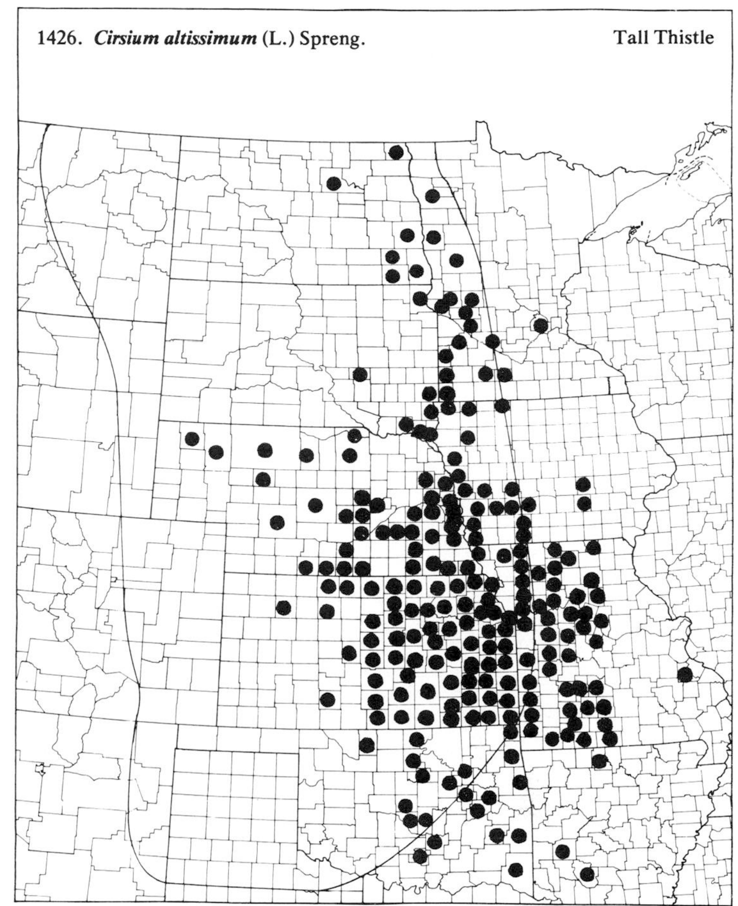
1426. ***Cirsium altissimum*** (L.) Spreng. Tall Thistle

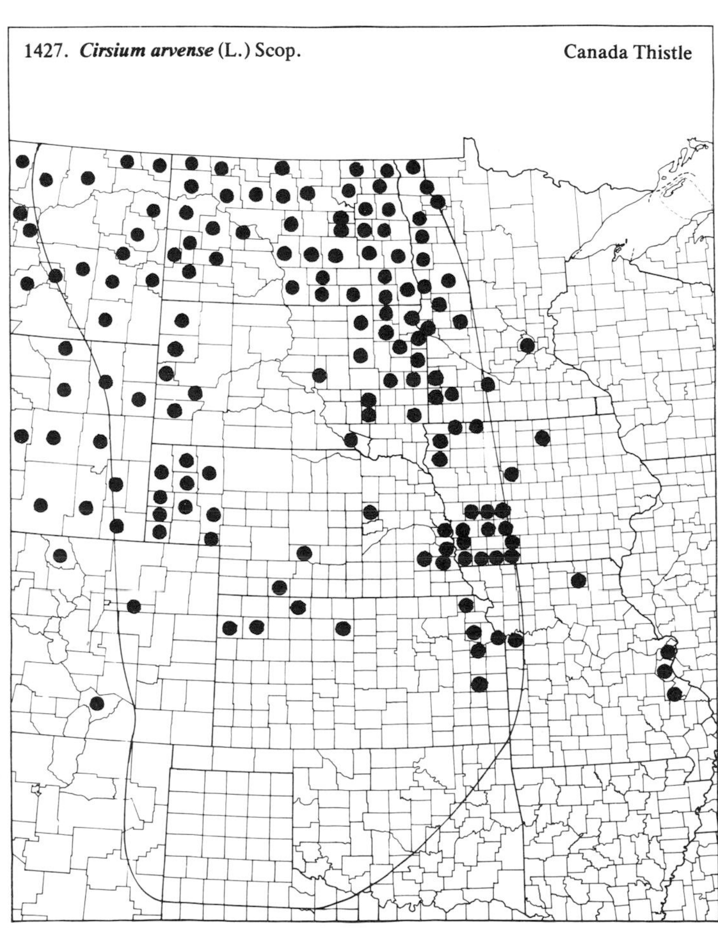
1427. ***Cirsium arvense*** (L.) Scop. Canada Thistle

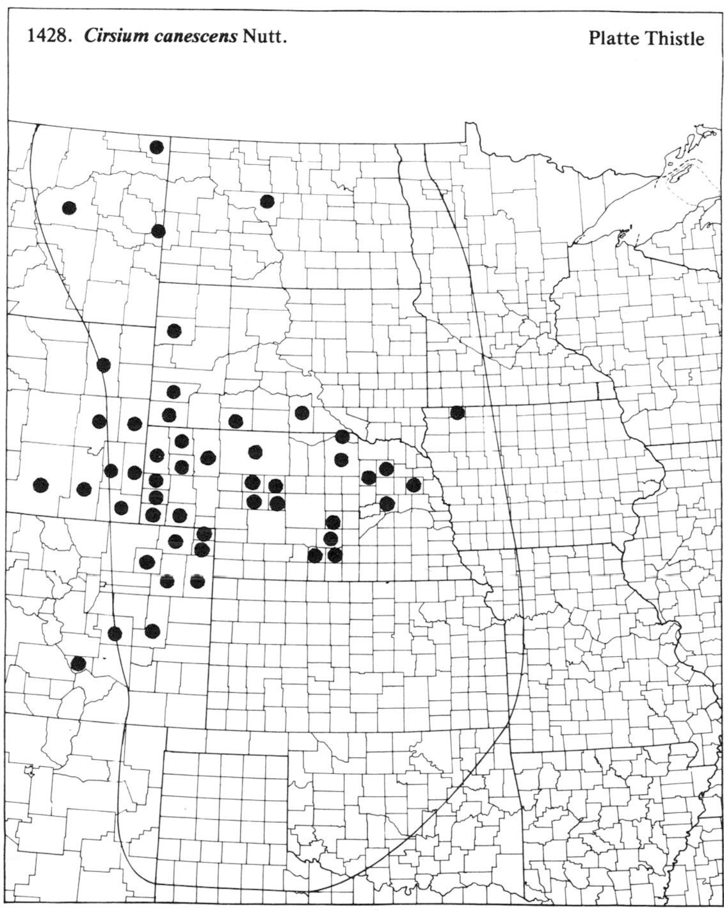
1428. ***Cirsium canescens*** Nutt. Platte Thistle

(Asteraceae)

1429. ***Cirsium discolor*** (Muhl.) Spreng. — Field Thistle

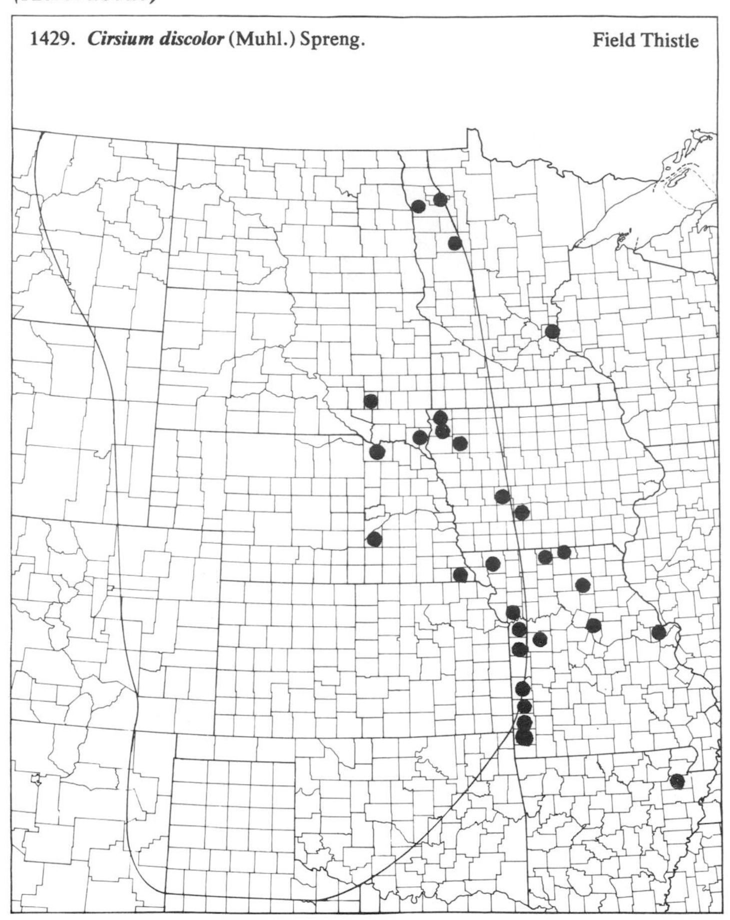

1430. ***Cirsium flodmani*** (Rydb.) Arthur — Prairie Thistle

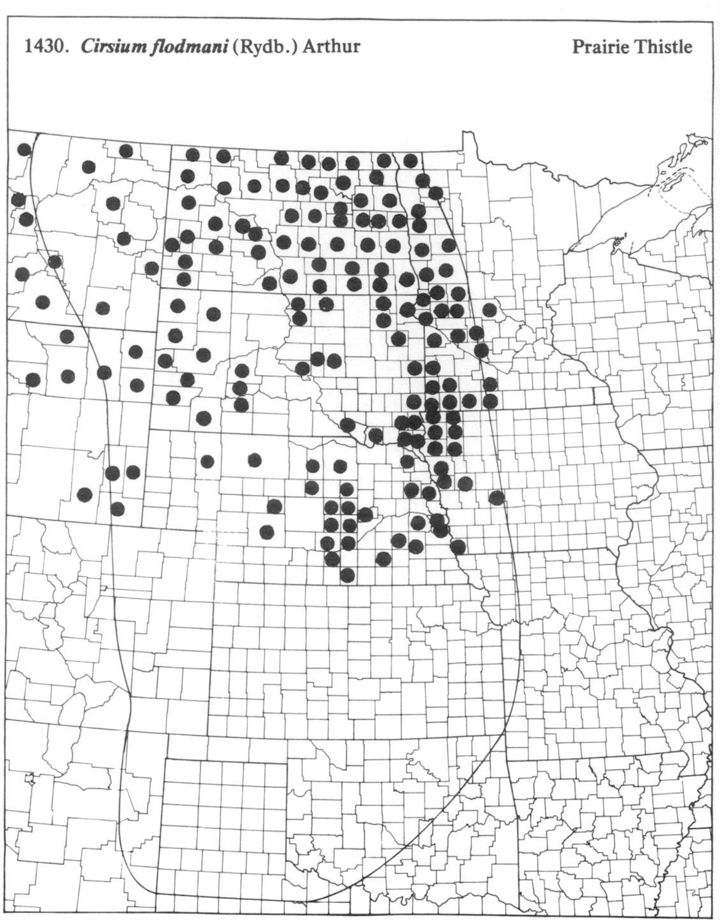

1431. ***Cirsium muticum*** Michx. — Swamp Thistle

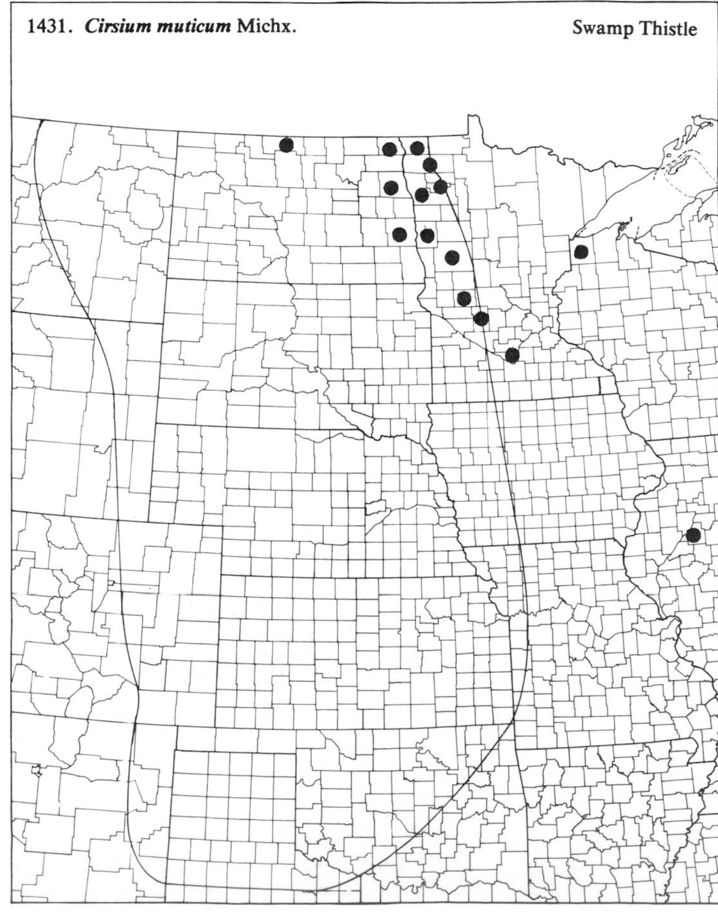

1432. ***Cirsium ochrocentrum*** Gray — Yellow Spine Thistle

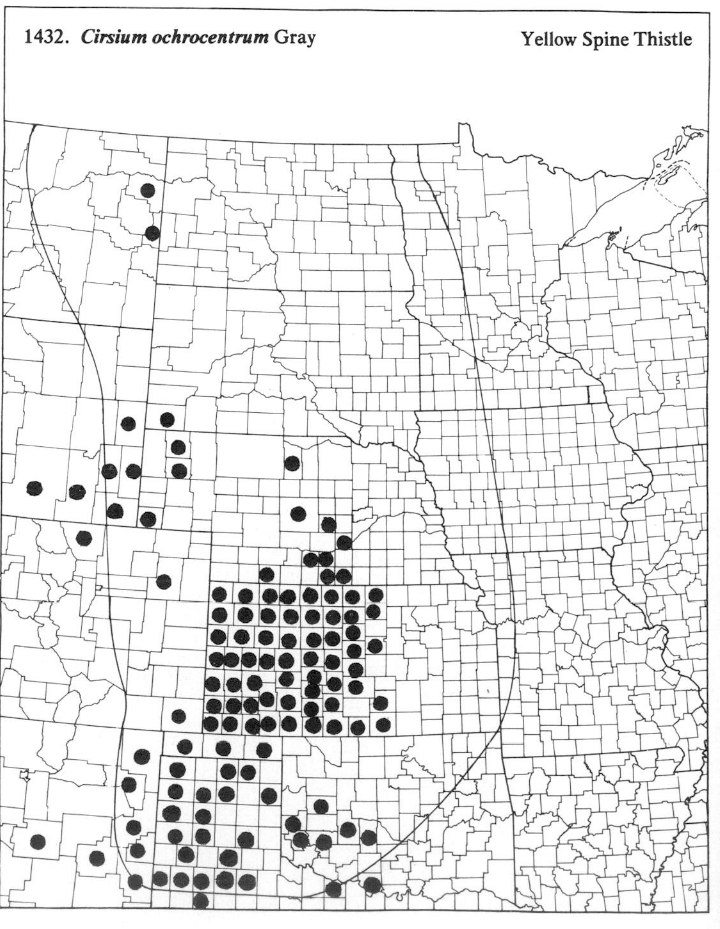

(Asteraceae)

1433. ***Cirsium undulatum*** (Nutt.) Spreng. — Wavyleaf Thistle

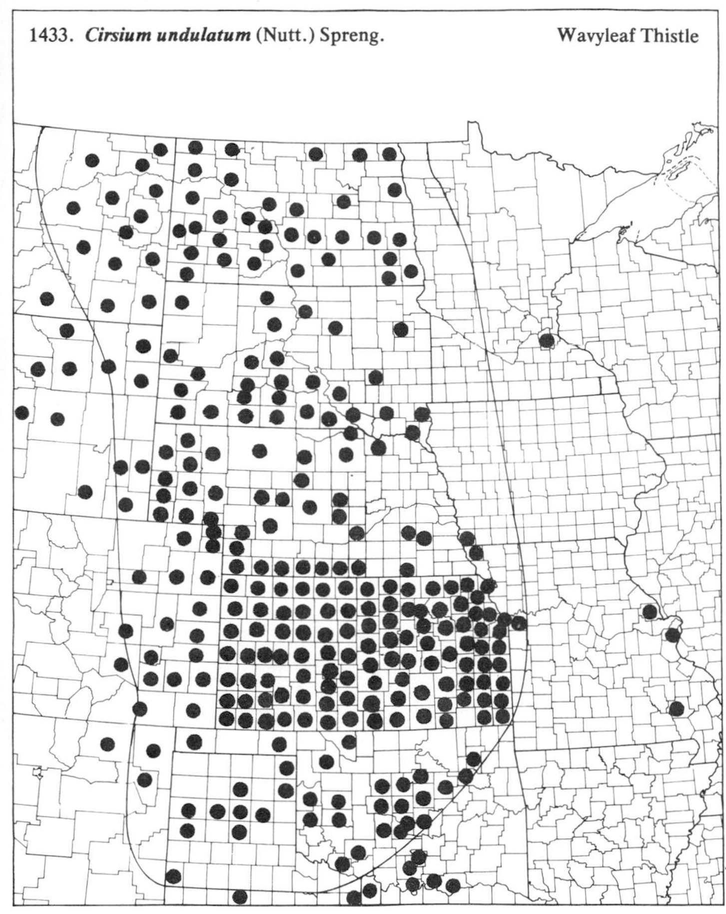

1434. ***Cirsium vulgare*** (Savi) Ten. — Bull Thistle

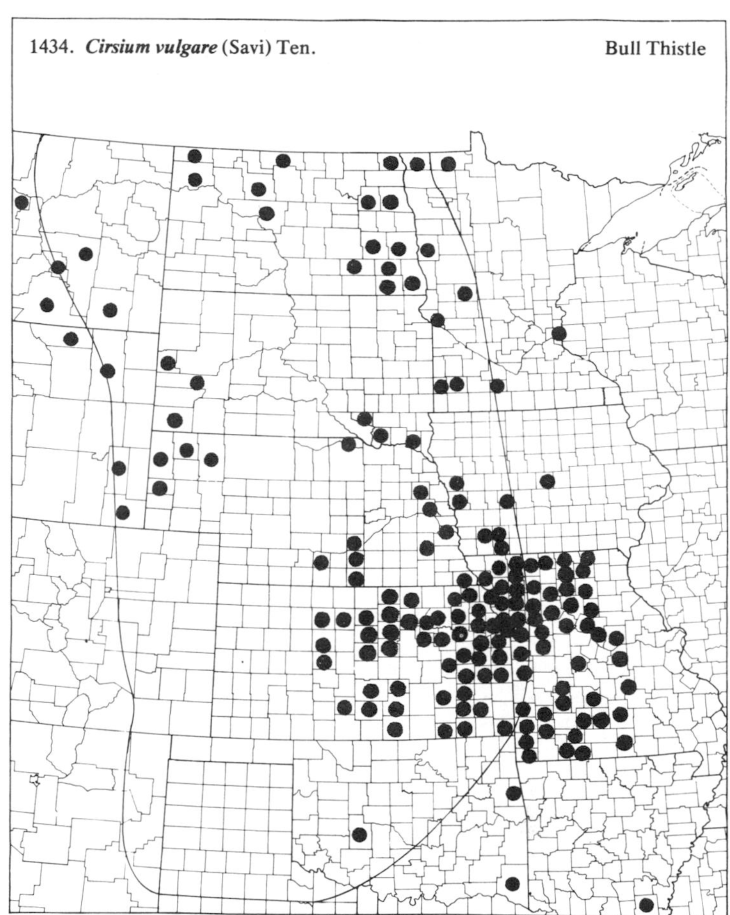

1435. ***Conyza canadensis*** (L.) Cronq. — Horseweed

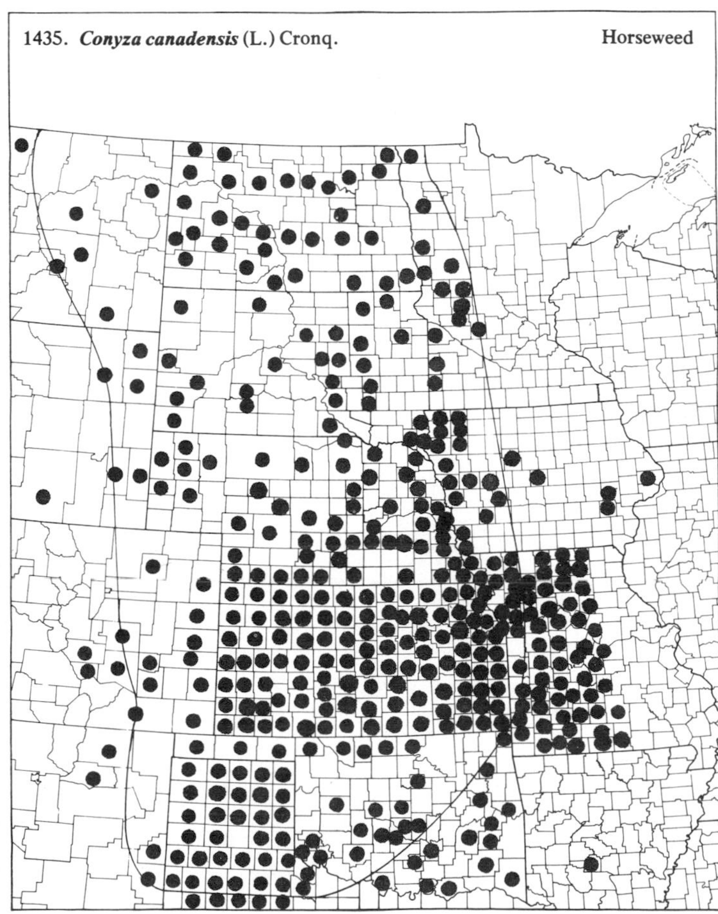

1436. ***Conyza ramosissima*** Cronq. — Spreading Fleabane

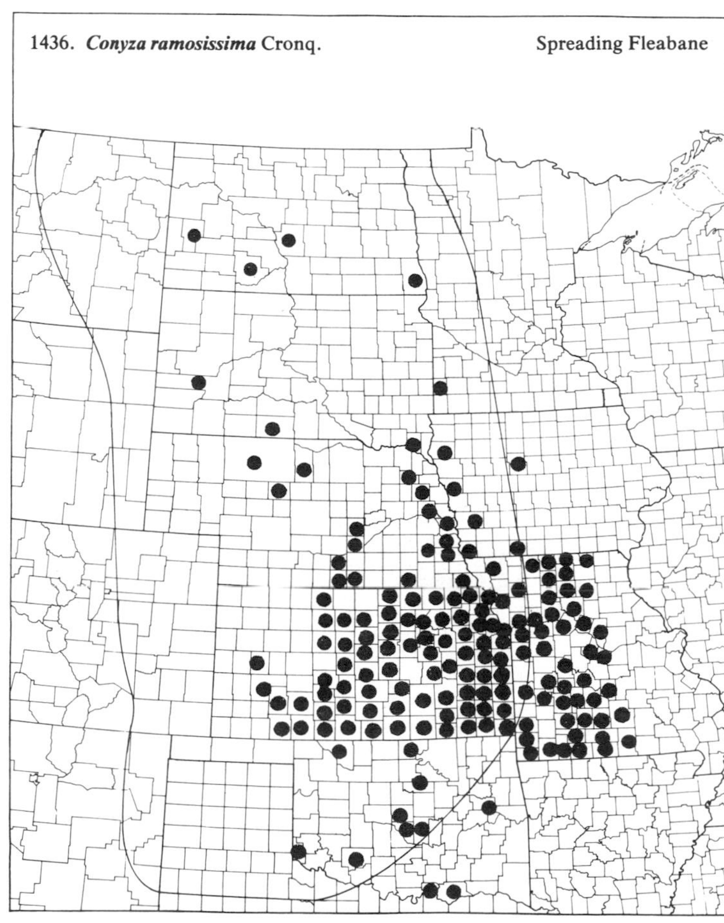

(Asteraceae)

1437. ***Coreopsis grandiflora*** Hogg — Bigflower Coreopsis

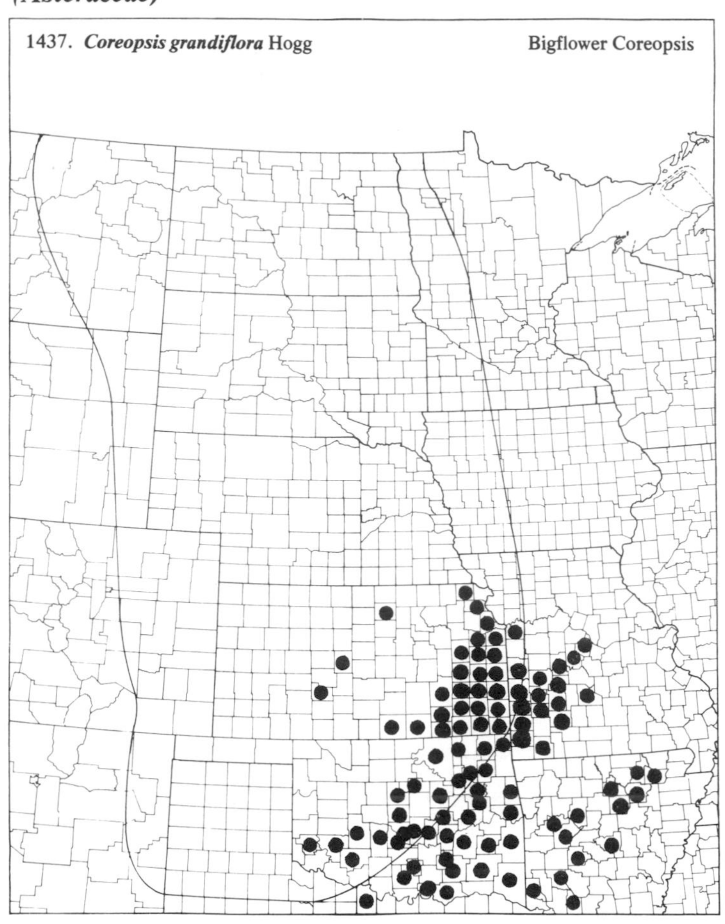

1438. ***Coreopsis palmata*** Nutt. — Finger Coreopsis

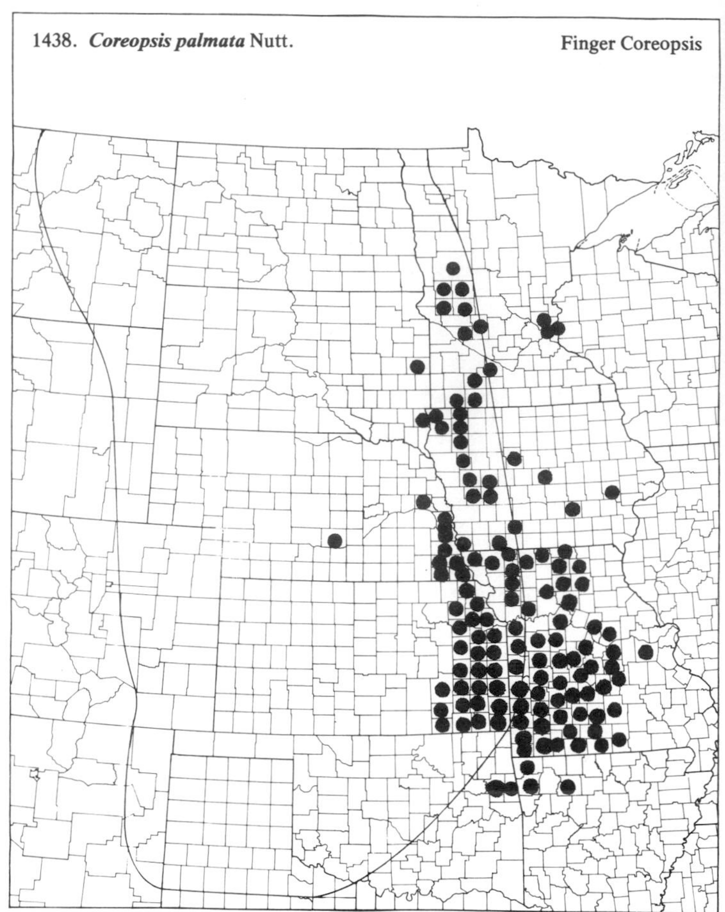

1439. ***Coreopsis tinctoria*** Nutt. — Plains Coreopsis

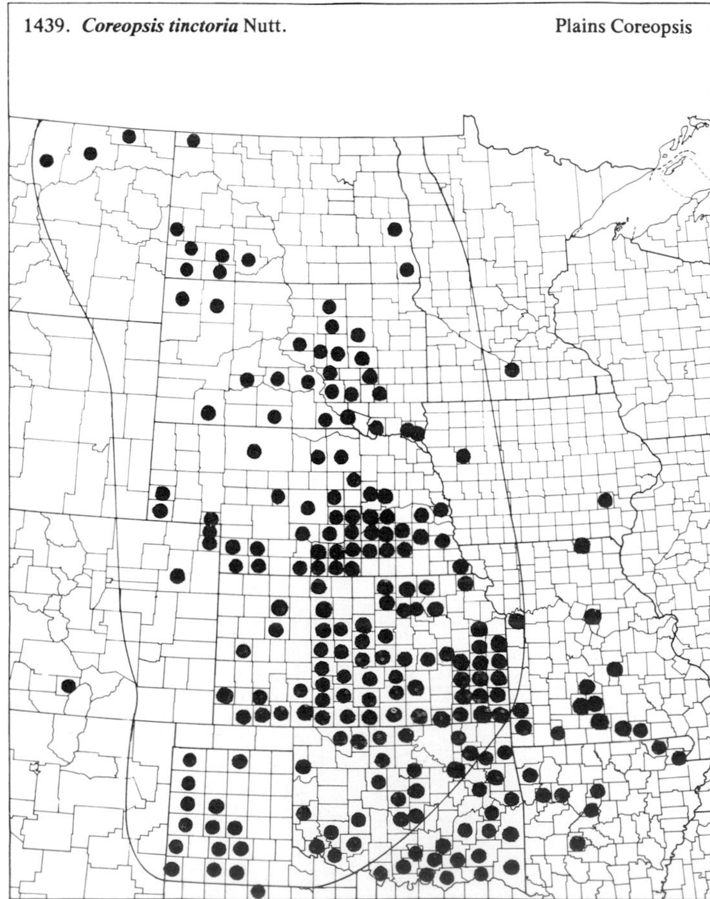

1440. ***Coreopsis tripteris*** L. — Tall Coreopsis

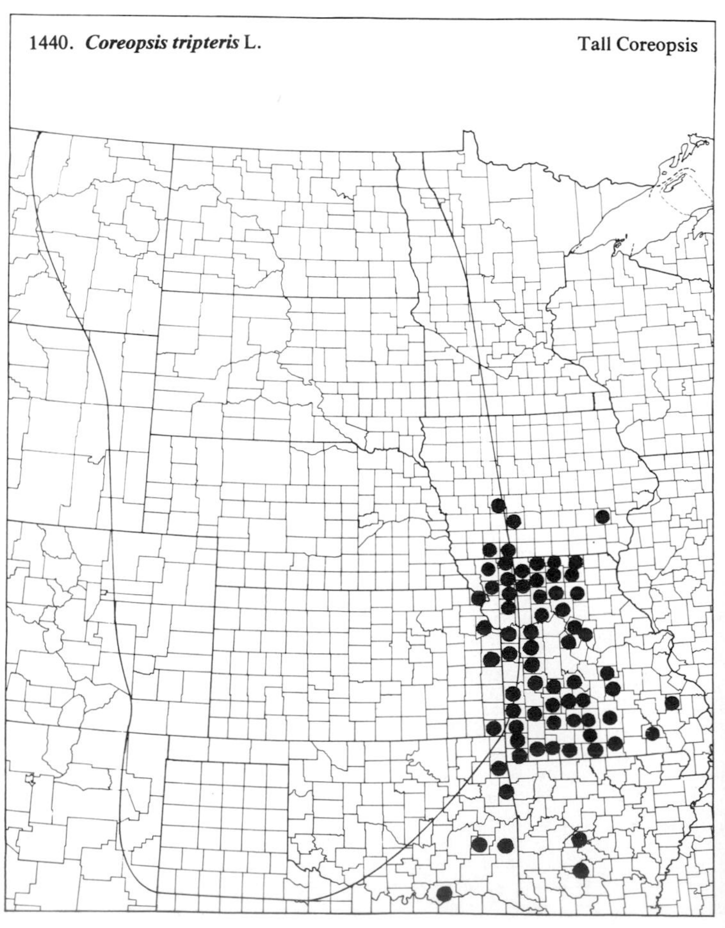

(Asteraceae)

1441. ***Crepis acuminata*** Nutt.

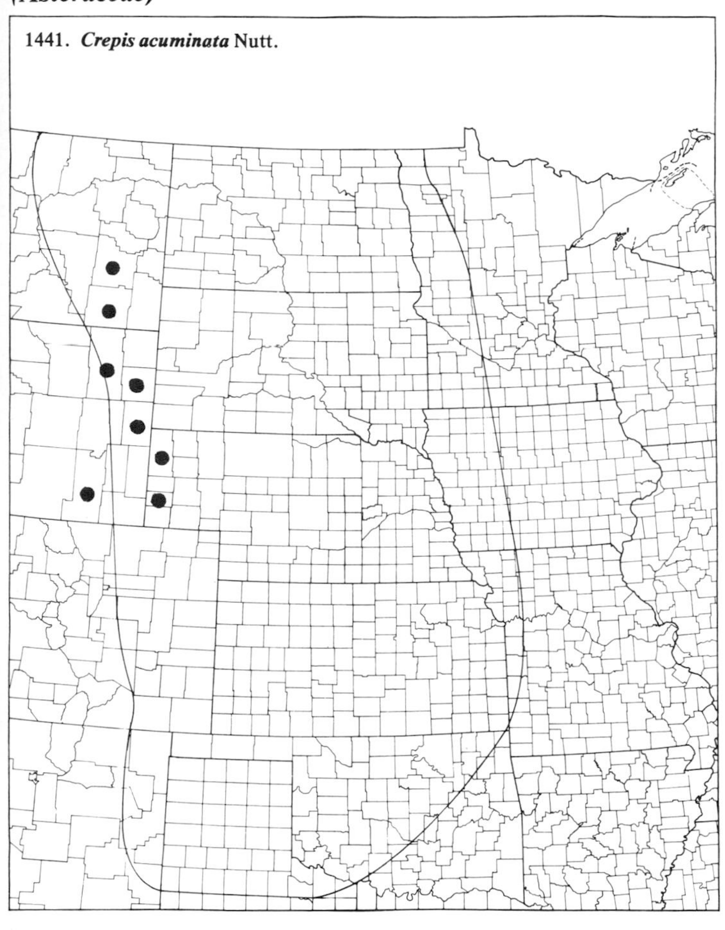

1442. ***Crepis occidentalis*** Nutt.

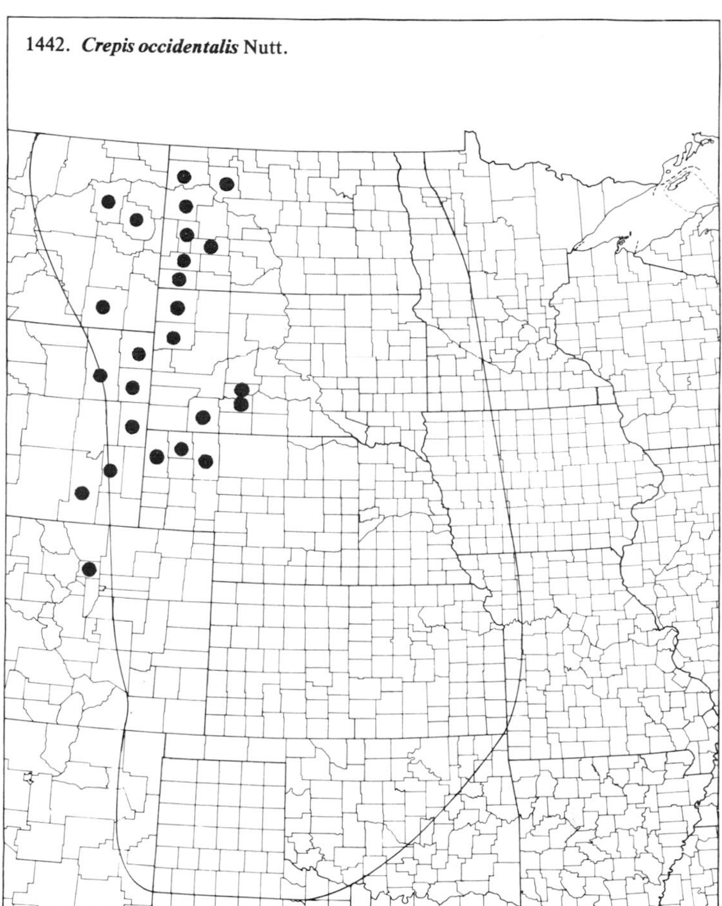

1443. ***Crepis runcinata*** (James) T. & G. Hawk's-beard

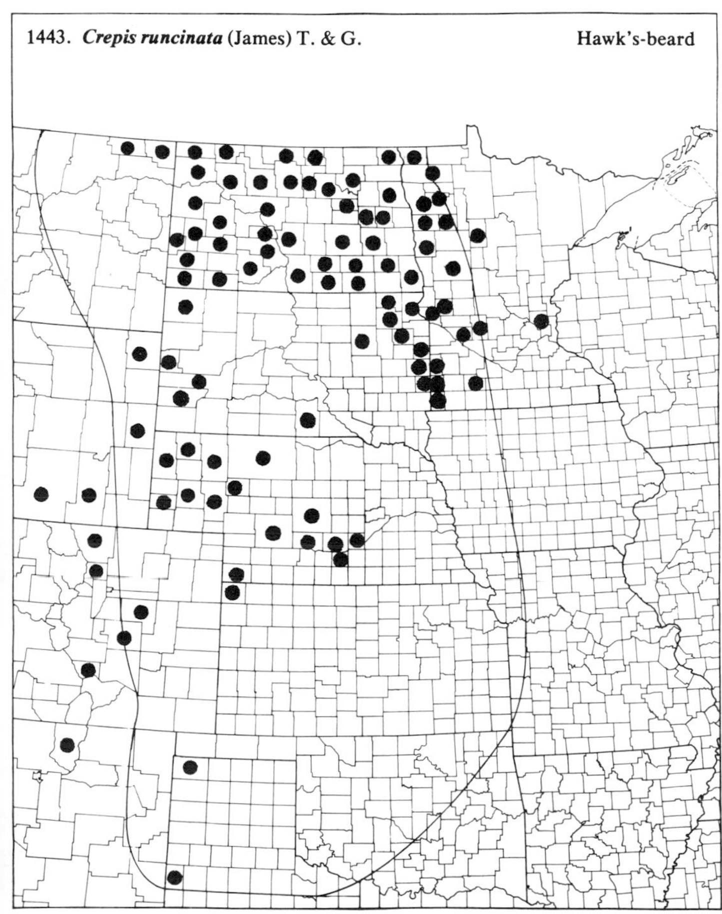

1444. ***Crepis tectorum*** L.

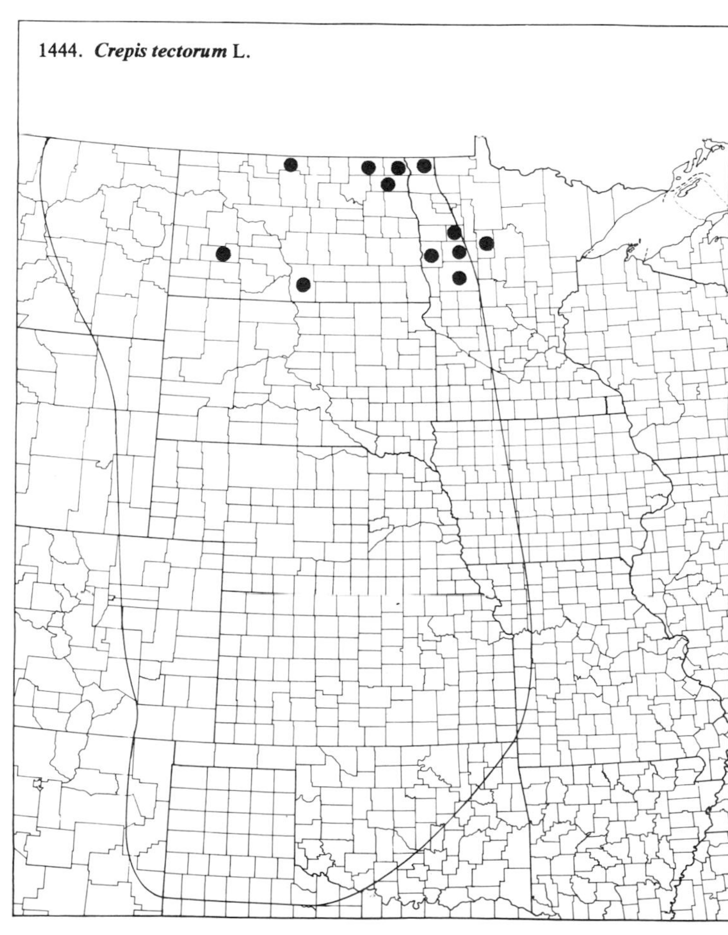

(Asteraceae)

1445. ***Dyssodia papposa*** (Vent.) Hitchc. Fetid Marigold

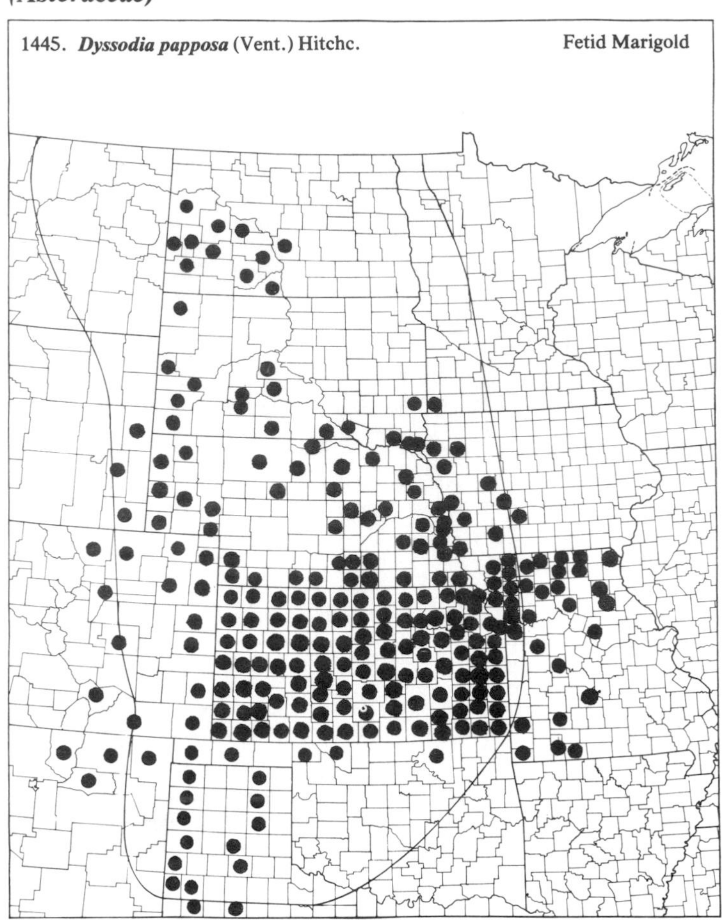

1446. ***Echinacea angustifolia*** DC. var. ***angustifolia*** Purple Coneflower

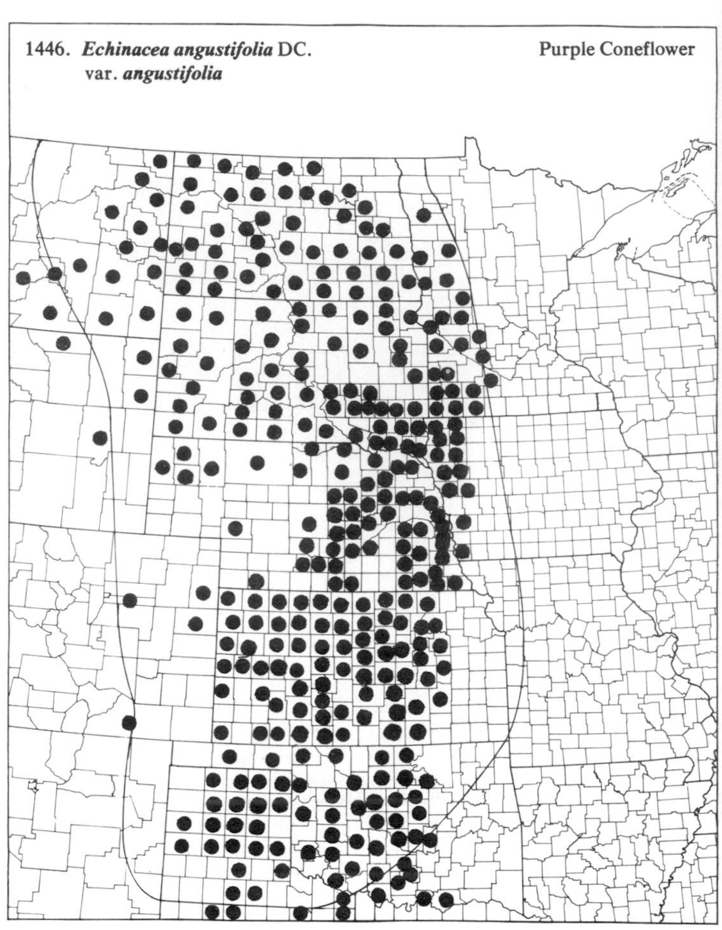

1447. ***Echinacea angustifolia*** DC. var. ***strigosa*** McGreg. Purple Coneflower

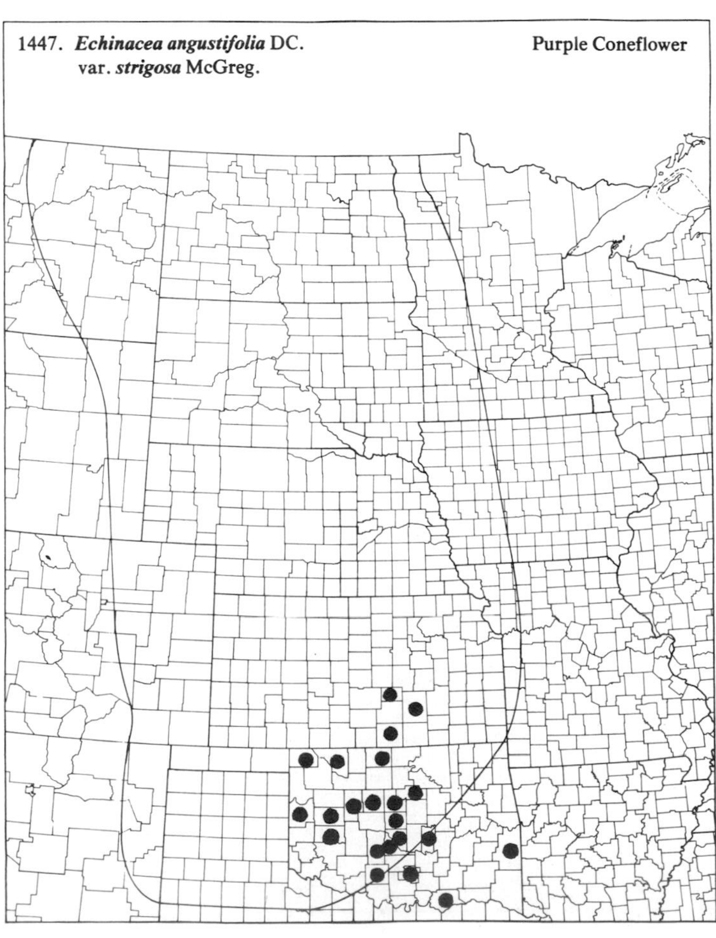

1448. ***Echinacea atrorubens*** Nutt. Yellow Echinacea

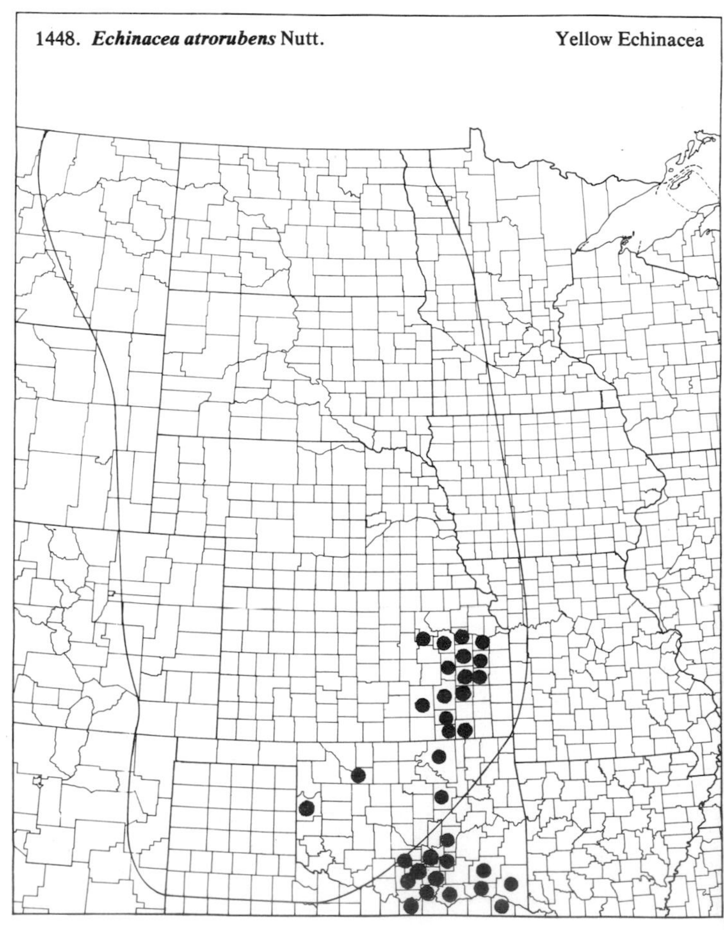

(Asteraceae)

1449. **Echinacea pallida** (Nutt.) Nutt. — Pale Echinacea

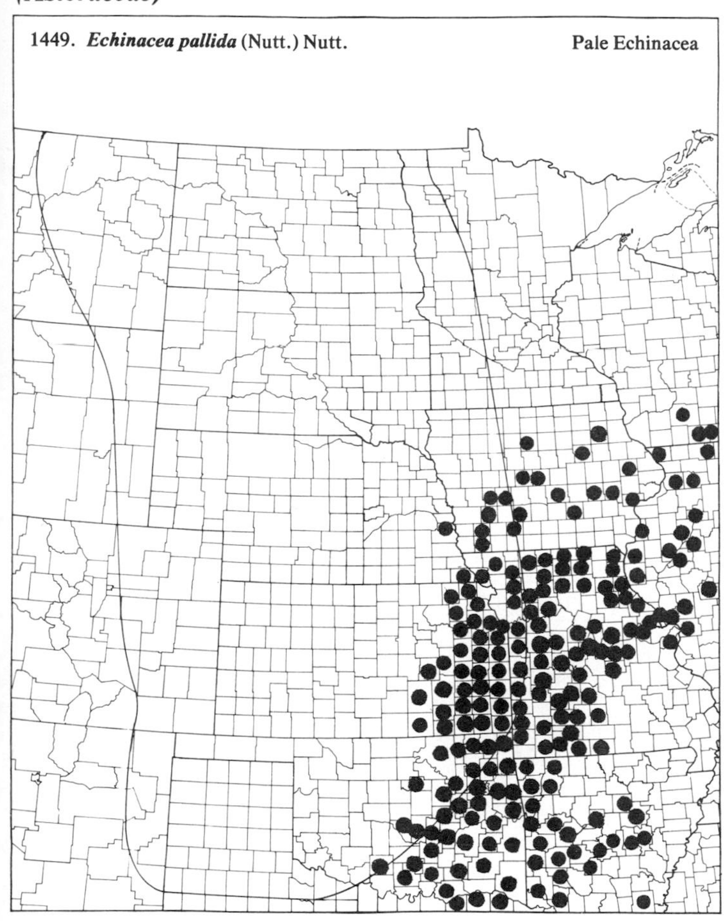

1450. **Eclipta alba** (L.) Hassk. — Yerba-de-Tajo

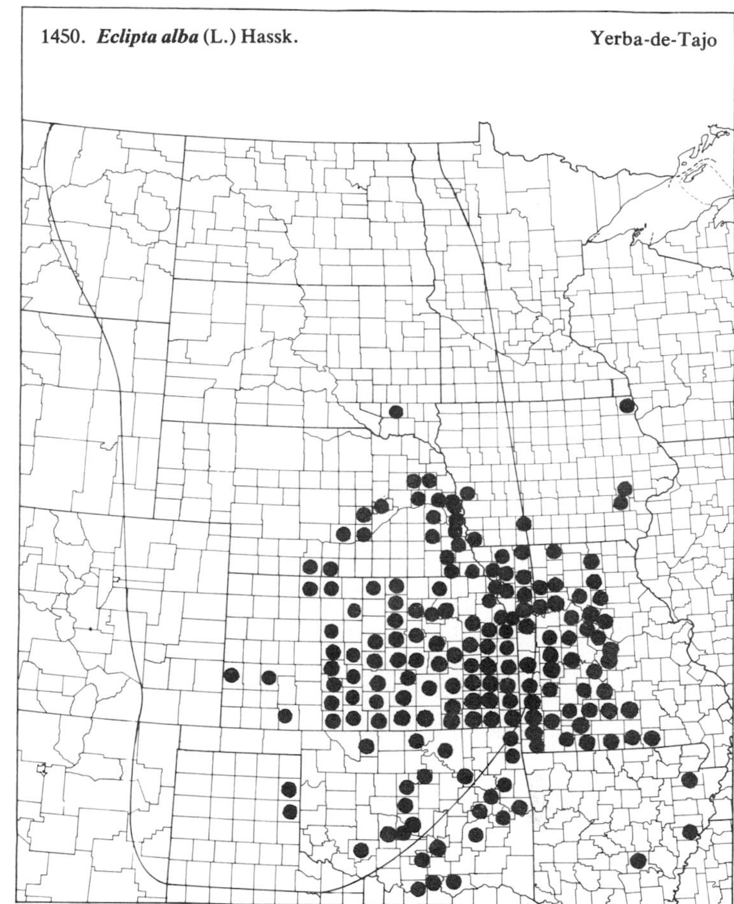

1451. **Elephantopus carolinianus** Willd. — Elephantsfoot

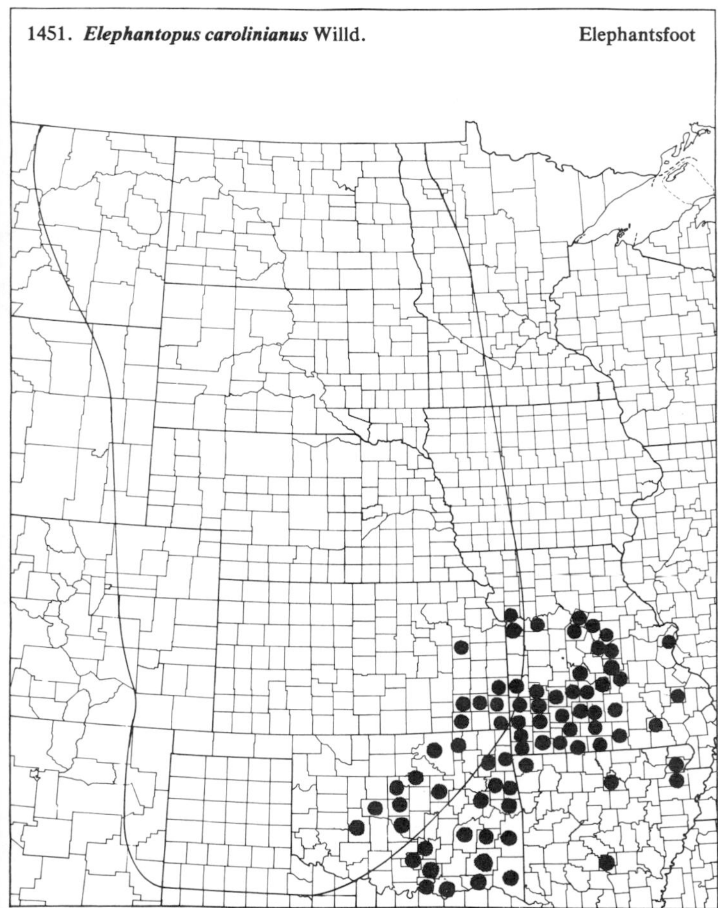

1452. **Engelmannia pinnatifida** T. & G. — Engelmann's Daisy

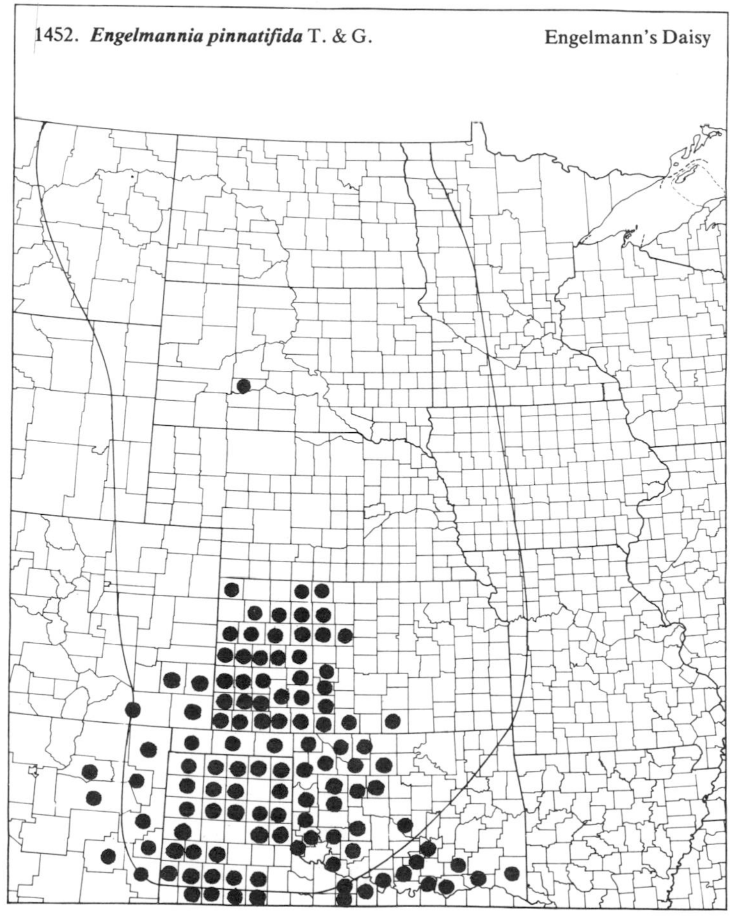

(Asteraceae)

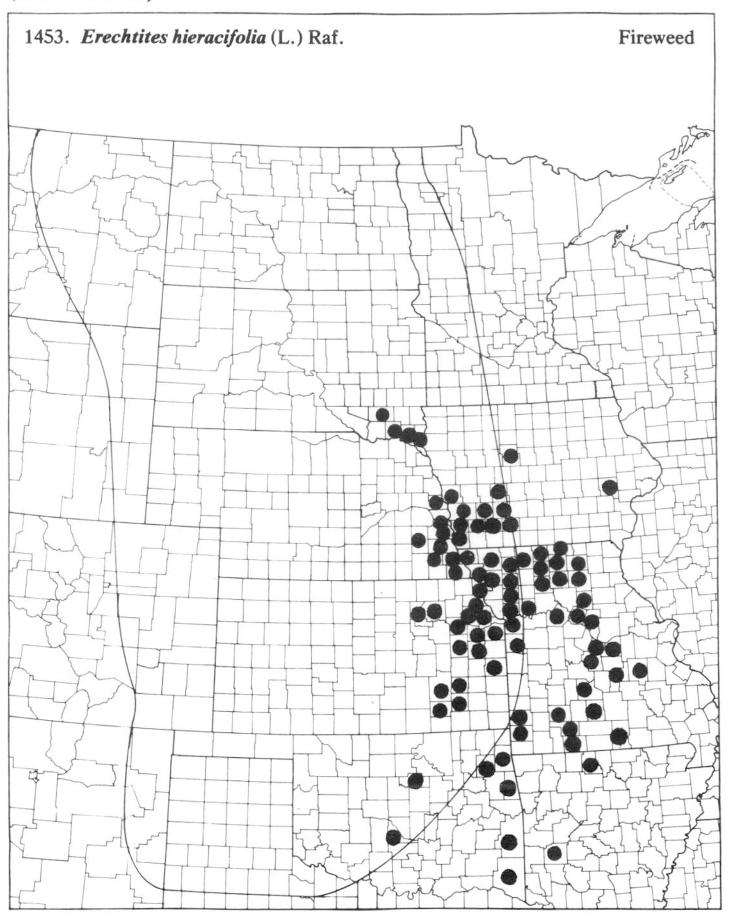
1453. *Erechtites hieracifolia* (L.) Raf.
Fireweed

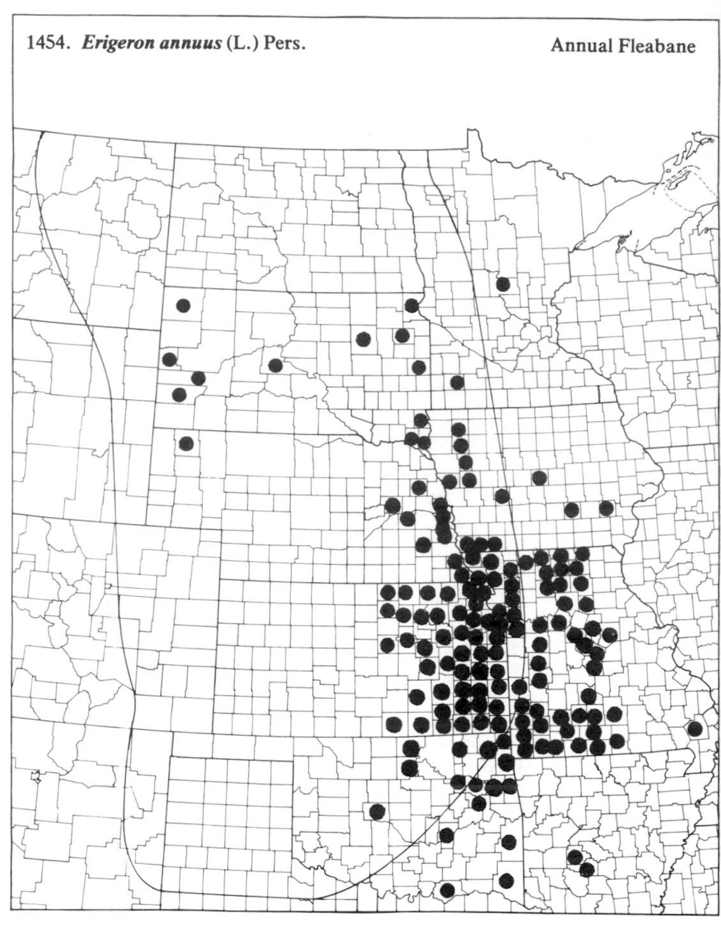
1454. *Erigeron annuus* (L.) Pers.
Annual Fleabane

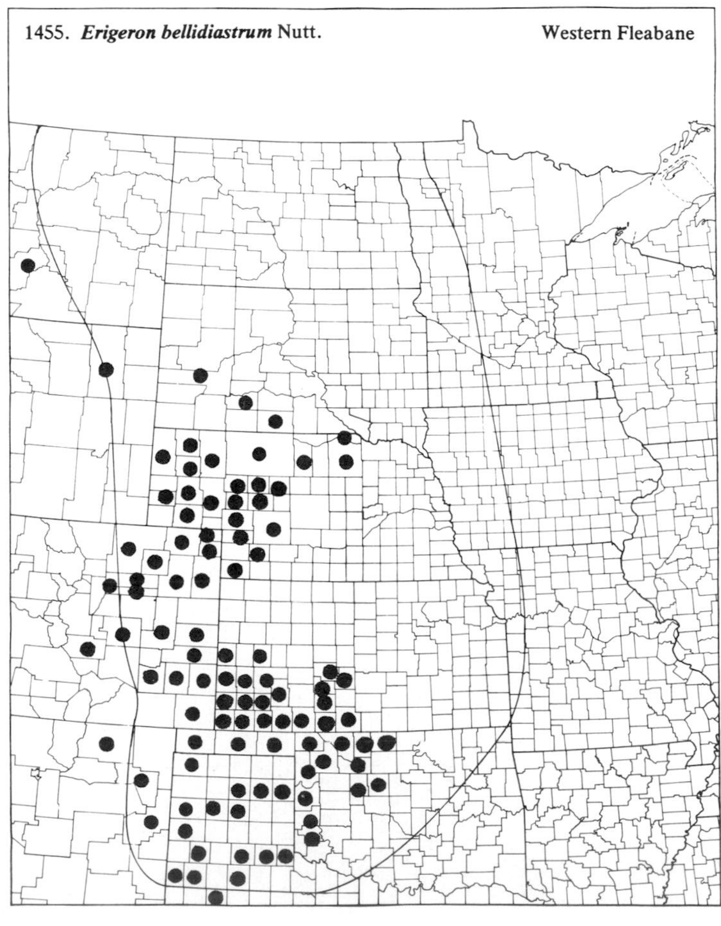
1455. *Erigeron bellidiastrum* Nutt.
Western Fleabane

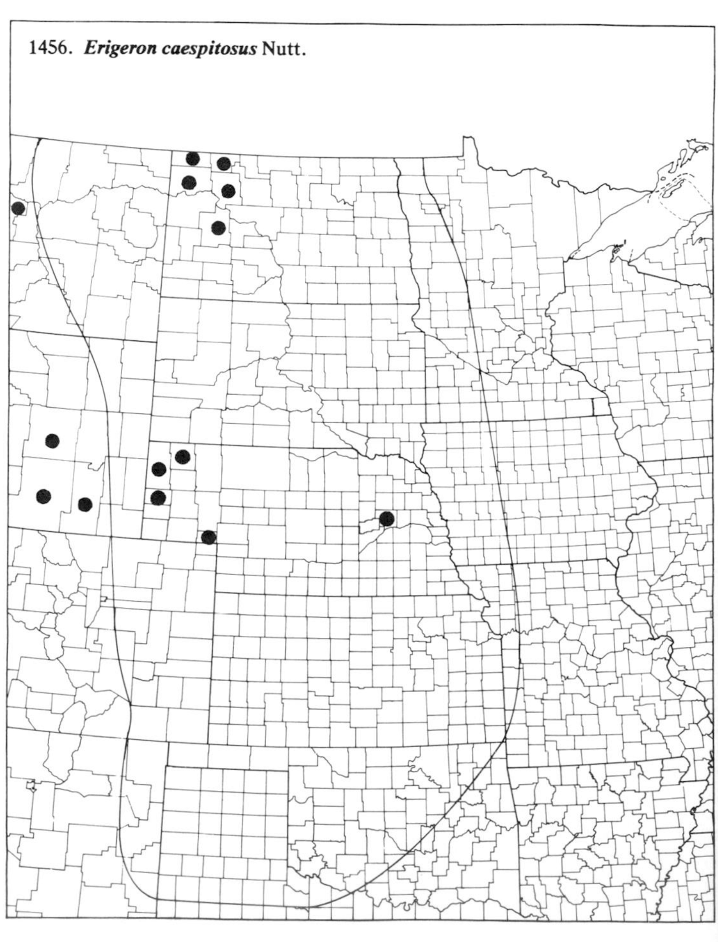
1456. *Erigeron caespitosus* Nutt.

(Asteraceae)

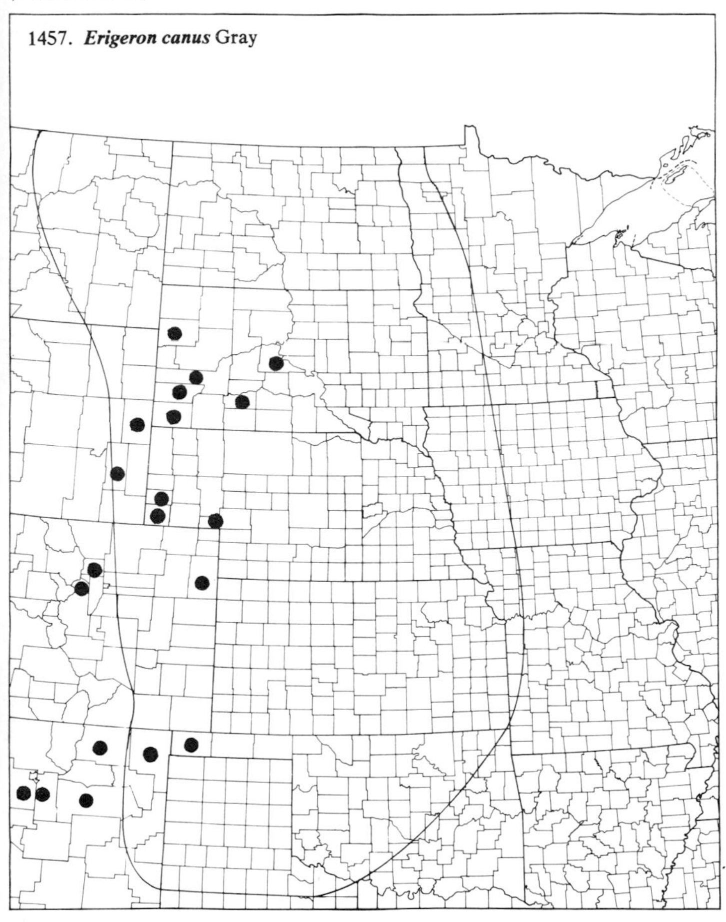
1457. ***Erigeron canus*** Gray

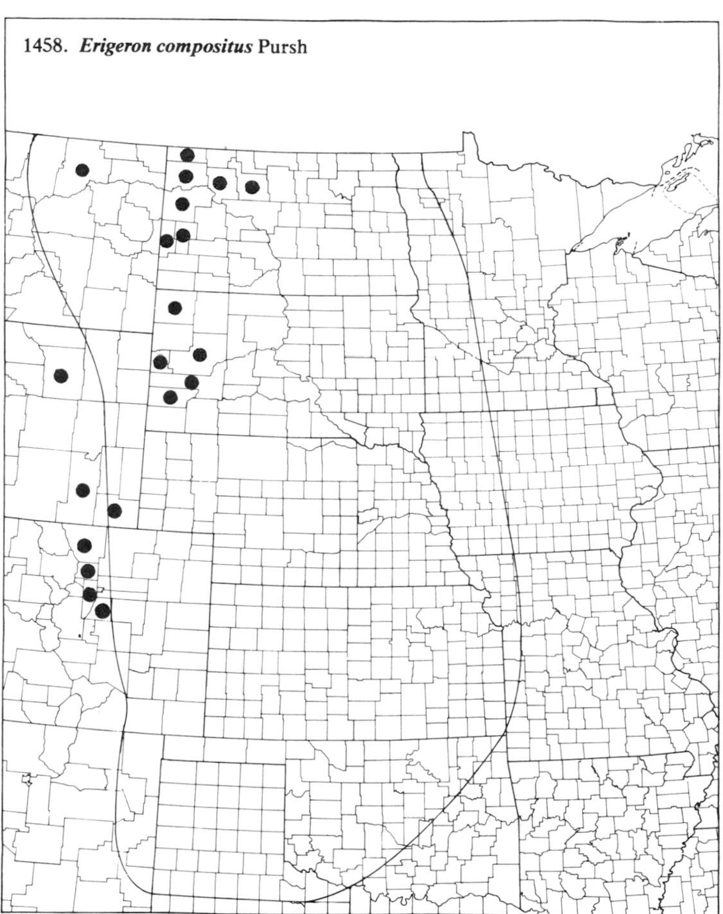
1458. ***Erigeron compositus*** Pursh

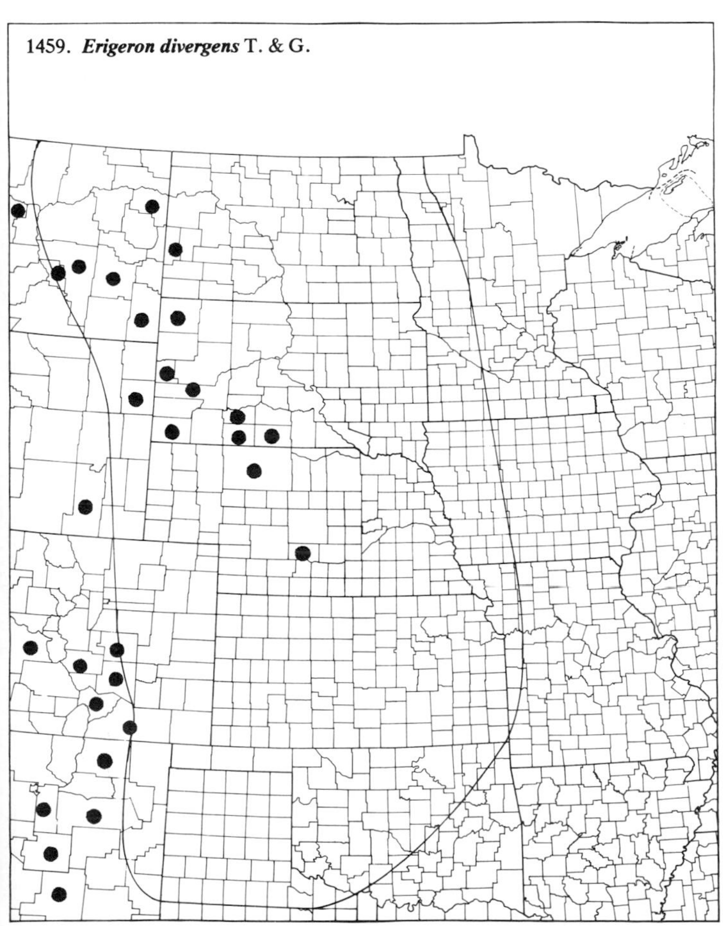
1459. ***Erigeron divergens*** T. & G.

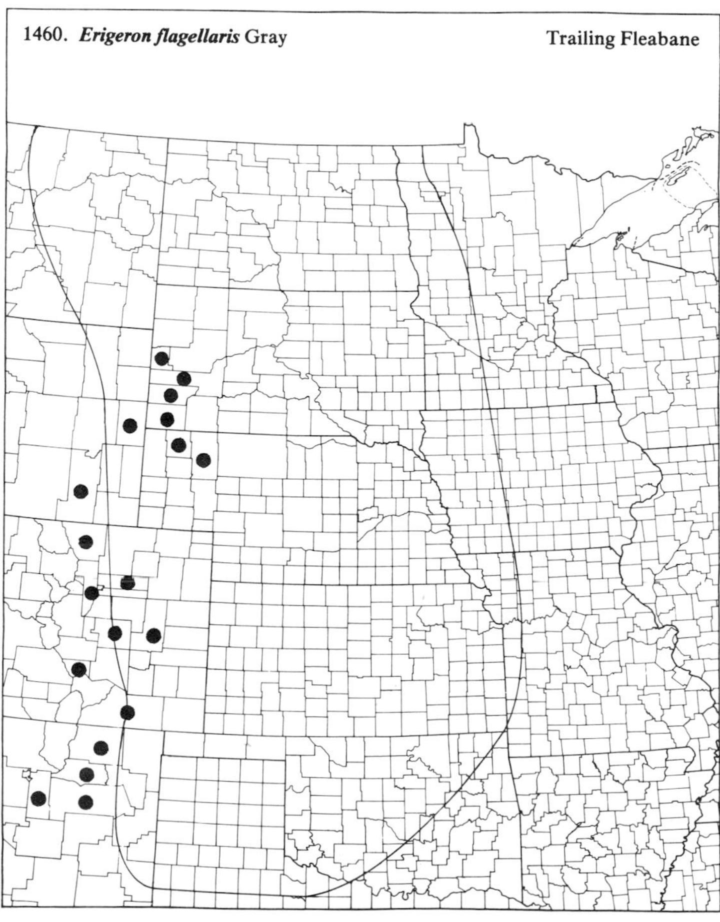
1460. ***Erigeron flagellaris*** Gray — Trailing Fleabane

(Asteraceae)

1461. ***Erigeron glabellus*** Nutt.

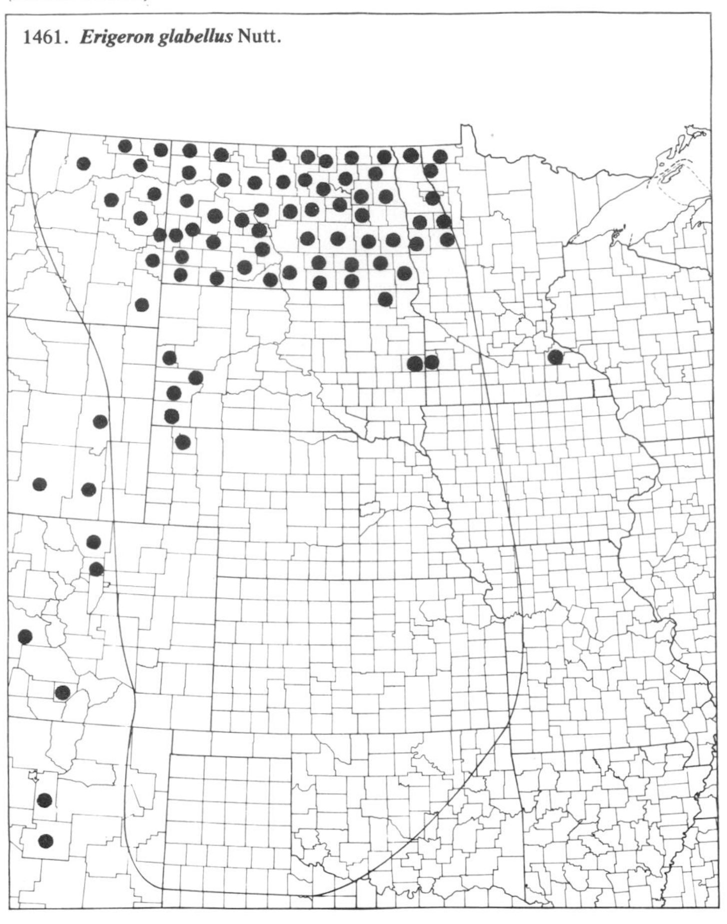

1462. ***Erigeron lonchophyllus*** Hook.

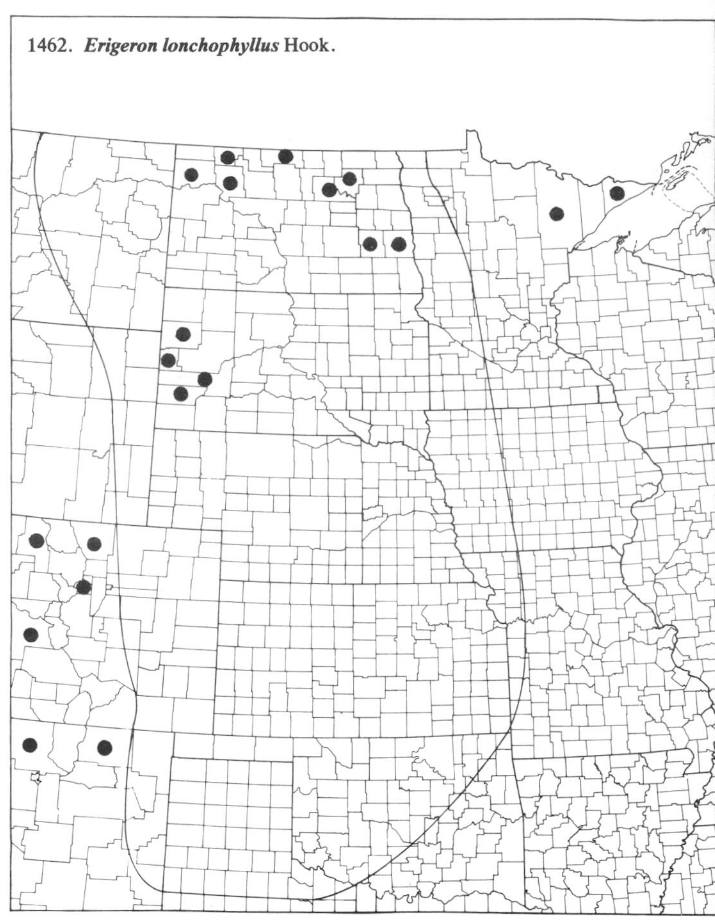

1463. ***Erigeron modestus*** Gray

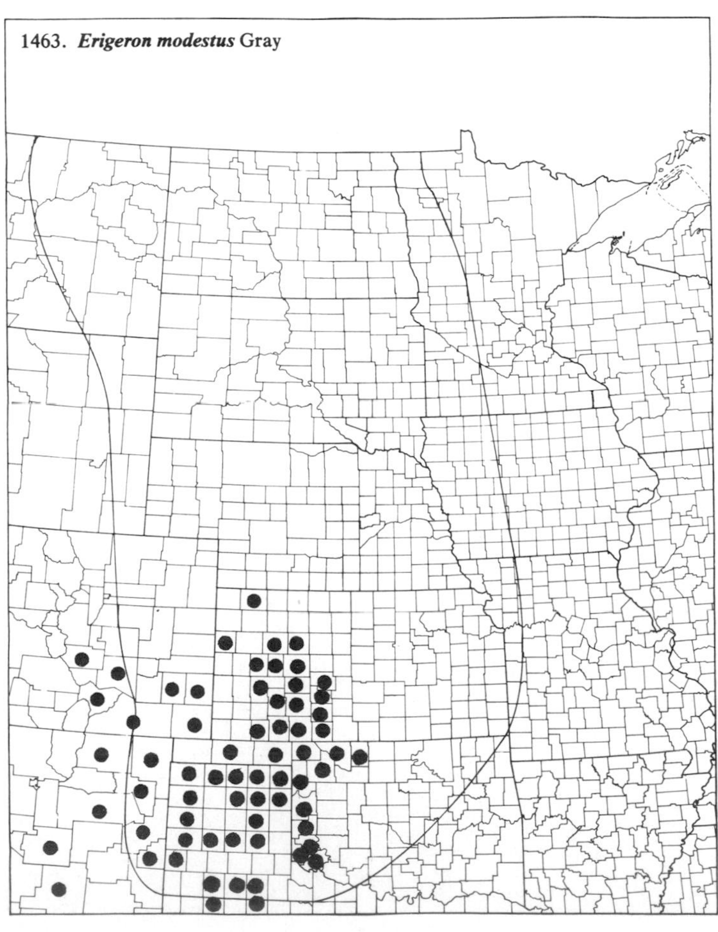

1464. ***Erigeron philadelphicus*** L. Philadelphia Fleabane

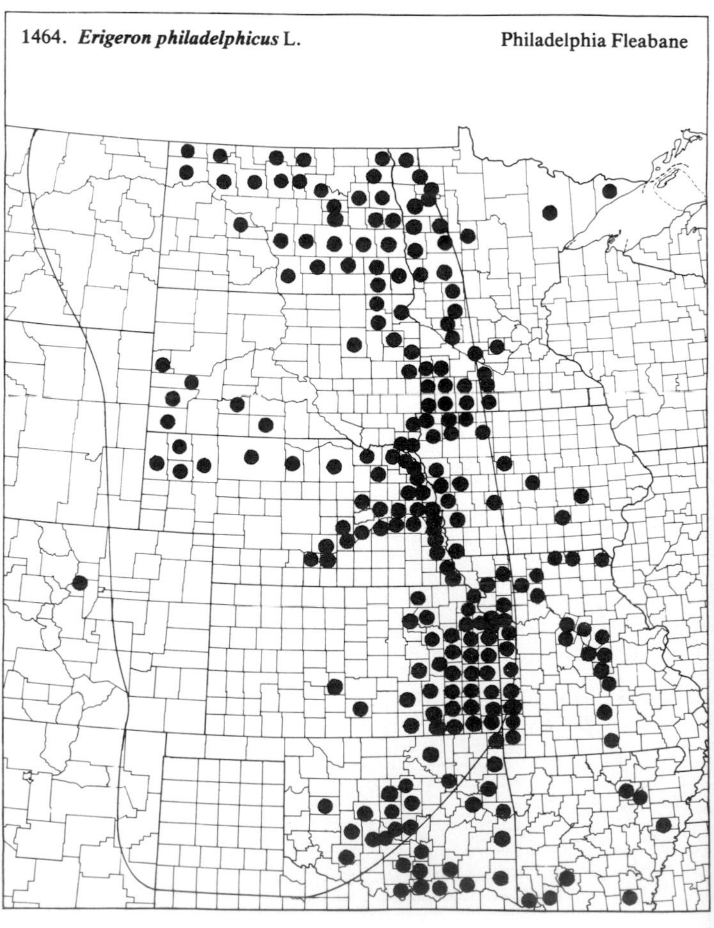

(Asteraceae)

1465. ***Erigeron pumilus*** Nutt. Low Fleabane

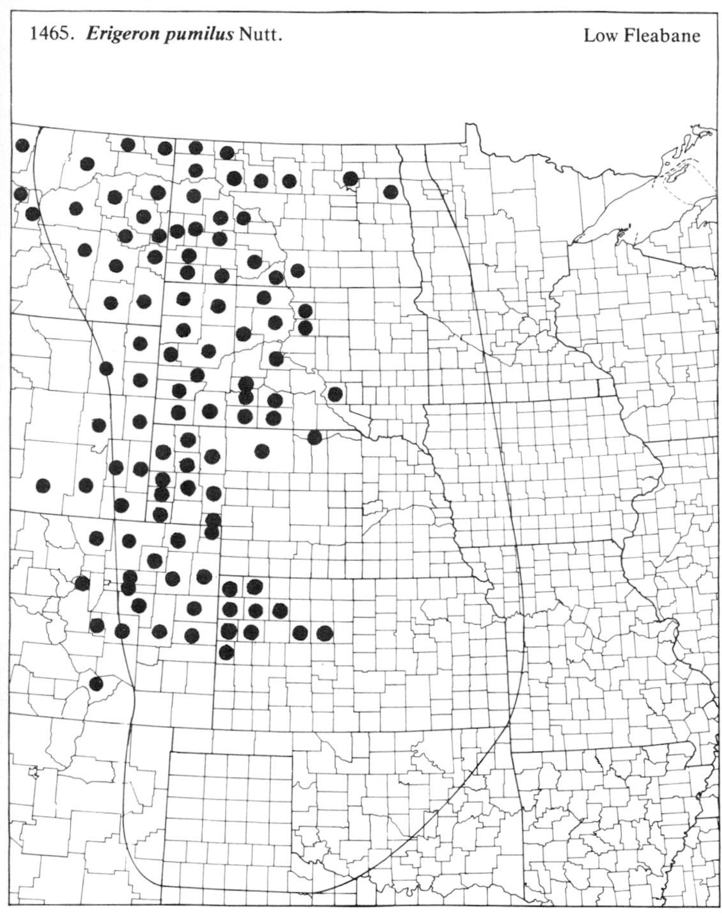

1466. ***Erigeron strigosus*** Muhl. Daisy Fleabane

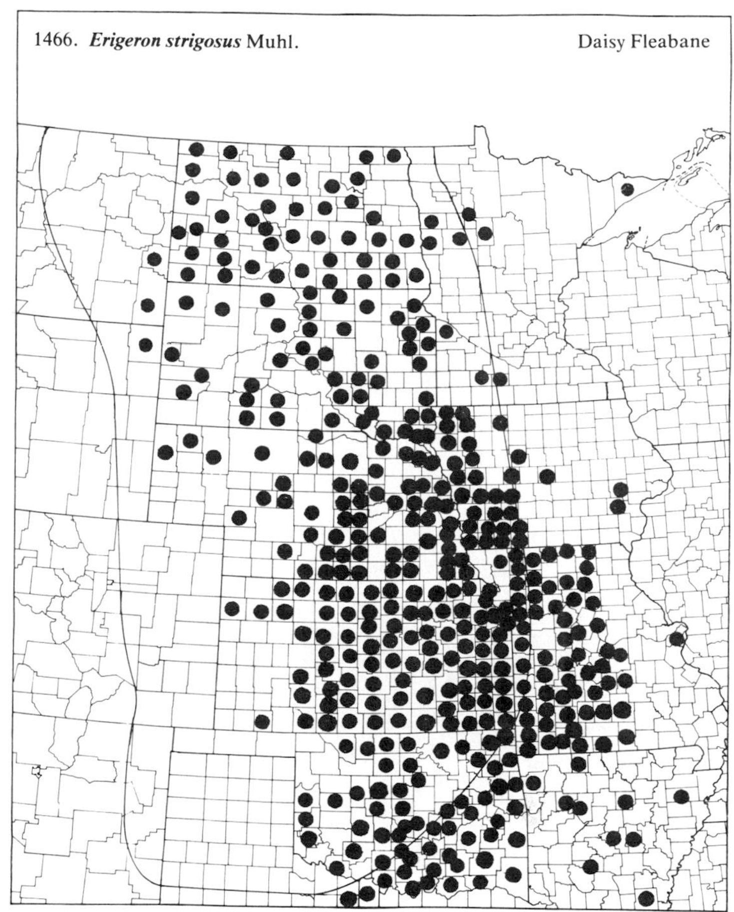

1467. ***Erigeron subtrinervis*** Rydb.

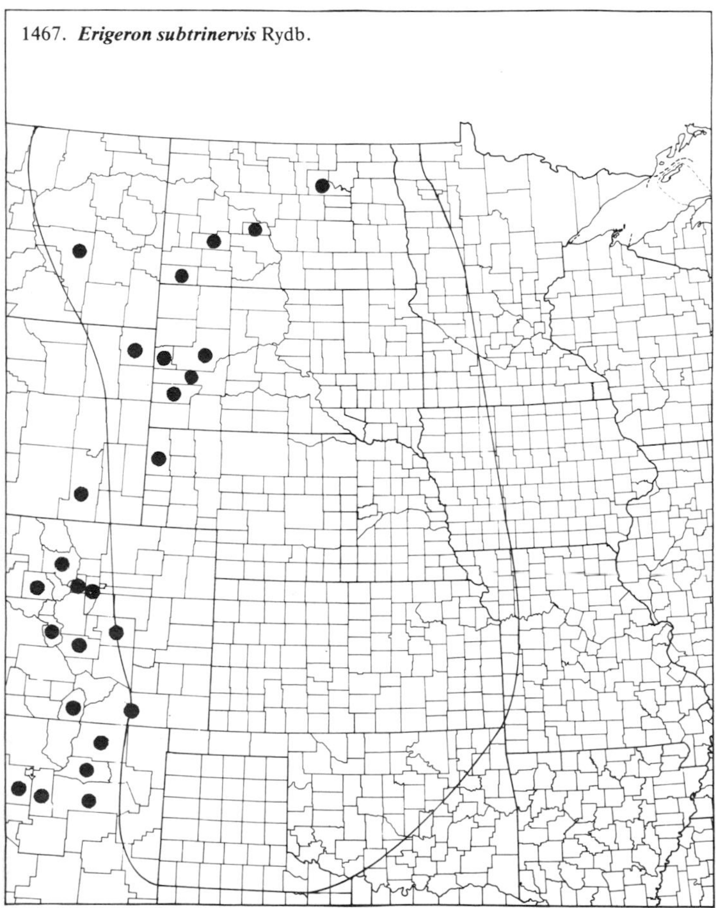

1468. ***Erigeron tenuis*** T. & G.

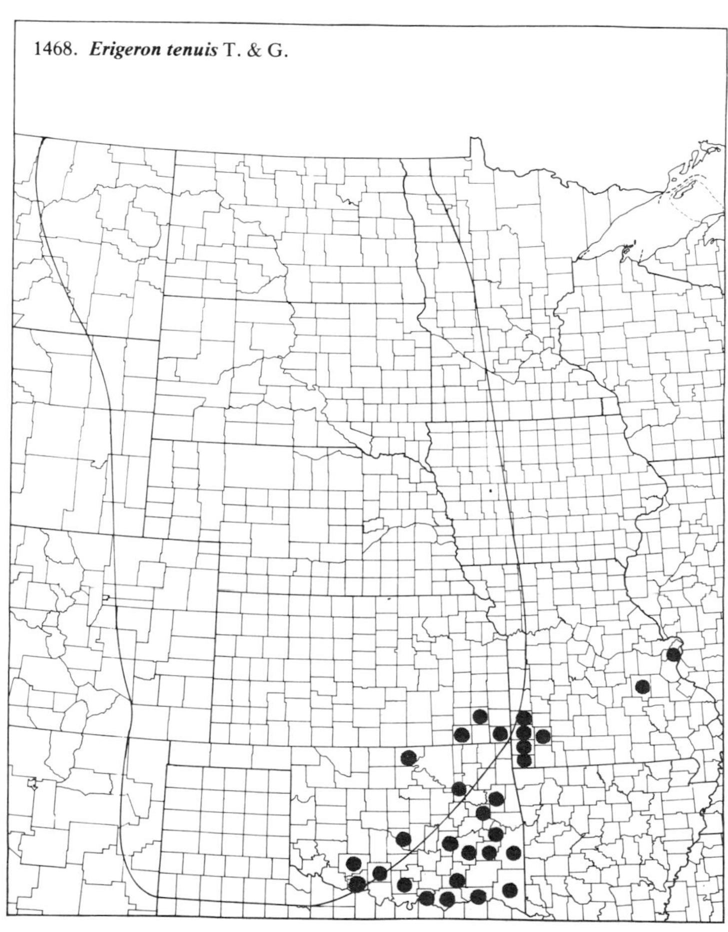

(Asteraceae)

1469. ***Eupatorium altissimum*** L. Tall Joe-Pye-weed

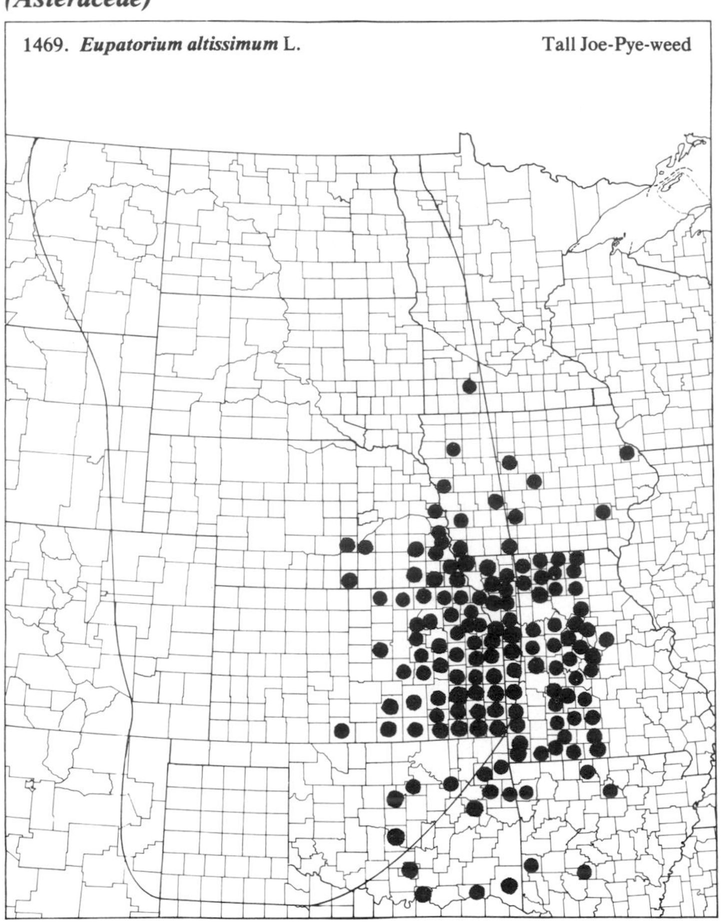

1470. ***Eupatorium coelestinum*** L. Mistflower

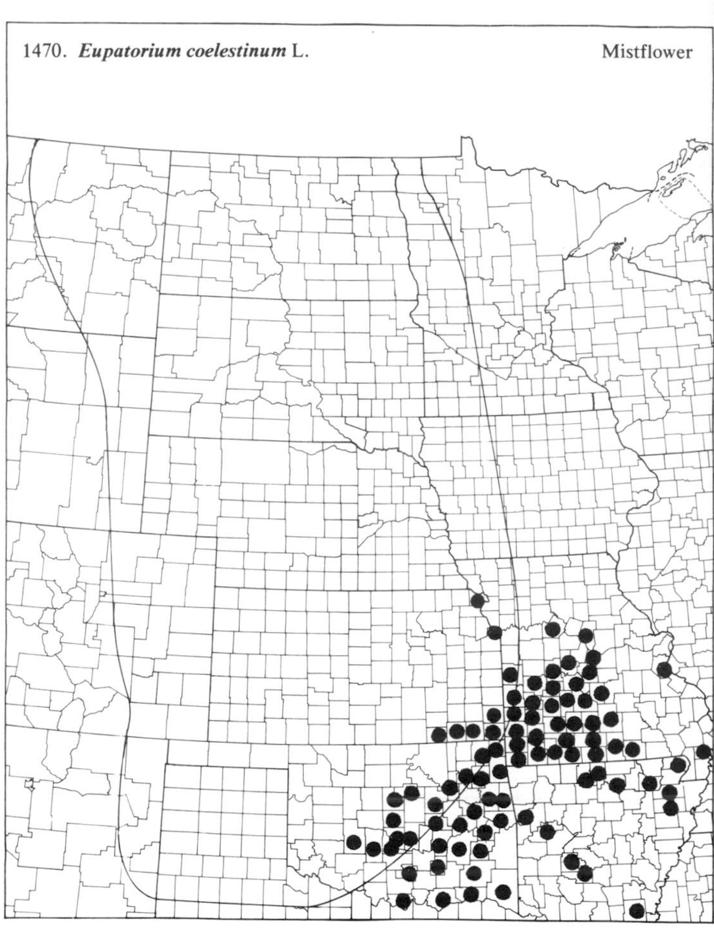

1471. ***Eupatorium maculatum*** L. var. ***bruneri*** (Gray) Breitung Spotted Joe-Pye-weed

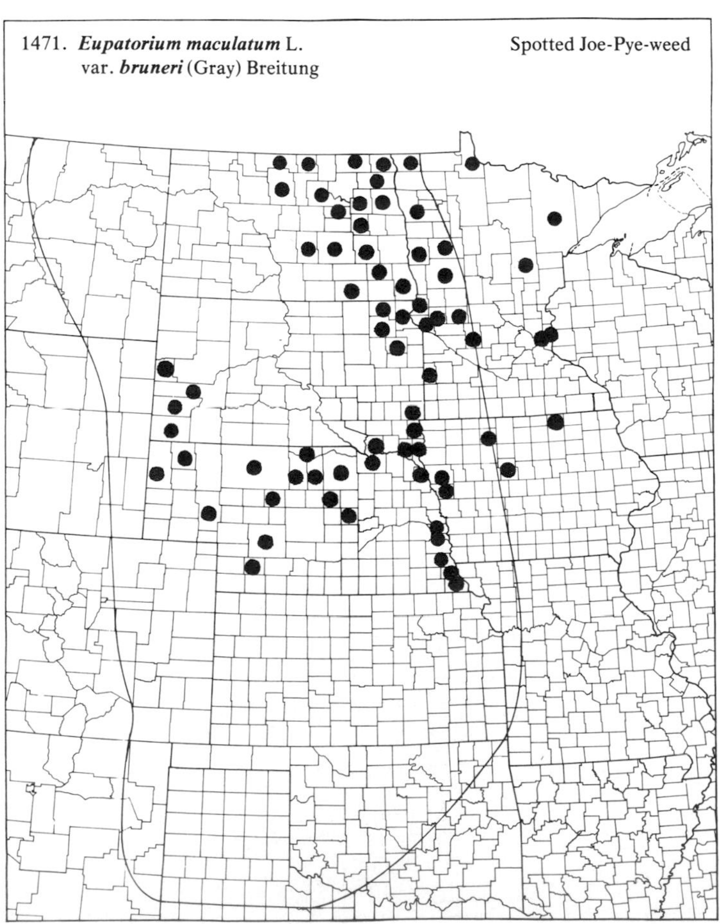

1472. ***Eupatorium perfoliatum*** L. Boneset

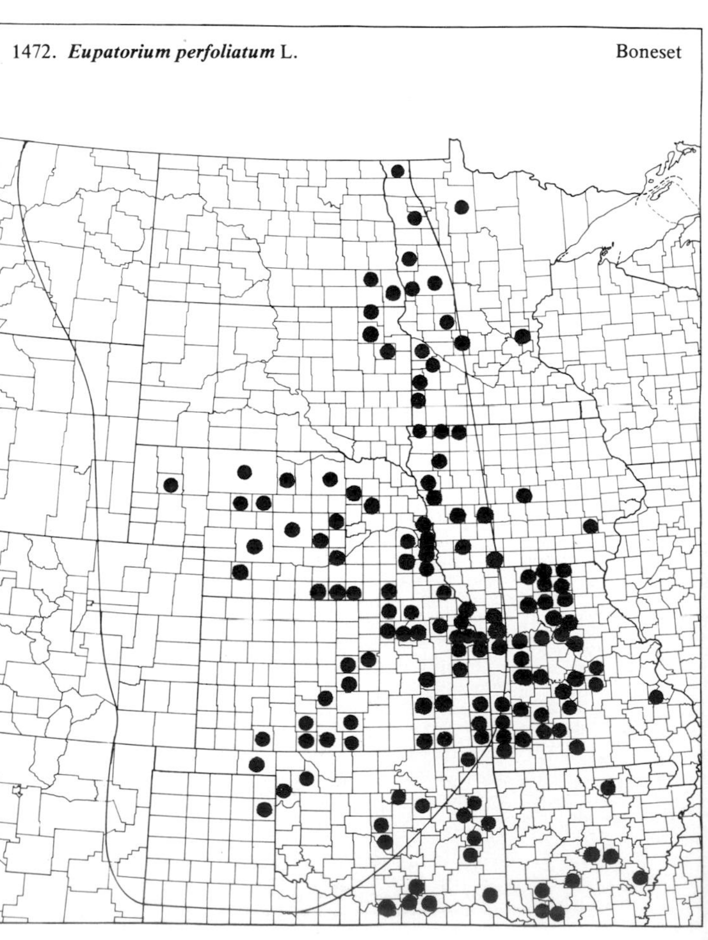

(Asteraceae)

1473. ***Eupatorium purpureum*** L. Sweet Joe-Pye-weed

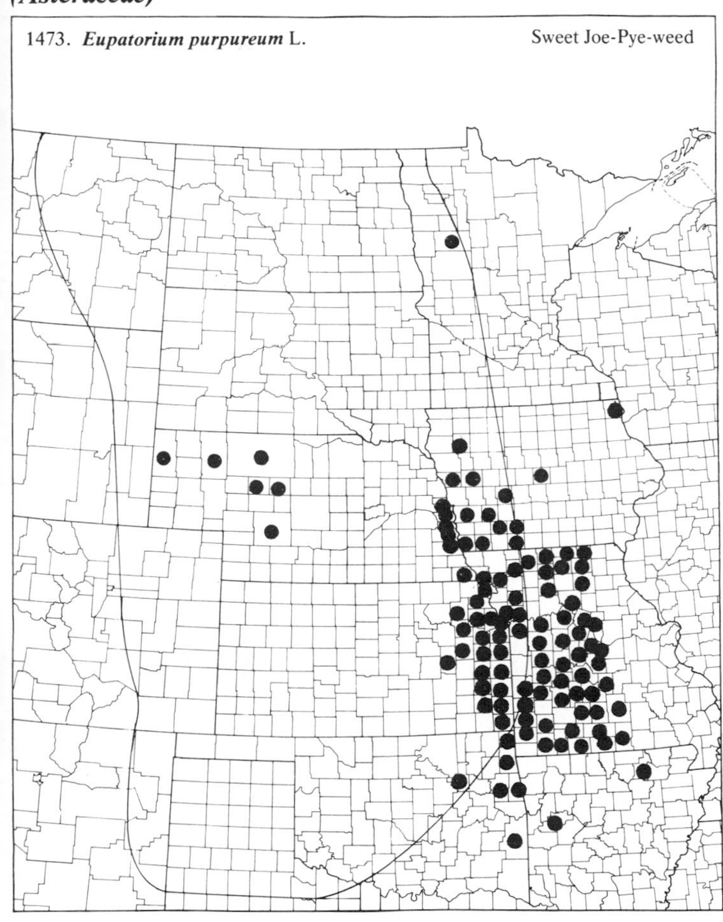

1474. ***Eupatorium rugosum*** Houtt. White Snakeroot

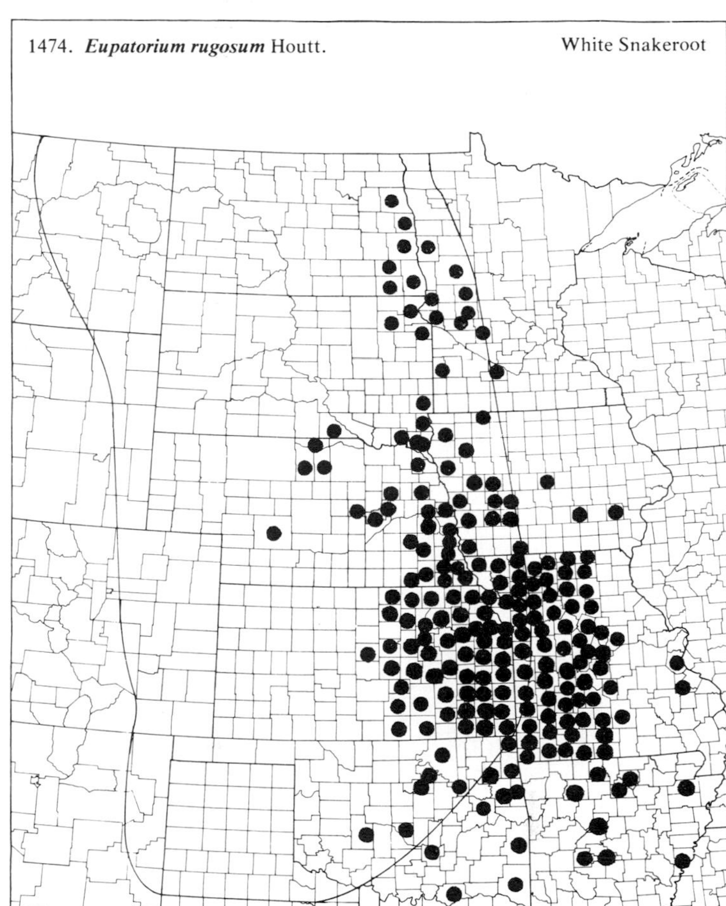

1475. ***Eupatorium serotinum*** Michx. Late Eupatorium

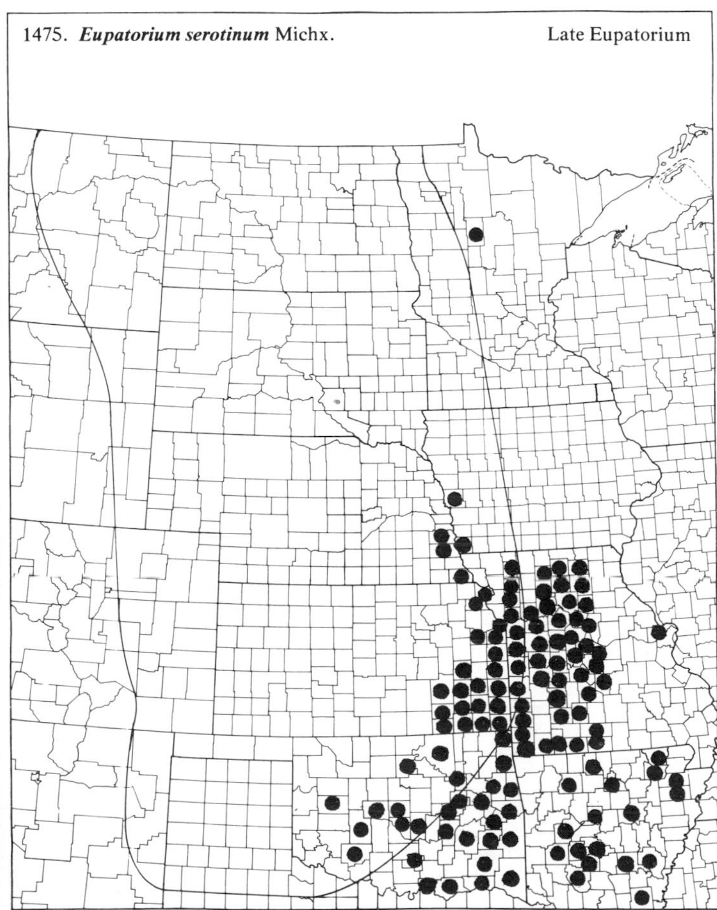

1476. ***Evax prolifera*** Nutt.

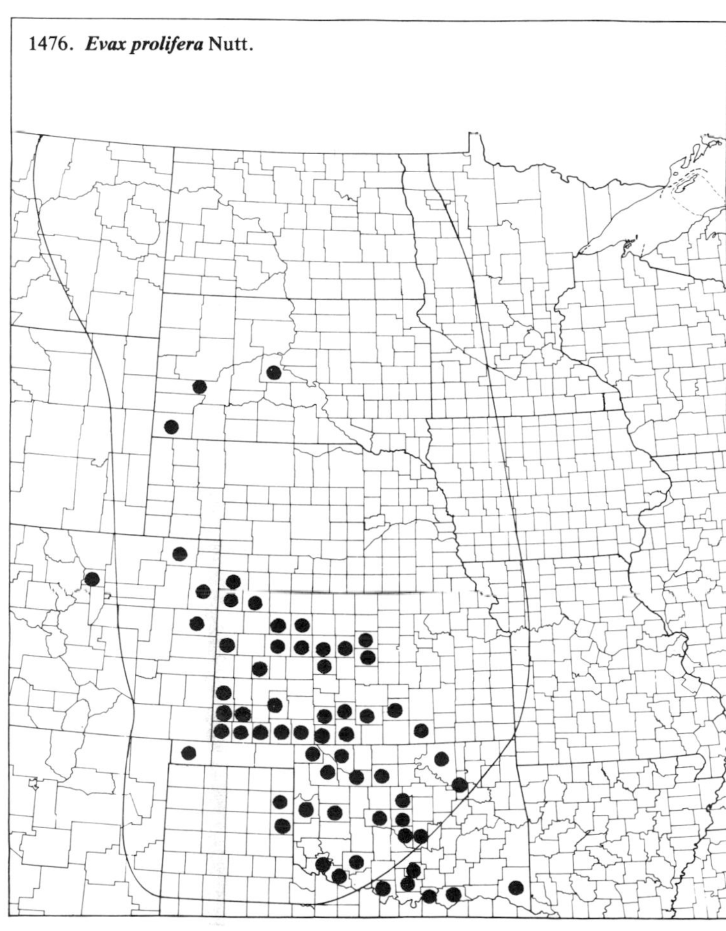

(Asteraceae)

1477. ***Flaveria campestris*** J.R. Johnst. Flaveria

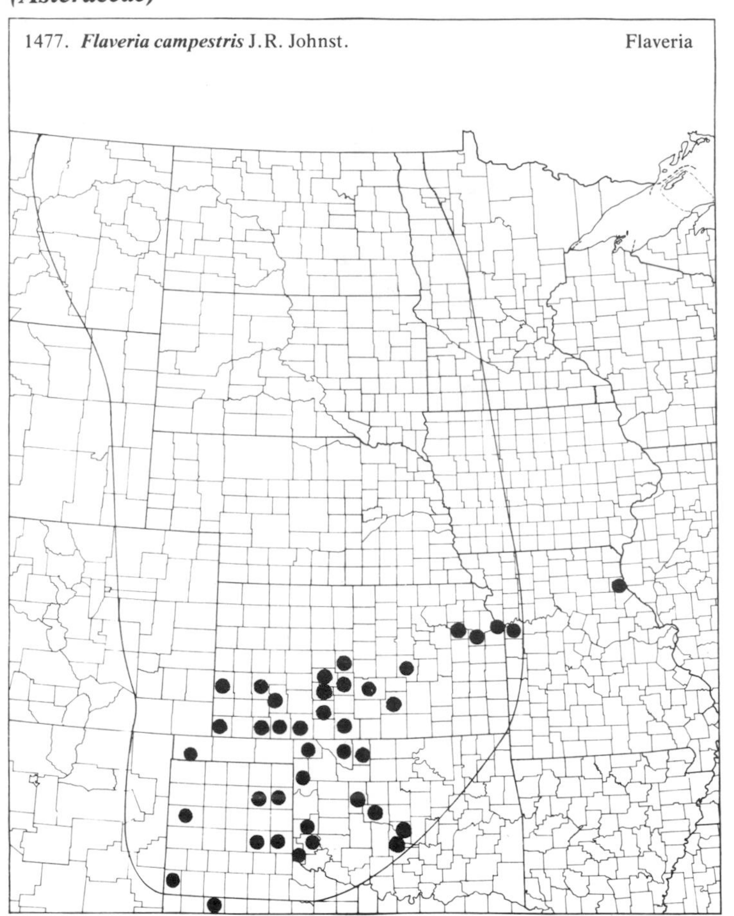

1478. ***Gaillardia aristata*** Pursh Blanketflower

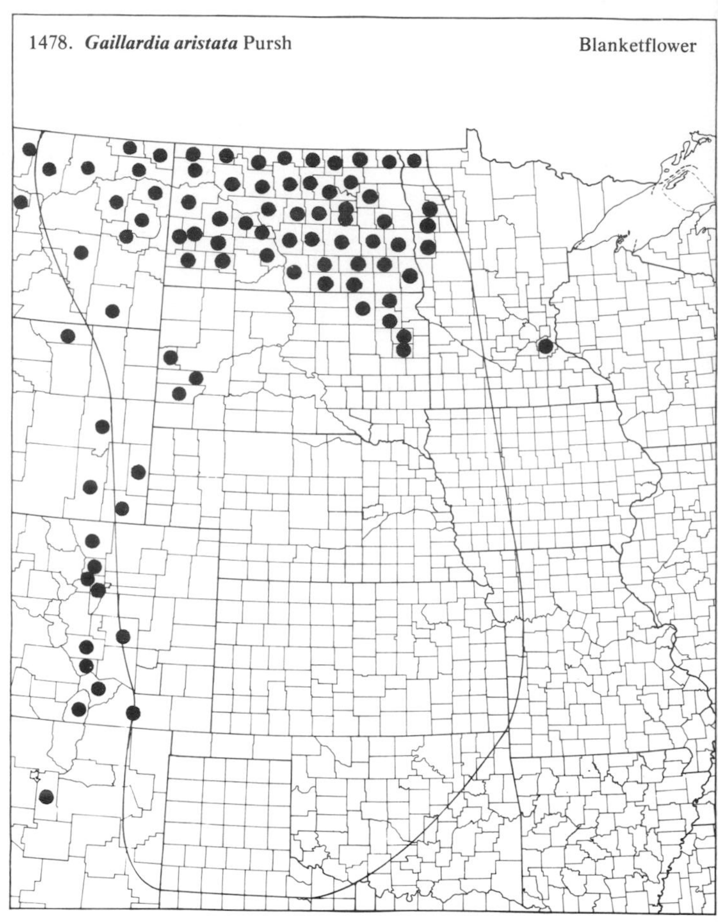

1479. ***Gaillardia fastigiata*** Greene Prairie Gaillardia

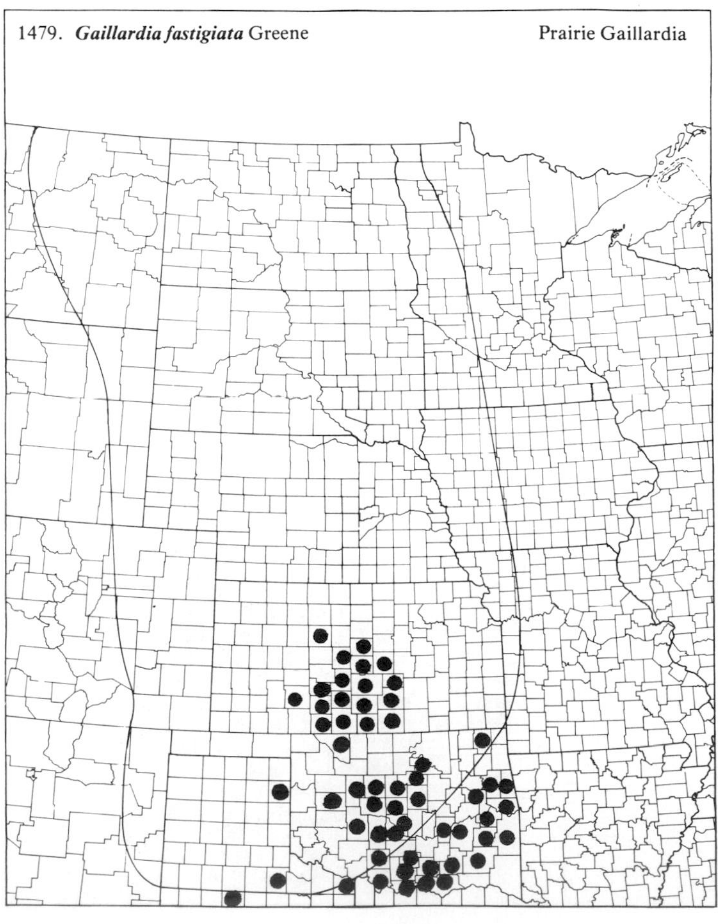

1480. ***Gaillardia pinnatifida*** Torr. Gaillardia

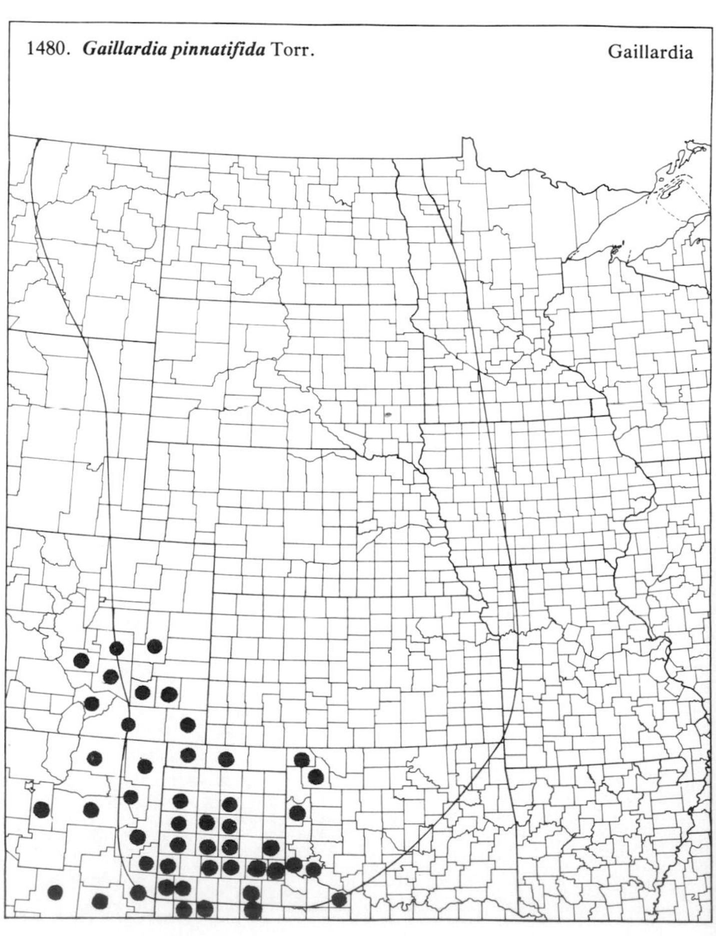

(Asteraceae)

1481. ***Gaillardia pulchella*** Foug. Rose-ring Gaillardia

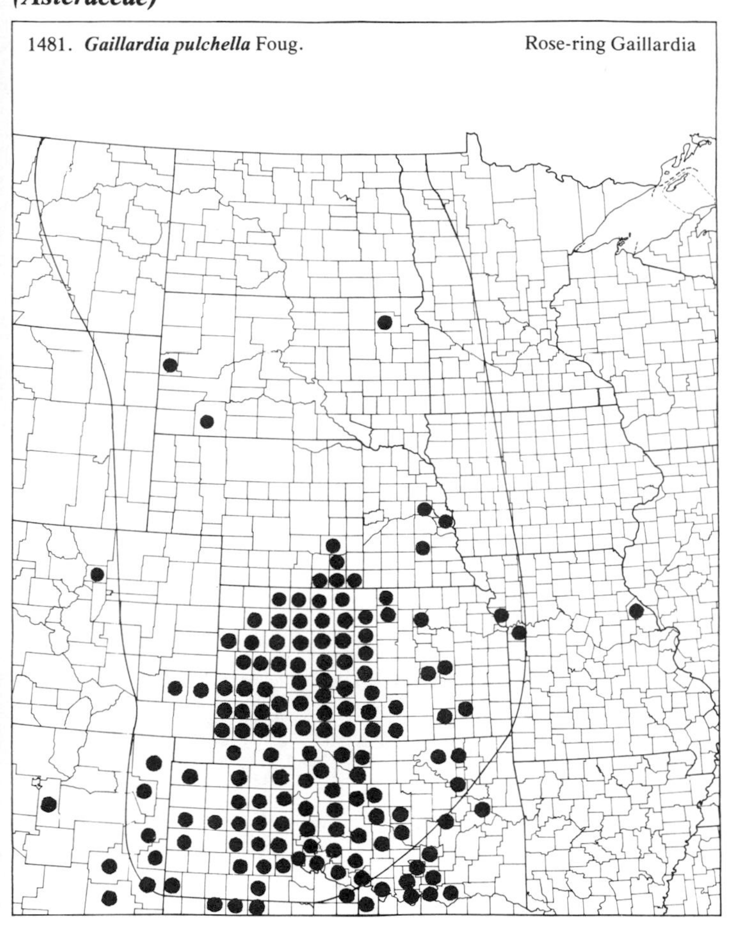

1482. ***Gaillardia suavis*** (Gray & Engelm.) Britt. & Rusby

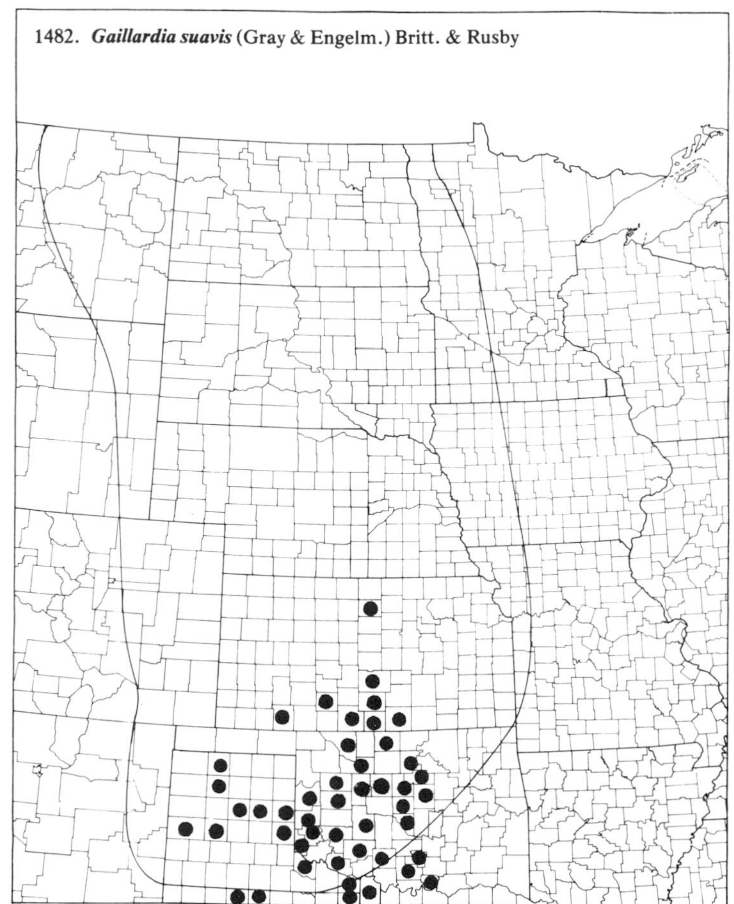

1483. ***Galinsoga ciliata*** (Raf.) Blake Fringed Quickweed

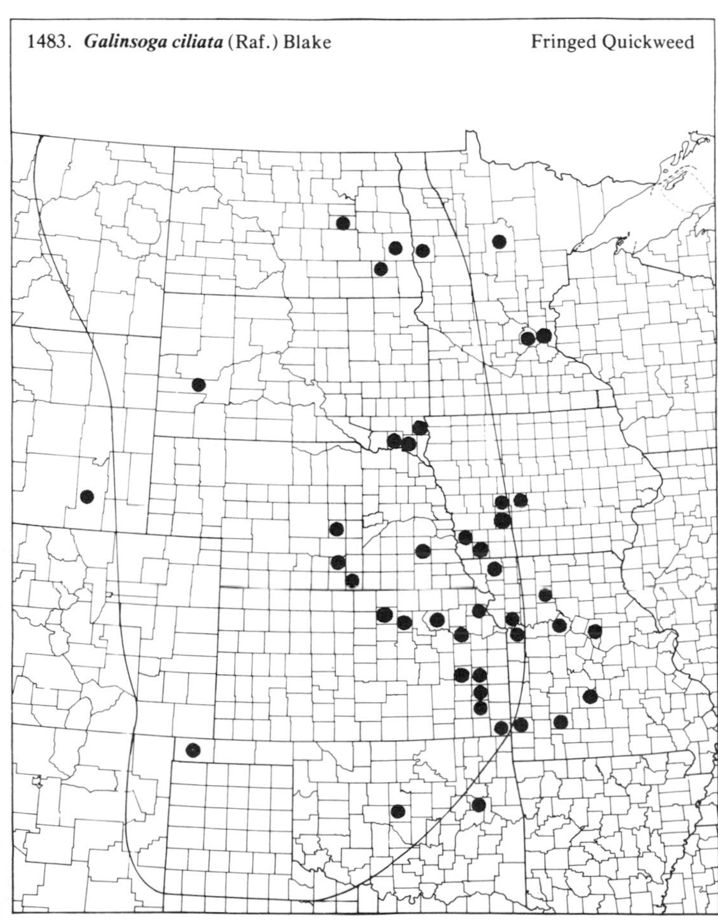

1484. ***Gnaphalium obtusifolium*** L. Fragrant Everlasting

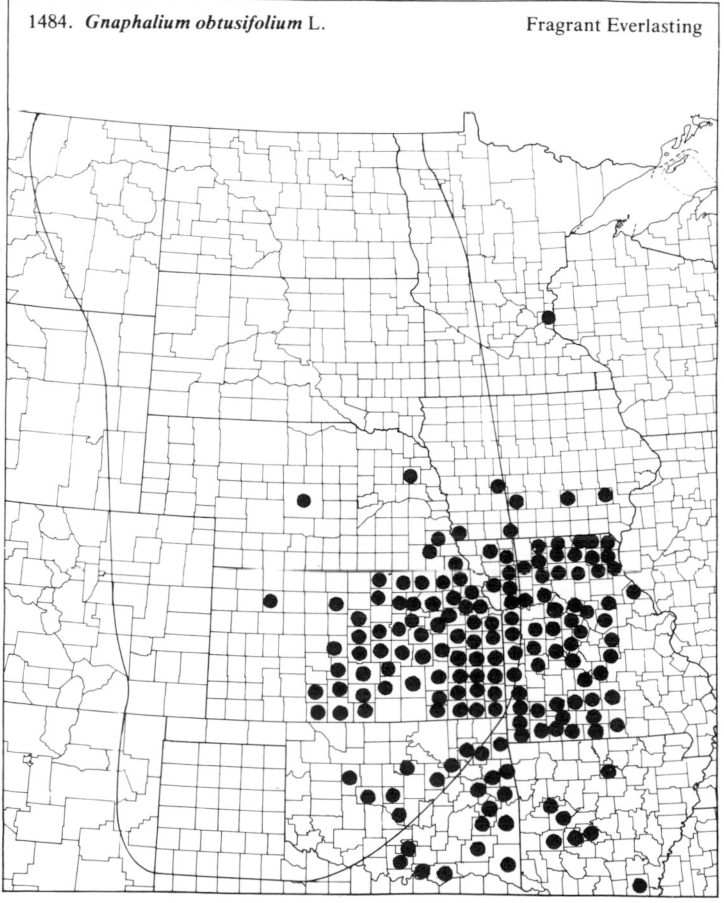

(Asteraceae)

1485. ***Gnaphalium palustre*** Nutt. Everlasting

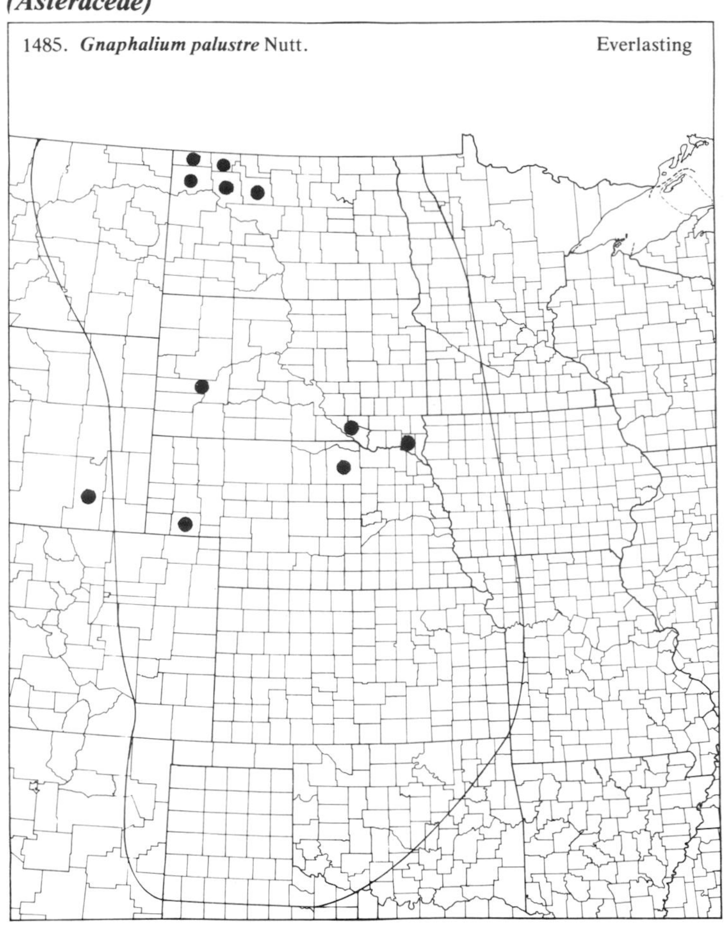

1486. ***Gnaphalium purpureum*** L. Purple Cudweed

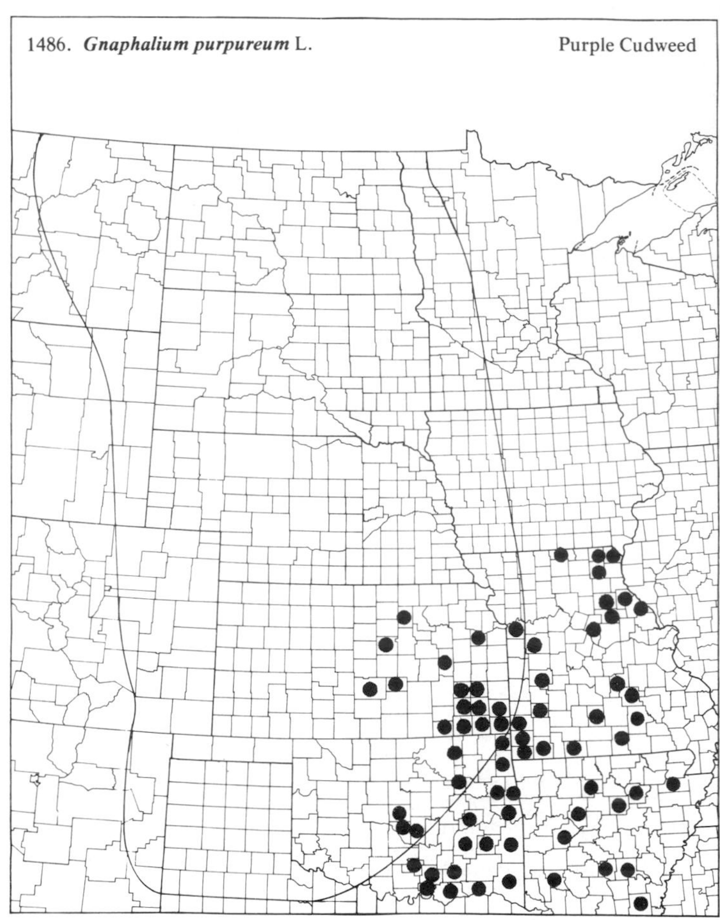

1487. ***Grindelia lanceolata*** Nutt. Spinytooth Gumweed

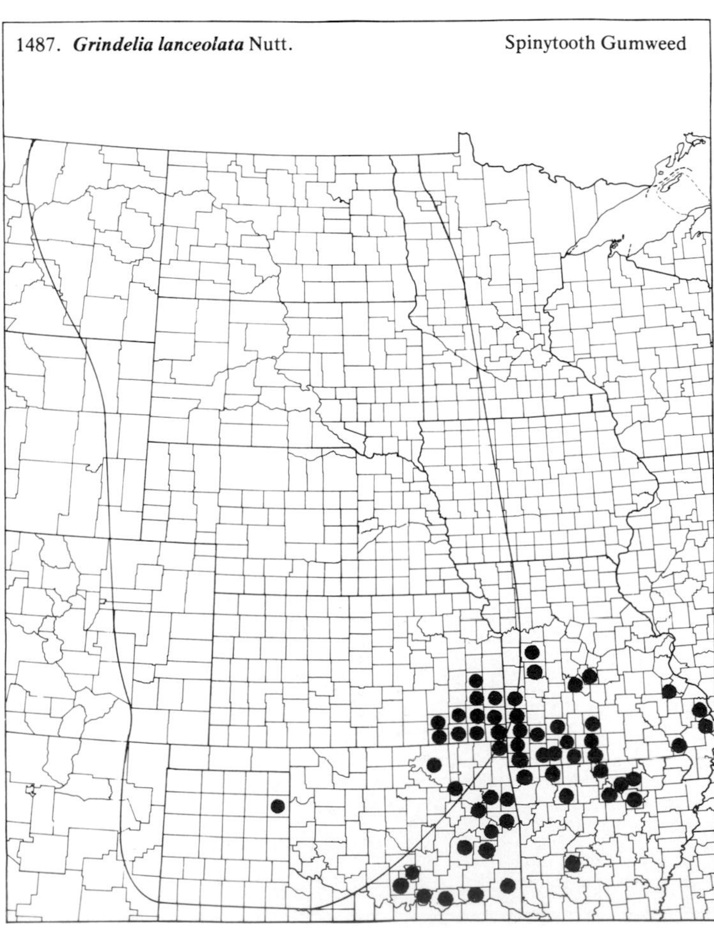

1488. ***Grindelia squarrosa*** (Pursh) Dun. var. ***squarrosa*** Curly-top Gumweed

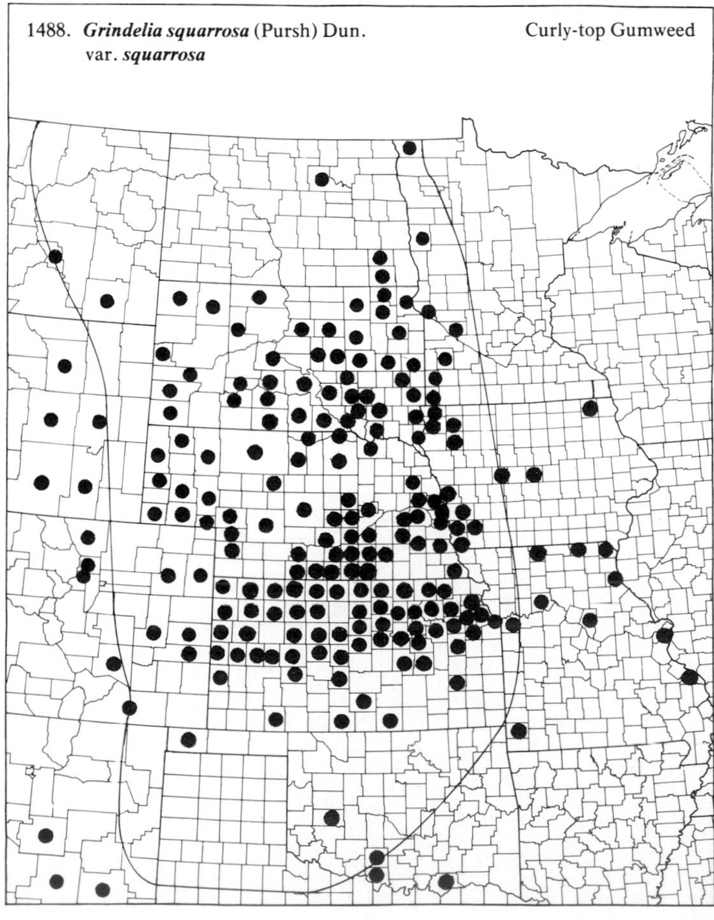

1489. ***Grindelia squarrosa*** (Pursh) Dun. var. ***nuda*** (Wood) Gray — Curly-top Gumweed

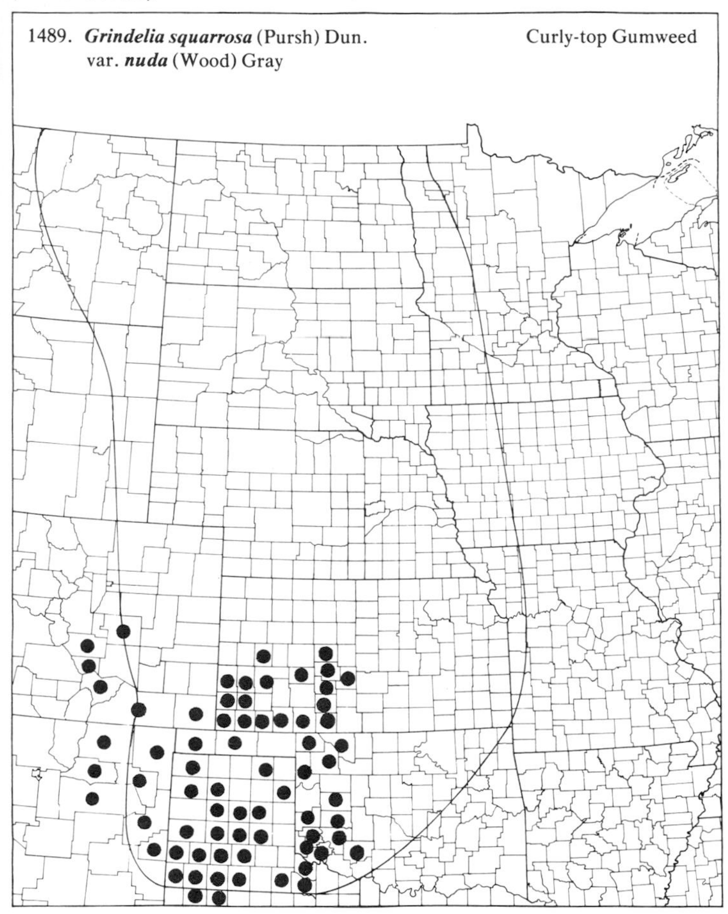

1490. ***Grindelia squarrosa*** (Pursh) Dun. var. ***quasiperennis*** Lunell — Curly-top Gumweed

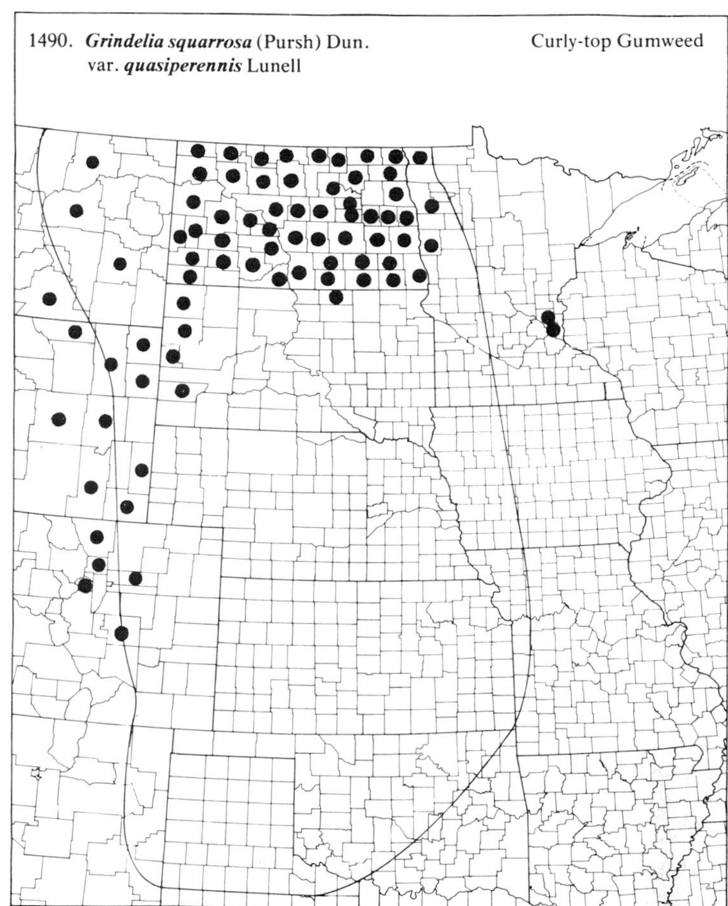

1491. ***Gutierrezia dracunculoides*** (DC.) Blake — Broomweed

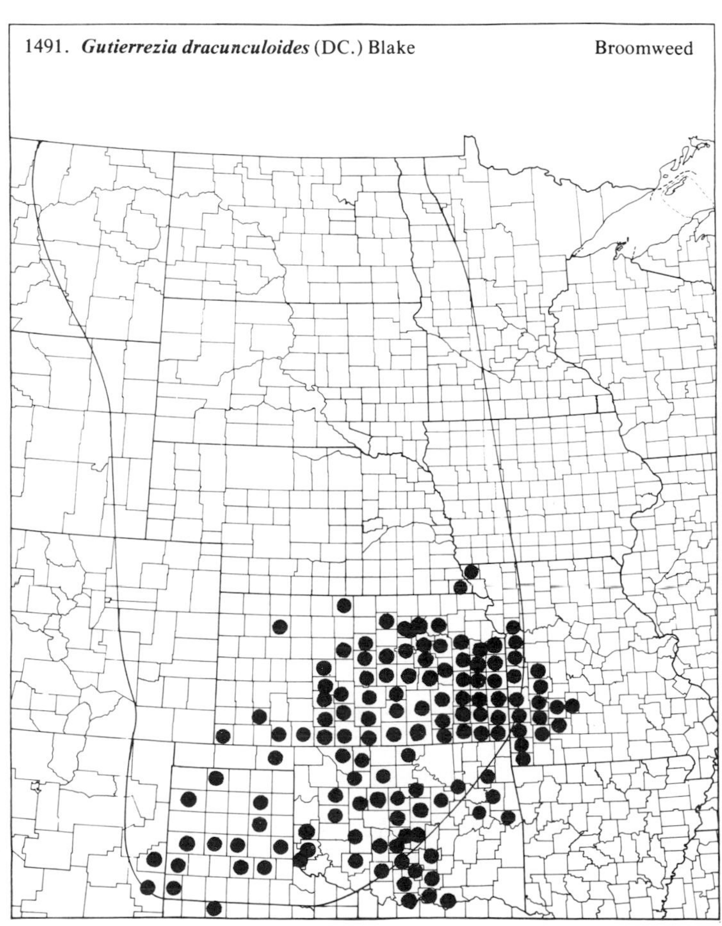

1492. ***Gutierrezia sarothrae*** (Pursh) Britt. & Rusby — Broom Snakeweed

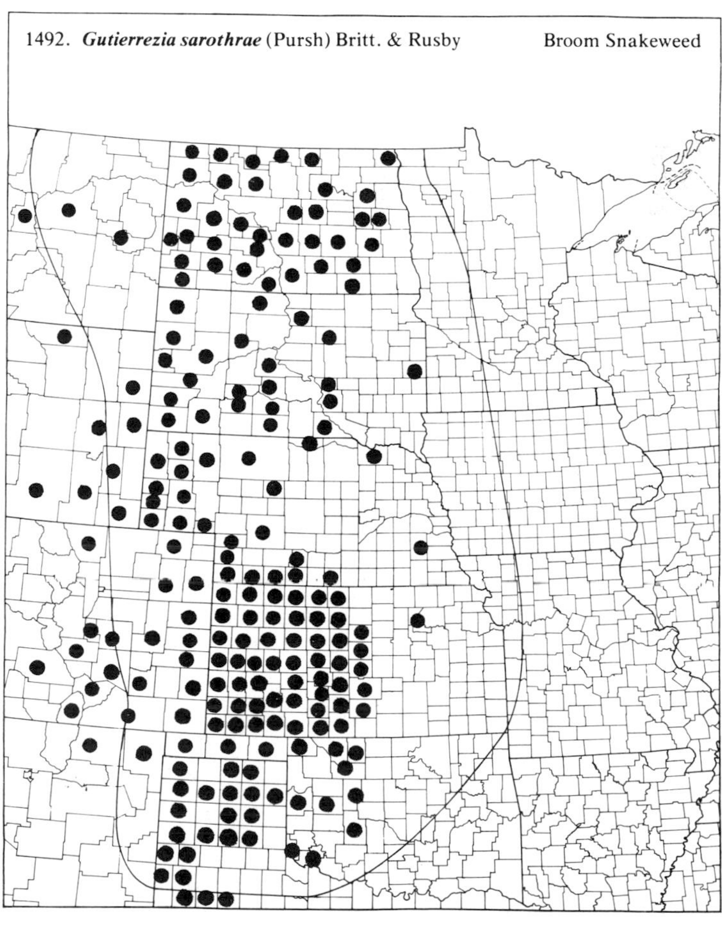

(Asteraceae)

1493. ***Haploesthes greggii*** Gray var. ***texana*** (Coult.) I.M. Johnst.

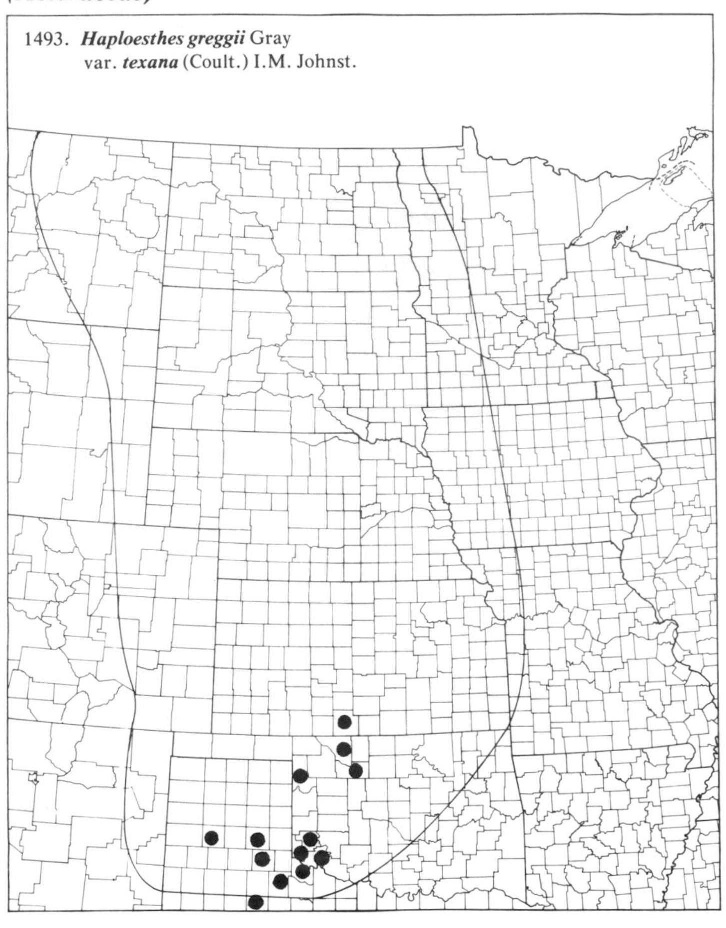

1494. ***Haplopappus annuus*** (Rydb.) Cory

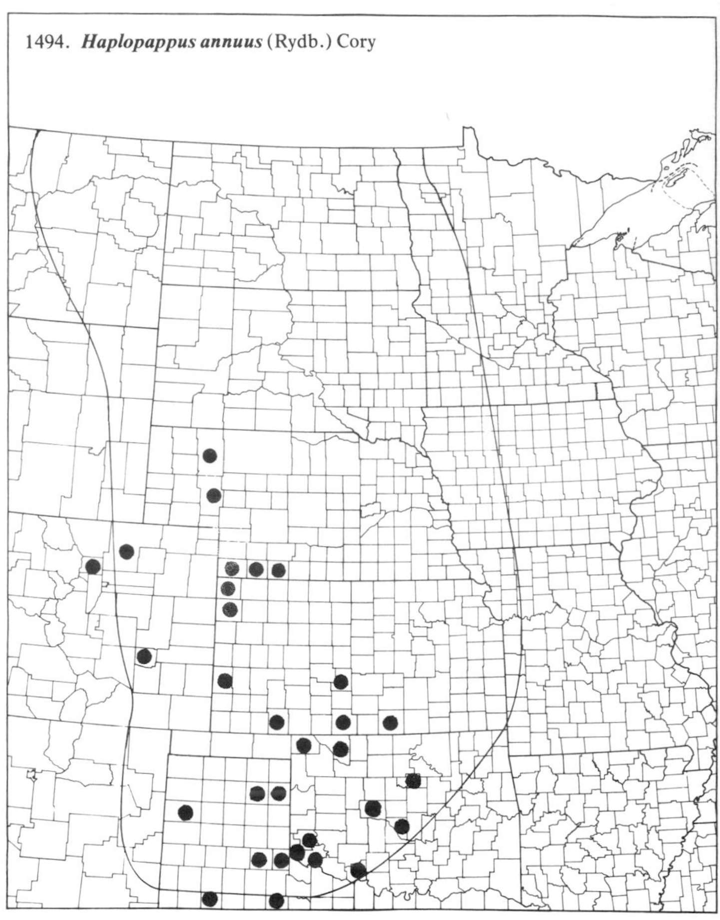

1495. ***Haplopappus armerioides*** (Nutt.) Gray

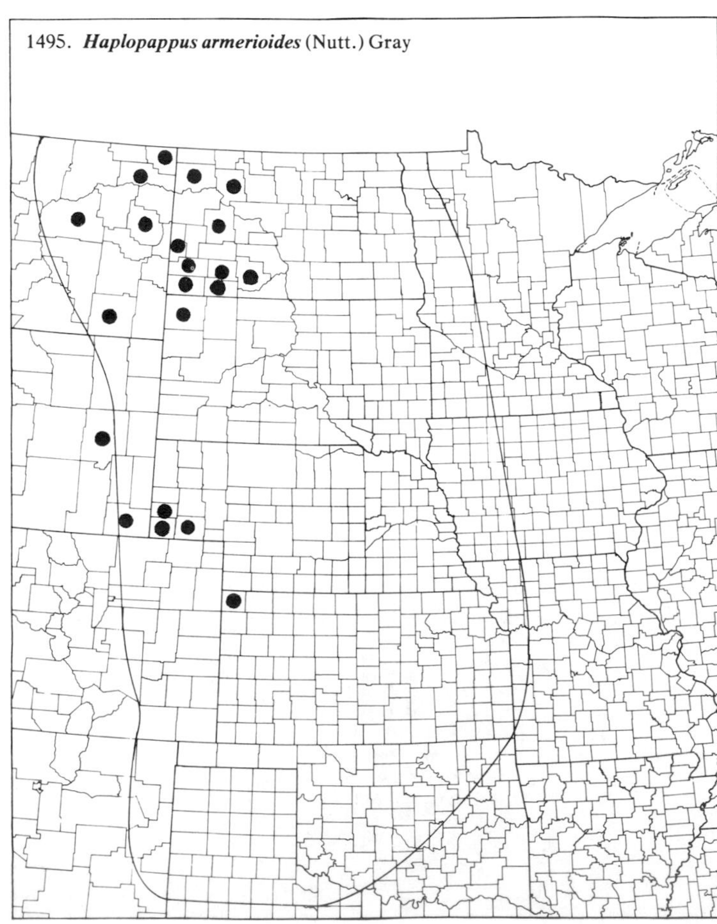

1496. ***Haplopappus ciliatus*** (Nutt.) DC. Wax Goldenweed

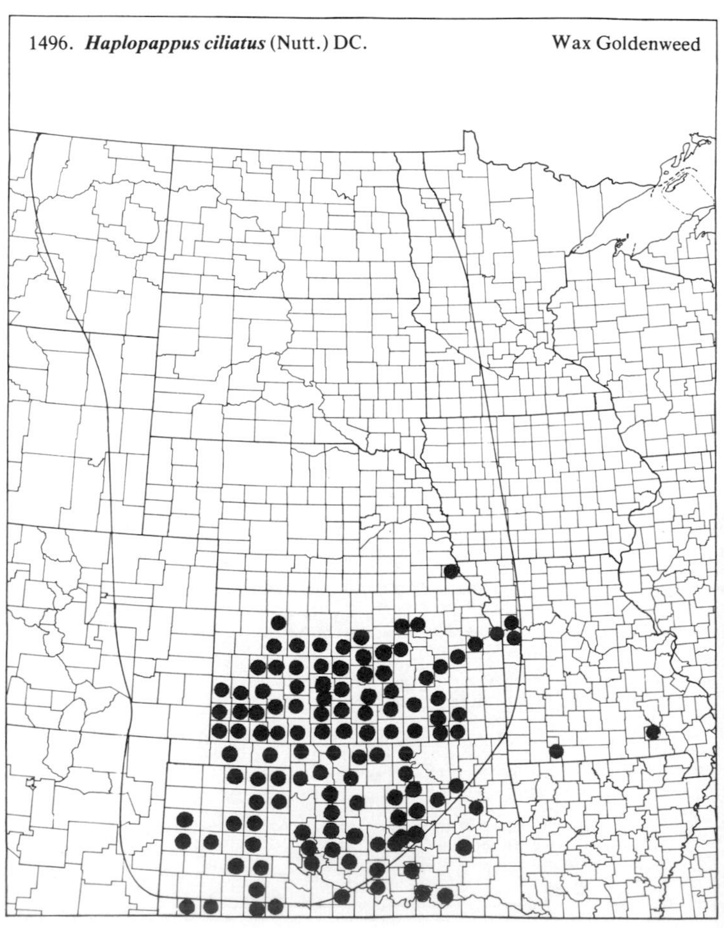

(Asteraceae)

1497. **Haplopappus heterophyllus** (Gray) Blake

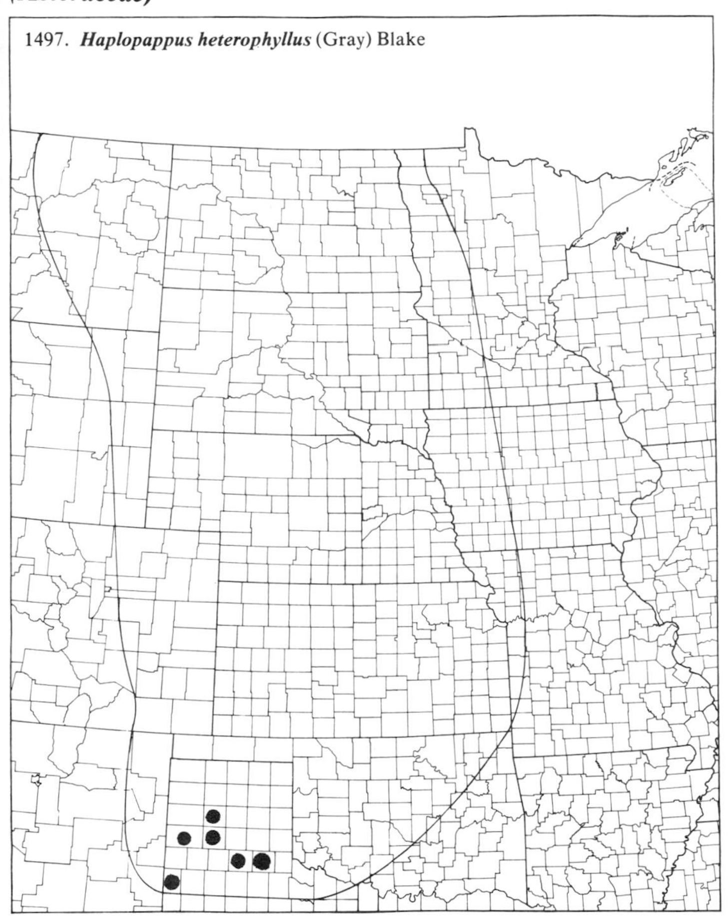

1498. **Haplopappus lanceolatus** (Hook.) T. & G.

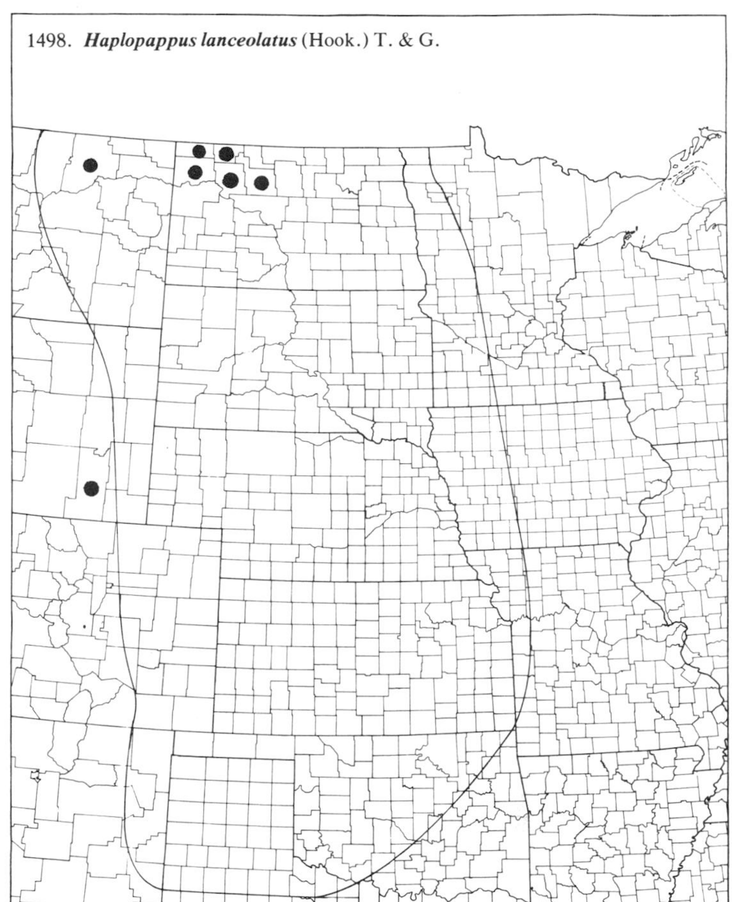

1499. **Haplopappus spinulosus** (Pursh) DC. Cutleaf Ironplant

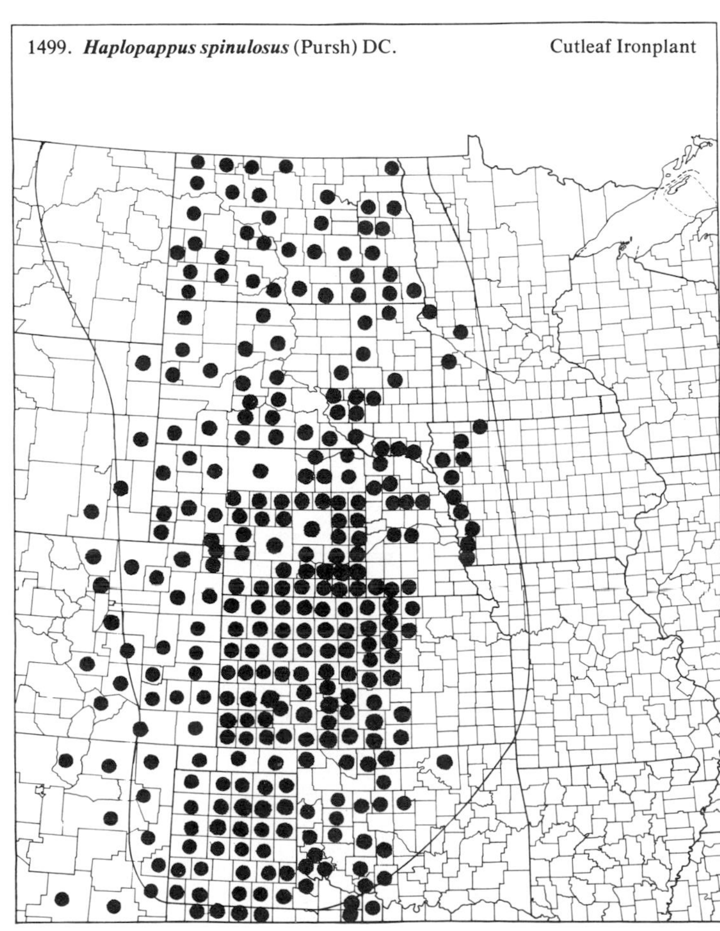

1500. **Haplopappus validus** (Rydb.) Cory

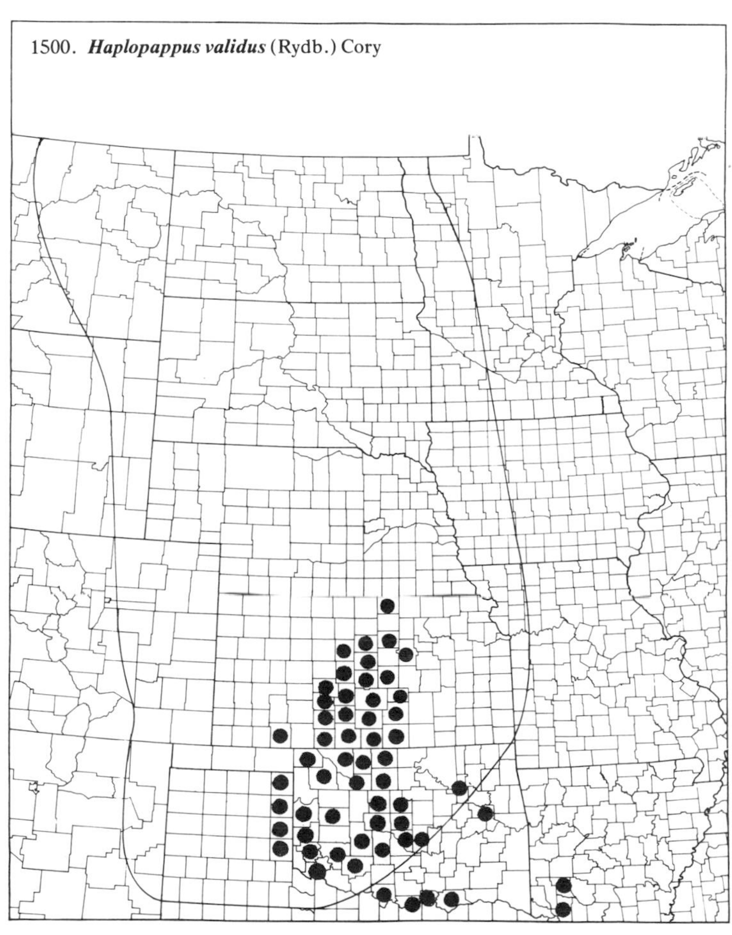

(Asteraceae)

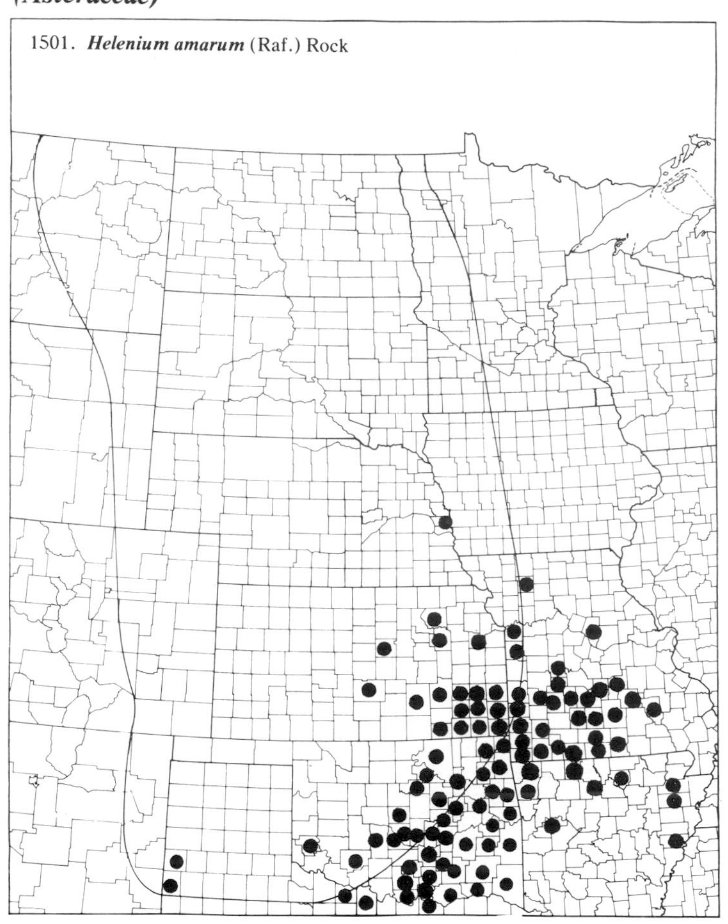
1501. **Helenium amarum** (Raf.) Rock

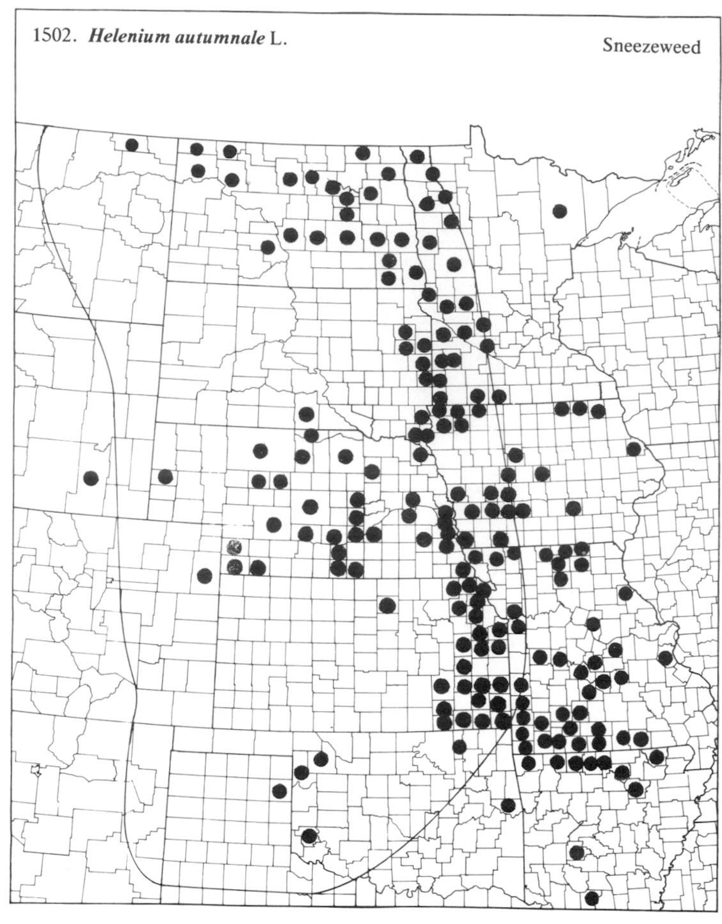
1502. **Helenium autumnale** L.
Sneezeweed

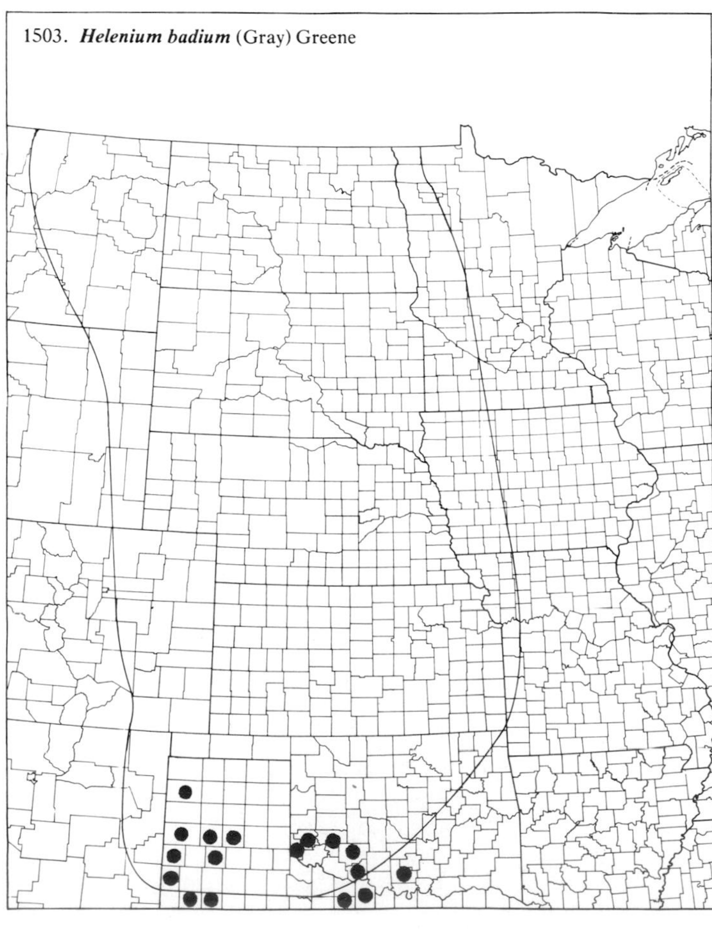
1503. **Helenium badium** (Gray) Greene

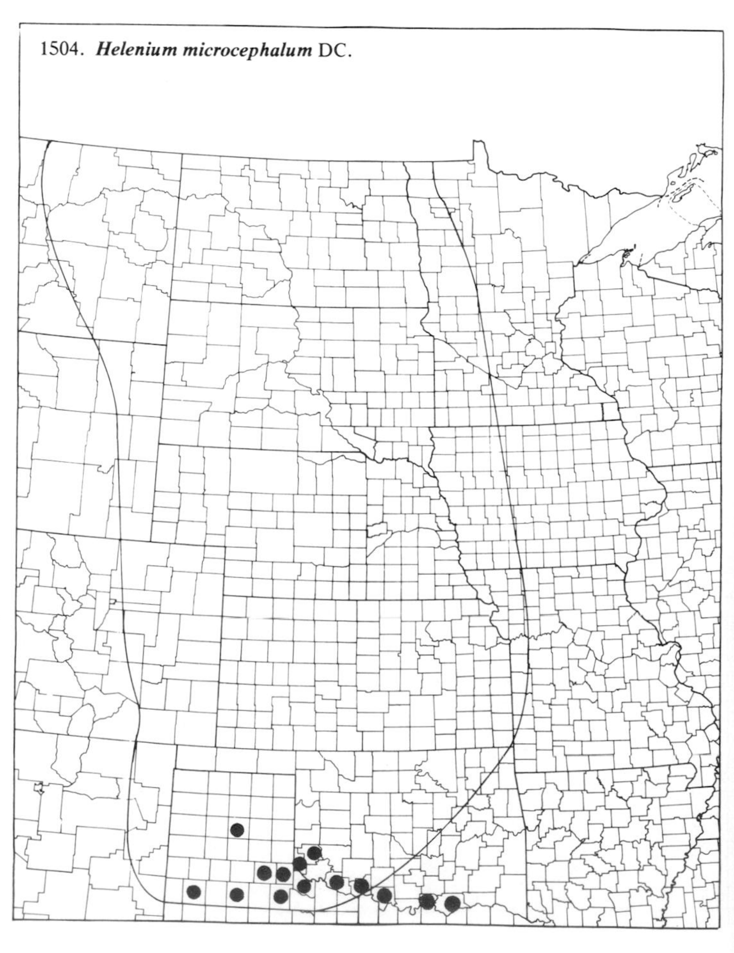
1504. **Helenium microcephalum** DC.

(Asteraceae)

1505. **Helianthus annuus** L. Common Sunflower

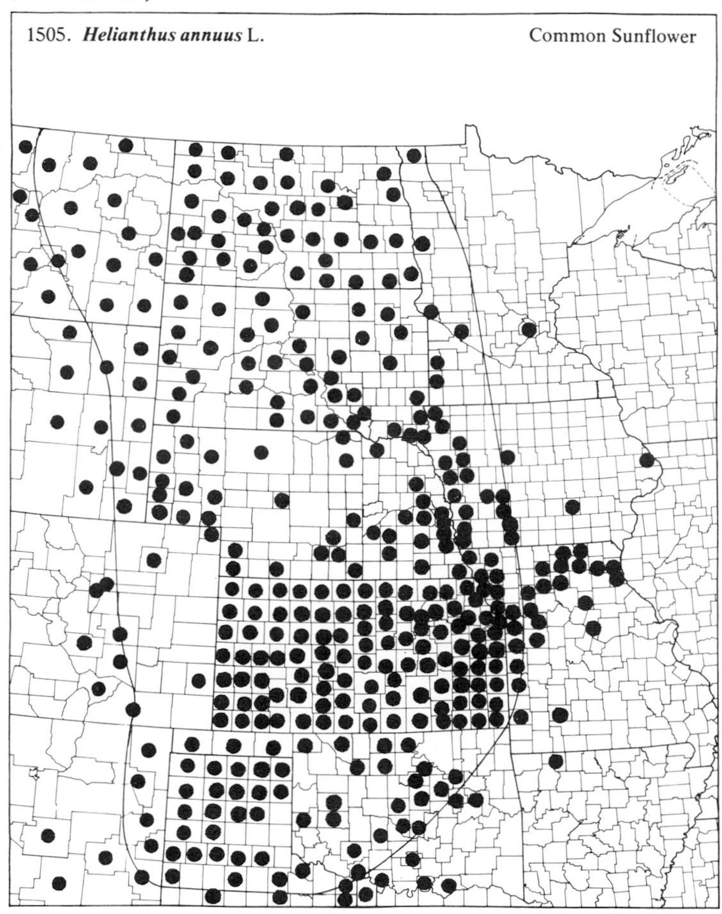

1506. **Helianthus ciliaris** DC. Texas Blueweed

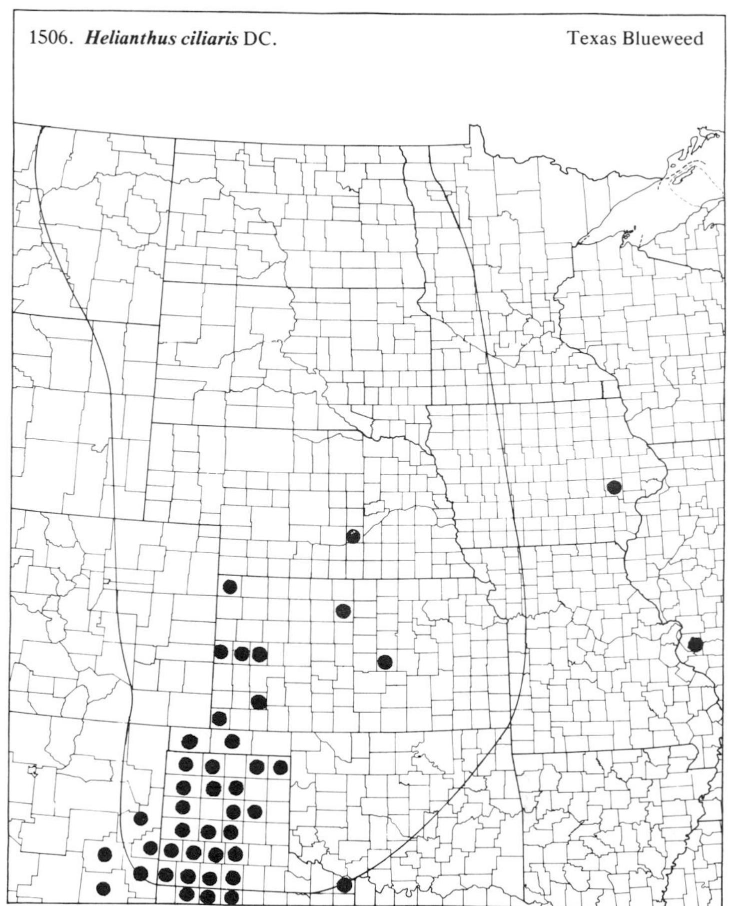

1507. **Helianthus grosseserratus** Martens Sawtooth Sunflower

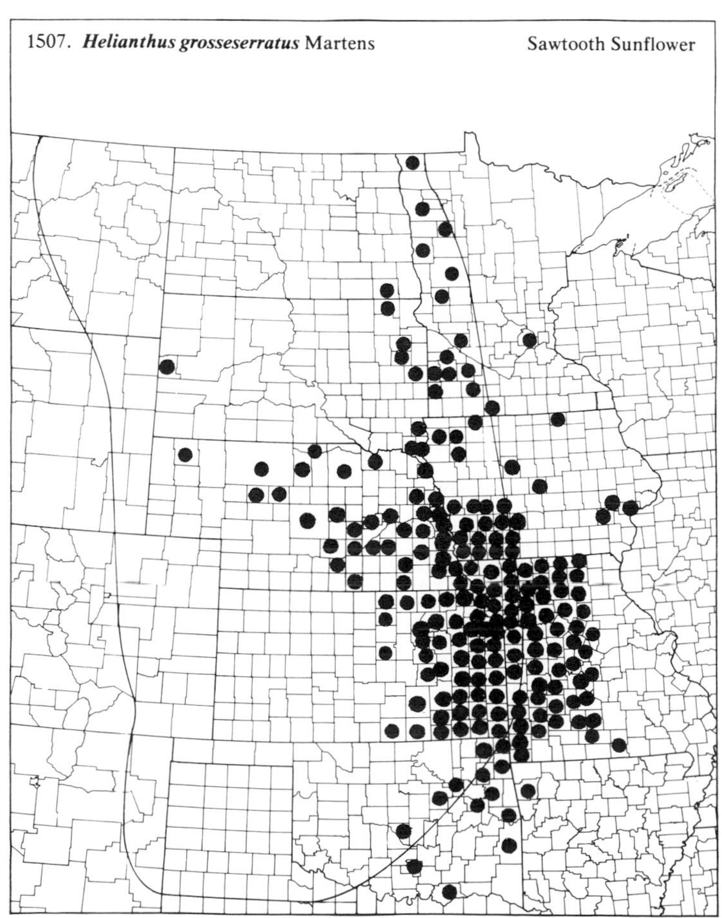

1508. **Helianthus hirsutus** Raf. Hairy Sunflower

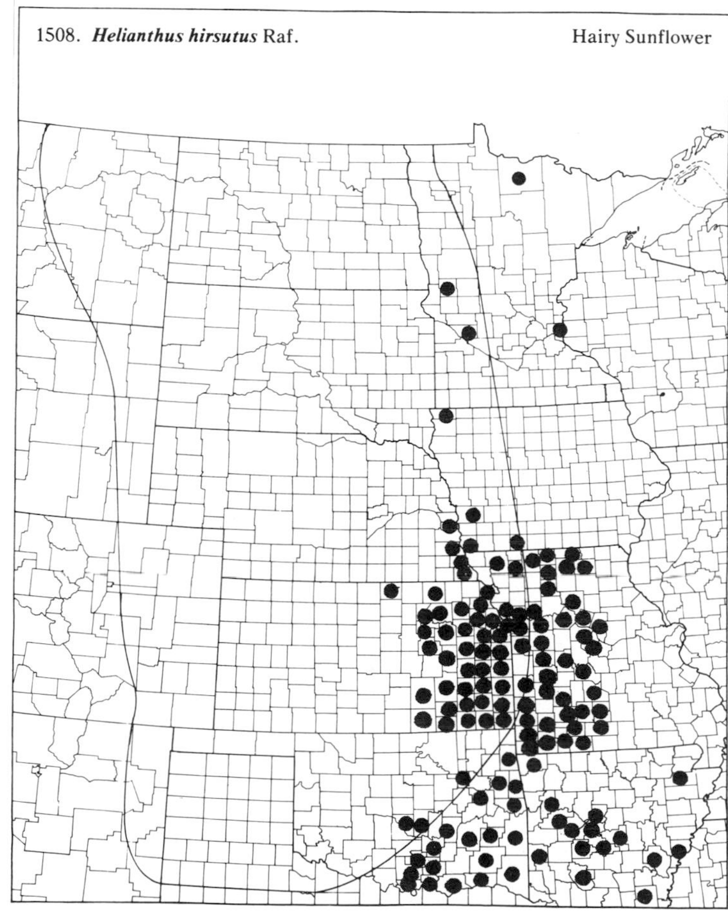

(Asteraceae)

1509. ***Helianthus maximiliani*** Schrad. Maximilian Sunflower

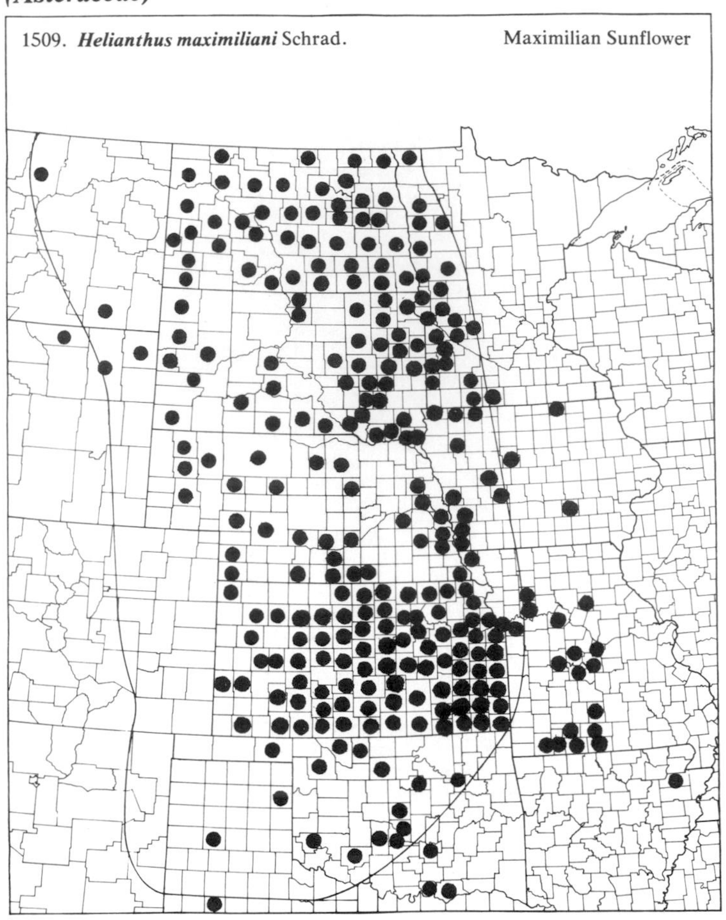

1510. ***Helianthus mollis*** Lam. Ashy Sunflower

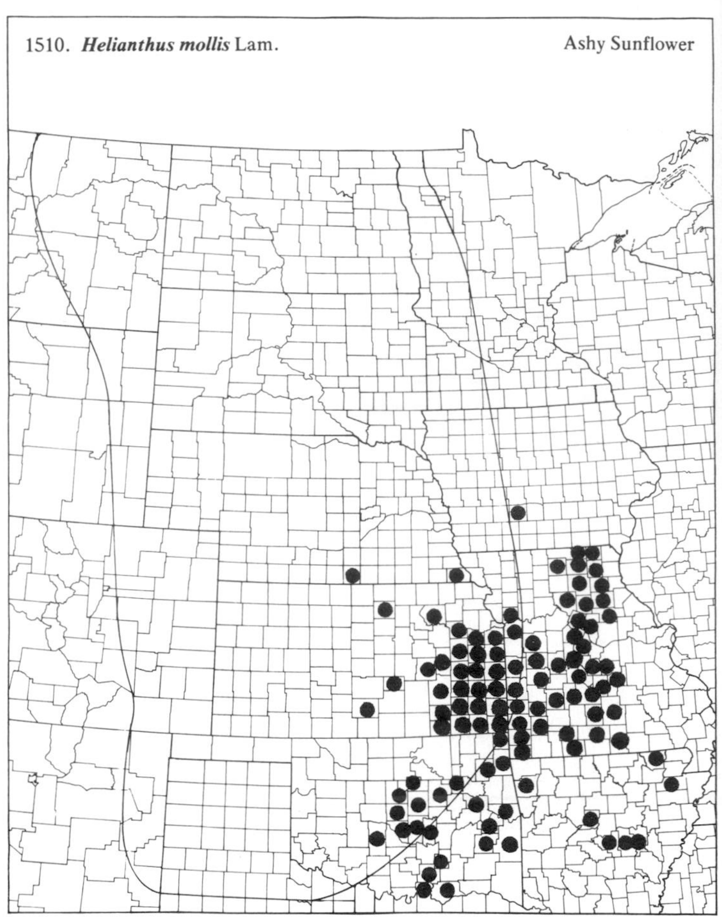

1511. ***Helianthus nuttallii*** T. & G. ssp. ***nuttallii*** Nuttall's Sunflower

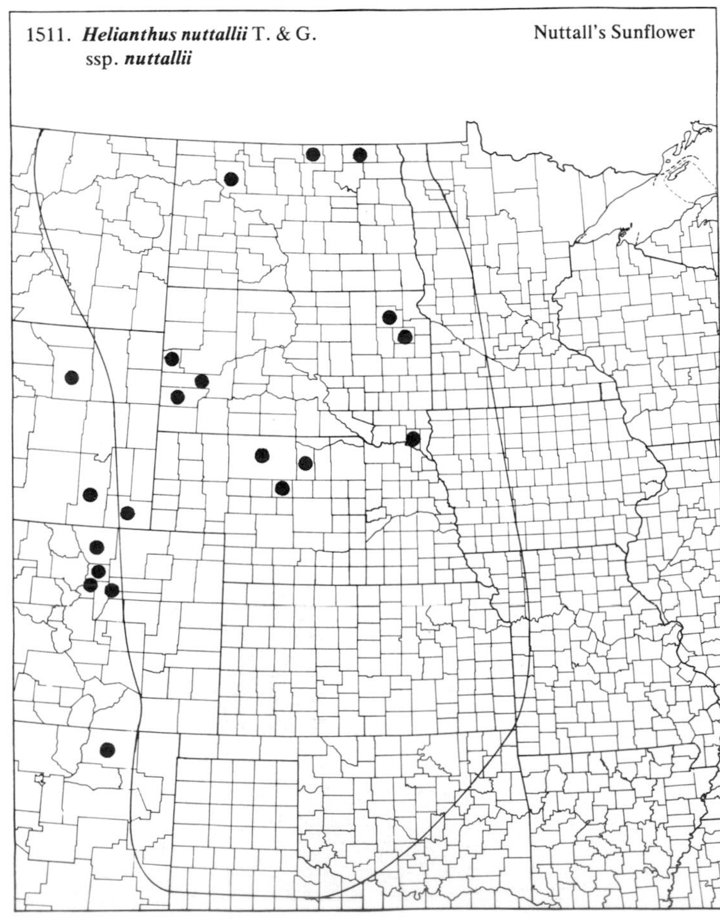

1512. ***Helianthus nuttallii*** T. & G. ssp. ***rydbergii*** (Britt.) Long Nuttall's Sunflower

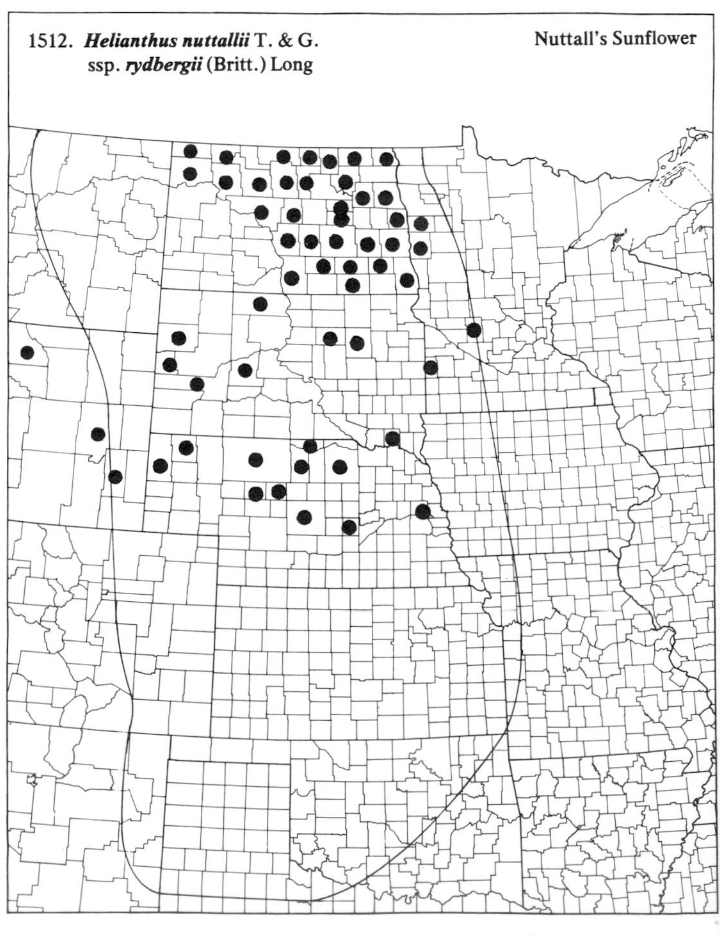

(Asteraceae)

1513. ***Helianthus petiolaris*** Nutt. — Plains Sunflower

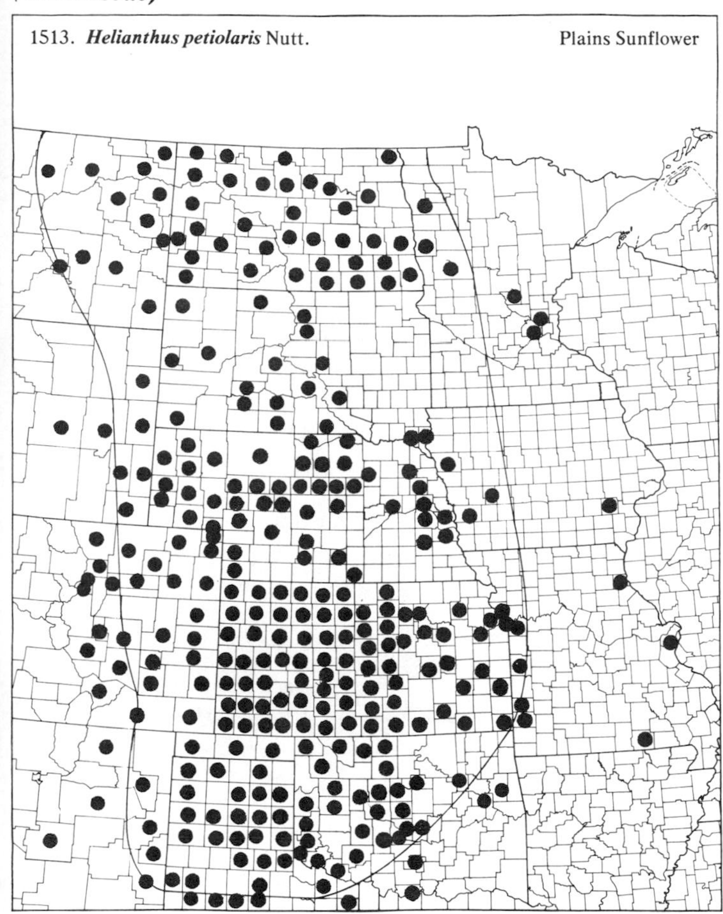

1514. ***Helianthus rigidus*** (Cass.) Desf. ssp. ***rigidus*** — Stiff Sunflower

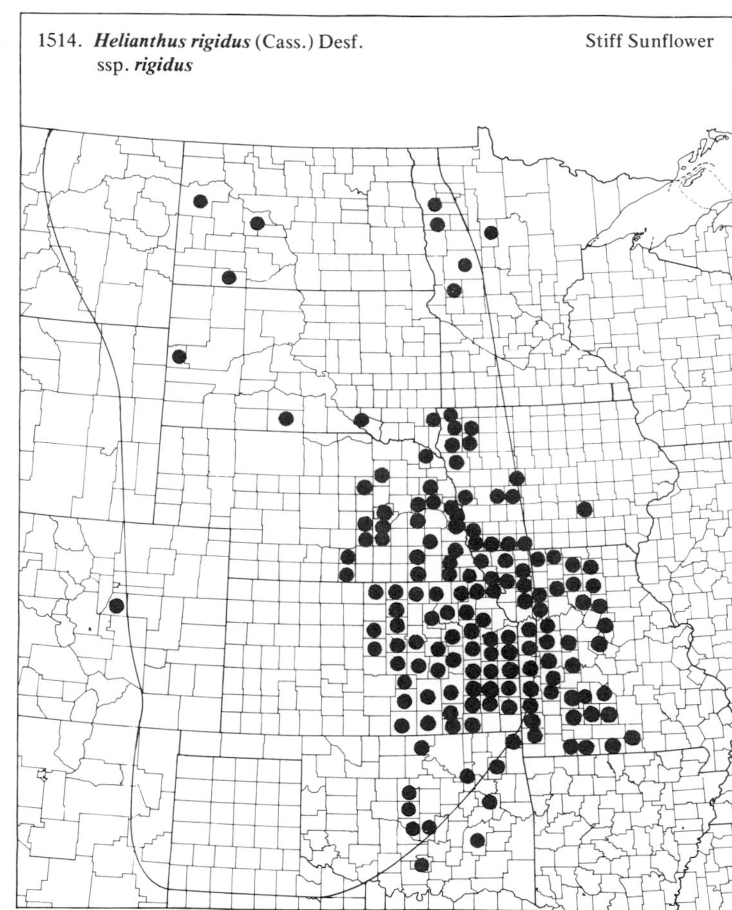

1515. ***Helianthus rigidus*** (Cass.) Desf. ssp. ***subrhomboideus*** (Rydb.) Heiser — Stiff Sunflower

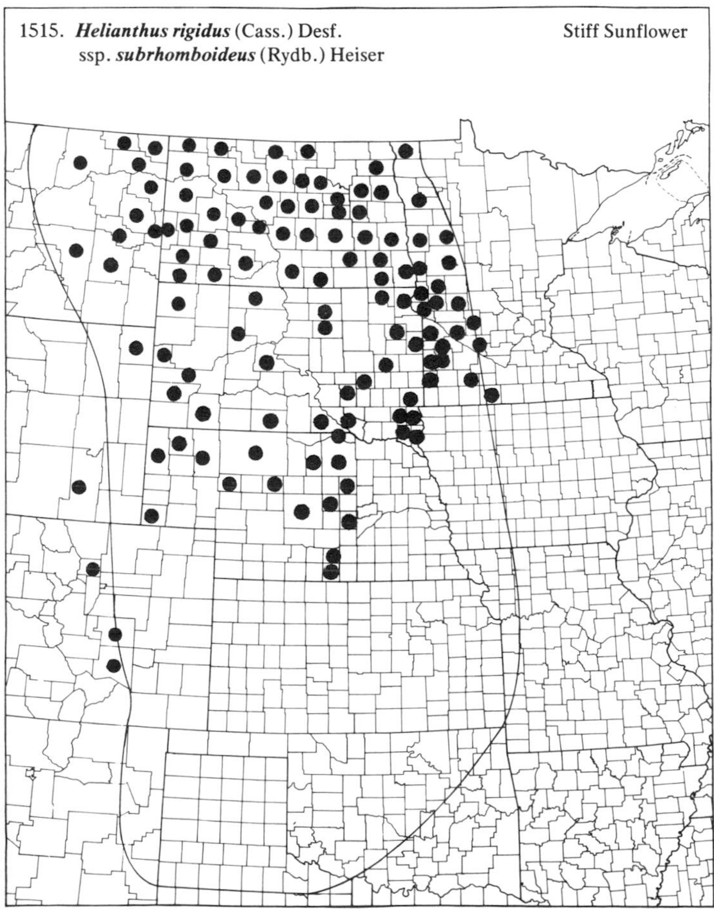

1516. ***Helianthus salicifolius*** A. Dietr. — Willow-leaved Sunflower

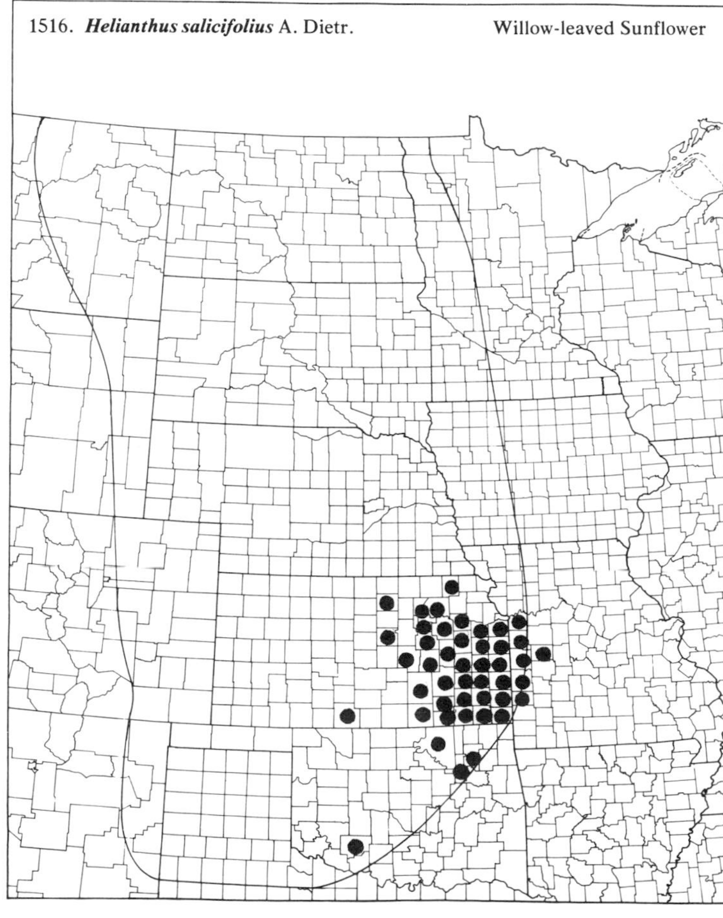

(Asteraceae)

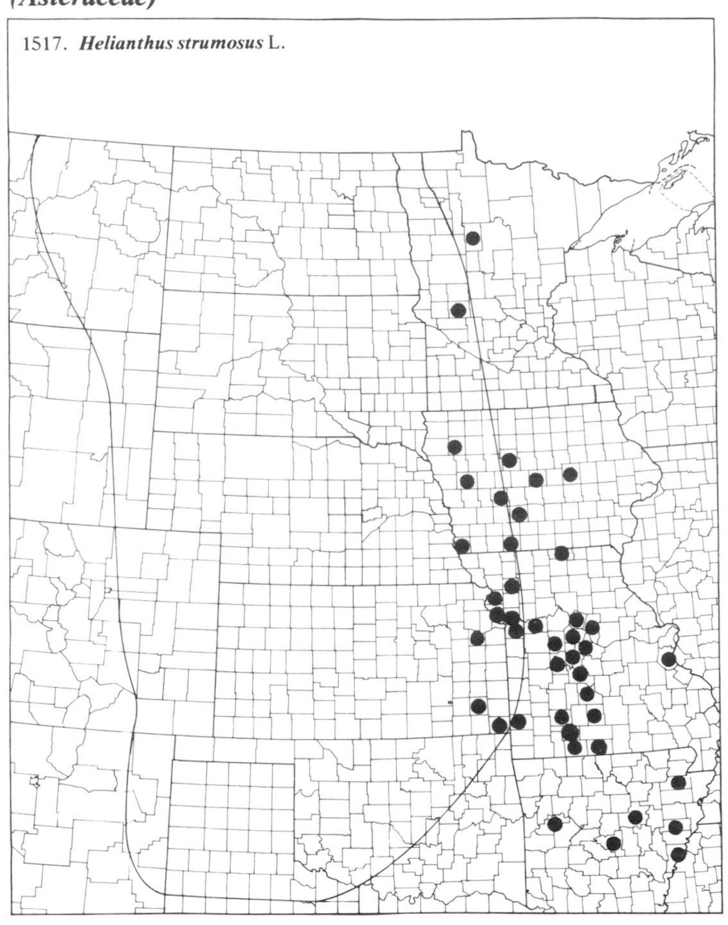
1517. ***Helianthus strumosus*** L.

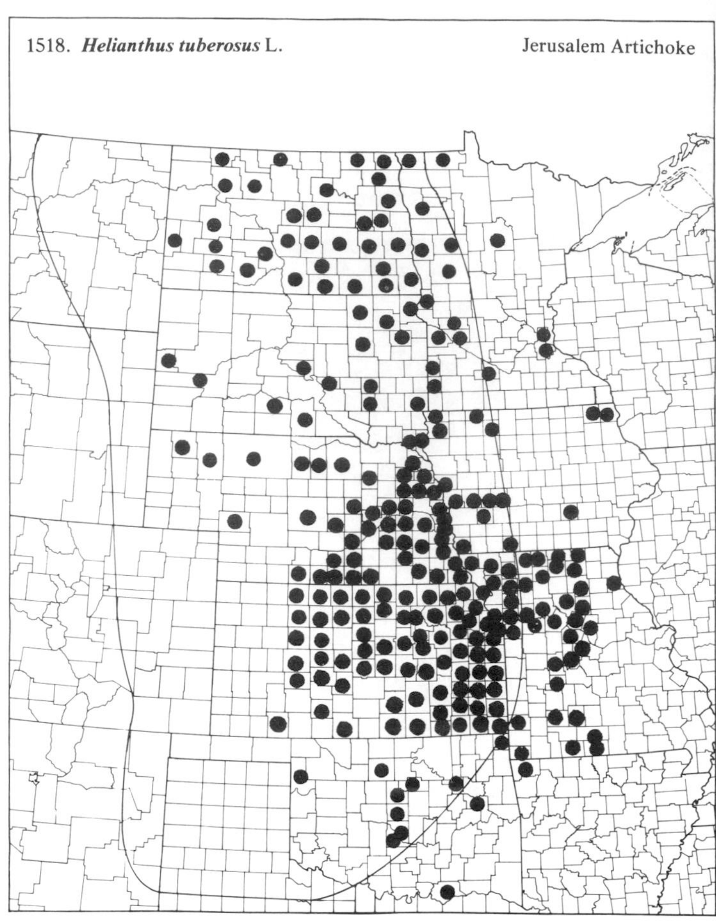
1518. ***Helianthus tuberosus*** L. Jerusalem Artichoke

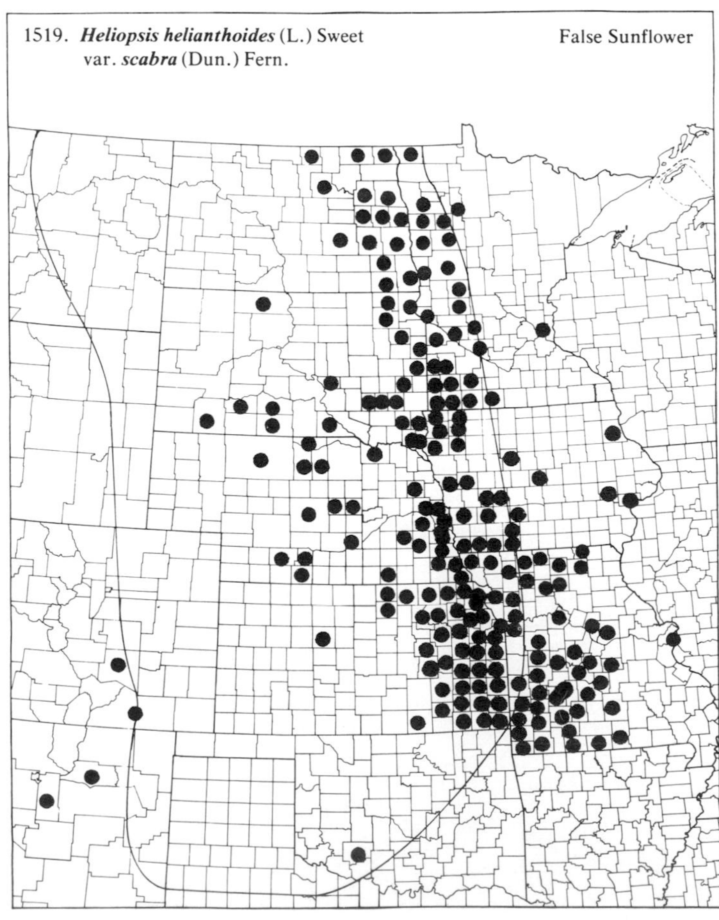
1519. ***Heliopsis helianthoides*** (L.) Sweet var. ***scabra*** (Dun.) Fern. False Sunflower

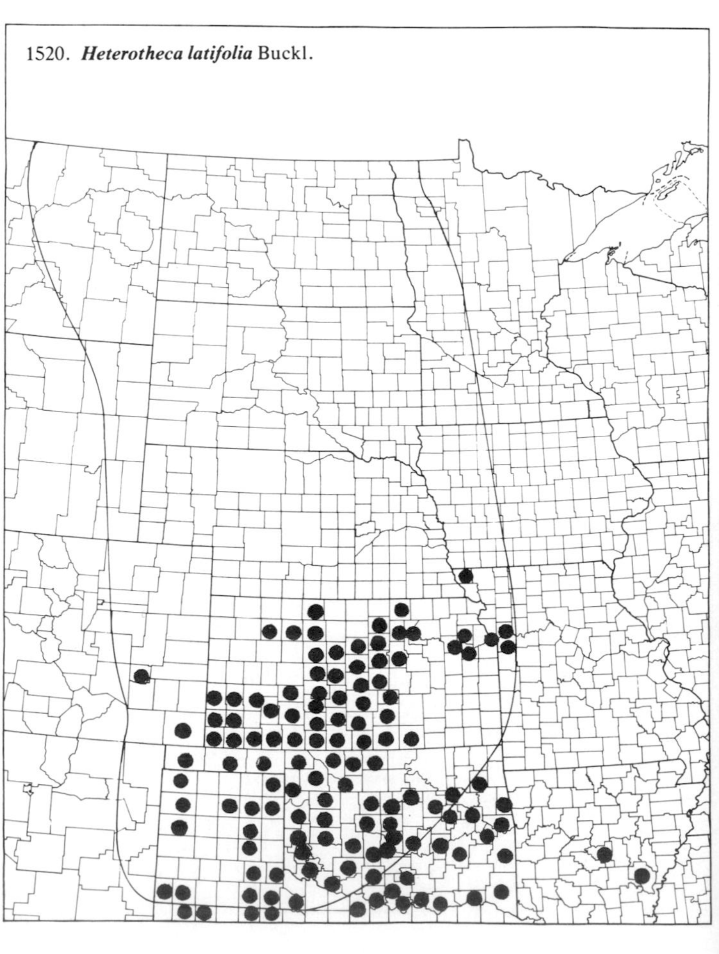
1520. ***Heterotheca latifolia*** Buckl.

(Asteraceae)

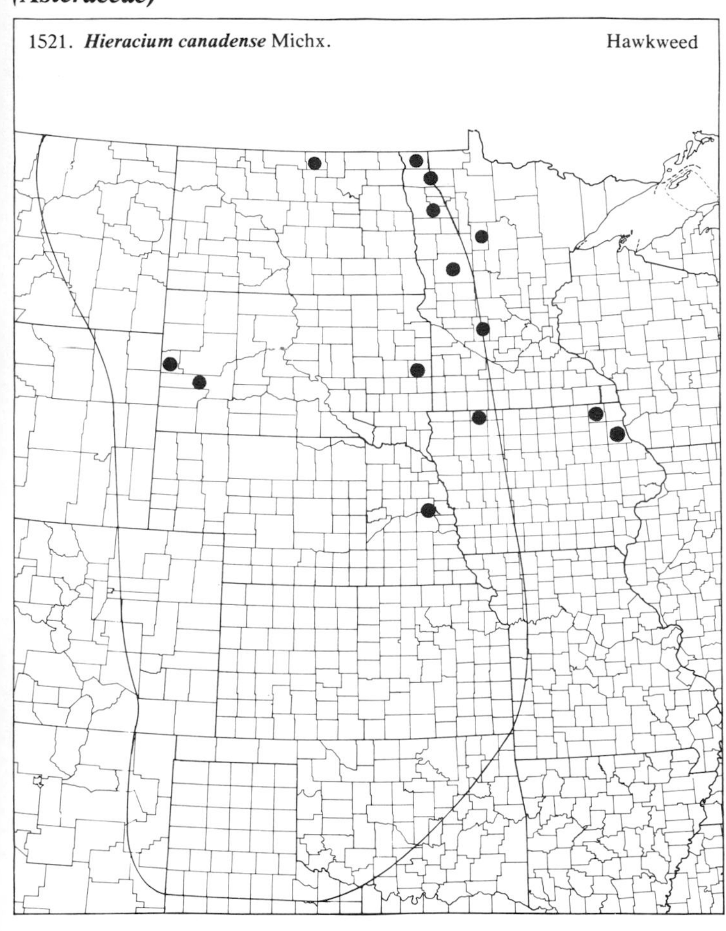
1521. **Hieracium canadense** Michx.
Hawkweed

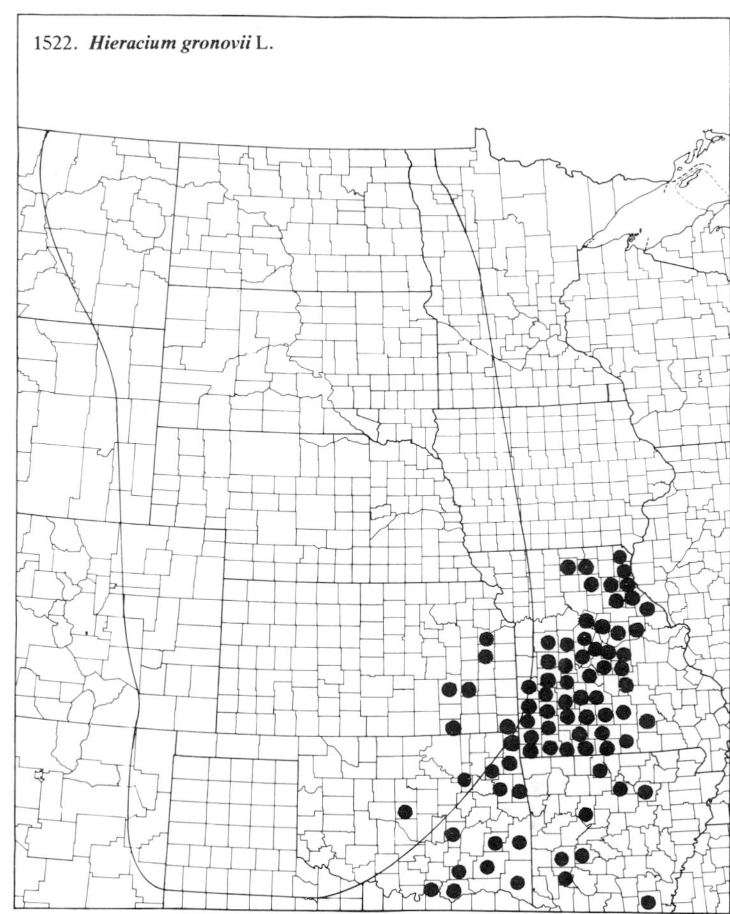
1522. **Hieracium gronovii** L.

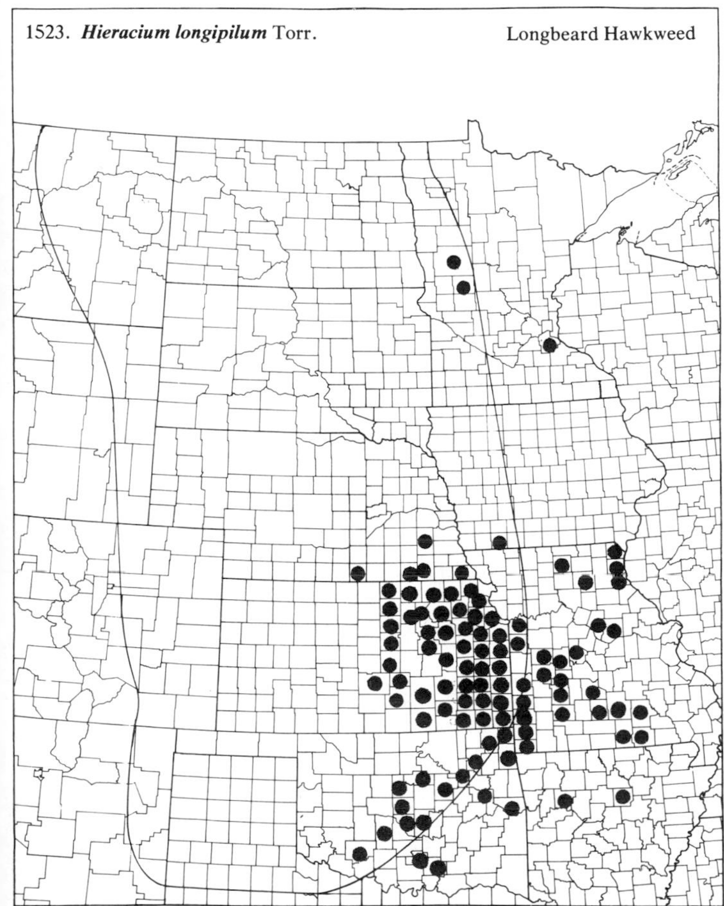
1523. **Hieracium longipilum** Torr.
Longbeard Hawkweed

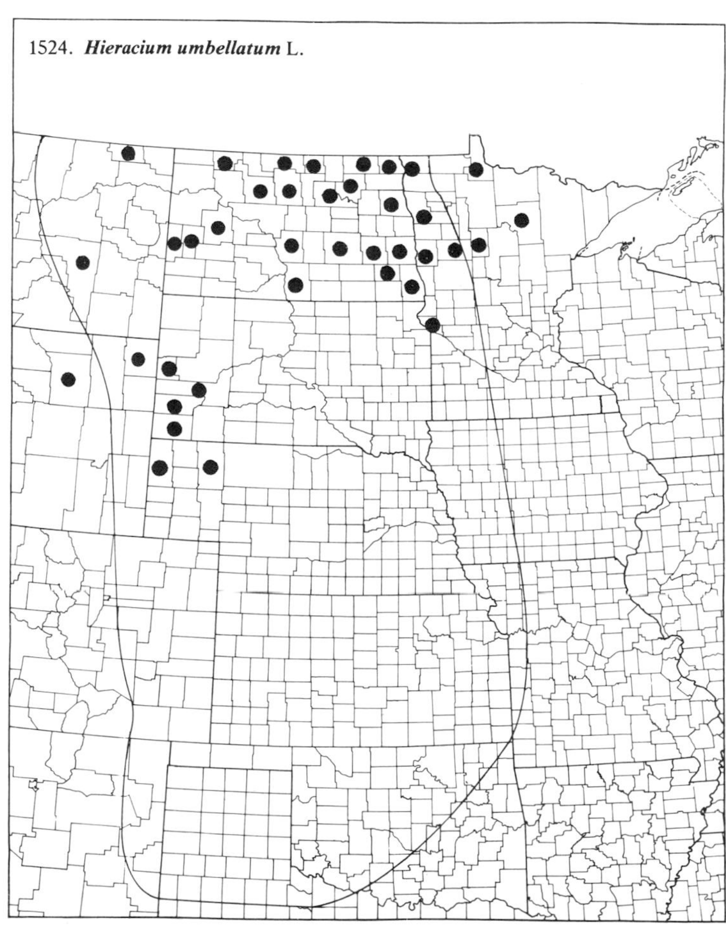
1524. **Hieracium umbellatum** L.

(Asteraceae)

1525. ***Hymenopappus filifolius*** Hook.
var. ***cinereus*** (Rydb.) I.M. Johnst.

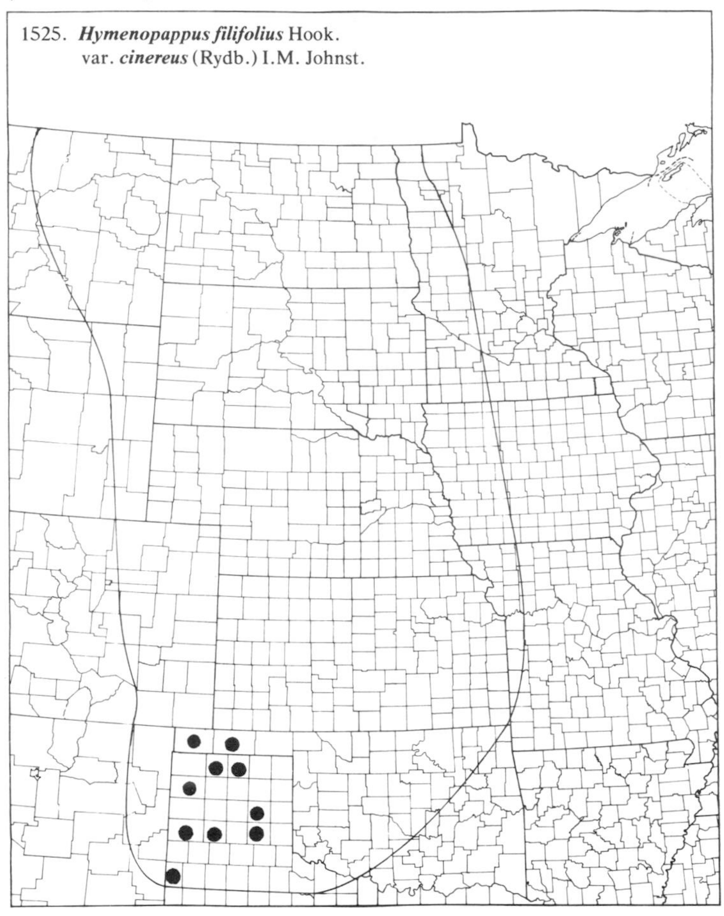

1526. ***Hymenopappus filifolius*** Hook.
var. ***polycephalus*** (Osterh.) B.L. Turner

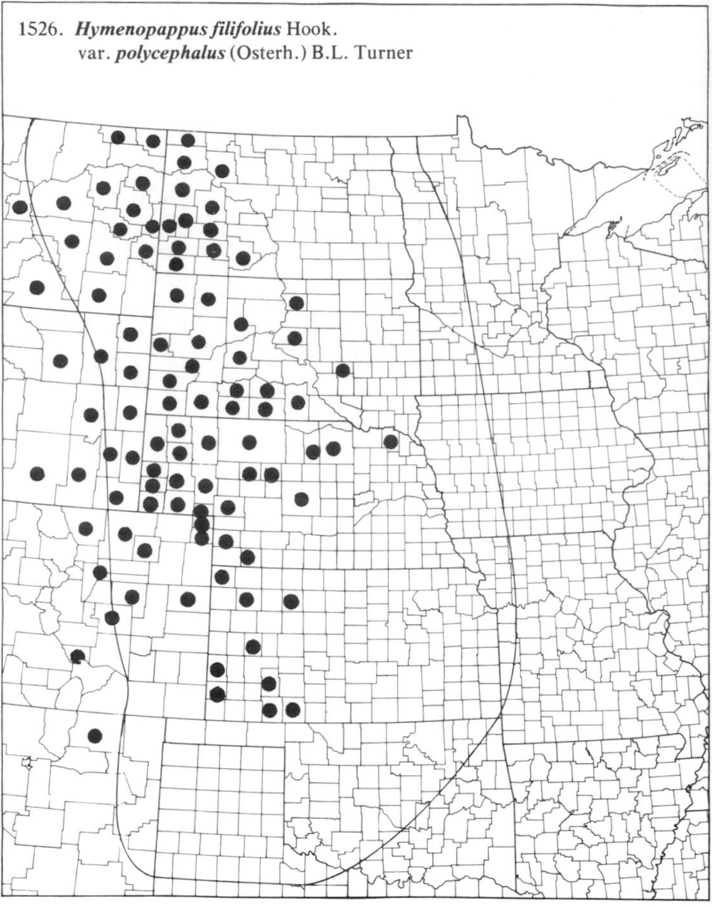

1527. ***Hymenopappus flavescens*** Gray

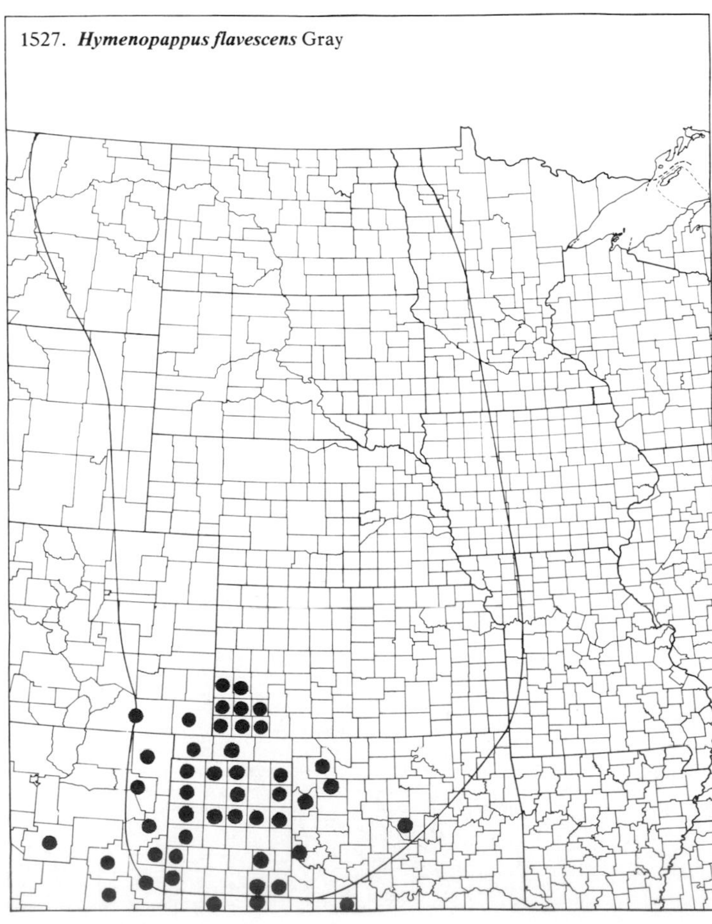

1528. ***Hymenopappus scabiosaeus*** L'Her.
var. ***corymbosus*** (T. & G.) B.L. Turner

Old Plainsman

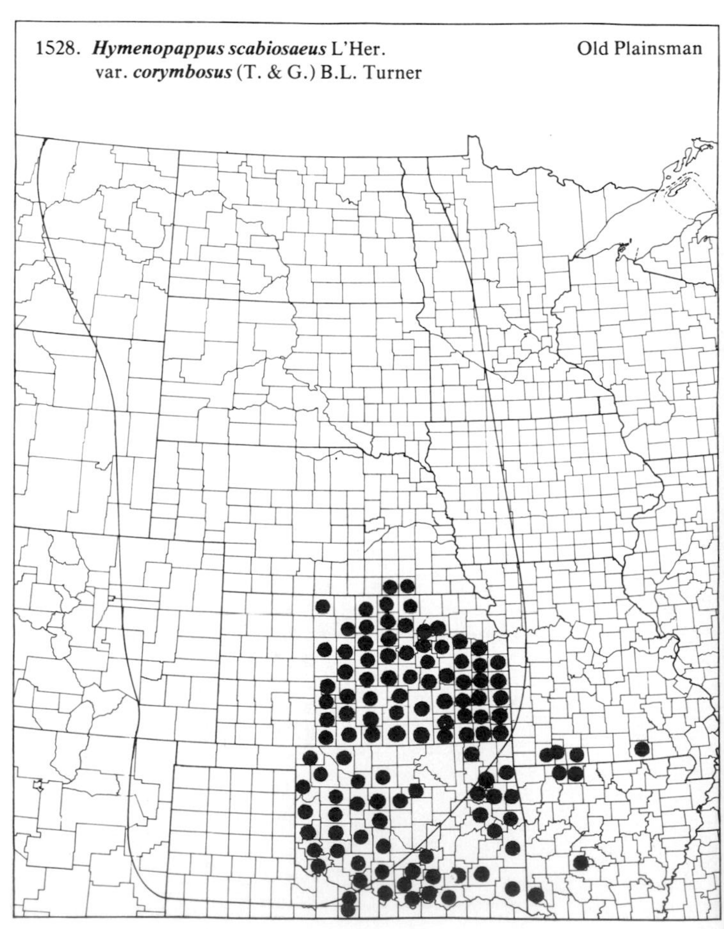

(Asteraceae)

1529. ***Hymenopappus tenuifolius*** Pursh — Slimleaf Hymenopappus

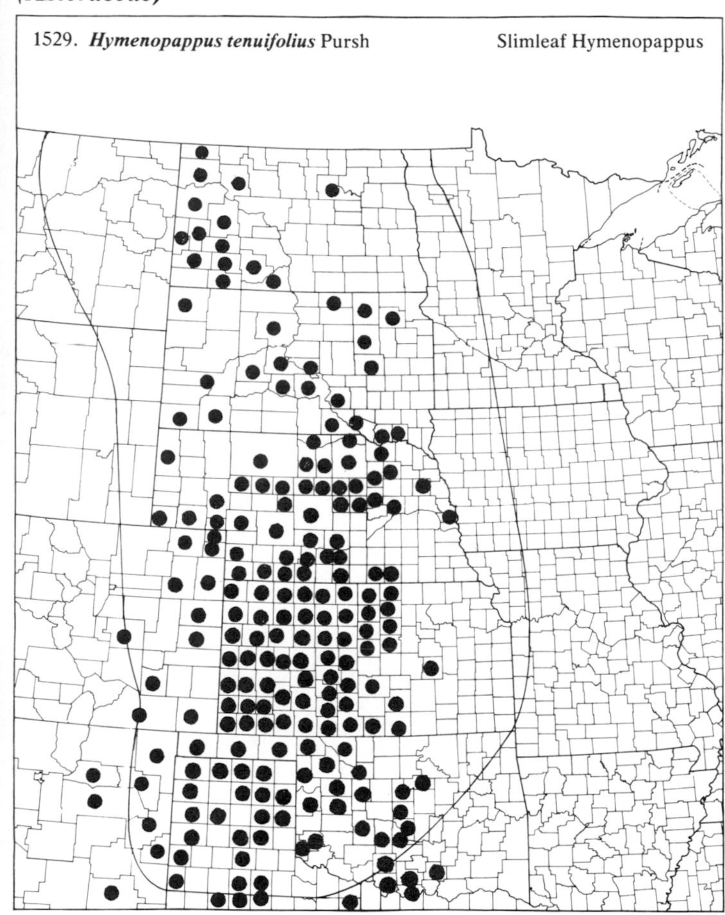

1530. ***Hymenoxys acaulis*** (Pursh) Parker — Stemless Hymenoxys

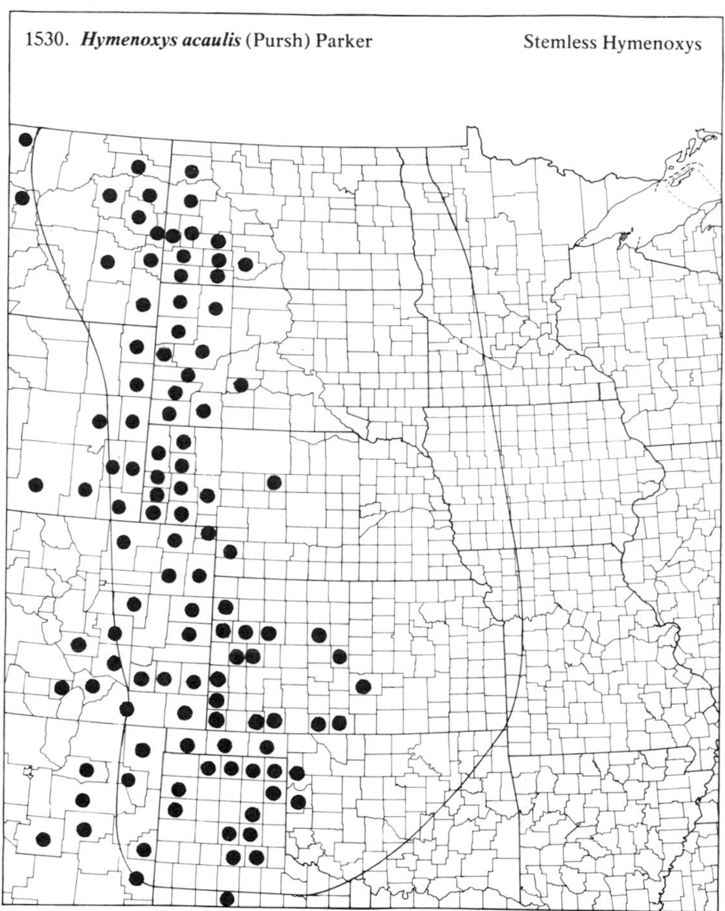

1531. ***Hymenoxys linearifolia*** Hook.

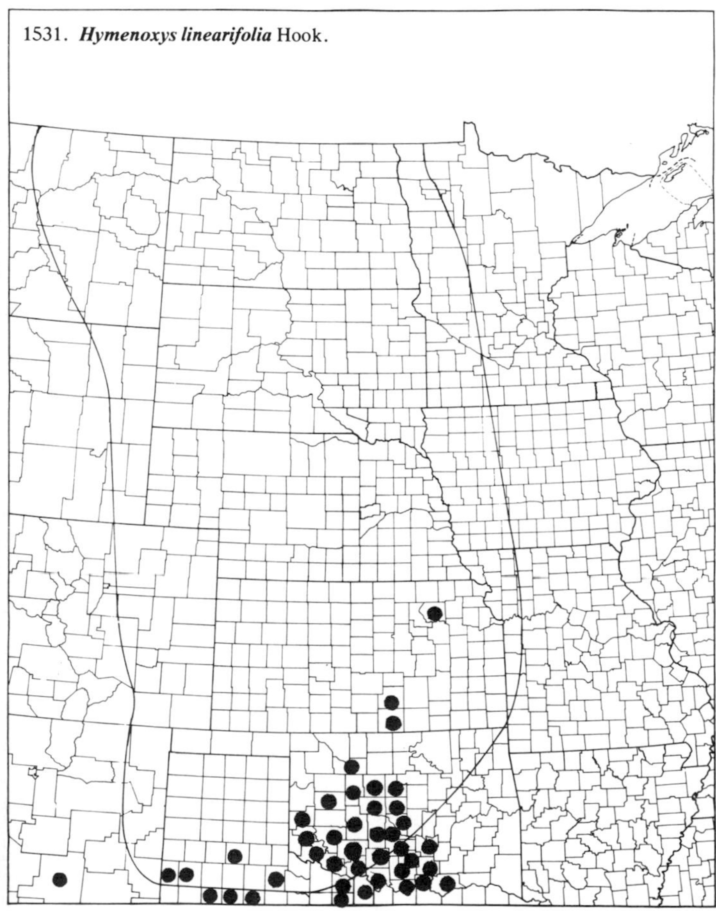

1532. ***Hymenoxys odorata*** DC. — Bitterweed

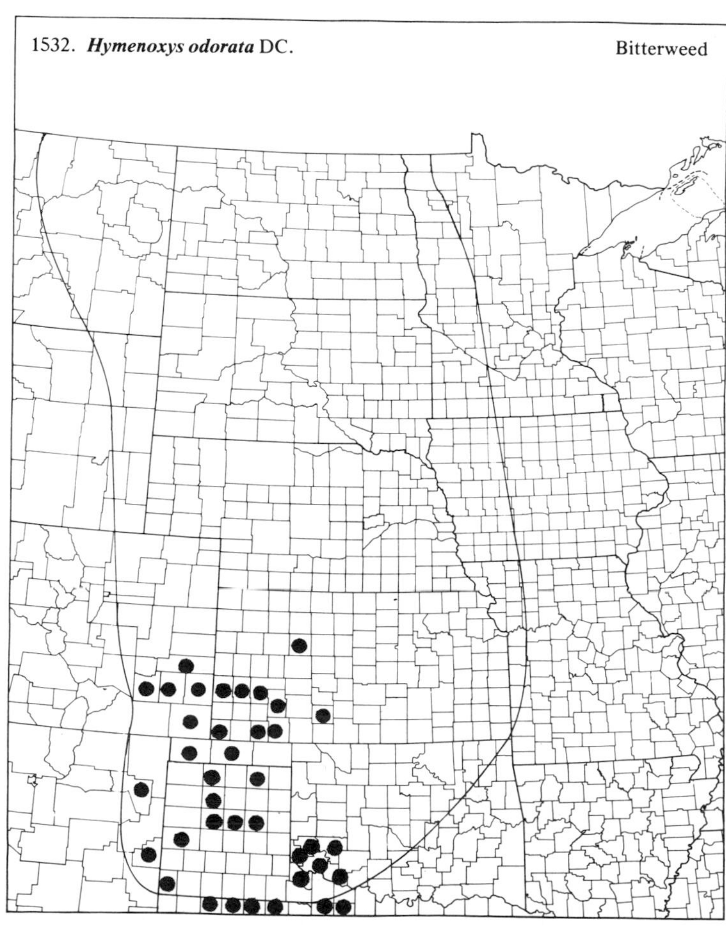

(Asteraceae)

1533. ***Hymenoxys richardsonii*** (Hook.) Cockll. Colorado Rubber Plant

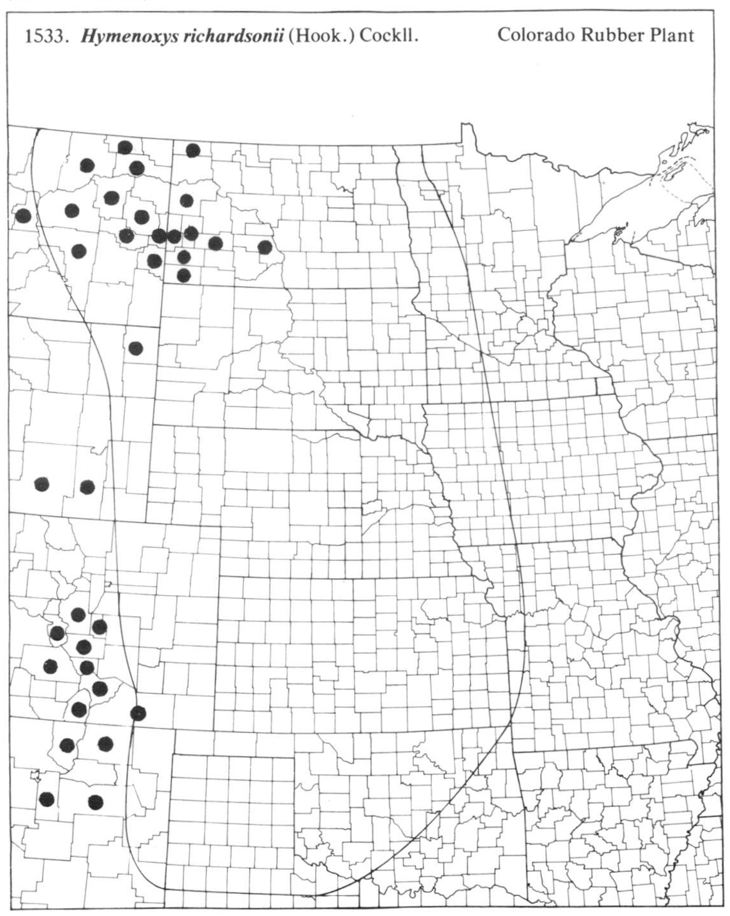

1534. ***Hymenoxys scaposa*** (DC.) Parker var. ***scaposa***

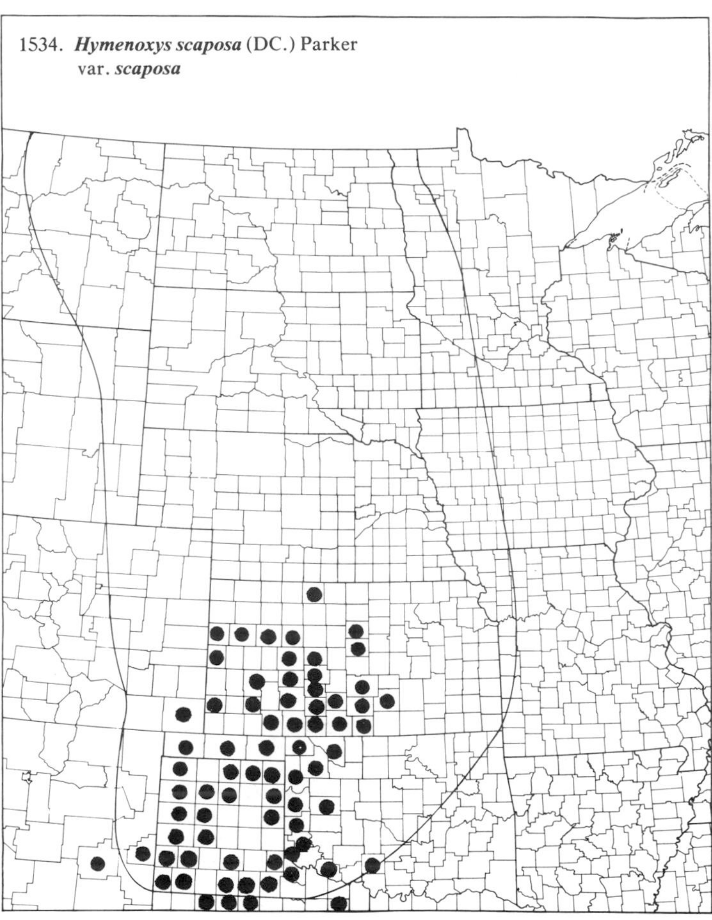

1535. ***Hymenoxys scaposa*** (DC.) Parker var. ***glabra*** (Nutt.) Parker

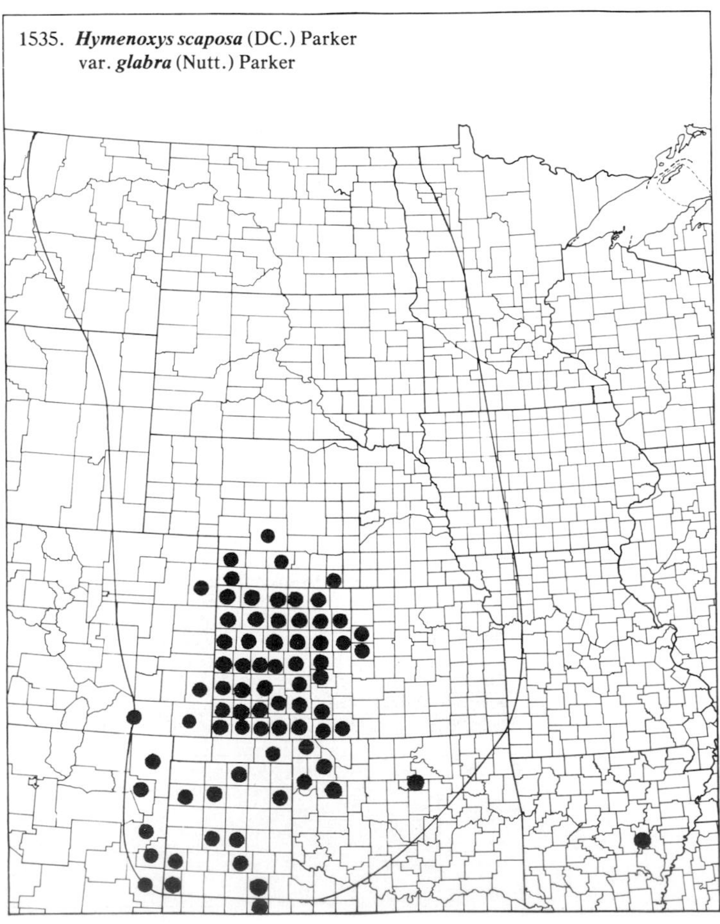

1536. ***Iva annua*** L.

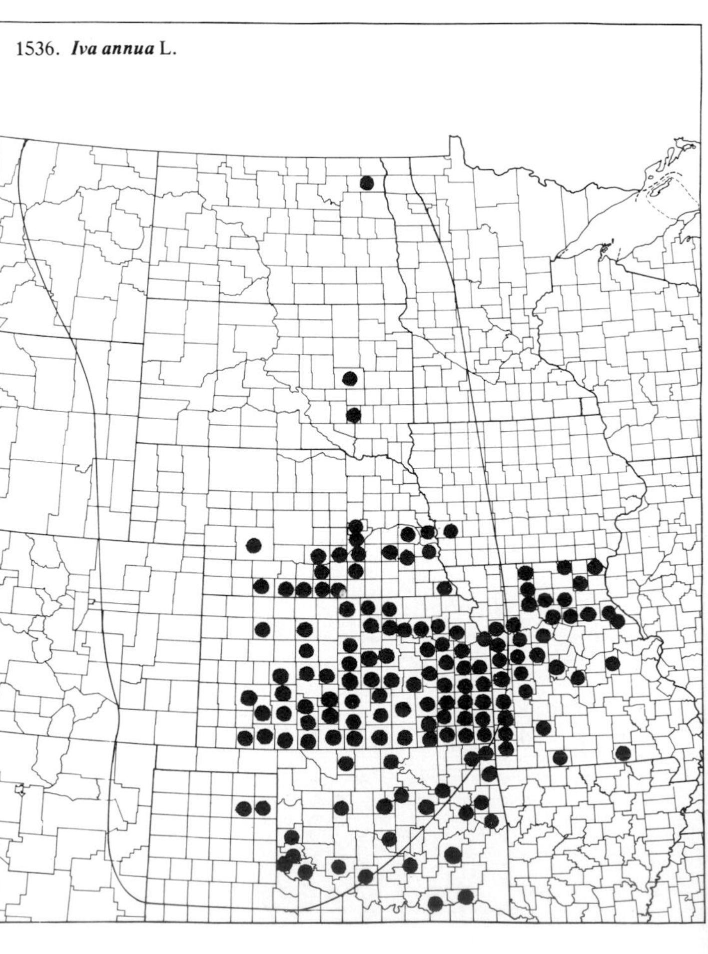

(Asteraceae)

1537. ***Iva axillaris*** Pursh — Povertyweed

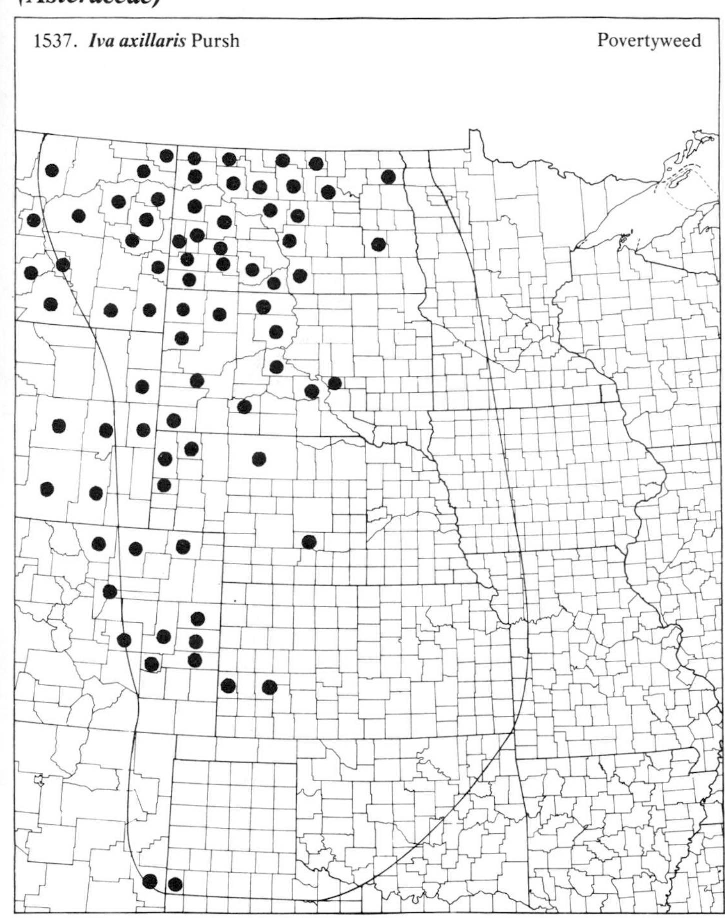

1538. ***Iva xanthifolia*** Nutt. — Marsh Elder

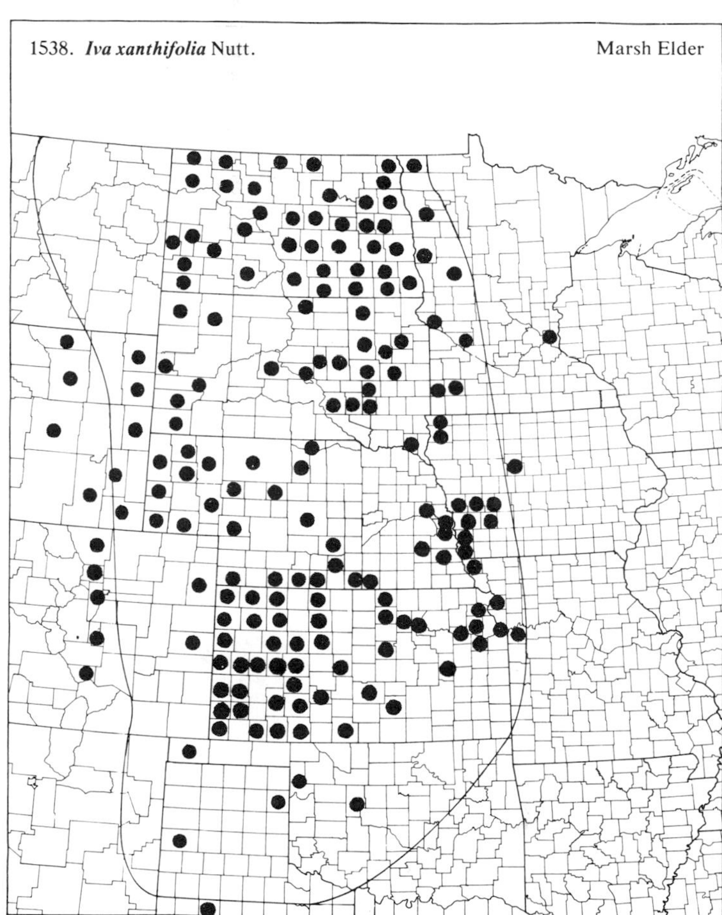

1539. ***Krigia dandelion*** (L.) Nutt. — Potato Dandelion

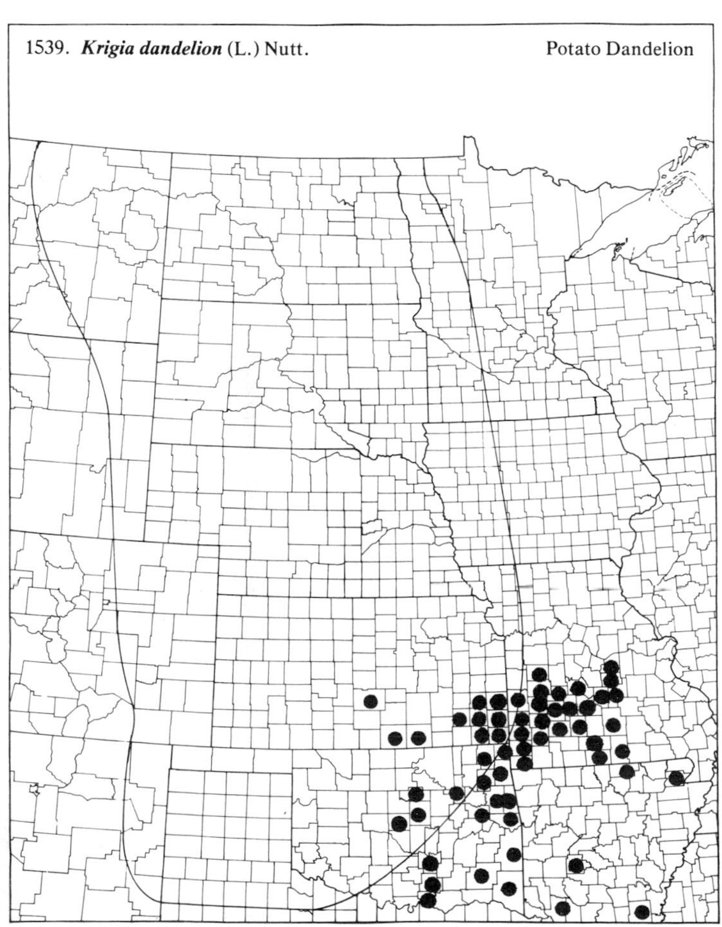

1540. ***Krigia occidentalis*** Nutt. — Western Dwarf Dandelion

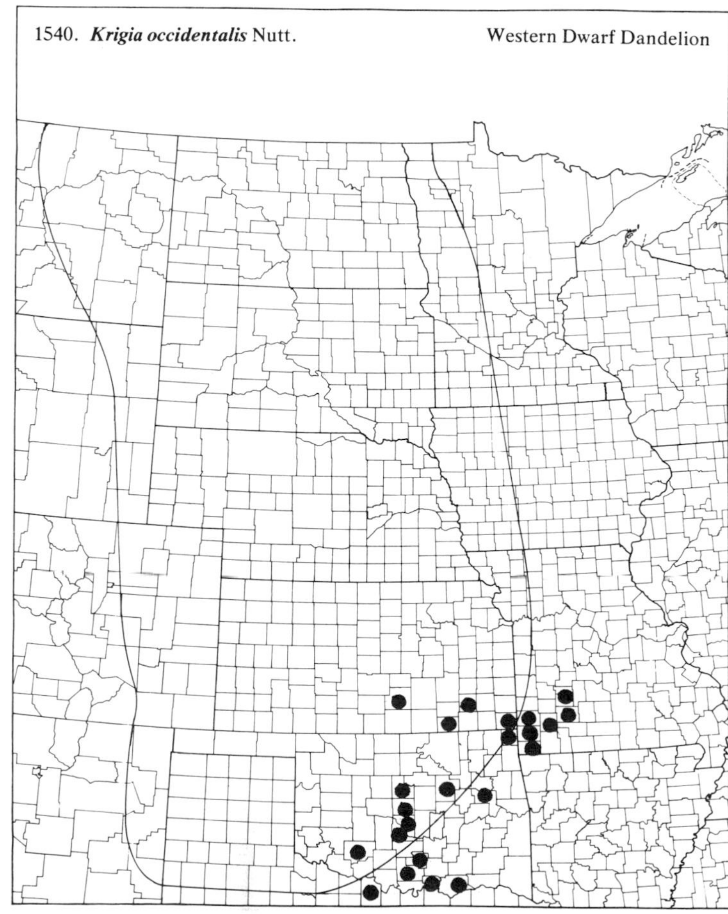

(Asteraceae)

1541. ***Krigia oppositifolia*** Raf. — Common Dwarf Dandelion

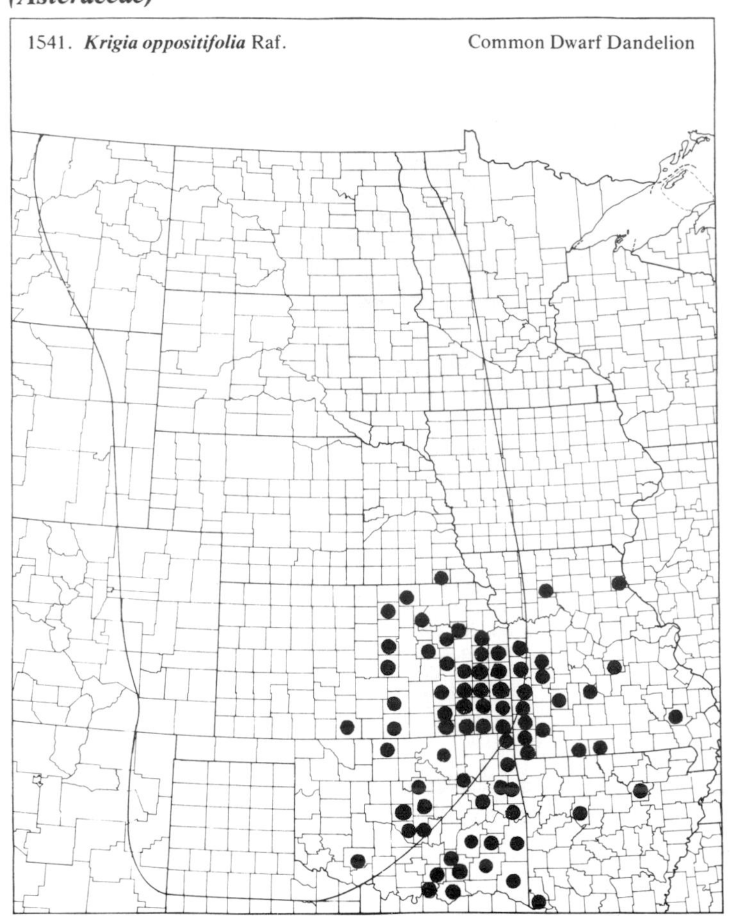

1542. ***Kuhnia eupatorioides*** L. var. ***corymbulosa*** T. & G. — False Boneset

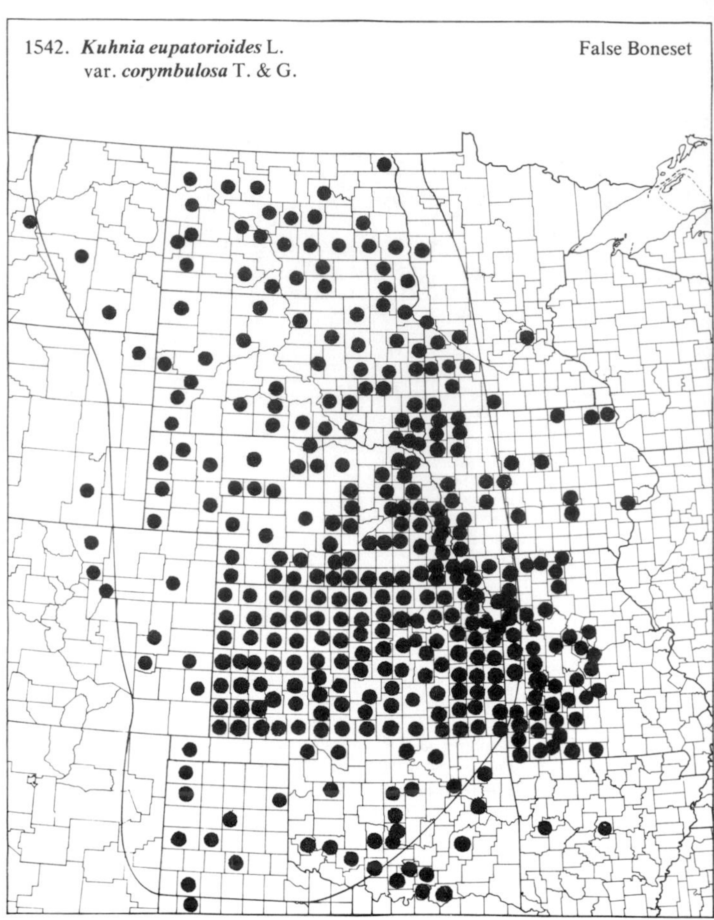

1543. ***Lactuca biennis*** (Moench) Fern. — BlueWood Lettuce

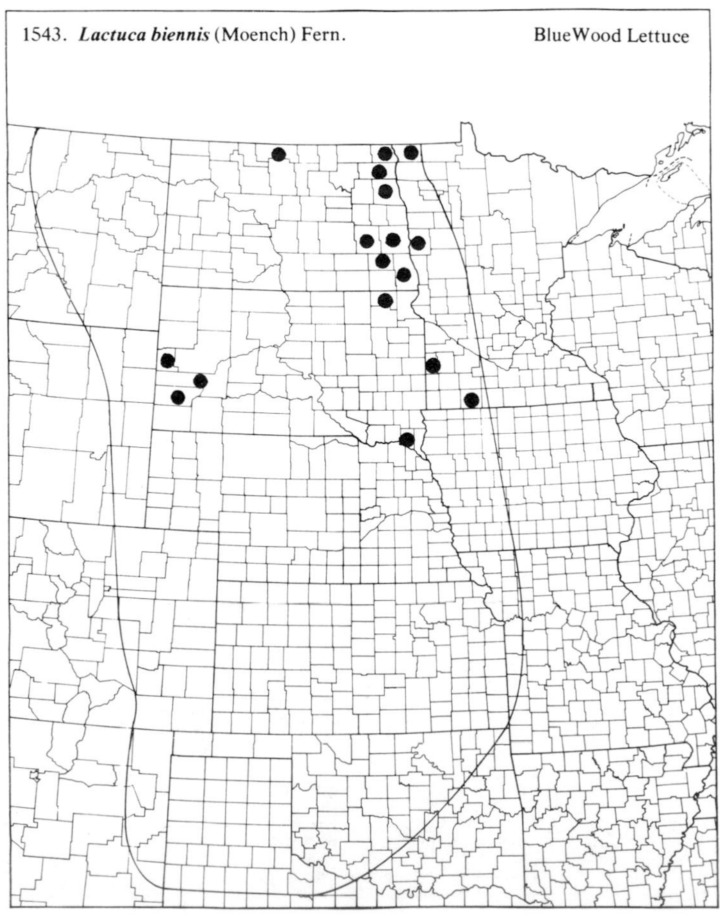

1544. ***Lactuca canadensis*** L. — Wild Lettuce

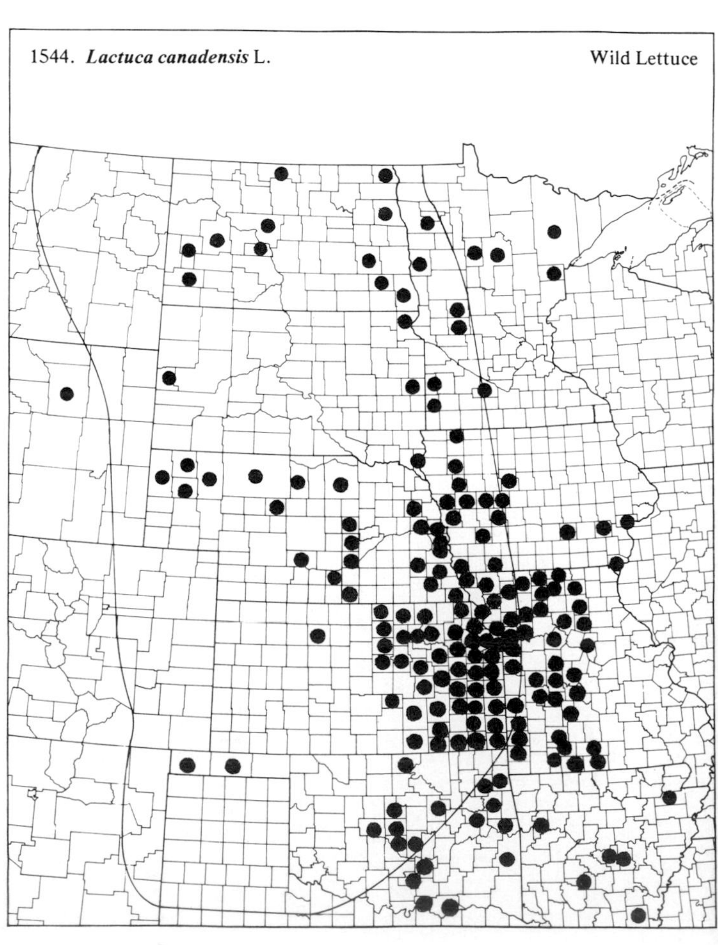

(Asteraceae)

1545. **Lactuca floridana** (L.) Gaertn. Florida Lettuce

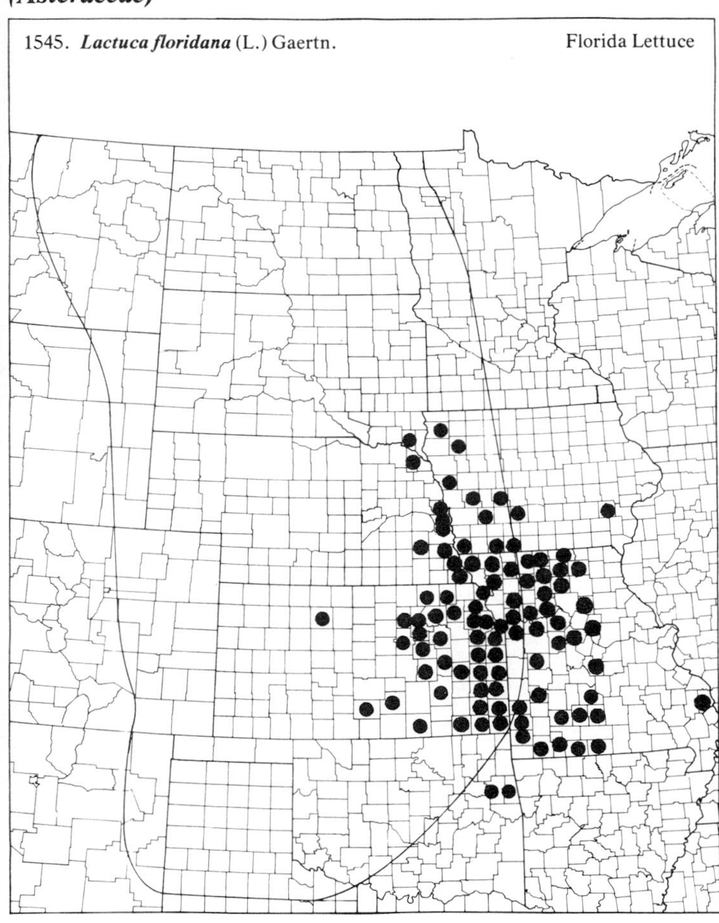

1546. **Lactuca ludoviciana** (Nutt.) DC. Western Wild Lettuce

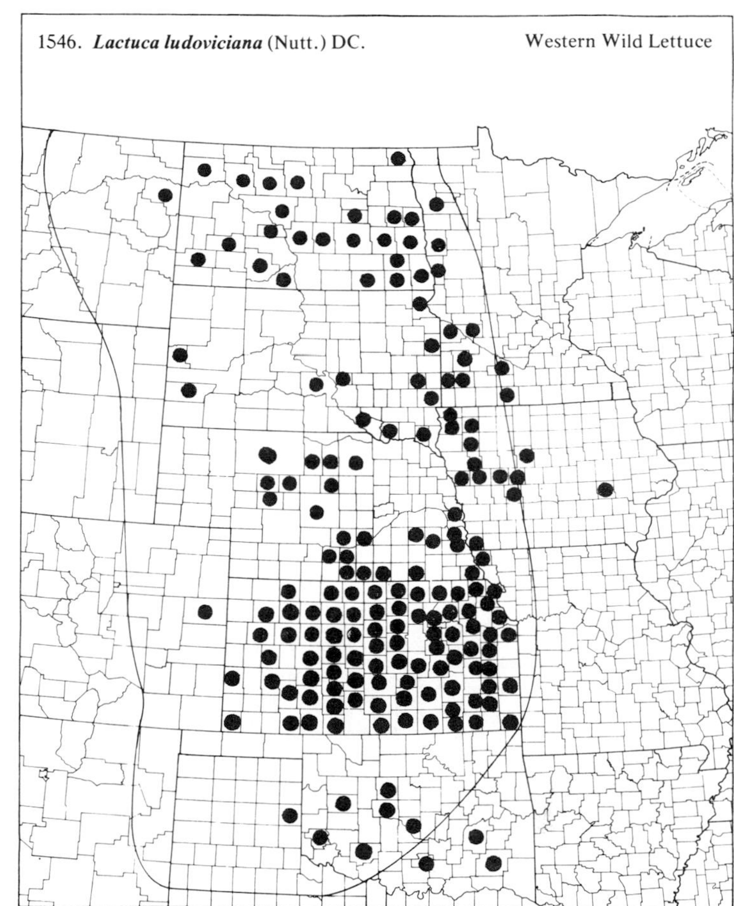

1547. **Lactuca oblongifolia** Nutt. Blue Lettuce

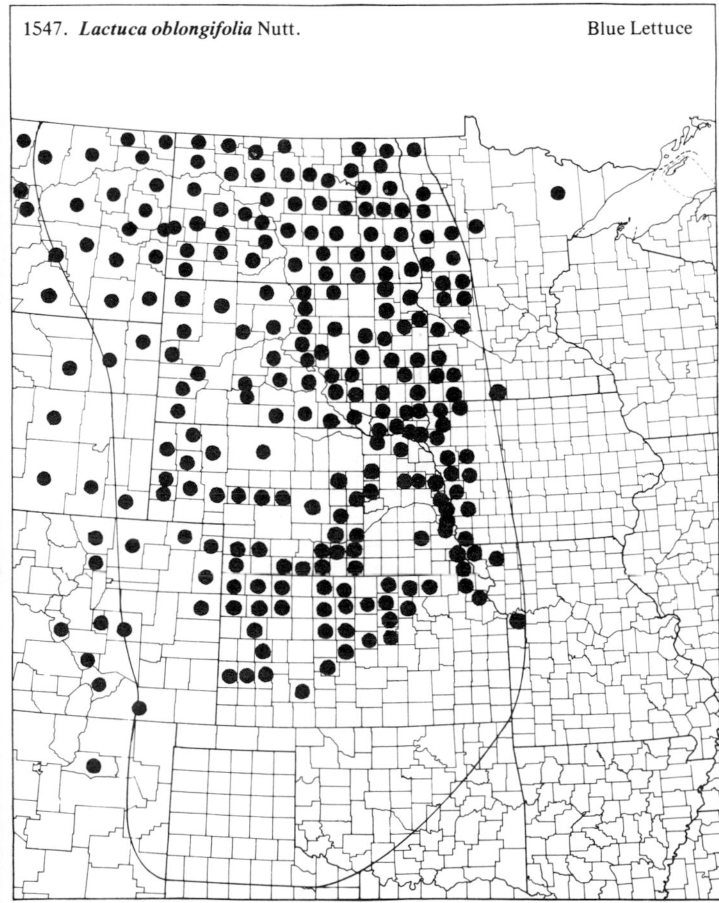

1548. **Lactuca saligna** L. Willow-leaved Lettuce

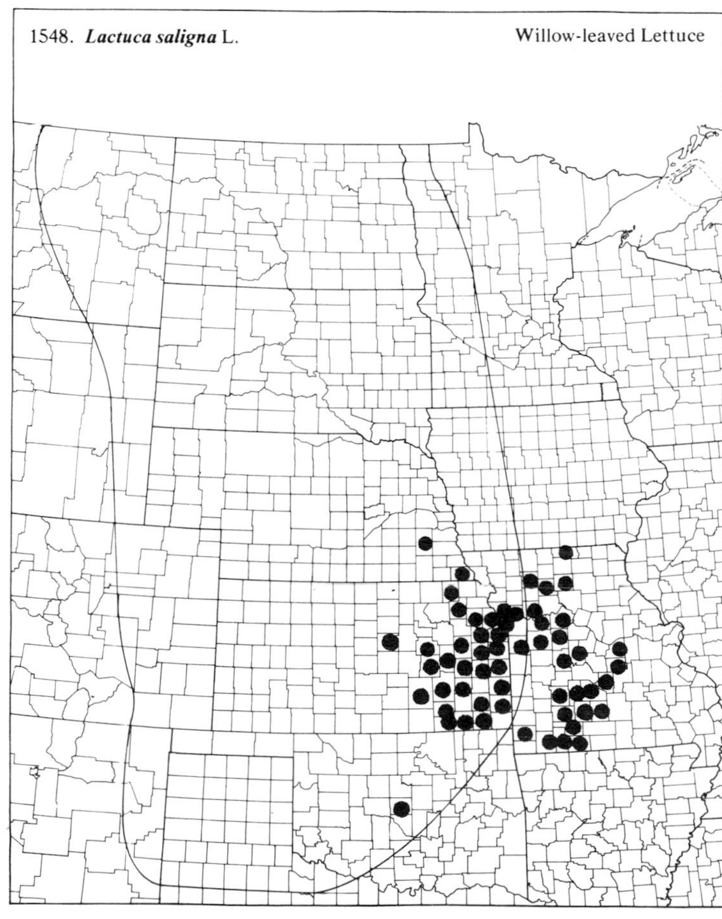

(Asteraceae)

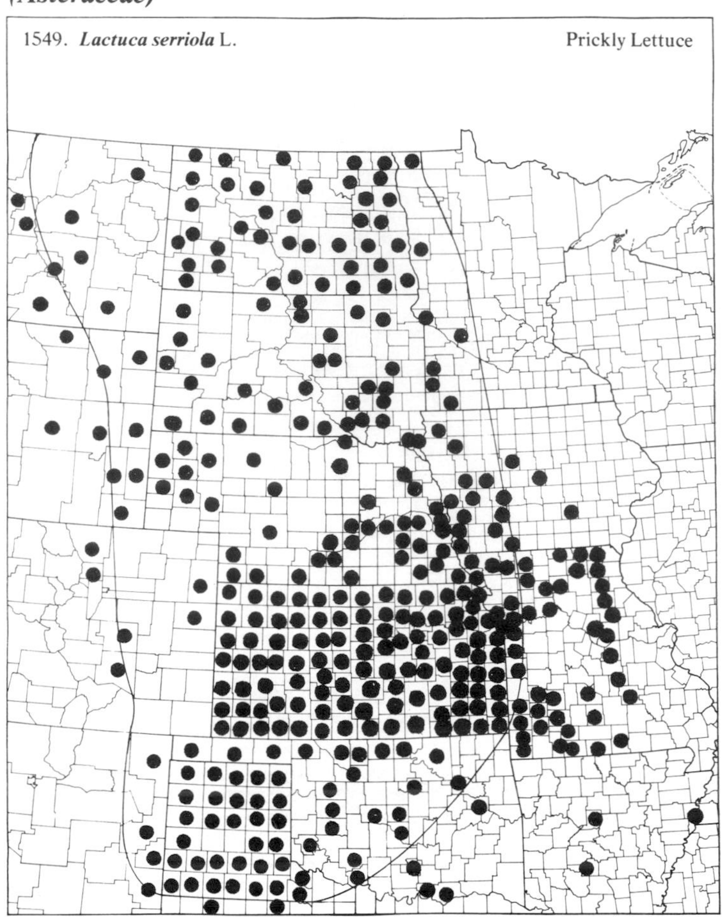

1549. ***Lactuca serriola*** L. Prickly Lettuce

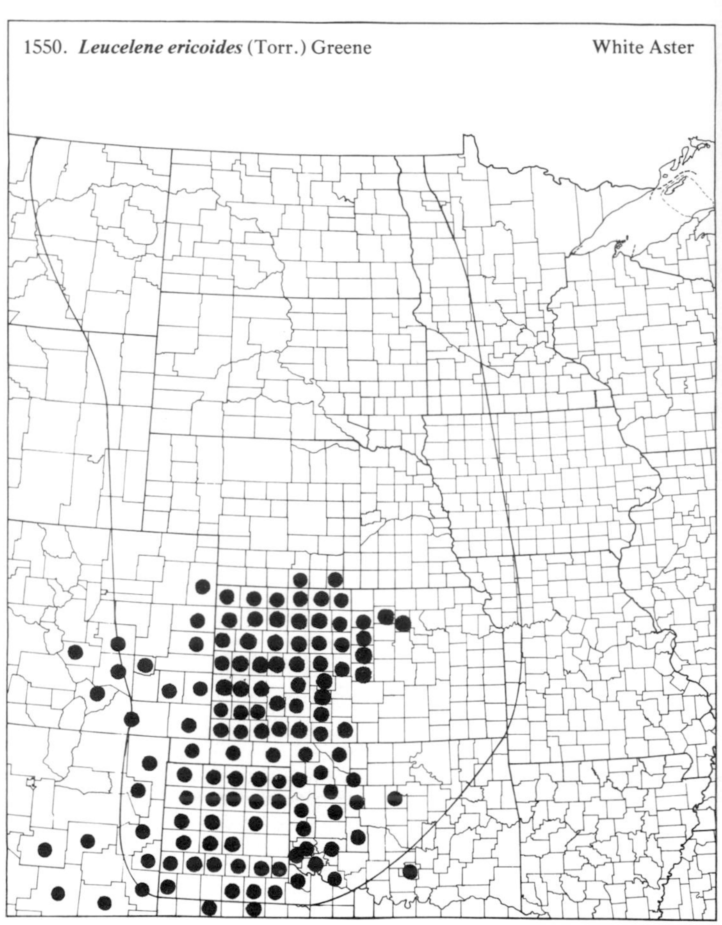

1550. ***Leucelene ericoides*** (Torr.) Greene White Aster

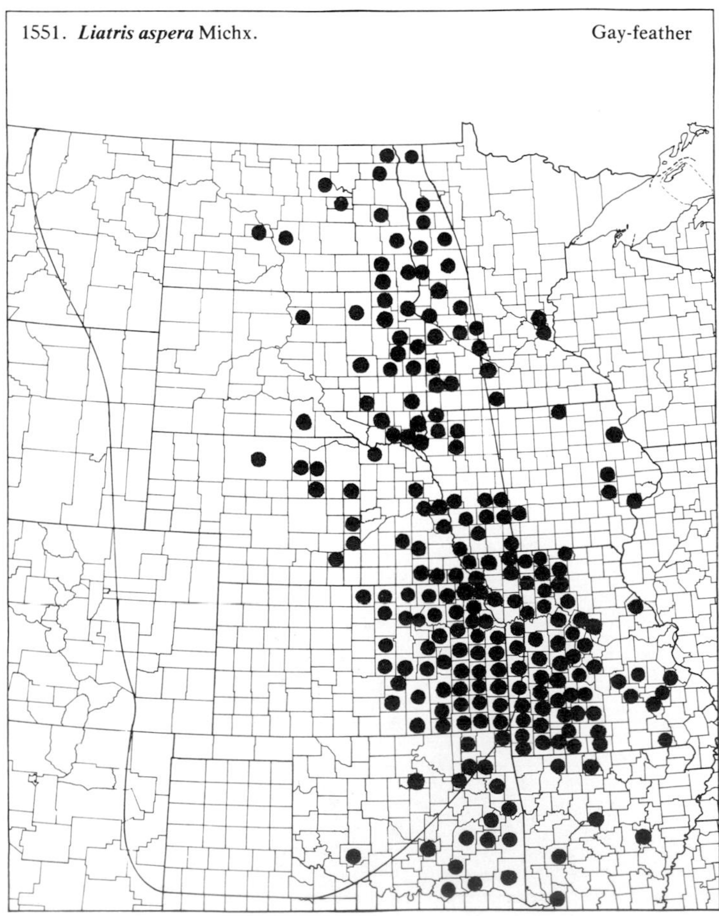

1551. ***Liatris aspera*** Michx. Gay-feather

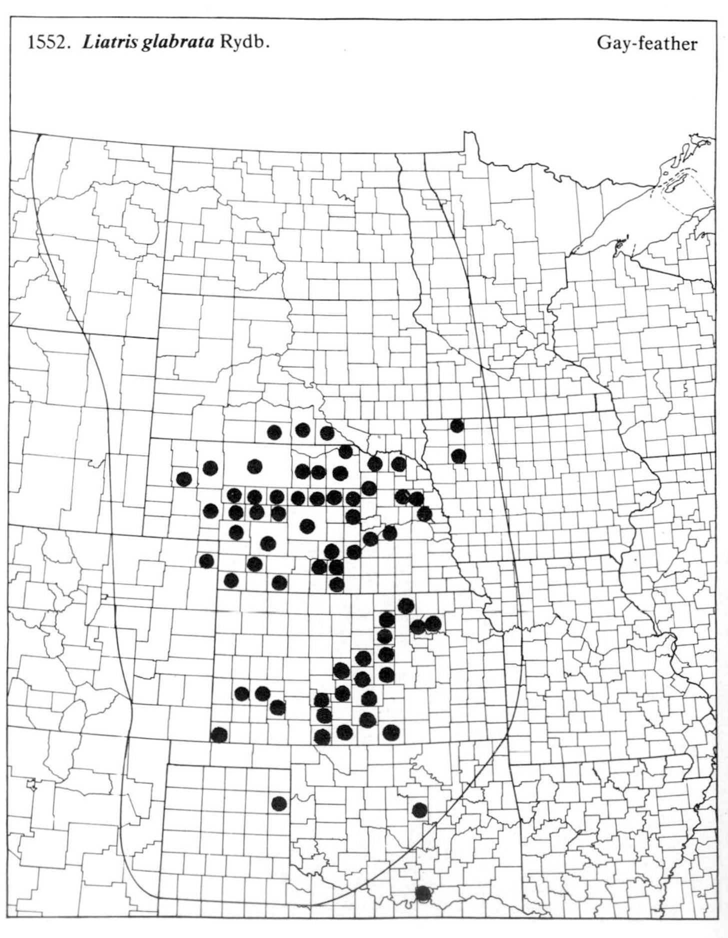

1552. ***Liatris glabrata*** Rydb. Gay-feather

(Asteraceae)

1553. ***Liatris hirsuta*** Rydb. Gay-feather

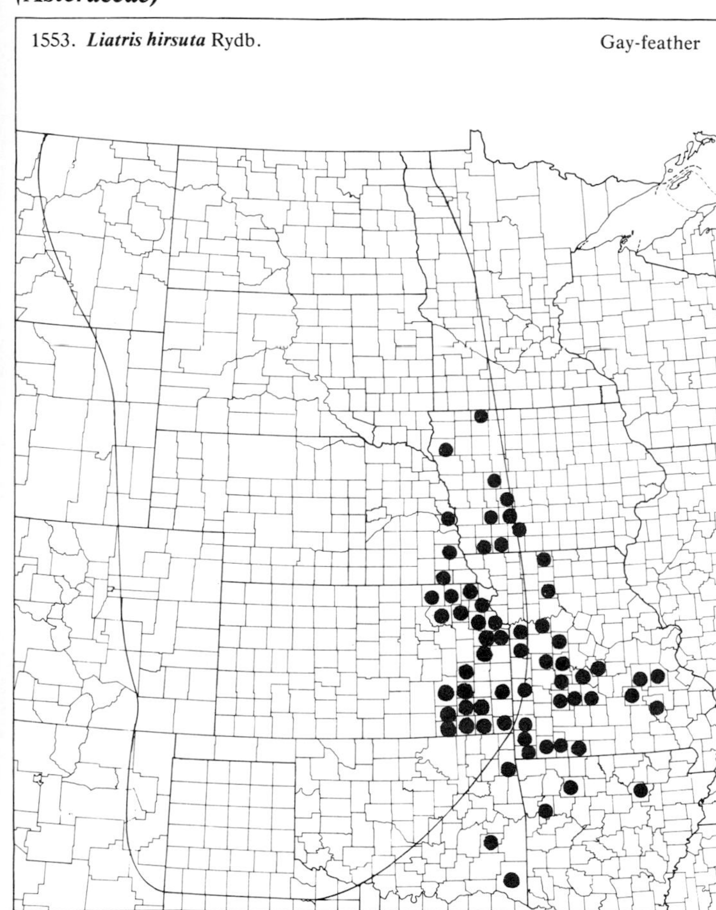

1554. ***Liatris lancifolia*** (Greene) Kittell Gay-feather

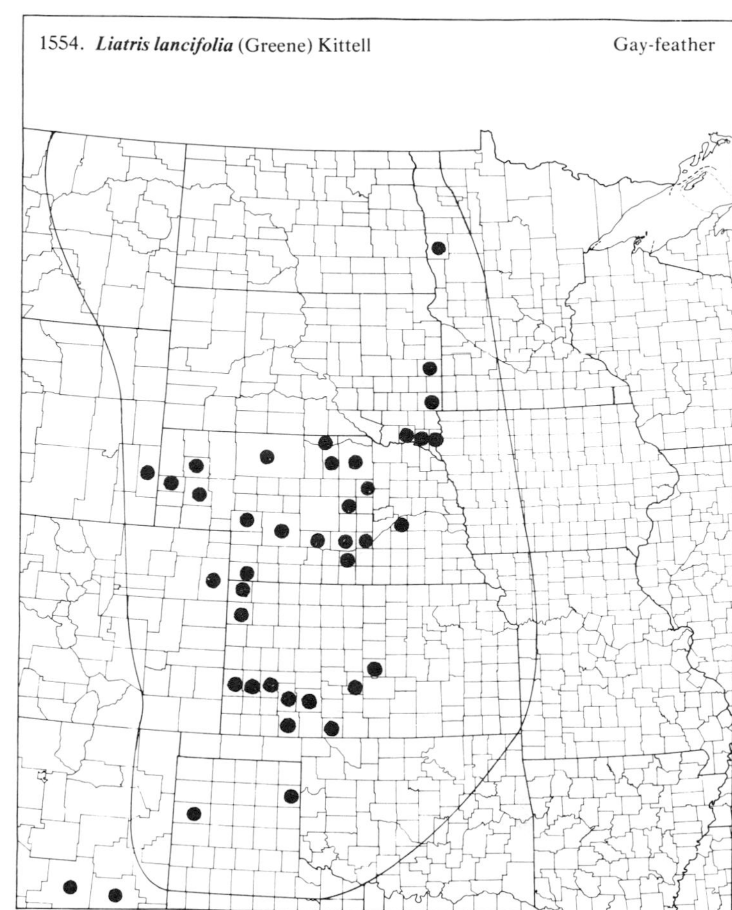

1555. ***Liatris ligulistylis*** (A. Nels.) K. Schum. Gay-feather

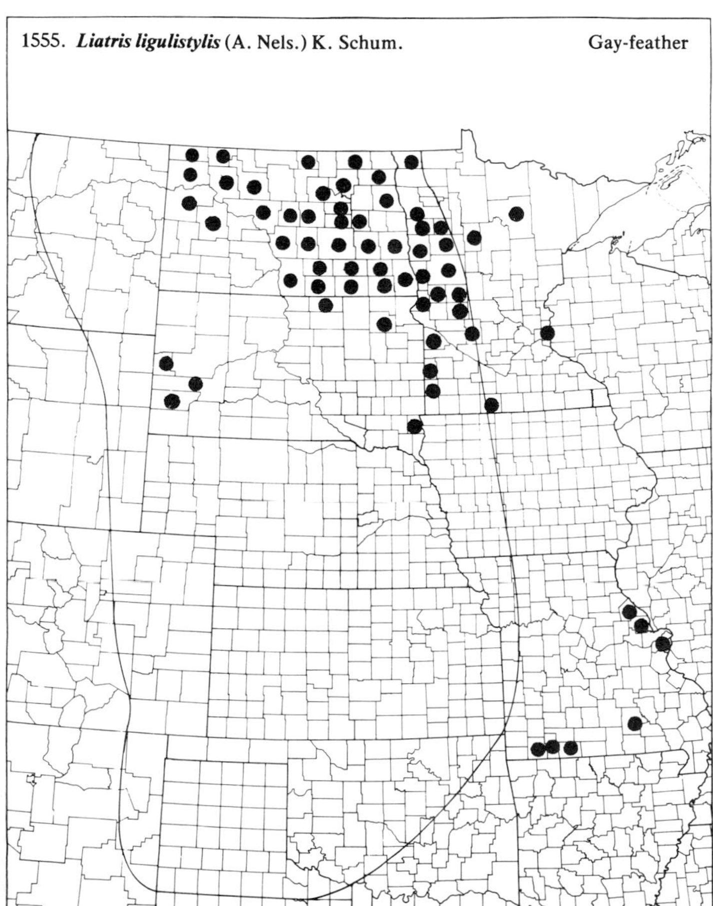

1556. ***Liatris mucronata*** DC. Gay-feather

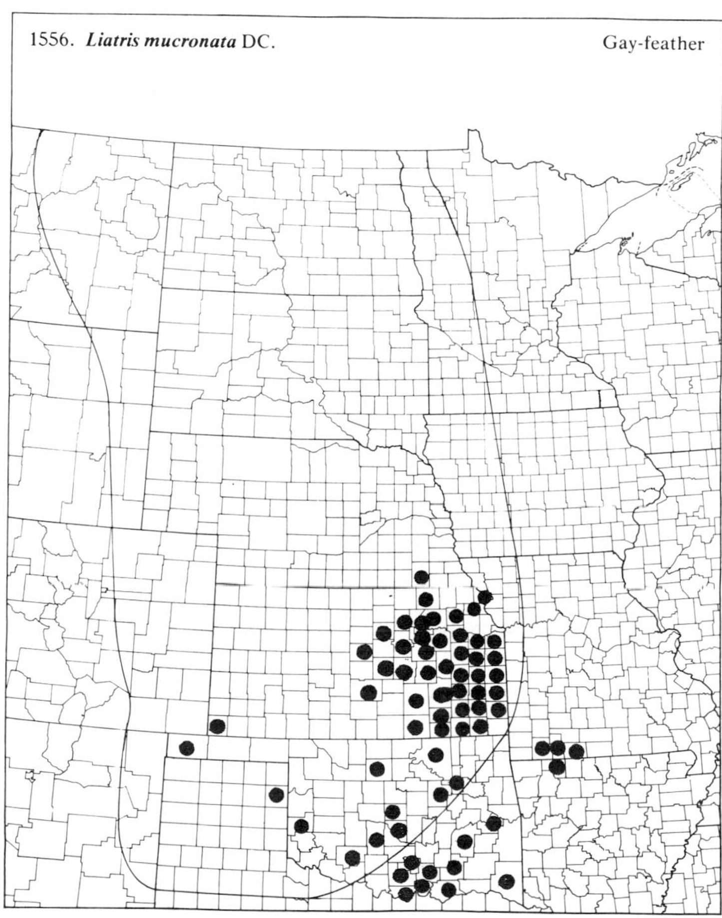

(Asteraceae)

1557. ***Liatris punctata*** Hook. — Blazing Star

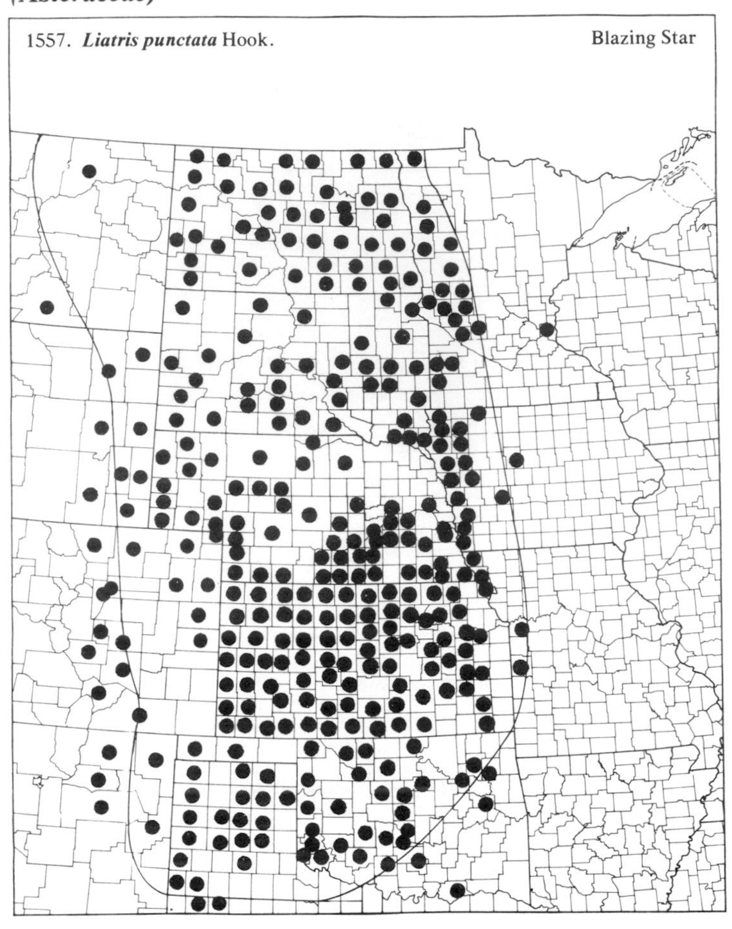

1558. ***Liatris pycnostachya*** Michx. — Tall Blazing Star

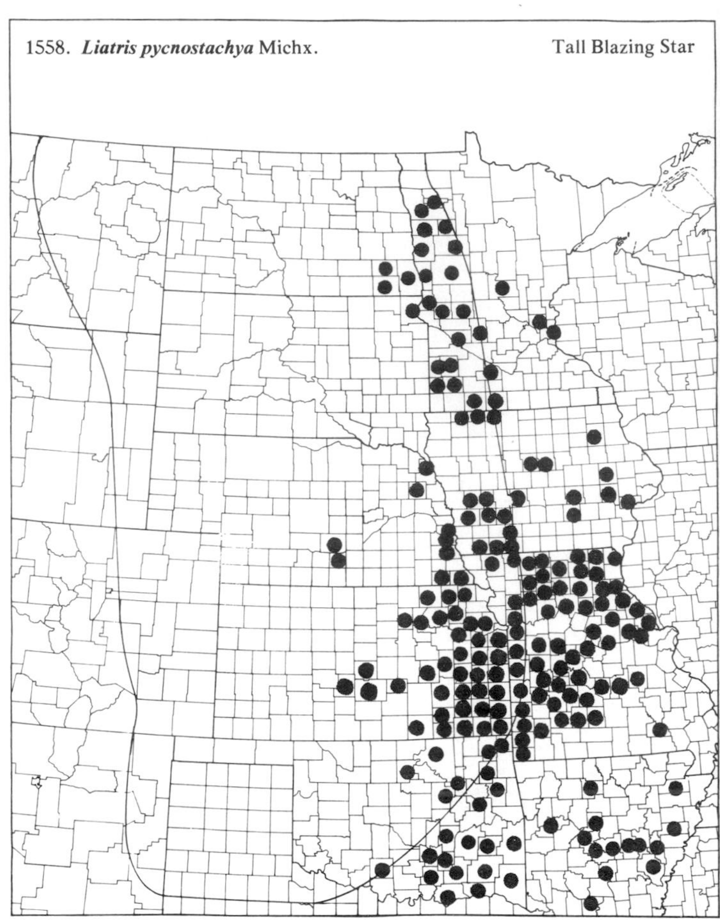

1559. ***Lygodesmia juncea*** (Pursh) Hook. — Skeleton-weed

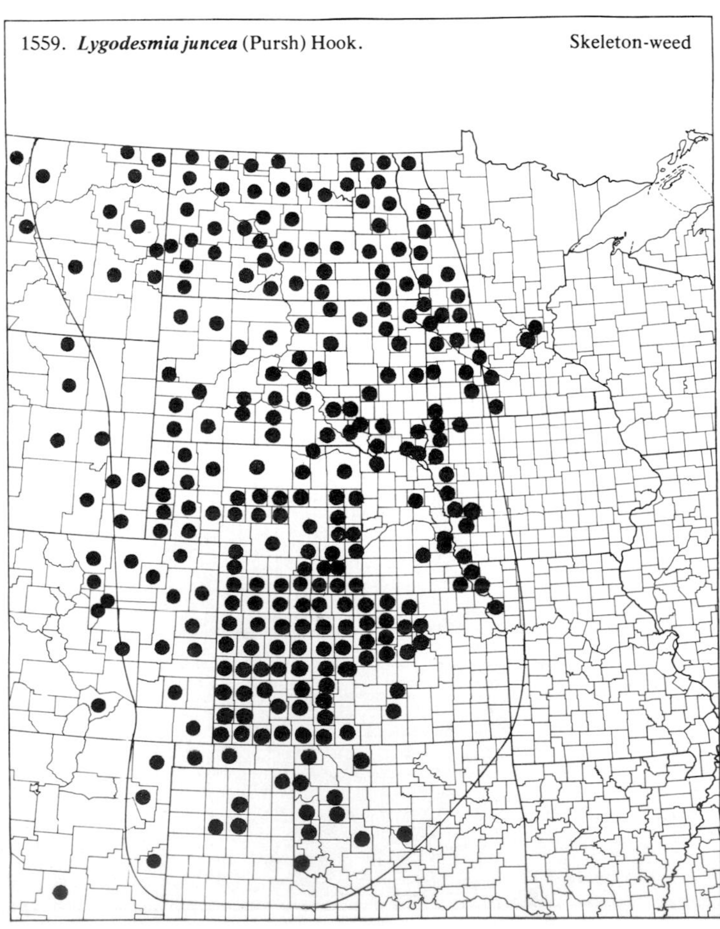

1560. ***Lygodesmia rostrata*** Gray — Annual Skeleton-weed

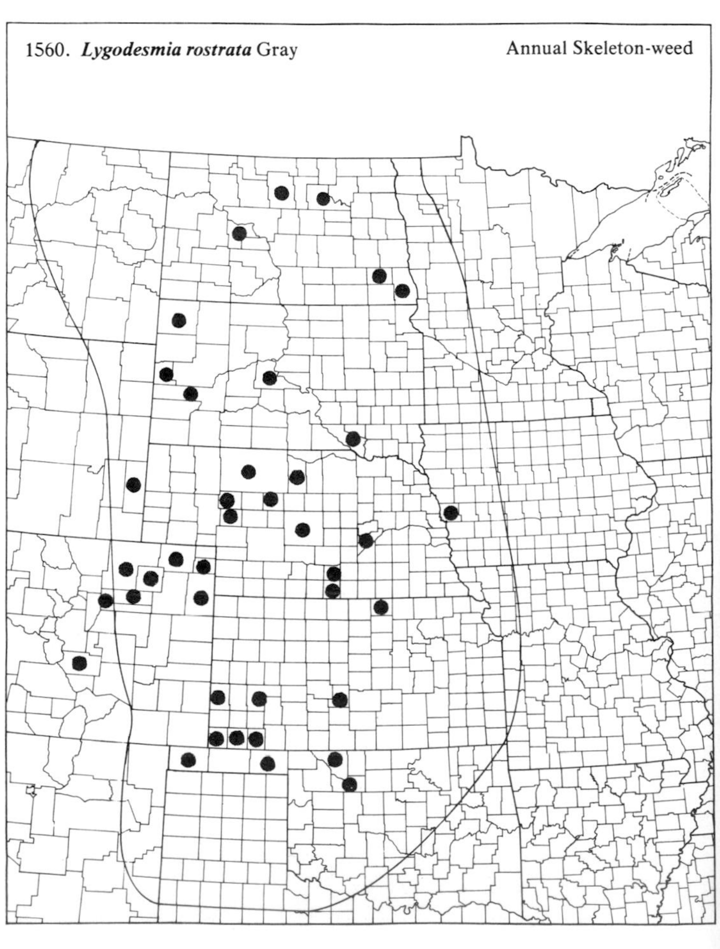

(Asteraceae)

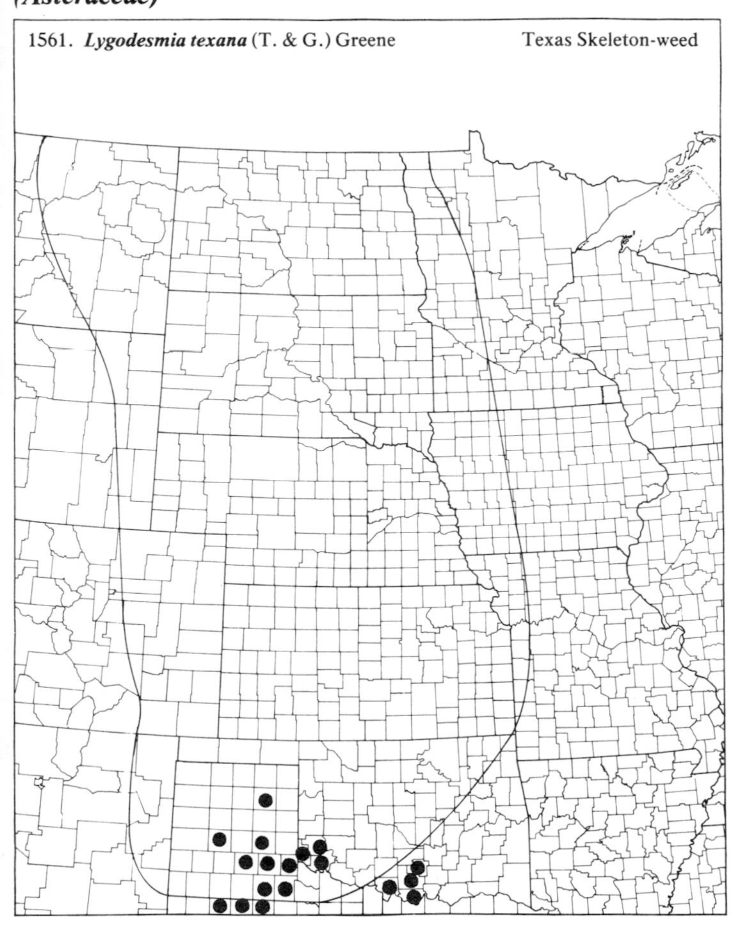
1561. *Lygodesmia texana* (T. & G.) Greene
Texas Skeleton-weed

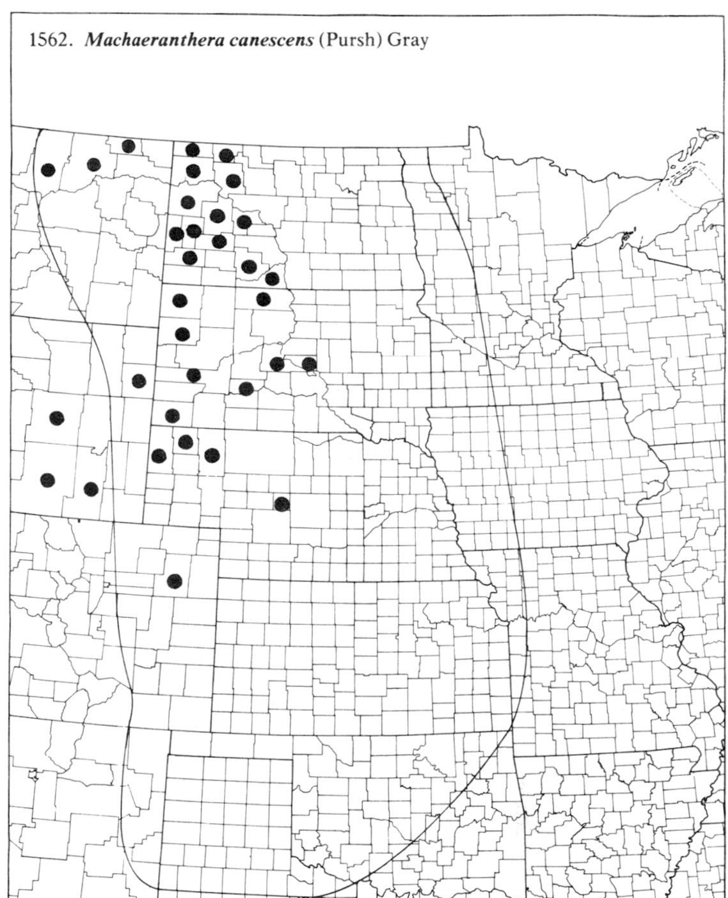
1562. *Machaeranthera canescens* (Pursh) Gray

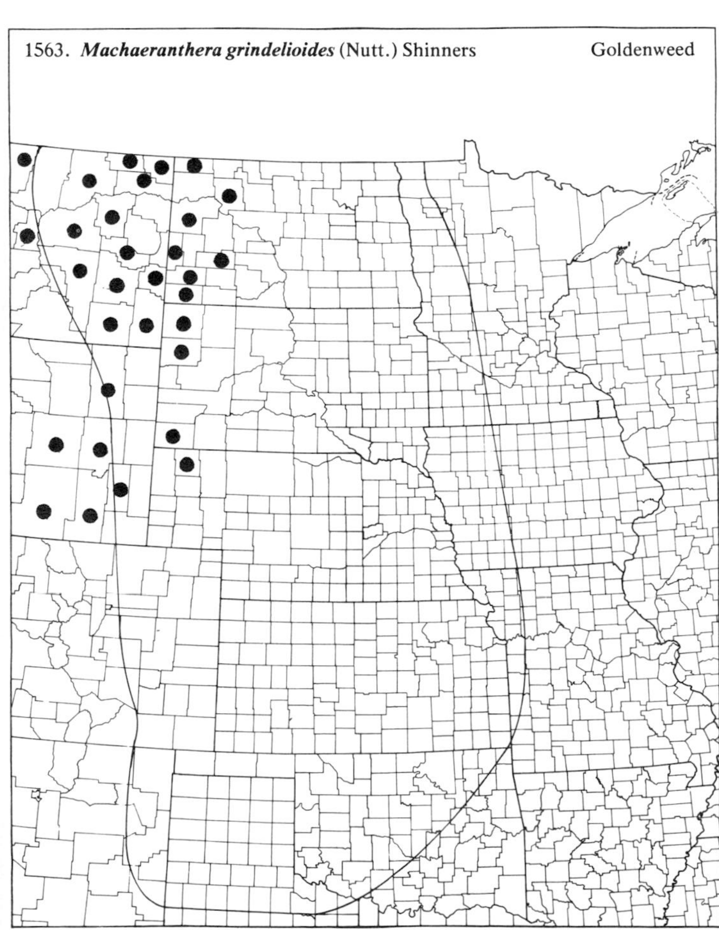
1563. *Machaeranthera grindelioides* (Nutt.) Shinners
Goldenweed

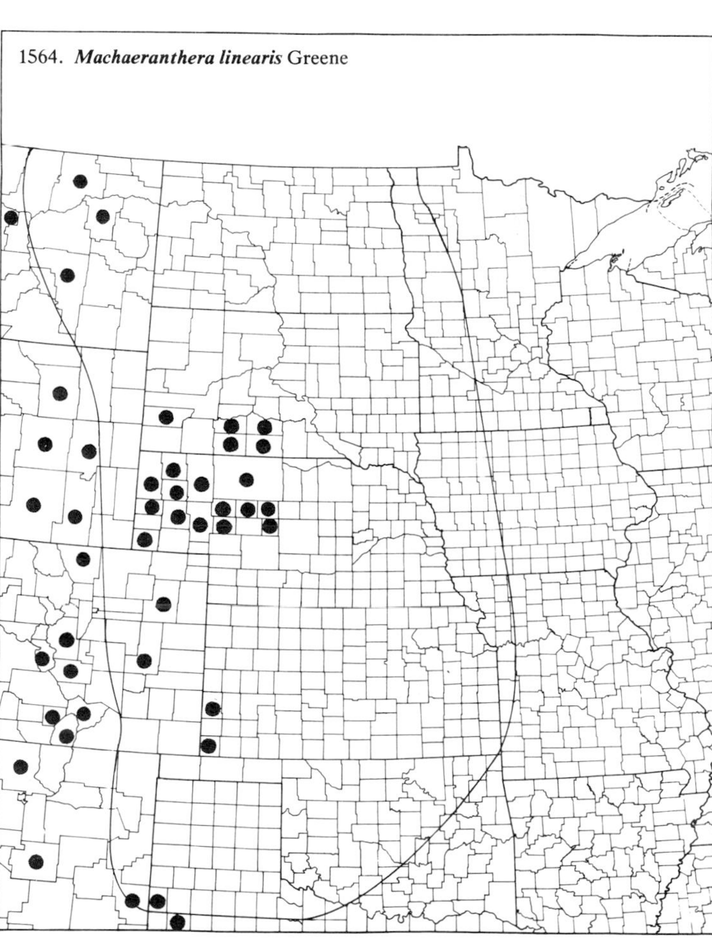
1564. *Machaeranthera linearis* Greene

(Asteraceae)

1565. ***Machaeranthera tanacetifolia*** (H.B.K.) Nees — Tansy Aster

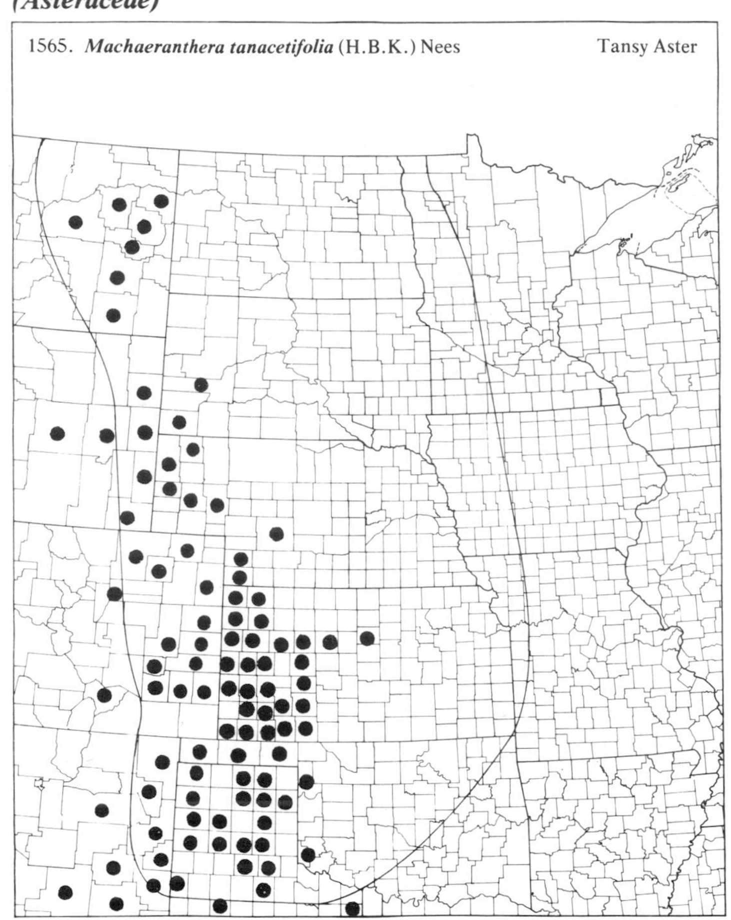

1566. ***Matricaria maritima*** L. — Wild Chamomile

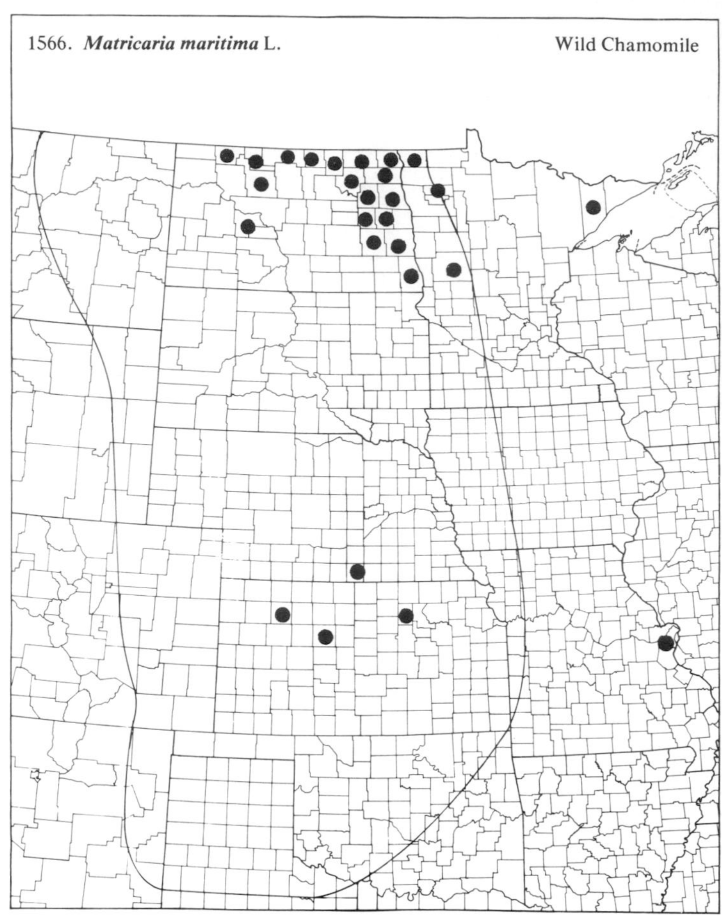

1567. ***Matricaria matricarioides*** (Less.) Porter — Pineapple-weed

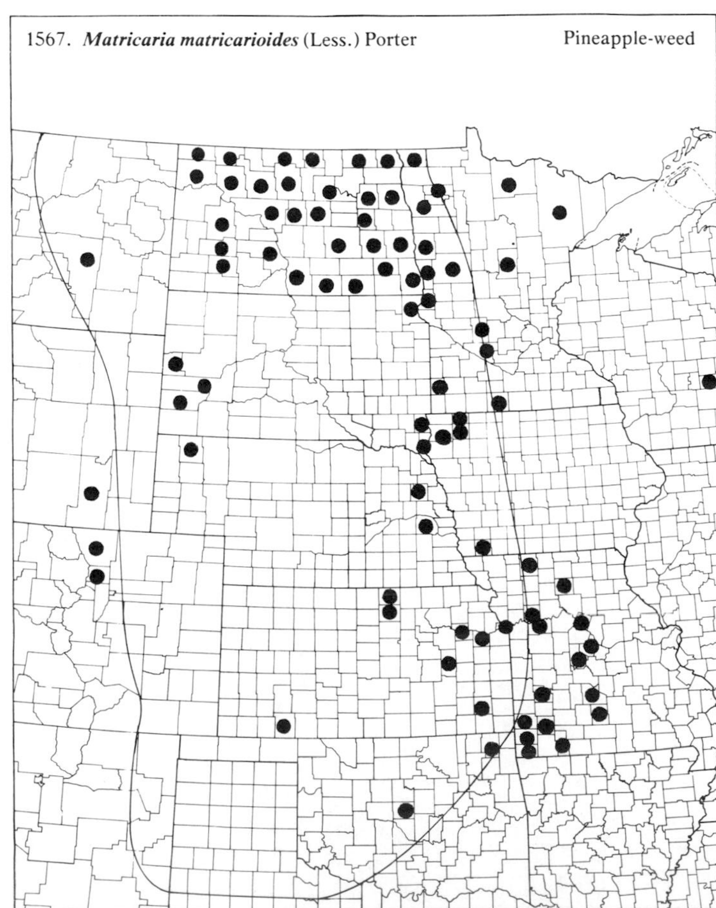

1568. ***Melampodium leucanthum*** T. & G.

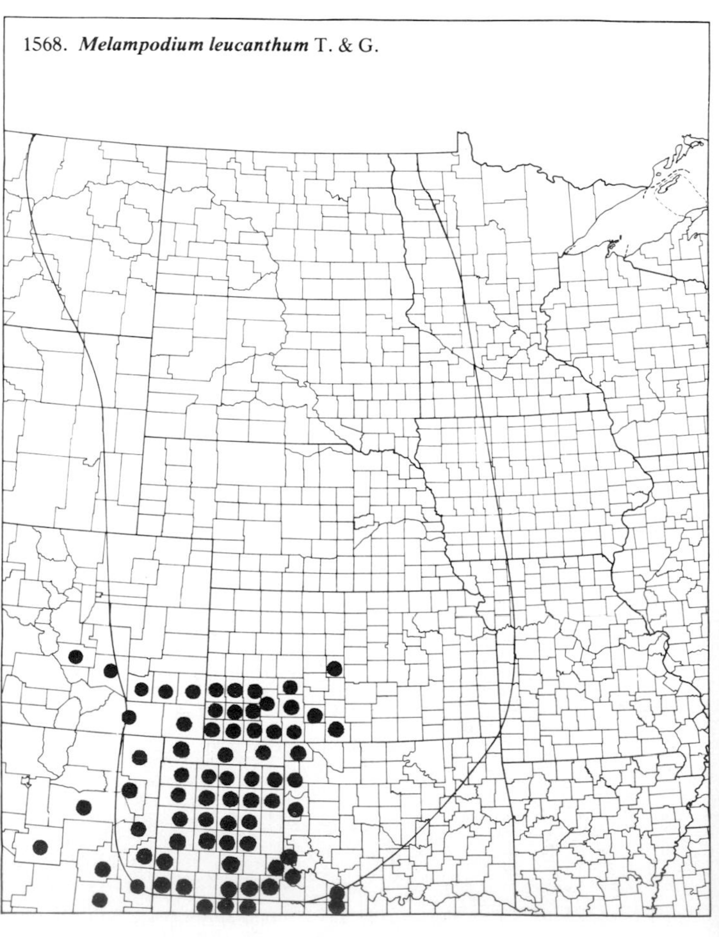

(Asteraceae)

1569. ***Microseris cuspidata*** (Pursh) Sch. Bip. False Dandelion

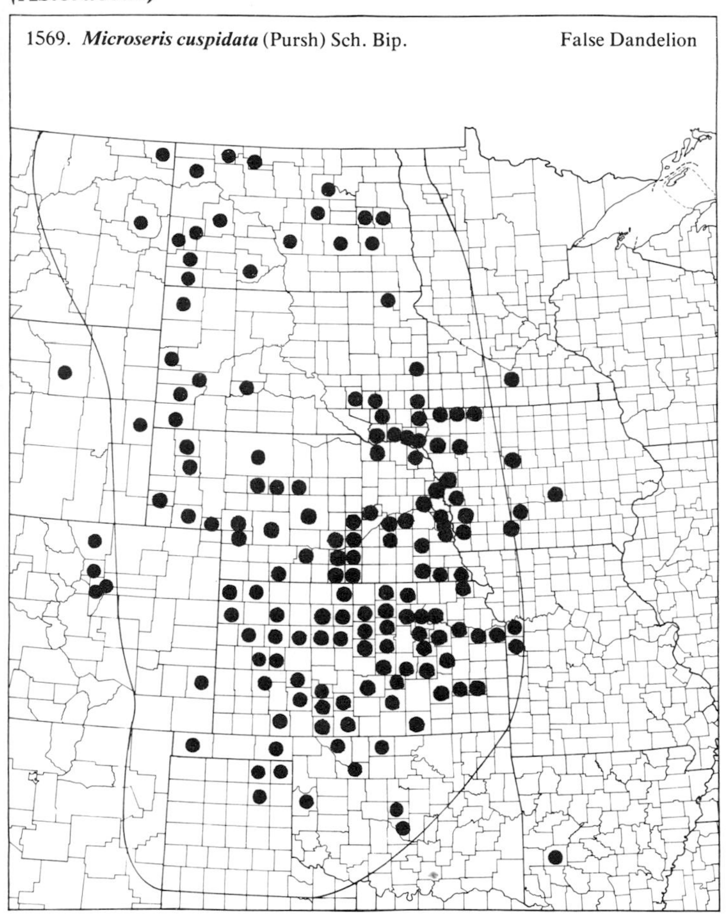

1570. ***Palafoxia rosea*** (Bush) Cory var. ***macrolepis*** Rydb.

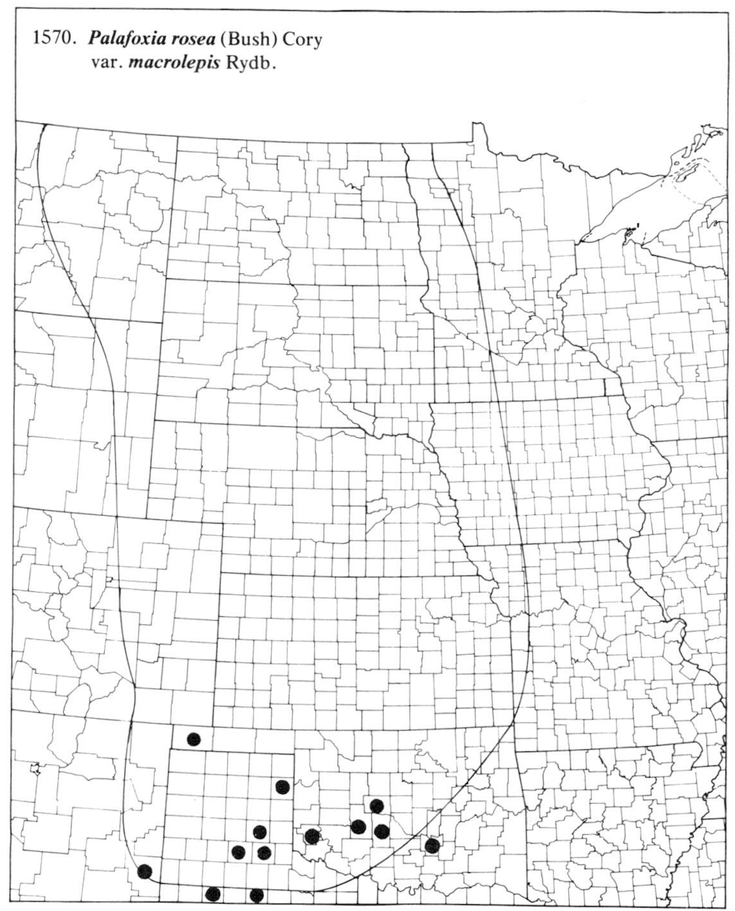

1571. ***Palafoxia sphacelata*** (Nutt.) Cory Othake

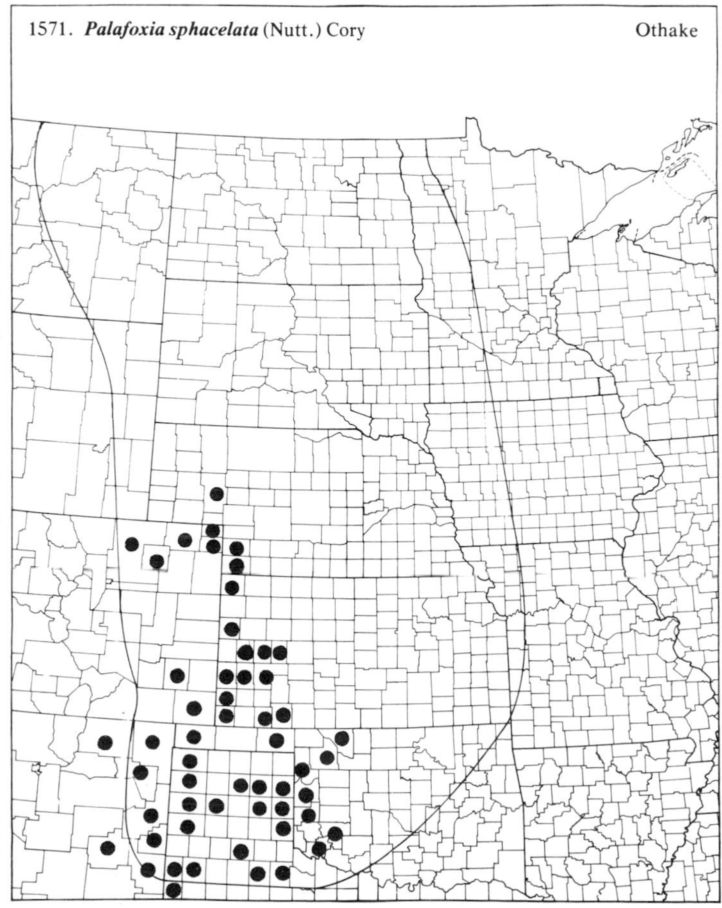

1572. ***Palafoxia texana*** DC.

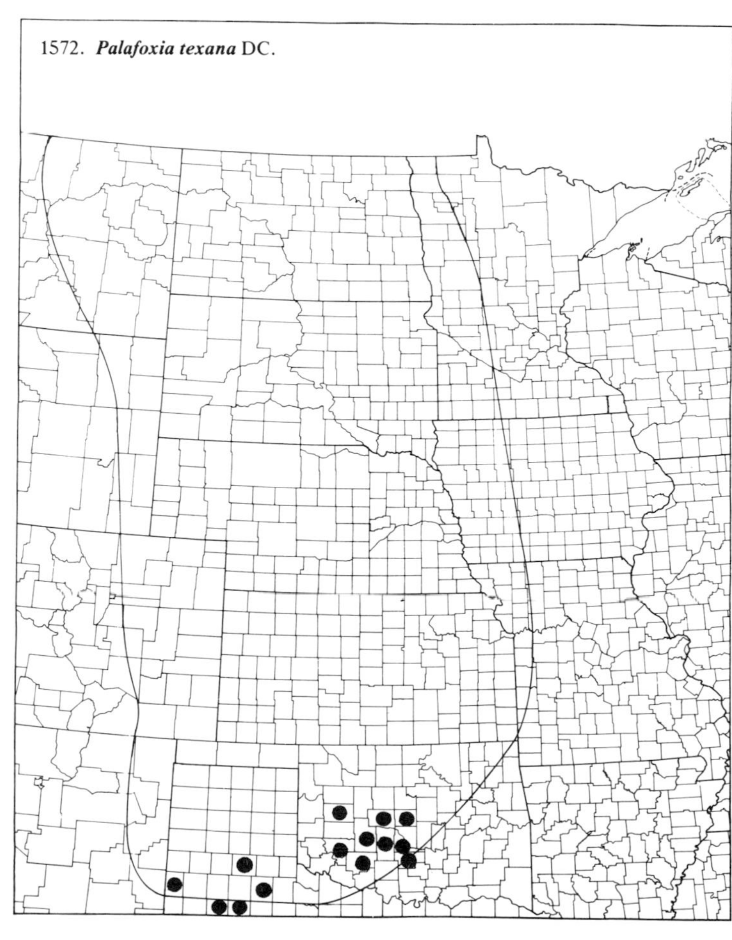

(Asteraceae)

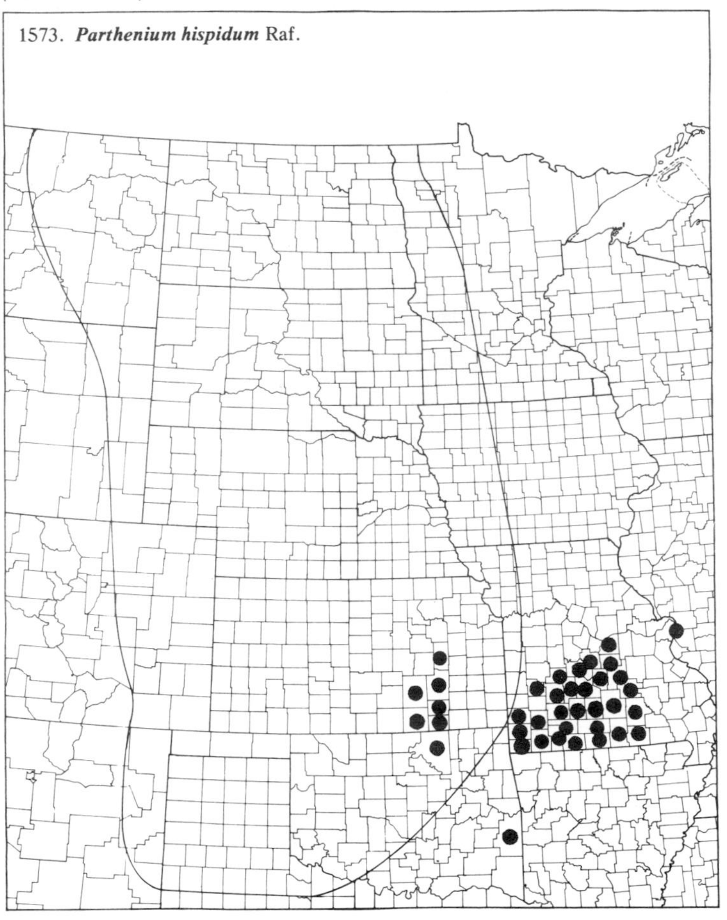
1573. ***Parthenium hispidum*** Raf.

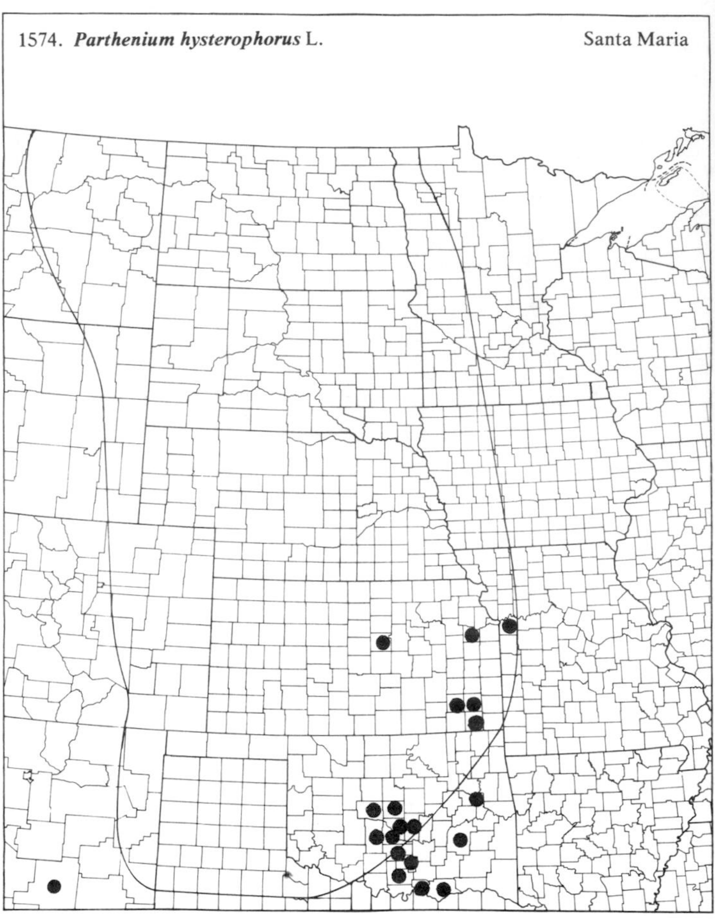
1574. ***Parthenium hysterophorus*** L. Santa Maria

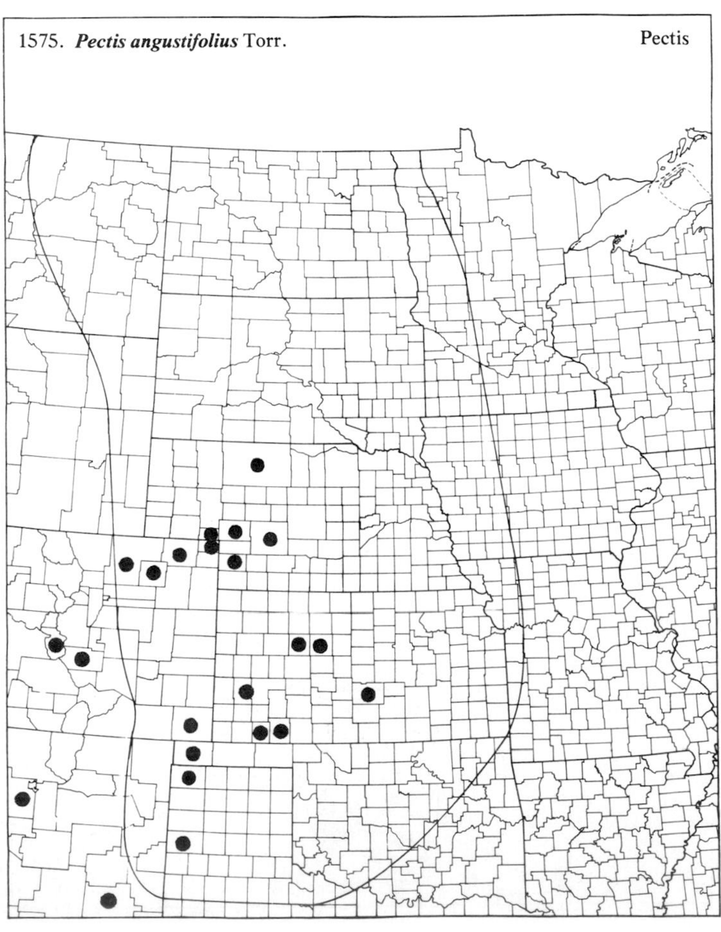
1575. ***Pectis angustifolius*** Torr. Pectis

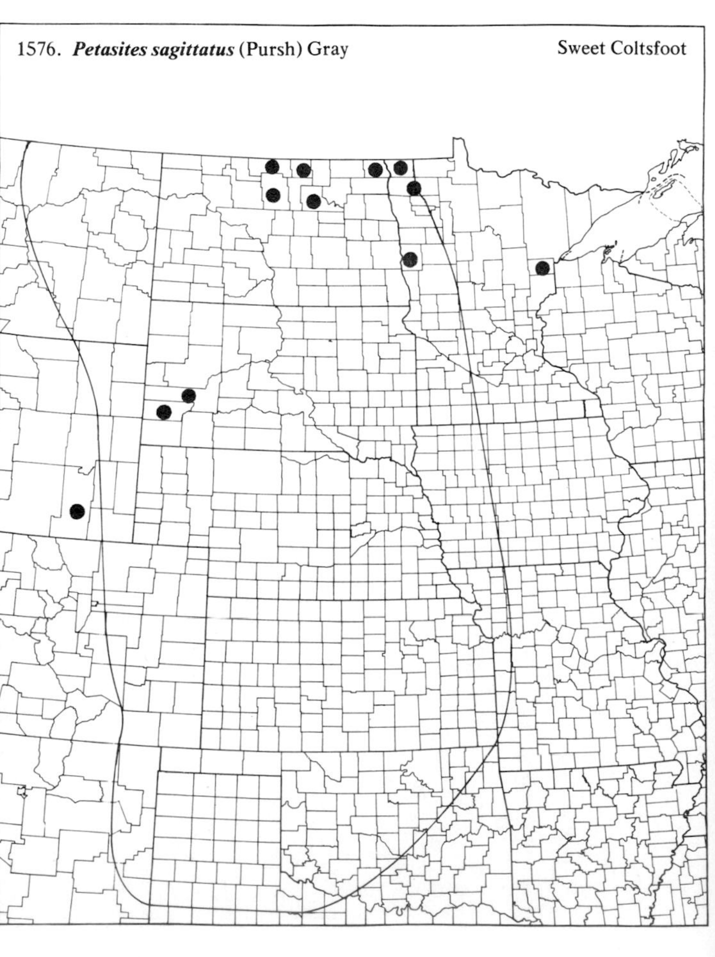
1576. ***Petasites sagittatus*** (Pursh) Gray Sweet Coltsfoot

(Asteraceae)

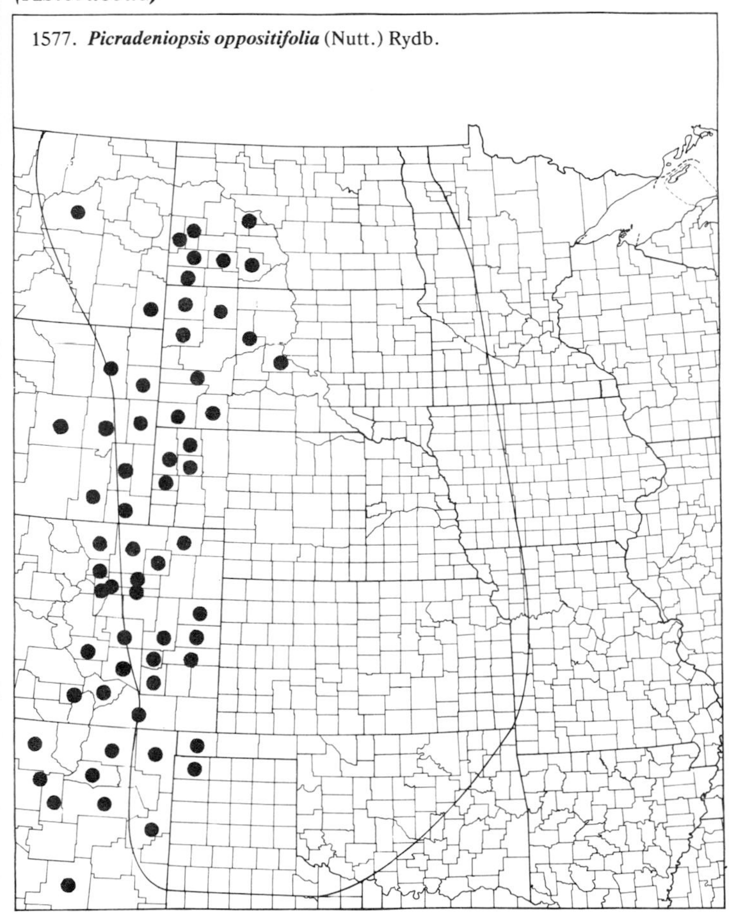
1577. ***Picradeniopsis oppositifolia*** (Nutt.) Rydb.

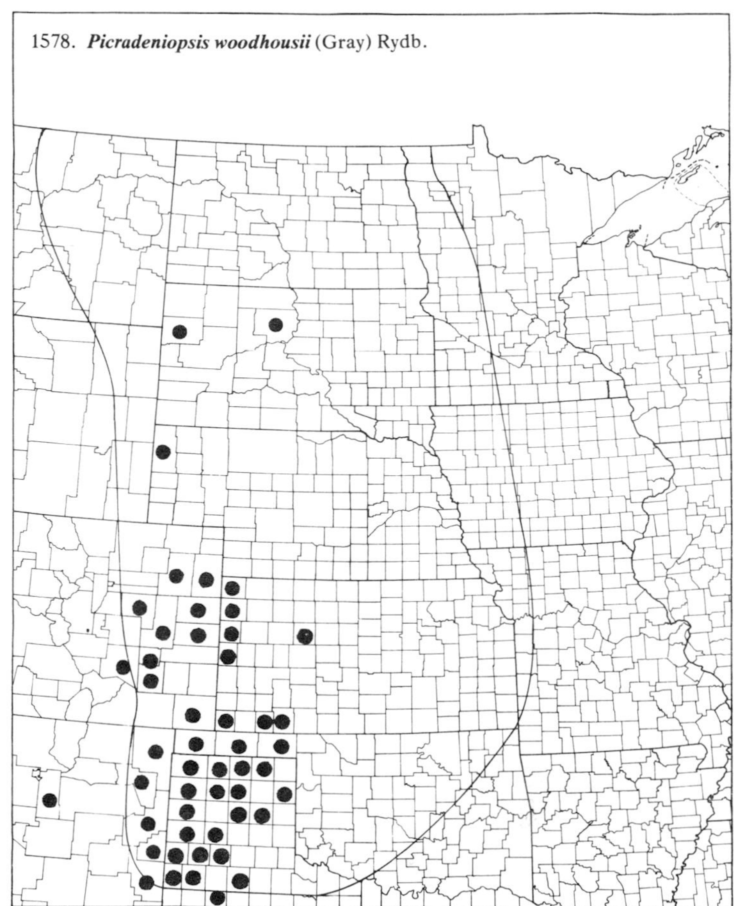
1578. ***Picradeniopsis woodhousii*** (Gray) Rydb.

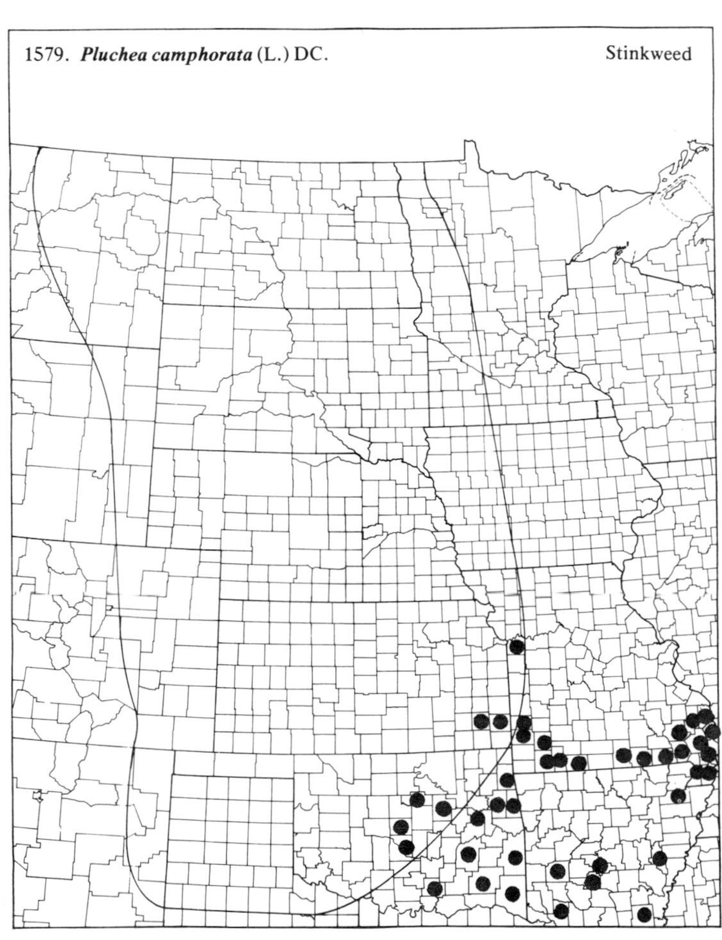
1579. ***Pluchea camphorata*** (L.) DC. Stinkweed

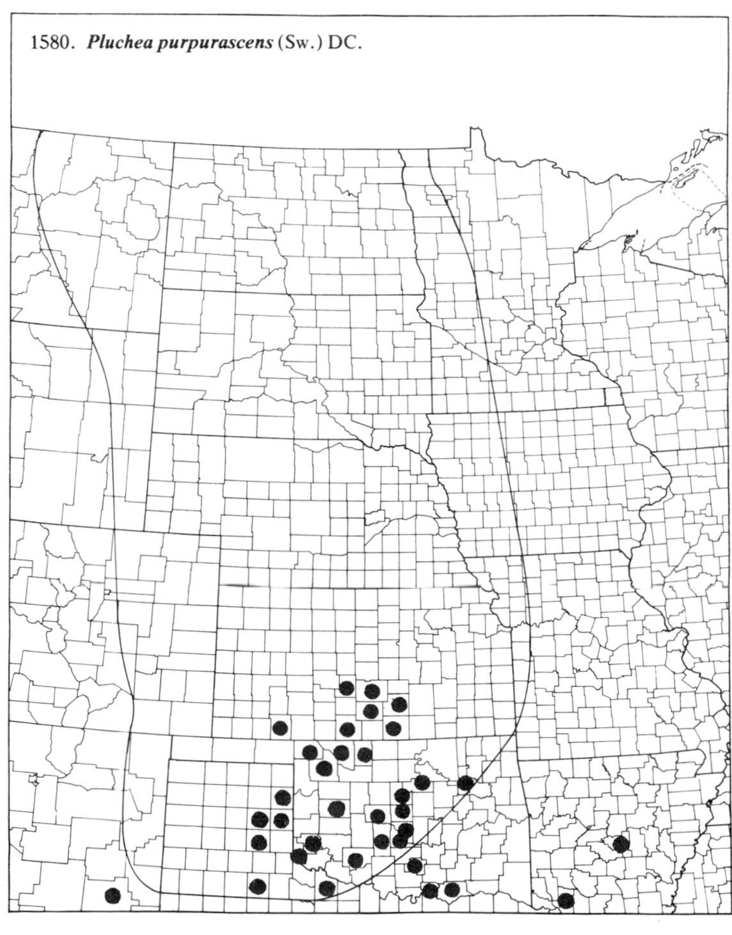
1580. ***Pluchea purpurascens*** (Sw.) DC.

(Asteraceae)

1581. ***Prenanthes alba*** L. White Rattlesnake-root

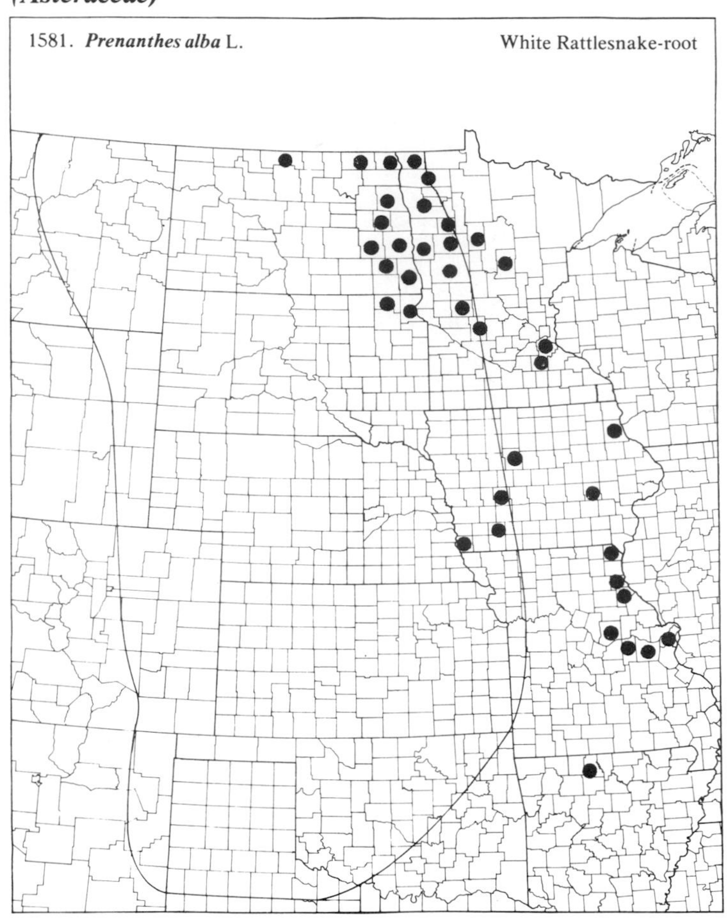

1582. ***Prenanthes aspera*** Michx. Rough Rattlesnake-root

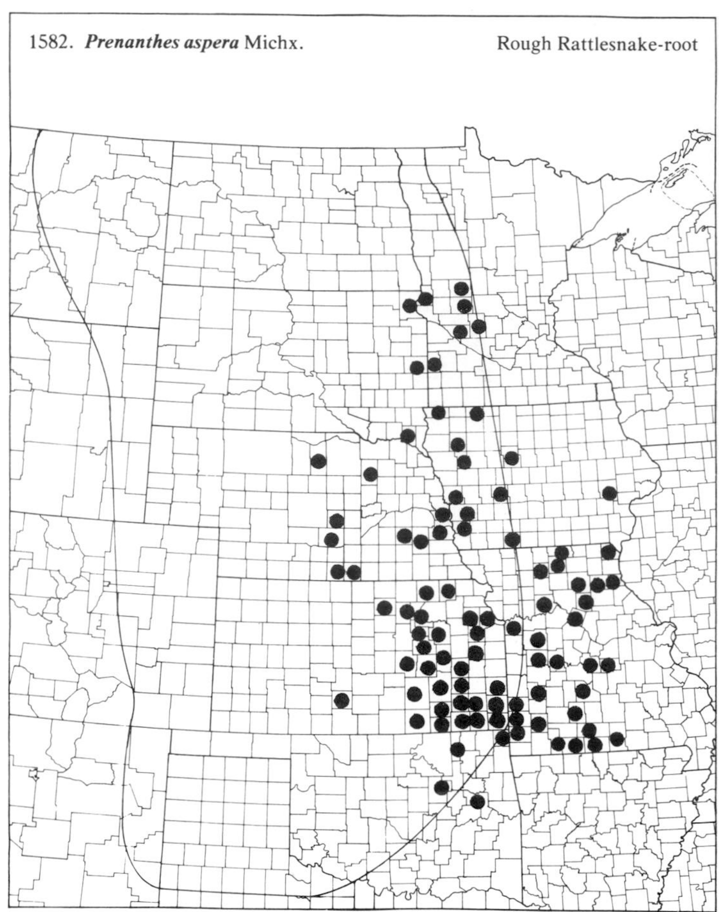

1583. ***Prenanthes racemosa*** Michx. ssp. ***multiflora*** Cronq.

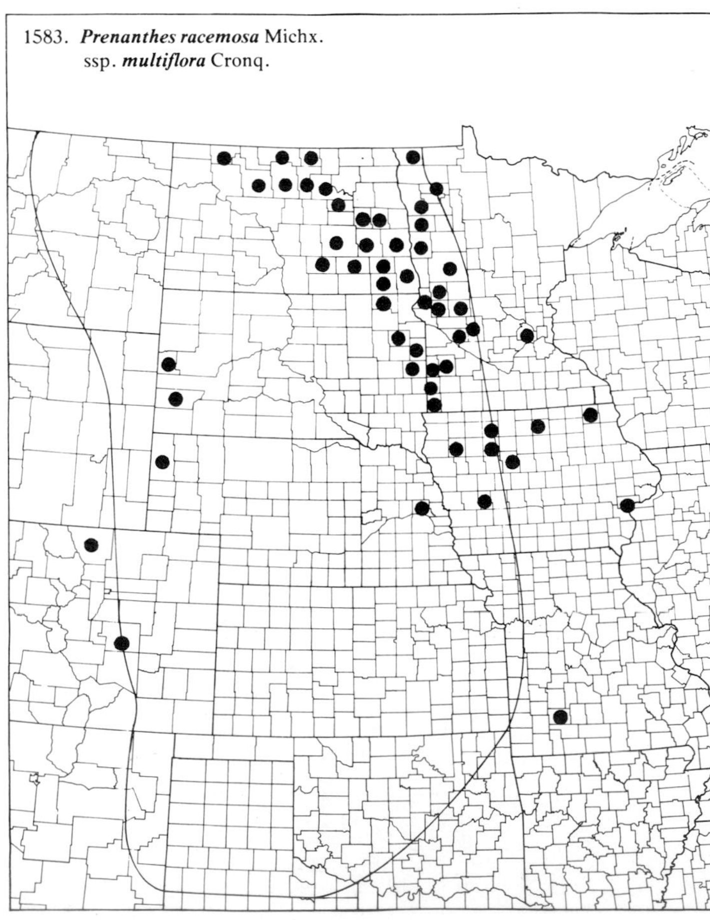

1584. ***Psilostrophe villosa*** Rydb. Paperflower

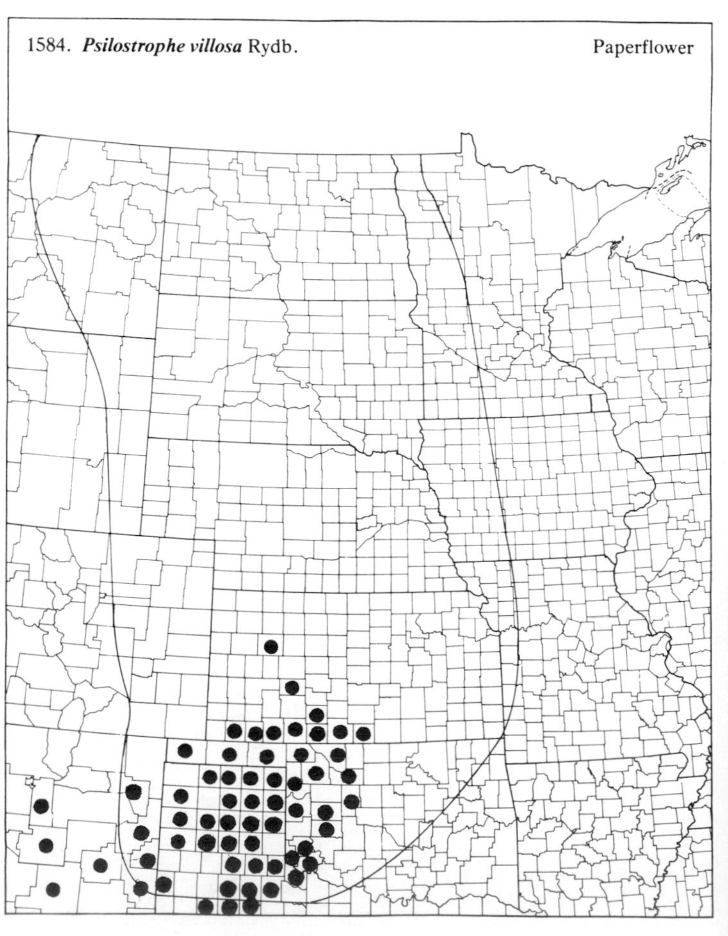

(Asteraceae)

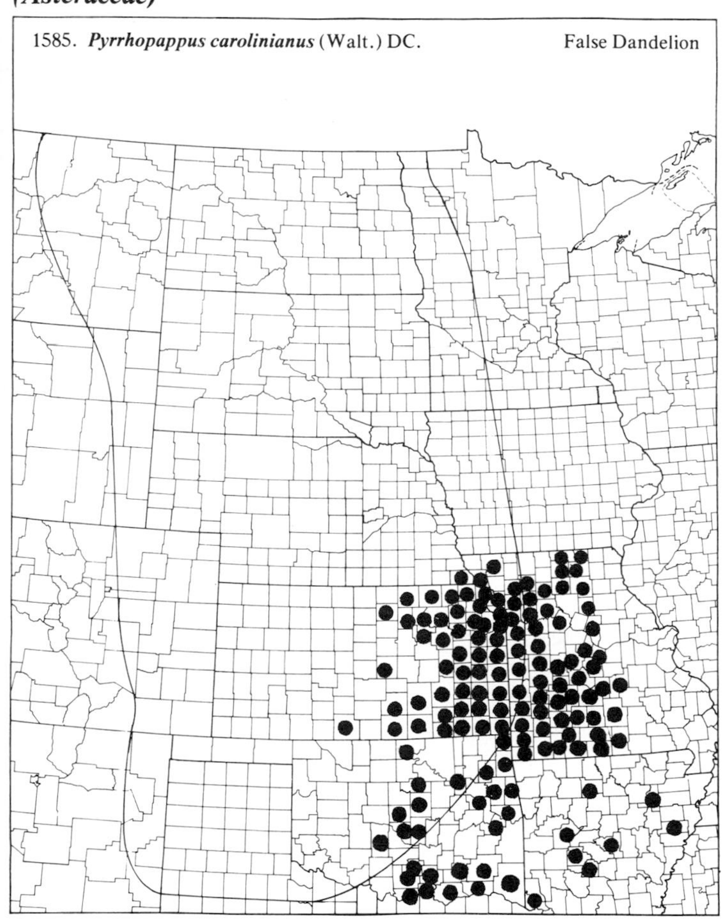
1585. ***Pyrrhopappus carolinianus*** (Walt.) DC.
False Dandelion

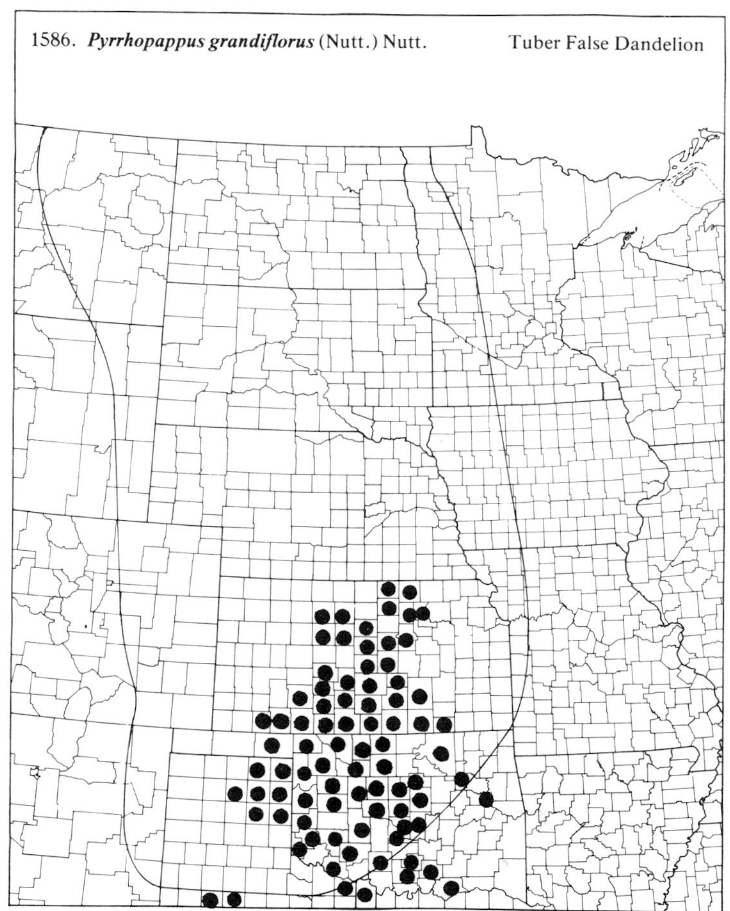
1586. ***Pyrrhopappus grandiflorus*** (Nutt.) Nutt.
Tuber False Dandelion

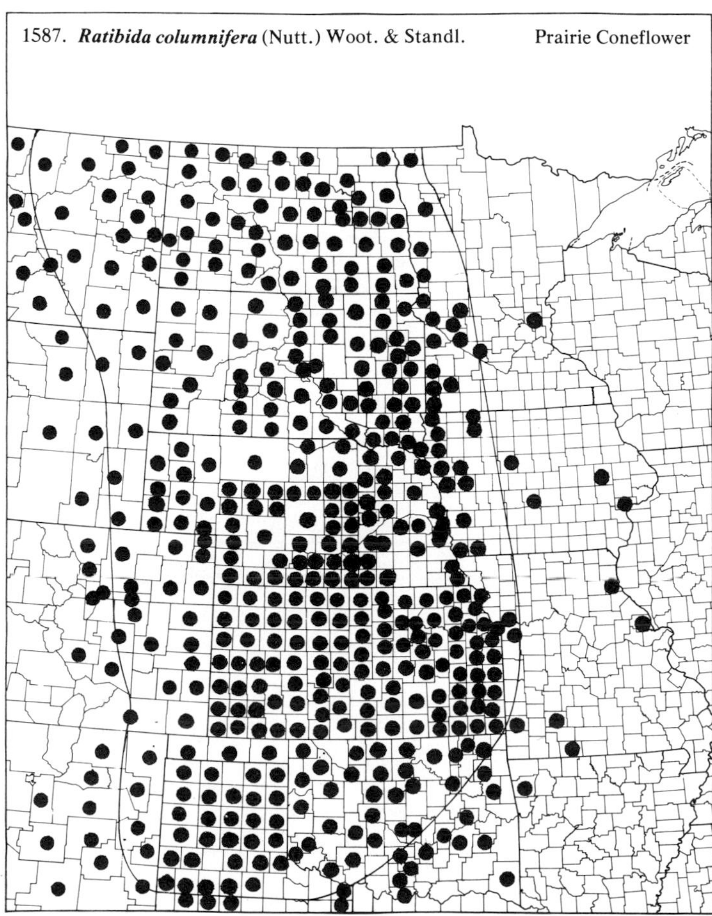
1587. ***Ratibida columnifera*** (Nutt.) Woot. & Standl.
Prairie Coneflower

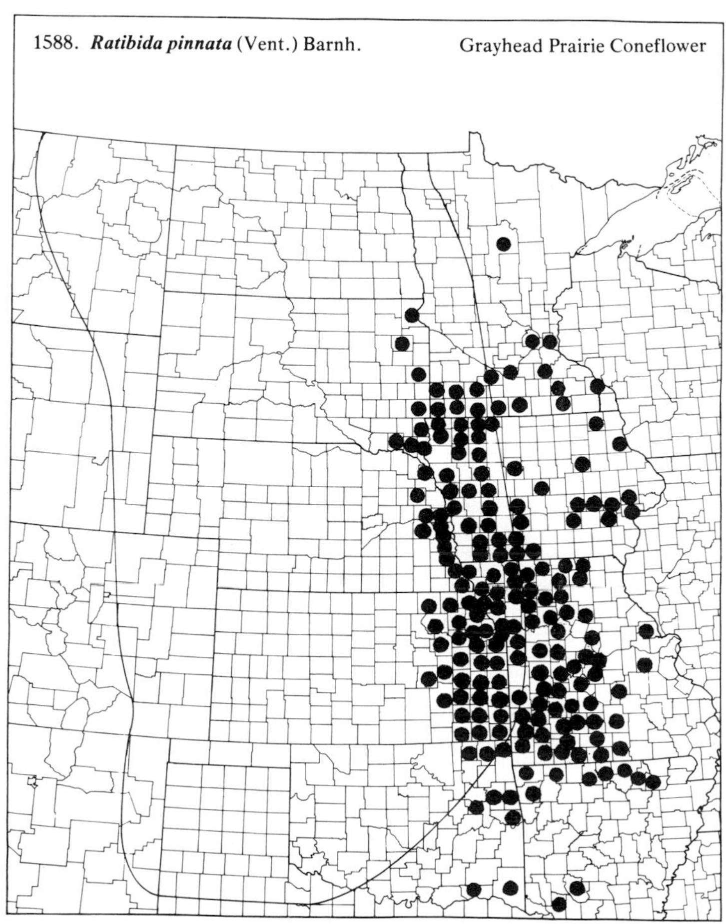
1588. ***Ratibida pinnata*** (Vent.) Barnh.
Grayhead Prairie Coneflower

(Asteraceae)

1589. ***Ratibida tagetes*** (James) Barnh. Shortray Prairie Coneflower

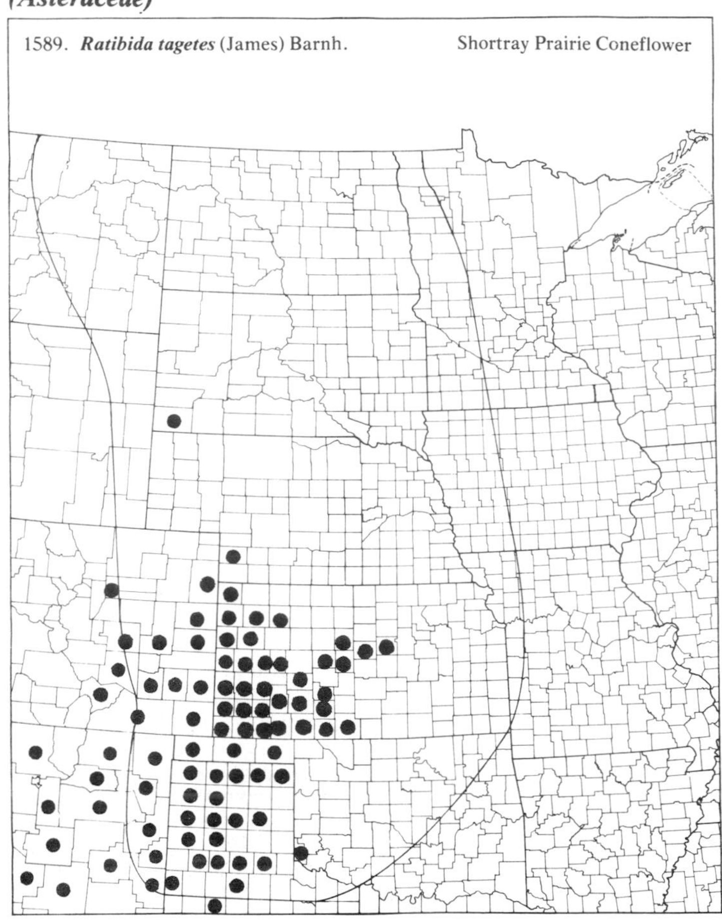

1590. ***Rudbeckia amplexicaulis*** Vahl

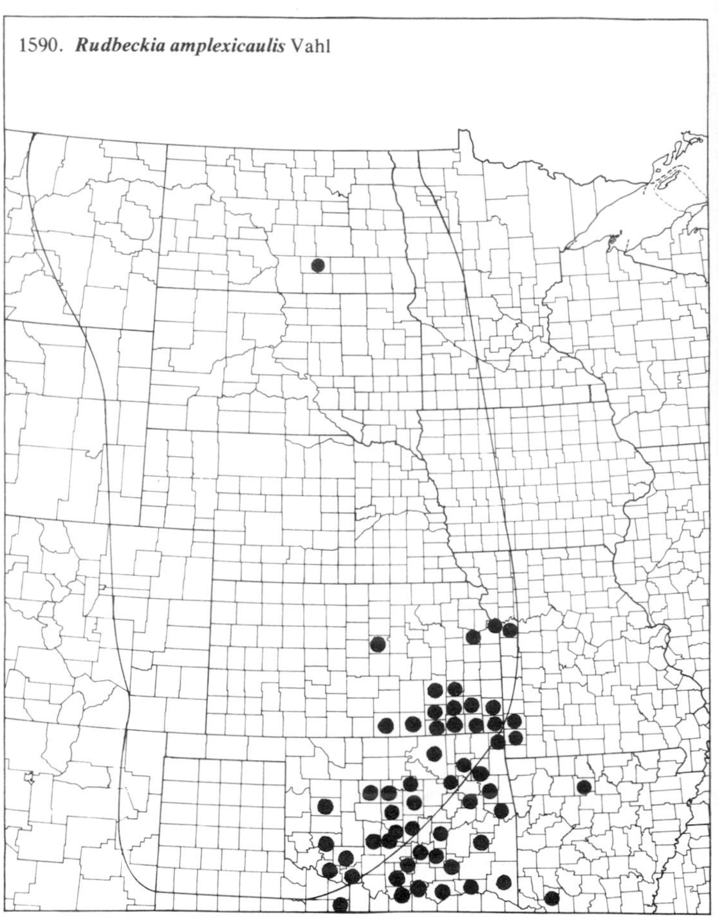

1591. ***Rudbeckia hirta*** L. Black-eyed Susan

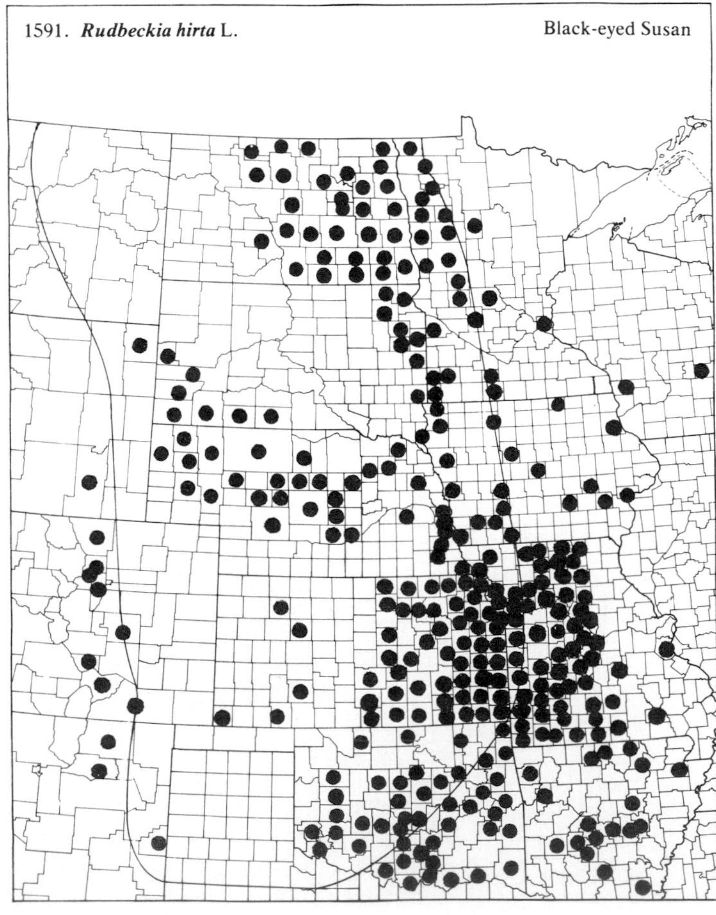

1592. ***Rudbeckia laciniata*** L. Goldenglow

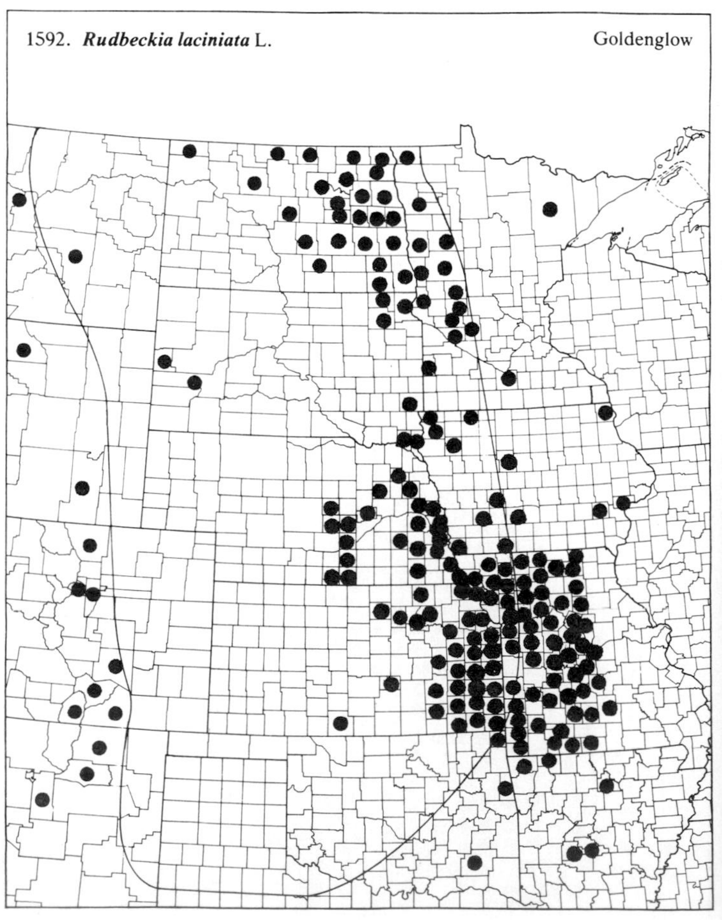

(Asteraceae)

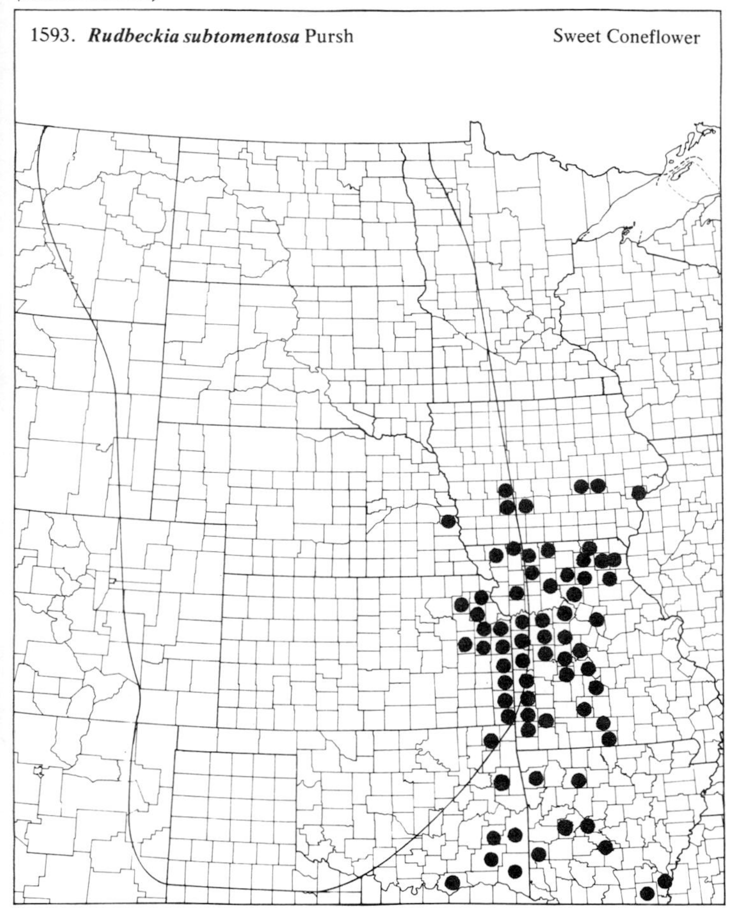
1593. **Rudbeckia subtomentosa** Pursh
Sweet Coneflower

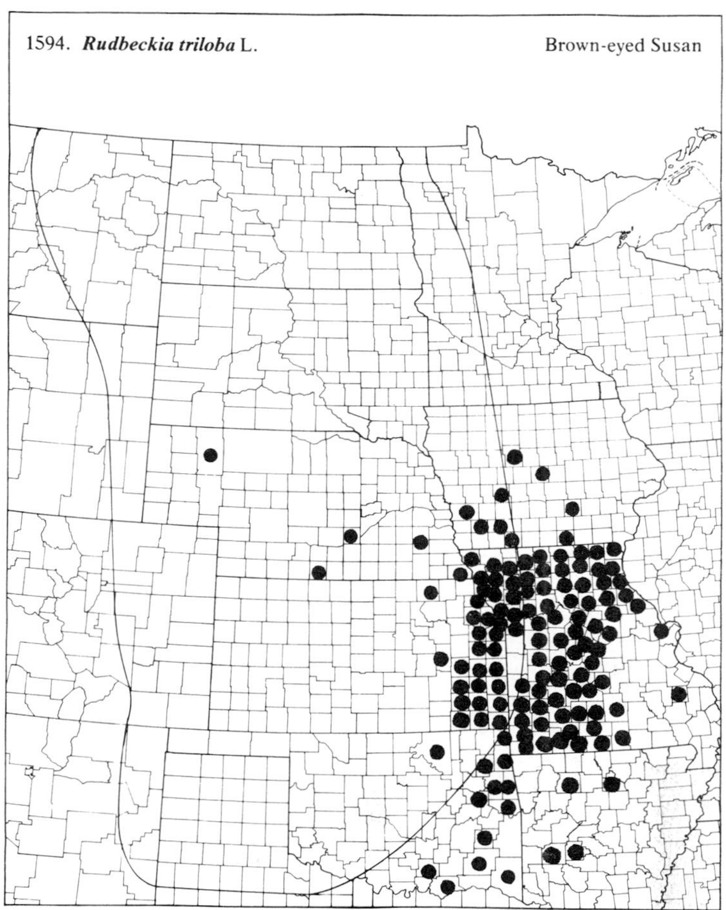
1594. **Rudbeckia triloba** L.
Brown-eyed Susan

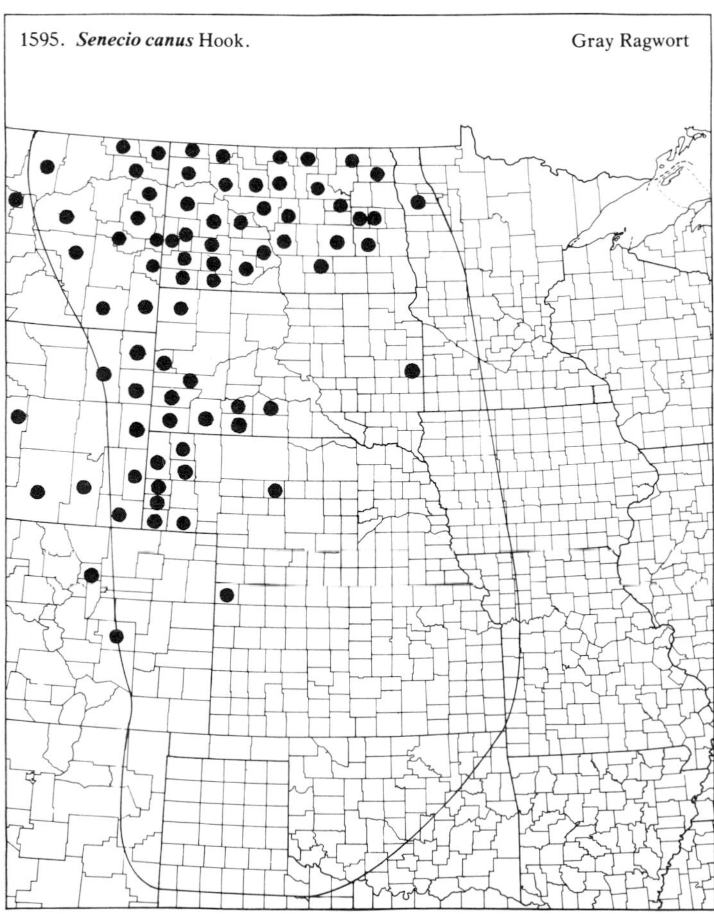
1595. **Senecio canus** Hook.
Gray Ragwort

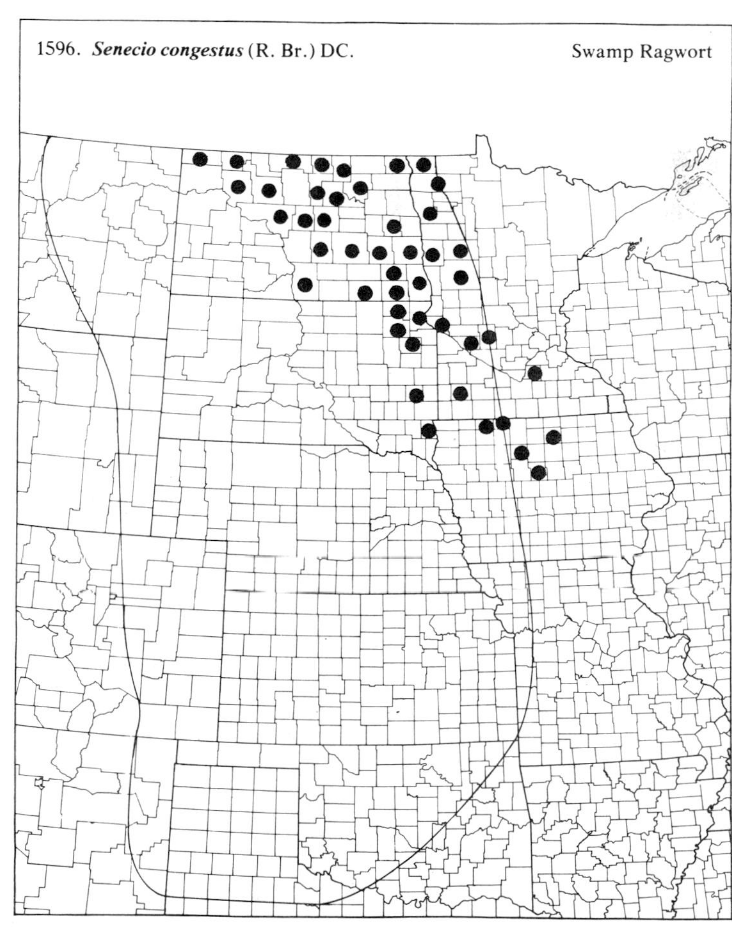
1596. **Senecio congestus** (R. Br.) DC.
Swamp Ragwort

(Asteraceae)

1597. ***Senecio douglasii*** DC.
var. ***longilobus*** (Benth.) L. Benson

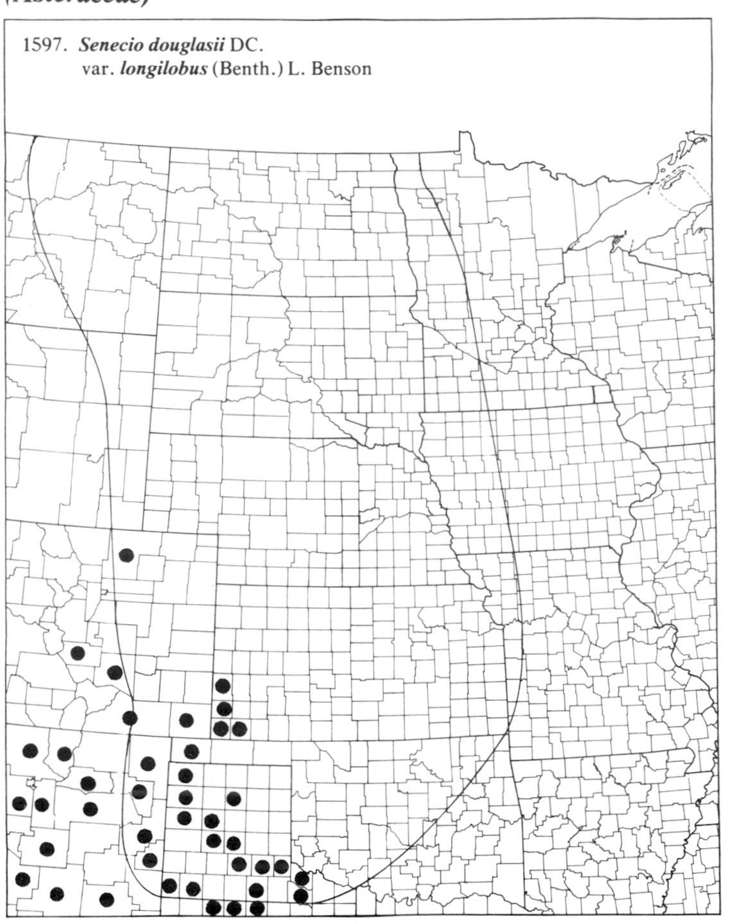

1598. ***Senecio glabellus*** Poir. Butterweed

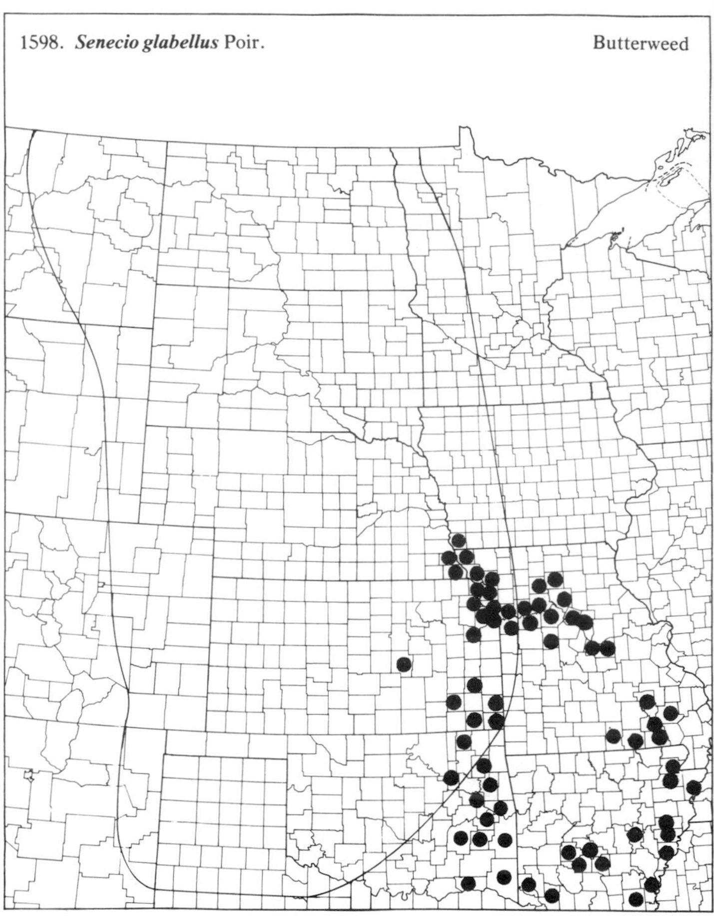

1599. ***Senecio imparipinnatus*** Klatt

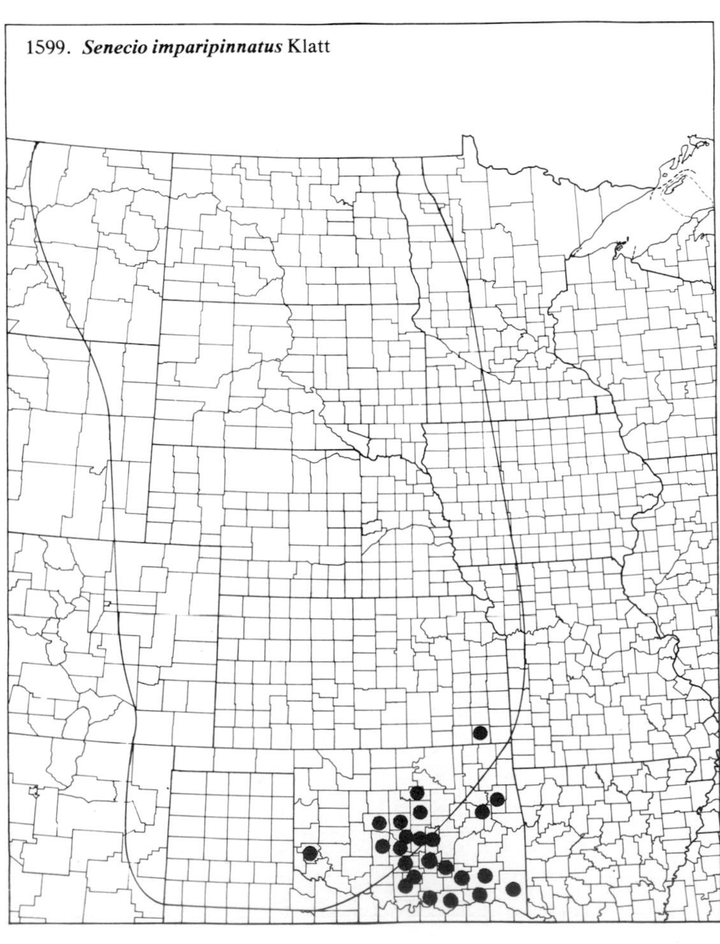

1600. ***Senecio integerrimus*** Nutt.

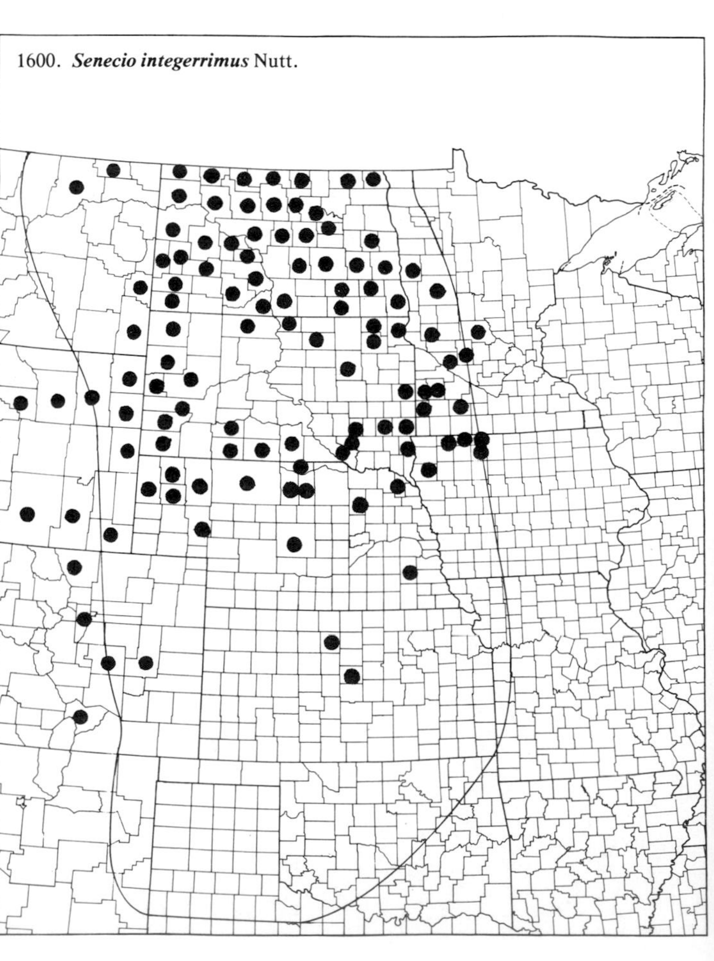

(Asteraceae)

1601. **Senecio obovatus** Muhl. Roundleaf Groundsel

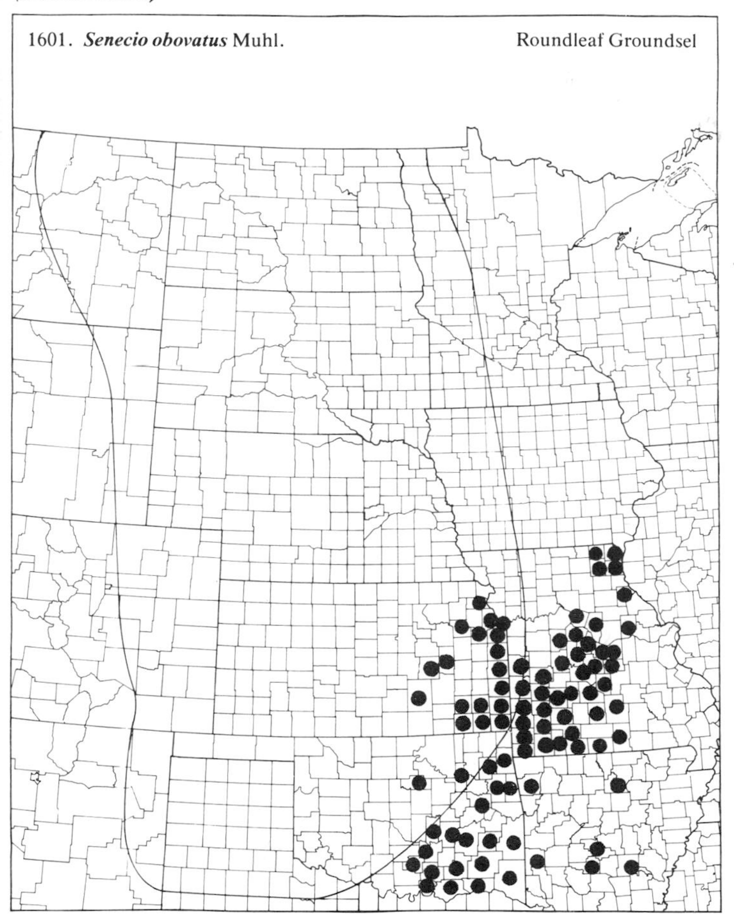

1602. **Senecio pauperculus** Michx. Balsam Groundsel

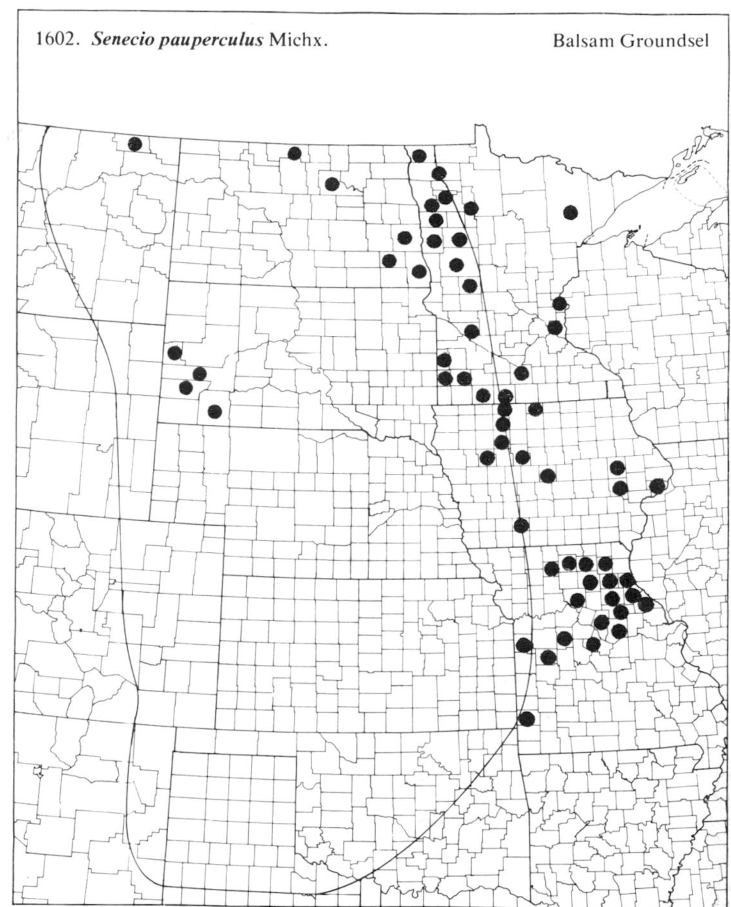

1603. **Senecio plattensis** Nutt. Prairie Ragwort

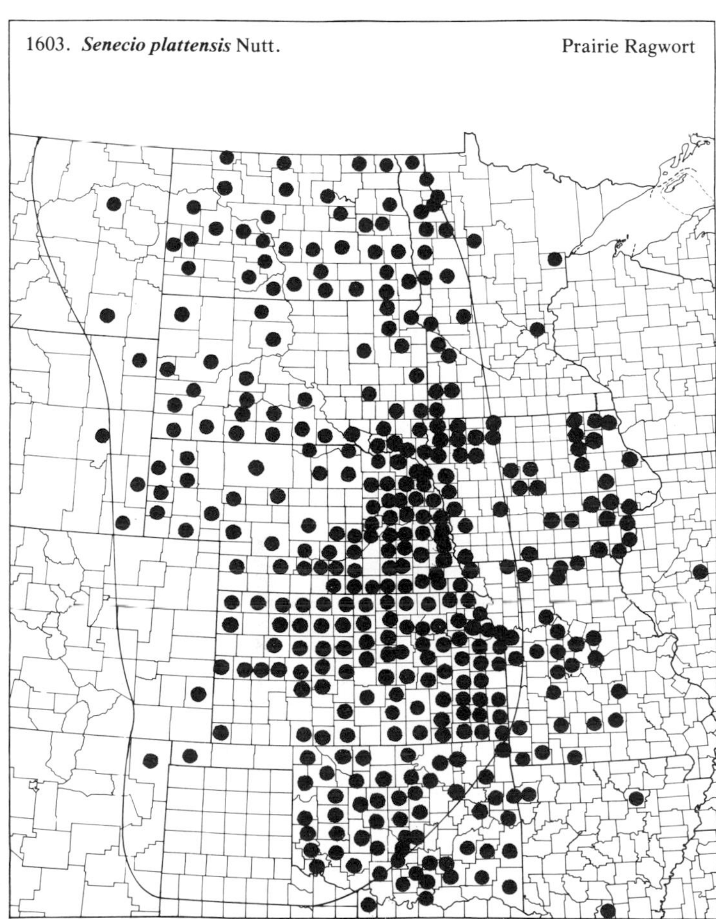

1604. **Senecio pseudaureus** Rydb. var. **semicordatus** (Mack. & Bush) T.M. Barkley

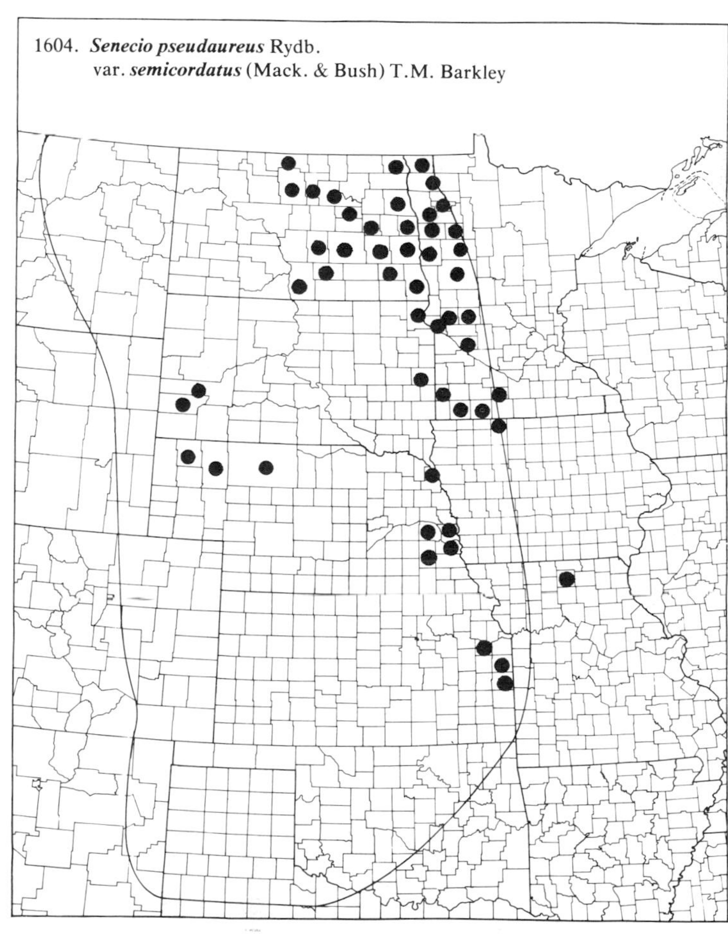

(Asteraceae)

1605. ***Senecio riddellii*** T. & G. Riddell Ragwort

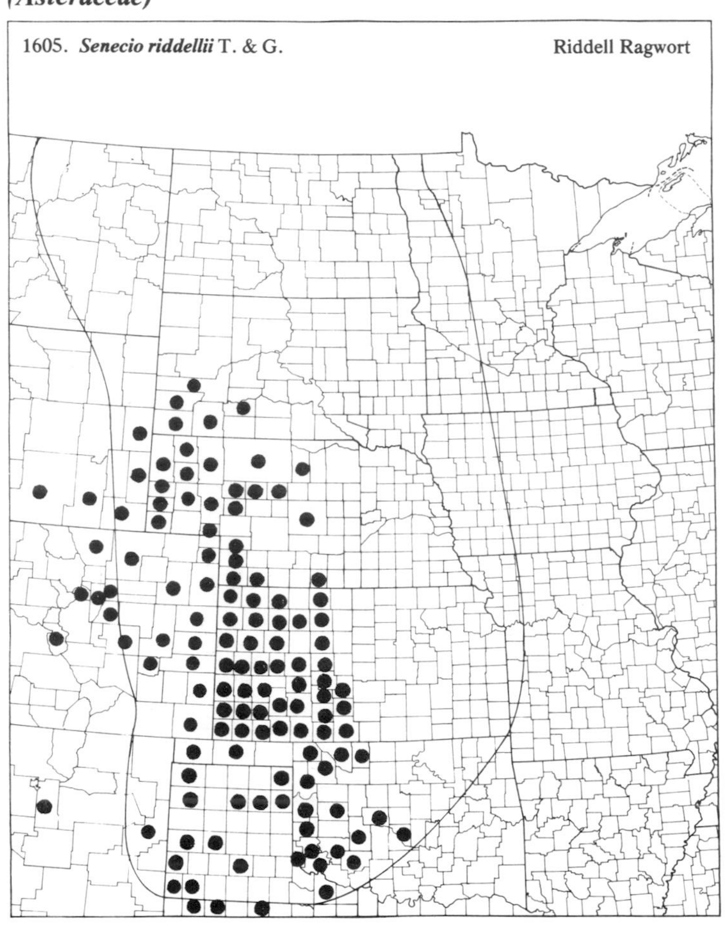

1606. ***Senecio spartioides*** T. & G.

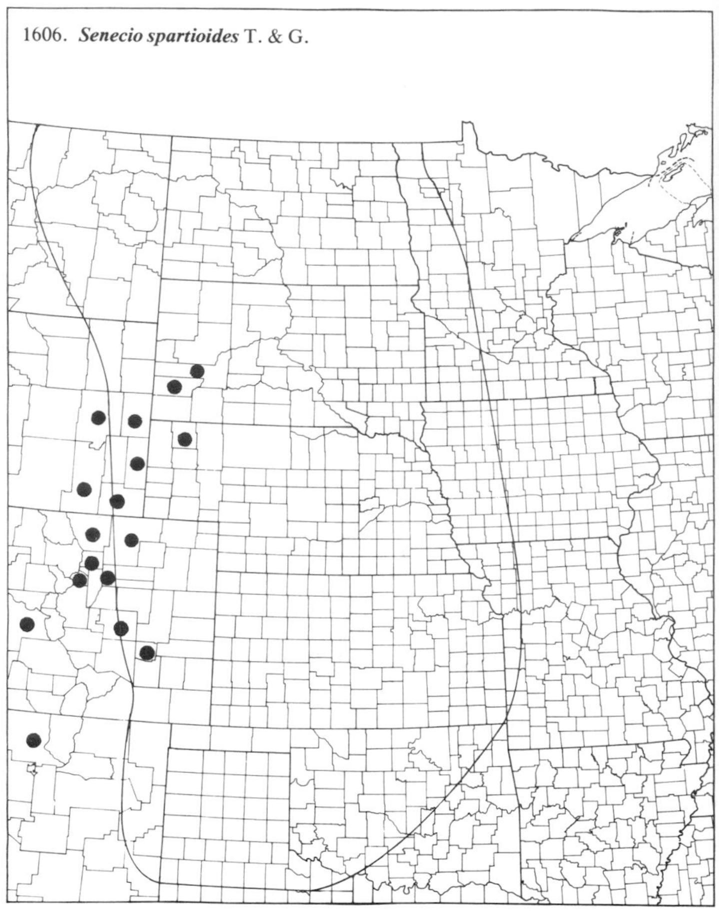

1607. ***Senecio tridenticulatus*** Rydb.

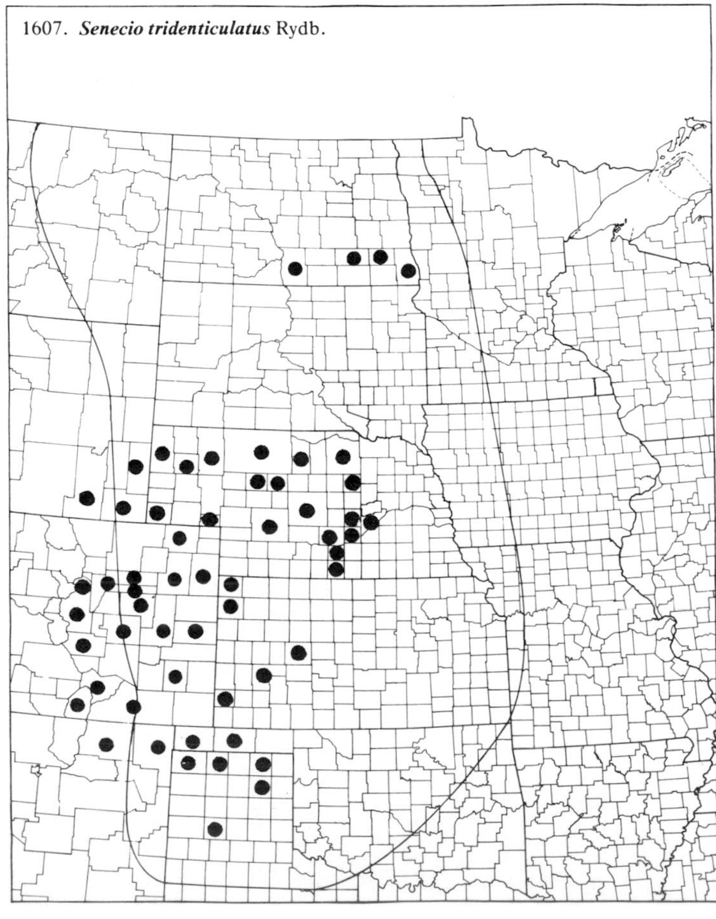

1608. ***Silphium integrifolium*** Michx. Wholeleaf Rosinweed

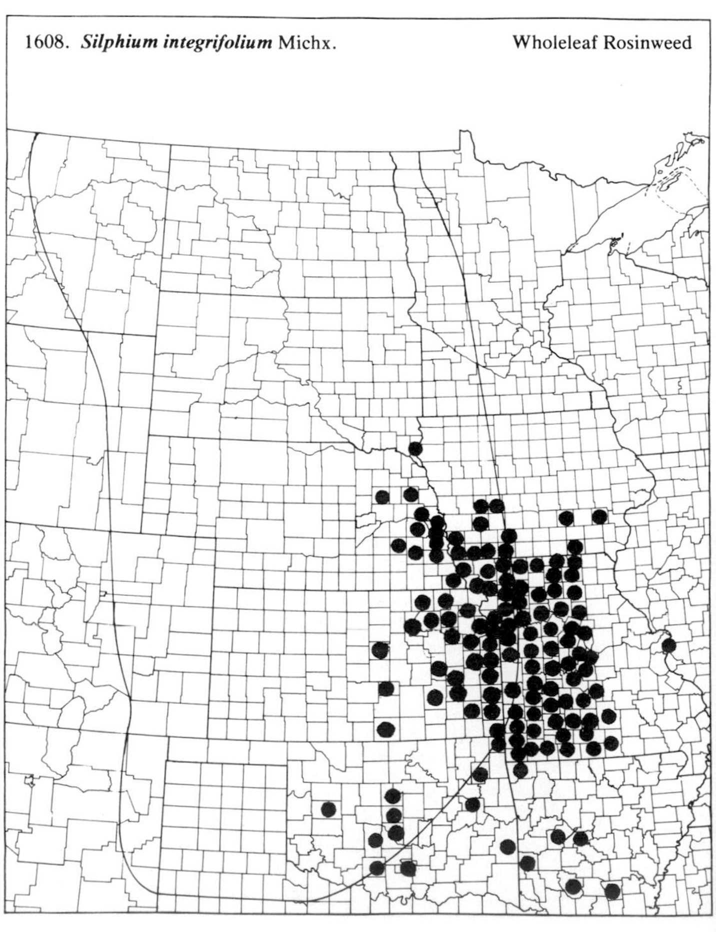

(Asteraceae)

1609. *Silphium laciniatum* L. Compass Plant

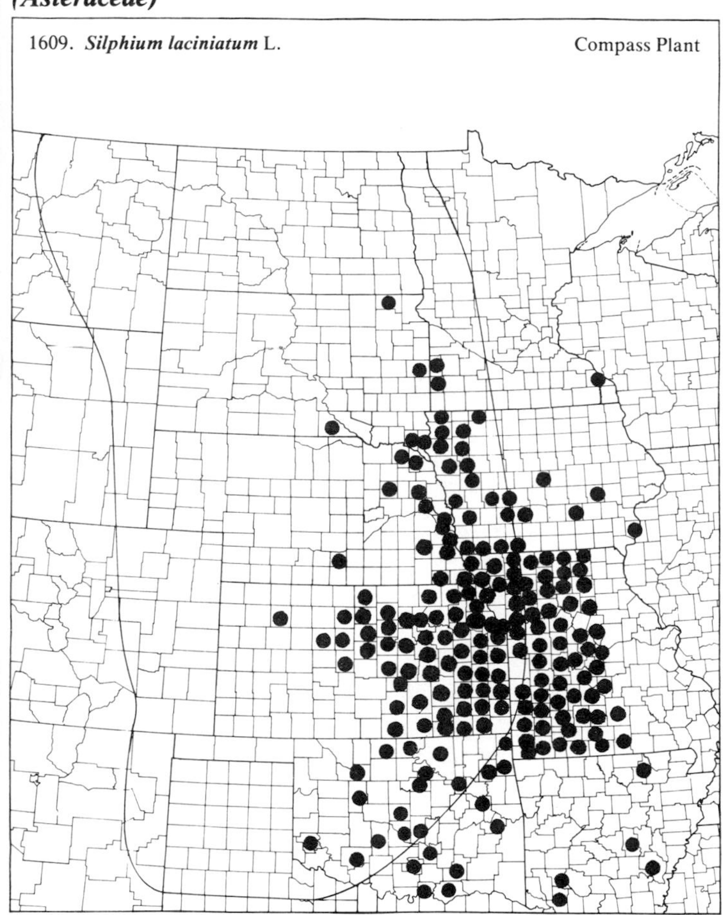

1610. *Silphium perfoliatum* L. Cup Plant

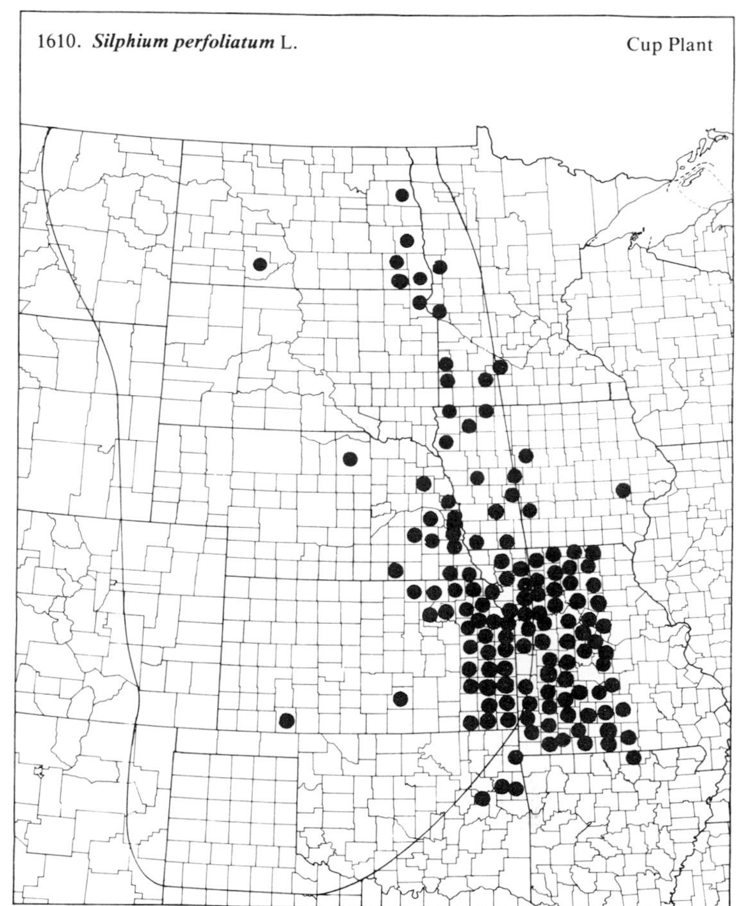

1611. *Silphium speciosum* Nutt. Showy Rosinweed

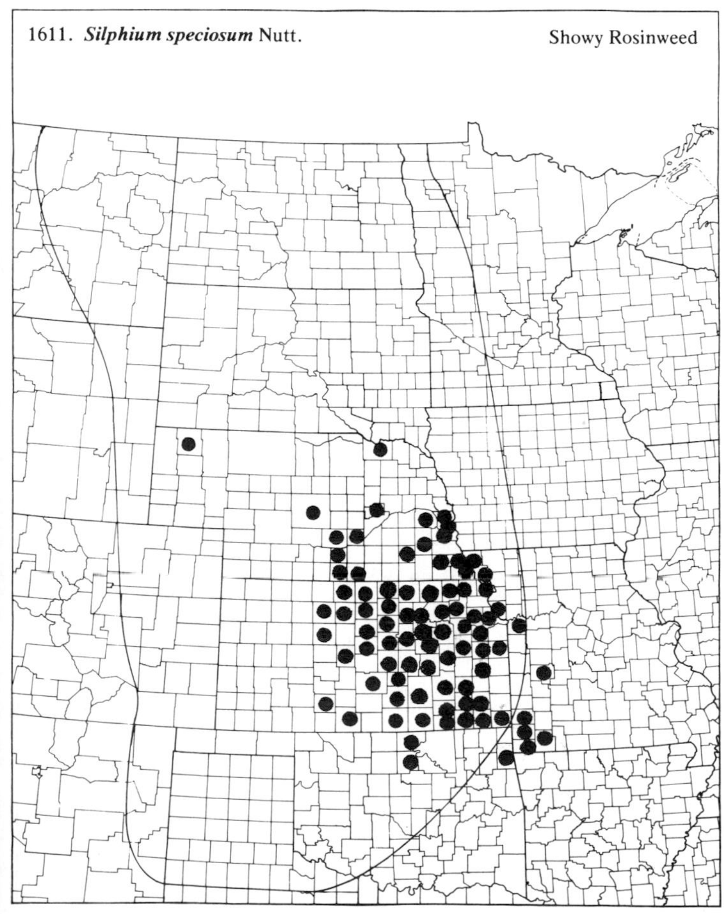

1612. *Solidago canadensis* L. var. *gilvocanescens* Rydb. Canada Goldenrod

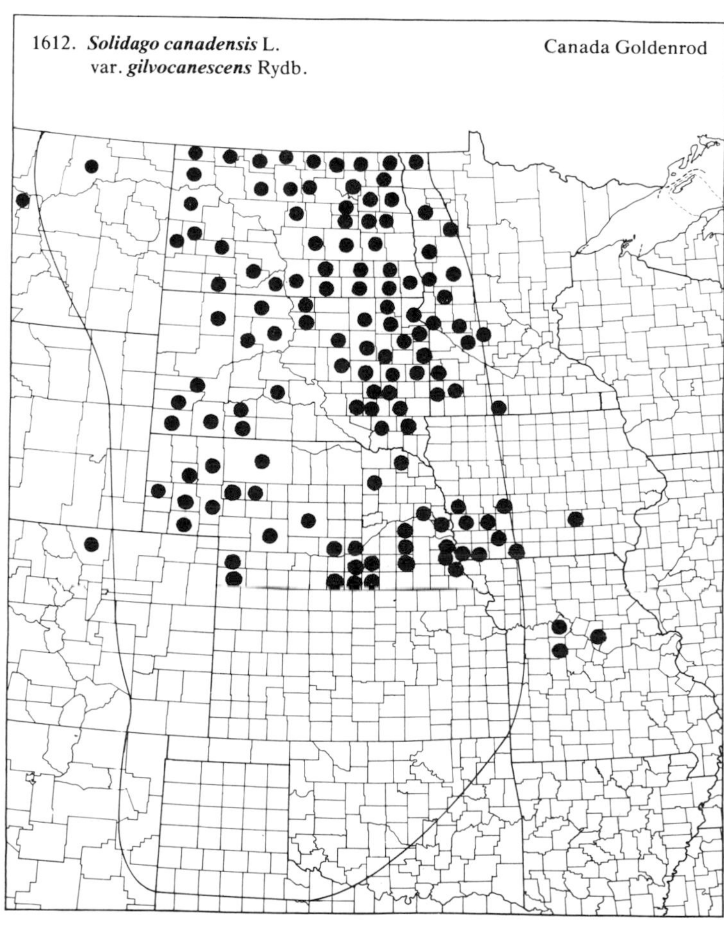

(Asteraceae)

1613. ***Solidago canadensis*** L. var. ***hargeri*** Fern. — Canada Goldenrod

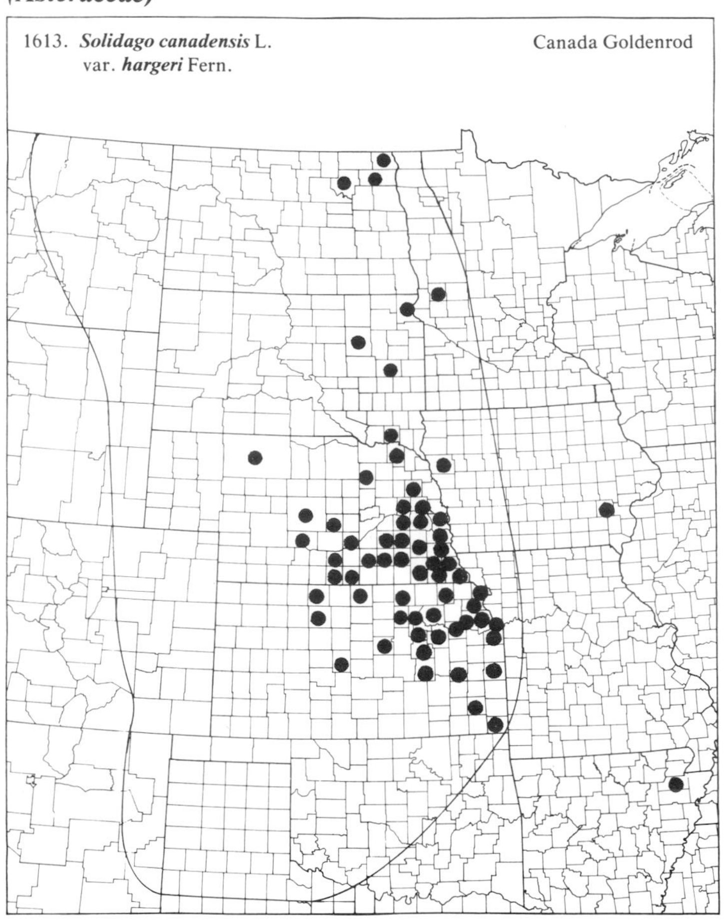

1614. ***Solidago canadensis*** L. var. ***scabra*** (Muhl.) T. & G. — Canada Goldenrod

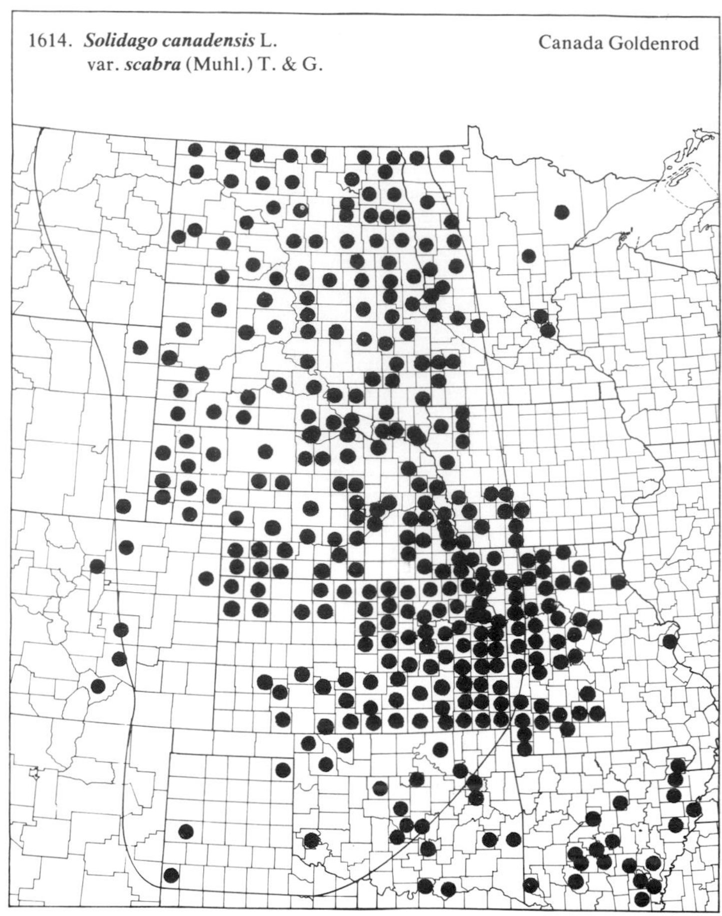

1615. ***Solidago flexicaulis*** L. — Broad-leaved Goldenrod

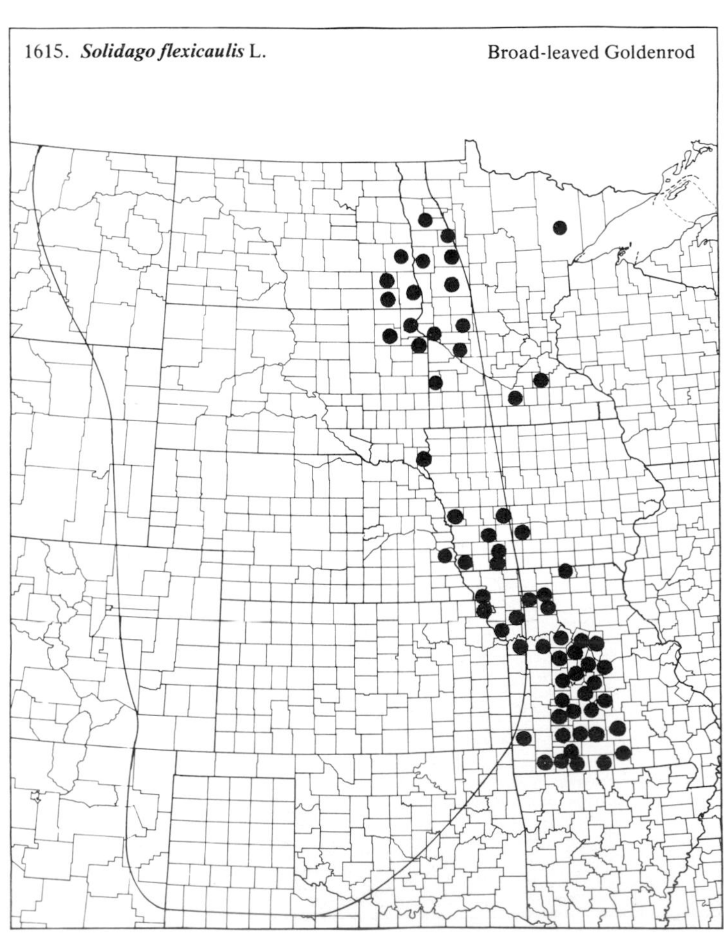

1616. ***Solidago gigantea*** Ait. — Late Goldenrod

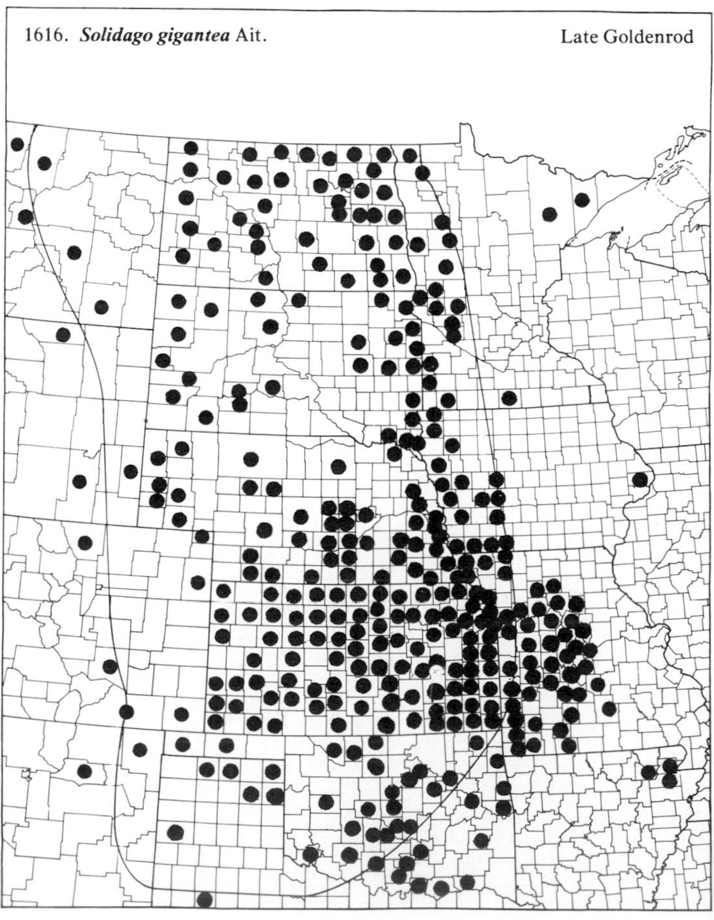

(Asteraceae)

1617. ***Solidago graminifolia*** (L.) Salisb. var. ***graminifolia*** — Narrow-leaved Goldenrod

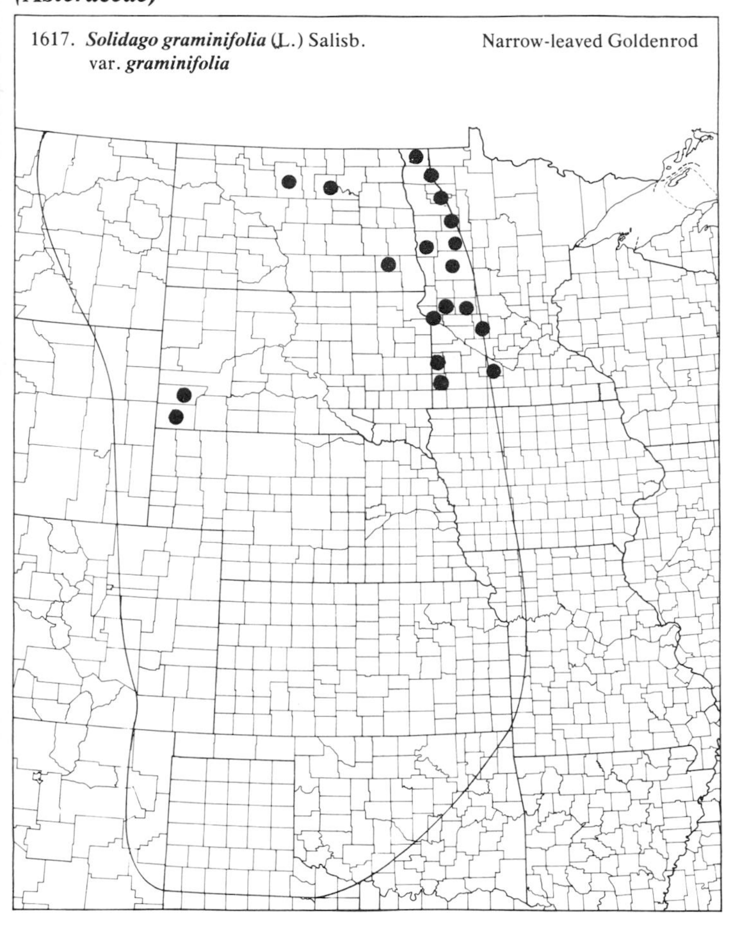

1618. ***Solidago graminifolia*** (L.) Salisb. var. ***gymnospermoides*** (Greene) Croat — Narrow-leaved Goldenrod

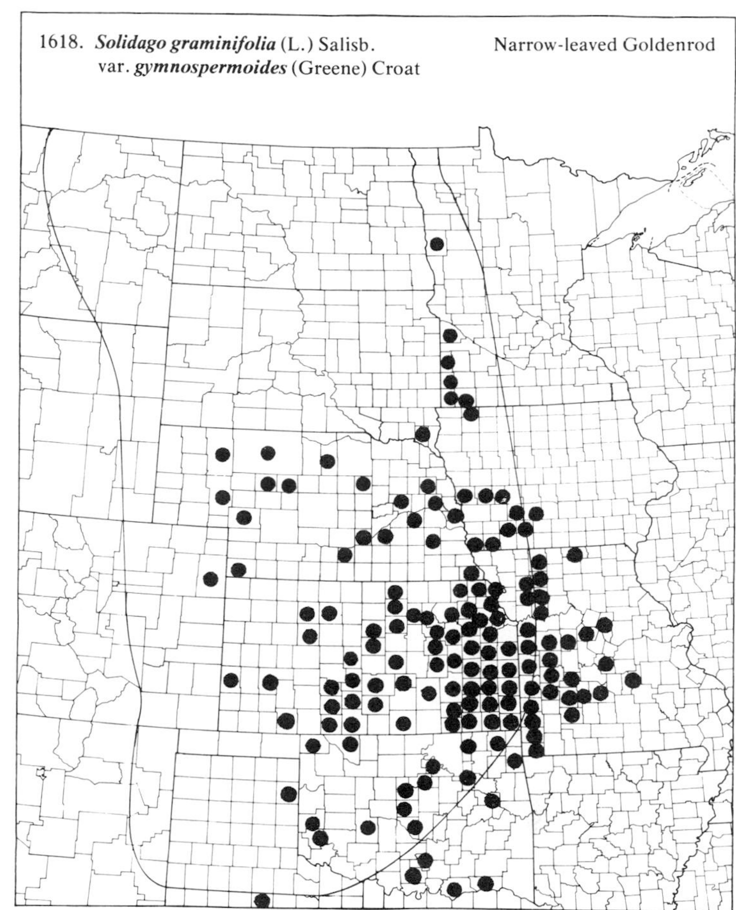

1619. ***Solidago graminifolia*** (L.) Salisb. var. ***major*** (Michx.) Fern. — Narrow-leaved Goldenrod

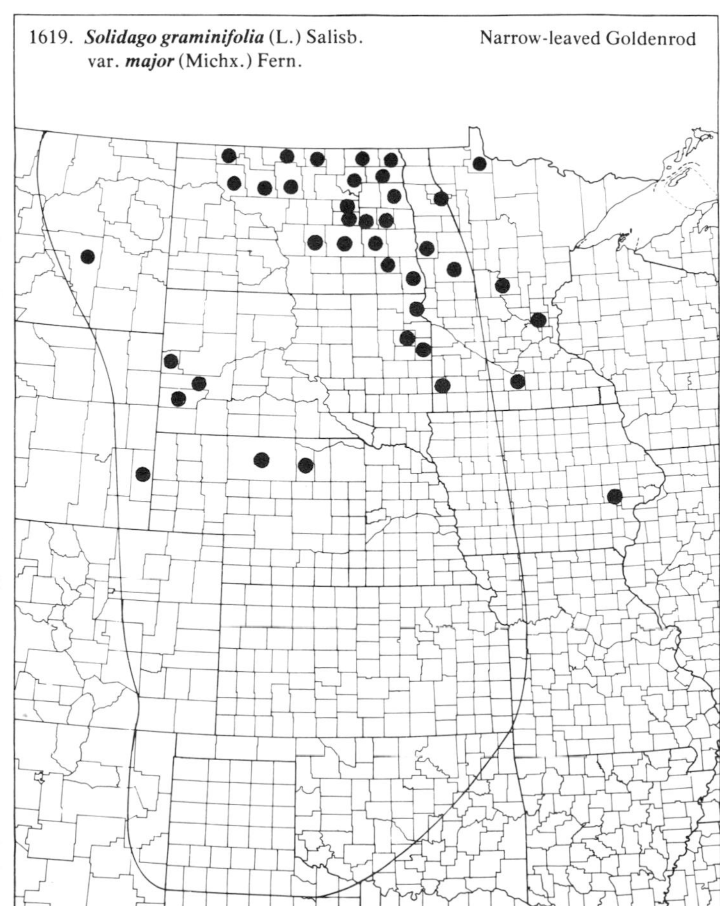

1620. ***Solidago graminifolia*** (L.) Salisb. var. ***media*** (Greene) Harris — Narrow-leaved Goldenrod

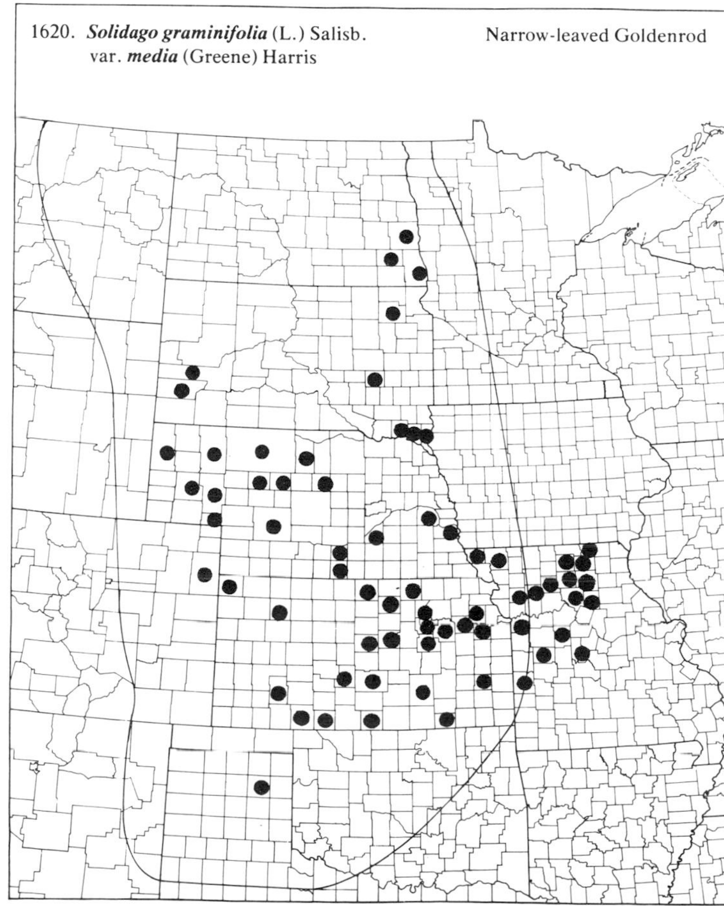

(Asteraceae)

1621. ***Solidago missouriensis*** Nutt. Prairie Goldenrod

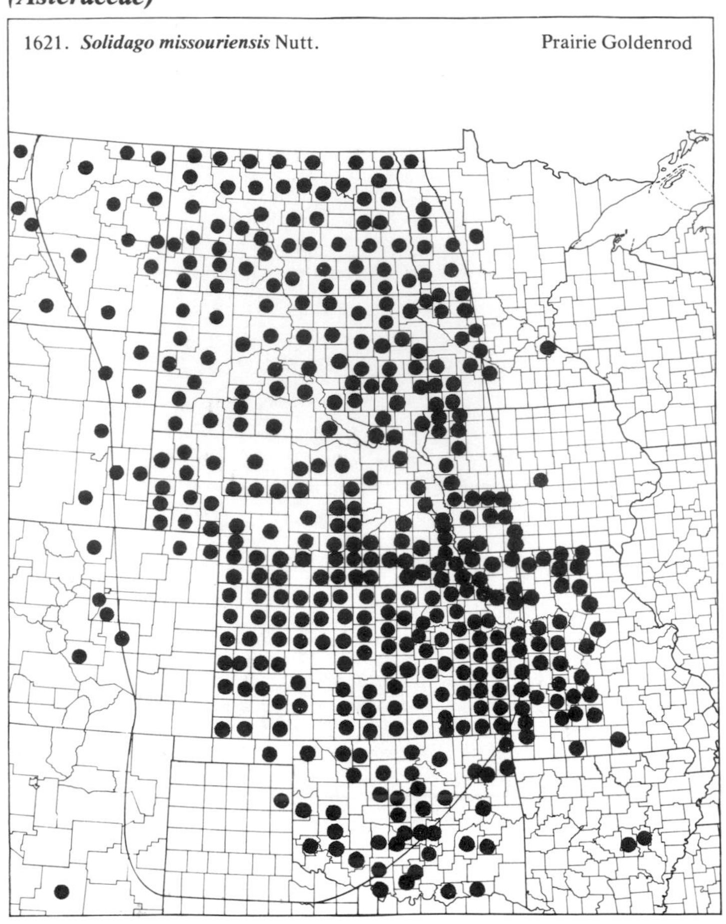

1622. ***Solidago mollis*** Bartl. Soft Goldenrod

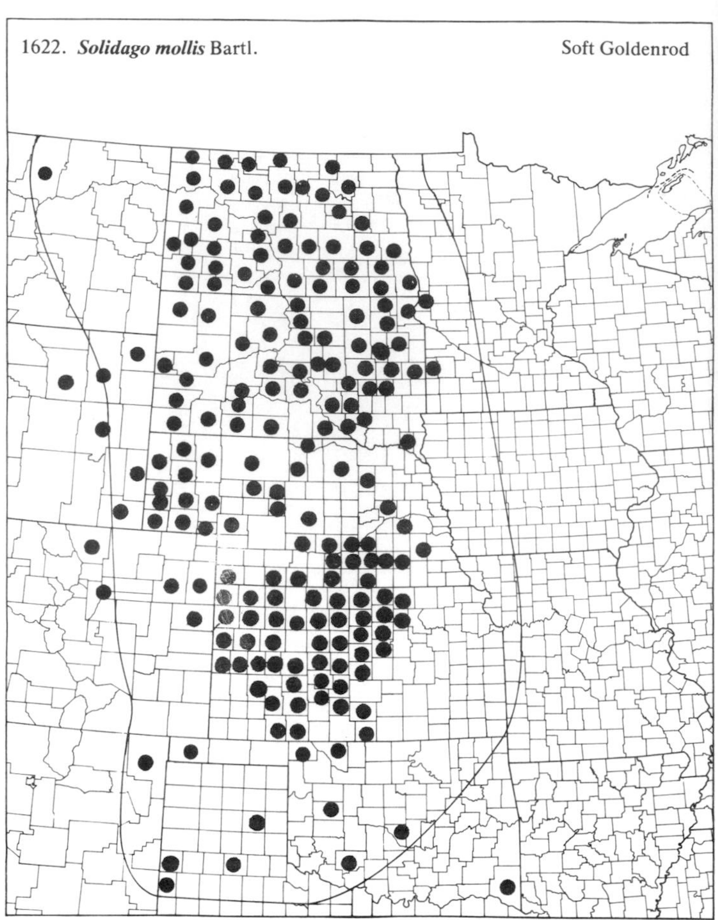

1623. ***Solidago nemoralis*** Ait. Gray Goldenrod

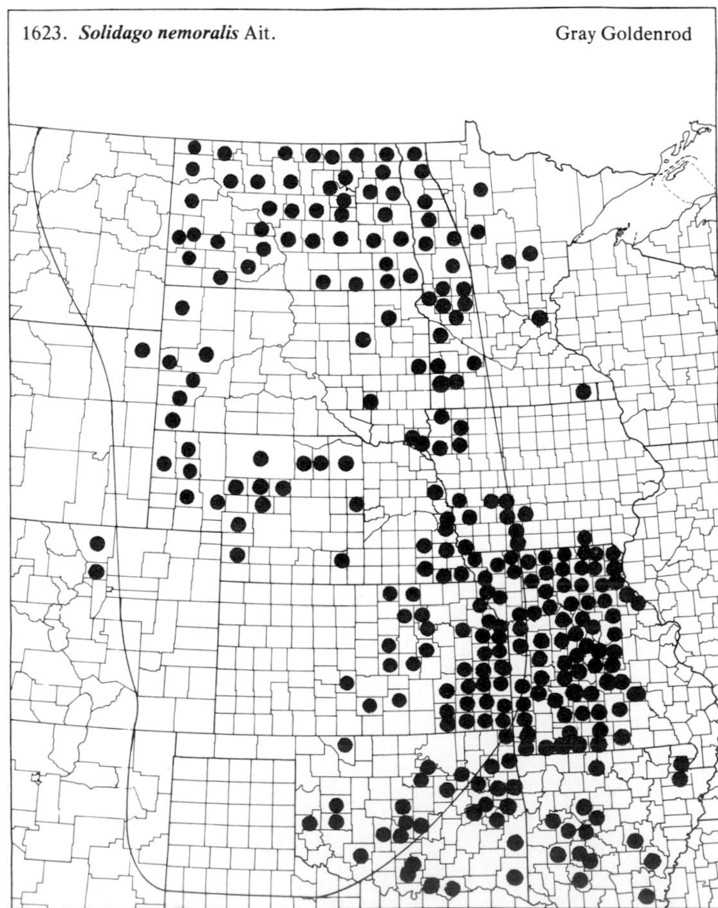

1624. ***Solidago petiolaris*** Ait. Downy Goldenrod

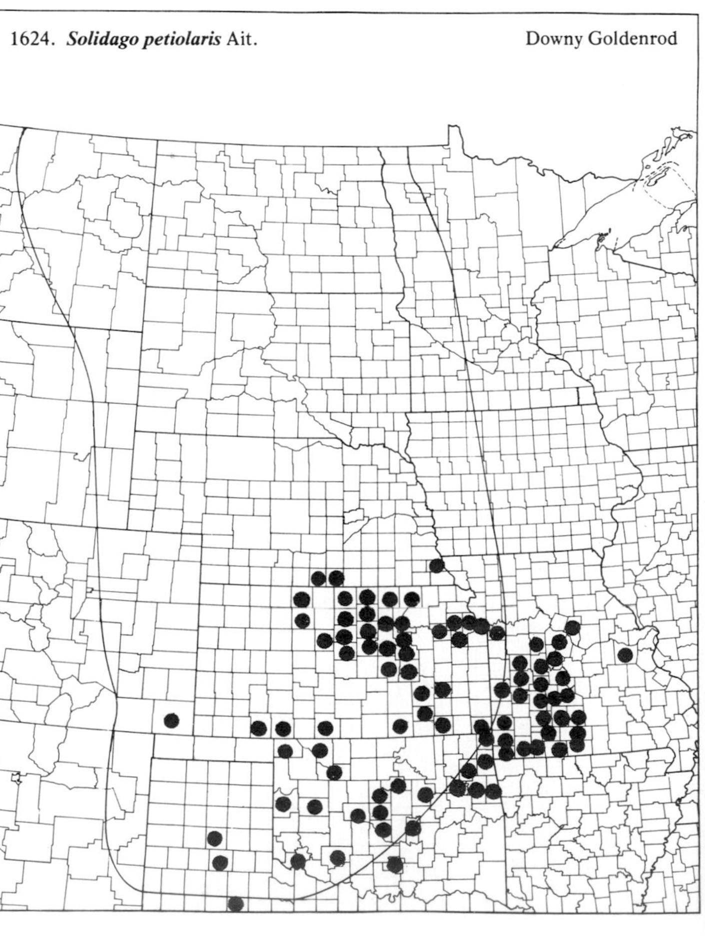

(Asteraceae)

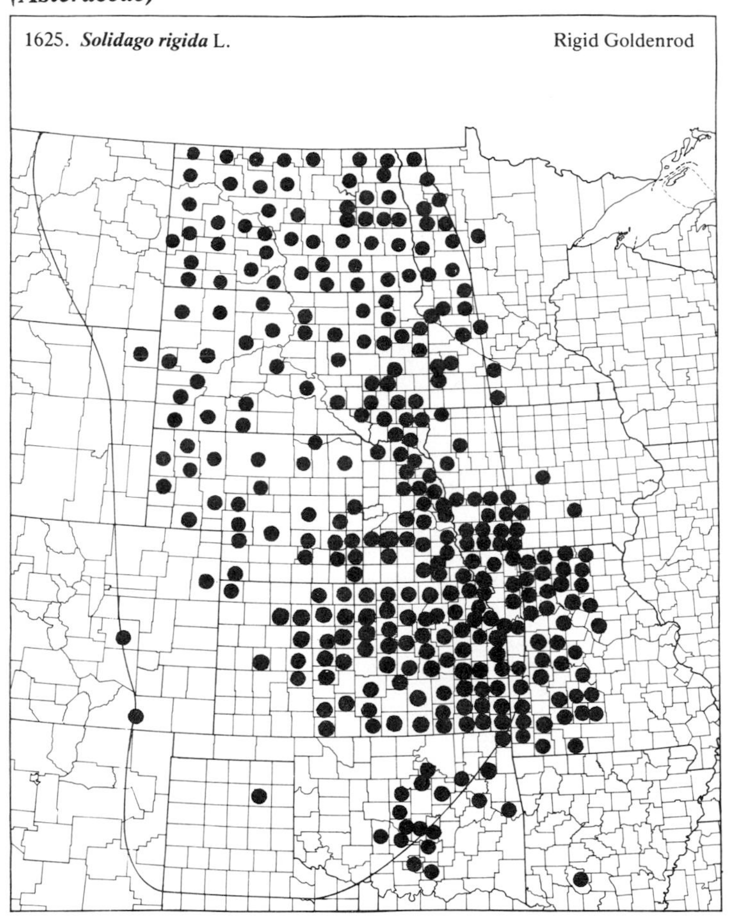
1625. *Solidago rigida* L.
Rigid Goldenrod

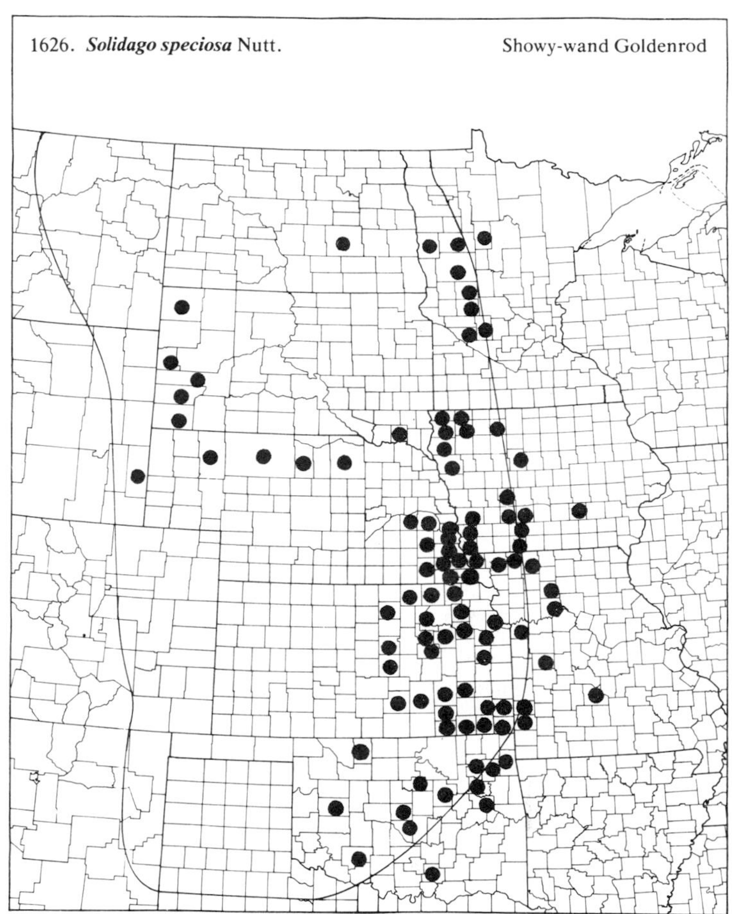
1626. *Solidago speciosa* Nutt.
Showy-wand Goldenrod

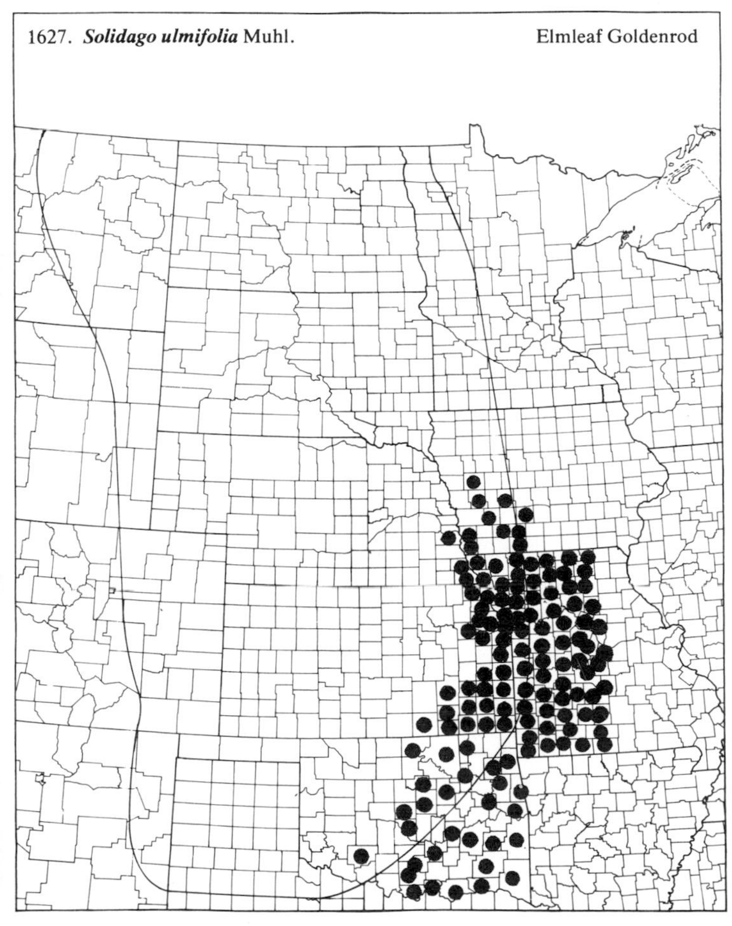
1627. *Solidago ulmifolia* Muhl.
Elmleaf Goldenrod

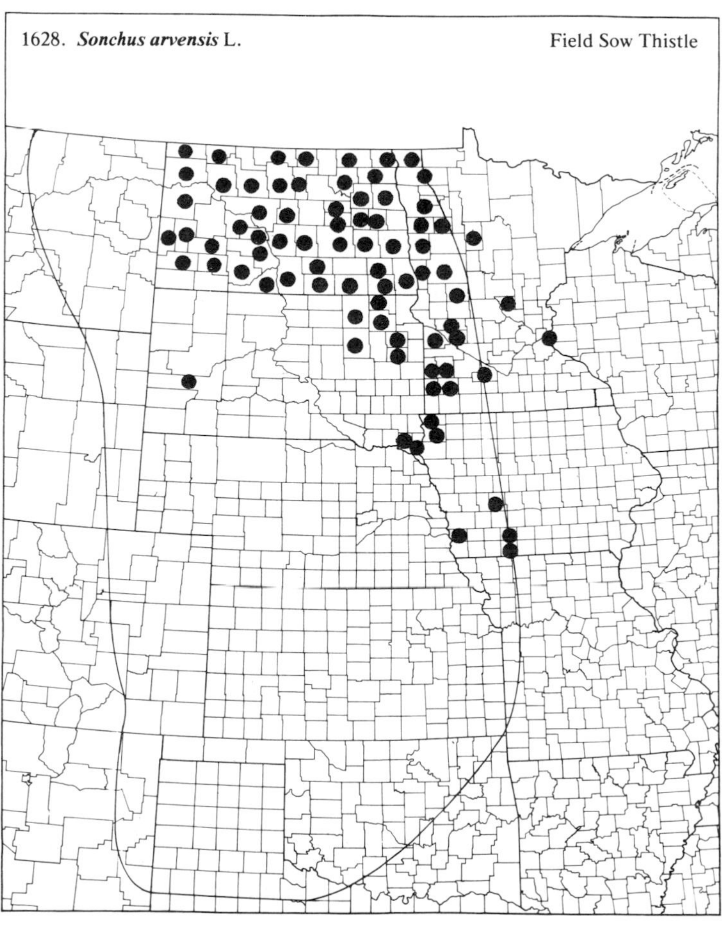
1628. *Sonchus arvensis* L.
Field Sow Thistle

(Asteraceae)

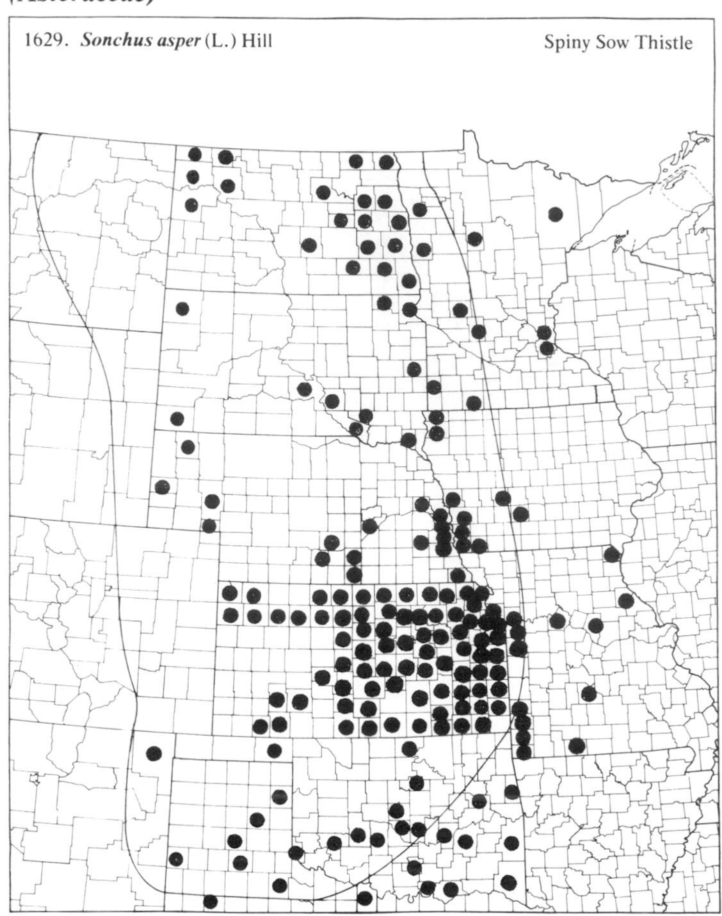
1629. *Sonchus asper* (L.) Hill
Spiny Sow Thistle

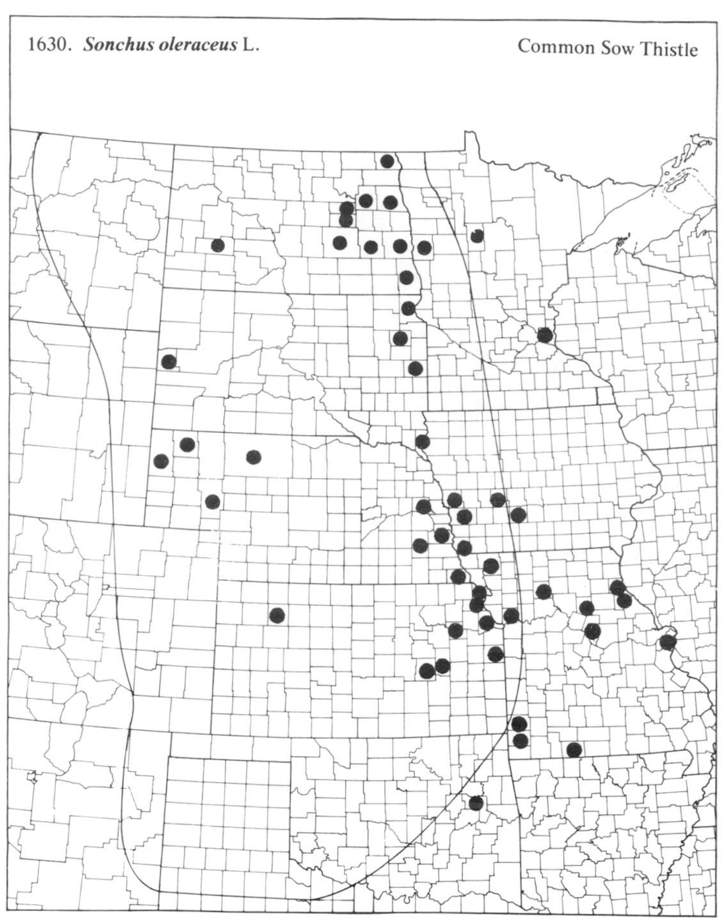
1630. *Sonchus oleraceus* L.
Common Sow Thistle

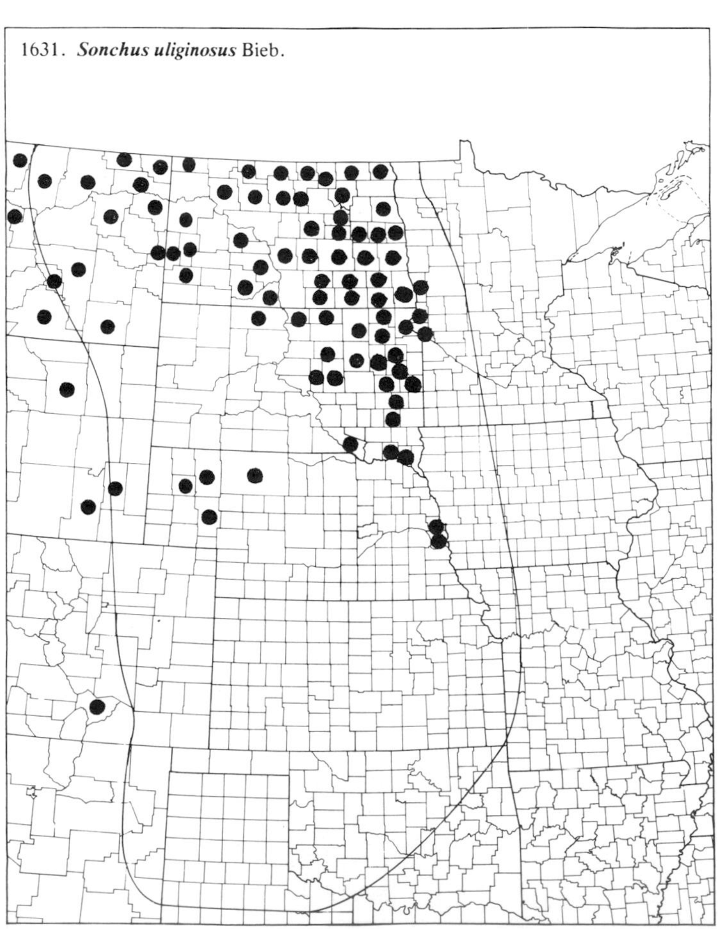
1631. *Sonchus uliginosus* Bieb.

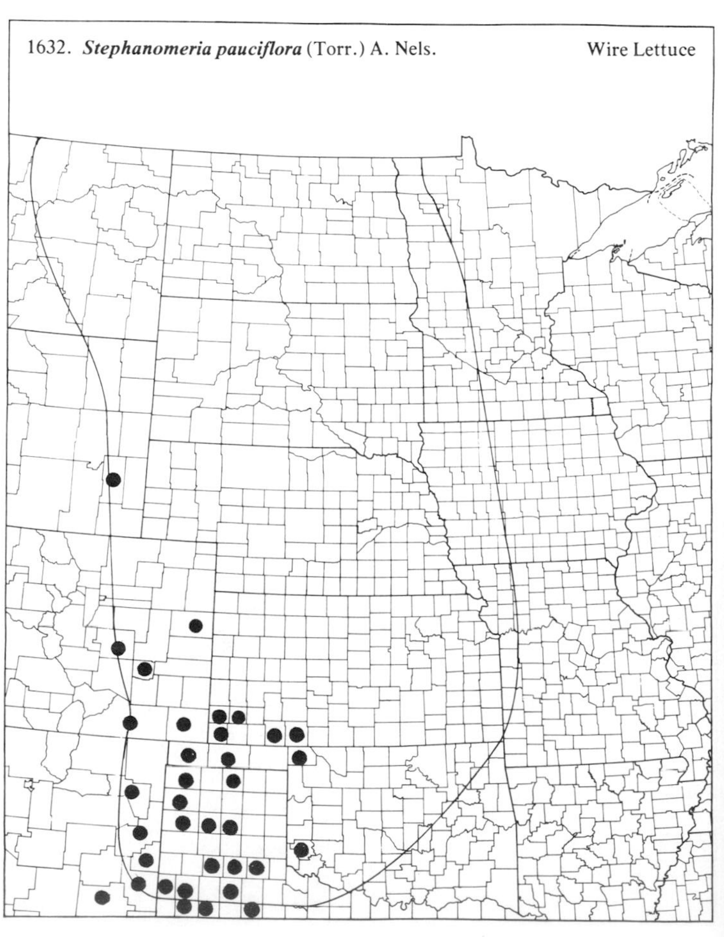
1632. *Stephanomeria pauciflora* (Torr.) A. Nels.
Wire Lettuce

(Asteraceae)

1633. ***Tanacetum vulgare*** L. Common Tansy

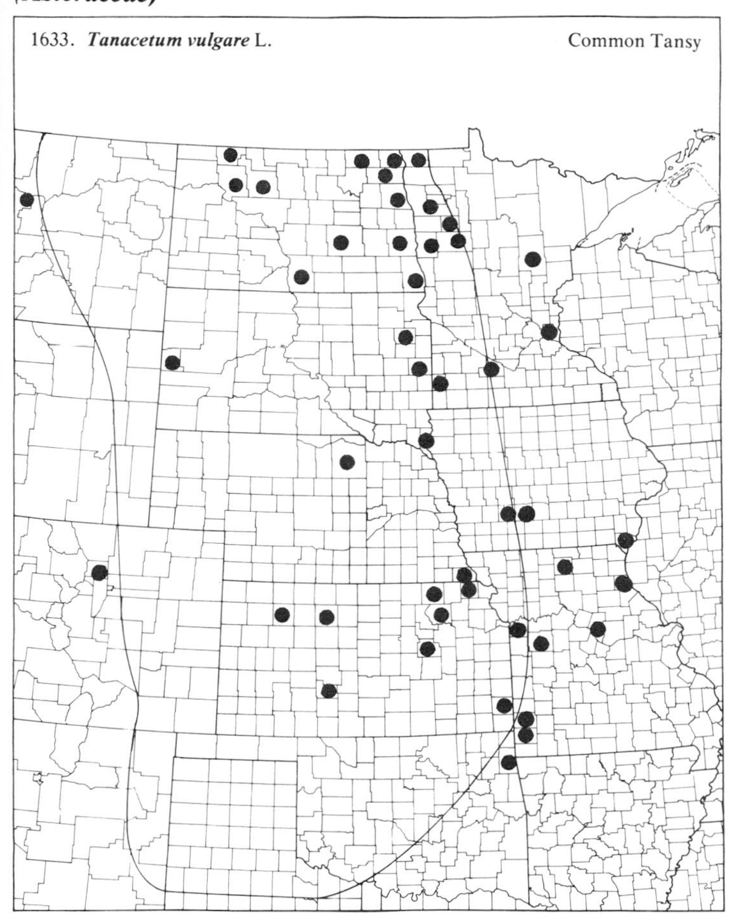

1634. ***Taraxacum laevigatum*** (Willd.) DC. Red-seeded Dandelion

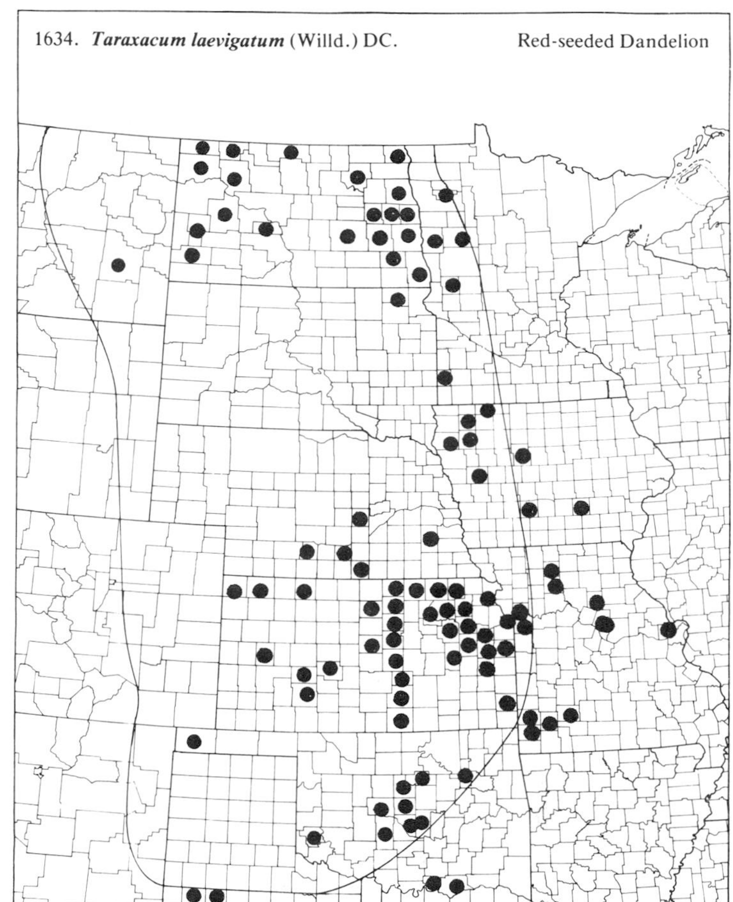

1635. ***Taraxacum officinale*** Weber Dandelion

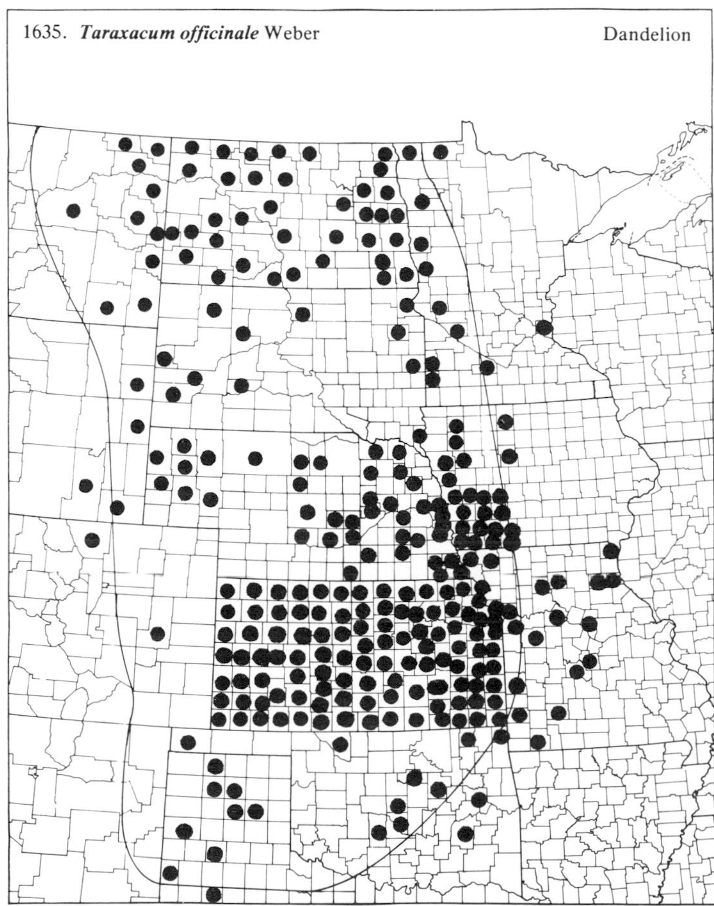

1636. ***Thelesperma filifolium*** (Hook.) Gray var. ***filifolium*** Greenthread

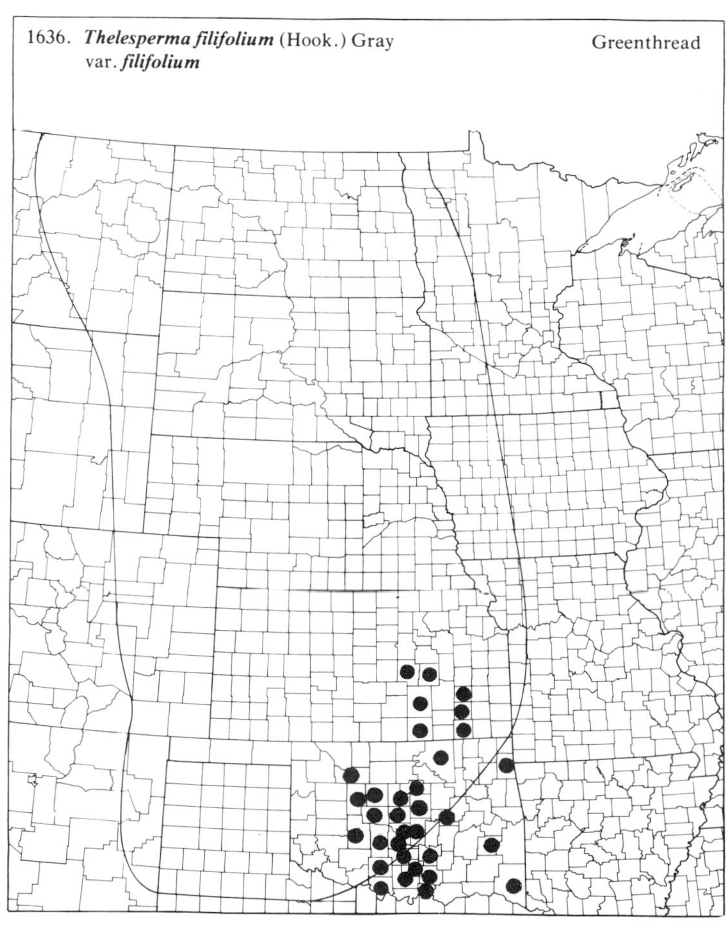

(Asteraceae)

1637. ***Thelesperma filifolium*** (Hook.) Gray var. ***intermedium*** (Rydb.) Shinners — Greenthread

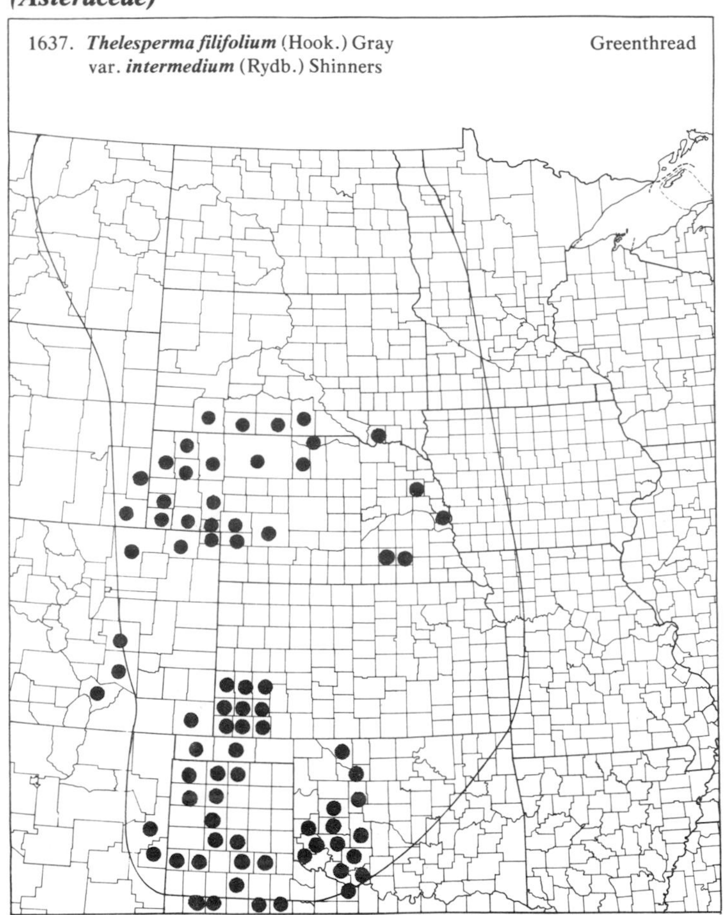

1638. ***Thelesperma megapotamicum*** (Spreng.) O. Ktze.

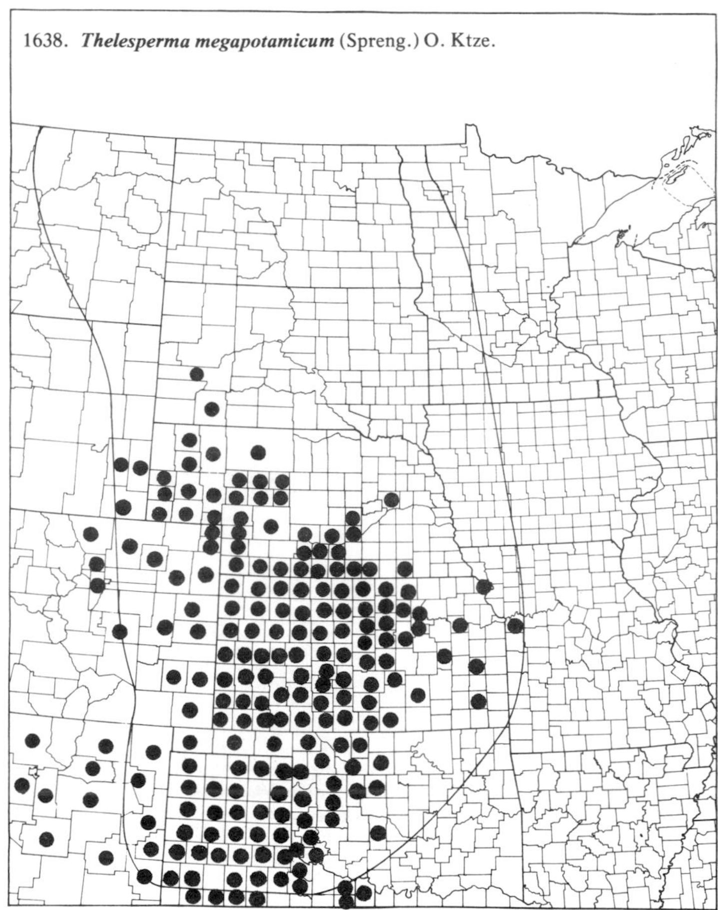

1639. ***Townsendia exscapa*** (Rich.) Porter — Easter Daisy

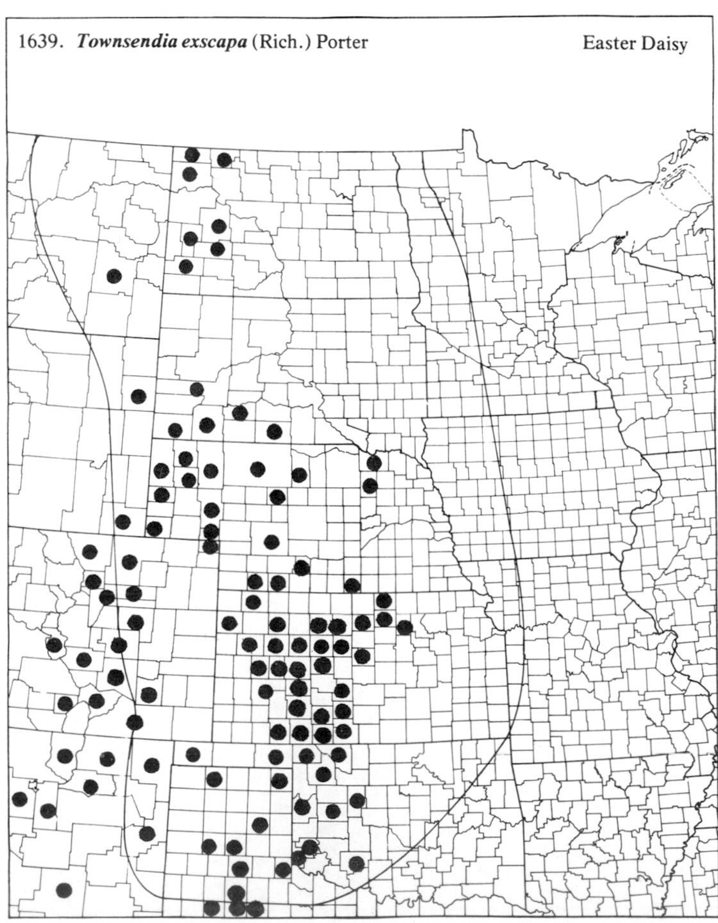

1640. ***Townsendia grandiflora*** Nutt.

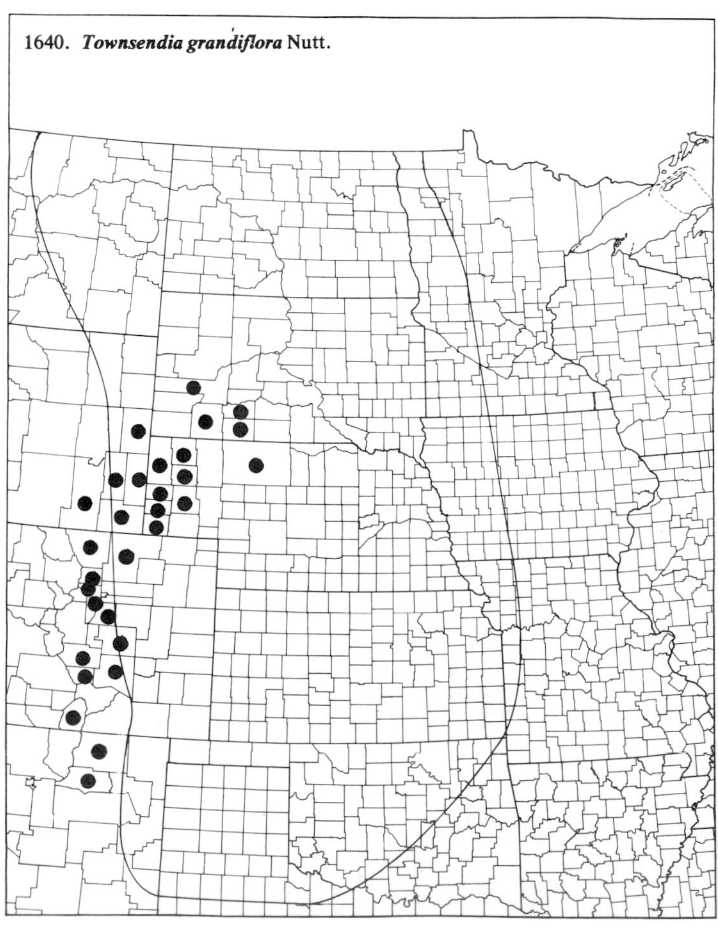

(Asteraceae)

1641. ***Townsendia hookeri*** Beaman

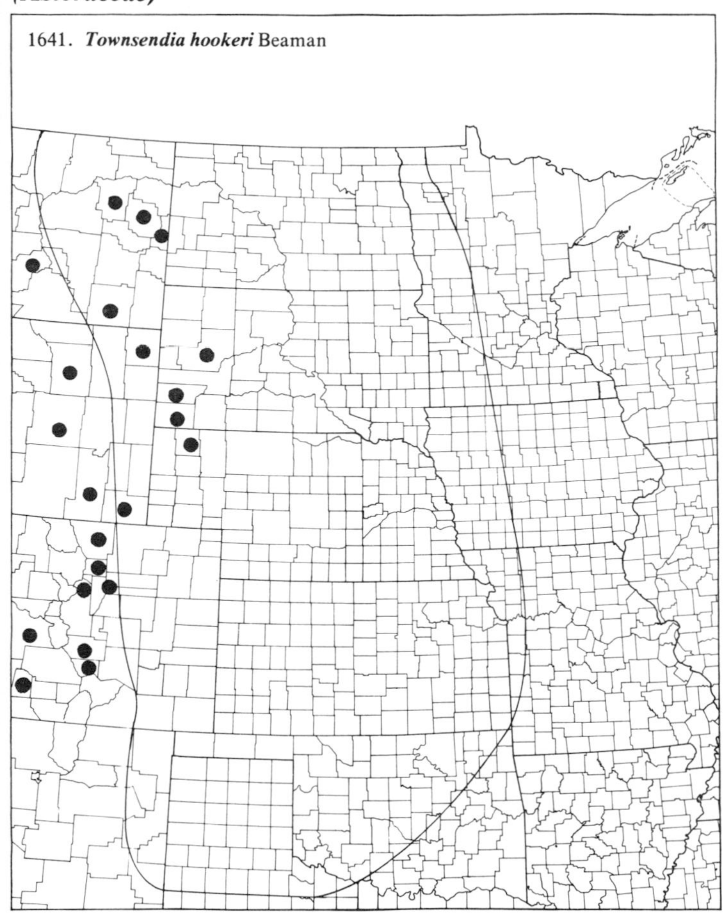

1642. ***Townsendia texensis*** Larsen

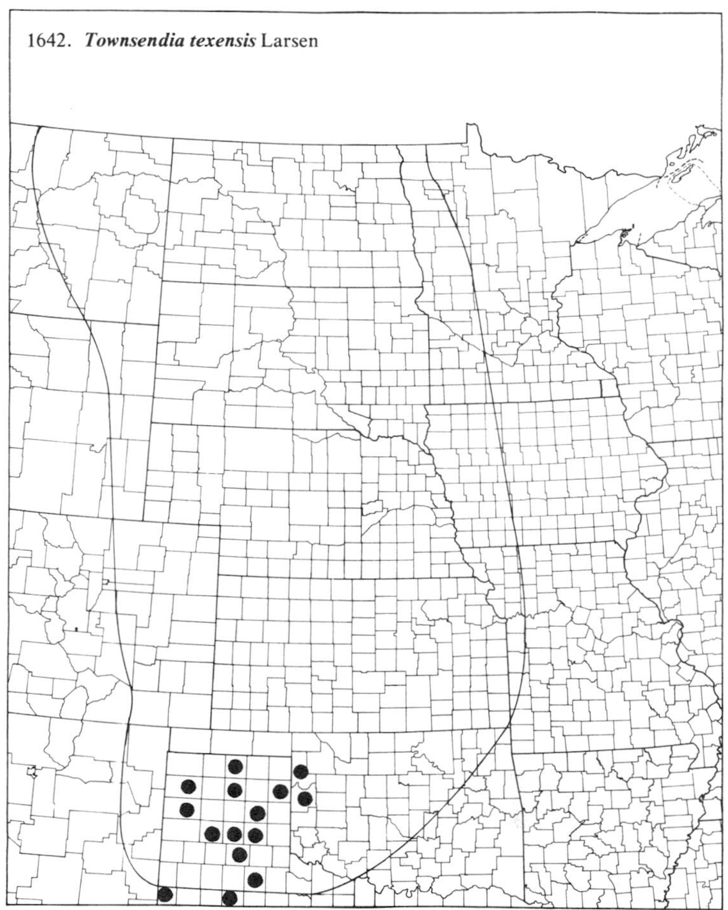

1643. ***Tragopogon dubius*** Scop. Goatsbeard

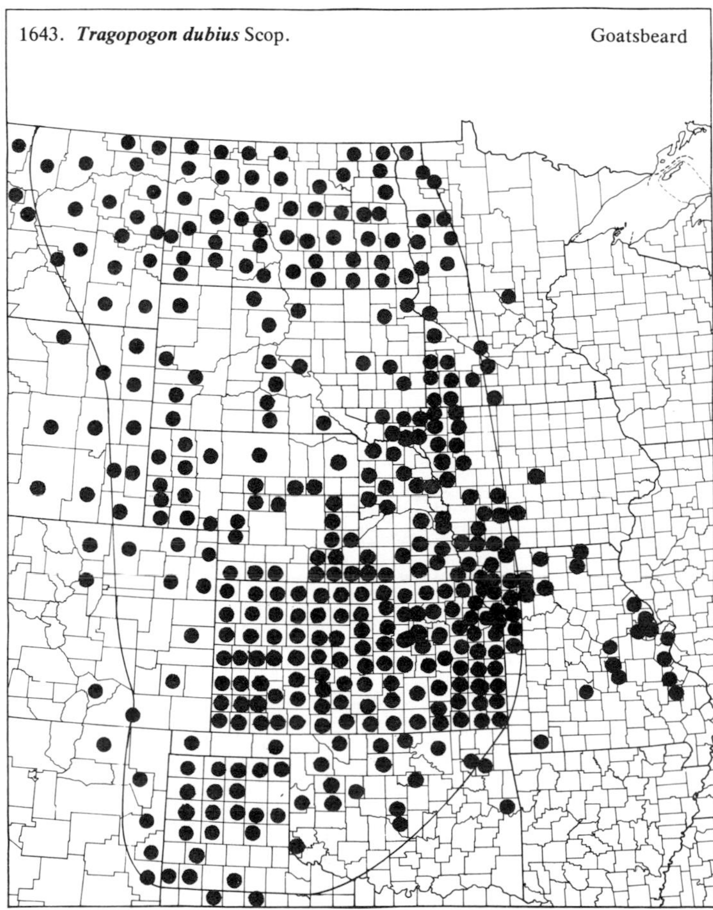

1644. ***Tragopogon porrifolius*** L. Salsify

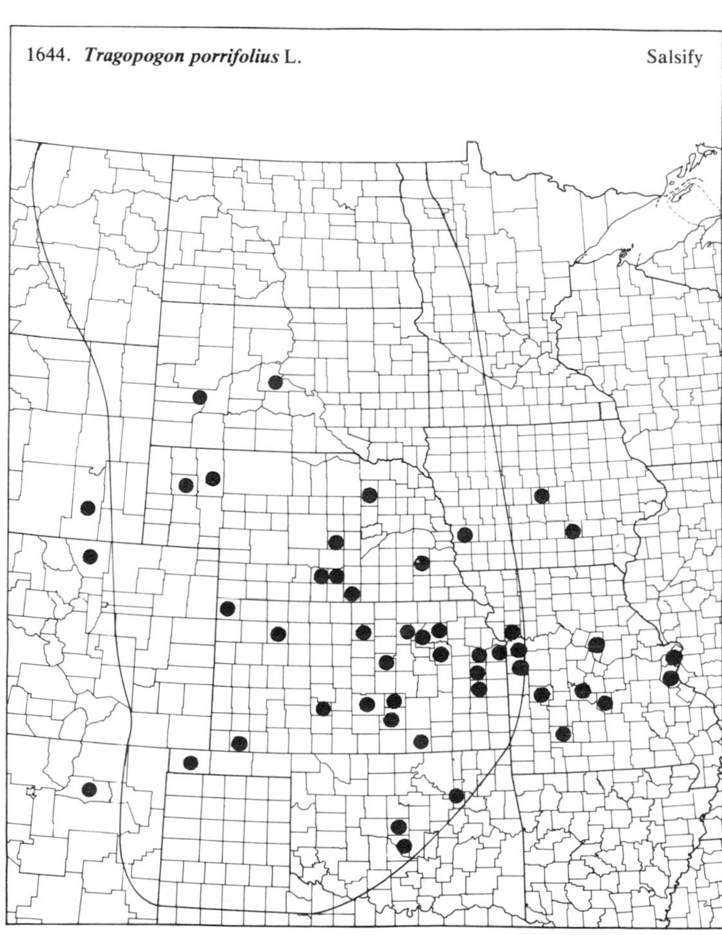

(Asteraceae)

1645. ***Verbesina alternifolia*** (L.) Britt. Wingstem

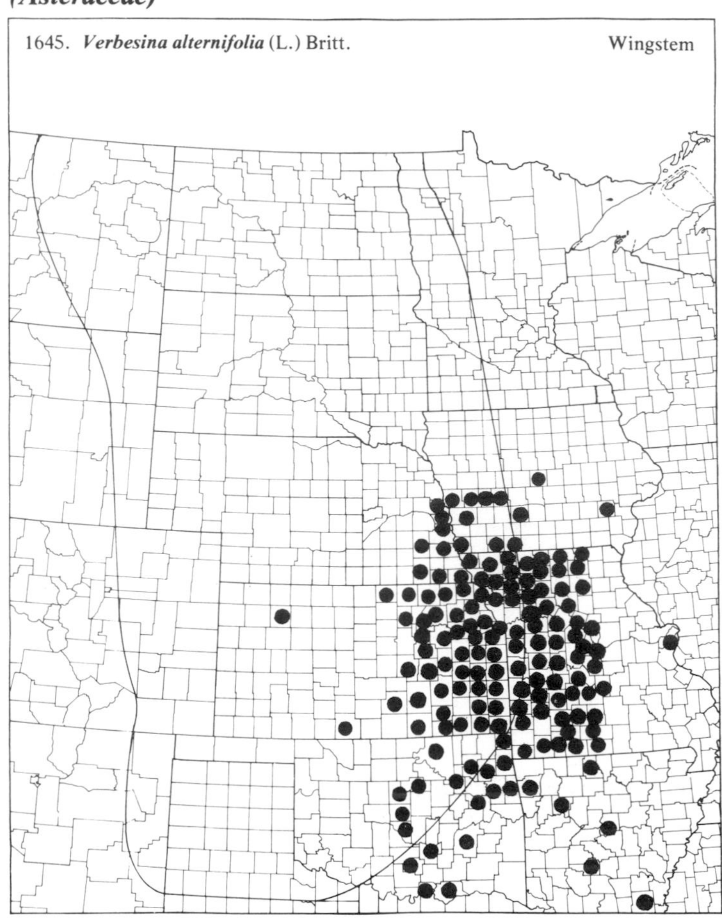

1646. ***Verbesina encelioides*** (Cav.) Babc. & Hall ssp. ***exauriculata*** (Robins. & Greene) Coleman Golden Crownbeard

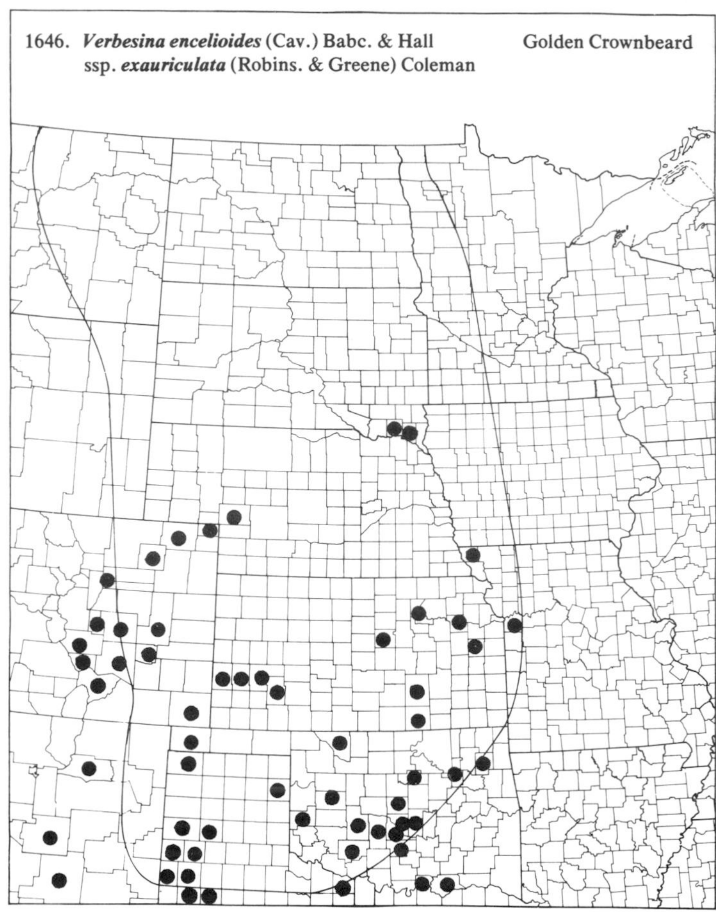

1647. ***Verbesina helianthoides*** Michx.

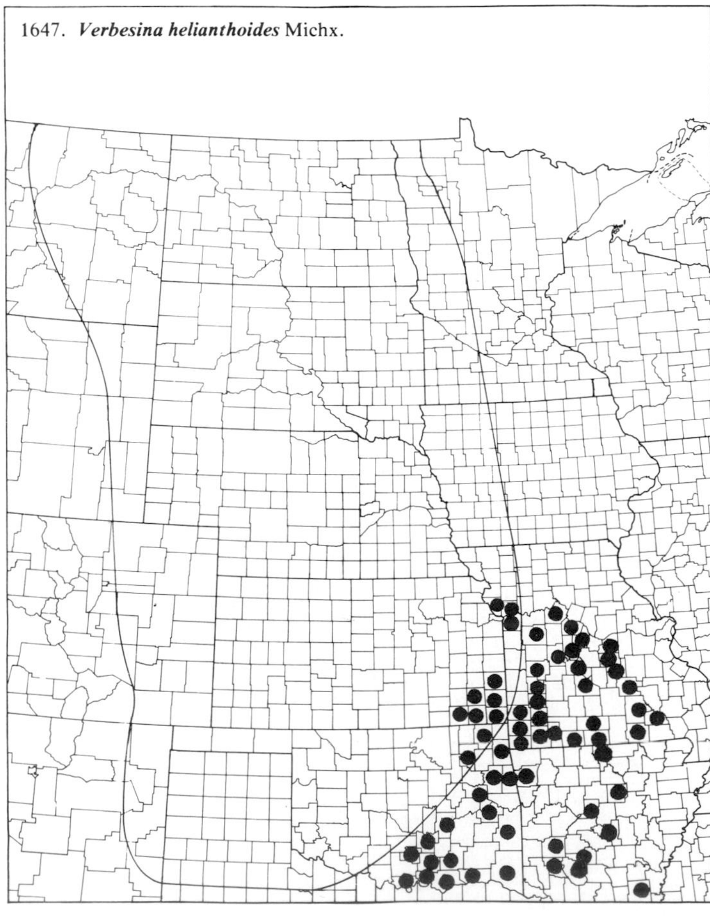

1648. ***Verbesina virginica*** L. Frostweed

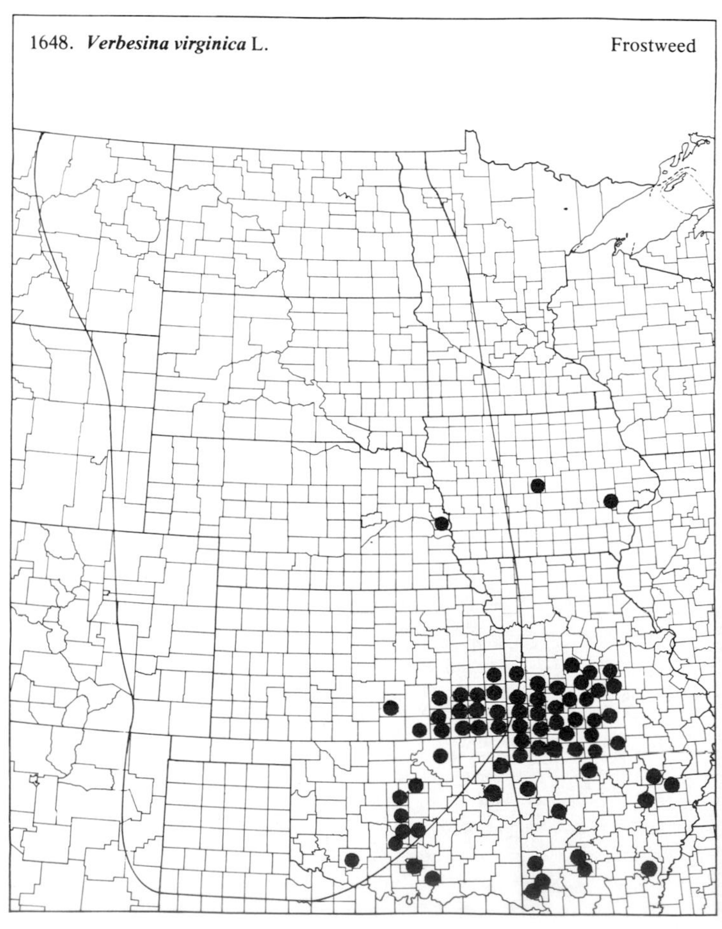

(Asteraceae)

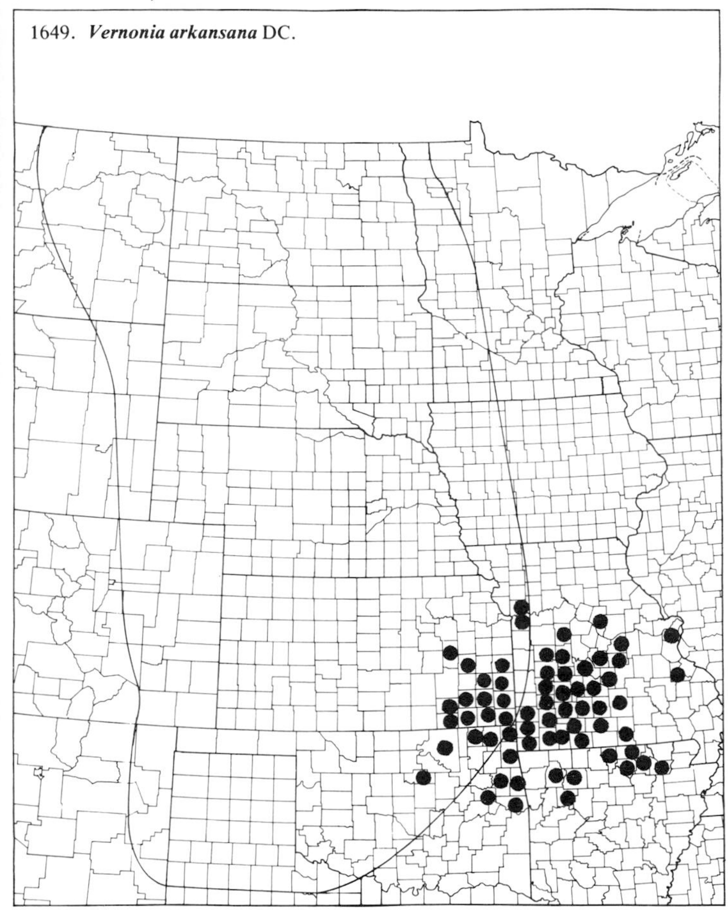
1649. Vernonia arkansana DC.

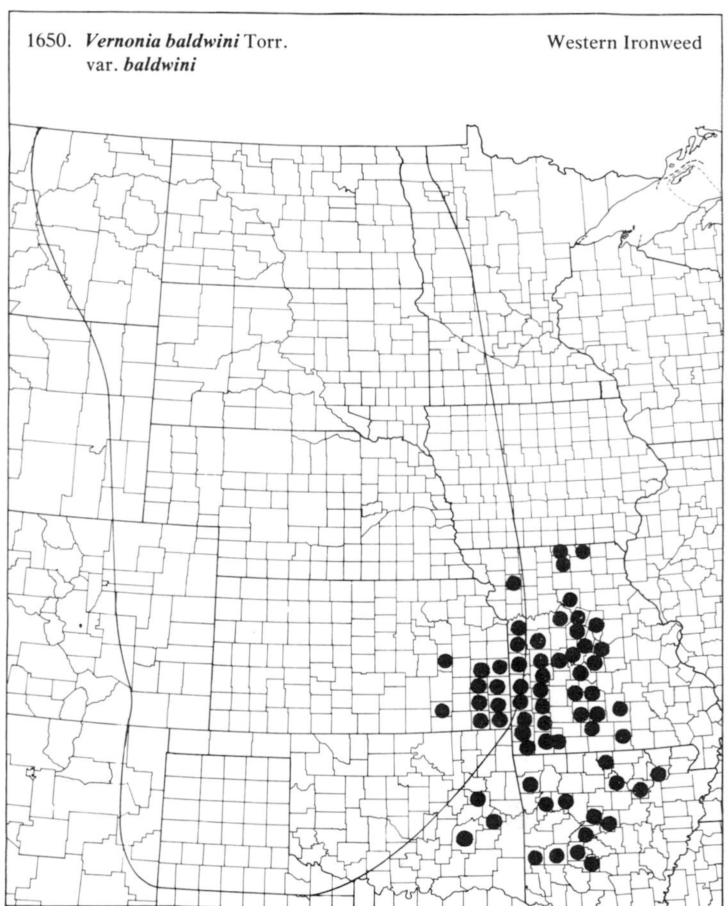
1650. Vernonia baldwini Torr.
var. baldwini
Western Ironweed

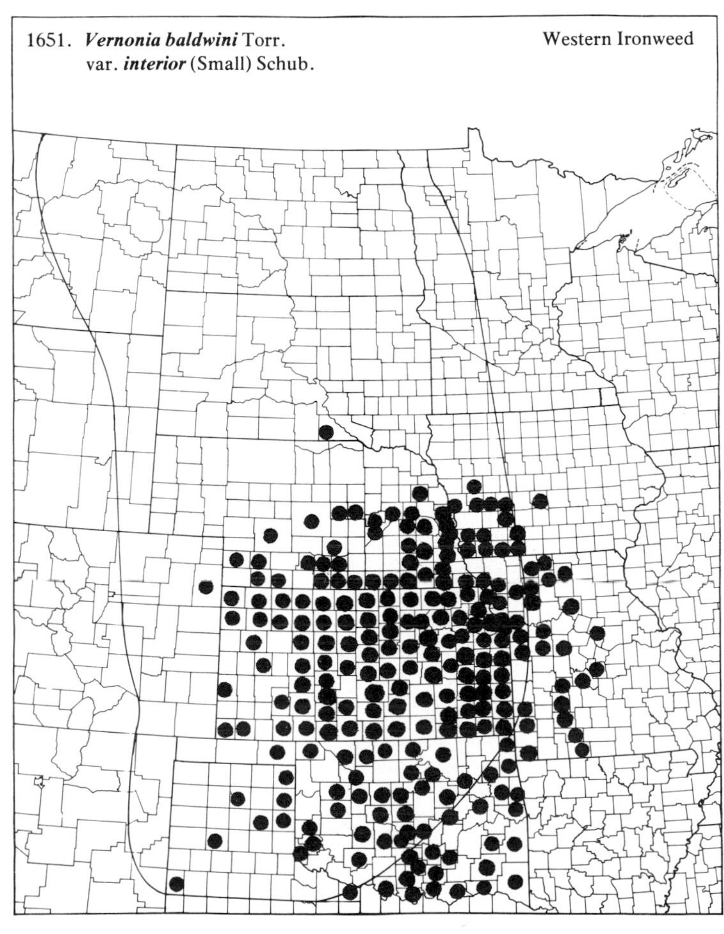
1651. Vernonia baldwini Torr.
var. interior (Small) Schub.
Western Ironweed

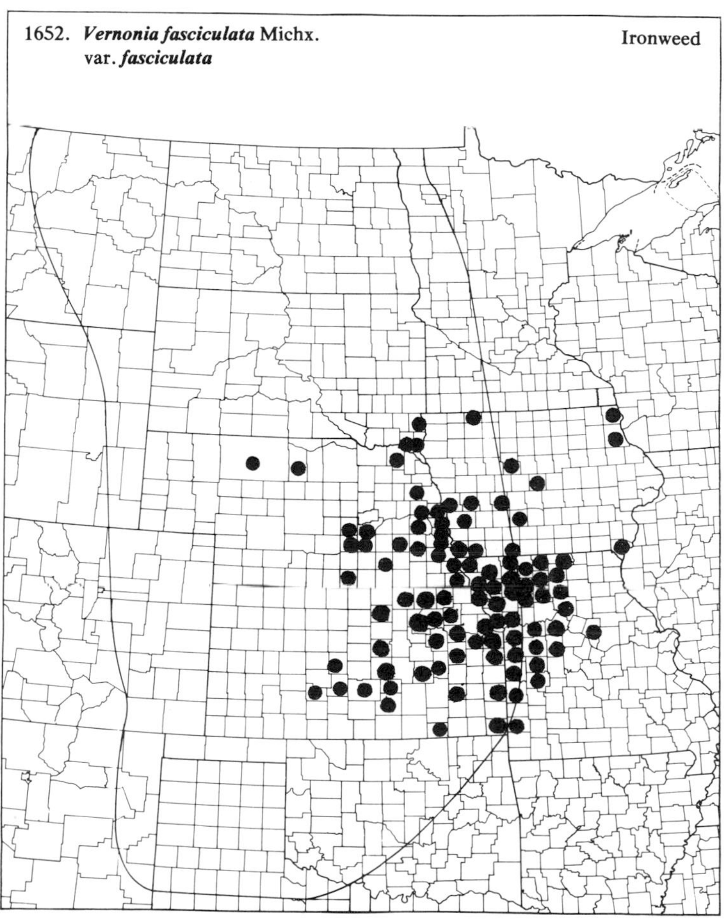
1652. Vernonia fasciculata Michx.
var. fasciculata
Ironweed

(Asteraceae)

1653. ***Vernonia fasciculata*** Michx. var. ***corymbosa*** (Schwein.) Schub. Ironweed

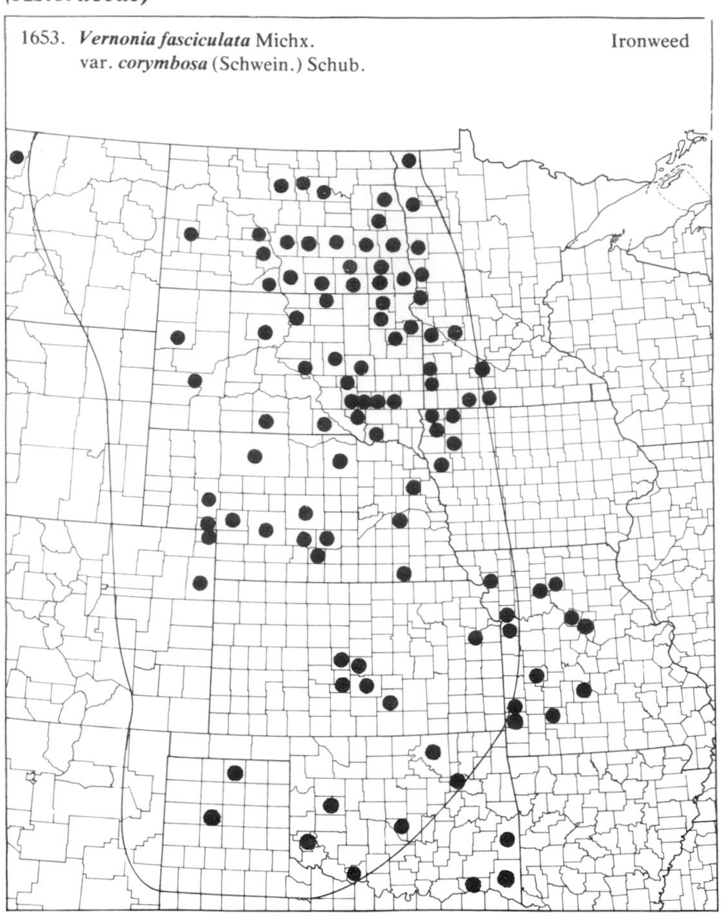

1654. ***Vernonia gigantea*** (Walt.) Trel.

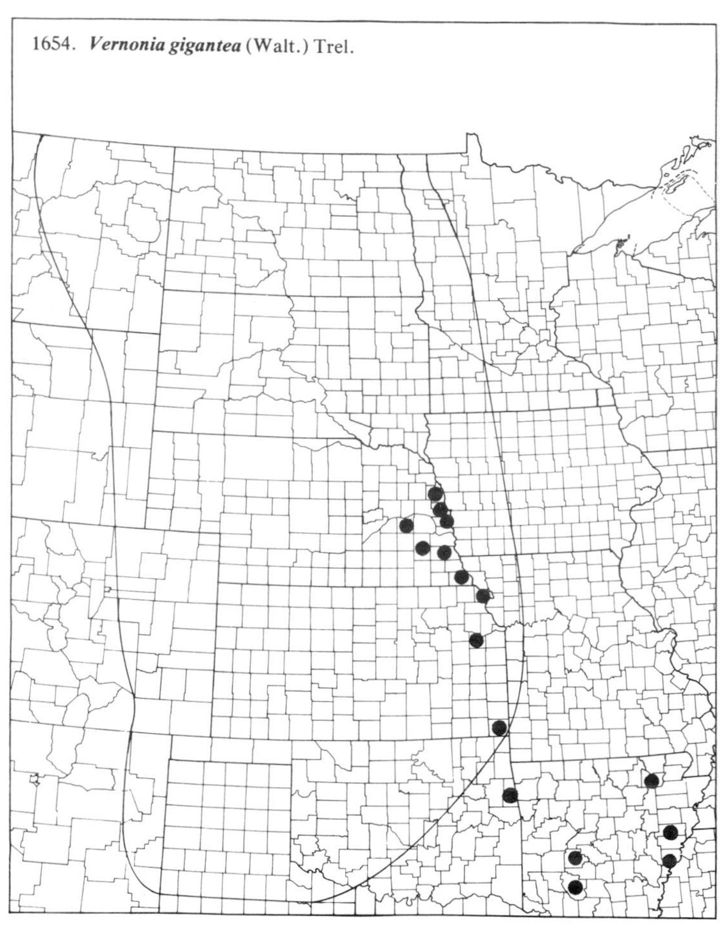

1655. ***Vernonia marginata*** (Torr.) Raf. Plains Ironweed

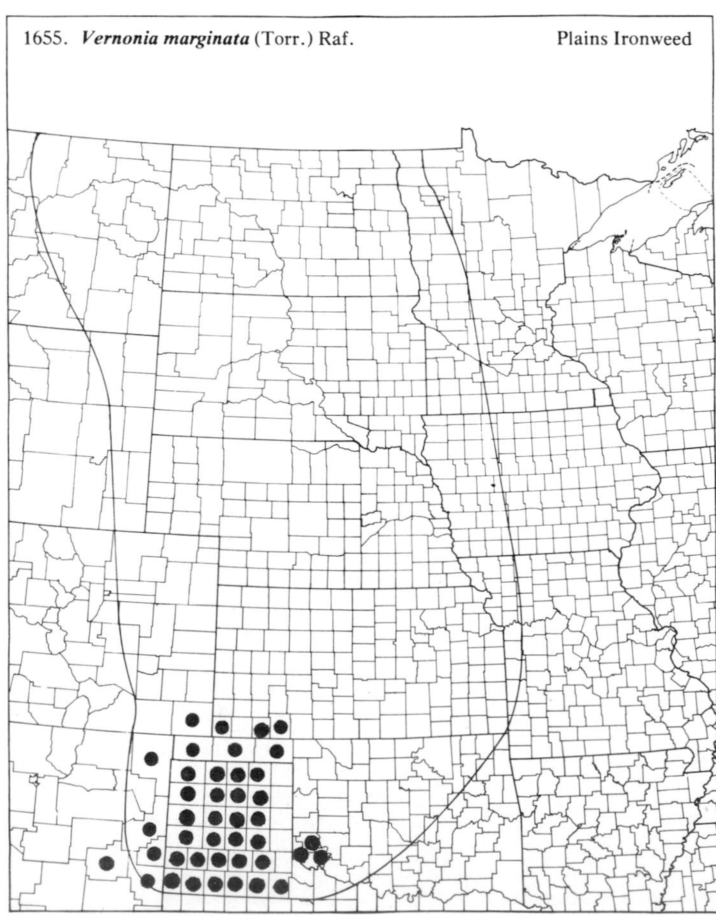

1656. ***Vernonia missurica*** Raf.

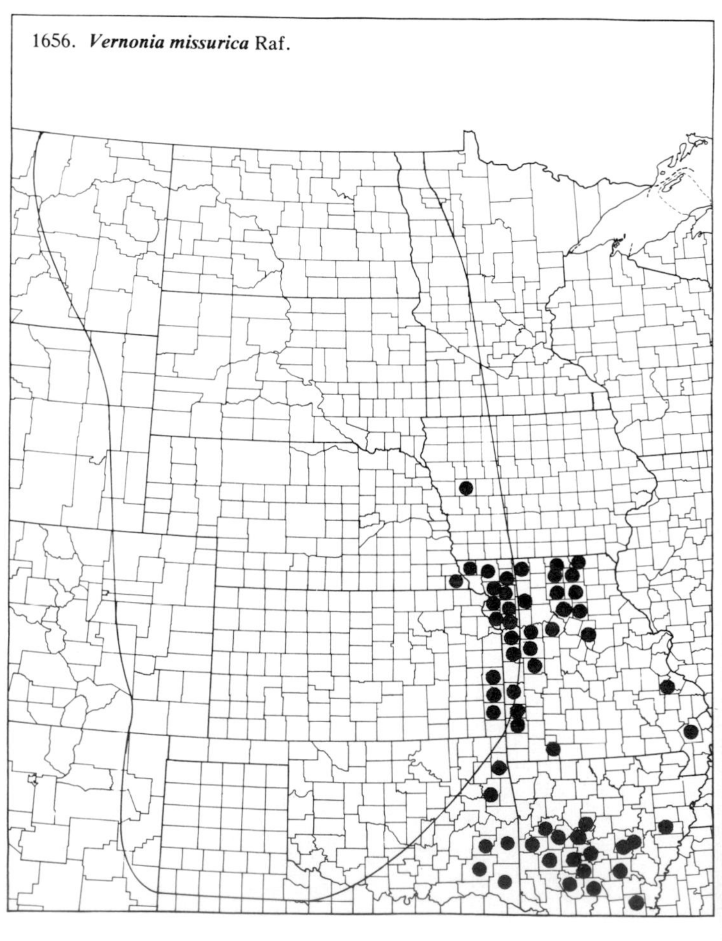

(Asteraceae)

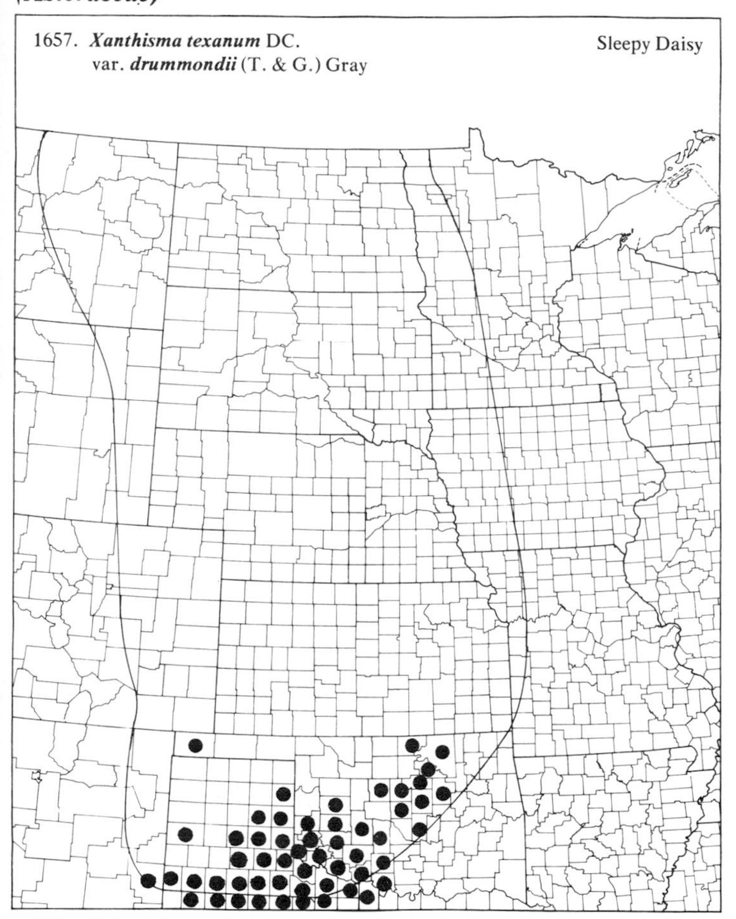

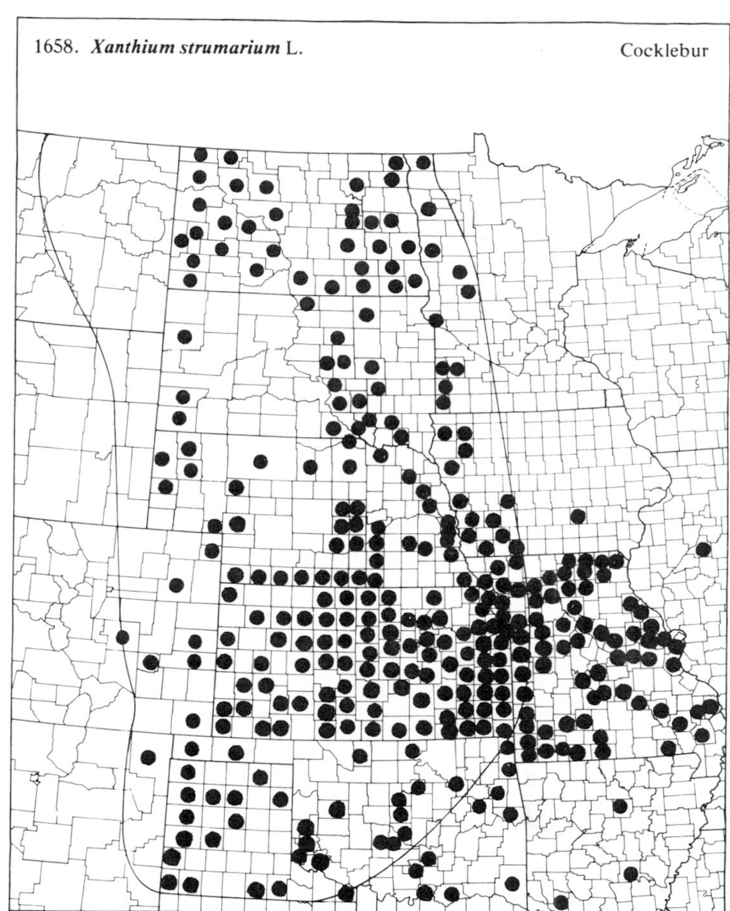

1659. *Zinnia grandiflora* Nutt. Rocky Mountain Zinnia

116. *Alismataceae* (Water Plantain Family)

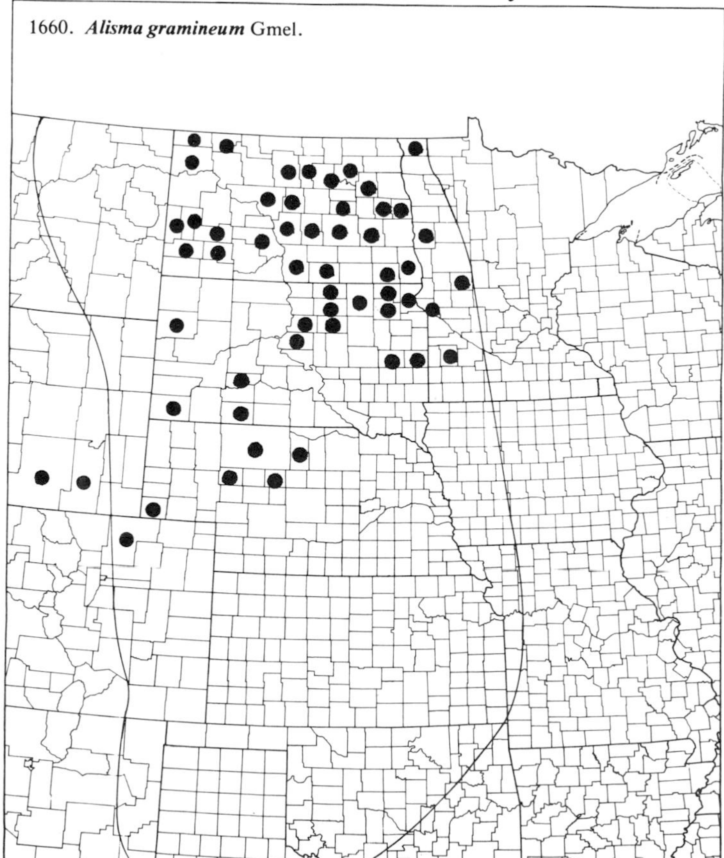

(Alismataceae)

1661. ***Alisma plantago-aquatica*** L. var. ***americanum*** R. & S. — Water Plantain

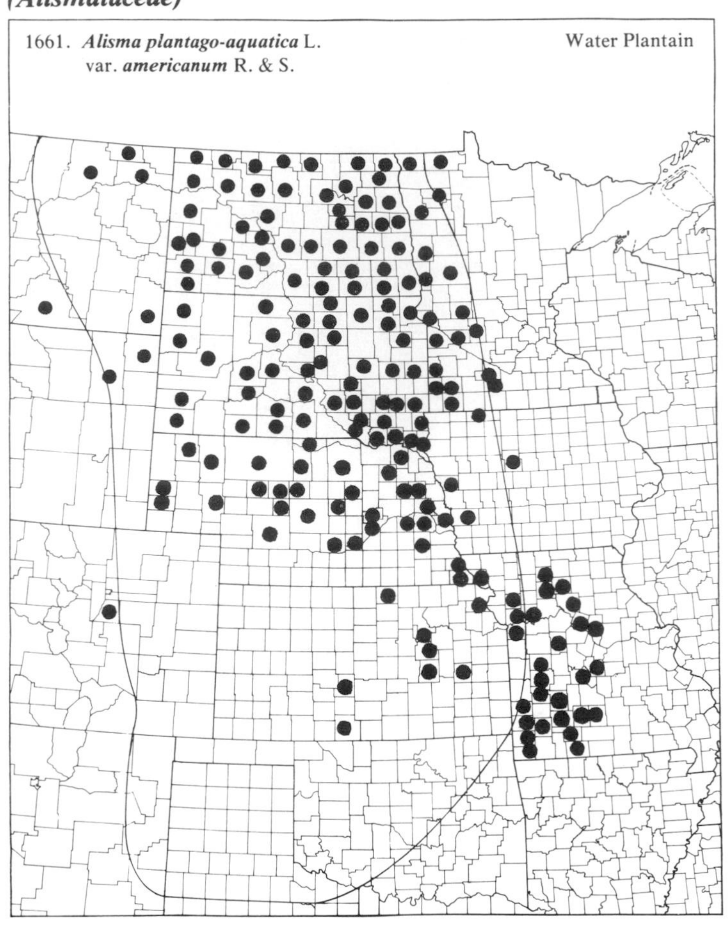

1662. ***Alisma subcordatum*** Raf. — Water Plantain

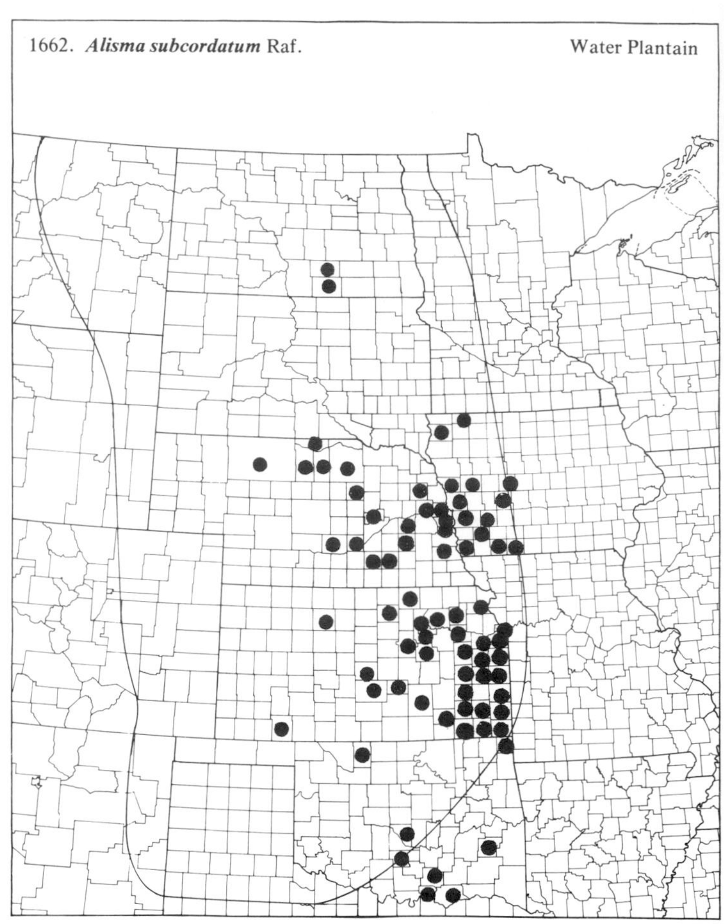

1663. ***Echinodorus cordifolius*** (L.) Griseb.

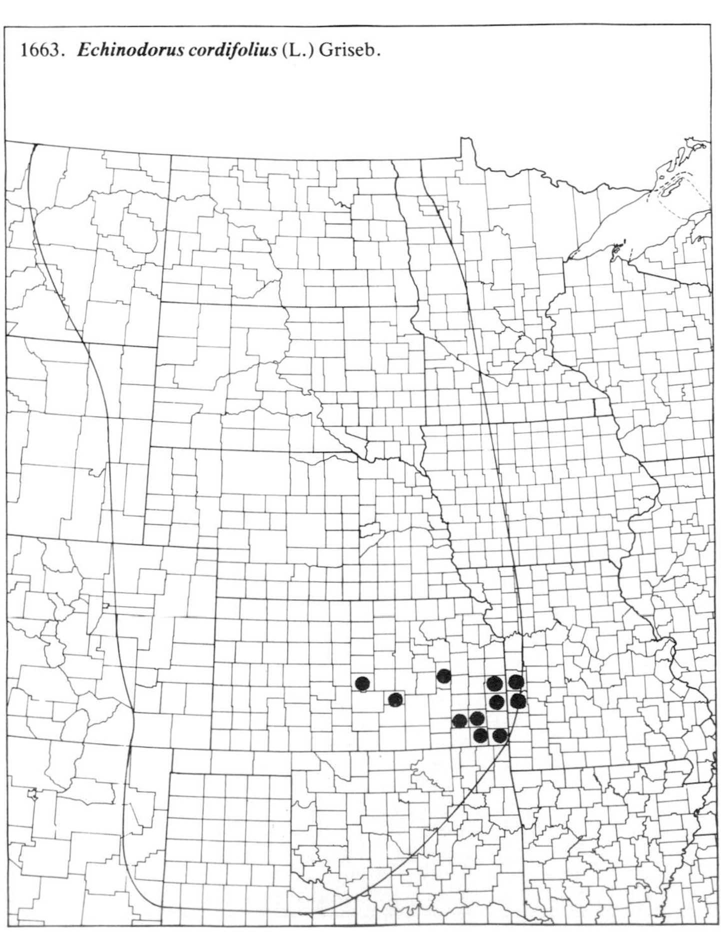

1664. ***Echinodorus rostratus*** (Nutt.) Engelm. — Burhead

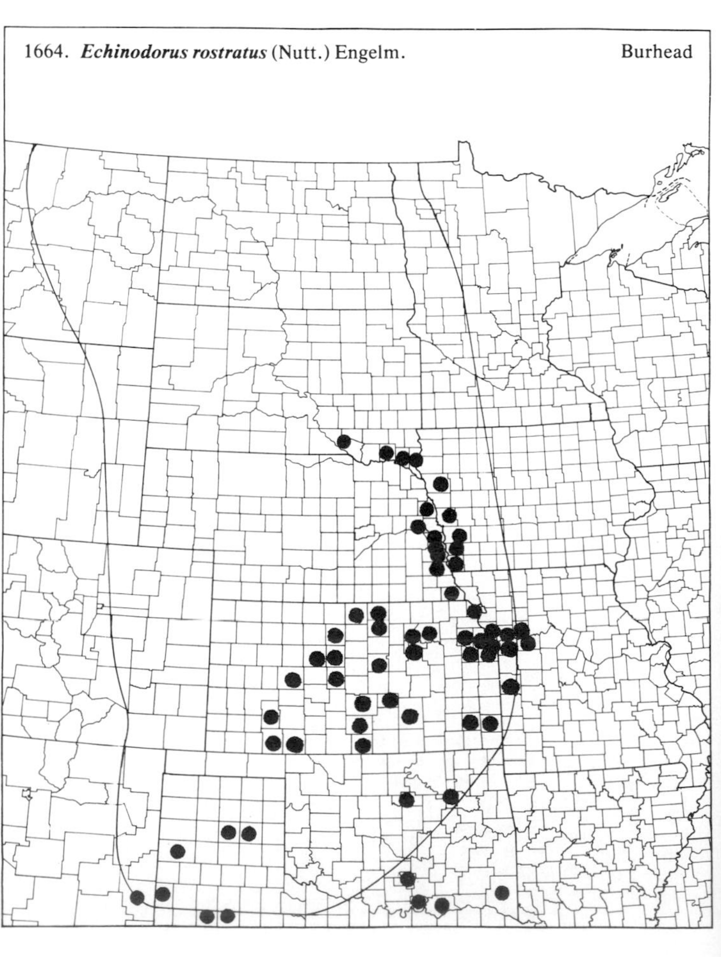

(Alismataceae)

1665. ***Sagittaria ambigua*** J.G. Sm. Arrowhead

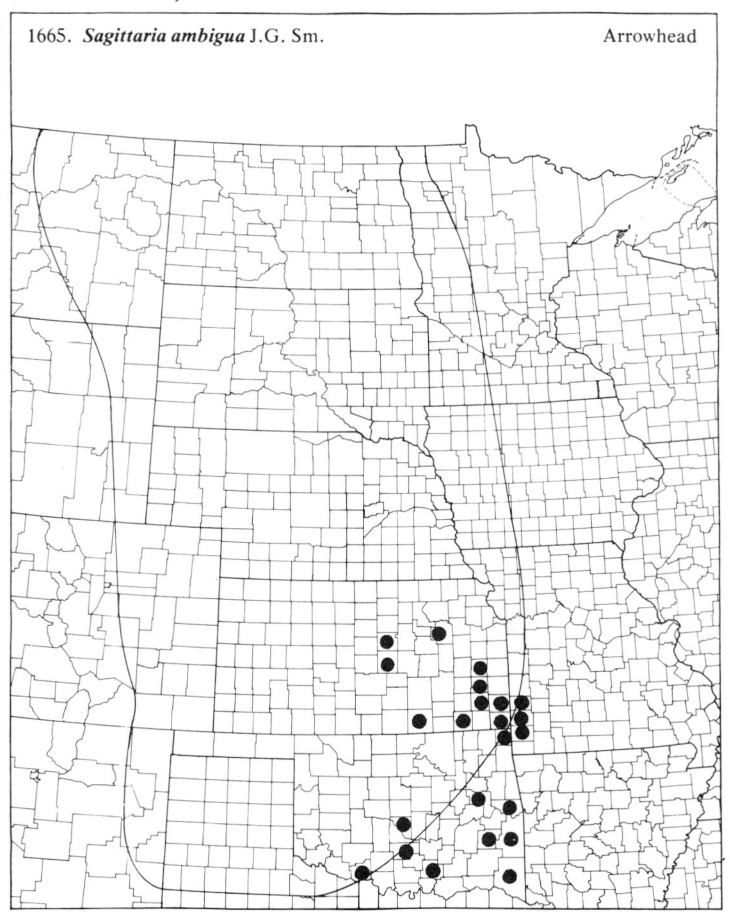

1666. ***Sagittaria cuneata*** Sheld. Duck Potato Arrowhead

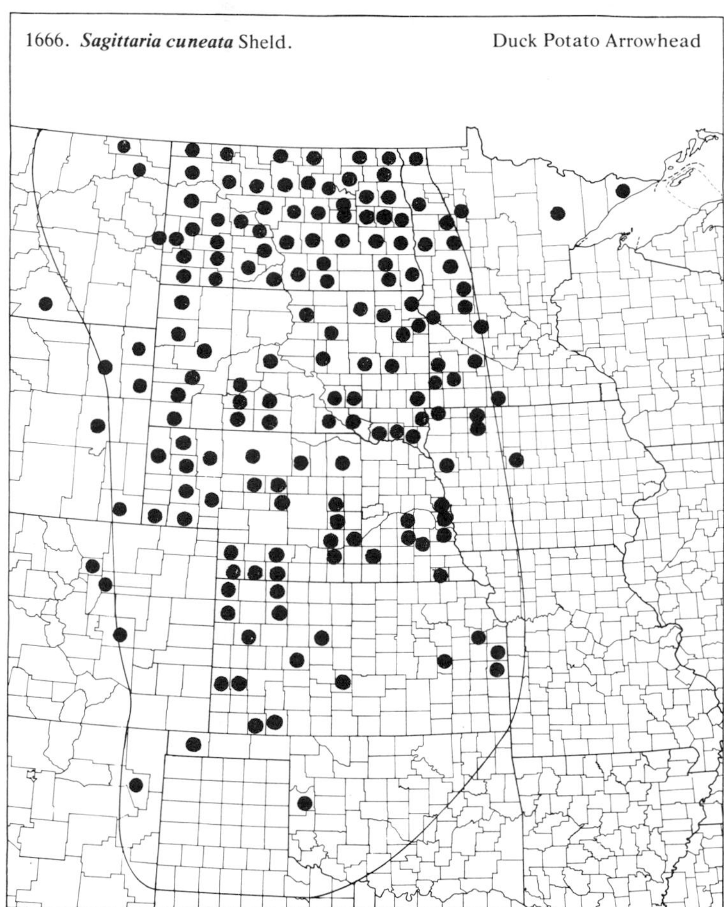

1667. ***Sagittaria engelmanniana*** J.G. Sm. ssp. ***brevirostra*** (Mack. & Bush) Bogin

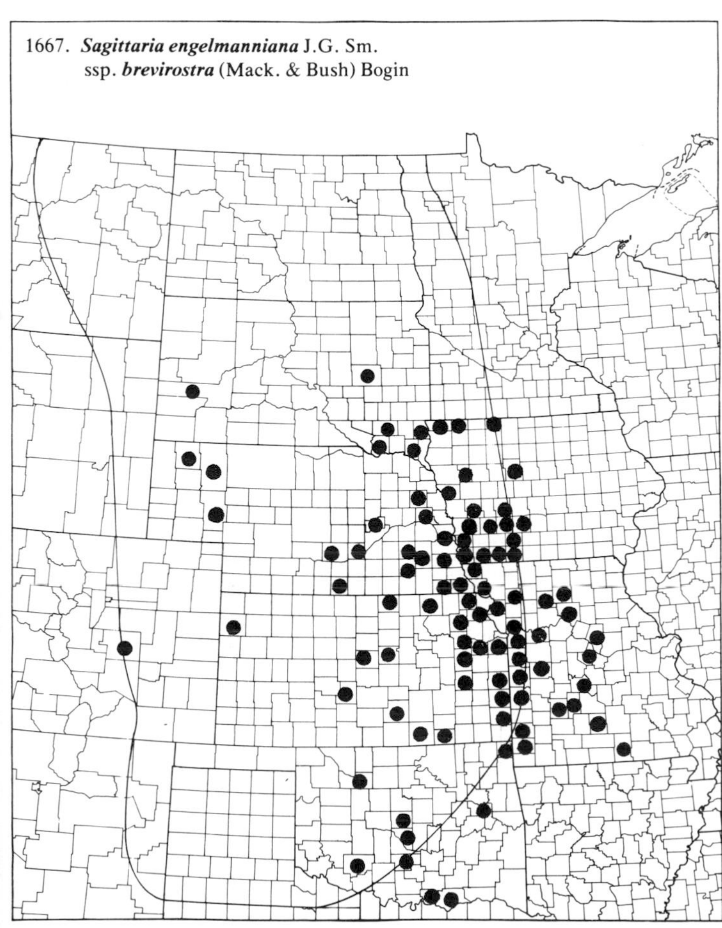

1668. ***Sagittaria graminea*** Michx.

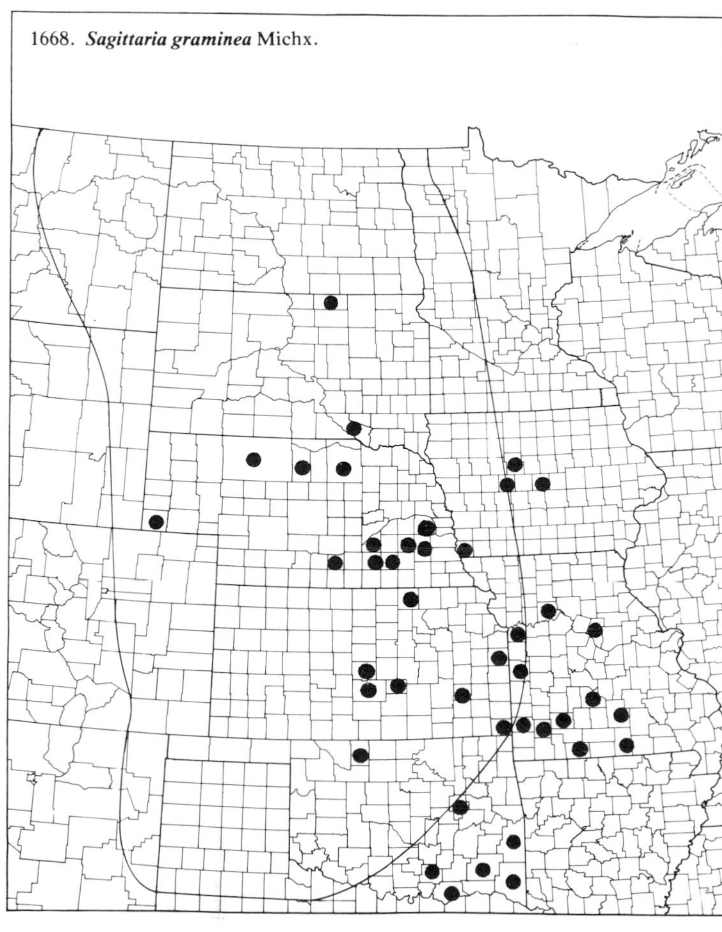

(Alismataceae)

1669. ***Sagittaria latifolia*** Willd. Common Arrowhead

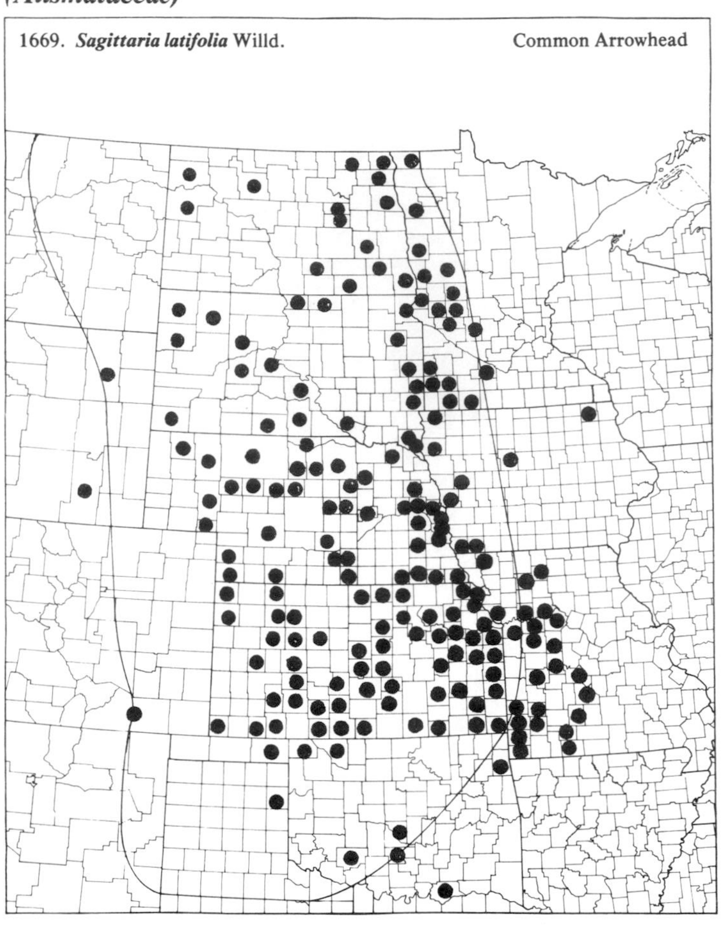

1670. ***Sagittaria longiloba*** Engelm. Longbarb Arrowhead

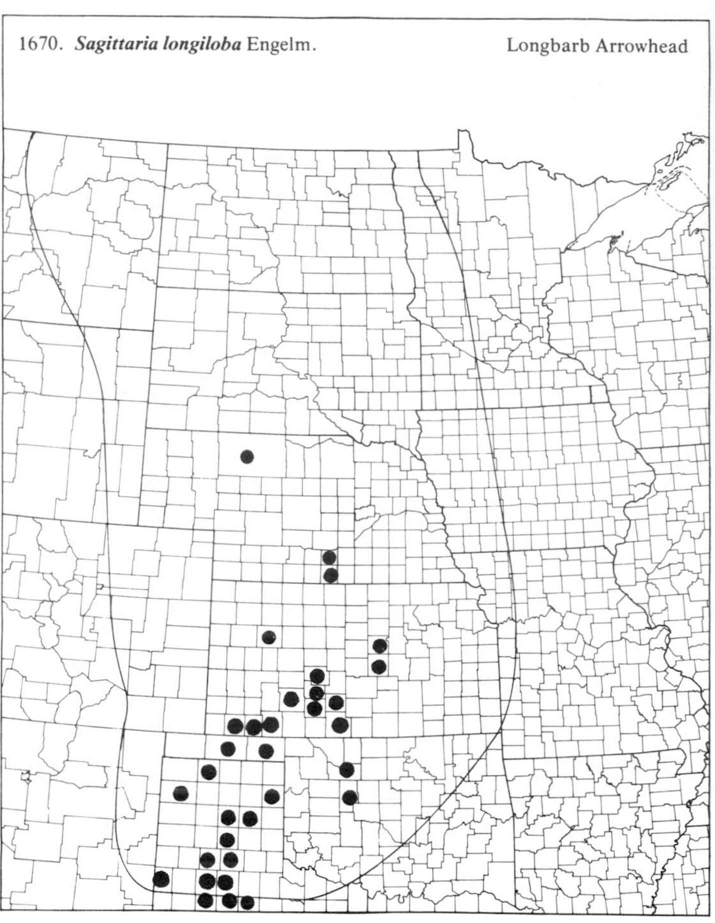

1671. ***Sagittaria montevidensis*** Schlect. & Cham. ssp. ***calycina*** (Engelm.) Bogin

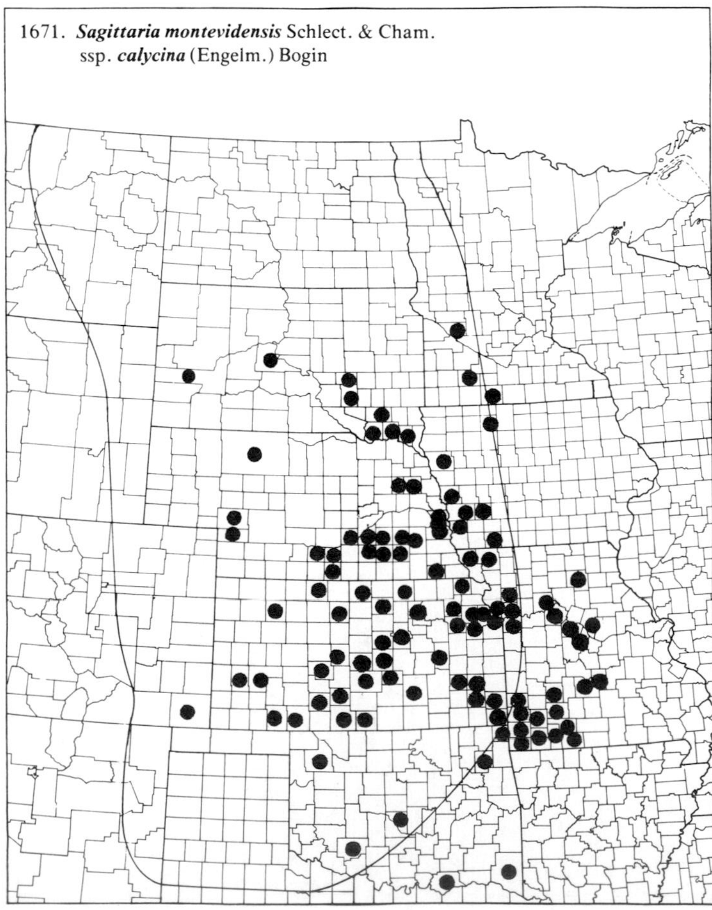

1672. ***Sagittaria rigida*** Pursh

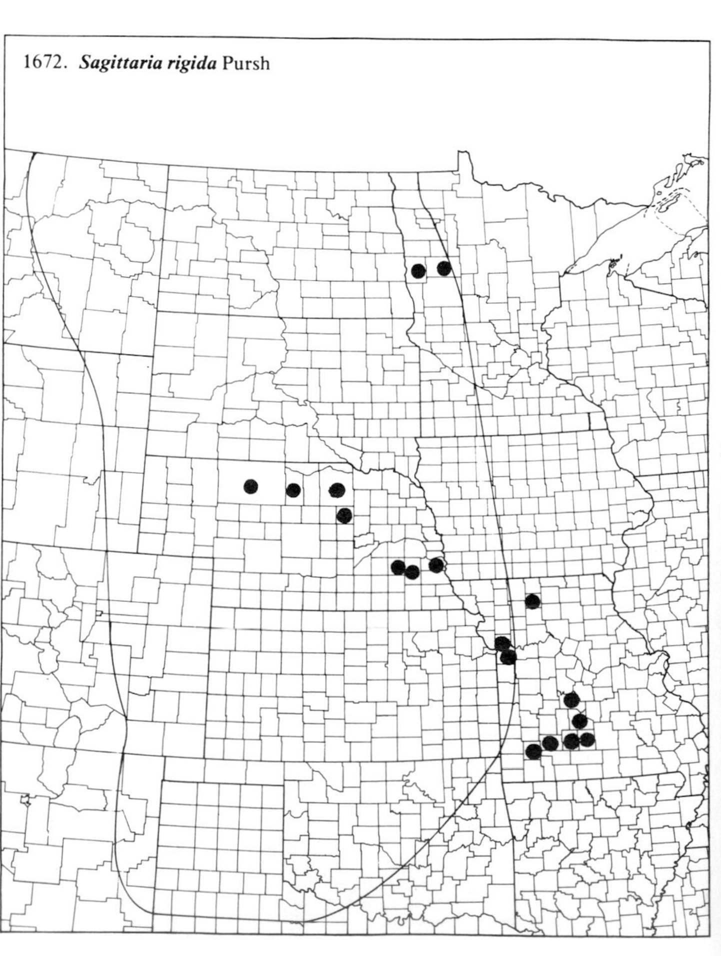

117. ***Hydrocharitaceae*** (Frog's-bit Family)

1673. ***Elodea canadensis*** Michx. Waterweed

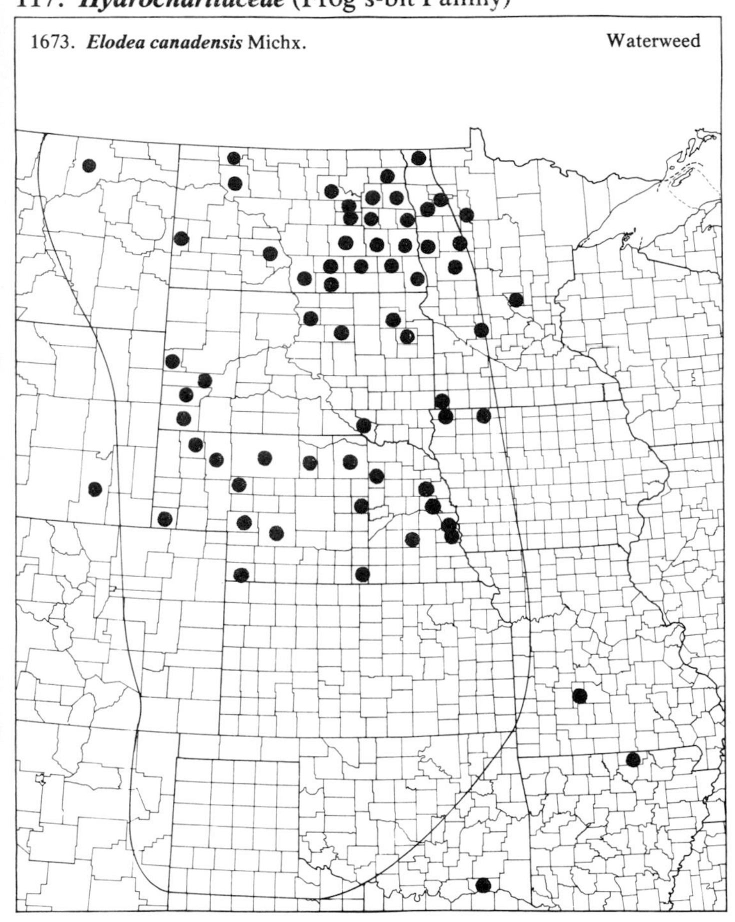

1674. ***Elodea nuttallii*** (Planch.) St. John Waterweed

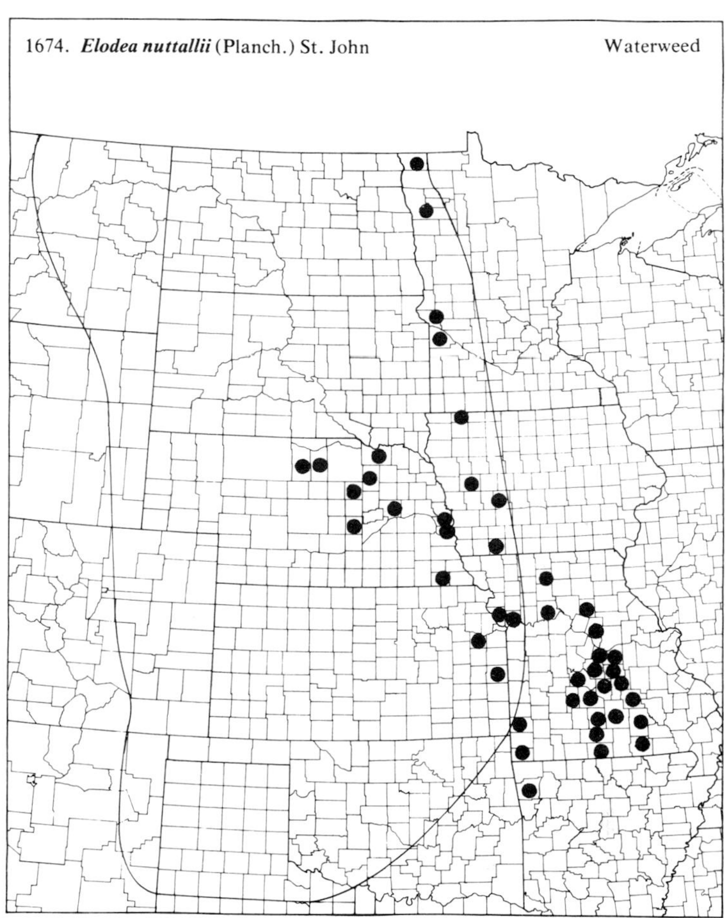

1675. ***Vallisneria americana*** Michx. Tapegrass

119. ***Juncaginaceae*** (Arrowgrass Family)

1676. ***Triglochin concinnum*** Davy
var. ***debile*** (M.E. Jones) J.T. Howell

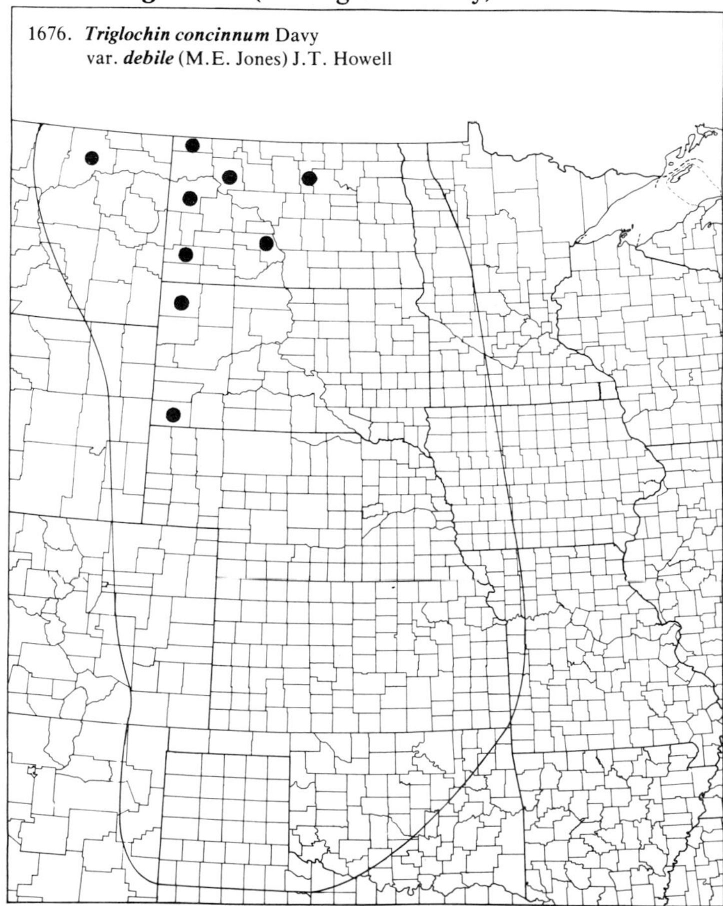

(Juncaginaceae)

1677. ***Triglochin maritimum*** L. Arrowgrass

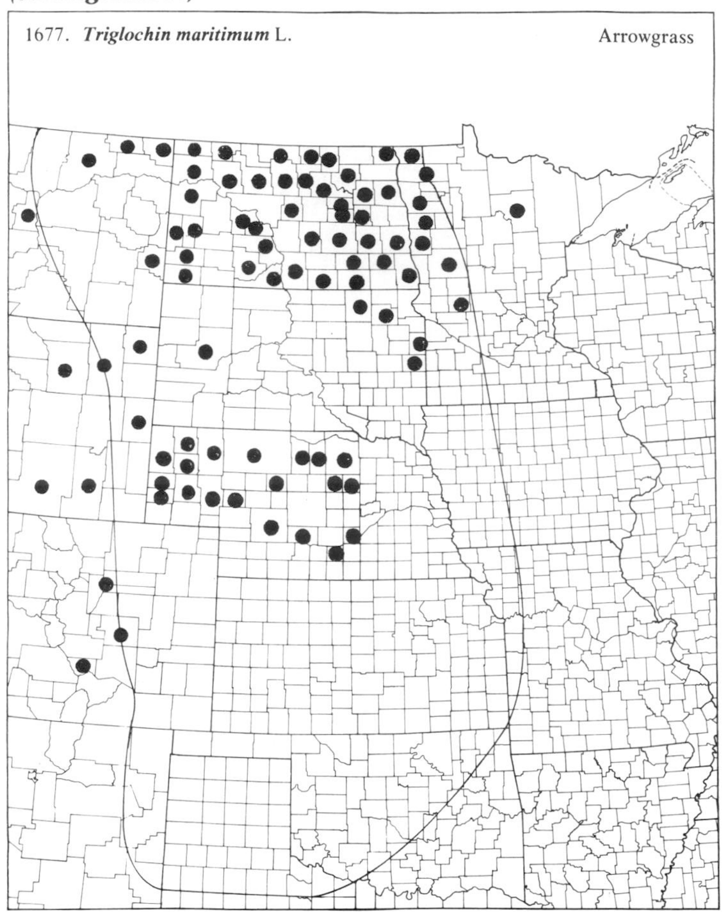

1678. ***Triglochin palustre*** L.

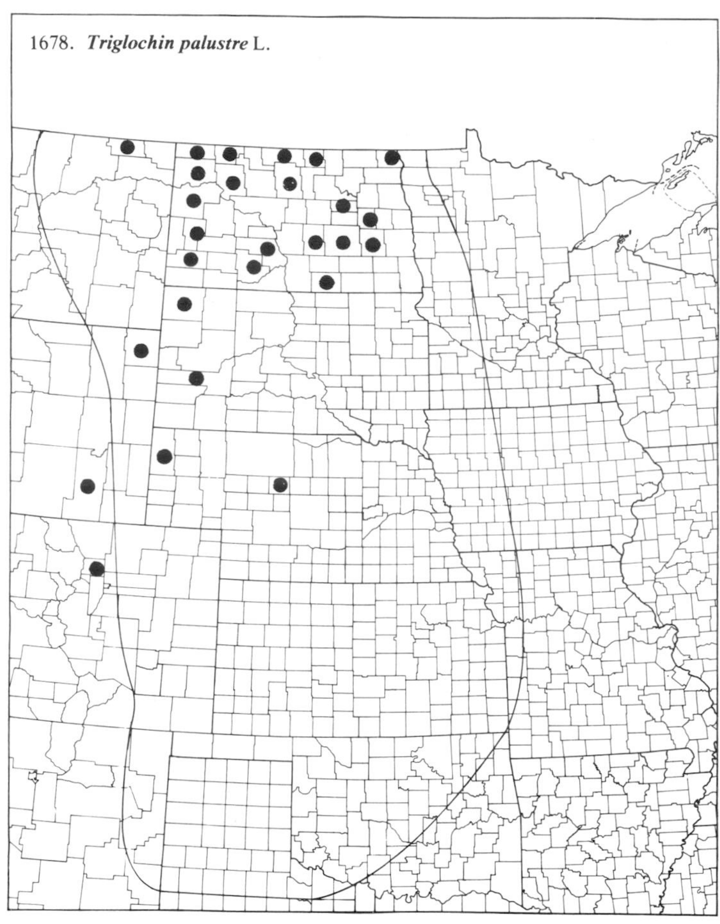

120. *Najadaceae* (Naiad Family)

1679. ***Najas flexilis*** (Willd.) Rostk. & Schmidt Naiad

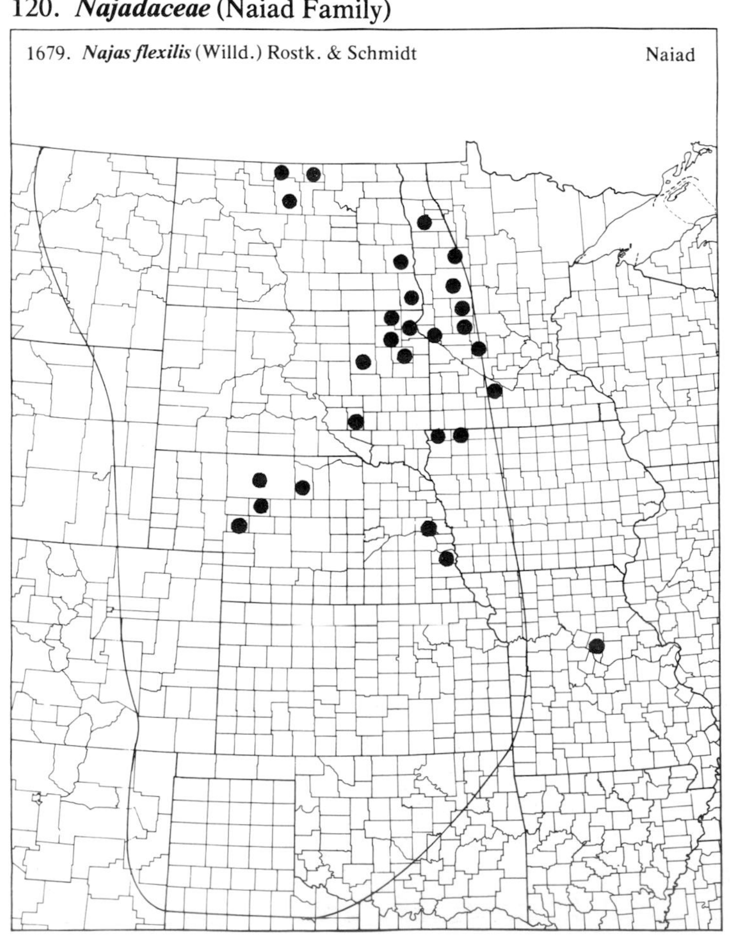

1680. ***Najas guadalupensis*** (Spreng.) Morong Naiad

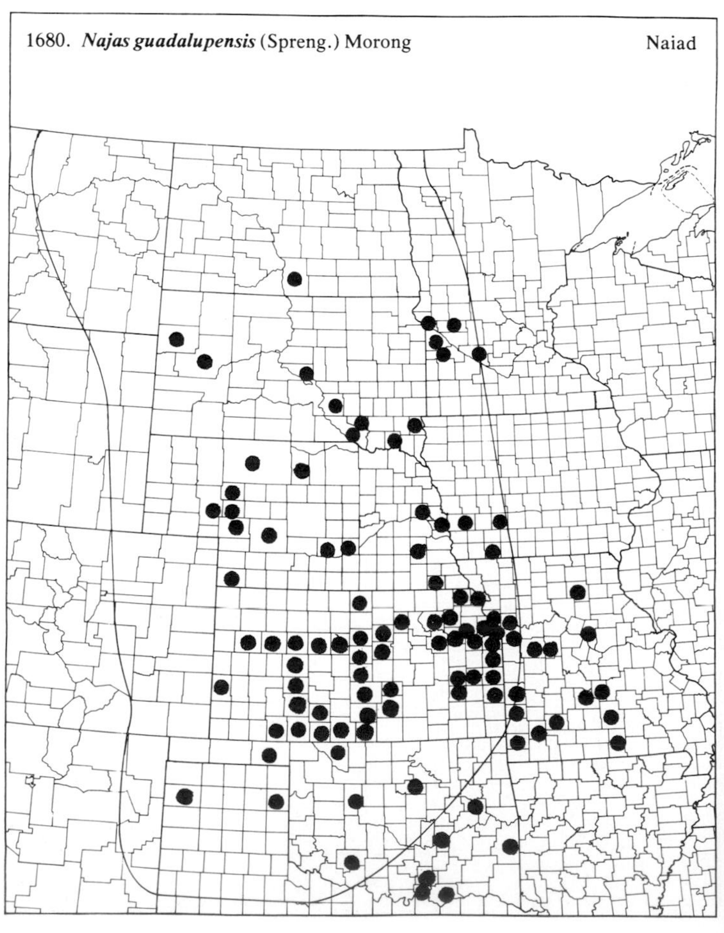

121. *Potamogetonaceae* (Pondweed Family)

1681. ***Potamogeton amplifolius*** Tuckerm.

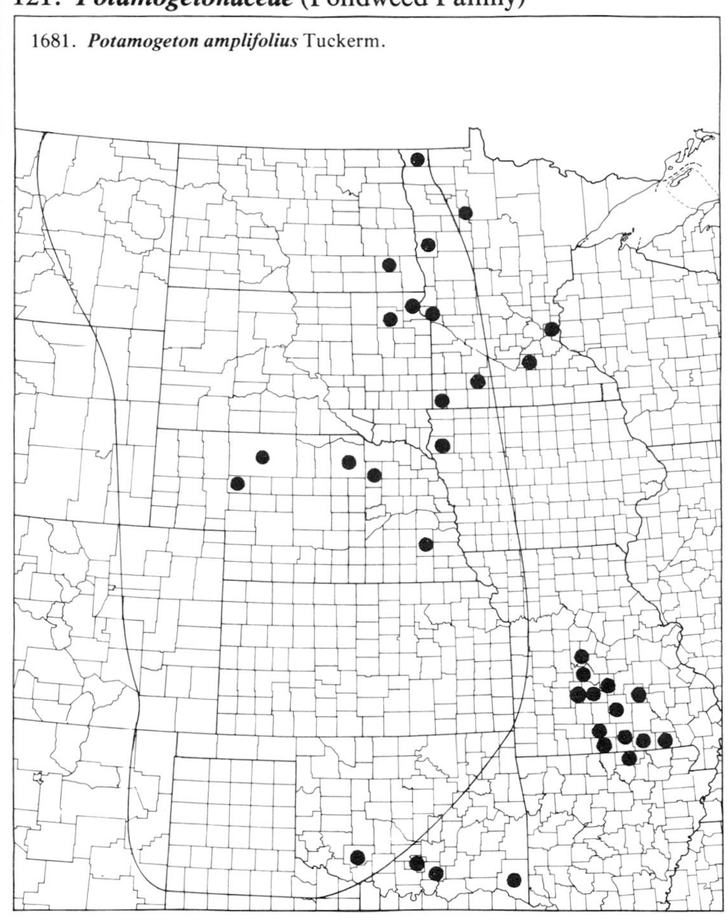

1682. ***Potamogeton crispus*** L.

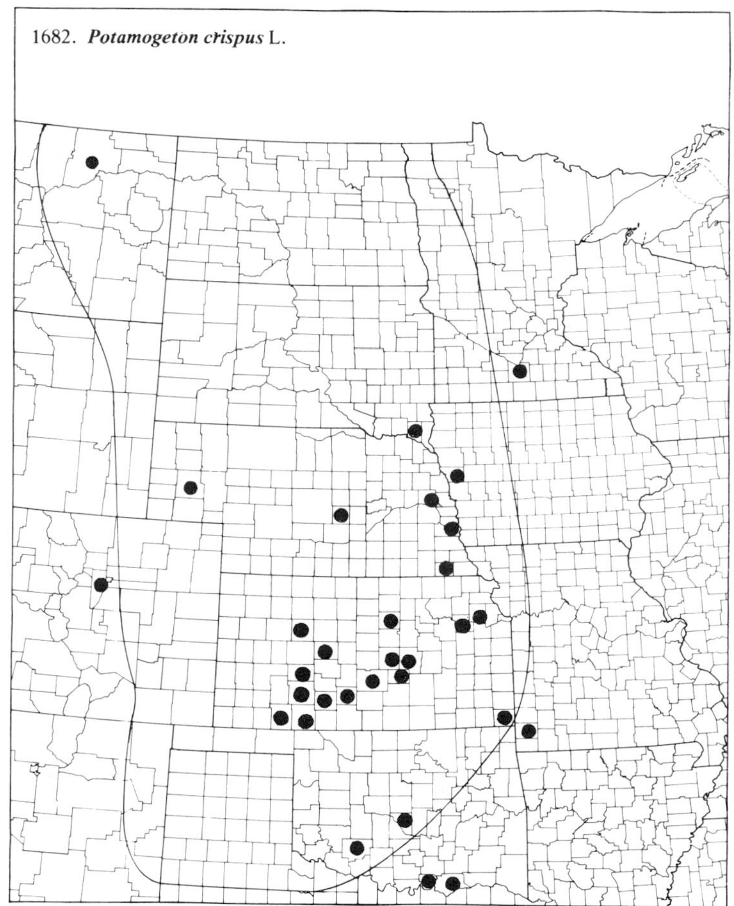

1683. ***Potamogeton diversifolius*** Raf. Water-thread Pondweed

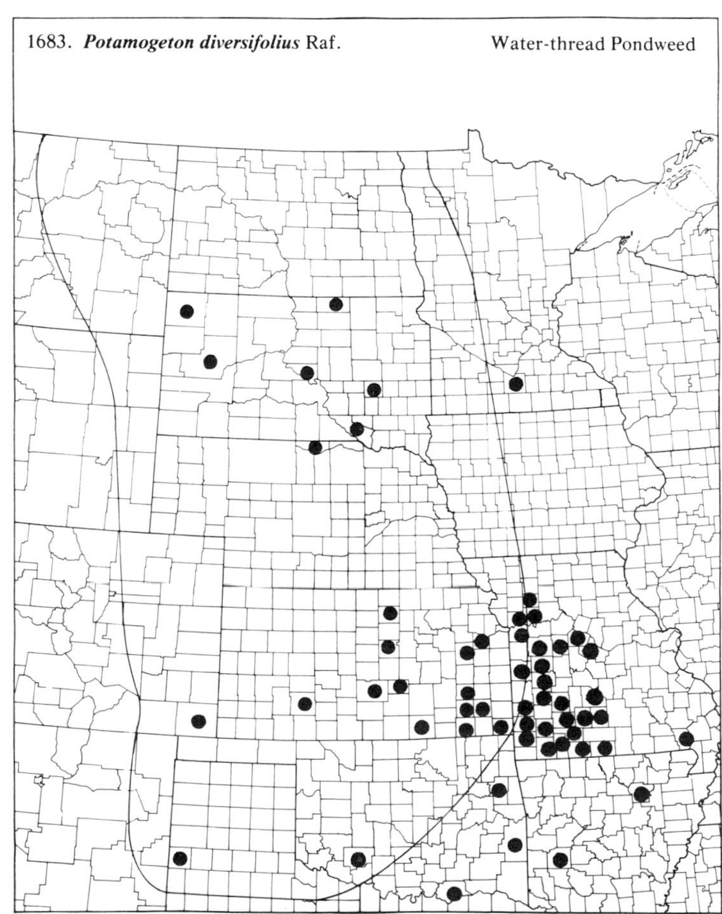

1684. ***Potamogeton filiformis*** Pers.

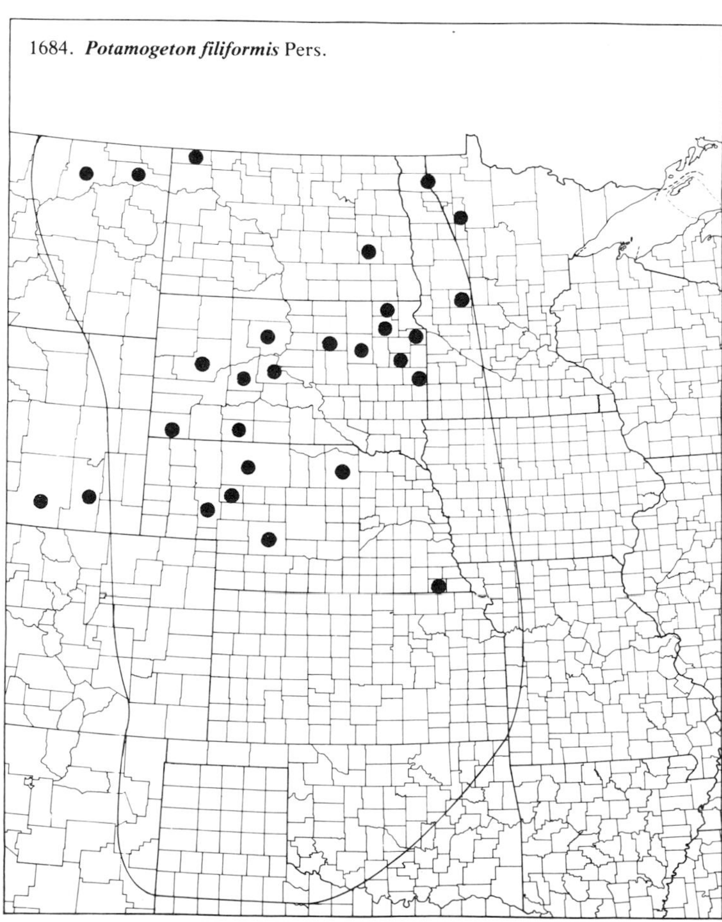

(Potamogetonaceae)

1685. ***Potamogeton foliosus*** Raf. Leafy Pondweed

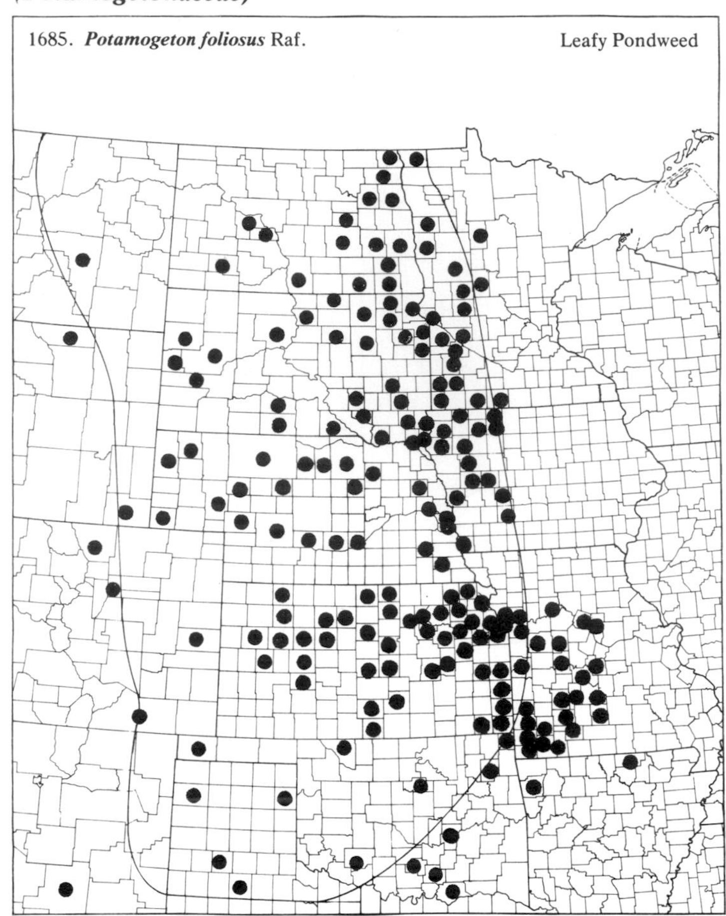

1686. ***Potamogeton friesii*** Rupr.

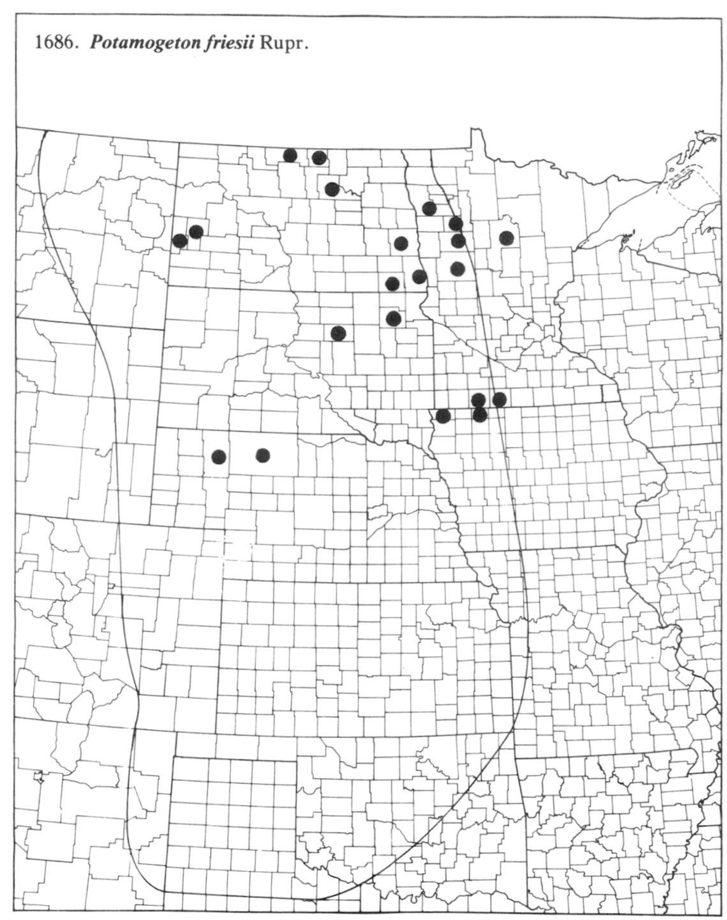

1687. ***Potamogeton gramineus*** L. Variable Pondweed

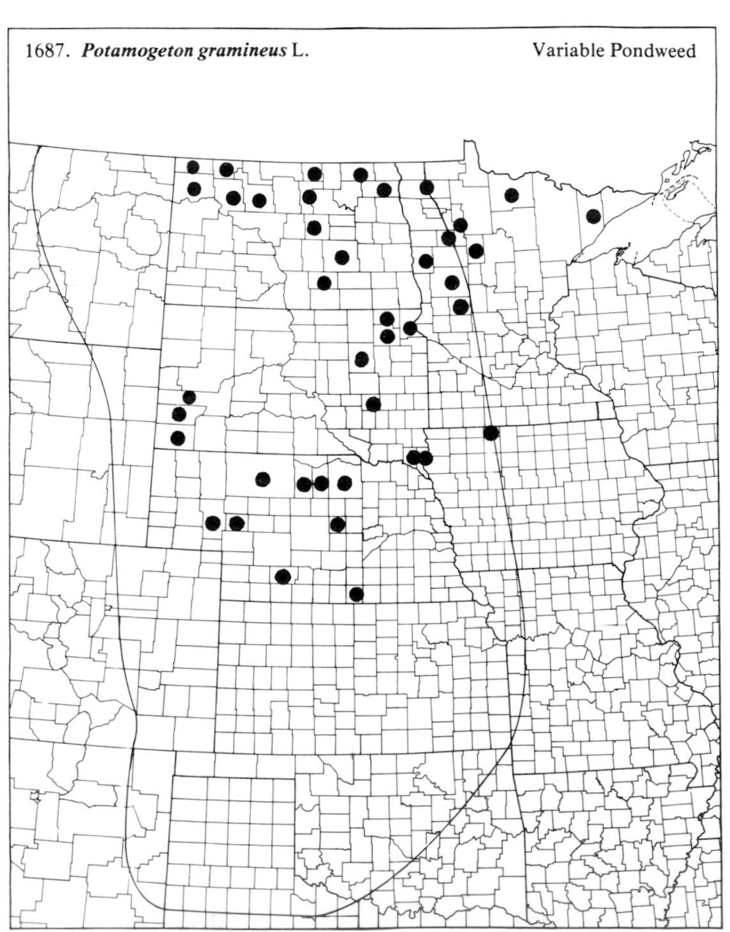

1688. ***Potamogeton illinoensis*** Morong Illinois Pondweed

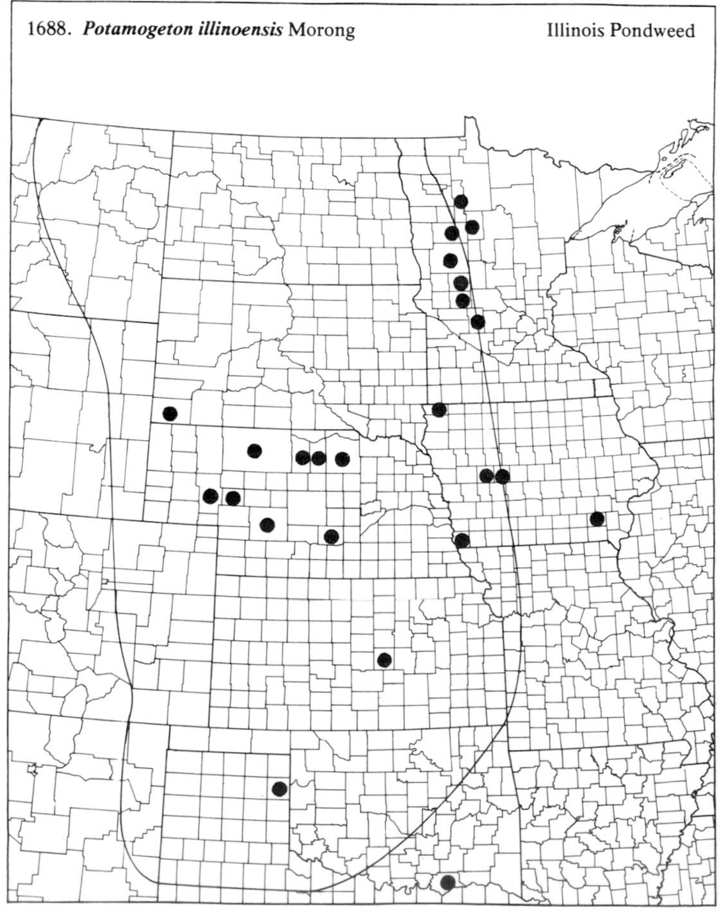

(Potamogetonaceae)

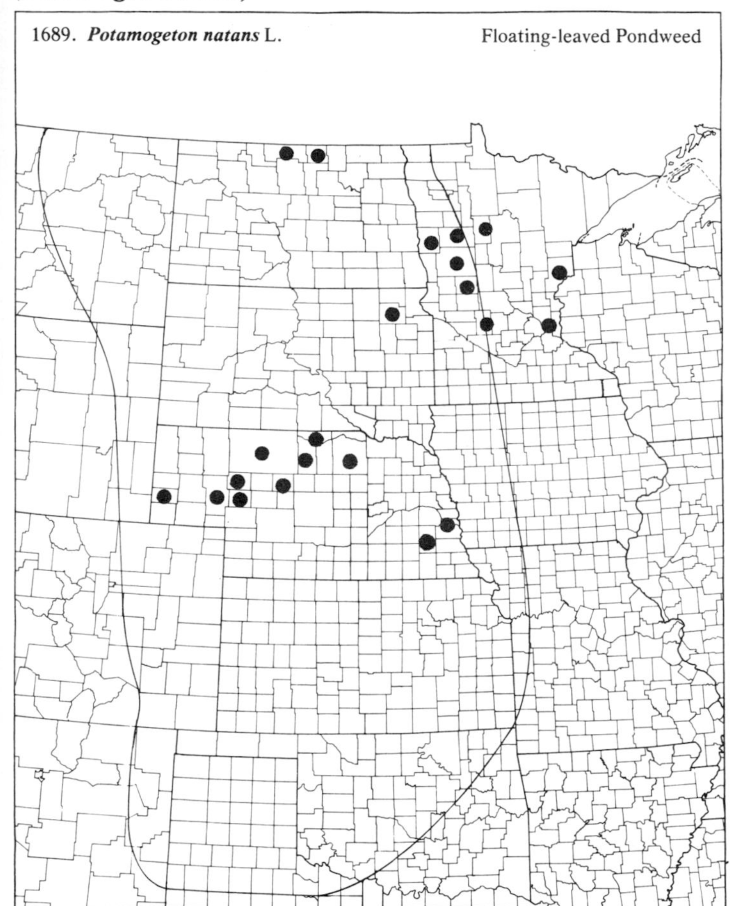
1689. ***Potamogeton natans*** L. Floating-leaved Pondweed

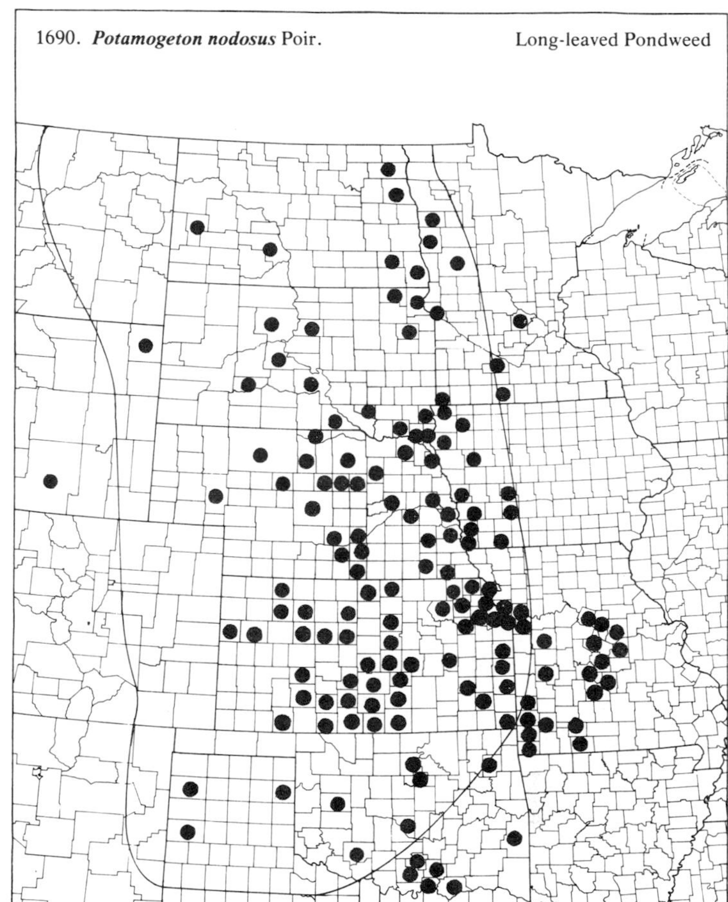
1690. ***Potamogeton nodosus*** Poir. Long-leaved Pondweed

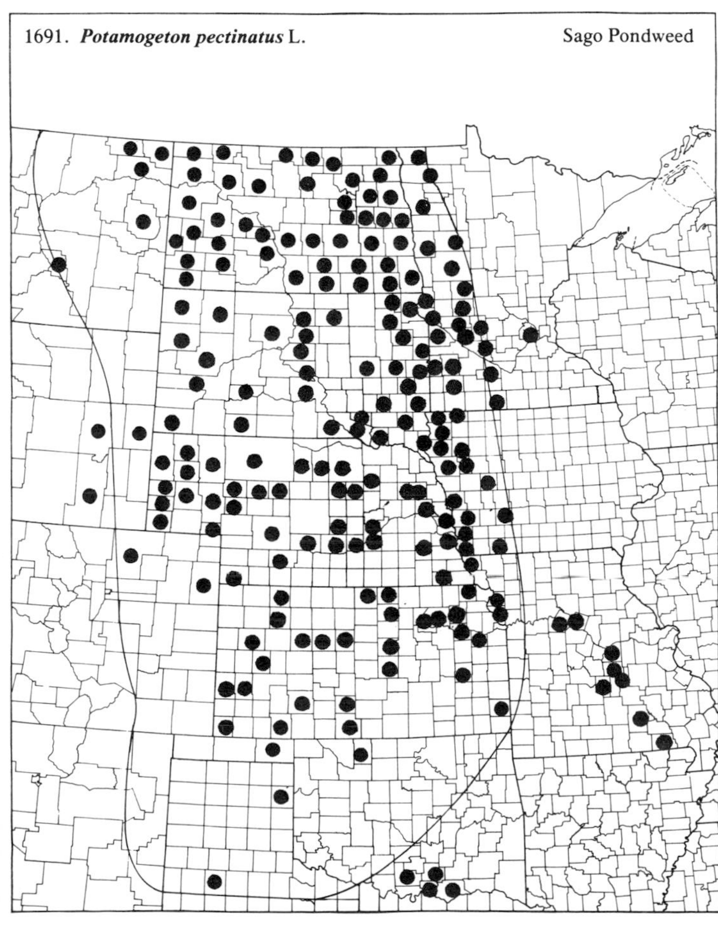
1691. ***Potamogeton pectinatus*** L. Sago Pondweed

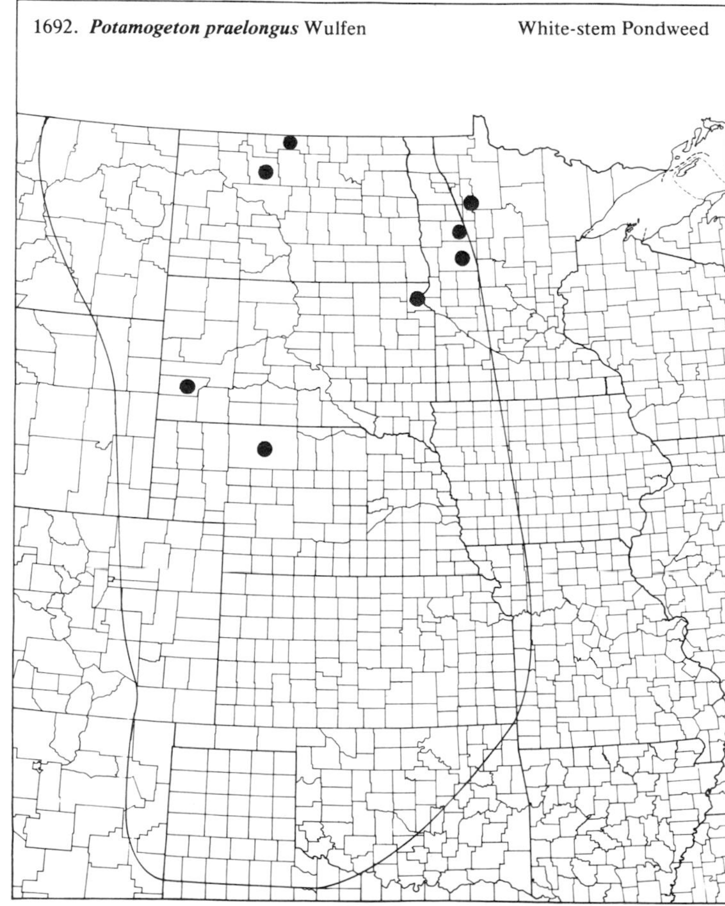
1692. ***Potamogeton praelongus*** Wulfen White-stem Pondweed

(Potamogetonaceae)

1693. ***Potamogeton pusillus*** L. var. ***pusillus*** — Baby Pondweed

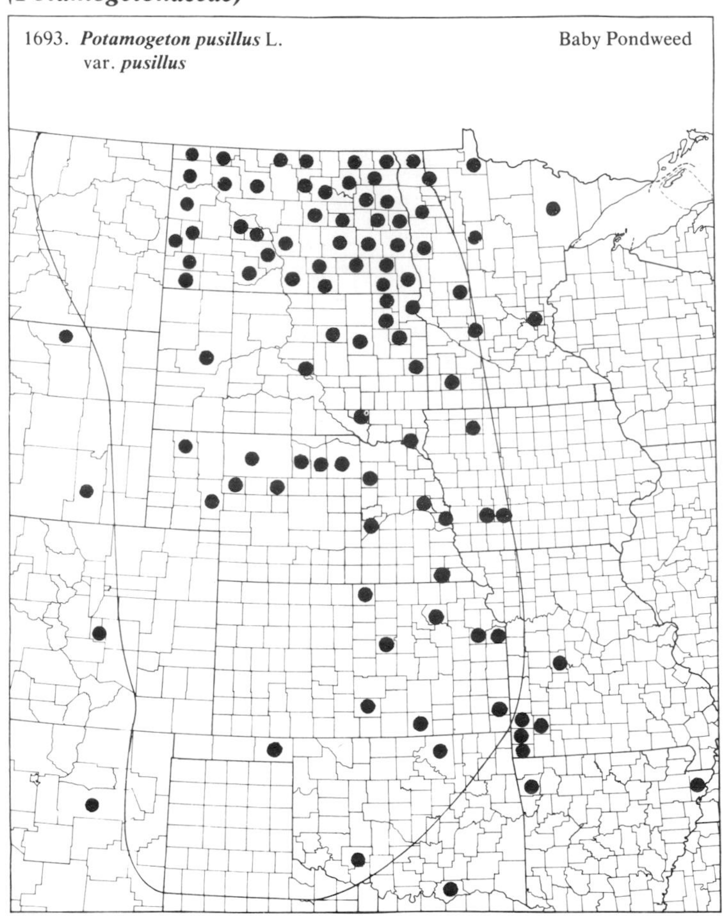

1694. ***Potamogeton richardsonii*** (Benn.) Rydb. — Red-head Pondweed

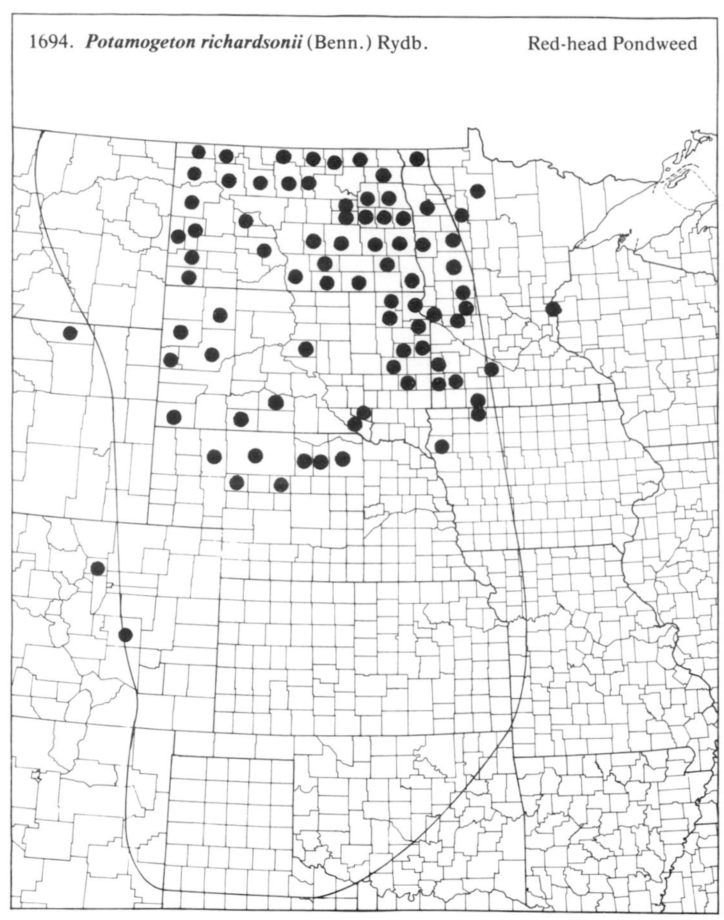

1695. ***Potamogeton strictifolius*** Benn.

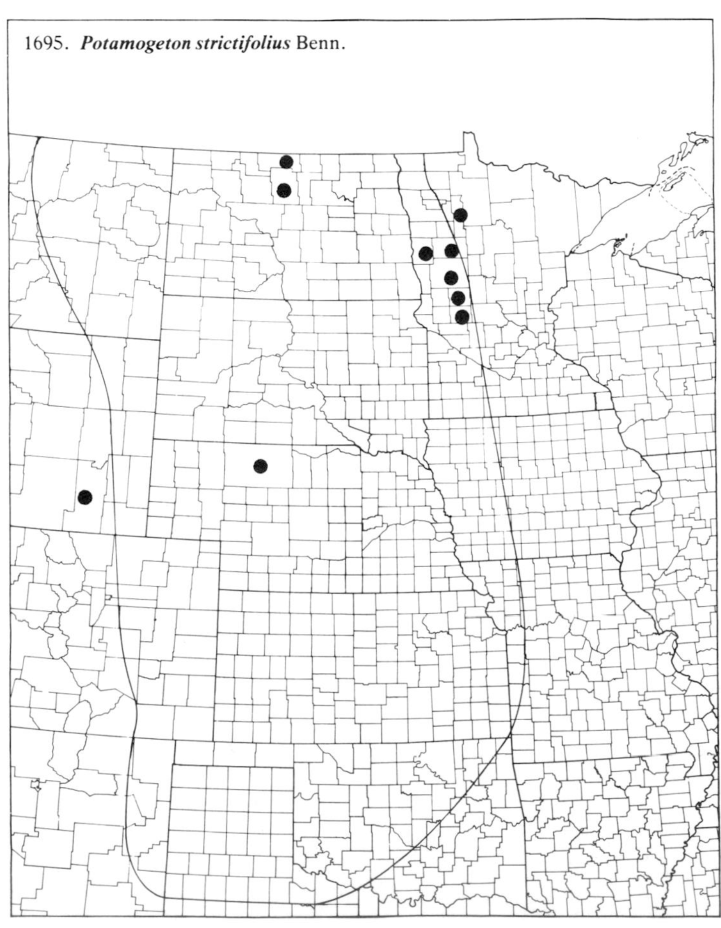

1696. ***Potamogeton zosteriformis*** Fern. — Flatstem Pondweed

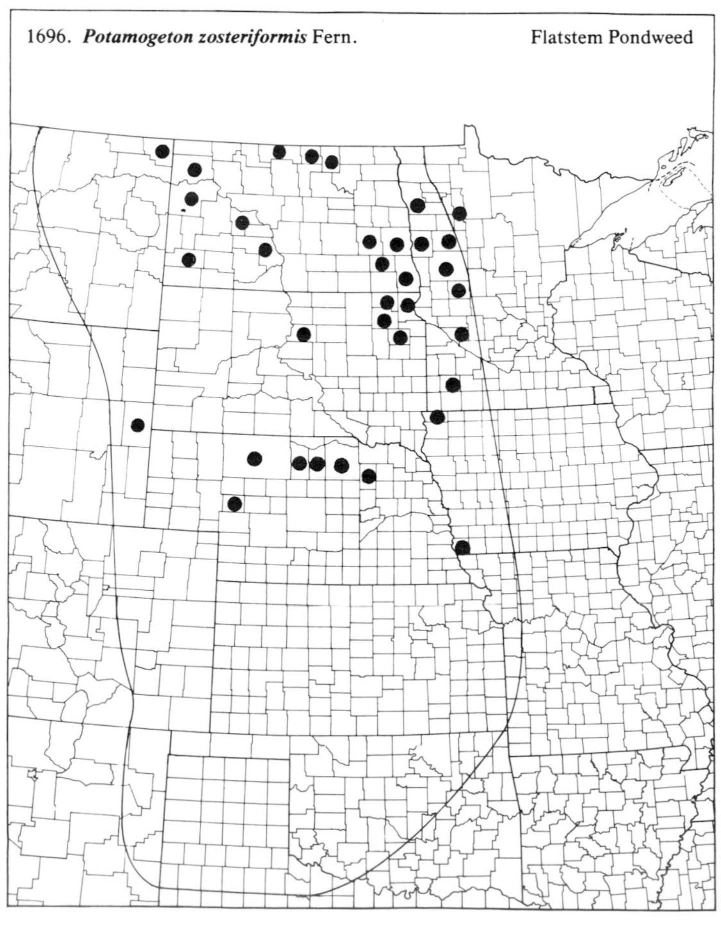

122. *Ruppiaceae* (Ditchgrass Family)

1697. *Ruppia maritima* L. Ditchgrass

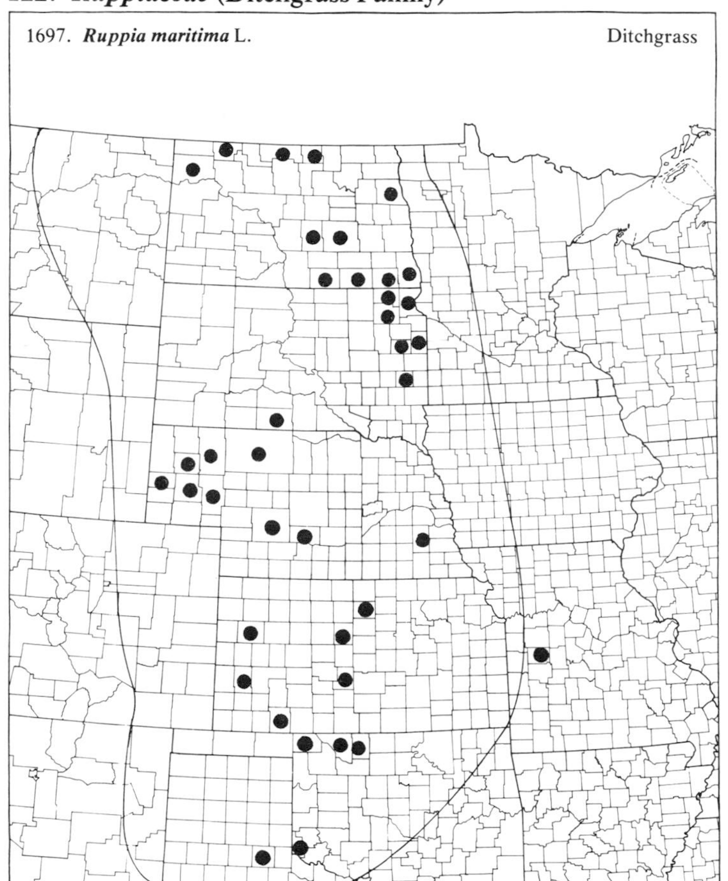

123. *Zannichelliaceae* (Horned Pondweed Family)

1698. *Zannichellia palustris* L. Horned Pondweed

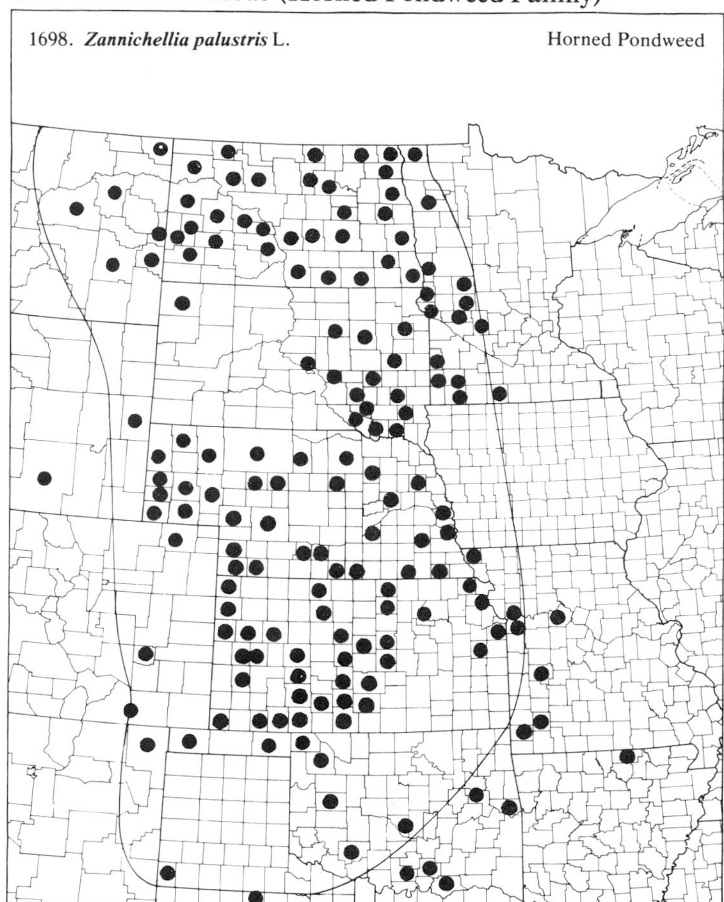

124. *Commelinaceae* (Spiderwort Family)

1699. *Commelina communis* L. Dayflower

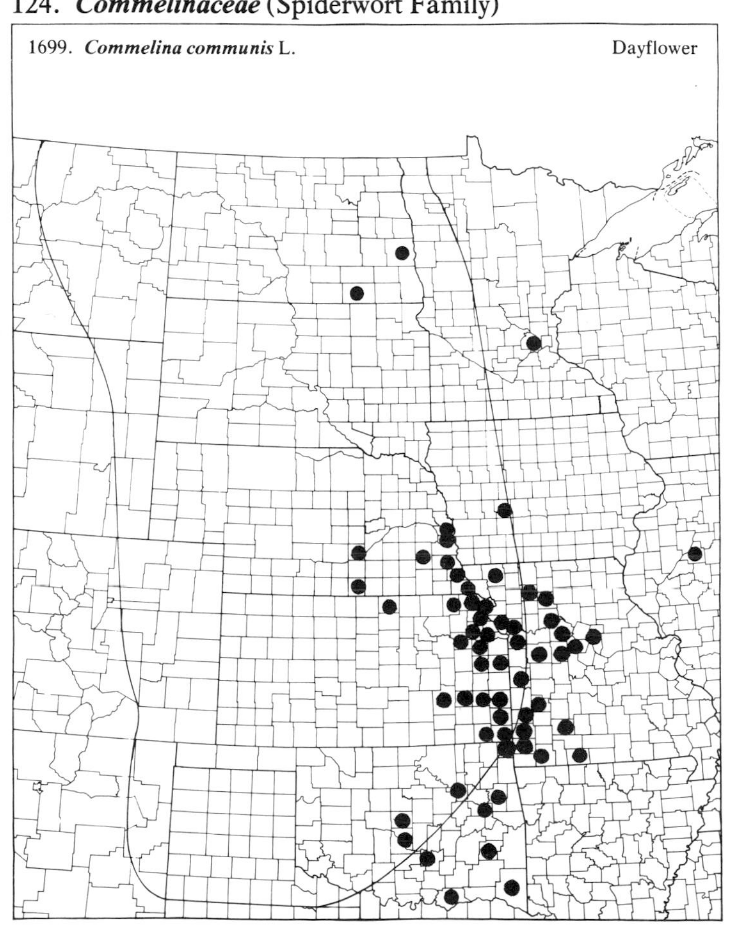

1700. *Commelina diffusa* Burm. f. Creeping Dayflower

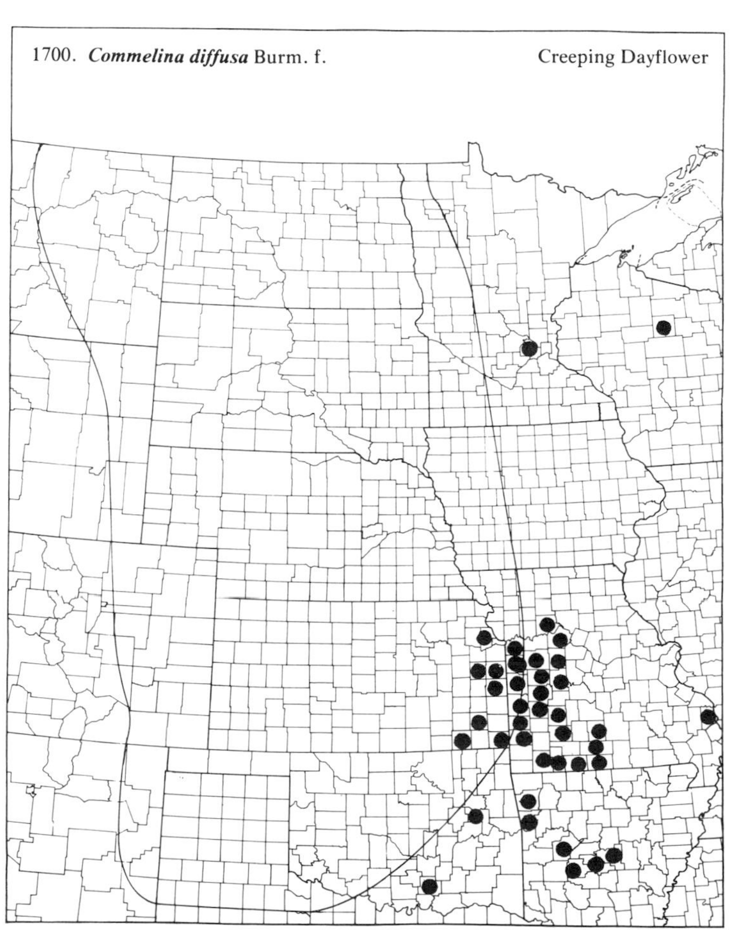

(Commelinaceae)

1701. ***Commelina erecta*** L. var. ***erecta*** — Erect Dayflower

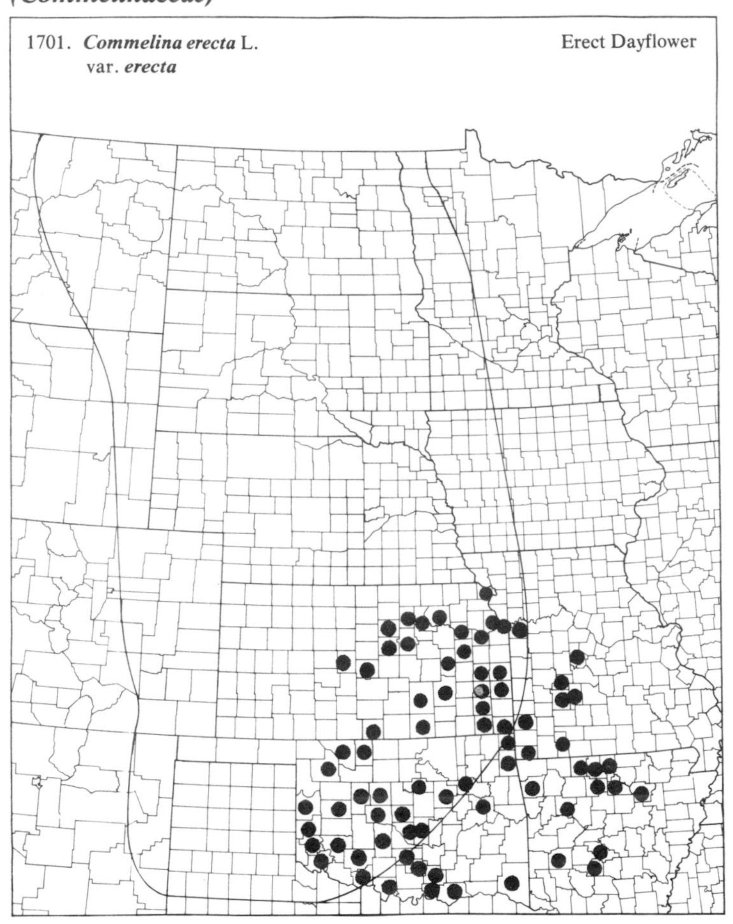

1702. ***Commelina erecta*** L. var. ***angustifolia*** (Michx.) Fern. — Erect Dayflower

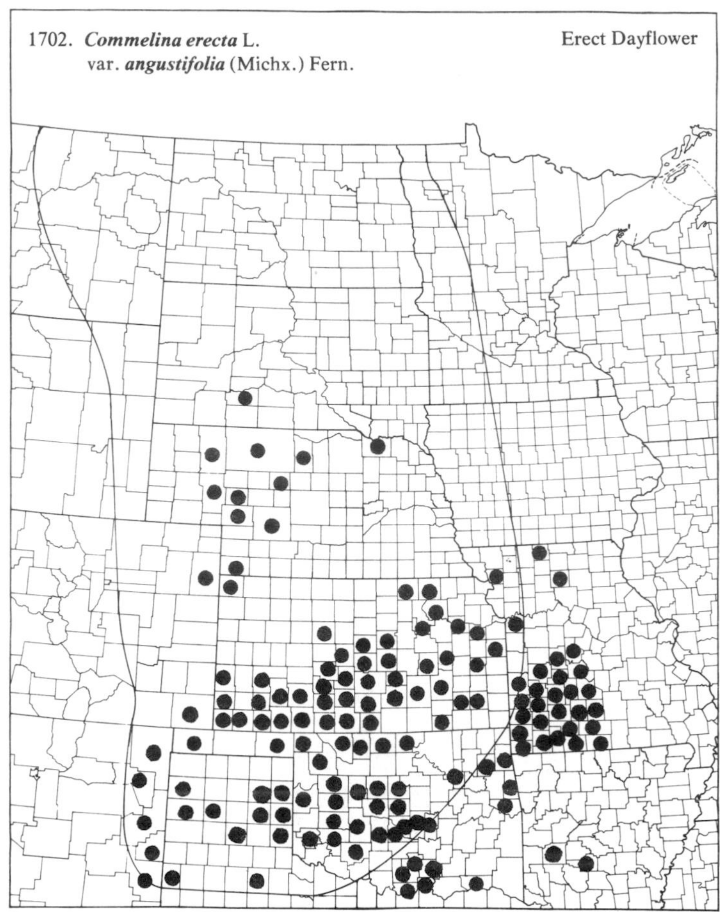

1703. ***Commelina virginica*** L.

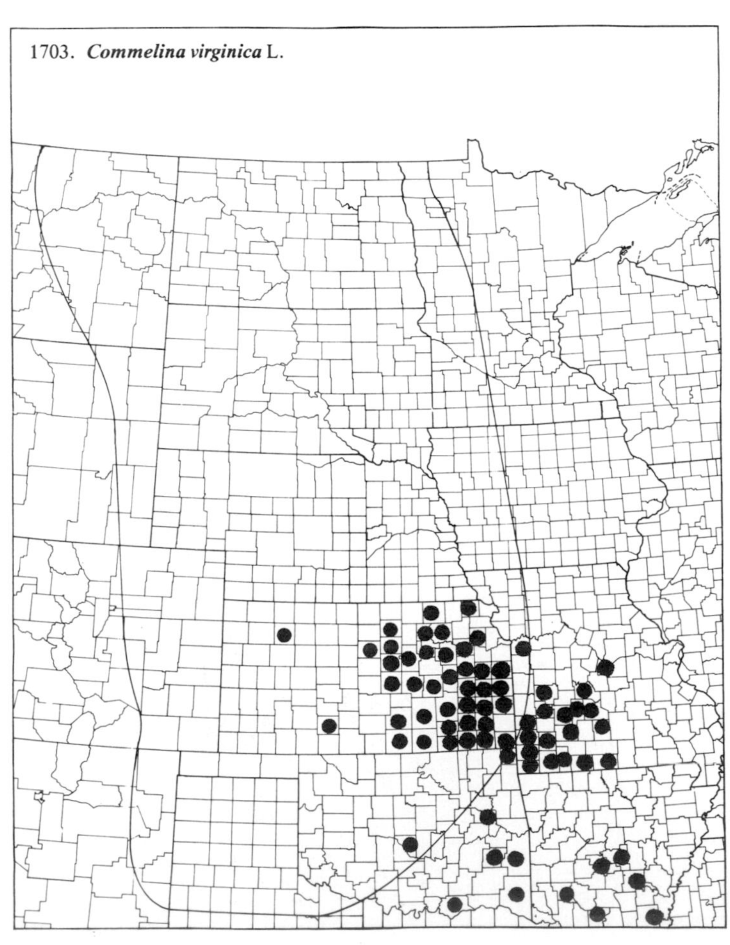

1704. ***Tradescantia bracteata*** Small — Spiderwort

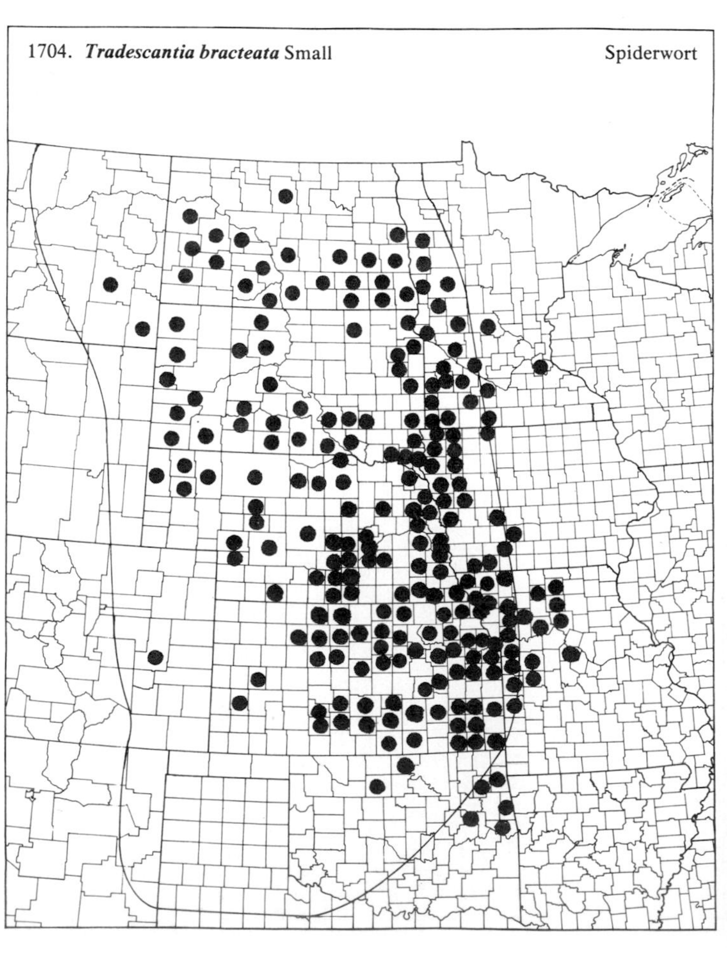

(Commelinaceae)

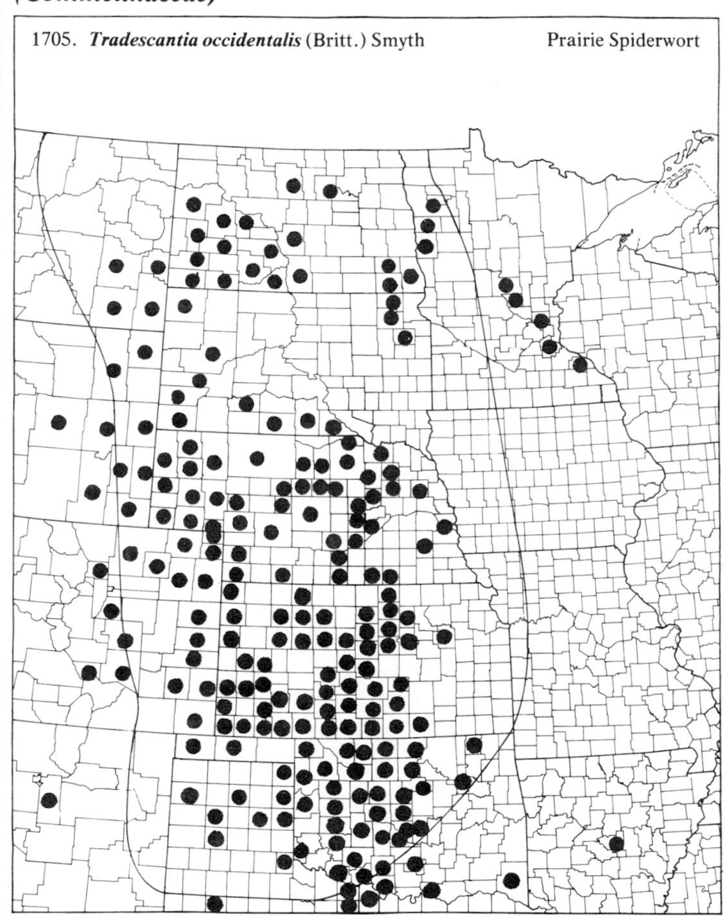

1705. ***Tradescantia occidentalis*** (Britt.) Smyth — Prairie Spiderwort

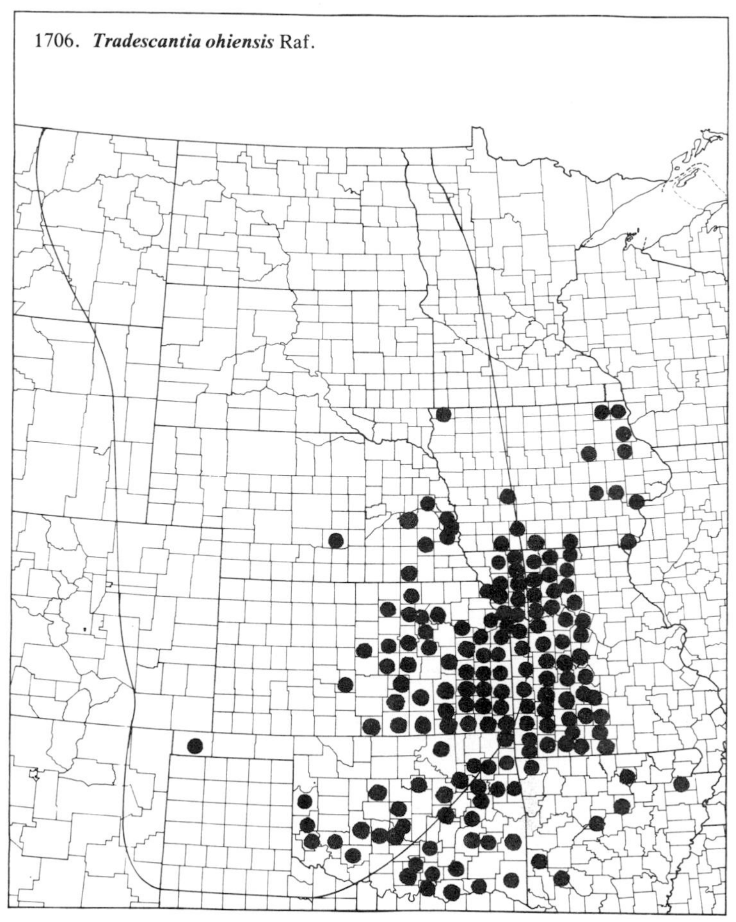

1706. ***Tradescantia ohiensis*** Raf.

1707. ***Tradescantia tharpii*** Anders. & Woods. — Tharp Spiderwort

125. *Juncaceae* (Rush Family)

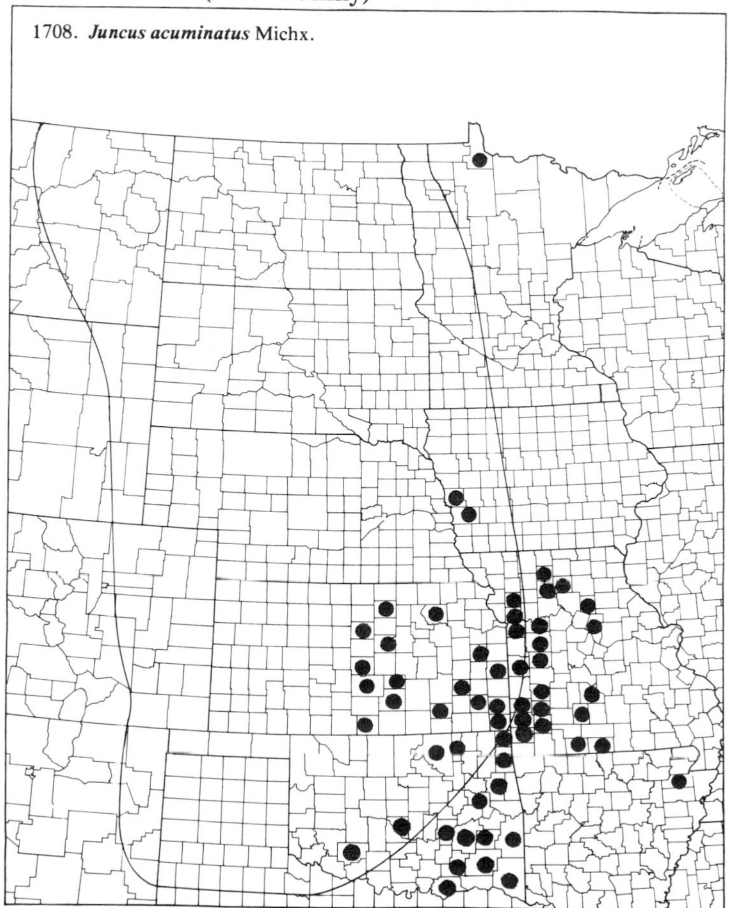

1708. ***Juncus acuminatus*** Michx.

(Juncaceae)

1709. ***Juncus alpinus*** Vill.

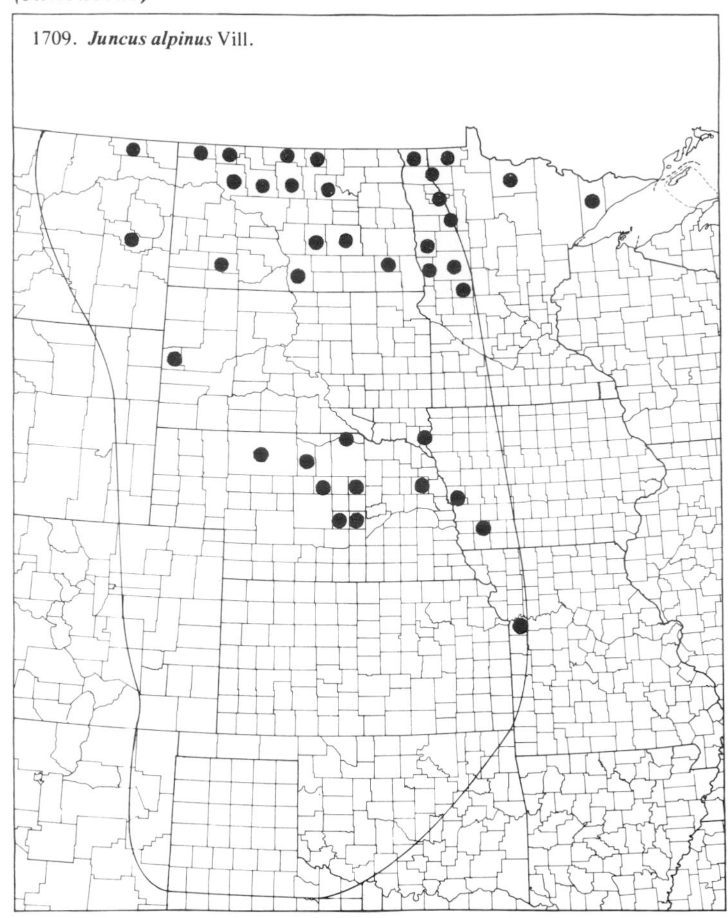

1710. ***Juncus balticus*** Willd. Baltic Rush

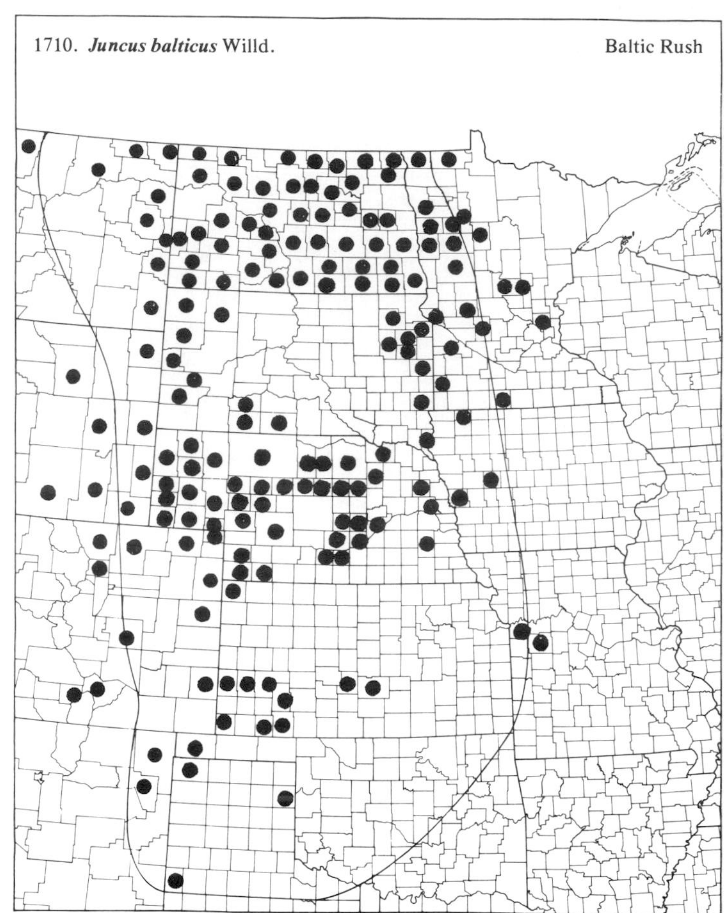

1711. ***Juncus brachyphyllus*** Wieg. Small-headed Rush

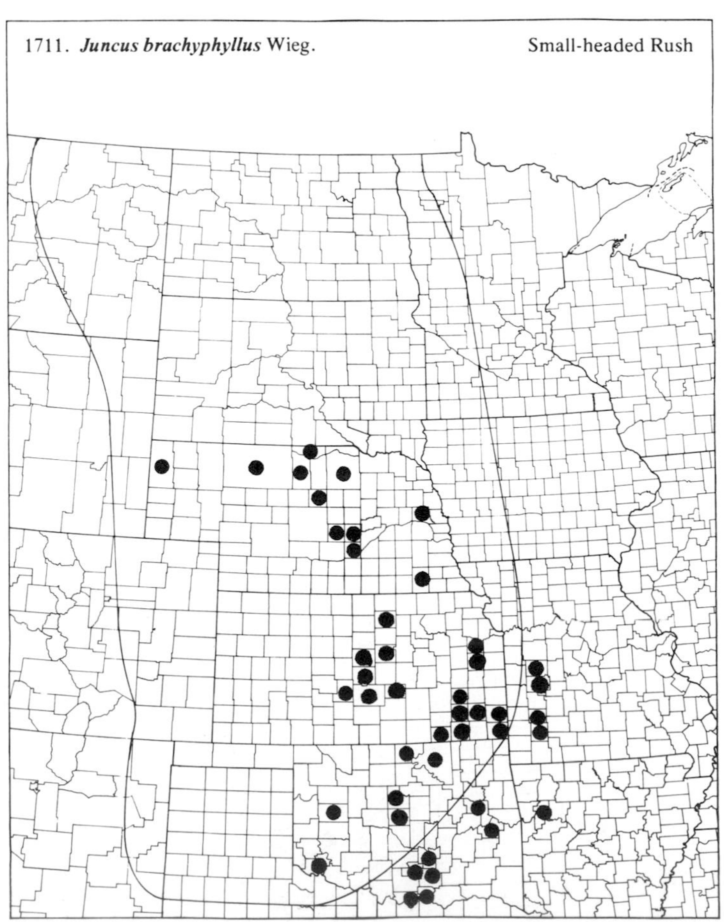

1712. ***Juncus bufonius*** L. Toad Rush

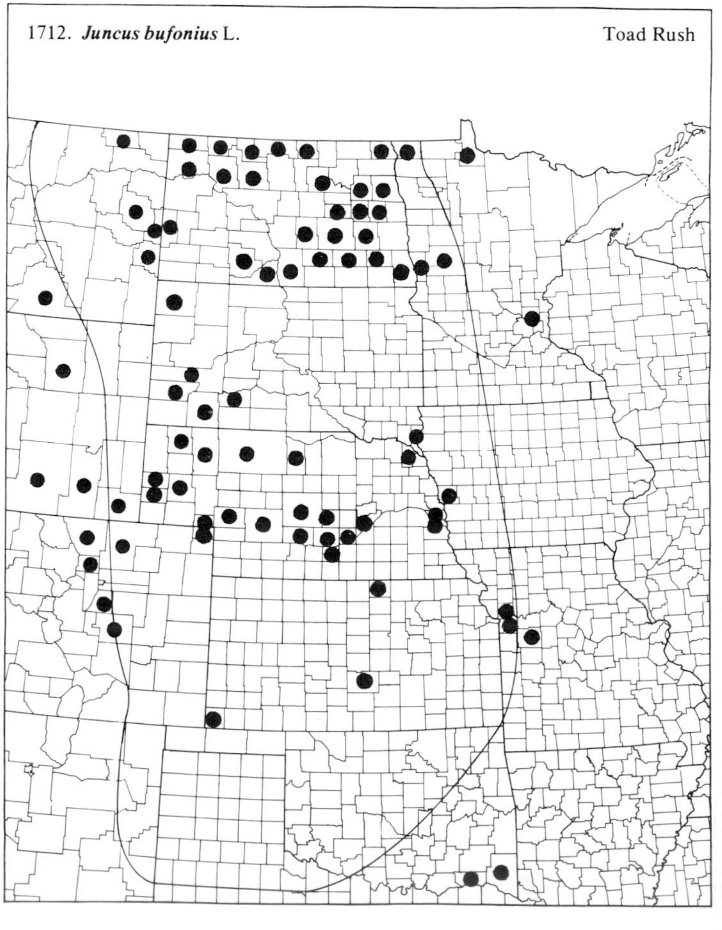

(Juncaceae)

1713. ***Juncus crassifolius*** Buch.

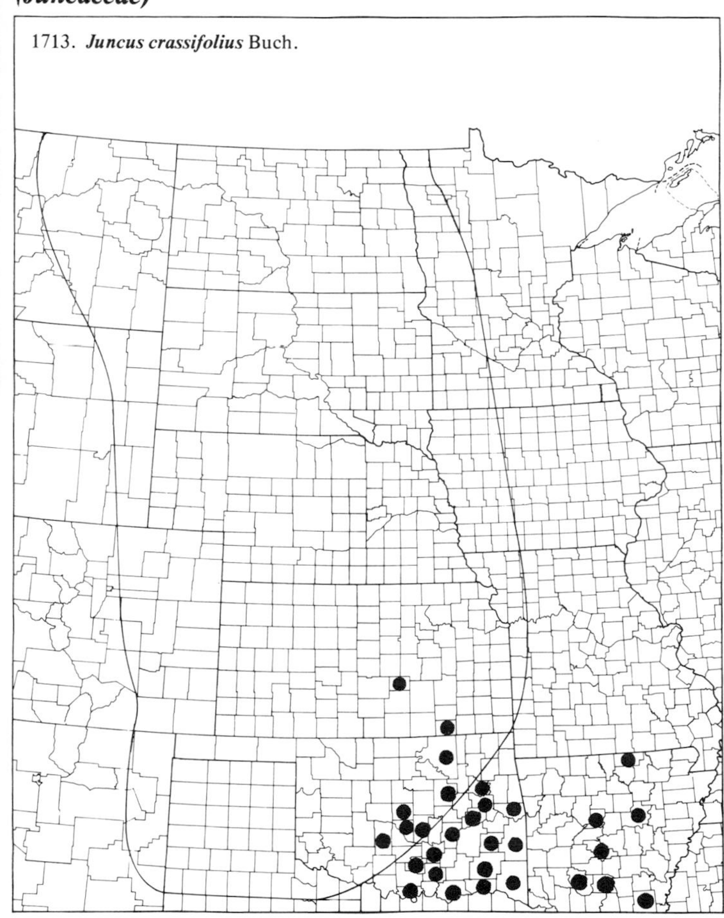

1714. ***Juncus diffusissimus*** Buckl. Slimpod Rush

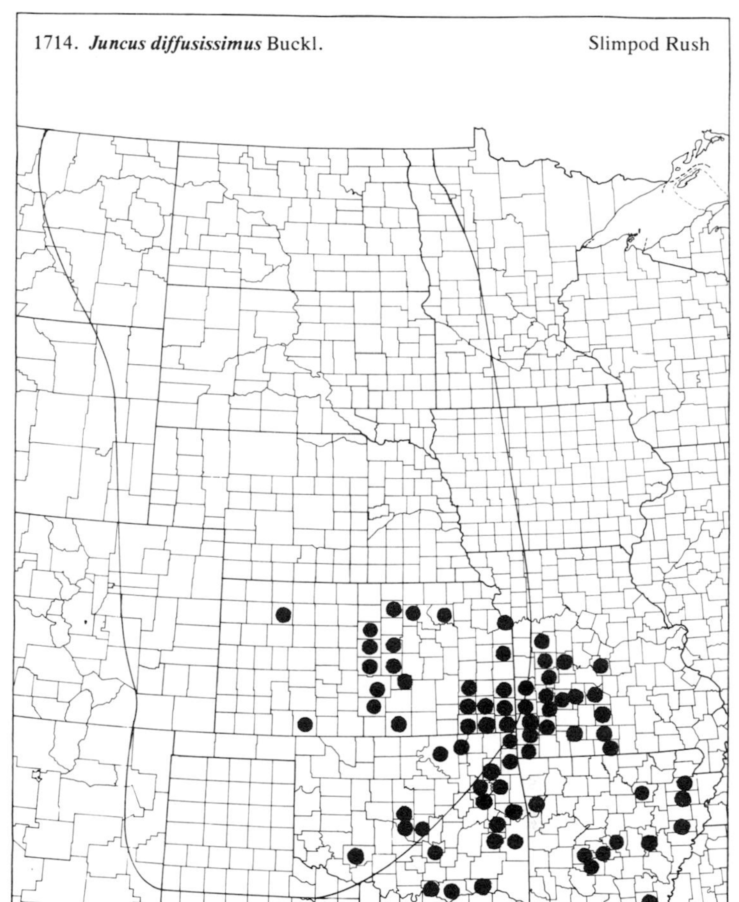

1715. ***Juncus dudleyi*** Wieg. Dudley Rush

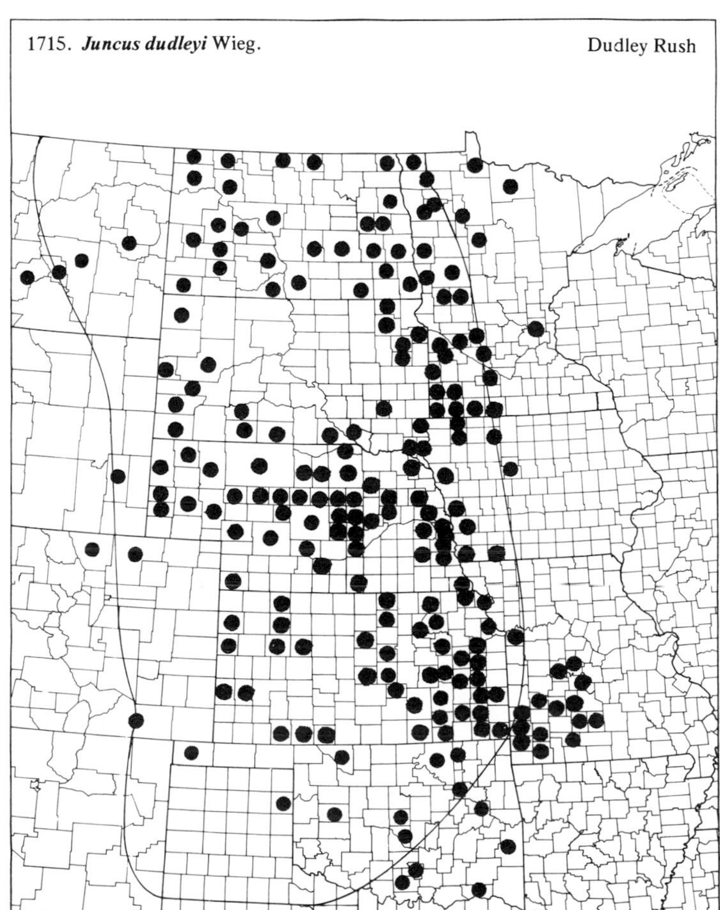

1716. ***Juncus effusus*** L. var. ***solutus*** Fern. & Wieg. Bog Rush

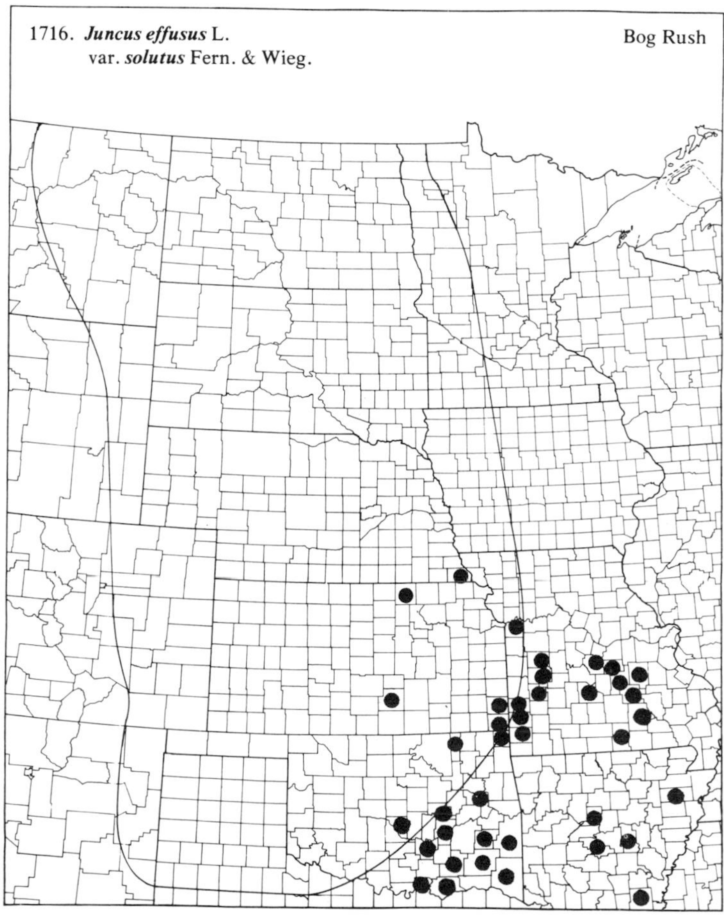

(Juncaceae)

1717. ***Juncus interior*** Wieg. Inland Rush

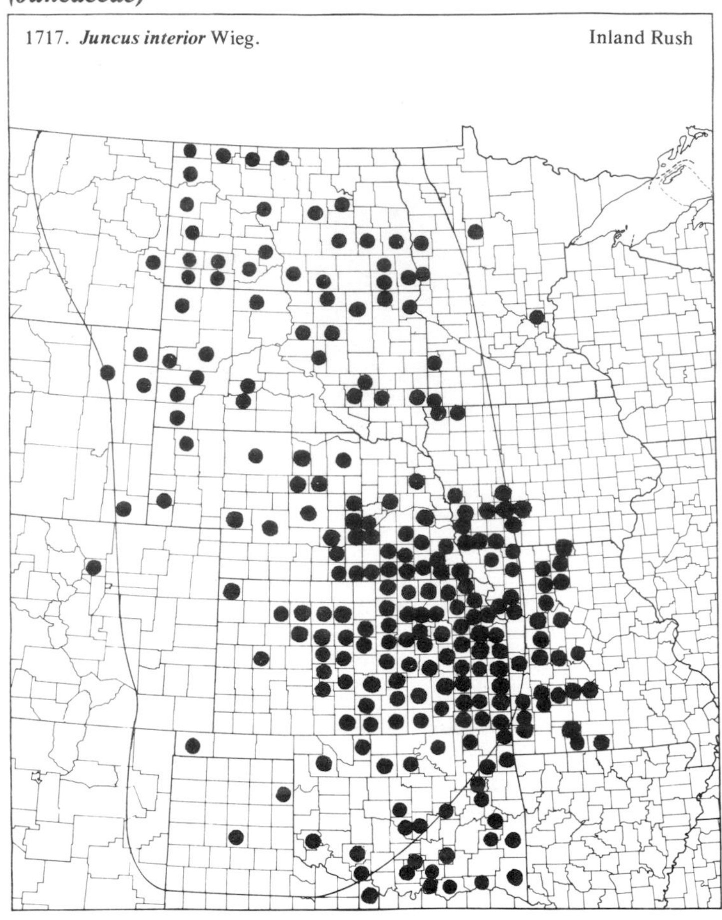

1718. ***Juncus longistylis*** Torr.

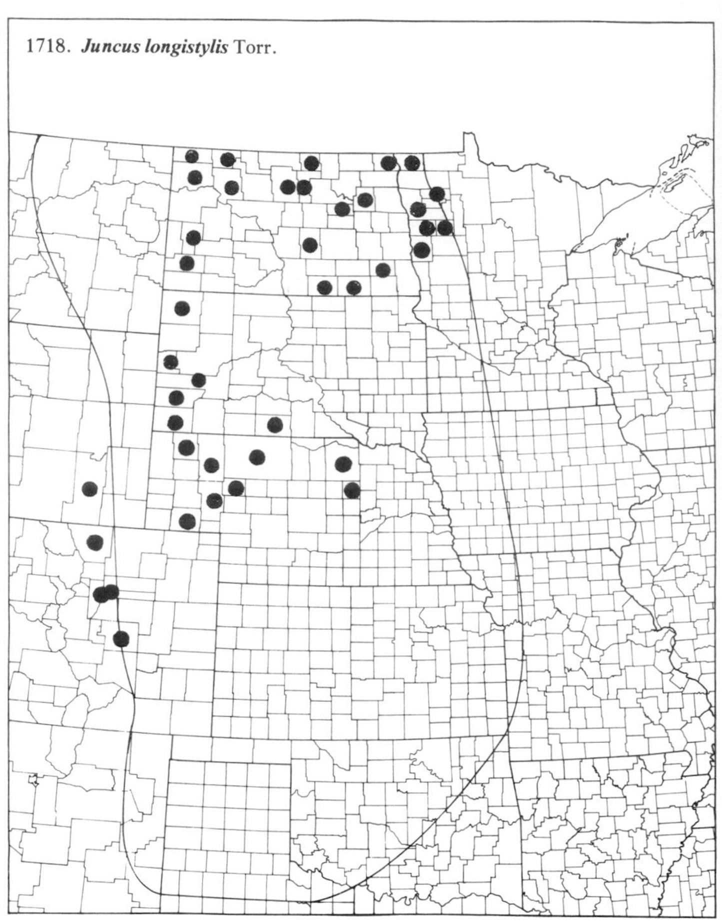

1719. ***Juncus marginatus*** Rostk. Grassleaf Rush

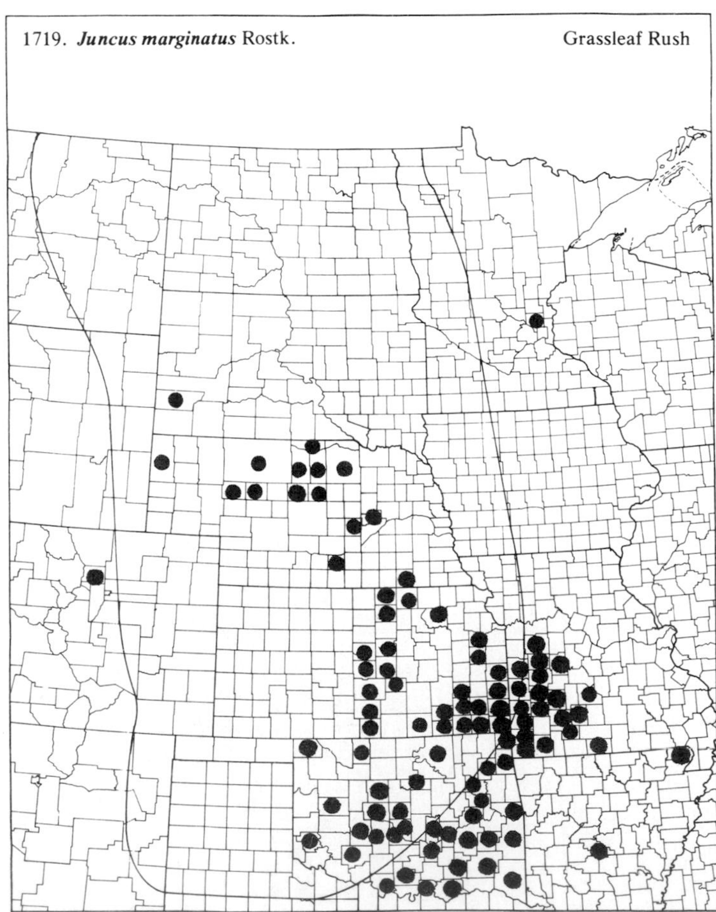

1720. ***Juncus nodatus*** Cov. Stout Rush

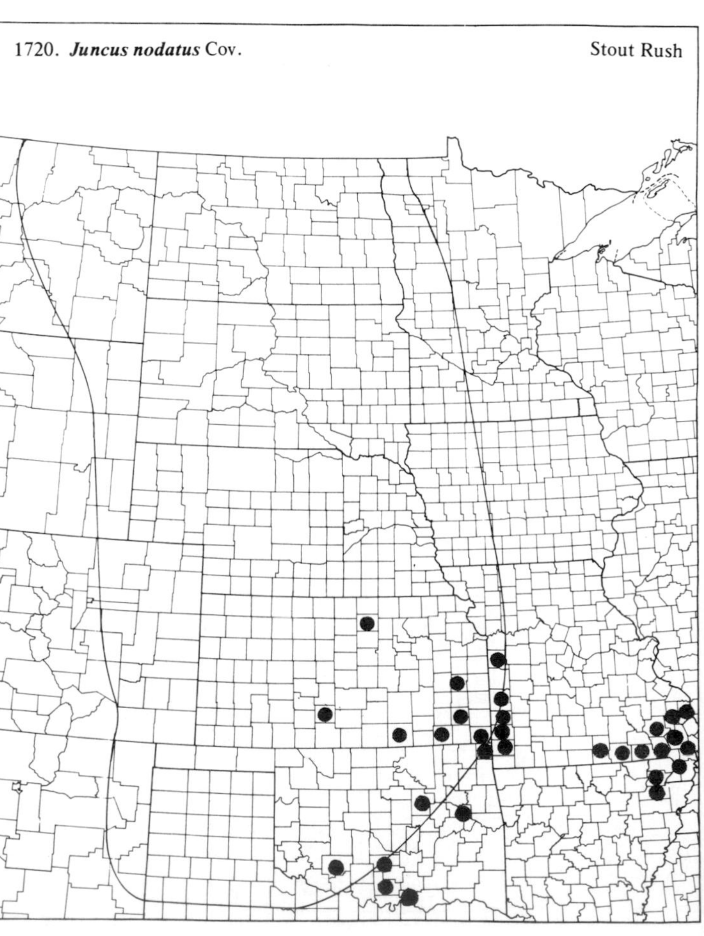

(Juncaceae)

1721. ***Juncus nodosus*** L. Knotted Rush

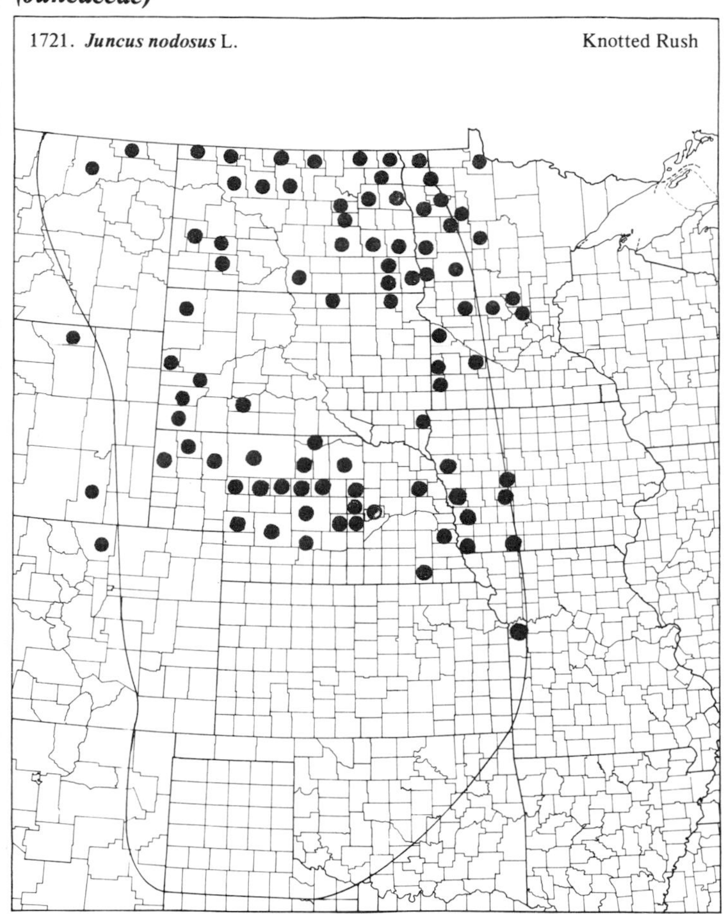

1722. ***Juncus scirpoides*** Lam.

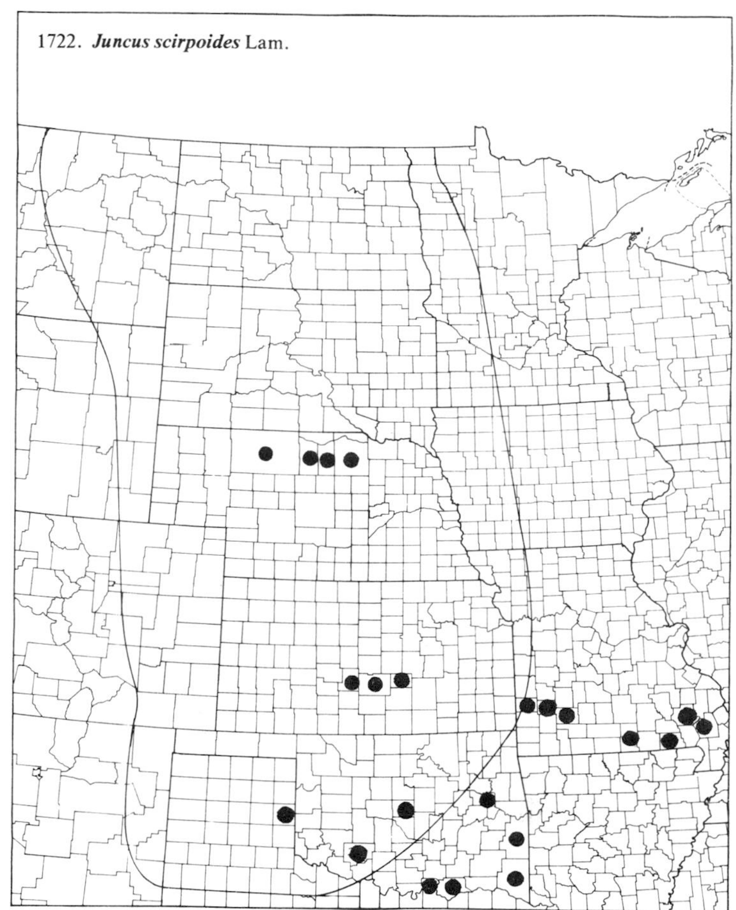

1723. ***Juncus tenuis*** Willd. Path Rush

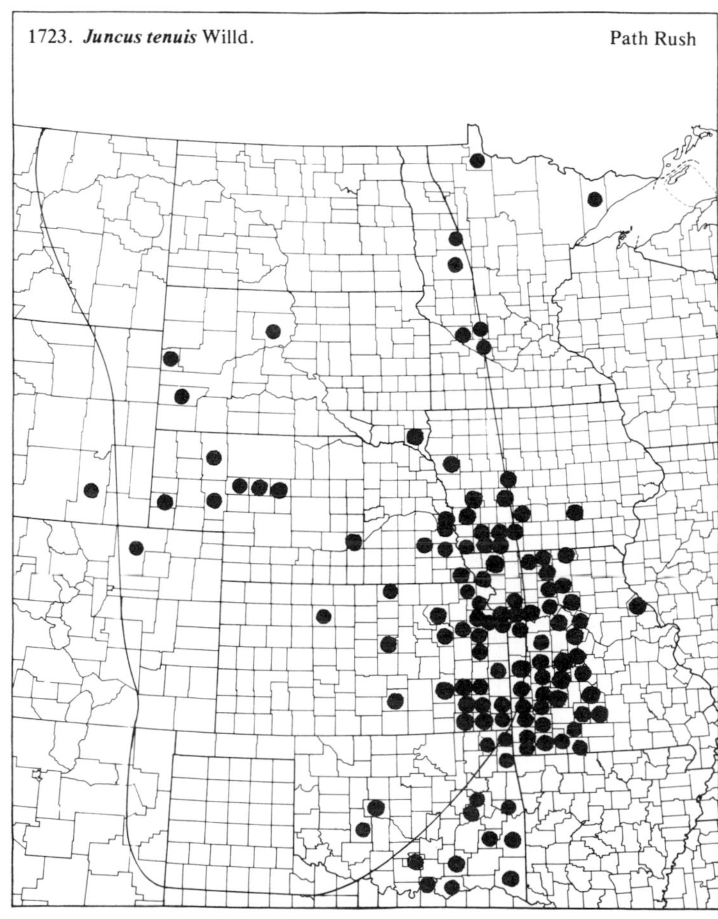

1724. ***Juncus torreyi*** Cov. Torrey's Rush

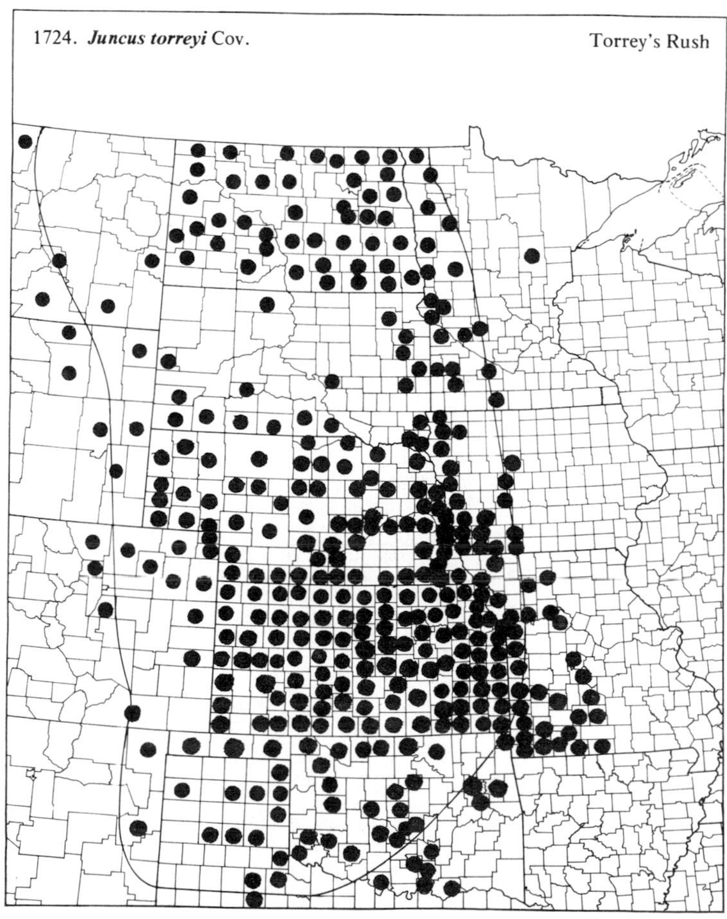

126. *Cyperaceae* (Sedge Family)

1725. ***Bulbostylis capillaris*** (L.) Clarke

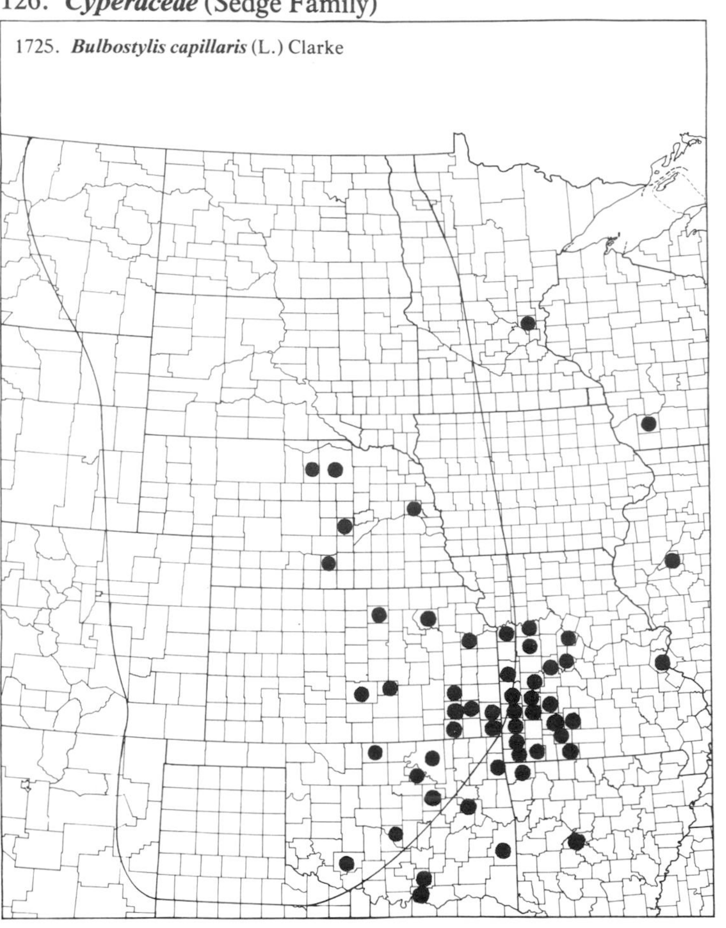

1726. ***Carex aenea*** Fern.

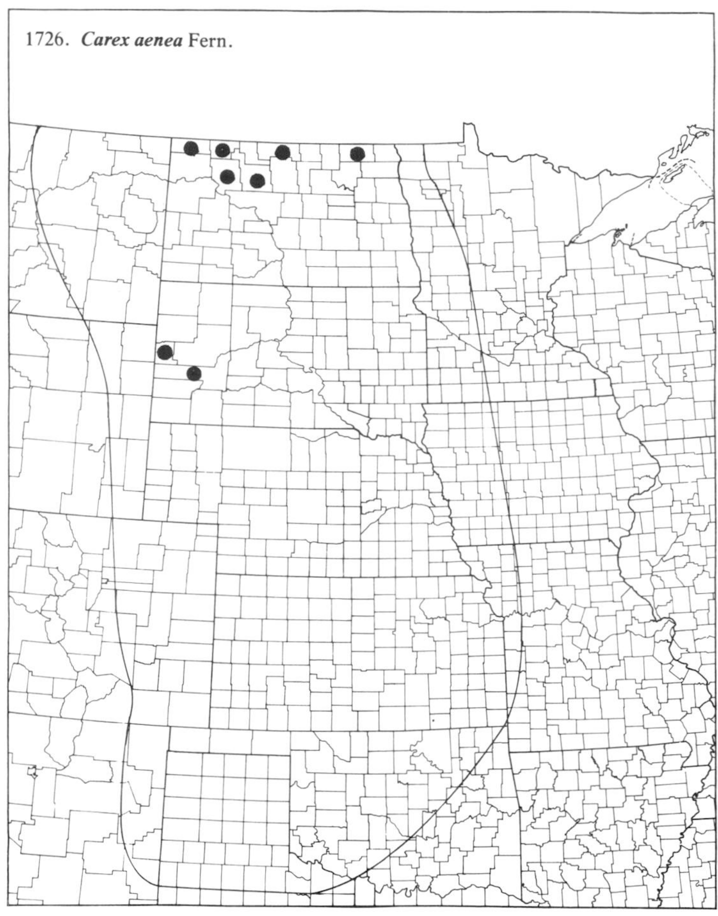

1727. ***Carex aggregata*** Mack. Glomerate Sedge

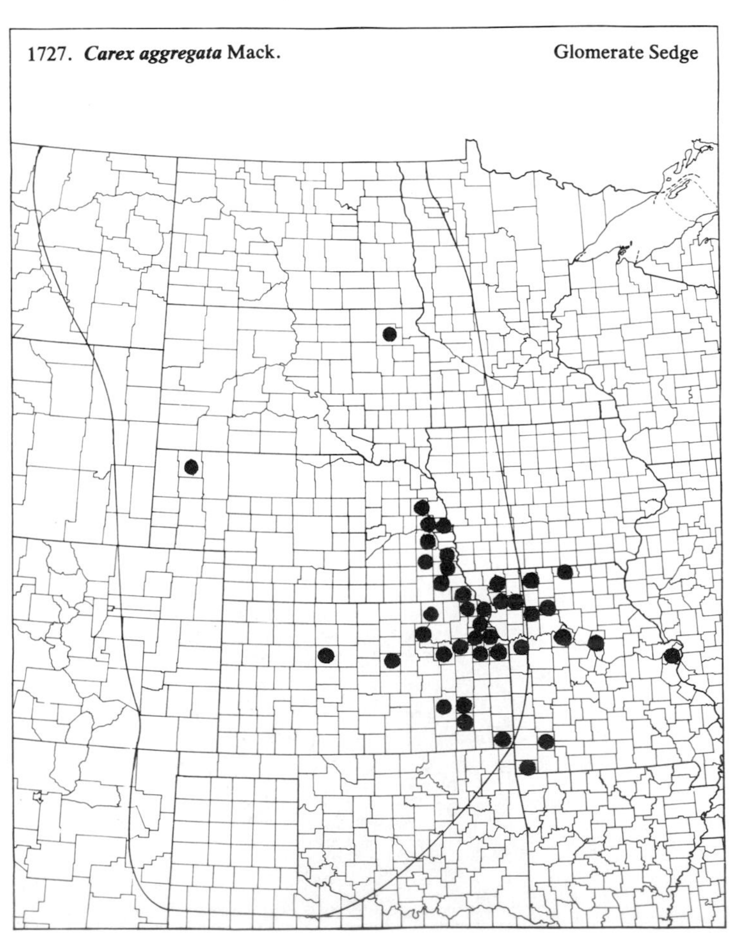

1728. ***Carex amphibola*** Steud. var. ***turgida*** Fern. Narrowleaf Sedge

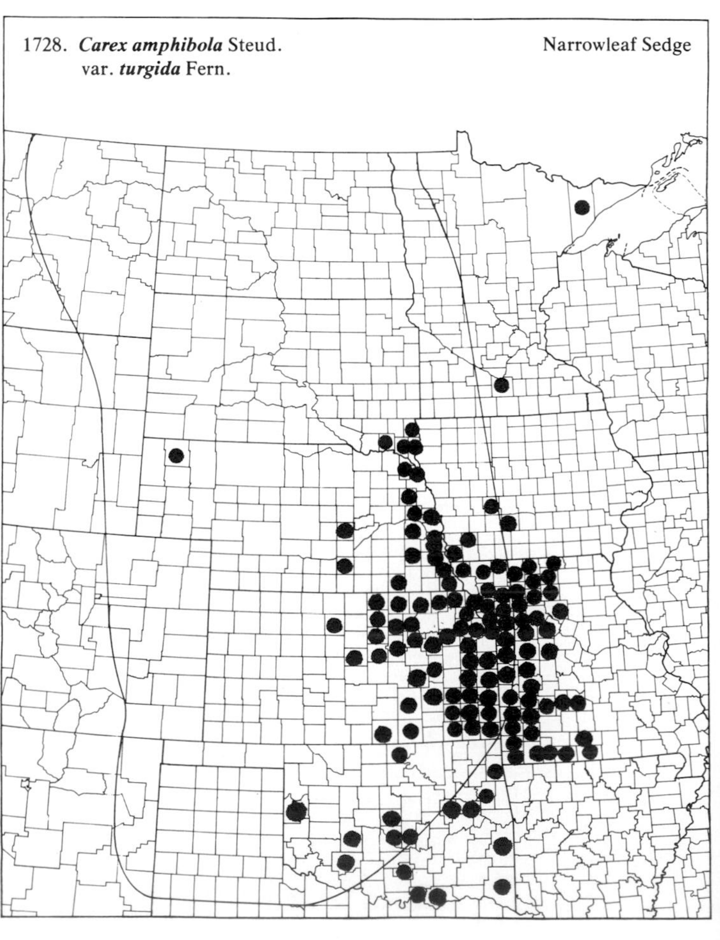

(Cyperaceae)

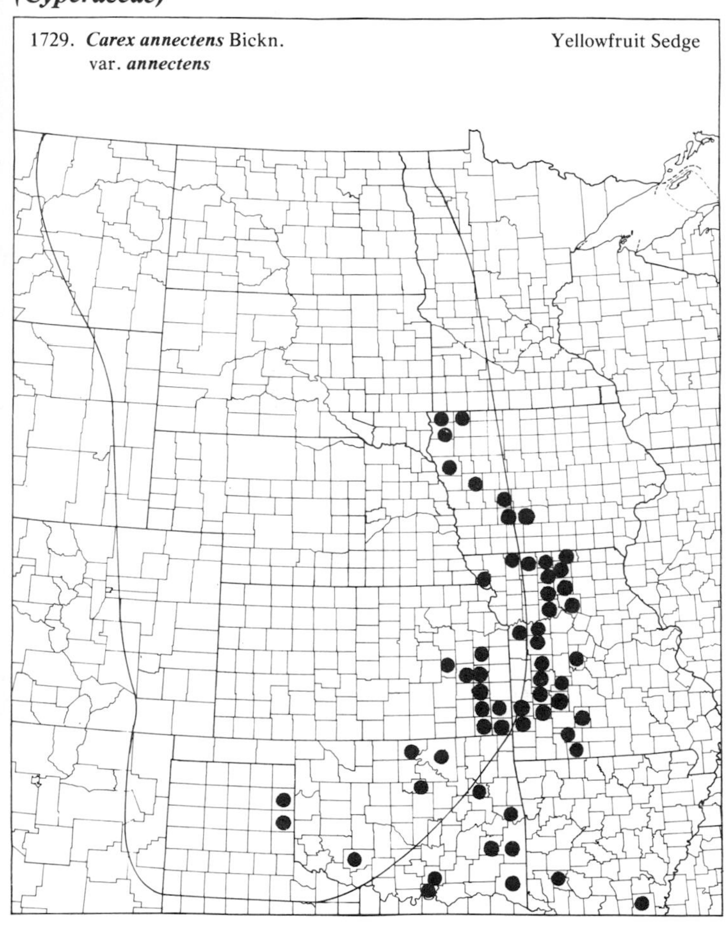
1729. *Carex annectens* Bickn.
var. *annectens*
Yellowfruit Sedge

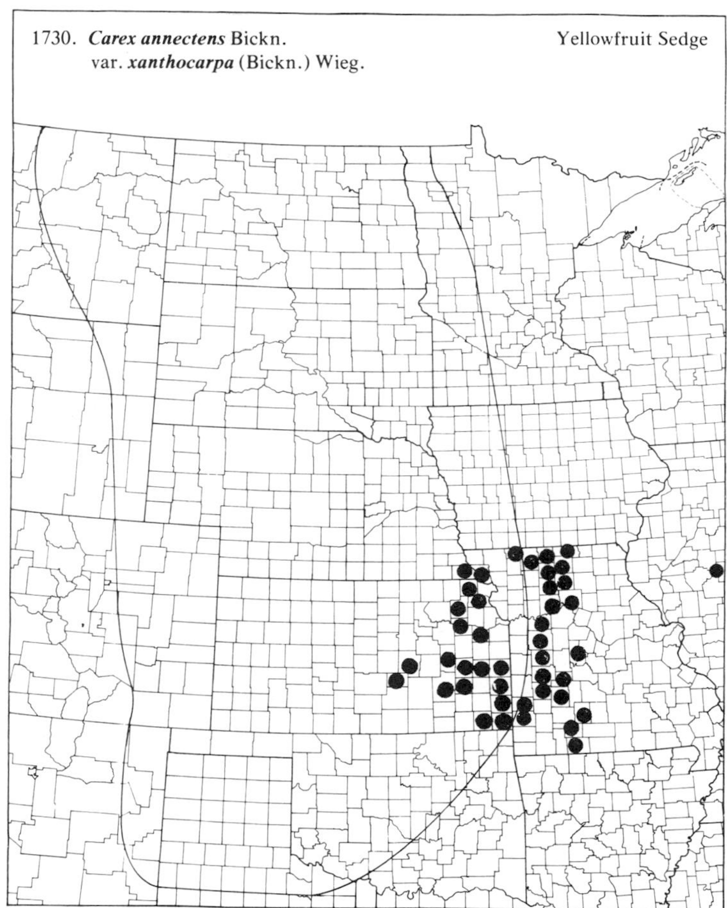
1730. *Carex annectens* Bickn.
var. *xanthocarpa* (Bickn.) Wieg.
Yellowfruit Sedge

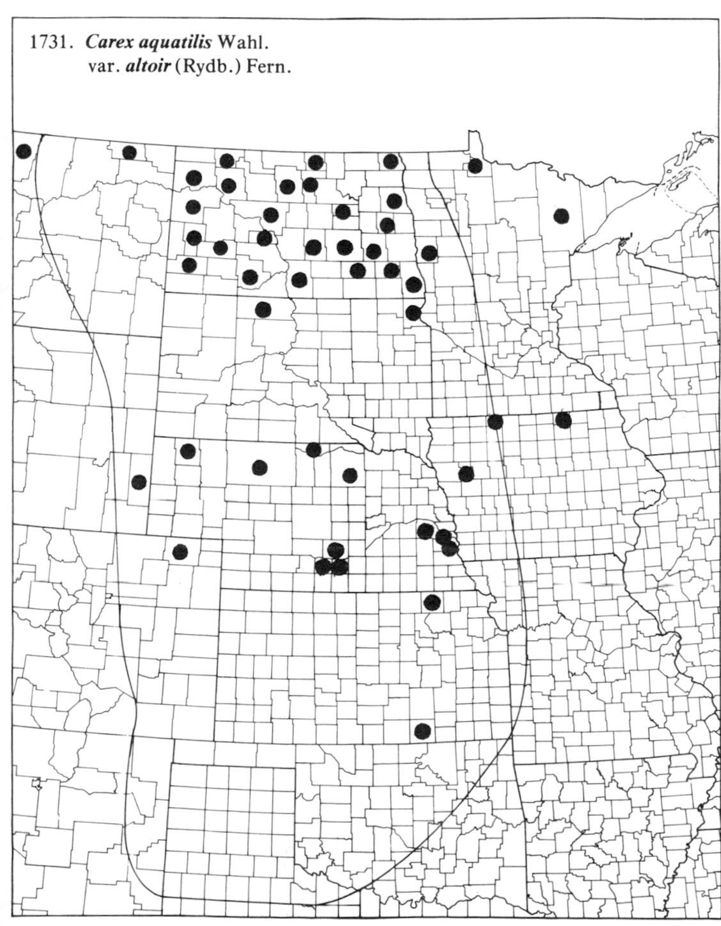
1731. *Carex aquatilis* Wahl.
var. *altoir* (Rydb.) Fern.

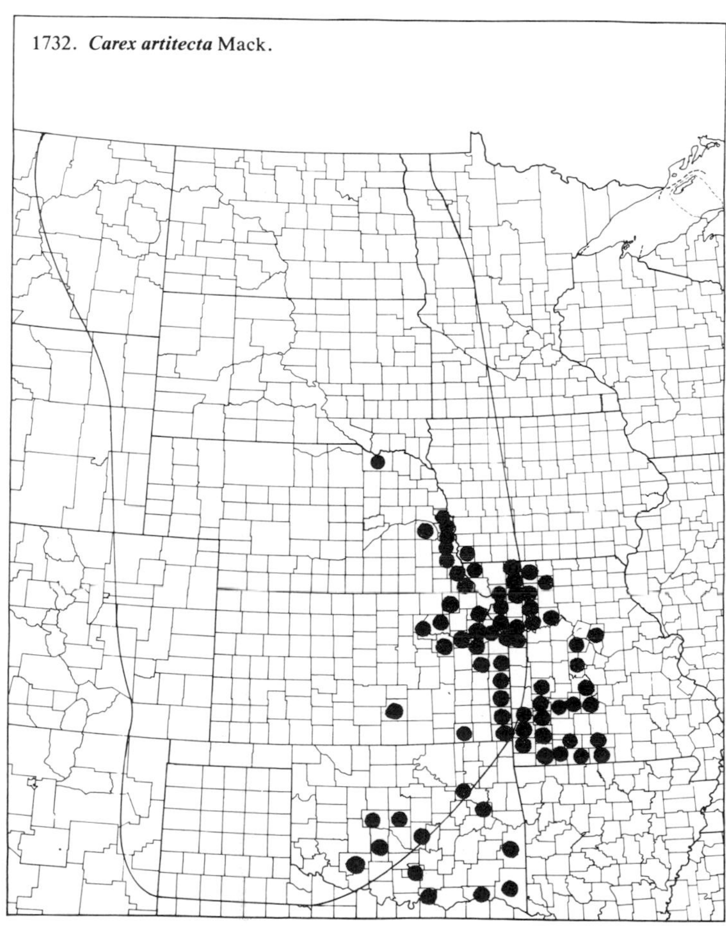
1732. *Carex artitecta* Mack.

(Cyperaceae)

1733. ***Carex assiniboinensis*** W. Boott

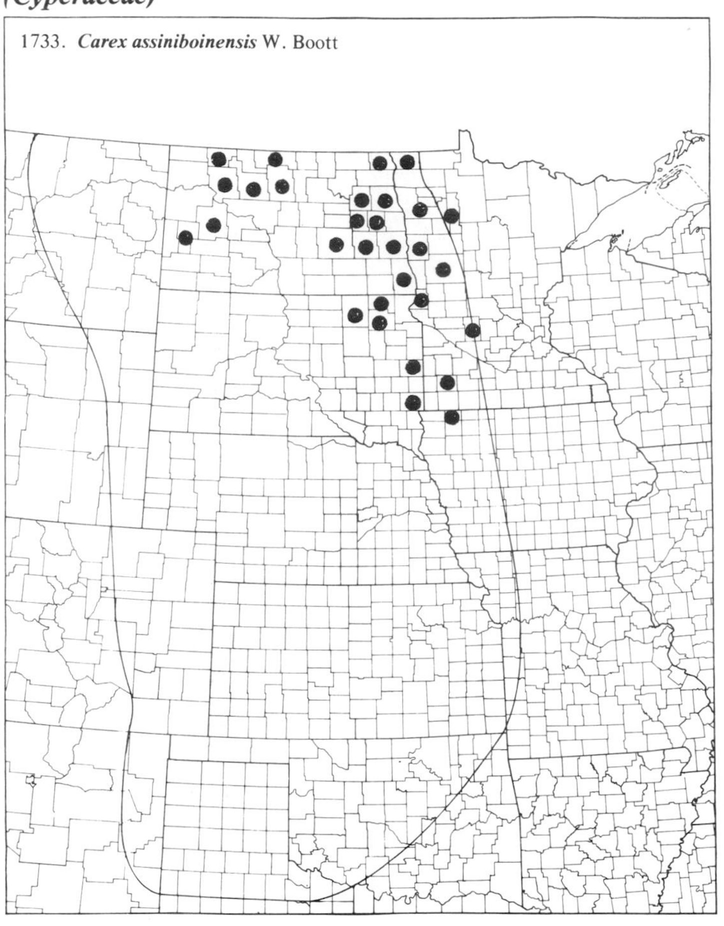

1734. ***Carex atherodes*** Spreng. Slough Sedge

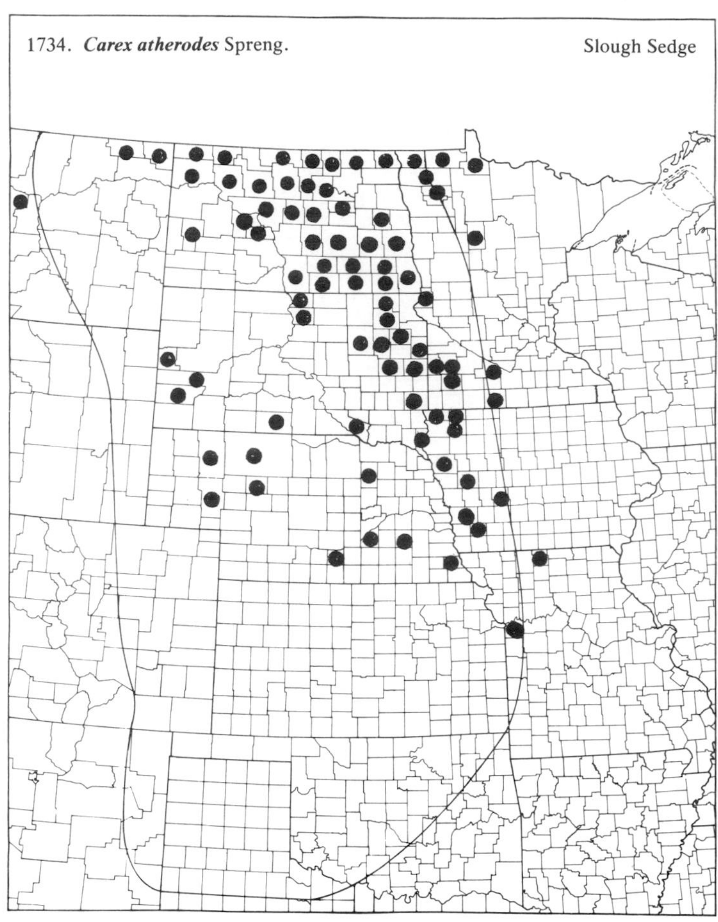

1735. ***Carex aurea*** Nutt. Golden Sedge

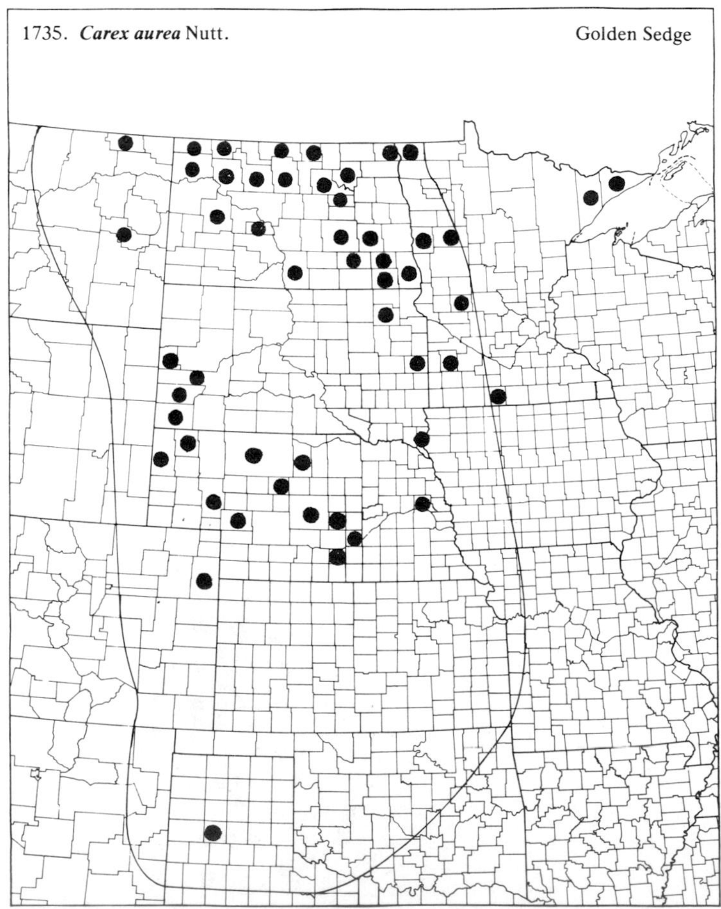

1736. ***Carex bebbii*** Olney ex Fern.

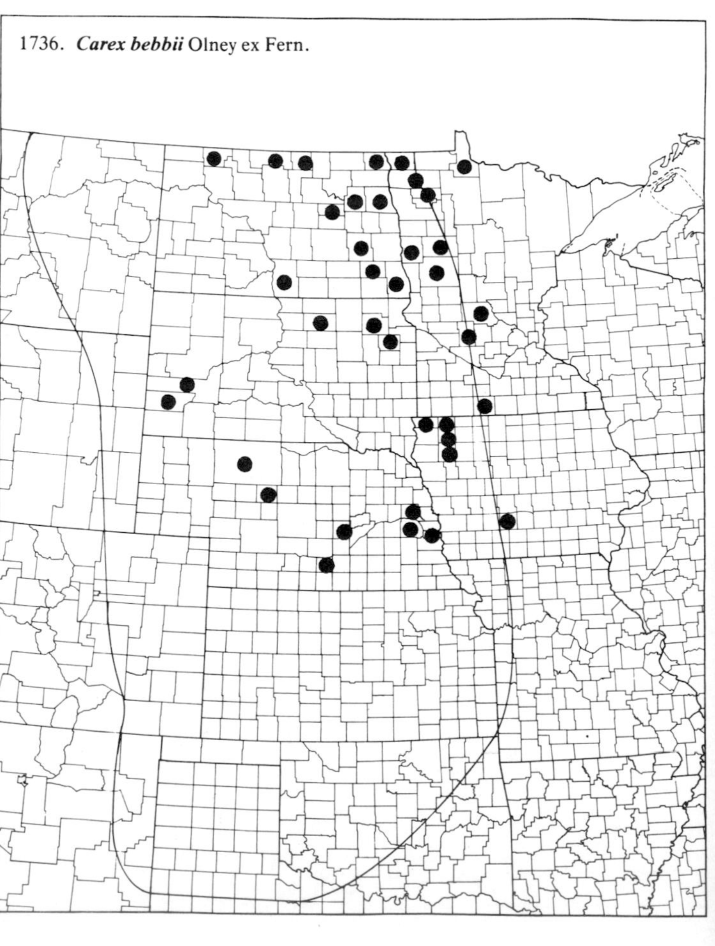

(Cyperaceae)

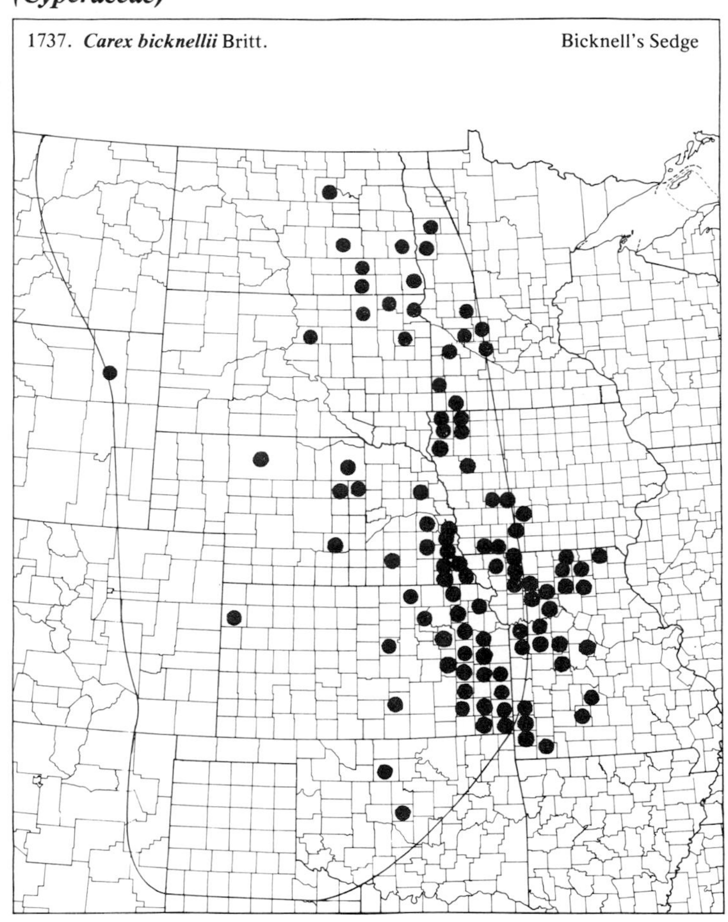
1737. Carex bicknellii Britt.
Bicknell's Sedge

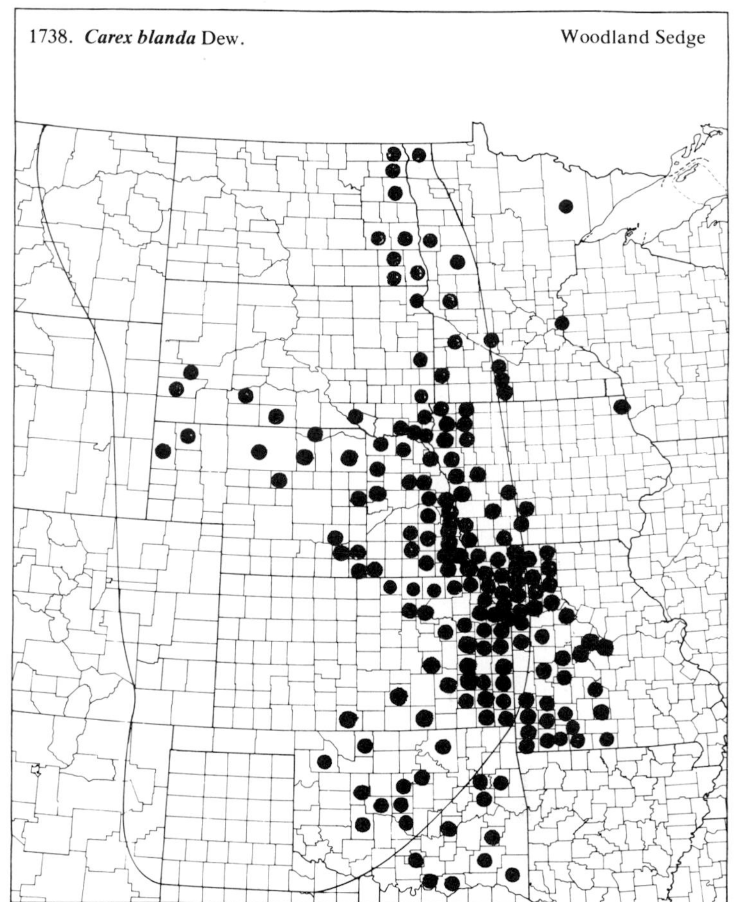
1738. Carex blanda Dew.
Woodland Sedge

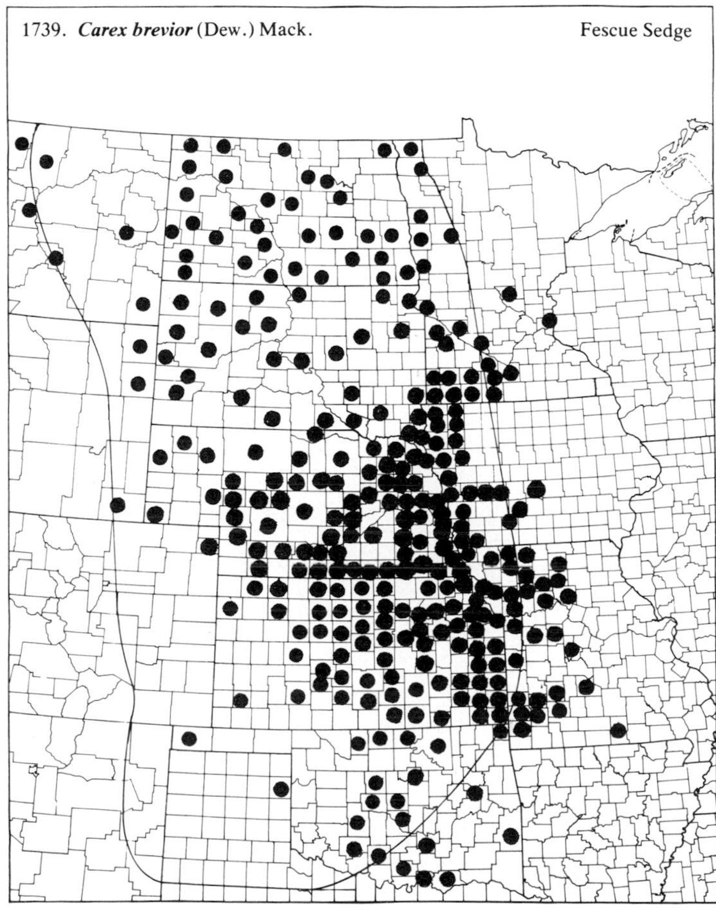
1739. Carex brevior (Dew.) Mack.
Fescue Sedge

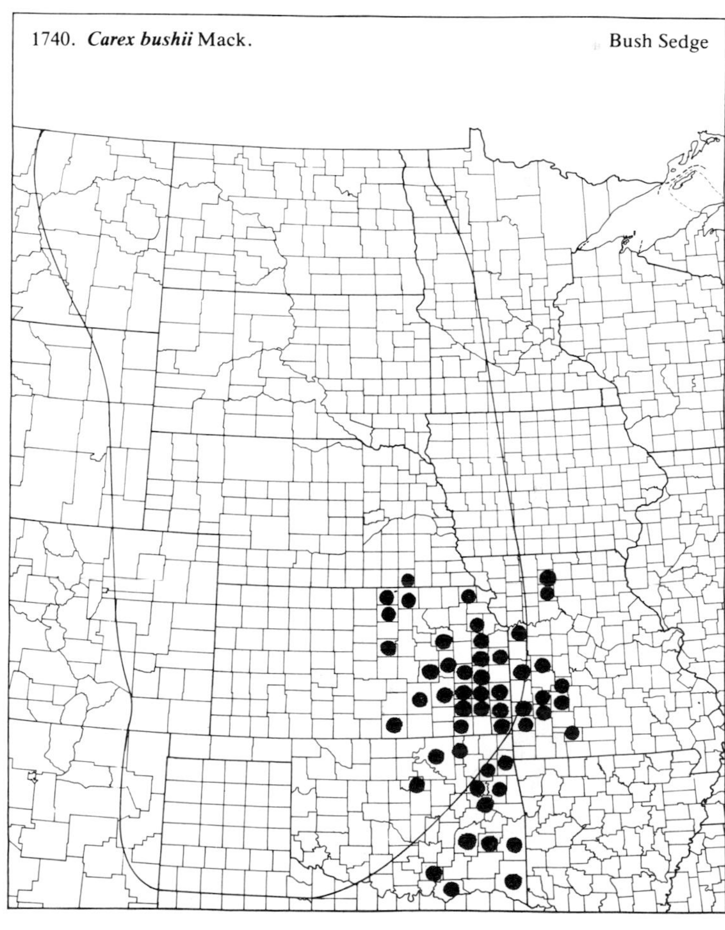
1740. Carex bushii Mack.
Bush Sedge

(Cyperaceae)

1741. *Carex buxbaumii* Wahl.

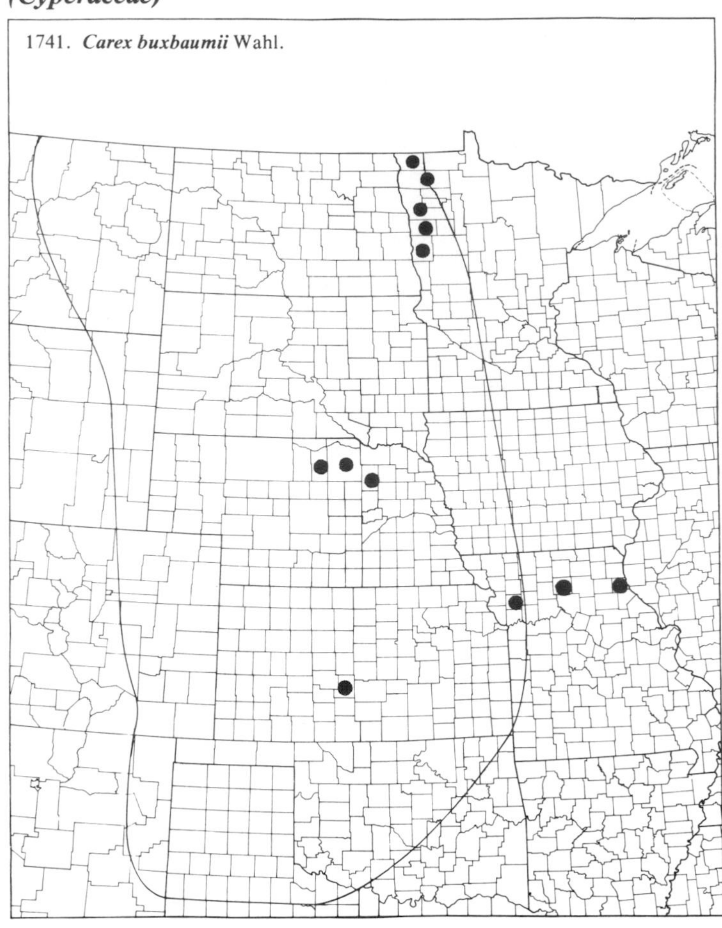

1742. *Carex cephalophora* Muhl. Woodbank Sedge

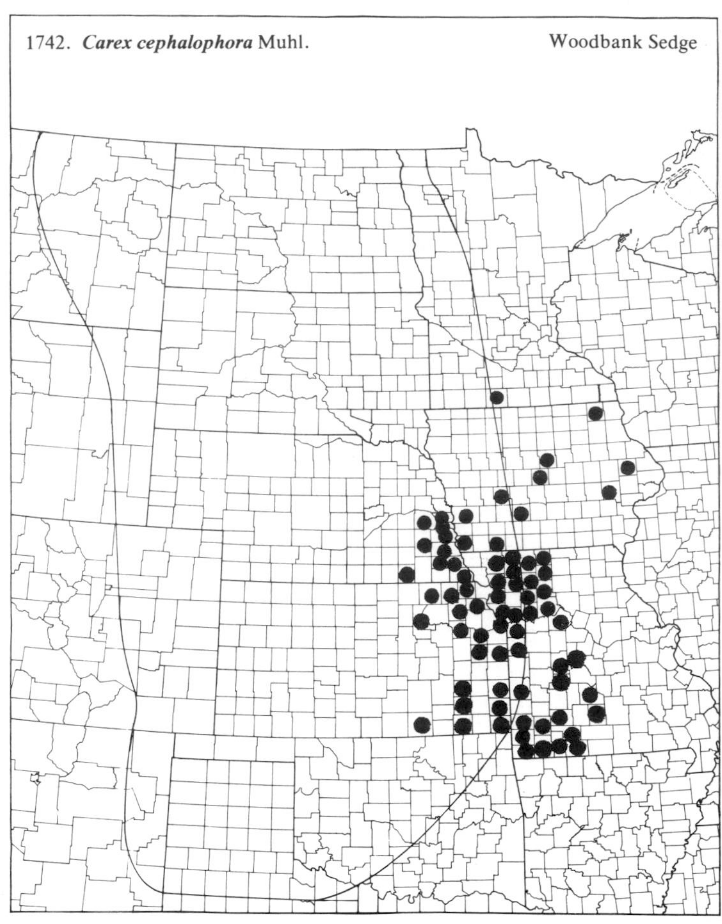

1743. *Carex comosa* W. Boott

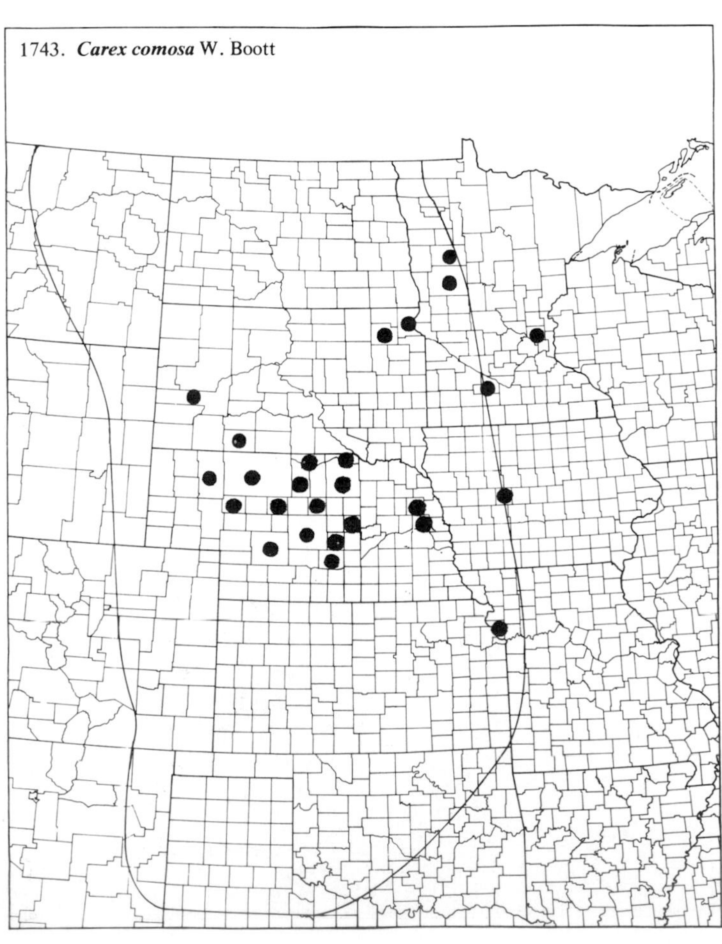

1744. *Carex conjuncta* W. Boott

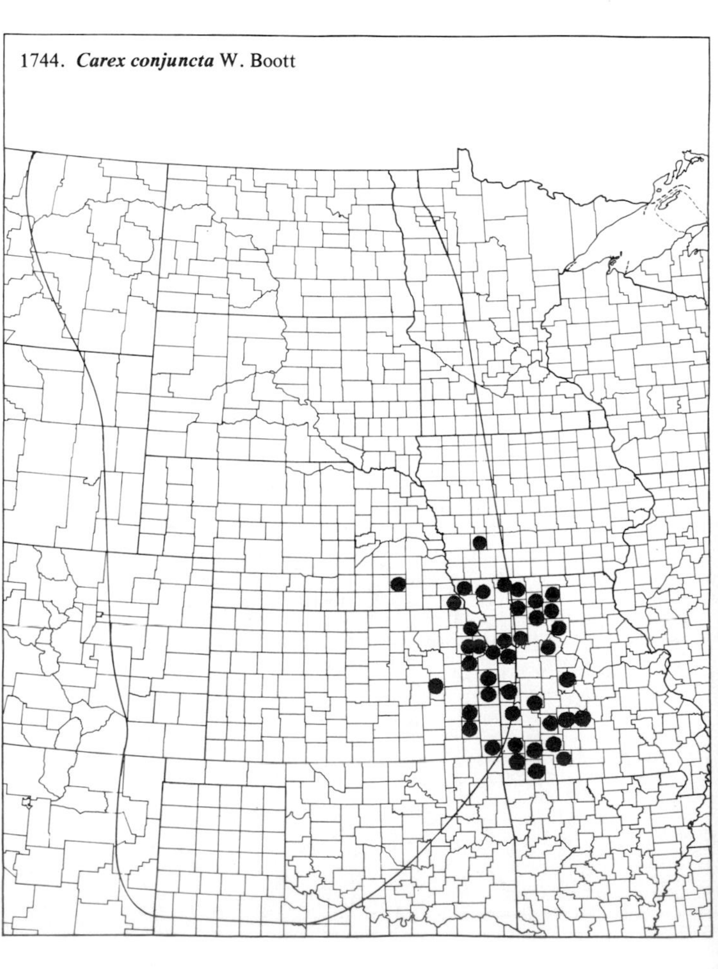

(Cyperaceae)

1745. ***Carex convoluta*** Mack.

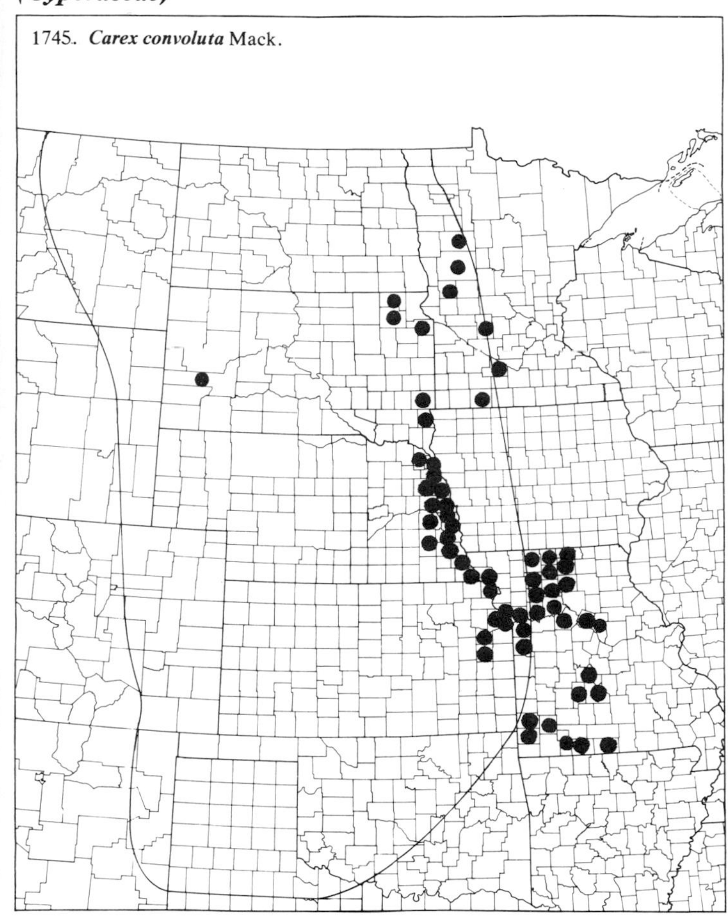

1746. ***Carex crawei*** Dew.

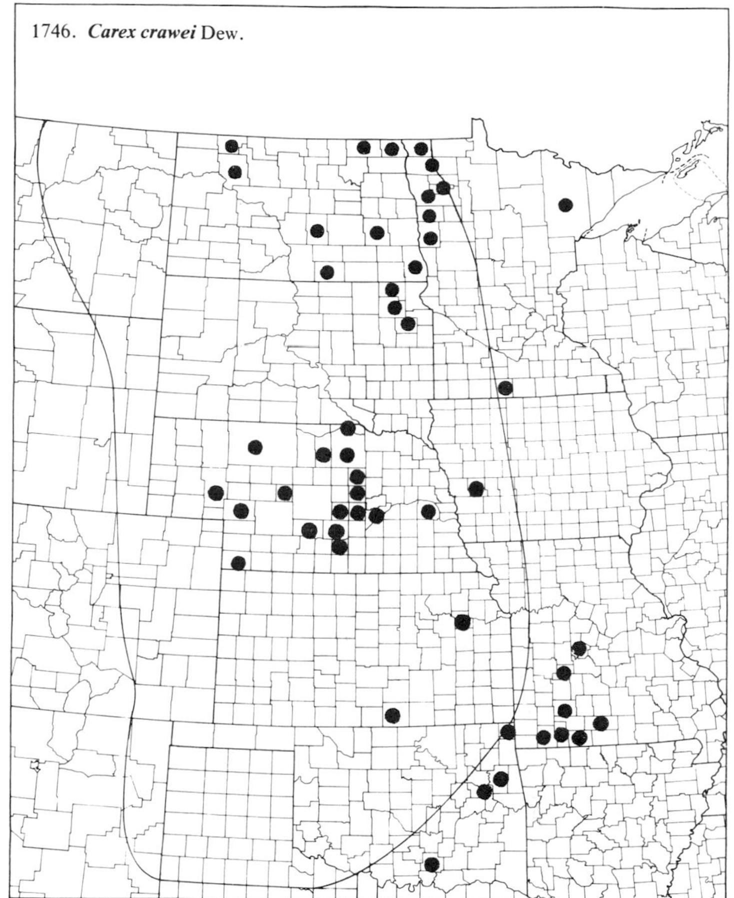

1747. ***Carex cristatella*** Britt.

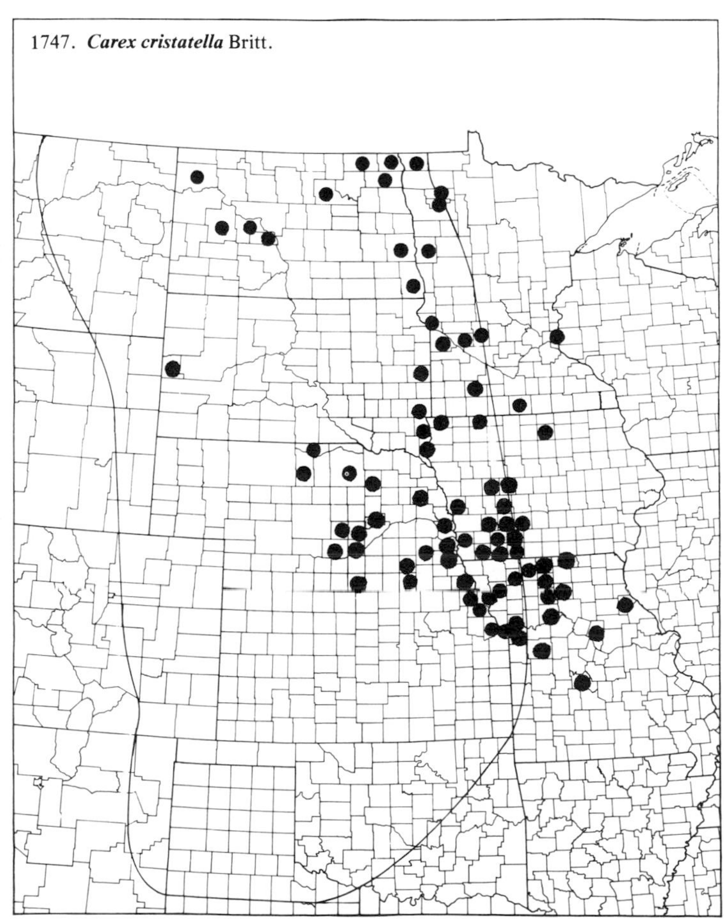

1748. ***Carex crus-corvi*** Shuttlew.

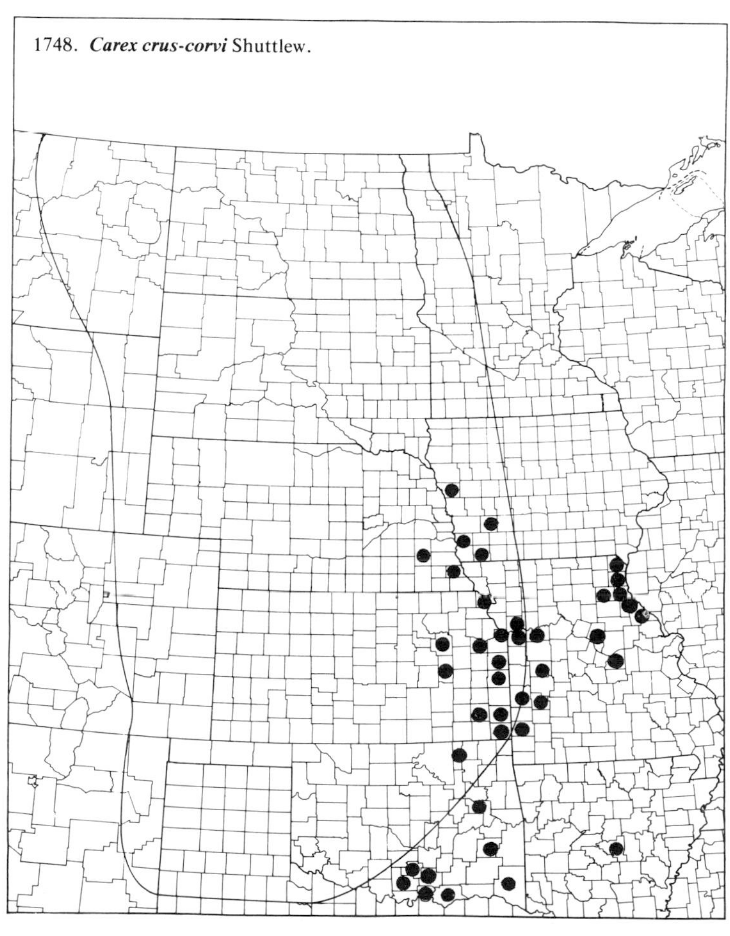

(Cyperaceae)

1749. ***Carex davisii*** Schwein. & Torr. Davis Sedge

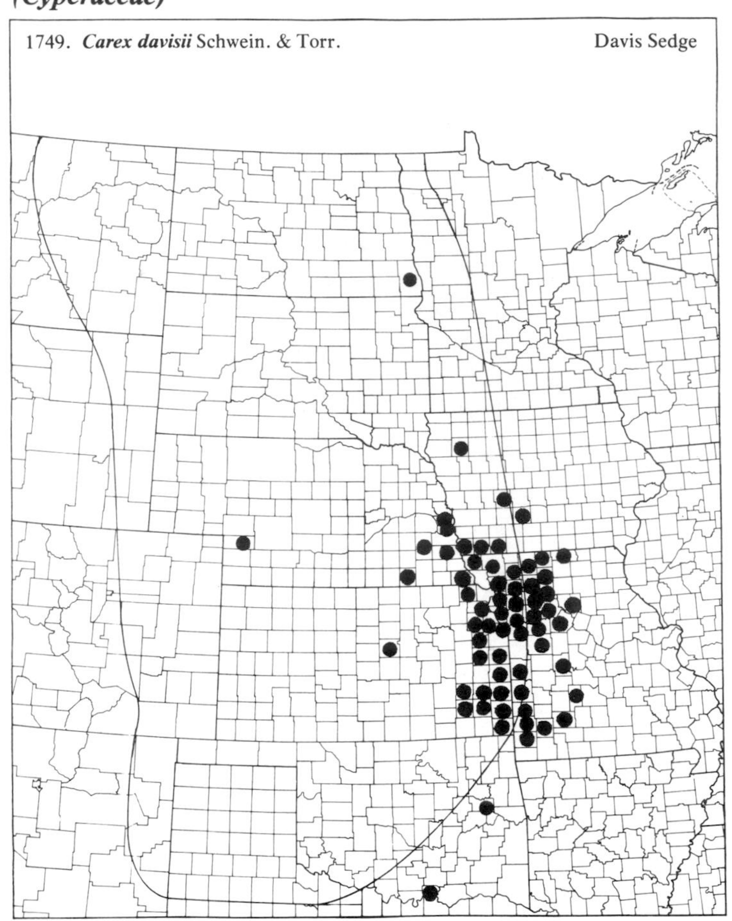

1750. ***Carex deweyana*** Schwein.

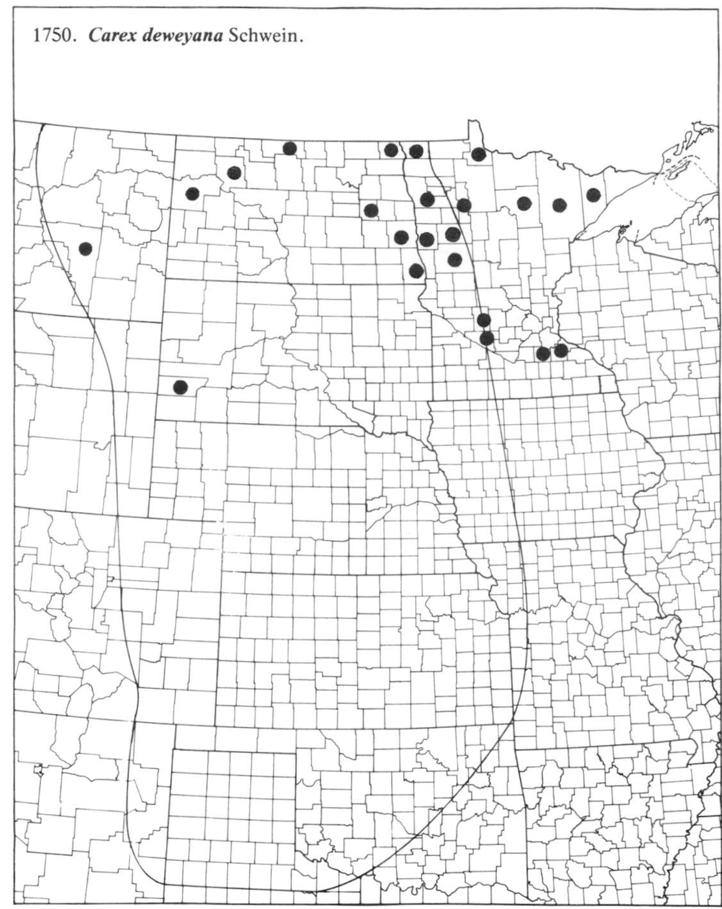

1751. ***Carex disperma*** Dew.

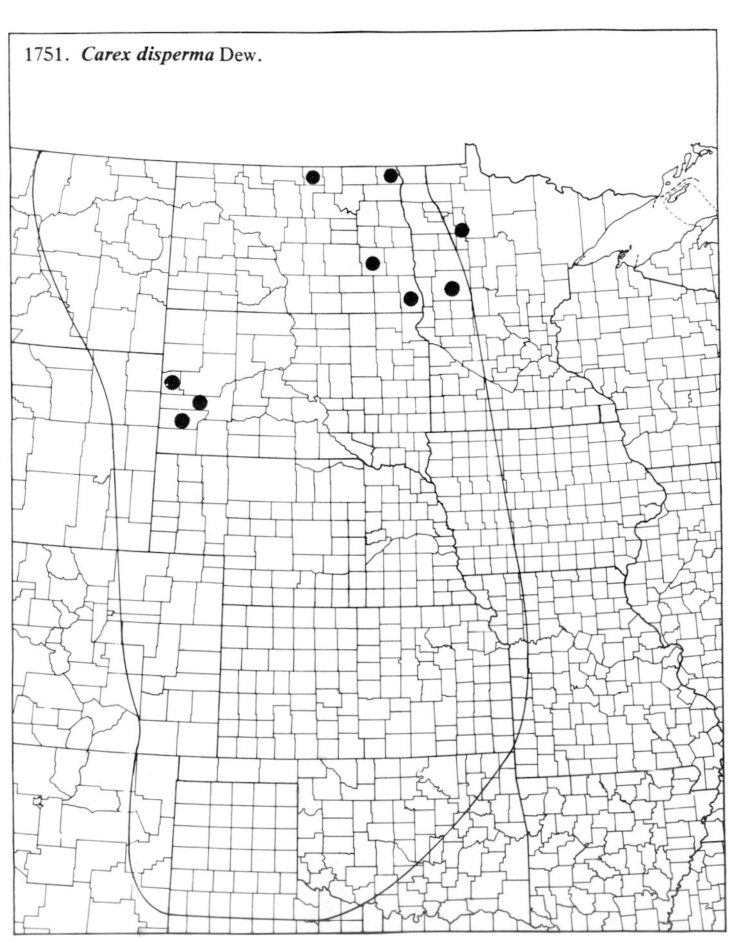

1752. ***Carex douglasii*** W. Boott

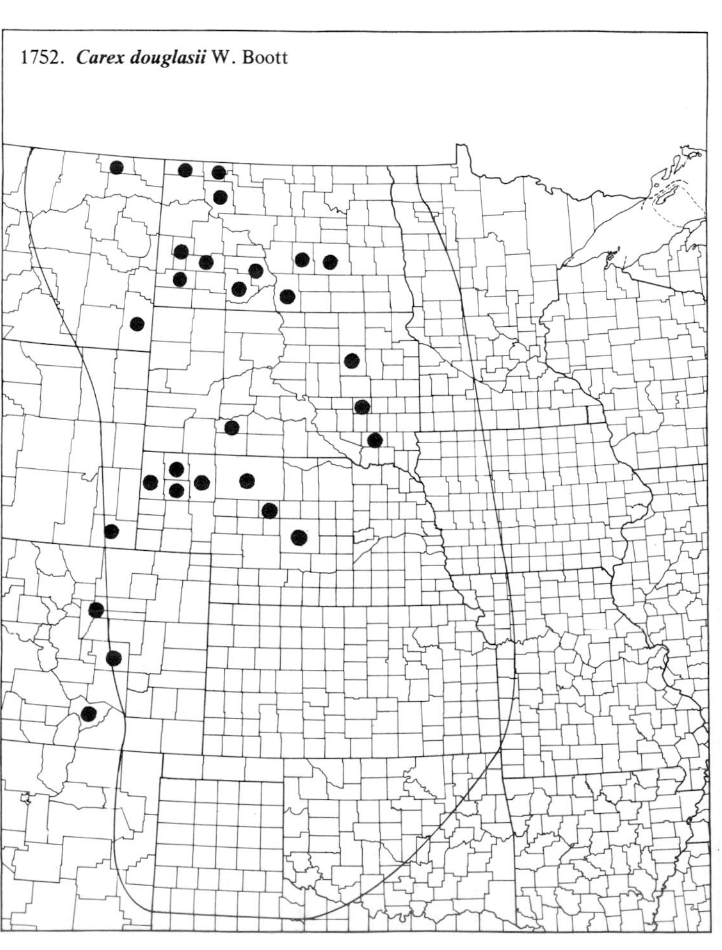

(Cyperaceae)

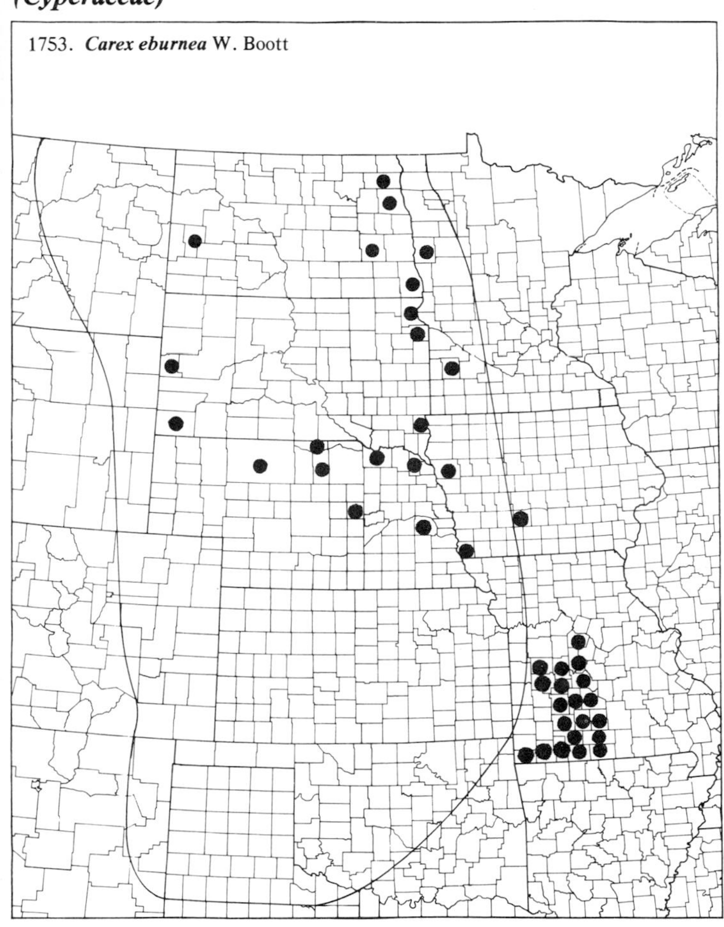
1753. Carex eburnea W. Boott

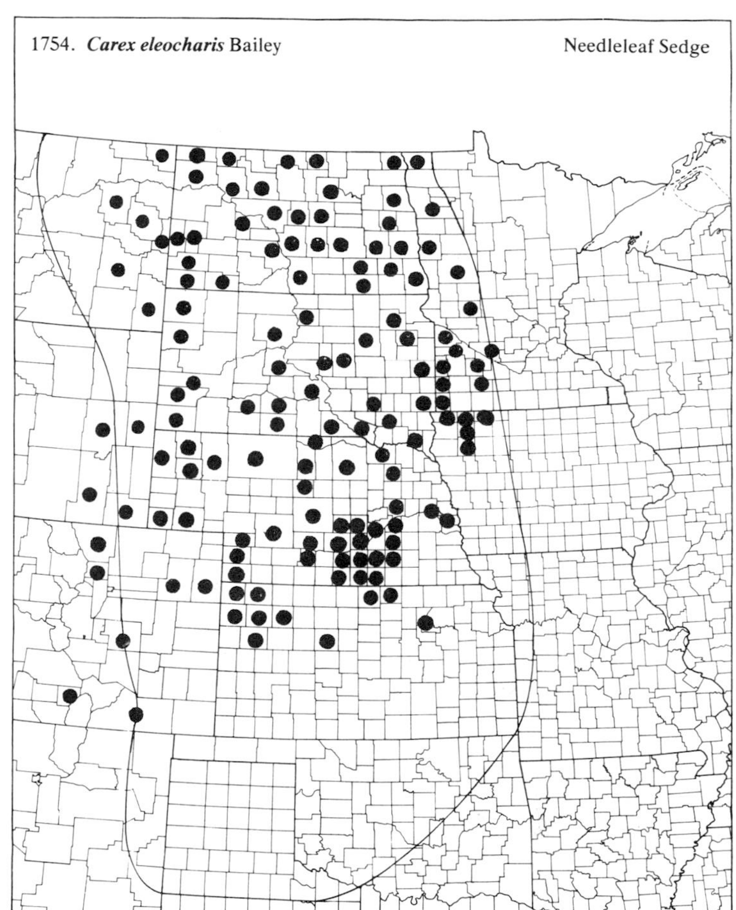
1754. Carex eleocharis Bailey
Needleleaf Sedge

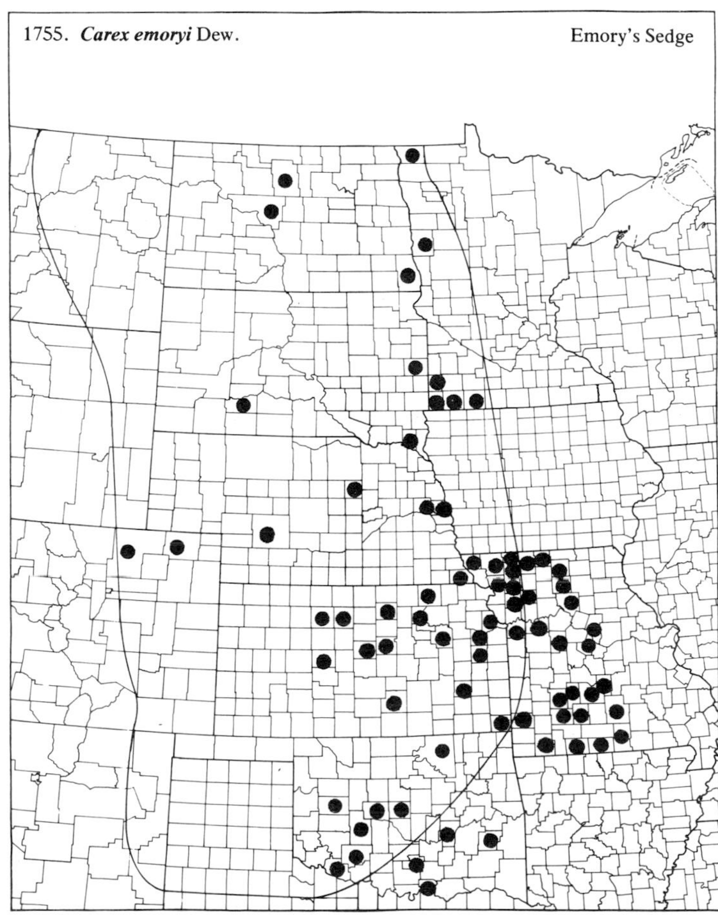
1755. Carex emoryi Dew.
Emory's Sedge

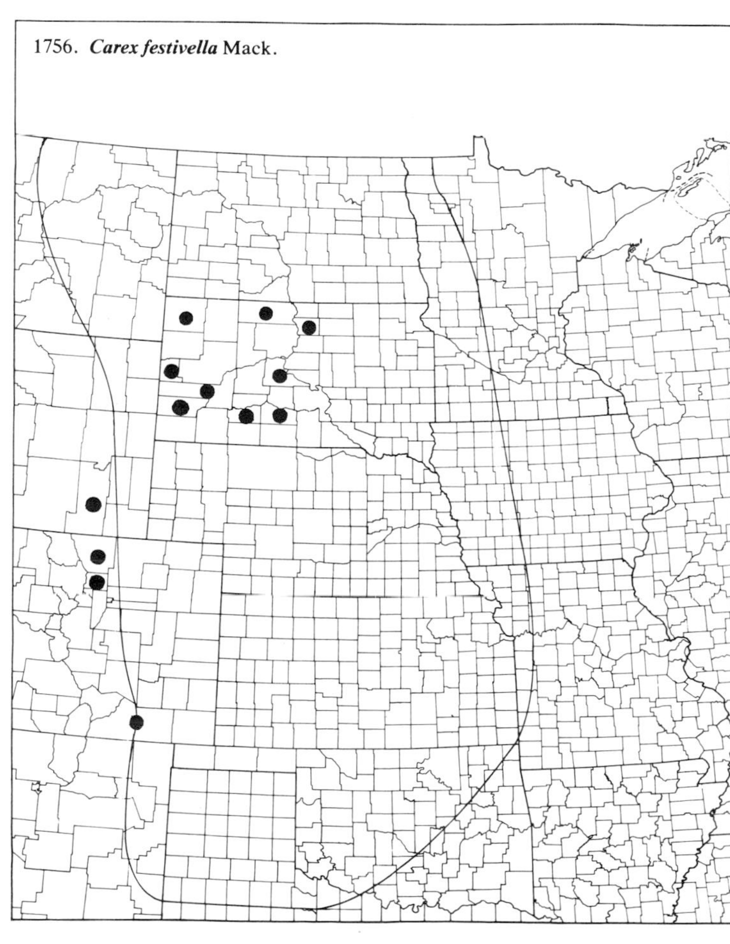
1756. Carex festivella Mack.

(Cyperaceae)

1757. ***Carex filifolia*** Nutt. Thread-leaved Sedge

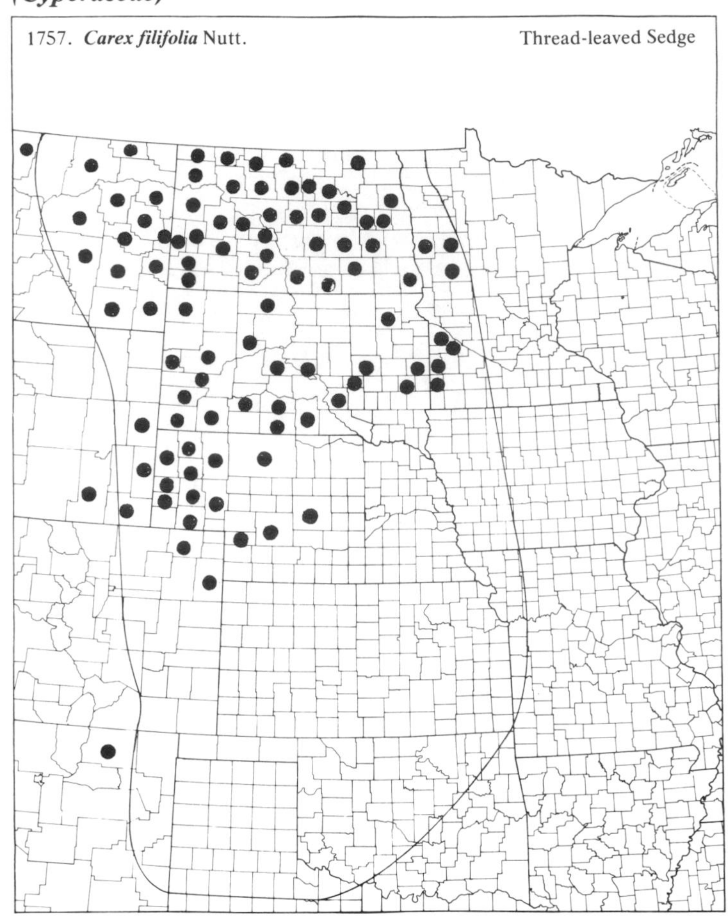

1758. ***Carex frankii*** Kunth Frank Sedge

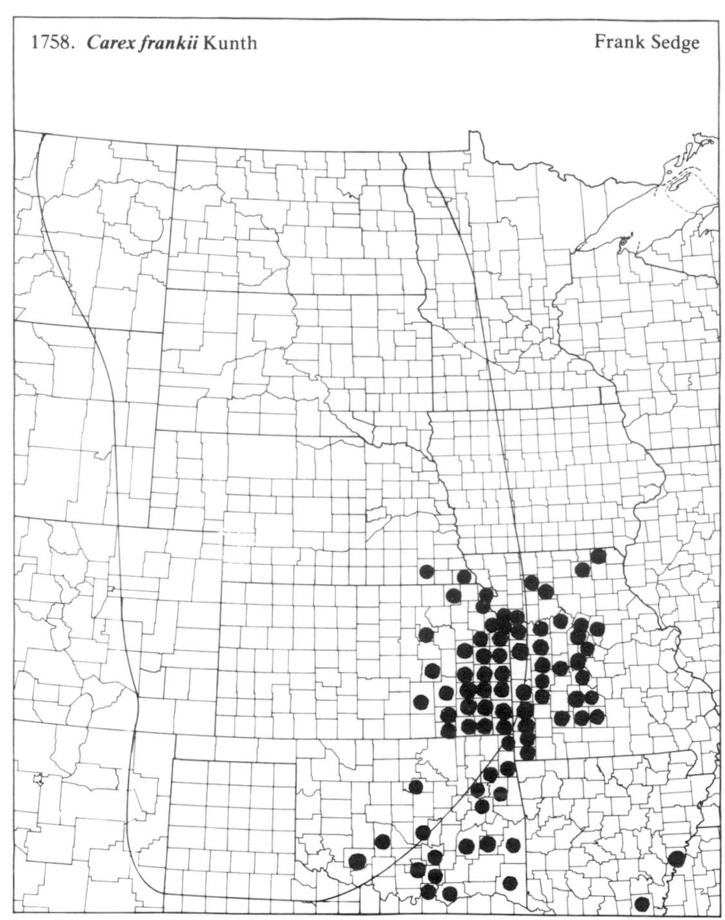

1759. ***Carex granularis*** Muhl. var. ***granularis*** Meadow Sedge

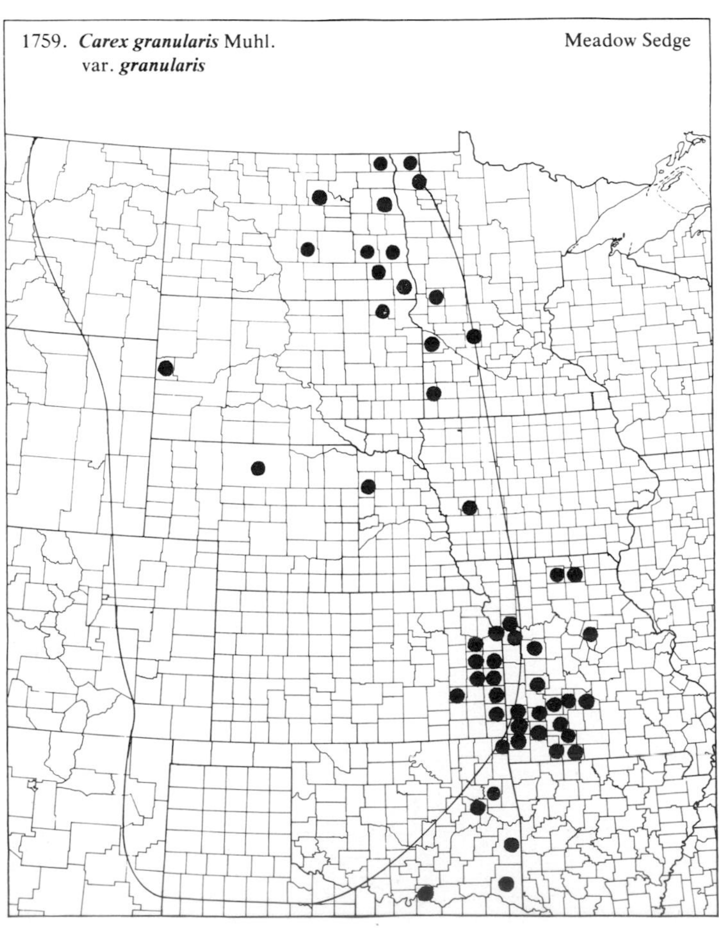

1760. ***Carex granularis*** Muhl. var. ***haleana*** (Olney) Porter Meadow Sedge

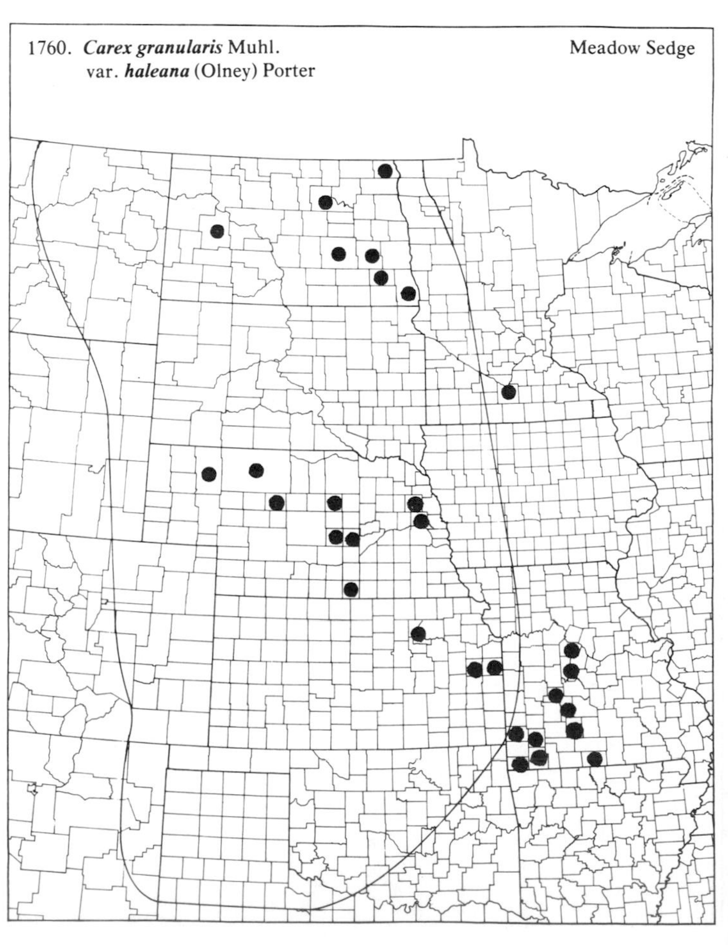

(Cyperaceae)

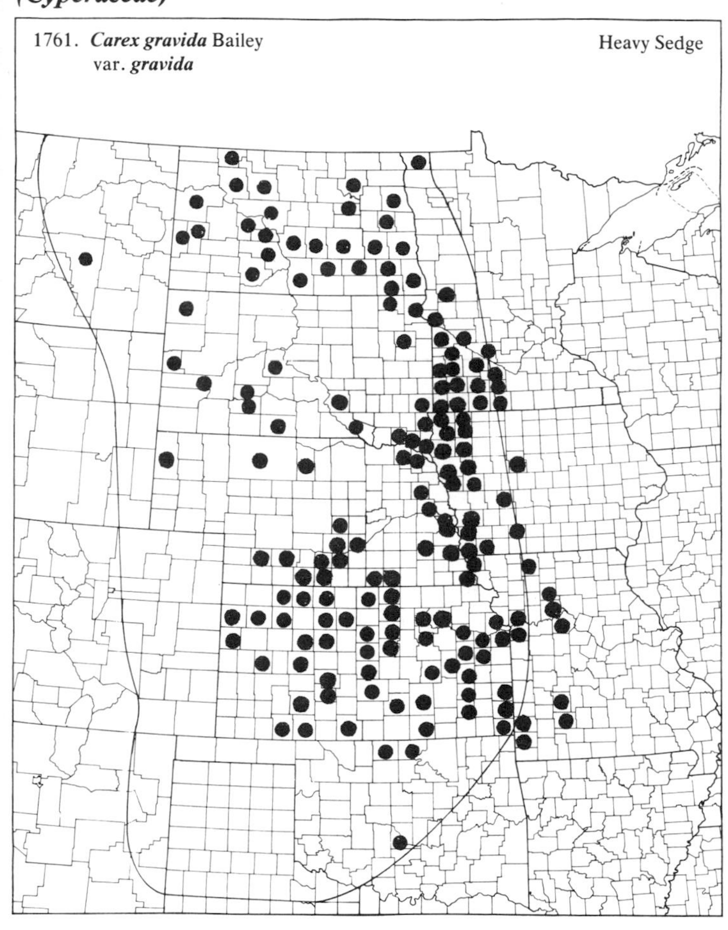

1761. ***Carex gravida*** Bailey var. ***gravida*** — Heavy Sedge

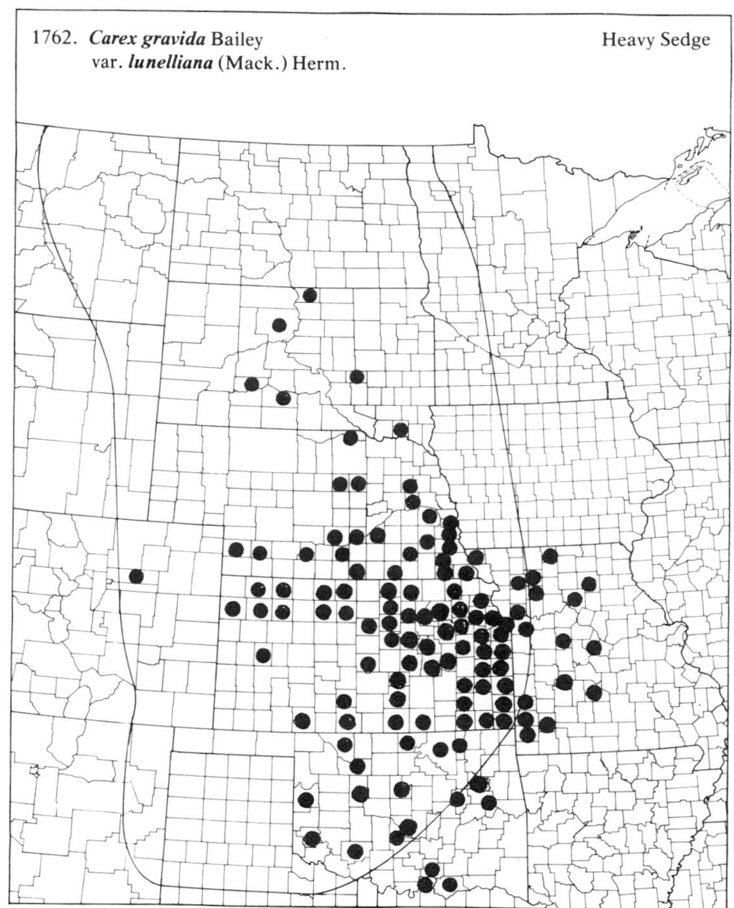

1762. ***Carex gravida*** Bailey var. ***lunelliana*** (Mack.) Herm. — Heavy Sedge

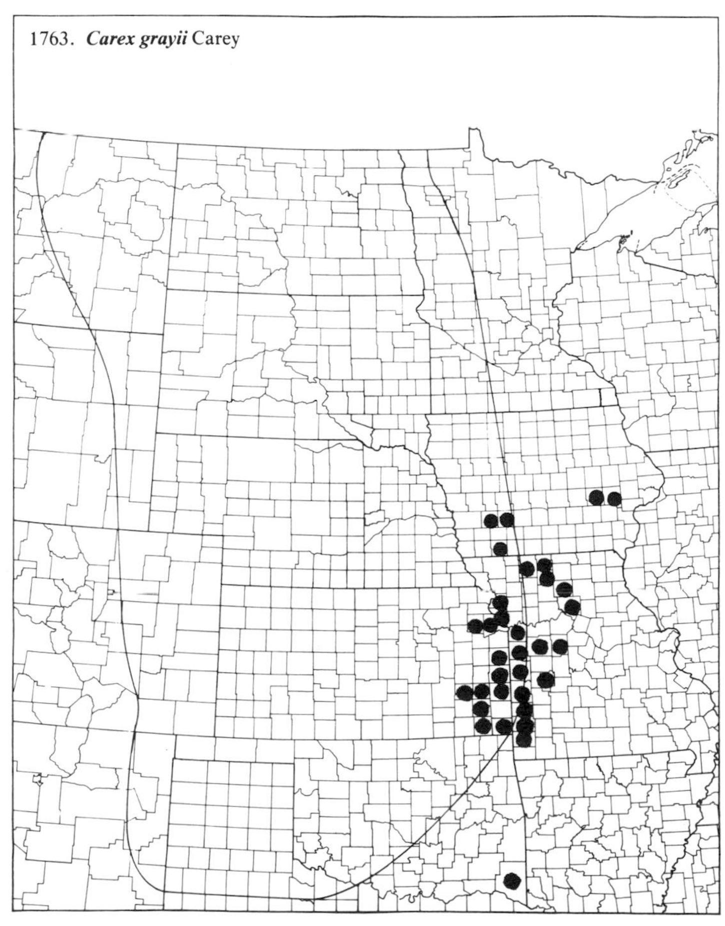

1763. ***Carex grayii*** Carey

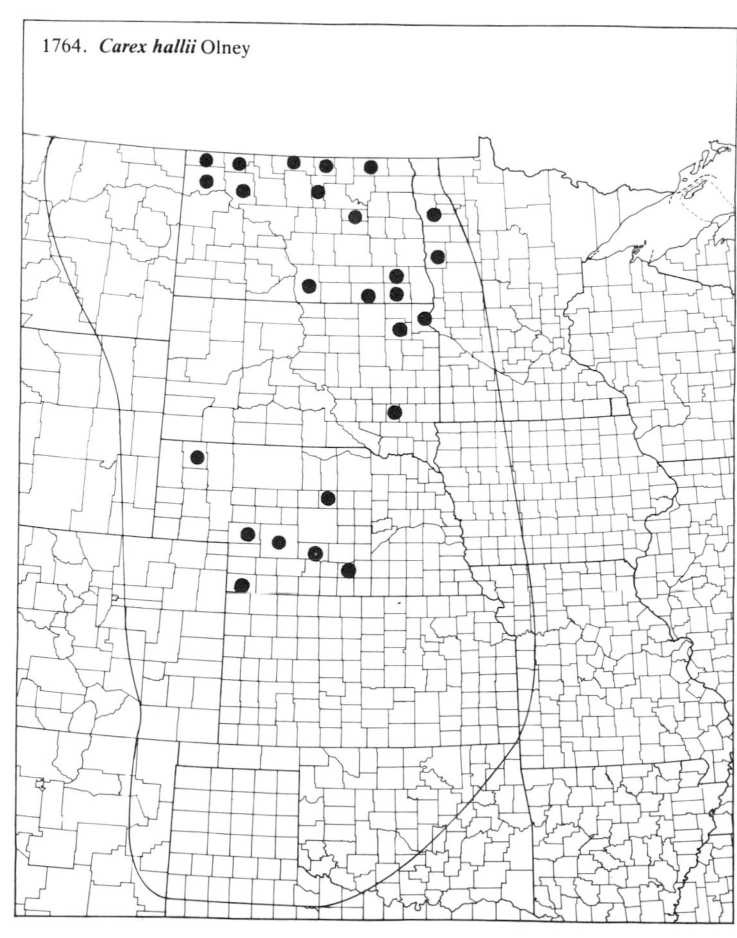

1764. ***Carex hallii*** Olney

(Cyperaceae)

1765. *Carex heliophila* Mack.

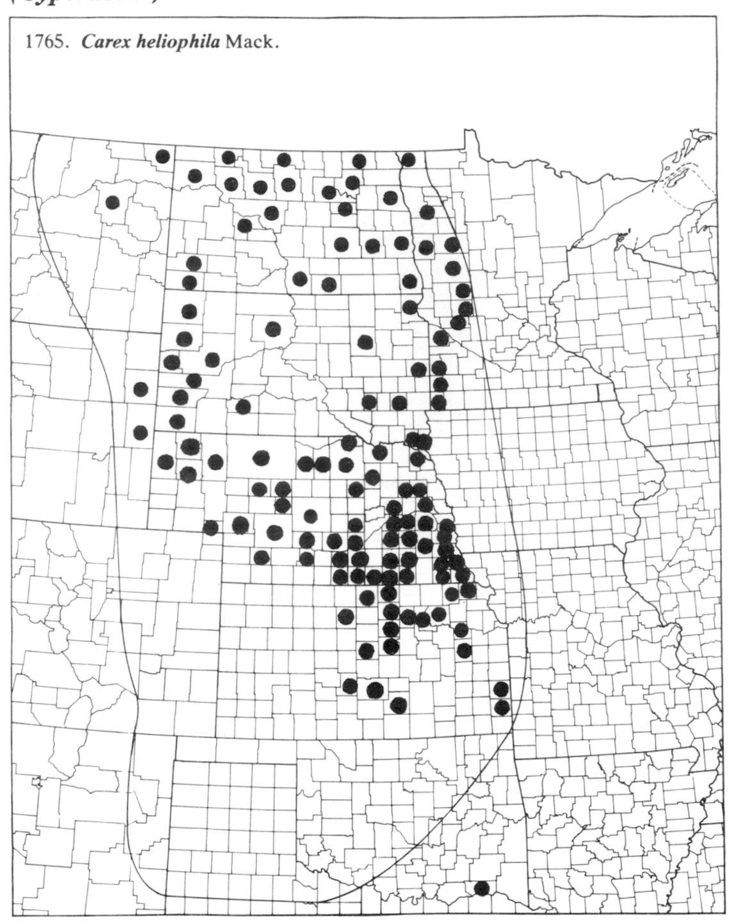

1766. *Carex hitchcockiana* Dew.

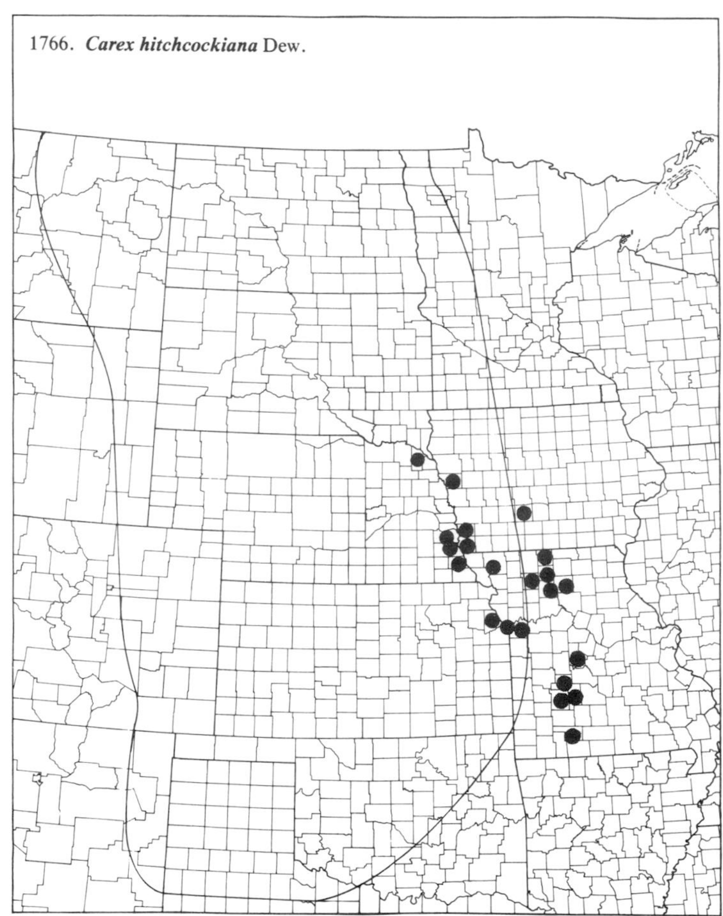

1767. *Carex hyalinolepis* Steud. Thin-scale Sedge

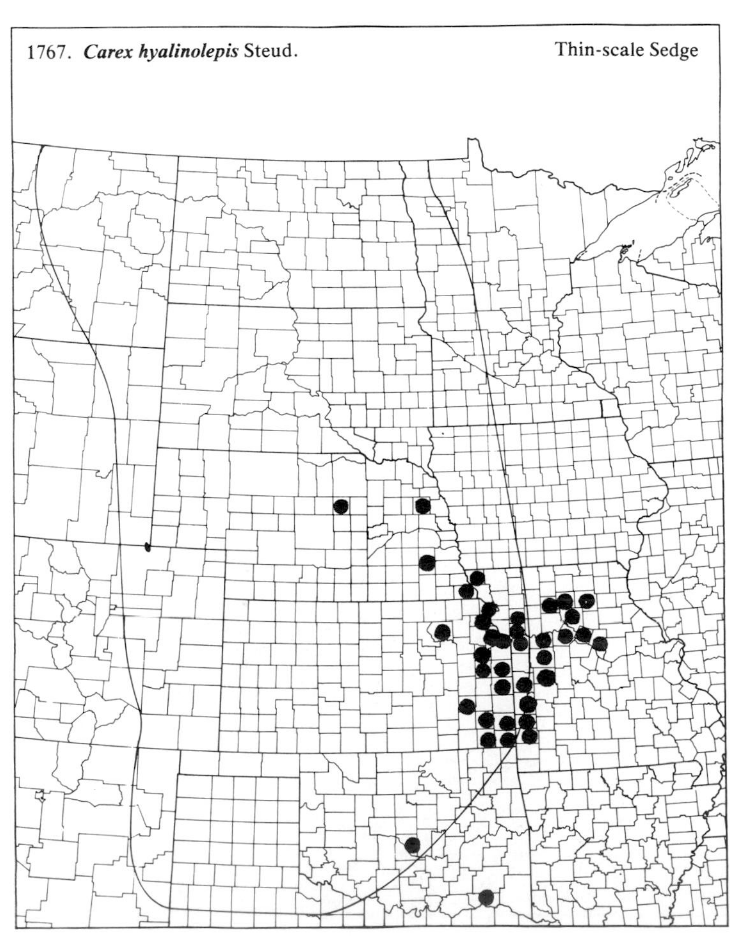

1768. *Carex hystericina* Muhl. Bottlebrush Sedge

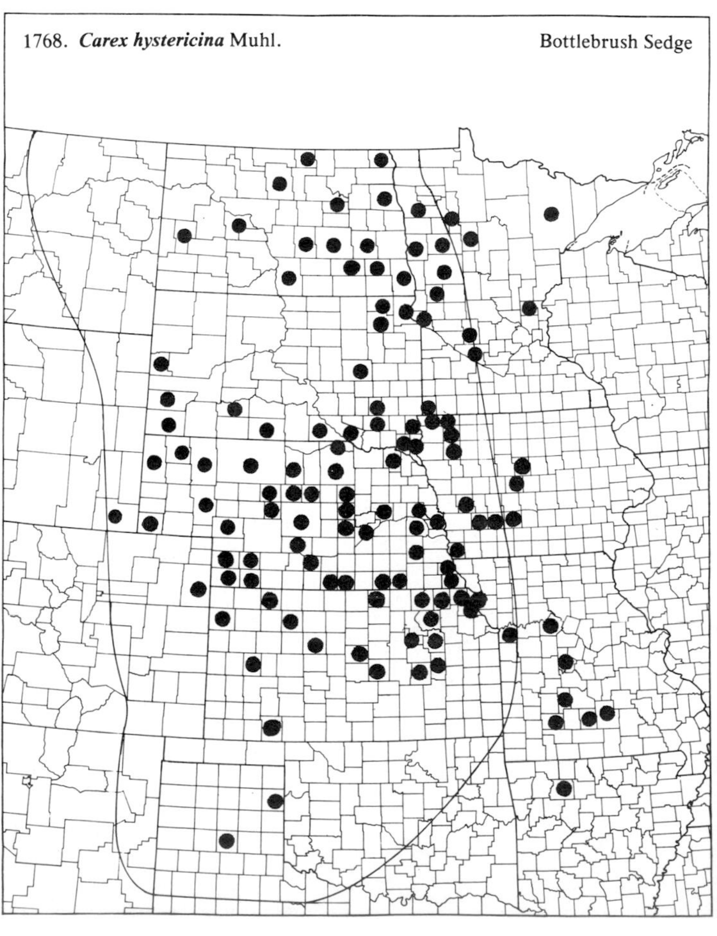

1769. ***Carex interior*** Bailey

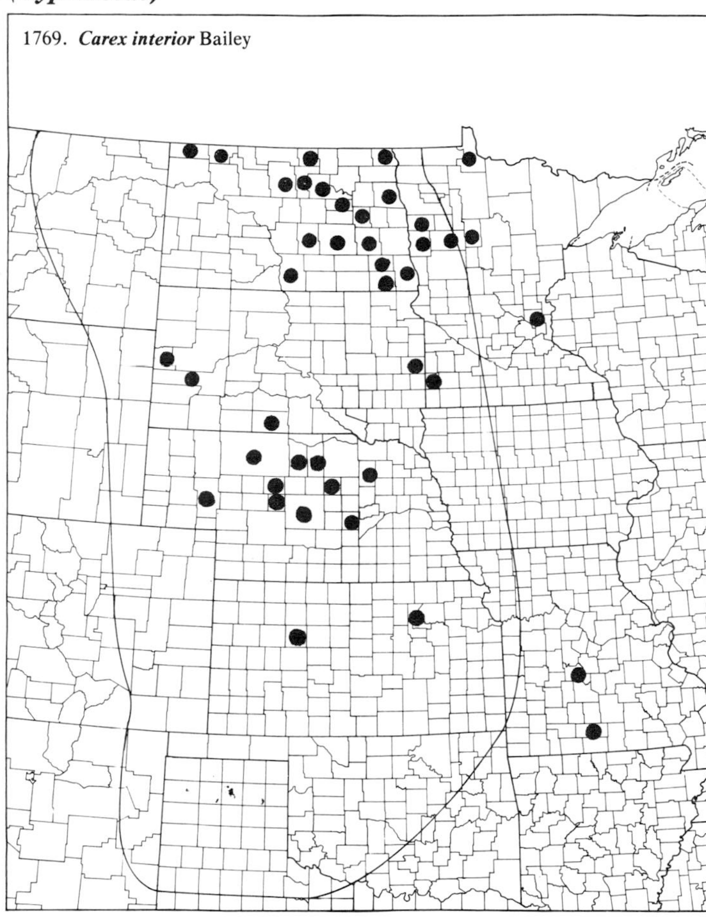

1770. ***Carex jamesii*** Schwein.

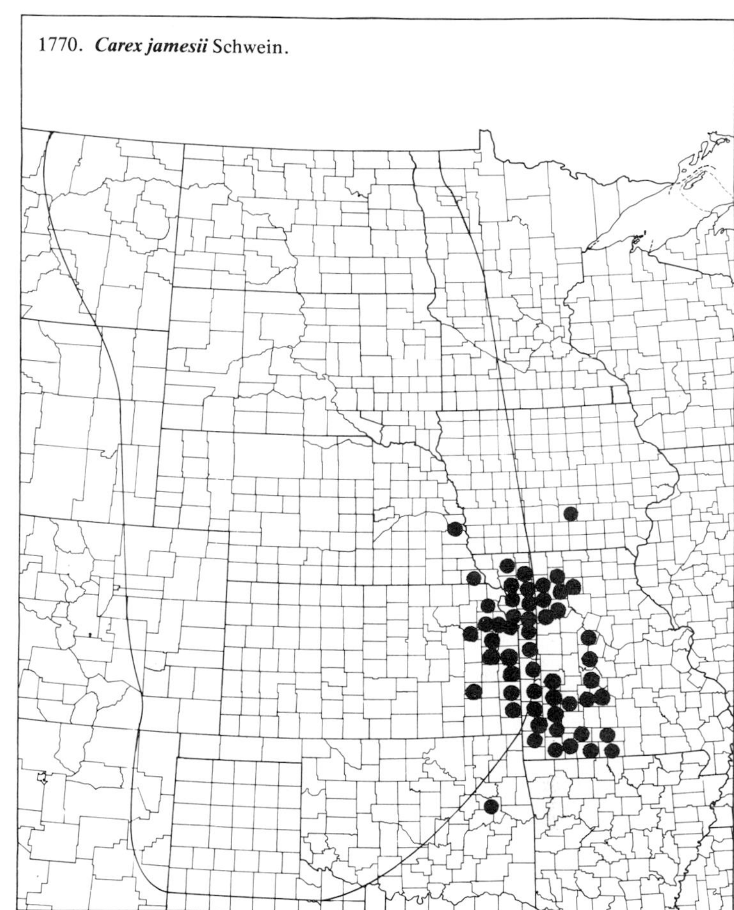

1771. ***Carex lacustris*** Willd.

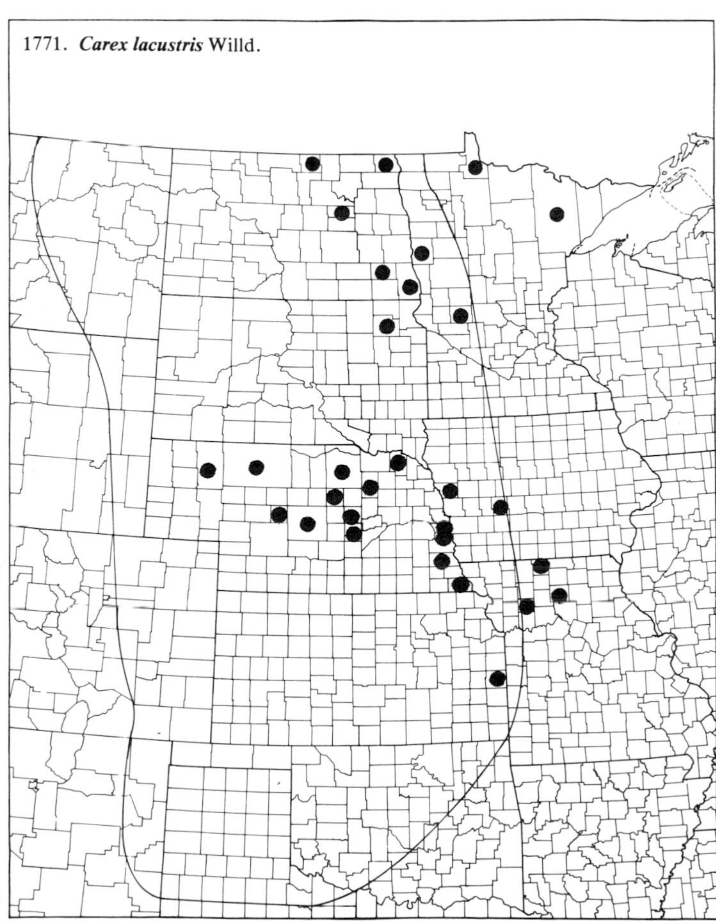

1772. ***Carex laeviconica*** Dew. Smoothcone Sedge

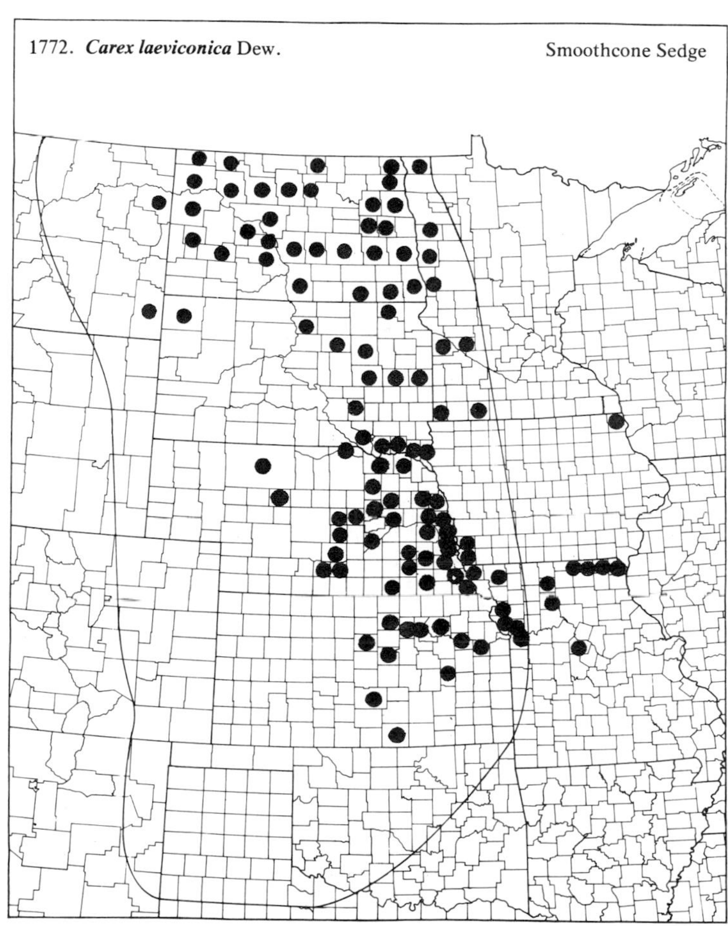

(Cyperaceae)

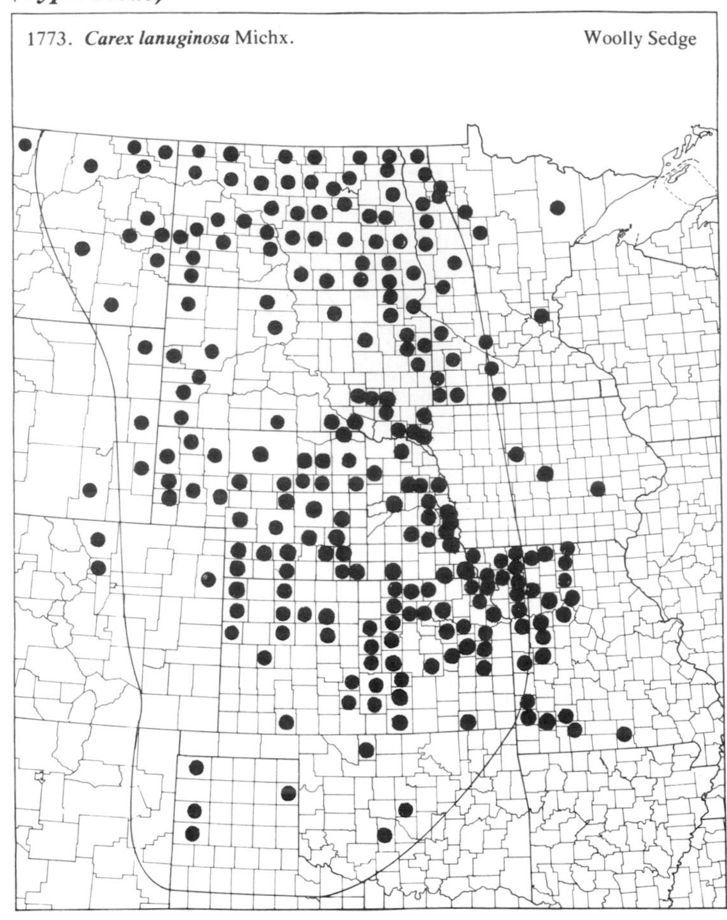

1773. ***Carex lanuginosa*** Michx. Woolly Sedge

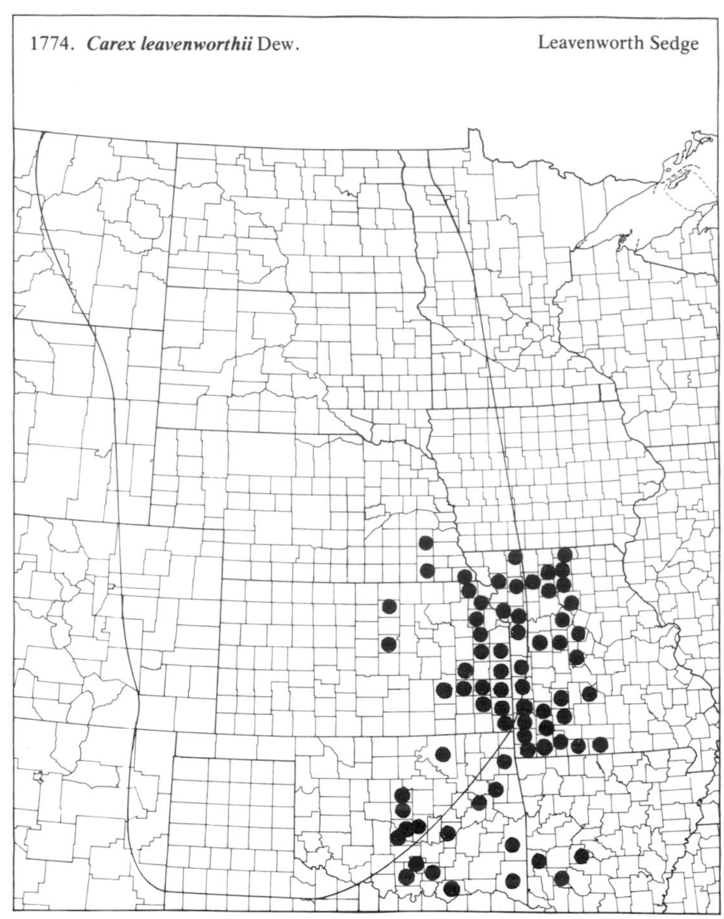

1774. ***Carex leavenworthii*** Dew. Leavenworth Sedge

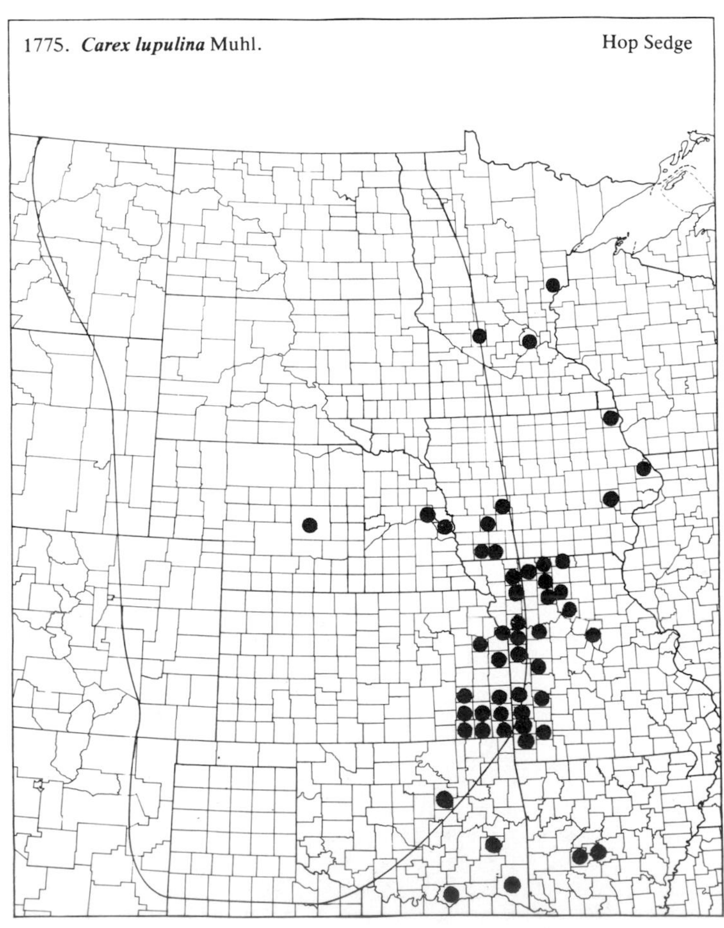

1775. ***Carex lupulina*** Muhl. Hop Sedge

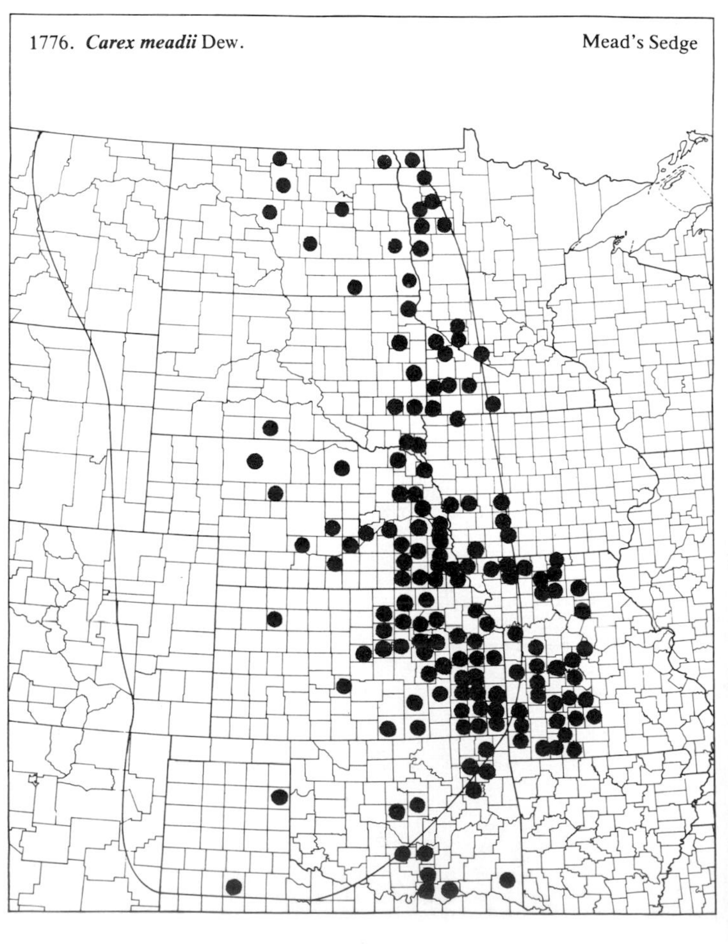

1776. ***Carex meadii*** Dew. Mead's Sedge

(Cyperaceae)

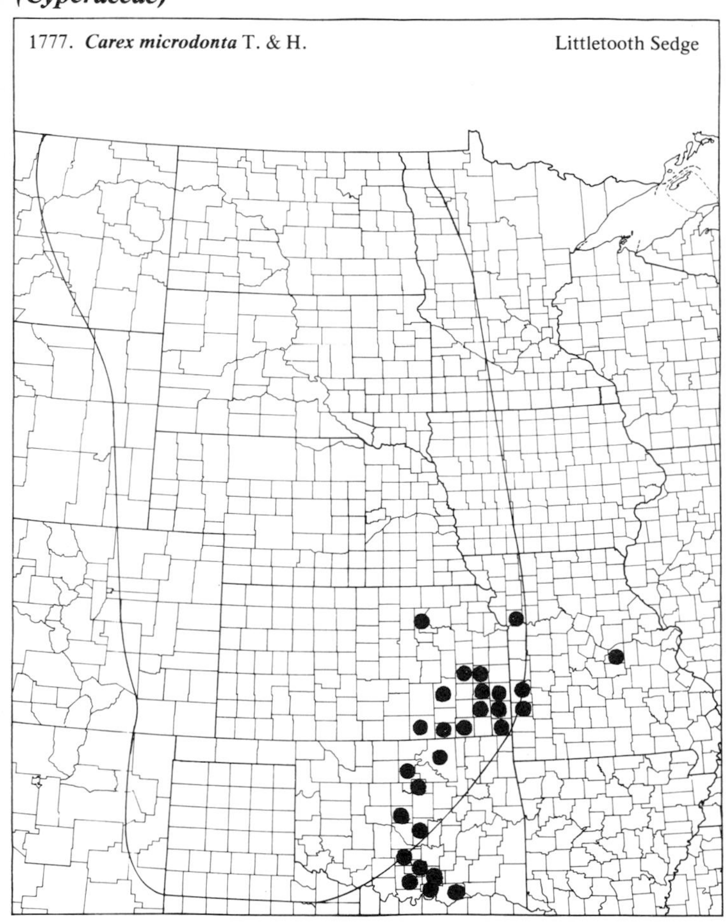

1777. *Carex microdonta* T. & H. Littletooth Sedge

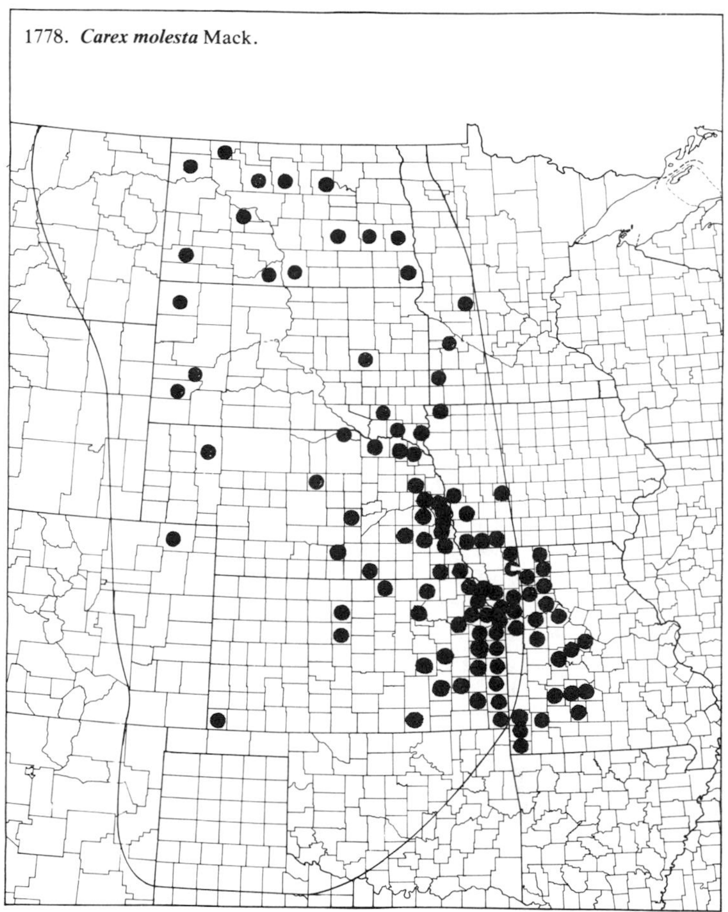

1778. *Carex molesta* Mack.

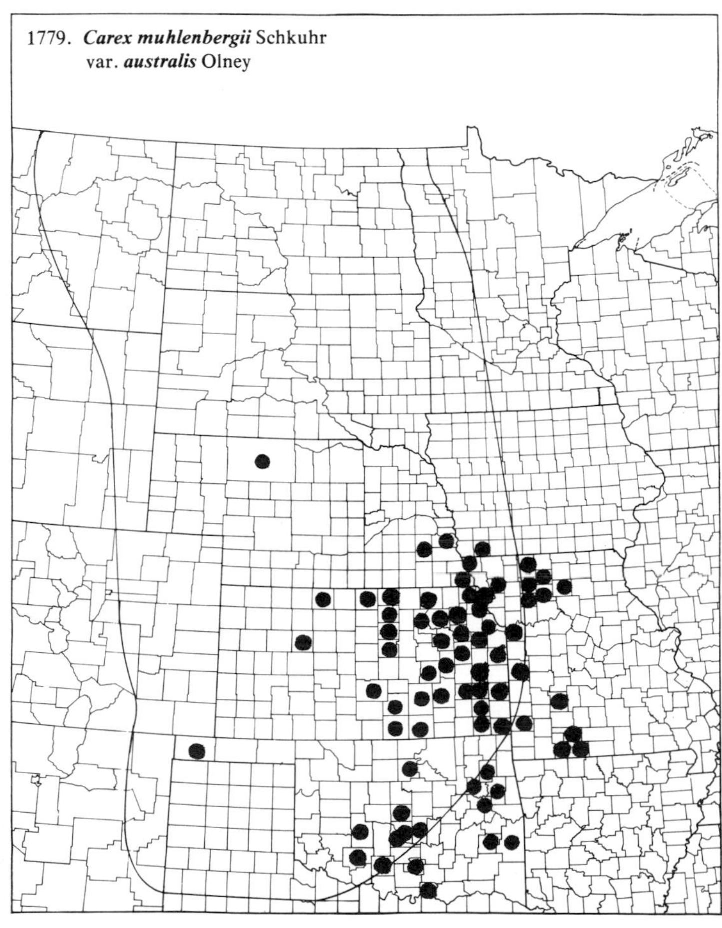

1779. *Carex muhlenbergii* Schkuhr var. *australis* Olney

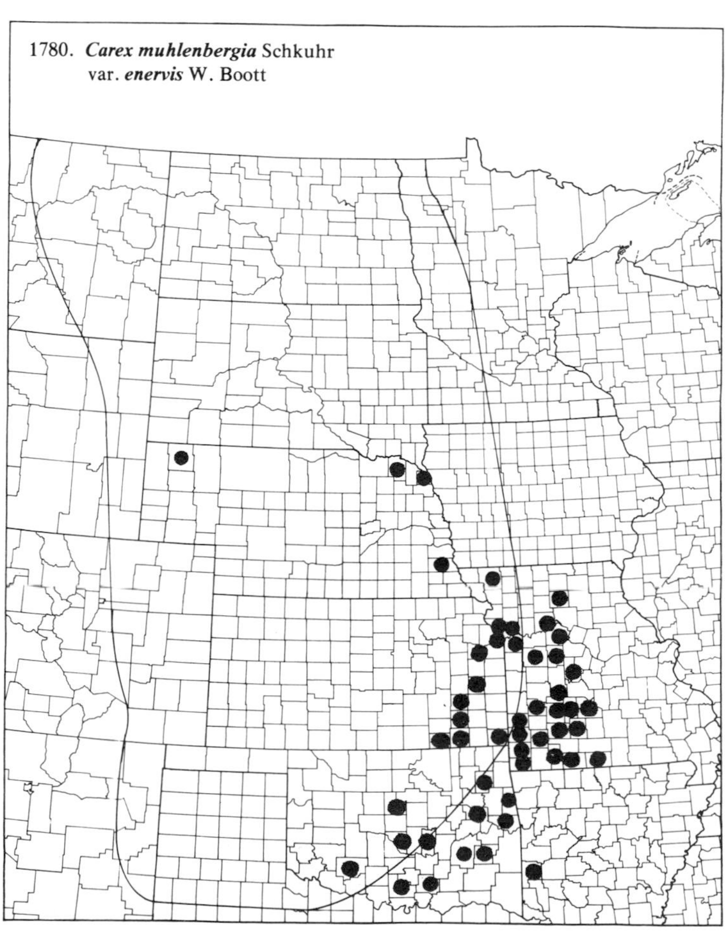

1780. *Carex muhlenbergia* Schkuhr var. *enervis* W. Boott

(Cyperaceae)

1781. ***Carex nebraskensis*** Dew.

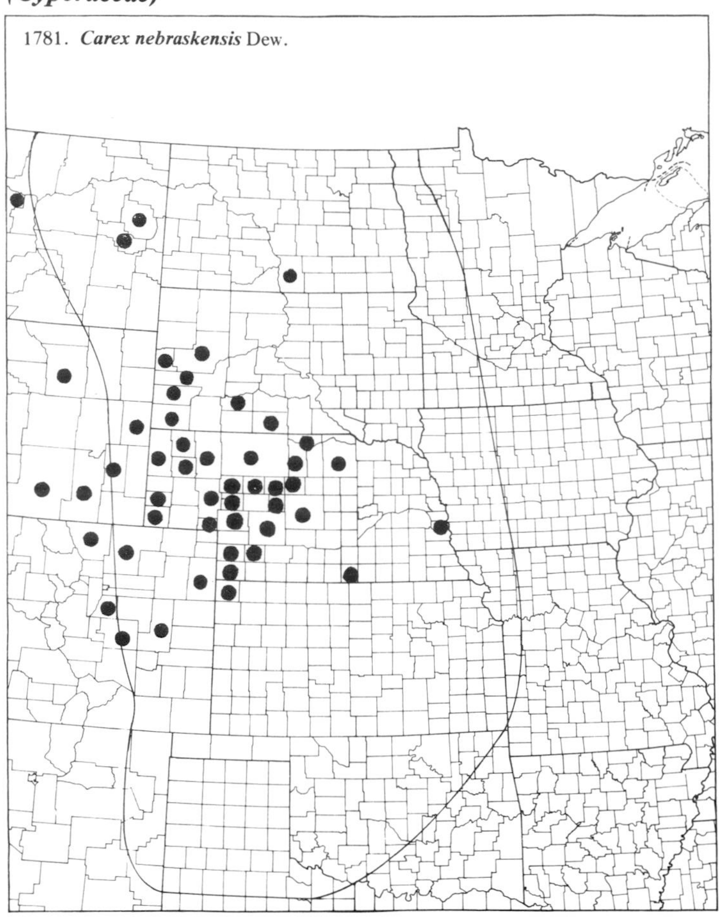

1782. ***Carex normalis*** Mack.

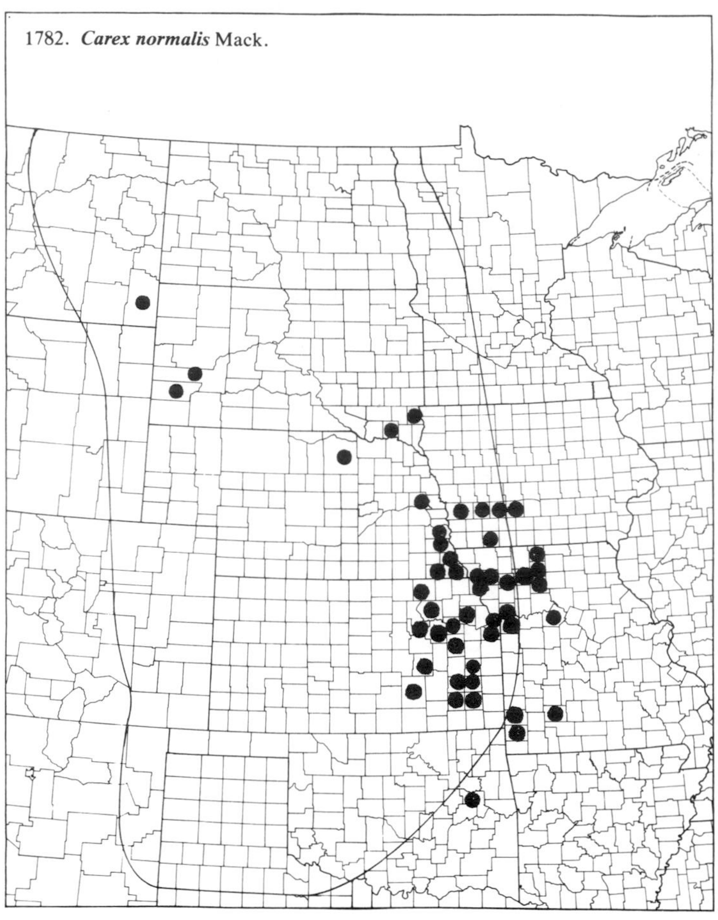

1783. ***Carex obtusata*** Lilj.

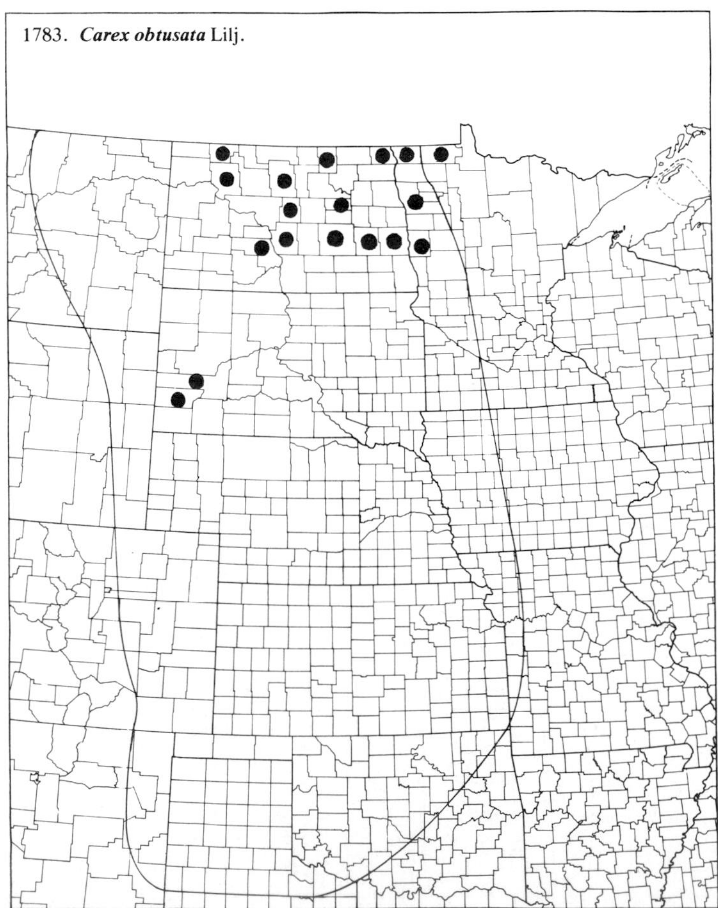

1784. ***Carex oligocarpa*** Schkuhr

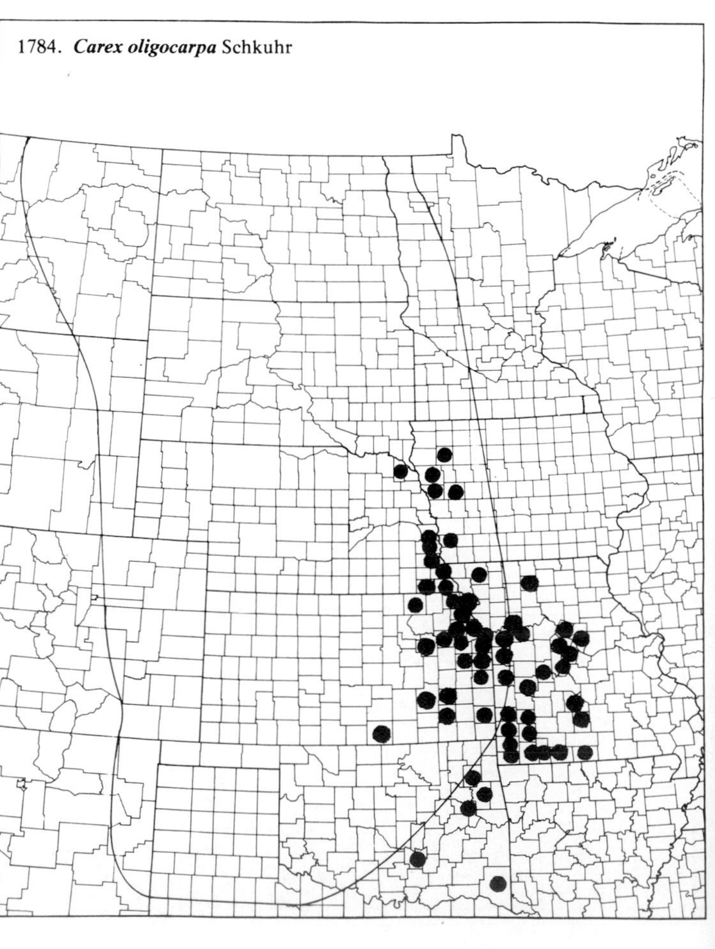

(Cyperaceae)

1785. **Carex peckii** Howe

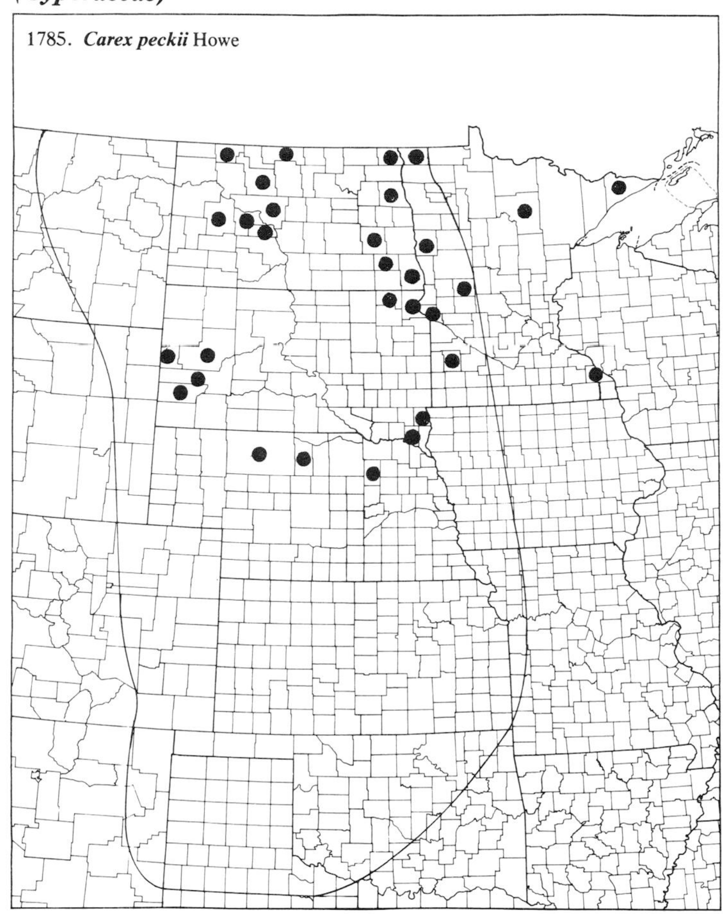

1786. **Carex pensylvanica** Lam. Pennsylvania Sedge

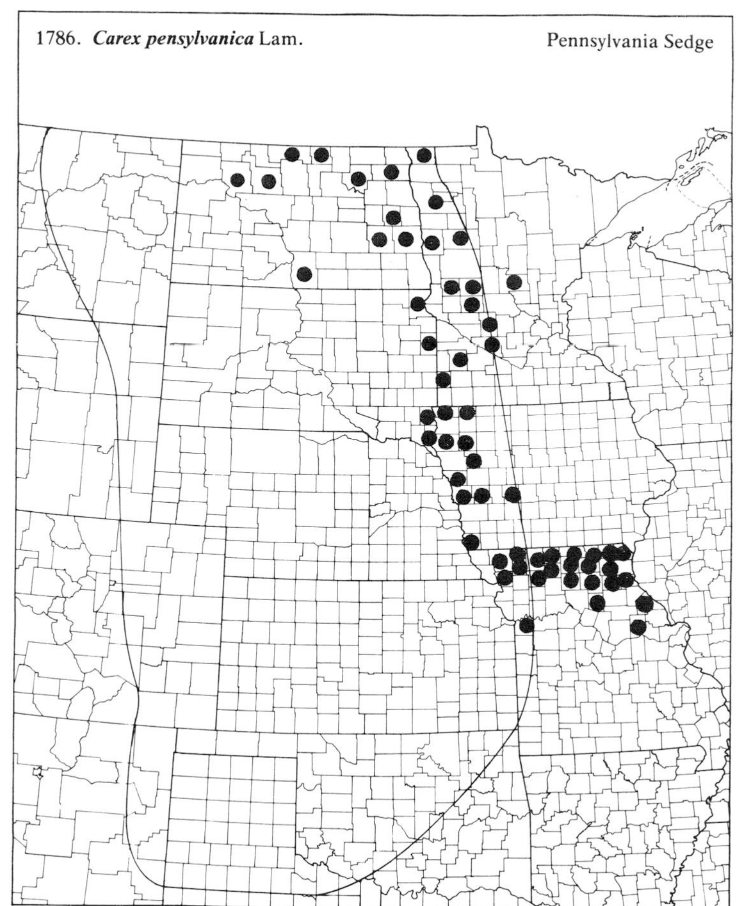

1787. **Carex praegracilis** W. Boott Clustered-field Sedge

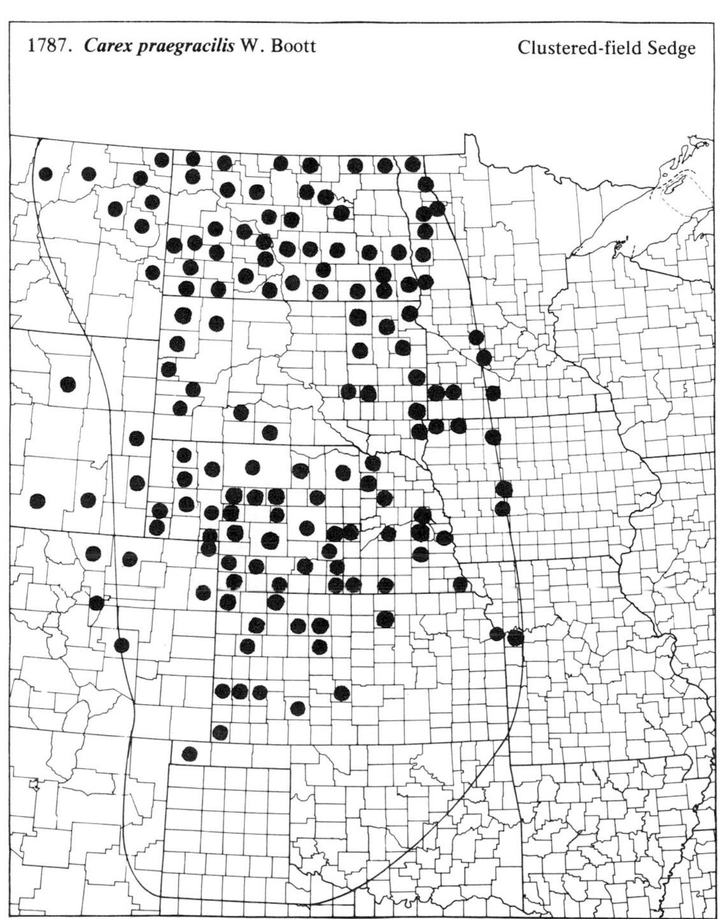

1788. **Carex prarisa** Dew.

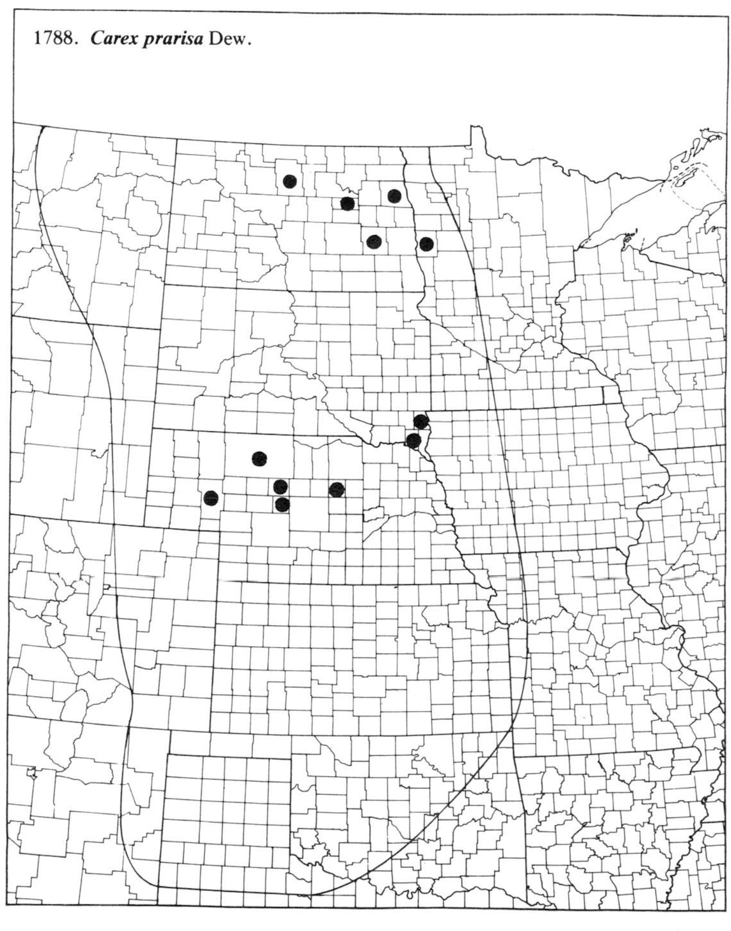

(Cyperaceae)

1789. *Carex retrorsa* Schwein.

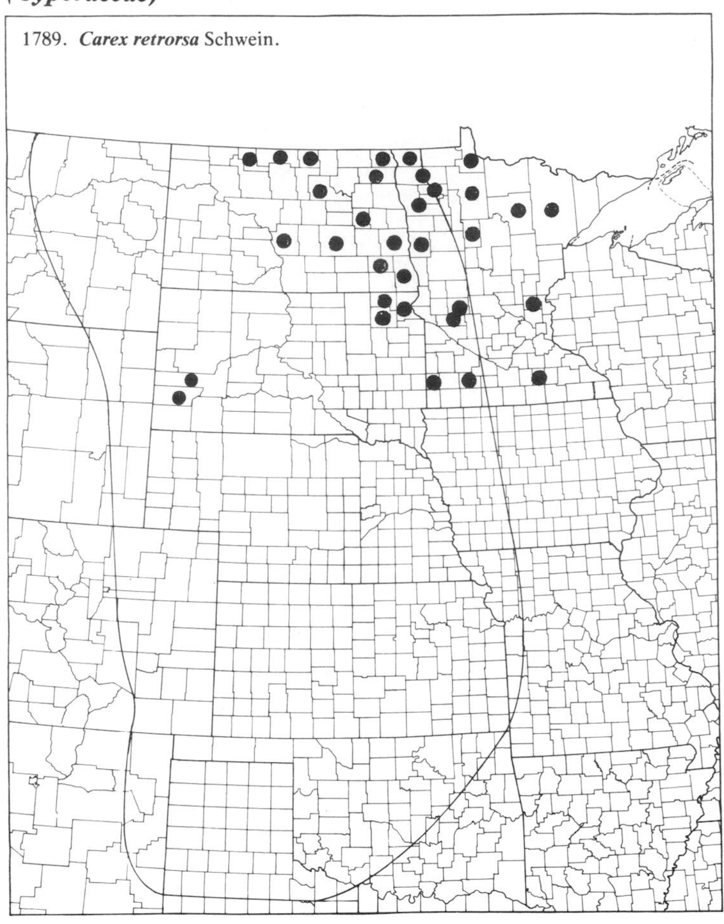

1790. *Carex rosea* Schkuhr

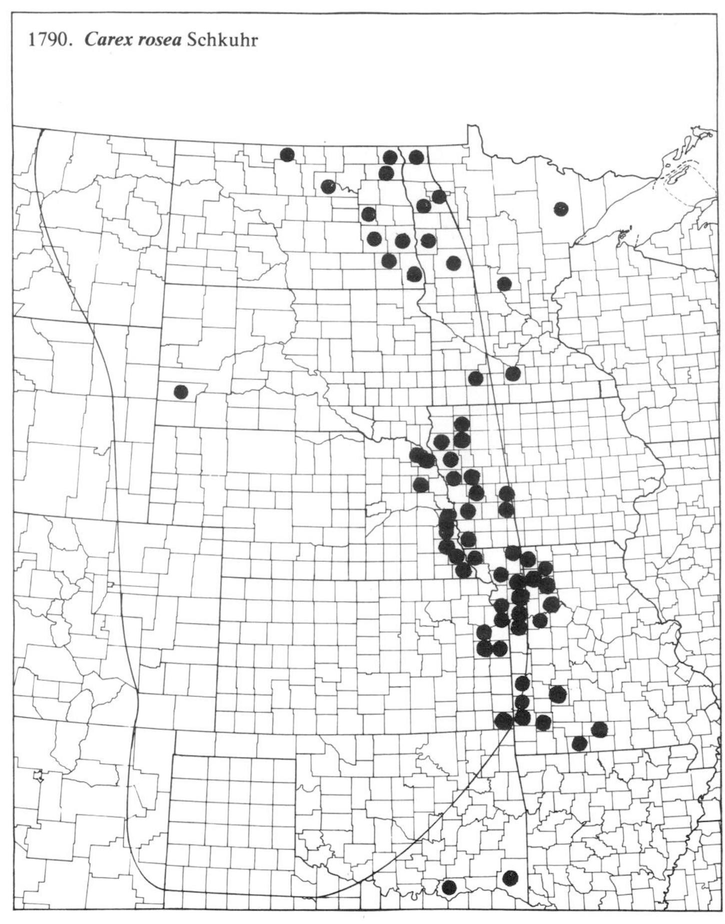

1791. *Carex rostrata* Stokes

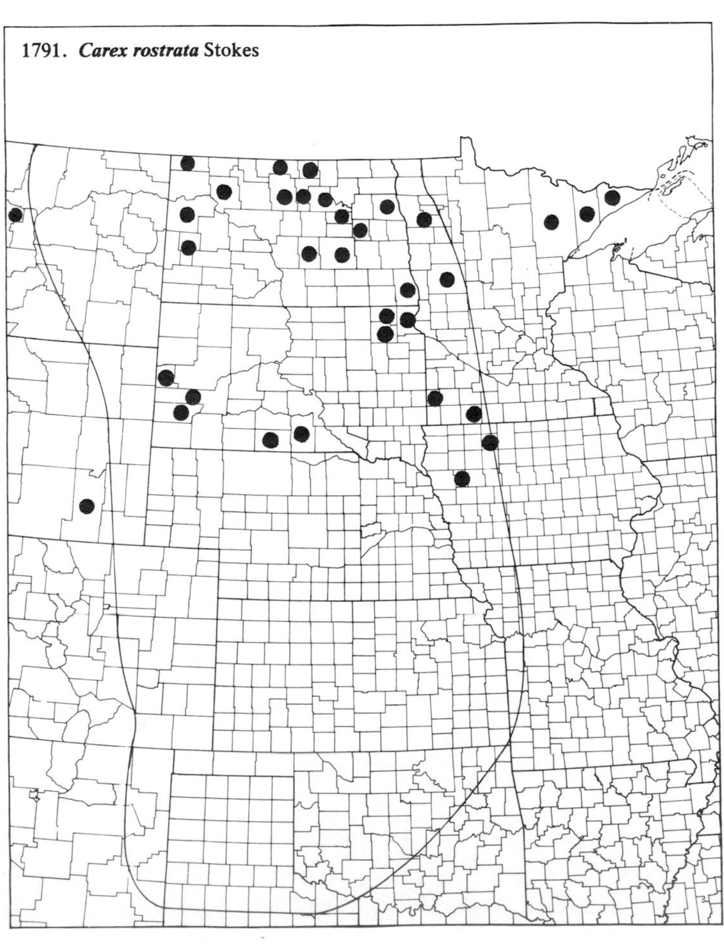

1792. *Carex sartwellii* Dew.

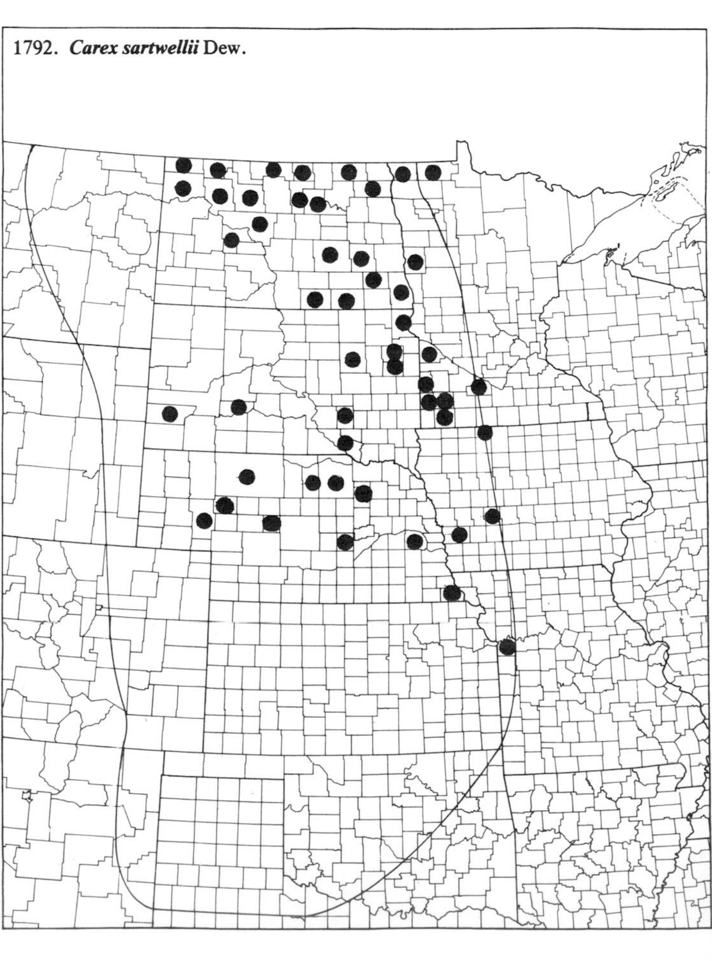

(Cyperaceae)

1793. ***Carex saximontana*** Mack.

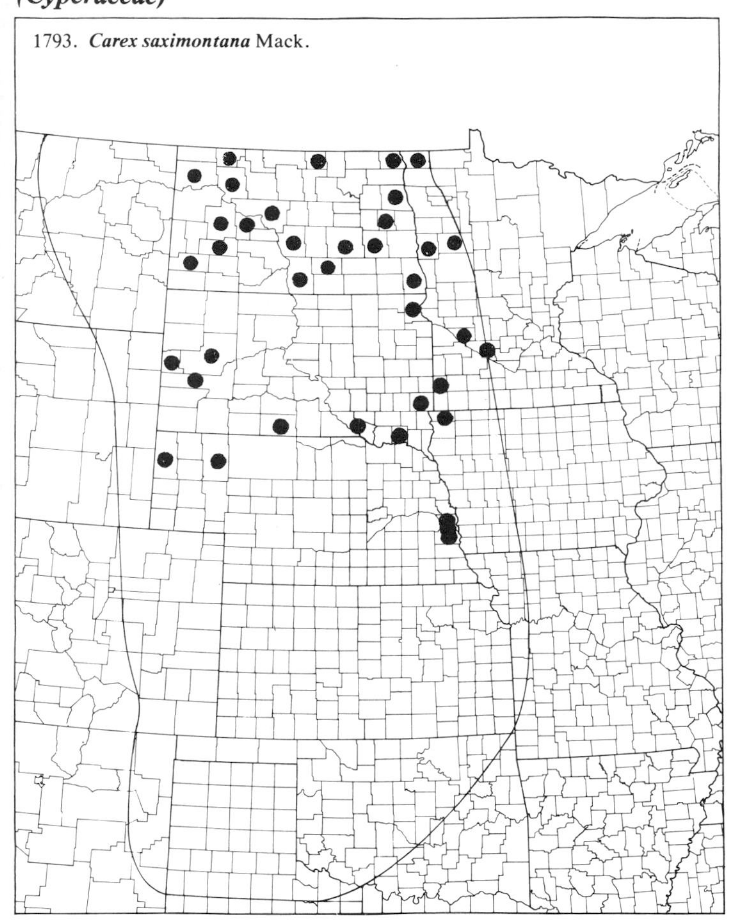

1794. ***Carex scoparia*** Schkuhr — Broom Sedge

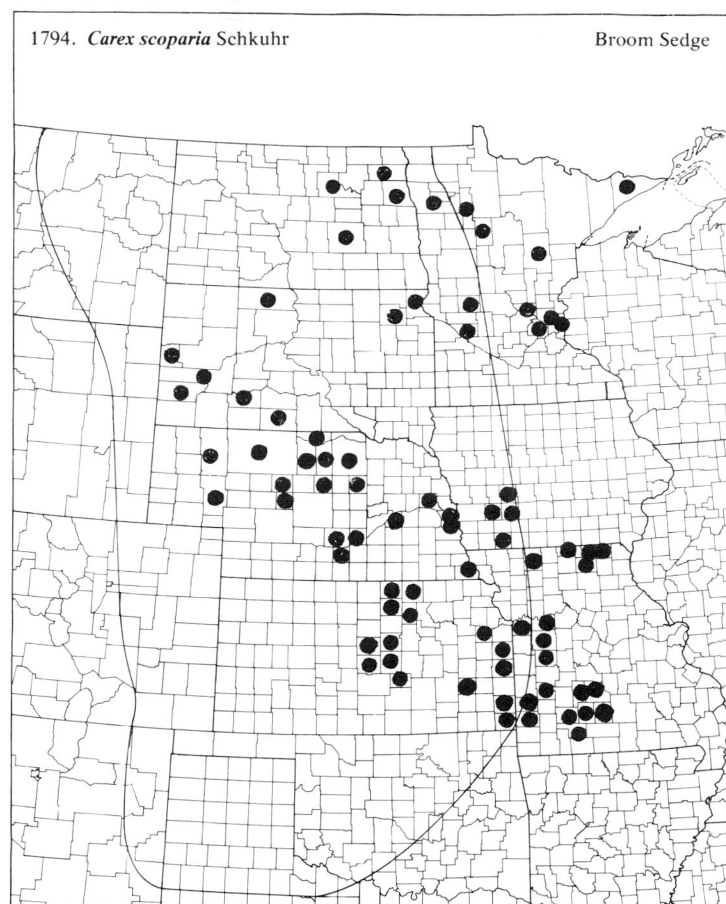

1795. ***Carex shortiana*** Dew.

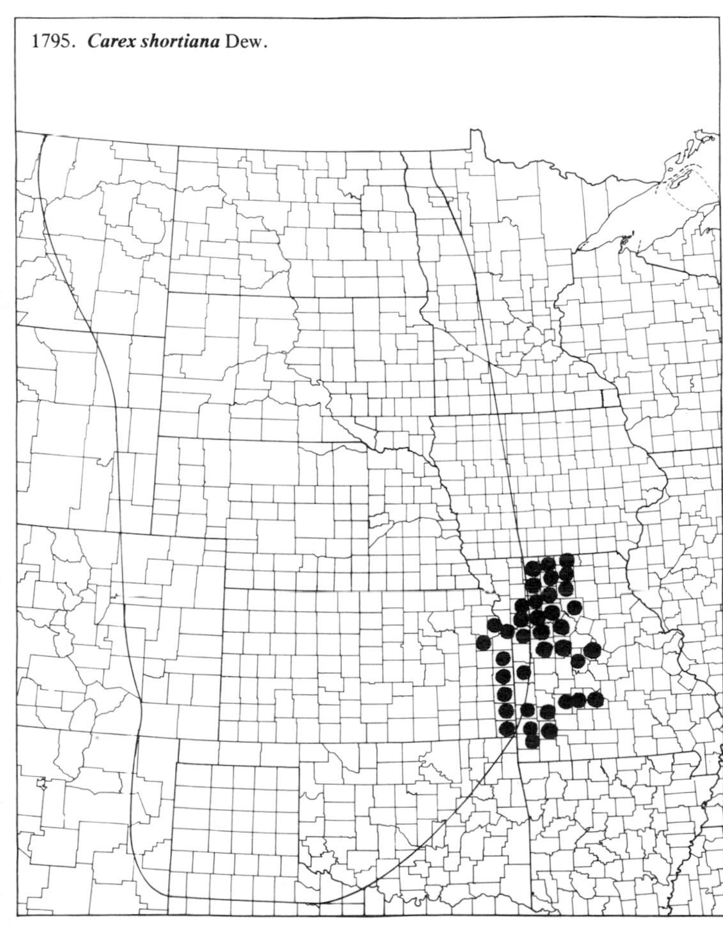

1796. ***Carex sparganioides*** Muhl.

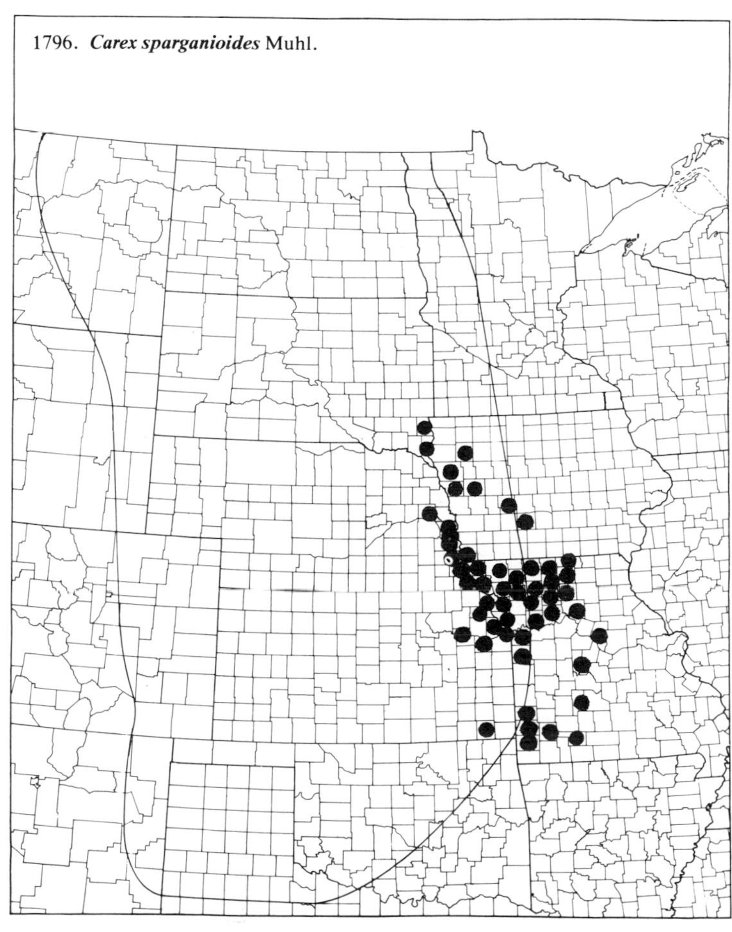

(Cyperaceae)

1797. ***Carex sprengelii*** Dew. Long-beaked Sedge

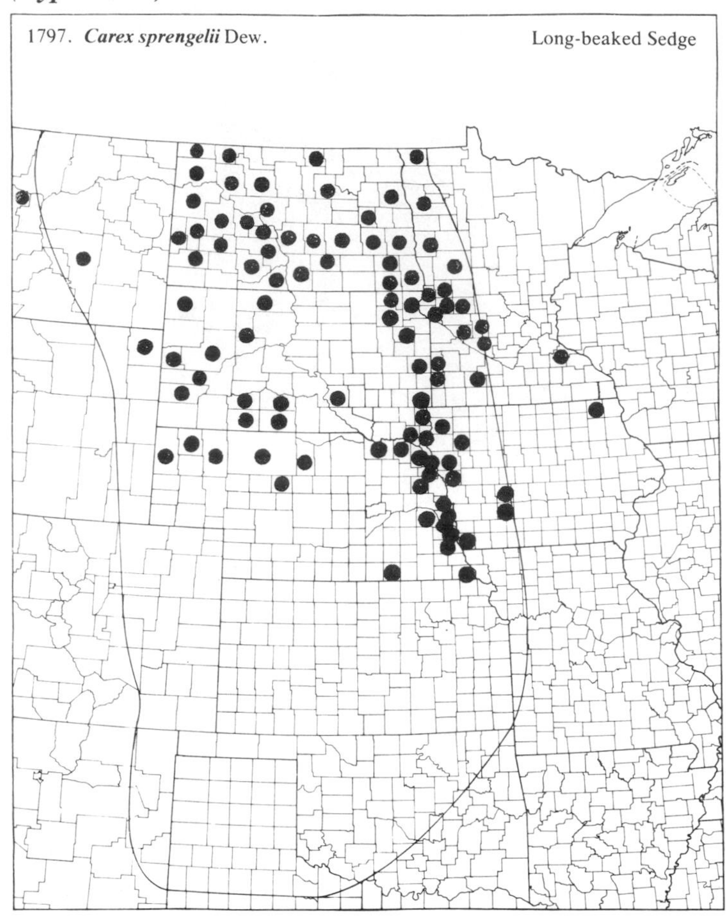

1798. ***Carex stipata*** Muhl. Saw-beak Sedge

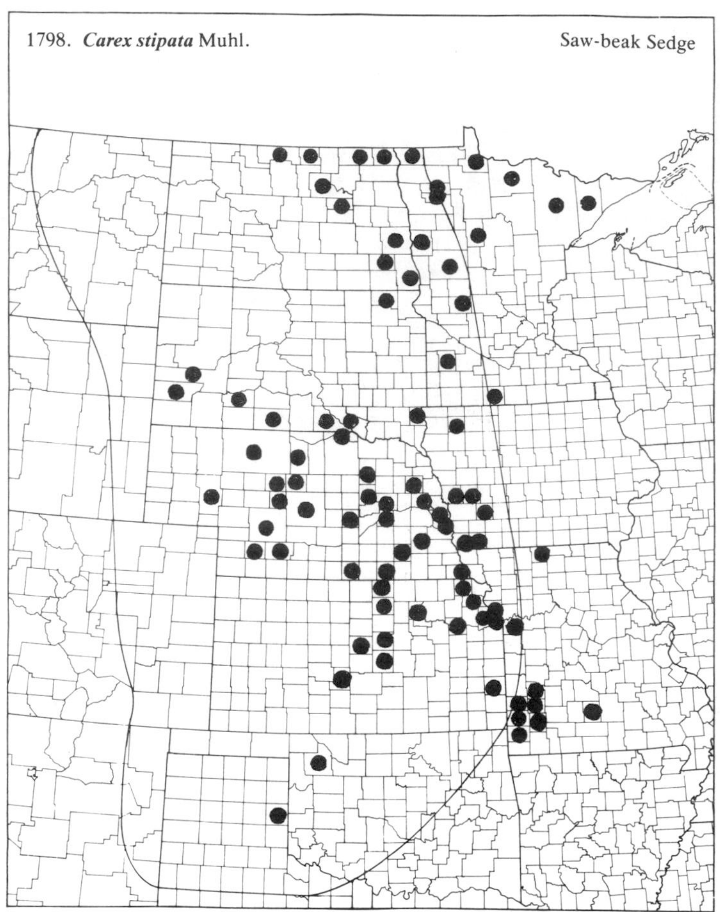

1799. ***Carex stricta*** Lam.

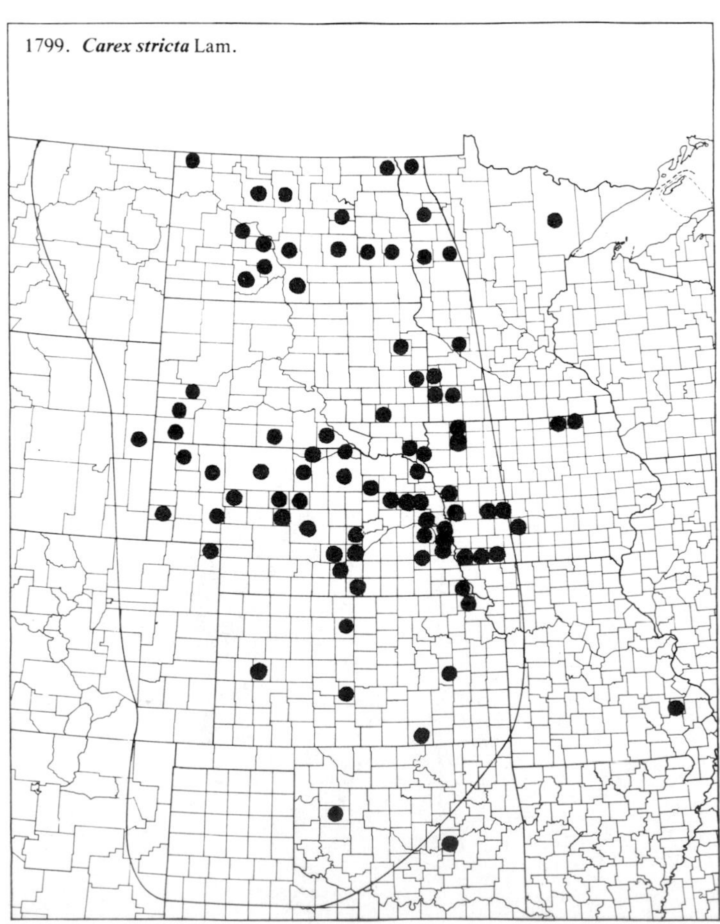

1800. ***Carex sychnocephala*** Carey

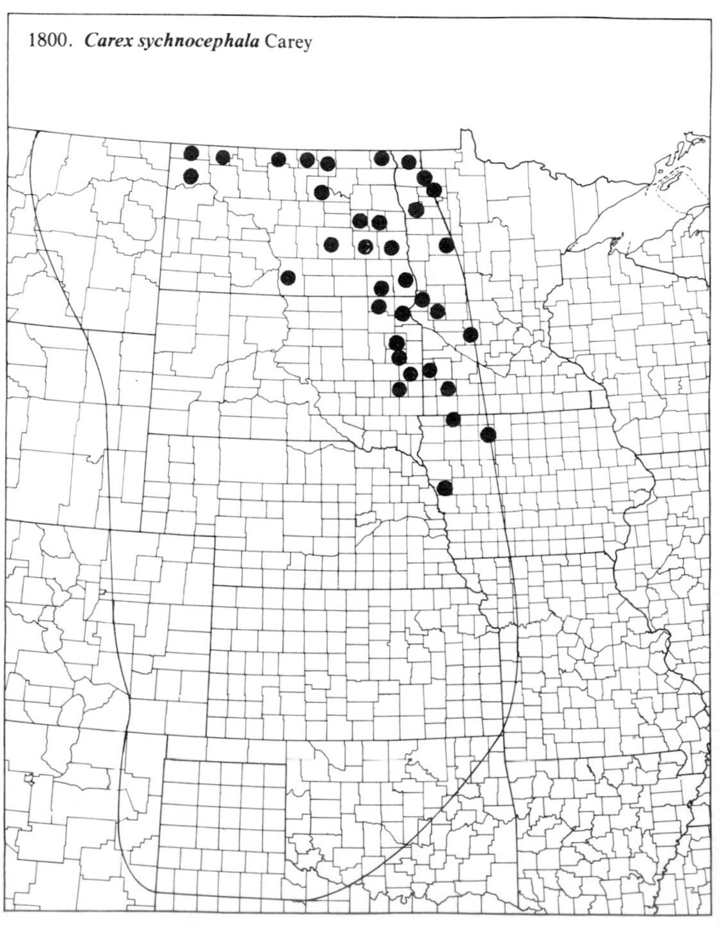

(Cyperaceae)

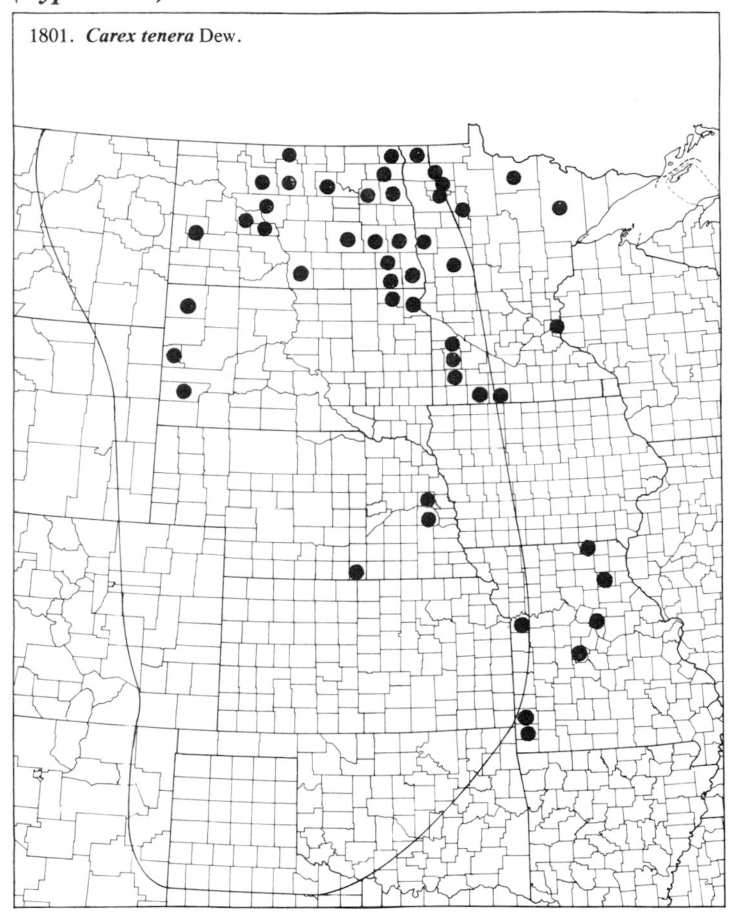

1801. *Carex tenera* Dew.

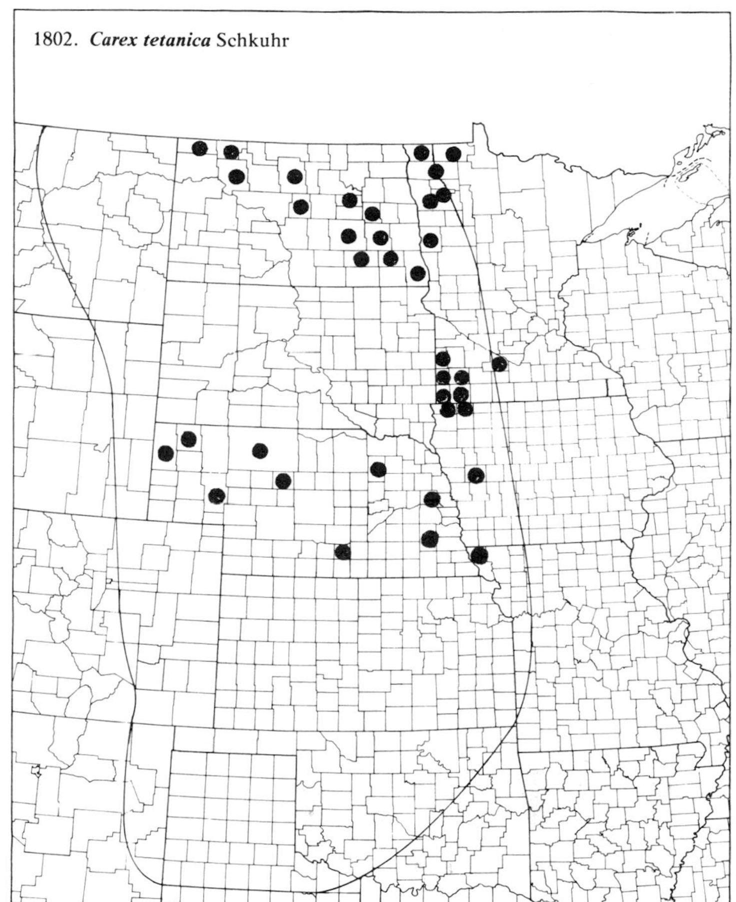

1802. *Carex tetanica* Schkuhr

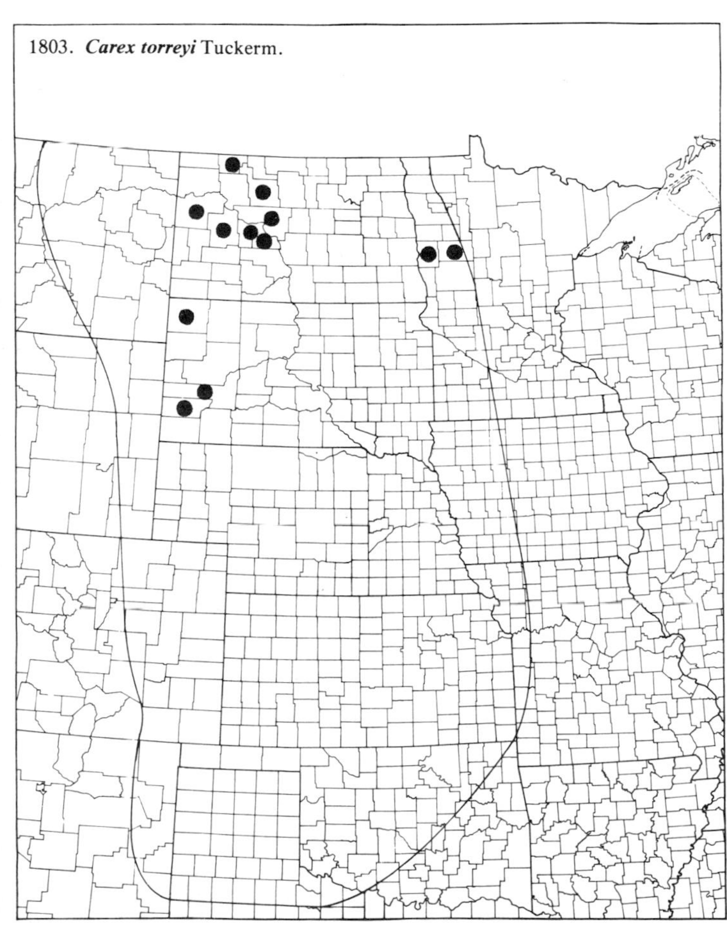

1803. *Carex torreyi* Tuckerm.

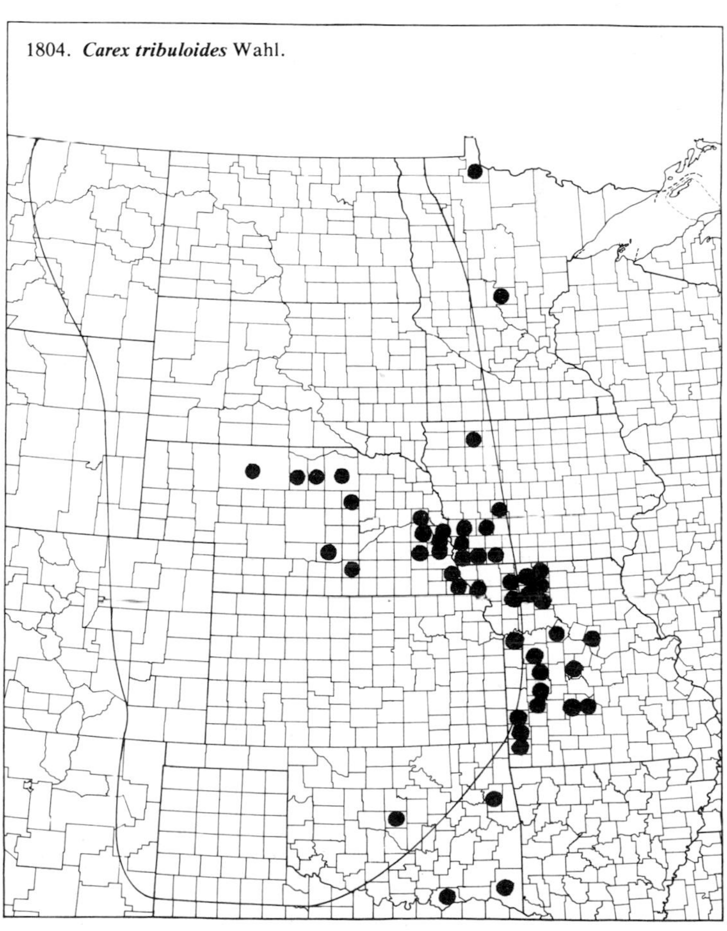

1804. *Carex tribuloides* Wahl.

(Cyperaceae)

1805. **Carex viridula** Michx.

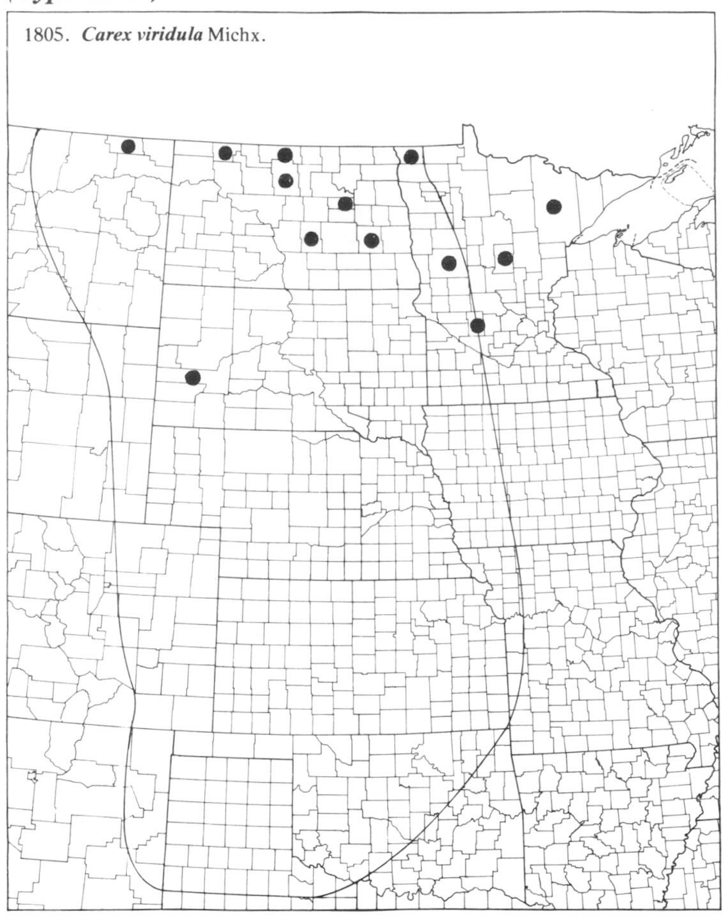

1806. **Carex vulpinoidea** Michx. Fox Sedge

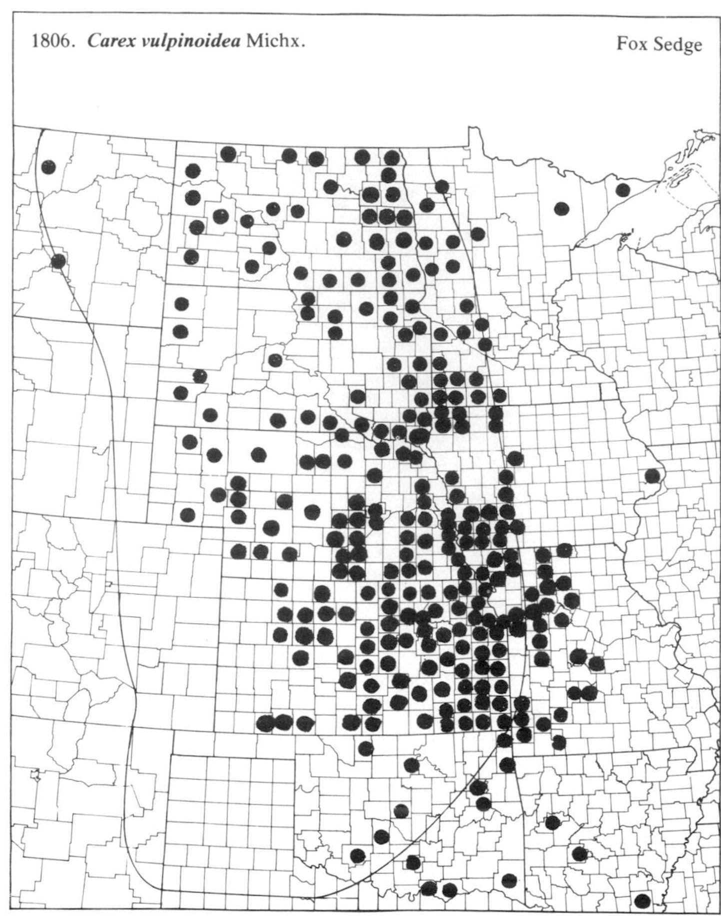

1807. **Carex xerantica** Bailey

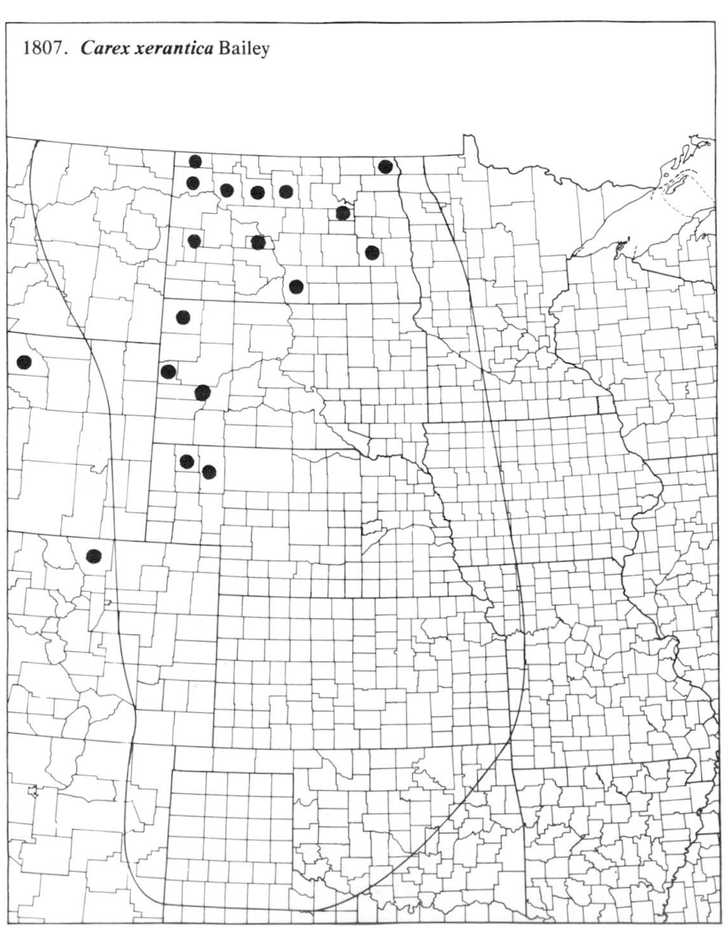

1808. **Cyperus acuminatus** T. & H. Tapeleaf Flatsedge

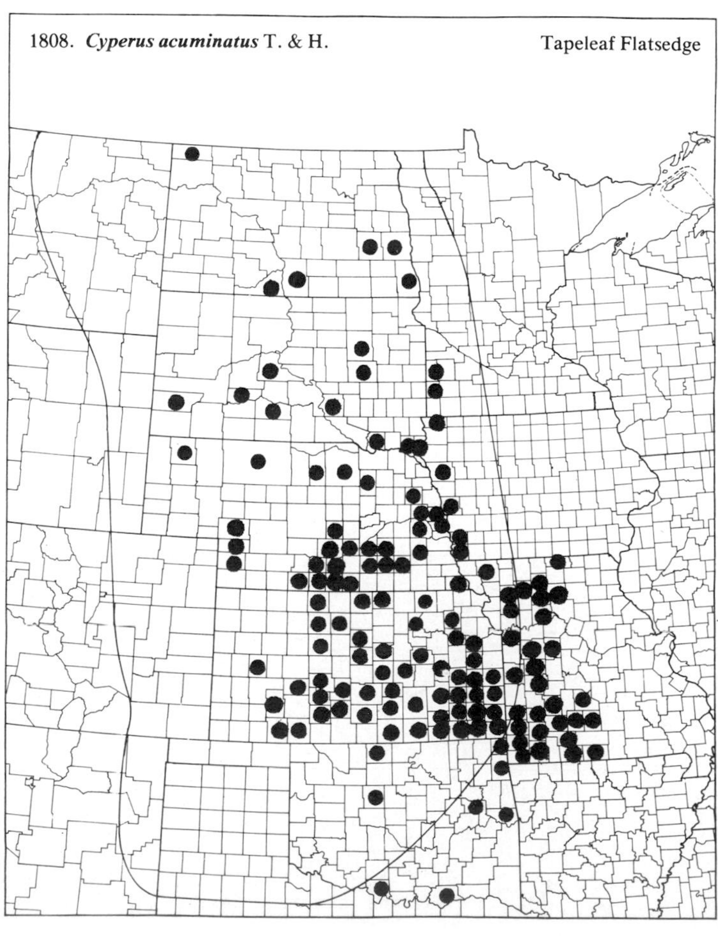

(Cyperaceae)

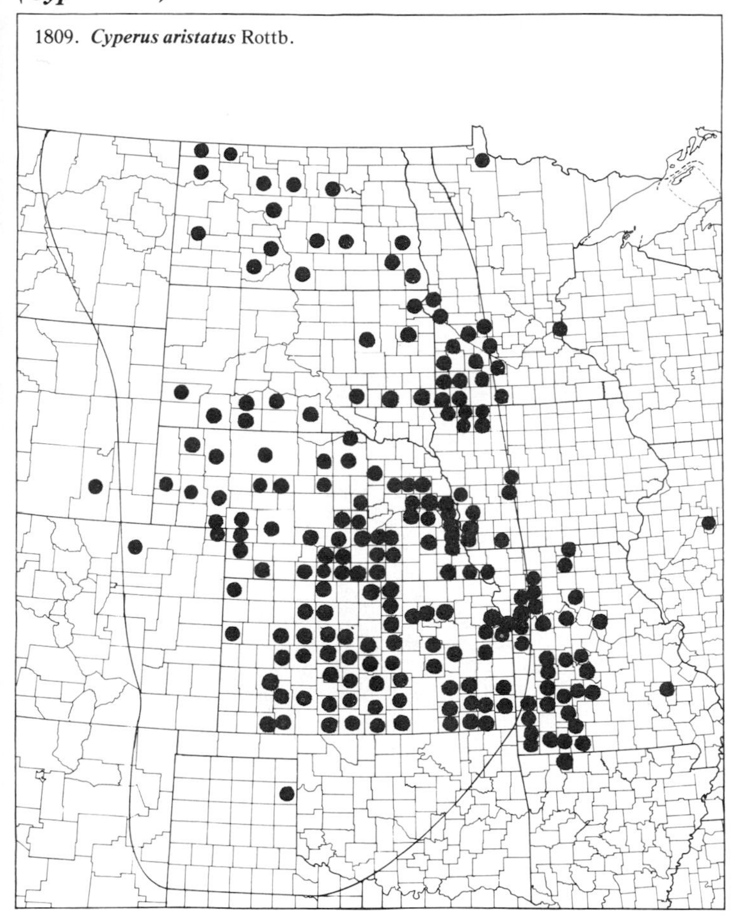
1809. Cyperus aristatus Rottb.

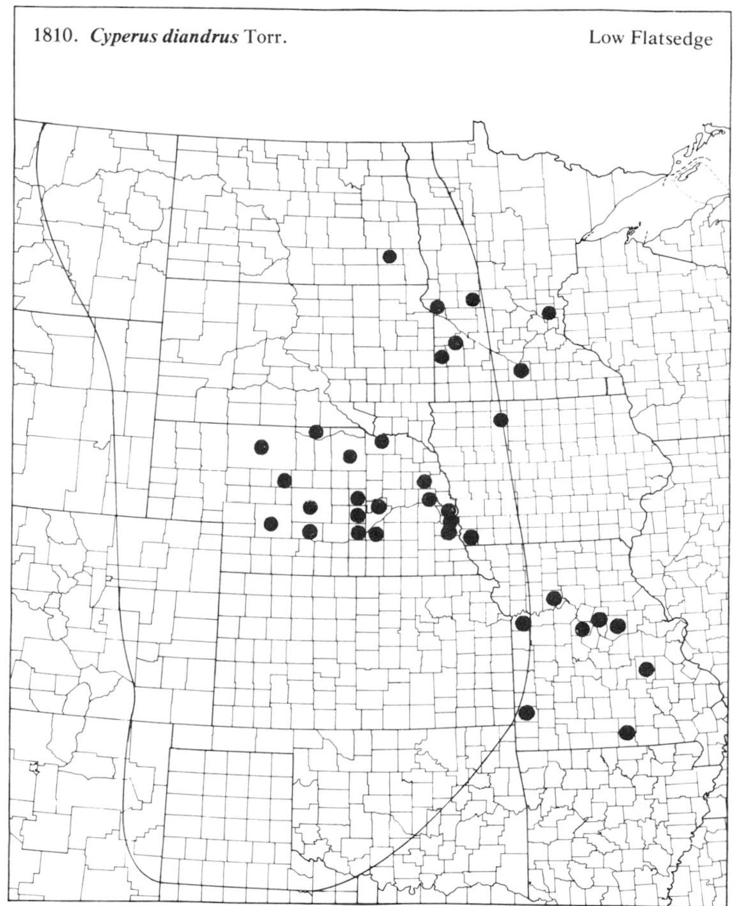
1810. Cyperus diandrus Torr.
Low Flatsedge

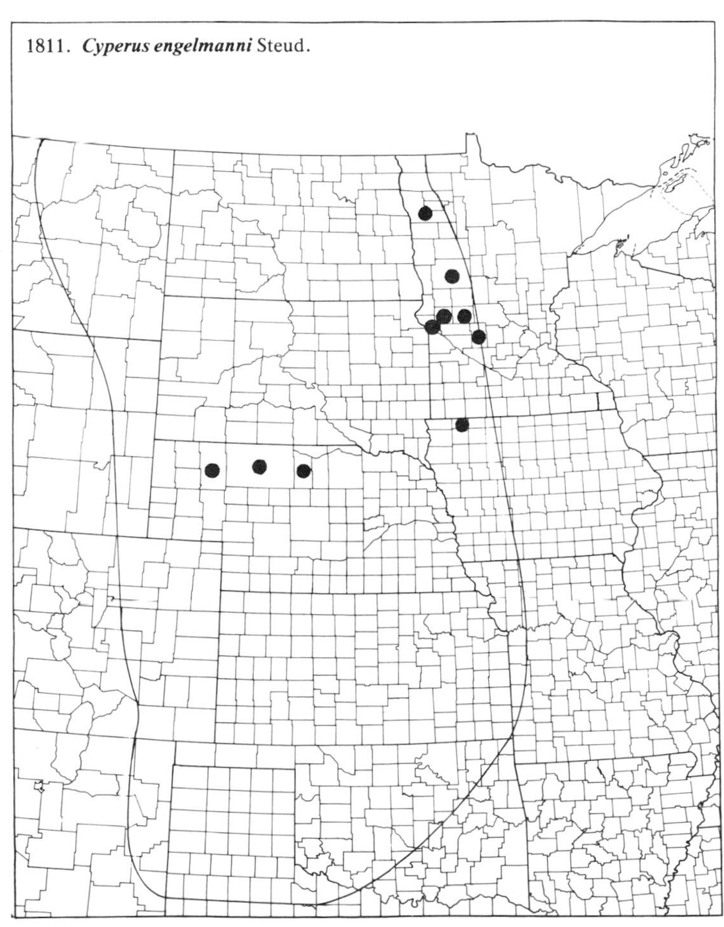
1811. Cyperus engelmanni Steud.

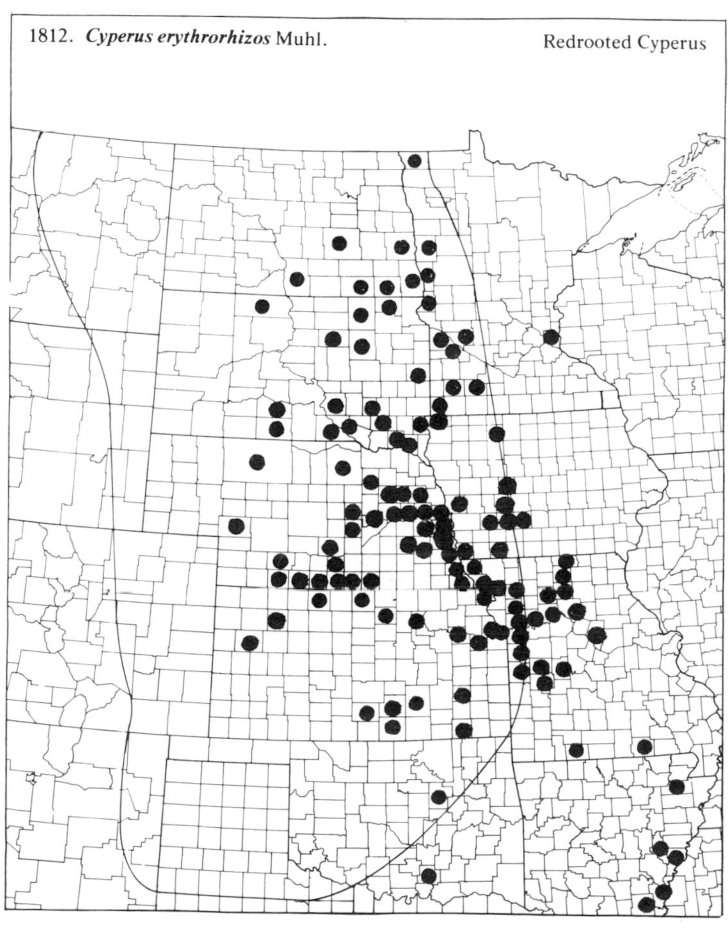
1812. Cyperus erythrorhizos Muhl.
Redrooted Cyperus

(Cyperaceae)

1813. *Cyperus esculentus* L. Yellow Nutsedge

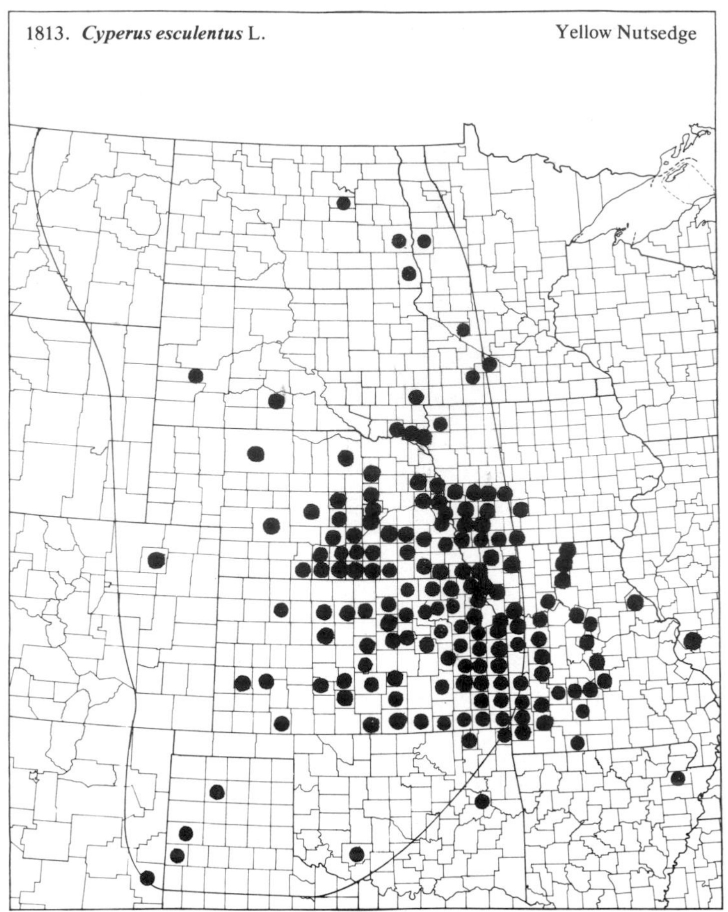

1814. *Cyperus ferruginescens* Boeck. Slender Flatsedge

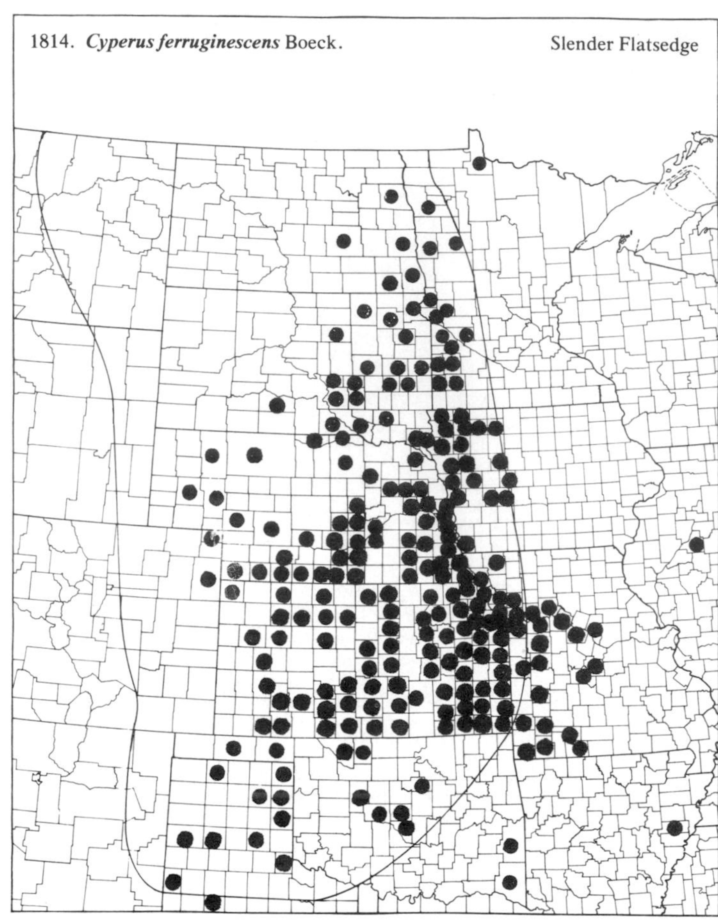

1815. *Cyperus filiculmis* Vahl Fern Flatsedge

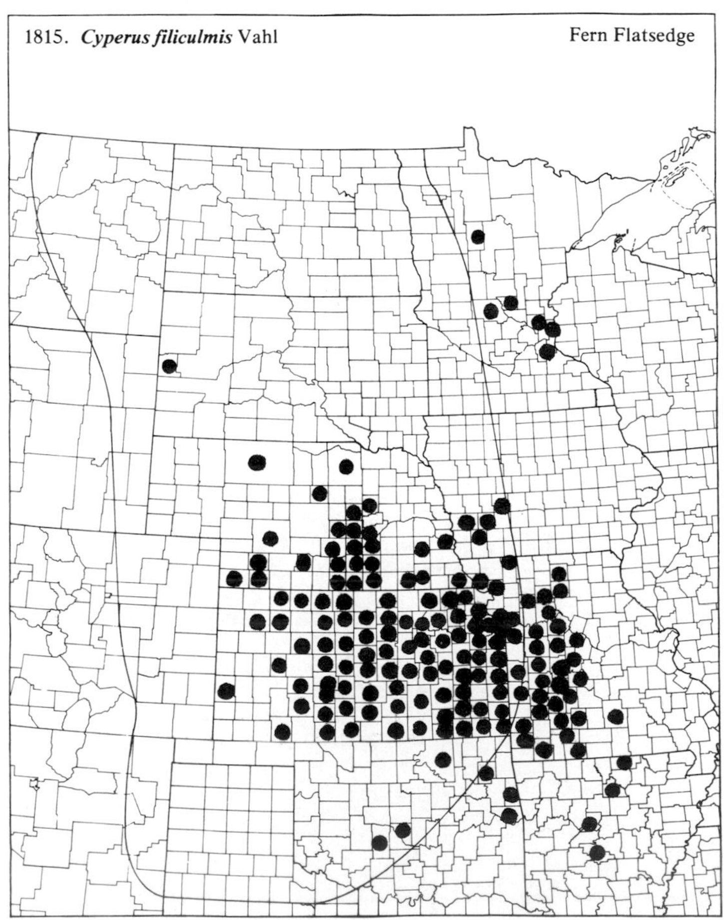

1816. *Cyperus lupulinus* (Spreng.) Marcks Houghton Flatsedge

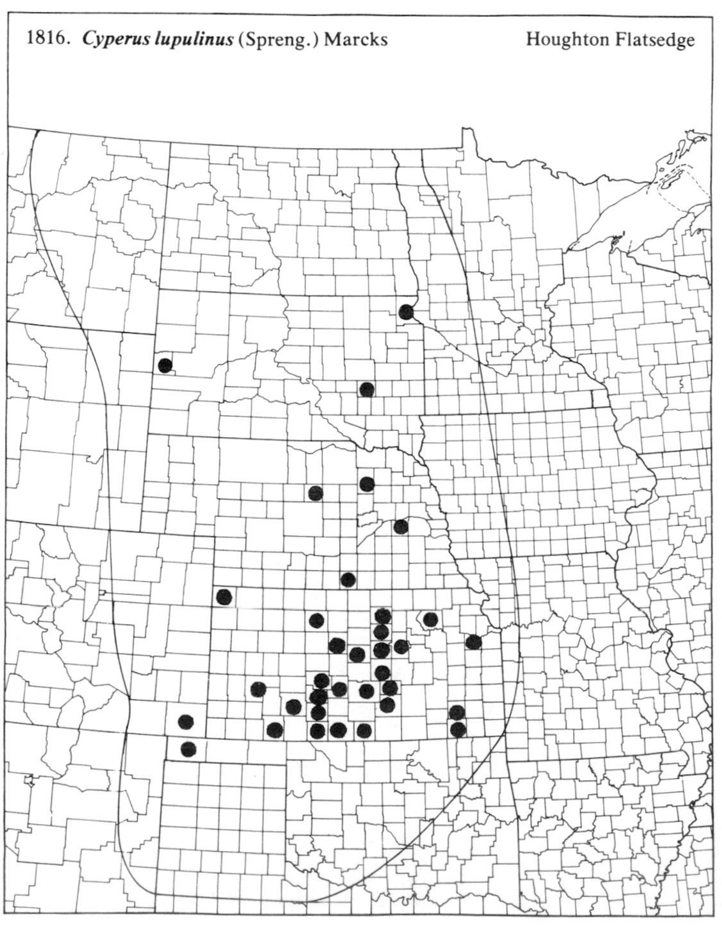

(Cyperaceae)

1817. ***Cyperus ovularis*** (Michx.) Torr. Globe Flatsedge

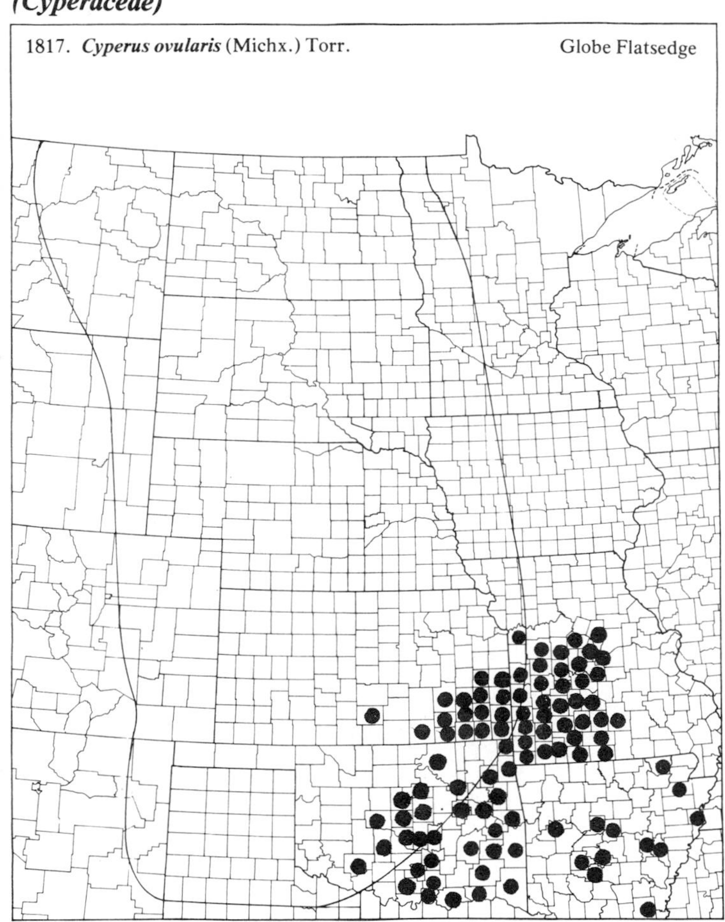

1818. ***Cyperus pseudovegetus*** Steud.

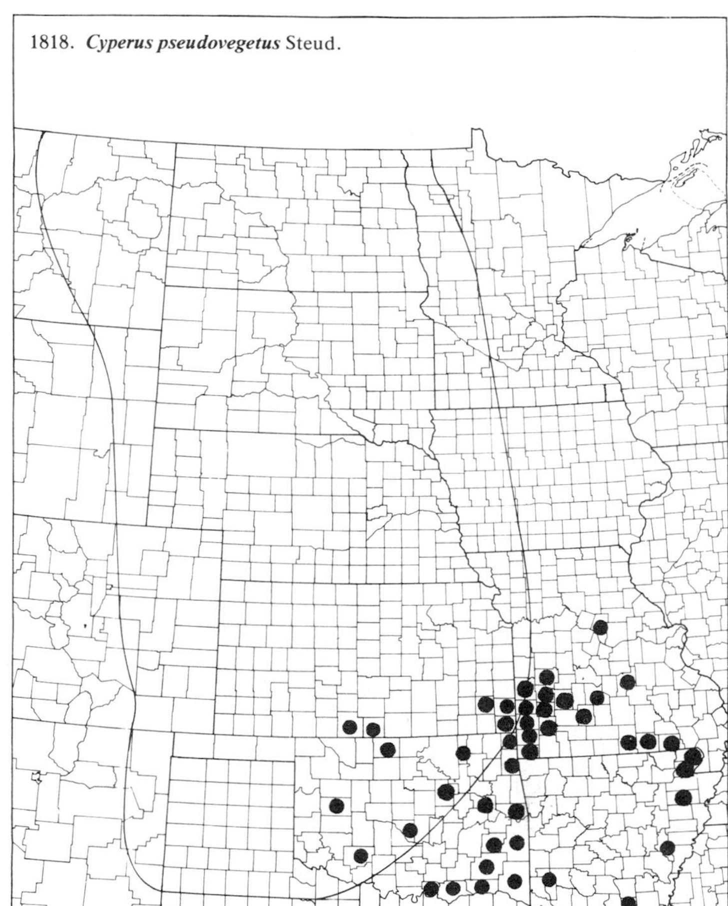

1819. ***Cyperus rivularis*** Kunth Brook Flatsedge

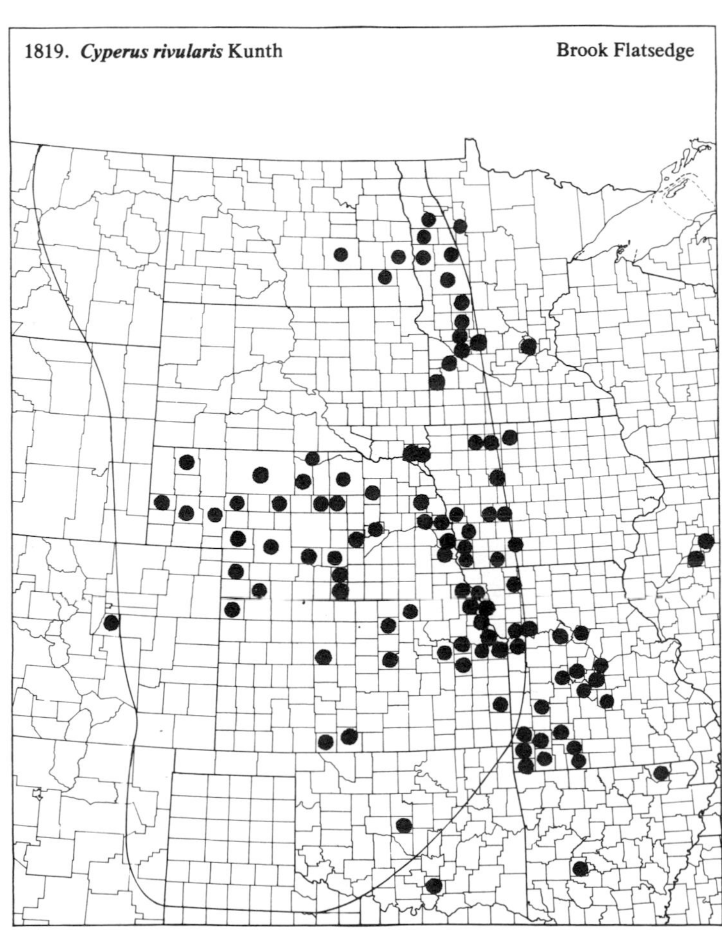

1820. ***Cyperus schweinitzii*** Torr. Schweinitz Flatsedge

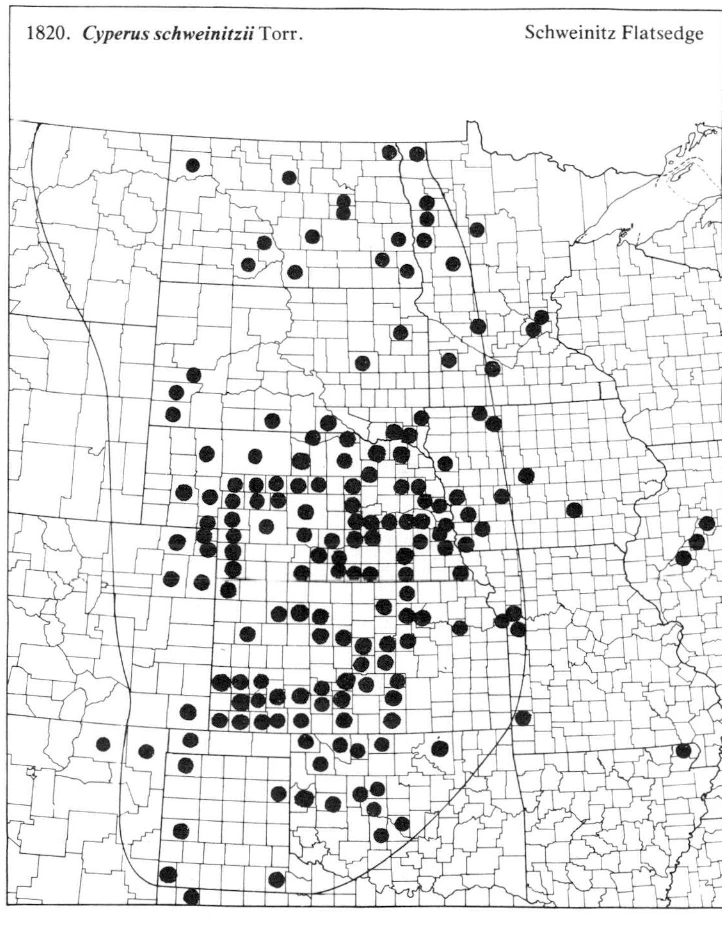

(Cyperaceae)

1821. ***Cyperus setigerus*** T. & H.

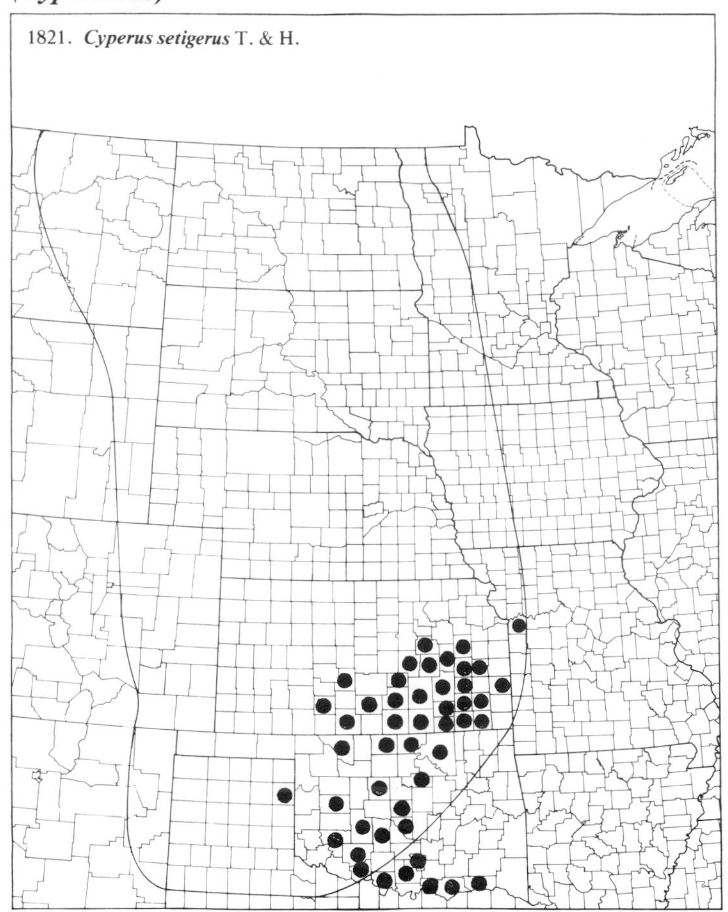

1822. ***Cyperus strigosus*** L. False Nutgrass

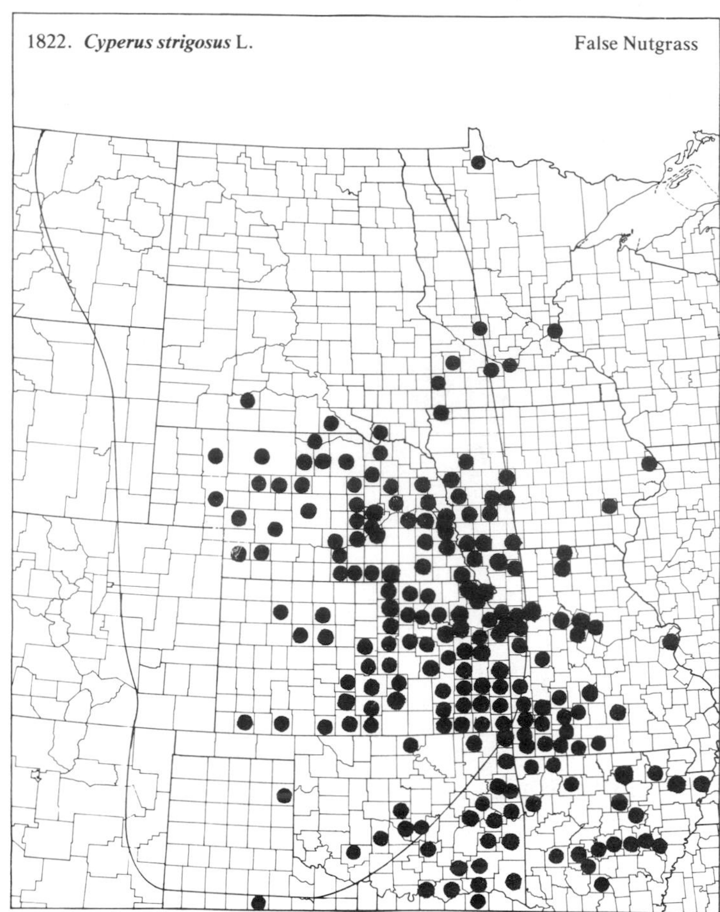

1823. ***Dulichium arundinaceum*** (L.) Britt. Three-way Sedge

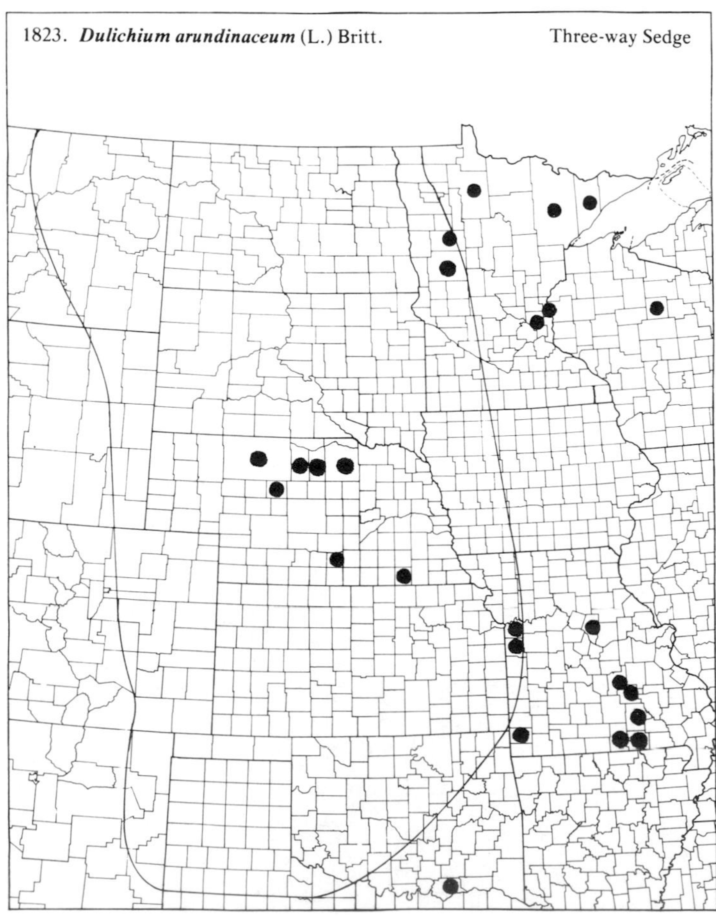

1824. ***Eleocharis acicularis*** (L.) R. & S. Needle Spikesedge

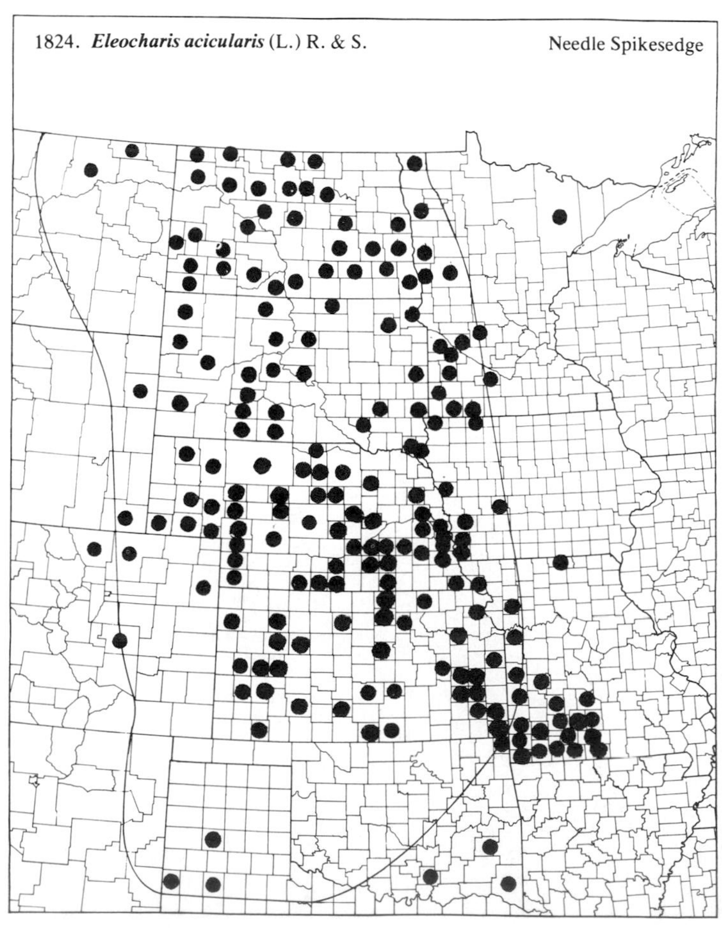

(Cyperaceae)

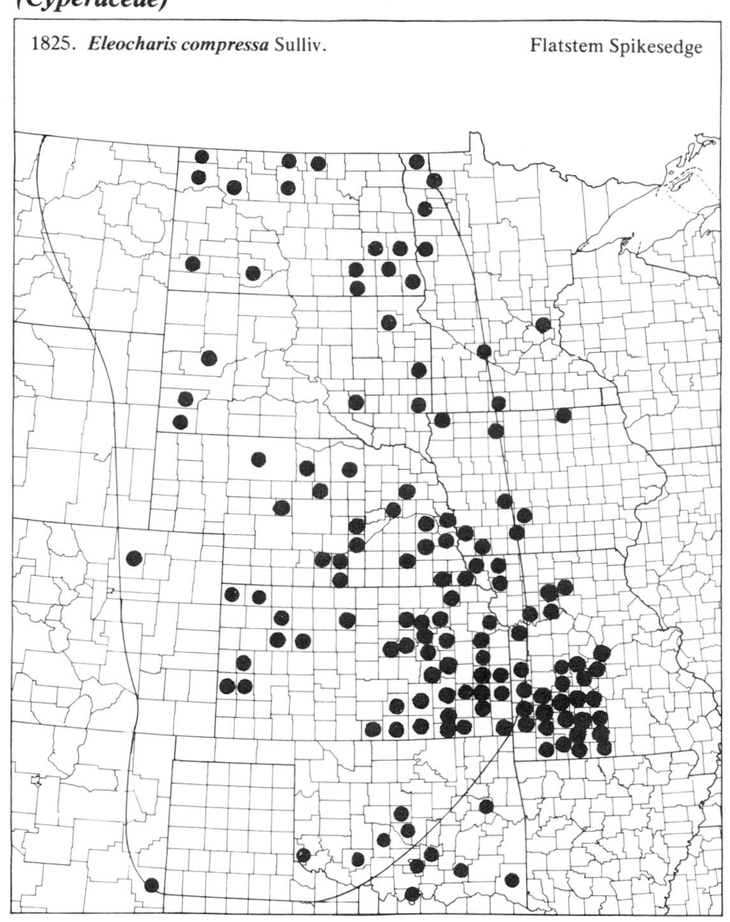
1825. ***Eleocharis compressa*** Sulliv. Flatstem Spikesedge

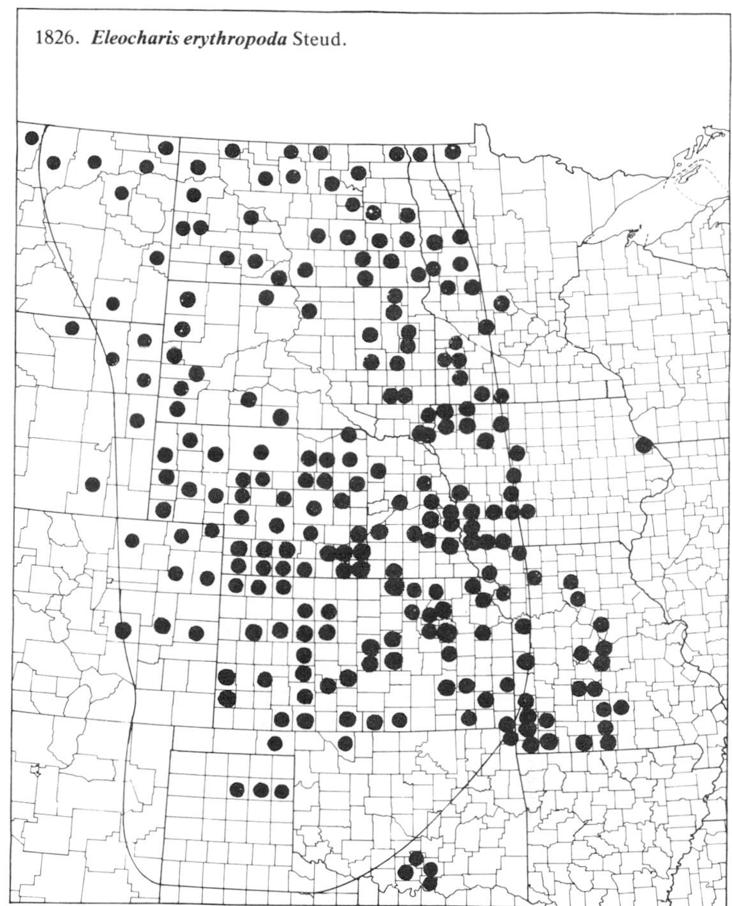
1826. ***Eleocharis erythropoda*** Steud.

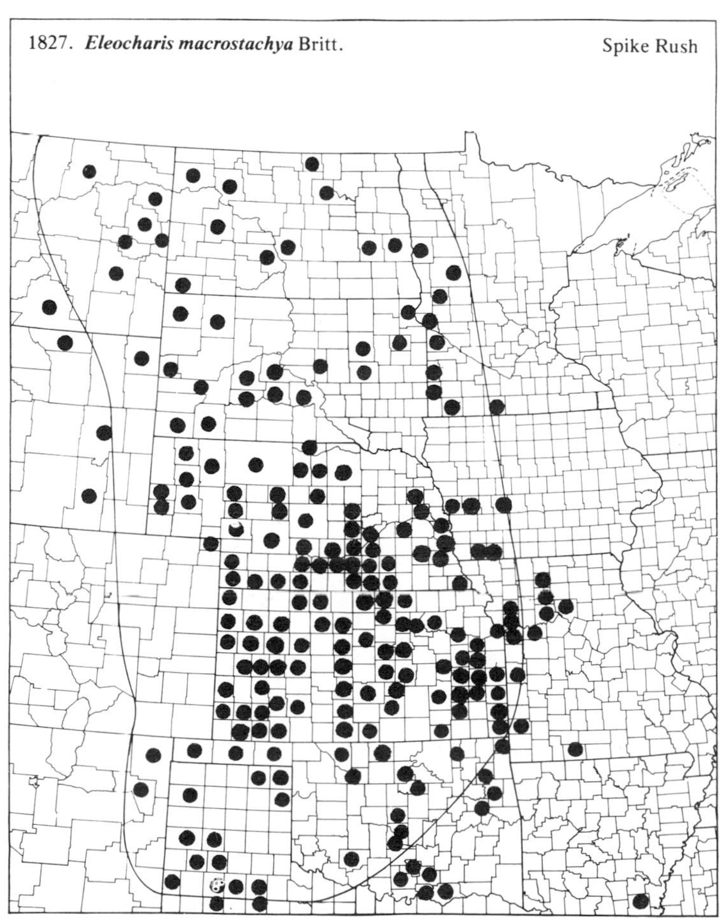
1827. ***Eleocharis macrostachya*** Britt. Spike Rush

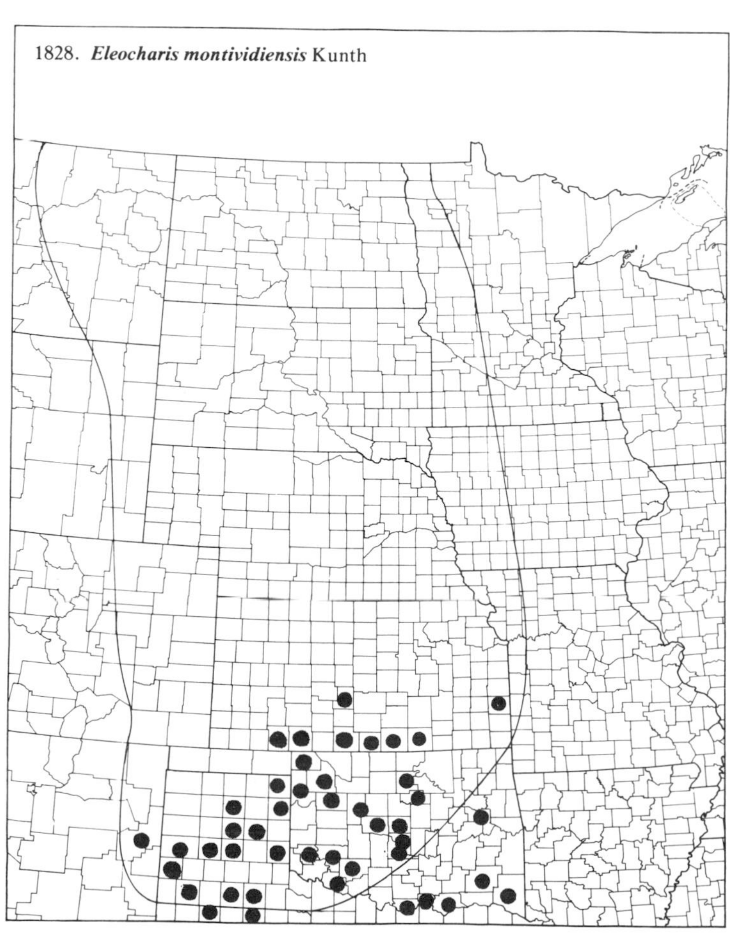
1828. ***Eleocharis montividiensis*** Kunth

(Cyperaceae)

1829. ***Eleocharis obtusa*** (Willd.) Schult. var. ***ovata*** (Roth) Drapalik & Mohlenbrock — Blunt Spikesedge

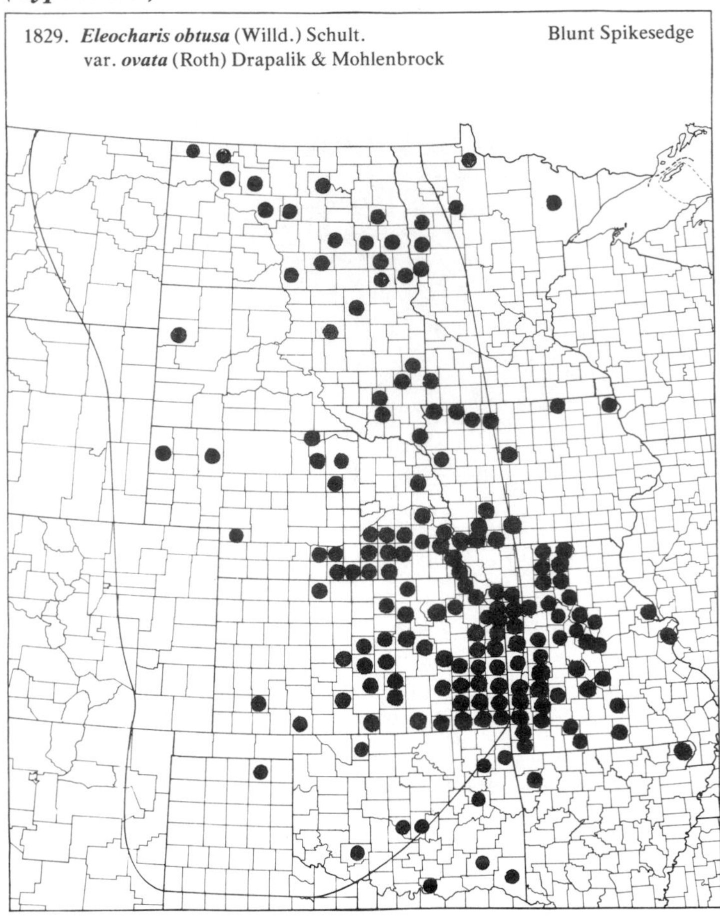

1830. ***Eleocharis parvula*** (R. & S.) Link var. ***anachaeta*** (Torr.) Svens.

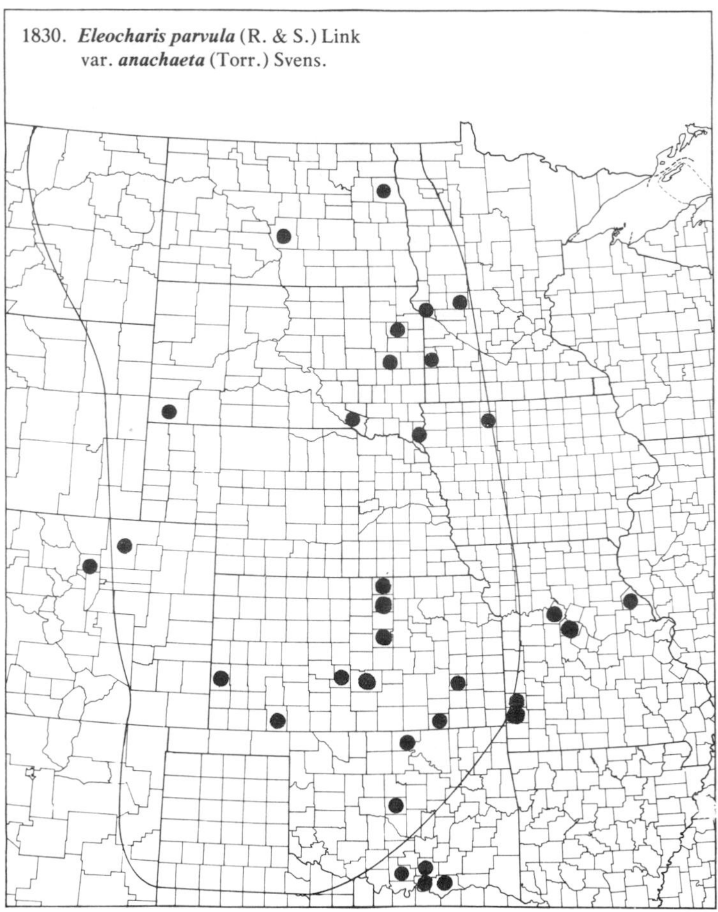

1831. ***Eleocharis rostellata*** Torr.

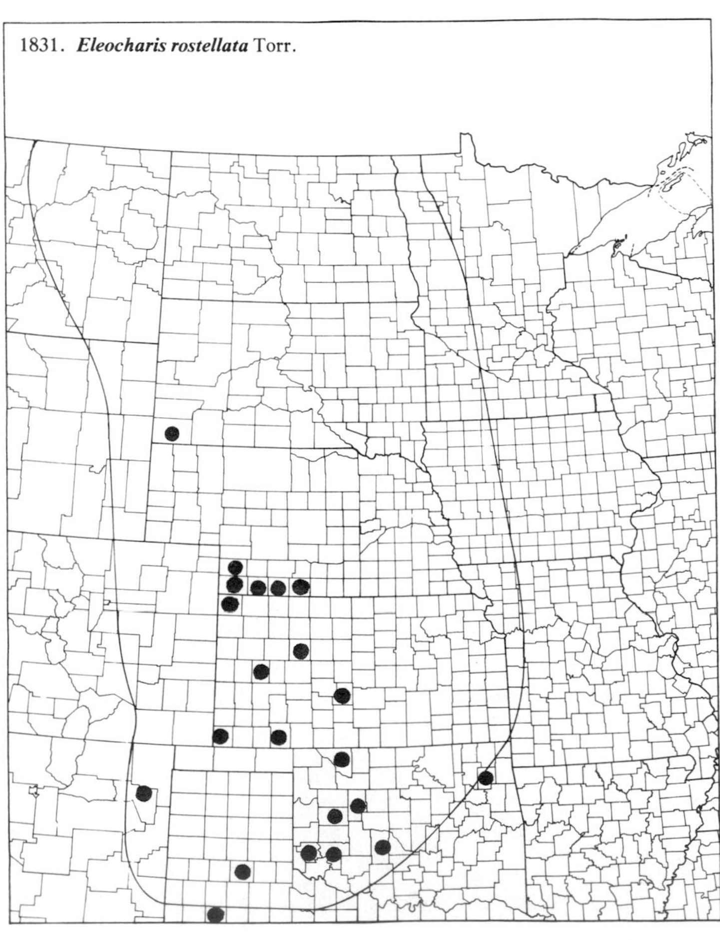

1832. ***Eleocharis smallii*** Britt.

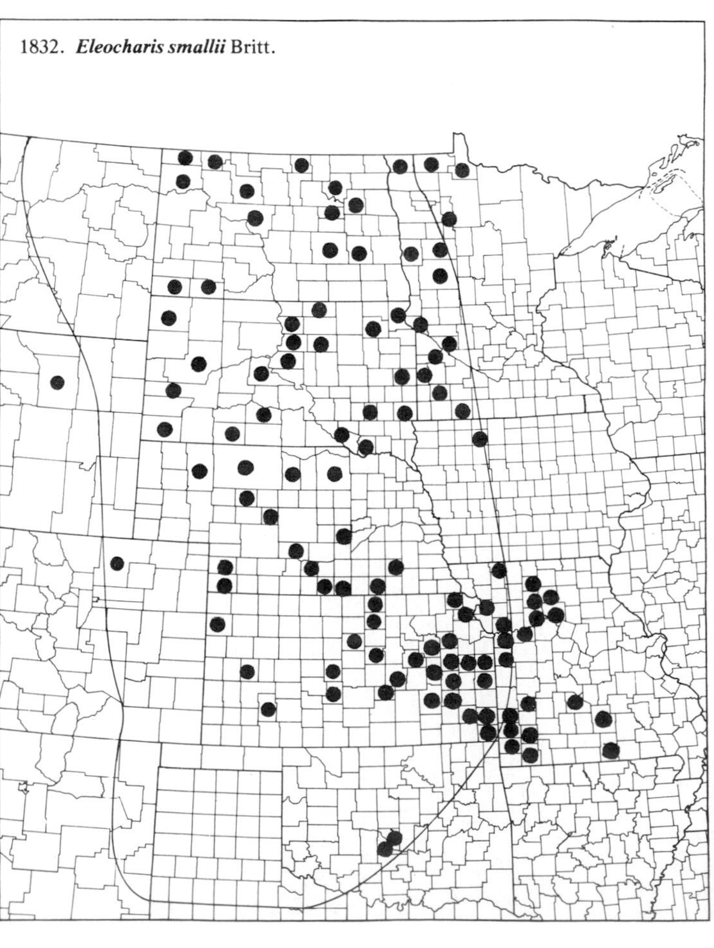

(Cyperaceae)

1833. ***Eleocharis tenuis*** (Willd.) Schult. var. ***verrucosa*** Svens.

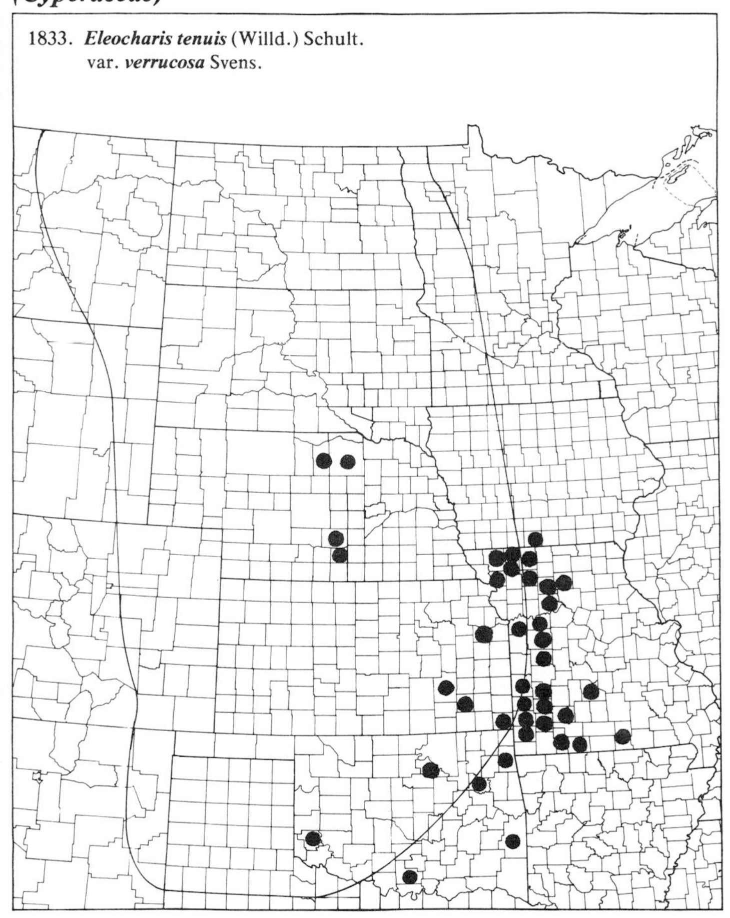

1834. ***Eleocharis wolfii*** Gray — Wolf Spikesedge

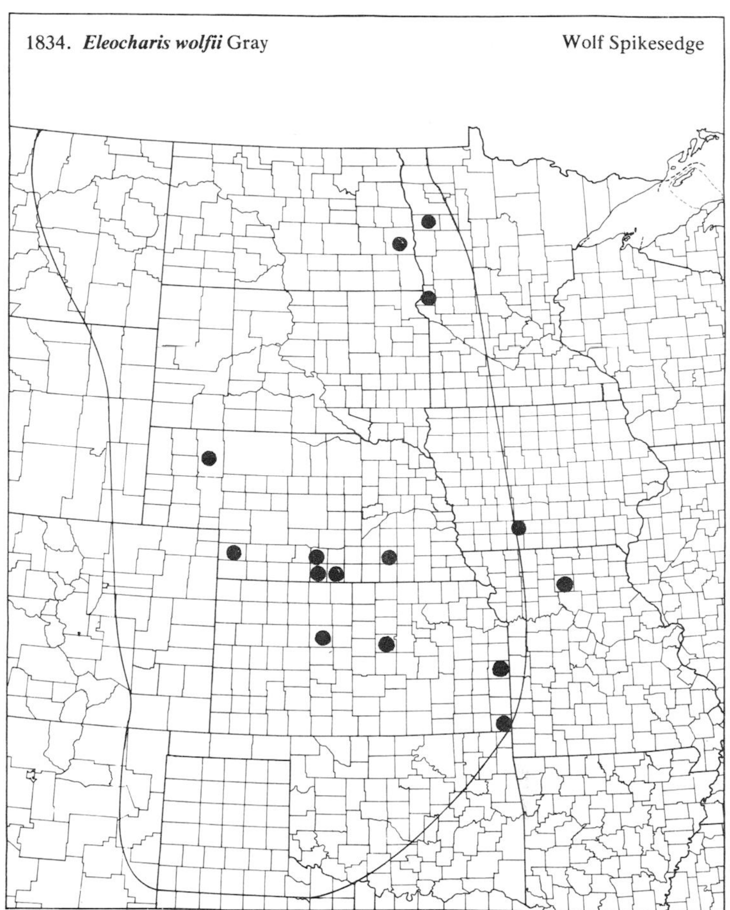

1835. ***Eleocharis xyridiformis*** Fern. & Brackett

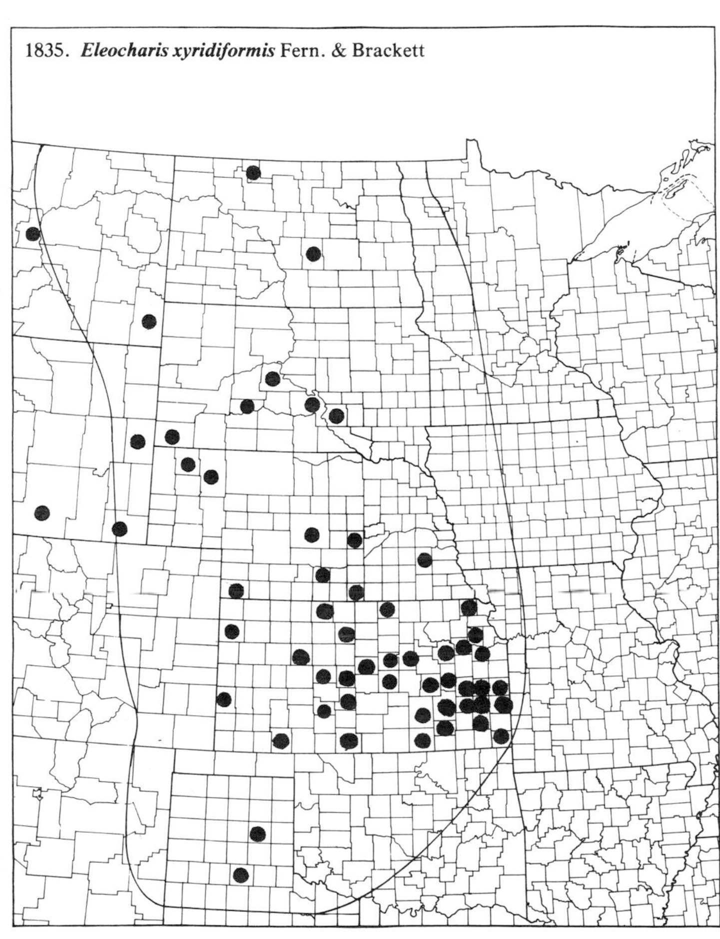

1836. ***Eriophorum polystachion*** L. — Cottongrass

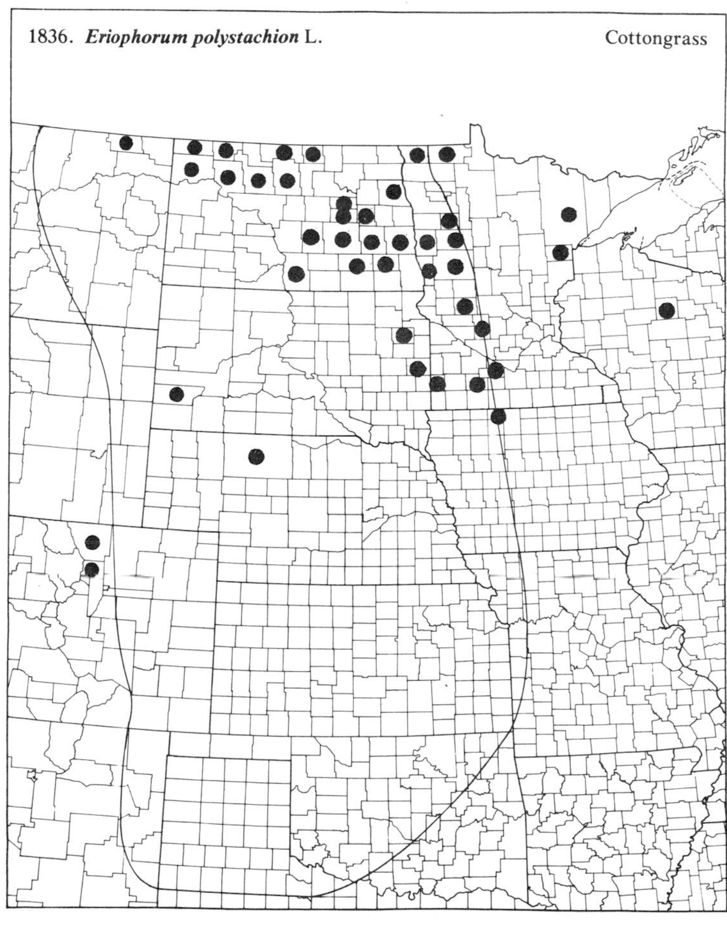

(Cyperaceae)

1837. ***Eriophorum viridicarinatum*** (Engelm.) Fern. Cottongrass

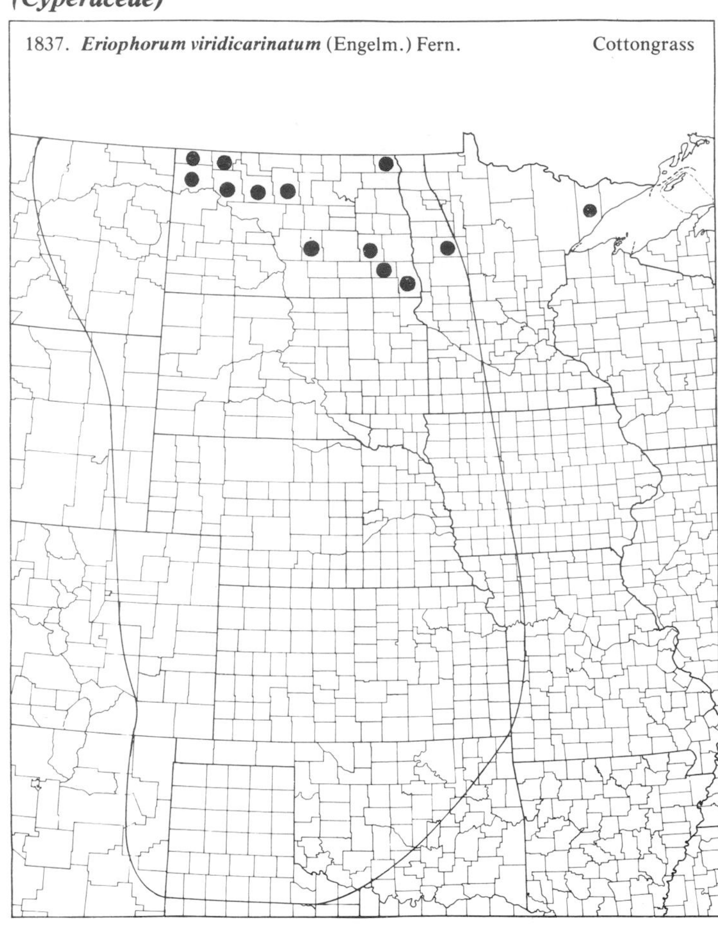

1838. ***Fimbristylis autumnalis*** (L.) R. & S.

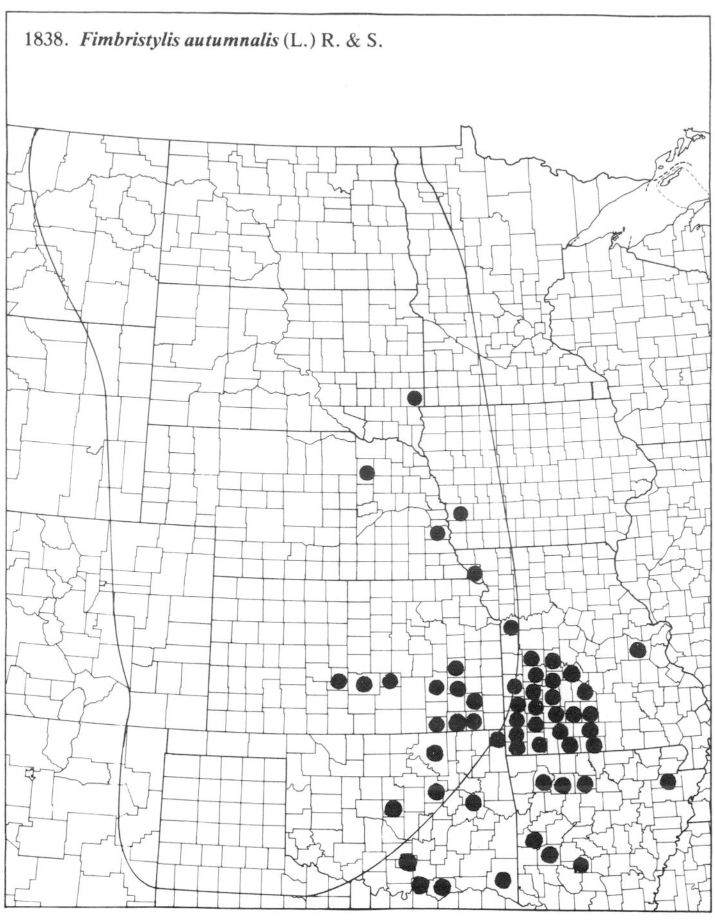

1839. ***Fimbristylis puberula*** (Michx.) Vahl var. ***puberula***

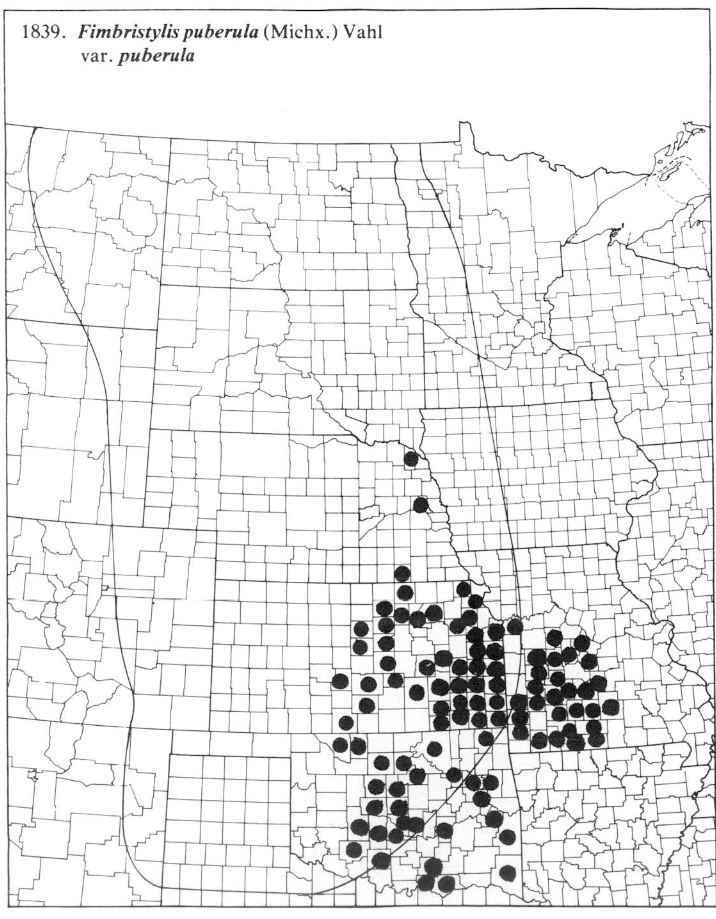

1840. ***Fimbristylis puberula*** (Michx.) Vahl var. ***interior*** (Britt.) Kral

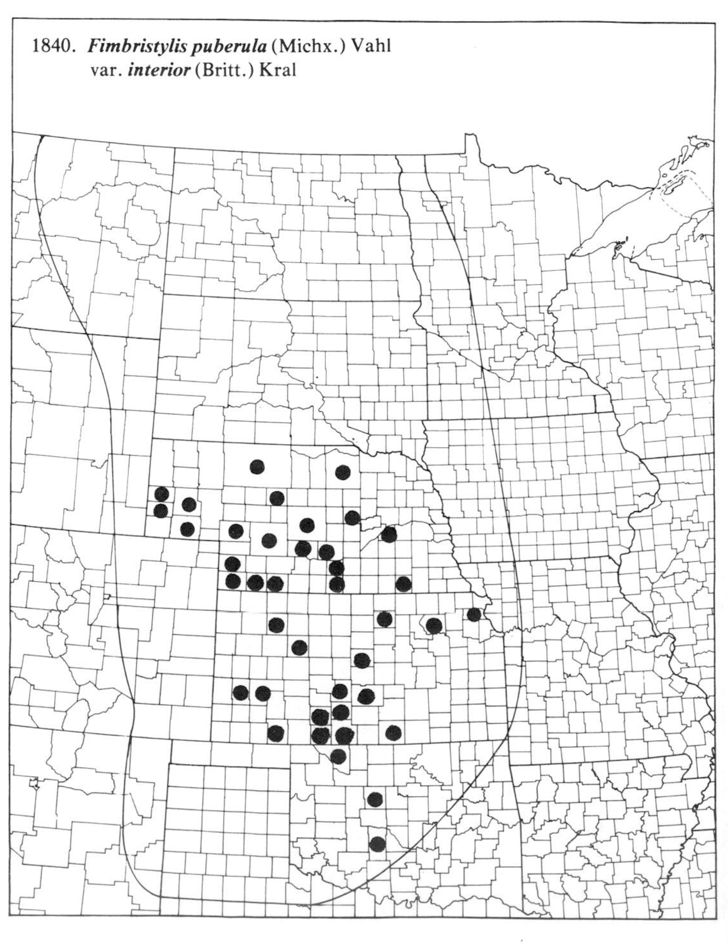

(Cyperaceae)

1841. ***Fimbristylis vahlii*** (Lam.) Link

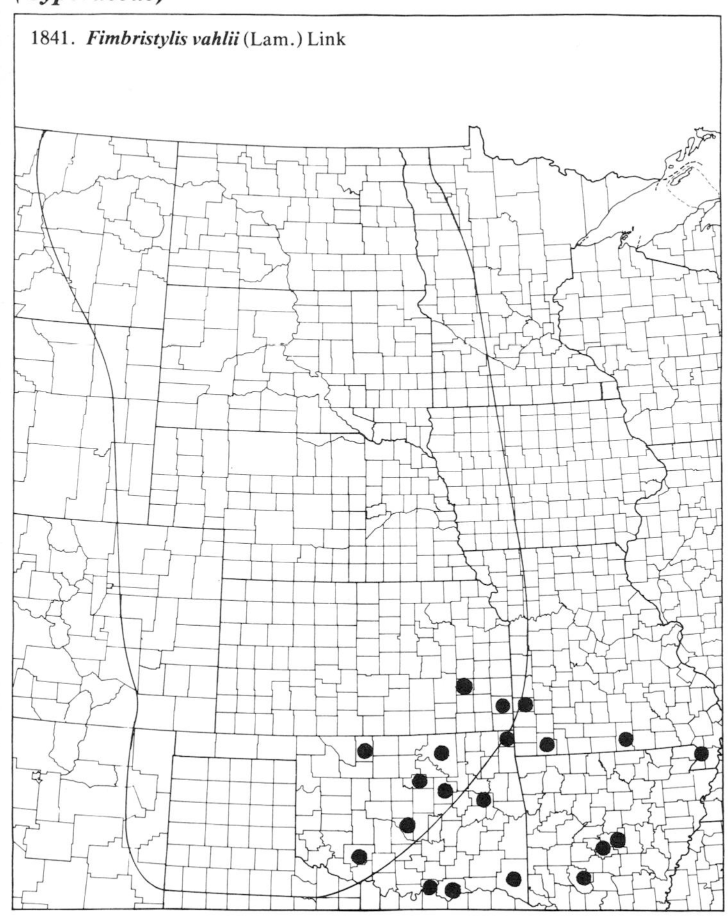

1842. ***Fuirena simplex*** Vahl — Fuirena

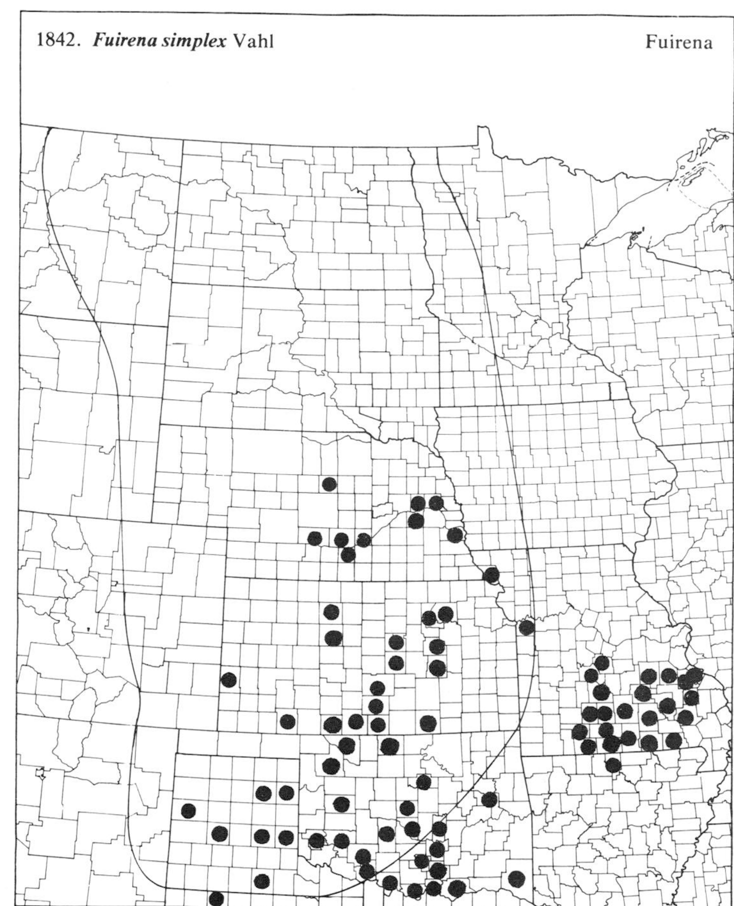

1843. ***Hemicarpha drummondii*** Nees — Drummond Hemicarpha

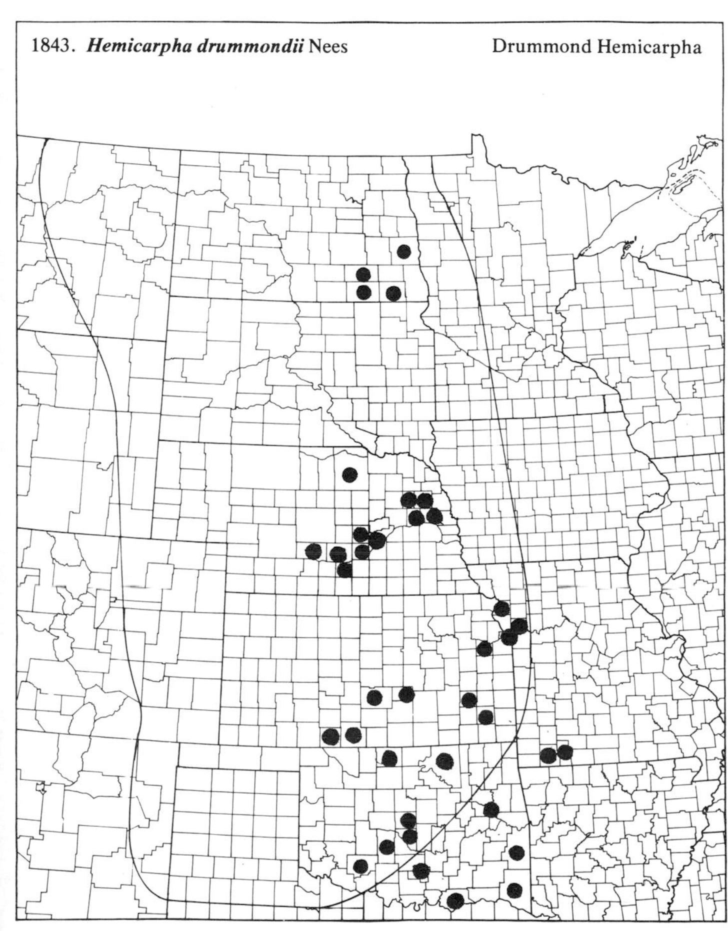

1844. ***Hemicarpha micrantha*** (Vahl) Pax — Common Hemicarpha

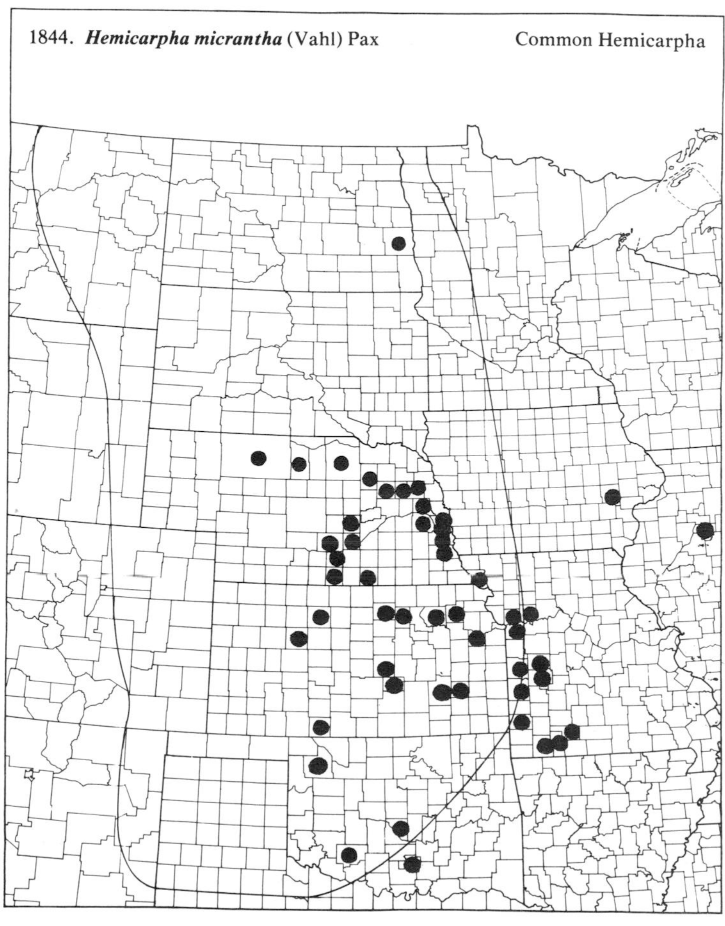

(Cyperaceae)

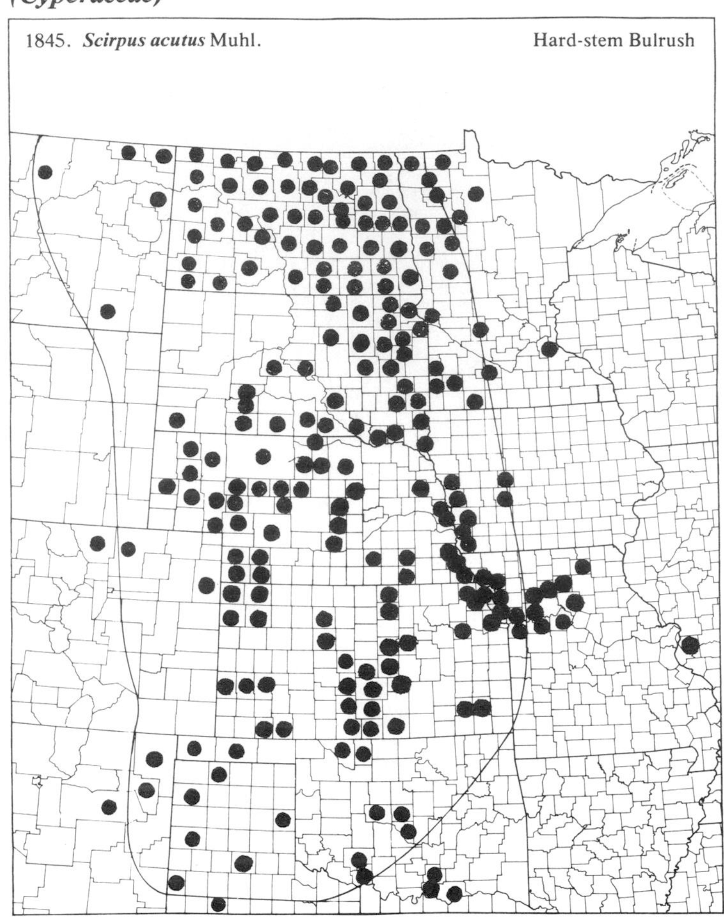

1845. *Scirpus acutus* Muhl. Hard-stem Bulrush

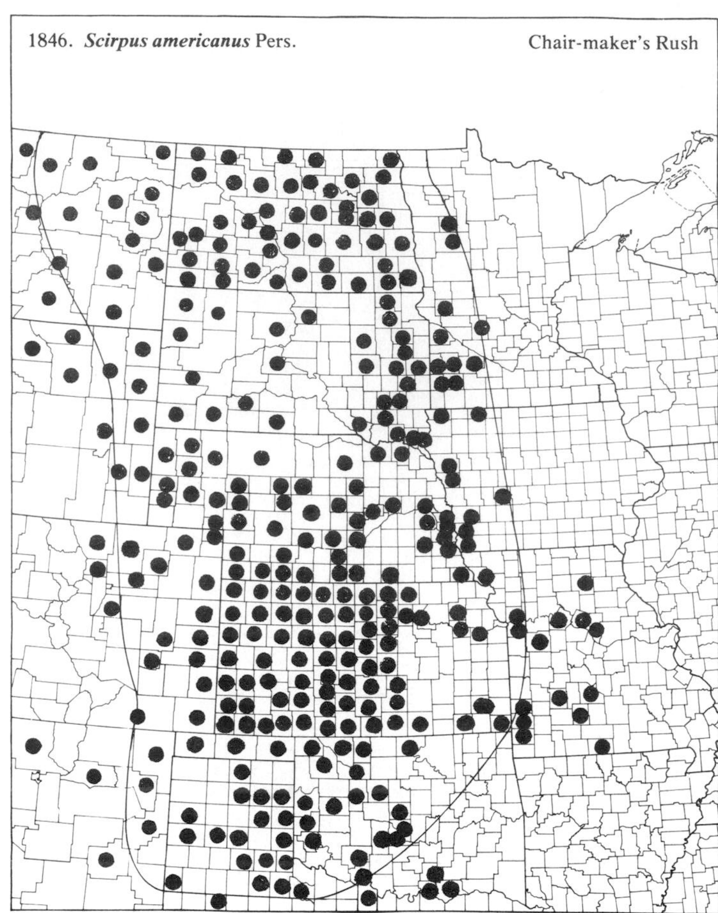

1846. *Scirpus americanus* Pers. Chair-maker's Rush

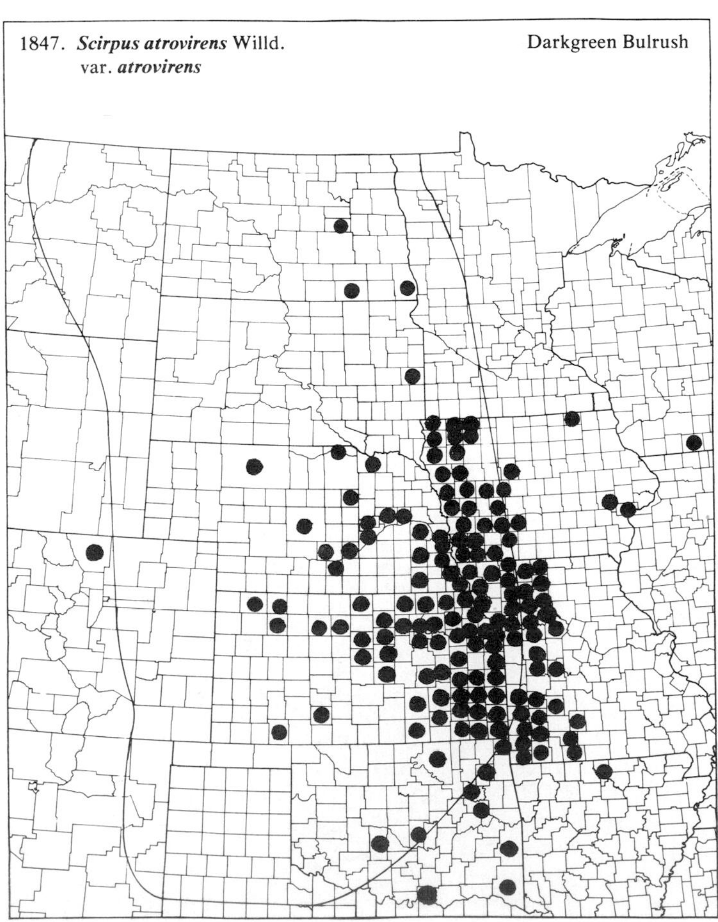

1847. *Scirpus atrovirens* Willd. var. *atrovirens* Darkgreen Bulrush

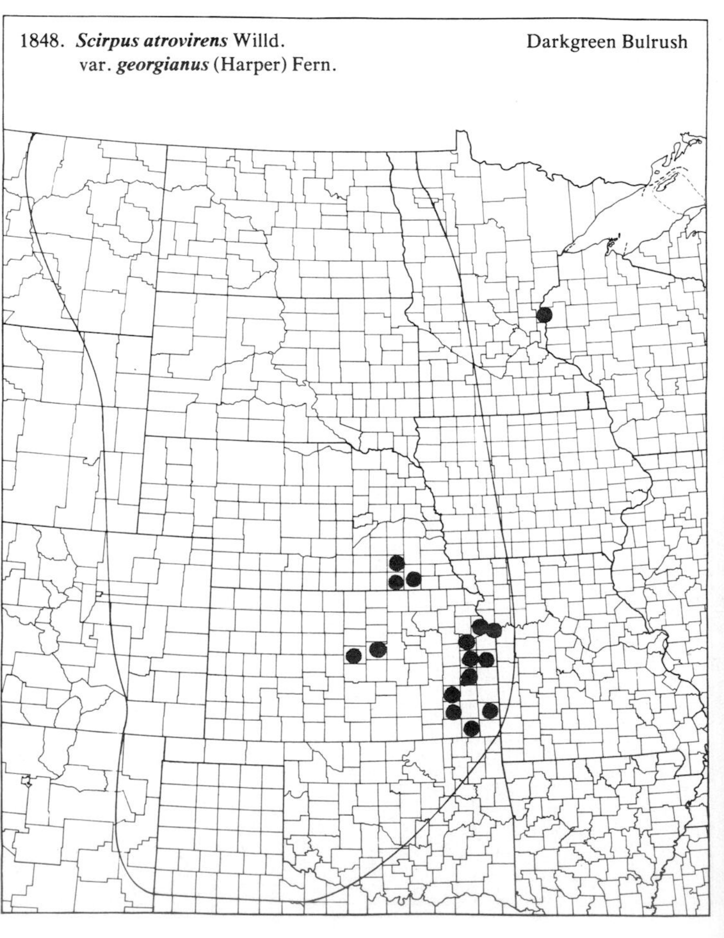

1848. *Scirpus atrovirens* Willd. var. *georgianus* (Harper) Fern. Darkgreen Bulrush

(Cyperaceae)

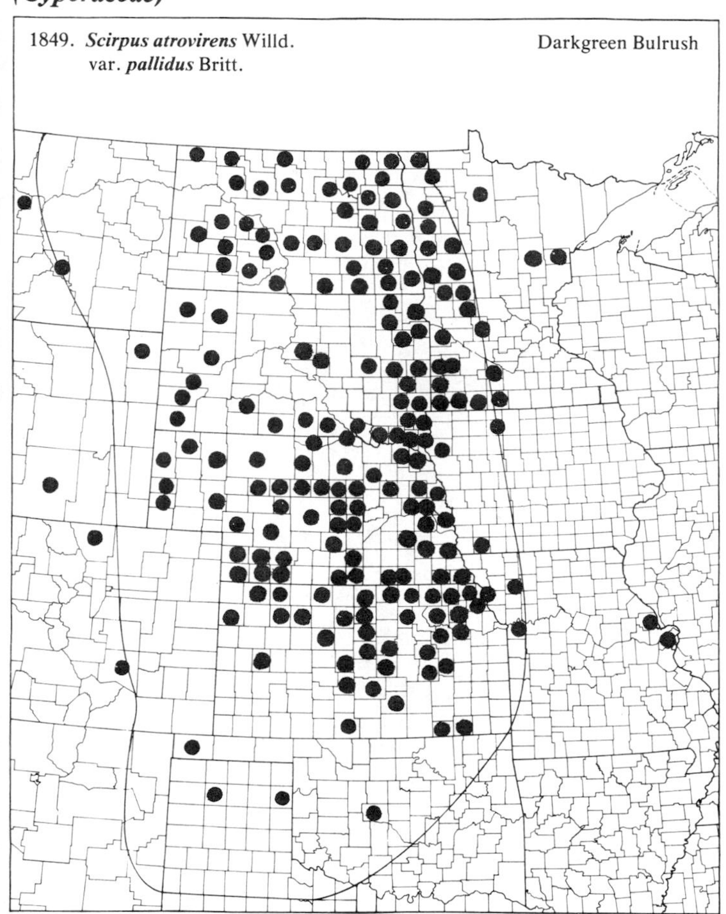
1849. ***Scirpus atrovirens*** Willd. var. ***pallidus*** Britt. Darkgreen Bulrush

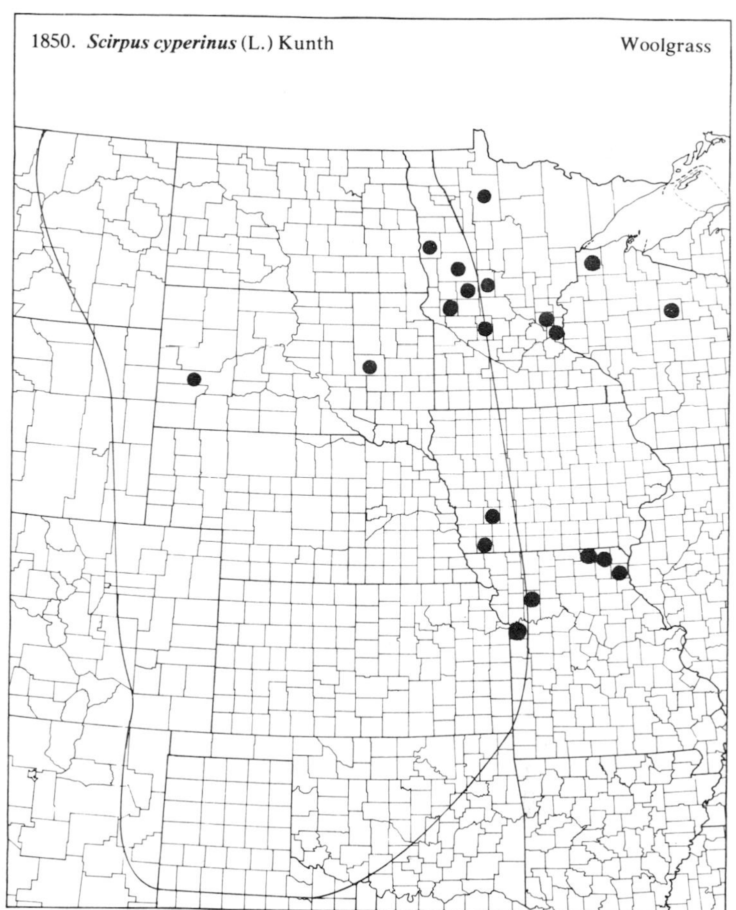
1850. ***Scirpus cyperinus*** (L.) Kunth Woolgrass

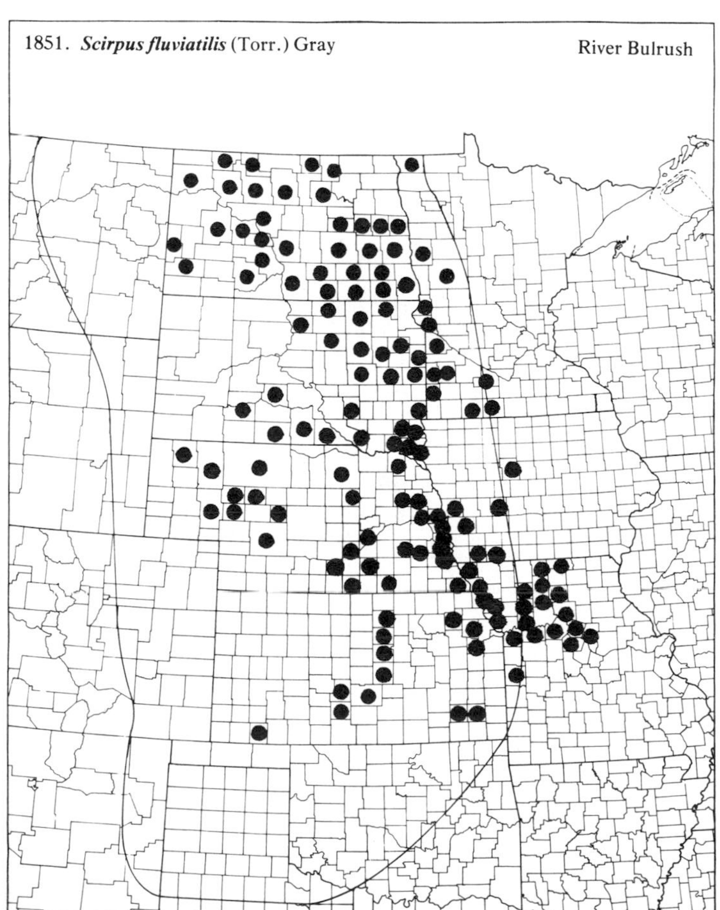
1851. ***Scirpus fluviatilis*** (Torr.) Gray River Bulrush

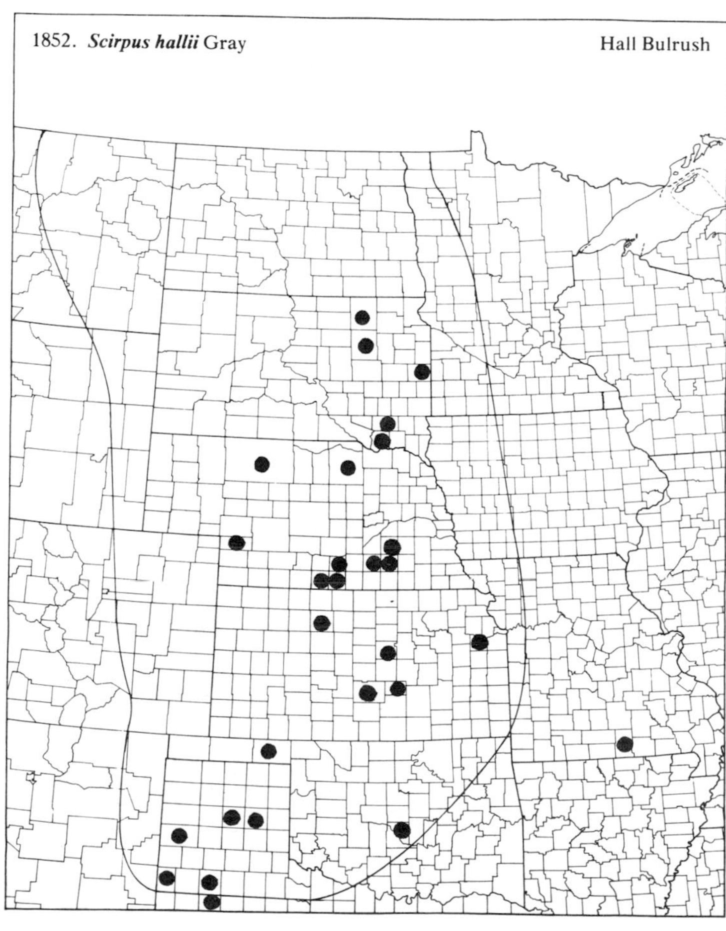
1852. ***Scirpus hallii*** Gray Hall Bulrush

(Cyperaceae)

1853. ***Scirpus heterochaetus*** Chase — Slender Bulrush

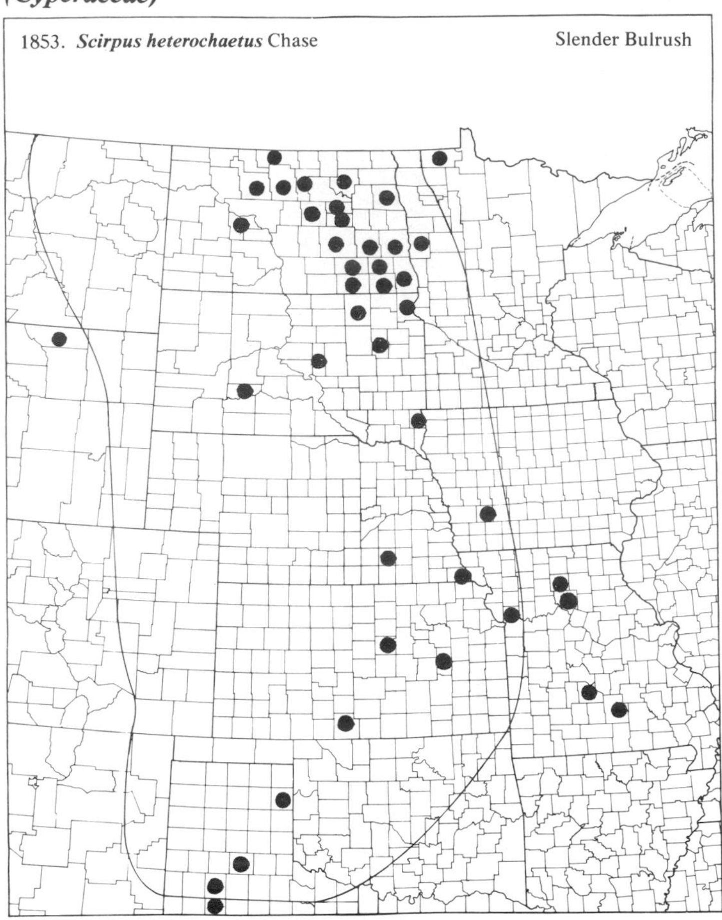

1854. ***Scirpus maritimus*** L. var. ***paludosus*** (A. Nels.) Kukenth. — Prairie Bulrush

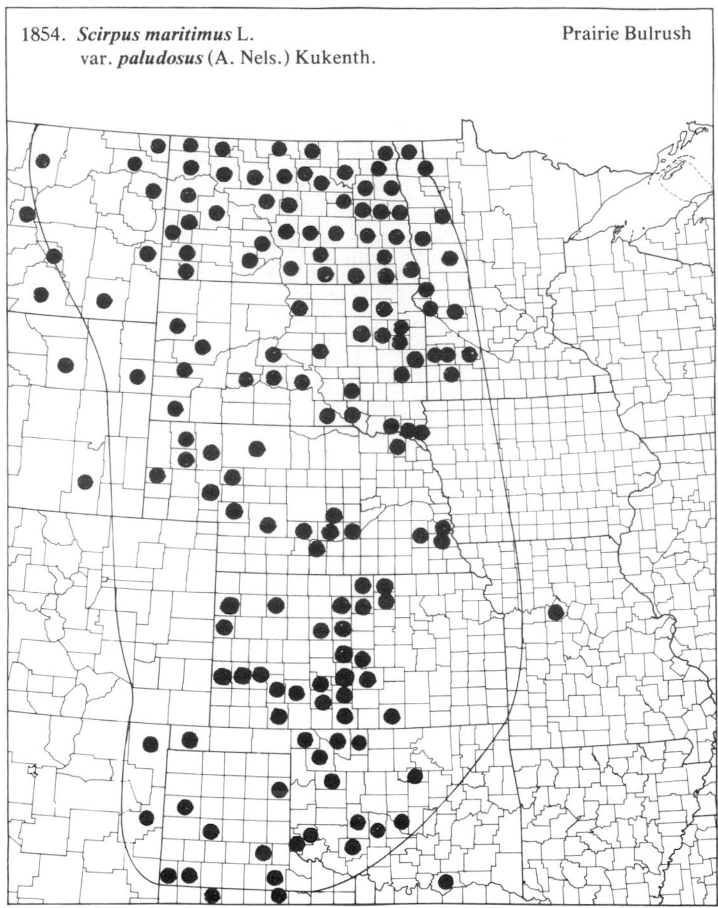

1855. ***Scirpus microcarpus*** Presl. var. ***rubrotinctus*** (Fern.) M.E. Jones

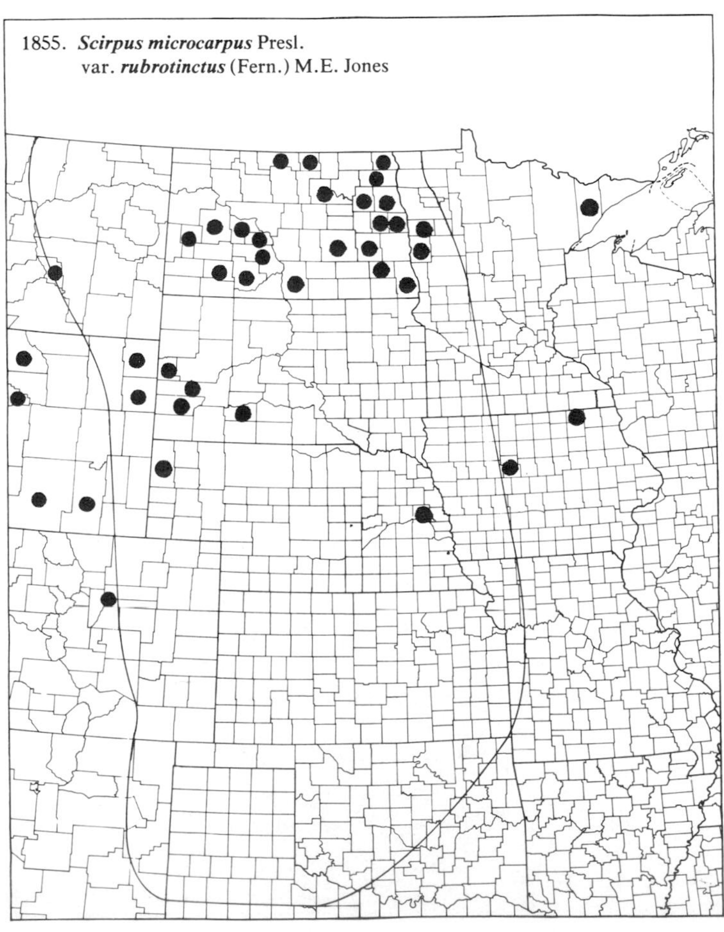

1856. ***Scirpus nevadensis*** Wats.

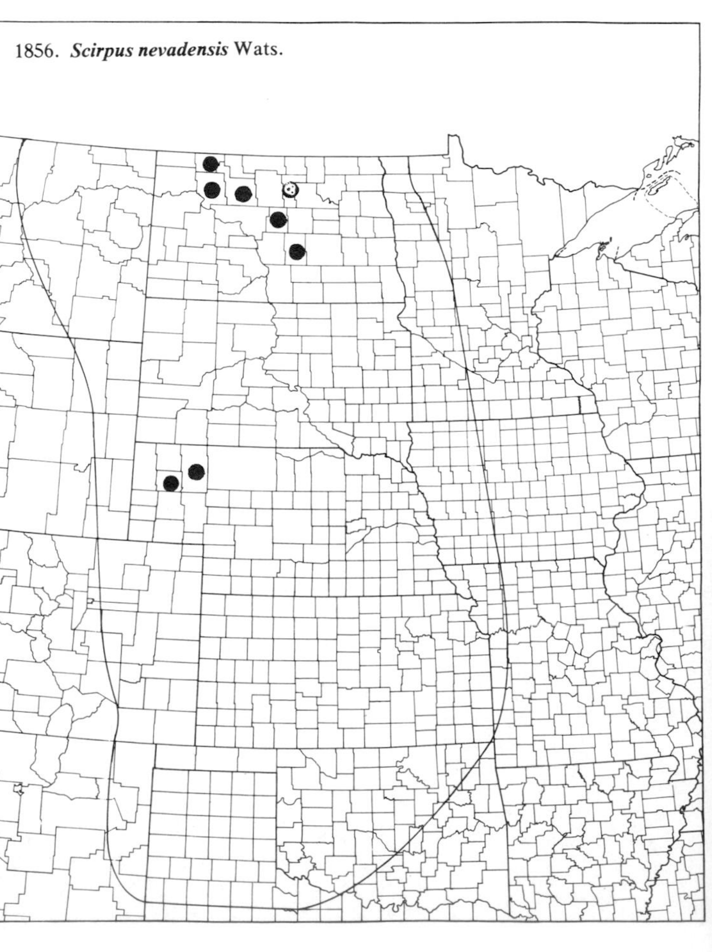

(Cyperaceae)

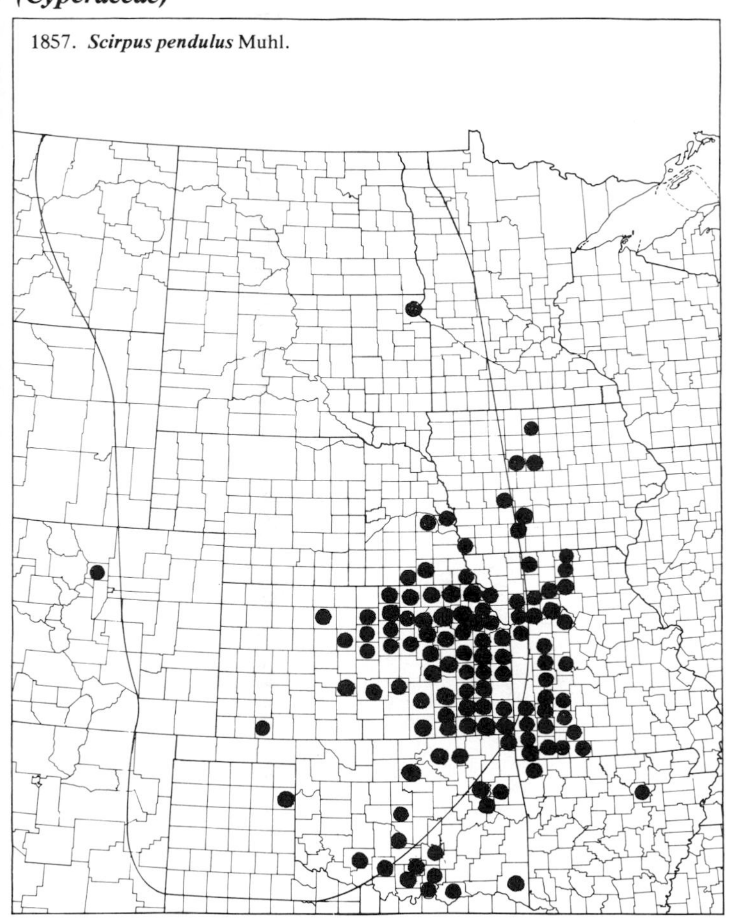

1857. ***Scirpus pendulus*** Muhl.

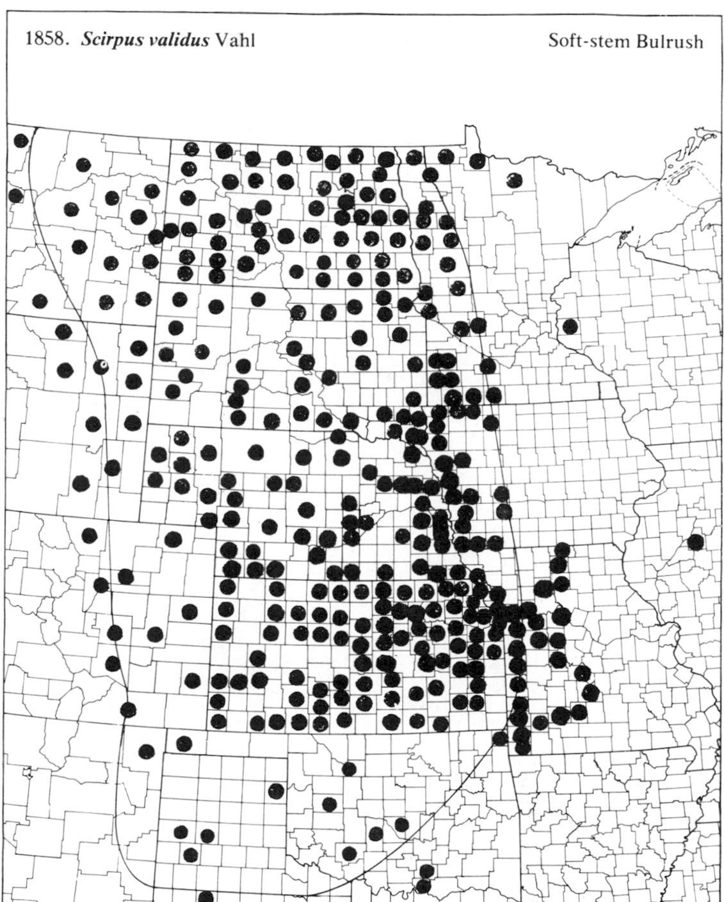

1858. ***Scirpus validus*** Vahl Soft-stem Bulrush

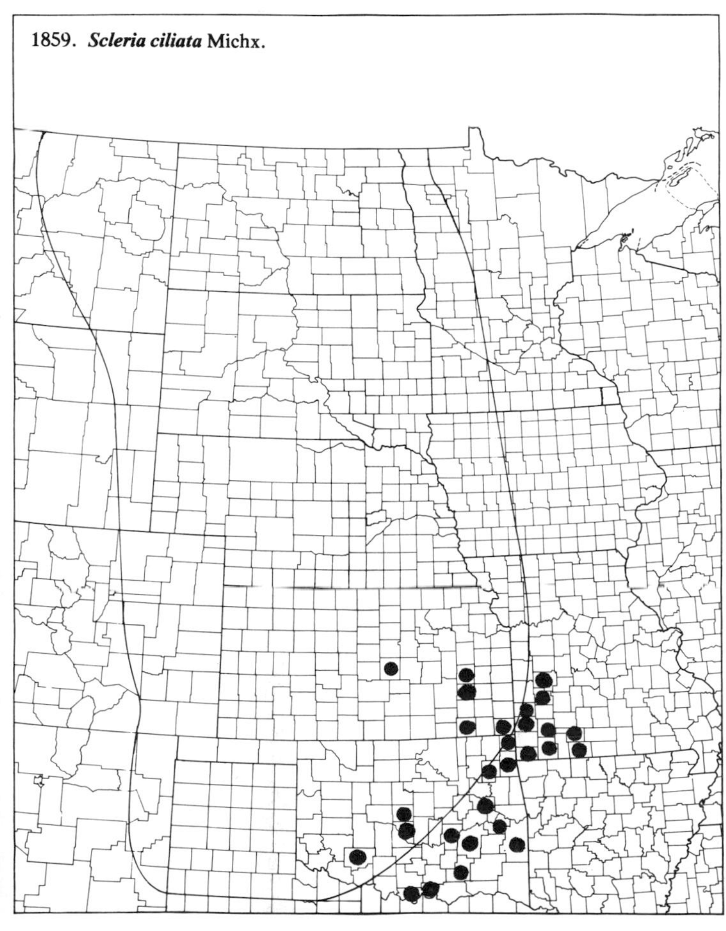

1859. ***Scleria ciliata*** Michx.

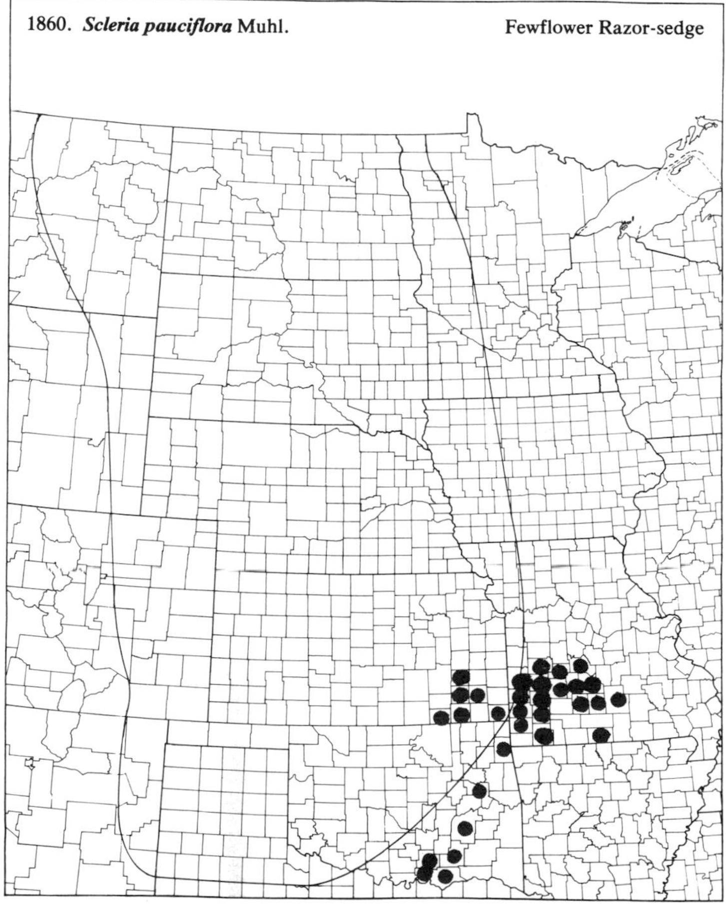

1860. ***Scleria pauciflora*** Muhl. Fewflower Razor-sedge

(Cyperaceae)

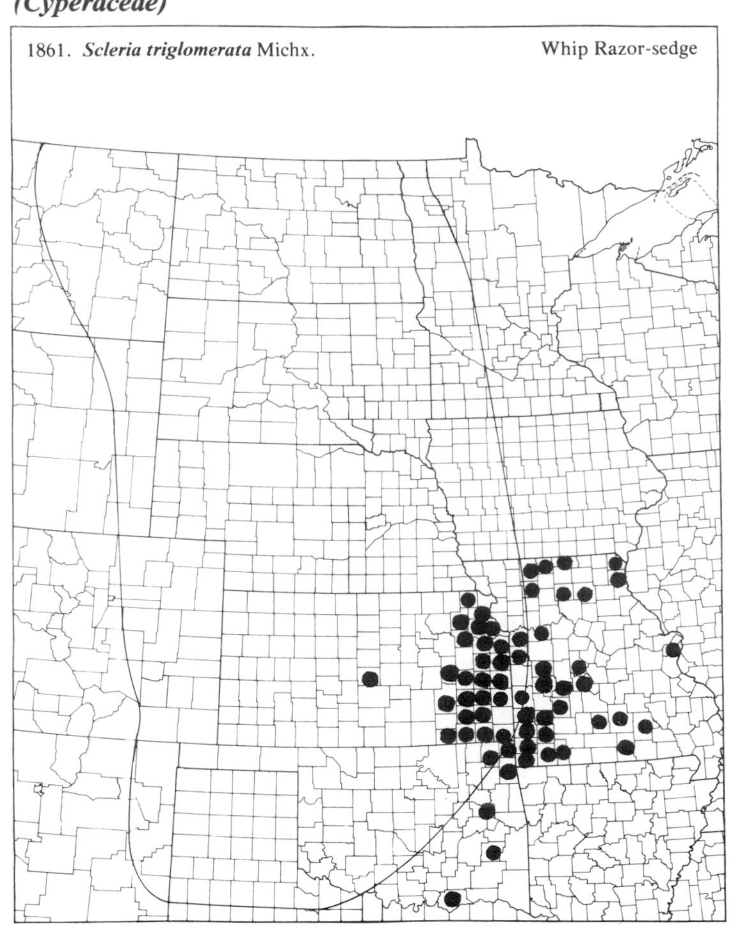

127. *Poaceae* (Grass Family)

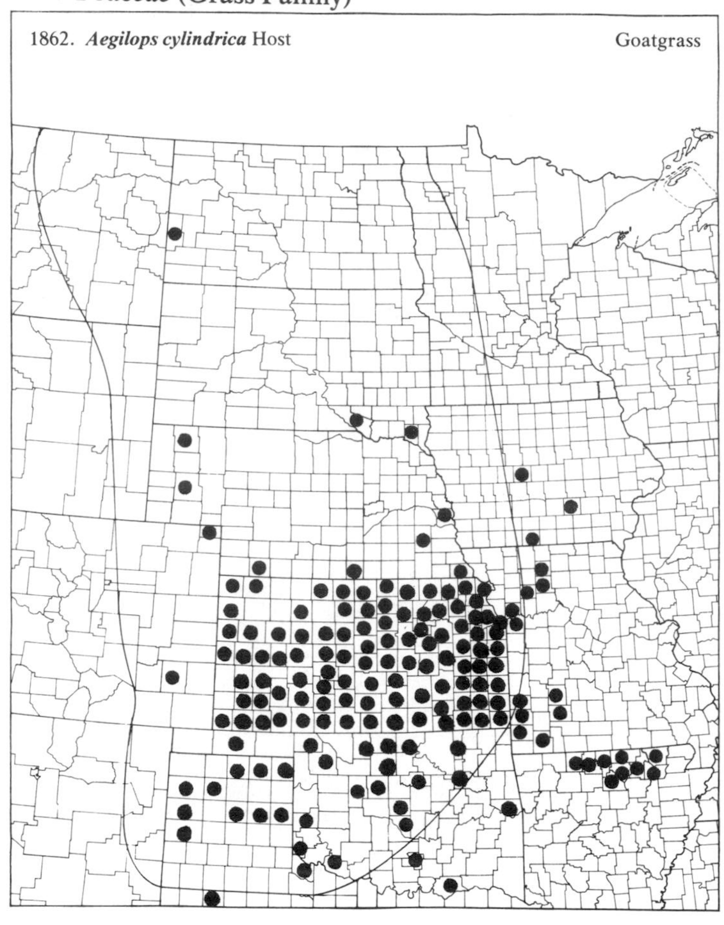

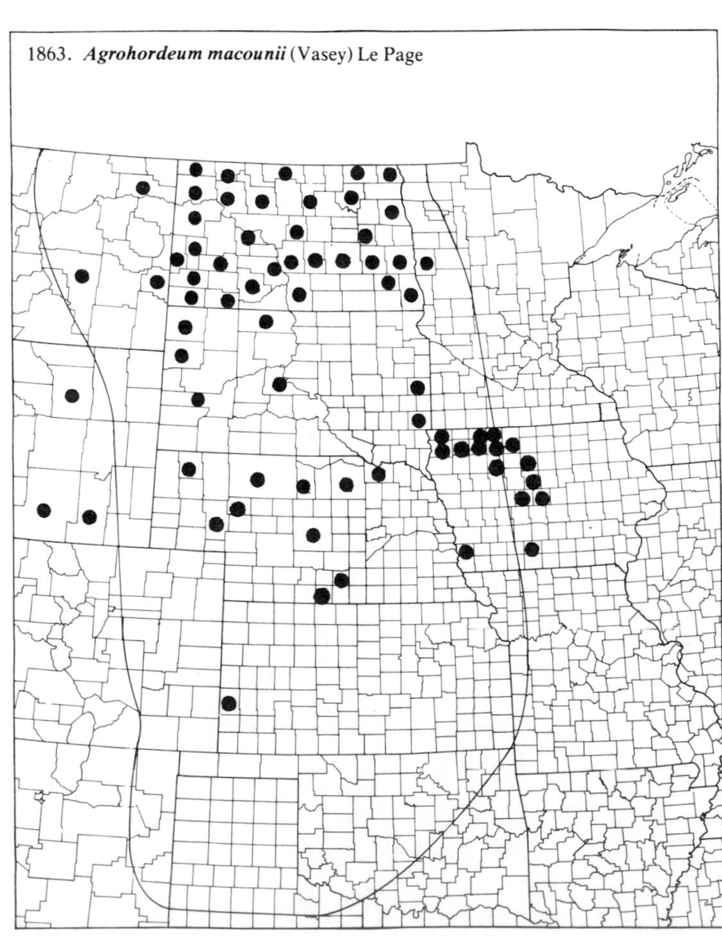

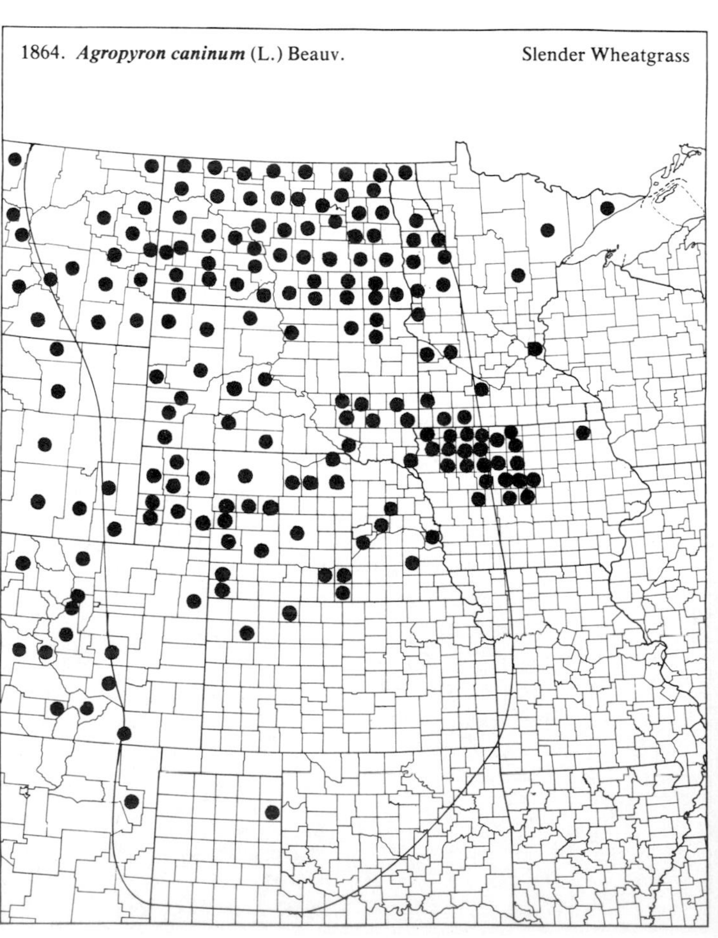

(Poaceae)

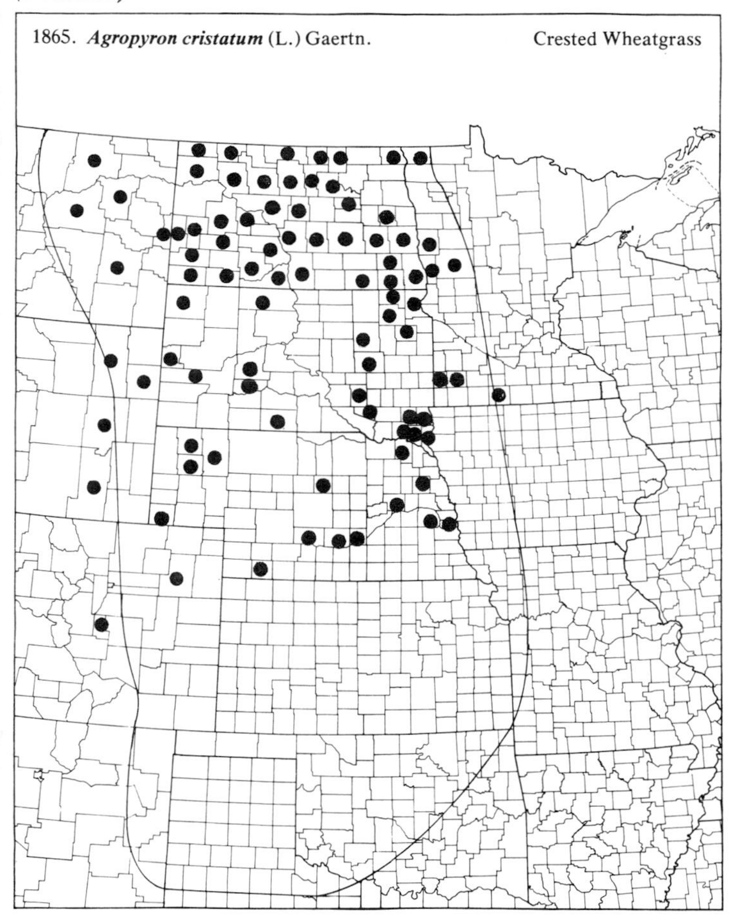
1865. ***Agropyron cristatum*** (L.) Gaertn. Crested Wheatgrass

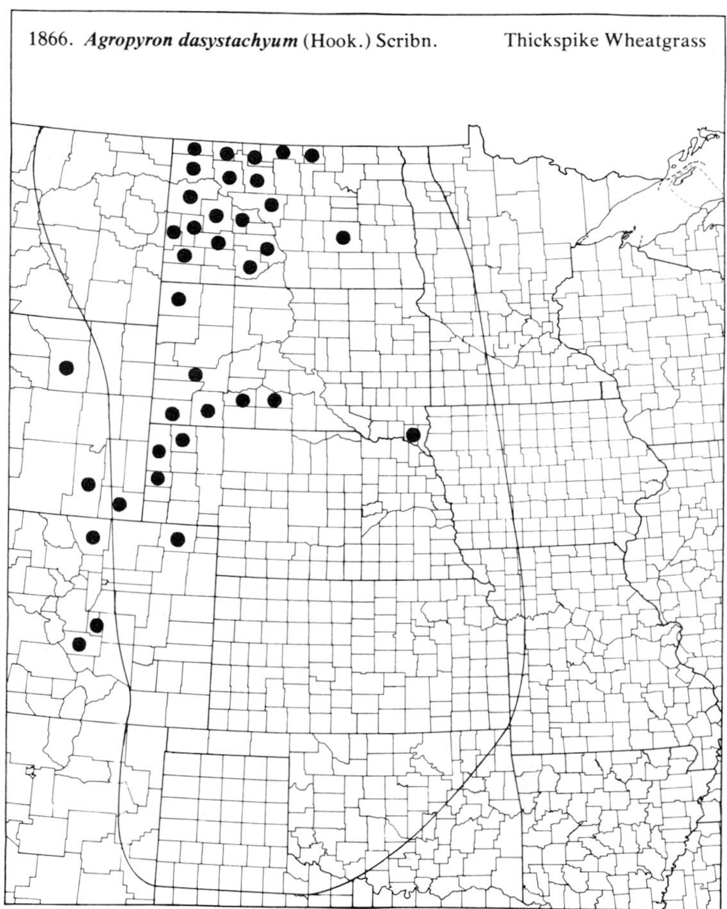
1866. ***Agropyron dasystachyum*** (Hook.) Scribn. Thickspike Wheatgrass

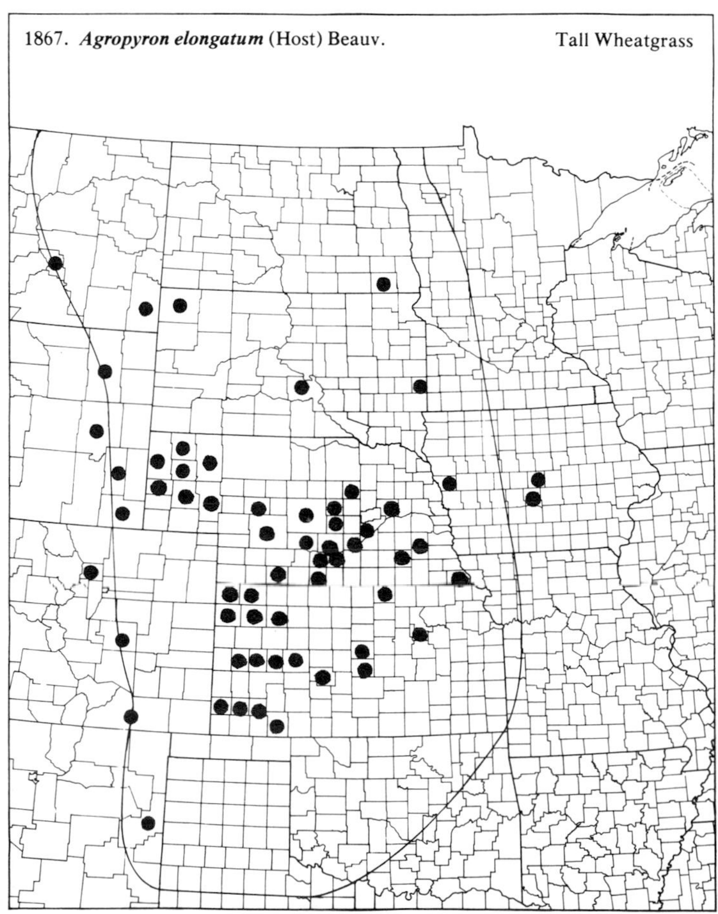
1867. ***Agropyron elongatum*** (Host) Beauv. Tall Wheatgrass

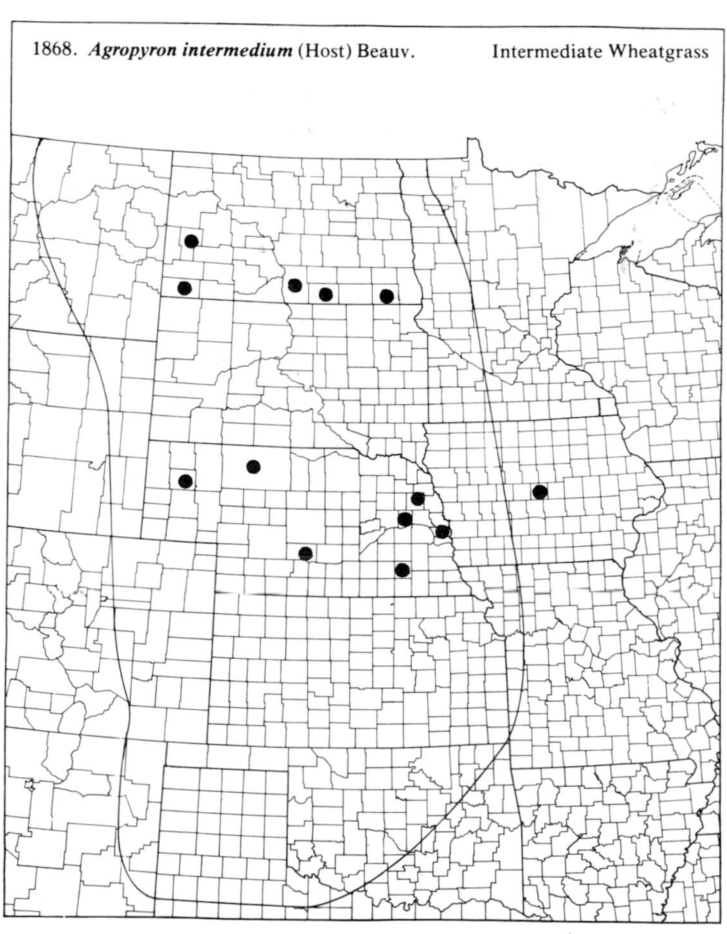
1868. ***Agropyron intermedium*** (Host) Beauv. Intermediate Wheatgrass

(Poaceae)

1869. ***Agropyron pectiniforme*** R. & S.

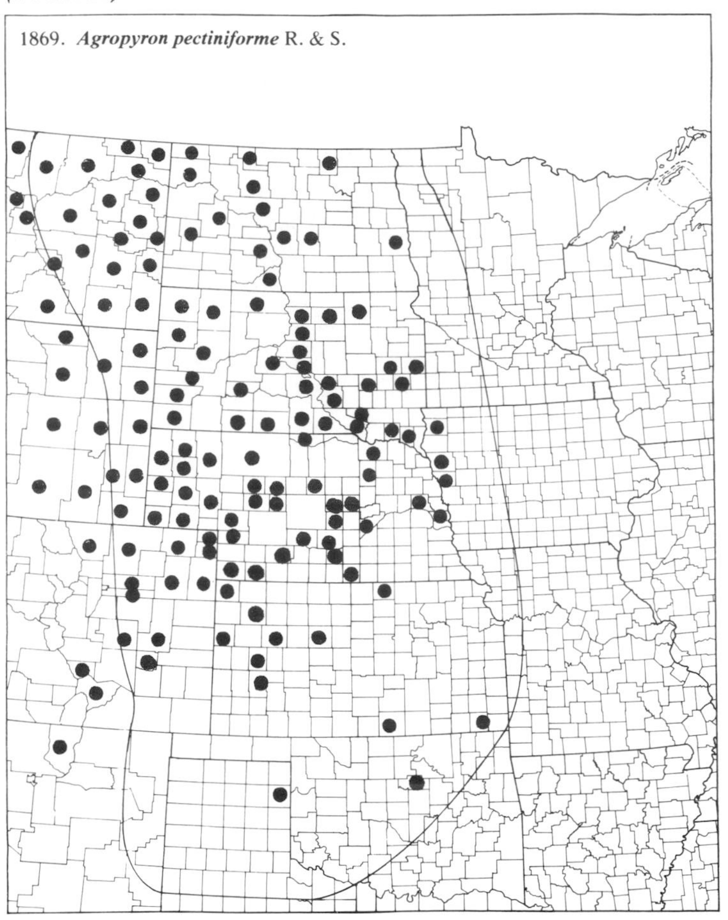

1870. ***Agropyron repens*** (L.) Beauv. Quackgrass

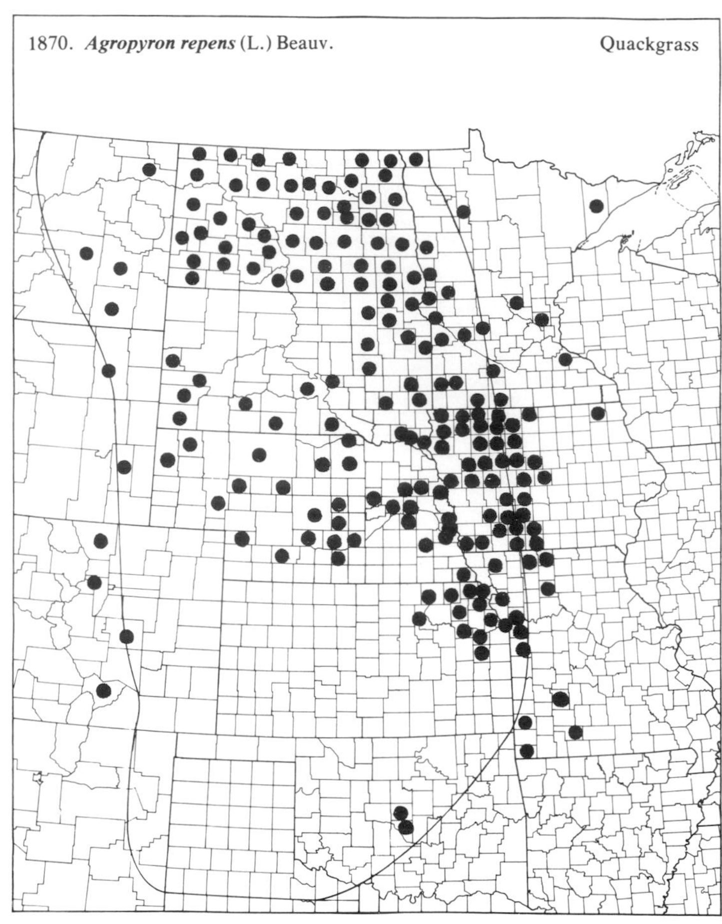

1871. ***Agropyron smithii*** Rydb. Western Wheatgrass

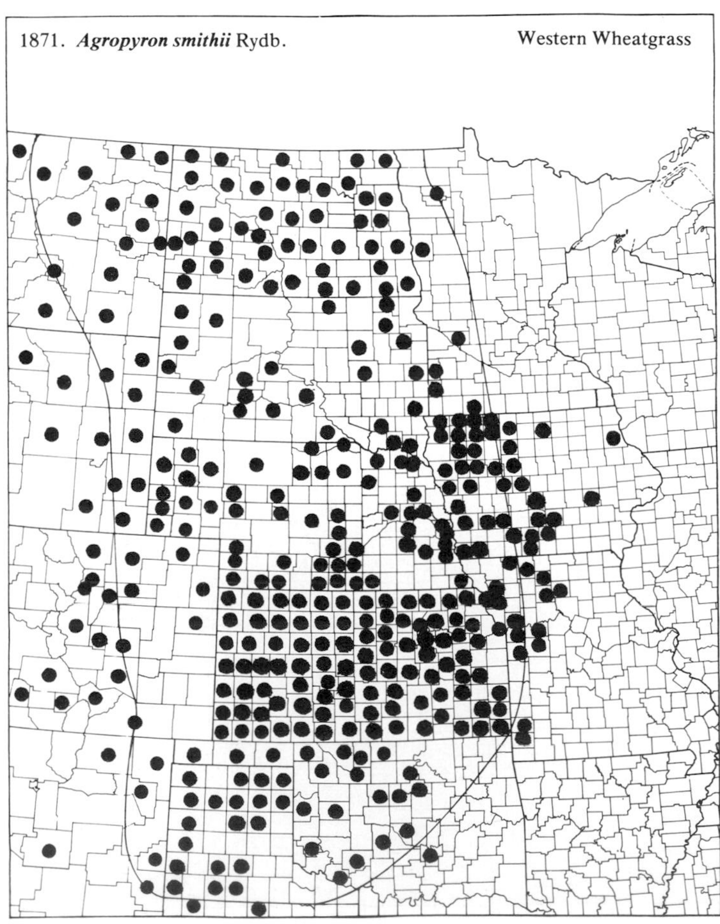

1872. ***Agropyron spicatum*** (Pursh) Scribn. & Sm. Bluebunch Wheatgrass

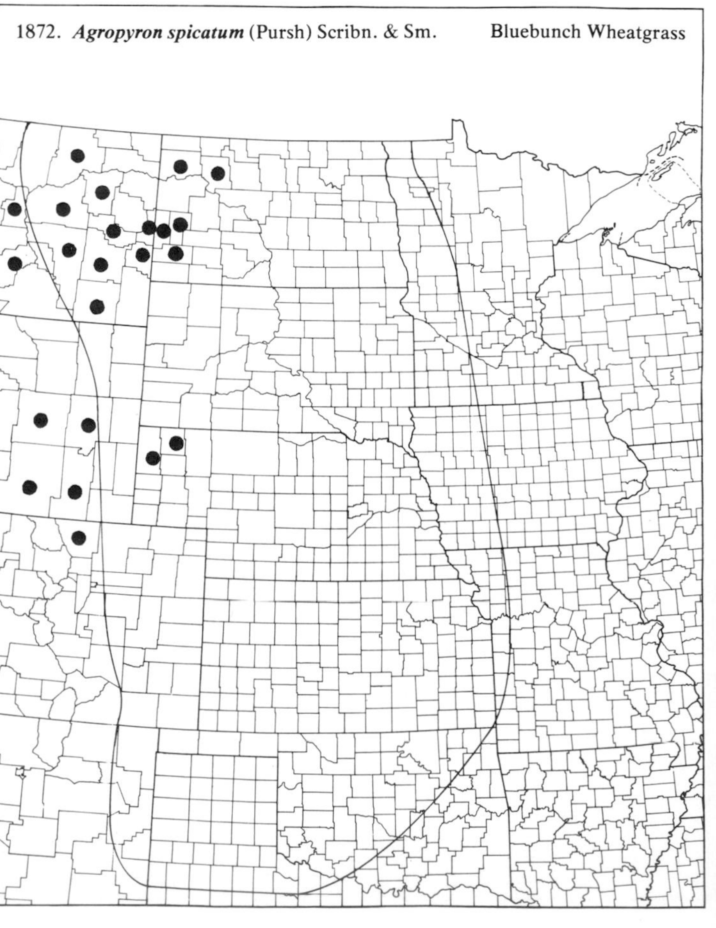

(Poaceae)

1873. ***Agrostis elliotiana*** Schult. Bentgrass

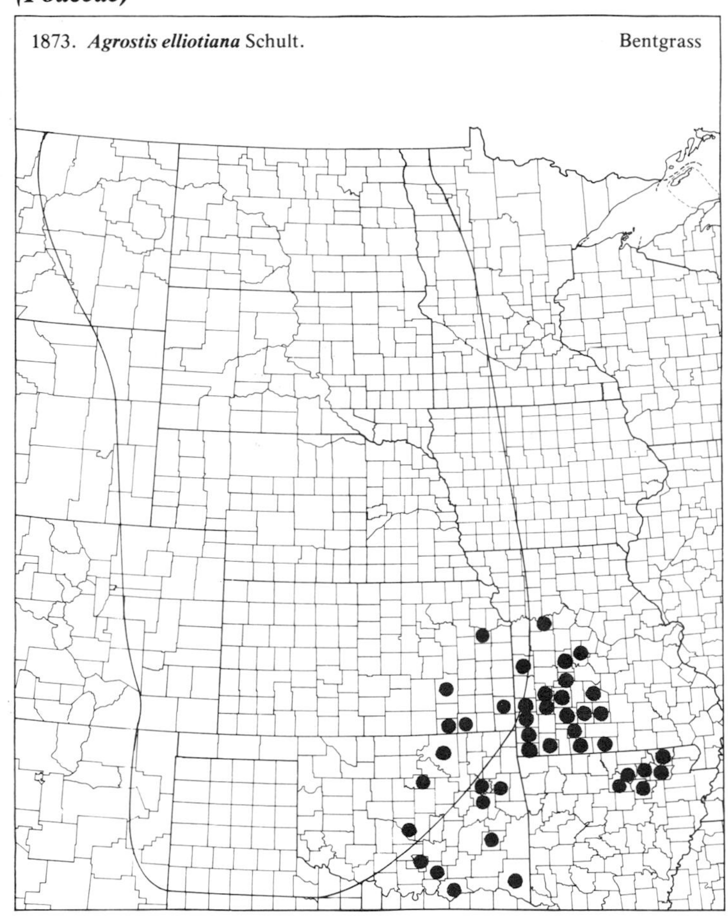

1874. ***Agrostis exarata*** Trin. ssp. ***minor*** (Hook.) C.L. Hitchc. Spikebent

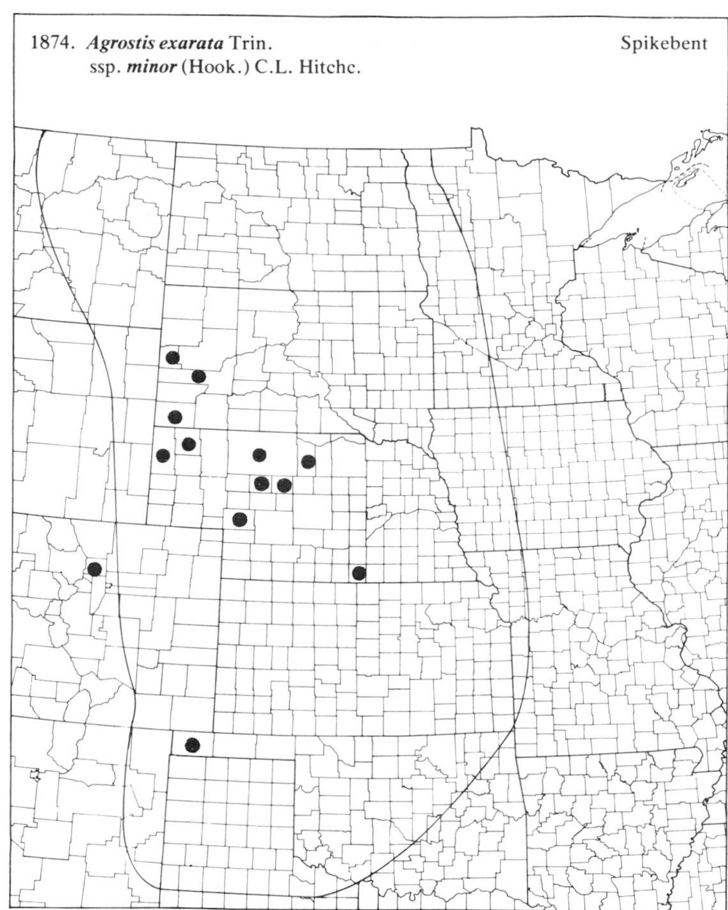

1875. ***Agrostis hyemalis*** (Walt.) B.S.P. var. ***hyemalis*** Ticklegrass

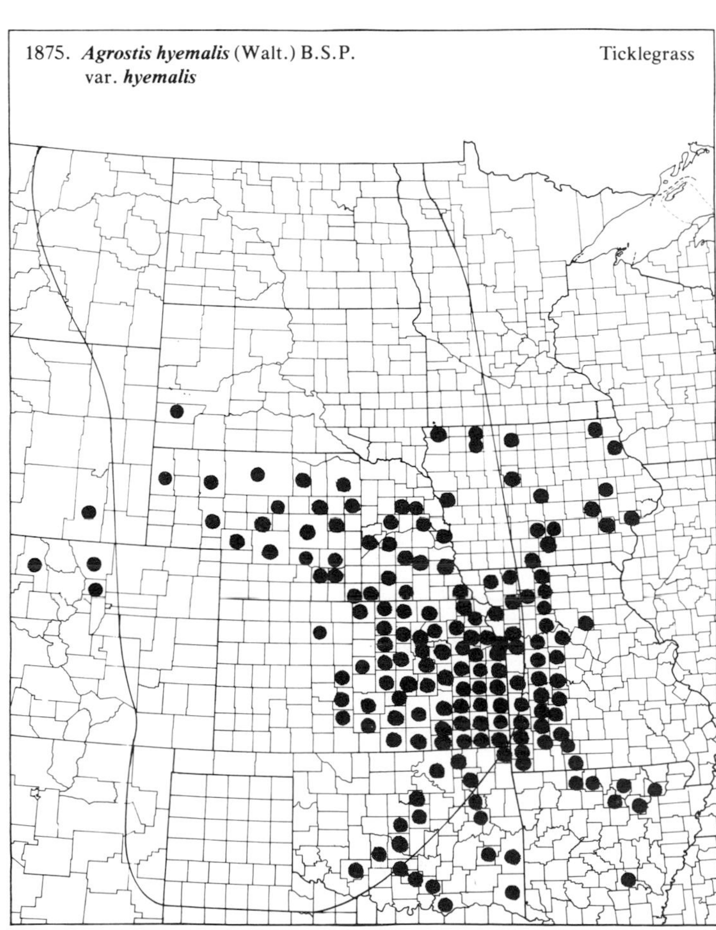

1876. ***Agrostis hyemalis*** (Walt.) B.S.P. var. ***tenuis*** (Tuckerm.) Gl. Ticklegrass

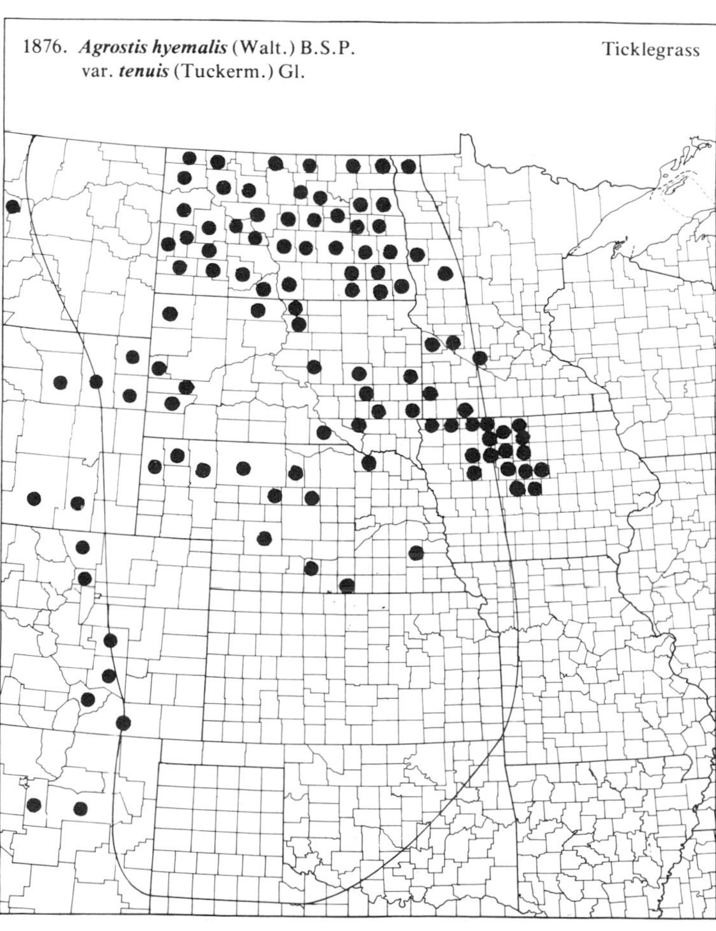

(Poaceae)

1877. ***Agrostis perennans*** (Walt.) Tuckerm. Autumn Bent

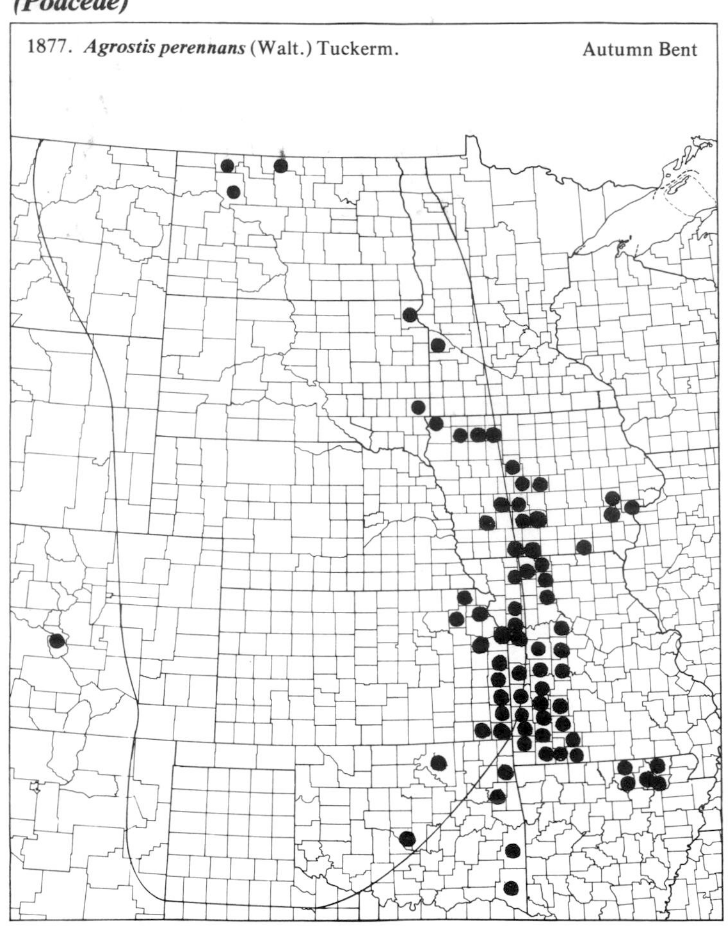

1878. ***Agrostis stolonifera*** L. Redtop

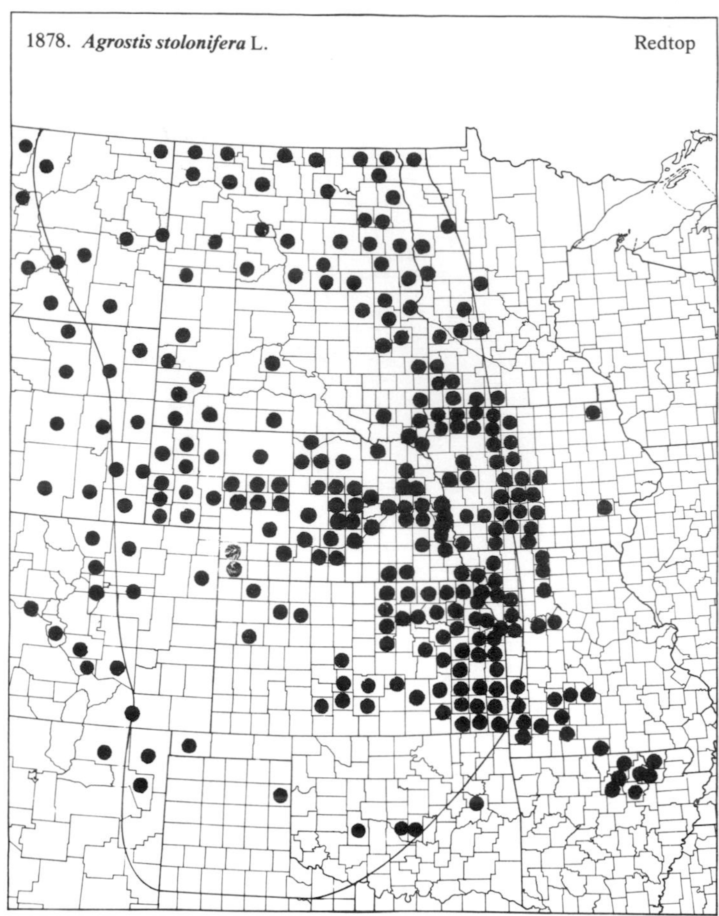

1879. ***Alopecurus aequalis*** Sobol. Shortawn Foxtail

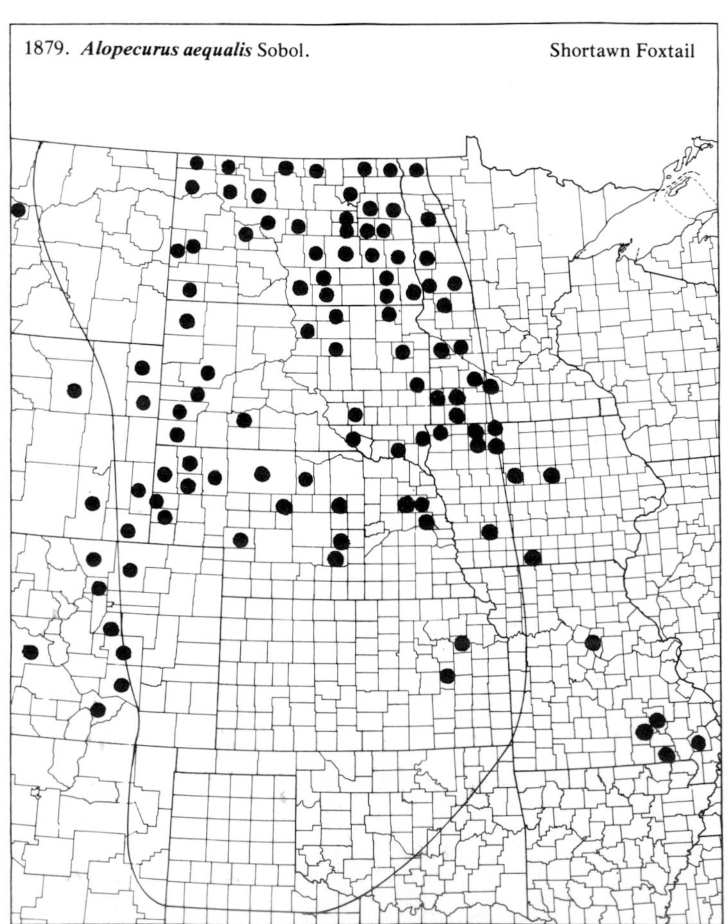

1880. ***Alopecurus carolinianus*** Walt. Carolina Foxtail

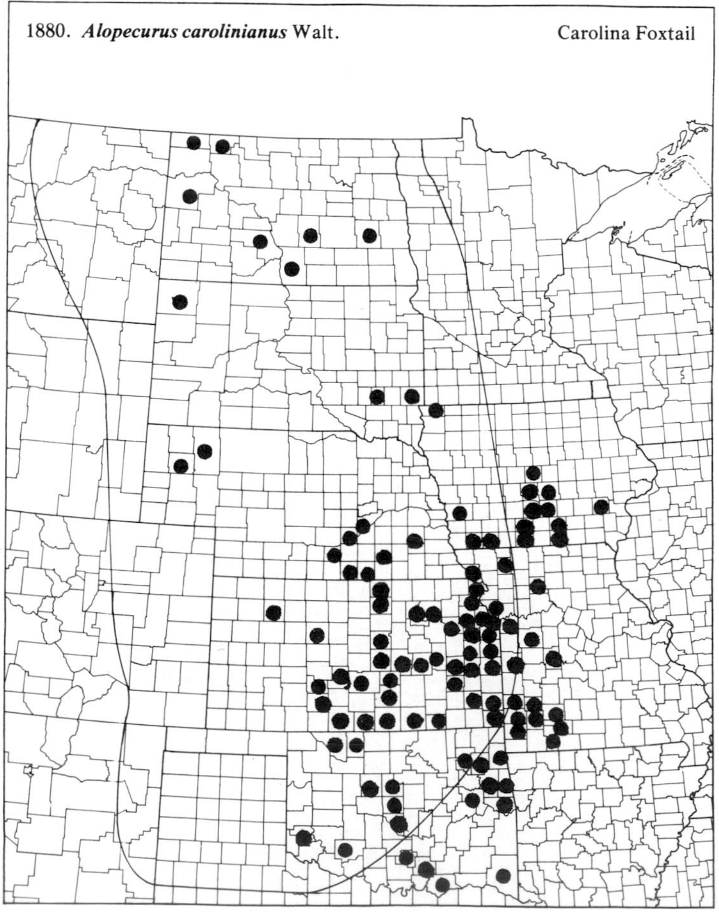

(Poaceae)

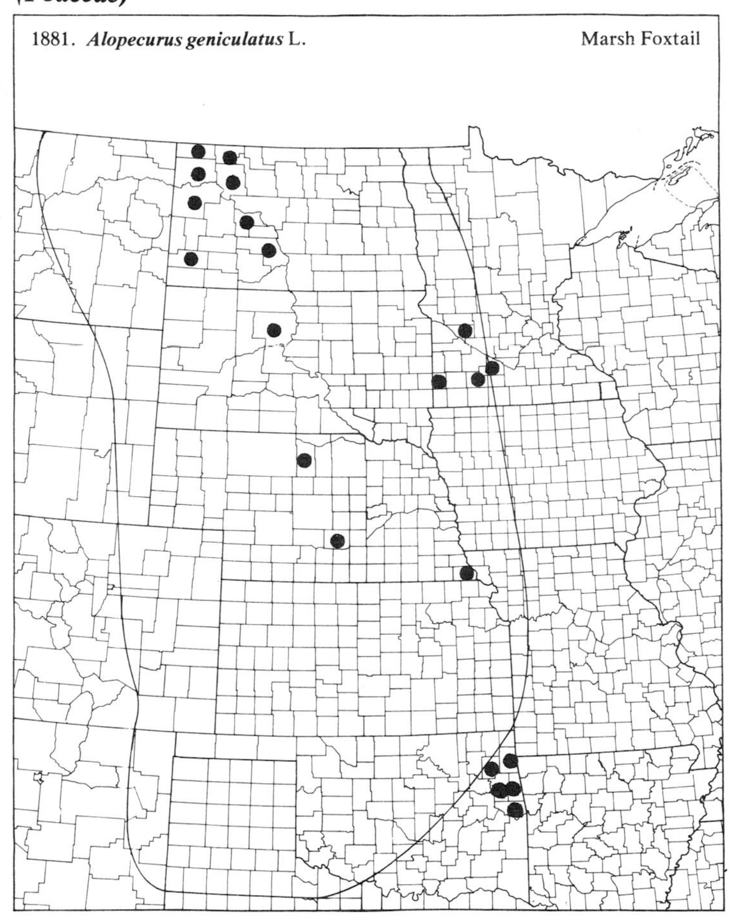
1881. *Alopecurus geniculatus* L. Marsh Foxtail

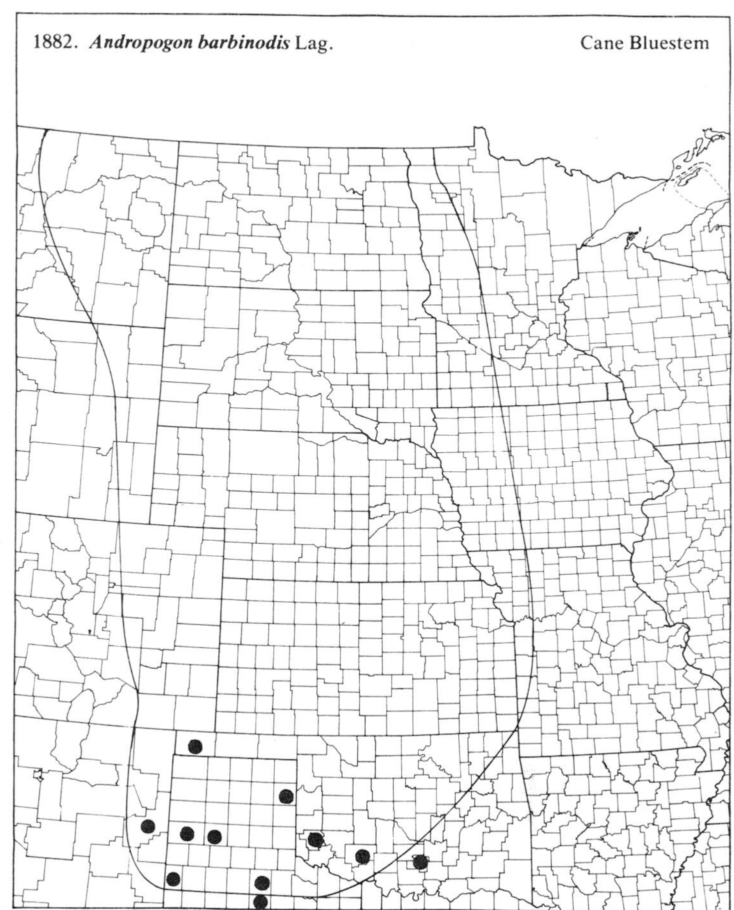
1882. *Andropogon barbinodis* Lag. Cane Bluestem

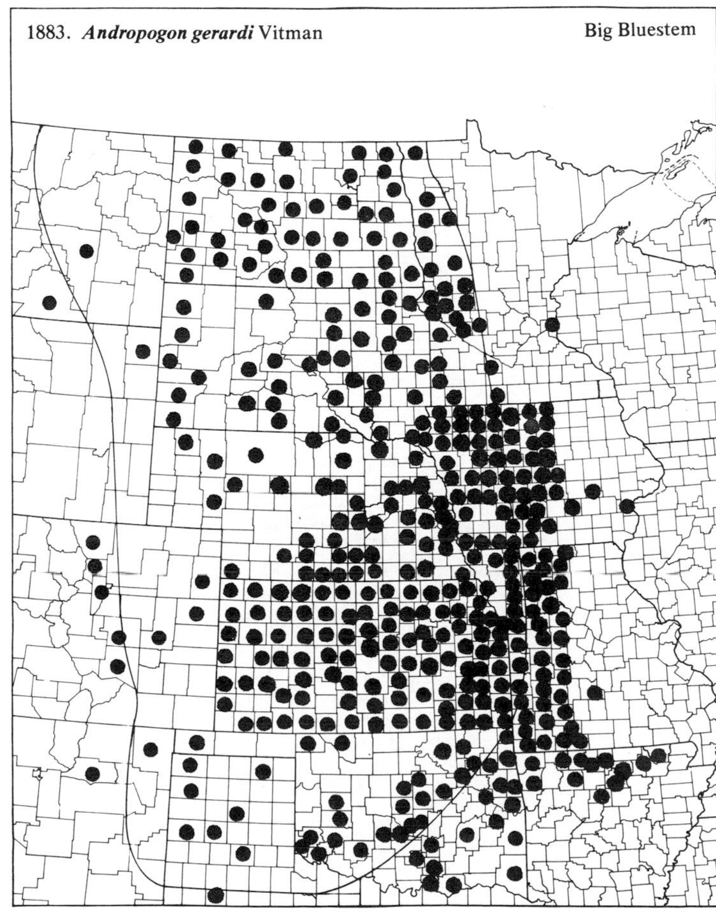
1883. *Andropogon gerardi* Vitman Big Bluestem

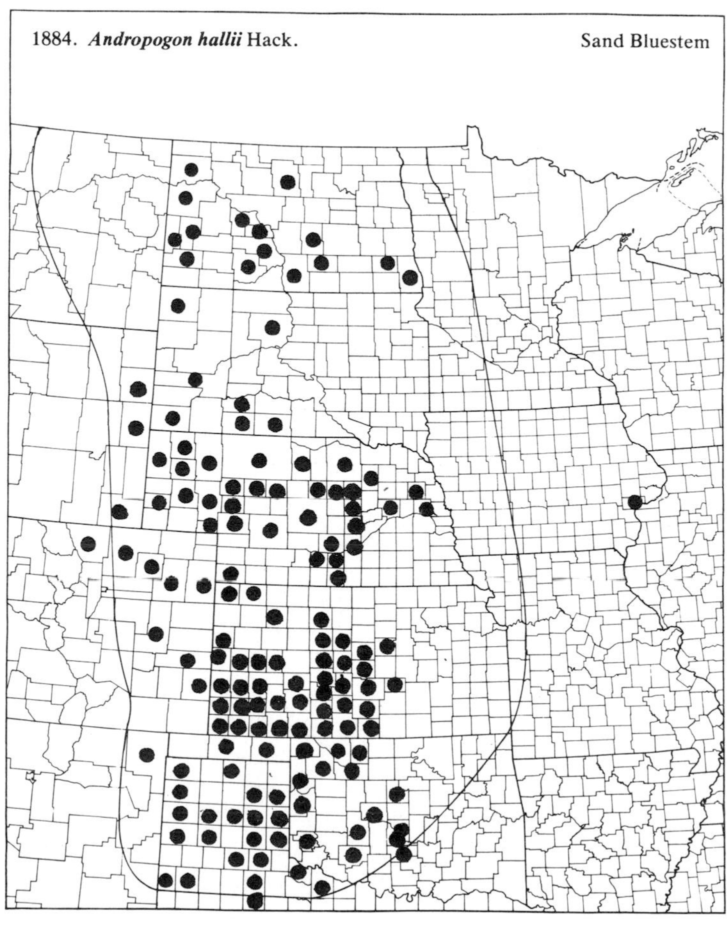
1884. *Andropogon hallii* Hack. Sand Bluestem

(Poaceae)

1885. ***Andropogon intermedius*** R. Br. Caucasian Bluestem

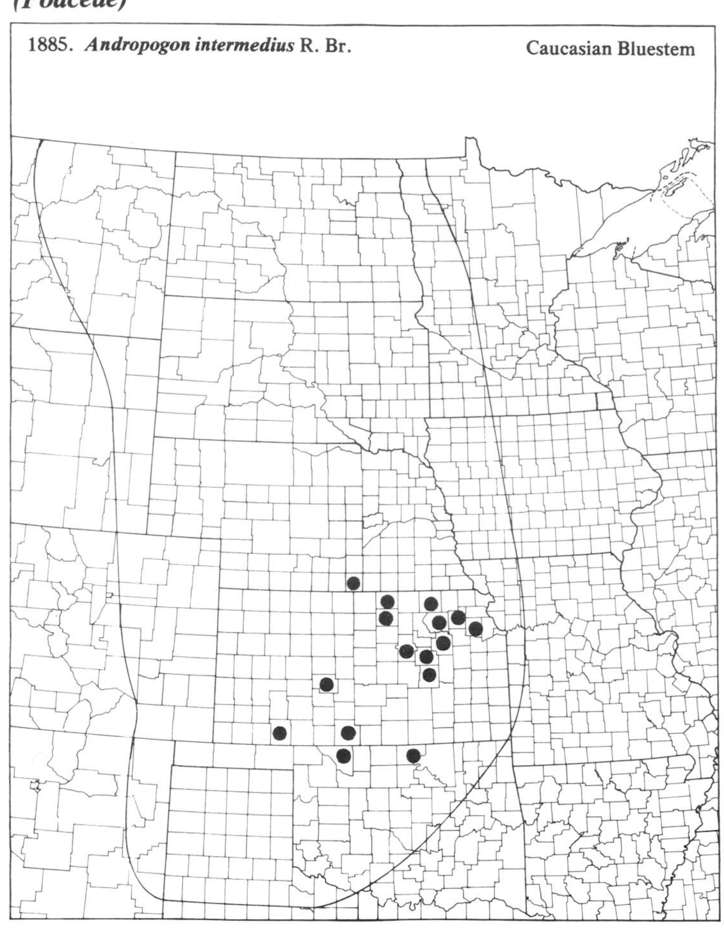

1886. ***Andropogon ischaemum*** L. var. ***songaricus*** Fisch. & Mey. Turkestan Bluestem

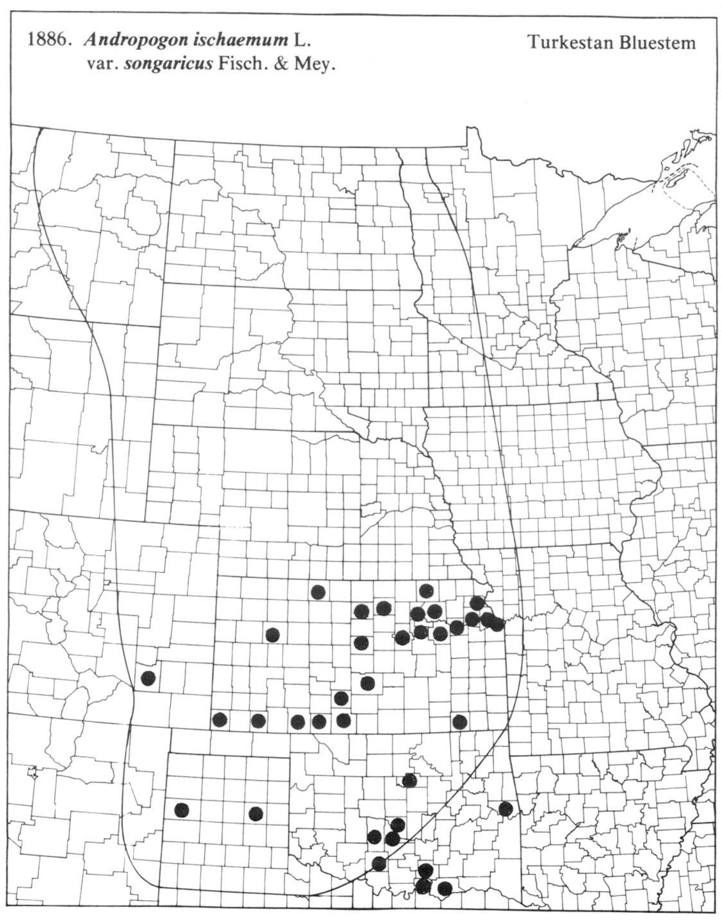

1887. ***Andropogon saccharoides*** Sw. Silver Bluestem

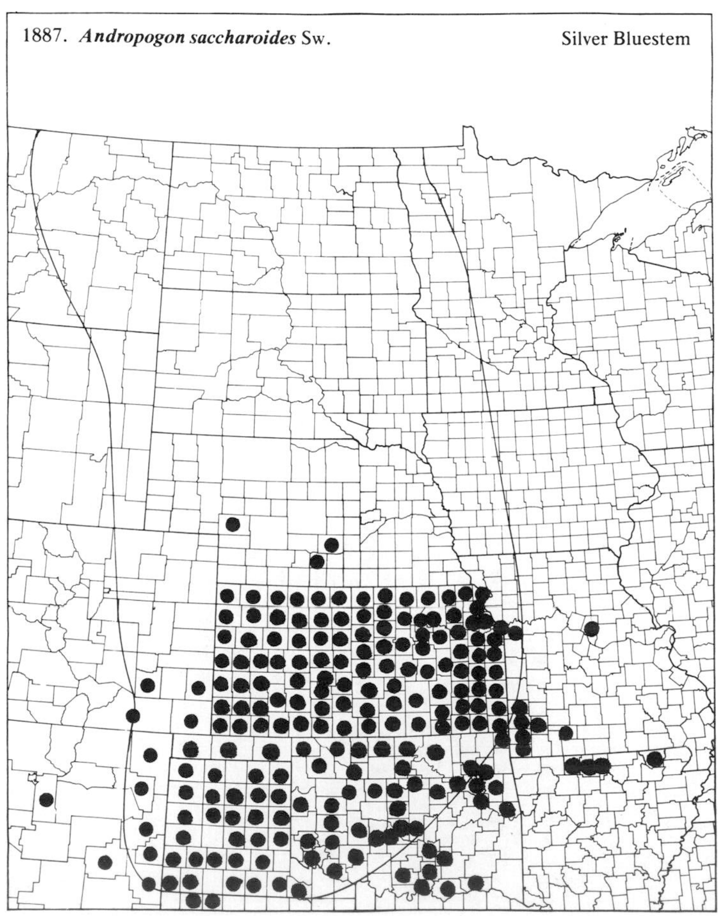

1888. ***Andropogon scoparius*** Michx. Little Bluestem

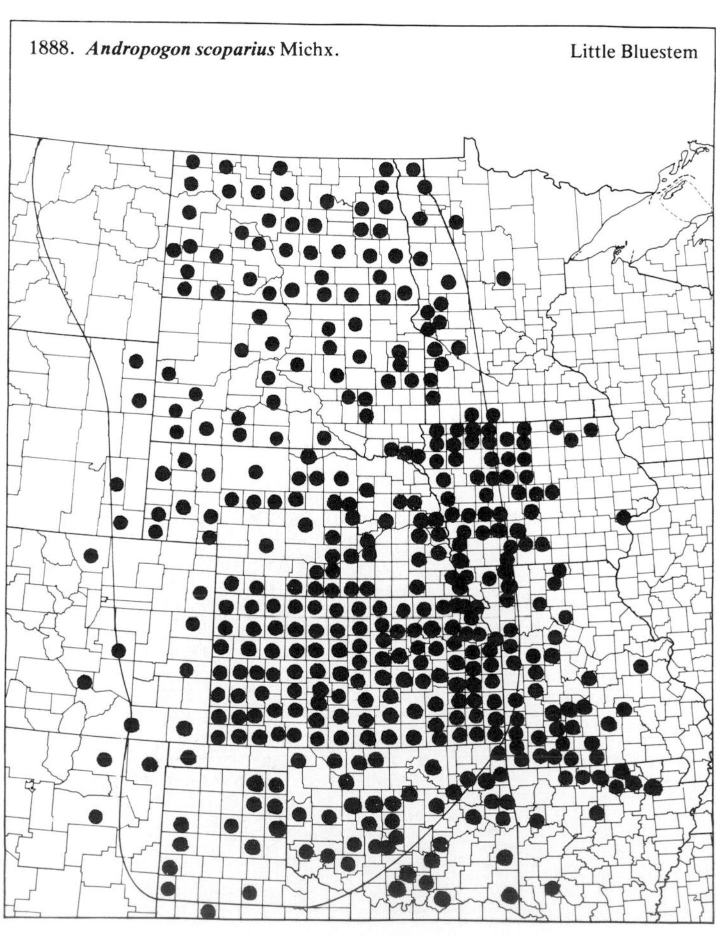

(Poaceae)

1889. ***Andropogon springfieldii*** Gould

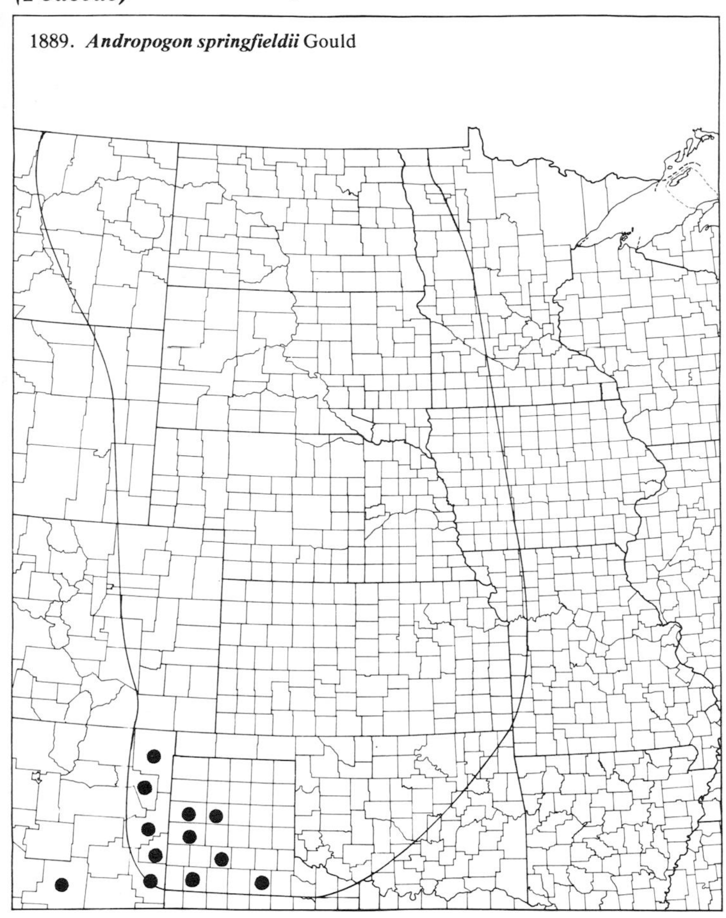

1890. ***Andropogon ternarius*** Michx. Splitbeard Bluestem

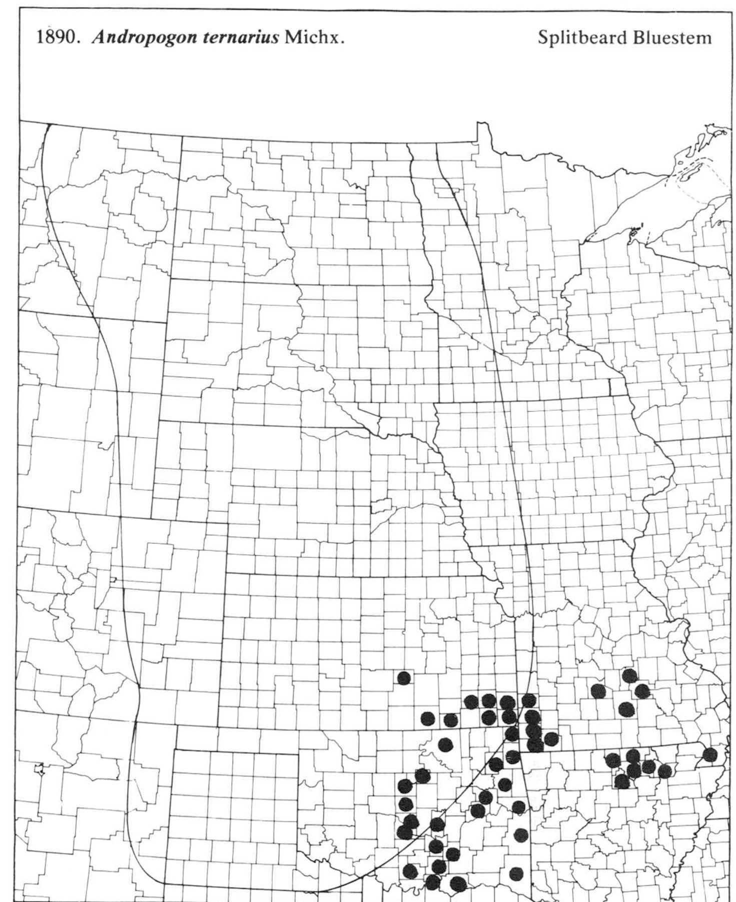

1891. ***Andropogon virginicus*** L. Broom Sedge

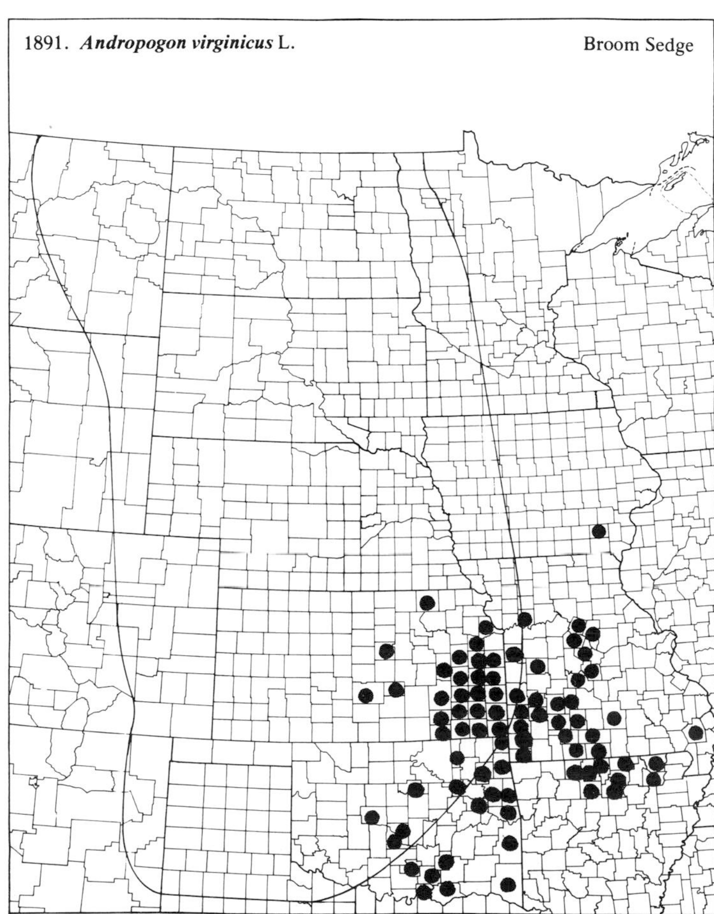

1892. ***Aristida adscensionis*** L. Six-weeks Threeawn

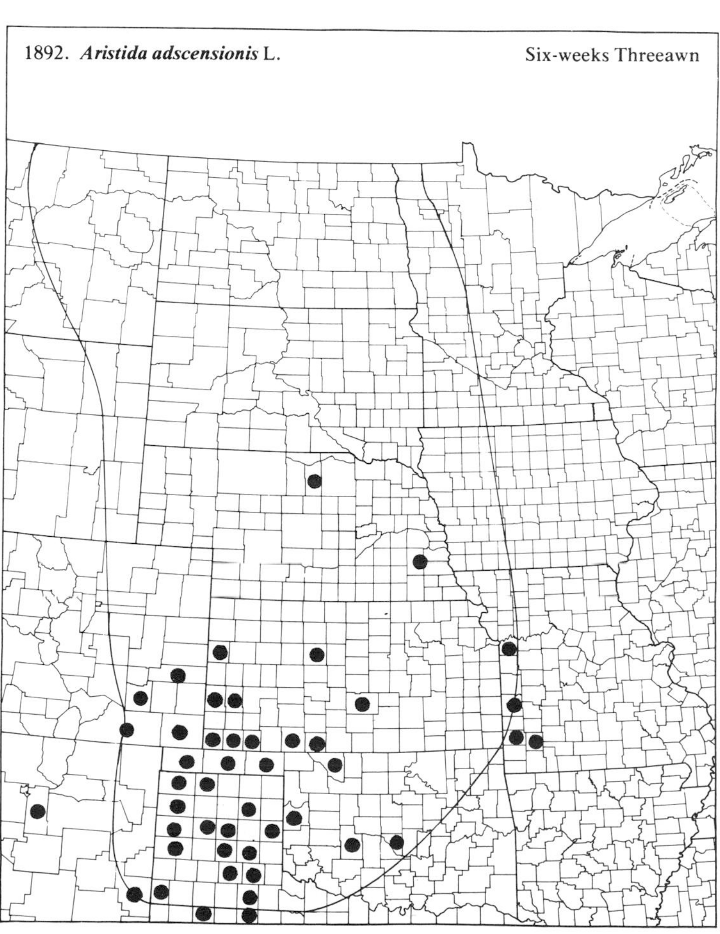

(Poaceae)

1893. ***Aristida basiramea*** Engelm. Forktip Threeawn

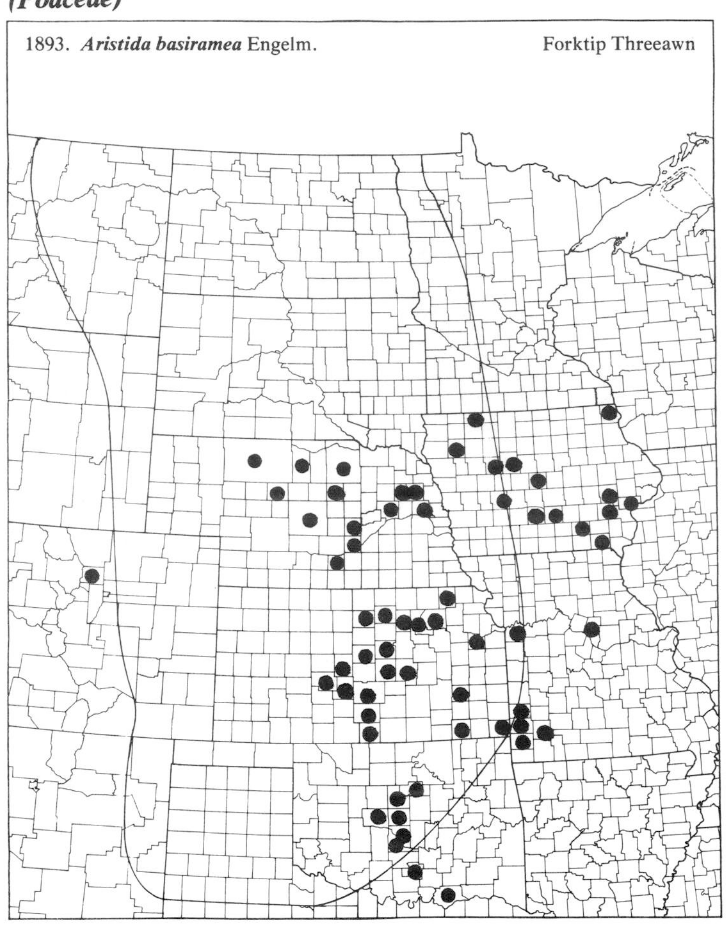

1894. ***Aristida dichotoma*** Michx. var. ***dichotoma*** Churchmouse Threeawn

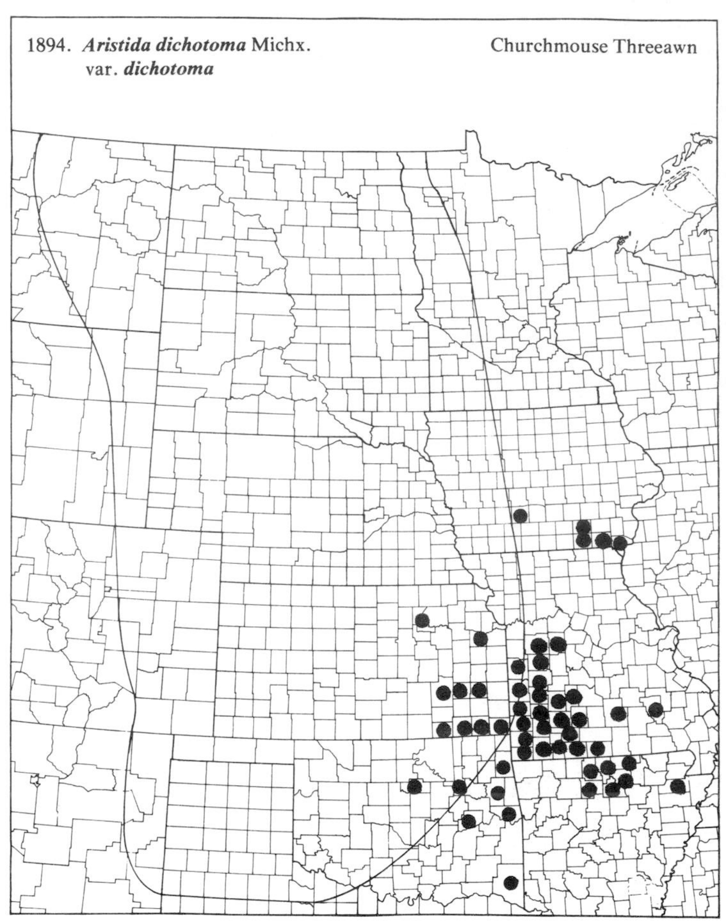

1895. ***Aristida dichotoma*** Michx. var. ***curtissii*** Gray Churchmouse Threeawn

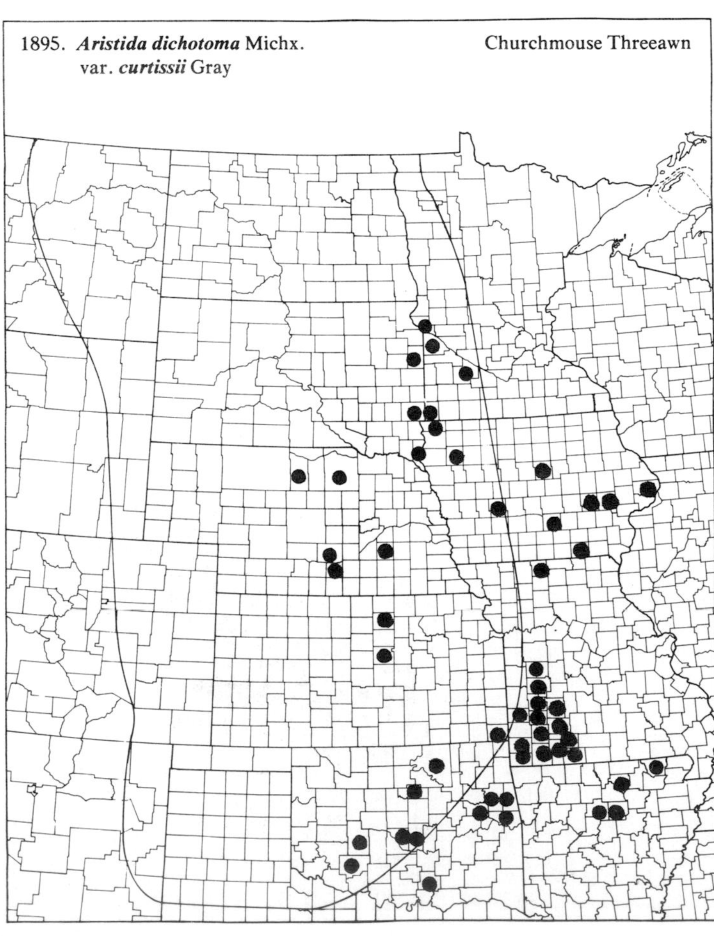

1896. ***Aristida divaricata*** H. & B. Poverty Threeawn

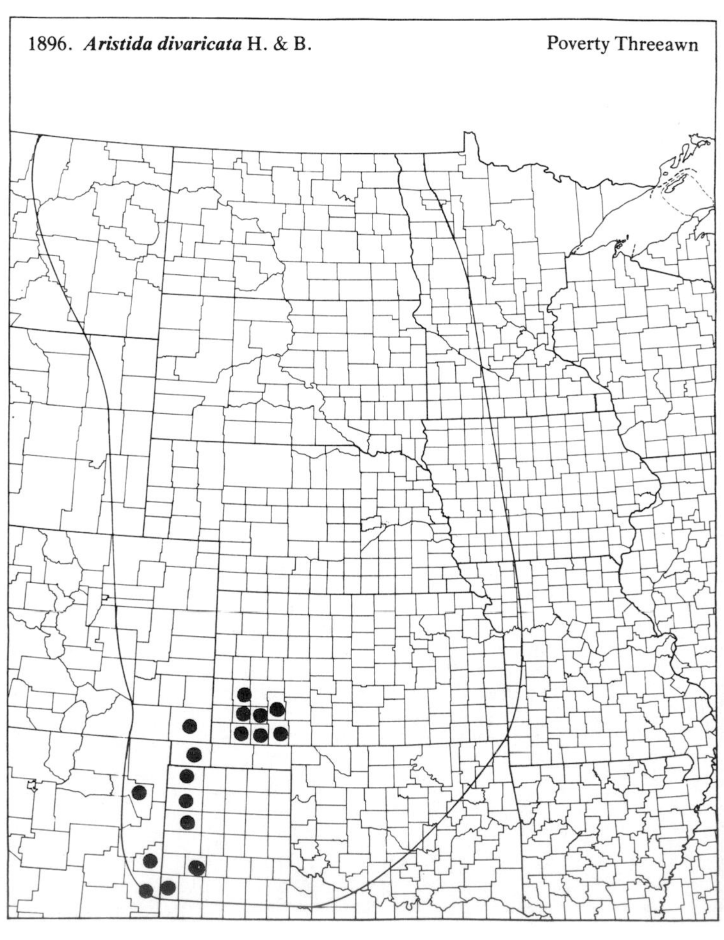

(Poaceae)

1897. ***Aristida glauca*** (Nees) Walp.

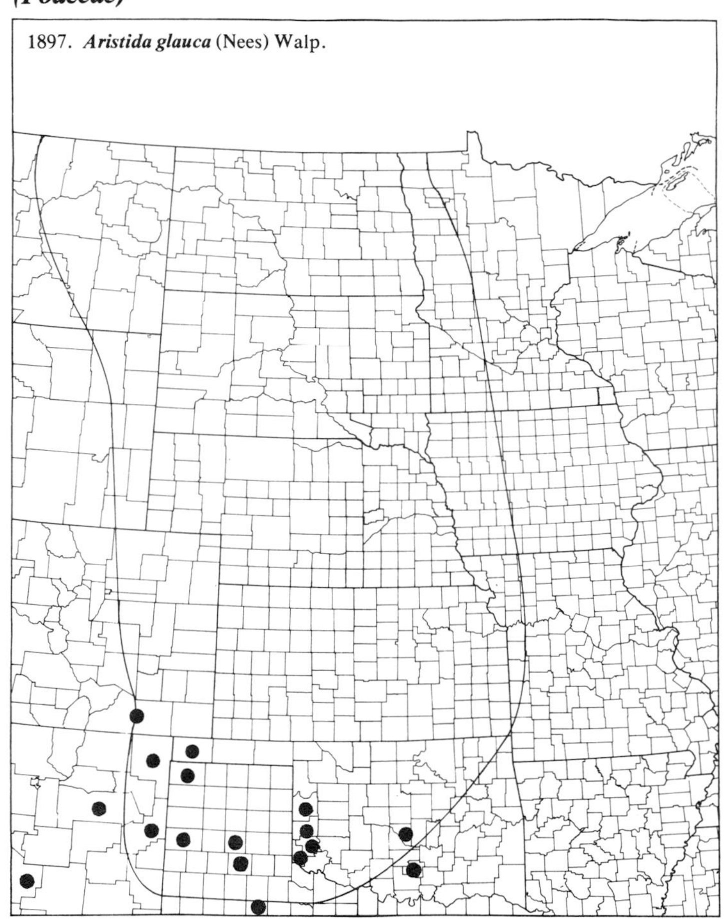

1898. ***Aristida longespica*** Poir. Slimspike Threeawn

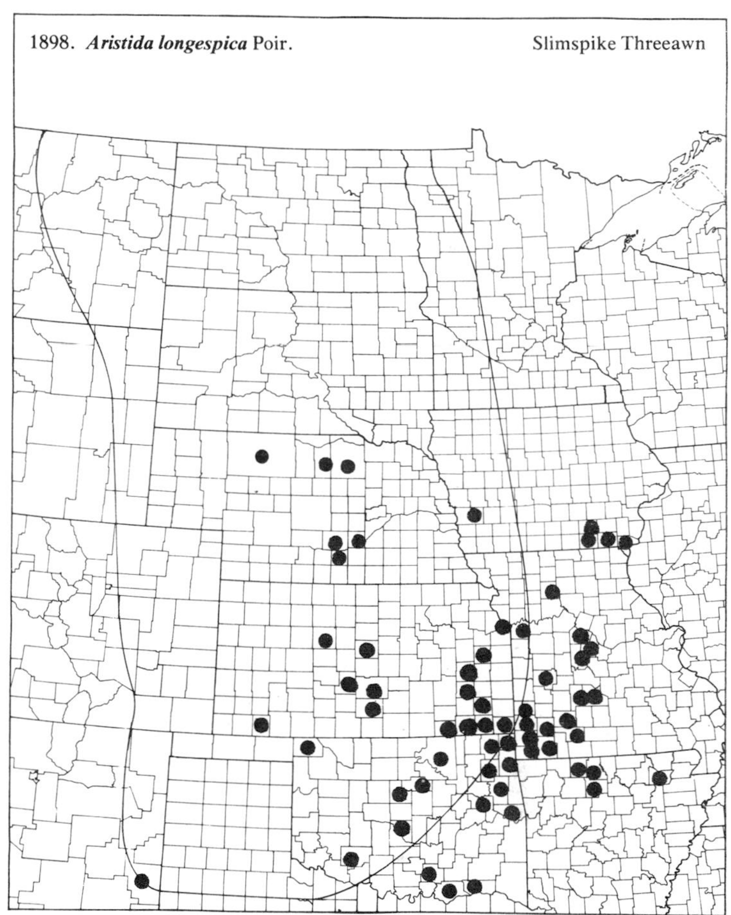

1899. ***Aristida longiseta*** Steud. Red Threeawn

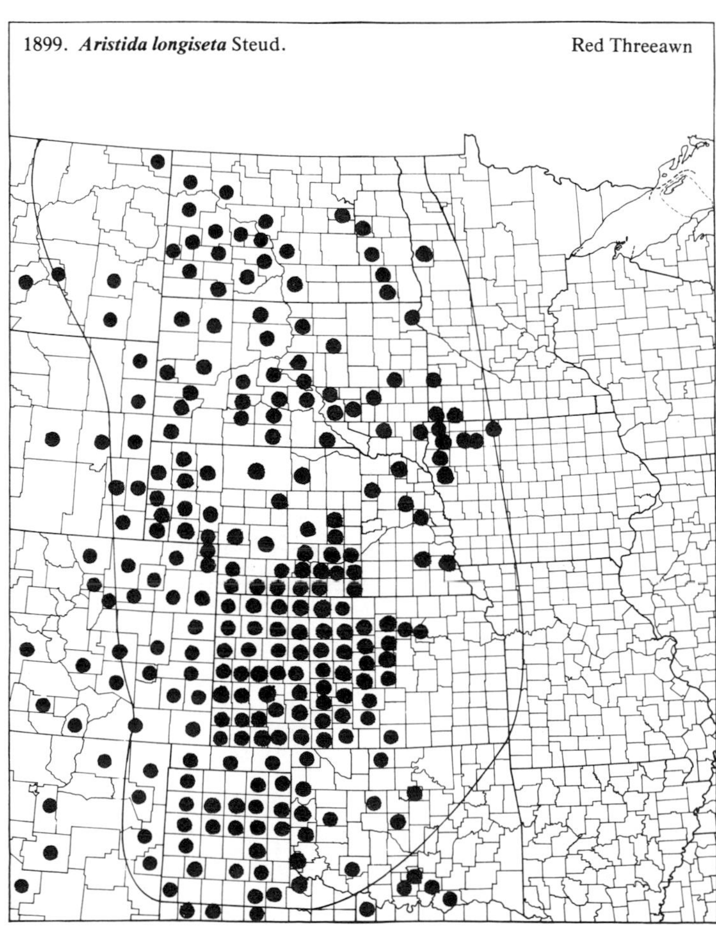

1900. ***Aristida oligantha*** Michx. Prairie Threeawn

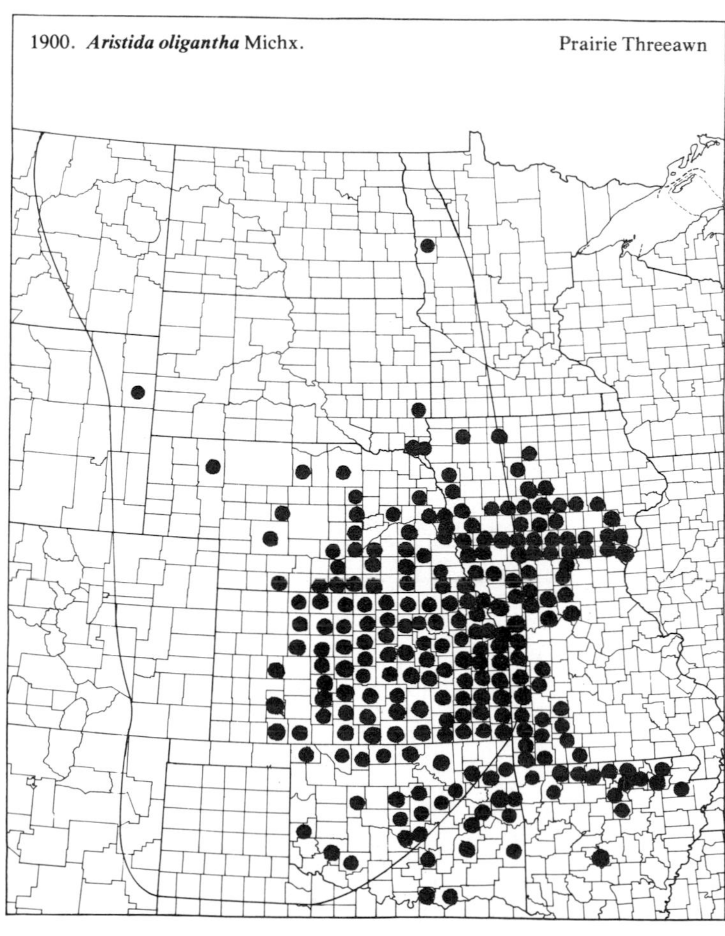

(Poaceae)

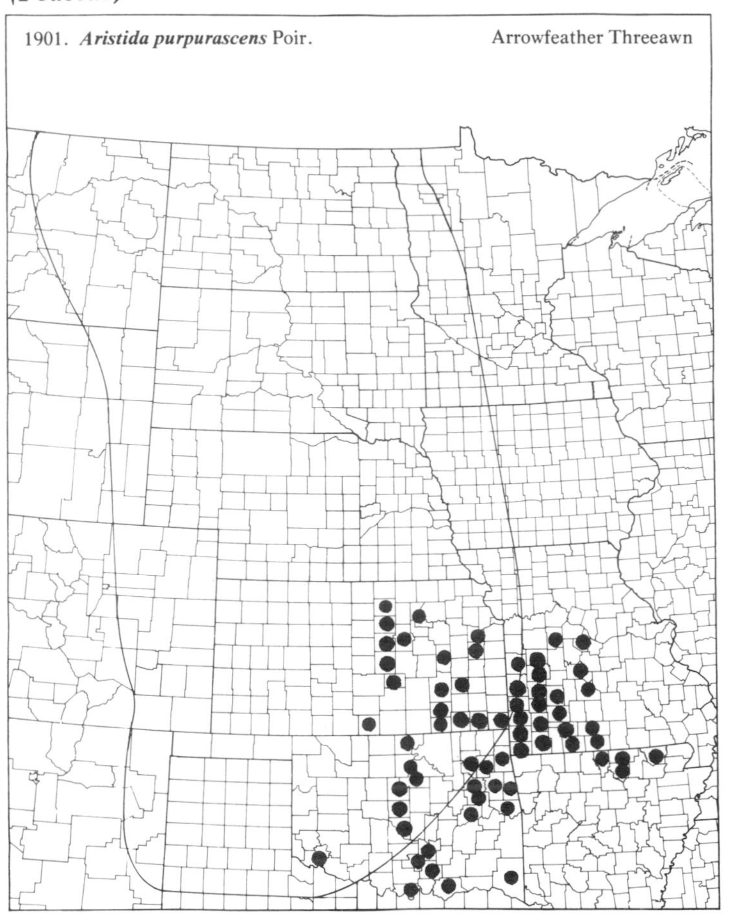
1901. *Aristida purpurascens* Poir.
Arrowfeather Threeawn

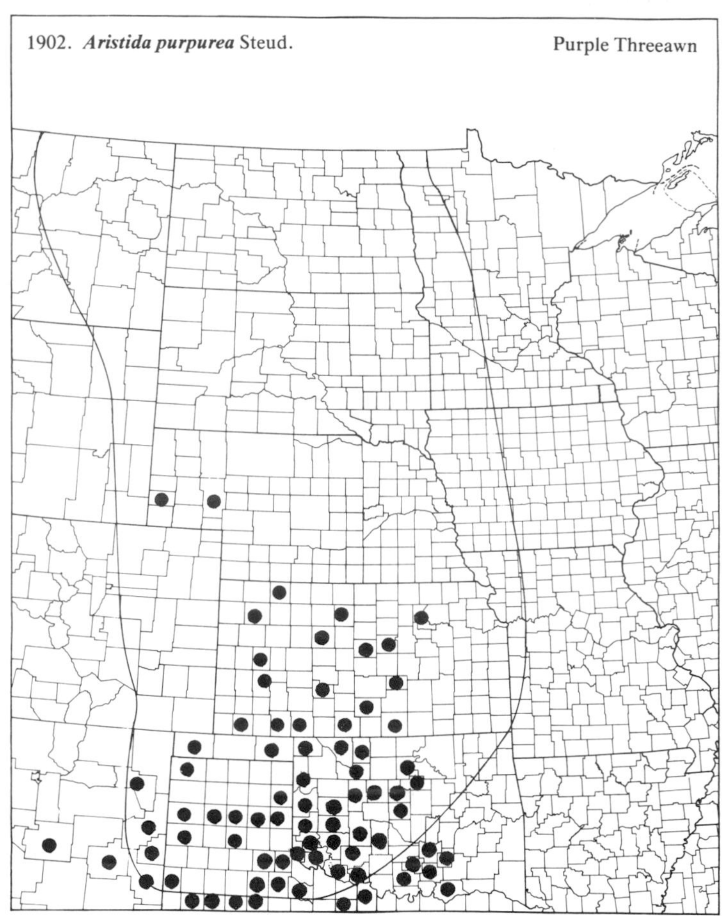
1902. *Aristida purpurea* Steud.
Purple Threeawn

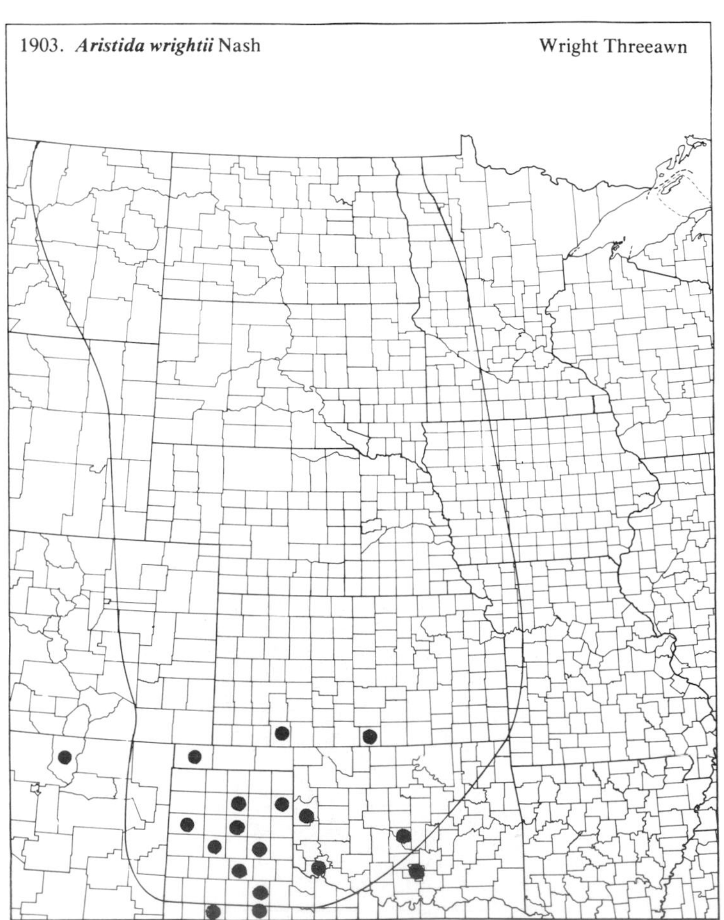
1903. *Aristida wrightii* Nash
Wright Threeawn

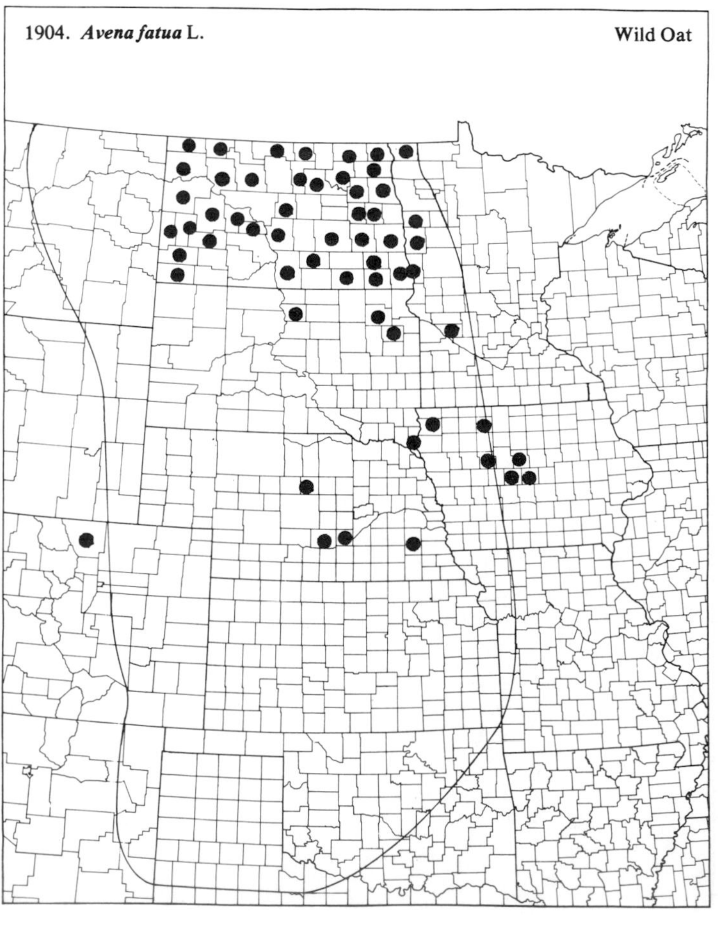
1904. *Avena fatua* L.
Wild Oat

(Poaceae)

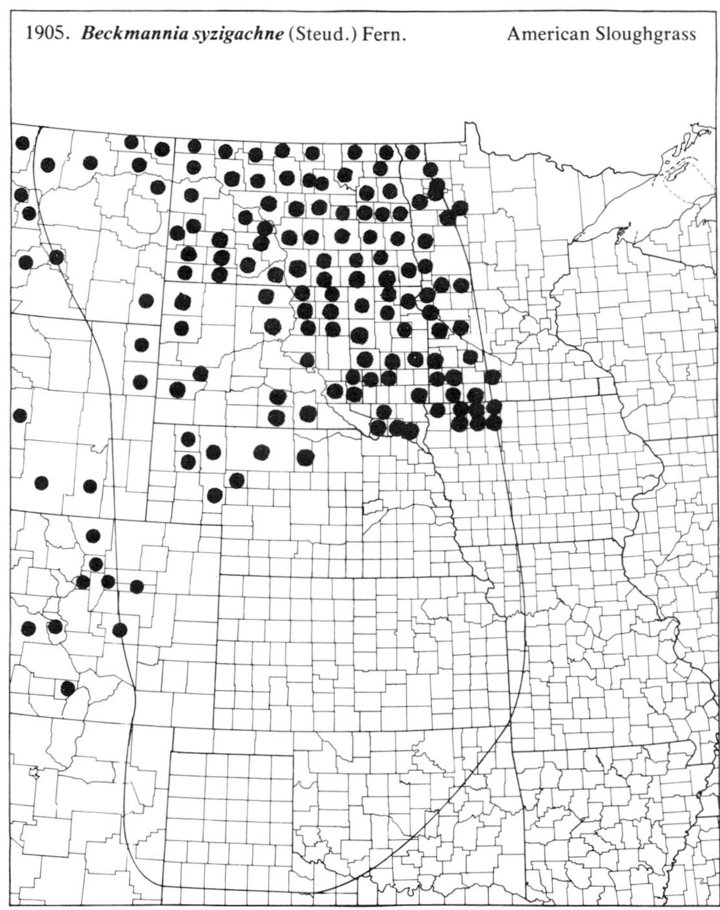
1905. Beckmannia syzigachne (Steud.) Fern.
American Sloughgrass

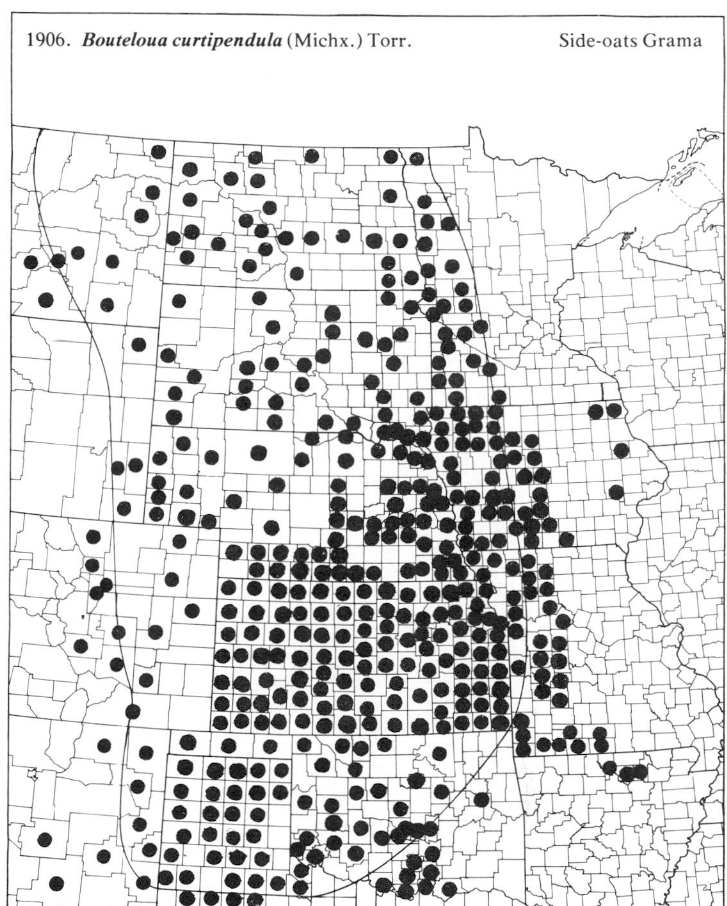
1906. Bouteloua curtipendula (Michx.) Torr.
Side-oats Grama

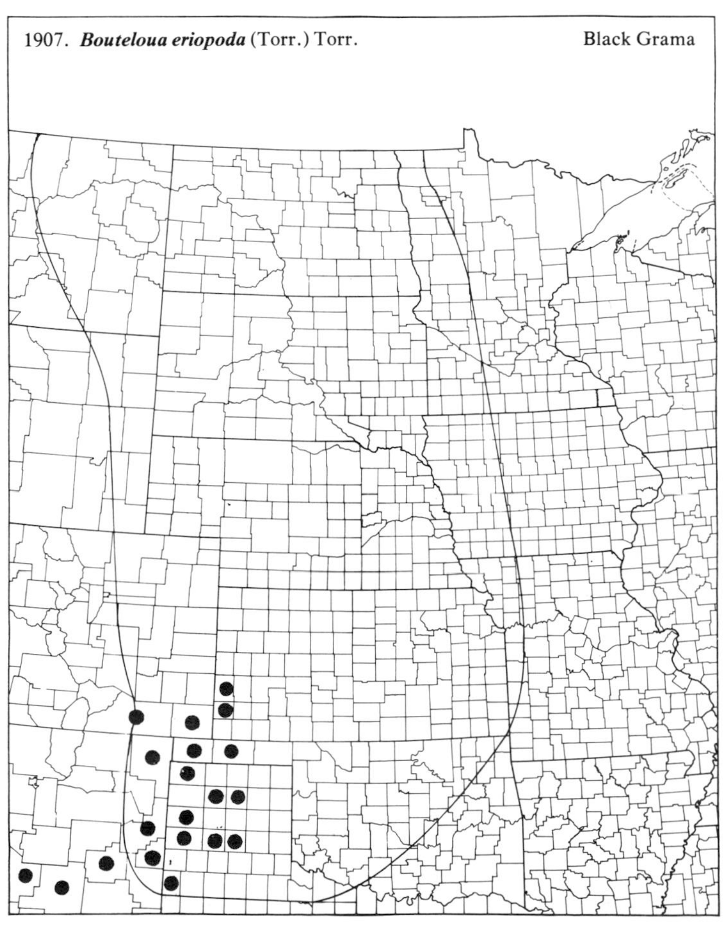
1907. Bouteloua eriopoda (Torr.) Torr.
Black Grama

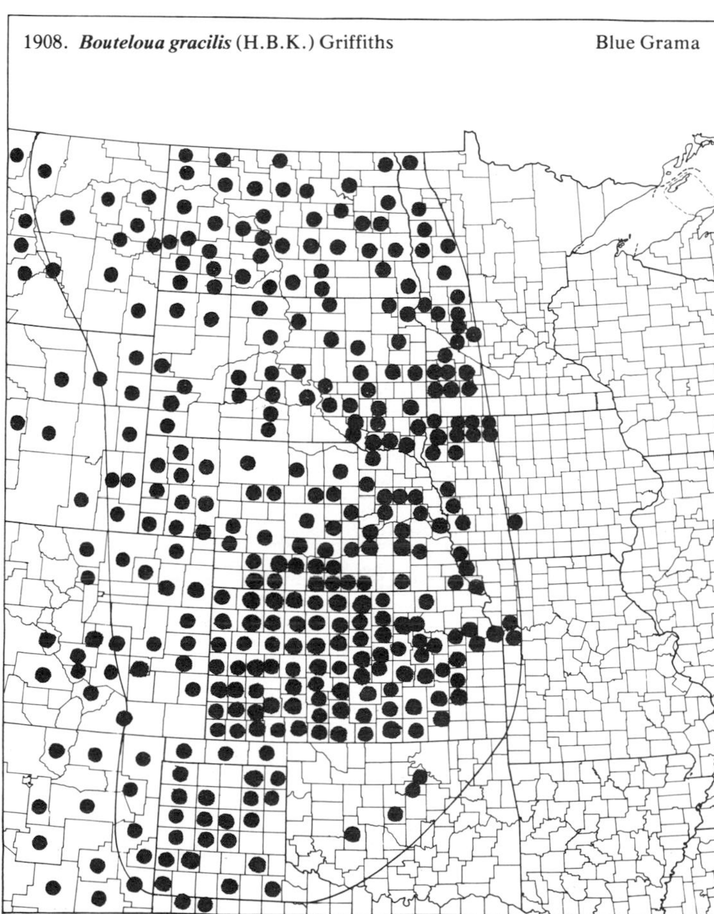
1908. Bouteloua gracilis (H.B.K.) Griffiths
Blue Grama

(Poaceae)

1909. ***Bouteloua hirsuta*** Lag. Hairy Grama

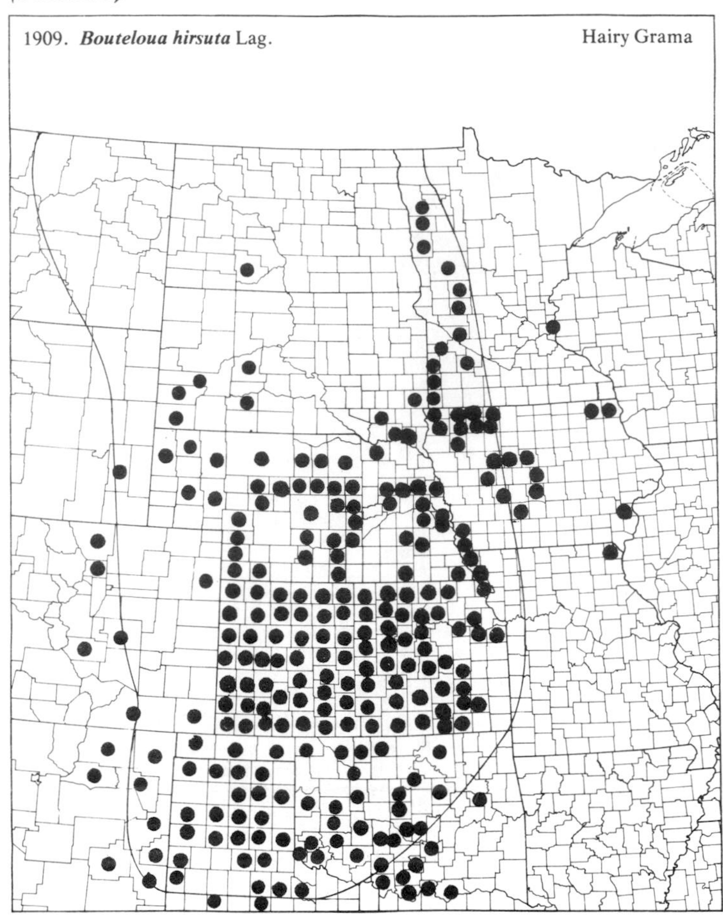

1910. ***Bouteloua rigidiseta*** (Steud.) Hitchc.

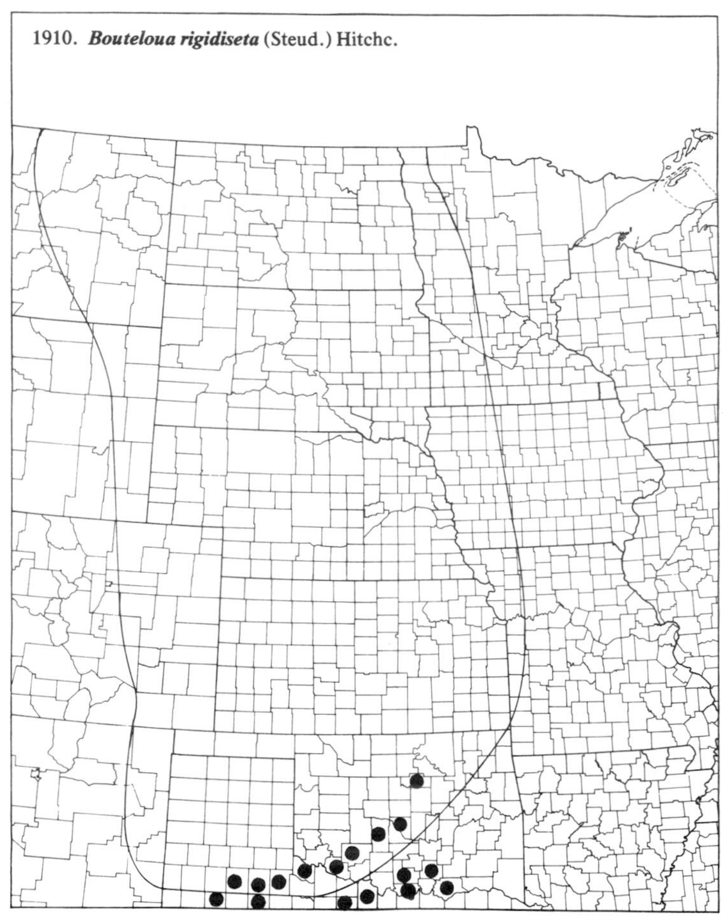

1911. ***Brachyelytrum erectum*** (Schreb.) Beauv.

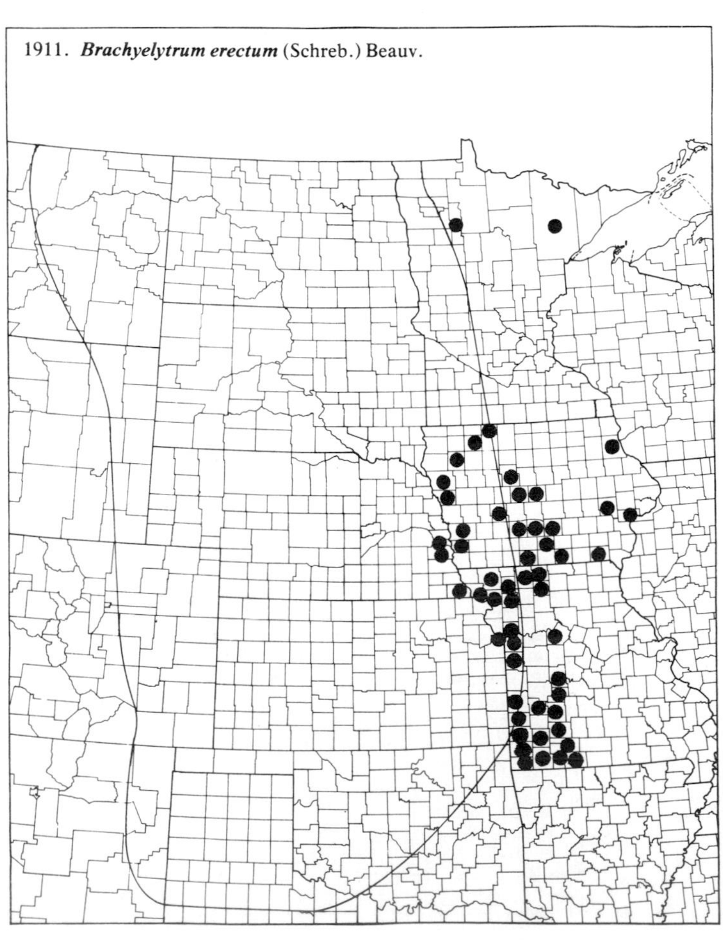

1912. ***Bromus ciliatus*** L. Fringed Brome

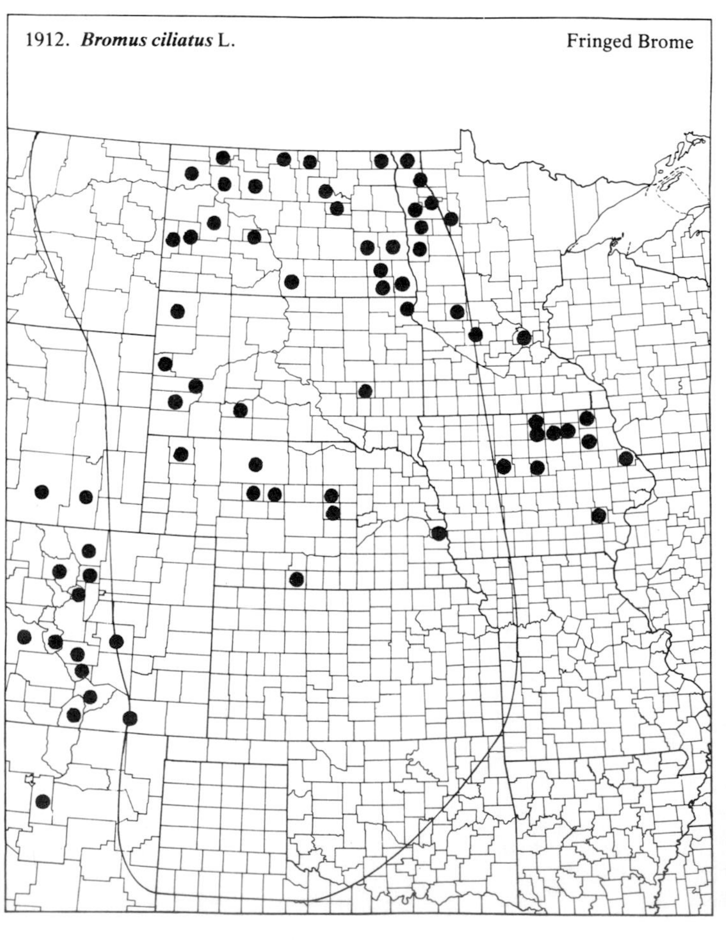

(Poaceae)

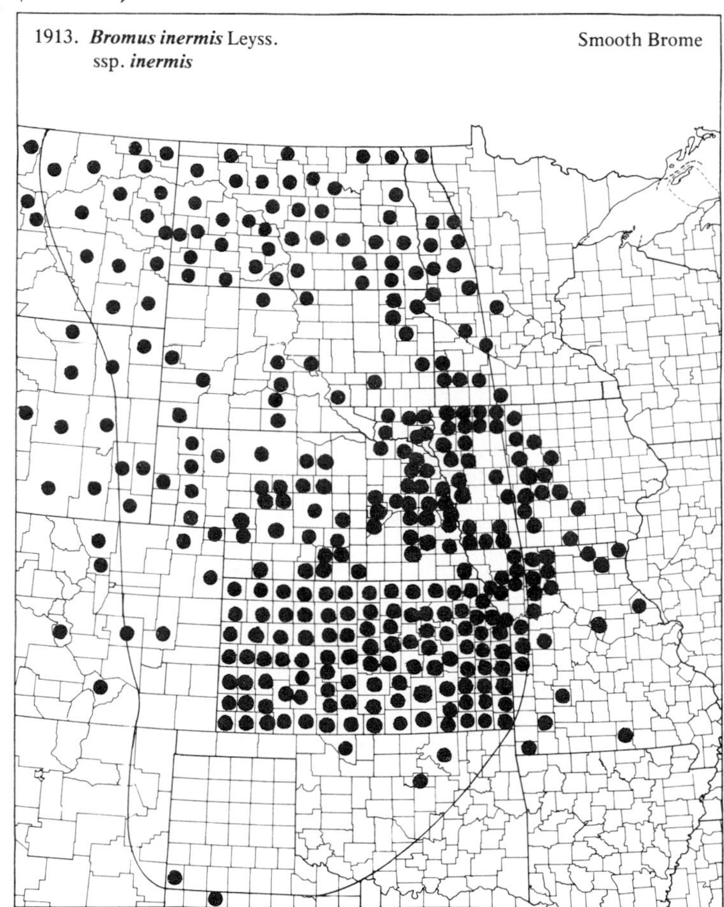
1913. ***Bromus inermis*** Leyss. ssp. ***inermis*** Smooth Brome

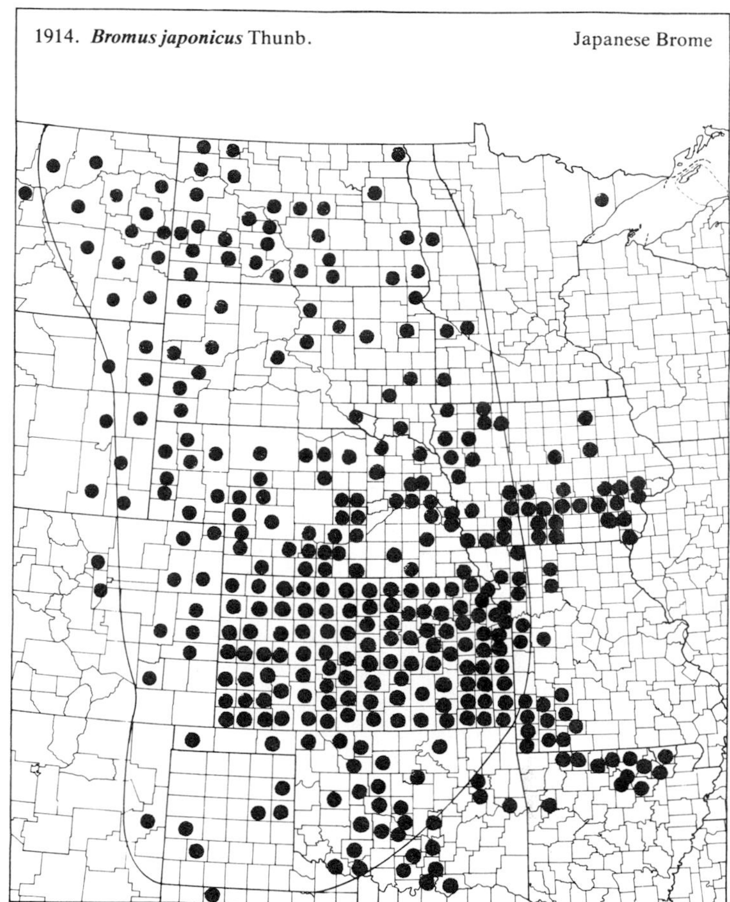
1914. ***Bromus japonicus*** Thunb. Japanese Brome

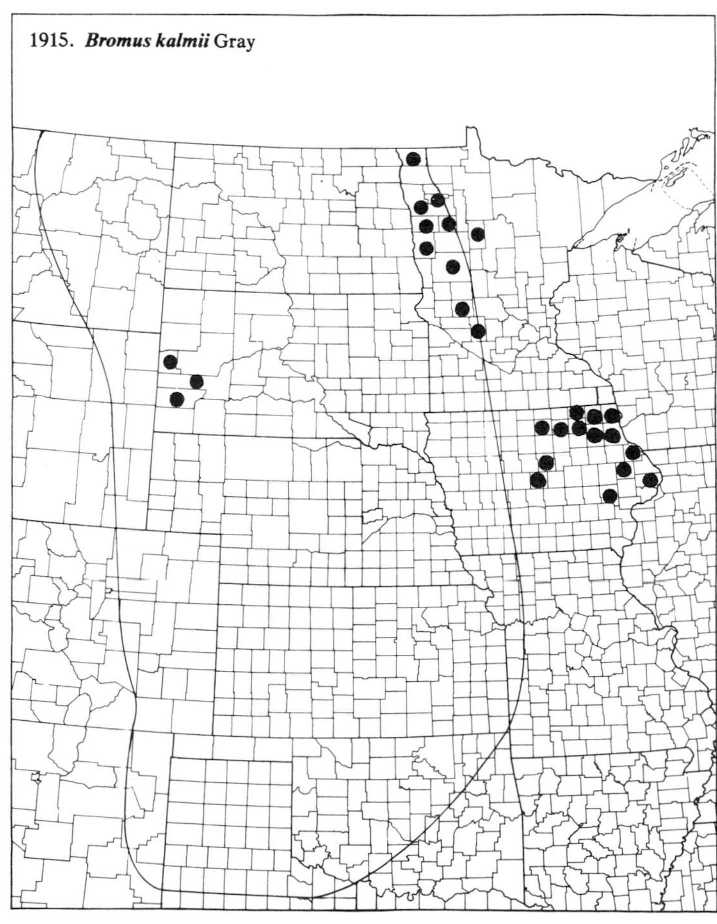
1915. ***Bromus kalmii*** Gray

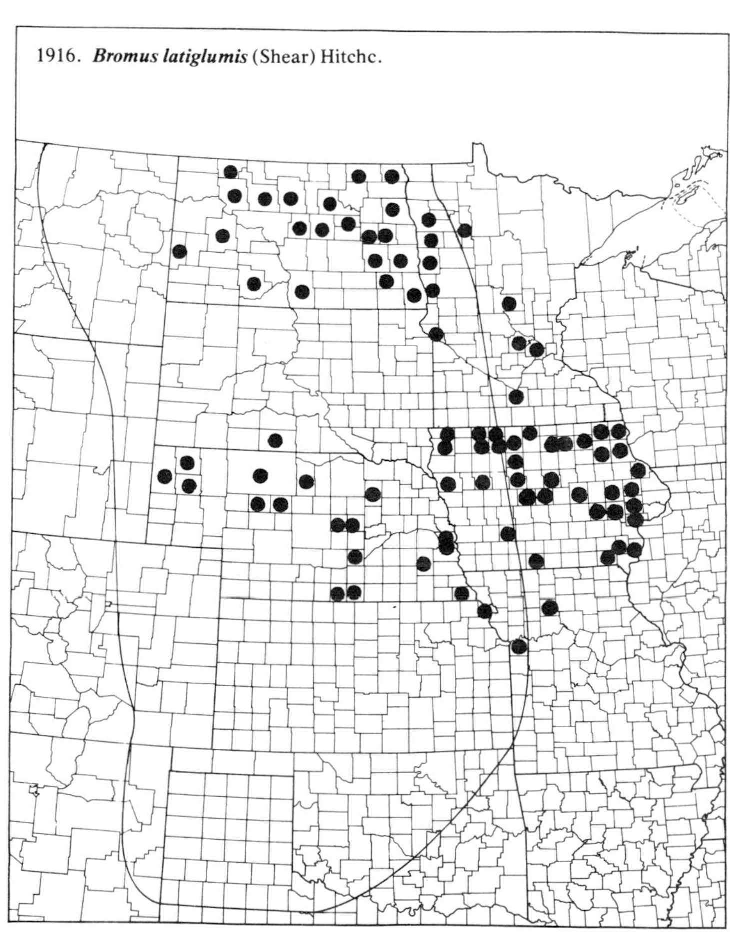
1916. ***Bromus latiglumis*** (Shear) Hitchc.

(Poaceae)

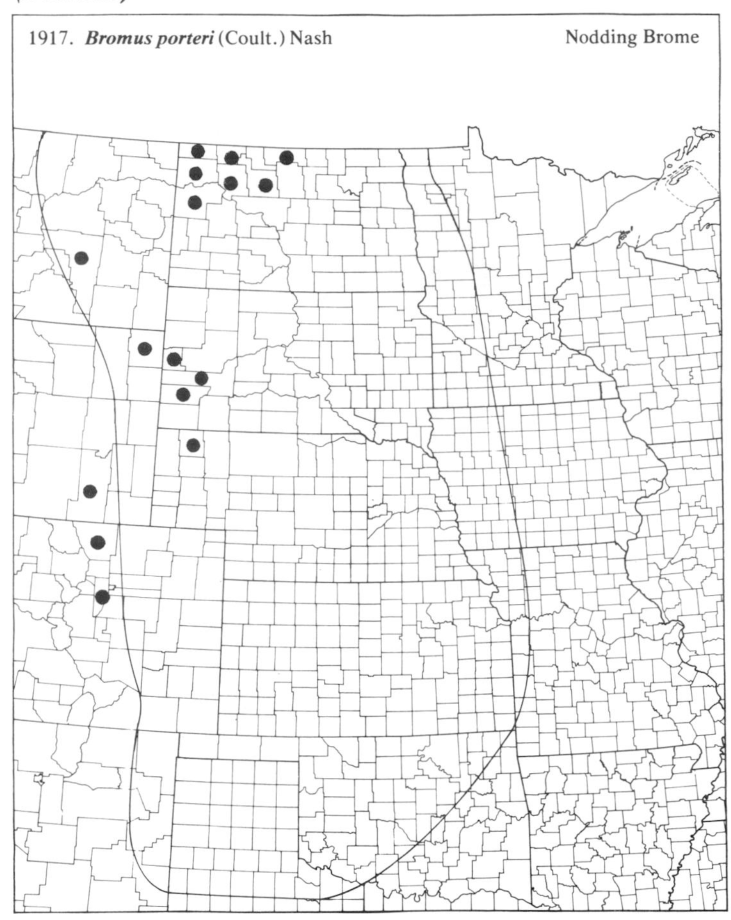
1917. **Bromus porteri** (Coult.) Nash
Nodding Brome

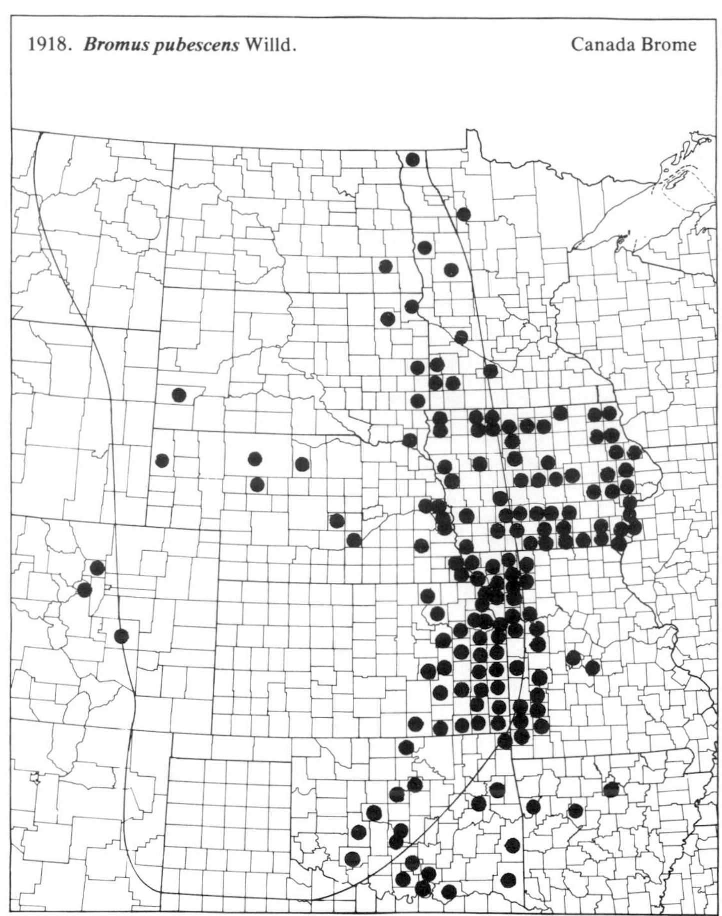
1918. **Bromus pubescens** Willd.
Canada Brome

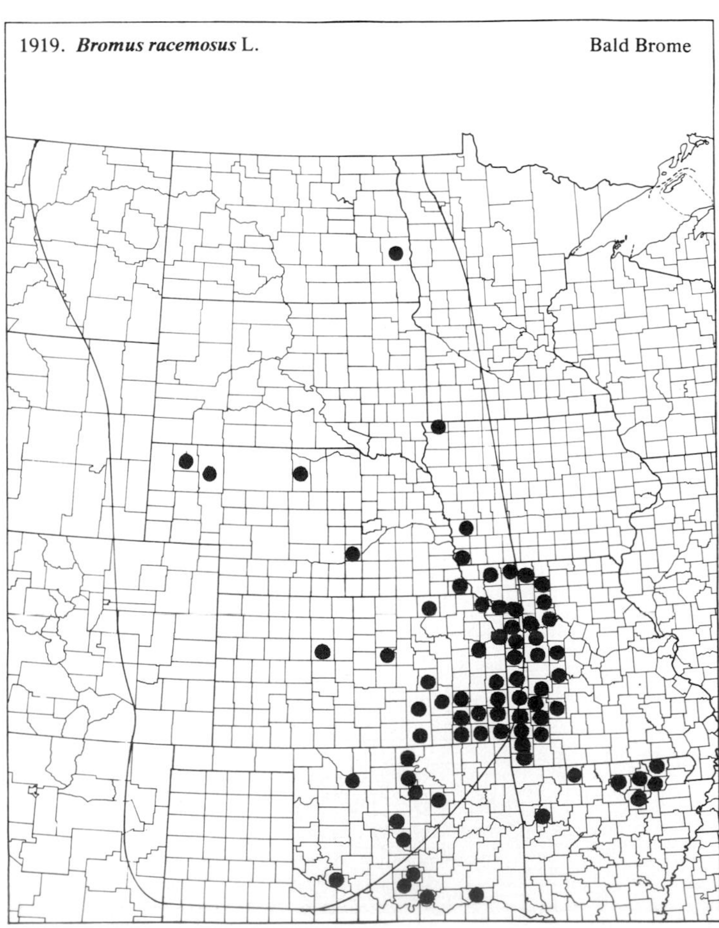
1919. **Bromus racemosus** L.
Bald Brome

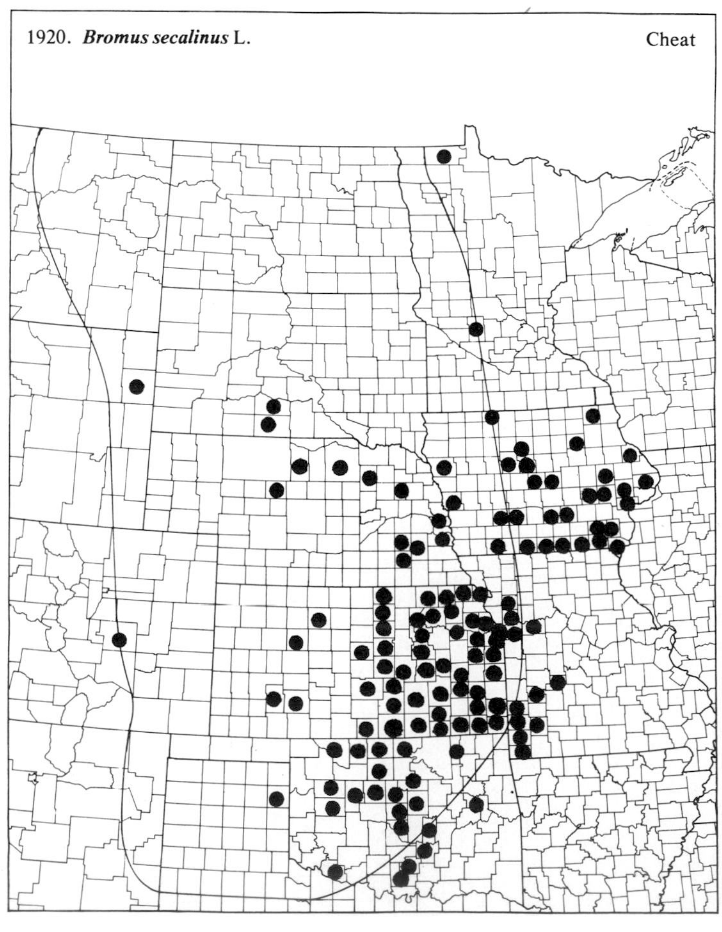
1920. **Bromus secalinus** L.
Cheat

(Poaceae)

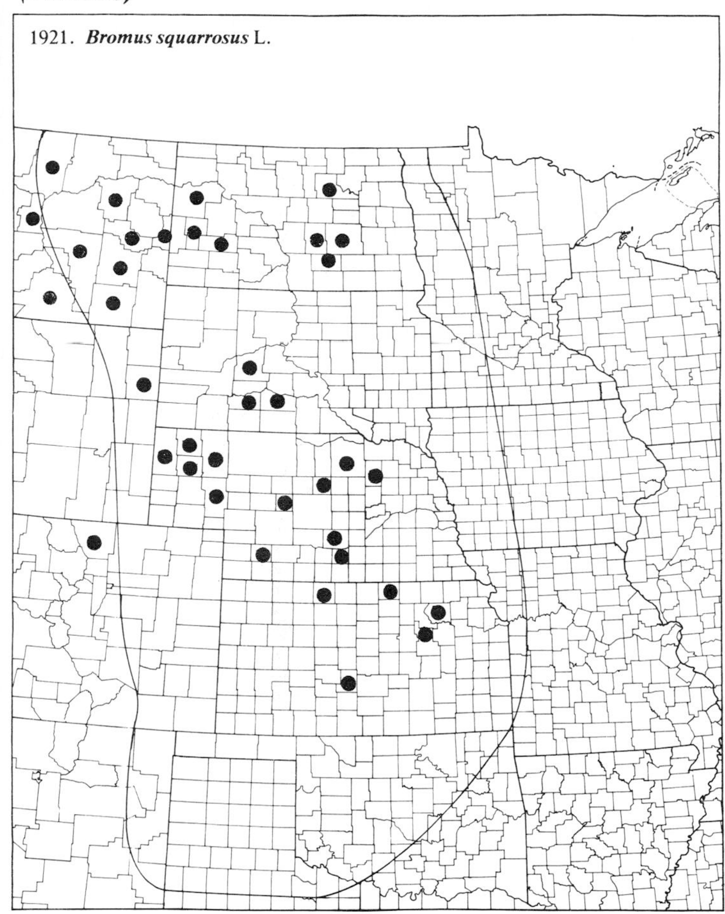

1921. ***Bromus squarrosus*** L.

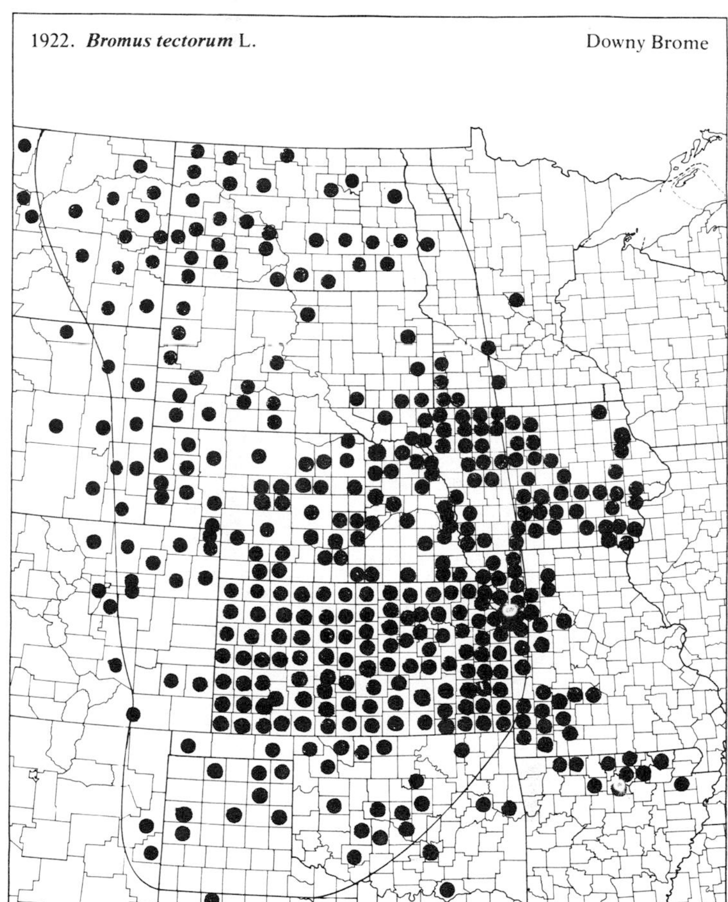

1922. ***Bromus tectorum*** L. Downy Brome

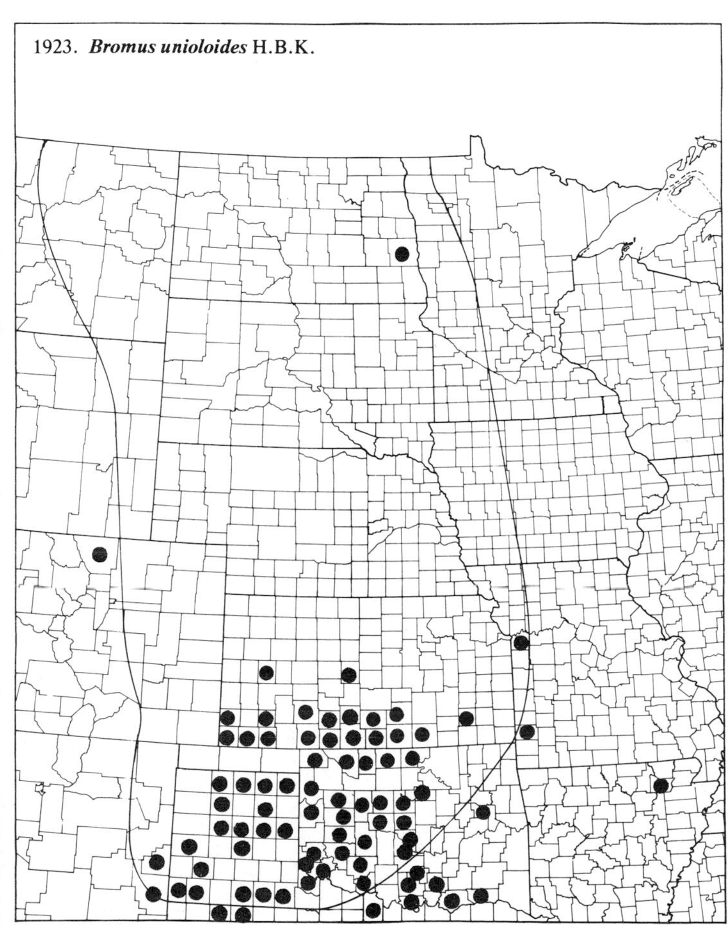

1923. ***Bromus unioloides*** H.B.K.

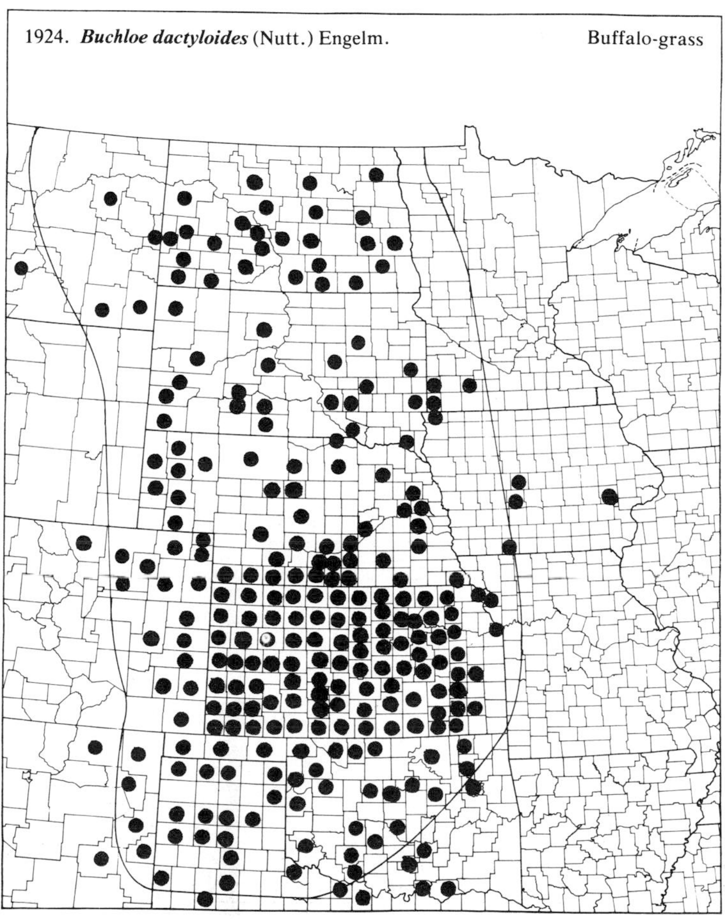

1924. ***Buchloe dactyloides*** (Nutt.) Engelm. Buffalo-grass

(Poaceae)

1925. ***Calamagrostis canadensis*** (Michx.) Beauv. Blue Joint

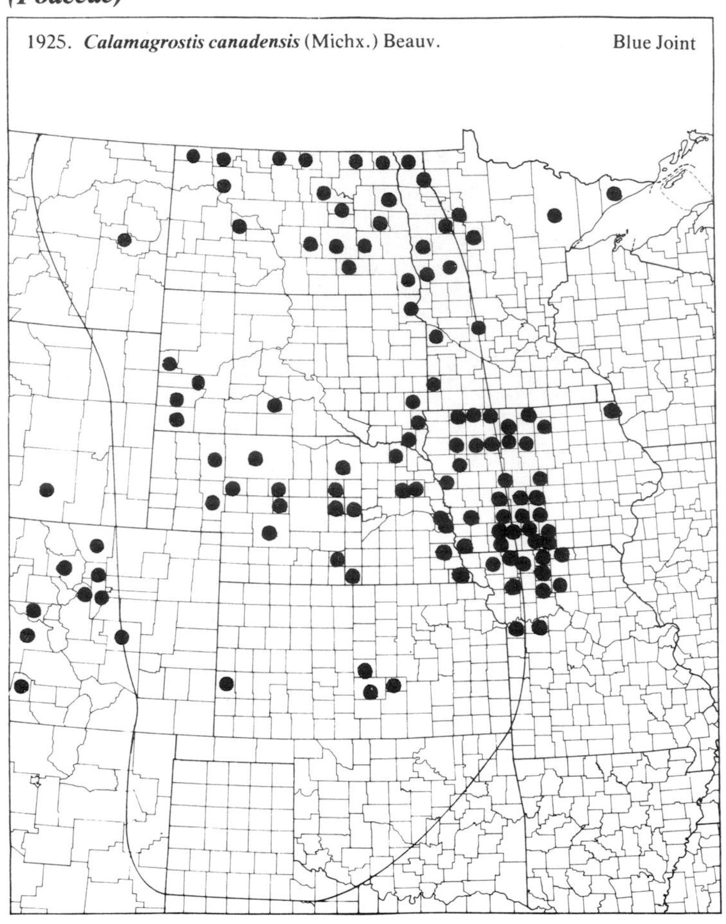

1926. ***Calamagrostis inexpansa*** Gray Northern Reedgrass

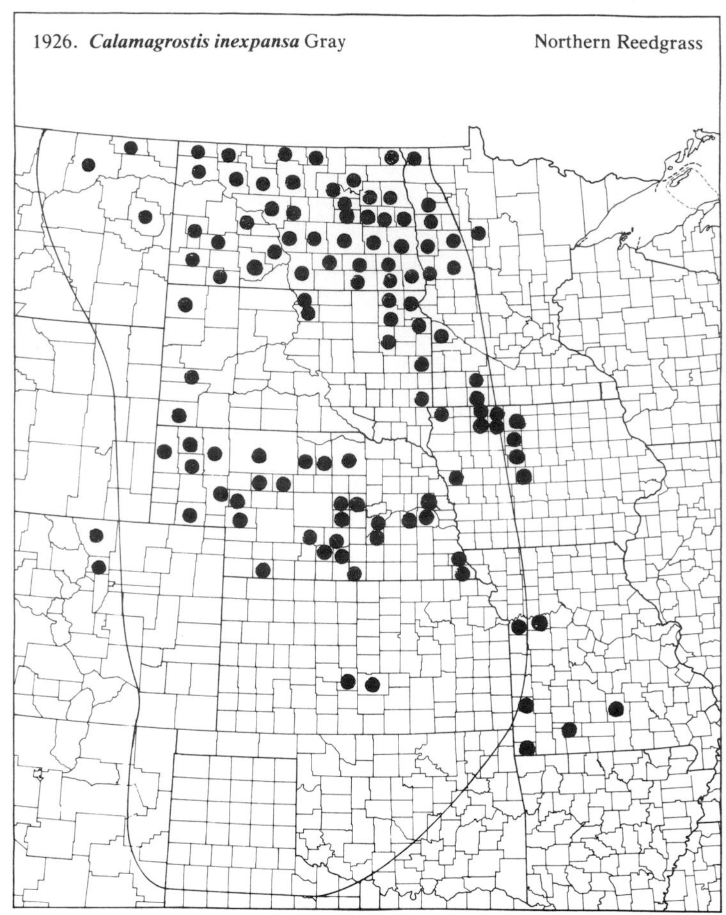

1927. ***Calamagrostis montanensis*** Scribn. Plains Reedgrass

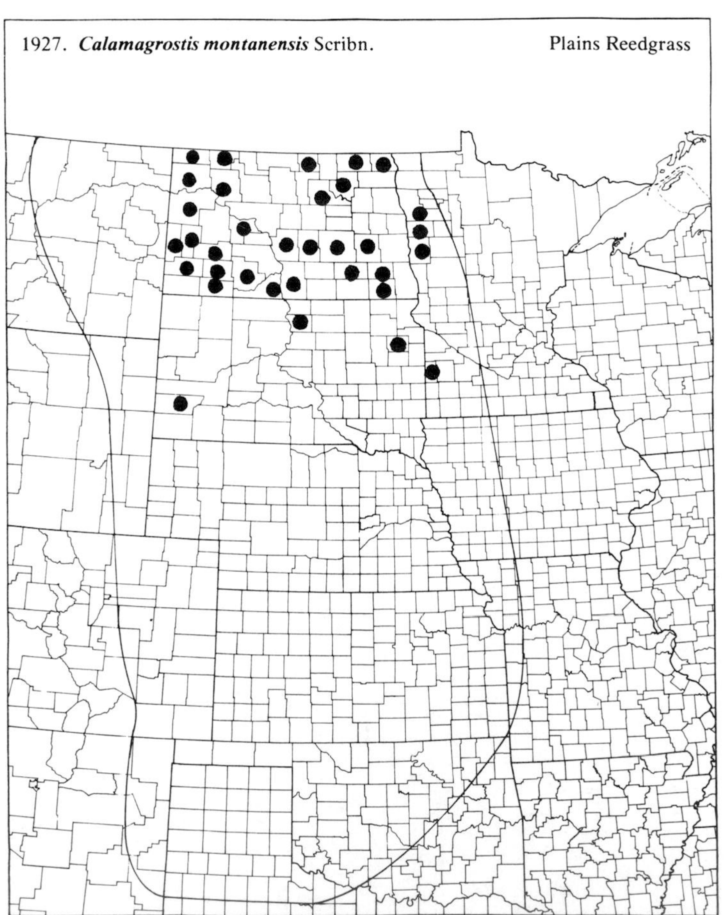

1928. ***Calamagrostis stricta*** (Timm.) Koel.

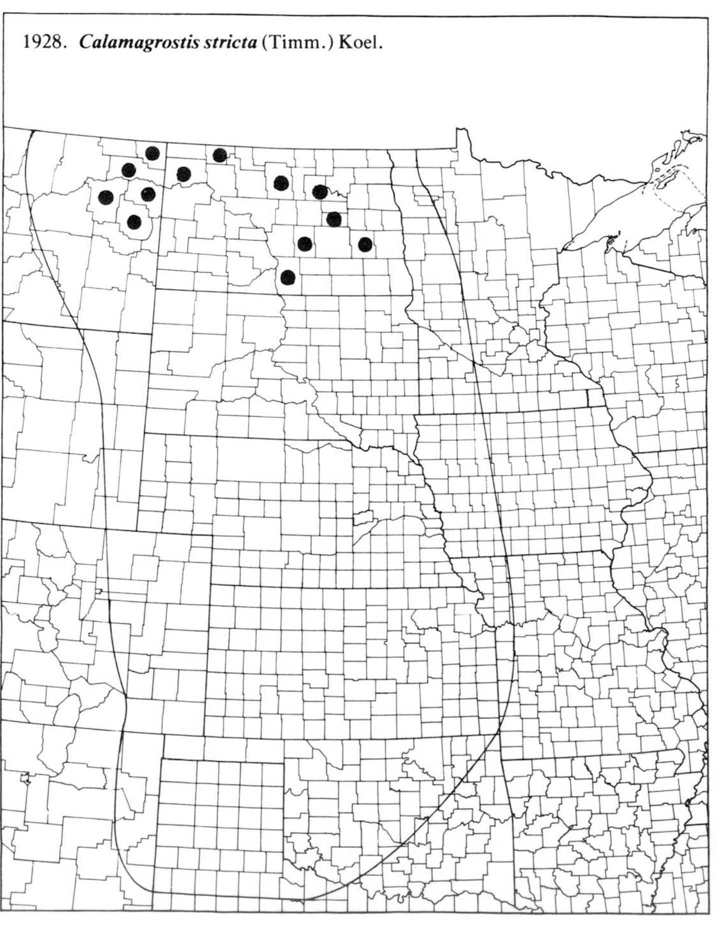

(Poaceae)

1929. ***Calamovilfa gigantea*** (Nutt.) Scribn. & Merr. — Big Sandreed

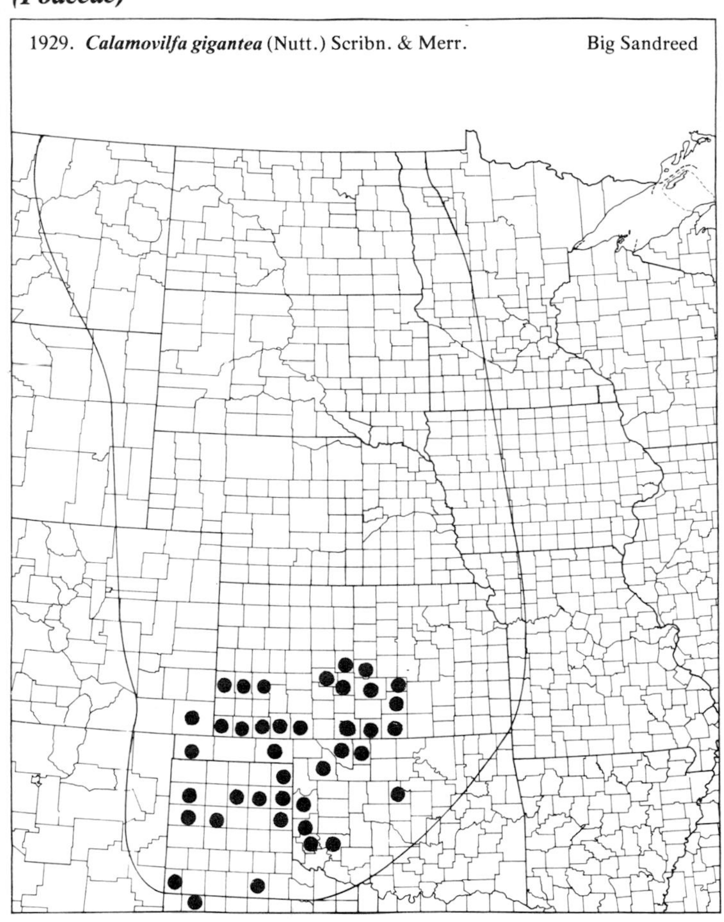

1930. ***Calamovilfa longifolia*** (Hook.) Scribn. — Prairie Sandreed

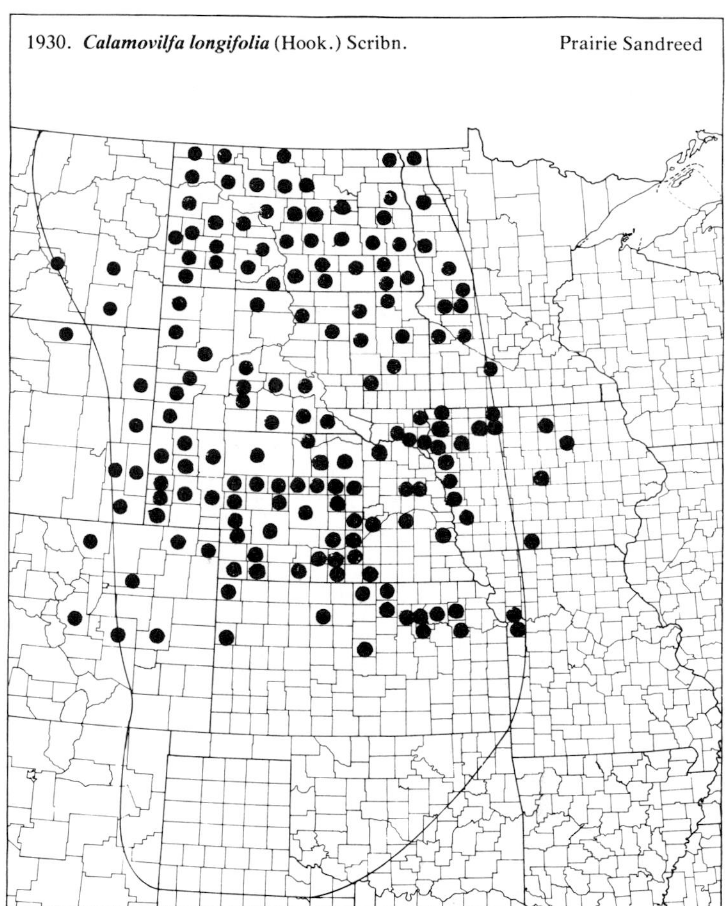

1931. ***Catabrosa aquatica*** (L.) Beauv. — Brookgrass

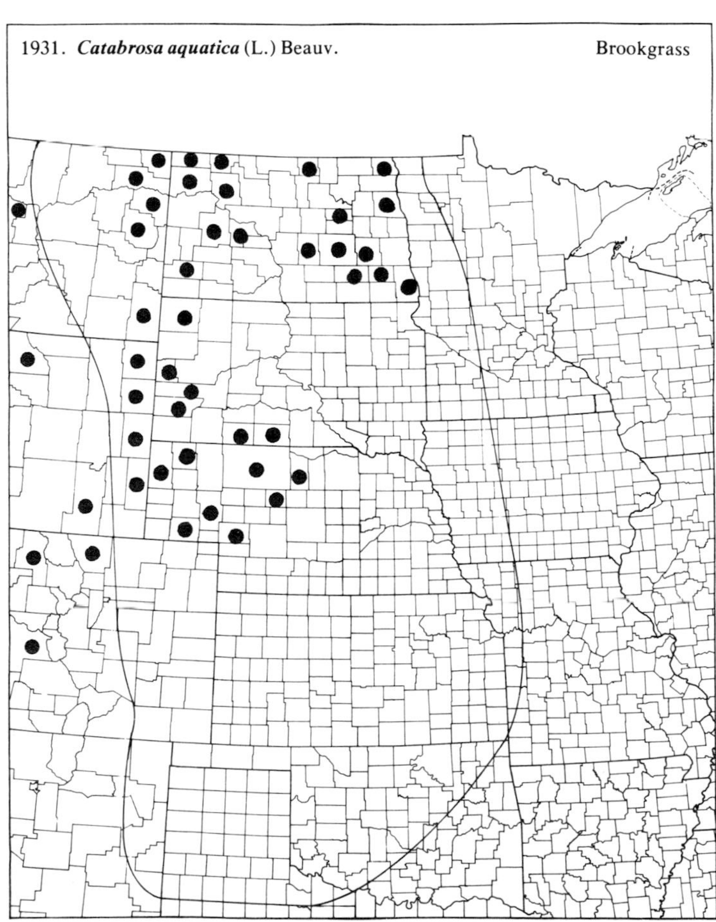

1932. ***Cenchrus incertus*** M.A. Curtis

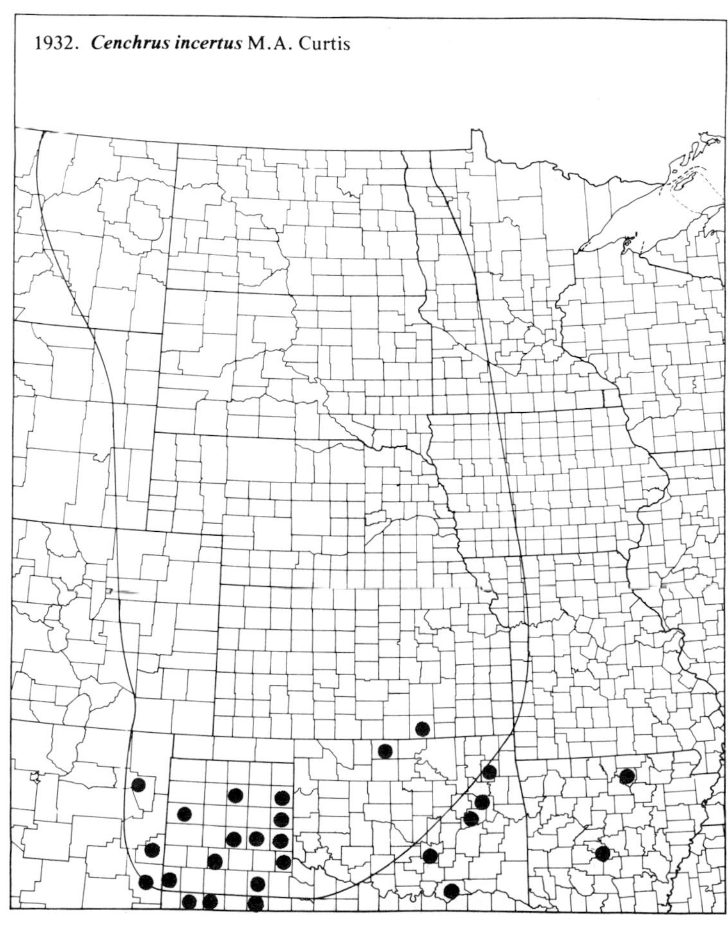

(Poaceae)

1933. ***Cenchrus longispinus*** (Hack.) Fern. Field Sandbur

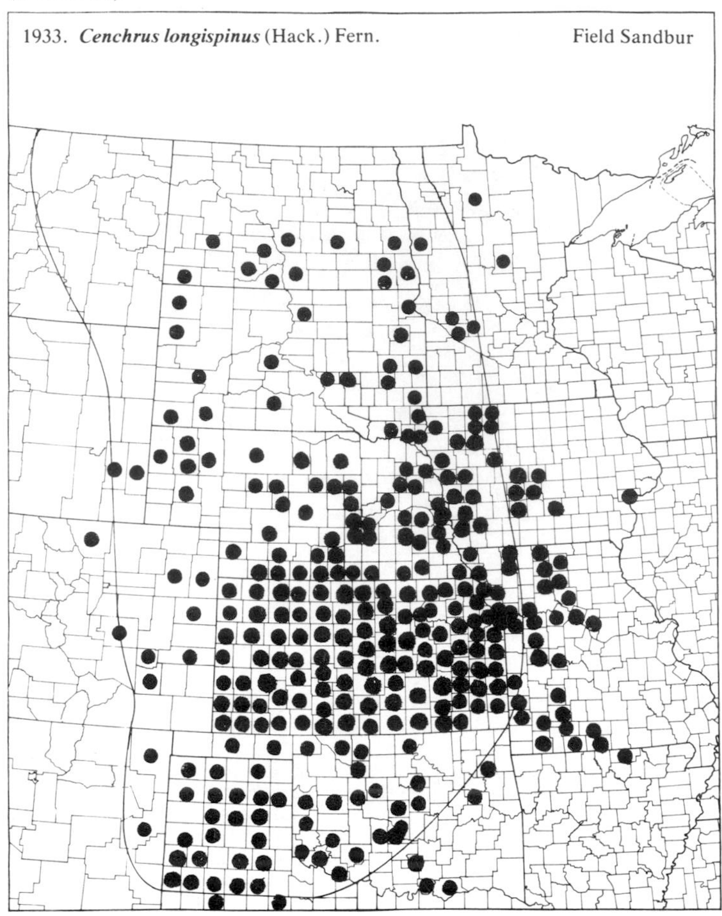

1934. ***Chloris cucullata*** Bisch.

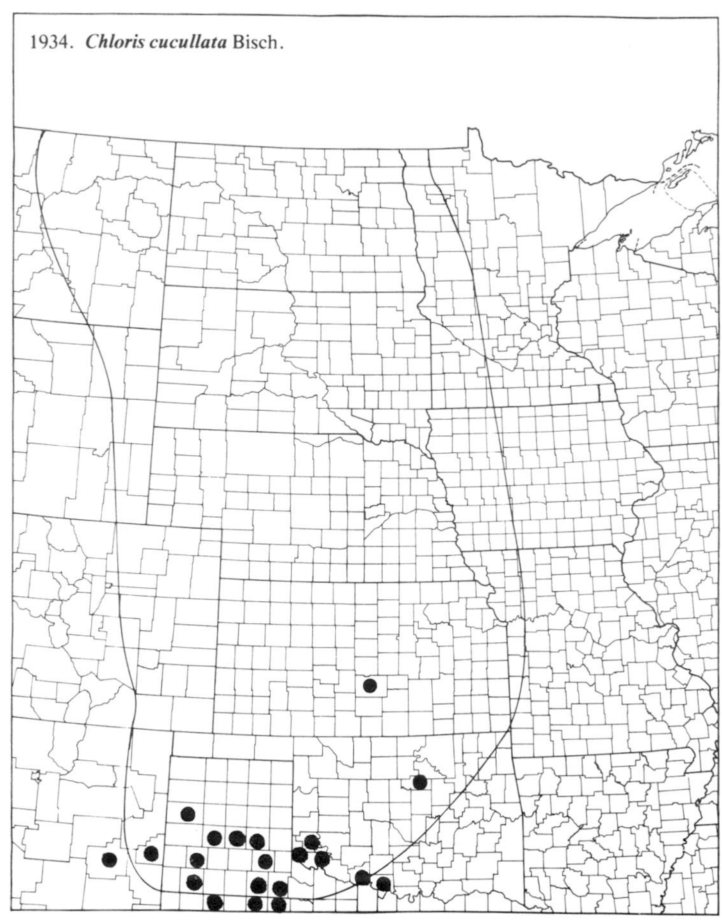

1935. ***Chloris verticillata*** Nutt. Windmillgrass

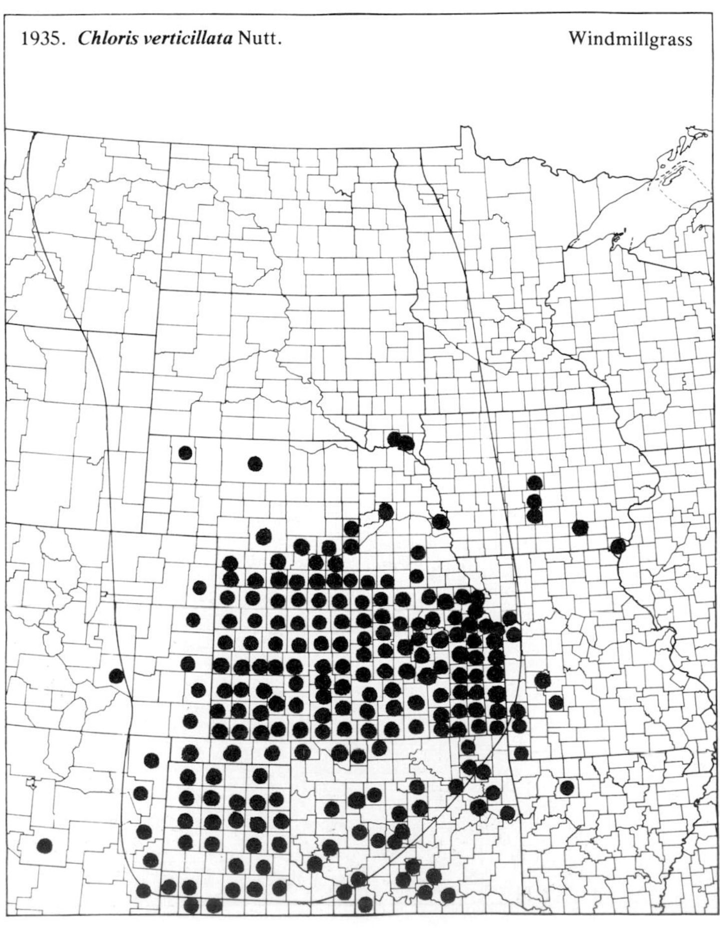

1936. ***Chloris virgata*** Sw. Showy Chloris

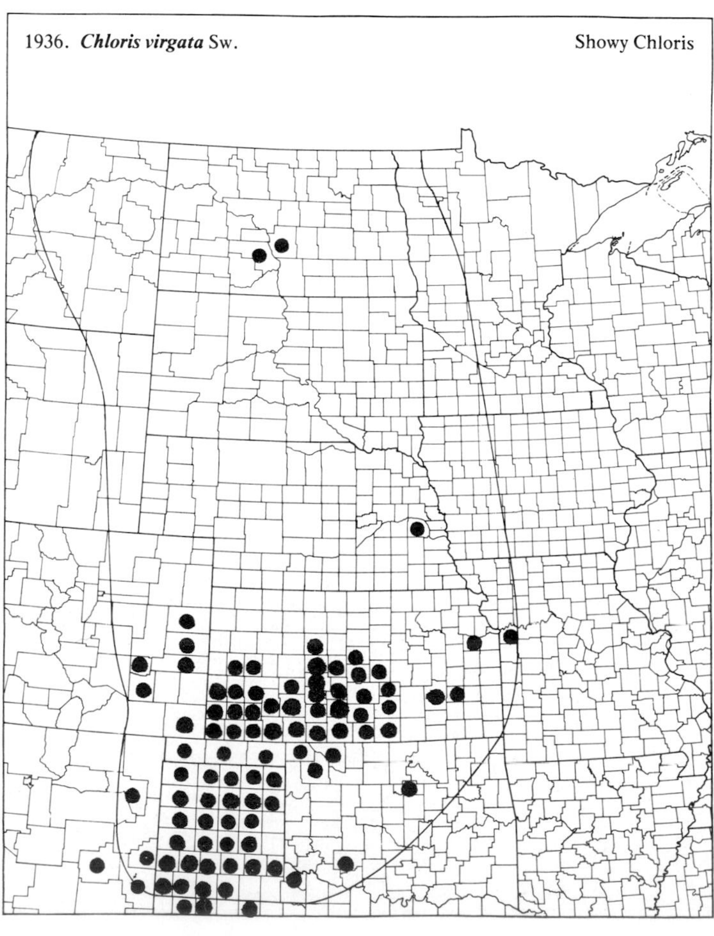

(Poaceae)

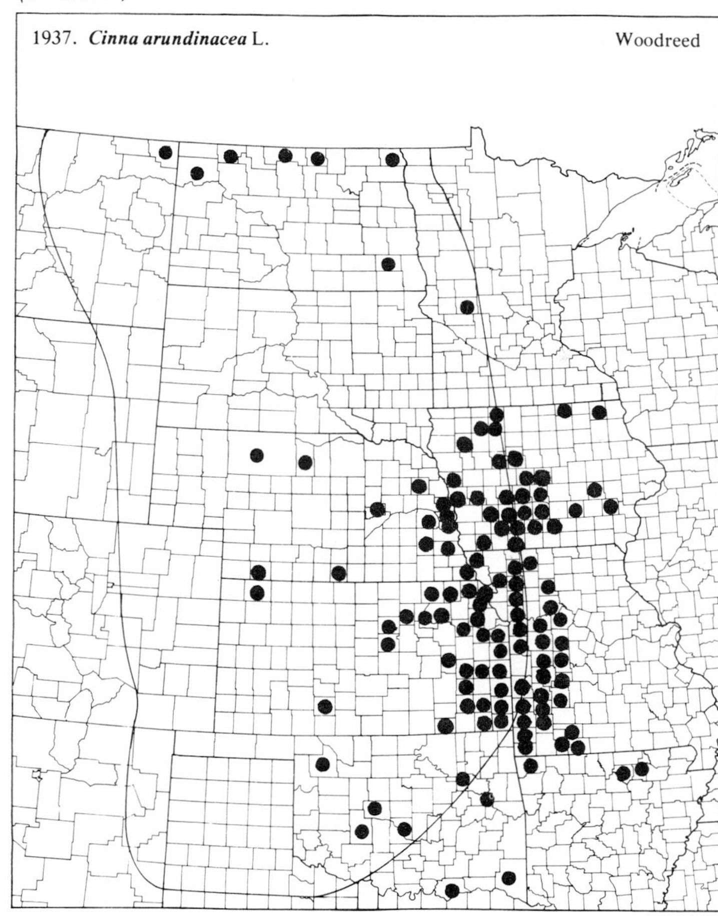
1937. *Cinna arundinacea* L.
Woodreed

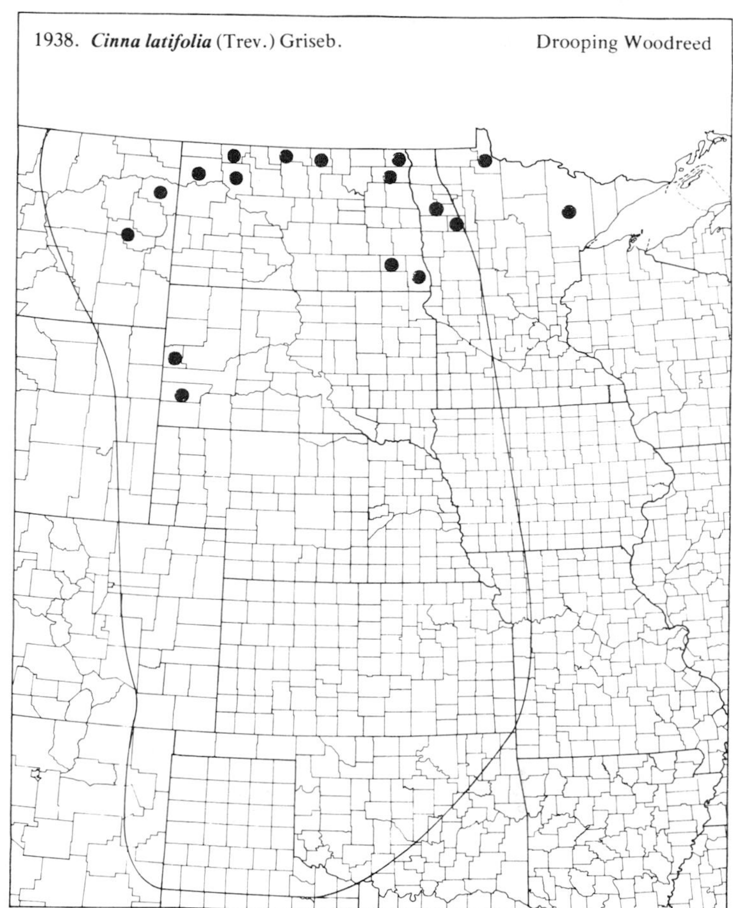
1938. *Cinna latifolia* (Trev.) Griseb.
Drooping Woodreed

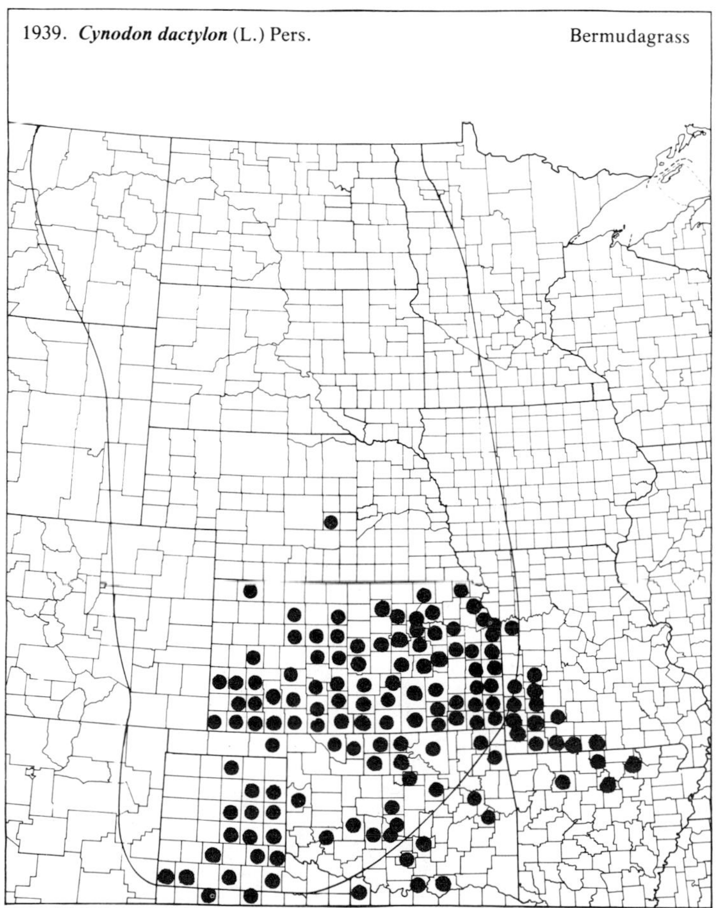
1939. *Cynodon dactylon* (L.) Pers.
Bermudagrass

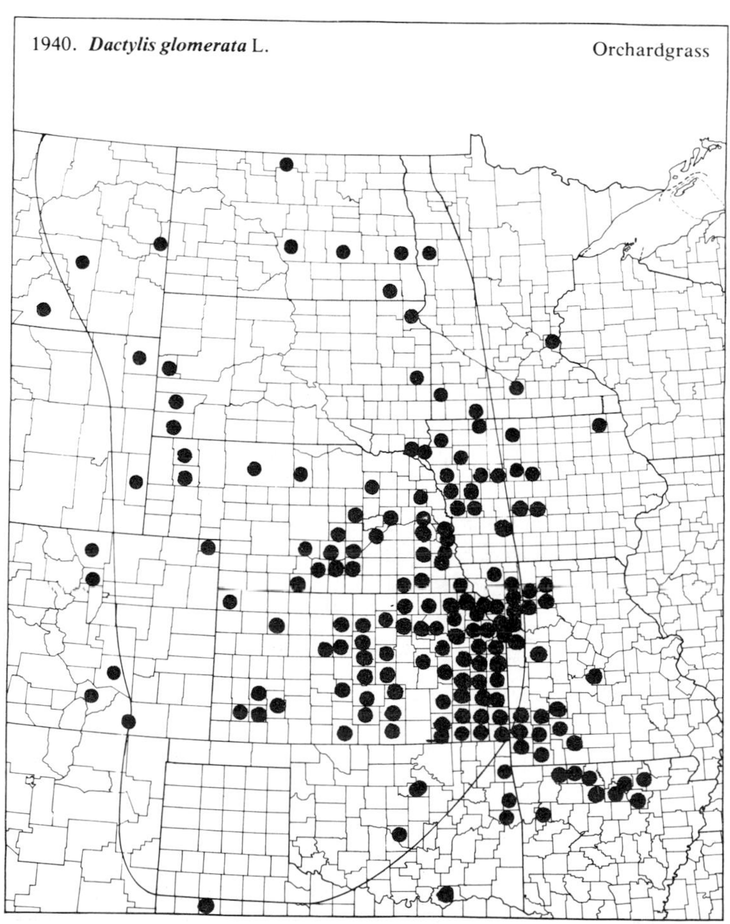
1940. *Dactylis glomerata* L.
Orchardgrass

(Poaceae)

1941. ***Danthonia spicata*** (L.) Beauv. Poverty Oatgrass

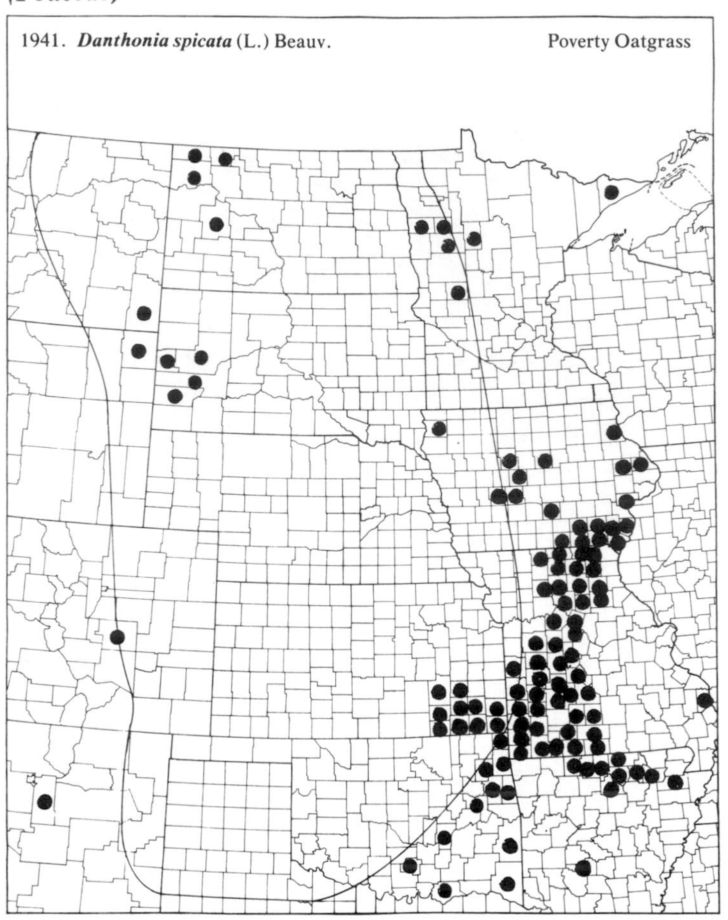

1942. ***Deschampsia caespitosa*** (L.) Beauv. Tufted Hairgrass

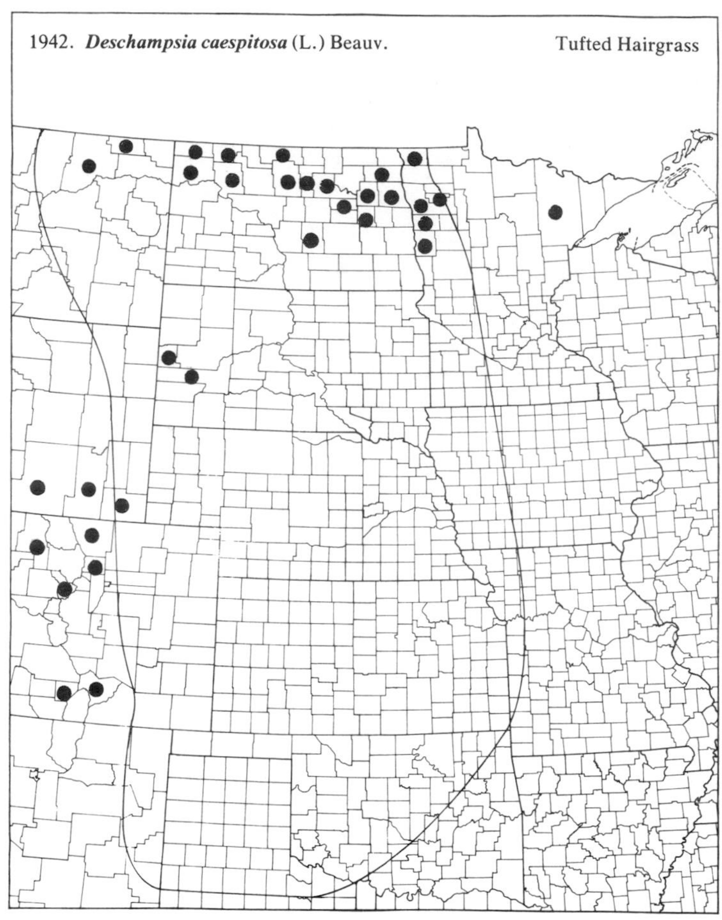

1943. ***Diarrhena americana*** Beauv. var. ***obovata*** Gl. American Beakgrain

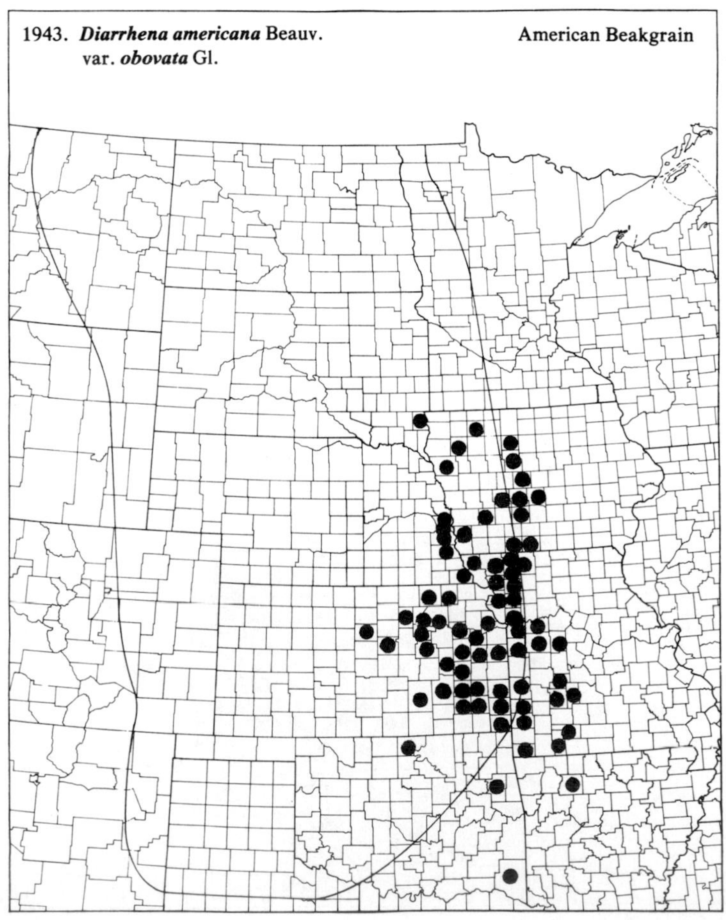

1944. ***Digitaria adscendens*** (H.B.K.) Henr.

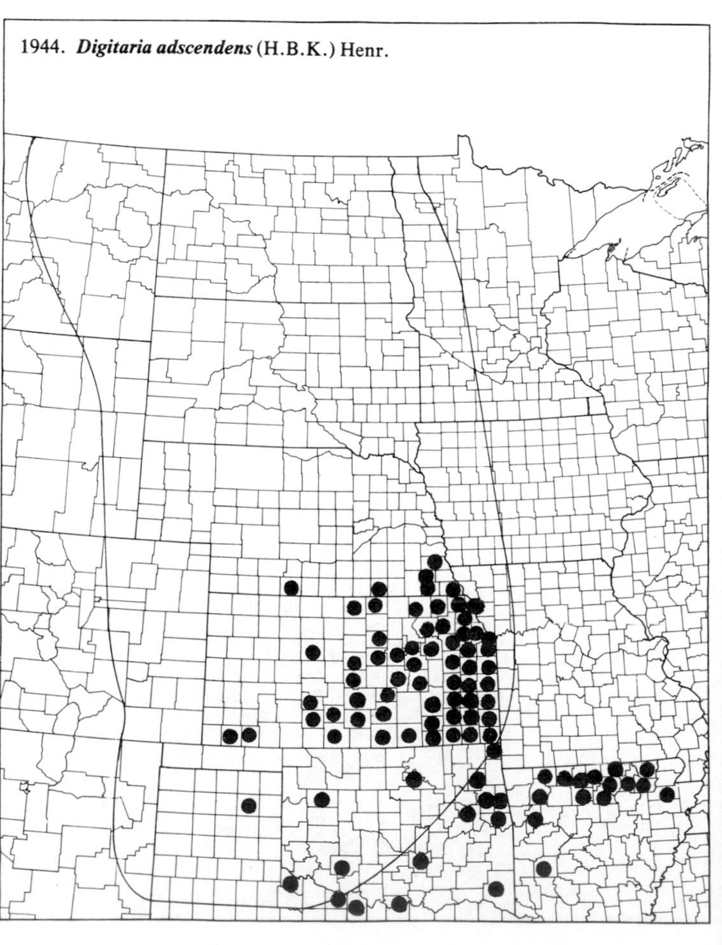

(Poaceae)

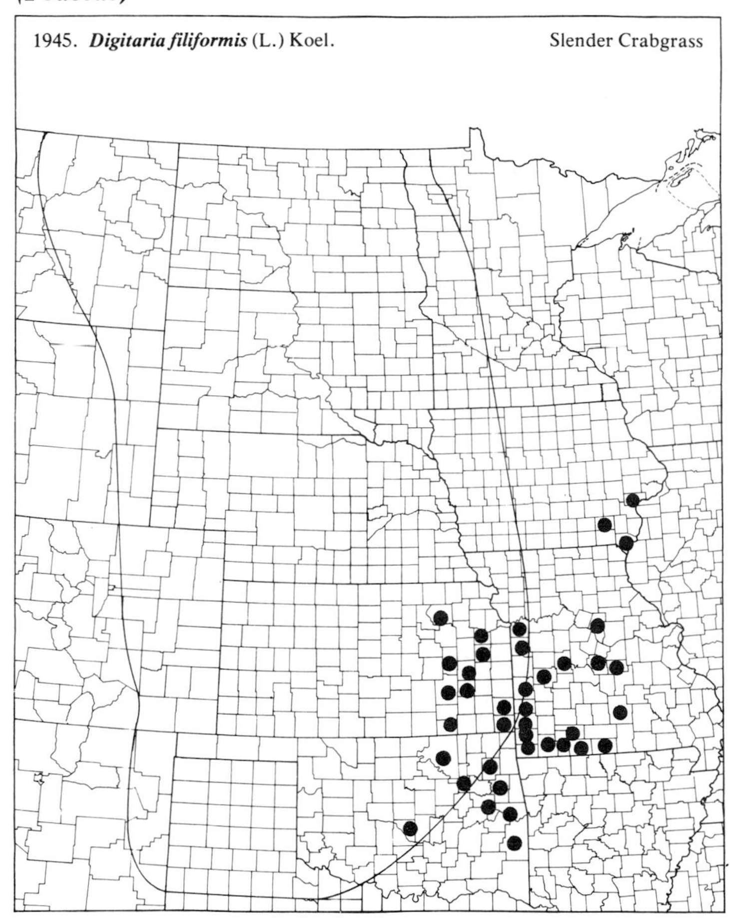
1945. ***Digitaria filiformis*** (L.) Koel.
Slender Crabgrass

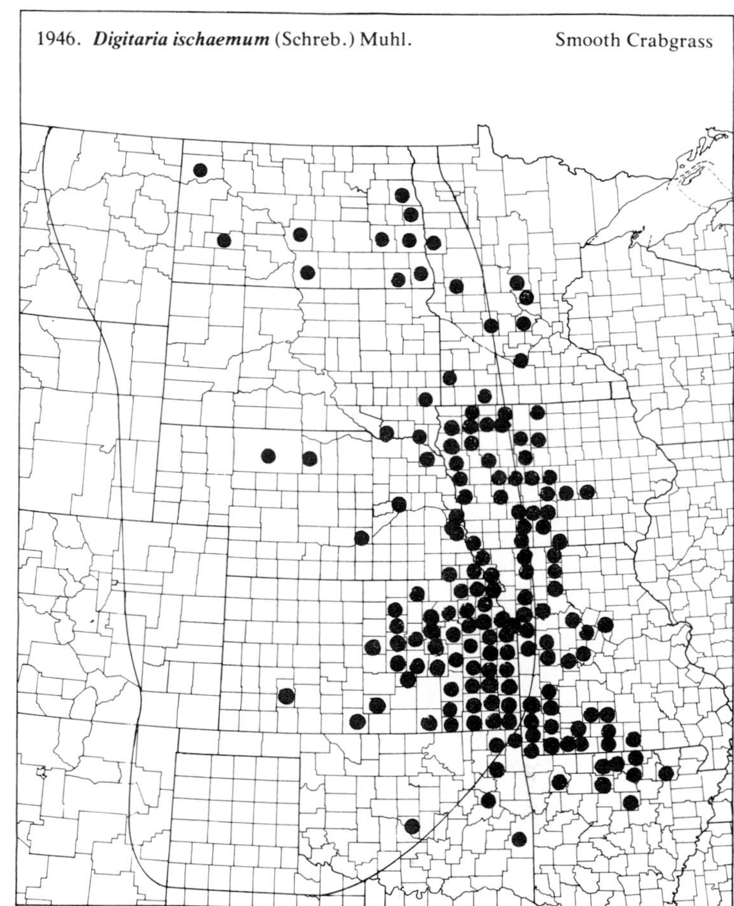
1946. ***Digitaria ischaemum*** (Schreb.) Muhl.
Smooth Crabgrass

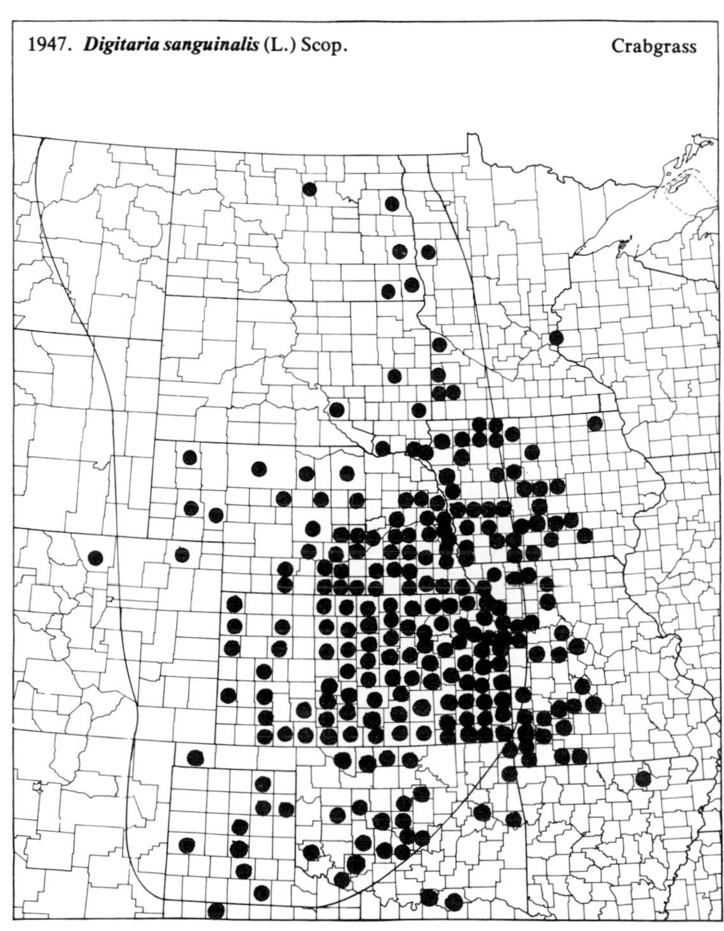
1947. ***Digitaria sanguinalis*** (L.) Scop.
Crabgrass

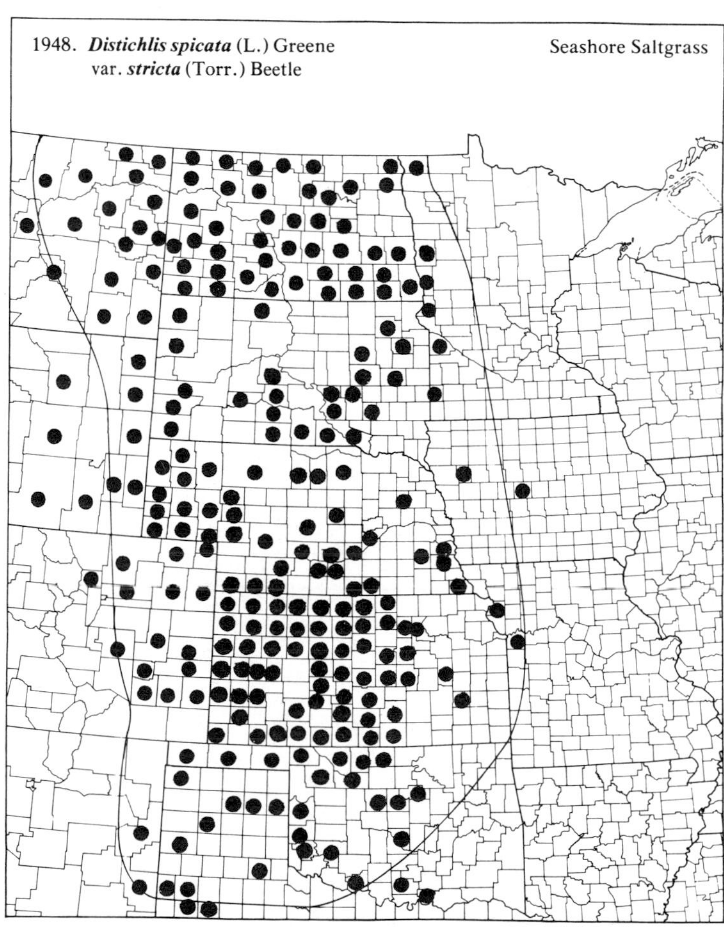
1948. ***Distichlis spicata*** (L.) Greene
var. ***stricta*** (Torr.) Beetle
Seashore Saltgrass

(Poaceae)

1949. *Echinochloa crusgalli* (L.) Beauv. Barnyardgrass

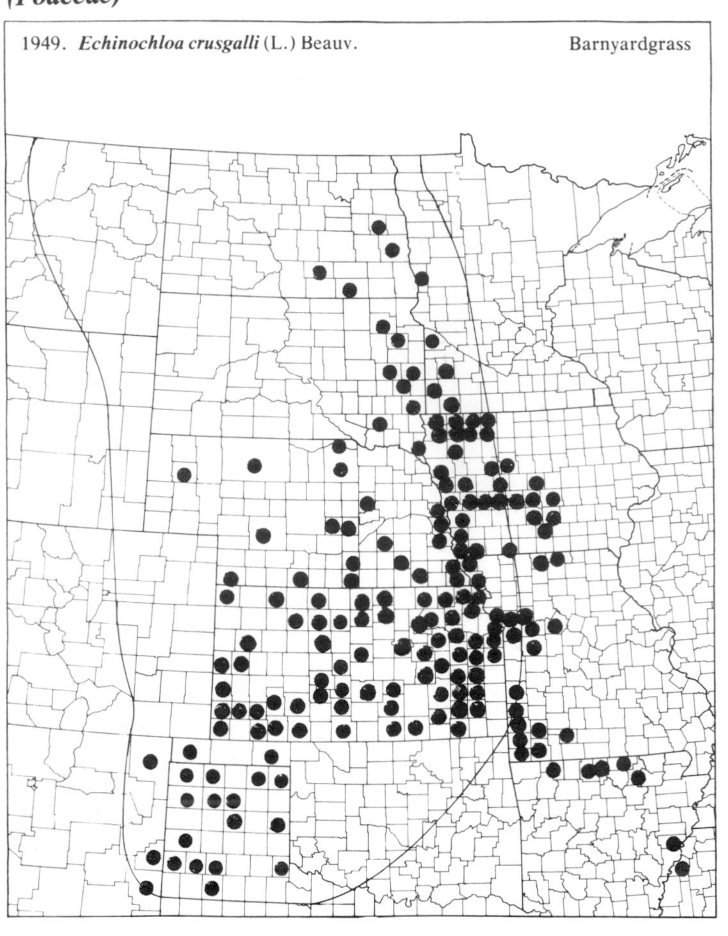

1950. *Echinochloa crus-pavonis* (H.B.K.) Schult. var. *macera* (Wieg.) Gould

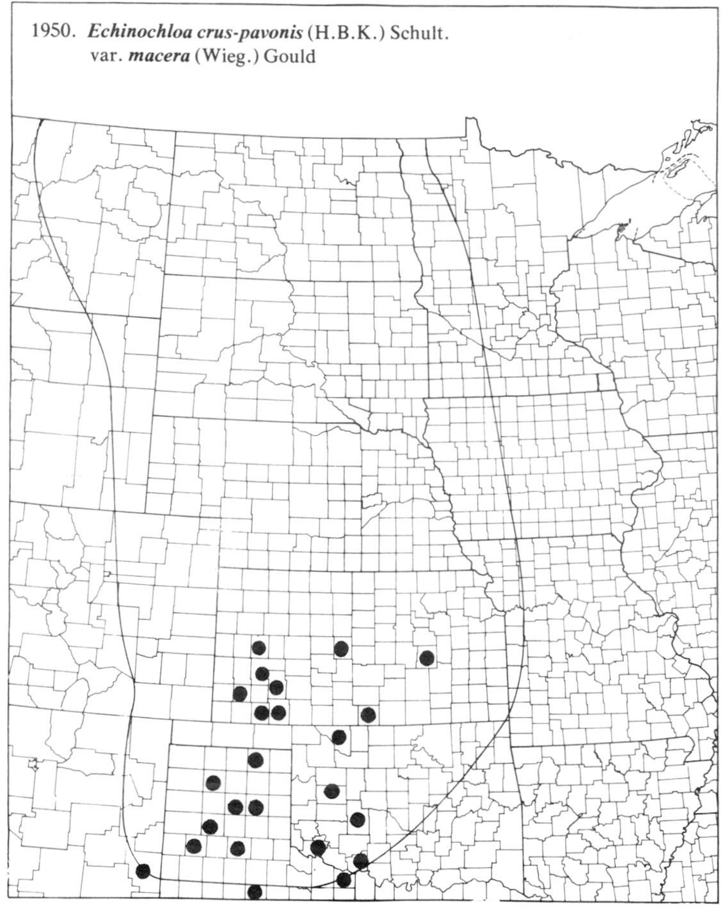

1951. *Echinochloa muricata* (Beauv.) Fern. var. *muricata*

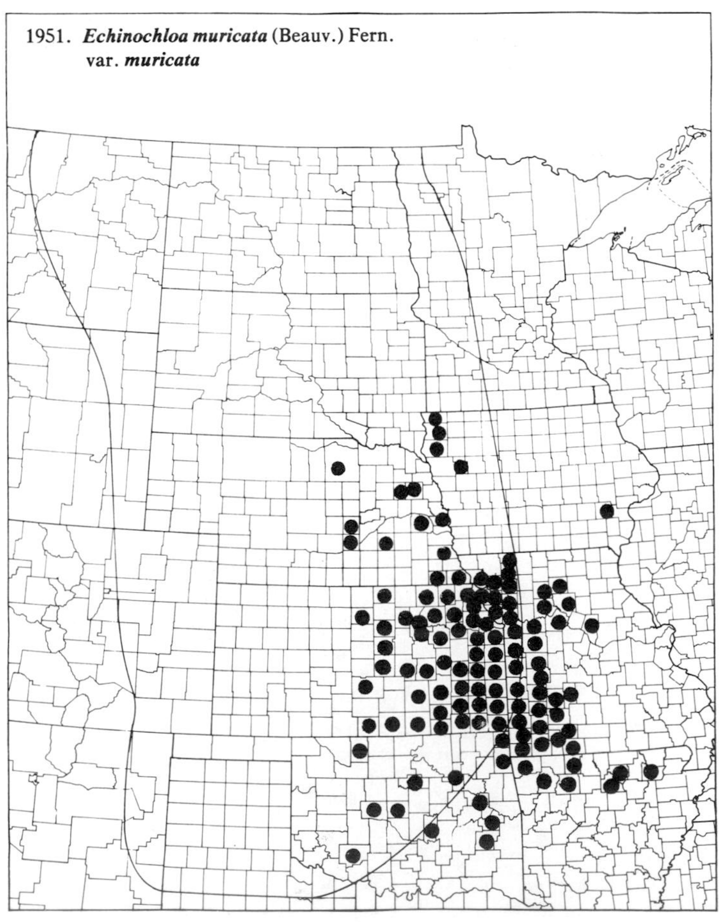

1952. *Echinochloa muricata* (Beauv.) Fern. var. *microstachya* Wieg.

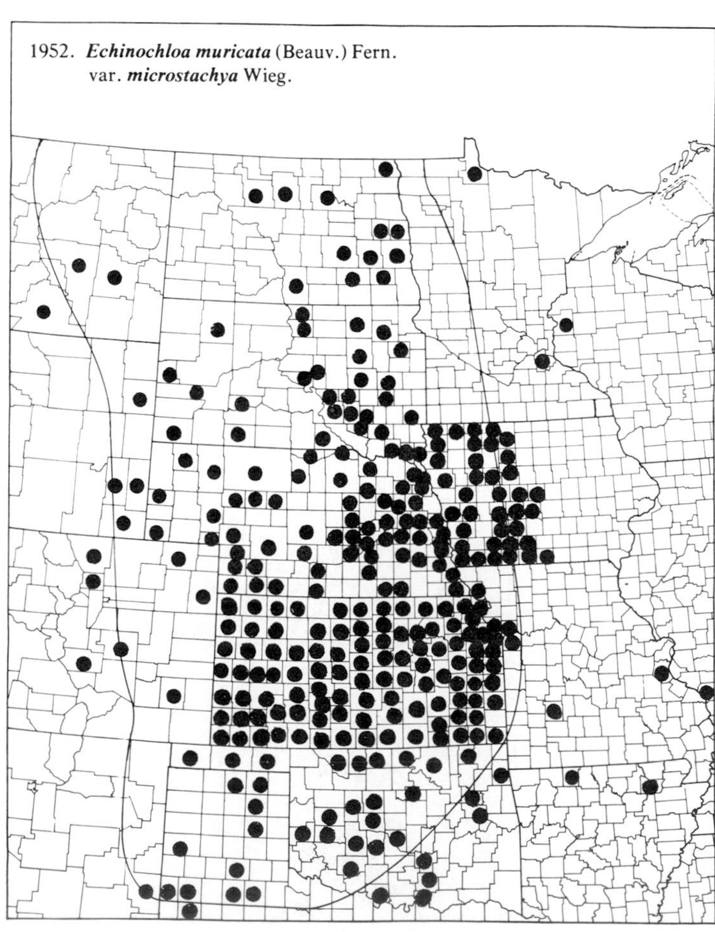

(Poaceae)

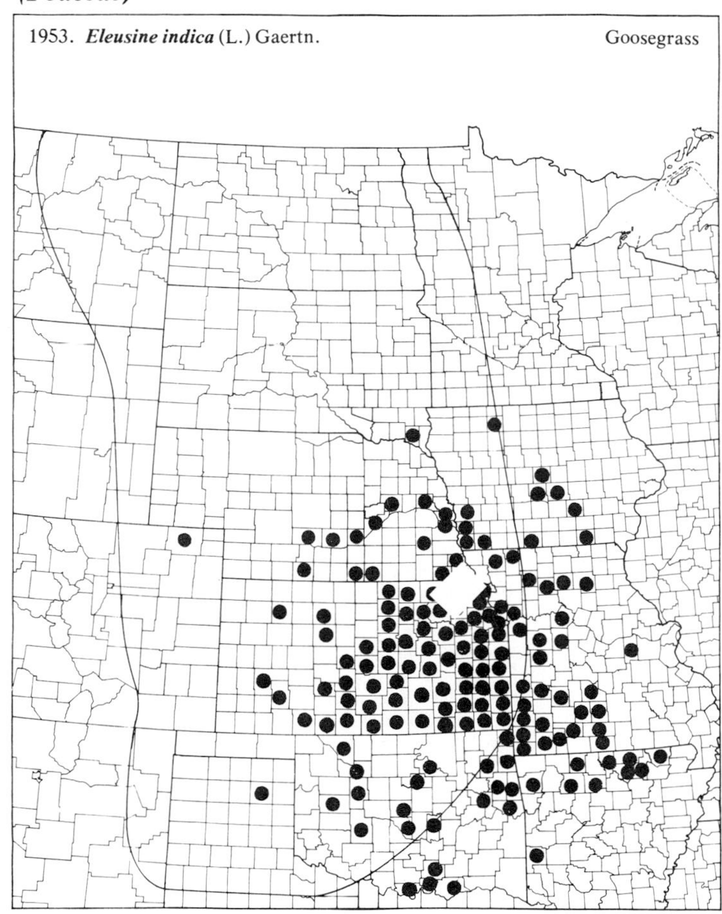
1953. Eleusine indica (L.) Gaertn.
Goosegrass

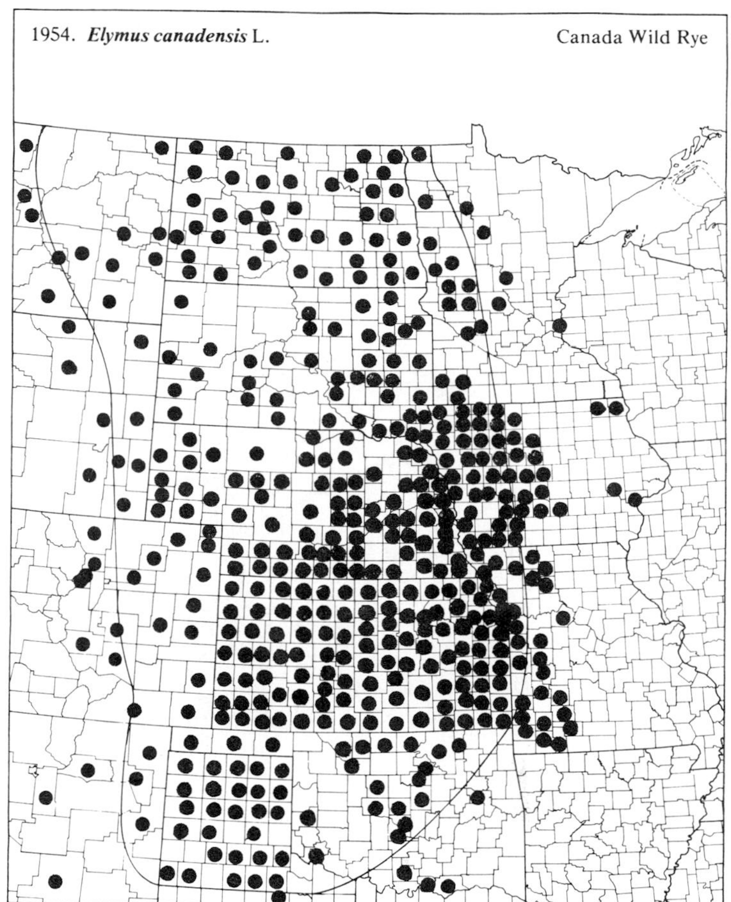
1954. Elymus canadensis L.
Canada Wild Rye

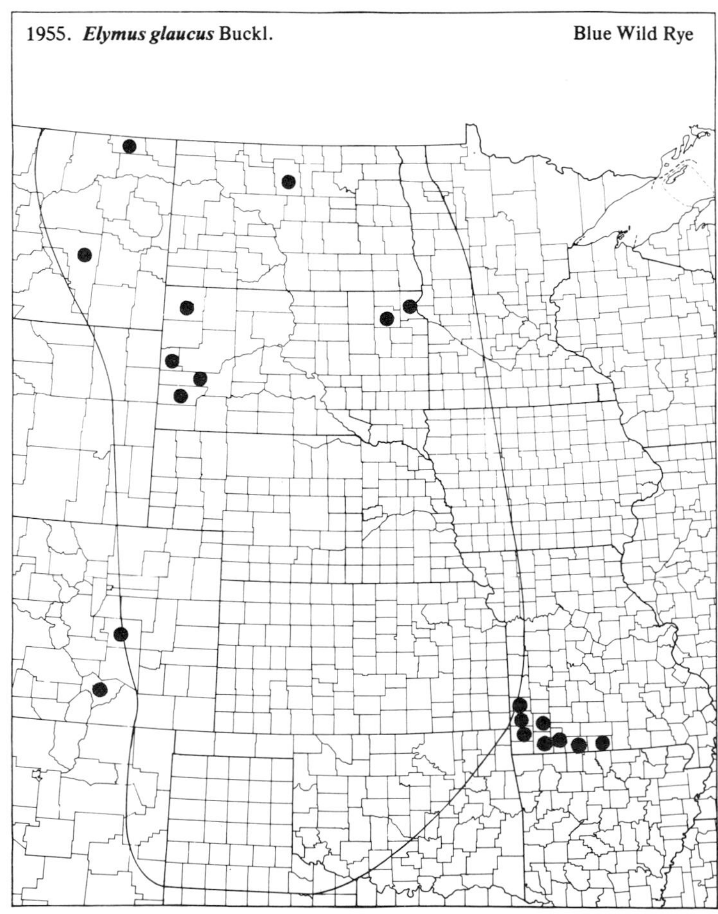
1955. Elymus glaucus Buckl.
Blue Wild Rye

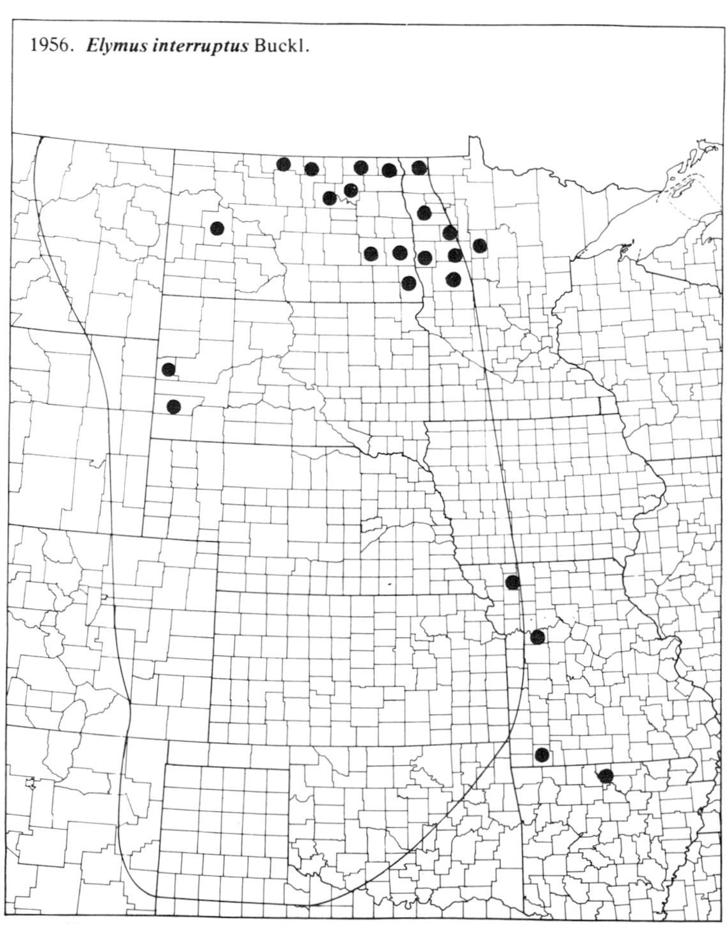
1956. Elymus interruptus Buckl.

(Poaceae)

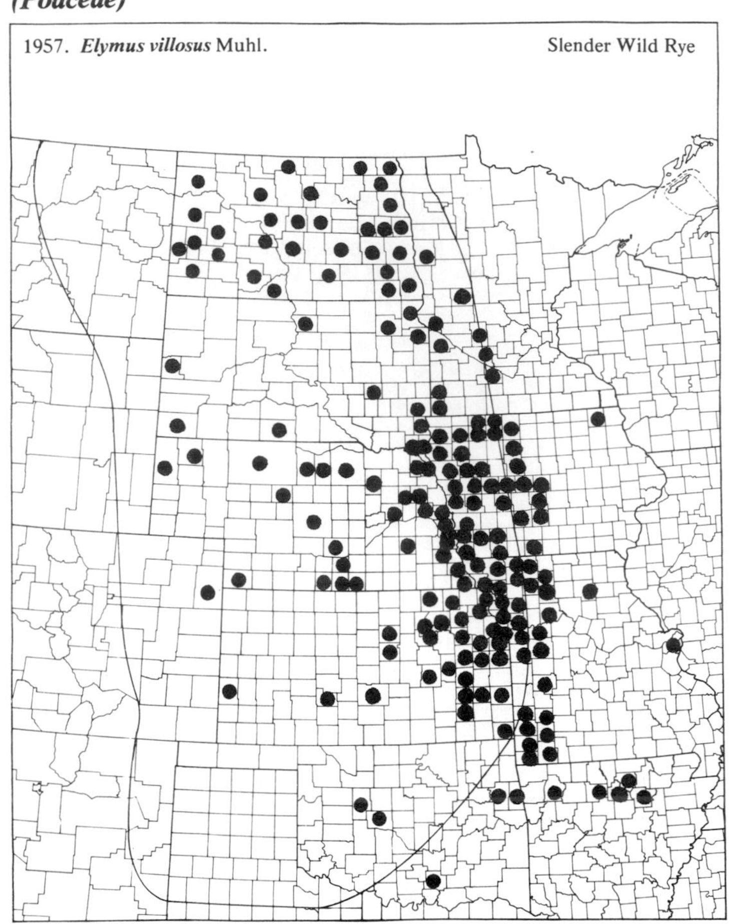
1957. *Elymus villosus* Muhl.
Slender Wild Rye

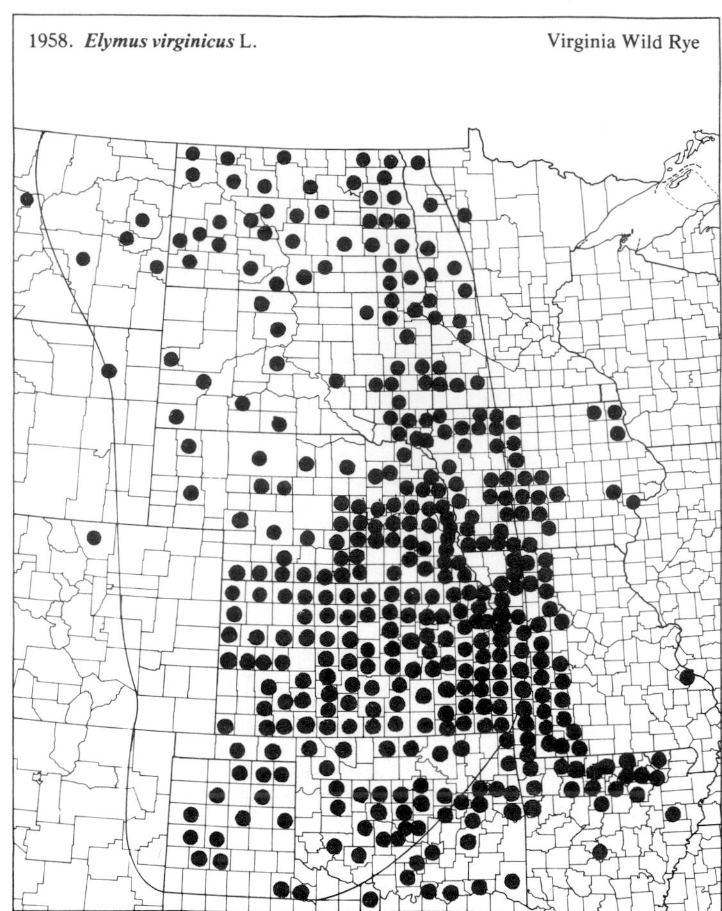
1958. *Elymus virginicus* L.
Virginia Wild Rye

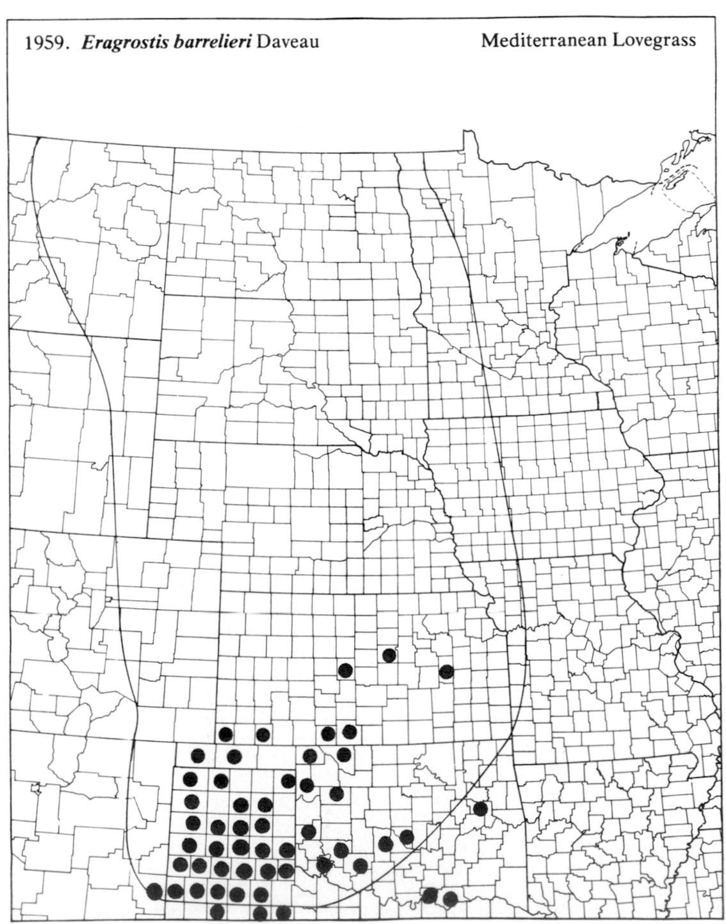
1959. *Eragrostis barrelieri* Daveau
Mediterranean Lovegrass

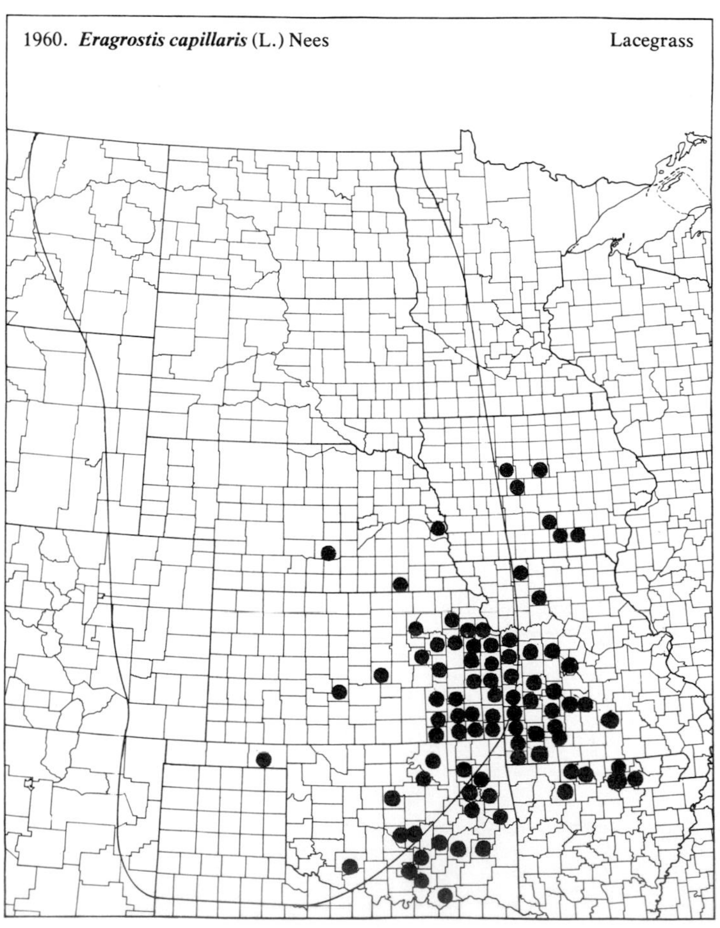
1960. *Eragrostis capillaris* (L.) Nees
Lacegrass

(Poaceae)

1961. ***Eragrostis cilianensis*** (All.) E. Mosher — Stinkgrass

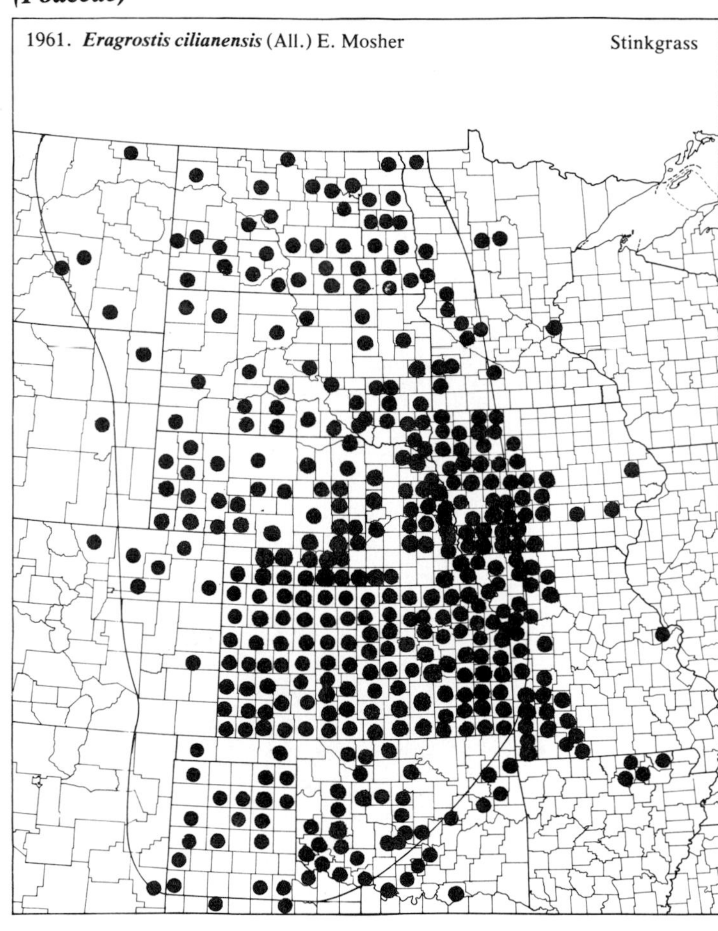

1962. ***Eragrostis curtipedicellata*** Buckl. — Gummy Lovegrass

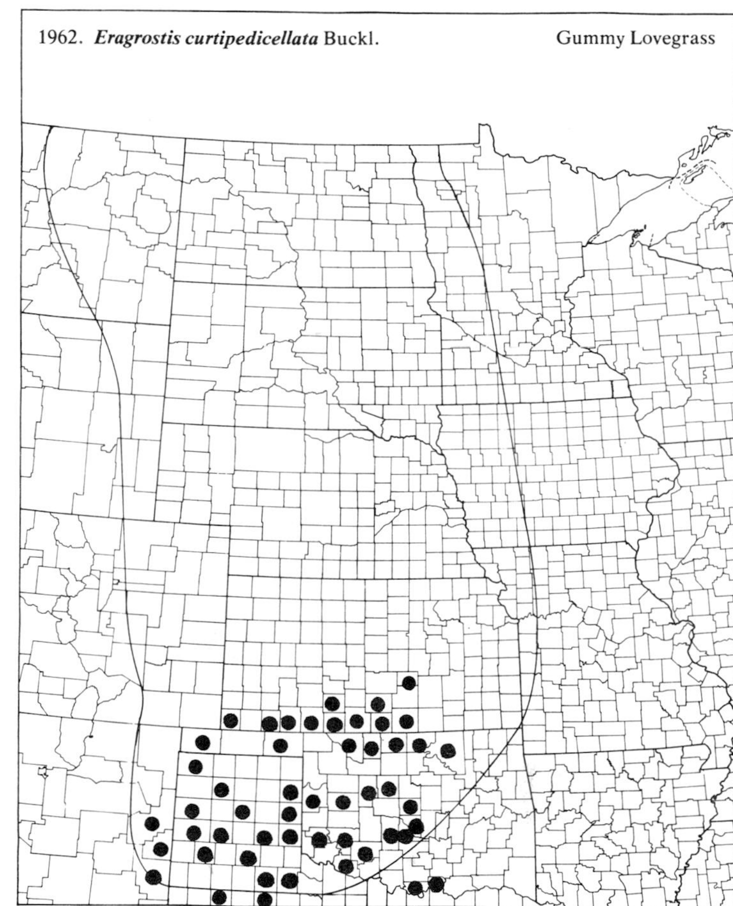

1963. ***Eragrostis curvula*** (Schrad.) Nees — Weeping Lovegrass

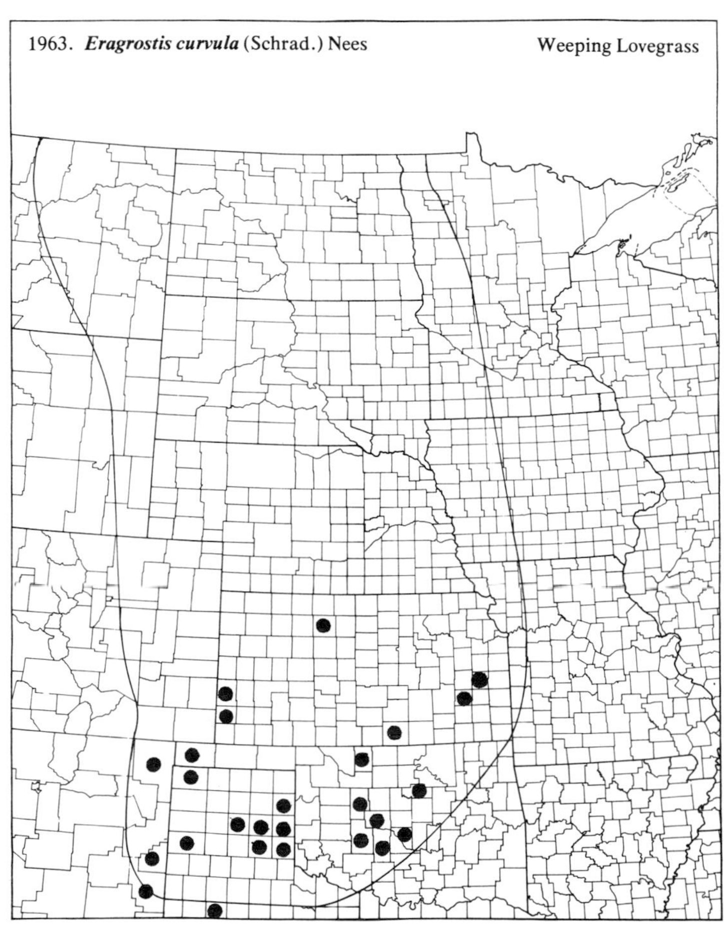

1964. ***Eragrostis frankii*** C.A. Mey. — Sandbar Lovegrass

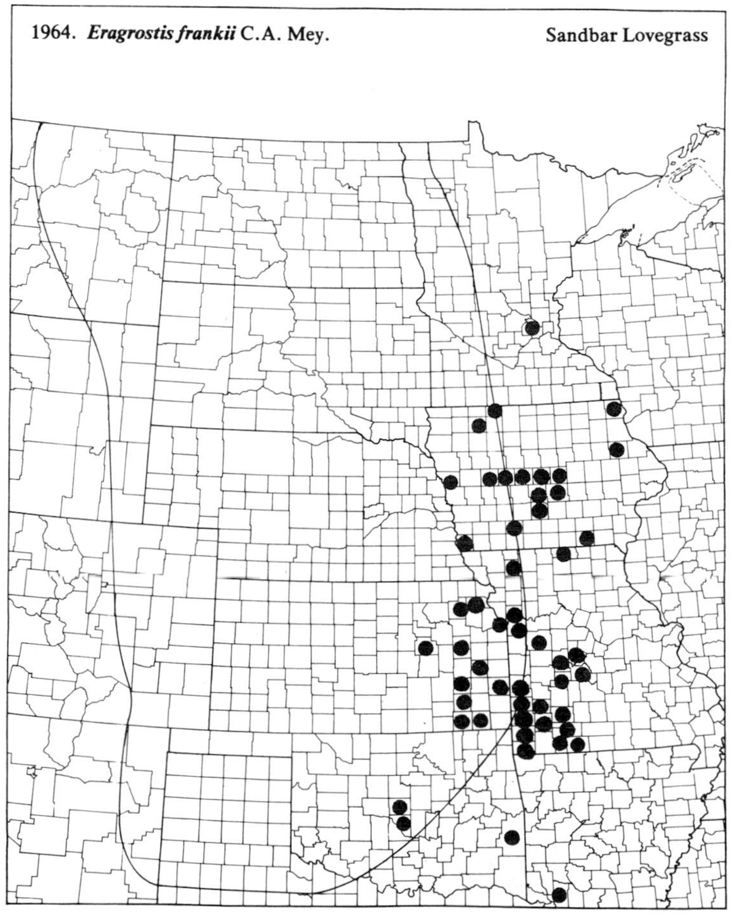

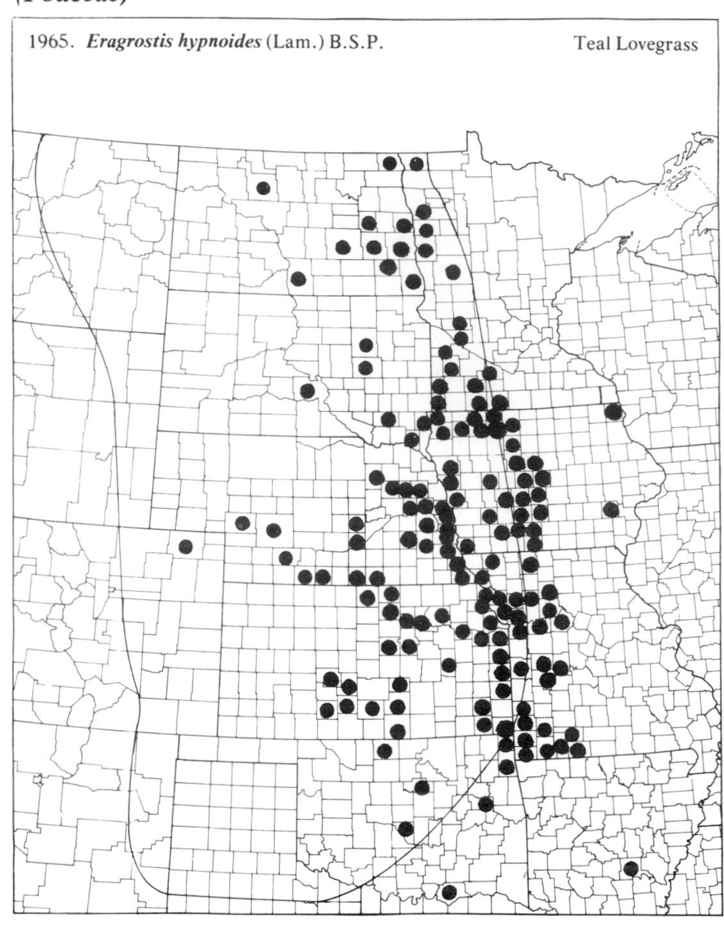

1965. ***Eragrostis hypnoides*** (Lam.) B.S.P. Teal Lovegrass

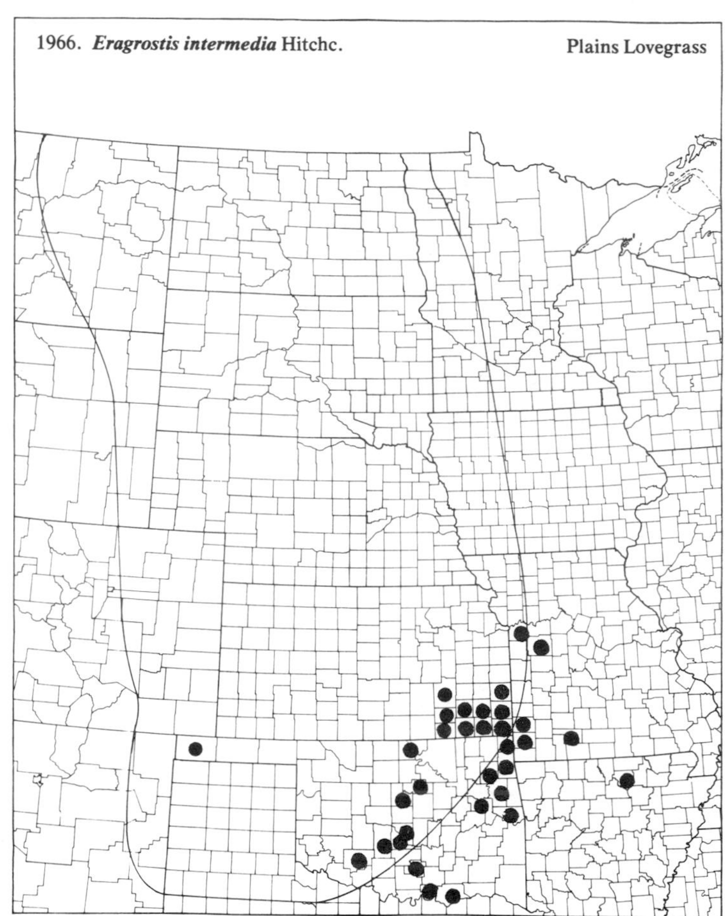

1966. ***Eragrostis intermedia*** Hitchc. Plains Lovegrass

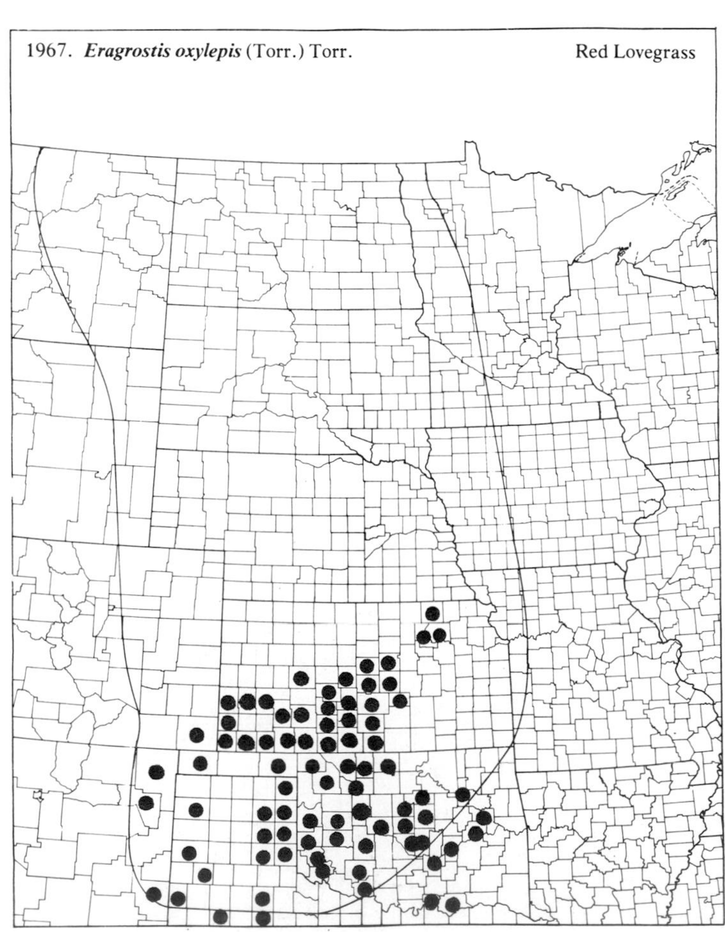

1967. ***Eragrostis oxylepis*** (Torr.) Torr. Red Lovegrass

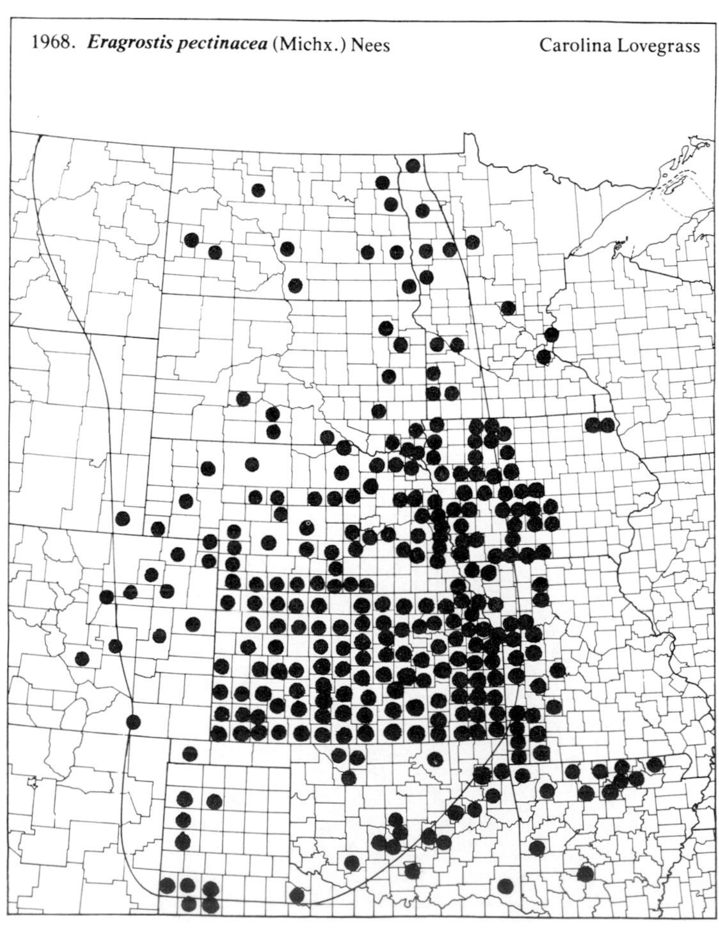

1968. ***Eragrostis pectinacea*** (Michx.) Nees Carolina Lovegrass

(Poaceae)

1969. ***Eragrostis reptans*** (Michx.) Nees

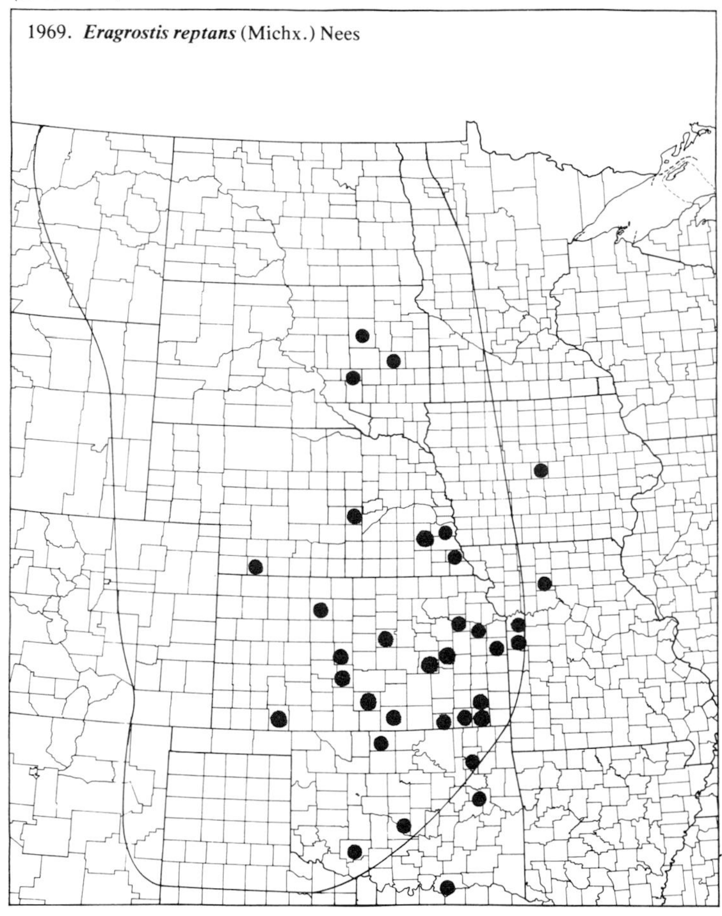

1970. ***Eragrostis sessilispica*** Buckl. Tumble Lovegrass

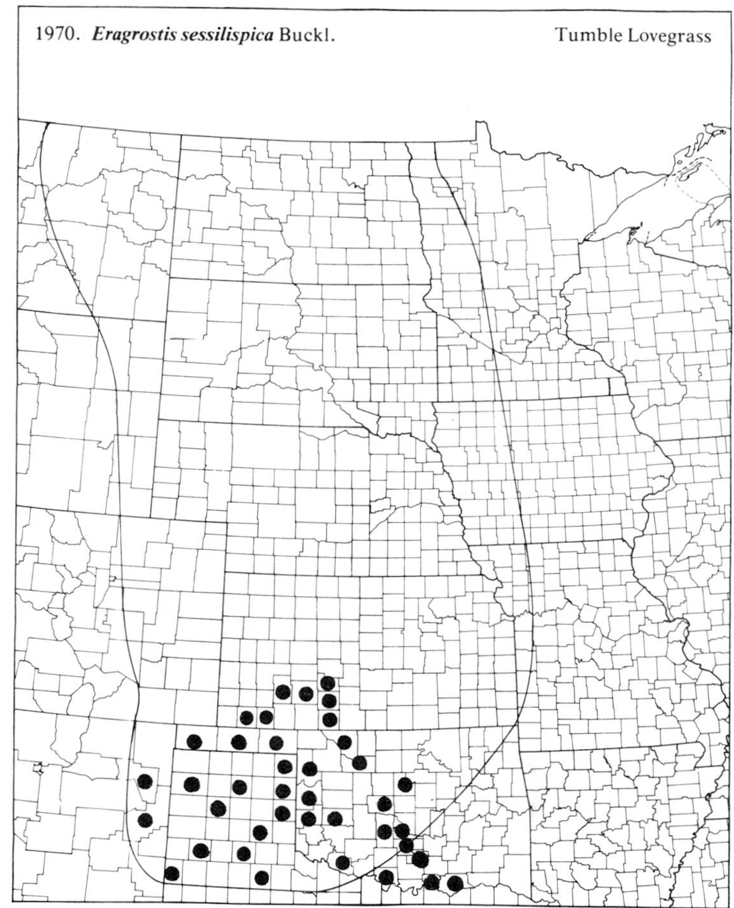

1971. ***Eragrostis spectabilis*** (Pursh) Steud. Purple Lovegrass

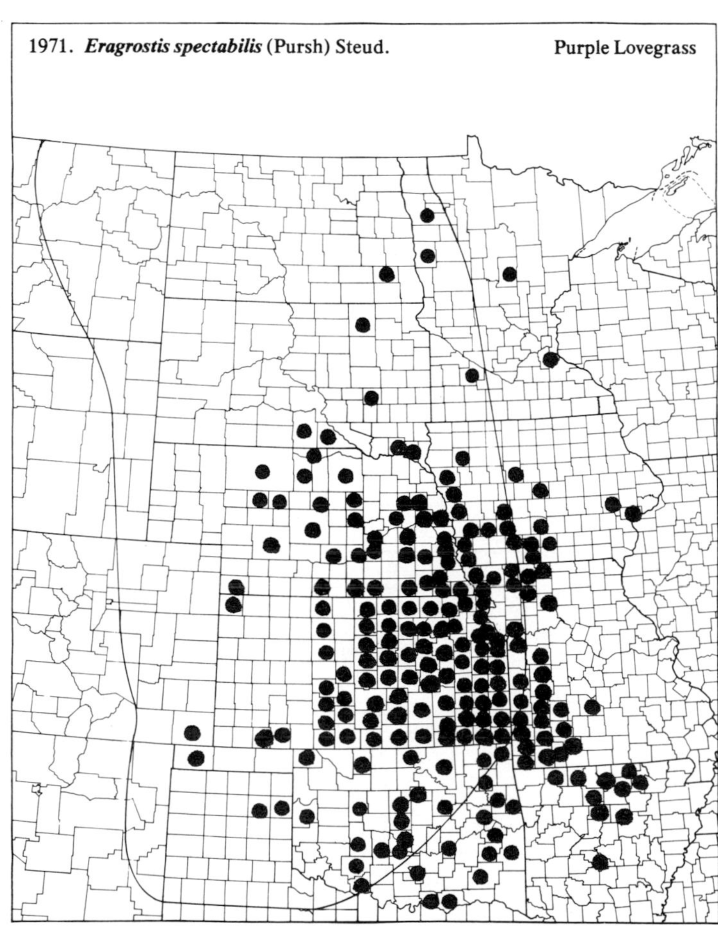

1972. ***Eragrostis trichodes*** (Nutt.) Wood Sand Lovegrass

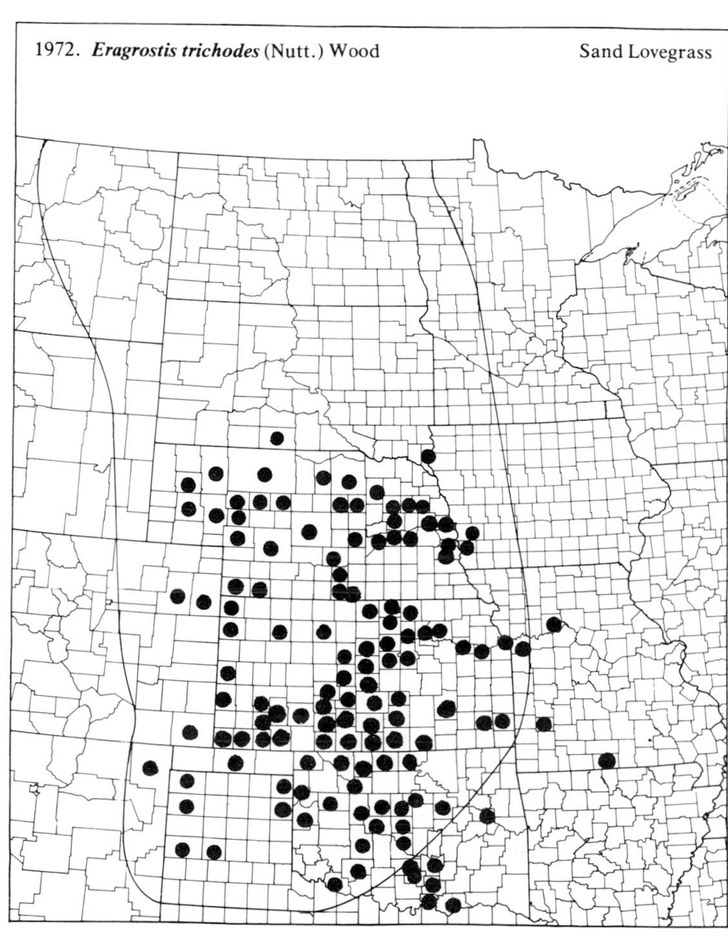

(Poaceae)

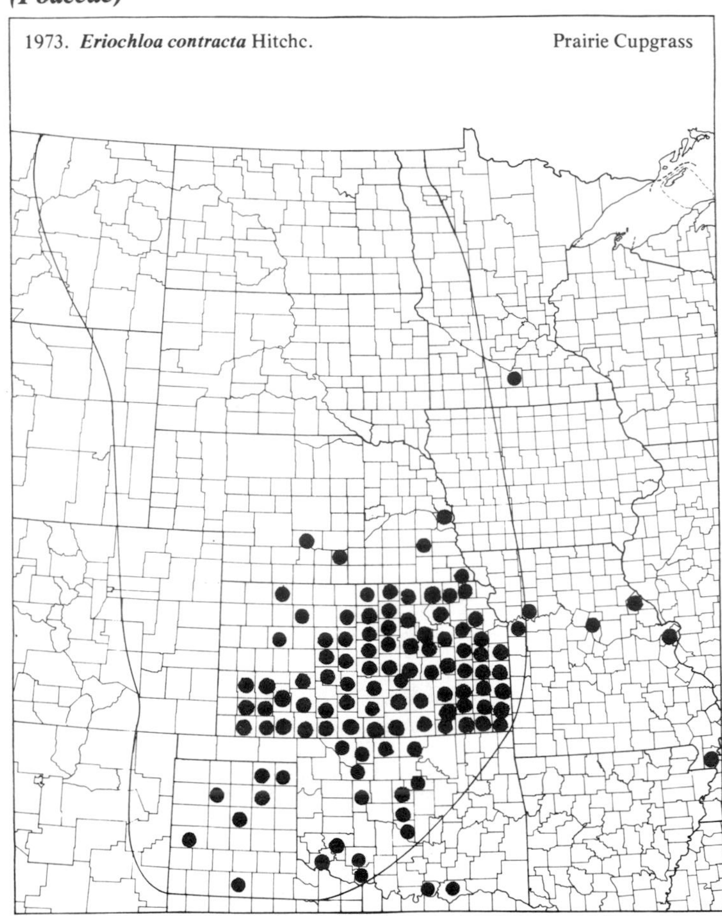
1973. *Eriochloa contracta* Hitchc.
Prairie Cupgrass

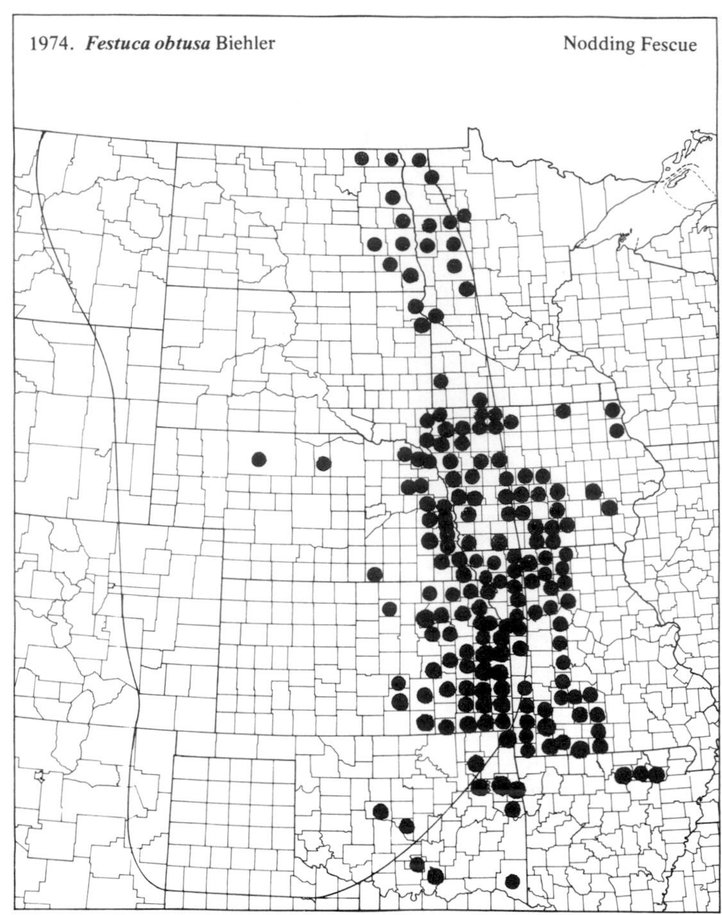
1974. *Festuca obtusa* Biehler
Nodding Fescue

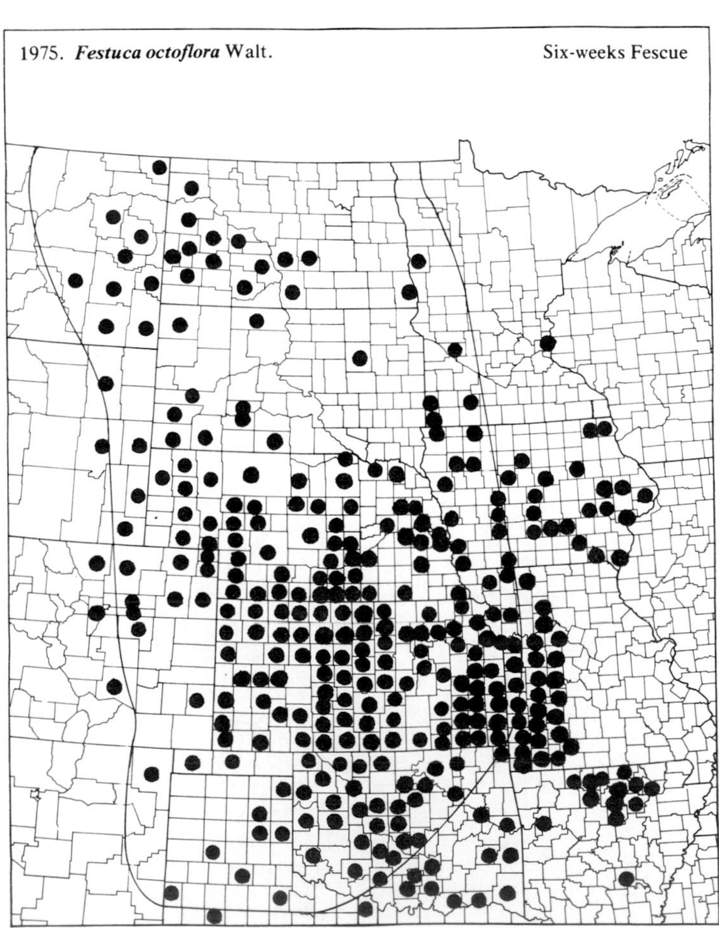
1975. *Festuca octoflora* Walt.
Six-weeks Fescue

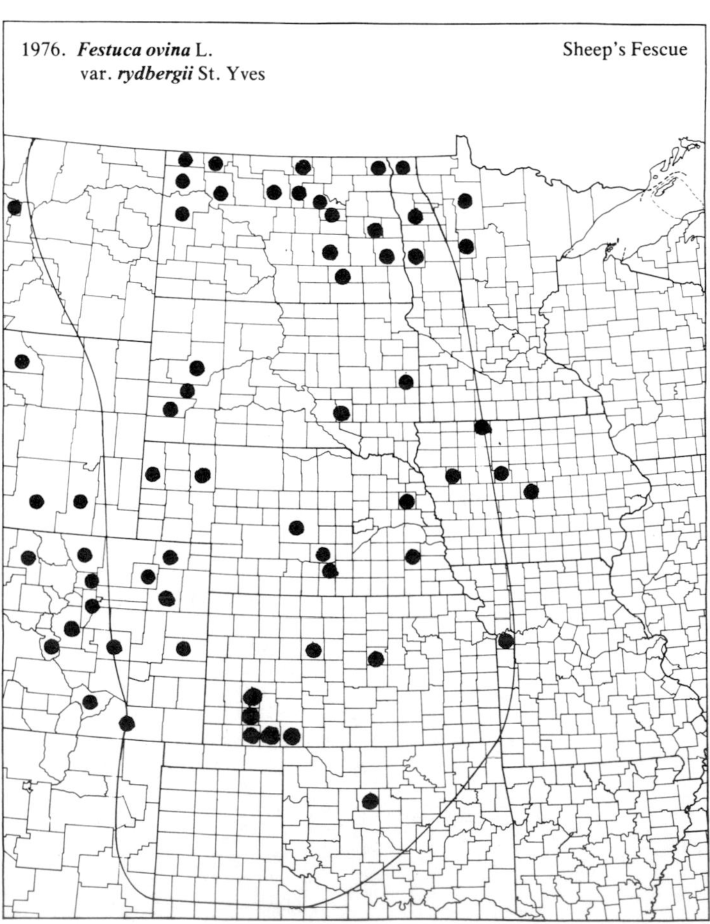
1976. *Festuca ovina* L.
var. *rydbergii* St. Yves
Sheep's Fescue

(Poaceae)

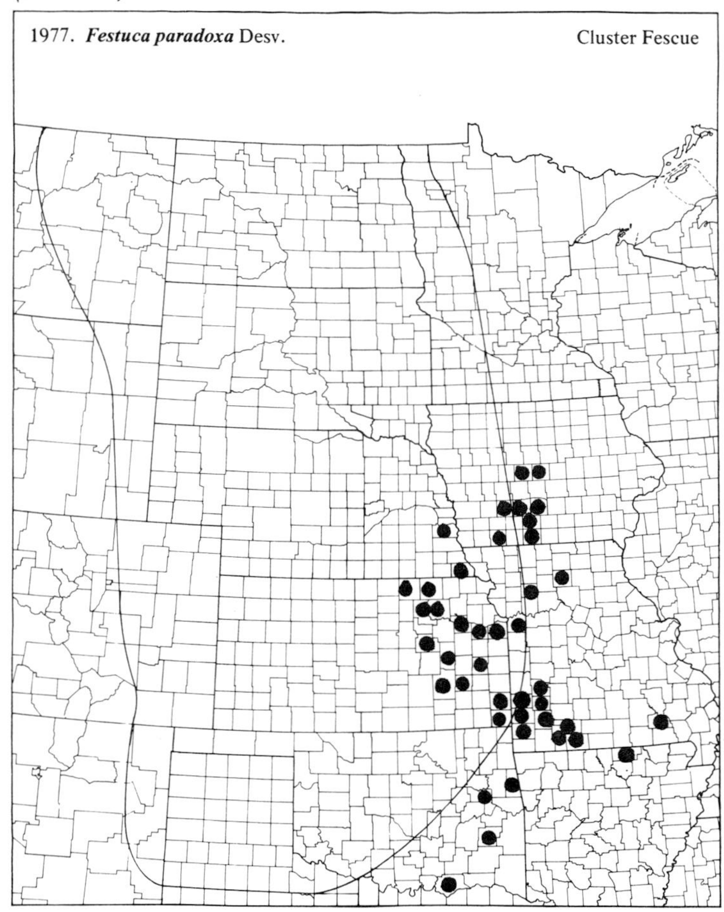
1977. **Festuca paradoxa** Desv. Cluster Fescue

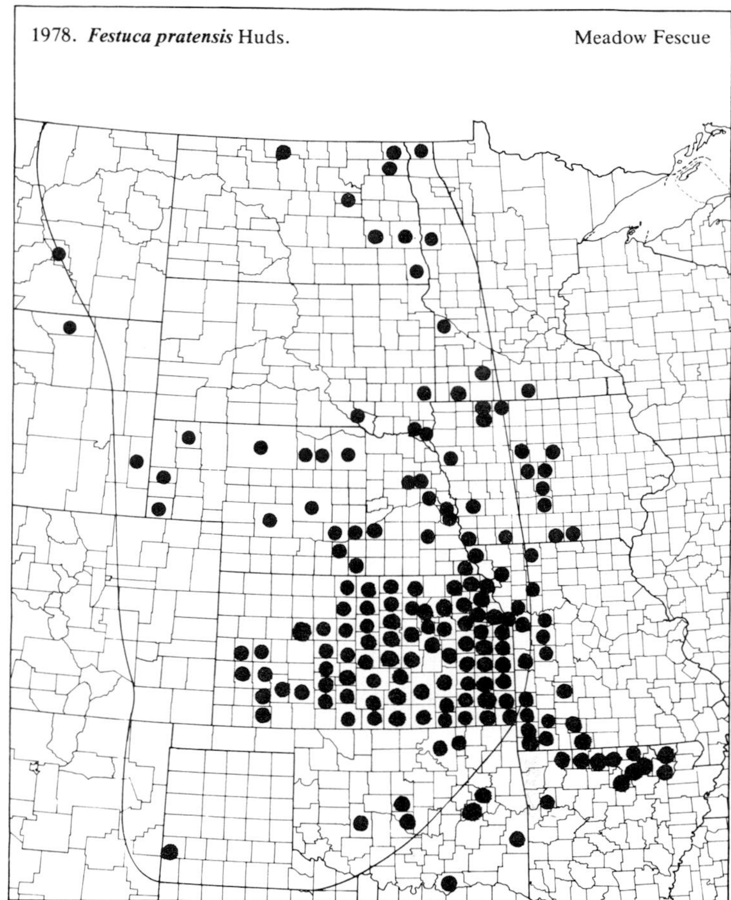
1978. **Festuca pratensis** Huds. Meadow Fescue

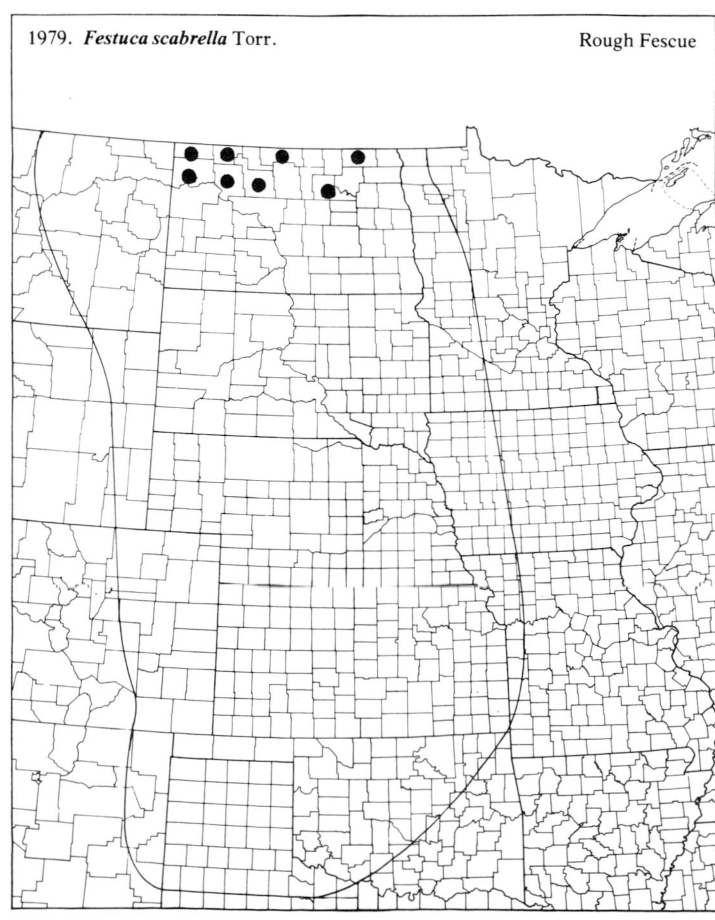
1979. **Festuca scabrella** Torr. Rough Fescue

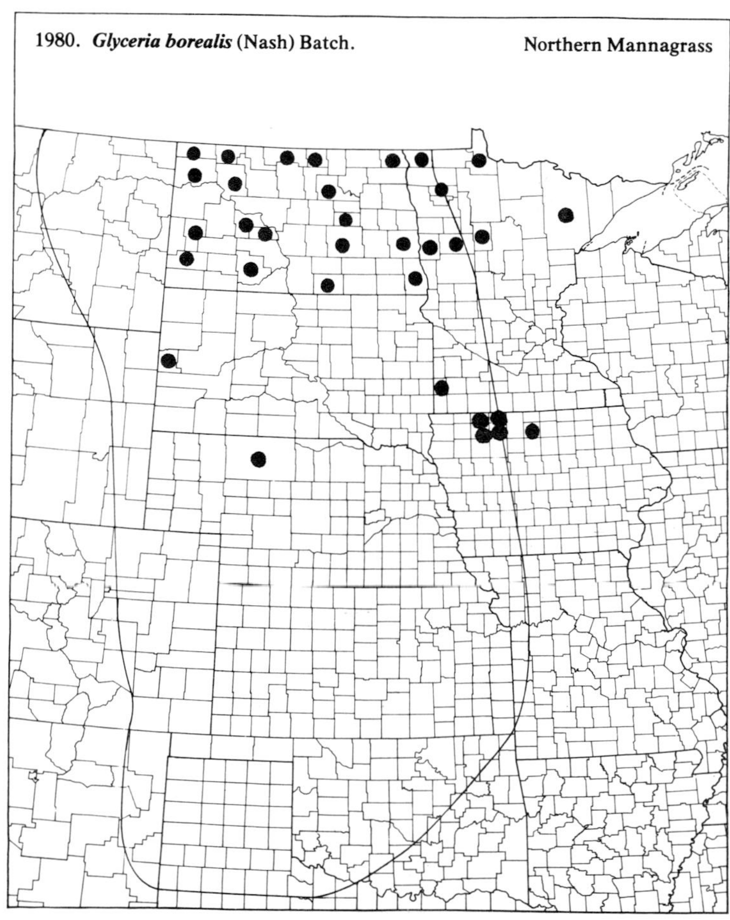
1980. **Glyceria borealis** (Nash) Batch. Northern Mannagrass

(Poaceae)

1981. ***Glyceria grandis*** Wats. Tall Mannagrass

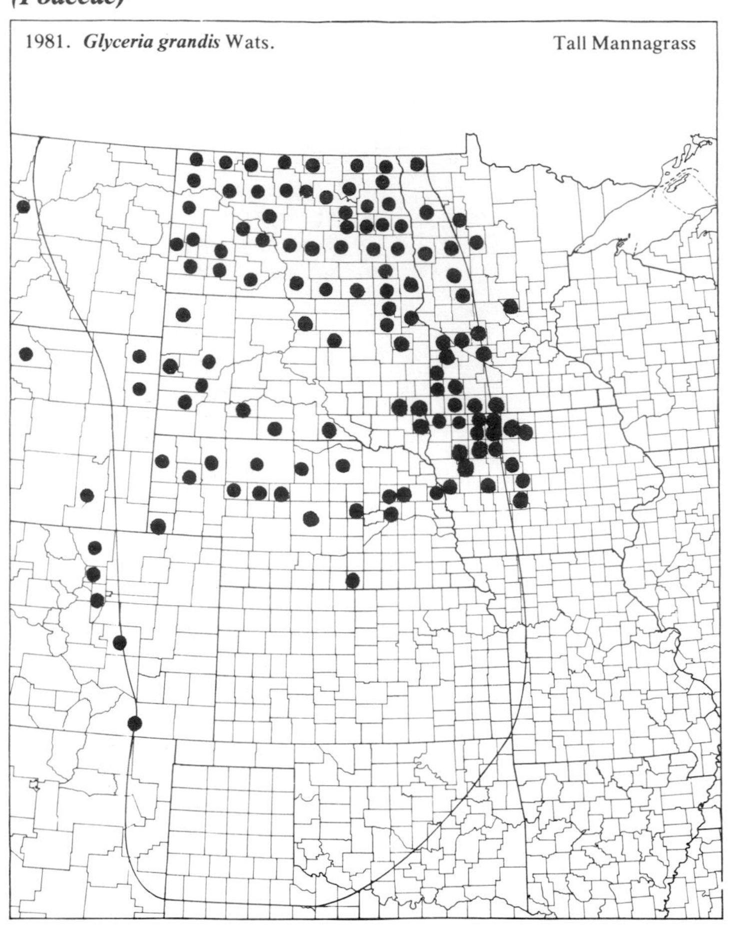

1982. ***Glyceria striata*** (Lam.) Hitchc. Fowl Mannagrass

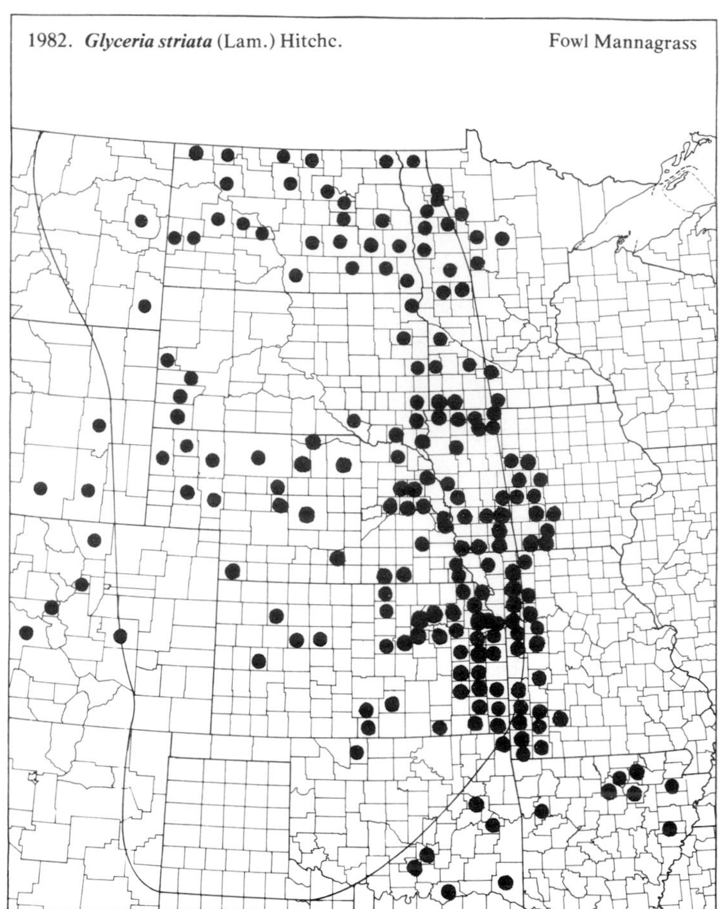

1983. ***Gymnopogon ambiguus*** (Michx.) B.S.P. Bearded Skeletongrass

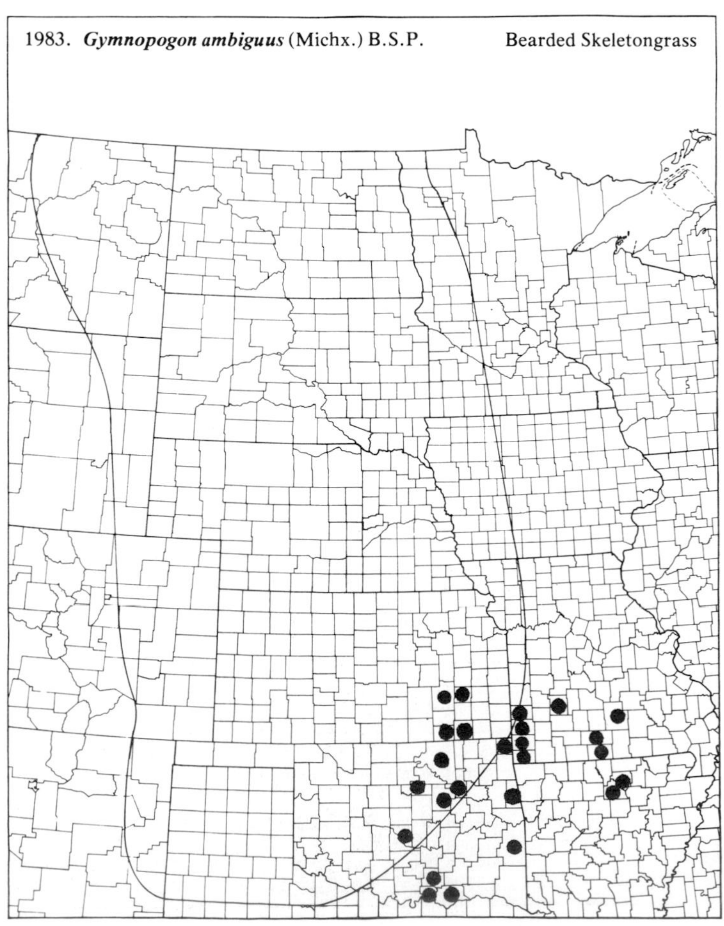

1984. ***Helictotrichon hookeri*** (Scribn.) Henr. Spike Oat

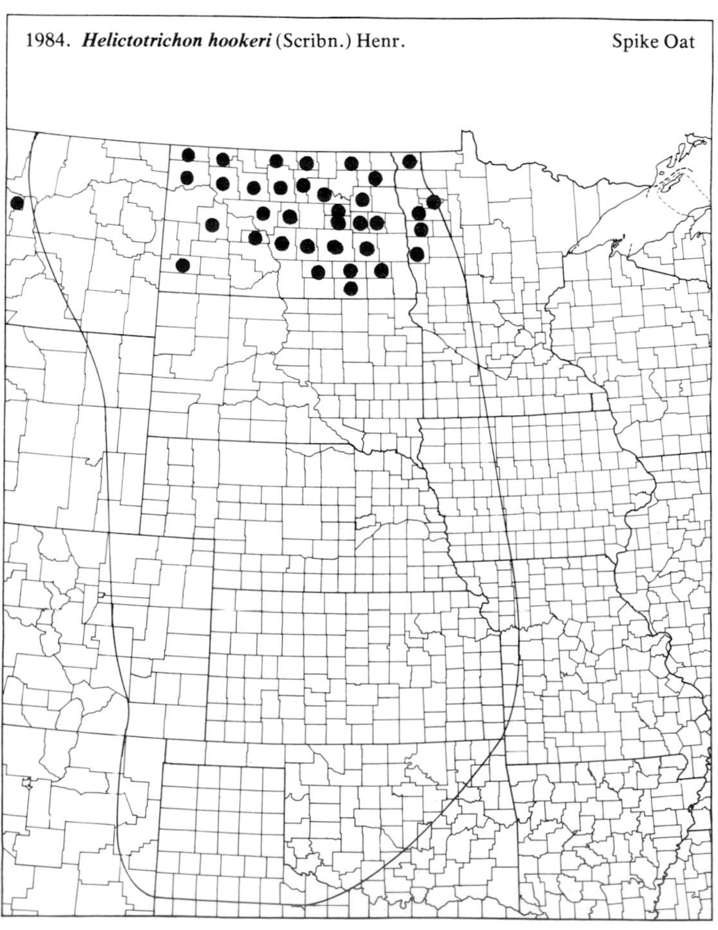

(Poaceae)

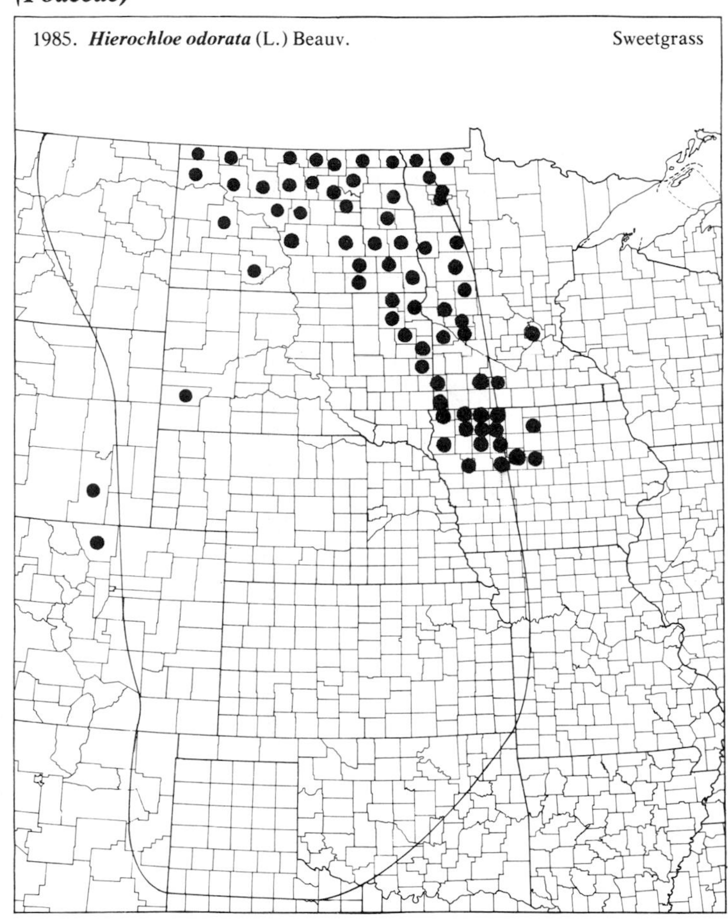
1985. Hierochloe odorata (L.) Beauv.
Sweetgrass

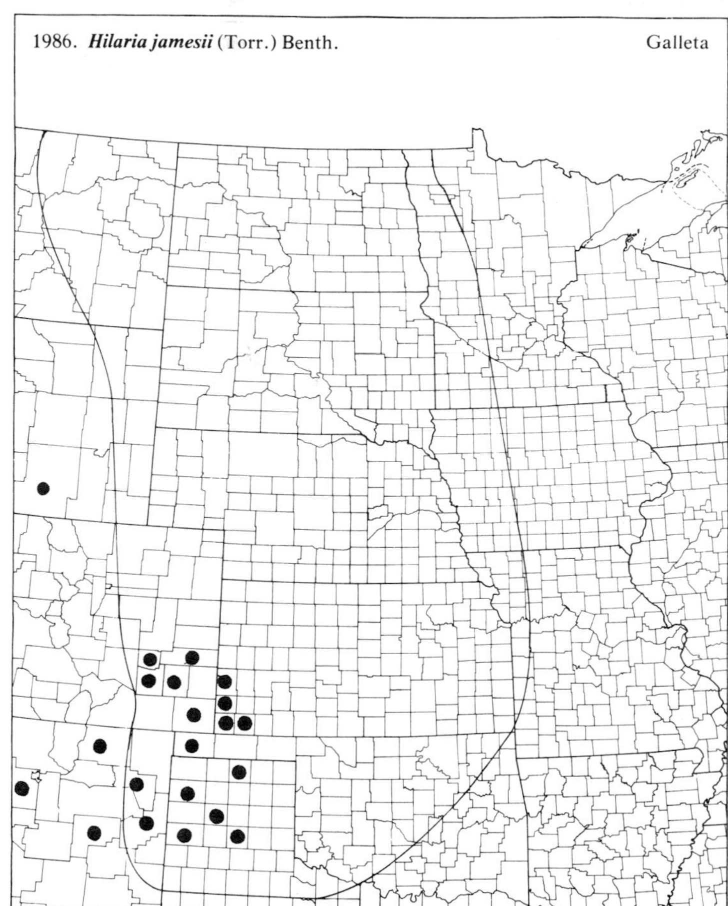
1986. Hilaria jamesii (Torr.) Benth.
Galleta

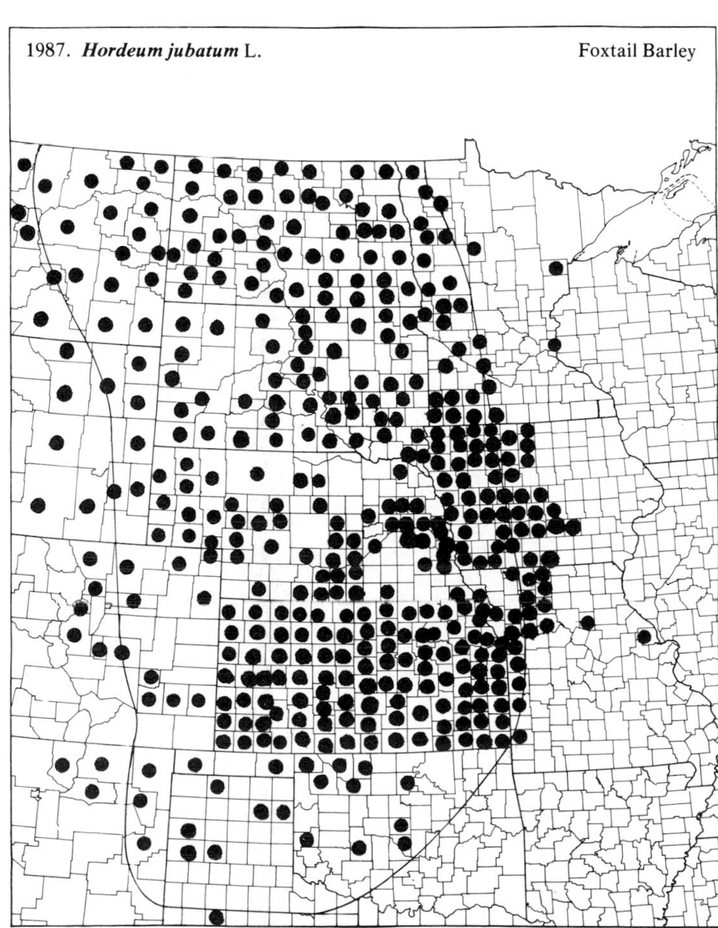
1987. Hordeum jubatum L.
Foxtail Barley

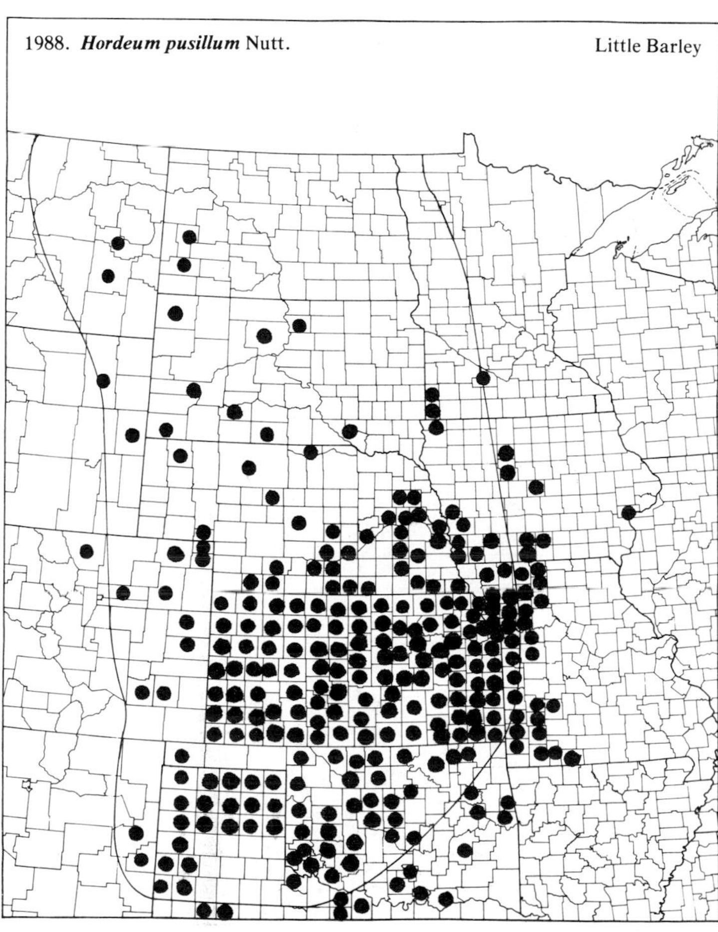
1988. Hordeum pusillum Nutt.
Little Barley

(Poaceae)

1989. ***Hystrix patula*** Moench — Bottlebrush-grass

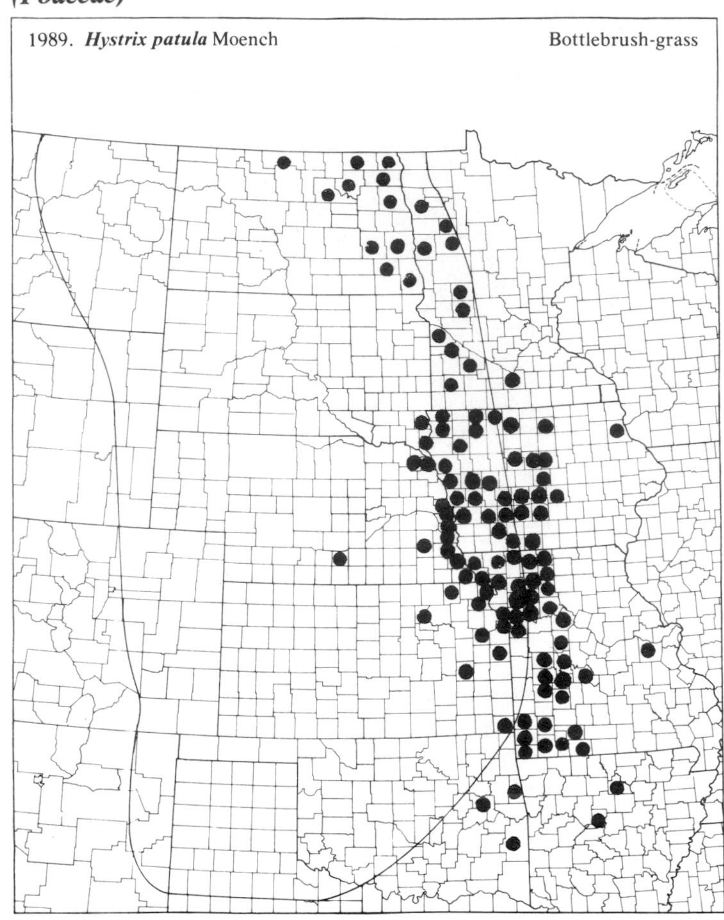

1990. ***Koeleria pyramidata*** (Lam.) Beauv. — Junegrass

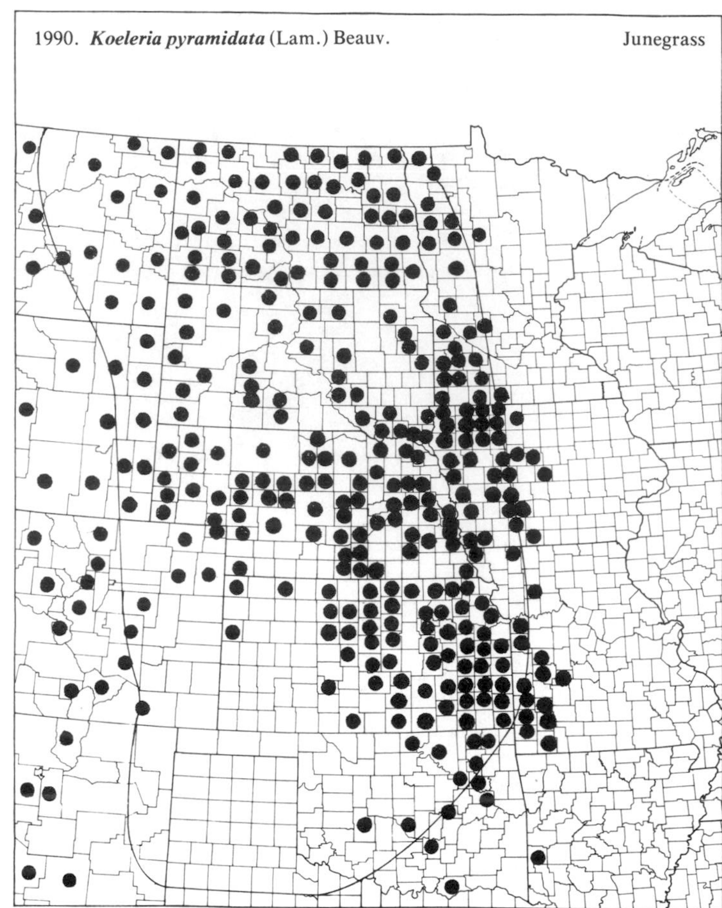

1991. ***Leersia oryzoides*** (L.) Sw. — Rice Cutgrass

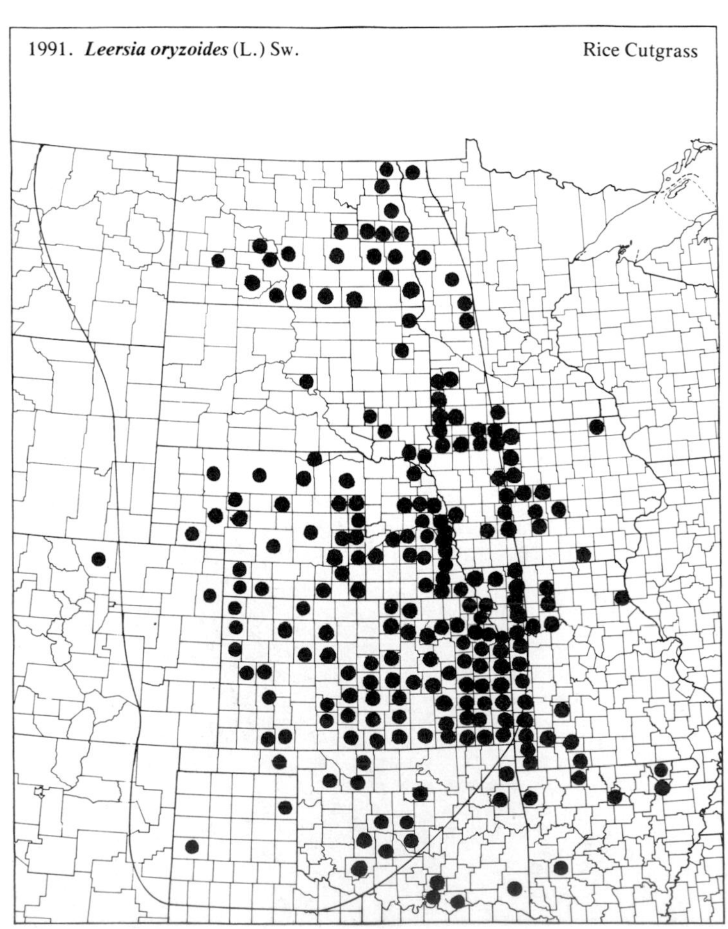

1992. ***Leersia virginica*** Willd. — Whitegrass

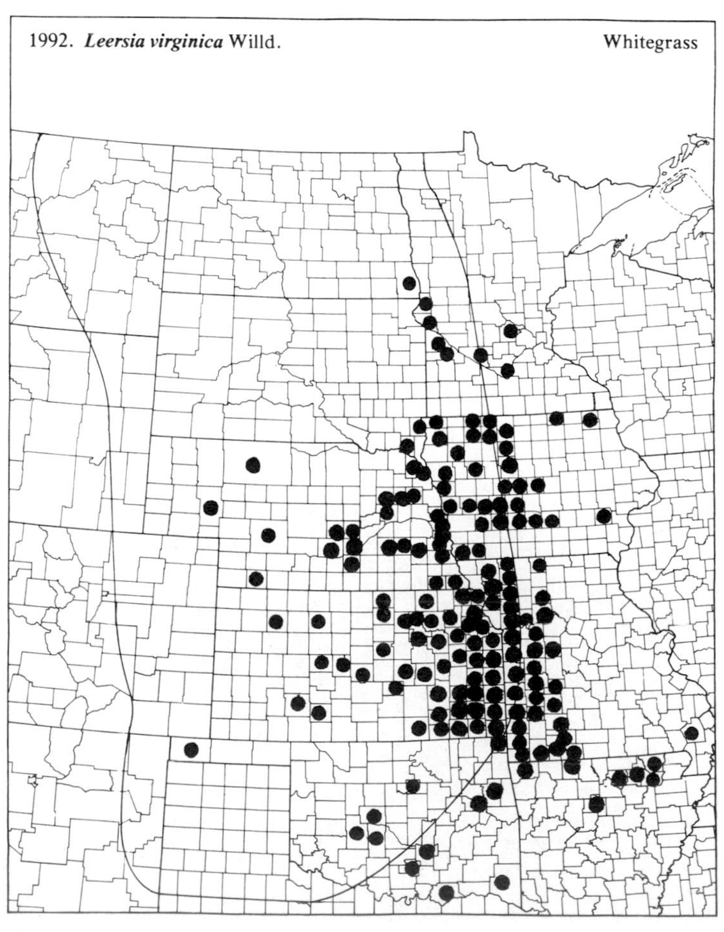

(Poaceae)

1993. ***Leptochloa fascicularis*** (Lam.) Gray — Bearded Sprangletop

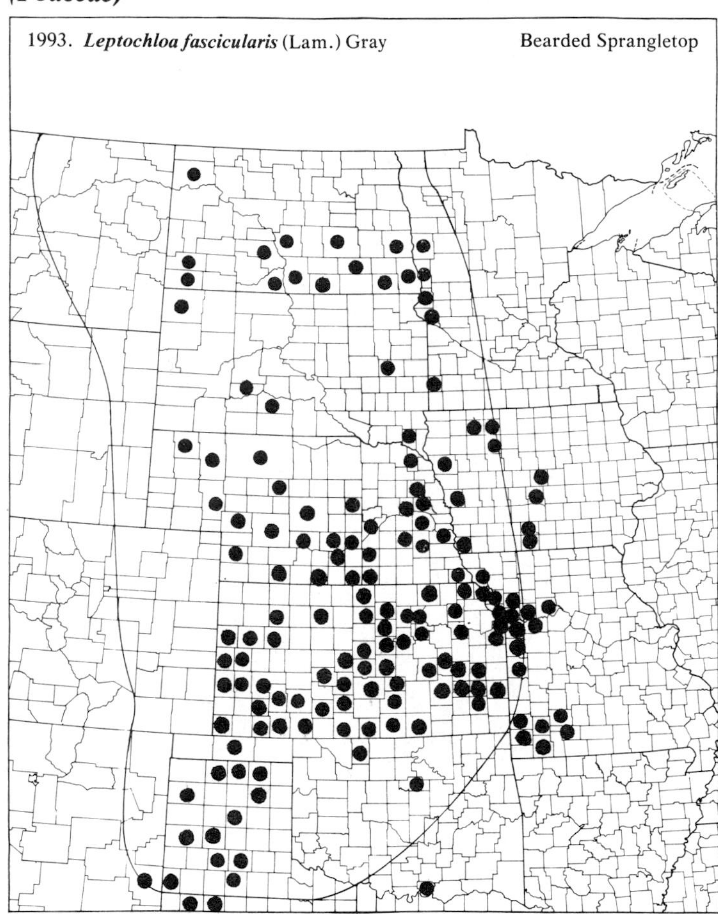

1994. ***Leptochloa filiformis*** (Lam.) Beauv. — Red Sprangletop

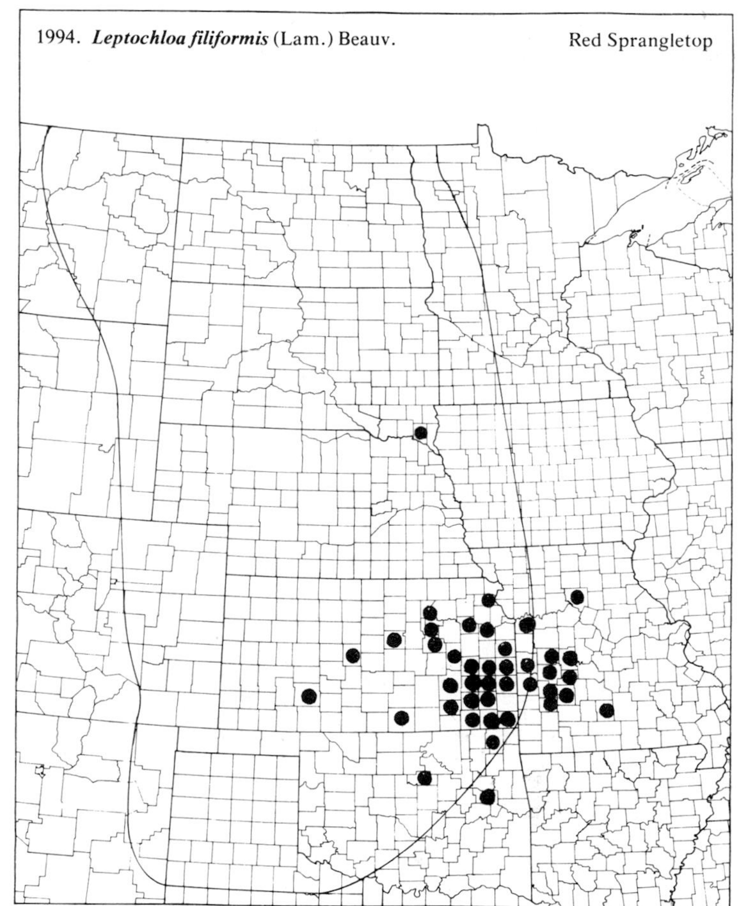

1995. ***Leptoloma cognatum*** (Schult.) Chase — Fall Witchgrass

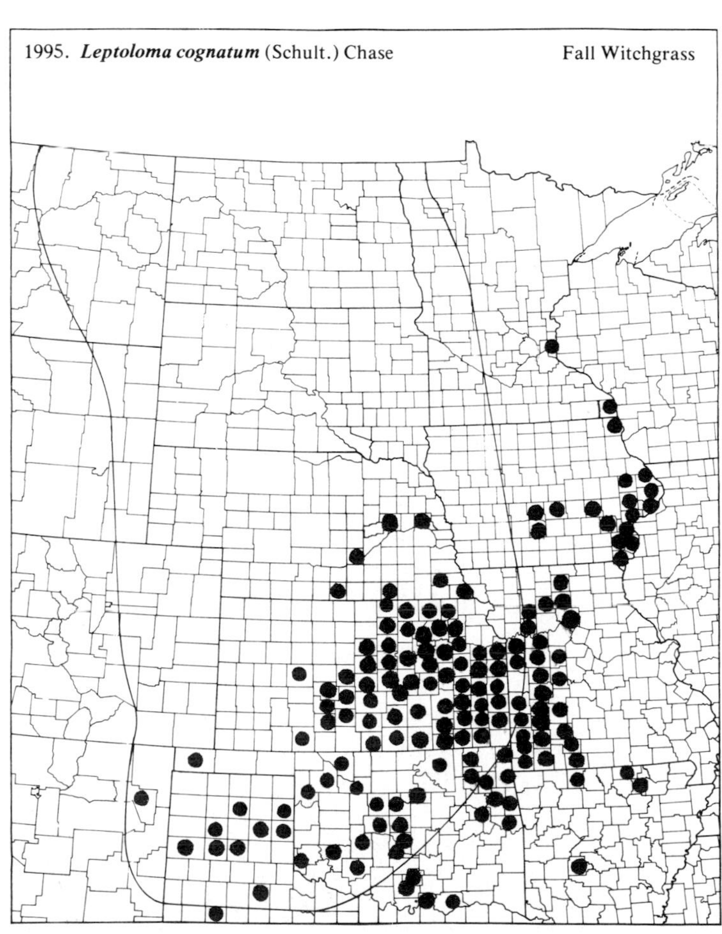

1996. ***Lolium perenne*** L. var. ***perenne*** — Perennial Ryegrass

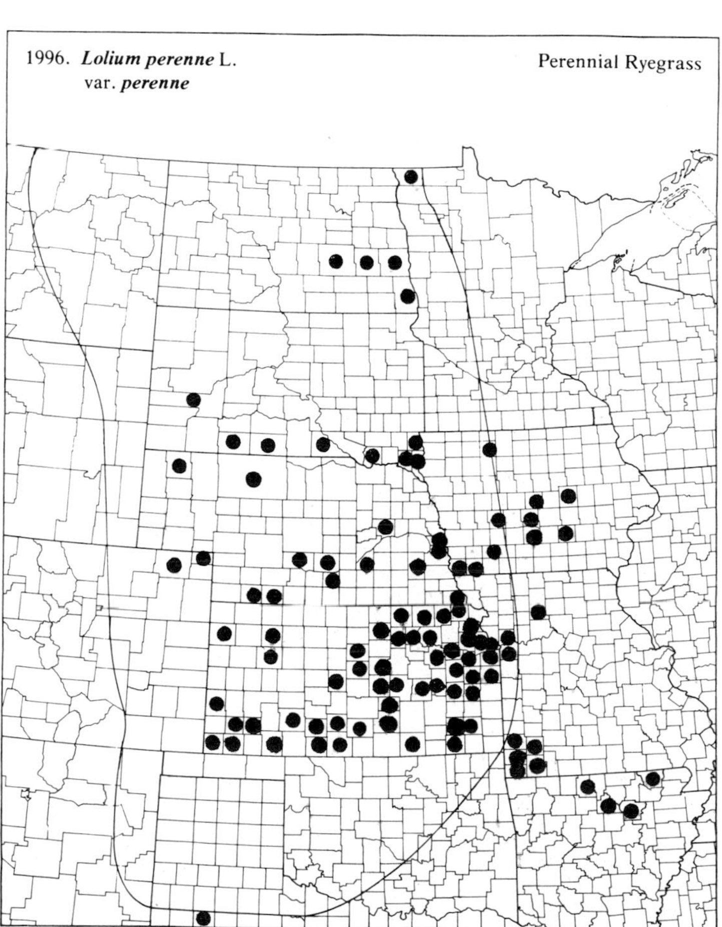

(Poaceae)

1997. ***Lolium perenne*** L. var. ***aristatum*** Willd. — Perennial Ryegrass

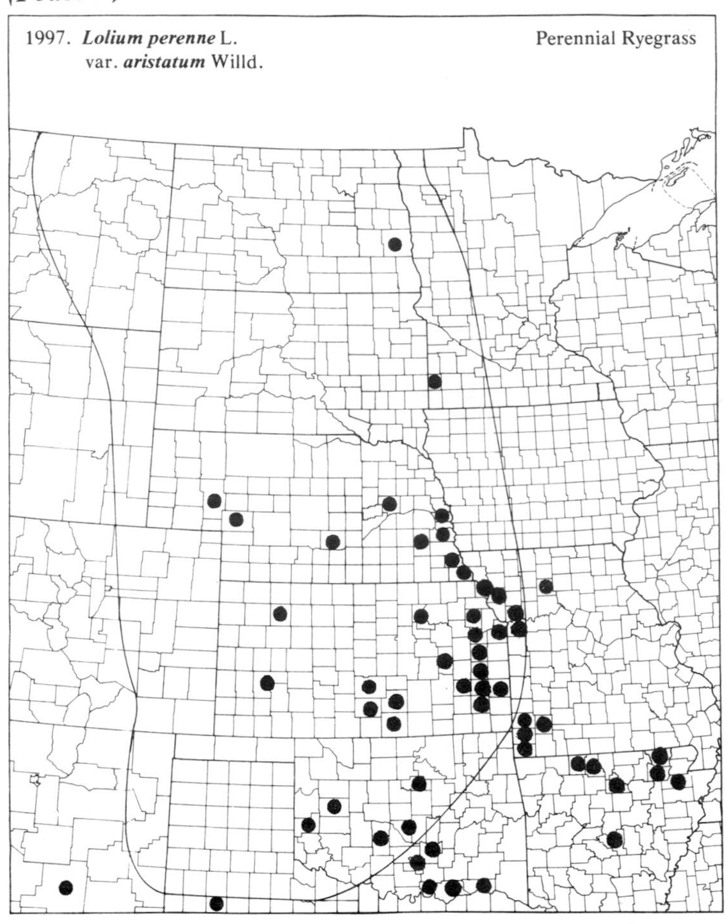

1998. ***Lolium persicum*** Bois & Hohen.

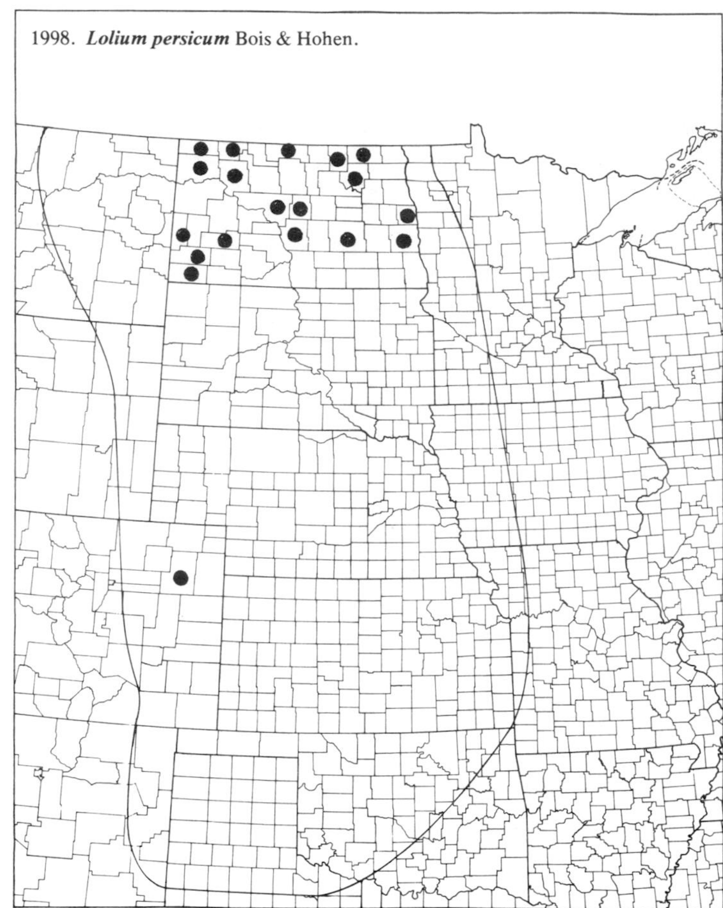

1999. ***Lycurus phleoides*** H.B.K. — Wolftail

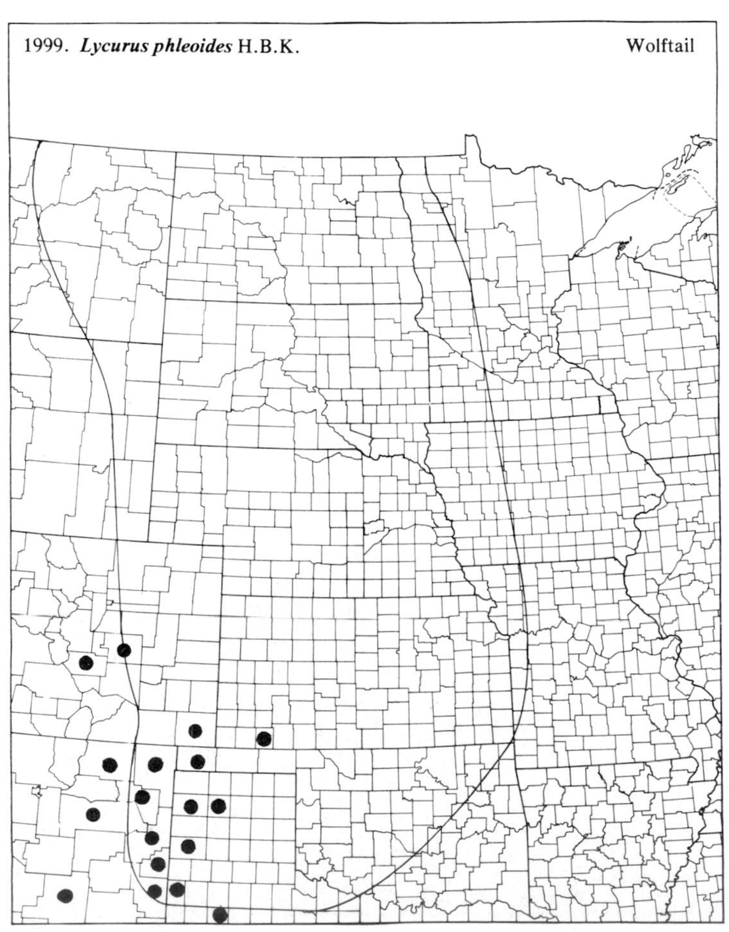

2000. ***Manisuris cylindrica*** (Michx.) O. Ktze. — Carolina Jointtail

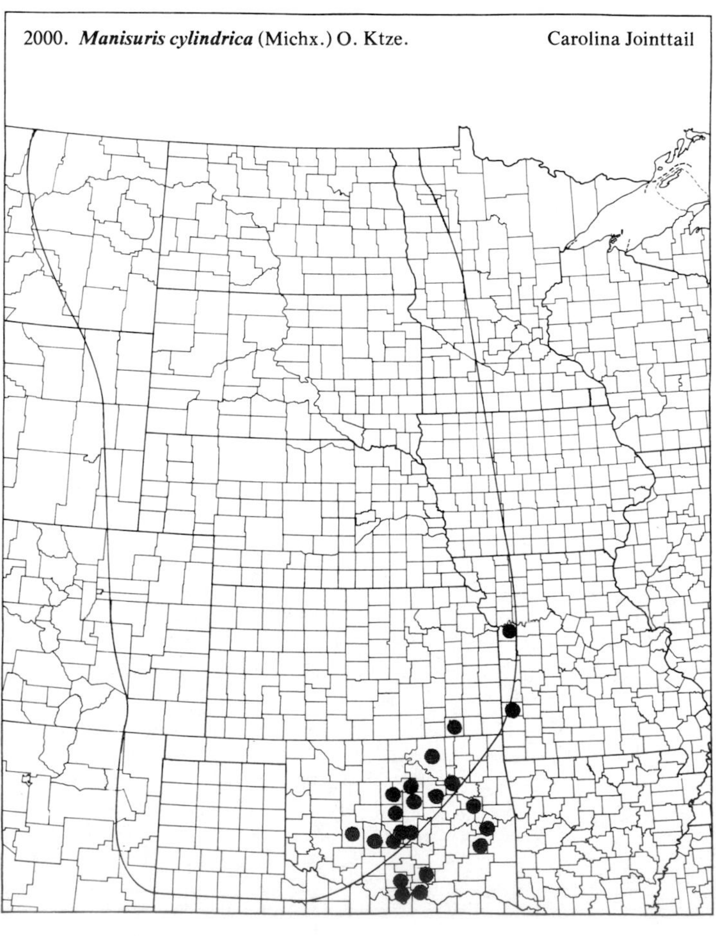

(Poaceae)

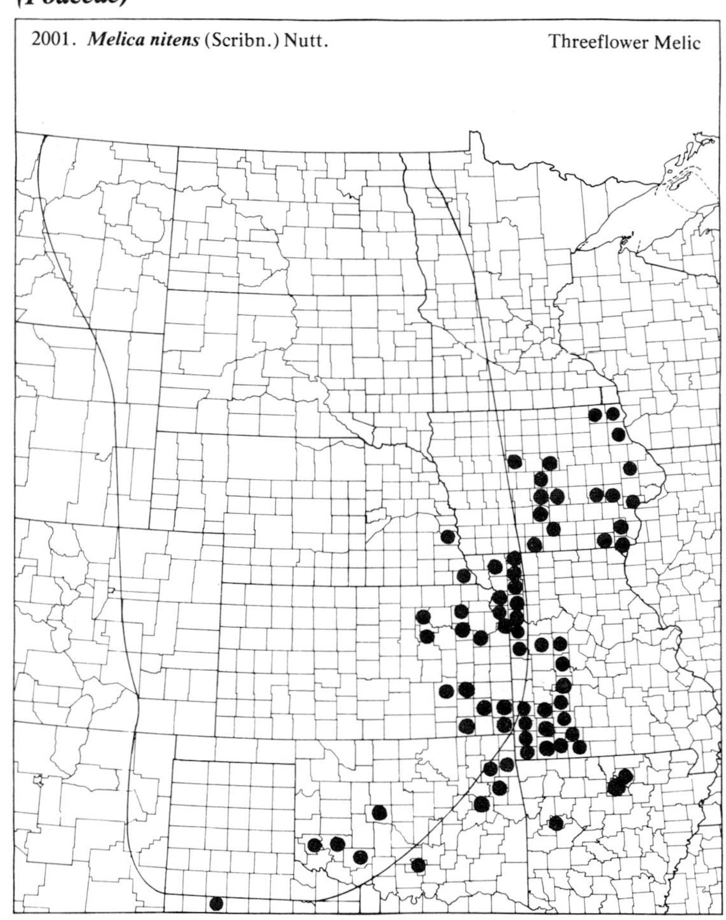
2001. Melica nitens (Scribn.) Nutt.
Threeflower Melic

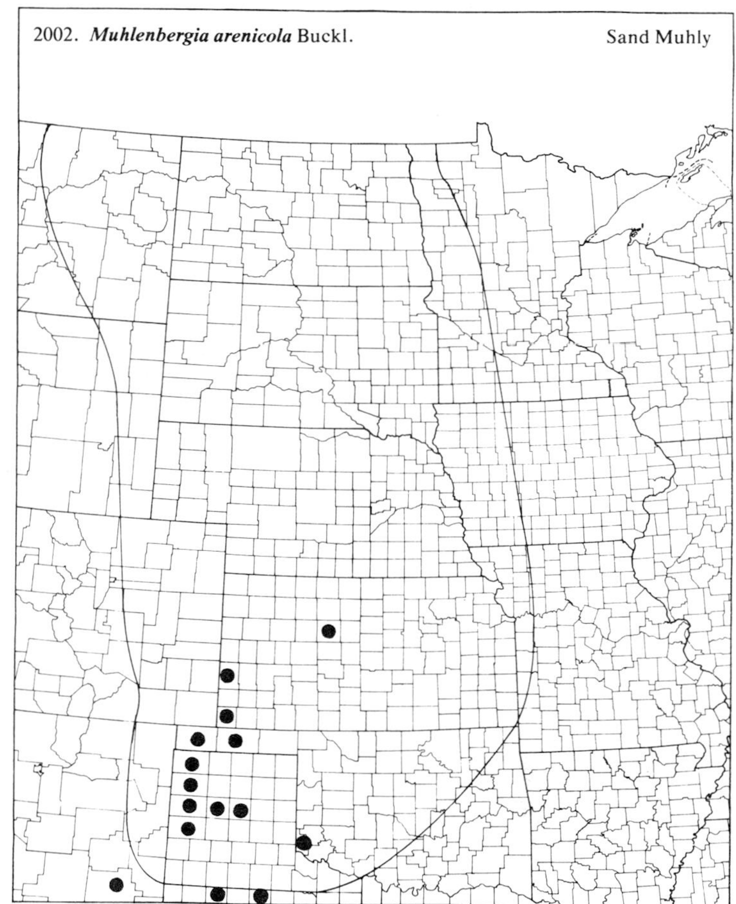
2002. Muhlenbergia arenicola Buckl.
Sand Muhly

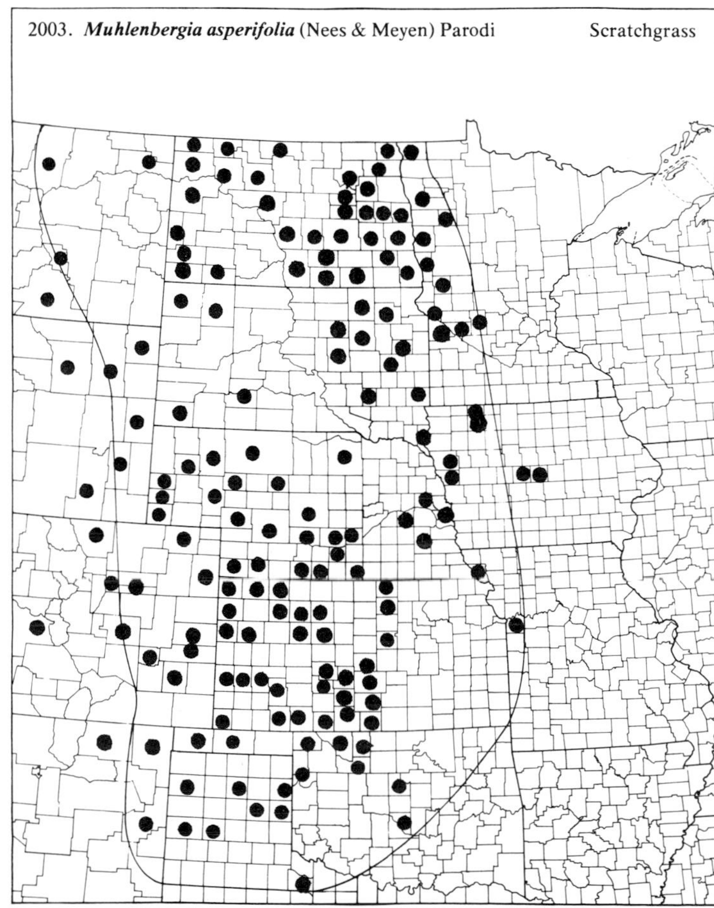
2003. Muhlenbergia asperifolia (Nees & Meyen) Parodi
Scratchgrass

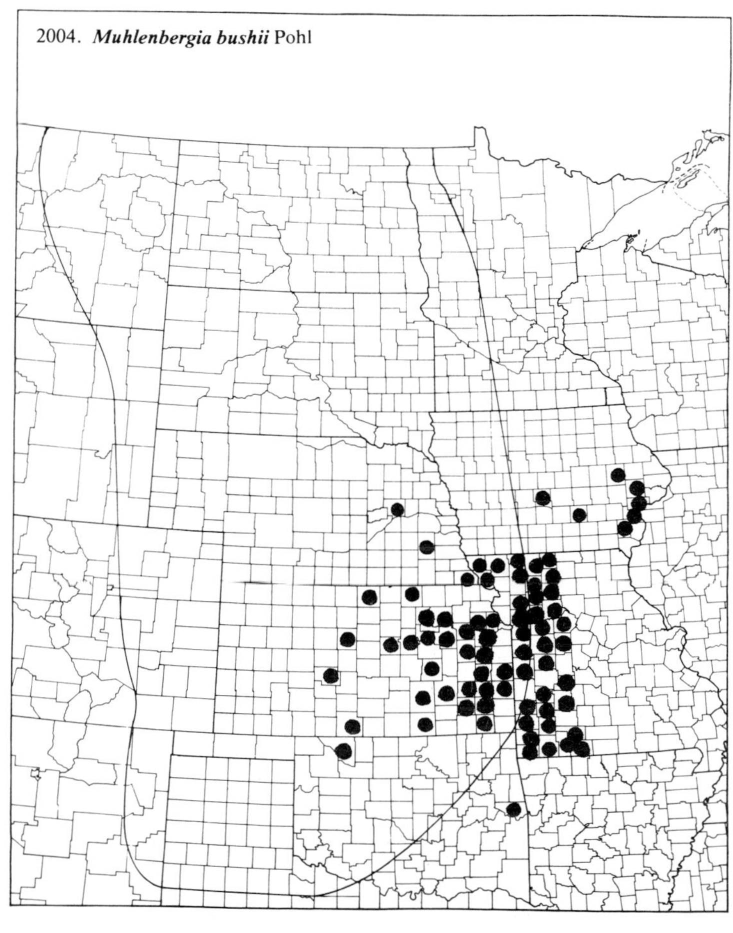
2004. Muhlenbergia bushii Pohl

(Poaceae)

2005. ***Muhlenbergia capillaris*** (Lam.) Trin. Hairgrass

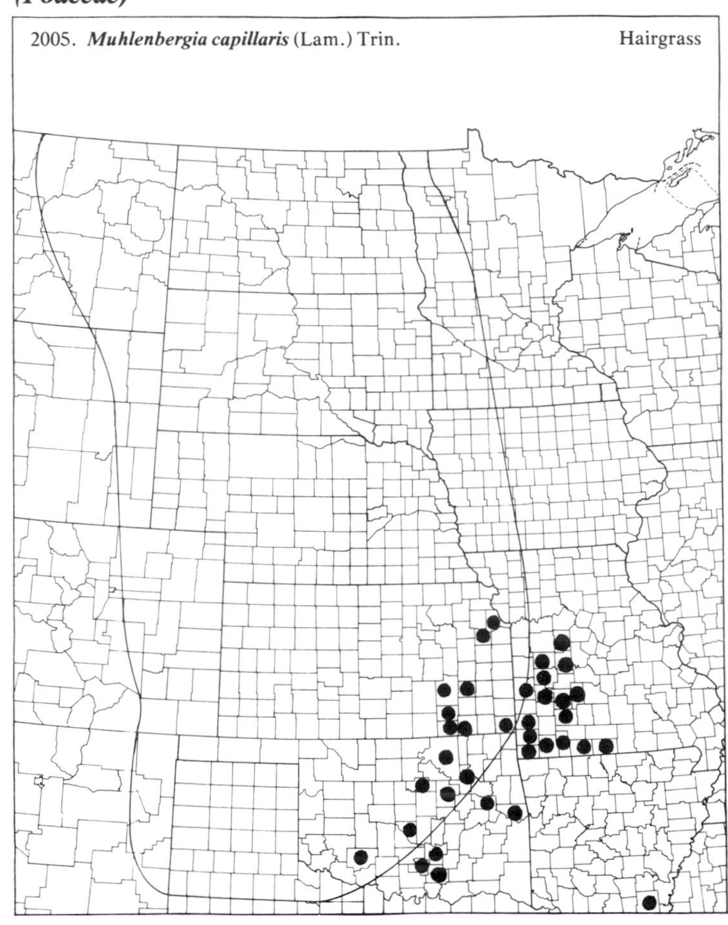

2006. ***Muhlenbergia cuspidata*** (Torr. in Hook.) Rydb. Plains Muhly

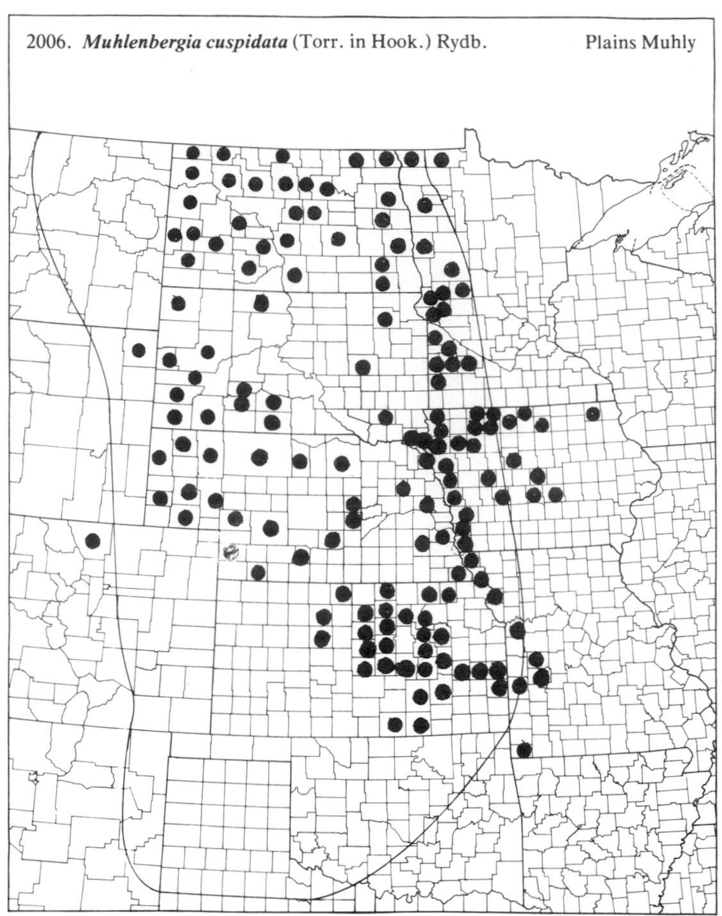

2007. ***Muhlenbergia frondosa*** (Poir.) Fern. Wirestem Muhly

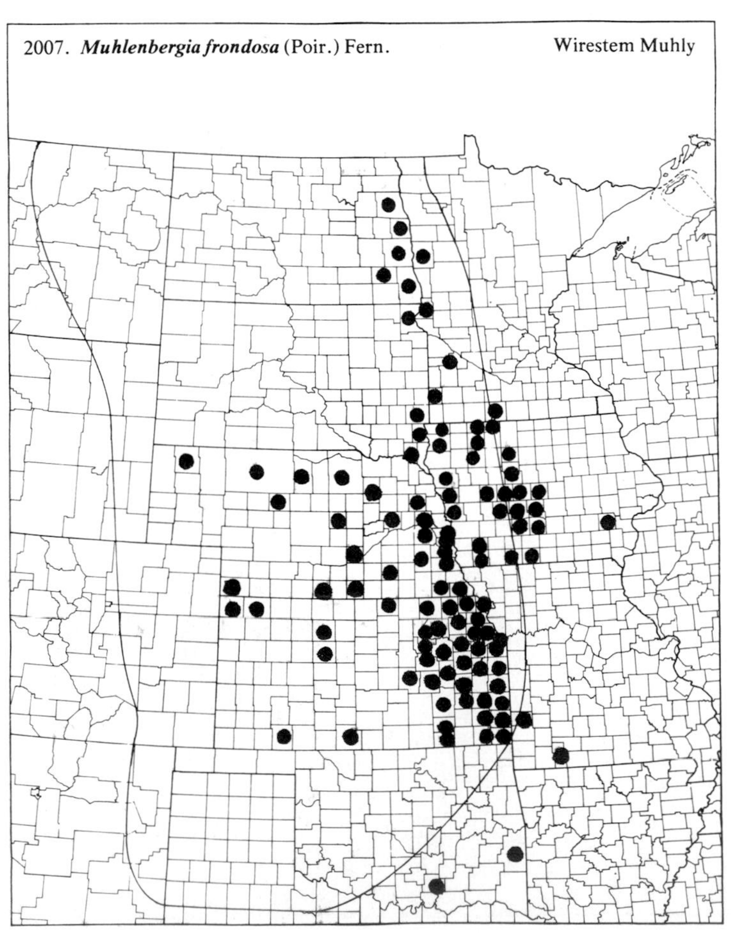

2008. ***Muhlenbergia glomerata*** (Willd.) Trin.

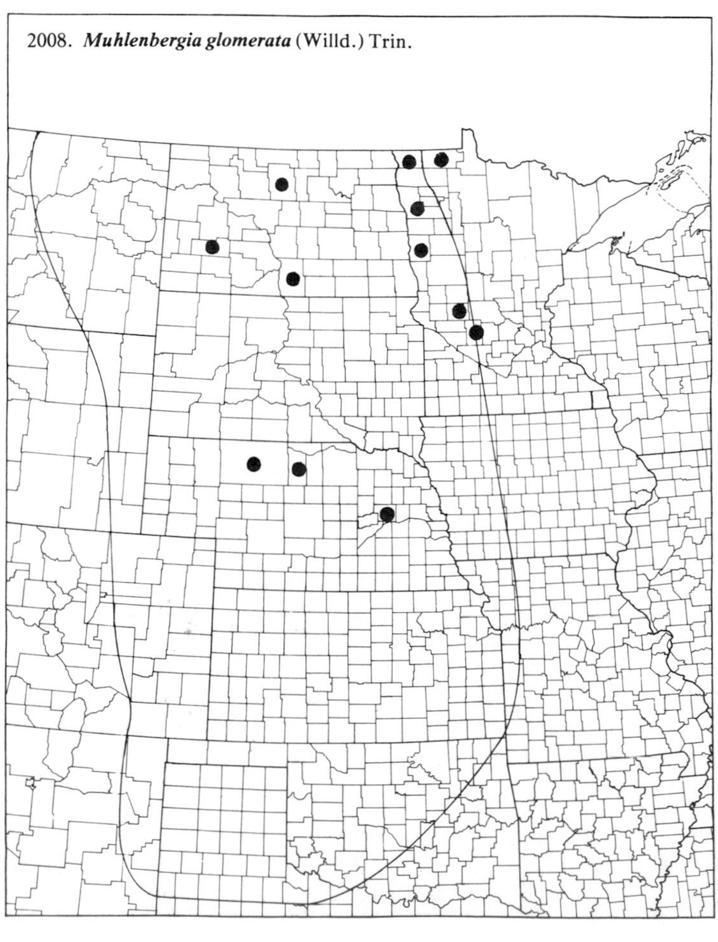

(Poaceae)

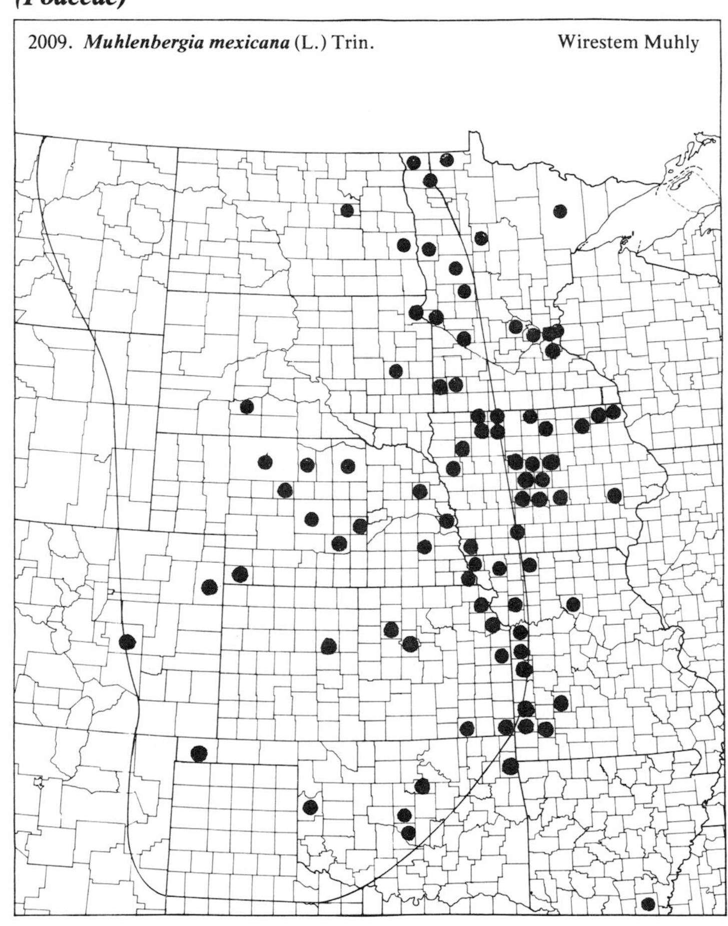

2009. ***Muhlenbergia mexicana*** (L.) Trin. Wirestem Muhly

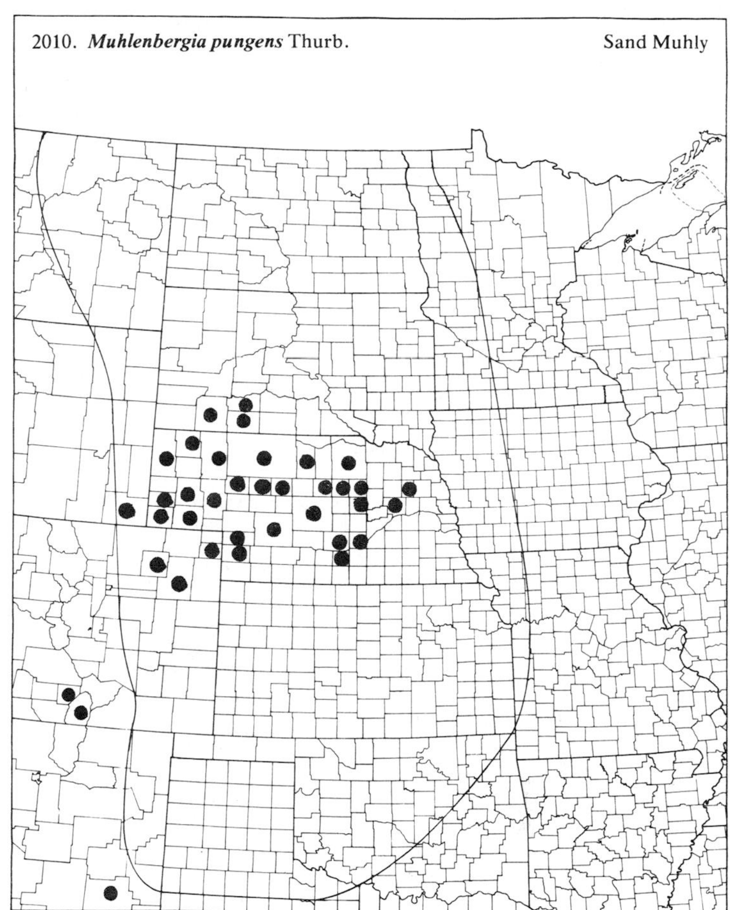

2010. ***Muhlenbergia pungens*** Thurb. Sand Muhly

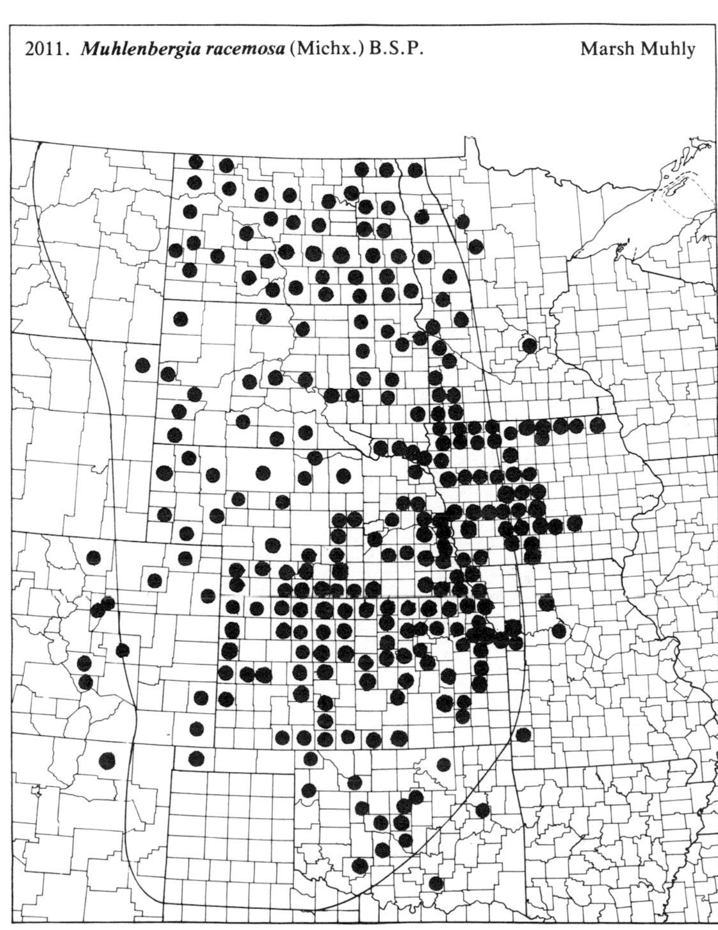

2011. ***Muhlenbergia racemosa*** (Michx.) B.S.P. Marsh Muhly

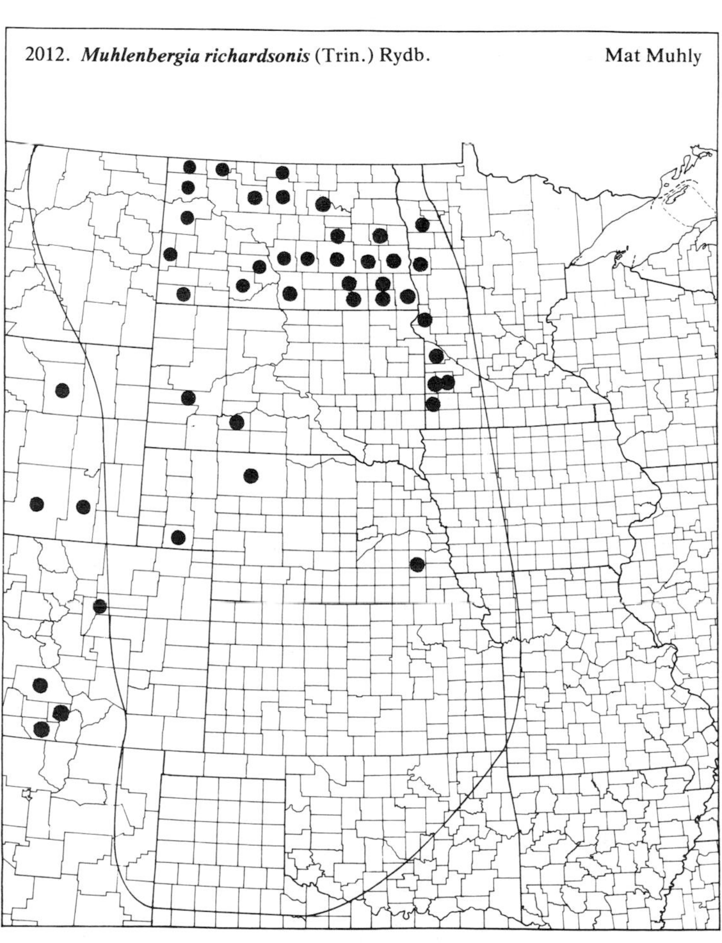

2012. ***Muhlenbergia richardsonis*** (Trin.) Rydb. Mat Muhly

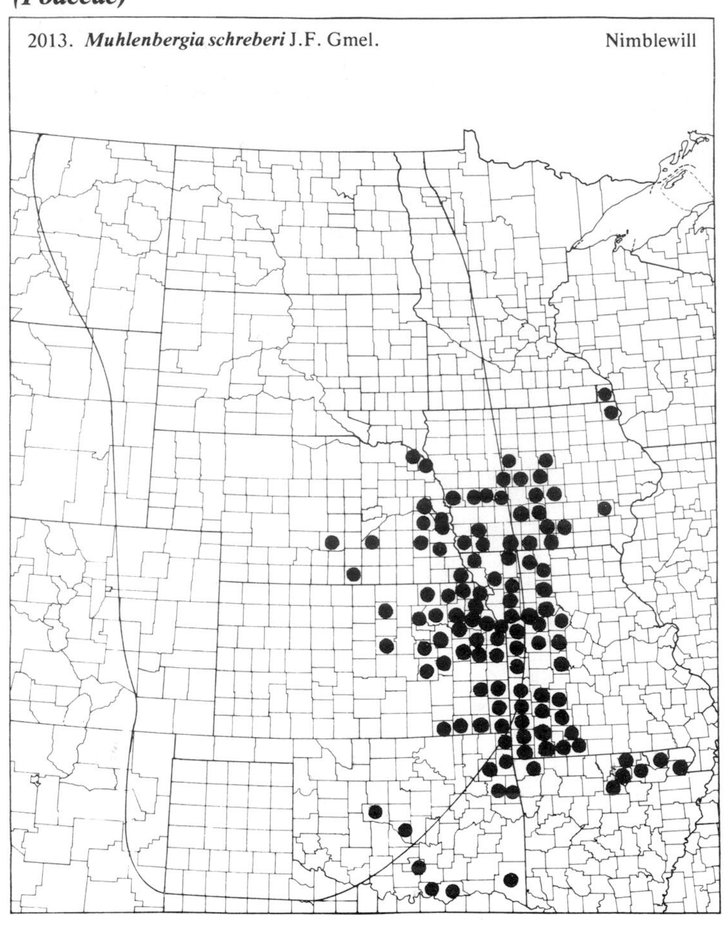
2013. *Muhlenbergia schreberi* J.F. Gmel.
Nimblewill

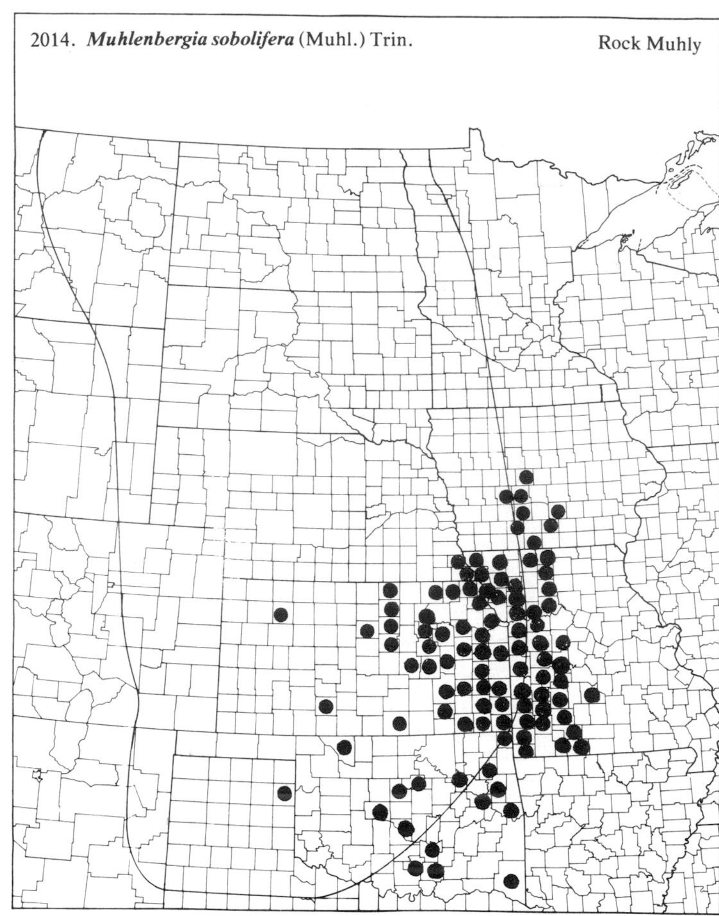
2014. *Muhlenbergia sobolifera* (Muhl.) Trin.
Rock Muhly

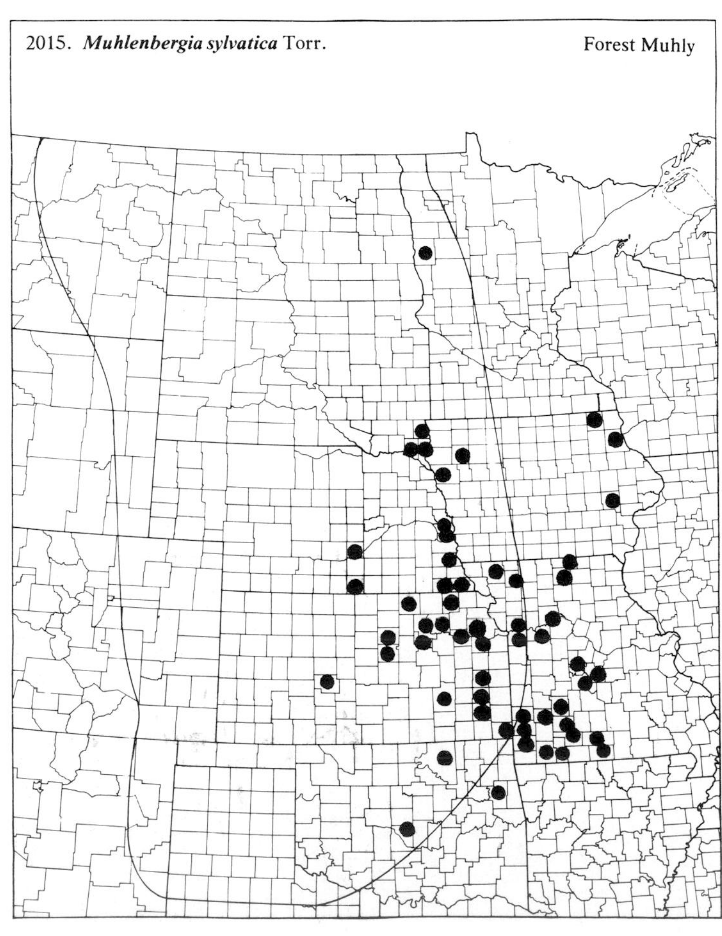
2015. *Muhlenbergia sylvatica* Torr.
Forest Muhly

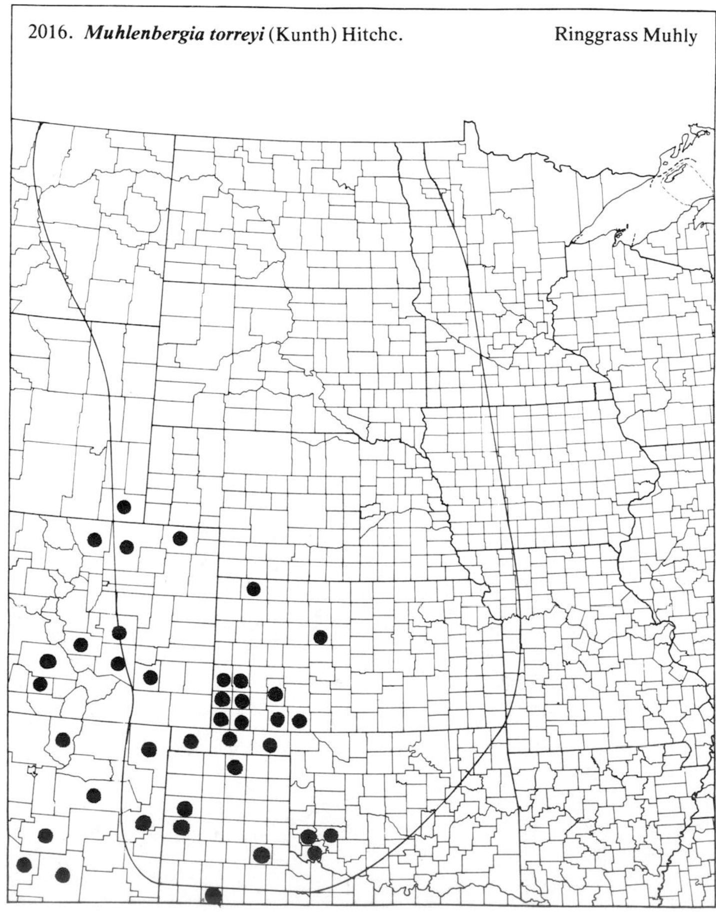
2016. *Muhlenbergia torreyi* (Kunth) Hitchc.
Ringgrass Muhly

(Poaceae)

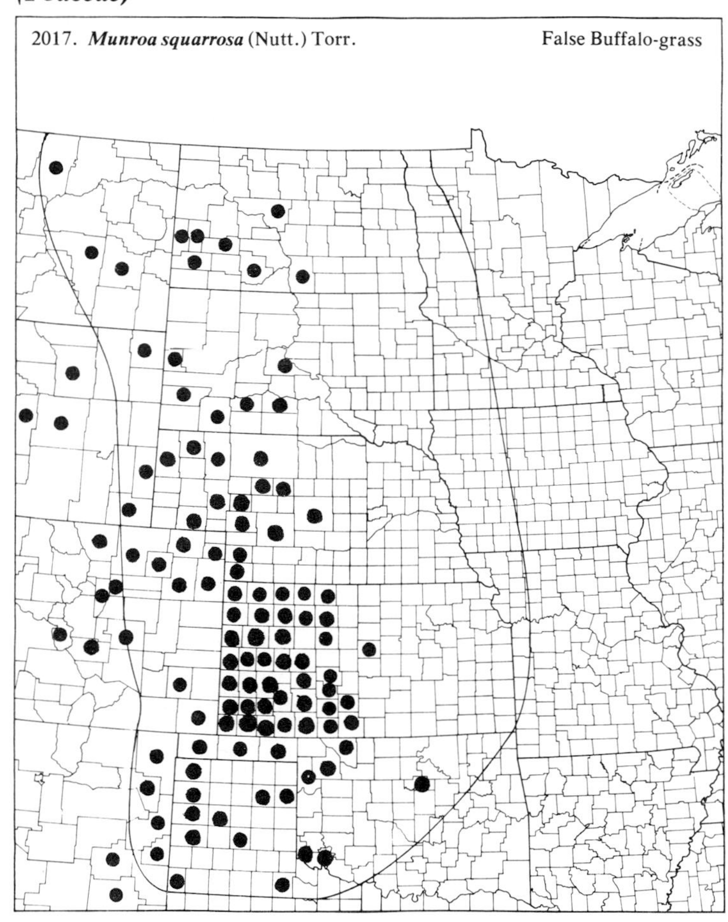
2017. ***Munroa squarrosa*** (Nutt.) Torr.
False Buffalo-grass

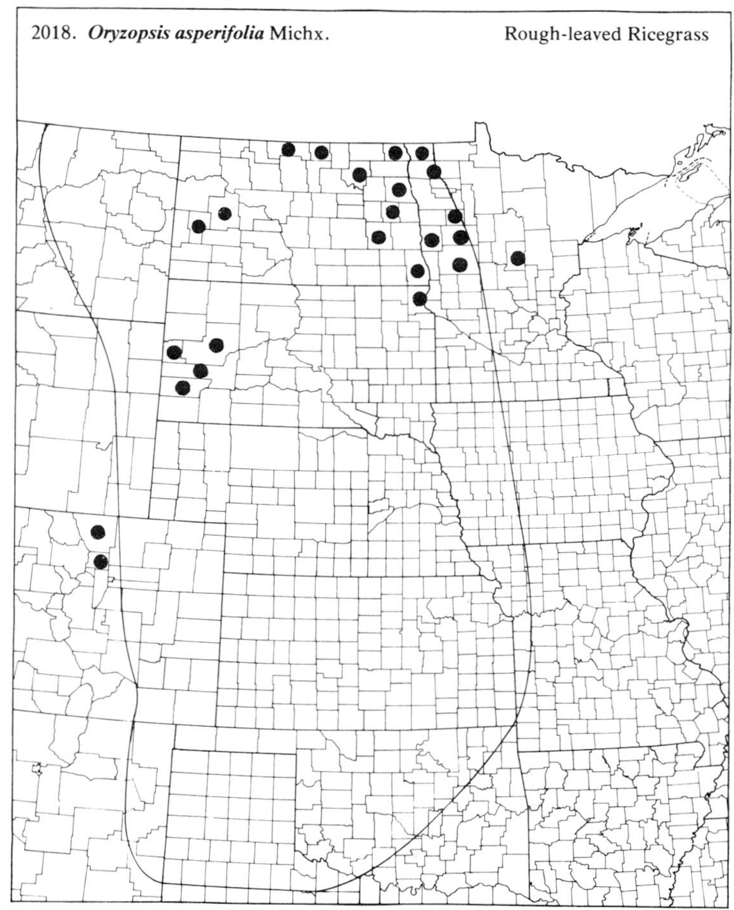
2018. ***Oryzopsis asperifolia*** Michx.
Rough-leaved Ricegrass

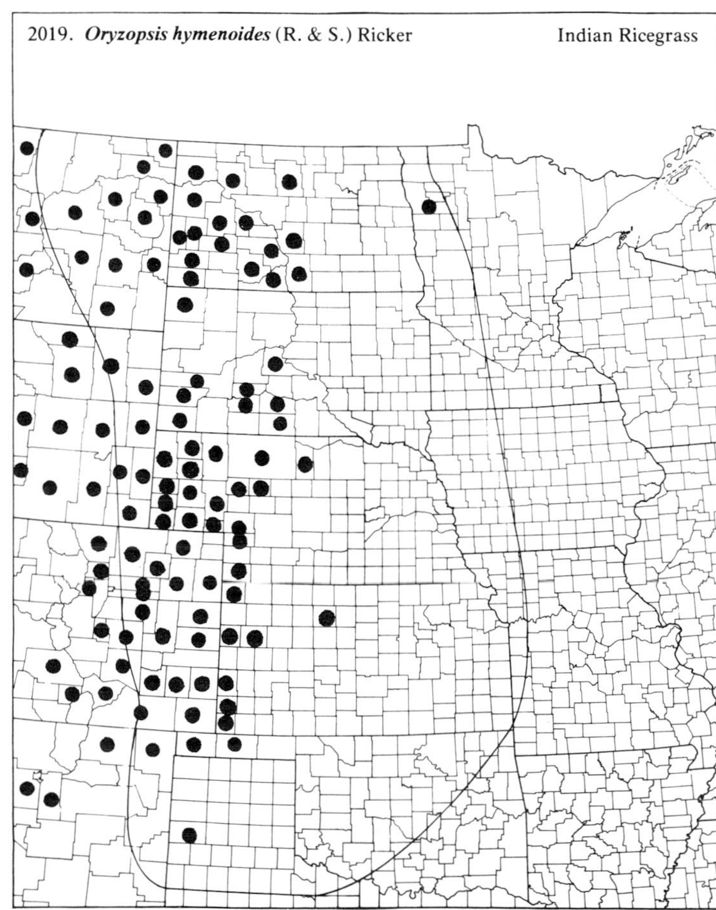
2019. ***Oryzopsis hymenoides*** (R. & S.) Ricker
Indian Ricegrass

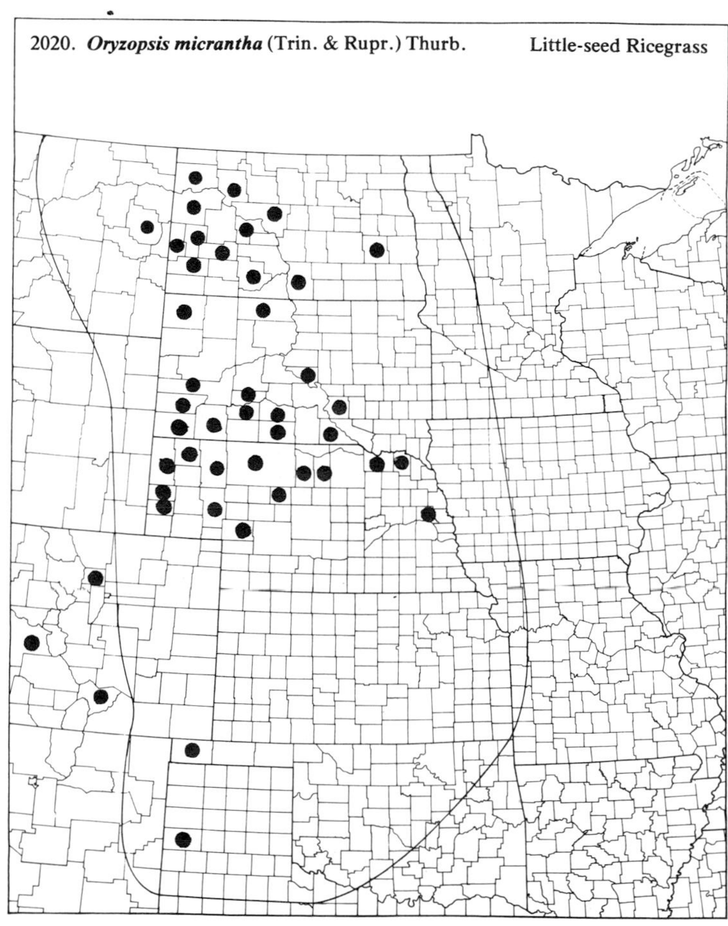
2020. ***Oryzopsis micrantha*** (Trin. & Rupr.) Thurb.
Little-seed Ricegrass

(Poaceae)

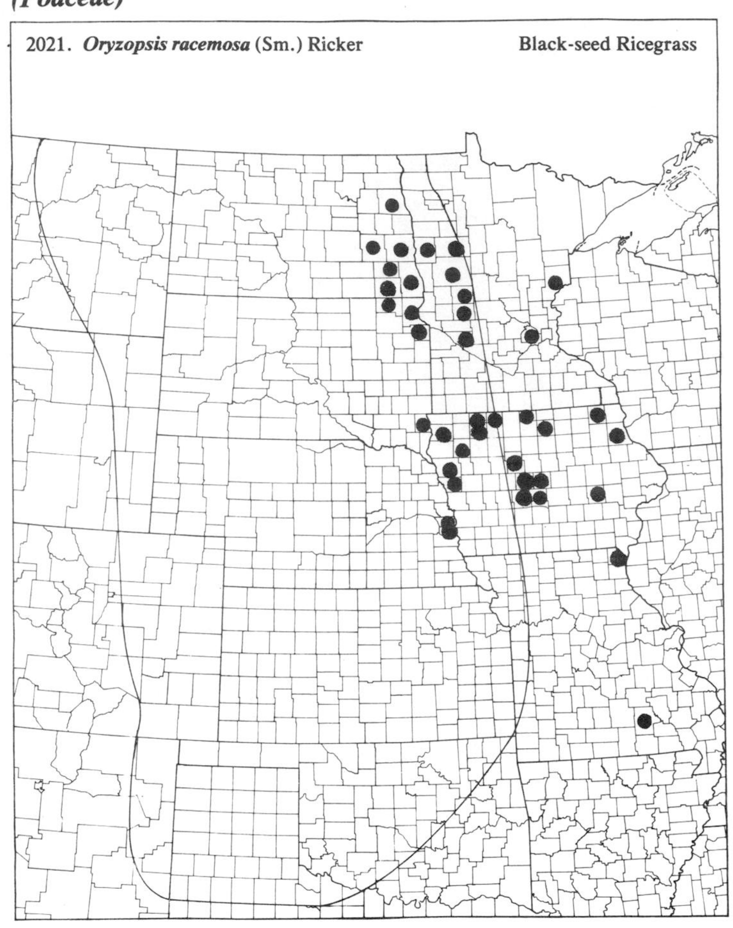

2021. ***Oryzopsis racemosa*** (Sm.) Ricker — Black-seed Ricegrass

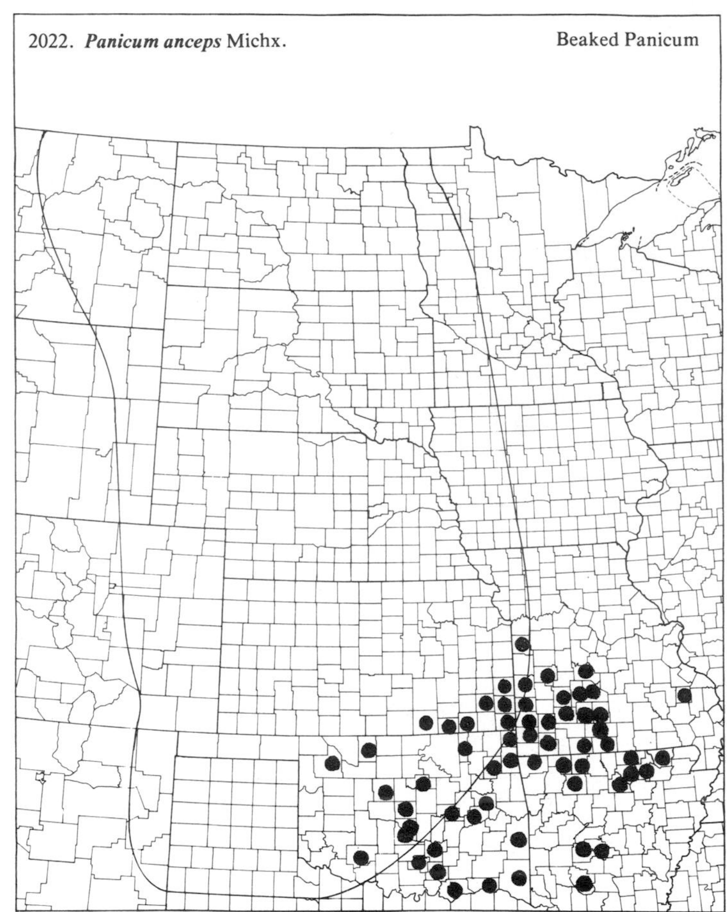

2022. ***Panicum anceps*** Michx. — Beaked Panicum

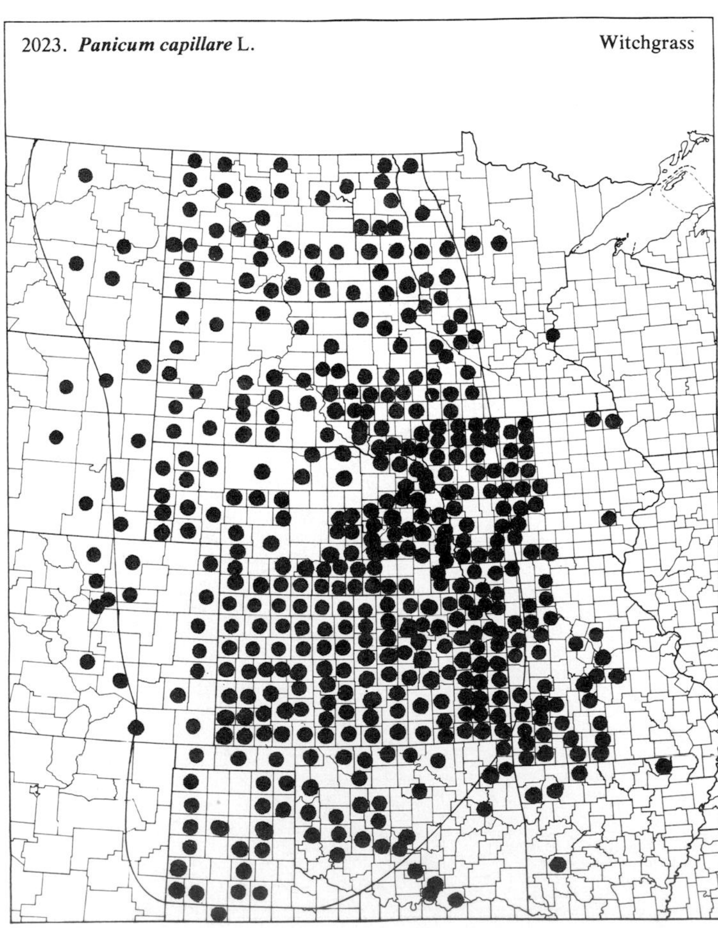

2023. ***Panicum capillare*** L. — Witchgrass

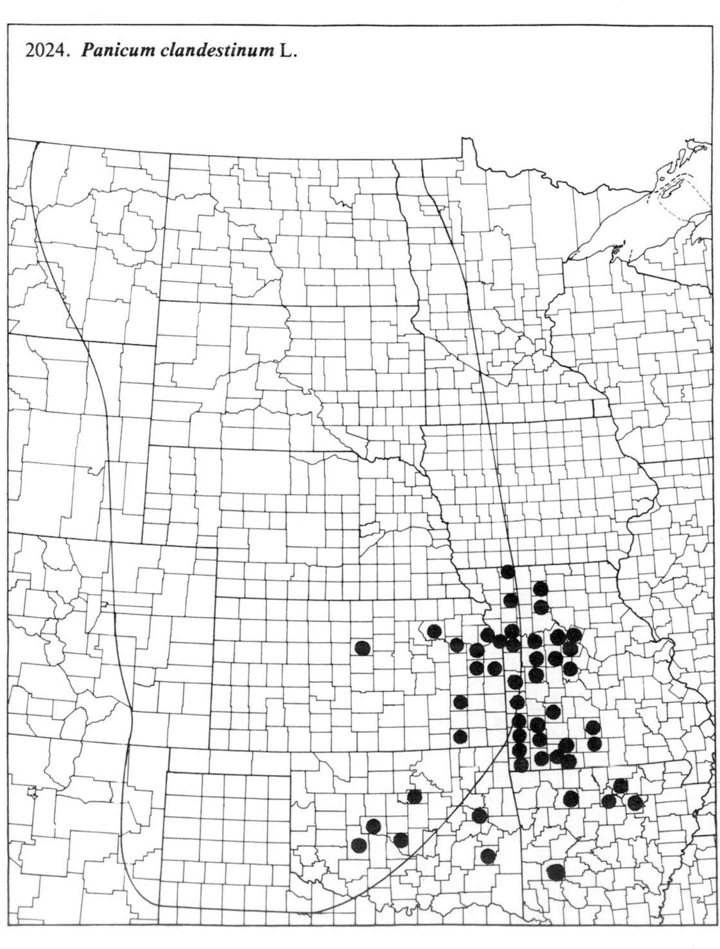

2024. ***Panicum clandestinum*** L.

(Poaceae)

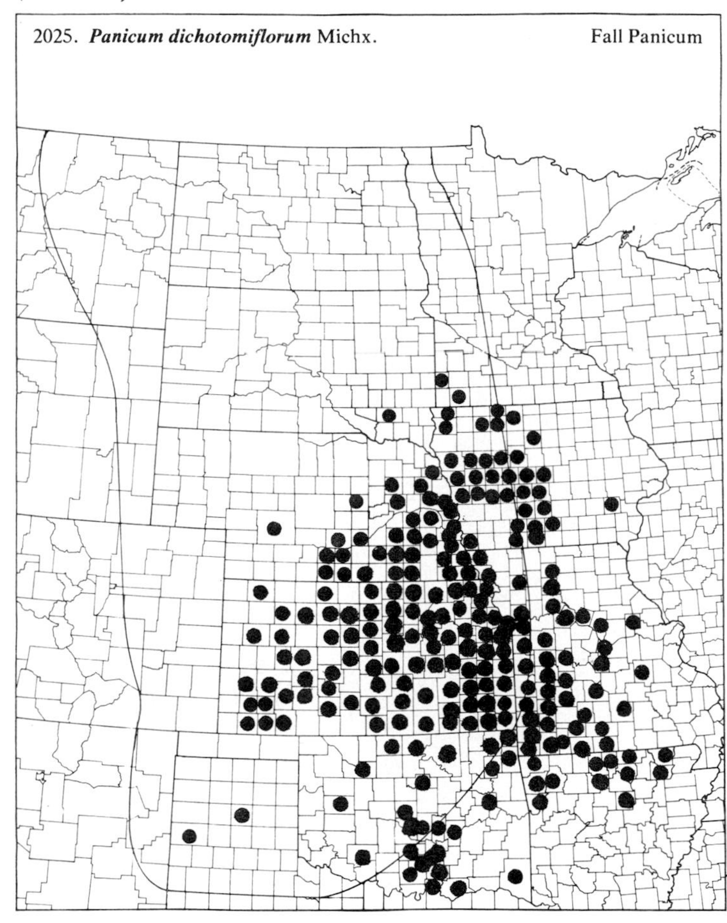
2025. ***Panicum dichotomiflorum*** Michx. Fall Panicum

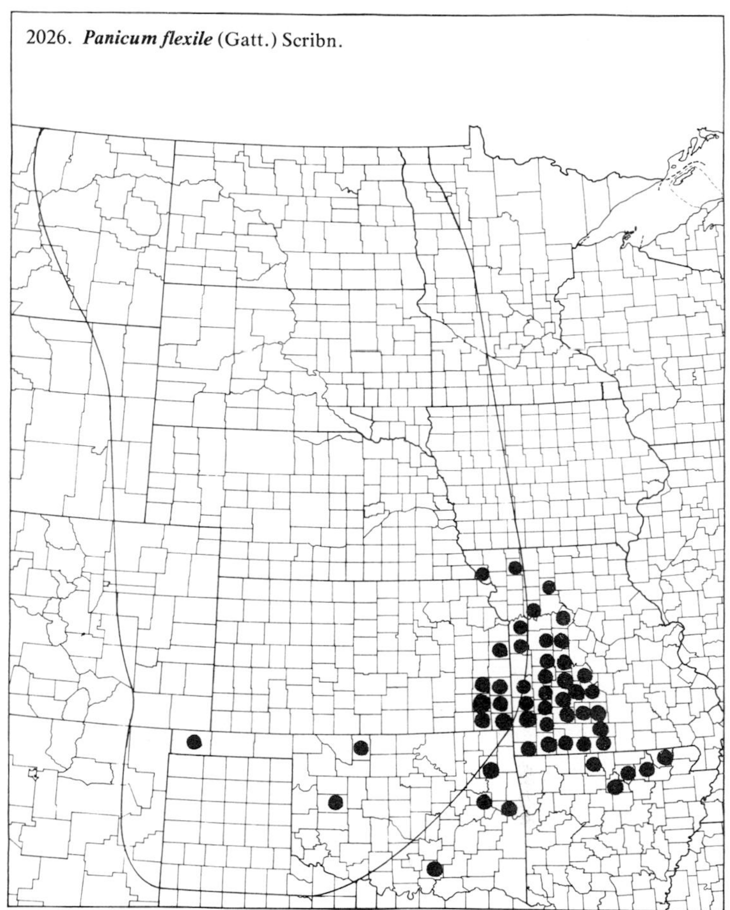
2026. ***Panicum flexile*** (Gatt.) Scribn.

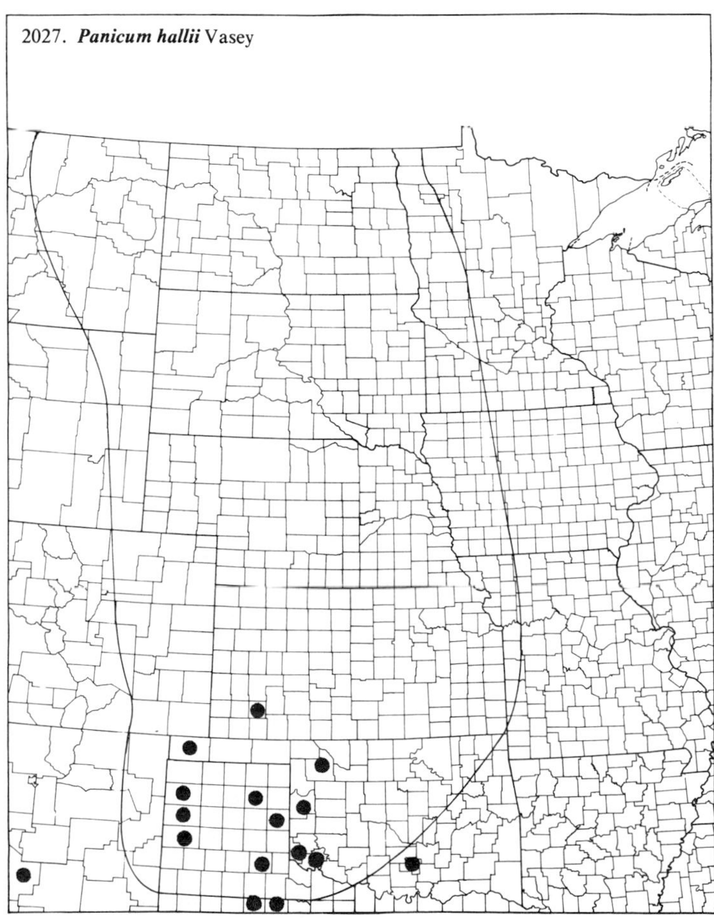
2027. ***Panicum hallii*** Vasey

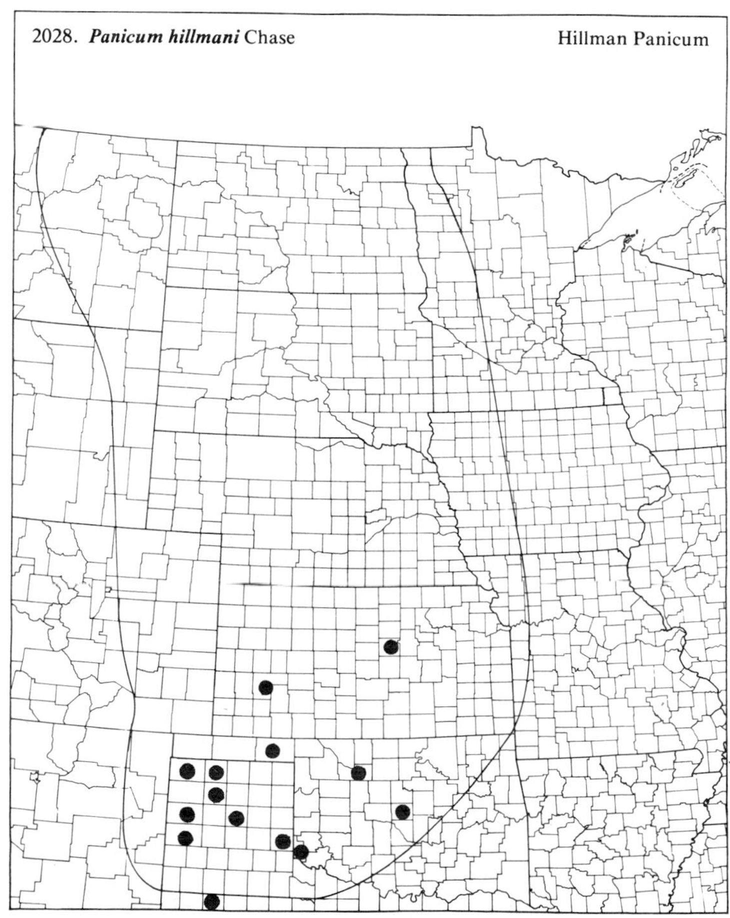
2028. ***Panicum hillmani*** Chase Hillman Panicum

(Poaceae)

2029. ***Panicum lanuginosum*** Ell. var. ***fasciculatum*** (Torr.) Fern.

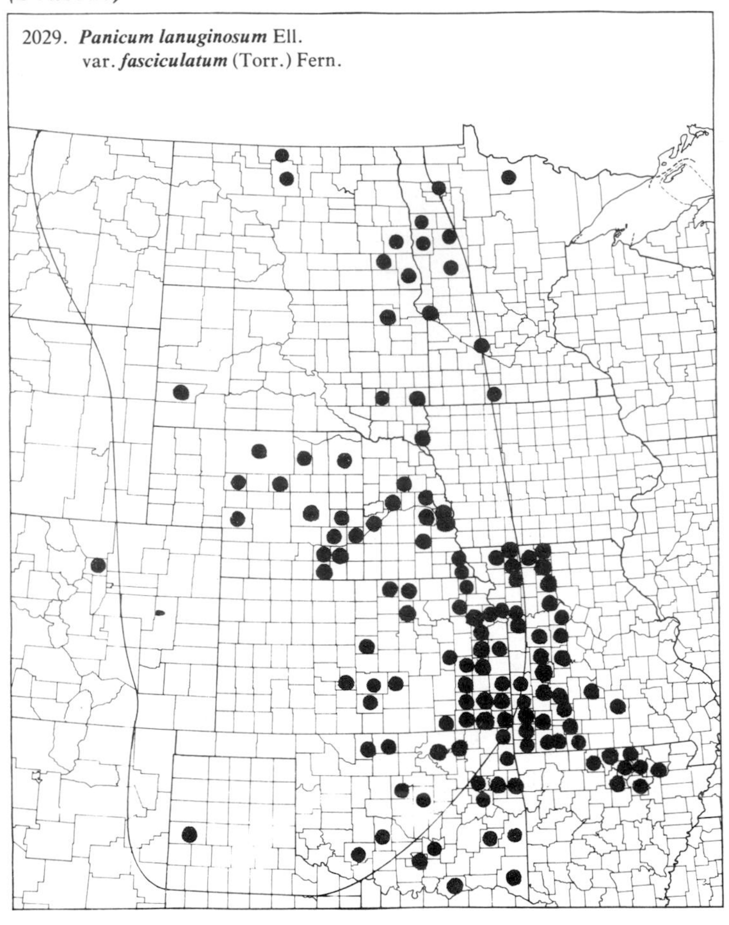

2030. ***Panicum lanuginosum*** Ell. var. ***lindheimeri*** (Nash) Fern.

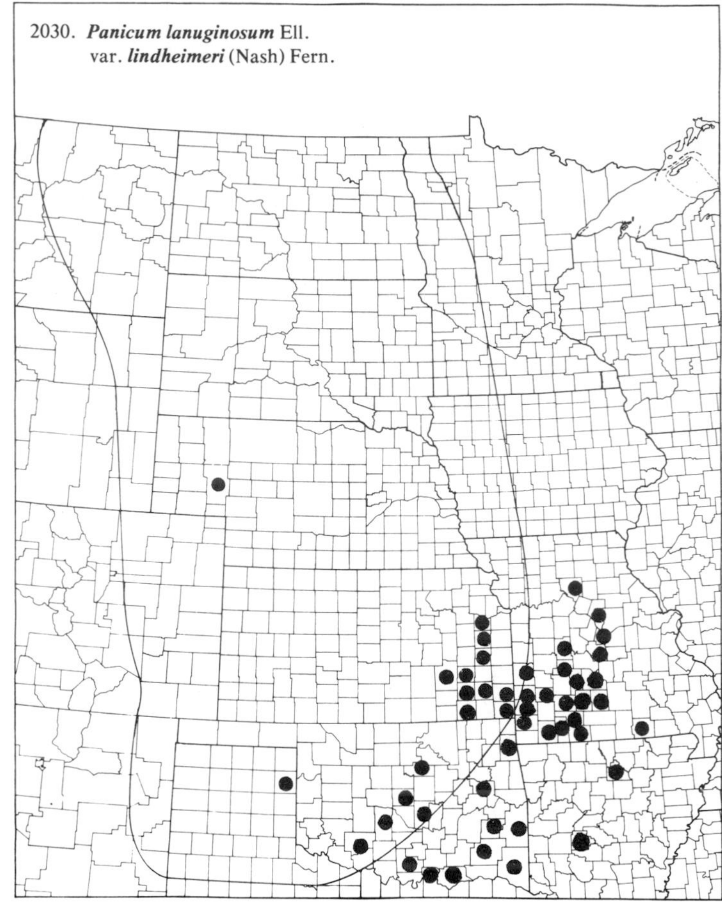

2031. ***Panicum latifolium*** L.

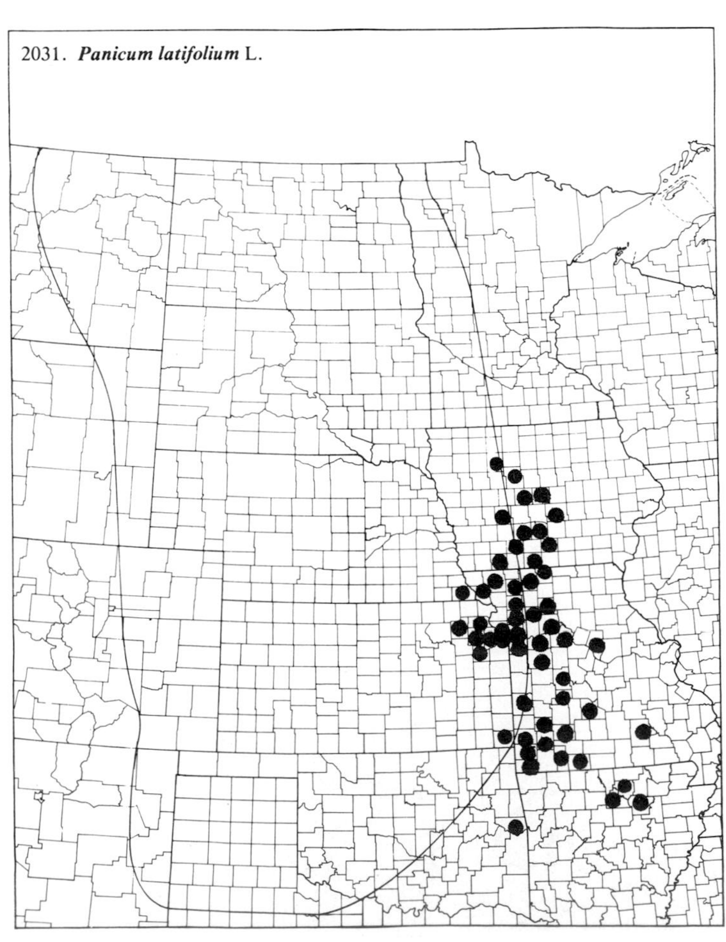

2032. ***Panicum leibergii*** (Vasey) Scribn. Leiberg Panicum

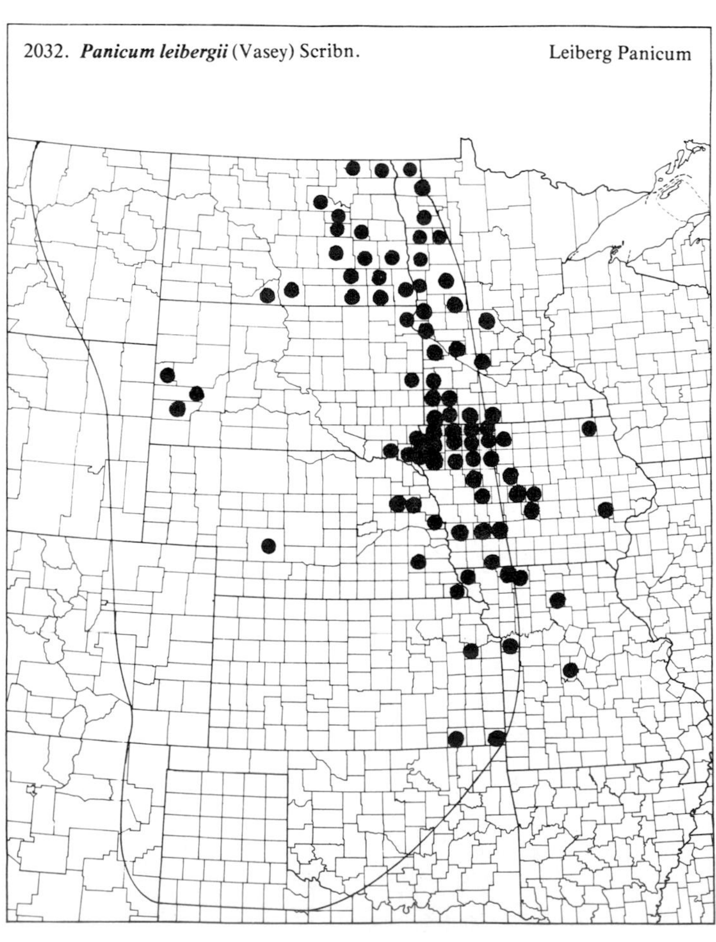

(Poaceae)

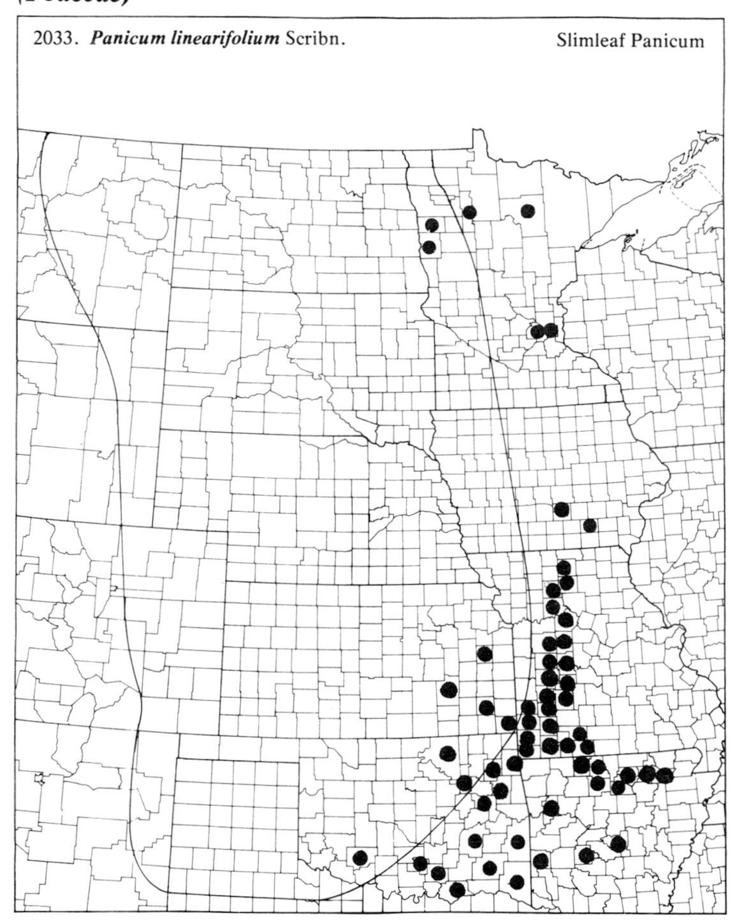
2033. ***Panicum linearifolium*** Scribn. Slimleaf Panicum

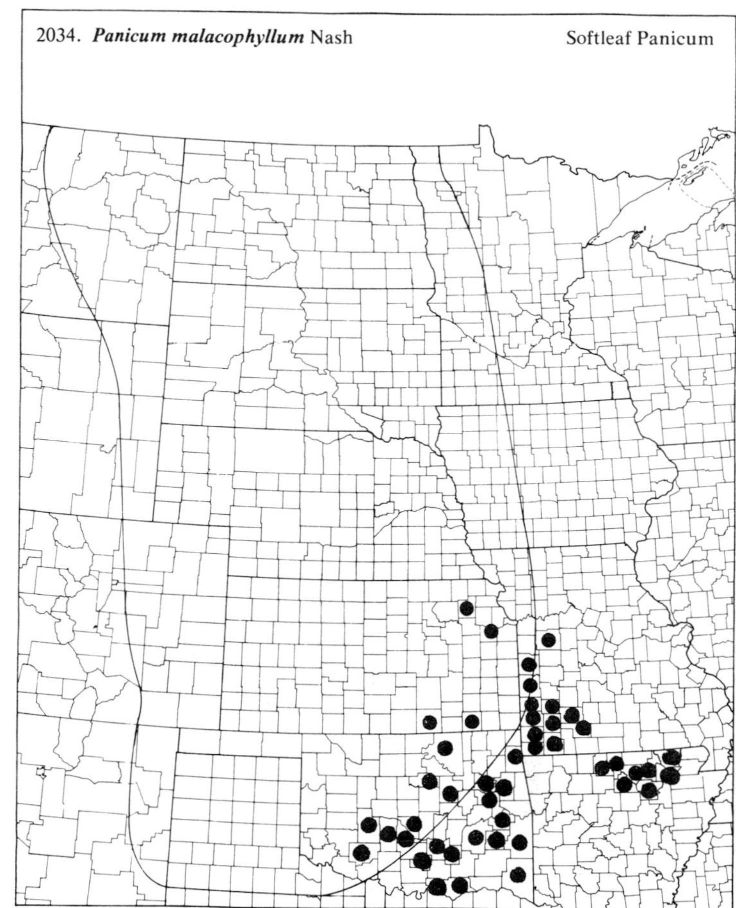
2034. ***Panicum malacophyllum*** Nash Softleaf Panicum

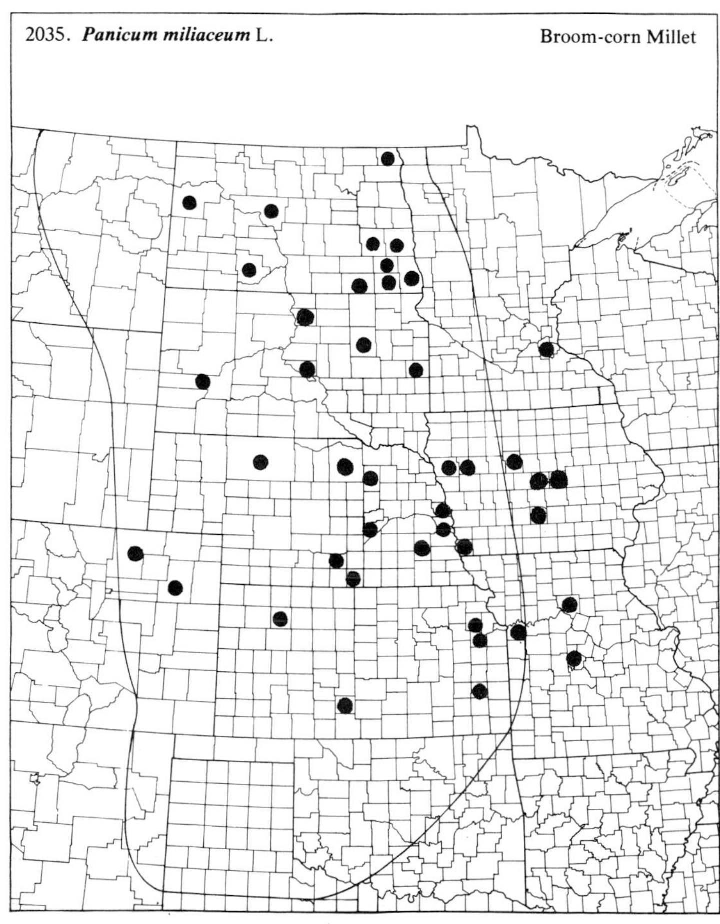
2035. ***Panicum miliaceum*** L. Broom-corn Millet

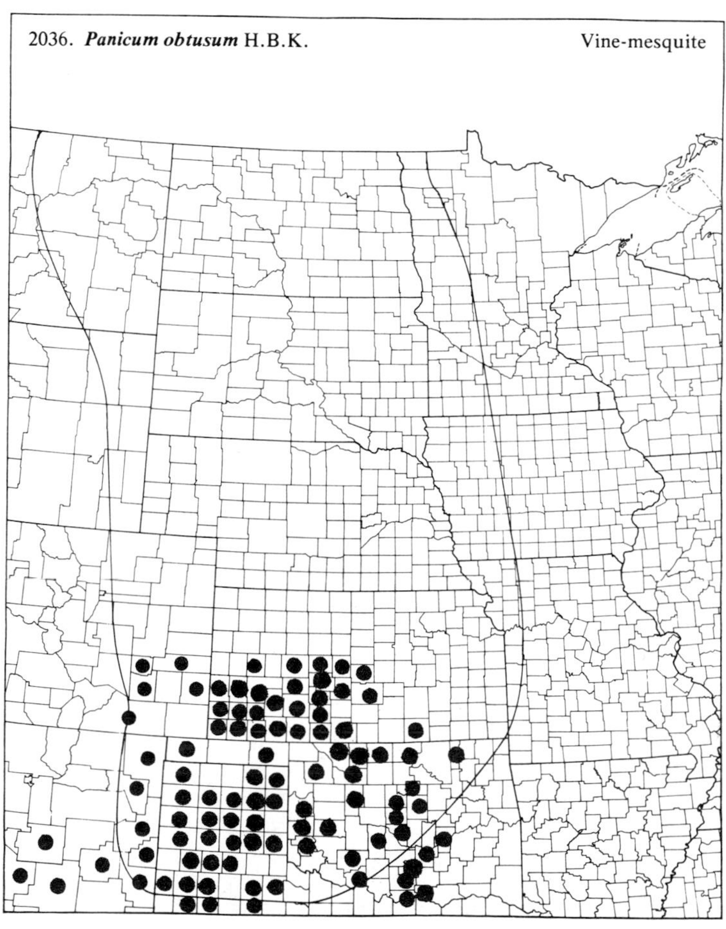
2036. ***Panicum obtusum*** H.B.K. Vine-mesquite

(Poaceae)

2037. ***Panicum oligosanthes*** Schult. var. ***scribnerianum*** (Nash) Fern. — Small Panicgrass

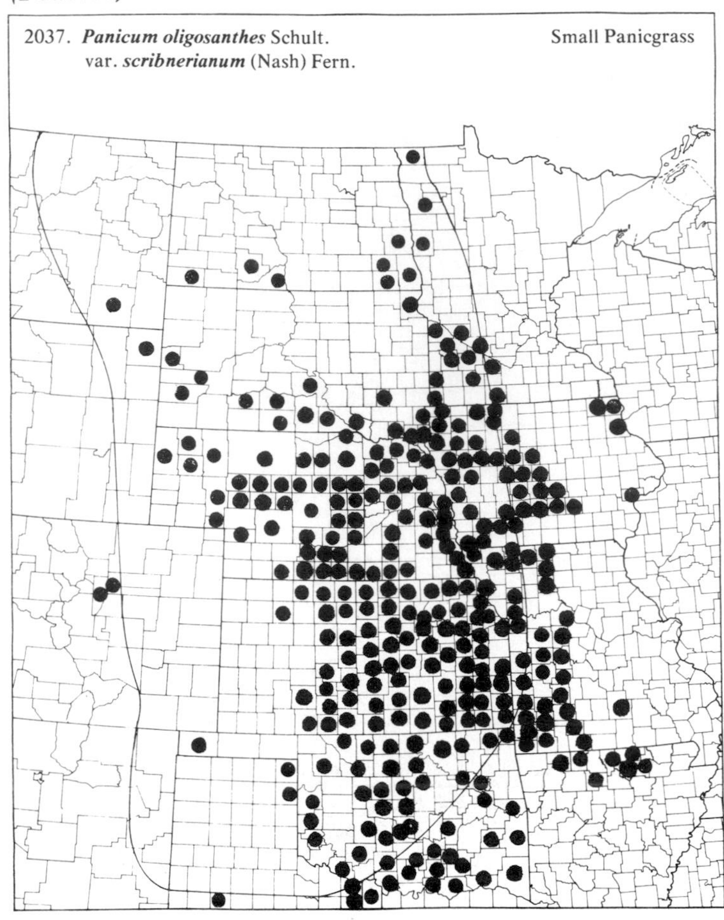

2038. ***Panicum perlongum*** Nash — Elongate Panicum

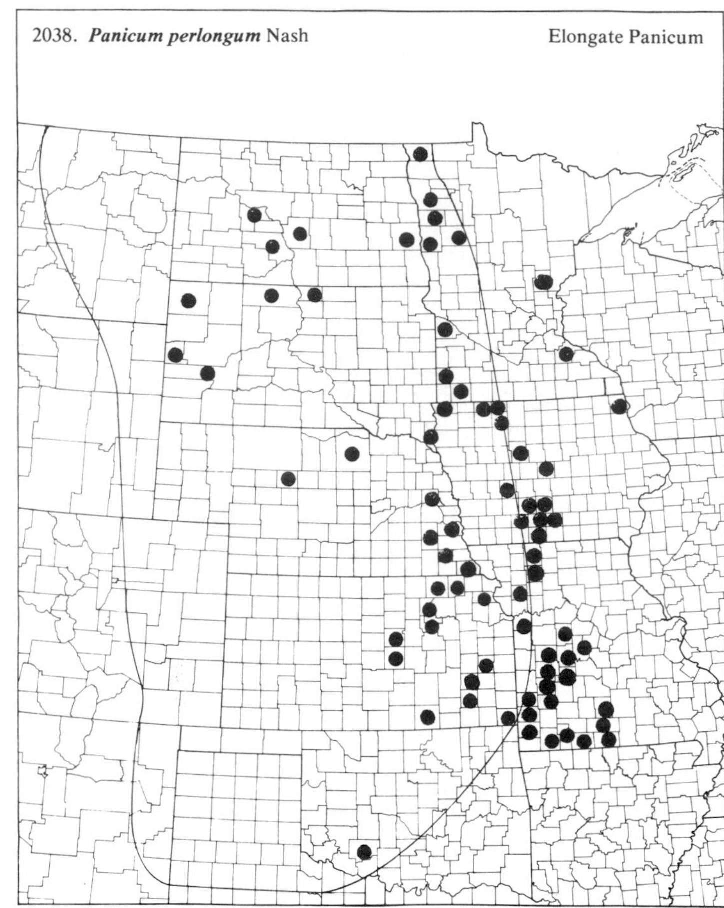

2039. ***Panicum philadelphicum*** Bernh.

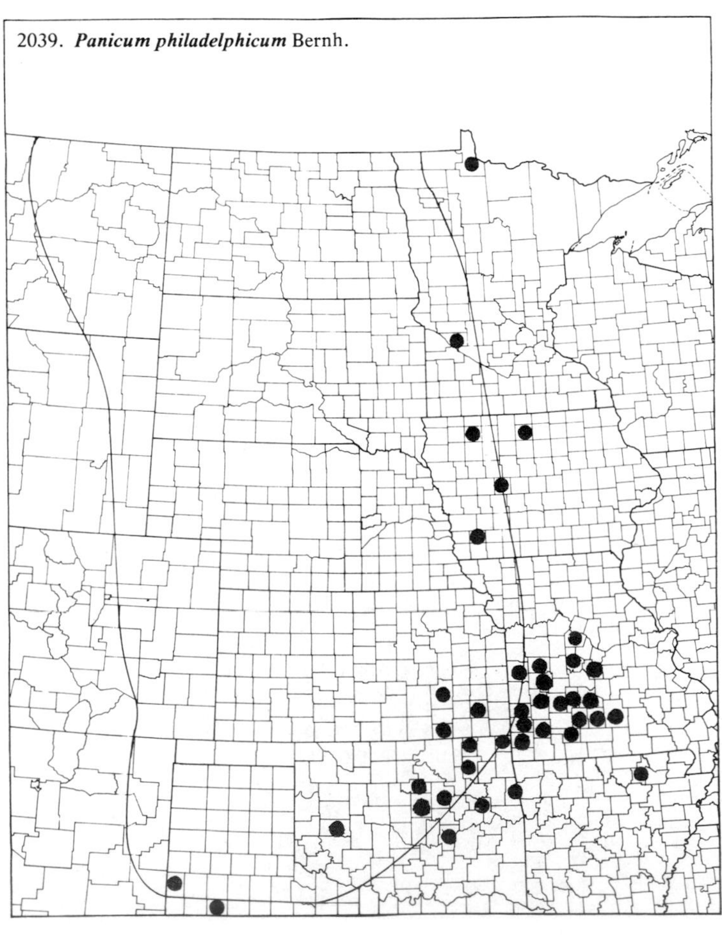

2040. ***Panicum praecocius*** Hitchc. & Chase — Early Panicum

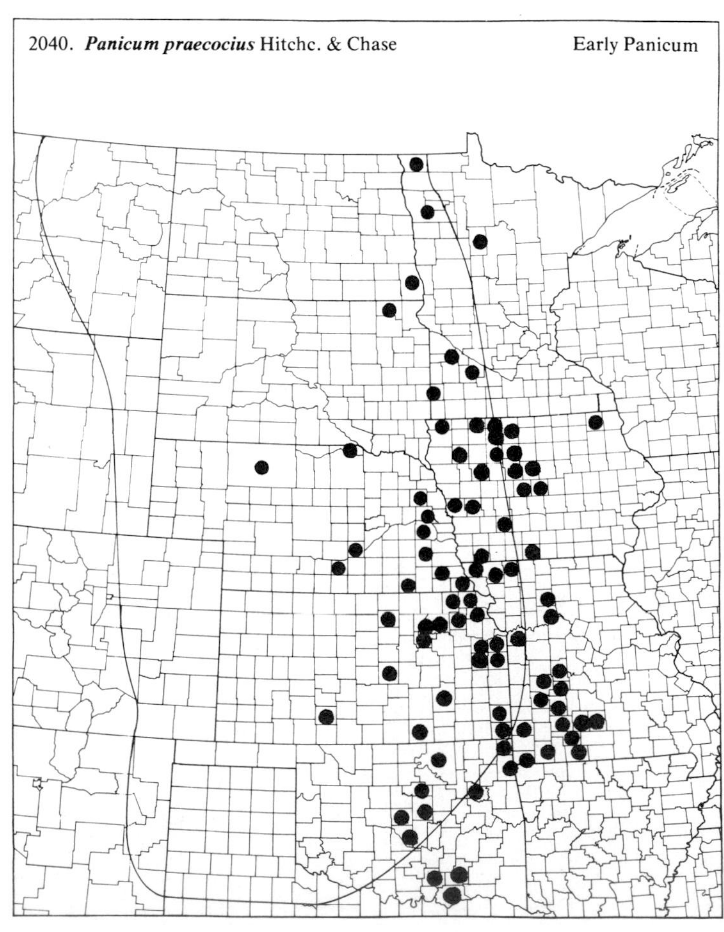

(Poaceae)

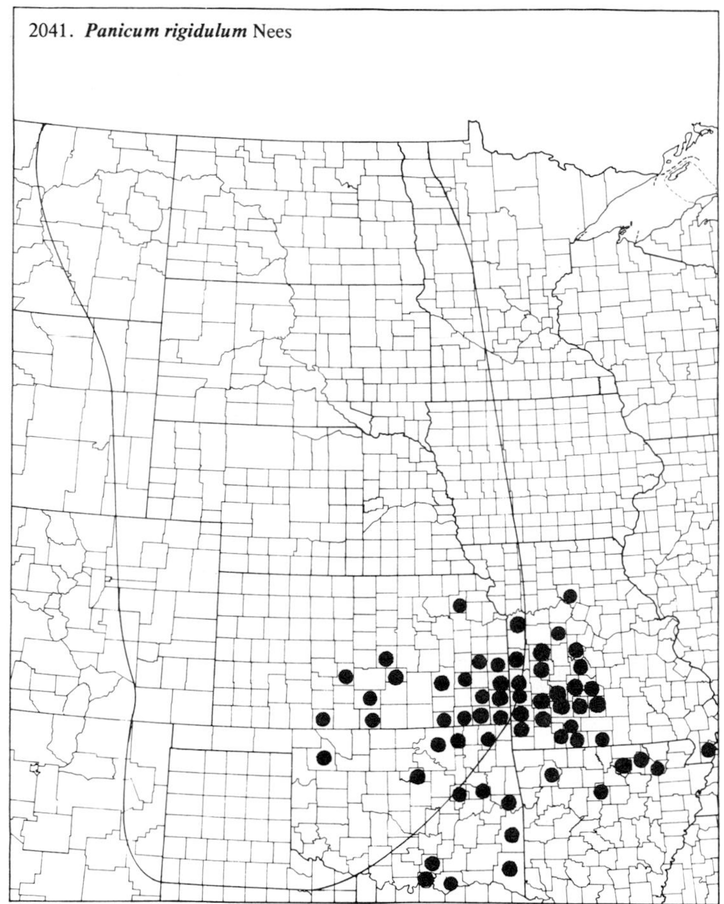
2041. Panicum rigidulum Nees

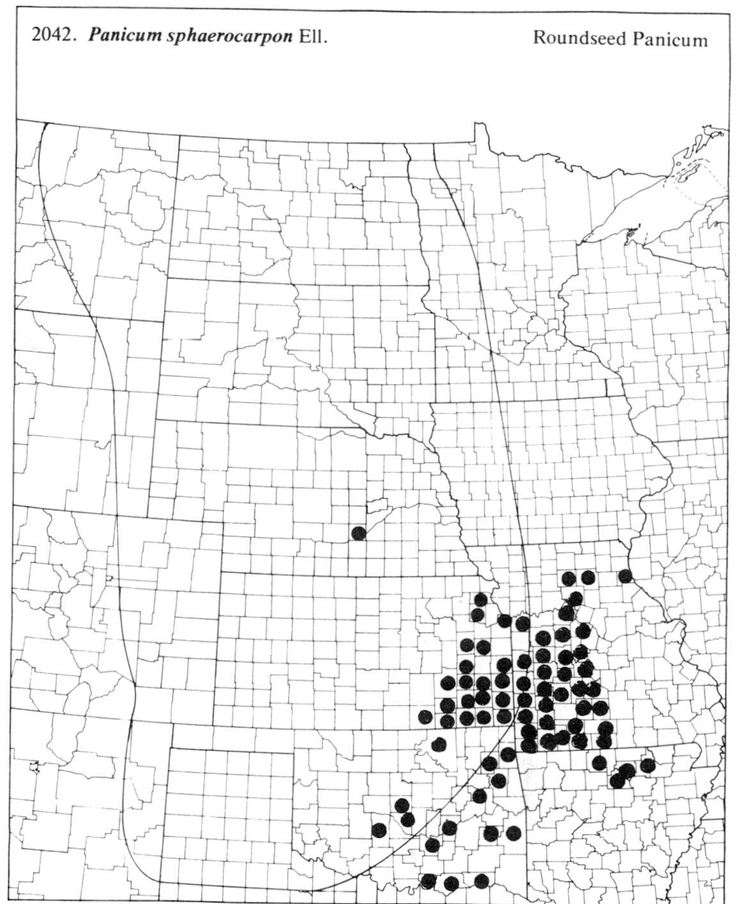
2042. Panicum sphaerocarpon Ell.
Roundseed Panicum

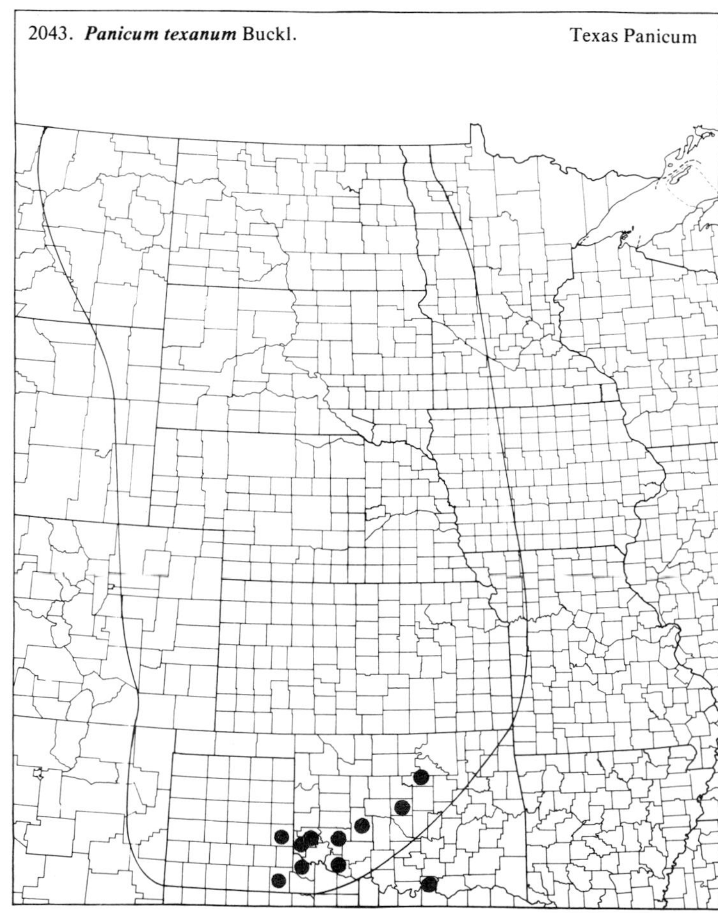
2043. Panicum texanum Buckl.
Texas Panicum

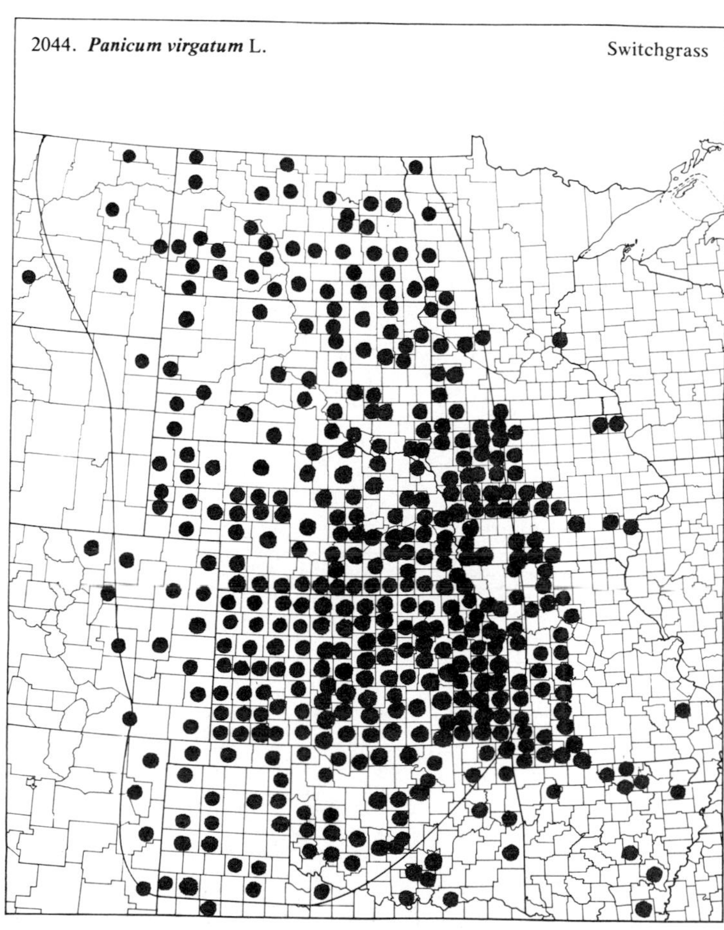
2044. Panicum virgatum L.
Switchgrass

(Poaceae)

2045. ***Panicum wilcoxianum*** Vasey — Wilcox Panicum

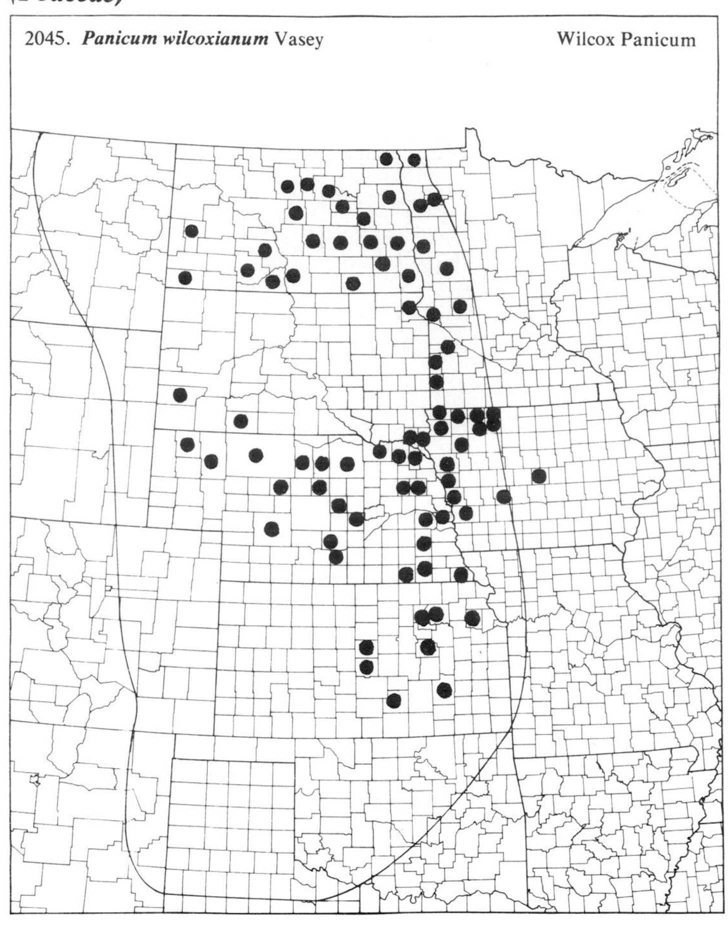

2046. ***Paspalum distichum*** L. — Knotgrass

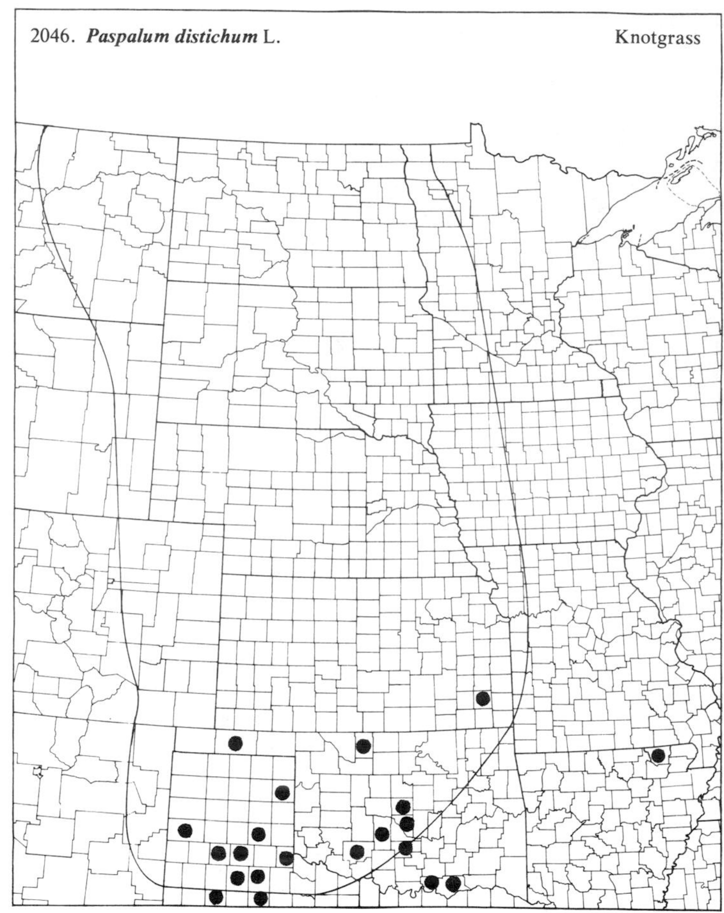

2047. ***Paspalum floridanum*** Michx. var. ***floridanum*** — Florida Paspalum

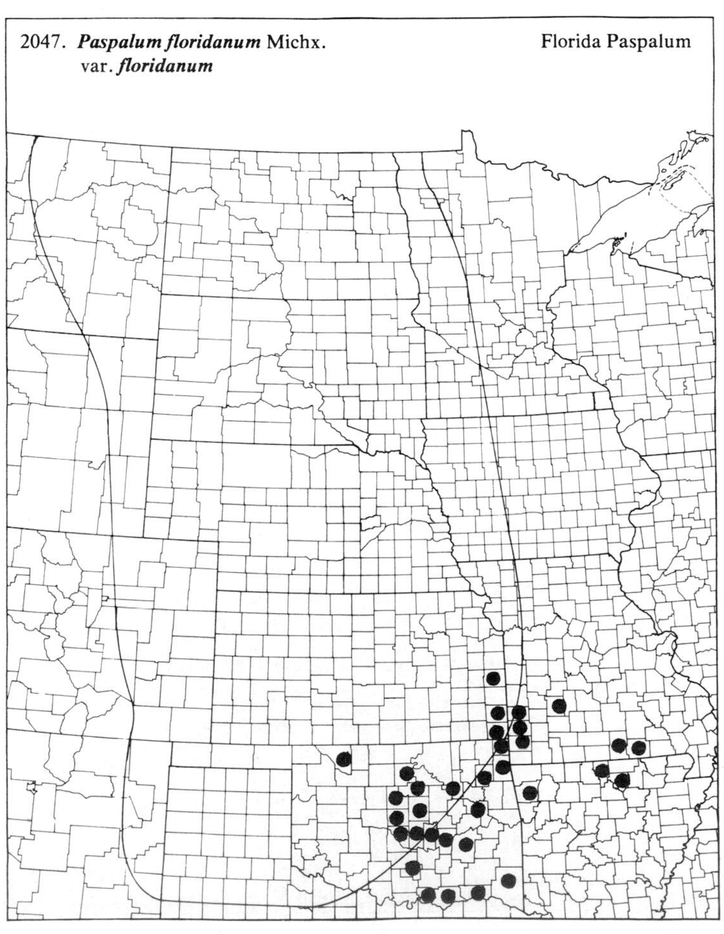

2048. ***Paspalum floridanum*** Michx. var. ***glabratum*** Engelm. — Smooth Florida Paspalum

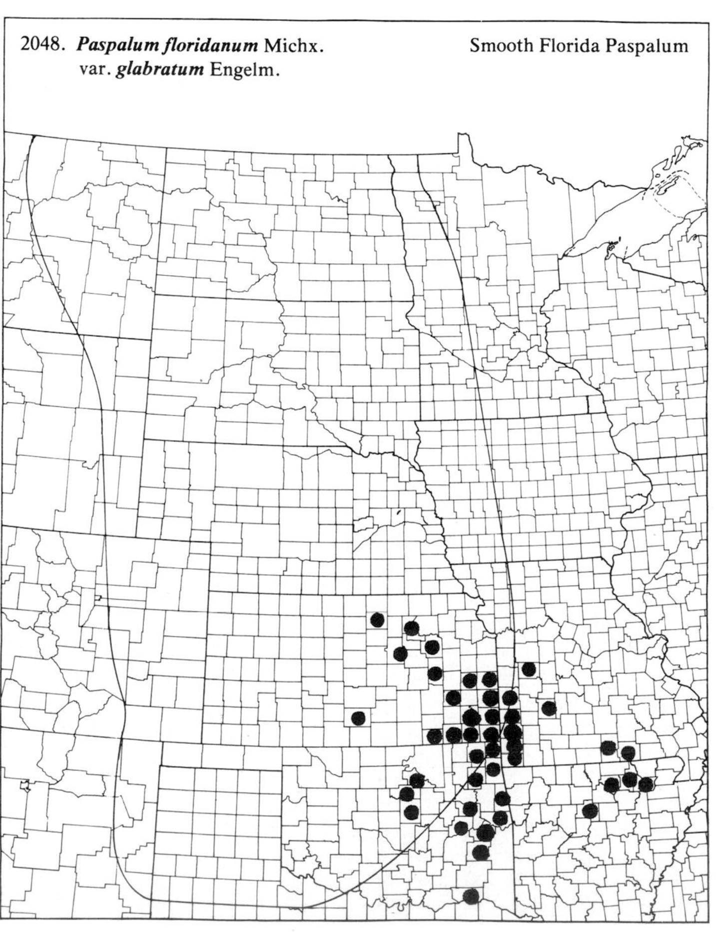

(Poaceae)

2049. **Paspalum fluitans** (Ell.) Kunth — Horsetail Paspalum

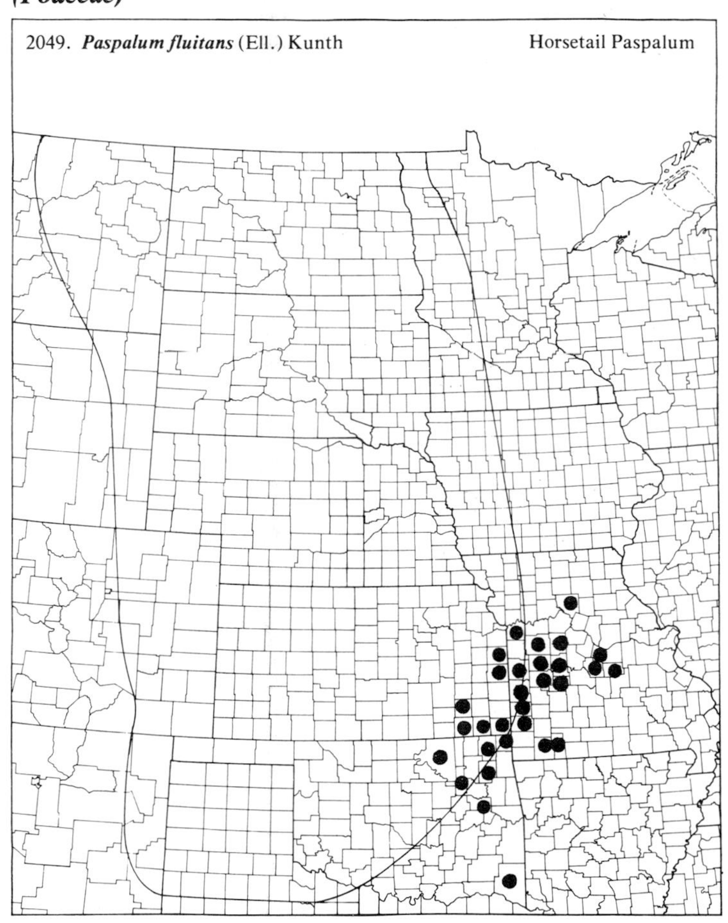

2050. **Paspalum laeve** Michx. var. **circulare** (Nash) Fern. — Field Paspalum

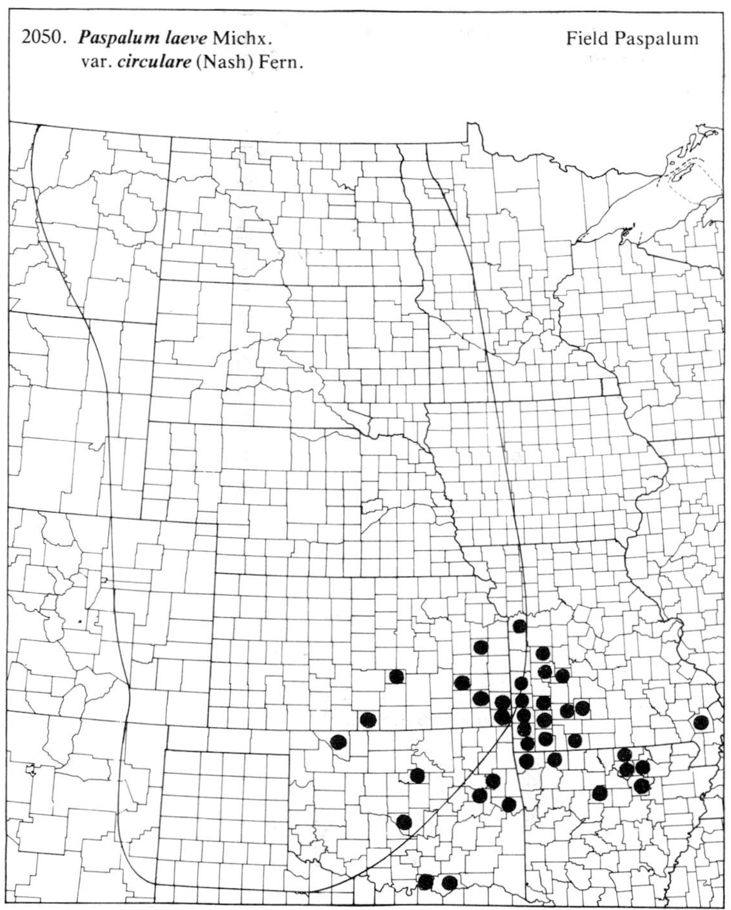

2051. **Paspalum pubiflorum** Rupr. var. **glabrum** Vasey — Hairyseed Paspalum

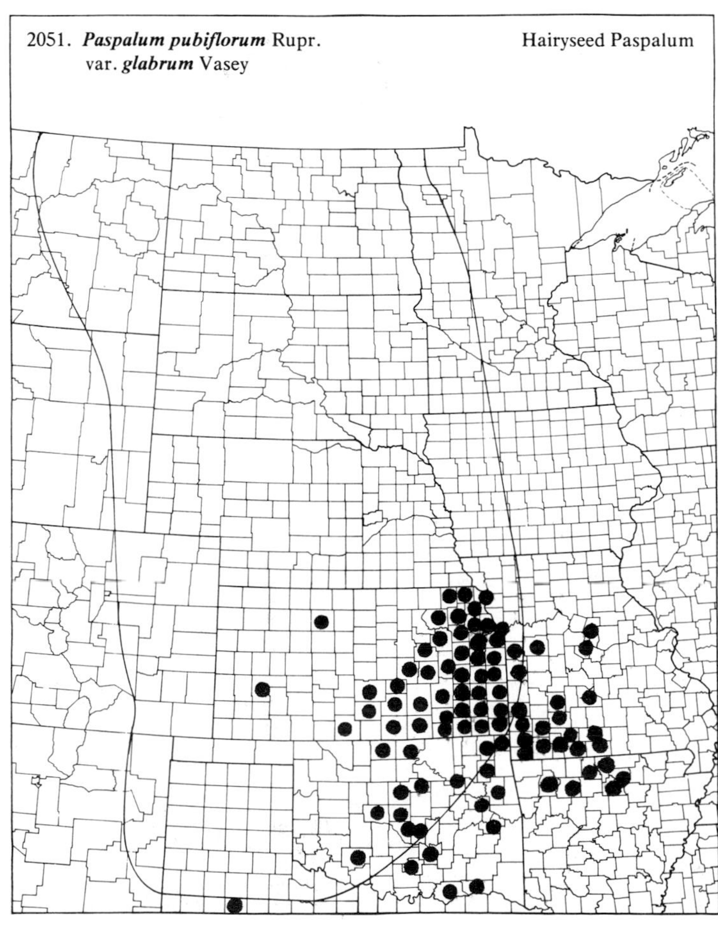

2052. **Paspalum setaceum** Michx. var. **muhlenbergii** (Nash) D. Banks

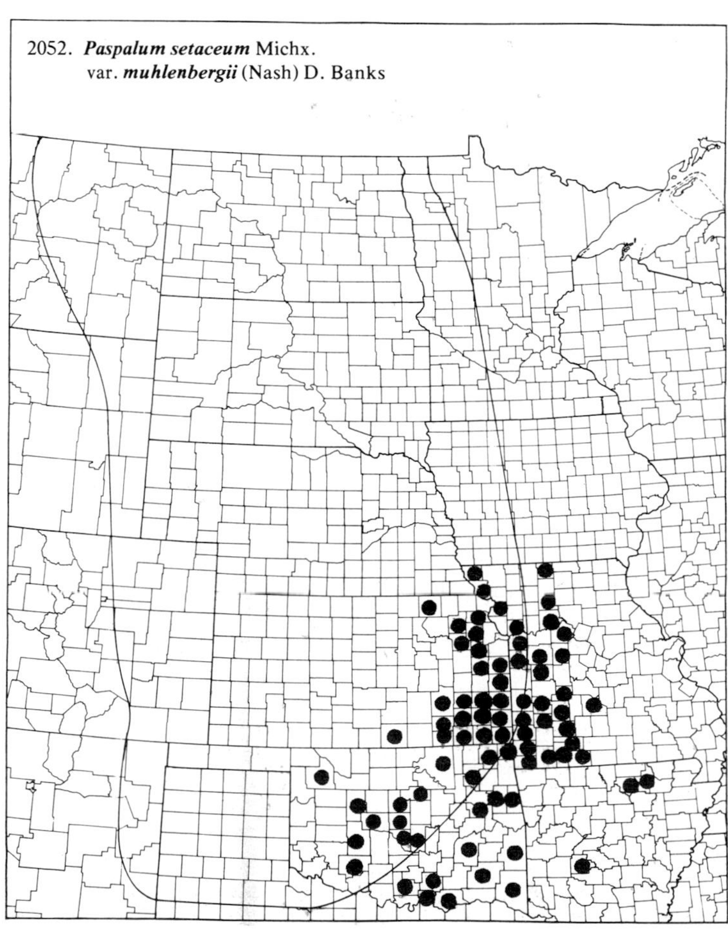

(Poaceae)

2053. ***Paspalum setaceum*** Michx. var. ***stramineum*** (Nash) D. Banks

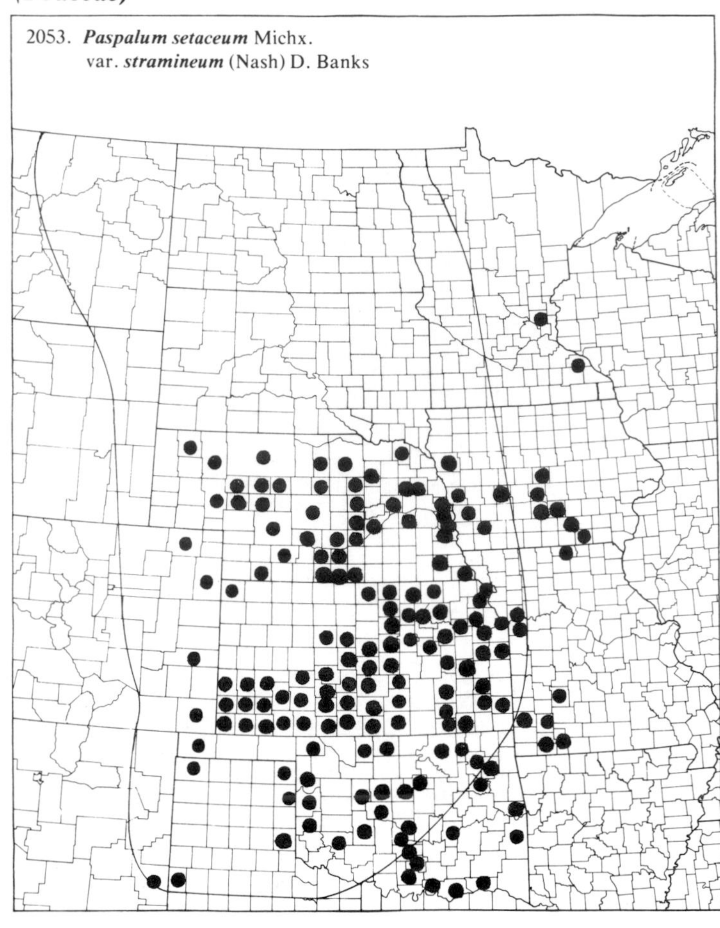

2054. ***Phalaris arundinacea*** L. Reed Canarygrass

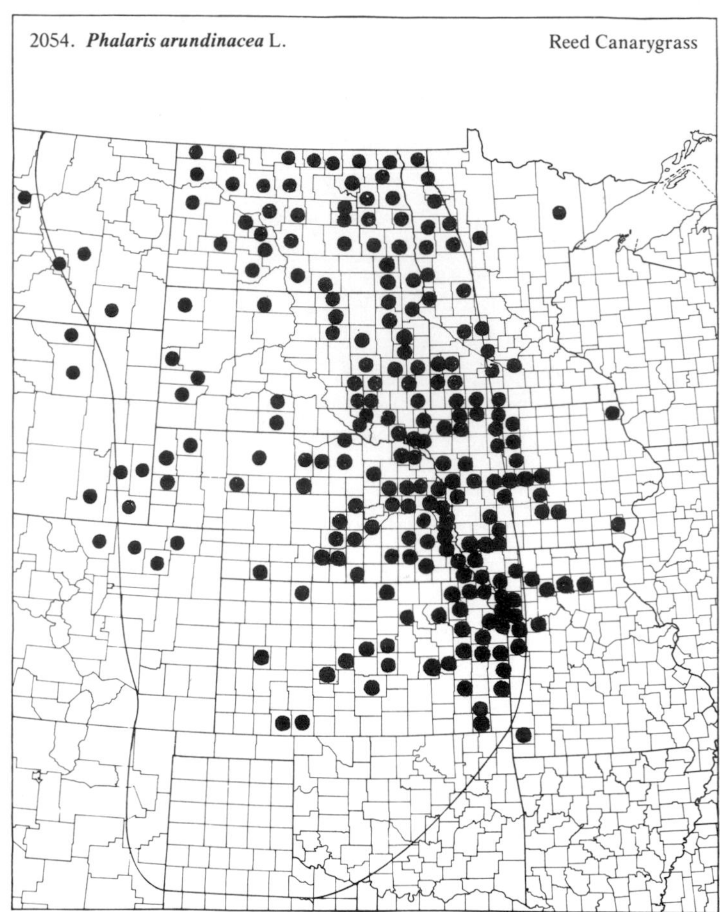

2055. ***Phalaris canariensis*** L. Canarygrass

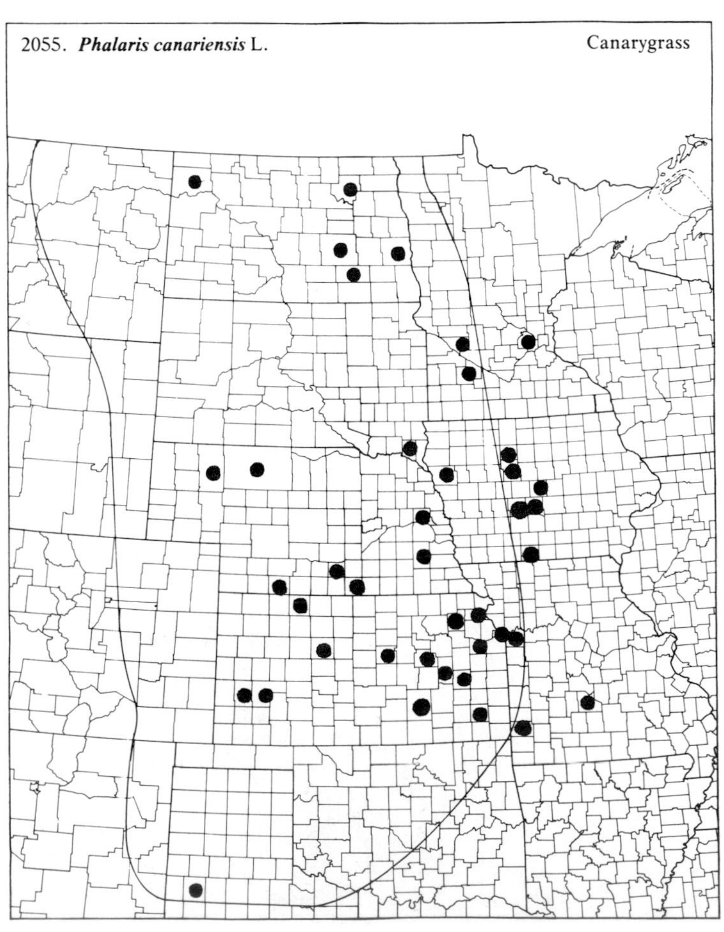

2056. ***Phalaris caroliniana*** Walt. Maygrass

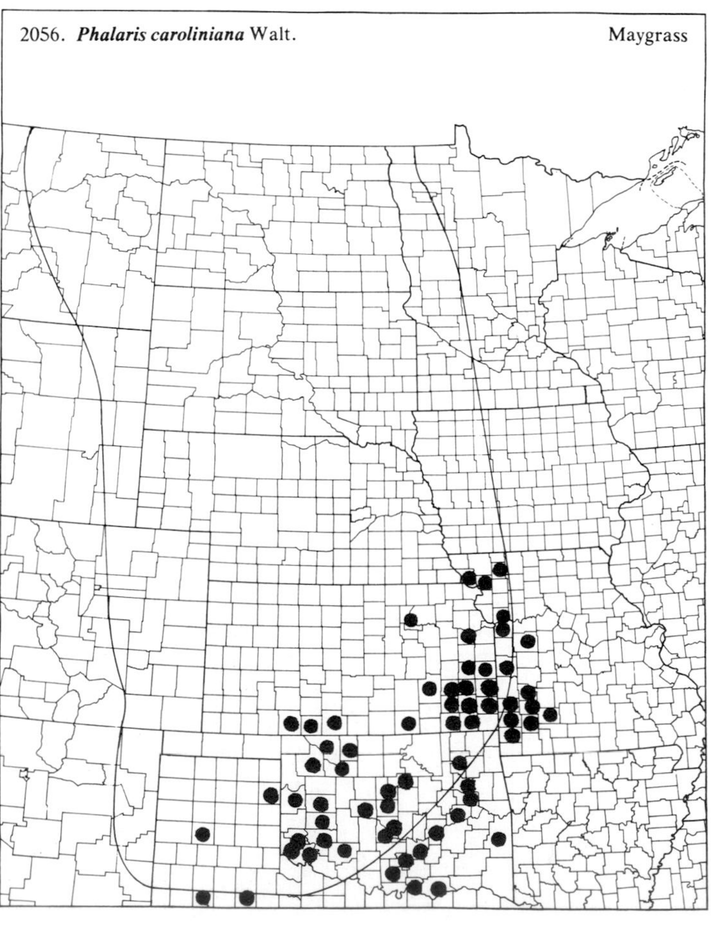

(Poaceae)

2057. ***Phleum pratense*** L. Timothy

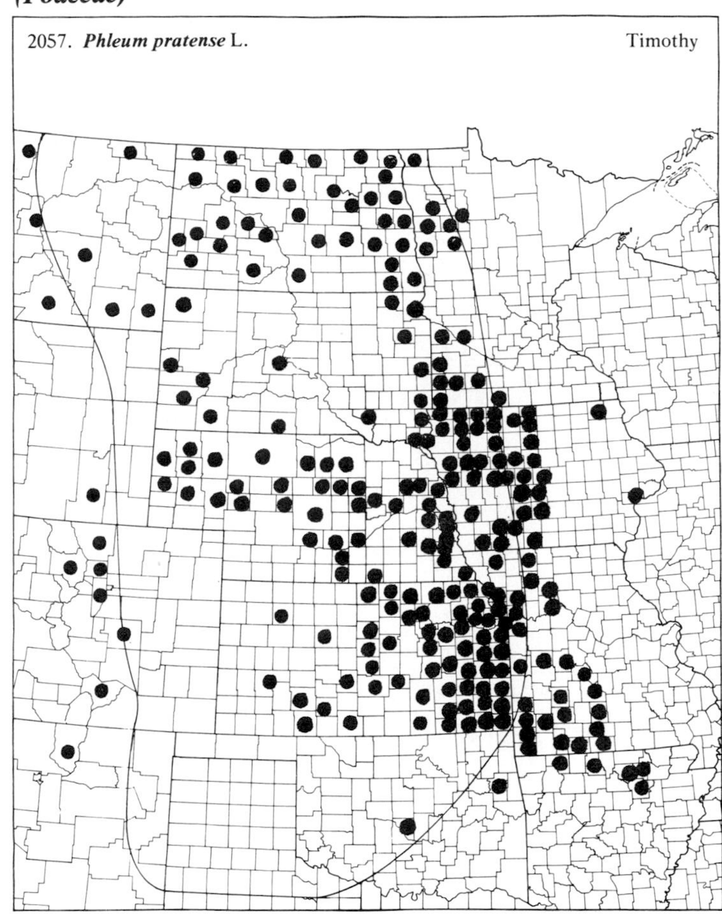

2058. ***Phragmites australis*** (Cav.) Trin. Common Reed

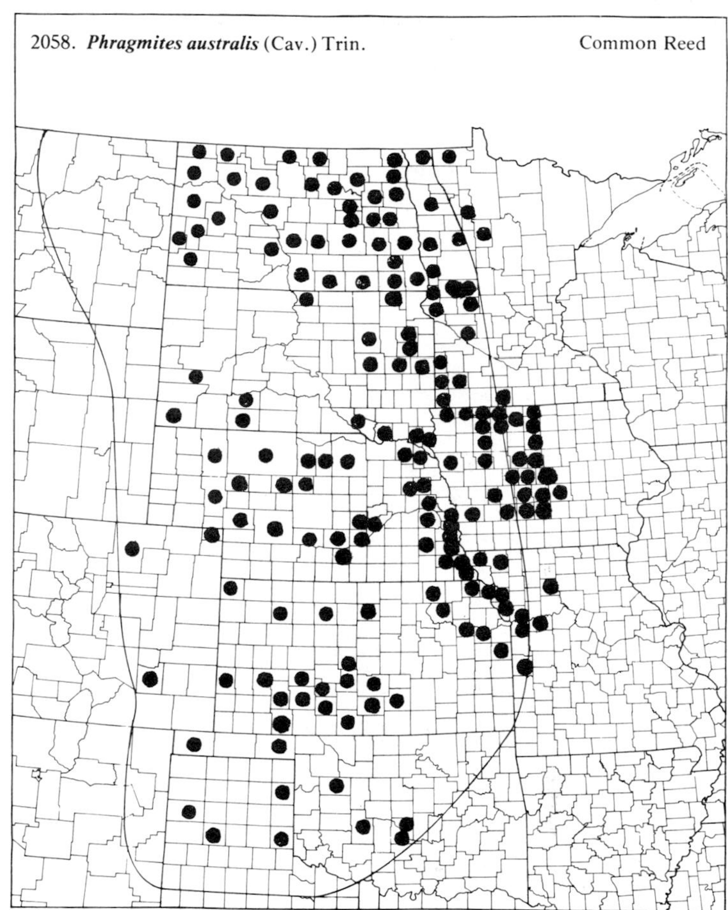

2059. ***Poa annua*** L. Annual Bluegrass

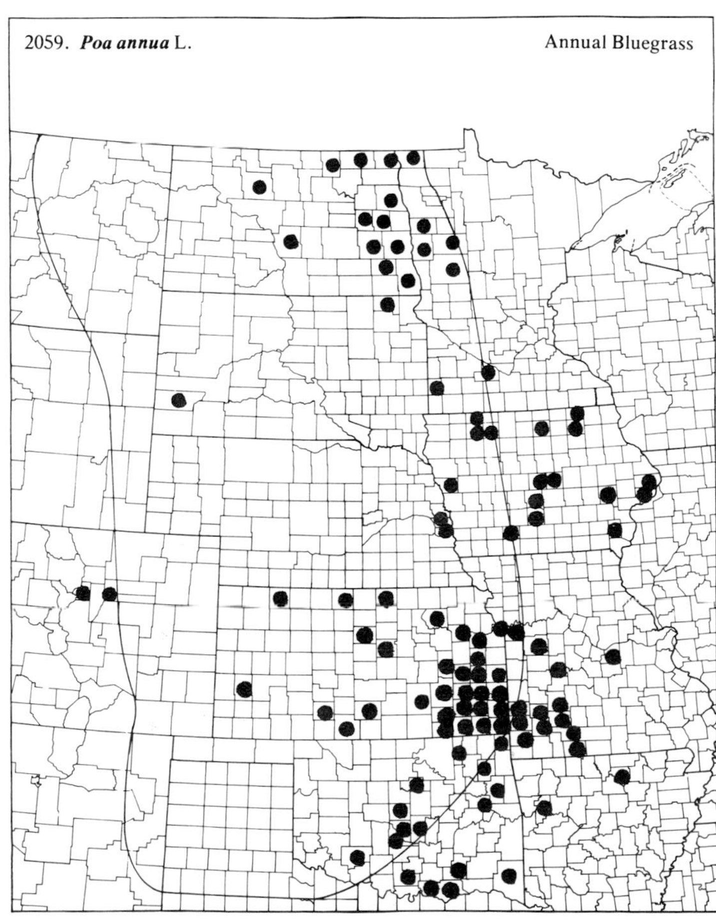

2060. ***Poa arachnifera*** Torr. Texas Bluegrass

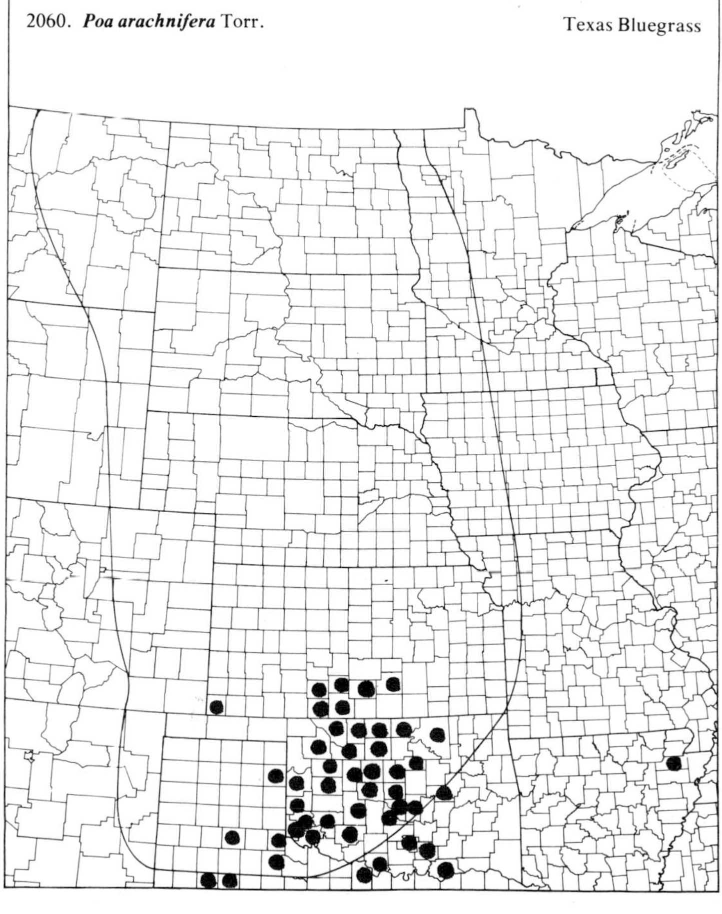

(Poaceae)

2061. ***Poa arida*** Vasey — Plains Bluegrass

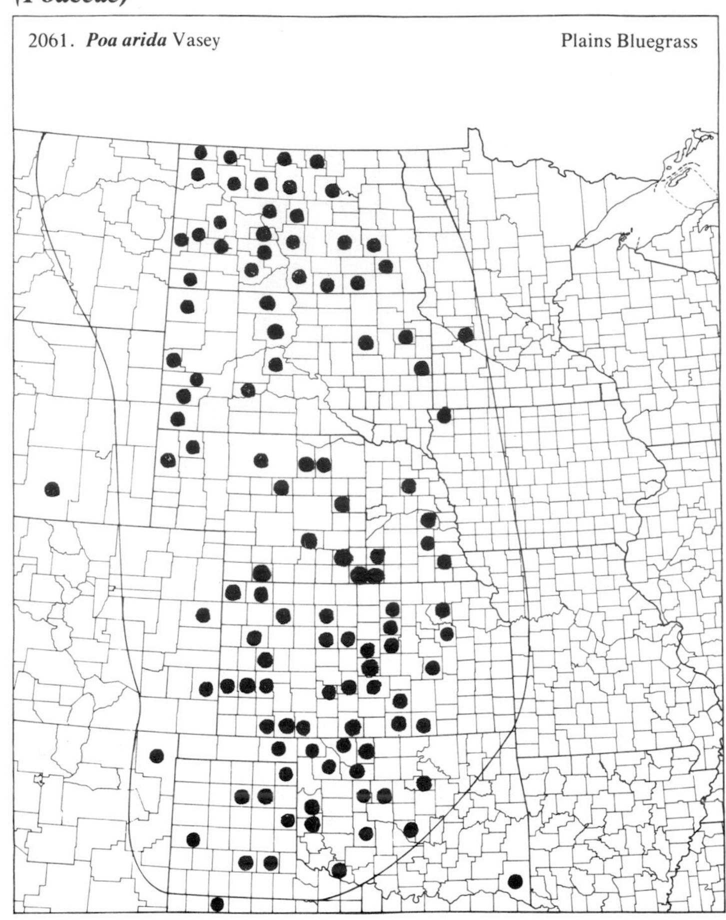

2062. ***Poa canbyi*** (Scribn.) Piper — Canby's Bluegrass

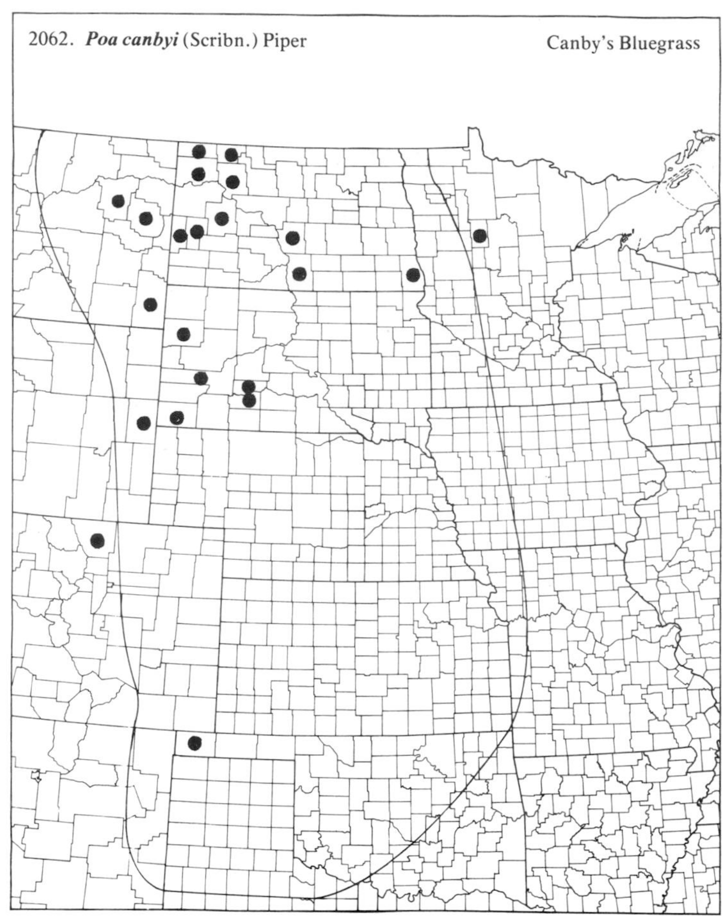

2063. ***Poa chapmaniana*** Scribn. — Chapman Bluegrass

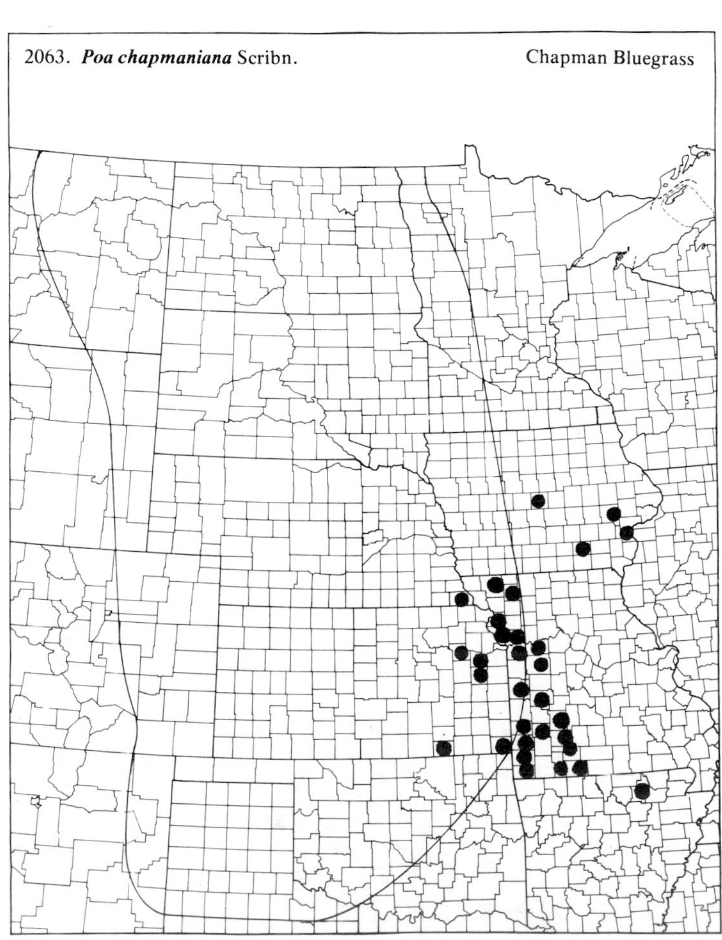

2064. ***Poa compressa*** L. — Canada Bluegrass

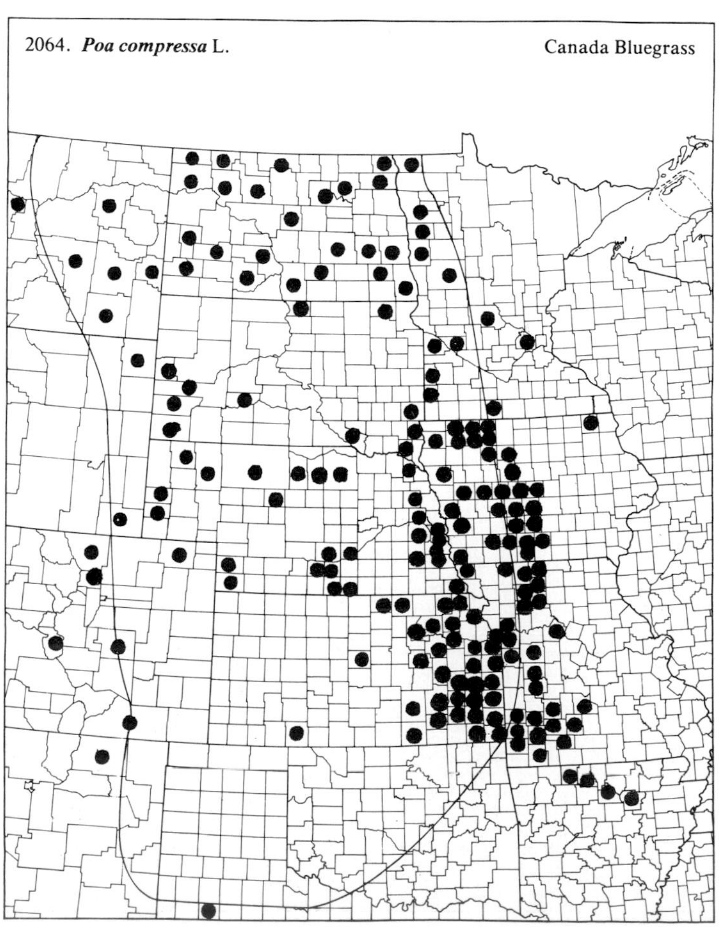

(Poaceae)

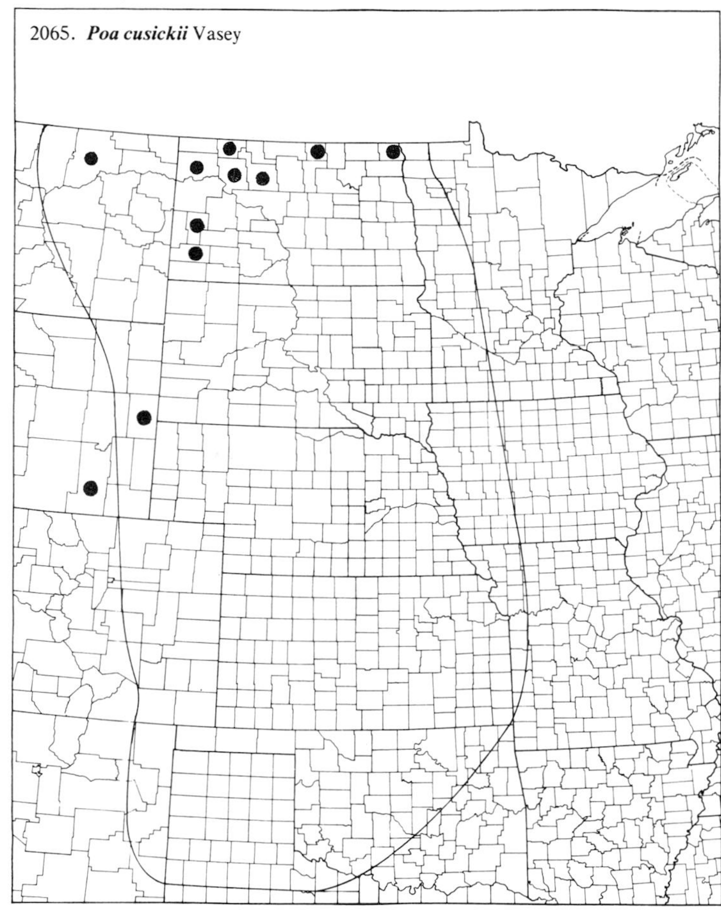
2065. **Poa cusickii** Vasey

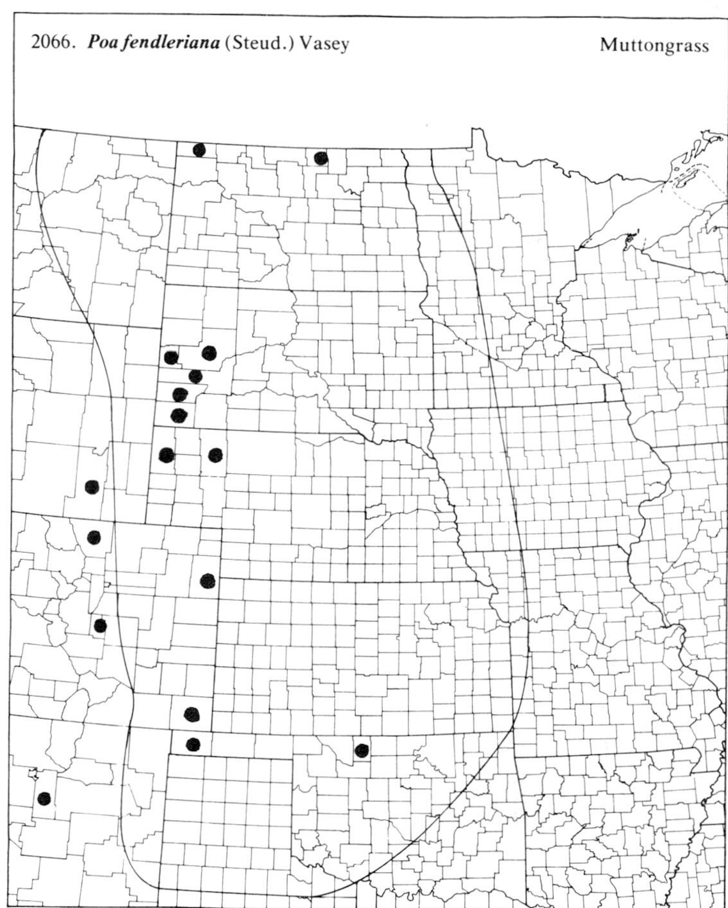
2066. **Poa fendleriana** (Steud.) Vasey — Muttongrass

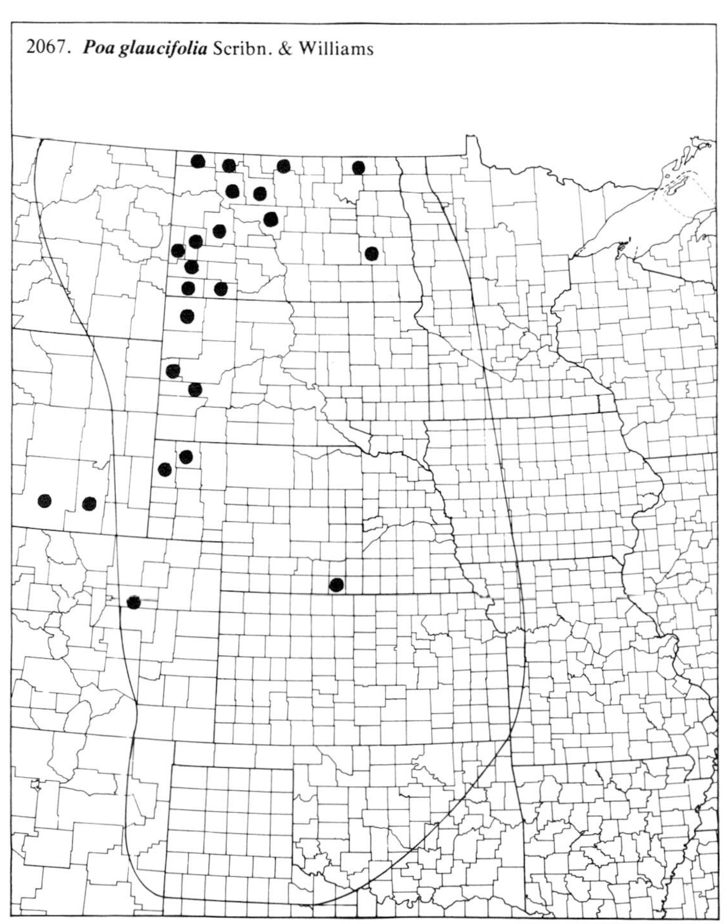
2067. **Poa glaucifolia** Scribn. & Williams

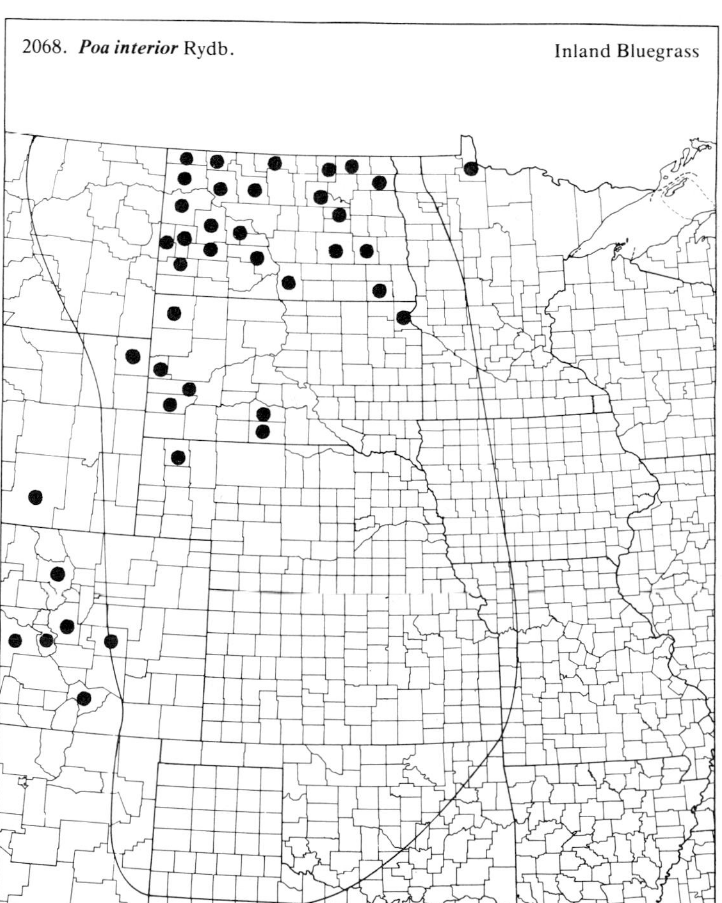
2068. **Poa interior** Rydb. — Inland Bluegrass

(Poaceae)

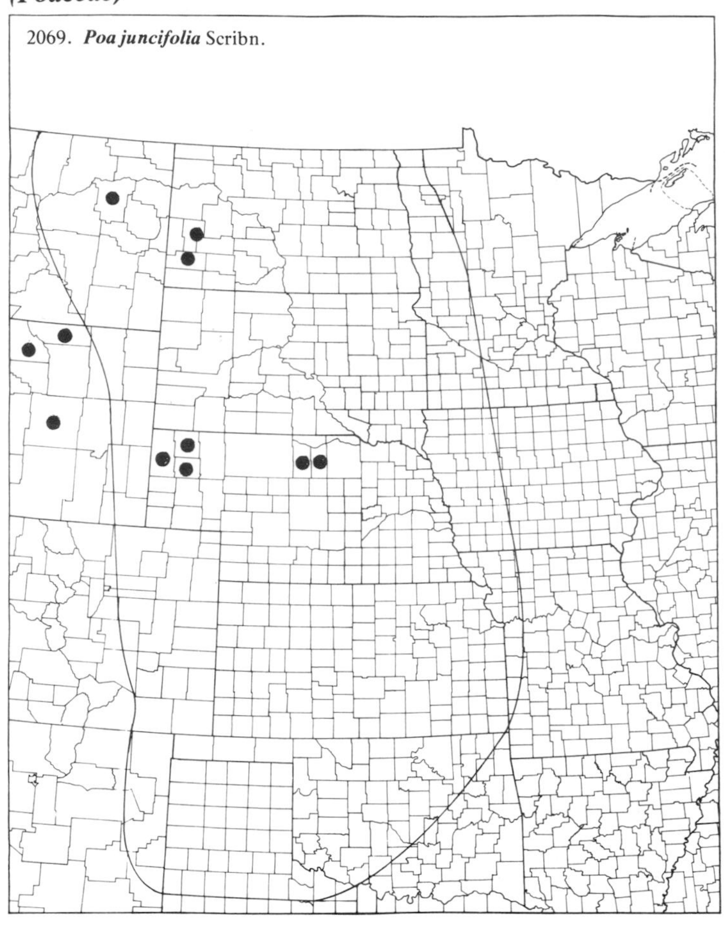
2069. *Poa juncifolia* Scribn.

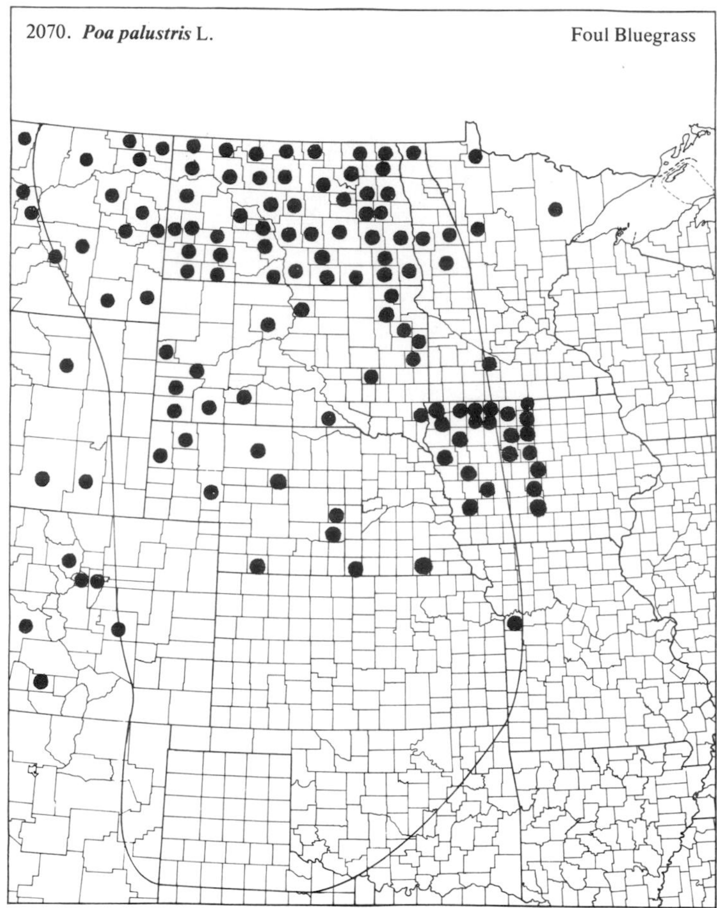
2070. *Poa palustris* L.
Foul Bluegrass

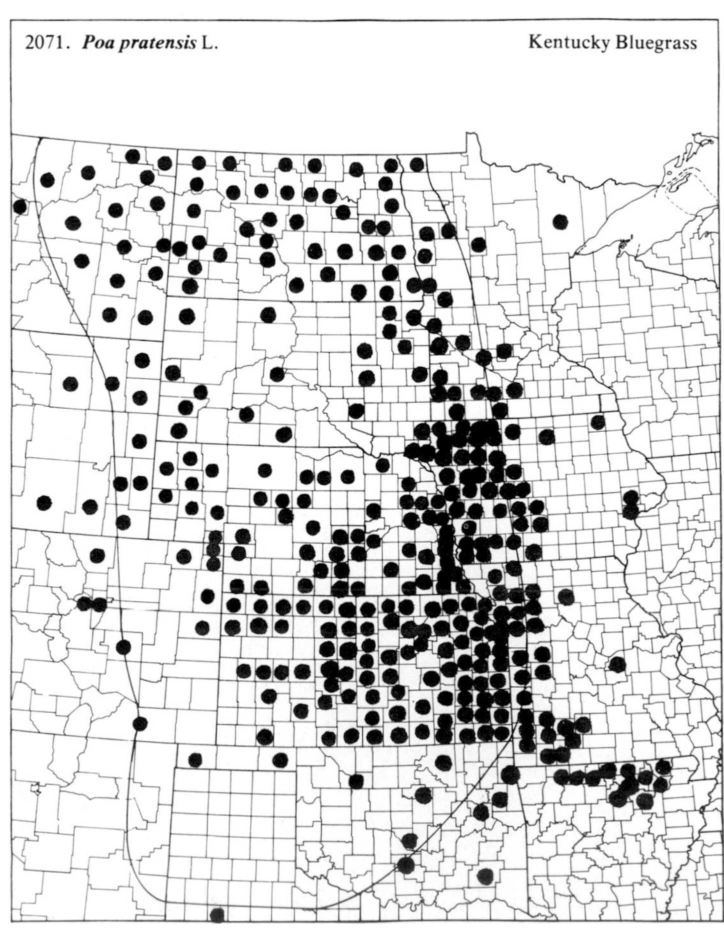
2071. *Poa pratensis* L.
Kentucky Bluegrass

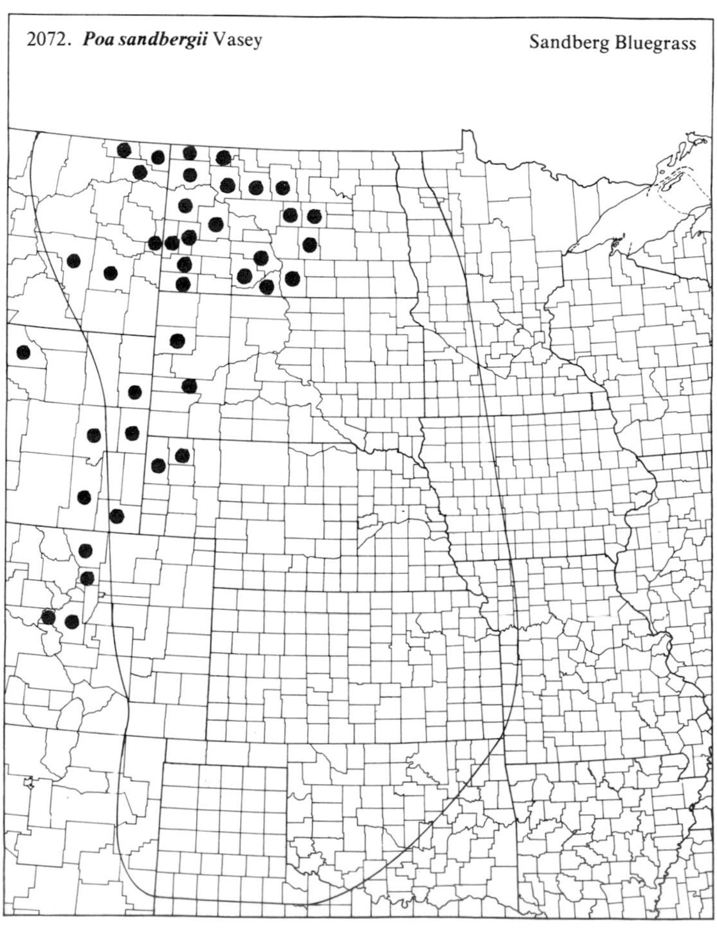
2072. *Poa sandbergii* Vasey
Sandberg Bluegrass

(Poaceae)

2073. ***Poa sylvestris*** Gray — Woodland Bluegrass

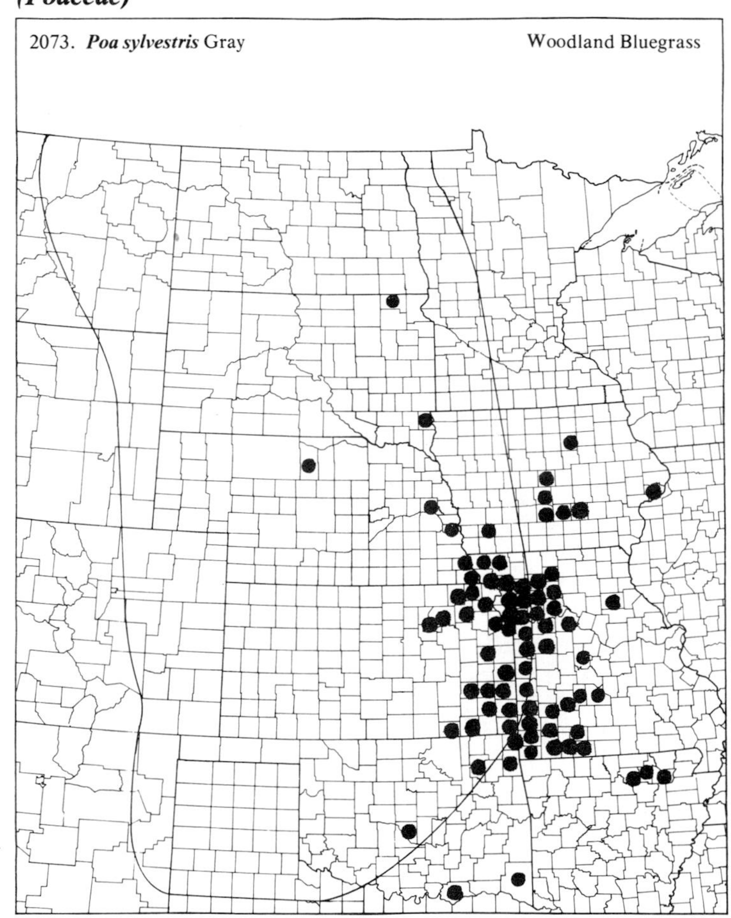

2074. ***Polypogon monspeliensis*** (L.) Desf. — Rabbitfoot-grass

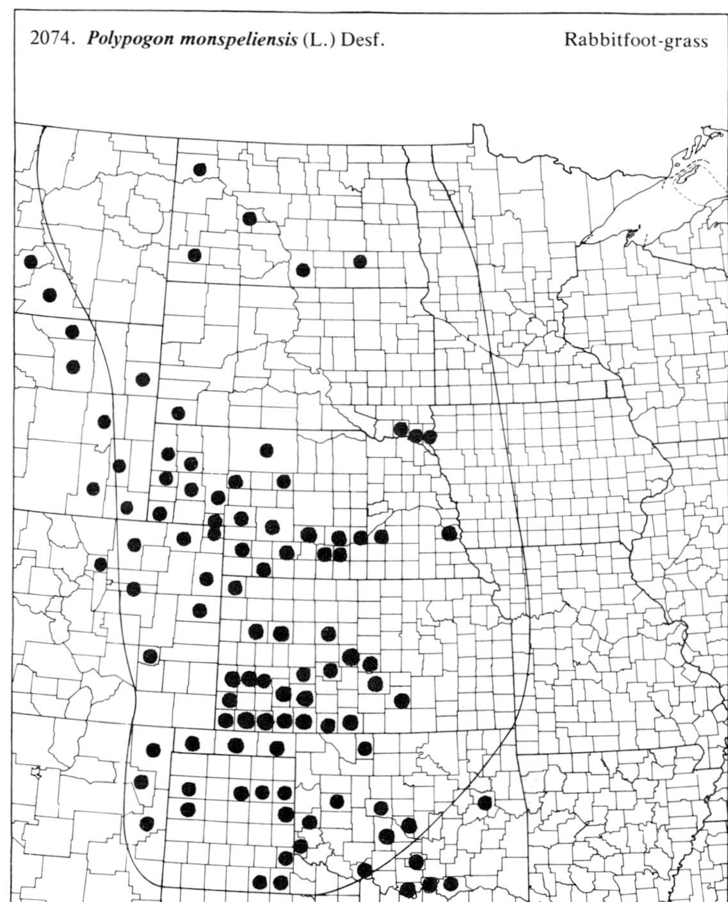

2075. ***Puccinellia cusickii*** Weath.

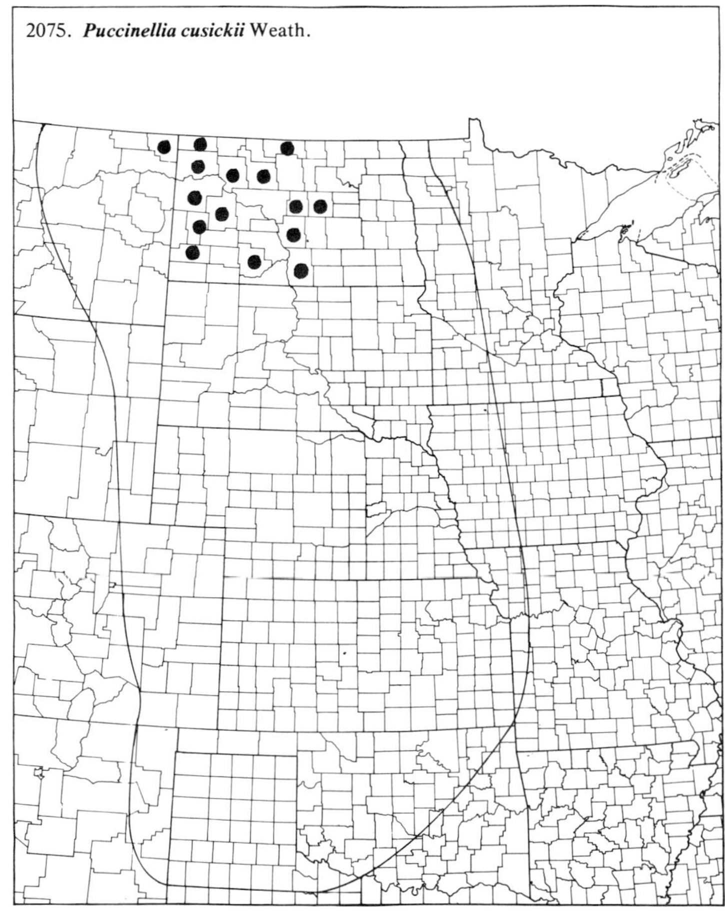

2076. ***Puccinellia distans*** (L.) Parl.

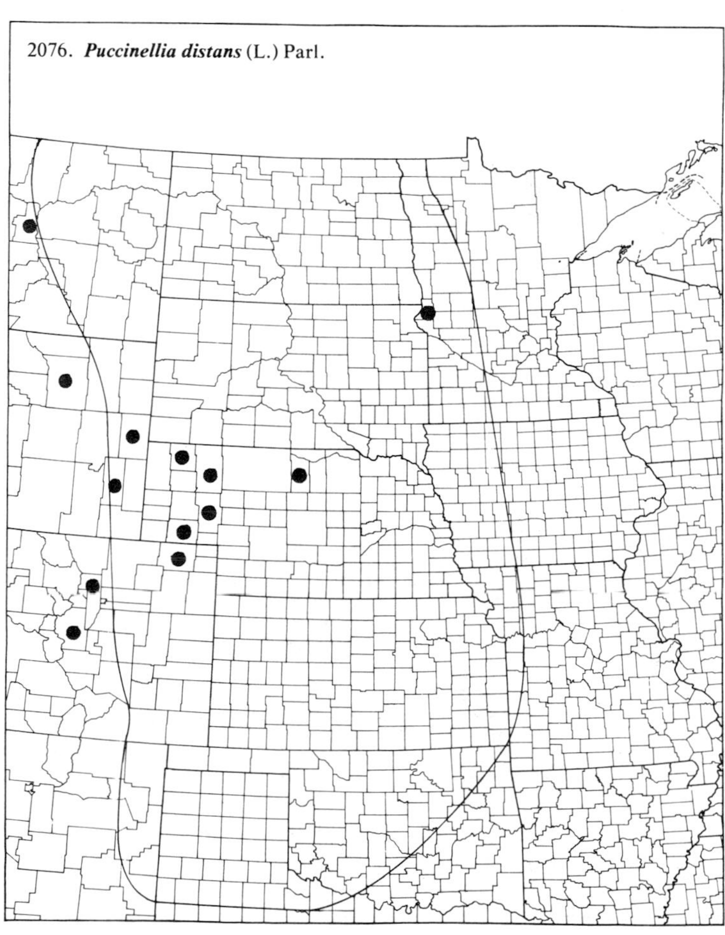

(Poaceae)

2077. ***Puccinellia nuttalliana*** (Schult.) Hitchc. Alkali-grass

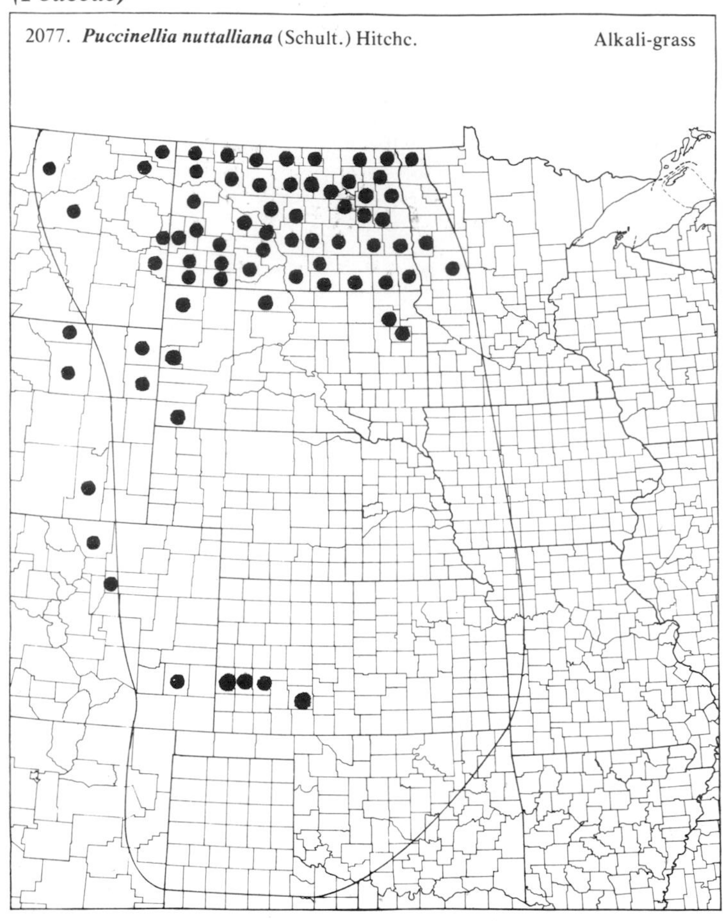

2078. ***Redfielda flexuosa*** (Thurb.) Vasey Blowout-grass

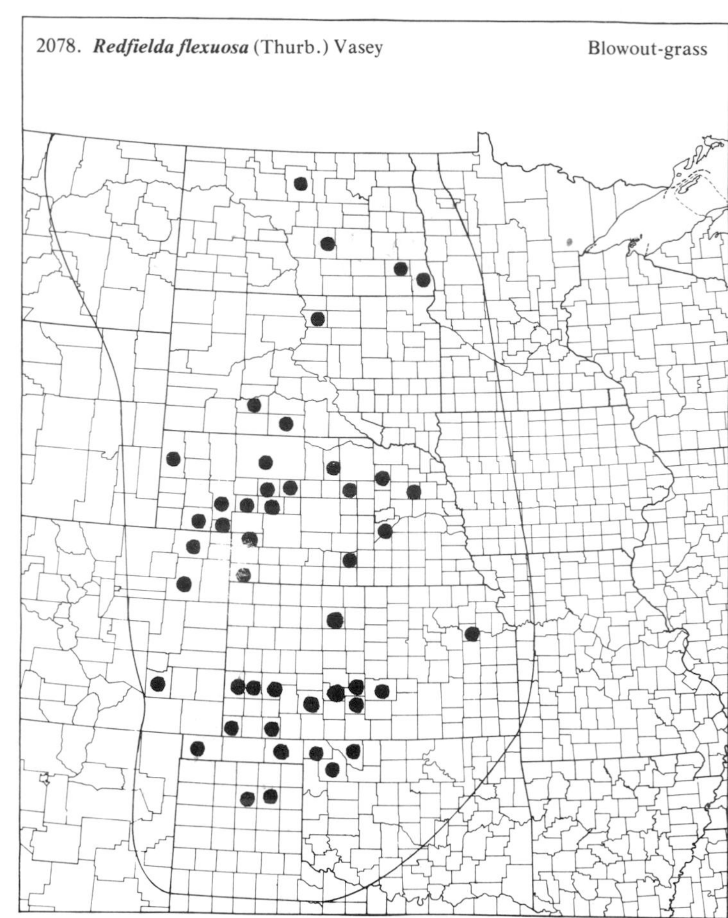

2079. ***Schedonnardus paniculatus*** (Nutt.) Trel. Tumblegrass

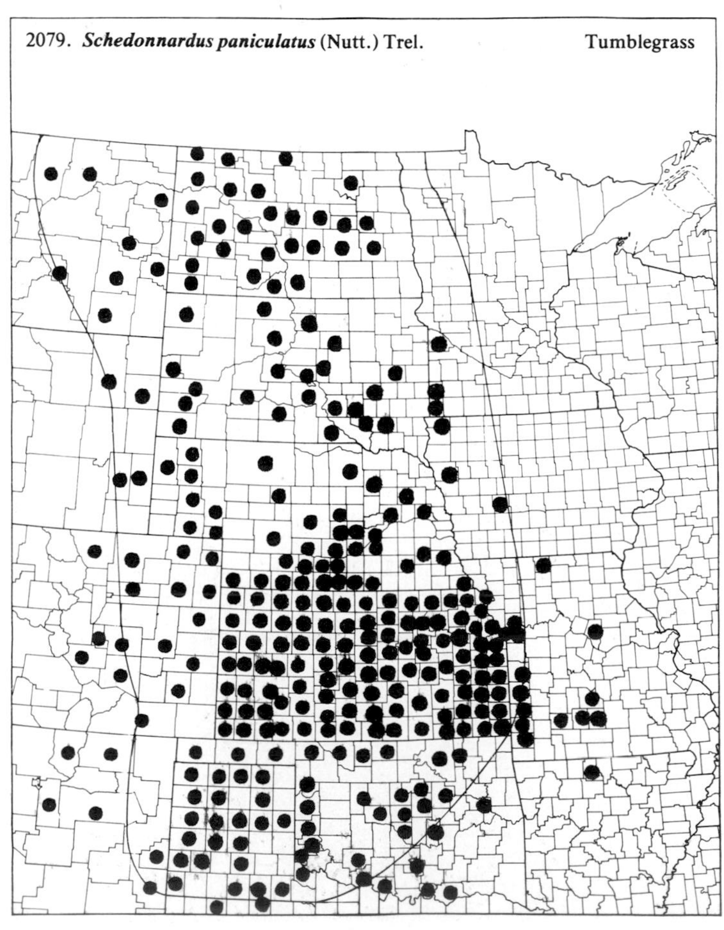

2080. ***Schizachne purpurascens*** (Torr.) Swall. False Melic

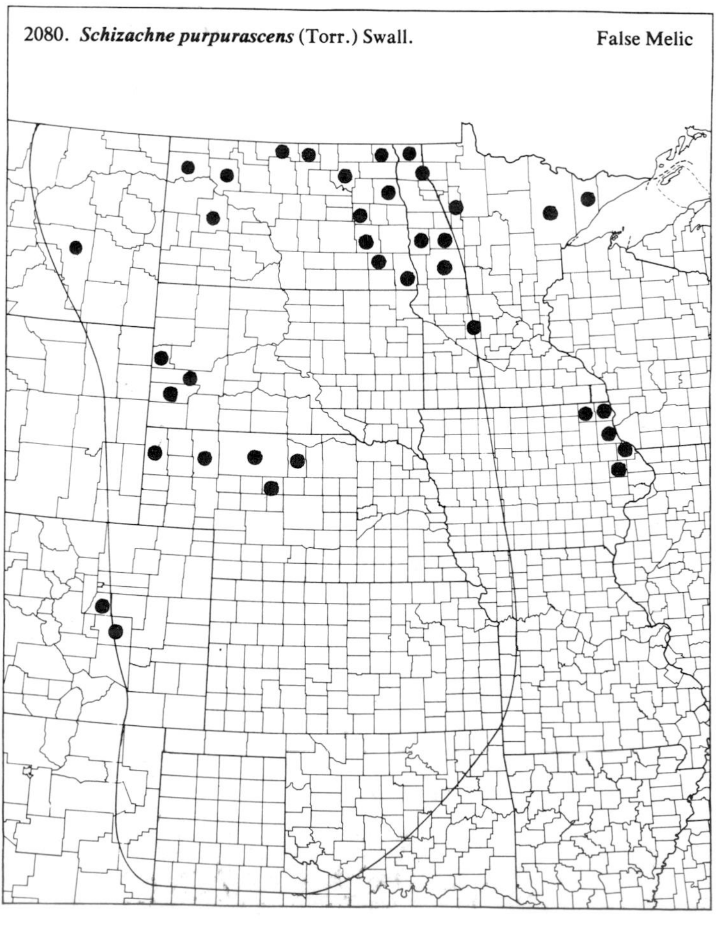

(Poaceae)

2081. ***Scolochloa festucacea*** (Willd.) Link — Sprangletop

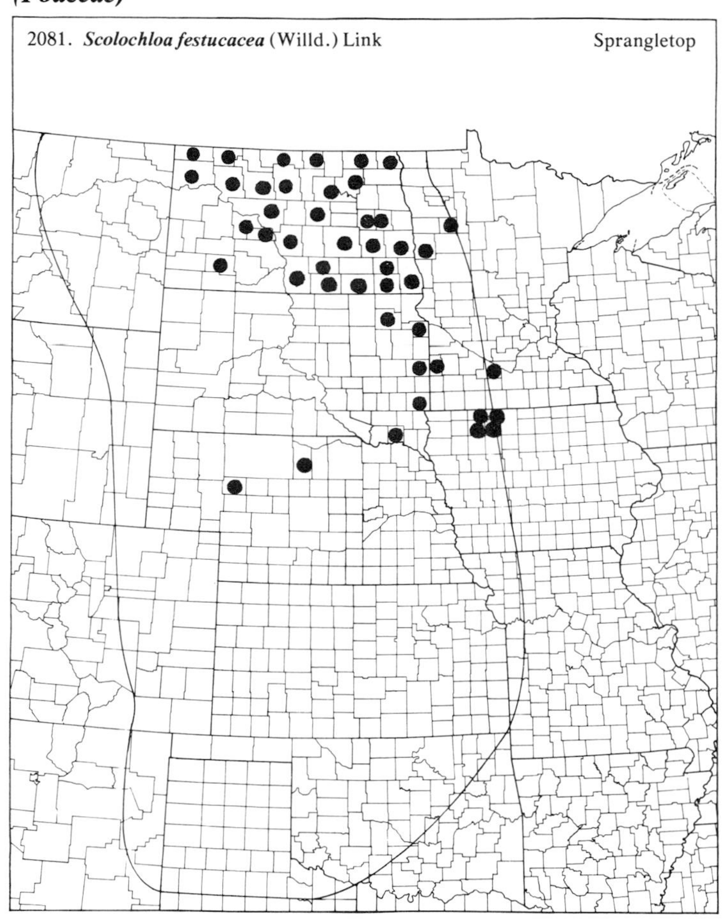

2082. ***Setaria faberii*** Herrm. — Chinese Foxtail

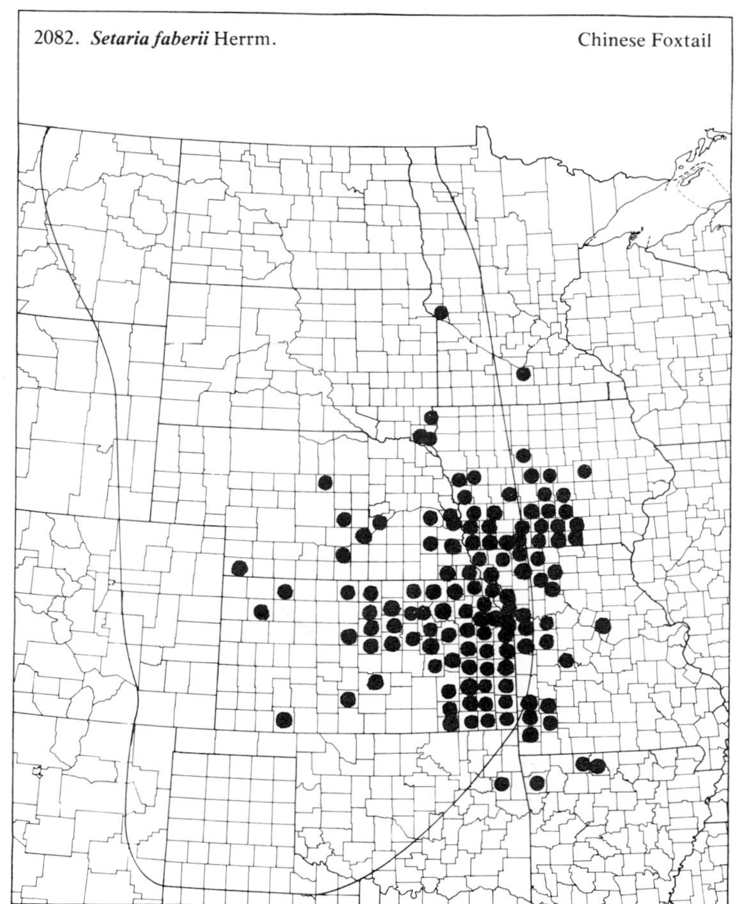

2083. ***Setaria geniculata*** (Lam.) Beauv. — Knotroot Bristlegrass

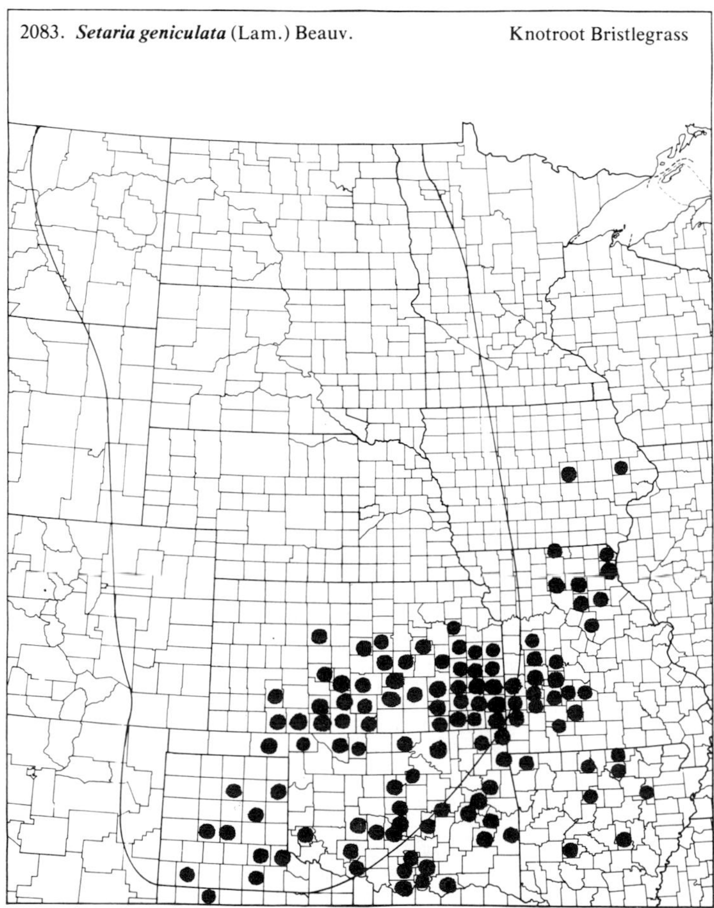

2084. ***Setaria glauca*** (L.) Beauv. — Yellow Foxtail

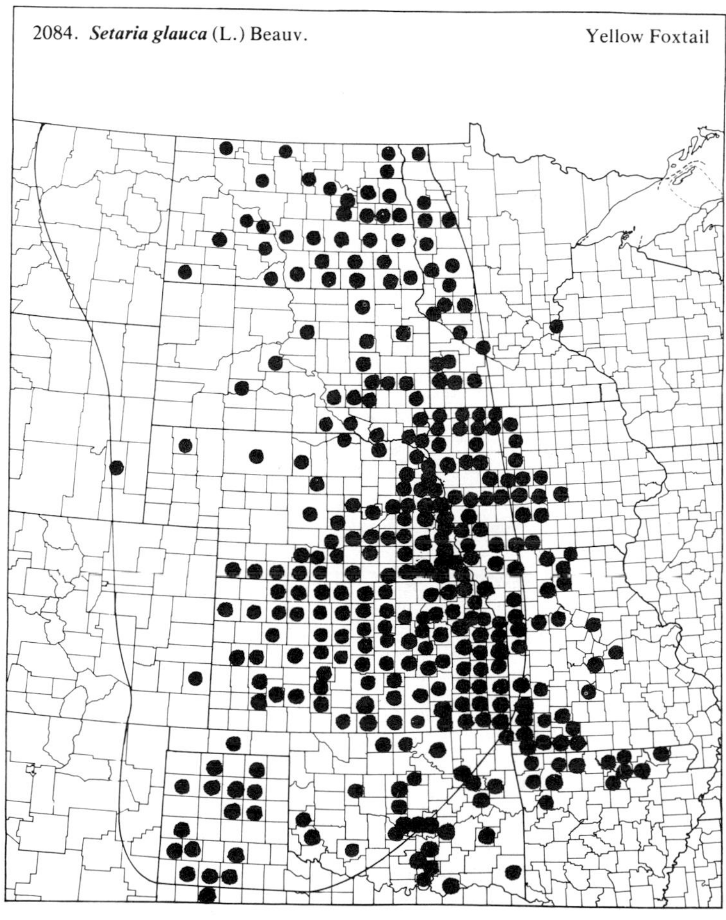

(Poaceae)

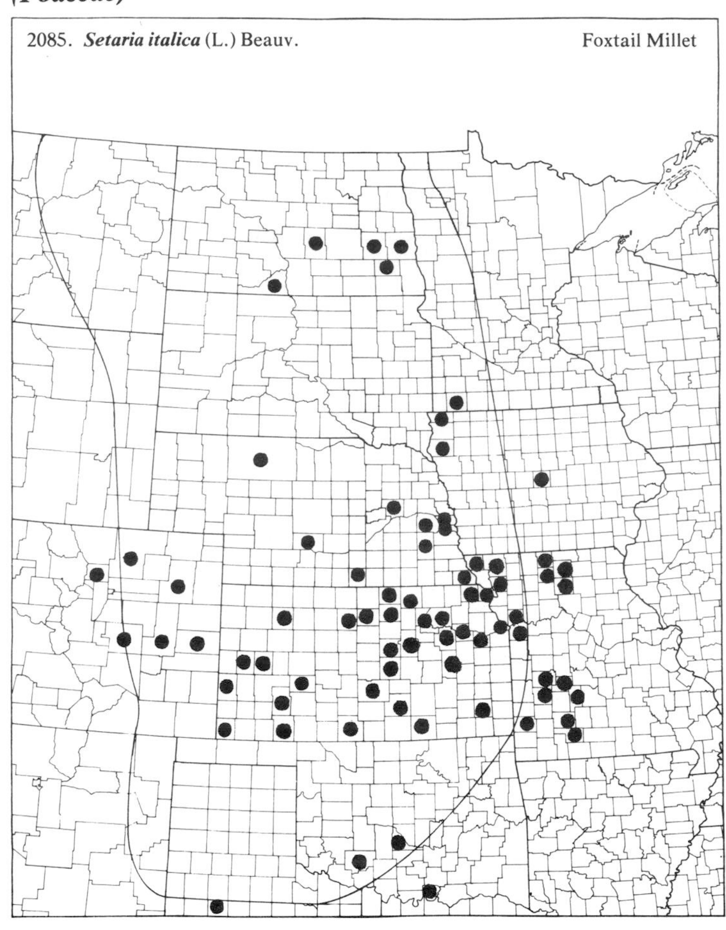
2085. Setaria italica (L.) Beauv.
Foxtail Millet

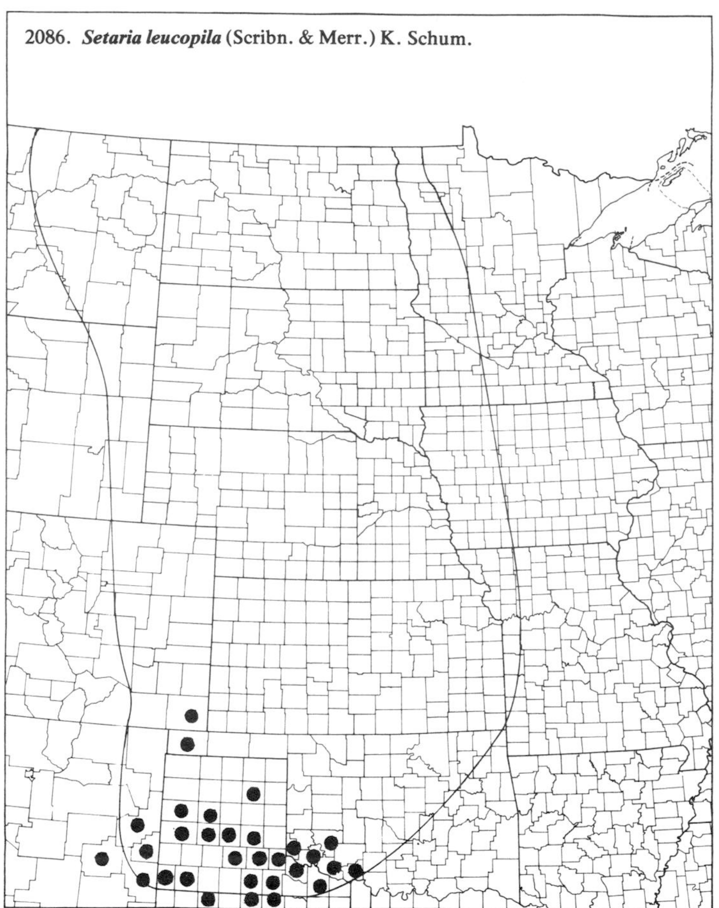
2086. Setaria leucopila (Scribn. & Merr.) K. Schum.

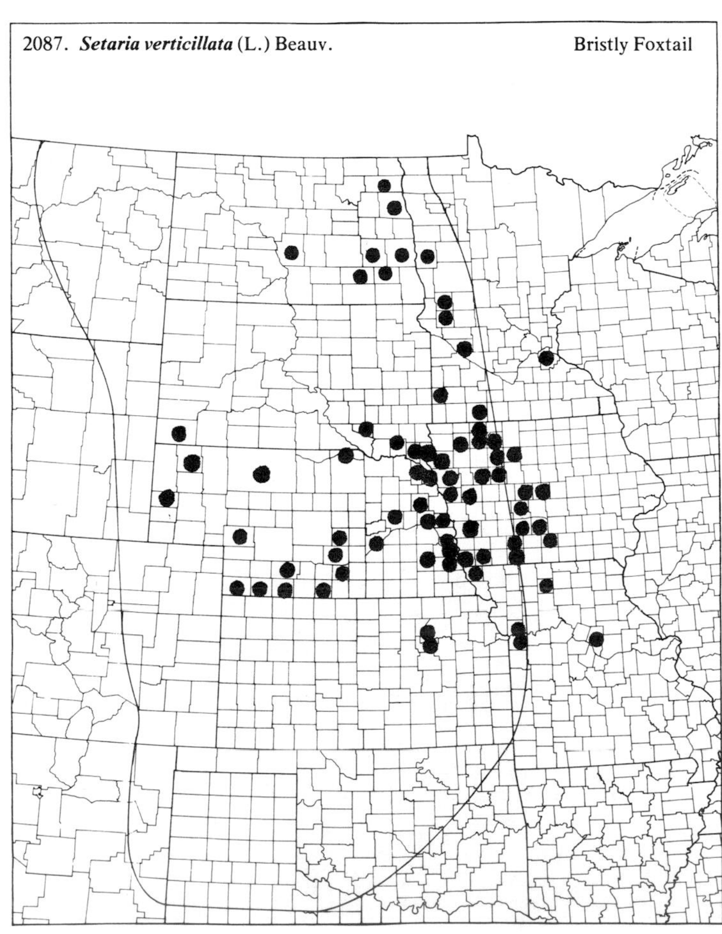
2087. Setaria verticillata (L.) Beauv.
Bristly Foxtail

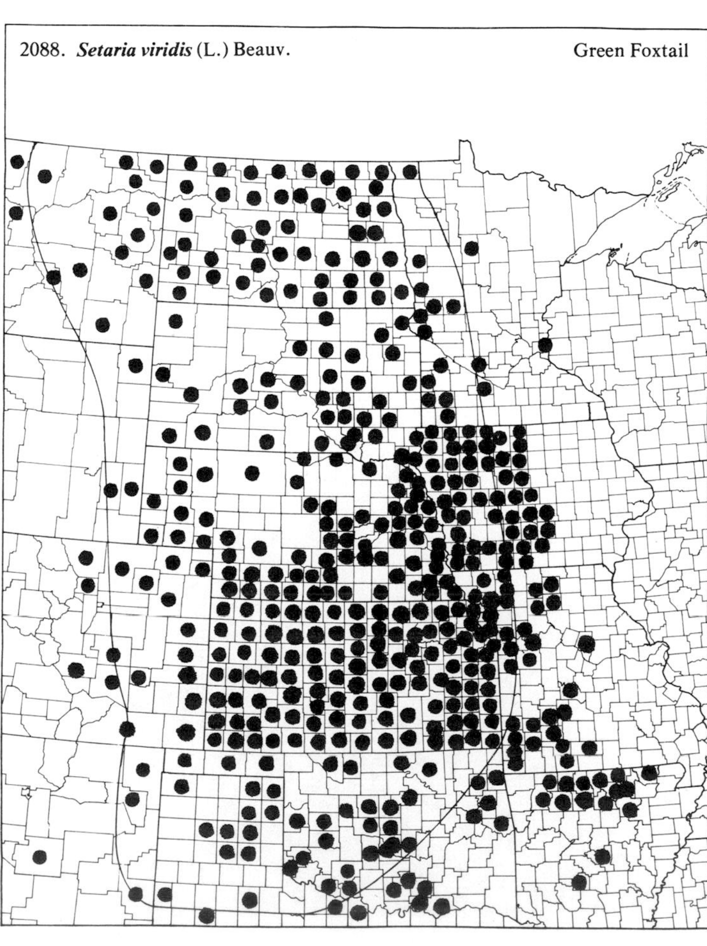
2088. Setaria viridis (L.) Beauv.
Green Foxtail

(Poaceae)

2089. ***Sitanion hystrix*** (Nutt.) Sm. var. ***brevifolium*** (Sm.) C.L. Hitchc. Squirreltail

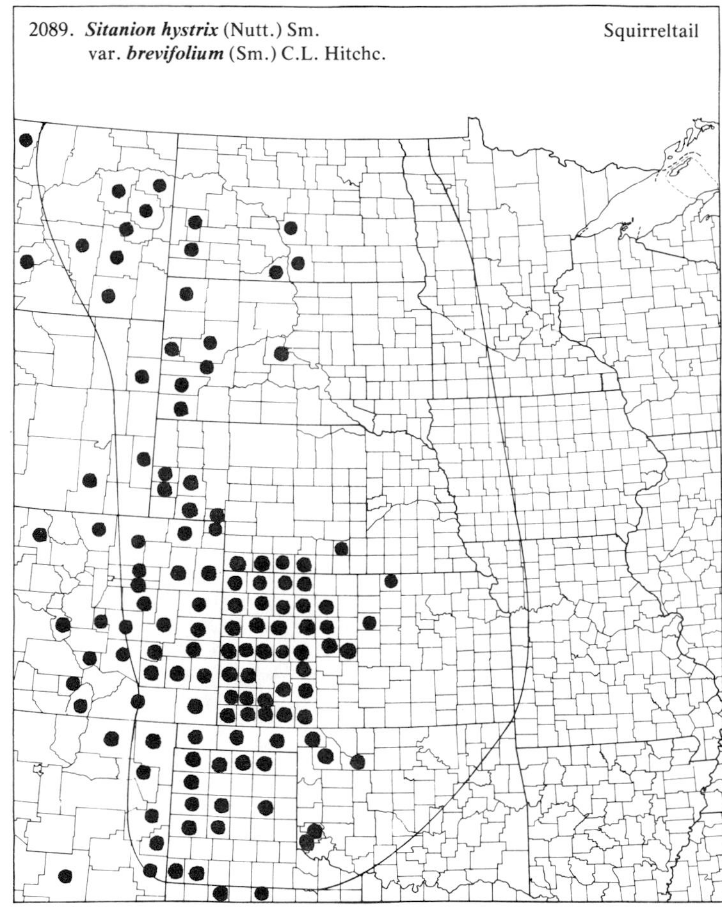

2090. ***Sorghastrum avenaceum*** (Michx.) Nash Indian-grass

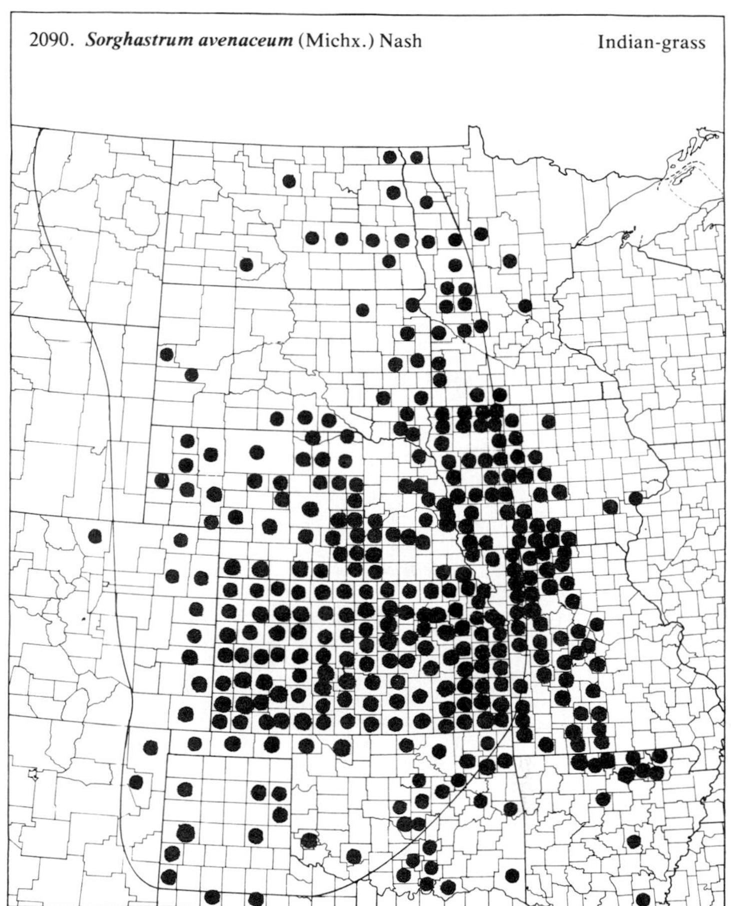

2091. ***Sorghum halepense*** (L.) Pers. Johnson-grass

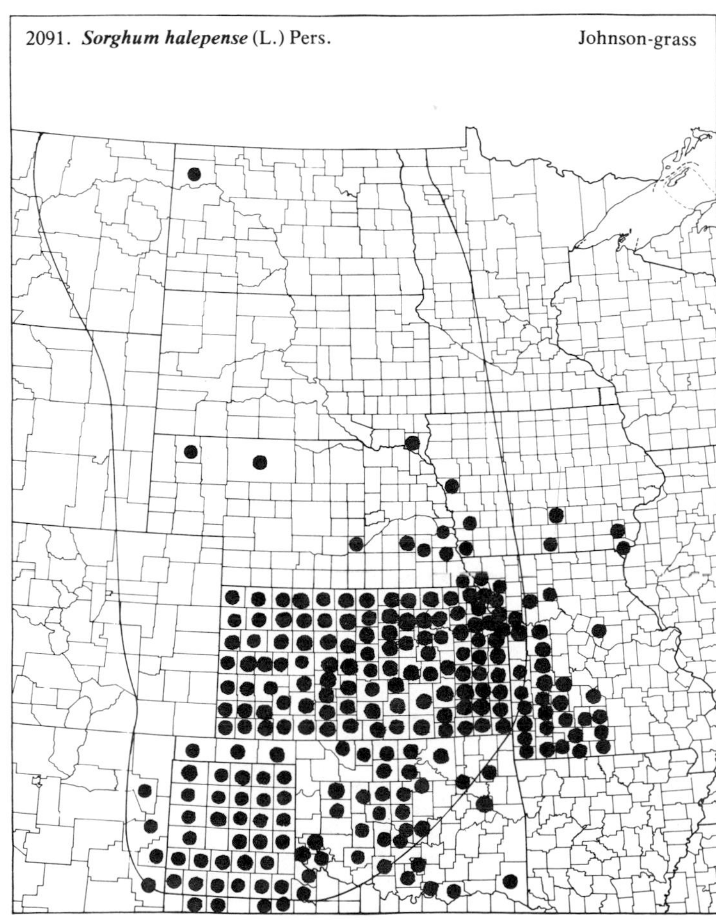

2092. ***Spartina gracilis*** Trin. Alkali Cordgrass

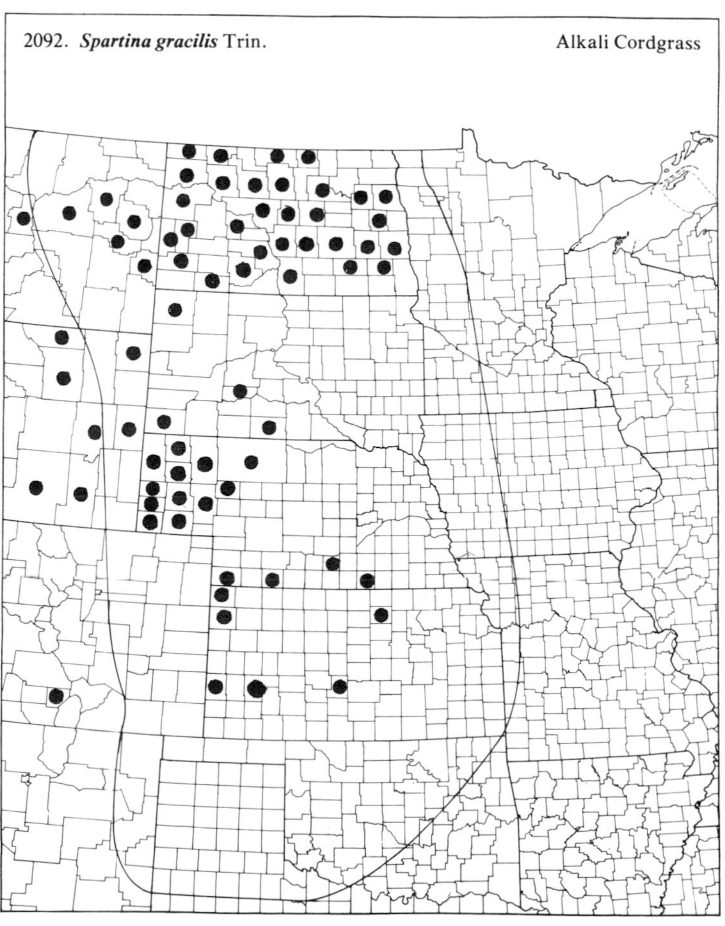

(Poaceae)

2093. ***Spartina pectinata*** Link — Prairie Cordgrass

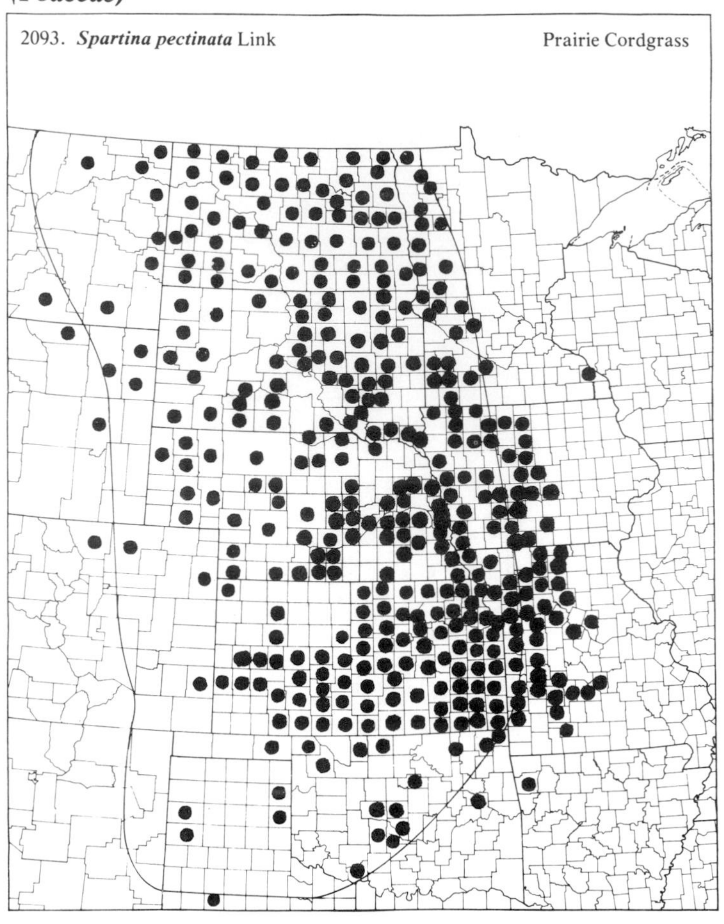

2094. ***Sphenopholis obtusata*** (Michx.) Scribn. var. ***obtusata*** — Prairie Wedgegrass

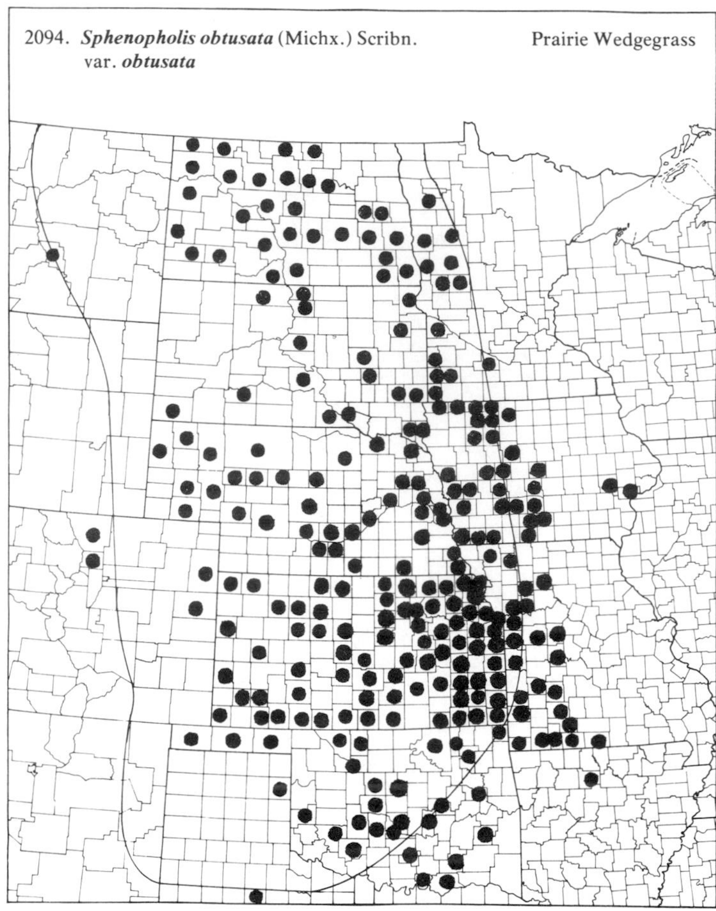

2095. ***Sphenopholis obtusata*** (Michx.) Scribn. var. ***major*** (Torr.) Erdman — Prairie Wedgegrass

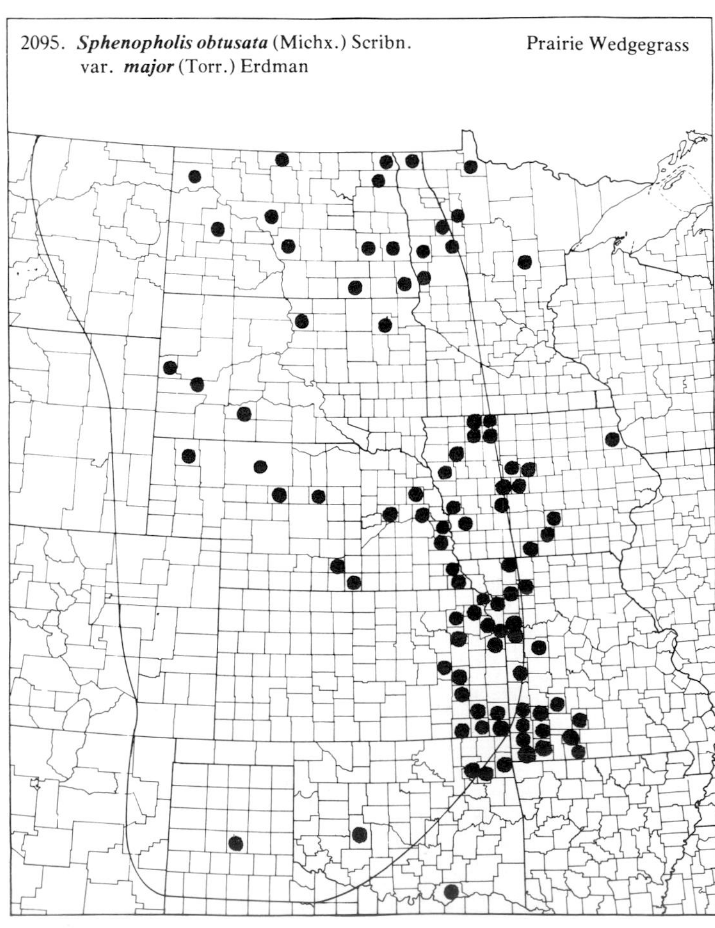

2096. ***Sporobolus airoides*** (Torr.) Torr. — Alkali Sacaton

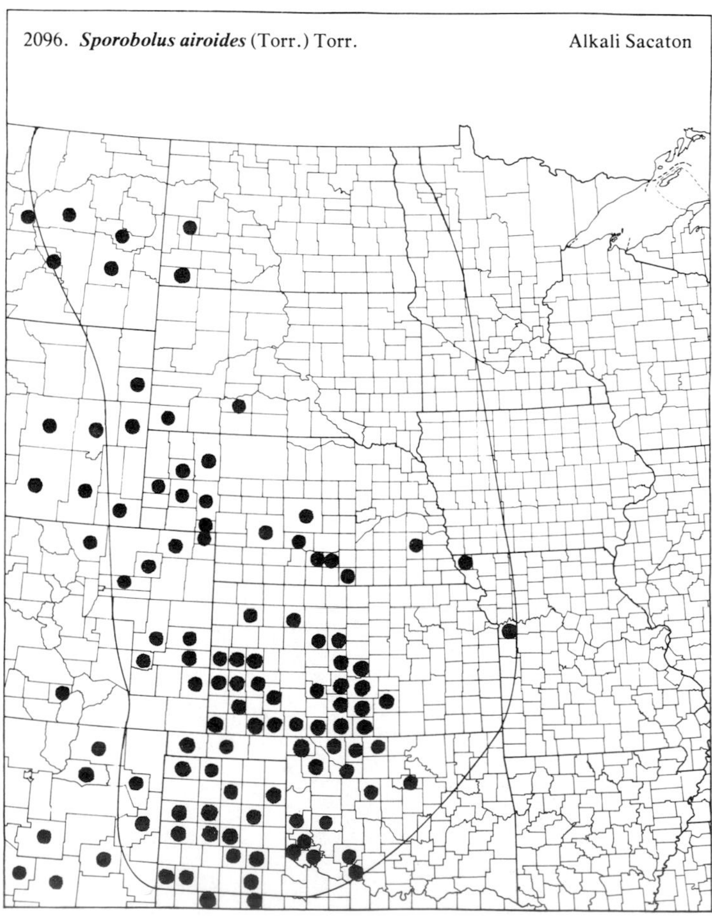

2097. ***Sporobolus asper*** (Michx.) Kunth var. ***asper*** — Rough Dropseed

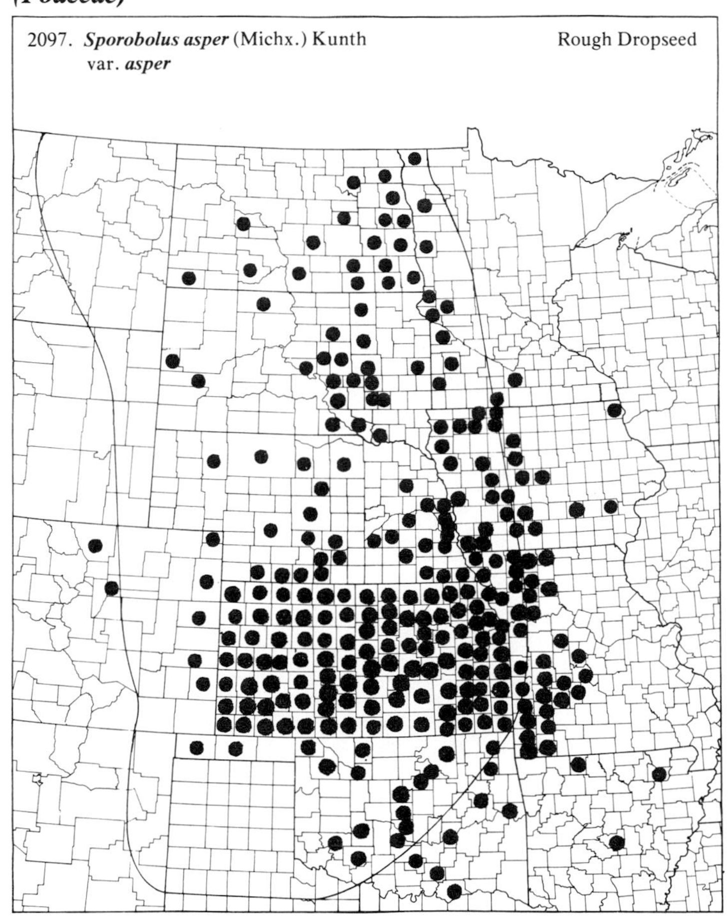

2098. ***Sporobolus asper*** (Michx.) Kunth var. ***clandestinus*** (Biehler) Shinners — Rough Dropseed

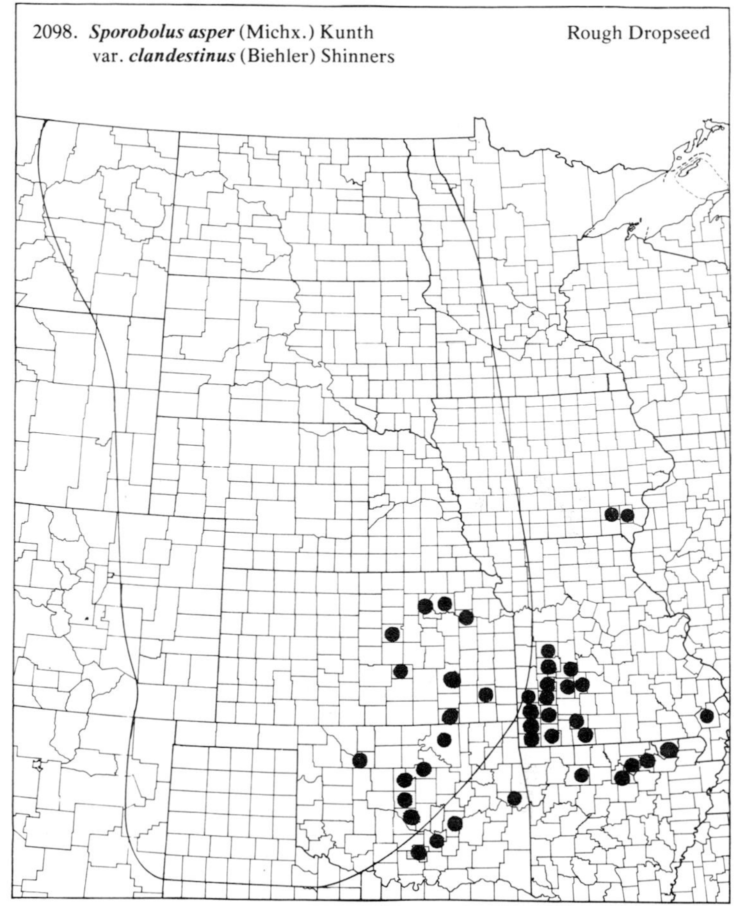

2099. ***Sporobolus asper*** (Michx.) Kunth var. ***hookeri*** (Trin.) Vasey — Rough Dropseed

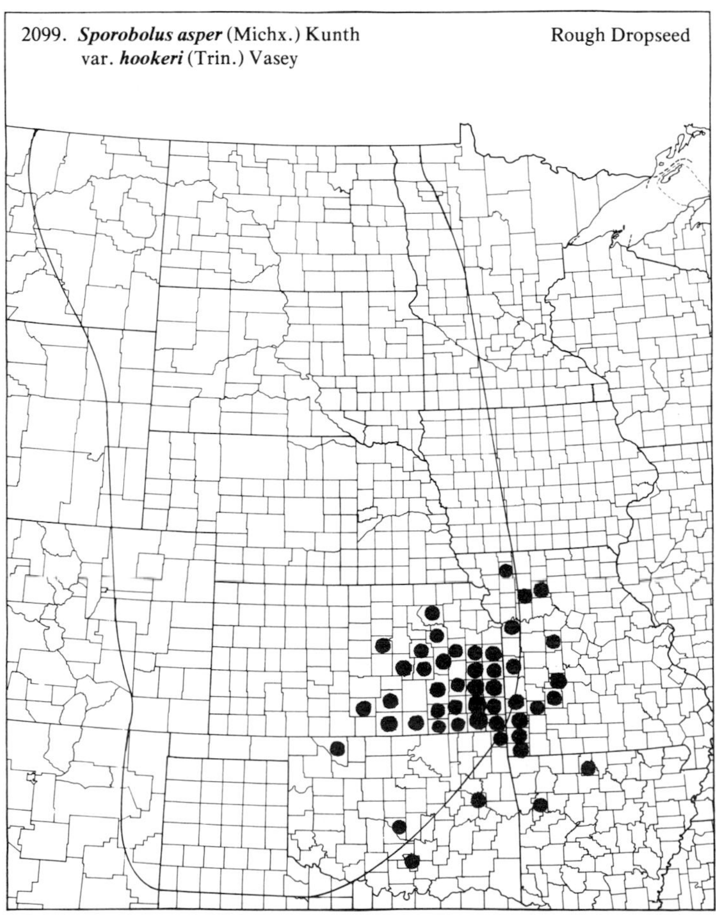

2100. ***Sporobolus asper*** (Michx.) Kunth var. ***pilosus*** (Vasey) Hitchc. — Rough Dropseed

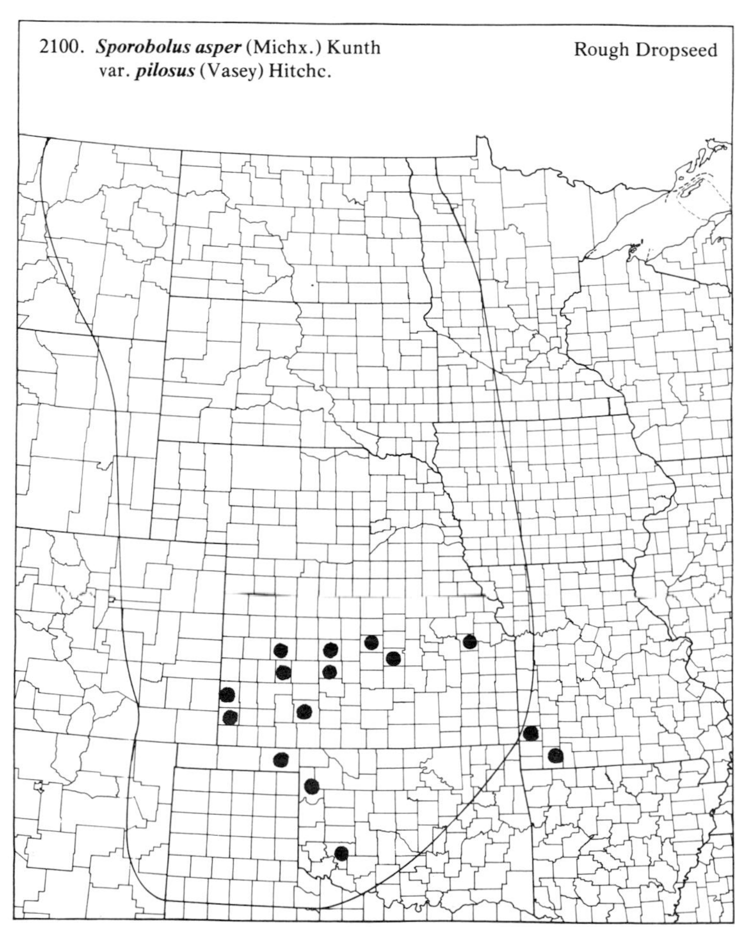

(Poaceae)

2101. ***Sporobolus cryptandrus*** (Torr.) Gray — Sand Dropseed

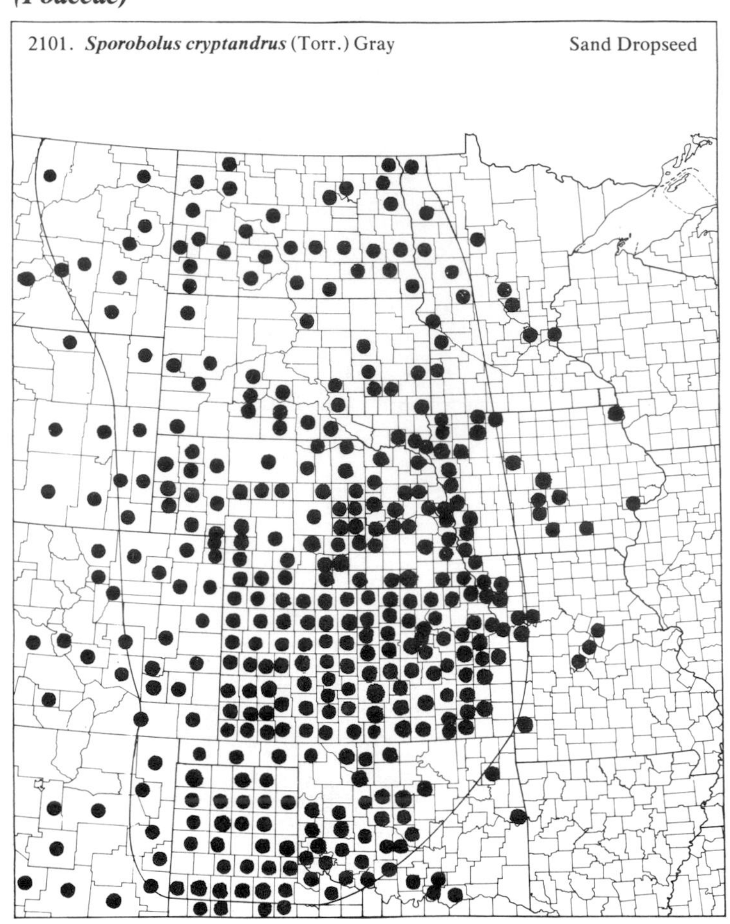

2102. ***Sporobolus giganteus*** Nash — Giant Dropseed

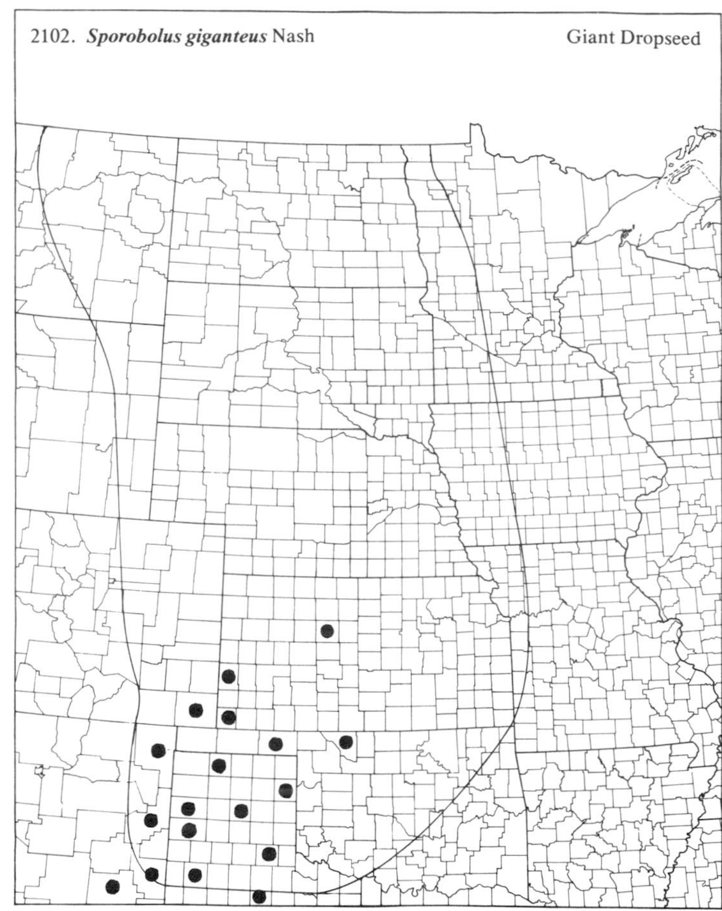

2103. ***Sporobolus heterolepis*** (Gray) Gray — Prairie Dropseed

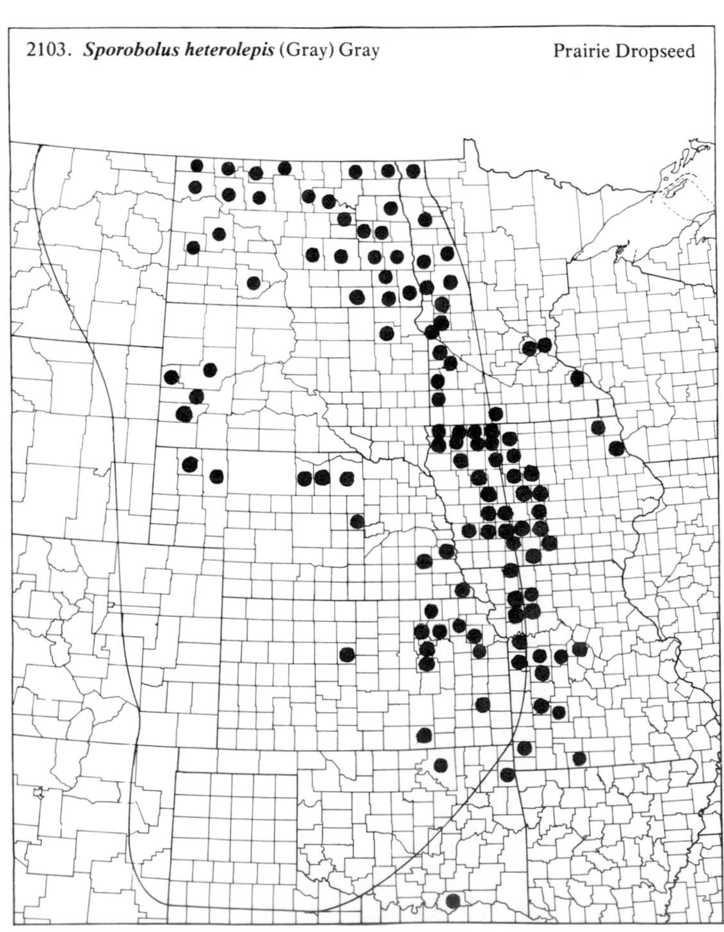

2104. ***Sporobolus pyramidatus*** (Lam.) Hitchc. — Whorled Dropseed

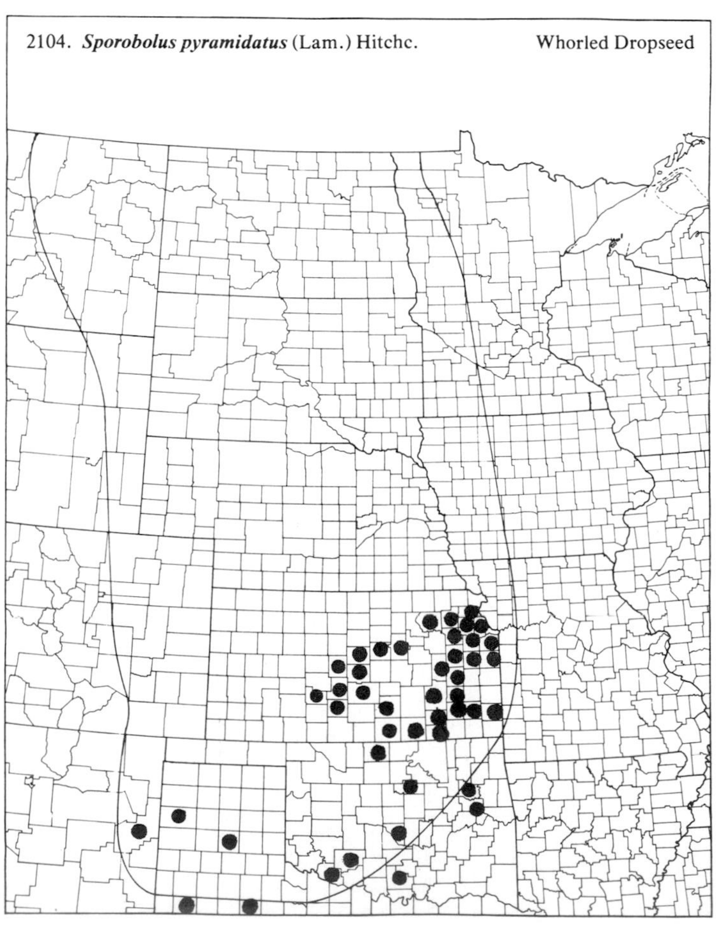

(Poaceae)

2105. ***Sporobolus texanus*** Vasey — Texas Dropseed

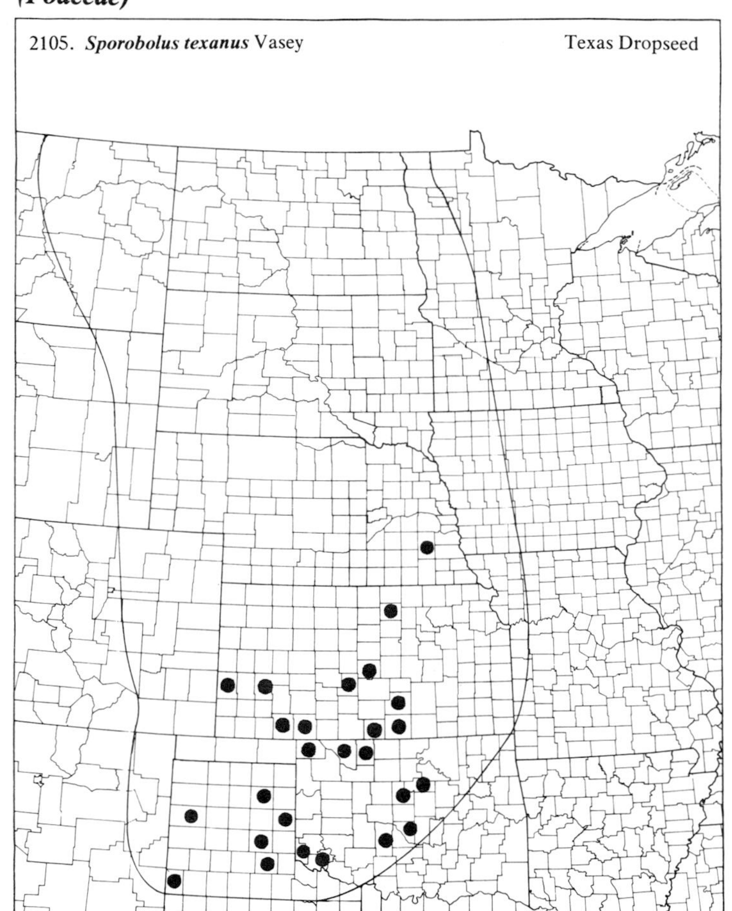

2106. ***Sporobolus vaginiflorus*** (Torr.) Wood var. ***vaginiflorus*** — Povertygrass

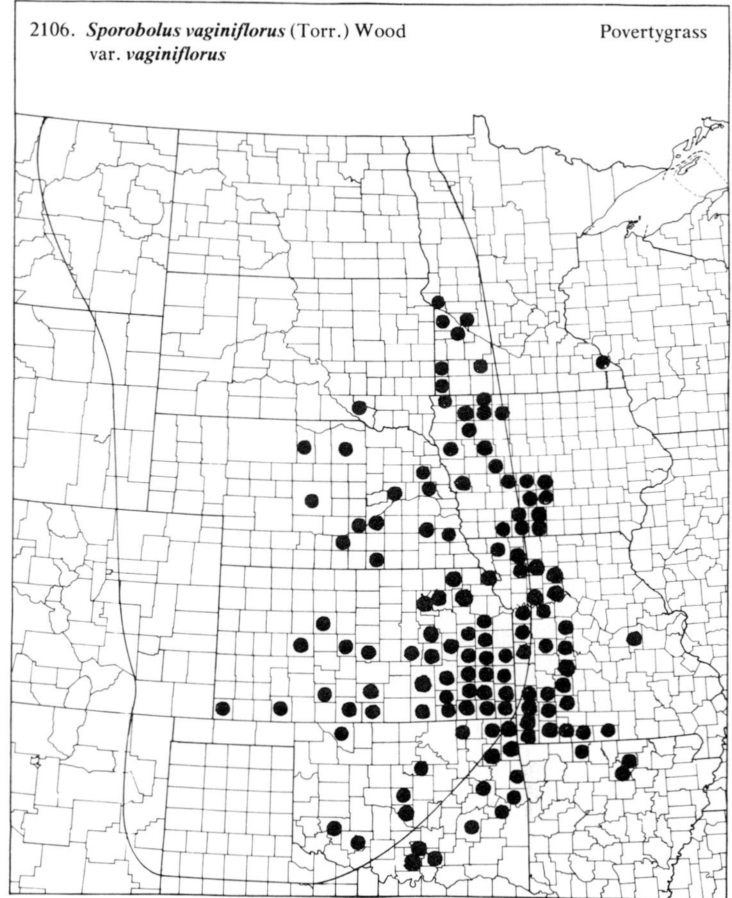

2107. ***Sporobolus vaginiflorus*** (Torr.) Wood var. ***neglectus*** (Nash) Scribn. — Povertygrass

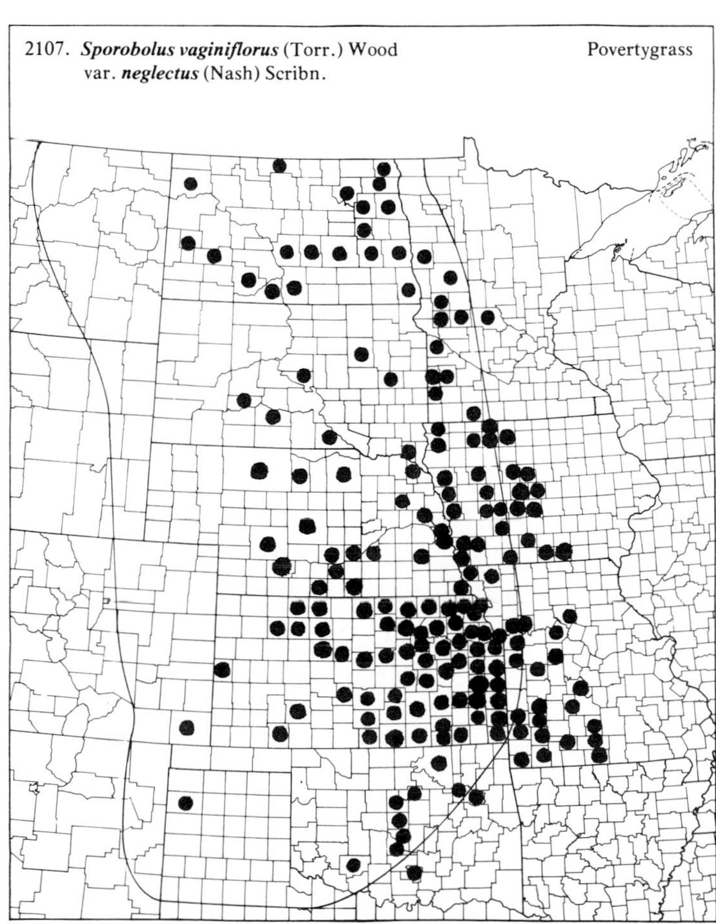

2108. ***Stipa comata*** Trin. & Rupr. — Needle-and-thread

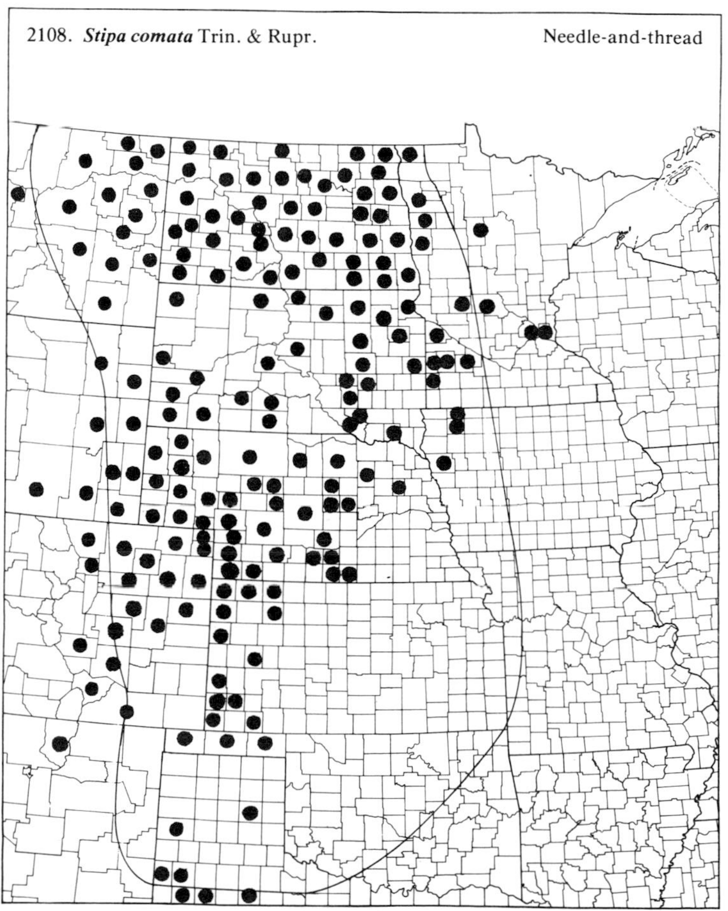

(Poaceae)

2109. ***Stipa spartea*** Trin. Porcupine-grass

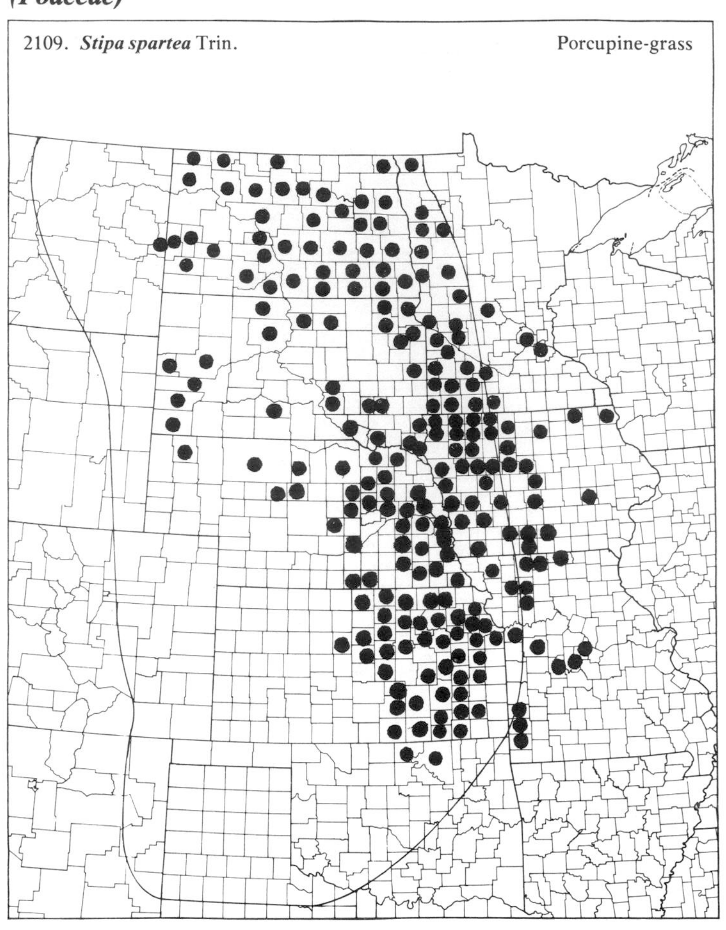

2110. ***Stipa viridula*** Trin. Green Needlegrass

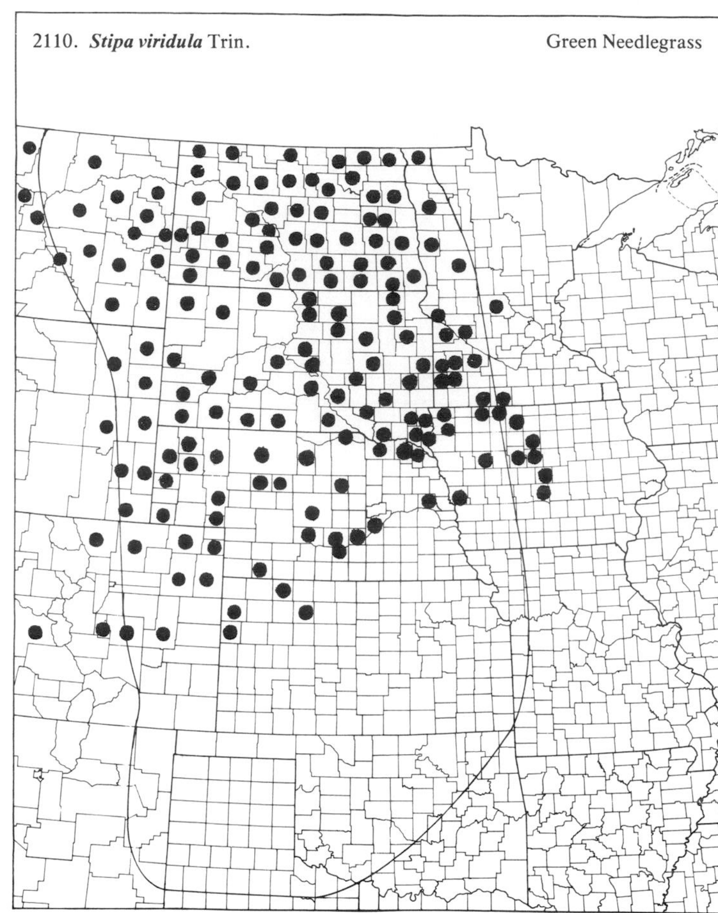

2111. ***Tridens albescens*** (Vasey) Woot. & Standl. White Tridens

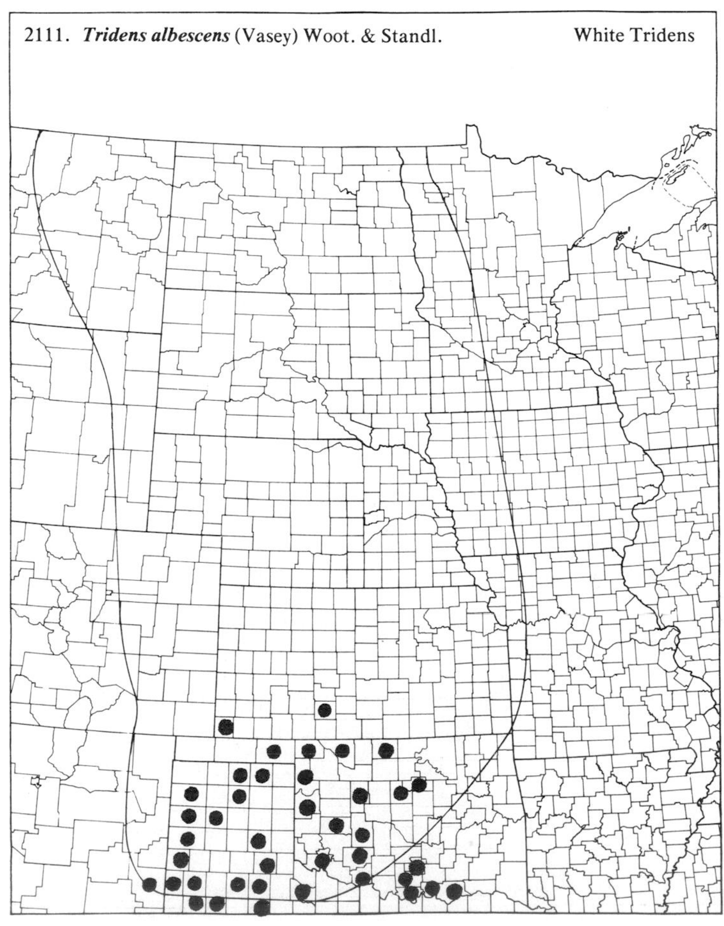

2112. ***Tridens flavus*** (L.) Hitchc. Purpletop

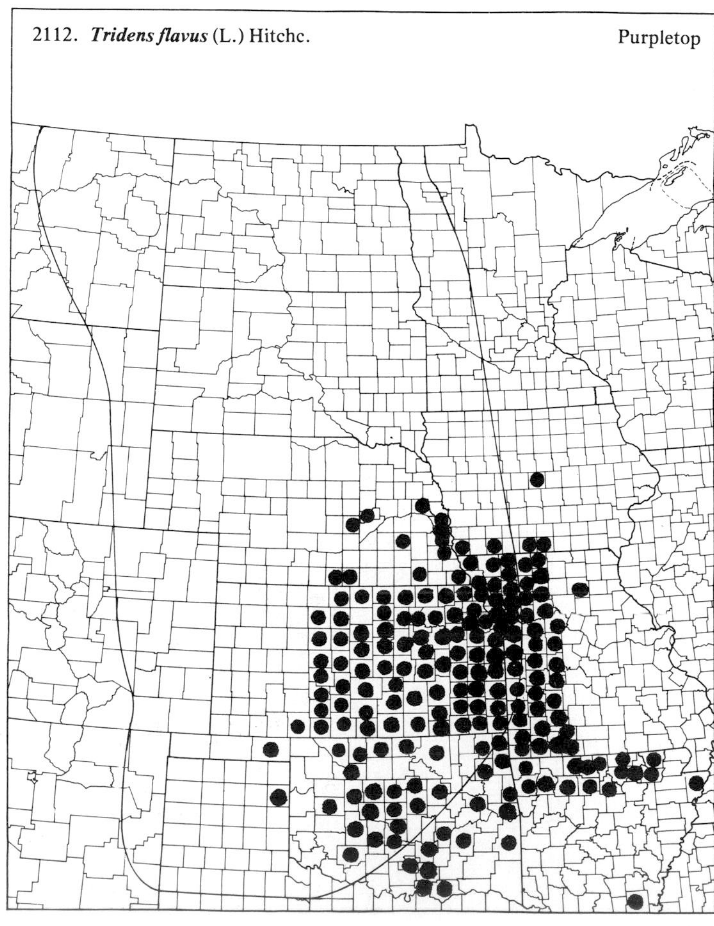

(Poaceae)

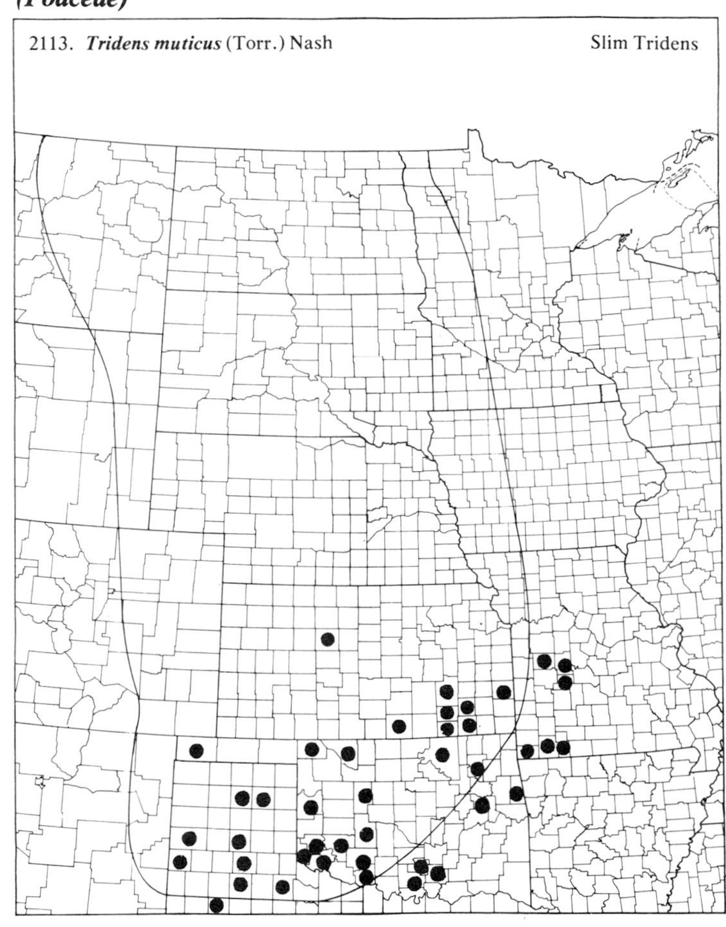
2113. *Tridens muticus* (Torr.) Nash
Slim Tridens

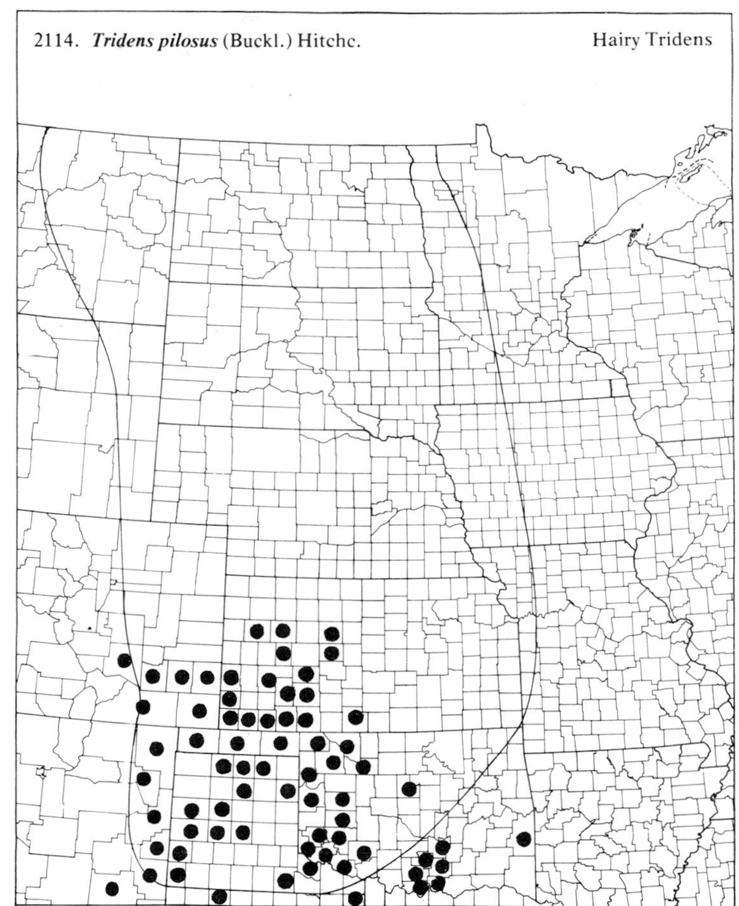
2114. *Tridens pilosus* (Buckl.) Hitchc.
Hairy Tridens

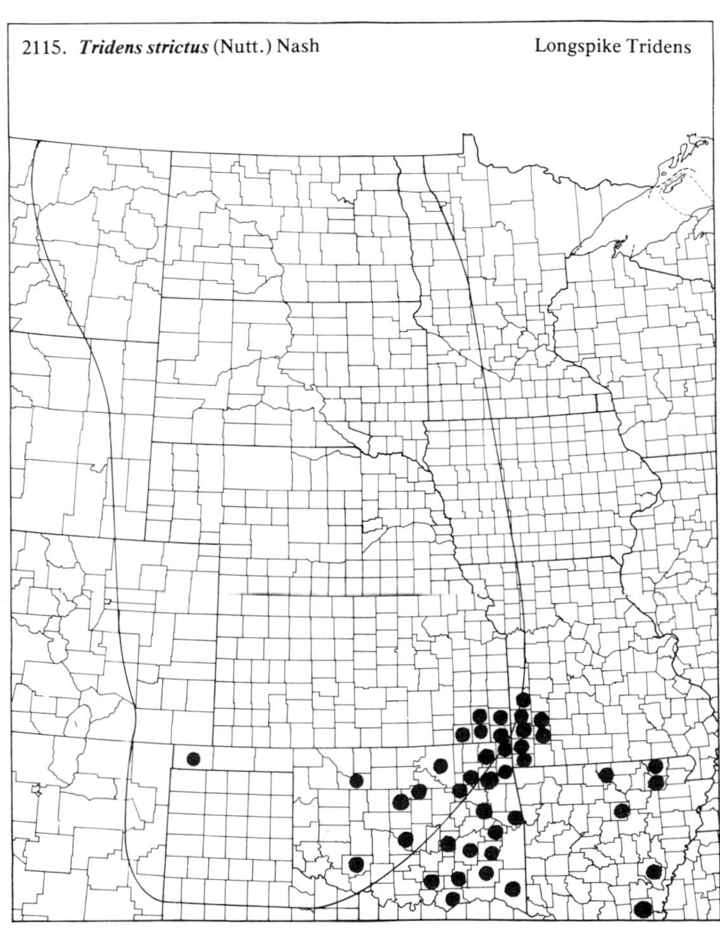
2115. *Tridens strictus* (Nutt.) Nash
Longspike Tridens

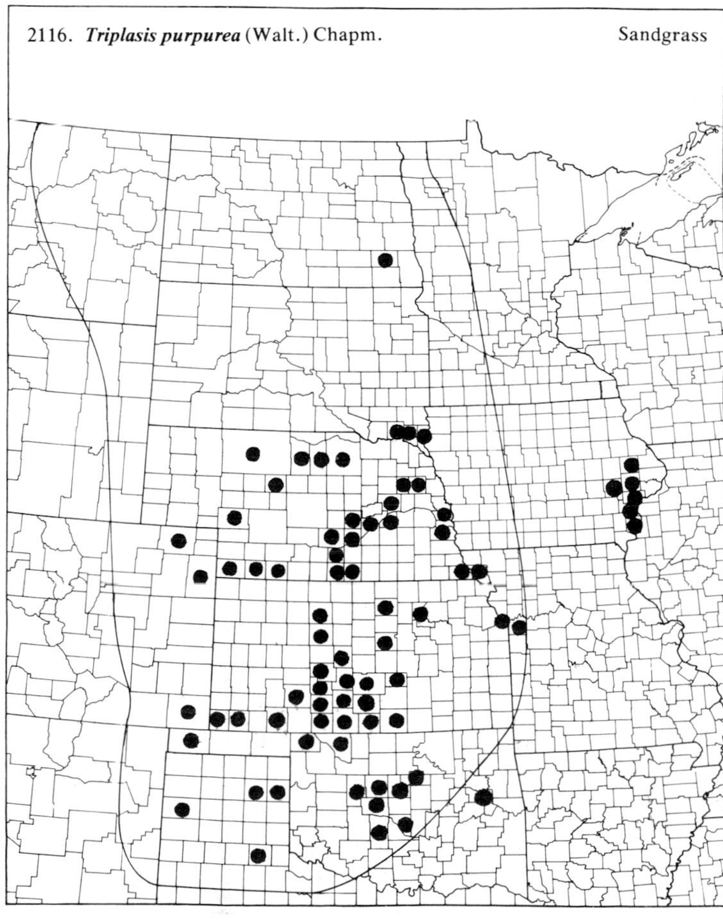
2116. *Triplasis purpurea* (Walt.) Chapm.
Sandgrass

(Poaceae)

2117. ***Tripsacum dactyloides*** L. Gamagrass

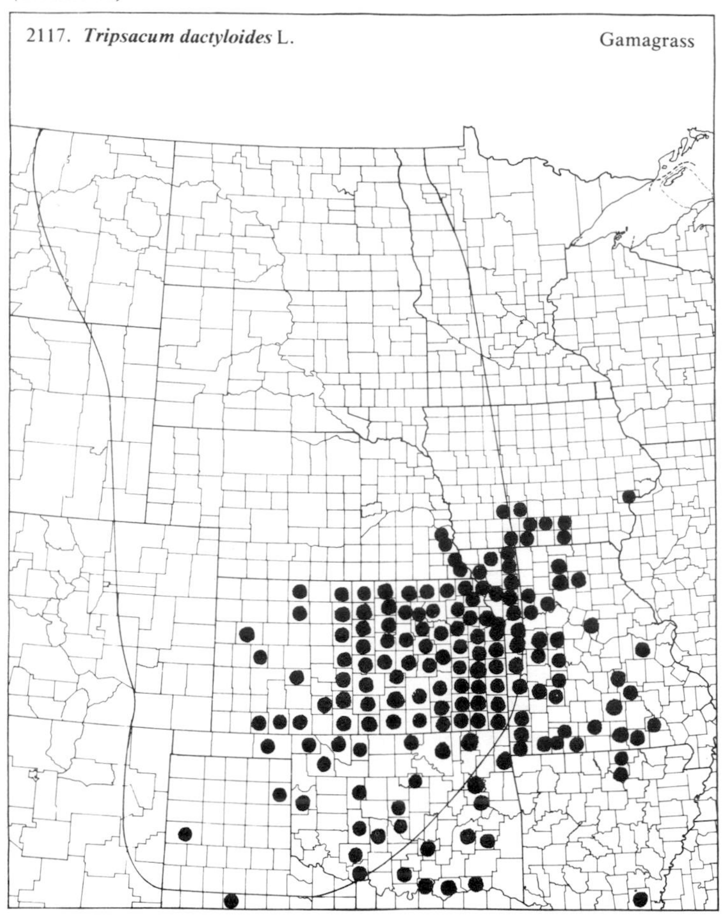

2118. ***Uniola latifolia*** Michx.

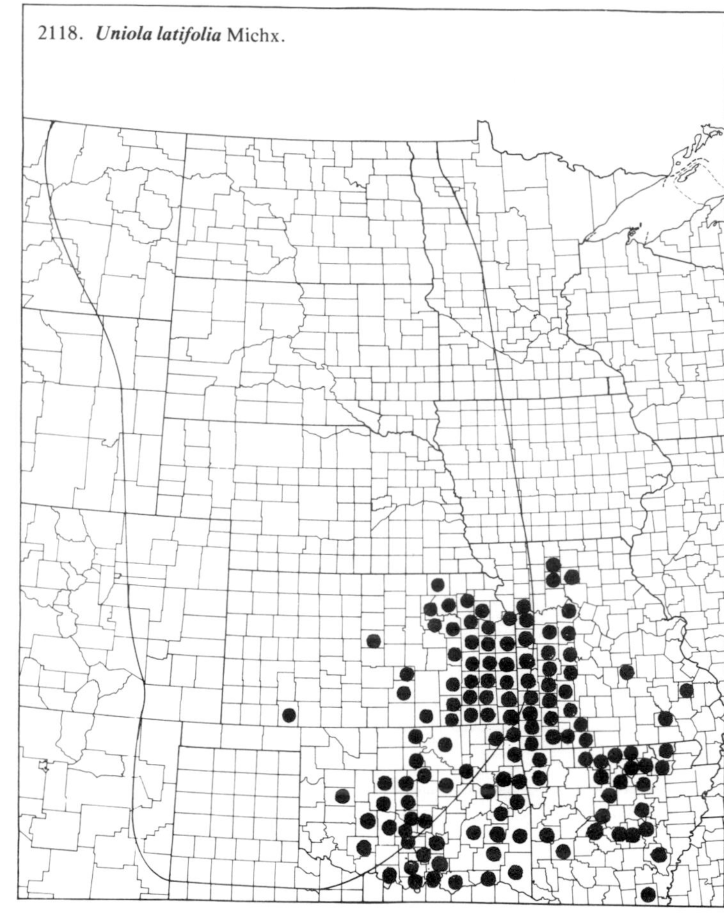

2119. ***Zizania aquatica*** L. Wild Rice

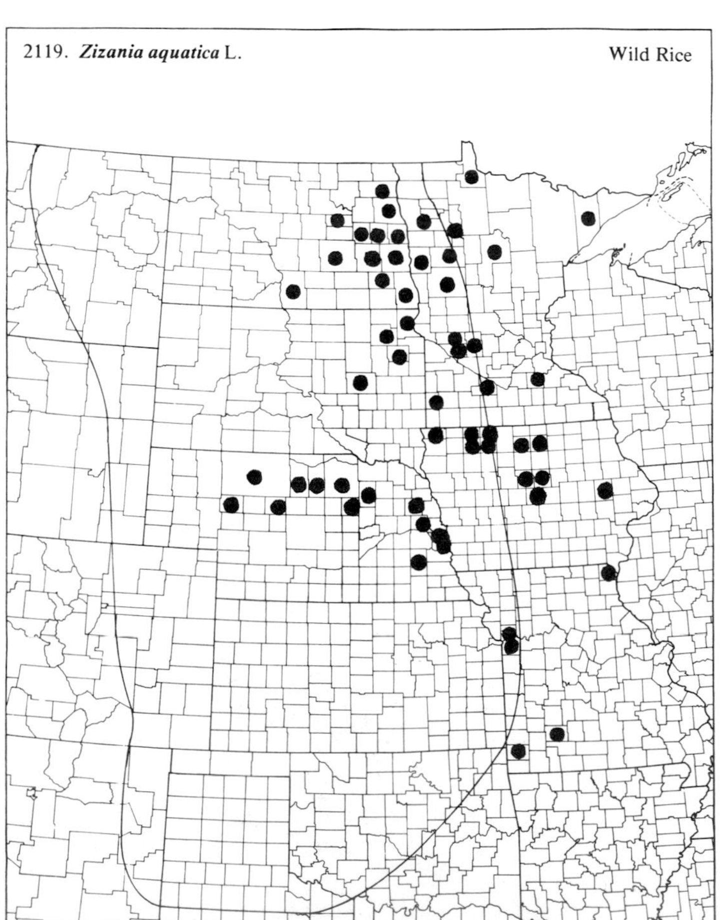

128. *Sparganiaceae* (Bur-reed Family)

2120. ***Sparganium chlorocarpum*** Rydb. Bur-reed

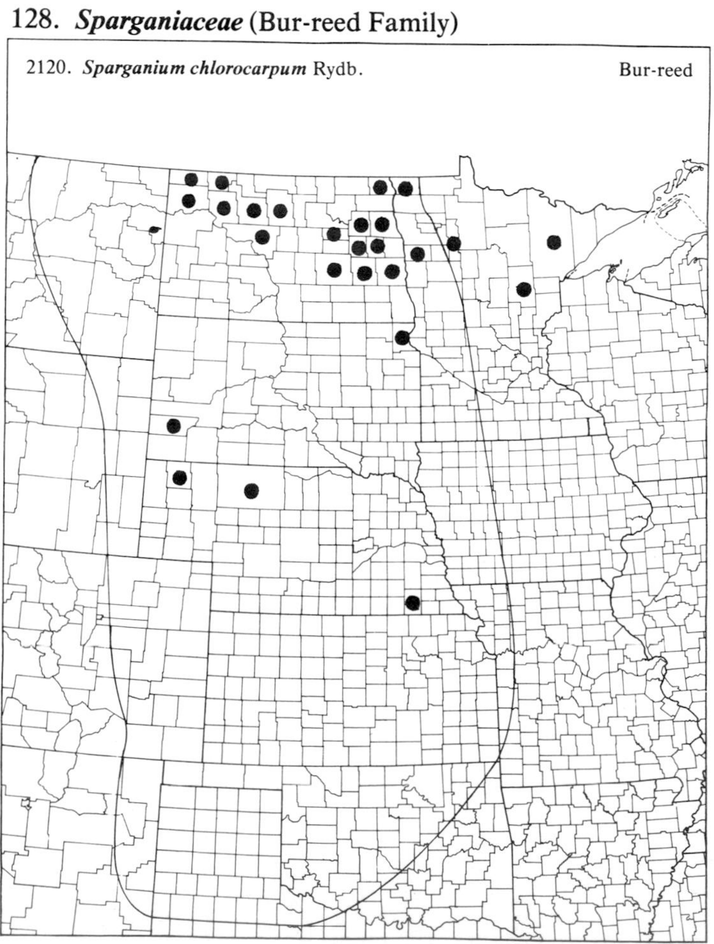

(Sparganiaceae)

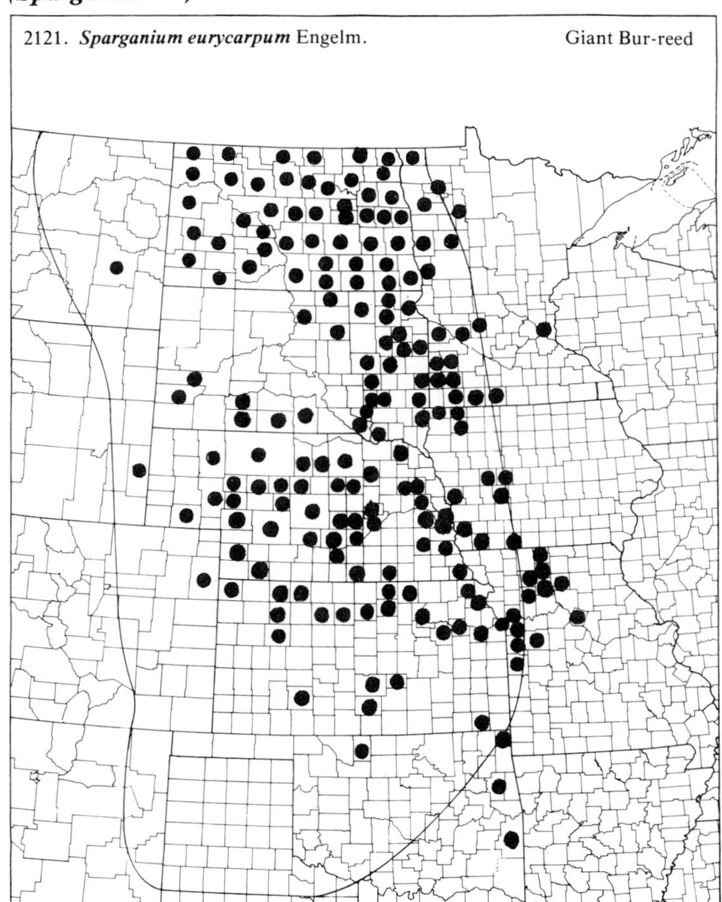
2121. ***Sparganium eurycarpum*** Engelm. Giant Bur-reed

129. *Typhaceae* (Cattail Family)

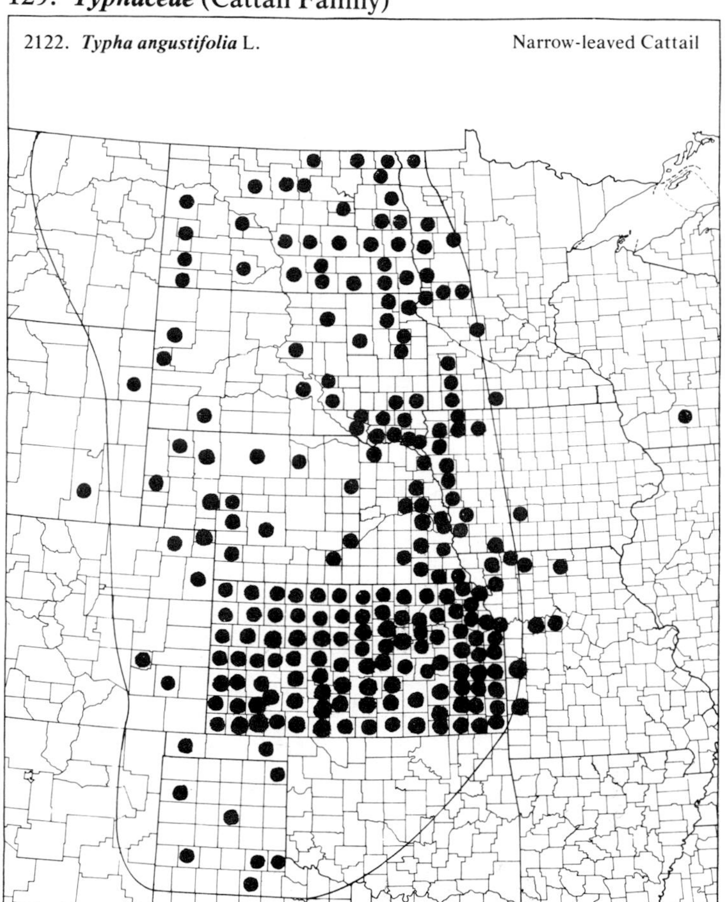
2122. ***Typha angustifolia*** L. Narrow-leaved Cattail

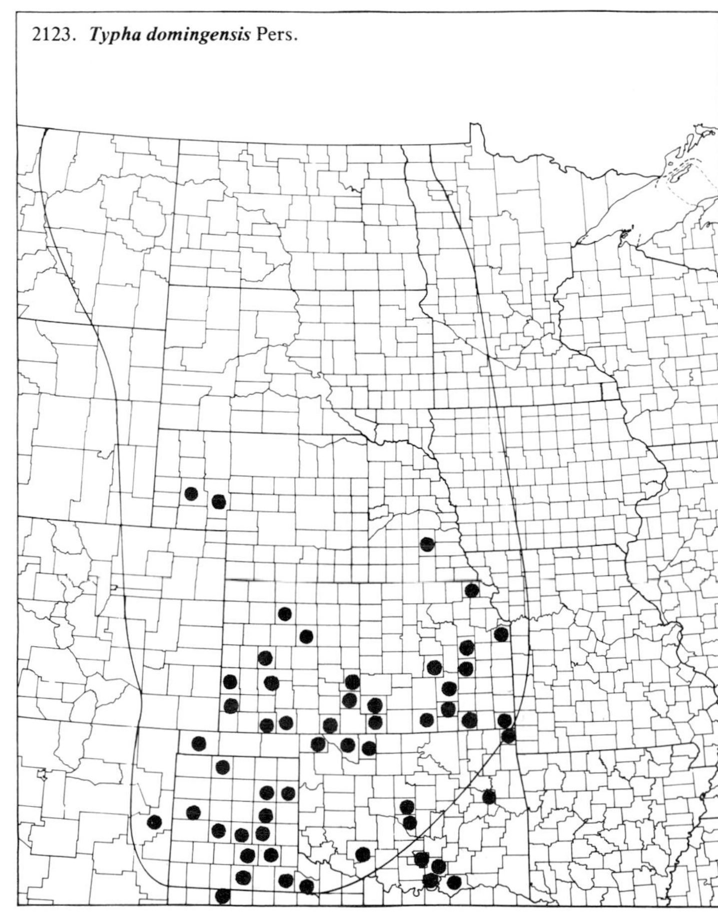
2123. ***Typha domingensis*** Pers.

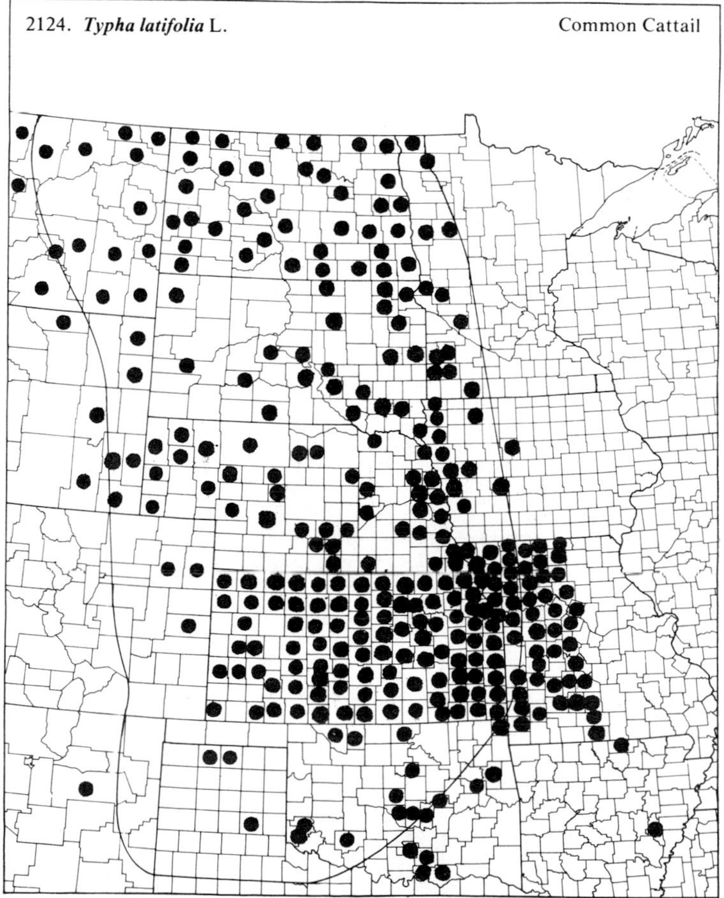
2124. ***Typha latifolia*** L. Common Cattail

130. *Araceae* (Arum Family)

2125. *Acorus calamus* L. — Sweet Flag

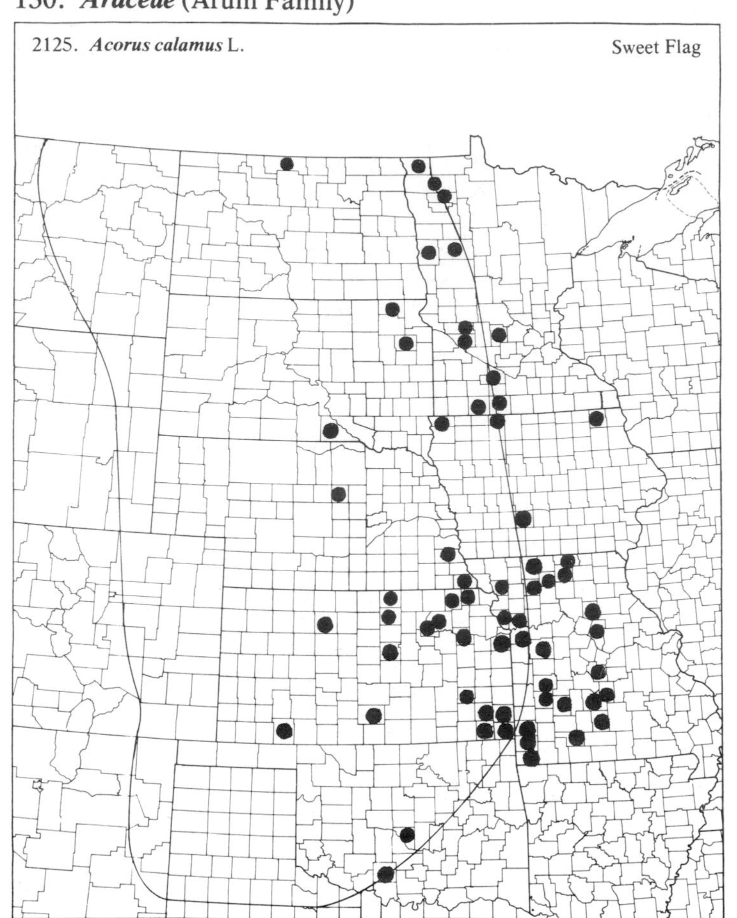

2126. *Arisaema dracontium* (L.) Schott — Dragonroot

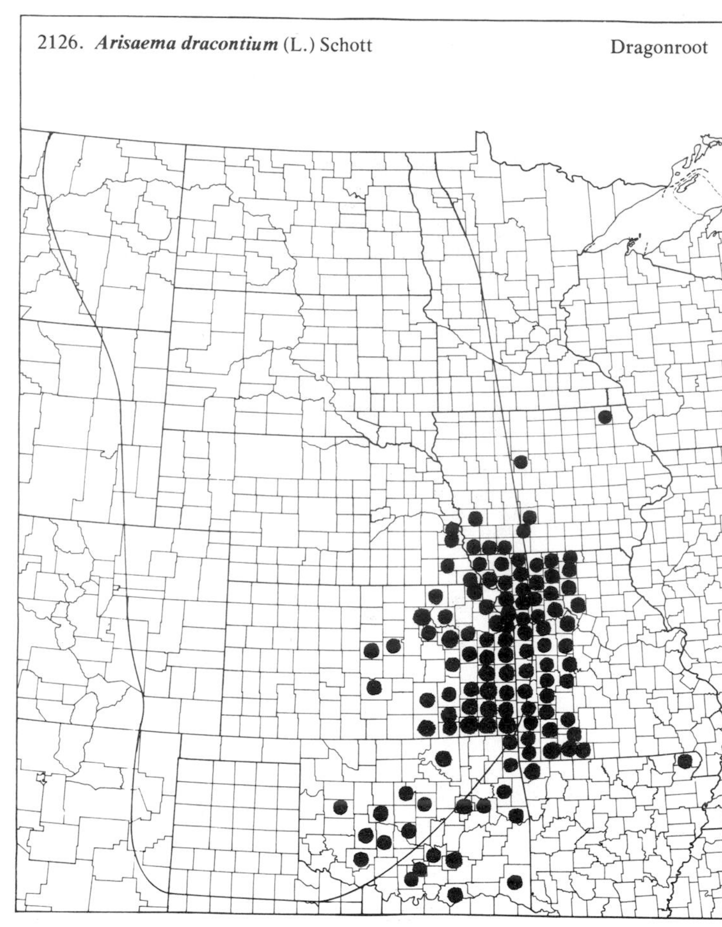

2127. *Arisaema triphyllum* (L.) Schott — Jack-in-the-pulpit

131. *Lemnaceae* (Duckweed Family)

2128. *Lemna gibba* L.

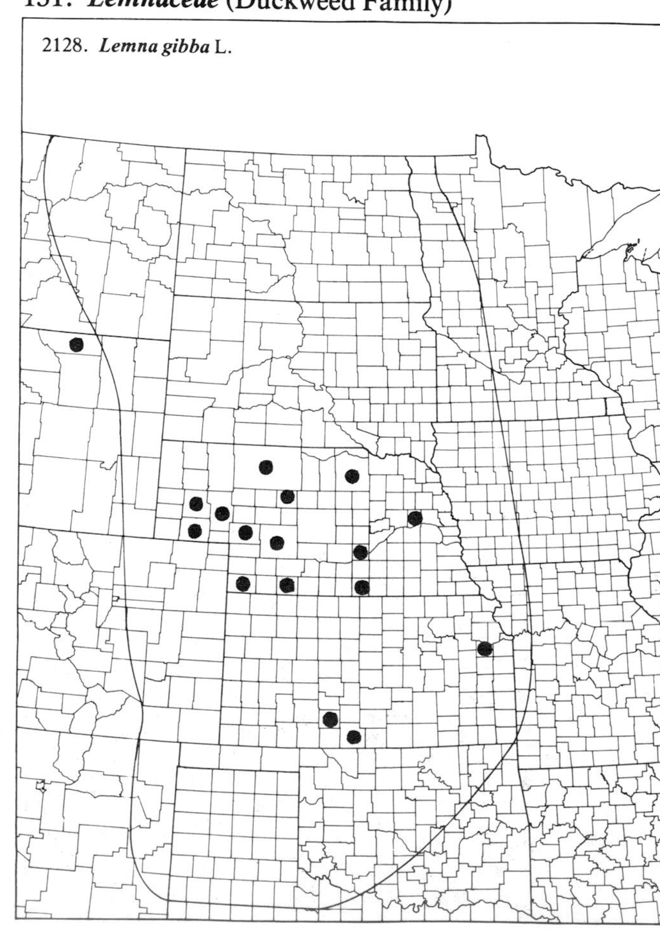

(Lemnaceae)

2129. ***Lemna minor*** L. Duckweed

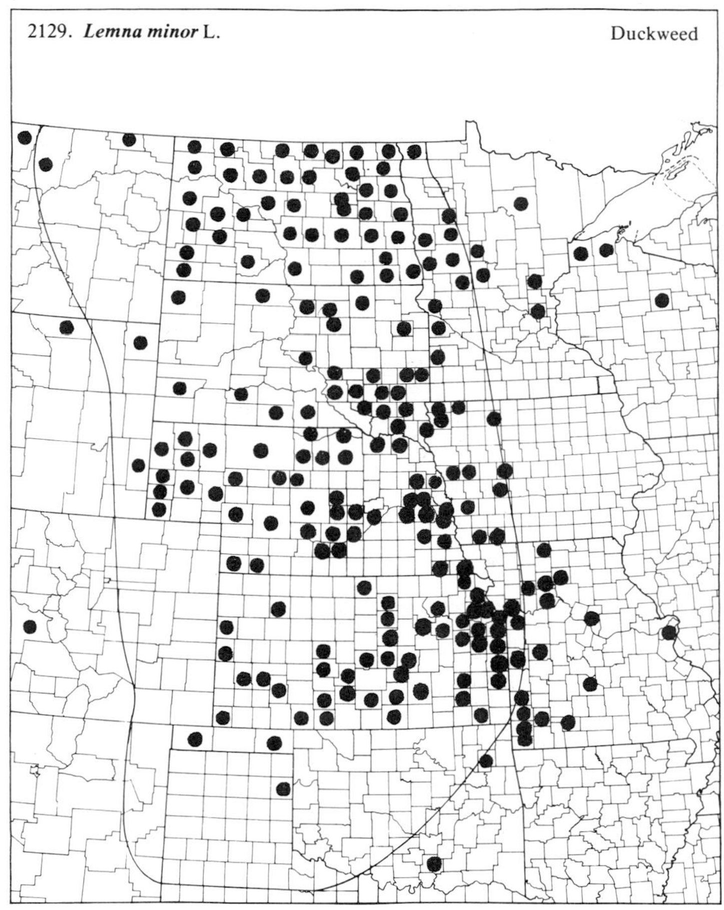

2130. ***Lemna perpusilla*** Torr. Minute Duckweed

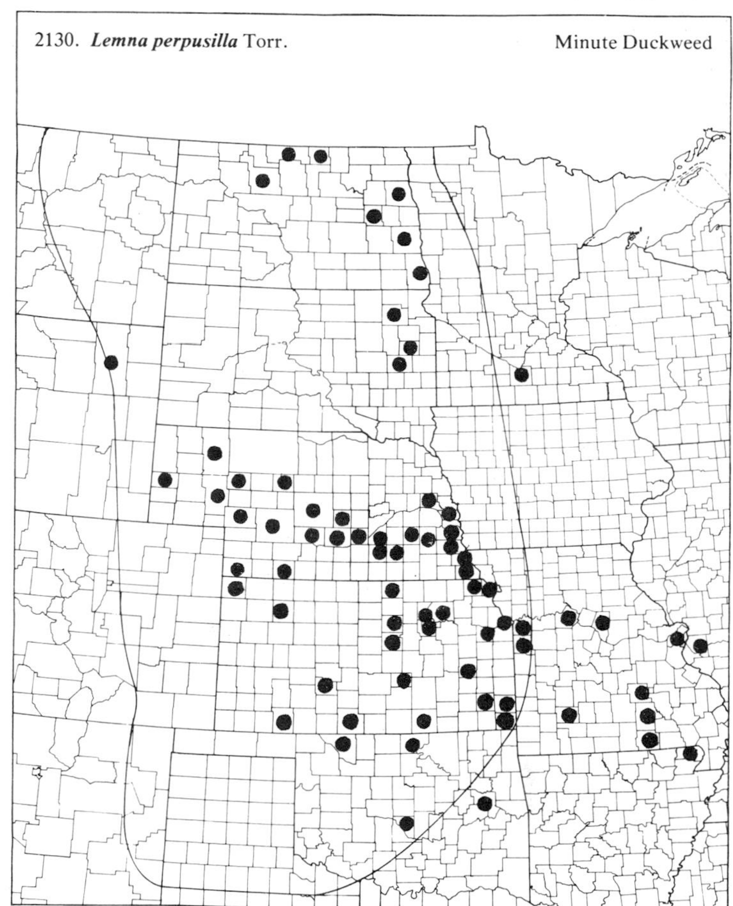

2131. ***Lemna trisulca*** L. Star Duckweed

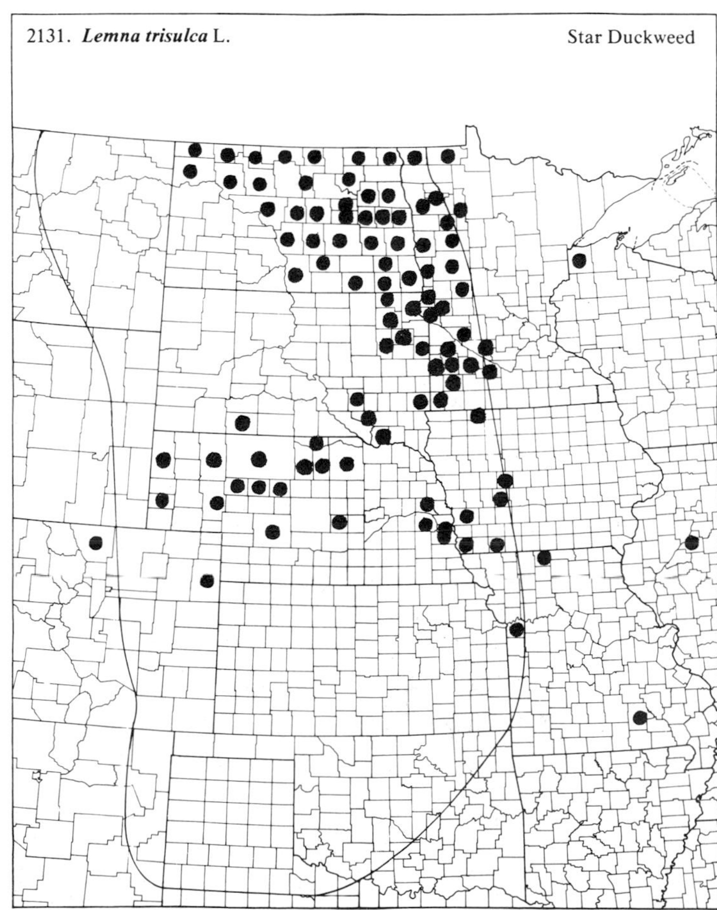

2132. ***Spirodela polyrhiza*** (L.) Schleid. Duck's Meat

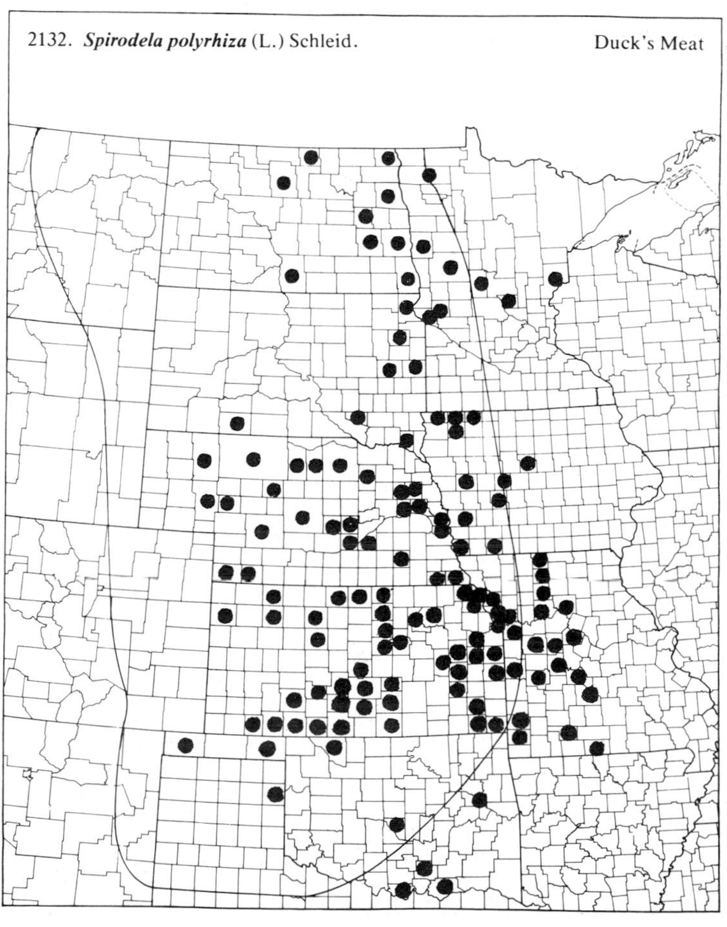

(Lemnaceae)

2133. ***Wolffia columbiana*** Karst.

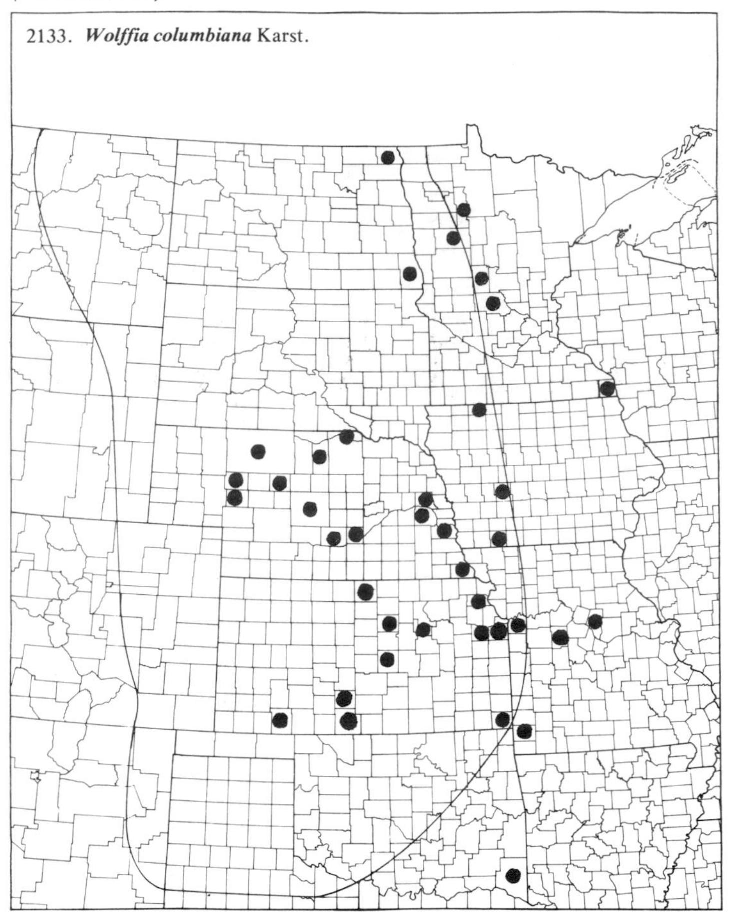

2134. ***Wolffia punctata*** Griseb.

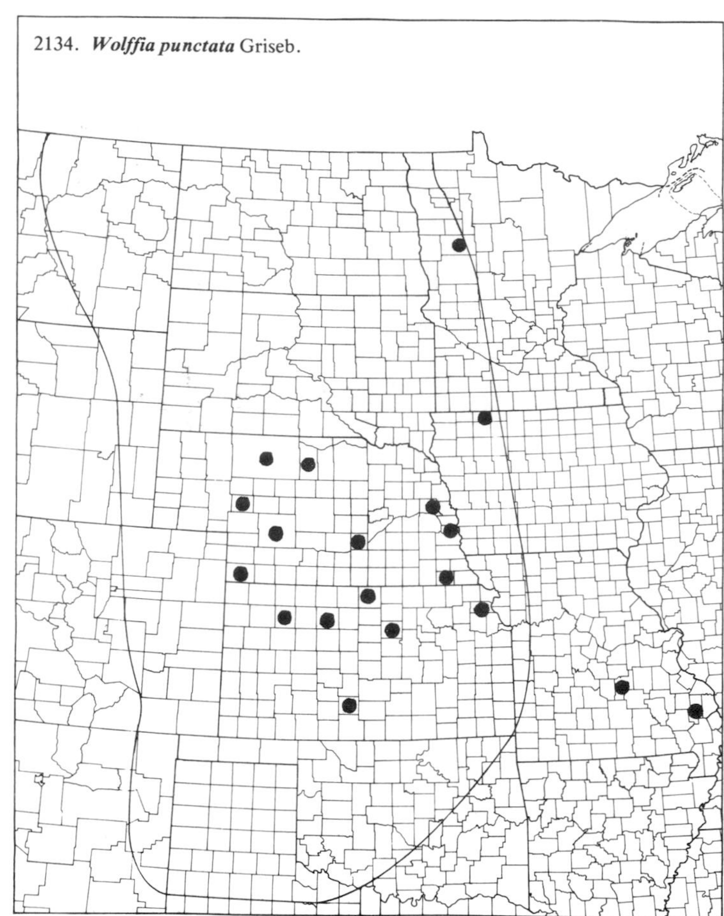

132. *Pontederiaceae* (Pickerelweed Family)

2135. ***Heteranthera dubia*** (Jacq.) MacM. — Water Stargrass

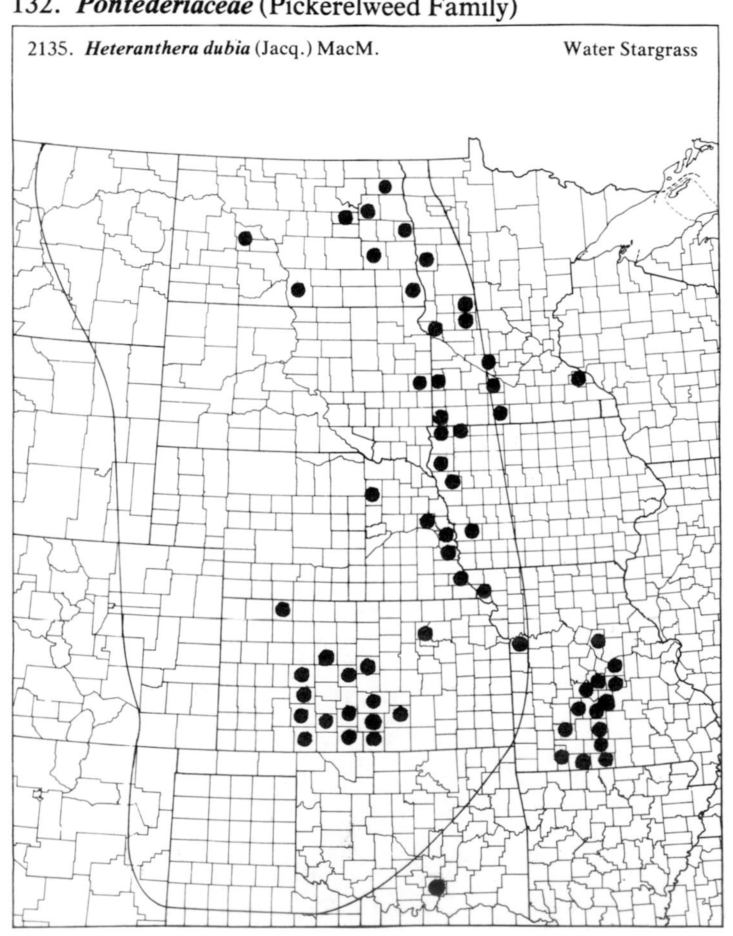

2136. ***Heteranthera limosa*** (Sw.) Willd. — Mud Plantain

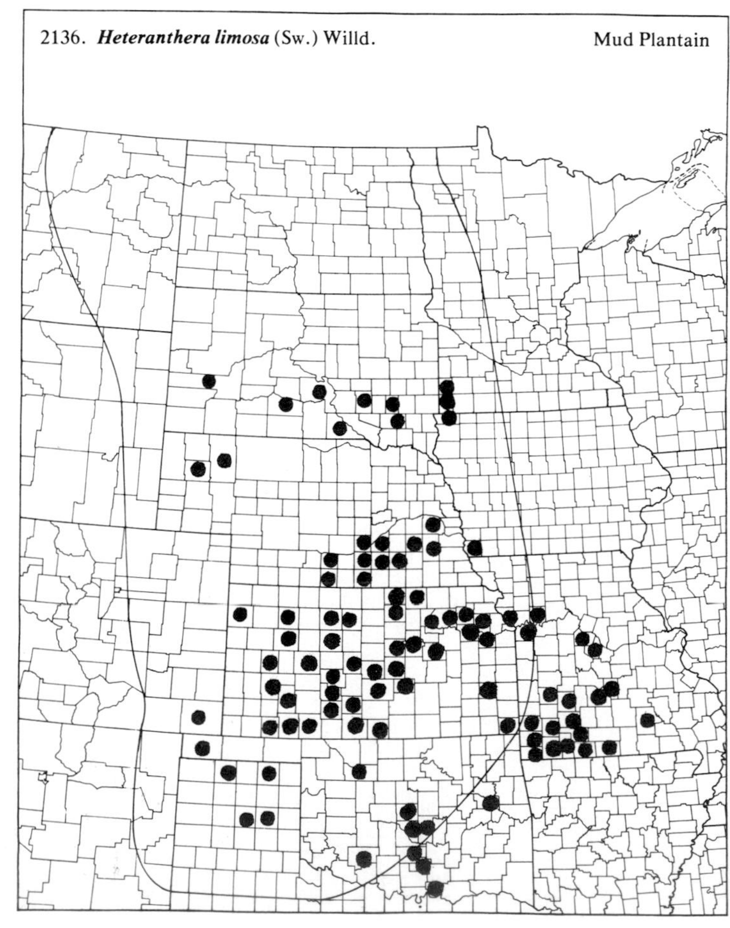

(Pontederiaceae)

2137. ***Heteranthera reniformis*** R & P.

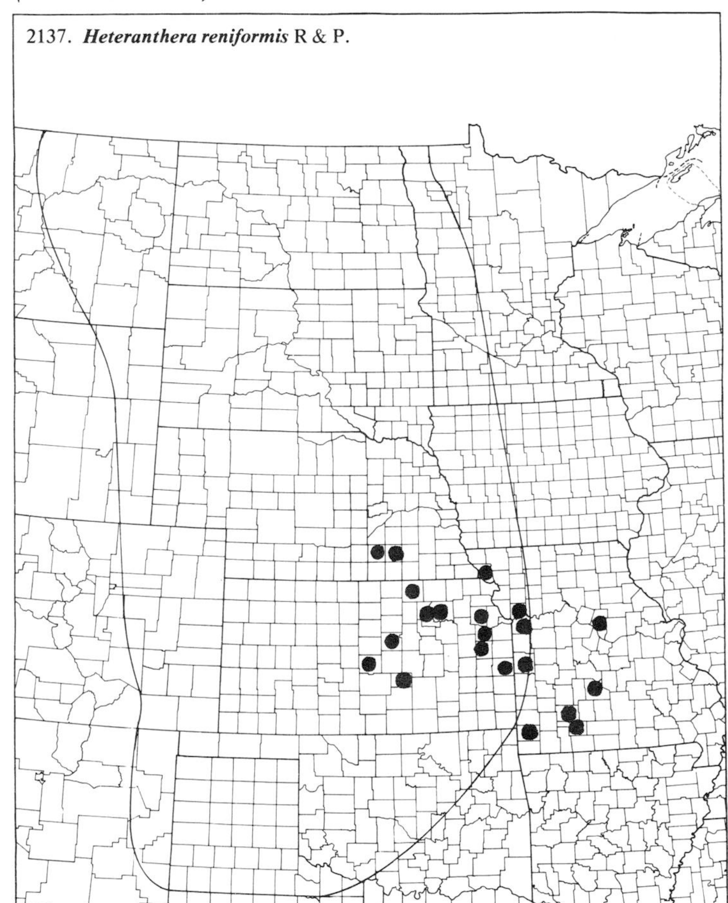

133. *Liliaceae* (Lily Family)

2138. ***Allium canadense*** L. var. ***canadense*** — Wild Onion

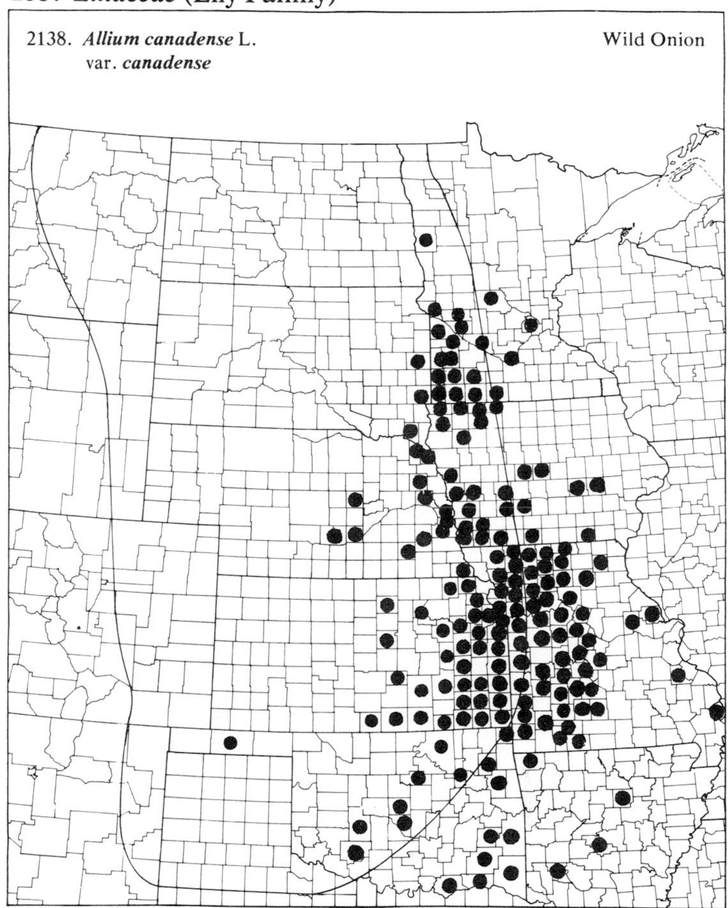

2139. ***Allium canadense*** L. var. ***fraseri*** M. Ownbey — Wild Onion

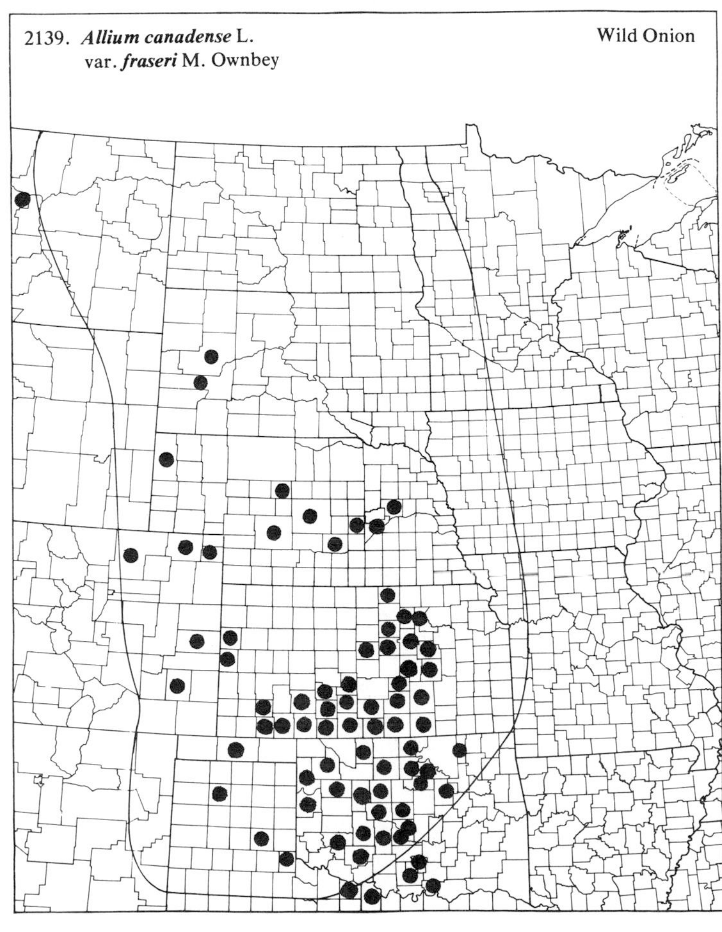

2140. ***Allium canadense*** L. var. ***hyacinthoides*** (Bush) M. Ownbey — Wild Onion

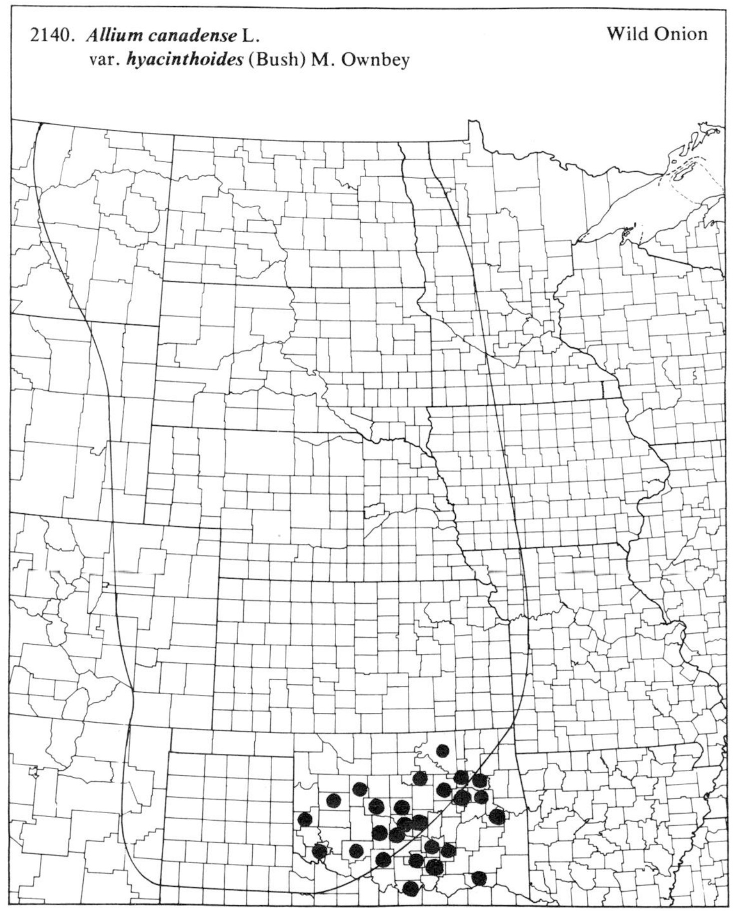

(Liliaceae)

2141. ***Allium canadense*** L. var. ***lavendulare*** (Bates) M. Ownbey & Aase — Wild Onion

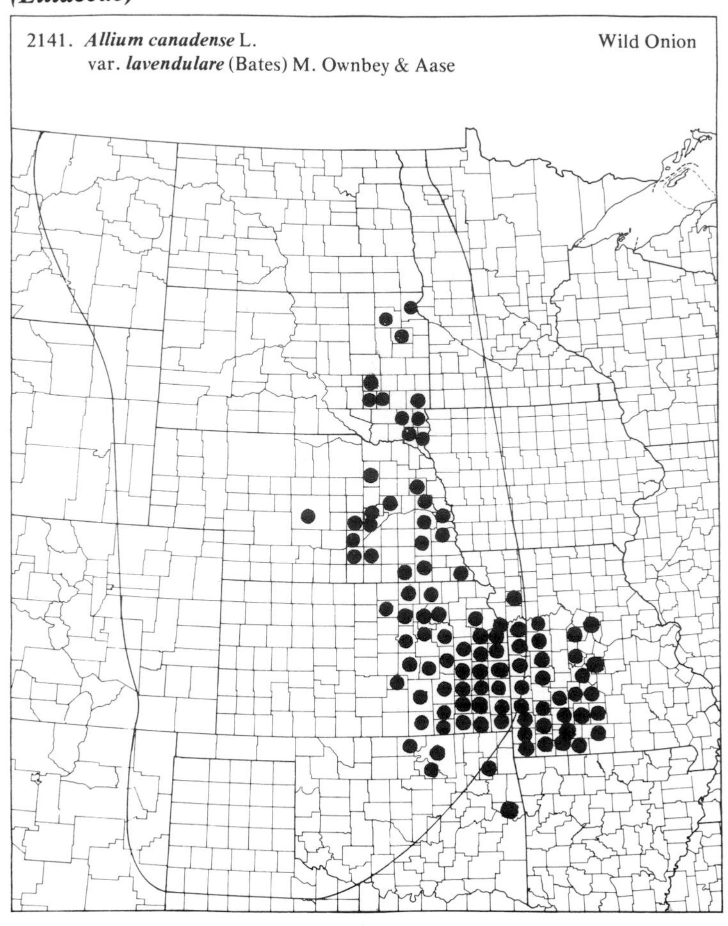

2142. ***Allium cernuum*** Roth — Wild Onion

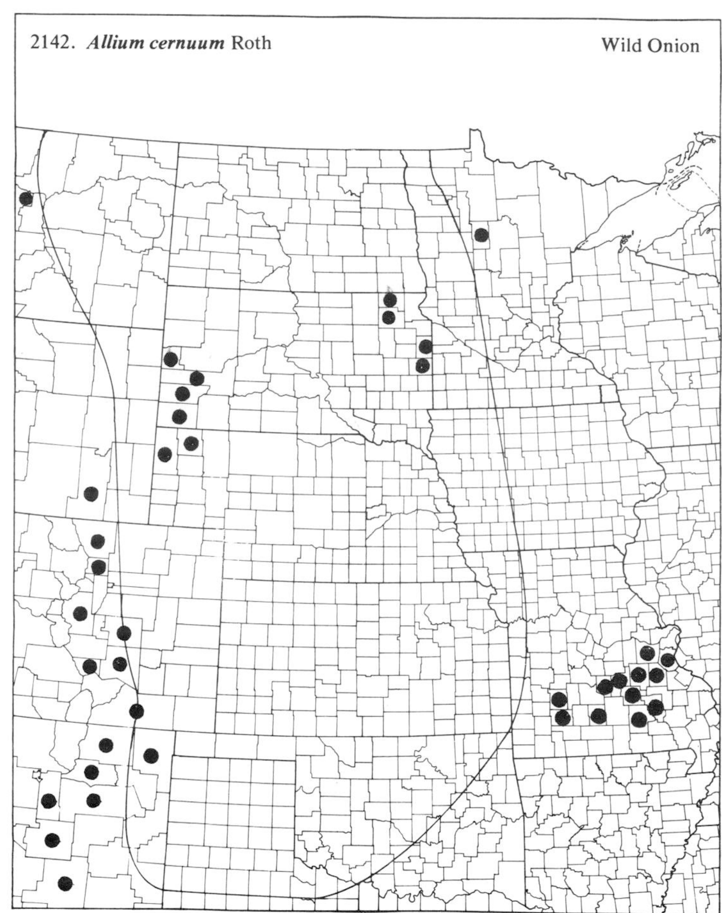

2143. ***Allium drummondii*** Regel — Wild Onion

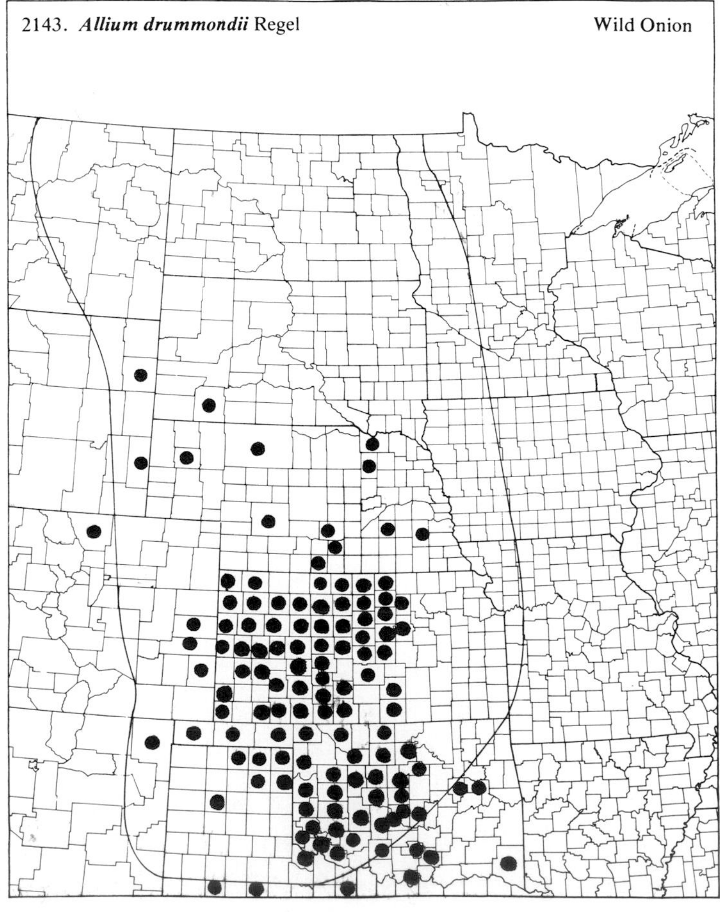

2144. ***Allium perdulce*** S.V. Fraser

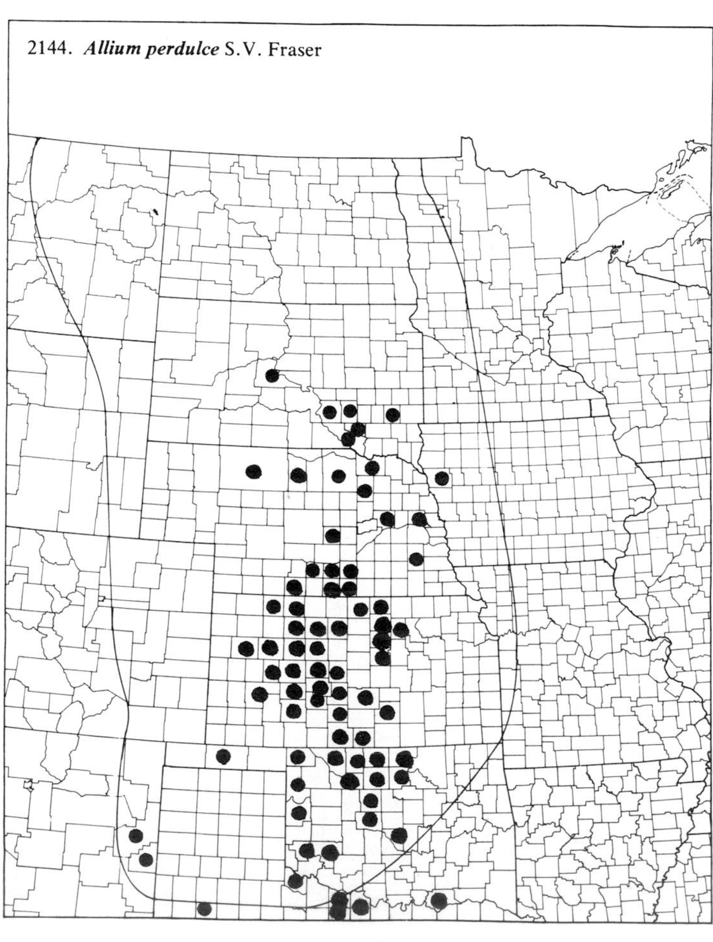

(Liliaceae)

2145. ***Allium stellatum*** Ker — Pink Wild Onion

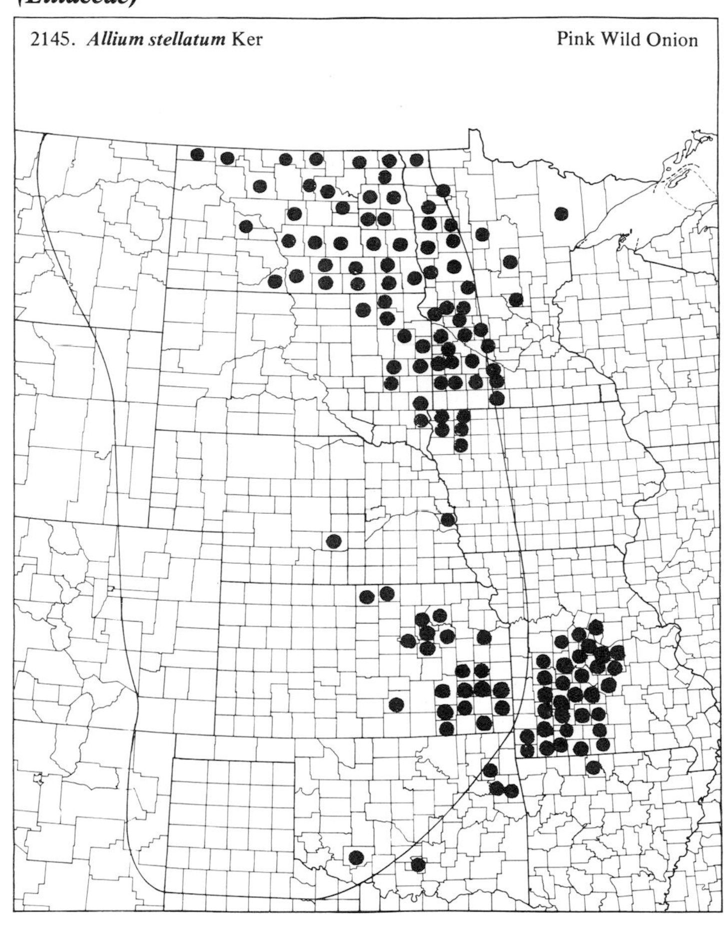

2146. ***Allium textile*** A. Nels. & Macbr. — White Wild Onion

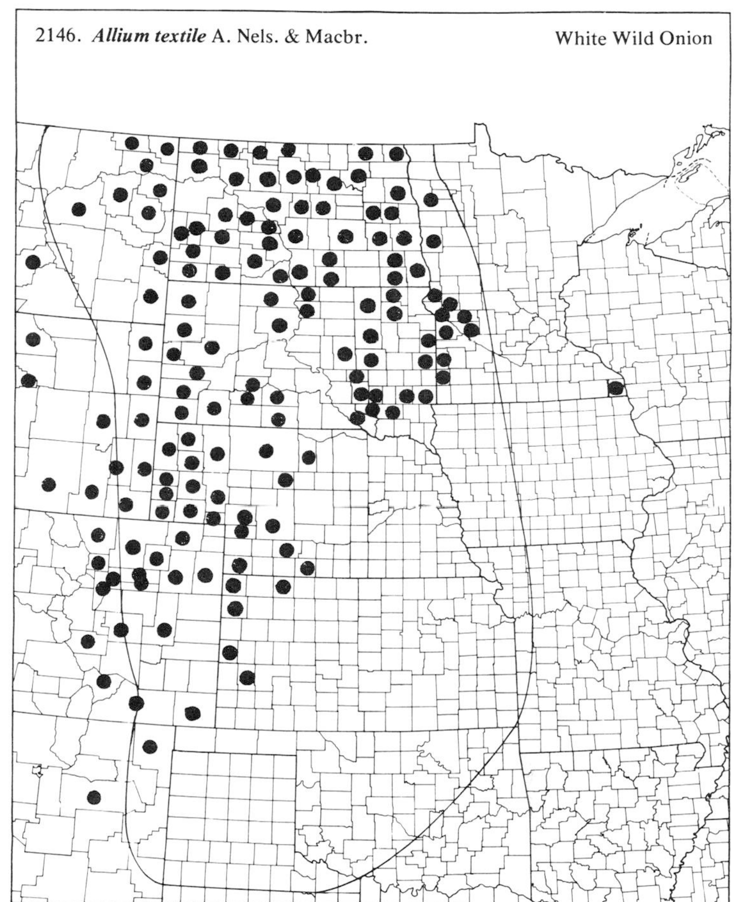

2147. ***Allium tricoccum*** Ait. — Wild Leek

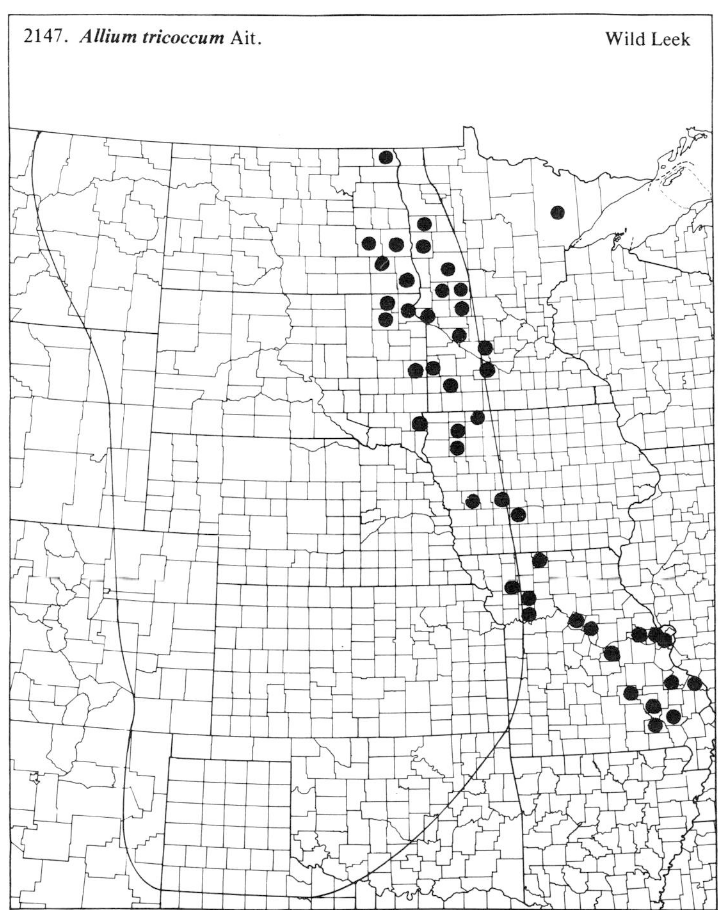

2148. ***Allium vineale*** L. — Wild Garlic

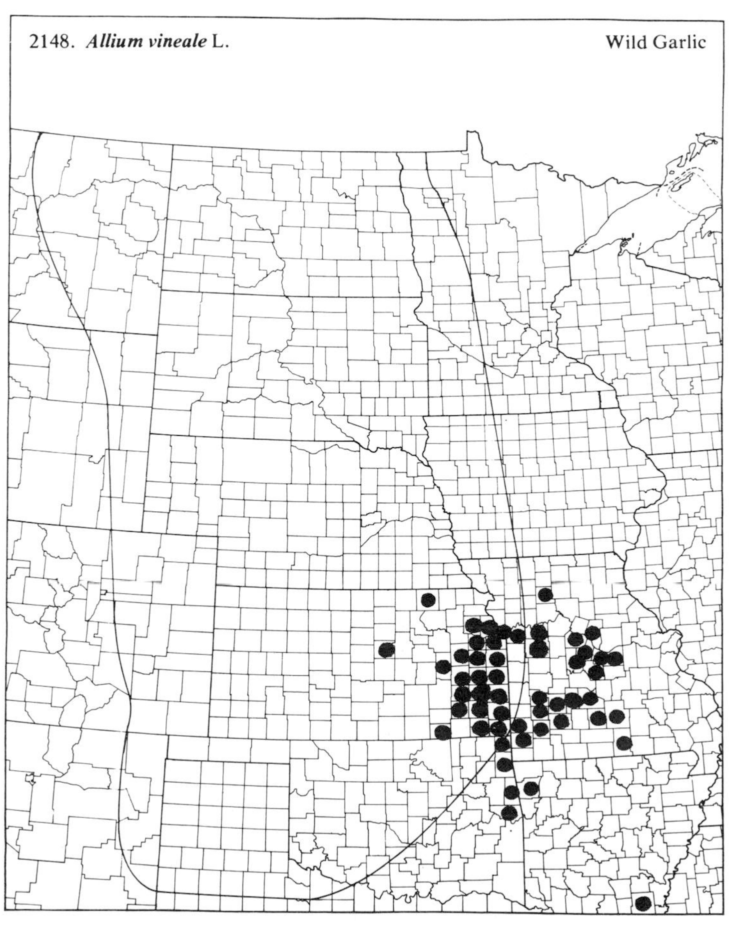

(Liliaceae)

2149. ***Androstephium caeruleum*** (Scheele) Greene

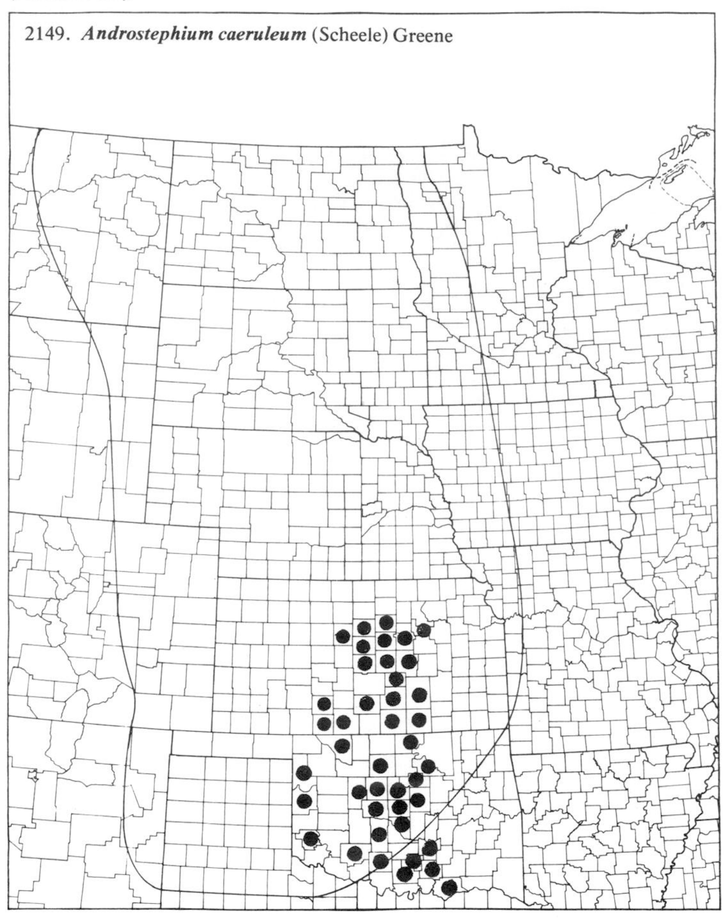

2150. ***Asparagus officinalis*** L. Asparagus

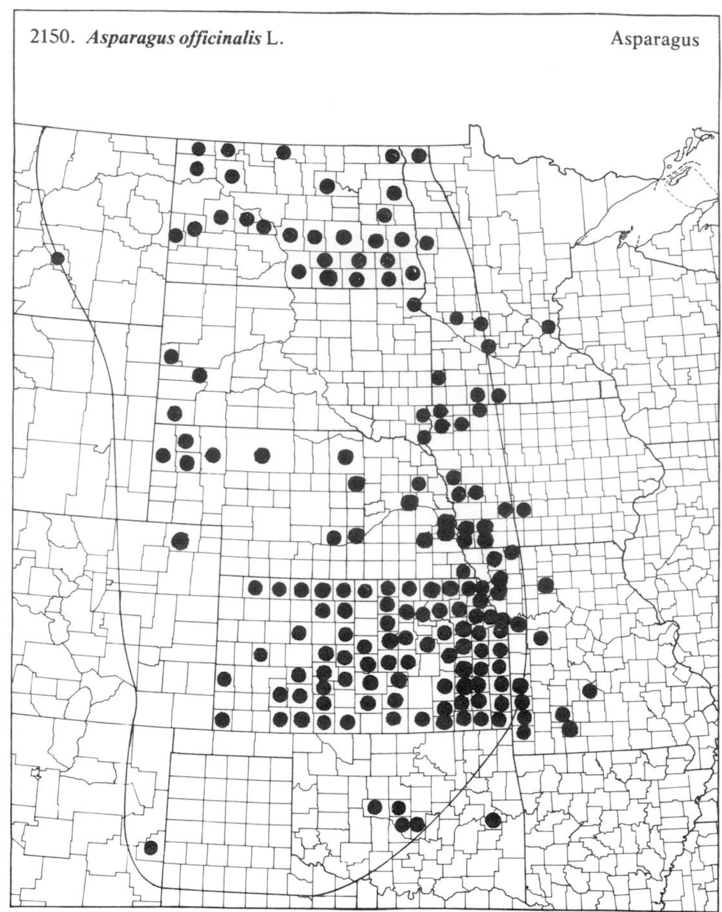

2151. ***Calochortus gunnisonii*** Wats. Sego Lily

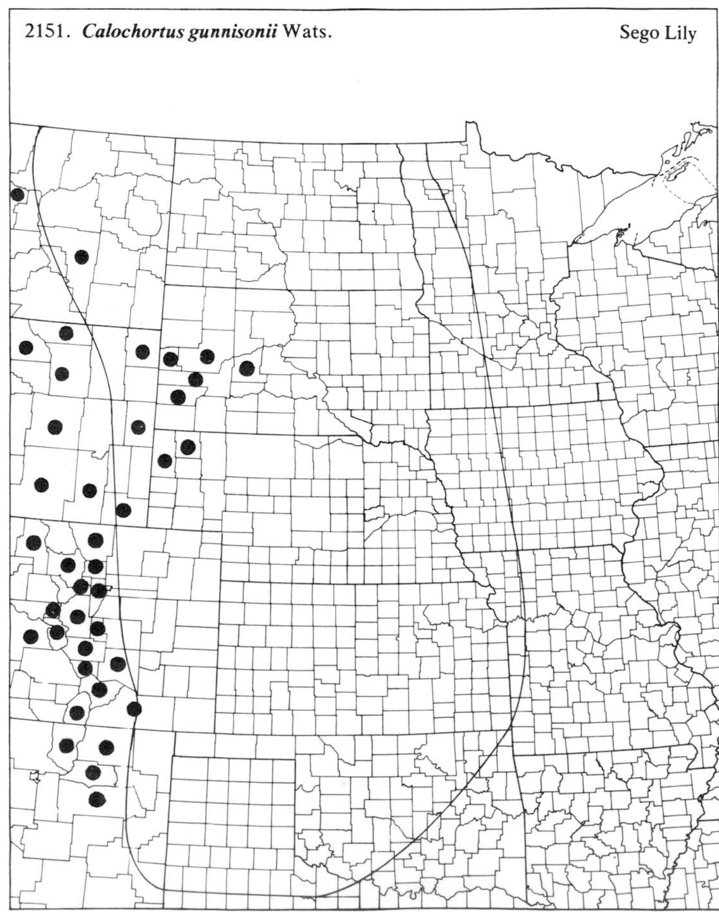

2152. ***Calochortus nuttallii*** T. & G. Marisposa Lily

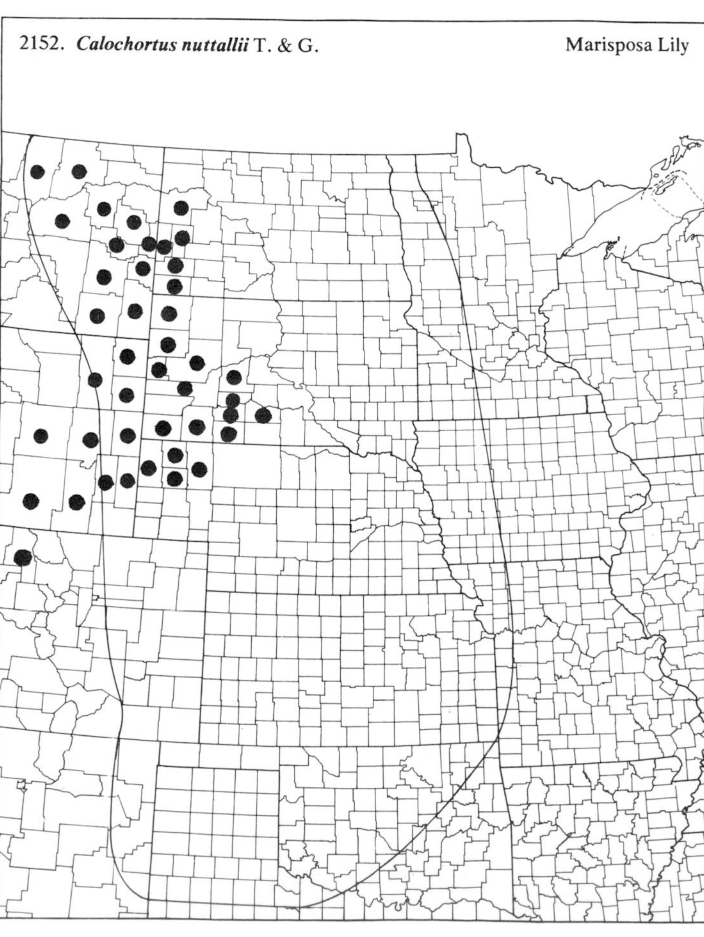

(Liliaceae)

2153. ***Camassia angusta*** (Engelm. & Gray) Blank.

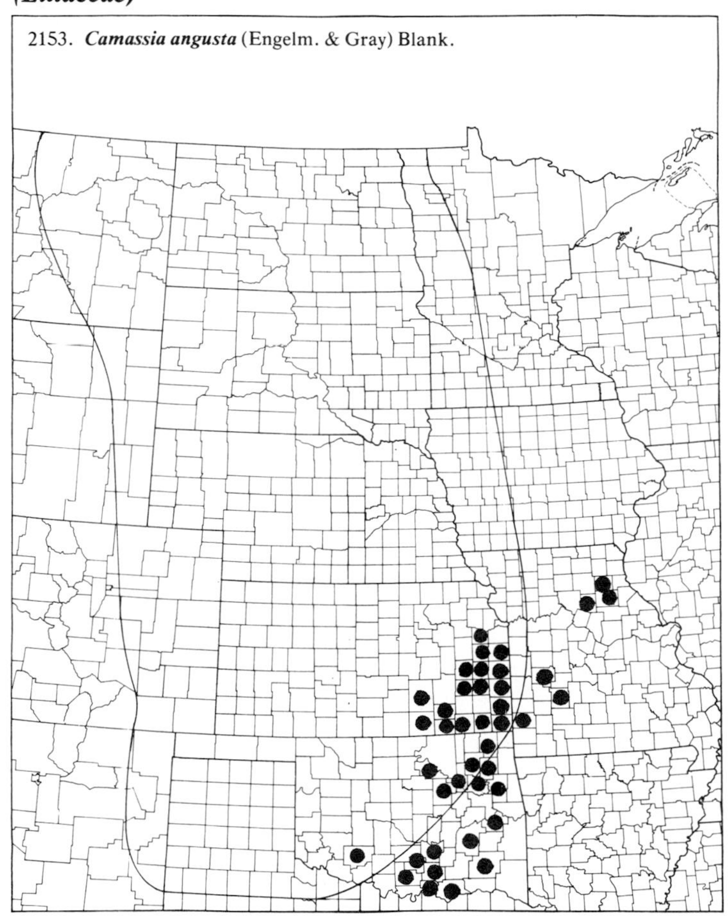

2154. ***Camassia scilloides*** (Raf.) Cory — Eastern Camass

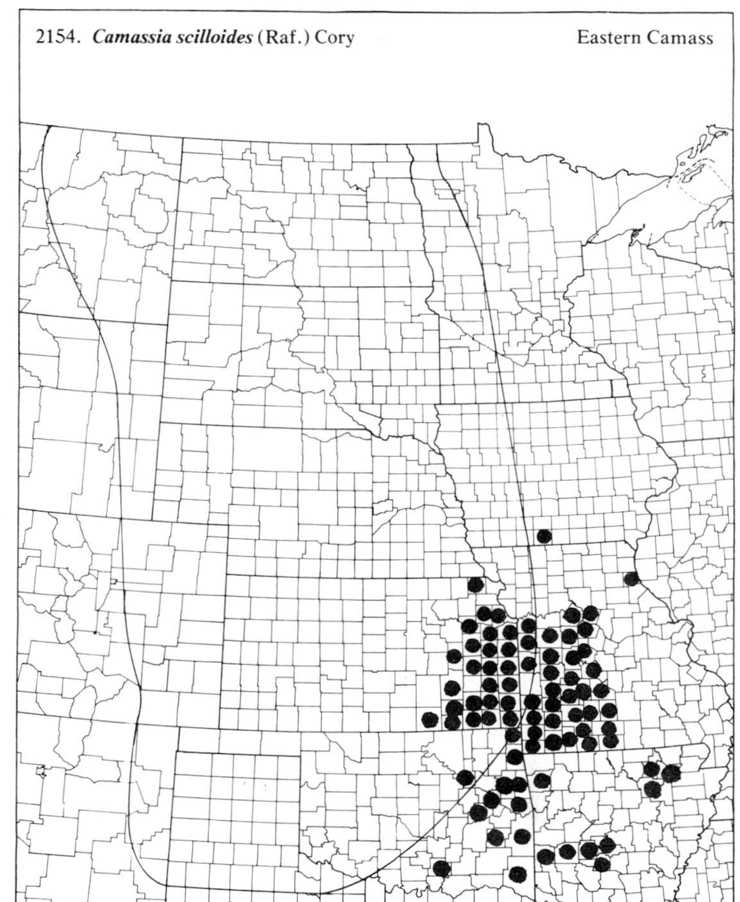

2155. ***Cooperia drummondii*** Herb.

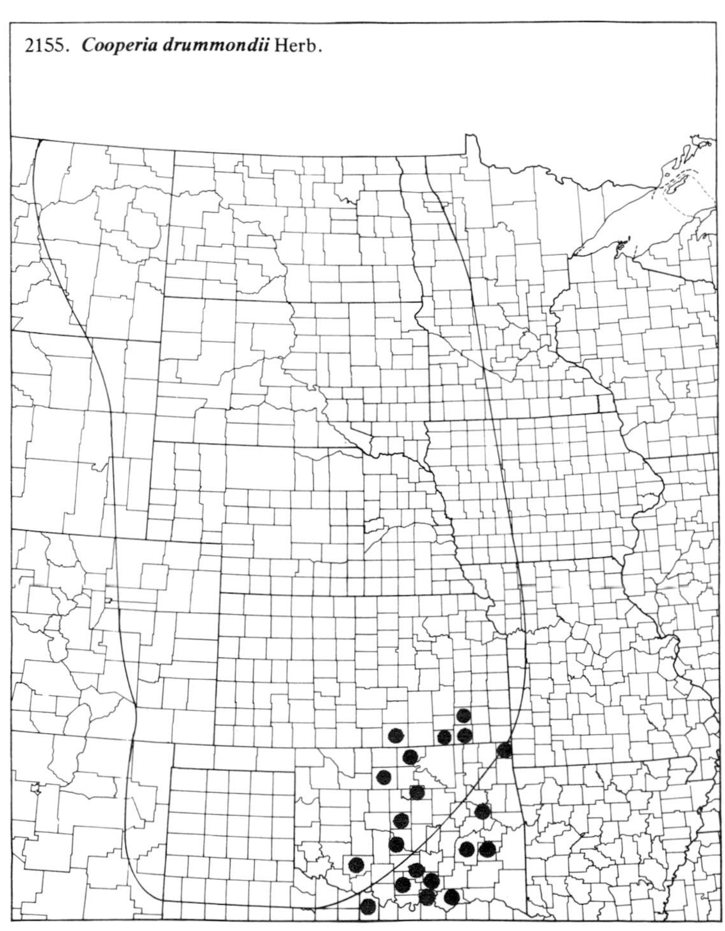

2156. ***Disporum trachycarpum*** (Wats.) Benth. & Hook. — Fairybells

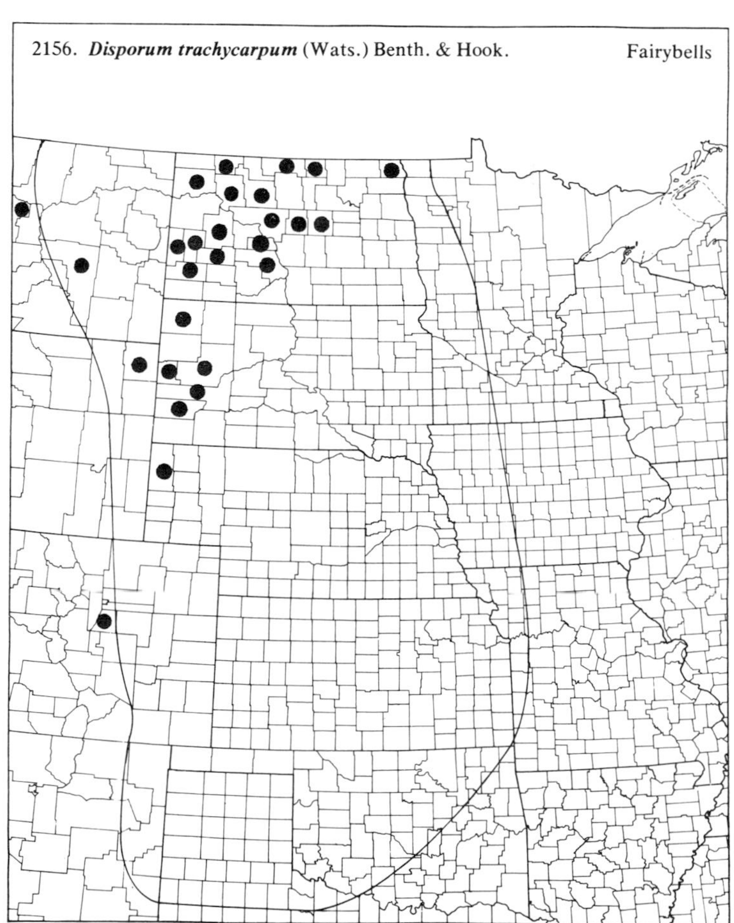

(Liliaceae)

2157. ***Erythronium albidum*** Nutt. White Dog's-tooth Violet

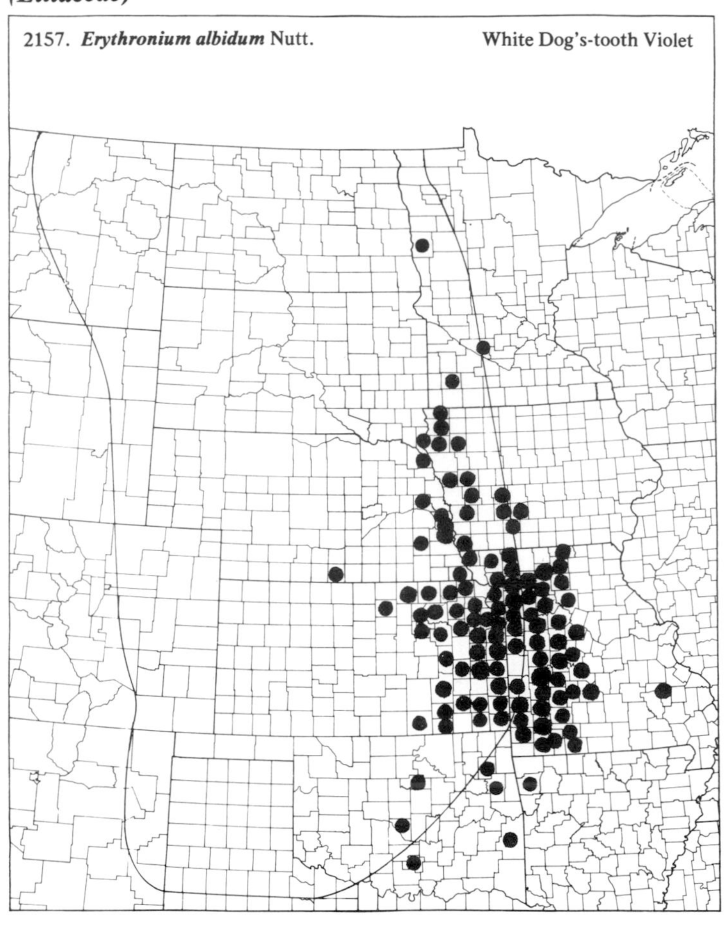

2158. ***Erythronium mesochoreum*** Knerr White Dog's-tooth Violet

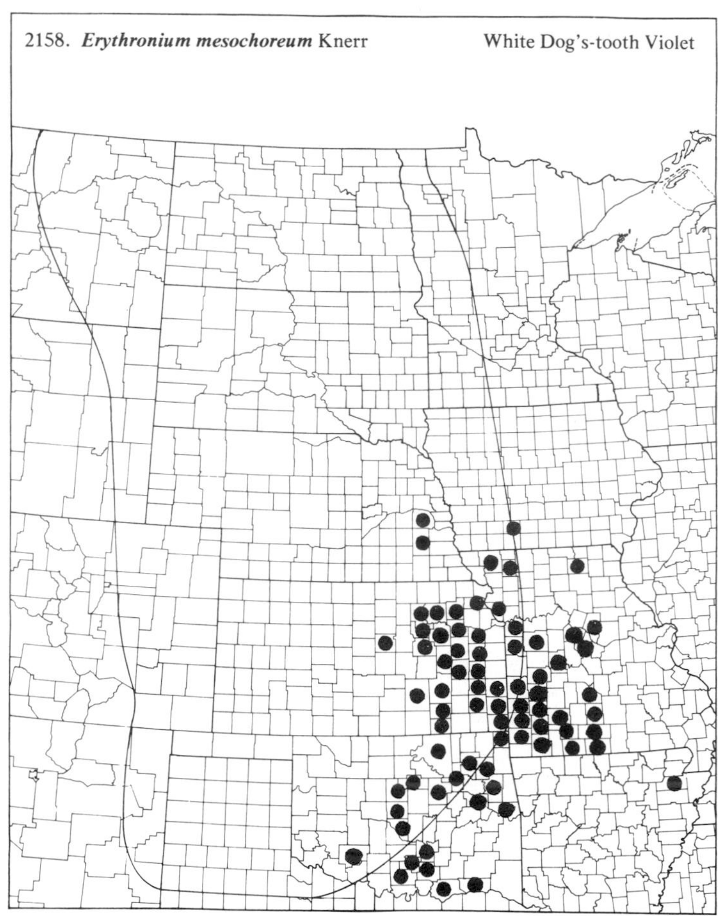

2159. ***Fritillaria atropurpurea*** Nutt. Leopard Lily

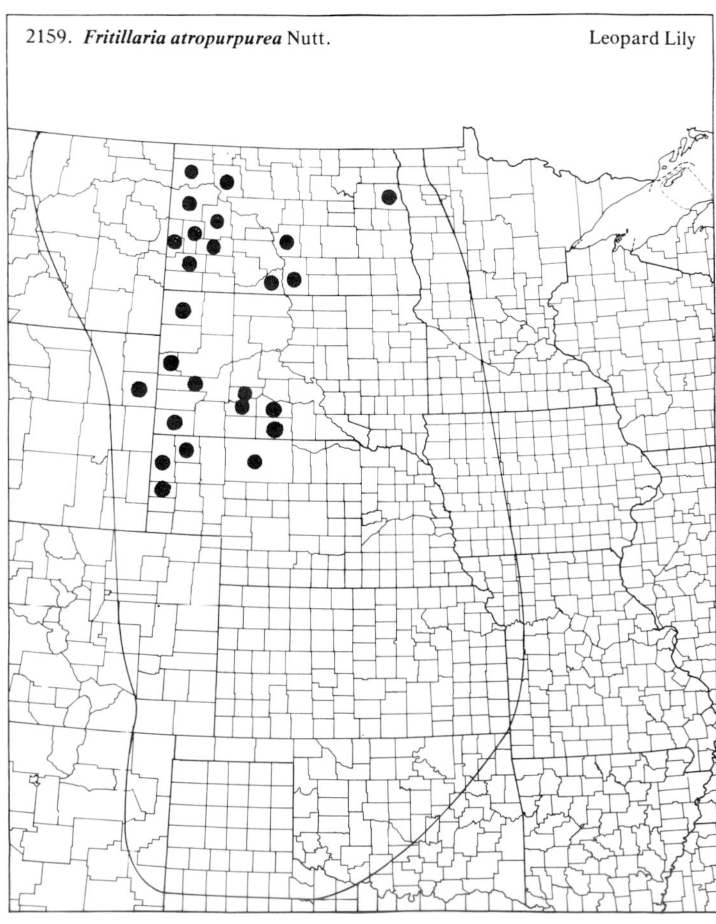

2160. ***Hemerocallis fulva*** (L.) L. Day Lily

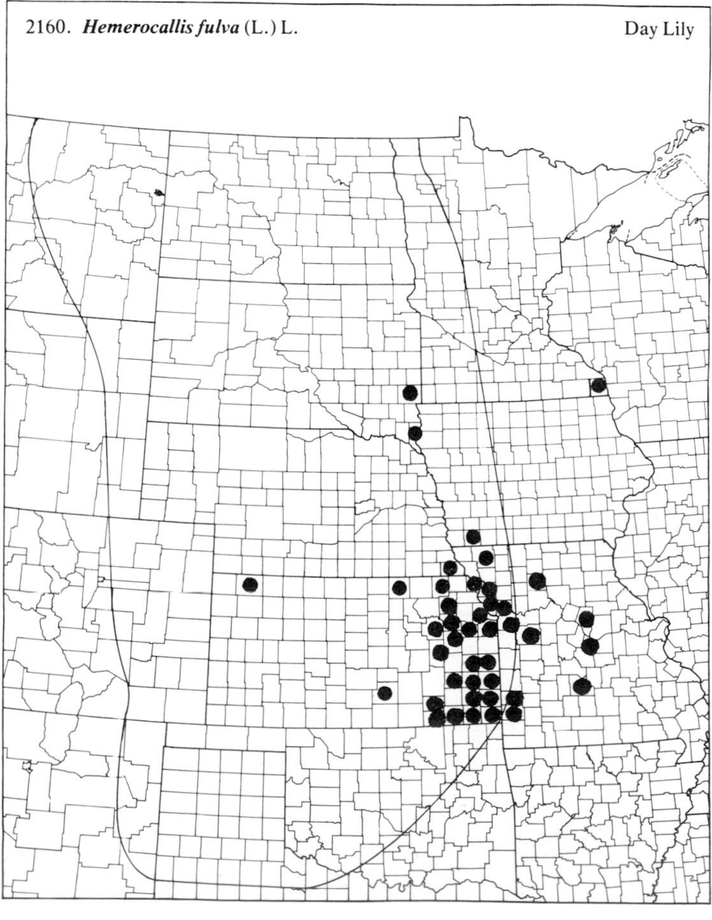

(Liliaceae)

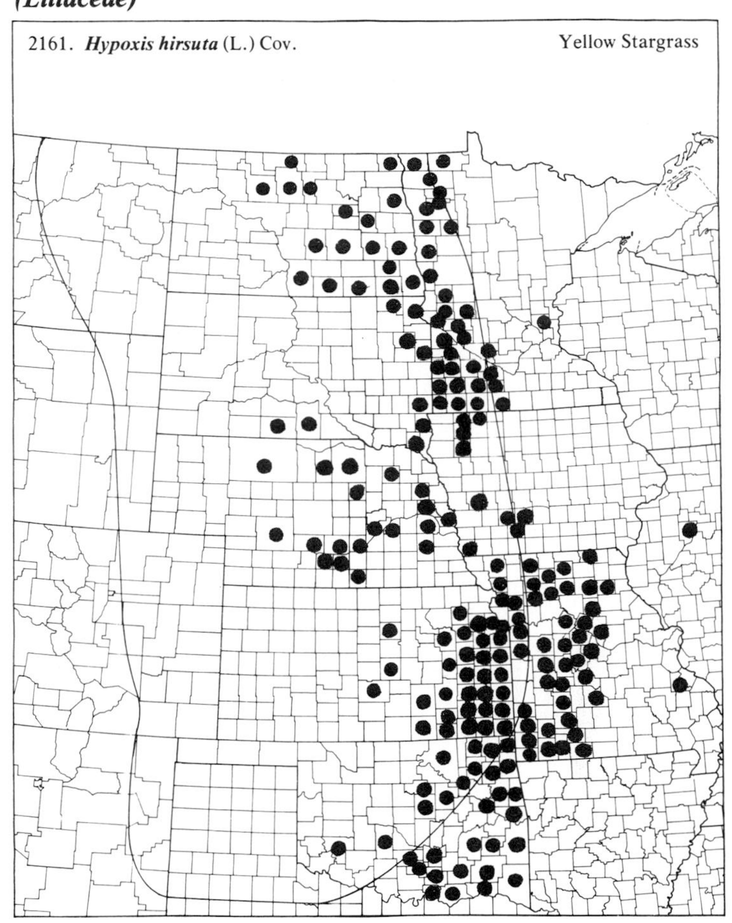
2161. Hypoxis hirsuta (L.) Cov.
Yellow Stargrass

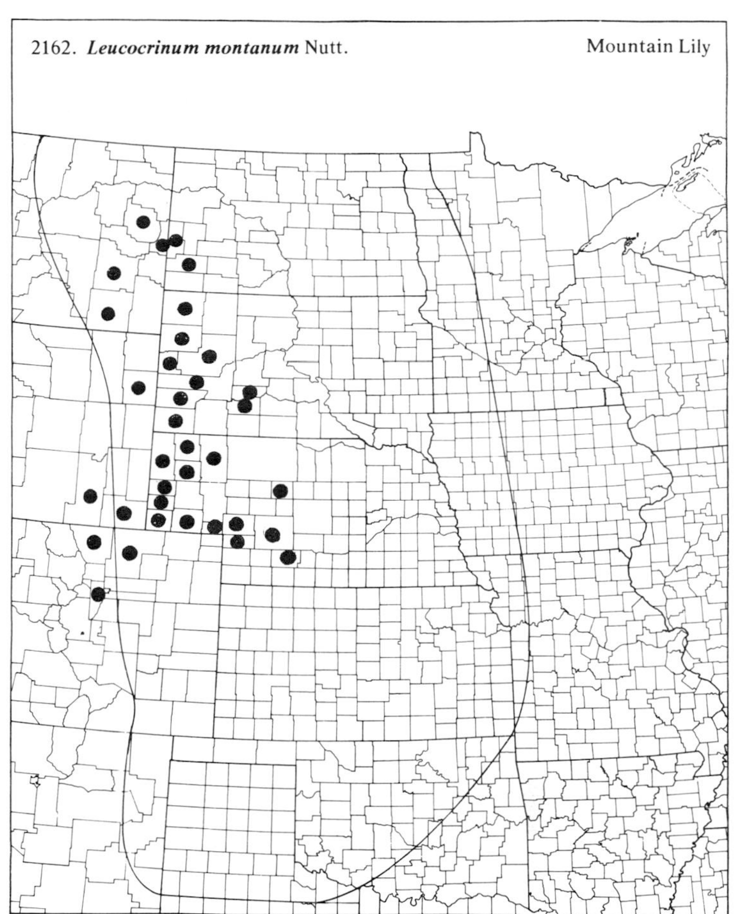
2162. Leucocrinum montanum Nutt.
Mountain Lily

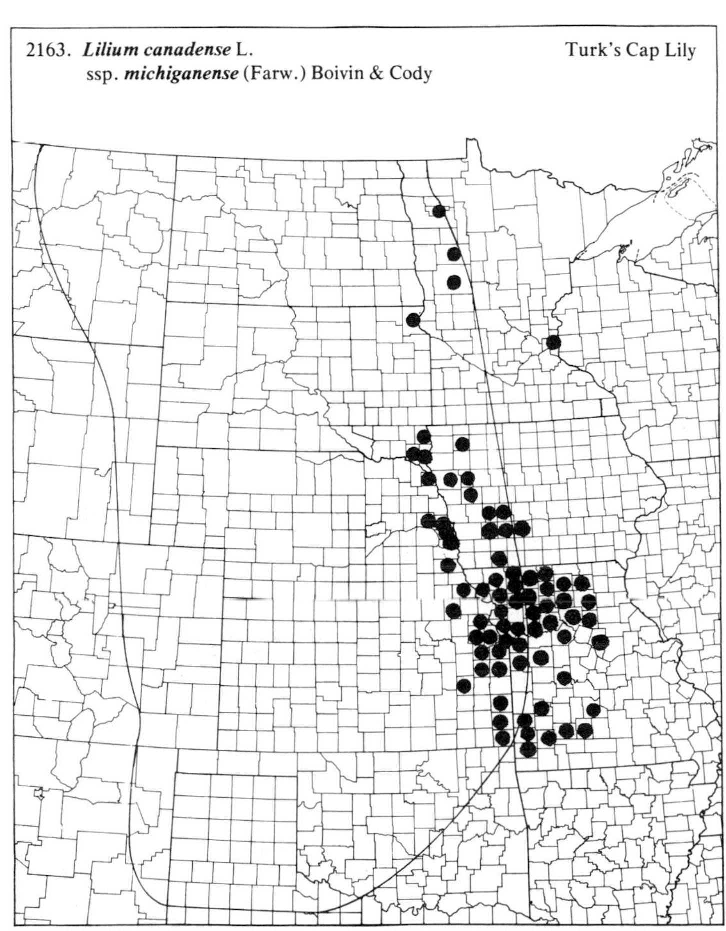
2163. Lilium canadense L.
ssp. michiganense (Farw.) Boivin & Cody
Turk's Cap Lily

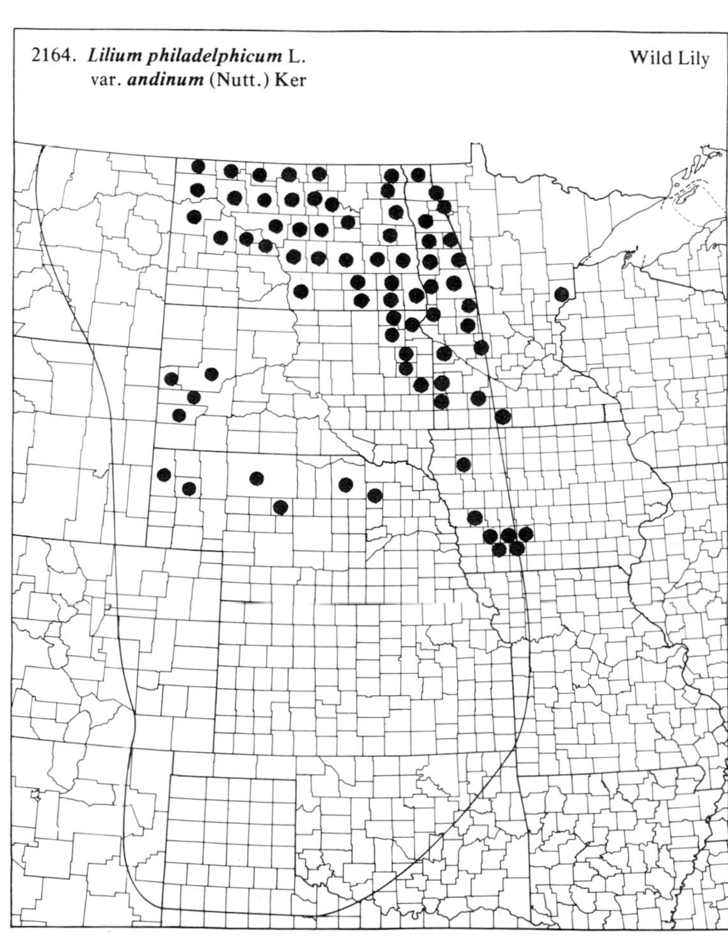
2164. Lilium philadelphicum L.
var. andinum (Nutt.) Ker
Wild Lily

(Liliaceae)

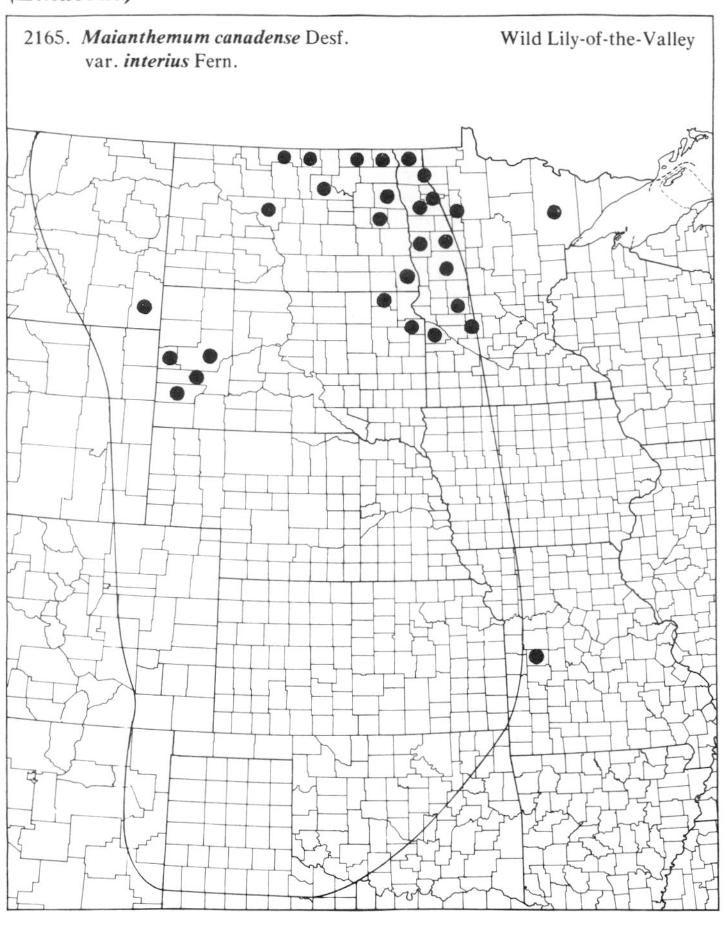

2165. ***Maianthemum canadense*** Desf. var. ***interius*** Fern. — Wild Lily-of-the-Valley

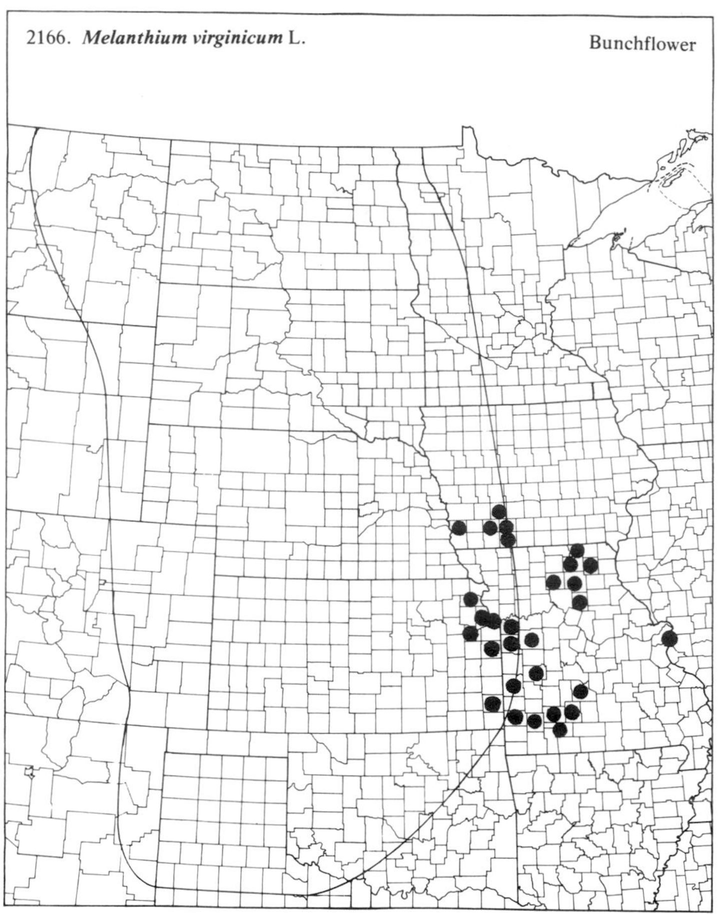

2166. ***Melanthium virginicum*** L. — Bunchflower

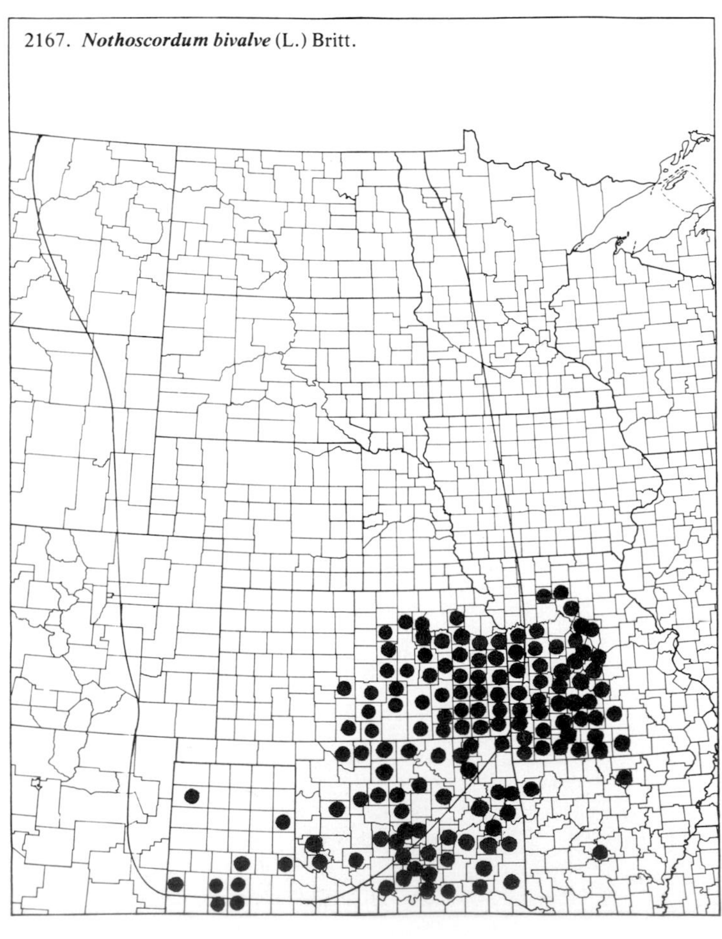

2167. ***Nothoscordum bivalve*** (L.) Britt.

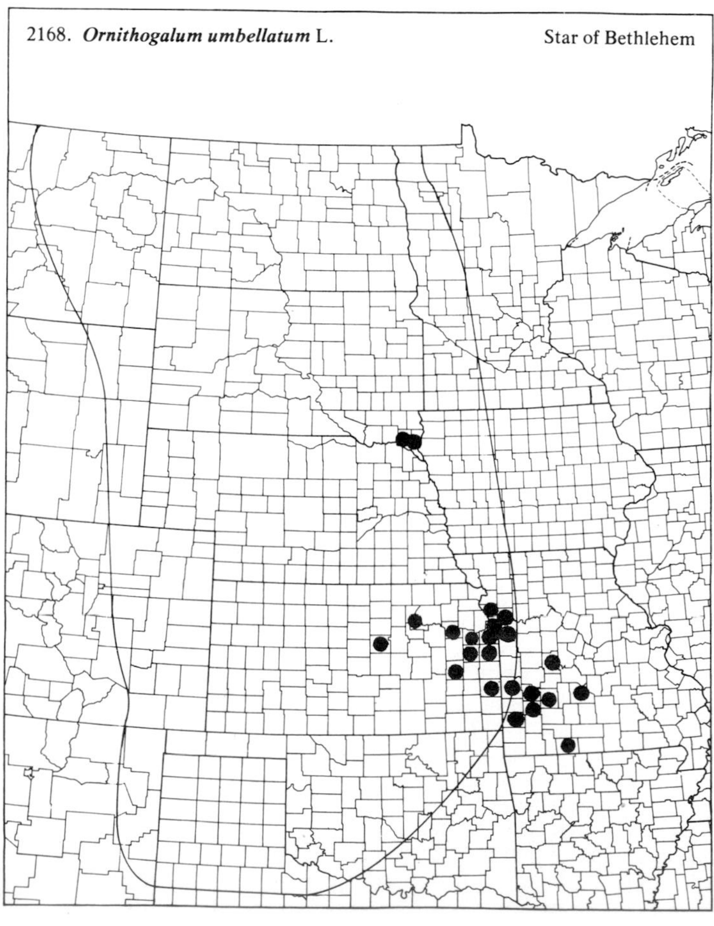

2168. ***Ornithogalum umbellatum*** L. — Star of Bethlehem

(Liliaceae)

2169. ***Polygonatum biflorum*** (Walt.) Ell. Solomon's Seal

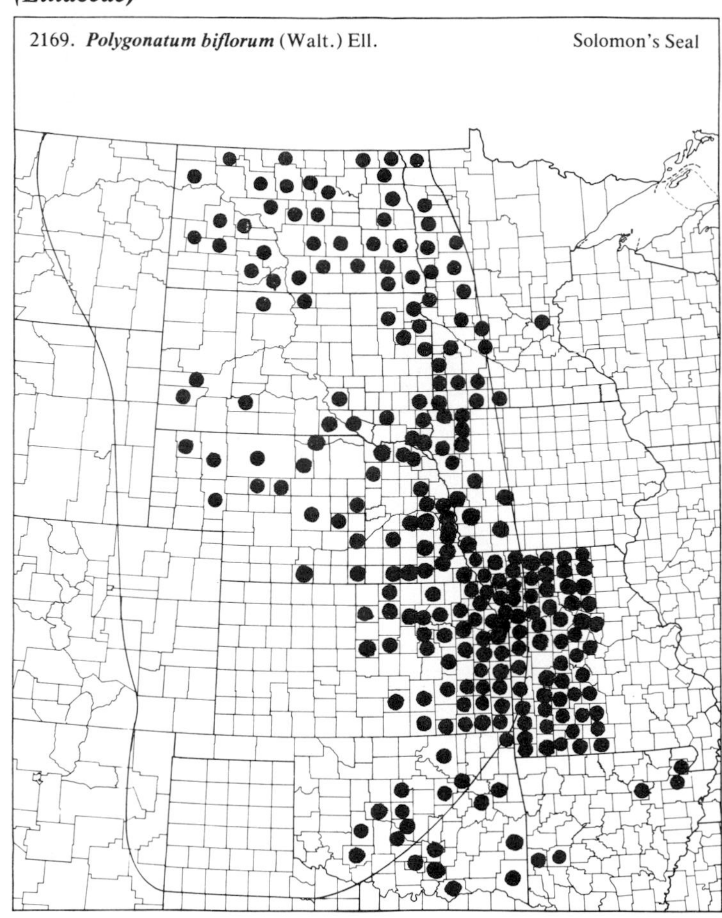

2170. ***Smilacina racemosa*** (L.) Desf. False Spikenard

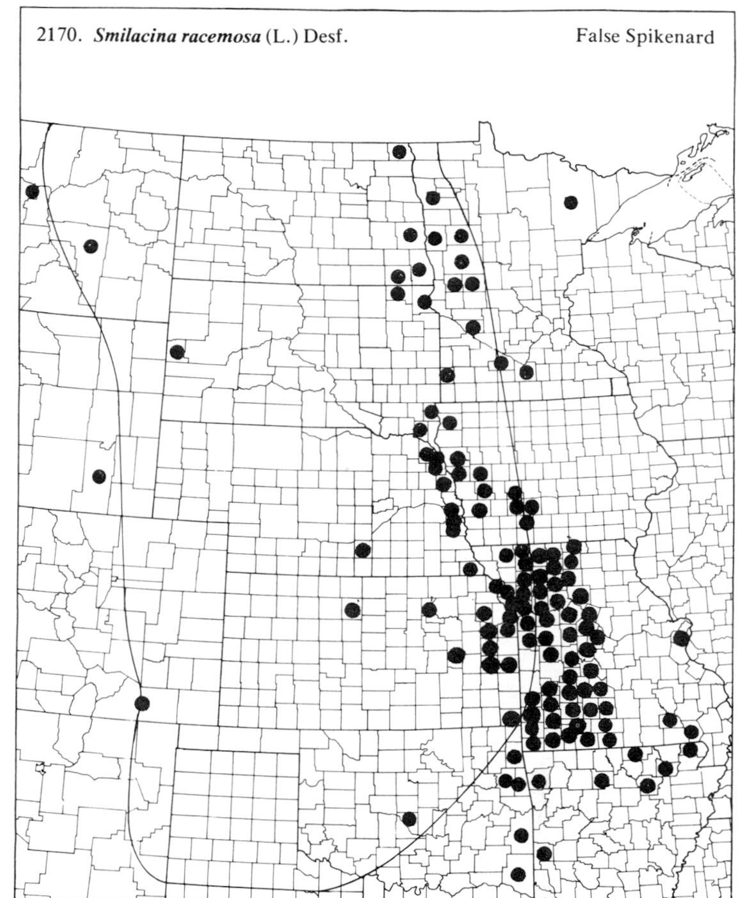

2171. ***Smilacina stellata*** (L.) Desf. Spikenard

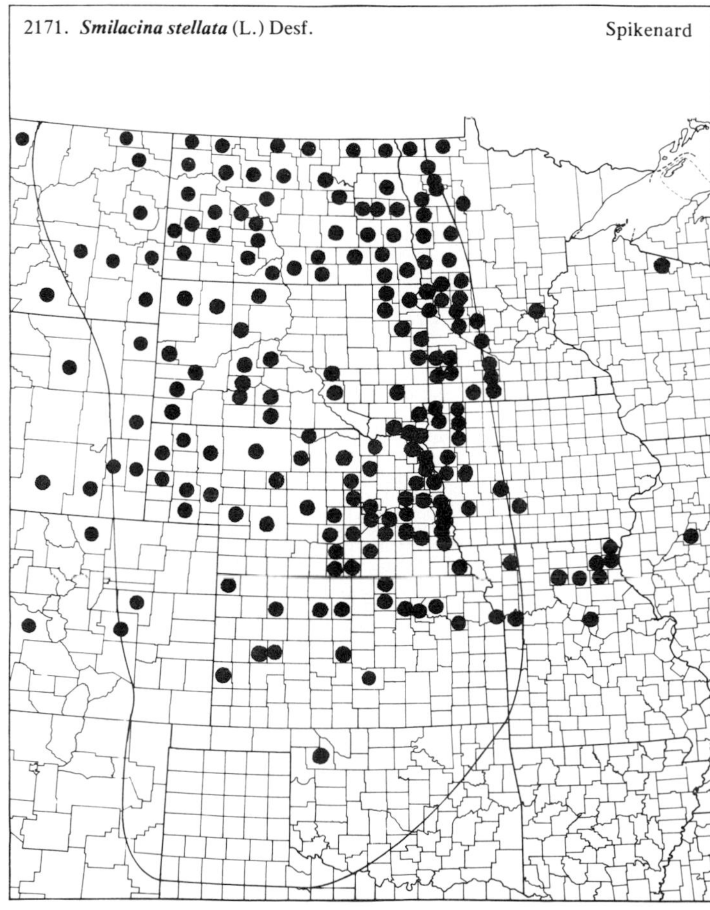

2172. ***Smilax bona-nox*** L. Greenbriar

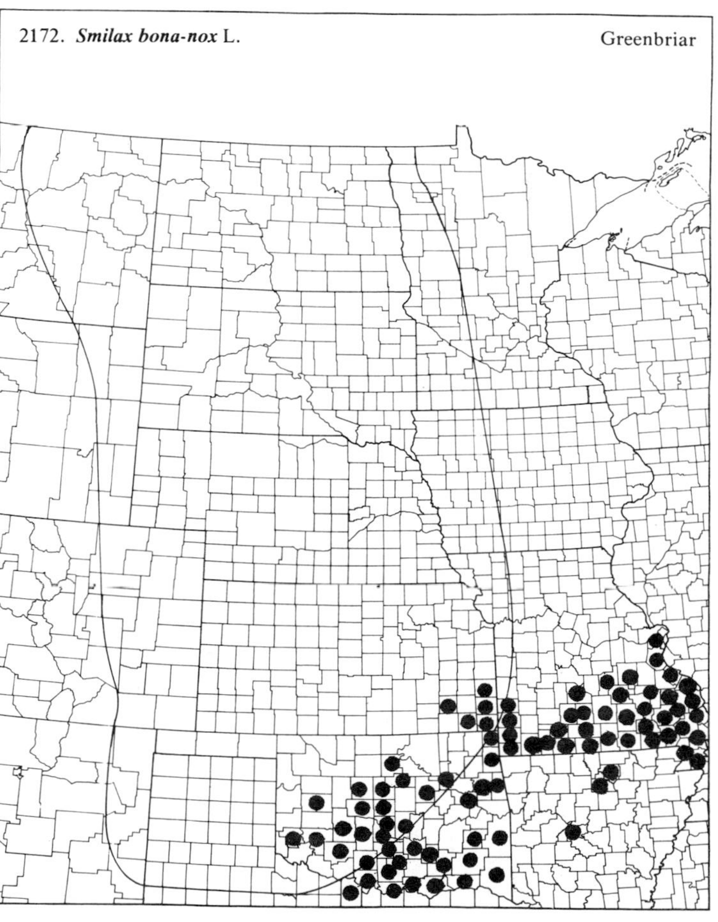

(Liliaceae)

2173. ***Smilax ecirrhata*** (Engelm. ex Kunth) Wats. Greenbriar

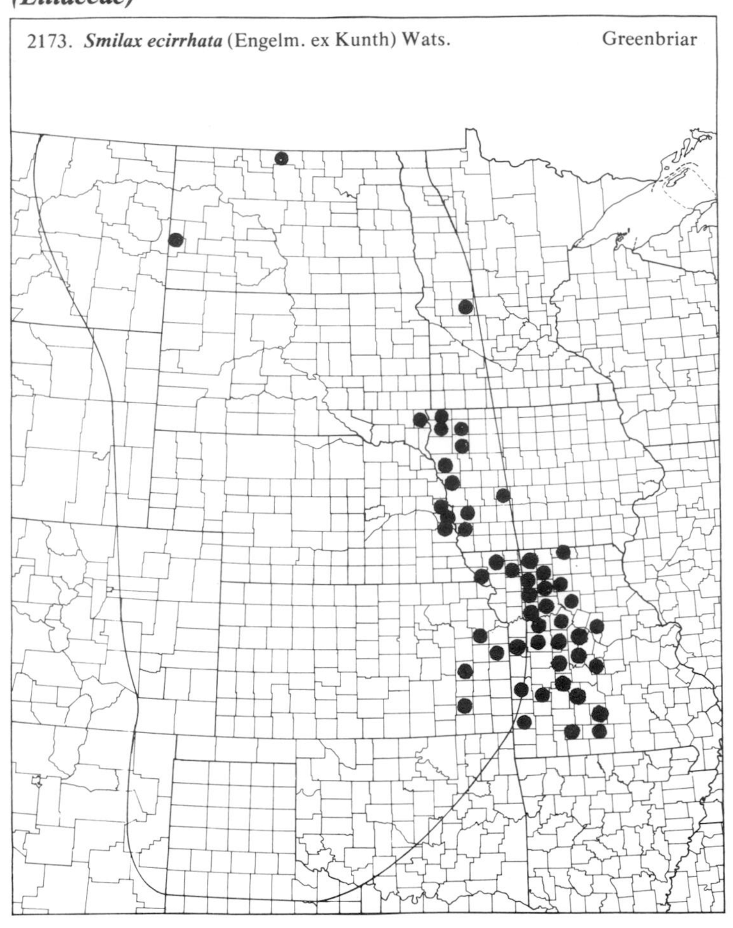

2174. ***Smilax herbacea*** L. var. ***herbacea*** Carrion-flower

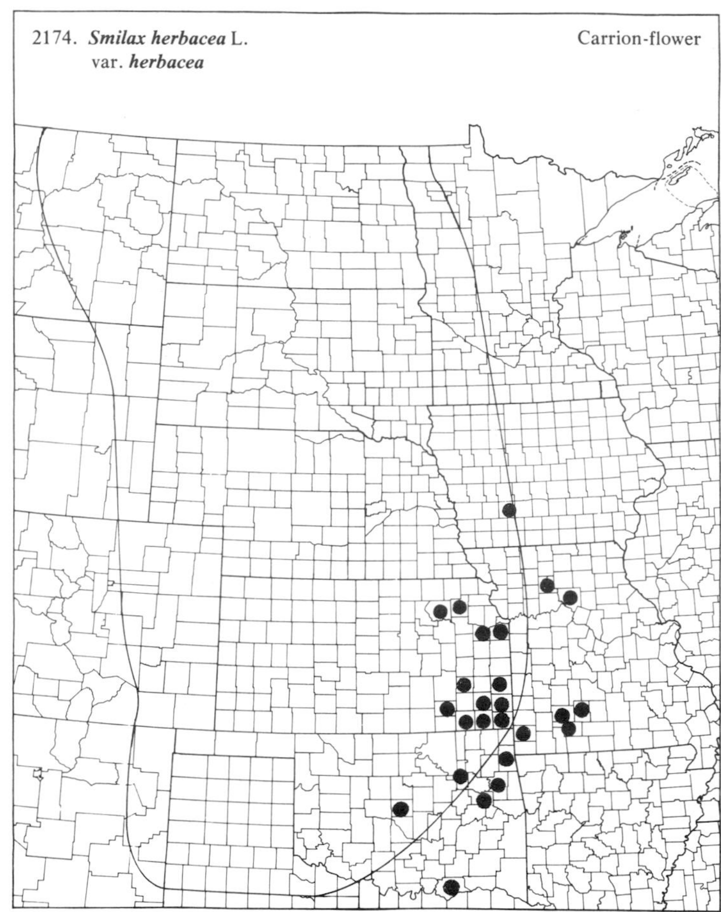

2175. ***Smilax herbacea*** L. var. ***lasioneuron*** (Small) Rydb. Carrion-flower

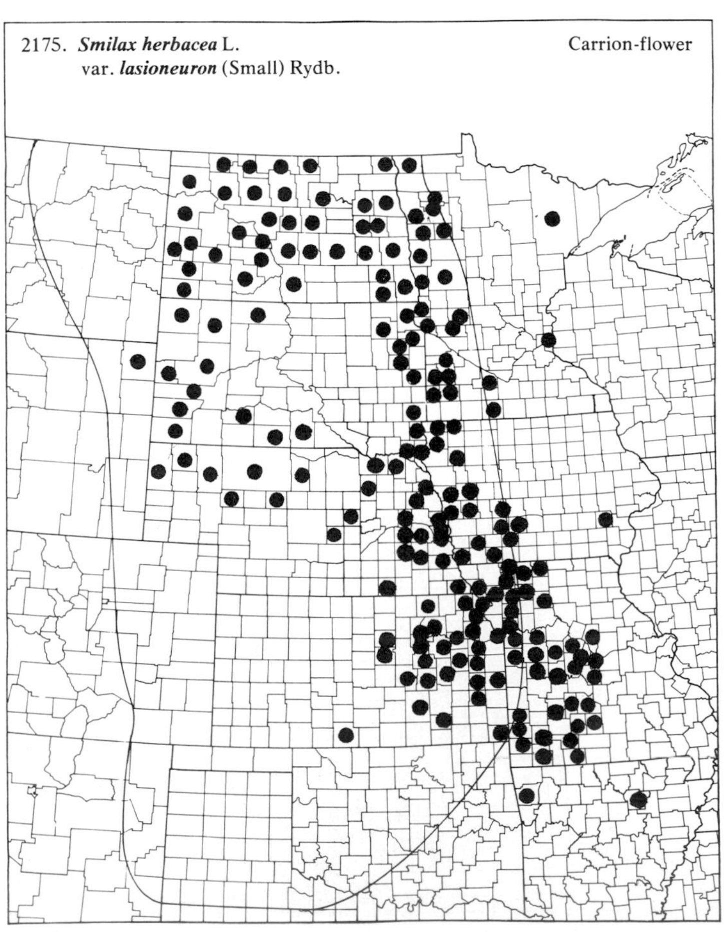

2176. ***Smilax hispida*** Muhl. Bristly Greenbriar

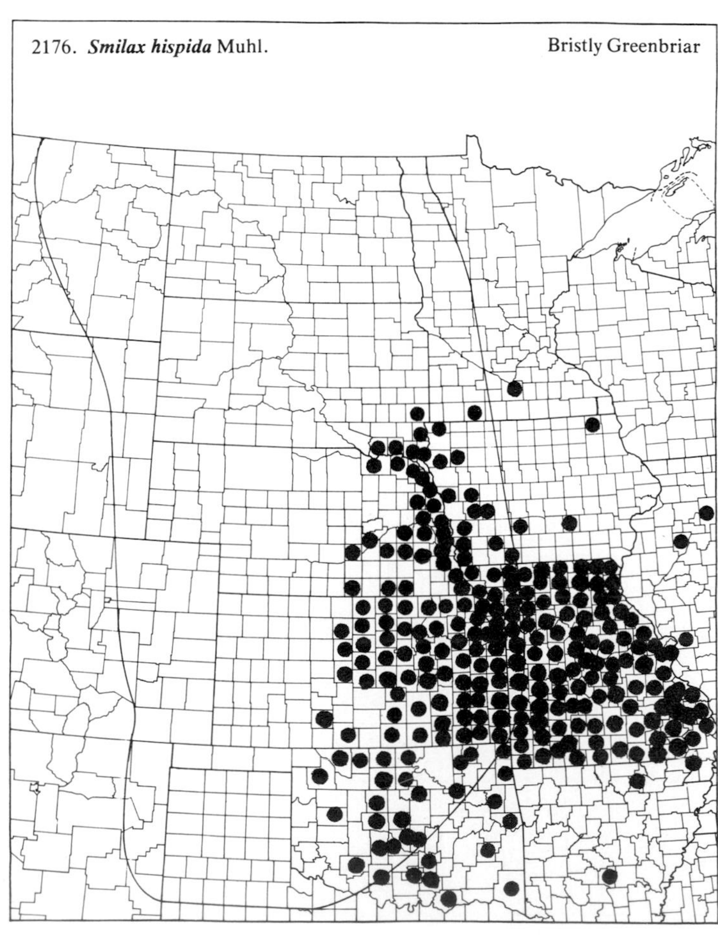

(Liliaceae)

2177. ***Trillium cernuum*** L. Nodding Trillium

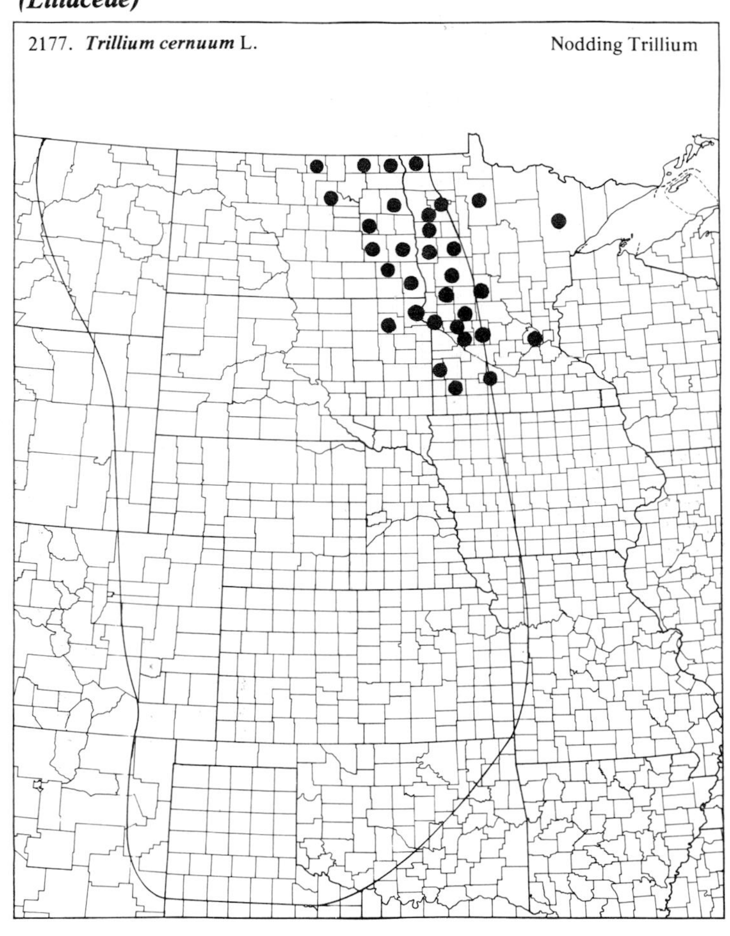

2178. ***Trillium sessile*** L. Toadshade

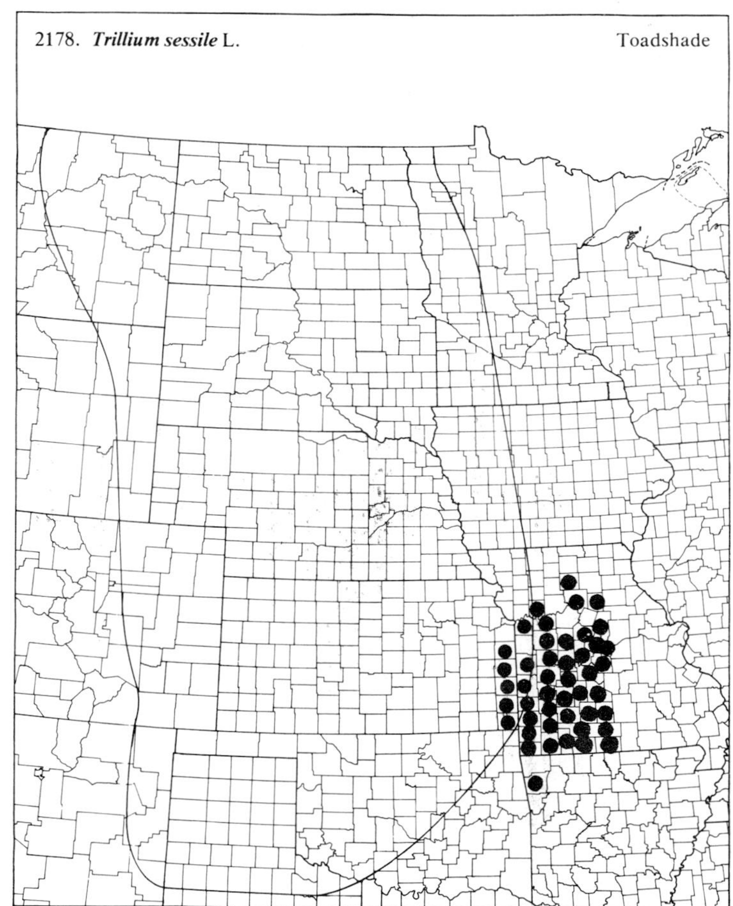

2179. ***Uvularia grandiflora*** Sm. Large Bellwort

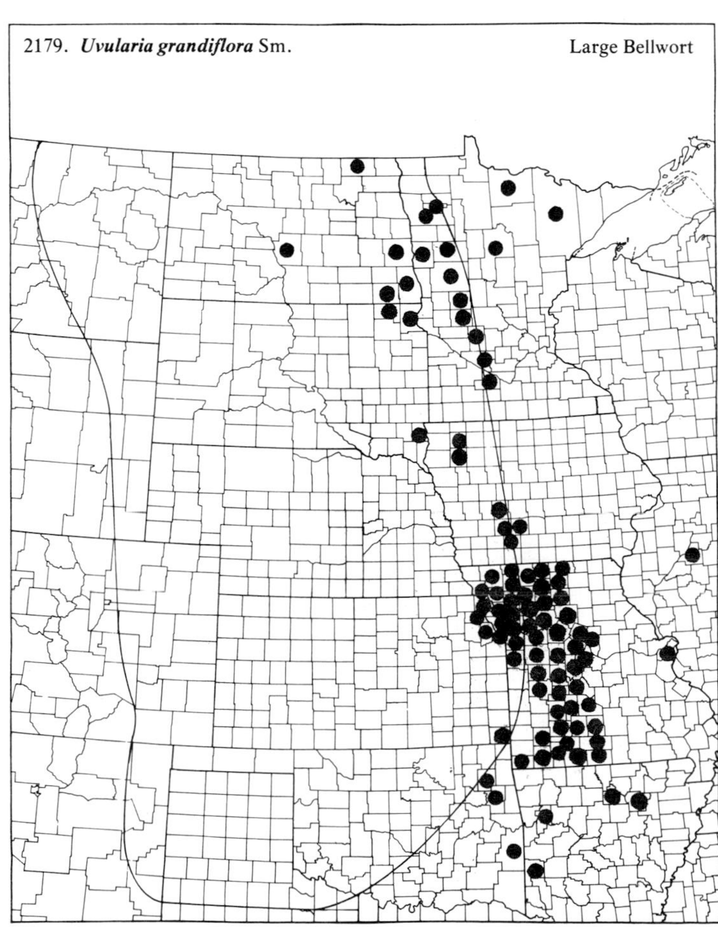

2180. ***Uvularia sessilifolia*** L. Small Bellwort

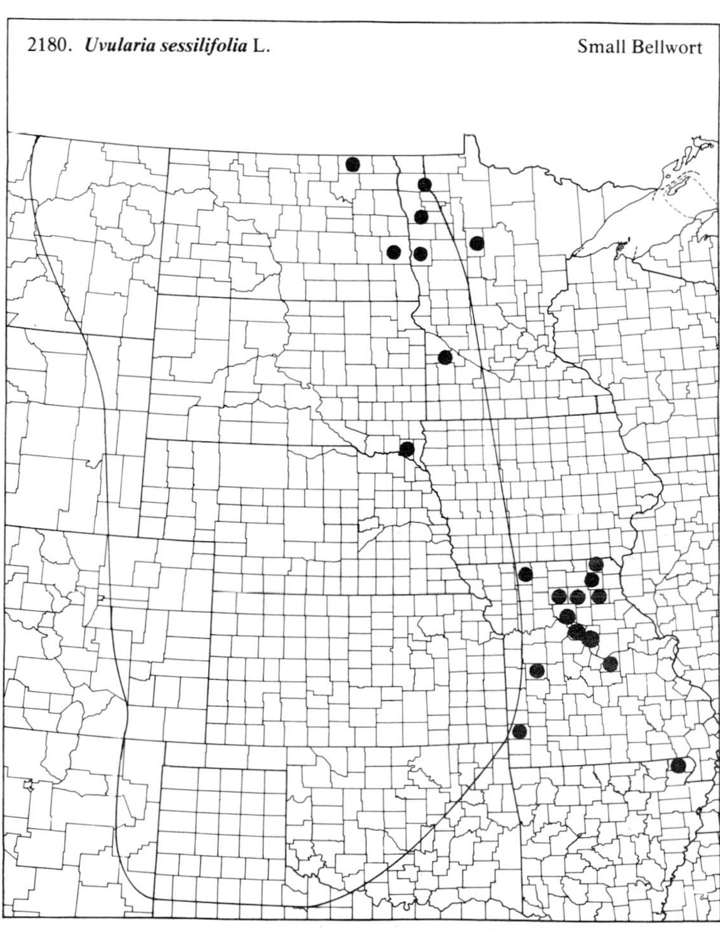

(Liliaceae)

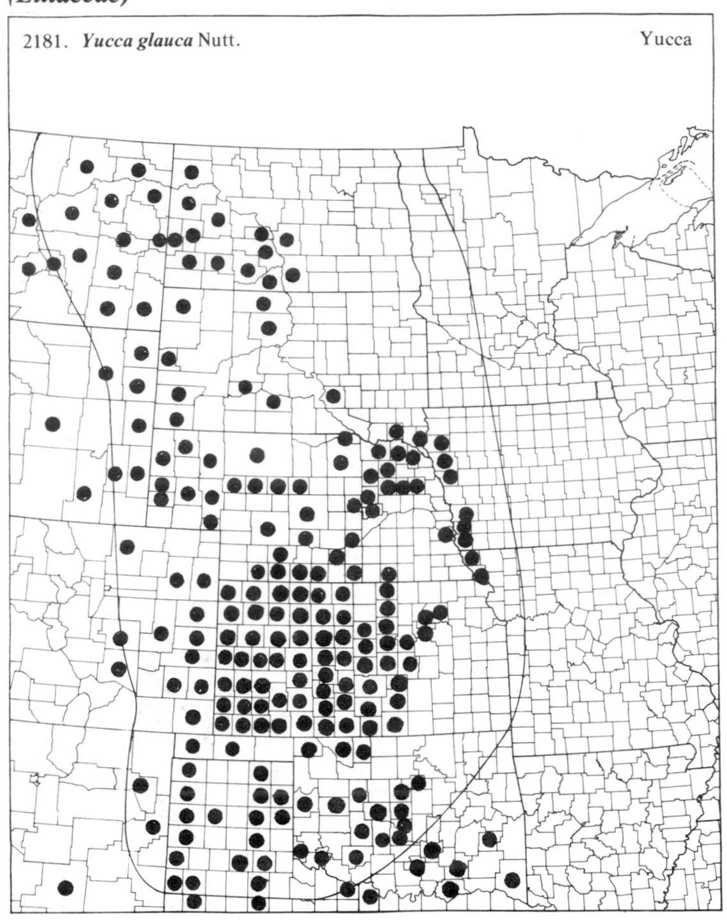

2181. ***Yucca glauca*** Nutt. Yucca

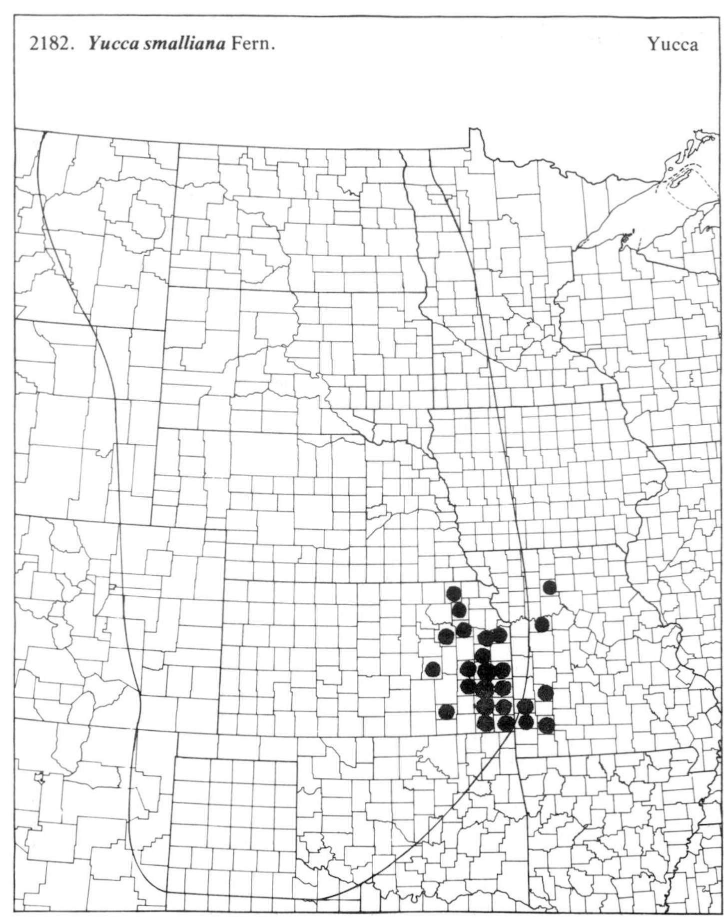

2182. ***Yucca smalliana*** Fern. Yucca

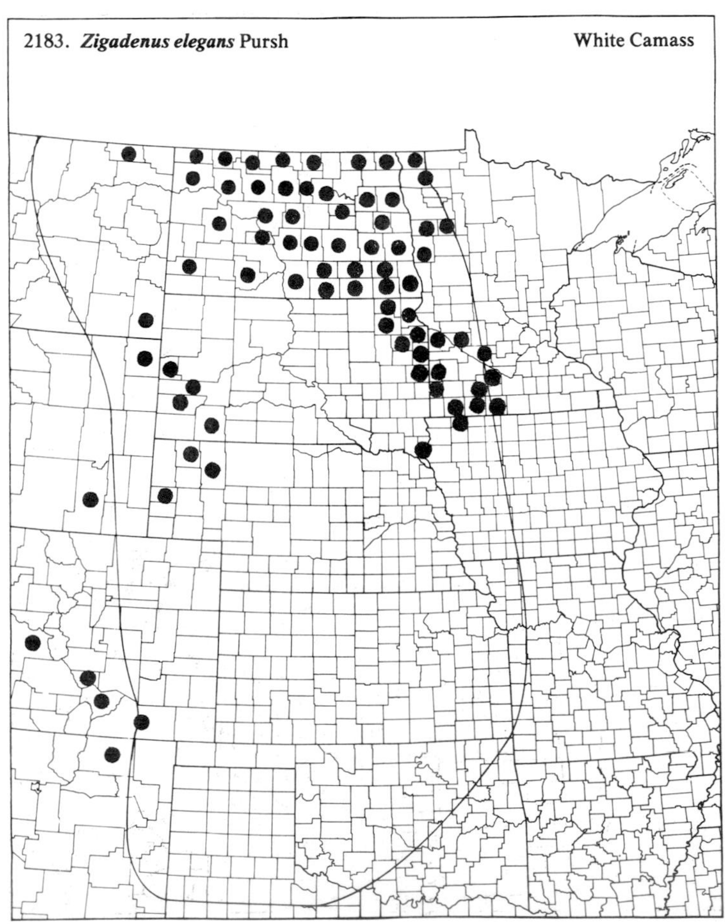

2183. ***Zigadenus elegans*** Pursh White Camass

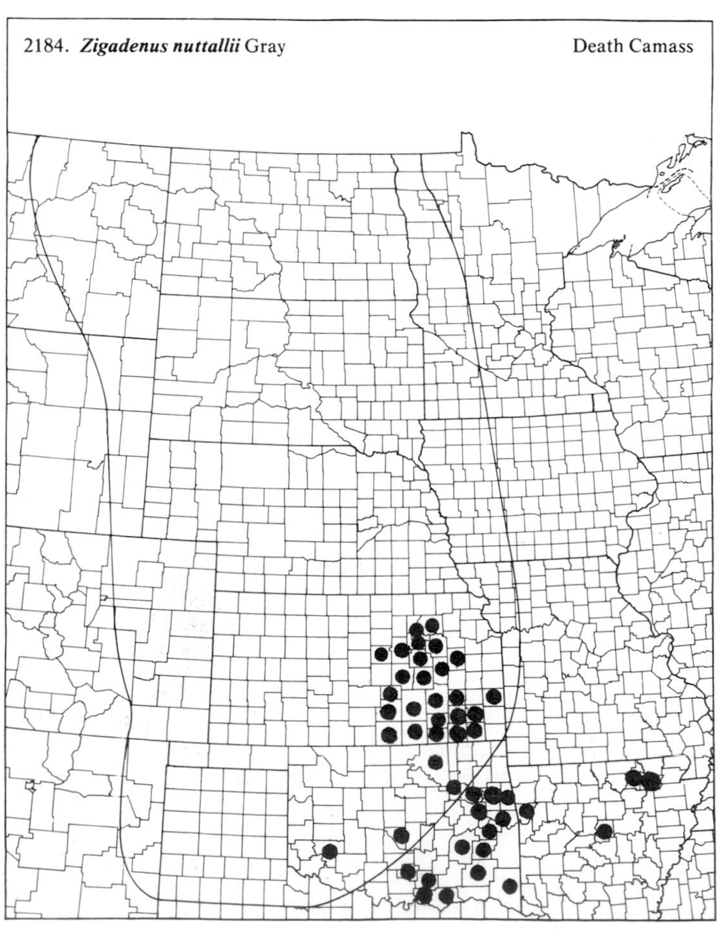

2184. ***Zigadenus nuttallii*** Gray Death Camass

(Liliaceae)

2185. ***Zigadenus venenosus*** Wats. var. ***gramineus*** (Rydb.) Walsh — Death Camass

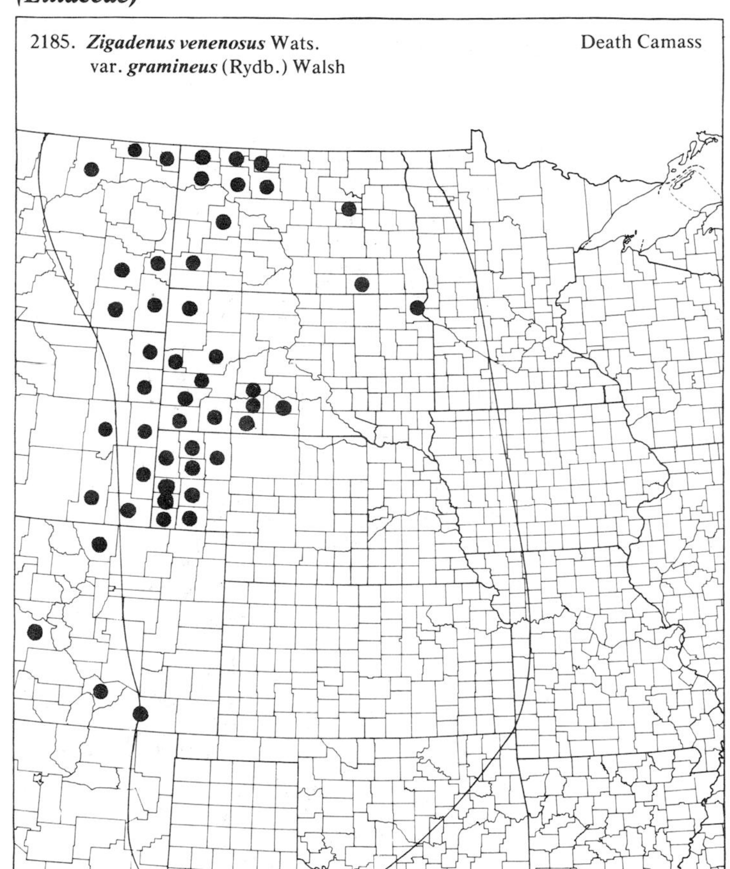

134. *Iridaceae* (Iris Family)

2186. ***Belamcanda chinensis*** (L.) DC. — Blackberry Lily

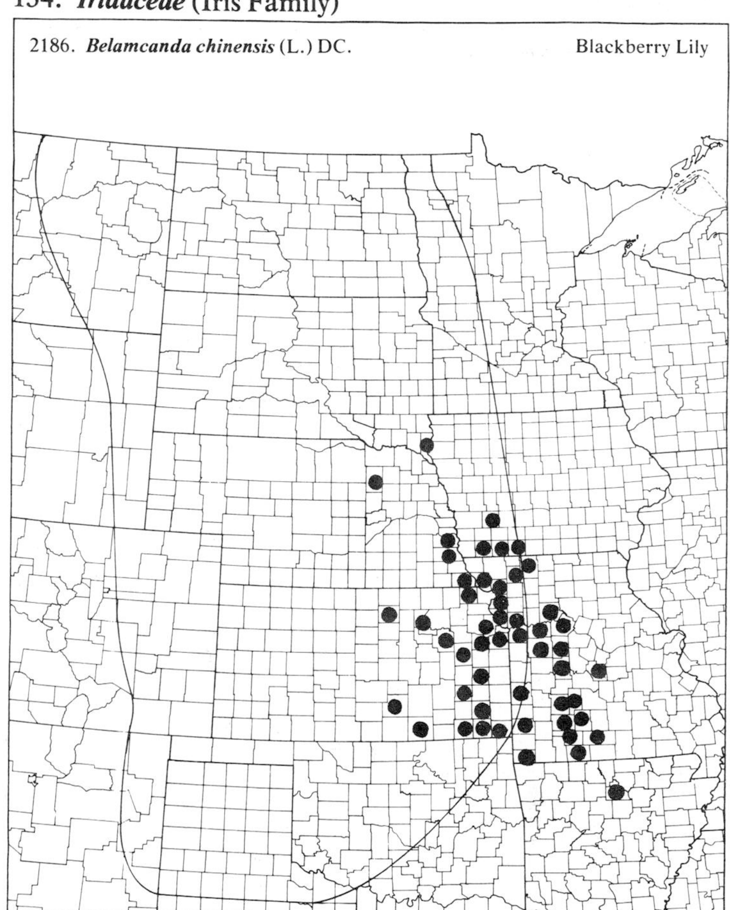

2187. ***Iris missouriensis*** Nutt. — Blue Flag

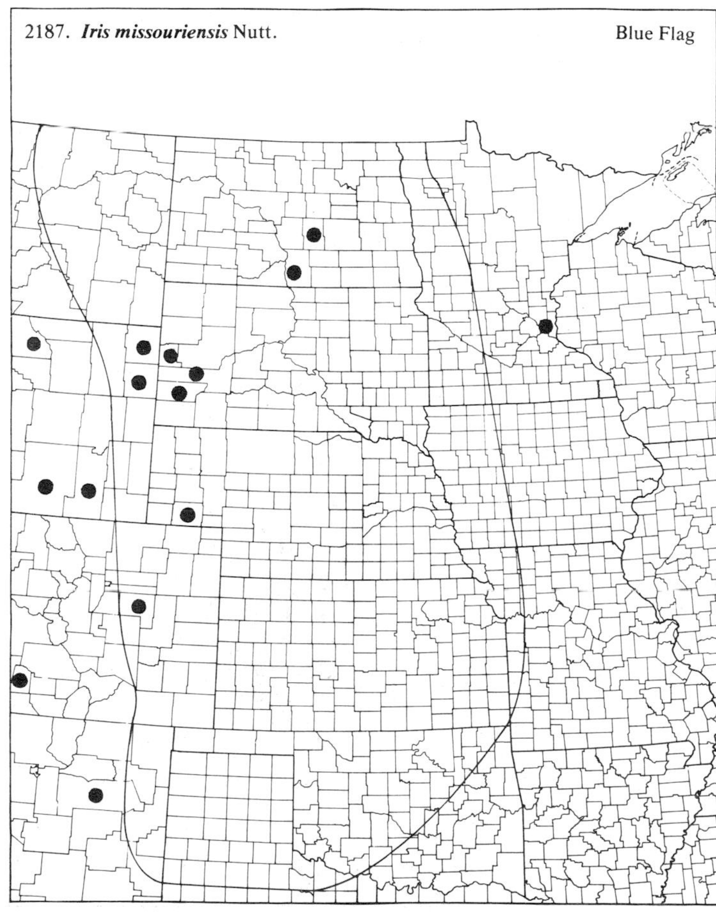

2188. ***Iris shrevei*** Small

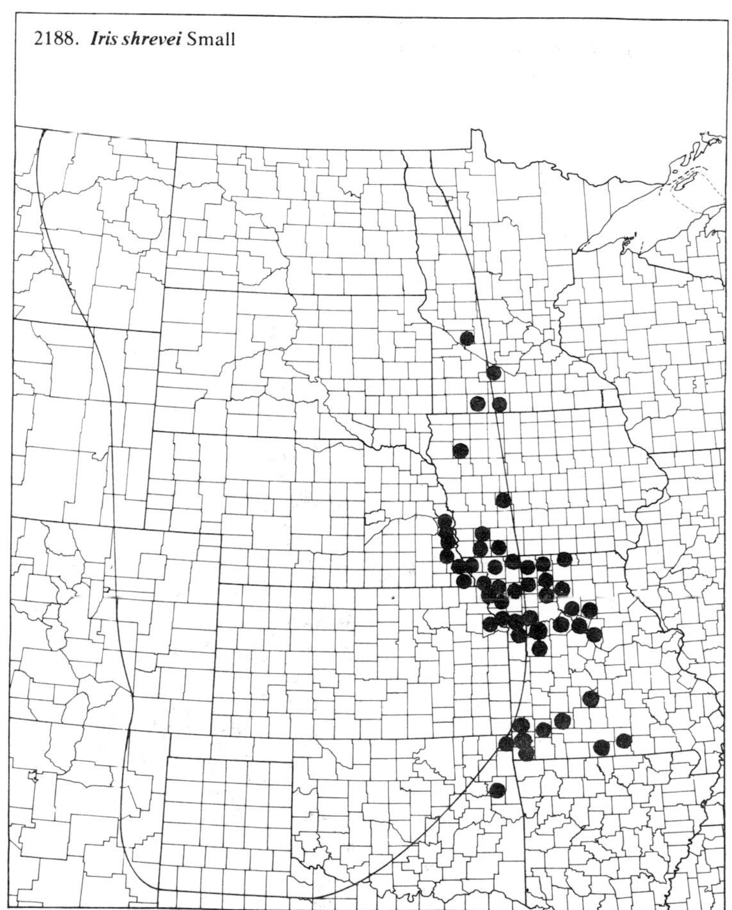

(Iridaceae)

2189. *Nemastylis geminiflora* Nutt. Prairie Iris

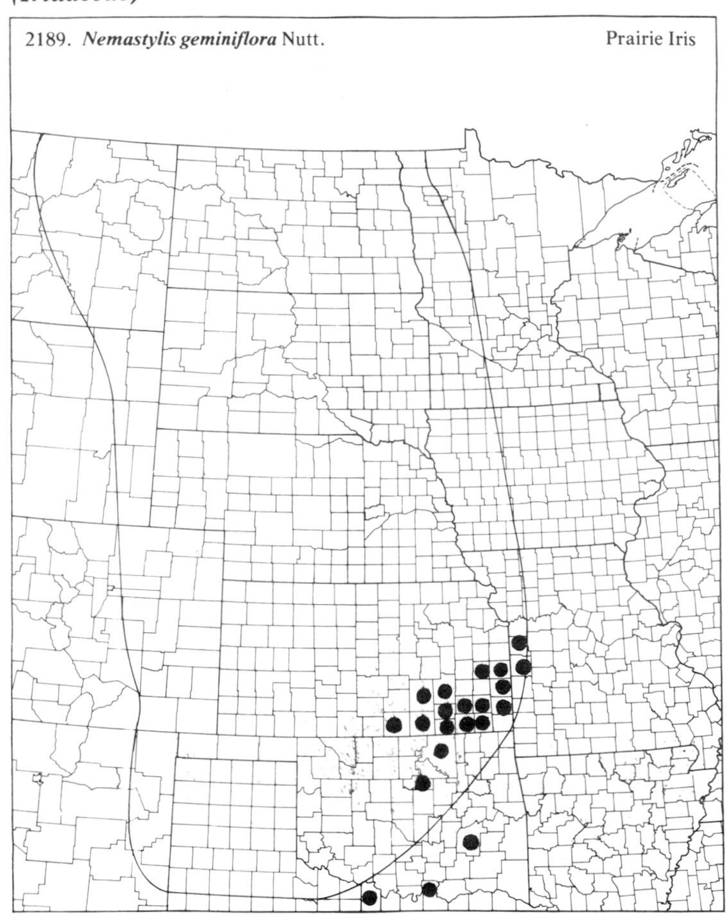

2190. *Sisyrinchium angustifolium* Miller Blue-eyed-grass

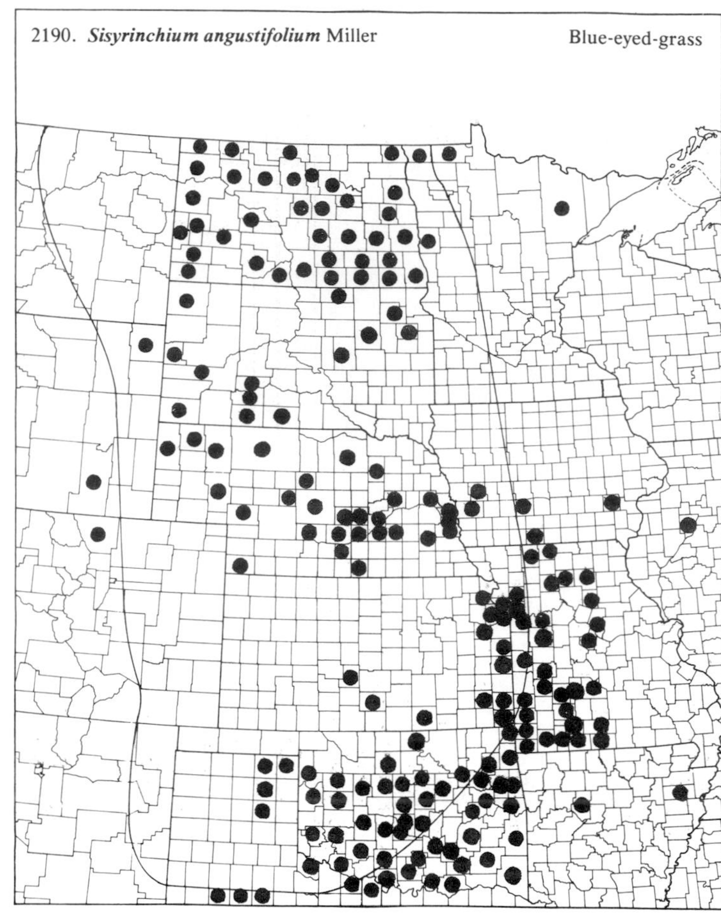

2191. *Sisyrinchium campestre* Bickn. var. *campestre* White-eyed-grass

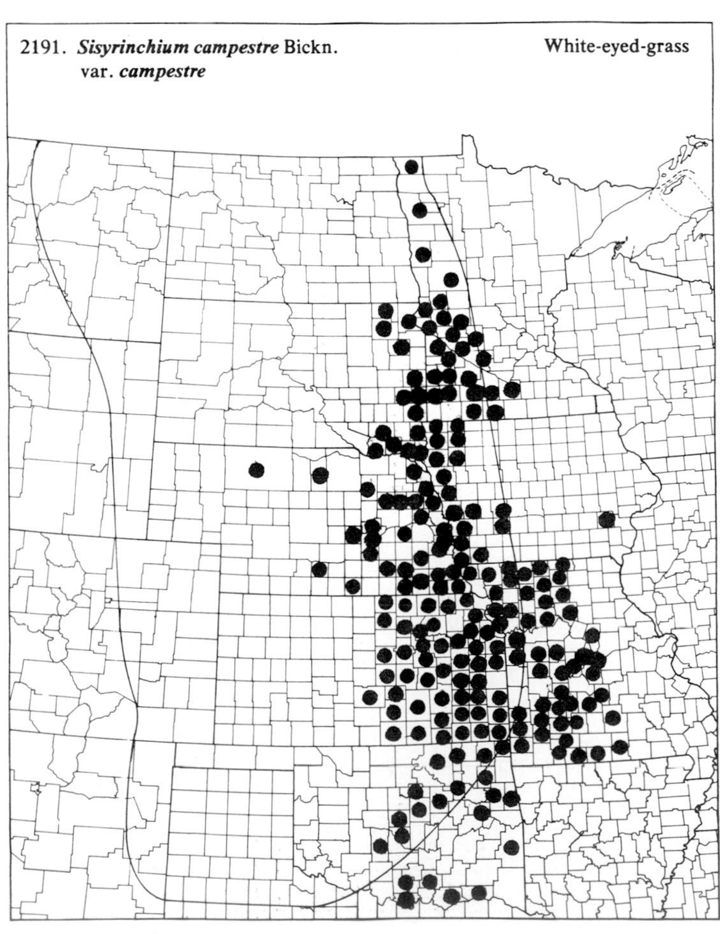

2192. *Sisyrinchium campestre* Bickn. var. *kansanum* Bickn. White-eyed-grass

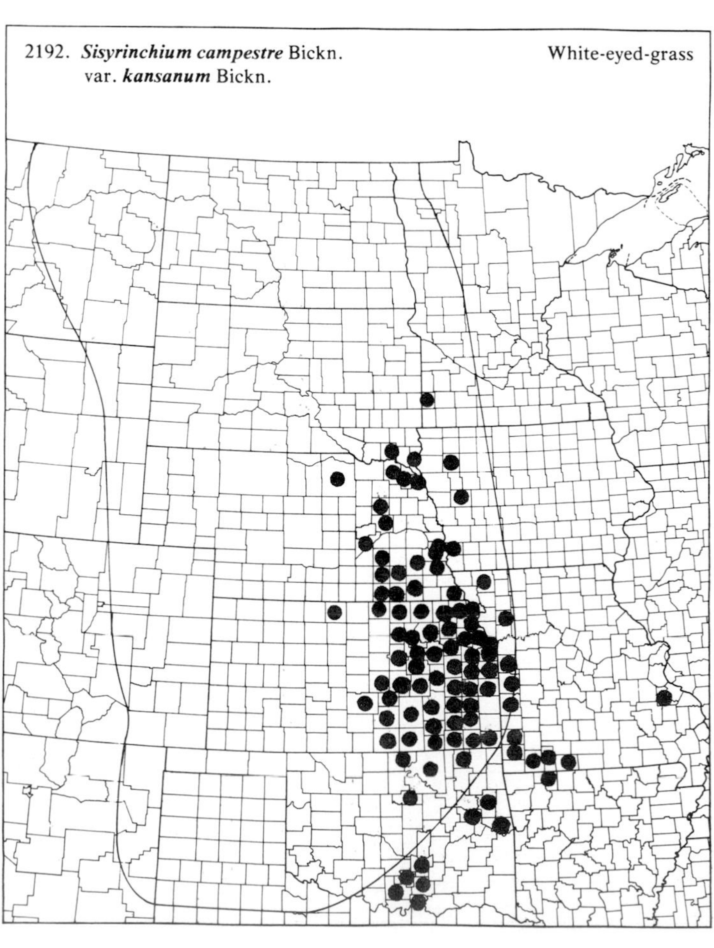

(Iridaceae)

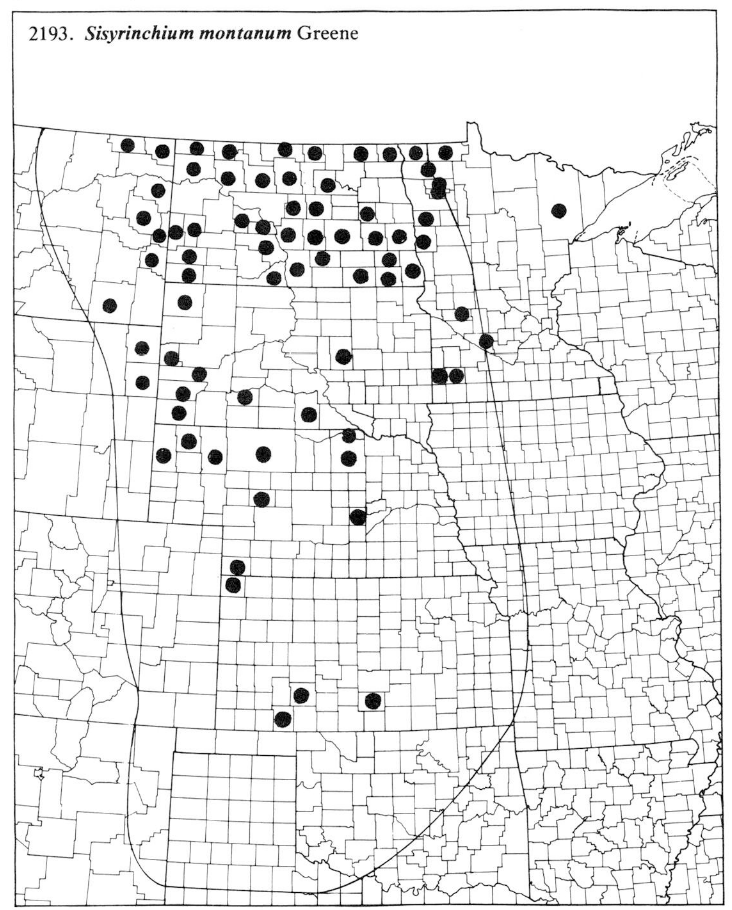

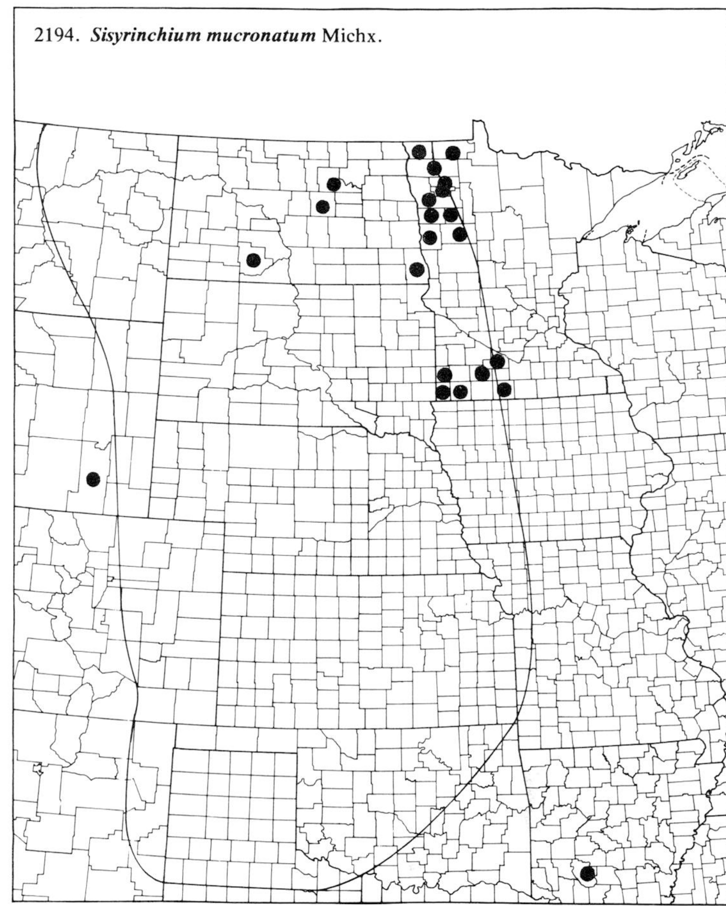

135. *Dioscoreaceae* (Yam Family)

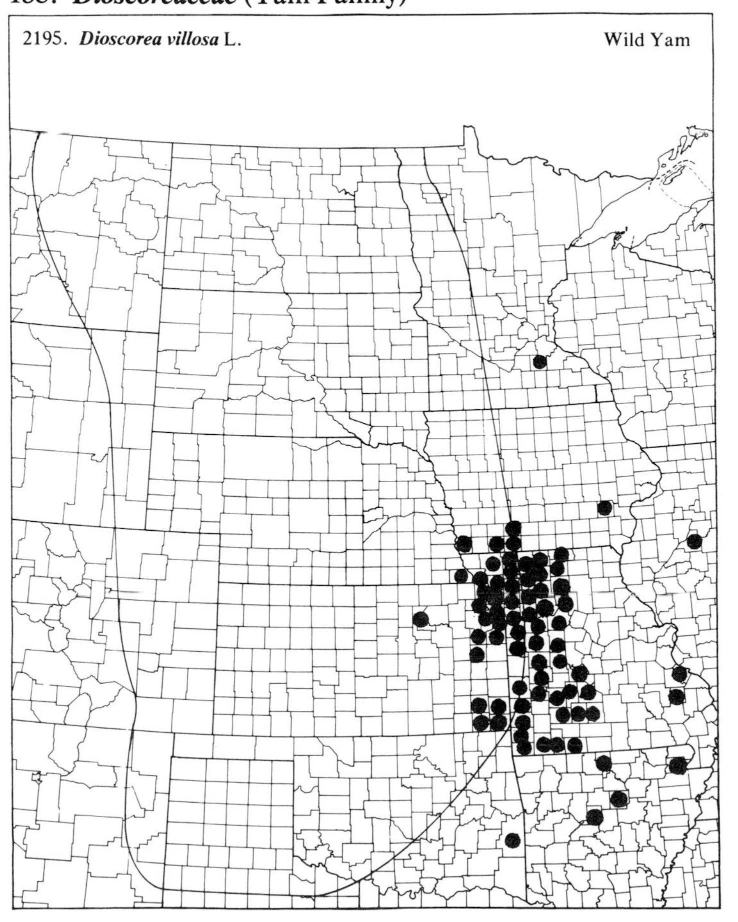

136. *Orchidaceae* (Orchid Family)

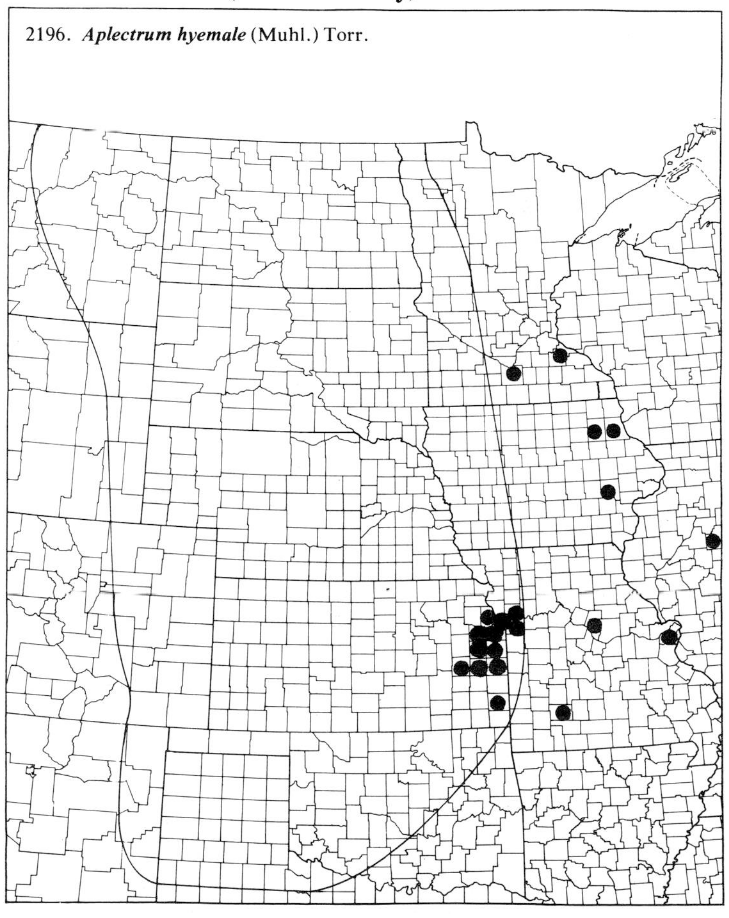

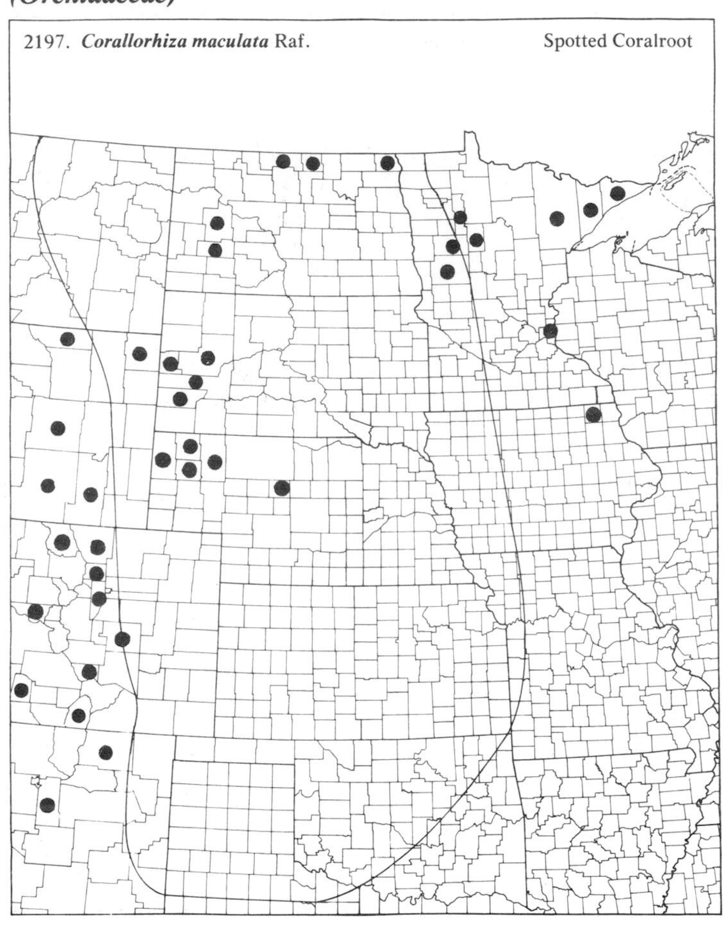
2197. ***Corallorhiza maculata*** Raf. Spotted Coralroot

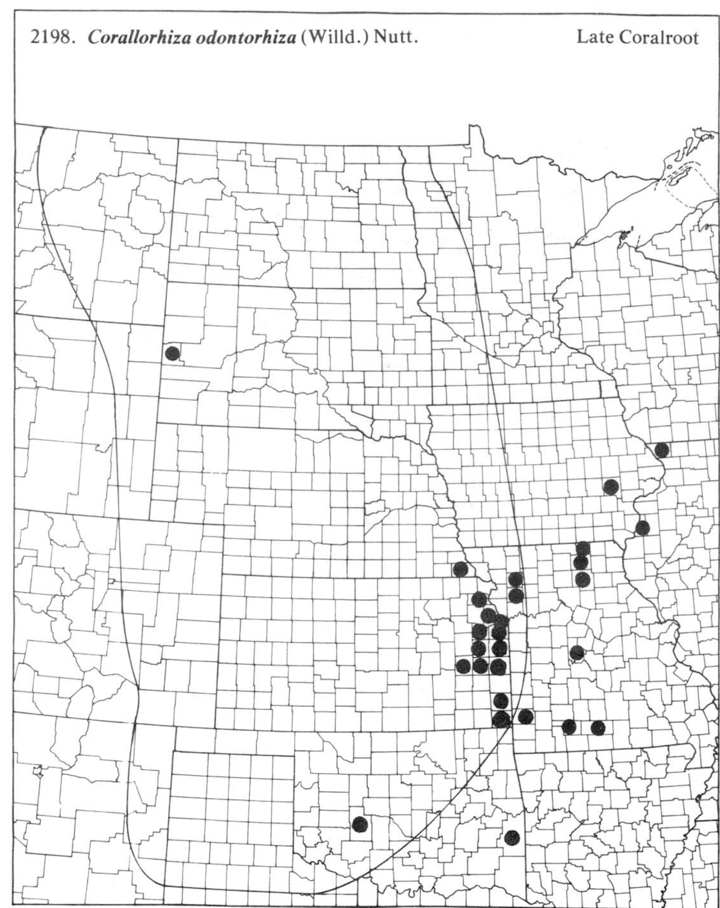
2198. ***Corallorhiza odontorhiza*** (Willd.) Nutt. Late Coralroot

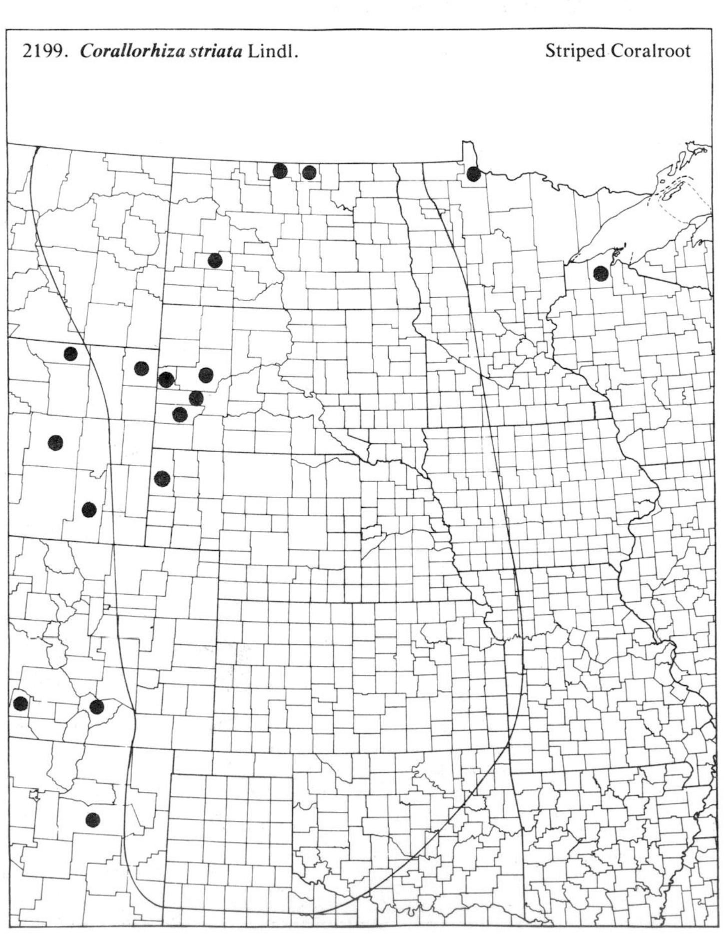
2199. ***Corallorhiza striata*** Lindl. Striped Coralroot

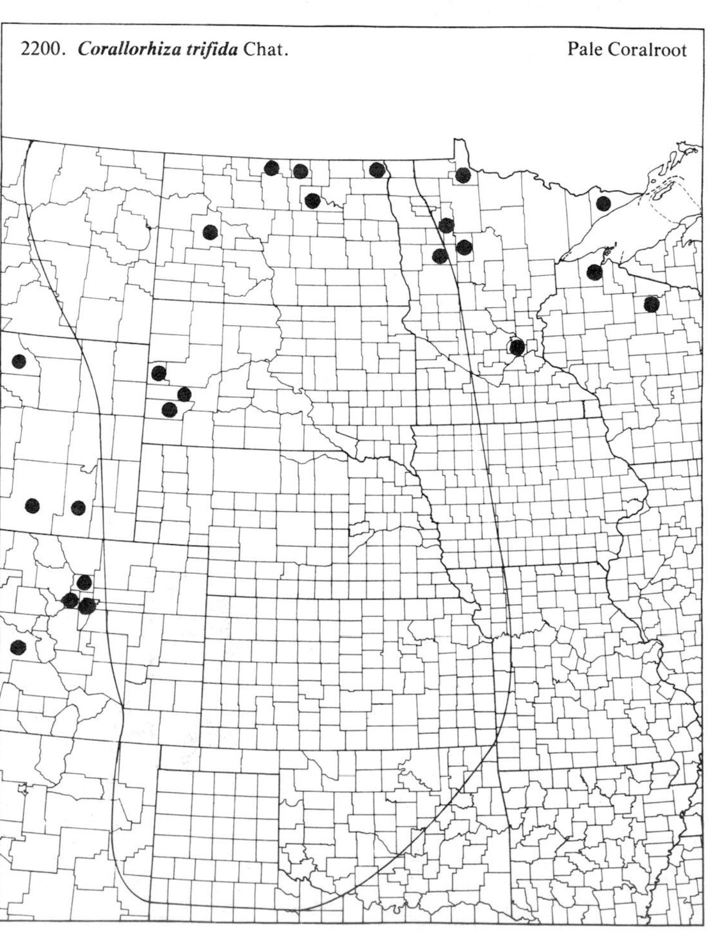
2200. ***Corallorhiza trifida*** Chat. Pale Coralroot

(Orchidaceae)

2201. **Corallorhiza wisteriana** Conrad — Wister's Coralroot

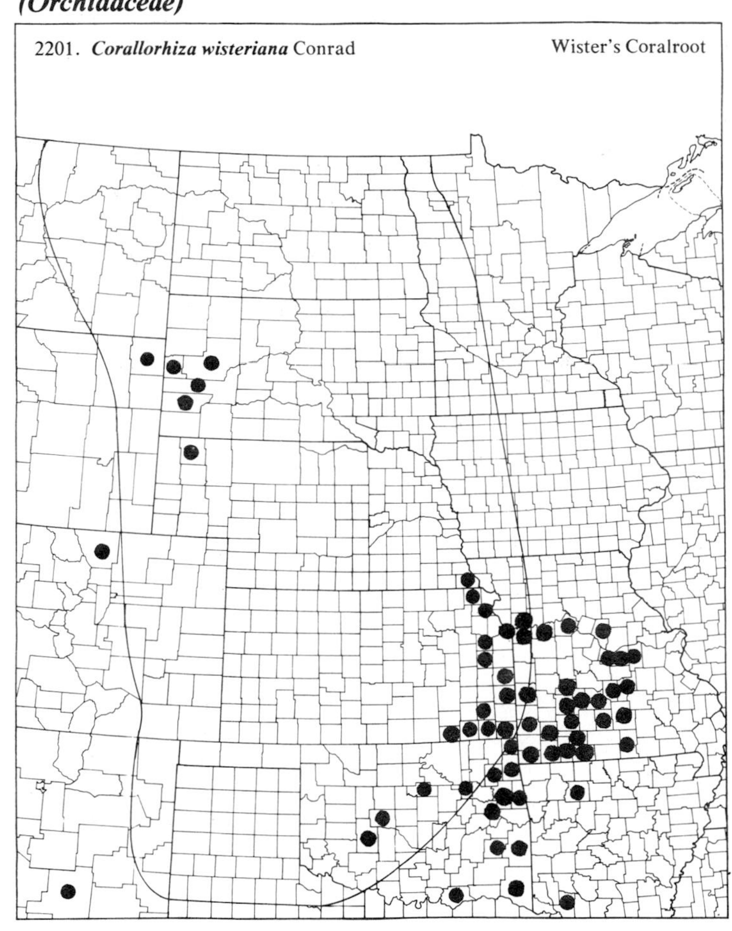

2202. **Cypripedium calceolus** L. var. **parviflorum** (Salisb.) Fern. — Yellow Lady's Slipper

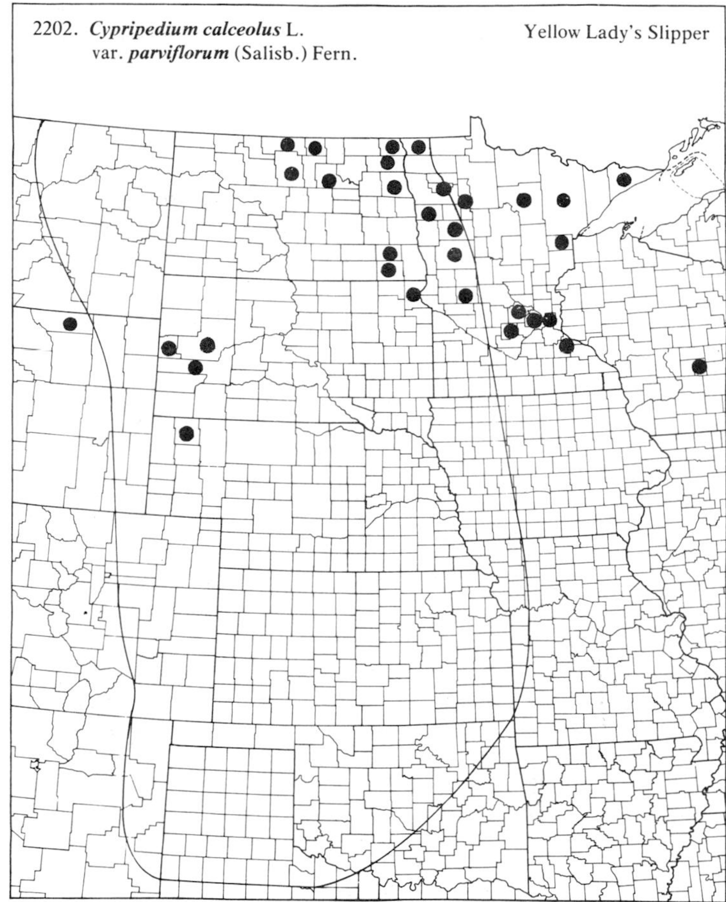

2203. **Cypripedium calceolus** L. var. **pubescens** (Willd.) Correll — Yellow Lady's Slipper

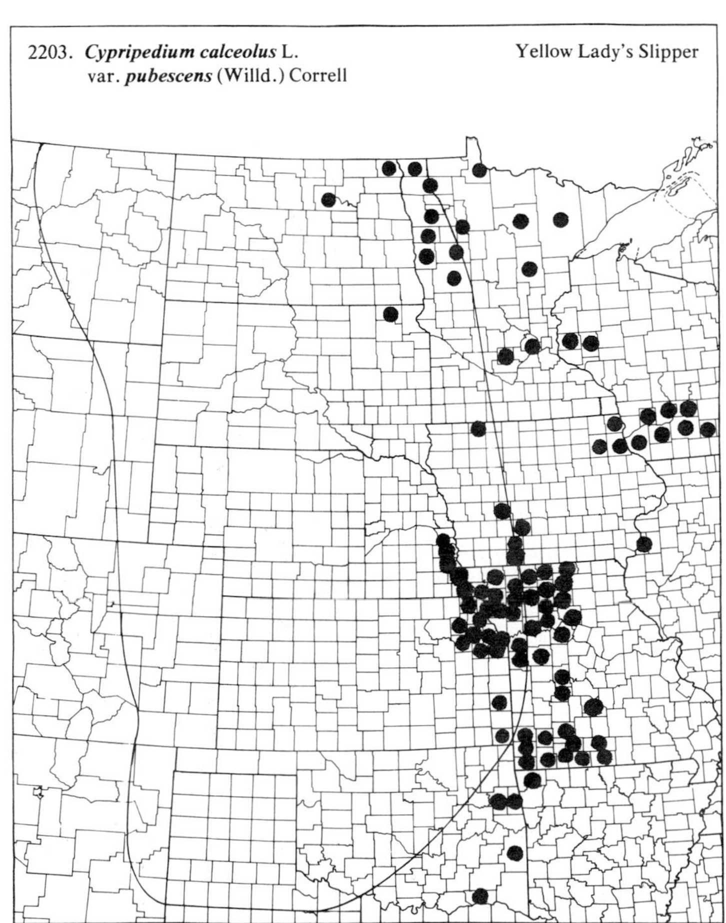

2204. **Cypripedium candidum** Muhl. — White Lady's Slipper

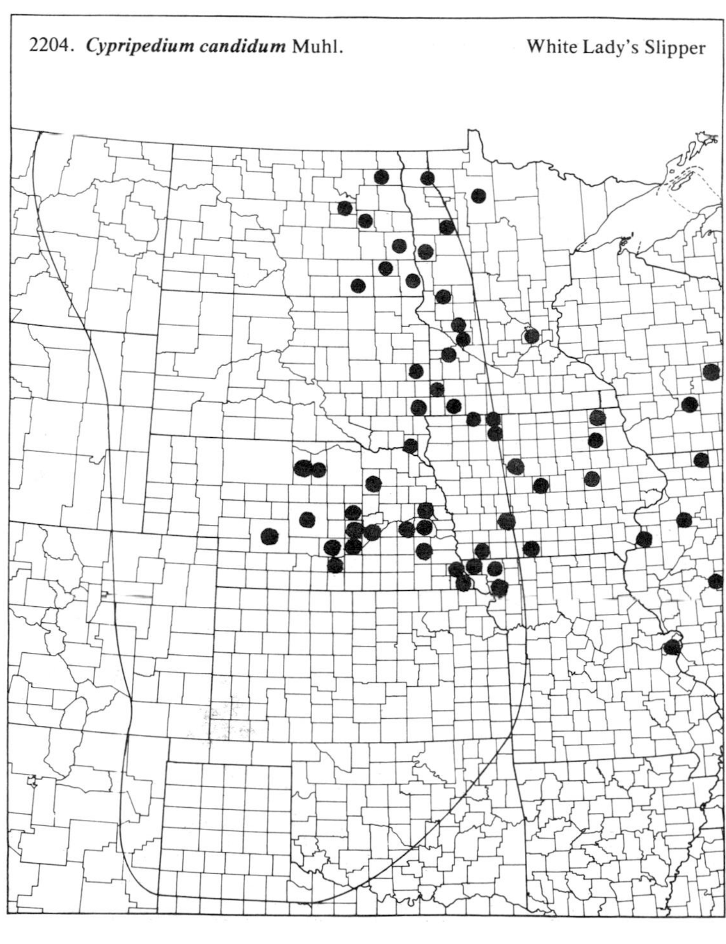

(Orchidaceae)

2205. ***Cypripedium reginae*** Walt. Showy Lady's Slipper

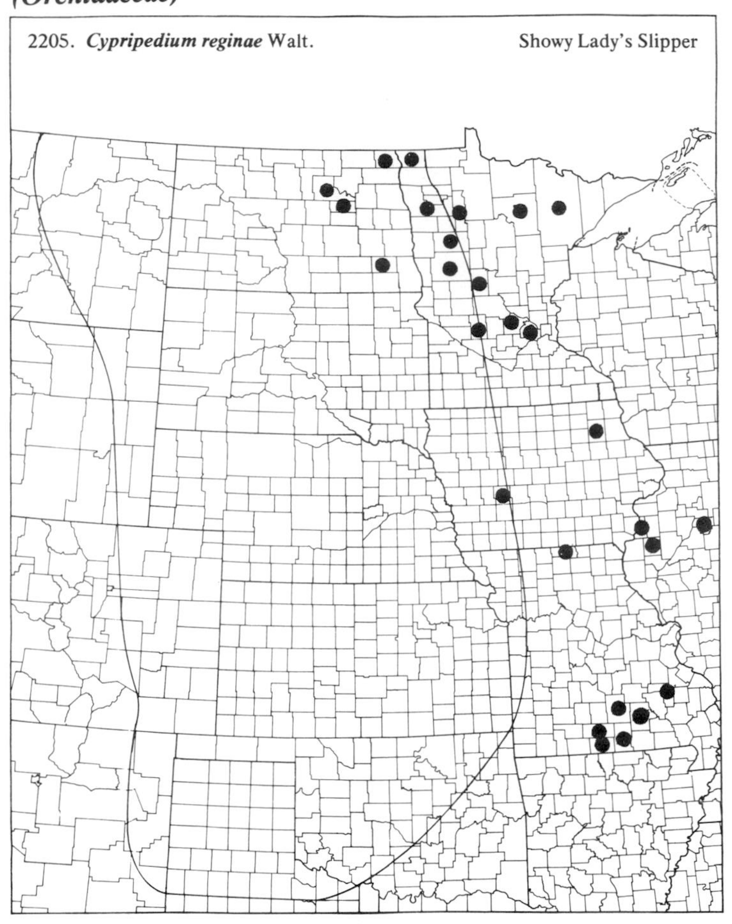

2206. ***Habenaria hyperborea*** (L.) R. Br. Northern Green Orchis

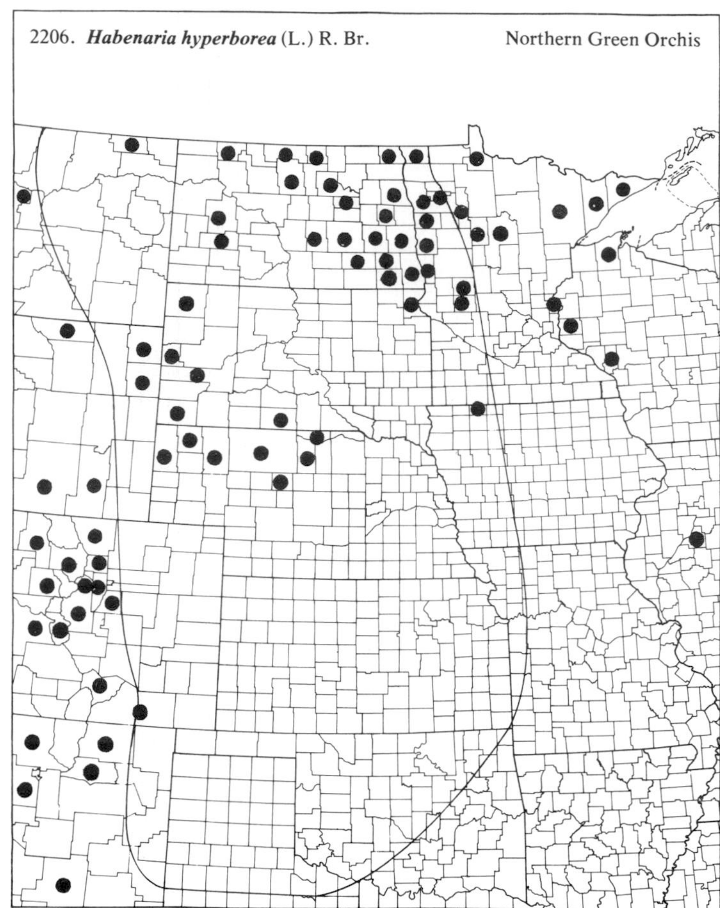

2207. ***Habenaria lacera*** (Michx.) Lodd. Ragged Orchis

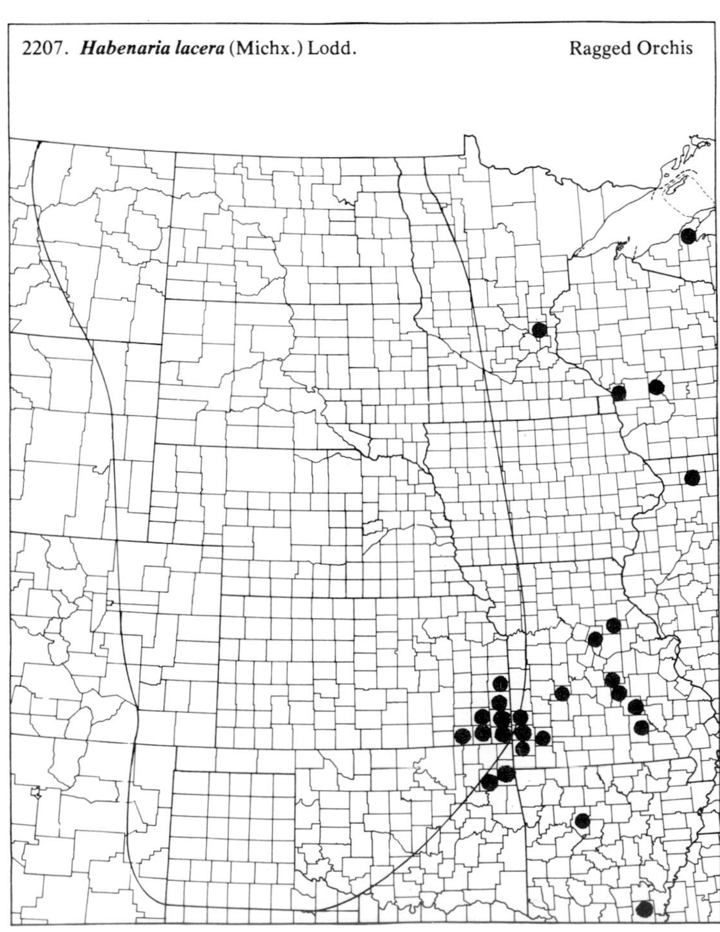

2208. ***Habenaria leucophaea*** (Nutt.) Gray Prairie Fringed Orchis

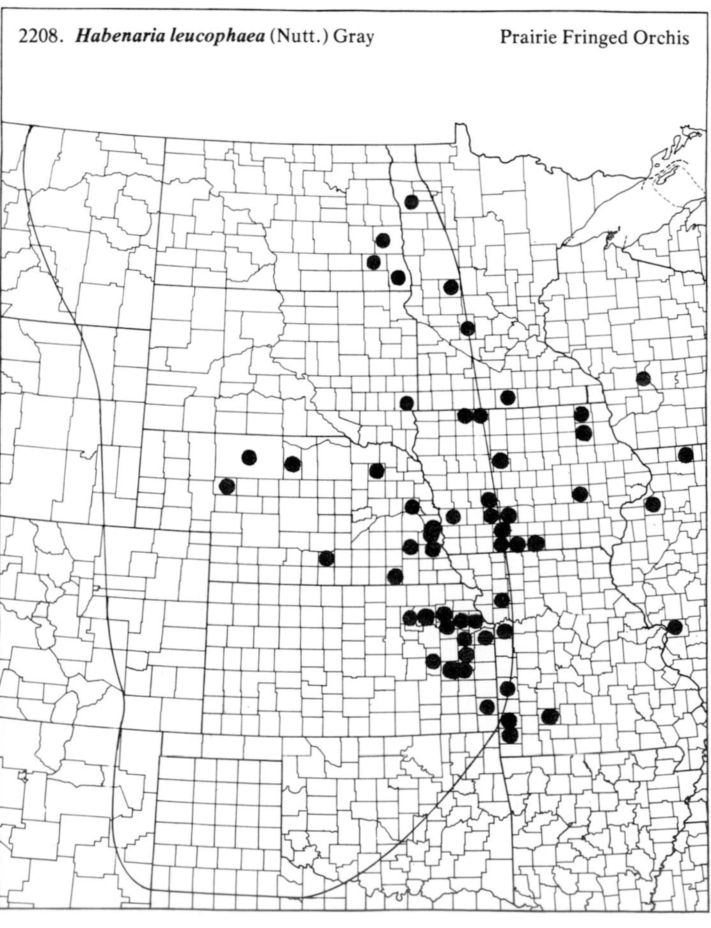

(Orchidaceae)

2209. ***Habenaria viridis*** (L.) R. Br. var. ***bracteata*** (Muhl.) Gray — Long-bracted Orchis

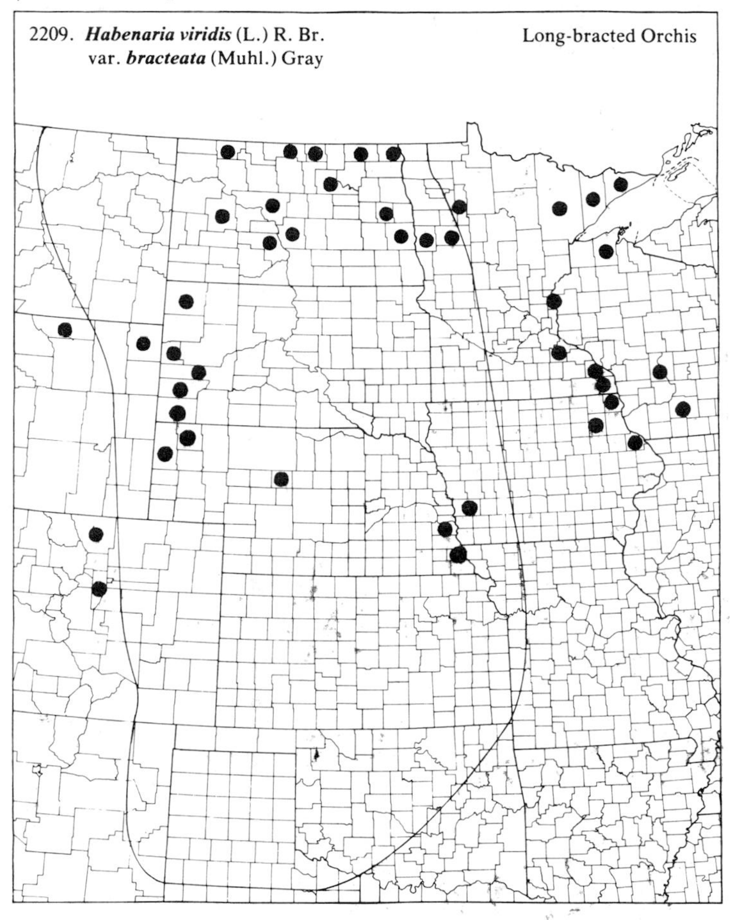

2210. ***Liparis loeselii*** (L.) Rich. — Loesel's Twayblade

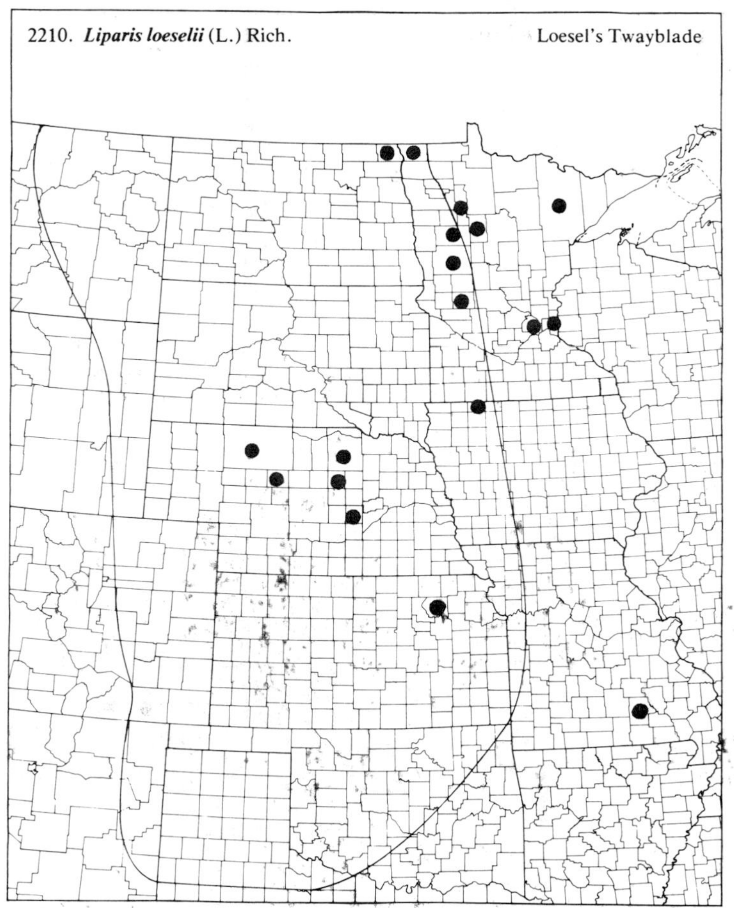

2211. ***Orchis spectabilis*** L. — Showy Orchis

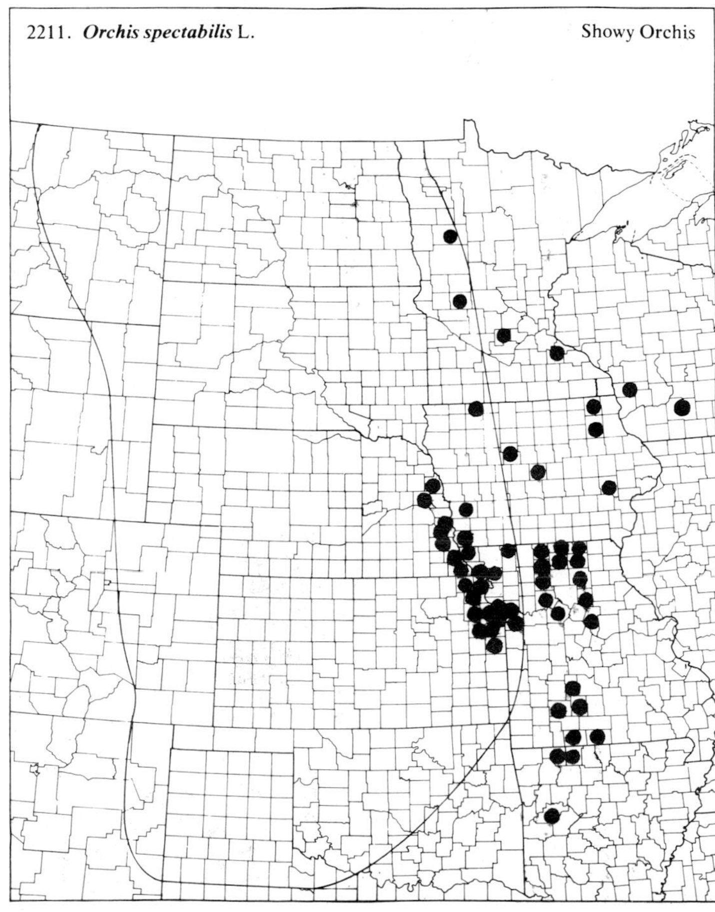

2212. ***Spiranthes cernua*** (L.) Rich. — Ladies'-tresses

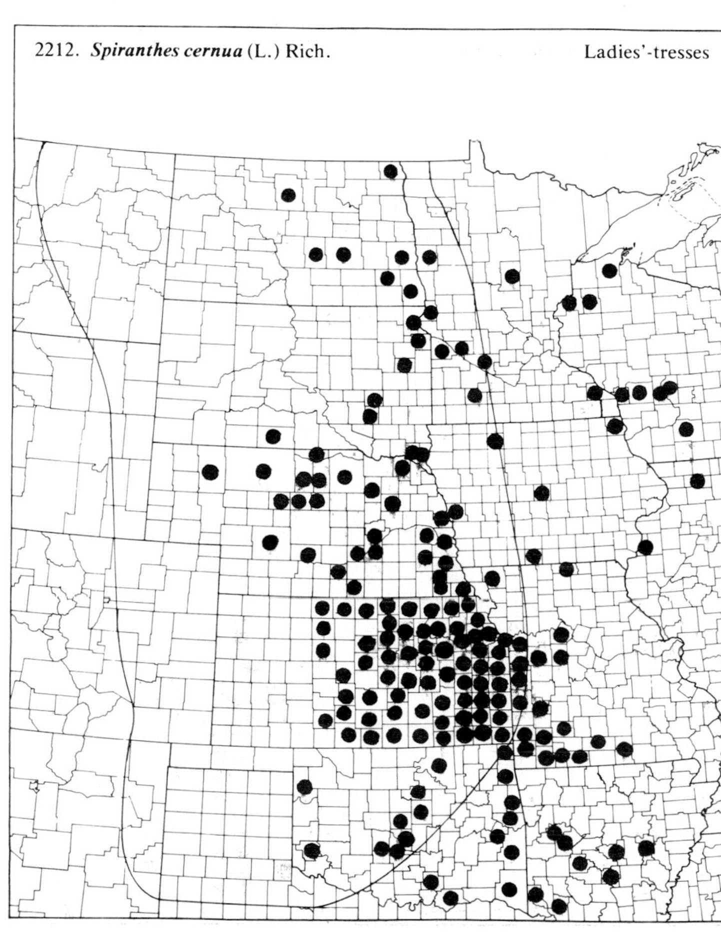

(Orchidaceae)

2213. **Spiranthes lacera** (Raf.) Raf. Slender Ladies'-tresses

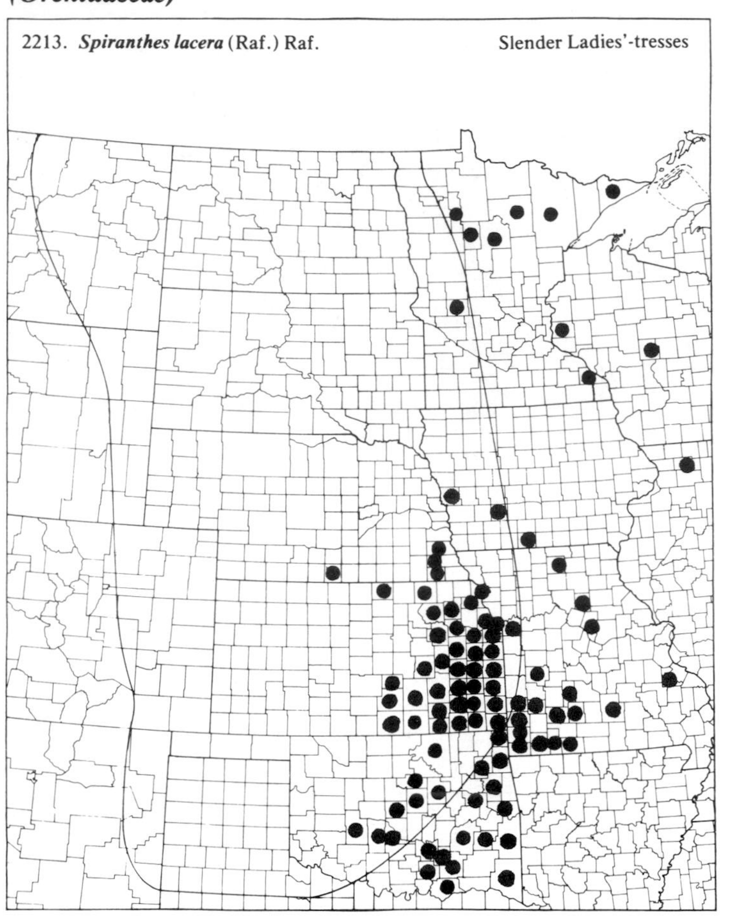

2214. **Spiranthes romanzoffiana** Cham. Hooded Ladies'-tresses

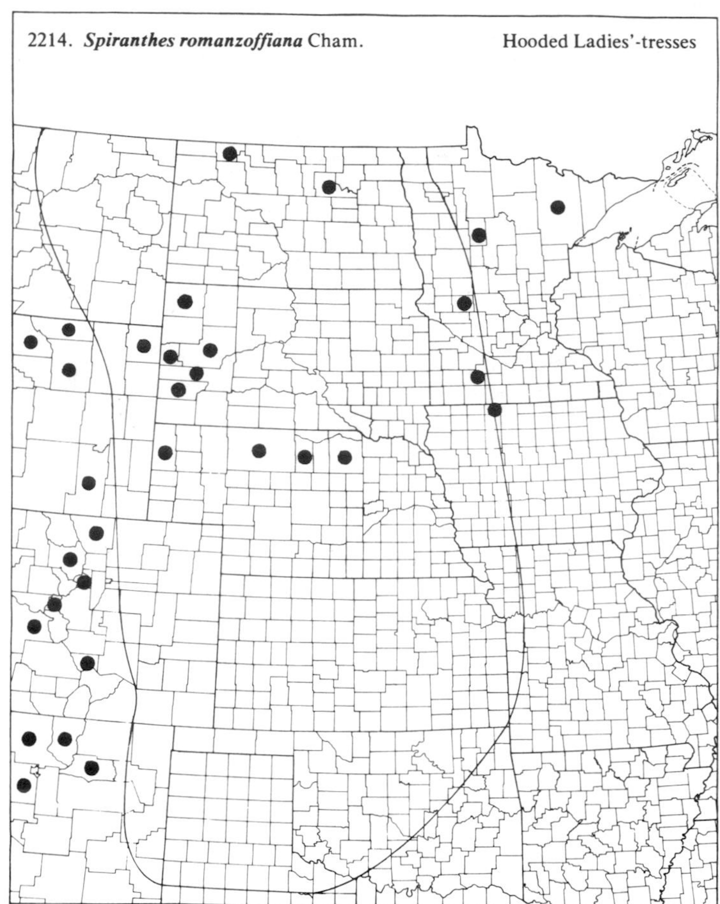

2215. **Spiranthes tuberosa** Raf. Little Ladies'-tresses

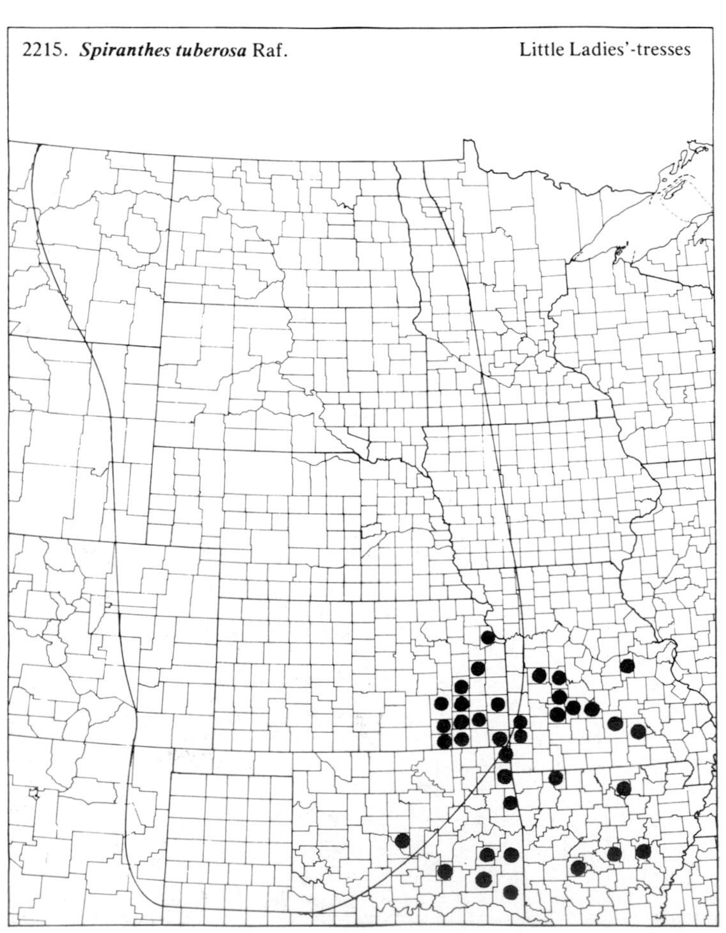

2216. **Spiranthes vernalis** Engelm. & Gray Twisted Ladies'-tresses

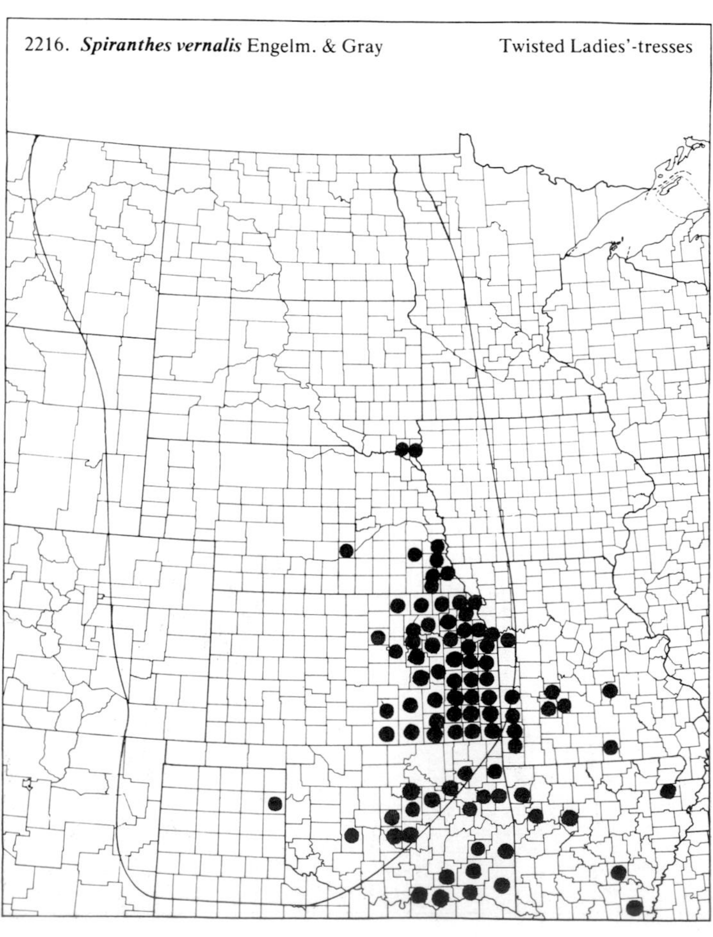

(Orchidaceae)

2217. ***Triphora trianthophora*** (Sw.) Rydb. in Britt. Nodding Pogonia

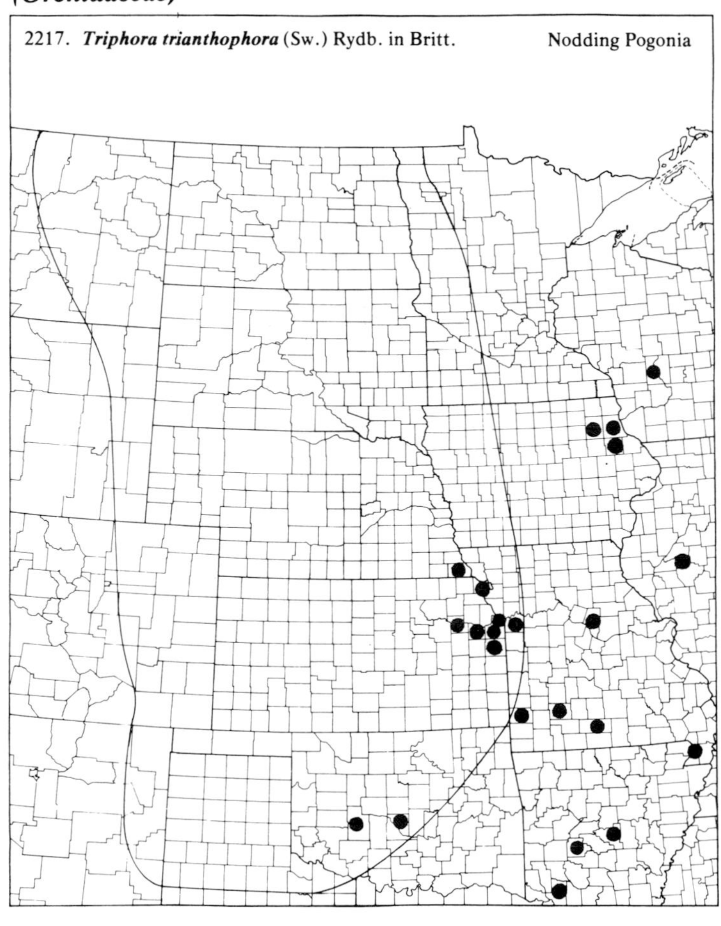

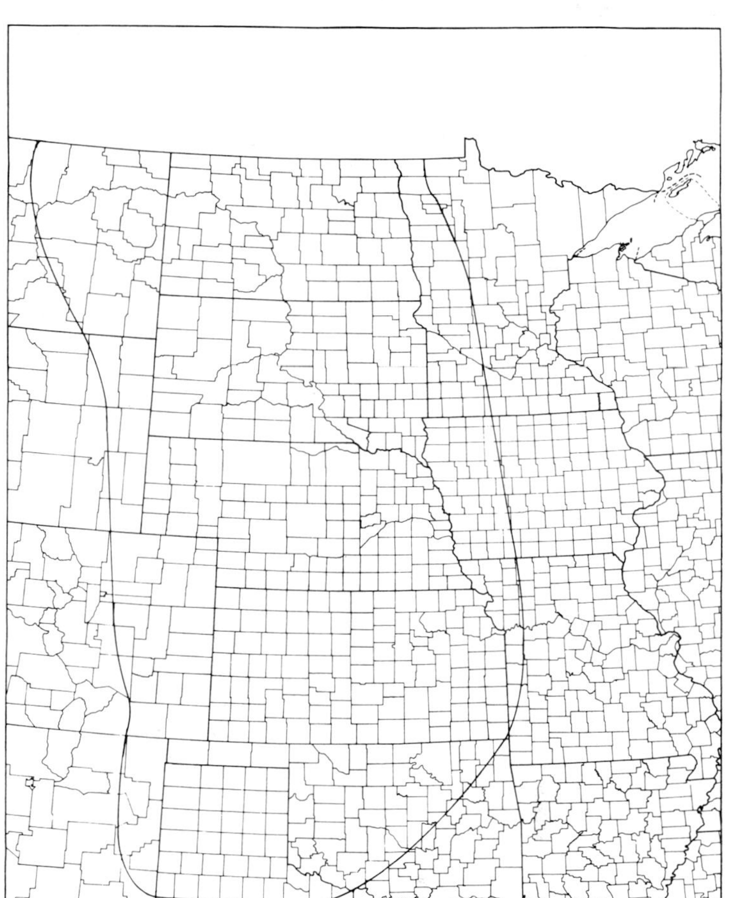

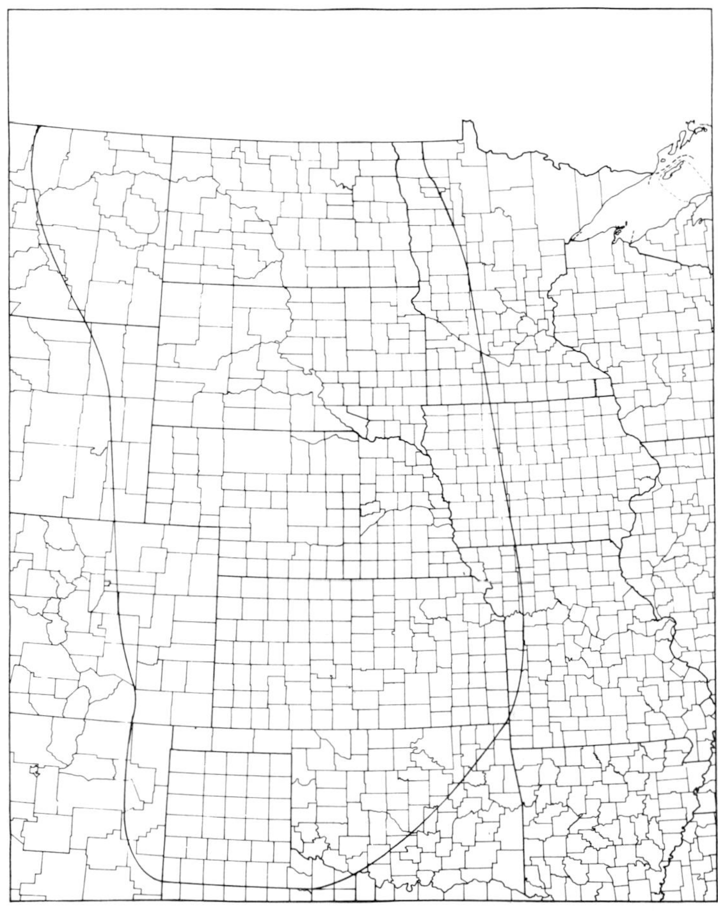

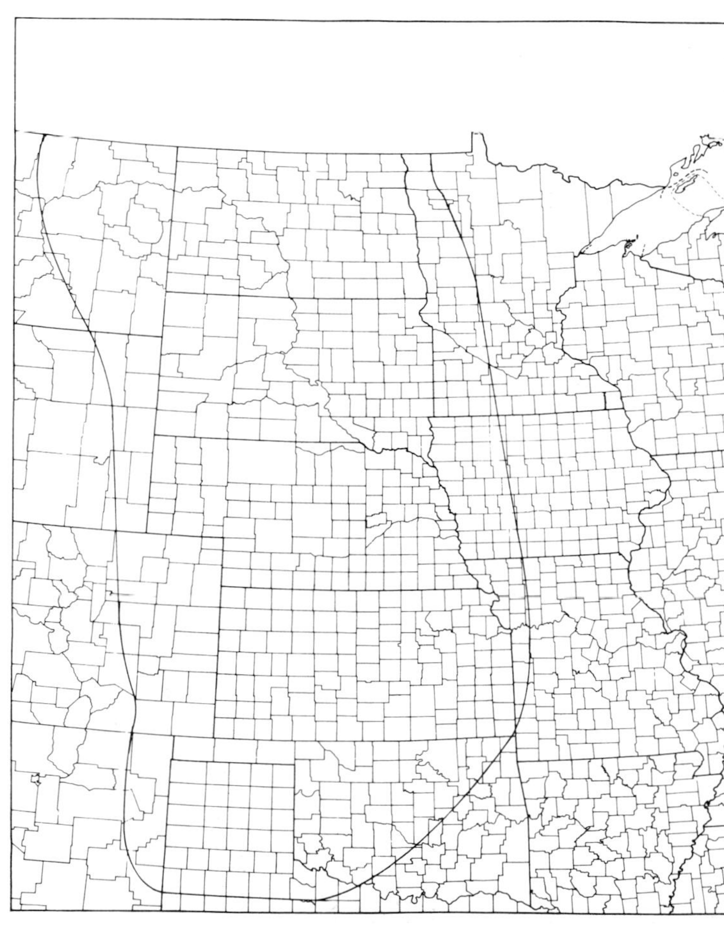

SECTION TWO

List of Infrequent Great Plains Taxa

THIS LIST includes some 850 taxa which are either rare or of restricted occurrence, but which are nonetheless members of the Great Plains flora. The list is divided into three sections in the manner of the distribution maps: Vascular Cryptogams, Conifers, and Flowering Plants. Families, genera, and species are in the same sequence as that of the distribution maps. Each taxon is followed by those states and counties in which it occurs. Postal service abbreviations for the states are used: Colorado—CO; Iowa—IA; Kansas—KS; Minnesota—MN; Missouri—MO; Montana—MT; Nebraska—NE; New Mexico—NM; North Dakota—ND; Oklahoma—OK; South Dakota—SD; Texas—TX; and Wyoming—WY. Custer, Lawrence, and Pennington counties in South Dakota include the Black Hills, thus accounting for the large number of taxa listed for these counties. Common names are included in parentheses where appropriate.

Vascular Cryptogams

LYCOPODIACEAE (Clubmoss Family)

Lycopodium annotinum L. (Bristly Clubmoss) NE: Cherry

Lycopodium obscurum L. (Ground Pine) MN: Grant; SD: Lawrence

SELAGINELLACEAE (Selaginella Family)

Selaginella peruviana (Milde) Hieron. NM: Union; OK: Caddo, Canadian, Cimarron, Comanche, Greer, Jackson, Kiowa

Selaginella underwoodii Hieron. CO: Baca; OK: Cimarron; WY: Laramie

ISOETACEAE (Quillwort Family)

Isoetes butleri Engelm. (Quillwort) KS: Cherokee; MO: Bates, Jasper; OK: Comanche

EQUISETACEAE (Horsetail Family)

Equisetum palustre L. (Meadow Horsetail) MN: Kittson; ND: Ransom, Richland

Equisetum scirpoides Michx. (Dwarf Scouring Rush) SD: Lawrence

Equisetum sylvaticum L. (Wood Horsetail) ND: Cavalier, Pembina; SD: Custer, Lawrence, Pennington, Roberts

Equisetum variegatum Schleich. (Variegated Horsetail) MN: Becker, Kittson, Mahnomen, Marshall, Red Lake

OPHIOGLOSSACEAE (Adder's-tongue Family)

Botrychium dissectum Spreng. var. *dissectum* KS: Cherokee

Botrychium lunaria (L.) Sw. MN: Norman; ND: Burke, McHenry

Botrychium matricariaefolium A. Br. MN: Lac Qui Parle

Botrychium multifidum (Gmel.) Rupr. (Leathery Grape Fern) MN: Otter Tail, Norman; SD: Pennington

Botrychium simplex E. Hitchc. SD: Custer, Lawrence

Ophioglossum vulgatum L. (Adder's-tongue) NE: Cherry; ND: Richland

OSMUNDACEAE (Osmunda Family)

Osmunda regalis L. (Royal Fern) KS: Wilson, Woodson; MO: Barton, Jackson, Vernon

POLYPODIACEAE (Fern Family)

Asplenium septentrionale (L.) Hoffm. CO: Baca; OK: Cimarron; SD: Custer, Pennington; WY: Laramie

Asplenium viride Huds. (Green Spleenwort) SD: Lawrence

Athyrium pycnocarpon (Spreng.) Tidest. (Glade Fern) KS: Leavenworth, Wyandotte; MO: Clay, Jackson

Cheilanthes alabamensis (Buckl.) Kunze, KS: Cherokee; OK: Ottawa

Cheilanthes fendleri Hook. OK: Comanche

Cheilanthes lindheimeri Hook. OK: Comanche
Cheilanthes tomentosa Link, KS: Cherokee; OK: Caddo, Comanche, Greer, Jackson, Kiowa, Osage
Cheilanthes wootonii Maxon, CO: Baca; NM: Union; OK: Cimarron, Greer, Kiowa
Cystopteris bulbifera (L.) Bernh. (Bulblet Bladder Fern) NE: Richardson; SD: Roberts
Gymnocarpium dryopteris (L.) Newm. (Oak Fern) MN: Mahnomen; SD: Custer, Lawrence
Notholaena fendleri Kunze, NM: Union; WY: Laramie
Notholaena sinuata (Lag.) Kaulf. var. *integerrima* Hook. NM: Union; OK: Cimarron; TX: Armstrong
Pellaea ternifolia (Cav.) Link var. *wrightiana* (Hook.) A. Tryon, OK: Cimarron, Comanche, Greer; TX: Hardeman
Polypodium hesperium Maxon, (Western Polypody) SD: Custer, Pennington, Shannon
Polypodium polypodioides (L.) Watt var. *michauxianum* Weath. (Resurrection Fern) KS: Chautauqua
Polypodium vulgare L. var. *virginianum* (L.) Eat. (Common Polypody) IA: Lyon; MN: Chippewa, Otter Tail, Redwood, Rock, Yellow Medicine; SD: Minnehaha
Polystichum minitum (Kaulf.) Presl. (Christmas Fern) SD: Pennington
Pteridium aquilinum (L.) Kuhn var. *pseudocaudatum* (Clute) Heller, (Bracken Fern) KS: Cherokee; MO: Jasper, Newton, Vernon
Thelypteris hexagonoptera (Michx.) Weath. KS: Cherokee; MO: Jasper, Newton; OK: Ottawa
Woodsia ilvensis (L.) R. Br. (Fragrant Woodsia) IA: Lyon; MN: Big Stone, Chippewa, Cottonwood, Redwood, Yellow Medicine
Woodsia mexicana Fee, CO: Las Animas; KS: Ellsworth; NM: Union; OK: Cimarron
Woodsia scopulina D. C. Eat. (Rocky Mountain Woodsia) SD: Corson, Custer, Fall River, Harding, Lawrence, Pennington; WY: Laramie
MARSILEACEAE (Marsilea Family)
Marsilea quadrifolia L. KS: Cherokee; MO: Jackson
Marsilea uncinata A. Br. TX: Potter
Pilularia americana A. Br. KS: Harvey, Reno; NE: Cherry; OK: Comanche

Conifers

PINACEAE (Pine Family)
Pinus contorta Dougl. (Lodgepole Pine) SD: Lawrence
EPHEDRACEAE (Ephedra Family)
Ephedra coryi E. L. Reed TX: Briscoe

Flowering Plants

2. LAURACEAE (Laurel Family)
Lindera benzoin (L.) Blume, (Spice Bush) KS: Cherokee, Crawford, Labette; MO: Barton, Jasper, Newton; OK: Craig, Osage, Ottawa
Sassafras albidum (Nutt.) Nees, (White Sassafras) KS: Cherokee, Labette; MO: Barton, Bates, Cass, Jasper, Newton; OK: Craig, Mayes, Ottawa
3. SAURURACEAE (Lizard's-tail Family)
Anemopsis californica (Nutt.) H. & A. KS: McPherson; TX: Hemphill
Saururus cernuus L. (Water Dragon) KS: Cherokee, Linn; MO: Barton, Jasper; OK: Mayes, Ottawa
4. ARISTOLOCHIACEAE (Birthwort Family)
Aristolochia serpentaria L. (Virginia Snakeroot) KS: Cherokee; MO: Cass, Jackson, Jasper
5. NYMPHAEACEAE (Water Lily Family)
Brasenia schreberi Gmel. (Purple Wen-Dock) KS: Johnson; MN: Otter Tail; MO: Barton, Jasper
Cabomba caroliniana Gray, (Fanwort) KS: Johnson
6. CERATOPHYLLACEAE (Hornwort Family)
Ceratophyllum echinatum Gray, KS: Lyon
7. RANUNCULACEAE (Buttercup Family)

Actaea alba (L.) Mill. IA: Adair; KS: Doniphan; MO: Atchison, Holt, Worth; NE: Richardson
Anemone heterophylla Nutt. TX: Briscoe, Floyd, Wheeler
Aquilegia brevistyla Hook. (Western Columbine) MN: Clay; SD: Custer, Lawrence, Pennington
Clematis drummondii T. & G. OK: Harmon; TX: Briscoe
Clematis hirsutissima Pursh, NE: Sheridan; SD: Fall River; WY: Niobrara
Clematis maximowicziana Levl. & Van. KS: Cherokee, Cowley, Crawford, Lyon, Neosho, Riley; MO: Jasper, Newton; NE: Furnas; OK: Ottawa
Clematis pseudoalpina (O. Ktze.) A. Nels. (Purple Virgin's Bower) ND: Dunn; SD: Lawrence, Pennington
Clematis tenuiloba (Gray) C. L. Hitchc. SD: Custer, Lawrence, Meade, Pennington
Myosurus aristatus Benth. MT: Daniels; ND: Slope, Ward, Williams
Ranunculus arvensis L. KS: Douglas
Ranunculus bulbosus L. (Bulbous Buttercup) KS: Cherokee, Lyon, Wyandotte; NE: Dawes, Franklin
Ranunculus cardiophyllus Hook. ND: McKenzie; SD: Custer, Lawrence, Pennington
Ranunculus inamoenus Greene, NE: Cheyenne; SD: Pennington
Ranunculus sardous Crantz, KS: Neosho
Ranunculus testiculatus Crantz, KS: Cheyenne, Decatur, Graham, Norton, Thomas; NE: Dawes; WY: Crook

8. BERBERIDACEAE (Barberry Family)
Berberis swaseyi Buckl. TX: Bailey
Berberis trifoliata Moric. TX: Armstrong, Bailey, Hardeman, Lamb

9. MENISPERMACEAE (Moonseed Family)
Calycocarpum lyoni (Pursh) Gray, KS: Cherokee, Johnson; MO: Barton, Bates, Cass, Jasper, Vernon; OK: Mayes, Tulsa

10. PAPAVERACEAE (Poppy Family)
Papaver rhoeas L. (Corn Poppy) KS: Chautauqua, Cherokee, Wabaunsee

13. ULMACEAE (Elm Family)
Ulmus alata Michx. (Small Elm) KS: Cherokee; OK: Creek, Ottawa, Tulsa

14. MORACEAE (Mulberry Family)
Broussonetia papyrifera (L.) Vent. (Paper Mulberry) KS: Leavenworth, Neosho, Wilson; NE: Richardson; OK: Grady
Morus microphylla Buckl. OK: Comanche; TX: Deaf Smith, Foard, Motley

15. CANNABACEAE (Hemp Family)
Humulus japonicus Sieb. & Zucc. (Japanese Hop) IA: Page; KS: Shawnee, Wabaunsee; MO: Jackson, Jasper; NE: Douglas, Nance

16. URTICACEAE (Nettle Family)
Pilea fontana (Lunell) Rydb. MN: Clay; NE: Cuming; ND: Richland, Stutsman; SD: Roberts
Urtica chamaedryoides Pursh, KS: Cherokee, Labette, Neosho; OK: Nowata, Ottawa, Tulsa

17. JUGLANDACEAE (Walnut Family)
Juglans cinerea L. (Butternut) IA: Fremont, Harrison, Pottawattamie

18. FAGACEAE (Beech Family)
Quercus undulata Torr. CO: Baca; NM: Harding, Union; OK: Cimarron
Quercus virginiana Mill. (Live Oak) OK: Comanche, Greer, Kiowa

19. BETULACEAE (Birch Family)
Alnus serrulata (Ait.) Willd. (Common Alder) KS: Cherokee; MO: Vernon

21. NYCTAGINACEAE (Four-O'Clock Family)
Allionia incarnata L. OK: Cimarron; TX: Armstrong, Deaf Smith, Randall
Boerhaavia erecta L. OK: Cleveland
Boerhaavia spicata Choisy, TX: Deaf Smith
Mirabilis jalapa L. (Four-O'Clock) TX: Hemphill
Mirabilis oxybaphoides (Gray) Gray, CO: El Paso; OK: Cimarron
Mirabilis pauciflora (Buckl.) Standl. OK: Greer, Kiowa
Selinocarpus diffusus Gray, OK: Harmon; TX: Oldham

22. CACTACEAE (Cactus Family)

Echinocactus texenis Hopffer, OK: Harmon; TX: Bailey, Cottle
Echinocereus baileyi Rose, OK: Comanche, Greer, Kiowa

23. AIZOACEAE (Carpetweed Family)
Glinus lotoides L. KS: Wilson; MO: Jackson, Vernon
Trianthema portulacastrum L. MO: Jackson, Jasper; OK: Cotton

24. CARYOPHYLLACEAE (Pink Family)
Arenaria drummondii Shinners, OK: Craig, Oklahoma, Tulsa
Arenaria rubella (Wahl.) Sm. SD: Lawrence, Pennington
Gypsophila elegans Bieb. MN: Kittson
Gypsophila muralis L. (Baby's Breath) MN: Otter Tail; SD: Custer, Lawrence, Pennington
Lychnis chalcedonica L. (Scarlet Lychnis) ND: Benson, Rolette, Sheridan
Scleranthus annuus L. KS: Cherokee, Crawford, Wilson; OK: Grady
Silene menziesii Hook. NE: Cherry, Sioux
Silene nivea (Nutt.) Otth. (Snowy Campion) NE: Buffalo, Dodge, Lancaster; SD: Custer, Lawrence, Pennington
Silene regia Sims, (Royal Catchfly) KS: Cherokee; MO: Jasper
Spergula arvensis L. (Corn Spurrey) KS: Riley; MO: Jackson, Jasper; ND: Cass
Spergularia marina (L.) Griseb. ND: Barnes, Burke, Divide, Grand Forks, La Moure
Spergularia rubra (L.) J. & C. Presl, ND: Barnes, Grand Forks, La Moure, Walsh, Ward
Stellaria aquatica (L.) Scop. KS: Doniphan; MN: Lac Qui Parle
Stellaria calycantha (Ledeb.) Bong. MN: Becker
Stellaria graminea L. (Common Stitchwort) ND: Cass

25. PORTULACACEAE (Purslane Family)
Portulaca umbraticola H.B.K. OK: Comanche
Talinum auranticum Engelm. TX: Armstrong
Talinum rugospermum Holz. KS: Harvey

26. CHENOPODIACEAE (Goosefoot Family)
Atriplex argentea Nutt. ssp. *expansa* (Wats.) Hall & Clem. (Silverscale Saltbush) TX: Childress
Atriplex powellii Wats. SD: Corson, Haakon, Jackson, Pennington
Atriplex wardii Standl. TX: Hall
Chenopodium capitatum (L.) Asch. (Strawberry Blite) MN: Becker; SD: Hutchinson, Lawrence, Pennington, Turner
Chenopodium foliosum (Moench) Asch. ND: Cavalier, Pembina
Chenopodium subglabrum (Wats.) A. Nels. MT: Custer; ND: Slope; NE: Cherry, Dawes, Scotts Bluff, Sheridan; SD: Mellette
Suaeda nigrescens I. M. Johnst. var. *glabra* I. M. Johnst. TX: Bailey
Suaeda suffrutescens Wats. OK: Major, Tillman; TX: Armstrong, Bailey, Childress, Hall
Suaeda torreyana Wats. TX: Bailey

27. AMARANTHACEAE (Pigweed Family)
Alternanthera caracasana H.B.K. OK: Tillman; TX: Cottle
Amaranthus powellii Wats. KS: Saline; WY: Goshen, Platte
Guilleminea densa (Willd.) Moq. OK: Cimarron; TX: Cottle, Motley
Guilleminea lanuginosa (Poir.) Hook. var. *sheldonii* (Uline & Gray) Mears, OK: Cleveland, Comanche, Harmon, Major
Guilleminea lanuginosa (Poir.) Hook. var. *tenuiflora* (Hook.) Mears, OK: Comanche, Creek, Osage

28. POLYGONACEAE (Buckwheat Family)
Eriogonum brevicaule Nutt. WY: Goshen
Eriogonum correllii Reveal, OK: Cimarron; TX: Armstrong, Briscoe, Floyd, Hardeman
Eriogonum gordonii Benth. NE: Sioux
Eriogonum tenellum Torr. CO: Baca, Bent; NM: Union; OK: Cimarron; TX: Bailey
Polygonella americana (Fisch. & Mey.) Small, TX: Lamb, Parmer
Polygonum cuspidatum Sieb. & Zucc. (Japanese Knotweed) KS: Cherokee, Crawford, Wyandotte; MO: Jackson
Polygonum densiflorum Meisn. KS: Cowley
Polygonum leptocarpum Robins. KS: Cowley; MT: Custer, Rosebud; SD: Fall River, Marshall, Stanley

Polygonum orientale L. (Kiss-me-over-the-garden-gate) IA: Adair; KS: Douglas, Leavenworth, Riley, Shawnee; MO: Barton, Cass, Clay, Jackson, Jasper; NE: Butler, Douglas

Polygonum persicarioides H.B.K. (Southwest Smartweed) KS: Riley, Saline, Washington; NE: Kearney; OK: Blaine, Cleveland, Oklahoma, Tulsa

Polygonum sawatchense Small, NE: Banner, Sioux; SD: Harding, Lawrence

Polygonum viviparum L. (Alpine Bistort) SD: Custer, Pennington; WY: Laramie

29. PLUMBAGINACEAE (Leadwort Family)

Limonium limbatum Small, OK: Cotton, Harmon, Jackson, Kingfisher, Major; TX: Bailey

31. HYPERICACEAE (St. John's-wort Family)

Hypericum canadense L. NE: Cherry, Hooker, Kearney, Thomas

Hypericum pyramidatum Ait. (Greater St. John's-wort) IA: Adair, Taylor; KS: Doniphan; MO: Atchison, Holt, Jackson; NE: Cass, Otoe, Richardson

Triadenum fraseri (Spach) Gl. (Marsh St. John's-wort) MN: Becker, Otter Tail, Polk; NE: Cherry, Holt, Thomas

33. MALVACEAE (Mallow Family)

Abutilon incanum (Link) Sweet, CO: Baca; OK: Comanche, Jackson

Abutilon parvulum Gray, TX: Deaf Smith

Althaea officinalis L. (Marshmallow) ND: Cass

Anoda cristata (L.) Schlecht. KS: Jefferson, Neosho

Callirhoe digitata Nutt. (Finger Poppy Mallow) KS: Cherokee; MO: Barton, Jasper; OK: Caddo, Ottawa, Tulsa

Callirhoe papaver (Cav.) Gray var. *bushii* (Fern.) Waterfall, KS: Neosho; OK: Tulsa

Malvella sagittifolia (Gray) Fryxell, TX: Bailey

Sida physocalyx Gray, OK: Harmon; TX: Bailey, Cottle, Hall

Sphaeralcea fendleri Gray, KS: Barber, Pawnee, Sedgwick; NM: Curry; TX: Oldham

34. DROSERACEAE (Sundew Family)

Drosera rotundifolia L. (Round-leaved Sundew) MN: Becker, Pope; ND: Bottineau

35. VIOLACEAE (Violet Family)

Viola arvensis Murr. (Wild Pansy) KS: Marshall, Saline; NE: Webster

Viola conspersa Reichb. (American Dog Violet) MN: Becker, Clay, Douglas, Kittson, Otter Tail; ND: Richland

Viola incognita Brainerd, MN: Becker, Douglas, Otter Tail; ND: Pembina

Viola lovelliana Brainerd, OK: Cleveland, Oklahoma

Viola renifolia Gray, (Kidney-leaved Violet) SD: Custer

Viola selkirkii Pursh, (Great Spurred Violet) SD: Pennington

Viola triloba Schwein. var. *dilatata* (Ell.) Brainerd, KS: Cherokee; MO: Jasper; OK: Ottawa

Viola villosa Walt. OK: Cleveland, McLain, Oklahoma

37. CISTACEAE (Rockrose Family)

Hudsonia tomentosa Nutt. (Poverty-grass) MN: Polk; ND: Ransom

Lechea intermedia Leggett, (Pinweed) SD: Pennington

38. TAMARICACEAE (Tamarisk Family)

Tamarix chinensis Lour. OK: Blaine, Greer, Oklahoma

39. LOASACEAE (Sand Lily Family)

Cevallia sinuata Lag. OK: Greer, Harmon, Jackson; TX: Cottle, Randall

Mentzelia albescens (Gill) Griseb. KS: Cherokee; MO: Jasper; OK: Comanche, Kiowa

Mentzelia albicaulis Dougl. CO: Washington, Weld; NE: Scotts Bluff; SD: Harding, Lawrence; WY: Campbell

Mentzelia chrysantha Engelm. NE: Morrill

Mentzelia dispersa Wats. MT: McCone; ND: Billings, Golden Valley, Morton, Slope; SD: Meade, Pennington

Mentzelia reverchonii (Urb. & Gilg.) Thomps. & Zavortink, TX: Dallam, Lamb

40. CUCURBITACEAE (Gourd Family)

Ibervillea lindheimeri (Gray) Greene, OK: Comanche, Greer, Harmon, Jackson; TX: Armstrong, Hardeman

41. SALICACEAE (Willow Family)

Populus angustifolia James, (Narrow-leaved Cottonwood) CO: El Paso; NM: Union; SD: Fall River, Lawrence; WY: Platte

Populus trichocarpa T. & G. ND: Billings
Salix fragilis L. (Crack Willow) IA: Crawford, O'Brien, Osceola, Woodbury; NE: Brown, Cherry, Cuming, Richardson, York; SD: Roberts
Salix monticola Bebb ex Coult. (Serviceberry Willow) SD: Lawrence, Pennington
Salix phylicifolia L. ssp. *planifolia* (Pursh) Hutonen, (Planeleaf Willow) SD: Custer, Lawrence, Pennington
Salix scouleriana Barr. (Western Pussy Willow) ND: Bottineau, Burke; SD: Custer, Lawrence, Meade, Pennington

42. CAPPARACEAE (Caper Family)
Cleome lutea Hook. NE: Lancaster

43. BRASSICACEAE (Mustard Family)
Arabidopsis thaliana (L.) Heynh. (Mouse-ear Cress) SD: Lawrence, Pennington
Arabis fendleri (Wats.) Greene, OK: Cimarron
Arabis laevigata (Muhl.) Poir. (Smooth Rock Cress) KS: Cherokee, Crawford, Douglas; MO: Buchanan, Jasper, Vernon; OK: Comanche, Ottawa
Barbarea orthoceras Ledeb. (Winter Cress) SD: Clay, Custer, Lawrence, Pennington
Cardaria pubescens (Mey.) Jarmolenko, (Whitetop) ND: Cass, McKenzie
Descurainia pinnata (Walt.) Britt. var. *intermedia* (Rydb.) C. L. Hitchc. (Tansy Mustard) MT: McCone, Sheridan; SD: Fall River, Lawrence; WY: Laramie, Niobrara
Diplotaxis muralis (L.) DC. (Sand Rocket) KS: Geary; NE: Douglas; SD: Clay
Draba aurea Vahl, SD: Custer, Pennington
Draba lanceolata Ledeb. SD: Lawrence, Pennington
Draba stenoloba Ledeb. SD: Lawrence, Pennington
Eruca sativa Lam. (Rocket-salad) IA: Mills, Pottawattamie; KS: Riley; MO: Jackson; ND: Cass, Pierce, Rolette; NE: Garfield
Lepidium austrinum Small, KS: Allen; OK: Blaine, Cleveland, Osage
Lepidium latifolium L. CO: Sedgwick; NE: Hooker
Lepidium sativum L. (Garden Cress) NE: Antelope, Valley
Lesquerella montana (Gray) Wats. NE: Kimball, Sioux; NM: Union; SD: Fall River; WY: Laramie, Platte
Lesquerella ovalifolia Rydb. ssp. *alba* (Goodm.) Roll. & Shaw, (Oval-leaf Bladderpod) OK: Caddo, Comanche, Custer, Jackson, Kiowa
Physaria didymocarpa (Hook.) Gray, (Double Bladderpod) ND: Billings, McKenzie, Slope; SD: Haakon, Harding, Jackson, Sheridan
Rorippa austriaca (Crantz) Bess. MN: Lincoln; ND: Cass, Pembina, Ransom, Richland; NE: Lancaster
Rorippa calycina (Engelm.) Rydb. MT: Custer; ND: McKenzie; NE: Scotts Bluff
Rorippa curvipes Greene, CO: Weld; KS: Ford; ND: Benson; SD: Brown, Faulk; WY: Platte
Rorippa palustris (L.) Bess. ssp. *glabra* (Schultz) Stuckey var. *glabrata* (Lunell) Stuckey, (Bog Yellow Cress) ND: Benson
Rorippa palustris (L.) Bess. ssp. *hispida* (Desv.) Jonsell var. *elongata* Stuckey, (Bog Yellow Cress) ND: Benson, Bottineau, Emmons; NE: Morrill, Wheeler; SD: Stanley, Minnehaha
Rorippa sylvestris (L.) Bess. (Creeping Yellow Cress) IA: Page; KS: Doniphan, Douglas, Franklin, Lyon; ND: Cass; NE: Dakota
Rorippa tenerrima Greene, CO: Weld; MO: Clay, Jackson; ND: McLean, Oliver; NE: Deuel; SD: Fall River
Sisymbrium irio L. CO: Logan; TX: Swisher
Streptanthus hyacinthoides Hook. KS: Barber, Comanche; OK: Caddo, Grady, Woods, Woodward
Thelypodium wrightii Gray, NM: Union; OK: Cimarron

44. RESEDACEAE (Reseda Family)
Reseda lutea L. IA: Harrison; NE: Dixon

45. ERICACEAE (Heath Family)
Vaccinium arboreum Marsh. (Sparkleberry) KS: Cherokee; MO: Jasper
Vaccinium membranaceum Dougl. (Mountain Huckleberry) SD: Lawrence

Vaccinium scoparium Leiberg, (Grouseberry) SD: Lawrence, Pennington
Vaccinium stamineum L. (Deerberry) OK: Craig, Ottawa
Vaccinium vacillans Torr. (Sugar Huckleberry) KS: Cherokee; MO: Barton, Jasper, Vernon

46. PYROLACEAE (Wintergreen Family)
Chimaphila umbellata (L.) Bart. (Prince's Pine) MN: Clay, Polk; SD: Lawrence
Pyrola picta Sm. SD: Lawrence
Pyrola rotundifolia L. var. *americana* (Sweet) Fern. (Wild Lily-of-the-Valley) MN: Beaver; SD: Custer, Lawrence, Pennington
Pyrola uniflora L. (One-flowered Wintergreen) SD: Lawrence, Pennington
Pyrola virens Schweigg. NE: Cherry, Sioux; SD: Lawrence, Pennington

47. MONOTROPACEAE (Indian Pipe Family)
Monotropa hypopithys L. (Pine Sap) KS: Cherokee, Douglas; MO: Jasper; NE: Box Butte, Richardson

50. PRIMULACEAE (Primrose Family)
Dodecatheon pulchellum (Raf.) Merr. SD: Custer, Fall River, Lawrence, Pennington
Lysimachia punctata L. (Garden Loosestrife) NE: Douglas
Primula incana M. E. Jones, MT: Daniels; ND: Burke, Mountrail
Samolus cuneatus Small, KS: Barber; TX: Gray, Hemphill, Wheeler

51. CRASSULACEAE (Stonecrop Family)
Sedum acre L. (Mossy Stonecrop) SD: Lawrence

52. SAXIFRAGACEAE (Saxifrage Family)
Hydrangea arborescens L. (Wild Hydrangea) KS: Cherokee; OK: Ottawa
Lithophragma bulbifera Rydb. (Woodland Star) SD: Lawrence
Lithophragma parviflora (Hook.) Nutt. (Prairie Star) NE: Dawes, Sioux; SD: Custer, Lawrence, Meade, Pennington; WY: Crook, Niobrara, Weston
Parnassia parviflora DC. (Grass-of-Parnassus) ND: Bottineau; SD: Lawrence, Pennington
Ribes lacustre (Pers.) Poir. (Swamp Currant) SD: Lawrence, Pennington
Saxifraga cernua L. (Saxifrage) SD: Lawrence
Saxifraga occidentalis Wats. SD: Lawrence
Saxifraga texana Buckl. KS: Chautauqua; MO: Barton, Jasper, Vernon; OK: Osage

53. ROSACEAE (Rose Family)
Crataegus berberifolia T. & G. KS: Cherokee, Neosho; MO: Barton, Jasper, Vernon
Crataegus coccinioides Ashe, KS: Riley; MO: Barton, Jasper
Crataegus collina Chapm. KS: Bourbon, Cherokee, Crawford, Neosho; MO: Barton, Jasper
Crataegus lanuginosa Sarg. KS: Cherokee; MO: Jasper
Crataegus palmeri Sarg. KS: Cherokee; MO: Barton, Bates, Jasper, Vernon
Crataegus pruinosa (Wendl.) K. Koch, KS: Cherokee, Crawford, Wabaunsee; MO: Andrew, Barton, Cass, Jackson, Jasper
Crataegus reverchoni Sarg. var. *discolor* (Sarg.) Palmer, KS: Cherokee, Franklin; MO: Jasper
Crataegus reverchoni Sarg. var. *stevensiana* (Sarg.) Palmer, KS: Cherokee
Fallugia paradoxa (D. Don) Endl. OK: Cimarron
Geum laciniatum Murr. var. *trichocarpon* Fern. IA: Adair, Cherokee, Fremont, Lyon, Page; KS: Anderson, Sedgwick; MO: Jackson; NE: Holt
Geum rivale L. (Purple Avens) MN: Otter Tail; ND: Pembina; SD: Day, Meade
Gillenia stipulata (Muhl.) Baill. (American Ipecac) KS: Cherokee, Crawford, Labette; MO: Barton, Jackson, Jasper, Vernon; OK: Craig, Ottawa
Petrophytum caespitosum (Nutt.) Rydb. (Rockplant Spiraea) SD: Lawrence
Physocarpus monogynus (Torr.) Coult. CO: Logan; OK: Cimarron; SD: Custer, Lawrence, Pennington
Potentilla biennis Greene, SD: Harding, Lawrence, Walworth
Potentilla diversifolia Lehm. ND: Billings, Slope; SD: Custer
Potentilla fissa Nutt. SD: Custer, Pennington; WY: Laramie
Potentilla glandulosa Lindl. SD: Codington, Custer, Lawrence, Meade, Pennington
Potentilla plattensis Rydb. ND: Burke, Divide, Mountrail, Williams; NE: Cheyenne
Potentilla tridentata Soland. (Three-toothed Cinquefoil) ND: Cavalier

Prunus nigra Ait. (Canada Plum) MN: Kittson, Otter Tail
Prunus rivalis Scheele, KS: Chautauqua, Cowley, Neosho; OK: Harper
Rosa eglanteria L. (Sweet Briar) KS: Jackson, Riley; MO: Jackson, Jasper
Rubus deliciosus Torr. CO: Baca, Elbert; NM: Union; OK: Cimarron
Rubus parviflorus Nutt. (Thimbleberry) SD: Lawrence, Pennington; WY: Crook
Rubus trivialis Michx. OK: Cleveland, Grady, Oklahoma
Sanguisorba annua Nutt. KS: Morton; OK: Blaine, Canadian, Kingfisher, Oklahoma
Sorbus scopulina Greene, (Western Mountain Ash) SD: Lawrence
Spiraea betulifolia Pall. var. *lucida* (Dougl.) C. L. Hitchc. (Mountain Meadowsweet) MT: Rosebud; SD: Brule, Custer, Lawrence, Pennington

54a. FABACEAE-Mimosoideae (Bean Family)
Mimosa biuncifera Benth. TX: Bailey, Floyd, Lamb, Randall

54b. FABACEAE-Caesalpinoideae (Bean Family)
Cassia roemeriana Scheele, NM: Roosevelt; OK: Jackson; TX: Bailey, Childress, Hardeman
Hoffmanseggia drepanocarpa Gray, NM: Curry; OK: Cimarron; TX: Bailey

54c. FABACEAE-Papilionoideae (Bean Family)
Anthyllis vulneraria L. (Kidney Vetch) ND: Cass
Astragalus alpinus L. SD: Custer, Lawrence, Pennington
Astragalus americanus (Hook.) M. E. Jones, SD: Lawrence, Pennington
Astragalus barrii Barneby, MT: Carter, Powder River; SD: Fall River, Pennington, Shannon
Astragalus bodini Sheld. WY: Laramie
Astragalus crassicarpus Nutt. var. *trichocalyx* (Nutt.) Barneby, KS: Cherokee; MO: Jackson, Jasper, Vernon; OK: Creek, Ottawa, Rogers, Tulsa
Astragalus miser Dougl. var. *hylophilus* (Rydb.) Barneby, SD: Pennington
Astragalus neglectus (T. & G.) Sheld. (Cooper's Milkvetch) MN: Marshall, Otter Tail, Polk; ND: Pembina
Astragalus praelongus Sheld. var. *ellisiae* (Rydb.) Barneby, NM: Roosevelt, Union; TX: Bailey, Deaf Smith, Motley
Astragalus punicens Osterh. CO: Las Animas; NM: Union; OK: Cimarron
Astragalus shortianus Nutt. NE: Banner, Kimball; WY: Laramie
Astragalus wootonii Sheld. TX: Hutchinson
Clitoria mariana L. KS: Cherokee; OK: Cleveland, Comanche, Creek, Payne, Tulsa
Dalea frutescens Gray, OK: Comanche; TX: Cottle, Foard
Desmodium marilandicum (L.) DC. KS: Cherokee, Montgomery; MO: Jackson, Jasper, Vernon; OK: Oklahoma
Desmodium nudiflorum (L.) DC. KS: Cherokee; MO: Barton, Jackson, Jasper, Vernon
Desmodium pauciflorum (Nutt.) DC. KS: Cherokee; MO: Jasper
Desmodium rigidum (Ell.) DC. KS: Chautauqua, Lyon, Woodson; MO: Barton, Jackson, Jasper, Vernon
Desmodium rotundifolium (Michx.) DC. KS: Cherokee; MO: Jasper
Galactia volubilis (L.) Britt. var. *mississippiensis* Vail, KS: Chautauqua, Cherokee; MO: Jasper; OK: Cleveland, Creek, Mayes, Oklahoma, Ottawa, Payne
Hedysarum alpinum L. var. *philoscina* (A. Nels.) Roll. SD: Custer, Lawrence, Pennington
Hedysarum occidentale Greene, SD: Pennington
Lathyrus latifolius L. (Everlasting Pea) KS: Atchison, Barber, Cowley, Coffey, Leavenworth, Lyon, Sedgwick; SD: Lawrence
Lathyrus pusillis Ell. (Low Peavine) KS: Cherokee, Crawford, Labette, Montgomery, Neosho; OK: Creek, Payne, Rogers, Tulsa
Lespedeza hirta (L.) Herm. (Hairy Lespedeza) KS: Cherokee; MO: Barton, Jasper, Vernon; OK: Grady
Lespedeza intermedia (Wats.) Britt. KS: Crawford, Harvey, Jackson, Linn, Lyon; MO: Jasper; OK: Alfalfa, Grady, Payne
Medicago minima L. (Bur Clover) KS: Cherokee, Clay, Cowley, Crawford, Labette, Lincoln, Montgomery, Neosho, Russell, Wilson, Woodson; OK: Caddo, Garfield, Grady, Oklahoma, Tillman
Neptunia lutea (Leavenw.) Benth. OK: Beckham, Cleveland, Comanche, Oklahoma, Payne

Oxytropis deflexa (Pall.) DC. var. *sericea* T. & G. ND: Bottineau, Cavalier, Pembina
Phaseolus polystachios (L.) B.S.P. (Wild Bean) MO: Jasper; NE: Sarpy, Seward; SD: Pennington
Pisum sativum L. var. *arvense* (L.) Poir. (Garden Pea) KS: Kingman, Sedgwick
Psoralea psoralioides (Walt.) Cory var. *eglandulosa* (Ell.) F. L. Freeman, (Sampson's Snakeroot) KS: Cherokee, Crawford, Labette, Neosho; MO: Barton, Bates, Jasper, Vernon; OK: Craig, Ottawa
Robinia hispida L. (Mossy Locust) KS: Coffey, Lyon, Osage, Woodson; NM: Union; OK: Cleveland, Logan
Sesbania macrocarpa Muhl. KS: Montgomery, Riley, Woodson, Wyandotte; OK: Tulsa
Thermopsis divaricata A. Nels. NM: Union
Trifolium arvense L. (Rabbit-foot Clover) KS: Jefferson, Lyon; MO: Buchanan, Jasper; ND: Cass
Trifolium beckwithii Brewer, SD: Deuel
Trifolium carolinianum Michx. (Carolina Clover) KS: Cherokee; MO: Barton, Jasper
Trifolium fragiferum L. (Strawberry-headed Clover) CO: Yuma; NE: Garden, Hall, Morrill; WY: Platte
Trifolium incarnatum L. (Crimson Clover) ND: Cass; NE: Lancaster; SD: Custer
Trifolium resupinatum L. (Reversed Clover) KS: Riley, Saline; ND: Cass
Trifolium stoloniferum Muhl. (Running Buffalo Clover) KS: Crawford, Linn, Miami, Rush; MO: Jasper; MT: Rosebud; NE: Cherry, Grant; WY: Weston
Vicia caroliniana Walt. (Wood Vetch) KS: Ellis, Lyon
Vicia cracca L. (Tufted Vetch) MN: Clay, Kittson; NE: Antelope, Buffalo, Cherry, Sheridan; SD: Lawrence
Vicia exigua T. & G. KS: Barber; NM: Union; OK: Cimarron; TX: Bailey
Vicia sativa L. (Spring Vetch) KS: Barber, Dickinson, Franklin, Wyandotte; MO: Jackson, Jasper; NE: Lancaster; SD: Brookings, Fall River

57. LYTHRACEAE (Loosestrife Family)
Didiplis diandra (Nutt.) Wood, (Water Purslane) KS: Cherokee, Jackson, Saline; MO: Barton, Bates, Cass, Clay, Jackson, Jasper

58. THYMELAEACEAE (Leatherwood Family)
Thymelaea passerina (L.) Coss. & Germ. KS: Cloud; NE: Cedar, Merrick, Morrill

59. ONAGRACEAE (Evening Primrose Family)
Boisduvalia glabella (Nutt.) Walp. ND: Billings, Hettinger; SD: Butte
Calylophus hartwegii (Benth.) Raven ssp. *filifolius* (Eastw.) Towner & Raven, OK: Ellis; TX: Lipscomb
Epilobium hornemanni Reichb. WY: Goshen
Epilobium palustre L. SD: Pennington
Gaura neomexicana Woot. ssp. *coloradensis* (Rydb.) Raven & Gregory, CO: Weld; WY: Laramie
Gayophytum diffusum T. & G. ssp. *parviflorum* Lewis & Szweykowski, (Baby's Breath) MT: Daniels; SD: Custer, Lawrence
Gayophytum ramosissimum T. & G. WY: Weston
Ludwigia glandulosa Walt. KS: Cherokee
Ludwigia repens Forst. OK: Alfalfa, Cleveland, Comanche; TX: Hemphill
Oenothera brachycarpa Gray, CO: El Paso, Otero
Oenothera parviflora L. MN: Kittson, Lincoln, Mahnomen, Pipestone

60. MELASTOMATACEAE (Melastome Family)
Rhexia mariana L. var. *interior* (Penn.) Kral & Bostick, KS: Cherokee; MO: Barton, Jasper, Vernon

62. CORNACEAE (Dogwood Family)
Cornus florida L. (Flowering Dogwood) KS: Cherokee; MO: Barton, Bates, Jasper, Vernon; OK: Craig, Ottawa

62a. GARRYACEAE (Garrya Family)
Garrya ovata Benth. TX: Briscoe

63. SANTALACEAE (Sandalwood Family)
Thesium linophyllum L. ND: Towner

68. EUPHORBIACEAE (Spurge Family)
Argythamnia humilis (Engelm. & Gray) Muell. Arg. var. *laevis* (Torr.) Shinners, OK: Cimarron
Croton capitatus Michx. var. *lindheimeri* (Engelm. & Gray) Muell. Arg. (Woolly Croton) KS: Miami
Croton dioicus Cav. TX: Briscoe, Floyd, Hardeman
Croton pottsii (Klotzsch) Muell. Arg. NM: Quay, Roosevelt; TX: Bailey, Briscoe, Deaf Smith
Euphorbia carunculata Waterfall, OK: Woods; TX: Hardeman, Wheeler
Euphorbia longicuris Scheele, OK: Comanche, Custer, Grady, Roger Mills, Washita
Euphorbia revoluta Engelm. CO: Baca
Euphorbia strictior Holz. NM: Curry, Harding; TX: Deaf Smith
Phyllanthus abnormis Baill. OK: Comanche, Custer
Phyllanthus caroliniensis Walt. KS: Cherokee, Crawford, Labette, Miami; MO: Barton, Bates, Jasper, Vernon
Phyllanthus polygonoides Nutt. OK: Caddo, Comanche, Jackson, Kiowa

69. RHAMNACEAE (Buckthorn Family)
Ceanothus fendleri Gray, CO: Las Animas; SD: Custer, Lawrence; TX: Childress, Cottle, Foard, Hardeman, Motley
Ceanothus velutinus Dougl. (Mountain Balm) SD: Lawrence, Pennington
Rhamnus davuricus Pall. ND: Burleigh, Oliver, Ransom, Richland; NE: Dawes; SD: Minnehaha
Ziziphus obtusifolia (T. & G.) Gray, OK: Cotton, Greer, Harmon, Jackson; TX: Briscoe, Hall, Randall

70. VITACEAE (Grape Family)
Vitis rupestris Scheele, (Sand Grape) OK: Cleveland, Comanche

73. HIPPOCASTANACEAE (Buckeye Family)
Aesculus glabra Willd. var. *glabra,* (Ohio Buckeye) IA: Audubon, Cass, Fremont, Mills, Taylor; KS: Wyandotte; MO: Andrew, Bates, Buchanan, De Kalb, Gentry, Jasper, Vernon

74. ACERACEAE (Maple Family)
Acer glabrum Torr. CO: Las Animas; NE: Sioux; WY: Converse, Laramie
Acer nigrum Michx. f. (Black Maple) MO: Andrew, Holt; NE: Sarpy; SD: Roberts

75. ANACARDIACEAE (Cashew Family)
Toxicodendron radicans (L.) O. Ktze. ssp. *pubens* (Engelm.) Gillis, (Poison Ivy) KS: Barton, Ellsworth, Ottawa, Shawnee; MO: Jackson
Toxicodendron radicans (L.) O. Ktze. ssp. *verrucosum* (Scheele) Gillis, (Poison Ivy) OK: Garfield, Woods
Toxicodendron toxicarium (Salisb.) Gillis, KS: Chautauqua, Cherokee; OK: Cleveland, Kingfisher, Oklahoma, Payne

77. RUTACEAE (Rue Family)
Thamnosma texana (Gray) Torr. TX: Bailey
Zanthoxylum hirsutum Buckl. OK: Cotton, Tillman

80. GERANIACEAE (Geranium Family)
Erodium texanum Gray, OK: Harmon, Jackson; TX: Hardeman
Geranium nervosum Rydb. CO: Larimer, Las Animas; SD: Lawrence, Pennington
Geranium richardsonii Fisch. & Trautv. SD: Custer, Lawrence, Pennington
Geranium viscosissimum Fisch. & Mey. MT: Rosebud; SD: Custer, Lawrence, Pennington; WY: Crook

82. LINACEAE (Flax Family)
Linum aristatum Engelm. TX: Hutchinson
Linum hudsonioides Planch. KS: Sedgwick; OK: Comanche, Greer
Linum medium (Planch.) Britt. var. *texanum* (Planch.) Fern. KS: Cherokee, Crawford; MO: Barton, Jasper, Vernon
Linum puberulum (Engelm.) Heller, CO: Baca, Bent, Weld; NE: Cheyenne; WY: Laramie

83. POLYGALACEAE (Milkwort Family)
Polygala tweedyi Britt. OK: Harmon; TX: Childress

85. APIACEAE (Parsley Family)
Ammoselinum butleri (Wats.) Coult. & Rose, KS: Crawford

Bupleurum rotundifolium L. KS: Coffey, Wabaunsee; MO: Clay
Cicuta douglasii (DC.) Coult. & Rose, MN: Clay; MT: Custer
Conioselinum chinense (L.) B.S.P. NE: Webster
Coriandrum sativum L. MO: Jackson; ND: Cass; OK: Oklahoma
Cymopterus macrorhizus Buckl. OK: Beckham, Custer; TX: Bailey, Childress, Cottle, Motley
Erigenia bulbosa (Michx.) Nutt. KS: Chautauqua, Cherokee, Crawford, Neosho; MO: Barton, Cass, Jasper
Eryngium diffusum Torr. OK: Alfalfa, Cotton, Grant, Greer, Tillman, Woodward
Eryngium planum L. SD: Lake
Eryngium prostratum Nutt. KS: Cherokee
Falcaria sioides (Wibel) Asch. IA: O'Brien, Sioux; KS: Atchison, Marshall; NE: Douglas, Lancaster; OK: Major
Foeniculum vulgare Mill. IA: Cass; KS: Neosho; MO: Jackson
Hydrocotyle ranunculoides L. f. KS: Kingman
Lomatium nuttallii (Gray) Macbr. NE: Scotts Bluff; SD: Custer
Osmorhiza chilensis H. & A. SD: Custer, Lawrence
Osmorhiza depauperata Phil. SD: Custer, Lawrence
Perideridia gairdneri (H. & A.) Math. (Squaw-root) SD: Lawrence, Pennington
Ptilimnium capillaceum (Michx.) Raf. (Mock Bishop-weed) KS: Crawford
Torilis nodosa (L.) Gaertn. OK: Custer

86. GENTIANACEAE (Gentian Family)
Centaurium exaltatum (Griseb.) Wight ex Piper, NE: Scotts Bluff
Centaurium texense (Griseb.) Fern. KS: Montgomery, Wilson; OK: Comanche
Gentianopsis procera (Holmgren) Ma ssp. *procera* MN: Otter Tail, Polk, Pope; ND: Kidder
Gentianopsis procera (Holmgren) Ma ssp. *macounii* (Holmgren) Iltis, MN: Clay, Kittson; NE: Eddy
Sabatia angularis (L.) Pursh, (Rose-Pink) KS: Cherokee, Crawford; MO: Barton, Jasper, Vernon; OK: Cleveland, Oklahoma
Swertia radiata (Kell.) O. Ktze. SD: Custer, Lawrence, Pennington; WY: Weston

87. APOCYNACEAE (Dogbane Family)
Amsonia ciliata Walt. var. *texana* (Gray) Coult. OK: Caddo, Comanche, Kingfisher; TX: Briscoe, Cottle
Amsonia illustris Woods. KS: Allen, Cherokee, Crawford; MO: Jackson, Jasper; OK: Ottawa
Amsonia tabernaemontana Walt. (Willow Amsonia) KS: Bourbon, Cherokee; MO: Jasper
Apocynum suksdorfii Greene var. *angustifolium* (Woot.) Woods. TX: Roberts
Vinca minor L. (Common Periwinkle) KS: Neosho; NE: Richardson

88. ASCLEPIADACEAE (Milkweed Family)
Asclepias brachystephana Torr. TX: Bailey, Deaf Smith, Randall
Asclepias involucrata Engelm. CO: Las Animas; KS: Stevens; NM: Curry; OK: Cimarron; TX: Sherman
Asclepias macrotis Torr. CO: Otero; NM: Curry; OK: Cimarron; TX: Deaf Smith
Asclepias oenotheroides Cham. & Schlecht. OK: Blaine, Major; TX: Bailey, Briscoe, Floyd
Asclepias quadrifolia Jacq. KS: Cherokee; MO: Barton, Jasper; OK: Ottawa, Tulsa
Asclepias tuberosa L. ssp. *terminalis* Woods. (Butterfly Milkweed) CO: Yuma; NM: Union; OK: Cimarron; SD: Fall River; TX: Hutchinson
Matelea biflora (Raf.) Woods. TX: Bailey, Hardeman
Matelea cynanchoides (Engelm.) Woods. OK: Canadian, Kiowa, Payne
Matelea decipiens (Alex.) Woods. KS: Cherokee; MO: Jasper
Matelea gonocarpa (Walt.) Shinners, KS: Cherokee; MO: Jasper
Periploca graeca L. KS: Greenwood
Sarcostemma crispum Benth. TX: Armstrong, Bailey, Deaf Smith, Hardeman
Sarcostemma cynanchoides Dcne. OK: Jackson; TX: Childress, Hall, Hemphill, Motley, Randall
Sarcostemma lobata Waterfall, OK: Cimarron

89. SOLANACEAE (Nightshade Family)
Chamaesaracha sordida (Dun.) Gray, TX: Briscoe, Cottle, Randall
Datura meteloides DC. OK: Cleveland

Datura quercifolia H.B.K. KS: Meade, Morton; OK: Cimarron; TX: Deaf Smith
Datura wrightii Regel, OK: Blaine; TX: Bailey, Motley
Lycium berlandieri Dun. OK: Cimarron, Harmon; TX: Randall
Lycium pallidum Miers, CO: Las Animas, Otero; NM: Union; OK: Cimarron
Nicotiana trigonophylla Dun. OK: Jackson
Physalis hederaefolia Gray var. *cordifolia* (Gray) Waterfall, OK: Cimarron
Solanum citrullifolium A. Br. NM: Harding; OK: Comanche
Solanum sarrachoides Sendt. KS: Chautauqua, Neosho; MO: Jasper; OK: Payne

90. CONVOLVULACEAE (Convolvulus Family)
Convolvulus pellitus Ledeb. KS: Bourbon, Jefferson, Montgomery, Shawnee; MO: Jackson, Platte
Cressa depressa Goodd. OK: Harmon; TX: Armstrong, Bailey
Ipomoea cristulata Hallier, KS: Hamilton, Neosho

91. CUSCUTACEAE (Dodder Family)
Cuscuta epithymum Murr. (Clover Dodder) ND: Bowman, Burke, Stutsman
Cuscuta glabrior (Engelm.) Yunck. OK: Washington; TX: Carson, Hansford
Cuscuta squamata Engelm. TX: Deaf Smith, Hemphill, Roberts
Cuscuta umbellata H.B.K. KS: Barber, Comanche, Hamilton, Kiowa; OK: Cimarron

92. MENYANTHACEAE (Buckbean Family)
Menyanthes trifoliata L. (Buckbean) MN: Becker, Otter Tail; ND: Ransom; SD: Brookings

93. POLEMONIACEAE (Polemonium Family)
Gilia calcarea M. E. Jones, CO: Lincoln; KS: Hamilton
Ipomopsis rubra (L.) Wherry, OK: Comanche, Kiowa, Roger Mills
Leptodactylon pungens (Torr.) Nutt. CO: Arapahoe, Morgan, Weld; NE: Scotts Bluff; WY: Campbell, Converse, Niobrara
Linanthus septentrionalis Mason, SD: Harding, Pennington
Phlox longipilosa Waterfall, OK: Comanche, Greer, Washita
Phlox muscoides Nutt. MT: Garfield, Richland, Rosebud, Sheridan; NE: Cheyenne, Kimball; WY: Laramie

94. HYDROPHYLLACEAE (Waterleaf Family)
Nemophila phacelioides Nutt. (Baby Blue Eyes) OK: Caddo
Phacelia congesta Hook. OK: Caddo, Comanche
Phacelia gilioides Brand. (Hairy Phacelia) KS: Crawford; MO: Barton, Jasper
Phacelia linearis (Pursh) Holz. MT: Carter, McCone, Powder River; SD: Pennington; WY: Campbell, Crook, Weston
Phacelia popei T. & G. TX: Armstrong, Bailey, Childress, Cottle
Phacelia strictiflora (Engelm. & Gray) Gray, OK: Blaine, Caddo, Canadian, Custer, Grady, Oklahoma

95. BORAGINACEAE (Borage Family)
Amsinckia intermedia Fisch. & Mey. (Fiddleneck) ND: Cass, Pembina, Pierce; NE: Hall, Webster
Amsinckia lycopsioides Lehm. NE: Hall; OK: Oklahoma
Asperugo procumbens L. KS: Douglas; ND: Dunn
Cryptantha macounii (Eastw.) Pays. ND: Burke, Divide, Mountrail, Williams
Cryptantha torreyana (Gray) Greene, ND: Bowman
Cynoglossum boreale Fern. (Northern Wild Comfrey) ND: Bottineau, Dunn, Rolette
Cynoglossum virginianum L. (Wild Comfrey) OK: Creek
Heliotropium indicum L. (Turnsole) KS: Cherokee, Crawford; MO: Barton, Bates, Jackson, Vernon; OK: Tulsa
Lycopsis arvensis L. (Small Bugloss) ND: Stark, Stutsman
Mertensia ciliata (James) D. Don, CO: Weld; MT: Wibaux; SD: Stanley; WY: Laramie
Myosotis scorpioides L. (True Forget-me-not) MO: Newton; SD: Lawrence
Myosotis sylvatica Hoffm. (Garden Forget-me-not) NE: Nemaha; SD: Custer, Lawrence, Pennington

97. VERBENACEAE (Vervain Family)
Phyla nodiflora (L.) Greene, OK: Alfalfa, Blaine, Cleveland, Major
Verbena ciliata Benth. CO: Baca; NM: Curry, Harding, Quay, Union; TX: Bailey, Dallam
Verbena halei Small, OK: Cotton

Verbena plicata Greene, OK: Cimarron, Harmon, Jackson; TX: Bailey, Childress, Cottle, Hall, Motley
Verbena wrightii Gray, NM: Harding, Quay, Union; OK: Cimarron, Harper
Vitex agnus-castus L. OK: Cleveland, Custer, Grady, Payne

99. LAMIACEAE (Mint Family)
Ajuga reptans L. KS: Cherokee, Crawford
Blephilia ciliata (L.) Benth. KS: Cherokee, Wilson, Wyandotte; MO: Jasper
Cunila origanoides (L.) Britt. (Common Dittany) KS: Cherokee; MO: Jasper; OK: Ottawa
Dracocephalum thymiflorum L. ND: Hettinger, Stark, Walsh
Leonurus marrubiastrum L. IA: Harrison; KS: Douglas; NE: Douglas, Nemaha, Richardson, Washington; SD: Hutchinson
Lycopus rubellus Moench, KS: Cherokee; MO: Jasper
Melissa officinalis L. (Common Balm) KS: Wyandotte; MO: Jackson, Jasper
Mentha cardiaca Gerarde, KS: Harper, Jackson, Jefferson; MO: Jasper
Mentha gentilis L. IA: Sioux; NE: Dawes, Dodge
Mentha longifolia L. (Horse Mint) KS: Sedgwick; MO: Clay, Jackson, Jasper
Mentha piperita L. (Peppermint) MO: Clay, Jackson, Jasper; NE: Brown, Sarpy; OK: Cleveland; TX: Wheeler
Mentha spicata L. (Spearmint) IA: Taylor; KS: Cloud, Kingman; MO: Clay, Jackson, Jasper; OK: Cleveland, Oklahoma
Monarda bradburiana Beck, KS: Cherokee, Greenwood; MO: Jasper; OK: Ottawa, Tulsa
Phlomis tuberosa L. ND: Foster
Physostegia angustifolia Fern. KS: Woodson; MO: Barton, Bates, Vernon; OK: Osage, Ottawa, Washington
Pycnanthemum albescens T. & G. KS: Cherokee; OK: Ottawa
Salvia lyrata L. (Cancer-weed) KS: Wilson
Salvia nemorosa L. SD: Clay, Gregory, Lincoln, Turner
Salvia sylvestris L. IA: Pottawattamie, Plymouth; KS: Sumner; ND: Cass, Logan, Slope; NE: Knox, Lancaster, Seward
Scutellaria incana Biehler, (Hoary Skullcap) KS: Cherokee; MO: Jasper

101. OLEACEAE (Olive Family)
Forestiera acuminata (Michx.) Poir. (Swamp Privet) KS: Cherokee, Labette; MO: Jasper
Forestiera pubescens Nutt. OK: Caddo; TX: Armstrong, Bailey, Briscoe, Motley, Randall, Roberts
Fraxinus nigra Marsh. (Black Ash) MN: Clay, Otter Tail; ND: Cass, Pembina

102. SCROPHULARIACEAE (Figwort Family)
Bacopa acuminata (Walt.) Robins. KS: Cherokee; MO: Jasper
Castilleja integra Gray, NM: Union; TX: Deaf Smith
Castilleja purpurea (Nutt.) G. Don var. *purpurea,* OK: Harper, Tulsa
Castilleja sulphurea Rydb. MT: Powder River; SD: Custer, Lawrence, Meade, Pennington
Collinsia verna Nutt. (Blue-eyed Mary) KS: Miami, Montgomery; MO: Cass, Clay, Jackson
Kickxia elatine (L.) Dum. (Canker-root) KS: Cherokee; MO: Clay, Jackson
Linaria genistifolia (L.) Mill. MN: Pope
Mimulus floribundus Dougl. SD: Custer, Pennington
Pedicularis grayi A. Nels. SD: Custer, Pennington
Penstemon haydeni Wats. NE: Cherry, Hooker, Thomas
Penstemon jamesii A. Nels. KS: Morton
Penstemon oklahomensis Penn. OK: Cleveland, Comanche, Lincoln, Oklahoma, Payne
Penstemon pallidus Small, (Pale Penstemon) KS: Douglas, Jefferson, Wilson; MO: Barton, Cass, Vernon
Penstemon procereus Dougl. MT: Daniels; ND: Burke
Veronica hederaefolia L. (Ivy-leaved Speedwell) NE: Douglas
Veronica longifolia L. MN: Otter Tail; ND: McHenry
Veronica officinalis L. (Common Speedwell) SD: Brookings
Veronica persica Poir. (Bird's-eye Speedwell) IA: Lyon, Sioux; SD: Lawrence
Veronica serpyllifolia L. (Thyme-leaved Speedwell) KS: Saline; NE: Lincoln
Veronica triphyllos L. KS: Cloud, Geary, Kingman, Saline; OK: Oklahoma, Woods

103. OROBANCHACEAE (Broomrape Family)
Orobanche riparia Collins, OK: Cimarron
104. BIGNONIACEAE (Bignonia Family)
Chilopsis linearis (Cav.) Sweet, TX: Bailey, Briscoe
105. ACANTHACEAE (Acanthus Family)
Dyschoriste linearis (T. & G.) O. Ktze. OK: Cotton, Jackson, Tillman; TX: Floyd
107. LENTIBULARIACEAE (Bladderwort Family)
Utricularia biflora Lam. OK: Cleveland, Comanche, Ottawa
108. CAMPANULACEAE (Bellflower Family)
Triodanis lamprosperma McVaugh, KS: Cherokee, Montgomery; MO: Jasper
109. RUBIACEAE (Madder Family)
Galium labradoricum Wieg. MN: Otter Tail, Pope; ND: Ransom
Galium texense Gray, OK: Comanche, Greer, Jackson, Kiowa
Hedyotis humifusa Gray, OK: Beckham, Cotton, Ellis, Greer, Jackson
Sherardia arvensis L. OK: Cleveland, Mayes
Sherardia orientalis Boiss. & Hohenack. ND: Cass
110. CAPRIFOLIACEAE (Honeysuckle Family)
Lonicera sempervirens L. (Trumpet Honeysuckle) IA: Adair; KS: Cherokee, Neosho; MO: Barton, Jasper; OK: Payne
Symphoricarpos palmeri G. N. Jones, OK: Cimarron
Triosteum angustifolium L. (Horse Gentian) KS: Cherokee
Viburnum edule (Michx.) Raf. (Squashberry) SD: Lawrence
111. ADOXACEAE (Adoxa Family)
Adoxa moschatellina L. (Moschatel) SD: Custer, Lawrence, Pennington
112. VALERIANACEAE (Valerian Family)
Valeriana acutiloba Rydb. SD: Custer, Lawrence, Pennington
Valeriana dioica L. SD: Custer, Lawrence, Pennington
Valeriana edulis Nutt. SD: Custer, Lawrence, Pennington
Valerianella amarella Krok, OK: Caddo, Cleveland, Grady, Lincoln, Oklahoma
113. DIPSACACEAE (Teasel Family)
Dipsacus laciniatus L. KS: Crawford
Knautia arvensis (L.) Duby, (Blue Buttons) MN: Clay; ND: Grand Forks, Pierce, Ward; SD: Day
114. ASTERACEAE (Sunflower Family)
Achillea sibirica Ledeb. MN: Roseau; ND: Bottineau, Cavalier, Pembina, Rolette
Agoseris aurantiaca (Hook.) Greene var. *purpurea* (Gray) Cronq. WY: Crook
Ambrosia linearis (Rydb.) Payne, CO: El Paso, Kiowa, Lincoln
Antennaria anaphaloides Rydb. WY: Albany, Niobrara
Antennaria dimorpha (Nutt.) T. & G. NE: Dawes, Sioux
Anthemis arvensis L. (Corn Chamomile) MO: Jackson, Jasper; ND: Grand Forks; NE: Douglas, Hamilton, Richardson
Anthemis tinctoria L. (Yellow Chamomile) ND: Bottineau, Pembina, Rolette
Aphanostephus riddellii T. & G. TX: Cottle, Crosby, Dickens, Donley, Lubbock, Motley
Arctium lappa L. (Great Burdock) NE: Custer, Sarpy; SD: Butte, Lawrence, Pennington, Union
Arctium tomentosum Mill. MN: Kittson; MO: Jackson; ND: Cass, Grand Forks, Pembina, Ransom; NE: Sherman; SD: Lawrence
Arnica cordifolia Hook. CO: Boulder, Larimer, Las Animas; MT: Rosebud; ND: Logan; SD: Custer, Lawrence; WY: Albany, Crook
Arnica lonchophylla Greene ssp. *arnoglossa* (Greene) Maguire, SD: Custer, Lawrence, Meade, Pennington
Arnica rydbergii Greene, SD: Custer, Pennington
Artemisia abrotanum L. (Southernwood) KS: Gove, Scott, Sheridan; MN: Kittson
Artemisia annua L. (Sweet Sagewort) KS: Anderson, Cowley, Doniphan, Johnson; MO: Jackson, Jasper
Artemisia bigelovi Gray, TX: Deaf Smith
Aster X *amethystinus* Nutt. ND: Logan; NE: Brown, Buffalo, Howard, Webster
Aster anomalus Engelm. (Manyray Aster) KS: Cherokee, Labette

Aster conspicuous Lindl. (Rough Aster) SD: Custer, Lawrence, Pennington; WY: Crook
Aster linariifolius L. KS: Cherokee
Aster lucidulus (Gray) Wieg. IA: Harrison; MN: Lincoln, Otter Tail, Pope, Red Lake; ND: Bottineau, Richland, Rolette
Aster paludosus Ait. ssp. *hemisphericus* (Alex.) Cronq. (Single-stemmed Bog Aster) KS: Cherokee, Crawford, Labette; MO: Barton, Jasper, Newton; OK: Delaware, Mayes, Ottawa
Aster sibiricus L. var. *meritus* (A. Nels.) Raup, SD: Custer, Lawrence, Pennington
Aster spinosus Benth. OK: Cleveland, Jackson, Payne; TX: Carson, Hemphill, Oldham
Aster turbinellus Lindl. KS: Cherokee; MO: Barton, Bates, Cass, Jasper; OK: Delaware, Mayes, Ottawa
Baccharis glutinosa (R. & P.) Pers. TX: Bailey
Baccharis texana (T. & G.) Gray, OK: Custer, Jackson
Balsamorhiza sagittata (Pursh) Nutt. (Balsam-root) MT: Rosebud; SD: Lawrence, Pennington; WY: Crook, Weston
Bellis perennis L. ND: Cass
Bidens bigelovii Gray, OK: Cimarron; TX: Carson
Brickellia brachyphylla (Gray) Gray, OK: Cimarron; TX: Deaf Smith
Brickellia californica (T. & G.) Gray, NM: Union; OK: Cimarron; TX: Deaf Smith
Carduus crispus L. (Welted Thistle) ND: Cass
Carthamus tinctorius L. (Safflower) KS: Clay, Sheridan, Stanton; ND: Slope, Stark
Centaurea scabiosa L. ND: Barnes, Golden Valley, Ransom
Chaenactis douglasii (Hook.) H. & A. MT: McCone; ND: Billings, Golden Valley; WY: Campbell, Crook, Niobrara, Weston
Chrysanthemum balsamita L. (Mint Geranium) KS: Ellsworth, Riley, Rooks; SD: Lawrence
Chrysothamnus parryi (Gray) Greene ssp. *howardii* (Parry) H. & C. NE: Kimball, Scotts Bluff; SD: Custer, Fall River
Chrysothamnus viscidiflorus (Hook.) Nutt. MT: Custer; NE: Sioux
Cirsium drummondii T. & G. SD: Custer, Lawrence, Pennington; WY: Weston
Cirsium pulcherrimum (Rydb.) K. Schum. NE: Scotts Bluff
Cirsium texanum Buckl. OK: Harmon, Jackson
Coreopsis pubescens Ell. KS: Cherokee; MO: Barton, Jasper
Crepis capillaris (L.) Wallr. ND: Cass, Mountrail
Dyssodia acerosa DC. TX: Deaf Smith
Dyssodia aurea (Gray) A. Nels. CO: Cheyenne, Crowley, Pueblo; KS: Hamilton
Dyssodia tagetoides T. & G. OK: Comanche, Custer, Kiowa, Tillman
Echinacea purpurea (L.) Moench, KS: Cherokee; MO: Clay, Jackson
Erigeron acris L. var. *asterioides* (Andrz.) DC. SD: Custer, Fall River, Pennington
Erigeron formosissimus Greene, SD: Custer, Lawrence, Pennington; WY: Weston
Erigeron ochroleucus Nutt. NE: Dawes; SD: Pennington; WY: Converse, Weston
Erigeron pulchellus Michx. (Robin's Plantain) KS: Cherokee; MN: Mahnomen; MO: Barton, Jasper, Vernon
Erigeron speciosus (Lindl.) DC. var. *macranthus* (Nutt.) Cronq. SD: Custer, Lawrence, Pennington
Erigeron vetensis Rydb. NE: Dawes; WY: Laramie
Evax verna Raf. OK: Greer, Harmon, Lincoln, Roger Mills; TX: Bailey, Hemphill
Galinsoga parviflora Cav. KS: Neosho; MO: Jackson, Jasper; ND: Cass, Foster
Gnaphalium chilense Spreng. OK: Comanche, Kiowa
Gnaphalium uliginosum L. (Low Cudweed) MN: Kittson; SD: Custer, Lawrence
Gnaphalium viscosum H.B.K. SD: Lawrence, Pennington
Gnaphalium wrightii Gray, CO: Baca; OK: Cimarron, Comanche
Gutierrezia texana (DC.) T. & G. OK: Comanche; TX: Hardeman
Haplopappus engelmannii (Gray) Hall, (Engelmann's Goldenweed) CO: Lincoln; KS: Hamilton
Haplopappus multicaulis (Nutt.) Gray, MT: Carter; SD: Fall River; WY: Converse, Niobrara, Weston
Helenium flexuosum Raf. KS: Cherokee, Crawford; MO: Barton, Jackson, Jasper, Vernon; OK: Ottawa

Helianthella quinquenervis (Hook.) Gray, SD: Custer, Lawrence, Pennington
Helianthus pumilus Nutt. CO: El Paso; MT: Valley; WY: Laramie
Hieracium albiflorum Hook. SD: Lawrence
Iva angustifolia Nutt. KS: Cherokee, Montgomery; OK: Ottawa
Krigia biflora (Walt.) Blake, KS: Cherokee; MN: Kittson, Mahnomen, Polk; MO: Jasper; OK: Ottawa
Kuhnia chlorolepis Woot. & Standl. OK: Cimarron, Texas; TX: Briscoe
Kuhnia eupatorioides L. var. *texana* Shinners, (False Boneset) OK: Cleveland, Oklahoma, Osage, Payne
Lapsana communis L. ND: Cass
Leontodon hastilis L. var. *vulgaris* Koch, KS: Douglas
Lindheimera texana Gray & Engelm. OK: Caddo, Comanche, Greer, Harmon, Jackson, Tillman
Machaeranthera glabriuscula (Nutt.) Cronq. & Keck, SD: Harding, Fall River; WY: Niobrara, Weston
Madia glomerata Hook. ND: Williams; SD: Lawrence, Pennington; WY: Crook
Marshallia caespitosa Nutt. KS: Montgomery; MO: Barton, Jasper
Matricaria chamomilla L. KS: Riley; ND: Burke, Emmons, Foster, Morton, Pembina, Ramsey
Onopordum acanthium L. (Scotch Thistle) KS: McPherson, Smith; MN: Rock: NE: Webster; WY: Converse
Parthenium incanum H.B.K. TX: Deaf Smith
Parthenium integrifolium L. (Wild Quinine) KS: Cherokee; MO: Barton, Cass, Jackson, Jasper, Vernon
Pericome caudata Gray, NM: Union; OK: Cimarron
Petasites frigidus (L.) Fries var. *palmatus* (Ait.) Cronq. ND: Bottineau
Picris echioides L. (Ox-tongue) ND: Cass, McLean
Pinaropappus roseus (Less.) Less. TX: Hemphill
Polymnia canadensis L. KS: Anderson, Cherokee, Miami; MO: Jackson
Polymnia uvedalia L. (Bearsfoot) KS: Cherokee; MO: Jasper; OK: Canadian, Cleveland, Oklahoma, Tulsa
Psilostrophe tagetina (Nutt.) Greene, NM: Curry; TX: Childress
Pyrrhopappus multicaulis DC. var. *geiseri* (Shinners) Northington, OK: Canadian, Cleveland, Comanche, Logan, McLain, Payne
Pyrrhopappus rothrockii Gray, TX: Hemphill, Moore, Potter
Rudbeckia grandiflora (D. Don) DC. (Rough Coneflower) KS: Cherokee, Crawford, Labette; OK: Craig
Senecio crassulus Gray, SD: Custer, Pennington
Senecio eremophilus Richards. ND: Bottineau, Rolette; SD: Custer, Lawrence, Pennington
Senecio hydrophilus Nutt. SD: Pennington
Senecio rapifolius Nutt. SD: Custer, Lawrence, Pennington
Senecio vulgaris L. (Common Groundsel) MN: Chippewa, Clay; ND: Cass, Stark; NE: Lancaster
Silphium asperrimum Hook. OK: Canadian, Comanche, Custer, Grady, Kiowa, Logan, Oklahoma, Ottawa, Payne
Solidago canadensis L. var. *canadensis,* (Canada Goldenrod) ND: Benson, Logan; SD: Custer, Grant, Lawrence
Solidago canadensis L. var. *salebrosa* (Piper) M. E. Jones, (Canada Goldenrod) ND: Bottineau, McKenzie, Rolette; SD: Custer, Fall River, Lawrence, Pennington, Todd
Solidago occidentalis (Nutt.) T. & G. NE: Hall, Kearney, Scotts Bluff
Solidago radula Nutt. KS: Cherokee; MO: Bates, Jasper, Vernon
Solidago riddellii Frank, MN: Chippewa, Clay, Norman; ND: Richland
Solidago sparciflora Gray, (Threenerve Goldenrod) OK: Cimarron; SD: Lawrence, Pennington; WY: Albany, Laramie
Stephanomeria runcinata Nutt. MT: McCone; NE: Scotts Bluff, Sioux; WY: Platte
Stephanomeria tenuifolia (Torr.) Hall, ND: Billings, Slope, Stark
Thelesperma ambiguum Gray, TX: Deaf Smith
Thelesperma marginatum Rydb. MT: Rosebud; ND: Divide, Williams

Viguiera stenoloba Blake, NM: Union
Xanthium spinosum L. (Spiny Cocklebur) KS: Cloud, Finney, Harvey, Kearny, Marion, Reno, Republic; NE: Douglas; TX: Deaf Smith

115. BUTOMACEAE (Butomus Family)
Butomus umbellatus L. (Flowering Rush) ND: Barnes, Bottineau; SD: Faulk

116. ALISMATACEAE (Water Plantain Family)
Echinodorus parvulus Engelm. KS: Harvey

117. HYDROCHARITACEAE (Frog's-bit Family)
Elodea densa (Planch.) Casp. KS: Douglas; NE: Hall

118. SCHEUCHZERIACEAE
Scheuchzeria palustris L. MN: Otter Tail, Pope

120. NAJADACEAE (Naiad Family)
Najas marina L. MN: Big Stone, Polk, Pope; ND: Richland

121. POTAMOGETONACEAE (Pondweed Family)
Potamogeton alpinus Balbis, NE: Cherry; SD: Lawrence, Roberts, Tripp
Potamogeton obtusifolius Mert. & Koch, (Bluntleaf Pondweed) KS: Harvey, Riley, Sheridan; MN: Kittson; NE: Cherry
Potamogeton pusillus L. var. *tenuissimus* Mert. & Koch, (Baby Pondweed) IA: Dickinson; MN: Douglas, Pipestone; ND: Stutsman; OK: Comanche
Potamogeton spirillus Tuckerm. MN: Lac Qui Parle; ND: Emmons; NE: Gage, Lancaster
Potamogeton vaginatus Turcz. MN: Otter Tail; ND: Bottineau, Rolette; SD: Pennington

125. JUNCACEAE (Rush Family)
Juncus articulatus L. SD: Roberts
Juncus brachycarpus Engelm. KS: Bourbon, Cherokee, Montgomery, Wilson; MO: Barton, Jasper, Vernon
Juncus brachycephalus (Engelm.) Buch. MN: Marshall
Juncus canadensis J. Gay, NE: Blaine, Brown, Cherry
Juncus ensifolius Wikst. var. *montanus* (Engelm.) C. L. Hitchc. MT: Custer; SD: Custer, Lawrence, Pennington; WY: Crook
Juncus gerardi Lois. (Blackgrass) MN: Clay, Kittson; ND: Cass, Richland
Juncus vaseyi Engelm. MN: Polk
Luzula campestris (L.) DC. var. *bulbosa* Wood, (Wood Rush) KS: Chautauqua, Cherokee, Neosho, Woodson; MO: Barton, Jasper, Vernon
Luzula campestris (L.) DC. var. *multiflora* (Ehrh.) Celak. (Wood Rush) MN: Kittson, Otter Tail; SD: Custer, Lawrence, Pennington; WY: Johnson

126. CYPERACEAE (Sedge Family)
Carex alopecoidea Tuckerm. MN: Kittson, Lac Qui Parle, Polk; ND: Barnes, Pembina, Richland
Carex arkansana Bailey, KS: Allen, Anderson, Cherokee, Labette, Neosho, Woodson; MO: Bates, Jasper
Carex athrostachya Olney, ND: Benson, Divide, Mountrail, Williams
Carex backii Boott, ND: Bottineau, Cavalier
Carex bella Bailey, SD: Custer
Carex brittoniana Bailey, OK: Cotton, Noble, Payne
Carex brunnescens (Pers.) Poir. MN: Kittson; SD: Custer
Carex capillaris L. var. *elongata* Olney, MN: Marshall; ND: Bottineau; SD: Lawrence
Carex caroliniana Schwein. KS: Cherokee, Linn, Neosho; MO: Barton, Jackson, Jasper
Carex concinna R. Br. SD: Lawrence, Meade
Carex crinita Lam. var. *brevicrinis* Fern. MO: Jackson, Jasper, Vernon; NE: Cass
Carex diandra Schrank, ND: Burke, Rolette; NE: Banner, Brown, Cherry, Rock, Thomas
Carex festucacea Schkuhr, IA: Lyon; MO: Barton, Holt, Jackson, Jasper; ND: Cass; NE: Richardson
Carex fissa Mack. KS: Cherokee, Neosho
Carex foenea Willd. MN: Clay, Polk; ND: Bottineau; NE: Loup; SD: Custer, Pennington
Carex garberi Fern. ND: Burke
Carex haydenii Dew. IA: Guthrie, Taylor; MO: Gentry, Jackson, Nodoway, Worth; ND: Dunn; NE: Lincoln; SD: Codington, Custer, Pennington

Carex hirsutella Mack. KS: Cherokee, Wilson, Woodson; MO: Barton, Clinton, Jackson
Carex hirtiflora Mack. KS: Wyandotte; MO: Clay, Jackson, Worth
Carex hoodii Boott, MT: Rosebud; SD: Lawrence; WY: Crook
Carex hookerana Dew. ND: Benson, Burke, Cavalier, Mountrail, Steele
Carex intumescens Rudge, MN: Becker, Pope; SD: Custer
Carex leptalea Wahl. ND: Pembina; SD: Custer
Carex limosa L. NE: Cherry
Carex melanostachya M. B. ex Willd. KS: Douglas
Carex microptera Mack. CO: Larimer; SD: Custer, Lawrence, Pennington
Carex microrhyncha Mack. OK: Caddo, Oklahoma
Carex muskingamensis Schwein. KS: Miami, Wyandotte; MO: Bates, Cass, Jackson, Vernon
Carex oklahomensis Mack. KS: Cherokee, Crawford, Douglas; MO: Henry, Jasper, Newton
Carex parryana Dew. ND: Burke, Ransom, Richland, Stutsman; WY: Crook
Carex pedunculata Muhl. MN: Becker, Clay, Otter Tail; SD: Roberts
Carex praticola Rydb. SD: Custer, Lawrence, Pennington
Carex pseudo-cyperus L. MN: Kittson, Marshall, Otter Tail, Polk; ND: Benson, Pembina
Carex retroflexa Muhl. KS: Cherokee, Crawford, Douglas, Miami; MO: Barton, Clay, Jasper; OK: Oklahoma
Carex richardsonii R. Br. ND: Cass, McHenry; SD: Custer, Pennington
Carex rossii Boott, ND: Slope; NE: Dawes, Sioux; SD: Pennington
Carex scirpiformis Mack. MN: Clay, Wilkin; ND: Dunn, McHenry
Carex squarrosa L. KS: Cherokee, Crawford; MO: Barton, Bates, Jasper; MN: Mahnomen; NE: Lancaster
Carex subimpressa Clokey, KS: Douglas
Carex texensis (Torr.) Bailey, KS: Atchison, Douglas, Wilson, Wyandotte; NE: Richardson
Carex triangularis Boeckl. KS: Allen, Cherokee, Neosho; TX: Deaf Smith, Hemphill, Wheeler
Carex vallicola Dew. SD: Custer, Harding, Lawrence
Carex vesicaria L. var. *monile* (Tuckerm.) Fern. IA: Fremont, Page; MN: Clay; SD: Marshall, Roberts
Cyperus flavescens L. var. *poaeformis* (Pursh) Fern. KS: Johnson, Wyandotte
Cyperus fuscus L. NE: Lincoln; SD: Tripp
Cyperus rotundus L. (Nutgrass) OK: Payne
Cyperus tenuifolius (Steud.) Dandy, KS: Cherokee, Douglas, Wyandotte; MO: Barton, Clay, Jackson, Jasper
Cyperus uniflorus T. & H. OK: Alfalfa, Blaine, Canadian, Custer; TX: Bailey, Cottle, Hockley
Dichromena nivea (Boeckl.) Britt. (White-top Sedge) OK: Grady, McLain
Eleocharis atropurpurea (Retz.) Kunth, KS: Harvey; NE: Franklin, Perkins, Polk; TX: Gray, Lubbock
Eleocharis caribaea (Rottb.) Blake, KS: Barber; OK: Cleveland, Oklahoma, Woodward
Eleocharis intermedia (Muhl.) Schult. MN: Norman
Eleocharis lanceolata Fern. KS: Cherokee, Morris; MO: Barton, Jasper
Eleocharis pauciflora (Lightf.) Link, MN: Norman; ND: McHenry; NE: Cherry
Eleocharis quadrangulata (Michx.) R. & S. var. *crassior* Fern. KS: Bourbon, Coffey, Neosho, Wilson; OK: Cleveland, Comanche, Woodward
Eriophorum chamissonis C. A. Mey. ND: Barnes, La Moure, Rolette
Eriophorum gracile Koch, CO: El Paso; NE: Cherry, Thomas
Fimbristylis annua (All.) R. & S. KS: Chautauqua, Crawford, Neosho, Wilson; MO: Barton, Jackson, Jasper, Vernon; OK: Ottawa
Rhynchospora capillacea Torr. (Beakrush) SD: Bennett
Rhynchospora globularis (Chapm.) Small var. *recognita* Gale, KS: Cherokee, Harvey; MO: Barton, Jasper, Vernon; OK: Creek, Oklahoma
Rhynchospora harveyi Boott, KS: Cherokee, Coffey, Harvey, Montgomery, Wilson, Woodson; OK: Harvey
Rhynchospora macrostachya Torr. (Horned Rush) KS: Allen, Anderson, Coffey, Woodson; MO: Barton; OK: Creek

Scirpus atrocinctus Fern. MN: Becker, Otter Tail
Scirpus koilolepis (Steud.) Gl. KS: Cherokee, Crawford, Labette, Montgomery, Neosho, Wilson; MO: Barton, Jasper, Newton; OK: Ottawa
Scirpus olneyi Gray, KS: Cherokee, Meade; OK: Beaver
Scirpus smithii Gray, NE: Sheridan
Scirpus torreyi Olney, NE: Dodge

127. POACEAE (Grass Family)

Agropyron X *pseudorepens* Scribn. & Sm. NE: Grant, Kearney
Agrostis semiverticillata (Forsk.) C. Chr. OK: Cimarron; TX: Armstrong, Hartley, Hutchinson
Alopecurus arundinaceus Poir. (Creeping Foxtail) ND: Burleigh, Cass, Emmons, Logan, McLean
Alopecurus myosuroides Huds. (Slender Foxtail) KS: Riley
Alopecurus pratensis L. (Meadow Foxtail) MN: Douglas; ND: Bottineau, McKenzie; NE: Cuming, Douglas, Gage, Lancaster, Sarpy
Andropogon glomeratus (Walt.) B.S.P. (Bushy Bluestem) OK: Caddo, Cleveland, McLain, Oklahoma
Aristida arizonica Vasey (Arizona Threeawn) OK: Cimarron, Roger Mills
Aristida barabata Fourn. CO: Baca; NM: Union; OK: Cimarron; TX: Dallam, Deaf Smith, Hemphill
Aristida roemeriana Scheele, TX: Hutchinson
Arrhenatherum elatius (L.) Presl. (Tall Oatgrass) CO: El Paso; IA: Dickinson, Page; MO: Jasper; NE: Box Butte, Douglas, Lancaster
Bouteloua barbata Lag. (Sixweeks Grama) KS: Harvey, Reno; OK: Harmon; TX: Deaf Smith
Bouteloua simplex Lag. NM: Harding; TX: Deaf Smith; WY: Laramie
Brachiaria ciliatissima (Buckl.) Chase, TX: Cottle
Bromus arvensis L. (Field Brome) ND: Pembina; OK: Comanche
Bromus brizaeformis Fisch. & Mey. (Rattlesnake Chess) ND: Kidder; NE: Scotts Bluff, Thomas; SD: Lawrence
Bromus carinatus H. & A. (California Brome) ND: Slope; NE: Dawes, Sheridan, Sioux; SD: Lawrence, Pennington
Bromus inermis Leyss. ssp. *pumpellianus* (Scribn.) Wagnon, (Smooth Brome) NE: Cherry, Sioux; SD: Custer, Day, Lawrence, Pennington; WY: Crook
Bromus lanatipes (Shear) Rydb. OK: Cimarron; TX: Deaf Smith
Bromus mollis L. (Soft Chess) KS: Kiowa, Riley; MO: Jackson; ND: Cass, Pierce; SD: Fall River
Calamagrostis purpurascens R. Br. (Purple Reedgrass) SD: Lawrence, Pennington
Danthonia intermedia Vasey, (Timber Oatgrass) SD: Lawrence, Pennington
Danthonia unispicata (Thurb.) Munro, WY: Crook, Weston
Echinochloa colonum (L.) Link, KS: Lyon; OK: Cleveland, Osage; TX: Briscoe, Deaf Smith, Gray
Elymus cinereus Scribn. & Merr. (Giant Wild Rye) MT: Rosebud; NE: Antelope, Dawes, Sioux; SD: Pennington, Walworth; WY: Albany, Converse, Johnson, Laramie, Sheridan, Weston
Elymus innovatus Beal, SD: Custer, Lawrence, Pennington, Shannon
Elymus junceus Fisch. (Russian Wild Rye) MT: McCone; ND: Sargent; NE: Thurston; SD: Shannon; WY: Goshen
Enneapogon desvauxii Beaux. OK: Cimarron; TX: Deaf Smith
Eragrostis hirsuta (Michx.) Nees, MO: Barton, Jasper; OK: Cimarron, Cleveland, Creek, Kay
Eragrostis neomexicana Vasey, TX: Briscoe, Gray, Roberts
Eragrostis pilosa (L.) Beauv. var. *perplexa* (Harvey) S. D. Koch, (India Lovegrass) CO: El Paso; KS: Edwards, Finney, Russell, Stanton, Wallace; NE: Hooker; SD: Mellette; TX: Dallam
Eragrostis pilosa (L.) Beauv. var. *pilosa,* (India Lovegrass) CO: Adams; MN: Lac Qui Parle; MO: Jackson; NE: Douglas; OK: Payne; SD: Clay; TX: Armstrong

Eragrostis poaeoides Beauv. (Little Lovegrass) KS: Douglas, Riley, Saline; MN: Lac Qui Parle; MO: Clay, Jackson; OK: Alfalfa, Cleveland

Eragrostis tephrosanthes Schult. CO: El Paso; TX: Carson, Castro, Deaf Smith, Gray, Hartley, Oldham, Potter

Eriochloa gracile (Fourn.) Hitchc. OK: Tillman; TX: Hale

Eriochloa sericea (Scheele) Munro, KS: Montgomery; OK: Canadian, Cleveland, Greer, Tillman

Festuca arundinacea Schreb. (Tall Fescue) IA: Plymouth; KS: Atchison, Bourbon, Lyon, Riley, Sumner; MN: Douglas; NE: Douglas, Lancaster, Sarpy, Washington; NM: Curry; TX: Swisher

Festuca idahoensis Elmer, (Blue Bunchgrass) CO: El Paso; SD: Pennington

Festuca myuros L. (Rattail Fescue) KS: Douglas; OK: Cleveland; TX: Carson

Festuca rubra L. (Red Fescue) NE: Douglas

Festuca sciurea Nutt. MO: Jasper; OK: Blaine, Kingfisher, Payne

Festuca subulata Trin. SD: Lawrence

Glyceria fluitans (L.) R. Br. (Mannagrass) SD: Custer

Hesperochloa kingii (Wats.) Rydb. KS: Marshall; NE: Dawes, Sioux; WY: Niobrara, Weston

Hilaria mutica (Buckl.) Benth. (Tobosa) OK: Harmon, Jackson; TX: Armstrong

Holcus lanatus L. (Velvetgrass) KS: Bourbon, Cherokee, Crawford, Douglas, Labette, Neosho; MO: Jackson, Jasper, Vernon

Hordeum brachyantherum Nevski, (Northern Meadow Barley) MT: Carter, Daniels, McCone, Powder River, Prairie, Rosebud

Hordeum X *montanense* Scribn. ND: Ward

Leersia lenticularis Michx. (Catchfly-grass) KS: Linn, Miami; MO: Bates, Clinton, Jasper, Vernon

Leptochloa dubia (H.B.K.) Nees, (Green Sprangletop) TX: Armstrong, Briscoe, Randall

Limnodea arkansana (Nutt.) L. H. Dewey, OK: Comanche, Greer, Jackson

Lolium temulentum L. var. *leptochaeton* A. Br. (Poison Darnel) KS: Cherokee, Riley, Wilson; MO: Jackson, Platte; ND: Richland, Stark, Walsh; OK: Payne

Melica mutica Walt. (Two-flowered Melic) OK: Canadian, Comanche, Kiowa

Melica porteri Scribn. (Porter's Melic) KS: Leavenworth, Montgomery

Melica smithii (Porter) Vasey, (Melicgrass) SD: Pennington

Miscanthus saccharifolius (Maxim.) Hack. IA: Ida, Montgomery, Shelby; NE: Douglas, Lancaster

Muhlenbergia filiformis (Thurb.) Rydb. (Pullup Muhly) KS: Finney; NE: Cherry, Thomas; SD: Pennington

Muhlenbergia minutissima (Steud.) Swall. NE: Scotts Bluff; SD: Pennington; WY: Weston

Muhlenbergia porteri Scribn. (Porter's Muhly) OK: Cimarron; TX: Deaf Smith

Muhlenbergia tenuiflora (Willd.) B.S.P. MO: Jackson, Jasper; NE: Douglas, Sarpy; OK: Comanche

Oryzopsis exigua Thurb. MT: Custer

Oryzopsis pungens (Torr.) Hitchc. (Mountain Ricegrass) SD: Lawrence

Panicum boscii Poir. KS: Montgomery; MO: Jackson, Jasper; OK: Ottawa

Panicum depauperatum Muhl. (Stunted Panicum) IA: Dickinson; KS: McPherson; MN: Norman, Polk; MO: Barton, Jasper

Panicum fasciculatum Sw. OK: Comanche, Grady; TX: Deaf Smith, Hardeman

Panicum hirticaule J. Presl. TX: Motley, Oldham

Panicum reverchonii Vasey, OK: Harmon, Jackson

Panicum scoparium Lam. (Velvet Panicum) KS: Cherokee, Douglas, Montgomery; MO: Barton, Jasper, Vernon; OK: Cleveland, Oklahoma, Osage

Panicum xanthophysum Gray, SD: Custer, Pennington

Paspalum dilatatum Poir. (Dallisgrass) OK: Cleveland, Custer, McLain

Paspalum laeve Michx. var. *pilosum* Scribn. (Field Paspalum) KS: Bourbon, Cherokee; MO: Barton, Jasper, Vernon

Phleum alpinum L. (Mountain Timothy) CO: Arapahoe; SD: Pennington; WY: Laramie

Poa bulbosa L. (Bulbous Bluegrass) KS: McPherson, Riley; ND: Stark; NE: Dawes, Sioux; OK: Cleveland; SD: Lawrence; WY: Crook
Poa trivialis L. (Rough-stalked Meadowgrass) KS: Shawnee; SD: Lawrence
Poa wolfii Scribn. NE: Brown
Sclerochloa dura (L.) Beauv. KS: Harvey, Mitchell; MO: Buchanan; OK: Cleveland, Payne
Scleropogon brevifolius Phil. (Burrograss) OK: Cimarron; TX: Bailey
Sitanion hystrix (Nutt.) J. G. Sm. var. *hystrix,* (Squirreltail) MT: Garfield, Prairie; ND: Sioux; SD: Pennington, Washabough
Sporobolus contractus Hitchc. OK: Cleveland; TX: Deaf Smith
Sporobolus flexuosus (Thurb.) Rydb. (Mesa Dropseed) KS: Seward; TX: Hall, Randall
Stipa neomexicana (Thurb.) Scribn. (New Mexico Feathergrass) CO: Otero; OK: Cimarron; TX: Armstrong, Bailey, Deaf Smith
Stipa occidentalis Thurb. ex Wats. var. *minor* (Vasey) C. L. Hitchc. (Columbia Needlegrass) SD: Custer, Lawrence, Pennington
Stipa richardsonii Link, (Richardson's Needlegrass) SD: Lawrence, Pennington
Stipa scribneri Vasey, OK: Cimarron
Trichachne californica (Benth.) Chase, (Arizona Cottontop) OK: Cimarron, Comanche, Greer, Kiowa; TX: Cottle, Motley
Trisetum flavescens (L.) Beauv. (Yellow Oats) KS: Riley; MO: Jackson
Trisetum interruptum Buckl. OK: Blaine, Cleveland, Roger Mills, Woods
Trisetum spicatum (L.) Richter, (Spike Trisetum) SD: Lawrence

128. SPARGANIACEAE (Bur-reed Family)
Sparganium americanum Nutt. KS: Douglas, Franklin, Osage; MO: Jackson, Jasper
Sparganium angustifolium Michx. SD: Marshall, Roberts
Sparganium multipedunculatum (Morong) Rydb. MT: Daniels; SD: Lawrence, Pennington, Todd

130. ARACEAE (Arum Family)
Calla palustris L. (Wild Calla) MN: Becker; ND: Pembina, Rolette

131. LEMNACEAE (Duckweed Family)
Lemna valdiviana Phil. KS: Douglas, Pratt; MO: Jackson; NE: Cherry; OK: Cleveland, Grady
Woffia papulifera Thomps. KS: Cherokee, Linn

132. PONTEDERIACEAE (Pickerelweed Family)
Eurystemon mexicanum (Wats.) Alex. TX: Carson
Pontederia cordata L. IA: Ringgold; KS: Cherokee, Lyon; MO: Barton, Bates, Clay, Jackson, Jasper, Ray; OK: Payne

133. LILIACEAE (Lily Family)
Allium canadense L. var. *mobilense* (Regel) Ownbey, (Wild Onion) KS: Cherokee; MO: Jasper; OK: Craig, Ottawa, Rogers
Allium geyeri Wats. CO: Las Animas; NM: Union; SD: Aurora, Corson, Custer, Fall River, Pennington; WY: Laramie
Allium porrum L. (Leek) IA: Mills; KS: Doniphan, Rush, Saline, Washington
Allium sativum L. (Garlic) KS: Allen, Butler, Coffey, Douglas, Elk, Greenwood, Neosho, Osage, Republic, Wabaunsee, Wilson; MO: Caldwell, Jasper, Platte, Vernon
Erythronium rostratum C. B. Wolf, KS: Cherokee; MO: Barton, Jasper
Fritillaria pudica (Pursh) Spreng. ND: Billings, Morton
Muscari botryoides (L.) Mills, KS: Douglas, Shawnee
Nolina texana Wats. NM: Harding, Quay; OK: Cimarron
Smilax herbacea L. var. *pulverulenta* (Michx.) Gray, (Carrion-flower) MN: Chippewa; MO: Jasper, Newton
Streptopus amplexifolius (L.) DC. (White Mandarin) SD: Custer, Lawrence, Pennington
Trillium gleasonii Fern. KS: Shawnee; MN: Clay; SD: Marshall, Roberts
Trillium nivale Ridd. (Dwarf White Trillium) IA: Cherokee, Lyon, O'Brien; MN: Douglas; MO: Daviess; NE: Cass; SD: Lincoln
Trillium viride Beck, KS: Cherokee; MO: Newton; OK: Ottawa

Yucca arkansana Trel. KS: Cowley, Montgomery, Wilson
Yucca baccata Torr. CO: Baca
Yucca neomexicana Woot. & Standl. CO: Baca; NM: Union; OK: Cimarron

134. IRIDACEAE (Iris Family)
Iris brevicaulis Raf. (Lamance Iris) KS: Leavenworth, Wyandotte; MO: Jackson
Iris pseudacorus L. (Yellow Iris) KS: Pottawatomie, Wabaunsee; NE: Seward
Iris versicolor L. (Blue Flag) MN: Becker, Cottonwood, Kittson, Pennington, Polk; NE: Sarpy
Sisyrinchium ensigereum Bickn. TX: Hemphill
Sisyrinchium pruinosum Bickn. KS: Harper; OK: Custer, Garfield, Grant, Kingfisher, Kiowa. Tulsa

136. ORCHIDACEAE (Orchid Family)
Calypso bulbosa (L.) Oakes, (Venus' Slipper) SD: Lawrence, Pennington; WY: Crook, Sheridan
Cypripedium X *andrewsii* A. M. Fuller, MN: Becker; ND: Nelson
Cypripedium calceolus L. var. *planipetalum* (Fern.) Vict. & Rousseau, (Yellow Lady's Slipper) ND: Ransom, Rolette
Epipactis gigantea Dougl. ex Hook. (Helleborine) SD: Fall River
Goodyera oblongifolia Raf. (Rattlesnake Plantain) SD: Custer, Lawrence; WY: Crook
Goodyera repens (L.) R. Br. (Dwarf Rattlesnake Plantain) MN: Becker, Nobles; SD: Custer, Lawrence, Pennington
Habenaria clavellata (Michx.) Spreng. (Green Woodland Orchis) ND: Grand Forks
Habenaria dilatata (Pursh) Hook. (White Orchis) MN: Becker, Pope; SD: Lawrence
Habenaria orbiculata (Pursh) Torr. (Round-leaved Orchis) MN: Becker; SD: Lawrence
Habenaria psycodes (L.) Spreng. (Soldier's Plume) NE: Thomas
Habenaria saccata Greene, SD: Lawrence
Habenaria unalascensis (Spreng.) Wats. (Alaskan Orchis) SD: Lawrence, Pennington; WY: Crook
Hexalectris spicata (Walt.) Barnh. (Crested Coralroot) KS: Montgomery; OK: Grady
Listera convallarioides (Sw.) Nutt. (Twayblade) SD: Lawrence; WY: Albany, Sheridan
Pogonia ophioglossoides (L.) Ker, ND: Grand Forks
Spiranthes lucida (H. H. Eaton) Ames, KS: Cloud; NE: Cass
Spiranthes ovalis Lindl. KS: Douglas, Linn, Miami; MO: Jackson; OK: Oklahoma

Index

[Family names appear in capital letters; generic names in italic type; and colloquial names in roman type. Numbers in parentheses refer to map numbers; numbers in roman type refer to pages in the map section. Italicized numbers indicate pages in the list of infrequent taxa]